Handbook *on* CLOSTRIDIA

Handbook *on* CLOSTRIDIA

Edited by

Peter Dürre

Taylor & Francis
Taylor & Francis Group

Boca Raton London New York Singapore

A CRC title, part of the Taylor & Francis imprint, a member of the Taylor & Francis Group, the academic division of T&F Informa plc.

Published in 2005 by
CRC Press
Taylor & Francis Group
6000 Broken Sound Parkway NW
Boca Raton, FL 33487-2742

Printed in the United States of America on acid-free paper
10 9 8 7 6 5 4 3 2 1

International Standard Book Number-10: 0-8493-1618-9 (Hardcover)
International Standard Book Number-13: 978-0-8493-1618-0 (Hardcover)
Library of Congress Card Number 2004064958

Library of Congress Cataloging-in-Publication Data

Handbook on clostridia / edited by Peter Dürre.
p. cm.
Includes bibliographical references and index.
ISBN 0-8493-1618-9 (alk. paper)
1. Clostridium--Handbooks, manuals, etc. I. Dürre, P. (Peter) II. Title.

QR82.B3H36 2005
579.3′64--dc22 2004064958

T&F informa

Taylor & Francis Group
is the Academic Division of T&F Informa plc.

Visit the Taylor & Francis Web site at
http://www.taylorandfrancis.com

and the CRC Press Web site at
http://www.crcpress.com

Preface

Clostridium is one of the largest bacterial genera, ranked second in size only to *Streptomyces*. Clostridia represent a heterogeneous taxonomic grouping, once only being united by rodlike morphology with a Gram-positive-type cell wall, anaerobic metabolism, ability to form endospores, and inability of dissimilatory sulfate reduction. Although intensive research has resulted in partial reclassification and creation of a number of new genus designations (e.g., *Caloramator, Filifactor, Moorella, Oxobacter, Oxalophagus, Paenibacillus, Syntrophospora, Thermoanaerobacter, Thermoanaerobacterium*), members of these taxonomic groups can still be considered descendants of a common, Gram-positive ancestor. Phylogenetically related genera are, e.g., *Anaerobacter, Epulopiscium, Ilyobacter,* and *Sporohalobacter*. This short list indicates the enormous metabolic potential of "the clostridia." Unfortunately, in the public eye these bacteria are notorious for causing severe diseases. While it is true that a few species produce extremely powerful toxins (botulinum is the most poisonous substance of biological origin known), their biotechnological importance and potential have often been overlooked. The acetone-butanol fermentation was the major production route to these solvents in the first half of the last century and, with respect to scale, represents one of the largest biotechnological processes ever performed. And some of the toxins have recently become indispensable tools in medical therapy. It is time to change the bad reputation of these microorganisms by pointing to the biotechnological and medical applications, in part still awaiting elucidation and exploitation, by providing a comprehensive overview on what is known on clostridia.

The *Handbook on Clostridia* summarizes aspects on methods, physiology, medical importance, regulation, ecosystems, genomics, as well as current and potential applications. The chapters are written by experts in the field. While in the past 15 years four books appeared featuring special clostridial topics, this is the first attempt to comprehensively cover the complete field of *Clostridium*. Almost all of the authors participated in one or both of the specialized international *Clostridium* conferences, which are regularly held in two- or three-year intervals (apathogenic, solventogenic: *Clostridium* I to VIII, from 1990 to 2004 [Salisbury, Blacksburg, Evanston, Ulm, Toulouse, Champaign/Urbana, Rostock, Edinburgh], *Clostridium* IX to be held in Houston, Texas, in 2006; pathogenic: ClostPath 1995 to 2003 [Rio Rico, Seillac, Chiba, Woods Hole], the fifth conference to be held in England in 2006). In addition, the European Commission has funded a specific Concerted Action on the genus *Clostridium*. Interested scientists and students are invited to check the respective Web sites for more information and future participation. All these activities underline the growing scientific interest in clostridia and their products. I therefore hope that not only specialists with a very specific research focus will read this book, but that the audience will include advanced undergraduate and graduate students as well as academic and industrial researchers, using it as a compilation of reviews, as a reference for teaching, and as a critical review on advantages of using clostridia for new applications — both in the medical and biotechnological fields.

It would have been impossible to complete this book without the efforts of numerous colleagues, many of whom became close friends during common research on clostridial species. Their enthusiasm, energy, and patience are gratefully acknowledged. Finally, I express my sincere thanks to Erika Dery of CRC Press who has guided me throughout the project. Her constant advice, encouragement, help, and patience were essential to publish the *Handbook on Clostridia*.

Peter Dürre

The Editor

Peter Dürre, Ph.D., professor of microbiology and head of the department of microbiology and biotechnology at the University of Ulm, Germany, received his undergraduate and Ph.D. degrees in microbiology in 1978 and 1981, respectively, from the University of Göttingen, Germany, Department of Microbiology. During that time, his research dealt with physiology and taxonomy of purine-degrading clostridia in the laboratory of Jan R. Andreesen.

From 1981 to 1983, he worked as a postdoctoral recipient of a DFG fellowship at the University of California, Berkeley, Department of Biochemistry, in the laboratory of Giovanna Ferro-Luzzi Ames, gaining research experience in molecular biology by studying the high-affinity histidine permease of *Salmonella typhimurium*. In 1983, he went back to the University of Göttingen, Germany, joined the Department of Microbiology, headed by Gerhard Gottschalk, and started his own research group. His major interests, physiology and molecular biology of the clostridia, with a special focus on regulation of solvent production by *Clostridium acetobutylicum*, are still pursued. Other recent topics are initiation of clostridial endospore formation and regulation of metabolic networks connected to this process. After having performed his "habilitation" in Göttingen in 1991, he accepted in 1995 the position of head of the department at the University of Ulm. In addition, in 2003, he became vice president for research at the university.

Dürre is a member of the Vereinigung für Allgemeine und Angewandte Mikrobiologie (VAAM), the Deutsche Gesellschaft für Hygiene und Mikrobiologie (DGHM), and the American Society of Microbiology (ASM). He serves in the Minerva Steering Committee (Beirat) of the Moshe Shilo Center for Marine Biogeochemistry, Hebrew University, Jerusalem, and in the ASIIN accrediting committee for biosciences, Frankfurt.

Dürre is not only editor of the *Handbook on Clostridia*, but has also collaborated on a number of other books including *Clostridia: Biotechnology and Medical Applications, Nucleic Acids Isolation Methods,* and *Regulatory Networks in Prokaryotes*. He is author and coauthor of numerous research articles and reviews. He is a member of the editorial boards of *Applied and Environmental Microbiology* and *Genome Letters*, was an editor of *FEMS Microbiology Reviews* from 1989 to 1998, and serves as editor of *Applied Microbiology and Biotechnology* and as senior editor of the *Journal of Molecular Microbiology and Biotechnology.*

Contributors

Farrukh Ahmad
Groundwater Service Inc.
Houston, Texas

Antoon D.L. Akkermans
Wageningen University
Wageningen, The Netherlands

Klaus Aktories
University of Freiburg
Freiburg, Germany

Jan R. Andreesen
Universitaet Halle
Halle, Germany

Jozef Anné
Katholieke Universiteit Leuven
Leuven, Belgium

Hubert Bahl
University of Rostock
Rostock, Germany

Sofie Barbé
Katholieke Universiteit Leuven
Leuven, Belgium

Holger Barth
University of Ulm
Ulm, Germany

George N. Bennett
Rice University
Houston, Texas

Hans P. Blaschek
University of Illinois
Urbana, Illinois

David W. Brett
Smith & Nephew, Inc.
Largo, Florida

Wolfgang Buckel
Philipps University
Marburg, Germany

Glen Carter
University of Nottingham
Nottingham, England

Jiann-Shin Chen
Virginia Polytechnic Institute and State University
Blacksburg, Virginia

Jackie K. Cheung
Monash University
Victoria, Australia

Anne Collignon
Université Paris-Sud
Châtenay-Malabry, France

Ian J. Davis
University of Nottingham
Nottingham, England

Sabrina Doß
University of Gießen
Gießen, Germany

Harold L. Drake
University of Bayreuth
Bayreuth, Germany

Bruno Dupuy
Institut Pasteur
Paris, France

Thaddeus C. Ezeji
University of Illinois
Urbana-Champaign, Illinois

Ralf-Jörg Fischer
University of Rostock
Rostock, Germany

Howard Goldfine
University of Pennsylvania
Philadelphia, Pennsylvania

Carolin Gröger
University of Gießen
Gießen, Germany

Jeralyn D. Haraldsen
Tufts University
Boston, Massachusetts

Joseph B. Hughes
Rice University
Houston, Texas

Sandra Hujer
University of Ulm
Ulm, Germany

Eric A. Johnson
University of Wisconsin
Madison, Wisconsin

Norah C. Johnston
University of Pennsylvania
Philadelphia, Pennsylvania

David T. Jones
University of Otago
Dunedin, New Zealand

Stefanie Keis
University of Otago
Dunedin, New Zealand

Tina Knauber
University of Gießen
Gießen, Germany

Sergey R. Konstantinov
Wageningen University
Wageningen, The Netherlands

Kirsten Küsel
University of Bayreuth
Bayreuth, Germany

Ronald G. Labbé
University of Massachusetts
Amherst, Massachusetts

Philippe Lambin
University of Maastricht
Maastricht, The Netherlands

William Landuyt
Katholieke Universiteit Leuven
Leuven, Belgium

Susan Leschine
University of Massachusetts
Amherst, Massachusetts

Dena Lyras
Monash University
Victoria, Australia

Paola Mastrantonio
Istituto Superiore di Sanitè
Rome, Italy

Bruce A. McClane
University of Pittsburgh
Pittsburgh, Pennsylvania

Sheena McGowan
Monash University
Victoria, Australia

Nigel P. Minton
University of Nottingham
Nottingham, England

Wilfrid J. Mitchell
Heriot-Watt University
Edinburgh, Scotland

Sandra Nuyts
University Hospital
Leuven, Belgium

Aharon Oren
The Hebrew University of Jerusalem
Jerusalem, Israel

Eleftherios T. Papoutsakis
Northwestern University
Evanston, Illinois

Oliver Pennington
University of Nottingham
Nottingham, England

Caroline M. Plugge
Wageningen University
Wageningen, The Netherlands

Michel R. Popoff
Insitut Pasteur
Paris, France

Nasib Qureshi
USDA NCAUR
Peoria, Illinois

Sharon J. Reid
University of Cape Town
Rondebosch, South Africa

Julian I. Rood
Monash University
Clayton, Australia

Steffen Schaffer
Degussa AG
Hanau-Wolfgang, Germany

Thorsten Selmer
Philipps University
Marburg, Germany

Hauke Smidt
Wageningen University
Wageningen, The Netherlands

Abraham L. Sonenshein
Tufts University
Boston, Massachusetts

J. Glenn Songer
University of Arizona
Tucson, Arizona

Alfons J.M. Stams
Wageningen University
Wageningen, The Netherlands

Bradley G. Stiles
U.S. Army Medical Research Institute
Fort Detrick, Maryland

Helen E. Stutz
University of Cape Town
Rondebosch, South Africa

Martin Tangney
Napier University
Edinburgh, Scotland

Jan Theys
University of Maastricht
Maastricht, The Netherlands

Kai Thormann
Stanford University
Stanford, California

Richard W. Titball
CBD Porton Down
Wiltshire, England

Christopher A. Tomas
Northwestern University
Evanston, Illinois

Anke Treuner-Lange
University of Gießen
Gießen, Germany

Seshu B. Tummala
Northwestern University
Evanston, Illinois

Rodney K. Tweten
University of Oklahoma
Oklahoma City, Oklahoma

Lieve Van Mellaert
Katholieke Universiteit Leuven
Leuven, Belgium

David E. Whitworth
University of Warwick
Coventry, England

Mike Young
Rockefeller University
New York, New York

Contents

PART I Methods

PART II Metabolic Pathways and Transport Systems

PART IV Regulatory Mechanisms

PART V Complex Regulatory Networks

PART VI Special Groups and Ecosystems

Part I

Methods

1 Species and Strain Identification Methods

David T. Jones and Stefanie Keis

CONTENTS

1.1 INTRODUCTION

One of the principal objectives of bacterial classification systems is to provide a reliable framework for the identification and classification of new isolates or unknown bacteria. An overarching goal in bacterial taxonomy has been the establishment of a natural phylogenetic classification that reflects the evolutionary pathways and history of individual groups of bacteria and provides insights with respect to their phylogeny and evolutionary relationships [1–3]. In practical terms, a phylogenetic-based classification system affords considerable advantage in providing greater stability for taxons, thereby reducing the often drastic rearrangements that have occurred with phenotypic-based systems used in the past.

The utilization of reliable molecular markers has been the key to the unravelling of evolutionary relationships and the establishment of a phylogenetic classification for bacteria. The 16S ribosomal RNA (rRNA) sequence has become established as the most generally accepted and widely used molecular marker for taxonomic studies in bacteria and has provided the basis for ascertaining phylogenetic relationships and delineating major taxonomic groupings [3]. The reason these nucleotide sequences have proved so valuable as markers is that the structural rRNA molecules are essential elements for protein synthesis in all cellular systems. Due to the highly conserved function of these molecules, the genomic sequences encoding them have changed more slowly during evolution than the bulk of the cell genome, so that some homology is retained by all organisms. The limits to divergence that have occurred means that these markers exhibit common conserved homologous sequences even between distantly related species. Individual lines of descent may therefore be inferred from rRNA sequences enabling them to be used as an evolutionary chronometer [1]. In addition, some segments of the molecule are more variable and have evolved more rapidly, allowing comparisons to be made among relatively closely related species. This has added to their value and usefulness for phylogenetic analysis. 16S rRNA sequence data has now reached the point where it is available for almost all validly described bacterial species, making it possible to obtain a comprehensive overview of evolutionary relationships among bacteria. The development of a phylogenetic classification has provided a unifying concept of the genus and higher ranked taxa.

The major impact that 16S rRNA sequence data has had on the elucidation of phylogenetic relationships can be no better exemplified than by the revolution that has occurred in the taxonomy of the clostridia. These developments are covered in a number of recent reviews [3–7]. The genus *Clostridium* was first described in 1880 by Prazmowski [8], and there are a number of excellent, comprehensive articles covering the discovery, recognition, and early history of the genus [4,9,10]. For the following 100 years the description of selected phenotypic characteristics remained the primary approach used to circumscribe the genus. During the first half of the previous century in particular, many new *Clostridium* species were described and named. However, many of these descriptions were incomplete and were often based on the investigation of just a single strain. Problems and inconsistencies also arose from the utilization of inappropriate methods or growth conditions and the contamination or subsequent loss of original strains [4]. Extensive patenting of industrial strains, and the problem of dealing with nonclostridial phenotypes, also contributed to the confused state of the classification of the genus and led to many species subsequently being declared illegitimate.

Even before the recent advances in our understanding, it became apparent that the genus *Clostridium* consisted of a heterogenous group of bacteria sharing only a small number of features in common [3,4,11]. These common properties are considered to represent ancient phylogenetic traits and include: rod-shaped morphology; Gram-positive staining cell walls containing a multilayer of peptidoglycan; formation of heat-stable endospores; anaerobic growth; fermentative metabolism without the capacity for dissimulatory sulphate reduction; and genomic DNA with a low %G+C content. Even these diagnostic features are not entirely reliable, as some groups show variations in morphology including the production of coccoid cells, as well as variations in physiology, including the ability to grow in air at atmospheric pressure. In many clostridial species, endospores are difficult to detect or are not produced at all under laboratory conditions. Although the ability to produce endospores appears to be monophyletic, the presence of endospores is not a phylogenetically coherent feature, as endospore formation occurs in species with widely divergent lines of descent. This is further complicated by the permanent loss of spore-forming ability and changes in morphology in many of the lineages.

The availability of 16S rRNA sequence data for all validly described clostridial species has had a radical impact on the taxonomy of the genus. Rather than constituting a single genus, spore-forming clostridia exhibit considerable genetic heterogeneity represented by a number of independent and deeply branching sublines within the *Clostridium*/*Bacillus* subdivision that also include many non-clostridial species. This led to the general acceptance that *Clostridium*, as originally defined, is not a phylogenetically coherent and homogenous taxon. Assuming constant rate of evolution, it would appear that the initial radiation into independent sublines occurred early on in the ancestry of these ancient bacteria, during the anaerobic phase of evolution [3]. The various sublines have continued to persist in niche environments, particularly those where the degradation of complex organic molecules occur under anaerobic conditions. The diverse sublines fall into around 20 clusters, with the genus *Clostridium* now restricted only to species that fall into the core clusters I and II [3,5]. These are equivalent to groups 1 and 2 originally proposed by Johnson and Francis [12].

1.2 DEFINITION OF A *CLOSTRIDIUM* SPECIES

The current concept of a bacterial species is that it consists of a phylogenetically coherent unit composed of closely related strains often forming clearly demarcated subgroups. As with other groups of bacteria, each *Clostridium* species has a designated type strain, along with all other strains that are viewed to be sufficiently similar to the type strain to be included within the species. The type strain is the strain selected to represent a permanent example of the species [1]. The strain selected as the type strain need not necessarily be the most typical or representative strain within the species but are usually strains that have been extensively described. Specimens of type strains are normally lodged with all of the major reference culture collections. To be included within a species, any new strain must be compared with the characteristics of the type strain.

It is now generally accepted that only strains that are truly closely related genetically should be included in the same species and that the strains comprising the species must form a genotypically rather than phenotypically coherent cluster or grouping. Genomic traits more fundamentally reflect the nature of the bacterial genotype than do phenotypic traits and provide a more accurate and stable basis for classification. This allows genetic homologies to be established first, before the phenotypic characteristics are compared. This has resulted in a decrease on the reliance upon traditional microbiological tests for gathering extensive phenotypic data for species definition.

Comparison of 16S rRNA sequence homology has now been widely accepted as providing the foundation for defining a *Clostridium* species. In many instances, the divergence in homology of 16S rRNA sequences is sufficient to clearly demarcate a species from the nearest related species, confirming a deep-rooted separate genetic lineage. A divergence in sequence homology of greater than 3% is generally considered sufficient to establish species identity [13]. However, as sequence data for clostridial species has accumulated, clusters of clostridial species were identified that share a high degree of 16S rRNA sequence homology. Stackebrandt and Hippe [3] list three clusters of closely related species that exhibit in excess of 98.5% homology, and identify additional closely related species that include *Clostridium haemolyticum* and *Clostridium botulinum* (neurotoxin types C and D), which share 99.3% sequence similarity and *Clostridium roseum* and *Clostridium acetobutylicum,* which share 99.8% similarity. A high level of 16S rRNA sequence homology has also been reported for three related species of solvent-producing clostridia, classified as *Clostridium beijerinckii*, *Clostridium saccharoperbutylacetonicum*, and *Clostridium saccharobutylicum* [14,15].

Where putative clostridial species exhibit more than 97% sequence homology, 16S rRNA sequence alone is not considered sufficient to confirm separate species status [13]. In such instances, the decisions that relate to the conferring of species status has to be based on information obtained by utilizing other techniques. A variety of other gene sequences, encoding for ubiquitously distributed, homologous housekeeping genes, have been assessed for suitability as molecular markers capable of elucidating phylogenetic relationships among bacteria [3]. Although some of these markers have been utilized for specific purposes, none approach the reliability and versatility provided by the 16S rRNA gene sequence.

The composition of chromosomal DNA is essentially unaffected by growth conditions, and although the amount might fluctuate under different environmental conditions, the nucleotide sequence remains essentially invariant. Chromosomal DNA may be analyzed at various levels that include: DNA base composition; DNA/DNA hybridization; genomic DNA fingerprinting and direct comparison of genomic DNA gene sequences. Used in combination, these techniques can provide a reliable approach to validating the boundaries of what appropriately constitutes a bacterial species. Hence, the more information that can be accumulated for a group, the more likely it is that various investigators will be in agreement as to which strain should legitimately be included within a species and which should be excluded.

1.3 MOLECULAR TECHNIQUES FOR SPECIES AND STRAIN IDENTIFICATION

There are a number of excellent practical handbooks and laboratory manuals that contain detailed step-by-step instructions for undertaking the various techniques discussed in this section [16–20]. Investigators wishing to undertake these procedures in the laboratory should consult these sources as well as descriptions of techniques involving *Clostridium* species reported in the current literature. The various techniques below are intended to provide an overview and guidelines for species and strain identification.

1.3.1 Techniques for Sequencing 16S Ribosomal RNA

A protocol for the direct sequencing of 16S rRNA utilizing reverse transcriptase was described by Lane et al. [21]. Although 16S rRNA contains base pairing combinations that can cause premature terminations, it proved to be a reliable, rapid, and relatively simple approach to obtaining comparative sequence data and superseded earlier techniques. The widespread use of nucleotide sequencing led to a rapid accumulation of sequence data for a wide range of bacterial species.

The development of PCR techniques provided investigators with a practical method for amplifying the 16S rRNA gene so that the DNA could either be sequenced directly or cloned and sequenced via standard DNA sequencing protocols. PCR techniques, coupled with automated sequencing of double-stranded (ds) DNA, have become the routine approach for obtaining nucleotide sequences. An alternative approach appropriate for some specific applications is to clone PCR-generated fragments in an appropriate plasmid or bacteriophage M13 cloning vector. Where a new taxonomic entity is to be characterized and described it is essential that the entire 16S rRNA gene sequence is accurately determined. However, 16S rRNA sequencing can also be used as a rapid screening procedure for the comparison of strains that are thought to share close nucleotide homology. For applications of this type, it can be sufficient to utilize specific internal PCR primers to amplify the variable regions of the gene. If strains exhibit a high degree of homology with these variable regions, their sequence homology in the conserved regions can be assumed to be invariant.

The reliability of the analysis of phylogenetic relationships between species and strains is dependent on the accuracy of the alignment of the sequence data and the nature of the algorithms used to generate the phylogenetic hierarchies [3]. It is essential that each nucleotide sequence is fully aligned with the common conserved ancestral sequences. The standard approach is to align the sequences that have been generated with respect to 16S rRNA sequence from *Escherichia coli*. Nucleotide sequences within the variable and highly variable regions can exhibit significant differences in length and can present difficulties with respect to accurate alignment.

Three approaches are used for nucleotide sequence analysis. These are maximum parsimony determination, maximum likelihood determination, and distance in pairs determinations. The maximum parsimony method is based on the assumption that evolution is reflected by the minimum number of changes in the positions of the aligned sequences. The algorithms search for the most parsimonious evolutionary trees from the various possible combinations. Normally the tree exhibiting the greatest consensus is selected from among 100 boot-strapped trees. The maximum likelihood method also evaluates sequence homology on a site-by-site basis. Trees are then generated, and the tree that best matches the congruence of the data is used to generate a maximum likelihood dendrogram.

The method that has been used most extensively for systematic studies of the clostridia is the distance method, in which a matrix of dissimilarity values is determined for the aligned sequences. These dissimilarity values are then converted into phylogenetic distances using various distance matrix and neighbor-joining algorithms. The most detailed phylogenetic analyses of the clostridia have been undertaken using this approach [3].

1.3.2 Determination of Guanine Plus Cytosine Content of DNA

The gross composition of bacterial chromosomal DNA is reflected by content of the four base pairs guanine plus cytosine (G+C) and adenine plus thymine (A+T). The mole% content of G+C in chromosomal DNA is constant for any specific strain, but the ratio can vary broadly among different groups of bacteria — ranging from a minimum of just below 25% to a maximum of around 75% [1]. Determination of the mole% content of G+C in chromosomal DNA can therefore provide an assessment of the degree of sequence homology shared between the chromosomal DNA from two strains of bacteria. The mole% content of G+C can be estimated by several different methods [20]. The thermal denaturation method is the most commonly used, as determinations utilizing this technique are rapid, inexpensive, and relatively insensitive to protein or RNA contamination of the DNA. The thermal denaturation of duplex DNA is measured by increased absorbance at 260 nm with the midpoint of the hyperchromic shift being taken as the melting temperature (Tm). Direct estimations of the nucleotide composition of DNA can also be determined by hydrolysis of DNA followed by separation and quantification of the products by high performance liquid chromatography. Determinations of G+C content have also been achieved using physiochemical approaches such as buoyant density equilibrium centrifugation in caesium chloride gradients, but this approach is seldom used these days. Detailed accounts of the procedures for undertaking the above techniques are described by Johnson [16,20]. All of the above procedures require a comparison of the unknown DNA against standard reference DNA.

Since DNA base composition does not take into consideration the linear sequence of the bases in the chromosomal DNA, overall base composition can only provide a crude measure of bacterial relatedness. It is possible for two strains of bacteria to have an identical base composition but be entirely different based on their sequence homology. However, the converse does not apply. If two organisms possess widely different base composition, they will have few DNA sequences in common and are likely to be only distantly related. In this regard, the mole% G+C DNA base composition can provide a useful measure of genetic heterogeneity, but can only be used to exclude strains from membership of a species. It is generally accepted that members of the same species should not differ by more than 4 to 5% mole% G+C. When deviation extends beyond this limit, significant genomic variation is indicated, warranting exclusion from the species. In making such assessments, it is important to take into consideration that mole% G+C determinations are seldom sufficiently accurate to distinguish reliably anything less than 2% in experimental variation [1].

Determination of DNA base composition also provides useful taxonomic information with relevance to the description of new species and is normally required for the complete description of the species. Within the genus *Clostridium* the mole% G+C of individual species is known to vary within a range of approximately 22 to 55% [22]. The majority of species form a cluster around 28 mole% G+C with other species forming smaller clusters in the region of 35, 45, and 52%.

1.3.3 DNA/DNA Hybridization

1.3.3.1 Techniques

DNA/DNA hybridization techniques provide a more useful and precise approach for obtaining estimations of relatedness of bacterial genotypes. The base composition and nucleotide sequence of chromosomal DNA for practical purposes remains invariant regardless of growth conditions. The use of DNA/DNA hybridization techniques allows the total genome of one bacterial strain to be compared with that of others. This enables elucidation of genomic relationships between relatively closely related groups of bacteria. The value and application of DNA hybridization techniques in taxonomic studies on the clostridia is exemplified by the pioneering studies undertaken by Johnson [23–25].

The methods for DNA reassociation have been comprehensively described in a variety of publications [17,20,26]. There are essentially two basic approaches to assessing DNA base sequence complementarity utilizing DNA/DNA hybridization techniques. Both approaches can be used to provide a quantitive estimate of DNA sequence homology between two species or strains. Similarities in base sequence homology between DNA from two organisms can be assessed by DNA reassociation methods in which DNA is first separated into single strands by thermal or alkali denaturation and subsequently allowed to anneal in the presence of a second single-stranded (ss) DNA molecule. If the nucleotide sequence of the two DNA samples exhibits significant homology, hybrid duplexes will be formed by base pairing. If the DNA share few sequences in common, there will be negligible hybridization formation. An alternative approach for assessing DNA sequence divergence is to compare stabilities of reassociated duplexes. Hybrid DNA duplexes that share lower base sequence homology will exhibit a lower melting temperature than either of the homologous parent molecules.

DNA/DNA reassociation techniques can be further divided into free-solution methods and membrane immobilization methods. When free-solution techniques are employed, the reassociation of denatured DNA fragments that have been radioactively labeled can be measured by adsorption to hydroxyapatite or by resistance to S1 nuclease. Alternatively, free-solution reassociation of denatured DNA can be measured optically without the need to radioactively label the DNA, by utilizing a recording spectrophotometer.

When undertaking free-solution reassociation assays with radioactively labeled DNA, an excess of DNA is first sheared to a constant molecular weight by sonication or extrusion through a French press. The DNA is then denatured and incubated with radioactive label (normally ^{14}C). Sheared ssDNA from the reference strain is added and renaturation is allowed to occur. Controls comprising the homologous reaction and labeled DNA with DNA known to share no homology with bacterial DNA (such as salmon sperm DNA) are required.

After renaturation has been allowed to occur, the re-annealed dsDNA has to be separated from the remaining ssDNA. This can be achieved by treating with S1 nuclease to hydrolyze the ssDNA but not dsDNA. The remaining intact dsDNA is precipitated with trichloroacetic acid, filtered, and the radioactivity in the precipitate is counted in a scintillation counter. The percentage homology values are obtained by dividing the counts per minute in heterologous resistant DNA by the activity of the homologous reaction and multiplying by 100. An alternative method for separating the dsDNA from ssDNA is the use of hydroxyapatite chromatography. In phosphate buffer, only the dsDNA will bind to hydroxyapatite, and the dsDNA can then be eluted with a high molarity buffer, precipitated, and counted in a scintillation counter.

Free-solution reassociation assays can also be undertaken without recourse to labeling the DNA. Reassociation of ssDNA is measured by a reduction in transmission at 260 nm. The initial rates of reassociation of an equimolar mixture of DNA from the two strains are compared with the initial rates of the two homologous reactions. The nonhomologous DNA reassociate more slowly than homologous molecules. This method has the advantage of being rapid, and labeled DNA is not required. Comparisons are made between paired strains rather than between reference and test strains. With appropriate care, this method is considered to produce results that are as reliable as those obtained using procedures involving radioactively labeled DNA [20,26].

Alternatives to free-solution techniques are membrane immobilization techniques. In immobilized DNA reassociation assays, the ssDNA is first bound to a solid support, such as nitrocellulose filters, so that ssDNA can form duplexes with complementary sequences. In the direct assay, filters containing immobilized ssDNA from the different strains are incubated with the labeled reference ssDNA, and renaturation is allowed to occur. Membranes are then washed, dried, and counted in a scintillation counter. The extent of homology is calculated in the same way as for free-solution assays with high-bound membrane counts indicating a high degree of homology.

To ensure reproducible results, DNA reassociation assays must be carefully standardized. The specificity of DNA reassociations can be influenced by the purity, concentration, and physical state of the DNA, including fragment size. The ionic strength of the buffer and the temperature at which the reassociation mix is incubated can also affect reassociation assays with optimal reassociation occurring at around 30°C [27].

A variation of the direct immobilized assay is the competition reassociation assay. The basis for this method is to immobilize reference ssDNA on a filter and add a small amount of labeled referenced DNA, along with an excess of heterologous competitor ssDNA. If the competitor DNA is complementary to the reference DNA, it will preferentially reassociate with the bound DNA because of its higher concentration, whereas if there is little or no homology shared between the competitor and immobilized DNA molecules, the labeled reference DNA will reassociate with the membrane-bound ssDNA. Low counts on the washed and dried membranes indicate high homology, and high counts indicate low homology. The advantage of this method is that only reference DNA is bound to the membranes. This provides greater accuracy and reduces the amount of work involved.

1.3.3.2 Applications

The formation of DNA-DNA duplexes do not normally occur when the base pair mismatches between two DNA molecules exceeds 10 to 20%. This tends to limit the usefulness of the techniques at the level of the genus. The deep branching in genomic homology between many of the *Clostridium* species means that DNA/DNA hybridization levels are usually too low to provide useful information about relatedness among divergent clostridial species.

The establishment of DNA homology is providing a unifying concept in defining a bacterial species. Among the clostridia, DNA/DNA reassociation analysis has provided a very effective approach to deducing evolutionary relatedness among closely related strains and species. As such, it can play a key role in validation of boundaries of what constitutes a legitimate species. One of the challenges of interpretation is that some species exhibit significantly greater genomic heterogeneity than others. The lower allowable limit for homology for the demarcation of bacterial species remains a controversial topic. A reassociation measurement of between 50 and 70% under optimal conditions seems to constitute the most generally accepted boundary. When more than one DNA/DNA homology grouping is identified among an assemblage of apparently similar strains, it can be taken as strong evidence that the strains constitute more than one species.

The main limitation to using these hybridization methodologies for taxonomic studies is that they do not readily lend themselves to simple, rapid, automated analyses. Due to the amount of work involved, it is unusual for full similarity matrices giving estimates of DNA homology between each and every strain under consideration. More commonly, matrices are restricted to more limited comparisons utilizing a limited number of selected reference strains and various groupings of test strains. These more limited comparisons can still provide very useful data and are generally reliable, provided the reference strain is representative of the species. Accordingly, DNA/DNA reassociation studies are most effectively undertaken in conjunction with other approaches.

DNA/DNA hybridization techniques utilizing free-solution reassociation, followed by digestion of radioactively labeled DNA using S1 nuclease hydrolysis, have been used in a number of studies of species differentiation and strain identification involving a variety of *Clostridium* species [24,25,28–30]. Used in combination with other approaches, DNA/DNA hybridization has provided a powerful and reliable technique for establishing genetic relationships between strains.

Although DNA/DNA hybridization techniques provide a valuable approach for assessing relatedness between bacterial genotypes, they provide no specific information with respect to the actual nucleotide sequence of the chromosomal DNA. As the number of fully sequenced bacterial genomes continues to expand, the options for making direct comparisons between the genomes of related species will increase and generate major advances in the understanding of evolutionary histories and phylogenetic relationships. Where the full genomic sequences have become available for clostridial species, these have provided valuable reference standards for comparisons with other closely related species and strains. It is, however, difficult to conceive of a time in the future where complete genomic sequences of bacterial species and strains become so prevalent that full DNA sequence comparison becomes a widely applied technique for taxonomic studies.

1.3.4 Pulsed-Field Gel Electrophoresis

1.3.4.1 Techniques

Pulsed-field gel electrophoresis (PFGE) provides a powerful technique that has many applications in bacterial genetics. Used in combination with rare-cutting restriction endonucleases, the technique can be utilized to provide a relatively quick and easy method for the physical analysis of bacterial genomes. The development of PFGE has facilitated the separation of large fragments of DNA ranging in size from 10 kb to more than 2500 kb [31,32]. There have been two main strategies developed to facilitate the electrophoretic separation of DNA fragments. The most extensively used technique is clamped homogenous electric field electrophoresis (CHEF) and an alternative approach is the use of transverse-alternating field electrophoresis (TAFE).

The standard approach for processing bacterial chromosomal DNA for use in PFGE studies involves the *in situ* preparation of DNA embedded in gel blocks or plugs using low-melting point agarose (LMPA). Cell cultures grown and harvested under the appropriate conditions are chilled on ice. Aliquots of cells are added to LMPA and cast into slabs or molds to facilitate the production of individual plugs suitable for incorporation into the wells of pulsed-field gels. The plugs containing the cells are then typically exposed to further treatment in a lysis buffer solution. Such solutions normally contain appropriate concentrations of lysozyme and detergents (e.g., Brij 58, deoxycholate, sodium lauryl sarcosine), RNase, and protease (e.g., proteinase K). Cells are incubated under appropriate conditions to achieve cell lysis and the *in situ* purification of chromosomal DNA. The resulting plugs containing the chromosomal DNA can then either be stored until required or exposed immediately to digestion with the appropriate restriction endonucleases.

When working with uncharacterized strains of *Clostridium*, preliminary trials using a range of restriction enzymes that have been used successfully in studies on related clostridial species should be undertaken. For most purposes, restriction enzymes that cleave the genome into around 5 to 20 fragments are the most useful. Restriction enzymes that generate a small number of restriction fragments can be useful for applications where very large chromosomal fragments are required and can be invaluable for undertaking double digests using combinations of restriction enzymes that cut the chromosome more frequently. Restriction enzymes that cleave the chromosomal DNA into a large number of smaller fragments can be useful for generating distinctive DNA fingerprint patterns, but the large number of resulting fragments can make the analysis of individual bands extremely difficult.

The most useful restriction endonucleases for cleaving clostridial genomic DNA are rare-cutting enzymes that recognize six or more base pairs containing a high proportion of G+C content in their recognition sites. Some of the most widely used restriction enzymes that have been employed in studies involving clostridial species are: *Apa*I, *Bss*HII, *Eag*I, *Ksp*I, *Mlu*I, *Nae*I, *Nar*I, *Nru*I, *Sst*II, *Xho*I, *Xma*I, *Rsr*II, *Sfi*I, and I-*Ceu*I.

Following electrophoresis, the gels are normally stained in ethidium bromide followed by destaining in deionized water and then viewed with a UV transillumintor. The sizes of fragments are estimated by comparing their migration rates with the migration rates of known fragment sizes such as those provided by commercially available molecular weight markers containing lambda ladders or yeast chromosomes.

Under appropriate conditions, the technique is capable of separating fragments ranging from as small as 10 kb to those that are up to about 2500 kb. However, it is not possible to achieve separation of fragments covering this range in one electrophoretic run. In order to achieve separation over this range, it is necessary to utilize a minimum of two settings — one resolving the upper range, the other resolving the lower range. Pulses can be controlled through ramping, and separation of different sized fragments can be achieved by varying the pulse times.

Fragments in excess of 2000 kb can be extremely difficult to separate by PFGE and often cluster as an unresolved band at the top of the gel or remain trapped in the cell debris in the well. Further differentiation of these large-sized fragments can sometimes be achieved by using double digests with a range of restriction enzymes. The existence of one or more chromosomal DNA fragments of very large size that remains unresolved has the potential to exert a huge impact on the calculation of the genome size of a particular strain and make it imperative that estimates of genome size are based on data generated from as many different restriction digests as possible.

Another phenomenon that may be encountered when analyzing chromosomal DNA fingerprint patterns is an uneven intensity of restriction fragments of a similar size present within the DNA fingerprint pattern. The presence of bands exhibiting double the intensity of the surrounding bands can be taken as strong evidence of the presence of two different restriction fragments that are very close in size that cannot be resolved as separate entities under the conditions used. In many instances, the presence of double bands can be confirmed by undertaking double digestions with a combination of two different restriction enzymes. Instances where multiple restriction fragments are superimposed to produce a single band can exacerbate difficulties in interpretation.

A further feature that can add to the complexity of interpreting chromosomal restriction digests is the over-representation of some macrorestriction fragments when compared with other fragments of a similar size [33,34]. A likely reason for this is that fragments located adjacent to the origin of replication are over-represented due to the presence of multiple replication forks in the DNA. This effect can be intensified when the genomic DNA under study has been isolated from rapidly replicating cells harvested during exponential growth.

The presence of plasmid(s) or extrachromosomal phage DNA can introduce a further source of difficulty in analyzing genomic DNA fingerprint patterns generated by PFGE. Normally, plasmids or extrachromosomal phage DNA produce fainter bands than those produced by the macrorestrictions fragments of chromosomal DNA. The presence of plasmid or extrachromosomal DNA can usually be identified by the electrophoresis of undigested genomic DNA or I-*Ceu*I-digested DNA that only cleaves chromosomal DNA [35]. This can determine whether plasmid bands are to be expected in the DNA fingerprint patterns of particular strains or isolates. One solution that has been applied to overcome the interference by plasmid DNA is to cut the plasmid into fragments of a small size with a frequent-cutting restriction enzyme.

One problem that has been encountered frequently when applying PFGE for the typing of clostridial strains is the presence of highly active and persistent endogenous DNase activity [36–42]. This activity associated with particular strains results in the shearing and degradation of chromosomal DNA, rendering it useless for PFGE analysis. Various approaches have been utilized in an attempt to overcome this problem with varying success. Methods that have been reported to try to overcome troublesome endogenous nuclease activity include: shortening of the lysis and DNA plug wash steps [36,43,44]; heating of harvested cells [41,42]; formaldehyde fixation of cells at the time of harvesting [43]; carrying out the plug preparation process under anaerobic conditions [14,45,46]; use of a lysis solution for resuspending of the cells prior to mixing with LMPA [44]; inclusion of hypertonic sucrose in the lysis solution [46]; and ultra short lysing steps during plug preparation [37].

A further technical difficulty that can affect the preparation of good quality DNA suitable for PFGE is the problem of obtaining effective cell lysis for the release of intact chromosomal DNA from some clostridial species. This is particularly marked in some strains of *C. botulinum* where the number of cells required to obtain sufficient genomic DNA to enable restriction analysis has been reported to be 5 to 10 times greater than for other species of *Clostridium* [37]. The main cause of difficulties with regard to strains that are recalcitrant to lysis appears to be associated with the characteristics of some Gram-positive cell walls that make them more resistant to the common agents used to achieve cell wall lysis. Growth conditions and the stage at which cell cultures are harvested can also exert an influence, with cells harvested in early-to-mid log phase producing the best results. Failure to achieve effective cell lysis can result in an excessive amount of the DNA being retained within the well. An additional technical difficulty that can be encountered with some clostridial strains is the failure of certain types of restriction enzymes to cut chromosomal DNA. This phenomenon appears to occur mainly as the result of specific CG methylation activity in some strains. Where this problem was encountered with strains that failed to be cleaved by *Sma*I, it proved possible to digest the DNA with its isoschizomer *Xma*I, which did not interfere with typing analysis [38].

1.3.4.2 Applications

PFGE has proved to be an excellent method for the typing of clostridial strains and is being used extensively in epidemiology and public health. Such studies include the analysis of the nosocomial pathogen *Clostridium difficile* [36,41,47,48], food-poisoning strains of *Clostridium perfringens* [49,50], and toxin-producing strains of *C. botulinum* [37,38], where it has provided a useful tool for characterizing type E strains, which exhibit marked differences in DNA fingerprint patterns. The ability to identify genomic differences can be used to monitor outbreaks and link isolates to incriminated food. Strain identification assists greatly with hazard analysis and the identification of critical control points. Other potential uses could also exist in legal and forensic applications.

Similar approaches have been utilized for the genotyping of various industrial and culture collection strains of solvent-producing clostridia that have been used for the commercial production of solvents [14,51]. The ability to identify distinct DNA fingerprint patterns in strains of solvent-producing clostridia played a key role in elucidating the origins, history, relationships, and characterization of this group of bacteria. This approach has the potential for application in the characterization of other industrial clostridial strains used for biomass degradation, enzyme production, and toxin production.

Another application of PFGE has been in the calculation of bacterial genome sizes, including certain clostridial strains [37,42,46,52]. An estimate of the size of the chromosome can be obtained by adding up the sizes assigned to each of the resolved restriction fragments. Such calculations should preferably always be based on estimates made from a number of different macrorestriction profiles for the same strain. In cases where the complete nucleotide sequences of the genome is not available, PFGE provides the most direct and accurate method available for determining the size of the chromosome [37,42]. When estimates of the genomic fragment sizes were compared, more accurate results were obtained by comparing the fragments in relation to the two closest marker fragments, as opposed to a calculation of fragment sizes based on measurements on the migration distance from the well, plotted on a standard curve [53]. Fragment size estimates generated through standard curves, even those based on a cubic spline formula were shown to be off by at least 10% at the top and bottom of the gel, where fragment migration is nonlinear [37]. Studies undertaken on pathogenic *C. botulinum* [37] and solvent-producing *C. beijerinckii* (unpublished data, Keis, S., Young, K.A., and Jones, D.T.) indicate that the genome sizes of related strains can vary quite considerably, suggesting that individual strains possess extensive genomic plasticity to enable them to cope with the varying requirements associated with distinct ecological niches. The application of PFGE has also enabled the construction of physical and genetic maps of the genome of a number of clostridial species including: *C. perfringens* [52,54], *C. beijerinckii* [33], and *C. saccharobutylicum* [34].

Other genome-based methods for strain characterization, such as plasmid profiling, restriction endonuclease analysis, ribotyping, and random amplification of polymorphic DNA, have not been used extensively to date in studies on *Clostridium* species. It is possible that some of these techniques could provide useful additional information for specific typing and identification applications with clostridial species.

1.4 CLASSICAL TECHNIQUES FOR SPECIES AND STRAIN IDENTIFICATION

A range of publications exist that provide detailed information relating to the use of classical microbiological phenotypic tests for the identification of *Clostridium* species [11,55–57].

1.4.1 Standard Phenotypic Characterization Techniques

Standard bacteriological phenotypic characterization tests to determine various morphological, biochemical, physiological, and metabolic properties continue to play an important role in clostridial systematics. The undertaking of these classical tests remains an essential part of formally describing a new species and supporting the development of more refined identification schemes. These tests also continue to play an important role in many routine testing and identification applications. A wide range of tests designed to undertake phenotypic characterization of clostridial strains are available and have been comprehensively described in a number of the publications listed below. The numerous classic bacteriological tests can be broadly divided into methods designed to assess the following characteristics:

- Colonial morphology on solid culture media and appearance in liquid culture media
- Cell morphology including cell size and shape, endospore morphology, motility and flagella morphology, attachment structures and extracellular capsule production, intracellular granules and staining reactions, and cell wall ultrastructure
- Growth characteristics including upper, lower, and optimum ranges for temperature, pH, and other growth parameters
- Degree of anaerobiosis and atmospheric requirements
- Nutrition, growth requirements, and biochemistry
- Sugar and amino acid uptake and utilization
- Ability to degrade complex macromolecules including carbohydrates and proteins
- Type of metabolism employed and end products produced
- Ability to utilize compounds as the sole sources of carbon for energy and growth
- Ability to utilize compounds as a source of nitrogen and nitrogen metabolism
- Inhibitory tests of various types
- Toxin and extracellular enzyme production

When performing phenotypic characterization tests, the inoculum for test media should be made from fresh subcultures that are in good physiological condition. Old cultures are not satisfactory as an inoculum, as they can give spurious results. The nature of the tests being undertaken tends to dictate whether broth or plate cultures are most appropriate. Ideally, phenotypic responses should be investigated under both rich and poor growth conditions in the presence and absence of carbohydrates.

The characteristics should be tested several times under the appropriate conditions, and the reliability of the test culture media and solutions must be confirmed by the inclusion of appropriate positive and negative controls. Tests should be performed under optimal growth conditions, and adaptations of standard test procedures need to be made to take into account pH, temperature, ionic conditions, and the gaseous atmospheric requirements of the particular strains that are being tested. This can be of particular importance with tests for carbohydrate utilization where nonspecific pH changes can result in false positives being recorded. It is sensible to select tests that are directly pertinent to the species or strain that is being investigated to avoid undertaking a large number of tests that will contribute little or no useful information. *Bergey's Manual of Systematic Bacteriology*

[11] provides a useful starting point for the selection and evaluation of phenotypic characteristics and identification keys.

The employment of commercially available kits can provide useful shortcuts for undertaking many routine identification tests. Such kits have been developed for testing a wide range of bacteria but must be carefully assessed with regard to suitability of each test to be performed. If there is any doubt about the reliability of results, they should be confirmed by independent testing methods. Many of the commercially available testing systems are designed to be used with dichotomous keys, diagnostic tables, or computer-based identification systems. Most of these kits are intended for specific purposes, such as the identification of common medically important bacteria, and may not be specifically suited to the identification of *Clostridium* species.

When classical phenotypic characterization tests are being undertaken for the purposes of describing or verifying a clostridial species, it is necessary to perform the full range of standard procedures. The results obtained with the new strain are then compared with either those of the type culture or some other firmly established and well-characterized strain. On the other hand, where phenotypic tests are being used for routine diagnostic or identification purposes, it is desirable that as few properties as practical should be tested to achieve a reliable outcome. The particular characteristics that are selected for testing should be stable, easy to test, common to all strains, and provide a high degree of discrimination. The availability of simple, rapid, reliable, and inexpensive tests that can provide the required degree of differentiation are invaluable for applications of this type.

1.4.2 Chemotaxonomic Techniques

The development of rapid and reliable analytical techniques in chemistry and molecular biology have provided a wealth of new information regarding the chemical composition of bacterial cells. This complementary knowledge has assisted in expanding the foundations on which modern bacterial classification is based. A wide range of analytical approaches have been employed in chemosystematics. At the level of the whole cell, mass pyrolysis coupled with analysis by gas liquid or mass spectrometry has been utilized. Analysis of cellular proteins include amino acid sequence determination, electrophoretic pattern and enzyme pattern analysis, along with serological comparisons. Analysis of bacterial cell wall composition include the determination of peptidoglycan structure as well as analysis of polysaccharides, teichoic acids, and amino acids. Determination of cellular fatty acids and lipids includes analysis of membrane fatty acids, polar lipids, mycolic acids, and isopreniod respiratory quinones. One of the limitations in utilizing chemosystematic data for classification purposes is the effect that variation in environmental conditions can have on the chemical constitution of the cell. Bacteria can modify the chemical composition of many classes of macromolecules to a significant extent in response to environmental fluctuations. A further factor that has detracted from widespread application of these techniques is the specialized equipment and expertise, and the time involved.

Within the genus *Clostridium* chemosystematic analysis has, in a number of instances, proved to be a valuable supplement to the traditional approaches used for classification and identification and has contributed to the establishment of a natural phylogenetic classification for members of the genus. The contribution that chemotaxonomic analysis has made to clostridial taxonomy has been mostly within the higher taxonomic groupings. Their usefulness as a discriminatory technique at the level of the species and strains is more limited, but such analyses have proven useful in certain specific applications. Pyrolysis mass spectrometry has been used for determining differences between certain strains of solvent-producing clostridia [58]. There have also been a number of taxonomic studies undertaken on various species of *Clostridium* employing the analysis of peptidoglycans [24,59–61]. The analysis of clostridial membrane phospholipids has also been extensively used in taxonomic studies [62–65]. For detailed information, refer to Chapter 15.

1.4.3 Biotyping Techniques

A number of other identification techniques have potential applications for species and strain identification. These include bacteriocin typing, bacteriophage typing, and serotyping, which have been developed and used with varying effectiveness for a variety of bacteria.

Two approaches have been used for undertaking bacteriocin typing. The first involves testing the ability of the strain under investigation to produce bacteriocins against a standard set of indicator strains. The second approach involves testing the susceptibility of the strain under investigation to a standard set of bacteriocins from known producer strains. Assays can be conducted in both liquid and solid culture. The most common technique used is to spot a small volume of a dilution series of broth containing the bacteriocin onto a lawn of the test strain plated out on solid culture media. Following incubation, the bacteriocin activity is assessed by inhibition of growth within the spot. An alternative option is to cut wells into the agar plate and add the bacteriocin dilution series to the wells and test by observing for zones of inhibition surrounding the wells. Plating out intersecting streaks of producer and sensitive strains onto agar plates and checking for inhibition also provides a simple method of checking for bacteriocin production. One of the main impediments to developing bacteriocin typing schemes is that bacteriocins tend to be active against a very broad range of strains, species, and in some cases even genera, which places limits on the discrimination that can be achieved using such typing schemes. Although bacteriocins are produced by individual strains belonging to many species of *Clostridium,* there have been few attempts to develop bacteriocin typing schemes for any clostridial species. A comprehensive coverage on this topic can be found in a review by Mahony [66].

Bacteriophage typing involves a similar strategy. The most commonly used techniques involve the use of lytic phages that are inoculated by dropping dilutions containing the phage onto freshly seeded bacterial lawns. Bacterial lawns can be produced by several methods including swabbing, inoculating with an excess of liquid culture, draining, and allowing the plates to dry, or adding cells to molten agar and using it to cover the surface of an agar plate with a thin overlay. Plates are then incubated for 15 to 48 hours, and a positive result is signified by the presence of zones of clearing in an otherwise uniform layer of bacterial growth. Although the overlay method is the method of choice for clostridial cells, problems can be encountered through the disruption of the agar overlay by vigorous gas production.

Phage typing schemes have been investigated for use with pathogenic species of clostridia such as *C. perfringens* [66] and *C. difficile* [67], as well as solvent-producing species [14]; however the reactions between bacteriophages and bacteria in most cases appeared to be extremely specific. In general, among the clostridia, individual phages commonly appear to only lyse strains belonging to the same DNA fingerprint group or taxospecies [14,66]. The narrow host range exhibited by most clostridial phages coupled with limited knowledge of their biology mitigates against the development of effective phage typing schemes for routine application with clostridial species. (Refer to Chapter 31 for further information.)

Serotyping techniques, including whole cell agglutinations, latex agglutination, ELISA, and direct and indirect fluorescent antibody tests utilizing both polyclonal and monoclonal antibodies, have been used extensively for the identification of bacterial cells or subcellular components. One of the main advantages provided by these serological methods is speed, but cross-reactions and other complexities can make results difficult to interpret. Serotyping has been used with some limited success for the differentiation of pathological clostridia (see review by Mahony [66]).

1.5 CONCLUSIONS

Before a basis for phylogenetic coherence was established, many clostridial species remained inadequately characterized and incorrectly classified, creating ongoing taxonomic difficulties. As new knowledge and understanding of evolutionary relationships based on 16S rRNA sequence homology has unfolded, it has become apparent there are many instances where genotypic properties of a group correlate relatively poorly with distribution of classical phenotypic properties used for identification. As such, phenotypic characteristics on their own are relatively poor discriminators of clostridial species. On the other hand, once genetic relatedness has been established, it is then possible to select with some confidence those phenotypic characteristics and tests that can best assist with rapid and reliable identification. This enables classical microbiological testing of phenotypic characteristics to be applied with much greater reliability. Where species identification cannot be determined by 16S rRNA sequence homology alone, then DNA/DNA hybridization — and genomic DNA fingerprinting in particular — has provided valuable assessment of the degree of similarity of genomic DNA. These techniques have proved to be very useful and effective for strain identification purposes.

REFERENCES

1. Austin, B. and Priest, F., *Modern Bacterial Taxonomy,* Van Nostrand Reinhold, Wokingham, UK, 1986.
2. Stackebrandt, E., Phylogenetic relationships vs. phenotypic diversity: how to achieve a phylogenetic classification system of the eubacteria, *Can. J. Microbiol.,* 34, 552, 1988.
3. Stackebrandt, E. and Hippe, H., Taxonomy and systematics, in *Clostridia. Biotechnology and Medical Applications*, Bahl, H. and Dürre, P., Eds., Wiley-VCH, Weinheim, Germany, 2001, 19.
4. Cato, E.P. and Stackebrandt, E., Taxonomy and phylogeny, in *Clostridia. Biotechnology Handbooks*, Minton, N.P. and Clarke, D.J., Eds., Plenum Press, New York, 1989, 1.
5. Collins, M.D., Lawson, P.A., Willems, A., Cordoba, J.J., Fernandez-Garayzabal, J., Garcia, P., Cai, J., Hippe, H., and Farrow, F.A.E., The phylogeny of the genus *Clostridium*: proposal of five new genera and eleven new species combinations, *Int. J. Syst. Bacteriol.,* 44, 812, 1994.
6. Stackebrandt, E. and Rainey, F.A., Phylogenetic relationships, in *The Clostridia: Molecular Biology and Pathogenesis*, Rood, J.I., McClane, B.A., Songer, J.G., and Titball, R.W., Eds., Academic Press, New York, 1997, 1.
7. Stackebrandt, E., Kramer, I., Swiderski, J., and Hippe, H., Phylogenetic basis for a taxonomic dissection of the genus *Clostridium, FEMS Immunol. Med. Microbiol.,* 24, 253, 1999.
8. Prazmowski, A., *Untersuchung über die Entwicklungsgeschichte und Fermentwirkung einiger Bacterien-Arten*, Inaugural Dissertation, 1880.
9. Morris, J.G., History and future potential of the clostridia in biotechnology, in *The Clostridia and Biotechnology,* Woods, D.R., Eds., Butterworth-Heinemann, Boston, 1993, 1.
10. Dürre, P., From Pandora's box to cornucopia: clostridia — a historical perspective, in *Clostridia. Biotechnology and Medical Applications*, Bahl, H. and Dürre, P., Eds., Wiley-VCH, Weinheim, Germany, 2001, 1.
11. Cato, E.P., George, W.L., and Finegold, S.M., Genus *Clostridium*, in *Bergey's Manual of Systematic Bacteriology,* Sneath, P.H.A., Mair, N.S., Sharpe, M.E., and Holt, J.G., Eds., Williams & Wilkins, Baltimore, 1986, 1141.
12. Johnson, J.L. and Francis, B.S., Taxonomy of the clostridia: ribosomal ribonucleic acid homologies among the species, *J. Gen. Microbiol.,* 88, 229, 1975.
13. Stackebrandt, E. and Goebel, B.M., Taxonomic note: a place for DNA-DNA reassociation and 16S rRNA sequence analysis in the present species definition in bacteriology, *Int. J. Syst. Bacteriol.,* 44, 846, 1994.
14. Keis, S., Bennett, C.F., Ward, V.K., and Jones, D.T., Taxonomy and phylogeny of industrial solvent-producing clostridia, *Int. J. Syst. Bacteriol.,* 45, 693, 1995.
15. Keis, S., Shaheen, R., and Jones, D.T., Emended descriptions of *Clostridium acetobutylicum* and *Clostridium beijerinckii*, and descriptions of *Clostridium saccharoperbutylacetonicum* sp. nov. and *Clostridium saccharobutylicum* sp. nov., *Int. J. Syst. Evol. Microbiol.,* 51, 2095, 2001.
16. Johnson, J.L., Determination of DNA base composition, in *Methods in Microbiology,* Gottschalk, G., Ed., Academic Press, London, 1985, 1.

17. Johnson, J.L., DNA reassociation and RNA hybridisation of bacterial nucleic acids, in *Methods in Microbiology*, Gottschalk, G., Ed., Academic Press, London, 1985, 34.
18. Johnson, J.L., Nucleic acid hybridization: principles and techniques, in *Nucleic Acids and Monoclonal Antibody Probes*, Swaminathan, B. and Prakash, G., Eds., Marcel Dekker, Inc., New York, 1989, 3.
19. Stackebrandt, E. and Goodfellow, M., *Nucleic Acid Techniques in Bacterial Systematics,* John Wiley & Sons Ltd., Chichester, UK, 1991.
20. Johnson, J.L., Similarity analysis of DNAs, in *Methods for General and Molecular Bacteriology*, Gerhardt, P., Ed., American Society for Microbiology, Washington, D.C., 1994, 655.
21. Lane, D.J., Pace, B., Olsen, G.J., Stahl, D.A., Sogin, M.L., and Pace, N.R., Rapid determination of 16S ribosomal RNA sequences for phylogenetic analysis, *Proc. Natl. Acad. Sci. U.S.A.*, 82, 6955, 1985.
22. Andreesen, J.R., Bahl, H., and Gottschalk, G., Introduction to the physiology and biochemistry of the genus *Clostridium*, in *Clostridia. Biotechnology Handbooks, Vol. 3*, Minton, N.P. and Clarke, D.J., Eds., Plenum Press, New York, 1989, 1.
23. Johnson, J.L., Relationship of deoxyribonucleic acid homologies to cell wall structure, *Int. J. Syst. Bacteriol.*, 20, 421, 1970.
24. Cummins, C.S. and Johnson, J.L., Taxonomy of the clostridia: wall composition and DNA homologies in *Clostridium butyricum* and other butyric acid-producing clostridia, *J. Gen. Microbiol.*, 67, 33, 1971.
25. Johnson, J.L., Use of nucleic-acid homologies in the taxonomy of anaerobic bacteria, *Int. J. Syst. Bacteriol.*, 23, 308, 1973.
26. Owen, R.J. and Pitcher, D., Current methods for estimating DNA base composition and levels of DNA-DNA hybridization, in *Chemical Methods in Bacterial Systematics*, Goodfellow, M. and Minnikin, D.E., Eds., Academic Press, London, 1985, 67.
27. Brenner, D.J., Fanning, G.R., Rake, A.V., and Johnson, K.E., Batch procedure for thermal elution of DNA from hydroxyapatite, *Anal. Biochem.*, 28, 447, 1969.
28. Matteuzzi, D., Hollaus, F., and Biavati, B., Proposal of neotype for *Clostridium thermohydrosulfuricum* and the merging of *Clostridium tartarivorum* with *Clostridium thermosaccharolyticum*, *Int. J. Syst. Bacteriol.*, 28, 528, 1978.
29. Kaneuchi, C., Miyazato, T., Shinjo, T., and Mitsuoka, T., Taxonomic study of helically coiled, sporeforming anaerobes isolated from the intestines of humans and other animals: *Clostridium cocleatum* sp. nov. and *Clostridium spiroforme* sp. nov., *Int. J. Syst. Bacteriol.*, 29, 1, 1979.
30. Johnson, J.L., Toth, J., Santiwatanakul, S., and Chen, J.-S., Cultures of "*Clostridium acetobutylicum*" from various collections comprise *Clostridium acetobutylicum*, *Clostridium beijerinckii*, and two other distinct types based on DNA-DNA reassociation, *Int. J. Syst. Bacteriol.*, 47, 420, 1997.
31. Smith, C.L. and Cantor, C.R., Preparation and manipulation of large DNA molecules, *Trends Biochem. Sci.*, 12, 284, 1987.
32. Smith, C.L. and Cantor, C.R., Purification, specific fragmentation and separation of large DNA molecules, *Methods Enzymol.*, 155, 449, 1987.
33. Wilkinson, S.R. and Young, M., Physical map of the *Clostridium beijerinckii* (formerly *Clostridium acetobutylicum*) NCIMB 8052 chromosome, *J. Bacteriol.*, 177, 439, 1995.
34. Keis, S., Sullivan, J.T., and Jones, D.T., Physical and genetic map of the *Clostridium saccharobutylicum* (formerly *Clostridium acetobutylicum*) NCP 262 chromosome, *Microbiology,* 147, 1909, 2001.
35. Katayama, S., Dupuy, B., Daube, G., China, B., and Cole, S.T., Genome mapping of *Clostridium perfringens* strains with I-*Ceu*I shows many virulence genes to be plasmid-borne, *Mol. Gen. Genet.*, 251, 720, 1996.
36. Chachaty, E., Saulnier, P., Martin, A., Mario, N., and Andremont, A., Comparison of ribotyping, pulsed-field gel electrophoresis and random amplified polymorphic DNA for typing *Clostridium difficile* strains, *FEMS Microbiol. Lett.*, 122, 61, 1994.
37. Hielm, S., Björkroth, J., Hyytiä, E., and Korkeala, H., Genomic analysis of *Clostridium botulinum* group II by pulsed-field gel electrophoresis, *Appl. Environ. Microbiol.*, 64, 703, 1998.
38. Hielm, S., Björkroth, J., Hyytiä, E., and Korkeala, H., Prevalence of *Clostridium botulinum* in Finnish trout farms: pulsed-field gel electrophoresis typing reveals extensive genetic diversity among type E isolates, *Appl. Environ. Microbiol.*, 64, 4161, 1998.
39. Samore, M.H., Kristjánsson, M., Venkataraman, L., DeGirolami, P.C., and Arbeit, R.D., Comparison of arbitrarily-primed polymerase chain reaction, restriction enzyme analysis and pulsed-field gel electrophoresis for typing *Clostridium difficile*, *J. Microbiol. Methods,* 25, 215, 1996.

40. Blaschek, H.P. and Klacik, M.A., Role of DNase in recovery of plasmid DNA from *Clostridium perfringens*, *Appl. Environ. Microbiol.,* 48, 178, 1984.
41. Kristjánsson, M., Samore, M.H., Gerding, D.N., DeGirolami, P.C., Bettin, K.M., Karchmer, A.W., and Arbeit, R.D., Comparison of restriction endonuclease analysis, ribotyping, and pulsed-field gel electrophoresis for molecular differentiation of *Clostridium difficile* strains, *J. Clin. Microbiol.,* 32, 1963, 1994.
42. Lin, W.-J. and Johnson, E.A., Genome analysis of *Clostridium botulinum* type A by pulsed-field gel electrophoresis, *Appl. Environ. Microbiol.,* 61, 4441, 1995.
43. Gibson, J.R., Sutherland, K., and Owen, R.J., Inhibition of DNAse activity in PFGE analysis of DNA from *Campylobacter jejuni*, *Lett. Appl. Microbiol.,* 19, 357, 1994.
44. Matushek, M.G., Bonten, M.J.M., and Hayden, M.K., Rapid preparation of bacterial DNA for pulsed-field gel electrophoresis, *J. Clin. Microbiol.,* 34, 2598, 1996.
45. Zappe, H., Jones, D.T., and Woods, D.R., Cloning and expression of *Clostridium acetobutylicum* endoglucanase, cellobiase and amino acid biosynthesis genes in *Escherichia coli*, *J. Gen. Microbiol.,* 132, 1367, 1986.
46. Wilkinson, S.R. and Young, M., Wide diversity of genome size among different strains of *Clostridium acetobutylicum*, *J. Gen. Microbiol.,* 139, 1069, 1993.
47. Kato, H., Kato, N., Watanabe, K., Ueno, K., Ushijima, H., Hashira, S., and Abe, T., Application of typing by pulsed-field gel electrophoresis to the study of *Clostridium difficile* in a neonatal intensive care unit, *J. Clin. Microbiol.,* 32, 2067, 1994.
48. Talon, D., Bailly, P., Delmee, M., Thouverez, M., Mulin, B., Iehl-Robert, M., Cailleaux, V., and Michel-Briand, Y., Use of pulsed-field gel electrophoresis for investigation of an outbreak of *Clostridium difficile* infection among geriatric patients, *Eur. J. Clin. Microbiol. Infect. Dis.,* 14, 987, 1995.
49. Ridell, J., Björkroth, J., Eisgrüber, H., Schalch, B., Stolle, A., and Korkeala, H., Prevalence of the enterotoxin gene and clonality of *Clostridium perfringens* strains associated with food poisoning outbreaks, *J. Food Prot.,* 61, 240, 1997.
50. Schalch, B., Sperner, B., Eisgrüber, H., and Stolle, A., Molecular methods for the analysis of *Clostridium perfringens* relevant to food hygiene, *FEMS Immunol. Med. Microbiol.,* 24, 281, 1999.
51. Montoya, D., Arévalo, C., Gonzales, S., Aristizabal, F., and Schwarz, W.H., New solvent-producing *Clostridium* sp. strains, hydrolyzing a wide range of polysaccharides, are closely related to *Clostridium butyricum*, *J. Ind. Microbiol. Biotechnol.,* 27, 329, 2001.
52. Canard, B. and Cole, S.T., Genome organization of the anaerobic pathogen *Clostridium perfringens*, *Proc. Natl. Acad. Sci. U.S.A.,* 86, 6676, 1989.
53. Tenover, F.C., Arbeit, R.D., Goering, R.V., Mickelsen, P.A., Murray, B.E., Persing, D.H., and Swaminathan, B., Interpreting chromosomal DNA restriction patterns produced by pulsed-field gel electrophoresis: criteria for bacterial strain typing, *J. Clin. Microbiol.,* 33, 2233, 1995.
54. Shimizu, T., Ohshima, S., Ohtani, K., Shimizu, T., and Hayashi, H., Genomic map of *Clostridium perfringens* strain 13, *Microbiol. Immunol.,* 45, 179, 2001.
55. Smith, L.D.S. and Hobbs, G., Genus III. *Clostridium* Prazmowski 1880, in *Bergey's Manual of Determinative Bacteriology*, Buchanan, R.E. and Gibbons, N.E., Eds., Williams & Wilkins, Baltimore, 1974, 551.
56. Holdeman, L.V., Cato, E.P., and Moore, W.E.C., *Anaerobe Laboratory Manual*, 4th ed., Virginia Polytechnic Institute and State University, Blacksburg, VA, 1977.
57. Smibert, R.M. and Krieg, N.R., Phenotypic characterisation, in *Methods for General and Molecular Bacteriology*, Gerhardt, P., Ed., American Society for Microbiology, Washington, D.C., 1994, 607.
58. Wilkinson, S.R., Young, M., Goodacre, R., Morris, J.G., Farrow, J.A.E., and Collins, M.D., Phenotypic and genotypic differences between certain strains of *Clostridium acetobutylicum*, *FEMS Microbiol. Lett.,* 125, 199, 1995.
59. Cummins, C.S., Cell wall composition in the classification of Gram-positive anaerobes, *Int. J. Syst. Bacteriol.,* 20, 413, 1970.
60. Schleifer, K.H. and Kandler, O., Peptidoglycan types of bacterial cell walls and their taxonomic implications, *Bacteriol. Rev.,* 36, 407, 1972.
61. Weiss, N., Schleifer, K.H., and Kandler, O., The peptidoglycan types of Gram-positive anaerobic bacteria and their taxonomic implications, *Rev. Inst. Pasteur (Lyon),* 14, 3, 1981.

62. Baumann, N.A., Hagen, P.-O., and Goldfine, H., Phospholipids of *Clostridium butyricum*. Studies on plasmalogen composition and biosynthesis, *J. Biol. Chem.,* 240, 1559, 1965.
63. Vollherbst-Schneck, K., Sands, J.A., and Montenecourt, B.S., Effect of butanol on lipid composition and fluidity of *Clostridium acetobutylicum* ATCC 824, *Appl. Environ. Microbiol.,* 47, 193, 1984.
64. Oulevey, J., Bahl, H., and Thiele, O.W., Novel alk-1-enyl ether lipids isolated from *Clostridium acetobutylicum*, *Arch. Microbiol.,* 144, 166, 1986.
65. Johnston, N.C. and Goldfine, H., Isolation and characterization of a novel four-chain ether lipid from *Clostridium butyricum*: the phosphatidylglycerol acetal of plasmenylethanolamine, *Biochim. Biophys. Acta,* 961, 1, 1988.
66. Mahony, D.E., Bacteriocin, bacteriophage and other epidemiological typing methods for the genus *Clostridium*, in *Methods in Microbiology*, Bergan, T. and Norris, J.R., Eds., Academic Press, London, 1979, 1.
67. Sell, T.L., Schaburg, D.R., and Fekety, D.R., Bacteriophage and bacteriocin typing scheme for *Clostridium difficile*, *J. Clin. Microbiol.,* 17, 1148, 1983.

2 Quantitative Proteome Analysis of Clostridia

Steffen Schaffer, Kai Thormann, Sandra Hujer, and Peter Dürre

CONTENTS

2.1 INTRODUCTION

Within the past decade molecular biology has gone through a transition from being an essentially reductionistic science focusing on one or few molecules at a time to a global science analyzing simultaneously whole classes of molecules, such as mRNA or protein pools. This paradigm shift has been initially triggered by the development of automated DNA sequencing methods, which allowed determination of the DNA sequence of entire bacterial genomes. Meanwhile, these techniques have been developed to a point, where deciphering eukaryal genomes is a feasible goal as has recently been demonstrated by sequencing the human genome.

The availability of complete genome sequences is an important prerequisite for being able to analyze the mRNA and protein pools of a given cell at a global scale. Analysis of the transcriptome, the sum of all mRNAs present in a cell at a given time, is usually performed by using DNA micro- or macroarrays containing thousands of oligonucleotides or PCR products representing all genes predicted to be present in a given organism. Simultaneous analysis of thousands of mRNAs is simplified by their similar physicochemical properties, which are relatively independent of sequence, and the fact that different mRNA species are thought to differ in abundance by not more than three orders of magnitude. Though many technical and scientific issues remain, gene expression profiling can be regarded as an established technique able of characterizing transcriptomes quite comprehensively.

In contrast, analysis of the proteome, the sum of all proteins present in a cell at a given time, is far more tedious as it requires analysis of thousands of different types of molecules with vastly different physicochemical properties and abundances (6 to 7 orders of magnitude; up to 10 orders of magnitude in human serum). Even the gene product derived from one given gene can come in a multitude of forms as alternative messenger splicing and post-translational modifications, most importantly phosphorylation and glycosylation events, are important regulatory tools. Moreover, as no amplification method for proteins analogous to polymerase chain reaction is available, researchers face the problem of limited amounts of sample material.

The workhorse of proteome analysis, if understood as global analysis of protein abundance, is two-dimensional gel electrophoresis. While the principles of this method are almost 30 years old [1], the method has been considerably improved in terms of reproducibility and ease of handling with the introduction of immobilized pH gradients (for a recent review see [2]). Even more progress has been made concerning the identification of proteins separated by 2D-PAGE. Mass spectrometry methods such as ESI-MS and MALDI-TOF-MS are modern high-throughput methods allowing the identification of proteins from femtomole amounts of sample. MALDI-TOF-MS-based peptide mass fingerprinting is the method of choice for high-throughput identification of proteins but *a priori* requires the knowledge of the genome sequence of the studied organism.

In this chapter, we will review proteomic technologies and applications of proteome analysis, most notably 2D-PAGE, for the analysis of clostridial physiology. Finally, some remarks will be made with respect to the scientific aims proteomic technology can be used for in the future and how the picture of proteomics, with respect to methodology and scientific applications, might change in the next few years.

2.2 MATERIALS AND METHODS

2.2.1 *In Vivo*-Labeling of Proteins with L-[^{35}S]Methionine

For labeling of newly synthesized proteins in *C. acetobutylicum*, 5 ml of culture are transferred to anaerobized 10-ml tubes and incubated for 3 min at 37°C in the presence of 10 μCi/ml L-[^{35}S]methionine (Amersham Pharmacia Biotech GmbH, Freiburg, Germany). Labeling is stopped by the addition of 1 ml stop solution (10 m*M* L-methionine, 1 mg/ml chloramphenicol in 10 m*M* Tris-HCl, 1 m*M* EDTA, pH 7.5) and setting the cells on wet ice. Cells are washed once with TE buffer (10 m*M* Tris-HCl, 1 m*M* EDTA, pH 7.5), suspended in 1.5 ml TE with 10 m*M* PMSF, and broken by sonication. Intact cells and cell debris is removed by centrifugation (20,000 × *g* for 30 min at 4°C) and aliquots of the cell-free extract containing $2 \cdot 10^6$ cpm are vacuum-dried and stored at –20°C [3].

2.2.2 Sample Preparation

For the preparation of unlabeled cell-free extracts, the cell suspensions with protease inhibitor (Complete Protease Inhibitor; Roche Diagnostics, Mannheim, Germany) and DNase/RNase solution (0.1 volume of DNase/RNase solution: 1 mg/ml DNase I, 0.25 mg/ml RNase A in 50 m*M* $MgCl_2$)

added are disrupted by five passages through a French pressure cell (SLM Instruments, Inc., Rochester, New York) at 12.5 Mpa, and the lysate is cleared by centrifugation (20,000 × *g* for 30 min at 4°C). Alternatively, cells can be broken by adding up to 2 ml of cell suspension to 1 g of zirconium-silica beads (diameter 0.2 mm; Roth, Karlsruhe, Germany) and three cycles of 30 sec bead-beating in a Silamat S5 device (Vivadent, Ellwangen, Germany) with intermittent cooling on ice for 2 min. Aliquots containing 100 and 500 µg protein for running analytical and preparative gels, respectively, are vacuum-dried and stored at –20°C. In case results obtained during IEF are unsatisfactory due to sample impurities (salts, metabolites, polymers, etc.), proteins can be precipitated by adding 9 volumes of acetone, overnight incubation at –20°C, centrifugation (15,000 × *g* for 10 min at 4°C), and air drying of the precipitated proteins prior to solubilization.

2.2.3 Isoelectric Focusing

Aliquots are solubilized in 400 µl rehydration buffer [8 *M* urea, 2 *M* thiourea, 1% (w/v) CHAPS, 20 m*M* DTT, 0.5% (v/v) Pharmalyte 3 to 10 (Amersham Pharmacia Biotech GmbH, Freiburg, Germany)] and used to rehydrate 18 cm Immobiline DryStrips with the desired p*I* range (Amersham Pharmacia Biotech GmbH, Freiburg, Germany), for 24 h. First-dimension electrophoresis is performed using a Multiphor II electrophoresis unit (Amersham Pharmacia Biotech GmbH, Freiburg, Germany) according to the recommendations of the manufacturer with the following electrophoresis parameters: 0 to 500 V gradient for 500 Vh, 500 V for 2500 Vh, 500 to 3500 V gradient for 10 kVh, 3500 V for 35 kVh. Alternatively, IEF can be conducted in an IPGphor IEF system (Amersham Pharmacia Biotech GmbH, Freiburg, Germany). In this case, only 350 µl of solubilized sample are required and rehydration of the Immobiline DryStrips is performed during the first two steps of electrophoresis. The total run has the following voltage profile: 6 h at 30 V, 6 h at 60 V, 1 h at 200 V, 1 h at 500 V, 1 h at 2000 V, 30 min gradient to 8000 V, 8000 V to a total of 70 to 80 kVh (wide pH gradients) or 150 to 180 kVh (narrow pH gradients). After finishing IEF, first-dimension gels can be stored in sealed Petri dishes at –70°C.

2.2.4 SDS-PAGE

In principle, both the use of homogeneous self-cast gels as well as of commercially available gradient gels lead to acceptable results. Each of the methods has its own advantages and drawbacks. While reproducibility of the separation pattern, as well as spot resolution and ease of handling, is superior when using precast gels, large proteins tend to be under-represented on such gels, at least if horizontal electrophoresis using the Multiphor II system is used. In contrast, large proteins can be detected on homemade gels, but there more care has to be taken to obtain reproducible spot patterns and spot resolution is lower.

When using self-cast gels, second-dimension electrophoresis is performed using a vertical electrophoresis apparatus fitting two homogeneous 12% polyacrylamide gels with dimensions 174 × 240 × 1 mm, the active separation length being reduced by a 4% stacking gel of 2 cm height (the use of a stacking gel increased spot resolution in our hands). To perform the second-dimension electrophoresis, the Immobiline DryStrip is equilibrated in equilibration solution A [50 m*M* Tris-HCl pH 6.8, 6 *M* urea, 30% (v/v) glycerol, 4% (w/v) SDS, 3.5 mg/ml DTT] and B (as equilibration solution A, except for omission of DTT and addition of 45 mg/ml iodoacetamide and a few grains of bromophenol blue) for exactly 15 min each. The equilibrated strip has to be trimmed to 16 cm length to fit the second-dimension gels (limiting the displayed p*I* range to 4 to 6.2) and is then placed on top of the second-dimension gels and held in place with embedding agarose [0.5% (w/v) agarose in 62.5 m*M* Tris-HCl pH 6.8, 4% (w/v) SDS]. After filling the electrophoresis apparatus with cold electrode buffer [50 m*M* Tris, 190 m*M* glycine, 0.1% (w/v) SDS], two gels are run in parallel at 4°C for 16 h with 2.5 W per gel.

Second-dimension separation using ready-made gradient gels is performed using Excel SDS gradient gels (12 to 14%) according to the recommendations of the manufacturer (Amersham Pharmacia Biotech GmbH, Freiburg, Germany).

2.2.5 Staining, Processing, and Analysis of Second-Dimension Gels

Nonradioactive analytical gels are silver-stained according to the protocol described by Blum et al. [4]. Nonradioactive preparative gels (homogeneous 12% polyacrylamide gels with 1 mm thickness) are stained with 0.2% Coomassie R-250 in an ethanol/glacial acetic acid/water mixture (70:25:5) and destained with the same solution without Coomassie.

Excel SDS gradient gels (12 to 14%) can alternatively be stained with colloidal Coomassie, resulting in lower detection limits (20 to 50 ng protein per spot). For unknown reasons, this method did not work with self-cast gels. Excel SDS gradient gels are fixed after electrophoresis for 1 h in freshly made fixing solution [1 part of methanol plus two parts of 25% (w/v) ammonium sulphate, 3% (v/v) phosphoric acid]. Subsequently, 1 ml Coomassie G-250 solution (10 mg/ml in water) is added, and the gels are stained overnight. The destaining procedure comprises three washes with water, 20 min each, a brief wash in 30% (v/v) methanol, and a final wash in water to remove residual methanol. Gels can then be stored in water or dried.

Radioactive gels are fixed in a methanol/glacial acetic acid/water mixture (50:10:40) and dried for 90 min at 70°C on a conventional gel dryer. Dried gels are exposed to BASIIIs Imaging Plates and analyzed on a BAS 1000 Phosphoimager [Fuji Photo Film (Europe) GmbH, Düsseldorf, Germany] using the Image Reader software, vers. 1.1, and MacBas software, vers. 2.3 [Fuji Photo Film (Europe) GmbH, Düsseldorf, Germany].

2.2.6 Concentration of Proteins from Second-Dimension Gels and N-Terminal Sequencing of Proteins

If protein spots from several gels have to be combined in order to acquire sufficient material for N-terminal microsequencing, this can be done according to the protocol of Rider et al. [5], except that the horizontal concentration effect by removing and reinserting spacers is omitted in our experiments. Transfer of concentrated protein bands onto Biotrace PVDF membranes (Gelman Sciences, Ann Arbor, Michigan) and subsequent Coomassie R-250-staining of PVDF membranes is performed following standard protocols and N-terminal microsequencing used in order to determine the sequences.

In case of clostridial species with their genome sequence determined and available in the public domain, protein identification is routinely performed by mass spectrometric methods. MALDI-TOF peptide mass fingerprinting requires only minute amounts of protein (femtomole to low picomole amounts) and can be adapted to high-throughput applications. Protein spots of interest are excised from destained Coomassie-stained gels with Pasteur pipettes and subjected to in-gel digestion [6]. Gel pieces are washed twice with 350 µl 0.1 *M* ammonium bicarbonate in 30% (v/v) acetonitrile, 10 min each wash. The destained and shrunken gel pieces are then vacuum-dried for 20 min in a conventional vacuum centrifuge and subsequently rehydrated with 1 µl 3 m*M* Tris-HCl, pH 8.8, containing 10 ng trypsin per µl. After 20 min, 2 µl 3 m*M* Tris-HCl, pH 8.8, without trypsin is added. Digestion is allowed to proceed overnight at room temperature. Peptides are then extracted by sequential addition of 6 µl water and 5 µl 0.1% (v/v) TFA in 30% (v/v) ACN. 0.5 µl of the resulting peptide solution is mixed on a stainless steel sample plate with 0.5 µl of a saturated α-cyano-4-hydroxy-*trans* cinnamic acid solution in 50% (v/v) ACN, 0.25% (v/v) TFA.

In cases where signal intensity of unpurified peptide mixes is unsatisfactory, the samples can be concentrated and desalted using ZipTip C_{18} (Millipore, Bedford, Massachusetts). Peptides are directly eluted into MALDI matrix solution.

2.2.7 Mass Spectrometry

Mass spectrometric analysis by peptide mass fingerprinting (PMF) and post-source decay (PSD) analysis is described as performed with a Voyager-DE™ STR MALDI-TOF mass spectrometer (Applied Biosystems, Weiterstadt, Germany).

100-well sample plates are calibrated using the OptiPlate tool of the Voyager Control Panel software, version 5.0 (Applied Biosystems, Weiterstadt, Germany). Calibration mixtures 1 and 2 of the Sequazyme Peptide Mass Standard Kit (Applied Biosystems, Weiterstadt, Germany) spotted on the reference position of the sample plate are used as calibrants. Samples can be analyzed manually or automatically in positive reflector mode with 20 kV accelerating voltage, 63% grid voltage, and the delay time set at 125 ns. Data acquisition and analysis is performed using the Voyager Control Panel software, version 5.0 and the Voyager Data Explorer software, version 3.5 (Applied Biosystems, Weiterstadt, Germany).

In order to assign the identity of individual peptides in cases where PMF led to ambiguous results, selected peptides can be subjected to PSD analysis. Typically, angiotensin is used as calibration substance, and peptides in the 1000 to 2000 Da mass range are isolated using a precursor ion selector window of +/– 10 Da. Fragmentation is induced using high laser energy, and fragment mass spectra are acquired in 12 segments with a decrement ratio of 0.75. The instrument parameters are as described for peptide mass fingerprints, except setting the grid voltage at 75% and delay time at 150 ns. The expected amino acid sequence of the analyzed peptide is then mapped onto the obtained PSD spectrum using the ion fragmentation calculator tool of the Voyager Data Explorer software.

2.2.8 Database Searches

The genome sequences of *C. acetobutylicum*, *C. perfringens*, and *C. tetani* are available in the public domain. Consequently, mass lists derived from PMF analyses can be used to search public databases. The most commonly used online tools for that purpose are MS-Fit (http://prospector.ucsf.edu/ucsfhtml4.0/msfit.htm), Peptident (http://www.expasy.ch/tools/peptident.html), and Profound (http://prowl. rockefeller.edu/cgi-bin/Profound). This subject has recently been reviewed by Fenyö [7].

2.3 ANALYZING CLOSTRIDIAL PHYSIOLOGY USING 2D-PAGE

2.3.1 General Remarks

Two-dimensional electrophoresis is still the most commonly used proteomic separation technology. However, it has to be kept in mind that it suffers from a variety of limitations. Due to the fact that only non- or zwitterionic detergents are compatible with isoelectric focusing, many hydrophobic proteins do not become solubilized during sample preparation or, if so, tend to precipitate during focusing, especially at pH values near p*I*, as the absence of charges favors hydrophobic interactions between proteins and between proteins and the acrylamide matrix. Therefore, these proteins are vastly under-represented on 2D gels [8]. The separation of proteins with a p*I* > 7 is also limited. This is due to a positive correlation between hydrophobicity and high p*I* and electroendoosmosis during isoelectric focusing, resulting in dehydration of the basic end of IEF gels and subsequent precipitation of basic proteins [9,10]. Finally, global analyses of protein abundance have shown that 2D-PAGE is only capable of displaying proteins simultaneously that differ in abundance by 2 to 3 orders of magnitude [3,11]. However, in the living cell, proteins are thought to differ in abundance by at least 6 or 7 orders of magnitude, implicating that it is very difficult, if not impossible, to display rare proteins (such as proteins involved in signal transduction) together with highly abundant proteins (such as enzymes of the central metabolic pathways or involved in protein biogenesis).

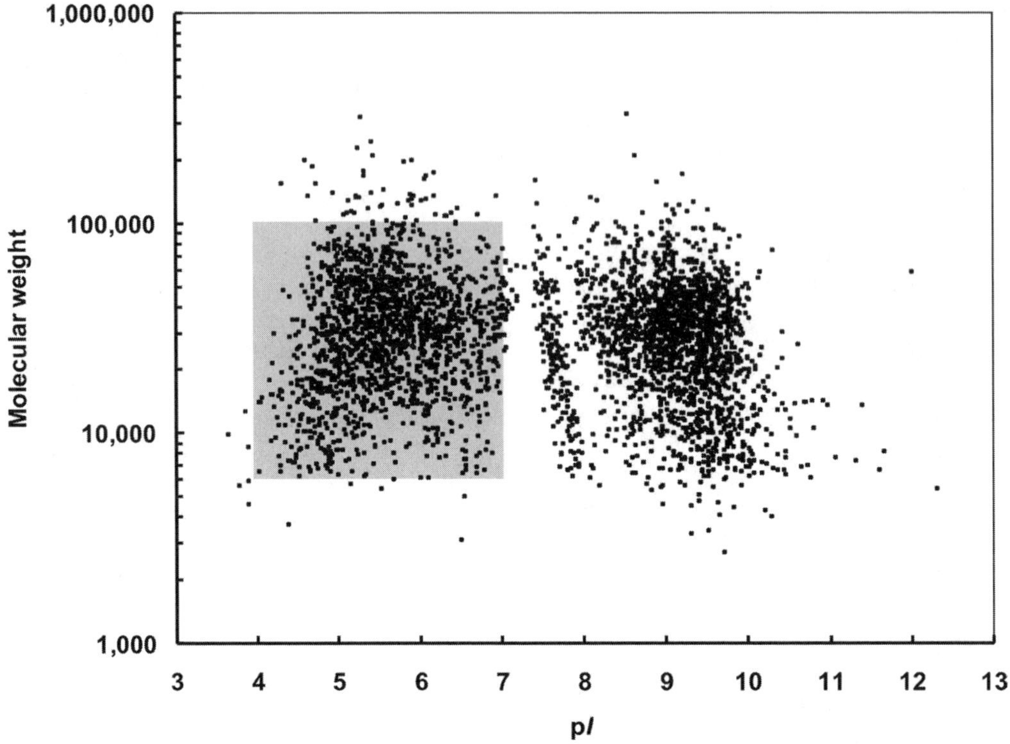

FIGURE 2.1 Conceptual proteome of *Clostridium acetobutylicum*. Molecular masses and p*I* values of all *C. acetobutylicum* proteins predicted by the EBI genome annotation (http://www.ebi.ac.uk/proteome) were calculated with the Compute pI/Mw tool available at the ExPASy server (http://www.expasy.org). The shaded area corresponds to proteins with molecular masses between 6 and 100 kDa and p*I* values ranging from 4 to 7 allowing their display when using the most common 2D-PAGE experimental regimes.

The genome sequence of three clostridial species, *C. acetobutylicum, C. perfringens,* and *C. tetani,* has been determined and made publicly available [12–14]. The chromosome-encoded proteome of these species is predicted to consist of 3676, 2660, and 2368 proteins, respectively, with 176, 63, and 61 additional gene products encoded by the megaplasmids pSOL1 (*C. acetobutylicum*), pCP13 (*C. perfringens*) and pE88 (*C. tetani*). Figure 2.1 shows the conceptual proteome of *C. acetobutylicum.* In contrast to other bacterial species, such as *Bacillus subtilis, Escherichia coli,* and *Mycobacterium tuberculosis,* most of the proteins of this organism (2303 out of 3852) are not within the most commonly used analysis window of p*I* 4 to 7 and MW 6 to 100 kDa [11,15]. Six proteins have p*I* values < 4, 2111 p*I* values > 7. Only few proteins have molecular masses higher than 100 kDa (153 proteins) or below 6 kDa (32 proteins). In principle, these findings also hold true for other clostridial species and further limit the applicability of 2D-PAGE for analyzing clostridial physiology.

The focus of the following sections is to review experimental work that has used two-dimensional electrophoresis as a major tool to analyze the physiology of both pathogenic and nonpathogenic clostridia.

2.3.2 Analysis of the Stress Response in Clostridia

The first proteome studies to be performed in clostridia were those by Terracciano et al. [16] and Pich et al. [17]. Both groups analyzed the response of *C. acetobutylicum* upon exposure to several environmental stresses, most notably to heat, and compared the results with changes in the protein synthesis patterns brought about by the metabolic shift from acidogenesis to solventogenesis.

Terracciano et al. [16] found eight proteins (designated Hsp16, Hsp18, Hsp22, Hsp28, Hsp49, Hsp68, Hsp74, and Hsp83, according to their molecular weight) with increased and two proteins (Hsp84 and Hsp100) with decreased synthesis rate upon heat shock. Time course experiments revealed that synthesis of most Hsps (with the exception of Hsp83, which is still synthesized with increased rates 4 hours after heat treatment) is only transiently increased. Synthesis of Hsp49 and Hsp68 had returned to preheat treatment levels in less than 1 hour, that of Hsp16 and Hsp74 after approximately 1 hour, and that of Hsp18, Hsp22, and Hsp28 after 4 hours. One additional 35 kDa protein appeared only after 1 hour of heat treatment. Many of these proteins were also synthesized with increased rates upon treating the cells with 11 m*M* butanol for 30 min (Hsp16, Hsp18, Hsp22, Hsp68, Hsp74, and Hsp83), even though the response was not as pronounced as after heat treatment. Synthesis of the 100 kDa protein was reduced after addition of butanol. Surprisingly, these effects could be observed regardless of the stage of growth at which butanol was added, even in late solventogenic phase when 48 mM butanol were already present in the culture supernatant. Finally, the effect of exposing *C. acetobutylicum* to air on synthesis of the Hsps was analyzed, and synthesis of Hsp22 and Hsp68 found to be induced, as well as that of the 100 kDa protein, decreased due to the applied environmental stress. Terracciano et al. [16] also characterized some of these proteins and could show that the Hsp18, Hsp49, Hsp74, and Hsp83 proteins are membrane-associated. Hsp74 was also shown to be immunoprecipitable with antiserum raised against *E. coli* DnaK, which makes the statements concerning cellular localization somewhat questionable.

The experiments performed by Pich et al. [17] were of similar nature. When shifting *C. acetobutylicum* cells from 30°C to 42°C for 8 min, the synthesis rate of at least 15 proteins (Hsp14, Hsp17, Hsp21, Hsp25, Hsp31, Hsp45, Hsp50, Hsp53, Hsp61, Hsp65a, Hsp65b, Hsp66, Hsp67, Hsp72, Hsp73) was increased significantly (4- to 20-fold). Again, Hsp72 was identified as DnaK, based on the observed cross-reactivity with antisera raised against *E. coli* DnaK, and Hsp67 was shown to cross-react with antisera directed against *E. coli* GroEL. The synthesis of most *C. acetobutylicum* Hsps reached its maximum at 10 to 15 min after the temperature shift. However, the synthesis of a subset of the Hsps (Hsp21, Hsp53, Hsp61) increased dramatically immediately after heat shock and peaked already at 6 min after shifting the cells to 42°C. It decreased rapidly thereafter, returning to synthesis rates comparable to those prior to heat shock 18 min after heat shock, while synthesis rates of the other Hsps remained at higher level (2-to 4-fold) 40 min after heat shock. Though most of the Hsps detected in these studies remain unidentified, and the molecular mechanisms underlying the observed changes in protein synthesis have yet to be elucidated (apart from DnaK and GroEL as well as Hsp17/Hsp18, which might be identical to the gene product of *hsp18* characterized by Sauer and Dürre [18]; see below), the observations make clear that heat-responsive gene expression in *C. acetobutylicum* is subject to complex regulation, as the response patterns of individual Hsps are remarkably different with respect to expression time course responsiveness toward environmental stimuli other than heat.

The same authors [19] have used 1D- and 2D-PAGE in order to reveal changes in the protein pattern of the thermophilic *Clostridium thermosulfurogenes* EM1 (now *Thermoanaerobacterium thermosulfurigenes* EM1) upon exposing the cell to 62°C. Surprisingly, they were able to observe changes in synthesis of only four proteins, designated Hsp21, Hsp66, Hsp70, and Hsp80 according to their molecular weight. Using heterologous antisera, they found Hsp71 to cross-react with α-*Mycobacterium tuberculosis* DnaK and Hsp66 with α-*E. coli* GroEL, suggesting that these proteins represent the *T. thermosulfurigenes* homologues. As the authors followed the synthesis rates of the individual Hsps by performing radioactive pulse-labeling at different time points after exposure to heat, they could demonstrate that in all four cases, the observed increase in synthesis rate is transient and peaks 10 to 15 min after transferring the cells to 62°C. Maximal induction of synthesis rates was 2-fold, 7.5-fold, 4-fold, and 9-fold for Hsp21, GroEL, DnaK, and Hsp80. In further experiments, the authors could show that in case of GroEL and DnaK the mRNA levels of the corresponding structural genes are also transiently increasing upon heat shock, reaching a maximum 15 min after heat shock, and the respective levels being increased 4-fold and 16-fold for the *dnaK* and *groEL* mRNAs, respectively. The authors noted that the number of proteins with increased synthesis is very low if compared with the extent of heat shock responses in mesophilic bacteria, for instance *C. acetobutylicum.*

Novak et al. [20] used differential scanning calorimetry (DSC) in order to identify cellular fractions of *C. perfringens,* which are damaged upon exposing the cells to heat. *C. perfringens* is a primary agent of foodborne illness caused by the increased consumer demand for low-preservative, minimally processed foods, and thermal treatment is used to make such products less perishable. They identified the ribosome-containing fraction as the only one and used 2D-PAGE to identify changes in this protein fraction upon cultivation at 28 and 46°C and found three protein spots up- and one spot down-regulated in the heat-shocked cells. Two of the up-regulated spots were identified and shown to represent glyceraldehyde-3-phosphate dehydrogenase and rubrerythrin. However, the physiological significance of these findings is not clear.

2.3.3 Comparison of Protein Patterns during Acidogenesis and Solventogenesis in *Clostridium acetobutylicum*

In first analyses, Terracciano et al. [16] and Pich et al. [17] determined to what extent Hsps they had previously identified as being induced in *C. acetobutylicum* during heat shock and other environmental stresses are also induced during solventogenesis. Terracciano et al. [16] noted that the synthesis ratios of Hsp22, Hsp68, DnaK, and Hsp84 are regulated in response to the growth phase. While the latter Hsp was most prominent during the switch phase, Hsp68 was synthesized during solvent production and Hsp22 as well as DnaK during late acidogenesis. Pich et al. [17] performed similar experiments but extended the analyses in order to dissect the influences of individual environmental parameters triggering induction of Hsp synthesis during the growth cycle. Cultivation of *C. acetobutylicum* took place in a phosphate-limited chemostat. In order to separate pH effects from solvent-induced effects, *C. acetobutylicum* was cultivated in a second set of experiments at lower phosphate concentrations, resulting in lower cell densities and concentrations of acids. Lowering the pH under these conditions does not lead to solvent production. Under these conditions the Hsp67, Hsp72, and Hsp73 proteins were synthesized with increased rates. In a third set of experiments, the switch from exponential growth to stationary phase was simulated by lowering the dilution rate while keeping the pH constant at 6.1, avoiding solvent formation. Under these conditions synthesis of Hsp21, Hsp67, and Hsp72 as well as of P34, P34.5, and P40 increased 2- to 3-fold. Taken together, these data show that synthesis of Hsps and other proteins is subject to complex regulation in response to the various stimuli experienced by the cell during the shift from acid to solvent production (pH, concentrations of acids and solvents, growth rate). These regulatory mechanisms act differentially on individual Hsps and their structural genes. For the most part, the molecular basis of the observed regulatory phenomena remains to be elucidated. However, expression of the *dnaK* and *groEL* genes is negatively regulated via the transcriptional repressors HrcA, a mechanism conserved in Gram-positive bacteria with low G+C content [21]. Expression of the gene coding for the small Hsp18 of *C. acetobutylicum* [18], on the other hand, seems to be controlled by the transcriptional repressor CtsR [22].

In order to identify proteins specifically synthesized during solventogenesis, Schaffer et al. [3] compared the protein synthesis and abundance pattern of acidogenic and solventogenic *C. acetobutylicum* cells. The cell material used for the analyses was isolated from a continuous culture experiment. The cells were cultivated in phosphate-limited mineral medium with a dilution rate of 0.08 at a pH of 6.0. Under equilibrium conditions, the concentration of the fermentation products butyrate, acetate, butanol, acetone, and ethanol was 59, 30, 2, 0, and 5 m*M*. In order to trigger solventogenesis, the pH set point was then lowered to 4.7 and samples were taken over the next 10 hours and analyzed with respect to concentration of fermentation products as well as expression of genes coding for solventogenic enzymes. Based on these analyses, cells taken 6 hours after the pH downshift were used as being representative of solventogenic cells. mRNA levels of the *sol* and *adc* operons (coding for butyraldehyde/butanol dehydrogenase E, CoA transferase subunits A and B, and acetoacetate decarboxylase) were maximal at this time point and synthesis of solvents had already started (concentration of fermentation products butyrate, acetate, butanol, acetone, and ethanol was 33, 21, 4, 2, and 3 m*M*). Comparison of the protein synthesis patterns in both metabolic states revealed global changes. Of the 130 most actively synthesized proteins, 52 proteins were synthesized with higher, 34 with lower rate during solventogenesis. Eleven of these proteins were identified by N-terminal sequencing. Adc and Hsp18 as well as DnaK and GroEL (identified by Pich et al. [17]) synthesis was significantly increased, as had been found in previous studies [17,18,23], with Adc being the protein with the highest induction rate (36-fold). However, in addition to that, several proteins were identified whose synthesis is also induced under solvent-producing conditions. Among these are three enzymes, phosphoglycerate dehydrogenase SerA, serine aminotransferase SerC, and seryl-tRNA synthetase SerS, with induction rates of about 2. This induction could be confirmed at the level of mRNA using Northern analyses. The genes are organized in a tetracistronic operon, *serCAXS*, which was shown to be transcribed from a typical σ^A-dependent promoter with –10 and –35 boxes identical to the respective consensus sequences, indicating that induction during solventogenesis is due to derepression. It is not clear, though, what is the physiological relevance of these observations, as the enzymes required for solventogenesis do not exhibit an unusual high serine content or unusual codon usage. Interestingly, *C. acetobutylicum* contains another gene coding for a seryl-tRNA synthetase, *serT*, which is organized monocistronically. Differential regulation of expression of two aminoacyl-tRNA synthetases with the same specificity has been shown for the *E. coli lysS* and *lysU* genes, coding for leucyl-tRNA synthetases. Here, *lysS* is expressed constitutively, while transcription of *lysU* is induced in response to heat, anaerobiosis, and changes in external pH [24,25].

Furthermore, 4-fold induction of synthesis of a protein, designated PdxY, which was recently shown to function in a novel pyridoxine biosynthesis pathway [26,27], was surprising. The protein is highly conserved in organisms from all three domains of life and was first identified in various organisms as being inducible by nutritional [28], oxidative [29], and general stresses [30], as well as being able to mediate resistance against reactive oxygen species [31,32]. None of the enzymes required for solvent formation is known to require pyridoxal phosphate as cofactor. However, it appears possible that solvent formation goes along with increased production of reactive oxygen species, which are known to be quenched by pyridoxine [33]. Finally, the Adc and Hsp18 proteins were each shown to exist in two isoforms with identical molecular weight but p*I* values differing by 0.19 and 0.24 pH units, respectively. This is an indication of covalent modification, which has not been described before for these proteins.

2.3.4 Analysis of Sulfonate Metabolism in Saccharolytic Clostridia

Experiments conducted by Bahl et al. [34] have shown that depletion of sulfur within the medium leads to an enhanced "shift" to solvent production and a higher overall yield of solvents in *C. acetobutylicum*. It is already known for aerobic bacteria, such as *E. coli*, *Pseudomonas aeruginosa*, *P. putida*, or *B. subtilis* [35–37], that the organisms induce a set of proteins under limitation of preferred sulfur sources like sulfates or cysteine to cope with the new conditions. Those gene products, sometimes referred to as "sulfur starvation induced" proteins (SSI proteins), are thought to be involved in acquisition of alternative sulfur sources, replacement of sulfur-rich proteins, and redistribution of intracellular sulfur stocks (reviewed in [38]). In contrast, knowledge about the reaction of anaerobic bacteria undergoing sulfur starvation is poor, and no reports are so far available about a system similar to that of aerobes [39]. An approach using the system of two-dimensional gel electrophoresis was used in studies by Thormann [40] and Hujer [41] to determine if such a system exists and how it might contribute to solvent production in saccharolytic clostridia.

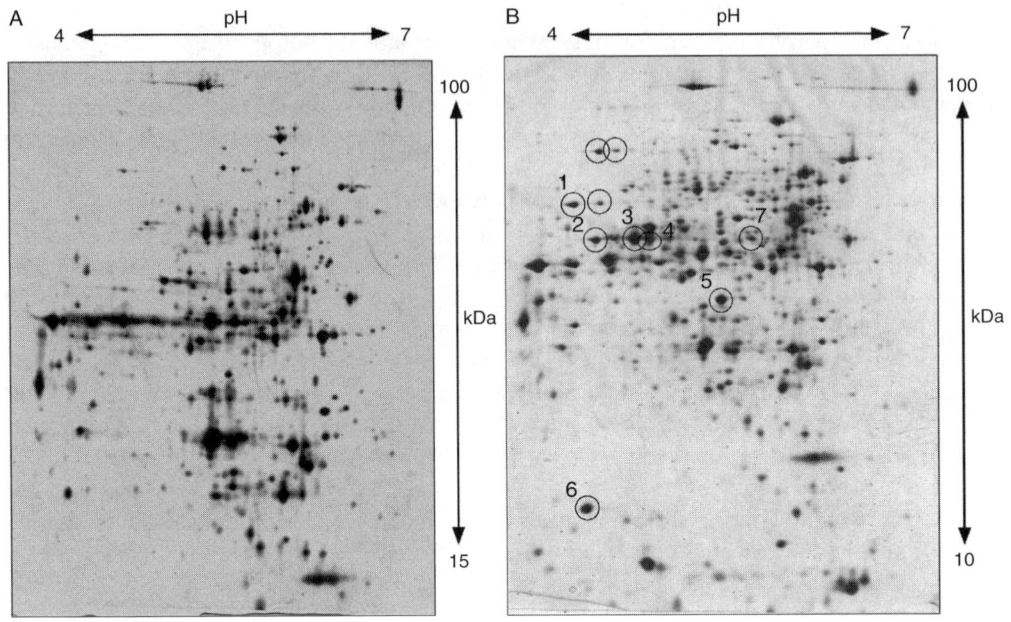

FIGURE 2.2 Protein pattern of *Clostridium beijerinckii* EV4 after growth with sulfate (A) and toluene sulfonate (B) as a sulfur source. Proteins (100 μg) were separated by 2D-gel electrophoresis and silver-stained. Spots with increased synthesis rates on toluene sulfonate are circled. N-termini of numbered spots have been determined.

The previous studies had shown that a switch from sulfur sources like cysteine or sulfate to sulfonates triggers the sulfur starvation response while at the same time providing a sulfur source for the bacteria to continue growth. Growth experiments with various sulfonates revealed that *C. acetobutylicum* DSM 792 is able to build up cell mass using taurine as sole sulfur source but exhibits a pronounced diminished growth rate. Accordingly, the differences in protein pattern obtained during the proteome analysis might not be exclusively attributed to sulfur starvation. Therefore, further studies were carried out using the closely related *C. beijerinckii* EV4, an isolate that had been shown to grow on arylsulfonates or taurine [42]. Comparison of protein expression patterns derived from exponentially growing cultures of *C. beijerinckii* EV4 led to the detection of 10 protein spots appearing exclusively or with increased abundance when sulfonate was the sole source of sulfur (Figure 2.2). Upon N-terminal sequencing of seven of these proteins, and comparing the amino acid sequence to those deduced from the known genome of *C. acetobutylicum* [12], one of the spots (5) could be identified. It appears to be a homologue of cysteine synthase (O-acetylserine sulfhydrolase, CysK), involved in basic sulfur metabolism by catalyzing the transfer of sulfide to O-acetyl-L-serine. CysK was also described to be among proteins induced under conditions of sulfur starvation in *E. coli* [43] and *B. subtilis* [36]; it was also identified among cold shock-induced proteins in the latter organism [44].

C. beijerinckii EV4 was found to degrade sulfonates exclusively when all other traces of sulfur sources like cysteine or sulfate were consumed completely [42], indicating induction on the transcriptional level. Taken together with the higher expression level of basic sulfur metabolism proteins during growth on sulfonates, there are strong implications that saccharolytic clostridia possess a system similar to the SSI-stimulon system, although it still remains obscure how this is linked to solventogenesis.

2.3.5 Identification of Abundant Cytosolic Proteins in *Clostridium pasteurianum*

In the course of a study focusing on clostridial iron-sulfur proteins and the purification of β-fructofuranosidase from *C. pasteurianum* Flengsrud and Skjeldal [45] have separated cell-free extracts of *C. pasteurianum* W5 and identified some of the most abundant proteins via N-terminal microsequencing. Of 20 proteins analyzed, only one was probably blocked N-terminally as no sequence was obtained. Of the remaining 19 N-terminal sequences, eight were found to match *C. pasteurianum* proteins and 15 to match clostridial database entries. Most of the identified proteins play vital roles in the cells' central metabolic pathways, such as glyceraldehyde-3-phosphate dehydrogenase, triosephosphate isomerase, acetyl-CoA acetyltransferase, acetoacetate decarboxylase, and butyrate kinase. Furthermore, several proteins that are part of the general stress response, such as DnaK, GroEL, GroES, and a putative thiol peroxidase, were found to be highly abundant under the applied cultivation conditions, indicating that they mean significant stress to the cells. Three proteins, pyruvate kinase, glyceraldehyde-3-phosphate dehydrogenase, and acetoacetate decarboxylase, were shown to be present at two positions on the 2D gel, respectively. In the case of glyceraldehyde-3-phosphate dehydrogenase and acetoacetate decarboxylase, the two spots have identical molecular weight and p*I* values differing by 0.2 pH units, respectively, indicating post-translational modification with a charged low-molecular-weight compound, such as phosphorylation. In case of pyruvate kinase, the two spots differ by p*I* ($\Delta = 0.3$) and molecular weight ($\Delta = 6.4$ kDa), indicating proteolytic processing either *in vivo* or during sample preparation.

2.3.6 Analysis of Toxin Production in *Clostridium difficile*

The genus *Clostridium* comprises several important pathogens, such as *C. botulinum*, *C. perfringens,* and *C. difficile.* Physiology and pathogenesis of some of these organisms have also been analyzed using proteomic technology.

Clostridium difficile-associated diarrhea (CDAD) has become a major hospital problem and is occasionally complicated by pseudomembranous colitis (PMC). Infections are usually caused by perturbations of the large-bowel microflora as a consequence of antibiotic treatment. This facilitates colonization of this ecosystem by *C. difficile* and overgrowth of the native microflora. The major virulence factors of *C. difficile* are two closely related toxins, A and B (ToxA and ToxB), which are hypothesized to be released from cells upon autolysis and cause glycosylation of small GTP-binding proteins of the host, such as Rho, Rac, and Cdc42, upon endocytosis be intestinal epithelial cells [46]. This leads to disruption of the cytoskeleton and results, ultimately, in cell death. Mukherjee et al. [47] have used 2D-PAGE in order to identify other proteins than toxins A and B that are also secreted into the culture supernatant and might play roles in pathogenesis of this organism. As it had been noticed in previous studies that toxin production is 20-fold lower upon cultivation in glucose-containing media, they compared the extracellular protein patterns of the high-toxin producing strain *C. difficile* VPI10463 after cultivation in glucose-free and glucose-containing medium in order to identify secreted proteins whose secretion follows a pattern similar to that of the toxins and which, therefore, might be related to pathogenesis as well. Of about 15 proteins detected in culture supernatants, four proteins were found to be more abundant in the absence of glucose and were subjected to N-terminal microsequencing. Two of the determined N-terminal sequences were highly similar to those of the FixA and FixB proteins from *E. coli*, which are known to be cytoplasmic proteins. Consequently, most *C. difficile* FixA and FixB was found in the cytoplasmic protein fraction (>99%), and the presence of these proteins in the culture supernatants can be attributed to partial cell lysis. Nevertheless, this was an interesting finding, as it is not known how ToxA and ToxB are released into the culture supernatant. The toxin proteins possess no signal sequences at their N-termini that would point at Sec-dependent translocation, and therefore, cell lysis was hypothesized to be responsible for toxin release. However, lysis of <1% of the cell population could not explain the efficient secretion of the toxins (almost half of total toxin in the culture supernatant) observed in the very same experiments. Another protein detected in the culture supernatants of strain VPI10463 showed homology to a gene product encoded by the *B. subtilis* prophage PBSX. A fourth protein (49 kDa) possesses a typical signal sequence for Sec-dependent translocation and is similar in its C-terminal part to outer-membrane efflux proteins of Gram-negative bacteria. In addition to differentially abundant proteins, the authors also attempted to identify the two most abundant extracellular proteins (50 and 36 kDa, respectively). In one case the obtained N-terminal sequence did not match any of the ORFs predicted to be encoded by the *C. difficile* genome. The second protein was identified as being part of the 72-kDa S-layer protein SlpA. The SlpA protein contains a signal peptide typical for proteins translocated via the plasma membrane in a Sec-dependent manner and was much more abundant in the cell pellet fraction, indicating that it associates with the cell wall. Fragments of this protein (24, 35, and 48 kDa) were shown to be represented by three more spots on the 2D gels. Two other spots (38 and 41 kDa) from the extracellular protein fraction were identified as representing fragments of a second S-layer protein, highly similar to SlpA.

ToxA and ToxB synthesis in *C. difficile* is also regulated in a temperature-dependent manner and is maximal at 37°C, demonstrating the result of the adaptation process of *C. difficile* to the temperature experienced in host animals. This control is exerted at the level of transcription and is mediated by the alternative σ factor TxeR [48–50]. In order to determine if the synthesis of proteins other than toxins A and B is subject to temperature-dependent control in *C. difficile*, Karlsson et al. [48] also compared the protein expression profiles of this organism in both rich medium and minimal medium without biotin grown at 22, 37, and 42°C. Biotin limitation is known to stimulate toxin production [51]. Therefore, the chosen growth conditions allowed studying temperature-controlled protein synthesis and conditions characterized by low and high toxin yields. 28 proteins were found to be most abundant at 37°C and seven of these were identified by N-terminal microsequencing. These proteins represented 3-hydroxybutyryl-CoA dehydrogenase, nitrate reductase, succinate-semialdehyde dehydrogenase, an NADH-dependent oxidase, a putative hydrolase, and a protein not matching any database entries. 3-hydroxybutyryl-CoA dehydrogenase was found to occur twice on the gels with the spots exhibiting a difference in molecular mass of 2 kDa but identical p*I*, indicating either C-terminal proteolytic processing (the experimentally determined N-terminal sequences were identical) or post-translational modification not affecting p*I*, such as glycosylation. Karlsson et al. [48] also compared the protein expression profiles of cells grown in defined medium under conditions of either biotin excess or limitation, which were subjected to thermal upshifts. Conditions known to be most favorable for toxin synthesis (biotin limitation and thermal upshift to 37°C) also led to maximal abundance of 12 proteins, two of which were identified by N-terminal microsequencing and shown to represent two isoforms of rubrerythrin, a protein apparently involved in protecting anaerobic bacteria from oxidative stress. Both spots exhibited identical molecular weight but differed in p*I* ($\Delta = 0.1$), indicating covalent modification of rubrerythrin with a charged compound.

2.3.7 Characterization of the VirR Regulon of *Clostridium perfringens*

Clostridium perfringens is the primary causative agent of clostridial myonecrosis, also known as gas gangrene, which is characterized by massive tissue destruction and necrosis due to the secretion of numerous toxins. Expression of the toxin structural genes is regulated by a two-component system, VirS-VirR. In order to characterize the VirR regulon globally, Shimizu et al. [52] have used 2D-PAGE for the comparison of the secretomes of *C. perfringens* wildtype and *virR* deletion strains. The authors used pH gradients 5-8 and 3-10 for first-dimension separation and homogeneous 10%

SDS-PAGE for second-dimension separation. Approximately 200 spots were detected in the culture supernatants of both strains, with many of them being proteins known to be localized in the cytoplasm and not having putative signal sequences directing them for secretion by the Sec machinery. Such proteins included ribokinase, β-hydroxybutyryl-coenzyme A dehydrogenase, fructose-bisphosphate aldolase, and elongation factor G. The abundance of eight proteins (ClpB, [Fe-S]-cofactor synthesis protein NifS, a probable cell-wall-binding protein, ribokinase, β-hydroxybutyryl-coenzyme A dehydrogenase, fructose-bisphosphate aldolase, elongation factor G, a conserved hypothetical protein with similarity to the p45 protein from *Listeria monocytogenes*) was increased, that of six proteins (2′,3′-cyclic nucleotide 2′-phosphodiesterase, perfringolysin O, a cysteine protease, collagenase, a probable surface protein, two proteins with hypothetical functions and similarity to YdaL from *B. subtilis* and chitinase A from *C. paraputrificum*, respectively) decreased in the *virR* mutant strain. In addition to the proteome approach, mRNA levels of the genes corresponding to the gene products putatively controlled by VirR, whether positively or negatively, were compared in wildtype and *virR* mutant strain. The mRNA levels of the structural genes coding for 2′,3′-cyclic nucleotide 2′-phosphodiesterase, perfringolysin O, the cysteine protease, collagenase, and YdaL were also reduced in the mutant, while those of the genes coding for a probable cell wall-binding protein and fructose-bisphosphate aldolase were increased in the *virR* strain, indicating that changes in abundance of only a subset of the identified proteins is due to altered transcription and, therefore, to possible VirR-dependent regulation.

As protease activity in the supernatant of *C. perfringens* had been reported to be positively regulated by the VirS-VirR two-component system, the authors also investigated the influence of adding various protease inhibitors to the culture supernatants on the protein pattern. In fact, they observed an increase in abundance of ClpB, NifS, ribokinase, fructose-bisphosphate aldolase, and the probable cell wall-binding protein upon adding benzamidine, leupeptin, or antipain to the culture supernatants, but no influence of PMSF, pepstatin, phosphoramidon, or E-64 on the wildtype protein pattern. A probable surface protein disappeared and the levels of perfringolysin and a cysteine protease decreased upon addition of benzamidine, PMSF, and phosphoramidon. These observations indicate that several proteases, inhibited by different compounds, are present in *C. perfringens* supernatants and that these proteases degrade some of the other proteins. Subsequently, the authors made use of the observation that the proteolytic activity in *C. perfringens* supernatants was inhibited by benzamidine, as they could purify the corresponding protease with a benzamidine-Sepharose column and showed that it is identical with the cysteine protease whose synthesis is positively regulated by VirR.

2.3.8 N-Terminal Processing of Proteins in Clostridia

As in most studies mentioned above, N-terminal microsequencing of proteins has been used to identify proteins of interest, some statements can be made concerning methionine aminopeptidase-dependent N-terminal processing of proteins in clostridia. The N-termini of 19 proteins from *C. pasteurianum*, 11 proteins from *C. acetobutylicum*, 2 proteins from *C. perfringens*, and 15 proteins from *C. difficile* have been determined. In those proteins where the penultimate amino acid residue was serine (2 proteins), alanine (4), or glycine (2), the initiator methionine had always been cleaved off by methionine aminopeptidase. If lysine (10), histidine (2), glutamate (1), aspartate (1), asparagine (1), leucine (3), isoleucine (1), or phenylalanine (1) was the penultimate residue, the N-terminus had not been processed. These data indicate that N-terminal processing in clostridia follows similar rules as in *E. coli* [11]. In *E. coli,* the N-terminal methionine is always cleaved when the penultimate amino acid residue is either serine or alanine. Occurrence of cleavage is variable with threonine, glycine, or proline being the penultimate residue and does not occur if other amino acids occupy the second position [11].

2.4 FUTURE PERSPECTIVES

Though 2D-PAGE is still the most widely used separation method in proteome analysis, a state that is not likely to change within the next decades, at least some of its limitations can probably be overcome only by gel-less separation systems. Currently, there are a number of methods in development that can be summarized by the term "shot-gun proteomics." They usually rely on the digestion of entire cell lysates using site-specific as well as unspecific proteases. Peptides are then identified by tandem mass spectrometry. Obviously, the most challenging task here is to analyze a sample with incredible complexity and to handle the generated data in a feasible manner. Several strategies have been derived to reduce the initial complexity of the proteolytically digested cell lysates, including multidimensional HPLC (MudPIT; [53]), chemical derivatization of sulfur-containing amino acids with subsequent reversed phase HPLC-based isolation of the modified peptide subpopulation (COFRADIC; [54,55]) and affinity labeling of certain amino acids (ICAT; [56]). In several proof-of-principle studies, these methods have already been demonstrated to be able to identify many more proteins within a given proteome than is possible by 2D-PAGE [57,58]. A severe drawback of gel-less separations is that mass spectrometric analysis per se does not allow for direct quantification of the identified peptides and, therefore, of the proteins these peptides are derived from. However, several strategies relying on isotope-coding of two samples to be compared and analyzed simultaneously have been developed. For instance, the ICAT reagent mentioned above comes in two forms, one carrying 8 deuterium atoms instead of hydrogen atoms resulting in a mass difference of 8 Da [56]. In consequence, a given cysteine-containing peptide representing the same protein species but derived from the two differentially labeled samples will result in two peaks in the mass spectrum separated by 8 Da. Therefore, the ratio of the proteins originally present in the samples can be quantified by calculating the corresponding peak areas. It will be interesting to follow how these methods perform in the future in comparison with the classical 2D approach.

REFERENCES

1. O'Farrell, P.H., High resolution two-dimensional electrophoresis of proteins, *J. Biol. Chem.,* 250, 4007, 1975.
2. Görg, A. et al., The current state of two-dimensional electrophoresis with immobilized pH gradients, *Electrophoresis,* 21, 1037, 2000.
3. Schaffer, S. et al., Changes in protein synthesis and identification of proteins specifically induced during solventogenesis in *Clostridium acetobutylicum*, *Electrophoresis,* 23, 110, 2002.
4. Blum, H., Beier, H., and Gross, H.J., Improved silver staining of plant proteins, RNA and DNA in polyacrylamide gels, *Electrophoresis,* 8, 93, 2000.
5. Rider, M.H. et al., An agarose-based gel-concentration system for microsequence and mass spectrometric characterization of proteins previously purified in polyacrylamide gels starting at low picomole levels, *Eur. J. Biochem.,* 230, 258, 1995.
6. Fountoulakis, M. and Langen, H., Identification of proteins by matrix-assisted laser desorption ionization-mass spectrometry following in-gel digestion in low-salt, nonvolatile buffer and simplified peptide recovery, *Anal. Biochem.,* 250, 153, 1997.
7. Fenyö, D., Identifying the proteome: software tools, *Curr. Opin. Biotechnol.,* 11, 391, 2000.
8. Santoni, V., Molloy, M., and Rabilloud, T., Membrane proteins and proteomics: un amour impossible?, *Electrophoresis,* 21, 1054, 2000.
9. Görg, A. et al., The current state of two-dimensional electrophoresis with immobilized pH gradients, *Electrophoresis,* 21, 1037, 2000.
10. Ohlmeier, S., Scharf, C., and Hecker, M., Alkaline proteins of *Bacillus subtilis*: first steps towards a two-dimensional alkaline master gel, *Electrophoresis,* 21, 3701, 2000.
11. Link, A.J., Robison, K., and Church, G.M., Comparing the predicted and observed properties of proteins encoded in the genome of *Escherichia coli* K-12, *Electrophoresis,* 18, 1259, 1997.

12. Nölling, J. et al., Genome sequence and comparative analysis of the solvent-producing bacterium *Clostridium acetobutylicum*, *J. Bacteriol.,* 183, 4823, 2001.
13. Shimizu, T. et al., Complete genome sequence of *Clostridium perfringens*, an anaerobic flesh-eater, *Proc. Natl. Acad. Sci. U.S.A.,* 99, 996, 2002.
14. Brüggemann, H. et al., The genome sequence of *Clostridium tetani,* the causative agent of tetanus disease, *Proc. Natl. Acad. Sci. U.S.A.,* 100, 1316, 2003.
15. Urquhart, B.L., Cordwell, S.J., and Humphery-Smith, I., Comparison of predicted and observed properties of proteins encoded in the genome of *Mycobacterium tuberculosis* H37Rv, *Biochem. Biophys. Res. Commun.,* 253, 70, 1998.
16. Terracciano, J.S., Rapaport, E., and Kashket, E.R., Stress- and growth phase-associated proteins of *Clostridium acetobutylicum*, *Appl. Environ. Microbiol.,* 54, 1989, 1988.
17. Pich, A., Narberhaus, F., and Bahl, H., Induction of heat shock proteins during initiation of solvent formation in *Clostridium acetobutylicum*, *Appl. Microbiol. Biotechnol.,* 33, 697, 1990.
18. Sauer, U. and Dürre, P., Sequence and molecular characterization of a DNA region encoding a small heat shock protein of *Clostridium acetobutylicum*, *J. Bacteriol.,* 175, 3394, 1993.
19. Narberhaus, F., Pich, A., and Bahl, H., Synthesis of heat shock proteins in *Thermoanaerobacterium thermosulfurigenes* EM1 (*Clostridium thermosulfurogenes* EM1), *Curr. Microbiol.,* 29, 13, 1994.
20. Novak, J.S., Tunick, M.H., and Juneja, V.K., Heat treatment adaptations in *Clostridium perfringens* vegetative cells, *J. Food Protect.,* 64, 1527, 2001.
21. Schulz, A. and Schumann, W., *hrcA*, the first gene of the *Bacillus subtilis dnaK* operon encodes a negative regulator of class I heat shock genes, *J. Bacteriol.,* 178, 1088, 1996.
22. Derre, I., Rapoport, G., and Msadek, T., CtsR, a novel regulator of stress and heat shock response, controls *clp* and molecular chaperone gene expression in Gram-positive bacteria, *Mol. Microbiol.,* 31, 117, 1999.
23. Gerischer, U. and Dürre, P., mRNA analysis of the *adc* gene region of *Clostridium acetobutylicum* during the shift to solventogenesis, *J. Bacteriol.,* 174, 426, 1992.
24. Kawakami, K., Ito, K., and Nakamura, Y., Differential regulation of two genes encoding lysyl-tRNA synthetases in *Escherichia coli*: *lysU*-constitutive mutations compensate for a *lysS* null mutation, *Mol. Microbiol.,* 6, 1739, 1992.
25. Leveque, F. et al., Control of *Escherichia coli* lysyl-tRNA synthetase expression by anaerobiosis, *J. Bacteriol.,* 173, 7903, 1991.
26. Ehrenshaft, M. et al., A highly conserved sequence is a novel gene involved in *de novo* vitamin B6 biosynthesis, *Proc. Natl. Acad. Sci. U.S.A.,* 96, 9374, 1999.
27. Osmani, A.H., May, G.S., and Osmani, S.A., The extremely conserved *pyroA* gene of *Aspergillus nidulans* is required for pyridoxine synthesis and is required indirectly for resistance to photosensitizers, *J. Biol. Chem.,* 274, 23565, 1999.
28. Padilla, P.A. et al., The highly conserved, coregulated SNO and SNZ gene families in *Saccharomyces cerevisiae* respond to nutrient limitation, *J. Bacteriol.,* 180, 5718, 1998.
29. Bernhardt, J. et al., Specific and general stress proteins in *Bacillus subtilis*-a two-dimensional protein electrophoresis study, *Microbiol. UK,* 143, 999, 1997.
30. Sivasubramaniam, S. et al., Characterisation of HEVER, a novel stress-induced gene from *Hevea brasiliensis*, *Plant Mol. Biol.,* 29, 173, 1995.
31. Ehrenshaft, M. et al., Functional characterization of SOR1, a gene required for resistance to photosensitizing toxins in the fungus *Cercospora nicotianae*, *Curr. Genet.,* 34, 478, 1999.
32. Ehrenshaft, M. et al., SOR1, a gene required for photosensitizer and singlet oxygen resistance in *Cercospora fungi*, is highly conserved in divergent organisms, *Mol. Cell,* 1, 603, 1998.
33. Bilski, P. et al., Vitamin B6 (pyridoxine) and its derivatives are efficient singlet oxygen quenchers and potential fungal antioxidants, *Photochem. Photobiol.,* 71, 129, 2000.
34. Bahl, H., Andersch, W., and Gottschalk, G., Continuous production of acetone and butanol by *Clostridium acetobutylicum* in a two-stage phosphate-limited chemostat, *Eur. J. Appl. Microbiol. Biotechnol.,* 15, 201, 1982.
35. Kertesz, M.A., Leisinger, T., and Cook, A.M., Proteins induced by sulfate limitation in *Escherichia coli*, *Pseudomonas putida* or *Staphylococcus aureus*, *J. Bacteriol.,* 175, 1187, 1993.
36. Coppée, J.Y. et al., Sulfur-limitation-regulated proteins in *Bacillus subtilis*: a two-dimensional gel electrophoresis study, *Microbiol. UK,* 147, 1631, 2001.

37. Hummerjohann, J. et al., Regulation of the sulfate starvation response in *Pseudomonas aeruginosa*: role of cysteine biosynthetic intermediates, *Microbiol. UK,* 144, 1375, 1998.
38. Kertesz, M.A., Riding the sulfur cycle-metabolism of sulfonates and sulfate esters in Gram-negative bacteria, *FEMS Microbiol. Rev.,* 24, 135, 2000.
39. Cook, A.M., Laue, H., and Junker, F., Microbial desulfonation, *FEMS Microbiol. Rev.,* 22, 399, 1998.
40. Thormann, K., Untersuchungen zur Regulation des *sol*-Operons von *Clostridium acetobutylicum*, Ph.D. thesis, University of Ulm, Germany, 2001.
41. Hujer, S., Proteomanalyse und genetische Systeme in anaeroben sulfonatabbauenden Bakterien, Diploma thesis, University of Ulm, Germany, 2001.
42. Denger, K. et al., Anaerobic desulfonation of 4-tolylsulfonate and 2-(4-sulfophenyl)butyrate by a *Clostridium* sp., *Appl. Environ. Microbiol.,* 62, 1526, 1996.
43. van der Ploeg, J.R. et al., Involvement of CysB and Cbl regulatory proteins in expression of the *tauABCD* operon and other sulfate starvation-inducible genes in *Escherichia coli*, *J. Bacteriol.,* 179, 7671, 1997.
44. Graumann, P. et al., Cold shock stress-induced proteins in *Bacillus subtilis*, *J. Bacteriol.,* 178, 4611, 1996.
45. Flengsrud, R. and Skjeldal, L., Two-dimensional gel electrophoresis separation and N-terminal sequence analysis of proteins from *Clostridium pasteurianum* W5, *Electrophoresis,* 19, 802, 1998.
46. McClane, B.A. and Rood, J.I., Clostridial toxins involved in human enteric and histotoxic infections, in *Clostridia: Biotechnology and Medical Applications,* Bahl, H. and Dürre, P., Eds., Wiley-VCH Verlag GmbH, Weinheim, Germany, 2001, 169.
47. Mukherjee, K. et al., Proteins released during high toxin production in *Clostridium difficile*, *Microbiol. UK,* 148, 2245, 2002.
48. Karlsson, S. et al., Expression of *Clostridium difficile* toxins A and B and their sigma factor TcdD is controlled by temperature, *Infect. Immun.,* 71, 1784, 2003.
49. Mani, N. and Dupuy, B., Regulation of toxin synthesis in *Clostridium difficile* by an alternative RNA polymerase sigma factor, *Proc. Natl. Acad. Sci. U.S.A.,* 98, 5844, 2001.
50. Mani, N. et al., Environmental response and autoregulation of *Clostridium difficile* TxeR, a sigma factor for toxin gene expression, *J. Bacteriol.,* 184, 5971, 2002.
51. Yamakawa, K. et al., Enhancement of *Clostridium difficile* toxin production in biotin-limited conditions, *J. Med. Microbiol.,* 44, 111, 1996.
52. Shimizu, T. et al., Proteome and transcriptome analysis of the virulence genes regulated by the VirR/VirS system in *Clostridium perfringens*, *J. Bacteriol.,* 184, 2587, 2002.
53. Washburn, M.P., Wolters, D., and Yates, J.R., 3rd, Large-scale analysis of the yeast proteome by multidimensional protein identification technology, *Nat. Biotechnol.,* 19, 242, 2001.
54. Gevaert, K. et al., Chromatographic isolation of methionine-containing peptides for gel-free proteome analysis: identification of more than 800 *Escherichia coli* proteins, *Mol. Cell Proteomics,* 1, 896, 2002.
55. Gevaert, K. et al., Exploring proteomes and analyzing protein processing by mass spectrometric identification of sorted N-terminal peptides, *Nat. Biotechnol.,* 21, 566, 2003.
56. Gygi, S.P. et al., Quantitative analysis of complex protein mixtures using isotope-coded affinity tags, *Nat. Biotechnol.,* 17, 994, 1999.
57. Koller, A. et al., Proteomic survey of metabolic pathways in rice, *Proc. Natl. Acad. Sci. U.S.A.,* 99, 11969, 2002.
58. Lipton, M.S. et al., Global analysis of the *Deinococcus radiodurans* proteome by using accurate mass tags, *Proc. Natl. Acad. Sci. U.S.A.,* 99, 11049, 2002.

3 Gene Cloning in Clostridia

Ian J. Davis, Glen Carter, Mike Young, and Nigel P. Minton

CONTENTS

3.1 INTRODUCTION

Clostridium is one of the largest prokaryotic genera, comprising a diverse assemblage of obligately anaerobic, Gram-positive, endospore-forming bacteria. As a whole, the genus exhibits extreme biocatalytic diversity. Some species such as *Clostridium acetobutylicum* are of considerable biotechnological interest as process organisms for the commercial production of chemical fuels and commodities from renewal biomass. However, the genus is best known as a consequence of the presence of organisms that cause human disease. Lethal intoxications are caused by *Clostridium botulinum* and *Clostridium tetani*, and to a lesser extent *Clostridium perfringens*, whereas debilitating infections are caused by *Clostridium difficile,* a major contributor to healthcare-associated illnesses within the U.K.

Following several years of intense activity, the genome sequences of representative strains of those species of greatest importance to the scientific community, namely *C. acetobutylicum*, *C. botulinum*, *C. tetani*, *C. perfringens*, and *C. difficile*, have now been determined; genome sequences for additional species, such as *Clostridium beijerinckii*, are in the pipeline. The availability of this information provides new opportunities for:

- Harnessing the biocatalytic potential of saccharolytic species in the commercial production of chemical fuels from renewable biomass
- Understanding the underlying mechanisms that contribute to the virulence of pathogenic species
- Assisting the development of improved methods for the detection and surveillance of the pathogenic organisms and of more effective countermeasures to prevent and treat infections

Effective means for genetically manipulating the organisms are required to maximize the potential utility of the wealth of available genome sequence information. In this chapter, we review the status of some of the genetic tools currently available, and we draw attention to the need for improved methods, where appropriate.

3.2 DNA ISOLATION

3.2.1 Genomic DNA

Protocols for the isolation of genomic DNA from different clostridia vary but tend to be based on the method first described by Marmur [1]. Examples include *C. acetobutylicum* [2], *C. thermocellum* [3], *C. botulinum* [4], and *C. cellulolyticum* [5]. This method involves an enzyme detergent lysis step, followed by extraction of cellular debris with phenol chloroform; the DNA is then precipitated out with ethanol. The method, described here in more detail, is taken from that used in *C. acetobutylicum* [2].

Cells are grown in clostridial basal media (CBM) [6] supplemented with glycine [0.4% (wt/vol)]. Pelleted cells (approx. 3 g wet weight) are then suspended in resuspension buffer [50 m*M* TRIS; 1 m*M* EDTA; 6.7% sucrose (wt/vol); pH 8.0] and incubated at 37°C for 15 min. Lysozyme solution (2.5 ml at 20 mg ml^{-1}) is then added, and the mixture is incubated for 10 min at 37°C. Clostridial DNAases can be inhibited at this stage by the addition of 1.25 ml of 0.5 *M* EDTA solution (pH 8.0). Lysis is achieved by the addition of 0.75 ml lysis buffer [20% SDS (wt/vol); 50 m*M* TRIS; 20 m*M* EDTA; pH 8.0] and incubation for 10 min at 37°C. Proteinase K solution (0.1 ml at 2.5 mg ml^{-1} in 10 m*M* TRIS; 1 m*M* EDTA) is added, and the mix incubated for 3 h at 37°C. Sodium perchlorate solution (5 *M*) is then added to a final concentration of 1 *M* and mixed. The emulsion is then extracted twice with chloroform/isoamylalcohol (24:1 vol/vol) by centrifugation for 5 min at 4000 to 10,000 g removing the aqueous phase. The DNA is then precipitated by the addition of 2 volumes of ethyl alcohol and can then be spooled out using a glass rod or pelleted via centrifugation. The DNA is then resuspended in TES (50 m*M* TRIS-HCl; 5 m*M* EDTA; 0.15 *M* NaCl). Treatment with Proteinase K and a further round of ethyl alcohol precipitation can be carried out to increase the purity of the DNA. Excess alcohol is removed by drying at room temperature and resuspension of the DNA in sterile water of TES.

It should be stated, however, that commercially available kits provide acceptable concentrations of purified genomic DNA for most applications and are less time-consuming to perform. They have been used successfully with many species, including *C. difficile* [7], *C. perfringens* [8], *C. acetobutylicum* [9] — and in our laboratory with *C. sporogenes* and *C. beijerinckii*. Anaerobic preincubation of the cell pellet in lysozyme (10 mg ml^{-1}) for 15 to 60 min at 37°C may be required to aid lysis with some commercial kits.

3.2.2 Plasmid DNA Isolation

Published procedures for isolating plasmid DNA from *C. acetobutylicum* (now *C. beijerinckii*) [10], and *C. cellulolyticum* [5] and *C. beijerinckii,* are based on the standard alkaline lysis method described by Sambrook et al. [11] with the addition of a prelysis step. All centrifugation steps are at 12,000 *g* at 4°C. A 1.5 ml aliquot of cells, taken from an overnight culture, is centrifuged. The pellet is then resuspended in 100 μl prelysis buffer [25% (w/v) sucrose, 25 m*M* Tris/HCl pH 8.0, 10 m*M* EDTA, 10 mg ml^{-1} lysozyme] and incubated anaerobically for 1 h at 37°C. A 200 μl aliquot of freshly prepared Solution II (0.2 N NaOH; 1% SDS) is added and mixed by inverting the tube followed by 150 μl of ice cold Solution III (5 *M* potassium acetate, 60 ml; glacial acetic acid, 11.5 ml; H_2O, 28.5 ml). The mixture is vortexed and held on ice for 3 to 5 min before centrifuging for 5 min. The supernatant is transferred to a fresh tube. An optional step is to add an equal volume of phenol chloroform and mix by vortexing prior to centrifuging for 2 min; the supernatant is transferred to a fresh tube. The DNA is precipitated by adding 2 volumes of ethanol and mixing by vortexing. The mixture is left at room temperature for 2 min then centrifuged for 5 min. The supernatant is removed by aspiration. The DNA is rinsed with 1 ml 70% ethanol (chilled at 4°C), which is aspirated off. The DNA is air dried for 10 min then resuspended in 50 μl TE (10 m*M* TRIS·Cl; 1 m*M* EDTA; pH 8.0).

Again, commercially available kits provide acceptable concentrations of plasmid DNA. They have been utilized with many species including *C. difficile*, *C. beijerinckii* [12], *C. botulinum* [13], *C. cellulolyticum* [5], and *C. perfringens* [14]. The addition of the prelysis step (incubation with lysozyme) described above is required, and large-scale kits may be required for low-copy number plasmids.

3.3 TRANSFORMATION OF CLOSTRIDIA

There are no reported examples of natural competence in clostridia. In order to promote DNA uptake, transformation procedures thus rely on physical alteration of the cell envelope. The conversion of cells to protoplasts prior to transformation is one such method. However, the optimization of protocols for protoplast production is complex and time-consuming, and electrotransformation of intact cells is much more commonly used. The concentrations of antibiotics used for selection in clostridia vary between species; typical concentrations used are: chloramphenicol, 5 to 35 μg ml^{-1}; thyamphenicol, 15 μg ml^{-1}; spectinomycin, 250 μg ml^{-1}; erythromycin, 10 to 50 μg ml^{-1}; tetracycline, 5 to 10 μg ml^{-1}.

3.3.1 Protoplast Production and Transformation

While this method has been largely superseded by electrotransformation, Reysset et al. [15] have developed an efficient protoplast transformation protocol for use with *C. acetobutylicum* NI-4081. This strain gave high transformation frequencies (up to 10^6 transformants μg^{-1} DNA) due to its reduced autolysin production (coupled with the use of autolysin inhibitors).

3.3.1.1 Protoplast Production

Bacteria are grown to mid-exponential phase (approximately 10^8 cells ml^{-1}) in T69 medium (per liter: 10 g glucose; 2 g ammonium acetate; 1 g yeast extract; 0.5 g casamino acids; 0.5 g bacto-tryptone; 0.5 g cysteine-HCl; 0.5 g KH_2PO_4; 0.3 g $MgSO_4.7H_2O$; 0.01 g $FeSO_4.7H_2O$; adjusted to pH 6.5 with NaOH). Solid sterile sucrose is added to a final concentration of 0.6 *M*. To remove the cell wall, lysozyme (100 μg ml^{-1}) and penicillin G (20 μg ml^{-1}) are added, and the cells incubated at 34°C for 1 h. The protoplasts are centrifuged at 3000 *g* and washed twice in protoplast wash buffer [T69 buffer supplemented with 0.6 *M* sucrose, 0.5% (w/v) bovine serum albumin (BSA) and 1 m*M* $CaCl_2$]. Protoplasts are then resuspended in protoplast buffer [T69 supplemented with

0.5 *M* xylose, 0.5% (w/v) BSA, 25 m*M* $MgCl_2$, and 25 m*M* $CaCl_2$]. The number of potential L-form cells can be estimated by diluting a sample of cells in T69C regeneration medium [T69 supplemented with 0.3 *M* sucrose, 0.25 *M* xylose, 0.5% (w/v) BSA and 1 m*M* $CaCl_2$] before plating on T69C solid medium [2.5% (w/v) agar]. Protoplast numbers can be estimated by phase-contrast light microscopy.

3.3.1.2 Protoplast Transformation

Plasmid DNA (50 to 800 ng), polyethylene glycol (PEG) 4000 [35% (w/v)], and 10^9 protoplasts are mixed and incubated at room temperature for 2 min. The mixture is then diluted 10-fold in T69 medium supplemented with 0.5 *M* xylose, 0.5% (w/v) BSA, 1 m*M* $CaCl_2$, and 4 mg choline ml^{-1}. The protoplasts are then centrifuged, washed, and resuspended in the same medium. Dilutions are then added to T69 top agar [T69 supplemented with 0.25 *M* xylose, 0.5% (w/v) BSA, 1 m*M* $CaCl_2$, 4 mg choline ml^{-1}, and 0.8% (w/v) agar]. This is then poured onto T69 agar [0.25 m*M* xylose and 2.5% (w/v) agar] and incubated at 34°C for 20 h. To select for plasmids with erythromycin resistance markers, a further 3 ml of top agar [T69 supplemented with 0.25 *M* xylose, 1 mg erythromycin ml^{-1}, and 0.8% (w/v) agar] is overlayed onto the plates and they are incubated for 4 to 6 days at 37°C.

3.3.2 Transformation of Thermophilic Clostridia

As previously reviewed by Mauchline et al. [16], procedures for the transformation of thermophilic clostridia remain poorly developed. *C. thermocellum* has not been reproducibly transformed. Reports of phenotypic conversion of transformed *C. thermocellum* protoplasts lack direct evidence of transformation [17]. The lack of reproducible transformation methods may be attributed to restriction systems such as that reported in *C. thermocellum* ATCC 27405 [18].

The biotechnologically important, ethanol-producing organism *C. thermohydrosulfuricum* has been successfully transformed [19]. Competence was induced by disruption of the surface S-layer and PEG-induced membrane permeabilization. The method was optimized for *C. thermohydrosulfuricum* DSM 568, which has low endogenous DNAase activity.

A fresh overnight culture is used to inoculate 50 ml of modified RCM broth (containing, per liter, 10 g tryptone; 4 g beef extract; 3 g yeast extract; 2.5 g glucose; 5 g NaCl; 0.5 g $Na_2S.9H_2O$; 0.5 g cysteine HCl; 0.5 g resazurin; 2 g K_2HPO_4; pH 7.2). Cells are grown at 68°C to stationary phase (OD_{578} = 0.45; approx. 1.5×10^9 cells ml^{-1}). The bacteria are centrifuged at 6000 *g* for 8 min, washed in wash buffer [50 m*M* Tris-HCl, pH 8.3, 0.05% (w/v) $Na_2S.9H_2O$, 0.05% (w/v) cysteine HCl], and then in 5 ml of electroporation buffer (wash buffer plus 0.35 *M* sucrose), and finally resuspended 0.2 ml of electroporation buffer. Plasmid DNA (2-5 µg) is added in TE buffer (25 m*M* Tris-HCl, pH 8.0, 10 m*M* EDTA) and incubated at 60°C for 5 min. A 1.5-ml sample of 40% PEG (mol. wt. 6000) is added, and the mixture is incubated for a further 60 min. A 2-ml sample of buffer [10 m*M* Tris-HCl, pH 8.3, 0.05% (w/v) $Na_2S.9H_2O$, 0.15 *M* NaCl, 0.05% (w/v) cysteine HCl] is added and the cells centrifuged before resuspending them in 1 ml of RCM broth. Bacteria are finally plated on RCM agar (2%, w/v), supplemented with selective antibiotic, as appropriate, and incubated at 55°C for between 4 and 6 days.

3.3.3 Electroporation Procedures

In this technique, cells are exposed to pulses of high-intensity electric fields that induce the transient formation of pores in the cell membrane, through which exogenous DNA can enter. The electroporation protocols are optimized for different species of clostridia by varying the method for cell preparation before electroporation; the electroporation parameters and the procedures for cell rescue post-electroporation. Numerous protocols have been documented. The most commonly used are described here. All of these procedures benefit from placing the electroporation chamber within an anaerobic workstation.

3.3.3.1 Electrotransformation of *C. beijerinckii* NCIMB 8052

The protocol for electrotransformation of *C. beijerinckii* NCIMB 8052 was first described by Oultram et al. [20]. A 10-ml culture in 2 × YTG (per liter: 16 g tryptone; 10 g yeast extract; 5 g glucose and 5 g NaCl) is inoculated and then serially diluted in further 10-ml cultures for incubation for 16 h (overnight) at 37°C. A culture in exponential phase is used to inoculate (1:10) 100 ml of 2 × YTG. When the culture reaches an OD_{600} = 0.6, it is cooled on ice and the cells pelleted by centrifugation at 2000 *g* for 10 min and washed in 10 ml precooled (4°C) electroporation buffer [270 m*M* sucrose, 1 m*M* $MgCl_2$ and 7 mM $NaHPO_4$ (pH7.4)]. The bacteria are then resuspended in 5 ml precooled electroporation buffer (4°C) and incubated on ice for 10 min before use.

Plasmid DNA (0.5 μg) is added to the prepared cells (0.8 ml), and the mixture is transferred to a precooled (4°C) electroporation cuvette (0.4 cm interelectrode distance). The mixture is incubated for 8 min on ice before electroporation. A 1.25 kV pulse (25 μF capacitance and 200 Ω resistance) is then applied using a Bio-Rad Gene Pulser. The cells are returned immediately to ice for 10 min. After dilution in 10 vol 2 × YTG and incubation at 37°C for 3 h, cells are concentrated by centrifugation at 2000 *g* for 10 min and resuspended in 100 μl 2 × YTG before plating on 2 × YTG plates containing the appropriate antibiotic. Plates are incubated at 37°C for 2 to 3 days.

3.3.3.2 Electrotransformation of *C. acetobutylicum* ATCC 824

C. acetobutylicum ATCC 824 contains a restriction system, Cac824I, that greatly reduces the electroporation efficiency of plasmid DNA lacking the appropriate methylation signature. To circumvent this, plasmid DNA may be methylated *in vivo* in *E. coli* containing the pAN1 plasmid, which encodes a bacteriophage methyltransferase [21]. The protocol described here (derived from Mermelstein et al. [22]) has been shown to produce 5×10^5 transformants μg^{-1} DNA, with a pIM13-based vector (4.8 kb). The electroporation parameters have since been optimized to produce a 10-fold increase in transformation frequency [23]. However, the equipment required to obtain optimal results is not readily available.

A culture is grown in 60 ml RCM (pH 5.2) to late exponential phase. The cells are then centrifuged, washed, and resuspended in 2.1 ml precooled electroporation buffer [272 m*M* sucrose (pH7.4), 5 m*M* NaH_2PO_4]. A sample of the cell suspension (0.7 ml) is added to a cuvette (0.4 cm interelectrode distance) and held on ice for 5 min. Plasmid DNA (0.1–10 μg) is then added and the suspension held on ice for 2 min before electroporation with a 2.0 kV pulse and 25 μF capacitance. RCM (10 ml) is added to the bacterial suspension, and it is incubated at 37°C for 4 h before plating on RCM agar (pH 5.8) containing the appropriate selective agents.

3.3.3.3 Electroporation of *C. botulinum*

Electroporation protocols have been optimized in examples of proteolytic (ATCC 3502) and nonproteolytic (ATCC 25765) strains of *C. botulinum* [13,24]. The only major difference between the protocols is the field strength used. Both strains yield comparable transformation frequencies when electroporated with plasmid pGK12 (4.4 kb). The nonproteolytic group II strain ATCC 25765 encodes a restriction modification system. This barrier is circumvented *in vivo* by methylation in an *E. coli* host containing a gene encoding the *B. subtilis* M.*Bsu*F1 methylase [13]. The protocol described below is for *C. botulinum* ATCC 25765, with alterations for ATCC 3502 given in square brackets.

A culture (20 ml) is grown for 16 h in TPGY medium (per liter: 20 g trypticase; 5 g yeast extract; 5 g peptone; 1 g glucose; 1 g cysteine-HCl) supplemented with glycine (1% w/v). A 300-ml

TPGY culture is then inoculated with this culture and grown to an OD_{660} = 0.8. The culture is equally divided, and the two aliquots are held on ice for 10 min before pelleting the cells by centrifugation at 6000 *g* for 10 min at 4°C. The supernatant is carefully removed and the bacterial pellet gently resuspended in 50 ml of precooled (4°C) electroporation buffer (10% PEG 6000 [PEG 8000 for ATCC 3502], 1 m*M* $MgCl_2$, 7 m*M* Na phosphate, pH 7.5). The cells are centrifuged as before and then resuspended in 3 ml precooled electroporation buffer. DNA (0.1–2.0 mg ml^{-1}) is added to a precooled electroporation cuvette (0.4 cm interelectrode distance) and mixed with 0.8 ml of cell suspension by gentle inversion 2 to 3 times. The bacteria are electroporated using a 2.0 kV pulse [2.5 kV for ATCC 3502] (25 μF capacitance and 400 Ω resistance). The cuvette is then immediately placed back on ice for 5 min. The bacterial suspension is added to 10 ml prewarmed (37°C) TPGY broth (supplemented with 25 m*M* $MgCl_2$), and the cells are left to recover for 5 h at 37°C. The bacterial suspension is then distributed into 6 eppendorf tubes, centrifuged for 3 min, the cells in each tube gently resuspended in 150 μl of TPGY broth and spread on individual TPGY agar (2.5%, w/v) plates containing the appropriate selective agent. Plates are incubated for between 24 and 48 h.

3.3.3.4 Electroporation of *C. cellulolyticum*

Electroporation of *C. cellulolyticum* ATCC 35319 was optimized using ATP leakage assays [25]. This strain encodes a restriction system, which is circumvented by *in vitro* methylation of vector DNA with *Msp*I methylase (1 U per μg of DNA incubated at 37°C for 3 h). Bacteria are grown on a defined medium containing (per liter): 6.7 g cellulose; 1.4 g KH_2PO_4; 2.9 g $K_2HPO_4.3H_2O$; 1 g $(NH_4)_2SO_4$; 0.1 g $MgCl_2.6H_2O$; 0·02 g $CaCl_2$; 25 μl 5% (w/v) $FeSO_4$ dissolved in 50 m*M* H_2SO_4; 1 ml trace element solution (described below); 10 ml vitamin solution (described below); 0.5 g Na_2S; and 0.5 ml 0.2% (w/v) resazurin [26].

The trace element solution contains (per liter): 5 g $FeSO_4{\cdot}7H_2O$; 1.44 g $ZnSO_4{\cdot}7H_2O$; 1.12 g $MnSO_4{\cdot}7H_2O$; 0.25 g $CuSO_4{\cdot}5H_2O$; 0.2 g $Na_2B_4O_7$; 1 g $(NH_4)_6(Mo)_7O_{24}{\cdot}4H_2O$; 0.04 g $NiCl_2$; 0.02 g $CoCl_2$; 0.03 g H_3BO_3; 0.02 g Na_2SeO_3; 50 ml HCl (10M).

The composition of the vitamin solution is (per 100 ml distilled water): 10 mg d-biotin; 25 mg p-aminobenzoic acid; 15 mg nicotinic acid; 25 mg ribofavin; 25 mg pantothenic acid; 25 mg thiamin; 10 mg cyanocobalamin. The vitamin solution is sterilized by passage through a 0.2-μm filter.

All incubations are at 34°C. Cells are grown to mid-log phase in 40 ml synthetic medium then washed with ice-cold electrotransformation buffer [270 m*M* sucrose, 5 m*M* K_2HPO_4 (pH 6.5)] and resuspended in 1.5 ml of the same buffer. A 0.5 ml aliquot of the cell suspension is added to 5 μl (0.5–1.0 μg) of methylated DNA in an electroporation cuvette (0.4 cm gap). The bacterial suspension is electroporated with a 2.0 kV pulse (25 μF capacitance and 1000 Ω resistance) and then immediately added to 5 ml prewarmed (34°C) synthetic medium for 6 h. Cells are then plated on selective medium (as above, but solidified with agar) and incubated for 3 to 5 days.

3.3.3.5 Electroporation of *C. perfringens*

Various methods have been devised for electroporation of *C. perfringens* [27,28]. The method most commonly used is based on that of Allen and Blaschek [28] and is given here.

Cells are incubated in TYG broth (3% biotrpticase, 2% yeast extract, 0.1% sodium thioglycollate, 0.5% glucose, pH7.4) for 16 h at 37°C. A 5-ml sample of cell suspension is centrifuged for 10 min at 5000 *g*. The cells are washed with 5 ml of electroporation buffer (0.272 *M* sucrose, 7 m*M* K_2HPO_4, and 1 m*M* $MgCl_2$) and resuspended in 1 ml of electroporation buffer. DNA (1 μg ml^{-1}) and 350 μl of the cell suspension are mixed in a cuvette (0.4 cm gap) and electroporated at 2.5 kV (25 μF capacitance and 200 resistance). The cuvette is incubated on ice for 5 min, after which 1 ml of YTG broth is added and incubation is at 37°C for 3 h. Cells are then spread on YTG agar (2%, w/v) containing the appropriate selective agent.

It should be noted that transformation frequency appears to be very strain-specific. Between 100- and 1000-fold differences have been reported between the transformation frequencies of different strains of this organism [28,29].

3.4 CONJUGATIVE TRANSFER

Conjugation is an important route for genetic exchange between bacteria — and indeed between bacteria and other organisms. It is thought to have played an important role in the spread of antibiotic resistance genes between organisms, particularly through the agency of conjugative transposons. The process requires close cell-to-cell contact and involves both a *cis*-acting nick site (*oriT*), as well as a number of *trans*-acting functions (most often plasmid encoded) that are necessary for mating pair formation as well as DNA processing and transfer of the conjugative plasmid to the recipient cell.

3.4.1 Conjugative Plasmids

Conjugative plasmids are widespread throughout the bacterial kingdom. However, in clostridia, naturally occurring conjugative plasmids appear to be relatively rare and, indeed, they have only been described in *C. perfringens*. In this species, all such plasmids are closely related, suggesting that they may have arisen from a common progenitor [30]. They encode two highly conserved antibiotic resistance genes. One of these, *tetP*, confers resistance to tetracycline via an efflux mechanism [31] and the other, *catP*, confers resistance to chloramphenicol [32] and resides on a transposable element [33].

These plasmids are readily transferred between different strains of *C. perfringens* [34], but despite this, they have so far proved to be of little use in the development of gene transfer technologies. To date, there is only a single report of the transfer of pIP401 to *C. difficile*, in which it was not stably inherited [35].

The conjugative plasmids from a number of other bacteria have very broad host ranges, and several examples from enterococci and streptococci, including pAMβ1 and pIP501, have been conjugated into *C. acetobutylicum* [36,37]. Plasmid pAMβ1 has also been conjugated into *C. butyricum* and *C. pasteurianum* [38]. The transfer efficiency of these plasmids is highly dependent on the donor organism and is typically most efficient when a Gram-positive species such as *Lactococcus lactis* is employed [36]. Transfer of nonconjugative plasmids from *B. subtilis* into *C. beijerinckii*, has also been achieved using the conjugative functions of pAMβ1. It relies on the formation of a co-integrate molecule, which is then transferred into the recipient cell [39]. This technique is inefficient and prone to the occurrence of large deletions affecting the conjugative functions of pAMβ1 in *B. subtilis* [40], and it has not been used widely.

3.4.2 Conjugative Transposons

In addition to conjugative plasmids, a number of Gram-positive organisms, including various clostridia, harbor conjugative transposons, such as Tn*916* of *E. faecalis* [41], Tn*1545* of *S. pneumoniae* [42], and Tn*4451* of *C. perfringens* [33]. They are large mobile genetic elements that generally reside within the bacterial chromosome and encode all of the functions necessary for their own transfer. Transposition of conjugative transposons relies on the formation of a covalently closed supercoiled circular DNA intermediate, which can then either reintegrate into the chromosome of the same cell or, following transfer by conjugation, insert into a recipient's genome. Insertion takes place without duplication of the target site, and they are replicated as part of the host chromosome [43]. They often possess an extremely wide host range [44] and have been transferred to a number of different clostridia, including *C. tetani* [45], *C. beijerinckii* [46],

C. acetobutylicum [2], *C. perfringens* [30], *C. botulinum* [4], and *C. difficile* [47,48]. These elements have been utilized as tools for the delivery of cloned gene fragments [49,50] and for transposon mutagenesis strategies (see Chapter 4).

3.4.3 Conjugative Mobilisation from *E. coli*

The conjugation strategy with perhaps the widest potential utility, however, is the transfer of plasmids from *E. coli* into clostridia. Indeed, the transfer of genetic material from *E. coli* into several species of clostridia has now been documented. These include *C. beijerinckii* (formerly known as *C. acetobutylicum*) [10], *C. perfringens* [51], *C. cellulolyticum* [5], *C. botulinum* [52], and *C. difficile* [12,53]. In all cases, transfer was reliant on components of the broad host range IncP family of plasmids.

A number of mobilizable *Clostridium/E. coli* shuttle vectors are available carrying the transfer origin (*oriT*) of an IncP plasmid, such as RP4 or RK2 (see Table 3.1). This region, in conjunction with several additional trans-acting functions (Tra functions), is absolutely required for the conjugative transfer of plasmid DNA to recipient cells. The Tra functions are provided by the *E. coli* (Tra^+) donor and may either be plasmid-encoded (e.g., carried by a IncP-type helper plasmid such as R702) or integrated into the chromosome, as is the case with *E. coli* strain SM10. In many instances, the transfer of IncP-containing vectors from *E. coli* (Tra^+) donors is remarkably efficient, particularly if a high donor-to-recipient ratio is used during conjugation. A typical protocol for conjugative plasmid transfer from *E. coli* to *Clostridium* is described below.

Inoculate the donor strain (e.g., *E. coli* HB101 containing R702, as well as the plasmid to be mobilized) into 5 ml of Brain Heart Infusion Broth (BHIB, Oxoid) containing the appropriate antibiotic selections, and incubate aerobically overnight at 37°C with shaking (200 rpm). Grow serial dilutions of the clostridial recipient in 5 ml aliquots of BHIB anaerobically overnight at 37°C. The following morning, dilute back one of the serial dilutions of the clostridial recipient that has not entered stationary phase to give an $OD_{600} = 0.05 - 0.1$ and grow anaerobically. Concurrently, dilute the *E. coli* donor culture to give an $OD_{600} = 0.01$ and grow aerobically. When both cultures have reached an $OD_{600} = 0.45 - 0.6$, place the donor culture into the anaerobic chamber, harvest the bacteria from 1 ml of culture, and wash them with 1 ml anaerobic phosphate buffered saline (PBS). Gently resuspend the bacterial pellet in 100 µl of the recipient clostridial culture (giving a ratio of 10:1, donor:recipient) and gently spread the mixture onto the surface of a 0.2 µm filter (Whatman) that has been placed on an anaerobic reinforced clostridial medium (RCM) agar plate. Incubate anaerobically overnight. Following this period of cocultivation, resuspend the bacterial growth from the filter in 1 ml anaerobic BHIB or PBS by vortex-mixing and spread 0.1 ml samples onto RCM agar containing the appropriate antibiotics to select for transconjugants (and counter-select against the donor and the recipient). To determine the transfer frequency, make serial dilutions of the donor-recipient mixture and spread onto appropriate agar plates to permit the separate growth of donor and recipient strains. Transconjugants are usually visible after anaerobic incubation at 37°C for 24 to 48 h. Most *E. coli* and *Bacillus subtilis* donors can be counter-selected using 10 µg trimethoprim ml^{-1}, and this is particularly efficient if the donor is a phage lambda *cI857* lysogen, since plating at 42°C provokes donor lysis, as a result of prophage induction. In the case of *C. difficile* recipients selection with D-cycloserine (250 µg ml^{-1}) and cefoxitin (8 µg ml^{-1}) will select against nonclostridial donors.

Gene transfer procedures based on conjugation are perhaps more labor-intensive than electrotransformation techniques. Despite this, they can offer distinct advantages. Most notably, conjugation may result in substantially higher plasmid transfer frequencies than can routinely be obtained with electrotransformation. Moreover, in some cases conjugation represents the only available option. Furthermore, it is not affected by extracellular nucleases because of the need for close cell-to-cell contact [69], and it does not require specialized and expensive equipment.

TABLE 3.1
Clostridial Plasmid Cloning Vectors

Plasmid	Size (kb)	Replicon[a]	Comments	Markers[b]	Ref.
pCAK1	11.6	CAK1 (*C. acetobutylicum*)	Phagemid ssDNA intermediate	Em^R [Ap^R]	54
pTYD101	4.0	pSC86 (*C. acetobutylicum*)	Deletion variant	Cm^R	55
pTYD104	7.6	pSC86 (*C. acetobutylicum*)	Deletion variant	Cm^R [Ap^R]	55
pCB3	7.03	pCB101 (*C. butyricum*)	ssDNA replication	Em^R [Ap^R]	56
pCTC511	7.85	pCB101 (*C. butyricum*)	IncP mobilizable	Em^R [Ap^R]	10
pMTL540E	5.23	pCB102 (*C. butyricum*)	Segregationally stable	Em^R [Ap^R]	57
pMTL540F	5.5	pCB102 (*C. butyricum*)	Expression vector	Em^R [Ap^R]	57
pMTL9401	4.6	pCB102 (*C. butyricum*)	IncP mobilizable	Em^R	12
pCB5	9.5	pCB103 (*C. butyricum*)	Uncharacterized replicon	Em^R [Ap^R]	56
pSYL2	8.7	pCBU2 (*C. butyricum*)		Em^R Tc^R	58
pMTL9301	7.1	pCD6 (*C. difficile*)	IncP mobilizable, transferred to *C. difficile* genome strain	Em^R	12
pAK201	8.0	pHB101 (*C. perfringens*)	Stable in *C. perfringens*	Cm^R	59
pJIR1456	6.8	pIP404 (*C. perfringens*)	IncP mobilizable	Cm^R	51
pJIR1457	6.2	pIP404 (*C. perfringens*)	IncP mobilizable, transferred to *C. botulinum* and *C. difficile*	Em^R	51
pJIR418	7.4	pIP404 (*C. perfringens*)	General purpose cloning vector	Em^R Cm^R	60
pMTL9611	5.5	pIP404 (*C. perfringens*)	IncP mobilizable	Em^R	12
pJU12	11.6	pJU121 (*C. perfringens*)		Tc^R	61
pHR106	7.9	pJU122 (*C. perfringens*)	Replicates in *C. beijerinckii*	Tc^R [Ap^R]	62
pRZE4	10.0	pJU122 (*C. perfringens*)	Replicates in *C. beijerinckii*	Em^R [Ap^R] Cm^R	62
pRZL3	10.8	pJU122 (*C. perfringens*)	Replicates in *C. beijerinckii*	Tc^R [Ap^R]	62
pSYL7	9.2	pJU122 (*C. perfringens*)		Em^R Tc^R	63
pBC16Δ1	2.8	pBC161 (*B. cereus*)	Unstable in *C. acetobutylicum*	Tc^R	63
pECII	4.5	pIM13 (*B. subtilis*)	Transferred to *C. cellulolyticum*	Em^R	5
pFNK1	2.4	pIM13 (*B. subtilis*)		Em^R	64
pIA		pIM13 (*B. subtilis*)	PACYC-based vector	Em^R	65
pIM13	2.3	pIM13 (*B. subtilis*)	ssDNA replication	Em^R	63
pKNT11	6.5	pIM13 (*B. subtilis*)	General purpose cloning vector	Em^R [Ap^R]	63
pKNT14	4.3	pIM13 (*B. subtilis*)	No *E. coli* replicon	Em^R	63
pKNT19	4.9	pIM13 (*B. subtilis*)	pUC19 polylinker	Em^R [Ap^R]	66
pSYL14	4.4	pIM13 (*B. subtilis*)		Em^R [Ap^R]	63
pXYLgusA		pIM13 (*B. subtilis*)	Xylose inducible expression vector	Em^R [Ap^R]	67
pAMβ1	25.5	pAMβ1 (*E. faecalis*)	Conjugative theta replicating	Em^R	36
pMTL500E	6.43	pAMβ1 (*E. faecalis*)	Broad host range cloning vector	Em^R [Ap^R]	20
pMTL500F	6.69	pAMβ1 (*E. faecalis*)	Expression vector	Em^R [Ap^R]	56
pMTL502E	7.52	pAMβ1 (*E. faecalis*)	Low copy version of pMTL500E	Em^R [Ap^R]	56
pMTL513	7.29	pAMβ1 (*E. faecalis*)	Stability cloning vector	Em^R [Ap^R]	56
pMU1328	7.5	pAMβ1 (*E. faecalis*)	Deletion variant in *C. acetobutylicum*	Em^R	56
pSYL9	8.9	pAMβ1 (*E. faecalis*)	Broad host range cloning vector	Em^R [Ap^R] Tc^R	63
pVA1	11.0	pAMβ1 (*E. faecalis*)	Broad host range cloning vector	Em^R	15
pIP501	35.0	pIP501 (*E. faecalis*)	Conjugative	Em^R Cm^R	37
pGK12	4.4	pWV01 (*L. lactis*)	Very broad host range vector	Em^R Cm^R	68
pT127	4.4	pT127 (*S. aureus*)	Unstable in *C. acetobutylicum*	Tc^R	63
pMK419	5.6	pUB110 (*S. aureus*)	Transferred to *C. thermocellum*	Cm^R [Ap^R]	17
pUB110	4.5	pUB110 (*S. aureus*)	ssDNA replication, thermostable	Km^R	19
pMTL30/31	4.36	None	IncP mobilizable integration vector	Em^R [Ap^R]	10

[a] Progenitor plasmid and host from which it was originally isolated.

[b] Em^R, erythromycin resistance; Tc^R, tetracycline resistance; Cm^R, chloramphenicol/thiamphenicol resistance; Km^R, kanamycin resistance; [Ap^R], ampicillin resistance in Gram-negative intermediate.

3.5 CIRCUMVENTION OF RESTRICTION BARRIERS

As indicated in Section 3.3, in many instances, the successful transfer of extrachromosomal elements, either by transformation or conjugation, has required the circumvention of the activity of endogenous restriction-modification (RM) systems. This is achieved through appropriate methylation of the vector DNA to be introduced. Organisms for which such a strategy has proven necessary include *C. acetobutylicum* ATCC 824 [21], *C. cellulolyticum* ATCC 35319 [5], *C. botulinum* ATCC 25765 [13], and *C. difficile* CD3 and CD6 [12]. In all cases, restriction activity was initially detected in bacterial lysates, after which the restriction and the methylation specificity of the RM system was determined. As more and more genome sequences become available, gene transfer strategies may be based on the characterization of cloned methylase genes identified by *in silico* approaches [70].

To isolate restriction enzyme activities, cell wall and associated nonspecific nucleases are enzymatically (lysozyme) removed in isotonic buffer before releasing the contents of the protoplasts formed by osmotic shock. A typical procedure, exemplified by the methodology employed in *C. difficile* [12], is as follows: Organisms are grown in 50 ml BHIB to an OD_{600} = 0.4 to 0.6, whereupon cells are harvested by centrifugation at 6000 g for 5 min and resuspended in 2 ml of 0.5 *M* sucrose/0.05 *M* Na phosphate buffer, pH7.0 containing lysozyme (10 mg ml^{-1}). This cell suspension is held anaerobically at 37°C for 1 to 4 h, during which protoplast formation occurs. Cells may be monitored under a light microscope following staining with safranin red and harvested by centrifugation once >90% of the cells are deemed to have formed protoplasts. The resulting pellets are resuspended in 500 µl of 0.05 *M* Na phosphate buffer, pH7.0, and lysis allowed to proceed for 5 to 10 min at room temperature. The cell debris may then be removed by centrifugation, and the resulting lysate stored at 4°C. Aliquots of a lysate (2 to 8 µl) are incubated with 0.5 µg of DNA of a plasmid of known nucleotide sequence in a range of restriction enzyme buffers for 4 h at 37°C, in a final volume of 20 µl, and the reaction products analyzed by agarose gel electrophoresis.

Using this type of approach, it has invariably proven possible to identify the point at which the marker plasmid is cleaved, most simply by comparison with the fragmentation pattern observed using commercially available restriction enzymes. Novel specificities require a more detailed analysis, in which fragmented DNA may be directly sequenced, or cloned and sequenced thereafter [12]. Having determined the specificity of the restriction enzyme, it is necessary to determine the methylation specificity of the RM system involved. This involves testing the ability of available methylase enzymes known to act on the identified recognition sequence. However, as the majority of clostridial methylases appear cytosine-specific, the methylation point may be determined directly, using a modification of the method of Feil et al. [71].

Approximately 2 µg of linearized plasmid DNA isolated from the strain under investigation is suspended in 100 µl of deionized water in a siliconized, 1.5-ml eppendorf tube and then denatured by adding 11 µl of 3 *M* NaOH and incubating at 37°C for 20 min. The tube is placed on ice, and 1.1 ml of 3.5 *M* $NaHSO_3$/1 mM hydroquinone, pH 5.0, is added. The solution is then overlayed with 150 µl of mineral oil and incubated in the dark for 24 h at 0°C. The sample is removed from beneath the mineral oil and transferred to a 1.5-ml siliconized eppendorf tube where the DNA is extracted using a Geneclean II kit (Stratech Scientific Ltd.) for 30 min at 4°C. The precipitated DNA is resuspended in 100 µl of deionized water and desulfonated by adding 11 µl of 2 *M* NaOH and incubating for 10 min at 20°C. This treatment converts all unmethylated cytosine bases to uracil. The modified DNA is precipitated by adding 5 *M* ammonium acetate, pH 7.0, to a final concentration of 3 *M*, and 3 volumes of ethanol. Following resuspension in 100 µl of deionized water, a 4 µl sample is used as a template in a PCR reaction using "modified" primers, which can only anneal if all of the targeted cytosine residues have been converted to uracil. The amplified region is then cloned and sequenced. Any cytosines present in the sequence obtained were methylated in the original DNA.

The application of this approach is reliant on being able to PCR-amplify a known sequence from the organism under investigation, which is likely (i.e., GC-rich DNA in an AT-rich organism), or experimentally known, to be cleaved by the endogenous enzyme. In the study of [13], this was made possible by the fact that a relatively GC-rich, foreign plasmid element (pGK12) could be transformed into *C. botulinum* at low efficiencies. In other cases, where transformation cannot be demonstrated, known genomic sequences may be targeted. Alternatively, it may prove possible to introduce foreign DNA using either a conjugative plasmid or a conjugative transposon. In this instance, the delivered element may be engineered to contain GC-rich DNA, either through the simple insertion of appropriate DNA fragments or by cointegration of a suitable plasmid vector with the conjugative element [47,72].

Prevention of DNA transfer by host RM systems is highly strain-specific. Indeed, there are many instances where restriction has not been a problem e.g., *C. beijerinckii* NCIMB 8052 [21], *C. perfringens* strain 13 [73], *C. difficile* strains CD37 [12] and CD630 [70], and *C. botulinum* ATCC 3502 [16]. Genome sequencing has shown that many of these organisms carry at least one type II methylase gene (most often more than one), but they lack genes encoding the cognate restriction enzymes. Thus, the recently completed genomes of *C. botulinum* ATCC 3502 and *C. perfringens* contain orphan copies of methylase genes (three and one, respectively), and are both readily transformable in the absence of any measures to circumvent restriction barriers [16,73]. In the case of *C. acetobutylicum,* a total of six methylase genes are present, two of which have an adjacent gene encoding a restriction enzyme; *C. thermocellum* has seven methylase genes, but only three restriction enzyme genes; while *C. tetani* has three methylase genes and a single restriction enzyme gene. The latter two organisms have yet to be transformed. On the other hand, *C. acetobutylicum* requires that the incoming DNA be protected from the activity of just one restriction endonuclease (*Cac*824I) for successful DNA transfer [21]. The circumvention of a single restriction activity is similarly required for transformation of *C. botulinum* ATCC 25765 [13], *C. cellulolyticum* [5], and *C. difficile* CD3 [12]. It now seems likely that the majority of type II methylase genes in clostridia plays no role in restriction/modification. To date, the potential effects of type I restriction systems on gene transfer in *Clostridium* spp. have not been analyzed.

3.6 AVAILABLE PLASMID CLONING VECTORS

Over the past two decades, genetic manipulation of several clostridia has been achieved, and a number of vectors are now available for use in these organisms. Despite this, the genus as a whole remains relatively recalcitrant to "laboratory based" genetic transfer, and the use of more genetically amenable organisms such as *E. coli* and *B. subtilis* as intermediates is necessary for undertaking plasmid constructions. While the available vectors are almost exclusively based on replicons derived from native clostridial plasmids, or from plasmids found in other Gram-positive hosts, they must also carry a second replication region, such as that of ColE1, which facilitates their maintenance in the intermediate host (normally *E. coli*).

3.6.1 Native Plasmids

Plasmids are widespread among the clostridia. Most remain cryptic, but in some cases functions have been ascribed. For example, pSOL1 of *C. acetobutylicum* ATCC 824 is a megaplasmid that encodes the genes needed for the production of acetone and butanol by this organism [74]. In addition, some important virulence factors of several pathogenic clostridia are plasmid-encoded, including the tetanus toxin of *C. tetani* [75], type G neurotoxin of certain strains of *C. botulinum* [76], and several of the toxins produced by some strains of *C. perfringens* [77].

3.6.2 Shuttle Vectors

The naturally occurring plasmids represent the most obvious candidates for use in the development of clostridial shuttle vectors, and it is not therefore surprising to find that several have been extensively characterized, including pIP404 from *C. perfringens* [78], pCB101 and pCB102 from *C. butyricum* [79], and, more recently, pCD6 from *C. difficile* [12]. As a result, a number of clostridial replicons have been identified and analyzed. The replication regions of pCB101 and pIP404 were identified by their ability to promote the stable maintenance of an otherwise nonreplicative vector in *B. subtilis* [80,81], while *C. beijerinckii* NCIMB 8052 was used to identify the replication regions of pCB102 [56] and pCD6 [12]. In addition, the use of sequence databases proved a valuable tool in predicting the location of the pCD6 replicon.

The replicon of pCB101 exhibits many of the characteristic features of plasmids that replicate via a rolling circle mechanism. One ORF is present, which shares homology with the replication proteins of other plasmids from Gram-positive bacteria that use this mode of replication [82]. The replication region of pCB102, on the other hand, shares no similarity at either the DNA or the protein level with any other characterized plasmid, and as such, the mechanism by which this plasmid replicates remains unknown. Likewise, it is not clear how pCD6 and pIP404 replicate. They do, however, possess several shared features, such as two large, distantly related, putative replication proteins and an extensive region of DNA repeats, which is often associated with plasmids that replicate by a theta mechanism [12]. In spite of our rudimentary understanding of the replication mechanisms of several of these plasmids, they have all been used successfully to develop clostridial cloning vectors (see Table 3.1).

An alternative strategy for vector construction uses replicons derived from other Gram-positive organisms. Plasmids such as pWV01 from *Lactococcus lactis* [83], pIM13 from *B. subtilis* [63], and pUB110 from *Staphylococcus* aureus [19] are good examples. The lack of specificity of the replication and partitioning systems of these particular plasmids, as well as the functionality of some of their associated antibiotic-resistance genes in clostridia, has permitted the development of many of the cloning vectors currently in use (see Table 3.1). Although many of these nonclostridial plasmids appear able to replicate in *Clostridium* spp. [5,10], none of them are segregationally stable (i.e., they tend to be lost in the absence of positive selection for bacteria that maintain them). With the notable exception of pAMβ1, which replicates via a unidirectional theta mechanism, the majority of these plasmids replicate using a rolling circle mechanism, involving a single stranded DNA (ssDNA) intermediate [84]. The highly recombinogenic ssDNA molecules may account for the structural instability of several of the recombinant vectors that have been constructed [56].

In addition to the creation of basic shuttle vectors, a number of specialized plasmids have been constructed, including expression vectors. These have been used to bring about heterologous gene expression, and of antisense RNA expression, in a number of species, including *C. acetobutylicum* [85–87] and *C. tetani* [88]. pMTL500F and pMTL540F are two examples of versatile expression vectors for use in clostridia. The former is based on the pAMβ1 replicon, while the latter utilizes the pCB102 replication region. Both carry an expression cassette that contains the promoter region of the *C. pasteurianum* ferredoxin (Fd) gene. They have been used in a number of situations, most prominently to bring about the production of prodrug converting enzymes useful in cancer therapy [56,57,89]. More recently, an inducible expression vector, pXYLgusA, which utilizes *xylR* and the *xylA* promoter-operator regions from the xylose operon of *Staphylococcus xylosus*, has been tested in *C. acetobutylicum* ATCC 824. This expression vector carries the *B. subtilis* pIM13 replication region and reportedly leads to a 17-fold increase in reporter gene activity when xylose, which functions as the inducer, is added to the system [67]. A more detailed description of such vectors, and other specialized vectors, is presented in other chapters in this volume.

3.7 CONCLUDING REMARKS

The basic gene transfer technologies described in this chapter have underpinned the recent development of strategies for targeted gene disruption via allelic replacement, generation of unmarked deletions, and antisense regulation of genes in several species of *Clostridium*. These approaches are documented elsewhere in this volume. They will permit exploration of the biological roles of the multitude of genes, whose presence has been revealed by genome sequencing projects, as well as the exploration of problems of fundamental biological interest, such as the contributions made by individual genes to metabolic regulation in response to environmental change. Nevertheless, considerable ingenuity will be required to perfect and further refine the tools already available for the small number of clostridia that we can now manipulate.

ACKNOWLEDGMENTS

The authors wish to acknowledge the financial support of the BBSRC (Grant No. 346/E13746) and the European Union (QLK3-CT-2001-01737).

REFERENCES

1. Marmur, J., A procedure for the isolation of deoxyribonucleic acid from micro-organisms, *J. Mol. Biol.,* 3, 208, 1961.
2. Bertram, J. and Dürre, P., Conjugal transfer and expression of streptococcal transposons in *Clostridium acetobutylicum*, *Archives of Microbiol.,* 151, 551–557, 1989.
3 Kataeva, I. et al., Cloning and sequence analysis of a new cellulase gene encoding CelK, a major cellulosome component of *Clostridium thermocellum*: evidence for gene duplication and recombination, *J. Bacteriol.,* 181, 5288, 1990.
4. Lin, W.J. and Johnson, E.A., Transposon Tn*916* mutagenesis in *Clostridium botulinum*, *Appl. Environ. Microbiol.,* 57, 2946, 1991.
5. Jennert, K.C. et al.,Gene transfer to *Clostridium cellulolyticum* ATCC 35319, *Microbiol.,* 146, 3071, 2000.
6. O'Brien, R.W. and Morris, J.G., Oxygen and the growth and metabolism of *Clostridium acetobutylicum*, *J. Gen. Microbiol.*, 68, 307, 1971.
7. Karlsson, S. et al., Expression of *Clostridium difficile* toxins A and B and their sigma factor TcdD is controlled by temperature, *Infect. Immun.,* 71, 1784, 2003.
8. Brynestad, S., Synstad, B., and Granum, P.E., The *Clostridium perfringens* enterotoxin gene is on a transposable element in type A human food poisoning strains, *Microbiol.,* 143, 2109, 1997.
9. Green, E.M. et al., Genetic manipulation of acid formation pathways by gene inactivation in *Clostridium acetobutylicum* ATCC 824, *Microbiol., 1*42, 2079, 1996.
10. Williams, D.R., Young, D.I., and Young, M., Conjugative plasmid transfer from *Escherichia coli* to *Clostridium acetobutylicum*, *J. Gen. Microbiol.,* 136, 819, 1990.
11. Sambrook, J., Fritsch, E.F., and Maniatis, T., *Molecular Cloning, A Laboratory Manual,* Cold Spring Harbor Laboratory Press, Cold Spring Harbor, NY, 1989.
12. Purdy, D. et al., Conjugative transfer of clostridial shuttle vectors from *Escherichia coli* to *Clostridium difficile* through circumvention of the restriction barrier, *Mol. Microbiol.,* 46, 439, 2002.
13. Davis, T.O. et al., Development of a transformation and gene reporter system for group II, non-proteolytic *Clostridium botulinum* type B strains, *J. Mol. Microbiol. Biotechnol.,* 2, 59, 2000.
14. Brynestad, S. and Granum, P.E., Evidence that Tn*5565*, which includes the enterotoxin gene in *Clostridium perfringens,* can have a circular form which may be a transposition intermediate, *FEMS Microbiol. Lett.,* 170, 281, 1999.
15. Reysset, G. et al., Transfection and transformation of *Clostridium acetobutylicum* strain N1-4081, *Biotechnol. Techniques*, 2, 199, 1988.
16. Mauchline, M.L. et al., Genetics of Clostridia, in *Manual of Industrial Microbiology and Biotechnology, 2nd Edition,* Demain, A.L. and Davies, J.E., Eds., ASM Press, Washington, D.C., 1999, 475–490.
17. Tsoi, T.V. et al., *Clostridium thermocellum* — a new object of genetic studies, *Mol. Gen. Mikrobiol. Virusol.*, 11, 18, 1987.
18. Klapatch, T.R., Demain, A.L., and Lynd, L.R., Restriction endonuclease activity in *Clostridium thermocellum* and *Clostridium thermosaccharolyticum, Appl. Microbiol. Biotechnol.,* 45, 127, 1996.

19. Soutschek-Bauer, E., Hartl, L., and Staudenbauer, W.L., Transformation of *Clostridium thermohydrosulfuricum* DSM 568 with plasmid DNA, *Biotechnol. Lett.*, 7, 705, 1985.
20. Oultram, J.D. et al., Introduction of plasmids into whole cells of *Clostridium acetobutylicum* by electroporation, *FEMS Microbiol. Letts.*, 56, 83, 1988.
21. Mermelstein, L.D. and Papoutsakis, E.T., *In vivo* methylation in *Escherichia coli* by the *Bacillus subtilis* phage phi 3T I methyltransferase to protect plasmids from restriction upon transformation of *Clostridium acetobutylicum* ATCC 824, *Appl. Environ. Microbiol.*, 59, 1077, 1993.
22. Mermelstein, L.D. et al., Expression of cloned homologous fermentative genes in *Clostridium acetobutylicum* ATCC 824, *Biotechnology (NY),* 10, 190, 1992.
23. Tyurin, M. et al., Electrotransformation of Clostridium acetobutylicum ATCC 824 using high-voltage radio frequency modulated square pulses, *J. Appl. Microbiol.*, 88, 220, 2000.
24. Zhou, Y. and Johnson, E.A., Genetic Transformation of *Clostridium botulinum* Hall A by electroporation, *Biotech. Letts.*, 15, 121, 1993.
25. Tardif, C. et al., Electrotransformation studies in *Clostridium cellulolyticum, J. Ind. Microbiol. Biotechnol.*, 27, 271, 2001.
26. Guedon, E. et al., Growth inhibition of *Clostridium cellulolyticum* by an inefficiently regulated carbon flow, *Microbiol.*, 145, 1831, 1999.
27. Phillips-Jones, M.K., Plasmid transformation of *Clostridium perfringens* by electroporation methods, *FEMS Microbiol. Lett.*, 54, 221, 1990.
28. Allen, S.P. and Blaschek, H.P., Factors involved in the electroporation-induced transformation of *Clostridium perfringens, FEMS Microbiol. Lett.,* 58, 217, 1990.
29. Zhao, Y. and Melville, S.B., Identification and characterization of sporulation-dependent promoters upstream of the enterotoxin gene (*cpe*) *of Clostridium perfringens*, *J. Bacteriol.*, 180, 136, 1998.
30. Rood, J.I. and Cole, S.T., Molecular genetics and pathogenesis of *Clostridium perfringens*, *Microbiol. Rev.*, 55, 621, 1991.
31. Abraham, L.J., Berryman, D.I., and Rood, J.I., Hybridization analysis of the class P tetracycline resistance determinant from the *Clostridium perfringens* R-plasmid, pCW3, *Plasmid,* 19, 113, 1998.
32. Abraham, L.J., Wales, A.J., and Rood, J.I., Worldwide distribution of the conjugative *Clostridium perfringens* tetracycline resistance plasmid, pCW3, *Plasmid,* 14, 37, 1985.
33. Abraham, L.J. and Rood, J.I., Identification of Tn*4451* and Tn*4452*, chloramphenicol resistance transposons from *Clostridium perfringens*, *J. Bacteriol.*, 169, 1579, 1987.
34. Rood, J.I., Scott, V.N., and Duncan, C.L., Identification of a transferable tetracycline resistance plasmid (pCW3) from *Clostridium perfringens*, *Plasmid,* 1, 563, 1978.
35. Ionesco, H., Transferable tetracycline resistance in "*Clostridium difficile*," *Ann. Microbiol.,* 131A, 171, 1980.
36. Oultram, J.D. and Young, M., Conjugal transfer of plasmid pAMβ1 from *Streptococcus lactis* and *Bacillus subtilis* to *Clostridium acetobutylicum*, *FEMS Microbiol. Letts*., 27, 129, 1985.
37. Reysset, G. and Sebald, M., Conjugal transfer of plasmid-mediated antibiotic resistance from streptococci to *Clostridium acetobutylicum, Ann. Inst. Pasteur Microbiol*., 136B, 275, 1985.
38. Young, M., Minton, N.P., and Staudenbauer, W.L., Recent advances in the genetics of the clostridia, *FEMS Microbiol. Rev.,* 5, 301, 1989.
39. Oultram, J.D., Davies, A., and Young, M., Conjugal transfer of a small plasmid from *Bacillus subtilis* to *Clostridium acetobutylicum* by cointegrate formation with plasmid pAMβ1, *FEMS Microbiol. Lett.*, 42, 113, 1987.
40. van der Lelie, D. and Venema, G., *Bacillus subtilis* generates a major specific deletion in pAMβ1, *Appl. Environ. Microbiol*., 53, 2458, 1987.
41. Franke, A.E. and Clewell, D.B., Evidence for conjugal transfer of a *Streptococcus faecalis* transposon (Tn916) from a chromosomal site in the absence of plasmid DNA, *Cold Spring Harb. Symp. Quant. Biol.,* 45, 77, 1991.
42. Courvalin, P. and Carlier, C., Transposable multiple antibiotic resistance in *Streptococcus pneumoniae, Mol. Gen. Genet*., 205, 291, 1996.
43. Scott, J.R., Sex and the single circle: conjugative transposition, *J. Bacteriol.* 174, 6005, 1992.
44. Clewell, D.B. and Gawron-Burke, C., Conjugative transposons and the dissemination of antibiotic resistance in streptococci, *Annu. Rev. Microbiol.,* 40, 635, 1986.

45. Volk, W.A. et al., Inter- and intrageneric transfer of Tn916 between *Streptococcus faecalis* and *Clostridium tetani, Plasmid*, 19, 255, 1988.
46. Woolley, R.C. et al., Transfer of Tn*1545* and Tn*916* to *Clostridium acetobutylicum, Plasmid,* 22, 169, 1989.
47. Mullany, P., Wilks, M., and Tabaqchali, S., Transfer of Tn*916* and Tn*916*ΔE into *Clostridium difficile*: demonstration of a hot-spot for these elements in *the C. difficile* genome, *FEMS Microbiol. Letts.,* 79, 191, 1991.
48. Haraldsen, J.D. and Sonenshein, A.L., Sporulation in *Clostridium difficile* requires disruption of the *sigmaK* gene, *Mol. Microbiol*., 48, 811, 2003.
49. Mullany, P. et al., Gene cloning in *Clostridium difficile* using Tn916 as a shuttle conjugative transposon, *Plasmid,* 31, 320, 1994.
50. Roberts, A.P. et al., Development of an integrative vector for the expression of antisense RNA in *Clostridium difficile, J. Microbiol. Methods*, 55, 617, 2003.
51. Lyras, D. and Rood, J.I., Conjugative transfer of RP4-oriT shuttle vectors from *Escherichia coli* to *Clostridium perfringens, Plasmid*, 39, 160, 1998.
52. Bradshaw, M., Goodnough, M.C., and Johnson, E.A., Conjugative transfer of the *Escherichia coli-Clostridium perfringens* shuttle vector pJIR1457 to *Clostridium botulinum* type A strains, *Plasmid*, 40, 233, 1998.
53. Mani, N. et al., Environmental and autoregulation of *Clostridium difficile* TxeR, a sigma factor for toxin gene expression, *J. Bacteriol*., 184, 5971, 2002.
54. Kim, A.Y. and Blaschek, H.P., Construction and characterization of a phage-plasmid hybrid (phagemid), pCAK1, containing the replicative form of viruslike particle CAK1 isolated from *Clostridium acetobutylicum* NCIB 6444, *J. Bacteriol.,* 175, 3838, 1993.
55. Yoshino, S. et al., Construction of shuttle vector plasmid between *Clostridium acetobutylicum* and *Escherichia coli, Agric. Biol. Chem.,* 54, 437, 1990.
56. Minton, N.P. et al., Clostridial cloning vectors, in *The Clostridia and Biotechnology,* Woods, D.R., Ed., Butterworth-Heinemann, Boston, 1993, 119–150.
57. Fox, M.E. et al., Anaerobic bacteria as a delivery system for cancer gene therapy: *in vitro* activation of 5-fluorocytosine by genetically engineered clostridia, *Gene Ther.*, 3, 173, 1996.
58. Lee, S.Y., Mermelstein, L.D., and Papoutsakis, E.T., Determination of plasmid copy number and stability in *Clostridium acetobutylicum* ATCC 824, *FEMS Microbiol. Lett*., 108, 319, 1993.
59. Kim, A.Y. and Blaschek, H.P., Construction of an *Escherichia coli-Clostridium perfringens* shuttle vector and plasmid transformation of *Clostridium perfringens, Appl. Environ. Microbiol.,* 55, 360, 1989.
60. Sloan, J. et al., Construction of a sequenced *Clostridium perfringens-Escherichia coli* shuttle plasmid, *Plasmid*, 27, 207, 1992.
61. Squires, C.H. et al., Shuttle plasmids for *Escherichia coli* and *Clostridium perfringens, J. Bacteriol*., 159, 465, 1984.
62. Birrer, G.A., Chesbro, W.R., and Zsigray, R.M., Electro-transformation of *Clostridium beijerinckii* NRRL B-592 with shuttle plasmid pHR106 and recombinant derivatives, *Appl. Microbiol. Biotechnol.,* 41, 32, 1994.
63. Truffaut, N., Hubert, J., and Reysset, G., Construction of shuttle vectors useful for transforming *Clostridium acetobutylicum, FEMS Microbiol. Letts.,* 49, 15, 1989.
64. Lee, S.Y. et al., Vector construction, transformation, and gene amplification in *Clostridium acetobutylicum* ATCC 824, *Ann. N. Y. Acad. Sci*., 665, 39, 1992.
65. Wong, J. and Bennett, G.N., Recombination-induced variants of *Clostridium acetobutylicum* ATCC 824 with increased solvent production, *Curr. Microbiol*., 32, 349, 1996.
66. Azeddoug, H., Hubert, J., and Reysset, G., Stable inheritance of shuttle vectors based on plasmid pIM13 in a mutant strain of *Clostridium acetobutylicum, J. Gen. Microbiol*., 138, 1371, 1992.
67. Girbal, L. et al., Development of a sensitive gene expression reporter system and an inducible promoter-repressor system for *Clostridium acetobutylicum, Appl. Environ. Microbiol.,* 69, 4985, 2003.
68. Kok, J., van der Vossen, J.M., and Venema, G., Construction of plasmid cloning vectors for lactic streptococci which also replicate in *Bacillus subtilis* and *Escherichia coli, Appl. Environ. Microbiol.,* 48, 726, 1984.

69. Mermelstein, L.D. et al., Genetic and metabolic engineering of *Clostridium acetobutylicum* ATCC 824, *Ann. NY Acad. Sci.,* 721, 54, 1994.
70. Herbert, M. et al., Gene transfer into *Clostridium difficile* CD630 and characterisation of its methylase genes, *FEMS Microbiol. Letts.,* 229, 103, 2003.
71. Feil, R. et al., Methylation analysis on individual chromosomes: improved protocol for bisulphite genomic sequencing, *Nucl. Acids Res.*, 22, 695, 1994.
72. Oultram, J.D. et al., Introduction of genes for leucine biosynthesis from *Clostridium pasteurianum* into *C. acetobutylicum* by cointegrate conjugal transfer, *Mol. Gen. Genet.,* 214, 177, 1988.
73. Scott, P.T. and Rood, J.I., Electroporation-mediated transformation of lysostaphin-treated *Clostridium perfringens, Gene,* 82, 327, 1989.
74. Cornillot, E. et al., The genes for butanol and acetone formation in *Clostridium acetobutylicum* ATCC 824 reside on a large plasmid whose loss leads to degeneration of the strain, *J. Bacteriol.*, 179, 5442, 1997.
75. Finn, C.W. et al., The structural gene for tetanus neurotoxin is on a plasmid, *Science.*, 224, 881, 1984.
76. Eklund, M.W. et al., Evidence for plasmid-mediated toxin and bacteriocin production in *Clostridium botulinum* type G, *Appl. Environ. Microbiol.*, 54, 1405, 1998.
77. Katayama, S. et al., Genome mapping of Clostridium perfringens strains I-CeuI shows many virulence genes to be plasmid-borne, *Mol. Gen. Genet.,* 251, 720, 1996.
78. Brefort, G. et al., Characterization and transferability of *Clostridium perfringens* plasmids, *Plasmid,* 1, 52, 1977.
79. Minton, N.P. and Morris, J.G., Isolation and partial characterisation of three cryptic plasmids from strains of *Clostridium butyricum, J. Gen. Microbiol.,* 127, 325, 1981.
80. Collins, M.E., Oultram, J.D., and Young, M., Identification of restriction fragments from two cryptic *Clostridium butyricum* plasmids that promote the establishment of a replication-defective plasmid in *Bacillus subtilis, J. Gen. Microbiol.,* 131, 2097, 1985.
81. Garnier, T. and Cole, S.T., Identification and molecular genetic analysis of replication functions of the bacteriocinogenic plasmid pIP404 from *Clostridium perfringens, Plasmid*, 19, 151, 1988.
82. Brehm, J.K. et al., Physical characterization of the replication origin of the cryptic plasmid pCB101 isolated from *Clostridium butyricum* NCIB 7423, *Plasmid,* 28, 1, 1992.
83. Vosman, B. and Venema, G., Introduction of a *Streptococcus cremoris* plasmid in *Bacillus subtilis, J. Bacteriol.*, 156, 920, 1983.
84. Gruss, A. and Ehrlich, S.D., The family of highly interrelated single-stranded deoxyribonucleic acid plasmids, *Microbiol. Rev.,* 53, 231, 1989.
85. Theys, J. et al., Stable *Escherichia coli-Clostridium acetobutylicum* shuttle vector for secretion of murine tumor necrosis factor alpha, *Appl. Environ. Microbiol.,* 65, 4295, 1999.
86. Tomas, C.A., Welker, N.E., and Papoutsakis, E.T., Overexpression of groESL in *Clostridium acetobutylicum* results in increased solvent production and tolerance, prolonged metabolism, and changes in the cell's transcriptional program, *Appl. Environ. Microbiol.,* 69, 4951, 2003.
87. Tummala, S.B., Junne, S.G., and Papoutsakis, E.T., Antisense RNA downregulation of coenzyme A transferase combined with alcohol-aldehyde dehydrogenase overexpression leads to predominantly alcohologenic *Clostridium acetobutylicum* fermentations, *J. Bacteriol.*, 185, 3644, 2003.
88. Marvaud, J.C. et al., TetR is a positive regulator of the tetanus toxin gene in *Clostridium tetani* and is homologous to botR, *Infect. Immun.*, 66, 5698, 1998.
89. Lemmon, M.J. et al., Anaerobic bacteria as a gene delivery system that is controlled by the tumor microenvironment, *Gene Ther.*, 4, 791, 1997.

4 Gene Analysis of Clostridia

Seshu B. Tummala, Christopher A. Tomas, and Eleftherios T. Papoutsakis

CONTENTS

4.1 INTRODUCTION

Recently, several advances in techniques to analyze genes and their functions for clostridia have been developed. In 1993, Mermelstein and Papoutsakis described an electroporation protocol for *C. acetobutylicum* that facilitated gene overexpression in *C. acetobutylicum* via plasmid DNA [1]. Another major advance involves the use of nonreplicative and replicative integrational plasmids to knock out specific clostridial genes by homologous recombination [2,3]. These developments in gene analysis techniques for clostridia have been reviewed extensively [4,5]. Three new exciting tools, gene expression reporter systems, antisense RNA, and large-scale transcriptional analyses using DNA microarrays have also been recently developed for the examination of gene function in clostridia and will be discussed in detail in this chapter. These tools will allow us to understand promoter regulation, assign functions to previously unknown genes, and develop new hypotheses for describing cellular processes.

4.2 GENE EXPRESSION REPORTER SYSTEMS

Gene expression reporter systems make it possible to study expression of both autologous and heterologous promoters in clostridia and to understand the regulation of these promoters. An improved understanding of promoter strength and regulation could lead to more effective clostridial expression vectors. Such vectors could eventually be used to augment expression of solventogenic pathway genes and could be coupled with antisense RNA strategies and knockout mutations to develop more complex metabolic engineering strategies in clostridia. In addition, gene expression reporter systems could be used to evaluate toxic gene expression patterns as affected by different environmental factors in pathogenic clostridia, and such studies would lead to a better understanding of pathogenesis and potential therapeutic interventions.

The first study to describe the use of a reporter gene in clostridia involved the use of the bioluminescent *luxA* and *luxB* genes of *Vibrio fischeri*. The bioluminescence produced by these genes was used to detect alpha-toxin gene (*cpe*) expression in *C. perfringens* [6,7]. Other gene expression reporter systems using the chloramphenicol acetyltransferase gene of *C. perfringens* (*catP*) [8] and a beta-glucuronidase from *E. coli* (*gusA*) [9,10] were also developed for examination of *cpe* gene expression in *C. perfringens*. Additionally, the *gusA* reporter gene was used to identify the promoter region of the epimerase gene (*nanE*) of *C. perfringens* [11]. The *gusA* system has also recently been adapted to examine gene expression in *C. acetobutylicum* [12]. In 1999, Tummala et al. developed a gene expression reporter system using the *lacZ* gene of *Thermoanaerobacterium sulfurogenes* EM1 to examine the expression patterns of acidogenic and solventogenic phase promoters in *C. acetobutylicum* [13,14]. Using this same gene as a reporter gene, Davis et al. developed a gene expression reporter system to examine toxic gene expression in group II, non-proteolytic *Clostridium botulinum* type B strains [15]. For *Clostridium beijerinckii*, a beta-1, 4-endoglucanase gene (*eglA*) has been used to develop a gene expression reporter system [16]. This reporter system has been used to examine gene expression from its endogenous promoter under different growth conditions, as well as from the glutamine synthetase (*glnA*) promoter region.

4.3 ANTISENSE RNA

4.3.1 OVERVIEW

Antisense RNA (asRNA) is now recognized as an efficient means of regulating gene expression post-transcriptionally. Generally, asRNA binds target RNA and prevents RNA translation by (1) hindering ribosome binding site interactions with the translational machinery of the cell (i.e., ribosomes) and/or (2) altering the structure of the target RNA such that ribonucleases can then degrade the target RNA [17]. Desai and Papoutsakis first demonstrated the effectiveness of asRNA strategies for the metabolic engineering of *C. acetobutylicum* [18]. They developed asRNA molecules against two genes involved in the primary metabolic network of *C. acetobutylicum*. One was directed toward the mRNA of the butyrate kinase (*buk*) gene, while the other was directed toward the mRNA of the phosphotransbutyrylase (*ptb*) gene. However, no rules or patterns were examined regarding the effectiveness of different asRNA designs, and the generality of these asRNA findings in downregulating other clostridial proteins and pathways still needed to be tested.

Most methods for designing effective asRNA molecules are based on the concept that by increasing the association rate between asRNA and target mRNA, inhibition of target gene expression can also be increased [19–22]. One method that was evaluated for designing asRNAs with higher annealing rates to the target mRNA is to incorporate stem-loop structures that are designed specifically for interacting with stem-loop structures in the target mRNA [23]. Experimental data, however, suggested that this approach does not necessarily work well (i.e., only 50% downregulation of target was achieved). Another method examined for generating asRNA candidates with high annealing rates to target mRNAs is *in vitro* selection of asRNA [24]. This technique involves using only asRNA candidates that bind with target mRNA with the highest association rate constants as determined from nondenaturing polyacrylamide gel electrophoresis [19,21]. However, only ca. 75% overall downregulation of target protein was achieved using this method [24]. The disadvantage of using *in vitro* selection is that *in vitro* conditions might not represent *in vivo* conditions accurately and, thus, may lead to inaccurate prediction of the ideal asRNA candidates. Finally, using theoretical analyses and computer algorithms that can provide information on structural elements that could be important for RNA duplex formation, a technique for developing asRNA based on structural properties has been investigated [25]. By examining several different structural features (including terminal unpaired nucleotides, components, and loop degree) of many different asRNAs designed to inhibit HIV-1 replication in human cells, Patzel and Sczakiel [25] showed a correlation between asRNA effectiveness and the number of terminal unpaired nucleotides.

Such approaches for asRNA design in prokaryotes have not been reported until recently. We have examined structural feature-based asRNA design [25] in *C. acetobutylicum* [26] using Mfold, a computational algorithm that predicts secondary structure, based on thermodynamics and structural information derived from studies of known RNA molecules [27–29]. Antisense RNAs directed against the acetoacetate decarboxylase gene (*adc*) were examined for secondary structure information that might explain differences in asRNA effectiveness and, thus, provide potential rules for the design of future asRNA constructs. Two criteria to assess asRNA structural differences were examined: free nucleotides and components [26]. Free nucleotides were defined as nucleotides in an asRNA molecule that are not involved in intramolecular bonding and, thus, are thought to provide potential sites with which the asRNA and target mRNA might interact (Figure 4.1) [25]. Components are structural features that contain regions of intramolecular binding called duplex RNA [30] that are at least two nucleotides long and may contain internal loops or bulges (Figure 4.1) [25]. The number of free nucleotides and components (normalized per 100 nucleotides of each asRNA molecule) were plotted separately for each asRNA against the percentage of overall protein downregulation of each asRNA's target enzyme. Using the data from the asRNA constructs developed in their study [26], as well as those of Desai and Papoutsakis [18] and the natural *glnA*-asRNA [31], only the normalized number of components correlated well with *in vivo* asRNA effectiveness with a correlation coefficient value of 0.87 for a second-order polynomial [26]. This value increases to 0.97 when only the *adc*-asRNAs developed in their study were considered. These results suggested

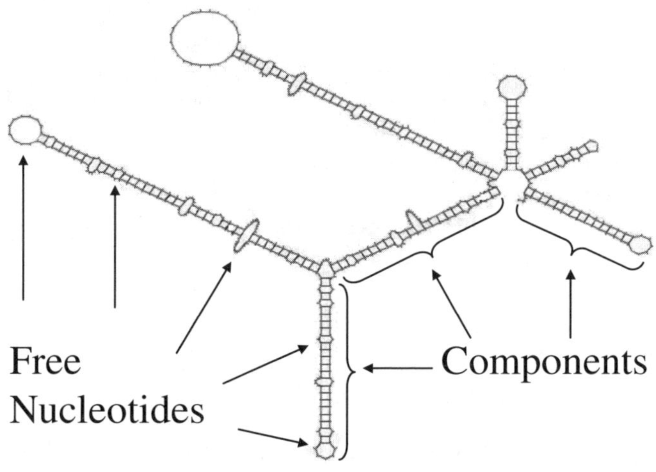

FIGURE 4.1 Free nucleotides and components. Examples of free nucleotides and components are shown for the predicted secondary structure of *ctfb1*-asRNA.

that the components/number of total nucleotides ratio was a useful predictor of asRNA effectiveness. To further corroborate the hypothesis that the components/nucleotides ratio can be used to design effective antisense RNA constructs in *C. acetobutylicum*, an antisense RNA targeting the CoA transferase subunit B gene (*ctfB1*-asRNA) with the components/nucleotides ratio as the main design criteria was introduced into *C. acetobutylicum*. *Ctfb1*-asRNA was the most effective asRNA seen in *C. acetobutylicum*, as shown by the highest percentage of overall protein downregulation compared to previous asRNA constructs, which further suggested that the components/nucleotide ratio is a good predictor of asRNA effectiveness.

Several other methods for exploiting the use of asRNA in *C. acetobutylicum* have been explored. The ability to express a single asRNA directed against two different proteins was examined [26]. The combined asRNA molecule was able to downregulate the first protein (CtfB) effectively, but its effects on the second protein (CtfA) were relatively small. A potential reason for this is that the *ctfA* mRNA might be embedded in the secondary structure of the polycistronic message (of which *ctfA* and *ctfB* are part [32]), which hinders the ability of the combined asRNA to bind its target region in the mRNA. Another potential reason might derive from the fact that the portion of combined asRNA targeted for CtfA downregulation is transcribed after the *ctfB* antisense portion. Since the *ctfB* antisense portion is transcribed first, it may bind its target and fix the combined asRNA's position on the polycistronic message and not allow the *ctfA* antisense portion either the time or the flexibility to bind the second intended target mRNA. The combination of asRNA and gene overexpression technologies has also been investigated for altering the metabolism of *C. acetobutylicum* [33,34]. Detailed product formation, metabolic flux, Western, and DNA-microarray analyses of a strain that overexpress an antisense RNA against the CoA Transferase subunit B gene (*ctfb1*-asRNA) and the alcohol-aldehyde dehydrogenase gene (*aad*) drastically altered the transcriptome and solvent production compared to the control strain (*ctfB1*-asRNA expression solely).

4.3.2 Design and Implementation of asRNA

For any gene that is to be downregulated using asRNA, the structural gene and the ribosome binding site must first be sequenced. Assuming the necessary DNA sequences for the target gene are known, potential antisense RNA constructs can then be generated and visualized using DNA representation software such as Gene Construction Kit2 (Textco, Inc., West Lebanon, NH). These constructs should contain in antisense orientation the DNA sequence of a portion of the gene's transcript, which includes the ribosome binding site and a fragment of the structural gene, with respect to an appropriate promoter in an expression vector. Although extensive experiments have not been performed to determine the optimal size of the portion of the structural gene's transcript used for asRNA production, anywhere between 33 and 100% of the target gene's transcript has been shown to be highly effective (>80% downregulation of the target gene) [26]. Furthermore, the end of the structural gene fragment should contain (or can be easily engineered to have) convenient restriction sites for subsequent cloning steps with the expression vector. A number of nucleotides (roughly up to 60 nucleotides) upstream of the ribosome binding site, which also contain (or can be easily engineered to have) convenient restriction sites, can also be included to facilitate directional cloning into the expression vector. Choice of promoter depends on the expression pattern of the target gene. Strong promoters that have similar gene expression patterns to that of the target gene are more likely to downregulate the target, because an excess of asRNA is expressed with respect to target mRNA during all growth phases, which increases the chance for asRNA to bind to the target mRNA and downregulate the target enzyme. Also, the promoter of the target gene can be used to drive expression of asRNA. In addition to having the same expression pattern as the target gene, using this promoter has the advantage of decreasing target gene expression via promoter dilution of potential regulatory proteins due to the additional copy of the promoter on the plasmid. It is important that the expression vector used for expression of asRNA also contain a stringent terminator sequence so that one can have confidence that asRNA transcription ends at the end of the expression cassette and does not continue into other regions of the construct, which might have deleterious consequences.

Potential asRNA construct candidates are further selected based on each asRNA secondary structure. Analysis of the secondary structures of RNA molecules is carried out using MFold (http://www.bioinfo.rpi.edu/~zukerm/rna/mfold-3.1.html). For quantitation of structural features, only the most energetically favorable structure as predicted by Mfold should be used. It is important to note that the RNA sequence that is entered into Mfold is derived from the DNA sequence that starts at the transcription start site of the promoter on the expression vector, continues through the ribosome binding site and structural gene portion of the asRNA, and ends at the last nucleotide of the terminator sequence on the expression plasmid. Inputting this sequence rather than just the antisense portions of the target gene would provide the most likely RNA sequence that is actually transcribed and, hence, leads to more accurate structural feature prediction. The number of components for the secondary structures of all of the potential asRNA molecules are enumerated and normalized based on total nucleotide length for each asRNA molecule. The DNA fragment that produced the asRNA candidate with the lowest components/nucleotides ratio is then cloned in the antisense orientation into the expression vector. Once the new construct is developed, it can be introduced into the host strain. The effectiveness of the asRNA molecule can then be ascertained by examination of the protein levels of the target gene. Protein levels, rather than mRNA levels, are a more accurate method for determining asRNA effectiveness, because both asRNA mechanisms (blocking of translation by asRNA binding, which does not affect mRNA levels and degradation of mRNA due to asRNA binding, which decreases mRNA levels) yield the desired effect of protein downregulation but have different effects on mRNA levels. Nonetheless, with the recent development of large-scale transcriptional analysis for *C. acetobutylicum* [35], the effects of the asRNA on the mRNA levels of the target gene as well as many other genes can be routinely tested to provide a more detailed picture of asRNA efficacy and its effects on different cellular programs [34].

4.3.3 Example

For instruction purposes, the process for the design of a construct that produces an asRNA against the CoAT subunit B gene (pCTFB1AS) is discussed. For the CoAT subunit B gene (*ctfB*), three potential antisense RNA constructs (pCTFB1AS, pCTFB2AS, and pCTFB3AS) were examined using Gene Construction Kit2. All three constructs included the DNA sequence of the ribosome binding site and a portion of the subunit's structural gene (ca. 39, 77, and 52% of structural gene, respectively) in an antisense orientation with respect to the thiolase (*thl*) promoter of *C. acetobutylicum* in an expression vector derived from pSOS95 (Figure 4.2) [26]. The *thl* promoter was chosen because it was thought to be a strong, constitutive-like promoter that would be able to strongly express the desired asRNA throughout the culture. This expression vector also has the advantage of containing the functional terminator sequence of the *adc* gene. All three of the fragments of the target genes used to create pCTFB1AS, pCTFB2AS, and pCTFB3AS in the DNA visualization software were modified to create mismatches to create *Bam*HI sites at the end of the structural gene portion of each asRNA. These sites were created for directional cloning purposes. Thus, cloning of the fragment that produced the best asRNA candidate could be easily performed.

Potential asRNA candidates were further selected by calculating a components/nucleotides ratio based on each asRNA's secondary structure. As described earlier, the RNA sequence entered into Mfold is derived from the DNA sequence that starts from the transcription start site of the *thl* promoter of the expression vector through the target gene sequences to the last nucleotide of the terminator sequence of the expression cassette. The results of the secondary structure predictions of Mfold for these asRNA candidates are shown in Figure 4.3. The number of components for each secondary structure is enumerated and normalized based on total length of each asRNA candidate

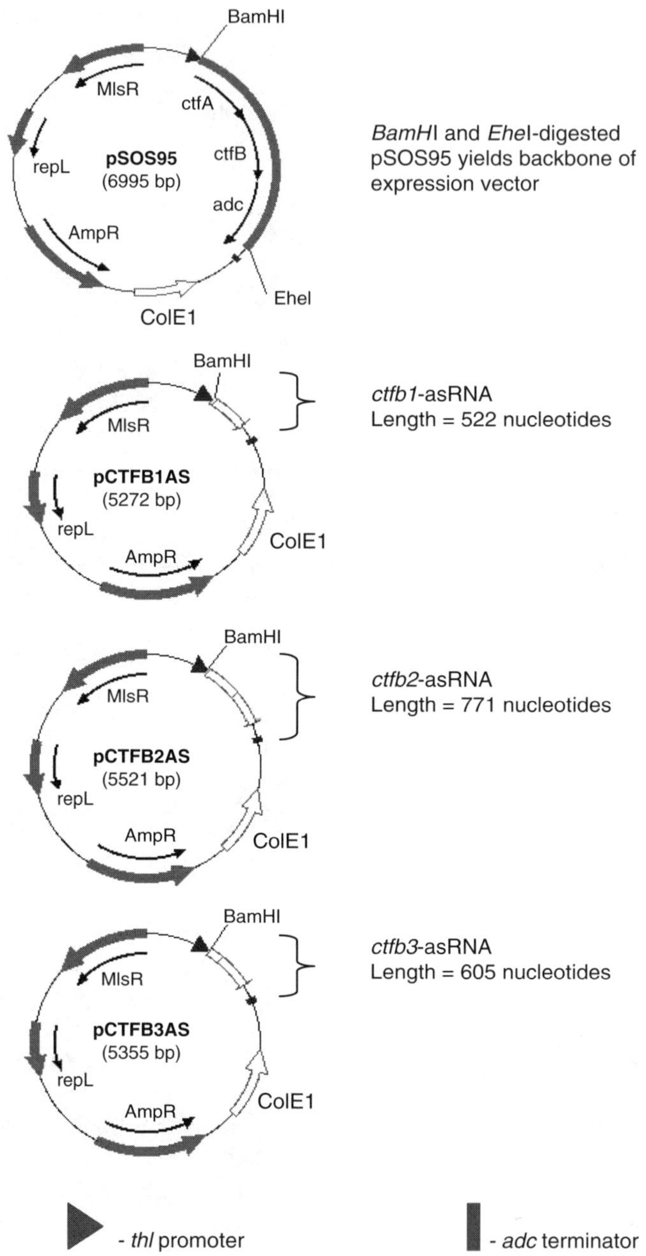

FIGURE 4.2 Antisense RNA constructs that express asRNA for downregulation of CtfB. For each plasmid, the locations and directions of transcription of relevant genes are indicated (arrows). Relevant restriction sites are shown. Abbreviations: *ctfA*, CoAT subunit A gene; *ctfB*, CoAT subunit B gene; *adc*, acetoacetate decarboxylase gene; MlsR, macrolide-lincosamide-streptogramin B resistance gene; repL, pIM13 origin of replication; AmpR, ampicillin resistance gene; ColE1, ColE1 origin of replication; *thl* promoter, promoter region for the thiolase gene of *C. acetobutylicum* ATCC 824; *adc* terminator, terminator sequence of the acetoacetate decarboxylase gene. All genes and plasmids are not drawn to scale.

[*ctfb1*-asRNA (1.34 components/nucleotides), *ctfb2*-asRNA (1.69 components/nucleotides), and *ctfb3*-asRNA (1.65 components/nucleotides)]. Since the asRNA that would be produced from pCTFB1AS had the lowest components/nucleotides ratio, the DNA fragment used to create pCTFB1AS in our

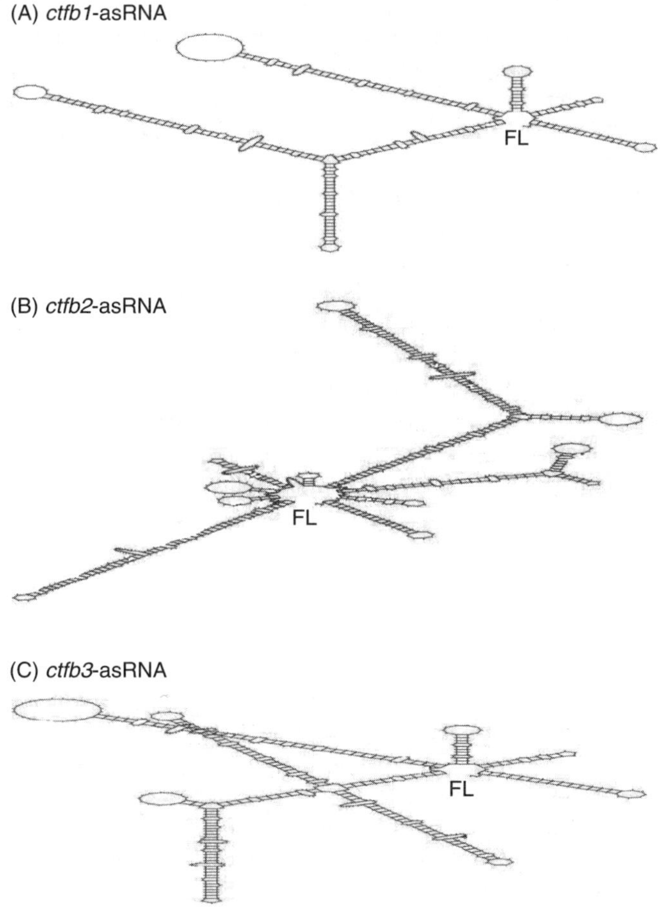

FIGURE 4.3 Predicted secondary structures of asRNA targeting *ctfB*. The first and last nucleotide of each asRNA molecule is designated F and L, respectively.

DNA visualization software was then PCR amplified to have phosphorylated blunt ends, digested with *Bam*HI, and cloned in the antisense orientation into *Bam*HI and *Ehe*I (a blunt-end cutter)-digested pSOS95 to form plasmid pCTFB1AS. To assess the effectiveness of asRNA downregulation of CtfB using the plasmid pCTFB1AS, Western blot analysis was performed with samples from static flask and bioreactor fermentations of strain *C. acetobutylicum* ATCC 824(pCTFB1AS) and the plasmid control strain, 824(pSOS95del). In both sets of experiments, strong downregulation of CtfB (>80% overall downregulation) occurred [26,33]. Large-scale transcriptional analysis of 824(pCTFB1AS) and 824(pSOS95del) also was performed and revealed that *ctfB* mRNA levels were lower in 824(pCTFB1AS) and that the expression of the asRNA resulted in changes to several cellular programs including sporulation, stress response, and motility [34].

4.3.4 Future Developments

The use of asRNA technology has so far proven very successful, but several improvements and developments would be desirable. One improvement is the development of a computational method that takes a target gene's sequence and generates the best potential asRNA candidate based on the components/nucleotide ratio. This development would significantly expedite the process of designing asRNA constructs. Another potential improvement to current asRNA constructs is to examine the possibility of downregulating multiple enzymes using different asRNA molecules on the same plasmid. Downregulation of multiple enzymes from a single plasmid would provide greater flexibility in altering key enzyme levels without having to introduce additional antibiotics in fermentations for selective pressure, which has been shown to be detrimental to solvent production [36]. Finally, by changing the promoters used to drive expression of asRNA, one may further improve current asRNA technologies. The ability to use different promoters to drive expression of asRNA would make it possible to better tailor the expression of asRNA to optimize downregulation of the target enzyme.

Antisense RNA also can be very useful for functional genomics studies. Numerous knockout strains of different species of clostridia need to be developed for functional genomic studies. Ideally, gene inactivation technologies using nonreplicative or replicative constructs should be used to create these knockout strains, but since these methods still remain a challenge and are not easily implementable in clostridia [2,3], another way to create these knockout strains is needed. By generating optimal asRNA for each open reading frame using a computational method that utilizes the components/nucleotide ratio as a design parameter, a library of strains that can downregulate all of the individual genes or operons of a clostridial species can be created. Then, molecular analyses (including microarray analyses) of all these strains can be performed to potentially elucidate the complete regulatory network.

4.4 DNA MICROARRAYS

4.4.1 OVERVIEW

Analysis of nucleic acids by hybridization has long been an accepted method of analysis in molecular biology. The established methods for gene expression analysis such as Northern blotting, reverse transcriptase (RT) PCR, and nuclease protection assays are only capable of analyzing small numbers of genes and samples in a single experiment. More advanced methods like differential display, subtractive hybridization, representational difference analysis, expressed sequence tags, cDNA fragment fingerprinting, and serial gene analysis of gene expression allow for discovery of differentially expressed genes. These methods are also limited by technical challenges and are only able to expand the analysis capability by a relatively small factor. The ability to simultaneously determine the expression levels of thousands of genes in one experiment can be achieved by reversing the Northern dot blotting principle so that the known sequences are immobilized (the probe for Northern dot blot analysis) and the labeled species is derived from an mRNA sample (deposited on a membrane for Northern dot blot analysis). Initial experiments, termed macroarrays, were also filter-based methods [37]. Filter-based methods have the disadvantages of requiring a relatively large membrane surface area (10^1 to 10^2 cm^2) and high membrane autofluorescence. Membrane autofluorescence greatly limits the ability to use multiplexed fluorescence techniques such as Cy-3 and Cy-5 based dyes. Fluorescent *in situ* hybridization (FISH) combines multiplexed fluorescence analysis with microscopy to allow for the detection of nucleic acids within cells and chromosomes. Immobilization of cells and chromosomes on microscope slides actually preceded filter-based gene expression analysis. The combination of macroarray and FISH technologies was made possible by advances in immobilizing nucleic acid sequences to a glass support. A procedure termed contact printing was developed to deposit purified nucleic acids onto a slide surface [38], while photolithography was used to synthesize oligonucleotides directly onto a glass support. The ability to spot DNA onto a glass support allowed for the miniaturization of the macroarray experiment. The first papers with the term microarray were published by Pat Brown (Stanford University) in 1995 and 1996 [39]. The method has quickly been adopted by the scientific community.

A variety of microarray platforms exist, all of which are miniaturized hybridization assays capable of examining gene expression for thousands of genes simultaneously. Despite differences in format, at their core, all microarrays contain four common features: (1) an array of immobilized nucleic acids or targets, (2) one or more labeled samples or probes, (3) hybridization of the probe to the array targets, and (4) a system for detection and quantitation of the hybridized signal.

4.4.1.1 The Nucleic Acid Targets

The spotted nucleic acid or target can be obtained from a number of different sources (e.g., synthesized oligonucleotides or cDNA from PCR reactions) and can be deposited onto the array surface by one of several different methods. Typical spot sizes range from 50 to 200 µm in diameter, allowing a single standard-sized glass slide (25 × 75 mm) to hold tens of thousands of spots. The use of cDNA clones as microarray targets has been enabled by DNA printing technologies and the development of modified glass surfaces with strong DNA binding characteristics. The use of cDNAs does not require one to know the entire DNA sequence or to have *a priori* knowledge about their expression. However, knowing the full DNA sequences allows for better design of cDNA targets by selecting regions of DNA with little or no homology to other target sequences, thereby minimizing nonspecific hybridization. DNA targets must be amplified and purified before being spotted. DNA obtained from PCR must be free of enzymes, nucleotides, and buffer components which interfere with DNA binding and microarray analysis. An important step in cDNA-based microarray production is verification that the products contain a single fragment of the correct size and that they are present in known concentrations. When handling such large numbers of cDNA fragments, it is critical to take special precautions to avoid contamination and mislabeling, which have been determined to be as high as 30% in some cases [40,41]. The optimal length of DNA targets is between 300 and 800 nucleotides. Shorter targets are not as easily bound to and retained on glass surfaces. However, longer targets require more DNA to guarantee the target DNA is present in excess relative to the probe.

4.4.1.2 The Labeled Probe

Labeling of samples to create probes typically consists of converting mRNA to a population of labeled cDNAs. The use of fluorescent dyes, primarily the cyanine dyes Cy3 and Cy5, to label samples has the advantage of allowing the detection of two signals on a single microarray. This allows for comparative analysis of multiple samples in a single experiment. The accuracy and throughput using this method is much greater than filter-based arrays on which a single radioactively labeled sample is analyzed at a time. There are several requirements for a successful labeling: (1) The labeling method should directly represent the original transcript population by equally labeling all transcripts, which vary in size, nucleotide sequences, and relative abundance. The relative abundance of transcripts varies widely, ranging from a few copies to thousands of copies per cell. (2) Labeled fragments should be 200 to 500 nucleotides long. Shorter fragments can lack specificity, while longer fragments take longer to hybridize with the target. (3) The amount of labeled probe must be sufficient to detect even the lowest level of mRNA, and is ideally sufficient for more than one hybridization. (4) The proper labeling density (the number of incorporated fluorescent molecules per nucleotide) is critical in maintaining the direct proportionality of the measured fluorescence to the number of transcripts. Two fluorescent molecules in close proximity will result in signal quenching and can dramatically reduce signal intensities (>90% in extreme cases). The nucleotide content of the transcripts has a large impact on labeling densities and must be considered when labeling. Sequences with A-rich content should be labeled with CyDye dCTP rather than CyDye dUTP. (5) Finally, the two dyes should label transcripts with the same efficiency. No single labeling protocol can produce an ideal labeling outcome, but methods have been developed to properly account for any shortcomings.

4.4.1.3 Hybridization of Probe to Target

Microarray hybridizations involve the formation of DNA duplexes consisting of labeled probe fragments and their complementary targets. For successful binding, the DNA must be single-stranded or denatured and readily accessible. The number of successful hybridizations is directly proportional to the number of each specific probe fragment in the sample, assuming the number of immobilized targets is in excess and not limited by hybridization kinetics. Microarray hybridizations are much more complex than any seen in most every other molecular biology tool. It is possible to have 10^4 different labeled fragments available to hybridize with upwards of 10^5 different targets. Therefore, it is critical that nonspecific hybridizations be minimized by maintaining stringent hybridization conditions.

4.4.1.4 Signal Detection and Quantitation

Once the labeled probes have been hybridized to their targets and any extraneous labeled material washed from the array, it is necessary to quantitate the amount of probe that has hybridized to each immobilized target. This can be achieved with a confocal microscope fitted with lasers. Most microarray scanners, such as the GSI Lumonics (PerkinElmer Life Sciences, Boston, MA) scanner, contain at least two lasers that emit light at different wavelengths corresponding to the excitation wavelength of the dyes used to prepare the labeled probes. The confocal microscope serves to record, with high resolution, the amount of light emitted from each microarray spot. When two samples are simultaneously hybridized on a single array, the probes from both samples compete for binding to the same targets. By comparing the intensities of the two fluorescent signals, one is able to determine the relative abundance of a given sequence. A variety of specialized software packages, such as ScanArray and QuantArray Microarray analysis software (PerkinElmer Life Sciences), is widely available to extract intensity data from scanned images. Additional software is required to normalize the array data to compensate for experimental variations. Finally, the data needs to be arranged and presented in a manner that allows investigators to confidently draw conclusions. This final step frequently utilizes a variety of clustering techniques including those developed by Eisen [42] and Tamayo [43].

4.4.2 Protocol

A cDNA microarray protocol has been developed and validated [35] using a targeted set of *C. acetobutylicum* genes. The same protocol would likely be directly applicable to other clostridia species. The first generation of *C. acetobutylicum* arrays was designed with spots representing 1019 open reading frames, approximately one-fourth of the genome. Full genome arrays are being constructed but have not yet been validated. The protocol outlined below has largely been adapted from the TIGR protocol and utilizes spotted cDNAs of a known sequence obtained from PCR [44]. The PCR products are purified, quantitated, and redissolved in a DMSO solution (for DNA annealing) and spotted on commercially available amine-modified glass slides. Labeled probes are obtained by reverse transcription reactions using fluorescently labeled (Cy-3 or Cy-5) nucleotides. After being hybridized with the labeled probes overnight, the slides are washed and analyzed with a microarray scanner. Finally, the data is normalized and genes exhibiting significant differences in expression are identified.

4.4.2.1 Design of DNA Targets

Several considerations must be made when designing a strategy to obtain DNA targets using PCR. PCR primers should be designed to amplify the target gene such that nonspecific hybridization between the bound target and labeled probes is minimized. Several genomics companies offer services to design PCR primers to minimize homologies between the amplified target and the rest of the genome. PCR primers for the first generation of *C. acetobutylicum* cDNA arrays were designed by Integrated Genomics (Chicago, IL) using a proprietary design algorithm. Alternatively, publicly available software such as PRIMEGENS is available for design of primers suitable for cDNA microarray production [45]. To avoid significant changes in the PCR amplification protocol, it is convenient to design PCR primers with melting temperatures that fall in a narrow range (2 to 3°C). After the primers have been designed, genes whose primers have the most similar melting temperatures can be grouped (for example, to fill a 96-well plate) for PCR to increase the likelihood of amplifying the desired products. PCR primers should be designed such that the amplified gene products are approximately 300 to 500 bp in length. Care should also be taken to minimize primer dimer formation. Primer dimers will result in an increased number of failed PCR reactions. PCR primers can be screened for primer dimer formation using publicly available software such as the Sigma Genosys Oligo Calculator (www.genosys.co.uk).

4.4.2.2 PCR Amplification and Product Purification

Two methods for PCR amplification have been used to generate DNA arrays for *C. acetobutylicum*. The first method, based on the TIGR protocol, utilizes a single 60 μl PCR reaction with approximately 4 μg chromosomal template DNA. The second method, based on Richmond et al. [46], utilizes two PCR reactions. The first reaction is performed with approximately 100 ng of chromosomal template DNA in a 50 μl reaction. The second reaction is performed using 1 μl of the first reaction as the template in a 100 μl reaction. The two-reaction method of Richmond et al. reduces the concentration of genomic DNA that is ultimately spotted with the target DNA. Genomic DNA is perfectly homologous to all possible labeled DNA probes, resulting in nonspecific binding. For either method, AmpliTaq Gold DNA Polymerase (Applied Biosystems, Foster City, CA) with the supplied buffer was found to provide the most consistent results with the highest yield. The resulting PCR products should be run on an agarose gel to verify the proper fragment size and that only a single DNA product was produced. Fast and efficient screening of PCR products for DNA arrays can be performed with a Ready-To-Run separation unit from Amersham Biosciences, or a similar apparatus. Reactions with missing or multiple bands should be repeated under less or more stringent PCR conditions, respectively, or an alternative set of PCR primers designed. After amplification, the PCR products are purified with a Montage PCR_{96} plate system and recovered with 116 μl of water. The products are run on a 200 ml, 96-well agarose gel (1.5%) to check for purity and to quantitate DNA yield. Plates with remaining primers or unreacted dNTPs (as indicated by a smear at the bottom of the gel) should be purified a second time. Impurities in the DNA probes result in decreased DNA binding to the glass slide surface and decreased hybridization efficiency. The DNA yield is determined using a digital image of the gel by comparing the intensity of the product band to the intensity of a similarly sized band in a standard of known concentration (for example, ϕX174 DNA – *HaeIII* digest). A DNA yield greater than 2.4 μg is necessary (if redissolved in 12 μl) to ensure the probe will not become saturated during hybridization. The purified products are then dried in a vacuum centrifuge and redissolved in 12 μl of 35% DMSO solution to yield a minimum target concentration of 0.2 mg/ml used for subsequent spotting. DNA solution volumes less than 12 μl can be used, but lower volumes increased the likelihood of failed pick-ups during the printing process.

4.4.2.3 DNA-Array Printing

To improve data reliability and reproducibility, all target DNA should be spotted at least in duplicate (if not triplicate). Genes of particular interest or genes thought to play a critical role for specific studies can be represented by more than three spots. For example, many genes involved in solventogenesis and sporulation were represented by as many as 12 spots on the *C. acetobutylicum* arrays. Each replicate spot should be spotted in a position away from the other replicate spots to minimize spatial biases. In addition to the target DNA, it is necessary to spot several negative control spots with no known homologies to the entire genome sequence. For the *C. acetobutylicum* arrays, 22 negative control genes [3 from *C. pasteurianum*, 9 from *Saccharomyces cerevisiae*, and 10 from *Arabidopsis thaliana* (SpotReport Array Validation System, Stratagene, LaJolla, CA)] with no known homologies to the *C. acetobutylicum* genome were used to correct for nonspecific hybridization. Genomic DNA should also be spotted as a positive control and can be useful in trouble shooting labeling and hybridization problems that may arise. When printing DNA in DMSO solutions, it is important to control the humidity and temperature [44]. Optimal spot morphology and subsequent hybridization were achieved when spotting was performed at 70 to 74 F and 40 to 45% relative humidity on Corning CMT-GAPS™ (Corning, NY) or TeleChem's (Sunnyvale, CA) ArrayIt™ Brand SuperAmine glass microarray slides. The first generation of *C. acetobutylicum* arrays was printed with a 125 μm spot size and 200 μm spacing. After spotting, the slides are UV cross-linked (Stratagene cross-linker) and baked in an oven at 80°C for 2 to 4 hours. Printed arrays should be stored in a desiccated container and kept in the dark. Array quality decreased considerably after 2 to 3 months with lower spot intensities and higher background.

4.4.2.4 RNA Sampling, Isolation, and Purification

High quality RNA is necessary to produce labeled cDNA probes used for DNA-array hybridizations. The following protocol produces high quality RNA from *C. acetobutylicum*. RNA quality is determined using spectrophotometer measurements ($A_{260}/A_{280} > 1.9$) and by running the RNA on an agarose gel to verify the absence of RNA degradation. Cell pellets from 2 to 5 ml of culture are collected by centrifugation at 4°C and 5000 × g for 10 minutes. Pellets are resuspended in 200 μl of SET buffer [25% sucrose, 50 m*M* Tris (pH 8), 50 m*M* EDTA (pH 8)] with 20 mg/ml lysozyme and the samples incubated at 37°C for 3 to 5 minutes [3]. One ml of cold TRIzol® Reagent (Invitrogen, Carlsbad, CA) is added and the samples vortexed for 30 seconds. The TRIzol® samples are immediately frozen at –85°C and the RNA isolated within one month to avoid degradation.

For isolation and purification, the TRIzol® samples are thawed at room temperature and diluted two-fold in ice cold TRIzol® up to 1 ml. Two hundred μl of chloroform is added to 1 ml of the diluted TRIzol® treated samples, vortexed, and allowed to stand for 2 minutes at room temperature. The samples are centrifuged at 12,000 × g for 15 minutes at 4°C and the aqueous phase transferred to a fresh tube. 0.5 ml of isopropanol is added, the tubes inverted several times, the samples allowed to stand for 3 minutes and then centrifuged at 12,000 × g for 10 minutes at 4°C. The resulting pellet is washed with 75% RNase-free ethanol and spun at 8000 × g for 4 minutes at 4°C. After drying for 10 minutes, the RNA is finally resuspended in RNase-free water and quantitated with a UV spectrophotometer (A260 and A280). Each sample is run on a 1.2% agarose gel to check for lack of RNA degradation. Purified RNA samples can be stored at –85°C but should be labeled within one week and the number of freeze-thaw cycles minimized.

4.4.2.5 cDNA Labeling

Labeled cDNA is synthesized by random hexamer-primed reverse transcription reactions in the presence of Cy3-dUTP or Cy5-dUTP using SuperScript II (Invitrogen) reverse transcriptase. Exposure of any Cy-dye material to light should be minimized whenever possible to avoid photobleaching. Twelve μg of RNA are mixed with 2.4 μg of random hexamer primers (Roche, Indianapolis, IN), heated to 70°C for 10 minutes and cooled on ice for one minute. Unlabeled dNTPs (0.60 m*M* dATP, 0.15 m*M* dTTP, and 0.40 m*M* dGTP and dCTP), 5X RT buffer (1X working concentration), DTT (1 m*M* working concentration), either Cy3 or Cy5 labeled dUTP (Amersham), and 400 U of reverse transcriptase, are added in the order stated to a final volume of 25 μl. The samples are incubated at 42°C for 2 hours. The reaction is stopped by addition of 20 m*M* EDTA and the RNA is subsequently degraded by addition of NaOH (30 m*M* final concentration) followed by incubation at 70°C for 10 minutes. The mixture is cooled on ice and neutralized by adding HCl (30 m*M* final concentration). The labeled probe is purified using a GFX purification kit (with two ethanol wash steps) and the DNA eluted with 50 μl TE (pH 8). The purified probe is then dried to near completion (<5 μl) in a rotary speedvac and stored at –20°C until use. Labeled cDNAs can be stored for several months without significant changes in hybridization results. Note that the most efficient labeling reactions are achieved when using fresh buffer components and recently opened (<1 week) Cy3- and Cy5-dUPTs.

4.4.2.6 Array Hybridization

All solutions used for array hybridizations should be of high quality and sterile filtered (0.2 μm) to ensure they are free of contaminants. For hybridizations, the spotted arrays are incubated in prehybridization buffer (5X SSC, 0.1% SDS, 1% bovine albumin) at 42°C for 45 minutes. The slides are then washed by dipping five times in Millipore water and twice in isopropanol. A lint-free tissue can be used to wipe excess isopropanol from the nonprinted areas of the slide and then allowed to air dry. Oppositely labeled dried probes are resuspended in 5 to 10 μl of sterile water and mixed. One microliter of sonicated salmon sperm DNA (10 mg/ml) is added, and the mixture is denatured at 95°C for 3 minutes. Salmon sperm DNA should be aliquoted to avoid repeated freeze thaws (which produces high background levels when used). An equal volume of 2X hybridization buffer (10X SSC, 50% formamide, 0.2% SDS) is added, and the sample is loaded onto the array under a LifterSlip (Erie Scientific, Portsmouth, NH). The slides are hybridized 12 to 18 hours at 42°C in Corning hybridization chambers with 100 μl of 10X SSC to maintain humidity. Longer hybridization times can increase the amount of bound labeled DNA. However, longer hybridization times result in drying of the sample under the cover slip to the slide surface, resulting in increased background fluorescence. After hybridization, the slides are washed with TeleChem ArrayIt™ DNA microarray wash buffer A at 42°C, followed by buffers B and C at room temperature, for 5 minutes in each buffer with vigorous agitation. The slides are quickly dipped in Millipore water, placed in a 50 ml conical centrifuge tube with lint-free tissue at the bottom, and dried by centrifugation for 5 minutes at 200 × g. The hybridized arrays are analyzed with a microarray scanner and software. Spot intensities can be quantitated with a wide variety of DNA array analysis software. Figure 4.4 shows several examples of scanned microarray slides with problems commonly encountered when first beginning microarray work.

4.4.2.7 Microarray Data Analysis

A significant percentage of genes analyzed in any given sample are likely to have spot intensities very close to the background intensities (or close to zero). Low-spot intensities have a high noise-to-signal ratio and can, therefore, lead to improper identification of differentially expressed genes. A proven method for minimizing such errors is use of a filtering criterion based on the spot intensity (signal) for a given channel (which has been background subtracted and corrected for nonspecific binding) and the standard deviation of the local background (noise). The criterion is described by the following:

$$\frac{x_{raw,i} - x_{bg,i} - \overline{x}_{neg}}{\beta \cdot SD_{bg,i}} > 1$$

where $x_{raw,i}$ is the raw channel intensity, $x_{bg,i}$ is the local background for the given channel, x_{neg} is the nonspecific binding, β is a constant, and $SD_{bg,i}$ is the standard deviation for the local channel background.

β can be chosen such that the channel intensity (background and nonspecific binding corrected) is greater than the noise of the background at a given confidence level. This serves as a very conservative measure for the reliability of the intensity data, particularly at lower intensities. Spots that fail to meet the above criterion for both channels should be ignored. Spots that fail to meet the criterion for one of the two channels can be subjected to a more strict criterion for the second channel (a β corresponding to a higher confidence level). Those that fail to meet the more stringent criterion for the second channel are also ignored. For spots that pass the more stringent criterion, one can estimate the fold-overexpression by dividing the intensity of the channel that exceeds the criterion by 1.96 times the standard deviation of the background for that channel. Data generated in this manner should be utilized only to fill in missing data, and not for identification of differentially expressed genes.

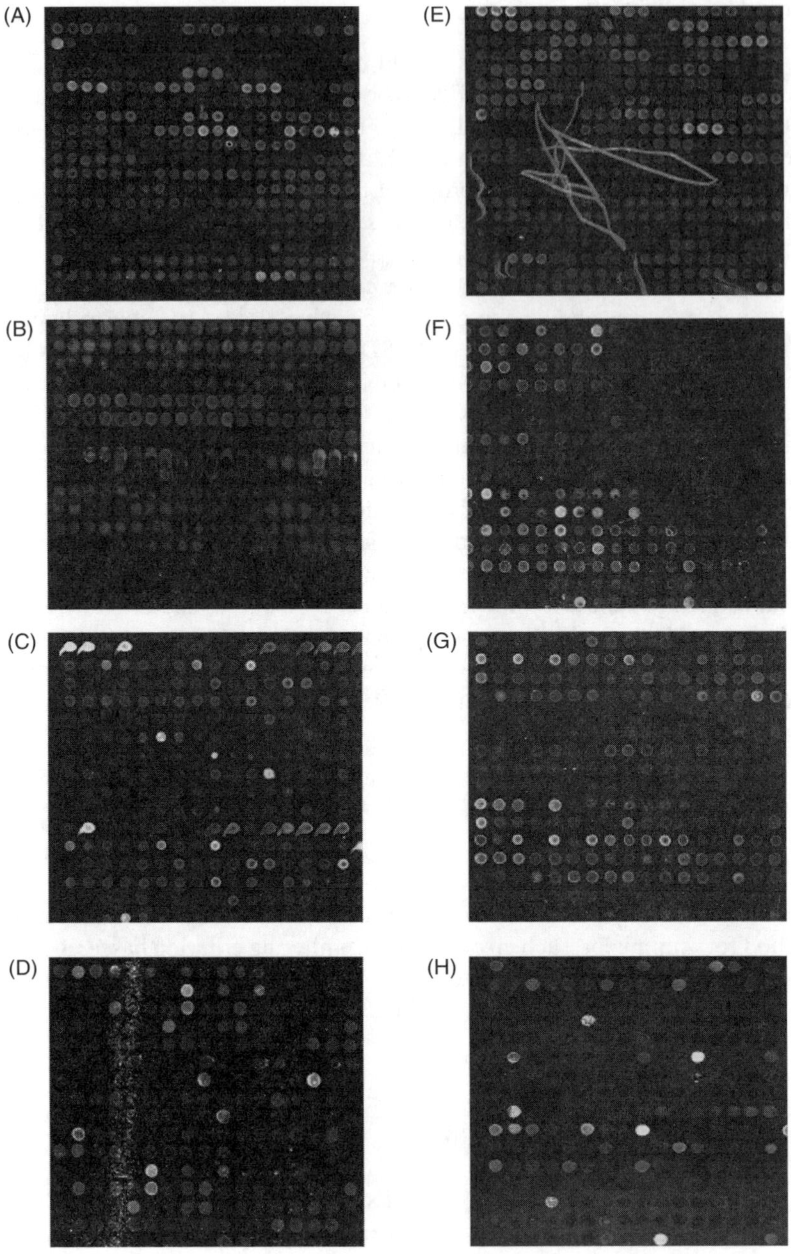

FIGURE 4.4 Common microarray slide pathologies. Examples of a slide (A) with good spot morphologies, low background, and high signal strength; (B) with smeared spots due to a printing robot error; (C) that contains spots with tails, likely due to a dusty printing pin; (D) that was not properly washed, as evidenced by streaking and colored specking; (E) that was scratched by the cover slip during the washing process; (F) that had a bubble under the cover slip during hybridization, leaving an area effectively unhybridized (upper right-hand corner); (G) has generally high background due to use of an expired slide or contaminated reagent; (H) with high background and reverse spotting, likely a result of sample evaporation during hybridization.

Variations in several factors and processes of array analysis result in two types of errors: random and system errors. Both types of errors affect the measured gene expression intensity. Random errors result from scanning errors, variations in the array surface, and spot-to-spot variations (on the same array) in the amount of deposited cDNA. These errors can generally be defined as noise with a mean of zero across all spots in an array. System errors can be defined as those resulting from variations in array quality (array surface and printing, array storage, and amount of DNA spotted), as well as in preparation between two labeled samples (RNA amount used, reverse transcription and labeling efficiencies, slide hybridization and washing). A normalization procedure can be applied to minimize the effects of such errors. The intuitive SNN-LERM normalization and gene identification method is based on segmentation of nearest neighbors (SNN) over the entire gene expression intensity range into a number of intensity intervals and determination of the mean (M) and standard deviation of the logarithms of expression ratios (LERs) for each interval [47]. After segmentation of the entire expression range into intensity intervals, the mean and standard deviation of the logarithm of expression ratios are calculated for each interval using the nearest neighbor genes. Genes with high differential expression are excluded from these calculations. Normalization is performed for each interval using the mean of the logarithm of expression ratios in the interval. The normalization program based on this method was written using a combination of Visual Basic and MATLAB (The MathWorks, Inc.). Compared to other normalization methods, this method delivered a drastic improvement in normalization for 22 *C. acetobutylicum* glass arrays. For identifying differentially expressed genes, upper and lower boundaries are constructed for each interval using the standard deviation of the expression-ratio logarithms. Using a *C. acetobutylicum* pSOL1 megaplasmid-deficient strain M5 [48], this method identified more "downregulated" pSOL1 genes with fewer misidentifications in a comparative array analysis of M5 versus the parent strain. A comparison of quantitative RT-PCR results with different gene identification methods also indicates that the proposed method is superior to other methods.

After normalization and gene identification, it is useful to group genes based on their common expression patterns. This can be achieved using a hierarchical clustering program like Cluster [42]. Alternatively, self organizing map (SOM) analysis can be used to identify gene clusters using GeneCluster (Whitehead Institute for Biomedical Research) [43]. Clusters identified by SOM analysis tend to be more stable than gene groups identified by hierarchical clustering. The hierarchically clustered genes and SOM gene clusters can be visualized using Tree View [42].

4.4.2.8 DNA-Array Validation

Two independent strategies were utilized for array validation. The first is based on the genes of the megaplasmid pSOL1, which is absent in strain M5 [49]. In a comparative analysis of WT and M5, any gene that resides on the pSOL1 megaplasmid should fall into one of two categories: significantly "upregulated" in WT or nondifferentially expressed. An Eisen plot of the 56 pSOL1 genes that are differentially expressed at the 95% confidence level for at least two time points revealed that the vast majority of pSOL1 genes had the expected higher expression in WT [42]. This demonstrated the validity of the DNA arrays and protocols used in the array analysis protocol that has been developed. For 132 pSOL1 genes (not including hypothetical proteins) analyzed on 17 arrays (17 arrays × 132 pSOL1 genes = 2244 classifications), 2231 were properly identified (99.4%). Analysis of the hypothetical proteins from the megaplasmid revealed that several genes had significant regions of homology with chromosomal genes, despite the effort to minimize such homologies when designing the cDNA probes for array spotting. This is likely due to the fact that many of the hypothetical proteins are short (<500 bp), making it more difficult to design probes without significant homologies to other genes. Of the 13 misidentified genes, only one was present on more than one slide. Nine of the misidentifications had fold ratios less than 1.9, while only one was greater than 2.5. These results suggest the array design, along with the array normalization and gene identification method, provide very accurate and reliable results.

A second validation method is based on direct comparison of array data to Northern analysis data for the strain pair WT and SKO1 (*spo0A* mutant). Ratios from Northern analysis of 824 and SKO1 samples were compared to microarray-generated ratios and previously published Northern data [3,35]. The relative mRNA levels generated by Northern analysis were normalized against the thiolase gene (*thlA*), which can be considered as a housekeeping gene in these experiments [3,14]. This was further verified by DNA-array analysis. Microarray ratios for the *aad-ctfA/B* and *ptb-buk* operon transcripts were generated by averaging the ratios calculated from spots consisting of cDNA for each individual gene. Because *spo0A* is disrupted in SKO1, Northern blots show up to a 38-fold decrease in transcript levels, where the microarrays showed up to a 3.5-fold decrease. The *sol* operon (*aad-ctfA/*B) has been shown to be directly regulated by Spo0A [3]. Northern blots indicate up to a 37-fold decrease in *aad-ctfA/*B transcript levels in SKO1, where the microarrays indicate a maximum 9.1-fold decrease at a 95% confidence level from transitional through mid-stationary growth phases. *adc* shows significant downregulation of expression during exponential growth to stationary phase. For both *aad-ctfA/B* and *spo0A* transcripts, the normalized ratios follow the downregulation trend but underestimate the overall ratio, a well-known characteristic of cDNA microarrays. This suggests that our DNA-array technology and computational tools are conservatively reliable for transcriptional analysis [47]. The *ptb-buk* transcript ratios show downregulation in SKO1 during exponential and transition phase followed by upregulation thereafter. The corresponding microarray data generally follow the same trend as in the Northern analysis, indicating that microarray analysis can be used to follow general transcriptional patterns of up- or downregulation.

4.4.3 Future Developments

Most fluorescently labeled cDNA targets used for microarray hybridizations are produced using the direct incorporation method [44,50]. The direct incorporation method utilizes bulky fluorescent dyes (Cy3 and Cy5), which are not efficiently incorporated by reverse transcriptases. Moreover, reverse transcriptases incorporate Cy3- and Cy5-labeled dNTPs at different rates. These problems can be avoided if an indirect labeling method is employed whereby aminoallyl-derivatized dUTP incorporation is followed by amino coupling of the fluorescent dyes using a simple chemical reaction [44,51,52]. In addition to overcoming the shortcomings of direct labeling, the indirect labeling method is significantly more cost effective (<50% less expensive per slide). An indirect labeling protocol should be tested and validated for use in future clostridial DNA-array work [44]. This protocol may provide increased sensitivity of detection and more accurate and reproducible results.

ACKNOWLEDGMENTS

This work was supported by grants BES-9911231 and BES-9905669 from the National Science Foundation and grant R828562 from the Environmental Protection Agency.

REFERENCES

1. Mermelstein, L.D. and Papoutsakis, E.T., In vivo methylation in *Escherichia coli* by the *Bacillus subtilis* phage phi 3T I methyltransferase to protect plasmids from restriction upon transformation of *Clostridium acetobutylicum* ATCC 824, *Appl. Environ. Microbiol.*, 59 (4), 107710, 1993.
2. Green, E.M. et al., Genetic manipulation of acid formation pathways by gene inactivation in *Clostridium acetobutylicum* ATCC 824, *Microbiol. UK*, 142, 2079, 1996.
3. Harris, L.M., Welker, N.E., and Papoutsakis, E.T., Northern, morphological, and fermentation analysis of *spo0A* inactivation and overexpression in *Clostridium acetobutylicum* ATCC 824, *J. Bacteriol.*, 184 (13), 3586, 2002.
4. Papoutsakis, E.T. and Bennett, G.N., Cloning, structure, and expression of acid and solvent pathway genes of *Clostridium acetobutylicum*, in *The Clostridia and Biotechnology*, Woods, D.R., Ed., Butterworth-Heinemann, Stoneham, MA, 1993, 157.

5. Tummala, S.B. et al., Genetic tools for solventogenic Clostridia, in *Clostridia: Biotechnology and Medical Applications*, Bahl, H. and Duerre, P., Eds., John Wiley & Sons, Inc., New York, 2001, 105.
6. Phillips-Jones, M.K., Bioluminescence (lux) expression in the anaerobe *Clostridium perfringens*, *FEMS Microbiol. Lett.,* 106 (3), 265, 1993.
7. Phillips-Jones, M.K., Use of a lux reporter system for monitoring rapid changes in alpha-toxin gene expression in *Clostridium perfringens* during growth, *FEMS Microbiol. Lett.,* 188 (1), 29, 2000.
8. Bullifent, H.L., Moir, A., and Titball, R.W., The construction of a reporter system and use for the investigation of *Clostridium perfringens* gene expression, *FEMS Microbiol. Lett.,* 131 (1), 99, 1995.
9. Melville, S.B., Labbe, R., and Sonenshein, A.L., Expression from the *Clostridium perfringens* cpe promoter in *C. perfringens* and *Bacillus subtilis*, *Infect Immun.,* 62 (12), 5550, 1994.
10. Zhao, Y. and Melville, S.B., Identification and characterization of sporulation-dependent promoters upstream of the enterotoxin gene (cpe) of *Clostridium perfringens*, *J. Bacteriol.,* 180 (1), 136, 1998.
11. Walters, D.M., Stirewalt, V.L., and Melville, S.B., Cloning, sequence, and transcriptional regulation of the operon encoding a putative N-acetylmannosamine-6-phosphate epimerase (nanE) and sialic acid lyase (nanA) in *Clostridium perfringens*, *J. Bacteriol.,* 181 (15), 4526, 1999.
12. Girbal, L. et al., Development of a sensitive gene expression reporter system and an inducible promoter-repressor system for *Clostridium acetobutylicum*, *Appl. Environ. Microbiol.,* 69 (8), 4985, 2003.
13. Durre, P. et al., Solventogenic enzymes of *Clostridium acetobutylicum*: catalytic properties, genetic organization, and transcriptional regulation, *FEMS Microbiol. Rev.,* 17 (3), 251, 1995.
14. Tummala, S. and Papoutsakis, E.T., Development and characterization of a gene-expression reporter system for *Clostridium acetobutylicum* ATCC 824, *Applied and Environ. Microbiol.,* 65, in press, 1999.
15. Davis, T.O. et al., Development of a transformation and gene reporter system for group II, non-proteolytic *Clostridium botulinum* type B strains, *J. Mol. Microbiol. Biotechnol.,* 2 (1), 59, 2000.
16. Quixley, K.W. and Reid, S.J., Construction of a reporter gene vector for *Clostridium beijerinckii* using a *Clostridium endoglucanase* gene, *J. Mol. Microbiol. Biotechnol.,* 2 (1), 53, 2000.
17. Wagner, E.G. and Simons, R.W., Antisense RNA control in bacteria, phages, and plasmids, *Ann. Rev. Microbiol.,* 48, 713, 1994.
18. Desai, R.P. and Papoutsakis, E.T., Antisense RNA strategies for metabolic engineering of *Clostridium acetobutylicum*, *Appl. Environ. Microbiol.,* 65 (3), 936, 1999.
19. Persson, C., Wagner, E.G., and Nordstrom, K., Control of replication of plasmid R1: kinetics of *in vitro* interaction between the antisense RNA, CopA, and its target, *CopT, Embo J.,* 7 (10), 3279, 1988.
20. Hjalt, T. and Wagner, E.G., The effect of loop size in antisense and target RNAs on the efficiency of antisense RNA control, *Nucleic Acids Res.,* 20 (24), 6723, 1992.
21. Rittner, K., Burmester, C., and Sczakiel, G., *In vitro* selection of fast-hybridizing and effective antisense RNAs directed against the human immunodeficiency virus type 1, *Nucleic Acids Res.,* 21 (6), 1381, 1993.
22. Hjalt, T.A. and Wagner, E.G., Bulged-out nucleotides in an antisense RNA are required for rapid target RNA binding *in vitro* and inhibition *in vivo*, *Nucleic Acids Res.,* 23 (4), 580, 1995.
23. Engdahl, H.M., Hjalt, T.A., and Wagner, E.G., A two unit antisense RNA cassette test system for silencing of target genes, *Nucleic Acids Res.,* 25 (16), 3218, 1997.
24. Patzel, V. and Sczakiel, G., *In vitro* selection supports the view of a kinetic control of antisense RNA-mediated inhibition of gene expression in mammalian cells, *Nucleic Acids Res.,* 28 (13), 2462, 2000.
25. Patzel, V. and Sczakiel, G., Theoretical design of antisense RNA structures substantially improves annealing kinetics and efficacy in human cells, *Nat. Biotechnol.,* 16 (1), 64, 1998.
26. Tummala, S.B., Welker, N.E., and Papoutsakis, E.T., Design of antisense RNA constructs for down-regulation of the acetone formation pathway of *Clostridium acetobutylicum*, *J. Bacteriol.,* 185 (6), 1923, 2003.
27. Jacobson, A.B. and Zuker, M., Structural analysis by energy dot plot of a large mRNA, *J. Mol. Biol.,* 233 (2), 261, 1993.
28. Zuker, M. and Stiegler, P., Optimal computer folding of large RNA sequences using thermodynamics and auxiliary information, *Nucleic Acids Res.,* 9 (1), 133, 1981.
29. Zuker, M., Calculating nucleic acid secondary structure, *Curr. Opin. Struc. Biol.,* 10 (3), 303, 2000.
30. Chastain, M. and Tinoco, I., Jr., Structural elements in RNA, *Prog. Nucleic Acid. Res. Mol. Biol.,* 41, 131, 1991.

31. Fierro-Monti, I.P., Reid, S.J., and Woods, D.R., Differential expression of a *Clostridium acetobutylicum* antisense RNA: implications for regulation of glutamine synthetase, *J. Bacteriol.*, 174, 7642, 1992.
32. Nair, R.V., Bennett, G.N., and Papoutsakis, E.T., Molecular characterization of an aldehyde/alcohol dehydrogenase gene from *Clostridium acetobutylicum* ATCC 824, *J. Bacteriol.*, 176 (3), 871, 1994.
33. Tummala, S.B., Junne, S.G., and Papoutsakis, E.T., Antisense RNA downregulation of coenzyme A transferase combined with alcohol-aldehyde dehydrogenase overexpression leads to predominantly alcohologenic *Clostridium acetobutylicum* fermentations, *J. Bacteriol.*, 185 (12), 3644, 2003.
34. Tummala, S.B. et al., Transcriptional analysis of product-concentration driven changes in cellular programs of recombinant *Clostridium acetobutylicum* strains, *Biotech. Bioeng.*, 84 (7), 842, 2003.
35. Tomas, C.A. et al., DNA array-based transcriptional analysis of asporogenous, nonsolventogenic *Clostridium acetobutylicum* Strains SKO1 and M5, *J. Bacteriol.*, 185 (15), 4539, 2003.
36. Harris, L.M. et al., Fermentation characterization and flux analysis of recombinant strains of *Clostridium acetobutylicum* with an inactivated *solR* gene, *J. Ind. Microbiol. Biotech.*, 27 (5), 322, 2001.
37. Baldwin, D., Crane, V., and Rice, D., A comparison of gel-based, nylon filter and microarray techniques to detect differential RNA expression in plants, *Curr. Opin. Plant Biol.*, 2 (2), 96, 1999.
38. Watson, A. et al., Technology for microarray analysis of gene expression, *Curr. Opin. Biotechnol.*, 9 (6), 609, 1998.
39. Schena, M. et al., Microarrays: biotechnology's discovery platform for functional genomics, *Trends Biotechnol.*, 16 (7), 301, 1998.
40. Knight, J., When the chips are down, *Nature*, 410 (6831), 860, 2001.
41. Taylor, E. et al., Sequence verification as quality-control step for production of cDNA microarrays, *Biotechniques*, 31 (1), 62, 2001.
42. Eisen, M.B. et al., Cluster analysis and display of genome-wide expression patterns, *Proc. Natl. Acad. Sci. USA*, 95 (25), 14863, 1998.
43. Tamayo, P. et al., Interpreting patterns of gene expression with self-organizing maps: methods and application to hematopoietic differentiation, *Proc. Natl. Acad. Sci. USA*, 96 (6), 2907, 1999.
44. Hegde, P. et al., A concise guide to cDNA microarray analysis, *Biotechniques*, 29 (3), 548, 2000.
45. Xu, D. et al., PRIMEGENS: robust and efficient design of gene-specific probes for microarray analysis, *Bioinformatics*, 18 (11), 1432, 2002.
46. Richmond, C.S. et al., Genome-wide expression profiling in *Escherichia coli* K-12, *Nucleic Acids Res.*, 27 (19), 3821, 1999.
47. Yang, H. et al., A segmental nearest neighbor normalization and gene identification method gives superior results for DNA-array analysis, *Proc. Natl. Acad. Sci. U.S.A.*, 100 (3), 1122, 2003.
48. Clark, S.W., Bennett, G.N., and Rudolph, F.B., Isolation and characterization of mutants of *Clostridium acetobutylicum* ATCC 824 deficient in acetoacetyl-Coenzyme A:acetate/butyrate:Coenzyme A transferase (EC 2.8.3.9) and in other solvent pathway enzymes, *App. Environ. Microb.*, 55, 970, 1989.
49. Cornillot, E. et al., The genes for butanol and acetone formation in *Clostridium acetobutylicum* ATCC 824 reside on a large plasmid whose loss leads to degeneration of the strain, *J. Bacteriol.*, 179 (17), 5442, 1997.
50. Fawcett, P. et al., The transcriptional profile of early to middle sporulation in *Bacillus subtilis*, *Proceedings of the National Academy of Sciences of the United States of America*, 97 (14), 8063, 2000.
51. Rosenow, C. et al., Prokaryotic RNA preparation methods useful for high density array analysis: comparison of two approaches, *Nucleic Acids Research*, 29 (22), 2001.
52. Yu, J.D. et al., Evaluation and optimization of procedures for target labeling and hybridization of cDNA microarrays, *Molecular Vision*, 8 (17), 130, 2002.

5 Anoxic Testing and Purification of Enzymes

Thorsten Selmer

CONTENTS

5.1 INTRODUCTION

The development of specific enzyme test systems, which allows the selective and quantitative measurement of an individual enzymatic reaction in crude samples like homogenized cells, is among the most challenging tasks in analytical biochemistry.

Though they have not been fully researched, enzyme testing and purification of the responsible enzymes are important analytical and preparative methods in microbiology. Enzyme assays are needed in order to compare the metabolic features of various species and can be used to follow metabolic changes during the growth of an organism. The knowledge of the key enzymes in the energy metabolism of microorganisms is a prerequisite for an understanding of bioenergetics and gross material fluxes in ecosystems. These tests are essential for the tracing of individual enzymes throughout a purification. Pure enzymes are essential starting materials for biochemical and structural studies on this important class of biomolecules. The knowledge of their functional properties can provide access to novel biotechnological processes or can play an important role in the development and progression of diseases.

This chapter is divided into four sections: General Remarks, Preparation of Anoxic Buffers and Solutions, Anoxic Testing of Enzymes, and Anoxic Purification of Enzymes. It aims to provide practical guidelines for enzyme testing and purification of oxygen-sensitive enzymes. The recommendations are based on practical experience and will likely differ from the solutions found in other laboratories. Economical constraints and the availability of equipment will enforce changes in the setups. The need to use sophisticated equipment, like anoxic glove boxes, will be essential for purification or analysis of extremely sensitive protein, while enzymatic testing of samples is frequently possible using simple equipment found in most microbiology laboratories.

5.2 GENERAL REMARKS

A living cell will easily contain several thousand enzymes. Some of these proteins are involved in the energy metabolism and abundant constituents of the cell. Others are involved in the biosynthesis of cofactors or secondary metabolites and will be present only in very low concentrations. While the former enzymes are frequently found at specific activities above 1 $\mu mol*mg^{-1}*min^{-1}$, the activities of the latter are frequently in the order of a few $nmol*mg^{-1}*min^{-1}$ or less. Obviously, the abundant enzymes are easily measured in short time continuous assays, while low activities will require longer incubation times and are measured most conveniently in stop time assays.

The situation is complicated by the promiscuity of some enzymes that exhibit broad natural substrate specificities. Abundant enzymes will, in general, render difficult or completely impede the measurement of a specific activity, which is present only at low levels, and it is a demanding task to find conditions that abolish or strongly suppress these unwanted reactions in an individual test system. Frequently, the aim to measure selectively one individual enzyme cannot be achieved by variation of the test composition (e.g., by changes in the pH, buffer composition, or selective inhibitors) and a fractionation of the crude sample will be required to succeed.

The most important point that determines the chance to measure an individual reaction is to keep the test system as simple as possible. An optimum system will only contain what is required to measure the desired activity. Any additional constituent can render troublesome testing, since additional compounds can support side reactions, interfere with the desired reaction, or cause unexpected side effects, including a loss of selectivity.

In an ideal case, the substrate consumption will perfectly match the product formation, and some spectral changes can be monitored directly. Many enzymatic systems are not that simple, and it will be required to use sophisticated analytical tools in order to measure substrate consumption and product formation simultaneously.

Moreover, the equilibrium of an individual reaction might strongly favor the substrate side in the physiological reaction. Such reactions are most conveniently measured starting with the product of the reaction in an inverted direction or must be coupled with a product consuming reaction in order to pull the net reaction in the desired direction.

Coupling of an individual reaction, which is not readily measurable, to a product consuming secondary reaction is a popular principle in order to facilitate enzymatic testing. In particular, dehydrogenases, which utilize nicotinamide dinucleotides as cosubstrates, are frequently used helper enzymes. The oxidation of NAD(P)H or the reduction of NAD(P) is easily followed by UV/Vis and will frequently allow to pull an unfavorable equilibrium into the desired direction. On the other hand, such enzymes may be useful in order to provide a substrate for a downstream reaction. In this case, the downstream reaction must pull the equilibrium in order to detect NAD(P)H oxidation or NAD(P) reduction. An example for the latter type of coupling is the malate dehydrogenase reaction, which is used to provide oxaloacetate. The equilibrium of this reaction strongly favors NAD and malate, but NADH is formed rapidly when oxaloacetate is consumed by downstream reactions like citrate synthase.

In practice, a coupling of desired enzymatic reactions to dehydrogenases may require several individual enzymes. In order to ensure the desired activity being measured, it is required to provide the helper enzymes in significant excess. It is generally recommended to provide a 10-fold excess of individual helper enzymes, as compared to the preceding activity in a sequence. The limits for the number of steps that can be employed following this rule are obvious, and it must be stated that the meaningful use of more than three helper enzymes in a test system must question the results in most cases.

Individual properties of the enzymes involved may completely abolish any coupling. For example, in our laboratory the reductive branch of the alanine fermentation in *Clostridium propionicum* is studied. The key enzyme in this pathway is (*R*)-lactoyl-CoA dehydratase (Lcd), catalyzing the reversible dehydration of (*R*)-lactoyl-CoA to acryloyl-CoA. The equilibrium of this reaction strongly favors lactoyl-CoA (99:1) and requires catalytic amounts of ATP and a low potential electron donor like sodium dithionite or titanium-(III)-citrate for *in vitro* testing. Acryloyl-CoA is subsequently reduced with NADH to propionyl-CoA by the acryloyl-CoA reductase complex (Acr) in *C. propionicum*. Since the latter reaction is essentially irreversible, it should theoretically provide a meaningful helper enzyme to detect Lcd activity. Indeed, it was very simple to detect the Lcd-dependent reduction of lactoyl-CoA to propionyl-CoA by mass spectrometry. However, the substrate reduction by Acr turned out to be decoupled from the oxidation of NADH under test conditions, and dithionite or titanium-(III)-citrate are used preferentially as electron supply for the reduction of acryloyl-CoA by Acr.

In addition to the general aspects that must be faced in order to establish specific enzyme test systems, the predominant strict anaerobic clostridial lifestyle has some consequences on enzymatic testing. During the evolution of clostridial enzymes, a selection toward a proper function in the presence of oxygen barely took place. The high concentration of this high potential electron acceptor in air might, therefore, cause massive problems in testing. Two basic problems must be accounted for when clostridial enzymes are investigated: the stability of the enzymes under air and possible side reactions in enzyme tests.

The response of clostridial enzymes toward exposure to molecular oxygen varies strongly among them (Table 5.1). Many cofactor-free enzymes are reasonably stable under air and can be handled without special precautions. However, long-term storage of these enzymes frequently requires the presence of thiols (e.g., DTT or DTE), in order to protect cysteinyl residues of the proteins from being slowly oxidized by oxygen, in particular in the presence of traces of heavy metal ions.

In contrast, several redox active enzymes from clostridial origin are highly sensitive to oxygen. These enzymes are adopted to function in an environment, which provides a low redox potential. Iron-sulfur clusters, flavines, or other cofactors in these proteins can easily react with molecular oxygen. In some cases, the oxygen acts as an artificial electron acceptor and gives rise to side reactions, which render difficult enzymatic testing. An example for the loss of selectivity frequently observed is the diasphorase activity of many clostridial flavoproteins; while these enzymes catalyze a selective reduction of substrates with NAD(P)H as electron source *in vivo*, the oxidation of NAD(P)H may be decoupled from the reduction of substrate under air, and hydrogenperoxide is formed. It is obvious that such side reactions render difficult enzymatic testing and, since owned by several different enzymes, impede tracing of an individual enzyme throughout a purification. Due to this property, the popular coupling of many enzymatic reactions to the oxidation of NAD(P)H or reduction of NAD(P) can cause problems when carried out in cell-free extracts under air; the diasphorase activity of many different enzymes will yield an overestimation of a substrate specific NAD(P)H oxidation, while the NAD(P)-dependent substrate oxidation is frequently underestimated due to the reoxidation of NAD(P)H.

The reaction of molecular oxygen with essential cofactors of enzymes can also yield an irreversible loss of activity. A variety of cofactors are irreversibly destroyed by this reaction. Cluster destruction in iron-sulfur proteins and polypeptide cleavage of glycyl radical enzymes exposed to

TABLE 5.1

Group	Testing	Purification
Extremely sensitive enzymes $t_{1/2}$ under air below 1 minute Glycyl radical enzymes and some iron-sulfur proteins	Strict anoxia for sample preparation and testing Anoxic glove boxes or physical shielding in combination with oxygen scavenging systems are essentially required	Strict anoxia for all handling steps Anoxic glove boxes are essentially required
Highly sensitive enzymes $t_{1/2}$ under air below 1 hour Many redox-active iron-sulfur and flavoproteins	Strict anoxia required for convenient sample preparation and testing Anoxic glove boxes or oxygen scavenging systems are highly recommended Initial testing is frequently possible presupposing the direct contact with air is avoided	Strict anoxia for all handling steps Anoxic glove boxes are essentially required
Moderately sensitive enzymes $t_{1/2}$ under air below 1 day Many redox-active iron-sulfur and flavoproteins Proteins containing oxygen sensitive thiols	Initial testing frequently possible under air, exclusion of oxygen will increase activity yield Reliable testing is frequently possible under air using anoxic solutions and physical shielding of the samples	Reduction of oxygen exposure is necessary Anoxic glove boxes are recommended, but convenient activity yield can be obtained using anoxic buffers and precautions to exclude air
Slightly sensitive enzymes $t_{1/2}$ under air around 1 week Proteins containing oxygen sensitive thiols	Testing is frequently possible under air Oxygen-induced side reactions may occur	Purification is frequently possible under air Reduction of the oxygen exposure may positively affect activity yield Anoxic storage is recommended
Insensitive enzymes Stable under air Many cofactor-free enzymes	Testing is frequently possible under air Oxygen-induced side reactions may occur	Purification and storage under air

air are well-studied examples. However, many moderately oxygen-sensitive proteins are measured successfully in cell extracts but will not survive purification steps under air. This is predominantly due to a reasonable protection of sensitive enzymes by high concentrations of reduced flavines and iron-sulfur proteins in cell extracts of clostridia.

In addition to ill effects on the enzymes or the loss of specificity, the presence of oxygen may also disable enzymatic testing of redox reactions using redox dye indicators. The spontaneous chemical reaction of oxygen with the dyes employed as electron donors, acceptors, or carriers will enforce an anoxic testing of otherwise oxygen-stable enzymes.

5.3 PREPARATION OF ANOXIC BUFFERS AND SOLUTIONS

5.3.1 General

Anoxic testing and purification of enzymes require methods and technical equipment in order to provide anoxic solutions and to maintain anaerobic conditions throughout all required handling steps. The choice of suitable methods will strongly depend not only on resources and equipment available, but also on the oxygen sensitivity of the enzymes. It is crucial to define the objectives before getting started. An initial testing of cell-free extracts in order to establish the oxygen sensitivity of an enzyme of interest requires less sophisticated methods than the purification or crystallization of extremely oxygen-sensitive proteins.

While the preparation of anoxic solutions can be achieved easily and reliably at various scales, the transfers of samples under anoxic conditions must be considered a more demanding problem. Anoxic solutions in tightly closed containers are well protected against the flood in, but all sample transfers under air own a high risk of oxygen to force its way back in the samples and to interfere with the downstream analysis. This problem is more easily controlled in systems where either chemical oxygen scavengers or enzymatic reactions are used for continuous removal of trace oxygen than in systems where these constituents must be avoided.

5.3.2 Physical Barriers

Physical barriers are the first line of defense against the flood in of oxygen from the surrounding environment. The maximum solubility of oxygen in water is very low as compared to 21% (v/v) oxygen in air (e.g., an air-saturated aqueous solution contains about 235 μM oxygen at 25°C). A proper shielding of solutions in tightly closed containers will reliably avoid the contact of sample with oxygen, while diffusion barriers like cushions of heavy gas (e.g., argon) or layers of mineral oil on top of solutions can slow down the rate at which oxygen forces its way back into the solution. While tightly closed containers will allow long-term storage but are more difficult to handle, diffusion barriers protect the samples for short-time applications only but will facilitate sample handling.

The simplest physical barrier is a tightly sealed bottle. Filled and tightly closed under inert gas, solutions in a container are reliably protected against oxygen. Setups of bottles sealed with rubber stoppers are frequently used to prepare and store anoxic solutions. The risk of oxygen forcing its way back into the bottle is further minimized when the solutions are kept slightly overpressured, since the inert gas will flow out rather than oxygen flooding in. Similar situations are enzymatic tests in glass or quartz cuvettes sealed with rubber stoppers. Like bottles, these containments can be tightly closed. Since the stoppers are easily pierced with hypodermic needles, reactions can be started by adding a missing constituent via gas-tight volumetric syringes.

While glass containers are very useful for many applications, the use of the popular disposable polypropylen test tubes in anoxic experiments should be considered difficult. Like other plastics, the containment wall will contain significant amounts of oxygen, which will slowly diffuse into the solution. The permeability of containment walls for molccular oxygen should be carefully considered, since these materials exhibit frequently low diffusion resistance against oxygen.

The most sophisticated setups will employ anoxic glove boxes (Figure 5.1). The glove boxes provide an inert gas atmosphere walled against the surrounding air by either flexible PVC foils or rigid polyacrylate walls. Gloves accessible from the outside will allow all the required manipulations to be performed in an oxygen-free atmosphere. Samples and equipment are introduced through a lock system, which is evacuated and purged with inert gas in order to minimize air flood in.

The glove boxes are filled with inert gases such as nitrogen, argon, or a mixture of nitrogen (95%) and hydrogen (5%). The latter gas mixture is particularly useful in maintaining strict anoxic conditions, since it is not yet explosive in the presence of air, and the palladium catalysts in the glove boxes will catalyze the reaction of molecular oxygen with hydrogen to yield water, thereby continuously removing trace oxygen flooding in. The presence of hydrogen within the glove boxes will also allow for sensing the oxygen content in a ppm range, an essential analytical tool for the handling and analysis of extremely oxygen-sensitive enzymes.

Though anoxic glove boxes will facilitate most of the required anoxic manipulations on samples, some drawbacks in this technology should be considered. The first problem is limited space within the glove boxes, as well as constraints in size for equipment that can pass the lock. The introduction of very long chromatographic columns or voluminous equipment will frequently require a laborious disassembly and reassembly of the glove box. The requirement of about 24 to 48 h in order to reestablish strict anoxic conditions afterward will slow down the progress of research significantly and renders difficult-to-adopt analytical schemes according to recent results.

(a)

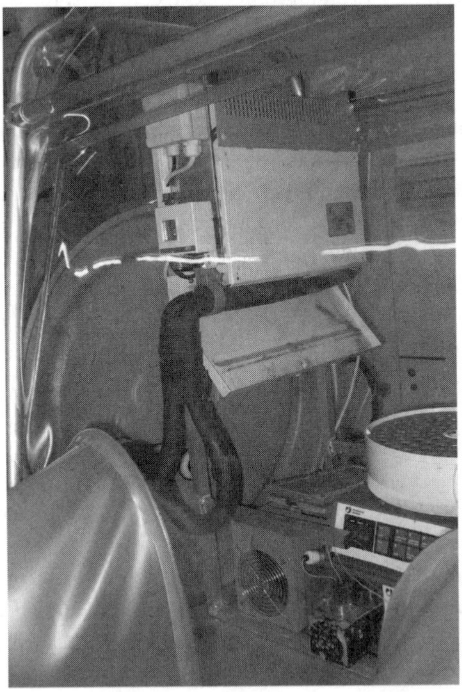

(b)

FIGURE 5.1 Anoxic glove boxes. The glove box provides a physical barrier against the flood in of molecular oxygen (a) and allows sample handling in an anoxic environment (N_2/H_2 95/5%). Samples and equipment are introduced via the lock at the right. Trace oxygen is constantly removed by palladium catalysts, and heat is drawn off using heat exchangers connected to a cold water circle (b). Water and hydrogensulfide are continuously removed, pumping the atmosphere over solid sodium hydroxide (c). Heat-sensitive samples are stored in liquid-chilled aluminium blocks (d), which are connected to external cold-water lines with Luer-adaptors. Both hydrogen and oxygen in the atmosphere are continuously monitored (e).

(c)

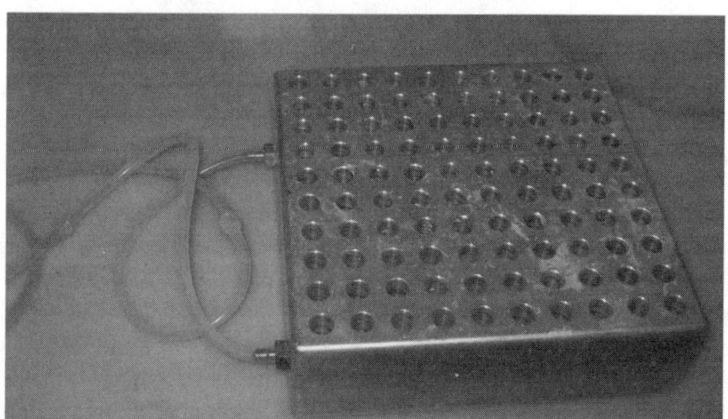

(d)

FIGURE 5.1 (continued).

Additional serious problems arise from the fact that the glove boxes (must) provide a closed system. Whatever is produced inside the wall will stay there. The most problematic products inside anoxic glove boxes are heat, water, and hydrogensulfide.

The most important sources of heat in the glove boxes are electrical devices like stirrers, heat blocks, and photometers. A single deuterium lamp in a photometer will produce about 500 Wh^{-1} heat. The wall of the glove box will not allow a rapid heat transmission, and the temperatures inside glove boxes with operating equipment will be significantly higher than the ambient temperature

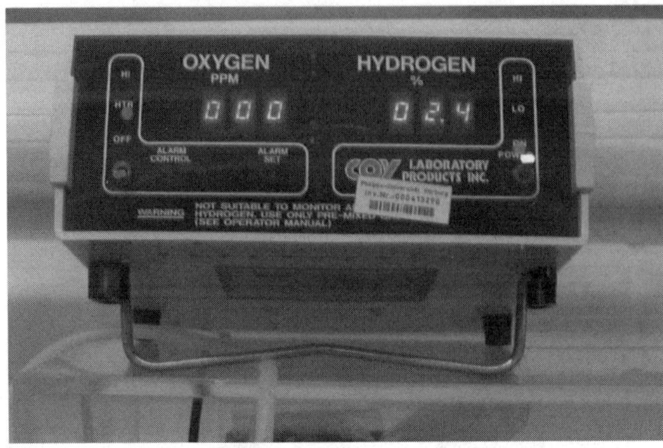

(e)

FIGURE 5.1 (continued).

(usually between 5 and 15°C). It is, therefore, recommended to switch off all devices that are not actually needed. However, a certain buildup of heat is inevitable and should be considered problematic when sensitive enzymes must be handled. As shown in the close-up of Figure 5.1, a simple system consisting of a cooling water cycle and commercially available ventilating heat exchangers inside the glove box will lead out the produced heat in a very efficient way. At ambient temperatures averaging 25°C and a cooling water temperature of 10°C, the temperature inside the glove box averages 15°C. The temperature can be lowered when the temperature of the cooling water is reduced.

It is obvious that ice water, frequently used to store sensitive samples on the bench, is not the first choice for cooling samples inside of glove boxes; during the melting process, small bubbles of air in the ice will release oxygen. We experienced that a single container with about 1 kg of ice will transiently yield up to 15 ppm oxygen in a 5-m^3 glove box. A refrigerating device outside the glove box can be used to provide a solution of 20% glycerol in water to chill aluminium blocks inside the glove box; at cooling liquid temperatures of –2 to –4°C, the cold blocks are cooled below 4°C within about 1 h.

The second major concern in the operation of glove boxes is the water vapor saturation of the atmosphere. Though only small amounts of water are generated by the catalytic removal of trace oxygen under normal operating conditions, significant water will evaporate from solutions handled inside the glove boxes. Consequently, without special precautions, the vapor saturation of the atmosphere will be reached quickly. Due to periodical temperature changes, condensation of water may give rise to dangerous (and expensive) damage of electrical equipment as well as inactivate the palladium catalysts. While the latter is reversible by heating at 180°C, the former must be avoided from the very beginning. Indeed, there are several possibilities to remove the water from the atmosphere. Usually, water-binding compounds like blue gel can be used to bind water. However, the compound has only a very limited capacity and requires frequent exchange and regeneration. An efficient and inexpensive way to eliminate water from the atmosphere is to provide a constant passage of the atmosphere over solid sodium or potassium hydroxide or lime. Placed in sealed bottles, this setup can reduce the vapor saturation far below 50%. Simple air pumps with a performance of about 500 Lh^{-1} for 5 m^3 glove boxes are sufficient to remove water, continuously keeping the vapor saturation below 50%. It is worth noting that more than 500 ml of water are collected by this setup within one week under operating conditions. Another advantage of sodium hydroxide is the capability to efficiently bind hydrogensulfide. Though this chemical is rarely intentionally used in the glove box, significant amounts will be released from bacterial cell-free extracts, in particular when acidic conditions are required. The compound is a well-established catalyst poison, which will irreversibly inactivate the palladium catalysts and cause expensive damages on the oxygen and hydrogen sensor units.

5.3.3 Preparation and Handling of Anoxic Solutions

Molecular oxygen is a major constituent of air, but its solubility in water is very low. The solubility of gases in solutions is a function of the partial pressure in the gas phase above, and a convenient way to reduce the concentration of dissolved oxygen is the storage under inert gases (e.g., nitrogen, helium, or argon). However, the diffusion of oxygen in solution is very slow. In order to accelerate the degassing process, purging of the solutions with gas or vigorous stirring under an inert atmosphere is recommended.

Larger volumes of buffers and solutions are satisfactorily degassed by boiling and subsequent exchange of the headspace atmosphere with inert gases; oxygen is almost insoluble in boiling water, and water vapor is heavier than air. The solution is boiled in a rubber-stoppered bottle for about 15 to 30 min. A hypodermic needle allows a pressure equilibrium during heating. Water vapor replaces the air above the boiling solution, and a subsequent cooling under slight overpressure of inert gas will yield an anoxic solution protected by an inert gas atmosphere. Obviously, the headspace above the solution should be small. Storage slightly overpressured (50 to 100 mbar) will avoid oxygen from flooding in. An alternative method, which is used in many laboratories, is boiling and subsequent cooling of the solutions under vacuum. Once the solutions are chilled down, the headspace is filled with inert gas. However, small leakages in the system during the cooling period under vacuum will allow oxygen to force its way back into the system.

Though useful to provide anoxic buffers or other solutions of heat stable compounds, these methods are not applicable for the preparation of solutions of heat-labile coenzymes, substrates, or enzymes. The preparation of such solutions strongly depends on the compounds to be dissolved. Solids are most satisfactorily dissolved in serum bottles in anoxic solutions. The air in the container is removed by purging of the system with inert gas, and the desired volume of anoxic solution is added using gas-tight volumetric syringes.

Solutions of volatile liquids or enzyme solutions cannot be prepared easily, since the former will evaporate during gas-purging steps, while for the latter solutions usually only small volumes of the stock solutions are required, which will dry, giving rise to the inactivation of sensitive enzymes. These solutions are best prepared in inert gas-purged containers of suitable volume; the desired volume of the anoxic buffer is transferred into the container, and the required volume of liquid or enzyme solution is added. The amount of oxygen introduced from the original liquids or solutions is minimized when previously stored under anoxic conditions.

5.3.4 Continuous Removal of Trace Oxygen

Though careful degassing and storage of anoxic solutions in tightly sealed containers will protect the solutions from oxygen, the transfer of solutions under air can cause oxygen to force its way back into the solutions. Physical methods are generally slow, damaging to biological molecules, and do not provide continuing protection against oxygen. Because proper handling steps are inevitable in order to carry out enzymatic tests or sample preparations, methods that allow a continuous removal of trace oxygen are required. These methods either employ enzymatic reactions or excess of chemical oxygen scavengers. Such systems are frequently suitable to maintain anoxia for reasonable periods of time, since the removal of oxygen is fast and continuously achieved.

Biological systems provide a wide range of oxygen-reducing systems that can be used as described above. The most frequently employed enzymatic method for the removal of trace oxygen is the glucose oxidase/catalase system [1].

Glucose oxidase (EC 1.1.3.4) catalyzes the reaction of two moles of glucose with two moles of oxygen (O_2), yielding two moles of hydrogenperoxide (H_2O_2) and gluconolactone, the latter spontaneously hydrolyzing to gluconic acid. Two moles of H_2O_2 are subsequently decomposed by catalase, regenerating one mole of oxygen and two moles of water. Consequently, two moles of gluconic acid are formed per mole oxygen reduced.

Glucose oxidase has been isolated from various sources, in particular from fungi (*Aspergillus niger* and several *Penicillium species*), and studied in detail. The enzymes from *A. niger, P. notatum,* and *P. amagasakiense* are commercially available. Glucose oxidase is a flavoprotein and is readily inactivated by 8-hydroxyquinoline, heavy metal ions (Ag^+, Cu^{2+}, Hg^{2+}), $NaHSO_3$, *o*-phthalate, putrescine, semicarbazide, and $NaNO_3$. The enzyme is partially inhibited by hydrazines, dimedone and hydroxylamine, and competitively inhibited by halogenide anions (F^-, Cl^-, and Br^-) at low pH. The enzymes are generally specific for β-D-glucose (K_m 30–110 m*M*), but 2-deoxy-D-glucose is an alternative substrate (K_m ~ 25 m*M*, relative activity rate about 6%). The K_m-value for oxygen varies between 0.18 m*M* (*A. niger*) and 0.95 m*M* (*Phanerochaete chrysosporium*), but alternative hydrogen acceptors such as 2,6-dichlorophenol indophenol are used in the absence of oxygen. The turnover number of the homodimeric enzyme (about 160,000 Da) from *A. niger* is >20,000 s^{-1} (>250 U/mg) at 25°C. The enzyme has a temperature optimum of 40 to 45°C (maximum about 55°C). It is stable between pH 3.4 to 7.5 with a pH-optimum at about 6.0 and >90% of the maximum activity retained pH 4 or 7, respectively. The enzyme is stable in the presence of up to 5% SDS, sodium azide (1.5 m*M*), phenol (10 m*M*), heparin, citrate, EDTA, and freezing/thawing. Stored in solution at –20°C or lyophilized at 4°C, the enzyme is stable for >1 year, dissolved at 4°C for several weeks.

Heme catalases (E.C. 1.11.1.6) are widespread in nature and found in most aerobes and aerotolerant organisms. These enzymes catalyze the disproportioning of two molecules of hydrogenperoxide to oxygen and water. Several other compounds including β-(3,4-dihydroxyphenyl)-L-alanine, catechol, diaminobenzidine, guajacol, methanol, *o*-dianisidine, or pyrogallol may serve as alternative hydrogen donors. The K_m-value (H_2O_2) for enzymes from different organisms varies from 1.9 to 19.3 m*M*, but also with the pH. The bovine enzyme is commercially available at specific activities >65,000 U/mg. Though enzymes from different organisms vary significantly in their pH-dependent activity profiles, catalases, in general, exhibit a rather broad pH-dependence between pH 5 and 10. The enzymes are active up to 55°C with a temperature optimum of about 40°C. The enzymes are usually large, homotetrameric proteins with a native molecular mass of >200,000 Da. Catalases are, in general, inactivated by aminotriazoles, azide, fluoride, and cyanide, while the sensitivity toward thiols (β-mercaptoethanol or dithiothreitol) varies significantly between the individual source organisms; the bovine enzyme is not affected significantly, while the enzymes from *Zea mays* are sensitive.

The enzymes are prepared as filter-sterilized stock solutions (e.g., 20 mg/ml glucose oxidase and 2 mg/ml catalase, respectively) and may be stored at 4°C for several weeks. The enzymatic test mixture is supplemented with 0.3% β-D-glucose. The enzymes are added to the test mixtures yielding final concentrations of 20 μg/ml glucose oxidase and 2 μg/ml catalase, respectively, and will quickly remove the trace oxygen present. However, in cases of significant flood in of oxygen, it is advisable to increase the concentration of the enzymes in order to maintain lower steady-state oxygen concentrations.

The enzymatic removal of trace oxygen is a very useful method in initial testing for enzymatic activities. The method is simple to use even in laboratories without sophisticated equipment. However, the presence of glucose can cause problems in some test systems. Though the protocatechuate dioxygenase system [2] can be used in such cases, a true alternative for trace oxygen removal is provided by chemical oxygen scavengers like sodium dithionite or titanium-(III)-citrate. These very low potential electron donors rapidly react with oxygen and, when present in excess, will continuously remove oxygen forcing its way back into the sample.

Sodium dithionite is a versatile, powerful, and commercially available reagent and is widely used as a reducing agent in biochemistry. Despite the large variety of applications described in the literature, there exists a considerable ignorance concerning the mode of action of this reagent [3,4]. It has been stated that this is due in part to the difficulties in the preparation of standardized solutions. It is worth noting that the best commercially available $Na_2S_2O_4$ preparations contain only about 80 to 85% dithionite. This value will decrease dramatically upon aerobic storage in the laboratory over a lengthy period of time. Purer preparations of sodium dithionite can be prepared by repeated, strictly anoxic recrystallization of the product from alkaline solution with methanol. Therefore, 60 g of sodium dithionite are dissolved in 200 ml of 100 m*M* sodium hydroxide, and then 440 ml 100 m*M* NaOH/methanol (20:80) are slowly added. The white, crystalline precipitate is harvested on a fritted glass filter funnel (F porosity), and the crystallization process is repeated. The quality of the final product can be determined by photometric measurement of an anoxic solution at 315 nm using an ε_{315nm} of 8.0 mM^{-1}cm^{-1} and purities >99 can be achieved by repeated recrystallization.

The reduction of molecular oxygen by dithionite is a two-step process. In the first, rapid reaction, one mole of dithionite reduces oxygen to hydrogenperoxide (Equation 5.1). In a second, much slower step, the latter product is further reduced to water by a second mole of dithionite (Equation 5.2). After addition of dithionite, it is recommended to allow the reaction to proceed for about 30 min at room temperature before the buffers are used.

$$S_2O_4^{2-} + 2\ H_2O + O_2 + 2\ HSO_3^- + H_2O_2 \tag{5.1}$$

$$S_2O_4^{2-} + H_2O_2 + 2\ HSO_3^- \tag{5.2}$$

Though the reduction of oxygen by dithionite can be used to keep the oxygen concentration in samples extremely low, it is not advisable to use excess of dithionite to prepare anoxic solution from aerated water. The reduction of one mole of oxygen will yield four moles of bisulfide. This compound is a rather strong oxidant and forms (reversible) adducts with N-5 of flavins, C-4 of nicotinamides and with carbonyls, which may impair enzyme testing or even destroy enzymes. Other problems associated with the use of this compound are the high UV-absorbance, which is a problem in UV measurements, and the low stability of dithionite in acid solutions. It is generally recommended to prepare dithionite solution with pH-values >7.5, since the compound breaks down with a rate of 2% per minute at pH 6.5, and much faster at lower pH.

An alternative oxygen scavenger, titanium-(III)-citrate, was introduced in 1976 by Zehnder and Wuhrmann [5]. The Ti^{3+} ion is a well-known, strong reducing agent in inorganic chemistry, but forms rapidly hydroxide precipitates in physiological solutions (the solubility product of $Ti(OH)_3$ is about 10^{-40}). Hence, a chelating agent such as citrate, tartrate, or oxalate is required to form soluble complexes. The oxidation of Ti^{3+} to Ti^{4+} eliminates oxygen from aqueous solution with a first order rate constant of $11.4 * 10^{-5}\ s^{-1}$ at 25°C.

Titanium-(III)-citrate stock solutions (50 m*M*) are prepared adding 5 ml of a 15% solution of $TiCl_3$ in 10% HCl to 50 ml 200 m*M* sodium citrate under anoxic conditions. The solution is neutralized with saturated sodium carbonate solution and adjusted to 100 ml with water.

The reduction of oxygen by Ti^{3+} proceeds in a number of consecutive one-electron reductions steps, giving rise to some level of highly reactive oxygen species, including superoxide, hydrogen-peroxide, and hydroxyl radical species. Thus, this reaction is very useful to deoxygenate buffers and solutions initially, but is not recommended in the presence of sensitive biomolecules. The presence of excess of Ti^{3+} will also help to keep the oxygen concentration extremely low.

Though very attractive in order to maintain strict anoxia, excess of chemical oxygen scavengers should be considered problematic in enzymatic testing of redox active enzymes. The low potential electron donors (see below) are able to reduce cofactors in iron-sulfur and flavoproteins and will, therefore, inhibit enzymes that require oxidized cofactors in order to initiate the catalytic cycle. Moreover, natural and artificial electron carriers usually employed in studies of such enzymes are chemically reduced by excess of either dithionite or Ti^{3+}.

5.3.5 Control of the Redox Potential

The preparation of anoxic media using physical methods enables the laboratory scientist to work under conditions that will avoid oxygen damage of sensitive proteins and suppress disturbing side reactions caused by the nonphysiological electron acceptor. However, these properties will not per se enable measurements of redox reactions. It is practically impossible to prepare solutions of a low redox potential only by removing traces of oxygen. Since most of the redox-active enzymes found in clostridial species are adapted to the low physiological potential in the living cell, it is essentially required to control this parameter in many enzymatic tests.

Many enzymes involved in the redox chemistry within the living cell require electron donors, carriers, or acceptors that are required for their function. These factors are either not known for many enzymatic reactions or do not provide a direct UV/Vis detection of the redox process *in vitro*. Hence, a variety of artificial electron acceptors, carriers, and donors have been used in order to facilitate the detection of such enzymes.

Sodium dithionite and titanium-(III)-citrate can serve as very low potential electron donors for a variety of enzymes and can be used to prepare the reduced forms of many redox dyes (see below). They can provide redox potentials of –660 mV (dithionite) or –480 mV (Ti^{3+}).

The redox properties of sodium dithionite have been studied in detail by Mayhew (1978) [3]. Since kinetic studies suggested that the sulfoxylate radical anion (SO_2^-) rather than the dimer ($S_2O_4^{2-}$), is the electron donating species, a midpoint potential E' of –660 mV is reached only at low concentrations (<10 n*M*), and the redox potential becomes less negative ($\Delta E_m/\Delta \log[S_2O_4^{2-}]$ = +0.29 mV) at higher dithionite concentrations (at 1 *M* dithionite and 2 *M* hydrogensulfite the potential is –386 mV at pH 7.0 and 25°C). The midpoint potential is also affected by pH and temperature. The pH-effect on the redox potential is controlled by the pK_s of the hydrogensulfite at 6.9: the slope $\Delta E'/\Delta$pH is –59 mV below and –118 mV above the pK. The temperature changes the midpoint potential by –1.6 mV/C between 2 and 40°C.

Most artificial electron carriers are synthetic organic dyes, which change the spectroscopic properties with changes in the redox state. Benzyl viologen (1,1′-dibenzyl-4,4′-bipyridinium dichloride) is an example. The oxidized compound is colorless to pale yellow and can serve as a low potential electron acceptor (E_0' = –359 mV at 30°C, independent of pH). The compound is readily reduced to yield a blue to violet radical cation using either dithionite, borohydride or Ti^{3+}. This form can serve as an electron donor in a variety of enzymatic reactions. Since the reduced compound is readily oxidized by molecular oxygen, it is useful to monitor anoxia. The reduced compound has an absorbance maximum of 555 nm at room temperature, which diminishes at a higher temperature, while a peak at 598 nm appears. At 80°C, the latter is the only visible one. Some properties of useful redox indicators and mediators are summarized in Table 5.2.

5.4 ANOXIC TESTING OF ENZYMES

5.4.1 General

In the previous section, some methods have been established that can provide anoxic solutions. In the following section we will try to give some handling guidelines for the anoxic preparation of solutions and samples for anoxic testing.

An initial qualitative testing of individual enzymes involved in a metabolic pathway will frequently yield first indications for oxygen-sensitive enzymes; such enzymes will usually give rise

TABLE 5.2

Compound	Properties	Remarks
Benzyl viologen (1,1′-dibenzyl-4,4′-bipyridylium dichloride)	409.4 g/mole soluble in water and MeOH $E_0' = -359$ mV at 30°C, independent of pH *Reduced compound* λmax 555 nm at room temperature ε = mM^{-1}cm^{-2} λmax 598 nm at 80°C ε = mM^{-1}cm^{-2}	Low potential electron acceptor and carrier. Oxidized form colorless to pale yellow. Readily reduced by dithionite, borohydride or titanium-(III)-citrate. Reduced compound blue to violet, readily oxidized by O_2 and used as electron donor. The extinction coefficients at 555 nm and 598 nm vary strongly with the temperature. The maximum at 555 nm diminishes with increasing temperature and the maximum at 598 nm appears. At 80°C, only the peak at 598 nm is visible.
Brilliant cresyl blue (7-(diethylamino)-3-imino-8-methyl-3*H*-phenoxazine hydrochloride)	335.8 g/mole soluble in water, EtOH and HAc $E_0' = +47$ mV at 30°C, pH 7.0 *Oxidized compound* λmax 625 nm at 40 mg/L ε = mM^{-1}cm^{-2}	Electron acceptor. Oxidized form blue, reduced form colorless. Reduced by flavoenzymes and by dithionite. Unstable in aqueous solution in particular in alkaline solutions. The reduced compound is re-oxidized by air at pH 7 but not at pH 6.4. Reoxidized by ferricyanide at al pH. With increasing concentration the absorbance maximum shifts to 575 nm. The compound is recrystallized from petrol ether. The commercial product is the diethylamino compound but older literature frequently refers to the dimethylamino compound, which has similar properties.
2,6-Dichlorophenol-indophenol (DCPIP, DPIP, N-(p-hydroxyphenyl-2,6-dichloro-p-benzoquinoneimine) Na salt	290.1 g/mole soluble in water and EtOH $E_0' = +217$ mV at 30°C, pH 7.0 *Oxidized compound* λmax 600 nm ε = 22.0 mM^{-1}cm^{-2} at pH 7.0	Electron acceptor. Oxidized form blue at neutral, pink at acidic pH. Reduced by ascorbic acid, dithionite, borohydride, phenazine methosulfate, ubiquinones, etc. Reduced compound colorless. Only slowly reoxidized by air and the exclusion of oxygen is not required. The compound is reduced by flavoenzymes and nicotinamide dinucleotide-linked dehydrogenases using phenazine methosulfate as mediator.
Diethylsafranin (3-(diethylamino)-7-amino-5-phenyl-phenazinium chloride	378.9 g/mole soluble in water $E_0' = +251$ mV at pH 7.0 *Oxidized compound* λmax 555 nm ε = 49.0 mM^{-1}cm^{-2} at pH 7.0	Hydrogen donor for fumarate reductase and similar reactions. The oxidized compound is red. Reduction by dithionite to the colorless hydrogen donor. Readily re-oxidized by air. Unstable in acid and under air; prepare fresh daily. Generally impure.

TABLE 5.2 (continued)

Compound	Properties	Remarks
Iodonitrotetrazolium chloride (INT, 2-p-Iodophenyl-3-p-nitrophenyl-5-phenyl-2H-tetrazolium chloride)	505.7 g/mole Slightly soluble in water, reduced compound insoluble *Reduced compound* λmax 500 nm ε = mM^{-1}cm^{-2} at pH 7.0	Electron acceptor. Oxidized form colorless, reduced form reddish purple. Readily reduced by flavoenzymes. Used in assay of dehydrogenases by non-enzymatic coupling to NAD(P)H with phenazine methosulfate. Reduced by reducing sugars, ascorbate, NH_2OH in alkaline solution, by reduced phenazine methosulfate, ubiquinones and vitamin K at neutral pH. Reduced compound (formazan) requires detergent for solubility. Not oxidized by O_2.
Menaquinone (Menadione, vitamin K3, menaphthone, 2-methyl-1,4-naphtoquinone)	172.2 g/mole Insoluble in water, soluble in EtOH or ether E'_0 = 0 mV at pH 7.0	Electron carrier. Reduced by flavoenzymes. Frequently used as electron carrier to O_2 or non-autooxidizable acceptors such as indophenols or tetrazolium salts in enzymatic tests of membranes. Solutions stable under air, but decomposition by sunlight, reducing agents or alkalis. Readily reoxidized by air at neutral or alkaline pH, only very slowly at pH 6.3 and below.
Methylene blue (methylthionine chloride, 3,7-bis(dimethylamino-phenazathionium chloride)	319.9 g/mole (anhydr.) Soluble in water and EtOH E'_0 = +11 mV at pH 7.0, 30°C *Oxidized compound* λmax nm ε = mM^{-1}cm^{-2} at pH 7.0	Electron acceptor. Oxidized form blue, reduced form colorless. Reduced by flavoenzymes, rate often increased by mediators (phenazine methosulfate). Spectrum depends on concentration. Reduced by dithionite, by reduced flavins, phenazine methosulfate and quinines. The reduced compound tends to stick to glass surfaces and is readily reoxidized by air or anaerobically by light; oxidation rates increase with pH. Stable under nitrogen.
Methyl viologen (1,1′-dimethyl-4,4′-bipyridylium dichloride, Paraquat)	257.2 g/mole Soluble in water and MeOH E'_0 = –446 mV at 30°C, independent of pH *Reduced compound* λmax 601 nm ε = 13.4 mM^{-1}cm^{-2} at pH 7.0	Very low potential electron acceptor and carrier. Oxidized form colorless. Readily reduced by dithionite, borohydride or titanium-(III)-citrate. Stable reduced form blue to violet, readily oxidized by O_2. The reduced compound is an electron donor for, e.g., hydrogenases.
Phenazine methosulfate (PMS, N-methyl-phenazonium methylsulfate)	306.3 g/mole Soluble in water, reduced compound very slightly E'_0 = +80 mV at 30°C, pH 7.0 *Oxidized compound* λmax 387 nm ε = 26.3 mM^{-1}cm^{-2} at pH 2–8 $\Delta\varepsilon_{(ox\text{-}red)}$ = 25.0 mM^{-1}cm^{-2}	Electron acceptor and carrier. Oxidized form yellow, reduced form almost colorless, green semiquinone sometimes observed. Rapidly reduced by flavoenzymes and non-enzymatically by NAD(P)H. Frequently used as electron carrier to less readily autooxidizable electron acceptors like indophenols or tetrazolium salts. H_2O_2 is formed by autooxidation of $PMSH_2$ under air.

TABLE 5.2 (continued)

Compound	Properties	Remarks
		Catalase is completely inhibited by PMS and other provision to remove peroxide may be necessary (CN^- or 8-hydroxyquinoline). The compound is reduced by dithionite, borohydride, NAD(P)H, and ascorbic acid. Reacts with thiols including enzyme bound cysteins.
Potassium ferricyanide	329.3 g/mole Soluble in water $E_0' = +420$ mV at 25°C, pH > 3.5 λmax 420 nm ε = 1.02 mM^{-1}cm^{-2} at pH 2–8	High potential electron acceptor and mild oxidizing agent. Slow decomposition in solution by light. Oxidizes thiols to disulfides.
Phycocyanine	210.2 g/mole Slightly soluble in water $E_0' = -34$ mV at 30°C, pH 7.0 Oxidized form λmax 700 nm ε= 3.4 mM^{-1}cm^{-2} at pH 7.0	Electron acceptor and carrier. Oxidized form blue, reduced form colorless. Reduced by dithionite, borohydride and in alkaline solutions by glucose. Very rapidly re-oxidized by O_2. The reduced compound is used as electron donor for membrane enzymes.
2,3′,6-Trichlorophenolindophenol (2,3′,6-Trichlorindophenol, o-chlorophenolindo-2,6-dichlorophenol) Na salt	324.5 g/mole Soluble in water $E_0' = +219$ mV at 25°C, pH 7.0 Reduced form λmax 645 nm ε = 27.0 mM^{-1}cm^{-2} at pH 7.3	Electron acceptor. Oxidized in neutral solutions blue, reduced colorless. The reduced compound is used as electron donor. The reoxidation under air is very slow. Aqueous solutions decompose, in particular in acidic solution and in light. Reduced by ascorbic acid, dithionite, and reduced phenazine methosulfate.

to a characteristic accumulation of the intermediates up- and downstream of the oxygen-sensitive reaction, which will diminish in anoxic assays. However, moderate oxygen-sensitive enzymes (with half-lifetimes in cell-free extracts in the minute-to-hour range under air) will frequently work well in concentrated cell-free extracts, while enzymatic testing in dilution will fail.

All manipulations of oxygen sensitive samples are most conveniently performed in an anoxic environment established using anoxic glove boxes. However, glove boxes are rather expensive and sophisticated tools that are not available in all laboratories. Economical constraints and the need to use large equipment that cannot be placed inside a glove box will frequently require anoxic sample transfer under air.

A number of standard biochemical procedures are available to brake cells, remove cell debris, prepare membrane preparations, or to concentrate samples. However, despite the ease of use under air, many of these methods will become very difficult, or useless, when they have to be applied on samples sensitive to oxygen. The next crucial steps in anoxic testing are sample clarification, dilutions, and transfer into test containers. Though more difficult to operate, some simple precautions can reliably maintain anoxia throughout these manipulations. A proper definition of the objectives is essential in order to keep the procedure as simple as possible — but as developed as required — in order to obtain reliable results.

5.4.2 Sample Preparation

The sample preparation for enzymatic testing should be preceded by preliminary experiments establishing high activity levels in the harvested cells. These data are easily accessible, correlating bacterial growth and substrate consumption or product formation, respectively, for various growth conditions. From such plots, optimum growth conditions and the optimum growth phase for cell harvest can be established.

Once the optimum conditions are known, cells for enzymatic testing can be prepared and analyzed. Whenever strict anoxia must be maintained, it is highly recommended to start with sufficiently high cell mass. The cells are usually precipitated by centrifugation at low g-values. The compact pellets thus formed will normally provide a high level of protection, since the solubility of oxygen in water is very low and its diffusion is very slow; reduced constituents of the upper cell layers of the pellet will protect sensitive components in deeper layers for extended periods of time when stored frozen. Wet packed cell masses of >10 g are easily handled without special precautions. These cells originating from 2 to 10 l of medium are most conveniently harvested by flow-through centrifugation from the culture flask slightly overpressurized with N_2. The bacterial cell slurry thus obtained is transferred carefully into suitable beakers and centrifuged. The medium is discarded and the solid cell pellets are immediately used or frozen at –80°C.

Though more difficult to handle, smaller culture volumes may be used. However, the less voluminous cell pellets thus obtained require special precautions in order to avoid significant damage of sensitive enzymes. Smaller cultures of 50 to 500 ml are cultivated in infusion bottles of 100 ml or 500 ml volume. These bottles can be centrifuged at low g-values in suitable adaptors in suitable rotors. The settled cells are separated from the medium; a slight overpressure (about 100 mbar) of N_2 is applied from a gas line, and the bottle is inverted. A hypodermic needle connected by a suitable tubing to a waste container is used to pierce the stopper and allows the medium to be replaced by nitrogen. It is highly recommended to use small amounts of cells immediately, since freezing in this setup will cause underpressure in the bottles, allowing oxygen to force its way into the container.

A washing step of the cells harvested from complex media is frequently recommended in order to avoid inhibition of enzymes by medium constituents or to remove compounds that interfere with downstream analysis. A washing step should be carried out prior to freezing of cells since freeze-thawing cycles can give rise to breaking of cells and, therefore, can cause losses of material. Since efficient washing steps will require a resuspension of the cell pellet, oxygen in the washing buffers will have access to the individual cells causing significant damage, in particular when repeated washing steps are needed. These steps are most conveniently carried out in infusion bottles and will work for comparable large quantities of cells, as well as for small quantities. The desired volume of washing buffer is introduced from tightly sealed, slightly overpressured bottles using either gas tight glass syringes or transfer lines. It is important to check the oxygen resistance of the connecting tubing in transfer lines prior to use. In order to ensure extremely low oxygen levels, the buffers may be supplemented with 0.3% glucose and 20 µg/ml each glucose oxidase and catalase, or chemical oxygen scavengers. Finally, the cells are resuspended in a buffer suitable for braking the cells and transferred into a smaller container (e.g., a tightly closed serum bottle purged with nitrogen).

All subsequent preparation steps (washing and braking of the cells, centrifugation, and concentration) will require anoxia in order to avoid possible damage of oxygen-sensitive enzymes. Moreover, it is highly recommended to perform the individual steps at low temperature (<4°C). These steps can be performed most conveniently in an anoxic environment provided by a glove box. In this environment, almost no restrictions in using equipment available on a laboratory bench must be considered when tightly closed containers (e.g., beakers for centrifuges, sample flasks, etc.) are used for outside handling steps, or equipment (e.g., sonicators, concentrators, photometers, etc.) can be placed inside the glove box.

A variety of methods are available to open cells for enzymatic testing. Most frequently used are French-pressure cells, sonication, freeze-thawing, and lysozyme in combination with either freeze-thawing or hypo-osmotic shock. However, for anoxic disintegration, using sonication is practically restricted to anoxic environments provided by glove boxes, while the other methods are readily adopted to the needs of anoxia. Freeze-thawing, enzymatic degradation of the cell wall, and osmotic shock are readily performed in closed containers and provide an efficient way to open cells of many species. In order to use the French press, some precautions must be taken to maintain anoxia during the procedure. First, the outlet of the pressure cell is connected to a nitrogen line, and the whole cell is purged for several times with nitrogen. Then, a hypodermic needle is connected to the nitrogen-filled cell, and the air in the needle and the connective line is removed, pressing down the piston. The needle is then used to pierce the rubber stopper of the serum bottle, and the sample is transferred into the cell. The only difference to the use of the cell under air is the repeated release of the suspension into a closed bottle.

The clarification of crude extracts is among the most challenging tasks in sample preparation for anoxic enzyme testing. Ultracentrifugation of crude cell extracts is used frequently in order to provide clear solutions — or in subcellular fractionations on a laboratory scale. However, the filling of the tubes under air must be considered difficult if strict anoxia has to be maintained, and the comparable long centrifugation times will expose oxygen-sensitive enzymes for a long time. The ill effects of oxygen will be almost negligible for large volumes of concentrated solutions protected by oxygen scavenging systems but will become devastating when small volumes of diluted proteins must be handled without protection against oxygen. While the concentrated, voluminous samples can be successfully handled under a constant stream of protecting gas and are sufficiently protected by a cushion of mineral oil supplemented with either pyrogallol or hydroquinone, using anoxic gloves boxes will be required when small volumes of diluted samples must be handled.

5.4.3 Assay Preparation

In the simplest cases, the enzyme to be measured has been described from various sources and needs to be characterized for another organism. The catalytic properties, including substrate specificity, cofactor requirements, pH-, and temperature-dependence of such enzymes are frequently available in the literature or in databases and can be adopted without problems to anoxic assay conditions. For other enzymes, the reaction catalyzed resembles reactions known from other enzymes. In such cases, it is in generally recommended to follow adopted protocols close to the procedures described for similar enzymes. However, many clostridial enzymes do not have counterparts and are not yet studied. Moreover, several substrates are utilized by different species yielding the same products, despite the fact that the metabolic pathways and the enzymatic reactions involved are entirely different. The same holds true for anabolic pathways providing similar products starting from very different sources. It is not possible to discuss all the individual points that must be considered in order to provide robust and reliable test systems for novel enzymes. Only a few points of general importance can be summarized here. In all cases where entirely new metabolic properties of an organism are the subject of research, initial characterization of intermediates accumulating transiently in the medium using chromatography-, radiochemical-, NMR- or mass spectrometry-based methods can provide information about individual reactions likely to be involved, but these data cannot exclude several individual steps that do not accumulate intermediates and, therefore, escape detection. In such cases, the development of a specific test is a challenging task since neither the substrates nor the products are known.

Once an individual reaction has been detected in cell-free extracts or membrane preparations, it will become possible to establish an enzymatic test system. However, the activities measured initially will likely not match the demand of the production rates observed *in vivo*. In the simplest cases, this discrepancy turns out to be due to parameters that do not match the needs of one individual enzyme (for example, buffer composition or pH, ionic strength, cofactor requirement, etc.). In more complicated situations, the substrate or the enzyme needs an activation by accessory enzymes, which can be limiting in the assay. These auxiliary components can become limiting in the assay, in particular during purification steps, and a supplementation will be required in order to restore full activity. And last but not least, oxygen may cause ill effects on the desired enzyme.

Obviously, even small contamination of buffers with oxygen will cause failure of enzymatic testing in diluted samples. Therefore, it is frequently recommended to initially add a concentrated solution of substrate(s) to the extract in order to monitor activity. Later, proper dilutions can be made in order to obtain quantitative data. The reason for this sequence of steps is easily exemplified by some theoretical calculations. Assuming a specific activity of 100 U/mg for an enzyme of 50 kDa, the product release in cultures allowed an estimate for the specific activity of 50 mU/mg protein *in vivo*. For a cell-free extract with a protein concentration of 10 mg/ml, one can expect 500 mU/ml or 5 μg/ml of target protein. Using the molecular mass, a concentration of 100 n*M* is calculated for the protein. Assuming an oxygen concentration of 5 n*M* in our buffers, we can expect almost 95% activity yield for the assay in cell extracts. However, a 10-fold dilution of the sample will cause more than 50% activity loss while a 100-fold dilution will yield no residual activity. Since such dilutions are frequently used in order to obtain relative activity rates of 5–100 μM*min^{-1} in enzymatic tests, the problems relating to the handling and measurement of such enzymes becomes apparent. Undoubtedly, such enzymes will require the most sophisticated equipment, including glove boxes, to be studied successfully.

However, most enzymes are by far less sensitive, in particular in cell-free extracts where reduced iron-sulfur proteins and flavoproteins provide a certain degree of protection. Using the precautions outlined, reliable measurements can be obtained using simple setups, like rubber stoppered cuvettes or small, septum closed reaction vessels (for example HPLC sample flasks) that are purged with inert gas prior to use. The transfer of assay components and samples is most conveniently achieved using gas-tight volumetric syringes for sample transfer. Since hypodermic needles can easily pierce rubber stoppers or septa, few precautions are required. However, it is recommended to purge the syringes first with headspace atmosphere of the storage bottles and subsequently with buffer or sample in order to minimize oxygen transfer. To reduce the risk of oxygen contamination further, it is recommended to apply slight overpressure to all containments and to overfill the syringes with buffer or sample for transfer. Once taken from the storage bottles, a small droplet of buffer will appear at the tip of the needle, which will provide a diffusion barrier for oxygen. When the seal of the test container is touched, the needle tip is placed on the surface and the piston is adjusted to the desired volume. Again, a droplet of buffer or sample is sacrificed in order to provide a small but effective diffusion barrier prior to piercing the seal.

5.5 ANOXIC PURIFICATION OF ENZYMES

5.5.1 General

Once a specific and reliable test system for an individual enzyme is available, it is possible to trace its activity throughout a purification. The goal of each purification is to provide sufficient amounts of the interesting enzyme suitable for the downstream analysis planed, but it is barely the aim of research by itself. The requirements in terms of yield, purity, and speed of an individual purification will vary significantly for various needs, and it is highly recommended to define these objectives before getting started.

An initial characterization of the target protein and of the most disturbing contaminants will allow a systematic method development. Many questions concerning the stability of the target and its binding properties to the available chromatographic media should be answered in order to allow a proper choice of chromatographic media and alternative purification methods, and finally, to combine the individual steps to an optimum purification strategy.

Finally, we must consider special needs of highly oxygen-sensitive samples. The requirements of anoxia should be defined prior to starting a purification. In the easiest case, it will be possible to do the majority of the required manipulations inside a glove box. However, if a glove box is not available for purification, it is required to check the chromatographic system and all its individual components with regard to the oxygen diffusion resistance. In particular, the tubing materials, connections, pumps, seals, and columns should be checked carefully against a flood in of oxygen. Many individual constituents of the chromatographic systems available have not been designed to exclude oxygen. In particular, the thin-walled Teflon tubing frequently used to connect pumps, columns, detectors, and other chromatographic equipment have a very low oxygen diffusion resistance. In some cases, this tubing can be replaced by thicker-walled PEEK-capillaries or wrapped with heat-shrinking PVC tubes in order to allow the purification of moderately sensitive proteins. However, if extremely oxygen-sensitive proteins or small quantities of sensitive proteins must be handled, it might become recommended to establish cooperation with better equipped laboratories rather than to adopt a system that is not suitable for this purpose.

5.5.2 Testing Equipment

The purification of an oxygen-sensitive enzyme is, in general, more demanding in terms of anoxia than the testing of the enzymatic activity in cell-free extracts. This is due to the fact that the enzyme is frequently bound to chromatographic media and the columns are washed and developed using comparable large volumes of buffers. These large volumes will allow oxygen to react with substantial quantities of the target enzyme, though it might be present in concentrations that do not significantly affect testing. Moreover, throughout the purification, other proteins that can partially protect the target in cell-free extract will be removed, and the damaging effect of oxygen is established.

When the purification is performed in anoxic glove boxes, the oxygen concentration is kept extremely low, minimizing these problems. However, many moderately sensitive proteins are successfully purified using chromatographic systems operated under air. In order to minimize oxygen to flood in during the procedure, it is necessary to establish a sufficient protection of the protein in the loading, operating, and fractionation steps.

In order to establish problems prior to getting started, methods are required to allow detection of oxygen forcing its way back into the system. The flood in of oxygen is most conveniently tested using anoxic solutions of benzyl- or methyl viologen, partially reduced with dithionite or titanium-(III)-citrate. The reduced dyes are easily monitored at 555 nm or 604 nm, respectively, and will react immediately with oxygen giving rise to bleaching of the dye. In order to check the oxygen resistance of a chromatographic system, an anoxic solution of the dye (about 100 μM) is prepared and kept under inert gas. The reducing agent is added to establish an initial absorbance of about 0.5. Then, the reservoir is connected to the system, and the pumps are vigorously purged with the solution. The solution is pumped directly through the detector at the highest possible flow rates until a stable absorbance value is reached. Then, the flow rate is reduced to about 10%, and the absorbance is further monitored using five-to-ten operating volumes of the system. In a gas-tight setup, no changes in absorbance due to the flow rates should be visible, while a decrease with reduced flow will indicate significant amounts of oxygen forcing their way in. Tiny leakages are detected stopping the pumps for defined periods of time. Subsequent restart allows the detection of decrease in absorbance caused by small amounts of oxygen flooding in. In a similar approach, loading and fractionation can be optimized.

A small gas leakage toward the surrounding air is frequently not avoidable, and it is an important to answer the question: How severely will this residual oxygen effect the purification prior to getting started? However, tiny leakages remaining are frequently covered using low concentrations of chemical oxygen scavengers, presupposing compatibility with the enzyme of interest.

Sample handling steps, including sample load and fractionation, will obviously own a high risk of oxygen contamination when done under air, and working in a glove box will facilitate these steps. However, voluminous chromatographic equipment will occupy valuable space inside the glove boxes. Using the methods for optimizing chromatographic systems, it will be possible to place only the most sensitive components (sample loading devices, columns, fraction collectors, and buffer reservoirs) inside the glove box, while space-consuming parts like pumps, mixers, columns, detectors, and controllers can be placed outside. Such setups will also facilitate technical service and overcome some restrictions owned by the glove box dimensions.

5.5.3 Define the Objectives

An individual protein purification strategy can be placed in a simple three-dimensional scheme characterized by the parameters purity, speed, and yield. The protein preparation for one particular downstream application frequently causes demands that will enforce one or two individual parameters to be optimized. Since the individual parameters are interconnected, it will hardly become possible to optimize one parameter on cost of the other without affecting the results in a sometimes nonpredictable way.

In an ideal case, purification is achieved in a single, fast step, and a pure preparation at almost quantitative yield will be obtained. This is hardly possible, and we will have to find a compromise that will supply the desired amounts of a protein at sufficient quality in reasonable time. The analysis planed with an individual preparation will shift the requirement predominated by either purity, speed, or yield.

Proteins needed for either therapeutic use or for *in vivo* studies must be extremely pure (>99%), while moderate purity (<95%) will be required when used as antigen for antibody production or when N-terminal sequencing is the aim of the purification. Most biochemical characterization methods fall in their demand in between these two extremes (95 to 99% purity). The purification of highly unstable proteins or the analysis of multiple samples (for example, in induction studies) requires rapid purification strategies, while many other applications are not directly affected by this parameter. Finally, the activity yield is of the highest importance if the enzymes are required in large quantities for biosynthetic purposes or material demanding analytical methods (for example, many spectroscopic applications) but may be less crucial in cases where the starting material is readily available or the product is needed only in small quantities.

It is frequently recommended to define the requirements according to the use of the final product, but it may be helpful to spend time on a proper choice of the source; a 5-fold higher activity in the starting material will reduce the purification factor required in order to obtain pure protein by a factor of 5, too, and the use of an organism whose genomic DNA-sequence is known will tremendously facilitate identification and subsequent cloning of the encoding genes.

Stating a high purity ("the protein was homogenous by Coomassie or silver-stained SDS-PAGE") is far from guarantee that the purification is sufficient for an intended application, since trace contaminants that either degrade or inactivate the target protein will strongly affect storage stability and all downstream applications. On the other hand, a low purity will not interfere with selective tracing of individual properties, assuming the key contaminants exhibiting similar properties are reliably removed. For example, EPR-studies on glycyl radical enzymes or many iron-sulfur proteins but also activity measurements are frequently possible in comparable crude samples depleted from interfering activities. Consequently, it might be recommended to define key contaminants, which must be removed throughout a purification. These components must be distinguished from constituents not affecting the intended application and can be tolerated, in particular in high throughput applications. It is quite obvious that this requires reliable testing for the target activity, as well as testing of the crucial contaminants.

Although the number of purification steps should be minimized since the product yield will strongly suffer with increasing numbers of steps, the quality of the end products should not be compromised. It is normally much better to overpurify an enzyme than to underpurify it. Subsequent results might be questioned when the sample purity is low and impurities are unknown.

5.5.4 Initial Characterization of the Target Enzyme

Trial and error approaches in method development will frequently achieve the goal to obtain pure enzymes, but these strategies will be costly in terms of time and money. A proper definition of the properties of the target protein and of critical impurities will allow a structured way for method development that pays off, in particular when multiple schemes for different downstream applications are needed.

The initial characterization of an enzyme aims to determine the stability window that will define the outer framework for optimization and method selection avoiding protein inactivation during the purification. A variety of information is needed for the target protein and the critical impurities and must be collected for an individual source in order to establish optimum conditions. Additional information from the literature will provide a rough guide to adopt existing protocols.

These data required must define the resistance of an enzyme activity toward changes of the environmental conditions used throughout a purification. The most important factors affecting protein stability are temperature, pH, and buffer composition, but a variety of other factors must be considered important. They include: resistance to organic solvents or antichaotropic salts, the requirement of detergents, cofactors or metal ions and the sensitivity to proteases, metal ions, or oxygen, respectively. Finally, it is highly recommended to establish proper storage conditions at the earliest possible stage; it makes no sense to invest lots of resources and time in the workout of a demanding purification scheme and expect activity losses downstream of those efforts.

The buffer composition will strongly affect the stability and the chromatographic behavior of individual proteins: While some proteins require phosphate or particular ions during the purification, others will suffer ill effects on the activity. Moreover, several chromatographic and precipitation procedures take advantage of salt or organic solvent concentration to achieve binding and release of a protein, and it is highly recommended to determine the stability and solubility window for the target protein and the crucial impurities with a proper selection of salts and solvents. In addition, some proteins contain loosely bound cofactors, metal ions, or require detergents, which are consecutively lost during the purification and must be present in the buffer to allow high activity yields, while others may have detrimental effects and should be removed by chelating agents. Cell-free extracts from various sources may contain highly active proteases that will have strong damaging effects on the target protein, and the use of protease inhibitors might be necessary. The redox stability window might require protection of the proteins from oxidation but might also require strict anoxia or the presence of strongly reducing agents.

Once the stability window of the target and the most important impurities have been established, the framework in which a purification must be achieved is known, and it is worth thinking about the choice of chromatographic media and the conditions for loading and recovery of the samples. Though sophisticated chromatographic systems are available that facilitate an optimization of the separation process by automated procedures, the need to analyze the samples following each individual run will render useless their application on crude samples. In contrast, larger numbers of starting conditions are easily checked in batch approaches using small aliquots of pre-equilibrated chromatographic media mixed with the sample. When the material pelleted, simple enzymatic detection in the supernatant will give indication for binding. Once these conditions have been established, the same method can be used to establish the conditions required to release the activity. Finally, this method can be used to establish the capacity for an individual chromatographic medium for the target enzyme under experimental conditions.

The data obtained from such preliminary experiments can be obtained for a variety of media in a reasonably short time and will allow the development of a reasonable purification strategy based on knowledge of the binding and release properties of the target. The properties of crucial impurities are readily collected by additional testing of the experimental samples in additional enzyme assays. Once collected, these data will allow the development of purification strategies for very different purposes and undesired surprises, including the loss of activity, are much more unlikely to occur.

5.5.5 The Choice of Methods

Only a small number of proteins can be purified according to individual properties using rather specialized methods. The waste majority of proteins is purified using chromatographic or precipitation procedures that will separate the individual molecules according to physicochemical properties such as affinity, charge, hydrophobicity, size, solubility, or stability. The different methods take advantage of individual properties or of a combination thereof and will differ remarkably in terms of capacity, resolution, recovery, and speed. Consequently, the individual methods will have advantages, disadvantages, and limits. Since the knowledge of these properties is essential for successful applications, they will be discussed here.

The stability of proteins in solution is strongly influenced by the temperature and the pH. Many proteins will denature when the individual limits are exceeded and those properties are useful in order to remove bulk impurities. In other cases, the target protein will denature, but will — in contrast to the impurities — readily refold when suitable conditions are restored. Though a powerful method for the enrichment of enzymes with comparable high tolerance toward heat, acid, or base, attention must be paid to the fact that coprecipitation will frequently cause remarkable losses, in particular when applied on concentrated samples, and that a defined heating of large volumes is very difficult to achieve.

The solubility of enzymes in solution is mediated by water shells surrounding the native molecules. When these water molecules are partially removed using either antichaotrophic salts (ammonium sulfate, potassium phosphate, etc.) or organic solvents (ethanol, acetone, etc.), the individual molecules can interact with each other and will start to precipitate. Since the hydrate shell is restored upon removal of the causative agent, the precipitation is frequently reversible. However, in particular, organic solvents pose a high risk to denature the proteins, leaving behind inactive enzymes. The precipitation process is concentration and time dependent, and remarkable losses will be inevitable when applied on highly diluted samples for short times. Though individual properties (solubility at very high or insolubility at very low concentrations of the precipitants) will cause a remarkable enrichment for a particular activity, the majority of proteins will precipitate in a similar concentration range, and the enrichment factor for these enzymes will be low. However, the method allows a reasonably fast concentration of moderately diluted samples.

The size of individual proteins can vary over a wide range. Small electron transferring proteins like ferredoxins are small, monomeric proteins with native molecular masses of about 5 kDa, while large hetero-oligomeric complexes can exceed masses of 1 million Da. Methods taking advantage of the different sizes in order to separate proteins are dialysis, ultrafiltration, and size exclusion chromatography (SEC, gel filtration). However, dialysis is a time-consuming process and is most frequently employed in order to replace buffers rather than in the removal of impurities. Ultrafiltration is most frequently used for sample concentration or buffer exchange. In order to separate individual proteins, the size exclusion chromatography provides a powerful tool for the separation of molecules according to their sizes. A variety of different media are available, which cover the range from 100 Da up to several 100,000 Da. Using high-resolution media of suitable separation range, globular molecules differing in their molecular masses by only about 10% are usually at least partially separated, while baseline separation is frequently seen for molecules with mass differences of >30%. Though very useful to determine the molecular mass of globular proteins, the separation is sometimes compromised by shape factors or by nonspecific interactions with the column material. Due to the low capacity in terms of application volumes (frequently <1% of the column volume), sample concentration (usually <10 mg/mL), and the low lateral flow rates required for optimum separation (frequently in the range <30 mL*cm^{-2}*h^{-1}), the method is slow and restricted to prepurified starting material. However, it should be noted that a group separation in order to achieve desalting is conveniently achieved for samples up to 20% of the column volume at lateral flow rates up to 200 mL*cm^{-2}*h^{-1}, which allows a rapid buffer exchange between individual chromatographic steps.

The separation of enzymes according to the charge is most conveniently achieved by ion exchange chromatography (IEX) and is strongly affected by the net charge of an individual molecule at a given pH, but also by the charge distribution on the surface, by sterical factors, and last but not least by nonspecific interactions with the carrier materials (for example, dextran-based materials will provide hydroxyl groups for additional hydrogen bonds, while styrene-based materials will support hydrophobic interactions). Since many of these factors are strongly influenced by the pH of the buffer system used, it will be frequently possible to separate proteins co-eluting under certain conditions when the pH is changed. Since the net charge of a molecule is zero at the isoelectric point, gradual changes of this parameter will also provide a powerful separation technique. The capacity of IEX materials in terms of binding and applicable sample volume is very high and a high resolution is frequently given. Since a strong binding is achieved at low ionic strength, the gradual increase of salt concentration will allow for concentration of samples during operation. The major disadvantage of IEX is the limited capability to predict the final buffer composition in terms of ionic strength and content of low molecular mass constituents derived from either the sample or the buffers.

In particular, when the solvent shells of dissolved proteins are partially removed by antichaotrophic salts (ammonium sulfate, potassium phosphate, etc.), many proteins will interact strongly with hydrophobic surfaces (for example phenyl-, isopropyl-, or ethylether groups) provided by hydrophobic interaction chromatography (HIC) media. Using a gradual decrease in the concentration of the salt, individual proteins are released according to their "hydrophobicity." The pH will also significantly affect HIC. The capacity, the resolution, and the concentrating effect of HIC materials are similar to IEX and the difficulties to predict the final sample composition are essentially the same. Moreover, some proteins will not bind to HIC but precipitate, while others will bind very strongly and will not be released in solely aqueous solutions. Elution of such proteins will frequently require the presence of ethyleneglykol or other organic solvents, which can cause ill effects on enzyme activity. The reversed-phase HPLC (RPC) can be considered as a HIC method employing strongly hydrophobic materials and predominantly denaturing solvent systems in order to separate protein subunits and peptides.

Specific binding of ligands owned by individual enzymes will provide an extremely selective binding and, therefore, allow the purification of the target molecule in ideal cases in a single chromatographic step. Such specific binding properties are the underlying principle in affinity chromatography (AC). The binding to individual materials might be due to not-yet-understood properties of the targets and owned by many enzymes (for example, the binding to minerals like hydroxylapatite). Other supports contain covalently linked organic dyes, which will mimic cofactors, substrates, or competitive inhibitors or, at the high end, monoclonal antibodies of moderate affinity or specific inhibitors directed against an individual target. These materials will selectively recruit individual targets or comparable small groups of proteins from rather crude and frequently also highly diluted samples. The recruitment of a very small fraction of the molecule from the bulk protein will provide, in general, a rather high capacity. Nonbinding constituents are easily removed and specific changes in either ionic strength, pH, or by ligand replacement will allow for recovery of the targets, frequently in a concentrated form. However, downstream removal of ligands can be troublesome, in particular when inhibitors, substrates, or cofactors are present and the final preparation is required for kinetic or crystallization studies.

The capacity of columns employed in an individual step of a purification deserves some attention. Large column dimensions will frequently provide high capacity. However, when used for small amounts of targets, the sample will be recovered in a very diluted form, owning a high risk of activity and material losses. On the other hand, loading large quantities of proteins to small columns will cause weakly bound proteins to be replaced. In extremes, a target known to bind under certain conditions will pass the column quantitatively at high sample concentrations. In particular, in crude samples containing the desired activity in very small concentrations relative to total protein content, it is highly recommended to optimize the capacity for the target prior to getting started. Anion exchange chromatography may serve as an example of how a method is optimized in terms of capacity; most suppliers will recommend high pH values (<9.0) in order to ensure complete binding. However, the pI-values of many clostridial enzymes are below 7.0, and it is reasonable to expect a certain fraction of proteins to bind at significantly lower pH (e.g., 6.0). The capacity for an individual target at pH 6.0 will be much higher than compared to pH 9.0, since many other proteins will not bind at the lower pH. Increasing the salt concentration in the loading buffer will further decrease the fraction of total protein bound and increase the capacity for the target, presupposing that the target will still bind. An increase in the capacity by a factor of 5 to 10 is often possible for many columns, and a factor of 100 is likely to become established for proteins with special properties (e.g., for acidic proteins in anion exchange). A column with high target capacity but low total protein binding will find its extreme in affinity chromatography using immobilized monoclonal antibodies, which will recruit at highest selectivity the target protein. In addition to the increased capacity, the target will be released by rather small changes of the buffer composition and it is self-explanatory that this will yield a dramatic reduction in separation time and increase speed and activity yield for sensitive enzymes.

5.5.6 Development of Purification Strategies

The cytosol of a living cell will easily contain several thousand individual proteins and a huge number of other biomolecules. Moreover, less than 5% of the total number of proteins will account for more than half of the protein molecules present, while others are present in only a few copies per cell. The purification of a target protein in a single step will be hard to achieve. The situation is further complicated by the fact that high intracellular turnover rates and regulatory properties of enzymes may cause a remarkable high heterogeneity of an individual target, which is further increased by the capability of proteins to form transient or stable aggregates with each other, or with other macromolecules in the cell. Considering that such interactions might be essential for or might interfere with function, it is a challenging task to release a protein or to stabilize these structures for purification.

Though the preliminary testing can give general guidance throughout the subsequent purification, it is far from a guarantee to achieve the particular goal, based on the conditions established. Often, it will be required to adopt the preliminary scheme according to the latest results. Throughout a purification, the proteins are exposed to variations of the physicochemical environment, which can enforce complex formation and disassembly. The dynamic nature of these processes will account for fundamentally different results obtained from purifications using the same columns and running conditions for individual steps but changing just the order in which those steps are performed. In other words, a contaminant easily removed in one order of steps will be persistent and hard to remove when this order is changed.

In order to achieve the objective, it is highly recommended to consider particular needs of subsequent purification phases and a kind of logical order exists for the combination of individual steps. This order is not only largely dictated by the properties of the individual methods employed, but also strongly affected by the source from which the enzyme needs to be purified. A size exclusion chromatography can provide an excellent method to remove trace contaminants, structural variants, and to obtain a final preparation in a defined buffer, but it will not be suitable for fractionation of voluminous, highly concentrated (and viscous) crude samples. From this example, it is apparent that the objectives for an individual step in a protein purification will vary.

TABLE 5.3

Component	Recommended Changes
Buffer reservoirs	Use tightly closed Schott bottles equipped with a suitable multicanal head-set to prepare anoxic solutions. Purge reservoirs constantly with inert gas, preferentially helium. Helium as carrier gas will conveniently remove oxygen from the solutions and its very low solubility in aqueous solutions will avoid "ghost peak" generated by air bubbles in the detector.
Solvent transfer lines and connections	Replace thin-walled Teflon tubing by thick-walled PEEK capillaries of large inner diameter. Replace Teflon seals and connectors by rubber seals and PEEK connectors. When a replacement is not possible, coat lines and connection with heat shrinking PVC tubing of suitable inner diameter and thick walls. Upon heating with a hot air blower, ensure tight contact between the inner Teflon and the outer PVC parts.
Pumps and mixer	Check the transfer line in- and outlets for Teflon seals and replace by rubber seals if possible. Replace the purging solution behind the pistons by a freshly prepared, anoxic solution containing low concentrations of oxygen suitable scavengers.
Sample loading	Load large samples directly from tightly sealed bottles using a transfer line operated by a peristaltic pump. Make sure that a pressure equilibrium with inert gas is possible. Small samples are most conveniently applied with gas tight volumetric syringes of suitable volume. Make sure that the sample loop is either from thick-walled PEEK tubing or coated by heat shrinking PVC tubing. Wash the sample loop with multiple volumes of anoxic solution, preferentially containing a low concentration of oxygen scavenger, immediately followed by the sample load. Apply sample immediately after loading into the loop.
Columns	Cover Teflon tubing with tightly fitting rubber tubes. Do not use heat shrinking PVC material since many columns will contain or are composed of heat sensitive material! Alternatively, small columns and the attached tubes are dipped in a solution containing a glucose oxidase/catalase oxygen scavenging system.
Fraction collectors	Using automated fraction collection is not recommended when oxygen-sensitive samples are handled under air. Oxygen is reliably excluded collecting samples in long test tube pre-filled with argon or about 2 cm mineral oil. Make sure that the outlet capillary reaches the bottom of the flask when the sample is collected and protect against convection or draft by suitable caps. Once the target enzyme containing fractions is known, it is highly recommended to transfer the sample into a well-protected, sealed container using either transfer lines or gas tight volumetric syringes.

Once the source of an enzyme has been defined, proper conditions for the braking of cells and the provision of starting material will be required. Almost quantitative yield in the release of the desired activity will be beneficial, but in practice, a quantitative release will frequently need methods, which detrimentally affect stability. Consequently, the result is frequently a compromise in terms of overall yield and specific activity. Occasionally, a change of the source can become required at this point; high specific activity in one source may be accompanied by a low product stability due to constituents causing ill effects on the target enzyme (e.g., proteases, inactivating activities, inhibitors), while sources with lower specific activity may provide rather stable preparations. The opportunity to obtain clear samples of low viscosity will be another point deserving attention. Due to the difficulties to clarify highly viscous samples, it will be required to remove, in particular, nucleic acids by nucleases or precipitating agents (e.g., protamine sulfate or polyethyleneimine).

The choice of the initial step in a purification scheme is among the most important decisions to be made in order to achieve the final goal. Highly active and homogeneous final preparations will frequently depend on the ability to remove all constituents causing ill effects on the desired activity. It is almost self-explanatory that proteases, inhibitors, and inactivating enzymes present in the original source will cause losses in activity yield and damaging effects in a time-dependent manner. A second major achievement desired is an almost complete removal of constituents that interfere with the detection of the target activity and, therefore, impede tracing during the subsequent purification steps. Once these contaminants have been removed, it is often possible to use facilitated enzyme test systems downstream of the first column. The third aim of the initial purification is a concentration of the sample. The observation that diluted but otherwise stable proteins at concentrations below 100 to 500 n*M* will often suffer severe losses of activity upon storage, is frequently due to the nonspecific and under physiological condition hardly reversible adherence to various containment surfaces. Since these losses will become negligible at higher concentrations, the importance of this factor is obvious.

A proper characterization of the sample composition after the first column is highly recommended. Analysis of the sample by two-dimensional gel electrophoresis will provide additional and very useful information; a combination of IEP and SDS-PAGE will provide evidence concerning the isoelectric points and the individual polypeptide composition, while a combination of native electrophoresis and SDS-PAGE will provide information on native molecular masses and subunit composition of the major constituents. N-terminal sequencing of selected polypeptides will frequently allow identification of the target and the major impurities at this early point, providing testing capabilities for subsequent separation. This identification is strongly facilitated in organisms whose genome is known: peptide mapping of individual polypeptides, followed by mass spectrometric readout, and database research will rapidly establish the sample composition.

The stabilized preparations obtained from the first purification phase will be subjected in the second phase to other methods aiming the removal of bulk contaminants. This goal is most likely achieved using highly selective chromatographic media, which will preferentially bind either the target enzyme or the major impurities. However, for separation problems lacking these specific binding properties, it is highly recommended to choose a method based on different properties than the previous one in order to maximize separation effects. The proper characterization of individual constituents of the product obtained from the initial phase will frequently allow choices of methods according to the known properties of either the target or the impurities.

In the last phase of a purification, the minor contaminants and structural variants of the target enzyme will be removed. Since it is recommended to establish proper buffer conditions for subsequent storage or applications in this polishing step, size exclusion chromatography on high resolution materials is frequently the method of choice. However, some downstream applications will be most conveniently performed using denatured protein rather than the native structure. The latter is particularly true for N-terminal sequencing, amino acid analysis, and peptide mapping of the individual polypeptides. Since many enzymes are hetero-oligomers, denaturing separations like SDS-PAGE and blotting on PVDF-membranes or reversed-phase HPLC are most conveniently employed.

Individual methods based on very different properties of the enzymes are most meaningfully used in a particular phase of a purification, and a proper arrangement of the individual steps will strongly improve the scheme. However, we should keep in mind that various additional handling steps might be required in order to connect the individual columns. These steps will frequently employ concentration and buffer exchange in order to match the requirements of downstream methods. Since such handling steps carry a considerable risk of sample or activity loss and are frequently time-consuming, all efforts must be made to restrict these steps to an essentially required minimum. Indeed, a proper arrangement of individual purification steps in the purification scheme can avoid many additional steps or replace these steps by less hazardous procedures.

5.5.7 Optimizing Purification Schemes for Downstream Applications

Once a purification scheme for the preparation of pure enzymes of high quality has been established, it will be desired to adopt the scheme for individual needs of downstream applications. Assuming a high-quality product is obtained by reasonable efforts in good activity yield, comparable small changes will be required to provide material for very different analytical methods. Considering the material demand, an upscale will often be required, but applications for a developed scheme to the analysis of various samples available in much smaller amounts might also require a downscale. An adaptation of individual columns employed in a purification is frequently possible by changes of the column dimensions accordingly, but can become very difficult or costly if either large quantities of a specific affinity material must be synthesized or expensive materials are needed, in particular when the properties of the source material will own a high inherent risk for column inactivation (e.g., proteases vs. monoclonal antibodies). Moreover, many methods employed for sample preparation and clarification in the laboratory scale will rapidly come to their limits when applied to either very large or small quantities of sample. The popular cell opening by mechanical methods will frequently not be the first choice for downscaled schemes and must be replaced, for example, by enzymatic steps and compatible detergents. Centrifugation steps frequently used for sample clarification or in precipitation steps are frequently replaced by filtration or solid phase extraction procedures. In other cases, the sample clarification is completely omitted and sophisticated column technologies (expanded bed columns) or batch approaches are used in the initial step.

An individual purification scheme will provide material for the aimed analysis but will fail to provide convenient material when other objectives are defined. For downstream applications, which will require a high product purity in small amounts, it is frequently recommended to apply only the purest fraction of an individual separation to the subsequent purification step, and it is obvious that the activity yield will suffer from this strategy. When a high product yield is needed, it often is required to add additional purification steps in order to achieve a similar result. On the other hand, the requirements in terms of purity can be lower, and it might become possible to work with only partially purified enzyme. The latter situation will apply, in particular, on systems that are either very unstable and, therefore, will suffer severe losses of activity when "over purified," or for systems that are composed of several individual components and will exhibit a loss of function when the individual constituents are separated. However, it should be mentioned that the composition of such partially purified samples should be known in order to avoid questioning of the results later.

5.5.8 Anoxic Purifications

It is hardly difficult to understand that all the problems in the handling of oxygen-sensitive samples will increase with decreasing amounts of sample, and it is obvious that the tolerance limit in terms of the total amount of oxygen (from trace oxygen in buffers to oxygen flooding in from the surrounding air during handling) strongly depends on the nature or the degree of oxygen sensitivity of the target, the amount of starting material available, the relative concentration of the target in the starting material, and on the degree of anoxia that can be achieved and maintained throughout a purification. It will also strongly depend on the desired application for which the enzyme is prepared. Consequently, a logical limit exists, enforcing the use of a strict anoxic environment for the successful handling of even moderately oxygen-sensitive enzymes.

The guidelines given in the previous sections also apply to the purification of oxygen-sensitive proteins. When anoxic glove boxes will provide a protected environment, most purifications are set up and operated in a similar fashion as purifications under air. Nevertheless, anoxic purification in these boxes generally is more difficult due to the given restrictions owned by the glove box technology. The physical barrier will render difficult many operations and the replacement of spare parts, or the provision of consumables will frequently require additional time. Finally, the heat problem will render difficult the handling of enzymes, which are oxygen-sensitive and heat-labile.

Under less protected conditions, additional problems may occur: Such purifications will always have a risk of activity losses due to a nonrecognized exposure of the target toward oxygen. Therefore, it is recommended to monitor anoxia. The easiest way to monitor (and maintain) anoxia is achieved by redox dyes (for example, reduced viologens). These compounds are added at low concentrations and provide a clearly visible indication for anoxia. Moreover, the reduced dye will provide a certain degree of protection, since oxygen flooding in will preferentially react with these molecules present in high concentrations, as compared to the target enzyme. However, using dyes is restricted to enzymes not being affected, and will have to be removed at some point, since these compounds might interfere with downstream applications.

Most protein purifications are performed in the laboratory at μmole-scales of starting material. For extremely oxygen-sensitive enzymes, an anoxic testing is frequently possible due to the fact that highly concentrated cell-free extracts will provide a certain degree of protection by other oxygen-reducing compounds, and even highly sensitive enzymes will suffer almost negligible oxygen damage when the oxygen concentration is reduced below 1 μ*M*. However, sensitive enzymes will often react with oxygen in a time-dependent manner. We must expect about 1% losses when 10 μmole of enzyme are dissolved in 100 mL buffer containing 1 μ*M* oxygen, but that 10% losses will be unavoidable when dissolved in 1 L of the same buffer. On the other hand, we must face the fact that operating chromatographic columns will frequently employ binding, washing, and elution. Bound to the column, the enzyme will react with trace oxygen in the buffers, and large volumes of buffers will cause substantial damage, even at very low oxygen concentrations. When we further consider that the oxygen-caused damage is time-dependent, the relative damage at a given flow rate will strongly depend on the capacity of the column for the target enzyme. A rapid elution after binding is always desired in order to minimize oxygen-induced damage. From these considerations it is obvious that a successful anoxic purification will highly depend on our ability to ensure that each individual step is performed in the lowest possible volume and aims to keep the target concentration as high as possible. Gradient elution, frequently employed in protein purifications under air, will hardly match this objective, and it is often recommended to employ preferential step elution procedures. This goal is most conveniently achieved using highly selective chromatographic media, and therefore, small columns operating close to the capacity limit will be the method of choice.

REFERENCES

1. Englander, S.W., Calhoun, D.B., and Englander, J.J., Biochemistry without oxygen, *Anal Biochem.*, 161, 300–306, 1987.
2. Patil, P.V. and Ballou, D.P., The use of protocatechuate dioxygenase for maintaining anaerobic conditions in biochemical experiments, *Anal Biochem.*, 286, 187–192, 2000.
3. Mayhew, S.G., The redox potential of dithionite and SO^{-2} from equilibrium reactions with flavodoxins, methyl viologen and hydrogen plus hydrogenase, *Eur. J. Biochem.*, 85, 535–547, 1978.
4. Mayhew, S.G., Abels, R., and Platenkamp, R., The production of dithionite and SO^{2} — by chemical reaction of (BI) sulphite with methyl viologen semiquinone, *Biochem. Biophys. Res. Commun.*, 77, 1397–1403, 1977.
5. Zehnder, A.J. and Wuhrmann, K., Titanium (III) citrate as a nontoxic oxidation-reduction buffering system for the culture of obligate anaerobes, *Science*, 194, 1165–1166, 1976.

Part II

Metabolic Pathways and Transport Systems

6 Degradation of Polymers: Cellulose, Xylan, Pectin, Starch

Susan Leschine

CONTENTS

6.1 INTRODUCTION

Many species of *Clostridium* grow and persist in a wide variety of environments that are rich in decaying plant material. For example, free-living clostridia are commonly present in relatively high numbers in terrestrial environments, such as soils and sediments. Also, clostridia inhabit the gastrointestinal tracts of herbivorous and wood-eating animals as members of complex host-associated microbial communities, and clostridia may be isolated directly from decaying plants and composts. Often, the clostridial species isolated from these environments produce the complex enzymatic machinery necessary to depolymerize plant polysaccharides, and usually these clostridia are able to utilize the decomposition products as fermentable substrates.

One of the most important features of plant polymers as a substrate for microbial growth is their insolubility [1]. Clostridial degradation of these polymers occurs exocellularly, either in association with the outer cell envelope layer or extracellularly. This suggests that assembly of enzyme systems, which may be extremely complex, also occurs exocellularly. Presumably, in their exocellular environments, these enzymes are stable and reasonably resistant to proteolytic attack. Hydrolysis products may be available as carbon and energy sources for other microbes that inhabit environments in which plant polymers are biodegraded, forming the basis of numerous interactions among diverse microbes and their products in these anaerobic conditions. Some of these interactions may affect polymer decomposition and the metabolism of polymer-degrading clostridia [2].

As growth substrates, plant polysaccharides occur as extremely complex networks of polymers. Plants invest enormous resources into producing these polymeric networks, in part as a deterrent to microbial attack. In spite of their complexity, numerous different clostridia have evolved effective strategies for decomposing these molecules. This chapter focuses on the known diversity of polymer-decomposing clostridia and their enzyme systems within the context of the remarkable complexity of their polysaccharide substrates.

6.2 PLANT POLYMERS AS GROWTH SUBSTRATES

On a global scale, the photosynthetic fixation of CO_2 yields enormous quantities of plant material, primarily polysaccharides (sugar chains or glycans), including cellulose, xylan, pectin, and starch. Plants are the master glycan producers of the planet, and most of this material is contained in their cell walls or extracellular matrix. The plant cell wall is a highly organized network of cellulose and cross-linked glycans embedded in a gel matrix of pectic substances and reinforced with structural proteins and aromatic substances [3,4]. The molecular composition and arrangements of wall polymers differ among plant species, among tissues within a single species, among individual cells, and even among regions of the wall of an individual plant cell [5].

Cellulose serves as the primary structural component of plant cell walls, accounting for up to 40% of the dry mass of all primary cell walls and an even larger percentage of secondary walls. Thus, cellulose comprises almost half of all biomass produced on the planet [6]. Cellulose is a homopolymer consisting of D-glucose units joined by β(1,4)-bonds. The disaccharide cellobiose is regarded as the repeating unit in cellulose, since each D-glucose unit is rotated by 180 relative to its neighbor. The size (degree of polymerization) of cellulose molecules varies from 500 glucose units per molecule in primary cell walls of plants to 7000 to 14,000 glucose moieties per molecule in secondary walls, although a degree of polymerization of up to 23,000 glucose units per cellulose molecule has been measured in certain algal species [7–10].

Cellulose molecules are strongly associated through inter- and intramolecular hydrogen bonding and van der Waals forces, forming microfibrils, which combine to form fibers. Within microfibrils, cellulose molecules are present in a parallel orientation, that is, with reducing ends of adjacent glucan chains located at the same end. Plant microfibrils are 36 molecules thick, on average, and although each cellulose molecule is just 2 to 3 micrometers in length, microfibrils may reach lengths of hundreds of micrometers because individual cellulose molecules begin and end at different places within the microfibril. Apparently, cellulose chain initiation and termination occur many times during the synthesis of one microfibril by the cellulose synthase complexes ("rosettes") of plants [8]. Cellulose molecules in microfibrils form highly ordered crystalline domains interspersed by more disordered, amorphous regions. The degree of crystallinity in native cellulose is 60 to 90%. Although cellulose can take on at least four different crystalline forms, as determined by X-ray crystallography, one form, cellulose I, predominates in native cellulose. Generally, the abundance of crystalline forms of cellulose within a plant cell wall varies according to cell type and developmental stage. Secondary cell wall microfibrils have a higher cellulose I crystallinity and may be thicker than primary wall microfibrils [8,10,11].

Cellulose almost never occurs alone in nature. For example, cellulose constitutes about 20 to 40% of wall dry weight in growing primary walls and increases to 40 to 60% in secondary walls. The secondary walls of cottonseed hairs, which are nearly 100% cellulose, are exceptional. Usually, cellulose is associated with other molecules, which may dramatically affect its degradation. Cellulose fibrils are associated by hydrogen bonds with cross-linking glycans forming a network that is embedded in a matrix of pectic polysaccharides and an additional network of structural proteins or phenylpropanoid substances [5]. Cellulose imparts tensile strength to the wall to resist turgor pressure. High compression strengths are achieved when lignin (a complex aromatic polymer, classified by the abundance of certain cinnamoyl alcohols) replaces water in the matrix of cell walls. Lignification greatly increases bonding within the wall and produces rigid, woody tissues able to withstand the compressive force of gravity [10]. The process of lignification is of commercial significance because it contributes to a loss in the nutritive quality of forage grasses, reduces pulping efficiency of wood products, and generally affects the degradability of plant biomass [5].

Primary cell walls of plants are made up of structurally independent but interacting networks around a scaffold of cellulose and tightly bound cross-linking glycans. Traditionally, these glycans were termed "hemicellulose" for the fraction extracted from plant materials with dilute alkali, irrespective of their structure. Most flowering plants (dicots and noncommelinoid monocots) have "Type I" walls [4], which contain approximately equal amounts of cellulose and cross-linking xyloglucans, with smaller amounts of arabinoxylans, glucomannans, and galacto-glucomannans. The xyloglucans consist of linear chains of β(1,4)-D-glucose with numerous α-D-xylose units linked at regular sites to the O-6 position of the glucosyl units. Some of the xylosyl units are substituted with α-L-arabinose or β-D-galactose, depending on the species, and sometimes the galactosyl groups are further substituted with α-L-fucose. Grasses and other commelinoid monocots have "Type II" walls [4,12] with cellulose microfibrils of the same structure as those of Type I walls, but with glucuronoarabinoxylans as the major cross-linking polymers and only minor amounts of xyloglucans. The attachment of arabinose and glucuronic acid side-groups to the backbone polymer of β(1,4)-D-xylopyranosyl units affects the ability of the xylans to hydrogen bond to cellulose and, thus, the extent of cross-linking. In addition to these xylans, a third type of cross-linking glycan, the mixed linkage β(1,3), (1,4)-D-glucans, accumulates in cell walls when grass cells begin to elongate. These unbranched polymers consist of cellotriose and cellotetraose units connected by β(1,3)-linkages and take on a corkscrew-like shape [5]. The mixed linkage β(1,3), (1,4)-D-glucans from barley and oats are nearly identical to lichenan [13], a readily available polysaccharide from Iceland moss (*Cetraria islandica*) that is widely used as a substrate for enzymes from polymer-degrading clostridia. Often, microbes isolated from malted barley or other grains, or from environments rich in decaying grass, such as strains of *Clostridium acetobutylicum* [14], *Clostridium cellulovorans* [15], and *Clostridium thermocellum* [16], produce enzymes that attack lichenan.

An intimate association of native cellulose with tightly bound cross-linking glycans may explain, at least in part, the richness of the diversity of glycoside hydrolases produced by cellulolytic clostridia and other microbes. Presumably, some hydrolases function to break down glycans encasing cellulose microfibrils and thereby increase accessibility of the cellulose to attack by endo- and exoglucanases. Occasionally, a microbe is unable to use the products of glycan hydrolysis as growth substrates, suggesting that cellulose is the primary target of the overall enzymatic attack. For example, *C. thermocellum* possesses genes that encode xylanases and produces xylanase activity, but is unable to ferment xylan, xylose, or other five-carbon sugars [17–19].

The cross-linked cellulose microfibril–glycan network is embedded in a matrix of pectins. Pectins are highly hydrated, branched, heterogeneous polysaccharides rich in D-galacturonic acid (GalA), and often contain the methylpentose L-rhamnose (Rha). These complex carbohydrates are believed to be involved in controlling wall porosity and charge density, as well as microfibril spacing and wall extension. Interactions with other macromolecular systems of the cell wall and smaller entities, such as calcium, borate, polyamines, and phenolic compounds, contribute to the multifunctional properties of the pectin matrix [20].

Two major pectins are homogalacturonans and rhamnogalacturonans. Homogalacturonans are α(1,4)-linked polymers containing as many as 200 galacturonic acid units that may be partially methyl esterified at the C-6 carboxyl group and O-acetylated at the C-3 or C-2 position. Before they are deposited in the cell wall, homogalacturonans are methyl esterified. Removal of methyl ester groups within the wall matrix results in cross-linking of homogalacturonan molecules by calcium and formation of supramolecular assemblies and gels. Rhamnogalacturonans include the RG-1 class of pectins, complex polysaccharides with a backbone of a repeating α(1,4)-linked disaccharide [α-D-GalA-(1,2)-α-L-Rha]. Branched and linear oligosaccharides composed predominantly of α-D-arabinofuranosyl and β-D-galactopyranosyl residues are linked to C4 of some of the Rha residues, which may also be O-acetylated at C2 or C3. Another class of rhamnogalacturonans, RGII, includes structurally modified homogalacturonans. Although RGII pectins form a relatively minor component of cell walls, they include molecules with the richest diversity of sugars and linkage structures found among plant polysaccharides. These very complex RGII molecules can dimerize as boron diesters. Pectins are largely restricted to primary cell walls, and, in general, pectins of Type I and Type II walls are similar, although grasses have relatively low amounts of these polysaccharides. Also, in Type I walls the pectic network may be covalently cross-linked to xyloglucans and certain structural proteins [5,12,20–22].

The framework of the plant cell wall is largely carbohydrate, although structural proteins may also form networks in the wall. As mentioned above, components of the pectin matrix in some Type I walls may interact with some of the several types of structural proteins through covalent bonds. A class of structural proteins, termed "arabinogalactan proteins" may contain more than 95% carbohydrate, and thus are more aptly described as proteoglycans. These proteins are attached to large, highly branched galactan chains and are subsequently decorated with arabinose units. In contrast with Type I walls, the cell walls of grasses have very little structural protein; rather, they have extensive interconnecting networks of phenylpropanoids. Most of these aromatics consist of hydroxycinnamic acids, such as ferulic and *p*-coumaric acids, which are attached as carboxyl esters to the O-5 position of some arabinose units in glucuronoarabinoxylans. A portion of the ferulic acid units of neighboring xylans may cross-link by phenyl–phenyl or phenyl–ether linkages to interconnect the xylans into a large network [5].

In general, the end of cell wall expansion often is marked by the formation of covalent cross-links, which results in irreversible associations between wall polymers [23]. Covalent cross-linking dramatically affects the mechanical properties of walls and also impacts the microbial degradation of wall polymers. In addition to the cross-linking of polysaccharides to various other polymers in nonlignified walls described above, direct covalent cross-linking also occurs in lignified walls. Mature cells of plant tissues usually lack protoplasts, and the wall is the only significant cellular structure present. Typically, these walls consist of a thin primary layer, a thicker, multilaminate secondary wall, and sometimes a tertiary layer. The secondary wall is rich in cellulose, with qualitative differences in noncellulosic polysaccharides as compared to primary cell walls, and, in many plants, these thickened walls also are lignified. Hydrophobic lignin, made from phenylpropanoid amino acids, replaces water in the wall during the lignification process and encrusts the cellulose microfibril-glycan and protein networks. Lignification increases the tensile strength and hydrophobicity of walls. At the interface between the hydrophobic surfaces of the lignin deposits and the polysaccharide and protein networks, several types of covalent cross-linkages occur. For example, ester linkages form between uronic acids on glucuronoxylans or rhamnogalacturonans and hydroxyl groups on lignin surfaces. Ether linkages also may occur, as well as glycosidic linkages to terminal phenolic or side-chain hydroxyls on lignin. Bifunctional ferulic acid may form covalent ester–ether bridges between polysaccharides and lignins, with feruloyl–arabinose esters likely contributing to these cross-links [23,24].

Covalent cross-links between lignin and glycans greatly complicate the microbial degradation of plant polysaccharides [23,25]. As mentioned above, some clostridia produce enzymes ("hemicellulases") that depolymerize xylans and other glycans encasing cellulose microfibrils, presumably as a means of gaining access to cellulose, inasmuch as these microbes may be unable to ferment xylan, xylose, or other five-carbon sugar hydrolysis products. Enzymes that attack cross-linking polymers have been commercialized for numerous applications and are discussed at the end of this chapter. For example, mixtures of enzymes, such as xylanases, mannanase, α-galactosidase, have been shown to be effective in biobleaching of paper pulp, a process that helps remove lignin and produces brighter pulp [26,27]. Although the mechanisms by which xylanases and other enzymes facilitate bleaching (lignin removal) are not fully understood, it is clear that enzyme treatment opens the structure of pulp fibers by degrading glycan polymers and thereby disrupting covalent glycan-lignin linkages [26,28].

While cellulose is the primary structural component of plants, starch serves as the principal food reserve. Starch is synthesized and stored in plastids, either temporarily in chloroplasts or for longer periods in amyloplasts in storage tissues such as tubers or seeds. Starch is deposited in plant cells as insoluble grains composed of α-amylose and amylopectin. α-Amylose is a linear polymer of several thousand glucose residues linked by α(1,4) bonds, which forms an irregularly aggregating helically coiled conformation. Amylopectin is a branched molecule consisting of α(1,4)-linked glucose residues with α(1,6)-linked branches every 20 to 30 glucose residues on average. Amylopectin molecules are among the largest molecules occurring in nature, with up to 10^6 glucose residues per molecule. Starch grains contain both amylose and amylopectin, with amylose accounting for approximately 30% of the total starch, although the percentage depends on the species and the plant tissue used for starch storage. Also, starch grains range in size from less than 1 μm in diameter to greater than 100 μm, and they vary in structure, as do the properties of the starches they contain. Amylopectin molecules take on a crystalline nature in starch grains due to hydrogen bonding between hydroxyl groups of glucose residues on adjacent chains. Upon heating, starch grains absorb water and swell, amylose separates from amylopectin, and, when cooled, amylose can form hydrogen bonds with other amylose molecules and with amylopectin forming a gelatinous network [29].

The chemical complexity of plant polymers is daunting. Within each plant species, the composition and arrangements of polymers vary among individual cells and among regions within cells. An even greater diversity of molecular composition is found among plant species. The complexity and variety of plant polymers is reflected in the diversity of clostridia able to attack these polymers and utilize them as growth substrates. Some clostridia are generalists, able to decompose a broad range of plant polymers. For example, *C. cellulovorans*, isolated from a woody biomass digestor, decomposes and ferments a broad range of plant polysaccharides, including cellulose, xylan, and pectin [30]. A forest soil microbe, *Clostridium phytofermentans*, ferments these cell wall polymers and also starch, the main carbon and energy storage polymer of plants [31]. Other clostridia are specialists. *Clostridium methylpentosum*, a human intestinal bacterium, has been found to utilize only four sugar products of the enzymatic depolymerization of dietary pectin, two methylpentoses (L-rhamnose and L-fucose) and two pentoses (L-lyxose and D-arabinose) [32]. The known *Clostridium* species involved in the decomposition of plant polymers comprise a large and diverse group. Considering the complexity of plant polymers found in nature and the range of anaerobic environments in which these molecules may be degraded, such diversity is not surprising.

6.3 CELLULOSE-FERMENTING CLOSTRIDIA

Cellulolytic clostridia have been isolated from a broad range of environments, including soils and sediments from geographically widely separated locations. In addition, these microbes are found in anaerobic digestors, sewage sludge, and composts, as well as the gastrointestinal tracts of herbivorous and wood-eating animals (Table 6.1). Typically, cellulolytic clostridia are members of complex polysaccharide-decomposing microbial communities that thrive in environments where plant materials accumulate. The products of cellulose decomposition are available as carbon and energy sources for other microbes, forming the basis of many microbial interactions that occur among members of these communities [2]. Light microscopic observations of cellulose-decomposing cultures of cellulolytic clostridial strains often reveal motile cells swimming in liquid as well as cells intertwined within cellulose fibers or adhering end-on to fibers (Figure 6.1A). In addition, cellulolytic clostridia may grow as biofilms on the surface of their insoluble substrate. For example, *C. phytofermentans* colonizes the surface of dialysis tubing ("regenerated cellulose") and forms structures with architectural features typical of single-species biofilms (Figure 6.1B) [33]. Generally, in biofilms, cell aggregates are embedded in an extracellular matrix and appear as mushroom-like structures, separated by channels and irregularly shaped openings [34].

TABLE 6.1
Cellulolytic Clostridia and Close Relatives

Phylogenetic Cluster[a]	Species	Source	Optimum Growth Temperature (°C)	Growth Substrates[b] Plant Wall Components	Starch	Fermentation Products[c]	Ref.
I	*Clostridium cellulovorans*	Anaerobic digestor	37	Broad	–	A, B, F, L	30
I	*Clostridium chartatabidum*	Rumen	38–42	Broad (xylan –)	–	A, B, E	39
I	*"Clostridium longisporum"*	Rumen	35–42	Broad (xylan –)	–	F, B, A, E	41
III	*Acetivibrio cellulolyticus*[d]	Sewage sludge	35	C, CB, salicin	–	A, E (tr)	42
III	*Bacteroides cellulosolvens*[d]	Sewage sludge	42	C, CB	–	A, E, L	43
III	*Clostridium aldrichii*	Anaerobic digestor	35	C, CB, xylan	–	A, P, B, IB, IV, L, S	231
III	*Clostridium cellobioparum*	Rumen	30–37	Broad	–	A, E, F, L	153
III	*Clostridium cellulolyticum*	Composted grass	32–35	C, CB, G, ⊕ (xylan, NR)	–	A, E, L, F	232
III	*Clostridium hungatei*	Soil	30–40	C, CB, G, ⊕	–	A, E, L, F	233
III	*Clostridium josui*	Compost	45	Broad	–	A, E, B (tr), P (tr)	234
III	*Clostridium papyrosolvens*	Sediment	25–32	C, CB, G, ⊕	–	A, E, L	148,235
III	*Clostridium stercorarium*	Compost	65	Broad	+	A, E, L	236
III	*Clostridium termitidis*	Termite gut	37	Broad (xylan, NR)	–	A, E	237
III	*Clostridium thermocellum*	Anaerobic digestor	60–64	C, CB (G, with lag)	–	A, E, L	17, 238
IV	*Clostridium cellulosi*	Composted manure	55–60	Broad (xylan, NR)	+	E, A	45
V	*Clostridium thermocopriae*	Soil, compost, animal feces, hot spring	60	Broad	+/–	E, A, B, L	46
XIVa	*Clostridium celerecrescens*	Cow manure	30–37	Broad (xylan, NR)	–	E, A, F, B, L, IB, IV, C, S	239
XIVa	*Clostridium herbivorans*	Pig intestine	39–42	C, CB, maltose	+	B, F, E	240
XIVa	*Clostridium phytofermentans*	Soil	35–37	Broad	+	E, A, F, L	31
XIVa	*Clostridium polysaccharolyticum*	Rumen	30–38	C, CB, xylan	+	F, B, E (tr)	241,242
XIVa	*Clostridium populeti*	Anaerobic digestor	35	Broad	–	B, L, A	243
XIVb	*Clostridium lenticellum*	River sediment	40	Broad	+	A, E, L (tr)	244

TABLE 6.1 (continued)
Cellulolytic Clostridia and Close Relatives

[a] Individual lineages containing *Clostridium* species as defined by Collins et al. [35], amended by the addition of *A. cellulolyticus* and *B. cellulosolvens* [44], *C. hungatei* [233], and *C. josui* [87], to cluster III; and *C. herbivorans* [240] and *C. phytofermentans* [31] to cluster XIVa.

[b] All strains use cellulose (C) and cellobiose (CB) as growth substrates. Some species also use glucose and other 6- and 5-carbon sugars, and xylan (C, CB, G, ⊕). Species with a "Broad" substrate range use additional substrates such as maltose, other disaccharides, and/or pectin. Abbreviations: +, substrate fermented; –, substrate not fermented; +/–, substrate fermented by some strains; NR, not reported.

[c] All strains that have been examined produce CO_2 and H_2 during cellulose fermentation. Products are listed in the order of their predominance, when known. Abbreviations: A, acetate; B, butyrate; C, caproate; E, ethanol; F, formate; IB, isobutyrate; IV, isovalerate; L, lactate; P, propionate; S, succinate; (tr), trace.

[d] Due to their close phylogenetic relationship with cellulolytic *Clostridium* species in cluster III, *A. cellulolyticus* and *B. cellulosolvens* are included for comparison.

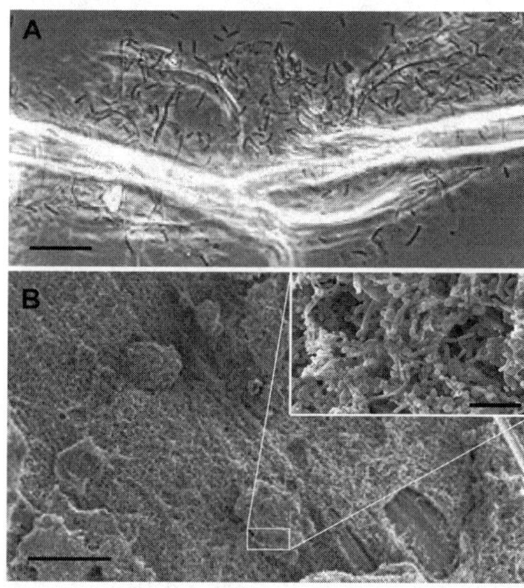

FIGURE 6.1 Light and scanning electron micrographs of cellulose-decomposing cultures of clostridial species. (A) Phase-contrast micrograph of an exponential-phase culture of *Clostridium hungatei* strain B3B grown with ball-milled filter paper [148] as carbon and energy source showing cells entangled in cellulose fibers. Bar, 10 µm. (B) Scanning electron micrograph of a *Clostridium phytofermentans* biofilm showing large cell aggregates on the surface of partially degraded dialysis tubing ("regenerated cellulose"). Bar, 50 µm. The higher magnification inset shows individual cells embedded in a stringy extracellular matrix, presumably dehydrated extracellular polysaccharide. Bar, 5 µm.

As described elsewhere in this book (Chapter 1), the genus *Clostridium* is extremely heterogeneous with many species phylogenetically intermixed with other genera [35,36]. A detailed phylogenetic analysis based on 16S rRNA gene sequences has revealed considerable phylogenetic diversity and has led to the identification of about 20 individual lineages or "clusters" [35]. Almost half of the clostridial species, including *Clostridium butyricum*, the type species of the genus, belong to a phylogenetically well-defined cluster, designated cluster I, equivalent to the rRNA homology group I of Johnson and Francis [37]. The remaining *Clostridium* species are dispersed among the numerous other clusters with separate lines of descent.

Despite sharing a common physiological trait, the cellulolytic clostridia display great phylogenetic diversity, with representatives in several of the major clusters defined by Collins et al. [35,38] (Table 6.1). *C. cellulovorans* from a woody biomass digestor [30], along with two rumen cellulolytic bacteria, *Clostridium chartatabidum* [39] and "*Clostridium longisporum*" [40,41], are members of cluster I (Table 6.1), which includes the type species and many other noncellulolytic *Clostridium* species.

The extensively investigated thermophilic species, *C. thermocellum*, isolated from sewage digestor sludge [17], along with other thermophiles and several mesophilic species from soils, sediments, compost, sewage sludge, and anaerobic digestors form a phylogenetically coherent group, cluster III [35] (Table 6.1), corresponding to group E of Rainey et al. [38]. Thus far, cluster III is defined solely by cellulolytic species. Two species from sewage sludge, *Acetivibrio cellulolyticus* [42] and *Bacteroides cellulosolvens* [43], also are members of cluster III [44] (Table 6.1) but were not classified as clostridia when originally described because the cells of both species stain Gram-negative and spores were not observed. However, many cellulolytic clostridia have a negative Gram stain reaction and, in some strains, spores are observed infrequently. Evidently, *A. cellulolyticus* and *B. cellulosolvens* are closely related to the cellulolytic clostridial members of cluster III, and they can share physiological properties, such as those related to cellulose decomposition. Similarities in the cellulase systems of *A. cellulolyticus*, *B. cellulosolvens*, and cellulolytic clostridia are discussed below.

Although cluster III contains the largest number of known cellulose-decomposing species of clostridia, several other individual lineages also have cellulolytic members (Table 6.1). For example, clusters IV and V each include at least one cellulolytic microbe, *Clostridium cellulosi* [45] and *Clostridium thermocopriae* [46], respectively. Cluster XIV, a group exhibiting considerable phylogenetic depth with two major subgroups, subclusters XIVa and XIVb [35], includes several cellulolytic *Clostridium* species isolated from a wide range of environments (Table 6.1). Interestingly, *Clostridium lentocellum*, the only known cellulolytic member of subcluster XIVb, apparently is closely related to the very large, not-yet-cultured symbiont from the intestine of surgeonfish, "*Epulopiscium*" sp. [35,47].

Cellulolytic members of cluster XIV are physiologically diverse; however, they appear to share certain characteristics that differentiate them in fundamental ways from the microbes in cluster III. Generally, in addition to cellulose, cluster XIV cellulolytic microbes decompose and ferment various polysaccharides and disaccharides derived from plant cell walls such as pectin, xylan, cellobiose, and gentiobiose. Many also ferment sucrose and maltose, as well as starch, a trait that is very uncommon among the cellulolytic clostridia in cluster III (Table 6.1). At least one cellulolytic member of cluster XIV, *C. celerecrescens*, is capable of reducing cinnamic acids, widely distributed aromatic compounds involved in cross-linking arabinoxylans and in polysaccharide-lignin linkages in plant cell walls [48]. Furthermore, cellulolytic members of cluster XIVa produce ethanol, butyrate or formate as their major nongaseous fermentation product, whereas cluster III microbes generally form predominantly acetate (Table 6.1), suggesting fundamental differences in the fermentation pathways used by microbes in these two groups.

6.4 ENZYMES THAT DEGRADE PLANT POLYMERS — GENERAL CONSIDERATIONS

The complexity and variety of plant polymers is matched not only by the diversity of clostridia able to attack these molecules, but also by the wide array of enzymes that catalyze their decomposition. Traditionally, these enzymes have been classified according to their substrate and the manner by which they interact with their substrate. Thus, enzymes active on cellulose, xylan, and mannan have been grouped in different categories. Likewise, endo-acting enzymes, which cleave polysaccharides in the middle of chains, and exo-acting enzymes, which cut polymers at their ends, also have been placed in different groups. However, it has become apparent that this classification scheme does not present a clear picture of the diversity that exists among enzymes that degrade plant polymers. For example, some endoglucanases exhibit processivity, with endoglucanolytic cleavage of internal glycosidic bonds and exoglucanolytic degradation of resulting ends of glucan chains. Presumably, the intricacy of the plant cell wall gives rise to a very large number of subtly different substrates, and an equally large set of enzymes has evolved that match these substrates.

The two major classes of enzymes that are active against plant polysaccharides, glycoside hydrolases and glycosyltransferases, are modular proteins. [49] Each module, or domain, comprises a contiguous amino acid sequence on the polypeptide chain and forms a functionally and structurally discrete unit that folds independently from the rest of the protein. [50,51,52] A new widely used classification system, which is designed to integrate both structural and mechanistic features of carbohydrate-active enzymes, groups modules into families of related amino acid sequences [52–54]. The catalytic mechanism and relative position of the catalytic residues are conserved within each family. Thus, once the stereochemical mechanism and catalytic residues are established for one member of a family, they may be extended to other members of that family [49]. The various families of enzymes are cataloged on the CAZy (Carbohydrate-Active Enzymes) Web site (http://afmb.cnrs-mrs.fr/CAZY/acc.html) [55], which also provides excellent background information. In general, clostridia produce a wide range of complementary enzymes containing catalytic modules from different families with various specificities to degrade a particular set of plant polymers.

Glycoside hydrolases (GHs) contain at least one catalytic module, which usually provides the basis for classification of the enzyme. Cellulases, and many other GHs, often have at least two sites that bind the enzyme to the substrate: the active site of the catalytic domain and a separately folded, functionally independent carbohydrate-binding module (CBM). These modules are usually attached to each other by way of a linker region. CBMs have been shown to play an essential role in catalysis [56–60]. For example, the catalytic domain of cellobiohydrolase CBH1 from *Trichoderma reesei* has very limited activity on cellulose without its CBM, and the binding specificity and adsorption constant are greatly reduced [61]. Studies have indicated that the various modules of carbohydrate-active enzymes may have different evolutionary histories [62]. Thus, enzymes that degrade plant polymers may evolve over time by "mixing and matching" of various modules, such that the activities of these enzymes are affected in subtle ways.

6.5 CELLULASE SYSTEMS — BACKGROUND

The enzymatic mechanism of cellulose degradation has been an area of active investigation for more than 50 years. Initially, the impetus for these studies was a persistent problem caused by the fungal attack on cotton clothing and tents of troops stationed in Southeast Asia during World War II. Research aimed at developing strategies to inhibit fungal cellulases was carried out at the U.S. Army Research and Development Command in Natick, Massachusetts. Under the direction of Elwyn Reese and Mary Mandels, this work ultimately resulted in the development of seminal concepts related to the mechanism of cellulose degradation, including the role of synergism among components of cellulase systems [63–65].

Consequently, early studies of cellulases involved enzyme systems produced by fungi, particularly *T. reesei*, which has served as a model. This enzyme system comprises three main activities: endoglucanases, which randomly hydrolyze β(1,4) bonds within cellulose molecules, thereby producing reducing and nonreducing ends; exoglucanases, which cleave cellobiose units from the nonreducing ends of cellulose polymers; and β-glucosidases, which hydrolyze cellobiose and low-molecular-weight cellodextrins, thereby yielding glucose. The enzymatic components act synergistically in the hydrolysis of crystalline cellulose. A proposed explanation for synergism has involved the idea that endoglucanases attack amorphous regions of cellulose fibers, forming sites for exoglucanases that are then able to hydrolyze cellobiose units from more crystalline regions of fibers.

Finally, β-glucosidases, by hydrolyzing cellobiose, prevent the accumulation of this disaccharide, which is an inhibitor of exoglucanase activity. Clearly, this model for the mechanism of action for fungal cellulase systems is an oversimplification because it does not explain all types of observed synergism [66]. Also, microbial cellulase systems can include proteins with activity similar to that of expansins, plant cell wall proteins that are thought to disrupt hydrogen bonding between cell wall polysaccharides without hydrolyzing them [67]. The discovery in *T. reesei* of swollenin, a protein that exhibits disruption activity on cellulosic materials, and shares sequence similarity with plant expansins [68], indicates the complexity of cellulase systems and variety of activities involved in cellulose decomposition.

The insoluble crystalline nature of cellulose, and its intimate association with other polymers, makes it an unusually recalcitrant substrate for hydrolytic enzymes. Generally, single enzymes are unable to hydrolyze crystalline cellulose. Degradation occurs by simultaneously interacting enzymes, or complexes of enzymes, in cooperation with noncatalytic, specific substrate-binding modules. The mechanism of cellulose hydrolysis can be viewed as a dual process involving the mechanical and structural "preparation" of the insoluble crystalline substrate, followed by hydrolytic cleavage [69].

6.6 CELLULOSOME SYSTEMS

Results of early work on the cellulases of clostridia revealed that the enzymatic machinery produced by these and other anaerobic microbes was fundamentally different from that of *T. reesei*. It is now known that, in anaerobic microorganisms, most enzymes active on cellulose are organized into large multiprotein complexes with an intricate and sophisticated architecture (Figure 6.2). These enzymes have been termed "cellulosomes" by Bayer, Lamed, and co-workers [70–72]. In general, cellulosomes efficiently catalyze the decomposition of cellulose as long as they retain their integrity, but even partial disassociation of the complexes, as occurs under relatively mild conditions, causes

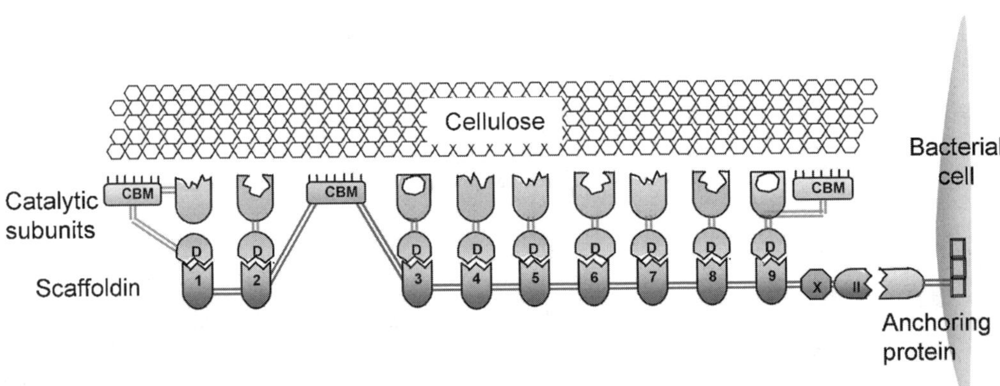

FIGURE 6.2 Schematic representation of the *Clostridium thermocellum* cellulosome and an associated cell-surface anchoring protein (adapted from Bayer et al. [114]). Cellulosomal subunits are modular proteins with various domains often connected by distinct linker sequences (=). The catalytic subunits are organized around the large nonhydrolytic scaffoldin subunit, which has nine internally repeated sequences (known as hydrophobic domains or type-1 cohesins) numbered 1 to 9, according to their position relative to the amino terminus. Integration of the catalytic subunits into the cellulosome structure occurs through interactions of the scaffoldin cohesins with the type-1 dockerin domains (D) of the catalytic subunits. The cellulosome binds to cellulose by way of a CBM-3a within the scaffoldin and also through the CBMs of some of the catalytic subunits. The scaffoldin contains a type-II dockerin domain (II) that interacts with the type-II cohesin of an anchoring protein, which is thought to connect the cellulosome to the cell surface by way of its surface-layer homology domain (rectangular boxes). In *C. thermocellum*, the scaffoldin also contains a hydrophilic domain (X) of unknown function.

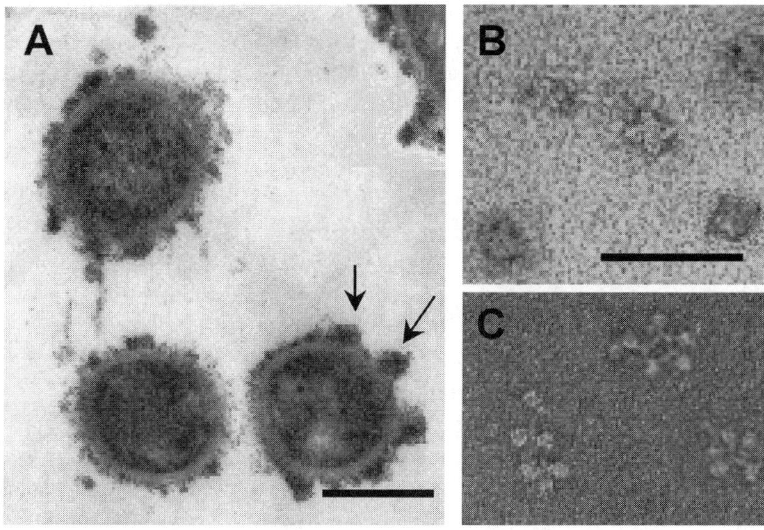

FIGURE 6.3 Transmission electron micrographs of cellulosome-producing *Clostridium thermocellum* cells (A) and negatively stained multiprotein cellulosomal complexes of *Clostridium papyrosolvens* (B, C). Protuberant structures (cellulosomes) on cell surfaces (A), indicated by arrows, were visualized in thin sections of *C. thermocellum* YS cells stained with cationized ferritin. [122] Bar, 0.5 μm. Xylanosome particles (B) and cellulosome particles (C) present in xylanase-active fractions and cellulase-active fractions, respectively, were purified from the cellulosomal system of *C. papyrosolvens* C7. [151] Bar, 50 nm.

loss of most activity against crystalline forms of cellulose. Cellulosomes were first discovered during attempts to identify the adherence factor linking cells to cellulose [73,74; reviewed in 19]. On the cell surface, these multiprotein enzymes appear as polycellulosomal aggregates (Figure 6.3A), and apparently promote adherence of the bacterium to cellulose [75–82].

Most of the protein subunits of cellulosomes have catalytic activity, such as endoglucanase, cellobiohydrolase, or xylanase activity, while others do not appear to have catalytic activity but rather serve a structural role. The main structural subunit of a cellulosome is a large glycoprotein that functions as a "cellulosome-integrating" protein or scaffoldin, and may also be involved in attachment of the cellulosome to the substrate. The catalytic subunits of cellulosomes are organized around the scaffoldin, which has internally repeated sequences (cohesin domains) that bind the catalytic subunits by way of their conserved, duplicated sequences, known as dockerin domains [72, 83–97; reviewed in 19,69,77,98]. It is believed that the large glycoprotein (scaffoldin) functions as a scaffold in the assembly of the cellulosome and that it concentrates and aligns the catalytic subunits for efficient cellulose hydrolysis [70]. Possibly, the cellulosome catalyzes multiple, nearly simultaneous cuttings of the glucan chain [81,99].

6.6.1 The Cellulosome System of *Clostridium thermocellum*

The thoroughly investigated cellulase system produced by *C. thermocellum* has served as a model for studies of multiprotein, multifunctional cellulosome systems produced by clostridia and other microbes. Cellulosomes of *C. thermocellum* strains range in molecular weight from 2.0 to 6.5 × 10^6 [81,96] and comprise 14 to 26 or more polypeptide subunits, depending on the strain and growth conditions [74,100]. The largest subunit, the cellulosome-integrating protein CipA (also designated S1 or S_L), is a glycoprotein with a mass of 210 to 250 kDa that functions as a scaffoldin in the *C. thermocellum* cellulosome. [72,95,96,101] It is a modular protein comprising a CBM-3a, with a broad binding specificity for crystalline cellulose, a hydrophilic module of unknown function (termed an "X" module), a modified dockerin domain (type-II), and nine internal repeated sequences, or cohesin domains, that bind the catalytic subunits (Figure 6.2) [94,102]. Synergism between a catalytic subunit (CelD) and the CBM of the scaffoldin has been demonstrated [103]. Cellulosomes are glycosylated, containing 6 to 13% carbohydrate. Although some catalytic components may be glycosylated, most carbohydrate is associated with the scaffoldin. Glycosidic residues are O-linked via galactopyranose to threonine residues in the proline-threonine-rich linker regions connecting the cohesin modules of the scaffoldin [104].

The type-II dockerin attached to the C-terminus of the *C. thermocellum* scaffoldin does not bind to scaffoldin type-I cohesins. Rather, it binds specifically to type-II cohesins of the cell surface-associated "anchoring" proteins SbdA, Orf2P, and OlpB [105–108]. These proteins contain an S-layer homology (SLH) module for associating with the cell surface, and thereby anchor cellulosomes to the cell through interaction of their type-II cohesin with the scaffoldin type-II dockerin (Figure 6.2). On the surface of *C. thermocellum* cells, cellulosomes appear as polycellulosomal aggregates (Figure 6.3A) and promote adherence of the bacterium to cellulose [75,79–81]. Products of cellulolysis may pass through fibrous "contact corridors" [19,99] that are observed between cells and cellulose. However, the observation that noncellulolytic commensal bacteria are able to grow in cellulose-decomposing cocultures with *C. thermocellum* [109] indicates that at least some soluble sugars are released from fibrous connectors or are otherwise available as growth substrate for other microorganisms.

Twenty-three genes for cellulosomal components, each containing a dockerin domain, have been detected in *C. thermocellum* [summarized in 69]. Four of the 10 proteins identified in cellulosomes are cellobiohydrolases (exoglucanases), and 9 are endoglucanases. In addition, a lichenase [β(1,3), (1,4)-D-endoglucanase)], a chitinase, a mannanase, and five xylanases have been identified [16,69,110–112]. Two of the xylanases also contain xylan esterase modules that remove feruloyl residues from native xylan [113]. As mentioned above, *C. thermocellum* is unable to ferment xylan, xylose, or other five-carbon sugars, and presumably these enzymes function to break down glycans encasing cellulose microfibrils and thereby increase accessibility of the cellulose to attack by cellulosomes. Several cellulosomal subunits are large proteins containing multiple catalytic domains and other functional modules as detailed by Bayer et al. [76]. In addition, several carbohydrate-active enzymes, which are not associated with cellulosomes, have been identified [114], including two endoglucanases that lack an SLH module, CelI [115], a GH-9 with two CBMs, and CelC [116], a GH-5 that lacks a CBM. *C. thermocellum* produces at least two other noncellulosomal carbohydrate-active enzymes: XynX, a modular xylanase including an N-terminal thermo-stabilizing domain, a GH-10 catalytic domain, a CBM-9, and an SLH module [117]; and Lic16A, a lichenase with β(1,3)-endoglucanase activity that also incorporates an SLH module but has four modules of CBM4a [118].

6.6.2 Other Cellulosome Systems

Cellulosomes are produced by many diverse anaerobes in addition to *C. thermocellum,* and their presence has been demonstrated in many of the cellulolytic clostridia and their close relatives listed in Table 6.1, including *A. cellulolyticus*, *B. cellulosolvens*, *C. cellobioparum*, *C. cellulolyticum*, *C. cellulovorans*, *C. josui*, and *C. papyrosolvens* [79,83,87,119,120–123]. Analyses of the modular structure of scaffoldins from these microbes have revealed interesting similarities and differences in comparisons with *C. thermocellum* CipA (Figure 6.4). Scaffoldins from *A. cellulolyticus* and *B. cellulosolvens* each contain a type-II dockerin domain at their C terminus and an internal CBM [124,125] as does CipA. Based on these characteristics, these three scaffoldins have been grouped together as "class I scaffoldins" in a scheme proposed by Bayer et al. [114]. CipV from *A. cellulolyticus*, which possesses a GH-9 module at its N-terminus, is the only known scaffoldin with a catalytic domain.

The cellulolytic clostridia *C. cellulovorans*, *C. cellulolyticum*, and *C. josui* produce class II scaffoldins, which lack a type-II dockerin domain and include a CBM at their N-terminus (Figure 6.4).

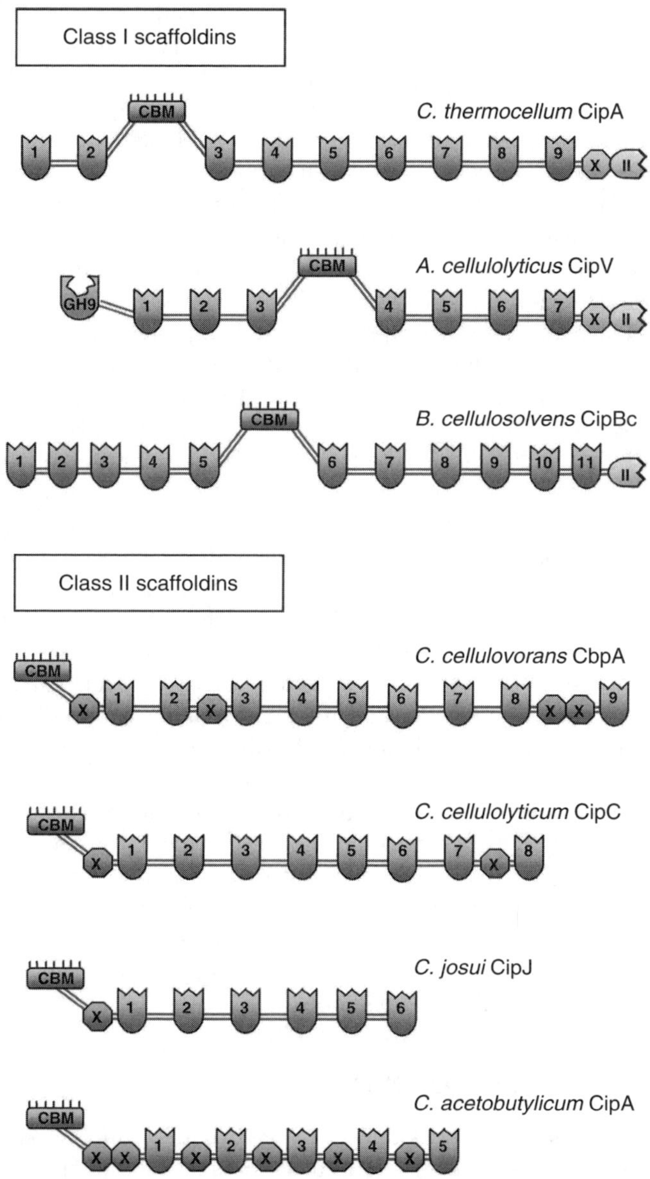

FIGURE 6.4 Modular organization of the known scaffoldins from clostridia and close relatives, organized according to a scheme proposed by Bayer et al. [114]. Class I scaffoldins have a C-terminal type-II dockerin domain (II) and an internal CBM, while Class II scaffoldins lack a dockerin domain and include a CBM at their N-terminus. All known scaffoldins are modular proteins containing internally repeated sequences (cohesins), numbered according to their position relative to the amino terminus. Most include one or more hydrophilic domains (X) of unknown function. The scaffoldin of *Acetivibrio cellulolyticus* has a glycoside hydrolase family 9 (GH-9) catalytic module at its N-terminus. In most scaffoldins, the cohesins are connected by short, proline–threonine-rich linkers sequences (=). However, in the *Clostridium acetobutylicum* scaffoldin, the cohesins are joined to each other by long hydrophilic domains.

Presumably, these microbes would not produce anchoring proteins that contain type-II cohesins, as have been found in *C. thermocellum* (described above), and cell-cellulosome interactions, if they occur, would be accomplished through an alternative mechanism. A possible role in cell surface attachment for the four hydrophilic domains (X modules) of CbpA, the *C. cellulovorans* scaffoldin, has been proposed [77,126]. These modules do not bind to cellulose or to enzyme subunits, nor do they bind to cohesins or other X modules [127]. The observed similarity between X modules and surface layer proteins indicates that X modules of CbpA might be SLH modules, and they might play a role in binding the cellulosome to the cell surface [126]. Furthermore, EngE, a major component of *C. cellulovorans* cellulosomes, potentially could bind to a cohesin of scaffoldin CbpA through its dockerin sequence, as well as to the cell surface by way of its three SLH modules, and thus play a role in binding the cellulosome to the cell surface [126]. Evidence in support of this hypothesis has been presented using recombinant EngE (rEngE) proteins with and without SLH modules [78]. The *C. cellulovorans* rEngE protein bound cell wall fragments to a mini-CbpA scaffoldin only when the rEngE protein included SLH modules. Possibly, an even more complex system of cell-surface-anchoring and adaptor scaffoldin-like proteins may be involved in the cell-associated supramolecular architecture of the *A. cellulolyticus* cellulosome [128].

Based on genome sequence analysis, a gene predicted to encode a scaffoldin has been identified as part of a large gene cluster in *C. acetobutylicum* ATCC 824, a solvent-producing bacterium not known to degrade cellulose [129]. Genes in the operon-like cluster encoding the scaffoldin (CipA) and other cellulosome components are ordered similarly to those in the mesophilic species *C. cellulolyticum* and *C. cellulovorans*, in contrast to the more dispersed arrangement of these genes in the chromosome of the thermophile *C. thermocellum* [129–134]. The scaffoldin of *C. acetobutylicum* contains an N-terminal CBM-3a, six hydrophilic X modules, and only five cohesins in a unique arrangement (Figure 6.4) [130,135]. The cohesins are joined to each other by long hydrophilic X modules, rather than by short, proline–threonine-rich linkers sequences as found in other scaffoldins (Figure 6.4). Although cellulase activity is not detectable in cultures of *C. acetobutylicum* ATCC 824, evidence for the production of a high-molecular-weight cellulosomal complex has been reported [130]. Also, a mini-CipA polypeptide, consisting of a CBM-3a and two cohesin domains, has been expressed in *C. acetobutylicum* and has been shown to result in the formation of a minicellulosome [135]. Although recombinant cellulosomes have been constructed *in vitro* [136], this is the first report of recombinant cellulosome assembly *in vivo*. Furthermore, CelG, a dockerin-containing GH-9 protein encoded by the *C. acetobutylicum* cellulosome gene cluster and exhibiting activity with carboxymethylcellulose or lichenan (a mixed linkage β(1,3),(1,4)-D-glucan) as substrate, is expressed when *C. acetobutylicum* cells are cultured on lichenan or xylose, but not on cellobiose or glucose, indicating that putative cellulosomal subunits are expressed and active under some growth conditions [14].

6.7 CELLULASE SYSTEMS OF CLOSTRIDIA

The cellulosome of *C. cellulolyticum* is smaller than that of *C. thermocellum* and consists of 600-kDa particles that form 16-MDa aggregates that contain 14 different proteins, ranging in size from 20 to 160 kDa [120]. The major subunits of the *C. cellulolyticum* cellulosome are the scaffoldin protein CipC, and two cellobiohydrolases, CelE and CelF, which are encoded by genes that form a large cluster along with genes for some noncellulosomal GHs [131,137]. Interestingly, the major catalytic subunits of cellulosomes in *C. cellulolyticum*, and also in *C. thermocellum* and *C. cellulovorans*, include two cellobiohydrolases that belong to GH families with opposite processivity [69,138,139]. Cellulosomal subunit composition and cellulose adhesion apparently are regulated in *C. cellulolyticum* by growth substrate. When cells are cultured with xylan as substrate, cellulosomes contain xylanases in addition to cellulases and are not cell-associated [140].

Cellulosomes of *C. cellulovorans* are intermediate in size between those of *C. thermocellum* and *C. cellulolyticum*, with a molecular mass of 1 MDa [121], and they comprise at least 10 different catalytic subunits [84,77]. Cellulosomes produced by *C. cellulovorans* can be separated into two fractions differing in activity [141]. The two fractions contain different relative amounts of the same subunits, suggesting that the quantitative composition of cellulosomes affects activity. As discussed above, cellulosomes bind to cells by way of the SLH modules of the scaffoldin CbpA and the associated catalytic subunits [78,126]. In addition to the scaffoldin CbpA, *C. cellulovorans* cellulosomes contain two major proteins, an endoglucanase EngE and an exoglucanase ExgS, and several minor subunits with molecular masses of 40 to 100 kDa, including a mannanase [142], a xylanase [143], and a pectate lyase [144]. The cellulase system of *C. cellulovorans* also comprises at least four noncellulosomal proteins [84,145]. Clearly, *C. cellulovorans* produces a battery of carbohydrate-active enzymes that degrade a wide range of polymers, suggesting that this bacterium may effectively decompose plant cell walls. Consistent with this notion, cellulosomes from *C. cellulovorans* have been shown to release reducing sugars from corn cell walls [141] and form protoplasts from cultured tobacco and *Arabidopsis* cells [146]. At least in part, the efficacy of the *C. cellulovorans* cellulase system in degrading plant polymers is due to synergism between cellulosomal xylanases and cellulases [147] and between cellulosomal and noncellulosomal subunits [145].

C. papyrosolvens strain C7, a mesophile isolated from freshwater sediment [148], lacks cellulosome clusters on its surface and cells do not adhere to cellulose fibers [122]. The cellulase system of this clostridium is present in culture supernatant fluids and comprises at least seven distinct cellulosomes with molecular weights ranging from 500,000 to 660,000, each with different polypeptide composition, enzymatic and structural properties (Figure 6.3B and Figure 6.3C) [149–151]. Two of the cellulosome complexes have xylanase activity (e.g., Figure 6.3B), and three others, Avicelase activity (e.g., Figure 6.3C). Moreover, hydrolysis of crystalline cellulose by this "multicellulosome" system involves synergistic interactions among its multiprotein components [149]. In many respects, the cellulosome system of *C. papyrosolvens* resembles that of *C. thermocellum*. Both require Ca^{2+} and a thiol reducing agent for activity, and some of the *C. papyrosolvens* cellulosome complexes are similar in ultrastructure to, although smaller than, the cellulosomes of *C. thermocellum*. However, discrete cellulosomes of different composition have not been isolated from *C. thermocellum* cellulosome preparations. Also, components of the cellulosome system of *C. papyrosolvens* are never found associated with cells, an uncommon property among cellulolytic clostridia. As pointed out by Forsberg et al. [152], cell-free cellulase complexes may be able to penetrate small spaces in cellulose fibers that the cell itself cannot enter, and thus, the complexes gain greater access to the cellulose substrate. Cells of *C. papyrosolvens* do not adhere to cellulose fibers, rather they accumulate near this substrate [148], a behavior observed by Hungate [153] in cellulolytic clostridia he had isolated. Hsing and Canale-Parola [154] proposed that this behavior results from a chemotactic response toward products of cellulose hydrolysis. Chemotactic behavior of this sort could play an important role in the overall process of cellulose degradation.

Generally, in clostridia, cellulase activity is due to the production of cellulosomes, while noncellulosomal enzymes do not have significant activity against crystalline forms of cellulose. However, *C. stercorarium* (member of cluster III, Table 6.1) produces a fundamentally different cellulase system, in spite of its close phylogenetic relationship to *C. thermocellum*. In this bacterium, there is neither biochemical nor genetic evidence for the production of cellulosomes [155,156]. The cellulase system of *C. stercorarium* includes two enzymes originally designated Avicelase I and II [156]. (Avicelase activity refers to the ability of cellulase preparations to hydrolyze Avicel, a microcrystalline form of cellulose, and often is considered a measure of exoglucanase activity.) Purified Avicelase I (CelZ) from *C. stercorarium* is a monomeric protein with a molecular weight of 100,000 that catalyzes the hydrolysis of crystalline cellulose and does not require additional proteins or cofactors. Apparently, Avicelase I (CelZ) is a processive endoglucanase, exhibiting endoglucanolytic cleavage of internal glycosidic bonds and exoglucanolytic degradation of resulting ends of glucan chains, and it differs from endoglucanases produced by other cellulolytic anaerobes by its ability to hydrolyze crystalline cellulose [156,157]. Avicelase II (CelY) is a novel type of exoglucanase that releases cellotetraose, cellotriose, or cellobiose from the nonreducing end of cellulose chains, and also hydrolyzes Avicel [158]. In combination, Avicelase I and II act synergistically in the degradation of microcrystalline cellulose [159]. The genes encoding these enzymes are located in close proximity on the genome of *C. stercorarium*, and both enzymes are modular [155]. Avicelase I consists of an N-terminal GH-9 catalytic domain, an adjacent CBM-3c, required for catalytic activity and thermostability, and a C-terminal CBM-3b, while Avicelase II has an N-terminal GH-48 catalytic domain and a C-terminal CBM-3 [155,160]. In *C. stercorarium*, these two enzymes comprise an active cellulase system of considerably lower complexity than those of other cellulolytic clostridia studied thus far.

6.8 DEGRADATION OF XYLAN AND OTHER GLYCANS (HEMICELLULOSE)

As discussed above, most clostridial cellulase systems include enzymes that degrade xylan and other plant cell wall cross-linking glycans, collectively known as hemicelluloses. The diversity of these enzymes in cellulolytic clostridia, even in those microbes that do not ferment hemicellulose breakdown products, is remarkable. For example, *C. thermocellum* produces a mannanase [111], two lichenases [16,118], and at least five xylanases [69,114]. Moreover, three of the xylanases also contain carbohydrate esterase modules [58,113]. This impressive array of enzymes is produced by a microbe that is unable to ferment xylans, mannans, xylose, or other five-carbon sugars, and clearly illustrates the importance of enzymes that decompose the glycans encasing cellulose microfibrils. As a result, cellulases gain access to cellulose, the primary carbon and energy source for *C. thermocellum*. The cellulosomal enzyme system of *C. thermocellum* also includes a chitinase [110], even though chitin does not serve as a growth substrate. The ability to degrade chitin is a widespread property of the cellulolytic clostridia [161]. In soils and sediments, cellulose is present in proximity to chitin, an insoluble polymer of β-(1,4)-linked *N*-acetylglucosamine residues that functions as a structural polysaccharide in the exoskeletons of arthropods and the cell walls of most fungi [162]. Possibly, cellulolytic clostridia use chitin as a nitrogen source, inasmuch as environments where cellulose accumulates often are deficient in combined nitrogen [161].

Enzymes active on the cross-linking glycans of plant cell walls have been found in nearly all cellulolytic clostridia that have been examined. These microbes produce a cassette of enzymes that mount an effective multipronged attack on the varied linkages that exist within glycan molecules and between glycans and other polymers of plant cell walls. As mentioned above, *C. thermocellum* produces at least five xylanases with GH family GH-10 or GH-11 modules, typical of xylanases [114]. Also, a xylanase-active GH-44 module is found in the endoglucanase, CelJ [163]. Two of the xylanases, XynY and XynZ, contain carbohydrate esterase modules that exhibit feruloyl esterase activity and remove feruloyl residues from arabinoxylans [113]. Presumably, these enzymes could break feruloyl-linkages between glycan chains. They also could attack feruloyl-arabinose ester bonds that covalently cross-link glycans to lignins in some plant cell walls and thereby help to release polysaccharides from lignin. Direct ester linkages between these polymers could be cleaved by the carbohydrate esterase activities exhibited by CelE or XynA (XynV), which remove acetyl groups from acetylated xylan [58]. The endoglucanase CelH of *C. thermocellum* contains a mannanase catalytic module of family GH-26, which could attack another class of cross-linking glycans in the network that encases cellulose fibrils [111]. The importance of mannanase activity in enzyme systems of polysaccharide-degrading clostridia is signaled by the presence of a gene for a family GH-5 mannanase in each of the large cellulosomal gene clusters found in *C. acetobutylicum*, *C. cellulolyticum*, and *C. cellulovorans* [129–132]. Finally, *C. thermocellum* produces two lichenases, LicA and LicB, which could assist in the decomposition of additional cross-linking glycans, the β(1,3)-glucans and mixed linkage β(1,3), (1,4)-glucans [16,118].

C. cellulovorans also produces a battery of enzymes that enables it to effectively decompose the cross-linking glycans of plant cell walls. In contrast to *C. thermocellum*, however, *C. cellulovorans* utilizes many of these polysaccharides as growth substrates. When *C. cellulovorans* is cultured with xylan as substrate, at least two major and three minor xylanase-active proteins are present in both cellulosomal and noncellulosomal fractions of culture supernatant fluids, and levels of xylanase activity are higher than when cells are cultured on cellulose or cellobiose. Thus, cells regulate the expression of xylanase activity and the composition of cellulosomes depending on growth substrate [164]. XylA, a major cellulosomal xylanase, contains a NodB domain that exhibits acetyl xylan esterase activity, in addition to a GH-11 catalytic domain. These two modules of XylA act synergistically in the hydrolysis of acetylated xylan [143]. Two noncellulosomal enzymes, which may play a role in the degradation of plant cell wall cross-linking glycans by *C. cellulovorans*, have been characterized. ArfA is an α-L-arabinofuranosidase that decomposes arabinan and arabinoxylans. BgaA is a β-galactosidase that has a higher affinity for α-L-arabinopyranose residues than β-D-galactopyranose residues. Interestingly, the noncellulosomal proteins ArfA and BgaA act synergistically with the cellulosomal enzyme XylA in degrading plant cell walls [145].

Several additional clostridia form enzymes that degrade xylans and other cross-linking glycans of plant cell walls. For example, *C. acetobutylicum* produces two endoxylanases and a β-xylosidase [165,166], *C. cellulolyticum* produces a β-xylosidase active on *p*-nitrophenylxylose and xylodextrins [167], and *C. josui* elicits a cellulosomal α-galactosidase, which includes a family GH-27 and shows a preference for highly polymeric galactomannan [168]. *C. josui* also produces a xylanase, Xyn10A, which is present in culture supernatant fluids and also is cell-associated [169]. Similar to other clostridial xylanases, Xyn10A is a multimodular protein, and includes a GH-10, a CBM-9, two SLHs and two thermostabilizing domains, which may be involved in binding the enzyme to cellulose. The hemicellulases of *C. stercorarium* are of special interest due to their notable thermostability. *C. stercorarium*, which forms a noncellulosomal cellulase system (described above), produces an extremely thermostable multimodular xylanase, Xyn10B, that is very similar in structure and activity to Xyn10A of *C. josui* [170,171]. Xyn10B is a cellulose- and xylan-binding protein that appears to bind cells to these insoluble polymers [56,171]. *C. stercorarium* produces additional thermostable enzymes active on cross-linking glycans and their breakdown products, including XylA, a β-xylosidase that exhibits α-arabinofuranosidase activity, and ArfB, a major α-arabinofuranosidase, and RamA [172,173].

Xylanases that are free of cellulase activity are of special interest because of their use in the pulp and paper industries to treat pulp prior to bleaching in order to minimize the use of chlorine [174]. Conventional processes for bleaching pulp result in the formation of chlorinated phenolic compounds such as polychlorinated biphenyls (PCBs), which are known to cause health problems when they enter the environment. Pretreatment of pulp with xylanases facilitates the release of lignin during the bleaching process and reduces the amount of chlorine required. Xylanases that are stable and active at high temperature and alkalinity are particularly useful since the Kraft pulping process is carried out under these conditions. In this regard, the cellulose-free thermostable xylanases produced by *Clostridium absonum* are noteworthy [175].

6.9 PECTIN DEGRADATION

In plant cell walls, the cross-linked cellulose microfibril-glycan network is embedded in a matrix of pectins, as described above. Pectins are hydrated, branched, and heterogeneous polysaccharides rich in D-galacturonic acid. They consist of homogalacturonans, α(1,4)-linked polymers of galacturonic acid units that usually are methyl esterified, and rhamnogalacturonans, extremely complex polysaccharides often containing a galacturonic acid–rhamnose backbone. Pectin molecules may be covalently cross-linked to xyloglucans and certain structural proteins of plant walls. Fermentation of pectin by clostridia and other microbes results in dissolution of the middle lamella of the cell wall and can cause plant tissue maceration. Pectin degradation forms the basis of retting, a traditional process used to extract fibers from plants, such as flax, coconut, and hemp, in the preparation of linen, coir, and other natural fibers. Clostridia are the most commonly isolated microbes from water-retted material [176]. A study by Tamburini et al. [176] demonstrated that isolates with high polygalacturonase activity belong to two phylogenetic clusters related to *Clostridium saccharobutylicum* [177] and the *C. acetobutylicum-Clostridium felsineum* group [178]. *C. felsineum*, along with *Clostridium flavum* and *Clostridium laniganii*, are pectinolytic species long associated with retting of plant material [179,180]. Generally, pectin-fermenting clostridia are widespread in soils, sediments, and decaying plant materials [181–183]. Also, they can play a role in plant pathogenesis, for example, in the development of "wetwood" of living trees [184]. In addition, pectinolytic clostridia are inhabitants of the gastrointestinal tracts of humans and other herbivorous and omnivorous animals [185–187].

Pectin degradation usually involves the combined action of a battery of enzymes that differ in substrate specificity and reaction mechanism. Pectin methylesterases remove methoxyl groups from pectin, releasing methanol and converting methylgalacturonate residues to galacturonate residues, or pectin to pectate. Due to the action of these enzymes, methanol is a major end product of pectin decomposition [188]. Depolymerases include lyases and hydrolases that cleave bonds between galacturonate or galacturonate–rhamnose units. Pectate lyases cleave by β-elimination generating oligosaccharides with a 4,5-unsaturated residue at the nonreducing end. Polygalacturonase and exopolygalacturonosidase cleave by hydrolysis [189].

Some of the cellulolytic species included in Table 6.1 are able to ferment pectin, although many do not. *C. cellulovorans*, a cellulolytic microbe that also ferments pectin, produces a pectate lyase, PelA, as part of its cellulosomal enzyme system [144]. This multimodular protein comprises a truncated N-terminal pectate lyase (PL) family 1 module, a CBM, a PL-4 catalytic domain, and a dockerin, which indicates that PelA is a cellulosomal subunit. A recombinant form of PelA cleaves polygalacturonic acid to trigalacturonic and digalacturonic acid but does not further degrade either of these products. *C. cellulovorans* cellulosome preparations, especially those with high levels of pectate lyase activity, are effective in forming protoplasts from cultured tobacco and *Arabidopsis* cells [146]. A cellulosome-associated rhamnogalacturonan lyase Rgl11Y has been identified in *C. cellulolyticum* [190]. This enzyme, which includes a PL-11 catalytic domain and a dockerin, cleaves α(1,4) bonds between L-Rha and D-GalA units in the backbone of rhamnogalacturonan I (RGI) via a β-elimination mechanism. Rgl11Y is more active with potato pectic galactan as substrate than rhamnogalacturonan, suggesting that it requires galactan side chains on the RGI backbone for activity. The noncellulosomal enzyme system of the cellulolytic thermophile *C. stercorarium* includes an α-rhamnosidase, RamA, a novel enzyme with an inverting mechanism and an unusual GH catalytic domain [191].

The pectinases produced by *Clostridium thermosaccharolyticum* (presently classified as *Thermoanaerobacterium thermosaccharolyticum* [35]), a thermophilic, noncellulolytic, pectinolytic microbe that causes spoilage of canned foods, also have been investigated. Under some growth conditions, this microbe produces an extracellular glycosylated protein complex that exhibits pectin methylesterase activity, releasing methanol, and exopolygalacturonosidase activity, resulting in the production of digalacturonate from the nonreducing end of pectin chains. These two enzymes act on pectin in a coordinated manner in the protein complex and release arabinose, rhamnose, and galactose, in addition to methanol, digalacturonate, and trigalacturonate [192,193]. *Clostridium multifermentans* produces an unusual enzyme complex that exhibits coordinated pectin methylesterase and pectate lyase activities, releasing unsaturated digalacturonic acid residues from the reducing end of pectin chains [194]. A pectinolytic system with similar activities is elicited by the human intestinal bacterium, "*Clostridium butyricum-beijerinckii*" [186]. This enzyme system, which includes a pectin methylesterase, an exopectate lyase, and also an endopectate lyase, may play a role in fiber digestion in humans [186].

6.10 STARCH DEGRADATION

As the principal storage compound of plants, starch is an abundantly produced polymer. As detailed above, starch is deposited in plant cells as insoluble grains of α-amylose, a linear polymer of glucose residues linked by α(1,4) bonds, and amylopectin, a branched molecule consisting of α(1,4)-linked glucose residues with α(1,6)-linked branches. Endo-acting (α-amylase) and exo-acting (β-amylase and glucoamylase) enzymes cleave α(1,4) bonds in starch molecules. α-Amylases hydrolyze linkages in the interior of starch polymers in a random fashion, leading to the formation of linear and branched oligosaccharides. Exo-acting hydrolases attack the substrate from the nonreducing end, producing maltose (β-amylases) or glucose (glucoamylases). Pullulanases ("debranching" enzymes) hydrolyze α(1,6) bonds of amylopectin or pullulan, a fungal glucan containing both α(1,6) and α(1,4) linkages. On the basis of substrate specificity and product formation, pullulanases are divided into two groups. Pullulanase type I enzymes hydrolyze the α(1,6) bonds in amylopectin and branched oligosaccharides, but are unable to attack α(1,4) linkages in starch or other glucans. Pullulanase type II enzymes, also known as "amylopullulanases," attack both α(1,6) linkages in amylopectin and α(1,4) bonds in linear and branched polysaccharides. Maltose and short oligosaccharides produced through the action of the above-mentioned enzymes are hydrolyzed by α-glucosidases [195].

Starch-degrading enzymes are widespread among the saccharolytic clostridia. Several cellulolytic clostridial members of the phylogenetically defined cluster XIV are able to ferment starch (Table 6.1). However, none of the known cellulosome-producing strains have been found to degrade starch, and cellulosomal enzymes active on starch have not been reported in clostridia.

The biotechnological potential of thermostable enzymes has stimulated investigations of the starch-degrading enzyme systems of thermophilic clostridial strains. Novel type II pullulanases, which cleave both α(1,6) and α(1,4) glucosyl linkages, have been isolated from strains of *Clostridium thermohydrosulfuricum* (reclassified as either *Thermoanaerobacter ethanolicus* or *Thermoanaerobacter thermohydrosulfuricus*) and from *Clostridium thermosulfurogenes* (now known as *Thermoanaerobacterium thermosulfurigenes*) [196–199]. Most of these enzymes are predominantly extracellular. However, a pullulanase produced by *T. thermosulfurigenes* strain EM1 appears to be cell bound, anchored to the cell surface by way of three SLH domains at the C-terminus of the protein [200]. During growth, the enzyme can be released into the growth medium, resulting in a change in the cell envelope ultrastructure [201,202]. *T. thermosulfurigenes* also produces a β-amylase [203,204] and an α-amylase that act synergistically with pullulanase in the degradation of native starch [205,206].

Thermophilic clostridia and close relatives also have proven to be good sources for glucoamylases, which have been purified and characterized from *T. ethanolicus* (*C. thermohydrosulfuricum*) strain 39E [207] and from strains of *T. thermosaccharolyticum* (*C. thermosaccharolyticum*) [208–210]. Structural analysis of a glucoamylase from *T. thermosaccharolyticum* has revealed a multimodular protein with a C-terminal GH-15 catalytic domain connected by a linker domain to an N-terminal domain "super-β-sandwich" of unknown function, which is composed of 18 antiparallel strands arranged in β-sheets [211].

The starch-hydrolyzing enzymes of *C. acetobutylicum* and *C. beijerinckii* are of interest because of the role of these microbes in industrial processes such as solvent production. *C. acetobutylicum* ATCC 824 produces a 61-kDa α-amylase that is more active with high-molecular-weight substrates than with low-molecular-weight maltooligosaccharides [212,213]. Also, in *C. acetobutylicum* NCIB 8052, two enzymes with α-glucosidase activity have been identified that are active with maltose, maltotriose, or isomaltose, but not starch, as substrates [214]. Starch-hydrolyzing clostridia that associate with humans also have received special attention. For example, a 76-kDa α-amylase produced by *Clostridium perfringens* has been characterized [215]. In addition, the role of clostridia in starch decomposition in the human colon has been considered. Depending on the type of starch ingested, a portion escapes digestion in the human small intestine and reaches the colon, where it is fermented by the resident microbial population. Butyrate, which is formed as a product of this fermentation, is important for normal development of colonic epithelial cells, inhibits growth of colon cancer cells, and may serve to protect individuals from colonic disease [216–218]. Pryde et al. [219] found that the most numerous butyrate-producing bacteria from human colonic contents are oxygen-sensitive anaerobes belonging to the clostridial clusters IV and XIV, which are known to include starch-decomposing microbes (see Table 6.1).

6.11 APPLICATIONS FOR POLYMER-DEGRADING CLOSTRIDIA AND THEIR ENZYME SYSTEMS

Cellulose and other components of biomass represent an abundant and valuable renewable natural resource that may be put to a wide range of uses, e.g., as a source of food, fiber, chemicals, energy, etc. Those microbes and their enzymes, which are active on components of biomass, are extremely useful as catalysts, effecting the changes necessary to convert biomass polymers into more valuable products. Although several chapters of this book are focused on applications involving clostridia, a few specific processes employing polymer-degrading clostridia and their enzymes deserve mention here.

Clostridia produce a broad range of industrially important enzymes. A major advantage of processes that employ microbial enzymes is that enzymatically driven reactions often are far less harmful to the environment than alternative methods. For example, as mentioned above, microbial xylanases are used commercially in the pulp and paper industry in biobleaching processes. Pulp is pretreated with xylanases and other enzymes to help remove lignin, which produces a brighter pulp and minimizes the use of harsh chemicals in the subsequent treatment stages of the Kraft pulping process [174,220]. Apparently, enzyme treatment opens the structure of pulp fibers by degrading cross-linking glycan polymers and disrupting covalent glycan-lignin linkages [26,28]. Mixtures of enzymes, such as xylanases, α-galactosidase, and clostridial mannanase have been shown to be effective in biobleaching of paper pulp [26,27]. Cellulase-free thermostable xylanases, such as those produced by *C. absonum*, may be particularly useful in these processes [174,175,221]. Xylanases also are used as additives to improve nutritional properties of silage and grain feed, in wheat flour to improve dough handling and quality of baked products, in extraction processes for production of coffee, plant oils, and starch, and, in combination with pectinases and cellulases, for the clarification of fruit juices [26].

In addition to xylanases, clostridia produce several other enzymes that are presently used, or have potential applications, in industry. Enzymes, such as lichenases, that hydrolyze cross-linking glycans other than xylans (e.g., the mixed linkage β(1,3), (1,4)-D-glucans from barley and oats) are potentially useful in the mashing process of beer brewing [222]. Starch-hydrolyzing enzymes, such as amylases, pullulanases, and glucoamylases, especially those with increased thermostability, serve important functions in food, chemical, and pharmaceutical industries [195]. Pectinases may be used to facilitate the extraction of fibers from plants in procedures analogous to traditional retting processes [223,224].

Although bacterial cellulases are not yet widely produced industrially, clostridial cellulosome preparations efficiently hydrolyze crystalline cellulose and are good candidates for commercial development. Various potential applications of cellulosomes and cellulosomal components are being explored. Bayer and colleagues [101] suggested that "designer cellulosomes" might be constructed with specific capabilities for effective waste treatment and commercial utilization of cellulosic materials. Active cellulosome chimeras that exhibit synergistic action on crystalline cellulose have been constructed using protein modules from the cellulosomes of *C. cellulolyticum* and *C. thermocellum* [225,226]. Also, Doi and co-workers have assembled cellulosomes from *C. cellulovorans* protein modules and demonstrated that certain combinations of "minicellulosomes" act synergistically in the degradation of corn cell walls [136,147,227]. Results of these studies suggest a potentially valuable strategy for designing recombinant cellulosomes with improved properties for specific applications.

Certainly, the most important commercial process employing clostridia is the acetone-butanol-ethanol (ABE) fermentation of *C. acetobutylicum*. Although production of solvents by this process has dropped off dramatically in recent years due to rising substrate (molasses) costs and the availability of inexpensive crude oil, several improvements in technology and clostridial strains suggest that the ABE fermentation may now be economically viable [228]. For example, Qureshi and Blaschek [229] have reported the development of a strain of *C. beijerinckii* that efficiently ferments cornstarch, a potentially less expensive substrate, with enhanced solvent yield. The discovery of cellulosomes in *C. acetobutylicum* ATCC 824 [129,130] and the ability to construct recombinant cellulosomes *in vivo* [135] improve the prospects for the biological production of solvents from cellulose, the ultimate cheap substrate. Also noteworthy is the recently reported isolation from soil of 13 new *Clostridium* strains, closely related to *C. acetobutylicum*, that produce high amounts of solvents and hydrolyze a wide variety of α- and β-glycans, including starch, xylan, pectin, inulin, and cellulose [230]. Considering the complexity of plant polymers in nature and the range of anoxic environments in which these molecules may be degraded, surely we have barely scratched the surface of understanding the natural diversity of potentially useful polymer-degrading microbes in the biosphere.

ACKNOWLEDGMENTS

This chapter is dedicated to University of Massachusetts Professor Emeritus Ercole Canale-Parola, who inspired my interest in the subject of this chapter, and who, during more than three decades as an active scientist and dedicated, well-loved teacher, sparked the curiosity of many students and associates to comprehend the amazing world of polymer-decomposing microbial communities. I wish to express my gratitude to all members of my laboratory for stimulating discussions on polymer-degrading clostridia. Research in my laboratory was supported by U.S. Department of Energy grants DE-FG02-88ER13898 and DE-FG02-02ER15330.

REFERENCES

1. Salyers, A.A., Reeves, A., and Delia, J., Solving the problem of how to eat something as big as yourself: diverse bacterial strategies for degrading polysaccharides, *J. Ind. Microbiol. Biotechnol.*, 17, 470, 1996.
2. Leschine, S.B., Cellulose degradation in anaerobic environments, *Annu. Rev. Microbiol.*, 49, 399, 1995.
3. Carpita, N., McCann, M., and Griffing, L.R., The plant extracellular matrix: news from the cell's frontier, *Plant Cell,* 8, 1451, 1996.
4. Carpita, N.C. and Gibeaut, D.M., Structural models of primary-cell walls in flowering plants — consistency of molecular-structure with the physical-properties of the walls during growth, *Plant, J.* 3, 1, 1993.
5. Carpita, N.C. and McCann, M.C., The cell wall, in *Biochemistry and Molecular Biology of Plants*, Buchanan, B.B., Gruissem, W., and Jones, R.L., Eds., American Society of Plant Physiologists, Rockville, MD, 2000, 52.
6. Eriksson, K.-E.L., Blanchette, R.A., and Ander, P., *Microbial and Enzymatic Degradation of Wood and Wood Components,* Springer-Verlag, Berlin, 1990.
7. Brown, R.M., The biosynthesis of cellulose, *J. Macromol. Sci.-Pure Appl. Chem.,* A33, 1345, 1996.
8. Delmer, D.P., Cellulose biosynthesis: exciting times for a difficult field of study, *Annu. Rev. Plant Physiol. Plant Mol. Biol.,* 50, 245, 1999.
9. Ljungdahl, L.G. and Eriksson, K.-E., Ecology of microbial cellulose degradation, in *Adv. Microbial Ecol.*, Marshall, K.C., Ed., Plenum, New York/London, 1985, 237.
10. Richmond, P.A., Occurrence and functions of native cellulose, in *Biosynthesis and Biodegradation of Cellulose*, Haigler, C.H. and Weimer, P.J., Eds., Marcel Dekker, Inc., New York, 1991, 5.
11. Marchessault, R.H. and Sundararajan, P.R., Cellulose, in *The Polysaccharides, Vol. 2*, Aspinall, G.O., Ed., Academic, New York, 1983, 11.
12. Carpita, N.C., Structure and biogenesis of the cell walls of grasses, *Annu. Rev. Plant Physiol. Plant Molec. Biol.,* 47, 445, 1996.
13. Dais, P. and Perlin, A.S., High-field, C-13-NMR spectroscopy of beta-D-glucans, amylopectin, and glycogen, *Carbohyd. Res.,* 100, 103, 1982.
14. Lopez-Contreras, A.M. et al., Production by *Clostridium acetobutylicum* ATCC 824 of CelG, a cellulosomal glycoside hydrolase belonging to family 9, *Appl. Environ. Microbiol.,* 69, 869, 2003.

15. Foong, F. et al., Nucleotide sequence and characteristics of endoglucanase gene Engb from *Clostridium cellulovorans*, *J. Gen. Microbiol.*, 137, 1729, 1991.
16. Zverlov, V.V. et al., Purification and cellulosomal localization of *Clostridium thermocellum* mixed linkage beta-glucanase LicB (1,3-1,4-beta-D-glucanase), *Biotechnol. Lett.*, 16, 29, 1994.
17. Ng, T.K., Weimer, P.J., and Zeikus, J.G., Cellulolytic and physiological properties of *Clostridium thermocellum*, *Arch. Microbiol.* 114, 1, 1977.
18. Shinmyo, A., Garcia-Martinez, D.V., and Demain, A.L., Studies on the extracellular cellulolytic enzyme complex produced by *Clostridium thermocellum*, *J. Appl. Biochem.*, 1, 202, 1979.
19. Shoham, Y., Lamed, R., and Bayer, E.A., The cellulosome concept as an efficient microbial strategy for the degradation of insoluble polysaccharides, *Trends Microbiol.*, 7, 275, 1999.
20. Willats, W.G.T. et al., Pectin: cell biology and prospects for functional analysis, *Plant Mol. Biol.*, 47, 9, 2001.
21. Carpita, N.C. and McCann, M.C., The functions of cell wall polysaccharides in composition and architecture revealed through mutations, *Plant Soil*, 247, 71, 2002.
22. Ridley, B.L., O'Neill, M.A., and Mohnen, D.A., Pectins: structure, biosynthesis, and oligogalacturonide-related signaling, *Phytochemistry*, 57, 929, 2001.
23. Iiyama, K., Lam, T.B.T., and Stone, B.A., Covalent cross-links in the cell-Wall, *Plant Physiol.*, 104, 315, 1994.
24. Scalbert, A. et al., Ether linkage between phenolic acids and lignin fractions from wheat straw, *Phytochemistry*, 24, 1359, 1985.
25. Hartley, R.D. and Ford, C.W., Phenolic constituents of plant cell walls and wall biodegradability, in *Plant Cell Wall Polymers, Biogenesis and Biodegradation, ACS Symp. Ser. 399*, Lewis, N.G. and Paice, Eds., M.G. American Chemical Society, Washington, D.C., 1989, 135.
26. Beg, Q.K. et al., Microbial xylanases and their industrial applications: a review, *Appl. Microbiol. Biotechnol.*, 56, 326, 2001.
27. Clarke, J.H. et al., A comparison of enzyme-aided bleaching of softwood paper pulp using combinations of xylanase, mannanase and alpha-galactosidase, *Appl. Microbiol. Biotechnol.*, 53, 661, 2000.
28. Beg, Q.K. et al., Enhanced production of a thermostable xylanase from Streptomyces sp. QG-11-3 and its application in biobleaching of eucalyptus kraft pulp, *Enzyme Microb. Technol.*, 27, 459, 2000.
29. Dennis, D.T. and Blakeley, S.D., Carbohydrate metabolism, in *Biochemistry and Molecular Biology of Plants*, Buchanan, B.B., Gruissem, W., and Jones, R.L., Eds., American Society of Plant Physiologists, Rockville, MD, 2000, 630.
30. Sleat, R., Mah, R.A., and Robinson, R., Isolation and characterization of an anaerobic, cellulolytic bacterium *Clostridium cellulovorans* sp. nov., *Appl. Environ. Microbiol.* 48, 88, 1984.
31. Warnick, T.A., Methé, B.A., and Leschine, S.B., *Clostridium phytofermentans* sp. nov., a cellulolytic mesophile from forest soil, *Int. J. Syst. Evol. Microbiol.*, 52, 1155, 2002.
32. Himelbloom, B.H. and Canale-Parola, E., *Clostridium methylpentosum* sp. nov.: a ring-shaped intestinal bacterium that ferments only methylpentoses and pentoses, *Arch. Microbiol.*, 151, 287, 1989.
33. Wittbold III, W.M., Rennie, J., and Leschine, S.B., unpublished data, 2003.
34. Stoodley, P. et al., Biofilms as complex differentiated communities, *Annu. Rev. Microbiol.*, 56, 187, 2002.
35. Collins, M.D. et al., The phylogeny of the genus *Clostridium*: proposal of five new genera and eleven new species combinations, *Int. J. System. Bacteriol.*, 44, 812, 1994.
36. Stackebrandt, E. and Hippe, H., Taxonomy and systematics, in *Clostridia: Biotechnology and Medical Applications*, Bahl, H. and Dürre, P., Eds., Wiley-VHC, Weinheim, Germany, 2001, 19.
37. Johnson, J.L. and Francis, B.S., Taxonomy of the clostridia: ribosomal ribonucleic acid homologies among the species, *J. Gen. Microbiol.*, 88, 229, 1975.
38. Rainey, F.A. and Stackebrandt, E., 16S rDNA analysis reveals phylogenetic diversity among the polysaccharolytic clostridia, *FEMS Microbiol. Lett.*, 113, 125, 1993.
39. Kelly, W.J., Asmundson, R.V., and Hopcroft, D.H., Isolation and characterization of a strictly anaerobic, cellulolytic spore former: *Clostridium chartatabidum* sp. nov., *Arch. Microbiol.*, 147, 169, 1987.
40. Hungate, R.E., *The Rumen and Its Microbes*, Academic, New York, 1966.
41. Varel, V.H., Reisolation and characterization of *Clostridium longisporum*, a ruminal sporeforming cellulolytic anaerobe, *Arch. Microbiol.*, 152, 209, 1989.
42. Patel, G.B. et al., Isolation and characterization of an anaerobic, cellulolytic microorganism, *Acetivibrio cellulolyticus* gen. nov., sp. nov., *Int. J. Syst. Bacteriol.*, 30, 179, 1980.

43. Murray, W.D., Sowden, L.C., and Colvin, J.R., *Bacteroides cellulosolvens* sp. nov., a cellulolytic species from sewage sludge, *Int. J. Syst. Bacteriol.,* 34, 185, 1984.
44. Lin, C.Z., Urbance, J.W., and Stahl, D.A., *Acetivibrio cellulolyticus* and *Bacteroides cellulosolvens* are members of the greater clostridial assemblage, *FEMS Microbiol. Lett.,* 124, 151, 1994.
45. He, Y.L., Ding, Y.F., and Long, Y.Q., Two cellulolytic *Clostridium* species: *Clostridium cellulosi* sp. nov. and *Clostridium cellulofermentans* sp. nov., *Int. J. Syst. Bacteriol.,* 41, 306, 1991.
46. Jin, F.X., Yamasato, K., and Toda, K., *Clostridium thermocopriae* sp. nov., a cellulolytic thermophile from animal feces, compost, soil, and a hot-spring in Japan, *Int. J. Syst. Bacteriol.,* 38, 279, 1988.
47. Angert, E.R., Brooks, A.E., and Pace, N.R., Phylogenetic analysis of *Metabacterium polyspora*: clues to the evolutionary origin of daughter cell production in *Epulopiscium* species, the largest bacteria, *J. Bacteriol.,* 178, 1451, 1996.
48. Chamkha, M., Garcia, J.L., and Labat, M., Metabolism of cinnamic acids by some *Clostridiales* and emendation of the descriptions of *Clostridium aerotolerans*, *Clostridium celerecrescens* and *Clostridium xylanolyticum*, *Int. J. Syst. Evol. Microbiol.,* 51, 2105, 2001.
49. Henrissat, B. and Davies, G.J., Glycoside hydrolases and glycosyltransferases. Families, modules, and implications for genomics, *Plant Physiol.,* 124, 1515, 2000.
50. Corpet, F., Gouzy, J., and Kahn, D., Recent improvements of the ProDom database of protein domain families, *Nucleic Acids Res.,* 27, 263, 1999.
51. Henrissat, B. and Bairoch, A., Updating the sequence-based classification of glycosyl hydrolases, *Biochem. J.,* 316, 695, 1996.
52. Henrissat, B., Teeri, T.T., and Warren, R.A.J., A scheme for designating enzymes that hydrolyse the polysaccharides in the cell walls of plants, *FEBS Lett.,* 425, 352, 1998.
53. Coutinho, P.M. and Henrissat, B., The modular structure of cellulases and other carbohydrate-active enzymes: an integrated database approach, in *Genetics, Biochemistry and Ecology of Cellulose Degradation*, Ohmiya, K. et al., Eds., Uni Publishers Co., Tokyo, 1999, 15.
54. Coutinho, P.M. and Henrissat, B., Carbohydrate-active enzymes: an integrated database approach, in *Recent Advances in Carbohydrate Bioengineering*, Gilbert, H.J. et al., Eds., The Royal Society of Chemistry, Cambridge, 1999, 3.
55. Coutinho, P.M. and Henrissat, B., Carbohydrate-Active Enzymes: http://afmb.cnrs-mrs.fr/~cazy/CAZY/index.html, 1999.
56. Ali, M.K. et al., Importance of the carbohydrate-binding module of *Clostridium stercorarium* Xyn10B to xylan hydrolysis, *Biosci. Biotechnol. Biochem.,* 65, 41, 2001.
57. Arai, T. et al., Characterization of a cellulase containing a family 30 carbohydrate-binding module (CBM) derived from *Clostridium thermocellum* CelJ: importance of the CBM to cellulose hydrolysis, *J. Bacteriol.,* 185, 504, 2003.
58. Fernandes, A.C. et al., Homologous xylanases from *Clostridium thermocellum*: evidence for bi-functional activity, synergism between xylanase catalytic modules and the presence of xylan-binding domains in enzyme complexes, *Biochem. J.,* 342, 105, 1999.
59. Sun, J.L. et al., Adsorption of *Clostridium stercorarium* xylanase A to insoluble xylan and the importance of the CBDs to xylan hydrolysis, *J. Ferment. Bioeng.,* 85, 63, 1998.
60. Tomme, P. et al., Cellulose-binding domains: classification and properties, in *Enzymatic Degradation of Insoluble Polysaccharides*, Saddler, J.N. and Penner, M., Eds., American Chemical Society, Washington, DC, 1995, 142.
61. Stahlberg, J., Johansson, G., and Pettersson, G., A new model for enzymatic hydrolysis of cellulose based on the 2-domain structure of cellobiohydrolase-I, *BioTechnology,* 9, 286, 1991.
62. Gilkes, N.R. et al., Domains in microbial β-1,4-glucanases: sequence conservation, function, and enzyme families, *Microbiol. Rev.,* 55, 303, 1991.
63. Gilligan, W. and Reese, E.T., Evidence for multiple components in microbial cellulases, *Can. J. Microbiol.,* 1, 90, 1954.
64. Reese, E.T. and Mandels, M., Enzymatic degradation, in *Cellulose and Cellulose Derivatives, Part V,* Bikales, N.M. and Segal, L., Eds., Wiley-Interscience, New York, 1971, 1079.
65. Reese, E.T., Siu, R.G.H., and Levinson, H.S., The biological degradation of soluble cellulose derivatives and its relationship to the mechanism of cellulose hydrolysis, *J. Bacteriol.,* 59, 485, 1950.
66. Klyosov, A.A., Trends in biochemistry and enzymology of cellulose degradation, *Biochemistry,* 29, 10577, 1990.

67. Cosgrove, D.J., Loosening of plant cell walls by expansins, *Nature,* 407, 321, 2000.
68. Saloheimo, M. et al., Swollenin, a *Trichoderma reesei* protein with sequence similarity to the plant expansins, exhibits disruption activity on cellulosic materials, *Eur. J. Biochem.,* 269, 4202, 2002.
69. Schwarz, W.H., The cellulosome and cellulose degradation by anaerobic bacteria, *Appl. Microbiol. Biotechnol.,* 56, 634, 2001.
70. Lamed, R. and Bayer, E.A., The cellulosome concept: exocellular/extracellular enzyme reactor centers for efficient binding and cellulolysis, in *Biochemistry and Genetics of Cellulose Degradation. FEMS Symp. No. 43,* Aubert, J.-P., Béguin, P., and Millet, J., Eds., Academic, London, 1988, 101.
71. Lamed, R. and Bayer, E.A., The cellulosome of *Clostridium thermocellum, Adv. Appl. Microbiol.,* 33, 1, 1988.
72. Lamed, R. et al., The cellulosome — a discrete cell surface organelle of *Clostridium thermocellum* which exhibits separate antigenic, cellulose-binding and various catalytic activities, *Biotechnol. Bioeng. Symp.,* 13, 163, 1983.
73. Bayer, E.A., Kenig, R., and Lamed, R., Adherence of *Clostridium thermocellum* to cellulose, *J. Bacteriol.,* 156, 818, 1983.
74. Lamed, R., Setter, E., and Bayer, E.A., Characterization of a cellulose-binding, cellulase-containing complex in *Clostridium thermocellum, J. Bacteriol.,* 156, 828, 1983.
75. Bayer, E.A. and Lamed, R., Ultrastructure of the cell surface cellulosome of *Clostridium thermocellum* and its interaction with cellulose, *J. Bacteriol.,* 167, 828, 1986.
76. Bayer, E.A. et al., Cellulosomes–structure and ultrastructure, *J. Struct. Biol.,* 124, 221, 1998.
77. Doi, R.H. and Tamaru, Y., The *Clostridium cellulovorans* cellulosome: an enzyme complex with plant cell wall degrading activity, *Chem. Rec.,* 1, 24, 2001.
78. Kosugi, A. et al., Cell-surface-anchoring role of N-terminal surface layer homology domains of *Clostridium cellulovorans* EngE, *J. Bacteriol.,* 184, 884, 2002.
79. Lamed, R. et al., Specialized cell surface structures in cellulolytic bacteria, *J. Bacteriol.,* 169, 3792, 1987.
80. Mayer, F., Cellulolysis: ultrastructural aspects of bacterial systems, *Electron Microsc. Rev.,* 1, 69, 1988.
81. Mayer, F. et al., Macromolecular organization of the cellulolytic enzyme complex of *Clostridium thermocellum* as revealed by electron microscopy, *Appl. Environ. Microbiol.,* 53, 2785, 1987.
82. Wiegel, J. and Dykstra, M., *Clostridium thermocellum*: adhesion and sporulation while adhered to cellulose and hemicellulose, *Appl. Microbiol. Biotechnol.,* 20, 59, 1984.
83. Doi, R.H. et al., The *Clostridium cellulovorans* cellulosome, *Crit. Rev. Microbiol.,* 20, 87, 1994.
84. Doi, R.H. et al., Cellulosome and noncellulosomal cellulases of *Clostridium cellulovorans, Extremophiles,* 2, 53, 1998.
85. Fujino, T., Béguin, P., and Aubert, J.-P., Cloning of a *Clostridium thermocellum* DNA fragment encoding polypeptides that bind the catalytic components of the cellulosome, *FEMS Microbiol. Lett.,* 94, 165, 1992.
86. Gerngross, U.T. et al., Sequencing of a *Clostridium thermocellum* gene (*cipA*) encoding the cellulosomal S_L-protein reveals an unusual degree of internal homology, *Molec. Microbiol.,* 8, 325, 1993.
87. Kakiuchi, M. et al., Cloning and DNA sequencing of the genes encoding *Clostridium josui* scaffolding protein CipA and cellulase CelD and identification of their gene products as major components of the cellulosome, *J. Bacteriol.,* 180, 4303, 1998.
88. Lytle, B. et al., Interactions of the CelS binding ligand with various receptor domains of the *Clostridium thermocellum* cellulosomal scaffolding protein, CipA, *J. Bacteriol.,* 178, 1200, 1996.
89. Pagès, S. et al., Sequence analysis of scaffolding protein CipC and ORFXp, a new cohesin-containing protein in *Clostridium cellulolyticum*: comparison of various cohesin domains and subcellular localization of ORFXp, *J. Bacteriol.,* 181, 1801, 1999.
90. Park, J.S., Matano, Y., and Doi, R.H., Cohesin-dockerin interactions of cellulosomal subunits of *Clostridium cellulovorans, J. Bacteriol.,* 183, 5431, 2001.
91. Poole, D.M. et al., Identification of the cellulose binding domain of the cellulosome subunit S1 from *Clostridium thermocellum* YS, *FEMS Microbiol. Lett.,* 99, 181, 1992.
92. Salamitou, S. et al., Involvement of separate domains of the cellulosomal protein S1 of *Clostridium thermocellum* in binding to cellulose and in anchoring of catalytic subunits to the cellulosome, *FEBS Lett.,* 304, 89, 1992.

93. Takagi, M. et al., The hydrophobic repeated domain of the *Clostridium cellulovorans* cellulose-binding protein (CbpA) has specific interactions with endoglucanases, *J. Bacteriol.,* 175, 7119, 1993.
94. Tokatlidis, K. et al., Interaction of the duplicated segment carried by *Clostridium thermocellum* cellulases with cellulosome components, *FEBS Lett.,* 291, 185, 1991.
95. Wu, J.H.D. and Demain, A.L., Proteins of the *Clostridium thermocellum* cellulase complex responsible for degradation of crystalline cellulose, in *Biochemistry and Genetics of Cellulose Degradation. FEMS Symp. No. 43*, Aubert, J.-P., Béguin, P., and Millet, J., Eds., Academic, London, 1988, 117.
96. Wu, J.H.D., Orme-Johnson, W.H., and Demain, A.L., Two components of an extracellular protein aggregate of *Clostridium thermocellum* together degrade crystalline cellulose, *Biochemistry,* 27, 1703, 1988.
97. Yaron, S. et al., Expression, purification and crystallization of a cohesin domain from the cellulosome of *Clostridium thermocellum, J. Biotechnol.,* 51, 243, 1996.
98. Bayer, E.A. et al., Cellulose, cellulases and cellulosomes, *Curr. Opin. Struct. Biol.,* 8, 548, 1998.
99. Felix, C.R. and Ljungdahl, L.G., The cellulosome: the exocellular organelle of *Clostridium, Annu. Rev. Microbiol.,* 47, 791, 1993.
100. Kohring, S., Wiegel, J., and Mayer, F., Subunit composition and glycosidic activities of the cellulase complex from *Clostridium thermocellum* JW20, *Appl. Environ. Microbiol.,* 56, 3798, 1990.
101. Bayer, E.A., Morag, E., and Lamed, R., The cellulosome — a treasure-trove for biotechnology, *Trends Biotechnol.,* 12, 379, 1994.
102. Salamitou, S. et al., Recognition specificity of the duplicated segments present in *Clostridium thermocellum* endoglucanase CelD and in the cellulosome-integrating protein CipA, *J. Bacteriol.,* 176, 2822, 1994.
103. Kataeva, I., Guglielmi, G., and Beguin, P., Interaction between *Clostridium thermocellu*m endoglucanase CelD and polypeptides derived from the cellulosome-integrating protein CipA: stoichiometry and cellulolytic activity of the complexes, *Biochem. J.,* 326, 617, 1997.
104. Gerwig, G.J. et al., The Nature of the carbohydrate-peptide linkage region in glycoproteins from the cellulosomes of *Clostridium thermocellum* and *Bacteroides cellulosolvens, J. Biol. Chem.,* 268, 26956, 1993.
105. Leibovitz, E. and Béguin, P., A new type of cohesin domain that specifically binds the dockerin domain of the *Clostridium thermocellum* cellulosome-integrating protein CipA, *J. Bacteriol.,* 178, 3077, 1996.
106. Leibovitz, E. et al., Characterization and subcellular localization of the *Clostridium thermocellum* scaffoldin dockerin binding protein SdbA, *J. Bacteriol.,* 179, 2519, 1997.
107. Lemaire, M. et al., OlpB, a new outer layer protein of *Clostridium thermocellum*, and binding of its S-layer-like domains to components of the cell envelope, *J. Bacteriol.,* 177, 2451, 1995.
108. Salamitou, S. et al., Subcellular localization of *Clostridium thermocellum* ORF3p, a protein carrying a receptor for the docking sequence borne by the catalytic components of the cellulosome, *J. Bacteriol.,* 176, 2828, 1994.
109. Pohlschröder, M., Leschine, S., and Canale-Parola, E., *Spirochaeta caldaria* sp. nov., a thermophilic bacterium that enhances cellulose degradation by *Clostridium thermocellum, Arch. Microbiol.,* 161, 17, 1994.
110. Zverlov, V.V., Fuchs, K.P., and Schwarz, W.H., Chi18A, the endochitinase in the cellulosome of the thermophilic, cellulolytic bacterium *Clostridium thermocellum, Appl. Environ. Microbiol.,* 68, 3176, 2002.
111. Halstead, J.R. et al., A family 26 mannanase produced by *Clostridium thermocellum* as a component of the cellulosome contains a domain which is conserved in mannanases from anaerobic fungi, *Microbiology,* 145, 3101, 1999.
112. Hayashi, H. et al., Nucleotide sequences of two contiguous and highly homologous xylanase genes xynA and xynB and characterization of XynA from *Clostridium thermocellum, Appl. Microbiol. Biotechnol.,* 51, 348, 1999.
113. Blum, D.L. et al., Feruloyl esterase activity of the *Clostridium thermocellum* cellulosome can be attributed to previously unknown domains of XynY and XynZ, *J. Bacteriol.,* 182, 1346, 2000.
114. Bayer, E.A., Shoham, Y., and Lamed, R., Cellulose-decomposing bacteria and their enzyme systems, in *The Prokaryotes: An Evolving Electronic Resource for the Microbiological Community, 3rd Edition (latest update release 3.7, April 2001)*, Dworkin, M. et al., Eds., Springer-Verlag, New York, 2000.

115. Hazlewood, G.P. et al., Gene sequence and properties of CelI, a family E endoglucanase from *Clostridium thermocellum*, *J. Gen. Microbiol.,* 139, 307, 1993.
116. Schwarz, W.H. et al., Nucleotide sequence of the CelC gene encoding endoglucanase C of *Clostridium thermocellum*, *Gene,* 63, 23, 1988.
117. Kim, H., Jung, K.H., and Pack, M.Y., Molecular characterization of *xynX*, a gene encoding a multidomain xylanase with a thermostabilizing domain from *Clostridium thermocellum*, *Appl. Microbiol. Biotechnol.,* 54, 521, 2000.
118. Fuchs, K.P. et al., Lic16A of *Clostridium thermocellum*, a non-cellulosomal, highly complex endo-beta-1, 3-glucanase bound to the outer cell surface, *Microbiology,* 149, 1021, 2003.
119. Lamed, R. et al., Cellulosome-like entities in *Bacteroides cellulosolvens*, *Current Microbiol.,* 22, 27, 1991.
120. Gal, L. et al., Characterization of the cellulolytic complex (cellulosome) produced by *Clostridium cellulolyticum*, *Appl. Environ. Microbiol.,* 63, 903, 1997.
121. Shoseyov, O. and Doi, R.H., Essential 170-kDa subunit for degradation of crystalline cellulose by *Clostridium cellulovorans* cellulase, *Proc. Natl. Acad. Sci. U.S.A.,* 87, 2192, 1990.
122. Cavedon, K., Leschine, S.B., and Canale-Parola, E., Cellulase system of a free-living, mesophilic clostridium (strain C7), *J. Bacteriol.,* 172, 4222, 1990.
123. Cavedon, K., Leschine, S.B., and Canale-Parola, E., Characterization of the extracellular cellulase from a mesophilic clostridium (strain C7), *J. Bacteriol.,* 172, 4231, 1990.
124. Ding, S.Y. et al., A novel cellulosomal scaffoldin from *Acetivibrio cellulolyticus* that contains a family 9 glycosyl hydrolase, *J. Bacteriol.,* 181, 6720, 1999.
125. Ding, S.Y. et al., A scaffoldin of the *Bacteroides cellulosolvens* cellulosome that contains 11 type II cohesins, *J. Bacteriol.,* 182, 4915, 2000.
126. Tamaru, W. and Doi, R.H., Three surface layer homology domains at the N terminus of the *Clostridium cellulovorans* major cellulosomal subunit EngE, *J. Bacteriol.,* 181, 3270, 1999.
127. Liu, C.C. and Doi, R.H., Properties of *exgS*, a gene for a major subunit of the *Clostridium cellulovorans* cellulosome, *Gene,* 211, 39, 1998.
128. Xu, Q. et al., The cellulosome system of *Acetivibrio cellulolyticus* includes a novel type of adaptor protein and a cell surface anchoring protein, *J. Bacteriol.,* 185, 4548, 2003.
129. Nolling, J. et al., Genome sequence and comparative analysis of the solvent-producing bacterium *Clostridium acetobutylicum*, *J. Bacteriol.,* 183, 4823, 2001.
130. Sabathe, F., Belaich, A., and Soucaille, P., Characterization of the cellulolytic complex (cellulosome) of *Clostridium acetobutylicum*, *FEMS Microbiol. Lett.,* 217, 15, 2002.
131. Bagnara-Tardif, C. et al., Sequence analysis of a gene cluster encoding cellulases from *Clostridium cellulolyticum*, *Gene,* 119, 17, 1992.
132. Tamaru, Y. et al., A large gene cluster for the *Clostridium cellulovorans* cellulosome, *J. Bacteriol.,* 182, 5906, 2000.
133. Fujino, T., Béguin, P., and Aubert, J.-P., Organization of a *Clostridium thermocellum* gene cluster encoding the cellulosomal scaffolding protein CipA and protein possibly involved in attachment of the cellulosome to the cell surface, *J. Bacteriol.,* 175, 1891, 1993.
134. Guglielmi, G. and Béguin, P., Cellulase and hemicellulase genes of *Clostridium thermocellum* from five independent collections contain few overlaps and are widely scattered across the chromosome, *FEMS Microbiol. Lett.,* 161, 209, 1998.
135. Sabathe, F. and Soucaille, P., Characterization of the CipA scaffolding protein and *in vivo* production of a minicellulosome in *Clostridium acetobutylicum*, *J. Bacteriol.,* 185, 1092, 2003.
136. Murashima, K. et al., Heterologous production of *Clostridium cellulovorans engB*, using protease-deficient *Bacillus subtilis*, and preparation of active recombinant cellulosomes, *J. Bacteriol.,* 184, 76, 2002.
137. Bélaich, J.-P. et al., The cellulolytic system of *Clostridium cellulolyticum*, *J. Biotechnol.,* 57, 3, 1997.
138. Gaudin, C. et al., CelE, a multidomain cellulase from *Clostridium cellulolyticum*: a key enzyme in the cellulosome?, *J. Bacteriol.,* 182, 1910, 2000.
139. Parsiegla, G. et al., The crystal structure of the processive endocellulase CelF of *Clostridium cellulolyticum* in complex with a thiooligosaccharide inhibitor at 2.0 angstrom resolution, *EMBO J.,* 17, 5551, 1998.

140. Mohand-Oussaid, O. et al., The extracellular xylan degradative system in *Clostridium cellulolyticum* cultivated on xylan: evidence for cell-free cellulosome production, *J. Bacteriol.,* 181, 4035, 1999.
141. Murashima, K., Kosugi, A., and Doi, R.H., Determination of subunit composition of *Clostridium cellulovorans* cellulosomes that degrade plant cell walls, *Appl. Environ. Microbiol.,* 68, 1610, 2002.
142. Tamaru, Y. and Doi, R.H., The *engL* gene cluster of *Clostridium cellulovorans* contains a gene for cellulosomal ManA, *J. Bacteriol.,* 182, 244, 2000.
143. Kosugi, A., Murashima, K., and Doi, R.H., Xylanase and acetyl xylan esterase activities of XynA, a key subunit of the *Clostridium cellulovorans* cellulosome for xylan degradation, *Appl. Environ. Microbiol.,* 68, 6399, 2002.
144. Tamaru, Y. and Doi, R.H., Pectate lyase A, an enzymatic subunit of the *Clostridium cellulovorans* cellulosome, *Proc. Natl. Acad. Sci. USA,* 98, 4125, 2001.
145. Kosugi, A., Murashima, K., and Doi, R.H., Characterization of two noncellulosomal subunits, ArfA and BgaA, from *Clostridium cellulovorans* that cooperate with the cellulosome in plant cell wall degradation, *J. Bacteriol.,* 184, 6859, 2002.
146. Tamaru, Y. et al., Formation of protoplasts from cultured tobacco cells and *Arabidopsis thaliana* by the action of cellulosomes and pectate lyase from *Clostridium cellulovorans, App. Environ. Microbiol.,* 68, in press, 2002.
147. Murashima, K., Kosugi, A., and Doi, R.H., Synergistic effects of cellulosomal xylanase and cellulases from *Clostridium cellulovorans* on plant cell wall degradation, *J. Bacteriol.,* 185, 1518, 2003.
148. Leschine, S.B. and Canale-Parola, E., Mesophilic cellulolytic clostridia from freshwater environments, *Appl. Environ. Microbiol.,* 46, 728, 1983.
149. Pohlschröder, M., Leschine, S.B., and Canale-Parola, E., Multicomplex cellulase-xylanase system of *Clostridium papyrosolvens* C7, *J. Bacteriol.,* 176, 70, 1994.
150. Pohlschröder, M., Leschine, S., and Canale-Parola, E., Regulation of the multicomplex cellulase-xylanase system of *Clostridium papyrosolvens*, in *Genetics, Biochemistry and Ecology of Lignocellulose Degradation*, Shimada, K. et al., Eds., Uni, Tokyo, 1994, 86.
151. Pohlschröder, M., Canale-Parola, E., and Leschine, S., Ultrastructural diversity of the cellulase complexes of *Clostridium papyrosolvens* C7, *J. Bacteriol.,* 177, 6625, 1995.
152. Forsberg, C.W., Beveridge, T.J., and Hellstrom, A., Cellulase and xylanase release from *Bacteroides succinogenes* and its importance in the rumen environment, *Appl. Environ. Microbiol.,* 42, 886, 1981.
153. Hungate, R.E., Studies on cellulose fermentation. I. The culture and physiology of an anaerobic cellulose-digesting bacterium, *J. Bacteriol.,* 48, 499, 1944.
154. Hsing, W. and Canale-Parola, E., Cellobiose chemotaxis by the cellulolytic bacterium *Cellulomonas gelida, J. Bacteriol.,* 174, 7996, 1992.
155. Bronnenmeier, K. et al., Structure of the *Clostridium stercorarium* gene *celY* encoding the exo-1,4-beta-glucanase Avicelase II, *Microbiology,* 143, 891, 1997.
156. Bronnenmeier, K. and Staudenbauer, W.L., Cellulose hydrolysis by a highly thermostable endo-1,4-β-glucanase (Avicelase I) from *Clostridium stercorarium, Enzyme Microb. Technol.,* 12, 431, 1990.
157. Riedel, K. and Bronnenmeier, K., Active-site mutations which change the substrate specificity of the *Clostridium stercorarium* cellulase CelZ — Implications for synergism, *Eur. J. Biochem.,* 262, 218, 1999.
158. Bronnenmeier, K., Rucknagel, K.P., and Staudenbauer, W.L., Purification and properties of a novel type of exo-1,4-beta-glucanase (Avicelase-II) from the cellulolytic thermophile *Clostridium stercorarium, Eur. J. Biochem.,* 200, 379, 1991.
159. Riedel, K., Ritter, J., and Bronnenmeier, K., Synergistic interaction of the *Clostridium stercorarium* cellulases Avicelase I (CelZ) and Avicelase II (CelY) in the degradation of microcrystalline cellulose, *FEMS Microbiol. Lett.,* 147, 239, 1997.
160. Riedel, K. et al., The modular cellulase CelZ of the thermophilic bacterium *Clostridium stercorarium* contains a thermostabilizing domain, *FEMS Microbiol. Lett.,* 164, 261, 1998.
161. Reguera, G. and Leschine, S.B., Chitin degradation by cellulolytic anaerobes and facultative aerobes from soils and sediments, *FEMS Microbiol. Lett.,* 204, 367, 2001.
162. Muzzarelli, R.A.A., *Chitin,* Pergamon Press, Oxford, 1977.
163. Ahsan, M.M. et al., Cloning, DNA sequencing, and expression of the gene encoding *Clostridium thermocellum* cellulase CelJ, the largest catalytic component of the cellulosome, *J. Bacteriol.,* 178, 5732, 1996.

164. Kosugi, A., Murashima, K., and Doi, R.H., Characterization of xylanolytic enzymes in *Clostridium cellulovorans*: expression of xylanase activity dependent on growth substrates, *J. Bacteriol.,* 183, 7037, 2001.
165. Lee, S.F., Forsberg, C.W., and Rattray, J.B., Purification and characterization of two endoxylanases from *Clostridium acetobutylicum* ATCC 824, *Appl. Environ. Microbiol.,* 53, 644, 1987.
166. Lee, S.F. and Forsberg, C.W., Isolation and some properties of a beta-D-xylosidase from *Clostridium acetobutylicum* ATCC 824, *Appl. Environ. Microbiol.,* 53, 651, 1987.
167. Saxena, S. et al., Biochemical properties of a beta-xylosidase from *Clostridium cellulolyticum, Appl. Environ. Microbiol.,* 61, 3509, 1995.
168. Jindou, S. et al., Alpha-galactosidase Aga27A, an enzymatic component of the *Clostridium josui* cellulosome, *J. Bacteriol.,* 184, 600, 2002.
169. Feng, J.X. et al., Cloning, sequencing, and expression of the gene encoding a cell-bound multi-domain xylanase from *Clostridium josui,* and characterization of the translated product, *Biosci. Biotechnol. Biochem.,* 64, 2614, 2000.
170. Fukumura, M. et al., Nucleotide sequence of the *Clostridium stercorarium xynB* gene encoding an extremely thermostable xylanase, and characterization of the translated product, *Biosci. Biotechnol. Biochem.,* 59, 40, 1995.
171. Ali, M.K. et al., The multidomain xylanase Xyn10B as a cellulose-binding protein in *Clostridium stercorarium, FEMS Microbiol. Lett.,* 198, 79, 2001.
172. Sakka, K. et al., Nucleotide sequence of the *Clostridium stercorarium xylA* gene encoding a bifunctional protein with beta-D-xylosidase and alpha-L-arabinofuranosidase activities, and properties of the translated product, *Biosci. Biotechnol. Biochem.,* 57, 268, 1993.
173. Schwarz, W.H. et al., Debranching of arabinoxylan — properties of the thermoactive recombinant alpha-L-arabinofuranosidase from *Clostridium stercorarium* (ArfB), *Appl. Microbiol. Biotechnol.,* 43, 856, 1995.
174. Subramaniyan, S. and Prema, P., Cellulase-free xylanases from *Bacillus* and other microorganisms, *FEMS Microbiol. Lett.,* 183, 1, 2000.
175. Rani, D.S. and Nand, K., Purification and characterisation of xylanolytic enzymes of a cellulase-free thermophilic strain of *Clostridium absonum* CFR-702, *Anaerobe,* 7, 45, 2001.
176. Tamburini, E. et al., Characterization of bacterial pectinolytic strains involved in the water retting process, *Environ. Microbiol.,* 5, 730, 2003.
177. Keis, S., Shaheen, R., and Jones, D.T., Emended descriptions of *Clostridium acetobutylicum* and *Clostridium beijerinckii*, and descriptions of *Clostridium saccharoperbutylacetonicum* sp. nov. and *Clostridium saccharobutylicum* sp. nov., *Int. J. Syst. Evol. Microbiol.,* 51, 2095, 2001.
178. Tamburini, E. et al., *Clostridium felsineum* and *Clostridium acetobutylicum* are two distinct species that are phylogenetically closely related, *Int. J. Syst. Evol. Microbiol.,* 51, 963, 2001.
179. Potter, L.F. and McCoy, E., The fermentation of pectin and pectic acid by *Clostridium felsineum, J. Bacteriol.,* 64, 701, 1952.
180. Lanigan, G.W., Studies on the pectinolytic anaerobes *Clostridium flavum* and *Clostridium laniganii, J. Bacteriol.,* 77, 1, 1959.
181. Chin, K.J. et al., Methanogenic degradation of polysaccharides and the characterization of polysaccharolytic clostridia from anoxic rice field soil, *Syst. Appl. Microbiol.,* 21, 185, 1998.
182. Ng, H. and Vaughn, R.H., *Clostridium rubrum* sp. nov. and other pectinolytic clostridia from soil, *J. Bacteriol.,* 85, 1104, 1963.
183. Schink, B. and Zeikus, J.G., Microbial ecology of pectin decomposition in anoxic lake sediments, *J. Gen. Microbiol.,* 128, 393, 1982.
184. Schink, B., Ward, J.C., and Zeikus, J.G., Microbiology of wetwood — importance of pectin degradation and *Clostridium* species in living trees, *Appl. Environ. Microbiol.,* 42, 526, 1981.
185. Heinrichova, K. and Stachova, M., Pectinolytic enzymes of *Clostridium lochheadii, Biologia,* 49, 813, 1994.
186. Nakajima, N. et al., Degradation of pectic substances by two pectate lyases from a human intestinal bacterium, *Clostridium butyricum-beijerinckii* group, *J. Biosci. Bioeng.,* 88, 331, 1999.
187. Varel, V.H. and Pond, W.G., Characteristics of a new cellulolytic *Clostridium* sp. isolated from pig intestinal tract, *Appl. Environ. Microbiol.,* 58, 1645, 1992.

188. Schink, B. and Zeikus, J.G., Microbial methanol formation — a major end product of pectin metabolism, *Curr. Microbiol.*, 4, 387, 1980.
189. Collmer, A., Ried, J.L., and Mount, M.S., Assay methods for pectic enzymes, in *Methods in Enzymology, Volume 161. Biomass, Part B: Lignin, Pectin, and Chitin*, Wood, W.A. and Kellogg, S.T., Eds., Academic Press, San Diego, 1988, 329.
190. Pagès, S. et al., A rhamnogalacturonan lyase in the *Clostridium cellulolyticum* cellulosome, *J. Bacteriol.*, 185, 4727, 2003.
191. Zverlov, V.V. et al., The thermostable alpha-L-rhamnosidase RamA of *Clostridium stercorarium*: biochemical characterization and primary structure of a bacterial alpha-L-rhamnoside hydrolase, a new type of inverting glycoside hydrolase, *Mol. Microbiol.*, 35, 173, 2000.
192. vanRijssel, M., Gerwig, G.J., and Hansen, T.A., Isolation and characterization of an extracellular glycosylated protein complex from *Clostridium thermosaccharolyticum* with pectin methylesterase and polygalacturonate hydrolase activity, *Appl. Environ. Microbiol.*, 59, 828, 1993.
193. vanRijssel, M. et al., Involvement of an intracellular oligogalacturonate hydrolase in metabolism of pectin by *Clostridium thermosaccharolyticum*, *Appl. Environ. Microbiol.*, 59, 837, 1993.
194. Sheiman, M.I. et al., Coordinated action of pectinesterase and polygalacturonate lyase complex of *Clostridium multifermentans*, *Eur. J. Biochem.*, 64, 565, 1976.
195. Bertoldo, C. and Antranikian, G., Starch-hydrolyzing enzymes from thermophilic archaea and bacteria, *Curr. Opin. Chem. Biol.*, 6, 151, 2002.
196. Lee, Y.E. et al., Taxonomic distinction of saccharolytic thermophilic anaerobes: description of *Thermoanaerobacterium xylanolyticum* gen. nov., sp. nov., and *Thermoanaerobacterium saccharolyticum* gen. nov., sp. nov.; reclassification of *Thermoanaerobium brockii*, *Clostridium thermosulfurogenes*, and *Clostridium thermohydrosulfuricum* E100-69 as *Thermoanaerobacter brockii* comb. nov., *Thermoanaerobacterium thermosulfurigenes* comb. nov., and *Thermoanaerobacter thermohydrosulfuricus* comb. nov, respectively; and transfer of *Clostridium thermohydrosulfuricum* 39E to *Thermoanaerobacter ethanolicus*, *Int. J. Syst. Bacteriol.*, 43, 41, 1993.
197. Melasniemi, H., Purification and some properties of the extracellular alpha-amylase pullulanase produced by *Clostridium thermohydrosulfuricum*, *Biochem. J.*, 250, 813, 1988.
198. Saha, B.C., Mathupala, S.P., and Zeikus, J.G., Purification and characterization of a highly thermostable novel pullulanase from *Clostridium thermohydrosulfuricum*, *Biochem. J.*, 252, 343, 1988.
199. Spreinat, A. and Antranikian, G., Purification and properties of a thermostable pullulanase from *Clostridium thermosulfurogenes* EM1 which hydrolyzes both alpha-1,6 and alpha-1,4-glycosidic linkages, *Appl. Microbiol. Biotechnol.*, 33, 511, 1990.
200. Matuschek, M. et al., Pullulanase of *Thermoanaerobacterium thermosulfurigenes* EM1 (*Clostridium thermosulfurogenes*): molecular analysis of the gene, composite structure of the enzyme, and a common model for its attachment to the cell surface, *J. Bacteriol.*, 176, 3295, 1994.
201. Antranikian, G., Herzberg, C., and Gottschalk, G., Production of thermostable alpha-amylase, pullulanase, and alpha-glucosidase in continuous culture by a new *Clostridium* isolate, *Appl. Environ. Microbiol.*, 53, 1668, 1987.
202. Antranikian, G. et al., Changes in the cell envelope structure of *Clostridium* sp. strain EM1 during massive production of alpha-amylase and pullulanase, *FEMS Microbiol. Lett.*, 41, 193, 1987.
203. Reddy, P.R.M., Swamy, M.V., and Seenayya, G., Purification and characterization of thermostable beta-amylase and pullulanase from high-yielding *Clostridium thermosulfurogenes* SV2, *World J. Microbiol. Biotechnol.*, 14, 89, 1998.
204. Shen, G.J. et al., Purification and characterization of a novel thermostable beta-amylase from *Clostridium thermosulfurogenes*, *Biochem. J.*, 254, 835, 1988.
205. Bahl, H. et al., Alpha-amylase of *Clostridium thermosulfurogenes* EM1: nucleotide sequence of the gene, processing of the enzyme, and comparison to other alpha-amylases, *Appl. Environ. Microbiol.*, 57, 1554, 1991.
206. Spreinat, A. and Antranikian, G., Analysis of the amylolytic enzyme system of *Clostridium thermosulfurogenes* EM1: purification and synergistic action of pullulanases and maltohexaose forming alpha-amylase, *Starch-Starke*, 44, 305, 1992.
207. Hyun, H.H. and Zeikus, J.G., General biochemical characterization of thermostable pullulanase and glucoamylase from *Clostridium thermohydrosulfuricum*, *Appl. Environ. Microbiol.*, 49, 1168, 1985.

208. Specka, U., Mayer, F., and Antranikian, G., Purification and properties of a thermoactive glucoamylase from *Clostridium thermosaccharolyticum*, *Appl. Environ. Microbiol.*, 57, 2317, 1991.
209. Ganghofner, D. et al., Purification and properties of an amylopullulanase, a glucoamylase, and an alpha-glucosidase in the amylolytic enzyme system of *Thermoanaerobacterium thermosaccharolyticum*, *Biosci. Biotechnol. Biochem.*, 62, 302, 1998.
210. Feng, P.H. et al., Production, purification, and characterization of *Thermoanaerobacterium thermosaccharolyticum* glucoamylase, *Starch-Starke,* 54, 328, 2002.
211. Aleshin, A.E. et al., Crystal structure and evolution of a prokaryotic glucoamylase, *J. Mol. Biol.,* 327, 61, 2003.
212. Annous, B.A. and Blaschek, H.P., Isolation and characterization of alpha-amylase derived from starch-grown *Clostridium acetobutylicum* ATCC 824, *J. Indust. Microbiol.,* 13, 10, 1994.
213. Paquet, V. et al., Purification and characterization of the extracellular alpha-amylase from *Clostridium acetobutylicum* ATCC-824, *Appl. Environ. Microbiol.,* 57, 212, 1991.
214. Albasheri, K.A. and Mitchell, W.J., Identification of 2 alpha-glucosidase activities in *Clostridium acetobutylicum* NCIB 8052, *J. Appl. Bacteriol.,* 78, 149, 1995.
215. Shih, N.J. and Labbe, R.G., Purification and characterization of an extracellular alpha-amylase from *Clostridium perfringens* type-A, *Appl. Environ. Microbiol.,* 61, 1776, 1995.
216. Wang, X. et al., *In vitro* utilization of amylopectin and high-amylose maize (amylomaize) starch granules by human colonic bacteria, *Appl. Environ. Microbiol.,* 65, 4848, 1999.
217. Weaver, G.A. et al., Acarbose enhances human colonic butyrate production, *J. Nutr.,* 127, 717, 1997.
218. Wolin, M.J. et al., Changes of fermentation pathways of fecal microbial communities associated with a drug treatment that increases dietary starch in the human colon, *Appl. Environ. Microbiol.,* 65, 2807, 1999.
219. Pryde, S.E. et al., The microbiology of butyrate formation in the human colon, *FEMS Microbiol. Lett.,* 217, 133, 2002.
220. Kulkarni, N., Shendye, A., and Rao, M., Molecular and biotechnological aspects of xylanases, *FEMS Microbiol. Rev.,* 23, 411, 1999.
221. Rani, D.S. and Nand, K., Production of thermostable cellulase-free xylanase by *Clostridium absonum* CFR-702, *Process Biochem.,* 36, 355, 2000.
222. Golovchenko, N.P. et al., Isolation and characterization of a lichenan-degrading hydrophobic endoglucanase of *Clostridium thermocellum*, *Appl. Microbiol. Biotechnol.,* 39, 74, 1993.
223. Akin, D.E., Himmelsbach, D.S., and Morrison, W.H., Biobased fiber production: enzyme retting for flax/linen fibers, *J. Polym. Environ.,* 8, 103, 2000.
224. Zhang, J., Henriksson, G., and Johansson, G., Polygalacturonase is the key component in enzymatic retting of flax, *J. Biotechnol.,* 81, 85, 2000.
225. Fierobe, H.P. et al., Design and production of active cellulosome chimeras — selective incorporation of dockerin-containing enzymes into defined functional complexes, *J. Biol. Chem.,* 276, 21257, 2001.
226. Fierobe, H.P. et al., Degradation of cellulose substrates by cellulosome chimeras — substrate targeting versus proximity of enzyme components, *J. Biol. Chem.,* 277, 49621, 2002.
227. Murashima, K., Kosugi, A., and Doi, R.H., Solubilization of cellulosomal cellulases by fusion with cellulose-binding domain of noncellulosomal cellulase EngD from *Clostridium cellulovorans*, *Proteins,* 50, 620, 2003.
228. Dürre, P., From Pandora's box to cornucopia: clostridia — a historical perspective, in *Clostridia: Biotechnology and Medical Applications*, Bahl, H. and Dürre, P., Eds., Wiley-VCH, Weinheim, Germany, 2001, 1.
229. Qureshi, N. and Blaschek, H.P., Recent advances in ABE fermentation: hyper-butanol producing *Clostridium beijerinckii* BA101, *J. Ind. Microbiol. Biotechnol.,* 27, 287, 2001.
230. Montoya, D. et al., New solvent-producing *Clostridium* sp. strains, hydrolyzing a wide range of polysaccharides, are closely related to *Clostridium butyricum*, *J. Ind. Microbiol. Biotechnol.,* 27, 329, 2001.
231. Yang, J.C. et al., *Clostridium aldrichii* sp. nov., a cellulolytic mesophile inhabiting a wood-fermenting anaerobic digester, *Int. J. Syst. Bacteriol.,* 40, 268, 1990.
232. Petitdemange, E. et al., *Clostridium cellulolyticum* sp. nov., a cellulolytic mesophilic species from decayed grass, *Int. J. Syst. Bacteriol.,* 34, 155, 1984.

233. Monserrate, E., Leschine, S.B., and Canale-Parola, E., *Clostridium hungatei* sp. nov., a mesophilic, N_2-fixing cellulolytic bacterium isolated from soil, *Int. J. Syst. Evol. Microbiol.,* 51, 123, 2001.
234. Sukhumavasi, J. et al., *Clostridium josui* sp. nov., a cellulolytic, moderate thermophilic species from Thai compost, *Int. J. Syst. Bacteriol.,* 38, 179, 1988.
235. Madden, R.H., Bryder, M.J., and Poole, N.J., Isolation and characterization of an anaerobic, cellulolytic bacterium, *Clostridium papyrosolvens* sp. nov., *Int. J. Syst. Bacteriol.,* 32, 87, 1982.
236. Madden, R.H., Isolation and characterization of *Clostridium stercorarium* sp. nov., cellulolytic thermophile, *Int. J. Syst. Bacteriol.,* 33, 837, 1983.
237. Hethener, P., Brauman, A., and Garcia, J.L., *Clostridium termitidis* sp. nov., a cellulolytic bacterium from the gut of the wood-feeding termite, *Nasutitermes lujae*, *Syst. Appl. Microbiol.,* 15, 52, 1992.
238. Wiegel, J., Formation of ethanol by bacteria: a pledge for the use of extreme thermophilic anaerobic bacteria in industrial ethanol fermentation processes, *Experientia,* 36, 1434, 1980.
239. Palop, M.L.L. et al., Isolation and characterization of an anaerobic, cellulolytic bacterium, *Clostridium celerecrescens* sp. nov., *Int. J. Syst. Bacteriol.,* 39, 68, 1989.
240. Varel, V.H., Tanner, R.S., and Woese, C.R., *Clostridium herbivorans* sp. nov., a cellulolytic anaerobe from the pig intestine, *Int. J. Syst. Bacteriol.,* 45, 490, 1995.
241. van Gylswyk, N.O., *Fusobacterium polysaccharolyticum* sp. nov., a gram-negative rod from the rumen that produces butyrate and fements cellulose and starch, *J. Gen. Microbiol.,* 116l, 157, 1980.
242. van Gylswyk, N.O., Morris, E.J., and Els, H.J., Sporulation and cell wall structure of *Clostridium polysaccharolyticum* comb. nov. (formerly *Fusobacterium polysaccharolyticum*), *J. Gen. Microbiol.,* 121, 491, 1980.
243. Sleat, R. and Mah, R.A., *Clostridium populeti* sp. nov., a cellulolytic species from a woody-biomass digestor, *Int. J. Syst. Bacteriol.,* 35, 160, 1985.
244. Murray, W.D. et al., *Clostridium lentocellum* sp. nov., a cellulolytic species from river sediment containing paper-mill waste, *Syst. Appl. Microbiol.,* 8, 181, 1986.

7 Genetic Organization and Regulation of Hexose and Pentose Utilization in the Clostridia

Sharon J. Reid

CONTENTS

7.1 CARBOHYDRATE UTILIZATION BY CLOSTRIDIA

The clostridia are anaerobic, Gram-positive, endospore-forming bacilli and are comprised of a large number of species, most of which were originally assigned to this group because they exhibit these three characteristics. This method of strain classification has resulted in the assignation of a heterogenous group of bacteria to the genus *Clostridium*. Bacteria from this group are therefore capable of degrading a large range of carbohydrate substrates and of exhibiting a wide variety of metabolic activities [1]. The classification of this genus is still under review, particularly since molecular taxonomic techniques have been used to re-examine the status of some of these strains [2]. The saccharolytic clostridia are a good example of recent, more rational reorganization of species assignment, and the group of bacteria previously designated as *Clostridium acetobutylicum* has now been divided into four different taxonomic clusters [3–5]. Although studies on clostridia have contributed greatly to our understanding of many metabolic pathways [6], knowledge of these activities is far from complete. However, as more molecular data regarding these processes become available, it is useful to compare the genetic structure and organization of the pathways found in different *Clostridium* species, so that common regulatory patterns become more apparent and evolutionary history can be deduced.

The saccharolytic clostridia have received much attention over the past years, particularly those that have biotechnological application and are capable of producing industrially important chemicals [7,8]. A number of bacterial strains, originally classified as *Clostridium acetobutylicum,* were isolated in different parts of the world from soil or vegetable material for the purpose of converting carbohydrate material into the industrially important acetone and butanol [7]. The soil environment contains a variety of macromolecular material, including a variety of carbohydrates derived from plant, animal, and microbial material. *Clostridium* species are able to utilize many of these carbohydrate sources and produce a battery of enzymes to facilitate this process. Many of these are extracellular enzymes that are excreted into the environment to undertake the breakdown of the polymeric carbohydrates into smaller entities that can be taken up by the bacterium and metabolized [9,10].

The oligo-, di-, or monosaccharides produced in this way are transported into the cell via specific membrane-bound transport systems [11], and the carbohydrates are subsequently catabolized via glycolysis or the pentose phosphate pathway. The monosaccharides consist of numerous hexose sugars, including hexitols and hexuronic acids, and several pentoses, namely arabinose, xylose, and ribose, while the major disaccharides are comprised mainly of sucrose, lactose, maltose, and cellobiose. Genomic sequence data released recently indicated that the genome of *C. acetobutylicum* has more than 90 genes that are involved in the degradation of carbohydrate substances, including at least 14 distinct families of glycosyl hydrolases [12]. Likewise, *Clostridium perfringens* has the genetic potential to utilize a wide variety of carbohydrate substrates [13]. Carbohydrates are transported into the cell by specific membrane transport proteins, which can consist of one or more protein components but typically have one region of the protein that spans the membrane with multiple transmembrane helical segments. Usually, a large number of these transport systems are present in any one bacterium. For instance, *Bacillus subtilis* has a total of about 50 different carbohydrate transporters that have been reported or identified from the genome sequence [14], and it is possible that not all of the permeases in *B. subtilis* have been identified. Similarly, the genome sequence of *Bifidobacterium lactis* showed that 8.5% of the genes identified were concerned with carbohydrate transport [15].

The bacterial cell accumulates mono- and disaccharides via different transport mechanisms. The majority of reports are of transport by the phosphoenolpyruvate-dependent phosphotransferase system (PTS) [11,16,17]. During the translocation of the carbohydrate from the exterior of the cell to the interior, it is chemically modified by phosphorylation, effectively performing two functions simultaneously. In some cases, the phosphorylated substrate that accumulates is immediately able to enter the glycolytic pathway, while in others, additional conversions have to take place before the metabolite can be incorporated. The occurrence of these transport systems in *Clostridium* is reviewed in Chapter 8. There are other mechanisms for carbohydrate transport that result in substrate accumulation without chemical modification, and these include ATP-driven transporters and other non-PTS permeases [11]. Although there is evidence of both PTS and non-PTS transport systems in the genome sequence data from *C. acetobutylicum* 824 [12], the molecular mechanisms of transport in *Clostridium* are still not well characterized.

The genes comprising carbohydrate catabolic operons are usually carefully regulated by the cell, so they are only expressed when their substrate is present and when preferred carbon sources are depleted. This is designed to maximize efficiency and to save cellular energy. Physiological studies have shown that both *Clostridium beijerinckii* NCIMB 8052 and *Clostridium saccharobutylicum* NCP262 metabolize glucose in preference to disaccharides. In *C. beijerinckii*, when glucose was present with either cellobiose, lactose, or maltose, the utilization of the disaccharide was completely prevented until the glucose levels had dropped sufficiently [18]. Similarly, the sucrose operon of *C. beijerinckii* was not expressed in the absence of sucrose, and sucrose was not utilized in the presence of glucose until the glucose was depleted [19]. In *C. saccharobutylicum* NCP262, a similar pattern emerges, and glucose repression of galactose utilization [20] and of cellobiose uptake and utilization [21] has been reported. The mechanism that prevails to facilitate this induction and repression is mediated by the operon-specific transcriptional repressors, in conjunction with the global carbon catabolite repression proteins. In many of the Gram-positive bacteria, this mechanism is implemented by the CcpA protein [22]. The CcpA protein represses transcription of catabolic operons by binding to palindromic sequences called CRE sites, usually situated in the promoter region upstream of the structural genes themselves. The *ccpA* gene (previously called *regA*), has been isolated from both *C. saccharobutylicum* [23] and *C. beijerinckii* [24]. The *C. saccharobutylicum ccpA* gene was capable of complementing a *B. subtilis ccpA* mutant, and restored amylase repression and acetoin production to the mutant [23]. Negative transdominance studies, such as those performed in *B. subtilis* [25], involving overexpression of the *C. beijerinckii ccpA* gene in a wildtype strain, led to an increase in glucose repression of sucrose utilization in a manner analogous to the *B. subtilis ccpA* studies [24].

Most of the research on carbohydrate substrate utilization in *Clostridium* has been focused on the solventogenic species, because of their biotechnological application. The substrates that have been investigated are mainly those classified as renewable resources, such as cellulose, hemicellulose, and starch [9,10]. However, less is known about the low molecular weight carbohydrate compounds, which are accumulated in the cell and are catabolized for cell growth and energy. It is important to understand both the physiology and the genetic regulation of the carbohydrate utilization pathways in order to design improved strains for optimal substrate conversion to useful products. Clostridial carbohydrate metabolism, and the cloning of several carbohydrate catabolic genes from solventogenic clostridial strains, has been previously reviewed [10,26,27]. This review will focus on the current knowledge of the genetic basis of carbohydrate utilization, using, in particular, the examples that have proven functional analysis for comparative studies with DNA sequence data from genome sequencing projects [12,13].

7.2 UTILIZATION OF MONOSACCHARIDES

In general, the saccharolytic bacteria, such as *Clostridium acetobutylicum,* can degrade most of the sugars occurring in plant biomass. However, hexoses and especially glucose are the most efficient substrates for the production of cellular energy and growth. The metabolic pathways by which the saccharolytic clostridia convert monosaccharides to acids and solvents have been extensively reviewed [7,28–30].

After their entry into the cell, all carbohydrates are converted into an intermediate of one of the central carbohydrate-degrading pathways: the glycolytic or Embden-Meyerhof-Parnas pathway, the pentose-phosphate pathway, or the Entner-Doudoroff (KDPG) pathway [13,31]. Depending on how a carbohydrate enters the cycle, different enzymes can be required to convert it to a glycolytic intermediate. Glycolysis, which results in the production of pyruvate, one of the key metabolic intermediates in bacterial cells, is one of the most conserved pathways in living organisms [32]. This has facilitated the identification of many components of the pathways from genomic DNA sequence data from different bacteria. Functionality of the proteins has been determined either experimentally or has been deduced by comparison to similar enzymes from other bacteria. Genes encoding enzymes from the EMP pathway have been cloned and characterized from *C. acetobutylicum* ATCC 824 [33]. These authors show that the genes encoding the phosphofructokinase and the pyruvate kinase are organized as an operon on the chromosome and appear to be cotranscribed. A cluster of genes encoding glycolytic enzymes has been isolated from *C. acetobutylicum* DSM 792 [34]. DNA sequencing of the *gap* (glyceraldehyde 3-phosphate dehydrogenase, GAPDH) gene and its flanking regions revealed that the genes encoding the phosphoglycerate kinase (*pgk*), triosephosphate isomerase (*tpi*), and phosphoglycerate mutase (*pgm*) lie downstream of the *gap* gene, in the order *gap-pgk-tpi-pgm*. They are all orientated in the same direction and, as there are no typical rho-independent terminator sequences in the intergenic regions, are probably cotranscribed. The *C. acetobutylicum* GAPDH protein has extremely high sequence homology to the protein from *C. pasteurianum* (95%), as well as significant homology to the deduced GAPDH proteins from *C. perfringens* (76%) and *C. tetani* (54%). In agreement with the observations of Dandekar et al. [32], the genetic organization of *gap-pgk-tpi* in *C. acetobutylicum* is found to be completely conserved in both *C. perfringens* and *C. tetani*, as well as in *B. subtilis*, *B. megaterium*, and the more distantly related strains, *Corynebacterium glutamicum* and *Lactobacillus delbrueckii* [34].

Hexitols, such as glucitol and mannitol, have been studied in different strains of *Clostridium*, although glucose is still the preferred substrate in mixtures of sugars [17]. The *gutB* and *gutD* genes were cloned from *C. pasteurianum*, and confirmed to encode the $EIIA^{glucitol}$ subunit of the PTS and the glucitol 6-phosphate dehydrogenase enzyme, respectively [26]. *C. beijerinckii* NCIMB 8052 was shown to possess the same gene order of *gutB-gutD*, which were found to be located in a gene cluster of glucitol PTS genes, in the order *gutA1-gutA2-orfX-gutB-gutD*. The function of *orfX* is unknown, but all of the *gut* genes were demonstrated to be induced by glucitol and repressed by glucose [36].

7.3 UTILIZATION OF DISACCHARIDES

7.3.1 Sucrose Degradation

Sucrose, because of its origin in higher plant tissues, is generally the disaccharide most commonly available for bacterial degradation. It is accumulated by bacterial cells by means of sucrose-specific transport systems. The predominant mechanism facilitating sucrose uptake in Gram-positive bacteria is the PEP-dependent sucrose PTS, but other sugar permeases have been reported recently, including ABC-transporters, which have been shown in *Streptococcus* to transport sugars such as maltose [11], and solute:cation symport systems, such as those encoded by the *scrT* gene in the sucrose operon in *Bifidobacterium lactis* [36] and the *lacS* gene in the lactose operon of *Streptococcus thermophilus* [37]. The PTS is composed of two energy-coupling proteins, Enzyme I and HPr, and several sugar-specific Enzyme II proteins [11]. Enzyme II proteins typically consist of three protein domains, EIIA, EIIB, and EIIC, and transport and phosphorylation of the sugar occurs when a phosphoryl group donated by PEP is passed via EI, HPr, EIIA, EIIB, and EIIC to the incoming sugar. The translocation of the sugar through the membrane is facilitated by the integral membrane domain $EIIC^{sugar}$. The organization of the EII domains varies, and may consist of a single fused protein or different fused and unfused domains, but it is the complex of the three enzyme domains or proteins, functioning together, that brings about the transport and phosphorylation of the sugar substrate.

Depending on which type of transport system is present in a particular bacterium, the sucrose will either accumulate inside the cell in the form of sucrose-6-P or, if the mechanism was non-PTS, as unmodified sucrose. A number of different enzymes catalyze the hydrolysis of intracellular sucrose. The β-fructofuranosidases of the glucosyl hydrolase family 32 constitutes possibly the largest group of enzymes capable of sucrose hydrolysis, and included are proteins from archaebacteria, eubacteria, and eukaryotes [38–40]. These enzymes catalyze the hydrolysis of terminal fructose moieties from fructose-containing disaccharides, trisaccharides, and polymers, with varying specificity. Some examples, such as levanases and inulinases, have increased specificity against high molecular weight fructose polymers, while others specifically hydrolyze low molecular weight fructose-containing sugars such as sucrose and raffinose. Within this group are sucrose-6-phosphate hydrolases associated with PTS-dependent sucrose utilization systems [41,42].

Sucrose phosphorylases also catalyze the hydrolysis of sucrose and simultaneously phosphorylate the glucose residue using inorganic phosphate to yield glucose-1-phosphate and fructose.

Examples of sucrose phosphorylases have been identified in several bacteria, including *Streptococcus mutans, Leuconostoc mesenteroides,* and *Bifidobacterium lactis* [36,43,44]. In both cases, the fructose resulting from the hydrolysis of either sucrose or sucrose-6-P is phosphorylated by ATP-dependent fructokinases to give fructose-6-P, which can be incorporated into the glycolytic pathway. Regulatory proteins, typically of the GalR-LacI family of transcriptional regulators, are usually associated with the sucrose utilization genes in order to induce or repress the structural genes, depending on the substrates present.

7.3.1.1 Genetic Organization of Sucrose Utilization in *C. beijerinckii*

In the *Clostridium* group, the first molecular analysis of genes involved in sucrose utilization was undertaken in *C. beijerinckii* NCIMB 8052 [19]. A genomic library of *C. beijerinckii* chromosomal DNA was screened in *E. coli* for clones able to utilize sucrose, indicating that the sucrose hydrolase was functional in the Gram-negative heterologous host. The presence of the *scr* genes on a plasmid conferred both sucrose hydrolase and fructokinase enzyme activities on the recombinant *E. coli* cells and showed that the sucrose hydrolase was able to catalyze the hydrolysis of sucrose and raffinose but not melezitose, inulin, and levan. The sucrase-containing region of the chromosome was sequenced and found to be comprised of four genes, the *scrARBK* genes (Figure 7.1).

DNA sequence analysis revealed that the most closely related proteins in every case were the deduced proteins from an analogous operon identified from the genome sequence of *C. perfringens*

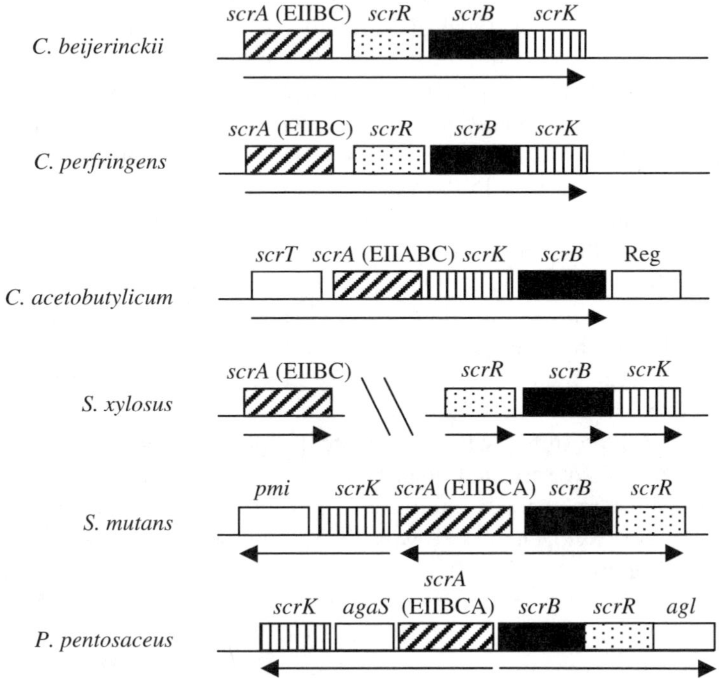

FIGURE 7.1 Comparison of the gene arrangement of the sucrose utilization locus found in different Gram-positive bacteria: *C. beijerinckii* NCIMB 8052 [19], *C. perfringens* strain 13 (NCBI: NC 003366.1) [13], *C. acetobutylicum* ATCC 824 (NCBI: NC 003030.1) [12,56], *S. xylosus* [51], *S. mutans* [65], and *P. pentosaceus* [66]. Genes have been assigned identities based either on their gene names or according to the probable gene products: *scrA,* PTS EII transport protein; *scrR,* sucrose regulator; *scrB,* sucrose-6-P hydrolase; *scrK,* fructokinase; *scrT,* putative transcriptional regulator with antitermination activity; Reg, predicted transcriptional regulator; *pmi,* mannose-6-P isomerase; *agaS,* α-galactosidase; *agl,* α-glucosidase.

TABLE 7.1
Sequence Comparison of the Deduced Proteins of the Sucrose Utilization Operons from Three Closely Related Clostridium Species

Gene	Amino Acid Sequence Identity to Corresponding *C. beijerinckii scr* Proteins (%)	
	C. perfringens	*C. acetobutylicum*
scrA	65	34[a]
scrR	58	—
scrB	50	37
scrK	63	53

[a] % identity over the corresponding EIIBC region of protein only.

(Acc. No: NC 003366.1), which ranged from 50 to 65% amino acid identity (Table 7.1). The *C. beijerinckii scrA* gene specifies a deduced protein of 451aa with strong homology to PTS $EIIBC^{scr}$ proteins, with the highest amino acid sequence identity to a number of *Streptococcus* species, including *S. sobrinus* ScrA (55.1% identity over the corresponding EIIBC domains). Several particularly well-conserved regions have been identified for $EIIBC^{scr}$ proteins, which are all present in the *C. beijerinckii* ScrA [45]. The putative *C. beijerinckii* ScrB protein showed homology to β-fructofuranosidases of the glucosyl hydrolase family 32 [39], with highest identity to a group of sucrose-6-phosphate hydrolases from *B. subtilis, B. anthracis,* and *B. cereus* (36%) [46,47], *Staphylococcus xylosus* (35%) [48], and *Vibrio alginolyticus* (35.5%) [49]. This would suggest that the product of the *scrB* gene normally hydrolyzes sucrose-6-phosphate rather than sucrose. The deduced ScrK protein from *C. beijerinckii* showed significant homology to several prokaryotic ATP/Mg^{2+}-dependent fructokinases [50] from the low G+C Gram-positive bacteria, including *B. cereus*, *B. anthracis*, and *Staphylococcus aureus* (47 to 55% identity), as well as to a group of eukaryotic fructokinase proteins from *Zea mays* and *Arabidopsis thaliana* (30% aa identity).

The deduced ScrR protein showed relatively low but significant homology to members of the GalR-LacI family of transcriptional regulators: 42% identity to *B. cereus* and *B. anthracis,* 31.8% identity to *S. xylosus,* and 26% identity to *Pediococcus pentosaceous* ScrR proteins, respectively. The most highly conserved region of the regulatory protein was the N-terminal helix-turn-helix DNA binding motif, but in addition, the *C. beijerinckii* and *S. xylosus* ScrR proteins share several other conserved domains, presumably the sucrose-6-phosphate binding regions [51]. Phylogenetic analyses of the *C. beijerinckii* ScrR and other GalR-LacI family members showed that within the GalR-LacI family, sucrose regulatory proteins have evolved separately and are represented on four different branches of the phylogenetic tree with the *C. beijerinckii, S. xylosus,* and *C. perfringens* proteins indicating a common origin [19].

The *scrARBK* gene cluster would therefore contain all of the genes required for the transport of sucrose into the cell and the subsequent hydrolysis of the substrate to assimilable products, as well as regulation of the operon. The *scrA* gene encodes only the $EIIBC^{sucrose}$ domains of the sucrose PTS, but could possibly be complemented by the $EIIA^{glu}$ protein as has been demonstrated in *B. subtilis*, where the $EIIA^{glu}$ protein supplements the $EIIBC^{sucrose}$ for the transport of sucrose across the membrane [52]. The hydrolysis of the sucrose-6-P would be accomplished by the ScrB protein, and the fructokinase presumably would be responsible for phosphorylation of the fructose generated, prior to it being incorporated into the glycolytic pathway.

7.3.1.2 Transcriptional Regulation of the *C. beijerinckii scr* Operon

The functionality of the regulatory gene, *scrR,* and the mechanism of regulation of the sucrose gene cluster, was clarified by RNA and enzyme analysis [19]. In *C. beijerinckii,* both sucrose hydrolase and fructokinase enzyme activities were very low in the presence of glucose, fructose, or maltose and were only induced by sucrose. Disruption of either the *scrR* or *scrB* genes by targeted plasmid integration resulted in strains unable to utilize sucrose, indicating that this is the only operon for the utilization of sucrose in this strain of *Clostridium.* RNA analysis confirmed that the genes of the *scr* operon were cotranscribed on a 5 kb mRNA transcript and that transcription was induced by sucrose. The role of the *scrR* gene was elucidated by Northern blot analysis of the *scrR* mutant, which clearly showed constitutive transcription of the upstream *scrA* gene, confirming that *scrR* encodes a transcriptional repressor, which probably acts by binding to an operator sequence in the promoter region upstream of *scrA.*

The transcriptional start site of the *scrARBK* gene cluster was located at an adenine residue 44 bp upstream of the *scrA* ATG start codon [19]. Examination of the sequence upstream of the *scr* genes (Figure 7.2) revealed possible –35 and –10 promoter sequences (5′-TTGACA-N_{18}-TATCAT), which lie between –51 and –80 bp upstream of the *scrA* initiation codon, and show considerable similarity to *E. coli* σ^{70} and clostridial consensus promoter sequences [53]. These regions including the promoter sequences show an exceptional degree of conservation between the two species, perhaps indicating the importance of this region in regulation of the operon. An imperfect palindrome (5′-GAAAAC/G′G′T′T′T′C′-3′) was identified by Reid et al. [19], between the putative *scrA* promoter and ATG initiation codon (Figure 7.2), which appears to be typical of the G/C centred operator sites at which a transcriptional regulator of the GalR-LacI family could bind [54], and could also represent a CRE site. However, this G/C centered inverted repeat sequence is not present in the *C. perfringens* sequence. A second putative CRE element was identified 139 bp upstream of the transcriptional start site in *C. beijerinckii* [24], and another region of dyad symmetry can be seen in the highly conserved region of the promoters of both *C. beijerinckii* and *C. perfringens* (Figure 7.2). Whether these regions represent the binding sites of one or more regulatory proteins still remains to be elucidated. The fact that one of the potential CRE boxes and the target site of the *scrR* repressor coincides might serve to ensure that the operon is repressed both in the absence of sucrose and in the presence of preferred carbon sources, such as glucose.

The location of the gene for the ScrR-negative regulator within the inducible *scrARBK* operon is somewhat unusual because *scrR* would be transcribed as part of a 5 kb mRNA transcript. This

```
                                                -35                -10
Cbeij.  TATAATTTATTATTTAGGTAAAAAACGTATTGACACATTATATATGGAACTGGTATCATATA  64
        || ||| ||   ||||             |||||| |   |||||||||||| || ||||||
Cperf.  TAAAATATAAAGTTTAAAGC.........TTGACAGAAACTATATGGAACTGATACCATATA  52

Cbeij.  TAACATGA...AAACGGTTTCCGAGGTTAATTG.AATTAGGAGGGGACATTAATGAGGAACA  117
           || ||   ||| | | ||  | |   || |  |||||||||    |  |||||| ||
Cperf.  ATTCAAGAAGTAAAGGATATCAAAAGAGGATAGGTTTTAGGAGGTTTTA.AAATGAGTAAAG  111
                   CRE                        RBS
```

FIGURE 7.2 Sequence alignment of the promoter region upstream of the *scrARBK* operons in *C. beijerinckii* NCIMB 8052 [19] and *C. perfringens* strain 13 [13]. The position of the transcriptional start site at an adenine residue of *C. beijerinckii* is shown in bold, and the putative promoter regions (–10 and –35 sites) are indicated. The ribosome-binding sites (RBS) and the ATG start codons of the genes are shown in bold letters. The bold arrows indicate the region of dyad symmetry, which overlaps the promoter region, and possibly represents a regulatory protein binding site, while the light arrows indicate the position of the CRE [19].

would imply that it is maximally expressed when the other genes are expressed and would, therefore, be negatively autoregulated. In other sucrose-utilization regulons, with the exception of *L. lactis* [55], the corresponding *scrR* genes are independently transcribed, and in *S. xylosus,* inactivation of the *scrR* gene leads to constitutive expression of the downstream *scrB* gene [51]. In *C. beijerinckii,* it would be unnecessary for large amounts of ScrR to be synthesized, as *scr* operon transcripts accumulate in the presence of sucrose in order to hydrolyze the substrate. There must therefore be other regulatory mechanisms that limit translation of *scrR* in order to allow for optimal expression of the *scr* operon and the build up of gene products for sucrose utilization while the inducer is present.

7.3.1.3 Sucrose Metabolism in Other *Clostridium* Species

Analysis of the genome sequence data for *C. perfringens* demonstrated that the sucrose locus has the identical gene arrangement of *scrARBK* that has been reported in *C. beijerinckii* [19] (Figure 7.1). The deduced amino acid sequences of all the genes is highly conserved and exhibit very high identities of between 50 and 65%, to the orthologues in *C. beijerinckii* (Table 7.1). The genes are all orientated in the same direction and presumably are transcribed as an operon on a single mRNA transcript from a promoter upstream of *scrA*, as is the case in *C. beijerinckii.* Comparison of the DNA sequence of the promoter regions of these two operons shows that the nucleic acid sequence is highly conserved over the –10 and –35 regions. A region of dyad symmetry is also found in this highly conserved stretch of DNA, possibly corresponding to a binding site for a regulator protein (Figure 7.2).

In contrast to the two *Clostridium* species described above, the genome sequence of *C. acetobutylicum* ATCC 824 showed a very different arrangement of sucrose utilization genes [56]. These authors established that sucrose metabolism in *C. acetobutylicum* was conducted via the same pathways as in *C. beijerinckii* [57] and that sucrose was transported into the cell by a sucrose-specific PTS. Examination of the *C. acetobutylicum* genome sequence data revealed that the structural genes were clustered in a similar way to *C. beijerinckii* [19], but in the order *scrAKB* [56] (Figure 7.1). The *scrB* gene (designated *sacA* by Nolling et al. [12]), encoding a sucrose-6-P hydrolase, is not highly homologous to the *scrB* genes of the other clostridia and shows only 40% aa identity to *B. subtilis,* and 39% and 37% identity to *C. perfringens* and *C. beijerinckii,* respectively (Table 7.1). The *scrB* gene is found in the same frame as a sucrose PTS-IIABC gene (*scrA*), and the *scrK* gene, encoding the fructokinase (53% I to both *C. perfringens* and *C. beijerinckii).* The *scrA* gene from *C. acetobutylicum* encodes a protein of 627 amino acids, compared to corresponding proteins of 451 aa from *C. beijerinckii* and *C. perfringens*, which comprises only the enzyme IIBC component of the sucrose-specific PTS protein. The *C. acetobutylicum* protein has homology over its entire length to the *scrA* gene from *Streptococcus agalactiae* (40%), which encodes the larger protein of 620 aa for the IIABC component. The three *C. acetobutylicum* sucrose genes are flanked by two regulatory genes (Figure 7.1). Upstream of the *scrA* gene and in the same orientation lies the *scrT* gene, which has 46% aa identity to the *licT* gene of *B. subtilis* [58]. Downstream of the *scr* operon lies a second regulator gene, which is a member of the AraC/XylS family, and shows 39% aa identity to AraC from *B. subtilis* [59]. Northern blot analysis demonstrated that the *scrTAKB* genes are expressed on a single transcript, and so would be regulated as a operon [56].

The *scrT* gene is a member of the BglG family of antitermination proteins [58], which not only has homology to *licT* from *B. subtilis,* but also to *sacT* and *sacY* (41%), the sucrose antitermination proteins from *B. subtilis* [58]. It also has 42% identity to the regulatory protein in the operon containing genes for aryl-β-glucoside utilization from the ruminal *Clostridium longisporum* [60]. The deduced amino acid sequence of ScrT was found to contain all the conserved histidine residues that are phosphorylated in LicT in *B. subtilis* [61] and BglG in *E. coli* [58,62]. Regulatory sequences with dyad symmetry, resembling intrinsic terminator structures, can be identified in both the leader region upstream of *scrT,* as well as in the intergenic region of 144 bp between the last codon of the *licT* gene and the ATG start codon of the *scrA* gene [56]. Partially overlapping these terminators are typical RAT sequences, characteristic of operons controlled by antiterminators of the Bgl family [64]. These features would imply that the *scr* genes of *C. acetobutylicum* are regulated by the ScrT protein, which acts via an antitermination mechanism. In addition to these features, a CRE site was identified overlapping the promoter of the *scrT* gene [56]. As the enzymes of the sucrose operon have been shown to be repressed in the presence of glucose, it is highly likely that this operon is, like the *C. beijerinckii* operon, under the control of the CcpA protein.

There are no genes in *C. tetani* with detectable homology to either the *C. acetobutylicum sacA* or the *C. beijerinckii scrARBK* genes, although homologues of the *scrT* gene can be identified in *C. tetani* (41% aa identity) and *C. thermocellum* (38% aa identity), but not in *C. beijerinckii* or *C. perfringens*. However, as these genes are not clustered with the other sucrose utilization genes, their actual function may not be sucrose regulation.

It has been suggested that bacterial sucrose utilization systems evolved by modular evolution, where individual genes sharing a common origin became independently associated into regulons and operons [45]. A comparison of the arrangement of the genes within the sucrose locus of several members of the low G+C Gram-positive bacteria shows interesting genome conservation (Figure 7.1). The *C. beijerinckii* and *C. perfringens scr* gene clusters have a number of similarities to the sucrose PTS regulons of other low G+C Gram-positive bacteria, including *S. xylosus* [51], *S. mutans* [65], and *Pedicoccus pentosaceous* [66] (Figure 7.1). The *S. xylosus* system has been reported to consist of an independent *scrA* and the *scrRBK* gene cluster [51,67], which is precisely the same order as the equivalent genes of the two clostridial sucrose systems. In contrast, the system of regulation of the *C. acetobutylicum* sucrose utilization genes would appear to be completely different, and the *scrA* gene encodes an enzyme IIABC component of the PTS, more like *S. mutans* [65] and *P. pentosaceous* [66]. The similarity in the molecular organization of the two sucrose operons from *C. beijerinckii* and *C. perfringens* suggests a very close evolutionary relationship between these two species. The significant degree of sequence homology and gene clustering between the clostridia and the lactic acid bacteria would also indicate a common evolutionary origin for these operons, although these have been altered and obscured by modular shuffling to different extents (Figure 7.1).

7.3.2 Lactose and Galactose Degradation in Clostridia

Lactose utilization by clostridia was studied intensively in some of the solventogenic species in the 1980s and 1990s, because of the opportunities afforded by the availability of large quantities of whey, a high-lactose waste product of the dairy industry. Several *C. acetobutylicum* strains have successfully been shown to use whey as a substrate for solvent production [68,69]. When tested for enzyme production under inducing conditions, all strains produced two different enzymes capable of hydrolyzing lactose, a phospho-β-galactosidase and a β-galactosidase. Both enzymes were induced during growth on whey and were repressed by glucose. Interestingly, the expression of these genes was found to be developmentally regulated, with the phospho-β-galactosidase highly expressed in acidogenic phase while the β-galactosidase expression was optimally expressed during solventogenesis and the early stages of sporulation [69]. The reports of both types of enzymes imply the presence of both a lactose-specific PTS as well as a nonphosphorylating transport system for lactose. Recently, a PTS-dependent lactose utilization system has been reported in *C. saccharobutylicum* NCP262 [20]. Only one β-galactosidase gene has been cloned and characterized from *Clostridium* species to date, specifically from *C. acetobutylicum* NCIMB 2951[70].

In general, the Gram-positive bacteria utilize lactose by two different pathways (Figure 7.3). The pathway reported in *S. aureus* involves lactose transport by a lactose PTS, and the subsequent hydrolysis by a phospho-β-galactosidase to yield glucose and galactose-6-phosphate [31]. The other pathway, initially described in the lactic acid bacteria, *Lactobacillus dulbrueckii* [71,72] and *Streptococcus thermophilus* [73,74], involves the transport of lactose by a permease protein (*lacS*), and then hydrolysis by β-galactosidase (*lacZ*) to glucose and galactose. The glucose generated is

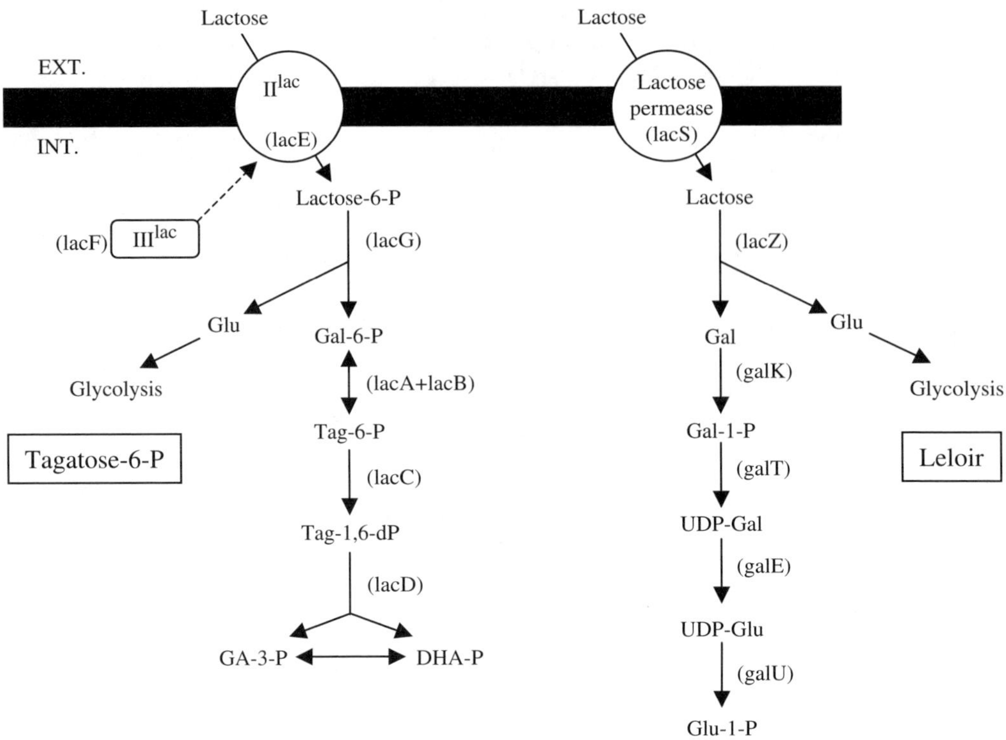

FIGURE 7.3 Pathways indicating the different mechanisms of transport and catabolism of lactose in the Gram-positive bacteria: a PTS-dependent lactose system results in the accumulation of lactose-6-phosphate, which would be hydrolyzed by a phospho-β-galactosidase (*lacG*), and a non-PTS permease transports the lactose into the cell unmodified, and this is cleaved by a β-galactosidase (*lacZ*). The products are metabolized by the tagatose-6-P [77] and the Leloir pathways [75], respectively. The genes encoding the enzymes of the pathways are indicated in brackets.

phosphorylated and incorporated into the glycolytic pathway, and galactose is metabolized by the Leloir pathway [75]. This pathway is common to many bacterial species and has been well-characterized in *E. coli* [76]. Here, galactose is converted to glucose-1-P in five steps, involving enzymes encoded by the *galK, galT, galE, galU,* and *pgm* genes (Figure 7.3). In contrast, the gal-6-P generated by the action of phospho-β-galactosidase is degraded to triose phosphates by the less common tagatose-6-phosphate pathway [77], which is encoded by four genes, the *lacABCD* genes. These enzymes are responsible for the conversion of gal-6-P to tag-6-P, and then to two molecules of triose-phosphate and glyceraldehydes (Figure 7.3).

7.3.2.1 PTS-Dependent Lactose Utilization in *Clostridium*

Recent releases of genomic DNA sequence from *C. acetobutylicum, C. perfringens,* and *C. tetani* have allowed us to build a better picture of lactose utilization in *Clostridium* and to substantiate the physiological findings previously reported [12,13; Genbank Acc. No. NC 004557.1]. In *C. acetobutylicum* 824, a unique arrangement of genes for lactose and galactose utilization is found. These genes form a large continuous cluster, 20 kb in length, on the *C. acetobutylicum* chromosome, in which the PTS-*lac* genes are arranged in the opposite orientation to the *gal* genes immediately downstream, and these are followed by the tagatose genes (Figure 7.4A and Figure 7.4B). The PTS lactose cluster consists of the *lacR* gene, a transcriptional PTS-repressor, *lacF* and *lacE*, encoding the lactose-specific PTS proteins, enzyme IIA and enzyme IIBC, and the *lacG* gene, coding for the

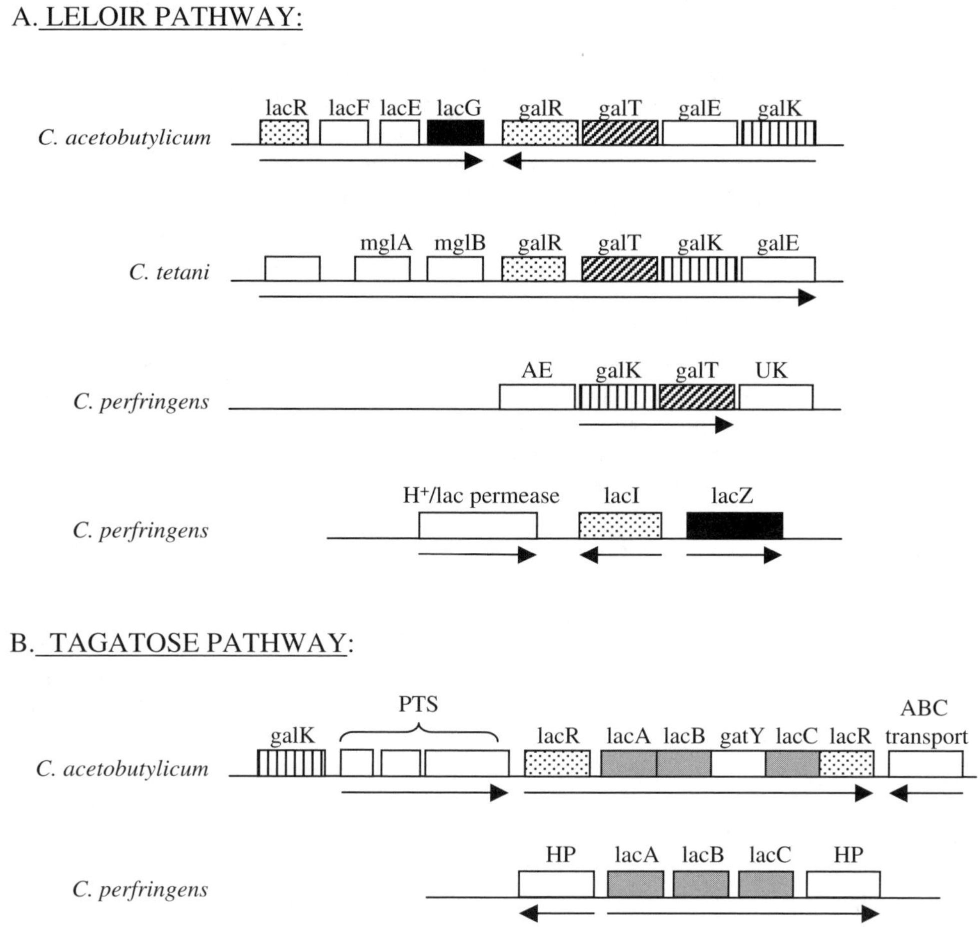

FIGURE 7.4 Organization of the genes encoding proteins involved in lactose and galactose utilization from the three *Clostridium* genomes sequenced to date: *C. acetobutylicum* ATCC 824 (NCBI: NC 003030.1), *C. perfringens* strain 13 (NCBI: NC 003366.1) and *C. tetani* E88 (NCBI: NC 004557.1). (A) Lactose transport and catabolism, and the Leloir pathway. (B) Galactose catabolic operons, and the tagatose-6-P pathway. Gene names are as given in Figure 7.3 or have been named according to their probable gene products: *mglA* and *mglB*, galactose transport; AE, aldose epimerase; UK, uridine kinase; HP, hypothetical protein. Note that the two regions of the *C. acetobutylicum* genome illustrated are continuous and are joined at the *galK* gene, forming a 20 kb cluster of lactose-galactose catabolic genes.

phospho-β-galactosidase enzyme. The individual genes show a close relationship to the corresponding genes from other low G+C Gram-positive bacteria, showing between 52 and 58% identity at the amino acid level to genes from *Enterobacterium faecium*, *S. aureus,* and a number of *Lactobacillus* species. The staphylococcal lactose operon consists of 7 genes that are inducible by lactose [31]. The order of the *lac* genes in *S. aureus* is the same as in *C. acetobutylicum*, except that four genes from the tagatose-6-P pathway are inserted between the *lacR* gene and the rest of the lactose PTS genes. A very similar gene arrangement exists in *Lactobacillus lactis* [31]. In contrast, *C. acetobutylicum* has an operon that constitutes the Leloir pathway downstream of the *lacRFEG* genes.

7.3.2.2 Non-PTS-Dependent Lactose-Galactose Systems in *Clostridium*

As previously mentioned, there is evidence of a non-PTS transport system for lactose in *Clostridium* [69]. Again, although physiological studies on galactose uptake in clostridia are not extensive, there is evidence that in *C. pasteurianum*, galactose uptake was found to be via an H^+-symport system [78]. In *C. saccharobutylicum,* treatment of the cells with an ATPase inhibitor or an energy uncoupler slowed the utilization of galactose and indicated that galactose uptake takes place by means of a proton-dependent system [20]. These authors also demonstrated that galactose utilization was repressed by glucose.

The non-PTS β-galactosidase enzymes from *Clostridium* appear to be a very diverse group of proteins, with little sequence homology to each other. The active *E. coli* β-galactosidase is comprised of four identical subunits of 1021 aa [79], and five major regions of high sequence identity have been identified as common to all β-galactosidases [71]. These conserved regions have been identified in the β-galactosidases of clostridia, and other Gram-positive bacteria, and correspond to the active site domains of *E. coli* and other well-characterized β-galactosidases. However, the three β-galactosidases from clostridia discussed below are distinctly different enzymes and only distantly related to each other. This might be because they are used in a different context and might also reflect very different origins.

Analysis of the genome sequence data from *C. perfringens* shows that it has a β-galactosidase gene that lies in a cluster of lactose genes, encoding a H^+/sugar symport protein, followed by the *lacI* and *lacZ* genes (Figure 7.4). The LacZ protein shows high sequence homology to LacZ proteins from *C. thermocellum* (46% aa identity) and *S. pyogenes* (37% aa identity), but very little conserved sequence homology to the well-known *E. coli* or *B. subtilis* proteins.

The first β-galactosidase gene cluster to be functionally characterized was from *C. acetobutylicum* NCIMB 2951 [70]. Two genes occur in this cluster: the *cbgA* gene, which encodes a β-galactosidase, and downstream, the *cbgR* gene. The deduced CbgA protein was found to be closely related to proteins found in the lactic acid bacteria, *Streptococcus* and *Lactobacillus* (between 50 and 54% aa identity), as well as in *B. megaterium* (44% I). CbgA has a MW of 105 kDa and has several of the conserved regions identified in the *Lactobacillus* β-galactosidases by Schmidt et al. [71]; however, the C-terminal region corresponding to region VIII is missing in CbgA [9]. The *cbgR* gene is 0.4 kb and encodes a sequence closely related to the missing C-terminal region VIII, which is highly conserved in most members of the β-glycosidase C family. Transposon mutagenesis indicated that this region was essential for the expression of β-galactosidase, and the authors propose that it is translationally coupled to the *cbgA* gene, which results in the expression of a functional enzyme [70]. Support for this model comes from reports of β-galactosidase enzymes from *Leuconstoc lactis,* [80], and *Lactobacillus sake* [81], where the enzymes are encoded by two genes, the *lacL* and *lacM* genes, which overlap slightly. The genes are transcribed as a single transcript, and the expression of both genes is required for functional β-galactosidase production [81]. It is, therefore, very likely that the same system of translational coupling occurs in *C. acetobutylicum* NCIMB 2951.

The promoter region of *cbgA* contains two regions of dyad symmetry, which may suggest regulation by an antitermination mechanism similar to the sucrose operon of *B. subtilis* [82]. However, as the deduced protein from *cbgR* does not have any homology with published antiterminator proteins, it is more probable that the inverted repeat sequences represent the binding sites of regulator proteins.

The *lacA* gene from *Thermoanaerobacter thermosulfurogenes* (formerly *Clostridium thermosulfurogenes*) [83] encodes a protein with a calculated Mr of 83.8 kDa, which appears to be only distantly related to other β-galactosidase proteins. Sequence comparisons showed that, in most cases, the homology was restricted to the N-terminal region of the gene, which includes the conserved active site of the enzyme [84]. However, the LacA protein does show amino acid similarity over the whole length of the molecule (56% identical residues) to a second putative β-galactosidase protein in the *C. perfringens* genome data, and it might therefore represent a new class of β-galactosidase enzymes in *Clostridium.*

No orthologues of either *lacZ, cbgA,* or *lacA* can be identified in the genomes of either *C. acetobutylicum* or *C. tetani*, although genes with deduced protein homology (26% identity) to the CbgA protein can be detected in both the *C. perfringens* and *C. thermocellum* genomes. It is also possible that the *cbgAR* genes from *C. acetobutylicum* NCIMB 2951 are part of a larger operon, and that additional *lac* genes would be found upstream of the *cbgA* genes. Two other β-galactosidase genes can be identified in the *C. perfringens* genome, but as these occur in different gene clusters together with genes such as xylanase or sialidase genes, these genes may have a more specialized function in carbohydrate metabolism.

7.3.2.3 Leloir and Tagatose Pathways

Genome sequence data from the *Clostridium* group show a great diversity in the pathways capable of metabolizing galactose. In *C. acetobutylicum,* for example, the classical Leloir pathway is found, and a cluster of four genes, the *galKETR* genes, have been identified (Figure 7.4A). These genes encode enzymes of this pathway; galactokinase, UDP-gal-4-epimerase, gal-1-P-uridyltransferase, and a transcriptional regulator, GalR. The proteins deduced from the DNA sequence are highly conserved, demonstrating between 49 to 56% identity at the amino acid level. In most cases, the closest relatives to these proteins are *E. faecium*, *S. thermophilus,* and *B. subtilis*. The four genes are possibly transcriptionally linked, as the genes virtually overlap each other. In *S. thermophilus*, the *galKTE* genes are transcribed together as a single transcript of 3.7 kb in the presence of lactose [74], while in *L. lactis*, the *gal* cluster is interrupted by the *lacA* and *lacZ* genes, giving a final arrangement of *galKT-lacAZ-galE* [85]. The gene arrangement in *C. acetobutylicum* is different to that of the streptococci, although in this group of lactic acid bacteria, the gene order is highly conserved [74]. In a similar arrangement, *C. tetani* possesses a cluster of seven genes involved in the utilization of galactose, all oriented in the same direction, although the order is slightly different. These include the *mglA-mglB* genes, which make up a galactose transporter and the *galR* regulator, followed by the *galTKE* genes (Figure 7.4A). Although orthologues of the *galT, galK,* and *galE* genes are also found in the genome of *C. perfringens* and have 50, 54, and 76% identity to *C. acetobutylicum,* respectively, only the *galK* and *galT* genes are linked on the chromosome.

In both *C. acetobutylicum* and *C. perfringens*, galactose is metabolized via the more unusual tagatose-6-P pathway (Figure 7.3). This pathway was first described in *S. aureus* [77] and *Lactococcus lactis* [86]. The *lacABCD* genes produce the enzymes, galactose-6-phosphate isomerase and tagatose-6-P kinase, which convert Gal-6-P into two triose-phosphate molecules. Three of these genes, the *lacABC* genes, can be identified in the genome of *C. perfringens* (Figure 7.4B). The putative *lacB* gene product, one of the subunits of the gal-6-P isomerase, is more similar to the corresponding proteins from *Enterococcus faecium* (67% identity) and *S. aureus* (58% identity) than to the gal-6-P isomerase from *C. acetobutylicum* 824 (49% identity).

Most of the deduced proteins from these *Clostridium* pathways are closely related to proteins of the lactic acid bacteria, rather than soil bacteria such as *B. subtilis*. It is interesting that, although most bacteria have been reported to possess either the genes encoding enzymes of the Leloir pathway [75], or the more unusual tagatose-6-P pathway [31], *C. acetobutylicum* has the genes for both of these clustered in a contiguous stretch of genomic DNA together with the phospho-β-galactosidase operon — a total of ca. 20 kb (Figure 7.4A and Figure 7.4B). In fact, the genes of the Leloir pathway would appear to have been acquired by *C. acetobutylicum* from the lactic acid bacteria by means of horizontal gene transfer. They became incorporated into the other operons involved in lactose and galactose catabolism in *C. acetobutylicum,* making this bacterium highly suited to a galactose-rich environment.

7.4 UTILIZATION OF PENTOSES

The most common pentose sugars, which are generated by the degradation of different carbohydrate polymers, include arabinose, xylose, and ribose. These sugars are metabolized intracellularly by a combination of phosphorylation, isomerization, and epimerization, and the resulting intermediates are converted to fructose-6-phosphate and glyceraldehyde-6-phosphate [87]. In *Clostridium,* a number of enzymes and the corresponding genes have been reported from different species, most of them concerning xylose metabolism in connection with degradation of hemicellulose-containing biomass [83,88–91]. Ounine et al. [92] have demonstrated that *C. saccharobutylicum* P262 is capable of fermenting xylose to solvents with a yield close to that obtained from glucose, although the growth rate was slightly slower. Xylose uptake was shown to be repressed by glucose in *C. acetobutylicum*, [90,91]; however, this is not the case for all clostridia, and *C. thermoaceticum* has been reported to use xylose preferentially to glucose and fructose [93]. Enzymes involved in ribose dissimilation have been reported from *C. perfringens* [88].

In *Bacillus subtilis,* these operons are not expressed unless the specific substrate is present, and each operon is regulated by transcriptional repressors for that regulon [31]. The transcriptional regulators of arabinose, xylose, and ribose are controlled by AraR, XylR, and RbsR, respectively, in conjunction with the global catabolite repressor protein, CcpA [94–96]. A comparative analysis of the pentose utilization genes occurring in the *Bacillus/Clostridium* group and other Gram-positive bacteria was performed by Rodionov et al. [97], who used the *B. subtilis* regulatory genes to identify orthologous genes and their regulons in other *Bacillus* spp., as well as in *C. acetobutylicum* and *C. difficile.* This may represent a more relevant way to find previously unidentified carbohydrate regulons in *Clostridium.* These authors also identified CRE sequences upstream of the different genes and operons and were able to identify which ones are likely to be under the control of the catabolite repressor protein, CcpA. The majority of genes for the utilization of pentoses, with the exception of *xylAB* of *C. difficile* and *xynCB* of *C. acetobutylicum,* have a predicted CRE box and, therefore, would be regulated by CcpA [97]. This comparative analysis has been extended here to include the genome sequence data from *C. perfringens* strain 13 (NCBI Acc. No: NC 003366.1) and *C. tetani* E88 (NCBI Acc. No: NC 004557.1).

7.4.1 The Arabinose Regulon

In *B. subtilis*, the utilization of arabinose is controlled by the repressor protein AraR, which belongs to the LacI/GalR family of repressors. In the absence of inducer, AraR binds to the operator site upstream of the *araABDLMNPG, araE,* and *araR* operons [94], which are located in different regions of the bacterial chromosome. The *araABD* genes encode three enzymes required for arabinose catabolism, namely, arabinose isomerase, ribulokinase, and ribulose-5-phosphate epimerase. The *araE* gene encodes a proton symporter for the transport of arabinose into the cell, and is organized as a divergent transcriptional unit with *araR* [94].

In *C. acetobutylicum,* a large cluster of genes involved in arabinose utilization is found in the chromosome. The *araR, araDA,* and *araE* genes are present in the cluster, but the arrangement of the genes is somewhat different, and both *araE* and *araA* are duplicated in the gene cluster, with *araR* transcribed divergently to the other *ara* genes. No orthologues of *araB*, the ribulose kinase gene, are found in *C. acetobutylicum*; however, a gene similar to several sugar kinases but not clustering on the phylogenetic tree with any of the known kinases was identified and designated *araK*. Rodionov et al. [97] propose that the *araK* gene has replaced *araB* in both the *C. acetobutylicum* and *Enterococcus faecium* genomes. No orthologues of the arabinose genes could be identified in the genome sequence of *C. perfringens* or *C. tetani,* possibly because these genes are much more diverse in structure and do not show much conservation across the different Gram-positive genera. CRE sites were identified upstream of both copies of the *araE* gene in *C. acetobutylicum,* indicating that the arabinose genes are under control of the CcpA protein and would be repressed by preferred carbon sources such as glucose.

7.4.2 The Ribose Regulon

The ribose transport genes from *B. subtilis* form an operon consisting of the structural genes for the ABC transporter system, *rbsABCD*, the *rbsK* gene coding for the ribokinase, and the regulator

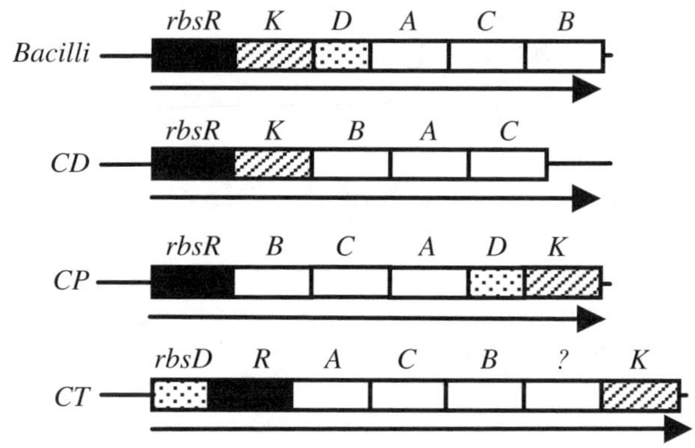

FIGURE 7.5 Organization of the ribose operons in various Gram-positive bacteria: Bacilli, *B. subtilis, B. halodurans, B. stearothermophilus,* and *B. anthracis* [97]; CD, *C. difficile* [97]; CP, *C. perfringens* strain 13 (NCBI: NC 003366.1) and CT, *C. tetani* E88 (NCBI: NC 004557.1). Gene names: *rbsR*, ribose regulator; *rbsK*, ribokinase; *rbsABCD*, components of the ribose transporter system.

gene, *rbsR* [95]. The genes are all orientated in the same direction, and would be transcribed from a promoter region upstream of *rbsR* (Figure 7.5). In *B. subtilis,* this operon is induced by ribose and is subject to catabolite repression [95]. The promoter region contains a CRE (CcpA binding site), so presumably the operon is under the control of the catabolite repressor protein, CcpA. This gene organization is conserved in other *Bacillus* spp. such as *B. halodurans*, *B. stearothermophilus,* and *B. anthracis,* as well as in *Lactococcus lactis* [97].

Interestingly, *C. difficile* has a very similar gene arrangement for ribose transport, but lacks both the *rbsD* gene (the transmembrane component of the ribose transporter) from the middle of the operon as well as the *rbsB* gene from the end of the cluster (Figure 7.5). *C. perfringens* has the full complement of genes arranged in the order *rbsR-B-C-A-D-K*, while *C. tetani* has the *rbs* genes clustered more like the Bacilli, *rbsD-R-A-C-B-?-K*. It is surprising that no similar ribose-uptake genes could be identified in the *C. acetobutylicum* genome, since this gene cluster shows a particularly high degree of conservation across the different genera. A search for potential CRE boxes, and for the target site of the *rbsR* repressor, revealed that in *C. difficile*, the two sites coincide [97]. This is similar to the regulatory region of the sucrose operon of *C. beijerinckii* [19], and can serve to ensure that the operon is repressed both in the absence of ribose and in the presence of preferred carbon sources.

7.4.3 The Xylose Regulon

The genes for xylose utilization in *B. subtilis* and *B. megaterium* form an operon, *xylAB*, and are negatively controlled by *xylR* repressor [96,98], which binds to the operator sites upstream of *xylAB*. XylR also regulates the expression of the *xynBC* operon, which encodes β-xyloside permease and β-xylosidase [98]. These two operons are adjacent to each other and are transcribed in the same direction but are physically separated by the *xylR* gene, which is divergently transcribed. The *xylR*/*xylAB* divergon (genes separated by a region containing the two divergent promoters) is conserved in all the *Bacillus* strains, but the adjacent genes are very different. Interestingly, the *B. subtilis* gene arrangement is more conserved in the distant relative, *C. acetobutylicum,* than in the other *Bacillus* strains [97]. *C. acetobutylicum* has precisely the same gene arrangement to that of *B. subtilis,* with the divergent *xylR* gene, except that *xylA* gene is not in the same cluster. The putative xylose repressor, XylR, does not closely resemble the *Bacillus* XylR proteins, however a common 23bp consensus sequence was identifed upstream of the *xynBC, xylB,* and *xylR* genes in *C. acetobutylicum,* which possibly is the target of XylR binding [97]. The *C. difficile* xylose genes are differently arranged. The *xylR*/*xylAB* divergon is conserved, while upstream lies a cluster of five genes, *xylS*-pts1-2-3-4, encoding a putative xylosidase and a probable xylose PTS transport system, respectively. Target sites for XylR were located both in the common regulatory sequence containing the divergent promoters and upstream of the *xylS* gene. *E. faecalis* has the same cluster of orthologous genes upstream of and part of the *xylAB* operon as *C. difficile* [97]. Comparison of the xylose utilization genes, therefore, shows that the genes from these two clostridium species are more closely related to other genera than to members of the *Clostridium* genus. A CRE sequence can be seen in the promoter region of the *xylB* gene of *C. acetobutylicum* [97], which presumably is responsible for the glucose repression of xylose utilization observed in *C. acetobutylicum* [90,91].

A similar *xylA-xylB* gene cluster has been reported from the thermophilic species, *C. thermosaccharolyticum* [89]. Here, the *xylAB* operon was reported to be induced by xylose, but although the genes were shown to be preceded by an ORF, which may encode the XylR repressor protein, the ORF had no amino acid sequence identity to known XylR proteins. Xylose isomerase genes, (*xylA*), have also been characterized from other thermophilic anaerobes, namely, *C. thermosulfurogenes* (renamed *Thermoanaerobacter*) and *C. thermohydrosulfuricum* [83,99]. These proteins are all very closely related, sharing amino acid identities of up to 93% [89]. Again, no *xylAB* or *xynBC* gene clusters have been identified in the genome sequences of *C. perfringens* and *C. tetani*; however, these genomes are not yet completely annotated.

7.5 CONCLUDING REMARKS

The comparative analysis of *Clostridium* genes discussed here gives an indication of the many and varied genes and operons concerned with carbohydrate utilization in this group of bacteria. On the whole, these systems demonstrate a very strong degree of conservation, particularly, as predicted, with genes associated with glycolysis [32]. This would indicate that the genes were present in a common ancestral prokaryote. However, there is frequently unexpected diversity in gene arrangement between species, where a gene or genes have obviously been inserted into an otherwise highly conserved group, and this is clearly the result of horizontal gene transfer (HGT) or of modular shuffling of the components of a regulon. There have not been any reports of regions of the *Clostridium* genome, other than the rRNA operons, which have a different G+C ratio [12], which normally indicates the acquisition of genes from another source. However, the close relationship of some of the lactose/galactose genes to their counterparts in the lactic acid bacteria could suggest such acquisitions. HGT presumably could involve whole operons rather than single genes, as this would be the most efficient way of expanding a bacterium's arsenal of carbohydrate utilization enzymes. In *C. acetobutylicum,* the genes of the Leloir pathway appear to have been inserted into an existing gene cluster, primarily concerned with lactose and galactose catabolism via the tagatose-6-P pathway. This would suggest that *C. acetobutylicum* acquired these genes from the lactic acid bacteria by means of horizontal gene transfer. In fact, the large 20 kb "island" of genes involved in galactose catabolism, which has been described in *C. acetobutylicum,* might have been a simultaneous acquisition, which allowed *C. acetobutylicum* to utilize the most abundant disaccharide in mammalian environments.

In order to maximize efficient utilization of nutrients, bacteria require global regulatory mechanisms, which integrate carbon metabolism with other essential metabolites, such as nitrogen. The CcpA protein in *B. subtilis* has been shown to do just this, and the number of pathways reported to be under the control of this protein is growing [14]. In *Clostridium,* however, relatively little is known of the regulation of the soluble carbohydrate substrates. More detailed studies need to be undertaken to elucidate the regulatory mechanisms controlling each regulon, so that an integrated picture is obtained. The isolation of the *ccpA* gene from *C. beijerinckii* and *C. saccharobutylicum* [23,24], and the characterization of operons such as sucrose, glucitol, and possibly glutamine synthetase, which possess CRE sequences and may be under the control of this protein, affords an excellent opportunity to clarify these global control mechanisms in *Clostridium.*

REFERENCES

1. Hippe, H., Andreeson, J.R., and Gottschalk, G., The genus *Clostridium*-nonmedical, in *The Prokaryotes*, Balows, A., et al., Eds., Springer-Verlag, New York, 1992, 1800.
2. Collins, M.D. et al., Taxonomic studies on a psychrophilic *Clostridium* from vacuum-packed beef: description of *Clostridium estertheticum* sp. nov., *FEMS Microbiol. Lett.*, 96, 235, 1992.
3. Johnson, J.L. and Chen, J.-S., Taxonomic relationships among strains of *Clostridium acetobutylicum* and other phenotypically similar organisms, *FEMS Microbiol. Lett.*, 17, 233, 1995.
4. Keis, S. et al., Taxonomy and phylogeny of industrial solvent-producing clostridia, *Int. J. Syst. Bacteriol.*, 45, 693, 1995.
5. Keis, S., Shaheen, R., and Jones, D.T., Emended descriptions of *Clostridium acetobutylicum* and *Clostridium beijerinckii*, and descriptions of *Clostridium saccharoperbutylacetonicum* sp. nov. and *Clostridium saccharobutylicum* sp. nov., *Int. J. Syst. Evol. Microbiol.*, 51, 2095, 2001.
6. Morris, J.G., History and future potential of the clostridia in biotechnology, in: *The Clostridia*, Woods, D.R., Ed., Butterworth-Heinemann, Stoneham, MA, 1993, 1.
7. Jones, D.T. and Woods, D.R., Acetone-butanol fermentation revisited, *Microbiol. Rev.*, 50, 484, 1986.
8. Jones, D.T. and Keis, S., Origins and relationships of industrial sovent-producing clostridial strains, *FEMS Microbiol. Rev.*, 17, 223, 1995.
9. Bronnenmeier, K. and Staudenbauer, W.L., Molecular biology and genetics of substrate utilization in clostridia, in *The Clostridia and Biotechnology*, Woods, D.R., Ed., Butterworth-Heinemann, Stoneham, MA, 1993, 261.
10. Verhasselt, P. and Vanderleyden, J., Cloning and expression of *Clostridium acetobutylicum* genes involved in carbohydrate utilization, in *Genetics and Molecular Biology of Anaerobic Bacteria*, Sebald, M., Ed., Springer-Verlag, New York, 1993, 301.
11. Saier, M. et al., Transport mechanisms, in *Bacillus Subtilis and Other Gram-Positive Bacteria: Bochemistry, Physiology and Molecular Genetics*, Sonenshein, A.L., Hoch, J.A., and Losick, R., Eds., ASM Press, Washington, DC, 1993, 133.
12. Nolling, J. et al., Genome sequence and comparative analysis of the solvent-producing bacterium *Clostridium acetobutylicum*, *J. Bacteriol.*, 183, 4823, 2001.
13. Shimizu, T. et al., Complete genome sequence of *Clostridium perfringens*, an anaerobic flesh-eater, *Proc. Natl. Acad. Sci. USA*, 99, 996, 2002.
14. Deutscher, J., Galinier, A., and Martin-Verstraete, I., Carbohydrate uptake and metabolism, in *Bacillus Subtilis and Its Closest Relatives: From Genes to Cells*, Sonenshein, A.L., Hoch, J.A., and Losick, R., Eds., ASM Press, Washington, DC, 2002, 129.
15. Schell, M.A. et al., The genome sequence of *Bifidobacterium longum* reflects its adaptation to the human gastrointestinal tract, *Proc. Natl. Acad. Sci. USA*, 99, 14422, 2002.
16. Lengeler, J.W., Jahreis, K., and Wehmeier, U.F., Enzymes II of the phosphoeonolpyrate-dependent phosphotransferase systems: their structure and function in carbohydrate transport, *Biochem. Biophis. Acta*, 1188, 1, 1994.
17. Mitchell, W.J., Carbohydrate uptake and utilization by *Clostridium beijerinckii* NCIMB 8052, *Anaerobe*, 2, 379, 1996.
18. Mitchell, W.J., Albasheri, K.A., and Yazadanian, M., Factors affecting utilization of carbohydrates by clostridia, *FEMS Microbiol. Rev.*, 17, 317, 1995.
19. Reid, S.J., Rafudeen, M.S., and Leat, N.G., The genes controlling sucrose utilization in *Clostridium beijerinckii* NCIMB 8052 constitute an operon, *Microbiol.*, 145, 1461, 1999.
20. Guttierrez, N.A. and Maddox, I.S., Galactose transport in *Clostridium acetobutylicum* P262, *Lett. in Appl. Microbiol.*, 23, 97, 1996.
21. Awang, G.M., Ingledew, W.M., and Jones, G.A., The effect of fermentable carbohydrate on sporulation and butanol production by *Clostridium acetobutylicum* P262, *Appl. Microbiol. Biotechnol.*, 38, 12, 1992.
22. Saier, M. et al., Protein phosphorylation and regulation of carbon metabolism in Gram-negative versus Gram-positive bacteria, *Trends Biochem. Sci.*, 20, 267, 1995.
23. Davison, S.P. et al., A *Clostridium acetobutylicum* regulator gene (*regA*) affecting amylase production in *Bacillus subtilis*, *Microbiol.*, 141, 996, 1995.

24. Rafudeen, M.S., Investigation of carbon catabolite repression in *Clostridium beijerinckii* NCIMB 8052, Ph.D. Thesis, University of Cape Town, South Africa, 2002.
25. Kraus, A. and Hillen, W., Analysis of CcpA mutations defective in carbon catabolite repression in *Bacillus megaterium, FEMS Microbiol. Lett.*, 153, 221, 1997.
26. Mitchell, W.J., Physiology of carbohydrate to solvent conversion by Clostridia, *Adv. Microbial Physiol.*, 39, 31, 1998.
27. Mitchell, W.J., Biology and physiology, in *Clostridia: Biotechnology and Medical Applications*, Bahl, H. and Dürre, P., Eds., Wiley-VCH Verlag GmbH, Weinheim, Germany, 2001, 49.
28. Bahl, H. and Dürre, P., Clostridia, in: *Biotechnology Vol. 1*, Sahm, H., Ed., VCH, Weinheim, Germany, 1993, 285.
29. Rogers, P., Genetics and biochemistry of *Clostridium* relevant to development of fermentation processes, *Adv. Appl. Microbiol.*, 31, 1, 1986.
30. Rogers, P. and Gottschalk, G., Biochemistry and regulation of acid and solvent production in clostridia, in *The Clostridia and Biotechnology*, Woods, D.R., Ed., Butterworth-Heinemann, Stoneham, MA, 25, 1993.
31. Steinmetz, M., Carbohydrate catabolism: pathways, enzymes, genetic regulation, and evolution, in *Bacillus Subtilis and Other Gram-Positive Bacteria: Biochemistry, Physiology and Molecular Genetics*, Sonenshein, A.L., Hoch, J.A., and Losick, R., Eds., ASM Press, Washington, DC, 1993, 157.
32. Dandekar, T. et al., Pathway alignment: application to the comparative analysis of glycolytic enzymes, *Biochem J.*, 343, 115, 1999.
33. Belouski, E., Watson, D.E., and Bennett, G.N., Cloning, sequence, and expression of the phosphofructokinase gene of *Clostridium acetobutylicum* ATCC 824 in *Escherichia coli*, *Curr. Microbiol.*, 37, 17, 1998.
34. Schreiber, W. and Dürre, P., The glyceraldehyde-3-phosphate dehydrogenase of *Clostridium acetobutylicum*: isolation and purification of the enzyme, and sequencing and localization of the *gap* gene within a cluster of other glycolytic genes, *Microbiology*, 145, 1999, 1839.
35. Tangney, M. et al., A gene system for glucitol transport and metabolism in *Clostridium beijerinckii* NCIMB 8052, *Appl. Environ. Microbiol.*, 64, 1612, 1998.
36. Trindade, M.I., Abratt, V.R., and Reid, S.J., Induction of the sucrose utilisation genes from *Bifidobacterium lactis* by sucrose and raffinose, *Appl. Environ. Microbiol.*, 69, 24, 2003.
37. Hutkins, R.W. and Ponne, C., Lactose uptake driven by galactose efflux in *Streptococcus thermophilus*: evidence for a galactose-lactose antiporter, *Appl. Environ. Microbiol.*, 57, 941, 1991.
38. Henrissat, B., A classification of glycosyl hydrolases based on amino acid sequence similarities, *Biochem. J.*, 280, 309, 1991.
39. Henrissat, B. and Bairoch, A., New families in the classification of glycosyl hydrolases based on amino acid sequence similarities, *Biochem. J.*, 293, 781, 1993.
40. Henrissat, B. and Bairoch, A., Updating the sequence-based classification of glycosyl hydrolases, *Biochem. J.*, 316, 695, 1996.
41. Lunsford, R.D. and Macrina, F.L., Molecular cloning and characterization of *scrB*: the structural gene for *Streptococcus mutans* PTS sucrose-6-phosphate hydrolase, *J. Bacteriol.*, 166, 426, 1986.
42. Thompson, J. et al., Transposon-encoded sucrose metabolism in *Lactococcus lactis*. Purification of sucrose-6-phosphate hydrolase and genetic linkage to N-(L-1-carboxyethyl)-L-ornithine synthase in strain K1, *J. Biol. Chem.*, 266, 14573, 1991.
43. Russell, R.R. et al., *Streptococcus mutans gtfA* Gene Specifies Sucrose Phosphorylase, *Infect. Immun.*, 56, 2763, 1988.
44. Koga, T. et al., Purification and some properties of sucrose phosphorylase from *Leuconostoc mesenteroides*, *Agric. Biol. Chem.*, 55, 1805, 1991.
45. Titgemeyer, F. et al., Molecular analysis of the *scrA* and *scrB* genes from *Klebsiella pneumoniae* and plasmid pUR400, which encode the sucrose transport protein Enzyme ll(Scr) of the phosphotransferase system and a sucrose-6-phosphate invertase, *Molec. Gen. Genet.*, 250, 197, 1996.
46. Fouet, A. et al., Cloning and expression in *Escherichia coli* of the sucrase gene from *Bacillus subtilis*, *Molec. Gen. Genet.*, 186, 399, 1982.
47. Fouet, A., Klier, A.F., and Rapoport, G., Nucleotide sequence of the sucrase gene of *Bacillus subtilis*, *Gene*, 45, 221, 1986.

48. Bruckner, R., Wagner, E., and Gotz, F., Characterization of a sucrase gene from *Staphylococcus xylosus*, *J. Bacteriol.*, 175, 851, 1993.
49. Blatch, G.L. and Woods, D.R., Nucleotide sequence and analysis of the *Vibrio alginolyticus scr* repressor-encoding gene (*scrR*), *Gene*, 101, 45, 1991.
50. Bork, P., Sander, C., and Valencia, A., Convergent evolution of similar enzymatic function on different protein folds: the hexokinase, ribokinase, and galactokinase families of sugar kinases, *Protein Science*, 2, 31, 1993.
51. Gering, M. and Bruckner, R., Transcriptional regulation of the sucrase gene of *Staphylococcus xylosus* by the repressor ScrR, *J. Bacteriol.*, 178, 462, 1996.
52. Sutrina, S.I. et al., The glucose permease of *Bacillus subtilis* is a single polypeptide chain that functions to energize the sucrose permease, *J. Biol. Chem.*, 265, 18581, 1990.
53. Young, M., Minton, N.P., and Staudenbauer, W.L., Recent advances in the genetics of the clostridia, *FEMS Microbiol. Rev.*, 63, 301, 1989.
54. Weickert, M.J. and Adhya, S., The family of bacterial regulators homologous to Gal and Lac repressors, *J. Biol. Chem.*, 267, 15869, 1992.
55. Rauch, P.J.G. and de Vos, W.M., Transcriptional regulation of the Tn5276-located *Lactococcus lactis* sucrose operon and characterisation of the *sacA* gene encoding sucrose-6-phosphate hydrolase, *Gene*, 121, 55, 1992.
56. Tangney, M. and Mitchell, W.J., Analysis of a catabolic operon for sucrose transport and metabolism in *Clostridium acetobutylicum* ATCC 824, *J. Mol. Microbiol. Biotechnol.*, 2, 71, 2000.
57. Tangney, M. et al., Sucrose transport and metabolism in *Clostridium beijerinckii* NCIMB 8052, *J. Appl. Microbiol.*, 84, 914, 1998.
58. Tortosa, P. et al., Sites of positive and negative regulation in the *Bacillus subtilis* antiterminators LicT and SacY, *Mol. Microbiol.*, 41, 1381, 2001.
59. Sa-Nogueira, I., Paveia, H., and de Lancastre, H., Isolation of constitutive mutants for L-arabinose utilization in *Bacillus subtilis*, *J. Bacteriol.*, 170, 2855, 1988.
60. Brown, G.D. and Thomson, J.A., Isolation and characterisation of an aryl-β-D-glucoside uptake and utilisation system (*abg*) from the gram-positive ruminal *Clostridium* species, *C. longisporum*, *Mol. Gen. Genet.*, 257, 213, 1998.
61. Lindner, C. et al., Regulation of the activity of the *Bacillus subtilis* antiterminator LicT by multiple PEP-dependent, Enzyme I- and HPr-catalysed phosphorylation, *Mol. Microbiol.*, 31, 995, 1999.
62. Chen, Q. et al., BglF, the sensor of the *Escherichia coli bgl* system, uses the same site to phosphorylate both a sugar and a regulatory protein, *EMBO J.*, 16, 4617, 1997.
63. Schnetz, K., Toloczyki, C., and Rak, B., Beta-glucoside (*bgl*) operon of *Escherichia coli* K12 — nucleotide sequence, genetic organization, and possible evolutionary relationship between regulatory components of two *Bacillus subtilis* genes, *J. Bacteriol.*, 169, 2579, 1987.
64. Houman, F., Diaz-Torres, M. R., and Wright, A., Transcriptional antitermination in the *bgl* operon of *E. coli* is modulated by a specific RNA binding protein, *Cell*, 62, 1153, 1990.
65. Sato, Y. et al., Isolation, characterization and sequence analysis of the *scrK* gene encoding fructokinase of *Streptococcus mutans*, *J. Gen. Microbiol.*, 139, 921, 1993.
66. Naumoff, D.G. and Livshits, V.A., Molecular structure of the *Lactobacillus plantarum* sucrose utilization locus: comparison with *Pediococcus pentosaceus*, *Mol. Biol.*, 35, 15, 2001.
67. Wagner, E., Gotz, F., and Bruckner, R., Cloning and characterization of the *scrA* gene encoding the sucrose-specific Enzyme II of the phosphotransferase system from *Staphylococcus xylosus*, *Mol. Gen. Genet.*, 241, 33, 1993.
68. Ennis, B.M. and Maddox, I.S., The effect of pH and lactose concentration on solvent production from whey permeate using *Clostridium acetobutylicum*, *Biotechnol. Bioeng.*, 29, 329, 1987.
69. Yu, P.-L., Smart, J.B., and Ennis, B.M., Differential induction of β-galactosidase and phospho-β-galactosidase activities in the fermentation of whey permeate by *Clostridium acetobutylicum*, *Appl. Microbiol. Biotech.*, 26, 254, 1987.
70. Hancock, K.R. et al., Expression and nucleotide sequence of the *Clostridium acetobutylicum* β-galactosidase gene cloned in *Escherichia coli*, *J. Bacteriol.*, 173, 3084, 1991.
71. Schmidt, B.F. et al., Expression and nucleotide sequence of the *Lactobacillus bulgaricus* β-galactosidase gene cloned in *Escherichia coli*, *J. Bacteriol.*, 171, 625, 1989.

72. Leong-Morgenthaler, P., Zwahlen, M.C., and Hottinger, H., Lactose metabolism in *Lactobacillus bulgaricus*: analysis of the primary structure and expression of the genes involved, *J. Bacteriol.*, 173, 1951, 1991.
73. de Vos, W.M., Boerrigter, I., and Van Rooyen, R.J., Characterization of the lactose-specific enzymes of the phosphotransferase system in *Lactococcus lactis*, *J. Biol. Chem.*, 265, 22554, 1990.
74. Vaillancourt, K., Moineau, S., and Frenette, M., Galactose and lactose genes from the galactose-positive bacterium *Streptococcus salivarius* and the phylogenetically related galactose-negative bacterium *Streptococcus thermophilus*: organization, sequence, transcription, and activity of the *gal* gene products, *J. Bacteriol.*, 184, 785, 2002.
75. Lin, E.C.C., Dissimilatory pathways for sugars, polyols and carboxylates, in *Escherichia coli and Salmonella typhimurium: Cellular and Molecular Biology*, Neidhardt, F.C., Ed., ASM Press, Washington, DC, 1987, 244.
76. Adhya, S., The galactose operon, in *Escherichia coli and Salmonella typhimurium: Cellular and Molecular Biology*, Neidhardt, F.C., Ed., ASM Press, Washington, DC, 1987, 1503.
77. Rosey, E.L., Oskouian, B., and Stewarts, G.C., Lactose metabolism by *Staphylococcus aureus*: characterization of *lacABCD*, the structural genes of the tagatose 6-phosphate pathway, *J. Bacteriol.*, 173, 5992, 1991.
78. Booth, I.R. and Morris, J.G., Proton-motive force in the obligately anaerobic bacterium *Clostridium pasteurianum:* a role in galactose and gluconate uptake, *FEBS Lett.*, 59, 153, 1975.
79. Kalnins, A. et al., Sequence of the *lacZ* gene of *Escherichia coli*, *EMBO J.*, 2, 593, 1985.
80. David, S., Stevens, H., and van Riel, M., *Leuconostoc lactis* β-galactosidase is encoded by two overlapping genes, *J. Bacteriol.*, 174, 4475, 1992.
81. Obst, M., Meding, E.R., and Vogel, R.F., Two genes encoding the β-galactosidase of *Lactobacillus sake*, *Microbiol.*, 141, 3059, 1995.
82. Debarbouille, M. et al., The *sacT* gene regulating the *sacPA* operon in *Bacillus subtilis* shares strong homology with transcriptional antiterminators, *J. Bacteriol.*, 172, 3966, 1990.
83. Lee, C. et al., Catalytic mechanism of xylose isomerase from *Clostridium thermosulfurogenes.* Characterization of the structural gene and function of the active site histidine, *J. Biol. Chem.*, 265, 19082, 1990.
84. Burchhardt, G. and Bahl, H., Cloning and analysis of the β-galactosidase-encoding gene from *Clostridium thermosulfurogenes* EM1, *Gene*, 106, 13, 1991.
85. Vaughan, E.E., Pridmore, R.D. and Mollett, B., Transcriptional regulation and evolution of lactose genes in the galactose-lactose operon of *Lactococcus lactis* NCDO2054, *J. Bacteriol.*, 180, 4893, 1998.
86. Van Rooijen, R.J., Vanschalkwijk, S., and de Vos, W.M., Molecular cloning, characterization and nucleotide sequence of the tagatose-6-phosphate pathway gene cluster of the lactose operon of *Lactococcus lactis*, *J. Biol. Chem.*, 266, 7182, 1991.
87. Rosenberg, S.L., Fermentation of pentose sugars to ethanol and other neutral products by microorganisms, *Enzyme Microbiol. Technol.*, 2, 185, 1980.
88. Cynkin, M.A. and Delwiche, E.A., Metabolism of pentoses by clostridia. I. Enzymes of ribose dissimilation in extracts of *Clostridium perfringens*, *J. Bacteriol.*, 75, 331, 1958.
89. Meaden, P.G. et al., The xylose isomerase-encoding gene (*xylA*) of *Clostridium thermosaccharolyticum*: cloning, sequencing and phylogeny of XylA enzymes, *Gene*, 141, 97, 1994.
90. Fond, O. et al., The acetone butanol fermentation on glucose and xylose. I. Regulation and kinetics in batch cultures, *Biotechnol. Bioeng.*, 28, 160, 1986.
91. Fond, O. et al., The acetone butanol fermentation on glucose and xylose. II. Regulation and kinetics in fed-batch cultures, *Biotechnol. Bioeng.*, 28, 167, 1986.
92. Ounine, K. et al., Regulation and butanol inhibition of D-xylose and D-glucose uptake in *Clostridium acetobutylicum*, *Appl. Environ. Microbiol.*, *49,* 874, 1985.
93. Andreeson, J.R. et al., Fermentation of glucose, fructose and xylose by *Clostridium thermoaceticum*: effects of metals on growth yields, enzymes and the synthesis of acetate from CO_2, *J. Bacteriol.*, 114, 743, 1973.
94. Mota, L.J., Tavares, P., and Sa-Nogueira, I., Mode of action of AraR, the key regulator of L-arabinose metabolism in *Bacillus subtilis*, *Mol. Microbiol.*, 33, 476, 1999.
95. Woodson, K. and Devine, K., Analysis of a ribose transport operon from *Bacillus subtilis*, *Microbiol.*, 140, 1829, 1994.

96. Schmiedel, D. et al., Regulation of expression, genetic organisation and substrate specificity of xylose uptake in *Bacillus megaterium*, *Mol. Microbiol.*, 23, 1053, 1997.
97. Rodionov, D.A., Mironov, A.A., and Gelfand, M.S., Transcriptional regulation of pentose utilisation systems in the *Bacillus/Clostridium* group of bacteria, *FEMS Microbiol. Lett.*, 205, 305, 2001.
98. Lindner, C., Stulke, J., and Hecker, M., Regulation of xylanolytic enzymes in *Bacillus subtilis*, *Microbiol.*, 140, 753, 1994.
99. Dekker, K. et al., Xylose (glucose) isomerase gene from the thermophile *Clostridium themohydrosulfuricum*: cloning, sequencing, and expression in *E. coli*, *Agric. Biol. Chem.*, 55, 221, 1991.

8 Carbohydrate Uptake by the Phosphotransferase System and Other Mechanisms

Wilfrid J. Mitchell and Martin Tangney

CONTENTS

8.1 INTRODUCTION

With the exception of a few species that are capable of autotrophic growth, clostridia lead a heterotrophic existence relying on organic molecules as a source of carbon and energy [1]. The acquisition of these molecules from the surrounding environment is a prerequisite for growth and development, and it is therefore unsurprising that specific mechanisms exist to catalyze and control their uptake. The majority of uptake, or transport, processes in bacteria utilize metabolic energy to enable concentration of the substrate in the cytoplasm at a level that can support metabolism. These active transport processes are classified according to the form of energy involved; either high-energy phosphate bonds (ATP or PEP) or ion gradients (H^+ or Na^+). Examples of each of these categories have been found in clostridia, and while the transport systems have in many cases not been characterized in detail, it is clear that clostridia employ typical bacterial mechanisms to support nutrient uptake. This chapter will review the current understanding of uptake of carbohydrates by clostridial species; strains that have been reclassified as members of the genera *Thermoanaerobacter* and *Thermoanaerobacterium* [2] have not been included.

8.2 THE BACTERIAL PHOSPHOENOLPYRUVATE (PEP)-DEPENDENT PHOSPHOTRANSFERASE SYSTEM

In common with many other obligately and facultatively anaerobic bacteria, clostridia rely heavily, but not exclusively, on the PEP-dependent phosphotransferase system (PTS) as a means of accumulating sugars and sugar derivatives [3]. The unique feature of this system is the catalysis of both

TABLE 8.1
Phosphotransferase Activities in Clostridia[a]

Strain	PTS Substrates[a]	Refs.
C. acetobutylicum	Cellobiose, fructose, glucose, lactose, maltose, mannitol, sucrose	4–9
C. beijerinckii [b]	Fructose, glucitol,[d] glucose, lactose, mannitol, sucrose	10–14
C. butyricum	Fructose	15
C. difficile	Mannitol	16
C. longisporum	Arbutin, salicin [e]	17
C. pasteurianum	Fructose, glucitol,[d] glucose, mannitol, sucrose	15,18–20
C. perfringens	Glucose, mannose [f]	21
C. roseum	Fructose	15
C. rubrum	Fructose	15
C. saccharobutylicum [c]	Galactose, glucose, lactose	22,23
C. thermocellum	Fructose, mannitol	24

[a] PTS activities demonstrated in either permeabilized cells or cell-free extracts, except where indicated otherwise.
[b] Strain NCIMB 8052, formerly *C. acetobutylicum.*
[c] Strain NCP 262 (P 262), formerly *C. acetobutylicum.*
[d] Also known as sorbitol.
[e] Evidence from interaction with the *E. coli* PTS *in vivo.*
[f] Evidence from demonstration of intracellular pool of sugar phosphate.

the uptake and the initial step of metabolism of the substrate, defined as group translocation. Thus, the substrate enters the cytoplasm as a phosphorylated derivative, the phosphate being derived from PEP. The economy of achieving two energy-dependent reactions in a single step provides a rationale for the preponderance of this mechanism in anaerobic, fermentative bacteria.

The operation of a PTS is usually demonstrated directly by monitoring PEP-dependent phosphorylation of substrate in permeabilized cells or cell-free extracts. Alternative assays have also been developed that involve coupling formation of pyruvate (the product of dephosphorylation of PEP) to NADH oxidation, or glucose 6-P (the product of the glucose PTS) to $NADP^+$ reduction using lactate dehydrogenase and glucose 6-P dehydrogenase, respectively. Principally by these criteria, a broad spectrum of phosphotransferases for monosaccharides, disaccharides and sugar alcohols has been shown to exist in clostridia (Table 8.1). Among the strains in which uptake of several of these substrates has been examined, it is clear that the PTS is the predominant mechanism in *C. acetobutylicum* and *C. beijerinckii,* but it appears to be less important in *C. thermocellum.* PTS activities for fructose and mannitol were found in *C. thermocellum* 651 [24]; however, despite observing a significant rate of PEP-dependent fructose phosphorylation in extracts, a later study with strain ATCC 27405 concluded that uptake of fructose was not PTS-dependent on the basis of assays of individual components of the system [25]. It is possible that both PTS and non-PTS mechanisms for fructose uptake operate in this species.

The PTS typically consists of a multiprotein phosphoryl transfer chain, which carries phosphate between PEP and the substrate (Figure 8.1). On the basis of sequence comparisons, a uniform nomenclature for the component proteins, or protein domains, of the system was proposed and is now universally used [26]. The first two proteins or domains, I and H (enzyme I and heat stable, histidine-phosphorylatable protein HPr encoded by the genes *ptsI* and *ptsH,* respectively), are hydrophilic and are usually recovered in the soluble fraction of cell extracts; they are referred to as the general PTS proteins, since, with few exceptions, they are common to all the phosphotransferases in the cell. The other domains constitute the substrate-specific part of the system ("enzyme II"). A typical enzyme II complex incorporates three domains, termed IIA, IIB, and IIC, while a small

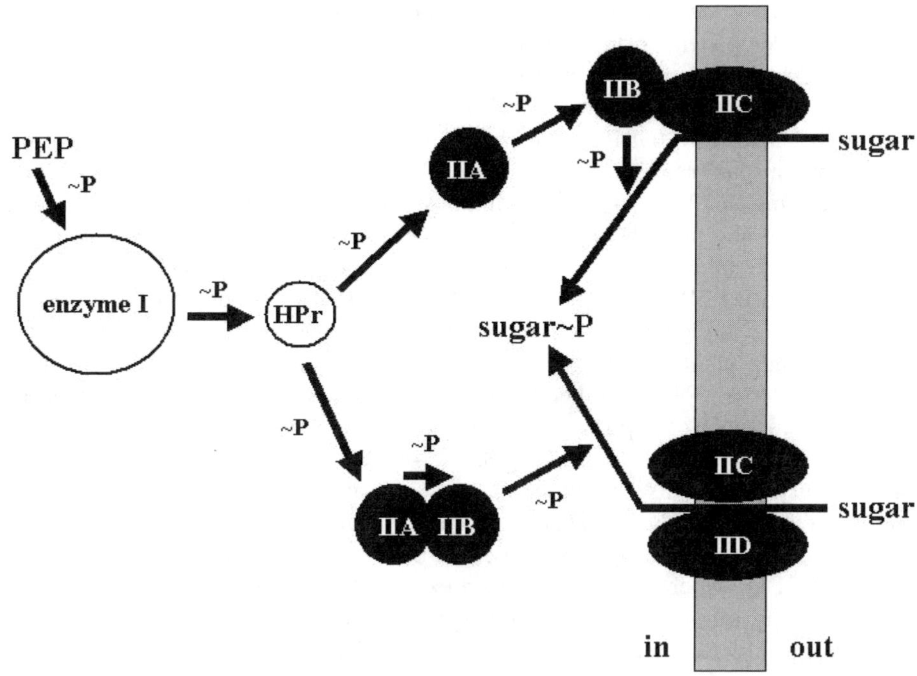

FIGURE 8.1 The bacterial phosphoenolpyruvate-dependent phosphotransferase system. The system comprises a phosphoryl transfer chain between the donor (PEP) and the substrate, which is phosphorylated as it enters the cell. Both general (enzyme I, HPr) and substrate-specific proteins are involved. The substrate-specific part typically incorporates three domains (IIA, IIB, and IIC), although some systems have an additional domain (IID). The molecular architecture of the IIABC domains is variable; they can be carried on separate proteins or fused together in different combinations. The examples shown represent the domain structure of the glucose PTS (top) and mannose PTS (bottom) of *Escherichia coli* [27].

number include an additional domain termed IID. The IIA and IIB domains are hydrophilic, typically around 100 residues in length, and contain sites for phosphorylation. On the other hand, IIC and IID domains are found within integral membrane proteins that function to translocate the substrate, and there is no evidence that they are phosphorylated. IIC proteins are considerably more hydrophobic than IID proteins and, in general, are predicted to comprise a greater number of transmembrane helices. Despite the fact that all phosphotransferases have a similar domain structure, there is considerable variation in their molecular architecture in that the domains can reside within separate proteins, or be fused together in different combinations [26–28]. This has been interpreted as evidence for the occurrence of extensive genetic rearrangements throughout evolution.

8.2.1 The Clostridial PTS

8.2.1.1 Functional Characterization

The clostridial PTS includes both soluble and membrane-bound components, as has been demonstrated by means of fractionation and reconstitution of cell extracts. This would be expected based on the cellular distribution of proteins shown in Figure 8.1. That the clostridial system comprises functional domains equivalent to those in other bacteria became evident from the results of *in vitro* heterologous sugar phosphorylation assays. Thus, the soluble fraction of *C. beijerinckii* (formerly *C. acetobutylicum*) NCIMB 8052 complemented membranes prepared from *Escherichia coli* and *Bacillus subtilis* (and also *C. pasteurianum*) for glucose phosphorylation, and vice-versa [10]. Furthermore, PTS activity was restored to extracts prepared from *ptsI* and *ptsH* mutants of *E. coli*,

B. subtilis, and *Staphylococcus aureus* by addition of clostridial extract [10,14]. Functional interactions between PTS components have also been shown to occur *in vivo*; thus a cloned β-glucoside PTS enzyme IIBC from *C. longisporum* enabled *E. coli* to grow on arbutin and salicin [17], while complementation of an *E. coli ptsH* mutant by *C. acetobutylicum* HPr has been reported recently [29].

The affinity of the clostridial PTS for glucose (K_m in the range 12 to 34 μ*M*) is similar to corresponding systems in other bacteria [10,21,30]. Competition studies suggested that glucose and mannose were accumulated by the same PTS in both *C. perfringens* and *C. beijerinckii* [10,14,21], while glucose uptake and phosphorylation by both *C. beijerinckii* and *C. acetobutylicum* ATCC 4259 were found to be inhibited by the analog 2-deoxyglucose (2DG), which itself was shown to be phosphorylated in a PEP-dependent manner [4,8,10,14]. On the other hand, neither methyl α-glucoside nor 3-O-methylglucose was found to be an effective inhibitor in the latter two strains [4,10]. These findings suggest that these glucose phosphotransferases have a similar substrate specificity to the glucose PTS in other Gram-positive bacteria, particularly streptococci [31]. Conversely, the glucose PTS of *C. pasteurianum* phosphorylates methyl α-glucoside and is inhibited by 2DG but not significantly by mannose [18,19], suggesting that it may belong to a different family, or subfamily, of permeases. Methyl α-glucoside is also phosphorylated by the glucose PTS of *C. acetobutylicum* ATCC 824 [8]. This strain has, in addition, been shown to accumulate 3-O-methylglucose, although no evidence was presented that the glucose PTS was responsible [32,33].

Resistance to 2DG, which is toxic, has been exploited as a means of selecting mutants with impaired PTS activity following chemical mutagenesis. One class of *C. pasteurianum* mutants exhibited a pleiotropic defect in PTS activity, consistent with loss of function of enzyme I or HPr [18]. The 2DG-resistant strain of *C. beijerinckii*, BA101, also appears to have a pleiotropic defect in PTS activity, with substrate phosphorylation in cell extracts reduced by around 50% [14]. This strain has attracted interest due to the fact that it exhibits more complete utilization of glucose and greater production of butanol than the parent strain [34,35]. Since glucose utilization was enhanced despite reduced PTS activity, it has been suggested that an alternative, non-PTS transport system may contribute to glucose utilization [14].

The distribution of the general PTS proteins between membrane and soluble fractions in both *C. pasteurianum* [19] and *C. beijerinckii* [10] was found to be atypical in that a proportion of enzyme I was firmly associated with the membrane. As a result, a combination of membranes and soluble HPr was sufficient to reconstitute a functional PTS. Although the distribution of enzyme I between fractions could simply be a consequence of the method of cell disruption and extract preparation, the membrane-bound enzyme could not be removed by salt treatment. It is therefore likely that this close association with the membrane reflects the physicochemical properties of the enzyme I protein. The *E. coli* enzyme I has been shown to exhibit hydrophobic characteristics [36], and there is evidence that the soluble PTS domains in this bacterium form a membrane-associated complex in the intact cell [37]. Furthermore, in the case of the photosynthetic bacteria *Rhodospirillum rubrum* and *Rhodopseudomonas sphaeroides* and the prosthecate, freshwater bacterium *Ancalomicrobium adetum*, the entire PTS complex was isolated in the membrane fraction [38–40]. The observed distribution of the general PTS proteins in clostridia is, therefore, not unique but does represent a variation on previously identified patterns.

The molecular architecture of a number of clostridial phosphotransferases has been investigated by means of assays with fractionated cell extracts. The principle of these assays, dependent on inducible synthesis of specific PTS permeases, is illustrated in Figure 8.2. Uninduced cells, grown on glucose (Figure 8.2A), will contain only the general PTS components enzyme I and HPr (which are required for glucose PTS activity), while induced cells will, in addition, contain the substrate-specific domains. As a result, PTS activity for the substrate in a combination of uninduced soluble fraction and induced membranes will be dependent on the domain organization within the enzyme II part of the system. If activity is detected, all substrate-specific domains must be membrane-associated (Figure 8.2B). On the other hand, lack of activity indicates that at least one substrate-specific domain is soluble (Figure 8.2C). The latter result is normally interpreted as evidence for

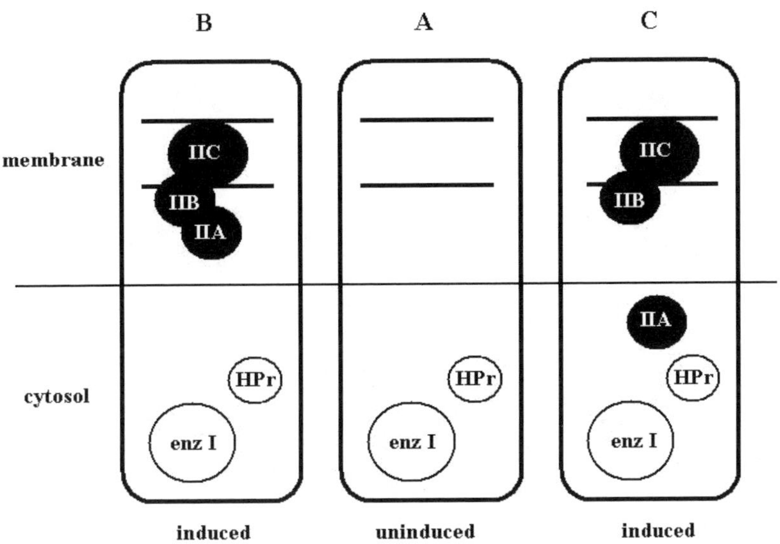

FIGURE 8.2 Determination of the molecular architecture of a phosphotransferase system. Cells that are uninduced for a specific permease contain only the general PTS proteins enzyme I and HPr (A). Following induction, the domains of the specific permease are either all associated with the membrane (B) or distributed between the membrane and cytosol (C). The combination of uninduced cytosol (A) and induced membrane (B) will give PTS activity, whereas the combination of uninduced cytosol (A) and induced membrane (C) will be inactive since the IIA domain is absent.

a soluble IIA protein, since no phosphotransferases are known in which the A and C domains are fused together with the B domain in a separate protein [41]. However, the possibility that both IIA and IIB are soluble cannot be excluded. This experimental approach has provided evidence for a soluble, substrate-specific component of the glucitol PTS in *C. beijer*inckii [12] and the mannitol PTS in *C. acetobutylicum* [6] while no such component was detected for the sucrose PTS in either *C. beijerinckii* [13] or *C. acetobutylicum* [5], nor the maltose PTS in *C. acetobutylicum* [7]. In each case, the biochemical data have been found to be consistent with the genetic structure of the system (see below).

8.2.1.2 The General PTS Domains

A minimal amount of characterization of clostridial enzyme I and HPr proteins has been reported. The two proteins from *C. beijerinckii* were separated by gel filtration chromatography and shown to be of a similar size to their counterparts in other bacteria [10], while phosphorylation of proteins from both *C. beijerinckii* and *C. acetobutylicum* has been demonstrated directly following incubation of extracts with ^{32}PEP [8]. Genome sequencing data have recently provided more detailed information on the structure of these proteins and their relationships to enzyme I and HPr in other bacteria.

Each of the clostridial genomes that has been completely sequenced contains a single gene encoding a putative enzyme I-like protein. Phylogenetic analysis indicates that the relationships between proteins of the enzyme I family broadly reflect the taxonomic relationships of the bacteria, and the clostridial proteins clearly form a separate cluster (Figure 8.3). These proteins exhibit between 50 and 70% identity with each other, and all contain the conserved histidine residue that is phosphorylated by PEP (His189 in *E. coli*) [27], as well as other residues that have been implicated as important in the enzyme's function in enteric bacteria [42–44]. The proteins from *C. acetobutylicum* and *C. perfringens*, however, lack approximately 30 amino acids at the C-terminal end.

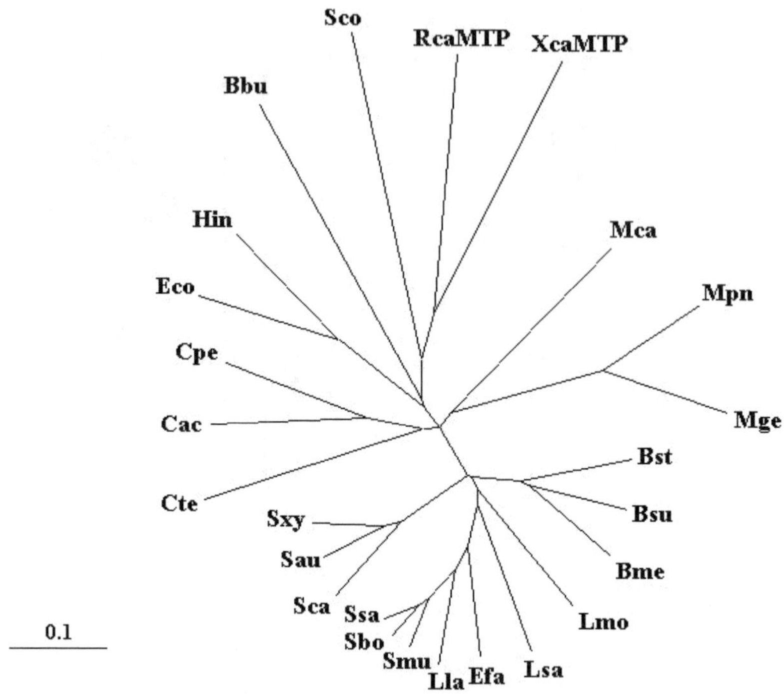

FIGURE 8.3 Phylogenetic analysis of enzyme I proteins and domains. An unrooted phylogenetic tree is presented showing clostridial enzyme I proteins in relation to other members of the enzyme I family. Abbreviations (for enzyme I proteins, unless otherwise stated) with accession numbers in parentheses are as follows: Bbu, *Borrelia burgdorferi* (NP_212692); Bme, *Bacillus megaterium* (O69251); Bst, *Bacillus stearothermophilus* (P42014); Bsu, *Bacillus subtilis* (P08838); Cac, *Clostridium acetobutylicum* (NP_349687); Cpe, *Clostridium perfringens* (NP_563273); Cte, *Clostridium tetani* (NP_782363); Eco, *Escherichia coli* (P08839); Efa, *Enterococcus faecalis* (NP_814461); Hin, *Haemophilus influenzae* (NP_439854); Lla, *Lactococcus lactis* (NP_266274); Lmo, *Listeria monocytogenes* (NP_464528); Lsa, *Lactobacillus sakei* (O07126); Mca, *Mycoplasma capricolum* (P45617); Mge, *Mycoplasma genitalium* (NP_073100); Mpn, *Mycoplasma pneumoniae* (NP_110316); RcaMTP, *Rhodobacter campestris* MTP enzyme I domain (P23388); Sau, *Staphylococcus aureus* (NP_371608); Sbo, *Streptococcus bovis* (Q9WXK9); Sca, *Staphylococcus carnosus* (P23533); Sco, *Streptomyces coelicolor* (NP_625674); Smu, *Streptococcus mutans* (AAA91093); Ssa, *Streptococcus salivarius* (P30299); Sxy, *Staphylococcus xylosus* (AAG38584); XcaMTP, *Xanthomonas campestris* MTP enzyme I domain (P45597).

Analysis reveals the presence of a single gene encoding HPr in the genomes of *C. acetobutylicum* and *C. perfringens*, while a putative HPr has also been identified in *C. tetani* [45]. The deduced HPr proteins of *C. acetobutylicum*, *C. perfringens,* and *C. tetani* share a minimum 70% identity with each other and form an independent cluster in the phylogenetic tree, separate from those of other low-GC Gram-positive bacteria (Figure 8.4). Nevertheless, both groups of proteins show almost complete conservation of amino acid sequence around the catalytic phosphorylated residue His15, and also around residue Ser46, phosphorylation of which by an ATP-dependent, metabolite-activated HPr kinase/phosphorylase is the key determinant of carbon catabolite repression in low GC Gram-positive bacteria [46] (see Chapter 25). There are no additional HPr homologues in the clostridia. This is of particular interest in light of the fact that *B. subtilis* has been found to contain a homologue of HPr termed Crh, which has an exclusively regulatory role in the cell [47] (see Chapter 25).

In the majority of eubacteria, the *ptsH* and *ptsI* genes encoding HPr and enzyme I are located in an operon and are co-expressed [27]. In the sequenced clostridial genomes, these genes are not adjacent and appear to be monocistronic. Nothing is known about control of their expression. However, upstream and divergent to *ptsI* in *C. acetobutylicum* is a gene encoding a putative

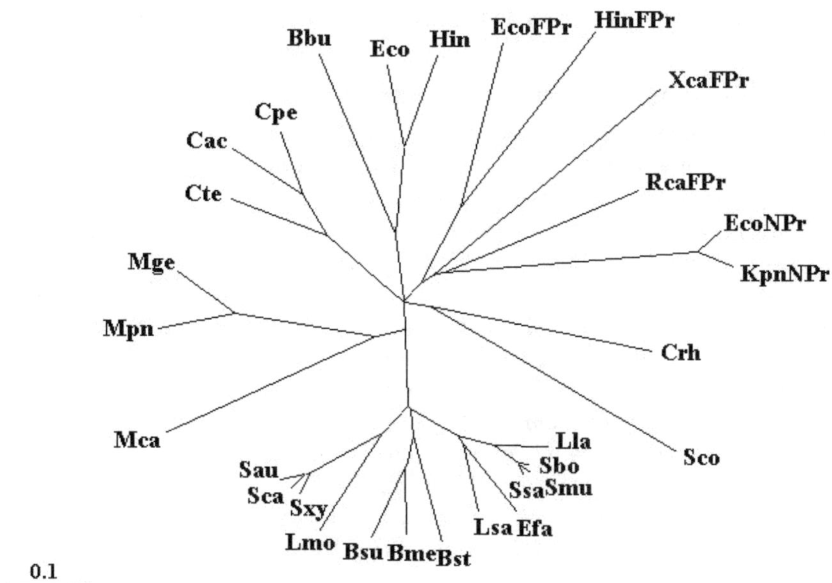

FIGURE 8.4 Phylogenetic analysis of HPr proteins and related PTS domains. An unrooted phylogenetic tree is presented showing clostridial HPr proteins in relation to other members of the HPr family. Abbreviations (for HPr proteins, unless otherwise stated) with accession numbers in parentheses are as follows: Bbu, *Borrelia burgdorferi* (AE001157); Bme, *Bacillus megaterium* (AJ005075); Bst, *Bacillus stearothermophilus* (U12340); Bsu, *Bacillus subtilis* (A46238); Cac, *Clostridium acetobutylicum* (AY196477); Cpe, *Clostridium perfringens* (NP_562585); Crh, *Bacillus subtilis* catabolite repression HPr-like protein (Z94043); Cte, *Clostridium tetani* (DAA02130); Eco, *Escherichia coli* (P07006); EcoFPr, *Escherichia coli* FPr (P24217); EcoNPr, *Escherichia coli* NPr (P33996); Efa, *Enterococcus faecalis* (A25053); Hin, *Haemophilus influenzae* (P43921); HinFPr, *Haemophilus influenzae* FPr (P44715); KpnNPr, *Klebsiella pneumoniae* NPr (P51185); Lmo, *Listeria monocytogenes* (AF030824); Lsa, *Lactobacillus sakei* (AF172726); Lla, *Lactococcus lactis* (Q9CJ83); Mca, *Mycoplasma capricolum* (P07515); Mge, *Mycoplasma genitalium* (P47287); Mpn, *Mycoplasma pneumoniae* (P75061); RcaFPr, *Rhodobacter campestris* FPr (P23388); Sau, *Staphylococcus aureus* (X93205); Sbo, *Streptococcus bovis* (AB027569); Sca, *Staphylococcus carnosus* (X60766); Sco, *Streptomyces coelicolor* (AL009204); Smu, *Streptococcus mutans* (A44562); Ssa, *Streptococcus salivarius* (Z17217); Sxy, *Staphylococcus xylosus* (AF316496); XcaFPr, *Xanthomonas campestris* FPr (P45597).

multidomain regulator (HRRP) comprising an N-terminal HPr-like domain and domains showing similarity to the NtrC family of response regulators [48]. Since HRRP contains a σ^N–activating domain, and *ptsI* is associated with a characteristic σ^N promoter, it was suggested that HRRP might be involved in regulation of *ptsI* expression in response to phosphorylation of the HPr domain by enzyme I [48]. Interestingly, the same gene arrangement is found in *C. perfringens*. The potential role of HRRP in regulation of PTS activity in clostridia is, therefore, worthy of further investigation.

8.2.1.3 The Enzyme II Domains

There are six families of PTS permeases [41]. Principally as a result of genome sequencing projects [49–51], a considerable number of clostridial phosphotransferases have been identified at the sequence level, and these include representatives of all six families. It is clear that clostridial systems comprise the same domains as the PTS in other bacteria, consistent with biochemical characterization described above, while a similar variability in gene organization and domain architecture is apparent. Sequence comparisons and assignment to a PTS family can provide some indication of the potential substrate of a permease, and, since many phosphotransferase genes are located in operons together with genes encoding enzymes for mobilization of the substrate within the cell, the identity of adjacent genes can also be informative. At this stage, however, the function of many systems has yet to be resolved.

8.2.1.3.1 The PTS Complement of C. acetobutylicum *ATCC 824*

Of the strains whose genomes have been completely sequenced, *C. acetobutylicum* has the greatest complement of phosphotransferases with 13 systems, including one encoded on the large pSOL1 plasmid that carries the genes for acetone and butanol formation [52]. An additional glucose-specific IIA domain (IIA^{glc}) is also present. The domain structure of these phosphotransferases is summarized in Figure 8.5. Of the 13 systems, six belong to the glucose-glucoside (Glc) family, two each to the fructose-mannitol (Fru) family, the lactose-*N,N′*-diacetylchitobiose-β-glucoside (Lac) family, and the mannose-fructose-sorbose (Man) family, and one to the galactitiol (Gat) family.

The Glc family includes permeases specific for glucose, N-acetylglucosamine, and a variety of α- and β-glucosides. Although PTS activity for a number of these substrates has been demonstrated in extracts of *C. acetobutylicum* ATCC 824 (Table 8.1), assignment of substrates to the six permeases within this group is incomplete, and a considerable amount of functional characterization of the phosphotransferases is still required. The sucrose PTS has been studied in this organism. Analysis of the sequences of PTS and associated genes encoding sucrose 6-P hydrolase and fructokinase, supported by measurements of enzyme activities and gene expression, clearly identified one of the "β-glucoside" systems as the sucrose PTS [5]. This IIBCA permease exhibits the BC domain arrangement and the motif –HCXTRLR- (around the phosphorylated Cys residue in the B domain), which are characteristic of the sucrose subfamily [28]. A second system has been implicated in the metabolism of maltose [7]. The gene products of the putative maltose operon, a

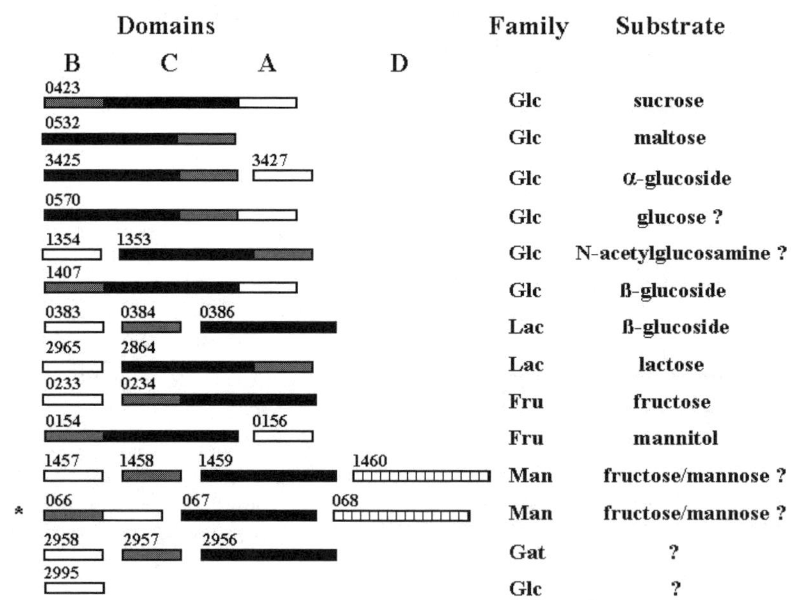

FIGURE 8.5 The phosphotransferase complement of *Clostridium acetobutylicum* ATCC 824. All systems are encoded on the chromosome with the exception of one member of the Man family, indicated by *, which is encoded on the plasmid pSOL1. CAC ORF numbers (or CAP ORF, in the case of the plasmid-borne genes) are shown above proteins, which are represented by horizontal bars. Architecture of each PTS is indicated by organization of domains within and between proteins, with the N-terminus at the left. The order of proteins reflects the order of genes within the respective operons, but in some cases genes encoding the proteins of the PTS are separated by metabolic or regulatory genes. The genes encoding the putative N-acetylglucosamine PTS are divergently transcribed. For each system, the family to which it is assigned by phylogenetic analysis, and the probable substrate, are indicated.

IICB PTS permease (MalP) and maltose 6-P hydrolase (MalH), show close identity to equivalent proteins in *B. subtilis* (67 and 75%, respectively), *Fusobacterium mortiferum* (61 and 81%), and *Klebsiella pneumoniae* (58 and 76%). A *malP* mutant of *B. subtilis* was shown to have a specific defect for growth on maltose [53], while the MalH (also known as GlvA or AglB) enzymes have been shown to be phospho-α-glucosides capable of hydrolyzing maltose 6-P [54–56]. The conclusion that the *malPH* operon in *C. acetobutylicum* is concerned with maltose metabolism is therefore compelling. Interestingly, recent studies have suggested that the maltose/α-glucoside phosphotransferases and associated hydrolases in *F. mortiferum* and *K. pneumoniae* are involved in uptake of isomers of sucrose [56,57], and we have found that ability of *C. acetobutylicum* to metabolize the isomers leucrose, palatinose, and turanose is similarly induced by maltose but not by sucrose [58]. As also observed for several other phosphotransferases of the Glc family, the putative *mal* operon of *C. acetobutylicum* does not encode a IIA domain. There is evidence, including for *B. subtilis* MalP, that the missing function is provided by the glucose-specific IIA^{glc} [53,59–62]. It is therefore likely that IIA^{glc} participates in maltose uptake in *C. acetobutylicum.*

A second putative α-glucoside operon also encodes a IICB permease of the Glc family and a phospho-α-glucosidase. The protein products show, respectively, 33 and 50% identity to MalP and MalH, but compared to these, they are much less closely related to the corresponding proteins in *F. mortiferum* and *K. pneumoniae*. This operon differs from the putative *mal* operon in encoding a discrete IIA component of the PTS, but biochemical analysis has shown that a soluble, substrate-specific IIA domain is not required for maltose phosphorylation [7]. Thus, although this system may have some affinity for maltose, it can be concluded that it is not essential for maltose metabolism, and its substrates remain to be identified.

On the basis of sequence homology and phylogeny, two further permeases that are not associated with metabolic genes seem likely to be involved in uptake of glucose (IICBA) and N-acetylglucosamine (IIA/IICB). However, it should be noted that the splitting of the putative N-acetylglucosamine PTS into two polypeptide chains is unusual, as is the divergent transcription of the corresponding genes. Furthermore, the genes are not associated with those for metabolism of N-acetylglucosamine 6-P that lie elsewhere on the chromosome. The final member of the Glc family is a IIBCA β-glucoside permease, which, according to the domain order and sequence around the phosphorylated Cys, belongs to the sucrose subfamily. This protein, closely related (65% identity across its entire length) to the aryl-β-D-glucoside PTS of *C. longisporum,* which has been shown to transport arbutin and salicin [17], is one of two PTS permeases in *C. acetobutylicum* associated with genes encoding phospho-β-glucosidase. The other is a tripartite IIA/IIB/IIC member of the Lac family, which is known to include permeases that transport cellobiose and other β-glucosides including lichenan degradation products and *N,N′*-diacetylchitobiose [63–66]. *C. acetobutylicum* does exhibit PTS activity for cellobiose [8], but whether one or both of these β-glucoside permeases is responsible is not known, and the spectrum of substrates accommodated by each remains to be resolved. The other permease that belongs to the Lac family is a lactose PTS with IIA and IICB components. This system is expressed in an operon with a phospho-β-galactosidase enzyme, and its synthesis is induced by lactose and galactose [9].

It seems clear that the two permeases of the Fru family are concerned with the uptake of fructose and mannitol, respectively. The bipartite IIA/IIBC fructose permease is encoded in an operon together with 1-phosphofructokinase; the implication is that, like other fructose permeases in this family [11,67,68], it generates fructose 1-P, which is then phosphorylated to fructose 1,6-bisP. The presence of a separate IIA protein is unusual, although not unique, in a fructose-1-P forming PTS. Similarly, the IICB/IIA mannitol PTS is associated with mannitol 1-P dehydrogenase. Expression of the equivalent *mtl* operon in the closely related strain *C. acetobutylicum* DSM 792, which encodes virtually identical proteins, has been shown to be induced by mannitol [6].

Two phosphotransferases in *C. acetobutylicum* ATCC 824 incorporate IID domains; one encoded on the chromosome and the other on the pSOL1 plasmid. Unsurprisingly, these belong to the Man family, whose members are the only PTS permeases that exhibit this trait. These systems also share the property that the IIB domain is phosphorylated on a His rather than a Cys residue [69], and the candidate His is conserved in the putative clostridial proteins. The proteins of the two systems are homologous but not closely related, with identities of the corresponding components ranging from 22 to 30%. An obvious difference is that in the plasmid-encoded system, the IIA and IIB domains are fused within a single protein. The PTS permeases of the Man family often exhibit a broad specificity, and as such, the function of the two systems can only be speculated upon at present. The chromosomal genes are associated with a gene encoding a putative levanase/invertase, indicating that together they may form the counterpart of the *lev* operon in *B. subtilis*, which encodes a PTS that accumulates fructose (as fructose 6-P) and mannose [70,71]. *C. acetobutylicum* grows on mannose, but we have found that this strain will not grow on sorbose [8]. This in itself does not rule out the possibility that sorbose may be transported and phosphorylated by one or both of the Man family phosphotransferases, since genes encoding enzymes for further metabolism of the sugar (sorbose 1-P reductase and glucitol 6-P dehydrogenase) are not present in the genome. It does, however, suggest that sorbose uptake is not their primary function.

The final PTS in *C. acetobutylicum* ATCC 824 is a member of the Gat family. The corresponding genes lie upstream of genes encoding enzymes of the tagatose 6-P pathway, similar to the gene arrangement in enteric bacteria [72]. It is noticeable that galactose 6-P isomerase, rather than galactitol 1-P dehydrogenase, is encoded within this gene cluster, suggesting that it is not concerned with metabolism of galactitol. A possible function could be the uptake and metabolism of galactose, but there is no evidence that PTS permeases of the Gat family can transport this sugar [73]. Furthermore, extracts of *C. acetobutylicum* grown on galactose were shown to have negligible PTS activity for this substrate [8]. A more likely role is metabolism of the galactose moiety of lactose, which, following PTS-mediated uptake of the disaccharide and subsequent hydrolysis, would appear in the cell as galactose 6-P. Indeed, equivalent genes directing metabolism of galactose 6-P are part of the lactose PTS operons of *Lactococcus lactis* [74], *Staph. aureus* [75] and *Streptococcus mutans* [76], and while this is not the case in *C. acetobutylicum* we have shown that expression of the galactose 6-P isomerase is induced by growth on lactose [9]. Therefore, the identity of the substrates recognized by the *C. acetobutyicum* "galactitol" PTS remains to be established.

In addition to the 13 complete phosphotransferase systems, the *C. acetobutylicum* genome contains an isolated gene encoding a putative IIA^{glc} domain. The deduced protein is not closely related to other IIA^{glc} domains, and it is possible that it is a redundant vestige of evolution. Alternatively, it may have evolved to play a specific regulatory role in the cell. It is well established that IIA^{glc} is a key regulatory protein in carbon catabolite control in enteric bacteria [27], although there is currently no evidence for a similar significance in Gram-positive bacteria.

In summary, *C. acetobutylicum* ATCC 824 possesses a spectrum of phosphotransferase systems, which enable it to accumulate and grow on a range of hexoses, disaccharides, and derivatives. The functions of many of these systems can be deduced with a degree of confidence, but more detailed physiological analysis will further increase our appreciation of the importance of the PTS in this industrially significant organism.

8.2.1.3.2 The PTS Complement of C. perfringens *and* C. tetani

There is a solitary report of PTS activity for glucose and mannose in *C. perfringens*, with the sugars apparently phosphorylated by the same system [21]. The genome sequence of strain 13 has uncovered several phosphotransferases that could potentially account for this activity. It appears that there are four complete IIA/IIB/IIC/IID permeases of the Man family. Genes encoding one of these systems are adjacent to a gene encoding 6-phosphofructokinase, indicative of a fructose 6-P forming PTS. A second system is closely related to the PTS encoded on pSOL1 of *C. acetobutylicum* with the same molecular architecture (i.e., A and B domains within a single protein) and is not linked to relevant metabolic genes. The other two systems are associated with genes involved in sugar or sugar phosphate metabolism, but their functions are not immediately apparent. One of these has the unusual gene order *BCDA* rather than *ABCD*, and the B domain is much less closely related to the Man family than are the others. This system might, therefore, have arisen by gene rearrangement following divergence of the PTS families.

C. perfringens also contains phosphotransferases, which, on the basis of phylogenetic relationships and associated metabolic genes, can be predicted to be involved in uptake of fructose (forming fructose 1-P), glucose, N-acetylglucosamine, and the disaccharides maltose, sucrose, and trehalose. The putative N-acetylglucosamine and maltose permeases are closely related to their counterparts in *C. acetobutylicum*, and the sucrose permease to that of *C. beijerinckii* (see below). The only IIA domain belonging to the Glc family is encoded as part of the putative N-acetylglucosamine PTS; as for the corresponding system in *C. acetobutylicum*, the A domain is transcribed divergently to the C and B domains, but the latter are separate proteins in this case. Whether this single IIA domain is involved in phosphoryl transfer between HPr and all the permeases of the Glc family is a matter of speculation. The *C. perfringens* genome also contains isolated genes encoding, respectively, a IIA domain and a IIB domain of the Lac family, but these must be of questionable significance in terms of PTS activity.

In keeping with a lifestyle that is less dependent on carbohydrate metabolism, the *C. tetani* genome has a lower content of PTS permeases than other clostridia. In fact, the genome sequence reveals just one complete permease, a IICBA protein that has been designated as a glucose PTS and is closely related (66% identity) to the putative glucose PTS of *C. acetobutylicum*. We are not aware of any reports of PTS activity in *C. tetani*.

8.2.1.3.3 The PTS in Other Clostridia

PTS activity for a number of substrates has been demonstrated in *C. pasteurianum* and *C. beijerinckii* (Table 8.1), but relatively little information is available at the molecular level. Two PTS operons, concerned with metabolism of glucitol and sucrose, respectively, have been cloned from *C. beijerinckii*. The glucitol operon, which was isolated via complementation of an *E. coli gutD* mutant lacking glucitol 6-P dehydrogenase, encodes a glucitol PTS and the dehydrogenase, together with an additional protein OrfX [12]. This protein is homologous to transaldolase enzymes, suggesting that it may be involved in metabolism of fructose-6-P, which is produced by oxidation of the PTS product glucitol 6-P; however, no aldolase-related activities could be demonstrated in the overexpressed and purified protein [8], and its function in glucitol metabolism therefore remains unknown. Expression of the operon and synthesis of the PTS and dehydrogenase was shown to be induced by glucitol, thus clearly implicating it in glucitol metabolism [12]. The gene order in the *gut* operon is *gutA1gutA2orfXgutBgutD*. The PTS is encoded by the *gutA1* and *gutA2* (IICB) and *gutB* (IIA) genes and belongs to the Gut family, which includes only glucitol PTS permeases. These permeases are unusual in that the IIB domain is encoded between two segments of the IIC domain; thus *gutA1* encodes a hydrophobic protein that is the first part of the IIC domain, while *gutA2* encodes the hydrophilic IIB domain fused to the second hydrophobic segment of IIC. An alignment of the *E. coli* and *C. beijerinckii* glucitol permeases was used to propose a topological model, which predicted that each segment of the IIC domain contains four transmembrane regions [77]. The *E. coli* permease was originally believed to be a single polypeptide chain [78], but a re-examination of the *gutA* sequence revealed that it too was divided [79]. The glucitol permeases of both *Erwinia amylovora* [80] and *Strep. mutans* [81] have also been shown to have the same structure, while additional, similar members of the Gut family have been revealed by a number of genome sequencing projects.

The sucrose PTS operon of *C. beijerinckii* was isolated by screening of a plasmid gene library in *E. coli* for a sucrose fermenting phenotype. The operon comprises genes *scrARBK*, encoding a IIBC PTS permease, a transcriptional regulator, sucrose 6-P hydrolase, and fructokinase, respectively [82]. Disruption of either the *scrR* or *scrB* genes by targeted integration prevented growth on sucrose, indicating that this operon was solely responsible for sucrose metabolism in *C. beijerinckii*. The gene arrangement is identical to that in the putative sucrose PTS operon of *C. perfringens*, even to the extent that the *scrB* and *scrK* genes overlap by eight nucleotides. On the other hand, the corresponding operon in *C. acetobutylicum* ATCC 824 encodes a sucrose PTS permease that includes a IIA domain and exhibits differences in gene order and regulation [5]. These differences are considered in detail in Chapter 25.

A PTS operon concerned with uptake and metabolism of aryl-β-glucosides has been isolated from the ruminal strain, *C. longisporum*. The operon was obtained from a gene library in *E. coli* by screening for hydrolysis of the fluorogenic substrate methylumbelliferyl β-D cellobioside (MUC) [17]. MUC hydrolysis in *E. coli* was found to be dependent on the presence of both enzyme I and HPr, indicating that a PTS was involved in uptake and hydrolysis of MUC and that components of the *E. coli* and clostridial PTS could interact productively *in vivo*. The genes of the *C. longisporum* operon, *abgG*, *abgF*, and *abgA*, encode, respectively, a transcriptional antiterminator, a IIBC PTS permease belonging to the Glc family, and a phospho-β-glucosidase. When expressed in an *E. coli* strain from which the resident cryptic *bgl* (β-glucoside) and *cel* (cellobiose) operons had been deleted, the *abgF* and *abgG* genes conferred the ability to grow on arbutin and salicin, but not cellobiose. Further analysis employing *E. coli* mutants specifically lacking either β-glucoside PTS or phospho-β-glucosidase activities confirmed the functions of these genes and the specificity of the encoded proteins [17].

Genome sequencing projects for *C. botulinum*, *C. difficile,* and *C. thermocellum* are in progress. PTS activities have been demonstrated in both *C. difficile* and *C. thermocellum* (Table 8.1), but there do not appear to have been any attempts to do likewise for *C. botulinum*. All three genomes encode proteins with similarity to general PTS proteins and PTS permeases, but analysis of the PTS complement and its significance in these bacteria must await completion of genome sequencing.

8.3 NON-PTS TRANSPORT MECHANISMS

In general, lack of PEP-dependent phosphorylation of a substrate by permeabilized cells or cell extracts is taken as indicative of the absence of a PTS, implying that accumulation by the cell must be by an alternative mechanism. Based on this criterion alone, it is apparent that some non-PTS transport systems must contribute to carbohydrate uptake in clostridia. Indeed, the plethora of putative ATP- and ion-dependent transport systems encoded by the sequenced genomes is testament to the general importance of non-PTS transport processes in clostridial physiology. There have, however, been few detailed studies of non-PTS permeases, and it is not clear at this stage how many are involved in carbohydrate accumulation as opposed to uptake of other solutes, or efflux of molecules from the cell.

In fermentative bacteria, ATP is generated by substrate level phosphorylation reactions incorporated into degradative metabolic pathways; hydrolysis of ATP via a membrane-bound ATPase then results in generation of a transmembrane ion gradient. The majority of bacteria contains a H^+-translocating F_1F_0-type ATPase, which establishes a primary H^+ gradient across the membrane (Figure 8.6). Exchange of H^+ and Na^+ via a Na^+/H^+ antiporter, driven by the H^+ gradient, can then generate a secondary gradient of Na^+ ions. On the other hand, a V-type ATPase may directly couple Na^+ translocation to ATP hydrolysis to generate a primary gradient of Na^+ ions. Although there have been few detailed studies of the energetics of heterotrophic clostridia, there is evidence that the requisite components for energy transductions are present. The ATPase complex in *C. pasteurianum* has been shown to translocate H^+ [83], and while no biochemical characterization has been carried out in *C. acetobutylicum* or *C. perfringens,* the presence of genes encoding a F_1F_0-type ATPase suggests a similar mechanism for establishment of a transmembrane H^+ gradient. *C. acetobutylicum* and *C. perfringens* also encode putative Na^+/H^+ antiporters, although a study of Na^+ translocation in the former failed to detect any antiport activity [84]. In contrast, the peptidolytic thermophile *C. fervidus*, having a V-type ATPase [85], apparently relies wholly on Na^+ ion circulation [86], and genome analysis suggests that the same is true for *C. tetani* [51]. On the basis of the available evidence, it appears that Na^+ ions have a greater significance for uptake of amino acids, rather than sugars and other carbohydrates.

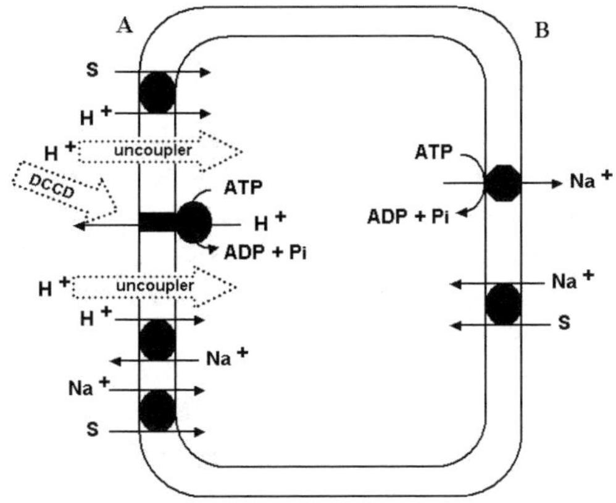

FIGURE 8.6 Generation and utilization of ion gradients in anaerobic bacteria. (A) Generation of primary H^+ gradient and secondary Na^+ gradient by F_1F_0-ATPase and Na^+/H^+ antiport, and use of both ion gradients to support solute uptake. Inhibitory action of uncouplers and DCCD is indicated. (B) Generation of primary Na^+ gradient by V-type ATPase, and use of Na^+ gradient to support solute uptake. S, solute.

The most comprehensive study of ion-dependent carbohydrate accumulation in clostridia to date has been the analysis of galactose and gluconate uptake by *C. pasteurianum* [87]. Accumulation of these substrates is inhibited by the proton-conducting uncouplers tetrachlorosalicylanilide (TCS) and carbonylcyanide *m*-chlorophenylhydrazone (CCCP), as well as by N,N′-dicyclohexylcarbodiimide (DCCD), an inhibitor of the membrane-bound ATPase, suggesting that they are taken up via a H^+-symport mechanism. This conclusion was supported by the demonstration that uptake is accompanied by alkalinization of the medium and can be driven by an artificially generated proton gradient. Galactose uptake was driven by both the charge (membrane potential) and chemical (pH) components of the gradient, whereas only the latter was found to be able to support uptake of gluconate (which carries a negative charge, thus resulting in an electroneutral H^+-gluconate translocation process). This would be expected if the H^+:substrate stoichiometry was 1:1. The subsequent observations that a *pts* mutant could transport and grow normally on both galactose and gluconate, and that phosphorylation of galactose by permeabilized cells is 10-fold greater in the presence of ATP compared to PEP, further verified that the PTS is not involved in uptake of these substrates [18]. Gluconate is, in fact, metabolized by *C. pasteurianum* without initial phosphorylation [88], while galactose is phosphorylated by galactokinase and then presumably metabolized via the Leloir pathway [89].

In extracts of *C. beijerinckii*, ATP-dependent galactokinase activity was again found to be more than 10-fold greater than the PEP-dependent phosphorylation rate [11], consistent with the conclusion that galactose is accumulated in an unaltered form by a non-PTS mechanism. The apparent PTS activity was extremely low, but could represent a galactose PTS. However, it seems more likely that it reflects an affinity for galactose of another system, most probably a lactose PTS. A similar pattern of galactose phosphorylation is exhibited by extracts of *C. acetobutylicum* ATCC 824 [8], the genome of which encodes the enzymes of the Leloir pathway. In *Staph. aureus* and some lactic streptococci, galactose is known to be a substrate of the lactose PTS [90,91], while galactose 6-P is an inducer of lactose utilization [92,93]. We have evidence that the same may be true in *C. acetobutylicum* [9]. The tagatose 6-P pathway (see above), which we have shown is expressed in *C. acetobutylicum* growing on lactose [9], would then constitute a potential second route for galactose metabolism, as well as being the principal mechanism for utilization of the galactose moiety of lactose [94].

Direct experimental demonstration of the existence of a galactose PTS appears to be restricted to *Streptococcus lactis* and *Lactobacillus casei* [95–97]. Nevertheless, a study of galactose transport in *C. saccharobutylicum* (formerly *C. acetobutylicum*) P262 concluded that two separate systems, a PTS and a non-PTS, are involved in uptake of the sugar [22]. The inhibitory effects of CCCP and DCCD on utilization of galactose, but not glucose, by energy-starved cells were consistent with the operation of a H^+-symport, while permeabilized cells showed an equivalent PEP-dependent PTS activity for the two sugars. These data indicate that galactose can be phosphorylated by a PTS, but it was not established that this activity was due to a galactose-specific permease.

There was no evidence for maltose PTS activity in extracts of *C. beijerinckii* NCIMB 8052 [98]. The uncoupler TCS caused complete inhibition of uptake of maltose by whole cells, demonstrating that accumulation of the sugar is energy-dependent. Although these findings are consistent with a H^+-dependent process, they are not definitive. It has been shown that TCS can cause a dramatic decrease in the cellular ATP pool in anaerobic bacteria, presumably as a result of hydrolysis of ATP via the membrane-bound ATPase as the cells attempt to maintain a transmembrane proton gradient [99]. A role for ATP in supporting maltose accumulation therefore cannot be excluded, and the mechanism of uptake remains to be established.

ATP-dependent accumulation of substrates has been shown to be of widespread significance in bacteria, and the genomes of *C. acetobutylicum* and *C. perfringens* reveal the presence of a large complement of members of the ABC (ATP binding cassette) family of transporters [49,50]. However, experimental evidence linking ATP to sugar accumulation has been presented only in the case of glucose and cellobiose in *C. thermocellum*. PTS activity for these substrates was not detected in cell extracts [25,100]. Strong inhibition of uptake of the sugars by whole cells was exerted by arsenate and fluoride, while DCCD and the uncouplers CCCP and dinitrophenol had a minimal effect [101,102]. In support of the conclusion that accumulation is ATP-dependent, collapse of the transmembrane proton gradient using ionophores was found not to significantly affect transport activity or ATP content of the cells, while sugar uptake following treatment with arsenate, fluoride, or iodoacetate could be correlated with depletion of the cellular ATP pool [103]. The cellobiose transport system also apparently mediates uptake of cellodextrins, emphasizing its importance in assimilation of cellulose degradation products to support growth of the bacterium on cellulose [103]. It is reasonable to assume that the two permeases are members of the ABC family, although no further characterization has been reported.

Given the potential importance of pentoses and pentose-containing polymers as fermentation substrates for production of fuels and solvents, it is surprising that there have been no detailed studies of the mechanism of uptake of these sugars in ethanologenic or solventogenic clostridia. Both *C. acetobutylicum* and *C. thermosaccharolyticum* have been shown to be capable of accumulating xylose [104,105], but in neither case was the mechanism established. Pentose sugars are not known to be substrates of the PTS in bacteria, and consequently a non-PTS uptake mechanism is to be expected.

8.4 REGULATION OF CARBOHYDRATE ACCUMULATION

Solute accumulation is a key regulatory step in metabolism, with the uptake of metabolizable carbon sources being adjusted to satisfy the cell's requirements for the support of growth and metabolism under a variety of conditions. Genes encoding bacterial transport systems, both PTS enzyme II domains and non-PTS permeases, are generally contained within operons that also encode enzymes involved in the initial steps of metabolism of the specific substrate. It is evident that the gene products should be required only when the substrate is present and represents the best available source of carbon and energy for growth. This is reflected in the fact that expression of the operons is generally induced in the presence of the substrate of the system and repressed in the presence of a more energetically favorable alternative. The heterotrophic clostridia in most cases behave in typical fashion, recognizing glucose as the preferred substrate that prevents metabolism of other carbon sources. It appears that the underlying mechanisms of transcriptional regulation may be similar to those that operate in other low-GC Gram-positive bacteria (see Chapter 25). Some observations have, however, indicated that glucose is not always the favored carbon source. *C. thermoaceticum* was shown to metabolize xylose in preference to fructose and glucose [106], while in *C. thermocellum,* synthesis of glucose permease and glucokinase, and consequently utilization of glucose, was prevented in the presence of cellobiose [102]. *C. thermocellum* metabolizes cellobiose via an energy-conserving phosphorolytic cleavage, resulting in a higher growth rate and yield on the disaccharide relative to glucose [102]. Although the mechanism of repression of glucose metabolism by cellobiose has not been established, this pattern is consistent with the notion that bacteria show a preference for the substrate that is most favorable in energy terms.

Control of *activity* of transport systems is also an important feature of metabolic regulation, both in terms of controlling the rate of substrate utilization and influencing gene expression by modulating the intracellular concentration of inducers of transcription. In a study of sugar utilization by *C. beijerinckii*, cultures were grown on one of several disaccharides (thus ensuring full induction of the respective uptake and metabolic systems) and subsequently inoculated into medium containing both glucose and the original growth substrate. In the case of lactose and cellobiose, little or no disaccharide was utilized until glucose was exhausted from the medium, but in contrast, utilization of maltose and sucrose continued in the presence of glucose [107]. An immediate inhibition of lactose utilization by *C. acetobutylicum* and xylose utilization by *C. thermosaccharolyticum* has also been observed when glucose is added to cultures growing on these sugars [9,105]. It is therefore apparent that glucose can exert an inhibitory effect on uptake of some alternative substrates. A potential mechanism for control, where the second substrate is accumulated via a PTS, is through competition for PEP, which is required to support uptake of both available sugars. Such competition has been shown to have physiological relevance in control of uptake of sorbitol (glucitol) in *C. pasteurianum* [20], and is likely to occur at the level of interaction between phospho-HPr and the enzyme IIA domains of the respective phosphotransferase systems. In addition, glucose 6-P, the product of the glucose PTS, was found to inhibit the glucitol PTS in permeabilized cells when present at a high concentration [20], and this may also be of general significance for control of sugar uptake by intact cells. An alternative regulatory mechanism, potentially affecting both PTS and non-PTS uptake systems, may involve the phosphorylation of HPr on residue Ser46 by metabolite-activated, ATP-dependent kinase/phosphorylase (see Chapter 25). In low-GC Gram-positive bacteria, this phosphorylation of HPr has been implicated both in regulation of transcription of catabolic genes and in control of carbohydrate accumulation. Phosphorylation of the Ser residue has a dramatic inhibitory effect on phosphorylation of His15 by enzyme I, and the rate of uptake of PTS substrates is therefore reduced [108–110], while HPr(ser)P has been shown to act as an inhibitor of accumulation of non-PTS substrates in *Lactobacillus brevis*, *Lact. casei,* and *L. lactis* [111–115]. There is currently no evidence that HPr(ser)P plays such a direct role in regulating transport processes in clostridia, but the demonstration that kinase-dependent phosphorylation of HPr occurs in *C. acetobutylicum* [29], the conservation of sequence around Ser46 in clostridial HPr proteins, and the presence of genes encoding HPr kinase/phosphorylase enzymes in sequenced genomes is indicative of the potential significance of HPr phosphorylation within the genus.

The regulatory phosphorylation of HPr was discovered as a result of investigation of the phenomenon of inducer expulsion, in which preaccumulated, phosphorylated PTS substrate is rapidly lost to the outside when cells encounter the more favorable carbon source glucose. All low-GC Gram-positive bacteria that exhibit this behavior have been found to contain a sugar-P phosphatase that is stimulated by HPr(ser)P [116]. Therefore a mechanism was proposed, supported by *in vitro* studies, in which formation of HPr(ser)P in the presence of glucose would trigger dephosphorylation of the preaccumulated alternative substrate and subsequent loss of free sugar from the cell. *C. saccharobutylicum* P262 has been shown to exhibit expulsion of the lactose analog thiomethylgalactoside (TMG) when exposed to glucose, suggesting that HPr(ser)P may be of relevance in control of carbohydrate accumulation in this strain [23]. However, the results of recent *in vivo* studies employing *ptsH* and *hprK* (HPr kinase) mutants of *Lact. casei* and *L. lactis* have cast doubt on the role of HPr(ser)P in inducer expulsion, since strains that could not phosphorylate HPr retained the ability to expel preaccumulated solutes [114,115]. Clearly, further investigation is required in order to understand the molecular basis of this phenomenon, both in clostridia and other bacteria.

8.5 CONCLUSION

Nutrient accumulation is an integral part of metabolism and a key regulatory step in the conversion of external solutes by actively growing cells, and it is clear that a thorough understanding of the processes involved is critical for a full appreciation of an organism's physiology. Many heterotrophic clostridia have the ability to utilize a range of carbohydrates to support growth. Although a small number of substrates have been shown to be accumulated by alternative mechanisms, they behave as typical anaerobes in making extensive use of the multiprotein PEP-dependent phosphotransferase system (PTS), which is predominant. The clostridial PTS exhibits strong similarities to the PTS in other bacteria, as shown by amino acid sequence comparisons and functional interactions, both *in vitro* and *in vivo*, between components from different organisms.

Genome sequencing projects are currently providing an unprecedented information resource that can direct experimental studies aimed at increasing our understanding of clostridial physiology and metabolism, and considerable advances can be expected in the future. In addition to prediction and analysis of the metabolic capabilities of the bacteria, it can be expected that the principles of metabolic regulation will be identified and subjected to experimental scrutiny, and this will be critical in enabling strategic control and exploitation of metabolic activities. The bacterial PTS has been shown to be intimately involved in regulation of metabolism [27,116,117], and its potential significance in clostridia is beginning to be revealed. Our current understanding of the regulation of carbohydrate uptake and metabolism in clostridia is described in Chapter 25.

REFERENCES

1. Mitchell, W.J., Biology and physiology, in *Clostridia: Biotechnology and Medical Applications*, Bahl, H. and Dürre, P., Eds., Wiley-VCH Verlag GmbH, Weinheim, Germany, 2001, 49.
2. Lee, Y-E. et al., Taxonomic distinction of saccharolytic thermophilic anaerobes: description of *Thermoanaerobacterium xylanolyticum* gen. nov., sp. nov., and *Thermoanaerobacterium saccharolyticum* gen. nov., sp. nov.; reclassification of *Thermoanaerobium brockii*, *Clostridium thermosulfurogenes*, and *Clostridium thermohydrosulfuricum* E100-69 as *Thermoanaerobacterium thermosulfurigenes* comb. nov., and *Thermoanaerobacter thermohydrosulfuricus*, respectively; and transfer of *Clostridium thermohydrosulfuricum* 39E to *Thermoanaerobacter ethanolicus*, *Int. J. Syst. Bacteriol.*, 43, 41, 1993.
3. Mitchell, W.J., Physiology of carbohydrate to solvent conversion by clostridia, *Adv. Mic. Physiol.*, 39, 31, 1998.
4. Hutkins, R.W. and Kashket, E.R., Phosphotransferase activity in *Clostridium acetobutylicum* from acidogenic and solventogenic phases of growth, *Appl. Environ. Microbiol.*, 51, 1121, 1986.
5. Tangney, M. and Mitchell, W.J., Analysis of a catabolic operon for sucrose transport and metabolism in *Clostridium acetobutylicum* ATCC 824, *J. Mol. Microbiol. Biotechnol.*, 2, 71, 2000.
6. Behrens, S., Mitchell, W.J., and Bahl, H., Molecular analysis of the mannitol operon of *Clostridium acetobutylicum* encoding a phosphotransferase system and a putative PTS-modulated regulator, *Microbiol.*, 147, 75, 2001.
7. Tangney, M., Winters, G.T., and Mitchell, W.J., Characterization of a maltose transport system in *Clostridium acetobutylicum* ATCC 824, *J. Ind. Microbiol. Biotechnol.*, 27, 298, 2001.
8. Mitchell, W.J. and Tangney, M., unpublished data.
9. Yu, Y. et al., Mechanism and regulation of lactose metabolism in *Clostridium acetobutylicum*, manuscript in preparation.
10. Mitchell, W.J., Shaw, J.E., and Andrews, L., Properties of the glucose phosphotransferase system of *Clostridium acetobutylicum* NCIB 8052, *Appl. Environ. Microbiol.*, 57, 2354, 1991.

11. Mitchell, W.J., Carbohydrate uptake and utilization by *Clostridium beijerinckii* NCIMB 8052, *Anaerobe*, 2, 379, 1996.
12. Tangney, M. et al., A gene system for glucitol transport and metabolism in *Clostridium beijerinckii* NCIMB 8052, *Appl. Environ. Microbiol.*, 64, 1612, 1998.
13. Tangney, M. et al., Sucrose transport and metabolism in *Clostridium beijerinckii* NCIMB 8052, *J. Appl. Microbiol.*, 84, 914, 1998.
14. Lee, J., Mitchell, W.J., and Blaschek, H.P., Glucose uptake in *Clostridium beijerinckii* NCIMB 8052 and the solvent-hyperproducing mutant BA101, *Appl. Environ. Microbiol.*, 67, 5025, 2001 (and *Appl. Environ. Microbiol.*, 68, 3181, 2002).
15. Hugo, H.V. and Gottschalk, G., Distribution of 1-phosphofructokinase and PEP:fructose phosphotransferase activity in clostridia, *FEBS Lett.*, 46, 106, 1974.
16. Kazamias, M.T. and Sperry, J.F., Enhanced fermentation of mannitol and release of cytotoxin by *Clostridium difficile* in alkaline culture media, *Appl. Environ. Microbiol.*, 61, 2425, 1995.
17. Brown, G.G. and Thomson, J.A., Isolation and characterisation of an aryl-β-D-glucoside uptake and utilisation system (*abg*) from the gram-positive ruminal *Clostridium* species *C. longisporum*, *Mol. Gen. Genet.*, 257, 213, 1998.
18. Booth, I.R. and Morris, J.G., Carbohydrate transport in *Clostridium pasteurianum*, *Biosci. Rep.*, 2, 47, 1982.
19. Mitchell, W.J. and Booth, I.R., Characterization of the *Clostridium pasteurianum* phosphotransferase system, *J. Gen. Microbiol.*, 130, 2193, 1982.
20. Roohi, M.S. and Mitchell, W.J., Regulation of sorbitol metabolism by glucose in *Clostridium pasteurianum*: a role for inducer exclusion, *J. Gen. Microbiol.*, 133, 2207, 1987.
21. Groves, D.J. and Gronlund, A.F., Carbohydrate transport in *Clostridium perfringens* type A, *J. Bacteriol.*, 100, 1256, 1969.
22. Gutierrez, N.A. and Maddox, I.S., Galactose transport in *Clostridium acetobutylicum* P262, *Lett. Appl. Microbiol.*, 23, 97, 1996.
23. Diez-Gonzalez, F. and Russell, J.B., The regulation of thiomethylgalactoside transport in *Clostridium acetobutylicum* P262 by inducer exclusion and inducer expulsion mechanisms, *FEMS Microbiol. Lett.*, 136, 123, 1996.
24. Patni, N.J. and Alexander, J.K., Catabolism of fructose and mannitol in *Clostridium thermocellum*: presence of phosphoenolpyruvate: fructose phosphotransferase, fructose 1-phosphate kinase, phosphoenolpyruvate: mannitol phosphotransferase and mannitol 1-phosphate dehydrogenase in cell extracts, *J. Bacteriol.*, 105, 226, 1971.
25. Nochur, S.V. et al., Mode of sugar phosphorylation in *Clostridium thermocellum*, *Appl. Biochem. Biotechnol.*, 33, 33, 1992.
26. Saier, M.H. Jr. and Reizer, J., Proposed uniform nomenclature for the proteins and protein domains of the bacterial phosphoenolpyruvate:sugar phosphotransferase system, *J. Bacteriol.*, 174, 1433, 1992.
27. Postma, P.W., Jacobson, G.R., and Lengeler, J.W., Phosphoenolpyruvate: carbohydrate phosphotransferase systems of bacteria, *Microbiol. Rev.*, 57, 543, 1993.
28. Lengeler, J.W., Jahreis, K., and Wehmeier, U.F., Enzymes II of the phosphoenolpyruvate-dependent phosphotransferase systems: their structure and function in carbohydrate transport, *Biochim. Biophys. Acta*, 1188, 1, 1994.
29. Tangney, M. et al., Analysis of the elements of catabolite repression in *Clostridium acetobutylicum* ATCC 824, *J. Mol. Microbiol. Biotechnol.*, 6, 6, 2003.
30. Ounine, K. et al., Regulation and butanol inhibition of D-xylose and D-glucose uptake in *Clostridium acetobutylicum*, *Appl. Env. Microbiol.*, 49, 874, 1985.
31. Meadow, N.D., Fox, D.K., and Roseman, S., The bacterial phosphoenolpyruvate: glycose phosphotransferase system, *Ann. Rev. Biochem.*, 59, 497, 1990.
32. Moreira, A.R., Ulmer, D.C., and Linden, J.C., Butanol toxicity in the butylic fermentation, *Biotech. Bioeng. Symp.*, 11, 567, 1981.
33. Yerushalmi, L., Volesky, B., and Votruba, J., Systems analysis of the culture physiology in acetone-butanol fermentation, *Biotech. Bioeng.*, 28, 1334, 1986.
34. Annous, B.A. and Blaschek, H.P., Isolation and characterization of *Clostridium acetobutylicum* mutants with enhanced amylolytic activity, *Appl. Environ. Microbiol.*, 57, 2544, 1991.

35. Formanek, J., Mackie, R., and Blaschek, H.P., Enhanced butanol production by *Clostridium beijerinckii* BA101 grown in semidefined P2 medium containing 6 percent maltodextrin or glucose, *Appl. Environ. Microbiol.,* 63, 2306, 1997.
36. Robillard, G.T., Dooijewaard, G., and Lolkema, J., *Escherichia coli* phosphoenolpyruvate dependent phosphotransferase system. Complete purification of enzyme I by hydrophobic interaction chromatography, *Biochem.,* 18, 2984, 1979.
37. Saier, M.H. Jr. et al., Evidence for the functional association of enzyme I and HPr of the phosphoenolpyruvate-sugar phosphotransferase system with the membrane in sealed vesicles of *Escherichia coli, J. Cell. Biochem.,* 18, 231, 1982.
38. Saier, M.H. Jr., Feucht, B.U., and Roseman, S., Phosphoenolpyruvate-dependent fructose phosphorylation in photosynthetic bacteria, *J. Biol. Chem.,* 246, 7819, 1971.
39. Brouwer, M., Elferink, M.G.L., and Robillard, G.T., Phosphoenolpyruvate-dependent fructose phosphotransferase system of *Rhodopseudomonas sphaeroides*: purification and physicochemical and immunochemical characterization of a membrane-associated enzyme I, *Biochem.,* 21, 82, 1982.
40. Saier, M.H. Jr. and Staley, J.T., Phosphoenolpyruvate:sugar phosphotransferase system in *Ancalomicrobium adetum, J. Bacteriol.,* 131, 716, 1977.
41. Saier, M.H. Jr. and Tseng, T-T., Evolutionary origins of transmembrane transport systems, in *Transport of Molecules Across Microbial Membranes*, Broome-Smith, J.K. et al., Eds., University Press, Cambridge, 1999, 252.
42. Seok, Y-J. et al., Importance of the region around glycine-338 for the activity of enzyme I of the *Escherichia coli* phosphoenolpyruvate:sugar phosphotransferase system, *Biochem.,* 35, 236, 1996.
43. Brokx, S.J. et al., Enzyme I of the phosphoenolpyruvate:sugar phosphotransferase system. *In vitro* intragenic complementation: the roles of arg126 in phosphoryl transfer and the C-terminal domain in dimerization, *Biochem.,* 38, 3624, 2000.
44. García-Alles, L.F. et al., Mechanism-based inhibition of enzyme I of the *Escherichia coli* phosphotransferase system. Cysteine 502 is an essential residue, *J. Biol. Chem.*, 277, 6934, 2002.
45. Tangney, M. and Mitchell, W.J., *Clostridium tetani* encodes a phosphocarrier protein HPr, *Microbiol.*, 150, 525, 2004.
46. Deutscher, J., Galinier, A., and Martin-Verstraete, I., Carbohydrate uptake and metabolism, in *Bacillus subtilis* and Its Closest Relatives, Sonenshein, A.L. et al., Eds., American Society for Microbiology, Washington, DC, 2002, 129.
47. Galinier, A. et al., The *Bacillus subtilis crh* gene encodes a HPr-like protein involved in carbon catabolite repression, *Proc. Natl. Acad. Sci. U.S.A.*, 94, 8439, 1997.
48. Reizer, J. et al., A hybrid response regulator possessing a PEP-dependent phosphorylation domain, *Microbiol.,* 145, 987, 1999.
49. Nölling, J. et al., Genome sequencing and comparative analysis of the solvent-producing bacterium *Clostridium acetobutylicum, J. Bacteriol.,* 183, 4823, 2001.
50. Shimizu, T. et al., Complete genome sequence of *Clostridium perfringens*, an anaerobic flesh-eater, *Proc. Natl. Acad. Sci. U.S.A.,* 99, 996, 2002.
51. Brüggemann, H. et al., The genome sequence of *Clostridium tetani*, the causative agent of tetanus disease, *Proc. Natl. Acad. Sci. U.S.A.,* 100, 1316, 2003.
52. Cornillot, E. et al., The genes for butanol and acetone formation in *Clostridium acetobutylicum* ATCC 824 reside on a large plasmid whose loss leads to degeneration of the strain, *J. Bacteriol.,* 179, 5442, 1997.
53. Reizer, J. et al., Novel phosphotransferase system genes revealed by genome analysis — the complete complement of PTS proteins encoded within the genome of *Bacillus subtilis*, *Microbiol.,* 145, 3419, 1999.
54. Bouma, C.L. et al., 6-Phospho-α-glucosidase from *Fusobacterium mortiferum*: cloning, expression and assignment to family 4 of the glycosylhydrolases, *J. Bacteriol.,* 179, 4129, 1997.
55. Thompson, J. et al., The gene *glvA* of *Bacillus subtilis* 168 encodes a metal-requiring, NAD(H)-dependent 6-phospho-α-glucosidase. Assignment to family 4 of the glycosylhydrolase superfamily, *J. Biol. Chem.*, 273, 27347, 1998.
56. Thompson, J. et al., Metabolism of sucrose and its five linkage-isomeric α-D-glucosyl-D-fructoses by *Klebsiella pneumoniae*. Participation and properties of sucrose-6-phosphate hydrolase and phospho-α-glucosidase, *J. Biol. Chem.,* 276, 37415, 2001.

57. Pikis, A. et al., Metabolism of sucrose and its five isomers by *Fusobacterium mortiferum*, *Microbiol.*, 148, 843, 2002.
58. Berepiki, A. and Mitchell, W.J., unpublished data.
59. Lengeler, J.W., Mayer, R.J., and Schmid, K., Phosphoenolpyruvate-dependent phosphotransferase system Enzyme III and plasmid-encoded sucrose transport in *Escherichia coli* K-12, *J. Bacteriol.*, 151, 468, 1982.
60. Boos, W. et al., Trehalose transport and metabolism in *Escherichia coli*, *J. Bacteriol.*, 172, 3450, 1990.
61. Sutrina, S.L. et al., The glucose permease of *Bacillus subtilis* is a single polypeptide chain that functions to energise the sucrose permease, *J. Biol. Chem.*, 265, 18581, 1990.
62. Dahl, M.K., Enzyme II^{Glc} contributes to trehalose metabolism in *Bacillus subtilis*, *FEMS Microbiol. Lett.*, 148, 233, 1997.
63. Parker, L.L. and Hall, B.G., Characterization and nucleotide sequence of the cryptic *cel* operon of *Escherichia coli* K12, *Genetics*, 124, 455, 1990.
64. Reizer, J., Reizer, A., and Saier, M.H. Jr., The cellobiose permease of *Escherichia coli* consists of three proteins and is homologous to the lactose permease of *Staphylococcus aureus*, *Res. Microbiol.*, 141, 1061, 1990.
65. Tobish, S. et al., Identification and characterization of a new β-glucoside system in *Bacillus subtilis*, *J. Bacteriol.*, 179, 496, 1997.
66. Keyhani, N.O. and Roseman, S., Wild-type *Escherichia coli* grows on the chitin disaccharide, *N,N′*-diacetylchitobiose, by expressing the *cel* operon, *Proc. Nat. Acad. Sci. U.S.A.*, 94, 14367, 1997.
67. Mitchell, W.J. et al., Identification of a phosphoenolpyruvate:fructose phosphotransferase system (fructose-1-phosphate forming) in *Listeria monocytogenes*, *J. Bacteriol.*, 175, 2758, 1993.
68. Fraenkel, D.G., The phosphoenolpyruvate-initiated pathway of fructose metabolism in *Escherichia coli*, *J. Biol. Chem.*, 243, 6458, 1968.
69. Erni, B. et al., Mannose permease of *Escherichia coli*. Domain structure and function of the phosphorylating subunit, *J. Biol. Chem.*, 264, 18733, 1989.
70. Martin-Verstraete, I. et al., Levanase operon of *Bacillus subtilis* includes a fructose-specific phosphotransferase system regulating the expression of the operon, *J. Mol. Biol.*, 214, 657, 1990.
71. Martin-Verstraete, I., Michel, V., and Charbit, A., The levanase operon of *Bacillus subtilis* expressed in *Escherichia coli* can substitute for the mannose permease in mannose uptake and bacteriophage lambda infection, *J. Bacteriol.*, 178, 7112, 1996.
72. Nobelmann, B. and Lengeler, J.W., Molecular analysis of the *gat* genes from *Escherichia coli* and of their roles in galactitol transport and metabolism, *J. Bacteriol.*, 178, 6790, 1996.
73. Lengeler, J.W., personal communication.
74. van Rooijen, R.J., van Schalkwijk, S., and de Vos, W.M., Molecular cloning, characterization, and nucleotide sequence of the tagatose 6-phosphate pathway gene cluster of the lactose operon of *Lactococcus lactis*, *J. Biol. Chem.*, 266, 7176, 1991.
75. Rosey, E.L., Oskoulian, B., and Stewart, G.C., Lactose metabolism by *Staphylococcus aureus*: characterization of *lacABCD*, the structural genes of the tagatose 6-phosphate pathway, *J. Bacteriol.*, 173, 5992, 1991.
76. Rosey, E.L. and Stewart, G.C., Nucleotide and deduced amino acid sequences of the *lacR*, *lacABCD*, and *lacFE* genes encoding the repressor, tagatose 6-phosphate gene cluster, and sugar-specific phosphotransferase system components of the lactose operon of *Streptococcus mutans*, *J. Bacteriol.*, 174, 6159, 1992.
77. Reizer, J. et al., Proposed topology of the glucitol permeases of *Escherichia coli* and *Clostridium acetobutylicum*, *Curr. Microbiol.*, 33, 331, 1996.
78. Yamada, M. and Saier, M.H. Jr., Glucitol-specific enzymes of the phosphotransferase system in *Escherichia coli*. Nucleotide sequence of the *gut* operon, *J. Biol. Chem.*, 262, 5455, 1987.
79. Reizer, J. et al., The glucitol permease of *Escherichia coli*: a tripartite permease of the phosphotransferase system, *Microbiol.*, 144, 1463, 1998.
80. Aldridge, P., Metzger, M., and Geider, K., Genetics of sorbitol metabolism in *Erwinia amylovora* and its influence on bacterial virulence, *Mol. Gen. Genet.*, 256, 611, 1997.
81. Boyd, D.A. et al., Identification of the operon for the sorbitol (glucitol) phosphoenolpyruvate:sugar phosphotransferase system in *Streptococcus mutans*, *Infect. Immun.*, 68, 925, 2000.

82. Reid, S., Rafudeen, M.S., and Leat, N.G., The genes controlling sucrose utilization in *Clostridium beijerinckii* NCIMB 8052 constitute an operon, *Microbiol.,* 145, 1461, 1999.
83. Clarke, D.J., Fuller, F.M., and Morris, J.G., The proton-translocating adenosine triphosphatase of the obligately anaerobic bacterium *Clostridium pasteurianum*, 1. ATP phosphohydrolase activity, *Eur. J. Biochem.,* 98, 597, 1979.
84. Terracciano, J.S., Schreurs, W.J.A., and Kashket, E.R., Membrane conductance of *Clostridium thermoaceticum* and *Clostridium acetobutylicum*: evidence for electrogenic Na^+/H^+ antiport in *Clostridium thermoaceticum, Appl. Environ. Microbiol.,* 53, 782, 1987.
85. Höner Zu Bentrup, K. et al., A Na^+-pumping V_1V_0 –ATPase complex in the thermophilic bacterium *Clostridium fervidus, J. Bacteriol.,* 179, 1274, 1997.
86. Speelmans, G. et al., Energy transduction in the thermophilic anaerobic bacterium *Clostridium fervidus* is exclusively coupled to sodium ions, *Proc. Natl. Acad. Sci. U.S.A.*, 90, 7975, 1993.
87. Booth, I.R. and Morris, J.G., Proton-motive force in the obligately anaerobic bacterium *Clostridium pasteurianum*: a role in galactose and gluconate uptake, *FEBS Lett.*, 59, 153, 1975.
88. Bender, R., Andreesen, J.R., and Gottschalk, G., 2-Keto-3-deoxygluconate, an intermediate in the fermentation of gluconate by clostridia, *J. Bacteriol.,* 107, 570, 1971.
89. Daldal, F. and Applebaum, J., Cloning and expression of *Clostridium pasteurianum* galactokinase gene in *Escherichia coli* K-12 and nucleotide sequence analysis of a region affecting the amount of the enzyme, *J. Mol. Biol.,* 186, 533, 1985.
90. Simoni, R.D. et al., Sugar transport. IV. Isolation and characterization of the lactose phosphotransferase system in *Staphylococcus aureus, J. Biol. Chem.,* 248, 932, 1973.
91. LeBlanc, D.J. et al., Influence of the lactose plasmid on the metabolism of galactose by *Streptococcus lactis, J. Bacteriol.,* 137, 878, 1979.
92. Morse, M.L. et al., Metabolism of lactose by *Staphylococcus aureus* and its genetic basis, *J. Bacteriol.,* 95, 2270, 1968.
93. Cords, B.R. and McKay, L.L., Characterization of lactose-fermenting revertants from lactose-negative *Streptococcus lactis* C2 mutants, *J. Bacteriol.,* 119, 830, 1974.
94. Thomas, T.D., Turner, K.W., and Crow, V.L., Galactose fermentation by *Streptococcus lactis* and *Streptococcus cremoris*: pathway, products and regulation, *J. Bacteriol.,* 144, 672, 1980.
95. Thompson, J., Galactose transport systems in *Streptococcus lactis, J. Bacteriol.,* 144, 683, 1980.
96. Park, Y.H., and McKay, L.L., Distinct galactose phosphoenolpyruvate-dependent phosphotransferase system in *Streptococcus lactis, J. Bacteriol.,* 149, 420, 1982.
97. Chassy, B.M., and Thompson, J., Regulation and characterization of the galactose-phosphoenolpyruvate-dependent phosphotransferase system in *Lactobacillus casei, J. Bacteriol.,* 154, 1204, 1983.
98. Albasheri, K.A., and Mitchell, W.J., Identification of two α-glucosidase activities in *Clostridium acetobutylicum* NCIB 8052, *J. Appl. Bacteriol.,* 78, 149, 1985.
99. Sahm, K. et al., Molecular analysis of the *amy* gene locus of *Thermoanaerobacterium thermosulfurigenes* EM1 encoding starch-degrading enzymes and a binding protein-dependent maltose transport system, *J. Bacteriol.,* 178, 1039, 1996.
100. Patni, N.J. and Alexander, J.K., Utilization of glucose by *Clostridium thermocellum*: presence of glucokinase and other glycolytic enzymes in cell extracts, *J. Bacteriol.,* 105, 220, 1971.
101. Hernandez, P.E., Transport of D-glucose in *Clostridium thermocellum* ATCC 27405, *J. Gen. Appl. Microbiol.,* 28, 469, 1982.
102. Ng, T. and Zeikus, J.G., Differential metabolism of cellobiose and glucose by *Clostridium thermocellum* and *Clostridium thermohydrosulfuricum, J. Bacteriol.,* 150, 1391, 1982.
103. Strobel, H.J. et al., Carbohydrate transport by the anaerobic thermophile *Clostridium thermocellum* LRQ1, *Appl. Environ. Microbiol.,* 61, 4012, 1995.
104. Ounine, K. et al., Regulation and butanol inhibition of D-xylose and D-glucose uptake in *Clostridium acetobutylicum, Appl. Environ. Microbiol.,* 49, 874, 1985.
105. Aduse-Opoku, J. and Mitchell, W.J., Diauxic growth of *Clostridium thermosaccharolyticum* on glucose and xylose, *FEMS Microbiol. Lett.,* 50, 45, 1988.
106. Andreesen, J.R. et al., Fermentation of glucose, fructose and xylose by *Clostridium thermoaceticum*: effects of metals on growth yields, enzymes and the synthesis of acetate from CO_2, *J. Bacteriol.,* 114, 743, 1973.

107. Mitchell, W.J., Albasheri, K.A., and Yazdanian, M., Factors affecting utilization of carbohydrates by clostridia, *FEMS Microbiol. Rev.,* 17, 317, 1995.
108. Deutscher J. et al., Bacterial phosphoenolpyruvate-dependent phosphotransferase system: P-ser-HPr and its possible regulatory function, *Biochem.,* 23, 4455, 1984.
109. Deutscher, J., Sossna, G., and Gonzy-Treboul, G., Regulatory function of the phosphocarrier protein HPr of the phosphoenolpyruvate-dependent phosphotransferase system in Gram-positive bacteria, *FEMS Microbiol. Rev.,* 63, 167, 1989.
110. Reizer, J. et al., Mechanistic and physiological consequences of HPr(ser) phosphorylation on the activities of the phosphoenolpyruvate:sugar phosphotransferase system in Gram-positive bacteria: studies with site-specific mutants of HPr, *EMBO J.,* 8, 2111, 1989.
111. Ye, J.J. et al., ATP-dependent phosphorylation of serine-46 in the phosphocarrier protein HPr regulates lactose/H^+ symport in *Lactobacillus brevis*, *Proc. Natl. Acad. Sci. U.S.A.,* 91, 3102, 1994.
112. Ye, J. J. et al., Regulation of the glucose:H^+ symporter by metabolite-activated ATP-dependent phosphorylation of HPr in *Lactobacillus brevis*, *J. Bacteriol.,* 176, 3484, 1994.
113. Viana, R. et al., Enzyme I and HPr from *Lactobacillus casei*: their role in sugar transport, carbon catabolite repression and inducer exclusion, *Mol. Microbiol.,* 36, 570, 2000.
114. Dossonnet, V. et al., Phosphorylation of HPr by the bifunctional HPr kinase/P-ser-HPr phosphatase from *Lactobacillus casei* controls catabolite repression and inducer exclusion but not inducer expulsion, *J. Bacteriol.,* 182, 2582, 2000.
115. Monedero, V. et al., Regulatory functions of serine-46-phosphorylated HPr in *Lactococcus lactis*, *J. Bacteriol.,* 183, 3391, 2001.
116. Saier. M.H. Jr. et al., Catabolite repression and inducer control in Gram-positive bacteria, *Microbiol.*, 142, 217, 1996.
117. Stülke, J. and Hillen, W., Coupling physiology and gene regulation in bacteria: the phosphotransferase sugar uptake system delivers the signals, *Naturwissen.,* 85, 583, 1998.

9 Special Clostridial Enzymes and Fermentation Pathways

Wolfgang Buckel

CONTENTS

9.1 INTRODUCTION

Clostridia exhibit special metabolic routes, which are not common in other organisms. They are able to produce short-chain fatty acids and hydrogen from carbohydrates and amino acids. Especially, the fermentation of all 20 proteinogenic amino acids represents a unique feature of this anaerobic bacterial group. Therefore, clostridia require special enzymes, which are the topic of this review. We define clostridia as members of the "Clostridia," class I of the phylum *Firmicutes* (BXIII), which include the family "Fusobactericeae" (Fusobacteria) of the phylum *Fusobacteria* (BXXI) [1]. The recent availability of whole bacterial genomes, however, has shown that several enzymes once thought to be unique to clostridia are also present in the phylum *Proteobacteria* (BXII) or in the phylum *Actinobacteria* (BXIV). An example is the conversion of glutamate to pyruvate and acetate discovered in *Clostridium tetanomorphum* [2], and related clostridia [3,4], which is part of the methylaspartate pathway leading to ammonia, acetate, CO_2, butyrate, and hydrogen (Figure 9.1). The first two enzymes of this pathway, the coenzyme B_{12}-dependent glutamate mutase and methylasparate ammonia lyase, have been detected in various enterobacteria — noteworthy also in two pathogenic strains, *Escherichia coli* O157:H7 Sakai [5] and *E. coli* O157:H7 EDL933 [6], but not in *E. coli* K-12. Glutamate mutase has also been shown to be involved in the biosynthesis of the antibiotic friulimicin in *Actinoplanes friuliensis* [7]. The third enzyme, mesaconate hydratase, for which fumarate is a better substrate, is related to the iron-sulfur-cluster-containing fumarase A from *E. coli*. The product of the hydration of mesaconate, (*S*)-citramalate,

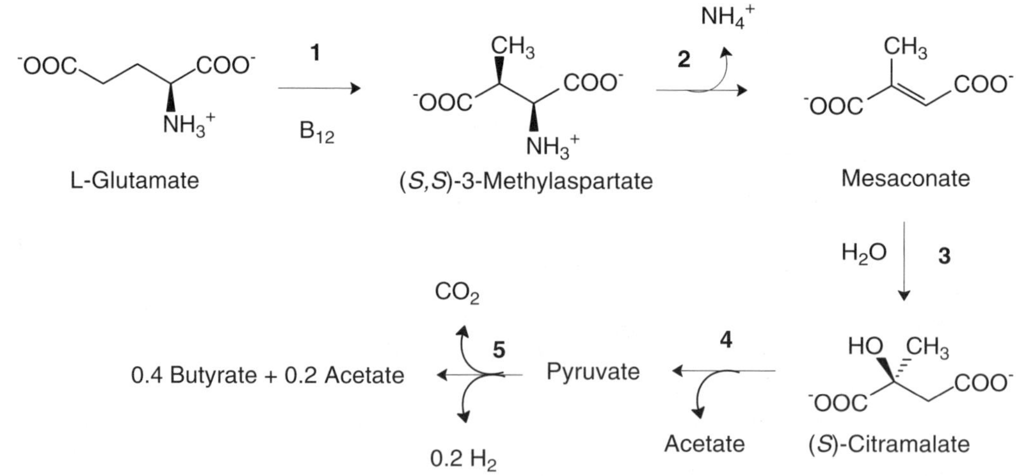

FIGURE 9.1 Fermentation of glutamate via 3-methylaspartate. Enzymes: (1) coenzyme B_{12}-dependent glutamate mutase (Mut/GlmELS); (2) methylaspartate ammonia lyase (MalA); (3) citramalate dehydratase (MesAB); (4) citramalate lyase (ClmABC) (see also Figure 9.11); (5) enzymes converting pyruvate via acetyl-CoA to butyrate (for details, see Figure 9.5)

is cleaved into acetate and pyruvate. This reaction is catalyzed by citramalate lyase, which closely resembles citrate lyase from *Klebsiella pneumoniae* and some clostridia. Nevertheless, we will include glutamate mutase and the consecutive enzymes in this chapter, since this pathway plays a major role in the energy metabolism of various clostridia. A reason for this unexpected occurrence of glutamate mutase in enterobacteria could be due to lateral gene transfer, which also may have happened between clostridia and *Euryarchaeota*, especially *Archaeoglobus fulgidus*.

We divide this chapter into 10 sections, which together comprise about 50 different enzymes. In each part, the enzymes are either metabolically or mechanistically related. Besides their typical occurrence in clostridia, the enzymes are selected due to their role in metabolism, their interesting mechanism, or their unusual catalyzed reaction. In this review, some biotechnologically important but mechanistically less interesting enzymes are omitted, especially hydrolytic enzymes such as cellulases, collagenases, and amylases, which are found in other chapters of this book.

9.2 ACETOGENESIS

The theory of the chemoautotrophic origin of life postulates the formation of organic compounds by reduction of carbon dioxide with hydrogen generated "in statu nascendi" from the reaction of ferrous sulfide and hydrogen sulfide to yield pyrite [8]. In extant Bacteria, there are four different pathways known by which CO_2 is reductively fixed into organic acids: the Calvin-Benson-Bassham cycle, the reverse citric acid cycle [9], the 3-hydroxypropionate pathway [10], and acetogenesis [11]. Whereas the first three pathways require as acceptors ribulose-1,5-bisphosphate, succinyl-CoA, and acetyl-CoA, respectively, in acetogenesis, acetate is formed directly by reduction of CO_2, preferentially with hydrogen as reducing agent, similar to the reaction proposed by Huber and Wächtershäuser [12].

9.2.1 ACETYL-COA SYNTHASE/CO DEHYDROGENASE

In acetogenic organisms like *Acetobacterium woodii* or *Moorella thermoacetica* (formerly called *Clostridium thermoaceticum*), both of which are members of class I "Clostridia," the key metabolite acetyl-CoA is synthesized from carbon monoxide, a methyl corrinoid, and CoASH, catalyzed by acetyl-CoA synthase (CO-methylating acetyl-CoA synthase; EC 2.3.1.169) (Figure 9.2). The enzyme is commonly associated with carbon-monoxide dehydrogenase (ferredoxin as acceptor; EC 1.2.7.4) and, therefore, called bifunctional CO dehydrogenase/acetyl-CoA synthase (CODH/ACS) (Equation 9.1 and Equation 9.2). Note that the enzyme is different from acetyl-CoA synthetase (acetate:CoA ligase; EC 6.2.2.1), which catalyzes the activation of acetate to acetyl-CoA, whereby ATP is hydrolyzed to AMP and diphosphate.

$$CO_2 + 2\ H^+ + 2\ e^- = CO + H_2O \tag{9.1}$$

$$CO + CH_3\text{-CFeSP} + \text{CoASH} = CH_3\text{-CO-SCoA} + H^+ + \text{CFeSP}^- \tag{9.2}$$

The reducing agents for the CO dehydrogenase are *in vivo* ferredoxin or flavodoxin; *in vitro* reduced methylviologen or dithionite are used. CFeSP is a corrinoid iron-sulfur protein composed of two subunits; the small subunit (33 kDa) contains the corrinoid (a derivative of vitamin B_{12}) and the large subunit (55 kDa) contains a [4Fe-4S] cluster, which keeps the corrinoid in the Co(I) state ready to accept a methyl group from methyltetrahydrofolate (CH_3-H_4F) [13]. CODH/ACS from *M. thermoacetica* is composed of two different subunits, α (AcsB, 78 kDa) and β (AcsA, 71 kDa), which form a heterotetramer $\alpha_2\beta_2$. The enzyme contains up to 4 Ni, 24 Fe, 28 S^{2-} and 2 Zn per 298 kDa [14,15]. The metal ions and the sulfur form three different clusters, A, B, and C, whose structures have been resolved recently. Cluster A, the active site of acetyl-CoA synthase is composed of a [4Fe-4S]-X-Ni cluster. In the first published structure, X was determined predominantly as Cu

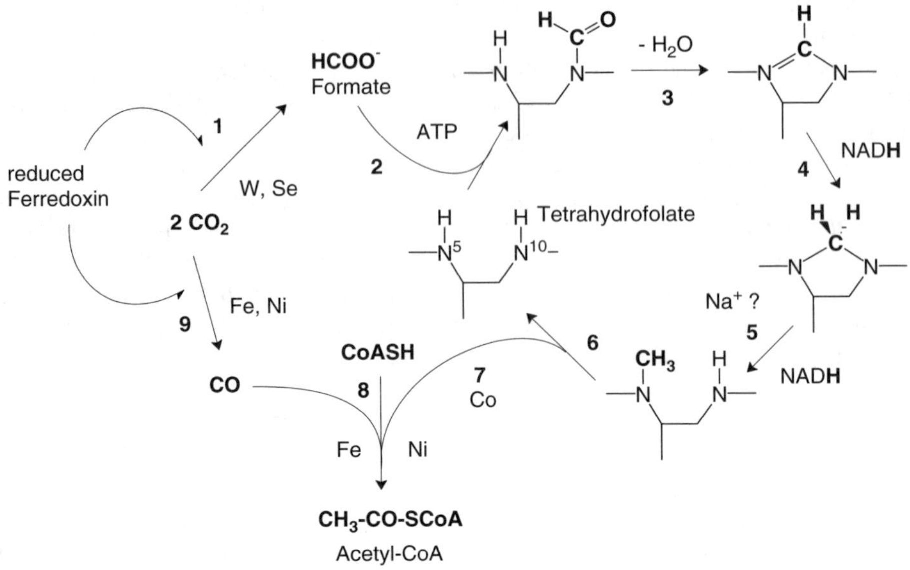

FIGURE 9.2 De novo synthesis of acetyl-CoA from CO_2. Enzymes: (1) Formate dehydrogenase; (2) formate-tetrahydrofolate ligase (EC 6.3.4.3); (3) Methenyltetrahydrofolate cyclodehydrase (EC3.5.4.9); (4) Methylene-tetrahydrofolate dehydrogenase (NAD^+, EC 1.5.1.15); (5) Methyltetrahydrofolate dehydrogenase; (6) Methyltetrahydrofolate methyltransferase; (7) CFeSP, a corrinoid iron-sulfur protein; (8) acetyl-CoA synthase (EC 2.3.1.169); (9) CO-dehydrogenase (EC 1.2.7.4).

[16]; the second structure exhibited two conformations, one with X = Zn and the other with X = Ni [17]. Finally, the most active acetyl-CoA synthase from the carboxytrophic organism *Carboxydothermus hydrogenoformans* ("Clostridia"), which is monofunctional without CO-dehydrogenase activity, revealed X = Ni [18]. These differences may arise from the fact that the second Ni (X) is labile and can easily be replaced by Zn or Cu [19].

Cluster C represents the active site of CO dehydrogenase. Again, the same authors have put forward three different structures. The most likely structure stems from *C. hydrogenoformans*, which has revealed a novel [4Fe-Ni-5S] cluster. In *M. thermoacetica,* in the vicinity of cluster C, is cluster B, a standard [4Fe-4S] cluster, which probably is involved in electron transport. The monofunctional CO-dehydrogenase of *C. hydrogenoformans* contains four such B-clusters [20]. In *M. thermoacetica,* the two active sites are connected by a channel, in which CO may diffuse from the C-cluster, where it is generated by reduction of CO_2, to the A-cluster, where it is combined with CH_3^+ and $CoAS^-$ to yield CH_3-CO-SCoA. It has been shown that the stable Ni of the A-cluster is methylated by the corrinoid-iron-sulfur-protein [21]. Then CO binds to the labile Ni, where CoASH also coordinates. The methyl group is connected with CO to yield an acetyl group bound at the labile Ni. Finally, the acetyl group is taken over by CoASH. The last two partial reactions can be measured directly by isotope exchange (Equation 9.3 and Equation 9.4):

$$CH_3\text{-}\mathbf{CO}\text{-SCoA} + CO = CH_3\text{-CO-SCoA} + \mathbf{CO} \tag{9.3}$$

$$CH_3\text{-CO-}\mathbf{S}\text{CoA} + HSCoA = CH_3\text{-CO-SCoA} + H\mathbf{S}CoA \tag{9.4}$$

Using this mechanism of acetyl-CoA synthesis as a guideline, Huber and Wächtershäuser combined FeS and NiS together with CO, H_2S, and methylmercaptane (CH_3-SH) for the synthesis of acetyl-S-CH_3 at 100°C. Although the thiol ester hydrolyzed to acetate under these conditions, its transient generation could be established by trapping with aniline as acetanilide. Furthermore,

CH_3-SH and acetate were formed solely from CO and H_2S by adding selenite to this primordial mixture [12].

9.2.2 Formate Dehydrogenase

The methyl group of acetyl-CoA also stems from CO_2, which is reduced to formate and activated in an ATP-dependent reaction to N^{10}-formyltetrahydrofolate (formyl-H_4F). Cyclization yields methenyl-H_4F, which is further reduced via methylene-H_4F to methyl-H_4F (Figure 9.2). The most interesting enzyme of this pathway is formate dehydrogenase from *M. thermoaceticum* and *Eubacterium acidaminophilum* (clostridia), which contains seleonocysteine and tungsten bound to molybdopterin [22,23]. There are many different formate dehydrogenases containing molybdenum or no metal, which are involved in the oxidation of formate to CO_2. It has been speculated that those that are able to catalyze the back reaction require tungsten in the active center. *In vitro* reduction of CO_2 is measured spectrophotometrically with dithionite-reduced methylviologen [24]; *in vivo* ferredoxin or flavodoxin serve as reducing agents. Both electron carriers in turn are reduced either by the oxidation of pyruvate catalyzed by pyruvate:ferredoxin (flavodoxin) oxidoreductase (Section 9.4.2) or by molecular hydrogen (Section 9.3). A reduction of ferredoxin by NADH is only possible by reversed electron transport (Section 9.3).

The enzymes, which catalyze the conversion of formate to methyl-H_4F, seem to be similar to those found in Eukarya and Bacteria. However, one of these enzymes should generate $\Delta\mu H^+$, as in the cytochrome b containing *M. thermoacetica* [25], or $\Delta\mu Na^+$, as in *Acetobacterium woodii* (clostridia) [26], from which ATP could be synthesized. Substrate level phosphorylation via acetylphosphate occurs with acetyl-CoA synthesized by acetyl-CoA synthase, but the derived ATP is consumed in the activation of formate to formyl-H_4F.

9.3 FORMATION OF HYDROGEN

Beside its consumption in acetogenesis, molecular hydrogen is a frequent end product of clostridial fermentations. In these reactions, usually part of the substrate is oxidized to an energy-rich compound, from which ATP is formed via substrate level phosphorylation, whereas the reducing equivalents are disposed on the other part of the substrate. Examples are the fermentations leading to crotonyl-CoA, which disproportionates to butyrate and two acetates, whereby 1 ATP/2 crotonyl-CoA is conserved. If, however, the reducing equivalents generated by the oxidation to two acetates would be used to reduce protons to molecular hydrogen, the ATP yield/crotonyl-CoA would be doubled. This "substrate-serving effect" might be the reason why hydrogen is produced whenever thermodynamically possible.

The enzymes that catalyze the reversible cleavage of molecular hydrogen into two protons and two electrons (Equation 9.5), are called hydrogenases; for a recent review, see Frey [27].

$$2\ H^+ + 2\ e^- = H_2 \tag{9.5}$$

9.3.1 Iron Only Hydrogenases

Clostridia contain FeFe or "iron only" type hydrogenases, which are 10 times more active hydrogen producers ($k_{cat} \leq 9000\ s^{-1}$) and 40 times faster consumers ($k_{cat} = 28{,}000\ s^{-1}$) than the NiFe type present in many bacteria and methanogenic archaea (Figure 9.3). On the other hand, the NiFe hydrogenases have a much lower K_m for hydrogen and are less sensitive toward oxygen. As indicated by the name, the crystal structures of the FeFe hydrogenases of *Desulfovibrio vulgaris* (Deltaproteobacteria) and of *Clostridium pasteurianum* revealed a binuclear Fe1-Fe2 cluster. The hydrogenase from *C. pasteurianum* contains 20 Fe/mol enzyme, which form the Fe1-Fe2 cluster, 3 [4Fe-4S] and 2 [3Fe-4S] clusters. The active site is located in the Fe1-Fe2 cluster, of which Fe1 is connected

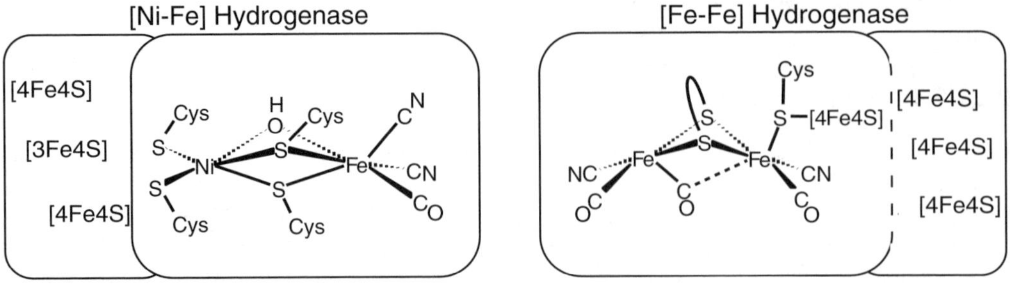

FIGURE 9.3 Active sites of the FeFe ("iron only") and the NiFe hydrogenases.

by the sulfur of a cysteine to one of the [4Fe-4S] clusters [28]. One CN^- and two CO coordinate to each Fe atom of the Fe1-Fe2 cluster. The two Fe are bridged by a fifth CO and a novel HS-CH_2-NH-CH_2-SH ligand, di(thiomethyl)amine (DTA). Hence, six ligands coordinate Fe1 and five ligands coordinate Fe2. Hydrogen binds probably to the sixth coordination site of Fe2, located at the end of a hydrophobic channel, through which the gas can diffuse in either direction. At Fe2, the hydrogen molecule is cleaved into a proton, which is removed by an adjacent basic amino acid residue and a hydride transiently bound to Fe2. Finally, the cluster withdraws two electrons from the hydride to yield the second proton. The electrons are further transferred, like in a wire, by the other four iron-sulfur clusters to the acceptor ferredoxin. Hydrogen formation follows just the reverse of this mechanism. Notably, the structure of the active site of the NiFe hydrogenases is very similar [29] (Figure 9.3).

9.3.2 NADH-Ferredoxin Oxidoreductase

Acidaminococcus fermentans ("Clostridia," "Acidaminococcaceae") produces hydrogen from glutamate, in addition to ammonia, CO_2, acetate, and butyrate. Initially, glutamate is converted via (*R*)-2-hydroxyglutaryl-CoA and glutaconyl-CoA to CO_2 and crotonyl-CoA, whose oxidation to two acetates leads to the reduction of NAD^+ (Figure 9.4 and Figure 9.5). The formed NADH ($E_0' =$ −320 mV) is reoxidized by acetoacetyl-CoA and crotonyl-CoA to yield butyrate (80%), and by

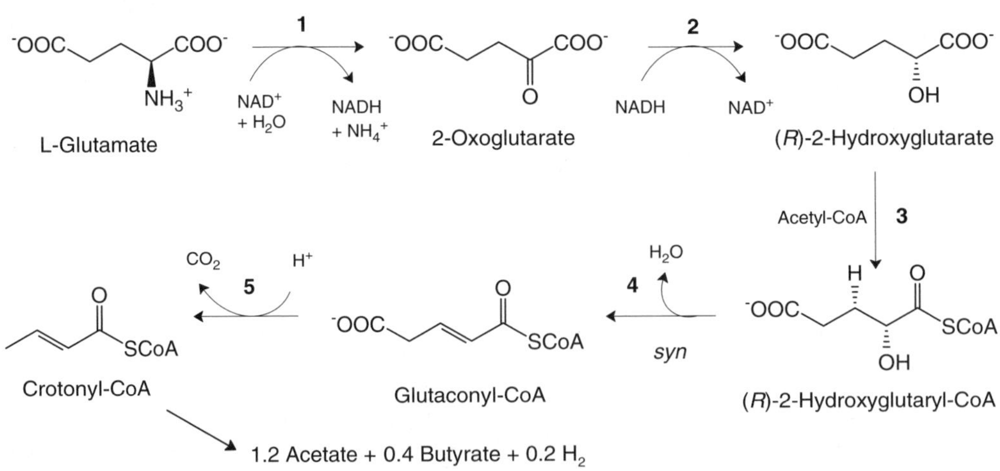

FIGURE 9.4 Fermentation of glutamate via (*R*)-2-hydroxyglutarate by *A. fermentans*. Enzymes: (1) Glutamate dehydrogenase; (2) 2-hydroxyglutarate dehydrogenase; (3) glutaconate/2-hydroxyglutarate CoA-transferase; (4) 2-hydroxyglutaryl-CoA dehydratase; (5) glutaconyl-CoA decarboxylase (see Figure 9.9). The disproportionation of crotonyl-CoA to acetate, butyrate and H_2 is described in Figure 9.5.

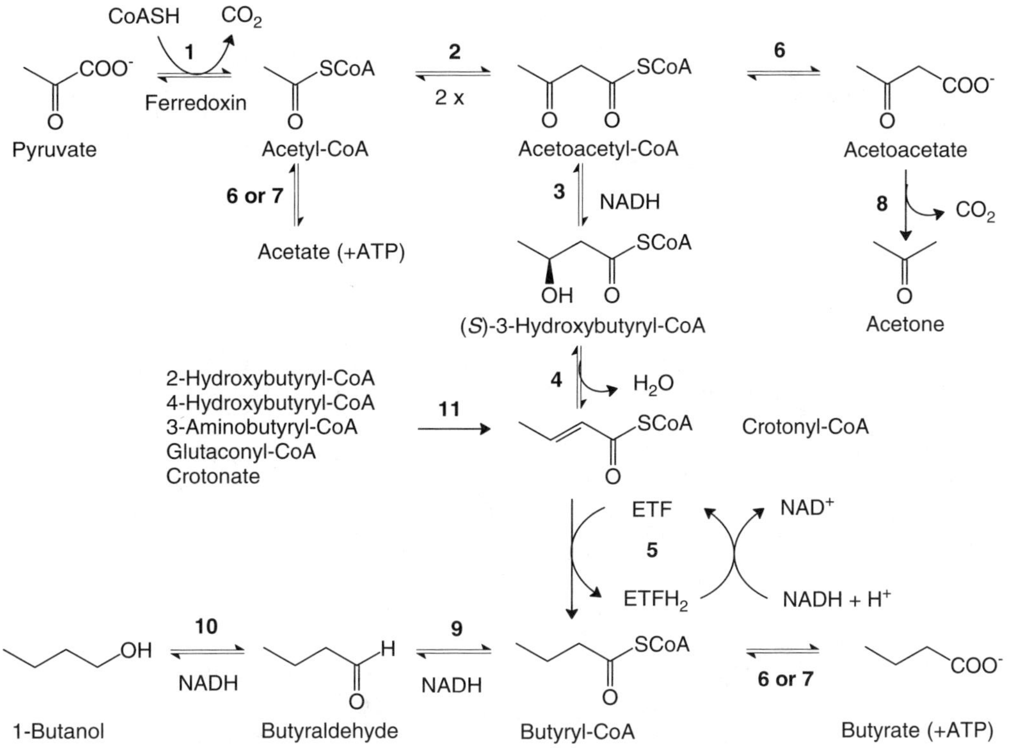

FIGURE 9.5 Synthesis of acetate, butyrate, acetone and 1-butanol. Enzymes: (1) Pyruvate:ferredoxin oxidoreductase; (2) thiolase, acetyl-CoA acetyl-transferase; (3) 3-hydroxybutyryl-CoA dehydrogenase; (4) crotonase, butenoyl-CoA hydratase; (5) butyryl-CoA dehydrogenase and electron transferring flavoprotein (ETF); (6) CoA-transferase; (7) acetate/butyrate kinase and phosphate acetyl/butyryl-transferase; (8) acetoacetate decarboxylase; (9) butyraldehyde dehydrogenase, CoA acylating; (10) 1-butanol (alcohol) dehydrogenase; (11) five other enzymes involved in crotonyl-CoA formation.

protons to yield hydrogen (20%; $E_0' = -420$ mV). Hence, a reversed electron transport must decrease the redox potential of NADH by 100 mV in order to reach that of hydrogen [30]. This could be achieved by a NADH:ferredoxin oxidoreductase integrated in the cytoplasmic membrane and driven by $\Delta\mu Na^+$, which is generated by the decarboxylation of glutaconyl-CoA to crotonyl-CoA (see Section 9.6). A very active NADH dehydrogenase has been detected in membranes of *A. fermentans* as measured by the reduction of ferricyanide, specific activity ca. 10 μmol × min^{-1} × mg^{-1} protein (J. Bresser and W. Buckel, unpublished data).

9.4 FORMATION OF SHORT-CHAIN FATTY ACIDS AND THEIR CoA-DERIVATIVES

9.4.1 Sources of Acetyl-CoA

There are several other ways to acetyl-CoA, whose de novo synthesis has already been described in the above. Clostridia usually activate acetate via acetylphosphate to acetyl-CoA, catalyzed by the enzymes acetate kinase (EC 2.7.2.1) and phosphate acetyltransferase (EC 2.3.1.8). In general, however, this pathway is used in the reverse direction for ATP-synthesis by substrate level phosphorylation. At the expense of another CoA-thioester, acetyl-CoA can be obtained by transfer of CoASH catalyzed by many CoA-transferases (EC 2.8.3. See next section). In *Clostridium kluyveri*, which ferments ethanol and acetate to butyrate, caproate (hexanoate), and hydrogen, acetyl-CoA

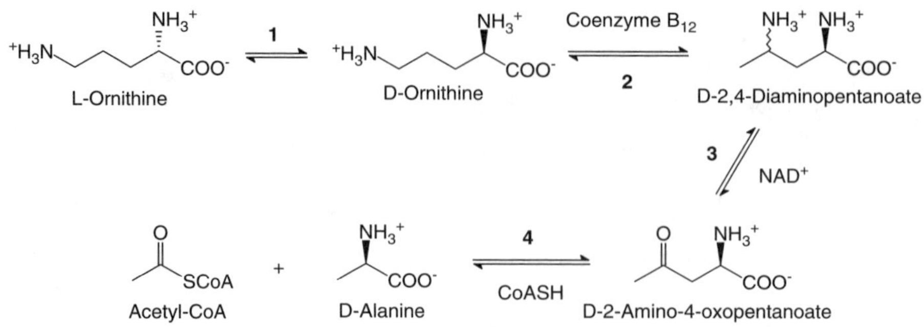

FIGURE 9.6 Oxidative branch of ornithine fermentation by *C. sticklandii*. (1) Ornithine racemase; (2) D-ornithine mutase; (3) D-2,4-aminopentanoate dehydrogenase; (4) D-2-amino-4-oxopentanoate cleavage enzyme.

is synthesized by the NAD^+-dependent oxidation of ethanol via acetaldehyde catalyzed by alcohol dehydrogenase (EC 1.1.1.1) and acetaldehyde dehydrogenase (CoASH acetylating; EC 1.2.1.10) [31]. In the fermentations of sugars and amino acids, acetyl-CoA is generated from pyruvate, either by oxidative decarboxylation catalyzed by pyruvate-ferredoxin oxidoreductase (pyruvate synthase; EC 1.2.7.1) (Figure 9.5) or by cleavage to formate mediated by pyruvate formate-lyase (EC 2.3.1.54). The latter enzyme is a member of the *S*-adenosylmethionine-dependent radical enzymes, which will be described in Section 9.10.2. A frequent source of acetyl-CoA is acetoacetyl-CoA, which is cleaved by CoASH into two acetyl-CoA catalyzed by thiolase (acetyl-CoA C-acetyltransferase; EC 2.3.1.9). Acetoacetyl-CoA is generated from crotonyl-CoA by hydration and NAD^+-dependent oxidation mediated by crotonase (3-hydroxybutyryl-CoA dehydratase; EC 4.2.1.55) and 3-hydroxybutyryl-CoA dehydrogenase (EC 1.1.1.157), respectively (Figure 9.5). In *Clostridium sticklandii,* a special source of acetyl-CoA is 2-amino-4-oxopentanoate derived from ornithine, which is cleaved by a pyridoxal-5′-phosphate dependent enzyme to alanine and acetyl-CoA [32] (Figure 9.6).

9.4.2 Pyruvate-Ferredoxin Oxidoreductase

In aerobic organisms, the oxidant of pyruvate is NAD^+ leading to acetyl-CoA, CO_2 and NADH. Due to the relatively high redox potential of NAD^+ ($E_0' = -320$ mV) as compared to that of pyruvate (E_0' ca. –500 mV), the reaction is irreversible. Anaerobes use oxygen-sensitive ferredoxins and flavodoxins as oxidants (E_0' ca. –450 mV), whose redox potentials are closer to that of pyruvate, which make the reactions reversible. Therefore, the corresponding enzyme is called pyruvate synthase. Indeed some autotrophic organisms as are the methanogenic Archaea, phylum Euryarchaeota (AII) [33], and the green sulfur bacteria *Chlorobium*, phylum Chlorobi (BXI) [9], fix part of their CO_2 by aid of this enzyme. With the exception of *C. kluyveri*, clostridia use pyruvate synthase predominantly in the reverse direction yielding acetyl-CoA and reduced ferredoxin, from which hydrogen can be formed directly. Although pyruvate synthase is common in clostridia, initial studies on the mechanism were performed with pyruvate synthase from *Halobacterium salinarium* (formerly called *H. halobium*, Euryarchaeota), leading to the detection of one of the first stable organic radicals in enzymes [34]. This exciting result was probably due to the fact that the enzyme from the facultative anaerobe *H. salinarium* is much more oxygen tolerant than those from the strict anaerobic clostridia. Also, the first crystal structure was obtained from a nonclostridial enzyme; the pyruvate synthase was isolated from *Desulfovibrio africanus* (Deltaproteobacteria). After soaking with pyruvate, an organic radical was formed in the crystalline enzyme [35]. The pyruvate reacted with the prosthetic group thiamine diphosphate in the carbanion form (Figure 9.7) and became decarboxylated (Figure 9.8). From the resulting hydroxyethyl thiamine diphosphate (enamine), one electron was removed by an adjacent [4Fe-4S] cluster leading to a radical cation (σ/n

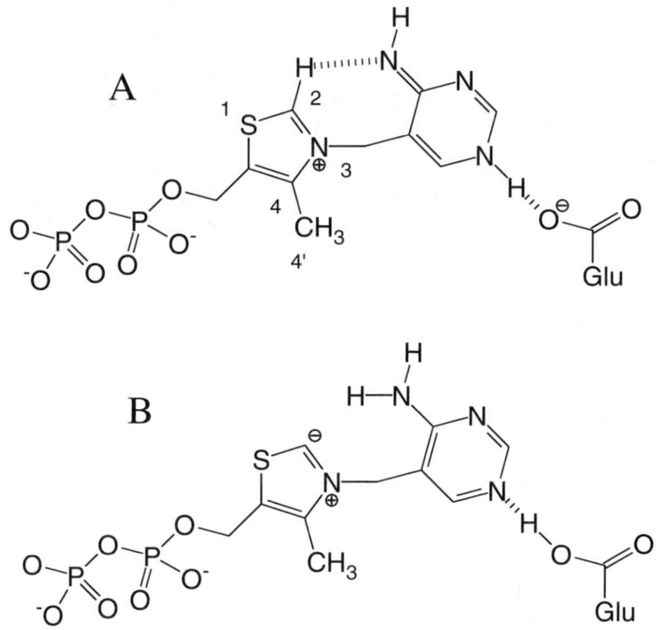

FIGURE 9.7 Thiamindiphosphate in the protonated (A) and reactive carbanion form (B). The prosthetic group is shown in the conformation found in the crystal structure [35]. This conformation has been predicted by A. Schellenberger 30 years ago [187].

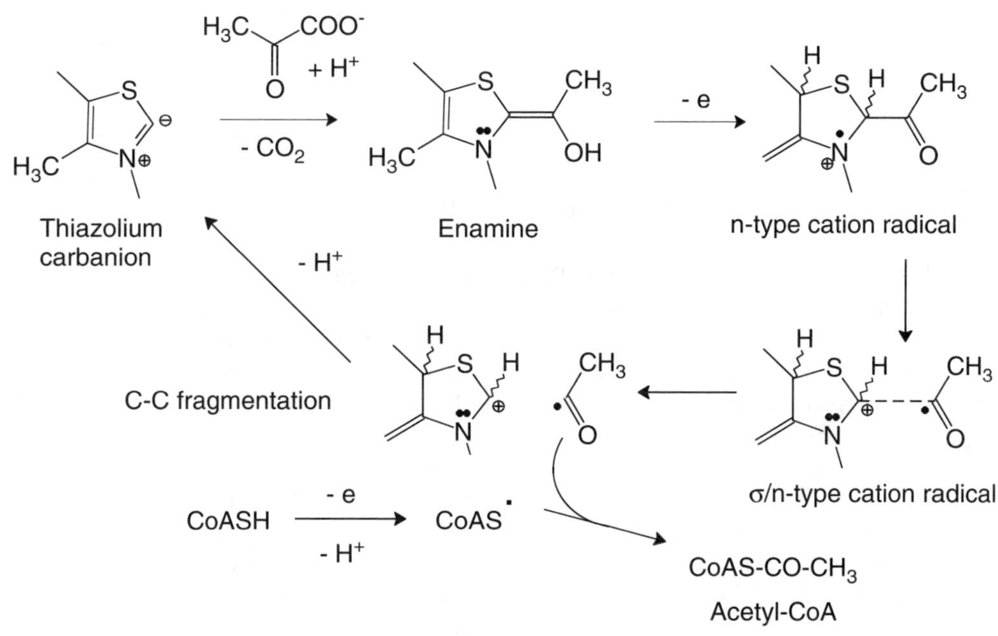

FIGURE 9.8 Proposed mechanism for the oxidative decarboxylation of pyruvate catalyzed by pyruvate ferredoxin oxidoreductase. The two electrons are removed by ferredoxin via the [4Fe-4S] clusters of the enzyme.

radical cation) localized in the unusual long bond between C2 of the thiazole ring and C2′ of the acetyl group (1.75–1.95 Å, as compared to a normal C-C-bond of 1.54 Å). Unexpectedly, the thiazol ring is slightly bent along a line connecting N3 to C5. Hence, the ring lost its aromaticity, which was thought to stabilize the radical. This bending is only possible if the double bond is shifted from C4=C5 to C4=C4′, showing for the first time that the methyl group C4′ (numbering in Figure 9.7) has an importance in catalysis. Only in the presence of an electron accepting ferredoxin, the reaction proceeds further by breaking the long C-C bond into an acetyl radical and the thiazole cation, which isomerizes to the ground state. Concomitantly, CoASH is oxidized by the [4Fe-4S] cluster to a thiyl radical, which recombines with the acetyl radical to yield acetyl-CoA.

9.4.3 Butyrate and Caproate Synthesis in *Clostridium kluyveri*

C. kluyveri ferments ethanol and acetate to butyrate and caproate [36] and molecular hydrogen, approximately according to Equation 9.6, whereby one ATP/2 H_2 is conserved [37,38].

$$\begin{aligned}&6\text{ Ethanol} + 3\text{ acetate}^- \rightarrow 3\text{ butyrate}^- + \text{caproate}^- + H^+ + 2\ H_2 + 4\ H_2O;\\&\Delta G^{\circ\prime} = -178\text{ kJ/mol caproate}\end{aligned} \quad (9.6)$$

Each ethanol is oxidized by two NAD^+ via acetaldehyde to acetyl-CoA in a thermodynamically reversible reaction (Equation 9.7).

$$\text{Ethanol} + 2\ NAD^+ \rightarrow \text{acetyl-CoA} + 2\ NADH + 2\ H^+;\ \Delta G^{\circ\prime} = +3\text{ kJ/mol ethanol} \quad (9.7)$$

The enzymes are alcohol dehydrogenase (EC 1.1.1.1) and aldehyde dehydrogenase (CoASH acetylating, EC 1.2.1.10). From one acetyl-CoA, ATP is generated via acetylphosphate (phosphotransacetylase, phosphate acetyl-transferase, EC 2.3.1.8, and acetate kinase, EC 2.7.2.1) (Equation 9.8), and two NADH are used for hydrogen formation, most likely via a membrane-bound ferredoxin–NAD^+ reductase (EC 1.18.1.3) and an only-iron hydrogenase (ferredoxin hydrogenase, EC 1.12.7.2) (Equation 9.9).

$$\text{Acetyl-CoA} + HPO_4^{2-} + ADP^{3-} \rightarrow \text{acetate}^- + ATP^{4-} + \text{CoASH};\ \Delta G^{\circ\prime} = 0\text{ kJ/mol} \quad (9.8)$$

$$2\ NADH + 2\ H^+ \rightarrow \mathbf{2\ H_2} + 2\ NAD^+;\ \Delta G^{\circ\prime} = +\ 39\text{ kJ/2 }H_2 \quad (9.9)$$

It has been shown that membranes from the related *Clostridium tetanomorphum* contain an Rnf-like NADH-ferredoxin oxidoreductase (EC 1.18.1.3), which may be driven by $\Delta\mu Na^+$ [39] (C.D. Boiangiu and W. Buckel, unpublished data). There is genetic evidence that the Rnf enzyme complex from *Rhodobacter capsulatus* provides reduced ferredoxin for nitrogen fixation by using NADH as reductant (Rnf = *Rhodobacter* nitrogen fixation) [40–42]. The required energy for hydrogen formation ($\Delta\mu Na^+$ or $\Delta\mu H^+$) is probably provided by the subsequent synthesis of butyrate and caproate from acetyl-CoA and acetate via acetoacetyl-CoA (acetyl-CoA C-acetyl-transferase, EC 2.3.1.9), (*S*)-3-hydroxybutyryl-CoA (NAD-dependent 3-hydroxyacyl-CoA dehydrogenase, EC 1.1.1.35), and butyryl-CoA (butyryl-CoA dehydrogenase, EC 1.3.99.2). The latter enzyme gets the electrons from NADH via the electron-transferring flavoprotein (ETF). Thus, the reductive condensation of two acetyl-CoA leads to butyryl-CoA by the reverse β-oxidation pathway (Equation 9.10); similarly, reductive condensation of acetyl-CoA and butyryl-CoA give rise to caproyl-CoA (Equation 9.11).

$$\begin{aligned}&2\text{ Acetyl-CoA} + 2\text{ NADH} + 2\ H^+ \rightarrow \text{Butyryl} - \text{CoA} + 2\ NAD^+ + H_2O + \text{CoASH};\\&\Delta G^{\circ\prime} = -42\text{ kJ/mol}.\end{aligned} \quad (9.10)$$

$$\text{Butyryl-CoA} + \text{acetyl-CoA} + 2\ \text{NADH} + 2\ \text{H}^+ \rightarrow \text{caproyl-CoA} + 2\ \text{NAD}^+ + \text{H}_2\text{O} + \text{CoASH}; \quad \Delta G^{\circ\prime} = -42\ \text{kJ/mol}. \tag{9.11}$$

The free fatty acids are formed by the action of a CoA-transferase with acetate as CoA-acceptor (butyryl-CoA: acetate CoA-transferase, EC 2.8.3.-) (Equation 9.12 and Equation 9.13).

$$\text{Butyryl-CoA} + \text{acetate} \rightarrow \text{butyrate} + \text{acetyl-CoA};\ \Delta G^{\circ\prime} = 0\ \text{kJ/mol} \tag{9.12}$$

$$\text{Caproyl-CoA} + \text{acetate} \rightarrow \text{caproate} + \text{acetyl-CoA};\ \Delta G^{\circ\prime} = 0\ \text{kJ/mol} \tag{9.13}$$

By these conversions, NADH is finally completely reoxidized to NAD^+, but nothing is known about how $\Delta\mu Na^+$ or $\Delta\mu H^+$ is generated by the fatty acid synthesis. Probably the fatty acid synthesis is located in the membrane [43]; especially the NADH-dependent reduction of crotonyl-CoA to butyryl-CoA mediated by ETF could give rise to an electrochemical ion gradient. Analysis of the deduced amino acid sequences of the enzymes catalyzing the synthesis of butyrate from acetyl-CoA and NADH in *C. tetani,* as well as activity measurements in cell-free extracts of *C. tetanomorphum,* revealed no integral membrane protein. Since *C. tetani* and *C. tetanomorphum* are closely related to *C. kluyveri* [44], it appears unlikely that butyrate and caproate syntheses are membrane processes in the latter organism. Therefore, coupling of fatty acid synthesis to hydrogen and ATP formation in *C. kluyveri* remains an enigma (C.D. Boiangiu, G. Herrmann, and W. Buckel, unpublished data).

9.4.4 Solvent Production in *Clostridium acetobutylicum*

Like many clostridia, *C. acetobutylicum* oxidizes glucose via the Emden-Meyerhof pathway to pyruvate, which is further oxidized to acetyl-CoA mediated by pyruvate:ferredoxin oxidoreductase. Usually, the reduced ferredoxin is a source of hydrogen, whereas the NADH generated in glycolysis is used for butyrate synthesis from two acetyl-CoA. Under certain conditions, however, *C. acetobutylicum* avoids the formation of too much acid. Therefore, two precursors of butyryate, acetoacetyl-CoA and butyryl-CoA, are partially converted to the neutral products acetone and 1-butanol, respectively (Figure 9.5). The CoA-moiety from acetoacetyl-CoA is transferred to acetate, and the resulting acetoacetate is decarboxylated to acetone. Acetoacetate decarboxylase contains in the active site a lysine residue with an extremely low pK = 5, which forms a Schiff base with acetoacetate in order to facilitate decarboxylation [45]. Butanol stems from the two-step NADH-dependent reduction of butyryl-CoA via butyraldehyde. The regulation and technical importance of this solvent-producing pathway will be discussed in other chapters of this book.

9.4.5 Propionate Synthesis in *Clostridium propionicum*

C. propionicum and *Megasphaera elsdenii* are able to ferment alanine, serine, and cysteine to propionate + acetate via pyruvate, as well as threonine to butyrate + propionate via 2-oxobutyrate [2,46–48]. Both 2-oxoacids are reduced to (*R*)-2-hydroxy acids as well oxidized to acetyl-CoA or propionyl-CoA, from which ATP is formed. Activation of the (*R*)-2-hydroxy acids to the CoA-derivatives by propionate CoA-transferase (EC 2.8.3.1) [49] is followed by unusual dehydrations to acryloyl-CoA or crotonyl-CoA, respectively [50–53]. Finally, the enoyl-CoAs are reduced to acyl-CoAs, from which the fatty acids are liberated by propionate CoA-transferase (Figure 9.9). Thus, three alanines give rise to one acetate + CO_2 and two propionates. The most interesting enzyme of this pathway, the lactyl-CoA dehydratase, will be discussed in Section 9.11. An FAD-containing enzyme complex composed of an ETF and an acyl-CoA dehydrogenase catalyzes the consecutive irreversible reduction of the product acryloyl-CoA to propionyl-CoA by NADH. Thus

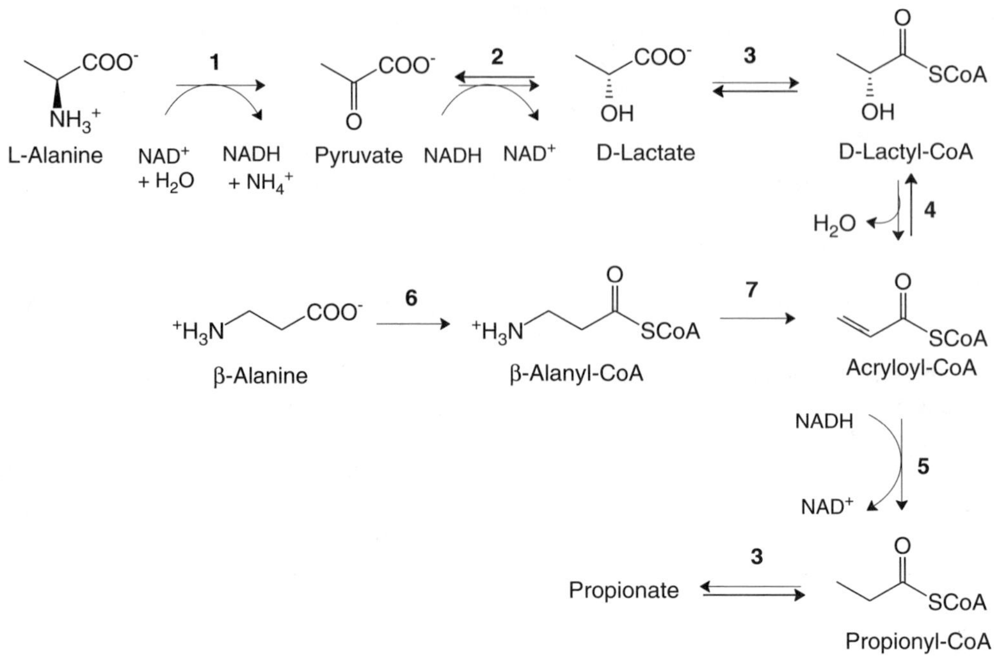

FIGURE 9.9 Reduction of α- and β-alanine to propionate in *C. propionicum*. Enzymes: (1) Alanine:2-oxoglutarate aminotransferase and glutamate dehydrogenase; (2) D-lactate dehydrogenase; (3) propionate CoA-transferase; (4) lactyl-CoA dehydratase; (5) acryloyl-CoA reductase; (6) β-alanine CoA-transferase; (7) β-alanyl-CoA ammonia lyase. In the complete fermentation, pyruvate is also oxidized to acetyl-CoA, from which ATP is conserved via acetylphosphate. Due to the reversibility of reactions 2, 3, and 4, β-alanine can be oxidized to pyruvate and further to acetyl-CoA [35].

NADH reduces FAD, which then transfers two electrons to acryloyl-CoA [54]. *C. propionicum* is also able to ferment 3 β-alanine to acetate, CO_2, and 2 propionate by the same stoichiometry as α-alanine. β-Alanine is converted to β-alanyl-CoA catalyzed by a specific CoA-transferase, followed by deamination to acryloyl-CoA [55], which enters the pathway of α-alanine fermentation [188]. Currently, enzymes of these pathways are used to develop a biotechnological process for the synthesis of 3-hydroxypropionate from glucose [56].

9.5 CoA-TRANSFERASES AND CLAISEN CONDENSATIONS

CoA-transferases catalyze the reversible transfer of CoASH from a donor acyl-CoA to an acceptor acid. Thus, a new CoA-thiol ester is formed without consumption of ATP. Due to this property, the enzymes are more common in anaerobic organisms, which thrive on minimal amounts of ATP. Although never observed directly, there is convincing evidence that the CoA-transfer proceeds via mixed anhydrides generated by the attack of a carboxylate anion at a thiol ester carbonyl [57] (Figure 9.10). This reaction is very similar to the Claisen ester condensation, in which an enolate anion of an acyl-CoA attacks a thiol ester carbonyl as in thiolase (acetyl-CoA acetyl-transferase, EC 2.3.1.9). Therefore, this section deals with both types of reactions. Since Ludwig Claisen also discovered the condensation of esters with aldehydes or ketones, sometimes these reactions are also called Claisen condensations [58]; enzymatic examples are malate synthase (EC 2.3.3.9), citrate synthase (EC 2.3.3.1), citrate lyase, and citramalate lyase (see next).

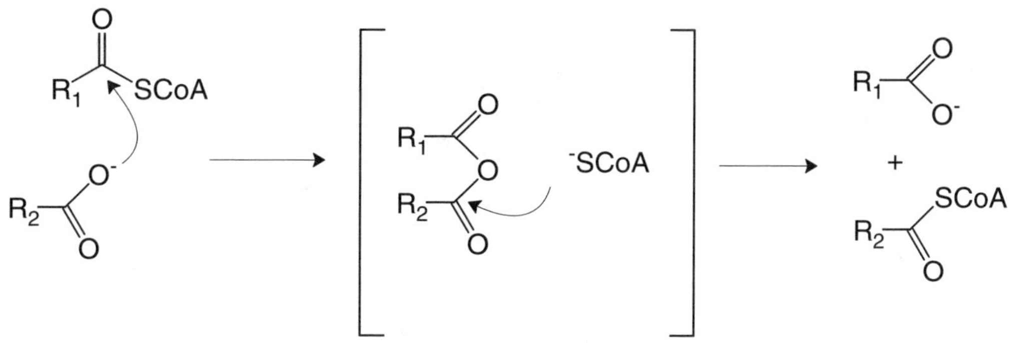

FIGURE 9.10 Mechanism of CoA-transferases. Intially the CoA-thiol ester is attacked by a carboxylate anion of the second substrate or of the enzyme (conserved glutamate residue), depending on the class of CoA-transferase. Subsequently, the formed mixed anhydride reacts with the transiently released CoA thiolate anion to form either the products or an enzyme-CoA-thiol ester intermediate, whereby the carboxylate product is generated.

9.5.1 Three Classes of CoA-Transferases

CoA-transferases have been grouped into three unrelated classes as deduced from their amino acid sequences [59]. Class I comprises the enzymes, in which a CoA-thiol ester is formed with a specific glutamate residue as intermediate, whereas in classes II and III this intermediate has not been found. The enzyme-CoA-thiol ester can be detected by MALDI-TOF mass spectrometry [57], as well as chemically by reduction with borohydride to the alcohol or by incubation with hydroxylamine, whereby the hydroxamic acid of the glutamate residue is formed. Hence, borohydride or hydroxylamine inactivate the enzyme at pH 7 but only in the presence of an acyl-CoA substrate. Furthermore, by measuring the K_m-values for one substrate at different concentrations of the second substrate, parallel lines are obtained in a Lineweaver-Burk plot. These parallel lines are characteristics of a ping-pong mechanism, by which the second substrate enters the reaction after the first product has been formed. Clostridial examples of class I enzymes are butyryl-CoA-acetoacetate CoA-transferase (EC 2.8.3.9) involved in acetone production in *C. acetobutylicum* [60], glutaconate CoA-transferase (Gct, EC 2.8.3.12) from *A. fermentans* [61], and propionate CoA-transferase (Pct, EC 2.8.3.1) from *C. propionicum* [49]. The catalytic glutamate residue of glutaconate CoA-transferase has been exchanged to aspartate, which converted the enzyme to a specific thiol ester hydrolase. Most likely, a water molecule now occupies the space of the deleted methylene group [62].

Other well-characterized but not sequenced clostridial CoA-transferases are 5-hydroxybutyrate CoA-transferase from *Clostridium viride* (formerly called *C. aminovalericum*) fermenting 5-aminovalerate [63,64] and butyryl-CoA:acetoacetate CoA-transferase in the lysine fermenting *Clostridium subterminale* (formerly called *C.* SB4) [65]. Since *F. nucleatum* also ferments lysine by the same pathway as *C. subterminale* [66] and contains a putative class I butyryl-CoA:acetoacetate CoA-transferase [67], the *C. subterminale* enzyme probably also belongs to this class.

9.5.2 Class II CoA-Transferases: Citrate and Citramalate Lyases

Citrate lyase (CitFED, EC 4.1.3.6) occurring in Proteobacteria and several clostridia [68,69] or citramalate lyase (ClmABC; EC 4.1.3.22) from *C. tetanomorphum* [70] catalyze the Mg^{2+}-dependent cleavage of citrate into acetate and oxaloacetate or (*S*)-citramalate into acetate and pyruvate, respectively. The reaction, however, is a two-step process because the actual C-C-cleavage of a retro Claisen-type condensation occurs at the thiol ester level. Therefore, the enzymes are complexes of a CoA-transferase (class II-CoA-transferase, Cit F/CmlA) and a lyase (CitE/CmlB) mediated

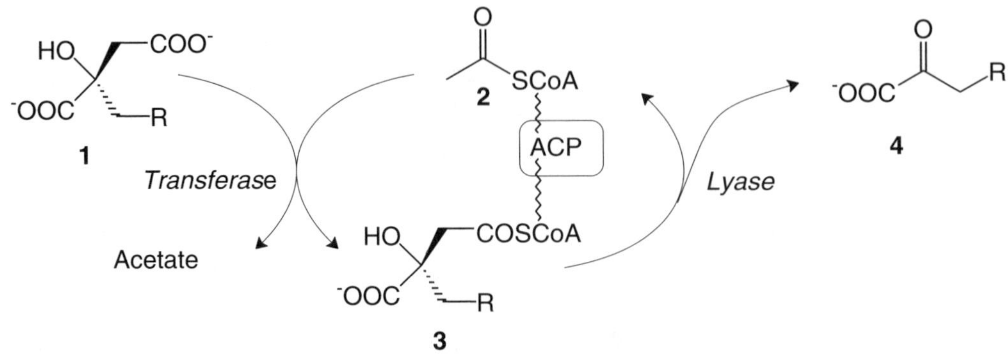

FIGURE 9.11 Mechanism of citrate and citramalate lyase. A, R = COO^-; B, R = H. 1A, Citrate; 1B, (S)-citramalate; 2, acetyl-CoA covalently bound to the acyl carrier protein (ACP); 3A, (S)-citryl-CoA-ACP; 3B, (S)-citramalyl-CoA-ACP; 4A, oxaloacetate; 4B, pyruvate.

by an acyl carrier protein (ACP, CitD/CmlC). The ACP contains a specific serine residue, to which acetyl-dephospho-CoA is covalently bound via ribose-5-phosphate [71]. The CoA-transferase, actually an acetyl-ACP:citrate/citramalate ACP-transferase, catalyzes the formation of (*S*)-citryl-ACP/(*S*)-citramalyl-ACP, which is cleaved to oxaloacetate/pyruvate and acetyl-ACP (Figure 9.11). Removal of the acetyl group from ACP inactivates both enzymes. Reactivation is achieved by acetylation with acetic anhydride [72] or by an ATP-dependent ligase (EC 6.2.1.22) [73]. The putative genes encoding the citramalate lyase complex from *C. tetani* (*cmlBCA*, 6 × 96 kDa), the ligase (*cmlL*), and the two enzymes (*cmlX* and *cmlG*) required for the biosynthesis of the prosthetic group of the ACP are located in this order at the end of a cluster starting with a regulatory gene for glutamate fermentation, followed by *mutS*, *mutL,* and *mutE* (*glm* in *Clostridium cochlearium*) encoding the E and S subunits of glutamate mutase and the repair enzyme MutL [74]. The consecutive genes *malA* and *mesAB* encode methylaspartate ammonia lyase (Mal) and the heterodimeric mesaconase (Mes), respectively [39] (A.J. Pierik, unpublished data) (Figure 9.1). In the biosynthesis of the prosthetic group, the ribose-triphosphate moiety of ATP is transferred by CmlG to dephospho-CoA, yielding 2′-(5″-triphosphoribosyl)-3′-dephospho-CoA, which is further transferred by ClmX to the specific serine residue of *apo*-ACP to give *holo*-ACP and diphosphate [71].

Acetyl-CoA-4-hydroxybutyrate CoA-transferase is involved in the fermentation of 4-aminobutyrate (γ-aminobutyrate, GABA) via 4-hydroxybutyryl-CoA and crotonyl-CoA to acetate and butyrate in *Clostridium aminobutyricum* [75] (Figure 9.12). The amino acid sequence shows 26% identities to CitF of citrate lyase [76]. Many related deduced proteins are found not only in other clostridia and *Fusobacterium nucleatum*, but also in *Eukarya* like *Drosophila melanogaster.* Some of these proteins are designated as acetyl-CoA hydrolases, which can be derived from class I CoA-transferases by exchange of the catalytic glutamate residue to aspartate [62]. Class II CoA-transferases appear to be a widespread, previously not well recognized and characterized family of enzymes.

9.5.3 The Emerging Class III CoA-Transferases

Class III is an emerging group of highly specific CoA-transferases found in Proteobacteria and Firmicutes; the only known clostridial examples are cinnamoyl-CoA:(*R*)-phenyllactate CoA-transferase from *Clostridium sporogenes* [77,78], and 2-hydroxyioscaproate CoA-transferase from *Clostridium difficile* [189]. The enzyme from *C. sporogenes* is part of the complex phenyllactate dehydratase, in which — like in citrate lyase — the substrate (*R*)-phenyllactate is activated by cinnamoyl-CoA to the thiol ester. Thereby (*E*)-cinnamate and (*R*)-phenyllactyl-CoA are formed.

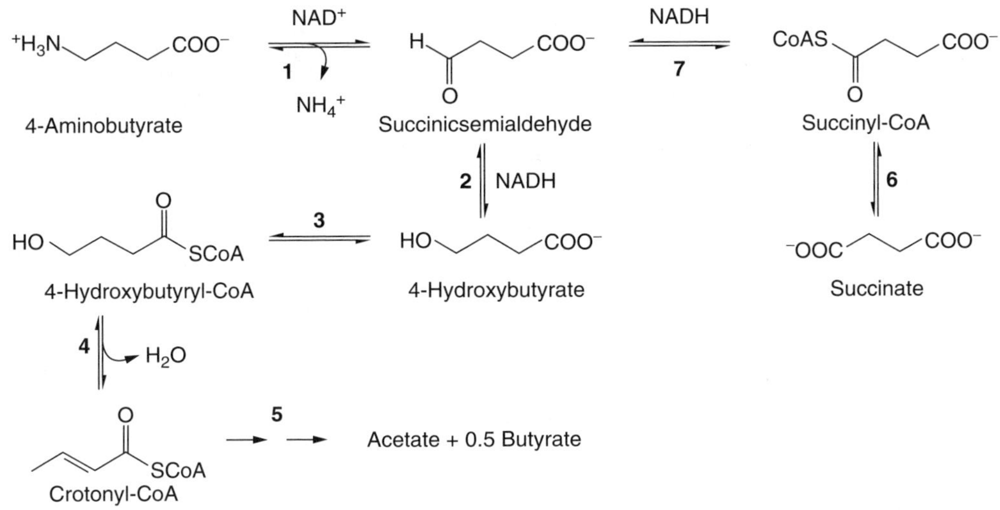

FIGURE 9.12 Fermentation of 4-aminobutyrate (GABA) by *C. aminobutyricum* and reduction of succinate to butyrate by *C. kluyveri*. Enzymes: (1) 4-Aminobutyrate aminotransferase and glutamate dehydrogenase; (2) 4-hydroxybutyrate dehydrogenase; (3) 4-hydroxybutyrate CoA-transferase; (4) 4-hydroxybutyryl-CoA dehydratase; (5) enzymes leading to acetate and butyrate (see Figure 9.5); (6) succinate CoA-transferase; (7) succinate semialdehyde dehydrogenase, CoA succinylating.

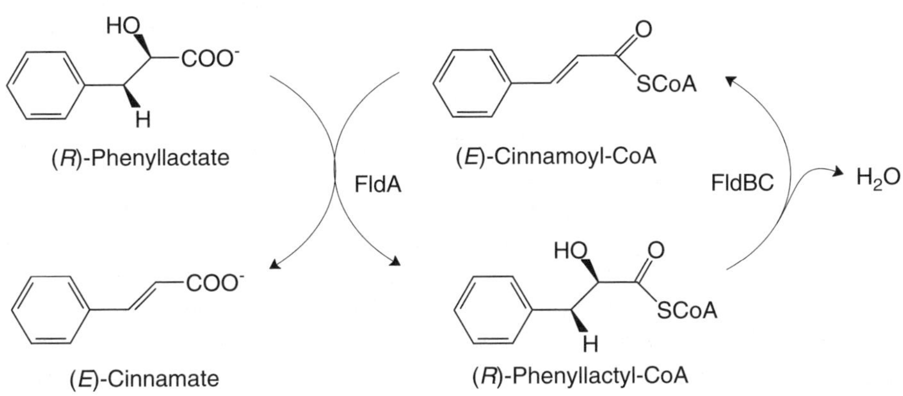

FIGURE 9.13 Reactions catalyzed by phenyllactate dehydratase from *C. sporogenes*. FldA, CoA-transferase; FldBC, phenyllactyl-CoA dehydratase.

The latter is dehydrated to cinnamoyl-CoA. Hence the overall reaction is the *syn*-dehydration of (*R*)-phenyllactate to (*E*)-cinnamate, mediated by catalytical amounts of cinnamoyl-CoA (Figure 9.13). Kinetic analysis of the CoA-transferase subunit indicated that a ternary complex between enzyme, phenyllactate, and cinnamoyl-CoA is formed. In the same way, 2-hydroxyioscaproate CoA-transferase generates (*R*)-2-hydroxyioscaproyl-CoA, which is dehydrated to isocaprenoyl-CoA, although the transferase seems not to be associated with the dehydratase.

9.5.4 A Single Reaction Comprising Two Claisen Condensations and One CoA-Transfer

The lysine fermenting *C. subterminale* and *F. nucleatum* contain enzymes that catalyze the acetyl-CoA-dependent cleavage of (*S*)-3-oxo-5-aminohexanoate to acetoacetate and (*S*)-3-aminobutyryl-CoA

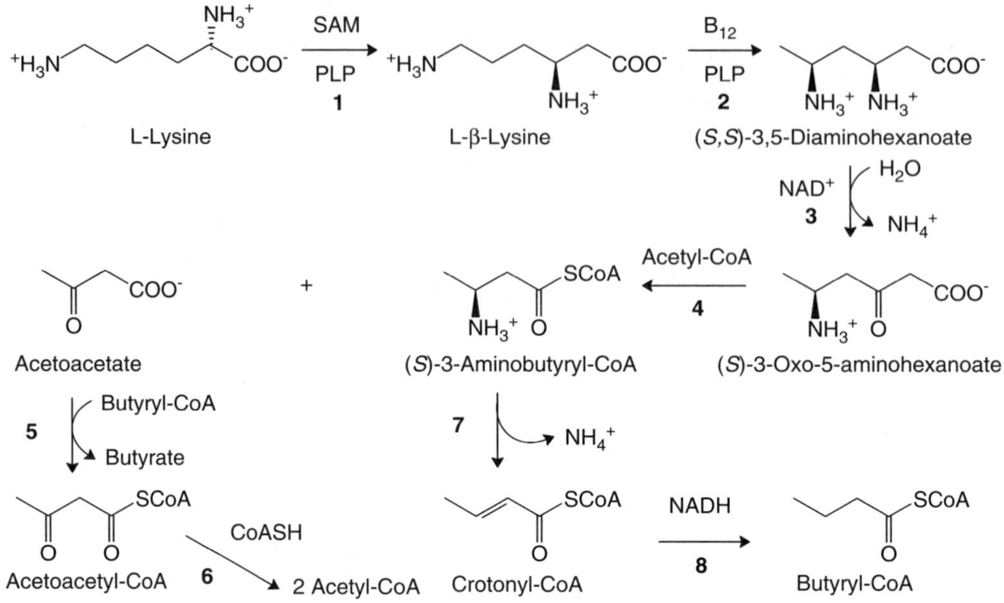

FIGURE 9.14 Fermentation of lysine by *C. subterminale* and *F. nucleatum*. SAM, S-adenosylmethionine; PLP, pyridoxal-5′-phosphate; B_{12}, coenzyme B_{12} or adenosylcobalamin; (1) lysine aminomutase; (2) β-lysine aminomutase; (3) 3,5-diaminohexanoate dehydrogenase; (4) 3-oxo-5-aminohexanoate cleavage enzyme; (5) acetoacetate CoA-transferase; (6) thiolase or acetyl-CoA acetyl-transferase; (7) 3-aminobutyryl-CoA ammonia lyase; (8) butyryl-CoA dehydrogenase. One mol acetyl-CoA is used for synthesis of ATP via acetylphosphate.

(3-oxo-5-aminohexanoate cleavage enzyme) [66,79] (Figure 9.14). An initial Claisen condensation of C2 of the enolate of the 3-oxo acid with acetyl-CoA could yield CoASH and an eight-carbon intermediate, 2,4-dioxo-3-carboxy-6-aminoheptane, which is cleaved again by CoASH at C4 into acetoacetate (C1-C3 + COO^-) and (*S*)-3-aminobutyryl-CoA. The eight-carbon intermediate can be regarded as a carba-analogue of the mixed anhydrides formed during catalysis of class II and class III CoA-transferases, since replacement of the CH_2-COO^- group between the two carbonyl groups by an oxygen atom leads to the anhydride between acetic acid and (*S*)-3-aminobutyric acid. Hence, the whole reaction is similar to a CoA-transferase, in which a carboxymethyl group, rather than an oxygen atom, is exchanged by CoA (Figure 9.15).

9.5.5 Citrate (*Re*)-Synthase

The well-known citrate synthase [citrate (*Si*)-synthase, EC 2.3.3.1], which occurs almost in every organism, catalyzes the enolization of acetyl-CoA, followed by an attack of the enolate anion at the *Si*-side of the carbonyl group of oxaloacetate. The resulting (*S*)-cityl-CoA is hydrolyzed to citrate and CoASH [80]. *C. kluyveri* and *Clostridium acidi-urici*, however, contain a citrate (*Re*)-synthase (EC 2.3.3.3) with (*R*)-cityl-CoA as intermediate [81–83]. Although the product citrate obtained with (*Re*)-citrate synthase is identical to that from (*Si*)-citrate synthase, with isotopically labeled acetate differently labeled products are obtained. Thus, incubation of cell-free extracts of most organisms with [1-^{14}C]acetate leads to (*S*)-[5-^{14}C]glutamate via (*S*)-[1-^{14}C]citrate, (*Z*)-[5-^{14}C]aconitate, (2*R*,3*S*)-[5-^{14}C]isocitrate and 2-oxo[5-^{14}C]glutarate. With cell-free extracts of *C. kluyveri*, however, (*S*)-[1-^{14}C]glutamate is formed from [1-^{14}C]acetate. Interestingly, (*Re*)-citrate synthase requires Mn^{2+} or Mg^{2+} ions for activity, whereas (*Si*)-citrate synthase does not and it is active in the presence of EDTA. Unfortunately, no sequence data of (*Re*)-citrate synthases are available.

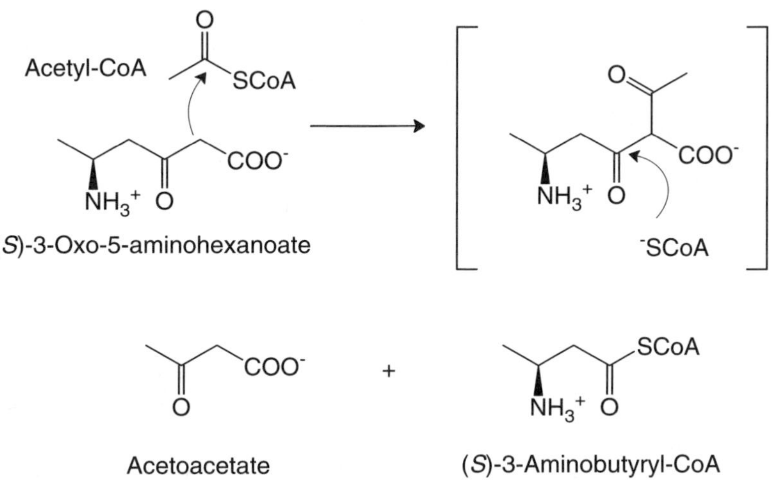

FIGURE 9.15 Proposed mechanism for the cleavage of (*S*)-3-oxo-5-aminohexanoate by acetyl-CoA. Initially the enolate anion of activated methylene group at C2 of the C6-substrate is C-acetylated by acetyl-CoA. The resulting C8-intermediate is thiolytically cleaved by the CoA thiolate anion into acetoacetate and (S)-3-aminobutyryl-CoA. Note the similar attack of a carboxylate at a thiol ester carbonyl in Figure 9.10.

9.6 Na⁺-DEPENDENT, BIOTIN-CONTAINING DECARBOXYLASES

9.6.1 Methylmalonyl-CoA and Glutaconyl-CoA Decarboxylases in Clostridia

There are three Na⁺-dependent biotin-containing decarboxylases, which catalyze the substitution of CO_2 by H^+ with retention of configuration: Oxaloacetate decarboxylase from Enterobacteria [84], methylmalonyl-CoA decarboxylase from *Veillonella parvula* (*Clostridiales,* "Acidaminococcaceae") [85] and *Propiogenium modestum* ("Fusobactericeae") [86], and glutaconyl-CoA decarboxylase from *Acidaminococcus fermentans* ("Acidaminococcaceae") [87] and *F. nucleatum* [88] (see Dimroth et al. [89] and Buckel [90]). The enzymes represent complexes of four functional domains or subunits, a carboxyltransferase (α), an alanine- and proline-rich biotin carrier (γ), a 9-11 membrane-spanning helices-containing Na⁺-dependent carboxybiotin decarboxylase (β), and a membrane anchor (δ). In the case of the fusobacterial glutaconyl-CoA decarboxylase, the molecular masses are 65 kDa (α), 33 kDa (β), 19 kDa (γ), and 16 kDa (δ). In the first catalytic step, the carboxyl group of the substrate is transferred by the α-subunit to biotin attached to the carrier (γ) and is thereby converted to a kinetically activated carboxylate in *N*-carboxybiotin. After swing-over of the carrier to the decarboxylase (β), *N*-carboxybiotin is decarboxylated, and the free energy of the decarboxylation ($\Delta G^{\circ\prime} = -30$ kJ/mol) is used to translocate 1-2 Na⁺ from the inside to the outside of the cytoplasma membrane, whereas the proton comes from the outside (Figure 9.16). The carboxyltransferase subunit of glutaconyl-CoA decarboxylase from *A. fermentans* has been crystallized, and the structure has been solved by X-ray crystallography at 2.2 Å resolution [91]. The enzyme is a homodimer, and each monomer comprises two related domains, which probably stem from gene duplication. The N-terminal domain of subunit 1 interacts with the C-terminal domain of subunit 2, and vice versa. The N-terminal domain provides the binding site of glutaconyl-CoA and C-terminal domain that for biotin attached to the carrier. Hence, the dimer is a functional unit, since the CO_2 is transferred from glutaconyl-CoA on subunit 1 to biotin on subunit 2. The enzyme belongs to the crotonase superfamily [92], in which two hydrogen bonds from backbone amides to the thiol ester carbonyl and the positive dipol of a α-helix stabilize the dienolate anion intermediate. The electrochemical Na⁺ gradient (ΔμNa⁺) formed by the Na⁺-dependent decarboxylase

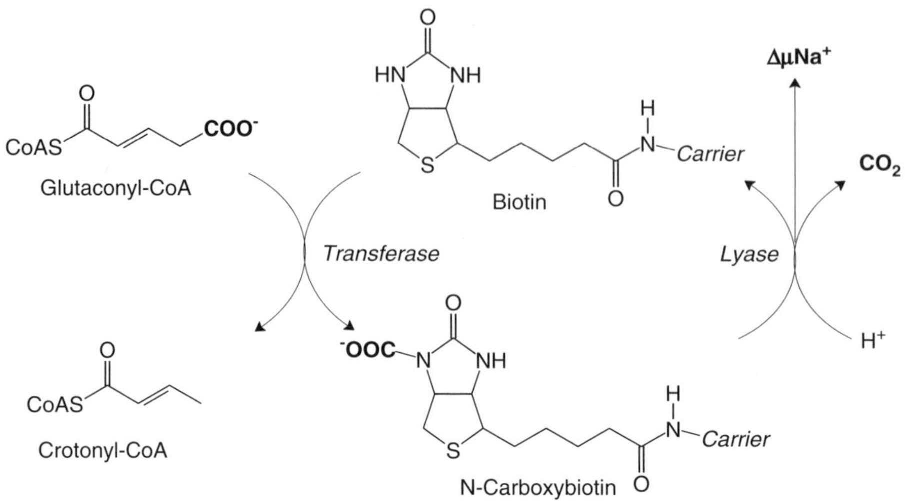

FIGURE 9.16 Mechanism of glutaconyl-CoA decarboxylase. The enzyme is composed of four different subunits, a hydrophilic carboxyl transferase, a hydrophobic decarboxylase, a biotin carrier, and a membrane anchor. The decarboxylase, an integral membrane protein, generates a sodium motive force.

is used for ATP-synthesis (*P. modestum*) and Na^+-driven symporters. The proposed Na^+-driven NADH-ferredoxin oxidoreductase has been described above (Section 9.3.2). It should be noted that the clostridial oxaloacetate decarboxylases appear to be soluble enzymes and are dependent on Mg^{2+} but independent on Na^+.

9.7 α-ELIMINATION OF AMINO GROUPS

Two different clostridial enzymes are known, glycine reductase and D-proline reductase (EC 1.21.3.3), which catalyze the α-elimination of amino groups (i.e., the substitution of an α-amino group by a hydrogen); similar reactions are mediated by sarcosine reductase and betaine reductase. The soluble enzyme systems are involved in the classical Stickland reaction, by which pairs of amino acids are fermented — one serves as electron donor and the other as acceptor. Among obligate donors are alanine, isoleucine, and valine, whereas glycine and proline are excellent acceptors. The donors are deaminated to the 2-oxo acids and oxidatively decarboxylated to the corresponding acyl-CoA, which is used for ATP conservation. The acceptor amino acids are reduced to fatty acids of the same carbon chain length. A typical fermentation is shown in Equation 9.14:

$$\text{Alanine} + 2\ \text{glycine} + 2\ H_2O \rightarrow 3\ \text{acetate}^- + 3\ NH_4^+ + CO_2;\ \Delta G^{\circ\prime} = -153\ \text{kJ/mol alanine} \quad (9.14)$$

According to Equation 9.14, both alanine and glycine are converted to acetate. The imino acid proline is reduced to 5-aminovalerate.

9.7.1 GLYCINE REDUCTASE

Glycine reductase has been discovered in *Clostridium sticklandii* [93,94]. The more recent work on this enzyme has been carried out with *Eubacterium acidaminophilum* (both *E. acidaminophilum* and *C. sticklandii* belong to *Clostridium* cluster XI) [95–98]. Glycine reductase comprises three protein complexes, A, B, and C, of which A and B contain selenocysteines. A pyruvoyl residue of the 47 kDa subunit GrdB of the heterodimeric protein B (GrdBB′) forms a Schiff base with glycine.

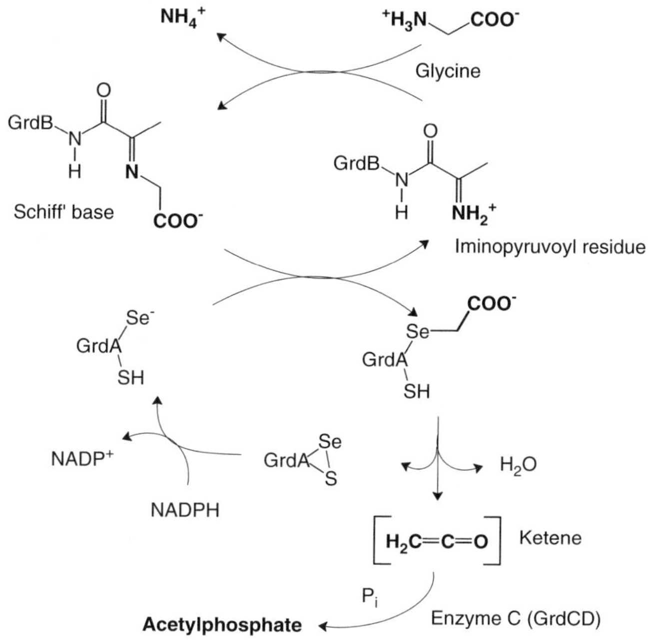

FIGURE 9.17 Reduction of glycine to ammonia and acetylphosphate. The scheme is a shortened version of that described in the text.

The selenocysteine residue of a small carbon- and redox-carrier protein A (17 kDa, GrdA) takes over the carboxymethyl group of the glycine, whereas the ammonia remains as imine on GrdB. The formed selenoether on protein A will be reductively split with the help of an adjacent SH-group to a hypothetical "ketene" intermediate [99] catalyzed by protein C. This leads to the formation of an oxidized S-Se-bond on protein A, which is reduced again by NADPH mediated by a thioredoxin reductase and thioredoxin. The "ketene" immediately reacts with a SH-group of one of the two subunits of protein C (GrdC and GrdD) to form an acetyl-thiol ester. The acetyl moiety is cleaved by phosphate to yield acetylphosphate as final product, catalyzed by GrdD. Hence, one mol ATP can be conserved per mol of glycine. Ammonia is released from the iminocarbonyl group of protein B either by hydrolysis or by the next incoming glycine (Figure 9.17).

9.7.2 D-Proline Reductase

The D-proline reductase system from *C. sticklandii* seems to be somewhat similar to the glycine reductase system due to the involvement of selenocysteine and an adjacent cysteine residue present on PrdB. The N-terminal pyruvoyl group formed by the cleavage of the proprotein PrdA is involved in forming a Schiff base with D-proline to facilitate the reductive cleavage to 5-aminovalerate as final product. NADH, not NADPH, is the natural electron donor. Genes coding for a supposedly electron-transferring protein PrdC and to homologues to both parts of the cleaved proprotein PrdA are present in both genomes of *C. sticklandii* and *C. difficile*. No ATP conservation has been observed so far [100,101].

According to Equation 9.7, the Stickland reaction between one alanine and two glycines should yield three ATP, one from acetyl-CoA derived by oxidation of alanine via pyruvate and the two others from glycine via acetylphosphate. Since $\Delta G^{\circ\prime}$ ca. 70 kJ/mol alanine is required to generate one ATP [102], the free energy of Equation 9.14 (–153 kJ) is not sufficient for 3 ATP. Probably some acetylphosphate has to be hydrolyzed to ensure optimal metabolic flux and hence growth.

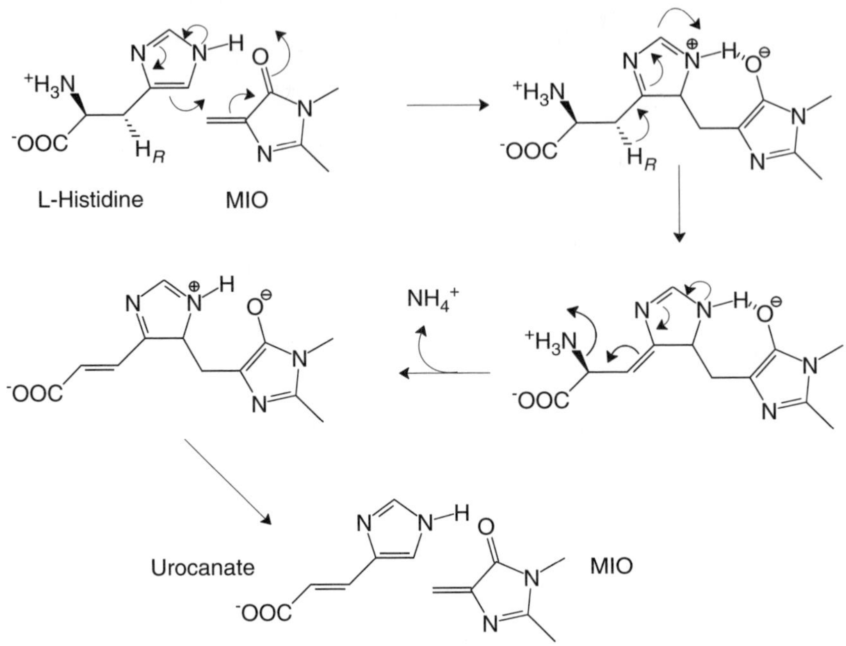

FIGURE 9.18 Mechanism of histidine ammonia lyase. The enzyme contains an electrophilic prosthetic group called MIO, 3,5-dihydro-5-methylidene-4H-imidazol-4-one. Electrophilic addition of this group to the aromatic imidazole ring of histidine lowers the p*K* of H_{Re} that a glutamate residue of the enzyme can remove the proton.

9.7.3 Histidine Ammonia α-Lyase

Clostridium tetanomorphum ferments histidine via glutamate to ammonia, CO_2, formate, acetate, butyrate, and hydrogen [2] (Figure 9.1). In the first step, histidine is deaminated to urocanate, whereby the α-amino group and the β-proton are eliminated. The very high p*K* of the β-proton is lowered by the electrophile methyleneimidazole (MIO), which transiently adds to the aromatic imidazole ring of histidine. MIO, which is related to the prosthetic group of the green fluorescent protein (Gfp), is formed by an autocatalytic posttranslational modification of the enzyme histidine ammonia lyase (histidase EC 4.3.1.3) [103] (Figure 9.18).

9.8 β-ELIMINATION OF AMINO GROUPS

9.8.1 Ammonia β-Lyases

In contrast to the α-elimination discussed above, the β-elimination of amino groups is a chemically facile process, for which the enzyme requires no prosthetic group apart from the fact that some of these deaminases need mono- and divalent metal ions for activity. The α-proton is activated by the carboxylate or thiol ester group and can be readily removed by a base from the enzyme. Nevertheless, there are different families of enzymes that catalyze these eliminations. Although the highly specific aspartase from many organisms [104] and the less specific 3-methylaspartase from *C. tetanomorphum* [105] (Figure 9.1) act on similar substrates, (*S*)-aspartate and (2*S*,3*S*)-3-methylaspartate, respectively, the enzymes belong to different families. Aspartase is related to class II fumarases, the argininosuccinate and adenylosuccinate lyases, whereas 3-methylaspartase is a member of the large enolase superfamily [106,107]. A third type of β-elimination of amino groups is catalyzed by 3-aminobutyryl-CoA ammonia lyase from the lysine-fermenting organisms

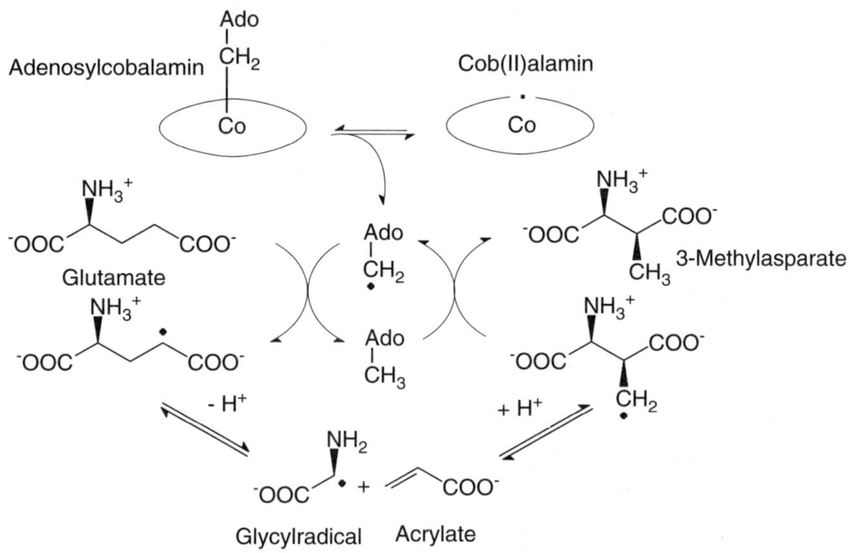

FIGURE 9.19 Proposed mechanism for glutamate mutase. Initially adenosylcobalamin (coenzyme B_{12}) is cleaved into cob(II)alamin and the 5′-deoxyadenosine radical (Ado-CH2•), which abstracts the 4-Si-hydrogen atom from glutamate. The 4-glutamyl radical fragments into 2-glycyl radical and acrylate, which recombine to 3-methyleneaspartate radical. Finally, the intermediate 5′-deoxyadenosine transfers a hydrogen atom to yield 3-methylaspartate.

C. subterminale [108] and *F. nucleatum* (Figure 9.14) [66], as well as by β-alanyl-CoA ammonia lyase from the α- and β-alanine fermenting *Clostridium propionicum* (Figure 9.9) [55]. Thus, (*S*)-3-aminobutyryl-CoA is reversibly deaminated to crotonyl-CoA and β-alanyl-CoA to acryloyl-CoA. The crystal structure of β-alanyl-CoA ammonia lyase (G. Herrmann, W. Buckel, T. Selmer, and K. Reuter, unpublished data) indicates that the β-aminoacyl-CoA ammonia lyases are related to (*R*)-3-hydroxybutyryl-CoA dehydratase from *Aeromonas caviae* [109].

A very different ammonia β-lyase represents ethanolamine ammonia lyase, which catalyzes the overall β-elimination of ammonia from ethanolamine, yielding acetic aldehyde. The enzyme, which was discovered in *Clostridium* sp., but purified from Enterobacteria and *Rhodococcus erythropolis*, contains coenzyme B_{12}, which mediates the migration of a hydrogen atom from the hydroxyl carbon 2 to carbon 1, whereas the amino group moves in the opposite direction. This mechanism is similar to that discussed for glutamate mutase in the following section (Figure 9.19). The resulting 1-aminoethane-1-ol dissociates into ammonia and acetic aldehyde [110].

9.9 B_{12}-ENZYMES

9.9.1 GLUTAMATE MUTASE (METHYLASPARTATE MUTASE)

C. tetanomorphum is the organism in which the first coenzyme form of vitamin B_{12} has been discovered [111]. The bacterium readily ferments glutamate to ammonia, CO_2, H_2, acetate, and butyrate approximately according to Equation 9.15.

$$\begin{aligned}&5\ \text{Glutamate}^- + 6\ H_2O + 2\ H^+ \rightarrow 5\ NH_4^+ + 5\ CO_2 + 6\ \text{Acetate}^- + 2\ \text{Butyrate}^- + H_2\\&\Delta G^{\circ\prime} = -62.6\ \text{kJ/mol glutamate, whereby 3 ATP can be obtained by substrate}\\&\qquad\text{level phosphorylation (104 kJ/mol ATP) [30].}\end{aligned} \tag{9.15}$$

Initially, glutamate is degraded to ammonia, acetate and pyruvate (Figure 9.1). The oxo acid is oxidatively decarboxylated to acetyl-CoA, catalyzed by pyruvate-ferredoxin oxidoreductase. The reduced ferredoxin is reoxidized by acetyl-CoA and protons leading to butyrate and hydrogen, respectively (Figure 9.5). The physiological task of glutamate mutase is the conversion of an α-amino acid to a β-amino acid, which is readily deaminated to mesaconate, as discussed in the previous section. Related conversions in the pathway of lysine fermentation catalyzed by lysine-2,3-aminomutase and β-lysine 5,6-aminomutase are shown below.

While studying the enzymes involved in the early steps of glutamate fermentation, H.A. Barker realized that cell-free extracts lost their activity upon irradiation with light or by treatment with charcoal. This was the starting point of the isolation of pseudocoenzyme B_{12}, the coenzyme of glutamate mutase, which catalyzes the reversible carbon-skeleton rearrangement of (*S*)-glutamate to (2*S*,3*S*)-3-methylaspartate (see also Section 9.1). Chemical [111] as well as X-ray analyses [112] established the structure of the coenzyme as a derivative of vitamin B_{12}, in which the axial base was adenine (pseudocoenzyme B_{12}) rather than 5,6-dimethybenzimidazole (coenzyme B_{12}, also called adenosylcobalamin), and, most importantly, 5′-deoxyadenosin was connected via its 5′-carbon to the cobalt by a Co-C-bond, a light sensitive metallo-organic bond, the only thermally stable one known in nature [113]. The enzyme glutamate mutase is a complex with the composition $\sigma_2\varepsilon_2$(pseudocoenzyme $B_{12})_2$. The subunits σ (component S from supernatant) and ε_2 (component E from eluate) have been purified separately from the related *C. cochlearium* and form an active and stable complex only in the presence of pseudocoenzyme B_{12} or coenzyme B_{12} [74]. The genes encoding the two subunits, *glmE* and *glmS*, have been cloned and separately expressed in *E. coli*. The crystal structure of glutamate mutase from *C. cochlearium* has been solved [114] and confirmed that the σ-subunit binds the coenzyme by replacing the "lower" adenine base from the Co-atom by a conserved histidine residue ("base-off, his-on") [115]. The "upper face" of the coenzyme is directed toward the ε-subunit, with the 5′-deoxyadenosyl moiety sticking into the active site cavity of ε. Hence, the coenzyme can be regarded as the glue holding the subunits together. The most important feature of the structure was the detection of (2*S*,3*S*)-tartrate in the crystal bound between three arginines and one glutamate residue at a distance of about 6 Å apart from the Co-atom of the coenzyme [114]. Since the bound tartrate molecule closely resembles the substrate (2*S*,3*S*)-3-methylaspartate, its location inside the ε-subunit could be the substrate binding site. Therefore, this arrangement of the three positive charges and the possibly protonated glutamate residue has been termed "arginine claw" [74].

The initial step in the carbon skeleton rearrangement is the homolysis of the carbon cobalt bond. Due to binding of the coenzyme to the enzyme, the bond is weakened and an equilibrium between the closed and open (biradical) form of the bond is established [116]. The substrate reacts with the open form, and 4-H_{Si} of glutamate is abstracted by the 5′-deoxadenosyl radical to yield cob(II)alamin, 5′-deoxyadenosine, and 4-glutamyl radical (Figure 9.19). The latter radical rearranges to 3-methyleneaspartate radical, which abstracts the hydrogen atom from 5′-deoxyadenosine to obtain the product 3-methylaspartate and regenerate adenosylcobalamin [117]. The rearrangement has been described by the Buckel–Golding mechanism as fragmentation of the substrate-derived radical into acrylate and 2-glycyl radical followed by recombination to the product-related radical [118]. Experimental evidence for this mechanism has been obtained during steady state of the glutamate mutase catalyzed reaction through the concerted inhibition by glycine + acrylate [118], the participation of C-2 of glutamate in the EPR spectrum showing cob(II)alamin interacting with substrate derived carbon-centred radicals at a distance of 6.5 Å [119] and quenching by mecaptoethanol, followed by identification of acrylate and glycine by HPLC [120].

9.9.2 2-Methyleneglutarate Mutase (3-Methylitaconate Mutase)

Eubacterium barkeri (formerly called *Clostridium barkeri*, family "Eubacteriaceae," order Clostridiales) is the only organism known to contain 2-methyleneglutarate mutase (3-methylitaconate

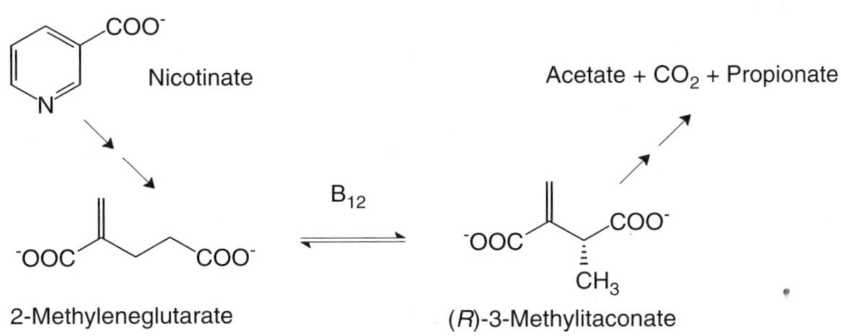

FIGURE 9.20 Reaction catalyzed by 2-methyleneglutarate mutase in the pathway of nicotinate fermentation by *E. barkeri*. B_{12}, coenzyme B_{12} or adenosylcobalamin. The complete pathway is described elsewhere [74].

mutase; EC 5.4.99.4). The enzyme catalyzes the coenzyme B_{12}-dependent carbon skeleton rearrangement of 2-methyleneglutarate (2-methylideneglutarate) to (*R*)-3-methylitaconate (2-methylidene-3-methylsuccinate) [121–123]. Thereby, the 4-H_{Re} is removed by the 5′-deoxyadenosine radical and added to carbon 3 to become the racemic methyl group of (*R*)-3-methylitaconate as analyzed using all three isotopes of hydrogen [124]. The mutase is involved in the fermentation of nicotinate to ammonia, propionate, acetate, and CO_2 [74] (Figure 9.20).

$$\text{Nicotinate}^- + 4\ H_2O \rightarrow NH_4^+ + \text{Propionate}^- + \text{Acetate}^- + CO_2$$

In this pathway, the carbon skeleton of nicotinate is retained until 2-methyleneglutarate, in which the hydrogen at C3 is not activated for further isomerizations (e.g., to 2-methylglutaconate). Upon rearrangement, however, the resulting (*R*)-3-methylitaconate readily isomerises to 2,3-dimethylmaleate. This symmetrical molecule is hydrated to (2*R*,3*S*)-2,3-dimethylmaleate and cleaved to propionate and pyruvate [125]. Recently, the gene encoding dimethylmaleate lyase of *E. barkeri* has been cloned, sequenced, and expressed as active enzyme in *E. coli*. The deduced amino acid sequence is related to those of isocitrate lyases, which catalyze the analogous cleavage of (2*R*,3*S*)-isocitrate to succinate and glyoxylate and confirms earlier biochemical observations of a common mechanism [190].

The gene encoding 2-methyleneglutarate mutase (*mgm*) has been cloned and overexpressed in *E. coli*. The purified *apo*-enzyme from *E. coli* is a homotetramer (4 × 67 kDa) without any activity. But on addition of coenzyme B_{12}, immediately full activity is obtained and two adenosylcobalmins are bound per tetramer. Unexpectedly, there is no sequence similarity of Mgm with GlmE of glutamate mutase, and only the C-terminus of Mgm contains the conserved DXHXXG motif of coenzyme binding of GlmS ("base-off his-on"). For glutamate mutase, it has been shown that this histidine coordinates to the Co-atom of the coenzyme. Unlike glutamate mutase from *C. cochlearium,* the natural coenzyme of 2-methylenglutarate mutase is adenosylcobalamin [74]. Until now, all attempts to crystallize the enzyme have failed (C. Kratky, Universität Graz, Austria, personal communication).

In contrast to glutamate mutase, for which only a fragmentation mechanism can be envisaged, the initially generated 2-methylene-4-glutaryl radical could intramolecularly add to the methylene double bond and form a 1-methylenecyclopropane-1,2-dicarboxylate radical intermediate, which could eliminate to the 3-methyleneitaconate radical (addition-elimination mechanism). All experimental evidence indicates, however, that the carbon-skeleton rearrangement catalyzed by 2-methyleneglutarate mutase follows the Buckel–Golding mechanism by fragmentation of the substrate-derived 2-methylene-4-glutaryl radical into 2-acrylyl radical and acrylate, which recombine to the product-related 3-methyleneitaconate radical [118]. Analogous to glutamate mutase, the EPR spectrum is slightly changed only if carbons 2 or 4 of 2-methyleneglutarate are ^{13}C-labelled [126]. Furthermore, the mutase catalyzes the isomerization of the exomethylene hydrogens of 2-methyleneglutarate and

(*R*)-3-methylitaconate. Thus, adding (*Z*)-3-methyl[*methylene*-$^2H_1$]itaconate to the active enzyme, the (*E*)-isomer and both (*E*) and (*Z*)-2-[*methylene*-$^2H_1$]methyleneglutarates are formed at a rate comparable to turnover. At equilibrium, 3% of each 3-methylitaconate isomer and 47% of each 2-methyleneglutarate isomer are present. Before reaching equilibrium, however, the concentration of the *E*-isomer of 3-methylitaconate rises up to 10%. This "*E*-overshoot" can only explained by the fragmentation mechanism, if one assumes an additional energy barrier between an arrangement of acryl radical + acrylate ready to recombine back to the 3-methyleneitaconate radical and ready to recombine forward to the 2-methylene-4-glutaryl radical [127].

9.9.3 D-Lysine/L-β-Lysine 5,6-Aminomutase

The fermentation of L-lysine [L-α-lysine, (2*S*)-2,6-diaminohexanoate] by *Clostridium subterminale*, *Clostridium sticklandii,* and *F. nucleatum* proceeds via L-β-lysine [(3*S*)-3,6-diaminohexanoate], which in the second step is converted to (3*S*,5*S*)-3,5-diaminohexanoate by a coenzyme B_{12}-dependent enzyme (Figure 9.14). Since this mutase also mediates the isomerization of D-lysine to (2*R*,5*S*)-2,5-diaminohexanoate, the enzyme has been named D-lysine/L-β-lysine 5,6-aminomutase (EC 5.4.3.3) [128,129]. Interestingly, the formation of L-β-lysine from L-lysine also involves an intramolecular 1,2-migration of an amino group, but the responsible enzyme L-lysine 2,3-aminomutase contains a [4Fe-4S] cluster and requires S-adenosylmethionine (SAM) as cofactor [129] (see Section 9.10).

D-Lysine/L-β-lysine 5,6-aminomutase from *C. sticklandii* is composed of two different subunits (α, 57 kDa; β, 29 kDa) forming a heteroteramer ($\alpha_2\beta_2$, ca. 170 kDa) also designated as E_1. Whereas the sequence of the α-subunit shares no similarity to other proteins, that of the β-subunit exhibits the motif of coenzyme B_{12}-binding as found in other "base off, his on" mutases (vide supra). Hence, the overall structure of the amino mutase is similar to that of glutamate mutase, in which the smaller subunit binds coenzyme B_{12}. In order to get activity, the enzyme requires, in addition to coenzyme B_{12}, pyridoxal-5′-phosphate, NH_4^+ or K^+, and β-mercaptoethanol. Remarkably, during turnover the enzyme becomes inactivated with a half-life of a few minutes [130]. *C. sticklandii* contains an E_2 protein (90 kDa), which in the presence of ATP and Mg^{2+} reactivates E_1 [129].

The proposed mechanism of D-lysine/L-β-lysine 5,6-aminomutase is initiated by the formation of an external aldimin (Schiff base) of the ε-amino group of the substrate with pyridoxal-5′-phosphate. The migration of the aldimine from C6 to C5 involves the general scheme as outlined for the carbon-skeleton rearranging mutases (Figure 9.19). The 5′-deoxyadenosylradical generated from coenzyme B_{12} abstracts one hydrogen atom at C5 to yield the substrate-derived radical, and cob(II)alamin. The aldimine migrates from C6 to C5 via a hypothetical azacyclopropylcarbinyl radical. Finally 5′-deoxyadenosine donates a hydrogen atom to the resulting product-related radical at C6 to yield the methyl group of 2,5/3,5-diaminohexanoate. The inactivation during turnover ("suicide") has been shown to proceed by electron transfer from the intermediate cob(II)alamin either to the substrate-derived or to the product-related radical, followed by protonation of the carbanion. Thereby 5′-deoxyadenosine, cob(III)alamin and substrate or product are generated [131].

9.9.4 D-Ornithine 4,5-Aminomutase

Washed cells of *C. sticklandii* ferment L-ornithine to ammonia, acetate, alanine, and 5-aminovalerate approximately according to Equation 9.16 [132].

$$2\ \text{L-Ornithine}^+ + 2\ H_2O \rightarrow 2\ NH_4^+ + \text{Acetate}^- + \text{D,L-Alanine} + \text{5-Aminovalerate} + H^+ \quad (9.16)$$

In the reductive branch, L-ornithine is converted to proline and reduced to 5-aminovalerate (Section 9.7.2). Therefore, 2-oxoglutarate transaminates L-ornithine to glutamate and glutamate semialdehyde (ornithine aminotransferase, EC 2.6.1.68), which after nonenzymatic cyclization to Δ^1-pyrrolin-5-carboxylate is reduced proline. The formed glutamate is oxidized to 2-oxoglutarate and ammonia catalyzed by glutamate dehydrogenase (EC 1.4.1.2). In *C. sporogenes,* this overall deaminating cyclization of ornithine to proline is catalyzed by only one enzyme, ornithine cyclodeaminase (EC 4.3.1.12). In the oxidative branch of the fermentation (Figure 9.6), L-ornithine is epimerized, and the δ-amino group of D-ornithine is shifted to the γ-position in a coenzyme B_{12}-dependent reaction to yield (2*R*,4*S*)-2,4-diaminovalerate [133]. Oxidative deamination with NAD^+ at C4 is followed by a pyridoxal-5′-phosphate cleavage to acetyl-CoA (C4 + C5) and D-alanine (C1-C3) [32]; the latter racemizes. D-Ornithine 4,5-aminomutase (EC 5.4.3.5) has been purified and characterized as an enzyme similar to D-lysine 5,6-aminomutase but with tightly bound coenzyme B_{12} [134].

9.10 *S*-ADENOSYLMETHIONINE-DEPENDENT RADICAL ENZYMES

C. subterminale and *F. nucleatum* ferment lysine to ammonia, acetate, and butyrate according to Equation 9.17 and Figure 9.14:

$$\text{L-Lysine}^+ + 2\ H_2O \rightarrow 2\ NH_4^+ + \text{Acetate}^- + \text{Butyrate}^- + H^+ \quad (9.17)$$

In the first step, the removal of the α-amino group is elegantly solved by lysine 2,3-aminomutase (EC 5.4.3.2), which catalyzes the intramolecular migration of the α-amino group to the β-position, yielding L-β-lysine, as well as the inter- or intramolecular migration of one β-hydrogen to the α-position. The oxygen-sensitive enzyme from *C. subterminale* contains a $[4Fe\text{-}4S]^{2+/1+}$cluster and requires pyridoxal 5′-phosphate (PLP) and *S*-adenosylmethionine (SAM) rather than coenzyme B_{12} for activity [135]. H.A. Barker, who discovered lysine 2,3-aminomutase, correctly assumed that the function of SAM is similar to that of coenzyme B_{12}. Due to its simpler structure, as compared to adenosylcobalamin, Barker called it the "poor man's coenzyme B_{12}" [cited in 136]. Today, however, about 600 SAM-dependent radical enzymes have been deduced from the genome sequences [137], and not much more than a dozen coenzyme B_{12}-dependent enzymes are known. This might be the reason why Perry A. Frey recently called SAM "a wolf in sheep's clothing or a rich man's coenzyme B_{12}" [138].

9.10.1 LYSINE-2,3-AMINOMUTASE

The enzyme from *C. subterminale* (formerly called *C.* SB4) catalyzes the exchange $3H_{Re}$ of (2*S*)-α-lysine with the α-amino group under inversion of the configurations at C2 and C3. $2H_{Re}$ is the newly introduced hydrogen of the product (3*S*)-β-lysine [139]. Like in coenzyme B_{12}-dependent reactions, there is no exchange of the migrating hydrogen with the solvent. The enzyme is a homohexamer (ca. 260 kDa) composed of 47 kDa subunits. It contains 1 pyridoxal-5′-phosphate and 1 Zn^{2+} or Co^{2+} per subunit as well as 3 $[4Fe\text{-}4S]^{2+}$ clusters per hexamer. Probably each cluster is bound between two subunits. SAM is not tightly bound to the enzyme but required as cofactor. With respect of amino group migration, the mechanism of lysine 2,3-aminomutase is similar to that of lysine 5,6-aminomutase but differs in the formation of the 5′-deoxyadenosine radical. Since in the active enzyme the [4Fe-4S] cluster needs to be in the +1 oxidation state, it appears likely that the cluster delivers the electron necessary to split SAM into methionine and 5′-deoxyadenosine radical.

The most interesting aspect of lysine 2,3-aminomutase is formation of the kinetic competent product-related radical (at C2), which is stabilized by the adjacent carboxylate. In contrast to a similar radical, the substrate-derived 4-glutamyl radical, observed with glutamate mutase, the radical of the aminomutase lies not in the "shade of cob(II)alamin"; it is the only radical visible by EPR during turnover. Deuterium- or ^{13}C-labelling of L-lysine, especially at C2, gave rise to huge changes in the EPR spectrum, but only minor effects were observed with glutamate mutase and

[4-^{13}C]glutamate or perdeuterated glutamate (see Section 9.9.1). Therefore, the radical at C2 of lysine could be much better characterized. In particular, the close neighborhood of the hydrogen at C4′ of pyridoxal-5′-phosphate could be established, whereas the classic test, the reduction of the imine double bond by sodium borohydride, failed. The concentration of the substrate-derived radical of lysine 2,3-aminomutase (radical at C3) is probably too low to be seen during steady state or equilibrium. If, however, carbon 3 of lysine is exchanged by sulfur, then a substrate analogue is obtained (4-thia-L-lysine or S-aminoethyl-L-cysteine) in which the radical at C3 is stabilized by the sulfur and can be observed by EPR. Again, by introducing ^{13}C or deuterium at C3, the EPR spectrum changes completely its shape [129].

9.10.2 Glycyl Radical Enzymes

Lysine 2,3-aminomutase was the first enzyme in which the novel function of SAM as a generator of 5′-deoxadenosine radicals was discovered. All other SAM-dependent radical enzymes, with the exception of the spore photoproduct lyase, use the 5′-deoxadenosine radical not for the reaction itself but to abstract a hydrogen atom from a specific glycine residue of the protein. This procedure converts the inactive proenzyme into a catalytically active glycyl radical enzyme. Examples of this type in clostridia are pyruvate-formate lyase, glycerol-dehydratase, and *p*-hydroxyphenylacetate decarboxylase. In addition the anaerobic ribonucleotide reductase from *E. coli* [140] and benzyl-succinate synthase from *Thauera aromatica* ("Betaproteobacteria") [141,142] belong to this class of radical enzymes. In general, glycyl radical enzymes comprise two proteins, a large homodimeric proprotein (2 × ca. 80 kDa) and an activating protein containing [4Fe-4S] clusters. In order to get activity, SAM and a strong reducing agent are required. The [4Fe-4S] clusters of the activating protein seem to have the same function as the clusters in lysine 2,3-aminomutase, the generation of 5′-deoxyadenosine radicals. The conversion of the dimeric proprotein to the active enzyme involves the conversion of only one of the two critical glycines into a glycine radical. This is readily seen by SDS-PAGE of the active enzyme under oxic conditions. Exposure to oxygen cleaves the enzyme at the site of the glycine radical, which sits near the C-terminus. Thus, a double band is observed, the uncleaved α-subunit and the truncated α′-subunit. Another diagnostic tool is the EPR spectrum of the glycine radical, which sometimes is seen already with whole cells. The observed doublet signal is due to the coupling of the radical with the remaining hydrogen of the glycine. In D_2O, this hydrogen slowly exchanges with deuterium, and a singlet is observed in the EPR. The catalytically active radical, however, is a thiyl radical formed by hydrogen atom transfer from a conserved cysteine residue to the glycyl radical.

The best-characterized glycyl radical enzyme is pyruvate formate lyase from *E. coli* [143–145]. In clostridia, the enzyme has been detected in *C. kluyveri*, *C. pasteurianum,* and *C. butyricum* [146,147]. There seems to exist no essential difference between the *E. coli* and clostridial pyruvate formate lyases. Hence, this is another example of a relationship between clostridia and "Gammaproteobacteria" at the enzyme level. Until recently, glycerol dehydratase and the related propane-1,2-diol dehydratases have been considered exclusively as coenzyme B_{12}-dependent enzymes. The only exception was a not-very-well-characterized B_{12}-independent diol dehydratase from membranes of *Clostridium glycolicum* [148], which gave an EPR signal at $g = 2.02$, indicating a stable organic radical. In 2003, the glycerol dehydratase from *Clostridium butyricum* was identified as a glycyl radical enzyme [149]. The enzyme might replace the coenzyme B_{12}-dependent glycerol dehydratase in the biotechnological production of 1,3-propanediol via glycerol and 3-hydroxypropionaldehyde. The disadvantage of the B_{12} enzyme is the rapid suicidal inactivation of the coenzyme, which has to be regenerated in an ATP-dependent process.

An unusual decarboxylase from *Clostridium difficile* has been characterized as a novel glycyl radical enzyme, which catalyzes the decarboxylation of *p*-hydroxyphenylacetate to *p*-cresol [150]. Chemical considerations suggested that the *p*-hydroxyl group destabilizes the carbanion resulting from CO_2 elimination. Therefore, it was assumed that oxidation of the phenolate to a phenoxyradical

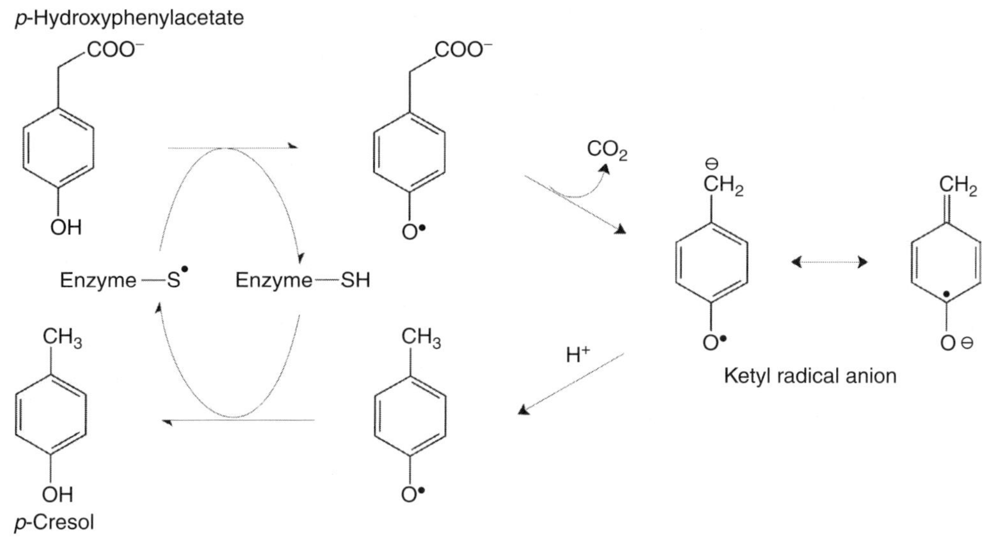

FIGURE 9.21 Radical reaction catalyzed by *p*-hydroxyphenylacetate decarboxylase from *C. difficile*. The enzyme contains a stable glycyl radical, which, upon addition of substrate, abstracts a hydrogen atom from a thiol at the active site. The formed thiyl radical initiates the decarboxylation.

would lead to a more stable ketyl radical anion [151]. The identification of *p*-hydroxyphenylacetate decarboxylase as glycyl radical enzyme revealed, however, that, most likely, the thiyl radical of the enzyme (vide supra) abstracts a hydrogen atom from the *p*-hydroxyl group. The resulting *p*-phenoxyacetate radical then decarboxylates to a ketyl radical anion, which is protonated to *p*-phenoxymethylene radical followed by hydrogen atom back-transfer from the enzyme to yield *p*-cresol (Figure 9.21). It has been assumed that the bactericidal *p*-cresol might help the pathogenic organism to overgrow competitors in the human intestine.

9.11 HYDROXYACID DEHYDRATASES

Hydroxy acids are a common substrate for enzymes catalyzing the elimination of water to yield α,β-unsaturated acids or derivatives thereof. These dehydratases occur in almost every metabolic pathway, such as glycolysis, citric acid cycle, fatty acid biosynthesis and degradation (β-oxidation), and in amino acid metabolism. There are two types of hydroxyacid dehydration: either the free acid or the CoA-derivative is dehydrated. Two rules are proposed to explain this difference. The first rule states that simple β-hydroxy acids (3-hydroxy acids) without any other functional group require activation to the CoA-thiol ester prior to dehydration. The second rule states that β-hydroxyacids, with either a second carboxylate in the β-position or a second substituent in the α-position, are dehydrated at the free acid level. These rules are also valid for β-amino acids. Examples for the first rule, i.e., requirement of thiol esters, are the dehydrations of β-hydroxyacyl-CoA to enoyl-CoAs (Figure 9.5), 3-hydroxyglutaryl-CoA to glutaconyl-CoA, and (*S*)-3-hydroxy-3-methylglutaryl-CoA to 3-methylglutaconyl-CoA [152], as well as the deaminations of β-aminobutyryl-CoA to crotonyl-CoA (Figure 9.14) and β-alanyl-CoA to acryloyl-CoA (Figure 9.9). The dehydrations of 2-, 4- and 5-hydroxyacids, which are derived from the corresponding amino acids, also occur at the CoA-thiol ester level. The reason for this is different and will be explained below. Examples of the second rule, i.e., thiol ester independent eliminations, are the dehydration of (*S*)-malate to fumarate, citrate and (3*R*,2*S*)-isocitrate to *cis*-aconitate, and (*S*)-citramalate to mesaconate (methylfumarate) (Figure 9.1), as well as of sugar acids such as 6-phosphogluconate to 2-dehydro-3-deoxy-6-phosphogluconate (also called KDPG), 2,3-dihydroxy acids to 2-oxoacids, and 2-phosphoglycerate to phosphoenolpyruvate.

The deaminations of (*S*)-aspartate to fumarate and (2*S*,3*S*)-3-methylaspartate to mesaconate also follow the second rule (Figure 9.11). The requirement for CoA-thiol esters in the dehydration and deamination of simple β-hydroxy and β-amino acids can be easily explained by the lowering of the p*K* of the α-proton from about 30 to 21. The p*K* is further lowered by hydrogen bonding to backbone amides or certain OH-groups. It is not clear, however, why substrates with an additional electron-withdrawing substituent, as indicated above, do not need to form a CoA-thiol ester prior to dehydration or deamination.

The free hydroxyamino acids serine and threonine are dehydrated to the corresponding enamino acids, which are hydrolyzed to pyruvate and 2-oxobutyrate, respectively. Although these eliminations appear to fit the second rule, all known threonine dehydratases and many serine dehydratases contain pyridoxal-5′-phosphate, which, like CoA-thiol esters, lowers the p*K* of the α-proton and, thus, activates the hydroxyamino acid.

9.11.1 L-Serine Dehydratase

In bacteria, notably in *Peptostreptococcus asaccharolyticus* ("Peptostreptococcaceae," Clostridiales), the L-enantiomer specific serine dehydratase is a [4Fe-4S] cluster-containing and pyridoxal-5′-phosphate-independent enzyme catalyzing the elimination of water to yield an enamino acid, which tautomerizes to an imine, followed by hydrolysis to ammonia and pyruvate. Since, under air, the fourth iron of the cluster is lost concomitant with activity, it has been assumed that the mechanism is similar to that of aconitase, in which the hydroxyl group of the substrate is coordinated to the labile iron [153,154]. Both genes coding for the heterodimeric L-serine dehydratase have been cloned and sequenced [155]. The derived amino acid sequences together show significant identities with both monomeric L-serine dehydratases from *E. coli* [156,157] and those of deduced L-serine dehydratases from the genomes of many bacteria. L-Serine and L-threonine dehydratase have also been purified from *C. propionicum*. Whereas L-serine is dehydrated by a [4Fe-4S] cluster-containing enzyme, and thus fits the second rule, the L-threonine dehydratase has pyridoxal-5′-phosphate as prosthetic group [48].

9.11.2 The Dehydration of (*R*)-2-Hydroxy Acids: 2-Hydroxyacyl-CoA Dehydratases

The majority of α-amino acids are much more resistant toward β-elimination than their β-isomers. This may be one reason why nature has chosen α-amino acids as building blocks of proteins. β-Elimination of a nucleophile (NH_3 or OH^-) adjacent to the electron-withdrawing α-carboxylate is difficult to achieve, since the β-proton is not activated (p*K* = 40). Nevertheless, several clostridia and *F. nucleatum* are able to ferment 12 of the 20 proteinogenous α-amino acids via their corresponding (*R*)-2-hydroxy acids, which are activated to (*R*)-2-hydroxyacyl-CoA and dehydrated to enoyl-CoA (Figure 9.22). (*R*)-Lactate is formed from alanine by amino transfer to 2-oxoglutarate, followed by an NADH-dependent reduction (Figure 9.9). In the same manner, phenylalanine yields (*R*)-3-phenyllactate, leucine (*R*)-2-hydroxyisocaproate, tryptophan (*R*)-indollactate, and tyrosine (*R*)-4-hydroxyphenyllactate. Direct oxidation of glutamate by NAD^+, followed by an NADH-dependent reduction, leads to (*R*)-2-hydroxyglutarate. Degradation of histidine and glutamine via glutamate also gives rise to (*R*)-2-hydroxyglutarate. Elimination of water from serine and threonine yields pyruvate and 2-oxobutyrate, which are reduced to (*R*)-lactate and (*R*)-2-hydroxybutyrate, respectively. The same two acids are formed from cysteine and methionine after β- and γ-elimination, respectively, and reduction. Prior to dehydration, all these (*R*)-2-hydroxy acids are converted to the (*R*)-2-hydroxyacyl-CoA derivatives using specific CoA-transferases (Section 9.5.1).

It has been proposed that the dehydration of (*R*)-2-hydroxyacyl-CoA to enoyl-CoA can only be achieved by conversion of the electrophilic thiol ester carbonyl into a nucleophile, a process called "Umpolung" (charge reversal). The thiol ester carbonyl has properties of a ketone, which

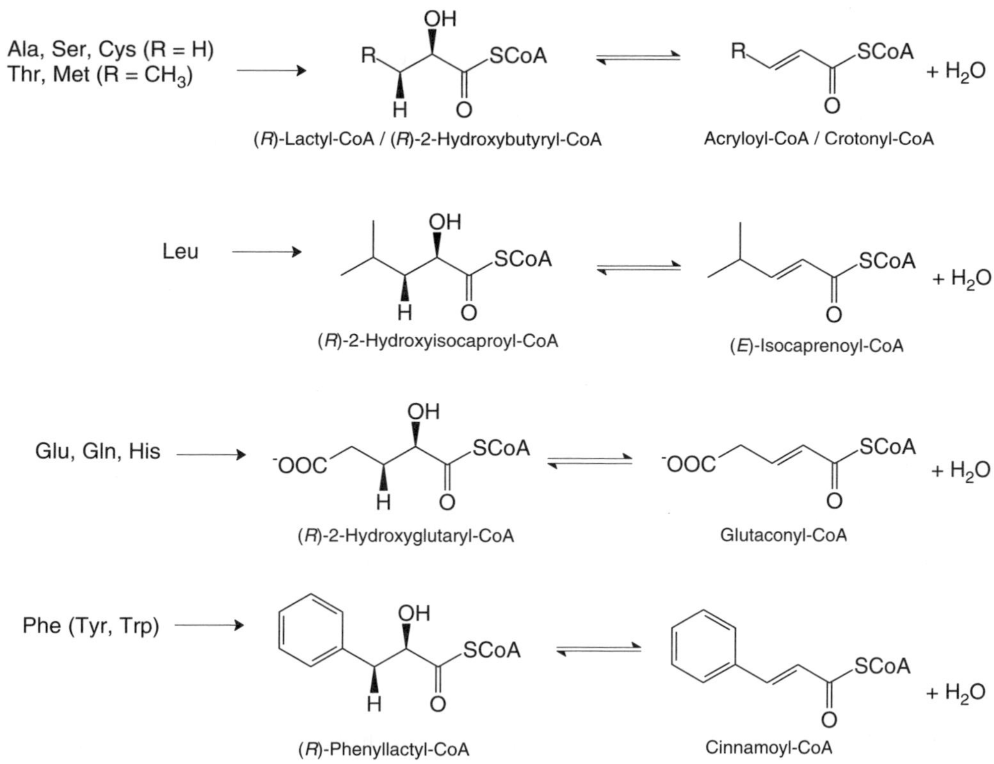

FIGURE 9.22 Fermentation of 12 proteinogenous α-amino acids via (*R*)-2-hydroxy acids. Only the reversible syn-dehydrations of (*R*)-2-hydroxyacyl-CoA to enoyl-CoA are shown. The (*R*)-2-hydroxyacyl-CoA derivatives are formed via the NADH-dependent reduction of the 2-oxoacids followed by CoA-transfer. The formed enoyl-CoAs are reduced to the saturated acycl-CoA, from which the fatty acids are liberated by CoA-transfer. An exception is glutaconyl-CoA, which is decarboxylated to crotonyl-CoA prior to reduction to butyrate or oxidation to acetate.

can be reduced by one electron to a nucleophilic ketyl radical anion. Hence, reduction of 2-hydroxy-acyl-CoA to its ketyl radical anion would facilitate the elimination of the hydroxyl group to yield an enoxy radical, which can be deprotonated to the ketyl radical anion of the product enoyl-CoA. Oxidation of the latter to the unsaturated product by the next incoming substrate would complete the catalytic cycle [158,159] (Figure 9.23). It has been calculated that the p*K* of the enoxy radical has been lowered to 14, about 26 units less than the p*K* of the β-proton of the 2-hydroxy acid [160]. Similar to β-hydroxyacyl-CoA, the p*K* of the enoxyradical may further be lowered by hydrogen bonding.

Until now, six different 2-hydroxyacyl-CoA dehydratases have been purified and characterized: lactyl-CoA dehydratase from *C. propionicum* [47,50–53], 2-hydroxyglutaryl-CoA dehydratase from *A. fermentans* [159,161–164], *C. symbiosum* [165] and *F. nucleatum* [166], phenyllactate dehydratase from *C. sporogenes* [77,78] and 2-hydroxyisocaproyl-CoA dehydratase from *C. difficile* [189]. All dehydratases are enzyme systems composed of two separable components, an extremely oxygen-sensitive activator or initiator (component A) and a moderate oxygen-sensitive component D, the actual dehydratase.

The activator, initiator, or component A from *A. fermentans* produced in *E. coli* [163] is a homodimeric enzyme with one [4Fe-4S] cluster between the two subunits (2 × 27 kDa) [167]. The most remarkable feature of component A are two helices, each from one subunit pointing with their N-termini toward the [4Fe-4S] cluster, forming a helix-cluster-helix angle of 105°. A similar architecture

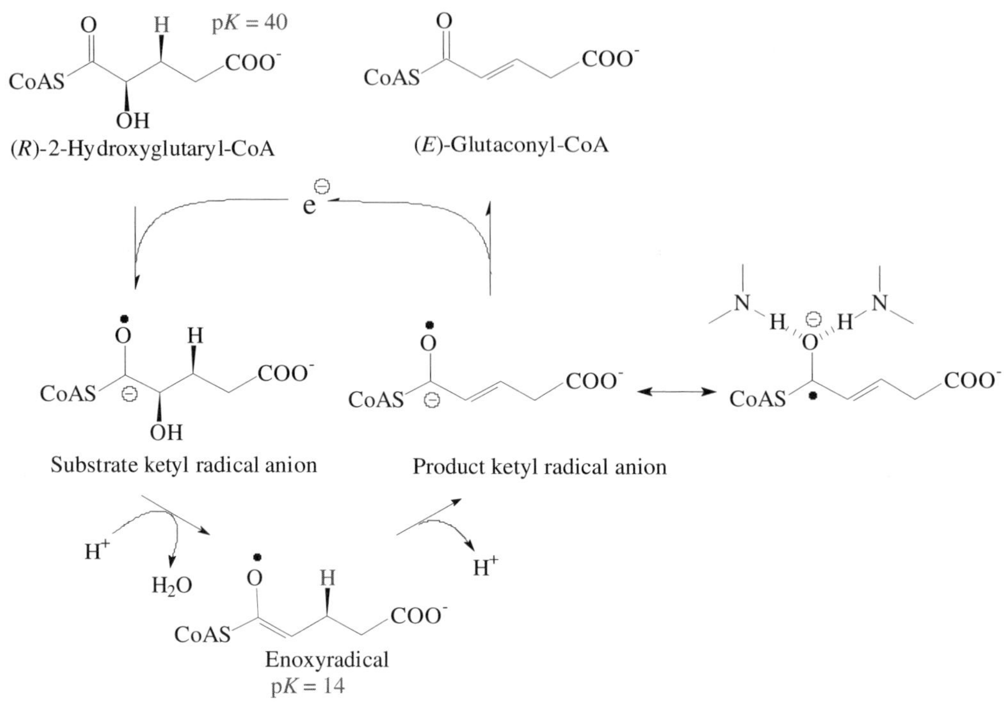

FIGURE 9.23 Proposed mechanism for the dehydration of (*R*)-2-hydroxyglutaryl-CoA to glutaconyl-CoA. Component A introduces into component D an energized electron, which is further transferred into the thiol ester carbonyl group. The thus obtained substrate-derived ketyl radical anion expels the adjacent hydroxyl group. The β-hydrogen of the resulting enoxy radical is now activated (shift from p*K* 40 to 14) and can be abstracted as a proton yielding the product-related ketyl radical anion. Finally, the electron is handed over to the next incoming substrate and the product is formed. In order to demonstrate how the ketyl radical anion could be stabilized, one resonance structure is given.

is found in the phylogenetically unrelated iron protein of nitrogenase from *Azotobacter vinelandii,* with a helix-cluster-helix angle of 150°. Upon binding to component D, the angle probably opens to 180°, as observed in the complex of nitrogenase iron protein with molybdenum-iron protein in the presence of ADP-AlF_4^- [168]. Component A has a low ATPase activity (ca. 1 s^{-1}) but only in the reduced $[4Fe\text{-}4S]^+$ state. In the presence of an equal amount of component D, the ATPase activity increases about 50-fold [189]. The structure of component A also revealed that the [4Fe-4S] cluster is easily accessible from the solvent. This may be the reason for the extreme oxygen sensitivity. The redox potential of component A could not be measured, but the cluster becomes almost completely reduced by flavodoxin (E_0' ca. –420 mV) or ferredoxin (E_0' = –405 mV) [169], indicating a potential of about –350 mV or even higher. The closely related components A from *C. sporogenes* [78], *F. nucleatum* [170], and *C. difficile* [189] have also been purified in the same way and revealed almost identical properties. Each known genome of an anaerobic bacterium (including *E. coli*) or archaeon contains at least one deduced homologue of this exciting protein. There are even four homologous genes of component A in the genome of *Clostridium acetobutylicum* [171].

Component D of *A. fermentans* has been characterized as a heterodimeric enzyme (54 + 42 kDa) containing one [4Fe-4S] cluster, one riboflavin-5′-phosphate (FMN), and about 0.1 riboflavin. Molybdenum has also been found in this protein, but the content of 0.1 mol/heterodimer appears to be too low to be significant. Furthermore, the same amount of Mo has been detected in component D from *C. symbiosum*, but this metal is absent in 2-hydroxyglutaryl-CoA dehydratase from *F. nucleatum* and 2-hydroxyisocapryloyl-CoA dehydratase from *C. difficile*. Interestingly, component

D from *C. symbiosum* contains two [4Fe-4S] clusters. The smaller β-subunit of this protein could be crystallized, and its crystal structure has been determined. It revealed one [4Fe-4S] cluster and one FMN at a distance of 17 Å. Unexpectedly, only three iron atoms of the cluster are coordinated by cysteines, whereas the fourth has a not-well-conserved tyrosine as ligand (H. Dobbek and B. Martins, unpublished data). The 2-hydroxyglutaryl-CoA dehydratase from *F. nucleatum* is unique, since it is composed of three different subunits [67,166,170]. The third subunit seems not to be related to any other protein. Component D of phenyllactate dehydratase from *C. sporogenes* is also a trimeric protein. The third and largest subunit (46 kDa), however, has been characterized as a cinnamoyl-CoA:phenyllactate CoA-transferase, which catalyzes the formation of (*R*)-phenyllactyl-CoA. The other two subunits are homologues of the α- and β-subunits (45 and 37 kDa) of components D from *C. symbiosum* and *A. fermentans*. Hence, activity of phenyllactate dehydratase requires, in addition to (*R*)-phenyllactate, ATP, $MgCl_2$, a reducing agent (see below), and component A, also catalytic amounts of cinnamoyl-CoA. The mechanism comprises a combination of the mechanisms of citrate lyase and 2-hydroxyglutaryl-CoA dehydratase. Initially (*R*)-phenyllactyl-CoA and the final product (*E*)-cinnamate are generated from cinnamoyl-CoA and (*R*)-phenyllactate (Figure 9.13). In the next step, cinnamoyl-CoA is regenerated by dehydration of (*R*)-phenyllactyl-CoA. This enzyme complex clearly shows that formation of the thiol ester substrate is a prerequisite for the dehydration of 2-hydroxyacids. In the fermentation of glutamate via 2-hydroxyglutarate, the dehydration at the thiol ester level could be due to the participation of the CoA-ester in the consecutive decarboxylation step [77,78].

The reductive activation of component D of the dehydratases requires component A, ATP, $MgCl_2$, and a reducing agent. *In vitro* dithionite or Ti(III)citrate are suitable one-electron donors, whereas *in vivo* a clostridial-type, two [4Fe-4S] cluster-containing ferredoxin [169] or a flavodoxin [164] serve for this purpose. The further fate of the electron in the activation process remains unclear. Whereas the Mössbauer spectrum revealed oxidation of component A during activation, the concomitant reduction of the cluster of component D could not be observed by this method. The active component D, however, exhibited an EPR signal ($g < 2.0$), which has been interpreted as that of Mo(V) [164]. The just-detected tyrosine-coordination of the [4Fe-4S] cluster by x-ray crystallography might lead to another speculation. This noninnocent ligand could be reduced to a radical anion stabilized by the cluster.

The putative mechanism of activation and dehydration can now be described in the following way: The cluster of component A, to which two ADP are bound, is reduced by ferredoxin or flavodoxin with one electron to $[4Fe\text{-}4S]^{+}$. Then, ADP is exchanged by ATP, which causes the helix-cluster-helix angle to open from 105° to 180°. This conformational change enables component A to dock at component D, and the electron is transferred from A to D with concomitant hydrolysis of two ATP. The electron transfer becomes irreversible and component A returns to its "ground state" with two ADP and oxidized $[4Fe\text{-}4S]^{2+}$. Upon addition of (*R*)-2-hydroxyacyl-CoA to the reduced component D, the electron is further transferred to the substrate to form the ketyl radical anion, which initiates the dehydration as proposed above. Afterward, the electron is returned to component D and transferred further to the next incoming substrate [164]. Thus, multiple turnovers are possible without additional consumption of ATP. Only if the electron is lost by oxidation, another activation with hydrolysis of two ATP becomes necessary. If each turnover would require hydrolysis of ATP, the organism would be unable to thrive from glutamate, since the fermentation only yields 0.6 mol ATP/mol glutamate. There are still several weak steps in this proposed mechanism that need to be clarified. The major problem comprises the localization of the electron in component D and the verification of the ketyl radical anion and enoxy radical intermediates. None of our experiments performed so far was able to disprove this proposal. The electron transfer from component A to D driven by ATP hydrolysis is comparable to an archer shooting arrows. Furthermore, the conformational changes of the string of his bow during shooting are similar to the proposed opening of the helix-cluster-helix angle from 105° to 180°. Hence, all enzymes related to component A have been called Archerases [172].

9.11.3 Dehydration of 4-Hydroxybutyrate: 4-Hydroxybutyryl-CoA Dehydratase

Like the dehydration of 2- and 3-hydroxyacyl-CoA, the dehydration of 4-hydroxybutyrate also occurs at the CoA-ester level. Formation of 4-hydroxybutyryl-CoA is mediated by a specific CoA-transferase with acetyl-CoA or butyryl-CoA as donor [75]. The reversible dehydration of 4-hydroxybutyryl-CoA to crotonyl-CoA is catalyzed by a homotetrameric moderate oxygen-sensitive enzyme (4×65 kDa) comprising one $[4Fe\text{-}4S]^{2+}$ cluster and one noncovalently bound FAD per subunit [173,174]. 4-Hydroxybutyrate is an intermediate in the fermentation of 4-aminobutyrate (γ-aminobutyrate, GABA) by *Clostridium aminobutyricum* (Figure 9.12). The γ-amino acid is converted in the usual way to 4-hydroxybutyrate by amino transfer to 2-oxoglutarate, yielding glutamate and succinate semialdehyde (EC 2.6.1.19) followed by an NADH-dependent reduction to 4-hydroxybutyrate (EC 1.1.1.61). During the further course of the fermentation, crotonyl-CoA disproportionates to butyryl-CoA and acetyl-CoA. Finally 1 ATP is formed from acetyl-CoA per 2 mol 4-aminobutyrate fermented (Equation 9.18).

$$2\times 4\text{-Aminobutyrate} + 2\ H_2O = 2\ NH_4^+ + 2\ \text{Acetate}^- + \text{Butyrate}^- + H^+$$
$$\Delta G^{\circ\prime} = -96\ \text{kJ/mol butyrate; 1 ATP}/2\times 4\text{-aminobutyrate.} \tag{9.18}$$

$$\text{Glutamate}^- + H^+ = 4\text{-Aminobutyrate};\ \Delta G^{\circ\prime}\ \text{ca.} -30\ \text{kJ/mol} \tag{9.19}$$

The value of –96 kJ/mol, which fits Thauer's rule for the minimal requirement of –70 kJ/mol ATP, has been calculated using ΔG° data for the formation from the elements [102]. The free energy of formation of 4-aminobutyrate from the elements ($\Delta G^{\circ}f = -312$ kJ/mol) has been obtained from the decarboxylation of glutamate assuming a $\Delta G^{\circ\prime}$ value similar to that of several other decarboxylations (Equation 9.19).

Another pathway involving the dehydration of 4-hydroxybutyryl-CoA comprises the reduction of succinate to butyryate in *C. kluyveri* [174,175] (Figure 9.12). As discussed above, the organism thrives from ethanol and acetate, whereby the thermodynamically unfavorable oxidation of the alcohol to acetate and H_2 is driven by butyrate and caproate synthesis. Succinate can act as electron acceptor, alternative to protons. It is reduced to butyrate via succinyl-CoA, succinate semialdehyde, 4-hydroxybutyrate, 4-hydroxybutyryl-CoA, and crotonyl-CoA (Equation 9.20). 4-Hydroxybutyryl-CoA dehydratase from *C. kluyveri* has been characterized as an enzyme very similar to that from *C. aminobutyricum*.

$$3\ \text{Ethanol} + 2\ \text{Succinate}^{2-} = 3\ \text{Acetate}^- + 2\ \text{Butyrate}^- + H_2O + H^+$$
$$\Delta G^{\circ\prime} = -171\ \text{kJ/3 mol ethanol; 1 ATP/3 ethanol} \tag{9.20}$$

Like in the ethanol-acetate fermentation (vide supra), the ATP yield is very low as compared to the free energy of the reaction. In this case, 3 ATP are formed from 3 acetyl-CoA obtained by oxidation of 3 ethanol, but 2 ATP are wasted for the activation of 2 succinate to 2 succinyl-CoA.

The genes encoding 4-hydroxybutyrate CoA-transferase and 4-hydroxybutyryl-CoA dehydratase (Figure 9.12) have been cloned from *C. aminobutyricum*. Deduced amino acid sequences related to that of the dehydratase are found in several genomes of anaerobic bacteria and archaea. Part of the sequence is related to those of some FAD-containing mono-oxygenases [76]. Recent X-ray crystallographic studies of 4-hydroxybutyryl-CoA dehydratase showed, however, that the fold of the protein is related to that of butyryl-CoA dehydrogenase. The $[4Fe\text{-}4S]^{2+}$ cluster and the FAD are located close together at a distance of 6 Å. Interestingly, the cluster is only coordinated by three cysteines; the fourth iron is connected via a histidine to the protein. Between both prosthetic groups, there is a cleft in which the substrate can be fitted. There are also glutamate residues, which may act as acid-base catalysts, and a tyrosine residue close to the $[4Fe\text{-}4S]^{2+}$ cluster [176].

4-Hydroxybutyryl-CoA dehydratase as isolated contains a stable, neutral flavin semiquinone radical [177]. Interestingly, upon exposure of the enzyme to air, the dehydratase activity is increased within the first minute, whereas upon longer incubation the enzyme becomes inactive, but most of the vinylacetyl-CoA isomerase activity remains. A gentler oxidation under anoxic conditions can be achieved with thionine without further inactivation. EPR spectroscopy at 77 K revealed that thereby the flavin semiquinone is oxidized to the quinone. Upon addition of the substrate — either 4-hydroxybutyryl-CoA or crotonyl-CoA — the flavin semiquinone appears again, which can also be detected by UV/visible spectroscopy around 500 nm. EPR spectroscopy at 4 K revealed the resonance of the semiquinone at $g = 2.004$, but also further resonances appeared at higher and lower g values, which could be due to the radical partner of the flavin semiquinone at the substrate or at an amino acid residue of the protein. Addition of 4-hydroxy[U-^{13}C]butyryl-CoA, however, did not change the EPR signals, indicating the absence of a substrate-based radical, the signal of which should have become broadened by coupling to ^{13}C. Therefore a protein-based radical appears likely (e.g., a tyrosine radical coupling with the cluster) [178]. EPR spectroscopy further revealed that the $[4Fe\text{-}4S]^{2+}$ cluster is only partial reduced by dithionite to the EPR-detectable $[4Fe\text{-}4S]^{1+}$; almost complete reduction can be achieved in the presence of substrate at pH >8 (I. Çinkaya and W. Buckel, unpublished data).

The dehydration of 4-hydroxybutryl-CoA involves all three prochiral methylene groups; two of them are converted to methine groups and the third to a methyl group. Using (*R*)- and (*S*)-4-hydroxy[3-$^{2}H_1$]butyryl-CoA, it could be established that the 3-*Si*-hydrogen is removed during catalysis [179]. Interestingly, this corresponds to the 3-*Re*-hydrogen of butyryl-CoA, which is transferred to FAD during the oxidation catalyzed by butyryl-CoA dehydrogenase [180].

The dehydration of 4-hydroxybutyryl-CoA to crotonyl-CoA poses similar mechanistic problems as those discussed with 2-hydroxyacyl-CoA. Again, the β-hydrogen, which has to be removed as a proton, is not acidic (pK = 40). It has been proposed, therefore, that the reaction is initiated by removal of the acidic α-proton (pK = 8 in acyl-CoA dehydrogenase [181]) followed by one-electron oxidation to the enoxy radical. The now-facilitated removal of the β-proton yields a ketyl radical anion able to eliminate the vinylogous α-hydroxyl group. The obtained dienoxy radical is reduced to the dienolate and protonated at the γ-carbon to the final product crotonyl-CoA [151,160]. The deprotonation at the α-carbon and protonation at the γ-carbon also explain the isomerization of vinylacetyl-CoA to crotonyl-CoA. Since this activity remains stable in the presence of air, the $[4Fe\text{-}4S]^{2+}$ cluster seems to be involved only in the dehydration. Most likely the iron atom coordinated by the histidine participates in the abstraction of the hydroxyl group (Figure 9.24).

9.11.4 Dehydration of 5-Hydroxy Acids: 5-Hydroxyvaleryl-CoA Dehydratase

During the Stickland reaction, proline is reduced to 5-aminovalerate, which is excreted by the organism. Other clostridia, notably *Clostridium aminovalericum*, ferment this δ-amino acid to ammonia, acetate, propionate, and valerate [182]. Later, H.A. Barker observed that the type-strain was unable to ferment the amino acid and isolated a new one [183], which was designated as *Clostridium viride*, because of the green color of several flavoproteins [64]. In his seminal paper, Barker reported that the organism converts the δ-amino acid in the usual way to the δ-hydroxy acid: amino transfer and NADH-dependent reduction of the resulting glutarsemialdehyde. He further observed that cell-free extracts converted 4-pentenoyl-CoA to 5-hydroxyvaleryl-CoA. This unusual reaction was of great interest, since in the reverse direction, the γ-hydrogen of 5-hydroxyvaleryl-CoA is too far away from the thiol ester to be activated. The solution to this problem would be the

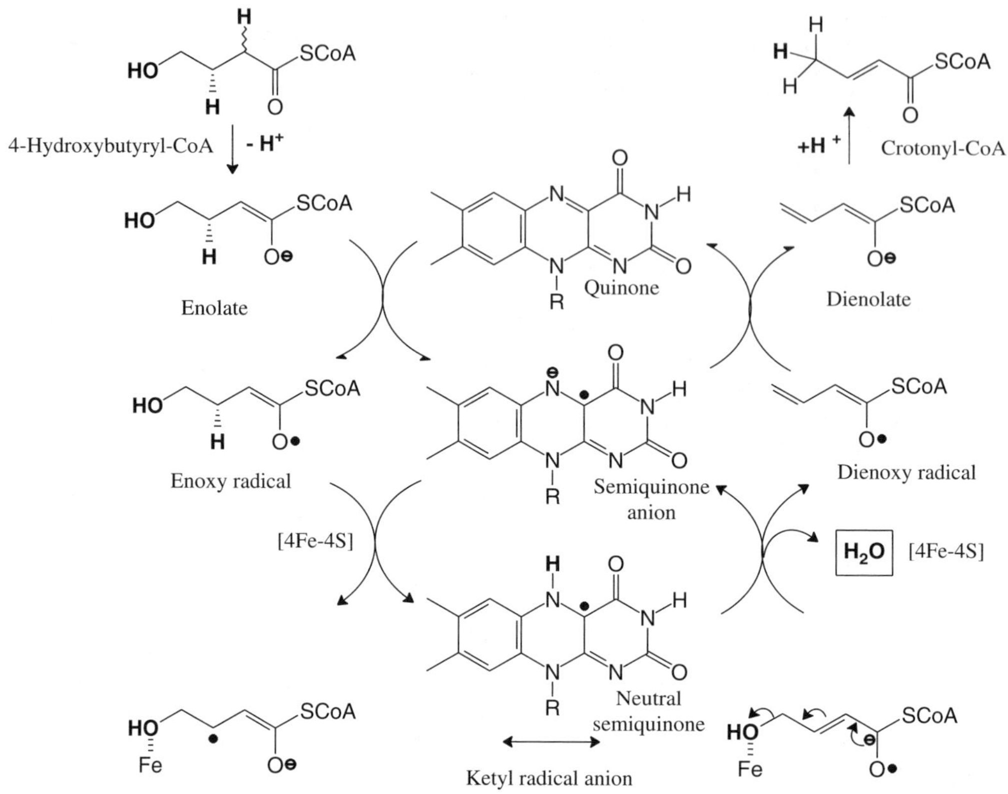

FIGURE 9.24 Proposed mechanism of 4-hydroxybutyryl-CoA dehydratase. Initially, a base removes the acidic α-proton of 4-hydroxybutyryl-CoA and FAD oxidizes the resulting enolate to the enoxy radical and the semiquinone anion of FAD, which is neutralized by the now acidic β-proton of the enoxy radical (pK = 14). The resulting ketyl radical anion expels the hydroxyl group assisted by the [4Fe-4S] cluster (symbolized by Fe). Water is formed from the hydroxyl group and the proton from the neutral semiquinone. In the final steps, an electron is transferred from the semiquinone anion to the dienoxy radical followed by protonation at C4 to yield the methyl group of the product crotonyl-CoA.

transient insertion of an α,β-double bond prior to hydration or dehydration. Indeed, a green FAD-containing 5-hydroxyvaleryl-CoA dehydrogenase could be isolated, which also catalyzed the elimination of the hydroxyl group. Hence, 5-hydroxyvaleryl-CoA is oxidized by FAD to 5-hydroxy-2-pentenoyl-CoA and dehydrated 2,4-pentadienoyl-CoA. Subsequent reduction of the α,β-double bond by $FADH_2$ yields *in vitro* 4-pentenoyl-CoA [184]. *In vivo*, however, this product is probably not formed, since the intermediate 2,4-pentadienoyl-CoA is much faster reduced to 3-pentenoyl-CoA, catalysed by a different green reductase. Thus, 5-hydroxyvaleryl-CoA is oxidatively dehydrated to 2,4-pentadienoyl-CoA, followed by reduction to 3-pentenoyl-CoA and isomerization to 2-pentenoyl-CoA which disproportionates to valerate, propionate, and acetate [183,185]. The electron acceptor of 5-hydroxyvaleryl-CoA dehydrogenase/dehydratase and the donor of 2,4-pentadienoyl-CoA reductase, which are probably the same molecule but different from NAD^+/NADH, have not yet been identified (Figure 9.25).

9.12 CONCLUSIONS

The previous 11 sections comprise the most exciting clostridial enzymes, which have been selected by the author. A complete coverage of all clostridial enzymes would have resulted in a textbook

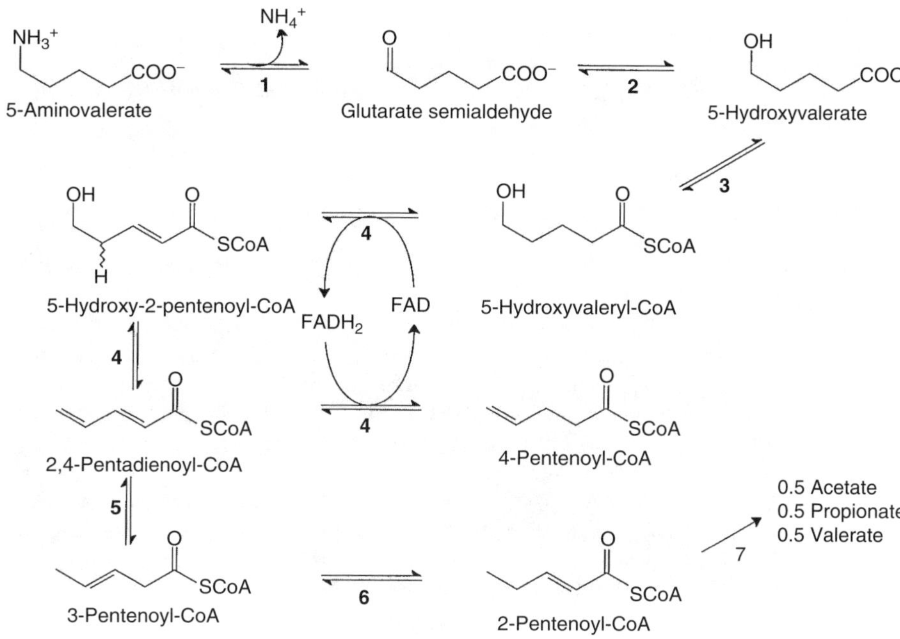

FIGURE 9.25 Fermentation of 5-aminovalerate by *C. viride*. Enzymes: (1) 5-Aminovalerate aminotransferase and glutamate dehydrogenase; (2) NAD^+-dependent 5-hydroxyvalerate dehydrogenase; (3) 5-hydroxyvalerate CoA-transferase; (4) 5-hydroxyvaleryl-CoA dehydratase/dehydrogenase; (5) NADH-dependent 2,4-pentadienoyl-CoA reductase (the electron transfer from 4 to 5 remains to be established); (6) 3-pentenoyl-CoA Δ-isomerase; (7) enzymes analogous to those of crotonyl-CoA disproportionation (Figure 9.5).

in several volumes. Most of the described enzymes also occur in other organisms but, as stated in the introduction, seem to be characteristic of clostridia. Besides glutamate mutase, among these enzymes are the oxygen sensitive acetyl-CoA synthase, pyruvate ferredoxin oxidoreductase, iron only hydrogenase, and pyruvate formate lyase. Those that seem to be restricted to clostridia and Fusobacteria are (*Re*)-citrate synthase and enzymes of the amino acid fermentation pathways, as are the glycine and proline reductases, the SAM and B_{12}-dependent aminomutases, *p*-hydroxyphenylacetate decarboxylase, and the three different hydroxyacyl-CoA dehydratases, which transiently reduce or oxidize their substrates in order to enable the dehydrations. The apparently very simple synthesis of butyrate from two acetyl-CoA is also a clostridial specialty, and seems to be restricted to the family Clostridiaceae. Thus, the formation of butyrate from acetyl-CoA has never been observed in cell-free extracts from *C. tetanomorphum* (Figure 9.1 and Figure 9.5, H.A. Barker, personal communication), whereas the disproportionation of crotonyl-CoA to acetate and butyrate can be readily studied in comparable preparations from *A. fermentans* (Figure 9.4 and Figure 9.5) [3,186].

ACKNOWLEDGMENTS

The work was supported by grants from the Deutsche Forschungsgemeinschaft (DFG, German research foundation), the European Commission, and the Fonds der Chemischen Industrie. The author thanks Professor Bernard T. Golding for helpful discussions. We are indebted to Erica Lyon (Max-Planck-Institut für terrestrische Mikrobiologie, Marburg, Germany) for drawing Figure 9.3.

REFERENCES

1. Garrity, G.M., Boone, D.R., and Castenholz, R.W., *Bergey's Manual of Systematic Bacteriology, 2nd Edition., Vol. 1: The Archaea and the Deeply Branching and Phototrophic Bacteria,* Springer, New York, Berlin, Heidelberg, 2001.
2. Barker, H.A., Fermentations of nitrogenous organic compounds in *The Bacteria*, Gunsalus, I.C., Ed., Academic Press Inc., New York, 1961, 151–207.
3. Buckel, W. and Barker, H.A., Two pathways of glutamate fermentation by anaerobic bacteria, *J. Bacteriol.,* 117, 1248–1260, 1974.
4. Buckel, W., Unusual enzymes involved in five pathways of glutamate fermentation, *Appl. Microbiol. Biotechnol.,* 57, 263–273, 2001.
5. Hayashi, T., Makino, K., Ohnishi, M., Kurokawa, K., Ishii, K., Yokoyama, K., Han, C.G., Ohtsubo, E., Nakayama, K., Murata, T., Tanaka, M., Tobe, T., Iida, T., Takami, H., Honda, T., Sasakawa, C., Ogasawara, N., Yasunaga, T., Kuhara, S., Shiba, T., Hattori, M., and Shinagawa, H., Complete genome sequence of enterohemorrhagic *Escherichia coli* O157:H7 and genomic comparison with a laboratory strain K-12, *DNA Res.,* 8, 11–22, 2001.
6. Perna, N.T., Plunkett, G. III, Burland, V., Mau, B., Glasner, J.D., Rose, D.J., Mayhew, G.F., Evans, P.S., Gregor, J., Kirkpatrick, H.A., Posfai, G., Hackett, J., Klink, S., Boutin, A., Shao, Y., Miller, L., Grotbeck, E.J., Davis, N.W., Lim, A., Dimalanta, E.T., Potamousis, K.D., Apodaca, J., Anantharaman, T.S., Lin, J., Yen, G., Schwartz, D.C., Welch, R.A. and Blattner, F.R, Genome sequence of enterohaemorrhagic *Escherichia coli* O157:H7, *Nature,* 409, 529–533, 2001.
7. Heinzelmann, E., Berger, S., Puk, O., Reichenstein, B., Wohlleben, W., and Schwartz, D., A glutamate mutase is involved in the biosynthesis of the lipopeptide antibiotic friulimicin in actinoplanes friuliensis, *Antimicrob. Agents Chemother.,* 47, 447–457, 2003.
8. Wächtershäuser, G., Groundworks for an evolutionary biochemistry: the iron-sulfur world, *Prog. Biophys. Mol. Biol.,* 58, 85–201, 1992.
9. Evans, M.C., Buchanan, B.B., and Arnon, D.I., A new ferredoxin-dependent carbon reduction cycle in a photosynthetic bacterium, *Proc. Natl. Acad. Sci. USA,* 55, 928–934, 1966.
10. Herter, S., Busch, A., and Fuchs, G., L-Malyl-coenzyme A lyase/beta-methylmalyl-coenzyme A lyase from Chloroflexus aurantiacus, a bifunctional enzyme involved in autotrophic CO_2 fixation, *J. Bacteriol.,* 184, 5999–6006, 2002.
11. Drake, H., *Acetogenesis,* Chapman & Hall, New York, London, 1994.
12. Huber, C. and Wächtershäuser, G., Activated acetic acid by carbon fixation on (Fe,Ni)S under primordial conditions, *Science,* 276, 245–247, 1997.
13. Menon, S. and Ragsdale, S.W., Role of the [4Fe-4S] cluster in reductive activation of the cobalt center of the corrinoid iron-sulfur protein from *Clostridium thermoaceticum* during acetate biosynthesis, *Biochemistry,* 37, 5689–5698, 1998.
14. Ragsdale, S.W. and Kumar, M., Nickel-containing carbon monoxide dehydrogenase/acetyl-CoA synthase, *Chem. Rev.,* 96, 2515–2539, 1996.
15. Loke, H.K., Bennett, G.N. and Lindahl, P.A., Active acetyl-CoA synthase from *Clostridium thermoaceticum* obtained by cloning and heterologous expression of acsAB in *Escherichia coli, Proc. Natl. Acad. Sci. USA,* 97, 12530–12535, 2000.
16. Doukov, T.I., Iverson, T.M., Seravalli, J., Ragsdale, S.W., and Drennan, C.L., A Ni-Fe-Cu center in a bifunctional carbon monoxide dehydrogenase/acetyl-CoA synthase, *Science,* 298, 567–572, 2002.
17. Darnault, C., Volbeda, A., Kim, E.J., Legrand, P., Vernede, X., Lindahl, P.A., and Fontecilla-Camps, J.C., Ni-Zn-[4Fe-4S] and Ni-Ni-[4Fe-4S] clusters in closed and open alpha subunits of acetyl-CoA synthase/carbon monoxide dehydrogenase, *Nat. Struct. Biol.,* 10, 271–279, 2003.
18. Svetlitchnyi, V., Dobbek, H., Meyer-Klaucke, W., Meins, T., Thiele, B., Romer, P., Huber, R., and Meyer, O., A functional Ni-Ni-[4Fe-4S] cluster in the monomeric acetyl-CoA synthase from Carboxydothermus hydrogenoformans, *Proc. Natl. Acad. Sci. USA,* 101, 446–451, 2004.
19. Seravalli, J., Xiao, Y., Gu, W., Cramer, S.P., Antholine, W.E., Krymov, V., Gerfen, G.J., and Ragsdale, S.W., Evidence that NiNi acetyl-CoA synthase is active and that the CuNi enzyme is not, *Biochemistry,* 43, 3944–3955, 2004.

20. Dobbek, H., Svetlitchnyi, V., Gremer, L., Huber, R., and Meyer, O., Crystal structure of a carbon monoxide dehydrogenase reveals a [Ni-4Fe-5S] cluster, *Science,* 293, 1281–1285, 2001.
21. Menon, S. and Ragsdale, S.W., The role of an iron-sulfur cluster in an enzymatic methylation reaction. Methylation of CO dehydrogenase/acetyl-CoA synthase by the methylated corrinoid iron-sulfur protein, *J. Biol. Chem.,* 274, 11513–11518, 1999.
22. Ljungdahl, L.G. and Andreesen, J.R. Tungsten, a component of active formate dehydrogenase from *Clostridium thermoacetium, FEBS Lett.,* 54, 279–282, 1975.
23. Gräntzdörffer, A., Rauh, D., Pich, A., and Andreesen, J.R., Molecular and biochemical characterization of two tungsten- and selenium-containing formate dehydrogenases from *Eubacterium acidaminophilum* that are associated with components of an iron-only hydrogenase, *Arch. Microbiol.,* 179, 116–130, 2003.
24. de Bok, F.A., Hagedoorn, P.L., Silva, P.J., Hagen, W.R., Schiltz, E., Fritsche, K., and Stams, A.J., Two W-containing formate dehydrogenases (CO_2-reductases) involved in syntrophic propionate oxidation by *Syntrophobacter fumaroxidans, Eur. J. Biochem.,* 270, 2476–2485, 2003.
25. Hugenholtz, J. and Ljungdahl, L.G., Metabolism and energy generation in homoacetogenic clostridia, *FEMS Microbiol. Rev,.* 7, 383–389, 1990.
26. Heise, R., Müller, V., and Gottschalk, G., Presence of a sodium-translocating ATPase in membrane vesicles of the homoacetogenic bacterium *Acetobacterium woodii, Eur. J. Biochem.,* 206, 553–557, 1992.
27. Frey, M., Hydrogenases: hydrogen-activating enzymes, *Chembiochem,* 3, 153–160, 2002.
28. Peters, J.W., Lanzilotta, W.N., Lemon, B.J., and Seefeldt, L.C., X-ray crystal structure of the Fe-only hydrogenase (CpI) from *Clostridium pasteurianum* to 1.8 angstrom resolution, *Science,* 282, 1853–1858, 1998.
29. Happe, R.P., Roseboom, W., Pierik, A.J., Albracht, S.P., and Bagley, K.A., Biological activation of hydrogen, *Nature,* 385, 126, 1997.
30. Härtel, U. and Buckel, W., Sodium ion-dependent hydrogen production in *Acidaminococcus fermentans, Arch. Microbiol.,* 166, 350–356, 1996.
31. Burton, R.M. and Stadtman, E.R., The oxidation of acetaldehyde to acetyl coenzyme A, *J. Biol. Chem.,* 202, 837, 1953.
32. Jeng, I.M., Somack, R., and Barker, H.A., Ornithine degradation in *Clostridium sticklandii*; pyridoxal phosphate and coenzyme A dependent thiolytic cleavage of 2-amino-4-ketopentanoate to alanine and acetyl coenzyme A, *Biochemistry,* 13, 2898–2903, 1974.
33. Shieh, J. and Whitman, W.B., Autotrophic acetyl coenzyme A biosynthesis in *Methanococcus maripaludis, J. Bacteriol.,* 170, 3072–3079, 1988.
34. Kerscher, L. and Oesterhelt, D., The catalytic mechanism of 2-oxoacid:ferredoxin oxidoreductases from *Halobacterium halobium*. One-electron transfer at two distinct steps of the catalytic cycle, *Eur. J. Biochem.,* 116, 595–600, 1981.
35. Chabrière, E., Vernède, X., Guigliarelli, B., Charon, M.-H., Hatchikian, E.C., and Fontecilla-Camps, J.C., Crystal structure of the free radical intermediate of pyruvate:ferredoxin oxidoreductase, *Science,* 294, 2559–2563, 2001.
36. Bornstein, B.T. and Barker, H.A., The energy metabolism of *Clostridium kluyveri* and the synthesis of fatty acids, *J. Biol. Chem.,* 172, 659, 1948.
37. Thauer, R.K., Jungermann, K., Henninger, H., Wenning, J., and Decker, K., The energy metabolism of *Clostridium kluyveri, Eur. J. Biochem.,* 4, 173–180, 1968.
38. Schoberth, S. and Gottschalk, G., Considerations on the energy metabolism of *Clostridium kluyveri, Arch. Mikrobiol.,* 65, 318–328, 1969.
39. Brüggemann, H., Bäumer, S., Fricke, W.F., Wiezer, A., Liesegang, H., Decker, I., Herzberg, C., Martinez-Arias, R., Merkl, R., Henne, A., and Gottschalk, G., The genome sequence of *Clostridium tetani*, the causative agent of tetanus disease, *Proc. Natl. Acad. Sci. USA,* 100, 1316–1321, 2003.
40. Schmehl, M., Jahn, A., Meyer zu Vilsendorf, A., Hennecke, S., Masepohl, B., Schuppler, M., Marxer, M., Oelze, J., and Klipp, W., Identification of a new class of nitrogen fixation genes in *Rhodobacter capsulatus*: a putative membrane complex involved in electron transport to nitrogenase, *Mol. Gen. Genet.,* 241, 602–615, 1993.

41. Kumagai, H., Fujiwara, T., Matsubara, H., and Saeki, K., Membrane localization, topology, and mutual stabilization of the rnfABC gene products in *Rhodobacter capsulatus* and implications for a new family of energy-coupling NADH oxidoreductases, *Biochemistry,* 36, 5509–5521, 1997.
42. Saeki, K. and Kumagai, H., The rnf gene products in *Rhodobacter capsulatus* play an essential role in nitrogen fixation during anaerobic DMSO-dependent growth in the dark, *Arch. Microbiol.,* 169, 464–467, 1998.
43. Hillmer, P. and Gottschalk, G., Particulate nature of enzymes involved in the fermentation of ethanol and acetate by *Clostridium kluyveri*, *FEBS Lett.,* 21, 351–354, 1972.
44. Collins, M.D., Lawson, P.A., Willems, A., Cordoba, J.J., Fernandez-Garayzabal, J., Garcia, P., Cai, J., Hippe, H., and Farrow, J.A., The phylogeny of the genus *Clostridium*: proposal of five new genera and eleven new species combinations, *Int. J. Syst. Bacteriol.,* 44, 812–826, 1994.
45. Schmidt, D.E. Jr. and Westheimer, F.H., pK of the lysine amino group at the active site of acetoacetate decarboxylase, *Biochemistry,* 10, 1249–1253, 1971.
46. Cardon, B.P. and Barker, H.A., Amino acid fermentations by *Clostridium propionicum* and *Diplococcus glycinophilus*, *Arch. Biochem. Biophys.,* 12, 165, 1947.
47. Schweiger, G. and Buckel, W., On the dehydration of (R)-lactate in the fermentation of alanine to propionate by *Clostridium propionicum*, *FEBS Lett.,* 171, 79–84, 1984.
48. Hofmeister, A.E., Grabowski, R., Linder, D., and Buckel, W., L-serine and L-threonine dehydratase from *Clostridium propionicum*. Two enzymes with different prosthetic groups, *Eur. J. Biochem.,* 215, 341–349, 1993.
49. Selmer, T., Willanzheimer, A., and Hetzel, M., Propionate CoA-transferase from *Clostridium propionicum*. Cloning of gene and identification of glutamate 324 at the active site, *Eur. J. Biochem.,* 269, 372–380, 2002.
50. Schweiger, G. and Buckel, W., Identification of acrylate, the product of the dehydration of (R)-lactate catalysed by cell-free extracts from *Clostridium propionicum*, *FEBS Lett.,* 185, 253–256, 1985.
51. Kuchta, R.D. and Abeles, R.H., Lactate reduction in *Clostridium propionicum*. Purification and properties of lactyl-CoA dehydratase, *J. Biol. Chem.,* 260, 13181–13189, 1985.
52. Kuchta, R.D., Hanson, G.R., Holmquist, B., and Abeles, R.H., Fe-S centers in lactyl-CoA dehydratase, *Biochemistry,* 25, 7301–7307, 1986.
53. Hofmeister, A.E. and Buckel, W., (R)-lactyl-CoA dehydratase from *Clostridium propionicum*. Stereochemistry of the dehydration of (R)-2-hydroxybutyryl-CoA to crotonyl-CoA, *Eur. J. Biochem.,* 206, 547–552, 1992.
54. Hetzel, M., Brock, M., Selmer, T., Pierik, A.J., Golding, B.T., and Buckel, W., Acryloyl-CoA reductase from *Clostridium propionicum*. An enzyme complex of propionyl-CoA dehydrogenase and electron-transferring flavoprotein, *Eur. J. Biochem.,* 270, 902–910, 2003.
55. Vagelos, P.R., Earl, J.M., and Stadtman, E.R., Propionic acid metabolism. I. The purification and properties of acrylyl coenzyme A aminase, *J. Biol. Chem.,* 234, 490–497, 1959.
56. Buckel, W., Gokarn, R.R., Gort, S.J., Jessen, H., Selifonova, O.V., and Selmer, T., Novel polypeptides having lactyl-CoA dehydratase, E1 activator, 3-hydroxypropionyl-CoA dehydratase or malonyl-CoA reductase activity, useful in producing 3-hydroxypropionic acid and other organic compounds, in Patent 014806513; WPI Acc No: 2002-627219/200267, Cargill, Inc. (CRGI), Minneapolis, 2002.
57. Selmer, T., and Buckel, W., Oxygen exchange between acetate and the catalytic glutamate residue in glutaconate CoA-transferase from *Acidaminococcus fermentans*. Implications for the mechanism of CoA-ester hydrolysis, *J. Biol. Chem.,* 274, 20772–20778, 1999.
58. March, J., *Advanced Organic Chemistry, 3rd Edition,* John Wiley & Sons, Inc., New York, 1985.
59. Heider, J., A new family of CoA-transferases, *FEBS Lett.,* 509, 345–349, 2001.
60. Cary, J.W., Petersen, D.J., Papoutsakis, E.T., and Bennett, G.N., Cloning and expression of *Clostridium acetobutylicum* ATCC 824 acetoacetyl-coenzyme A:acetate/butyrate:coenzyme A-transferase in *Escherichia coli*, *Appl. Environ. Microbiol.,* 56, 1576–1583, 1990.
61. Buckel, W., Dorn, U., and Semmler, R., Glutaconate CoA-transferase from *Acidaminococcus fermentans*, *Eur. J. Biochem.,* 118, 315–321, 1981.
62. Mack, M. and Buckel, W., Conversion of glutaconate CoA-transferase from *Acidaminococcus fermentans* into an acyl-CoA hydrolase by site-directed mutagenesis, *FEBS Lett.,* 405, 209–212, 1997.

63. Eikmanns, U. and Buckel, W., Properties of 5-hydroxyvalerate CoA-transferase from *Clostridium aminovalericum*, *Biol. Chem. Hoppe Seyler*, 371, 1077–1082, 1990.
64. Buckel, W., Janssen, P., Schumann, A., Eikmanns, U., Messner, P., Sleytr, U.B., and Liesack, W., *Clostridium viride* sp. nov., a strictly anaerobic bacterium using 5-aminovalerate as growth substrate, previously assigned to *Clostridium aminovalericum*, *Arch. Microbiol.,* 162, 387–394, 1994.
65. Barker, H.A., Jeng, I.M., Neff, N., Robertson, J.M., Tam, F.K., and Hosaka, S., Butyryl-CoA:acetoacetate CoA-transferase from a lysine-fermenting clostridium, *J. Biol. Chem.,* 253, 1219–1225, 1978.
66. Barker, H.A., Kahn, J.M., and Hedrick, L., Pathway of lysine degradation in *Fusobacterium nucleatum*, *J. Bacteriol.,* 152, 201–207, 1982.
67. Kapatral, V., Anderson, I., Ivanova, N., Reznik, G., Los, T., Lykidis, A., Bhattacharyya, A., Bartman, A., Gardner, W., Grechkin, G., Zhu, L., Vasieva, O., Chu, L., Kogan, Y., Chaga, O., Goltsman, E., Bernal, A., Larsen, N., D'Souza, M., Walunas, T., Pusch, G., Haselkorn, R., Fonstein, M., Kyrpides, N., and Overbeek, R., Genome sequence and analysis of the oral bacterium *Fusobacterium nucleatum* strain ATCC 25586, *J. Bacteriol.,* 184, 2005–2018, 2002.
68. Buckel, W., Buschmeier, V., and Eggerer, H., The action mechanism of citrate lyase from *Klebsiella aerogenes*, *Hoppe Seyler's Z. Physiol. Chem.,* 352, 1195–1205, 1971.
69. Antranikian, G., Friese, C., Quentmeier, H., Hippe, H., and Gottschalk, G., Distribution of the ability for citrate uitlisation amongst clostridia., *Arch. Microbiol.,* 138, 179–182, 1984.
70. Buckel, W. and Bobi, A., The enzyme complex citramalate lyase from *Clostridium tetanomorphum*, *Eur. J. Biochem.,* 64, 255–262, 1976.
71. Schneider, K., Dimroth, P., and Bott, M., Biosynthesis of the prosthetic group of citrate lyase, *Biochemistry,* 39, 9438–9450, 2000.
72. Buckel, W., Acetic anhydride: an intermediate analogue in the acyl-exchange reaction of citramalate lyase, *Eur. J. Biochem.,* 64, 263–267, 1976.
73. Schmellenkamp, H. and Eggerer, H., Mechanism of enzymic acetylation of des-acetyl citrate lyase, *Proc. Natl. Acad. Sci. USA,* 71, 1987–1991, 1974.
74. Buckel, W., Bröker, G., Bothe, H., and Pierik, A.J., Glutamate mutase and 2-methyleneglutarate mutase, in *Chemistry and Biochemistry of* B_{12}, Banerjee, R., Ed., John Wiley & Sons, Inc., New York, 1999, 757–781.
75. Scherf, U. and Buckel, W., Purification and properties of 4-hydroxybutyrate coenzyme A transferase from *Clostridium aminobutyricum*, *Appl. Environ. Microbiol.,* 57, 2699–2702, 1991.
76. Gerhardt, A., Cinkaya, I., Linder, D., Huisman, G., and Buckel, W., Fermentation of 4-aminobutyrate by *Clostridium aminobutyricum*: cloning of two genes involved in the formation and dehydration of 4-hydroxybutyryl-CoA, *Arch. Microbiol.,* 174, 189–199, 2000.
77. Dickert, S., Pierik, A.J., Linder, D., and Buckel, W., The involvement of coenzyme A esters in the dehydration of (R)-phenyllactate to (E)-cinnamate by *Clostridium sporogenes*, *Eur. J. Biochem.,* 267, 3874–3884, 2000.
78. Dickert, S., Pierik, A.J., and Buckel, W., Molecular characterization of phenyllactate dehydratase and its initiator from *Clostridium sporogenes*, *Mol. Microbiol.,* 44, 49–60, 2002.
79. Yorifuji, T., Jeng, I.M., and Barker, H.A., Purification and properties of 3-keto-5-aminohexanoate cleavage enzyme from a lysine-fermenting Clostridium, *J. Biol. Chem.,* 252, 20–31, 1977.
80. Buckel, W. and Eggerer, H., Intramolecular nucleophilic catalysis on the hydrolysis of citryl-CoA, *Hoppe Seyler's Z. Physiol. Chem.,* 350, 1367–1376, 1969.
81. Gottschalk, G. and Barker, H.A., Synthesis of glutamate and citrate by *Clostridium kluyveri*. A new type of citrate synthase, *Biochemistry,* 5, 1125–1133, 1966.
82. Gottschalk, G., Partial purification and some properties of the (R)-citrate synthase from *Clostridium acidi-urici*, *Eur. J. Biochem.,* 7, 301–306, 1969.
83. Gottschalk, G., Dittbrenner, S., Lenz, H., and Eggerer, H., Studies on the re-citrate synthase reaction, *Eur. J. Biochem.,* 26, 455–461, 1972.
84. Dimroth, P., A new sodium-transport system energized by the decarboxylation of oxaloacetate, *FEBS Lett.,* 122, 234–236, 1980.
85. Hilpert, W. and Dimroth, P., Conversion of the chemical energy of methylmalonyl-CoA decarboxylation into a Na^+ gradient, *Nature,* 296, 584–585, 1982.

86. Bott, M., Pfister, K., Burda, P., Kalbermatter, O., Woehlke, G., and Dimroth, P., Methylmalonyl-CoA decarboxylase from *Propionigenium modestum* — cloning and sequencing of the structural genes and purification of the enzyme complex, *Eur. J. Biochem.,* 250, 590–599, 1997.
87. Buckel, W. and Semmler, R., A biotin-dependent sodium pump: glutaconyl-CoA decarboxylase from *Acidaminococcus fermentans, FEBS Lett.,* 148, 35–38, 1982.
88. Beatrix, B., Bendrat, K., Rospert, S., and Buckel, W., The biotin-dependent sodium ion pump glutaconyl-CoA decarboxylase from *Fusobacterium nucleatum* (subsp. nucleatum). Comparison with the glutaconyl-CoA decarboxylases from gram-positive bacteria, *Arch. Microbiol.,* 154, 362–369, 1990.
89. Dimroth, P., Jockel, P., and Schmid, M., Coupling mechanism of the oxaloacetate decarboxylase Na^+ pump, *Biochim. Biophys. Acta,* 1505, 1–14, 2001.
90. Buckel, W., Sodium ion-translocating decarboxylases, *Biochim. Biophys. Acta,* 1505, 15–27, 2001.
91. Wendt, K.S., Schall, I., Huber, R., Buckel, W., and Jacob, U., Crystal structure of the carboxyltransferase subunit of the bacterial sodium ion pump glutaconyl-coenzyme A decarboxylase, *EMBO J.,* 22, 3493–3502, 2003.
92. Holden, H.M., Benning, M.M., Haller, T., and Gerlt, J.A., The crotonase superfamily: divergently related enzymes that catalyze different reactions involving acyl coenzyme a thioesters, *Acc. Chem. Res.,* 34, 145–157, 2001.
93. Turner, D.C. and Stadtman, T.C., Purification of protein components of the clostridial glycine reductase system and characterization of protein A as a selenoprotein, *Arch. Biochem. Biophys.,* 154, 366–381, 1973.
94. Stadtman, T.C., Selenium-dependent enzymes, *Annu. Rev. Biochem.,* 49, 93–110, 1980.
95. Andreesen, J.R., Glycine metabolism in anaerobes, *Antonie Van Leeuwenhoek,* 66, 223–237, 1994.
96. Andreesen, J.R., Wagner, M., Sonntag, D., Kohlstock, M., Harms, C., Gursinsky, T., Jäger, J., Parther, T., Kabisch, U., Gräntzdörffer, A., Pich, A., and Söhling, B., Various functions of selenols and thiols in anaerobic gram-positive, amino acids-utilizing bacteria, *Biofactors,* 10, 263–270, 1999.
97. Wagner, M., Sonntag, D., Grimm, R., Pich, A., Eckerskorn, C., Söhling, B., and Andreesen, J.R., Substrate-specific selenoprotein B of glycine reductase from *Eubacterium acidaminophilum.* Biochemical and molecular analysis, *Eur. J. Biochem.,* 260, 38–49, 1999.
98. Kohlstock, U.M., Rücknagel, K.P., Reuter, M., Schierhorn, A., Andreesen, J.R., and Söhling, B., Cys359 of GrdD is the active-site thiol that catalyses the final step of acetyl phosphate formation by glycine reductase from *Eubacterium acidaminophilum, Eur. J. Biochem.,* 268, 6417–6425, 2001.
99. Buckel, W., Amino acid fermentation: Coenzyme B_{12}-dependent and -independent pathways, in *The Molecular Basis of Bacterial Metabolism 41. Colloquium — Mosbach,* Hauska, G. and Thauer, R.K., Eds., Springer-Verlag, Heidelberg, 1990, 21–30.
100. Kabisch, U.C., Gräntzdörffer, A., Schierhorn, A., Rücknagel, K.P., Andreesen, J.R., and Pich, A., Identification of D-proline reductase from *Clostridium sticklandii* as a selenoenzyme and indications for a catalytically active pyruvoyl group derived from a cysteine residue by cleavage of a proprotein, *J. Biol. Chem.,* 274, 8445–8454, 1999.
101. Bednarski, B., Andreesen, J.R., and Pich, A., In vitro processing of the proproteins GrdE of protein B of glycine reductase and PrdA of D-proline reductase from *Clostridium sticklandii*: formation of a pyruvoyl group from a cysteine residue, *Eur. J. Biochem.,* 268, 3538–3544, 2001.
102. Thauer, R.K., Jungermann, K., and Decker, K., Energy conservation in chemotrophic anaerobic bacteria, *Bacteriol. Rev.,* 41, 100–180, 1977.
103. Rétey, J., Discovery and role of methylidene imidazolone, a highly electrophilic prosthetic group, *Biochim. Biophys. Acta,* 1647, 179–184, 2003.
104. Viola, R.E., L-aspartase: new tricks from an old enzyme, *Adv. Enzymol. Relat. Areas Mol. Biol.,* 74, 295–341, 2000.
105. Barker, H.A., Smyth, R.M., Wilson, R.M., and Weissbach, H., The purification and properties of beta-methylaspartase, *J. Biol. Chem.,* 234, 320–328, 1959.
106. Asuncion, M., Blankenfeldt, W., Barlow, J.N., Gani, D., and Naismith, J.H., The structure of 3-methylaspartase from *Clostridium tetanomorphum* functions via the common enolase chemical step, *J. Biol. Chem.,* 277, 8306–8311, 2002.
107. Levy, C.W., Buckley, P.A., Sedelnikova, S., Kato, Y., Asano, Y., Rice, D.W., Baker, P.J., Insights into enzyme evolution revealed by the structure of methylaspartate ammonia lyase, *Structure (Camb),* 10, 105–113, 2002.

108. Jeng, I.M. and Barker, H.A., Purification and properties of L-3-aminobutyryl coenzyme A deaminase from a lysine-fermenting clostridium, *J. Biol. Chem.,* 249, 6578–6584, 1974.
109. Hisano, T., Tsuge, T., Fukui, T., Iwata, T., Miki, K., and Doi, Y., Crystal structure of the (R)-specific enoyl-CoA hydratase from *Aeromonas caviae* involved in polyhydroxyalkanoate biosynthesis, *J. Biol. Chem.,* 278, 617–624, 2003.
110. Bandarian, V. and Reed, G.H., Ethanolamine ammonia-lyase; in *Chemistry and Biochemistry of* B_{12}, Banerjee, R., Ed., John Wiley & Sons, Inc., New York, 1999, 811–833.
111. Barker, H.A., Weissbach, H., and Smyth, R.D., A coenzyme containing pseudovitamin B_{12}, *Proc. Natl. Acad. Sci. U.S.A.,* 44, 1093–1097, 1958.
112. Lenhert, P.G. and Hodgkin, D.C., Structure of the 5,6-dimethylbenzimidazolylcobamide coenzyme, *Nature (London),* 192, 937–938, 1961.
113. Kratky, C. and Kräutler, B., x-ray crystallography of B_{12}; in *Chemistry and Biochemistry of* B_{12}, Banerjee, R., Ed., John Wiley & Sons, Inc., New York, 1999, 9–42.
114. Reitzer, R., Gruber, K., Jogl, G., Wagner, U.G., Bothe, H., Buckel, W., and Kratky, C., Glutamate mutase from *Clostridium cochlearium*: the structure of a coenzyme B_{12}-dependent enzyme provides new mechanistic insights, *Structure Fold. Des.,* 7, 891–902, 1999.
115. Zelder, O., Beatrix, B., Kroll, F., and Buckel, W., Coordination of a histidine residue of the protein-component S to the cobalt atom in coenzyme B_{12}-dependent glutamate mutase from *Clostridium cochlearium, FEBS Lett.,* 369, 252–254, 1995.
116. Chen, D., Abend, A., Stubbe, J., and Frey, P.A., Epimerization at carbon-5′ of (5′R)-[5′-^{2}H]adenosylcobalamin by ribonucleoside triphosphate reductase: cysteine 408-independent cleavage of the Co-C5′ bond, *Biochemistry,* 42, 4578–4584, 2003.
117. Golding, B.T. and Buckel, W., Corrin-dependent reactions; in *Comprehensive Biological Catalysis,* Sinnott, M.L., Ed., Academic Press, 1998, 239–259.
118. Beatrix, B., Zelder, O., Kroll, F., Örlygsson, G., Golding, B.T., and Buckel, W., Evidence for a mechanism involving transient fragmentation in coenzyme B_{12}-dependent carbon skeleton rearrangements, *Angew. Chem. Int. Edn. Engl.,* 34, 2398–2401, 1995.
119. Bothe, H., Darley, D.J., Albracht, S.P., Gerfen, G.J., Golding, B.T., and Buckel, W., Identification of the 4-glutamyl radical as an intermediate in the carbon skeleton rearrangement catalyzed by coenzyme B_{12}-dependent glutamate mutase from *Clostridium cochlearium, Biochemistry,* 37, 4105–4113, 1998.
120. Chih, H.W. and Marsh, E.N.G., Mechanism of glutamate mutase: Identification and kinetic competence of acrylate and glycyl radical as intermediates in the rearrangement of glutamate to methylaspartate, *J. Am. Chem. Soc.,* 122, 10732–10733, 2000.
121. Kung, H.F., Cederbaum, S., Tsai, L., and Stadtman, T.C., Nicotinic acid metabolism. V. A cobamide coenzyme-dependent conversion of alpha-methyleneglutaric acid to dimethylmaleic acid, *Proc. Natl. Acad. Sci. USA,* 65, 978–984, 1970.
122. Kung, H. and Tsai, L., Nicotinic acid metabolism. VII. Mechanisms of action of clostridial alpha-methyleneglutarate mutase (B_{12}-dependent) and methylitaconate isomerase, *J. Biol. Chem.,* 246, 6436–6443, 1971.
123. Hartrampf, G. and Buckel, W., On the steric course of the adenosylcobalamin-dependent 2-methyleneglutarate mutase reaction in *Clostridium barkeri, Eur. J. Biochem.,* 156, 301–304, 1986.
124. Ciceri, D., Pierik, A.J., Hartrampf, G., Bröker, G., Speranza, G., Buckel, W., Cornforth, J., and Golding, B.T., Stereochemistry of the methyl group in (R)-3-methylitaconate derived by rearrangement of 2-methylideneglutarate catalysed by a coenzyme B_{12}-dependent mutase, *Helv. Chim. Acta,* 83, 2550–2561, 2000.
125. Eggerer, H., Completion of the degradation scheme for nicotinic acid by *Clostridium barkeri, Curr. Top. Cell. Regul.,* 26, 411–418, 1985.
126. Pierik, A.J., Ciceri, D., Lopez, R.F., Kroll, F., Bröker, G., Beatrix, B., Buckel, W., and Golding, B.T., Searching for intermediates in the carbon skeleton rearrangement of 2-methyleneglutarate to (R)-3-methylitaconate catalysed by coenzyme B_{12}-dependent 2-methyleneglutarate mutase from *Eubacterium barkeri,* in preparation.
127. Pierik, A.J., Ciceri, D., Broker, G., Edwards, C.H., McFarlane, W., Winter, J., Buckel, W., and Golding, B.T., Rotation of the exo-methylene group of (R)-3-methylitaconate catalyzed by coenzyme B_{12}-dependent 2-methyleneglutarate mutase from *Eubacterium barkeri, J. Am. Chem. Soc.,* 124, 14039–14048, 2002.

128. Baker, J.J., van der Drift, C., and Stadtman, T.C., Purification and properties of beta-lysine mutase, a pyridoxal phosphate and B_{12} coenzyme dependent enzyme, *Biochemistry,* 12, 1054–1063, 1973.
129. Frey, P.A. and Chang, C.H., Aminomutases; in *Chemistry and Biochemistry of B_{12}*, Banerjee, R., Ed., John Wiley & Sons, Inc., New York, 1999, 835–857.
130. Chang, C.H. and Frey, P.A., Cloning, sequencing, heterologous expression, purification, and characterization of adenosylcobalamin-dependent D-lysine 5, 6-aminomutase from *Clostridium sticklandii*, *J. Biol. Chem.,* 275, 106–114, 2000.
131. Tang, K.H., Chang, C.H., and Frey, P.A., Electron transfer in the substrate-dependent suicide inactivation of lysine 5,6-aminomutase, *Biochemistry,* 40, 5190–5199, 2001.
132. Dyer, J.K. and Costilow, R.N., Fermentation of ornithine by *Clostridium sticklandii*, *J. Bacteriol.,* 96, 1617–1622, 1968.
133. Dyer, J.K. and Costilow, R.N., 2,4-Diaminovaleric acid: an intermediate in the anaerobic oxidation of ornithine by *Clostridium sticklandii*, *J. Bacteriol.,* 101, 77–83, 1970.
134. Tsuda, Y. and Friedmann, H.C., Ornithine metabolism by *Clostridium sticklandii.* Oxidation of ornithine to 2-amino-4-ketopentanoic acid via 2,4-diaminopentanoic acid; participation of B_{12} coenzyme, pyridoxal phosphate, and pyridine nucleotide, *J. Biol. Chem.,* 245, 5914–5926, 1970.
135. Chirpich, T.P., Zappia, V., Costilow, R.N. and Barker, H.A., Lysine 2,3-aminomutase. Purification and properties of a pyridoxal phosphate and S-adenosylmethionine-activated enzyme, *J. Biol. Chem.,* 245, 1778–1789, 1970.
136. Baker, J.J. and Stadtman, T.C., Aminomutases, in *B_{12}: Biochemistry and Medicine,* Dolphin, D., Ed., Wiley-Interscience, New York, 1982, 203–232.
137. Sofia, H.J., Chen, G., Hetzler, B.G., Reyes-Spindola, J.F., and Miller, N.E., Radical SAM, a novel protein superfamily linking unresolved steps in familiar biosynthetic pathways with radical mechanisms: functional characterization using new analysis and information visualization methods, *Nucleic Acids Res.,* 29, 1097–1106, 2001.
138. Frey, P.A. and Magnusson, O.T., S-Adenosylmethionine: a wolf in sheep's clothing, or a rich man's adenosylcobalamin?, *Chem. Rev.,* 103, 2129–2148, 2003.
139. Aberhart, D.J., Gould, S.J., Lin, H.J., Thiruvengadam, T.K., and Weiller, B.H., *J. Amer. Chem. Soc.,* 105, 5461–5470, 1983.
140. Eliasson, R., Fontecave, M., Jornvall, H., Krook, M., Pontis, E., and Reichard, P., The anaerobic ribonucleoside triphosphate reductase from *Escherichia coli* requires S-adenosylmethionine as a cofactor, *Proc. Natl. Acad. Sci. USA,* 87, 3314–3318, 1990.
141. Leuthner, B., Leutwein, C., Schulz, H., Horth, P., Haehnel, W., Schiltz, E., Schägger, H., and Heider, J., Biochemical and genetic characterization of benzylsuccinate synthase from *Thauera aromatica*: a new glycyl radical enzyme catalysing the first step in anaerobic toluene metabolism, *Mol. Microbiol.,* 28, 615–628, 1998.
142. Beller, H.R. and Spormann, A.M., Analysis of the novel benzylsuccinate synthase reaction for anaerobic toluene activation based on structural studies of the product, *J. Bacteriol.,* 180, 5454–5457, 1998.
143. Knappe, J., Neugebauer, F.A., Blaschkowski, H.P., and Gänzler, M., Post-translational activation introduces a free radical into pyruvate formate-lyase, *Proc. Natl. Acad. Sci. USA,* 81, 1332–1335, 1984.
144. Knappe, J., Bohnert, E., and Brummer, W., S-Adenosyl-L-methionine, a component of the clastic dissimilation of pyruvate in *Escherichia coli*, *Biochim. Biophys. Acta,* 107, 603–605, 1965.
145. Becker, A., Fritz-Wolf, K., Kabsch, W., Knappe, J., Schultz, S., and Volker Wagner, A.F., Structure and mechanism of the glycyl radical enzyme pyruvate formate-lyase, *Nat. Struct. Biol.,* 6, 969–975, 1999.
146. Wood, N.P. and Jungermann, K., Inactivation of the pyruvate formate lyase of *Clostridium butyricum*, *FEBS Lett.,* 27, 49–52, 1972.
147. Weidner, G. and Sawers, G., Molecular characterization of the genes encoding pyruvate formate-lyase and its activating enzyme of *Clostridium pasteurianum*, *J. Bacteriol.,* 178, 2440–2444, 1996.
148. Hartmanis, M.G. and Stadtman, T.C., Solubilization of a membrane-bound diol dehydratase with retention of EPR g = 2.02 signal by using 2-(N-cyclohexylamino)ethanesulfonic acid buffer, *Proc. Natl. Acad. Sci. USA,* 84, 76–79, 1987.
149. Raynaud, C., Sarçabal, P., Meynial-Salles, I., Croux, C., and Soucaille, P., Molecular characterization of the 1,3-propanediol (1,3-PD) operon of *Clostridium butyricum*, *Proc. Natl. Acad. Sci. USA,* 100, 5010–5015, 2003.

150. Selmer, T. and Andrei, P.I., p-Hydroxyphenylacetate decarboxylase from *Clostridium difficile*. A novel glycyl radical enzyme catalysing the formation of p-cresol, *Eur. J. Biochem.*, 268, 1363–1372, 2001.
151. Buckel, W. and Golding, B.T., Radical species in the catalytic pathways of enzymes from anaerobes, *FEMS Microbiol. Rev.*, 22, 523–541, 1999.
152. Liesert, M., Zschocke, J., Linder, D., and Buckel, W., Substrate assisted catalysis in 3-methylglutaconyl-CoA hydratase from Acinetobacter sp., in preparation.
153. Grabowski, R., Hofmeister, A.E., and Buckel, W., Bacterial L-serine dehydratases: a new family of enzymes containing iron-sulfur clusters, *Trends Biochem. Sci.*, 18, 297–300, 1993.
154. Hofmeister, A.E., Albracht, S.P., and Buckel, W., Iron-sulfur cluster-containing L-serine dehydratase from *Peptostreptococcus asaccharolyticus*: correlation of the cluster type with enzymatic activity, *FEBS Lett.*, 351, 416–418, 1994.
155. Hofmeister, A.E., Textor, S., and Buckel, W., Cloning and expression of the two genes coding for L-serine dehydratase from *Peptostreptococcus asaccharolyticus*: relationship of the iron-sulfur protein to both L-serine dehydratases from *Escherichia coli, J. Bacteriol.*, 179, 4937–4941, 1997.
156. Su, H.S., Lang, B.F., and Newman, E.B., L-Serine degradation in *Escherichia coli* K-12: cloning and sequencing of the sdaA gene, *J. Bacteriol.*, 171, 5095–5102, 1989.
157. Su, H. and Newman, E.B., A novel L-serine deaminase activity in *Escherichia coli* K-12, *J. Bacteriol.*, 173, 2473–2480, 1991.
158. Buckel, W. and Keese, R., One electron redox reactions of CoASH esters in anaerobic bacteria. A mechanistic proposal, *Angew. Chem. Int. Edn. Engl.*, 34, 1502–1506, 1995.
159. Müller, U. and Buckel, W., Activation of (R)-2-hydroxyglutaryl-CoA dehydratase from *Acidaminococcus fermentans*, *Eur. J. Biochem.*, 230, 698–704, 1995.
160. Smith, D.M., Buckel, W., and Zipse, H., Deprotonation of Enoxy Radicals: Theoretical Validation of a 50-Year-Old Mechanistic Proposal, *Angew. Chem. Int. Edn. Engl.*, 42, 1867–1870, 2003.
161. Schweiger, G., Dutscho, R., and Buckel, W., Purification of 2-hydroxyglutaryl-CoA dehydratase from *Acidaminococcus fermentans*. An iron-sulfur protein, *Eur. J. Biochem.*, 169, 441–448, 1987.
162. Dutscho, R., Wohlfarth, G., Buckel, P., and Buckel, W., Cloning and sequencing of the genes of 2-hydoxyglutaryl-CoA dehydratase from *Acidaminococcus fermentans*, *Eur. J. Biochem.*, 181, 741–746, 1989.
163. Hans, M., Buckel, W., and Bill, E., The iron-sulfur clusters in 2-hydroxyglutaryl-CoA dehydratase from *Acidaminococcus fermentans*. Biochemical and spectroscopic investigations, *Eur. J. Biochem.*, 267, 7082–7093, 2000.
164. Hans, M., Bill, E., Cirpus, I., Pierik, A.J., Hetzel, M., Alber, D., and Buckel, W., Adenosine triphosphate-induced electron transfer in 2-hydroxyglutaryl-CoA dehydratase from *Acidaminococcus fermentans*, *Biochemistry*, 41, 5873–5882, 2002.
165. Hans, M., Sievers, J., Müller, U., Bill, E., Vorholt, J.A., Linder, D., and Buckel, W., 2-Hydroxyglutaryl-CoA dehydratase from *Clostridium symbiosum*, *Eur. J. Biochem.*, 265, 404–414, 1999.
166. Klees, A.G., Linder, D., and Buckel, W., 2-Hydroxyglutaryl-CoA dehydratase from *Fusobacterium nucleatum* (subsp. nucleatum): an iron-sulfur flavoprotein, *Arch. Microbiol.*, 158, 294–301, 1992.
167. Locher, K.P., Hans, M., Yeh, A.P., Schmid, B., Buckel, W., and Rees, D.C., Crystal structure of the *Acidaminococcus fermentans* 2-hydroxyglutaryl-CoA dehydratase component A, *J. Mol. Biol.*, 307, 297–308, 2001.
168. Schindelin, H., Kisker, C., Schlessman, J.L., Howard, J.B., and Rees, D.C., Structure of ADP × AIF_4^--stabilized nitrogenase complex and its implications for signal transduction, *Nature*, 387, 370–376, 1997.
169. Thamer, W., Cirpus, I., Hans, M., Pierik, A.J., Selmer, T., Bill, E., Linder, D., and Buckel, W., A two [4Fe-4S]-cluster-containing ferredoxin as an alternative electron donor for 2-hydroxyglutaryl-CoA dehydratase from *Acidaminococcus fermentans*, *Arch. Microbiol.*, 179, 197–204, 2003.
170. Tamannaei, A., Untersuchungen zum Katalysemechnismus der 2-Hydroxyglutaryl-CoA-Dehydratase aus *Fusobacterium nucleatum*, Diploma Thesis, Philipps-Universität Marburg, 2003.
171. Nölling, J., Breton, G., Omelchenko, M.V., Makarova, K.S., Zeng, Q., Gibson, R., Lee, H.M., Dubois, J., Qiu, D., Hitti, J., Wolf, Y.I., Tatusov, R.L., Sabathe, F., Doucette-Stamm, L., Soucaille, P., Daly, M.J., Bennett, G.N., Koonin, E.V., and Smith, D.R., Genome sequence and comparative analysis of the solvent-producing bacterium *Clostridium acetobutylicum*, *J. Bacteriol.*, 183, 4823–4838, 2001.

172. Kim, J., Hetzel, M., Boiangiu, C.D., and Buckel, W., Dehydration of (R)-2-hydroxyacyl-CoA to enoyl-CoA in the fermentation of alpha-amino acids by anaerobic bacteria, *FEMS Microbiol.*, Reviews in press, 2004.
173. Scherf, U. and Buckel, W., Purification and properties of an iron-sulfur and FAD-containing 4-hydroxybutyryl-CoA dehydratase/vinylacetyl-CoA delta 3-delta 2-isomerase from *Clostridium aminobutyricum*, *Eur. J. Biochem.*, 215, 421–429, 1993.
174. Scherf, U., Söhling, B., Gottschalk, G., Linder, D., and Buckel, W., Succinate-ethanol fermentation in *Clostridium kluyveri*: purification and characterisation of 4-hydroxybutyryl-CoA dehydratase/vinylacetyl-CoA delta 3-delta 2-isomerase, *Arch. Microbiol.*, 161, 239–245, 1994.
175. Söhling, B. and Gottschalk, G., Molecular analysis of the anaerobic succinate degradation pathway in *Clostridium kluyveri*, *J. Bacteriol.*, 178, 871–880, 1996.
176. Martins, B.M., Dobbek, H., Çinkaya, I., Buckel, W., and Messerschmidt, A., Crystal structure of 4-hydroxybutyryl-CoA dehydratase: Radical catalysis involving a [4Fe-4S] cluster and flavin, *Proc. Natl. Acad. Sci. U.S.A.*, 101, 15645–15649, 2004.
177. Çinkaya, I., Buckel, W., Medina, M., Gomez-Moreno, C., and Cammack, R., Electron-nuclear double resonance spectroscopy investigation of 4-hydroxybutyryl-CoA dehydratase from *Clostridium aminobutyricum*: comparison with other flavin radical enzymes, *Biol. Chem. Hoppe Seyler*, 378, 843–849, 1997.
178. Näser, U., Pierik, A.J., Scott, R., Çinkaya, I., Golding, B.T., and Buckel, W., Synthesis of ^{13}C-labelled gamma-butyrolactones and EPR study of 4-hydroxybutyryl-CoA dehydratase., in preparation, 2004.
179. Scott, R., Näser, U., Friedrich, P., Çinkaya, I., Buckel, W., and Golding, B.T., Stereospecific elimination of the 3-Si hydrogen of 4-hydroxybutyryl-CoA during catalysis of 4-hydroxybutyryl-CoA dehydratase from *Clostridium aminobutyricum*, *Chem. Com. (London)*, 2004, 1210–1211, 2004.
180. Bucklers, L., Umani-Ronchi, A., Rétey, J., and Arigoni, D., Stereochemistry of the enzymatic dehydration of butyryl-CoA, *Experientia*, 26, 931–933, 1970.
181. Engst, S., Vock, P., Wang, M., Kim, J.J., and Ghisla, S., Mechanism of activation of acyl-CoA substrates by medium chain acyl-CoA dehydrogenase: interaction of the thioester carbonyl with the flavin adenine dinucleotide ribityl side chain, *Biochemistry*, 38, 257–267, 1999.
182. Hardman, J.K. and Stadtman, T.C., Metabolism of omega-acids. II. Fermentation of delta-aminovaleric acid by *Clostridium aminovalericum* n. sp., *J. Bacteriol.*, 79, 549–552, 1960.
183. Barker, H.A., D'Ari, L. and Kahn, J., Enzymatic reactions in the degradation of 5-aminovalerate by *Clostridium aminovalericum*, *J. Biol. Chem.*, 262, 8994–9003, 1987.
184. Eikmanns, U. and Buckel, W., Crystalline green 5-hydroxyvaleryl-CoA dehydratase from *Clostridium aminovalericum*, *Eur. J. Biochem.*, 197, 661–668, 1991.
185. Eikmanns, U. and Buckel, W., A green 2,4-pentadienoyl-CoA reductase from *Clostridium aminovalericum*, *Eur. J. Biochem.*, 198, 263–266, 1991.
186. Buckel, W., The reversible dehydration of (R)-2-hydroxyglutarate to (E)-glutaconate, *Eur. J. Biochem.*, 106, 439–447, 1980.
187. Schellenberger, A., Sixty years of thiamin diphosphate biochemistry, *Biochim. Biophys. Acta*, 1385, 177–186, 1998.
188. Herrmann, G., Selmer, T., Jessen, H.J., Gokarn, R.R., Selifonova, O., Gort, S.J., and Buckel, W., Two beta-alanine ammonia lyases in *Clostridium propionicum*, *FEBS Journal*, in press (2005).
189. Kim, J., Darley, D. and Buckel, W., 2-Hydroxyisocaproyl-CoA dehydratase and its activator from *Clostridium difficile*, *FEBS Journal*, in press (2005).
190. Wagener, N., (2R,3S)-2,3-Dimethylmalate lyase from *Eubacterium barkeri*, Diploma Thesis, Marburg, 2003.
191. Martins, B.M., Macedo-Ribeiro, S., Bresser, J., Buckel, W., and Messerschmidt, A., Structural basis for stereo-specific catalysis in NAD$^+$-dependent (*R*)-2-hydroxyglutarate dehydrogenase from *Acidaminococcus fermentans*, *FEBS Journal*, in press (2005).

10 Degradation of Heterocyclic Compounds

Jan R. Andreesen

CONTENTS

10.1 "CLOSTRIDIA" COVERED IN THIS REVIEW

This short review will cover the metabolism of heterocyclic, mainly N-containing, compounds by bacteria that would belong to the genus *Clostridium* according to the old terms [1], although it has become clear that only species of cluster I can be correctly assigned to the genus *Clostridium* [2]. However, at the time the Biotechnology Handbook on clostridia [3] and the second edition of *The Prokaryotes* [4] were published, this discrimination could not be made due to the lack of 16S rDNA sequence data. Today, many organisms are still associated with their old names, although it has become clear that they should be renamed. Such a reclassification has been done quite cautiously, e.g., for the nicotinic acid-fermenting organism *Clostridium barkeri* of cluster XV, now known as *Eubacterium barkeri* [2]. Also, many members of the former genus *Peptostreptococcus* (*Clostridium* cluster XIII) have already been renamed, such as the purinolytic specialist *Psc. barnesae* [5] to *Gallicola barnesae* [6] or as the related glycine-utilizing *Psc. magnus* and *Psc. micros* to *Finegoldia magna* and *Micromonas micros* [7,8], both showing relationships to purine-utilizing clostridia.

However, some species such as the purinolytic specialists *C. acidiurici*, *C. cylindrosporum*, and *C. purinolyticum* (cluster XII) remain to be renamed, as they still carry the name *"Clostridium,"* although they belong to the same group as, e.g., *Tissierella creatinophila*, a specialist for degradation of creatine and similar compounds [9].

10.2 HETEROCYCLIC COMPOUNDS ARE MORE ABUNDANT THAN THOUGHT

Heterocyclic compounds such as purines and pyrimidines constitute a high percentage of the biomass due to their presence in RNA and DNA. They also play an important role in coenzymes involved in energy metabolism (adenine nucleotides), activation, synthesis or regulation of cellular compounds (coenzyme A, uracil/thymine/cytosine/guanine nucleotides), or in electron transport [NAD(P), FMN, FAD, cytochromes, pyrroloquinoline quinone (PQQ)], and in cofactors such as thiamin, pyridoxal phosphate, riboflavin, biotin, folic acid, and corrinoids. The biosynthesis of the amino acid histidine exhibits with its imidazole ring system strong relationships to that of purines, whereas its degradation by clostridia ends up in most studied cases in a common path with glutamate [10]. The benzopyrrole ring system of the amino acid tryptophan is generally formed via anthranilate, which is also an intermediate in the anaerobic degradation of the indole ring system (in the only studied case of *Desulfobacterium indolicum*). In that degradative path, two hydroxylation steps are followed by hydrolytic cleavage of isatin and a subsequent decarboxylation reaction [11]. However, the degradation of amino acids with a heterocyclic side chain as well as of the imino acid proline are traditionally included in the section dealing with amino acid metabolism.

A totally neglected field comprises the anaerobic transformation of alkaloids, including a vast variety of very different, physiologically highly active heterocyclic compounds [12,13] of which caffeine and nicotine are commonly known. One of the few studied examples represents the toxic compound mimosine, a tyrosine analogue present in a tropical forage legume, which is converted by some anaerobic bacteria in ruminants to 2,3,-dihydroxypyridine [14]. (Iso)Flavonoids are decomposed by anaerobic bacteria of the gut, including a *Clostridium* sp., to polyphenols [15–17].

Despite the fact that heterocyclic compounds comprise a high percentage of industrially produced compounds [18], and especially of the biomass of a prokaryotic cell (RNA and DNA make up to 25% of its dry mass), these compounds do not attract as much attention as sugars [only 8% of the dry weight is present in glycogen, peptidoglycan (murein) and lipopolysaccharides], whereas amino acids dominate, followed by fats (55% as protein and 9% as lipids, respectively) [19]. This might be a reflection of the human mind and of eating habits, preferring polysaccharides (rice, potatoes, pastas, marmelades, etc.) or meat and milk products. On the other hand, single-cell protein of bacteria, yeasts, or algae turned out to become a general flop as a nutritional substitute for conventional food in the 1970s, partly due to its high content of nucleic acids favoring gout and other diseases in experimental animals. Heterocyclic compounds formed during conventional cooking (due to ring closure and dehydration reactions between aldehydes of sugars and amino groups of proteins) contribute extensively to its flavor. Nothing is known about the fate of these compounds as they pass the anaerobic parts of the gut. This again demonstrates a necessity to restart the investigation of heterocyclic compounds by anaerobic bacteria.

Scanning relevant reviews regarding the degradation of heterocyclic compounds, it becomes obvious that most articles deal with their aerobic decomposition, and only a few indicate in their title different redox or anaerobic conditions employed [18,20–23]. The most comprehensive basic review regarding purine and pyrimidine degradation by defined cultures was compiled by Vogels and van der Drift [24], whereas Bollag and Kaiser [21] give the longest list containing the different heterocyclic compounds for which a transformation has been reported under anaerobic conditions.

10.3 INITIAL COMPARATIVE STUDIES USING CLOSTRIDIA

The capacity of many different species of the "clostridia" to metabolize (aromatic) amino acids or pyrimidines has been screened by members of the ARC at Norwich, U.K. [25–29], whereas the degradation of purines by highly specialized clostridial species was initially studied in the laboratories of H.A. Barker [30] and J.C. Rabinowitz [31] at Berkeley, CA, and later continued by our group in Göttingen, Germany [10,23,32,32a]. Anaerobic metabolism of nicotinic acid by *Clostridium barkeri*, now *Eubacterium barkeri* [2], was initially studied in the labs of both, Earl R. and Thressa C. Stadtman [33–35], at NIH, Bethesda, MD, showing the involvement of a B_{12}-coenzyme dependent 2-methyleneglutarate mutase reaction [36], still a main focus to be studied with this organism [37]. Later the involvement of molybdenum and a labile selenium moiety in nicotinic acid hydroxylase [38] and xanthine dehydrogenase [39] became a point of interest as well.

10.4 PYRIMIDINE DEGRADATION

Orotic acid is the general biosynthetic precursor of pyrimidines [40]. Its degradation by clostridia has only been reported for *C. oroticum* [41] and involves a (reversible) NADH-dependent reduction to dihydroorotate via an iron-sulfur cluster plus FAD- and FMN-containing 62 kDa enzyme. Due to these cofactors, it also catalyzes diaphorase activities (summarized in Argyrou et al. [42]). The reversible hydrolytic cleavage of 5,6-dihydroorotate is catalyzed by the homodimeric enzyme dihydroorotase that involves four zinc in its catalytic mechanism and can be partially replaced by Co^{2+} [43–46]. This enzyme has been studied only for very few prokaryotes [47] in contrast to amidases involved in a more general hydrolysis of cyclic imides [48]. The biotechnologically important hydantoinases belong to the latter group and have been found in many species of clostridia and related organisms [49]. The N-carbamoyl-aspartate formed is later hydrolytically cleaved by an ureidosuccinase to ammonia, CO_2, and aspartate, probably without intermediate formation of carbamoylphosphate [24].

Uracil is the decarboxylation product of orotic acid. Therefore, its degradation scheme is quite similar to the latter, except that β-alanine instead of aspartate is the intermediate product, utilized afterward by e.g., *C. propionicum* [50]. A uracil-converting organism was isolated, *C. uracilicum* [51]. However, although uracil was reduced in an NADH-dependent manner, it did not stimulate growth. The original isolate was lost, and new enrichment cultures with uracil yielded strains that resembled *C. glycolicum*. However, these strains lacked a characteristic ability to utilize ethylene glycol, and most used glycerol instead [29]. *C. glycolicum* is known to be mainly saccharolytic, whereas only *C. sporogenes* and closely related strains of *C. botulinum* types A and B metabolize uracil to dihydrouracil from the tested proteolytic species. Extracts of *C. sporogenes* catalyze the same reaction if supplied with NADPH [27]. Cytosine is deaminated to uracil and ammonia before being reduced. The enzyme cytosine deaminase is only formed by prokaryotes and fungi as part of the pyrimidine salvage pathway; thus, being a potential drug target [52]. Thymine and uracil are also reduced by resting cells of *C. sticklandii*, forming first the dihydro form. Dihydrothymine is cleaved to N-carbamoyl-β-aminoisobutyrate (β-ureidoisobutyrate), but no $^{14}CO_2$ is released from the respective C-2-labeled pyrimidine [53]. Therefore, both compounds serve as alternative electron acceptors in Stickland reactions.

As has been emphasized before [24], the reductive degradation of pyrimidines is more often employed than the oxidative pathway that forms barbiturate by introducing a third oxo-group that will be cleaved to malonic acid and urea.

10.5 METABOLISM OF PURINES AND ITS DERIVATIVES

Compared to pyrimidines, purines have attracted more interest, probably due to the fact that uric acid is a four nitrogen-containing excretion product of birds, reptiles, mollusks, and partly of primates, such as humans. Thus, guano of birds is a good fertilizer. However, uric acid precipitation in bone joints causes gout, whereas its presence in blood plasma might be useful as an antioxidant defense against radicals and aging [54]. Some insects store uric acid as a nitrogen source in special fat bodies and rely on its degradation by anaerobic bacteria of the gut, being present in high numbers [5,55]. Recently, uric acid — in its crystal form — has been found to act as a "danger signal" to alert damaged tissue cells if invaded by microorganisms [56].

The poorly soluble guanine is part, as nucleoside, of the iridescent colors in scales of fish and reptiles. Adenine is toxic for many organisms due to its impact on regulatory networks and salvage pathways, in addition to its role of being a potential substrate for xanthine dehydrogenase, thus leading to different highly toxic and insoluble hydroxylated adenines [57,58]. In both cases, the amino function present in adenine and guanine has to be replaced first by the action of specific deaminases, which convert these bases to hypoxanthine and xanthine, respectively. The latter are good substrates for a variety of highly specialized purinolytic bacteria.

In the studies on purinolytic bacteria, a major breakthrough was achieved when the requirement for special trace element ions such as tungstate, molybdate, and selenite — for (better) utilization of purines and its intermediary products, formate and glycine — was recognized [57,59–61]. Uric acid is predominantly fermented to acetate, formate, glycine, CO_2 and ammonia if no selenium is added, or if it is present due to a contamination of chemicals, glassware, or water. In such a case, a cleavage of the imidazole ring to 4,5-diaminouracil is observed [62]. If the mentioned trace elements are no longer limiting, xanthine dehydrogenase, formate dehydrogenase, and glycine reductase will be formed as highly active enzymes. This allows for better growth and changes the former typical product pattern of *C. cylindrosporum*, traditionally separated from *C. acidiurici* by forming, without a selenite supplementation, a much higher amount of formate and glycine besides acetate [30,31], compared to later studies performed in their presence [63]. Probably, these species have a different ability to take up traces of these anions being present as contaminants in chemicals [64]. Subsequently, by supplementing media with these trace elements, so-far-unrecognized anaerobic purinolytic species were isolated and described, such as *C. purinolyticum*, also utilizing adenine [57]; *Eubacterium angustum*, restricted to guanine, xanthine, and uric acid, but growing exceedingly fast with a t_d of 45 min [65] (being related by 16S rDNA to the former mentioned purinolytic clostridia belonging to cluster XII [2]); and *Peptostreptococcus barnesae*, utilizing also adenine and glycine [5] (now reclassified as *Gallicola barnesae* [6]). It could be demonstrated that *Peptococcus glycinophilus* (similar to *Pc. micros*, now *Micromonas micros* [7]) and *Pc. variabilis* (similar to *Pc. magnus*, now *Finegoldia magna* [7]) formed an active glycine reductase after being supplemented with selenite, thus reducing it now directly to acetate and avoiding the previously described transformation via serine and pyruvate as is also shown by studies using radioactive glycine [66]. However, the above-mentioned strains were unable to utilize different purine compounds under the conditions tested [67], although these species belong to the same group of anaerobic cocci as the strict purinolytic *G. barnesae*, at least according to the 16S rDNA homology [6]. This inability was further investigated for a large number of strains and species of pepto(strepto)cocci and anaerobic cocci belonging to different Hare groups [68]. However, only those strains that are now reclassified as *Peptinophilus asaccharolyicus*, *Pp. indolicus,* and *Pp. harei* [6] were found to ferment a variety of purines but not glycine [68].

In extracts of three "classical" strictly purinolytic clostridial species studied [69,70], xanthine is first hydrolytically cleaved via known intermediates to ammonia, CO_2, and formiminoglycine, followed by a transfer of the formyl group to tetrahydrofolate (THF), yielding formyl-THF, ammonia, and glycine. Formyl-THF is transformed to formate and ATP by the reversible reaction of formyl-THF synthetase. Thus, these bacteria are able to conserve energy by a pathway involving no redox reaction up to this step. However, the energetic situation would be much improved by oxidizing formate to CO_2 and by reducing glycine to acetyl phosphate; the latter can be transformed by acetate kinase to acetate and ATP [71–74].

10.5.1 Glycine as an Additional Substrate Due to the Presence of Selenium

Glycine utilization as sole or main substrate by purinolytic clostridia was first obtained for *C. purinolyticum*. Even di- or triglycine and some glycine-containing dipeptides serve as a substrate in contrast to serine and pyruvate [57]. However, if *C. purinolyticum* is grown in larger culture volumes, a requirement for 1 m*M* adenine as sort of a "vitamin" becomes obvious by a significant decrease in growth rate [75]. Similar results were obtained for different strains of *C. acidiurici* and *C. cylindrosporum*. All strains are actually able to grow on glycine; however, all require the presence of a purine compound that differs individually. In general, this purine should be utilized slowly [76]. This could reflect the specialized metabolism of purinolytic organisms that might have no obvious need to synthesize purine compounds *de novo*. This notion might be further supported by the observed general requirement for thiamin as a vitamin. One part of thiamin is commonly synthesized from 4-aminoimidazoleriboside (AIR), which should be formed from 4-aminoimidazole as the branching intermediate during xanthine hydrolysis to formiminoglycine, CO_2, and ammonia by all these specialists [69,70].

The only difference to former degradation schemes for purines (see details in Dürre and Andreesen, 1983 [58]) resides in the direct reduction of glycine by the glycine reductase system [75] containing two selenocysteine-containing proteins, GrdA [72] and GrdB [77,78]. The electrons necessary for this reduction can be delivered by formate dehydrogenase, a selenocysteine- and tungsten-containing enzyme, in most of these anaerobes [60,61,64]. Formate dehydrogenase has been investigated in more detail for *Eubacterium acidaminophilum* [79] and *C. sticklandii* [80] (both organisms belong to clostridial cluster XI). Interestingly, the purinolytic bacteria are able to discriminate specifically between the ions tungstate and molybdate: the former is incorporated into formate dehydrogenase-, but not into xanthine dehydrogenase, a classical molybdoenzyme in *C. acidiurici* and *C. purinolyticum* [63,64,81,82]. Molybdate and tungstate are very similar in their chemical and physical properties, and therefore, the incorporation of molybdate or tungstate into the incorrect enzyme is a danger that has to be prevented somehow in organisms that use both trace elements in their cofactors [83]. So far, the tungstate-binding lipoprotein TupA of the respective ABC transport system from *E. acidaminophilum* (oriented to the outside) is the only studied example for a highly tungstate-specific protein [84]. In contrast, the small cytoplasmic Mop protein does not show such discrimination between tungstate and molybdate [85]. Other components that are responsible for the final selective incorporation into the respective pyranopterin cofactor remain to be identified and characterized.

10.5.2 Xanthine Dehydrogenase and Purine Hydroxylase, Central Enzymes for Purine Interconversions Contain Labile Selenium

The involvement of selenium in promoting xanthine dehydrogenase activity was discovered during studies on formate dehydrogenase [60,61]. Molybdenum, iron-sulfur clusters, and flavins are known redox-active components of xanthine dehydrogenase from all three purinolytic clostridia [57,61,81,82] as well as from *Eubacterium (Clostridium) barkeri* [39,86]. The xanthine dehydrogenase from *C. cylindrosporum* has been purified quite early [87] and contains molybdenum, iron, and FAD. The hydroxylation pattern of different purines (e.g., position 8 is favored), as well as kinetic constants, was of interest and differs from those of the enzyme from milk. The purified homogeneous enzyme from *C. acidiurici* is rather unstable and might partially cleave during purification, resulting in (unusual) five subunits (110, 83, 56/53, and 26 kDa) of the purified enzyme [82]. The determination of cofactors indicates that parts of molybdenum, iron, acid labile sulfur, and FAD are lacking to give a full complement. The content of selenium is very low (0.13 g-atom per dimer), indicating its presence in a labile form distinct from selenocysteine. A labile selenium moiety is also known for the nicotinic acid hydroxylase of *E.(C.) barkeri* [88] and has later been fully substantiated for the latter enzyme [38]. The substrate specificity of the xanthine dehydrogenase of *C. acidiurici* changes during purification [82]. Perhaps an acetaldehyde-oxidizing activity is removed, which could be part of a second enzyme, such as aldehyde oxidases, that is known to be inactive toward xanthine but not toward other purines [89,90].

The presence of a second, different purine-converting enzyme, purine hydroxylase, is substantiated for *C. purinolyticum* [81,91,92], which — like an aldehyde oxidase — is inactive with xanthine but active with other purines. Both enzymes from *C. purinolyticum* differ in the numbers and sizes of subunits [three vs. four: 80, 35, and 16 kDa vs. 54, 42 (if both would be cleavage products, they would add to 96), 34, and 20 kDa] in the N-terminal sequence and sensitivity toward cyanide. Both enzymes contain the same cofactors, including a labile form of selenium. The xanthine dehydrogenase is rather insensitive toward oxygen [81] and its level in activity increases after growth on xanthine or uric acid, whereas that of purine hydroxylase is high with purine or adenine as growth substrate [91]. Thus, the purine interconversions measured before by an optimized HPLC separation [69] using cell extracts [58] of these three clostridial purinolytic species might reflect the action of two similar, but somewhat different, enzymes.

In *C. purinolyticum*, adenine is first deaminated by an adenine deaminase to hypoxanthine before it is oxidized to xanthine, whereas the other two species produce the more insoluble and, thus, dangerous 6-amino-8-hydroxypurine, but neither hypoxanthine nor xanthine [58]. In accord with this, the substrate specificity of both purine-converting enzymes from *C. purinolyticum* exhibit no activity with adenine [81], whereas adenine is a poor substrate for the purified xanthine dehydrogenase of *C. cylindrosporum* [87]. The activity with adenine becomes undetectable during purification of the enzyme from *C. acidiurici* [82]. No natural electron carrier for any of the enzymes described above could be identified so far.

10.5.3 The Natural Electron Carrier of Xanthine Dehydrogenases of Purinolytic Clostridia Is Still Unknown

2,6-Dichlorophenolindophenol (DCPIP) was used as acceptor during purification of the enzymes from *C. cylindrosporum* and *C. purinolyticum*, whereas DCPIP (also in combination with other carriers) is only up to about 10% as effective as methyl viologen. The specific activity obtained for the purified enzyme is the highest in the case of *C. acidiurici*. The enzyme is able to reduce uric acid to xanthine by reduced methyl viologen [82]. Ferredoxin seems to be an ideal electron carrier to couple formate oxidation with uric acid reduction as only observed in crude extracts of *C. cylindrosporum* [93]. However, ferredoxin (from *C. pasteurianum*) and NAD(P) do not function as electron acceptors in cases of the enzymes from *C. acidiurici* and *C. purinolyticum*.

The inability of these natural carriers to obtain electrons from xanthine dehydrogenases seems to be surprising, as oxygen (as in xanthine oxidases) cannot be a natural acceptor for anaerobic bacteria. The three different domains generally found in eukaryotic xanthine oxidase/dehydrogenase or subunits of bacterial enzymes correspond to a small 20 kDa (N-terminal) Fe/S domain, a 40 kDa FAD-binding domain (in the middle), and an 85 kDa (C-terminal) domain carrying the substrate-converting catalytically active molybdopterin moiety [94], which transfers the electrons from the substrate via the molybdopterin and the two 2Fe-2S clusters to the flavins. The electrons at the (partially) reduced flavins interact with pyridine nucleotides, generally NAD, or, if the flavin is exposed at the protein surface, also with oxygen. In case of the enzyme from *C. acidiurici*, some interaction with added free flavins can be observed [82]. However, the low redox potential of the uric acid/xanthine couple of –440 mV [90] would favor ferredoxin. A specialized, unique form of ferredoxin is, perhaps, present in these organisms [31] and might act as the natural carrier.

10.5.4 NADP-Dependent Xanthine Dehydrogenase from *E. barkeri*

The xanthine dehydrogenase of *E.* (*C.*) *barkeri* uses NADP(H) as electron carrier [39], as also reported before for the similar enzyme nicotinic acid hydroxylase of that organism. Nicotinate has served as selective substrate for the isolation of that organism [33,95]. *E. barkeri* was tested to use purines as growth substrates after discovering a selenium requirement for obtaining a very active nicotinic acid hydroxylase [59,96] that exhibits, now, a low side activity with some purines. The effect of selenium is much more pronounced and selective as observed for xanthine dehydrogenase of purinolytic clostridia [82]. Hypoxanthine, xanthine, guanine, and uric acid are utilized by *E. barkeri* in contrast to the related organisms *E. limosum* and *Acetobacterium woodii*. However, no growth occurred with a variety of other pyridine carboxylates and pyrimidines tested. The xanthine dehydrogenase activity is selectively induced by growth of *E. barkeri* on purines [96] and differs significantly from nicotinic acid hydroxylase in its native mass and mobility during electrophoresis after a respective activity stain [39]. The specific acitivity with the natural acceptor NADP is 160 μmoles min^{-1} mg^{-1} protein with xanthine, thus, being much higher than of the nicotinic acid hydroxylase for its substrate nicotinate (18 μmoles min^{-1} mg^{-1}) [38]. The subunit composition of xanthine dehydrogenase is 81, 30, and 17.5 kDa, and the molecular mass of the native enzyme is 530 kDa, indicating a dedecameric structure, also by electron microscopy [86] being clearly different from nicotinic acid hydroxylase (50, 37, 33, and 23 kDa subunits, making up to a 400 kDa protein) [38]. The cofactor content represents with about 18 Fe/S, 2.3 Mo, 1.1 W, 2.8 FAD and 0.95 Se per mol holoenzyme no full complement [39], especially for selenium, which can be removed by cyanide. Reconstitution experiments are successful using selenide and reducing conditions. Thus, selenide might be present analogous to the cyanolyzable sulfur group at the molybdopterin cofactor [97] or as selanylcysteine residue close to the latter, similar to the proposed situation for a CO dehydrogenase [98]. However, later studies of the CO dehydrogenase from *Oligotropha carboxidovorans* now indicate a [CuSMo] cluster at the active site instead of the selanylcysteine group [99].

The amount of cross-reacting material toward xanthine dehydrogenase is much less affected by selenite compared to molybdate deficiency or a control without a supplementation of *E. barkeri*, especially if it is compared to the strong influence of selenite supplementation on the expression of a high xanthine dehydrogenase activity [86]. This effect could be explained by a post-translational incorporation of "selenium" at a late step. The antibodies raised against the xanthine dehydrogenase show confluent as well as unique bands after rocket immunoelectrophoresis, using a mixture of xanthine dehydrogenase with a partially purified nicotinic acid hydroxylase. This points to an immunological relationship of both enzymes.

10.5.5 Purine Utilization by *C. sticklandii*

C. sticklandii is known for its ability to utilize amino acids as electron donors or acceptors [34,100,101]. Glycine and proline are excellent electron acceptors. If cells of *C. sticklandii* are starved by one of these acceptors, NADH fluorescence decreases substantially (if, e.g., xanthine, uric acid, or even adenine is added, indicating a function as electron acceptor [102]). In further studies, a catabolism of purines (1 to 4 m*M*) by cell suspensions is documented. Adenine is first converted to hypoxanthine and some new UV-absorbing compounds. Adenine-8-^{14}C is degraded to ^{14}C-formate and some xanthine [103]. The authors discuss a cleavage in the imidazole ring, as detected later for *C. purinolyticum* in selenium-deprived cells [62].

10.6 Ecological Considerations Regarding Organisms Involved in Purine Degradation

Feces of birds contain about 3 mmol uric acid per 100 g, sufficient to allow growth of all tested purinolytic organisms, even without the addition of the previously mentioned trace elements. About 10% of the total culturable organisms obtained from the feces of chickens is able to utilize uric acid, compared to 2% for human, and none for rat feces. The formation of halos around colonies clearing precipitated uric acid is an easy indication of this capacity [55]. Enrichments for spore-formers led to the isolation of strains belonging to the three known species, *C. acidiurici*, *C. cylindrosporum,* and *C. purinolyticum,* whereas strains of *G. barnesae* were obtained without a preceding pasteurization step, requiring the mentioned trace elements if they were cultured in defined media [5]. When all available strains of clostridial purinolytic species were analyzed, they could unambiguously be assigned to one of these three species before using DNA/DNA homology studies as the "gold standard" [63], now also supported by 16S rDNA homologies. Similar results were obtained previously by Champion and Rabinowitz [31], using immunological methods. *C. cylindrosporum* is validly described [104], although the simple method to differentiate it from *C. acidiurici* by the main products formed or the metal ion requirement (for obtaining a highly active formate dehydrogenase) is no longer valid [63]. Interestingly, all these clostridial strains show a rather high temperature optimum (42 to 45°C), pointing to a natural habitat in birds, which have a higher body temperature than mammals.

An unusual pattern of uric acid utilization was observed for strain 2-41 from feces of birds, which later was shown to be similar to *Propionibacterium acnes*. This strain forms halos around colonies grown on uric acid-containing solid media, but crusts of crystalline uric acid are formed within the colonies [55]. The organism utilizes about half of the uric acid supplied. Its consumption is improved by adding glucose as cosubstrate, preferentially in the stationary phase.

10.7 NICOTINIC ACID METABOLISM BY *E. BARKERI*

This pyridine 3-carboxylate is the only pyridine compound whose utilization by a strict anaerobe has been investigated in detail [21,22,105]. It acts as a selective substrate for *E. (C.) barkeri* [33], a member of clostridial cluster XV [2]. Its transformation to propionate, acetate, CO_2, and ammonia has been reviewed recently [106,107].

The degradation of nicotinate is initiated by formation of 6-hydroxynicotinate via an NADP-dependent enzyme, originally described as an FAD-containing iron-sulfur protein of about 300 kDa, which is very substrate specific except for pyrazine-carboxylate [95]. An NADPH oxidase and NADPH diaphorase activity is also associated with the enzyme. This can be attributed by the cofactors mentioned. However, these unspecific side reactions, the presence of molybdate during isolation of the organism, the already known cofactors, the reported molecular mass, and especially the anaerobically occurring hydroxylation reaction are very similar to xanthine dehydrogenases. These considerations resulted, finally, in the detection of two further constituents.

The addition of 0.1 μM selenite enhances the enzymatic activity 16-fold. The further addition of molybdate adds 10% to that high activity, whereas the presence of tungstate is not inhibitory [59]. Unexpectedly, the obtained highly active preparation exhibited — contrary to the original description — was a side activity with some purines. This observation served as the basis for studying the similar, but distinct, enzyme xanthine dehydrogenase in that organism [39,59,86]. Using ^{75}Se, the radioactivity copurified with nicotinic acid hydroxylase activity, although the specific activity and radioactivity did not increase, pointing to a labile compound [88]. A labile selenium-containing compound was released from these preparations by heat or chaotropic reagent treatment. After alkylation, dialkylselenide seems to be a product. At least, selenocysteine can be excluded as a selenium-carrying moiety. A similar copurification is also noticed for iron, molybdenum, and a pterin cofactor [108].

Oligotropha (Pseudomonas) carboxydovorans also contains a complex enzyme being a molybdenum-containing iron-sulfur flavoprotein with a labile selenium bound to the protein [109,110]. By crystallographic analysis, selenium was first interpreted as a S-selanylcysteine moiety closely associated with the molybdenum cofactor [98]. However, the selenium ligation has been questioned again [99], as discussed above, for xanthine dehydrogenase of *E. barkeri*. The nicotinic acid hydroxylase of *E. barkeri* is reported to be composed of four subunits (50, 33, 30, and 23 kDa) [38,111], in contrast to the three subunits that constitute the above described xanthine dehydrogenase (81, 30, and 17.5 kDa) of the same organism. The N-terminus of the 50 kDa subunit seems to be post-translationally processed, as it does not start with a Met and is unusually small within the class of molybdenum hydroxylases. Perhaps the finding of 50 and 33 kDa subunits in nicotinic acid hydroxylase is due to a cleavage that might not affect activity, a phenomenon that is not unusual for this type of protein [90,94,106]. The final specific activity reported is about one order of magnitude lower than observed for the xanthine dehydrogenase. The presence of a labile selenium moiety is verified by different means. According to EPR studies, it seems to be directly coordinated with the molybdenum of the molybdopterin. Surprisingly, the enzyme is very resistant toward cyanide, even up to 100 m*M*, and cannot be reactivated by e.g., selenide [38,111]. The content of the cofactors is (per 160 kDa) about 1, 1, 7, and 0.55 for FAD, molybdenum, iron, and selenium, respectively. The iron-sulfur centers are of the [2Fe-2S] type as is usual for molybdenum-containing hydroxylases. The molybdopterin is present as a dinucleotide [38].

The product of the nicotinate hydroxylase, 6-hydroxynicotinate, acts as inhibitor of the hydroxylase, but the latter compound is converted by a reductase to 6-oxo-1,4,5,6-tetrahydronicotinate using reduced ferredoxin. This reductase is quite oxygen-labile, sensitive toward iron-chelating agents, and catalyzes a reversible reaction [95]. In contrast to 2-formylglutarate and the successively formed 2-hydroxymethylglutarate, 2-methyleneglutarate is a well-characterized intermediate [35]. The involvement of a cobamide coenzyme in the mutase reaction forming 3-methylitaconate (follwed by an isomerase yielding 2,3-dimethylmaleate) [36] represents a profoundly studied example for B_{12}-coenzymes. These studies became possible after obtaining both enzymes in a homogeneous form [112]. The homolytic cleavage of the carbon-cobalt bond of the B_{12}-coenzyme initiates the formation of a 2-methyleneglutar-4-yl radical that undergoes a "fragmentation-recombination" reaction by which a 2-acrylyl radical detaches from C-3 to become bonded to C-4. After fragmentation, the latter represents the α-carbon of an additionally formed acrylate. After religation, the radical product abstracts the hydrogen from the corrinoid cofactor to give (R)-3-methylitaconate as final product of the 2-methyleneglutarate mutase reaction [37]. Although an "addition-elimination" mechanism like in methylmalonyl-CoA mutase reaction seems to be energetically favorable, the use of deuterated substrates strongly points to the "fragmentation-recombination" mechanism.

10.8 INDIGO FORMATION

Indigo is the only natural source of blue color and has to be solubilized before being applied as a pigment for dyeing. In medieval times, a woat vat was used where leaves of the woad plant *Isatis tinctoria* ("Färberwaid") were fermented at pH 9 (by adding lime or potash) and a temperature around 50°C. This results in a hydrolysis of the plants indoxyl-5-oxofuranone-D-glucoside (indican) to 3-hydroxyindole (indoxyl) and a decoloration of indigo by the low redox potential to the leuco form (concommitant with a noxious odor due to formation of, e.g., dimethylsulfide from the cruciferous plant). The textiles immersed in this "brew" are dyed to the blue color of indigo in the presence of air [113]. The responsible aerotolerant organism for indigo reduction has been isolated and described as *Clostridium isatidis* [114], belonging to cluster I of the clostridia. The aerotolerant *C. carnis* is closely related. The organism utilizes a variety of sugars forming acetic, lactic and formic acid, besides H_2 and CO_2, however, it does not produce indole from tryptophan. It seems that *C. isatidis* reduces indigo only as a side reaction. No further degradation is reported, e.g., of oxindole or isatin, as is the case for *Desulfobacterium indolicum* [11].

10.9 CONVERSION OF SECONDARY PLANT METABOLITES

Mimosine is a toxic amino acid (3,4-dihydroxy-phenylalanine) derived from *Leucaena leucocephala* herbage fed to ruminants affecting the retention of ammonia. 2,3-Dihydroxypyridine and 3-hydroxy-4-pyridone are intermediates in the degradation by a saccharolytic *Clostridium* spec., thus, toxic symptoms do not occur in sheep containing this organism in the rumen [14,115]. Ruminal organisms like *C. sticklandii* and *C. aminophilum* are inhibited in the deaminase reaction by 1-[(E)-2-(2-methyl-4-nitrophenyl)diaz-1-enyl]pyrrolidine-2-carboxylate, whereas diphenyliodonium chloride does not inhibit these hyper-ammonia-producing bacteria [116]. Retention of the amino acids would be beneficial for the ruminants.

Flavonoids represent a high portion within the group of secondary plant metabolites generally formed via the initial chalcon synthase reaction. *Eubacterium ramulus* (*Clostridium* cluster XIVa [2]) is able to split the aromatic ring system of many flavonoids, thereby producing phenolic acids. These transformations require glucose as a primary energy source [117]. *C. orbiscidens* is another important organism in the anaerobic degradation of flavonoids such as quercitin and taxifolin to 3,4-dihydroxyphenylacetic acid, CO_2, and phloroglucinol, whereas the respective glucosides are not utilized by this asaccharolytic organism [17]. Phloroglucinol is anaerobically degraded by the delta-proteobacterium *Pelobacter acidigallici* to three acetates [118]. *C. scidens*, *E. desmolans*, and *Clostridium* strain HGH136 are other anaerobic organisms of the gut able to convert flavonoids and isoflavonoids [15,16,119] as well as steroidal compounds [120].

10.10 CONVERSION OF HETEROCYCLIC COMPOUNDS LIKE TRINITROTOLUENE AND TRINITROTRIAZINE

2,4,6-trinitrotoluene (TNT), hexahydro-1,3,5-trinitro-1,3,5-triazine (RDX), and related compounds are widely used and powerful explosives, similar to the well-known trinitroglycerol and 2,4,6-trinitrophenol (picric acid) [121]. This chemical property is a result of the presence of strong oxidizing groups (e.g.,-NO_2) attached to an oxidizable moiety within the same molecule. The decomposition of the pure compounds requires an initiation but then is highly exothermic, resulting in an explosion. The aerobic bacterial decomposition of such nitro compounds has been studied in detail [122] and exhibits many interesting reactions [123]. However, the anaerobic transformations are rarely studied by pure cultures. Thus, TNT becomes first reduced via a possible monoamino-dinitro-compound to 2,4-diamino-6-nitrotoluene (DANT) followed via the 6-hydroxylamino-derivative (DAHAT) to triaminotoluene (TAT) [124]. These authors suggest that the reduction of DANT to DAHAT can be unspecifically carried out by pyruvate ferredoxin oxidoreductase or CO dehydrogenase/acetyl-CoA synthase of anaerobic or acetogenic bacteria, whereas the reduction of the hydroxylamino group of DAHAT might be catalyzed by a sulfite reductase. At least, such reduction reactions are catalyzed as a side reaction by many enzymes [125]. The amino-compounds are of higher toxicity than the original substrates. Therefore, these derivatives should be fixed during bioremediation into the organic matrix of the soil to become insolubilized [126,127].

The trinitrotriazines, such as RDX or HMX, are anaerobically reduced by *C. bifermentans* and extracts of *C. acetobutylicum* via different routes involving different nitroso compounds and their removal as nitrite [128,129]. *C. bifermentans* is able to remove RDX and its nitroso metabolites completely to unaccountable products, methanol, formaldehyde, CO_2, and N_2O. The proposed pathway for RDX decomposition by *C. acetobutylicum* differs substantially, as triaminotriazine is the final product that can be formed by extracts of that organism.

10.11 OUTLOOK

The enormous chemical variety of secondary metabolites — known to be produced especially by bacteria, plants, and animals — involves oxygen-, sulfur-, and nitrogen-containing compounds that are often cyclized to rings and aromatic heterocyclic structures [12,13,130,131]. All these compounds have to and will be remineralized by prokaryotes. Anaerobic conditions represent the first atmosphere for living cells on earth that nowadays still suits organisms like clostridia and other anaerobes.

In the last years, compounds like saturated and aromatic hydrocarbons, methane, and ammonia, whose degradation was formerly believed to strictly require molecular oxygen for the initial attack, are now known to be converted also in the absence of oxygen [132–134]. The extreme diversity of microorganisms will take care of any biologically produced natural compound to become remineralized. The bacteria represent the highest biomass on earth, but so far only about 1% of the species are described or cultivated by man [135]. Up to now, mostly organisms are known for which "extreme" conditions have been used for their isolation, such as a high substrate concentration, to come up with "visible" numbers. Very recently, the community structure of an extremely specialized, acid- and metal ion-containing environment could be resolved using a metagenomic approach [136]. However, under more natural conditions, such as very low substrate concentrations, symbiosis and cometabolism become important factors [137]. Thus, the rumen and the gut represent important natural places where, e.g., ingested heterocyclic alcaloids — although perhaps "designed" for a protection against feeding — are finally transformed by bacteria to simple organic and inorganic compounds.

ACKNOWLEDGMENTS

The experimental work reported from the author's laboratorium has been supported in part by the former BMBF–Schwerpunkt in Göttingen and the Deutsche Forschungsgemeinschaft. I thank Thomas Brüser for critically reading the manuscript.

REFERENCES

1. Hippe, H., Andreesen, J.R., and Gottschalk, G., The genus Clostridium-nonmedical, in *The Prokaryotes — A Handbook on the Biology of Bacteria: Ecophysiology, Isolation, Identification, Applications,* Balows, A., Trüper, H.G., Dworkin, M., Harder, W., and Schleifer, K.-H., Eds., Springer-Verlag, New York, 1991, 1800–1866.
2. Collins, M.D., Lawson, P.A., Willems, A., Cordoba, J.J., Fernandez-Garayzabal, J., Garcia, P., Cai, J., Hippe, H., and Farrow, J.A.E., The phylogeny of the genus *Clostridium*: Proposal of five new genera and eleven new species combinations, *Int. J. Syst. Bacteriol.,* 44, 812–826, 1994.
3. Minton, N.P. and Clarke, D.J., *Clostridia,* Plenum Press, New York, 1989.
4. Balows, A., Trüper, H.G., Dworkin, M., Harder, W., and Schleifer, K.-H., *The Prokaryotes — A Handbook on the Biology of Bacteria: Ecophysiology, Isolation, Identification, Applications,* Balows, A., Trüper, H.G., Dworkin, M., Harder, W., and Schleifer, K.-H., Eds., Springer-Verlag, New York, 1991.
5. Schiefer-Ullrich, H. and Andreesen, J.R., *Peptostreptococcus barnesae* sp. nov., a gram-positive, anaerobic, obligately purine utilizing coccus from chicken feces, *Arch. Microbiol.,* 143, 26–31, 1985.
6. Ezaki, T., Kawamura, Y., Li, N., Li, Z.-Y., Zhao, L., and Shu, S., Proposal of the genera *Anaerococcus* gen. nov., *Peptoniphilus* gen. nov. and *Gallicola* gen. nov., for members of the genus *Peptostreptococcus, Int. J. Syst. Evol. Microbiol.,* 51, 1521–1528, 2001.
7. Murdoch, D.A., Collins, M.D., Willems, A., Hardie, J.M., Young, K.A., and Magee, J.T., Description of three new species of the genus *Peptostreptococcus* from human clinical specimens: *Peptostreptococcus harei* sp. nov., *Peptostreptococcus ivorii* sp. nov., and *Peptostreptococcus octavius* sp. nov., *Int. J. Syst. Bacteriol.,* 47, 781–787, 1997.
8. Song, Y., Liu, C., McTeague, M., Vu, A., Liu, J.Y., and Finegold, S.M., Rapid identification of Gram-positive anaerobic coccal species originally classified in the genus *Peptostreptococcus* by multiplex PCR assays using genus- and species-specific primers, *Microbiology,* 149, 1719–1727, 2003.
9. Harms, C., Schleicher, A., Collins, M.D., and Andreesen, J.R., *Tissierella creatinophila* sp. nov., a gram-positive, anaerobic, non-spore-forming, creatine-fermenting organism, *Int. J. Syst. Bacteriol.,* 48, 983–993, 1998.

10. Andreesen, J.R., Bahl, H., and Gottschalk, G., Introduction to the physiology and biochemistry of the genus *Clostridium*, in *Clostridia,* Minton, N.P. and Clarke, D.J., Eds., Plenum Press, New York, 1989, 27–62.
11. Johansen, S.S., Licht, D., Arvin, E., Mosbaek, H., and Hansen, A.B., Metabolic pathways of quinoline, indole and their methylated analogs by *Desulfobacterium indolicum* (DSM 3383), *Appl. Microbiol. Biotechnol.,* 47, 292–300, 1997.
12. Luckner, M., *Secondary Metabolism in Microorganism, Plants, and Animals,* VEB Gustav Fischer Verlag, Jena, Germany, 1990.
13. Croteau, R., Kutchan, T.M., and Lewis, N.G., Natural Products (Secondary Metabolites), in *Biochemistry & Molecular Biology of Plants,* Buchanan, B.B., Gruissem, W., Jones, R.L., Eds., American Society of Plant Physiologists, Rockville, MD, 2000, 1250–1318.
14. Dominguez-Bello, M.G. and Stewart, C.S., Degradation of mimosine, 2,3-dihydroxy pyridine and 3-hydroxy-4(1H)-pyridone by bacteria from the rumen of sheep in Venezuela, *FEMS Microbiol. Ecology,* 73, 283–290, 1990.
15. Blaut, M., Schoefer, L., and Braune, A., Transformation of flavonoids by intestinal microorganisms, *Int. J. Vitam. Nutr. Res.,* 73, 79–87, 2003.
16. Hur, H.-G., Beger, R.D., Heinze, T.M., Lay, J.O., Freeman, J.P., Dore, J., and Rafi, F., Isolation of an anaerobic intestinal bacterium capable of cleaving the C-ring of the isoflavonoid daidzein, *Arch. Microbiol.,* 178, 8–12, 2002.
17. Schoefer, L., Mohan, R., Schwiertz, A., Braune, A., and Blaut, M., Anaerobic degradation of flavonoids by *Clostridium orbiscindens, Appl. Environ. Microbiol.,* 69, 5849–5854, 2003.
18. Fetzner, S., Bacterial degradation of pyridine, indole, quinoline, and their derivatives under different redox conditions, *Appl. Microbiol. Biotechnol.,* 49, 237–250, 1998.
19. Neidhardt, F.C., Ingraham, J.L., and Schaechter, M., *Physiology of the Bacterial Cell — A Molecular Approach,* Sinauer Associates Publishers, Sunderland, MA, 1990.
20. Berry, D.F., Francis, A.J., and Bollag, J.-M., Microbial metabolism of homocyclic and heterocyclic aromatic compounds under anaerobic conditions, *Microbiol. Rev.,* 51, 43–59, 1987.
21. Bollag, J.-M. and Kaiser, J.-P., The transformation of heterocyclic aromatic compounds and their derivatives under anaerobic conditions, *Crit. Rev. Environ. Control,* 21, 297–329, 1991.
22. Kaiser, J.-P., Feng, Y., and Bollag, J.-M., Microbial metabolism of pyridine, acridine, and their derivates under aerobic and anaerobic conditions, *Microbiol. Rev.,* 60, 483–498, 1996.
23. Koenig, K. and Andreesen, J.R., Aerober und anaerober Abbau von heterocyclischen, aromatischen Verbindungen durch Bakterien, *BioEngineering,* 8(2), 78–84, 1992.
24. Vogels, G.D. and van der Drift, C., Degradation of purines and pyrimidines by microorganisms, *Bacteriol. Rev.,* 40, 403–468, 1976.
25. Elsden, S.R. and Hilton, M.G., Amino acid utilization patterns in Clostridial taxonomy, *Arch. Microbiol.,* 123, 137–141, 1979.
26. Elsden, S.R., Hilton, M.G., and Waller, J.M., The end products of the metabolism of aromatic amino acids by clostridia, *Arch. Microbiol.,* 107, 283–288, 1976.
27. Hilton, M.G., Mead, G.C., and Elsden, S.R., The metabolism of pyrimidines by proteolytic clostridia, *Arch. Microbiol.,* 102, 145–149, 1975.
28. Mead, G.C., The amino acid-fermenting clostridia, *J. Gen. Microbiol.,* 67, 47–56, 1971.
29. Mead, G.C., Adams, B.W., Hilton, M.G., and Lord, P.G., Isolation and characterization of uracil-degrading clostridia from soil, *J. Appl. Bacteriol.,* 46, 465–472, 1979.
30. Barker, H.A., Fermentations of nitrogenous compounds, in *Bacterial Fermentations,* Barker, H.A., Ed., New York, John Wiley & Sons, 1956, 57–90.
31. Champion, A.B. and Rabinowitz, J.C., Ferredoxin and formyltetrahydrofolate synthetase: comparative studies with *Clostridium acidiurici, Clostridium cylindrosporum,* and newly isolated anaerobic uric acid-fermenting strains, *J. Bacteriol.,* 132, 1003–1020, 1977.
32. Andreesen, J.R., Glycine metabolism in anaerobes, *Antonie van Leeuwenhoek,* 66, 223–237, 1994.

32a. Andreesen, J.R., Glycine reductase mechanism, *Curr. Opin. Biol.,* 8, 454–461, 2004.

33. Stadtman, E.R., Stadtman, T.C., Pastan, I., and Smith, L.D., *Clostridium barkeri* sp. n., *J. Bact.,* 110, 758–760, 1972.
34. Stadtman, T.C., A gold mine of fascinating enzymes: those remarkable, strictly anaerobic bacteria, *Methanococcus vannielii* and *Clostridium sticklandii, J. Biol. Chem.,* 277, 49091–49100, 2002.

35. Tsai, L. and Stadtman, E.R., Anaerobic degradation of nicotinic acid, *Methods Enzymol.*, 18B, 233–249, 1971.
36. Kung, H.F., Cederbaum, S., Tsai, L., and Stadtman, T.C., Nicotinic acid metabolism, V. A cobamide coenzyme-dependent conversion of α-methyleneglutaric acid to dimethylmaleic acid, *Proc. Natl. Acad. Sci, U.S.A.*, 65, 978–984, 1970.
37. Pierik, A.J., Ciceri, D., Bröker, G., Edwards, C.H., McFarlane, W., Winter, J., Buckel, W., and Golding, B.T., Rotation of the exo-methylene Group of (R)-3-Methylitaconate catalyzed by coenzyme B_{12}-dependent 2-methyleneglutarate mutase from *Eubacterium barkeri, J. Am. Chem. Soc.*, 124, 14039–14048, 2002.
38. Gladyshev, V.N., Khangulov, S.V., and Stadtman, T.C., Properties of the selenium- and molybdenum-containing nicotinic acid hydroxylase from *Clostridium barkeri*, *Biochem.*, 35, 212–223, 1996.
39. Schräder, T., Rienhöfer, A., and Andreesen, J.R., Selenium-containing xanthine dehydrogenase from *Eubacterium barkeri*, *Eur. J. Biochem.*, 264, 862–871, 1999.
40. Wasternack, C., Degradation of pyrimidines — enzymes, localization and role in metabolism, *Biochem. Physiol. Pflanzen.*, 173, 467–499, 1978.
41. Wachsman, J.T. and Barker, H.A., Characterization of an orotic acid fermenting bacterium, *Zymobacterium oroticum*, nov. gen., nov. spec., *J. Bacteriol.*, 68, 400–404, 1954.
42. Argyrou, A., Washabaugh, M.W., and Pickart, C.M., Dihydroorotate dehydrogenase from *Clostridium oroticum* is a class 1B enzyme and utilizes a concerted mechanism of catalysis, *Biochem.*, 39, 10373–10384, 2000.
43. Lipscomb, W.N. and Sträter, N., Recent advances in zinc enzymology, *Chem. Rev.*, 96, 2375–2433, 1996.
44. Pettigrew, D.W., Bidigare, R.R., Mehta, B.J., Williams, M.I., and Sander, E.G., Dihydro-orotase from *Clostridium oroticum*. Purification and reversible removal of essential zinc., *Biochem. J.*, 230, 101–108, 1985.
45. Pettigrew, D.W., Mehta, B.J., Bidigare, R.R., Choudhury, R.R., Scheffler, J.E., and Sander, E.G., Enzyme elements involved in the interconversion of L-carbamylaspartate and L-dihydroorotate by dihydroorotase from *Clostridium oroticum*, *Arch. Biochem. Biophys.*, 243, 447–453, 1985.
46. Taylor, W.H., Taylor, M.L., Balch, W.E., and Gilchrist, P.S., Purification and properties of dihydro-orotase, a zinc-containing metalloenzyme in *Clostridium oroticum*, *J. Bacteriol.*, 127, 863–873, 1976.
47. Ogawa, J. and Shimizu, S., Purification and characterization of dihydroorotase from *Pseudomonas putida*, *Arch. Microbiol.*, 164, 353–357, 1995.
48. Soong, C.-L., Ogawa, J., and Shimizu, S., A novel amidase (half-amidase) for half-amide hydrolysis involved in the bacterial metabolism of cyclic imides, *Appl. Microbiol. Biotechnol.*, 66, 1947–1952, 2000.
49. Morin, A., Touzel, J.-P., Lafond, A., and Leblanc, D., D-hydantoinase from anaerobic microorganisms, *Appl. Microbiol. Biotechnol.*, 35, 536–540, 1991.
50. Cardon, B.P. and Barker, H.A., Amino acid fermentations by *Clostridium propionicum* and *Diplococcus glycinophilus*, *Arch. Biochem.*, 12, 165–180, 1947.
51. Campbell, L.L., Reductive degradation of pyrimidines. I. The isolation and characterization of a uracil fermenting bacterium, *Clostridium uracilicum* nov. spec., *J. Bacteriol.*, 73, 220–224, 1957.
52. Ireton, G.H., Black, M.E., and Stoddard, B.L., The 1.14 A crystal structure of yeast cytosine deaminase: evolution of nucleotide salvage enzymes and implications for genetic chemotherapy, *Structure (Camb.)*, 11, 961–972, 2003.
53. Schäfer, R. and Schwartz, A.C., Degradation of pyrimidine bases in *Clostridium sticklandii*, *Arch. Microbiol.*, 124, 111–114, 1980.
54. Ames, B.N., Cathcart, R., Schwiers, E., and Hochstein, P., Uric acid provides an antioxidant defense in humans against oxidant- and radical-caused aging and cancer: a hypothesis, *Proc. Natl. Acad. Sci. U.S.A.*, 78, 6858–6862, 1981.
55. Schiefer-Ullrich, H., Harnsäureabbau im Hühnerdarm durch anaerobe Bakterien, Ph.D. Thesis, Georg-August-Universität, Göttingen, Germany, 1984.
56. Heath, W.R. and Carbone, F.R., Dangerous liaisons, *Nature*, 425, 460–461, 2003.
57. Dürre, P., Andersch, W., and Andreesen, J.R., Isolation and characterization of an adenine-utilizing, anaerobic sporeformer, *Clostridium purinolyticum* sp. nov., *Int. J. Syst. Bacteriol.*, 31, 184–194, 1981.
58. Dürre, P. and Andreesen, J.R., Purine and glycine metabolism by purinolytic clostridia, *J. Bacteriol.*, 154, 192–199, 1983.

59. Imhoff, D. and Andreesen, J.R., Nicotinic acid hydroxylase from *Clostridium barkeri*: Selenium-dependent formation of active enzyme, *FEMS Microbiol. Lett.,* 5, 155–158, 1979.
60. Wagner, R. and Andreesen, J.R., Differentiation between *Clostridium acidiurici* and *Clostridium cylindrosporum* on the basis of specific metal requirements for formate dehydrogenase formation, *Arch. Microbiol.,* 114, 219–224, 1977.
61. Wagner, R. and Andreesen, J.R., Selenium requirement for active xanthine dehydrogenase from *Clostridium acidiurici* and *Clostridium cylindrosporum*, *Arch. Microbiol.,* 121, 255–260, 1979.
62. Dürre, P. and Andreesen, J.R., Anaerobic degradation of uric acid via pyrimidine derivatives by selenium-starved cells of *Clostridium purinolyticum*, *Arch. Microbiol.,* 131, 255–260, 1982.
63. Schiefer-Ullrich, H., Wagner, R., Dürre, P., and Andreesen, J.R., Comparative studies on physiology and taxonomy of obligately purinolytic clostridia, *Arch. Microbiol.,* 138, 345–353, 1984.
64. Wagner, R. and Andreesen, J.R., Accumulation and incorporation of ^{185}W-tungsten into proteins of *Clostridium acidiurici* and *Clostridium cylindrosporum*, *Arch. Microbiol.,* 147, 295–299, 1987.
65. Beuscher, H.U. and Andreesen, J.R., *Eubacterium angustum* sp. nov., a Gram-positive anaerobic, non-sporeforming, obligate purine fermenting organism, *Arch. Microbiol.,* 140, 2–8, 1984.
66. Dürre, P., Spahr, R., and Andreesen, J.R., Glycine fermentation via a glycine reductase in *Peptococcus glycinophilus* and *Peptococcus magnus*, *Arch. Microbiol.,* 134, 127–135, 1983.
67. Dürre, P., Selenabhängige Vergärung von Purinen und Glycin durch *Clostridium purinolyticum* und *Peptococcus glycinophilus*, Ph.D. Thesis, Georg-August-Universität, Göttingen, Germany, 1981.
68. Tziaka, C., Untersuchungen zur Charakterisierung von Stämmen der Gattungen Peptococcus und Peptostreptococcus, Ph.D. Thesis, Georg-August-Universität, Göttingen, Germany, 1987.
69. Dürre, P. and Andreesen, J.R., Separation and quantitation of purines and their anaerobic and aerobic degradation products by high-pressure liquid chromatography, *Anal. Biochem.,* 123, 32–40, 1982.
70. Rabinowitz, J.C., Intermediates in purine breakdown, *Meth. Enzymol.,* 6, 703–713, 1963.
71. Arkowitz, R.A. and Abeles, R.H., Mechanism of action of clostridial glycine reductase: isolation and characterization of a covalent acetyl enzyme intermediate, *Biochem.,* 30, 4090–4097, 1991.
72. Garcia, G.E. and Stadtman, T.C., Selenoprotein A component of the glycine reductase complex from *Clostridium purinolyticum*: nucleotide sequence of the gene shows that selenocysteine is encoded by UGA, *J. Bacteriol.,* 173, 2093–2098, 1991.
73. Kohlstock, U.-M., Rücknagel, K.P., Reuter, M., Schierhorn, A., Andreesen, J.R., and Söhling, B., Cys359 of GrdD is the active-site thiol that catalyses the final step of acetyl phosphate formation by glycine reductase from *Eubacterium acidaminophilum*, *Eur. J. Biochem.,* 268, 6417–6425, 2001.
74. Schräder, T. and Andreesen, J.R., Purification and characterization of parotein P_C, a component of glycine reductase from *Eubacterium acidaminophilum*, *Eur. J. Biochem.,* 206, 79–85, 1992.
75. Dürre, P. and Andreesen, J.R., Selenium-dependent growth and glycine fermentation by *Clostridium purinolyticum*, *J. Gen. Microbiol.,* 128, 1457–1466, 1982.
76. Lebertz, H., Selenabhängiger Glycin-Stoffwechsel bei anaeroben Bakterien und vergleichende Untersuchungen zur Glycin-Reduktase und zur Glycin-Decarboxylase, Ph.D. Thesis, Georg-August-Universität, Göttingen, Germany, 1984.
77. Kreimer, S. and Andreesen, J.R., Glycine reductase of *Clostridium litorale* — cloning, sequencing, and molecular analysis of the grdAB operon that contains two in-frame TGA codons for selenium incorporation, *Eur. J. Biochem.,* 234, 192–199, 1995.
78. Wagner, M., Sonntag, D., Grimm, R., Pich, A., Eckerskorn, C., Söhling, B., and Andreesen, J.R., Substrate-specific selenoprotein B of glycine reductase from *Eubacterium acidaminophilum*. Biochemical and molecular analysis, *Eur. J. Biochem.,* 260 (1), 38–49, 1999.
79. Graentzdoerffer, A., Rauh, D., Pich, A., and Andreesen, J.R., Molecular and biochemical characterization of two tungsten- and selenium-containing formate dehydrogenases from *Eubacterium acidaminophilum* that are associated with components of an iron-only hydrogenase, *Arch. Microbiol.,* 179, 116–130, 2003.
80. Gräntzdörffer, A., Formiat-Stoffwechsel in *Eubacterium acidaminophilum*: Molekulare und biochemische Charakterisierung der Wolfram- und Selen-haltigen Formiat-Dehydrogenasen sowie einer Eisen-Hydrogenase, Ph.D. Thesis, Martin-Luther-Universität, Halle, Germany, 2000.
81. Self, W.T. and Stadtman, T.C., Selenium-dependent metabolism of purines: a selenium-dependent purine hydroxylase and xanthine dehydrogenase were purified from *Clostridium purinolyticum* and characterized, *Proc. Natl. Acad. Sci. U.S.A.,* 97, 7208–7213, 2000.

82. Wagner, R., Cammack, R., and Andreesen, J.R., Purification and characterization of xanthine dehydrogenase from *Clostridium acidiurici* grown in the presence of selenium, *Biochim. Biophys. Acta*, 791, 63–74, 1984.
83. Pau, R.N. and Lawson, P.A., Transport, homeostasis, regulation, and binding of molybdate and tungstate to proteins, in *Metal Ions in Biological Systems,* Sigel, A. and Sigel, H., Eds., Marcel Dekker, New York, 2002, 31–74.
84. Makdessi, K., Andreesen, J.R., and Pich, A., Tungstate uptake by a highly specific ABC transporter in *Eubacterium acidaminophilum, J. Biol. Chem.,* 276, 24557–24564, 2001.
85. Makdessi, K., Fritsche, K., Pich, A., and Andreesen, J.R., Identification and characterization of the cytoplasmid tungstate/molybdate-binding protein (Mop) from *Eubacterium acidaminophilum, Arch. Microbiol.,* 181, 45–51, 2004.
86. Rienhöfer, A., Strukturelle und immunologische Untersuchungen zur Xanthin-Dehydrogenase aus Butyribacterium barkeri, ein Seleno-Molybdo-Eisen-Schwefel-Flavoprotein, Ph.D. Thesis, Georg-August-Universität, Göttingen, Germany, 1985.
87. Bradshaw, W.H. and Barker, H.A., Purification and properties of xanthine dehydrogenase from *Clostridium cylindrosporum, J. Biol. Chem.,* 235, 3620–3629, 1960.
88. Dilworth, G.L., Properties of the selenium-containing moiety of nicotinic acid hydroxylase from *Clostridium barkeri, Arch. Biochem. Biophys.,* 219, 30–38, 1982.
89. Krenitsky, T.A., Neil, S.M., Elion, G.B., and Hitchings, G.H., A comparison of the specificities of xanthine oxidase and aldehyde oxidase, *Arch. Biochem. Biophys.,* 150, 585–599, 1972.
90. Coughlan, M.P., Aldehyde oxidase, xanthine oxidase and xanthine dehydrogenase: hydroxylases containing molybdenum, iron-sulfur and flavin, in *Molybdenum and Molybdenum-Containing Enzymes,* Coughlan, M.P., Ed., Pergamon Press, Oxford, 1980, 119–185.
91. Self, W.T., Regulation of purine hydroxylase and xanthine dehydrogenase from *Clostridium purinolyticum* in response to purines, selenium, and molybdenum, *J. Bacteriol.,* 184, 2039–2044, 2002.
92. Self, W.T., Selenium-dependent enzymes from clostridia, in *Biochemistry and Physiology of Anaerobic Bacteria,* Ljungdahl, L.G., Adams, M.W., Barton, L.L., Ferry, J.G., and Johnson, M.K., Eds., Springer, New York, 2003, 157–170.
93. Bradshaw, W.H. and Reeder, D.J., Ferredoxin coupling of formate oxidation to urate reduction in extracts of *Clostridium cylindrosporum, Bacteriol. Proc.,* Abstract P 124, 110, 1964.
94. Pai, E.F. and Nishino, T., The molybdenum-containing xanthine oxidoreductases and picolinate dehydrogenases, in *Metal Ions in Biological Systems,* Sigel, A. and Sigel, H., Eds., Marcel Dekker, New York, 2002, 431–454.
95. Holcenberg, J.S. and Stadtman, E.R., Nicotinic acid metabolism. III. Purification and properties of a nicotinic acid hydroxylase, *J. Biol. Chem.,* 244, 1194–1203, 1969.
96. Imhoff, D., *Clostridium barkeri*: Selenabhängiger Nikotinsäure- und Purinabbau sowie taxonomische Neueinordnung, Ph.D. Thesis, Georg-August-Universiät, Göttingen, Germany, 1981.
97. Hille, R., Molybdenum and tungsten in biology, *Trends Biochem. Sci.,* 27, 360–367, 2002.
98. Dobbek, H., Gremer, L., Meyer, O., and Huber, R., Crystal structure and mechanism of CO dehydrogenase, a molybdo iron-sulfur flavoprotein containing S-selanylcysteine, *Proc. Natl. Acad. Sci. U.S.A.,* 96, 8884–8889, 1999.
99. Gnida, M., Ferner, R., Gremer, L., Meyer, O., and Meyer-Klaucke, W., A novel binuclear (CuSMo) cluster at the active site of carbon monoxide dehydrogenase: Characterization by X-ray absorption spectroscopy, *Biochem.,* 42, 222–230, 2003.
100. Graentzdoerffer, A., Pich, A., and Andreesen, J.R., Molecular analysis of the grd operon coding for genes of the glycine reductase and of the thioredoxin system from *Clostridium sticklandii, Arch. Microbiol.,* 175, 8–18, 2001.
101. Kabisch, U.C., Graentzdoerffer, A., Schierhorn, A., Ruecknagel, K.P., Andreesen, J.R., and Pich, A., Identification of D-proline reductase from *Clostridium sticklandii* as a selenoenzyme and indications for a catalytically active pyruvoyl group derived from a cysteine residue by cleavage of a proprotein, *J. Biol. Chem.,* 274, 8445–8454, 1999.
102. Schwartz, A.C. and Schäfer, R., New amino acids, and heterocyclic compounds participating in the Stickland reaction of *Clostridium sticklandii, Arch. Microbiol.,* 93, 267–276, 1973.
103. Schäfer, R. and Schwartz, A.C., Catabolism of purines in *Clostridium sticklandii, Zbl. Bakt. Hyg.,* I. Abt. Orig. A., 235, 165–172, 1976.

104. Andreesen, J.R., Zindel, U., and Dürre, P., *Clostridium cylindrosporum* (ex Barker and Beck 1942) nom. rev., *Int. J. Syst. Bacteriol.,* 35, 206–208, 1985.
105. Eggerer, H., Completion of the degradation scheme for nicotinic acid by *Clostridium barkeri*, *Curr. Top. Cell. Reg.,* 26, 411–418, 1985.
106. Andreesen, J.R. and Fetzner, S., The molybdenum-containing hydroxylases of nicotinate, isonicotinate, and nicotine, in *Metal Ions in Biological Systems,* Sigel, A. and Sigel, H., Eds., Marcel Dekker, New York, 2002, 405–430.
107. Buckel, W., Anaerobic energy metabolism, in *Biology of the Prokaryotes,* Lengeler, J.W., Drews, G., and Schlegel, H.G., Eds., Thieme, Stuttgart, Germany, 1999, 278–326.
108. Dilworth, G.L., Occurrence of molybdenum in the nicotinic acid hydroxylase from *Clostridium barkeri*, *Arch. Biochem. Biophys.,* 221, 565–569, 1983.
109. Meyer, O., Chemical and spectral properties of carbon monoxide: Methylene blue oxidoreductase, *J. Biol. Chem.,* 257, 1333–1341, 1982.
110. Meyer, O. and Rajagopalan, K.V., Selenite binding to carbon monoxide oxidase from *Pseudomonas carboxydovorans*. Selenium binds covalently to the protein and activates specifically the CO-methylene blue reaction., *J. Biol. Chem.,* 259, 5612–5617, 1984.
111. Gladyshev, V.N., Khangulov, S.V., and Stadtman, T.C., Nicotinic acid hydroxylase from *Clostridium barkeri*: electron paramagnetic resonance studies show that selenium is coordinated with molybdenum in the catalytically active selenium-dependent enzyme, *Proc. Natl. Acad. Sci. U.S.A.,* 91, 232–236, 1994.
112. Michel, C., Hartrampf, G., and Buckel, W., Assay and purification of the adenosylcobalamin-dependent 2-methyleneglutarate mutase from *Clostridium barkeri*, *Eur. J. Biochem.,* 184, 103–107, 1989.
113. Padden, A.N., Dillon, V.M., John, P., Edmonds, J., Collins, M.D., and Alvarez, N., Clostridium used in mediaeval dyeing, *Nature,* 396, 225, 1998.
114. Padden, A.N., Dillon, V.M., Edmonds, J., Collins, M.D., Alvarez, N., and John, P., An indigo-reducing moderate thermophile from a woad vat, *Clostridium isatidis* sp. nov., *Int. J. Syst. Bacteriol.,* 49, 1025–1031, 1999.
115. Dominguez-Bello, M.G. and Stewart, C.S., Characteristics of a rumen Clostridium capable of degrading mimosine, 3(OH)-4-(1H)-pyridone and 2,3 dihydroxypyridine, System. *Appl. Microbiol.,* 14, 67–71, 1991.
116. Floret, F., Chaudhary, L.C., Ellis, W.C., El Hassan, S., McKain, N., Newbold, C. J., and Wallace, R.J., Influence of 1[(E)-2-(2-methyl-4-nitrophenyl)diaz-1-enyl] pyrrolidine-2-carboxylic acid and diphenyliodonium chloride on ruminal protein metabolism and ruminal microorganisms, *Appl. Environ. Microbiol.,* 65, 3258–3260, 1999.
117. Schneider, H. and Blaut, M., Anaerobic degradation of flavonoids by *Eubacterium ramulus*, *Arch. Microbiol.,* 173, 71–75, 2000.
118. Brune, A. and Schink, B., Phloroglucinol pathway in the strictly anaerobic *Pelobacter acidigallici*: fermentation of trihydroxybenzenes to acetate via triacetic acid, *Arch. Microbiol.,* 157, 417–424, 1992.
119. Winter, J., Moore, L.H., Dowell, V.R., and Bokkenheuser, V.D., C-ring cleavage of flavonoids by human intestinal bacteria, *Appl. Environ. Microbiol.,* 55, 1203–1208, 1989.
120. Baron, S.F. and Hylemon, P.B., Biotransformation of bile acids, cholesterol, and steroid hormones, in *Gastrointestinal Microbiology,* Macki, R.I. and White, B.A., Eds., Chapman & Hall, New York, 1997, 470–510.
121. Gorontzy, T., Drzyzga, O., Kahl, M.W., Bruns-Nagel, D., Breitung, J., von Loew, E., and Blotevogel, K.-H., Microbial degradation of explosives and related compounds, *Crit. Rev. Microbiol.,* 20, 265–284, 1994.
122. Johnson, G.R. and Spain, J.C., Evolution of catabolic pathways for synthetic compounds: bacterial pathways for degradation of 2,4-dinitrotoluene and nitrobenzene, *Appl. Microbiol. Biotechnol.,* 62, 110–123, 2003.
123. Nadeau, L.J., He, Z., and Spain, J.C., Bacterial conversion of hydroxylamino aromatic compounds by both lyase and mutase enzymes involves intramolecular transfer of hydroxyl groups, *Appl. Environ. Microbiol.,* 69, 2786–2793, 2003.
124. Preuss, A., Fimpel, J., and Diekert, G., Anaerobic transformation of 2,4,6-trinitrotoluene (TNT), *Arch. Microbiol.,* 159, 345–353, 1993.

125. Watrous, M.M., Clark, S., Kutty, R., Huang, S., Rudolph, F.B., Hughes, J.B., and Bennet, G.N., 2,4,6-Trinitrotoluene reduction by an Fe-only hydrogenase in *Clostridium acetobutylicum*, *Appl. Environ. Microbiol.,* 69, 1542–1547, 2003.
126. Drzyzga, O., Bruns-Nagel, D., Gorontzy, T., Blotevogel, K.-H., and von Löw, E., Anaerobic incorporation of the radiolabeled explosive TNT and metabolites into the organic soil matrix of contaminted soil after different treatment procedures, *Chemosphere,* 38, 2081–2095, 1999.
127. Hawari, J., Beaudet, S., Halasz, A., Thiboutot, S., and Ampleman, G., Microbial degradation of explosives: biotransformation versus mineralization, *Appl. Microbiol. Biotechnol.,* 54, 605–618, 2000.
128. Zhang, C. and Hughes, J.B., Biodegradation pathways of hexahydro-1,3,5-trinitro-1,3,5-triazine (RDX) by *Clostridium acetobutylicum* cell-free extract, *Chemosphere,* 50, 665–671, 2003.
129. Zhao, J.-S., Paquet, L., Halasz, A., and Hawari, J., Metabolism of hexahydro-1,3,5-trinitro-1,3,5-triazine through initial reduction to hexahydro-1-nitroso-3,5-dinitro-1,3,5-triazine followed by denitration in *Clostridium bifermentans* HAW-1, *Appl. Microbiol. Biotechnol.,* 63, 187–193, 2003.
130. Crosa, H.J. and Walsh, C.T., Genetics and assembly line enzymology of siderophore biosynthesis in bacteria, *Microbiol. Mol. Biol. Rev.,* 66, 223–249, 2002.
131. Kwon, H.-J., Smith, W.C., Scharon, A.J., Hwang, S.H., Kurth, M.J., and Shen, B., C - O bond formation by polyketide synthases, *Science,* 297, 1327–1330, 2002.
132. Jetten, M.S.M., Sliekers, O., Kuypers, M., Dalsgaard, T., van Niftrik, L., Cirpus, I., van de Pas-Schoonen, K., Lavik, G., Thramdrup, B., Le Paslier, D., Op den Camp, H.J.M., Hulth, S., Nielsen, L.P., Abma, W., Third, K., Engström, P., Kuenen, J.G., Jorgensen, B.B., Canfield, D.E., Sinninghe Damsté, J.S., Revsbech, N.P., Fuerst, J., Weissenbach, J., Wagner, M., Schmidt, I., Schmid, M., and Strous, M., Anaerobic ammonium oxidation by marine and freshwater planctomycete-like bacteria, *Appl. Microbiol. Biotechnol.,* 63, 107–114, 2003.
133. Nauhaus, K., Boetius, A., Krüger, M., and Widdel, F., *In vitro* demonstration of anaerobic oxidation of methane coupled to sulphate reduction in sediment from a marine gas hydrate area, *Environ. Microbiol.,* 4, 296–305, 2002.
134. Widdel, F. and Rabus, R., Anaerobic biodegradation of saturated and aromatic hydrocarbons, *Curr. Opinion Biotechnol.,* 12, 259–276, 2001.
135. Schleifer, K.-H., Microbial diversity: Facts, problems and prospects, *System. Appl. Microbiol.,* 27, 3–9, 2004.
136. Tyson, G.W., Chapmann, J., Hugenholtz, P., Allen, E.E., Ram, R.J., Richardson, P.M., Solovyev, V.V., Rubin, E.M., Rokhsar, D.S., and Banfield, J.F., Community structure and metabolism through reconstruction of microbial genomes from the environment, *Nature,* 428, 37–43, 2004.
137. Schink, B., Energetics of syntrophic cooperation in methanogenic degradation, *Microbiol. Mol. Biol. Rev.,* 61, 262–280, 1997.

11 Nitrogen Assimilation in Clostridia

Sharon J. Reid and Helen E. Stutz

CONTENTS

11.1 INTRODUCTION

The industrial process using solvent-producing *Clostridium* species traditionally has been a batch process, characterized by the initial growth or acidogenic phase, followed by the stationary or solventogenic growth phase, which culminates ultimately in sporulation [1]. The regulation and factors involved in triggering the transitions from the early phase to the solventogenic phase, as well as the factors that affect sporulation, are complex and have been the subject of much research [1,2]. When molasses, which is deficient in nitrogen, replaced the historical corn mash, the industrial process became more complex because nitrogenous compounds had to be added to the sugar-based medium [3]. Solvent yield was affected by the nitrogen source; however, the effect is difficult to analyze in a complex medium. The use of soluble, chemically defined media for laboratory studies is complicated by pH control issues because the buffering capacity of the medium is significantly reduced when ammonia salts or amino acids replace less metabolizable proteins and peptides.

Varying results have been reported on the effect of nitrogen (ammonia) limitation on solvent production, and this has been further complicated by the reclassification of a number of strains formerly all classified as *Clostridium acetobutylicum* [4]. In a chemically defined medium, under conditions of high glucose and low ammonia concentration, *C. acetobutylicum* ATCC 824 produced a high concentration of solvents at pH 5.0 [5,6]. Similarly, Andersch et al. [7] found that strain DSM1731 produced both acids and solvents at low pH under ammonia limitation. The ratio between the carbon and nitrogen substrates, as well as the culture pH, has been reported to have a significant effect on solvent productivity [8]. On the other hand, *C. beijerinckii* NCIMB 8052 and the closely related *C. saccharobutylicum* NCP262 (formerly *C. acetobutylicum* NCP262) did not produce solvents in ammonia-limited cultures [9,10]. Under these nitrogen-limited conditions, the motile rod-shaped NCP262 cells did not differentiate into clostridial stage cells, and sporulation did not occur. This is consistent with the finding that both *C. acetobutylicum* and *C. beijerinckii* require partially hydrolyzed proteins or amino acids for solvent production, presumably reflecting a preference for these amino acids as nitrogen sources [3,9]. Thus, a number of variables may combine to influence the effectiveness of solvent formation under ammonia-limited conditions by the saccharolytic clostridia, and a systematic analysis of nitrogen assimilation would facilitate the understanding of these pathways.

11.2 NITROGEN ASSIMILATION PATHWAYS

Bacteria can utilize a wide variety of nitrogen compounds as sources of cellular nitrogen. These range from simple inorganic compounds, such as diatomic nitrogen and nitrate, to more complex compounds including urea, amino acids, amino sugars, and nucleosides. Endogenous ammonia is an absolute requirement for the biosynthesis of at least some essential nitrogen-containing compounds. Thus, the growth rate of bacteria on different nitrogen sources is generally limited by the rate at which ammonia ions can be generated from them. It is not surprising, then, that in the *Enterobacteriaceae* and many other bacteria, ammonia supports the fastest cell growth rate. However, this is not always the case, and in *Bacillus subtilis* and *Corynebacterium callunae,* glutamine and glutamate are the preferred nitrogen sources, respectively [11,12]. The expression of the metabolic pathways involved in nitrogen metabolism are generally regulated in response to nitrogen availability. These include pathways required for nitrogen fixation, for the transport, catabolism, and conversion of nitrogenous compounds to ammonium ions and other components, and for the assimilation of ammonia. Under conditions of nitrogen excess, the cells generally utilize the preferred nitrogen source and repress the pathways involved in the utilization of alternative nitrogen sources. However, when ammonium ions become limiting, these alternative routes can be activated. The combination of pathways available, and the regulatory systems controlling the preferential use of nitrogen compounds, depends on the particular organism. Such a system of nitrogen control has been studied most comprehensively in the enteric bacteria, although significant progress is being made in other bacterial groups.

In most bacteria, the major route for the assimilation of ammonium ions is via the amino acids glutamine and glutamate. These two compounds are key metabolites in bacterial metabolism, serving as principal precursors for nearly all other nitrogen-containing compounds. Two major pathways control the biosynthesis of these two amino acids, the most important being the energy dependent glutamine synthetase-glutamate synthase (GS-GOGAT) pathway, which is ubiquitous in bacteria. The enzyme glutamine synthetase (GS: EC 6.3.1.2) catalyzes the amidation of endogenous glutamate to form glutamine (reaction 1), the only pathway for the biosynthesis of glutamine from ammonia:

Reaction 1: NH_3 + glutamate + ATP $\rightarrow$ glutamine + ADP + Pi

Glutamate is synthesized by glutamate synthase (designated GOGAT: EC 1.4.1.13). This enzyme catalyzes the NAD(P)H-dependent transfer of the amide group from glutamine to 2-ketoglutarate to produce two molecules of glutamate (reaction 2). The reactions can take place independently, but can also be understood as a cycle in which every turn results in the net synthesis of one molecule of glutamate from ammonia and 2-ketoglutarate (reaction 3). The consumption of energy drives these coupled reactions strongly in a forward direction, making this essentially a physiologically irreversible pathway.

Reaction 2: glutamine + 2-ketoglutarate + NADPH + $H^+ \rightarrow$ 2-glutamate + $NADP^+$

Reaction 3: NH_3 + 2-ketoglutarate + NADPH + $H^+ \rightarrow$ glutamate + $NADP^+$

The second route for ammonia assimilation is via glutamate dehydrogenase (GDH: EC 1.4.1.4). GDH catalyzes the reductive amination of 2-ketoglutarate to produce glutamate in a reversible reaction in the presence of NAD(P)H (reaction 4):

Reaction 4: NH_3 + 2-ketoglutarate + NAD(P)H + $H^+ \leftrightarrow$ L-glutamate + $NAD(P)^+$

In the majority of microorganisms both the GDH and the GS-GOGAT pathways operate, but the relative contribution of these pathways to ammonia assimilation and glutamate synthesis varies greatly depending on the organism and the nitrogen conditions. In *C. kluyveri* and *C. butyricum,* for example, the GDH pathway predominates in ammonia-grown cells but was found to play a minor role relative to the GS/GOGAT pathway in cells grown under nitrogen-fixing conditions [13]. In some microorganisms, just one of the pathways can be detected. The assimilation of ammonia occurs solely by the coupled GS-GOGAT pathway under all nitrogen conditions in *Clostridium pasteurianum* [14], *C. thermoautotrophicum* [15], and *C. saccharobutylicum* [16], as well as in *B. subtilis* [17] and *Streptomyces clavuligerus* [18]. On the other hand, GDH appears to be the sole enzyme in ammonia assimilation in the ruminal bacterium *Streptococcus bovis* [19] and is the primary enzyme involved in glutamate formation in *Bacteroides fragilis* [20] and the nitrogen-fixing bacterium *Bacillus macerans* [21]. While the biosynthesis of glutamine occurs solely via the action of GS and absolutely requires ammonia, glutamate can be produced directly from the degradation of a number of amino acids including glutamine, histidine, arginine, and proline, or indirectly via a transamination reaction involving the transfer of the α-amino group from one amino acid, e.g., aspartate, to the keto group of 2-ketoglutarate [22,23]. For the purpose of this review, we will be focusing on the GS-GOGAT pathway in relation to nitrogen metabolism in clostridia.

11.3 THE GS AND GOGAT ENZYMES

Although the reactions catalyzed by the GS and GOGAT pathways are tightly coupled processes and are highly conserved in bacteria from diverse ecological niches, the genes encoding these enzymes are not physically linked, and no coordinate regulation has been documented. Furthermore, the structure and regulation of these key enzymes of nitrogen assimilation vary considerably in different bacteria belonging to the same genus, presumably reflecting different metabolic needs. Recent developments in the field of genomics have allowed the use of databases to search for homologues of different genes. A number of nitrogen-related genes have been identified in the three *Clostridium* genomes sequenced to date, viz. *C. acetobutylicum* 824, *C. perfringens* strain 13, and *C. tetani* E88 (Table 11.1).

11.3.1 Glutamine Synthetase

The DNA sequence of many of the genes encoding GS enzymes has by now been determined, and these *glnA* genes have been shown to be highly conserved in different species. Three main classes of GS enzymes have been described in bacteria [16]. GSI enzymes are the most ubiquitous type of bacterial GS and have been extensively studied in the *Enterobacteriaceae*, but are also present

TABLE 11.1
Homologues of Nitrogen Genes in Clostridia

Gene/Protein	*C. saccharobutylicum* and *C. beijerinckii*[1]	*C. acetobutylicum* 824[2]	*C. perfringens* Strain 13[3]	*C. tetani* E88[4]
glnA (GS)	*glnA-nitR-gltAB*	*glnA*	*glnA*	*glnA*
	GS type I	GS typeIII	GS type I	GS type I
gltAB (GOGAT)	*glnA-nitR-gltAB*	*gltAB*	*gltB* (?)	*gltB* (?)
gdhA (GDH)	No GDH activity detected	*gdhA*	*gdhA*	ND[5]
σ^L		σ^L	σ^L	σ^L
GltT (Na/glutamate symporter)		*gltT*	*gltT*	ND[5]
glnB (P_{II})	2 *glnB*-like genes in *nif* gene cluster[6]	2 *glnB*-like genes in *nif* gene cluster[6]	ND	ND
glnD (UTase) and *glnE* (ATase)		ND[5]	ND	ND
glnR, tnrA, amtR, nrgAB		*nrgAB*	ND	ND

1. [41, 80, 83].
2. [32]; (GenBank Acc. No: NC 003030.1)
3. GenBank Acc. No: NC 003366.1
4. GenBank Acc. No: NC 004557.1
5. ND: Not detected in database searches or not annotated.
6. [44]

in the Gram-positive organisms *Bacillus, Clostridium, Streptomyces* [24], and *Corynebacterium* [25]. The GSI enzyme is a dodecamer composed of 12 identical subunits, ranging between 44 and 59 kDa, which are organized in two superimposed hexagonal rings. This GS group can be divided into two further groups depending on whether they are regulated post-translationally by covalent modification, resulting in a reduction of GS activity. GSII enzymes are typical of eukaryotes but are also found, together with GSI, in many symbiotic nitrogen-fixing bacteria including *Agrobacterium, Rhizobium, Bradyrhizobium,* and *Frankia* species [16], and also in the free-living *Streptomyces* [26]. The GSII enzyme is characteristically a thermolabile octomer of identical subunits arranged in two discs of four subunits [27]. The smaller GSII subunit (~36 kDa) lacks the C-terminal portion of the GSI subunit, which includes the adenylylation site involved in post-translational control of activity. Although the overall amino acid similarity between the GSI and GSII subunits is low (~15% identity), five regions associated with the GS active sites remain well conserved [28]. GSIII was first identified in the anaerobic intestinal bacterium *Bacteroides fragilis* [29]. It is a hexamer of six large identical subunits (75 kDa) and is quite distinct from either GSI or GSII at both the amino acid and enzyme levels [16]. Homologous genes have since been identified in *Butyrivibrio* [30], *Ruminococcus* [31], *Clostridium acetobutylicum* 824 [32], and *C. thermocellum* (GenBank Acc: ZP00061653), as well as the cyanobacteria *Synechocystis* (GenBank Acc: BAA18510) and *Synecococcus* (GenBank Acc: AAF91344). It is interesting that *C. acetobutylicum* should have the gene encoding a GSIII enzyme [32], while the *glnA* genes from both *C. saccharobutylicum* and *C. beijerinckii* code for the more common GSI enzyme [33].

11.3.2 Glutamate Synthase (GOGAT)

Numerous GOGAT structural genes have been cloned and sequenced, and a number of GOGAT enzymes have been purified to homogeneity from a variety of microorganisms, fungi, and plants

[34]. Analyses of their combined structural, biochemical, and sequence characteristics have led to the identification of three classes of GOGAT enzymes.

The bacterial class of the enzyme is dependent on reduced pyridine nucleotides (NAD(P)H) for their reducing equivalents (NAD(P)H-GOGAT) and is composed of two dissimilar subunits, the large (α) subunit conserved around ~150 kDa, and the small (β) subunit conserved at ~50 kDa, that together form the active $\alpha\beta$ protomer (~200 kDa) [34]. The second class of enzyme depends on reduced ferredoxin as its electron donor (Fd-GOGAT; EC 1.4.7.1) and is found in photosynthetic cyanobacteria [35], and in the photosynthetic tissues of plants [36,37]. Molecular characterization of these enzymes has revealed that they appear to exist as homodimers composed of a single polypeptide chain varying in size from 125 to 180 kDa. The eukaryotic pyridine-dependent forms of the enzyme constitute the third class of GOGAT (NADH-GOGAT; EC 1.4.1.14), found in yeast, fungi, and nonphotosynthetic tissues (nodules) of plants [38]. They are composed of a single high molecular weight polypeptide of approximately 200 kDa.

In bacteria, the structural genes for the α and β subunit polypeptides are encoded by the *gltB* and *gltD* genes, respectively, except in *B. subtilis,* where the corresponding genes have been designated *gltA* and *gltB* [39]. In all cases so far, the genes are genetically linked, with the α subunit generally encoded upstream of the β subunit, except in *Azospirillum brasilense,* where the gene order is uniquely reversed [40]. Comparative analysis of the amino acid sequences of GOGAT from eukaryotes and bacteria has revealed highly conserved regions, both within and between each class of enzyme [34]. Specifically, the α subunit of bacterial GOGAT is very similar to the single polypeptide chain of Fd-dependent GOGAT, and to the amino terminal three quarters of eukaryotic NADH-dependent GOGAT, while the bacterial β subunit shares significant homology with the C-terminal region of eukaryotic NADH-dependent GOGAT. The eukaryotic pyridine dependent form of the enzyme appears to derive from the fusion of polypeptides corresponding to the α and β subunits of bacterial GOGAT, linked by a short nonconserved polypeptide chain. The Fd-GOGATs lack a region corresponding to the bacterial β subunit [34].

In *C. saccharobutylicum* and *C. beijerinckii,* the *gltAB* gene cluster encodes the two subunits of a NADH-dependent GOGAT enzyme with the normal arrangement of the gene for the large α subunit preceding the smaller β subunit [41]. The same gene arrangement is conserved in *C. acetobutylicum* 824, but cannot be detected in either *C. perfringens* or *C. tetani* (Figure 11.1). Two β subunit-like genes can be identified in *C. perfringens* and one in *C. tetani*; however, it is likely that these are not true GOGAT genes as they do not have the *gltA* gene upstream. A similar β subunit-like gene (*gltX*) has been characterized in *C. saccharobutylicum* but was found not to be regulated by nitrogen. The *gltX* gene, therefore, is probably not the structural gene for a GOGAT protein but is more likely to belong to a larger group of β subunit-like genes with oxidoreductase activity [41].

11.4 REGULATION OF NITROGEN ASSIMILATION

Since GS and GOGAT are key enzymes in nitrogen assimilation in bacteria, their activities are generally tightly controlled in response to the levels of nitrogen source available. Thus, cells growing with preferred nitrogen sources generally contain low levels of GS and GOGAT, whereas high levels of these enzymes are present during nitrogen-limited growth to ensure adequate supplies of glutamine and glutamate. The most comprehensive analysis of the regulation of nitrogen sensing and assimilation has been performed in the *Enterobacteriaceae*, so that this has become the paradigm for this field. Genomics has demonstrated the extent to which proteins involved in this process are conserved, with the GS enzyme one of the most highly conserved examples across widely diverse groups of organisms. However, recent reports on the regulation of these genes illustrate how different and novel each particular mechanism is. The first part of this review will therefore provide a brief overview of the different elements and features of nitrogen regulation in the *Enterobacteriaceae* and the Gram-positive bacteria for comparative purposes.

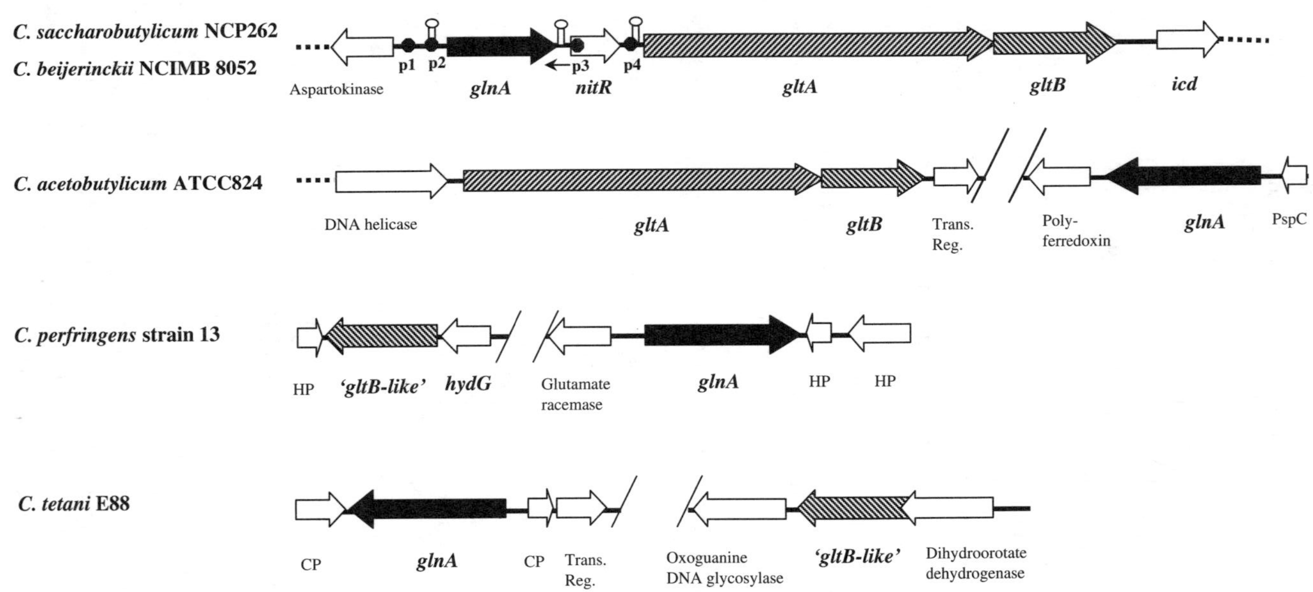

FIGURE 11.1 Comparison of the arrangement of the nitrogen assimilatory genes in *C. saccharobutylicum* and *C. beijerinckii* with the three *Clostridium* genomes sequenced to date: *C. acetobutylicum* ATCC 824 (NCBI: NC 003030.1), *C. perfringens* strain 13 (NCBI: NC 003366.1), and *C. tetani* E88 (NCBI: NC 004557.1). Genes have been assigned identities based on their gene names: *glnA,* glutamine synthetase; *gltA* and *gltB,* NAD(P)H-dependent large and small subunits of GOGAT, respectively; *icd,* isocitrate dehydrogenase and *hydG,* probable hydrogenase gamma chain — or according to the probable gene products; PspC, putative stress-responsive transcriptional regulator; Trans. Reg, predicted transcriptional regulator; HP, hypothetical protein or CP, conserved protein. No GOGAT large subunit could be identified in the genomes of either *C. perfringens* or *C. tetani*. As discussed in the text, the four promoters (● P1-P4) and the positions of the terminator structures (⍡) are indicated. Dotted lines signify truncated gene sequences.

11.5 GLOBAL NITROGEN REGULATION IN THE *ENTEROBACTERIACEAE*

The regulation of GS activity has been extensively studied in *E. coli*, *Salmonella typhimurium,* and *Klebsiella aerogenes* [23,42,43], and as the most researched bacterial system, it often serves as the model against which other systems are assessed. The regulation of nitrogen assimilation in the enteric bacteria is complex and involves mechanisms for the regulation of GS activity that are interlinked and act at both the level of transcription and the activity of the enzyme. In *E. coli*, the *glnA* gene is part of a complex operon (the *glnA-ntrBC* operon). The *ntrB* gene encodes the histidine kinase or NR_I protein and the *ntrC* gene encodes the NR_{II} protein, which together form the two-component nitrogen regulatory system. NR_{II} is the response regulator of transcription of the *glnA* as well as the other *ntr* and *nif* genes. Depending on the nitrogen conditions, the *glnA-ntrBC* operon is expressed from either one of two tandem promoters — one being a consensus σ^{70} promoter and the other requires a minor σ factor, σ^N, (formerly σ^{54}), in conjunction with the enhancer protein, phosphorylated NR_I.

The transcriptional and post-translational regulation of GS activity is controlled via the Ntr system, which consists of a series of proteins able to recognize the nitrogen status of the cell [23]. The key protein is the product of the *glnB* gene, P_{II}, which is present in two states, uridylylated (P_{II}-UMP) or deuridylylated (P_{II}). The modification of P_{II} is catalyzed by the uridylyl-transferase/uridylyl-removing activity of the UT/UR enzyme (encoded by *glnD*), which responds to the nitrogen levels in the cell. In turn, P_{II} controls the adenylylation state and, hence, the activity of the GS enzyme, by regulating the ATase enzyme. As well as affecting GS activity, P_{II} affects the ability of the *ntrB* gene product, NR_I, to phosphorylate the *ntrC* gene product NR_{II}. Under conditions of nitrogen excess, P_{II} stimulates the dephosphorylation activity of NR_{II}, resulting in dephosphorylated NR_I, which is unable to activate transcription by σ^N at the second *glnA* promoter. P_{II} therefore controls the appropriate responses required under conditions of nitrogen excess, namely, inactivation of GS activity and the inhibition of *glnA* transcription. Under conditions of nitrogen limitation, P_{II} controls the activation of GS activity and the transcription of *glnA*. The proteins UT/UR and P_{II} link the two cascades: one involving UT/UR, P_{II}, ATase and GS, which controls GS activity; the other involving UT/UR, P_{II}, NR_{II} and NR_I, which controls *glnA* and other Ntr-regulated operons.

The only gene for which homologous genes can be detected in clostridia, is the *glnB* gene, which, in the enteric bacteria, encodes the P_{II} protein (Table 11.1). Two *glnB*-like genes have been isolated in the *nif* gene clusters of both *C. acetobutylicum* and *C. beijerinckii* [44], but the specific function of these genes remains unknown. No homologues of *glnD* (UTase) or *glnE* (ATase) are found in any of the low-G+C Gram-positive bacteria [45], which agrees with the evidence that the GS from *C. saccharobutylicum* is not subject to post-transcriptional modification such as adenylylation as in the *Enterobacteriacae* [33].

11.6 NITROGEN CONTROL IN GRAM-POSITIVE BACTERIA

The regulation of nitrogen metabolism in Gram-positive bacteria has not been as well characterized as in the enteric bacteria. Evidence for a global regulatory network equivalent to the Ntr system is lacking, although homologues of some of the components of the enteric system have been identified by DNA sequence homology in *B. subtilis* [46], *Streptomyces* [26], and *Corynebacterium* [47,48]. There is still not much information available on the control of enzymes involved in the utilization of nitrogen sources by different bacteria. There is also no evidence, to date, that a universal global regulatory system exists in either the low-G+C or the high-G+C Gram-positive bacteria, which will be characteristic of that particular group.

11.7 NITROGEN REGULATION IN *BACILLUS SUBTILIS*

In *B. subtilis,* amino acids (glutamine, arginine, and possibly others), rather than ammonia, are the preferred nitrogen source in that they support the fastest growth rate [11]. While the synthesis of amino acid degradative enzymes is substrate inducible, with the exception of asparaginase, their expression is generally not regulated in response to nitrogen availability [11,24]. However, nitrogen limitation stimulates the expression of other enzymes involved in nitrogen metabolism including urease, asparaginase, GSI, and enzymes involved in nitrate assimilation [11]. Only recently has a global system of nitrogen control begun to emerge in *B. subtilis* [49]. Although the mechanism by which nitrogen levels are sensed is not known, the GS protein plays an important role in transmitting information about nitrogen availability, as well as in controlling its own synthesis [50]. At least three independent global regulatory proteins are involved in nitrogen regulation: GlnR, a member of the dicistronic *glnRA* operon; TnrA; and CodY. These regulators function under different nutritional conditions, and their activities support a system of control in which they direct the cell toward adaptive vegetative growth rather than toward sporulation, when nitrogen becomes limiting [51]. In addition, σ^L, a homologue of σ^{54}, plays a role in recognizing promoters of genes involved in nitrogen metabolism.

The regulation of GS activity in *B. subtilis* has been reviewed by Schreier [24] and Belitsky [45]. The activity of the GS enzyme is controlled primarily at the level of transcription in response to nitrogen availability. The *glnRA* operon is transcribed by the vegetative (σ^A-dependent) form of RNA polymerase [24]. During growth with excess nitrogen, the *glnR* product, GlnR, a small (135-amino acid) dimeric protein, represses expression of the *glnRA* operon [52]. TnrA also negatively regulates *glnRA* expression by binding to the *glnRA*O2 operator, which overlaps the –35 region of the promoter [49]. The binding of GlnR to the operators requires GS, but the exact nature of the interaction between these two proteins is not fully understood, and the mechanisms by which nitrogen levels are sensed in *B. subtilis* are not yet known.

GlnR and TnrA are related proteins belonging to the MerR family of transcriptional regulators [51]. In response to nitrogen limitation, TnrA activates the expression of asparaginase, γ-aminobutyric acid permease (*gabP*), urease (*ureABC*), the *nrgAB* operon encoding a putative ammonia permease (NrgA), a homologue of the *E. coli* P_{II} protein (NrgB), and the nitrate assimilatory enzymes (*nasABCDEF*). In addition, TnrA positively regulates its own synthesis and represses the expression of *glnRA* and the genes encoding GOGAT [49]. In contrast, GlnR represses the expression of *glnRA, tnrA,* and *ureABC* in cells growing with excess (preferred) nitrogen sources. These related proteins are DNA-binding regulatory proteins, which have almost identical N-terminal DNA-binding domains, and bind to similar DNA sequences (TGTNAN$_7$TNACA), the GlnA/TnrA site under the different nutritional conditions. Thus, GlnR and TnrA not only regulate their own synthesis, but also cross-regulate each other's expression, and both negatively regulate GS activity.

CodY-dependent regulation responds to the total nutritional status of the cell. It represses the expression of several genes involved in nitrogen metabolism, including the histidine degradative operon (*hut*), the dipeptide transport operon *dpp*, the isoleucine and valine degradative operon (*bkd*), *ureABC,* and *gabP,* as well as genes involved in competence and acetate metabolism [49]. The highest levels of CodY-dependent repression occurred in cells growing rapidly in a medium rich in amino acids. While the signal regulating CodY activity is unknown, it is thought that the mechanism of transcriptional regulation involves binding of CodY to a structure formed by AT-rich DNA sequences [51]. When carbon or nitrogen become limiting, CodY-dependent regulation is relieved.

There is no evidence that a nitrogen-regulatory system analogous to that of *B. subtilis* is present in any other Gram-positive bacteria, although the genetic organization and GlnR-dependent regulation of the *B. subtilis glnRA* operon is preserved in a number of *Bacillus* and *Streptococcus* spp., as well as in *Staphylococcus aureus* and *Lactococcus lactis* [45]. This arrangement is not present in any of the three *Clostridium* genomes, and genes for other *Bacillus* regulatory elements, such as TnrA, could not be identified. There is, however, evidence of CodY and σ^L genes in the *C. acetobutylicum, C. perfringens,* and *C. tetani* genomes, and the *nrgAB* gene cluster, coding for an ammonium transporter and a P_{II}-like nitrogen regulatory protein, respectively, is detectable in *C. acetobutylicum* (Table 11.1).

11.8 NITROGEN REGULATION IN THE HIGH G-C GRAM-POSITIVE BACTERIA

In contrast to other Gram-positive prokaryotes, *Streptomyces* species synthesize both GSI and GSII enzymes, and the GSI activity, encoded by *glnA*, is regulated at both the transcriptional and post-translational levels in response to nitrogen availability [53,54]. At present, nothing is known about the regulation of the *glnII* genes encoding GSII subunits in *Streptomyces* spp., and the respective roles of the two *gln* genes remains unclear. Neither *glnA* nor *glnII* single mutants appear to require glutamine for growth, and in *S. viridochromogenes*, no differential regulation of the expression of GSI and GSII in response to different nitrogen sources was observed, although GSI activity was always dominant [54]. In contrast to the *E. coli glnAntrBC* and *B. subtilis glnRA* operons, the *S. coelicolor glnA* transcription unit is monocistronic and does not contain any potential regulatory genes [53]. Although the mechanisms regulating the expression of GSI in *Streptomyces* in response to nitrogen availability are not known, transcription of *glnA* in *S. coelicolor* requires a positive regulatory gene, *glnR,* which encodes a 29-kDa polypeptide with significant homology to other response regulator proteins that are known to act as transcriptional activators [53].

A *glnE* homologue encoding a putative adenylyltransferase and a *glnB* (P_{II}) homologue were characterized from *S. coelicolor* and were found to be localized between the *glnA* and *glnII* genes [55]. This is similar to the closely related species, *Mycobacterium tuberculosis,* where the *glnA, glnE,* and *glnII* genes are adjacent [55], but is different from the situation in *E. coli,* where *glnE* is not physically linked to the gene encoding GS [56].

Control of nitrogen assimilation in *Corynebacterium* appears to be very similar to the *Streptomycetes,* in that the *glnA* gene is not clustered with other nitrogen genes and is monocistronically transcribed [25,47]. Like the *Streptomycetes,* the *Corynebacterium* GS can be inactivated by adenylylation and is controlled by the post-translational modification of P_{II} by the UR/UTase enzyme. Homologues of *glnK* and *glnD,* which encode the signal transducing protein, P_{II}, and the uridylyltransferase enzyme, respectively, have been identified in both *C. glutamicum* and *C. diphtheria* [48,57]. The GOGAT genes are found in their normal operon arrangement, *gltBD*, with a histidine kinase gene, *hkm*, upstream of the *gltB* gene. Expression of the *hkm* gene was found to be required for optimal GOGAT expression, indicating that the histidine kinase may be involved in transmitting signals of the nitrogen status of the cell [25].

Recent reports indicate that the regulatory protein AmtR is a global regulator in *Corynebacterium* [48,58]. When *C. glutamicum* is grown in nitrogen-rich conditions, AmtR binds to a specific sequence in the promoter regions of the *glnA* and *amt* genes, as well as the *amtB-glnK-glnD* and *gltBD* operons, and prevents transcription of these genes. When nitrogen becomes limiting, however, AmtR dissociates from its binding sites, resulting in the transcription of the nitrogen genes. By virtue of the global AmtR repressor, the regulatory system in *C. glutamicum* differs significantly from that in other Gram-positive bacteria, such as *B. subtilis* [49] and *S. coelicolor* [53], and is perhaps more similar to that of *E. coli*. However, what makes nitrogen control in *C. glutamicum* unique is the fact that a single regulatory protein controls ammonium uptake, as well as the glutamate and glutamine biosynthetic pathways. Furthermore, the global nature of AmtR is demonstrated by its ability to control GS at two levels. AmtR regulates transcription of firstly, the *glnA* gene, which affects cellular GS concentrations, and secondly, of *glnK* and *glnD,* which encode proteins involved in the post-translational regulation of the GS enzyme. There is no evidence of AmtR homologues in any of the *Clostridium* genomes (Table 11.1).

11.9 ANTISENSE RNA AS A REGULATORY MECHANISM

Since the first reports of natural antisense RNAs (asRNA) in the 1980s [59], the number of examples of small RNA molecules that play a regulatory role in cell metabolism has increased vastly. Regulatory RNAs, sometimes referred to as riboregulators, are small complementary RNA molecules that act by binding to complementary sequences on target mRNA molecules, resulting in the post-transcriptional regulation of the target gene function. These riboregulators add an additional dimension to the regulation of the target genes, often acting to fine-tune cellular responses to environmental signals. An interesting feature of asRNA regulation is that a small RNA molecule can affect more than one target gene. Most of the examples of natural asRNA regulated systems occur in the prokaryotes, and the majority of these have been detected in plasmids, phages, or transposons [60,61]. In these cases, the RNA-encoding gene overlaps the target gene (*cis* encoded), while the few chromosomally located examples of asRNA, such as MicF, OxyS, and DsrA [62,63], are usually *trans* encoded, at genetic loci distant to their target genes. It is highly likely that, as the genomic analysis of more diverse bacteria is completed, many more examples of asRNA regulation will be described. RNA interference (RNAi) is widely found in many eukaryotes as a gene silencing technique and involves complementary base pairing between the mRNA and the asRNA, followed by mRNA degradation of the duplex molecule [64].

Antisense RNAs vary in the mechanisms by which they effect the required regulation [64], and although they generally function to repress the target gene function, there are examples of where they serve as activators, such as the *Staphylococcus agr* system [65]. One of the best-characterized systems is the replication control of plasmid ColE1, in which the asRNA inhibits the formation of the primer for replication [59]. In other plasmid systems, such as those of the *E. coli* plasmid R1 [60,61] and the streptococcal plasmid pMV158, the asRNA blocks the initiation of translation and prevents ribosome binding to the replicase gene [66]. Another mechanism by which asRNA regulates gene expression is by transcriptional attenuation such as that first described for the *Staphylococcus* plasmid pTI181 [67] and for the streptococcal plasmids pIP501 [68] and pAM1 [69]. This mechanism would appear to be unique to Gram-positive bacteria.

The chromosomal asRNA systems are mostly encoded *in trans* and, therefore, do not necessarily form perfect complementary duplexes. This would not only facilitate the recognition of a broader range of asRNA binding sites on multiple targets, but also would allow for an additional control mechanism, in that the bulges caused by misprimed bases prevent rapid degradation by the cellular RNAase III [70]. The 93-bp MicF-RNA is induced by osmotic, redox, or temperature stress and serves to repress the OmpF porin of the *E. coli* outer membrane [71]. The binding of MicF RNA both inhibits OmpF translation and promotes OmpF-mRNA degradation. OxyS-RNA is 109 bp, is induced by osmotic stress, and has been shown to regulate up to 40 genes by inhibiting translation of RpoS, as well as blocking the RBS of other genes [72,73].

The presence of natural antisense riboregulators in *Clostridium* was first demonstrated in the regulation of the glutamine synthetase gene in *C. saccharobutylium* NCP262. A 43-bp asRNA has been identified in the *glnA* operon and has been implicated in the down-regulation of the GS enzyme [74,75]. asRNA can be used with considerable success to inhibit gene expression in prokaryotes [76], and it has become a powerful laboratory tool particularly in organisms that have a limited number of genetic tools available. In this approach, the target gene is cloned in the reverse orientation behind a promoter on a plasmid. The resulting asRNA is assumed to bind to the target mRNA and prevent translation by either blocking the ribosome binding site or by making the double-stranded molecule more susceptible to attack by RNA-degrading enzymes [76]. Evidence that asRNA can be effectively employed to influence gene expression in *Clostridium* has been provided by the construction of *C. acetobutylicum* 824 strains with altered metabolic flux patterns during solventogenesis [77,78]. Plasmid-borne asRNA constructs to two of the enzymes of the acetone pathway, acetoacetate decarboxylase (AADC) and coenzyme A-transferase (CoAT), were introduced into *C. acetobutylicum* [79]. Although down-regulation of the AADC protein was evident from Western blot analysis, no significant reduction in acetone levels was observed. Similar experiments with the CoAT asRNA constructs showed substantially lower acetone and butanol levels and greatly increased the amount of ethanol produced. This demonstrates that asRNA technology can be effectively used to make beneficial alterations to various metabolic pathways of *Clostridium.*

11.10 NITROGEN ASSIMILATION IN *C. SACCHAROBUTYLICUM* AND *C. BEIJERINCKII*

Very few studies of the genes and enzymes involved in nitrogen assimilation have been carried out in clostridia, other than to show that the nitrogen content of the fermentation medium is important for both optimal solvent production and successful sporulation [9,10]. Although evidence has been provided for these pathways in diverse *Clostridium* species [13–15], the only molecular analysis of nitrogen assimilation has been done in *C. saccharobutylicum* and *C. beijerinckii* [80,81].

11.10.1 Structure of *glnA* Region from *C. saccharobutylicum* and *C. beijerinckii*

The region encoding the *glnA* gene from *C. saccharobutylicum* NCP262 (formerly known as *C. acetobutylicum* P262) was cloned by complementation of an *E. coli glnAntrBC* mutant, which lacks any GS activity and is therefore unable to grow on inorganic nitrogen sources [33]. The enzyme activity produced by the cloned *Clostridium glnA* gene was not regulated by adenylylation. DNA sequence analysis revealed that the structural gene encoded a GS I protein of 444 amino acids, which was expressed from either of two promoter sequences, P1 or P2, at positions –124 to –95 and –65 to –37 respectively, upstream from the *glnA* start codon [74,82]. In addition, an extensive inverted repeat sequence (158 nucleotides), typical of an intrinsic terminator, was located immediately downstream of *glnA*. Four regions exhibiting dyad symmetry are located upstream of *glnA*, three upstream of P1, and one overlapping P2 [80]. These features are summarized in Figure 11.1.

Subsequent DNA analysis of the region downstream of the *glnA* gene [80] revealed an ORF of 566 nucleotides present in the same orientation as the *glnA* gene and separated from it by an intergenic region of 200 bp (Figure 11.1). The deduced amino acid sequence of this ORF showed limited sequence homology to very few proteins in the databases, all of which were members of the response regulator family of two-component signal transduction systems. The gene was, therefore, designated *nitR.*

Downstream from the *nitR* gene and separated by an intergenic region of 108 bp, lie two genes coding for the α and β subunits of GOGAT, *gltA* (4554 bp) and *gltB* (1473 bp), respectively, which are present in the same orientation as *glnA* and *nitR* [41]. The two ORFs are separated by 12 bp, and their arrangement is typical of that found in other bacteria, in which the α subunit precedes the β subunit. Downstream from *gltB* and separated by an intergenic region of 866 bases, is a truncated gene that has been identified as the *icd* gene, coding for the isocitrate dehydrogenase enzyme. Promoter regions were identified upstream of the *gltA* and *icd* genes, which showed homology to the clostridial extended consensus promoter sequences [41]. Two major inverted repeat sequences, with the potential to form stable stem-loop structures, were identified in this region. The first region of dyad symmetry is located between the putative *gltA* promoter and the *gltA* structural gene, while the second is located 18 to 62 bp downstream of the *gltB* termination codon and forms a typical transcriptional terminator (Figure 11.1).

A 3.8 kb DNA fragment was identified and cloned from *C. beijerinckii* NCIMB 8052, using internal fragments from the *C. saccharobutylicum glnA*, *nitR,* and *gltA* genes as probes for DNA hybridization. The recombinant plasmid was able to complement glutamine auxotrophy in the *E. coli glnAntrBC* mutant, demonstrating a functional GS enzyme. DNA sequence analysis identified homologues of these nitrogen genes in this chromosomal fragment. Thus, *C. beijerinckii* and the closely related strain *C. saccharobutylicum* have the identical unique arrangement of genes encoding the two key enzymes of nitrogen assimilation clustered in one region of the chromosome [83]. Unexpectedly, this gene arrangement does not occur in *C. acetobutylicum* 824, where the GS encoded by the *glnA* gene belongs to the group III class of GS enzymes. It is also not found in *C. perfringens* or *C. tetani*, where the *glnA* genes are not adjacent to other nitrogen genes on the chromosome, and the GOGAT genes have not been identified as yet (Figure 11.1).

11.10.2 Antisense RNA

An interesting and novel regulatory element has been identified for nitrogen assimilation in *C. saccharobutylicum.* There is evidence to support the involvement of an antisense mRNA (asRNA) in the down-regulation of *glnA* gene expression [74,75]. A functional promoter, P3, situated 227 to 257 nucleotides from the 3′ end of the *glnA* gene, and in the opposite orientation to the *glnA* gene (Figure 11.1), has been shown to direct the synthesis of a short 43 bp RNA transcript in *C. saccharobutylicum.* This RNA transcript is complementary to a region at the start of both the *glnA* and the *gltA* mRNA, including the ribosome binding sites and the AUG initiation codons of both genes (Figure 11.2). An up-promoter mutation in P3 resulted in reduced levels of GS activity when the *glnA* region was expressed in *E. coli* [74]. In addition, the production of this asRNA was differentially regulated by the nitrogen source, such that under nitrogen-limiting conditions when *glnA* mRNA transcription was induced and GS activity was elevated, the expression of the *as*-RNA was repressed; levels of *glnA* mRNA exceeded that of the asRNA by 5-fold. Under nitrogen rich conditions that repress GS activity, the situation was reversed; there was ~1.6-fold more *as*-RNA transcripts over *glnA* mRNA transcripts. Thus, it was proposed that the asRNA hybridized to the complementary regions on the *glnA* mRNA (Figure 11.2), thereby blocking the ribosome binding site and the translational start site, and inhibiting translation of nascent GS during nitrogen rich conditions [75]. The presence of a similar complementary sequence upstream of the *gltAB* genes would result in the blocking of translation of *gltAB* mRNA under the same conditions.

11.10.3 Identity of the Regulatory Gene *nitR*

Database searches based on amino acid sequence homologies suggested that the deduced product of *nitR* represented a response regulator protein [80]. The deduced amino terminal domain showed sequence homology to the conserved amino terminal domains of known response regulatory proteins, which are members of the family of two-component signal transduction systems. These proteins function by phosphorylation of the response regulator by the protein kinase component in response to environmental changes such as nutrient limitations. The phosphorylated regulator protein is thus activated and can carry out its regulatory function. NitR shares significant homology over its entire length with two probable response regulators, one from *S. coelicolor* (GenBank Acc: T35758) and the other from *M. tuberculosis* (Genebank Acc: H70558), as well as to the response regulator protein NasT, from *Azotobacter vinelandii* (27% identity). NasT has been shown to be required for the expression of the assimilatory nitrite-nitrate reductase operon (*nasAB*) [84], but the other two proteins have only been identified as a result of genome sequencing programs. The predicted carboxy-terminus of the *nitR* gene shares similarity with the carboxy-terminal domains of several regulatory proteins with antitermination activity. These include NasT, as previously mentioned, the aliphatic amidase regulator, AmiR, of *Pseudomonas aeruginosa* [85], and NasR from *Klebsiella oxytoca* [86]. AmiR positively controls the expression of the amidase encoded *amiE* gene via an antitermination mechanism by allowing RNA polymerase to read through the rho-independent terminator present in the region between the *amiE* promoter and the *amiE* structural gene [87,88]. The carboxy-terminal region of NitR showed significant homology to the carboxy-terminal region of NasR (28 identical residues over a region of 70 residues), the nitrite-nitrate responsive positive regulator of the *nasFEDCBA* operon encoding the assimilatory nitrite and nitrate reductases of *K. oxytoca* [86]. Interestingly, the expression of this operon in *K. oxytoca* (M5al) is

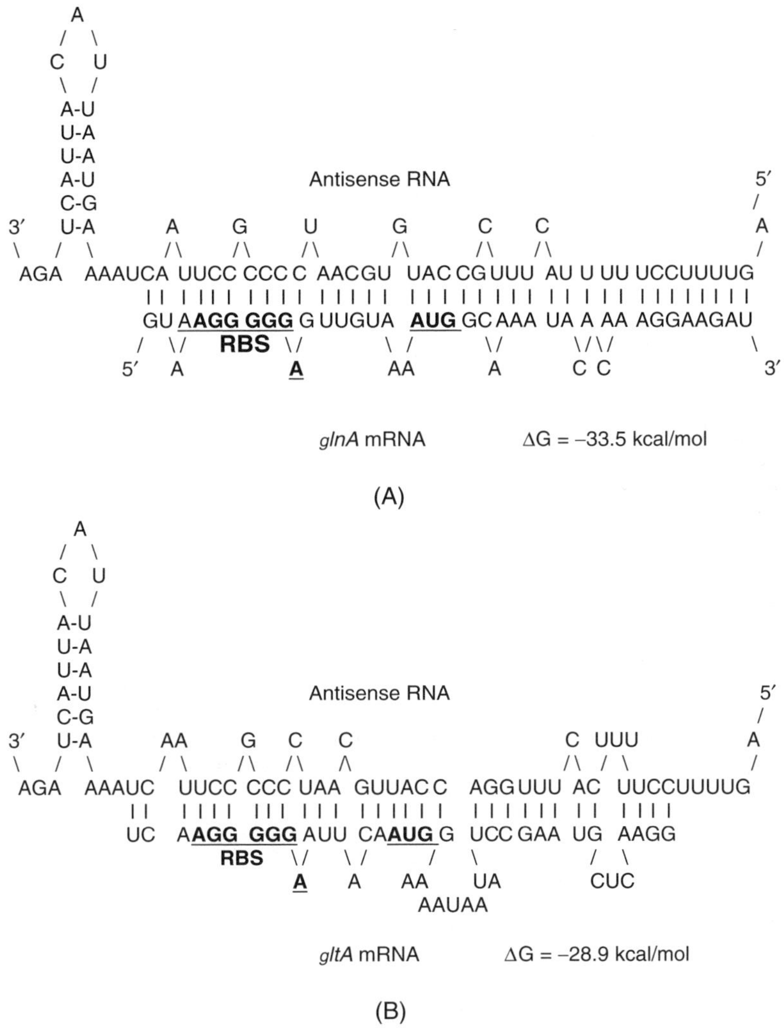

FIGURE 11.2 Proposed RNA-RNA hybrid formation between (A) *glnA* mRNA and asRNA and (B) *gltA* mRNA and asRNA, blocking the ribosome binding sites (RBS) and the AUG start codons of the genes.

also controlled by a transcriptional antitermination mechanism. In this case, it is hypothesized that the NasR protein binds to a stem-loop structure in the *nasF* leader region in response to nitrate or nitrite, thereby mediating transcriptional readthrough of an adjacent downstream factor-independent terminator [89,90].

It, therefore, seems likely that in both *C. saccharobutylicum* and *C. beijerinckii,* NitR represents a response regulator protein that may respond to nitrogen conditions and positively control *glnA* expression via an antitermination mechanism. This is consistent with the presence of a long leader region in the mRNA, between promoter P1 and the *glnA* structural gene, with the potential to form stem-loop structures (Figure 11.3 and Figure 11.4).

11.10.4 Transcriptional Regulation of the Nitrogen Assimilation Genes

RNA analysis showed pronounced induction and repression in the transcription levels of all four genes under defined nitrogen-limiting and nitrogen-rich conditions, respectively [41]. Probes specific for the *glnA* and the *nitR* genes detected a single band of approximately 2.3 kb, corresponding

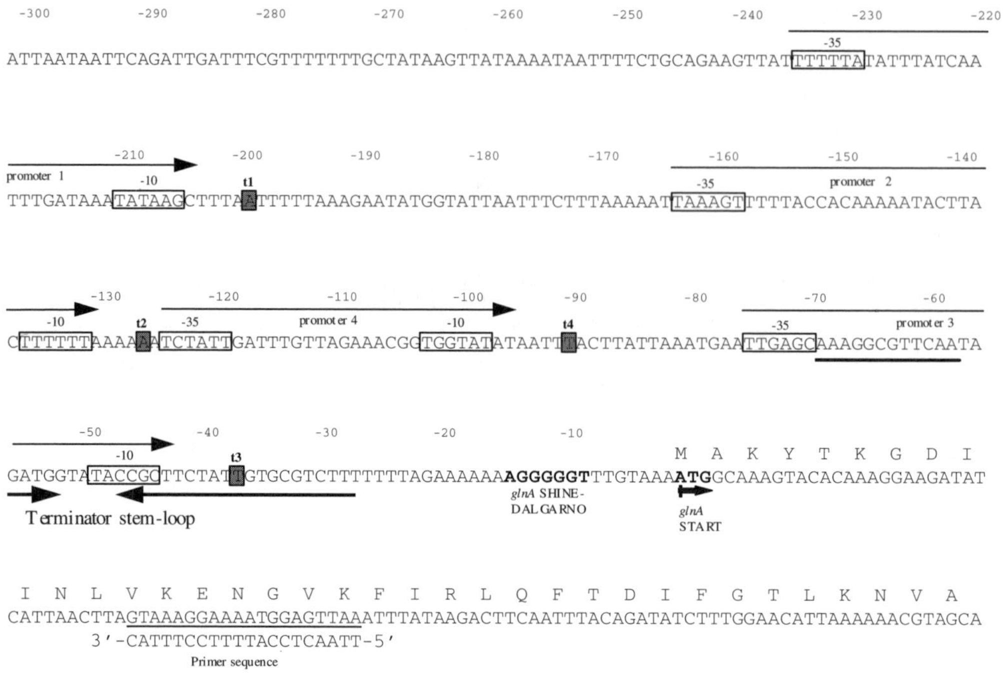

FIGURE 11.3 The 5′ terminus and region upstream of *glnA* in *C. beijerinckii* NCIMB 8052. The positions of the four transcriptional start points are shown, as well as putative promoter regions (–10 and –35 sites are boxed and labeled). Also shown is the stem of the putative terminator structure (bold arrows) and the position of the reverse primer for primer extension experiments.

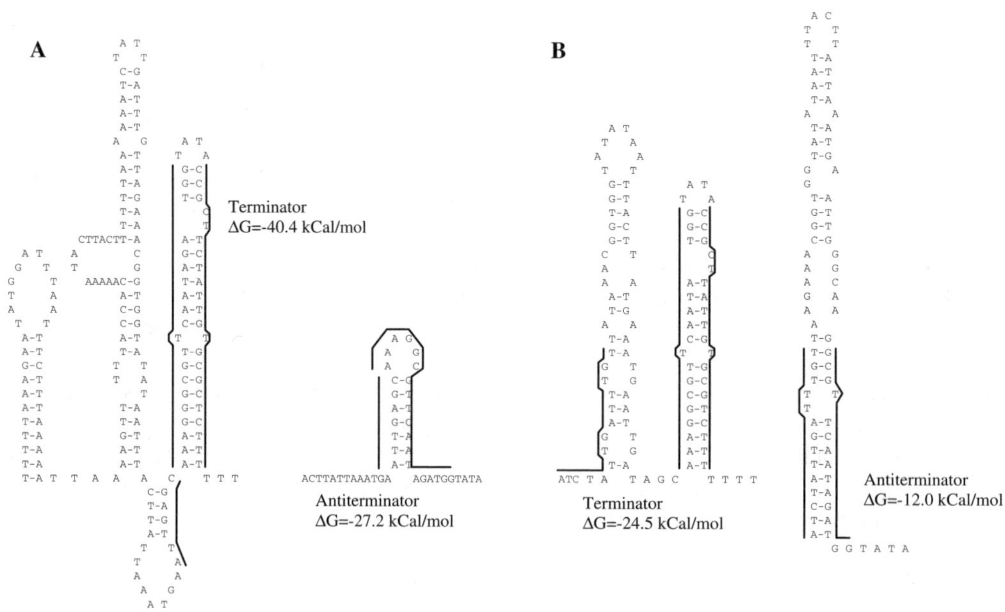

FIGURE 11.4 Putative secondary structures formed by the *C. beijerinckii glnA* leader transcripts t1 (A) and t2 (B). Black lines indicate regions involved in terminator or antiterminator formation.

to the 1.3 kb *glnA* gene, the 0.57 kb *nitR,* and the 200 bp leader and intergenic regions. These results imply that under nitrogen-limiting conditions, *glnA* and *nitR* are transcribed as an operon and that the inverted repeat sequence identified downstream of *glnA* [82] does not function as a transcriptional terminator. It is possible that under nitrogen-rich conditions, there may be evidence of a smaller transcript terminating at this sequence. However, under these conditions, gene expression is repressed, and RNA levels are extremely low. It follows that, under nitrogen-limiting conditions, expression of the asRNA would be repressed as the divergent as promoters would not be available for RNA polymerase binding. Northern blots also confirmed that the α and β subunits of GOGAT are cotranscribed, and a single, common 6.2 kb hybridization signal was detected for both the *gltA*- and *gltB*-specific probes [41]. Again, this is the expected size for the two genes and the inverted repeat sequence downstream of the *gltB* gene. The two genes are separated by a 12 bp intergenic region, which reinforces the polycistronic nature of the genes and helps to ensure that ribosomes do not dissociate at the end of the first gene before re-initiating at the AUG of the second gene [91]. Cotranscription and translational coupling of the *gltA* and *gltB* genes would ensure that equimolar amounts of the two subunits are synthesized.

The relative levels of *glnA*, *nitR*, *gltA,* and *gltB* gene expression in response to different nitrogen conditions were determined by quantitating the signal intensities generated from Northern dot blots [41]. At early exponential phase, there was a 7- to 8-fold induction of the *glnA* and *nitR* genes under nitrogen-limiting conditions, and the two genes were cotranscribed. Similarly, expression levels for the *gltA* and *gltB* genes were induced 6-fold and were cotranscribed under all conditions. By late exponential phase, there was a drop in the induced expression levels of *glnA* and *nitR* of approximately 1.6- and 1.7-fold, respectively, whereas expression of the *gltA* and *gltB* genes remained relatively constant. By comparing these results with the GS and GOGAT activity profiles, it was evident that the changes in the *glnA*, *gltA,* and *gltB* mRNA levels in relation to the nitrogen source was reflected in the corresponding levels of GS and GOGAT activities. These results confirm that GS and GOGAT activities are regulated by the nitrogen source primarily at the level of transcription.

While there was no indication to suggest that GOGAT was regulated at the post-transcriptional or post-translational level, these results suggest that GS activity was regulated post-transcriptionally to some degree. Specifically, at late exponential phase, the drop in the level of GS activity in the inducing media (~3.5 fold) was not reflected in the corresponding levels of RNA (which decreased by 1.6-fold). This may reflect post-transcriptional regulation by asRNA, since it is induced under these conditions. Antisense RNA was shown to be differentially regulated by nitrogen conditions in comparison with the *glnA* mRNA and GS activity [75].

11.10.5 Transcriptional Initiation Site Determination

The promoter region essential for transcription initiation of the *glnA* gene in *C. beijerinckii* 8052 was identified by using a reporter plasmid based on a promoter-less endoglucanase gene *(eglA)* from *C. saccharobutylicum* [92,93]. A transcriptional fusion was created using the *C. beijerinckii glnA* promoter region extending from position –300 to the translational start site of the *glnA* gene [81]. High levels of endoglucanase activity could be measured under different growth conditions from this fusion plasmid, and levels were increased as much as 8-fold over either the native *eglA* promoter or the promoter of the *C. beijerinckii* sucrose operon [94], indicating that the *glnA* promoter is particularly efficient.

In order to determine the start of transcription of *glnA* in *C. beijerinckii* NCIMB 8052 under conditions of nitrogen excess and nitrogen limitation, primer extension experiments were performed on equal amounts of RNA extracted from cultures grown under these conditions [83]. Two major and two minor transcriptional start sites were identified under both nitrogen excess and limitation (Figure 11.3). The major start sites, t_1 and t_2, occurred at positions –200 and –125, respectively, with minor sites at positions –37 and –90. These latter sites may represent spurious start sites due to the processivity of the reverse transcriptase enzyme in regions of high secondary structure. In all cases, approximately 2- to 3-fold more primer extension product was obtained under nitrogen-limiting conditions, indicating that transcription from all four promoter sites is increased under conditions of nitrogen limitation, in agreement with the results obtained from Northern blot analysis (see above).

The existence of a 200 bp leader region in the mRNA of the *glnA* gene, with four apparent transcriptional start points upstream of *glnA*, suggests a model for the regulation of the operon. Secondary structures formed by transcripts initiated at each of the four start sites were predicted using RNA structure v3.71 [94]. The proposed structures are shown in Figure 11.4. The intrinsic terminator structure occurs between positions –70 and –28 and would, therefore, be present in all but the shortest transcripts. However, possible antiterminator structures, which would sequester residues necessary for the intrinsic terminator formation, are present in both major leader transcripts. These different antiterminator structures may be involved in the recognition and binding of an antiterminator protein, such as NitR, which would result in the prevention of terminator formation, and read-through by the RNA polymerase into the structural genes would occur.

11.10.6 Model for Regulation of Nitrogen Assimilation in *C. saccharobutylicum* and *C. beijerinckii*

In light of these results, a simple working model was proposed to account for the adaptive response to nitrogen conditions (Figure 11.5). DNA sequence analysis suggests that NitR represents a response regulator component that is able to respond to changes in the nitrogen environment, as well as a regulatory protein that acts by an antitermination mechanism to control gene transcription. The 200 bp leader region upstream of the *glnA* gene has extensive stretches of inverted repeat

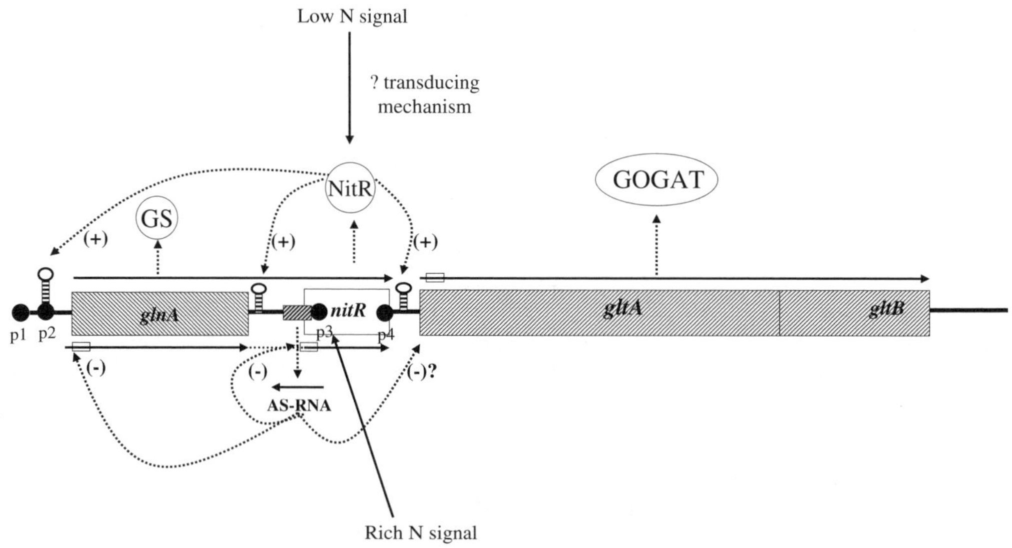

FIGURE 11.5 Model for the regulation of the *C. saccharobutylicum* P262 *glnA, nitR, gltA* and *gltB* genes by nitrogen content of the medium (N). The two *glnA* promoter sequences are indicated by P1 and P2. The putative *gltAB* promoter is indicated by P4. Promoter P3 is situated in the 5′ region of the *nitR* gene and directs the synthesis of the antisense RNA (AS-RNA), which has the potential to bind the 5′ regions of the *glnA*, *nitR* and *gltA* mRNA transcripts, reducing expression of the respective genes in response to a nitrogen rich signal. NitR may act as a transcriptional antiterminator by interacting with the terminator-like structures, indicated by ⚲, in response to a low nitrogen signal transduction mechanism.

sequences, and both the *nitR* gene and the *gltAB* operon have at least one potential terminator structure upstream of the structural genes. NitR could, therefore, positively control *glnA* transcription via an antitermination mechanism by binding to the putative termination structure present in the leader region between promoter 1 (p1) and the *glnA* initiation codon. Furthermore, the position of a putative transcription terminator between the likely *gltA* promoter sequence and the *gltA* structural gene suggests that, in addition to acting as a transcriptional terminator for *nitR*, this secondary structure may influence the expression of the *gltAB* operon. The *glnA* and *nitR* genes are transcribed together under nitrogen-limiting conditions, implying that, under these conditions, the inverted repeat sequence identified downstream of *glnA* [82] does not function as a transcriptional terminator [41]. The most obvious explanation is that NitR regulates transcriptional read-through of this terminator-like structure, thus positively regulating its own expression during nitrogen limitation. A low level of constitutive *glnA* and *nitR* mRNA expression is detected under all conditions, presumably ensuring that sufficient NitR protein was always available should nitrogen become limiting. It is envisaged that under nitrogen-limiting conditions, a signal transduction mechanism activates residual NitR protein to promote the transcriptional induction of the *glnAnitR* and *gltAB* operons, by allowing RNA polymerase to read through the terminator-like structures.

Regulation by antitermination is effected by a number of different mechanisms. The regulatory protein either binds to the target RNA to stabilize an alternative antiterminator structure and, hence, prevent the formation of the terminator structure as in the *E. coli bgl* operon [96], or may act directly by blocking terminator formation, similar to the *K. oxytoca nas* operon [89]. In some cases, an additional protein factor was found to be essential for the antitermination mechanism, as is the case with AmiC in *Pseudomonas* [86] and NasS in *Azotobacter* [84]. There are also different mechanisms by which the antiterminator protein interacts with its specific ligand. Some RNA-binding proteins that regulate by this mechanism, like the TRAP protein in *B. subtilis* [97], require the presence of their respective ligand before binding to the target RNA. There are, therefore, a number of questions that still have to be answered with respect to the regulation in *C. saccharobutylicum* and *C. beijerinckii*, such as the nature of the signal to which NitR responds, and the actual mechanism by which the proposed antitermination of transcription is effected.

The regulation by asRNA would appear to represent a fine tuning system of the nitrogen response. Under nitrogen-rich conditions, *glnA* asRNA has been implicated in the negative regulation of GS expression by binding to a 43 bp complementary sequence spanning the ribosome binding site and start codons of the *glnA* mRNA [75]. In the same way, it would also inhibit translation of the *nitR* gene and the *gltAB* operon. In addition, since it is transcribed off the opposite DNA strand in the 5′ start region of the *nitR* gene, it would interfere with the transcription of *nitR*. This is consistent with the model in which expression of asRNA would be preferentially achieved under nitrogen-rich conditions, when *glnA* and *nitR* expression is repressed. It would appear that the asRNA is a rapid mechanism for down-regulating the expression of existing *glnA* and *gltAB* mRNA, once the GS and GOGAT are no longer required, and may reflect the importance of energy saving in *Clostridium*, as the GS-GOGAT pathway requires ATP. This is reminiscent of the down-regulation of *E. coli* GS by adenylylation, which progressively inactivates the enzyme already present in the cells when it is no longer required.

In conclusion, the two saccharolytic clostridial species, *C. saccharobutylicum* and *C. beijerinckii,* show a unique pattern of gene arrangement of the nitrogen assimilation genes, where the genes encoding the GS and GOGAT enzymes, as well as a putative regulatory gene, are clustered on the chromosome. This clustering of genes and, presumably, their coordinated regulation, would ensure that both essential enzymes required for the cycling of glutamine would be produced under the same nutritional conditions. It is extremely interesting that *C. acetobutylicum* 824 does not have a similar gene cluster — the *glnA* gene in this case does not appear to be genetically linked to other nitrogen assimilation regulatory genes. It is also intriguing that the *glnA* gene from *C. acetobutylicum* encodes a GSIII enzyme and not a GSI enzyme as in *C. saccharobutylicum* and *C. beijerinckii*, giving credence to the fact that they are not very closely related and supporting their assignment to different species. The other *Clostridium* strain, which produces a GSIII enzyme, is the cellulolytic *C. thermocellum.* Other bacterial genera reported to possess this form of the GS enzyme are either anaerobic rumen or intestinal bacteria, such as *Butyrivibrio*, *Bacteroides,* or *Ruminococcus* species [16], or members of the cyanobacteria.

Finally, this unique cluster of nitrogen assimilation genes, found only in *C. saccharobutylicum* and *C. beijerinckii* to date, would appear to be controlled at a number of different regulatory levels, involving novel regulatory mechanisms such as transcriptional antitermination and antisense regulation. However, the other elements of global nitrogen regulation are still to be identified in *Clostridium,* and a number of hypotheses remain to be tested, before a final model of nitrogen gene expression is obtained, which coordinate other nitrogen metabolic pathways such as nitrogen fixation [44].

REFERENCES

1. Jones, D.T. and Woods, D.R., Acetone-butanol fermentation revisited, *Microbiol. Rev.*, 50, 484, 1986.
2. Rogers, P. and Gottschalk, G., Biochemistry and regulation of acid and solvent production in Clostridia, In *The Clostridia and Biotechnology*, Woods, D.R., Ed., Butterworth-Heinemann, Reading, MA, 25, 1993.
3. Abou, Z.A., Fouad, M., and Yassein, M., Microbiological production of acetone-butanol by *Clostridium acetobutylicum, Zentralbl. Bakteriol. Naturwiss,* 133, 125, 1978.
4. Keis, S. et al., Taxonomy and phylogeny of industrial solvent-producing clostridia, *Int. J. Syst. Bacteriol.*, 45, 693, 1995.
5. Monot, F. et al., Acetone and butanol production by *Clostridium acetobutylicum* in a synthetic medium, *Appl. Environ. Microbiol.*, 44, 1318, 1982.
6. Monot, F. and Engasser, J.M., Production of acetone and butanol by batch and continuous culture of *Clostridium acetobutylicum* under nitrogen limitation, *Biotech. Lett.,* 5, 213, 1983.
7. Andersch, W., Bahl, H., and Gottschalk, G., Acetone-butanol production by *Clostridium acetobutylicum* in an ammonium-limited chemostat at low pH Values, *Biotechnology Letters,* 41, 29, 1982.
8. Roos, J.W., McLaughlin, J.K., and Papoutsakis, E.T., The effect of pH on nitrogen supply, cell lysis and solvent production in fermentations of *Clostridium acetobutylicum*, *Biotechnol. Bioeng.,* 27, S401, 1985.
9. Gottschal, J. and Morris, J.G., Non-production of actone and butanol by *Clostridium acetobutylicum* during glucose- and ammonium-limitation in continuous culture, *Biotech. Lett.*, 3, 525, 1981.
10. Long, S., Jones, D.T. and Woods, D.R., Initiation of solvent production, clostridial stage and endospore formation in *Clostridium acetobutylicum* P262, *Appl. Environ. Microbiol.*, 20, 256, 1984.
11. Atkinson, M.R. and Fisher, S.H, Identification of genes and gene products whose expression is activated during nitrogen-limited growth in *Bacillus subtilis, J. Bacteriol.*, 173, 23, 1991.
12. Ertan, H., The effect of various culture conditions on the levels of ammonia assimilatory enzymes of *Corynebacterium callunae*, *Arch. Microbiol.*, 158, 42, 1992.
13. Kanamori, K., Weiss, R.L., and Roberts, J.D, Ammonia assimilation pathways in nitrogen-fixing *Clostridium kluyverii* and *Clostridium butyricum, J. Bacteriol.,* 171, 2148, 1989.
14. Dainty, R.H., Glutamate biosynthesis in *Clostridium pasteurianum* and its significance in nitrogen metabolism, *Biochem. J.,* 126, 1055, 1972.
15. Bogdahn, M. and Kleiner, D., N_2 fixation and NH_4^+ assimilation in the thermophilic anaerobes *Clostridium thermosaccharolyticum* and *Clostridium thermoautotrophicum*, *Arch. Microbiol.*, 144, 102, 1986.
16. Woods, D.R. and Reid, S.J., Recent developments on the regulation and structure of glutamine synthetase enzymes from selected bacterial groups, *FEMS Microbiol. Rev.,* 11, 273, 1993.
17. Fisher, S.H. and Sonenshein, A.L., Control of carbon and nitrogen metabolism in *Bacillus subtilis*, *Annu. Rev. Microbiol.,* 45, 107, 1991.
18. Brana, A.F., Wolfe, S., and Demain, A.L, Relationship between nitrogen assimilation and cephalosporin synthesis in *Streptomyces clavuligerus*, *Arch. Microbiol.,* 146, 46, 1986.

19. Chen, G.J. and Russell, J.B, Transport of glutamine by *Streptococcus bovis* and conversion of glutamine to pyroglutamic acid and ammonia, *J. Bacteriol.,* 171, 2981, 1989.
20. Yamamoto, I., Abe, A., and Ishimoto, M., Properties of glutamate dehydrogenase purified from *Bacteroides fragilis, J. Biochem. (Tokyo),* 101, 1391, 1984.
21. Kanamori, K., Weiss, R.L. and Roberts, J.D, Role of glutamate dehydrogenase in ammonia assimilation in nitrogen-fixing *Bacillus macerans, J. Bacteriol.,* 169, 4692, 1987.
22. Fisher, S.H., Utilization of amino acids and other nitrogen-containing compounds, in *Bacillus subtilis and Other Gram-Positive Bacteria,* A.L. Sonenshein, J.A. Hoch, and Losick R., Eds., ASM Press, Washington, DC, 1993, 221.
23. Reitzer, L.J., Ammonia assimilation and the biosynthesis of glutamine, glutamate, aspartate, asparagine, L-alanine and D-alanine, in *Escherichia coli and Salmonella typhimurium: Cellular and Molecular Biology,* Neidhardt, F.C. et al., Eds., ASM Press, Washington, DC, 1996, 391.
24. Schreier, H.J., Biosynthesis of glutamine and glutamate and the assimilation of ammonia, in *Bacillus subtilis and Other Gram-positive Bacteria*, A.L. Sonenshein, J.A. Hoch, and Losick R., Eds., ASM Press, Washington, DC, 1993, 281.
25. Schulz, A.A., Collett, H.J., and Reid, S.J., Nitrogen and carbon regulation of glutamine synthetase and glutamate synthase in *Corynebacterium glutamicum* ATCC 13032, *FEMS Microbiol. Lett.*, 205, 361, 2001.
26. Fisher, S.H., Glutamine synthesis in *Streptomyces* — a review, Gene, 115, 13, 1992.
27. Carlson, T.A. and Chelm, B.K., Apparent eukaryotic origins of glutamine synthetase II from the bacterium *Bradyrhizobium japonicum, Nature,* 322, 568, 1986.
28. Rawlings, D.E. et al., Nucleotide sequence of the glutamine synthetase gene and its controlling region from the acidophilic autotroph *Thiobacillus ferrooxidans, Gene,* 53, 211, 1987.
29. Southern, J.A., Parker, J.R., and Woods, D.R., Novel structure, properties and inactivation of glutamine synthetase cloned from *Bacteroides fragilis, J. Gen. Microbiol.*, 133, 2437, 1989.
30. Goodman, H.J. and Woods, D.R., Cloning and nucleotide sequence of the *Butyrivibrio fibrisolvens* gene encoding a type III glutamine synthetase, *J. Gen. Microbiol.*, 139, 1487, 1993.
31. Reyes, J.C. and Florencio, F.J., A new type of glutamine synthetase in cyanobacteria: the protein encoded by the glnN gene supports nitrogen assimilation in *Synechocystis* sp. strain PCC 6803, *J. Bacteriol.,* 176, 1260, 1994.
32. Nolling, J. et al., Genome sequence and comparative analysis of the solvent-producing bacterium *Clostridium acetobutylicum, J. Bacteriol.*, 183, 4823, 2001.
33. Usdin, K.P. et al., Cloning, expression and purification of glutamine synthetase from *Clostridium acetobutylicum, Appl. Environ. Microbiol.*, 52, 413, 1986.
34. Vanoni, M.A. and Curti, B., Glutamate synthase: a complex iron-sulfur flavoprotein, *Cell. Mol. Life Sci.*, 55, 617, 1999.
35. Marques, S., Florencio, F.J., and Candau, P., Purification and characterization of the ferredoxin-glutamate synthase from the unicellular cyanobacterium *Synechococcus* sp. PCC 6301, *Eur. J. Biochem.,* 206, 69, 1992.
36. Knaff, D.B. and Hirasawa, M., Ferredoxin-dependent chloroplast enzymes, *Biochim. Biophys. Acta,* 1056, 93, 1995.
37. Avila, C. et al., Cloning and sequence analysis of a cDNA for barley ferredoxin-dependent glutamate synthase and molecular analysis of photorespiratory mutants deficient in the enzyme, *Planta*, 189, 475, 1993.
38. Gregerson, R.G. et al., Molecular characterization of NADH-dependent glutamate synthase from alfalfa nodules, *Plant Cell*, 5, 215, 1993.
39. Bohannon, D.E., Rosenkrantz, M.S., and Sonenshein, A.L., Regulation of *Bacillus subtilis* glutamate synthase genes by the nitrogen source, *J. Bacteriol.*, 163, 957, 1985.
40. Pelanda, R. et al., Glutamate synthase genes of the diazotroph *Azospirillum brasilense*. Cloning, sequencing, and analysis of functional domains, *J. Biol. Chem.*, 268, 3099, 1993
41. Stutz, H., Regulation and characterization of the nitrogen assimilatory gene cluster in *Clostridium acetobutylicum* P262, Ph.D. Thesis, University of Cape Town, South Africa, 2000.
42. Reitzer, L.J. and Magasanik, B., Ammonia assimilation and the biosynthesis of glutamine, glutamate, aspartate, asparagine, L-alanine and D-alanine, in: *Escherichia coli and Salmonella typhimurium: Cellular and Molecular Biology,* Neidhardt, F.C. et al., Eds., ASM Press, Washington, DC, 1987, 302.

43. Merrick, M.J. and Edwards, R.A., Nitrogen control in bacteria, *Microbiol. Rev.,* 59, 604, 1995.
44. Chen, J.-S., Toth, J., and Kasap, M., Nitrogen-fixation genes and nitrogenase activity in *Clostridium acetobutylicum* and *Clostridium beijerinckii, J. Ind. Microbiol. Biotechnol.*, 27, 281, 2001.
45. Belitsky, B.R., Biosynthesis of amino acids of the glutamate and aspartate families, alanine and polyamines, in *Bacillus subtilis and Its Closest Relatives: From Genes to Cells*, Sonenshein, A.L., Hoch, J.A., and Losick, R., Eds., ASM Press, Washington, DC, 2002, 203.
46. Wray, L.V.J., Atkinson, M.R., and Fisher, S.H., The nitrogen-regulated *Bacillus subtilis nrgAB* operon encodes a membrane protein and a protein highly similar to the *Escherichia coli glnB*-encoded PII protein, *J. Bacteriol.*, 176, 108, 1994.
47. Jakoby, M., Kramer, R., and Burkovski, A., Nitrogen regulation in *Corynebacterium glutamicum*: isolation of genes involved and biochemical characterization of corresponding proteins, *FEMS Microbiol. Lett.*, 173, 303, 1999.
48. Nolden, L. et al., Sensing nitrogen limitation in *Corynebacterium glutamicum*: the role of *glnK* and *glnD, Mol. Microbiol.,* 42, 1281, 2001.
49. Fisher, S.H., Regulation of nitrogen metabolism in *Bacillus subtilis*: vive la difference!, *Molec. Microbiol.,* 32, 223, 1999.
50. Schreier, H.J. and Sonenshein, A.L, Altered regulation of the *glnA* gene in glutamine synthetase mutants of *Bacillus subtilis, J. Bacteriol.,* 167, 35, 1986.
51. Fisher, S.H. and Debarbouille, M., Nitrogen source utilization and its regulation., in *Bacillus subtilis and Its Closest Relatives: From Genes to Cells*, Sonenshein, A.L., Hoch, J.A., and Losick, R., Eds., ASM Press, Washington, DC, 2002, 181.
52. Schreier, H.J. et al., Regulation of *Bacillus subtilis* glutamine synthetase gene expression by the product of the *glnR* gene, [published erratum appears in *J. Mol. Biol.*, 226, 571, 1992], *J. Mol. Biol.,* 210, 51, 1989.
53. Fisher, S.H. and Wray, L.V.J, Regulation of glutamine synthetase in *Streptomyces coelicolor, J. Bacteriol.,* 171, 2378, 1989.
54. Hillemann, D. et al., Genetic and biochemical characterization of the two glutamine synthetases GSI and GSII of the phosphinothricyl-alanyl-alanine producer, *Streptomyces viridochromogenes* Tu494, *J. Gen. Microbiol.,* 139, 1773, 1993.
55. Fink, D. et al., Nitrogen metabolism in *Streptomyces coelicolor* A3(2): modification of glutamine synthetase I by an adenylyltransferase, *Microbiology,* 145, 2313, 1999.
56. van Heeswijk, W.C. et al., The genes of the glutamine synthetase adenylylation cascade are not regulated by nitrogen in *Escherichia coli, Mol. Microbiol.,* 9, 443, 1993.
57. Nolden, L., Beckers, G., and Burkovski, A., Nitrogen assimilation in *Corynebacterium diphtheriae*: pathways and regulatory cascades, *FEMS Microbiol. Lett.*, 208, 287, 2002.
58. Jakoby, M. et al., AmtR, a global repressor in the nitrogen regulation system of *Corynebacterium glutamicum, Mol. Microbiol*., 37, 964, 2000.
59. Eguchi, Y., Itoh, T., and Tomizawa, J., Antisense RNA, *Annu. Rev. Biochem*., 60, 631, 1991.
60. Wagner, E.G. and Simons, R.W., Antisense RNA control in bacteria, phages, and plasmids, *Annu. Rev. Microbiol.,* 48, 713, 1994.
61. Wagner, E.G., Altuvia, S. and Romby, P., Antisense RNAs in bacteria and their genetic elements, *Adv. Genetics,* 46, 361, 2002.
62. Soderbom, F. et al., Regulation of plasmid R1 replication: PcnB and RNase E expedite the decay of the antisense RNA, CopA, *Mol. Microbiol*., 26, 493, 1997.
63. Xu, F., Lin-Chao, S., and Cohen, S. N., The *Escherichia coli pcnB* gene promotes adenylylation of antisense RNAI of ColE1-type plasmids *in vivo* and degradation of RNAI decay intermediates, *Proc. Natl. Acad. Sci. USA*, 90, 6756, 1993.
64. Brantl, S., Antisense-RNA regulation and RNA interference. *Biochim. Biophys. Acta,* 1575, 15, 2002.
65. Morfeldt, E. et al., Activation of alpha-toxin translation in *Staphylococcus aureus* by the trans-encoded antisense RNA, RNAIII, *EMBO J.,* 14, 4569, 1995.
66. del Solar, G. and Espinosa, M., Plasmid copy number control: an ever-growing story, *Mol. Microbiol.,* 37, 492, 2000.
67. Novick, R.P. et al., pT181 plasmid replication is regulated by a countertranscript-driven transcriptional attenuator, *Cell,* 59, 395, 1989.

68. Brantl, S., Birch-Hirschfeld, E., and Behnke, D, RepR protein expression on plasmid pIP501 is controlled by an antisense RNA-mediated transcription attenuation mechanism, *J. Bacteriol.,* 175, 4052, 1993.
69. Le Chatelier, E., Ehrlich, S., and Janniere, L., Countertranscript-driven attenuation system of the pAMβ1 *repE* gene, *Mol. Microbiol.,* 20, 1099, 1996.
70. Kolb, F.A. et al., Bulged residues promote the progression of a loop-loop interaction to a stable and inhibitory antisense-target RNA complex, *Nucleic Acids Res.,* 29, 3145, 2001.
71. Delihas, N. and Forst, S., MicF: an antisense RNA gene involved in response of *Escherichia coli* to global stress factors, *J. Mol. Biol.*, 313, 1, 2001.
72. Altuvia, S. et al., A small, stable RNA induced by oxidative stress: role as a pleiotropic regulator and antimutator, *Cell,* 90, 43, 1997.
73. Argaman, L. and Altuvia, S., *fhlA* repression by OxyS RNA: kissing complex formation at two sites results in a stable antisense-target RNA complex, *J. Mol. Biol.*, 300, 1101, 2000.
74. Janssen, P.J., Jones, D.T., and Woods, D.R., Studies on *Clostridium acetobutylicum glnA* promoters and antisense RNA, *Mol. Microbiol.,* 4, 1575, 1990.
75. Fierro-Monti, I.P., Reid, S.J., and Woods, D.R., Differential expression of a *Clostridium acetobutylicum* antisense RNA: implications for regulation of glutamine synthetase, *J. Bacteriol.*, 174, 7642, 1992.
76. Inouye, M., Antisense RNA: its functions and applications in gene regulation — a review, *Gene,* 72, 25, 1988.
77. Desai, R.P. et al., Metabolic flux analysis elucidates the importance of the acid-formation pathways in regulating solvent production by *Clostridium acetobutylicum, Metab. Eng.*, 1, 206, 1999.
78. Tummala, S., Welker, N., and Papoutsakis, E.T., Design of antisense RNA constructs for downregulation of the acetone formation pathway of *Clostridium acetobutylicum*, *J. Bacteriol.,* 185, 1923, 2003.
79. Tummala, S.B., Junne, S.G., and Papoutsakis, E.T., Antisense RNA downregulation of coenzyme A transferase combined with alcohol-aldehyde dehydrogenase overexpression leads to predominantly alcohologenic *Clostridium acetobutylicum* fermentations, *J. Bacteriol.*, 185, 3644, 2003.
80. Woods, D.R. and Reid, S.J., Regulation of nitrogen metabolism, starch utilisation and the *beta-hbd-adh1* gene cluster in *Clostridium acetobutylicum*, *FEMS Microbiol. Rev.,* 17, 299, 1995.
81. Quixley, K.W. and Reid, S.J., Construction of a reporter gene vector for *Clostridium beijerinckii* using a *Clostridium* endoglucanase gene, *J. Mol. Microbiol. Biotechnol.*, 2, 53, 2000.
82. Janssen, P.J. et al., Molecular analysis and regulation of the *glnA* gene of the gram-positive anaerobe *Clostridium acetobutylicum* [published errata appeared in *J. Bacteriol.* 170, 3319, 1988], *J. Bacteriol.*, 170, 400, 1988.
83. Quixley, K.W., Studies on the regulation of nitrogen metabolism in *Clostridium acetobutylicum* NCP262 and *C. beijerinckii* NCIMB 8052, M.Sc. Thesis, University of Cape Town, South Africa, 1999.
84. Guttierrez, J.C. et al., *nasST,* two genes involved in the induction of the assimilatory nitrite-nitrate reductase operon (*nasAB*) of *Azotobacter vinelandii, Mol. Microbiol.*, 18, 579, 1995.
85. Lowe, N., Rice, P.M., and Drew, R., Nucleotide sequence of the aliphatic amidase regulator gene (*amiR*) of *Pseudomonas aeruginosa*, *FEBS Lett.*, 246, 39, 1989.
86. Lin, J.T., Goldman, B.S., and Stewart, V., The *nasFEDCBA* operon for nitrate and nitrite assimilation in *Klebsiella pneumoniae* M5al, *J. Bacteriol.,* 176, 2551, 1994.
87. Wilson, S.A. et al., Antitermination of amidase expression in *Pseudomonas aeruginosa* is controlled by a novel cytoplasmic amide-binding protein, *EMBO J.,* 12, 3637, 1993.
88. O'Hara, B.P. et al., Crystal structure and induction mechanism of AmiC-AmiR: a ligand-regulated transcription antitermination complex, *EMBO J.*, 18, 5175, 1999.
89. Chai, W. and Stewart, V., NasR, a novel RNA-binding protein, mediates nitrate-responsive transcription termination of the *Klebsiella oxytoca* M5al *nasF* operon leader *in vitro, J. Mol. Biol.*, 283, 339, 1998.
90. Lin, J.T. and Stewart, V., Nitrate assimilation by bacteria, *Adv. Microb. Physiol.,* 39, 1, 1998.
91. Goodrich, J.A., Cutler, G., and Tjian, R., Contacts in context: promoter specificity and macromolecular interactions in transcription, *Cell,* 84, 825, 1996.
92. Zappe, H., Jones, D.T., and Woods, D.R., Cloning and expression of *Clostridium acetobutylicum* endoglucanase, cellobiase and amino acid biosynthesis genes in *Escherichia coli*, *J. Gen. Microbiol.,* 132, 1367, 1986.
93. Abratt, V.R., Zappe, H., and Woods, D.R., A reporter gene vector to investigate the regulation of glutamine synthetase in *Bacteroides fragilis* Bf1, *J. Gen. Microbiol.,* 139, 59, 1993.

94. Reid, S.J., Rafudeen, M.S., and Leat, N.G., The genes controlling sucrose utilisation in *Clostridium beijerinckii* NCIMB 8052 constitute an operon, *Microbiol.,* 145, 1461, 1999.
95. Mathews, D. et al., Expanded sequence dependence of thermodynamic parameters improves prediction of RNA secondary structure, *J. Mol. Biol.*, 288, 911, 1999.
96. Aymerich, S. and Steinmetz, M., Specificity determinants and structural features in the RNA target of the bacterial antiterminator proteins of the Bgl/SacY family, *Proc. Natl. Acad. Sci. U.S.A.,* 89, 10410, 1992.
97. Babitzke, P., Regulation of tryptophan biosynthesis: Trp-ing the TRAP or how *Bacillus subtilis* invented the wheel, *Mol. Microbiol.*, 26, 1, 1997.

12 Nitrogen Fixation

Jiann-Shin Chen

CONTENTS

12.1 INTRODUCTION

Nitrogen is an essential element for all life forms. Although nitrogen gas (N_2 or dinitrogen) constitutes about 80% of the Earth's atmosphere, plants and animals and most microbes cannot use N_2 as a nitrogen source because of the chemical stability of the N_2 molecule. The usable form of nitrogen is "fixed" nitrogen, where nitrogen has been transformed into part of a nitrogenous compound such as ammonia or nitrate. In nature, there are two main routes of transforming or fixing nitrogen. One is by lightning, and it contributes, at most, about 10% of the total annual yield of fixed nitrogen [1]. By far, the most important route of nitrogen fixation is from the activity of certain soil bacteria, which contributes about 65% of the total annual yield of fixed nitrogen, with the balance (25%) coming from industrial production using the Haber-Bosch process.

Biological nitrogen fixation is a metabolic capability found only in the prokaryotes, including both bacteria and archaea. However, symbiotic and associative nitrogen fixation involves both bacteria and the host plants, and nitrogen-fixing bacteria also colonize animals, where nitrogen fixation can be a significant source of nitrogen for wood-infesting termites and shipworms [1]. Nitrogen-fixing bacteria are found across the taxonomic groups, including aerobic, anaerobic, facultatively anaerobic, and photosynthetic bacteria, as well as cyanobacteria [2]. Among the nitrogen-fixing organisms, *Clostridium* occupies a very special place. The species *Clostridium pasteurianum* is the first known free-living nitrogen-fixing bacterium, and it has been studied in the laboratory ever since its isolation by S. Winogradsky more than 100 years ago [3]. After the ^{15}N

tracer technique became available, a number of *Clostridium* species, including *C. acetobutylicum*, *C. beijerinckii*, and *C. butyricum*, were found to be capable of nitrogen fixation [4].

Biochemical studies conducted with *C. pasteurianum* between the late 1950s and early 1970s led to pivotal findings that are the foundation of our current knowledge of the enzymology of biological nitrogen fixation. These crucial events include the preparation of consistently active nitrogen-fixing cell-free extracts [5,6], the discovery of the low-redox potential iron-sulfur protein ferredoxin as the natural electron donor for nitrogenase [7–9], the discovery that ATP, in addition to ferredoxin, is required for nitrogenase activity [10], and the separation of nitrogenase into two metalloproteins [11,12].

In retrospect, the use of an obligate anaerobe in early nitrogen-fixation research had a profound effect on the progress of this field. Because of the O_2-sensitivity of the source organism, anaerobic conditions were used in the preparation of cell-free extracts and pyruvate was used in the assay to support the *in vitro* nitrogen-fixing activity. This combination of circumstances facilitated the realization that nitrogenase is extremely O_2-sensitive and that reduced ferredoxin and ATP, both of which are produced during the phosphoroclastic cleavage of pyruvate, are necessary for nitrogenase-catalyzed reduction of dinitrogen to ammonia.

Although nitrogenase is a highly conserved enzyme in terms of both the structure and the enzymic properties, the nitrogenase of clostridia exhibits distinct properties. Most notably, the nitrogenase component proteins of *C. pasteurianum* do not form an active complex in heterologous combinations with a component protein from *A. vinelandii* or *K. pneumoniae* [13]. This is consistent with the characteristic structural features of the Fe and MoFe proteins of *C. pasteurianum*, and such features are also present in the corresponding proteins of *C. acetobutylicum* and *C. beijerinckii* (see Section 12.7). At the DNA level, novel *nif* genes, including the fused *nifN-B* gene and the split *nifV*ω and *nifV*α genes, have so far only been observed in *C. acetobutylicum*, *C. beijerinckii*, and *C. pasteurianum* (see Section 12.8). In addition, the control elements for transcriptional activation of the *nif* genes in these three clostridial species also appear different from the extensively studied NifA system (see Section 12.10). The clostridia therefore possess some unique genetic elements for nitrogen fixation, and these elements expand the choices of properties for an engineered *nif* system.

12.2 ISOLATION OF THE FIRST FREE-LIVING NITROGEN-FIXING BACTERIUM — *CLOSTRIDIUM PASTEURIANUM*

Winogradsky [14] first presented evidence for the assimilation of nitrogen gas by a butyric acid-producing and spore-forming bacillus. It was followed by the description of a nitrogen-fixing *Clostridium* that was isolated from soil in 1892 [3]. This first free-living diazotroph that was studied in isolation was named *Clostridium pasteurianum* in honor of Louis Pasteur [3; p. 330]:

> Nous l'appellerons en honneur du grand savant, créateur de notre science: *Clostridium pasteurianum*.

In a subsequent report, Winogradsky presented a careful sketch of *C. pasteurianum* cells undergoing the sporulation and germination processes [15], and photomicrographs were used in the following report to illustrate the characteristic shape of sporulating *C. pasteurianum* cells [16]. It is interesting to note that some authors thought the name initially given by Winogradsky was *Clostridium pastorianum* to indicate that the spindle-shaped organism was from the field [e.g., 17]. In fact, for some unknown reason, the name *Clostridium pastorianum* was used in the 1902 publications [15,16] but was rarely used after. The culture of *C. pasteurianum* from Winogradsky was subjected to single-cell isolation in the laboratory of McCoy at the University of Wisconsin, and this further purification did not alter the nitrogen-fixing activity or efficiency of the culture [18]. *C. pasteurianum* strain W5 (= ATCC 6013), which was from the laboratory of McCoy, represents the organism isolated by Winogradsky. Among the nitrogen-fixing clostridia, *C. pasteurianum* is the most thoroughly characterized in terms of the nitrogen-fixing system.

12.3 NITROGEN-FIXING CLOSTRIDIA

The genus *Clostridium* is a broad collection of rod-shaped, anaerobic bacteria. It has been proposed that the traditional genus *Clostridium* be rearranged into different genera [19]. The proposed scheme will separate the currently known nitrogen-fixing clostridia into different genera. However, *C. acetobutylicum*, *C. beijerinckii*, *C. butyricum*, and *C. pasteurianum* will remain in the redefined genus *Clostridium*, which is reserved for species belonging to the rRNA group I of Johnson and Francis [20].

Following the isolation and characterization of *C. pasteurianum* as a nitrogen fixer by earlier workers, Rosenblum and Wilson [4] reported the fixation of $^{15}N_2$ by several clostridia, including *C. aceticum*, *C. acetobutylicum*, *C. beijerinckii*, *C. butylicum*, *C. butyricum*, *C. felsineum*, *C. kluyverii*, *C. lactoacetophilum*, *C. madisoni*, *C. pasteurianum*, *C. pectinovorum*, and *C. tetanomorphum*. Among these species, *C. butylicum* [21], *C. lactoacetophilum* [22], and *C. madisoni* [23] have been reclassified as *C. beijerinckii*. Nitrogen fixation by *C. acetobutylicum* [24], *C. beijerinckii* [24], *C. butyricum* [25], and *C. kluyverii* [25] has been confirmed. *C. aceticum* did not grow in a yeast extract-containing medium that supports nitrogen-fixing growth of *C. pasteurianum* [26]. It should be noted that the medium used by Rosenblum and Wilson [4] contained low concentrations of yeast extract and tryptone (as carryover from the inoculum). Besides these mesophilic and saccharolytic species, the nitrogen-fixing clostridia include other species that belong to other physiological groups: the acetogenic *C. formicoaceticum* [27], the cellulytic *C. hungatei* [28], and the acid-tolerant *C. akagii* and *C. acidisoli* [29]. The nitrogen-fixing thermophile *C. thermosaccharolyticum* [30] has been renamed as *Thermoanaerobacterium thermosaccharolyticum* [19].

The genus *Paenibacillus* contains nitrogen-fixing species that form spores under anaerobic conditions [31]. The former *Clostridium durum* (or *Paenibacillus durum*), the dominant organism found in a sediment core from the Black Sea, has been reclassified as a member of the species *Paenibacillus azotofixans* [31]. Criteria used in the reclassification include the DNA relatedness at the genome level as measured by the DNA–DNA reassociation technique. Therefore, some of the spore-forming, nitrogen-fixing rods do not belong to the genus *Clostridium*.

Among the nitrogen-fixing clostridia, *C. pasteurianum* is the most frequently used species for the study of biochemistry and physiology of nitrogen fixation. Accordingly, *C. pasteurianum* was first chosen for a study of the *nif* genes from an obligate anaerobe [32]. At present, the nucleotide sequences of the *nif* genes encoding the nitrogenase component proteins and those genes required for the synthesis of the iron-molybdenum cofactor (FeMo cofactor) have been determined, either completely or partially, in *C. acetobutylicum*, *C. beijerinckii*, *C. pasteurianum*, and *C. hungatei*. In addition, the complete genome sequence of *C. acetobutylicum* has been determined, which allows a search for nitrogen fixation-related genes outside the *nif* cluster of an obligate anaerobe.

12.4 GROWTH MEDIA FOR THE NITROGEN-FIXING CLOSTRIDIA

Both chemically defined and complex growth media are available for the study of nitrogen fixation by clostridia. Among the known nitrogen-fixing enzymes, the most widely distributed, and also the best characterized, is the molybdenum-containing nitrogenase, consisting of two iron-sulfur proteins — the iron protein and the molybdenum-iron protein. Because of the composition of the nitrogenase component proteins, a growth medium supporting nitrogen fixation is expected to contain adequate amounts of molybdenum, iron, and sulfur compounds. Molybdenum was shown to stimulate nitrogen fixation by *C. butyricum* [33]. Similarly, iron stimulates nitrogen fixation by *C. pasteurianum* [34]. It is interesting to note that when the isolation of *C. pasteurianum* was reported [3], the growth medium did not contain added molybdenum. (The medium consisted of a mineral solution of the following composition: distilled water, 1000 cc; potassium phosphate, 1 gm; magnesium sulfate, 0.5 gm; sodium chloride, 0.01 to 0.02 gm; iron sulfate, 0.01 to 0.02 gm; manganese sulfate, 0.01 to 0.02 gm. To the mineral solution, 2 to 4% of glucose and a small amount of calcium carbonate were added.) The level of magnesium sulfate was later lowered to 0.2 gm per liter of the mineral solution [15]. The term "Winogradsky's medium" was introduced by McCoy and co-workers in 1928 [18], and the composition was given as K_2HPO_4, 1 gm; $MgSO_4.7H_2O$, 0.2 gm; NaCl, 0.01 gm; $FeSO_4.7H_2O$, 0.01 gm; $MnSO_4.4H_2O$, 0.01 gm; glucose, 20 gm; distilled water, 1000 cc; calcium carbonate, 30 gm, to be added after sterilization. The synthetic medium used by McCoy et al. [18] also did not contain any added molybdenum or vitamins. It might be presumed that in the early studies, either a sufficient amount of molybdenum was present as an impurity in the media or a molybdenum-independent nitrogenase was responsible for the observed nitrogen-fixing activity. Genes for a putative molybdenum- and vanadium-independent nitrogenase are present in *C. pasteurianum* W5 (see Section 12.8.2).

Modified versions of Winogradsky's medium have been widely used in the study of nitrogen fixation by clostridia. A medium containing $NaMoO_4$, $CoCl_2$, $CaCl_2$, $FeCl_3$, $CuSO_4$, $ZnSO_4$, $MnSO_4$, $MgSO_4$, K_2HPO_4, Na_2SO_4, biotin, *p*-aminobenzoic acid, and sucrose was developed for *C. pasteurianum* [35]. However, the need for all of the elements in the medium has not been established. On the other hand, a synthetic medium containing only K_2HPO_4, KH_2PO_4, $MgSO_4$, $CaCl_2$, $FeSO_4$, $NaMoO_3$, biotin, and sucrose was also routinely used for nitrogen-fixing cultures of *C. pasteurianum* [36–38]. Nitrogen-fixing *C. pasteurianum* was also grown in a continuous culture (at a dilution rate of 0.2) using a medium containing sucrose, ferric citrate, $MgSO_4$, Na_2MoO_4, $MnSO_4$, KH_2PO_4, and biotin [39]. Based on a comparison of six commonly used nitrogen-fixing media, a medium containing K_2HPO_4, KH_2PO_4, $MgSO_4$, $CaCl_2$, NaCl, Na_2MoO_4, Na_2Fe-EDTA, biotin, *p*-aminobezoic acid, yeast extract, and three carbon substrates (sucrose, mannitol, sodium lactate) was developed, and it supported good growth of *C. pasteurianum* [40].

A nitrogen-fixing culture is always provided with a sufficient amount of N_2 as the nitrogen source. The amount and variety of nitrogenous compounds (to provide amino acids, nucleotides, and vitamins) to be included in a nitrogen-fixing medium, when necessary, must be determined with each organism so that the synthesis of nitrogenase is not repressed. When Rosenblum and Wilson [4] surveyed 15 strains of clostridia for their nitrogen-fixing activity, the stock was grown on a modified Winogradsky medium, which contained, in addition to the salts and glucose, 1 gm tryptone, 1 g agar, and 2 gm $CaCO_3$ per liter. Nitrogen fixation was tested in a N-free medium in which the tryptone was omitted and 1 ml of yeast water (1 mg N per ml) was added as a source of growth factor. Because 1 ml of the stock culture was added to 15 ml of the N-free medium, a certain concentration of the nitrogenous compounds from tryptone could be present in the N-free medium. The presence of these nitrogenous compounds apparently did not prevent the strains from fixing nitrogen. It is also possible that the nitrogen-fixing system was synthesized after the fixed N in the medium had been consumed [41].

Either biotin or biotin plus *p*-aminobenzoic acid is required for maximal growth for clostridial species including *C. acetobutylicum*, *C. beijerinckii* (formerly *C. butylicum*), and *C. felsineum* [42]. Growth of *C. pasteurianum* under nitrogen-fixing conditions is noticeably stimulated by biotin [34]. Although *p*-aminobenzoic acid is often included in the growth medium for the clostridia, it is not required by nitrogen-fixing *C. pasteurianum* [43,44]. Besides biotin and *p*-aminobenzoic acid, other vitamins [28,45] or yeast extract [46] have been added to the growth medium for nitrogen-fixing clostridia, but a requirement for the other vitamins has not been proven.

Members of the genus *Clostridium* are metabolically diverse. A broad range of compounds can serve as the carbon source. Therefore, besides sucrose and glucose, other fermentable substrates, including polymers such as cellulose, may be used as the carbon source in a nitrogen-fixing medium [28]. The combined-carbon medium (containing sucrose, mannitol, and lactate) of Rennie [40] was successfully used, after replacing yeast extract with a trace-element solution, to demonstrate nitrogen fixation with *C. akagii* and *C. acidisoli* [29].

12.5 PREPARATION OF CELL-FREE EXTRACTS AND MEASUREMENT OF NITROGEN-FIXING ACTIVITY

Because of the oxygen-sensitivity of nitrogenase and possible cold-lability of the nitrogenase Fe protein, the preparation of cell-free nitrogen-fixing extracts from a new organism requires special techniques. The methods for the preparation of active extracts from *C. pasteurianum* may be a useful guide for works involving other clostridia. These methods encompass the production, storage, and lysis of nitrogen-fixing cells and the handling and storage of cell-free extracts.

Batch cultures of butyric acid-producing clostridia require adequate pH control to allow sufficient growth. $CaCO_3$ is an effective buffering agent in growth media for nitrogen-fixing, butyric acid-producing clostridia, and the first consistently active nitrogen-fixing cell extracts were prepared from cells grown in a medium containing $CaCO_3$ [6]. However, the unused $CaCO_3$ in a culture will sediment with cells during centrifugation, and its presence may interfere with cell lysis and the ensuing analytical procedures. The addition of KOH (8–10 N) through a pH controller is an effective way of maintaining the culture pH [41]. Nitrogen-fixing cells may be collected by centrifugation or by filtration and centrifugation. The residual sugar is removed by washing the sedimented cells with an anaerobic buffer so that excessive H_2 and acid production are prevented when cell lysis is carried out in a closed vessel at room temperature. Excessive production of acetic and butyric acids during cell lysis may cause a drop in pH and inactivation of nitrogenase. Paste of nitrogen-fixing cells may be rapidly frozen in liquid nitrogen and then stored in closed containers (closed after the liquid nitrogen has been evaporated) below –20°C. Cell paste of *C. pasteurianum* can also be dried under a vacuum in rotary evaporators that are kept at 30 to 40°C in water bath to prevent the cell paste from freezing [6; see reference 47 for a photograph of the setup for drying cells]. Dried cell paste is collected as flakes. When moisture and air are excluded, dried cells can be stored in sealed bottles at –15°C or below for months without loss of nitrogen-fixing activity [6,48].

Autolysis of dried *C. pasteurianum* cells occurs readily in a buffer without the addition of lysozyme [6]. However, the addition of lysozyme [48] and DNase [49] to the buffer facilitates the lysis of *C. pasteurianum* cells, especially when frozen cells are used for the preparation of cell-free extracts. Exclusion of air is crucial during the preparation of cell-free extracts or the nitrogenase activity will be lost. A vacuum manifold system is most practical for the preparation of oxygen-free buffers and for the preparation of nitrogen-fixing cell extracts.

There are several assays for the measurement of nitrogenase activity. The acetylene-reduction assay, which measures the nitrogenase-catalyzed formation of ethylene from acetylene by gas chromatography, is the most convenient and, hence, the most commonly used assay for nitrogen fixation [50]. The acetylene-reduction assay is suitable for the measurement of nitrogen-fixing activity in samples ranging from soils, whole cells, to purified nitrogenase component proteins. Several methods are available for the measurement of N_2 fixation [50]. $^{15}N_2$ is a sensitive and specific tracer for nitrogen fixation, but the measurement of $^{15}N_2$ requires mass spectrometry. Ammonia, the physiological product of the nitrogenase-catalyzed reaction, can be measured with the Nessler's reagent [50] or the phenol/hypochlorite method [51]. When ammonia formation by crude extracts of *C. pasteurianum* is measured, the incorporation of ammonia into glutamine by glutamine synthetase in the extracts can introduce an appreciable error [50].

The nitrogen-fixation reaction requires a source of electron and a source of energy. Ferredoxin or flavodoxin is the usual electron donor for nitrogenase, whereas ATP hydrolysis provides the energy. The amount of ATP consumed during the transfer of a pair of electrons from the Fe protein to the MoFe protein depends on the reaction conditions. Under optimal conditions, the nitrogenase-catalyzed reaction has the following stoichiometry:

$$N_2 + 8\ e^- + 8\ H^+ + 16\ \text{Mg-ATP} \rightarrow 2\ NH_3 + H_2 + 16\ \text{Mg-ADP} + 16\ \text{Pi}$$

The formation of H_2 is an obligatory side reaction when N_2 is reduced to NH_3, and it cannot be suppressed by increasing the N_2 concentration [52]. In *in vitro* assays, sodium dithionite is an effective reductant for nitrogenase, and a separate electron donor is not required. The physiological electron donor for the clostridial nitrogenase is ferredoxin [10]. In nitrogen-fixing clostridia, reduced ferredoxin can be generated by pyruvate:ferredoxin oxidoreductase [8] or hydrogenase [7]. Therefore, when the nitrogen-fixing activity of a cell-free extract is measured, either pyruvate or H_2, besides dithionite, can be the source of the reducing power. When pyruvate is used as the reductant, it can also provide ATP via the action of phosphotransacetylase (Acetyl-CoA + Pi → acetyl phosphate + CoA) and acetokinase (acetyl phosphate + ADP → acetate + ATP).

12.6 PURIFICATION OF THE MOLYBDENUM-NITROGENASE FROM *C. PASTEURIANUM*

The molybdenum-nitrogenase consists of two component proteins, the Fe protein and the MoFe protein, which differ in size and net charge. After a cell-free extract of *C. pasteurianum* has been treated with either protamine sulfate [48,53] or polyethylene glycol [38] to precipitate out nucleic acids, the two component proteins of nitrogenase are resolved by gel filtration [53] or by anion-exchange chromatography [38]. The Fe and MoFe proteins are further purified by column chromatography. The purification steps must be carried out in an O_2-free environment because the Fe protein and MoFe proteins, respectively, have a half-life of about 45 seconds and 10 minutes in air [54]. Techniques for maintaining and monitoring anaerobic conditions for biochemical work can be found in Mortenson, 1972; Eady, 1980; and Beinert et al., 1978 [48,54,55]. The Fe protein of *C. pasteurianum* is unstable at temperatures near 0°C, so its purification, including the centrifugation steps, is performed at 15 to 25°C [38,48]. Both Fe and MoFe proteins can be stored as frozen droplets in liquid nitrogen for months or longer [48,54].

12.7 PROPERTIES OF THE *C. PASTEURIANUM* MOLYBDENUM-NITROGENASE

The molybdenum-nitrogenase consists of the Fe protein and the MoFe protein. The polypeptides that constitute the Fe and the MoFe proteins are highly conserved among all known diazotrophs [56]. During catalysis, the Fe and the MoFe proteins associate and dissociate when electrons are transferred from the Fe protein to the MoFe protein with the concomitant hydrolysis of ATP. The Fe protein is thus the specific electron donor for the MoFe protein. Substrate reduction occurs only after a sufficient number of electrons have been accumulated on the MoFe protein [57].

Because of the high degree of conservation at the protein level, an active, heterologous nitrogenase complex can form between component proteins isolated from different species [13]. However, heterologous combinations involving a component protein of *C. pasteurianum* generally do not yield an active nitrogenase complex. The Fe protein of *C. pasteurianum* and the MoFe protein of *A. vinelandii* form a tight-binding, inactive heterocomplex [58]. The Mo-nitrogenase of *C. pasteurianum* has distinct enzymic properties. Whereas nitrogenase is generally sensitive to H_2 inhibition, the Mo-nitrogenase of *C. pasteurianum* is less sensitive to H_2 as an inhibitor than nitrogenases from other organisms [59]. Compared with nitrogenases from *A. vinelandii* and *K. pneumoniae*, the *C. pasteurianum* nitrogenase has a higher specificity for nucleotides [60]. These functional differences can be attributed to the characteristic amino acid sequences of the Fe and MoFe proteins of *C. pasteurianum*, and the crystallographic structures revealed distinct regions in the Fe and MoFe proteins of *C. pasteurianum* [61–63].

12.7.1 Fe Protein

The Fe protein of *C. pasteurianum* is a homodimer with a mol wt of about 60,000, which is lighter than the mol wt (about 65,000) of the Fe proteins from many other diazotrophs. The polypeptide is 273 amino acids in length and with an estimated *pI* of 4.85, making it the most acidic of Nif proteins of *C. pasteurianum* [32]. The amino acid sequence was first determined from the purified Fe protein [64], and the sequence corresponds to that encoded by *nifH1* [65]. The crystallographic structure of the *C. pasteurianum* Fe protein has an overall similarity to the Fe protein of *A. vinelandii* [63]. A single 4Fe:4S cluster covalently bridges the two monomers through the thio ligands of Cys94 and Cys129 of each monomer. The Fe protein provides the binding sites for MgATP and MgADP.

Using free *A. vinelandii* Fe protein and the Fe protein with nucleotides bound, it was shown that there is significant conformational change at the dimer interface following the binding of the nucleotides [66]. Earlier studies with the *C. pasteurianum* Fe protein revealed that the binding of MgATP to the Fe protein increases the exposure of its FeS cluster to solvent [67] and lowers the redox potential of the protein by 110 mV [68]. Despite an overall similarity, there are discernible structural differences between the Fe proteins of *C. pasteurianum* and *A. vinelandii*. The greatest structural diversity is observed in the connecting loops between secondary structural elements and the C-terminus. The shorter C-terminal sequence of the *C. pasteurianum* NifH does not allow it to extend onto the opposing subunit to create substantial interaction, which may have a negative effect on the overall stabilization of the dimer [63]. The deduced NifH amino acid sequences of *C. acetobutylicum* ATCC 824 and *C. beijerinckii* strains NRRL B593 and NCIMB 8052 share a high degree of sequence identity with the *C. pasteurianum* NifH1 throughout the entire length of the sequence, including the characteristically short C-terminal region [24,69]. It suggests that the Fe protein of *C. acetobutylicum*, *C. beijerinckii*, and *C. pasteurianum* share similar molecular properties.

12.7.2 MoFe Protein

The MoFe protein is a tetramer of the $\alpha_2\beta_2$ structure, which may be viewed as two $\alpha\beta$ subunit pairs. The *C. pasteurianum* MoFe protein has a mol wt of about 218,000. The α-subunit is encoded by the *nifD* gene and is 533 amino acids in length, whereas the β-subunit is encoded by the *nifK* gene and is 458 amino acids in length. Each $\alpha\beta$ subunit-pair contains two metalloclusters, an iron-sulfur cluster known as the P-cluster and a molybdenum-iron-sulfur cluster known as the iron-molybdenum cofactor (or FeMo-cofactor). The nitrogen-reduction site is believed to reside in the FeMo-cofactor, and electron transfer from the Fe protein to the FeMo-cofactor is mediated by the P-cluster.

For the *C. pasteurianum* MoFe protein, the α-subunit (NifD) is longer and the β-subunit (NifK) is shorter than the corresponding proteins of other diazotrophs [70]. The longer NifD and shorter NifK are also observed in *C. acetobutylicum* and *C. beijerinckii* [24]. There is an extra stretch of about 50 amino acids occurring between residues 375 and 430 in NifD and a shorter N-terminal region (by about 50 residues) in NifK. The overall crystallographic structure is similar between the MoFe proteins of *C. pasteurianum* and *A. vinelandii*, especially the tertiary structure around the metal centers [61,62]. However, significant differences also exist. For example, because of the ~50-residue insertion, the general shape of the *C. pasteurianum* MoFe protein is slightly elongated along one axis when compared with that of the *A. vinelandii* MoFe protein [62]. Also, the FeMo-cofactor is buried more deeply in the *C. pasteurianum* MoFe protein than in the *A. vinelandii* MoFe protein, because of the loop involving residues 383 to 397, which is characteristic of the *C. pasteurianum* α subunit (NifD). The structure of the MoFe protein of *C. acetobutylicum* and *C. beijerinckii* may be expected to resemble that of *C. pasteurianum* because the amino acid sequences of NifD and NifK of these three species are highly conserved [69].

Clostridium acetobutylicum

***nifH* ▸ *nifI$_1$* ▸ *nifI$_2$* ▸ *nifD* ▸ *nifK* ▸ *nifE* ▸ *nifN-B* ▸ *nifVω* ▸ *nifVα* ▸**

Clostridium beijerinckii

***nifH* ▸ *nifI$_1$* ▸ *nifI$_2$* ▸ *nifD* ▸ *nifK* ▸ *nifE* ▸ *nifN-B* ▸ *fdxA* ▸ *nirJ1* ▸ *nirJ2* ▸ *nirD* ▸ *nirH* ▸ *nifVω* ▸ *nifVα* ▸**

Clostridium pasteurianum

***nifH2* ▸ *nifH1* ▸ *nifD* ▸ *nifK* ▸ *nifE* ▸ *nifN-B* ▸ *modA* ▸ *modB* ▸ *nifVω* ▸ *nifVα* ▸**

FIGURE 12.1 Organization of the *nif* cluster of *C. acetobutylicum* ATCC 824, *C. beijerinckii* NRRL B593, and *C. pasteurianum* W5. Besides *nif*-specific genes, each *nif* cluster also contains genes that may play a role in nitrogen regulation or nitrogenase synthesis. The direction of translation is indicated by arrowheads.

12.8 THE NITROGEN-FIXATION GENES OF THE CLOSTRIDIA

Those genes that are required for the activity of the molybdenum-nitrogenase are designated the *nif* genes. Most of the *nif* genes typically occur in a cluster, but additional *nif* genes may occur outside the main cluster. A cluster of *nif* genes has been identified in three species of clostridia: *C. acetobutylicum*, *C. beijerinckii*, and *C. pasteurianum* (Figure 12.1). The nucleotide sequence of the *nif* clusters of *C. acetobutylicum* ATCC 824, *C. beijerinckii* NRRL B593, and *C. pasteurianum* W5 has been determined. The organization of the *nif* genes is conserved in these three species. The proposed gene products or functions of the *nif* genes are listed in Table 12.1. The presumed boundaries for the *nif* cluster are defined by long intergenic regions bordering the cluster and the presence of flanking genes that are unrelated to nitrogen fixation or nitrogen metabolism. In addition to the *nif* genes, the clostridia also have putative *anf* genes, which encode an alternative nitrogen-fixing system that does not involve molybdenum or vanadium.

12.8.1 THE MOLYBDENUM-DEPENDENT NITROGEN-FIXATION (*NIF*) GENES

The *nif* cluster of *C. acetobutylicum* and *C. pasteurianum* is each composed of nine genes, whereas the *nif* cluster of *C. beijerinckii* is composed of 14 genes. Seven of these genes are common for the three species, and they are the *nif*-specific genes arranged in the order of *H* ▸ *D* ▸ *K* ▸ *E* ▸ *N-B* ▸ *Vω* ▸ *Vα* ▸. These seven *nif* genes are conceivably the minimum required for nitrogen fixation. The fused *nifN-B* gene and the split $nifV_\omega$ and $nifV_\alpha$ genes are characteristics of these three *Clostridium* species. The intergenic region preceding the *nif* cluster of these three *Clostridium* species is between 459 and 553 bp. The length of this intergenic region seems to mark a clear upstream boundary for the *nif* cluster.

The genes in the *nif* cluster of *C. acetobutylicum*, *C. beijerinckii*, and *C. pasteurianum* have two other general features. One is an overlap between the *nifD* and *nifK* genes [69,70]. The other is a biased codon usage pattern [32]. The third position of the codons for each amino acid is predominantly or exclusively A and U for the *nifHDK* genes, whereas the other *nif* genes and the *nir* genes use codons with G or C at the third position more frequently than the *nifHDK* genes. The very biased codon usage pattern for the *nifHDK* genes probably reflects the high level of synthesis of nitrogenase component proteins but may also indicate a different origin for these genes.

The *nif* clusters of *C. acetobutylicum*, *C. beijerinckii*, and *C. pasteurianum* can be differentiated by the non-*nif* genes that are present in each cluster (Figure 12.1). The *nif* cluster of *C. acetobutylicum* does not contain any non-*nif* gene and is, thus, the simplest. The *nif* cluster of

TABLE 12.1
Composition of the *nif* Cluster of *C. acetobutylicum* ATCC 824, *C. beijerinckii* NRRL B593, and *C. pasteurianum* W5 and the Proposed Product or Function of the Genes

C. acetobutylicum	*C. beijerinckii*	*C. pasteurianum*	Proposed Gene Product or Function	Ref.
		nifH2	Homologue of nitrogenase Fe protein	65
nifH	*nifH*	*nifH1*	Nitrogenase Fe protein; synthesis of FeMo cofactor	24,65
nifI$_1$	*nifI1*		Regulation of nitrogenase activity (switch-off)	24
nifI$_2$	*nifI2*		Regulation of nitrogenase activity (switch-off)	24
nifD	*nifD*	*nifD*	Nitrogenase MoFe protein, α subunit	24,70
nifK	*nifK*	*nifK*	Nitrogenase MoFe protein, β subunit	24,70
nifE	*nifE*	*nifE*	Synthesis of FeMo cofactor	79
nifN-B	*nifN-B*	*nifN-B*	Synthesis of FeMo cofactor	32
		modA	Molybdate transport	24
		modB	Molybdate transport	24,80
	fdx		2Fe-2S ferredoxin	69
	nirJ1		Synthesis of heme d_1 or coenzyme PQQ[a]	69
	nirJ2		Synthesis of heme d_1 or coenzyme PQQ	69
	nirD		Synthesis of heme d_1	69
	nirH		Synthesis of heme d_1	69
nifVω	*nifVω*	*nifVω*	Homocitrate synthase; synthesis of FeMo cofactor	69,81
nifVα	*nifVα*	*nifVα*	Homocitrate synthase; synthesis of FeMo cofactor	69,81

[a] PQQ: pyrroloquinoline quinone.

C. pasteurianum does not have the *nifI* genes but has the *modA* and *modB* genes, instead. The *nif* cluster of *C. beijerinckii* has the *nifI* genes, and instead of the *mod* genes, it has five other non-*nif* genes (*fdxA* and *nirJ1, J2, D, and H*) occupying the location of the *mod* genes in *C. pasteurianum.*

Although the concise *C. acetobutylicum nif* cluster lacks the *mod* genes and the *nir* genes, these genes are present in the *C. acetobutylicum* genome. They are CAC0281 (*modA*), CAC0280 (*modB*), CAC2796 (*nirJ1*), CAC2795 (*nirJ2*), CAC2794 (*nirD*), and CAC2793 (*nirH*). A gene similar to the *fdxA* gene of *C. beijerinckii* is not found in *C. acetobutylicum*; however, *C. pasteurianum* has a homologous gene that encodes a [2Fe-2S] ferredoxin that is synthesized in nitrogen-fixing cells but not in ammonia-grown cells.

The organization of *nif* and non-*nif* genes in the *nif* cluster of *C. acetobutylicum, C. beijerinckii,* and *C. pasteurianum* suggests that the seven core *nif* genes have existed in these species as a cluster in the present order prior to the insertion of other genes into the cluster. The insertion of the *mod* genes would result in the *nif* cluster of *C. pasteurianum*, whereas the insertion of the *nifI* genes would result in the *nif* cluster of *C. acetobutylicum.* The insertion of the five genes (*fdxA* to *nirH*) into the *nif* cluster of *C. acetobutylicum* would then result in the *nif* cluster of *C. beijerinckii.*

A brief description of the *nif* genes of *C. acetobutylicum*, *C. beijerinckii*, and *C. pasteurianum* is given below:

Clostridium pasteurianum. The organization of *nif* genes in *C. pasteurianum* has been reviewed [24,32]. The major *nif* cluster (Figure 12.1) spans a region of 13.4 kb (between *nifH2* and *nifVα*), and it is 12.2 kb between *nifH1* and *nifVα.* The orientation of all genes in the cluster is from *nifH2* toward *nifVα.* The *nifH1* gene encodes the purified nitrogenase Fe protein [65], whereas the *nifH2* gene encodes a polypeptide of 272 amino acids, which differs from the *nifH1* product in 23 amino acids (8%).

The presence of the *modA* and *modB* genes within the *nif* cluster is rare. Similar *mod* genes are present in other nitrogen-fixing organisms [71], but they are located either elsewhere in the genome or at the boundary of the *nif* cluster, such as in *Acetobacter diazotrophicus* [71]. The proposed function for the putative *modA* and *modB* genes is molybdate transport.

In addition to *nifH1*, *nifH2*, and *nifH3* (*anfH*; see next section), *C. pasteurianum* has two unlinked *nifH*-like genes, *nifH4* and *nifH5* [44,71a]. The length of the polypeptides encoded by *nifH1* through *nifH6* ranges from 272 to 275 amino acids. Except for NifH3, which is about 65% identical to NifH1 at the amino acid level, the other five NifH polypeptides are between 91.6 and 99.6% identical among themselves.

A recent study detected the presence, in nitrogen-fixing cells of *C. pasteurianum*, of a polypeptide that is the product of the *nifH2* gene [45]. The presence of this polypeptide in nitrogen-fixing cells but not in ammonia-grown cells suggests a role related to nitrogen fixation for this protein; however, its actual function remains to be determined.

Clostridium acetobutylicum. The *nif* cluster of *C. acetobutylicum* ATCC 824 spans 10.7 kb [24,72] and is the shortest *nif* cluster that has been described to date (Figure 12.1). The orientation of all genes in the cluster is from *nifH* toward *nifV*α. The assigned gene number is as follows:

nifH, CAC0253; *nifI*$_1$, CAC0254; *nifI*$_2$, CAC255; *nifD*, CAC256, *nifK*, CAC0257; *nifE*, CAC0258; *nifN-B*, CAC0259; *nifV*ω, CAC0260; *nifV*α, CAC0261

The *nifI*$_1$ and *nifI*$_2$ genes correspond to the homologous genes that are present at the same location in the *nif* cluster of several methanogens, and these genes were previously designated as the *glnB*$_1$ and *glnB*$_2$ genes, respectively, for their sequence relatedness to the nitrogen-regulatory gene *glnB* encoding the P_{II} protein [73]. The *nifI*$_1$ and *nifI*$_2$ genes of *Methanococcus maripaludis* are required for the ammonia switch-off of nitrogen fixation [74].

Outside the *nif* cluster, an open-reading frame (CAC2229) that corresponds to the *nifJ* gene (encoding pyruvate:ferredoxin/flavodoxin oxidoreductase) of *Klebsiella pneumoniae* can be found on the chromosome of *C. acetobutylicum* ATCC824. The ferredoxin/flavodoxin-linked pyruvate dehydrogenase is a hallmark of the saccharolytic clostridia and is not specific for nitrogen fixation.

In *C. acetobutylicum* ATCC 824, *nifS*-like genes are found outside the *nif* cluster as CAC2234 and CAC2972 [69]. The *nifS* and the related *iscS* genes encode a cysteine desulfurase that is required for the formation of iron-sulfur clusters, as well as the synthesis of other compounds [75]. Open-reading frames similar to *nifS* are also found in the genomes of *C. perfringens* and *C. tetani*, although the genomes of these two clostridia do not contain any other *nif* genes [69]. It suggests that the "*nifS*" genes found in these *Clostridium* species are not *nif*-specific.

Clostridium beijerinckii. The *nif* cluster of *C. beijerinckii* NRRL B593 (Figure 12.1) resembles those of *C. acetobutylicum* and *C. pasteurianum*, both in the structure and order of the *nif* genes and the location where non-*nif* genes are present in the cluster. However, the identity of the non-*nif* genes within the *nif* cluster is species-specific for these three clostridial species. The direction of transcription for the genes in the *C. beijerinckii nif* cluster is from *nifH* toward *nifV*α.

The five non-*nif* genes, from *fdxA* to *nirH*, are situated between *nifN-B* and *nifV*ω, the same location where the *modA* and *modB* genes are situated in *C. pasteurianum*. The *C. acetobutylicum* genome does not have a gene similar to *fdxA* of *C. beijerinckii*. On the other hand, the [2Fe-2S] ferredoxin of *C. pasteurianum* [76] is highly related to the *fdxA*-encoded sequence, although the gene encoding the *C. pasteurianum* [2Fe-2S] ferredoxin is not part of the *nif* cluster. In *C. pasteurianum*, the level of the [2Fe-2S] ferredoxin is highly increased in nitrogen-fixing cells than in ammonia-grown cells, but there is no evidence that the [2Fe-2S] ferredoxin is involved in nitrogen fixation.

The gene clusters CAC2796 through CAC2793 of *C. acetobutylicum* correspond to *nirJ1*, *nirJ2*, *nirD*, and *nirH*, respectively, of *C. beijerinckii*. The *nir* cluster of *C. acetobutylicum* is about 1.3 Mb away (across the proposed origin of replication) from the *nif* cluster [69]. The *nirJ, D, H* genes are postulated to play a role in the synthesis of heme d_1 or coenzyme PQQ (pyrroloquinoline quinone), but it is not known if heme d_1 or coenzyme PQQ is present in the clostridial cell.

12.8.2 Alternative Nitrogen-Fixation (anf) Genes

Besides the *nif* genes encoding the molybdenum-nitrogenase system, putative *anf* genes, which encode the iron-only nitrogenase, have been identified in two nitrogen-fixing clostridia. Separate from the major *nif* cluster, *C. pasteurianum* has a putative *anf* cluster consisting of five open reading frames: ***anfH* ► ◄ *ORF anfD* ► *anfG* ► *anfK*►** [77]. The *C. pasteurianum anfH* gene is synonymous to the previously reported *C. pasteurianum nifH3* gene [65]. The ORF between *anfH* and *anfD* is in an opposite orientation relative to the *anf* genes [44,77]. The separation of the *anfH* and *anfD* genes by an ORF in *C. pasteurianum* resembles the organization of the *vnf* genes in *Azotobacter* [78]. A cluster of *anfHDGK* genes is present in the nitrogen-fixing, cellulytic *Clostridium hungatei*; however, the *anfH* and *anfD* of *C. hungatei* are not separated by an ORF (GenBank accession number U59415).

On the basis of 16S rRNA sequences, *C. pasteurianum* and *C. hungatei* are phylogenetically distant. *C. pasteurianum* (with a G+C content of 26 to 28%) belongs to cluster I of the clostridia [19], whereas *C. hungatei* (with a G+C content of 40 to 42%) is associated with species belonging to cluster III of the clostridia [28]. Further sequence analysis of the putative *anf* genes and the flanking regions in the two clostridia may provide clues on the path of propagation of the *anf* genes.

12.9 NITROGEN FIXATION-RELATED GENES OF CLOSTRIDIA

The nitrogenase proteins are highly sensitive to oxidative damage by molecular oxygen, and the nitrogen-fixation process depends on an adequate supply of ATP and a reductant with a low redox potential. It is, thus, natural that the nitrogen-fixation activity is tightly regulated in response to both the redox and the nitrogen status of the cell [73]. For clostridia, the regulatory genes for nitrogen fixation are yet to be identified. However, the discovery of the *nifH*-linked *nifI* genes in some clostridia should accelerate the elucidation of this important aspect of nitrogen fixation in this group of diazotrophs.

In the proteobacteria, the expression and activity of the transcriptional regulator (NifA) for the *nif* genes are generally modulated by proteins that respond to the nitrogen status (NtrB, NtrC, and the P_{II} protein) and the oxygen level (NifL). In addition to its role in the transcriptional control of nitrogen fixation, the P_{II} protein may also regulate the nitrogenase activity at a post-translational level in response to ammonia and other fixed nitrogen sources. The P_{II} protein is encoded by the *glnB* gene and represents a family of conserved signal-transduction proteins that play a significant role in the coordination of nitrogen metabolism in a wide variety of bacteria.

Because clostridia are obligately anaerobic bacteria and are evolutionarily ancient, one may not expect to find a *nifL*-like gene in the clostridia, especially if the *nif* genes or their ancestral forms have existed in the clostridia before the earth's atmosphere became aerobic. Indeed, a *nifL*-like gene has not been reported in clostridia. In addition, a gene similar to *nifA* is not present in the *nif* cluster of *C. acetobutylicum*, *C. beijerinckii*, or *C. pasteurianum* or elsewhere in the genome of *C. acetobutylicum*. The presumed promoter regions for the *nif* genes of *C. pasteurianum* do not have the motif of the *nifA*-regulated promoters [44]. Therefore, the *nif*-specific transcriptional regulators of clostridia may differ from the well-characterized proteins of proteobacteria.

Interestingly, two *glnB-like* genes are present between the *nifH* and *nifD* genes in *C. acetobutylicum* and *C. beijerinckii*. These two *glnB*-like genes are referred to as the $nifI_1$ and $nifI_2$ genes here because the gene designation has been proposed for similar genes found in *Methanococcus maripaludis* [74]. The $nifI_1$ and $nifI_2$ genes are required for ammonia switch-off of nitrogen fixation in *M. maripaludis*; this switch-off is reversible and does not seem to involve ADP-ribosylation or any other covalent modification of the Fe protein of nitrogenase. Also, the switch-off does not affect *nif* gene transcription, *nifH* mRNA stability, or the stability of the Fe protein [74].

A search of the genome of two nondiazotrophic clostridia, *C. perfringens* and *C. tetani*, did not reveal any open-reading frames related to the *nifI* gene, which further suggests a *nif*-specific role for the *nifI* genes of *C. acetobutylicum* and *C. beijerinckii* [69]. The *nif* cluster of *C. pasteurianum* does not have any *glnB*-like genes, and the *in vivo* nitrogenase activity in *C. pasteurianum* does not show switch-off by the addition of ammonia, although the synthesis of nitrogenase is repressed by ammonia [41]. In contrast, the *in vivo* nitrogenase activity of *C. beijeirinckii* falls rapidly following the addition of ammonia, but the *in vitro* nitrogenase activity shows a much smaller drop. Furthermore, there is no change in the mobility of the *C. beijerinckii* Fe protein on SDS-PAGE during the period that the *in vivo* nitrogenase activity exhibits a drop [45]. The nature of the switch-off of nitrogenase activity by ammonia in *C. beijerinckii* is yet to be determined.

12.10 REGULATION OF NITROGEN FIXATION IN CLOSTRIDIA

The expression of the *nif* genes is tightly controlled in all diazotrophs in which this property has been examined. The *nif* cluster of *C. acetobutylicum*, *C. beijerinckii*, and *C. pasteurianum* does not contain an apparent regulatory gene. Therefore, the genetic elements for the transcriptional control of the *nif* genes have not yet been found in these species. However, there is evidence that the expression of the *nif* genes in clostridia is also tightly controlled. When a sufficient amount of ammonia is present in the growth medium, nitrogenase is not synthesized by *C. pasteurianum* [82], *C. thermosaccharolyticum* [30], or *C. formicoaceticum* [27]. When provided with a limiting amount of ammonia, a culture of *C. pasteurianum* switches from ammonia-dependent growth to nitrogen-fixing growth, after a diauxic lag following the exhaustion of ammonia in the growth medium [83].

When ammonia was added to a nitrogen-fixing culture of *C. pasteurianum*, synthesis of nitrogenase immediately stopped [39,41,45]. The already present nitrogenase, however, remains active through several cell divisions, which results in a decrease in nitrogenase activity per cell due to dilution of the enzyme. It is interesting that the addition of ammonia had different effects on the *in vivo* activity of existing nitrogenase in *C. pasteurianum* and in *C. beijerinckii* [45]. Whereas addition of ammonia does not affect the *in vivo* activity of existing nitrogenase in *C. pasteurianum*, similar additions caused an immediate drop in *in vivo* nitrogenase activity in *C. beijerinckii* [45]. However, there was no similar drop in *in vitro* nitrogenase activity following addition of ammonia to a nitrogen-fixing culture of *C. beijerinckii*. It may be speculated that products of the *nifI*$_1$ and *nifI*$_2$ genes play a role in regulating the activity of nitrogenase in response to the concentration of ammonia in *C. beijerinckii*, assuming *nifI*-like genes are not present in the genome (outside the *nif* cluster) of *C. pasteurianum*.

When sporulation was induced in batch cultures of nitrogen-fixing *C. pasteurianum*, either by the addition of calcium acetate or by an increase in pH, partial inactivation of nitrogenase and excretion of ammonia were observed [84]. A drastic drop (>90%) in *in vivo* nitrogenase activity occurred immediately after the culture pH was changed from 6.0 to 8.4; however, about 40% of the activity recovered in 2 h when the pH was maintained at 8.4. At the time of the pH change, the *in vitro* nitrogenase activity fell about 65%, and the activity remained at that level for at least 2 h. It appears that the pH change stopped nitrogenase synthesis, caused partial inactivation of existing nitrogenase, and generated an initial physiological condition that did not support full nitrogenase activity.

To investigate the mechanism of transcriptional control of the *nif* genes, the transcription start site for six *nif* transcription units (including those of *nifH*-like genes) has been determined in *C. pasteurianum* [44]. Transcription of *nifH1*, which encodes the characterized Fe protein, starts 62 base pairs preceding the translation start site. An analysis of the upstream sequences revealed similarities in the –130 region of these transcription units, and the consensus sequence (ATCAATAT-N_{6-10}-ATGGATTC) is different from the TGT-N_{10}-ACA consensus that is present in the –100 region of the *nif* transcription units of *K. pneumoniae* and *A. vinelandii*. Because a *nifA*-like gene is not present in the *nif* clusters of *C. acetobutylicum*, *C. beijerinckii*, and *C. pasteurianum* or elsewhere in the genome of *C. acetobutylicum*, the promoter and upstream sequences for the *nif* genes of these *Clostridium* species can be expected to differ from those involving NifA.

12.11 CONCLUDING REMARKS

Nitrogen fixation has been demonstrated in clostridial species belonging to different physiological groups. Up until now, most of the biochemical and physiological data have been obtained with *C. pasteurianum*, whereas genetic data have been obtained with *C. acetobutylicum*, *C. beijerinckii*, and *C. pasteurianum*. For growth, these species can use a number of carbon-substrates, including sucrose, cellobiose (except *C. pasteurianum*) and starch (except *C. pasteurianum*). Cellulose, an abundant carbon-substrate, supports growth and nitrogen fixation of some cellulytic clostridia. There has not been a systematic search among the more than 170 species of *Clostridium* for the nitrogen-fixing capability. A comprehensive examination of this large genus could yield many more nitrogen-fixing species that some of them may possess an as-yet-unknown nitrogen-fixing enzyme or a different control mechanism for their nitrogen-fixing systems. The broad substrate range of the clostridia suggests the possibility of finding nitrogen-fixing species in diverse environments, which could range from soils to the digestive tract of ruminant animals [1]. Despite the high nitrogen-fixing activity of *C. pasteurianum*, the contribution of nitrogen-fixing clostridia to the cycling of nitrogen in nature is yet to be determined.

Our knowledge of the genetics of nitrogen fixation in clostridia is primarily from studies involving *C. acetobutylicum*, *C. beijerinckii*, and *C. pasteurianum*. In these species, both the structure and the organization of the *nif* genes display distinctive features, including the fused *nifN-B* gene. The *nif* clusters of *C. acetobutylicum*, *C. beijerinckii*, and *C. pasteurianum* are more concise than all other known *nif* clusters, and they may represent the minimal set of *nif*-specific genes that are required for diazotrophic growth in the absence of O_2. At present, the *nif* cluster (nine genes) of *C. acetobutylicum* ATCC 824 is the most concise. To understand the regulation of expression of *nif* genes in clostridia and to elucidate the precise function of the *nifB* gene product in relation to the *nifE* and *nifN* gene products, the *nif* system of *C. acetobutylicum* is attractive as a model system, despite the fact that the biochemical properties of the nitrogen-fixing system of *C. pasteurianum* have been more thoroughly investigated. The availability of the genome sequence of *C. acetobutylicum* ATCC 824 and the number of useable genetic tools for this strain could facilitate the acquisition of useful new information on several frontiers of nitrogen-fixation research.

REFERENCES

1. Fisher, K. and Newton, W.E., Nitrogen fixation — a general overview, in *Nitrogen Fixation at the Millennium*, Leigh, G.J., Ed., Elsevier, Amsterdam, 2002, Chapter 1.
2. Sprent, J.I. and Sprent, P., *Nitrogen Fixing Organisms*, Chapman & Hall, London, 1990, Chapter 1.
3. Winogradsky, S., Recherches sur l'assimilation de l'azote libre de l'atmosphère par les microbes, *Archives des Sciences Biologiques*, 3, 297, 1895.
4. Rosenblum, E.D. and Wilson, P.W., Fixation of isotopic nitrogen by *Clostridium*, *J. Bacteriol.*, 57, 413, 1949.
5. Carnahan, J.E. et al., Nitrogen fixation in cell-free extracts of *Clostridium pasteurianum*, *Biochim. Biophys. Acta*, 38, 188, 1960.
6. Carnahan, J.E. et al., Nitrogen fixation in cell-free extracts of *Clostridium pasteurianum*, *Biochim. Biophys. Acta*, 44, 520, 1960.
7. Mortenson, L.E., Valentine. R.C., and Carnahan, J.E., An electron transport factor from *Clostridium pasteurianum*, *Biochem. Biophys. Res. Commun.*, 7, 448, 1962.
8. Mortenson, L.E., Valentine, R.C., and Carnahan, J.E., Ferredoxin in the phosphoroclastic reaction of pyruvic acid and its relation to nitrogen fixation in *Clostridium pasteurianum*, *J. Biol. Chem.*, 238, 794, 1963.
9. Mortenson, L.E., Ferredoxin requirement for nitrogen fixation by extracts of *Clostridium pasteurianum*, *Biochim. Biophys. Acta,* 81, 473, 1963.
10. Mortenson, L.E., Ferredoxin and ATP, requirements for nitrogen fixation in cell-free extracts of *Clostridium pasteurianum*, *Proc. Natl. Acad. Sci. U.S.A.*, 52, 272, 1964.

11. Mortenson, L.E., Nitrogen fixation in extracts of *Clostridium pasteurianum*, in *Non-Heme Iron Proteins: Role in Energy Conversion*, San Pietro, A., Ed., The Antioch Press, Yellow Springs, OH, 1965, 243.
12. Mortenson, L.E., Components of cell-free extracts of *Clostridium pasteurianum* required for ATP-dependent H_2 evolution from dithionite and for N_2 fixation, *Biochim. Biophys. Acta,* 127, 18, 1966.
13. Emerich, D.W., Hageman, R.V., and Burris, R.H., Interactions of dinitrogenase and dinitrogenase reductase, *Adv. Enzymol.*, 52, 1, 1981.
14. Winogradsky, S., Sur L'assimilation de l'azite gazeux de l'atmosphère par les microbes, *Comp. rend. hebd. Sénc. Acad. Sci. (Paris)*, 116, 1385, 1893.
15. Winogradsky, S., *Clostridium Pastorianum*, seine Morphologie und seine Eigenschaften als Buttersäuereferment, *Zentrabl. Bakteriol. Parasitenk. Infektionskr. Hyg. Abt. II*, 9, 43, 1902.
16. Winogradsky, S., *Clostridium Pastorianum*, seine Morphologie und seine Eigenschaften als Buttersäuereferment, *Zentrabl. Bakteriol. Parasitenk. Infektionskr. Hyg. Abt. II*, 9, 107, 1902.
17. Postgate, J., *Nitrogen Fixation*, 3rd Ed., Cambridge University Press, Cambridge, 1998, 49.
18. McCoy, E., Higby, W.M., and Fred, E.B., The assimilation of nitrogen by pure cultures of *Clostridium pasteurianum* and related organisms, *Centralbl. Bakteriol. Parasit. II,* 76, 314, 1928.
19. Collins, M.D. et al., The phylogeny of the genus *Clostridium*: proposal of five new genera and eleven new species combinations, *Int. J. Syst. Microbiol.*, 44, 812, 1994.
20. Johnson, J.L. and Francis, B.S., Taxonomy of the clostridia: ribosomal ribonucleic acid homologies among the species, *J. Gen. Microbiol.,* 88, 229, 1975.
21. George, H.A. et al., Acetone, isopropanol, and butanol production by *Clostridium beijerinckii* (syn. *Clostridium butylicum*) and *Clostridium aurantibutyricum*, *Appl. Environ. Microbiol.*, 45, 1160, 1983.
22. Cummins, C. and Johnson, J.L., Taxonomy of the clostridia: wall composition and DNA homologies in *Clostridium butyricum* and other butyric acid-producing clostridia, *J. Gen. Microbiol.*, 67, 33, 1971.
23. Keis, S., Shaheen, R., and Jones, D.T., Emended description of *Clostridium acetobutylicum*, *Clostridium beijerinckii*, and descriptions of *Clostridium saccharobutylacetonicum* sp. nov. and *Clostridium saccharobutylicum* sp. nov., *Int. J. Syst. Evol. Microbiol.*, 51, 2095, 2001.
24. Chen, J.-S., Toth, J., and Kasap, M., Nitrogen-fixation genes and nitrogenase activity in *Clostridium acetobutylicum* and *Clostridium beijerinckii*, *J. Ind. Microbiol. Biotechnol.*, 27, 281, 2001.
25. Kanamori, K., Weiss, R.L., and Roberts, J.D., Ammonia assimilation pathways in nitrogen-fixing *Clostridium kluyverii* and *Clostridium butyricum*, *J. Bacteriol.*, 171, 2148, 1989.
26. Szech, U., Braun, M., and Kleiner, D., Uptake and excretion of amino acids by saccharolytic clostridia, *FEMS Microbiol. Lett.*, 58, 11, 1989.
27. Bogdahn, M., Andreesen, J.R., and Kleiner, D., Pathways and regulation of N_2, ammonium and glutamate assimilation by *Clostridium formicoaceticum*, *Arch. Microbiol.*, 134, 167, 1983.
28. Monserrate, E., Leschine, S.B., and Canale-Parola, E., *Clostridium hungatei* sp. nov., a mesophilic, N_2-fixing cellulytic bacterium isolated from soil, *Int. J. Syst. Evol. Microbiol.*, 51, 123, 2001.
29. Kuhner, C.H. et al., *Clostridium akagii* sp. nov. and *Clostridium acidisoli* sp. nov.: acid-tolerant, N_2-fixing clostridia isolated from acidic forest soil and litter, *Int. J. Syst. Evol. Microbiol.*, 50, 873, 2000.
30. Bogdahn, M. and Kleiner, D., N_2 fixation and NH_4^+ assimilation in the thermophilic anaerobes *Clostridium thermosaccharolyticum* and *Clostridium thermoautotrophicum*, *Arch. Microbiol.*, 144, 102, 1986.
31. Rosado, A.S., van Elsas, J.D., and Seldin, L., Reclassification of *Paenibacillus durun* (formerly *Clostridium durum* Smith and Cato 1974) Collins et al. 1994 as a member of the species *P. azotofixans* (formerly *Bacillus azotofixans* Seldin et al. 1984) Ash et al. 1994, *Int. J. Syst. Bacteriol.*, 47, 569, 1997.
32. Chen, J.-S. and Johnson, J.L., Molecular biology of nitrogen fixation in the clostridia, in *The Clostridia and Biotechnology*, Woods, D.R., Ed., Boston: Butterworth-Heinemann, Boston, 1993, 371.
33. Jensen, H.L. and Spencer, D., The influence of molybdenum and vanadium on nitrogen fixation by *Clostridium butyricum* and related organisms, *Proc. Linnean Soc. N. S. Wales,* 72, 73, 1947.
34. Carnahan, J.E., and Castle, J.E., Some requirements of biological nitrogen fixation, *J. Bacteriol.,* 75, 121, 1958.
35. Bui, P.T., Mechanism of the enzymic reduction of N_2, Ph.D. Thesis, Purdue University, West Lafayette, IN, 1968.
36. Westlake, D.W.S. and Wilson, P.W., Molecular hydrogenase and nitrogen fixation by *Clostridium pasteurianum, Can. J. Microbiol.,* 5, 617, 1959.

37. Vandecasteele, J.-P. And Burris, R.H., Purification and properties of the constituents of thc nitrogenase complex from *Clostridium pasteurianum, J. Bacteriol.*, 101, 794, 1970.
38. Tso, M.-Y.W., Ljones, T., and Burris, R.H., Purification of the nitrogenase proteins from *Clostridium pasteurianum, Biochim. Biophys. Acta,* 267, 600, 1972.
39. Kleiner, D., Regulation of ammonium uptake and metabolism by nitrogen fixing bacteria III. *Clostridium pasteurianum, Arch. Microbiol.*, 120, 263, 1979.
40. Rennie, R.J., A single medium for the isolation of acetylene-reducing (dinitrogen-fixing) bacteria from soils, *Can. J. Microbiol.*, 27, 8, 1981.
41. Daesch, G. and Mortenson, L.E., Effect of ammonia on the synthesis and function of the N_2-fixing enzyme system in *Clostridium pasteurianum, J. Bacteriol.,* 110, 103, 1972.
42. Lampen, J.O. and Peterson, W.H., Growth factor requirements of clostridia, *Arch. Biochem.,* 2, 443, 1943.
43. Daesch, G. and Mortenson, L.E., Sucrose catabolism in *Clostridium pasteurianum* and its relation to N_2 fixation, *J. Bacteriol.*, 96, 346, 1968.
44. Wang, S.-Z., Chen, J.-S., and Johnson, J.L., The presence of five *nifH*-like sequences in *Clostridium pasteurianum*: sequence divergence and transcription properties, *Nucleic Acid Res.*, 16, 439, 1988.
45. Kasap, M., Nitrogen metabolism and solvent production in *Clostridium beijerinckii* NRRL B593, Ph.D. Thesis, Virginia Polytechnic Institute and State University, Blacksburg, 2002.
46. Jordan, D.C. and McNicol, P.J., A new nitrogen-fixing *Clostridium* species from a high Arctic ecosystem, *Can. J. Microbiol.,* 25, 947, 1979.
47. Benemann, J.R. and Valentine, R.C., High-energy electrons in bacteria, *Adv. Microbial Physiol.*, 5, 135, 1971.
48. Mortenson, L.E., Purification of nitrogenase from *Clostridium pasteurianum, Methods Enzymol.*, 24, 446, 1972.
49. Chen, J.-S. and Blanchard, D.K., Purification and properties of the H_2-oxidizing (uptake) hydrogenase of the N_2-fixing anaerobe *Clostridium pasteurianum, Biochem. Biophys. Res. Commun.,* 122, 9, 1984.
50. Burris, R.H., Nitrogen fixation — assay methods and techniques, *Methods Enzymol.*, 24, 415, 1972.
51. Dilworth, M.J., Eldridge, M.E., and Eady, R.R., Correction for creatine interference with the direct indophenol measurement of NH_3 in steady-state nitrogenase assays, *Anal. Biochem.*, 206, 6, 1992.
52. Simpson, F. and Burris, R.H., A nitrogen pressure of 50 atmosphees does not prevent evolution of hydrogen by nitrogenase, *Science*, 224, 1095, 1984.
53. Zumft, W.G. and Mortenson, L.E., Evidence for a catalytic-centre heterogeneity of molybdoferredoxin from *Clostridium pasteurianum, Eur. J. Biochem.,* 35, 401, 1973.
54. Eady, R.R., Isolation and characterization of various nitrogenases, *Methods Enzymol.*, 69, 753, 1980.
55. Beinert, H., Orme-Johnson, W., and Palmer, G., Special techniques for the preparation of samples for low-temperature EPR spectroscopy, *Methods Enzymol.*, 54, 111, 1978.
56. Dean, D.R. and Jacobson, M.R., Biochemical genetics of nitrogenase, in *Biological Nitrogen Fixation*, Stacey, G., Burris, R.H., and Evans, H.J., Eds., Chapman & Hall, New York, 1992, Chapter 20.
57. Henderson, R.A., Advances towards the mechanism of nitrogenases, in *Nitrogen Fixation at the Millennium*, Leigh, G.J., Ed., Elsevier, Amsterdam, 2002, Chapter 9.
58. Emerich, D.W., Ljones, T., and Burris, R.H., Nitrogenase: properties of the catalytically inactive complex between the *Azotobacter vinalandii* MoFe protein and the *Clostridium pasteurianum* Fe protein, *Biochim. Biophys. Acta,* 527, 359, 1978.
59. Guth, J.H. and Burris, R.H., Inhibition of nitrogenase-catalyzed NH_3 formation by H_2, *Biochemistry*, 22, 5111, 1983.
60. Weston, M.F., Kotake, S., and Davis, L.C., Interaction of nitrogenase with nucleotide analogs of ATP and ADP and the effect of metal ions on ADP inhibition, *Arch. Biochem. Biophys.,* 225, 809, 1983.
61. Bolin, J.T. et al., The unusual metal clusters of nitrogenase: structural features revealed by x-ray anomalous diffraction studies of the MoFe protein from *Clostridium pasteurianum, Proc. Natl. Acad. Sci. U.S.A.*, 90, 1078, 1993.
62. Kim, J., Woo, D., and Rees, D.C., X-ray crystal structure of the nitrogenase molybdenum-iron protein from *Clostridium pasteurianum* at 3.0-Å resolution, *Biochemistry*, 32, 7104, 1993.
63. Schlessman, J L. et al., Conformational variability in structures of the nitrogenase iron proteins from *Azotobacter vinelandii* and *Clostridium pasteurianum, J. Mol. Biol.*, 280, 669, 1998.
64. Tanaka, M. et al., The amino acid sequence of *Clostridium pasteurianum* iron protein, a component of nitrogenase, *J. Biol. Chem.*, 252, 7093, 1977.

65. Chen, K.C.-K., Chen, J.-S., and Johnson, J.L., Structural features of multiple *nifH*-like sequences and very biased codon usage in nitrogenase genes of *Clostridium pasteurianum*, *J. Bacteriol.*, 166, 162, 1986.
66. Benton, P.M.C., Sen, S., and Peters, J.W., Nitrogenase structure, in *Nitrogen Fixation at the Millennium*, Leigh, G.J., Ed., Elsevier, Amsterdam, 2002, Chapter 2.
67. Walker, G.A. and Mortenson, L.E., Effect of magnesium adenosine 5′-triphosphate on the accessibility of the iron of clostridial azoferredoxin, a component of nitrogenase, *Biochemistry*, 13, 2382, 1974.
68. Zumft, W.G., Mortenson, L.E., and Palmer, G., Eletron-paramagnetic-studies on nitrogenase — investigation of the oxidation-reduction behavior of azoferredoxin and molybdoferredoxin with potentiometric and rapid-freeze techniques, *Eur. J. Biochem.*, 46, 525, 1974.
69. Toth, J. and Chen, J.-S., Unpublished data, 2002.
70. Wang, S.-Z., Chen, J.-S., and Johnson, J.L., Distinct structural features of the α and β subunits of nitrogenase molybdenum-iron protein of *Clostridium pasteurianum*: an analysis of amino acid sequences, *Biochemistry*, 27, 2800, 1988.
71. Lee, S. et al., Characterization of a major cluster of *nif, fix,* and associated genes in a sugarcane endophyte, *Acetobacter diazotrophicus*, *J. Bacteriol.*, 182, 7088, 2000.
71a. Tollin, C., Tran, A., Toth, J., and Chen, J.-S., Unpublished data, 2004.
72. Nölling, J. et al., Genome sequence and comparative analysis of the solvent-producing bacterium *Clostridium acetobutylicum*, *J. Bacteriol.*, 183, 4823, 2001.
73. Arcondéguy, T., Jack, R., and Merrick, M., P_{II} signal transduction proteins, pivotal players in microbial nitrogen control, *Microbiol. Mol. Biol. Rev.*, 65, 80, 2001.
74. Kessler, P.S., Daniel, C., and Leigh, J.A., Ammonia switch-off of nitrogen fixation in the methanogenic archeon *Methanococcus maripaludis*: mechanistic features and requirement for the novel GlnB homologues, $NifI_1$ and $NifI_2$, *J. Bacteriol.*, 183, 882, 2001.
75. Mihara, H. and Esaki, N., Bacterial cysteine desulfurase: their function and mechanisms, *Appl. Microbiol. Biotechnol.*, 60, 12, 2002.
76. Meyer, J., Cloning and sequencing of the gene encoding the [2Fe-2S] ferredoxin from *Clostridium pasteurianum*, *Biochim. Biophys. Acta.* 1174, 108, 1993.
77. Zinoni, F., Robson, R.M., and Robson, R.L., Organization of potential alternative nitrogenase genes from *Clostridium pasteurianum*, *Biochim. Biophys. Acta,* 1174, 83, 1993.
78. Bishop, P.E. and Premakumar, R., Alternative nitrogen fixation systems, in *Biological Nitrogen Fixation*, Stacey, G., Burris, R.H. and Evans, H.J., Eds., Chapman & Hall, New York, 1992, 736.
79. Wang, S.-Z., Chen, J.-S., and Johnson, J.L., Nucleotide and deduced amino acid sequences of the *nifE* from *Clostridium pasteurianum*, *Nucleic Acid Res.*, 17, 3299, 1989.
80. Wang, S.-Z., Chen, J.-S., and Johnson, J.L., A nitrogen-fixation gene (*nifC*) in *Clostridium pasteurianum* with sequence similarity to *chlJ* of *Escherichia coli*, *Biochem. Biophys. Res. Commun.*, 169, 1122, 1990.
81. Wang, S.-Z. et al., The N-terminal and C-terminal portions of NifV are encoded by two different genes in *C. pasteurianum*, *J. Bacteriol.*, 173, 3041, 1991.
82. Mortenson, L.E., Morris, J., and Jeng, D.Y., Purification, metal composition and properties of molybdoferredoxin and azoferredoxin, two of the components of the nitrogen-fixing system of *Clostridium pasteurianum*, *Biochim. Biophys. Acta*, 141, 516, 1967.
83. Mortenson, L.E., Regulation of nitrogen fixation, *Curr. Topics Cell. Reg.*, 13, 179, 1978.
84. Vallespinós, F. and Kleiner, D., Selective inactivation of nitrogenase and glutamate synthase during sporulation of *Clostridium pasteurianum*, *J. Gen. Microbiol.*, 117, 543, 1980.

13 Clostridial Potassium Transport Systems

Anke Treuner-Lange and Peter Dürre

CONTENTS

13.1 INTRODUCTION

The intracellular milieu of any organism must remain relatively constant with regard to pH, metabolite levels, and ionic composition to ensure an active metabolism. This cytoplasmic homeostasis is essential for the growth and survival of bacteria. K^+-transport is believed to play an essential role in maintaining homeostasis by regulating cell turgor and the cytoplasmic pH. Potassium is also required for the activity of several enzymes, therefore it is not surprising that bacteria developed several uptake and efflux systems for potassium. Few experimental data on potassium metabolism in clostridia are available. An early report describes potassium to be an essential factor in the fermentation of maize meal by *Clostridium acetobutylicum* [1], although the underlying metabolic requirement is still unknown. This chapter will focus on the different known potassium transport systems, in part experimentally characterized in *C. acetobutylicum* and detected in the complete genomes of *C. acetobutylicum, C. perfringens,* and *C. tetani,* as well as those found in the still not completely sequenced genomes of *C. thermocellum, C. botulinum,* and *C. diffcile.*

13.2 THE TRK SYSTEM

In *Escherichia coli,* the Trk system is a high-rate, but low-affinity multidomain K^+-uptake system. It consists of at least three proteins: an integral membrane protein (TrkH or TrkG), an NADH-binding peripheral membrane protein TrkA, and one or two ATP-binding proteins (TrkE or SapD). K^+-uptake via the Trk system depends on $\Delta_{\mu H+}$, and ATP and is believed to occur in a symport with protons [2]. The transmembrane proton motive force supposedly drives the K^+-transport [3], and ATP is believed to activate the system [4]. Most K-12 strains contain two of these systems (TrkG and TrkH), whereas other bacteria have been shown to comprise only one system. Since the *trkG* gene is located within the prophage rac region, the *trkH* gene is considered as the intrinsic *E. coli* gene (Figure 13.1) [5].

Searching the already mentioned clostridial genomes for TrkH or TrkG-encoding genes, clostridial homologous proteins could be deduced and identified. With two exceptions, all species

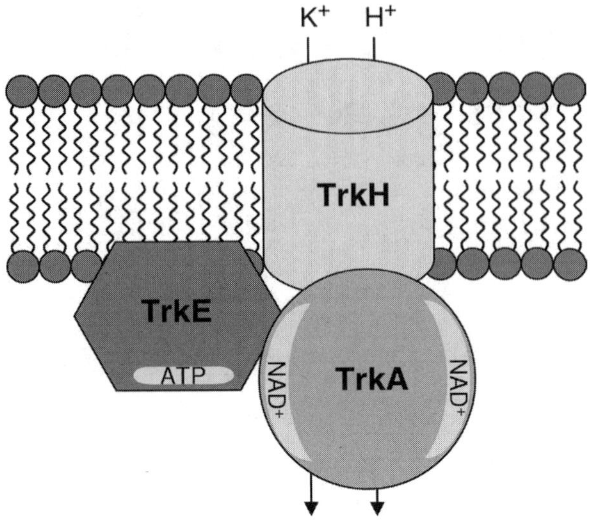

FIGURE 13.1 Model of the enterobacterial Trk-system.

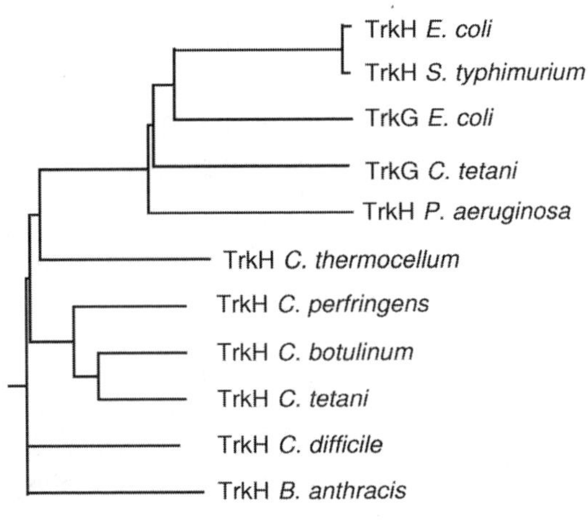

FIGURE 13.2 ClustalW dendogram generated by Phylip's Drawgram with following sequences: TrkH *E. coli* sp_P21166, TrkH *S. typhimurium* sp_Q9L6L2, TrkG *E. coli* sp_P23849, TrkG *C. tetani* NP_780946.1, TrkH *P. aeruginosa* gi_9949330, TrkH *C. thermocellum* gi_23021113, TrkH *C. perfringens* NP 5628071, *C. botulinum* gnl_SANGER_36826_Cbot440b12.q1c, TrkH *C. tetani* NP_782824.1, TrkH *C. difficile* gnl_Sanger_1496_Cd183h6.p1t, TrkH *B. anthracis* gi_21399306.

investigated comprise a putative Trk system. We suggest designating the corresponding genes *trkH* to mark them as the intrinsic genes coding for the integral membrane protein of the system. No such system could be found in *C. acetobutylicum,* whereas *C. tetani* comprises two systems. A ClustalW analysis of the clostridial TrkH proteins with similar proteins from other bacteria indicates that the clostridial proteins cluster together, whereas the second Trk(H) protein from *C. tetani,* rather, clusters with a group of proteins from Gram-negative bacteria (Figure 13.2). We therefore designated that gene *trkG* (*C. tetani* NP_780946.1). TrkH and TrkG from *E. coli* are assumed to span the membrane at least 10 times [2]. All clostridial TrkH(G) proteins reveal very similar Kyte-Doolittle hydropathy profiles so they can be assumed to be membrane proteins, too (data not shown).

The *trkA* gene product is required for the activity of the Trk system from *E. coli* [2]. TrkA is described as a peripheral membrane protein that is attached to the inner side of the cytoplasmic

```
                                                * *  *
TrkA  C. botulinum   ------------------MGKKQFVVIGLGRFGTSVAKTLYTLGNDVLAI
TrkAg C. tetani      ------------------MANKQFVVIGLGRFGTSVAQTLYSLGNDVLAI
TrkA  C. perfringens ------------------MSSKQFVIIGLGRFGSSVAKTLYALGHDVLAI
TrkA  C. difficile   --------------------MKQYIVIGCGRFGSSVASTMHLLGHQVMAI
TrkA  C. thermocellum--------------------MKSFLVIGIGRFGKHLSRKLVELGNEVMIV
TrkAh C. tetani      MLGIILGPYFFTYFLNQEMIKMKVIIVGVGKLGYKLAEAMLNEDIDVTLV
TrkA  E. coli        ---------------------MKIIILGAGQVGGTLAENLVGENNDITVV

                     *                           *        *
TrkA  C. botulinum   DSSDDIVQSISDSVTH-SVQMNATDENSLRALGIRNFDVAVITIGSDIQA
TrkAg C. tetani      DSDEDRVQSISENVTH-AIQADATDENSLRSIGVRNFDVAVVTIGSDLQA
TrkA  C. perfringens DSNEDLVQEISDSVTH-AVQMDATDENALRTLGLRNFDVAVVTIGANIQA
TrkA  C. difficile   DKNEDSVQSISDKVTH-SLIVDVTDEQALRSLGLGNFDVAVVAIGSDIRA
TrkA  C. thermocellumDKNEENIRELLPIVTN-AQIGDCTKEEVLRSLGIDNYDVCFVCVENDFQS
TrkAh C. tetani      DSNPKVLERINDHLDVLTVTANGIEMDVLKELSIETYDLLVAATSSDETN
TrkA  E. coli        DTNGERLRTLQDKFDLRVVQGHGSHPRVLREAGADDADMLVAVTSSDETN

                                     *                       *    **
TrkA  C. botulinum   STMATLLVKEMGVKYIIAKANTEIHAKVLYK------IGADRVVFPERDM
TrkAg C. tetani      SVMATLLVKELGVGYTIAKANSELHAKVLYK------IGADKVVLPERDM
TrkA  C. perfringens SVMATLLVKDMGIKYIIAKGNSDLHAKVLYK------IGADRVILPEKDM
TrkA  C. difficile   SIMATLIAKEMGVELIICKAKDELQAKVLYK------IGADRVVFPERDM
TrkA  C. thermocellumSLEITNLLKDFGAKYVISKSSRDIQTKFLLR------NGADEVIDPERNI
TrkAh C. tetani      TIICALAKKVG-CKKTIARIRNPEYMKQLD--FVKAEMGIDHIINPELAT
TrkA  E. coli        MVACQVAYSLFNTPNRIARIRSPDYVRDADKLFHSDAVPIDHLIAPEQLV

                                                           *
TrkA  C. botulinum   GVRVAHNLVSTNILDYIELSP---NYSIAEIVIPKIWHGKTLNELNIRAN
TrkAg C. tetani      GVRVAHNLVSTNILDYIELSP---DFSIAEVISPKEWYGKNLEDLSIRAN
TrkA  C. perfringens GVRVAHNLVSSSILDYIELSP---DYSIIEIESPKEWYGKSMKELSLRSK
TrkA  C. difficile   GVRVAHNLVSDNILDHIELDP---EYSIVEIVTPNSWVGKTLIELELRAR
TrkA  C. thermocellumAEKLAVRCSANHVFDYIELTN---DISIYEIPPAKSWIGRSIKEINFRVR
TrkAh C. tetani      ANAIEKYLLKNYNFYSGDFASGKVSMVDFNIGNMNNFVGKKIEDIDDLD-
TrkA  E. coli        IDNIYRLIEYPGALQVVNFAEGKVSLAVVKAYYGGPLIGNALSTMREHMP

                                      *              *
TrkA  C. botulinum   YGINVVALKRGEEINVSPVAEDTIESGDIIVAIGSEEDLTKVEI----LN
TrkAg C. tetani      YGINIVAIKKKEEINVSPTAEDVIEEGDIIVAIGGTDELNRLET----LV
TrkA  C. perfringens YGINVMAIKRNNEVNISPDADDVINKDDIVVAIGSAEDLTKLEG----KI
TrkA  C. difficile   YEITVLAIKTGKNINVTPSPDEELTAGSILVIIGQNTSITAITSGNKGII
TrkA  C. thermocellumYHVSILATKAGNDISPLPVADYIFKPEEHLLVIGRRKDVEKLLK----HI
TrkAh C. tetani      --GLLITAISRNGTIIIPYGSIELVENDVIHIIGQSKNIEKLNSKLDVNI
TrkA  E. coli        HIDTRVAAIFRHDRPIRPQGSTIVEAGDEVFFIAASQHIRAVMSELQR--

                       *  *     ***  *  ** **      *  **        * *** *
TrkAh C. tetani      NKKHVKNAMILGGGKIGYYLAQRLTSYNINVVIIEQDKKRCEYLSEKLNN
TrkA  E. coli        LEKPYKRIMLVGGGNIGAGLARRLEKD-YSVKLIERNQQRAAELAEKLQN

                          **     ** **     *  *  *  ** *      **  *  *
TrkAh C. tetani      VLIIHGDGTDMNLLEEENLSSMDAFIGVTGYDEQNLLMALMAKQSGVNKT
TrkA  E. coli        TIVFFGDASDQELLAEEHIDQVDLFIAVTNDDEANIMSAMLAKRMGAKKV

                        * *  **          *      * *  *        * ** *  *
TrkAh C. tetani      IAKISRPSYVHIIDKLGLDVALNPVNITASNILKYIRGGKVVSVSLLLDG
TrkA  E. coli        MVLIQRRAYVDLVQGSVIDIAISPQQATISALLSHVRKADIVGVSSLRRG

                       *  *  *         **       ** * **** *      * *
TrkAh C. tetani      QGEVTEIIASEN---TPIVGKPLSKLGLPKGIIIGAIVHEENVIIPNGDS
TrkA  E. coli        VAEAIEAVAHGDESTSRVVGRVIDEIKLPPGTIIGAVVRGNDVMIANDNL

                      *   *  *
TrkAh C. tetani      IIHPNDRIIVFCLTSDLPSLNMFIKSHKGGMFSELWNHNQGTGKYIKR
TrkA  E. coli        RIEQGDHVIMFLTDKKFITDVERLFQPSPFFL----------------
```

FIGURE 13.3 ClustalW alignment of the clostridial TrkA proteins with EcTrkA. The areas corresponding to the two NAD$^+$-binding domains of EcTrkA are boxed [7]. Fully conserved residues are labeled with an asterisk. The depicted TrkA sequences are the following: *C. botulinum* gnl_SANGER_36826_Cbot440b12.q1c, *C. tetani* gi 28211879, *C. perfringens* gi 18310872, *C. difficile* gnl_Sanger_1496_Cd183h6.p1t, *C. thermocellum* ZP_00060801.1, GI:23021114, *C. tetani* NP_780947.1, gi 2821003, *E. coli* gi 16131169.

membrane via TrkH or TrkG [6]. EcTrkA consists of two similar halves, each one containing a putative NAD$^+$-binding site (Figure 13.3).

Every listed clostridial *trkH* and *trkG* is coorganized with a putative *trkA* gene. That indicates that the second Trk system of *C. tetani* was most likely inherited as a complete system, whereas in *E. coli* TrkH and TrkG are sharing the same TrkA protein. Interestingly, all putative TrkA proteins, except for one, are only half the size of EcTrkA. The exception is the second TrkA protein of *C. tetani*, which gene we suggest to call *trkAg* to mark it as the gene coorganized with *trkG* (Figure 13.3). This observation raised the question: are these smaller TrkA proteins specific for clostridia? A databank search for TrkA proteins from other bacterial species indicated these small TrkA proteins do occur in other bacterial species like *Thermoanaerobacter tengcongensis* (gi 20515179), *Listeria monocytogenes* (gi 16410425), *Bacillus halodurans* (gi 15615226), *Enterococcus faecalis* (gi29377374). The double-sized TrkA proteins occur, e.g., in *E. coli* (gi 16131169), *Salmonella typhimurium* (gi 16766698), *Yersinia pestis* (gi 16120577), and *Vibrio alginolyticus* (gi 541661).

Since the two halves of EcTrkA are very similar, it might be speculated that the corresponding clostridial one-half proteins are fulfilling the same function working as dimers.

The two EcTrk systems depend differently on a third gene product TrkE. EcTrk^{H} requires TrkE, whereas EcTrk^{G} shows residual activity without TrkE [8]. However, a Trk system of *Vibrio alginolyticus* is fully active in an *E. coli trkE* strain [9]. In *E. coli* and *S. typhimurium trkE* maps inside the *sapABCDF* operon, which encodes a putative peptide-uptake ATP-binding cassette transporter of unknown function [10,11].

More recent data suggest the ATP-binding SapD protein to represent *TrkE* [12]. At that point, the limit of a genome survey is reached. Looking for *trkE*-like encoding genes in the clostridial genomes, several homologues can be found in each genome, even in the genome of *C. acetobutylicum,* which is obviously lacking a Trk system. To answer the question whether the clostridial Trk systems depend on further proteins or not, experimental data are necessary.

13.3 KCSA POTASSIUM CHANNEL

In 1995, the identification and electrophysiological characterization of the first prokaryotic potassium channel was described [13]. The corresponding *kcsA* gene encodes a protein with two N-terminal transmembrane helices with a deduced molecular mass of 17.6 kDa. Although the KcsA K^+-channel contains two membrane-spanning regions, it resembles, rather, a six membrane-spanning K^+-channel from eukaryotes. Both types of channels are working as tetramers of mostly identical subunits [14]. An interesting pmf dependent tetramerization of KcsA was suggested by van Dalen et al [15]. There are many studies concerning the biophysical features, the channel architecture, and the structure of KcsA, and interested readers are referred to further literature [14,16–20].

The pore sequence of KcsA is located between the two transmembrane helices and contains the K^+-channel signature sequence [14]. This signature was already found in a deduced protein from *C. acetobutylicum* in 1998 [14]. Indeed, this protein is now annotated as a potassium transport protein in the genome of *C. acetobutylicum* (NP 347946.1). In only one other clostridial genome could a similar sequence be found. One gene (position 2921494-2922549 on contig gnl|Sanger_1496|Cd183h6.p1t) of *C. difficile* encodes a slightly similar sequence.

Although these signature sequences are obviously well conserved (Figure 13.4), the proteins differ in their lengths (KcsA 160 aa, *C. acetobutylicum* 256 aa, *C. difficile* 352 aa) and their

65-ALWW-SVETAT**TVGYG**DLYPVTLWGR- 89 *S. lividans KcsA*
163-ALWW-SFVTIT**TVGYG**DISPSTPFGR-187 *C. acetobutylicum*
230-VEKFFDQIARY**T**T**GYG**EKPWNILLVI-245 *C. difficile I*
300-DLWYFSMATFS**TVGYG**MVATS-LIGK-325 *C. difficile II*

FIGURE 13.4 Amino acid alignment around the K^+-signature sequence from KcsA of *S. lividans*, with the putative KcsA sequences of *C. acetobutylicum* and *C. difficile*.

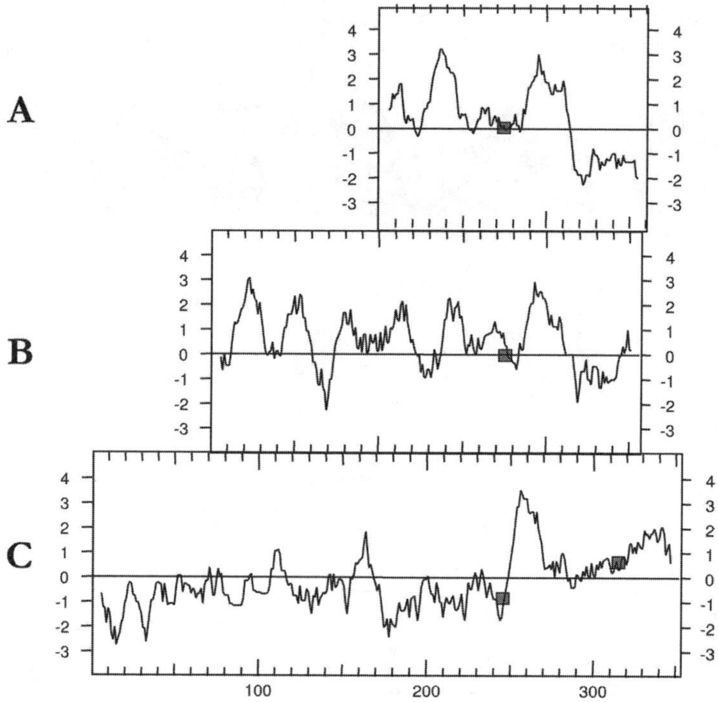

FIGURE 13.5 Kyte-Doolittle hydropathy profiles from KcsA in *S. lividans* (A), Orf in *C. acetobutylicum* (B), and Orf in *C. difficile* (C). The small grey boxes indicate the position of the K^+-signature sequences.

hydropathy profiles (Figure 13.5). Whereas the gene product from *C. acetobutylicum* shows the same architecture as KcsA with the pore sequence located between the two C-terminal transmembrane helices, the *C. difficile* gene product looks different (Figure 13.5). It lacks similarity around the first transmembrane region of KcsA but has two potential K^+-signature sequences.

The gene product from *C. acetobutylicum* seems to comprise six transmembrane helices. Nevertheless, it is probably another bacterial K^+-channel, as the two transmembrane-spanning K^+-channels, such as KcsA, have been described to be only shorter versions of the six transmembrane-spanning K^+-channels [14].

13.4 KDP

The *kdp* system represents, in most bacteria, a regulated, high-affinity, emergency K^+-translocating P-type ATPase. The system normally comprises at least three ATPase subunits, KdpA, KdpB, and KdpC, as well as a two-component signal transduction system with KdpD representing the membrane sensor histidine kinase and KdpE the response regulator. The increasing number of sequenced bacterial genomes shows that, although there is significant amino acid sequence conservation within the proteins, there is quite a variety in the operon structures. The best studied *kdp* system from *E. coli* comprises the genes *kdpFABCDE*, organized in two transcriptional units (for a review see [21,22]). The EcKdpF subunit is a 3-kDa polypeptide, fulfilling a stabilizing function for the ATPase [23].

Parts of a clostridial *kdp* system were identified and characterized in 1996 [24], showing some differences to the enterobacterial *kdp* system. There was another *orf* between *kdpC* and *kdpD*, which

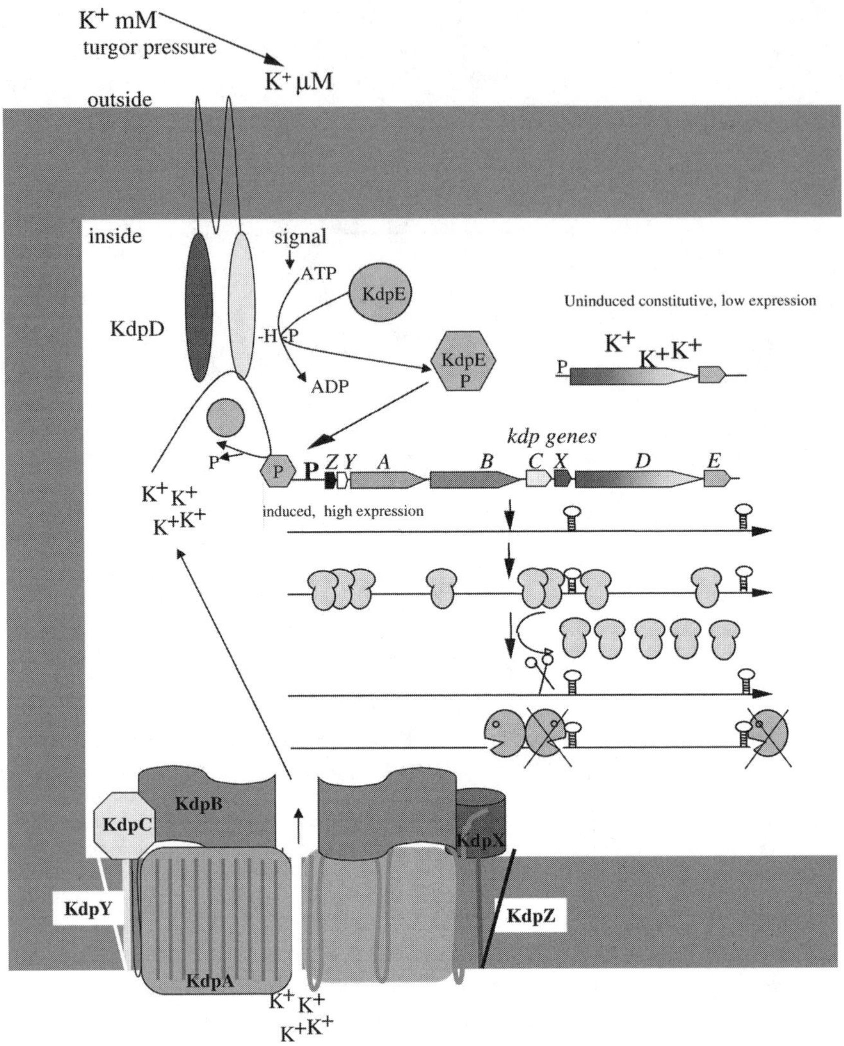

FIGURE 13.6 Current model of the *kdp*-system in *C. acetobutylicum.*

are adjacent in the *Eckdp* system. The gene was called *kdpX,* and its gene product shows no significant similarity to other proteins. The analysis of the 5′ end of the clostridial *kdp* operon was reported in 1997 [25]. Batch fermentation experiments with continously decreasing concentrations of potassium showed induction of the *kdp* system when the potassium concentration in the medium became growth limiting. Maximal induction was observed between 50 and 170 µ*M* of potassium. Further analysis of the *kdp* mRNA transcripts led to the current model (Figure 13.6):

- Under growth-supporting K^+-concentration there is a low and constitutive expression of the *kdpDE* operon, starting from the *kdpDE* promoter.
- When the potassium concentration becomes growth limiting, the autokinase activity of KdpD is activated, causing a phosphorylation of KdpD resulting in phosphotransfer to KdpE.
- KdpE in its phosphorylated form binds to the *kdpZYABCXDE* promoter region and activates transcription [26].

- A full-length *kdpZYABCXDE* transcript is produced, but not maintained. A putative RNase cleavage splits the full-length transcript into an unstable *kdpZYABCX* transcript and a protected *kdpDE* transcript.
- Translation of the *kdpZYABCX* transcript results in the synthesis of the K^+-transporting ATPase, and internal potassium concentration increases at the expense of ATP.
- At increasing potassium concentration, the phosphatase activity of KdpD causes dephosphorylation of KdpE, and transcription of the *kdp* operon stops.

So far, only indirect evidence indicates that *kdpX* encodes a KdpX protein necessary for the clostridial *kdp* system, as a functional Kdp-ATPase complex was not yet purified from *C. acetobutylicum* cells. RT-PCR experiments indicated a whole length *kdpZYABCXDE* mRNA transcript under K^+ limitation [25], also *in vitro* translation experiments with different *kdpX* carrying plasmids showed a protein of the expected size [25]. Complementation studies using *kdp* mutants from *E. coli* transformed with plasmids encoding either Ca*kdpZYABCX* or Ca*kdpZYABC* genes showed the mutants to grow much slower under K^+-limitation, when *kdpX* was missing [27]. Membranes of similarly prepared *E. coli kdp* mutants showed different substrates specificity, different stimulation profiles, and decreased transport capabilities of the clostridial Kdp-ATPase without the putative KdpX protein [28].

The promoter region upstream of the *C. acetobutylicum kdpABCXDE* operon, deduced from primer extension experiments [25], contains several T-tracts, some of which were found to be involved in recognition by purified and phosphorylated KdpE from *C. acetobutylicum*. The binding motif is similar, but not identical, to that of *E. coli* [26]. While EcKdpE~P is able to bind to the clostridial sequence, the CaKdpE proved to be much more specific and did not bind in unphosphorylated form or to other similar DNA sequences from *C. acetobutylicum* or *E. coli* [26].

Another difference between the enterobacterial and clostridial *kdp* system concerns the KdpF subunit of *E. coli*. The determined clostridial *kdpABCXDE* promoter includes a region upstream of *kdpA*, which encodes two little KdpF-like peptides designated KdpZ and KdpY [25]. Since the EcKdp-ATPase complex seems to work in a $KdpF_2A_2B_2C_2$ conformation, it was hypothesized that the clostridial Kdp-ATPase works either as a $KdpZYA_2B_2CX$ (Figure 13.6) or as a $KdpZ_2Y_2A_2B_2C_2X_2$ [25,27]. This hypothesis is interesting to test by searching for other clostridial *kdp* systems. In the available clostridial genomes, only two other *kdp* systems could be found, namely in *C. difficile* (contig gnl|Sanger_1496|Cd183h6.p1t) and *C. perfringens* (gnl|TIGR_19510 |contig:4304). These systems show some interesting differences to the Ca*kdp*-system (Figure 13.7).

First of all, the *kdp* system of *C. difficile* encodes two putative peptides similar to KdpZ and KdpY of *C. acetobutylicum*. But it lacks a *kdpX*-like gene, and the putative *kdpDE* operon is separated from the *kdpZYABC* operon by more than 266 kbp. Neither upstream of the *kdpDE* operon nor downstream of the *kdpZYABC* operon could *kdpX*-like regions be found. Also, in the *kdp* region

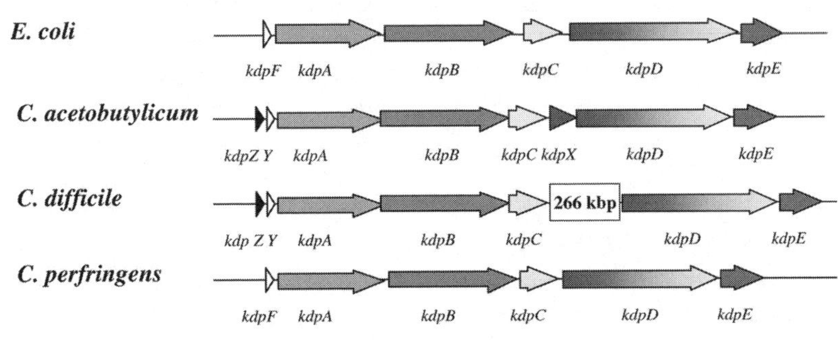

FIGURE 13.7 *kdp* operon organization in selected bacteria.

```
                                  +      +  +* ++++++*++
KdpF E. coli            MSAGVITGVLL---VFLLLGYLVYALINAEAF
KdpY C. acetobutylicum         MILLAIIIIELFIYLCYALFNPEKF
KdpY C. difficile         MEGLPMWFLAVVIIMLIIYLVYALLNPEKF
KdpF C. perfringens        MDLILIITGLTFLILFIYLWYILLRGERNE
KdpF S. aureus                MITLLAVVVIALILFLFYALIWSEKF
KdpF S. spec.           M-57-YALGLLVIIVLGLAVYLAMVIIEPERF
KdpY B. anthracis          MTIALSIIVAAITVYLVYLVYALLNPEKF
KdpY B. cereus                MMIALSIIVAAITVYLVYALLNPEKF

                        +++  ++   +    ++  ++ *  +   + *
KdpZ C. acetobutylicum  MLDVIFLVLIVLGFLFLRYFINWCEG-TINKK
KdpZ C. difficile       MLDIVMIALLFIGFVSMKLLADWCGK-QTEKK
KdpZ B. anthracis            MIGIFVVLVASMVSLASWSDKVVKEGKQS
KdpZ B. cereus          MLDVVMIGIFVVLVASMASLASWSDKVVKEGKQS
```

FIGURE 13.8 Amino acid alignment of KdpZ- and KdpF/Y-like peptides. The following sequences have been used: KdpF of *E. coli* P3693; KdpY and KdpZ of *C. acetobutylicum* U44892; KdpY and KdpZ of *C. difficile* gnl|Sanger 1496|Cd183h6.p1t; KdpY and KdpZ of *B. cereus* gnl TIGR 222523 contig:1754:b 10987; KdpF of *C. perfringens* (gnl|TIGR_19510 |contig:4304; KdpY and KdpZ of *B. anthracis* NC003995.1; KdpF of *S. aureus* NC002745.2; KdpF of *Synecchocystis* sp. BAA17930.1. Fully conserved residues are labeled by an asterisk (*), and conserved residues in more than 50% of the sequences are labeled by a plus sign (+).

of *C. perfringens,* no *kdpX*-like gene could be identified, and the genes are organized as the *kdp* operon from *E. coli*. In summary, these genome comparisons cannot shed any light on the importance or significance of having *kdp* systems with one or two putative membrane anchors, or with or without a *kdpX*-like gene.

Figure 13.8 shows an amino acid alignment of KdpZ and KdpF/KdpY-like sequences in some bacteria. So far, systems with a single small peptide contain gene products resembling KdpF. Systems with two peptides are encoded by operons, carrying a *kdpF/Y* gene, usually downstream of the *kdpZ* gene. The conservation of certain residues within the respective little peptides becomes more obvious with an increasing number of sequences. The C-terminal PEKF-motif in 50% of the KdpF/KdpY peptides might be involved in the putative ATPase-stabilizing function. Regarding the KdpY sequence of *C. perfringens* we are somewhat suspicious whether the available DNA sequence is correct. Assuming only one frame shift, this peptide would end with GEKK instead of GERNE. Also, the former peptide would be translationally coupled with the following *kdpA* gene in that the TGA stop codon overlaps with the ATG of the *kdpA* gene. The other peptide would end within the third codon of *kdpA*. Within the *kdp* system of *Synecchocystis* sp., a small orf exists between *kdpB* and *kdpC* [29,30]. This orf encodes a little protein, which shows similarity to KdpF/KdpY peptides in its C-terminus (Figure 13.8). We consider that protein to be an extended version of KdpF/KdpY peptides, possibly evolved by fusion of *kdpF/Y* and *kdpZ* genes. The role of KdpZ and KdpY for the Kdp-ATPase of *C. acetobutylicum* has been analyzed by complementation of certain *E. coli* mutants. These studies revealed that a lack of KdpZ or KdpY changes the activity and also the substrate specificity of the clostridial ATPase [28]. Thus, these peptides might be more important for the functionality of the corresponding Kdp-ATPases, as has been appreciated in the past.

13.5 KUP

The Kup system is another K^+-uptake system known from *E. coli*. The *kup* gene is expressed constitutively and encodes a two-domain protein. The N terminus probably spans the membrane 12 times, whereas the hydrophilic domain is supposed to be involved in regulation of transport activity. The system was shown to be the major K^+-transport system at acidic pH, working supposedly like a H^+/K^+-symporter [31]. In the genome of *C. perfringens,* a putative Kup-encoding

TABLE 13.1
Occurrence of Potassium-Transport System in Different Clostridial Species

	TrkH	TrkG	KcsA	Kdp	Kup
C. acetobutylicum	-	-	+	+	-
C. perfringens	+	-	-	+	+
C. thermocellum	+	-	-	-	-
C. botulinum	+	-	-	-	-
C. difficile	+	-	(+)	+	-
C. tetani	+	+	-	-	-

gene could be detected (TIGR 195103 contig:4304). The deduced gene product is 37.5% identical to EcKup (gi 16131615) and shows a similar hydropathy profile.

13.6 SUMMARY

The following potassium transport systems could be found at least in one of the investigated clostridial genomes (Table 13.1): The TrkH system is obviously used by most of the investigated organisms. *C. difficile* and *C. perfringens* have the highest number of different transport systems, whereas *C. thermocellum* and *C. botulinum* seem to achieve potassium transport, and most likely osmoadaption, by using just the TrkH system. Nevertheless, we cannot rule out the existence of so far unknown potassium transport systems in these organisms.

REFERENCES

1. Davies, R., Studies on the acetone-butyl alcohol fermentation. 2. Intermediates in the fermentation of glucose by *Cl. acetobutylicum*. 3. Potassium as an essential factor in the fermentation of maize meal by *Cl. acetobutylicum* (BY), *Biochem. J.*, 36, 582, 1942.
2. Bakker, E.P., Low-affinity K^+ uptake systems, in *Alkali Cation Transport Systems in Prokaryotes*, Bakker, E.P., Ed., CRC Press, Boca Raton, FL, 1993, 253.
3. Stumpe, S. et al., K^+ circulation across the prokaryotic cell membrane: K^+-uptake systems, in *Handbook of Biological Physics*, Konings, W.N., Kaback, H.R., and Lolkema, J.S., Eds., Elsevier, Amsterdam, 1996, 473.
4. Stewart, L.M., Bakker, E.P., and Booth, I.R., Energy coupling to K^+ uptake via the Trk system in *Escherichia coli*: the role of ATP, *J. Gen. Microbiol.*, 131, 77, 1985.
5. Wood, J.M., Osmosensing by bacteria: signals and membrane-based sensors, *Microbiol. Mol. Biol. Rev.*, 63, 230, 1999.
6. Bossemeyer, D. et al., K^+-transport protein TrkA of *Escherichia coli* is a peripheral membrane protein that requires other *trk* gene products for attachment to the cytoplasmic membrane, *J. Biol. Chem.*, 264, 16403, 1989.
7. Schlosser, A. et al., NAD^+ binding to the *Escherichia coli* K($^+$)-uptake protein TrkA and sequence similarity between TrkA and domains of a family of dehydrogenases suggest a role for NAD^+ in bacterial transport, *Mol. Microbiol.*, 9, 533, 1993.
8. Dosch, D.C. et al., Genetic analysis of potassium transport loci in *Escherichia coli*: evidence for three constitutive systems mediating uptake potassium, *J. Bacteriol.*, 173, 687, 1991.
9. Nakamura, T. et al., Cloning of the *trkAH* gene cluster and characterization of the Trk K($^+$)-uptake system of *Vibrio alginolyticus*, *Microbiology*, 144, 2281, 1998.
10. Parra-Lopez, C., Baer, M.T., and Groisman, E.A., Molecular genetic analysis of a locus required for resistance to antimicrobial peptides in *Salmonella typhimurium*, *Embo J.*, 12, 4053, 1993.

11. Linton, K.J. and Higgins, C.F., The *Escherichia coli* ATP-binding cassette (ABC) proteins, *Mol. Microbiol.,* 28, 5, 1998.
12. Harms, C. et al., Identification of the ABC protein SapD as the subunit that confers ATP dependence to the K^+-uptake systems Trk(H) and Trk(G) from *Escherichia coli* K-12, *Microbiology,* 147, 2991, 2001.
13. Schrempf, H. et al., A prokaryotic potassium ion channel with two predicted transmembrane segments from *Streptomyces lividans*, *Embo J.,* 14, 5170, 1995.
14. Doyle, D.A. et al., The structure of the potassium channel: molecular basis of K^+ conduction and selectivity, *Science,* 280, 69, 1998.
15. van Dalen, A. et al., Efficient membrane assembly of the KcsA potassium channel in *Escherichia coli* requires the protonmotive force, *EMBO Rep.,* 1, 340, 2000.
16. Holyoake, J. et al., KcsA closed and open: modelling and simulation studies, *Eur. Biophys. J.*, 2003 (Epub).
17. Garofoli, S. and Jordan, P.C., Modeling permeation energetics in the KcsA potassium channel, *Biophys. J.,* 84, 2814, 2003.
18. Guidoni, L. and Carloni, P., Potassium permeation through the KcsA channel: a density functional study, *Biochim. Biophys. Acta,* 1563, 1, 2002.
19. LeMasurier, M., Heginbotham, L., and Miller, C., KcsA: it's a potassium channel, *J. Gen. Physiol.,* 118, 303, 2001.
20. Miller, C., Ion channels: doing hard chemistry with hard ions, *Curr. Opin. Chem. Biol.,* 4, 148, 2000.
21. Altendorf, K. and Epstein, W., The Kdp-ATPase of *Escherichia coli*, in *Biomembranes (ATPases)*, Lee, A., Ed., JAI Press Inc., Greenwich, CT, 1996, 403.
22. Altendorf, K. et al., Structure and function of the Kdp-ATPase of *Escherichia coli*, *Acta Physiol. Scand. Suppl.,* 643, 137, 1998.
23. Gassel, M. et al., The KdpF subunit is part of the K(+)-translocating Kdp complex of *Escherichia coli* and is responsible for stabilization of the complex *in vitro*, *J. Biol. Chem.,* 274, 37901, 1999.
24. Treuner-Lange, A. and Dürre, P., Molecular biological analysis of *kdpD/E*, a sensor histidine kinase/response regulator system in *Clostridium acetobutylicum*, *Anaerobe,* 2, 351, 1996.
25. Treuner-Lange, A., Kuhn, A., and Dürre, P., The *kdp* system of *Clostridium acetobutylicum*: cloning, sequencing, and transcriptional regulation in response to potassium concentration, *J. Bacteriol.,* 179, 4501, 1997.
26. Behrens, M. and Dürre, P., KdpE of *Clostridium acetobutylicum* is a highly specific response regulator controlling only the expression of the *kdp* operon, *J. Mol. Microbiol. Biotechnol.,* 2, 45, 2000.
27. Treuner, A., Das *kdp*-System von *Clostridium acetobutylicum*: Klonierung, Sequenzierung, Induktion und molekularbiologische Charakterisierung, Ph.D. Thesis, Georg-August-Universität Göttingen, Germany, 1996.
28. Behrens, M., Schreiber, W., and Dürre, P., The high-affinity K^+-translocating ATPase complex from *Clostridium acetobutylicum* consists of six subunits, *Anaerobe,* 7, 159, 2001.
29. Flemisch, S.E., Klonierung, Untersuchung zur Transkription und Analyse der Sequenzdaten des *kdp*-Systems von *Synechocystis* sp. *PCC6803*, Diploma Thesis, University of Ulm, Germany, 1998.
30. Kaneko, T. et al., Sequence analysis of the genome of the unicellular cyanobacterium *Synechocystis* sp. *strain PCC6803*. II. Sequence determination of the entire genome and assignment of potential protein-coding regions (supplement), *DNA Res.,* 3, 185, 1996.
31. Zakharyan, E. and Trchounian, A., K^+ influx by Kup in *Escherichia coli* is accompanied by a decrease in H^+ efflux, *FEMS Microbiol. Lett.,* 204, 61, 2001.

14 Transport of Phosphate

Ralf-Jörg Fischer and Hubert Bahl

CONTENTS

14.1 INTRODUCTION

Phosphorus is an essential nutrient for prokaryotes. It belongs to the macronutrients [1] or major bioelements [2] and makes up to 3% of the cell mass [3]. Phosphorus is primarily a constituent of nucleic acids and phospholipids but is also present in teichoic acids and nucleotides such as ATP, GTP, NAD^+, and FAD.

Generally, major bioelements are required by the bacteria in relatively high concentrations of above 0.1 m*M* [2]. Cells satisfy their demand for phosphorus by uptake of inorganic (P_i, HPO_4^{2-}) or organic phosphate compounds. However, in most soils, soluble phosphate is present at concentrations of 0.1 to 10 μ*M* only [4]. These low concentrations result from the formation of essentially insoluble metallophosphate compounds, which are produced upon addition of P_i to soil. Thus, especially in soil, inorganic phosphate is often expected to be the critical growth-limiting factor [5]. It can be assumed that bacteria in nature predominantly have to cope with phosphate limitation. Hence, the ability to respond and to adapt to phosphate limitation is critical for bacterial survival in nature. Due to the occurrence of phosphorus in various different cell compounds, it is not a surprise that a complex regulatory network is required enabling a crosstalk between the different branches of metabolism for their P_i demand. Only a tight interaction of regulatory cascades guarantees an optimal distribution of this valuable nutrient among the competing anabolic pathways. Furthermore, the cells themselves display a variety of structural alterations. The best-understood systems are the ones of *Escherichia coli* and *Bacillus subtilis,* which have been studied for almost 50 years [6,7] (reviewed by Wanner [8] and Hulett [5]).

14.2 UPTAKE OF PHOSPHATE

14.2.1 High-Affinity Systems

Generally, as soon as the extracellular availability of P_i drops below a threshold concentration, one basic important cellular reaction is the *de novo* expression of a highly efficient phosphate uptake system. In most cases, a periplasmic protein-dependent phosphate-specific transport system (Pst) is synthesized. These phosphate uptake systems belong to the superfamily of ABC (ATP-binding cassette) transporters [9,10]. They are composed of a membrane-associated complex of four proteins or protein domains. Two of them, PstC and PstA, are integral membrane polypeptides spanning the cytoplasmic membrane. They build the translocase and are often denoted as permease components. Two other proteins, PstB, frequently act as a dimer and are associated at the cytoplasmic side of the complex. PstB proteins bind ATP and provide the energy for the transport via its hydrolysis. Thus, they are described as ATP-binding components, or traffic ATPases. Bacterial ABC-type importers usually contain one additional component located outside of the cytoplasmic membrane, (periplasmic) binding proteins. In phosphate-specific uptake systems, these components are often known as PstS proteins. In Gram-negative bacteria, PstS polypeptides are freely moving in the periplasm. Equivalent proteins of Gram-positive bacteria are lipoproteins, which are anchored in the cytoplasmic membrane via fatty-acylated cysteine residues at the N-terminus in the mature protein [11,12].

Nearly nothing is known about the response of clostridia to phosphate starvation on the molecular level. In contrast, effects of phosphate limitation on the metabolism and development of *C. acetobutylicum* were described more than 20 years ago [13,14, see below]. The clostridial genome sequencing projects provide a basis for progress in the understanding of clostridial biology in general and of phosphate metabolism in particular. Three of them have been finished [15] — first *C. acetobutylicum* ATCC 824 in 2001 [16], a nonpathogenic, harmless soil-living solvent producer; secondly, in 2002, *C. perfringens* str. 13 [17], the causative agent of gas gangrene; and finally, *C. tetani* E88 in 2003 [18], the producer of tetanus toxin, the second most poisonous substance known. Another five clostridial genome sequencing projects are in progress [19] — *C. botulinum* A (NC_003223), *C. difficile* (NC_002933), *C. perfringens* ATCC 13124 (NC_003913), *C. thermocellum* ATCC 27405 (NZ_AABG00000000), and *C.* sp. BC1.

First steps of the analysis of the clostridial response to phosphate starvation on the molecular level have been taken using *C. acetobutylicum*, the model organism of nonpathogenic clostridia. BLAST-searches [20] with the amino acid sequence of the phosphate-specific binding protein PstS of *B. subtilis* (300 aa, gi903303 [21]) against the genome of *C. acetobutylicum* predicted that the open reading frame CAC1705 (298 aa, gi15894982) encodes a phosphate-specific binding protein as it reveals striking similarities of at least 47% identical and 66% similar amino acid residues [22,23]. The putative *pstS* gene is followed by four genes sharing significant similarities with the corresponding members of typical phosphate-specific ABC transport systems, as described above. Accordingly, the *pst* operon of *C. acetobutylicum* comprises five genes, *pstS*, *pstC*, *pstA*, *pstB,* and *phoU,* in the same order as *E. coli* (Figure 14.1 [24]). Experimental data proved that the *pst* operon of *C. acetobutylicum* is only expressed under P_i limitation, with concentrations below 0.1 m*M* in the external medium [23].

The genomic organization of the *pst* operons of *C. acetobutylicum, C. perfringens, C. tetani, E. coli,* and *B. subtilis* is schematically drawn in Figure 14.1. Nearly identical operon structures can be predicted in *C. perfringens* and *C. acetobutylicum*, with the exception that the *pstS* gene seems to be duplicated in the former one (61% identical, 80% similar amino acid residues). The *pst* operons in both organisms contain a *phoU* gene. In *E. coli* PhoU is not directly involved in the uptake process, but is proposed to be part of a "repression complex" in the membrane, which consists of the phosphate-specific ABC transporter and the two-component regulatory system PhoP/PhoR [8]. Interestingly, PhoU is missing in *B. subtilis* [21,25], the model organism of aerobic Gram-positive bacteria.

Based on Figure 14.1, it can be concluded that clostridia likely possess *pst*-type high-affinity phosphate-specific uptake systems. The membrane-spanning parts, as well as the ATP binding components, seem to be encoded by different genes and not to be organized in the form of protein domains. Interestingly, in *C. perfringens,* two tandemly arranged *pstS* genes can be predicted. Similar duplications of *pstS* genes were found, for example, in the genomes of *Methanobacterium thermoautotrophicum* ΔH [26] or *Lactococcus lactis* ssp. *lactis* IL1403 [27].

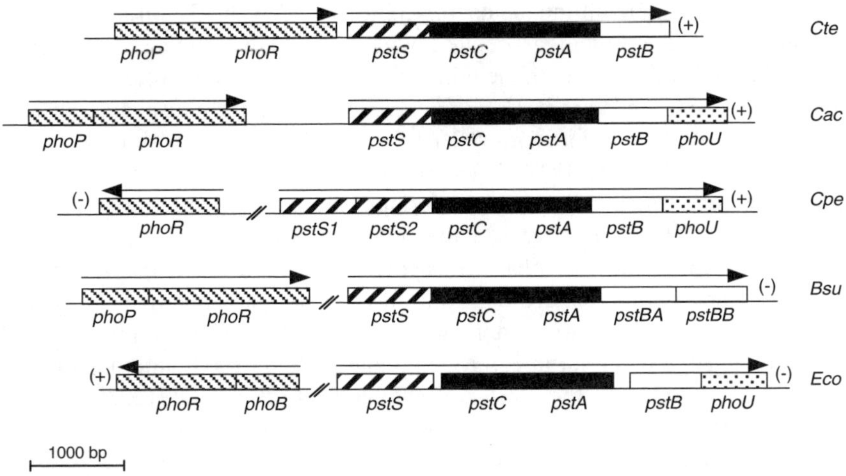

FIGURE 14.1 Genetic organizations of *pst* operons and relative locations of the major pho regulon regulatory two-component systems in *C. tetani* (*Cte*), *C. acetobutylicum* (*Cac*), *C. perfringens* (*Cpe*), *B. subtilis* (*Bsu*), and *E. coli* (*Eco*). The arrangement of structural genes is shown in boxes, and their direction of transcription is indicated by arrows. The function of the gene products is indicated by the shading of the boxes: roughly hatched boxes, binding proteins; dark boxes, inner membrane components; light boxes, ATPase; dotted boxes, genes that do not encode proteins directly involved in substrate transport across the inner membrane; finely hatched boxes, major phosphate-specific two-component regulatory protein systems. Plus or minus signs in parentheses symbolize the leading or lagging strand of the chromosome. Genes are: *C. tetani*, CTC01130 to CTC01135 (*phoPR* and *pst* operons); *C. acetobutylicum*, CAC1700, CAC1701 (PhoP/PhoR) and CAC 1705 to CAC 1709 (*pst* operon); *C. perfringens*, CPE1757 (*phoR*), CPE0637 to CPE0641 (*pst* operon); *B. subtilis*, PhoP, PhoR, PstS, PstC, PstA, PstBA, PstBB; *E. coli*, PhoB, PhoR, PstS, PstC, PstA, PstB, PhoU.

The *pst* operons of clostridia obviously show a nonuniform composition. Thus far, the genome of *C. tetani* harbors the smallest *pst* operon, comprising solely the four genes encoding the phosphate transport complex, *pstS*, *pstC*, *pstA*, and *pstB* but not *phoU*.

14.2.2 Low-Affinity Systems

Bacteria usually contain at least two major P_i uptake machineries. Besides the high-affinity, low-velocity Pst-like system (see above), which is expressed only under phosphate starvation, members of the inorganic phosphate transporter (PiT) family are widespread, as they were found in Gram-negative and Gram-positive bacteria, archaea, yeast, fungi, plants, and animals [28]. PitA and PitB of *E. coli* represent low-affinity P_i transporters [29,30] ensuring the supply of phosphate when extracellular P_i is in excess. Unlike Pst, Pit is a single-component transporter, not part of the Pho regulon, and appears to be synthesized constitutively in *E. coli*, whereas the expression of proteins with similar proposed function in other organisms could be repressed under conditions of P_i limitation (e.g., Pit of *Rhizobium meliloti* [31]). Furthermore, P_i uptake via Pit in *E. coli* and, for instance, *Acinetobacter johnsonii* depends on the presence of divalent cations (Mg^{2+}, Ca^{2+}, Co^{2+}, or Mn^{2+}). They form a soluble, neutral, metal-phosphate chelate ($MeHPO_4$) complex, which is driven by the proton motive force, symported together with one proton per mol, thus indicating a cotransport of divalent cations, P_i, and H^+ [32–34]. Other functionally characterized members of the PiT family appear to catalyze inorganic phosphate uptake by Na^+ symport, like Pho4 of the filamentous fungi *Neurospora crassa*, which is described as a P_i -repressible P_i:Na^+ symporter [35].

The knowledge about low-affinity inorganic phosphate uptake systems in clostridia is limited to similarity data of deduced proteins from genome sequences. *C. acetobutylicum* reveals a putative "phosphate permease" of 330 amino acids (gi15896344, CAC3093) directly preceded by an "uncharacterized conserved protein associated with phosphate permease" (210 amino acids, gi15896345, CAC3094), both located on the minus strand. Their derived protein products are highly similar to the same-sized pendants of *C. perfringens*, a "probable phosphate transport protein" CPE0583 (gi18144242) of 332 amino acids (62% identical, 78% similar residues) and a "conserved hypothetical protein" CPE0582 (gi18144241, 210 amino acids, 43% identical, 71% similar residues). The open reading frames indicate 8 bp overlaps, suggesting that both genes are transcribed as a single transcript. Based on similarity data, the phosphate permeases belong to the Pho4 subclass (pfam01384, [36]) of inorganic phosphate transporters mentioned above, indicating a sodium-dependent phosphate uptake process. In the genome of *C. tetani,* these putative phosphate permeases are not present. However, a "sodium-dependent phosphate transport protein" (CTC00877, gi28210590) is predicted, a potential integral membrane protein of 548 amino acids. The deduced protein seems to be a member of the cluster of orthologous genes COG1283 [37], revealing the signature Pfam Pf02690 [36] of the protein families database. This is a family of mainly mammalian type II renal Na^+/P_i -cotransporters with other related sequences from lower eukaryotes and bacteria, some of which are also Na^+/P_i -cotransporters. Interestingly, CTC00877 shares highest similarities (56% identity, 74% similarity) with the "conserved hypothetical protein" CPE2368 (537 aa, gi18146034) of *C. perfringens* and weaker relationship with putative membrane proteins of *C. thermocellum*. In *C. acetobutylicum* no such gene product seems to be present.

14.2.3 Pho Regulon

The genes of a bacterium that are regulated in response to phosphate starvation and under the control of a two-component regulatory system are referred to as the Pho regulon. In *E. coli* the Pho regulon comprises at least 38 proteins of 9 operons [8], and in *B. subtilis,* 31 different genes of 10 operons [5,38].

Investigations of clostridial Pho regulons are just at the beginning. To date, only two operons together with seven genes have been shown to be upregulated in *C. acetobutylicum* when the cells grow under phosphate limitation. The first one, the *pst* operon, includes five genes (see above), and the second operon shows two open reading frames of a putative two-component regulatory system [22,23]. The latter two deduced proteins are functionally similar to PhoB and PhoR of *E. coli* [39,40] or PhoP and PhoR of *B. subtilis* [41,42]. Thus, they might play a key role within the control of the Pho regulon, as they probably monitor the external P_i concentration and constitute the top of the hierarchical regulatory cascade initiated under phosphate starvation.

In *C. tetani,* the deduced *phoP/phoR* genes directly precede the *pst* genes (Figure 14.1) on the same strand and with identical orientation. This architecture, in close vicinity, can implicate functional relatedness of the gene products. A similar gene arrangement can be assumed in *C. acetobutylicum* because upstream of the *pst* operon, in a distance of only about 1000 bp, two open reading frames (CAC1700 and CAC1701) are located, which are predicted to encode a phosphate-specific response regulator and its sensor histidine kinase (see above). Thus, this chromosomal area harboring both operons was named the "*phoPR-pst*" gene region [22]. In *C. perfringens,* the situation seems to be different, as only a conserved *phoR*-like gene product can be predicted, an orphan sensor histidine kinase (CPE1757) sharing highest similarities (47% identical, 67% similar amino acid residues) with PhoR of *C. acetobuylicum.* (For further information about two-component regulatory systems of clostridia refer to Chapter 24.)

If we compare these genetic organizations with the situations in *E. coli* and *B. subtilis,* various differences are obvious. For example, in *C. perfringens*, the gene loci of the Pho regulon regulators are not located in the neighborhood of their *pst* operons. The *phoPR* and the *pst*-operon genes of *B. subtilis* are separated by at least 400 kilobase pairs, and, in contrast to the situation in *C. tetani* and *C. acetobutylicum*, both gene loci are found on the complementary strand of the DNA [43]. In *E. coli,* the distance between the equivalent operons is more than one megabase pairs, as the *phoBR* operon resides at 9 min, and the *pst* operon is located at about 84 min on the chromosome. Furthermore, it is remarkable that they are contrarily orientated — *phoBR* is located on the plus strand and the *pst* locus on the minus strand.

A common regulatory principle of Pho regulon genes is based on the existence of "Pho boxes," which are often found in different numbers in their promoter regions [5,8]. Pho boxes are conserved DNA motives of at least two repeated AT-rich hexamer DNA motifs, which are separated by five variable nucleotides. They are recognized and bound by, normally phosphorylated, PhoP- or PhoB-like response regulator proteins, a prerequisite for transcription initiation.

In clostridia, Pho boxes have not been experimentally identified. Due to the high AT content (>68 mol%) of the chromosomal DNA of *C. acetobutylicum,* motifs with at least partial similarity to the Pho boxes of *B. subtilis* or *E. coli* can be found, e.g., in front of the *pst* operon. However, whether they function as regulatory sites remains to be determined. On the other hand, it is possible that Pho boxes in clostridia are quite different from those in *E. coli* or *B. subtilis*.

Up to now, Pho regulons in clostridia have not been well defined. More proteins need to be identified which are expressed after the depletion of phosphate. Our current projects focus on proteome/secretome analysis as well as on DNA macroarray-based transcriptional analysis of *C. acetobutylicum* during growth under high- and low-phosphate concentrations.

14.2.4 Effects of Phosphate Limitation

The effect of phosphate limitation on the metabolism of clostridia has been extensively studied in *C. acetobutylicum*. The developmental cycle of *C. acetobutylicum* is characterized by a metabolic switch from acid to solvent production and the production of endospores. Both events are initiated when the environmental conditions do not allow further exponential growth. So far, no single parameter has been identified that is able to induce solvent formation. In fact, a combination of several parameters, resulting in stressful growth conditions, seems to be important. As an example, an excess of fermentable carbohydrates, low pH values (<5), and high concentration of butyrate result in a slow growth rate and the metabolic shift to solvent production. Interestingly, phosphate in growth-limiting concentrations (<1 m*M*) also favors solvent formation. Thus, a phosphate-limited chemostat allows the continuous production of acetone and butanol over a long period of time [13]. Also in batch culture, low phosphate concentrations increase the substrate turnover and the solvent yield (Table 14.1).

Phosphate limitation is not only an important factor for a reliable metabolic switch of *C. acetobutylicum*, but also for the initiation of sporulation in clostridia. Depending on the organism or the strain, sporulation cannot always be observed in laboratory cultures. In our experience, cultures of *C. acetobutylicum, C. pasteurianum, C. tyrobutyricum,* and others that encounter phosphate limitation during a certain growth phase initiate sporulation in the majority of cells. This always occurs during the setup of a phosphate-limited chemostat, after the inflow of fresh phosphate-limited medium to the growing start-up batch culture (with higher phosphate concentrations) has begun [14,44].

Note that the conditions of sporulation and early regulatory steps of this developmental process differ in clostridia and the aerobic model *B. subtilis*. Saccharolytic clostridia, as *C. acetobutylicum,* do not sporulate when the fermentable carbohydrate is exhausted. Furthermore, several key components of the phosphorelay present in *B. subtilis* have not been found in the genome of *C. acetobutylicum.* How phosphate metabolism is involved in the regulation of sporulation and the metabolic switch at the molecular level remains to be elucidated.

TABLE 14.1
Substrate Consumption and Product Formation of *C. acetobutylicum* DSM 1731 in Batch Cultures at Low- and High-Phosphate Concentrations After 3 Days

	Culture 1	Culture 2
Initial substrate concentrations (m*M*)		
Phosphate	0.62	12.25
Glucose	303	303
Fermentation products after 3 days (m*M*)		
Butanol	175	47
Acetone	77	12
Ethanol	22	5
Butyrate	9	68
Acetate	18	42
Acetoin	9	8
Substrate concentrations after 3 days (m*M*)		
Glucose	0	129
Phosphate	0	11
Glucose consumption	100%	57%
Glucose converted to solvents	88%	36%

Source: After Bahl, H. et al., *Eur. J. Environ. Microbiol.,* 15, 201, 1982.

ACKNOWLEDGMENTS

Experimental work carried out in the authors' laboratories was supported by the Deutsche Forschungsgemeinschaft, the former Bundesministerium für Forschung und Technologie, and the Fonds der Chemischen Industrie.

REFERENCES

1. Madigan, M.T., Martinko, J.M., and Parker, J., *Brock Biology of Microorganisms*, 10th Edition, Prentice Hall, Pearson Education, Inc., Upper Saddle River, NJ, 2003, Chapter 4.
2. Gottschalk, G., *Bacterial Metabolism,* 2nd Edition, Springer, New York, 1986, Chapter 1.
3. Fuchs, G. and Kröger, A., Growth and nutrition, in *Biology of the Prokaryotes*, Lengeler, J.W., Drews, G., and Schlegel, H.G., Eds., Blackwell Science, Malden, MA, 1999, Chapter 6.
4. Bieleski, R.L., Phosphate pools, phosphate transport, and phosphate availability, *Annu. Rev. Plant Physiol.*, 24, 225, 1973.
5. Hulett, F.M., The pho regulon, in *Bacillus subtilis and Its Closest Relatives: From Genes to Cells*, Sonenshein, A.L., Hoch, J.A., and Losick, R., Eds., ASM Press, Washington, DC, 2002, Chapter 15.
6. Torriani, A., Influence of inorganic phosphate in the formation of phosphatases by *Escherichia coli*, *Biochem. Biophys. Acta,* 38, 460, 1960.
7. Torriani-Gorini, A., Ed., *Phosphate in Microorganisms: Cellular and Molecular Biology*, ASM Press, Washington, DC, 1994.
8. Wanner, B.L., Phosphorus assimilation and control of the phosphate regulon, in *Escherichia coli and Salmonella: Cellular and Molecular Biology*, 2nd Edition, Neidhardt, F.C., Ed., ASM Press, Washington, DC, 1996, Chapter 87.

9. Ames, G.F.-L., Bacterial periplasmic transport systems: structure, mechanism, and evolution, *Ann. Rev. Biochem.*, 55, 379, 1986.
10. Boos, W. and Lucht, J.M., Periplasmic binding protein-dependent ABC transporters, in *Escherichia coli and Salmonella: Cellular and Molecular Biology*, 2nd ed., Neidhardt, F.C., Ed., ASM Press, Washington, DC, 1996, Chapter 76.
11. Tjalsma, H. et al., Signal peptide-dependent protein transport in *Bacillus subtilis*: a genome-based survey of the secretome, *Microbiol. Mol. Biol. Rev.*, 64, 515, 2000.
12. Mazmanian, S.K. and Schneewind, O., Cell wall-anchored surface proteins and lipoproteins of Gram-positive bacteria, in *Bacillus subtilis and Its Closest Relatives: From Genes to Cells*, Sonenshein, A.L., Hoch, J.A., and Losick, R., Eds., ASM Press, Washington, DC, 2002, Chapter 6.
13. Bahl, H., Andersch, W., and Gottschalk, G., Continuous production of acetone and butanol by *Clostridium acetobutylicum* in a two-stage phosphate limited chemostat, *Eur. J. Appl. Environ. Microbiol.,* 15, 201, 1982.
14. Meinecke, B., Bahl, H., and Gottschalk, G., Selection of an asporogenous strain of *Clostridium acetobutylicum* in continuous culture under phosphate limitation, *Appl. Environ. Microbiol.*, 48, 1064, 1984.
15. National Center for Biotechnology Information, http://www.ncbi.nlm.nih.gov/genomes/MICROBES/Complete.html, Bethesda, MD.
16. Nolling, J. et al., Genome sequence and comparative analysis of the solvent-producing bacterium *Clostridium acetobutylicum*, *J. Bacteriol.*, 183, 4823, 2001.
17. Shimizu,T. et al., Complete genome sequence of *Clostridium perfringens*, an anaerobic flesh-eater, *Proc. Natl. Acad. Sci. USA*, 99, 996, 2002.
18. Brüggemann, H. et al., The genome sequence of *Clostridium tetani*, the causative agent of tetanus disease, *Proc. Natl. Acad. Sci. USA*, 100, 1316, 2003.
19. National Center for Biotechnology Information, http://www.ncbi.nlm.nih.gov/genomes/MICROBES/InProgress.html, Bethesda, MD.
20. Altschul, S.F. et al., Gapped BLAST and PSI-BLAST: a new generation of protein database programs, *Nucleic Acids Res.*, 25, 3389, 1997.
21. Takemura, K.-I., Mizuno, M., and Kobayashi, Y., A *Bacillus subtilis* gene cluster similar to the *Escherichia coli* phosphate-specific transport (*pst*) operon: evidence for a tandemly arranged *pstB* gene, *Microbiology*, 142, 2017, 1996.
22. Meyer, U., Analyse des ECF-Sigmafaktors SigX und Phosphat-regulierter Gene aus *Clostridium acetobutylicum*, Ph.D. Thesis, University of Rostock, Germany, 2001.
23. Fischer, R.-J. et al., Manuscript in preparation, 2005.
24. Blattner, F.R. et al., The complete genome sequence of *Escherichia coli* K-12, *Science*, 277, 1453, 1997.
25. Qi, Y., Kobayashi, Y., and Hulett, F.M., The *pst* operon of *Bacillus subtilis* has a phosphate-regulated promoter and is involved in phosphate transport but not in regulation of the *pho* regulon, *J. Bacteriol.*, 179, 2534, 1997.
26. Smith, D.R. et al., Complete genome sequence of *Methanobacterium thermoautotrophicum* ΔH: functional analysis and comparative genomics, *J. Bacteriol.*, 179, 7135, 1997.
27. Bolotin, A. et al., The complete genome sequence of the lactic acid bacterium *Lactococcus lactis* ssp. *lactis* IL1403, *Genome Res.,* 11, 731, 2001.
28. Transport Classification Database, http://tcdb.ucsd.edu/tcdb/family2.php?tc=2.A.20, 2004.
29. Willsky, G.R. and Malamy, M.H., Characterization of two genetically separable inorganic phosphate transport systems in *Escherichia coli*, *J. Bacteriol.*, 144, 356, 1980.
30. Harris, R.M. et al., Characterization of PitA and PitB from *Escherichia coli*, *J. Bacteriol.,* 183, 5008, 2001.
31. Bardin, S.D., Voegele, R.T., and Finan, T.M., Phosphate assimilation in *Rhizobium* (*Sinorhizobium*) *meliloti*: identification of a *pit*-like gene, *J. Bacteriol.*, 180, 4219, 1998.
32. Van Veen, H.W. et al., Translocation of metal phosphate via the phosphate inorganic transport system of *Escherichia coli*, *Biochemistry*, 33, 1766, 1994.
33. Van Veen, H.W. et al., Characterization of two phosphate transport systems in *Acinetobacter johnsonii* 210A, *J. Bacteriol.*, 175, 200, 1993.

34. Van Veen, H.W. et al., Mechanism and energetics of the secondary phosphate transport system of *Acinetobacter johnsonii* 210A, *J. Biol. Chem.*, 268, 19377, 1993.
35. Mann, B.J. et al., Nucleotide sequence of pho-4+, encoding a phosphate-repressible phosphate permease of *Neurospora crassa*, *Gene,* 83, 281, 1989.
36. Bateman, A. et al,, The Pfam protein families database, *Nucleic Acids Res.,* Database Issue 32:D138-D141 32, 2004.
37. Tatusov, R.L. et al., The COG database: an updated version includes eukaryotes, *BMC Bioinformatics,* 4, 41, 2003.
38. Antelmann, H., Scharf, C., and Hecker, M., Phosphate starvation-inducible proteins of *Bacillus subtilis*: proteomics approach and transcriptional analysis, *J. Bacteriol.,* 182, 4478, 2000.
39. Makino, K. et al., Nucleotide sequence of the phoB gene, the positive regulatory gene for the phosphate regulon of *Escherichia coli* K-12, *J. Mol. Biol.,* 190, 37, 1986.
40. Makino, K. et al., Nucleotide sequence of the *phoR* gene, a regulatory gene for the phosphate regulon of *Escherichia coli*, *J. Mol. Biol.,* 192, 549, 1986.
41. Lee, J.W. and Hulett, F.M., Nucleotide sequence of the *phoP* gene encoding PhoP, the response regulator of the phosphate regulon of *Bacillus subtilis*, *Nucleic Acids Res.*, 20, 5848, 1992.
42. Seki, T. et al., Nucleotide sequence of the *Bacillus subtilis phoR* gene, *J. Bacteriol.*, 170, 5935, 1988.
43. Kunst, F., et al., The complete genome sequence of the gram-positive bacterium *Bacillus subtilis*, *Nature*, 390, 249, 1997.
44. Dabrock, B., Bahl, H., and Gottschalk, G., Parameters affecting solvent production by *Clostridium pasteurianum*, *Appl. Environ. Microbiol.*, 58, 1233, 1992.

Part III

Cellular Components and Medical Aspects

15 Membrane Lipids of Clostridia

Howard Goldfine and Norah C. Johnston

CONTENTS

15.1 INTRODUCTION

Given the breadth and depth of the genus *Clostridium*, it is no surprise that the membrane lipids show considerable diversity. Indeed, it appears that their fatty acids show relationships to those found in both Gram-positive and Gram-negative bacteria. Unfortunately, it is difficult to make broad generalizations because the polar lipids have been studied in only a few of the more than 140 recognized species. Since the outer structures of bacteria show the greatest chemical diversity, an examination of their lipids provides useful data related to their taxonomy. As the genomes of more clostridia are sequenced, these relationships will undoubtedly be clarified and refined.

15.2 THE ACYL AND ALKENYL CHAINS

15.2.1 THE ACYL CHAINS

There is a wealth of published data on the acyl chains of *Clostridium.* These are presented in Table 15.1. Based on acyl chain composition, we have divided the genus *Clostridium* into four major groups (Table 15.2). Group A contains those organisms that have the most simple acyl chain composition, which consist predominantly of even-numbered saturated or monounsaturated straight chains ranging from C_{12} to C_{18}. Species in group B have similar acyl chains but have, in addition, cyclopropane fatty acids, which are derived from the monounsaturated chains. The acyl chains of groups A and B resemble those of many Gram-negative bacteria. Most organisms in group C have ≥5% branched-chain fatty acids of the *anteiso-* and *iso-*types, usually C_{13} to C_{18} in length (i.e., with a branched methyl group at either the penultimate or terminal carbon, respectively). These are present in addition to the straight-chain fatty acids found in group A and B species. The *iso* acids

TABLE 15.1
Fatty Acyl Chain Composition of *Clostridium* Species[a]

Species	12:0	14:0	16:0	16:1	18:0	18:1	17:cyc	19:cyc	Anteiso br	Iso-br	Straight-Chain Odd	Ref. Cited	Other Refs.
C. acetobutylicum		8	64.5	5.6	3	9	1	9	0	0	0	[18]	[56,57]
C. argentinense	2.5	4.2	9	3	1.1	35	0	0	0	0	0	[15]	
C. baratii[c]	1.4	6.5	14	4.8	2.5	18	0	2.7	0	0		[15]	
C. beijerinckii		4.5	51.5	20.6	2	4	10	4.5	0	0	0	[18]	
C. bifermentans	Tr[b]	4	27	3	8.5	5.7	11	0	6.7	20	24	[1]	[58]
C. botulinum B[c]	2.5	22.5	21	4.7	1.5	13	0	0	0	0	0	[15]	
C. botulinum C[c]	2	18	9.5	5	1	26	0	0	0	0	0	[15]	
C. botulinum D[c]	1	17	14.5	4.7	1.6	21	0	0	0	0	0	[15]	
C. botulinum E[c]	1.3	18	27	4	3	13	0	0	0	0	0	[15]	
C. botulinum F[c]	1.6	15	19	4	2.	21	0	0	0	0	0	[15]	
C. butyricum		11	42	14	2.3	22	0	8.4	0	0	0	[18]	
C. cadaveris	1.4	21	52	3.7	12	1.4	0	0	2	2	3	[1]	
C. caloritolerans	tr	17	27	10	3.3	3	0	0	1	7	29	[1]	
C. chauvoei	Similar to *C. histolyticum*											[58]	
C. cochlearium	4.7	47	34	2	3	2	0	0	0	0	6	[1]	[16]
C. difficile	3	3.3	37	8	12	1.5	0	0	2.3	24	7	[1]	[58]
C. esterthеticum[c]	0	28	8	22	0	0	4.3	0	0	0		[16]	
C. gasigenes	1	6	9	6	3	10	0	0	0	0	0	[10]	
C. fallax		7.5	45.5	20	1.5	17	2	3	0	0	0	[18]	
C. ghoni	2	8	35	16	5.7	9.5	0	0	3.3	10	8	[1]	
C. glycolicum	2	5.7	30	12	2.6	4.3	0	0	4.4	16	21	[1]	
C. histolyticum	6	33	20	10	2	4	0	0	0	0	9	[58]	[1]
C. innocuum[c]	3	10	26	13	14	17	0	0	0	0	0	[8]	
C. lentoputrescens	2	40	20	22	1.5	2.8	0	0	0	0	4	[1]	
C. limosum	2	28	39	17	2	2.5	0	0	0	0	7	[1]	[16]
C. lituseburense	2.4	11	35	7	11	4	0	0	1.6	7.4	19	[1]	
C. malenominatum	1	38	20	21	4	2.7	0	0	0	0	6	[1]	[16]
C. mangenotii	1.6	4	36	14	10	7.5	0	0	2	7.2	17	[1]	
C. pascui[c]	0	33	11	34	1	1	0	0	0	0		[16]	
C. pasteurianum	0	0	93	3.3	0	0	4	0	0	0	0	[18]	[59,60]
C. perfringens	29	21	8	3	8	3	0	0	5			[58]	[61]
C. propionicum	3.6	21	20	3.6	11	5.3	0	0	5	21	6	[1]	
C. pseudofallax		6.4	51	18	2	20	tr	tr	0	0		[18]	
C. putrefaciens	10	5	27	12	2	13	0	0	0	1	7	[1]	[58]
C. putrificum	1.2	27	23	8.4	3.4	1.8	0	0	1	4.7	36	[1]	[15]
C. saccharoper-butylacetonicum	tr	10	34	22.5	tr	5.9	0	0	3.9	13	6.3	[62]	
C. scatologenes	3.5	8	41	19	5.4	2.3	0	0	1	4	15	[1]	
C. septicum	Similar to *C. histolyticum*											[58]	
C. sordellii	tr	2	19	3	9.6	7.3	0	0	11	33	14	[1]	[58]
C. sporogenes	tr	18	27	11	3	3	0	0	1.2	7.8	26	[1]	[15,58]
C. sticklandii	2.4	17	45	11	2.5	3	0	0	1.5	8	7	[1]	
C. subterminale	6.5	22	39	8	3	2	0	0	5	7.4	2	[1]	[15,58]
C. symbiosum[d]	2.6	2	22	8.2	3.6	7	0	0	0	10	44	[63]	
C. tartarivorum[e]	tr	8.6	16	0	2	1.5	0	0	3	45	0	[60]	
C. tertium	Similar to *C. histolyticum*											[58]	

TABLE 15.1 (continued)
Fatty Acyl Chain Composition of *Clostridium* Species[a]

Species	12:0	14:0	16:0	16:1	18:0	18:1	17:cyc	19:cyc	Anteiso br	Iso-br	Straight-Chain Odd	Ref. Cited	Other Refs.
C. tetani	7.7	35	18	10	7.3	7.4	0	0	0	0	7	[1]	[16;58]
C. tetanomorphum	tr	26	18	17	5.6	26	0	0	0	0	2	[1]	[16]
C. thermocellum	tr	1	22	25[f]	1	tr	0	0	see[f]	38	tr	[64]	
C. thermosac-charolyticum[e]	1.5	14	10	0	1	tr	0	0	4.3	51	1.4	[60]	
C. vincentii		2.3	9	3	2	5	0	0	0	0	15	[10]	

[a] All data represent% of total acyl chains, except where otherwise noted.
[b] Tr = trace (<1%).
[c] % of total chains. For remainder of total chains see alk-1-enyl chains.
[d] Strain ATCC 14940, originally described as *Bacteroides symbiosus.*
[e] *C. tartarivorum* contains C_{13} (1.8%) and C_{15} (8.7%) saturated cyclopropane and 11% of unsaturated cyclopropane acyl chains. *C. thermosaccharolyticum* contains C_{13} (2.8%) and C_{15} (3.9%) saturated cyclopropane and 7.8% unsaturated cyclopropane acyl chains.
[f] Includes anteiso 17:0.

TABLE 15.2
Clostridia Grouped According to Fatty Acid Chain Composition

Group A	Group B	Group C	Group D
C. argentinense	*C. acetobutylicum*	*C. bifermentans*	*C. chauvoei*
C. botulinum	*C. baratii*	*C. cadaveris*	*C. cochlearium*
B, C, D, E, F	*C. beijerinckii*	*C. caloritolerans*	*C. histolyticum*
C. chauvoei	*C. butyricum*	*C. difficile*	*C. limosum*
C. gasigenes	*C. estertheticum*	*C. ghoni*	*C. malenominatum*
C. innocuum	*C. fallax*	*C. glycolicum*	*C. putrefaciens*
C. lentoputrescens	*C. pasteurianum*	*C. lituseburense*	*C. septicum*
C. pascui		*C. mangenotii*	*C. tetani*
C. perfringens[a]		*C. propionicum*	*C. vincentii*
C. pseudofallax		*C. putrificum*	
C. tertium		*C. saccharoperbutylacetonicum*	
C. tetanomorphum		*C. scatologenes*	
		C. sordellii	
		C. sporogenes	
		C. sticklandii	
		C. subterminale	
		C. symbiosum	
		C. tartarivorum	
		C. thermocellum	
		C. thermosaccharolyticum	

[a] Some strains of *C. perfringens* have ~5% branched-chain fatty acids.

have both even and odd numbers of carbon atoms, in which the terminal branch is largely derived by α-oxidation from valine and leucine, respectively. The *anteiso* acids all have odd numbers of carbon atoms, presumably derived from isoleucine [1]. These species may also have straight chains with odd-numbered acyl chains. Group D has moderate (>6% to <10%) or high (>10%) straight odd-numbered acyl chains but little or no branched chains.

In bacteria, unsaturated fatty acids usually contain one double bond, but unlike the predominant monounsaturated chains found in higher organisms, there may be more than one positional isomer. For example, the most common C_{18} unsaturated fatty acid in higher organisms is oleic acid, *cis*-9, but in bacteria these may be *cis*-9, *cis*-11 (*cis*-vaccenic acid), or even *cis*-13. Similarly, the C_{16} monounsaturated fatty acids can be *cis*-7 or *cis*-9. This arises from the different biosynthetic pathways in eukaryotes and prokaryotes. In the former, the double bond is inserted into a saturated fatty acid by a position-specific desaturase, whereas in bacteria the double bond is usually inserted during the process of chain elongation, and this may occur at different stages (Section 15.4.1) [2,3].

The acyl chain compositions presented in Table 15.1 are not fixed. Microorganisms have the ability to vary the compositions of their membrane lipids according to the growth temperature, which usually leads to increased levels of lower melting unsaturated or branched fatty acids at lower growth temperatures. Acyl chains may also vary with the phase of growth, and the composition and pH of the growth medium. The presence of alcohols or solvents may also affect the compositions of the acyl chains [4].

In recent phylogenies of the genus *Clostridium,* based on sequencing of 16S rRNA genes [5,6], most of the species in groups A and B have been placed in cluster I. Despite having a wide variety of phenotypes, including saccharolytic and proteolytic species, psychrophiles, mesophiles, and thermophiles, dendograms based 16S rRNA indicate strong relatedness among this cluster, which is equivalent to rRNA group I of Johnson and Francis [7]. An exception is *C. innocuum*, which has been placed in cluster XVI, a group of three loosely associated species, which falls within the mycoplasma supercluster [5]. The unique polar lipid composition of *C. innocuum* is consistent with its affinity to the genus *Clostridium* [8,9] (see Section 15.3). Recent phylogenies do not include *C. lentoputrescens*.

Many of the species in group C, characterized by moderate to high amounts of branched-chain fatty acids, are proteolytic. These organisms oxidize valine, leucine, and isoleucine to the corresponding branched-chain acyl-CoA derivatives, which serve as precursors of the carbon-terminals of branched fatty acids [1]. Most of these organisms have been included in cluster XI, with the exceptions of *C. putrificum, C. scatologenes, C. sporogenes,* and *C. subterminale*, which have been placed in cluster I but are in a distinct subgroup within this cluster [5]. *C. propionicum* has been placed in subcluster XIVb, consisting of six species, most of which are not closely related. *C. thermosaccharolyticum* is grouped in cluster VII, which contains organisms of the genus *Thermoanaerobacterium.* Its classification as *Clostridium* is currently doubtful [5].

Most of the species in group D, characterized by ≥5% straight-chain C-odd fatty acids were placed in cluster I [5,6,10]. The two exceptions, *C. histolyticum* and *C. limosum*, have been included in cluster II, which includes only one other species, *C. proteolyticum.*

We conclude that fatty acid chain composition correlates well with 16S rRNA sequences. As noted by Stackebrandt et al. [6], decisions of classifications "must await analysis of chemotaxonomic, biochemical, morphological and other more 'classical' taxonomic properties." Acyl chains are among the chemotaxonomic properties that have to be considered, although they might not be infallible.

15.2.2 The Alk-1-Enyl Chains

Alk-1-enyl chains are usually found in the *sn*-1 position of plasmalogens (Figure 15.1) and are usually isolated as either long-chain aldehydes after acid hydrolysis or dimethylacetal derivatives after acid methanolysis. They generally reflect the compositions of the acyl chains, indicating a

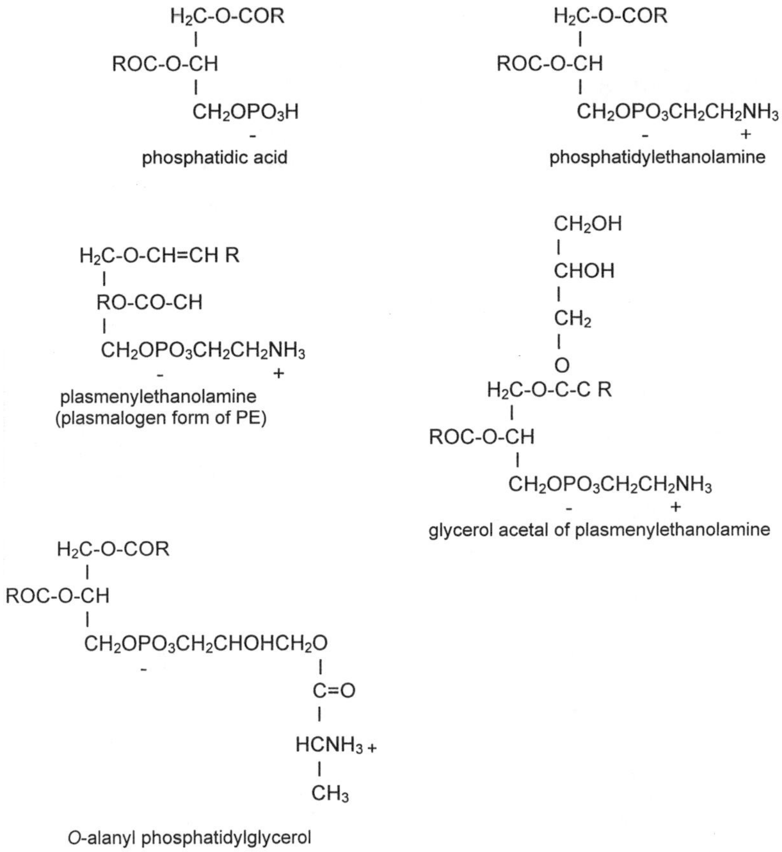

FIGURE 15.1 Structures of some common phospholipids in clostridia. Phosphatidic acid is an important intermediate usually found in trace amounts.

common origin [11,12]. However, in some cases, only a subset of the acyl chain structures are represented among the alk-1-enyl chains, and in some species the ratios of the alk-1-enyl chains are very different from those of the acyl chains (Table 15.1 and Table 15.3). Plasmalogens are found in a wide variety of anaerobic Gram-positive and Gram-negative bacteria [13,14]. Most of the species listed in Table 15.3 are in fatty acid groups A and B (Table 15.2) and are in *Clostridium* cluster I [5]. Relatively few of the proteolytic organisms that contain branched-chain fatty acids (group C, Table 15.2) were found to contain alk-1-enyl chains in an initial survey [1]; however, later studies on some of these species including *C. putrificum, C. sporogenes, C. subterminale,* and *C. tetani* revealed significant amounts of long-chain dimethylacetals derived from the alk-1-enyl chains [15,16]. These results suggest that other proteolytic species should be re-examined for the presence of plasmalogens.

15.3 THE POLAR LIPIDS

Information on the polar lipids of *Clostridium* is limited. Most studies have concentrated on a small group of related organisms in cluster I. Many of these organisms are capable of producing alcohols and solvents such as propanol, acetone, and butanol. This group includes *C. acetobutylicum, C. beijerinckii, C. butyricum, C. saccharobutylicum*, and *C. saccharoperbutylacetonicum* [17]. All of these species have phosphatidylethanolamine (PE) (Figure 15.1), phosphatidylglycerol (PG) and cardiolipin (diphosphatidylglycerol, DPG) as major phospholipids, and in each case the diacyl

TABLE 15.3
Alk-1-Enyl Chain Composition of *Clostridium* Species[a]

Species	12:0	14:0	16:0	16:1	18:0	18:1	17:cyc	19:cyc	Anteiso br	Iso-br	Ref. Cited	Other Refs.
C. acetobutylicum		2.3	17.5	7.7	5.7	15	5	46	0	0	[18]	
C. argentinense[b]		5.4	6.7	3.8	0	8.3	0	0	0	0	[15]	
C. baratii[b]	0	0	1	1	1.7	22	0	4.6	0	0	[15]	
C. beijerinckii		tr	52	16	4.2	4.2	15	7.5	0	0	[18]	
C. botulinum B[b]	0	0	5.8	2.1	tr	14	0	0	0	0	[15]	
C. botulinum C[b]	0	1.4	1	1.5	0	8	0	0	0	0	[15]	
C. botulinum D[b]	0	2.3	1.4	2	0	7	0	0	0	0	[15]	
C. botulinum E[b]	0	0	4.9	1.8	1.4	13	0	0	0	0	[15]	
C. botulinum F[b]	0	0	4.1	1.3	1	18	0	0	0	0	[15]	
C. butyricum		tr	18	21	4.8	44	tr	5.9	0	0	[18]	
C. cochlearium[b]	0	4	3	9							[16]	
C. estertheticum[b]	0	0	0	23			3.4				[16]	
C. fallax		tr	96.5	2.7	tr	tr	0	0	0	0	[18]	
C. innocuum[b]	0	0	9	0	6	0	0	0	0	0	[8]	
C. limosum[b]	0	5	2	10							[16]	
C. malenominatum[b]	0	0	0	3							[16]	
C. pascui	0	1.3	1.4	12							[16]	
C. pasteurianum		0	23.5	3	0	0	74	0	0	0	[18]	
C. pseudofallax		2.8	69	15	5.2	7.6	0	0	0	0	[18]	
C. putrificum[b]	0	5.7	1	3	0	3	0	0	0	0	[15]	
C. sporogenes[b]	0	6.3	1	3	0	4	0	0	0	0	[15]	
C. subterminale[b]	0	4	3.4	3	0	10	0	0	0	0	[15]	[16]
C. tetanib[b]	0	7.8	2.8	12							[16]	
C. tetanomorphum[b]	0	7	4	10			3.6				[16]	

[a] % of total alk-1-enyl chains, unless noted otherwise.
[b] As% of total acyl plus alk-1-enyl chains.

forms are accompanied by the corresponding plasmalogens. *C. beijerinckii* is readily distinguished from the others by having more phosphatidyl-N-methylethanolamine (PME) and the plasmalogen form (PlaME) than PE and PlaE (Figure 15.1) [18]. Unique glycerol acetals of PlaE and PlaME are also present in these species (Figure 15.1). A second novel class of lipids found in *C. butyricum* and *C. innocuum* are phosphatidylglycerol acetals of the corresponding plasmalogens; the first to be discovered was a PG acetal of plasmenylethanolamine [19] (Figure 15.2). *C. acetobutylicum* has, in addition to PE, PG, and DPG, large amounts of glycosyldiradylglycerols [20,21]. In strain DSM 1731, the principal glycolipids were found to be galactosyl-1-alk-1-enyl-2-acyl glycerol and β-glucosyl-galactosyl-1-alk-1-enyl-2-acyl glycerol [21,22], i.e., they are mainly in the form of plasmalogens.

C. saccharoperbutylacetonicum has the same phospholipids as *C. beijerinckii,* but the ratio of N-methylethanolamine lipids to ethanolamine lipids is much lower [23]. As noted above, N-methylethanolamine is the predominant nitrogenous head group in *C. beijerinckii*; thus these two species can be distinguished based on the ratio: (PME + PlaME)/(PE + PlaE). The polar lipids of *C. saccharobutylicum* are virtually indistinguishable from those of *C. butyricum* when examined by two-dimensional thin layer chromatography. Quantitative analysis has not been done. Recent studies on the taxonomy of this group based on comparative gene analysis have indicated that *C. saccharoperbutylacetonicum* is more closely related to *C. beijerinckii* than *C. saccharobutylicum* [24,25] (D.T. Jones, personal communication).

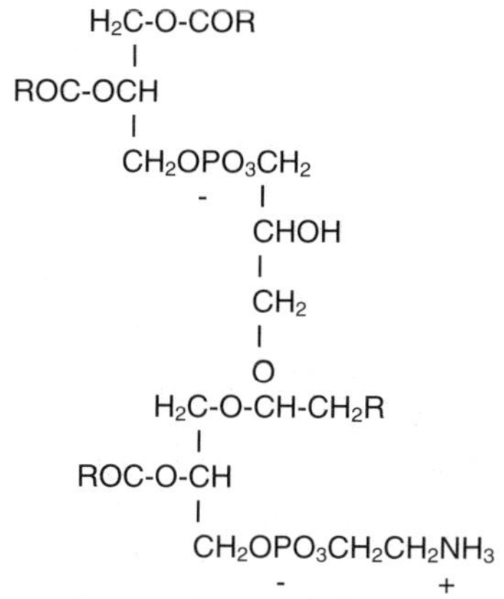

phosphatidylglycerol acetal of plasmenylethanolamine

FIGURE 15.2 Structure of the phosphatidylglycerol acetal of plasmenylethanolamine.

TABLE 15.4
Polar Lipid Composition of *Clostridium* Species[a]

Species	PE[b]	PME	GAs	PG	DPG	GDG	Aminoacyl PG	Refs.
C. acetobutylicum	46(ND)[c]	0	10	29(~100)	15(ND)	+(~100)	0	[20,21,57]
C. beijerinckii[d]	12(55)	34(78)	29	25(38)	+	tr	0	[18,65,66]
C. botulinum (NCIB 4270) grown at 30°C	28(0)	0	0	24(0)	21(0)		0	[67]
C. butyricum[e]	48(78)	0	16	21(~30%)	13(~30%)	tr	0	[18,68]
C. difficile	0	0	0	100	0	0	0	[31]
C. innocuum[f]	0	0	0	1	26	5[g]	0	[8,9,32]
C. perfringens[h]	++	0	0	+	+	tr	++	[27,30]
C. saccharobutylicum	++(ND)	0	+	+(ND)	++(ND)	+	0	[69]
C. saccharoper-butylacetonicum[h]	++(ND)	+(ND)	+	+(ND)	++(ND)	+	0	[62,69]

[a] Percentage of total lipid P, except where otherwise noted.
[b] Abbreviations: DPG, diphosphatidylglycerol (cardiolipin); GAs, glycerol acetals of the plasmalogen forms of PME and/or PE; GDG, glycosyldiradylglycerols; PE, phosphatidylethanolamine; PG, phosphatidylglycerol; PME, phosphatidyl-N-methylethanolamine.
[c] Numbers in parentheses represent % plasmalogen in lipid class, ND = Present, but % not determined.
[d] Strains of *C. beijerinckii* have, in addition, varying amounts of phosphatidylglycerolacetals of the plasmalogen forms of PE and PME [71].
[e] *C. butyricum* has, in addition, approximately 5% of a phosphatidylglycerolacetal of the plasmalogen form of PE [19].
[f] *C. innocuum* has, in addition: four phosphoglycolipids, which represent 72 mol% of the total lipid (58% of total lipid P), and phosphatidylglycerolacetals of the plasmalogen forms of DPG and lyso-DPG, which represent 4 mol% of the total lipid (8% of total lipid P).
[g] mol% of the total lipid.
[h] ++, major polar lipid; +, minor polar lipid; tr, trace.

The acyl chain and phospholipid compositions at various growth temperatures of a nonproteolytic strain of *C. botulinum* were determined by Evans et al. [26]. The phospholipid composition resembles that of other organisms in cluster I except for the relatively large proportion of phosphatidylserine (PS), which is usually found in trace amounts in clostridia.

C. perfringens, which is also in cluster I, is only distantly related to the solvent-producing species discussed above. It is readily distinguishable by its acyl chains and its polar lipids from those organisms. The acyl chains consist predominantly of the shorter, saturated straight chains C_{12} and C_{14}, with smaller amounts of the longer C_{16} and C_{18} species (Table 15.1). In the original report on its polar lipids, the major lipids described were PG, DPG, and *O*-amino acyl-PGs, with alanine as the predominant substituent on the head group glycerol (Figure 15.1). Lysine and ornithine were also found [27]. However, Carman and co-workers reported the presence of PS synthase and PS decarboxylase in *C. perfringens,* suggesting strongly that this organism makes PE [28,29]. Our recent studies on the polar lipids have confirmed the presence of PE as one of the major phospholipids. The major molecular species were found to have, as expected, C_{12}, C_{14}, and C_{16} saturated acyl chains [30].

C. difficile, which has been placed in cluster XI [5], is unique among clostridia in its polar lipid composition, since PG was the only phospholipid found [31]. Other Gram-positive organisms, including *Staphylococcus epidermidis* and *Micrococcus varians,* have PG as their predominant polar lipid [4].

C. innocuum has been placed in cluster XVI, a group of three loosely associated species, which falls within the mycoplasma supercluster [5]. The polar lipids consist mainly of a mixture of phospholipids and phosphoglycolipids [8,9]. The predominant phospholipids are PG, DPG, and a phosphatidylglycerol acetal of the plasmalogen form of DPG. The major phosphoglycolipids are derived from the membrane glycolipid D-Gal*p*(α1-2)D-Glc*p*(α1-3)radyl$_2$glycerol. Both compounds carry a 2-amino-1(3)-dihydroxypropane 3(1)phosphate residue attached to the O-6 of the galactopyranosyl residue. A less-polar compound is esterified with a third fatty acid at O-6 of the glucopyranosyl residue [8,32]. A comparison of the lipids of *C. innocuum* with those of *Acholeplasma laidlawii* revealed some similarities and major differences. Both species have PG and DPG in common and both lack phospholipids with a nitrogen-containing polar head group such as PE. Both have glycosyl and diglycosyldiradylglycerol lipids, but the structures of the diglycosyldiradylglycerols differ in having as the terminal sugar α-D-galactose in *C. innocuum* and α-D-glucose in *A. laidlawii.* The terminal α-D-glucose on diglycosyldiacylglycerol is found in many *Acholeplasma.* One exception is *A. axanthum*, which has the α-D-galactose terminal sugar [33]. The 2-amino-1(3)-dihydroxypropane 3(1)phosphate substituents are not found on the *A. laidlawii* glycolipids, nor are the plasmalogen forms of these glycolipids.

15.4 LIPID BIOSYNTHESIS

15.4.1 FATTY ACIDS

Although early work on the biosynthesis of fatty acids in bacteria was done with clostridia, the enzymes involved have not been investigated [2]. The pathways to saturated and unsaturated fatty acids appear to be similar, but not identical to those studied extensively in *E. coli.* In most prokaryotes, as distinct from eukaryotes, the enzymes involved in the condensation of C_2 units, and those involved in subsequent reductions and dehydration are separable and have been designated type II fatty acid synthase (FAS II), whereas in higher organisms the fatty acid synthase (FAS I) is a large, multifunctional protein of molecular mass exceeding 10^6 Daltons [34]. In bacteria, multiple cycles of condensation, reduction, dehydration, and reduction lead to saturated fatty acyl derivatives of the small acyl carrier protein (ACP). If initiated with branched-chain precursors such as isobutyryl-, isovaleryl-, or 2-methylbutyryl-CoA esters, the resulting long chain fatty acids will be of the *iso* or the *anteiso* class.

Unsaturated fatty acids are formed when an intermediate, for example, D-(-)-β-hydroxydecanoyl-S-ACP, undergoes reduction to *trans*-2-decenoyl-S-ACP followed by isomerization to the *cis*-3 compound or by direct β,γ-dehydration of the hydroxy intermediate [35]. Since this step can occur with intermediates of different chain length, depending on the organism, mixtures of monounsaturated fatty acids, for example, *cis*-7-16:1 and *cis*-9-16:1 and *cis*-9-18:1 and *cis*-11-18:1, can be produced, as was originally observed in clostridia [2]. It is important to note that most bacteria, even facultative and aerobic species, use a similar mechanism that differs from that found in eukaryotic organisms in which long-chain acyl-CoA esters are desaturated by an oxygen-dependent desaturase.

Cyclopropane fatty acids are formed by addition of a C_1 group derived from S-adenosylmethionine to the double bond of an unsaturated fatty acid in a preformed phospholipid [36,37]. In *C. beijerinckii* ATCC 6015 (formerly called *C. butyricum*), the cyclopropane synthase showed a strong preference for the $\omega - 7$ chains, especially with the C_{16} unsaturated acyl chains [38]. Studies with *E. coli* demonstrated similar positional specificity [39,40].

15.4.2 Polar Lipids

Phospholipid synthesis in clostridia has been studied in only a few species. As in *E. coli*, de novo phospholipid synthesis begins with the transfer of an acyl chain from ACP to *sn*-glycerol-3-phosphate, giving rise to monoacyl glycerol-3-P (Figure 15.3). A second acylation produces phosphatidic acid.

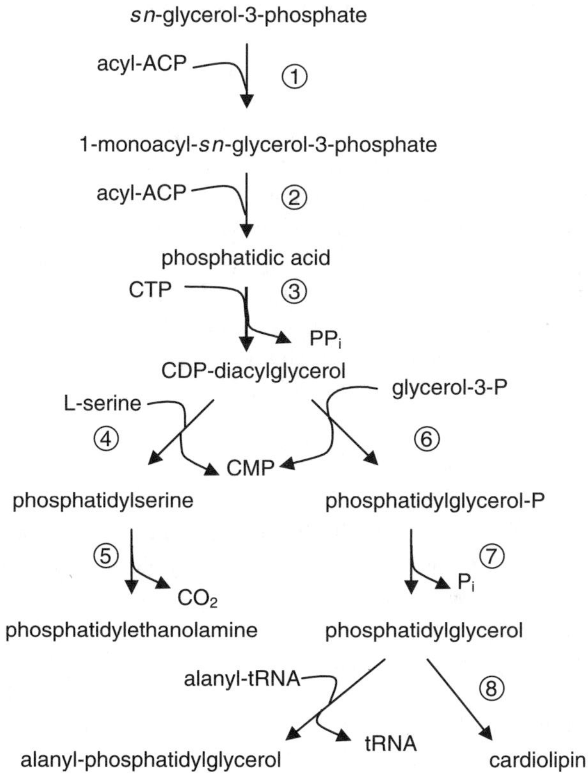

FIGURE 15.3 Pathways of phospholipid biosynthesis in *Clostridium beijerinckii*. Enzymes indicated in circled numbers are: (1) *sn*-glycerol-3-phosphate acyltransferase; (2) monoacyl-glycerol-3-phosphate acyltransferase; (3) phosphatidate cytidylyltransferase; (4) phosphatidylserine synthase; (5) phosphatidylserine decarboxylase; (6) phosphatidylglycerolphosphate synthase; (7) phosphatidylglycerolphosphate phosphatase; (8) cardiolipin synthase.

In vitro, with membrane particles from *C. beijerinckii*, the first acyltransferase was highly specific for acyl-ACP as acyl donor, but the second acyltransferase reaction was less specific in that both acyl-ACP and acyl-CoA served as acyl donors [41,42]. An acyltransferase was cloned from *C. butyricum,* and the gene was designated *plsD* [43].

After the formation of phosphatidic acid, the next step is the formation of CDP-diacylglycerol by reaction with CTP [44]. The pathway then bifurcates by reaction with either L-serine to yield phosphatidylserine (PS) or with *sn*-glycerol-3-P to give phosphatidylglycerol-P (PGP). PS is then decarboxylated to give PE, and PGP is hydrolyzed by a specific phosphatase to give PG [44,45]. Cardiolipin is formed by condensation of two molecules of PG [46]. As in *E. coli,* these enzymes in clostridia are membrane-bound and have not been purified. This pathway is the same as that found in *E. coli,* and whole-cell studies with *C. butyricum* have produced data consistent with this scheme [47,48]. Whole genome sequences for both *C. acetobutylicum* and *C. perfringens* (http://www.tigr.org/tigr-scripts/CMR2/CMRHomePage.spl) reveal the presence of genes that have been annotated for all the enzymes of this pathway with the exception of *pgpA and pgpB*, which encode PGP phosphatases in *E. coli*. Some doubt exists concerning the essential PGP phosphatase in *E. coli* [49]. Aminoacyl-PG is formed by the transfer of an amino acid from an aminoacyl-tRNA to phosphatidylglycerol. This reaction was demonstrated with membrane particles from *C. perfringens* (*C. welchii*) with alanyl-tRNA and lysyl-tRNA [50]. Studies of phospholipid biosynthesis in *C. perfringens* have provided evidence for the presence of PS synthase and PS decarboxylase and the requisite genes have been found in the recently published genome [28,29,51].

15.4.3 Plasmalogens

As is the case for the biosynthesis of unsaturated fatty acids, prokaryotes have evolved an anaerobic pathway for plasmalogen biosynthesis that differs fundamentally from the oxygen-dependent pathway found in higher organisms [52]. In the eukaryotic pathway, a saturated ether bond on carbon-1 of the glycerol-P backbone is desaturated by an oxygen-dependent cytochrome-containing enzyme. Prior to that step, *sn*-1-acyl dihydroxyacetone-P is reduced to 1-*O*-alkyl dihydroxyacetone-P, an intermediate that has been ruled out in *C. beijerinckii* by isotopic experiments [53]. Although considerable information is available on the origin of the alkenyl chains and the kinetics of formation of plasmalogens relative to the formation of the diacylphospholipids (reviewed in Goldfine, 1997, [54]), the mechanism of formation of the alkenyl bond in bacteria is still not known.

15.5 SUMMARY

The study of membrane lipids of clostridia began half a century ago with the seminal studies of Broquist and Snell on the requirement for biotin in fatty acid biosynthesis [55]. During the second half of the 20th century, much was learned about the mechanism of fatty acid and phospholipid biosynthesis, and the compositions of the membrane lipids of clostridia. However, many questions remain unanswered. For example, of the major pathogenic clostridia, the polar lipid compositions of only three, *C. difficile,* one strain of *C. botulinum*, and *C. perfringens*, have been determined, and only a few of the more than 100 known nonpathogenic species have been studied with respect to their membrane polar lipids. A complete database of fatty acids and aldehydes derived from the lipids is used in a commercial system for the identification of clostridia at the species level [15] and much of this information is available in the literature (Table 15.1 and Table 15.3). Beyond knowledge of the compositions of clostridial membranes, information on the mechanism of biosynthesis of the ether lipids is still incomplete. Current efforts to sequence the genomes of many bacteria, including clostridia, should eventually provide the foundation for an understanding of these metabolic processes.

REFERENCES

1. Elsden, S.R., Hilton, M.G., Parsley, K.R., and Self, R., The lipid fatty acids of proteolytic clostridia, *J. Gen. Microbiol.,* 118, 115–123, 1980.
2. Bloch, K., Baronowsky, P., Goldfine, H., Lennarz, W.J., Light, R., Norris, A.T., and Scheuerbrandt, G., Biosynthesis and metabolism of unsaturated fatty acids, *Federation Proc.*, 20, 921–927, 1961.
3. Bloch, K., Enzymatic synthesis of monounsaturated fatty acids, *Acc. Chem. Res.,* 2, 193–202, 1969.
4. O'Leary, W.M. and Wilkinson, S.G., Gram-positive bacteria, in *Microbial Lipids, Volume 1*, Ratledge, C. and Wilkinson, S.G., Eds., Academic Press, London, 1988, 117–201.
5. Collins, M.D., Lawson, P.A., Willems, A., Cordoba, J.J., Fernandez-Garayzabal, J., Garcia, P., Cai, J., Hippe, H., and Farrow, J.A.E., The phylogeny of the Genus *Clostridium*; proposal of five new genera and eleven new species combinations, *Int. J. Syst. Bacteriol.,* 44, 812–826, 1994.
6. Stackebrandt, E., Kramer, I., Swiderski, J., and Hippe, H., Phylogenetic basis for a taxonomic dissection of the genus *Clostridium*, *FEMS Immunol. Med. Microbiol.,* 24, 253–258, 1999.
7. Johnson, J.L. and Francis, B.S., Taxonomy of the Clostridia: ribosomal ribonucleic acid homologies among the species, *J. Gen. Microbiol.,* 88, 229–244, 1975.
8. Johnston, N.C., Goldfine, H., and Fischer, W., Novel polar lipid composition of *Clostridium innocuum* as the basis for an assessment of its taxonomic status, 140, 105–111, 1994.
9. Johnston, N.C. and Goldfine, H., Isolation and characterization of new phosphatidylglycerol acetals of plasmalogens — a family of ether lipids in clostridia, *Eur. J. Biochem.,* 223, 957–963, 1994.
10. Broda, D.M., Saul, D.J., Lawson, P.A., Bell, R.G., and Musgrave, D.R., *Clostridium gasigenes* sp. nov., a psychrophile causing spoilage of vacuum-packed meat, *Int. J. Syst. Evol. Microbiol.,* 50, 107–118, 2000.
11. Goldfine, H., Composition of the aldehydes of *Clostridium butyricum* plasmalogens: cyclopropane aldehydes, *J. Biol. Chem.,* 239, 2130–2134, 1964.
12. Baumann, N.A., Hagen, P.-O., and Goldfine, H., Phospholipids of *Clostridium butyricum*: studies on plasmalogen composition and biosynthesis, *J. Biol. Chem.,* 240, 1559–1567, 1965.
13. Goldfine, H. and Hagen, P.-O., Bacterial plasmalogens, in *Ether Lipids: Chemistry and Biology*, Snyder, F., Ed., Academic Press, New York, 1972, 329–350.
14. Horrocks, L.A. and Sharma, M., Plasmalogens and *O*-alkyl glycerophospholipids, in *Phospholipids*, Hawthorne, J.N. and Ansell, G.B., Eds., Elsevier Biomedical, Amsterdam, 1982, 51–93.
15. Ghanem, F.M., Ridpath, A.C., Moore, W.E.C., and Moore, L.V.H., Identification of *Clostridium botulinum, Clostridium argentinense,* and related organisms by cellular fatty acid analysis, *J. Clin. Microbiol.,* 29, 1114–1124, 1991.
16. Wilde, E., Collins, M.D., and Hippe, H., *Clostridium pascui* sp. nov., a new glutamate-fermenting sporeformer from a pasture in Pakistan, *Int. J. Syst. Bacteriol.,* 47, 164–170, 1997.
17. Keis, S., Shaheen, R., and Jones, D.T., Emended descriptions of *Clostridium acetobutylicum* and *Clostridium beijerinckii*, and descriptions of *Clostridium saccharoperbutylacetonicum* sp nov and *Clostridium saccharobutylicum* sp. nov., *Int. J. Syst. Evol. Microbiol.,* 51, 2095–2103, 2001.
18. Johnston, N.C. and Goldfine, H., Lipid composition in the classification of the butyric acid-producing clostridia, *J. Gen. Microbiol.,* 129, 1075–1081, 1983.
19. Johnston, N.C. and Goldfine, H., Isolation and characterization of a novel four chain phospholipid, the phosphatidylglycerol acetal of plasmenylethanolamine from *Clostridium butyricum, Biochim. Biophys. Acta,* 961, 1–12, 1988.
20. Johnston, N.C. and Goldfine, H., Replacement of the aliphatic chains of *Clostridium acetobutylicum* by exogenous fatty acids: regulation of phospholipid and glycolipid composition, *J. Bacteriol.,* 174, 1848–1853, 1992.
21. Oulevey, J., Bahl, H., and Thiele, O.W., Novel alk-1-enyl ether lipids isolated from *Clostridium acetobutylicum, Arch. Microbiol.,* 144, 166–168, 1986.
22. Thiele, O.W., Oulevey, J., and Bahl, H., Neuartige Alkenylether-lipide aus anaeroben Bakterien, *Fette Seifen Anstrichm.,* 87, 551–556, 1985.
23. Johnston, N.C. and Goldfine, H., Unpublished data, 2003.
24. Keis, S., Sullivan, J.T., and Jones, D.T., Physical and genetic map of the *Clostridium saccharobutylicum* (formerly *Clostridium acetobutylicum*) NCP 262 chromosome, *Microbiology-Sgm.,* 147, 1909–1922, 2001.

25. Keis, S., Bennett, C.F., Ward, V.K., and Jones, D.T., Taxonomy and phylogeny of industrial solvent-producing clostridia, *Intl. J. Syst. Bacteriol.*, 45, 693–705, 1995.
26. Evans, R.I., McClure, P.J., Gould, G.W., and Russel, N.J., The effect of growth temperature on the phospholipid and fatty acyl compositions of non-proteolytic *Clostridium botulinum, Int. J. Food Microbiol.,* 40, 159–167, 1998.
27. Macfarlane, M.G., Characterization of lipoamino-acids as O-amino-acid esters of phosphatidylglycerol, *Nature,* 196, 136–138, 1962.
28. Cousminer, J.J., Fischl, A.S., and Carman, G.M., Partial purification and properties of phosphatidylserine synthase from *Clostridium perfringens, J. Bacteriol.,* 151, 1372–1379, 1982.
29. Carman, G.M. and Wieczorek, D.S., Phosphatidylglycerophosphate synthase and phosphatidylserine synthase activities in *Clostridium perfringens, J. Bacteriol.,* 142, 262–267, 1980.
30. Johnston, N.C., Baker, J.K., and Goldfine, H., Phospholipids of *Clostridium perfringens*: A reexamination, *FEMS Microbiol. Lett.*, 233, 65–68, 2004.
31. Drucker, D.B., Wardle, H.M., and Boote, V., Phospholipid profiles of *Clostridium difficile, J. Bacteriol.,* 178, 5844–5846, 1996.
32. Fischer, W., Hartmann, R., Peter-Katalinic, J., and Egge, H., (*S*)-2-amino-1,3-propanediol-3-phosphate-carrying diradylglyceroglycolipids — novel major membrane lipids of *Clostridium innocuum, Eur. J. Biochem.,* 223, 879–892, 1994.
33. Smith, P.F., Archaebacteria and other specialized bacteria, in *Microbial Lipids Vol. 1,* Ratledge, C. and Wilkinson, S.G., Eds., Academic Press, London, 1988, 489–553.
34. Schweizer, E., Biosynthesis of fatty acids and related compounds, in *Microbial Lipids,* Ratledge, C. and Wilkinson, S.G., Eds., Academic Press, London, 1989, 13–50.
35. Rock, C.O. and Jackowski, S., Forty years of bacterial fatty acid synthesis, *Biochem. Biophys. Res. Commun.,* 292, 1155–1166, 2002.
36. Zalkin, H., Law, J.H., and Goldfine, H., Enzymatic synthesis of cyclopropane fatty acids catalyzed by bacterial extracts, *J. Biol. Chem.,* 238, 1242–1248, 1963.
37. Law, J.H., Biosynthesis of cyclopropane rings, *Acc. Chem. Res.,* 4, 199–203, 1971.
38. Goldfine, H. and Panos, C., Phospholipids of *Clostridium butyricum* IV. Analysis of the positional isomers of monounsaturated and cyclopropane fatty acids and alk-1′-enyl ethers by capillary column chromatography, *J. Lipid Res.,* 12, 214–220, 1971.
39. Marinari, L.A., Goldfine, H., and Panos, C., Specificity of cyclopropane fatty acid synthesis in *Escherichia coli.* Utilization of isomers of monounsaturated fatty acids, *Biochem,* 13, 1978–1983, 1974.
40. Ohlrogge, J.B., Gunstone, F.D., Ismail, I.A., and Lands, W.E.M., Positional specificity of cyclopropane ring formation from *cis*-octadecenoic acid isomers in *Escherichia coli, Biochim. Biophys. Acta,* 431, 257–267, 1976.
41. Goldfine, H., Ailhaud, G.P., and Vagelos, P.R., Involvement of acyl carrier protein in acylation of glycerol 3-phosphate in *Clostridium butyricum.* II. Evidence for the participation of acyl thioesters of acyl carrier protein, *J. Biol. Chem.,* 242, 4466–4475, 1967.
42. Goldfine, H. and Ailhaud, G.P., Fatty acyl-acyl carrier protein and fatty acyl-CoA in the biosynthesis of phosphatidic acid in *Clostridium butyricum, Biochem. Biophys. Res. Commun.,* 45, 1127–1133, 1971.
43. Heath, R.J., Goldfine, H., and Rock, C.O., A gene (*plsD*) from *Clostridium butyricum* that functionally substitutes for the *sn*-glycerol-3-phosphate acyltransferase gene (*plsB*) of *Escherichia coli, J. Bacteriol.,* 179, 7257–7263, 1997.
44. Silber, P., Borie, R.P., and Goldfine, H., The enzymes of phospholipid synthesis in *Clostridium butyricum, J. Lipid Res.,* 21, 1022–1031, 1980.
45. Verma, J.N. and Goldfine, H., Phosphatidylserine decarboxylase from *Clostridium butyricum, J. Lipid Res.,* 26, 610–616, 1985.
46. Walton, P.A. and Goldfine, H., Transphosphatidylation activity in *Clostridium butyricum.* Evidence for a secondary pathway by which membrane phospholipids may be synthesized and modified, *J. Biol. Chem.,* 262, 10355–10361, 1987.
47. Koga, Y. and Goldfine, H., Biosynthesis of phospholipids in *Clostridium butyricum*: the kinetics of synthesis of plasmalogens and the glycerol acetal of ethanolamine plasmalogen, *J. Bacteriol.,* 159, 597–604, 1984.

48. MacDonald, D.L. and Goldfine, H., Phosphatidylglycerol acetal of plasmenylethanolamine as an intermediate in ether lipid formation in *Clostridium butyricum, Biochem. Cell Biol.,* 68, 225–230, 1990.
49. Heath, R.J., Jackowski, S., and Rock, C.O., Fatty acid and phospholipid metabolism in prokaryotes. In *Biochemistry of Lipids, Lipoproteins and Membranes (4th Edition),* Vance, D.E. and Vance, J.E., Eds., Elsevier, Amsterdam, 2002, 55–93.
50. Gould, R.M. and Lennarz, W.J., Biosynthesis of aminoacylderivatives of phosphatidylglycerol, *Biochem. Biophys. Res. Commun.,* 26, 510–515, 1967.
51. Shimizu, T., Ohtani, K., Hirakawa, H., Ohshima, K., Yamashita, A., Shiba, T., Ogasawara, N., Hattori, M., Kuhara, S., and Hayashi, H., Complete genome sequence of *Clostridium perfringens*, an anaerobic flesh-eater, *Proc. Natl. Acad. Sci. U.S.A.*, 99, 996–1001, 2002.
52. Snyder, F., Lee, T.C., and Wykle, R.L., Ether-linked lipids and their bioactive species, in *Biochemistry of Lipids, Lipoproteins and Membranes,* Vance, D.E. and Vance, J.E., Eds., Elsevier, Amsterdam, 2002, 233–262.
53. Hill, E.E. and Lands, W.E.M., Formation of acyl and alkenyl glycerol derivatives in *Clostridium butyricum, Biochim. Biophys. Acta,* 202, 209–211, 1970.
54. Goldfine, H., Structure, biosynthesis, physical properties, and functions of the polar lipids of *Clostridium,* in *Advances in Lipobiology, Vol. 2,* Gross, R.W., Ed., JAI Press, Inc., Greenwich, CT, 1997, 109–142.
55. Broquist, H.P. and Snell, E.E., Biotin and bacterial growth. I. Relation to aspartate, oleate, and carbon dioxide, *J. Biol. Chem.,* 188, 431–444, 1951.
56. Vollherbst-Schneck, K., Sands, J.A., and Montenecourt, B.S., Effect of butanol on lipid composition and fluidity of *Clostridium acetobutylicum* ATCC 824, *Appl. Environ. Microbiol.,* 47, 193–194, 1984.
57. Lepage, C., Fayolle, F., Hermann, M., and Vandecasteele, J.-P., Changes in membrane lipid composition of *Clostridium acetobutylicum* during acetone-butanol fermentation: effects of solvents, growth temperature and pH, *J. Gen. Microbiol.,* 133, 103–110, 1987.
58. Moss, C.W. and Lewis, V.J., Characterization of clostridia by gas chromatography, *Appl. Microbiol.,* 15, 390–397, 1967.
59. Deo, Y.M., Bryant, R.D., and Laishley, E.J., Differences in cellular fatty acids and lipid composition in *Clostridium pasteurianum* under nitrogen- and non-nitrogen-fixing conditions, *Curr. Microbiol.,* 3, 55–58, 1979.
60. Chan, M., Himes, R.H., and Akagi, J.M., Fatty acid compostion of thermophilic, mesophilic, and psychrophilic clostridia, *J. Bacteriol.,* 106, 876, 1971.
61. Sinyak, K.M., Rudichenko, V.F., Yaroshenko, M.N., and Kruk, V.I., The regulation of fatty-acid composition of Cl-Perfringens type a by hydrogen and hydroxyl ions, *Izvestiya Akademii Nauk Sssr Seriya Biologicheskaya*, 393–401, 1981.
62. Ogata, S., Yoshino, S., Okuma, Y., and Hayashida, S., Chemical composition of autoplast membrane of *Clostridium saccharoperbutylacetonicum, J. Gen. Appl. Microbiol.,* 28, 293–301, 1982.
63. Ifkovits, R.W. and Ragheb, H.S., Cellular fatty acid composition and identification of rumen bacteria, *Appl. Microbiol.,* 16, 1406–1413, 1968.
64. Herrero, A.A., Gomez, R.F., and Roberts, M.F., Ethanol-induced changes in the membrane lipid composition of *Clostridium thermocellum, Biochim. Biophys. Acta,* 693, 195–204, 1982.
65. Khuller, G.K. and Goldfine, H., Phospholipids of *Clostridium butyricum.* V. Effects of growth temperature on fatty acid, alk-1-enyl group, and phospholipid composition, *J. Lipid Res.,* 15, 500–507, 1974.
66. Goldfine, H., Khuller, G.K., Borie, R.P., Silverman, B., Selick, H., Johnston, N.C., Vanderkooi, J.M., and Horwitz, A.F., Effects of growth temperature and supplementation with exogenous fatty acids on some physical properties of *Clostridium butyricum* phospholipids, *Biochim. Biophys. Acta,* 488, 341–352, 1977.
67. Evans, R.I., McClure, P.J., Gould, G.W., and Russel, N.J., The effect of growth temperature on the phospholipid and fatty acyl compositions of non-proteolytic *Clostridium botulinum, Int. J. Food Microbiol.,* 40, 159–167, 1998.
68. Goldfine, H., Johnston, N.C., and Bishop, D.G., Ether phospholipid asymmetry in *Clostridium butyricum, Biochem. Biophys. Res. Commun.,* 108, 1502–1507, 1982.
69. Johnston, N.C. and Goldfine, H., Unpublished data, 2003.

16 Adhesins

Paola Mastrantonio and Anne Collignon

CONTENTS

Clostridia, both in the human and animal body and in the environment, have to attach themselves firmly to the host cells through cell surface molecules to avoid being flushed away. Different from other bacteria, clostridia seem to adhere to the host cells through one of the most commonly used strategies, the afimbrial adhesins, which are bacterial surface proteins without a rodlike structure. In fact, the presence of fimbriae or pili has never been clearly associated with adhesive properties in these bacteria. The most studied proteins associated with adhesion activity, so far, have been the surface-layer (S-layer) proteins, mainly in *Clostridium difficile,* some proteins from cellulolytic clostridia that mediate attachment to substrates, and other cell surface proteins of *Clostridium difficile*.

16.1 INTRODUCTION TO S-LAYER

S-layer proteins are two-dimensional crystalline arrays formed by (glyco)protein subunits on the outer surface of many bacteria. The monomer subunits are frequently exported from the cytoplasm through the general secretory pathway by an N-terminal leader peptide and can be extracted from the cell surface by dissociating agents, such as 5M lithium chloride or EDTA, or by chaotropic denaturants, such as guanidine hydrochloride or urea. The isolated subunits can assemble spontaneously into regular arrays after removal of the disrupting agent, and this entropy-driven self-assembly forms lattices with the same geometric pattern of the original S-layer. The subunits can be aligned in lattices with oblique, square, or hexagonal symmetry (Figure 16.1). Center-to-center spacings of the morphological units vary from 2.5 to 35.0 nm, and about 5×10^5 subunits are required to cover the cell surface. The S-layer proteins are often the most abundant proteins, comprising up to 15% of the total cellular protein content. By sequence comparison, S-layer-homologous (SLH) motifs have been identified at the N-terminal part of many S-layer proteins and at the C-terminal of cell-associated exoenzymes and other exoproteins among different microorganisms. An SLH domain is usually composed of either a single or three repeating SLH motifs of approximately 50 to 60 residues, fairly divergent with an average identity of 27%. In Gram-positive

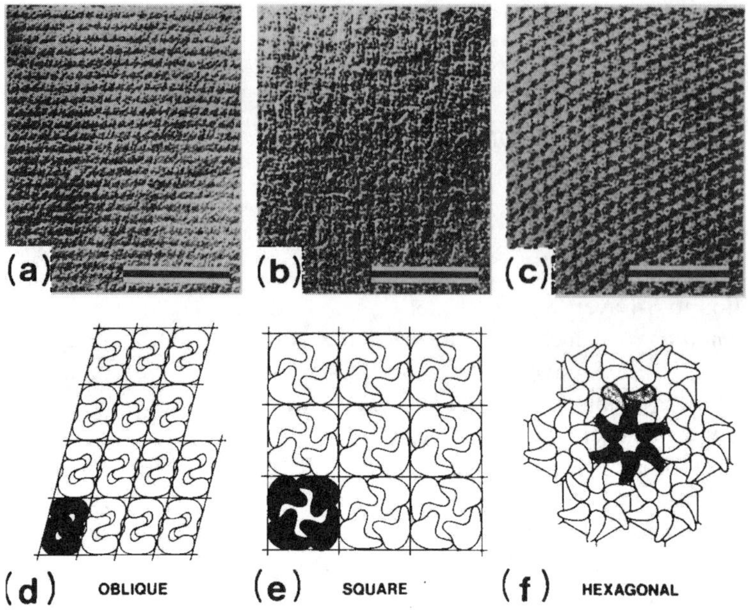

FIGURE 16.1 Electron micrographs (a-c) of freeze-etched preparations of intact cells and (d-f) schematic presentation of the corresponding lattice types of the S-layers. (a) and (d) Oblique lattice of *B. stearothermophilus* NRS2004/3a; (b) and (e) square lattice of *B. stearothermophilus* H4-65; (c) and (f) hexagonal lattice of *C. thermohydrosulfuricum* L111-69. Bar marker: 100 nm. (From Messner, P., Sleytr, U.B., in *Bacterial Cell Surface Techniques*, Hancock I.C. and Poxton I.R., Eds., John Wiley & Sons, Chichester, 1988, Chapter 4. With permission.)

bacteria, SLH motifs are suggested to anchor the proteins to the cell surface and in some microorganisms to mediate adhesion to the host cells. This short introduction to the S-layers intends only to introduce the section on *Clostridium difficile* surface structure and adhesion. For a more exhaustive picture, several recent reviews are suggested [1–4].

16.2 *CLOSTRIDIUM DIFFICILE* ADHESION

Clostridium difficile, the Gram-positive spore-forming anaerobe recognized as one of the major agents of antibiotic associated diarrhea, is a common cause of nosocomial infections in Europe and the U.S. [5–7]. It is primarily a disease of the elderly and is increasing in frequency concomitant with an aging European population. Clinical severity ranges from a self-limiting antibiotic associated diarrhea (AAD), through acute and severe AAD, to life-threatening clinical manifestations, such as pseudomembranous colitis [8,9]. Infection is facilitated by the use of antibiotics that perturb the large bowel microflora causing colonization and overgrowth by *C. difficile* [10], followed by the production of two large toxins (toxin A and B), which mediate the damage to the colonic mucosa [11–12].

Both toxins are very large, with molecular masses of 308 and 207 kDa, respectively. The toxins are endocytosed by intestinal epithelial cells and glycosylate small GTP-binding proteins (Rho, Rac, and Cde42), resulting in disruption of the actin cytoskeleton and, finally, cell death [13–14]. Besides the well-known virulence factors, toxin A and toxin B of *Clostridium difficile*, other factors play a fundamental role in the pathogenesis of disease, mainly those that can promote colonization of the host cells, such as flagella, degrading enzymes, and surface proteins that may mediate adhesion and also interaction with the immune system. In fact, in 1988, Peter Borriello [15] clearly

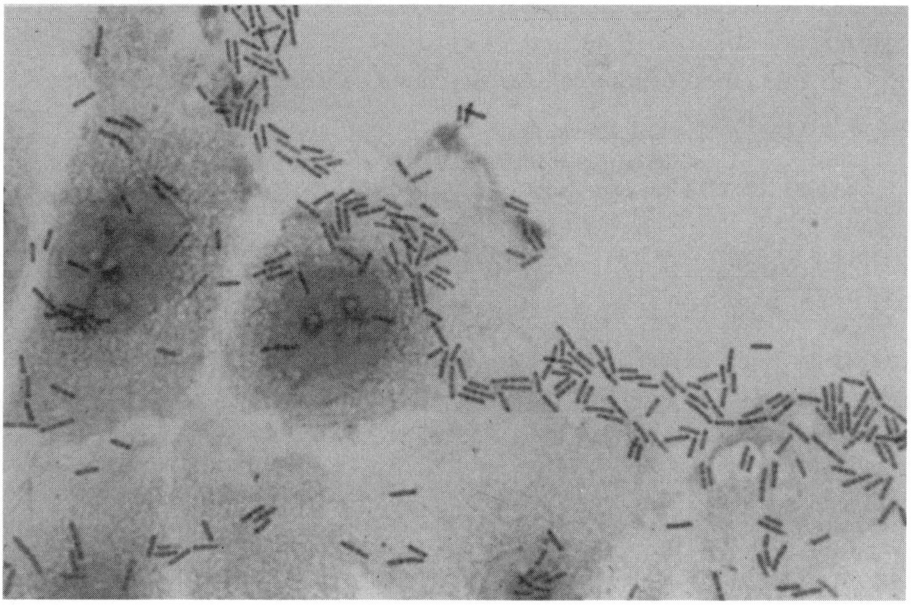

FIGURE 16.2 Caco-2 cells infected with *C. difficile* 253. Light micrograph of a Gram and Giemsa stained 3-day-old monolayer showing bacteria closely interacting with the periphery of an islet. Bar: 20 mm. (From Cerquetti, M. et al., *FEMS Immunol. Med. Microbiol.*, 32, 211, 2002. With permission.)

established that *C. difficile* can associate with intestinal mucosa in hamsters, and there was a positive correlation between virulence and mucosal adherence *in vivo*. But the exact sequence of events leading to *C. difficile* establishment on human intestinal mucosa has yet to be elucidated.

The adhesion of this microorganism to epithelial cells *in vitro,* by using different epithelial cell lines in particular, to human intestinal epithelial cell lines has been documented by various authors [16–18]. The human colonic epithelial cell line CaCo2 resembles small intestinal epithelial cells, and, grown *in vitro* under standard culture conditions, spontaneously exhibits signs of structural and functional differentiation and polarization characteristic of mature enterocytes. Polarization is characterized by the presence of both tight junctions and cell polarity; differentiation corresponds to a maturation process resulting in functionally differentiated brush-border microvilli. Differentiation is total at late confluency (14 to 21 days), but polarization occurs earlier and after confluency involves the whole monolayer [19].

Nonconfluent CaCo2 monolayers (3 to 4 days old) consist mainly of large islands containing several cells in which polarization and differentiation start from the central cells of the islands, whereas in polarized monolayers (15 days old) the cells are closely attached to each other and are only apically exposed to the outside environment.

Microscopy observations showed bacteria closely interacting with the periphery of the islets of cells in nonconfluent monolayer (Figure 16.2), whereas in the post-confluent monolayers, only a very small number of bacteria — or no bacteria at all — interacted with the cells [18]. This pattern of adherence resembled that of some enteroinvasive bacteria that enter through the basolateral surface of differentiated and nondifferentiated polarized CaCo2 cells [20–22]. To verify this possibility, Cerquetti et al. [18] treated post-confluent monolayers of Caco2 cells with EGTA as calcium chelator to disrupt the intercellular junctions and make the basolateral cell surface accessible to bacteria. They observed a dramatic increase of adherent bacteria and, in quantitative assays, about 20-fold increase in the adhesion level of *C. difficile* to CaCo2 cells (Table 16.1).

TABLE 16.1
Effect of Disruption of Intercellular Junctions on Adhesion of *C. difficile* to Caco-2 Cells

	Adherent Bacteria to 3-Day-Old Monolayers[a]		Adherent Bacteria to 15-Day-Old-Monolayers[a]	
Strain	No Pretreatment	Pretreatment with S-MEM +EGTA	No Pretreatment	Pretreatment with S-MEM+EGTA
C253	17.3 ± 5.50	31.5 ± 4.44[b]	1.7 ± 0.10	33.6 ± 1.52[c]
Cd79685	18.3 ± 5.95	22.1 ± 5.43	2.1 ± 0.30	11.9 ± 2.10[c]
ATCC 43597	24.7 ± 8.60	28.0 ± 6.55	1.8 ± 0.23	16.0 ± 2.64[c]

[a] Three- and 15-day-old-monolayers untreated or previously treated with S-MEM plus EGTA were infected with 1 × 10^8 CFU of *C. difficile*. After 1.30 h of incubation, adherent bacteria were determined on agar plates. Values are mean numbers of adherent bacteria per cell ± standard deviations of three separate experiments. Statistical analysis between values for EGTA-pretreated cells vs. those for untreated cells was performed by Student's t-test.

[b] $P < 0.05$

[c] $P < 0.01$.

Source: Cerquetti, M. et al., *FEMS Immunol. Med. Microbiol.*, 32, 211, 2002. With permission.

By confocal microscopy and immunofluorescence experiments, the CaCo2 cell surface area was explored by using the transferrin receptor as a marker of the basolateral area of polarized monolayers. In fact, only after treatment with EGTA could the transferrin receptor at the basolateral surface be detected. Bacteria visible in red and colocalization of the two signals (bacteria and transferrin) in yellow are present in focal planes just above those in which the transferrin receptor is most visible. This result strongly suggests *C. difficile* adhesion to the lateral cell surface. Also, by flow cytometric analysis it has been possible to quantify the cells of *C. difficile* adhered to the intestinal epithelial cell line CaCo2 and to human intestinal biopsies [17].

Scanning Electron Microscopy observations also confirmed the presence of adhering bacteria along the intercellular junctions in gaps between cells, mostly evident after EGTA treatment, but there were also several bacteria interacting directly with extracellular matrix fibers. ELISA performed using immobilized matrix components confirmed these data [18]. In particular, *C. difficile* was able to bind efficiently to collagen, fibronectin, and vitronectin. By extrapolating these results to the *in vivo* setting, a multistep process might be supposed: *C. difficile* might initially interact with the apical microvilli of the intestinal cells and begin to release toxins A and B, which disrupt epithelial barrier functions probably through Rho proteins on apical actin structure and on tight junctions organization. The basolateral pole of epithelial cells thus becomes accessible, and a larger number of bacteria might interact with a receptor therein located with the extracellular matrix proteins.

Different research groups have attempted to identify surface proteins of *C. difficile* with adhesive properties. Some studies on electrophoretic profiles of EDTA-extracted proteins, used for typing [23], have shown that a major protein of 36 kDa was present in many strains associated with the disease and that this protein was an antigen expressed *in vivo* in many patients with antibiotic-associated diarrhea [24]. Immunofluorescence assay revealed that the 36 kDa protein was exposed on the cell surface of *C. difficile* 253, a prototype of clinical isolates with the electrophoretic profile [25]. Some chemical-physical features of this protein common to the surface proteins involved in the formation of the crystalline surface layer (S-layer) in other bacteria [26,27] led to extended investigations on some toxinogenic and nontoxinogenic isolates that searched for the presence of a regular arranged S-layer on their outermost surface.

16.3 *CLOSTRIDIUM DIFFICILE* S-LAYER

A crystalline cell surface layer was first described by Kawata et al. in 1984 [28]. The presence of a regular array, squarely arranged, was demonstrated on the outer layer of the cell wall of about 10 strains of *C. difficile*, which possessed two major cell wall proteins with molecular weights ranging from 32 kDa to 48 kDa. Successively [29], the strain *C. difficile* GAI 4131 confirmed to have an S-layer composed by two proteins, and in 1991, Takumi et al. [30] demonstrated that the intact bacteria adhered significantly to HeLa cells, whereas after removal of the two S-layer proteins, the adherence was dramatically reduced. Moreover, Fab fragments of anti-S-layer protein antibodies were able to inhibit the adhesion of the bacterium to the cells, indicating a direct role of S-layer proteins in adhesion processes.

Other electron microscopy studies [31] demonstrated the presence of a crystalline regular array by a freeze etching technique on the cell surface of all *C. difficile* strains examined — either toxinogenic or not. To determine the symmetries of the crystalline arrays, a Fourier analysis (a digital analysis) was performed showing angles characteristic of the square symmetry model. In particular, the complete analysis indicated the presence of square ordered lattices with similar but not identical center-to-center spacing on the cell surface of the different strains examined. Examined by negative staining, the isolated cell wall fragments of the same strains, contrary to the square structure observed in the whole bacterial cells, showed an hexagonally ordered lattice. This result, similar to that observed by other authors [32], in *Bacillus brevis*, suggests the occurrence of two superimposed S-layers in each strain, and in the negative staining image the thicker inner S-layer apparently masks the much thinner outer S-layer, revealed only by freeze etching, which is a technique more sensitive to the outermost surface topology.

By treating the bacterial cells with 8 *M* urea, the complete removal of the S-layer was observed and the urea extracts of each strain, examined by SDS-PAGE, showed the presence of two main protein components. In fact, each strain carries a lower molecular mass protein and a higher molecular mass protein. The glycan/protein double labeling kit showed two distinct glycoprotein bands for each strain, indicating that, like many other S-layers, these proteins are glycosylated. Western immunoblot analysis with rabbit antisera against the 36 and 47 kDa proteins of the prototype strain C253 showed the antigenic relatedness of the high molecular mass S-layer proteins of different strains and the absence of antigenic relationships among the low molecular mass proteins [31]. Therefore, each strain carries on its surface one high molecular weight S-layer protein with antigenic similarities with high M.W. proteins of other strains and one low M.W. S-layer protein antigenically unique to each strain. The existence of two separate families of S-layer proteins was also confirmed by data obtained from N-terminal sequence analysis of different strains. Each family shared one common structured motif that had not been previously recognized in other bacterial S-layer proteins [31].

Purification and chemical characterization of the 36 and 47 kDa S-layer proteins of *C. difficile* C253 were performed by liquid chromatography coupled to electrospray ionization mass spectrometry (ESI-MS), matrix-assisted laser desorption/ionization time of flight (MALDI-TOF), and liquid chromatography coupled to diode array detector and ESI-MS [33]. The comparison of the molecular weight data obtained by SDS-PAGE, ESI-MS, and MALDI-TOF showed that the mass results from these two last techniques deviated from those obtained by gel electrophoresis, mainly for P47, but this discrepancy is probably due to a different electrophoretic behavior of glycated proteins.

By analyzing the enzymatic digestions of the proteins with protease V8 by liquid chromatography coupled to LC/DAD-ESI-MS, it was possible to characterize the peptide fingerprinting of the two proteins and obtain a preliminary identification of the glycosylated peptides. In order to search for sequence similarity to other proteins, the peptide sequences of *C. difficile* 253 were submitted to the *C. difficile* 630 genome database using TBLASTN, and all peptides could be mapped to the same contig. Karjalainen et al. [34] analyzed the sequence of this region that revealed a unique 2160 bp gene capable of coding for a protein of 73.41 kDa. PCR with oligonucleotides derived from *C. difficile* 630 also revealed the presence of this gene in the strain 253, and it was named *slpA* according to the nomenclature used for S-layer genes of other bacteria.

The encoded protein shows a characteristic prokaryotic signal sequence at the N-terminus and a two-domain structure. The presence of two S-layer proteins P36 and P47 in the cell surface suggests that the cleavage of the precursor occurred after translocation through the membrane. The post-translational cleavage is internally around position 350 and at the N-terminus to release the signal sequence. The availability of the N-terminal sequence of the two proteins indicated that the P47 corresponds to the C-terminal part of the precursor and the P36 to the N-terminal part.

Homology searches with other proteins showed that the C-terminal of SlpA has a significant homology with the family of autolysins from *B. subtilis* with an amidase activity and that at the N-terminus, 78 amino acids sequence shows a low homology to the SLH (S-layer homologous) domain present in most S-layer proteins.

Also the precursor to S-layer protein of another strain, *C. difficile* 79685 showed the same two-domain structure as in strains *C. difficile* 253 and 630, and only the C-terminal domain was well conserved among the three strains (77% identity). This domain also displayed sequence conservation with the cell wall anchoring domain (N-terminal) of the surface protein Cwp66 of *C. difficile* 79685, which has been characterized as an adhesin to Vero cells [35]. (See also Section 16.4.1.) Interestingly, in the recent genome sequence of *C. tetani* 19 homologues of Cwp66 [36] have been found, all of which possess multiple copies of the putative cell wall binding domain, identified in characterized SLPs.

Also, the 36 kDa protein of the S-layer in *C. difficile* 253 is involved in the adhesion process to CaCo2 cells. In fact, antiserum against the whole bacterium and monospecific antiserum against P36 significantly reduced the number of adherent bacteria per cell compared to the adherence of bacteria pretreated with nonimmune serum [34].

A 37 kb DNA fragment flanking *SlpA* and *Cwp66* genes has been examined, and 17 ORFs in the same orientation have been found (Figure 16.3). Eleven of these ORFs encode a domain present in either the amino or carboxy-terminal part of these putative secreted proteins showing homology to the Cwp autolysin of *B. subtilis,* and the authors suggest that all these proteins could be cell wall anchored through this domain. Each protein also has a variable domain and a signal peptide. Interestingly, the gene immediately downstream from slpA, ORF7, has a very significant homology with *SecA,* whose product is an essential component of the general secretory pathway.

Calabi et al. [37] published a paper with similar results but demonstrating also that the high MW SlpA proteins exhibited amidase activity in contrast to the low MW proteins in which this activity could not be detected by zymography. Moreover, there is a sequence conservation at 60% between the high MW SLPs in the amidase domain, indicating a strong conservation of this activity.

The same authors also investigated the direct binding of native SL proteins and recombinant subunits to human gastrointestinal tissues by both microscopic and biochemical approaches [38]. By using human colon biopsy specimens, either native or purified recombinant high molecular weight S-layer proteins showed a strong specific bond both to the epithelium and the lamina propria. The highest positivity was found in the apical cytoplasm, and particularly along the apical border, suggesting prevalent localization on the brush border.

In the lamina propria, staining was both widespread and finely fibrillar, indicating binding to components of the extracellular matrix. The low molecular weight subunits showed only a punctate staining in a subpopulation of epithelial cells both in the surface epithelium and in the necks of the glands. Thus, it is possible to hypothesize that the surface layer proteins of *C. difficile* might be multifunctional adhesins and that the binding may happen in two stages. First, S-layer proteins could mediate binding of *C. difficile* to brush border components, allowing the delivery of toxins A and B to enterocytes. Later, after epithelial damage, SLPs can bind *C. difficile* to the basolateral area of the cells and to extracellular matrix components.

In conclusion, the identification of the *Slp*A gene in *C. difficile* led also to the identification of a large family of related genes that cluster around the *slp*A locus, and future studies aimed at

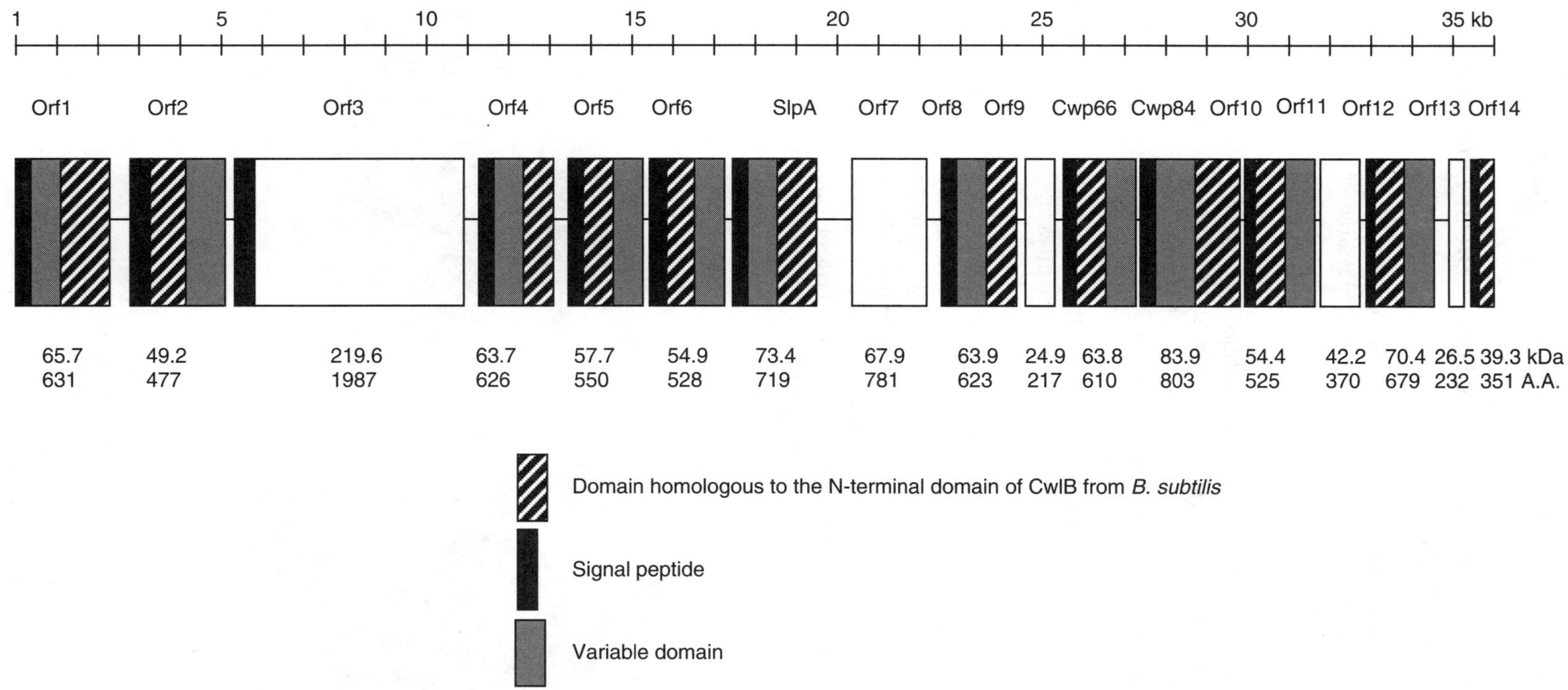

FIGURE 16.3 Genetic organization of a 37-kb DNA fragment carrying the *slpA* and *Cwp66* genes from the genome of *C. difficile* 630. (From Karjalainen, T. et al., *Infect. Immun.*, 69, 3442, 2001. With permission.)

characterizing gene products of this putative virulence gene cluster will greatly contribute to the comprehension of the colonization processes adopted by this microorganism.

16.4 OTHER SURFACE PROTEINS OF *CLOSTRIDIUM DIFFICILE* INVOLVED IN ADHERENCE

Clostridium difficile 79685 showed to adhere to cultured cells, and adherence is increased by various stresses such as heat shock, hyperosmolarity, low-iron concentration, and low pH [39].

16.4.1 Cwp66 Adhesin

By an immunoscreening of a gene library with antibodies prepared against heat-shocked bacteria in toto and adsorbed against non-heat-shocked *C. difficile*, a cell-surface associated protein, named Cwp66, was isolated [35]. Cwp66 is a 610-amino acid protein with molecular mass of 66,323 Da and contains two domains, one presenting homologies to the autolysin CwlB of *Bacillus subtilis*. The conservation of *cwp*66 gene region has been evaluated in different strains, and the N-terminal portion of the protein is well conserved (99.3% identity) at least with other two strains tested, whereas the C-terminal domain is more variable (58% identity).

Efforts to express the entire protein in *E. coli* failed; subsequently, the two domains were expressed and purified separately. Two bands, a 35-kDa and a 44-kDa band corresponding to Cwp66-N terminus and Cwp66-C terminus, respectively, were obtained and used to prepare rabbit polyclonal antibodies. By immunoelectron microscopy, the C-terminal domain was found to be cell surface exposed after heat shock, whereas the N-terminal part was inaccessible to the antibodies (Figure 16.4). Surface localization of Cwp66 suggested that it could play a role in adherence to cells. This was investigated by adherence inhibition assays of *C. difficile* to culture cells. Both antibodies (anti Cwp66-Nter and -Cter) and partially purified Cwp66-C and Cwp66-N proteins inhibited adherence of *C. difficile* to cultured cells confirming the protein as an adhesin.

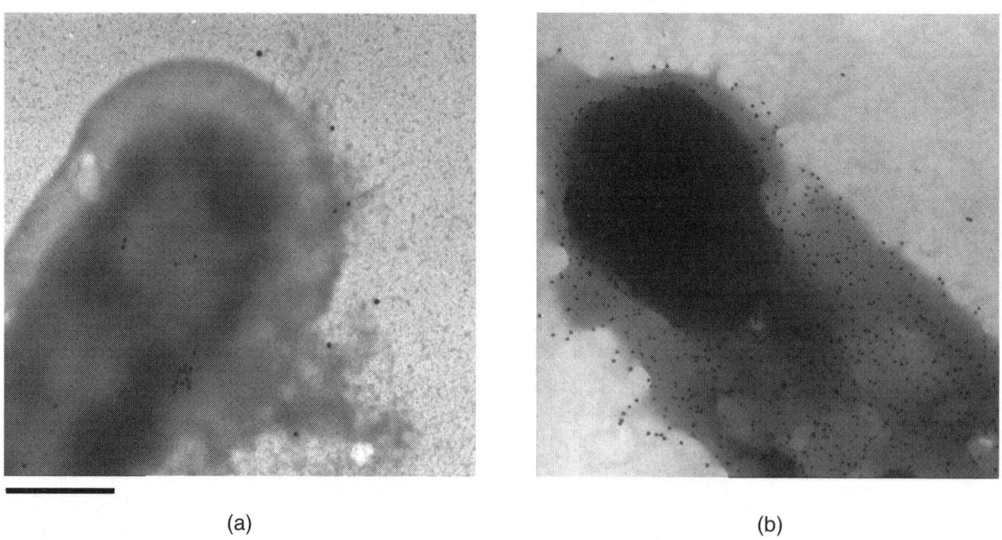

(a) (b)

FIGURE 16.4 Surface localization of *C. difficile* Cwp66 by immunoelectron microscopy. a and b: reaction of whole cells of *C. difficile* subjected to heat shock (60°C for 20 min) to polyclonal antibodies raised against Cwp66-Nter (a) and Cwp66-Cter (b), followed by protein A-colloidal gold labeling and negative staining. Bar, 600 nm. (Modified from Waligora, A.J. et al., *Infect. Immun.*, 69, 2144, 2001. With permission.)

Finally, the presence of the *cwp66* gene has been investigated in 36 strains of *C. difficile* belonging to different serogroups, and even if the gene was detected in all the strains, no correlation with the serogroups was found.

16.4.2 Heat Shock Protein GroEL

As the adherence of *C. difficile* is increased by heat shock, the role of heat shock proteins in the colonization process was suspected. The gene *groEL* was amplified by PCR using primers based on the *Clostridium perfringens groEL* gene and cloned into the expression vector pGEX-6P-1 in fusion with glutathione S-transferase. The 58 kDa protein GroEL was expressed in *E. coli* and purified by affinity chromatography. The *groEL* gene is highly conserved between various *C. difficile* isolates, as determined by PCR-RFLP and DNA sequencing. Cell fractionation experiments showed that the GroEL protein is found after heat shock mainly in the cytoplasmic membrane fractions and the supernatant, and weakly in the cell wall. Immunoelectron microscopy experiments demonstrated that GroEL is released extracellularly after heat shock; this protein could subsequently be absorbed to the surface of adjacent bacteria. In adherence inhibition assays, GroEL antibodies, as well as the recombinant protein, partially inhibited *C. difficile* cell attachment, suggesting a role for GroEL in adherence [40].

16.4.3 Fibronectin-Binding Protein: Fbp68

C. difficile binds to extracellular matrix proteins such as fibronectin, fibrinogen, collagen, and vitronectin [18]. It has been shown that the S-layer proteins of *C. difficile* bind to collagen and vitronectin but not to fibronectin [38]. Binding to fibronectin could be due to fibronectin-binding proteins (Fbps), which have been described in streptococci and staphylococci. The gene has been isolated by PCR using primers derived from the putative Fbp gene of *Bacillus subtilis* and the gene sequence of the *C. difficile* 630 strain. The one copy gene is highly conserved in *C. difficile* isolates. Like GroEL, the gene was cloned into the expression vector pGEX-6P-1 in fusion with glutathione S-transferase. The 68 kDa protein was expressed in *E. coli* and purified. Far-immuno dot-blotting demonstrated that Fbp68 binds fibronectin, fibrinogen, and also vitronectin. Indirect immunofluorescence assay and ELISA demonstrated that *C. difficile* bind both soluble and immobilized fibronectin. Cell fractionation and immuno-blot experiments showed that Fbp68 is located mainly in the cytoplasmic and, to a lesser extent, in the cell wall fraction. In adherence inhibition assays, the Fbp68-specific antibodies partially inhibited attachment of *C. difficile* to fibronectin and Vero cells. Thus, Fbp68 appears to be one of the several adhesins of *C. difficile* [41].

16.4.4 The Flagellar Proteins

In 1990, Delmee et al. [42] described the presence of uniformly distributed flagella in different strains of *C. difficile*. When purified by differential centrifugation, the SDS-PAGE revealed one band with an apparent molecular mass of 39 kDa. More recently, flagella were extracted from six *C. difficile* strains and the N-terminal sequences of the flagellin proteins were determined [43]. Using degenerate primers, the gene encoding the flagellum subunit, *fliC*, was isolated and sequenced. PCR-RFLP analysis of amplified flagellin gene products revealed interstrain heterogeneity.

To complete the study on the role of flagella in digestive colonization, the gene *fliD* encoding the flagellar cap protein was characterized. The flagellar cap protein is conserved among isolates of *C. difficile* [44]. Involvement of the flagellin filament protein (Fli C) and of the flagellar cap (FliD) in adherence of *C. difficile* to eukaryotic cells was investigated using antibodies raised against the purified proteins. No inhibition was detected when bacteria were preincubated with antibodies raised against FliC; in contrast FliD antibodies inhibited partial adherence, suggesting that the flagellar cap protein FliD appears to be one of the multiple adhesins of *C. difficile*.

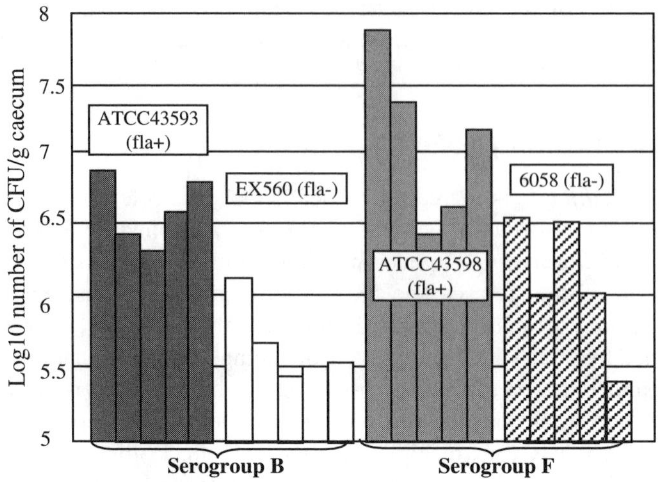

FIGURE 16.5 Adherence of flagellated and nonflagellated *C. difficile* strains to axenic mouse cecum. Adherence of flagellated and nonflagellated strains, belonging to the same serogroup, is expressed as log10 of number of bacteria per gram of cecum from five mice. Strains belonging to serogroup B are toxin A and B negative, whereas those belonging to serogroup F are toxin A negative. (From Tasteyre, A. et al., *Infect. Immun.*, 69, 7937, 2001. With permission.)

In vivo experiments using flagellated and nonflagellated strains belonging to the same serogroup confirmed the role of flagella in adherence and digestive colonization. Nonflagellated strains colonized axenic mouse intestine at the same rate as flagellated strains, but the adherence of nonflagellated strains to mouse ceca was significantly less than that of flagellated strains (Figure 16.5). Thus, flagella seem implicated in adherence to the mouse cecum *in vivo* [45].

From these studies, it appears that *C. difficile* possesses multiple adhesins responsible for adherence to mucus, enterocytes, and extracellular matrix components that are involved in the intestinal colonization process.

16.5 CLOSTRIDIAL CELLULOSOME AND ADHESION

The cellulolytic microbes occupy a broad range of habitats in the environment. The primary event in the degradation of cellulose is the tight adhesion of the bacterium to the substrate. The molecular mechanism of adhesion has been extensively studied in few cellulolytic clostridia, mainly in *Clostridium thermocellum* [46].

Clostridium thermocellum is a thermophilic clostridium whose growth is restricted to cellulose and its degradation products. The cellulose-binding factor has been characterized as a multienzyme complex, the cellulosome, capable of complete solubilization of cellulosic substrates. It is a 2-MDa conglomerate of many subunits packed into protuberance-like organelles, most of which are large enzymes (i.e., cellulase and xylanase). Characteristic of the clostridial cellulosomes is the presence of a large nonhydrolytic subunit, the scaffolding protein CipA, to which the catalytic subunits are attached. The aminoacid sequence of three scaffolding proteins showed multiple copies of hydrophobic domains, the cohesin domains. These domains interact with another complementary class of domains, the dockerin domains on the enzymatic subunit. One functional domain of the scaffolding protein is the cellulose-binding domain (CBD), which delivers complementary enzymes together with the entire cell to the cellulosic substrate. A clear adhesion property of this structure has been demonstrated so far only for *C. thermocellum* and *C. cellulolyticum.*

For a more detailed description of this structure and its functions see Reference 46.

REFERENCES

1. Sleytr, U.B. and Beveridge, T.J., Bacterial S-layers, *Trends Microbiol.,* 7, 253, 1999.
2. Sara, M. and Sleytr, U.B., S-Layer proteins, *J. Bacteriol.,* 182, 859, 2000.
3. Full Issue, *FEMS Microbiol. Rev.,* 20, Issue 1–2, 1997.
4. Sleytr, U.B., Messner, P., Pum, D., and Sara, M., *Crystalline Bacterial Cell Surface Proteins,* Landes R.G., Ed., Academic Press, Austin, 1996.
5. Bartlett, J.G., *Clostridium difficile*: clinical considerations, *Rev. Infect. Dis.*, 12 Suppl 2, S243, 1990.
6. Spencer, R.C., Clinical impact and associated costs of *Clostridium difficile*-associated disease, *J. Antimicrob. Chemother.,* 41 Suppl C, 5, 1998.
7. Barbut, F. and Petit, J.C., Epidemiology of *Clostridium difficile*-associated infections, *Clin. Microbiol. Infect.,* 7, 405, 2001.
8. Hurley, B.W. and Nguyen, C.C., The spectrum of pseudomembranous enterocolitis and antibiotic-associated diarrhea, *Arch. Intern. Med.,* 162, 2177, 2002.
9. Jabbar, A. and Wright, R.A., Gastroenteritis and antibiotic-associated diarrhea, *Prim. Care,* 30, 63, 2003.
10. Spencer, R.C., The role of antimicrobial agents in the aetiology of *Clostridium difficile*-associated disease, *J. Antimicrob. Chemother.,* 41 Suppl C, 21, 1998.
11. Nusrat, A. et al., *Clostridium difficile* toxins disrupt epithelial barrier function by altering membrane microdomain localization of tight junction proteins, *Infect. Immun.,* 69, 1329, 2001.
12. Savidge, T.C. et al., *Clostridium difficile* toxin B is an inflammatory enterotoxin in human intestine, *Gastroenterology,* 125, 413, 2003.
13. Just, I. et al., *Clostridium difficile* toxin B acts on the GTP-binding protein Rho, *J. Biol. Chem.,* 269, 10706, 1994.
14. Just, I. et al., The low molecular mass GTP-binding protein Rho is affected by toxin A from *Clostridium difficile, J. Clin. Invest,* 95, 1026, 1995.
15. Borriello, S.P. et al., Mucosal association by *Clostridium difficile* in the hamster gastrointestinal tract, *J. Med. Microbiol.,* 25, 191, 1988.
16. Eveillard, M. et al., Identification and characterization of adhesive factors of *Clostridium difficile* involved in adhesion to human colonic enterocyte-like Caco-2 and mucus-secreting HT29 cells in culture, *Mol. Microbiol.,* 7, 371, 1993.
17. Drudy, D. et al., Flow cytometric analysis of *Clostridium difficile* adherence to human intestinal epithelial cells, *J. Med. Microbiol.,* 50, 526, 2001.
18. Cerquetti, M. et al., Binding of *Clostridium difficile* to Caco-2 epithelial cell line and to extracellular matrix proteins, *FEMS Immunol. Med. Microbiol.,* 32, 211, 2002.
19. Zweibaum, A. et al., Enterocytic differentiation of a subpopulation of the human colon tumor cell line HT-29 selected for growth in sugar-free medium and its inhibition by glucose, *J. Cell Physiol.,* 122, 21, 1985.
20. Mounier, J. et al., *Shigella flexneri* enters human colonic Caco-2 epithelial cells through the basolateral pole, *Infect. Immun.,* 60, 237, 1992.
21. Gaillard, J.L. and Finlay, B.B., Effect of cell polarization and differentiation on entry of *Listeria monocytogenes* into the enterocyte-like Caco-2 cell line, *Infect. Immun.,* 64, 1299, 1996.
22. Hirakata, Y. et al., Adherence to and penetration of human intestinal Caco-2 epithelial cell monolayers by *Pseudomonas aeruginosa, Infect. Immun.,* 66, 1748, 1998.
23. Pantosti, A., Cerquetti, M., and Gianfrilli Mastrantonio, P., Electrophoretic characterization of *Clostridium difficile* strains isolated from antibiotic-associated colitis and other conditions, *J. Clin. Microbiol.,* 26, 540, 1988.
24. Pantosti, A. et al., Immunoblot analysis of serum immunoglobulin G response to surface proteins of *Clostridium difficile* in patients with antibiotic-associated diarrhea, *J. Clin. Microbiol.,* 27, 2594, 1989.
25. Cerquetti, M. et al., Purification and characterization of an immunodominant 36 kDa antigen present on the cell surface of *Clostridium difficile, Microb. Pathog.,* 13, 271, 1992.
26. Blaser, M.J. and Pei, Z., Pathogenesis of *Campylobacter fetus* infections: critical role of high-molecular-weight S-layer proteins in virulence, *J. Infect. Dis.,* 167, 372, 1993.
27. Munn, C.B. et al., Role of surface components in serum resistance of virulent *Aeromonas salmonicida, Infect. Immun.,* 36, 1069, 1982.

28. Kawata, T. et al., Demonstration and preliminary characterization of a regular array in the cell wall of *Clostridium difficile*, *Fems Microbiol. Lett.*, 24, 323, 1984.
29. Masuda, K., Itoh, M., and Kawata, T., Characterization and reassembly of a regular array in the cell wall of *Clostridium difficile* GAI 4131, *Microbiol. Immunol.*, 33, 287, 1989.
30. Takumi, K. et al., Self-assembly, adhesion, and chemical properties of tetragonally arrayed S-layer proteins of *Clostridium difficile*, *J. Gen. Appl. Microbiol.*, 37, 455, 1991.
31. Cerquetti, M. et al., Characterization of surface layer proteins from different *Clostridium difficile* clinical isolates, *Microb. Pathog.*, 28, 363, 2000.
32. Sara, M. et al., Characterization of S-layer from mesophilic bacillaceae and studies on their protective role towards muramidases, *Arch. Microbiol.*, 153, 209, 1990.
33. Mauri, P.L. et al., Characterization of surface layer proteins from *Clostridium difficile* by liquid chromatography/electrospray ionization mass spectrometry, *Rapid Commun. Mass Spectrom.*, 13, 695, 1999.
34. Karjalainen, T. et al., Molecular and genomic analysis of genes encoding surface-anchored proteins from *Clostridium difficile*, *Infect. Immun.*, 69, 3442, 2001.
35. Waligora, A.J. et al., Characterization of a cell surface protein of *Clostridium difficile* with adhesive properties, *Infect. Immun.*, 69, 2144, 2001.
36. Bruggemann, H. et al., The genome sequence of *Clostridium tetani,* the causative agent of tetanus disease, *Proc. Natl. Acad. Sci. U.S.A.*, 100, 1316, 2003.
37. Calabi, E. et al., Molecular characterization of the surface layer proteins from *Clostridium difficile*, *Mol. Microbiol.*, 40, 1187, 2001.
38. Calabi, E. et al., Binding of *Clostridium difficile* surface layer proteins to gastrointestinal tissues, *Infect. Immun.*, 70, 5770, 2002.
39. Walligora, A.J. et al., *Clostridium difficile* cell attachment is modified by environmental factors, *App. Environ. Microbiol.*, 65, 4234, 1999.
40. Hennequin, C. et al., GroEL (Hsp60) of *Clostridium difficile* is involved in cell adherence, *Microbiology,* 147, 87, 2001.
41. Hennequin, C. et al., Identification and characterization of a fibronectin-binding protein from *Clostridium difficile, Microbiology,* 149, 2779, 2003.
42. Delmee, M. et al., Characterization of flagella of *Clostridium difficile* and their role in serogrouping reactions, *J. Clin. Microbiol.*, 28, 2210, 1990.
43. Tasteyre, A. et al., A *Clostridium difficile* gene encoding flagellin, *Microbiology,* 146, 957, 2000.
44. Tasteyre, A. et al., Molecular characterization of fliD gene encoding flagellar cap protein and its expression among *Clostridium difficile* isolates from different serogroups, *J. Clin. Microbiol.,* 39, 1178, 2001.
45. Tasteyre, A. et al., Role of FliC and FliD flagellar proteins of *Clostridium difficile* in adherence and gut colonization, *Infect. Immunol.,* 69, 7937, 2001.
46. Bayer, E.A., Morag, E., Shoham, Y., Tormo, J., and Lamed, R., *The Cellulosome: A Cell Surface Organelle for the Adhesion to and Degradation of Cellulose in Bacterial Adhesion,* Fletcher M., Ed., Wiley-Liss, New York, 1996.

17 Clostridial Toxins vs. Other Bacterial Toxins

Michel R. Popoff and Bradley G. Stiles

CONTENTS

17.1 INTRODUCTION

Clostridia are widespread in the environment, and some of them are responsible for severe diseases in humans and animals, due to the production of highly potent protein toxins. Many other bacteria also produce toxins that directly compromise the host, thus promoting bacterial survival and subsequent invasion throughout host tissues. It is surprising that some *Clostridium* species, which are inherently environmental bacteria commonly found in soil and water, secrete so many target-specific toxins that affect various mammals. This chapter is a comparative review of various bacterial, and in particular clostridial, protein toxins regarding their genetics, structures, mechanisms of action, and elicited pathologies.

17.2 CLOSTRIDIAL AND OTHER BACTERIAL TOXINS

The genus *Clostridium* is a vast and heterogeneous group that contains more than 150 species. Among them, 35 species are considered pathogenic but only 15 produce potent protein toxins. Therefore, toxigenic strains represent less than 10% of the total *Clostridium* species described in the literature. However, these bacteria produce a total of 58 toxins that represent an impressive 18% of the 327 known bacterial toxins identified to date [1]. Indeed, *Clostridium* is regarded as the "most toxic" bacterial genus, as evidenced by synthesis of the highest number of known protein toxins (Figure 17.1). Approximately one third of all bacterial (35%), or clostridial (32%) toxins cause damage localized on targeted cell membranes. This raises an important question regarding why and how such few environmental bacterial species have acquired the ability to produce such potent toxins directed towards eukaryotic cells. Perhaps this phenomenon is linked to acquisition of toxin genes from other organisms and/or a dynamic evolutionary process in some clostridial species?

Based upon DNA alignment of ribosomal RNA genes, *Clostridium* species belong to 16 different clusters that further illustrate the phylogenetic heterogeneity of this genus. Most of the toxigenic *Clostridium* species (n=10) are classified into cluster I, which is considered the only "true" representative of the genus *Clostridium*, and the other toxigenic species are scattered into three other clusters [2] (Table 17.1). This indicates that all of the toxigenic *Clostridia* are not related phylogenetically.

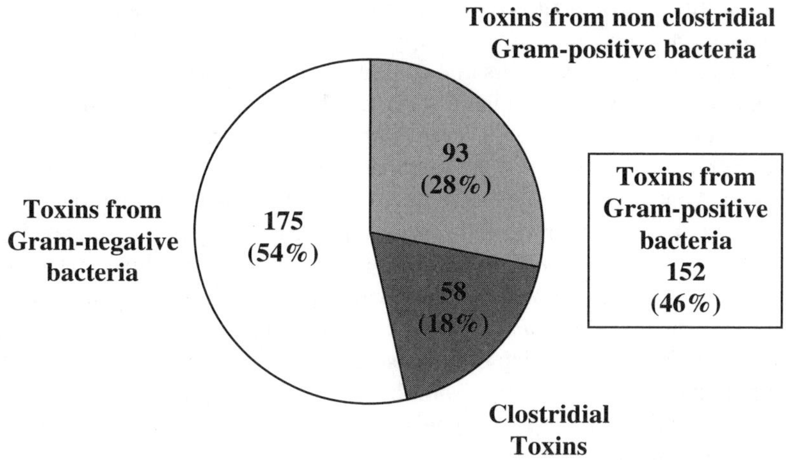

FIGURE 17.1 A comparison of the number of protein/peptide toxins produced by Gram-positive and Gram-negative bacteria relative to *Clostridium.*

TABLE 17.1
Toxigenic *Clostridium* Species, Number of Toxins and Repartition in Phylogenetic Clusters

Species	Toxins	16S rDNA Cluster
C. argentinense	1	I
C. baratii	2	I
C. bifermentans	3	XI
C. botulinum	3	I
C. butyricum	1	I
C. chauvoei	4	I
C. difficile	3	XI
C. haemolyticum	3	I
C. histolyticum	5	II
C. novyi	8	I
C. perfringens	14	I
C. septicum	4	I
C. sordellii	4	XI
C. spiroforme	1	XVIII
C. tetani	2	I
Total	58	

Note: The number of *C. chauvoei* toxins is not yet firmly established.

According to References 1 and 2.

17.3 TOXIGENIC CLOSTRIDIA ARE ENVIRONMENTAL BACTERIA THAT PRODUCE SPECIFIC TOXINS AND ENZYMES

Clostridia produce a myriad of potent toxins responsible for severe diseases in humans and animals. These anaerobic, fermentative spore formers are, however, generally regarded as better adapted to life in the environment as per the production of spores. Because of their ability to form spores, they can survive in the environment for very long periods in a dormant state. Most of the species are fermentative anaerobes that derive their energy from incomplete oxidation of organic molecules. Metabolically, *Clostridia* are quite versatile and degrade an extremely wide range of organic materials that include carbohydrates, organic acids, alcohols, aromatic compounds, peptides, amino acids, amines, purines, and pyrimidines. Thereby, they effectively participate in an important aspect of ecology involving biomass renewal. The vast majority of *Clostridium* species are saccharolytic, proteolytic, or both. Although synthetic media have been defined, *Clostridia* grow better in complex media with numerous sources of carbon and energy [3]. The recent genomic sequencing of *Clostridium acetobutylicum, Clostridium perfringens*, and *Clostridium tetani* show that these species are defective in many biosynthetic pathways for amino acids, thus underscoring their need for complex media necessary for growth. *C. acetobutylicum* and *C. perfringens* contain many genes for utilizing different sugars, unlike *C. tetani*, which only possesses genes for glucose utilization but has many for metabolizing amino acid [4–6]. Large carbohydrate polymers or monomers are fermented by various *Clostridium* species, but peptones generally represent preferred substrates for proteolytic species. Therefore, the *Clostridia* synthesize and secrete numerous hydrolytic enzymes that degrade organic molecules in the microenvironment into more easily "digested" or assimilated compounds. The resulting monomeric compounds required for growth are brought into their cytoplasm by various transporters. Perhaps the toxins produced by some *Clostridium* species evolved from "ancestral" hydrolytic enzymes by acquisition of novel specific properties that include pore formation, translocation across lipid membranes, and recognition of crucial eukaryotic targets. So, one might ask whether bacterial toxins, like those produced by the clostridia, represent an evolutionary protein "tool" derived from less toxic, food-gathering origins.

The term "toxin" etymologically refers to poison that was used on arrow heads during the ancient Greek period. This weapon could then be employed against an enemy or perhaps incapacitate prey for sustenance, thus aiding the protection and perpetuation of one's kind. Therefore, a toxin corresponds to a compound that induces disastrous effects at low concentration. Toxigenic bacteria act at a distance by secreting potent protein weapons, which directly interact with a targeted quarry (i.e., a cell). Many toxins possess an enzymatic property, and the specific activity (designated as activity per milligram of protein) is variable, according to the studied toxin and employed assay. Some of these molecules act at very low concentrations, like the botulinum neurotoxins, but others possess an enzymatic activity akin to typical enzymes. Thereby, some toxins have been classified as "major," or more potent, toxins while others are regarded as "minor," or less potent, toxins.

Historically, toxicity testing has involved lethality in mice upon injecting 0.5 to 1 ml of culture supernatant from a bacterial strain. Specific identification of the bacterium, via identification of a particular toxin, is classically made with antiserum and neutralization of the lethal effects. Indeed, toxins produced at a lethal concentration with such a volume of cell-free culture supernatant were considered major toxins, in contrast to those proteins produced in sublethal amounts, that are accordingly known as minor toxins. Actually, a progressive range of activity is observed between classical enzymes and potent toxins. Moreover, "active" molecules not transported into cells are considered as enzymes and only become cytotoxic when used at relatively high concentrations (10^{-6} *M* and above). A classical example is the exoenzyme C3 from *Clostridium botulinum*, which exhibits the same level of *in vitro* activity vs. other ADP-ribosylating toxins, but it does not enter cells. Biologically, it is peculiar why a bacterium would produce such a molecule that requires entry into a cell to elicit a biological effect, yet there is evidently no known "facilitator" for enabling an encounter between C3 and its substrate located within a cell. Clearly, we do not have all the answers to all the questions that microbes are perpetually posing in their own subtle way.

17.4 GENETIC ORGANIZATION OF BACTERIAL TOXIN GENES

Most bacteria contain a single circular chromosome and mobile extrachromosomal DNA elements, such as plasmids and phages, which can be integrated into the chromosome. Additional mobile

TABLE 17.2
Genomic Localization of the Main Bacterial Toxins

Toxin Gene Localization	Nonclostridial Toxins	Clostridial Toxins
Chromosome	*Pseudomonas aeruginosa* exotoxin A, exotoxin S	*C. perfringens* alpha-toxin, PFO, enterotoxin
	E. coli CNF1, α-hemolysin	*C. difficile* ToxA, ToxB, CDT
	Bordetella Pertussis Toxin	*C. spiroforme* toxin
	Shiga Toxin	*C. botulinum*: neurotoxin A, B, E, F
	Helicobacter pylori VacA Toxin	*C. septicum* α-toxin
	Aerolysin	
	Staphylococcus α-toxin, γ-hemolysin	
	Streptolysin O	
	Listeriolysin	
	Staphylococcus enterotoxins G, H, and I, TSST-1	
Plasmid	*E. coli* CNF2, heat stable toxin STa, STb, heat labile toxin, α-hemolysin	*C. perfringens* beta1, beta2, epsilon, iota toxins, enterotoxin
	B. anthracis toxin	*C. botulinum*: neurotoxin G
	Staphylococcus enterotoxins D and J	*C. tetani* Tetanus Toxin
Phage	Cholera Toxin	*C. novyi* α-toxin
	Diphtheria Toxin	*C. botulinum*: neurotoxin C1, D,
	Shiga Toxin	C2 toxin
	Streptococcus erythrogenic exotoxins A and C	*C. botulinum* C3 enzyme
	Staphylococcus enterotoxins A and E	

elements like transposons and transposable insertion sequences permit genetic transfer between these different DNA structures. DNA rearrangements are also supported by other mechanisms such as duplication and recombination. Therefore, bacterial genomes are highly dynamic structures ensuring genetic variations by horizontal transfer of DNA between bacteria from the same or different species, and by DNA rearrangements from one generation to another. Bacterial toxin genes are either located on chromosomes, plasmids, or phages (Table 17.2). In some bacteria, the toxin and virulence determinant genes, as well as other functionally related genes, are clustered in DNA segments called pathogenicity islands [7]. In *Clostridium,* the toxin genes are dispersed within the whole genome or DNA elements (pathogenicity islets) that are shorter than those found in toxigenic Gram-negative bacteria. Horizontal transfer plays an important role in disseminating toxin genes intra- and inter-species. Since *Clostridia* produce the greatest number of toxins, this genus possibly constitutes a reservoir of toxin genes that have spread to varying degrees throughout the bacterial world.

17.4.1 Bacteria Other Than *Clostridium* Species

Pathogenicity islands were first described in Gram-negative bacteria and are characteristically an unstable, large piece of chromosomal DNA (>30 kb) with a G+C content differing from the entire genome. Pathogenicity islands are flanked by direct repeats, tRNA genes, or insertion sequences that contain functional or cryptic genes involved in mobility, such as insertion sequences, integrases, transposases, and origins of plasmid replication. Pathogenicity islands can be lost by site-specific recombination between the flanking direct repeats and excision, thus resulting in reduced virulence of the host bacterium. Pathogenicity islands have been described in various Gram-negative species, and as an example, uropathogenic *Escherichia coli* contain several pathogenicity islands that carry the α-hemolysin gene. Other Gram-negative bacteria containing pathogenicity islands with important genes for virulence include *Yersinia* species (genes for iron acquisition), *Shigella* species (hemagglutinin and mucinase genes), *Vibrio cholerae* (genes encoding adhesin and receptor for toxin encoding phage), *Salmonella typhimurium* (genes for type III secretion system), *Helicobacter pylori* (genes for secreting toxin or other virulence factors), and *Bordetella pertussis* (*pertussis* toxin gene). Smaller pathogenicity islands (5–7 kb), also referred to as pathogenicity islets, that contain toxin or other virulence genes have been found in various bacteria such as *Hemophilus, Pseudomonas,* and *Salmonella*.

Gram-positive bacteria also possess pathogenicity islands. In *Staphylococcus aureus*, the toxic shock syndrome toxin-1 (TSST-1) gene is localized on a 15 kb pathogenicity island and mobilized by a phage. Strains which do not produce TSST-1 are defective in this pathogenicity island. In *Listeria monocytogenes*, virulence genes that encode listeriolysin, phospholipases, ActA, as well as regulatory proteins are also clustered on pathogenicity islands [7].

17.4.2 The Clostridia

The Clostridia are rather diverse regarding toxin gene organization. Clearly, the increasing number of completed genome sequences for this genus will permit more precise definition of the genome organization used by pathogenic *Clostridium*. DNA transfer between strains, as well as DNA rearrangements during growth, account for the diversity of toxigenic strains and emergence of new toxigenic variant strains.

17.4.2.1 *Clostridium perfringens*

As stated before, *C. perfringens* is the most prolific toxin producer among known microorganisms via toxin genes that appear either on the chromosome or plasmids. The genome sequence of a *C. perfringens* type A strain reveals a low G+C content (28.6%) equally distributed throughout the chromosome without any region exhibiting a remarkably higher — or lower — G+C content. The genes for toxins and extracellular enzymes do not form pathogenicity islands or possess insertion, transposon, or phage-related sequences with one known exception being the enterotoxin gene (see below) [5].

The gene for alpha-toxin, which is the main virulence factor responsible for a life-threatening form of myonecrosis often associated with soiled wounds and commonly known as gangrene, is by classic definition produced by all *C. perfringens* strains and localized at the same site on a variable region of the chromosome near the origin of replication [8]. Sequencing of the alpha-toxin gene reveals that it is highly conserved, and various studies have revealed few variations at the nucleotide and amino acid level. However, alpha-toxin from avian isolates possesses a greater degree of sequence divergence [9,10], but the reason for this animal-linked variation is still unknown. The genes of other toxins, such as perfringolysin O (also known as θ–toxin or PFO), collagenase (κ-toxin), and extracellular enzymes (i.e., hyaluronidase and neuraminidase), are also localized within a variable region near the alpha-toxin gene locus.

In contrast, the other major toxin genes (*beta1, beta2, epsilon,* and *iota*) are localized on plasmids of varying sizes. These plasmids can be lost or transferred, thus accounting for the various *C. perfringens* toxinotypes as classically defined by the production of one or more toxins. Therefore, *C. perfringens* type A that produce only one major toxin (alpha) represent the basic toxinotype for this species, which upon acquisition of a plasmid encoding for another specific toxin (beta, epsilon, or iota) yields another distinct toxinotype (B, C, D, or E) [11]. Toxin genes on a plasmid are commonly flanked by insertion sequences and indeed, IS*1151* lies upstream from the epsilon toxin gene, and immediately upstream of IS*1151,* there is a transposase-like gene. Intriguingly, insertion sequences related to *Staphylococcus* or *Lactococcus* transposases are also located downstream of the epsilon toxin gene (Figure 17.2) [8]. Perhaps such sequences are a hint of gene sharing among a broad array of bacteria.

Iota toxin, like the other clostridial binary toxins, consists of two independent proteins as exemplified by the enzymatic (Ia) and binding (Ib) components of iota. The two proteins are encoded

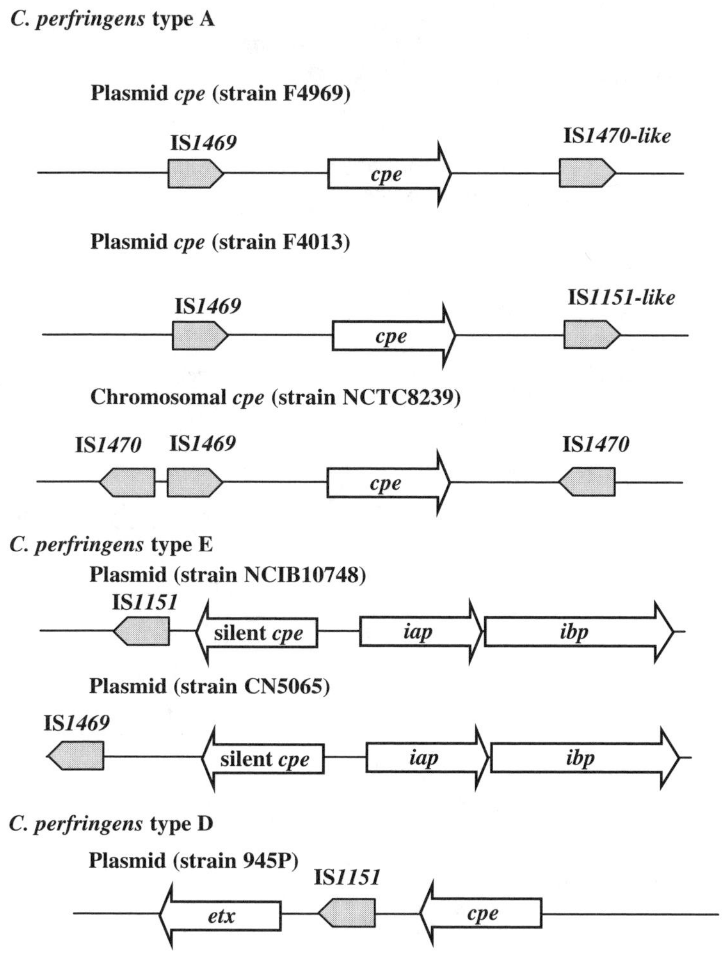

FIGURE 17.2 Genetic organization of the *Clostridium perfringens* enterotoxin (*cpe*) gene in various *C. perfringens* strains. *iap,* iota-toxin enzymatic component gene; *ibp*, iota-toxin binding component gene; *etx*, epsilon-toxin gene.

by distinct, yet adjacent, genes. The gene for Ia is located immediately upstream from the Ib gene, and both are regulated by the same promotor. The iota toxin genes of *C. perfringens* type E are localized on a large plasmid, whereas the highly similar binary toxin genes of *C. spiroforme* toxin (CST), *C. difficile* transferase (CDT), and the less related *C. botulinum* C2 toxin are on the chromosome [12].

Only 6% of *C. perfringens* isolates contain the *cpe* gene and produce the enterotoxin (CPE). *Cpe* is either located on the chromosome or a large 100 to 120 kb plasmid. Most of the enterotoxigenic *C. perfringens* type A strains involved in human food poisoning carry *cpe* on the chromosome, which is located on a 6.3 kb transposon-like structure (Tn*1565*) flanked by the insertion sequences IS*1470* and IS*1469* (Figure 17.2) [8]. In contrast, *C. perfringens* type A isolates from human cases of nonfoodborne disease, or those from animals, possess *cpe* on a plasmid, and this gene is also flanked by insertion sequences [13]. In *C. perfringens* type E, a silent *cpe* associated with IS*1151* and IS*1469* is located on the same plasmid, and in the same vicinity, as the iota toxin genes [14,15]. *Cpe* is also found on plasmids from other strains of *C. perfringens* types, such as D. The CPE amino acid sequence is relatively conserved among various strains, although the *cpe*

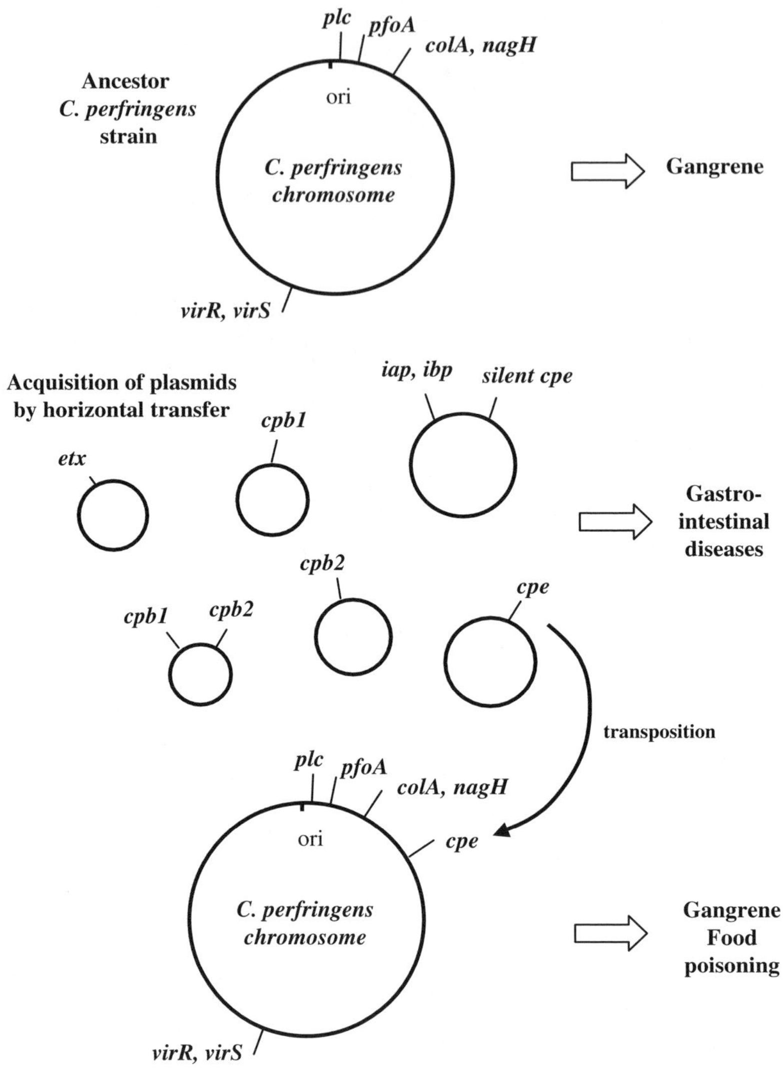

FIGURE 17.3 Chromosomal and plasmid localizations of toxin genes in *Clostridium perfringens*, as well as a hypothetical evolution of *C. perfringens* strains. *colA*, collagenase gene; *cpb1*, beta1 toxin gene; *cpb2*, beta-2 toxin gene; *cpe*, *C. perfringens* enterotoxin gene; *iap,* iota-toxin enzymatic component gene; *ibp*, iota-toxin binding component gene; *etx*, epsilon-toxin gene; *nagH*, neuraminidase gene; *plc*, alpha-toxin gene; *pfoA*, perfringolysin gene.

gene can be located on a chromosome or plasmid. This further suggests that *cpe* can be mobilized by transposition, and indeed, a circular transposon intermediate form has been discovered [16].

C. perfringens genes for toxins involved in gangrene are located on the chromosome, whereas the toxins responsible for gastrointestinal and foodborne diseases are plasmid encoded. It is evident that the *cpe* gene can be mobilized between a plasmid and chromosome. Strains with chromosomal *cpe* are mainly of human origin, and strains containing plasmidic *cpe* or other toxin genes (*beta1, beta2, epsilon,* and *iota*) are most often isolated from animals (Figure 17.3). Plasmids probably contain additional genes responsible for host specificity [11], which naturally raises the question: what are the selection pressures for maintaining plasmids and toxin genes in animal strains?

17.4.2.2 *Clostridium difficile*

C. difficile produces potent cytotoxins, toxin A (ToxA) and toxin B (ToxB), which are part of the large clostridial toxin group that also includes the *Clostridium sordellii* lethal (LT) and hemorrhagic (HT) toxins, as well as *Clostridium novyi* α-toxin. The *C. difficile* ToxA and ToxB genes are localized on the chromosome, whereas the α-toxin gene of *C. novyi* is harbored by a phage that, when cured, renders *C. novyi* nontoxigenic (see below) [17].

Genetic organization of the large clostridial toxin genes is probably conserved among the different *Clostridium* species, but to date this has been namely characterized in *C. difficile. ToxA* and *toxB* genes are clustered on a 19.6 kb DNA fragment, termed the pathogenicity locus (PaLoc). In nontoxigenic strains, the PaLoc is deleted and replaced by a short 115 bp sequence containing a direct repeat. The *C. difficile* chromosome exhibits only one single integration site for PaLoc that always occurs in the same orientation [18].

ToxA and toxB genes are similar in size (7 and 8 kb, respectively) and contain repetitive sequences within their 3′ ends. They are transcribed in the same orientation and separated by a small gene designated as tcdE (Figure 17.4). The toxin genes are flanked by two other small genes, tcdD or txeR in the 5′ part, and tcdC in the 3′ end, which is transcribed in an opposite direction [19]. TxeR, which is respectively homologous to botR and tetR in *C. botulinum* and *C. tetani* (see below), has been identified as an alternative sigma factor gene [20], and tcdC represents an inhibitory regulator of transcription that is perhaps an anti-sigma factor [21].

PaLoc possesses a great level of genetic variation, as toxigenic *C. difficile* strains produce either both toxins (A^+B^+), or only ToxB (A^-B^+). Deletions of varying lengths in the 3′ part of *toxA* gene account for the A^-B^+ strains. Several point mutations in *toxA* and *toxB* genes have been identified by restriction endonuclease analysis. Genetic variation extends also to the accessory genes and, in particular, nonsense mutations in the *tcdC* gene are accompanied by increased toxin production [21]. Therefore, *C. difficile* strains are classified into 17 types related to the previously characterized serogroups on the basis of bacterial surface antigens [22–28]. In addition, the insertion of a 1975 bp element with a combined feature of group I introns and insertion sequences containing two transposase genes has been evidenced in the *toxA* gene. This element has the ability to splice, but not impair, the translation of functional *toxA* [29].

17.4.2.3 *C. botulinum* and *C. tetani*

C. botulinum and *C. tetani* produce highly potent neurotoxins, known as the botulinum neurotoxins (BoNTs) and tetanus toxin (TeTx), respectively. The seven toxinotypes of BoNTs (A, B, C1, D, E, F, and G) are synthesized by various *Clostridium* species that form a rather heterogeneous class of bacteria:

- Group I: *C. botulinum* A, and proteolytic strains of *C. botulinum* B and F
- Group II: *C. botulinum* E, and glucidolytic strains of *C. botulinum* B and F
- Group III: *C. botulinum* C and D
- Group IV: *C. botulinum* G or *C. argentinense*
- Toxigenic *C. butyricum* strains
- Toxigenic *C. baratii* strains

BoNTs are physically associated with nontoxic proteins (ANTPs), which include the nontoxic, nonhemagglutinating (NTNH), and hemagglutinating components (HAs) that form large botulinum protein complexes of various sizes (230 to ~900 kDa), depending upon the toxinotypes and strains. The main role of ANTPs is supposedly to afford protection for the BoNTs from gastric acidity and digestive proteases, as well as to mediate BoNT absorption through the intestinal mucosa. In striking contrast, TeTx does not associate with other proteins to form a complex, and it does not readily translocate through the intestinal mucosa. Genetic organization of the BoNT and ANTP gene has been determined in representative strains of *C. botulinum* and TeTx gene in *C. tetani* [30].

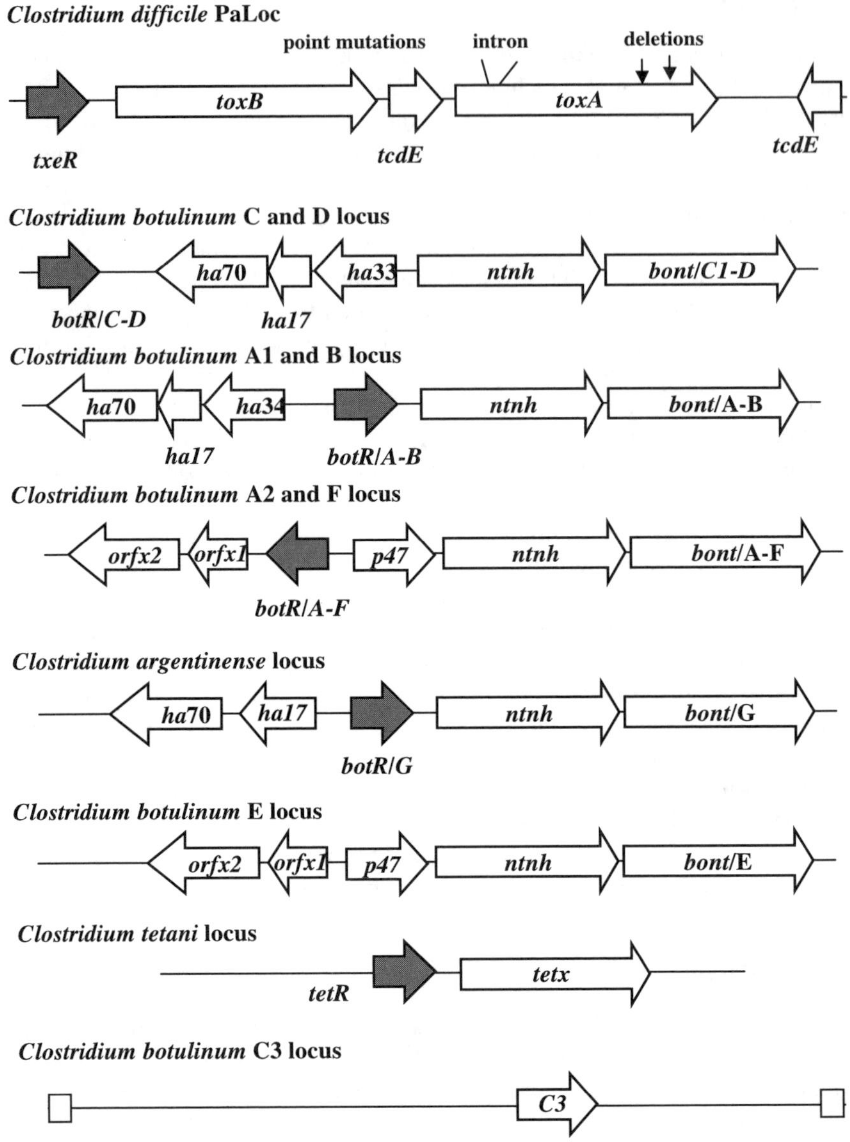

FIGURE 17.4 Pathogenicity islets in *Clostridium*: pathogenicity locus (PaLoc) in *C. difficile*, *C. botulinum*, *C. tetani*, and the C3 loci in neurotoxigenic *Clostridium*.

17.4.2.3.1 Genetic Organization

In all strains, the neurotoxin and ANTP genes are clustered in close vicinity and constitute a botulinum locus (Figure 17.4). This locus in *C. botulinum* A strain Hall is a 16 kb DNA fragment flanked by an insertion sequence and transposon gene [31,32]. In contrast to pathogenicity islands, the G+C content of the botulinum locus is similar to that found throughout the genome, and organization of the botulinum locus is conserved within the 3′ part, but differing slightly in the 5′ part among the different BoNT-producing toxinotypes. The botulinum neurotoxin genes (*bont*) are localized in the 3′ part of the locus and preceded by the genes for NTNH (*ntnh*). Both *ntnh* and *bont* genes are transcribed in the same orientation (Figure 17.4).

The hemagglutinin (HA) genes (*ha*33, *ha*17, and *ha*70) are upstream of the *ntnh-bont* genes and transcribed in an opposite orientation. The *ha* genes are missing in the nonhemagglutinating toxinotypes A2, E, and F. In *C. botulinum* G, the *ha* genes consist of only *ha*17 and *ha*70. In the toxinotypes A2, E, and F, a gene (*p*47) encoding a 47 kDa protein is immediately upstream of the *ntnh* gene, and both are transcribed in the same orientation. In addition, a gene (*orf*X1) encoding an 18 kDa protein not related to *ha*17, as well as the gene *orf*X2, both lie upstream of *p*47. A gene (*botR*) encoding for a 21 to 22 kDa protein that has regulatory features is localized in the 5′ part of the botulinum locus in *C. botulinum* C and D, whereas the same gene is positioned between the *ntnh-bont* and *ha* genes in *C. botulinum* types A1, B, and G (Figure 17.4).

Usually, one clostridial strain produces only one neurotoxin type, and the botulinum locus is evident as one copy on the genome, as suggested by Southern blotting of DNA fragments separated by pulse-field gel electrophoresis [33]. Rare strains synthesize two types of BoNT, as evidenced by isolates producing BoNT/A-BoNT/B, BoNT/A-BoNT/F, and BoNT/B-BoNT/F [33,34]. Investigations on an A-B strain show that it contains two *bont* genes related to those of *C. botulinum* A2 and proteolytic *C. botulinum* B, respectively [35].

It has been reported that some clostridial strains contain silent *bont* genes. Indeed, several *C. botulinum* A strains isolated from foodborne and infant botulism contain a silent *bont*B gene. Characterization of type A strain NCTC2916 shows that it has two loci, A and B, which are chromosomally localized but separated by ~40 kbp. The botulinum B locus consists of *bont*/B, *ntnh*, *botR/B*, *ha*33, and *ha*11 genes. The *bont*/A gene is identical to that of *C. botulinum* A1 strains, but organization of the botulinum A locus is similar to that of *C. botulinum* A2 and F strains. The *bont/B* nucleotide sequence found in NCTC2916 is related to that of other *C. botulinum* B strains (97% identity), but it has a stop mutation at position 128 and two base deletions (positions 2839 and 2944) that result in reading frameshifts and multiple stop codons [33]. Silent *bont*/B has also been evidenced in nontoxigenic *Clostridium subterminale* strains [36,37].

In *C. tetani*, one gene (*tet*R) equivalent to *bot*R was found upstream of the *tetx* gene (Figure 17.4), but no gene related to *antp* from *C. botulinum* has been identified in *C. tetani* [4].

17.4.2.3.2 Genomic Localization

In *C. tetani* and *C. argentinense*, the neurotoxin gene has been localized on a large plasmid (51 and 73 MDa, respectively). Plasmids of various sizes and bacteriophages have been found in *C. botulinum* A, B, E, and F, but toxigenicity has not been associated with these genetic elements. The neurotoxin genes of these toxin types have been cloned from chromosomal DNA, thus confirming a chromosomal localization (Table 17.2). In neurotoxigenic strains of *C. butyricum*, the *bont*/E gene has been found in a plasmid using polymerase chain reaction amplification, but further analysis by DNA/DNA hybridization more firmly suggests localization on the chromosome [30].

In *C. botulinum* C and D, it has been clearly evidenced that BoNT/C1 and BoNT/D are encoded by bacteriophages. *C. botulinum* C and D strains cured of their phages do not produce BoNT/C1 and BoNT/D, respectively, but they continue to produce the C2 enterotoxin. Such microorganisms could be converted into neurotoxigenic strains C or D by reinfection with phages obtained from toxigenic *C. botulinum* C or D strains. In addition, the BoNT/C1 and BoNT/D genes have been cloned and sequenced from purified phage DNA emanating from *C. botulinum* C-468 and *C. botulinum* D-1873 strains, respectively [30].

17.4.2.4 *C. botulinum* C3 Gene Locus

C. botulinum C and D produce an additional ADP-ribosyltransferase known as the C3 exoenzyme. The exoenzyme C3 gene is also harbored by the same bacteriophages carrying *bont*/C1 and *bont*/D genes, as further evidenced by genetic analysis of a mutant phage (CN) lacking a 21.5 kb fragment containing the C3 gene. This fragment was found in several bacteriophages for types C and D, and is delineated by an AAGGAG core motif. This motif is found in only one copy at the deletion junction in the CN-phage DNA. The DNA sequences from phages C and D, which flank the 21.5 kb fragment, are unrelated. However, the sequence on the left-end remains homologous throughout a 61 bp stretch upstream of the core motif, whereas the sequences diverge immediately downstream from the core motif on the right-end. The 21.5 kb fragment seems to be a mobile DNA element responsible for disseminating the C3 gene throughout *C. botulinum* types C and D, and has a similar feature to that of the site-specific transposon family of Tn1554 that includes; (i) asymmetric ends, (ii) absence of either inverted or terminal repeats, and (iii) presence of a 6 bp core motif at both insertion junctions and the insertion site [38].

17.5 TOXIN GENE MOBILITY AND EVOLUTION

Acquisition of novel genes by horizontal transfer (i.e., plasmid, phage), integration into the genome (i.e., transposition, recombination), and DNA rearrangements (i.e., deletion, insertion, duplication, mutation) represent the basis for genetic evolution in prokaryotes, and that certainly includes *Clostridium* [39,40]. However, the molecular mechanisms of gene transfer in *Clostridia* are still poorly understood. Toxin gene mobilization has been experimentally performed in only a few clostridial strains, and the intra-/inter-species transfer of toxin genes within *Clostridium,* and subsequent DNA rearrangements, are supported by genetic analysis and toxin sequence comparisons. For example, the similarity between BoNTs and TeTx, the fact that different *Clostridium* species can produce BoNT, and evidence that some strains contain several combinations of *bont* genes (A and B, A and silent B, A and F, B and F) [34,41], all strongly suggest that *bont* and *tetx* genes are derived from a common ancestor and have been transferred over time between various *Clostridium* strains. Nontoxigenic derivatives were evidenced in certain toxinotypes such as in *C. botulinum* B [42], thus indicating an instability of the DNA fragment harboring the *bont* genes.

17.5.1 PLASMIDS

Conjugation and mobilization of large plasmids in *Clostridium*, as that evidenced in *C. perfringens* have already been reported [43]. Since the main toxin genes of *C. perfringens* are distributed on several plasmids, the large diversity of *C. perfringens* toxinotypes results from plasmid exchange between strains. Therefore, it has been found that *cpe* [44] and etx genes (unpublished) are located on conjugative plasmids. Moreover, some strains like those producing β-toxin are unstable, and plasmid loss probably accounts for the emergence of nontoxic variants.

The *tetx* and *bont*/G genes are also localized on large plasmids in *C. tetani* and *C. argentinense,* respectively [45,46]. Their transfer from bacterium to bacterium could be achieved by mobilization of the corresponding plasmids. However, up to now, only nontoxigenic *C. argentinense* and *C. tetani* variants free of plasmids have been obtained.

17.5.2 BACTERIOPHAGES

In *C. botulinum* C and D, it has been clearly evidenced that bacteriophages mediate transfer of the *bont* genes. Bacteriophages that carry BoNT/C1 and BoNT/D genes can facilitate transfer to different host strains of *Clostridium.* The *C. botulinum* C or D strains free of bacteriophage are indistinguishable from *C. novyi* cured of bacteriophage. The nontoxigenic derivatives are indifferently transfected with *C. botulinum* C, D, or *C. novyi* bacteriophage yielding toxigenic strains. Thereby, within this group of bacteria, the bacteriophage type determines the resultant *Clostridium* species [47,49].

A pseudolysogenic relationship corresponding to the presence of bacteriophages free within the bacterial cytoplasm exists between these phages and hosts. Thus, variants free of bacteriophages can be obtained at high frequency via curing with acridine orange or ultraviolet light. Under laboratory culture conditions, a proportion of the bacteria that are dependent upon the strain and growth conditions (temperature, salinity, etc.) are lyzed and subsequently liberate free bacteriophage. Other bacteria lose their bacteriophages and can be reinfected with free bacteriophages. Such lysogeny and reinfection cycles likely occur throughout the environment (i.e., soil, intestinal tracts of birds and animals), thus probably accounting for the isolation of nontoxigenic or low toxin-producing variants [50].

In *C. botulinum* A and F, bacteriophage involvement has been suggested on the basis of a gene (*lyc*) identified in proximity to the *bont* genes. The *lyc*/A and *lyc*/F genes have been mapped approximately 1 kb downstream from the corresponding *bont* genes, and are partially related to bacteriophage genes encoding lytic enzymes in *Lactobacillus* and *Streptococcus pneumoniae* [34]. Since lytic enzymes are naturally involved in the bacteriophage life cycle, presence of *lyc* genes in the vicinity of *bont*/A gene in NCTC2916 could indicate that the botulinum neurotoxin locus is part of an integrated prophage.

17.5.3 Transposons and Recombination

Mobilizable transposons harboring antibiotic resistance genes have been characterized in various *Clostridium* species and include *C. perfringens* Tn*4451* and Tn*4452*, as well as the closely related Tn*4453a*, Tn*4453b*, and Tn*5398* from *C. difficile* [51]. Some clostridial toxin genes were evidently mobilized by transposition. This is seemingly the case for *C. perfringens cpe*, *C. difficile* PaLoc, and the *C. botulinum* C3 gene, which are localized on transposon-like elements (see above) that may have lost their mobility due to additional genetic modifications.

Transposable elements analogous to that encompassing the C3 gene could account for the different sites of the neurotoxin genes (i.e., chromosomal, plasmid, bacteriophage) and gene transfer between *Clostridium* strains. However, such elements have not yet been clearly identified but nucleotide sequence analysis of *C. botulinum* A suggests their possible existence. A 97 nucleotide stretch downstream of the stop codon of the *bont*/A gene is identical in *C. botulinum* A strains 62 and NCTC2916, whereas the following nucleotides are totally unrelated between both strains. The 97 nucleotide region could be part of a mobile DNA element encompassing the *bont*/A and *antp* genes. At least two sites of *bont*/A localization exist on the chromosome of *C. botulinum* A and reinforce the involvement of a mobile DNA element [30]. Similarly, the upstream region of *bont*/A and silent *bont*/B clusters in strain NCTC2916 share a marked homology indicating a similar evolutionary origin. In addition, insertion sequences with multiple internal mutations could be involved in gene transfer and have been identified downstream of the *bont*/A cluster [31].

Neurotoxigenic *C. butyricum* strains were probably derived from nontoxigenic *C. butyricum* strains by acquisition of the *bont*/E gene from *C. botulinum* E [52]. The *bont*/E gene and its flanking regions are absent in nontoxigenic *C. butyricum* strains, suggesting a possible gene mobilization by a mobile DNA element [53]. In the laboratory, this gene has been transferred from a neurotoxigenic *C. butyricum* strain to a nontoxigenic *C. botulinum* E strain by a protocol resembling transduction with a defective phage [54]. The precise method of molecular transfer has not been elucidated since DNA/DNA hybridization studies suggest the *bont*/E gene is localized on chromosomal, not phage, DNA in toxigenic *C. butyricum* [55].

In addition to DNA rearrangements that likely occur by homologous recombination, this also seems to have occurred between genes from the botulinum locus of different toxinotypes. This is supported by atypical strains that have genes with hybrid sequences from classical strains: mosaic *bont* gene from *C. botulinum* C and D [56], mosaic *ntnh* gene from *C. botulinum* A and C [57], or that from *C. botulinum* A and B [33].

17.6 TOXIN GENE REGULATION

In most bacteria, the transcription of toxin genes is a tightly regulated process in response to environmental factors. It is thought that most toxins are synthesized and transported across the bacterial membrane only when they are needed by the bacterium. Invasive bacterial genera, for example *Salmonella*, *Shigella*, and *Yersinia*, produce virulence factors only when they adhere to the target cell. Other toxigenic bacteria sense environmental factors and regulate the toxin production perhaps via quorum sensing, thus better enabling the bacteria to survive a hostile environment. Toxins could afford a benefit for the bacterium regarding escape of host defences and gaining access to essential nutrients. Some commonly identified environmental factors that trigger toxin production include temperature, osmolarity, nutrient availability, and iron concentration.

The production of toxins such as diphtheria toxin (DT), shiga toxin (ST), *Pseudomonas* exotoxin A (Exo A), *E. coli* hemolysin, and *H. pylori* VacA are all regulated by iron, which is also essential for bacterial growth. In a low iron concentration environment, these bacteria synthesize protein toxins that damage host tissue and cells, which subsequently liberates additional iron for bacterial growth. In contrast, an excess of iron can be toxic, and bacteria will then repress their toxin gene expression. Various iron-based regulation systems have been developed by bacteria, in which an iron-binding repressor protein plays an essential role [58].

The adaptive response of bacteria to environmental conditions involves a coordinated activation or repression of the transcriptional process for a specific gene. Many transcriptional regulatory proteins have been identified, including the two-component systems, quorum sensing factors, and molecules from LysR and AraC/XylS family. In several Gram-negative bacteria, the AraC proteins control, at the transcriptional level, stress responses, carbohydrate catabolism, and production of virulence determinants such as *Pseudomonas* exoenzyme S (ExoS), urease, and adhesion factors [59].

The primary control of gene transcription requires the RNA polymerase, which contains a dissociable element designated as sigma factor, that ensures promoter recognition. Primary sigma factors such as *E. coli* σ^{70} are required for rapid growth of bacteria. Alternative or secondary sigma factors that replace the primary sigma factor and redirect RNA polymerase to other gene promoters control growth transitions, such as passage from exponential to stationary growth phase, stress response, and morphological differentiation, which includes flagellar biosynthesis and sporulation. Although *Clostridium* sporulate, most of the toxins are produced during the exponential growth phase. However, there are sporulation-dependent toxins among the *Clostridia*. Therefore, transcription of these latter toxin genes are controlled by sigma factors involved in sporulation, which also play a role with certain *Bacillus* toxins. A recently identified group of alternative sigma factors (TxeR family, see below) control the production of some toxins in *C. difficile*, *C. perfringens*, *C. botulinum,* and *C. tetani* [60].

A key regulatory system in bacteria is based on the two-component system for signal transduction. This regulatory pathway includes two proteins, one being a membrane-associated histidine kinase (sensor), which, upon an external signal, autophosphorylates a conserved histidine and subsequently transfers the phosphate group to an aspartate residue of the second protein (regulator). The regulator stimulates or represses transcription of target genes. An advantage of two-component systems is amplification of the regulation signal, since a single sensor molecule can phosphorylate multiple regulator proteins. Two-component systems are adapted to coordinately control multiple genes localized through the bacterial genome (regulon). They control toxin synthesis in various bacteria, such as *B. pertussis* (BvgA/BvgS), *S. aureus* (AgrC/AgrA), and *Streptococcus pyogenes* (CovR/CovS) [61]. A related, but slightly different, system (ToxR, S, T) regulates the cholera toxin (CT) gene in *V. cholerae* [62]. Two-component systems play an important role in *Clostridium*.

17.6.1 Clostridial Toxin Genes under the Regulation of Alternative Sigma Factors

17.6.1.1 Clostridial Toxin Genes under the Control of Alternative Sigma Factors Involved in Sporulation

The CPE and *C. botulinum* C2 toxin are only produced during sporulation, and the molecular mechanism for regulating these toxins has been mainly investigated in *C. perfringens*. CPE is produced and accumulates in sporulating *C. perfringens*, thus forming large crystals visible by electron microscopy. The toxin is released into the external medium, such as the intestinal content, after sporangium lysis. CPE, like the C2 toxin, does not contain a signal peptide sequence and is not secreted from vegetative cells. The linkage between CPE production and sporulation occurs at the transcriptional level for both the chromosome or plasmid located *cpe* gene. The *cpe* promoter sequence is identical for both the chromosome or plasmid-borne gene. The *cpe* gene is transcribed as a monocistronic mRNA with a characteristic rho-independent transcriptional terminator [63]. Three transcription start sites (P1, P2, and P3) have been identified upstream of the coding region. The DNA sequence upstream of P1 is similar to consensus SigK-dependent promoters, whereas P2 and P3 are similar to consensus SigE-dependent promoters. SigK and SigE are also alternative sigma factors involved in sporulation of *Bacillus subtilis* [64]. In addition, Hpr-(hyperprotease producing factor) binding sequences have also been identified upstream and downstream of the *cpe* gene. Hpr negatively regulates expression of many genes during the exponential growth phase of *B. subtilis*, and an Hpr-like factor could repress the *cpe* gene during vegetative growth; however, an Hpr molecule in *C. perfringens* has not yet been identified [65].

17.6.1.2 Clostridial Toxin Genes under the Regulation of Alternative Sigma Factors of the TxeR Family

17.6.1.2.1 C. difficile

In *C. difficile*, the small *txeR* gene in the 5′ extremity of PaLoc encodes a 22 kDa basic protein with a DNA binding motif (helix-turn-helix). By using a reporter gene with the *toxA* and *toxB* gene promoters in *E. coli*, it was found that *txeR* is a positive regulator of *C. difficile* toxin gene expression [66]. *ToxA* and *toxB* genes are weakly expressed during the exponential growth phase of *C. difficile*, and mRNAs of the corresponding genes accumulate as the bacterium enters stationary phase. The transcription of *toxA* and *toxB* genes is blocked by glucose in the culture medium [67]. Since the sequences of the *toxA* and *toxB* gene promoters do not retain the consensus nucleotides of the canonical σ^{70} promoters of prokaryotes, it was suggested that these genes are transcribed by an uncommon RNA polymerase. The RNA polymerase holoenzyme from *C. difficile* does not bind to the *toxA* and *toxB* gene promoters, but it associates with the promoter of another gene (gdh) expressed during exponential growth. Purified TxeR alone also does not bind to *toxA* and *toxB* gene promoters, but the combination of TxeR with the core RNA polymerase from *C. difficile* or *E. coli* leads to functional RNA polymerase that associates with the toxin gene promoters. TxeR interacts with the RNA polymerase core enzyme in the absence of DNA, and the association of TxeR-RNA polymerase core enzyme is required for transcribing the toxin genes. Therefore, TxeR acts as an alternative sigma factor [20]. In recombinant *C. perfringens* with a reporter gene, TxeR positively regulates *toxA* and *toxB* gene promoters at the end of exponential growth and is inhibited by glucose. This correlates with the high production of *toxA* and *toxB* at the beginning of the stationary phase and depletion of nutrients like glucose. In addition, TxeR stimulates the transcription of its own gene [68], thus the model suggests that TxeR is repressed during exponential growth, but its expression occurs rapidly upon entry into stationary phase as a consequence of autoregulation. Subsequently, *toxA* and *toxB* genes are transcribed, and *tcdC* has been proposed to be the inhibitor or anti-sigma factor of TxeR [21].

17.6.1.2.2 C. botulinum *and* C. tetani

It was first reported in *C. botulinum* type C, that *botR*/C encodes a 22 kDa protein possessing the features of a DNA binding protein: basic pI (10.4) and a helix-turn-helix motif [69]. The *botR*/C is localized upstream of the *ha* genes (Figure 17.4) in *C. botulinum* A, but the homologous gene, *botR*/A, has a different localization and is inserted between the *ha* and *ntnh-bont* genes. This gene is also conserved in proteolytic and nonproteolytic *C. botulinum* B, D, F, and G, as well as *C. tetani* [70,71], but it has not yet been detected in *C. botulinum* E. In *C. tetani*, the gene *tet*R is localized immediately upstream of the *tetx* gene and is the only one from the botulinum locus.

BotRs from the different toxinotypes and TetR share 51 to 97% identity and are also related to other known regulatory proteins, which include: UviA (25 to 28% identity), which regulates the bacteriocin production in *C. perfringens*; Msmr protein (21 to 26%), which regulates sugar transport in *Streptococcus mutans;* and TxeR (20 to 24%). Interestingly, *txe*R and *uviA* have a similar localization as *botR* and *tetR*, since they lie upstream of the toxin and bacteriocin gene loci, respectively [72].

The function of *bot*R/A was analyzed *in vivo* by overexpressing this gene in *C. botulinum* A. *Bot*R/A was cloned in a high copy number shuttle vector that was then transferred into *C. botulinum* A by electroporation. A significant increase in BoNT/A and ANTPs was evident, with a corresponding increase in mRNA levels detected in the recombinant strain. In contrast, partial inhibition of the *botR*/A expression by antisense mRNA resulted in reduced production of BoNT/A and ANTPs [73]. Thus, it was concluded that BotR/A positively regulates the transcription of *bont* and *antp* genes. In the same way, TetR is a positive regulator of *tetx* gene expression, while BotR/A and, to a lesser extent, BotR/C are also functional in *C. tetani* [74]. This indicates that regulation of the *bont* and *tetx* genes is conserved among *C. botulinum* and *C. tetani*, and this constitutes further evidence that *bont* and *tetx* gene loci are derived from a common ancestor.

Like the scenario for TxeR, it has been found that BotR/A and TetR require the RNA polymerase core enzyme to interact with the target promoters. In *C. tetani*, TetR seems specific for the *tetx* gene, whereas BotR/A directs transcription of the two divergent operons, *ntnh-bont/A* and *ha* (Figure 17.4) (manuscript in preparation). On the basis of these results, it was concluded that BotR/A and TetR are alternative sigma factors. Sequence homologies at the amino acid level and similar functional properties suggest that BotR and TetR belong to the same family of alternative sigma factors already described in *C. difficile* (TxeR) and *C. perfringens* (UviA) [20,68] (Dupuy et al., in preparation). Moreover, they are functionally interchangeable (Dupuy et al., in preparation). Because they are unrelated to other sigma factors, these alternative sigma factors have been assigned to a new group of the extracytoplasmic function sigma 70 family (group 5 or TxeR subfamily) [60]. Interestingly, the large plasmid containing the *tetx* gene also codes for four other alternative sigma factor genes related to TetR, the function of which is still unknown [4].

17.6.2 Toxin Genes Regulated by Two-Component Systems

Many putative two-component system genes have been identified in the genome sequences of *C. perfringens* (28 sensor and 20 regulator genes), *C. botulinum* (34 two-component system genes), *C. tetani* (one two-component system on the plasmid harboring *tetx* gene, and 29 on the chromosome), and *C. difficile* (at least 20) [4,5] and sequences available on the Sanger Center.

In *Clostridia*, the best-characterized two-component system is the VirR/VirS system, which regulates various toxin genes in *C. perfringens* that code for alpha-toxin, theta-toxin, kappa-toxin, sialidase, protease, hemagglutinin, and various enzymes [75,76]. Proteome and transcriptome analysis indicate that seven other genes are regulated by VirR/VirS at the transcriptional level, and eight additional proteins are modified by a VirR/VirS-dependent protease [77]. Indeed, it was discovered that *C. perfringens* iota toxin is activated by a trypsin-like protease regulated by VirR/VirS in some *C. perfringens* strains [78]. VirR/VirS also controls differentially the expression of various genes located on the plasmid. The collagen adhesin gene is negatively regulated, whereas the *beta2* toxin gene is positively regulated by VirR/VirS. Both genes are harbored by a plasmid in strain CP13 [79]. VirR acts either by binding directly to the promoter of target genes, such as that for theta-toxin, or by promoting the transcription of a VR-RNA (VirR-regulated RNA) that positively or negatively controls gene expression [80].

Among 20 component systems tested in the Hall strain of *C. botulinum* by an antisense RNA method, two are directly or indirectly involved in the BoNT/A production (S. Raffestin, unpublished data). This indicates that factors other than BotR control the BoNT/A synthesis, as reported previously [81].

17.6.3 Environmental Factors That Control Expression of Toxin Genes

Little is known concerning the environmental and nutritional factors that regulate toxin production in *Clostridia*. Amino acids, biotin, carbon sources, and temperature can all regulate production of ToxA and ToxB in *C. difficile* [67,82–84]. By comparison, bicarbonate controls toxin synthesis in other sporulating, Gram-positive bacteria such as *Bacillus anthracis* [85], while in other Gram-positive or -negative bacilli, the synthesis of various toxins such as DT, ST, and ExoA are transcriptionally regulated by iron [86]. However, the environmental factors that govern toxin synthesis in *C. botulinum*, *C. perfringens,* and *C. tetani* are largely unknown. Nutrient availability, in particular carbon and nitrogen sources, is important for the toxinogenesis in *C. botulinum* and *C. tetani* [66,73,81]. High levels of arginine in a minimal medium represses the BoNT/A and protease production, whereas casein supplementation increases both toxin and protease synthesis [87]. Specific peptides from a casein pancreatic digest are essential for the toxinogenesis in *C. tetani* [88]. Indeed, hydrophobic peptides, consisting of four to eight amino acids in length and a Pro-aromatic-Pro motif, enhance TeTx production. It is possible that these peptides act as trypsin inhibitors that prevent toxin degradation, or perhaps they directly stimulate toxin synthesis via an unknown mechanism [88]. Overall, it is clear that the key extracellular factors controlling production of the most potent clostridial toxins largely remain a mystery.

17.7 TOXIN SECRETION AND PROCESSING

Protein toxins are virulence determinants either secreted into the external medium or directly injected into targeted cells via a protein complex that molecularly mimics a needle and syringe. Therefore, protein toxins must be exported from the host bacterium to have an effect on the eukaryotic host. Gram-negative bacteria exhibit a complex membrane structure with an inner cytoplasmic membrane, periplasmic space, and outer lipid leaflet. In contrast, Gram-positive bacteria, such as the *Clostridia*, are simpler and contain only a cytoplasmic membrane surrounded by the peptidoglycan wall. Thereby, the proteins that cross the cytoplasmic membrane of Gram-positive bacteria are directly released into the external medium.

17.7.1 The General Secretion Pathway

Translocation across the cytoplasmic membrane, known as the general secretory pathway, is conserved in both Gram-positive and Gram-negative bacteria. It consists of a membrane ATPase, SecA, and at least three transmembrane proteins (SecE, SecG, and SecY) that form an aqueous channel. Cytosolic chaperones, SecB in *E. coli* or CsaA in *B. subtilis*, prevent the tight folding and aggregation of the precursor proteins. The driving force of translocation is provided by ATP hydrolysis and proton motive gradient across the membrane. Proteins secreted by this mechanism contain an N-terminal leader or signal sequence consisting of 16 to 26 amino acids in Gram-negative bacteria, and 40 to 45 residues in Gram-positive bacteria. Signal peptides consist of a basic N-terminal domain, a hydrophobic central core, and a distal hydrophylic domain containing a proteolytic cleavage site. They ensure targeting to SecA and subsequent translocation. After protein transport across the membrane, the signal peptide is removed by a serine peptidase that generally cuts at an Ala-X site. Secreted proteins then acquire their correct conformation with the help of extracellular chaperones and metal ions, such as Ca^{++} in the cell wall of Gram-positive bacteria [89,90].

17.7.2 Toxin Secretion in Gram-Negative Bacteria

In Gram-negative bacteria, the export of physiological macromolecules across the complex membrane is accomplished by several types of secretion mechanisms (types I–V). Toxins use these secretion methods for release into the external medium or directly into a targeted cell.

The type I export mechanism requires three membrane proteins, including an ATPase, adaptor protein, and TolC that assemble into a trimeric structure which forms a pore through the outer membrane of Gram-negative bacteria. Type I exported proteins possess an uncleaved C-terminal signal sequence that does not retain a conserved amino acid sequence; however, they all possess a common structure. The best-characterized proteins secreted by this method include the *E. coli* α-hemolysin, as well as hemolysins, leukotoxins, and hydrolytic enzymes (i.e., Ipases, metalloproteases) from a myriad of Gram-negative bacteria.

The type II secretion system is widely used by various bacteria to export toxins (CT, ExoA, aerolysin) and hydrolytic enzymes. Proteins secreted by this route contain an N-terminal sequence and use the general secretion pathway to cross the inner membrane. After passage through the inner membrane, signal peptide is cleaved from the protein and released into the periplasmic space where the protein subsequently folds and, if it is part of a multimeric toxin like CT, properly assembles into a biologically active quarternary state. The toxin is then secreted via a second step across the outer membrane by means of a secretion apparatus encoded by the *eps* genes [91].

The type III secretion machinery consists of at least 14 proteins that assemble to form a pore through the inner and outer bacterial membrane, as well as through the eukaryotic membrane (translocon) coupled to an ATPase in the bacterial cytosol. Contact of the bacterium with a targeted eukaryotic cell represents the main initiation signal that controls synthesis of type III secretion proteins and injection of the effector proteins into target cells [92,93]. Although type III secretion is not used in Gram-positive bacteria, an equivalent was identified in *S. pyogenes,* as evidenced by NAD-glycohydrolase transport into targeted cells through the pore formed by the secreted cholesterol-dependent cytotoxin streptolysin O [94].

The type IV system, which probably evolves from conjugation pili, is used to transport DNA or proteins (i.e., Pertussis toxin) either directly into targeted cells, like that evidenced for the type III system, or into the extracellular medium. Architecture of the type IV system is reminiscent of a pilus containing at least 10 different proteins that possibly form a central pore connecting the bacterial cytoplasm with the extracellular medium or cytosol of a targeted cell [95].

In the type V secretion systems, proteins such as the vacuolating toxin from *H. pylori* transport themselves. They possess an N-terminal signal peptide that mediates export of the protein across the inner membrane through the general secretion apparatus, like that found in the type II system. The C-terminal part forms a β-barrel that inserts and generates a pore in the outer membrane, thus permitting passage of the molecule. The C-terminus is then cleaved, and the mature protein is released into the extracellular medium.

17.7.3 Toxin Secretion in *Clostridium*

Pathogenic Clostridia have not been shown to specifically bind to healthy eukaryotic cells and do not directly inject virulence determinants into cells. Clostridium species secrete large amounts of toxins into the external environment, which diffuse and act at a distance from the bacteria. Furthermore, genomic sequencing of toxigenic Clostridia such as *C. botulinum*, *C. difficile, C. perfringens,* and *C. tetani* does not reveal any type III secretion genes. Numerous clostridial toxins use the general secretion pathway, as genetic analysis reveals a characteristic N-terminal signal sequence in many of these proteins (Table 17.3). This is further supported by biochemical studies showing that the signal peptide is cleaved from the secreted toxins. However, several clostridial toxins do not contain a signal peptide, and their mode of secretion still remains mysterious. For example, cell lysis that occurs after sporulation seems to be the main means for mediating release of CPE from the bacterium [63]. However, in *C. difficile*, *C. botulinum,* and *C. tetani*, the toxins accumulate in the external medium during the end of exponential growth, without significant bacterial lysis. An exfoliation of the cell wall has been proposed for the export of BoNTs [96]. In *C. difficile*, toxins could be transported by a novel secretion system that includes a 47 kDa extracellular protein weakly related to TolC and a putative holin (TcdE) [97,98]. Interestingly, the

TABLE 17.3
Signal Peptides in Clostridial Toxins

Species	Toxins Containing a Signal Peptide	Signal Peptide Length	Toxins Devoid of Signal Peptide
C. perfringens	alpha-toxin	28	enterotoxin
	perfringolysin	27	
	beta1	27	
	beta2	30	
	epsilon	32	
	iota toxin Ia	41	
	Ib	39	
	ColA	39	
C. spiroforme	*C. spiroforme* toxin CSTa	44	
	CSTb	44	
C. difficile	*C. difficile* transferase CDTa	43	ToxA, ToxB
	CDTb	42	
C. sordellii			LT
C. novyi			α-toxin
C. septicum	α-toxin	32	
C. histolyticum	ColG	39	
	ColH	40	
C. botulinum	C3	40	C2 toxin C2-I, C2-II BoNTs
C. limosum	C3	45	
C. tetani			TeTx

C. tetani genome possesses a collection of conserved genes from the Sec system, as well as 10 genes related to types II and IV secretion systems [4]. Note that in contrast to the multicomponent toxins from Gram-negative bacteria like CT, which are assembled in a heterohexameric structure (A-B5) in the periplasmic space, the binary toxins from *Bacillus* and *Clostridium* are exported through the unique cell wall as independent proteins into the external medium. The binary toxin components subsequently associate when the corresponding binding component interacts with cell surface receptor.

17.7.4 Toxin Processing

Certain clostridial toxins are secreted as inactive forms (prototoxin) that are activated in the extracellular medium. Activation generally consists of a proteolytic release of an N-, C-, or both N- and C-terminal peptides (Table 17.4). Indeed, the binding components of clostridial binary toxins, as well as the lethal and edema toxins of *B. anthracis*, are activated by cleavage of a 20 kDa peptide from the N-terminus. The anthrax binding component (PA) is related at the amino acid sequence level to those of the clostridial binary toxins. Although the inactive forms of these binding components can recognize their specific cellular receptor, they do not undergo the oligomerization process that is a prerequisite step in toxin activity. For these binary toxins, production of inactive binding components that are subsequently activated by removing an N-terminal domain probably controls the correct assembly of monomers at the cell surface into "biologically active" heptamers. Clostridial binary toxins are activated by trypsin, α-chymotrypsin, or proteases produced by the bacterium or host intestinal tract prior to binding of target cells. In contrast to the clostridial binary toxins, the PA molecule possesses a furin cleavage site that is activated either in solution by serine-type proteases, or uniquely activated after binding to its cellular receptor by furin, a ubiquitous cell-surface protease [12,78,85].

TABLE 17.4
Proteolytic Processing of Clostridial Toxins

Species	Removing of an N-Terminal Peptide (Amino Acids)	Removing of an N-Terminal or C-Terminal Peptide	Proteolytic Cleavage and Formation of a Disulfide Bridge
C. perfringens	enterotoxin (45)	epsilon toxin	
	iota-toxin Ia (9–11)	(11–13 N-ter, and	
	Ib (173)	29 C-ter)	
	ColA (86)		
C. difficile	CDTb (167)		
C. spiroforme	CSTb (171)		
C. botulinum	C2-II (180)		BoNTs (L and H chains)
C. tetani			TeTx (L and H chains)
C. septicum		α-toxin (45 C-ter)	
C. histolyticum	ColG (110)		
	ColH (40)		

The pore-forming toxins such as *C. perfringens* epsilon toxin, CPE, *C. septicum* α-toxin, and *C. histolyticum* collagenases have all evolved an activation step that involves peptide removal from the N-terminus, C-terminus, or both (Table 17.4). Active clostridial neurotoxins result from a proteolytic cleavage between the L and H chains, as well as reduction of an interchain disulfide bridge. Toxin activation by proteolytic nicking and reduction of an intermolecular disulfide bridge is common in bacterial toxins active intracellularly (DT, CT, *E. coli* thermolabile and thermostable enterotoxins). Proteolytic nicking is localized between the enzymatic domain responsible for intracellular activity and the translocation domain required for internalization into cells. Separation of the enzymatic and translocation domains probably permits a partial unfolding of the enzymatic domain necessary for its passage across the endocytic vesicle membrane, whereas the translocation domain requires a conserved structure necessary for insertion into the membrane. The disulfide bridge essentially traps the enzymatic domain into the membrane via the translocation domain. Once inside the cytosol, the disulfide bridge is reduced and subsequently liberates a refolded, biologically active enzymatic domain [99].

Activating proteases are produced by the same toxigenic *Clostridium* strain, other resident bacteria, or the host, as evidenced by intestinal proteases like the cell-surface associated protease furin. Most of the clostridial toxins involved in gastrointestinal diseases are activated by intestinal proteases such as trypsin or α-chymotrypsin, as is the case for *C. perfringens* epsilon, iota, CPE, *C. septicum* α-toxin, *C. botulinum* C2 toxin, and BoNTs (Table 17.4). Other enterotoxins such as *C. difficile* ToxA, ToxB, as well as *C. sordellii* LT, do not require proteolytic activation, but they also are resistant to proteolytic degradation. However, in ironic contrast, the beta1 and beta2 toxins of *C. perfringens* are also enterotoxins, but they are rapidly degraded by intestinal proteases [100]. The clostridial toxins responsible for myonecrotic lesions are fully active without proteolysis and in some cases are actually destroyed by proteases (*C. perfringens* alpha-toxin), but exceptions include the *C. histolyticum* collagenases and *C. septicum* α-toxin.

Other types of post-translational modifications can occur during the processing of bacterial protein toxins. For example, acylation is used during the activation of some toxins such as hemolysins from the RTX family produced by Gram-negative bacteria, as well as the adenylcyclase from *B. pertussis*. Addition of one or two fatty acyl chains enhances hydrophobicity of a protein and subsequent binding to the target membrane. Furthermore, acyl chains are required for pore formation by RTX toxins, but acylation has not been found for any clostridial toxins [101].

17.8 STRUCTURE FUNCTION

Like other bacterial toxins, those from *Clostridia* exhibit either a pore-forming or enzymatic activity. Some enzymatic properties that clostridial and non-clostridial toxins share include ADP-ribosylation, proteolysis, and phospholipase activities. Glucosylation displayed by the large clostridial toxins is unique, but in contrast, the *Clostridia* do not produce toxins with glucosidase (ST), DNase (cytolethal distending toxins), or deamidase [*E. coli* cytotoxic necrotizing factor (CNF), *Bordetella* dermonecrotic toxins] activities. Certain toxins, like those employing type III secretion routes (i.e., ExoS) from Gram-negative bacteria, exhibit more than one enzymatic property, but clostridial toxins possess only a single enzymatic activity with one notable exception being *C. perfringens* alpha-toxin, which possesses phospholipase C and sphingomyelinase properties. Clostridial toxins with elucidated crystal structures have overall structures or, at least, active sites similar to those evident in other bacterial toxins or even eukaryotic enzyme counterparts. Many toxins mimic eukaryotic effectors by retaining a conserved region of interaction with their target, although they often have an overall different folding pattern [102].

17.8.1 Pore-Forming Toxins and Binding Components of Intracellularly Active Toxins

Almost one third of the clostridial toxins forms pores in cell membranes. Perhaps most representative of any of the clostridial pore-forming toxins are the cholesterol binding cytotoxins (CBC), the prototype of which is perfringolysin (also known as PFO or *C. perfringens* theta-toxin), which is related to streptolysin O, listeriolysin, etc. [103]. Perfringolysin has an unusual elongated rod shape rich in β-sheet structure, and it is hydrophilic without any significant hydrophobic patches evident on the toxin surface. Four domains can be distinguished in the PFO molecule. Domain 1 has a seven-stranded antiparallel β-sheet and is connected to domain 4 by the elongated domain 2. Domain 3 consists of β-sheets and α-helices. The C-terminal part (domain 4) folds into a separate and compact β-sandwich domain [104]. The most striking structural feature of PFO is that domain 3 within the water soluble monomers contains four β-strands that terminate into two bundles of three α-helices. During the oligomerization process and pore formation, the α-helices unfold and prolong the β-strands into two anti-parallel amphipathic β-strands. Two β-hairpins from each monomer assemble into a β-barrel [105], and 40 to 50 monomers associate to form large pores of 300 to 450 Å diameter [104–107].

A second group of pore-forming toxins consists of *Aeromonas hydrophila* aerolysin, *S. aureus* α-hemolysin as well as the binary staphylococcal leukocidins, such as LukF. The crystal structure of *Staphylococcus* α-hemolysin and LukF has been solved. Each monomer consists of a very hydrophilic sequence essentially arranged in β-sheets. Heptamerization leads to a mushroom-shaped homooligomer consisting of a cap, rim, and stem domains. The N-terminus detaches from the core monomer, unmasking a small hydrophobic surface and facilitating assembly with the corresponding domains of the neighboring monomers to form the cap. In contrast to PFO, only one antiparallel β-hairpin of each monomer unfolds and contributes to the stem formation, which consists of a 14-stranded-β-barrel. The resulting pores possess a small diameter ranging from 15 to 45 Å.

Like *Aeromonas* aerolysin, *C. septicum* α-toxin is secreted as a nontoxic molecule that undergoes proteolytic cleavage of the 45 C-terminal amino acids. *C. septicum* α-toxin shares 27% amino acid identity with *Staphylococcus* α-hemolysin, as well as aerolysin, and it is likely that these toxins probably share a similar structure and pore assembly mechanism [108,109]. The other clostridial pore-forming toxins probably belong to the same toxin family as the α-hemolysin and aerolysin. Indeed, *C. perfringens* epsilon toxin contains three domains that are mainly β-sheets with an overall conformation significantly related to that of aerolysin, although these toxins do not share significant sequence identity at the amino acid level [110]. *C. perfringens* beta1 toxin is also significantly related to *Staphylococcus* α-, and γ-hemolysins, as well as other binary hemolysins at the amino acid sequence level, and is thus likely to possess a related structure and mode of action [111]. In contrast to the *S. aureus* toxins, beta1-toxin is cytotoxic, but it is neither cytolytic nor hemolytic [111]. Beta1 toxin probably acts by forming multimeric pores in membranes, perhaps as reported with human umbilical vein endothelial cells [112]. Additionally, cation selective channels are also formed by beta1 toxin in artificial phospholipid bilayers [113]. *C. perfringens* beta2 toxin does not show any sequence homology with any other known protein. However, beta2 and beta1 toxins induce similar biological effects *in vitro* and *in vivo*, perhaps the result of a common conformation and mechanism of action between these two toxins, despite diverging amino acid sequences [100]. CPE also seems to have a conformation related to that of aerolysin (Colle, personal communication).

Interestingly, the binding components of clostridial binary toxins (C2-II, CDTb, CSTb, and Ib), *Bacillus* anthrax toxin (PA), as well as *Bacillus cereus* and *Bacillus thuringiensis* vegetative insecticidal protein (VIP-I) also form pores permitting the passage of enzymatic components into the cytosol of a targeted cell. Each binding component specifically mediates internalization of the corresponding enzymatic component. Thereby, PA from *B. anthracis* edema and lethal toxins internalizes only the corresponding enzymatic components edema factor (EF) and lethal factor (LF), while C2-II only internalizes the C2 enzymatic component (C2-I). However, binding components (CDTb, CSTb, and Ib) of the iota toxin family indifferently mediate translocation of the respective enzymatic components CDTa, CSTa, and Ia. At the amino acid sequence level, the binding components C2-II, CDTb, CSTb, Ib, PA, and VIP-I are significantly related (31 to 43% identity). The crystal structure of PA has been solved, and this basic conformation is likely conserved among the other binding components. Like PFO, PA is organized into four domains essentially composed of hydrophilic β-strands. The N-terminal domain contains a binding site for the enzymatic components (EF and LF), but EF and LF only bind to PA heptamers. The C-terminal domain (domain 4) is involved in recognizing the cell surface receptor. Domain 3, which is the smallest, likely plays a role in oligomerization [113a]. Domain 2 contains long β-strands and, in particular, an amphipathic flexible loop (amino acids 302–325) that forms a β-hairpin which plays a central role in pore formation [114–116]. The C-terminus of Ib and C2-II, corresponding to domain 4 of PA, has also been identified as the receptor binding domain [117,118]. The low amino-acid identity (less than 10%) between domain 4 suggests that each toxin recognizes distinct receptors [119]. However, the mechanisms for oligomerization and insertion into the membrane, as well as the shape of PA heptamers, are more similar to those of *Staphylococcus* α-hemolysin, LukF, and aerolysin than to those of *C. perfringens* PFO [114,120,121]. Indeed, the flexible loop of PA domain 2 partially detaches from the monomer and associates with the six other loops of the neighboring monomers to form a β-barrel that inserts into the membrane like *Staphylococcus* α-hemolysin and leukocidin. Interestingly, leukocidins are also binary toxins involved in pore formation but remain on the cell surface, whereas PA enables the translocation of partially unfolded enzymatic components into the cytosol [115,116,122,123].

This last point strongly suggests that the pore-forming toxins have evolved from a common ancestor. Two main functional classes of proteins can be considered; the hemolysins and leukocidins, as well as the binding components of binary toxins. According to their structure and mode of insertion into the membrane, these proteins can be divided into several groups (Figure 17.5). The binding components have retained a core structure also found in the β-sheet pore-forming toxins, and they have acquired the ability to recognize and translocate specific enzymatic components. However, in contrast to the hemolysins and leukocidins, they do not cause drastic cellular effects by pore formation. The binding components of binary toxins use pore formation as a means to inject a toxic protein into cells, and particularly the cytosol, whereas the pore-forming toxins more simply cause gross membrane damage that leads to loss of intracellular compounds.

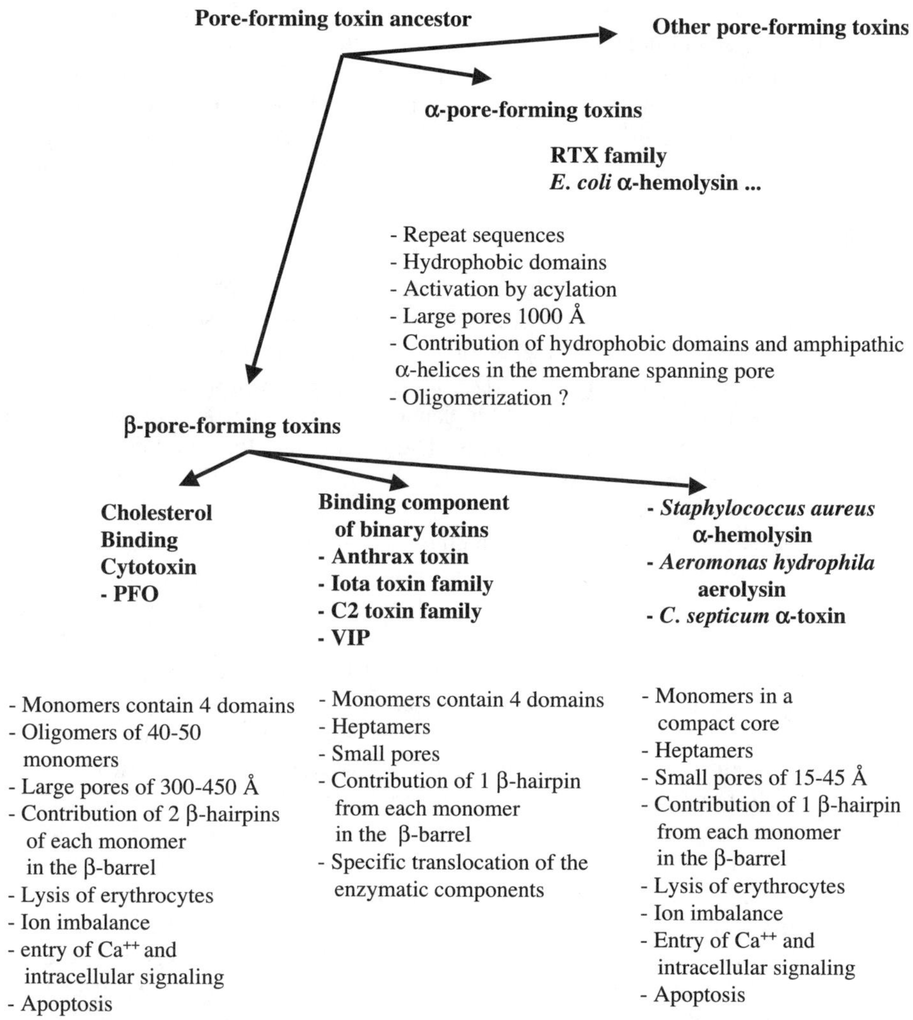

FIGURE 17.5 Bacterial pore-forming toxin families and a hypothetical route of evolution.

17.8.2 Phospholipases

C. perfringens alpha-toxin is the most thoroughly studied phospholipase C. It is composed of two domains that include an N-terminal domain primarily consisting of alpha-helices and an active site with three Zn^{++} binding sites, as well as a C-terminal domain essentially constituted of β-sheets (eight-stranded antiparallel β-sandwich). Since *C. perfringens* alpha-toxin is significantly related to the other *Bacillus* and *Clostridium* phospholipases C at the amino acid sequence level, these toxins probably share a common structure. The C-terminal domain of *C. perfringens* alpha-toxin, which contains a Ca^{++} binding site and recognizes cell surface receptor, is structurally similar to the C2 domains of eukaryotic proteins like pancreatic lipase, soybean lipoxygenase, synaptotagmin, phosphoinositide specific phospholipase C (PI-PLC) and arachidonate lipoxygenase. All of these proteins are involved in Ca^{++}-dependent binding of phospholipids such as phosphatidylcholine [124]. This raises the question of how *C. perfringens* alpha-toxin and related toxins have acquired a structural domain from a specific eukaryotic protein. The C-terminal domain of *C. perfringens* alpha-toxin is responsible for binding to membrane phospholipids, but it also possesses hemolytic, lethal, and sphingomyelinase activities [125]. Indeed, phosphatidylcholine-dependent phospholipase C

(PC-PLC) from *B. cereus*, which lacks a C-terminal domain and only contains an N-terminal domain, is neither lethal nor hemolytic. However, although *C. bifermentans* phospholipase C retains a C-terminal domain, this enzyme is nontoxic and only weakly hemolytic. Additionally residues Tyr-331 and Phe-334, which are essential for the interaction between phospholipase and membrane lipids, are missing in *C. bifermentans* phospholipase C [126]. Therefore, it seems apparent that the C-terminal domain greatly potentiates the biological activities of clostridial phospholipases.

17.8.3 ADP-Ribosylating Toxins

Many bacterial toxins are ADP-ribosyltransferases, and although primary amino acid sequences are not well conserved among these molecules, they do retain a conserved structure at the active site that includes three regions. The first region contains an α-helix bend on a β-strand (region 2) that ensures structural integrity of the NAD binding pocket, as well as a conserved STS motif between the α-helix and β-strand. Region 2 is flanked by a His or Arg in region 1, which plays an essential role in the correct orientation of an NAD molecule and the catalytic Glu found in region 3. The catalytic region 3 contains only a single Glu residue, or a biglutamic motif Glu-X-Glu, necessary for catalysis [127–129].

The clostridial ADP-ribosyltransferases include C3, whose substrate is the Rho protein, as well as the actin ADP-ribosylating toxins that are divided into two families: iota (*C. perfringens* iota toxin, CDT, CST) as well as C2. The actin ADP-ribosylating toxins are closely related to VIP from *B. cereus* and *B. thuringiensis*. In contrast to the other bacterial ADP-ribosyltransferases, the substrates of which are GTP binding proteins, the clostridial actin ADP-ribosylating toxins recognize ATP-binding proteins. It is noteworthy that the iota family toxins recognize all actin isoforms, whereas the substrate of C2 toxins is restricted to muscular actin. C3 modifies Rho, which is part of the small GTP binding protein superfamily, whereas the non-clostridial toxins recognize heterotrimeric G proteins (CT and pertussis toxin) or elongation factors for protein synthesis (DT and ExoA).

The crystal structures of C3, Ia, and VIP-II have been solved, and Ia and VIP-II are interestingly organized into two structurally similar subdomains. They consist of a perpendicular packing of five mixed β-sheet strands against a three-stranded antiparallel β-sheet flanked by four consecutive helices. The C-terminal subdomain contains the catalytic site, whereas the N-terminal subdomain interacts with the binding component and mediates toxin endocytosis. The two structural subdomains of Ia and VIP-II likely evolved by duplication of an ancestral gene [128,130]. One subdomain has acquired or retained the ADP-ribosylating activity, whereas the other subdomain has undergone modifications and adopted a novel function as evidenced by interactions with the binding component for translocation across the cell membrane. In contrast, C3 enzyme is constituted of only one subdomain highly homologous to the catalytic subdomain of Ia or VIP-II. Thus C3 enzyme does not possess a domain that interacts with a binding component, and a putative binding component for C3 has not correspondingly been found. Therefore, C3 is considered an enzyme and not a toxin, since the internalization machinery is lacking. C3 enzymes have been identified in *C. botulinum* C and D, *Clostridium limosum*, *B. cereus*, and *S. aureus* (epidermal cell differentiation inhibitor, EDIN), while the ADP-ribosylating binary toxins have been discovered in *C. botulinum* C and D, *C. difficile*, *C. perfringens*, *C. spiroforme*, *B. cereus*, and *B. thuringiensis*. The G+C content of these genes ranges from 27 to 32%, similar to that of the *Bacillus*, *Clostridium,* and *Staphylococcus* genomes. This suggests that these genes have been spread by horizontal transfer (see below), but their origin remains mysterious.

Genes for the binding and enzymatic components of the ADP-ribosylating binary toxins are closely located and organized in an operon. This supposes that a putative ancestral C3 gene has undergone a duplication and further modification to yield an enzymatic component gene, which has been mobilized near a binding component gene derived from an ancestral pore-forming toxin gene. Some *C. perfringens* and *C. difficile* strains only have an enzymatic or binding component gene. This indicates that the binary toxin genes also evolve independently. The C3 gene is the smallest bacterial ADP-ribosyltransferase gene discovered to date. As discussed above, it could be the ancestral "mother" of the various actin ADP-ribosylating component genes. Might the C3 gene also be the ancestor of other bacterial ADP-ribosyltransferases? When one considers the possible age of bacteria, *Clostridia* (and perhaps other anaerobes) may represent an archeal form of eubacteria that developed during a time period when earth was enshrouded by a less aerobic atmosphere. Although the enzymatic site is structurally conserved among the ADP-ribosylating toxins, their substrate specificity differs. This suggests that variations have preferentially occurred within this domain or among residues supporting substrate specificity. C3 enzymes and actin ADP-ribosylating toxins contain two adjacent protruding turns known as the ADP-ribosylating toxin turn turn (ARTT) motif, which are involved in substrate recognition. A Phe is conserved in turn 1, but C3 enzymes contain a Gln within turn 2, instead of a Glu, in actin ADP-ribosylating toxins. The ARTT motif is conserved in CT and the closely related *E. coli* thermolabile enterotoxin with a Glu found in the second turn [129], but the full molecular basis of substrate specificity still requires further characterization.

17.8.4 Glucosyltransferases

Large clostridial toxins are 250 to 300 kDa proteins that include *C. difficile* ToxA and ToxB, *C. sordellii* LT and HT, and *Clostridium novyi* α-toxin (α-novyi). ToxB and LT are highly related (76% amino acid sequence identity), whereas ToxA and α-novyi are more distant with only 48 to 60% sequence identity [131].

Large clostridial toxins are single-protein chains containing three functional domains. One third of the C-terminus exhibits multiple repeat sequences (7 class-I repeat units containing 30 amino acids each, and 31 class-II repeats of 20 to 21 residues) that are involved in recognizing cell surface receptor. These repeats possess a significant similarity with those of the ligand-binding domain of *Streptococcus* glucosyltransferases and muramidases, thus possibly indicating a common origin [132,133].

Large clostridial toxins specifically glucosylate Rho and Ras proteins by using UDP-glucose as a sugar donor, with an exception being α-novyi, which uses UDP-N-acetylglucosamine. The enzymatic site, characterized by the Asp-X-Asp motif surrounded by a hydrophobic region, as well as the substrate recognition domain, are both localized within the 546 N-terminal residues [134,135]. Aspartic residues seem to be involved in coordinating a divalent cation (mainly Mn^{++}), which increases the hydrolase activity and binding to UDP-glucose [136]. The region encompassing amino acids 133 to 517 is implicated in the nucleotide-sugar specificity [137], and a conserved hydrophobic residue (Trp-102) plays a critical role in binding of the cosubstrate UDP-glucose [138]. A very large number of eukaryotic and bacterial glucosyltransferases have been identified, and most of them retain the catalytic Asp-X-Asp site. The structures, which are known for only a few of these proteins, reveal that despite different amino acid sequences, folding of the catalytic domain is quite conserved [139,140]. Large clostridial toxins probably share a common evolution with other glycosyltransferases, but they have an unusually long sequence that possibly results from the fusion of a glucosylating domain with translocation and receptor binding domains.

17.8.5 Proteases

Several bacterial toxins exhibit a metalloprotease activity such as LF from *B. anthracis* lethal toxin, BoNTs, TeTx, *Bacteroides fragilis* enterotoxin (BFT), and collagenases from *C. histolyticum*, *C. perfringens*, and other *Clostridia*. The bacterial proteolytic toxins belong to a common class of zinc-dependent endopeptidases. However, these proteolytic toxins are unrelated at the amino acid sequence level and possess variable tertiary structures. All of these molecules do retain a conserved active site motif, His-Glu-X-X-His, involved in coordinating Zn^{++} (Figure 17.6).

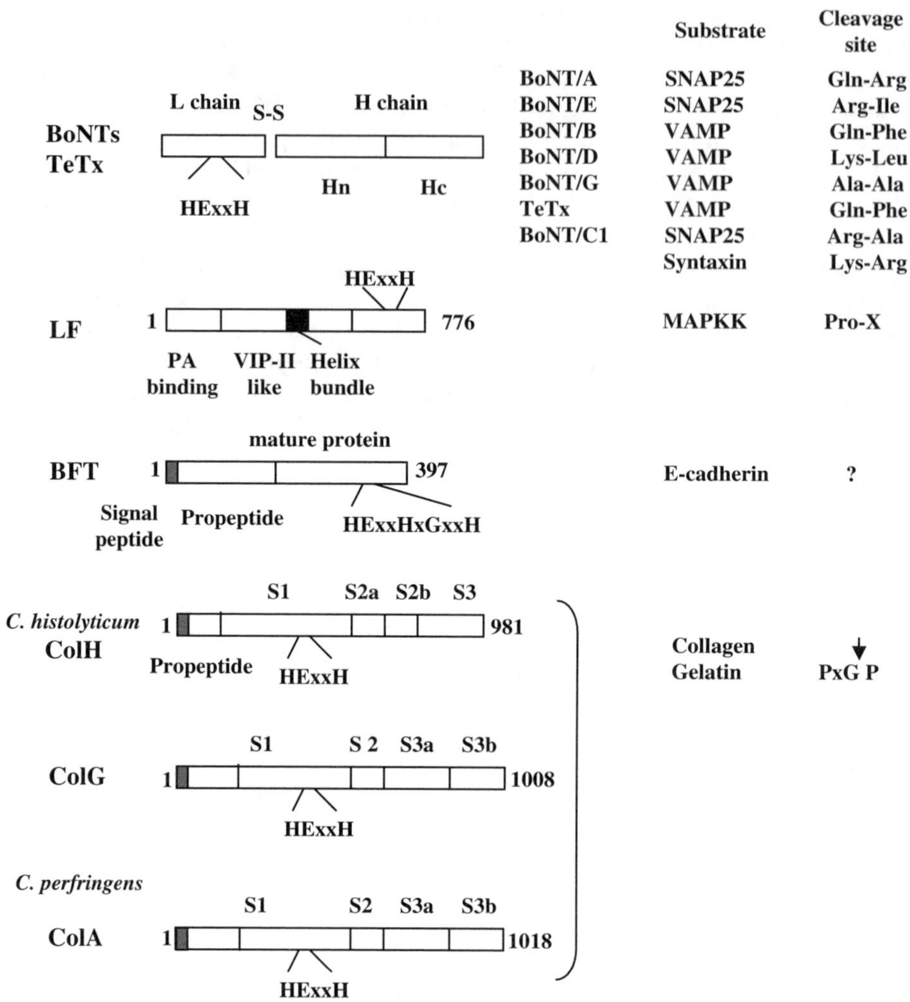

FIGURE 17.6 Domain organization, substrate, and specific cleavage sites of bacterial toxins that exhibit metalloprotease activity. BoNT, botulinum neurotoxin; TeTx, tetanus toxin; LF, anthrax lethal factor; BFT, *Bacteroides fragilis* enterotoxin; Col, collagenase; SNAP25, synaptosomal-associated protein; VAMP, vesicle-associated membrane protein; MAPKK, mitogen-activated protein kinase kinase.

The clostridial neurotoxins all share a common structure and are synthesized as precursor proteins (about 150 kDa) that are inactive or weakly active. The active molecule consists of a light chain (L, about 50 kDa) and heavy chain (H, about 100 kDa) linked by a disulfide bridge. Their structure possesses three distinct domains: (1) the L chain is formed of α-helices and a β-strand containing the catalytic zinc binding motif; (2) the N-terminal part of the H chain, which forms two unusually long and twisted α-helices; and (3) the C-terminal portion of the H chain, consisting of two distinct subdomains involved in receptor recognition [141–144]. The H chain mediates specific transport of the neurotoxin to a targeted neuron and facilitates subsequent internalization. The L chain is the intracellular active domain containing the His-Glu-X-X-His motif. BoNTs and TeTx cleave one of three members of the SNARE protein complex that play a key role in evoking neuroexocytosis. TeTx and BoNT/B, D, F, and G cleave synaptobrevin (or VAMP for vesicular associated membrane protein), BoNT/A and E cut SNAP25, and BoNT/C1 cleaves both SNAP25 and syntaxin. The cleavage sites differ for each neurotoxin, except BoNT/B and TeTx, which both cut synaptobrevin at the same site. The clostridial neurotoxins recognize their specific substrate by means of the SNARE motifs characterized by three acidic residues interspersed with hydrophobic and hydrophilic residues. SNARE motifs adopt a helical conformation with negatively charged residues exposed on a face that is flanked by a hydrophobic face. VAMP and syntaxin contain two SNARE motifs designated as V1 and V2, whereas SNAP25 possesses four motifs. The SNARE motifs are not equivalent for all neurotoxins, as V2 has a major role in recognition of VAMP by BoNT/B and BoNT/G, whereas V1 is implicated in interactions with TeTx, BoNT/D, and BoNT/F. In addition, a basic region flanking the cleavage site of TeTx and BoNT/B lies opposite V1 and is required for toxin binding and subsequent cleavage [145–147].

LF (90 kDa) from *B. anthracis* lethal toxin contains four domains. As for EF, domain 1 (amino acids 1–254) consists of a 12-helix bundle involved in interactions with PA. Curiously, the structure of domain 2 is similar to that of the catalytic domain of *B. cereus* VIP2 and *C. perfringens* iota toxin (see above). However, LF is devoid of ADP-ribosylating activity. Domain 3 forms a small helical bundle required for substrate recognition. Domain 4 (residues 552–776), consisting of a nine-helix bundle packed against a four-stranded β-sheet, contains the metalloprotease active site designated by the His-Glu-X-X-His motif. Domains 2, 3, and 4 form a broad deep groove, which represents the substrate binding pocket [148]. LF is a zinc-dependent protease that specifically cleaves mitogen-activated protein kinase kinase (MAPKK) 1 to 7, except MAPKK5, near the N-terminus, thereby yielding inactive molecules [149,150].

BFT is synthesized by *B. fragilis* as a 397 amino acid protein that is secreted by means of a signal peptide (18 amino acids) as a nonactive precursor. The 193 N-terminal residues (propeptide) are linked to the 186 C-terminal residues (mature protein) by a trypsin cleavage site. The mature protein contains a characteristic zinc-binding motif for metalloproteases (His-Glu-X-X-His-X-X-Gly-X-X-His). *In vitro,* BFT is able to proteolyze actin, gelatin, casein, and azocoll. Three highly related (92 to 96% identity) isoforms have been characterized, and BFT2, but not BFT1 or BFT3, contains an additional C-terminal extension of 20 amino acids that forms an amphipathic structure. This suggests that BFT2 could oligomerize and insert into cell membranes, thus forming a pore. BFT does not enter cells, but cleaves the extracellular domain of E-cadherin in an ATP-independent manner. This step is followed by degradation of the intracellular domain of E-cadherin, perhaps by ATP-dependent cellular proteases. To date, BFT is the only known bacterial toxin that modifies the actin cytoskeleton by cleaving a cell surface molecule. The BFT-dependent proteolysis of the extracellular domain of E-cadherin leads to loss of the intracellular domain and subsequent disorganization of the actin cytoskeleton [151].

C. histolyticum toxins are collagenases (also called α-clostripain) that exist in at least six different forms, with molecular masses ranging from 68 to 125 kDa. They are divided into two classes, ColG and ColH, based on amino acid sequence similarities and specificities toward peptide substrates. They are respectively encoded by two genes, *colG* and *colH*. Both genes, which probably derive from duplication of an ancestral gene, encode for a 116 kDa precursor that is processed within the C-terminus to yield multiple collagenase forms [152]. ColH consists of four segments, S1, S2a, S2b, and S3, where S2a and S2b are quite homologous. ColG also possesses four segments designated as S1, S2, S3a, and S3b. All collagenases contain within the N-terminus (segment 1) the consensus active site His-Glu-X-X-His of zinc metalloproteases (Figure 17.6) [153]. The other regions of these molecules are involved in substrate recognition and binding. Thus, S3 is required for binding to collagen [154,155]. *C. histolyticum* collagenases are specific for collagen and gelatin. They cleave peptide bonds on the amino side of a glycine residue in the Pro-X-Gly-Pro sequence, and they degrade their substrates into small dialyzable peptides. Collagenases play an important role in degrading connective tissue and facilitating dissemination of a pathogen from the initial infection site. *C. perfringens* also produces a collagenase, ColA or κ-toxin, which is related to ColG from *C. histolyticum* and contains the characteristic motif His-Glu-X-X-His [152,156].

The active site of the bacterial metalloproteases is similar to that of thermolysin and contains a structural Zn^{++} coordinated by the two His residues within the His-Glu-X-X-His motif, an additional Glu, and a water molecule interacting by an hydrogen bond with the Glu residue of the

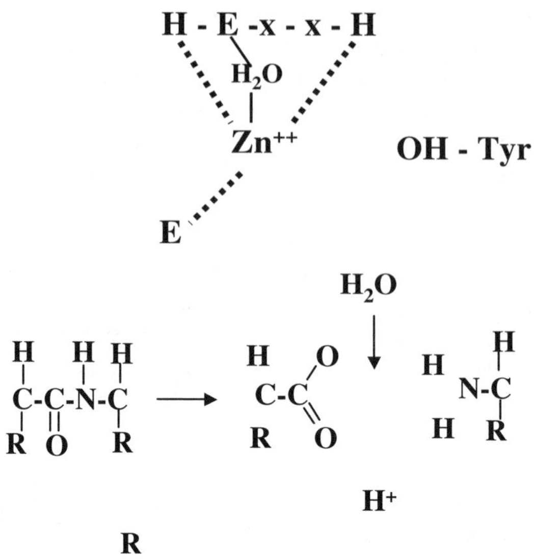

FIGURE 17.7 Schematic representation of the active site for zinc-dependent proteases. The zinc atom is coordinated by two His residues and Glu within the His-Glu-X-X-His motif. The catalytic water molecule is coordinated by a zinc atom and Glu found in the His-Glu-X-X-His motif. A conserved Tyr probably protonates the resulting amine group.

His-Glu-X-X-His motif. A water molecule is responsible for the hydrolytic reaction of proteolysis, and a conserved Tyr is probably involved in protonation of the resulting cleaved amine group (Figure 17.7).

17.9 TOXIN RECEPTORS

The first step of intoxication by any toxin involves recognition of a specific receptor on the cell surface and, more rarely, on the extracellular matrix (i.e., collagenases). The presence and abundance of specific receptors determine the relative sensitivity of cells as well as, ultimately, an animal species to any toxin. For example, neurotoxins are only active on certain neuronal cells, enterotoxins on intestinal cells, and leukotoxins on lymphocytes, but other toxins generically called cytotoxins recognize a wide range of cell types via ubiquitous receptors. As one specific example, mice do not possess specific DT receptors, and are resistant to DT, but humans are highly sensitive to DT and concomitantly possess high numbers of receptors on their cells. Clearly, effective interaction of a toxin with its receptor is a key process that dictates the subsequent intoxication steps, which include: (1) persistence of the toxin at the cell surface and thus permitting its local activity (pore formation or enzymatic degradation of certain membrane compounds); (2) internalization; and (3) subsequent intracellular trafficking.

Toxins interact with their cell receptor by a specific domain structurally distinct from the other functional domains involved in translocation and catalysis. The receptor binding domains generally possess a β-sheet rich structure that is variable among the various bacterial toxins (Figure 17.8). Within the single-chain toxins, the receptor binding domain is located at one extremity, which often encompasses the C-terminus, but it is in close proximity to the translocation domain in the intracellularly active toxins or oligomerization domain in pore-forming toxins. In the toxins that undergo proteolytic activation and reduction of a disulfide bridge (e.g., DT and clostridial neurotoxins), the receptor binding and translocation domains are separated from the catalytic domain as evidenced by two distinct protein chains. The binary toxins (anthrax and clostridial binary) have the translocation and receptor binding domains on a protein chain genetically distinct from that of

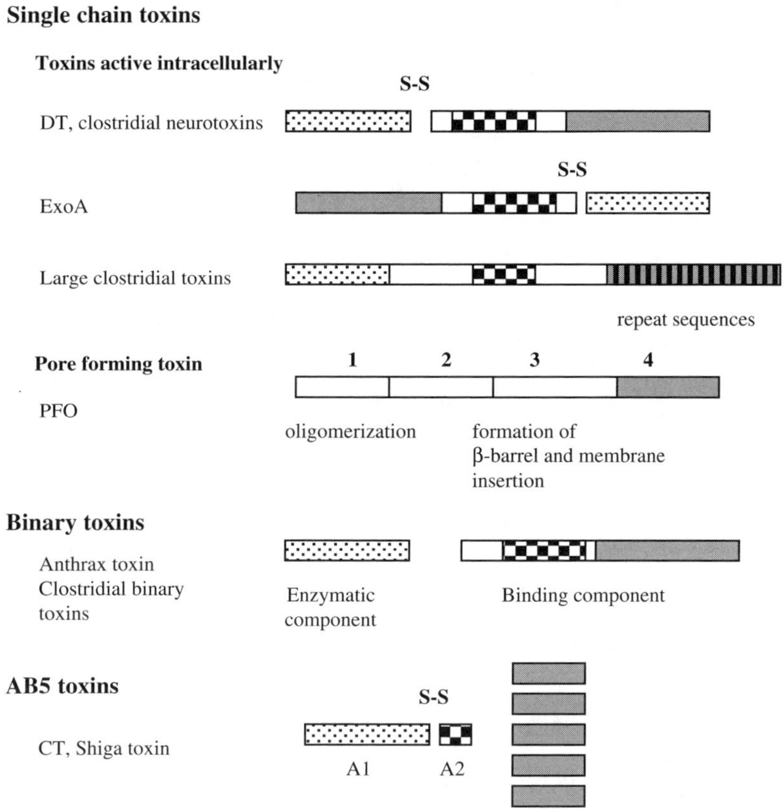

FIGURE 17.8 Localization of the receptor binding domain in selected bacterial toxins. DT, diphtheria toxin; CT, cholera toxin; ExoA, *Pseudomonas* exotoxin A; PFO, *Clostridium perfringens* perfringolysin: enzymatic domain, binding domain, translocation domain.

the enzymatic component. Organization of the receptor binding domains is more complex in the multicomponent toxins, also classified as AB5 toxins (CT, *E. coli* heat labile enterotoxin, ST, and Vero toxin). The receptor binding domains are synthesized as single proteins, which assemble within the bacterium into a homologous protein ring consisting of 5 identical monomers that form a central pore in a cell membrane. In contrast, and more complex, the pentamer of pertussis toxin consists of 5 heterologous B subunits (S2, S3, S4, S4, and S5). The translocation and enzymatic domains (A chain) interact by noncovalent bonds with the B subunits. These preformed, pentameric binding structures ensure high affinity binding toward cell receptors.

Identification of cell receptors for toxins represents one of the most challenging aspects in toxinology, and only a few receptors have been identified so far (Table 17.5). In particular, the receptors of clostridial toxins are very poorly understood. The receptors are diverse and belong to all classes of membrane compounds (i.e., carbohydrates, gangliosides, glycolipids, glycoprotein, phospholipids, and proteins). These molecules have a physiological function for cells, and toxins essentially hijack these receptors to exert their biological activities upon targeted cells. The few examples of clostridial toxin receptors described in the literature will now be addressed.

The general mode of action for pore-forming toxins includes the binding of toxin monomers to their cell surface receptors, lateral diffusion of the monomers bound to their receptors on the membrane surface which then facilitates monomer–monomer interactions, oligomerization, and subsequent insertion into the membrane. Indeed, this theory supposes that the receptors for pore-forming toxins are molecules having sufficient mobility on the membrane. CBC, such as PFO, recognize cholesterol as a surface receptor, and are concentrated into microdomains or lipid rafts,

TABLE 17.5
Cellular Receptors of the Main Bacterial Toxins (Reference)

Clostridial Toxin	Cellular Receptor(s)
C. botulinum Neurotoxins	Gangliosides (GT1b and GD1b) + protein (Synaptotagmin 2 for BoNT/B) (164–165)
C. tetani Tetanus Toxin	Ganglioside (GT1b) + GPI-anchored protein (161–163)
C. difficile ToxA	Carbohydrate (Gal-α1-3Gal-β1-4GlcNac motif) (136)
C. perfringens Enterotoxin	Proteins (Claudin and Occludin) (225, 286)
C. perfringens Perfringolysin	Cholesterol (103)
C. septicum α-toxin	GPI-anchored proteins (Thy-1 and Folate Receptor) (157)
Other Bacterial Toxins	**Cellular Receptor(s)**
Cholesterol Binding Cytolysins	Cholesterol (103)
A. hydrophila Aerolysin	GPI-anchored proteins (Thy-1 and Contactin) (287)
B. anthracis Lethal and Edema Toxins	Proteins (Tumor Endothelium Marker 8 and Human Capillary Gene 2 Protein) (288, 289)
B. pertussis Pertussis Toxin	Glycoprotein (NeuAcα(2, 6)-Galβ4GlcNAc) (290)
Bordetella Adenylcyclase	Protein ($\alpha_M\beta_2$ integrin) (291)
C. diphtheriae Diphtheria Toxin	Protein (Heparin binding Epidermal Growth Factor-like Growth Factor) (292) Sphingolipid (3′-Sulfogalactosyl-Ceramide) (293)
E. coli Heat Stable Toxin (STb)	Protein (Guanylate Cyclase C) (294)
E. coli Heat Stable Toxin (STa)	Protein (Tyrosine Phosphatase Protein Receptor beta) (295)
H. pylori Vacuolating Cytotoxin (VacA)	Protein (Low density lipoprotein-receptor-related protein (296) Globotriosides (GB3 and GB4) (297)
P. aeruginosa Exotoxin A	Ganglioside (GM1) (158)
S. dysenteriae Shiga Toxin and *E. coli* Shiga-Like Toxin	
V. cholerae Cholera Toxin	
Staphylococcal and streptococcal Superantigens	Proteins (Major Histocompatibility Complex Class II molecules and Vβ Specific T cell Receptors) (265)

that promote oligomerization of toxin monomers. However, cholesterol seems to have a more complex role than solely being a receptor for certain pore-forming toxins. *C. septicum* α-toxin and related toxins also interact with glycosylphosphatidylinositol (GPI)-anchored proteins that are also localized into membrane microdomains, which favors subsequent oligomerization of toxin monomers [105,157].

The receptor for PA from anthrax edema and lethal toxins involves at least two homologous membrane proteins (tumor endothelium marker 8 and human capillary morphogenesis protein) that possess an extracellular von Willebrand factor A domain. The current model for the anthrax toxins includes PA protoxin (PA83) binding to a cell surface receptor and subsequent cleavage by furin or other cell-associated proteases. Receptor-bound PA63 assembles into ring-shaped homoheptamers that are clustered into lipid-rafts and bind EF or LF [85,123] before endocytosis. A similar model is probably involved with binding or oligomerization of the binding components for the clostridial binary toxins. Proteolytically activated Ib from iota toxin binds to a protein membrane receptor and forms oligomers that are concentrated in lipid microdomains of Vero cells [157a]. Receptors for binding components of binary toxins are apparently not localized in microdomains under physiological conditions, as evidenced by the prototoxin form of Ib, which binds to cells, does not form oligomers, and is not localized into lipid microdomains. Somehow, the interaction of toxin monomers with receptor triggers localization into membrane microdomains and subsequent oligomerization.

The receptor of CT is the ganglioside GM1, which is also localized into lipid raft microdomains. The preformed, pentameric binding platform interacts with the GM1 receptors, at a stoichiometry of one GM1 molecule per binding subunit. The whole toxin is then endocytosed [158]. To date, the binding stoichiometry of the clostridial or *B. anthracis* binary toxins with receptor is unknown, but a recent report suggests a 1:1 mole ratio of PA63:receptor [158a].

Toxins that cause gross damage on the cell surface through a phospholipase C enzymatic activity, such as *C. perfringens* alpha-toxin and related proteins, bind to membrane phospholipids. This interaction with phospholipids was first elucidated with *C. perfringens* alpha-toxin. The C-terminal domain, which is structurally related to the C2 domain of similar eukaryotic proteins, mediates cell binding of this toxin (see Section 17.8.2) via charge–charge interactions with the phospholipid headgroups and calcium ions. The calcium ions confer positive charges on phospholipid membranes, thus facilitating the binding to a negatively charged toxin domain. Three calcium binding sites are partially coordinated by side-chains within the C-terminus and phospholipid phosphate groups [124,159]. It is assumed that the C-terminal domain of alpha-toxin preferentially binds to phospholipids in the intact membrane, subsequently triggering conformational change and activation of the enzyme. The active site evidently closes after detachment from the membrane. Therefore, alpha-toxin likely discriminates between phospholipids from intact and lyzed cells, resulting in increased cell lysis and release of nutrients for bacterial growth [160].

The large clostridial toxins possess a C-terminus rich in repeats, like that found in glucosidases and glucosyltransferases. This suggests that the specific receptor is a carbohydrate or glycoprotein, and, in fact, a trisaccharide (Gal-α1-3Gal-β1-4GlcNac) is the motif recognized by ToxA. The cell receptor in rabbit ileal brush border has been identified as the glycoprotein sucrase-isomaltase. However, this receptor is not expressed in the human colon, and therefore it is reasonable to surmise that the "real" receptor for this toxin still remains elusive [136].

TeTx and each serotype of the BoNTs recognize distinct specific receptors on demyelinated nerve endings. The current model is that clostridial neurotoxins interact with a dual receptor — one part being a ganglioside from the GD1b-GT1b series, and the other part represented by a protein or glycoprotein. The protein receptor part is supposedly highly specific for nerve cells, since these toxins only affect distinct neuronal populations. The protein part of the TeTx receptor has been identified as a 15 kDa GPI-anchored protein [161–163]. The localization of TeTx receptor on lipid rafts probably favors bound toxin mobility on the membrane and subsequent oligomerization before cell entry. TeTx has been found to form tetramers in lipid membranes. The high affinity of BoNTs and TeTx for presynaptic membranes is a likely result of multiple interactions with the ganglioside and protein portions of the receptor. Experimental data support that synaptotagmin isoforms, which are transmembrane proteins from the synaptic vesicles, represent a protein receptor of BoNTs. It has also been found that synaptotagmin II associated to ganglioside GT1b is involved in the binding of BoNT/B [164,165]. GT1b is required in neuroblastomas, which also express synaptotagmin I, for binding and subsequent activity of BoNT/A [166]. Polysialoganglioside binding sites have also been characterized within the C domain of the heavy chain of BoNT/A [141,144]. TeTx exhibits two carbohydrate binding sites, whereas BoNT/A and B possess only one [167].

17.10 CELLULAR TRAFFICKING OF TOXINS

When bound to their receptor, toxins are transported in the appropriate conformation to the correct cellular site to undertake their biological activities. For that, toxins use the inherent cell machinery. They hijack the normal physiological functions of target molecules to traffic, and ultimately exert, their toxic activity. As discussed above, toxins active on the cell surface first bind to their receptor, and move laterally on the membrane to either interact with the proper substrate or oligomerize with other toxin monomers, which then triggers pore formation. In particular, many bacterial toxins do not remain on the cell surface and must enter targeted cells to recognize and modify an intracellular target. Toxins secreted into the extracellular medium use receptor-mediated endocytosis to enter cells. One exception is the adenylcyclase from *B. pertussis*, which forms a pore in the plasma membrane and directly delivers its catalytic domain into the cytosol from the targeted cell surface. The toxins active intracellularly contain all the information required for their trafficking and activity, as certainly is the case for toxins produced by anaerobes. The *Clostridium* toxins active intracellularly belong to three classes: (1) neurotoxins; (2) large clostridial toxins; and (3) binary toxins. Another class of toxins or virulence factors secreted by type III and certain type IV systems are directly injected into the cytosol of the target cells. These toxins, generally produced by Gram-negative bacteria, lack the binding and translocation domains required for recognizing cell surface receptor and subsequent internalization. However, these type III molecules normally possess two enzymatic domains with two different enzymatic activities.

The intracellularly active toxin bound to its cell surface receptor is then embedded in endocytic vesicles generated by invagination of the cell membrane. Toxins such as DT and ExoA are internalized via clathrin-coated vesicles, which are generally used by a cell for the uptake of hormones or growth factors. This endocytic pathway is the best characterized and molecules, such as dynamin and Eps15, are required in this process. Dominant negative forms of dynamin or Eps15 block the cell entry of compounds that utilize this pathway. Other types of endocytic vesicles have been described, and among them are caveolae and nonclathrin-coated vesicles, which have been studied in the most detail. Caveolae are rich in detergent-insoluble lipids, such as glycosphingolipids and cholesterol, as well as certain proteins like caveolin. Their composition varies according to the cell types, and they constitute a subset of lipid rafts. GM1 ganglioside, the CT receptor, is clustered in lipid rafts, and CT enters cells via caveolae. However, toxins can use more than one endocytic pathway. For example, depending on the cell type, CT can enter a cell by clathrin-coated vesicles, caveolae, or both. Other toxins are internalized by different vesicles, which are nonclathrin and noncaveolae structures, such as *E. coli* CNF1 [168]. However, the endocytic pathway used by clostridial toxins has been poorly understood up until now. There are two main intracellular trafficking routes used by toxins, termed short and long pathways (Figure 17.9), as described below.

17.10.1 SHORT INTRACELLULAR PATHWAY

During the internalization process designated here as short trafficking, toxins migrate into the early or late endosomes, where translocation of the enzymatic domain into the cytosol occurs following a pH gradient (Figure 17.9). Acidification of the vesicles is a prerequisite for translocation, as it triggers a proteolytic processing and conformational change that facilitate a partial unfolding of the protein necessary for its translocation. The transmembrane pH gradient provides the driving force for translocation.

The best-characterized paradigm of a toxin short pathway is that taken by DT. DT bound to its receptor is endocytosed into clathrin-coated vesicles and is transported to early-late endosomes and lysosomes where degradation takes place. However, a fraction of DT translocates from early endosomes into the cytosol [169] following acidification of the endocytic vesicles (pH less than 6). Acidification triggers a conformational change of the translocation (T) domain, which facilitates the insertion of α-helices (in particular TH5-7 and TH8-9) into the membrane form to cation-selective

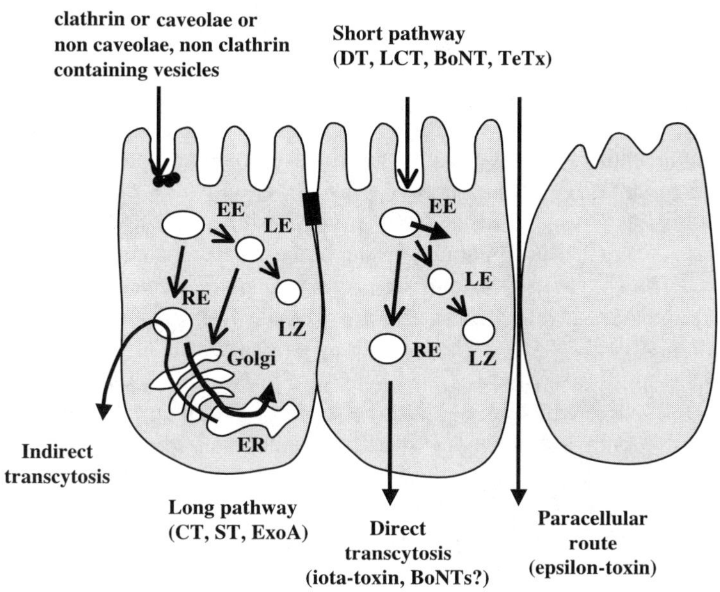

FIGURE 17.9 Schematic representation of the main intracellular trafficking patterns employed by bacterial toxins and passage of toxins across a cell barrier (direct and indirect transcytosis, paracellular route). EE, early endosome; LE, late endosome; LZ, lysosome; RE, recycling endosome; ER, endoplasmic reticulum; CT, cholera toxin; ExoA, *Pseudomonas* exotoxin A; ST, shiga toxin; BoNT, botulinum neurotoxin; LCT, large clostridial toxin; TeTx, tetanus toxin.

channels. Experiments with artificial membranes showed that the enzymatic domain (A) does not translocate through the hydrophilic environment of the channel formed by the T domain [170]. At low pH, the T domain interacts with membrane molten globule-like proteins that partially unfold and thus increase their relative hydrophobicity. The fragment A also partly unfolds and becomes hydrophobic in a molten globule-like conformation. Translocation could proceed by a series of transient association–dissociation events between the T domain and A chain. It has been proposed that the T domain may act like a transmembrane chaperone, forming a "sticky pore" that binds to the hydrophobic A-chain surface, thus maintaining it in an unfolded state during translocation [171,172]. In addition, cytosolic factors which include β-COP, heat shock protein 90, (Hsp90), thioredoxin reductase, and ATP are required for translocating the fragment A across a vesicle membrane [169,173]. The interchain disulfide bridge is reduced during the translocation process, and the A chain refolds in the neutral pH found in the cytosol.

All of the intracellularly active clostridial toxins, such as the neurotoxins, large clostridial toxins, and binary toxins, use a short intracellular trafficking route into targeted cells and require an acidification step of the endocytic vesicles for translocation. The translocation mechanism of the clostridial single-chain toxins, which include the neurotoxins and large clostridial toxins, is probably related to that of DT [145,147,174–179].

The intracellularly active binary toxins, like the clostridial binary toxins and anthrax lethal or edema toxins, also use a similar short pathway of cell entry as that employed by the single-chain toxins, but the translocation mechanism is different. Since the enzymatic and binding components are on distinct protein chains, each type of protein can assemble separately and form structures appropriate for an efficient translocation. Thus, the binding components heptamerize and form a transmembrane pore, which probably mediates the passage of partially unfolded enzymatic component through a pH gradient [180–183]. Cytosolic chaperones, such as Hsp90, seem to facilitate transport of C2-I across the vesicle membrane [184], possibly in a similar manner as that recently found for DT [173].

17.10.2 Long Intracellular Pathway and Transcytosis

The long intracellular pathway has been characterized by using the multicomponent toxins built on the classic AB5 model (CT, ST, and pertussis toxin), and also monomeric ExoA, from various Gram-negative pathogens (Figure 17.9). For example, vesicles containing CT bound to the GM1 ganglioside migrate to the Golgi apparatus through the early and late endosomes via a Rab-9-dependent pathway. The enzymatic A subunit is proteolytically activated by furin cleavage between the A1 (27.5 kDa) and A2 (4.5 kDa) fragments that remain linked by a disulfide bridge, probably in the trans-Golgi network [185]. CT then undergoes retrograde transport to the endoplasmic reticulum (ER) via coatomer I-coated vesicles. The KDEL C-terminal sequence of the A2 fragment is an ER retention sequence, which recognizes the receptor Erd2p and directs Golgi-ER trafficking [186,187]. The A2 fragment, consisting of an α-helix, mediates translocation of the A1 fragment across the ER membrane via the Sec61p complex [188,189]. ST uses a slightly different pathway, as it enters cells by clathrin-dependent as well as clathrin-independent pathways, and once in a cell, it is transported directly from early/recycling endosomes to the Golgi apparatus and finally the ER [190]. The ST receptor, Gb3 ganglioside, associates differently with membrane microdomains according to the cell type. ST undergoes retrograde transport only in cells where the B-subunits have associated with detergent-resistant membranes. In other cell types though, the toxin is predominantly routed to the lysosomes. Therefore, the association of B subunits to membrane microdomains is required for targeting into the retrograde route [191,192]. In contrast to CT, ST does not contain the KDEL retention signal for ER proteins, and its transport from Golgi to ER depends upon Rab6a [193].

Intracellular acting toxins must undergo an intracellular trafficking step in appropriate compartments not only to exert their biological activity, but also to cross a cell barrier and disseminate throughout the body and enter various organs to encounter a larger number of target cells. Some toxins are endocytosed on one side of the cell and exported to the other side (transcytosis). Transcytosis is used in physiological conditions to transport macromolecules such as immunoglobulins across a mucosal barrier. These molecules are endocytosed into early endosomes, sorted in the recycling endosomes, and then reexposed on the cell surface (direct transcytosis). CT, ST, and Vero toxins (VT or ST-like toxins) can transcytose across the epithelial cell barriers [194,195] by migrating to the Golgi, recycling to endosomes, and back again to the cell membrane. Although not proven, CT and ST may also move from the ER to the recycling compartment (indirect transcytosis). The GM1 ganglioside receptor localized in lipid rafts is a key determinant for intracellular trafficking and transcytosis of CT [196]. The Gb3 receptor could have a similar role for ST and VT, but VT can also transcytose across intestinal cells lacking Gb3 [195]. However, it has also been shown that VT-producing *E. coli* stimulate neutrophil recruitment to the intestinal lumen, and transmigration of neutrophils across the intestinal epithelium permits the paracellular passage of VT [197]. Transcytosis facilitates toxin dissemination through the circulatory system as evidenced in the hemolytic uremic syndrome caused by VT-producing *E. coli*, in which VT is transported to the kidney endothelium by migrating polymorphonuclear leukocytes via binding to low-affinity receptors on the leukocyte surface [198].

Some clostridial toxins also undergo a transcytotic transport. For example, *C. perfringens* iota toxin can transcytose across polarized intestinal epithelial cells, such as CaCo-2. Ib can bind to either the apical or basolateral surface, recycle to the opposite cell surface, and then dock as well as subsequently internalize Ia. It is intriguing that Ib remains on the cell surface up to 3 h and mediates endocytosis of several Ia molecules [199,200]. The long duration of Ib on the cell surface is advantageous, as it enhances the chances of Ia docking with the Ib heptamer and ultimately increases toxin efficiency and effects upon a cell. It has not been shown whether transcytosed Ib interacts with the neighboring cells or disseminates to other tissues by blood circulation. However, Sakurai showed that if Ib is injected intradermally and Ia is given systemically at a distal site in guinea pigs, a lesion forms at the Ib injection site [200a]. This suggests that Ia is able to "migrate" throughout the body and "locate" Ib. The exact pathway of Ib transcytosis is not fully known; however, since iota toxin is not transported by the Golgi apparatus, it likely undertakes a direct transcytotic route via early and recycling endosomes.

Botulism is normally acquired by the oral route through consumption of food containing preformed BoNT. However, as is the case for infection-borne botulism, BoNT is directly produced in the intestinal lumen during growth of neurotoxigenic *Clostridium*. BoNTs cross the intestinal mucosa and circulate throughout the body by the blood and lymphatic systems to target motoneuron endings, where they block the evoked release of acetylcholine that then leads to a flaccid paralysis. *In vitro*, BoNTs are transported across intestinal cell monolayers by a transcytotic mechanism. BoNTs do not perturb the integrity of the cell barrier or cause any toxic effect on epithelial cells. However, the exact transcytotic pathway taken by the BoNTs remains undetermined [201].

Other clostridial toxins produced in the intestine can also cross the intestinal barrier and diffuse through the bloodstream into various target organs. They generally use the paracellular route, after alteration of the barrier integrity and opening of the intracellular junctions caused by these toxins themselves (or others produced by *Clostridium)*. This is the case with *C. perfringens* epsilon toxin, which causes animal enterotoxemias. Toxin secreted in the small intestinal lumen increases mucosa permeability and diffuses through the blood circulation to the brain, kidneys, and other organs, thus causing dramatic physiological changes. Epsilon toxin tightly binds to specific cell surface receptors, oligomerizes, forms pores, and rapidly increases the permeability of polarized cell monolayers *in vitro,* as evidenced by swelling and cell lysis [202,203]. Since epsilon toxin does not undergo receptor-mediated endocytosis, it likely uses the paracellular route to access the bloodstream.

TeTx is secreted in wounds contaminated by *C. tetani* and subsequently diffuses to local extremities containing cholinergic neuronal cells. Like nerve growth factors, TeTx is endocytosed into nonacidified vesicles that contain the neurotrophin receptor $p75^{NTR}$ and then retrogradely transport toxin to the cell body in the central nervous system. Retrograde transport through the axon is microtubule- and actin cytoskeleton-dependent, as well as involves the activity of dynein and kinesin [204]. This transport is derived by the C-terminal fragment of TeTx, which also transports heterologous protein via a retroaxonal way [205,206]. Toxin is then delivered into the intersynaptic space entering the targeted neurons, which are inhibitory interneurons involved in regulating motoneurons. Toxin entry involves a receptor-mediated endocytosis into vesicles, which become acidified and thereby permit translocation of the L chain into the cytosol and subsequent intracellular activity [147,162,207]. Thus, TeTx undergoes a transcytotic pathway in the first neuron, subsequently entering the targeted neuron by an acidic short pathway. This raises a number of questions, including whether TeTx recognizes the same receptor on the first, and then targeted, neurons.

17.11 MOLECULAR BASIS OF *CLOSTRIDIUM* PATHOGENESIS

Pathogenic Clostridia secrete potent toxins in the extracellular medium that are responsible for all of the symptoms and lesions associated with clostridial diseases. The main characteristic of clostridial pathogenesis is bacterial growth at one site of the host organism, or outside (i.e., food), to a sufficient level that facilitates the production and secretion of toxin amounts that induce pathological modifications in host target cells and tissues.

Clostridia do not possess adherence pili and do not adhere to healthy eukaryotic cells, which could promote high colonization of the mucosal surface or other specific tissue like that described for various Gram-negative bacteria such as adherent *E. coli*, which readily colonize the digestive tract of various hosts. Potential adherence factors, such as pili and surface-anchored proteins which are expressed after exposure to stressful conditions, have been reported in *C. difficile*. These factors seem to mediate adherence to the extracellular matrix, such as mucus [208–210], but their relevance during the onset of *C. difficile* colitis has not been clearly established. Flagella have also been described in *Clostridium chauvoei*, and they seem important for the virulence of this bacterium, which causes blackleg in cattle [211,212]. Two genes that encode for the putative fibronectin binding proteins and genes coding proteins homologous to internalin A, which mediates the binding of *L. monocytogenes* to E-cadherin, and to SipA, a *Salmonella* adherence factor for host tissues, have been discovered in the genome of *C. tetani* [4]. The implication of these possible adherence factors in wound colonization still remains undetermined.

Additionally, *Clostridium* is not known to invade healthy eukaryotic cells. They do not contain a type III secretion system or the machinery for cell invasion. They are anaerobic bacteria, and they cannot survive in cells that normally maintain a high oxidoreduction potential. However, *C. perfringens* is perhaps an exception, as it is an aero-tolerant anaerobe that can survive in the cytoplasm of macrophages [213].

The Clostridia can enter an organism by two main routes that involve either the direct continuity with the environment, such as the oral route, and the tegument following a penetrative wound. Therefore, the Clostridia are rather versatile pathogens responsible for gastrointestinal or foodborne diseases, as well as gangrene and wound-associated intoxication (tetanus and botulism). Toxigenic bacteria, other than *Clostridium*, can also colonize multiple sites that include the digestive tract, breaches in the skin, and the respiratory tract (*Bordetella*, *Corynebacterium*, *Pseudomonas*, *Staphylococcus*, and *Streptococcus* species), as well as the urinary tract (*E. coli*), thus causing disease. These specific compartments within the body do not offer conditions that are suitable for *Clostridium* growth (aerobiosis, lack of appropriate nutrients, etc.).

Therefore, the toxins produced by *Clostridium* act at a distance from the bacterium. The toxins interact locally with host tissues, or they are transported to other target organs by crossing the mucosal barrier and disseminating through the circulatory system or neural axons. In essence, these bacterial toxins are like protein "bombs" released by the bacterium to "weaken" a perceived nemesis and to also obtain nutrients for itself and future progeny. The establishment of this biological "beach head" is critical for the survival and future dissemination of the microorganism, and in an opposite sense, the well being of the host also hinges upon how well the body manages the infection nidus.

17.11.1 Diseases Resulting from *Clostridium* Entry into Wounds

Accidental as well as surgically contaminated wounds can support *Clostridium* growth. The colonization of wounds by *Clostridium* requires various risk factors. Deep, penetrative wounds with a small opening through the skin can facilitate a sufficiently anaerobic condition into a protein-rich environment that can support the growth of proteolytic-type anaerobes. The presence of necrotic tissue and destruction of small blood vessels, which would normally bring needed oxygen and leucocytes into a traumatically injured site, also provide a favorable environment (anaerobiosis and nutrients) appropriate for *Clostridium* growth and dissemination.

17.11.1.1 Gangrene

The *Clostridium* species involved in gangrene and their toxins are summarized in Table 17.6, and are compared to other gangrene-linked toxins (Figure 17.10). In contaminated wounds and under suitable conditions, *Clostridium* will grow and secrete various necrosis-inducing toxins that diffuse locally and alter the neighboring healthy tissues. The resulting new zone of necrotic, relatively anaerobic tissue is rapidly invaded by *Clostridium*, which grows and synthesizes more toxin. Successively, the necrotic lesions rapidly progress along with bacterial growth and further dissemination of toxin by the bloodstream, which can induce systemic shock and eventual death. From a

TABLE 17.6
***Clostridium* Species and Their Main Toxins Involved in Gangrene and Wound-Acquired Diseases**[a]

Species	Main Toxin Involved in the Disease	Diseases
C. perfringens	alpha-toxin, Perfringolysin	Humans: gangrene, puerperal septicemia
C. septicum	α-septicum	Humans: gangrene, intestinal nontraumatic myonecrosis
C. histolyticum	collagenases	Animals: gangrene
		Humans, animals: gangrene
C. novyi (or *oedematiens*)	α-novyi	Humans, animals: gangrene
C. sordellii	LT, HT?	Humans, animals: gangrene
C. chauvoei	?	Cattle: blackleg
C. tetani	TeTx	Humans, animals: tetanus

[a] Note that *C. chauvoei* blackleg is a nontraumatic myonecrosis acquired from the passage of intestinal *C. chauvoei* into the blood circulation and then muscles.

historical and medical perspective, this disease has had a huge impact upon morbidity and mortality experienced during times of armed conflict. Prior to the antibiotic era, wounds incurred on the arms or legs were often dealt with swiftly via amputation, as the fear of gangrene was quite high. Unfortunately, due to less than antiseptic operating theaters (especially in the field near sites of armed conflict), and a basic lack of knowledge linked to disease and microbes, amputation could just as well lead to gangrene and death.

The toxins involved in gangrene are mainly membrane-damaging toxins (i.e., phospholipase C and pore-forming toxins) and those active on the extracellular matrix (i.e., collagenases and other hydrolytic enzymes). Gangrene due to *C. sordellii* and *C. novyi* implicate large clostridial toxins (i.e., *C. sordellii* LT and α-novyi), which enter cells and inactivate Rho and Ras proteins by glucosylation, thereby disrupting the actin cytoskeleton and inhibiting corresponding cell functions. The large clostridial toxins induce marked edema and necrosis of muscle cells.

The best-characterized model of gangrene is that caused by *C. perfringens*. Among the toxins produced by *C. perfringens*, alpha-toxin is the main toxin responsible for gangrene lesions. Alpha-toxin has phospholipase C and sphingomyelinase activities, and large amounts of toxin immediately lyze cells by degrading the membrane phospholipids [214]. However, cell lysis does not only result from the alpha-toxin phospholipase C activity, but also from subsequent activation of endogenous phospholipases A_2, C, and D [214–216]. The products of phospholipid hydrolysis (i.e., diacylglycerol and ceramide) also turn on the arachidonic cascade, thus resulting in release of leukotrienes, prostaglandins, and thromboxanes, which generate a local inflammation and vasoconstriction. This contributes to the anoxia of infected tissues, further promoting the local growth of *C. perfringens* [216]. In contrast to other forms of bacterial myonecrosis, *C. perfringens* gas gangrene is characterized by the absence of inflammatory cells in affected tissues and accumulation of leukocytes within vessels at the lesion periphery. At sublytic concentrations, as found at the periphery of necrotic lesions, alpha-toxin impairs leukocyte migration and promotes aggregation. The toxin strongly stimulates the expression of endothelial cell-leukocyte adherence molecule-1 (ELAM-1) and intercellular leukocyte adherence molecule-1 (ICAM-1), thus causing hyperadhesion of leukocytes to vessel endothelium. The recruitment of leukocytes coupled with toxin-induced hyperadhesion contributes to the evident leukostasis at the lesion periphery, and this too could enhance the respiratory burst activity and neutrophil-mediated vascular injury [217]. The rapid destruction of tissues also results from a drastic attenuation of local and regional blood flow. Alpha-toxin also promotes both platelet/platelet and platelet/neutrophil aggregates. The toxin activates the platelet

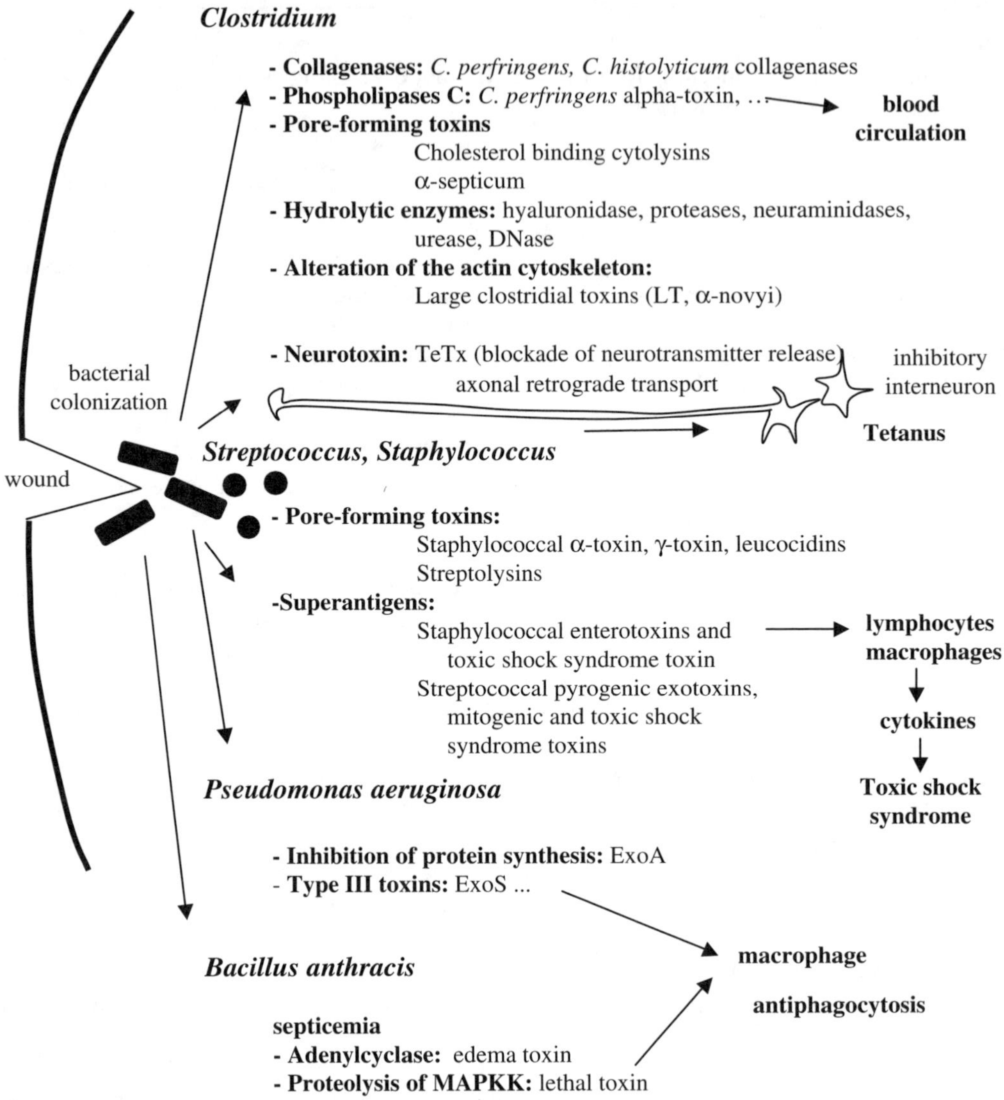

FIGURE 17.10 Main bacterial toxins involved in diseases acquired from wound contamination.

fibrinogen receptor (glycoprotein gpIIb/IIIa) and stimulates homotypic and heterotypic platelet aggregation in a soluble fibrinogen-dependent manner. Leukocyte aggregates are observed intravascularly and associate with the vascular endothelium. Free intravascular aggregates are responsible for the blockade of blood flow in small vessels, subsequently leading to anoxia and tissue necrosis [218–220].

17.11.1.2 Wound Tetanus and Botulism

C. tetani and proteolytic *C. botulinum* types A and B have the ability to grow in wounds. Conditions required for growth of the neurotoxigenic *Clostridium* in wounds (i.e., anaerobiosis, necrotic tissues, etc.) are the same as those for the gangrenous *Clostridium*. *C. tetani* and *C. botulinum* also produce CBCs (tetanolysin and botulinolysin, respectively), however their effects on clostridial colonization and growth in an infected wound is not known. The neurotoxigenic *Clostridia* do not produce any histological damage, but they synthesize TeTx and BoNT which diffuse to nerve endings (Figure 17.10). BoNT enters terminal motoneurons and inhibits release of acetylcholine, thus causing a flaccid paralysis, whereas TeTx is retrogradely transported (1–5 μm/s) through the cell body in the spinal cord and transynaptically to inhibitory interneurons, where it blocks the release of glycine or GABA (see Section 17.8.5). The reflexes are no longer controlled by the inhibitory interneurons leading to a spastic paralysis that is a classic characteristic of tetanus [145,147,179].

Tetanus is still rather common in humans (200,000–800,000 deaths per year), and neonatal forms resulting from contamination of the umbilical cord are most frequent [221]. Wound botulism is rare, but intravenous drug users represent a high-risk population because of repetitive injections with material that is occasionally contaminated by *C. botulinum.*

17.11.2 Gastrointestinal and Foodborne Diseases

The digestive tract, and mainly the small and large intestines which contain high amounts of nutrients, supports the growth of many facultative and strictly anaerobic bacteria such as *Clostridium.* Most pathogenic Clostridia are involved in gastrointestinal and foodborne diseases that represent not only an important problem for public health in humans, but these bacteria also pose a major cause of economical losses in domesticated animals (Table 17.7) [222].

Clostridial foodborne diseases can either arise from the ingestion of preformed toxin in food (intoxication) or intestinal colonization by *Clostridium* and subsequent production of toxin *in situ*

TABLE 17.7
***Clostridium* Species and Their Main Toxins Involved in Gastrointestinal Diseases and Food-Borne Acquired Diseases**

Species	Main Toxin Involved in the Disease	Diseases
C. perfringens	enterotoxin	Humans: foodborne poisoning, sporadic diarrhea, sudden infant death syndrome, antibiotic associated diarrhea
		Young animals (pigs): diarrhea
	alpha-toxin	Birds: necrotic enteritis
	beta-toxins	Humans: necrotic enteritis (PigBel)
		Young animals (mainly piglets): necrotic enteritis
		Horses: typhlocolitis
	epsilon-toxin	Sheep: enterotoxemia (struck)
	iota-toxin	Animals (sheep, goats, cattle): enterotoxemia
C. difficile	ToxA, ToxB	Humans: antibiotic associated diarrhea, pseudomembranous colitis
		Animals (pigs): enteritis, diarrhea
C. septicum	α-septicum	Humans: intestinal nontraumatic myonecrosis
		Animals (calves): enterotoxemia
		Animals (sheep): enterotoxemia (braxy)
C. sordellii	LT, HT	Animals: entritis, hemorrhagic enteritis
C. spiroforme	CST	Animals (rabbit): necrotic enteritis
C. botulinum A, B, E, F	BoNTA, B, E, F	Humans: botulism
		Animals (cattle, horse): botulism
C. botulinum C	BoNT/C1	Animals (birds, minks, cattle): botulism
C. botulinum D	BoNT/D	Animals (cattle): botulism
C. baratii	BaNT/F	Humans: botulism
C. butyricum	BuNT/E	Humans: botulism

(toxi-infection). Clostridia are environmental bacteria, and some foods constitute an appropriate site for colonization and toxin production. The main risk of clostridial intoxication is that due to botulism. Food storage conditions such as anaerobiosis, temperature [room temperature for proteolytic *C. botulinum* or low temperature (4 to 8°C) for nonproteolytic *C. botulinum* B and E, neutral pH, and low salt concentration permit *C. botulinum* growth, as well as toxin production [223,224]. The ingestion of preformed BoNT in food is responsible for most cases of botulism in adults.

Colonization of the intestinal lumen by *Clostridium* requires specific risk factors. Ingested bacteria are killed in great part by the acidic conditions of the stomach, but the spores can survive. In the intestine, resident microflora can exert a "barrier effect" against the bacteria ingested with food. Since *Clostridium* does not commonly exhibit adherence and invasive properties toward the healthy intestinal mucosa, they are directly in competition with the resident microflora for nutrient availability and are exposed to their "barrier effect" (i.e., bacteriocins, toxic metabolites, and lack of "land" to freely grow unencumbered by neighboring bacteria, etc.). Factors that perturb the physiological equilibrium of the intestines and resident microflora can subsequently permit colonization by *Clostridium*. The main risk factors involved in the onset of clostridial gastrointestinal diseases include:

- Antibiotic therapy that disturbs the resident microflora, thus facilitating growth of *C. difficile* or, more rarely, other *Clostridium* species, and the onset of diarrhea as well as pseudomembranous colitis
- Ingestion of a large number of enterotoxigenic *C. perfringens* (food containing at least 10^5 bacteria per g) that cause human foodborne poisoning
- Rapid change in diet from foraging to a rich food source consisting of readily fermentable carbohydrates and proteins, transport stress, rapid change of ambient temperature, and ingestion of frozen foods that are responsible for enterotoxemia in animals
- Young age with dysfunctional or incompletely established resident microflora, and clostridial necrotic enteritis in young babies and animals
- Intestinal carcinoma promoting colonization by *C. septicum* with an ensuing gangrene

Certain toxins, and specifically the BoNTs, only pass across the intestinal barrier and act at a distance on distinct target cells, whereas other clostridial toxins produced in the intestine induce alterations of the intestinal barrier with various levels of severity (Figure 17.11). *C. perfringens* food intoxication due to CPE consists of diarrhea, emesis, abdominal pain, and spontaneous recovery in 2 to 4 days, whereas *C. perfringens* beta1 and beta2 toxins, as well as large clostridial toxins, produce severe necrotic lesions and hemorrhaging of the intestinal mucosa accompanied by an intense inflammatory response. The clostridial toxins responsible for gastrointestinal diseases belong to two main classes: pore-forming toxins and those that modify the actin cytoskeleton (large clostridial toxins and clostridial binary toxins).

17.11.2.1 Clostridial Diseases Resulting from the Passage of Toxin through the Intestinal Mucosa and Dissemination to Other Target Cells: Botulism and Enterotoxemia

A specific neurological disease acquired by the oral route is botulism. BoNTs are preformed in food or produced in the intestine subsequent to the growth of *C. botulinum*. Neonates, young children with an incompletely established or partially functional resident microflora, and adults that have had abdominal surgery represent high-risk populations for intestinal colonization by *C. botulinum*. BoNTs cross the digestive mucosa by transcytosis without causing any damage to the intestinal cells (see Section 17.10.2), and they diffuse through the extracellular fluid and

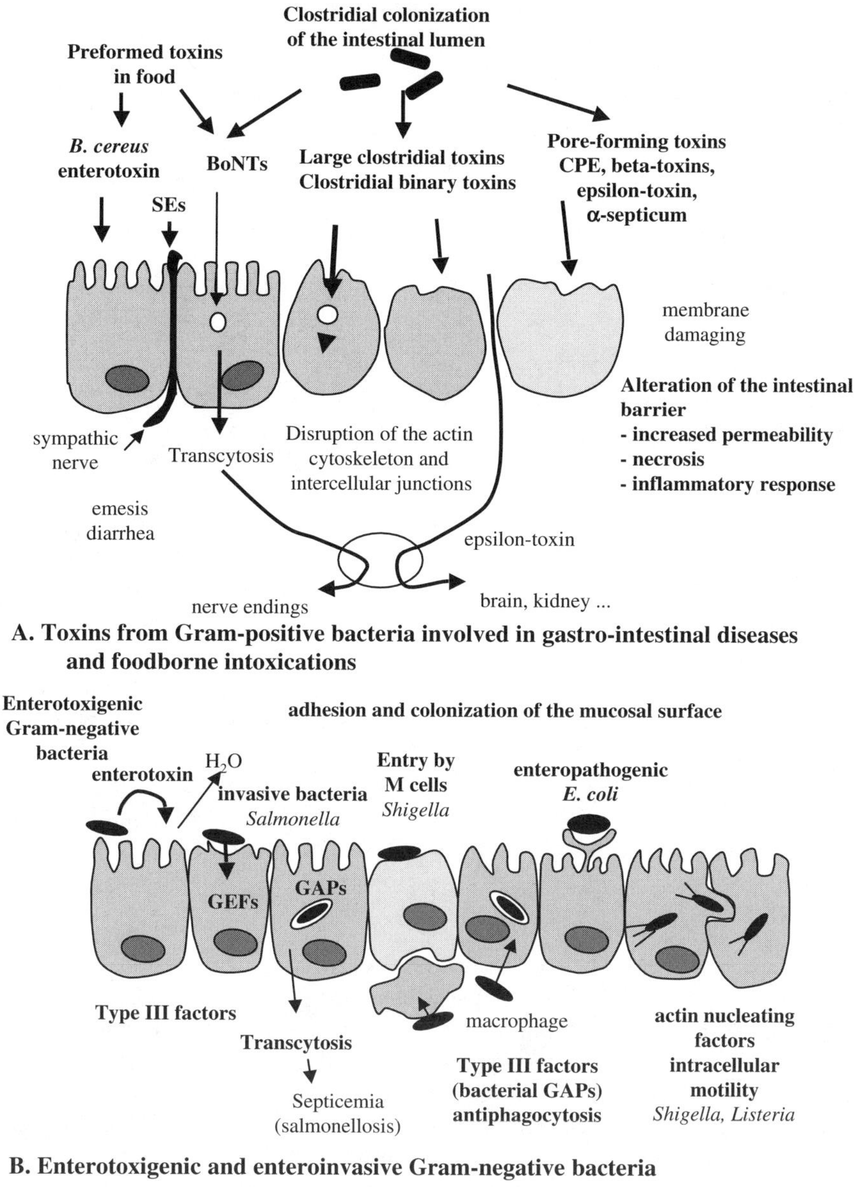

FIGURE 17.11 Schematic representation of the main bacterial toxins involved in gastrointestinal and foodborne diseases (A), in comparison with the enterotoxigenic and enteroinvasive Gram-negative bacteria (B). Clostridial toxins preformed in food or produced and secreted from the bacterium once in the intestinal lumen interact with the intestinal cells and alter the intestinal barrier via pore-formation, perturbation of the actin cytoskeleton, or disruption of intercellular junctions. *Bacillus cereus* enterotoxin and *Staphylococcus aureus* enterotoxins (SEs) are also preformed toxins present in contaminated food. Enterotoxigenic Gram-negative bacteria possess both adherent pili that promote mucosal colonization and secreted enterotoxins (cholera toxin, heat labile and heat stable enterotoxins) that alter cell homeostasis that leads to secretion of water and ions into the intestinal lumen, thus causing diarrhea. Invasive bacteria produce virulence factors or toxins, which also modify the actin cytoskeleton as clostridial toxins, but with a different mechanism. Invasive bacteria such as *Salmonella* inject into cells type III virulence factors that activate guanine nucleotide exchange factors (GEFs) and subsequently inactivate GTPase activating proteins (GAPs) like Rho-GTPases that permit internalization of bacteria into a cell. Once inside the cytosol, *Listeria* and *Shigella* lyse the phagocytic vesicle, move into the cytosol, and invade neighboring cells via actin nucleating factors and the host actin machinery. Some bacterial genera, such as *Yersinia* and *Pseudomonas*, escape phagocytosis by injecting type III factors into macrophages that subsequently depolymerize the actin cytoskeleton. Enteropathogenic *E. coli* promote specific cell membrane lesions characterized by microvilli loss and formation of actin pedestals.

bloodstream to the motoneuron endings. After internalization, the enzymatic L chain blocks the release of acetylcholine (see Section 17.8.5), which subsequently leads to a flaccid paralysis.

Enterotoxemias, which frequently occur in animals such as sheep and cattle, result in a profound shock syndrome. Sudden death occurs due to the rapid dissemination of toxins formed in the intestine which circulates throughout the body via the bloodstream into various organs that ultimately fail. The best-characterized enterotoxemia is that caused by *C. perfringens* epsilon-toxin (Table 17.7).

17.11.2.2 Clostridial Pore-Forming Toxins Involved in Gastrointestinal Diseases

The clostridial pore-forming toxins involved in gastrointestinal diseases include: CPE, beta1, beta2, epsilon, and α-septicum toxins.

17.11.2.2.1 C. perfringens *Enterotoxin (CPE)*

CPE, which is responsible for *C. perfringens* food intoxication, is released in the intestinal lumen when the sporulating cells lyze. CPE binds to a membrane receptor identified as claudin (isoforms 3, 4, 6, 7, 8, 14), an essential component of tight junctions existing between cells. This results in the formation of a 90 to 100 kDa complex (small complex) in membranes. A postbinding maturation subsequently occurs when cells are incubated at 37°C, and this consists of intermediate (135 kDa) and then large (160 or 200 kDa) complexes formed by association with membrane proteins. The large complex is significantly resistant to SDS and pronase, perhaps a result of CPE insertion into the membrane. Occludin, a major structural protein of tight junctions, is part of the 200 kDa complex [225]. CPE large complex increases the membrane permeability to small molecules (<200 Da), possibly by pore formation induced by the 160 kDa complex. The loss of small molecules (i.e., amino acids, ions, nucleotides, etc.) inhibits macromolecule synthesis and cytopathic effects leading to morphological changes, permeability alterations for large molecules, cell lysis, and of course diarrhea *in vivo*. In addition, the 200 kDa complex formed in polarized CaCo-2 cells results in the removal of occludin from tight junctions and increased paracellular permeability [226,227]. This mechanism is probably responsible for the intestinal fluid accumulation and epithelium desquamation observed *in vivo*.

17.11.2.2.2 C. perfringens *Beta Toxins*

C. perfringens beta toxins are involved in necrotic enteritis of young animals and humans (also called Pigbel and Darmbrand), as well as sheep enterotoxemia (Table 17.7). The disease is characterized by necrosis and inflammation of the intestinal mucosa with bleeding into the lumen. Beta1 toxin is very labile, as evidenced by protease degradation. For this reason, the beta1-induced pathology is only observed in particular circumstances such as that found in newborns and, in particular, within the digestive tract where protease activity is low. The risk factors involved in human disease include a low-protein diet that induces low tryptic activity in the intestine and consumption of sweet potatoes, which contain a trypsin inhibitor. The low-protease activity permits a high level of biologically active toxin in the intestinal lumen and, as discussed in Section 17.8.1, the beta1 and beta2 toxins probably disrupt intestinal cells by pore formation.

17.11.2.2.3 C. perfringens *Epsilon Toxin*

Epsilon toxin is the major toxin formed by *C. perfringens* types B and D, and unlike beta toxin, epsilon is activated by proteolysis. Fatal enterotoxemia, which is common in sheep, goats, and more rarely in cattle, is caused by *C. perfringens* type D. Overgrowth of *C. perfringens* type D in the intestine of susceptible animals, generally a consequence of overeating foods that contain a large starch or sugar content, concomitantly produce large amounts of epsilon-toxin. Toxin secreted in the intestine increases permeability of the mucosa and then spreads to different organs via the circulatory system, thus causing elevated blood pressure, increased vascular permeability, lung edema, kidney necrosis (pulpy kidney disease in lambs), as well as congestion and edema of the meninges.

Major pathological changes are observed in the brain, and they consist of congestion and edema of the meninges, perivascular and intercellular edema, as well as necrotic foci of the nervous tissue. Epsilon-toxin passes through the blood–brain barrier and accumulates specifically in the brain [228,229]. The neurological disorders (i.e., retraction of the head, opithotonus, convulsions, agonal struggling, hazardous roaming) seem to result from epsilon-toxin action on the hippocampus that leads to an excessive release of glutamate [230].

Mature epsilon-toxin is cytotoxic for MDCK (canine) cells and some murine kidney cell lines. The toxin preferentially binds to the apical cell surface through a specific receptor consisting of a 34 kDa membrane protein or glycoprotein, and subsequently forms large and stable membrane complexes that correspond to heptamerization and pore formation. In contrast to CPE and a heterotrimeric complex, epsilon-toxin seemingly interacts with only one class of membrane protein [203,231].

Epsilon cytotoxicity is associated with a rapid loss of intracellular K^+, but an increase in Cl^- and Na^+, whereas increased Ca^{++} levels occur later. In addition, the loss of viability also correlates with propidium iodide uptake, indicating that the epsilon-toxin forms large pores in cell membranes. Pore formation was also evidenced in artificial lipid bilayers. Epsilon-toxin induces water-filled channels permeable to hydrophilic solutes up to a molecular mass of 1000 Da, which likely represents molecule movement due to general diffusion pores slightly selective for anions [232]. In polarized MDCK cells, epsilon-toxin induces a rapid and dramatic increase in permeability. Pore formation in cell membranes is likely responsible for the permeability change evident in cell monolayers. The actin cytoskeleton and organization of tight and adherens junctions are not altered, and the paracellular permeability to macromolecules is not significantly increased upon epsilon-toxin treatment [202]. Thereby, epsilon-toxin is a very potent toxin that alters the permeability of cell monolayers of epithelium and endothelium origins, finally eliciting an edematous outcome.

Intravenous injection of epsilon-toxin in experimental animals leads to damage and increased permeability of the blood–brain barrier, accumulation of the toxin in the brain, and formation of perivascular edema [228,233–235]. Endothelial cells are likely the primary target, but epsilon-toxin also alters the blood–brain barrier and binds specifically to the brain tissue [228,233]. Epsilon-toxin could directly interact with hippocampus neurons, thus leading to an excessive release of glutamate likely responsible for the nervous symptoms of excitation [203,230,236,237]. However, to date, the precise mechanism of epsilon-toxin on neuronal cells still remains to be elucidated.

17.11.2.2.4 C. septicum *α-Toxin*

C. septicum is involved in nontraumatic myonecrosis of the intestinal mucosa, which occurs in patients with intestinal malignancy, neutropenia, leukemia, or diarrhea. This infection is accompanied by a fulminant and profound shock that is often fatal [238,239]. In animals, *C. septicum* is responsible for an enterotoxemia that is often referred to as bradsot or braxy.

C. septicum α-toxin is secreted as an inactive precursor that binds to a GPI-anchored membrane proteins such as *Aeromonas* aerolysin. After binding to receptor, the prototoxin is cleaved into an active form (41 kDa) by furin, and perhaps other cell proteases. Subsequently, the monomers bound to their receptor oligomerize into a hexamer or heptamer. An amphipathic β-strand of each monomer assembles with a neighboring monomer, and corresponding amphipathic β-strand structures form a β-barrel that inserts into the cell membrane to form a pore with a 10 to 20 Å diameter [109,157].

17.11.2.3 Clostridial Toxins That Modify the Actin Cytoskeleton and Are Involved in Gastrointestinal Diseases

17.11.2.3.1 Clostridial Binary Toxins

The clostridial binary toxins are involved in necrotizing enteritis as well as diarrhea of animals, and are occasionally implicated in human diseases. Iota toxin-producing *C. perfringens* (Type E) cause enterotoxemia in calves and other young animals, and *C. spiroforme* is responsible for a lethal enteritis in rabbits and, perhaps, humans. CDT possibly contributes to *C. difficile* pseudomembranous colitis in humans. C2 toxin induces hemorrhagic and necrotic intestinal lesions experimentally in mice and in naturally acquired botulism among birds [222,240].

The enzymatic components of binary toxins catalyze the mono-ADP-ribosylation of actin monomers specifically at Arg-177; however, polymerized F-actin is not readily modified by these enzymes. Residue Arg-177 is located in the actin–actin binding site that is buried upon formation of actin filaments. Enzymatic components of the iota family modify all actin isoforms that include cellular and muscular versions of the molecule, whereas C2 toxin only interacts with cytoplasmic and smooth muscle γ-actin [241–244]. Actin complexed to gelsolin also represents a substrate for ADP-ribosylation. A cumbersome ADP-ribose moiety on the actin binding site prevents nucleation and polymerization of ADP-ribosylated actin monomers. Moreover, ADP-ribosylated actin acts as a capping protein that binds to the barbed end of a growing actin filament and inhibits further addition of unmodified actin monomers. Depolymerization of actin filaments naturally occurs, and the subsequently released actin monomers are immediately ADP-ribosylated. In addition, ADP-ribosylation inhibits the intrinsic ATPase activity of actin [245]. Microinjection of cells with ADP-ribosylated actin monomers induces the same effects as C2 or iota toxin [245,246]. This results in a complete disassembly of actin filaments and accumulation of monomeric actin (Figure 17.12) [247–250].

The main effects induced by clostridial ADP-ribosylating toxins on cells include: (1) change in morphology (rounding); (2) inhibition of migration and activation of leukocytes [251]; (3) inhibition of smooth muscle contraction; and (4) impairment of endocytosis, exocytosis, as well as cytokinesis. Contraction of the ileum muscle, which is rich in β-actin, is inhibited by C2 toxin. Since this toxin specifically modifies monomeric actin, a dynamic equilibrium between monomeric and polymerized actin is required for smooth muscle contraction in these cell types [252].

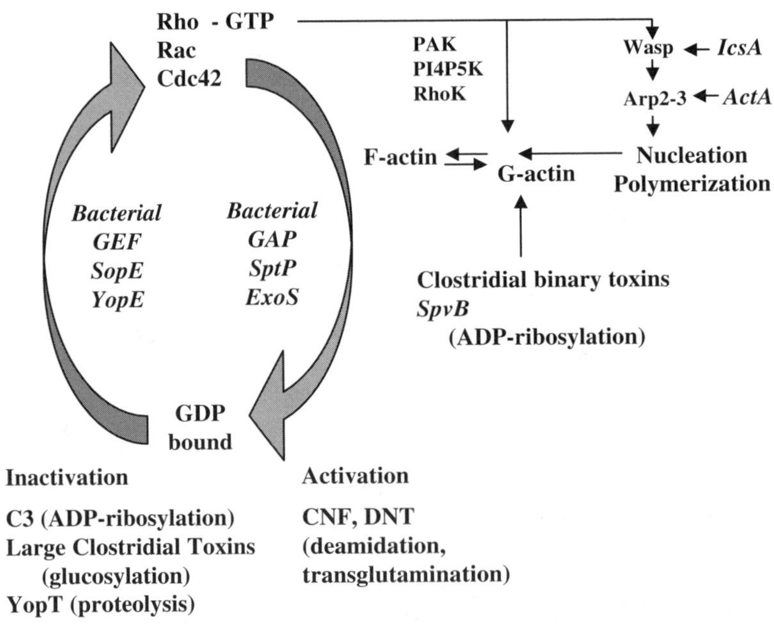

FIGURE 17.12 Bacterial toxins and type III virulence factors that modify the actin cytoskeleton. Bacterial toxins and type III virulence factors directly modify the actin monomers that lead to depolymerization of the actin filaments, induce actin nucleation, or alter the Rho-GTPase cycle. CNF, cytotoxic necrotizing factor from *E. coli*; DNT, dermonecrotic toxin from *B. pertussis*; SpvB, *Salmonella* plasmid virulence factor; Yop, *Yersinia* outer protein.

The depolymerization of actin cytoskeleton by iota-toxin induces a disorganization of tight and basolateral intercellular junctions as well as causes a subsequent increase in permeability of cultured intestinal cell monolayers [199]. In rabbit lungs, C2 toxin causes edema and alteration of the endothelial cell membranes [253]. The massive edema and hypotonic shock observed in animals injected with C2 toxin is a likely result from breakdown of the vascular endothelium barrier [247].

17.11.2.3.2 C3 Enzyme

C. botulinum types C and D produce an additional ADP-ribosyltransferase called C3, the substrate of which is Rho protein. ADP-ribosylation at Asn41 located in the switch I of Rho increases stability of the Rho-GDI (guanine dissociation inhibitor) complex, thus blocking translocation of Rho to the membrane, where it is physiologically activated and interacts with its effectors (Figure 17.12) [254–256]. Therefore, C3 inactivates Rho and depolymerizes the actin cytoskeleton. Curiously, C3 does not efficiently enter cells, and to date there has not been any identified binding (or vector) molecule to facilitate its entry into cells. Therefore, it is not surprising that there is no reported *in vivo* pathology due to C3.

17.11.2.3.3 Large Clostridial Toxins

Pseudomembranous colitis and postantibiotic therapy diarrhea caused by *C. difficile* are the most common nosocomial diseases in industrial countries. A close relative, *C. sordellii*, is associated with some hemorrhagic enteritis in cattle and enterotoxemia in animals.

The N-terminal part of the large clostridial toxins represents the fragment that is active intracellularly. Large clostridial toxins catalyze the glucosylation of 21 kDa G-proteins from UDP-glucose, except α-novyi, which uses UDP-*N*-acetylglucosamine as cosubstrate. ToxA and ToxB glucosylate Rho, Rac, and Cdc42 at Thr-37, whereas LT glucosylates Ras at Thr-35, Rap, Ral, and Rac. The large clostridial toxins cleave the cosubstrate and transfer the glucose moiety to the acceptor amino acid on Rho proteins (Figure 17.12) [174,257,258]. The conserved Thr that is glucosylated, is also located in switch I and Thr37/35 are involved in coordinating Mg^{++} and, subsequently, binding of the β and γ phosphates of GTP. The hydroxyl groups of Thr37/35 are exposed on the surface in its GDP-bound form, which is the only accessible substrate for glucosylating toxins. Glucosylation of Thr35 prevents the recognition of a downstream effector, blocking the G-protein in the inactive form [259]. In addition, glucosylation of GTPase slightly reduces the intrinsic GTPase activity, completely inhibiting hydrolysis by GAP-stimulated GTP [259], and leading to membrane accumulation of the GTP-bound form of Rho, where it is tightly linked [260].

Large clostridial toxins that modify Rho proteins induce cell rounding, with concomitant loss of actin stress fibers, reorganization of the cortical actin, and disruption of the intercellular junctions. ToxA and ToxB disrupt apical as well as basal actin filaments, and subsequently disorganize the ultrastructure and component distribution (ZO-1, ZO-2, occludin, claudin) of tight junctions, whereas E-cadherin junctions are minimally altered by these toxins [261,262].

ToxA and ToxB have been reported to induce apoptosis as a consequence of Rho glucosylation. In addition to the cytoskeletal effects, inactivation of Rho proteins impairs many other cellular functions such as endocytosis, exocytosis, lymphocyte activation, immunoglobulin-mediated phagocytosis in macrophages, NADPH oxidase regulation, smooth muscle contraction, phospholipase D activation, transcriptional activation mediated by JNK and p38 [136].

In monocytes, ToxA stimulates cytokine (TNF-α, IL-1β, IL-6 and IL-8) release and activation of p38 MAP kinase, whereas the activation of ERK and JNK is only transient [263,264]. The p38 activation is required for IL-8 production, IL-1β release, monocyte necrosis, and inflammation of the intestinal mucosa. ToxA-induced p38 activation could be mediated by toxin binding to a membrane receptor independently of Rho-GTPase glucosylation [264]. These effects could be involved in the marked inflammatory response of the intestinal mucosa observed during the natural disease.

17.11.2.4 Comparison with the Non-*Clostridium* Enteropathogenic Bacteria

The digestive tract is one of the main entry points for bacteria into a host organism, and the intestinal mucosa represents one of the first barriers to ensure sterility of host internal tissues. Bacteria have developed various strategies to productively interact with the intestinal mucosa in order to overcome host defenses and subsequently gain access to a hospitable environment well suited for their growth and survival (Figure 17.11). Like *Clostridium*, some bacteria produce potent enterotoxins directly in food, or they colonize the intestinal mucosa and secrete toxins that profoundly alter the enterocyte physiology. Coevolution of some pathogens with a host has led to bacterial adaptation into specific niches sheltered from the immune defenses. Such bacteria usually cause self-limiting disease, in contrast to the toxigenic pathogens that are accidentally encountered during life, which unfortunately can induce severe pathologies. A key pathogenic mechanism is the type III secretion system through which bacteria can inject virulence factors into a targeted cell, which modulates cellular functions. A common target for bacterial toxins and type III virulence factors is the actin cytoskeleton, which is involved in many cellular functions.

In contrast to clostridial toxins, which disrupt the actin filaments and markedly damage protective cellular barriers, invasive bacteria use virulence factors to hijack the actin cytoskeletal machinery for their own benefit that includes entry into cells, intracellular motility, as well as subsequent intercellular dissemination, and antiphagocytosis. Among the large number of molecules involved in controlling the actin cytoskeleton, the bacterial type III virulence factors, like clostridial toxins, target two main classes of proteins, which include actin monomers and Rho-GTPases. By comparison with clostridial toxins, the main paradigms of enteropathogenicity due to nonclostridial virulence factors will now be presented.

17.11.2.4.1 Intoxication via Preformed Toxin in Food

Bacteria other than *Clostridium* species, such as *S. aureus* and *B. cereus*, can grow and produce protein toxins in food. Staphylococcal enterotoxins (SEs), which are preformed in food improperly stored and contaminated by toxigenic *S. aureus*, are responsible for a major form of food poisoning in humans found throughout the world. They have a potent superantigenic activity that occurs at the cell surface by cross-linking receptor from Vβ-specific T lymphocytes with class II molecules of antigen-presenting cells. This results in a profound proliferation of T lymphocytes and a massive release of proinflammatory cytokines [265]. If present in large enough quantities, proinflammatory cytokines can induce an ill feeling that can include, in severe cases, toxic shock and potentially death. Remarkably, the role played by SE in enteritis is poorly understood, but these toxins seem to affect sympathic nerve endings in the intestinal mucosa that subsequently triggers emesis and diarrhea.

Enterotoxigenic *B. cereus* strains produce an emetic dodecapeptide and protein enterotoxins; the latter are still poorly characterized and include a tripartite hemolytic (BL), tripartite nonhemolytic (Nhe), and single-chain proteins [266]. It is known that the NheC component of Nhe is a metalloprotease related to *C. histolyticum* and *C. perfringens* collagenases [267].

17.11.2.4.2 Colonizing and Enterotoxigenic Gram-Negative Bacteria

Enterotoxigenic Gram-negative bacteria (i.e., *E. coli and V. cholerae*) possess adherence factors (pili) that facilitate colonization of the intestinal mucosal surface. Additionally, they also produce potent toxins such as CT, as well as heat labile and heat stable (STa and STb) enterotoxins that disturb the homeostasis of intestinal cells. Indeed, pathogens like these that produce both adherence factors and toxins represent formidable foes for both the medical and veterinary community.

CT as well as other heat labile toxins act intracellularly and, in the presence of the membrane factor ARF (ADP-ribosylating factor), ADP-ribosylate the active subunit Gsα of the heterotrimeric GTP-binding proteins that positively control adenylcyclase activity. ADP-ribosylated Gsα has a reduced GTPase activity and, therefore, remains in its active GTP-bound form. This results in a permanent adenylcyclase activity in cells, elevation of cAMP, and ultimately Cl^- and H_2O secretion [158].

Other enterotoxigenic *E. coli* secrete STa and STb, which are short compact peptides with three disulfide bridges. STa acts as a hormone (guanylin) and binds to the extracellular domain of the guanylate cyclase on the apical membrane of enterocytes. This induces a conformational change in the protein kinase-like domain, resulting in an uncontrolled increase of the guanylate cyclase activity, elevated cGMP levels, and a subsequent increase in Cl^- and H_2O secretion. These toxins are responsible for enteritis with a concomitant severe diarrhea [268,269].

In essence, these toxins modify a key function of enterocytes, which leads to a massive secretion of water and electrolytes by interacting with essential molecules intimately involved in cell homeostasis, signaling, and hormone recognition.

17.11.2.4.3 Enteroinvasive Bacteria and Modulation of the Actin Cytoskeleton

Enteroinvasive bacteria first specifically adhere to intestinal cells and ingeniously induce a local rearrangement of the actin cytoskeleton, which mediates bacterial engulfment into the cell. Two main mechanisms have been described for this process. The first involves a "triggering" process in which bacteria inject type III virulence factors into targeted cells that subsequently modulate the actin cytoskeleton, whereas the "zippering" mechanism results from the interaction of bacterial surface proteins with cell receptors (generally integrin), which directly transduce signals to the cell cytoskeleton [270]. Toxins and type III virulence factors of invasive bacteria act on the same intracellular target as those used by the clostridial toxins that modify the actin cytoskeleton (actin monomers and Rho-GTPases). These microbial proteins are involved in the invasion process and also prevent phagocytosis, ultimately enabling the bacterium to escape host defenses once they have crossed the intestinal barrier. Some bacteria use the actin machinery of the cell for entry (intracellular motility) or to develop particular membrane structures (actin pedestal). *Clostridium* species, as well as invasive and enteropathogenic Gram-negative bacteria, produce various toxins and virulence factors that commonly modulate the actin cytoskeleton, but this is accomplished by different mechanisms that lead to various enteropathogenic effects such as disruption of the cell barrier, cell invasion, antiphagocytosis, intracellular motility, and modification of cell membranes by enteroinvasive and enteropathogenic bacteria.

17.11.2.4.3.1 Toxins and Virulence Factors Involved in Bacterial Invasion of Cells

E. coli CNF is secreted by an unknown mechanism and enters cells by receptor-mediated endocytosis. CNF catalyzes the deamidation of Gln61/63 within switch II of Rho, Rac, and Cdc42 [271,272]. Therefore, the GTPase activity is abolished and Rho-GTPases are blocked in their active forms. Like the large clostridial toxins and C3 enzyme, CNF directly modifies Rho-GTPases, but this is done by a different enzymatic reaction (deamidation) that activates Rho-GTPases. Increased actin polymerization generates abundant stress fibers and lamellipodia structures that cause cellular dysfunctions such as higher cell motility, increased endocytosis, and modification of the intercellular junctions. However, deamidated Rho-GTPases are rapidly inactivated by ubiquitylation and degradation by the proteasome. Thereby, CNF induces a transient activation of Rho-GTPases that probably enhances bacterial internalization [273].

Salmonella provides the paradigm of "triggering" entry into epithelial cells, which involves type III virulence factors. These microbes increase and decrease the actin polymerization in a coordinated manner, thus permitting the assembly-disassembly of lamellipodia and subsequent endocytic entry of bacteria. When *Salmonella* adheres to intestinal cells, the bacteria inject into cells via the type III secretion protein SopE, which consists of two isoforms designated as SopE1 and SopE2. SopE exerts a GTP exchange factor (GEF) activity toward Rac and Cdc42, which stimulates actin polymerization and formation of membrane ruffling required for the engulfment of *Salmonella* [274,275]. Although SopE does not share any similar structure with eukaryotic GEFs, it uses a similar mechanism of action. SopE binds to switch I and II of Rho-GTPases and introduces a loop containing the GAGA motif, which reorientates switch I and II, thus releasing GDP; whereas the catalytic core of Rho-GEFs consists of an α-helix containing a critical Lys residue needed for enzymatic activity [276]. Rho-GTPases then spontaneously bind GTP, which is in higher concentration than GDP in the cytosol.

When in the endocytic vesicle, *Salmonella* reverses the activation of lamellipodia by injecting the type III *S. typhimurium* effector protein (Sptp). The Sptp N-terminus contains a GAP domain, which downregulates membrane ruffling by enhancing the GTPase activity of Rac and Cdc42, leading to the passage of Rho-GTPases in their GDP inactive form. Sptp is structurally unrelated to eukaryotic GAPs. However, Sptp uses an Arg finger located between the switch I and II of the Rho-GTPases to stabilize Gln61/63 in a favorable catalytic conformation. The catalytic Arg is localized on the top of an α-helix, which forms with three other α-helices a protruding bulge that interacts with the Rac active site. This is in contrast with eukaryotic GAPs in which the Arg residue is positioned on an exposed loop. In both cases, the conserved Arg plays a critical role in GAP activity [102,277].

Additionally, *Salmonella* secretes another type III factor, SpvB, which is an ADP-ribosyltransferase specific for nonmuscular actin. SpvB retains the critical residues involved in the activity of *Clostridium* and *Bacillus* ADP-ribosylating toxins, such as the Ser-Thr-Ser motif flanked by the downstream catalytic biglutamic and upstream Arg residues. Like clostridial ADP-ribosyltransferases, SpvB depolymerizes actin filaments. SptP and SpvB reverse the membrane ruffling and actin cytoskeletal reorganization mediated by bacterial entry, thus permitting the infected cells to regain their normal architecture after invasion [278].

17.11.2.4.3.2 Virulence Factors Targeting the Actin Cytoskeleton and Involved in Antiphagocytosis

Yersinia and *Pseudomonas* adhere to macrophages and inject type III GAP molecules, YopE and ExoS-ExoT, respectively. YopE is a 23 kDa protein that only possesses a GAP domain, whereas the other bacterial GAPs are bifunctional proteins. SptP, ExoS, and ExoT have in common an N-terminal GAP domain that is structurally related to that of YopE but possess an additional C-terminal domain, which is a tyrosine phosphatase in SptP and an ADP-ribosyltransferase domain in ExoS and ExoT. Structure of the α-helix bundle containing the catalytic Arg and their mode of interaction with the active site of Rho-GTPases is conserved in the bacterial GAPs. YopE, ExoS, and ExoT exhibit a GAP activity toward several proteins, including Rho, Rac, and Cdc42. They disorganize the actin filaments in macrophages, thus preventing bacterial engulfment. These type III bacterial GAPs are involved in the antiphagocytic effects of *Yersinia* and *Pseudomonas* [279–282].

In addition, *Yersinia* produce another type III factor, YopT, which is also implicated in downregulating of the actin cytoskeleton and antiphagocytic activity. YopT is a cysteine protease that specifically cleaves the prenylated Rho-GTPases near the C-terminus. Therefore, YopT prevents the distribution of Rho-GTPases to the membrane and their subsequent interaction with effector molecules [283]. YopT represents an additional model of toxin interaction with Rho-GTPases that include inactivation by proteolysis.

Intracellular bacteria such as *Shigella* and *Listeria* escape the endocytic vesicles, whereas *Salmonella* remain embedded in the vesicles and undergo a transcytosis into the bloodstream and, subsequently, to other tissues. *Listeria* uses a phospholipase C (listeriolysin) from the CBC family, which also encompasses clostridial hemolysins like PFO, to lyze the vesicle membrane.

17.11.2.4.3.3 Actin Nucleating Factors and Intracellular Motility

Inside the cytosol, *Shigella* and *Listeria* escape the phagocytic vesicle and spread from one cell to an adjacent neighbor by using the host actin cytoskeleton machinery. For accomplishing this feat, the bacteria express molecules on their surface, such as IcsA in *Shigella* and ActA in *Listeria*, which trigger actin nucleation, polymerization, and formation of an actin tail corresponding to the motile force of the bacteria. In physiological conditions, the nucleation of actin monomers is required for further polymerization of actin filaments. Two partners that assemble into a complex and have an essential role in the nucleation activity are N-Wasp, activated by binding to Cdc42, and the Arp2/3 complex. IcsA recruits the cellular nucleating factors activated by N-Wasp and

Arp2/3, thus promoting the formation of actin tails to one pole of the bacterium. ActA induces similar effects, but mimics N-Wasp and directly recruits the Arp2/3 complex [284].

17.11.2.4.3.4 A Particular Actin Structure Induced by Enteropathogenic E. coli

Enteropathogenic *E. coli* (EPEC) and enterohemorrhagic *E. coli* (EHEC) attach to intestinal cells and colonize the intestinal mucosa like enterotoxigenic Gram-negative bacteria. Moreover, EHEC secrete VT, which acts locally and systemically. Although EPEC and EHEC remain extracellularly, they produce modifications of the cell membrane that are closely related to those caused by invasive bacteria. They induce a loss of microvilli (attaching and effacing lesions) and subsequent remodeling of the cell membrane and underlying actin cytoskeleton, called the actin pedestal. The lesions correlate with intestinal colonization by the bacteria and onset of the clinical diarrheal enteritis. EPEC inject into host cells, through the type III secretion mechanism, a translocated intimin receptor (Tir) which localizes at the cell membrane and functions as a receptor for intimin found on the bacterial surface. In addition, Tir plays a central role in triggering formation of the actin pedestal. Tir phosphorylated on Tyr-474 binds to the adaptor protein Nck, which stimulates N-WASP and then activates the Arp2-3 complex independently of the Rho-GTPases. This promotes actin nucleation and polymerization required for pedestal formation with many other regulating proteins, focal adhesion, and lipid raft associated proteins whose functions are still unclear [285].

17.12 CONCLUDING REMARKS

Clostridium is the bacterial genus that produces the greatest number of toxins with numerous biological activities, as well as those that are most potent. However, only a few *Clostridium* species are toxigenic, which corresponds to the general property that the *Clostridia* are essentially environmental bacteria. Therefore, clostridial diseases are likely the result of an accidental encounter between a toxigenic *Clostridium* and a potential host that possesses specific risk factors.

Are clostridial toxins fundamentally different from other toxins produced by Gram-positive or Gram-negative bacteria? Like the other bacterial toxins, numerous clostridial toxins act on the cell surface by damaging membranes and, thus, are mainly involved in gangrene and necrotic enteritis. Most of these toxins are pore-forming proteins related to other bacterial pore-forming toxins. From the bacterium's perspective, pore-forming activity can be used to extract nutrients from cells required for bacterial growth, or these "holes" can be used to directly inject specific proteins into cells, which subsequently causes cell damage and lysis. The particular mode of action employed by many bacterial toxins is to cross the cell membrane and then elicit biological effects intracellularly.

The translocation machinery coupled to a receptor binding domain seems to derive from an ancestral pore-forming toxin. Perhaps some evidence of this is seen with the binary toxins, which are the result of a functional complementation between a binding component highly similar to *bona fide* pore-forming toxins and a separate enzymatic component. The single-chain intracellularly active toxins, such as clostridial neurotoxins, probably originated from a genetic fusion between binding and enzymatic proteins. Indeed, active clostridial neurotoxins require proteolytic cleavage and reduction of a cystine bond, which subsequently results in two chains that then possess a feature related to that of the binary toxins. A more complex fusion could have led to the intracellularly active large clostridial toxins.

Do clostridial toxins exhibit distinct enzymatic activities? Clostridial and non-clostridial toxins share ADP-ribosylation, phospholipase, and protease activities, but the clostridial toxins recognize different specific targets. For example, clostridial neurotoxins are zinc-dependent proteases with an active structural site similar to that of other metalloproteases. However, unlike the other proteases, these neurotoxins are highly specific regarding substrate, and only cleave specific types of neuronal proteins. The glucosyltransferase activity is also unique in clostridial toxins, since only the large clostridial toxins possess this activity. A myriad of intracellular compounds represents potential substrates for toxins; however, the actin cytoskeleton represents a privileged target for many toxins.

Among the numerous partners involved in regulating actin polymerization, the various toxins that modify the actin cytoskeleton only act at two essential levels, which include the actin monomers and the regulatory Rho-GTPases.

Several toxins can use the same cellular target yet elicit different effects upon a targeted cell. Bacteria such as Clostridia secrete actin-modifying exotoxins, that act from a distance by first entering cells via receptor-mediated endocytosis ultimately disrupting cell barriers and tissues, which then facilitates a massive bacterial colonization in necrotic tissues. In contrast, invasive bacteria use specific toxins or type III virulence factors that induce limited modulation of the actin cytoskeleton, resulting in bacterial invasion or escape from host defenses.

Finally, the multiple potencies of bacterial toxins open the way to various applications, which include therapy (i.e., use of botulinum neurotoxins and immunotoxins), cell biology tools useful for dissecting cell trafficking pathways, and transport of heterologous molecules into cells. Clearly, over time, there has been much gleaned by studying bacterial toxins, but it is clear that we do not possess all the answers, so we forge on.

REFERENCES

1. Alouf, J.E., Milestones in clostridial toxin research, in *Protein Toxins of the Genus Clostridium and Vaccination,* Duchesnes, C., Mainil, J., Popoff, M.R., and Titball, R., Eds., Presses de la Faculté de Médecine Vétérinaire, Liège, Belgium, 2003, 17.
2. Stackebrandt, E. and Hippe, H., Taxonomy and systematics, in *Clostridia,* Bahl, H. and Dürre, P., Eds., Willey-VCH, Weinheim, Germany, 2001, 19.
3. Mitchell, W.J., Biology and Physiology, in *Clostridia,* Bahl, H. and Dürre, P., Eds., Wiley-VCH, Weinheim, Germany, 2000, 49.
4. Brüggemann, H. et al., The genome sequence of *Clostridium tetani*, the causative agent of tetanus disease, *Proc. Natl. Acad. Sci. U.S.A.,* 100, 1316, 2003.
5. Shimizu, T. et al., Complete genome sequence of *Clostridium perfringens,* an anaerobic flesh-eater, *Proc. Natl. Acad. Sci. U.S.A.,* 99, 996, 2002.
6. Nölling, J. et al., Genome sequence and comparative analysis of the solvent-producing bacterium *Clostridium acetobutylicum*, *J. Bacteriol.,* 183, 4823, 2001.
7. Dobrindt, U. and Hacker, J., Plasmids, phages and pathogenicity islands: lessons on the evolution of bacterial toxins, in *The Comprehensive Source Book of Bacterial Protein Toxins,* Alouf, J.E. and Freer, J.H., Eds., Academic Press, San Diego, 1999, 3.
8. Rood, J.R., Virulence genes of *Clostridium perfringens*, *Annu. Rev. Microbiol.,* 52, 333, 1998.
9. Justin, N. et al., The first strain of *Clostridium perfringens* isolated from an avian source has an alpha-toxin with divergent structural and kinetics properties, *Biochem.,* 41, 6253, 2002.
10. Tsutsui, K. et al., Phylogenetic analysis of phospholipase C genes from *Clostridium perfringens* types A to E and *Clostridium novyi*, *J. Bacteriol.,* 177, 7164, 1995.
11. Petit, L. et al., *Clostridium perfringens*: toxinotype and genotype, *Trends Microbiol.,* 7, 104, 1999.
12. Popoff, M.R., Molecular Biology of Actin-ADP-Ribosylating Toxins, in *Bacterial Protein Toxins,* Aktories, K. and Just, I., Eds., Springer, Berlin, 2000, 275.
13. Miyamoto, K. et al., Organization of the plasmid *cpe* locus in *Clostridium perfringens* type A isolates, *Infect. Immun.,* 70, 4261, 2002.
14. Billington, S.J. et al., *Clostridium perfringens* type E animal enteritis isolates with highly conserved, silent enterotoxin gene sequences, *Infect. Immun.,* 66, 4531, 1998.
15. Gibert, M. et al., *Clostridium spiroforme* toxin genes are related to *C. perfringens* iota toxin genes but have a different genomic localization, *Syst. Appl. Microbiol.,* 20, 337, 1997.
16. Brynestadt, S. and Granum, P.E., Evidence that Tn5565, which includes the enterotoxin gene in *Clostridium perfringens*, can have a circular form, which may be a transposition intermediate, *FEMS Microbiol. Lett.,* 170, 281, 1999.
17. Schallehn, G. et al., Phage conversion of *Clostridium novyi* type A, *Zentralbl. Bakteriol.,* 247, 95, 1980.
18. Braun, M. et al., Definition of the single integration site of the pathogenicity locus in *Clostridium difficile*, *Gene,* 181, 29, 1996.

19. Thelestam, M. et al., Clostridial toxins acting on the cytoskeleton, in *The Comprehensive Sourcebook of Bacterial Protein Toxins,* Alouf, J.E. and Freer, J.H., Eds., Academic Press, London, 1999, 147.
20. Mani, N. and Dupuy, B., Regulation of toxin synthesis in *Clostridium difficile* by an alternative RNA polymerase sigma factor, *Proc. Ntl. Acad. Sci. U.S.A.,* 98, 5844, 2001.
21. Spigaglia, P. and Mastrantonio, P., Molecular analysis of the pathogenicity locus and polymorphism in the putative negative regulator of toxin production (TcdC) among *Clostridium difficile* clinical isolates, *J. Clin. Microbiol.,* 40, 3470, 2003.
22. Moncrief, J.S. et al., Genetic characterization of toxin A-negative, toxin B-positive *Clostridium difficile* isolates by PCR, *J. Clin. Microbiol.,* 38, 3072, 2000.
23. Barbut, F. et al., Prevalence and genetic characterization of toxin A variant strains of *Clostridium difficile* among adults and children with diarrhea in France, *J. Clin. Microbiol.,* 40, 2079, 2002.
24. Johnson, S. et al., International typing study of toxin A-negative, toxin B-positive *Clostridium difficile* variants, *J. Clin. Microbiol.,* 41, 1543, 2003.
25. Rupnik, M. et al., A novel toxinotyping scheme and correlation of toxinotypes with serogroups of *Clostridium difficile* isolates, *J. Clin. Microbiol.,* 36, 2240, 1998.
26. Rupnik, M. et al., Comparison of toxinotyping and PCR ribotyping of *Clostridium difficile* strains and description of novel toxinotypes, *Microbiol.,* 147, 439, 2001.
27. Rupnik, M. et al., New types of toxin A-negative, toxin B-positive strains among *Clostridium difficile* isolates from Asia, *J. Clin. Microbiol.,* 41, 1118, 2003.
28. Sambol, S.P. et al., Toxin gene analysis of a variant strain of *Clostridium difficile* that causes human clinical disease, *Infect. Immun.,* 68, 5480, 2000.
29. Braun, V. et al., A chimeric ribozyme in *Clostridium difficile* combines features of group I introns and insertion elements, *Mol. Microbiol.,* 36, 1447, 2000.
30. Popoff, M.R. and Marvaud, J.C., Structural and genomic features of clostridial neurotoxins, in *The Comprehensive Sourcebook of Bacterial Portein Toxins,* Alouf, J.E. and Freer, J.H., Eds., Academic Press, London, 1999, 174.
31. Dineen, S.S. et al., Neurotoxin gene clusters in *Clostridium botulinum* type A strains: sequence comparison and evolutionary implications, *Cur. Microbiol.,* 46, 342, 2003.
32. Bennik, M. et al., *The genome sequence of proteolytic Clostridium botulinum ATCC3502 (Hall): some highlights*, Presented at the Clostridia 03 Pathogenesis, Woods Hole, MA, 2003.
33. Hutson, R.A. et al., Genetic characterization of *Clostridium botulinum* type A containing silent type B neurotoxin gene sequences, *J. Biol. Chem.,* 271, 10786, 1996.
34. Henderson, I. et al., The genetic basis of toxin production in *Clostridium botulinum* and *Clostridium tetani*, in *The Clostridia: Molecular Biology and Pathogenesis,* Rood, I., Ed., Academic Press, New York, 1997, 261.
35. Fujinaga, Y. et al., Type A and B neurotoxin genes in a *Clostridium botulinum* type AB strain, *Biochem. Biophys. Res. Commun.,* 213, 737, 1995.
36. Franciosa, G. et al., Detection of type A, B, and E botulism neurotoxin genes in *Clostridium botulinum* and other *Clostridium* species by PCR: evidence of unexpressed type B toxin genes in type A toxigenic organisms, *J. Clin. Microbiol.,* 32, 1911, 1994.
37. Jovita, M.R. et al., Gene organization and sequence determination of the two botulinum neurotoxin gene clusters in *Clostridium botulinum, Cur. Microbiol.,* 36, 226, 1998.
38. Hauser, D. et al., Comparative analysis of C3 and botulinal neurotoxin genes and their environment in *Clostridium botulinum* C and D., *J. Bacteriol.,* 175, 7260, 1993.
39. Arber, W., Genetic variation: molecular mechanisms and impact on microbial evolution., *FEMS Microbiol. Rev.,* 24, 1, 2000.
40. Arber, W., Evolution of prokaryotic genomes, *Cur. Topics Microbiol. Immunol.,* 264, 1, 2002.
41. Franciosa, G. et al., Recovery of a strain of *Clostridium botulinum* producing both neurotoxin A and neurotoxin B from canned macrobiotic food, *Appl. Environ. Microbiol.,* 63, 1148, 1997.
42. Yamakawa, K. et al., Emergence of *Clostridium botulinum* type B-like nontoxigenic organisms in a patient with type B infant botulism, *J. Clin. Microbiol.,* 35, 2163, 1997.
43. Brefort, G. et al., Characterization and transferability of *Clostridium perfringens* plasmids., *Plasmid,* 1, 52, 1977.
44. Brynestad, S. et al., Enterotoxin plasmid from *Clostridium perfringens* is conjugative, *Infect. Immun.,* 69, 3483, 2001.

45. Eklund, M.W., Evidence for plasmid-mediated toxin and bacteriocin production in *Clostridium botulinum* type G, *Appl. Environ. Microbiol.*, 54, 1405, 1988.
46. Finn, C.W. et al., The structural gene for tetanus neurotoxin is on a plasmid, *Science,* 224, 881, 1984.
47. Eklund, M.W., Bacteriophage and the toxigenicity of *Clostridium botulinum* type C, *Science,* 172, 480, 1971.
48. Eklund, M.W. et al., Interspecies conversion of *Clostridium botulinum* type C to *Clostridium novyi* type A by bacteriophage, *Science,* 172, 480, 1974.
49. Eklund, M.W. and Poysky, F.T. Interconversion of type C and D strains of *Clostridium botulinum* by specific bacteriohages, *Appl. Microbiol.,* 27, 251, 1974.
50. Eklund, M.W. and Dowell, V.R., *Avian Botulism,* Charles C. Thomas, Springfield, IL, 1987.
51. Lyras, V.A. et al., The clostridial mobilisable transposons, *Cell. Mol. Life Sci.,* 59, 2033, 2002.
52. Poulet, S. et al., Sequences of the botulinal neurotoxin E derived from *Clostridium botulinum* type E (strain Beluga) and *Clostridium butyricum* (strains ATCC43181 and ATCC43755), *Biochem. Biophys. Res. Commun.,* 183, 107, 1992.
53. Hauser, D. et al., Plasmid localization of a type E botulinal neurotoxin gene homologue in toxigenic *Clostridium butyricum* strains, and absence of this gene in non-toxigenic *C. butyricum* strains, *FEMS Microbiol. Lett.,* 99, 251, 1992.
54. Zhou, Y. et al., Transfer of neurotoxigenicity from *Clostridium butyricum* to a nontoxigenic *Clostridium botulinum* type E-like strain, *Appl. Environ. Microbiol.,* 59, 3825, 1993.
55. Zhou, Y. and Johnson, E.A., Genetic transformation of *Clostridium botulinum* Hall A by electroporation, *Biotech. Lett.,* 15, 121, 1993.
56. Moriishi, K. et al., Mosaic structures of neurotoxins produced from *Clostridium botulinum* types C and D, *Biochem. Biophys. Acta,* 1307, 123, 1996.
57. Kubota, T. et al., Mosaic structure of the nontoxic-nonhemagglutinating component gene in *Clostridium botulinum* type A strain isolated from infant botulism in Japan, *Biochem. Biophys. Res. Commun.,* 224, 843, 1996.
58. Payne, D., Regulation of bacterial toxin synthesis by iron, in *Bacterial Protein Toxins,* Burns, D.L., Barbieri, J.T., Iglewski, B.H., and Rappuoli, R., Eds., ASM Press, Washington, DC, 2003, 25.
59. Frank, D.W. and Hunt, M.L., AraC family regulators and transcriptional control of bacteria virulence determinants, in *Bacterial Protein Toxins,* Burns, D.L., Barbieri, J.T., Iglewski, B.H., and Rappuoli, R., Eds., ASM Press, Washington, DC, 2003, 39.
60. Helmann, J.D., The extracytoplasmic function (ECF) sigma factors, *Adv. Microbial. Physiol.,* 46, 47, 2002.
61. Stibitz, S., Two-component systems, in *Bacterial Protein Toxins,* Burns, D.L., Barbieri, J.T., Iglewski, B.H., and Rappuoli, R., Eds., ASM Press, Washington, DC, 2003, 3.
62. Hirst, T., Cholera toxin and *Escherichia coli* heat-labile enterotoxin, in *The Comprehensive Sourcebook of Bacterial Protein Toxins,* Alouf, J.E. and Freer, J.H., Eds., Academic Press, London, 1999, 104.
63. McClane, B. and Rood, J.R., Clostridial toxins involved in human enteric and histotoxic infections, in *Clostridia,* Bahl, H. and Dürre, P., Eds., Willey-VCH, Weinheim, Germany, 2001, 169.
64. Zhao, Y. and Melville, S.B., Identification and characterization of sporulation-dependent promoters upstream of the enterotoxin gene (*cpe*) of *Clostridium perfringens, J. Bacteriol.,* 180, 136, 1998.
65. Brynestad, S. et al., A complex array of Hpr consensus DNA recognition sequences proximal to the enterotoxin gene in *Clostridium perfringens* type A, *Microbiol.,* 140, 97, 1994.
66. Moncrief, J.S. et al., Positive regulation of *Clostridium difficile* toxins, *Infect. Immun.,* 65, 1105, 1997.
67. Dupuy B. and Sonenshein, A.L., Regulated transcription of *Clostridium difficile* toxin genes, *Mol. Microbiol.,* 27, 107, 1998.
68. Mani, N. et al., Environmental response and autoregulation of *Clostridium difficile* TxeR, a sigma factor for toxin gene expression, *J. Bacteriol.,* 184, 5971, 2002.
69. Hauser, D. et al., Botulinal neurotoxin C1 complex, clostridial neurotoxin homology and genetic transfer in *Clostridium botulinum, Toxicon,* 33, 515, 1995.
70. Bhandari, M. et al., Molecular characterization of the clusters of genes encoding the botulinum neurotoxin complex in *Clostridium botulinum (Clostridium argentinense)* type G and nonproteolytic *Clostridium botulinum* type B, *Curr. Microbiol.,* 35, 207, 1997.

71. East, A.K. et al., Organization and phylogenetic interrelationships of genes encoding components of the botulinum toxin complex in proteolytic *Clostridium botulinum* types A, B, and F: evidence of chimeric sequences in the gene encoding the nontoxic nonhemagglutinin component, *Int. J. Syst. Bacteriol.,* 46, 1105, 1996.
72. Hundsberger, T. et al., Transcription analysis of the genes *tcdA-E* of the pathogenicity locus of *Clostridium difficile*, *Eur. J. Biochem.,* 244, 735, 1996.
73. Marvaud, J.C. et al., *botR* is a positive regulator of botulinum neurotoxin and associated non toxic protein genes in *Clostridium botulinum* A, *Mol. Microbiol.,* 29, 1009, 1998.
74. Marvaud, J.C. et al., *tetR* is a positive regulator of the tetanus toxin gene in *Clostridium tetani* and is homologous to *botR*, *Infect. Immun.,* 66, 5698, 1998.
75. Banu, S. et al., Identification of novel VirR/VirS-regulated genes in *Clostridium perfringens*, *Mol. Microbiol.,* 35, 854, 2000.
76. Rood, J.I. and Lyristis, M., Regulation of extracellular toxin production in *Clostridium perfringens*, *Trends Microbiol.,* 3, 192, 1995.
77. Shimizu, T. et al., Proteome and transcriptome analysis of the virulence genes regulates by the VirR/VirS system in *Clostridium perfringens*, *J. Bacteriol.,* 184, 2587, 2002.
78. Gibert, M. et al., *Clostridium perfringens* iota-toxin requires activation of both binding and enzymatic components for cytopathic activity, *Infect. Immun.,* 68, 3848, 2000.
79. Ohtani, K. et al., The VirR/VirS regulatory cascade affects transcription of plasmid-encoded putative virulence genes in *Clostridium perfringens* strain 13, *FEMS Microbiol. Lett.,* 222, 137, 2003.
80. Shimizu, T. et al., Clostridial VirR/VirS regulon involves a regulatory RNA molecule for expression of toxins, *Mol. Microbiol.,* 43, 257, 2002.
81. Johnson, E.A. and Bradshaw, M., *Clostridium botulinum* and its neurotoxin: a metabolic and cellular perspective, *Toxicon,* 39, 1703, 2001.
82. Karlsson, S. et al., Expression of *Clostridium difficile* toxins A and B and their sigma factor TcdD is controlled by temperature, *Infect. Immun.,* 71, 1784, 2003.
83. Karlsson, S. et al., Suppression of toxin production in *Clostridium difficile* VPI 10463 by amino acids, *Microbiol.,* 145, 1683, 1999.
84. Yamakawa, K. et al., Inhibition of enhanced toxin production by *Clostridium difficile* in biotin-limited conditions, *J. Med. Microbiol.,* 47, 767, 1998.
85. Leppla, S.A., Anthrax toxin, in *Bacterial Protein Toxins,* Aktories, K. and Just, I., Eds., Springer, Berlin, 2000, 445.
86. Payne, D., Regulation of bacterial toxin synthesis by iron, in *Bacterial Protein Toxins,* Burns, D.L., Barbieri, J.T., Iglewski, B.H., and Rappuoli, R., Eds., ASM Press, Washington, DC, 2003, 25.
87. Patterson-Curtis, S.I. and Johnson, E.A., Regulation of neurotoxin and protease formation in *Clostridium botulinum* Okra B and Hall by arginine, *Appl. Environ. Microbiol.,* 55, 1544, 1989.
88. Porfirio, Z. et al., Specific peptides of casein pancreatic digestion enhance the production of tetanus toxin, *J. Appl. Microbiol.,* 83, 678, 1997.
89. Pugsley, A., The complete general secretory pathway in Gram-negative bacteria, *Microbiol. Rev.,* 57, 50, 1993.
90. van Wely, K.H.M. et al., Translocation of proteins across the cell envelope of Gram-positive bacteria, *FEMS Microbiol. Lett.,* 25, 437, 2001.
91. Sandkvist, M., Biology of type II secretion, *Mol. Microbiol.,* 40, 271, 2001.
92. Hueck, C.J., Type III protein secretion systems in bacterial pathogens of animals and plants, *Microbiol. Mol. Biol. Rev.,* 62, 379, 1998.
93. Lee, C.A., Type III secretion systems: machines to deliver bacterial proteins into eukaryotic cells?, *Trends Microbiol.,* 5, 148, 1997.
94. Madden, J.C. et al., Cytolysin-mediated translocation (CMT): a functional equivalent of type III secretion in Gram-positive bacteria, *Cell,* 104, 143, 2001.
95. Christie, P.J. and Covacci, A., Bacterial type IV secretion systems: DNA conjugation machines adapted for export of virulence factors, in *Cellular Microbiology,* Cossart, P., Boquet, P., Normark, S., and Rappuoli, R., Eds., ASM Press, Washington, DC, 2000, 265.
96. Call, J.E. et al., *In situ* characterization of *Clostridium botulinum* neurotoxin synthesis and export, *J. Appl. Bacteriol.,* 79, 257, 1995.

97. Tan, K.S. et al., Evidence for holin function of *tcdE* gene in the pathogenicity of *Clostridium difficile*, *J. Med. Microbiol.,* 50, 61619, 2000.
98. Mukherjee, K. et al., Proteins released during high toxin production in *Clostridium difficile*, *Microbiol.,* 148, 2245, 2002.
99. Holmes, R.K. et al., The *E. coli/Vibrio cholerae* family of enterotoxins, in *Molecular Toxins in Foods and Feeds-Cellular and Molecular Modes of Action,* Pohland, E.E., Vulus, R.D., and Richard, J.L., Eds., Plenum, New York, 1991, 91.
100. Gibert, M. et al., Beta2 toxin, a novel toxin produced by *Clostridium perfringens*, *Gene,* 203, 65, 1997.
101. Ludwig, A. and Goebel, W., The family of the multigenic encoded RTX toxins, in *The Comprehensive Sourcebook of Bacterial Protein Toxins,* Alouf, J.E. and Freer, J.E., Eds., Academic Press, London, 1999, 330.
102. Stebbins, C.E. and Galan, J.E., Structural mimicry in bacterial virulence, *Nature,* 412, 701, 2001.
103. Alouf, J.E., Molecular features of the cytolytic pore forming bacterial protein toxins, *Fol. Microbiol.,* 48, 5, 2003.
104. Rossjohn, J. et al., Structure of a cholesterol-binding thiol-activated cytolysin and a model of its membrane form, *Cell,* 89, 685, 1997.
105. Heuck, A.P. et al., β-barrel pore-forming toxins: intriguing dimorphic proteins, *Biochemistry,* 40, 9065, 2001.
106. Shatursky, O. et al., The mechanism of membrane insertion of a cholesterol-dependent cytolysin: a novel paradigm for pore-forming toxins, *Cell,* 99, 293, 1999.
107. Ramachandran, R. et al., Structural insights into the membrane-anchoring mechanism of a cholesterol-dependent cytolysin, *Nat. Struct. Biol.,* 9, 823, 2002.
108. Tweten, R.K. and Sellman, B.R., *Clostridium septicum* pore-forming and lethal α-toxin, in *The Comprehensive Sourcebook of Bacterial Protein Toxins,* Alouf, J.E. and Freer, J.E., Eds., Academic Press, London, 1999, 435.
109. Tweten, R.K., *Clostridium perfringens* beta toxin and *Clostridium septicum* alpha toxin: their mechanisms and possible role in pathogenesis, *Vet. Microbiol.,* 82, 1, 2001.
110. Cole, S., Structural studies on epsilon toxin from *Clostridium perfringens*, in *Protein Toxins of the Genus Clostridium and Vaccination,* Duchesnes, C., Mainil, J., Popoff, M.R., and Titball, R., Eds., Presses de la Faculté de Médecine Vétérinaire, Liège, Belgium, 2003, 95.
111. Hunter, S.E. et al., Molecular genetic analysis of beta-toxin of *Clostridium perfringens* reveals sequence homology with alpha-toxin, gamma-toxin, and leukocidin of *Staphylococcus aureus*, *Infect. Immun.,* 61, 3958, 1993.
112. Steinthorsdottir, V. et al., *Clostridium perfringens* beta-toxin forms multimeric transmembrane pores in human endothelial cells, *Microb. Pathog.,* 28, 45, 2000.
113. Shatursky, O. et al., *Clostridium perfringens* beta-toxin forms potential-dependent, cation-selective channels in lipid bilayers, *Infect. Immun.,* 68, 5546, 2000.
113a. Mogridge, J. et al., Involvement of domain 3 in oligomerization by the protective antigen moiety of anthrax toxin, *J. Bacteriol.*, 183, 2111, 1999.
114. Petosa, C. et al., Crystal structure of the anthrax toxin protective antigen, *Nature (London),* 385, 833, 1997.
115. Cunningham, K. et al., Mapping the lethal factor and edema factor binding sites on oligomeric anthrax protective antigen, *Proc. Natl. Acad. Sci. U.S.A.,* 99, 7049, 2002.
116. Mogridge, J. et al., The lethal and edema factors of anthrax toxin bind only to oligomeric forms of the protective antigen, *Proc. Natl. Acad. Sci. U.S.A.,* 99, 7045, 2002.
117. Marvaud, J.C. et al., *Clostridium perfringens* iota-toxin:mapping of receptor binding and Ia docking domains on Ib, *Infect. Immun.,* 69, 2435, 2001.
118. Blöcker, D. et al., The C terminus of component C2II of *Clostridium botulinum* C2 toxin is essential for receptor binding, *Infect. Immun.,* 68, 4566, 2000.
119. Stiles, B.G. et al., *Clostridium perfringens* iota toxin: binding studies and characterization of cell surface receptor by fluorescence-activated cytometry, *Infect. Immun.,* 68, 3475, 2000.
120. Song, L. et al., Structure of staphylococcal α-hemolysin, a heptameric transmembrane pore, *Science,* 274, 1859, 1996.
121. Pedelacq, J.D. et al., The structure of a *Staphylococcus aureus* leucocidin component (LukF-PV) reveals the fold of the water soluble species of family of transmembrane pore-forming toxins, *Struct. Fold Des.,* 7, 277, 1999.

122. Mogridge, J. et al., Stoichiometry of anthrax toxin complexes, *Biochem.*, 41, 1079, 2002.
123. Abrami, L. et al., Anthrax toxin triggers endocytosis of its receptor via a lipid raft-mediated clathrin-dependent process, *J. Cell Biol.*, 160, 321, 2003.
124. Naylor, C.E. et al., Structure of the key toxin in gas gangrene, *Nature Struct. Biol.*, 5, 738, 1998.
125. Titball, R.W., Biochemical and immunological properties of the C-terminal domain of the alpha-toxin of *Clostridium perfringens*, *FEMS Microbiol. Lett.*, 110, 45, 1993.
126. Jepson, M.A., et al., Differences in the carboxy-terminal (putative phospholipid binding) domains of *Clostridium perfringens* and *Clostridium bifermentans* phospholipases C influence the hemolytic and lethal properties of these enzymes, *Infect. Immun.*, 67, 3297, 1999.
127. Pizza, M. et al., Molecular, functional and evolutionary aspects of ADP-ribosylating toxins, in *The Comprehensive Sourcebook of Bacterial Protein Toxins,* Alouf, J.E. and Freer, J.H., Eds., Academic Press, London, 1999, 45.
128. Han, S. et al., Evolution and mechanism from structures of an ADP-ribosylating toxin and NAD complex, *Nature Struct. Biol.*, 6, 932, 1999.
129. Han, S. et al., Crystal structure and novel recognition motif of Rho ADP-ribosylating C3 exoenzyme from *Clostridium botulinum*. Structural insights for recognition specificity and catalysis, *J. Mol. Biol.*, 305, 95, 2001.
130. Tsuge, H. et al., Crystal structure and site-directed mutagenesis of enzymatic components from *Clostridium perfringens* iota-toxin, *J. Mol. Biol.*, 325, 471, 2003.
131. Richard, J.F. et al., Bacterial toxins modifying the actin cytoskeleton, *Internatl. Microbiol.*, 2, 185, 1999.
132. Wren, B.W., A family of clostridial and streptococcal ligand-binding proteins with conserved C-terminal repeat sequences, *Mol. Microbiol.*, 5, 797, 1991.
133. Von Eichel-Streiber, C. et al., Evidence for a modular structure of the homologous repetitive C-terminal carbohydrate-binding sites of *Clostridium difficile* toxins and *Streptococcus mutans* glucosyltransferases, *J. Bacteriol.*, 174, 6707, 1992.
134. Hofmann, F. et al., Localization of the glucosyltransferase activity of *Clostridium difficile* Toxin B to the N-terminal part of the holotoxin, *J. Biol. Chem.*, 272, 11074, 1997.
135. Hofmann, F. et al., Chimeric clostridial cytotoxins: identification of the N-terminal region involved in protein substrate recognition, *Infect. Immun.*, 66, 1076, 1998.
136. Just, I. et al., Molecular mechanism of action of the large clostridial cytotoxins, in *Bacterial Protein Toxins,* Aktories, K. and Just, I., Eds., Springer, Berlin, 2000, 307.
137. Busch, C. et al., Characterization of the catalytic domain of *Clostridium novyi* alpha toxin, *Infect. Immun.*, 68, 6378, 2000.
138. Busch, C. et al., Involvement of a conserved tryptophan residue in the UDP-glucose binding of large clostridial cytotoxin glycosyltransferases, *J. Biol. Chem.*, 275, 13228, 2000.
139. Breton, C. and Imberty, A., Structure/function studies of glycosyltransferases, *Curr. Opin. Struct. Biol.*, 9, 563, 1999.
140. Ünligil and Rini, Glycosyltransferase structure and mechanism, *Curr. Opin. Struct. Biol.*, 10, 510, 2000.
141. Lacy, D.B. et al., Crystal structure of botulinum neurotoxin type A and implications for toxicity, *Nature Struct. Biol.*, 5, 898, 1998.
142. Lacy, D.B. and Stevens, R.C., Sequence homology and structural analysis of the clostridial neurotoxins, *J. Mol. Biol.*, 291, 1091, 1999.
143. Emsley, P. et al., The structures of the Hc fragment of tetanus toxin with carbohydrate subunit complexes provide insight into ganglioside binding, *J. Biol. Chem.*, 275, 8889, 2000.
144. Umland, T.C. et al., The structure of the receptor binding fragment Hc of tetanus neurotoxin, *Nature Struct. Biol.*, 4, 788, 1997.
145. Schiavo, G. et al., Neurotoxins affecting neuroexocytosis, *Physiol. Rev.*, 80, 717, 2000.
146. Meunier, F. et al., Molecular mechanism of action of botulinal neurotoxins and the synaptic remodeling they induce *in vivo* at the skeletal neuromuscular junction, in *Handbook of Neurotoxicology,* Massaro, J., Ed., Humana Press, Totowa, NJ, 2002, 305.
147. Humeau, Y. et al., How botulinum and tetanus neurotoxins block neurotransmitter release, *Biochimie,* 82, 427, 2000.
148. Pannifer, A.D. et al., Crystal structure of the anthrax lethal factor, *Nature (London),* 414, 229, 2001.

149. Duesbery, N.S. et al., Proteolytic inactivation of MAP-Kinase-Kinase by anthrax lethal factor, *Science,* 280, 734, 1998.
150. Vitale, N. et al., Susceptibility of mitogen-activated protein kinase kinase family members to proteolysis by anthrax lethal factor, *Biochem. J.,* 352, 739, 2000.
151. Sears, C.L., The toxins of *Bacteroides fragilis*, *Toxicon,* 39, 1737, 2001.
152. Matsushita, O. et al., Gene duplication and multiplicity of collagenases in *Clostridium histolyticum*, *J. Bacteriol.,* 181, 923, 1999.
153. Jung, M. et al., Identification of metal ligands in the *Clostridium histolyticum* ColH collagenase, *J. Bacteriol.,* 181, 2816, 1999.
154. Matsushita, O. et al., A study of the collagen-binding domain of a 116 kDa *Clostridium histolyticum* collagenase, *J. Biol. Chem.,* 273, 3643, 1998.
155. Matsushita, O. and Okabe, A., Clostridial hydrolytic enzymes degrading extracellular components, *Toxicon,* 39, 1769, 2001.
156. Matsushita, O. et al., Purification and characterization of a *Clostridium perfringens* 120-kilodalton collagenase and nucleotide sequence of the corresponding gene, *J. Bacteriol.,* 176, 149, 1994.
157. Gordon, V.M. et al., *Clostridium septicum* alpha-toxin uses glycosylphosphatidylinositol-anchored protein receptors, *J. Biol. Chem.,* 274, 27274, 1999.
157a. Hale, M. et al., Detergent-resistant membrane microdomains facilitate Ib oligomer formation and biological activity of *Clostridium perfringens* iota-toxin, *Infect. Immun.,* 72, 2186, 2004.
158. Holmes, R.K. et al., Cholera toxin and related enterotoxins of Gram-negative bacteria, in *Bacterial Toxins and Virulence Factors in Disease,* Moss, J., Iglewski, B., Vaughan, M., and Tu, A.T., Eds., Marcel Dekker, New York, 1995, 225.
158a. Wigelsworth D.J. et al., Binding stoichiometry and kinetics of the interaction of a human anthrax toxin receptor, CMG2, with protective antigen, *J. Biol. Chem.,* 279, 23349, 2004.
159. Guillouard, I. et al.,The carboxy-terminal C2-like domain of the α-toxin from *Clostridium perfringens* mediates calcium-dependent membrane recognition, *Mol. Microbiol.,* 26, 867, 1997.
160. Eaton, J.T. et al., Crystal structure of the *C. perfringens* alpha-toxin with the active site closed by a flexible loop region, *J. Mol. Biol.,* 319, 275, 2002.
161. Munro, P. et al., High sensitivity of mouse neuronal cells to tetanus toxin requires a GPI-anchored protein, *Biochem. Biophys. Res. Comm.,* 289, 623, 2001.
162. Herreros, J. et al., Lipid rafts act as specialized domains for tetanus toxin binding and internalization into neurons, *Mol. Biol. Cell,* 12, 2947, 2001.
163. Herreros, J. et al., Tetanus toxin fragment C binds to a protein present in neuronal cell lines and motoneurons, *J. Neurosci.,* 74, 1941, 2000.
164. Nishiki, T. et al., Identification of protein receptor for *Clostridium botulinum* type B neurotoxin in rat brain synaptosomes, *J. Biol. Chem.,* 269, 10498, 1994.
165. Nishiki, T. et al., The high-affinity of *Clostridium botulinum* type B neurotoxin to synaptotagmin II associated with gangliosides GT1b/GD1a, *FEBS Lett.,* 378, 253, 1996.
166. Yowler, B.C. et al., Botulinum neurotoxin A activity is dependent upon the presence of specific gangliosides in neuroblastoma cells expressing synaptotagmin I, *J. Biol. Chem.,* 27, 32815, 2002.
167. Rummel, A. et al., Two carbohydrate binding sites in the Hcc-domain of tetanus neurotoxin are required for toxicity, *J. Mol. Biol.,* 326, 835, 2003.
168. Boquet, P., The cytotoxic necrotizing factor 1 (CNF1) from *Escherichia coli*, *Toxicon,* 39, 1673, 2001.
169. Lemichez, E. et al., Membrane translocation of diphtheria toxin fragment A exploits early to late endosome trafficking machinery, *Mol. Microbiol.,* 23, 445, 1997.
170. Murphy, et al., Diphtheria toxin and related fusion proteins: autonomous systems for the delivery of proteins and peptides to the cytosol of eukaryotic cells, in *Bacterial Toxins and Virulence Factors in Disease,* Moss, J., Iglewski, B., Vaughan, M., and Tu, A.T., Eds., Marcel Dekker, New York, 1995, 23.
171. Ren, X.D. et al., Interaction of diphtheria toxin T domain with molten globule-like proteins and its implications for translocation, *Science,* 284, 955, 1999.
172. Hammond, K. et al., Interaction of the membrane-inserted diphtheria toxin T domain with peptides and its possible implications for chaperone-like T domain behavior, *Biochemistry,* 41, 3243, 2002.
173. Ratts, R. et al., The cytosolic entry of diphtheria toxin catalytic domain requires a host cell cytosolic translocation factor complex, *J. Cell Biol.,* 160, 1139, 2003.

174. Popoff, M.R. et al., Ras, Rap, and Rac small GTP-binding proteins are targets for *Clostridium sordellii* lethal toxin glucosylation, *J. Biol. Chem.,* 271, 10217, 1996.
175. Fiorentini, C. and Thelestam, M., *Clostridium difficile* toxin A and its effects on cells, *Toxicon,* 29, 543, 1991.
176. Barth, H. et al., Low pH-induced formation of ion channels by *Clostridium difficile* toxin B in target cells, *J. Biol. Chem.,* 276, 10670, 2001.
177. Qa'dan, M. et al., pH-induced conformational changes in *Clostridium difficile* toxin B, *Infect. Immun.,* 68, 2470, 2000.
178. Qa'dan, M. et al., pH-induced cytopathic effects of *Clostridium sordellii* lethal toxin, *Infect. Immun.,* 69, 5487, 2001.
179. Meunier, F. et al., Botulinum neurotoxins: from paralysis to recovery of functional neuromuscular transmission, *J. Physiol.,* 96, 105, 2002.
180. Barth, H. et al., Cellular uptake of *Clostridium botulinum* C2 toxin requires oligomerization and acidification, *J. Biol. Chem.,* 275, 18704, 2000.
181. Knapp, O. et al., Interaction of *Clostridium perfringens* iota-toxin with lipid bilayer membranes, *J. Biol. Chem.,* 277, 6143, 2002.
182. Schmid, A. et al., Interaction of *Clostridium botulinum* C2 toxin with lipid bilayer membranes, *J. Biol. Chem.,* 269, 16706, 1994.
183. Bachmeyer, C. et al., Interaction of *Clostridium botulinum* C2 toxin with lipid bilayer membranes and Vero cells: inhibition of channel function in chloroquine and related compounds *in vitro* and toxin action *in vivo*, *FASEB J.,* 15, 1658, 2001.
184. Haug, G. et al., The host cell chaperone Hsp90 is essential for translocation of the binary *Clostridium botulinum* C2 toxin into the cytosol, *J. Biol. Chem.,* 278, 32266, 2003.
185. Garred, O. et al., Furin-induced cleavage and activation of Shiga toxin, *J. Biol. Chem.,* 270, 10817, 1995.
186. Lencer, W.I. et al., Targeting of Cholera Toxin and *Escherichia coli* Heat Labile Toxin in polarized epithelia: role of COOH-terminal KDEL, *J. Cell Biol.,* 131, 951, 1995.
187. Majoul, I. et al., KDEL receptor (Erd2p)-mediated retrograde transport of the cholera toxin A subunit from the Golgi involves COPI, p23, and the COOH terminus of Erd2p, *J. Cell Biol.,* 143, 601, 1998.
188. Zhang, R.G. et al., The three-dimensional crystal structure of cholera toxin, *J. Mol. Biol.,* 251, 563, 1995.
189. Schmitz, A. et al., Cholera toxin is exported from microsomes by the Sec61p complex, *J. Cell Biol.,* 148, 1203, 2000.
190. Mallard, F. et al., Direct pathway from early/recycling endosomes to the Golgi apparatus revealed through the study of Shiga toxin B-fragment transport, *J. Cell Biol.,* 143, 973, 1998.
191. Falguières, T. et al., Targeting of Shiga toxin B-subunit to retrograde transport route in association with detergent-resistant membranes, *Mol. Biol. Cell,* 12, 2453, 2001.
192. Johannes, L. and Lamaze, C., Clathrin-dependent or not: is it still the question?, *Traffic,* 3, 443, 2002.
193. Johannes, L., The epithelial cell cytoseleton and intracellular trafficking. I. Shiga toxin B-subunit system: retrograde transport, intracellular vectorization, and more, *Am. J. Physiol. Gastrointest. Liver Physiol.,* 10, G1, 2002.
194. Lencer, W.I. et al., Transcytosis of cholera toxin subunits across model human intestinal epithelia, *Proc. Natl. Acad. Sci. U.S.A.,* 92, 10094, 1995.
195. Philpot, D.J. et al., Translocation of verotoxin-1 across T84 monolayers: mechanism of bacterial toxin penetration of epithelium, *Am. J. Physiol.,* 273, G1349, 1997.
196. Wolf, A.A. et al., Ganglioside structure dictates signal transduction by cholera toxin and association with caveolae-like membrane domains in polarized epithelia, *J. Cell Biol.,* 141, 917, 1998.
197. Hurley, B.P. et al., Shiga toxin translocation across intestinal epithelial cells is enhanced by neutrophil transmigration, *Infect. Immun.,* 69, 6148, 2001.
198. Te Loo, D.M. et al., Binding and transfer of verocytotoxin by polymorphonuclear leukocytes in hemolytic uremic syndrome, *Blood,* 95, 3396, 2000.
199. Richard, J.F. et al., Transcytosis of iota toxin across polarized CaCo-2 cell monolayers, *Mol. Microbiol.,* 43, 907, 2002.
200. Nagahama, M. et al., Binding component of *Clostridium perfringens* iota-toxin induces endocytosis in Vero cells, *Infect. Immun.,* 70, 1909, 2002.

200a. Sakurai, J. and Kobayashi, K., Lethal and dermonecrotic activities of *Clostridium perfringens* iota toxin: biological activities induced by cooperation of two nonlinked components, *Microbiol. Immunol.*, 39, 249, 1995.

201. Maksymowych, A.B. et al., Pure botulinum neurotoxin is absorbed from the stomach and small intestine and produces peripheral neuromuscular blockade, *Infect. Immun.,* 67, 4708, 1999.

202. Petit, L. et al., *Clostridium perfringens* Epsilon Toxin rapidly decreases membrane barrier permeability of polarized MDCK Cells, *Cell. Microbiol.,* 5, 155, 2003.

203. Petit, L. et al., *Clostridium perfringens* epsilon-toxin acts on MDCK cells by forming a large membrane complex, *J. Bacteriol.,* 179, 6480, 1997.

204. Lalli, G. et al., The journey of tetanus and botulinum neurotoxins in neurons, *Trends Microbiol.,* 11, 431, 2003.

205. Maslos, U. et al., Retrograde trans-synaptic transfer of green fluorescent protein allows the genetic mapping of neuronal circuits in transgenic mice, *Proc. Natl Acad. Sci. U.S.A.,* 99, 10120, 2002.

206. Li, X. et al., Recombinant forms of tetanus toxin engineered for examining and exploiting neuronal trafficking pathways, *J. Biol. Chem.,* 276, 31394, 2001.

207. Lalli, G. and Schiavo, G., Analysis of retrograde transport in motor neurons reveals common endocytic carriers for tetanus toxin and neutrophin receptor p75NTR, *J. Cell Biol.,* 156, 233, 2002.

208. Karjalainen, T. et al., Molecular and genomic analysis of genes encoding surface-anchored proteins from *Clostridium difficile*, *Infect. Immun.,* 69, 3442, 2001.

209. Tasteyre, A. et al., Role of FliC and FliD flagellar proteins of *Clostridium difficile* in adherence and gut colonization, *Infect. Immun.,* 69, 7937, 2001.

210. Waligora, A.J. et al., Characterization of a cell surface protein of *Clostridium difficile* with adhesive properties, *Infect. Immun.,* 69, 2144, 2001.

211. Kojima, A. et al., Cloning and expression of a gene encoding the flagellin of *Clostridium chauvoei*, *Vet. Microbiol.,* 76, 359, 2000.

212. Tamura, Y. et al., Antigenic analysis of *Clostridium chauvoei* flagella with protective and non-protective monoclonal antibodies, *J. Gen. Microbiol.,* 138, 537, 1992.

213. O'Brien, D.K. and Melville, S.B., The anaerobic pathogen *Clostridium perfringens* can escape the phagosome of macrophages under aerobic conditions, *Cell. Microbiol.,* 2, 505, 2000.

214. Sakurai, J. et al., Evidence for coupling of *Clostridium perfringens* alpha-toxin-induced hemolysis to stimulated phosphatidic acid formation in rabbit erythrocytes, *Infect. Immun.,* 61, 3711, 1993.

215. Ochi, S. et al., Phospholipid metabolism induced by *Clostridium perfringens* alpha-toxin elicits a hot cold type of hemolysis in rabbit erythrocytes, *Infect. Immun.,* 64, 3930, 1996.

216. Titball, R. et al., The *Clostridium perfringens* α-toxin, *Anaerobe,* 5, 51, 1999.

217. Bryant, A.E. and Stevens, D.L., Phospholipase C and perfringolysin O from *Clostridium perfringens* upregulate endothelial cell-leukocyte adherence molecule 1 and intercellular leukocyte adherence molecule 1 expression and induce interleukin-8 synthesis on cultured human umbilical vein endothelial cells, *Infect. Immun.,* 64, 358, 1996.

218. Bryant, A.E. et al., Clostridial gas gangrene. II. Phospholipase C-induced activation of platelet gpI-IbIIIa mediates vascular occlusion and myonecrosis in *Clostridium perfringens* gas gangrene, *J. Infect. Dis.,* 182, 808, 2000.

219. Bryant, A.E. et al., Clostridial gas gangrene. I. Cellular and molecular mechanisms of microvascular dysfunction induced by exotoxins of *Clostridium perfringens*, *J. Infect. Dis.,* 182, 799, 2000.

220. Bryant, A.E. et al., Activation of platelet gpIIIa by phospholipase C from *Clostridium perfringens* involves store-operated calcium entry, *J. Infect. Dis.,* 187, 408, 2003.

221. Vandelaer, J. et al., Tetanus in developing countries: an update on the maternal and neonatal tetanus elimination initiative, *Vaccine,* 21, 3442, 2003.

222. Songer, J.G. Clostridial enteric diseases of domestic animals, *Clin. Microbiol. Rev.,* 9, 216, 1996.

223. Popoff, M.R., Ecology of neurotoxigenic strains of Clostridia, in *Clostridial neurotoxins,* Montecucco, C., Ed., Springer-Verlag, Heidelberg, 1995, 1.

224. Lund, B.M. and Peck, M.W., *Clostridium botulinum*, in *The Microbiology Safety and Quality of Food,* Lund, B.M., Baird-Parker, T.C., and Gould, G.W., Eds., Aspen Publishers, Gaithersburg, MD, 2000, 1057.

225. Singh, U. et al., CaCo-2 cells treated with *Clostridium perfringens* enterotoxin form multiple complex species, one of which contains the tight junction protein occludin, *J. Biol. Chem.,* 275, 18407, 2000.

226. Singh, U. et al., Comparative biochemical and immunocytochemical studies reveal differences in the effects of *Clostridium perfringens* enterotoxin on polarized CaCo-2 cells versus Vero cells, *J. Biol. Chem.,* 276, 33402, 2001.
227. McClane, B.A., The complex interaction between *Clostridium perfringens* enterotoxin and epithelial tight junctions, *Toxicon,* 39, 1781, 2001.
228. Nagahama, M.Y. and Sakurai, J., Distribution of labeled *Clostridium perfringens* epsilon toxin in mice, *Toxicon,* 29, 211, 1991.
229. Nagahama, M.Y. and Sakurai, J., High-affinity binding of *Clostridium perfringens* epsilon-toxin to rat brain, *Infect. Immun.,* 60, 1237, 1992.
230. Miyamoto, O. et al., Neurotoxicity of *Clostridium perfringens* epsilon-toxin for the rat hipocampus via glutamanergic system, *Infect. Immun.,* 66, 2501, 1998.
231. Miyata, S. et al., *Clostridium perfringens* ε-toxin forms a heptameric pore within the detergent-insoluble microdomains of Madin-Darby Canine Kidney Cells and rat synaptosomes, *J. Biol. Chem.,* 277, 39463, 2002.
232. Petit, L. et al., *Clostridium perfringens* epsilon-toxin induces a rapid change in cell membrane permeability to ions and forms channels in artificial lipid bilayers, *J. Biol. Chem.,* 276, 15736, 2001.
233. Zhu, C. et al., *Clostridium perfringens* prototoxin-induced alteration of endothelial barrier antigen (EBA) immunoreactivity at the blood brain barrier (BBB), *Exp. Neurol.,* 169, 72, 2001.
234. Finnie, J.W. et al., Neuronal damage produced in rat brains by *Clostridium perfringens* type D epsilon-toxin, *J. Comp. Path.,* 120, 415, 1999.
235. Buxton, D., The use of an imunoperoxidase technique to investigate by light and electron microscopy the sites of binding of *Clostridium welchii* type D ε-toxin in mice, *J. Med. Microbiol.,* 11, 289, 1978.
236. Borrmann, E. et al., Effect of *Clostridium perfringens* epsilon toxin on MDCK cells, *FEMS Immunol. Med. Microbiol.,* 31, 85, 2001.
237. Miyamoto, O. et al., *Clostridium perfringens* epsilon toxin causes excessive release of glutamate in the mouse hippocampus, *FEMS Microbiol. Lett.,* 189, 109, 2000.
238. O'Rourke, F. et al., *Clostridium septicum*: a malignant pathogen, *J. Infect.,* 41, 286, 2000.
239. Chen, S.S. and Lubowski, D.Z., *Clostridium septicum* and malignancy, *ANZ J. Surg.,* 71, 647, 2001.
240. Stubbs, S. et al., Production of actin-specific ADP-ribosyltransferase (binary toxin) by strains of *Clostridium difficile*, *FEMS Microbiol. Lett.,* 186, 307, 2000.
241. Mauss, S. et al., ADP-ribosylation of actin isoforms by *Clostridium botulinum* C2 toxin and *Clostridium perfringens* iota toxin, *Eur. J. Biochem.,* 194, 237, 1990.
242. Vandekerckhove, J. et al., *Clostridium perfringens* iota toxin ADP-ribosylates skeletal muscle actin in Arg-177, *FEBS Lett.,* 255, 48, 1987.
243. Aktories, K. et al., Botulinum C2 toxin ADP-ribosylates actin, *Nature,* 322, 390, 1986.
244. Ohishi, I. and Tsuyama, S., ADP-ribosylation of nonmuscle actin with component I of C2 toxin, *Biochem. Biophys. Res. Commun.,* 136, 802, 1986.
245. Reuner, K.H. et al., Regulation of actin synthesis in rat hepatocytes by cytoskeleton rearrangements, *Eur. J. Cell Biol.,* 69, 189, 1996.
246. Kiefer, G. et al., Cytotoxic effects by microinjection of ADP-ribosylated skeletal muscle G-actin in PtK2 cells in the absence of *Clostridium perfringens* iota toxin, *Med. Microbiol. Immunol.,* 184, 175, 1996.
247. Aktories, K. and Koch, G., Modification of actin and Rho proteins by clostridial ADP-ribosylating toxins, in *Bacterial Toxins and Virulence Factors in Disease,* Moss, J., Iglewski, B., Vaughan, M., and Tu, A.T., Eds., Marcel Dekker, New York, 1995, 491.
248. Aktories, K. and Wegner, A., Mechanisms of the cytopathic action of actin-ADP-ribosylating toxins, *Mol. Microbiol.,* 6, 2905, 1992.
249. Aktories, K., Bacterial Protein Toxins as Tools in Cell Biology and Pharmacology, in *Cellular Microbiology,* Cossart, P., Boquet, P., Normark, S., and Rappuoli, R., Eds., ASM Press, Washington, DC, 2000, 221.
250. Aktories, K., Clostridial ADP-ribosylating toxins; effects on ATP and GTP-binding proteins, *Mol. Cell Biochem.,* 138, 167, 1994.
251. Vershueren, H. et al., Effects of *Clostridium botulinum* C2 toxin and cytochalasin D on *in vitro* invasiveness, motility and F-actin content of a murine T-lymphoma cell line, *Eur. J. Cell Biol.,* 66, 335, 1995.

252. Mauss, S. et al., Inhibition of the contraction of the isolated longitudinal muscle of the guinea-pig ileum by botulinum C2 toxin: evidence for a role of G/F-actin transition in smooth muscle contraction, *Naunyn-Schmiedebergs Arch. Pharmacol.,* 340, 345, 1989.
253. Ermert, L. et al., Ultrastructural changes of lung capillary endothelium in response to botulinum C2 toxin, *J. Appl. Physiol.,* 82, 382, 1997.
254. Genth, H. et al., Entrapment of Rho ADP-ribosylated by *Clostridium botulinum* C3 exoenzyme in the Rho-guanine nucleotide dissociation inhibitor-1 complex, *J. Biol. Chem.,* 278, 28523, 2003.
255. Sehr, P. et al., Glucosylation and ADP ribosylation of Rho proteins: effects on nucleotide binding, GTPase activity, and effector coupling, *Biochem.,* 37, 5296, 1998.
256. Fujihara, H. et al., Inhibition of RhoA translocation and calcium sensitization by *in vivo* ADP-ribosylation with the chimeric toxin DC3B, *Mol. Biol. Cell,* 8, 2437, 1997.
257. Just, I. et al., Glucosylation of Rho proteins by *Clostridium difficile* toxin B, *Nature (London),* 375, 500, 1995.
258. Just, I. et al., The enterotoxin from *Clostridium difficile* (ToxA) monoglucosylates the Rho proteins, *J. Biol. Chem.,* 270, 13932, 1995.
259. Hermann, C. et al., Functional consequences of monoglucosylation of Ha-Ras at effector domain amino acid threonine 35, *J. Biol. Chem.,* 273, 16134, 1998.
260. Genth, H. et al., Monoglucosylation of RhoA at threonine 37 blocks cytosol membrane recycling, *J. Biol. Chem.,* 274, 29050, 1999.
261. Nusrat, A. et al., *Clostridium difficile* toxins disrupt epithelial barrier function by altering membrane microdomain localization of tight junction proteins, *Infect. Immun.,* 69, 1329, 2001.
262. Chen, M.L. et al., Protein kinase C signaling regulates ZO-1 translocation and increased paracellular flux of T84 colonocytes exposed to *Clostridium difficile* toxin A, *J. Biol. Chem.,* 277, 4247, 2002.
263. Pothoulakis, C. and Lamont, J.T., Microbes and microbial toxins: paradigms for microbial-mucosa interactions II. The integrated response of the intestine to *Clostridium difficile* toxins, *Am. J. Physiol. Gastrointest. Liver Physiol.,* 280, G178, 2001.
264. Warny, M. et al., p38 MAP kinase activation by *Clostridium difficile* toxin A mediates monocyte necrosis, IL-8 production, and enteritis, *J. Clin. Invest.,* 105, 147, 2000.
265. Alouf, J.E. and Müller-Alouf, H., Staphylococcal and streptococcal superantigens: molecular, biological and clinical aspects, *Int. J. Med. Microbiol.,* 292, 429, 2003.
266. Gilmore, M.S. et al., *Enterococcus faecalis* cytolysin and *Bacillus cereus* bi- and tri-component haemolysins, in *The Comprehensive Sourcebook of Bacterial Protein Toxins,* Alouf, J.E. and Freer, J.H., Eds., Academic Press, London, 1999, 419.
267. Lund, T. and Granum, P.E., The 105-kDa protein component of *Bacillus cereus* non-haemolytic enterotoxin (Nhe) is a metalloprotease with gelatinolytic and collagenolytic activity, *FEMS Microbiol. Lett.,* 178, 355, 1999.
268. Rudner, X.L. et al., Regulation of cell signalling by the cytoplasmic domains of the heat-stable enterotoxin receptor: identification of autoinhibitory and activating motifs, *Proc. Natl. Acad. Sci. U.S.A.,* 92, 5169, 1995.
269. Hirayama, T. and Wada, A., Heat-stable enterotoxin of *Escherichia coli*, in *Bacterial Protein Toxins* Aktories, K. and Just, I., Eds., Springer, Berlin, 2000, 577.
270. Sansonetti, P., Host-pathogen interactions: the seduction of molecular cross talk, *Gut,* 50, ii2, 2002.
271. Flatau, G. et al., Toxin-induced activation of the G protein p21 Rho by deamidation of glutamine, *Nature (London),* 387, 729, 1997.
272. Schmidt, G. et al., Gln63 of rho is deaminated by *Escherichia coli* cytotoxic necrotizing factor-1, *Nature (London),* 387, 725, 1997.
273. Doye, A. et al., CNF1 exploits the ubiquitin-proteasome machinery to restrict Rho GTPase activation for bacterial host cell invasion, *Cell,* 111, 553, 2002.
274. Friebel, A. et al., SopE and SopE2 from *Salmonella typhimurium* activate different sets of RhoGTPases of the host cell, *J. Biol. Chem.,* 276, 34035, 2001.
275. Hardt, W.D. et al., *Salmonella typhimurium* encodes an activator of Rho GTPases that induces membrane ruffling and nuclear response in host cells, *Cell,* 93, 815, 1998.
276. Buchwald, G. et al., Structural basis for the reversible activation of a Rho protein by the bacterial toxin SopE, *EMBO J.,* 21, 3286, 2002.

277. Stebbins, C.E. and Galan, J.E., Modulation of host signaling by a bacterial mimic: structure of the *Salmonella* effector SptP bound to Rac1, *Mol. Cell,* 6, 1449, 2000.
278. Lesnick, M.L. et al., The *Salmonella spvB* virulence gene encodes an enzyme that ADP-ribosylates actin and destabilizes the cytoskeleton of eukaryotic cells, *Mol. Microbiol.,* 39, 1464, 2001.
279. Black D.S. and Bliska, J.B., The RhoGAP activity of the *Yersinia pseudotuberculosis* cytotoxin YopE is required for antiphagocytic function and virulence, *Mol. Microbiol.,* 37, 515, 2000.
280. Evdokimov, A. et al., Crystal structure of the *Yersinia pestis* GTPase activator YopE, *Prot. Sci.,* 11, 401, 2002.
281. Kazmierczak, B.I. and Engel, J.N., *Pseudommonas aeruginosa* ExoT acts *in vivo* as a GTPase-activating protein for RhoA, Rac1 and Cdc42, *Infect. Immun.,* 70, 2198, 2002.
282. Würtele, M. et al., How the *Pseudomonas aeruginosa* ExoS toxin downregulates Rac, *Nature Struct. Biol.,* 8, 23, 2001.
283. Shao, F. et al., A *Yersinia* effector and a *Pseudomonas* avirulence protein define a family of cysteine proteases functioning in bacterial pathogenesis, *Cell,* 109, 575, 2002.
284. Cossart, P. et al., Invasion of mammalian cells by *Listeria monocytogenes*: functional mimicry to subvert cellular functions, *Trends Cell Biol.,* 13, 23, 2003.
285. Campellone, K.G. and Leong, J.M., Tails of two Tirs: actin pedestal formation by enteropathogenic *E. coli* and enterohemorrhagic *E. coli* O157:H7, *Cur. Opin.,* 6, 82, 2003.
286. Sonoda, N. et al., *Clostridium perfringens* enterotoxin fragment removes specific claudin from tight junction: evidence for direct involvement of claudin in tight junction barrier, *J. Cell Biol.*, 147, 195, 1999.
287. Mackenzie, C.R. et al., Analysis of receptor binding by the channel-forming toxin aerolysin using surface plasmon resonance, *J. Biol. Chem.,* 274, 22604, 1999.
288. Bradley, K.A. et al., Identification of the cellular receptor for anthrax toxin, *Nature,* 414, 225, 2001.
289. Scobie, H.M. et al., Human capillary morphogenesis protein 2 functions as an anthrax toxin receptor, *Proc. Natl. Acad. Sci. U.S.A.,* 100, 5170, 2003.
290. Brennan, M.J. et al., Binding of pertussis toxin to a 165-kilodalton Chinese hamster ovary cell glycoprotein, *J. Biol. Chem.,* 263, 4895, 1988.
291. Guermonprez, P. et al., The adenylate cyclase toxin of *Bordetella pertussis* binds to target cells via the $\alpha_M\beta_2$integrin (CD11b/CD18), *J. Exp. Med.,* 193, 1035, 2001.
292. Naglich, J.G. et al., Expression cloning of a diphtheria toxin receptor: identity with heparin-binding EGF-like growth factor precursor, *Cell,* 69, 1051, 1992.
293. Labrie, V. et al., Binding of sulfatide and enterotoxicity of various *Escherichia coli* Stb mutants, *Microbiol.,* 147, 3141, 2001.
294. Vaandrager, A.B. et al., Guanylyl cyclase c is an *N*-linked glycoprotein receptor that accounts for multiple heat-stable enterotoxin-binding proteins in the intestine, *J. Biol. Chem.,* 268, 2174, 1993.
295. Kimura, B. et al., High molecular weight factor in FCS inhibits *Helicobacter pylori* VacA-binding to its receptor, RPTPbeta, on AZ-521, *Mol. Microbiol.,* 47, 105, 2003.
296. Kounnas, M.Z. et al., The α_2-macroglobulin receptor/low-density lipoprotein-receptor-related protein binds and internalizes *Pseudomonas* exotoxin A, *J. Biol. Chem.*, 267, 12420, 1992.
297. Sandvig, K. et al., Endocytosis from coated pits of shiga toxin: a glycolipid-binding protein from *Shigella dysenteriae1*, *J. Cell Biol.,* 108, 1331, 1989.

18 Clostridial Enterotoxins

Bruce A. McClane

CONTENTS

18.1 INTRODUCTION TO CLOSTRIDIAL ENTEROTOXINS

Several clostridial species (Table 18.1) exploit the anaerobic, nutrient-rich conditions of the mammalian gastrointestinal (GI) tract to cause enteric disease in humans and domestic animals. Clostridial enteric diseases are very common, often life-threatening, and impose significant economic burdens on medicine and animal husbandry. For example, prolonged hospitalizations from nosocomially acquired *C. difficile* infections annually drain billions of dollars from global healthcare systems.

The defining virulence attribute of all enteropathogenic clostridia is their production of toxins active on the GI tract. A list of clostridial toxins with known GI tract activity is shown in Table 18.1. Many of those toxins are described elsewhere in this book; this chapter will focus on two intestinally active toxins produced by certain *Clostridium perfringens* isolates, *C. perfringens* enterotoxin and *C. perfringens* beta2 toxin.

18.2 *CLOSTRIDIUM PERFRINGENS* ENTEROTOXIN

Clostridium perfringens enterotoxin (CPE) was first purified and characterized in the early 1970s (for review, see McDonel [1]). Those early studies identified the CPE protein as a 35 kDa polypeptide with a pI of 4.3. They also established that the biologic activity of this enterotoxin is heat- and pH-labile.

As shown in Table 18.2, *C. perfringens* isolates are commonly classified into types A–E, based upon expression of the four major typing toxins, i.e., alpha, beta, epsilon, and iota toxins [1]. Molecular studies [2–5] conducted during the past decade revealed that only a small fraction (~1 to 5%) of the global *C. perfringens* population carries the enterotoxin gene (*cpe*). While some *C. perfringens* type C and D isolates express the enterotoxin, the great majority of CPE-producing *C. perfringens* isolates classify as type A [2–5]. Interestingly, most (or all) *C. perfringens* type E isolates carry silent, defective *cpe* sequences ([6], further discussed in Section 18.2.2).

18.2.1 Role of CPE in Disease

18.2.1.1 *C. perfringens* Type A Food Poisoning

In the early 1970s, CPE-positive strains of *C. perfringens* type A became linked to *C. perfringens* type A food poisoning [1,7], which annually ranks as the second or third most common foodborne

TABLE 18.1
Clostridial Toxins with GI Tract Activity

Clostridium spp.	Toxin	Molecular Action	Associated GI Diseases
C. botulinum	C2 toxin	ADP ribosylates actin	?
C. difficile	Toxin A	Monoglucosylates rho family proteins	Pseudomembranous colitis and antibiotic-associated diarrhea (AAD)
	Toxin B	"	"
	CDT	ADP-ribosylates Actin	?
C. perfringens	α toxin	Phospholipase C	Animal enterotoxemias?
	β toxin	Pore-formation?	Necrotizing enteritis of humans (pig bel) and animal enterotoxemias
	β2 toxin	?	Animal enterotoxemias
	ι toxin	ADP-ribosylates Actin	Animal enterotoxemias
	ε toxin	Pore formation?	Animal enterotoxemias
	Enterotoxin (CPE)	Affects membrane permeability and tight junctions	*C. perfringens* type A food poisoning; nonfoodborne human GI diseases (e.g., AAD) and animal enterotoxemias
C. septicum	α toxin	Pore formation	Neutropenic enterocolitis in humans and animal enterotoxemias
C. spiroforme	α toxin	ADP ribosylates actin	Animal enterotoxemias

TABLE 18.2
***C. perfringens* Toxin Typing Scheme**

C. perfringens Type	Toxins Produced			
	Alpha	Beta	Epsilon	Iota
A	+	–	–	–
B	+	+	+	–
C	+	+	–	–
D	+	–	+	–
E	+	–	–	+

illness in the U.S. and U.K. [8,9]. In most cases of *C. perfringens* type A food poisoning, symptoms are usually limited to diarrhea and abdominal cramps. Typically, these GI symptoms develop 12 to 24 h after ingestion of a contaminated food and then persist for ~24 h before full recovery occurs. However, this food poisoning can have serious or even lethal outcomes, particularly in the elderly or debilitated. Although under-appreciated, the potential severity of *C. perfringens* type A food poisoning is evident from its rank as the second leading cause of foodborne deaths in the U.K. from 1990 to 2000, when this food poisoning claimed an average of >100 deaths per year in the U.K. [9].

C. perfringens type A food poisoning usually develops after ingestion of foods contaminated with large numbers (>10^6 bacteria/gram) of CPE-positive vegetative cells [7]. Such high *C. perfringens* food burdens often follow improper cooking or storage of foods, particularly meat products [7]. Vegetative cells and spores of CPE-positive *C. perfringens* type A food poisoning strains typically exhibit exceptional heat resistance, which favors their survival in improperly cooked or held foods [10]. After improper cooking or holding, vegetative cells (or vegetative cells arising from heat-germinated spores) then exploit the short doubling time (~10 to 15 min) of *C. perfringens* to rapidly reach pathogenic levels in contaminated food.

Once ingested, many *C. perfringens* vegetative cells die when exposed to the low pH of the stomach acid [7]. However, when a grossly contaminated food is ingested, some vegetative cells survive the stomach and then pass into the small intestines. After an initial multiplication, these bacteria commit to sporulation in the intestinal lumen. The trigger for this *in vivo* sporulation is not yet fully understood, but may involve exposure of *C. perfringens* vegetative cells to pH changes or bile salts while transiting through the GI tract [11]. During their *in vivo* sporulation, *C. perfringens* cells begin to express CPE (further discussed in Section 18.2.2). Following expression, the enterotoxin accumulates in the cytoplasm of the mother cell, i.e., CPE is not a secreted protein, but instead is released into the intestinal lumen when the mother cell lyses to release its mature endospore [7].

Considerable evidence directly implicates CPE as the virulence factor responsible for most, if not all, GI symptoms of *C. perfringens* type A food poisoning. For example, CPE is rarely (if ever) detectable in feces of healthy people, but is present in feces of virtually all *C. perfringens* type A food poisoning victims, often at levels known to produce pathophysiologic intestinal effects in animal models [12,13]. Additionally, human volunteer feeding experiments conducted in the 1970s demonstrated that ingestion of purified CPE is sufficient to reproduce the diarrheic and cramping symptoms of natural *C. perfringens* type A food poisoning [14]. Finally, molecular approaches recently demonstrated that enterotoxin expression is necessary for the enteric virulence of CPE-positive *C. perfingens* food poisoning isolates [15]. In those studies, lysates prepared from sporulating, but not vegetative, cultures of SM101 (a transformable derivative of the CPE-positive food poisoning isolate NCTC 8798 [16]) induced both histopathologic damage and fluid and electrolyte secretion in rabbit ileal loops (Figure 18.1). Since CPE is expressed only during sporulation [17], that observation is consistent with CPE being the SM101 virulence factor responsible for causing

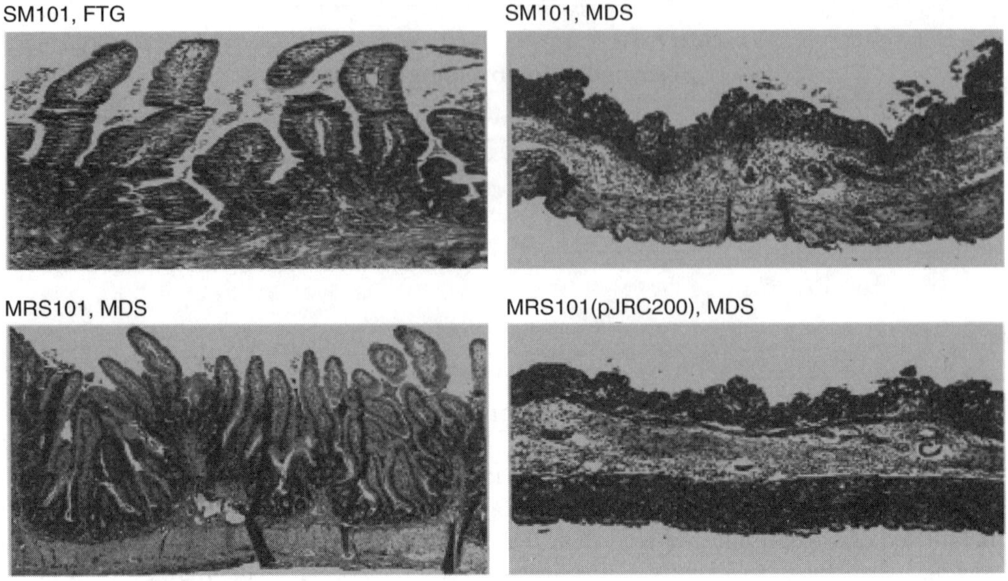

FIGURE 18.1 Histopathologic damage to rabbit ileal loops induced by sporulating culture lysates of *C. perfringens* isolate SM101 and derivatives. Tissue specimens shown were collected after ~18 hours of treatment with concentrated vegetative (FTG) or sporulating (MDS) culture lysates of wild-type isolate SM101, the isogenic *cpe* knock-out mutant MRS101, or the *cpe*-complementing strain MRS101(pJRC200). Note the extensive desquamation and villus shortening present in loops treated with MDS lysates of SM101 or MRS101(pJRC200). This damage is absent from loops treated with FTG lysates of SM101 [or MRS101 or MRS101(pJRC200), not shown] or MDS lysates of MRS101. (Reprinted with permission from Sarker, M.R. et al., *Molec. Microbiol.,* 33, 946, 1999.)

both intestinal damage and secretion. A conclusive role for CPE in causing the enteropathogenic effects of SM101 lysates was then established by specifically inactivating (by allelic exchange) the *cpe* gene of SM101. The resultant isogenic mutant was completely avirulent, i.e., sporulating culture lysates of this *cpe* knockout mutant no longer damaged or induced fluid and electrolyte secretion in rabbit ileal loops. However, complementing this *cpe* knockout mutant with a shuttle plasmid carrying the wild-type *cpe* gene fully restored enteric virulence. Therefore, the loss of enteric virulence observed with the isogenic mutant can be specifically attributed to *cpe* gene inactivation. Collectively, these observations fulfilling molecular Koch's postulates prove that CPE expression is required for the enteric virulence of SM101 in this model system.

18.2.1.2 CPE-Associated Non-Foodborne Human GI Diseases

In the mid-1980s, Borriello's group reported that CPE-positive *C. perfringens* type A isolates also cause such non-foodborne human GI diseases as antibiotic-associated diarrhea and sporadic diarrhea [18,19]. Later studies suggested these CPE-associated non-foodborne GI diseases represent between 1 and 20% of all non-foodborne human diarrhea cases [20–22]. The pathogenesis of CPE-associated non-foodborne GI diseases is not yet fully understood, but some evidence suggests these illnesses are true exogenous infections involving person-to-person transmission of *cpe*-positive isolates [18].

Several findings indicate that CPE also plays an important role in the pathogenesis of the non-foodborne human GI diseases caused by CPE-positive *C. pefringens* strains. For example, the enterotoxin is detectable, often at high levels, in the feces of many people suffering from these GI illnesses [21]. Furthermore, sporulating culture lysates prepared from an isogenic *cpe* knockout mutant of non-foodborne human GI isolate F4969 no longer damaged or induced fluid and electrolyte loss in rabbit ileal loops [15]. Complementation of that isogenic mutant with the wild-type *cpe* gene fully restored virulence, i.e., sporulating culture lysates of the complemented *cpe* knockout mutant once again caused histopathologic damage and fluid and electrolyte loss in rabbit ileal loops.

18.2.1.3 CPE-Associated Veterinary GI Diseases

During the past 10 to 15 years, CPE-positive type A isolates also became linked to several veterinary GI diseases, most convincingly equine and canine diarrheas [23]. These associations are primarily based upon the detection of CPE or CPE-positive isolates in feces from sick animals; no attempt has yet been made to inactivate the *cpe* gene of a veterinary disease isolate in order to experimentally demonstrate that CPE plays a role in these illnesses.

18.2.2 CPE Genetics and Expression

18.2.2.1 Introduction to the *cpe* Gene

Cloning studies revealed that the *cpe* gene encodes a 957 bp ORF whose sequence is identical in most (or all) *cpe*-positive type A isolates [17,24,25]. Primer extension analyses, RNase T2 protection assays, and deletion mutagenesis approaches identified three putative promoters immediately upstream of the *cpe* ORF [16]; these *cpe* promoters also appear to be conserved in many (or all) type A isolates [26]. Downstream of the *cpe* ORF is a stem-loop structure followed by an oligo-T tract [17] that may serve as a rho-independent transcriptional terminator for the *cpe* gene and may also influence CPE expression levels (see below).

18.2.2.2 Association of the *cpe* Gene with Mobile Genetic Elements

Like many bacterial toxin genes, the *cpe* gene appears to be associated with mobile genetic elements. Cole's group determined in the mid-1990s that the *cpe* gene can reside on either large plasmids or on the *C. perfringens* chromosome [25]. Even when chromosomally located, the *cpe* gene may still be associated with mobile genetic elements (i.e., Brynestad et al. proposed in 1997 that the chromosomal *cpe* gene is present on a putative 6.3 kb transposon named Tn*5565* [27]). This putative transposon, which has terminal IS*1470* repeats (see Figure 18.2), apparently inserted between the purine permease (*uabC*) and quinolinate phosphoribosyltransferase (*nadC*) housekeeping genes on the chromosome of *C. perfringens* strain NCTC8239 [27]. Brynestad et al. also detected excision of several possible Tn*5565* species from the NCTC8239 chromosome and demonstrated that those excised species form circular intermediates that could contribute to *cpe* gene movement [28].

Recently, Miyamoto et al. compared the organization of the chromosomal vs. plasmid *cpe* loci of type A isolates [26]. Those studies examining 19 *cpe*-positive type A isolates revealed that an ~3 kb DNA region (containing an upstream IS*1469* element, the *cpe* gene, and downstream flanking sequences) is identical in both the plasmid and chromosomal *cpe* locus. That finding is consistent with the chromosomal and plasmid *cpe* genes of type A isolates sharing a common origin. Beyond the conserved region immediately surrounding the *cpe* gene, Miyamoto et al. detected substantial differences between the plasmid vs. chromosomal *cpe* loci of type A isolates. Specifically, the type A plasmid *cpe* locus (Figure 18.2) lacks the IS*1470* element present upstream of the chromosomal *cpe* gene, instead carrying an upstream cytosine methylase (*dcm*) gene. Additionally, the IS*1470* element located downstream of the chromosomal *cpe* locus has been replaced in type A plasmid *cpe* loci by either an oppositely oriented and defective IS*1470*-like element, or by an IS*1151* element (Figure 18.2).

The plasmid *cpe* locus organizations depicted in Figure 18.2 suggest that the *cpe* plasmid of type A isolates might have originated from integration of a *cpe*-containing genetic element near the *dcm* sequences of a *C. perfringens* plasmid. For example, an excised Tn*5565* species containing both IS*1469* and the *cpe* gene has been detected in chromosomal *cpe* isolates [28]. A similar IS*1469*-*cpe* genetic element might have excised from the chromosome of a *C. perfingens* isolate containing

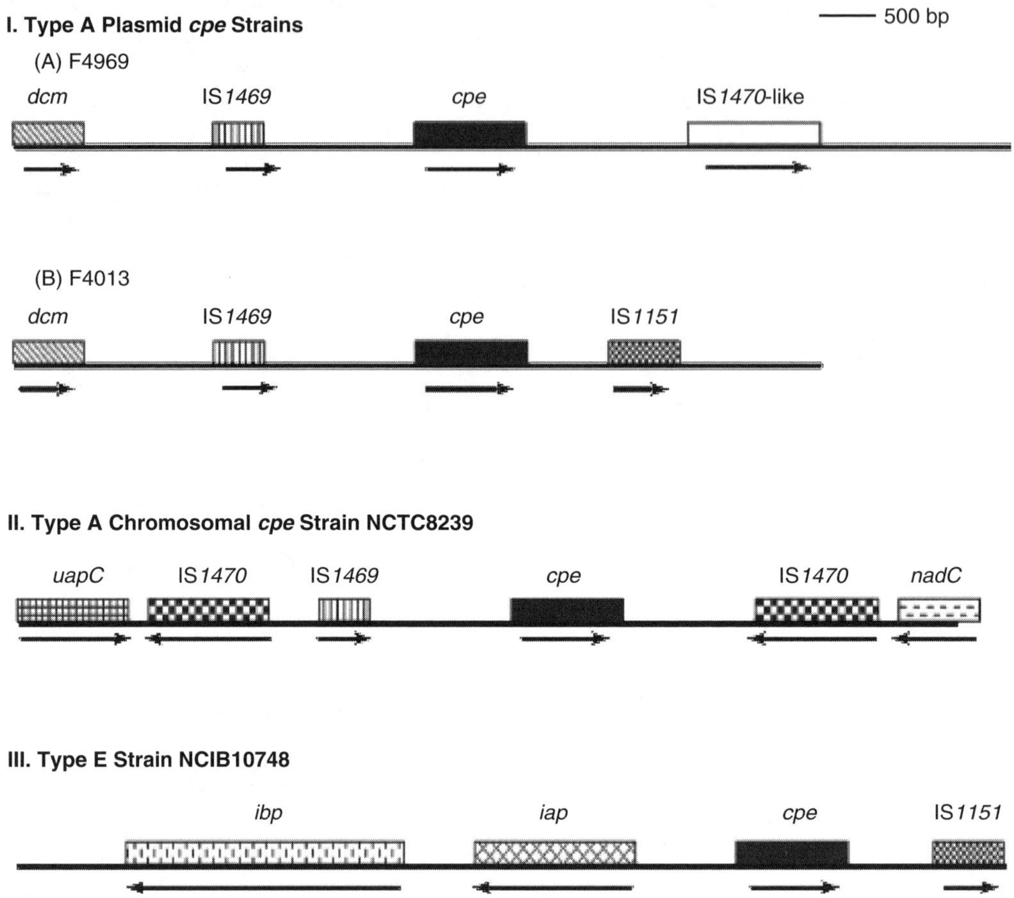

FIGURE 18.2 Genetic organization of the *cpe* locus in *C. perfringens* type A and E isolates. (I) Genetic organization of the *cpe* locus in two Type A isolates carrying a plasmid-borne, functional *cpe* gene, including: (a) isolate F4969 (note the presence of IS*1470*-like sequences downstream of the *cpe* gene in this isolate) and (b) isolate F4013 (note the presence of IS*1151* sequences downstream of the *cpe* gene in this isolate). (II) Genetic organization of the *cpe* locus in food poisoning strain NCTC8239, which carries its functional *cpe* gene on the chromosome (note the presence of IS*1470* sequences upstream and downstream of the *cpe* gene in this isolate). (III) Genetic organization of the *cpe* locus in type E isolate NCIB10748, which carries a defective *cpe* gene containing nonsense, missense, or frameshift mutations in its promoter region, ribosome binding site, and ORF. This figure is based upon results published in Billington et al., Miyamoto et al., and Brynestad et al. [6,26,27]. Gene abbreviations shown include: *dcm*, cytosine methyltransferase; *uapC*, purine permease; *nadC*, quinolinate phosphorylribosyltransferase; *ibp*, iota toxin B component; *iap*, iota toxin A component.

Tn*5565* and then inserted into some plasmid already present in that same isolate to create a progenitor *cpe* plasmid.

The detected similarity [26] of the plasmid *cpe* loci among type A isolates opens the possibility of horizontal transfer of the progenitor *cpe* plasmid to other type A isolates. This possibility receives direct support from recent *in vitro* studies demonstrating conjugative transfer of the *cpe* plasmid between *C. perfringens* type A isolates [29]. The precise mechanisms of conjugative plasmid transfer in *C. perfringens* awaits further study.

Interestingly, most (or all) *C. perfringens* type E isolates also carry *cpe* sequences [6], although those type E *cpe* sequences are defective and silent due to an accumulation of numerous nonsense, missense, and frame-shift mutations in their promoter region, ribosome binding site, and ORF [6].

As indicated in Figure 18.2, the plasmid-borne silent *cpe* sequences present in type E isolates are located near the iota toxin genes. This observation could suggest that the type E virulence plasmid originated by a mobile genetic element carrying iota toxin sequences inserting near the promoter of the plasmid *cpe* gene of a type A isolate. That putative insertion event may have inactivated the plasmid *cpe* gene (favoring the accumulation of additional mutations in the now defective type E *cpe* gene). It would also convert that type A isolate to type E. Putative interspecies transfer of iota toxin sequences to type E isolates is supported by the presence of iota toxin gene homologues in several other clostridial and *Bacillus* spp. [6].

The presence of nearly identical mutations in the silent *cpe* genes of most (or all) type E isolates, as well as the lack of chromosomal clonality among type E isolates [6], suggests that the type E plasmid carrying defective *cpe* sequences and functional iota toxin genes evolved fairly recently and then transferred between *C. perfringens* isolates [6]. If the type E plasmid is derived from the type A *cpe* plasmid (as discussed above), horizontal transfer of the type E plasmid carrying both silent *cpe* genes and iota toxin genes might be expected, given the recently demonstrated conjugative transfer of the type A *cpe* plasmid [29].

18.2.2.3 Relationship Between *cpe* Genotypes and GI Disease

To date, all examined *cpe*-positive, type A isolates carry either a plasmid-borne or chromosomal *cpe* gene, i.e., no *cpe*-positive type A isolate carrying *both* a chromosomal *cpe* gene and a plasmid *cpe* gene has yet been identified [26]. In early studies, Cole's group noted that *cpe*-positive isolates of European veterinary origin carried a plasmid *cpe* gene, while *cpe*-positive isolates associated with European cases of food poisoning had a chromosomal *cpe* gene [25]. The author's laboratory extended those initial findings by demonstrating that most (or all) North American food poisoning isolates also carry a chromosomal *cpe* gene [30]. More importantly, our group also determined that most (or all) CPE-positive non-foodborne human GI disease isolates, regardless of their geographic origin, have a plasmid-borne *cpe* gene [30]. These findings collectively demonstrate that particular *cpe* genotypes are strongly linked to specific CPE-associated human GI diseases. The discovery of *cpe* genotype–human GI disease associations suggests that *cpe* genotyping is a useful diagnostic and epidemiologic tool for investigating CPE-associated GI disease cases. In response, simple and rapid PCR assays for *cpe* genotyping of disease isolates have recently been developed ([31] and Figure 18.3).

The basis for the strong *cpe* genotype–human GI disease associations described above is not yet fully understood, although some clues are emerging. Both the vegetative cells and spores of food poisoning isolates carrying a chromosomal *cpe* gene were found to exhibit more heat resistance than cells and spores of other *C. perfringens* isolates, including type A plasmid *cpe* isolates [10]. Their greater heat resistance phenotype should favor the involvement of chromosomal *cpe* isolates in food poisoning, as this trait appears highly advantageous for a foodborne pathogen spread mainly through improperly cooked or held meats. The basis for the unusual heat resistance of food poisoning isolates carrying a chromosomal *cpe* gene is not yet known.

Potential explanations for the observed linkage between type A plasmid *cpe* isolates and CPE-associated non-foodborne GI diseases are also now becoming apparent [29]. Unlike *C. perfringens* food poisoning, which results from ingestion of massive numbers of *C. perfringens* cells growing in foods, the CPE-associated non-foodborne GI diseases usually result from ingestion or inhalation of a small dose of *C. perfringens* cells or spores present in the nosocomial environment [18,32]. Therefore, it should be helpful for the small number of infecting isolates responsible for causing CPE-associated non-foodborne human GI diseases to carry their *cpe* gene on a transferable plasmid (i.e., transfer of the *cpe* plasmid from these few infecting bacteria to the many *cpe*-negative *C. perfringens* isolates present in the normal intestinal flora should convert those recipient normal flora isolates to enteric virulence and, thus, help achieve the *cpe*-positive *C. perfringens* load needed to establish enteric disease). Presuming normal flora, *C. perfringens* isolates are under selective

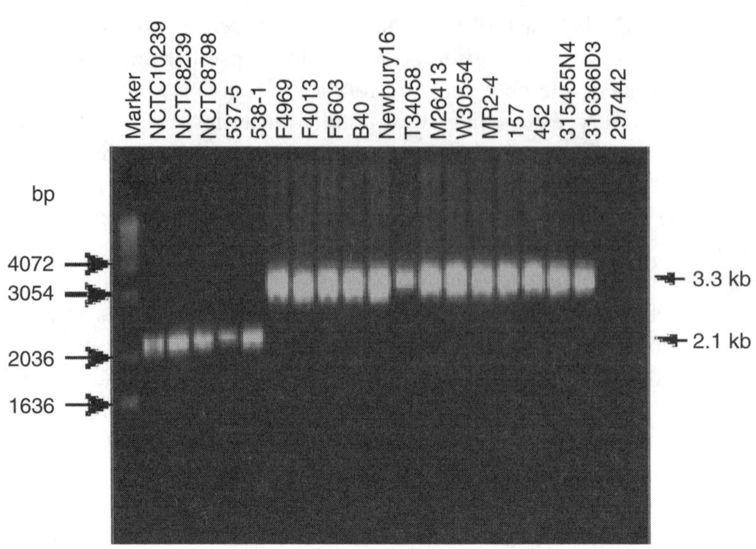

FIGURE 18.3 Duplex PCR genotyping assay for distinguishing *C. perfringens* type A isolates carrying a plasmid-borne *cpe* gene vs. a chromosomal *cpe* gene. Utilizing the type A *cpe* loci organizations shown in Figure 18.2, this duplex PCR produces a *dcm-cpe* PCR product (see Figure 18.2) of ~3.3 kb from plasmid *cpe* type A isolates and a *cpe*-IS*1470* PCR product of ~2.1 kb from chromosomal *cpe* type A isolates. Molecular size markers are present in lane 1, lanes 2 to 6 show duplex PCR results with type A isolates known to carry a chromosomal *cpe* gene, lanes 7 to 19 show duplex PCR results with type A isolates known to carry a plasmid *cpe* gene, and lane 20 shows duplex PCR results with a known *cpe*-negative type A isolate. (Reproduced with permission from Wen, Q. et al., *J. Clin. Microbiol.,* 41, 1494, 2003.)

pressure for their ability to colonize and persist in the GI tract; the putative *in vivo* virulence conversion of normal flora *C. perfringens* isolates could also increase the amount of time that *cpe*-positive isolates are present in the GI tract during CPE-associated non-foodborne human GI diseases. Such increased persistence might explain why symptoms of CPE-associated non-foodborne GI diseases are typically more prolonged and severe [33] than the symptoms of *C. perfringens* type A food poisoning, a disease caused by exogenous strains presumably not well adapted for colonization of the GI tract.

18.2.2.4 Regulation of CPE Expression

Quantitative Western blot analyses have demonstrated >1500-fold greater CPE expression by sporulating vs. vegetative cells of the *cpe*-positive, type A food poisoning strain NCTC8239 [17]. CPE expression appears to be similarly sporulation-associated in all other examined *cpe*-positive *C. perfringens* type A isolates, regardless of whether the *cpe* gene is chromosomal or plasmid-borne [24]. RNA slot blot and Northern blot studies revealed that this sporulation-associated CPE expression is regulated at the transcriptional level, with detectable *cpe* mRNA levels present during sporulation, but not during vegetative growth [16,34,35]. Those early Northern blot studies [34] also suggested that *cpe* is transcribed as a monocistronic message of ~1.2 kb, a result consistent with the identified location of upstream promoters and downstream transcriptional terminators (as described in Section 18.2.2.2).

An early study of CPE regulation [34] transformed *cpe*-negative *C. perfringens* type A, B, and C isolates with a shuttle plasmid carrying the *cpe* gene cloned from strain NCTC8239. All resultant transformants expressed CPE in a properly regulated manner (i.e., CPE was produced during sporulation but not during vegetative growth). Those observations suggested that CPE expression involves a positive, sporulation-associated regulator made by most *C. perfringens* isolates. The possible identity of this positive regulator was then provided by sequence analyses detecting homology between *cpe* transcriptional start sites and SigE- and SigK-dependent promoters [16]. This identified homology between *cpe* promoter sequences and SigE-/SigK-dependent promoters is probably important, since SigE and SigK-encoding genes are present on the *C. perfringens* chromosome [36] and are known to be active in *Bacillus cereus* mother cells during sporulation [16].

Other early studies suggested that CPE expression could also involve negative regulation during vegetative growth [37]. Brynestad et al. identified putative Hpr (hyperprotease-producing)-like binding sequences upstream and downstream of the *cpe* gene [37]. Furthermore, they reported that an *hpr*-specific probe can hybridize to DNA from several *C. perfringens* strains [37]. Since Hpr is known to negatively regulate expression of many proteins during the exponential growth of *B. subtilis*, the presence of Hpr-binding sequences near the *cpe* gene and the apparent presence of *hpr*-encoding sequences in *C. perfringens* isolates led Brynestad et al. to propose that Hpr negatively regulates CPE expression during vegetative growth [37]. This possibility remains unproven.

Besides being sporulation-associated, another interesting aspect of CPE expression is the exceptionally large amounts of CPE often produced by type A isolates during sporulation. CPE often accounts for >15% of total protein present inside sporulating cells of type A isolates [17,24]. In fact, so much CPE accumulates in the cytoplasm of many sporulating cells that the enterotoxin becomes localized in paracrystalline inclusion bodies [7,38]. This high-level CPE expression is not dependent on whether an isolate carries a plasmid-borne or chromosomal *cpe* gene and is also not attributable to a gene dosage effect (type A isolates generally carry only a single copy of the *cpe* gene [24]). One possible contributing factor to abundant CPE expression by many type A isolates could be message stability. An older study [39] reported that the *cpe* message has a half-life of ~58 min, which is exceptionally stable for a prokaryotic message. This putative *cpe* message stability could result, at least in part, from the presence of the previously mentioned stem-loop structure located downstream of the *cpe* gene, as similar stem-loop structures contribute to stability of other prokaryotic messages [17].

18.2.3 The Cellular Action of CPE

18.2.3.1 Early Events: CPE Binding and Formation of the Small CPE Complex

Recent studies [40] demonstrated that CPE can form channels in artificial lipid membranes, an observation indicating that protein receptors are not absolutely necessary for CPE action. However, this *in vitro* effect developed much more slowly than CPE effects on sensitive mammalian cells (see Section 18.2.3.3) and also required use of high (i.e., nonphysiologic) enterotoxin concentrations. Under pathophysiologic conditions, protein receptors appear to be critical for CPE-induced biologic activity. For example, Pronase pretreatment of naturally sensitive mammalian cell lines effectively eliminates the ability of those cells to specifically bind pathophysiologic CPE concentrations [41–43]. Furthermore, experiments have shown that when mouse L cells (which do not naturally bind or respond to the enterotoxin) are transfected to express claudin protein –3, –4, –6, –7, –8, or –14, those transfectants become very CPE-sensitive [44–46].

Transfection experiments demonstrating that expression of certain claudins allows mouse L cells to bind and respond to the enterotoxin also identifies those particular claudins as CPE receptor candidates. The claudin family currently consists of >20 small (~22 kDa) proteins found in the epithelial tight junction (TJ), where they serve important structural and functional roles [47,48]. Claudins differ among themselves primarily in their cytoplasmic, C-terminal tail [47,48]. Interestingly, expression of claudins 1, 2, 5, and 10 does not convert L cell transfectants to CPE sensitivity, suggesting that only certain claudins may function as CPE receptors [46].

The putative role of certain claudins as the physiologic CPE receptor still requires verification. For example, no study has yet demonstrated that CPE interacts with claudins in naturally CPE-sensitive cells, particularly in physiologically relevant target cells such as enterocytes or cultured

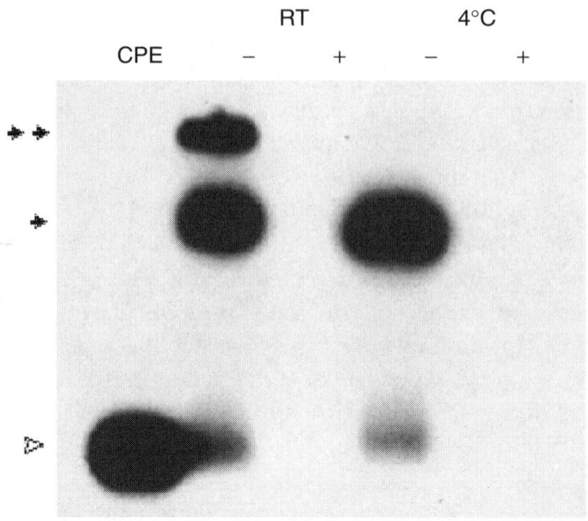

FIGURE 18.4 CPE small complex formation in rabbit intestinal brush border membranes (BBMs). Right panel: ^{125}I-CPE was incubated with BBMS at 4°C in the presence (+) or absence (-) of 50-fold excess unlabeled CPE. After washing away unbound toxin, the BBMs were extracted with Triton X-100 and analyzed by nondenaturing gel electrophoresis. The single closed arrow depicts the migration of the small complex formed under these conditions. Left panel: The same experiment performed in the right panel except it was performed at room temperature (RT) or 37°C (not shown) instead of at 4°C; note the additional band of higher M_r material (double closed arrow), which corresponds to the larger, SDS-resistant complexes discussed in the text. Similarly, if BBMs are CPE-treated at 4°C as described for the left panel, washed, and then warmed to 37°C, results identical to the right panel were obtained (not shown), indicating that formation of the small complex is a precursor for formation of the larger, SDS-resistant complexes. For comparison, migration of free CPE (no BBMs) in this gel system is noted by the single open arrow. (Reproduced with permission from Wieckowski, E.U. et al., *J. Biol. Chem.,* 269, 10838, 1994.)

enterocyte-like cell lines (e.g., CaCo-2 cells). A potential complication to certain claudins being the CPE receptor was provided by co-immunoprecipitation data [49], indicating that CPE closely interacts with an ~45 to 50 kDa protein upon binding to intestinal brush border membranes (the natural CPE target membrane) or CaCo-2 cells. At least four interpretations are possible for that co-immunoprecipitation result: (i) the ~45 to 50 kDa protein, rather than a claudin, is really the physiologic CPE receptor; (ii) there is more than one receptor capable of independently binding CPE; (iii) the ~45 to 50 kDa protein is simply a claudin aggregate, a phenomenon recently described by Anderson's laboratory [50]; or (iv) the ~45 to 50 kDa protein is a co-receptor that, in combination with certain claudins, binds CPE to sensitive cells.

CPE interacting with the ~45–50 kDa protein quickly becomes localized in a small complex of ~90 kDa ([49] and Figure 18.4). This CPE-containing small complex is SDS-sensitive and appears important for CPE activity since it forms in all naturally CPE-sensitive cells examined to date. However, formation of this small CPE complex is not sufficient to obtain toxicity. For example, several CPE point and deletion variants have been identified that can bind and form small complex in CPE-sensitive cells (e.g., Vero or CaCo-2 cells), yet remain nontoxic at 37°C [51,52]. Furthermore, CPE binds and forms the small complex at 4°C, but exerts no cytotoxic effect on sensitive cells at this low temperature [49].

18.2.3.2 Formation of Larger CPE-Containing Complexes

If Vero or CaCo-2 cells treated with CPE at 4°C (where all CPE localizes in the small complex, see Section 18.2.3.1) are subsequently warmed to 37°C, the enterotoxin rapidly associates with SDS-resistant material of >150 kDa [49]. Simultaneously with formation of that high M_r CPE-containing

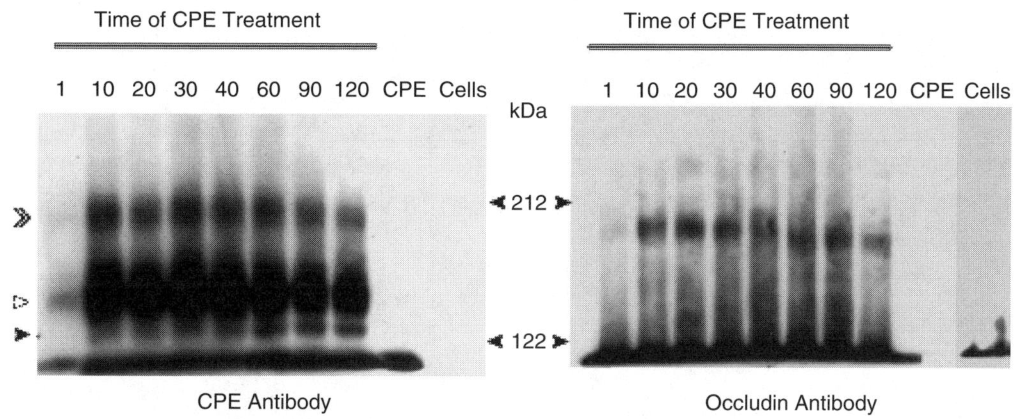

FIGURE 18.5 Large CPE complex formation kinetics in CaCo-2 cells treated in suspension with 1 μg/ml of CPE at 37°C. After the specified incubation times, the CPE-treated cell suspension was centrifuged, extracted with SDS, electrophoresed on 4% acrylamide gels containing SDS, and then Western immunoblotted with antibodies against either CPE (left panel) or occludin (right panel). The double, open, and closed arrows indicate the location of the ~200, ~155, and ~135 kDa large CPE complexes, respectively. (Reproduced with permission from Singh, U. et al., *J. Biol. Chem.,* 275, 18407, 2000.)

material, the warmed cells die. Those observations strongly suggest a cause-and-effect relationship between formation of the >150 kDa material and CPE-induced cytotoxicity. The importance of the >150 kDa material for CPE action is further supported by CPE point mutant and deletion fragment studies [51,52] that identified several CPE species capable of forming the small CPE complex, but not the >150 kDa CPE-containing material. Those CPE variants were found to be uniformly nontoxic. Furthermore, CPE deletion fragments lacking up to the first 53 N-terminal amino acids were shown to form more of the >150 kDa material than does the native enterotoxin, and those deletion fragments were also found to possess more cytotoxic activity than the native enterotoxin (see Section 18.2.7).

Singh et al. recently used SDS gels containing a low (4%) acrylamide content to better resolve the SDS-resistant, >150 kDa CPE-containing material [53]. Suspensions of enterocyte-like CaCo-2 cells treated with enterotoxin at 37°C, lysed with SDS, and run on those 4% acrylamide SDS gels (Figure 18.5) were shown to contain two major CPE-containing species, with sizes of ~155 kDa and ~200 kDa. A minor CPE species of ~135 kDa was also sometimes detectable in these experiments. Singh et al. also performed Western blot (Figure 18.5) and co-immunoprecipitations (not shown) demonstrating that the ~200 kDa CPE complex contains occludin, which (like claudins) is an important protein component of TJs (further discussion below in Section 18.2.3.3).

Which of the recently resolved SDS-resistant complexes is responsible for initiating CPE activity? To address that question, Singh et al. performed additional studies using CaCo-2 cells grown in Transwell® cultures, which permits those cells to polarize and form TJs. Those studies [54] revealed that after 15 min of CPE treatment (by which time cytotoxic effects are fully apparent, see Section 18.2.3.3), the ~155 kDa species is the only SDS-resistant CPE complex present in the Transwell® CaCo-2 cell cultures. In this Transwell® model system, ~200 kDa CPE complex formation was a slow-developing consequence of enterotoxin treatment, requiring at least 1 h before its appearance. These kinetic findings imply that formation of the ~155 kDa complex is sufficient to initiate CPE-induced cytotoxicity.

18.2.3.3 CPE-Induced Cytotoxicity

Formation of the ~155 kDa CPE complex apparently kills cells by triggering massive changes in plasma membrane permeability [55]. In all CPE-sensitive cells examined to date, these permeability

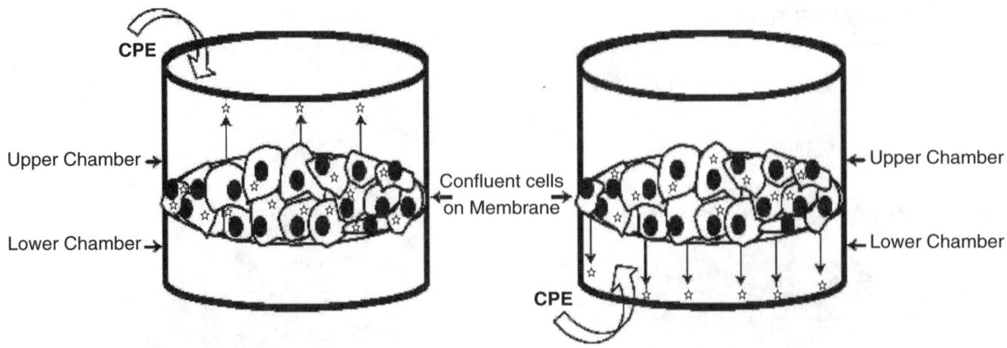

FIGURE 18.6 The cytotoxic effects of CPE on Transwell® cultures of CaCo-2 cells. CaCo-2 cultures were grown to confluency and then labeled with ^{86}RbCl. After the addition of CPE to the upper or lower chambers of the radiolabeled cultures (as indicated), culture supernatants in the upper and lower Transwell® chambers were collected separately and radioactivity in those samples was then determined using a gamma counter. After correction for spontaneous (background) radiolabel release, levels of CPE-induced ^{86}Rb-release are depicted by arrows (i.e., more arrows indicate greater radiolabel release). If no arrow is shown, no CPE-induced ^{86}Rb-release was detected into that chamber.

alterations are dose-dependent and can develop within 5 min of enterotoxin treatment [56]. Initially, the plasma membranes of CPE-treated cells become more permeable to small molecules of <200 Da. These CPE-induced small molecule permeability alterations are nonselective (i.e., influx and efflux of both ions and amino acids increase in CPE-treated cells). Those permeability changes damage the cellular colloid-osmotic equilibrium [57], which (at very high CPE concentrations) can lead to cell lysis. At lower CPE concentrations, these permeability changes alter cytoplasmic small molecule pools, thereby disrupting vital metabolic processes such as macromolecular synthesis and energy production [55].

Singh et al. also recently used [54] Transwell® culture systems to compare the effects of CPE treating apical vs. basolateral plasma membranes of polarized epithelial cells (such as CaCo-2 cells). Those studies (Figure 18.6) determined that CaCo-2 cells develop permeability alterations when CPE-treated on their apical surface, confirming that the apical membrane (which is the plasma membrane surface initially exposed to CPE during natural GI disease) is enterotoxin-sensitive. Interestingly, those studies also revealed that Transwell® CaCo-2 cultures exhibit even greater plasma membrane permeability changes when CPE-treated on their basal surface (i.e., the CPE sensitivity of basolateral membranes exceeds that of apical membranes). The relative CPE sensitivities of the apical vs. basolateral membranes of CaCo-2 cells grown in Transwell® cultures were found to correlate closely with the amounts of the ~155 kDa complex formed after apical vs. basolateral enterotoxin treatment of those cultures. This observation further supports the ~155 kDa CPE complex's importance for inducing the membrane permeability alterations responsible for cytotoxicity in the CPE-treated cell.

The mechanism by which ~155 kDa CPE complex formation affects plasma membrane permeability properties is not yet clear. However, increasing evidence suggests that, like many membrane-active bacterial toxins, this process might involve creation of channels/pores in plasma membranes. One observation supporting this hypothesis was already mentioned in Section 18.2.3.1 (i.e., at high concentrations, CPE can form channels in otherwise protein-free artificial membranes [40]). Furthermore, CPE present in SDS-resistant complexes (including the ~155 kDa complex) cannot dissociate from rabbit intestinal BBMs. Also, those CPE species are not degraded when present in BBMs treated with extracellular proteases [58]. This protease resistance requires the SDS-resistant complexes to be present in plasma membranes, since these CPE complexes are rapidly degraded if detergent-extracted from BBMs prior to protease-treatment. Collectively, these observations are consistent with CPE (and perhaps some eucaryotic complex proteins) inserting into plasma membranes when localized in SDS-resistant complexes; the inserted toxin might then exhibit channel/pore-forming activity.

Recent studies by Chakrabarti et al. examined the mechanism of cell death induced by CPE treatment [59]. Lower CPE doses (1 μg/ml) were found to induce a classical apoptosis effect involving mitochondrial membrane depolarization and caspase-3 activation. For example, CaCo-2 cells treated with a low CPE dose displayed a ladderlike DNA cleavage pattern that can be inhibited by either the broad-spectrum caspase inhibitor Z-VAD-FMK (Figure 18.7) or the caspase 3/7 inhibitor DEVD-CHO (not shown). However, neither the caspase 1 inhibitor YVAD-CHO (not shown) nor the oncosis inhibitor glycine (Figure 18.7) could affect this low CPE dose DNA cleavage. In contrast, high CPE doses (10 μg/ml) were found to cause oncosis, e.g., CaCo-2 cultures treated with the high CPE dose exhibited "smeared" DNA cleavage that could be inhibited by glycine (Figure 18.7) but not by the broad-spectrum caspase inhibitor Z-VAD-FMK. These findings have pathophysiologic relevance since both the low and high CPE concentrations used in these *in vitro* experiments have been detected in feces from patients sickened with CPE-associated GI disease [12,13] (i.e., both oncosis and apoptosis likely can occur during natural CPE-mediated GI disease). Possible contributions of CPE-induced apoptosis and oncosis to CPE-induced GI disease are described in Section 18.2.6.

18.2.4 CPE Interactions with Tight Junctions

If the ~155 kDa CPE complex initiates cytotoxicity, then what role might the ~200 kDa complex play in CPE action? As mentioned briefly in Section 18.2.3.2, Singh et al. demonstrated the presence of the TJ protein occludin in the ~200 kDa CPE complex formed by polarized CaCo-2 cells [53]. Those researchers later showed that, concurrent with ~200 kDa complex formation, occludin is removed from TJs and internalizes into the cytoplasm of CPE-treated CaCo-2 cells grown in Transwell® cultures [54]. This CPE-induced occludin redistribution is at least a somewhat specific effect as ZO-1, another TJ-associated protein, remains present at the TJs of CPE-treated CaCo-2 cell cultures. However, occludin may not be the only TJ protein whose cellular distribution is affected by CPE treatment. A nontoxic CPE fragment reportedly induced the removal of claudin 4 from the TJs into the cytoplasm of polarized MDCK cell cultures [60].

TJs serve as both a barrier (separating apical from basolateral membranes) and a gate (separating internal from external ions and fluids) in the polarized epithelium of the small intestine. Since occludin and claudins are both considered important TJ components, CPE-induced removal of these proteins from TJs likely has structural and functional consequences for the intestinal epithelium. For example, CPE-induced removal of occludin or claudins from TJs probably causes or contributes to the TJ structure breakdown and paracellular permeability alterations that have been observed when polarized epithelial models were treated with either native CPE or a nontoxic, binding-capable CPE fragment [61]. Similar TJ changes in the intestinal epithelium may contribute to CPE-mediated GI disease (see Section 18.2.6).

Interestingly, the presence or absence of TJs also affects how mammalian cell monolayers respond to CPE treatment. For example, CPE-treatment of ^{86}Rb-labeled Transwell® cultures of polarized CaCo-2 cells (Figure 18.6) or MDCK cells (not shown) causes radiolabel release only into the same Transwell® chamber to which the enterotoxin had been added. In contrast, CPE treatment of ^{86}Rb-labeled Transwell® cultures of Vero cells, which (unlike CaCo-2 and MDCK cells) lack TJs, induces release of radiolabel into both Transwell® chambers. The presence of TJs could be affecting monolayer responsiveness of polarized cells to CPE treatment in several ways. First, the barrier function of TJs might prohibit released ^{86}Rb from moving across a CPE-treated monolayer, at least prior to the development of overt cell damage. Second, the fence function of TJs separating apical from basolateral membranes might limit movement of the ~155 kDa CPE complex between the apical and basolateral membranes, thus preventing the development of permeability alterations in membrane surfaces not directly exposed to CPE.

18.2.5 Intestinal Effects of CPE

In vivo studies using a rabbit model confirmed that highly purified CPE is an enterotoxin (i.e., this toxin rapidly reverses small intestinal fluid and electrolyte transport from net absorption to net secretion [62–66]). All segments of the rabbit small intestine are CPE-sensitive, with the ileum being particularly responsive [64]. While CPE does not induce fluid and electrolyte changes in the rabbit colon [62], the sensitivity of the human colon to this enterotoxin remains unknown.

In the rabbit small intestines, purified CPE also produces histopathologic damage [63–66] that is indistinguishable from the damage caused by sporulating culture lysates of a CPE-positive isolate (see Figure 18.1). CPE-induced histopathologic damage is dose-dependent but can develop within 30 min of treatment [63–66]. This damage initially develops at villus tips, but later progresses to include villus shortening and epithelial desquamation [63–66]. In some cases (see Section 18.2.6) inflammatory cells infiltrate into intestinal regions damaged by CPE [63–66].

Several observations suggest the histopathologic damage described in the preceding paragraph initiates fluid and electrolyte transport alterations in the CPE-treated rabbit small intestine. For

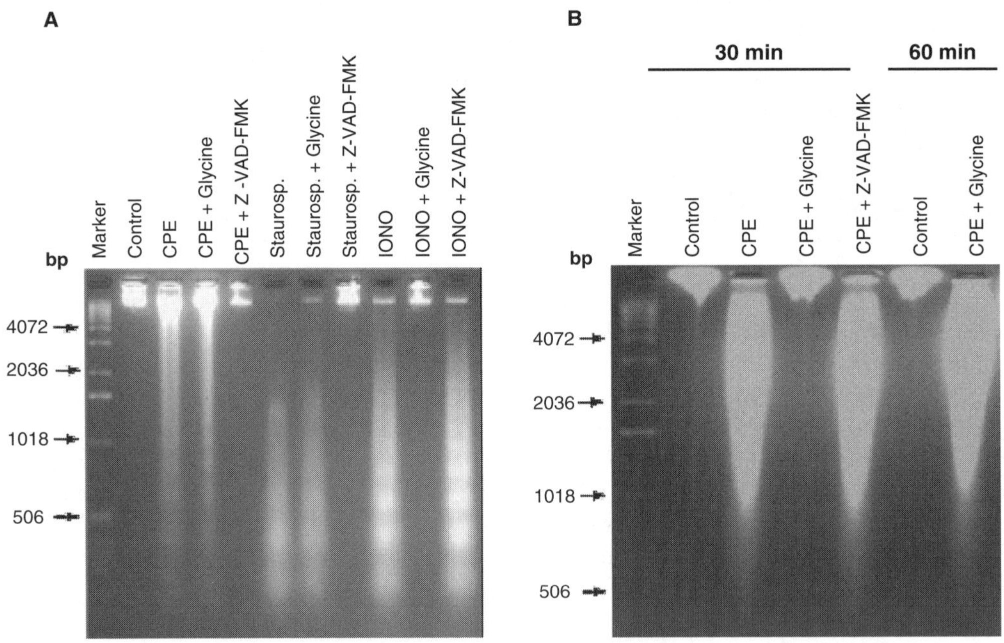

FIGURE 18.7 CPE-induced DNA cleavage in CaCo-2 cells. Confluent CaCo-2 cells were pretreated for 2 h at 37°C with Hanks balance salts solution (HBSS) that did or did not contain Z-VAD-FMK, a broad-spectrum capase inhibitor or the oncosis inhibitor glycine. Those cultures were then treated for: (A) 60 min at 37°C with HBSS containing 1 µg/ml of CPE or (B) 30 or 60 min (as indicated) at 37°C with HBSS containing 10 µg/ml of CPE. In all cases the same inhibitor (if any) was present during CPE treatment as had been applied during pretreatment. Also shown for panel A are CaCo-2 cells treated for 8 h at 37°C with 2 µ*M* staurosporine (Staurosp.) or 5 µ*M* ionomycin (IONO), which are positive controls for apoptosis and oncosis, respectively. Control lanes received only HBSS. After completion of the specified treatment incubation, DNA was extracted from each sample and run on 2% agarose gels. Those gels were then stained with ethidium bromide. (Reproduced with permission from Chakrabarti, G. et al., *Infect. Immun.*, 71, 4260, 2003.)

example, a close correlation exists between the onset of CPE-induced intestinal damage and the start of fluid and electrolyte transport changes [66]. Furthermore, only those CPE doses capable of causing intestinal histopathologic damage can trigger changes in intestinal fluid and electrolyte transport [63].

18.2.6 A Comprehensive Model for CPE's Intestinal Action

The data presented in Section 18.2.3 and Section 18.2.5 suggest the following comprehensive model for CPE action during GI disease (see Figure 18.8). After CPE is released into the intestinal lumen following lysis of the mother *C. perfringens* cell, the enterotoxin first binds to the apical surface of enterocytes, possibly via certain claudins or a 45 to 50 kDa protein. This binding results in formation of the ~90 kDa small CPE-containing complex. At physiologic temperatures, where membrane diffusion can occur, the small CPE complex rapidly interacts with other, still-unidentified proteins to form the ~155 kDa CPE-containing complex. Formation of this CPE complex disrupts the apical plasma membrane's small molecule permeability barrier, perhaps because CPE (and possibly other proteins) present in the 155 kDa complex inserts into the plasma membrane bilayer to form a channel/pore.

Some exposed enterocytes die directly from CPE-induced membrane permeability changes. When exposed to a high CPE dose, enterocytes form large amounts of the ~155 kDa complex and die rapidly from oncosis. Enterocytes exposed to a lower (but still lethal) CPE dose contain less ~155 kDa complex, and a classical caspase-3 mediated apoptosis is induced. Either the CPE-induced apoptotic or oncotic cell death pathway results in histopathologic damage that initiates intestinal fluid and electrolyte secretion changes. In regions of the small intestine exposed to high CPE concentrations such that substantial oncosis occurs, an inflammatory response might be triggered. That inflammatory response could contribute to intestinal fluid and electrolyte loss, particularly during the later stages of CPE-induced GI disease.

The onset of CPE-induced intestinal epithelial desquamation could expose the basolateral membranes of still-undamaged enterocytes, allowing CPE to then interact with previously inaccessible TJ proteins, such as occludin. This results in formation of the ~200 kDa CPE complex, which probably removes occludin, and possibly claudins, from TJs. This redistribution of TJ proteins damages the structure of the TJ and thereby induces paracellular permeability changes that contribute to CPE-induced intestinal fluid and electrolyte transport alterations.

18.2.7 The CPE Structure/Function Relationship

The 319 amino acid CPE polypeptide has a unique primary sequence, except for some limited homology (of unknown significance) with a non-neurotoxic protein made by some *Clostridium botulinum* strains [7]. The CPE amino acid sequence is identical in all CPE-positive type A isolates examined to date [24].

While the three-dimensional structure of CPE remains to be solved, extensive deletion and point mutagenesis studies have mapped functionally important regions of this enterotoxin [51,52,67]. Those mutagenesis studies revealed that, like most bacterial protein toxins, the receptor-binding and toxicity domains of CPE segregate to different regions. The extreme C-terminus of CPE contains a receptor-binding domain [52,67], while the N-terminal region of the native enterotoxin protein (i.e., amino acids 45–53) is essential for toxicity [51,52]. Those N-terminal sequences are necessary for biologic activity due to their involvement in formation of both the ~155 kDa and ~200 kDa CPE complexes [51–53].

Extreme N-terminal amino acids (i.e., residues 1–45) of native CPE play no apparent role in cytotoxicity. In fact, removing these residues activates enterotoxin activity by ~2- to 3-fold [52]. This activation may increase the exposure of the CPE amino acid 45–53 region and, thus, promote

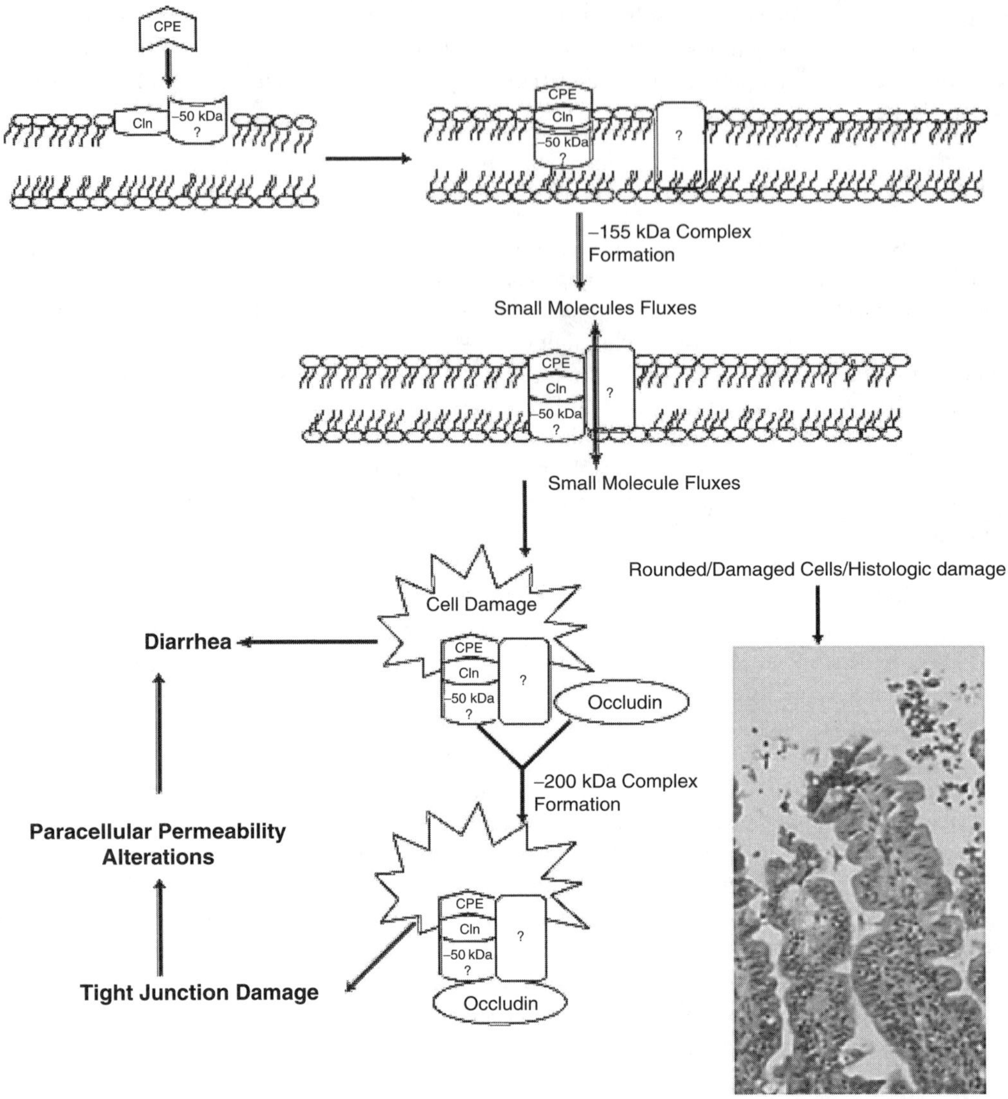

FIGURE 18.8 Current model for CPE action. CPE first binds to a claudin or an ~50 kDa protein to form the small (~90 kDa) CPE complex; that complex then interacts with additional proteins to form an ~155 kDa large CPE complex. Formation of that CPE species alters plasma membrane permeability, perhaps due to pore formation. These CPE-induced permeability changes produce morphologic damage that initiates CPE fluid and electrolyte transport alterations and also exposes the basolateral surface of enterocytes to CPE. That effect leads to additional ~155 kDa complex formation and also allows CPE to interact with occludin to form an ~200 kDa CPE complex. Formation of the ~200 kDa complex removes occludin and claudins from tight junctions (TJs). Consequently, these TJs become leaky, increasing paracellular permeability and contributing to CPE-induced intestinal fluid and electrolyte losses.

~155 kDa complex formation [52]. Interestingly, the intestinal proteases trypsin and chymotrypsin both remove extreme N-terminal CPE residues and also increase CPE activity *in vitro* [68–70]. Those findings suggest that CPE might undergo similar proteolytic activation during GI disease.

18.2.8 CPE Vaccines

Epitope mapping studies demonstrated the presence of a linear epitope in the extreme C-terminal sequences of CPE [67,68]. Furthermore, a monoclonal antibody reacting with that C-terminal epitope was shown to neutralize CPE cytotoxicity by blocking CPE binding [67,71], which further supports the localization of receptor-binding activity in the extreme C-terminus of CPE (see Section 18.2.7).

As C-terminal CPE fragments are noncytotoxic because they lack the amino acid 45–53 region necessary for ~155 kDa complex formation (see Section 18.2.7), the identification of a neutralizing linear epitope in C-terminal CPE fragments suggested the potential use of nontoxic C-terminal CPE fragments as vaccine candidates. This possibility was explored [72] by preparing a synthetic peptide to the extreme C-terminal sequence of CPE (i.e., the C-terminal 30 amino acids of the native enterotoxin). After that 30-mer peptide was chemically coupled to a carrier protein; the resultant conjugate was found to specifically induce high titers of CPE-neutralizing IgG antibodies when administered i.p. to mice. The mice showed no ill effects from this immunization, further supporting the potential usefulness of C-terminal CPE fragments for vaccination purposes. If further development of a CPE vaccine becomes warranted, it would be necessary to develop an immunization strategy to induce the strong intestinal IgA responses needed for protecting against CPE-mediated GI disease.

18.2.9 Potential Use of CPE for Anticancer Therapy

One of the most exciting recent CPE research developments has been provided by studies suggesting that the potent cytotoxic action of CPE might be harnessed for therapeutic use against certain solid tumors. One such study [73] suggests that claudin-4, which can serve as a functional CPE receptor, is over-expressed in most pancreatic cancer tissue and cell lines. Direct injection of CPE into human pancreatic tumor xenografts grown in mice was found to induce necrosis and shrinkage of those tumors. Since the mice carrying these tumors were unharmed by the enterotoxin injection they received, CPE appears to be a promising new treatment to specifically target claudin-4 expressing pancreatic tumors.

The potential antitumor activity of CPE does not appear limited to pancreatic tumors. Persistent high-level expression of claudin-3, another functional CPE receptor, has been detected in prostate adenocarcinomas and those tumor cells were also shown to be highly CPE-sensitive [74].

18.3 *C. PERFRINGENS* β2 TOXIN

In the 1980s Popoff's group sought to purify β toxin from *C. perfringens* type C strain CWC245 [75]. They obtained a purified protein of 28 kDa that caused cytotoxic effects in CHO cells and induced hemorrhagic necrosis in guinea pig ligated intestinal loops [75]. However, later cloning and sequencing studies indicated that the β toxin gene (*cpb*) of other *C. perfringens* type B and C strains encodes an ~35 kDa protein [75a]. To resolve the difference between the 28 kDa protein purified from stain CWC245 and the 35 kDa protein encoded by the *cpb* gene of other type B and C strains, Popoff's group then cloned and sequenced the CWC245 gene encoding their previously-purified ~28 kDa toxin [76]. Those studies demonstrated that the ~28 kDa protein encoded by CWC245 is not β toxin, but instead represents a distinct new toxin named beta2 (β2) toxin.

18.3.1 Role of β2 Toxin in Gastrointestinal Disease

To date, β2 toxin has been implicated as a possible virulence factor in GI diseases of several animals. Multiplex PCR and immunohistochemistry studies demonstrated the presence of *cpb*2-postive isolates or β2 toxin, respectively, in diarrheic piglets, horses with typhlocolitis, diarrheic dogs, an African elephant with ulcerative enteritis, and calves with enterotoxemia [77–83]. Recent studies have confirmed the ability of type A and C *cpb2*-positive porcine GI disease isolates to express β2 toxin [80]. A conclusive role for β2 toxin in veterinary disease awaits studies using *cpb2* knockout mutants. The contribution, if any, of β2 toxin to human GI disease has not yet been evaluated.

18.3.2 Genetics and Expression of β2 Toxin

18.3.2.1 The *cpb2* Gene

Cloning of the *cpb2* gene of strain CWC245 revealed an ORF encoding a 265 amino acid protein [76]. Amino acid sequencing analyses also performed in that study demonstrated that the 30 N-terminal amino acid region encoded by *cpb2* is removed from the mature β2 protein [76]. Presumably, this removed N-terminal region, which contains a signal peptidase recognition site, serves as a signal peptide facilitating β2 secretion. As mentioned earlier, β2 toxin shares no homology with β toxin or any other known proteins.

A consensus ribosome binding site is present immediately upstream of the *cpb2* initiation codon [76]. An inverted repeat capable of forming a hairpin loop lies downstream of the *cpb2* termination codon and may represent a Rho-independent transcriptional terminator [76]. The *cpb2* transcriptional start site has recently been mapped to 28 bases upstream of the start codon [84].

Multiplex PCR studies have detected the *cpb2* gene in all *C. perfringens* types; this toxin gene is present in nearly all type E isolates and appears to be least common in type D isolates [78,85]. Those studies also suggested a possible link between *cpe* and *cpb2* in some veterinary isolates.

18.3.2.2 Association of the *cpb2* Gene with Mobile Genetic Elements

Southern blot studies have demonstrated that the *cpb2* gene is present on large plasmids in *C. perfringens* strain CWC245, as well as several other *cpb2*-positive isolates associated with porcine diarrhea [76]. In addition, genome sequencing of *C. perfringens* strain 13 determined that type A strain carries its *cpb2* gene on a 54 kb plasmid, named pCP13 [36]. To date, there is no indication that the *cpb2* gene is carried on transposons.

18.3.2.3 Regulation of *cpb2* Expression

Unlike the sporulation-associated transcription of the *cpe* gene (Section 18.2.2.4), the *cpb2* gene is transcribed during vegetative growth, with maximal *cpb2* transcription occurring during the mid-exponential growth phase [76]. Shimizu's group has demonstrated that *cpb2* transcription is positively regulated by the VirR/VirS two-component regulatory system, which also regulates expression of other *C. perfringens* toxins, such as α toxin, κ toxin, and θ toxin [84]. This VirR/VirS regulation of *cpb2* transcription is mediated through the VR-RNA regulatory RNA [84]. Involvement of regulatory RNAs and two component regulatory systems in controlling virulence factor expression is an increasingly recognized strategy used by Gram-positive pathogens.

18.3.3 The Mechanism of Action of β2 Toxin

Limited information is currently available regarding the action of β2 toxin [76]. This toxin is lethal for mice (minimal lethal dose ~3 μg) and cytotoxic for CHO and I407 cells, causing rounding and inhibition of ^{3}H-thymidine incorporation. β2 toxin does not appear to damage the cytoskeleton. Finally, as mentioned earlier, β2 toxin can be considered an enterotoxin, since it causes fluid accumulation in guinea pig ileal loops [75]. In that animal model, the toxin also induces substantial necrosis.

18.3.4 β2 Vaccines

There have not yet been published attempts to develop a β2 toxin vaccine.

18.4 CONCLUDING REMARKS

Despite recent progress in understanding the action and genetics of CPE and some other enteric clostridial toxins listed in Table 18.1, there remains an urgent need for additional basic research on these toxins and a better evaluation of their role in pathogenesis. Only increased study will lead to improved vaccine and therapeutics against these toxins and will allow the harnessing of their potent activities for beneficial purposes, such as cancer therapy, or for use as probes to study normal mammalian cell function.

ACKNOWLEDGMENTS

Preparation of this chapter was supported, in part, by Public Health Service Grants AI19844 and AI56177 from the National Institute of Allergy and Infectious Diseases and by Grant 2001-0517 from the Ensuring Food Safety Research Program of the United States Department of Agriculture. The author thanks Dr. Ganes Chakrabarti for his assistance in preparing figures for this chapter.

REFERENCES

1. McDonel, J.L., Toxins of *Clostridium perfringens* types A, B, C, D, and E, in *Pharmacology of Bacterial Toxins*, Dorner, F. and Drews, H., Eds., Pergamon Press, Oxford, 1986, 477.
2. Kokai-Kun, J.F. et al., Comparison of Western immunoblots and gene detection assays for identification of potentially enterotoxigenic isolates of *Clostridium perfringens*, *J. Clin. Microbiol.,* 32, 2533, 1994.
3. Daube, G. et al., Hybridization of 2,659 *Clostridium perfringens* isolates with gene probes for seven toxins (α, β, ε, ι, θ, μ and enterotoxin) and for sialidase, *AJVR,* 57, 496, 1996.
4. Songer, J.G. and Meer, R.M., Genotyping of *Clostridium perfringens* by polymerase chain reaction is a useful adjunct to diagnosis of clostridial enteric disease in animals, *Anaerobe*, 2, 197, 1996.
5. Van Damme-Jongsten, M., et al., Synthetic DNA probes for detection of enterotoxigenic *Clostridium perfringens* strains isolated from outbreaks of food poisoning, *J. Clin. Microbiol.,* 28, 131, 1990.
6. Billington, S.J. et al., *Clostridium perfringens* type E animal enteritis isolates with highly conserved, silent enterotoxin sequences, *Infect. Immun.,* 66, 4531, 1998.
7. McClane, B.A., *Clostridium perfringens*, in *Food Microbiology: Fundamentals and Frontiers,* Doyle, M.P., Beuchat, L.R., and Montville, T.J., Eds, ASM Press, Washington, DC, 2001, 351.
8. Bean, N.H. et al., Surveillance for foodborne-disease outbreaks-United States, 1988–1982, *Morbid. Mortal. Weekly Reports,* 45, 1, 1996.
9. Adak, G.K., Long, S.M., and O'Brien, S.J., Trends in indigenous foodborne disease and deaths, England and Wales: 1992 to 2000, *Gut*, 51, 832, 2002.
10. Sarker, M.R. et al., Comparative experiments to examine the effects of heating on vegetative cells and spores of *Clostridium perfringens* isolates carrying plasmid versus chromosomal enterotoxin genes, *Appl. Environ. Microbiol.,* 66, 3234, 2000.
11. Heredia, N.L. et al., Growth, sporulation and enterotoxin production by *C. perfringens* type A in the presence of human bile salts, *FEMS Microbiol Lett.*, 84, 15, 1991.
12. Batholomew, B.A. et al., Development and application of an enzyme-linked immunosorbent assay for *Clostridium perfringens* type A enterotoxin, *J. Clin. Pathol.,* 38, 222, 1985.
13. Birkhead, G. et al., Characterization of an outbreak of *Clostridium perfringens* food poisoning by quantitative fecal culture and fecal enterotoxin measurement, *J. Clin. Microbiol.*, 26, 471, 1988.
14. Skjelkvale, R. and Uemura, T., Experimental diarrhea in human volunteers following oral administration of *Clostridium perfringens* enterotoxin, *J. Appl. Bacteriol.,* 46, 281, 1977.
15. Sarker, M.R., Carman, R.J., and McClane, B.A., Inactivation of the gene (*cpe*) encoding *Clostridium perfringens* enterotoxin eliminates the ability of two *cpe*-positive *C. perfringens* type A human gastrointestinal disease isolates to affect rabbit ileal loops, *Molec. Microbiol.*, 33, 946, 1999.

16. Zhao, Y. and Melville, S.B., Identification and characterization of sporulation-dependent promoters upstream of the enterotoxin gene (*cpe*) of *Clostridium perfringens*, *J. Bacteriol.*, *180*, 136, 1998.
17. Czeczulin, J.R., Hanna, P.C. and McClane, B.A., Cloning, nucleotide sequencing, and expression of the *Clostridium perfringens* enterotoxin gene in *Escherichia coli*, *Infect. Immun.*, 61, 3429, 1993.
18. Borriello, S.P. et al., Epidemiology of diarrhea caused by enterotoxigenic *Clostridium perfringens*, *J. Med. Microbiol.*, 20, 363, 1985.
19. Borriello, S.P. et al., Enterotoxigenic *Clostridium perfringens*: a possible cause of antibiotic-associated diarrhea, *Lancet*, 1, 305, 1984.
20. Mpamugo, O., Donovan, T., and Brett, M.M., Enterotoxigenic *Clostridium perfringens* as a cause of sporadic cases of diarrhea, *J. Med. Microbiol.*, 43, 442, 1995.
21. Abrahao, C., et al., Similar frequency of detection of *Clostridium perfringens* enterotoxin and *Clostridium difficile* toxins in patients with antibiotic-associated diarrhea, *Eur. J. Clin. Microbiol. Infect. Dis.*, 20, 676, 2001.
22. Brett, M.M. et al., Detection of *Clostridium perfringens* and its enterotoxin in cases of sporadic diarrhea, *J. Clin. Pathol.*, 45, 609, 1992.
23. Marks, S. et al., Genotypic and phenotypic characterization of *Clostridium perfringens* and *Clostridium difficile* in diarrheic and healthy dogs, *J. Vet. Intern. Med.*, 16, 533, 2002.
24. Collie, R.E., Kokai-Kun, J.F., and McClane, B.A., Phenotypic characterization of enterotoxigenic *Clostridium perfringens* isolates from non-foodborne human gastrointestinal diseases, *Anaerobe*, 4, 69, 1998.
25. Cornillot, E. et al., The enterotoxin gene (*cpe*) of *Clostridium perfringens* can be chromosomal or plasmid-borne, *Molec. Microbiol.*, 15, 639, 1995.
26. Miyamoto, K. et al., Organization of the plasmid *cpe* locus of *Clostridium perfringens* type A isolates, *Infect. Immun.*, 70, 4261, 2002.
27. Brynestad, S., Synstad, B., and Granum, P.E., The *Clostridium perfringens* enterotoxin gene is on a transposable element in type A human food poisoning strains, *Microbiology*, 143, 2109, 1997.
28. Brynestad, S. and Granum, P.E., Evidence that Tn*5565*, which includes the enterotoxin gene in *Clostridium perfringens*, can have a circular form which may be a transposition intermediate, *FEMS Microbiol. Lett.*, 170, 281, 1999.
29. Brynestad, S. et al., The enterotoxin (CPE) plasmid from *Clostridium perfringens* is conjugative, *Infect. Immun.*, 69, 3483, 2001.
30. Collie, R.E. and McClane, B.A., Evidence that the enterotoxin gene can be episomal in *Clostridium perfringens* isolates associated with nonfoodborne human gastrointestinal diseases, *J. Clin. Microbiol.*, 36, 30, 1998.
31. Wen, Q., Miyamoto, K., and McClane, B.A., Development of a duplex PCR genotyping assay for distinguishing *Clostridium perfringens* type A isolates carrying chromosomal enterotoxin (*cpe*) genes from those carrying plasmid-borne enterotoxin (*cpe*) genes, *J. Clin. Microbiol.*, 41, 1494, 2003.
32. Borriello, S.P., Clostridial diseases of the gut, *Clin. Infect. Dis.*, 20 (Suppl. 2), S242, 1995.
33. Carman, R.J., *Clostridium perfringens* in spontaneous and antibiotic-associated diarrhea of man and other animals, *Rev. Med. Microbiol.* 8, Suppl. 1, S43, 1997.
34. Czeczulin, J.R., Collie, R E., and McClane, B.A., Regulated expression of *Clostridium perfringens* enterotoxin in naturally *cpe*-negative type A, B, and C isolates of *C. perfringens*, *Infect. Immun.*, 64, 3301, 1996.
35. Melville, S.B., Labbe, R., and Sonenshein, A.L., Expression from the *Clostridium perfringens cpe* promoter in *C. perfringens* and *Bacillus subtilus*, *Infect. Immun.*, 62, 5550, 1994.
36. Shimizu, T. et al., Complete genome sequence of *Clostridium perfringens*, an anaerobic flesh-eater, *Proc. Natl. Acad. Sci. U.S.A..*, 99, 996, 2002.
37. Brynestad, S. et al., A complex array of Hpr consensus DNA recognition sequences proximal to the enterotoxin gene in *Clostridium perfringens* type A, *Microbiology*, 140, 97, 1994.
38. Labbe, R.G., *Clostridium perfringens*, in *Foodborne Bacterial Pathogens*, Doyle, M.P., Ed., Marcel Decker, New York, 1989, 192.
39. Labbe, R.G. and Duncan, C.L., Evidence for stable messenger ribonucleic acid during sporulation and enterotoxin synthesis by *Clostridium perfringens* type A, *J. Bacteriol.*, 129, 843, 1977.
40. Hardy, S.P. et al., Cationic currents induced by *Clostridium perfringens* type A enterotoxin in human intestinal Caco-2 cells, *J. Med. Microbiol.*, 48, 235, 1999.

41. McDonel, J.L., Binding of *Clostridium perfringens* 125I-enterotoxin to rabbit intestinal cells, *Biochem.,* 21, 4801, 1980.
42. Wnek, A.P. and McClane, B.A., Comparison of receptors for *Clostridium perfringens* type A and cholera enterotoxins in isolated rabbit intestinal brush border membranes, *Microb. Pathog.,* 1, 89, 1986.
43. McClane, B.A. et al., Divalent cation involvement in the action of *Clostridium perfringens* type A enterotoxin, *J. Biol. Chem.,* 263, 2423, 1988.
44. Katahira, J. et al., Molecular cloning and functional characterization of the receptor for *Clostridium perfringens* enterotoxin, *J. Cell Biol.*, 136, 1239, 1997.
45. Katahira, J. et al., *Clostridium perfringens* enterotoxin utilizes two structurally related membrane proteins as functional receptors *in vivo*, *J. Biol. Chem.*, 272, 26652, 1997.
46. Fujita, K. et al., *Clostridium perfringens* enterotoxin binds to the second extracellular loop of claudin-3, a tight junction membrane protein, *FEBS Lett.,* 476, 258, 2000.
47. Tsukita, S., Furuse, M., and Itoh, M., Multifunctional strands in tight junctions, *Nature Reviews: Molecular Cell Biology,* 2, 285, 2001.
48. Tsukita, S. and Furuse, M., Claudin-based barrier in simple and stratified cellular sheets, *Curr. Opin. Cell Biol.,* 14, 531, 2002.
49. Wieckowski, E.U., Wnek, A.P., and McClane, B.A., Evidence that an ~50 kDa mammalian plasma membrane protein with receptor-like properties mediates the amphiphilicity of specifically-bound *Clostridium perfringens* enterotoxin, *J. Biol. Chem.,* 269, 10838, 1994.
50. Mitic, L., Unger, V., and Anderson, J., Expression, solubilization, and biochemical characterization of the tight junction transmembrane protein claudin-4, *Protein Sci.,* 12, 218, 2003.
51. Kokai-Kun, J.F. et al., Identification of a *Clostridium perfringens* enterotoxin region required for large complex formation and cytotoxicity by random mutagenesis, *Infect. Immun.,* 67, 6534, 1999.
52. Kokai-Kun, J.F. and McClane, B.A., Deletion analysis of the *Clostridium perfringens* enterotoxin, *Infect. Immun.,* 65, 1014, 1997.
53. Singh, U. et al., CaCo-2 cells treated with *Clostridium perfringens* enterotoxin form multiple large complex species, one of which contains the tight junction protein occludin, *J. Biol. Chem.,* 275, 18407, 2000.
54. Singh, U. et al., Comparative biochemical and immunochemical studies reveal differences in the effects of *Clostridium perfringens* enterotoxin on polarized CaCo-2 cells versus Vero cells, *J. Biol. Chem.*, 276, 33402, 2001.
55. Hulkower, K.I., Wnek, A.P., and McClane, B.A., Evidence that alterations in small molecule permeability are involved in the *Clostridium perfringens* type-A enterotoxin-induced inhibition of macromolecular synthesis in Vero cells, *J. Cell. Physiol.*, 140, 498, 1989.
56. McClane, B.A., *Clostridium perfringens* enterotoxin acts by producing small molecule permeability alterations in plasma membranes, *Toxicology,* 87, 43, 1994.
57. McClane, B.A., Osmotic stabilizers differentially inhibit permeability alterations induced in Vero cells by *Clostridium perfringens* enterotoxin, *Biochim. Biophys. Acta,* 777, 99, 1984.
58. Wieckowski, E., Kokai-Kun, J.F., and McClane, B.A., Characterization of membrane-associated *Clostridium perfringens* enterotoxin following Pronase treatment, *Infect. Immun.,* 66, 5897, 1998.
59. Chakrabarti, G., Zhou, X., and McClane, B.A., Death pathways activated in CaCo-2 cells by *Clostridium perfringens* enterotoxin, *Infect. Immun.,* 71, 4260, 2003.
60. Sonoda, N. et al., *Clostridium perfringens* enterotoxin fragments removes specific claudins from tight junction strands: evidence for direct involvement of claudins in tight junction barrier, *J. Cell Biol.,* 147, 195, 1999.
61. Rahner, C., Mitic, L., and Anderson, J., Heterogeneity in expression and subcellular localization of Claudins 2, 3, 4, and 5 in the rat liver, pancreas, and gut, *Gastroenterology,* 120, 411, 2001.
62. McDonel, J.L. and Demers, G.W., *In vivo* effects of enterotoxin from *Clostridium perfringens* type A in rabbit colon: binding vs. biologic activity, *J. Infect. Dis.*, 145, 490, 1982.
63. McDonel, J.L. and Duncan, C.L., Histopathological effect of *Clostridium perfringens* enterotoxin in the rabbit ileum, *Infect. Immun.*, 12, 1214, 1975.
64. McDonel, J.L. and Duncan, C.L., Regional localization of activity of *Clostridium perfringens* type A enterotoxin in the rabbit ileum, jejunum and duodenum, *J. Infect. Dis.,* 136, 661, 1977.
65. McDonel, J.L. et al., The effects of *Clostridium perfringens* enterotoxin on rat and rabbit ileum: an electron microscopy study, *Lab. Invest.,* 39, 210, 1978.

66. Sherman, S., Klein, E., and McClane, B.A., *Clostridium perfringens* type A enterotoxin induces concurrent development of tissue damage and fluid accumulation in the rabbit ileum, *J. Diarrheal Dis. Res.,* 12, 200, 1994.
67. Hanna, P.C. et al., Localization of the receptor-binding region of *Clostridium perfringens* enterotoxin utilizing cloned toxin fragments and synthetic peptides. The 30 C-terminal amino acids define a functional binding region, *J. Biol. Chem.*, 266, 11037, 1991.
68. Hanna, P.C. et al., Mapping functional regions of *Clostridium perfringens* type A enterotoxin, *Infect. Immun.,* 60, 2110, 1992.
69. Granum, P.E. and Richardson, M., Chymotrypsin treatment increases the activity of *Clostridium perfringens* enterotoxin, *Toxicon,* 29, 445, 1991.
70. Granum, P.E., Whitaker, J.R., and Skjelkvale, R., Trypsin activation of enterotoxin from *Clostridium perfringens* type A, *Biochim. Biophys. Acta,* 668, 325,1981.
71. Wnek, A.P., Strouse, R.J., and McClane, B.A., Production and characterization of monoclonal antibodies against *Clostridium perfringens* type A enterotoxin, *Infect. Immun.,* 50, 442, 1985.
72. Mietzner, T.A. et al., A conjugated synthetic peptide corresponding to the C-terminal region of *Clostridium perfringens* type A enterotoxin elicits an enterotoxin-neutralizing antibody response in mice, *Infect. Immun.,* 60, 3947, 1992.
73. Michl, P. et al., Claudin-4: a new target for pancreatic cancer treatment using *Clostridium perfringens* enterotoxin, *Gastroenterology,* 121, 678, 2001.
74. Long, H. et al., Expression of *Clostridium perfringens* enterotoxin receptors claudin-3 and claudin-4 in prostate cancer epithelium, *Cancer Res.*, 61, 7878, 2001.
75. Jolivet-Reynaud, C. et al., Enteropathogenicity of *Clostridium perfringens* beta toxin and other clostridial toxins, *Zbl. Bakteriol.,* S15, 145, 1986.
75a. Hunter, S.E. et al., Molecular genetic analysis of beta-toxin of *Clostridium perfringens* reveals sequence homology with alpha-toxin, gamma-toxin, and leukocidin of *Staphylococcus aureus*, *Infect. Immun.,* 61, 3958, 1993.
76. Gibert, M., Jolivet-Reynaud, C., and Popoff, M.R., Beta2 toxin, a novel toxin produced by *Clostridium perfringens*, *Gene,* 203, 65, 1997.
77. Herholz, C. et al., Prevalence of β2-toxigenic *Clostridium perfringens* in horses with intestinal disorders, *J. Clin. Microbiol.,* 37, 358, 1999.
78. Garmory, H.S. et al., Occurrence of *Clostridium perfringens* β2-toxin amongst animals determined using genotyping and subtyping PCR assays, *Epidemiol. Infect.*, 124, 61, 2000.
79. Manteca, C. et al., The role of *Clostridium perfringens* beta2-toxin in bovine enterotoxaemia, *Vet. Microbiol.*, 86, 191, 2002.
80. Waters, M. et al., Genotyping and phenotyping of beta2-toxigenic *Clostridium perfringens* fecal isolates associated with gastrointestinal diseases in piglets, *J. Clin. Microbiol.*, 41, 3584, 2003.
81. Klassen, H.L.B.M. et al., Detection of the β2 toxin gene of *Clostridium perfringens* in diarrheic piglets in the Netherlands and Switzerland, *FEMS Immunol. Med. Microbiol.*, 24, 325, 1999.
82. Thiede, S., Goethe, R., and Amtsberg, G. Prevalence of beta2-toxin gene in *Clostridium perfringens* type A from diarrheic dogs, *Vet. Rec.*, 149, 273, 2001.
83. Bacciarni, L.N. et al., *Clostridium perfringens* beta2-toxin in an African elephant with ulcerative colitis, *Vet. Rec.*, 149, 618, 2001.
84. Ohtani, K. et al., The VirR/VirS regulatory cascade affects transcription of plasmid-encoded putative virulence genes in *Clostridium perfringens*, *FEMS Microbiol. Lett.,* 222, 137, 2003.
85. Bueschel, D.M. et al., Prevalence of *cpb2*, encoding beta2 toxin, in *Clostridium perfringens* field isolates: correlation of genotype with phenotype, *Vet. Microbiol.,* 94, 121, 2003.

19 Clostridial Cytotoxins

Holger Barth and Klaus Aktories

CONTENTS

Clostridia produce a large array of protein toxins. Besides the clostridial neurotoxins, various membrane active clostridial toxins and *Clostridium perfringens* enterotoxin, a group of clostridial cytotoxins has been the focus of intensive research during the past few years. This group of clostridial toxins includes *large clostridial cytotoxins, C3 exoenzymes,* and the *binary actin-ADP-ribosylating toxins*. Although diverse in properties, functions, and structures, these toxins all attack the actin cytoskeleton of eukaryotic target cells. Whereas the large clostridial cytotoxins and C3 exoenzymes accomplish this property by modification of Rho GTPases, which are master regulators of the cytoskeleton, the binary toxins directly target actin, which is one of the central elements of the cytoskeleton. Here, we review the biology and pharmacology of large clostridial cytotoxins, C3 exoenzymes, and binary actin-ADP-ribosylating toxins and focus on recent structure–function analyses of those toxins.

19.1 LARGE CLOSTRIDIAL CYTOTOXINS

The family of large clostridial cytotoxins consists of *Clostridium difficile* toxin A and toxin B, the hemorrhagic and the lethal toxin from *Clostridium sordellii,* and the α-toxin from *Clostridium novyi*. However, several isoforms have been described, extending this family of cytotoxins. All these toxins share sequence identities ranging from 36 to 90% and have molecular masses between 250 and 308 kDa [1,2].

19.1.1 *Clostridium difficile* Toxin A and Toxin B

Clostridium difficile plays a crucial role in the pathogenesis of antibiotic-associated diarrhea and pseudomembranous colitis often initiated by alteration of the normal intestinal flora during antibiotic therapy. Overgrowth of *Clostridium difficile* and release of toxin A and toxin B lead to inflammation and damage of the colonic mucosa [3]. Historically, the development in the field of the large clostridial cytotoxins is tightly connected to the discovery that pseudomembranous colitis induced by antibiotics is caused by *Clostridium difficile* and not, as earlier suggested, by *Staphylococcus aureus* or by viruses [4–6]. Of major importance was also the finding that *Clostridium sordellii* antitoxin neutralized the cytotoxic activity of a toxin isolated from patients with colitis [7]. Later, it was recognized that not only one but at least two toxins are produced by *Clostridium difficile* [8]. Further milestones in research on the toxins were the cloning and sequencing of the genes coding the toxins [9–12].

Because it was observed that toxin B was 100- to 1000-fold more toxic to cultured cells as compared to toxin A, it was designated as *cytotoxin*. On the other hand, enterotoxicity studied in animal models was only associated with toxin A, which was therefore named *enterotoxin* [13,14]. Later, it was recognized that toxin A-negative/toxin B-positive strains can also cause symptoms of pseudomembranous colitis, and recently enterotoxic properties were reported for toxin B as well [15,16]. The next major step in the understanding of the role of the toxins in disease was the finding that *Clostridium difficile* toxin A and toxin B transfer the glucose moiety of UDP-glucose to members of the Rho family of small GTPases [17,18], thereby, a novel and unique type of toxin action was identified. Thus, all toxins from the family of the large clostridial cytotoxins are glucosyltransferases.

19.1.1.1 Structure and Activity of Large Clostridial Cytotoxins

Having masses of 250 to 308 kDa, the large clostridial cytotoxins belong to the largest bacterial protein toxins known. The toxins are typical AB-toxins, harboring an enzyme domain and a binding/translocation domain (Figure 19.1). The glycosyltransferase activity is located at the N-terminus of the toxins, whereas receptor binding is suggested to be mediated by the C-terminus. In the middle of the proteins, a hydrophobic region is located, which is suggested to be involved in translocation of the proteins across cellular membranes (Figure 19.1).

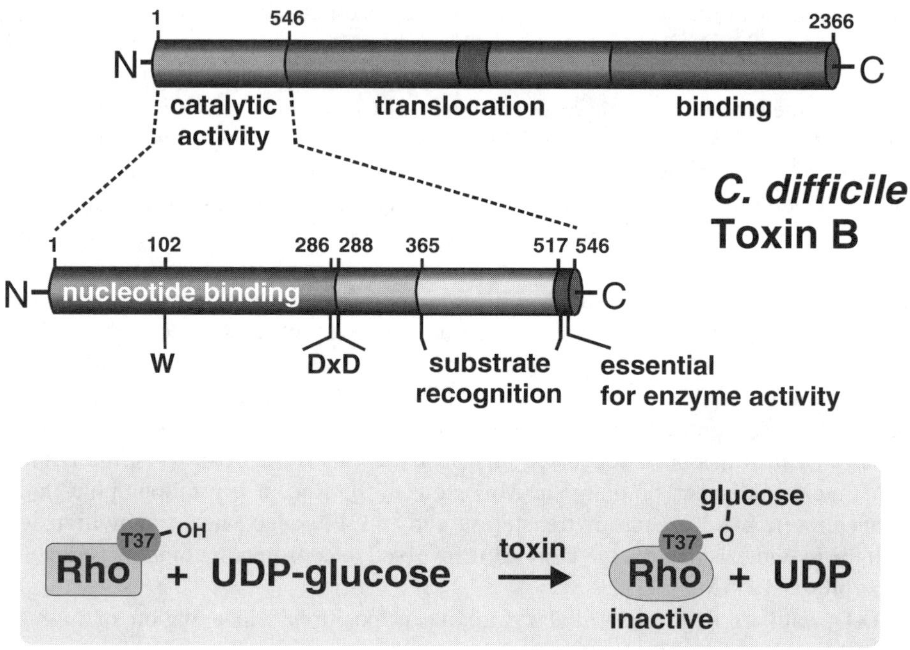

FIGURE 19.1 Structure of large clostridial cytotoxins. Large clostridial cytotoxins can be divided in three main domains. The catalytic domain is located at the N-terminus (for toxin B about residues 1–546), the C-terminus covers oligopeptide repeats (called CROPs) and is suggested to bind to carbohydrates. The middle part is suggested to be involved in translocation of toxins into the cytosol. The magnification of the enzyme domain gives conserved amino acid residues. The DXD motif is observed in several glycosyltransferases and most likely involved in divalent cation (Mn^{2+})-binding. A conserved tryptophan residue (W) is important for affinity toward the nucleotide sugar. The protein substrate recognition region seems to be located between residues 365 and 517. Deletion of residues 517–546 of toxin B blocks enzyme activity.

19.1.1.2 The N-terminal Enzyme Domain

Several findings indicate that the glycosyltransferase activity of the large clostridial cytotoxins is N-terminally located [19,20]. Microinjection of recombinant C-terminal truncations of various large clostridial cytotoxins caused the same morphological changes in eukaryotic cells as the holotoxins. *In vitro* glycosyltransferase activity of the truncations and of the native toxins were nearly identical. For *Clostridium difficile* toxin A and toxin B (and also for *Clostridium sordellii* lethal toxin) it was shown that the first 546 amino acids were sufficient for full glucosyltransferase activity. Fragments consisting of the first 516 amino acids were enzymatically inactive [19].

Structure–function analysis of the active fragments of *Clostridium sordellii* lethal toxin and the *Clostridium difficile* toxin B enzyme domains (amino acids 1–546), as well as *Clostridium novyi* α-toxin (amino acids 1–551), suggested that the substrate recognition site is located in the C-terminal region of the enzyme domains.

Importantly, these toxins differ in their substrate specificity from the prototypes *Clostridium difficile* toxin A and toxin B. Whereas toxins A and B, and also *Clostridium sordellii* hemorrhagic toxin and the α-toxin from *Clostridium novyi,* selectively modify Rho GTPases but not members of other subfamilies of small GTPases, *Clostridium sordellii* lethal toxin glucosylates Rac (to some extent Cdc42 *in vitro*) but not Rho. In addition, the lethal toxin accepts Ras, Ral, Rap, and R-Ras as protein substrates [21–24]. On the other hand, the α-toxin of *Clostridium novyi* is unique in respect to its second substrate. It uses preferentially UDP-GlcNAc as a cosubstrate, whereas all other toxins studied so far prefer UDP-glucose. *In vivo* modification of Cdc42 by α-toxin was demonstrated by specific [^{14}C]galactosylation of proteins bearing a N-acetylglucosamine moiety [25].

These differences in the substrate and cosubstrate specificity were exploited to study protein regions of the toxins involved in substrate recognition and interaction by constructing chimeras between the various members of the toxin family. Thus, studies on chimeric proteins of the active fragments assign the protein substrate recognition site of *Clostridium sordellii* lethal toxin to amino acids 365 to 516 [26]. The region involved in binding of activated sugar nucleotide sugar is less defined. Chimeras of the active fragments of *Clostridium sordellii* lethal toxin and *Clostridium novyi* α-toxin, which use UDP-glucose and UDP-N-acetylglucosamine as cosubstrate, respectively, suggest that the region involved in nucleotide-sugar specificity is localized between amino acids 133 and 517 of α-toxin [20].

Several highly conserved amino acid residues among the group of large clostridial cytotoxins have been recognized to be essential for enzyme activity. Best characterized is the "DXD-motif," which is located in the center of the enzyme domains. Change of one of the aspartate residues to alanine or asparagine inhibits enzyme activity [27]. Interestingly, the DXD-motif is conserved among many different families of glycosyltransferases [28,29]. Although its exact role in the enzyme reaction is not clear, it has been suggested that, in large clostridial cytotoxins, the DXD-motif is involved in nucleotide-sugar binding via Mn^{2+} ions [30]. Recent crystallographic data of two glycosyltransferases, β-1,3-glucuronyltransferase and α-1,4-N-acetyl-hexosaminyltransferase, are in line with the hypothesis that the DXD-motif is involved in coordinating binding of the nucleotide sugar in the presence of Mn^{2+} [31].

The DXD-motif of large clostridial cytotoxins is positioned in a region of high sequence homology [27]. Many different eukaryotic and prokaryotic glycosyltransferases share this "extended" DXD-motif [32]. Interestingly, almost all glycosyltransferases studied so far sharing the extended DXD-motif are α-retaining glycosyltransferases. This also appears to be true for the large clostridial cytotoxins, at least for *Clostridium sordellii* lethal toxin [33,34]. Mainly based on the above described sequence homology, the family of large clostridial cytotoxin has been designated glycosyltransferase family 44 as defined by Henrissat et al. (available at http://afmb.cnrs-mrs.fr/CAZY/GT_44.html). This family also includes genes from *Escherichia coli* [35] and *Chlamydia trachomatis* [36] coding for putative glycosyltransferases.

19.1.1.3 The Translocation and Binding Domains of Large Clostridial Cytotoxins

As mentioned above, the putative receptor-binding domain is located at the C-terminus of large clostridial cytotoxins. The domain is characterized by small repetitive sequence motifs, also called CROPS, clostridial repetitive oligopeptides. CROPS consist of 20 to 50 amino acids and are repeated 14 to 30 times [1,10,11]. A conserved motif of three aromatic residues YYF is typical for many CROPS. Several findings suggest that the C-terminus is involved in receptor binding. First, the C-terminus is structurally related to carbohydrate-binding domains of glycosyltransferases from *Streptococcus* species [37]. Second, recombinant fragments of *Clostridium difficile* toxin A and toxin B covering the putative binding domains block intoxication by the holotoxins [38]. Interestingly, it was shown recently that the whole repetitive region of toxin A is necessary for binding and endocytosis [39]. Finally, an antibody directed against the C-terminus of *Clostridium difficile* toxin A inhibits cytotoxicity [40]. However, Barroso and co-workers reported that deletion of the C-terminal repetitive domain of toxin B decreases the cytotoxicity only by a factor of 10; this finding is not in line with the hypothesis that only the CROPS define the binding site of the toxins [41].

19.1.1.4 Receptors of Large Clostridial Toxins

Our knowledge about the precise nature of membrane receptors for large clostridial cytotoxins is very limited. Most studies on receptor binding were performed with toxin A. Deduced from these binding studies and from the proposed properties of CROPs, it was suggested that toxin A binds in a lectinlike manner to Galα-1-3Galβ1-4GlcN structures [42,43]. From brush border cells of small intestines of infant hamsters, a 160 kDa galactose- and N-acetylglucosamine-containing glycoprotein was purified, which was suggested to function as a toxin receptor [44]. The same group showed that binding of toxin A is inhibited by lectin specific for Gal and GlcNAc and by immunoglobulin and nonimmunoglobulin components of human milk [45]. Furthermore, it was reported that toxin A binds to the membranous sucrose-isomaltase glycoprotein on rabbit cells [46]. However, this receptor is lacking in most of the toxin A sensitive cell lines. Also, a human glycosphingolipid was described to bind to toxin A, suggesting a lectin-like binding of the protein [47]. Recently it was shown that diethylpyrocarbonate, which specifically interacts with histidine residues, reduces ligand binding activity in a rabbit erythrocyte haemagglutination assay, suggesting that surface-exposed histidine residues are involved in binding [48]. The receptors for the other toxins are unknown.

19.1.1.5 Toxin Processing and Uptake

After receptor binding, uptake of the toxins follows an endosomal pathway [49,50]. Like diphtheria toxin and many other bacterial toxins, the large clostridial cytotoxins — or at least their enzyme domains — are released from early endosomes into the cytosol of target cells (see below). Evidence for this model was obtained by blocking endosomal acidification by bafilomycin A1, a drug inhibiting the vesicular H^+-ATPase of the endosomes. The pretreatment of eukaryotic cells with bafilomycin A1 effectively blocks the cytopathic effects of toxin B. Interestingly, the inhibitory effect of bafilomycin A1 can be circumvented by acidification of the culture medium (pH 5.2), which allows direct translocation of the toxins across the cell membrane [51,52].

The process of toxin translocation is still enigmatic. The question whether the complete holotoxin or only the enzymatic domain of the large clostridial cytotoxins is translocated into the cytosol was successfully addressed recently (Figure 19.2). Supported by different experimental approaches, it was reported that only the enzyme domain is delivered into the cytosol, whereas the translocation and receptor domains remain in the endosomes [53]. It was shown that after treatment of Vero cells

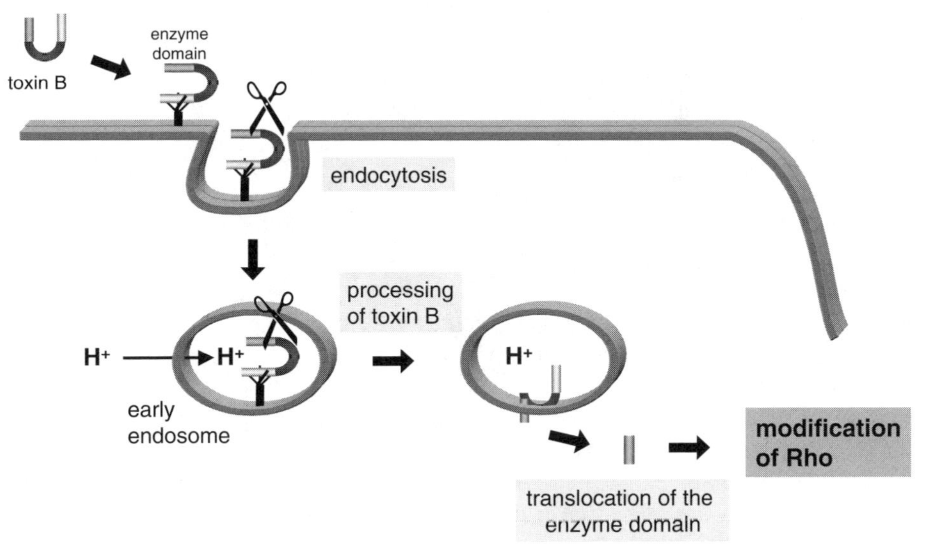

FIGURE 19.2 Processing of *C. difficile* toxin B. It has been suggested that toxin B is proteolytically processed because only the N-terminal catalytical domain appears in the cytosol and not the C-terminal binding and middle translocation domain. Where processing occurs is unclear. Possible locations are: (i) at the surface of the target cells, (ii) in the endosome after internalization and acidification, and (iii) during or after the translocation step.

with toxin B holotoxin, only an enzymatically active fragment but not the toxin B holotoxin is detectable in the cytosol by immunoblot analysis, fluorescence microscopy, and mass spectrometry. Fluorescence microscopy data suggested that the translocation/binding domain remains in the endosomes [53]. All these data indicate that the toxins are processed during uptake.

Whether proteolytic cleavage of toxin B takes place at the cell surface of the target cells or after toxin uptake is unclear. Proteolytic processing at the cell surface is one possibility, another is the cleavage in endosomes, but it is also conceivable that cleavage occurs after translocation of the catalytic domain across the endosomal membrane by a cytosolic protease.

A hydrophobic region is located in the central region of all large clostridial cytotoxins (amino acids 956–1128 in *Clostridium sordellii* lethal toxin), which is believed to be involved in translocation. As suggested from the uptake of other toxins, the acidic pH in the endosomes is believed to trigger a conformational change in the hydrophobic stretch, which allows membrane insertions and, eventually, translocation of the catalytic domain into the cytosol [52].

Clostridium difficile toxin B (amino acids 547–2366), as well as its binding and translocation domains, forms pores in intact membranes of Chinese hamster ovary cells [51]. By contrast, the catalytic domain is not able to form pores. In addition, toxin B induces acidic pH-dependent channels in artificial bilayers. However, the role of pore formation in translocation has not been clarified. So far, it can only be speculated that the enzyme domain of large clostridial cytotoxins is transported through this pore.

19.1.2 Other Members of the Family of Large Clostridial Cytotoxins

19.1.2.1 Toxins from *Clostridium sordellii*

Clostridium sordellii lethal toxin consists of 2364 amino acid residues with a mass of 270 kDa [54]. The toxin causes diarrhea and enterotoxemia in domestic animals and in wound infections in humans (e.g., gas gangrene) [55]. The toxin shares many biological properties with toxin A and B [56–58]. It is much less cytotoxic than toxin B and, in this respect, more similar to toxin A. Like toxin B, lethal toxin is not very effective as an enterotoxin in animal models. However, it is 10 times more lethal than toxins A and B in mice after intraperitoneal injection. As mentioned above, on the molecular level, lethal toxin shares glucosyltransferase activity with toxins A and B. However, one major difference between lethal toxin and *Clostridium difficile* toxins A and B is that lethal toxin differs in its substrate specificity and is able to glucosylate Rac, in addition to Ras subfamily GTPases, but not RhoA. *Clostridium sordellii* hemorrhagic toxin has a mass of 300 kDa. It is very similar to toxin A and is also able to elicit fluid response in ligated loop assays [57,58]. Concerning its enzyme activity, it shares the properties of toxin A [21].

19.1.2.2 *Clostridium novyi* α-Toxin

Clostridium novyi α-toxin has been implicated in gas gangrene in humans and wound infection in animals [59,60]. Edema formation induced by the α-toxin appears to be typical. The toxin has been cloned and sequenced by the group of von Eichel-Streiber [61]. It has a mass of about 250 kDa and consists of 2178 amino acid residues. α-toxin has biological properties very similar to that of toxins A and B [56,62–64]. The cytotoxic activity is similar to that of toxin B. The minimum mouse lethal dose was reported to be 5 ng. The toxin has no hemagglutination activity and has apparently no enterotoxic activity in the sense that it does not induce fluid secretion in ligated loop assays. Although sharing many properties with other toxins from the family of large clostridial cytotoxins, *Clostridium novyi* α-toxin is the only representative of this toxin family that prefers UDP-N-acetylglucosamine as a cosubstrate (see above) [25]. UDP-glucose is also utilized as a cosubstrate, but to a much lesser extent. In kinetic studies with an enzymatically active fragment (amino acids 1–551), a Michaelis-Menten constant (Km) of 17 μM was determined for UDP-N-acetylglucosamine and an about 20-fold higher Km value for UDP-glucose [20].

19.1.2.3 Toxin Variants

Recently, many toxin variants have been described [65,66]. Functional chimeras of *Clostridium difficile* toxin B (reference strain VPI 10463) and *Clostridium sordellii* lethal toxin exist, e.g., *Clostridium difficile* toxin B from strain 1470 and strain 8864. Their substrate specificities mainly resemble that of lethal toxin [67,68]. *Clostridium difficile* strain C34 produces a toxin B variant modifying Rho, Rac, Cdc42, as well as R-Ras, Ral, and Rap [69]. Interestingly, this toxin isoform is the only one reported, glucosylating RhoA and members of the Ras family in parallel.

19.1.3 Functional Consequences of Toxin-Mediated Glycosylation of the Small GTP-Binding Proteins

The targets of large clostridial cytotoxins are Rho and Ras proteins, which belong to the superfamily of low molecular mass GTPases. These proteins act as molecular switches in a broad spectrum of signaling pathways [70–72]. The switch proteins are regulated by a GTPase cycle, which is explained here for Rho GTPases (Figure 19.3). Rho proteins are inactive in the GDP-bound form and then localized in the cytosol. In the active, GTP-bound form, they are at the plasma membrane. In the cytoplasm, the Rho proteins are complexed with the guanine nucleotide dissociation inhibitor

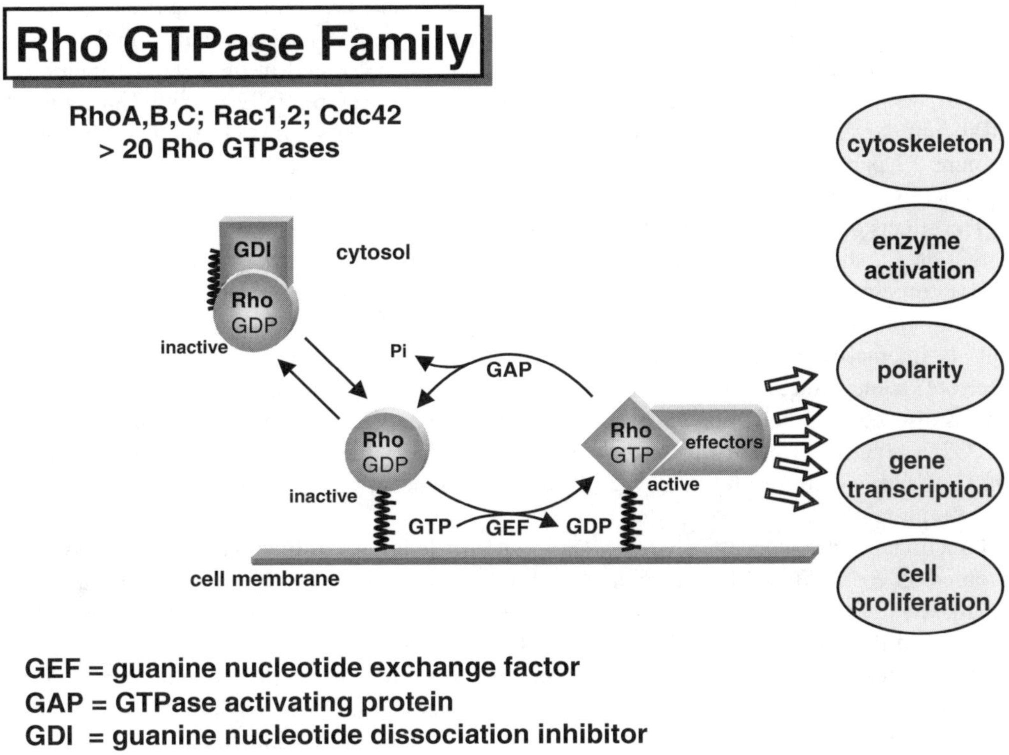

FIGURE 19.3 Regulation of Rho GTPases. Rho GTPases are inactive in the GDP-bound form. In the inactive form, they are associated with guanine nucleotide dissociation inhibitors (GDI), which keeps the GTPase in the cytosol. Activation of the GTPases starts with translocation to membranes, where guanine nucleotide exchange factors (GEFs) induce release of GDP. The active GTP-bound form of Rho proteins interacts with different effectors to control various signaling processing, including regulation of the cytoskeleton, enzyme activation (e.g., protein kinases, phospholipases), control of polarity, gene transcription, and cell proliferation. The active state is terminated by hydrolysis of bound GTP. GTP hydrolysis is facilitated by GTPase-activating proteins (GAP). Rho proteins are isoprenylated. This is important for GDI- and membrane-binding.

(GDI), which appears to prevent nucleotide exchange. Receptor-mediated activation triggers translocation of the GTPases to the plasma membrane and, in case of the Rho proteins, uncoupling from the GDIs. Exchange of GDP to GTP is catalyzed by guanine exchange factors (GEFs; note: more than 50 different GEFs may exist), resulting in the active GTP-bound form of the GTPases. Proteins of the family of GTPase-activating proteins (GAPs; note: again, more than 50 GAPs may exist) cause hydrolysis of GTP to GDP and, thereby, turn off the active state of small GTPases (see Figure 19.3).

Whereas Ras proteins are essential for the regulation of proliferation and cell differentiation [73], Rho proteins have important regulatory functions in the organization of the actin cytoskeleton [70]. Rho GTPases comprise a family of more than 15 related proteins, including Rho (A, B, C, D, G), Rac (1, 1b, 2, 3), Cdc42, G25K, Rnd (1, 2, 3), TTF, Rif, and TC10 (for review see Van Aelst and D'Souza-Schorey; Mackay and Hall; and Ridley [74–76]). Best studied are the RhoA, Rac, and Cdc42 subtypes. For example, RhoA induces formation of actin stress fibers and focal adhesions [77], Rac leads to formation of lamellipodia and membrane ruffles [78], and Cdc42 induces formation of microspikes/filopodia [79]. Therefore, Rho GTPases are crucial for cell migration, control of morphogenesis, and cell polarity. However, various other regulatory functions for the family of Rho GTPase are reported, like cell cycle control, activation of transcription, apoptosis, and transformation (for review see Etienne-Manneville and Hall; Bishop and Hall; and Takai et al. [70–72]).

All large clostridial cytotoxins glucosylate Rho/Ras GTPases at the same (equivalent) position. The toxins modify Rho A, B, and C at threonine 37 and Rac, Cdc42, or Ras at the corresponding position threonine 35 [17,18,23] (Figure 19.4). This target amino acid residue is highly conserved among the superfamily of small GTP-binding proteins. It is located in the switch-I region of the GTPases, which is involved in effector interaction. Rho is preferentially modified in its inactive, GDP-bound state, in which threonine 37 is directed toward the solvent. In the GTP-bound state, threonine 37 participates in Mg^{2+} and nucleotide binding [80] and is less accessible for the large clostridial cytotoxins [17].

The effects of toxin-catalyzed glucosylation on the functions of Rho proteins have been studied in great detail, showing diverse functional consequences for GTPases (Figure 19.5). Inhibition of effector coupling and subsequent blocking of signal transduction pathways is considered to be the most important consequence of glycosylation [81,82]. However, not all effectors do, exclusively or at all, interact with the switch-I region of the small GTPases where the glucosylation occurs. Therefore, some effectors might still interact with glycosylated proteins. Glycosylation of Rho proteins also blocks nucleotide exchange by GEFs [81] and inhibits the intrinsic and GAP-stimulated GTPase activity [82]. It was further reported that glucosylated Rho is no longer able to interact with GDI and is therefore found at the plasma membrane, where it interacts with an unidentified ~70 kDa protein [83].

Functional consequences of *Clostridium sordellii* lethal toxin-mediated Ras glucosylation at threonine 35 are similar. Whereas nucleotide binding is not affected, the intrinsic GTPase activity is markedly decreased, and the GAP-stimulated GTPase activity is completely blocked. Like glucosylated Rho proteins, the GEF-catalyzed GDP exchange to GTP is also reduced. However, it is suggested that the most important consequence of Ras glucosylation is the inhibition of coupling to the downstream effectors. It was shown that the interaction of glucosylated Ras with the Ras-binding domain (RBD) of the Raf kinase is completely blocked. Kinetic data revealed a K_d for the interaction of Ras-GTP and Raf-RBD of about 15 n*M*, whereas glucosylation of Ras increased the K_d to >1 m*M* [81]. Also, the crystal structure analysis of glucosylated and nonmodified Ras confirmed that glucosylation of Ras blocks effector interaction. Furthermore, crystallographic data suggests that, due to sterical hindrance, the glucose moiety prevents formation of the GTP conformation of the effector region, whereas binding to GTP itself is not disturbed [33]. Notably, threonine 35 of Ras, which is glucosylated by lethal toxin, is not directly involved in effector binding. As mentioned above, it has been suggested that the preferred form for glucosylation is the GDP-bound form of GTPases. In the active form of Ras, threonine 35 is directed into the molecule and is

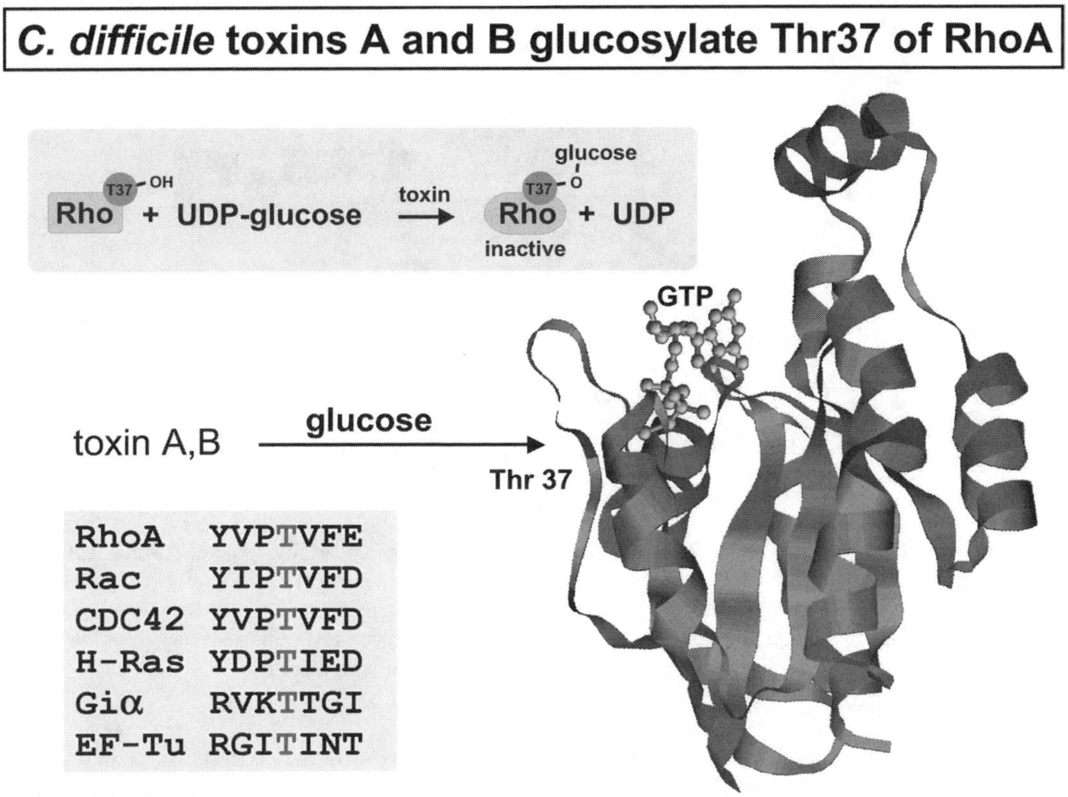

FIGURE 19.4 Glucosylation of RhoA by large clostridial cytotoxins. *C. difficile* toxins A and B glucosylate RhoA at Thr37. The toxins use UDP-glucose as a second substrate. Glucosylation occurs in the "effector region" of RhoA. Thr37 is highly conserved among small GTPases, including other Rho GTPases, Ras GTPases, and even alpha-subunits of heterotrimeric G-proteins. This threonine residue is important for magnesium and nucleotide binding.

involved in Mg^{2+} binding. Therefore, it was proposed that Ras glucosylation is solely possible in the GDP-bound form [81]. However, NMR analysis of soluble Ras bound to the GTP analogue GppNHp suggested that the effector loop exists in two distinct conformations that exchange rapidly [84]. It appears that only in one of these conformational states is threonine 35 involved in Mg^{2+} binding; in the other state, however, it is probably accessible for glucosylation. After glucosylation, the effector loop seems to be stabilized in the inactive state. Thus, NMR data suggest that glucosylation of GTP-bound Ras is possible [34]. NMR and crystal structure analysis support the view that glucose is bound in the α-anomeric form to the hydroxyl group of the threonine 35 side chain [33,34].

19.1.4 Effects of Large Clostridial Toxins on Cells

Glucosylation of Rho/Ras proteins by large clostridial toxins leads to drastic morphological changes in eukaryotic cells. The actin cytoskeleton of intoxicated cells largely redistributes, resulting in shrinking and rounding of cells, initially accompanied by formation of neurite-like retraction fibers. Finally, these fibers disappear and cells detach from the matrix [85–88]. Several other cellular responses to the inactivation of Rho and Ras proteins by large clostridial cytotoxins have been described, all explained by inhibition of the numerous functions of the small GTPases. Reports include inhibition of secretion [89], phospholipase D activity [90], calcium mobilization [91],

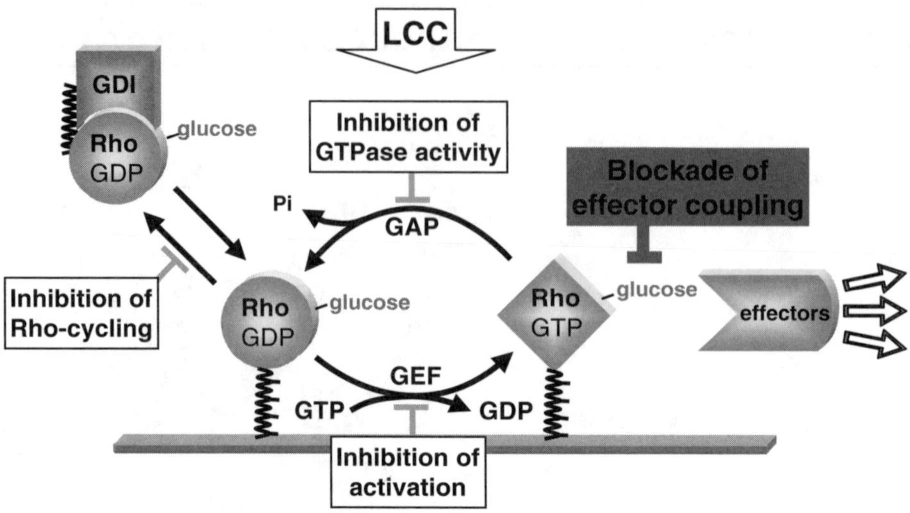

FIGURE 19.5 Functional consequences of glucosylation of Rho GTPases by large clostridial cytotoxins (LCC). LCC glucosylate Rho GTPases. The functional consequences are: (1) interactions with effectors are blocked, (2) activation by GEFs is blocked, (3) glucosylated Rho is membrane located and not "extracted" from the membrane by GDI, and finally, (4) GTP hydrolysis is blocked. Most important for inhibition of Rho signaling by glucosylation appears to be inhibition of the interaction with effectors.

muscarinic receptor signaling to focal adhesion kinase [92], as well as deregulation of neurotransmitter exocytosis [93], apoptosis [94], chemoattractant receptor signaling [95], neuronal axon formation [96], and phagocytosis [97].

19.1.5 Role of *Clostridium difficile* Toxins in Disease

Without question, our knowledge about large clostridial cytotoxins and especially about toxins A and B from *Clostridium difficile,* increased enormously during recent years. However, the precise pathogenetic pathways and sequences, which finally result in toxin-induced diarrhea and pseudomembranouse colitis, are still not clear. Concerning toxin-induced diarrhea, tissue damage and inhibition of the barrier function of the enterocytes might explain fluid response. In fact, numerous studies reported that the toxins have major effects on tight junctions, which were shown by decreased transepithelial resistance, increased paracellular bacterial migration, and changes in morphological features of tight junctions and associated proteins [16,98–105]. Particularly, it was found that F-actin restructuring was accompanied by dissociation of occluding ZO-1 and ZO-2 from lateral tight junction without affecting adherence junctions. Many of these data are in line with the view that Rho GTPases are crucially involved in tight junction regulation [98,106].

In addition to alteration of the barrier function of enterocytes, toxin A and B induce a pathological scenario in the gut, which can be summarized as a major inflammatory response. It was reported that the toxins induce massive neutrophil infiltration and the production and release of various inflammatory mediators, including prostaglandins, and leukotriens [107,108], Il-8 [15,108–111], and TNF-α [112]. Toxin A reportedly activates nuclear factor NF-κB, with subsequent release of various chemoattractants. In rat intestine macrophages are activated to release MIP-2 [113]. Furthermore, it was shown that the toxins activate intestinal nerves to release neuropeptides substance P and calcitonin gene-related peptide (CGRP) [114], which might have proinflammatory properties. Furthermore, a specific role is ascribed to mast cells, which are degranulated very early after toxin A exposure [115]. Although it is known that Rho GTPases are crucially involved in regulation of immune actions and transcriptional activation of immune cells, it was suggested that some of the above-mentioned responses observed with toxins A and B are independent of Rho GTPases. For example, mitochondrial damage and p38 mitogen activation was reported to be independent of Rho. Similarly, it was reported that toxin A causes ZO-1 translocation and increases paracellular flux via protein kinase C signal pathways in a process that occurs earlier than glucosylation of Rho proteins by the toxin [116]. Thus, it might be that large clostridial toxins harbor activities not recognized so far.

19.2 CLOSTRIDIAL C3 ADP-RIBOSYLTRANSFERASES

Clostridium botulinum C3 ADP-ribosyltransferase was serendipitously discovered in the 1980s during the course of screening for higher producer strains of the actin-ADP-ribosylating C2 toxin [117]. Because it was recognized that this protein consists only of the enzyme domain without having any binding and translocation domains, which are typical of AB toxins, C3 is often designated as an exoenzyme and not as a toxin. The ADP-ribosyltransferase C3 is produced together with the unrelated botulinum neurotoxins C1 or D, because the neurotoxin gene is located on the same phage as C3 exoenzyme [118,119]. It turned out that at least two isoforms of C3 (C3bot1 and C3bot2) were produced by *Clostridium botulinum* [120,121]. Later, another related C3 exoenzyme was found to be produced by *Clostridium limosum* (C3lim). This enzyme shares about 63% identical amino acid residues with that of C3 from *Clostridium botulinum* [122]. However, C3-like transferases are not only restricted to clostridia. A C3-like exoenzyme with similar activities as the clostridial exoenzymes were found in *Staphylococcus aureus* [123,124]. This enzyme was originally designated EDIN. To indicate the relationship of EDIN to C3 exoenzymes, we suggested recently the name C3stau1 for the C3-like enzyme from *S. aureus*. Wilde et al. [125] cloned a related ADP-ribosyltransferase from genomic DNA of *Staphylococcus aureus* strain HMI6 called C3stau2. A third enzyme (C3stau3) was reported recently [126]. Another C3-like ADP-ribosyltransferase has been identified in *Bacillus cereus* [127]. All C3-like exoenzymes are 23 to 28 kDa proteins with an isoelectric point greater than 9. Therefore, native proteins can be easily purified from culture supernatant by ion exchange chromatography [122,124,125,127–129]. In recent years, recombinant C3-like ADP-ribosyltransferases without the signal sequence often have been used for pharmacological and cell biological studies.

19.2.1 Structure of C3 Exoenzyme

Tainer and co-workers solved the crystal structure of C3 exoenzyme from *Clostridium botulinum* (C3bot1) [130]. Meanwhile, a second crystal structure analysis of C3bot was reported [131], and recently, the structure of the related C3stau was solved [132]. These analyses showed that C3 shares a fold typical for many ADP-ribosyltransferases. The enzyme is a mixed α/β-protein with a core consisting mainly of a five-stranded mixed β-sheet, which is positioned against a three-stranded antiparallel β-sheet. Four α-helices flank the three-stranded β-sheet. An additional α-helix flanks the five-stranded β-sheet. The catalytic pocket, including the NAD-binding site, is built by the β-sheet core and one α-helix (α3). The active site resembles the catalytic domains of other ADP-ribosylating toxins mentioned above, e.g., diphtheria toxin [133,134], *E. coli* heat-labile toxin [135], *Pseudomonas aeruginosa* exotoxin A [136], and pertussis toxin [137]. C3 has no "active site loop" that might participate in protein substrate recognition by diphtheria toxin. Instead, helix α3 fills this place in C3. The three-dimensional structure of C3bot1 shares an identical topology with the catalytic domain of the enzyme component VIP2 of the actin-ADP-ribosylating vegetative-insecticidal protein from *Bacillus cereus* [138]. It is noteworthy that the catalytic domain of VIP2 shares more than 30% sequence identity with C3bot1. The sequence identity between C3bot1 and C3stau1 is in the same range (34%). The structural similarity between C3-like transferases and C2 toxin is one explanation for the high efficiency of the chimeric toxin, consisting of the N-terminal part (C2IN) of the C2 toxin (adaptor part) and the C3lim exoenzyme (see below).

19.2.2 Structure-Function Analysis

C3bot1 shares a number of conserved amino acids in the active site with VIP2, and most of these residues are conserved among other members of C3-like ADP-ribosyltransferases. Glu174 (Glu214 including the signal sequence) is the *catalytic* glutamic acid residue. An equivalent glutamate is present in C3lim (Glu174). Changes of the catalytic Glu174 to Asp or Gln in C3lim or to any amino acid in C3bot reduces ADP-ribosyltransferase activity and leads to a decrease of the affinity for NAD [139,140]. The hydrophilic residues Ser134, Thr135, and Ser136 of C3bot1 are amino acids of the "STS"-motif (Figure 19.6), which is found in many ADP-ribosyltransferases including C2 toxin [141]. In C3bot1, Ser134 is involved in correct positioning of the catalytic Glu174. Moreover, the STS motif appears to stabilize the two perpendicular orientated β-sheets (e.g., β6 and β3). Further similarities between C3bot1 and the actin-ADP-ribosylating transferases can be found in the region involved in the binding of the adenine ring. In C3bot, Arg51, Tyr94, Arg127, Glu129, and Tyr183 form a pocket for the adenine ring. The two NAD phosphate groups appear to interact with the two conserved arginine residues, C3bot Arg88 and Arg146. Both residues can form hydrogen bonds with Asp90 and 91, respectively. Mutation of the equivalent residues Arg85 in C3stau2, Arg299 in C2 toxin, and Arg295 in iota toxin led to a drastic decrease in ADP-ribosyltransferase or cytotoxic activities [142–144].

All C3 exoenzymes ADP-ribosylate Rho GTPases (e.g., RhoA, B, and C) at asparagine-41 [145]. In Rho complexed with a GTP-analogue, the side chain of the acceptor amino acid Asn41 is solvent exposed [80,146]. Tainer and co-worker suggested that, in C3bot1, a motif termed "ARTT" (ADP-ribosylating-toxin-Turn-Turn-motif) is crucial for the recognition of the RhoA [130] (Figure 19.7). The ARTT-motif consists of two stretches covering residues Ser167-Ala170 (Turn1) and Gly171-Glu174 (Turn2) of C3bot. Both "turns" are located close to the N1 and the C1′ atoms of NAD, which form the N-glycosidic bond. One ARTT-motif, containing the solvent exposed Gln172, which is conserved among all Rho-ADP-ribosylating toxins, interacts with the carbonyl and amide group of Asn41 of RhoA. The second ARTT-motif, with residue Phe169 of C3bot (Phe in C3lim), is suggested to be involved in the recognition of RhoA by interacting with the hydrophobic, solvent-exposed patches around Rho Asn41.

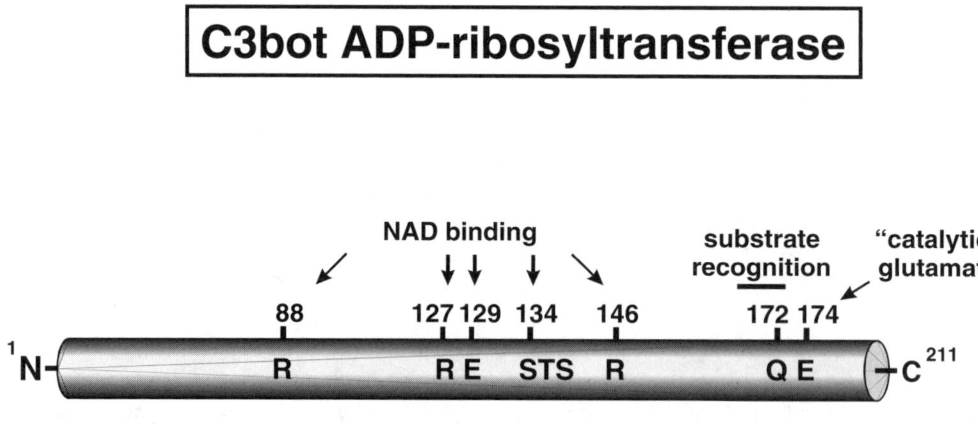

FIGURE 19.6 Structure of C3 from *Clostridium botulinum*. C3 consists only of the catalytic domain. The residue at position 174 (note, numbering is without signal sequence) is the "catalytic glutamate." Glutamine 172 is suggested to be involved in protein substrate recognition. The other residues indicated appear to be involved in NAD-binding.

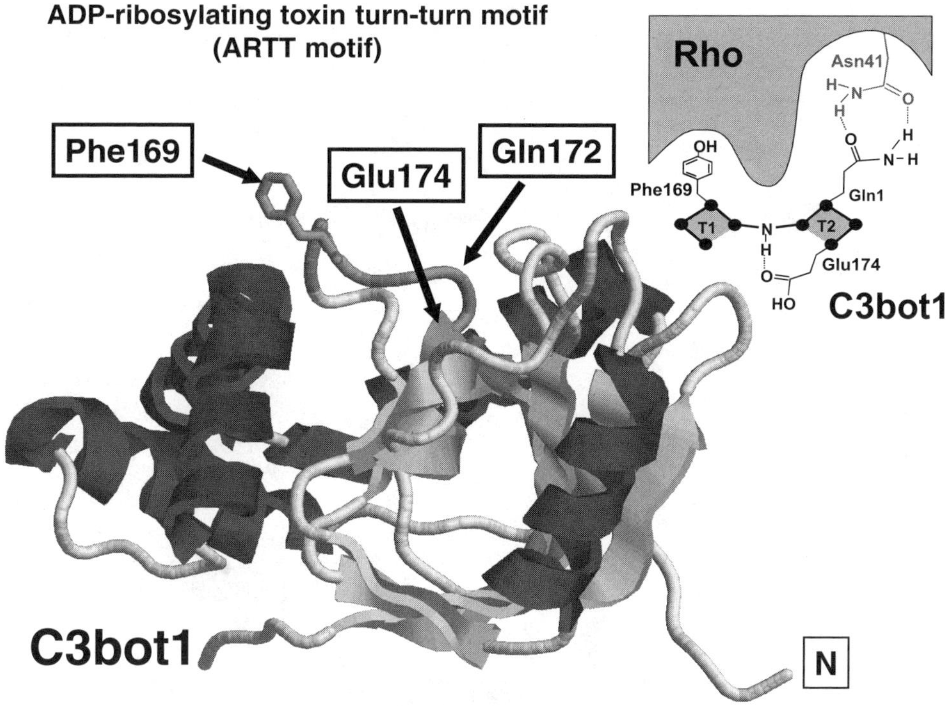

FIGURE 19.7 3D-Structure of C3. The 3D-Structure of C3bot1 is given according to the crystal structure reported by Tainer and co-workers [130]. It is proposed that the ADP-ribosylating toxin-turn-turn (ARTT)-motif is involved in substrate recognition (see text for details).

19.2.3 Functional Consequences of ADP-Ribosylation of Rho Proteins by C3 ADP-Ribosyltransferases

C3-like exoenzymes ADP-ribosylate RhoA, B, and C [147–149] and the modification of Rho occurs at asparagine-41 [145]. Although most other members of the Rho GTPase family contain an equivalent asparagine residue, they are poor *in vitro* substrates (e.g., Rac and see above) [122]. Asparagine-41, the acceptor amino acid for C3-catalyzed ADP-ribosylation is located in or near the "switch-1 region" (residues 28–41) of the GTPase [80,146]. The switch-1 region undergoes major conformational changes depending on the nucleotide bound, thereby playing a key role in signal transduction by the Rho GTPase. Note that Asn1 is not far away from Thr37, the residue of which is glucosylated *Clostridium difficile* toxins A and B. In fact, glucosylation at Thr37 blocks subsequent ADP-ribosylation. Conversely, ADP-ribosylation at Asn41 blocks subsequent glucosylation at Thr37.

Many studies have shown that ADP-ribosylation of RhoA renders the protein biologically inactive. Therefore, C3 exoenzymes cause a dramatic redistribution of the actin cytoskeleton typical for the inactivation of RhoA [149–151]. ADP-ribosylation has only minimal effects on nucleotide binding and GTPase activity of Rho [150]. Surprisingly, ADP-ribosylated RhoA is still able to bind to various effectors (e.g., protein kinase N) [82] (Figure 19.8). This is in contrast to Rho glucosylated at Thr37 by *Clostridium difficile* toxin B. Because binding of Rho to effector kinases appears to be sufficient for activation of the kinase activity, blockade of the Rho/effector interaction is not the mechanism underlying inactivation of Rho signaling by ADP-ribosylation. ADP-ribosylation of Rho may even increase the affinity toward its effector (e.g., phosphatidylinositol-4-phosphate-5-kinase) [152], possibly resulting in sequestering of the effector. However, further evidence for this

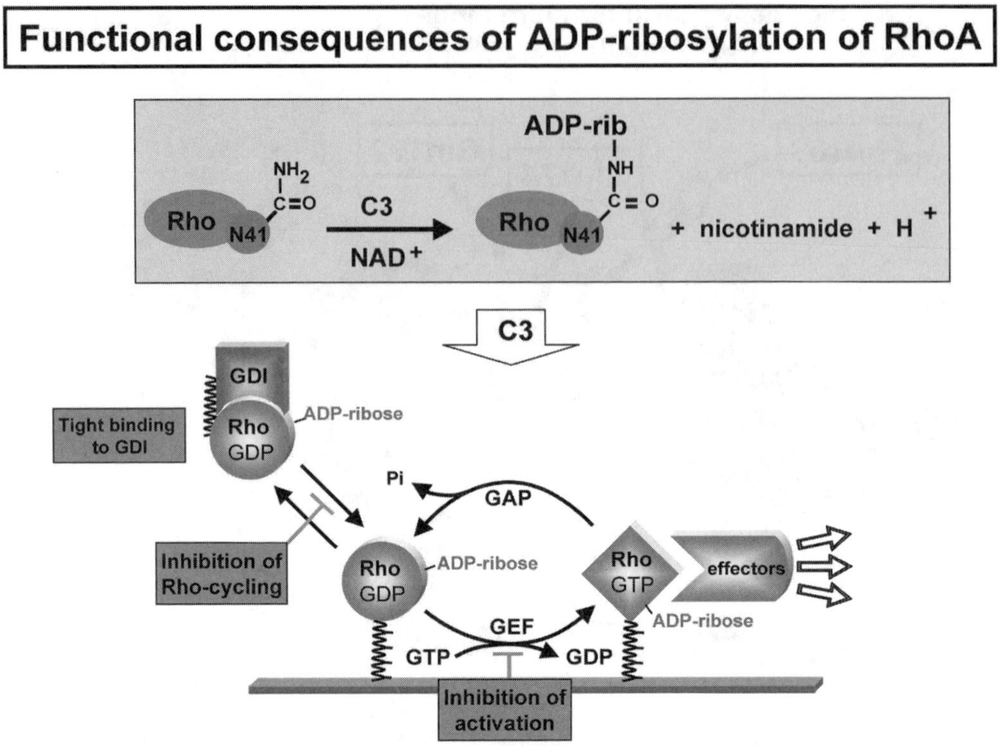

FIGURE 19.8 Functional consequences of ADP-ribosylation of RhoA by C3 exoenzymes. C3 exoenzymes are ADP-ribosyltransferases, which modify RhoA at asparagine 41. The functional consequences are as follows: (1) ADP-ribosylation blocks activation by guanine nucleotide exchange factors (GEFs), (2) ADP-ribosylated Rho has a higher affinity for GDI and is no longer at the membrane.

hypothesis is missing. It was suggested that ADP-ribosylation renders RhoA sensitive for degradation [153,154]. Clearly, ADP-ribosylation also inhibits the activation of RhoA by GEF (e.g., Lbc) [82]. Another mechanism might be particularly important. Rho requires a translocation from the cytoplasm to the membrane to become activated. ADP-ribosylated RhoA accumulates in the cytoplasm, and the release from GDI appears to be blocked [155,156]. Accordingly, translocation and activation of Rho will not occur.

19.2.4 C3 Exoenzymes as Tools

C3 is widely applied as a tool to inactivate RhoA, B, and C and to study the functional consequences of Rho inactivation on signaling processes. Because C3 transferase only consists of an enzyme domain and lacks a cell binding and transport domain, it is poorly accessible for cells. For studies with intact cells, high concentrations (e.g., 5 to 50 μg/ml) and long incubation times (up to 24 to 48 h) are necessary [151,157–159]. In earlier studies, the toxin was microinjected [77,150,160–163]. Other methods to introduce the toxin into cultured cells is permeabilization of cells by digitonin [164], streptolysin O [165], electropermeabilization [166,167], or by scrape loading [168]. The C3 gene was introduced into eukaryotic cells by transient and stable transfection with plasmids or by viral infection [97,153,169–172].

To improve cell accessibility, chimeras were constructed consisting of C3 ADP-ribosyltransferases and the cell binding/tanslocation domain of "complete" toxins. In one approach, C3bot was fused to the binding and translocation subunit of diphtheria toxin [173]. With the chimeric toxin,

600 pg/ml were sufficient to induce the redistribution of F-actin structures in Vero cells. Recently, the *Clostridium botulinum* C2 toxin was used to construct a chimeric fusion toxin (see below).

19.2.5 Cellular Effects of C3-Like ADP-Ribosyltransferases

A large number of studies on Rho functions were performed with C3 transferases. It is beyond the scope of this chapter to give a comprehensive review of these studies. Therefore, only some examples of studies are given, documenting the successful application of C3-like ADP-ribosyltransferases as pharmacological and cell biological tools.

In Vero cells, C3bot (5 μg/ml, for 12–24 h) induces morphological changes characterized by rounding up of the cells with concomitant destruction of stress fibers [149]. Similar observations were made with many cell types and with different C3 exoenzymes [77,125,150,161,174]. A common finding is that treated cells remain in contact via small extensions, in contrast to complete loss of cell contacts after treatment with the actin-ADP-ribosylating C2 toxin [151]. Actin-staining shows loss of stress fibers caused by inactivation of RhoA. Cortical actin appears to be more resistant toward C3 treatment. Usually, cells are viable after C3 treatment. Exchange of the medium to remove C3 reverses cell rounding after a few hours or days, probably due to neosynthesis of Rho [175]. After microinjection, the effects are more rapidly observed and occur within 10 to 15 min [150]. In the pioneering studies from the laboratory of Alan Hall, C3 was used to identify the role of Rho proteins in stress fiber formation induced by growth factor [77,164,176]. Processes that are mediated by Rac or Cdc42, like lamellipodia and microspike formation in fibroblasts, are not affected by C3, underlining the substrate specificity of the transferases [78,79,177]. It is also important to mention that C3 was successfully applied in studies on the role of Rho GTPases neurite outgrowth [178–181]. C3 was applied to study the role of Rho in endocytosis [182] and phagocytosis [97,182,183,184]. By using C3, the involvement of Rho proteins in the regulation of the phosphatidylinositol-4-phosphate-5-kinase (PI-4P-5-kinase) was studied [185]. Another effector which regulation via Rho was studied using C3 transferases is the phospholipase D (PLD) [90,153,186–188].

C3 was applied to study the role of Rho in signaling to the nucleus and in regulation of gene transcription [170,189]. Moreover, C3bot was successfully employed in delineation of the role of Rho in the signaling of various heptahelical receptors to the nucleus via heterotrimeric G proteins. C3bot inhibited the activation of SRF by α_1-adrenergic receptor signaling via $G\alpha_q$ in cardiomyocytes [190], by muscarinic m1 receptors via $G\alpha_{12}$ [191] or by thrombin, LPA, thromboxane, and endothelin receptors via $G\alpha_{12/13}$ [192,193].

19.2.6 C3 as a Virulence Factor

The large amount of information about the cellular functions of Rho GTPases, which is due to application of C3-like exoenzymes, contrasts our limited knowledge of the role of C3-like ADP-ribosyltransferases as potential virulence factors. As a virulence factor, a pathophysiological action of C3-like ADP-ribosyltransferases on the immune system of the target organism is likely. In line with this notion are many reports, showing that C3 affects immune cell functions [194–199]. Rho GTPases are reportedly involved in several processes that regulate the host immune response [195,200–202], and participate in the barrier functions of epithelial cells [98,203] and in wound healing [204]. The crucial role of Rho GTPases in these processes is probably the reason for their targeting by several bacterial toxins. The toxins inactivate (ADP-ribosylation and glucosylation; described here), or even activate, the proteins by deamidation (cytotoxic necrotizing factor CNF from *E. coli*; [205]) and transglutamination (dermonecrotizing toxin DNT from *Bordetella* species [206]), indicating that a fine-tuning of Rho GTPases is essential for cellular function. In line with the important role of Rho GTPases in pathogen-host interaction are recent findings that Rho GTPases are modulated or modified by various bacterial effector proteins that are introduced into the eukaryotic target cells by type-III secretion including exoenzyme S from *Pseudomonas aeruginosa*

[207], YopE and YopT from *Yersinia* [208,209], and SptP and SopE from *Salmonella* [210,211]. Taken together, a possible role of C3 as a virulence factor is conceivable. The question remains how C3-like transferases reach their cytosolic target proteins. Recently, two possible answers to this important question have been offered. It has been suggested for C3stau-producing *S. aureus* that they are able to invade cells and are located in the cytosol [212]. Release of the transferase would then occur at the place where Rho GTPases are located without further need for membrane crossing. On the other hand, it was proposed that membrane damaging bacterial toxins facilitate the cellular entry of other bacterial enzymes and effectors [213]. If this is also true for clostridia, it is feasible that C3 enters cells with the help of hemolysins or other pore-forming toxins, which are produced by many clostridia, including those producing C3.

19.3 BINARY ACTIN-ADP-RIBOSYLATING TOXINS FROM CLOSTRIDIA

Various bacterial toxins exhibit their cytotoxicity by acting on the actin cytoskeleton of eukaryotic cells as a target. A special type of actin-modifying toxin is represented by the family of actin ADP-ribosylating toxins, which has a binary structure. In contrast to most bacterial AB-type toxins, where the enzymatically active (A) and the binding/translocation (B) domains are localized on the same protein, the A- and B-domains of these binary toxins are localized on two individual and nonlinked proteins called components. The toxin components are secreted by the bacteria and have to assemble on the surface of eukaryotic cells to act cytotoxic. The binding/translocation components bind to specific cell surface receptors and mediate cell entry of the enzyme components. All toxins of this family act cytotoxic by ADP-ribosylating G-actin in the cytosol of eukaryotic cells. The enzyme components of the toxins transfer the ADP-ribose moiety of NAD onto the acceptor amino acid arginine-177 of their substrate protein G-actin [214,215]. The consequences of actin-ADP-ribosylation are due to altered properties of modified actin [216]. The family of actin ADP-ribosylating toxins consists of *Clostridium botulinum* C2 toxin [217], *Clostridium perfringens* iota toxin [218], *Clostridium spiroforme* toxin [219], *Clostridium difficile* ADP-ribosyltransferase CDT [220], and the vegetative insecticidal proteins (VIP) from *Bacillus cereus* [138].

19.3.1 *Clostridium botulinum* C2 toxin

Certain strains of *Clostridium botulinum* produce a variety of exotoxins, which are released from the bacteria and are involved in serious human diseases. Most famous are the extremely potent neurotoxins causing botulism [221,222]. Besides the neurotoxins, *Clostridium botulinum* type C and D strains also produce the actin-ADP-ribosylating C2 toxin. While the genes for the neurotoxins C1 and D are located on phage DNA, the gene for the C2 toxin is located in the bacterial chromosome, and phage curing of *Clostridium botulinum* results in strains that produce C2 toxin but no more neurotoxin [223]. It appears that production of C2 toxin is closely related to sporulation, because less toxin is synthesized during vegetative growth of the clostridia than during sporulation [224].

An involvement of C2 toxin in disease is still not clear, because all strains of *Clostridium botulinum* that produce C2 toxin also produce the neurotoxins. Because of their extremely high toxicity, the neurotoxins dominate the infection, while C2 toxin seems not to be involved in botulism. On the other hand, isolated C2 toxin acts lethal when it is applied to animals. The half lethal dose LD_{50} (i. v.) of C2 toxin for mice is less than 50 fmol and after application of 1 to 2 pmol C2 toxin, mice, rats, guinea pigs, and chickens died after about 1 hour [225]. C2 toxin is enterotoxic and causes necrotic and hemorrhagic lesions in the intestinal wall [226]. Simpson compared the pharmacological properties of botulinum neurotoxin type C1 with C2 toxin in detail and described that

C2 toxin causes hypotension, hemorrhaging into the lungs but not muscle relaxation. Hypotension is apparently not caused by vasodilatation but by volume depletion [225].

C2 toxin represents the prototype of clostridial binary actin ADP-ribosylating toxins. C2 toxin was the first toxin of this family, which was identified as a binary toxin. In 1980, C2 toxin was purified by Ohishi and co-workers, who reported that C2 toxin consists of two nonlinked proteins, which must assemble to act cytotoxic to eukaryotic cells [217]. When cultured cells are treated with both components of C2 toxin in combination, they round up as a consequence of the actin modification caused by C2 toxin [151,227,228]. Aktories and co-workers showed for C2 toxin, for the first time, that actin was ADP-ribosylated by a bacterial toxin [229]. The enzyme component C2I (~50 kDa) specifically ADP-ribosylates G-actin at the acceptor amino acid arginine-177 [215,229]. This observation introduced the new family of binary actin ADP-ribosylating toxins. The activated binding/translocation component C2II (80 kDa) mediates the transport of C2I into the cytosol of target cells [230]. The receptor for C2 toxin was the first identified cellular receptor for a toxin of this family [231]. The cellular uptake pathway for C2 toxin was identified, both components of the toxin were cloned, and their specific roles in cytotoxic action were investigated [232,233]. It was described for C2 toxin that the proteolytically activated binding/translocation component forms heptamers and pores in membranes [232], and later this was confirmed for other members of this toxin family [234]. Moreover, C2 toxin was the first binary toxin for which the dependence on a host cell chaperone for its uptake into eukaryotic cells was shown [235]. The other toxins of this family share the general properties of the components, the cytotoxic mode of action, and the cellular uptake pathway with C2 toxin.

19.3.1.1 The Enzyme Component C2I

The enzymatic component C2I of C2 toxin is an ADP-ribosyltransferase [236]. C2I was cloned and sequenced in 1996 [237] (Figure 19.9). Later, the protein was extensively characterized by truncation and mutation analyses. C2I (49.4 kDa) consists of 431 amino acids and shares significant sequence similarities with the other actin-ADP-ribosylating toxins (see below). This allowed to take the 3D-structure of the *Bacillus cereus* VIP2 toxin, which was solved by Tainer and co-workers [138], as a model for C2I and for other actin–ADP-ribosylating toxins. Meanwhile, the crystal structure of the enzyme domain of iota toxin has been reported [238], corroborating the close structural and functional relationship of the enzyme components of this toxin family. Deduced from these studies, we have major insights into the structural feature of C2I. All the toxins have a common structure of two domains. Both domains of the enzyme components of the toxins share the same folding typical for ADP-ribosyltransferases. However, whereas the C-terminal domain possesses the ADP-ribosyltransferase activity, the N-terminal domain functions as an adaptor part for the respective binding component. Thus, it has been suggested that the N-terminal domain evolved by gene duplication [138], which lost its enzyme activity and developed as an adaptor for interaction with the binding and translocation component.

According to the above-mentioned structure, the C-terminal domain of C2I harbors the ADP-ribosyltransferase activity. Bacterial mono-ADP-ribosyltransferases possess highly conserved catalytic domains, although their primary sequence identity is very low. The catalytic domain is formed by two antiparallel sheets flanked by two α-helices. This becomes obvious by analysis of the crystal structures of various bacterial ADP-ribosyltransferases including diphtheria toxin [134], *Pseudomonas aeruginosa* exotoxin A [239], *E. coli* heat-labile enterotoxin [135], and pertussis toxin [137]. Several highly conserved amino acid residues are located in the active site of most ADP-ribosyltransferases. A highly conserved glutamic acid residue is essential for transferase activity [240], and in all ADP-ribosyltransferases, which ADP-ribosylate arginine residues, a second essential glutamic acid residue is located two residues upstream of this "catalytic" glutamic acid [240–243]. Many ADP-ribosyltransferases are characterized by additional conserved residues, such as the

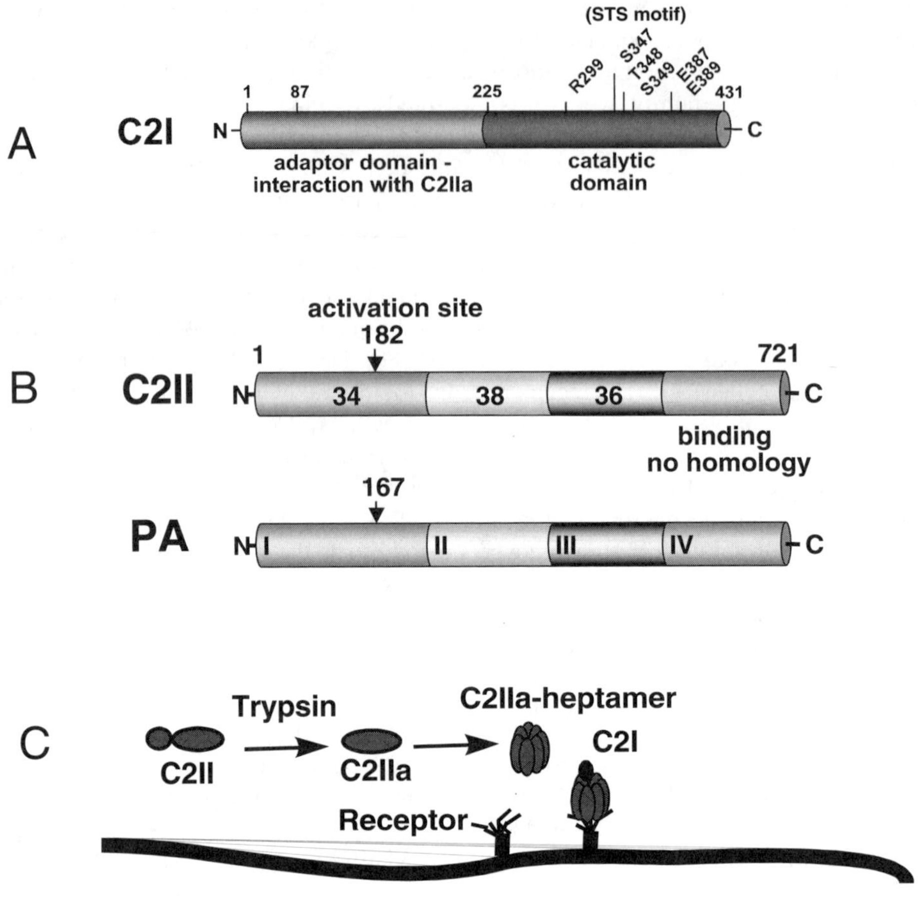

FIGURE 19.9 The binary *Clostridium botulinum* C2 toxin: Cytotoxic mode of action. The intracellular mode of action of the C2I enzyme component of C2 toxin is depicted schematically. C2I ADP-ribosylates G-actin at amino acid residue Arg-177. ADP-ribosylated G-actin molecules assemble to the fast-growing end of the actin filaments. There, they act as "capping proteins" and block the further association of actin monomers. This leads to depolymerization of actin filaments and to a complete breakdown of the actin cytoskeleton. For further details, see text.

serine-threonine-serine sequence ("STS" motif) upstream of the glutamic acids and an arginine residue upstream of the "STS" motif [244,245]. The catalytic site of C2I from *Clostridium botulinum* strain KZZ1577 (92-13) was characterized in detail by site-directed mutagenesis of the conserved amino acid residues [142]. The resulting proteins were assayed for ADP-ribosyltransferase activity and for glycohydrolase activity. Glu389 was identified as the "catalytic glutamic acid" in C2I, because exchange of Glu389 to glutamine blocked both ADP-ribosyltransferase and NAD-glycohydrolase activity of the C2I protein [142]. Exchange of Glu387 to glutamine blocked ADP-ribosyltransferase but not NAD-glycohydrolase activity [142]. The conserved amino acid residues of the catalytic site of C2I are depicted in Figure 19.9A.

As mentioned above, the N-terminal part of C2I (C2IN, amino acid residues 1–225) is enzymatically inactive and interacts with the binding/translocation component C2IIa and mediates translocation of C2I into the cytosol of eukaryotic cells [174]. To study the role of C2IN in more detail, a series of fusion toxins was prepared, containing various N-terminal fragments of C2I fused to C3 exoenzyme from *Clostridium limosum* as a reporter enzyme [246]. The C3 exoenzyme (~23 kDa) ADP-ribosylates and, thereby, inactivates the Rho-GTPase, but C3 exoenzyme is not able to enter cells readily on its own because this protein lacks a binding/translocation domain. The fusion toxins were used to determine the minimal essential part of C2IN, which is able to bind to C2IIa and to translocate into the cytosol. This was measured by the cytotoxic action of C3 exoenzyme in intact cells. Amino acid residues 1–87 of C2I are sufficient to mediate the uptake of a $C2I^{1\text{-}87}$-C3 fusion toxin into cells thus representing the minimal sequence necessary for interaction with C2IIa and for translocation of C2I into the cytosol. Truncation of the N-terminal residues 1-29 results in a $C2I^{30\text{-}225}$-C3 fusion toxin, which is not transported into cells by C2IIa but showed some residual binding to cells via C2IIa [246]. With respect to the 3D-structures available, the N-terminal 87 amino acid residues, which allowed interaction with the binding component and uptake into cells, correspond to the N-domain of VIP2 that contains the α-helices 1–4. The residues 12–29 of C2I correspond to the first α-helical structure (α1) in VIP2 [246]. In the crystal structure of VIP2, the N-terminal helices α1, α2, α3, and α4 (residues 1–87 in C2I) are exposed on the surface of the protein [138], which is in line with a putative role of this protein region as an adaptor unit.

19.3.1.2 The Binding Component C2II

The role of the binding component C2II for the transport of the enzyme component C2I was studied in detail in past years. C2II from *Clostridium botulinum* strain KZZ1577 (92-13) was cloned and sequenced, revealing a protein of 721 amino acid residues with a molecular mass of 80.5 kDa [247] (see Figure 19.9B). A detailed structure–function analysis was performed with recombinant C2II proteins because this method allowed the study of the consequences of amino acid exchanges or truncations in C2II. The protein is released from bacteria as a precursor, which has to become activated by proteolytic cleavage to mediate cytotoxic effects. The C2II precursor protein is neither able to interact with cellular receptors nor with C2I [232]. A peptide of ~20 kDa is cleaved from the N-terminus of C2II, thereby generating the active C2IIa fragment (59.8 kDa) [232]. As a consequence of proteolytic activation, C2IIa forms heat-labile but SDS-stable oligomers in solution [232] (Figure 19.9C). Analytical ultracentrifugation of C2IIa oligomers revealed that their molecular mass is about 420 kDa, indicating that C2IIa forms heptamers [232]. Electron microscopy proved that the heptamers are annular structures with an inner diameter of about 2 nm and an outer diameter of about 12 nm [232]. Interestingly, C2IIa, but not C2II, forms ion-permeable channels in artificial black lipid bilayer membranes and in the membranes of intact eukaryotic cells [232,248]. The pore formation by C2IIa in cell membranes was strictly dependent on acidic conditions [233]. Based on these observations, it became evident that C2IIa heptamers have two different functions in cellular uptake of C2 toxin. They act as a binding as well as a translocation component, depending on the specific cellular conditions. On the cell surface, the heptamers are in the "prepore state" and serve as docking platforms for C2I molecules and facilitate receptor-mediated endocytosis of the toxin complex [232]. Later, in endosomal compartments, acidification of the endosomal lumen triggers a conformational change of C2IIa, leading to membrane insertion and pore formation of the protein [233]. The role of the pores for cell entry of C2I is discussed later.

Sequencing of C2II revealed a remarkable homology not only to the binding components of the other members of the family of actin-ADP-ribosylating toxins, but also to the protective antigen (PA) from *Bacillus anthracis* [247,249] (Figure 19.9B). *Bacillus anthracis* produces two different binary toxins, which are the causative agents for the anthrax disease. PA (83 kDa) is the central transport component for two different enzyme components, the lethal factor (LF) and the edema factor (EF) (for review see Leppla [249]). LF is a metalloprotease, which cleaves MAP-kinases [250], whereas EF is an adenylyl cyclase [251,252]. The lethal toxin is the combination PA and LF and the edema toxin consists of PA and EF. There are several similarities between PA and C2II. As found for C2II, PA is proteolytically cleaved by furin and, thereby, activated. After removal of a 20 kDa fragment from the N-terminus, the resulting PA-63 represents the biologically active species. PA-63 binds to a recently identified cell receptor and forms heptamers [250,253,254]. The heptamers form ion-permeable channels in artificial lipid bilayer membranes and under acidic conditions also in cell membranes [255,256].

The knowledge about the structure of PA and its function for cellular uptake of the anthrax toxins increased rapidly in past years, and the crystal structure of PA was solved [255] (for review see Mourez et al.; and Collier and Young [257,258]. Based on the crystal structure, it became clear that PA is composed of four functional domains: the N-terminal domain 1 (activation site for the protease and interaction with enzyme components LF and EF), domain 2 (membrane insertion), domain 3 (unknown function, maybe oligomerization), and the C-terminal domain 4 (receptor binding). Alignment of C2II and PA reveals significant sequence identity within domains 1–3 (about 50%) but no homology in domain 4 [247]. This is not surprising because PA and C2II bind to different cellular receptors. The domain structure of C2II and PA is depicted schematically in Figure 19.9B.

The crystal structure of PA and the sequence alignment between PA and C2II were used to postulate a four-domain structure for the C2II protein [247]. N-terminal peptide sequencing of both proteins C2II and C2IIa revealed that the trypsin cleavage site is located between residues Lys181 and Ala182 [247]. Furthermore, it was suggested that the N-terminal domain of C2II (residues 1–181) might act as an intramolecular chaperone, important for proper folding of the protein prior to proteolytic activation. When truncated, the "genetically activated" $C2II^{182-721}$ (corresponding to C2IIa) was cloned and expressed as recombinant protein; it was neither able to bind to cells nor to mediate cytotoxic effects in the presence of C2I [247]. Moreover, the protein did not form pores in membranes [247]. Domain 2 of C2II is most likely essential for insertion of the protein into lipid bilayer membranes and for pore formation. In the case of the anthrax PA, the amino acid residues 325–356 most likely are responsible for membrane insertion of the protein and for pore formation [255]. These amino acid residues correspond to amino acid residues 303–331 in C2IIa [247]. This part has a conserved pattern of alternating hydrophilic and hydrophobic residues, which are suggested to be involved in membrane insertion. The C2IIa channel in artificial membranes is voltage-dependent and cation-selective [248]. When chloroquine and related compounds were added to the cis side of preformed C2IIa pores, the pores were closed for high negative voltages [259]. Chloroquine is positively charged and is most likely bound to negatively charged amino acid residues at the cis side of the C2IIa channel [259]. The precise role of domain 3 is still not known, but this part of the protein might be involved in oligomerization of C2IIa. The C-terminal domain 4 of C2IIa is responsible for the binding to its membrane receptor [247]. An antibody raised against the C-terminal domain of C2II (domain 4, amino acid residues 596–721) blocked binding of C2IIa to cells [247]. When C2IIa was C-terminally truncated, i.e., domain 4 was omitted, C2IIa was no longer able to bind to cells and to mediate cytotoxicity, but it still formed oligomers and channels [247]. Particularly, the 7 C-terminal residues appear to be involved in binding, because the deletion of these residues prevents binding of C2IIa to cells [247]. An antibody raised against domain 4 also blocked insertion of C2IIa channels into artificial lipid bilayer membranes [247]. Therefore, the C-terminus is involved in docking of C2IIa to the membrane, or it is important for the correct structure of the C2IIa channel. However, the antibody had no effect on the properties of existing channels when it was added after insertion of C2IIa into the membrane.

19.3.1.3 The Receptor for C2 Toxin

The first receptor identified for a binary actin-ADP-ribosylating toxin was the receptor for C2 toxin. Recently, it was reported that C2 toxin binds to asparagine-linked complex and hybrid carbohydrates on the surface of eukaryotic cells [231]. The structure of the receptor for C2 toxin is depicted schematically in Figure 19.10. To identify the receptor for C2 toxin, a mutant CHO cell line (CHO-RK14) was generated by chemical mutagenesis [260]. CHO-RK14 cells were resistant to C2 toxin but still sensitive to the similar binary *Clostridium perfringens* iota toxin [260]. Closer analysis of this cell line revealed that the cells were not able to bind C2IIa on their surface [260].

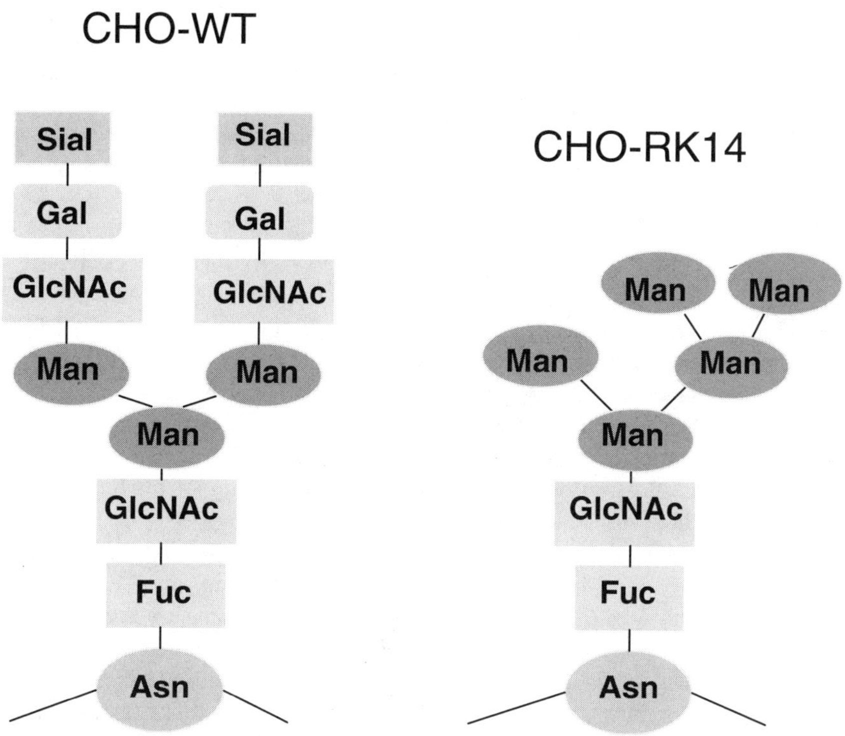

FIGURE 19.10 The components of C2 toxin. A. The enzyme component C2I has two functional domains. The N-terminal domain (residues 1–225) acts as an adaptor for interaction with the activated C2IIa binding component and mediates translocation of C2I across membranes. The C-terminal domain harbors the ADP-ribosyltransferase activity. The conserved amino acid residues in the catalytic domain of C2I are depicted. B. The binding/translocation component C2II has 4 functional domains, based on its sequence homology to the protective antigen (PA) from *Bacillus anthracis* (for details, see text). C2II is activated by trypsin cleavage at position 182 within the domain 1. Domain 2 seems to be involved in membrane insertion; domain 3 has a yet-unknown function; and domain 4 interacts with the cell receptor. C. Consequences of the C2II activation. Activated C2IIa forms ring-shaped heptamers ("pre-pores"), which represent the biologically active species of C2II. C2IIa binds to the cell receptor and interacts with C2I.

This finding gave evidence for a defect receptor for C2IIa, and, in more detailed studies, it became evident that CHO-RK14 cells lack complex and hybrid N-glycans, which are essential for C2 toxin. The reason is a single point mutation in the gene for N-acetylglucosaminyltransferase I [231]. Transfection of these mutated cells with the intact N-acetylglucosaminyltransferase I gene recovered sensitivity of the cells against C2 toxin and also the ability of the cells to bind C2IIa [231]. All vertebrate cell lines and primary cultured cells that have been tested so far were sensitive to C2 toxin, indicating that the receptor for C2 toxin is present on all eukaryotic cells, including fish cells [231]. Interestingly, insect cells, which have different carbohydrate structures on their cell surface, are resistant to C2 toxin [231]. This finding suggests that the carbohydrate structure might be sufficient for binding of C2IIa. However, so far, it cannot be excluded that specific proteins are involved in the endocytic uptake mechanism of C2 toxin.

19.3.1.4 Cellular Uptake of C2 Toxin

Because bacterial toxins modify substrates in the cytosol of their host cells, they must deliver at least their enzymatic active domains across cellular membranes. Therefore, the toxins developed sophisticated strategies, which allow the usage of the cellular protein trafficking routes to enter the cytosol. Based on their cellular uptake pathways, the toxins can be classically divided into two groups: (i) the "short trip toxins," which escape from early endosomal compartments into the cytosol, and (ii) the "long trip toxins," which take the "retrograde pathway" from early endosomes via late endosomes, the Golgi apparatus and the endoplasmic reticulum (ER). From the ER, these toxins, for which cholera toxin represents the prototype, translocate into the cytosol (for review see Sandvig and Van Deurs [261]).

The initial step for cellular uptake of C2 toxin is binding of C2IIa heptamers to the cellular receptor. After assembly of C2I, the receptor-toxin complexes are taken up via receptor-mediated endocytosis and reach early endosomal compartments. C2 toxin, as all binary actin ADP-ribosylating toxins, belongs to the "short trip toxins." Neither brefeldin A nor nocodazole, both inhibitors of the retrograde pathway, protects cells from intoxication with C2 toxin [232]. C2 toxin translocates from early acidic endosomes into the cytosol (see Figure 19.11). This step can be blocked by the drug bafilomycin A1, which inhibits the activity of the vacuolar-type H^+-ATPase (v-ATPase) [232]. This ATPase is located in the endosomal membranes and drives the acidification of endosomal compartments [262]. Acidification of the endosomal lumen is essential for inducing a conformational change of the C2IIa protein and leads to exposure of hydrophobic residues to the surface of the protein. As a consequence, membrane insertion and pore formation of C2IIa occurs, and C2I translocates into the cytosol [233] (see Figure 19.11). Pore formation by C2IIa and acidification are required for translocation of C2I. Based on this observation, there are some central questions arising about the mechanism of this translocation step. What is the precise role of the C2IIa pores? Does C2I translocate directly through the pore lumen? If so, it should be expected that C2I has to become unfolded during translocation because the dimensions of this 50 kDa protein are too extended to pass the pore (inner diameter ~2 nm) in a folded conformation. Finally, if C2I translocates in an unfolded conformation, how is its refolding in the cytosol achieved?

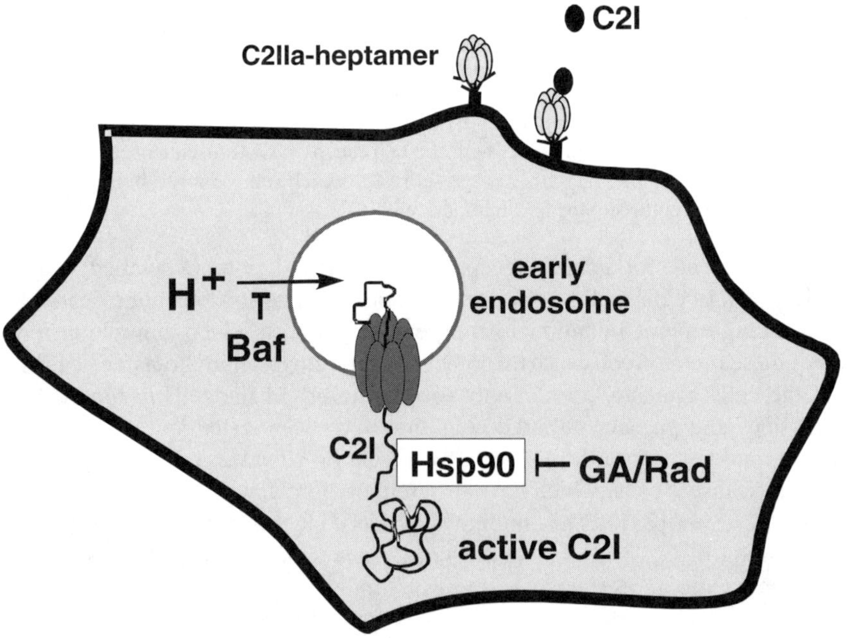

FIGURE 19.11 The cellular receptor for C2 toxin. C2 toxin binds to asparagine (Asn)-linked complex and hybrid carbohydrates. The receptor for C2 toxin of Chinese hamster ovary (CHO) wild type (WT) and of C2 toxin-resistant CHO-RK14 mutant cells is depicted.

The advantage of binary toxins for studying translocation processes is that the role of the single components for translocation can be assayed individually. The endosomal conditions for the translocation of C2 toxin into the cytosol can be experimentally mimicked on the surface of intact cells [232]. In brief, cells are incubated with C2IIa and C2I at 4°C to allow binding of the proteins to the cell surface but no endocytosis. Subsequently, warm and acidic medium (pH ≤5.5, 37°C) is added to the cells for 5 min to induce the direct translocation of C2I across the cell membrane into the cytosol. This method allows the monitoring of the single steps of translocation as membrane insertion and pore formation by C2IIa and translocation of C2I into the cytosol. Pore formation of C2IIa in cell membranes can be detected by measuring the efflux of radioactive rubidium-86 from preloaded cells [233]. C2IIa pores were only observed under acidic conditions and pore formation was dependent on binding of C2IIa to its receptor, indicating that pore formation is not due to an unspecific insertion of C2IIa heptamers into the cell membrane [233].

When C2I was delivered into the cytosol under the described experimental conditions, it became evident that very precise timing and coordination of the individual steps was essential. For translocation of C2I, it was essential that C2I was bound to C2IIa prior to acidification [233]. When first C2IIa pores were formed in the cell membranes and thereafter C2I was added into the medium, no intoxication of cells occurred [233]. This finding implies that C2IIa pores do not act as nonspecific channels that deliver "free" C2I molecules passively into the cytosol. Taken together, insertion of C2IIa into membranes, pore formation, and translocation of C2I is a precisely coordinated process, which is initiated by an acidic milieu. The C2IIa pores are essential for delivery of C2I into the cytosol. The central question is the role of the C2IIa pore for the translocation mechanism of C2I. If the C2IIa pores are directly involved in the membrane transport of C2I, an interaction of C2I with the lumen of C2IIa pores during the membrane translocation step should be postulated. In fact, a direct interaction of the enzyme component C2I with the C2IIa pores was experimentally demonstrated. The rubidium release assay was used to study the interaction between C2IIa pores and C2I. When C2I was present, the efflux of rubidium from intact cells was reduced [233]. This observation suggests that C2I interacts with the lumen of the C2IIa pore. Moreover, it was C2I interacted with the lumen of C2IIa pores in artificial black lipid bilayer membranes [233]. Although these findings are strong hints that C2I translocates directly through the C2IIa pores from acidic endosomes into the cytoplasm, the final direct proof for this hypothesis is still missing.

If C2I translocates directly through the C2IIa pore, the protein has to become unfolded for passage of the channel. This unfolding was studied with a recombinant fusion protein consisting of C2I and dihydrofolate reductase (DHFR) [263]. The DHFR/methotrexate (MTX) system can be used for studying unfolding of proteins during membrane translocation. MTX is a substrate for DHFR, and when MTX is bound to DHFR, the DHFR protein is stabilized in a tightly folded conformation and cannot become unfolded. DHFR was C-terminally fused to C2I, and the resulting C2I-DHFR fusion protein was active *in vitro* and was delivered by C2IIa into the cytosol of cells [263]. The presence of MTX inhibited the intoxication of cells with C2IIa/C2I-DHFR [263]. MTX bound to the DHFR domain of the C2I-DHFR fusion toxin and prevented its unfolding during translocation of the protein across endosomal membranes [263]. This observation speaks for an unfolding of the C2I protein during its translocation into the cytosol of cells.

19.3.1.5 The Heat Shock Protein Hsp90 Is Essential for Cellular Uptake of C2 Toxin

The recent finding that C2I translocates in an unfolded conformation from early endosomes into the cytosol of target cells implies that the ADP-ribosyltransferase activity must become recovered in the cytosol. One hypothesis is that host cell chaperones are involved in refolding of the C2I protein in the cytosol. To address this question, the heat shock protein Hsp90 was inactivated by the specific inhibitors geldanamycin and radicicol. Each of the inhibitors dramatically delayed the subsequent intoxication of cells by C2 toxin [235]. In the presence of geldanamycin or radicicol, C2I was trapped in endosomal compartments, and the translocation of C2I from early acidic endosomes into the cytosol was blocked [235]. The Hsp90 inhibitors had no influence on binding of C2 toxin to the cell surface, pore formation by C2IIa, and interaction between C2IIa and C2I [235]. The role of Hsp90 in cellular uptake of C2 toxin is depicted in Figure 19.11. This was the first report on an involvement of host cell chaperones in the cellular uptake mechanism of binary bacterial toxins. Recently, it was demonstrated that Hsp90 also inhibited intoxication of cells by iotalike toxins [264]. Therefore, cellular chaperones seem to be generally essential for cellular uptake of binary actin-ADP-ribosylating toxins. The precise role of Hsp90 in the translocation of these toxins, the involvement of further co-chaperones and the interaction with the toxin components will be central topics for future studies. Interestingly, Ratts and co-workers reported recently that Hsp90 is involved in the cellular uptake of diphtheria toxin [265]. Taken together, these findings suggest that the interplay between bacterial toxins and host cell chaperones during cellular uptake of the toxins might be a more general principle.

19.3.2 The Other Members of the Family of Actin ADP-Ribosylating Toxins

19.3.2.1 *Clostridium perfringens* Iota Toxin

The binding components of the iotalike toxins are interchangeable (i.e., the binding components are able to transport the enzyme components of other iotalike toxins into eukaryotic cells). In contrast, they do not transport the enzyme component of *Clostridium botulinum* C2 toxin [266,267]. Moreover, antibodies raised against *Clostridium perfringens* iota toxin prevent intoxication of cells with *Clostridium spiroforme* toxin, but not with C2 toxin. Iota toxin is exclusively produced by *Clostridium perfringens* type E strains, where the genes for iota toxin are located on a plasmid [268]. Stiles and co-workers identified the binary structure of iota toxin [218]. The binding/translocation component Ib mediates cell delivery of the enzyme component Ia. The ADP-ribosyltransferase Ia, specifically ADP-ribosylates G-actin at arginine-177 [214]. The toxin acts as an enterotoxin causing diarrhea in calves and lambs and is lethal for mice and dermonecrotic in guinea pigs [269].

19.3.2.2 *Clostridium spiroforme* Toxin

Clostridium spiroforme produces a binary actin-ADP-ribosylating toxin (CST) [219,266,270] that causes diarrhea and acts lethal in rabbits [271]. The components of CST, Sa (enzyme component) and Sb (binding/translocation component), show high homology to the components of iota toxin, Ia and Ib, respectively, and the CST proteins show cross-reactivity with *Clostridium perfringens* type E antiserum [266].

19.3.2.3 *Clostridium difficile* Transferase

The third member of the iotalike toxins is the *Clostridium difficile* toxin, CDT [272]. The components CDTa (enzyme component) and CDTb (binding/translocation component) share 81% and 84% amino acid sequence identity with *Clostridium perfringens* Ia and Ib, respectively [220,272]. As described above, *Clostridium difficile* causes antibiotic-associated diarrhea and pseudomembranous colitis in humans. Toxins A and B are the causative agents of this disease, but an involvement of CDT in pathogenesis is still not clear. Only a few (~6%) of the *Clostridium difficile* strains, which have been isolated from patients, produce this binary toxin. At present, it is being studied whether the binary toxin has any influence on the course or severity of the disease [273].

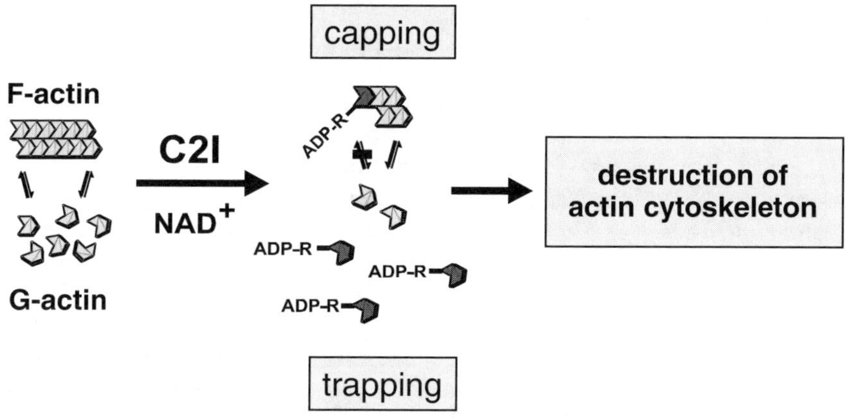

FIGURE 19.12 Cellular uptake mechanism of *Clostridium botulinum* C2 toxin. The actual model for the cellular uptake of C2 toxin is depicted schematically. C2IIa heptamers ("pre-pores"), the active species of the C2II component, bind to the receptor on the cell surface, as well as to the C2I enzyme components, and mediate uptake of the receptor/toxin complex into endosomal compartments. Following acidification of the endosomal lumen by the action of the v-type ATPase in the endosomal membrane, C2IIa heptamers insert into the endosomal membrane and thereby convert from "pre-pores" to pores. Finally, C2I is translocated into the cytosol, most likely through the pores. C2I must become unfolded for translocation. The inhibitor of the v-type ATPase, bafilomycin A1, prevents pore formation of C2IIa and translocation of C2I. The host cell chaperone Hsp90 is essential for translocation of C2I from early acidic endosomes into the cytosol. Inhibition of Hsp90 by its specific inhibitors radicicol (Rad) or geldanamycin (GA) prevents translocation of C2I into the cytosol.

19.3.3 ADP-Ribosylation of Actin by C2 Toxin

The cytotoxic mode of action of C2 toxin is depicted schematically in Figure 19.12. C2 toxin specifically ADP-ribosylates G-actin at arginine-177, while F-actin is not substrate for C2I [229,274,275]. C2I ADP-ribosylates only nonmuscle G-actin and smooth muscle actin isoforms but not α-actin isoforms [274,276]. ADP-ribosylation of G-actin inhibits assembly of actin monomers into filaments and leads to dissociation of the filaments [229,275]. As a consequence, C2 toxin leads to rounding up of cultured cells within a few hours. The molecular mechanisms involved in these actions include a capping function of ADP-ribosylated actin at the fast-growing ends (plus or barbed ends) of actin filaments without directly affecting the slow-growing end (minus or pointed end). This leads to an increase of the "critical concentration" for actin polymerization and tends to depolymerize F-actin [277,278]. The actin monomers released are substrate for ADP-ribosylation and trapped as monomers in their ADP-ribosylated form. Deduced from the 3D-structure of actin and the model of F-actin, it is suggested that the ADP-ribosylated at arginine-177 occurs in or near the contact site of actin monomers, thereby polymerization is prevented by sterical hindrance [279,280]. The ATPase activity of ADP-ribosylated actin is strongly inhibited even in the presence of cytochalasin [281]. ADP-ribosylation of actin affects its interaction with gelsolin by reducing the nucleation activity of the actin-gelsolin complex [282–284]. The interaction of ADP-ribosylated actin with the actin binding proteins deoxyribonuclease I, vitamin D-binding protein, thymosin beta(4), and cofilin is still possible [285]. It is reported that under special conditions a special type of actin filaments is even formed with ADP-ribosylated actin [285].

19.3.4 Recombinant Binary Fusion Toxins as Cellular Protein Delivery Systems: Tools for Cell Biology and Pharmacology

Binary bacterial toxins are good candidates to act as carriers for delivering foreign proteins into the cytosol of cells. As described above for the C2 toxin, there are adaptor domains within the enzyme components that have no enzyme activity, but interact with the binding/translocation components and are able to translocate across membranes. For C2I it was demonstrated that the N-terminal amino acid residues 1–225 are able to transport C3 transferases via the binding and translocation component C2IIa [174] (see Figure 19.13). C3 transferases, also called C3-like exoenzymes, are bacterial ADP-ribosyltransferases, which are produced by *Clostridium botulinum*, *Clostridium limosum*, *S. aureus*, and *B. cereus* (for an overview on C3 transferases, see earlier). C3 transferases modify and, thereby, inactivate the small GTPase Rho by ADP-ribosylation at residue asparagine-41. C3-like exoenzymes have neither binding nor transport components and are not taken up into eukaryotic cells in a specific manner. The proteins are most likely taken up by cells via nonspecific pinocytosis which usually requires high concentrations (>10 μg/ml medium) of C3

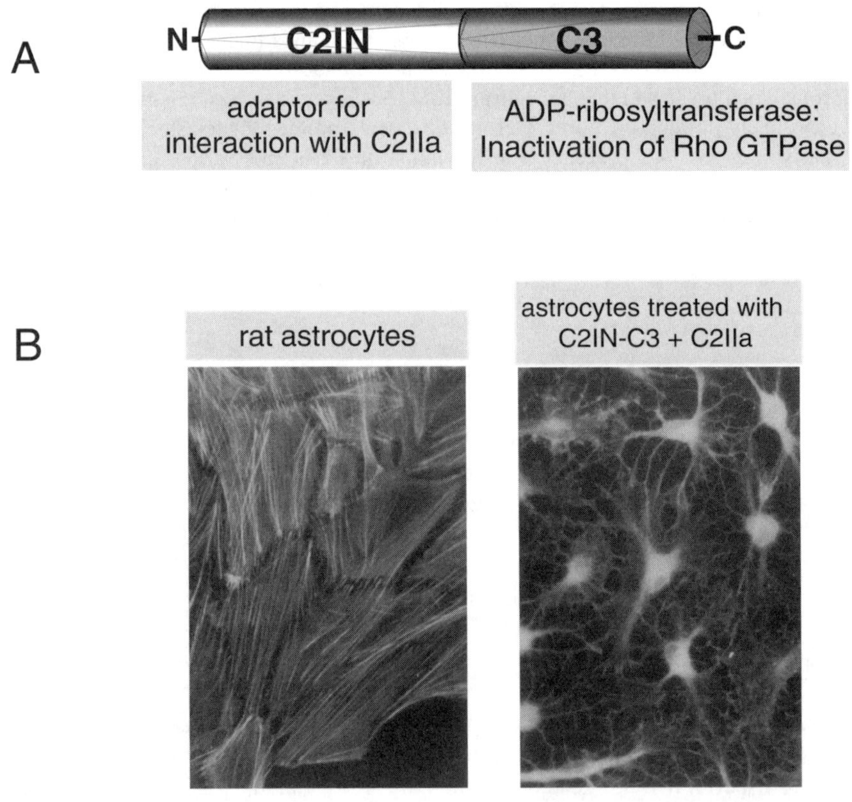

FIGURE 19.13 The binary C2 toxin as a cellular protein delivery system for foreign protein. (A) The C2IN-C3 fusion toxin is composed of the N-terminal domain of C2I (C2IN, amino acid residues 1–225) and the C3 transferase from *Clostridium limosum*. C3 transferase exclusively ADP-ribosylates the GTPase RhoA, B, and C but is not taken up into eukaryotic cells. The adaptor C2IN (amino acid residues 1–225 of C2I) is sufficient for interaction with C2IIa and for translocation of fusion proteins into the cytosol. (B) Cytotoxic effects of C2IN-C3/C2IIa on cells. Cells were incubated for 3h at 37°C with C2IN-C3 (100 ng/ml) together with C2IIa (200 ng/ml). For control, cells were incubated without toxin. The actin filaments of the cells were stained with rhodamine-phalloidin.

transferase for intoxication of cells. Moreover, incubation times for the intoxication of cells with C3 transferases are extended up to 24 h.

The C2IN-C3 fusion toxin, which consists of the amino acid residues 1–225 from C2I and C3 transferase from *Clostridium limosum,* represents the first recombinant fusion toxin based on C2 toxin [174]. This toxin is a cell permeable specific Rho inhibitor, which completely inactivates Rho in cultured cells at low concentrations (100 ng/ml medium) within 2 to 3 hours when it is applied together with the binding component C2IIa [174]. Morphological alterations of monolayer cells treated with C2IIa/C2IN-C3 start after 30 to 60 min, and after 2 to 3 hours, cells show the typical "C3-morphology," i.e., contraction of the cell body and formation of neurite-like protrusions (see Figure 19.13B). The analysis of time/concentration curves suggests that the potency of the C2IN-C3 fusion toxin increases more than 300-fold as compared to C3 exoenzymes [174]. ADP-ribosylation of Rho in intact cells can be monitored by gel shift analysis in immunoblots and by sequential ADP-ribosylation of Rho protein from cell lysates [174]. The cytopathic effects caused by the C2IN-C3 fusion toxin in cells were reversed within a few hours by removal of the toxin from the cell medium [175]. Cells reorganized their actin stress fibers, reconstituted their actin cytoskeleton, and recovered normal morphology. The rapid recovery of cells was due to a very small amount of newly synthesized Rho protein, and recovery was not observed when cells were pretreated with inhibitors of protein synthesis, or with the proteasome inhibitor lactacystin [175]. This suggests that both *de novo* synthesis of Rho as well as degradation of the C2IN-C3 fusion toxin at the proteasome complex are essential for the reconstitution of the actin cytoskeleton induced by Rho. This reversibility of cytopathic effects seems to be specific for the recombinant C2IN-C3 fusion toxin because the C3 exoenzyme from *Clostridium limosum* did not show these effects after removal of the toxin.

Smaller parts of C2I (residues 1–87) are sufficient for the transport of C3 transferase into cells but the prototype $C2I^{1\text{-}225}$-C3 is the most efficient fusion toxin [246]. The position of the C2IN adaptor within the fusion toxin is not limiting the transport ability of the protein. The C3-$C2I^{1\text{-}225}$ fusion toxin with the adaptor located C-terminally and the GST-$C2I^{1\text{-}225}$-C3 fusion toxin, where the C2IN adaptor is placed in the middle of the protein, are also transported into the cytosol by C2IIa [246].

In past years, several fusion toxins consisting of C2IN and various C3 transferases were constructed. Because the various C3 enzymes differ with respect to their substrate specificity toward Rho, the individual fusion toxins can be used to study the role of members of the Rho subfamily (i.e., RhoE is modified by C3stau from *S. aureus*) in intact cells. Wilde and co-workers reported recently that a fusion toxin consisting of $C2I^{1\text{-}225}$ and the C3-like exoenzyme from *S. aureus* is also efficiently transported by C2IIa into the cytosol of eukaryotic cells [125]. One reason for the high transport capacity of the C2-system for C3 transferases might be that C3 exoenzymes, like C2I, are ADP-ribosyltransferases. Because C3 proteins are also structurally highly similar to the catalytic domain of C2I, the fusion toxin mimics the original C2I protein in size and in its overall structure. In the past, only bacterial proteins were used as fusion partners to generate C2 toxin-based chimeras. It will be interesting to use this system for the delivery of eukaryotic proteins, as well as for nucleic acids into the cytosol of cells.

The "natural" actin-ADP-ribosylating toxins, as C2 toxin, and also the C2 toxin-based "artificial" recombinant fusion toxins, are ideal tools to study the role of their substrate proteins actin and Rho in various processes in intact eukaryotic cells. Table 19.1 summarizes the studies in which C2 toxin was used for cell biological and pharmacological studies. The C2IN-C3 fusion toxin is a specific cell permeable and reversible inhibitor of Rho A, B, and C. These features make this toxin valuable to study kinetics and functional consequences of Rho-ADP-ribosylation in intact cells. Studies that used the recombinant Rho-inactivating C2IN-C3 fusion toxin as a tool are listed in Table 19.2.

TABLE 19.1
The C2 Toxin as a Tool in Cell Biology and Pharmacology: Studies Using C2 Toxin to Test the Involvement of the Actin Cytoskeleton in Cellular Processes

	Refs.
Activation and peptide receptor dynamics of neutrophils	[286–290]
Activation of B lymphocytes	[291–294]
Apoptosis in B lymphocytes	[295]
Exocytosis in PC12	[296]
Exocytosis in chromaffin cells	[297]
Exocytosis in mast cells	[89,298,299]
Exocytosis of insulin	[300]
Steroid output from adrenal Y-1 cells	[301]
Intracellular calcium mobilization	[91]
Insulin signaling in L6 myoblasts	[302]
Golgi-ER transport	[303]
Glucose transport in 3T3LI	[304]
Actin expression	[305–307]
Cell cycle control	[308]
Metastasis and invasion	[309]
Endothelial permeability	[310–314]

TABLE 19.2
The C2IN-C3 Fusion Toxin as a Tool in Cell Biology and Pharmacology: Studies on the Involvement of Rho in Cellular Processes

	Refs.
Activation of anion channels	[315]
Secretion of van Willebrand factor	[316]
Ephrin-A5-induced neuronal growth cone collapse	[181]
Uptake and degradation of lipoproteins by macrophages	[317]
IL-1 receptor induced signaling	[318]
Aquaporin2 translocation	[319]
Cerebellar granule neuronal survival	[320]
Cyclooxygenase-2-gene expression	[321]
Monocyte transendothelial migration	[322]
Volume-regulated anion channels in vascular endothelial cells	[323]
Angiotensin II-induced vasoconstriction	[324]
Mechanism of formation of branched dendrites	[325]

REFERENCES

1. Von Eichel-Streiber, C., Boquet, P., Sauerborn, M., and Thelestam, M., Large clostridial cytotoxins — a family of glycosyltransferases modifying small GTP-binding proteins, *Trends Microbiol.*, 4, 375, 1996.
2. Busch, C. and Aktories, K., Microbial toxins and the glucosylation of Rho family GTPases, *Curr. Opin. Struct. Biol.*, 10, 528, 2000.
3. Kelly, C.P. and LaMont, J.T., *Clostridium difficile* infection, *Annu. Rev. Med.*, 49, 375, 1998.

4. Bartlett, J.G., Moon, N., Chang, T.W., Taylor, N., and Onderdonk, A.B., Role of *Clostridium difficile* in antibiotic-associated pseudomembranous colitits, *Gastroenterology*, 75, 778, 1978.
5. Larson, H.E. and Proce, A.B., Pseudomembranous colitis; presence of clostridial toxin, *Lancet*, II, 1312, 1977.
6. Larson, H.E., Price, A.B., Honour, P., and Borriello, S.P., *Clostridium difficile* and the aetiology of pseudomembranous colitis, *Lancet*, 20, 1063, 1978.
7. Rifkin, G.D., Fekety, F.R., Silva, J., and Sack, R.B., Antibiotic-induced colitis. Implication of a toxin neutralised by *Clostridium sordellii* antitoxin, *Lancet*, II, 1103, 1977.
8. Taylor, N.S., Thorne, G.M., and Bartlett, J.G., Comparison of two toxins produced by *Clostridium difficile*, *Infect. Immun.*, 34, No.3, 1036, 1981.
9. Barroso, L.A., Wang, S.-Z., Phelps, C.J., Johnson, J.L., and Wilkins, T.D., Nucleotide sequence of *Clostridium difficile* toxin B gene, *Nucl. Acids Res.*, 18, No.13, 4004, 1990.
10. Dove, C.H., Wang, S.Z., Price, S.B., Phelps, C.J., Lyerly, D.M., Wilkins, T.D., and Johnson, J.L., Molecular characterization of the *Clostridium difficile* toxin A gene, *Infect. Immun.*, 58, 480, 1990.
11. Von Eichel-Streiber, C., Laufenberg-Feldmann, R., Sartingen, S., Schulze, J., and Sauerborn, M., Comparative sequence analysis of the *Clostridium difficile* toxins A and B, *Mol. Gen. Genet.*, 233, 260, 1992.
12. Sauerborn, M. and Von Eichel-Streiber, C., Nucleotide sequence of *Clostridium difficile* toxin A, *Nucleic Acids Res.*, 18, 1629, 1990.
13. Lyerly, D.M., Krivan, H.C., and Wilkins, T.D., *Clostridium difficile*: its disease and toxins, *Clin. Microbiol. Rev.*, 1, 1, 1988.
14. Wilkins, T.D., Role of *Clostridium difficile* toxins in disease, *Gastroenterology*, 93, No.2, 389, 1987.
15. Savidge, T.C., Pan, W.-H., Newman, P., O'Brien, M., Anton, P.M., and Pothoulakis, C., *Clostridium difficile* toxin B is an inflammatory enterotoxin in human intestine, *Gastroenterology*, 125, 413, 2003.
16. Riegler, M., Sedivy, R., Pothoulakis, C., Hamilton, G., Zacheri, J., Bischof, G., Cosentini, E., Feil, W., Schiessel, R., LaMont, J.T., and Wenzl, E., *Clostridium difficile* toxin B is more potent than toxin A in damaging human colonic epithelium *in vitro*, *J. Clin. Invest.*, 95, 2004, 1995.
17. Just, I., Selzer, J., Wilm, M., Von Eichel-Streiber, C., Mann, M., and Aktories, K., Glucosylation of Rho proteins by *Clostridium difficile* toxin B, *Nature*, 375, 500, 1995.
18. Just, I., Wilm, M., Selzer, J., Rex, G., Von Eichel-Streiber, C., Mann, M., and Aktories, K., The enterotoxin from *Clostridium difficile* (ToxA) monoglucosylates the Rho proteins, *J. Biol. Chem.*, 270, 13932, 1995.
19. Hofmann, F., Busch, C., Prepens, U., Just, I., and Aktories, K., Localization of the glucosyltransferase activity of *Clostridium difficile* toxin B to the N-terminal part of the holotoxin, *J. Biol. Chem.*, 272, 11074, 1997.
20. Busch, C., Schömig, K., Hofmann, F., and Aktories, K., Characterization of the catalytic domain of *Clostridium novyi* alpha-toxin, *Infect. Immun.*, 68, 6378, 2000.
21. Genth, H., Hofmann, F., Selzer, J., Rex, G., Aktories, K., and Just, I., Difference in protein substrate specificity between hemorrhagic toxin and lethal toxin from *Clostridium sordellii*, *Biochem. Biophys. Res. Commun.*, 229, 370, 1996.
22. Hofmann, F., Rex, G., Aktories, K., and Just, I., The Ras-related protein Ral is monoglucosylated by *Clostridium sordellii* lethal toxin, *Biochem. Biophys. Res. Commun.*, 227, 77, 1996.
23. Just, I., Selzer, J., Hofmann, F., Green, G.A., and Aktories, K., Inactivation of Ras by *Clostridium sordellii* lethal toxin-catalyzed glucosylation, *J. Biol. Chem.*, 271, 10149, 1996.
24. Popoff, M.R., Chaves, O.E., Lemichez, E., Von Eichel-Streiber, C., Thelestam, M., Chardin, P., Cussac, D., Chavrier, P., Flatau, G., Giry, M., Gunzburg, J., and Boquet, P., Ras, Rap, and Rac small GTP-binding proteins are targets for *Clostridium sordellii* lethal toxin glucosylation, *J. Biol. Chem.*, 271, 10217, 1996.
25. Selzer, J., Hofmann, F., Rex, G., Wilm, M., Mann, M., Just, I., and Aktories, K., *Clostridium novyi* α-toxin-catalyzed incorporation of GlcNAc into Rho subfamily proteins, *J. Biol. Chem.*, 271, 25173, 1996.
26. Hofmann, F., Busch, C., and Aktories, K., Chimeric clostridial cytotoxins: identification of the N-terminal region involved in protein substrate recognition, *Infect. Immun.*, 66, 1076, 1998.
27. Busch, C., Hofmann, F., Selzer, J., Munro, J., Jeckel, D., and Aktories, K., A common motif of eukaryotic glycosyltransferases is essential for the enzyme activity of large clostridial cytotoxins, *J. Biol. Chem.*, 273, 19566, 1998.

28. Breton, C. and Imberty, A., Structure/function studies of glycosyltransferases, *Curr. Opin. Struct. Biol.*, 9, 563, 1999.
29. Wiggins, C.A.R. and Munro, S., Activity of the yeast *MNN1* α-1,3-mannosyltransferase requires a motif conserved in many other families of glycosyltransferases, *Proc. Natl. Acad. Sci. U.S.A.*, 95, 7945, 1998.
30. Busch, C., Hofmann, F., Gerhard, R., and Aktories, K., Involvement of a conserved tryptophan residue in the UDP-glucose binding of large clostridial cytotoxin glycosyltransferases, *J. Biol. Chem.*, 275, 13228, 2000.
31. Negishi, M., Dong, J., Darden, T.A., Pedersen, L.G., and Pedersen, L.C., Glucosaminylglycan biosynthesis: what we can learn from the X-ray crystal structures of glycosyltransferases GlcAT1 and EXTL2, *Biochem. Biophys. Res. Commun.*, 303, 393, 2003.
32. Keusch, J., Manzella, S.M., Nyame, K.A., Cummings, R.D., and Baenziger, J.U., Cloning of Gb_3 synthase, the key enzyme in globo-series glycosphingolipid synthesis, predicts a family of α1,4 glycosyltransferases conserved in plants, insects and mammals, *J. Biol. Chem.*, 275, in press, 2000.
33. Vetter, I.R., Hofmann, F., Wohlgemuth, S., Herrmann, C., and Just, I., Structural consequences of mono-glucosylation of Ha-Ras by *Clostridium sordellii* lethal toxin, *J. Mol. Biol.*, 301, 1091, 2000.
34. Geyer, M., Wilde, C., Selzer, J., Aktories, K., and Kalbitzer, H.R., Glucosylation of Ras by *Clostridium sordellii* lethal toxin: Consequences for the effector loop conformations observed by NMR spectroscopy, *Biochemistry*, 42, 11951, 2003.
35. Burland, V., Shao, Y., Perna, N.T., Plunkett, G., Sofia, H.J., and Blattner, F.R., The complete DNA sequence and analysis of the large virulence plasmid of *Escherichia coli* O157:H7, *Nucl. Acids Res.*, 26, 4196, 1998.
36. Stephens, R.S., Kalman, S.L.C., Fan, J., Marathe, R., Aravind, L., Mitchell, W., Olinger, L., Tatusov, R.L., Zhao, Q., Koonin, E.V., and Davis, R.W., Genome sequence of an obligate intracellular pathogen of humans: *Chlamydia trachomatis*, *Science*, 282, 754, 1998.
37. Von Eichel-Streiber, C., Sauerborn, M., and Kuramitsu, H.K., Evidence for a modular structure of the homologous repetitive C-terminal carbohydrate-binding sites of *Clostridium difficile* toxins and *Streptcoccus mutans* glucosyltransferases, *J. Bacteriol.*, 174, 6707, 1992.
38. Sauerborn, M., Leukel, P., and Von Eichel-Streiber, C., The C-terminal ligand-binding domain of *Clostridium difficile* toxin A (TcdA) abrogates TcdA-specific binding to cells and prevents mouse lethality, *FEMS Microbiol. Lett.*, 155, 45, 1997.
39. Frisch, C., Gerhard, R., Aktories, K., Hofmann, F., and Just, I., The complete receptor-binding domain of *Clostridium difficile* toxin A is required for endocytosis, *Biochem. Biophys. Res. Commun.*, 300, 706, 2003.
40. Frey, S.M. and Wilkins, T.D., Localization of two epitopes recognized by monoclonal antibody PCG-4 on *Clostridium difficile* toxin A., *Infect. Immun.*, 60, 2488, 1992.
41. Barroso, L.A., Moncrief, J.S., Lyerly, D.M., and Wilkins, T.D., Mutagenesis of the *Clostridium difficile* toxin B gene and effect on cytotoxic activity, *Microb. Pathog.*, 16, 297, 1994.
42. Krivan, H.C., Clark, G.F., Smith, D.F., and Wilkins, T.D., Cell surface binding site for *Clostridium difficile* enterotoxin: evidence for a glycoconjugate containing the sequence Galα1-3Galβ1-4GlcNAc, *Infect. Immun.*, 53, 573, 1986.
43. Tucker, K.D. and Wilkins, T.D., Toxin A of *Clostridium difficile* binds to the human carbohydrate antigens I, X, and Y., *Infect. Immun.*, 59, 73, 1991.
44. Rolfe, R.D. and Song, W., Purification of a functional receptor for *Clostridium difficile* toxin A from intestinal brush border membranes of infant hamsters, *Clin. Infect. Dis.*, 16, 219, 1993.
45. Rolfe, R.D. and Song, W., Immunoglobulin and non-immunoglobulin components of human milk inhibit *Clostridium difficile* toxin A-receptor binding, *J. Med. Microbiol.*, 42, 10, 1995.
46. Pothoulakis, C., Gilbert, R.J., Cladaras, C., Castagliuolo, I., Semenza, G., Hitti, Y., Montcrief, J.S., Linevsky, J., Kelly, C.P., Nikulasson, S., Desai, H.P., Wilkins, T.D., and LaMont, J.T., Rabbit sucrase-isomaltase contains a functional intestinal receptor for *Clostridium difficile* toxin A, *J. Clin. Invest.*, 98, 641, 1996.
47. Teneberg, S., Lönnroth, I., López, J.F.T., Galili, U., Halvarsson, M.Ö., Ångström, J., and Karlsson, K.A., Molecular mimicry in the recognition of glycosphingolipids by Galα3Galβ4GlcNAcβ-binding *Clostridium difficile* toxin A, human natural anti α-galactosyl IgG and the monoclonal antibody Gal-13: characterization of a binding-active human glycosphingolipid, non-identical with the animal receptor, *Glycobiology*, 6, 599, 1996.

48. Roberts, A.K. and Shone, C.C., Modification of surface histidine residues abolishes the cytotoxic activity of *Clostridium difficile* toxin A, *Toxicon*, 39, 325, 2001.
49. Florin, I. and Thelestam, M., Internalization of *Clostridium difficile* cytotoxin into cultured human lung fibroblasts, *Biochim. Biophys. Acta*, 763, 383, 1983.
50. Henriques, B., Florin, I., and Thelestam, M., Cellular internalisation of *Clostridium difficile* toxin A, *Microb. Pathogen.*, 2, 455, 1987.
51. Barth, H., Pfeifer, G., Hofmann, F., Maier, E., Benz, R., and Aktories, K., Low pH-induced formation of ion channels by *Clostridium difficile* toxin B in target cells, *J. Biol. Chem.*, 276, 10670, 2001.
52. Qa'Dan, M., Spyres, L.M., and Ballard, J.D., pH-induced conformational changes in *Clostridium difficile* toxin B, *Infect. Immun.*, 68, 2470, 2000.
53. Pfeifer, G., Schirmer, J., Leemhuis, J., Busch, C., Meyer, D.K., Aktories, K., and Barth, H., Cellular uptake of *Clostridium difficile* toxin B: translocation of the N-terminal catalytic domain into the cytosol of eukaryotic cells, *J. Biol. Chem.*, 278, 44535, 2003.
54. Green, G.A., Schué, V., and Monteil, H., Cloning and characterization of the cytotoxin L-encoding gene of *Clostridium sordellii*: Homology with *Clostridium difficile* cytotoxin B, *Gene*, 161, 57, 1995.
55. Hatheway, C.L., Toxigenic clostridia, *Clin. Microbiol. Rev.*, 3, 66, 1990.
56. Bette, P., Oksche, A., Mauler, F., Von Eichel-Streiber, C., Popoff, M.R., and Habermann, E., A comparative biochemical, pharmacological and immunological study of *Clostridium novyi* α-toxin, *C. difficile* toxin B and *C. sordellii* lethal toxin, *Toxicon*, 29, 877, 1991.
57. Martinez, R.D. and Wilkins, T.D., Comparison of *Clostridium sordellii* toxins HT and LT with toxins A and B of *C. difficile*, *J. Med. Microbiol.*, 36, 30, 1992.
58. Martinez, R.D. and Wilkins, T.D., Purification and characterization of *Clostridium sodellii* hemorrhagic toxin and cross-reactivity with *Clostridium difficile* toxin A (enterotoxin), *Infect. Immun.*, 56, 1215, 1988.
59. Songer, J.G., Clostridial enteric diseases of domestic animals, *Clin. Microbiol. Rev.*, 9, 216, 1996.
60. Songer, J.G., Clostridial diseases of animals, in *The Clostridia: Molecular Biology and Pathogenesis,* Rood, J.I., McClane, B.A., Songer, J.G., and Titball, R.W., Eds., Academic Press, San Diego, 1997, 153.
61. Hofmann, F., Herrmann, A., Habermann, E., and Von Eichel-Streiber, C., Sequencing and analysis of the gene encoding the α-toxin of *Clostridium novyi* proves its homology to toxins A and B of *Clostridium difficile*, *Mol. Gen. Genet.*, 247, 670, 1995.
62. Ball, D.W., Van Tassell, R.L., Denton Roberts, M., Hahn, P.E., Lyerly, D.M., and Wilkins, T.D., Purification and characterization of alpha-toxin produced by *Clostridium novyi* type A, *Infect. Immun.*, 61, 2912, 1993.
63. Bette, P., Frevert, J., Mauler, F., Suttorp, N., and Habermann, E., Pharmacological and biochemical studies of cytotoxity of *Clostridium novyi* type A alpha-toxin, *Infect. Immun.*, 57, 2507, 1989.
64. Oksche, A., Nakov, R., and Habermann, E., Morphological and biochemical study of cytoskeletal changes in cultured cells after extracellular application of *Clostridium novyi* alpha-toxin, *Infect. Immun.*, 60, 3002, 1992.
65. Rupnik, M., Avesani, V., Janc, M., Von Eichel-Streiber, C., and Delmée, M., A novel toxinotyping scheme and correlation of toxinotypes with serogroups of *Clostridium difficile* isolates, *J. Clin. Microbiol.*, 36, 2240, 1998.
66. Rupnik, M., Braun, V., Soehn, F., Janc, M., Hofstetter, M., Laufenberg-Feldmann, R., and Von Eichel-Streiber, C., Characterization of polymorphisms in the toxin A and B genes of *Clostridium difficile*, *FEMS Microbiol. Lett.*, 148, 197, 1997.
67. Chaves-Olarte, E., Löw, P., Freer, E., Norlin, T., Weidmann, M., Von Eichel-Streiber, C., and Thelestam, M., A novel cytotoxin from *Clostridium difficile* serogroup F is a functional hybrid between two other large clostridial cytotoxins, *J. Biol. Chem.*, 274, 11046, 1999.
68. Soehn, F., Wagenknecht-Wiesner, A., Leukel, P., Kohl, M., Weidmann, M., Von Eichel-Streiber, C., and Braun, V., Genetic rearrangements in the pathogenicity locus of *Clostridium difficile* strain 8864: implications for transcription, expression and enzymatic activity of toxins A and B, *Molecular & General Genetics*, 258, 222, 1998.
69. Mehlig, M., Moos, M., Braun, V., Kalt, B., Mahony, D.E., and Von Eichel-Streiber, C., Variant toxin B and a functional toxin A produced by *Clostridium difficile* C34., *FEMS Microbiol. Lett.*, 198, 171, 2001.

70. Etienne-Manneville, S. and Hall, A., Rho GTPases in cell biology, *Nature*, 420, 629, 2002.
71. Bishop, A.L. and Hall, A., Rho GTPases and their effector proteins, *Biochem. J.*, 348, 241, 2000.
72. Takai, Y., Sasaki, T., and Matozaki, T., Small GTP-binding proteins, *Physiol. Rev.*, 81, 153, 2001.
73. Wittinghofer, A., Signal transduction via Ras, *Biol. Chem.*, 379, 933, 1998.
74. Van Aelst, L. and D'Souza-Schorey, C., Rho GTPases and signaling networks, *Genes & Development*, 11, 2295, 1997.
75. Mackay, D.J.G. and Hall, A., Rho GTPases, *J. Biol. Chem.*, 273, 20685, 1998.
76. Ridley, A., Rho, in *GTPases*, Hall, A., Ed., Oxford University Press, Oxford, 2000, 89.
77. Ridley, A.J. and Hall, A., The small GTP-binding protein rho regulates the assembly of focal adhesions and actin stress fibers in response to growth factors, *Cell*, 70, 389, 1992.
78. Ridley, A.J., Paterson, H.F., Johnston, C.L., Diekmann, D., and Hall, A., The small GTP-binding protein rac regulates growth factor-induced membrane ruffling, *Cell*, 70, 401, 1992.
79. Kozma, R., Ahmed, S., Best, A., and Lim, L., The Ras-related protein Cdc42Hs and bradykinin promote formation of peripheral actin microspikes and filopodia in Swiss 3T3 fibroblasts, *Mol. Cell. Biol.*, 15, 1942, 1995.
80. Ihara, K., Muraguchi, S., Kato, M., Shimizu, T., Shirakawa, M., Kuroda, S., Kaibuchi, K., and Hakoshima, T., Crystal structure of human RhoA in a dominantly active form complexed with a GTP analogue, *J. Biol. Chem.*, 273, 9656, 1998.
81. Herrmann, C., Ahmadian, M.R., Hofmann, F., and Just, I., Functional consequences of monoglucosylation of H-Ras at effector domain amino acid threonine-35, *J. Biol. Chem.*, 273, 16134, 1998.
82. Sehr, P., Joseph, G., Genth, H., Just, I., Pick, E., and Aktories, K., Glucosylation and ADP-ribosylation of Rho proteins — effects on nucleotide binding, GTPase activity, and effector-coupling, *Biochemistry*, 37, 5296, 1998.
83. Genth, H., Aktories, K., and Just, I., Monoglucosylation of RhoA at Threonine-37 blocks cytosol-membrane cycling, *J. Biol. Chem.*, 274, 29050, 1999.
84. Geyer, M., Schweins, T., Herrmann, C., Prisner, T., Wittinghofer, A., and Kalbitzer, H.R., Conformational transitions in $p21^{ras}$ and in its complexes with the effector protein Raf-RBD and the GTPase activating protein GAP, *Biochemistry*, 35, 10308, 1996.
85. Ottlinger, M.E. and Lin, S., *Clostridium difficile* toxin B induces reorganization of actin, vinculin, and talin in cultures cells, *Exp. Cell Res.*, 174, 215, 1988.
86. Fiorentini, C., Arancia, G., Paradisi, S., Donelli, G., Giuliano, M., Piemonto, F., and Mastrantonio, P., Effects of *Clostridium difficile* toxins A and B on cytoskeleton organization in HEp-2 cells: a comparative morphological study, *Toxicon*, 27, 1209, 1989.
87. Malorni, W., Fiorentini, C., Paradisi, S., Giuliano, M., Mastrantonio, P., and Donelli, G., Surface blebbing and cytoskeletal changes induced *in vitro* by toxin B from, *Exp. Mol. Pathol.*, 52, 340, 1990.
88. Thelestam, M. and Chaves-Olarte, E., Cytotoxic effects of the *Clostridium difficile* toxins, *Curr. Top. Microbiol. Immunol.*, 250, 85–96, 2000.
89. Prepens, U., Just, I., Von Eichel-Streiber, C., and Aktories, K., Inhibition of FcεRI-mediated activation of rat basophilic leukemia cells by *Clostridium difficile* toxin B (monoglucosyltransferase), *J. Biol. Chem.*, 271, 7324, 1996.
90. Schmidt, M., Rümenapp, U., Bienek, C., Keller, J., Von Eichel-Streiber, C., and Jakobs, K.H., Inhibition of receptor signaling to phospholipase D by *Clostridium difficile* toxin B — role of Rho proteins, *J. Biol. Chem.*, 271, 2422, 1996.
91. Djouder, N., Prepens, U., Aktories, K., and Cavalié, A., Inhibition of calcium release-activated calcium current by Rac/Cdc42-inactivating clostridial cytotoxins in RBL cells, *J. Biol. Chem.*, 275, 18732, 2000.
92. Linseman, D.A., Hofmann, F., and Fisher, S.K., A role for the small molecular weight GTPases, Rho and Cdc42, in muscarinic receptor signaling to focal adhesion kinase, *J. Neurochem.*, 74, 2010, 2000.
93. Doussau, F., Gasman, S., Humeau, Y., Vitiello, F., Popoff, M., Boquet, P., Bader, M.-F., and Poulain, B., A Rho-related GTPase is involved in Ca^{2+}-dependent neurotransmitter exocytosis, *J. Biol. Chem.*, 275, 7764, 2000.
94. Subauste, M.C., Von Herrath, M., Benard, V., Chamberlain, C.E., Chuang, T.H., Chu, K., Bokoch, G.M., and Hahn, K.M., Rho family proteins modulate rapid apoptosis induced by cytotoxic T lymphocytes and Fas., *J. Biol. Chem.*, 275, 9725, 2000.

95. Servant, G., Weiner, O.D., Herzmark, P., Balla, T., Sedat, J.W., and Bourne, H.R., Polarization of chemoattractant receptor signaling during neutrophil chemotaxis, *Science*, 287, 1037, 2000.
96. Bradke, F. and Dotto, G.P., The role of local actin instability in axon formation, *Science*, 283, 1931, 1999.
97. Caron, E. and Hall, A., Identification of two distinct mechanisms of phagocytosis controlled by different Rho GTPases, *Science*, 282, 1717, 1998.
98. Nusrat, A., Giry, M., Turner, J.R., Colgan, S.P., Parkos, C.A., Carnes, D., Lemichez, E., Boquet, P., and Madara, J.L., Rho protein regulates tight junctions and perijunctional actin organization in polarized epithelia, *Proc. Natl. Acad. Sci. U.S.A.*, 92, 10629, 1995.
99. Nusrat, A., Von Eichel-Streiber, C., Turner, J.R., Verkade, P., Madara, J.L., and Parkos, C.A., *Clostridium difficile* toxins disrupt epithelial barrier function by altering membrane microdomain localization of tight junction proteins, *Infect. Immun.*, 69, 1329, 2001.
100. Liu, T.S., Musch, M.W., Sugi, K., Walsh-Reitz, M.M., Ropeleski, M.J., Hendrickson, B.A., Pothoulakis, C., LaMont, J.T., and Chang, E.B., Protective role of HSP72 against *Clostridium difficile* toxin A-induced intestinal epithelial cell dysfunction, *Am. J. Physiol. Cell Physiol.*, 284, C1073, 2003.
101. Moore, R., Pothoulakis, C., LaMont, J.T., Carlson, S., and Madara, J.L., C. difficile toxin A increases intestinal permeability and induces Cl-, *Am. J. Physiol.*, 259, G165, 1990.
102. Feltis, B.A., Wiesner, S.M., Kim, A.S., Erlandsen, S.L., Lyerly, D.L., Wilkins, T.D., and Wells, C.L., *Clostridium difficile* toxins A and B can alter epithelial permeability and promote bacterial paracellular migration through HT-29 enterocytes, *Shock*, 14, 629, 2000.
103. Gerhard, R., Schmidt, G., Hofmann, F., and Aktories, K., Activation of Rho GTPases by *Escherichia coli* cytotoxic necrotizing factor 1 increases intestinal permeability in Caco-2 cells, *Infect. Immun.*, 66, 5125, 1998.
104. Hecht, G., Pothoulakis, C., LaMont, J.T., and Madara, J.L., *Clostridium difficile* toxin A perturbs cytoskeletal structure and tight junction permeability of cultured human intestinal epithelial monolayers, *J. Clin. Invest.*, 82, 1516, 1988.
105. Hecht, G., Koutsouris, A., Pothoulakis, C., LaMont, J.T., and Madara, J.L., *Clostridium difficile* toxin B disrupts the barrier function of T_{84} monolayers, *Gastroenterology*, 102, 416, 1992.
106. Jou, T.-S., Schneeberger, E.E., and Nelson, W.J., Structural and functional regulation of tight junctions by RhoA and Rac1 small GTPases, *J. Cell Biol.*, 142, 101, 1998.
107. Pothoulakis, C., Sullivan, R., Melnick, D.A., Triadafilopoulos, G., Gadenne, A.-S., Meshulam, T., and LaMont, J.T., *Clostridium difficile* toxin A stimulates intracellular calcium release and chemotactic response in human granulocytes, *J. Clin. Invest.*, 81, 1741, 1988.
108. Mahida, Y.R., Makh, S., Hyde, S., Gray, T., and Borriello, S.P., Effect of *Clostridium difficile* toxin A on human intestinal epithelial cells: induction of interleukin 8 production and apoptosis after cell detachment, *Gut*, 38, 337, 1996.
109. Warny, M., Keates, A.C., Keates, S., Castagliuolo, I., Zacks, J.K., Aboudola, S., Qamar, A., Pothoulakis, C., LaMont, J.T., and Kelly, C.P., p^{38} MAP kinase activation by *Clostridium difficile* toxin A mediates monocyte necrosis, IL-8 production, and enteritis, *J. Clin. Invest.*, 105, 1147, 2000.
110. Jefferson, K.K., Smith, M.F. Jr., and Bobak, D.A., Roles of intracellular calcium and NF-κB in the *Clostridium difficile* toxin A-induced up-regulation and secretion of IL-8 from human monocytes, *J. Immunol.*, 163, 5183, 1999.
111. He, D., Sougioultzis, S., Hagen, S., Liu, J., Keates, S., Keates, A.C., Pothoulakis, C., and LaMont, J.T., *Clostridium difficile* toxin triggers human colonocyte IL-8 release via mitochondrial oxygen radical generation, *Gastroenterology*, 122, 1048, 2002.
112. Ishida, Y., Maegawa, T., Kondo, T., Kimura, A., Iwakura, Y., Nakamura, S., and Mukaida, N., Essential involvement of IFN-gamma in *Clostridium difficile* toxin A-induced enteritis, *J. Immunol.*, 172, 3018, 2004.
113. Pothoulakis, C. and LaMont, J.T., Microbes and microbial toxins: paradigms for microbial-mucosal interactions. II. The integrated response of the intestine to *Clostridium difficile* toxins, *Am. J. Physiol. Gastrointest. Liver Physiol.*, 280, G178, 2001.
114. Pothoulakis, C., Castagliuolo, I., LaMont, J.T., Jaffer, A., O'Keane, J.C., Snider, R.M., and Leeman, S.E., CP-96,345, a substance P antagonist, inhibits rat intestinal responses to *Clostridium difficile* toxin A but not cholera toxin, *Proc. Natl. Acad. of Sci. U.S.A.*, 91, 947, 1994.

115. Wershil, B.K., Castagliuolo, I., and Pothoulakis, C., Direct evidence of mast cell involvement in *Clostridium difficile* toxin A-induced enteritis in mice, *Gastroenterology*, 114, 956, 1998.
116. Chen, M.L., Pothoulakis, C., and LaMont, J.T., Protein kinase C signaling regulates ZO-1 translocation and increased paracellular flux of T84 coloncytes exposed to *Clostridium difficile* toxin A., *J. Biol. Chem.*, 277, 4247, 2002.
117. Aktories, K., Weller, U., and Chhatwal, G.S., Clostridium botulinum type C produces a novel ADP-ribosyltransferase distinct from botulinum C2 toxin, *FEBS Lett.*, 212, 109, 1987.
118. Hauser, D., Gibert, M., Eklund, M.W., Boquet, P., and Popoff, M.R., Comparative analysis of C3 and botulinal neurotoxin genes and their environment in *Clostridium botulinum* types C and D, *J. Bacteriol.*, 175, 7260, 1993.
119. Popoff, M.R., Hauser, D., Boquet, P., Eklund, M.W., and Gill, D.M., Characterization of the C3 gene of *Clostridium botulinum* types C and D and its expression in *Escherichia coli*, *Infect. Immun.*, 59, 3673, 1991.
120. Nemoto, Y., Namba, T., Kozaki, S., and Narumiya, S., *Clostridium botulinum* C3 ADP-ribosyltransferase gene, *J. Biol. Chem.*, 266, 19312, 1991.
121. Moriishi, K., Syuto, B., Saito, M., Oguma, K., Fujii, N., Abe, N., and Naiki, M., Two different types of ADP-ribosyltransferase C3 from *Clostridium botulinum* type D lysogenized organisms, *Infect. Immun.*, 61, 5309, 1993.
122. Just, I., Mohr, C., Schallehn, G., Menard, L., Didsbury, J.R., Vandekerckhove, J., van Damme, J., and Aktories, K., Purification and characterization of an ADP-ribosyltransferase produced by *Clostridium limosum*, *J. Biol. Chem.*, 267, 10274, 1992.
123. Inoue, S., Sugai, M., Murooka, Y., Paik, S.-Y., Hong, Y.-M., Ohgai, H., and Suginaka, H., Molecular cloning and sequencing of the epidermal cell differentiation inhibitor gene from *Staphylococcus aureus*, *Biochem. Biophys. Res. Commun.*, 174, 459, 1991.
124. Sugai, M., Enomoto, T., Hashimoto, K., Matsumoto, K., Matsuo, Y., Ohgai, H., Hong, Y.-M., Inoue, S., Yoshikawa, K., and Suginaka, H., A novel epidermal cell differentiation inhibitor (EDIN): Purification and characterization from *Staphylococcus aureus*, *Biochem. Biophys. Res. Commun.*, 173, 92, 1990.
125. Wilde, C., Chhatwal, G.S., Schmalzing, G., Aktories, K., and Just, I., A novel C3-like ADP-ribosyltransferase from *Staphylococcus aureus* modifying RhoE and Rnd3, *J. Biol. Chem.*, 276, 9537, 2001.
126. Yamaguchi, T., Hayashi, T., Takami, H., Ohnishi, M., Murata, T., Nakayama, K., Asakawa, K., Ohara, M., Komatsuzawa, H., and Sugai, M., Complete nucleotide sequence of a *Staphylococcus aureus* exfoliative toxin B plasmid and identification of a novel ADP-ribsyltransferase, EDIN-C., *Infect. Immun.*, 69, 7760, 2001.
127. Just, I., Selzer, J., Jung, M., van Damme, J., Vandekerckhove, J., and Aktories, K., Rho-ADP-ribosylating exoenzyme from *Bacillus cereus* — purification, characterization and identification of the NAD-binding site, *Biochemistry*, 34, 334, 1995.
128. Aktories, K., Rösener, S., Blaschke, U., and Chhatwal, G.S., Botulinum ADP-ribosyltransferase C3. Purification of the enzyme and characterization of the ADP-ribosylation reaction in platelet membranes, *Eur. J. Biochem.*, 172, 445, 1988.
129. Moriishi, K., Syuto, B., Yokosawa, N., Oguma, K., and Saito, M., Purification and characterization of ADP-ribosyltransferases (Exoenzyme C3) of *Clostridium botulinum* type C and D strains, *J. Bacteriol.*, 173, 6025, 1991.
130. Han, S., Arvai, A.S., Clancy, S.B., and Tainer, J.A., Crystal structure and novel recognition motif of Rho ADP-ribosylating C3 exoenzyme from *Clostridium botulinum*: structural insights for recognition specificity and catalysis, *J. Mol. Biol.*, 305, 95, 2001.
131. Ménétrey, J., Flatau, G., Stura, E.A., Charbonnier, J.-B., Gas, F., Teulon, J.-M., Le Du, M.-H., Boquet, P., and Ménez, A., NAD binding induces conformational changes in Rho ADP-ribosylating *Clostridium botulinum* C3 exoenzyme, *J. Biol. Chem.*, 277, 30950, 2002.
132. Evans, H.R., Sutton, J.M., Holloway, D.E., Ayriss, J., Shone, C.C., and Acharya, K.R., The crystal structure of C3stau2 from *Staphylococcus aureus* and its complex with NAD, *J. Biol. Chem.*, 278, 45924, 2003.
133. Bell, C.E. and Eisenberg, D., Crystal structure of diphtheria toxin bound to nicotinamide adenine dinucleotide, *Biochemistry*, 35, 1137, 1996.

134. Choe, S., Bennett, M.J., Fujii, G., Curmi, P.M.G., Kantardjieff, K.A., Collier, R.J., and Eisenberg, D., The crystal structure of diphtheria toxin, *Nature*, 357, 216, 1992.
135. Sixma, T.K., Pronk, S.E., Kalk, K.H., Wartna, E.S., van Zanten, B.A.M., Witholt, B., and Hol, W.G.J., Crystal structure of a cholera toxin-related heat-labile enterotoxin from *E. coli*, *Nature*, 351, 371, 1991.
136. Allured, V.S., Collier, R.J., Carroll, S.F., and McKay, D.B., Structure of exotoxin A of *Pseudomonas aeruginosa* at 3,0-Angström resolution, *Proc. Natl. Acad. Sci. U.S.A.*, 83, 1320, 1986.
137. Stein, P.E., Boodhoo, A., Armstrong, G.D., Cockle, S.A., Klein, M.H., and Read, R.J., The crystal structure of pertussis toxin, *Structure*, 2, 45, 1994.
138. Han, S., Craig, J.A., Putnam, C.D., Carozzi, N.B., and Tainer, J.A., Evolution and mechanism from structures of an ADP-ribosylating toxin and NAD complex, *Nature Struct. Biol.*, 6, 932, 1999.
139. Böhmer, J., Jung, M., Sehr, P., Fritz, G., Popoff, M., Just, I., and Aktories, K., Active site mutation of the C3-like ADP-ribosyltransferase from *Clostridium limosum* — Analysis of glutamic acid 174, *Biochemistry*, 35, 282, 1996.
140. Saito, Y., Nemoto, Y., Ishizaki, T., Watanabe, N., Morii, N., and Narumiya, S., Identification of Glu^{173} as the critical amino acid residue for the ADP-ribosyltransferase activity of *Clostridium botulinum* C3 exoenzyme, *FEBS Lett.*, 371, 105, 1995.
141. Domenighini, M. and Rappuoli, R., Three conserved consensus sequences identify the NAD-binding site of ADP-ribosylating enzymes, expressed by eukaryotes, bacteria and T-even bacteriophages, *Molecular Microbiology*, 21, 667, 1996.
142. Barth, H., Preiss, J.C., Hofmann, F., and Aktories, K., Characterization of the catalytic site of the ADP-ribosyltransferase *Clostridium botulinum* C2 toxin by site-directed mutagenesis, *J. Biol. Chem.*, 273, 29506, 1998.
143. Perelle, S., Domenighini, M., and Popoff, M.R., Evidence that Arg-295, Glu-378, and Glu-380 are active-site residues of the ADP-ribosyltransferase activity of iota toxin, *FEBS Lett.*, 395, 191, 1996.
144. Wilde, C., Just, I., and Aktories, K., Structure-function analysis of the Rho-ADP-ribosylating exoenzyme C3stau2 from *Staphylococcus aureus*, *Biochemistry*, 41, 1539, 2002.
145. Sekine, A., Fujiwara, M., and Narumiya, S., Asparagine residue in the rho gene product is the modification site for botulinum ADP-ribosyltransferase, *J. Biol. Chem.*, 264, 8602, 1989.
146. Wei, Y., Zhang, Y., Derewenda, U., Liu, X., Minor, W., Nakamoto, R.K., Somlyo, A.V., Somlyo, A.P., and Derewenda, Z.S., Crystal structure of RhoA-GDP and its functional implications, *Nature Struct. Biol.*, 4, 699, 1997.
147. Braun, U., Habermann, B., Just, I., Aktories, K., and Vandekerckhove, J., Purification of the 22 kDa protein substrate of botulinum ADP-ribosyltransferase C3 from porcine brain cytosol and its characterization as a GTP-binding protein highly homologous to the rho gene product, *FEBS Lett.*, 243, 70, 1989.
148. Aktories, K., Braun, U., Rösener, S., Just, I., and Hall, A., The rho gene product expressed in *E. coli* is a substrate of botulinum ADP-ribosyltransferase C3, *Biochem. Biophys. Res. Commun.*, 158, 209, 1989.
149. Chardin, P., Boquet, P., Madaule, P., Popoff, M.R., Rubin, E.J., and Gill, D.M., The mammalian G protein rho C is ADP- ribosylated by *Clostridium botulinum* exoenzyme C3 and affects actin microfilament in Vero cells, *EMBO J.*, 8, 1087, 1989.
150. Paterson, H.F., Self, A.J., Garrett, M.D., Just, I., Aktories, K., and Hall, A., Microinjection of recombinant $p21^{rho}$ induces rapid changes in cell morphology, *J. Cell Biol.*, 111, 1001, 1990.
151. Wiegers, W., Just, I., Müller, H., Hellwig, A., Traub, P., and Aktories, K., Alteration of the cytoskeleton of mammalian cells cultured *in vitro* by *Clostridium botulinum* C2 toxin and C3 ADP-ribosyltransferase, *Eur. J. Cell Biol.*, 54, 237, 1991.
152. Ren, X.-D., Bokoch, G.M., Traynor-Kaplan, A., Jenkins, G.H., Anderson, R.A., and Schwartz, M.A., Physical association of the small GTPase Rho with a 68-kDa phosphatidylinositol 4-phosphate 5-kinase in swiss 3T3 cells, *Mol. Biol. Cell*, 7, 435, 1996.
153. Meacci, E., Vasta, V., Moorman, J.P., Bobak, D.A., Bruni, P., Moss, J., and Vaughan, M., Effect of Rho and ADP-ribosylation factor GTPases on phospholipase D activity in intact human adenocarcinoma A549 cells, *J. Biol. Chem.*, 274, 18605, 1999.
154. Malcolm, K.C., Elliott, C.M., and Exton, J.H., Evidence for Rho-mediated agonist stimulation of phospholipase D in Rat1 fibroblasts — effects of *Clostridium botulinum* C3 exoenzyme, *J. Biol. Chem.*, 271, 13135, 1996.

155. Fujihara, H., Walker, L.A., Gong, M.C., Lemichez, E., Boquet, P., Somlyo, A.V., and Somlyo, A.P., Inhibition of RhoA translocation and calcium sensitization by *in vivo* ADP-ribosylation with the chimeric toxin DC3B, *Mol. Biol. Cell*, 8, 2437, 1997.
156. Genth, H., Gerhard, R., Maeda, A., Amano, M., Kaibuchi, K., Aktories, K., and Just, I., Entrapment of Rho ADP-ribosylated by *Clostridium botulinum* C3 exoenzyme in the Rho-GDI-1 complex., *J. Biol. Chem.*, in press, 2003.
157. Morii, N. and Narumiya, S., Preparation of native and recombinant *Clostridium botulinum* C3 ADP-ribosyltransferase and identification of Rho proteins by ADP-ribosylation, in *Methods in Enzymology,* Balch, W.E., Der, C.J., and Hall, A., Eds., Academic Press, San Diego, 1995, 196.
158. Verschueren, H., De Baetselier, P., De Braekeleer, J., Dewit, J., Aktories, K., and Just, I., ADP-ribosylation of Rho-proteins with botulinum C3 exoenzyme inhibits invasion and shape changes of T-lymphoma cells, *Eur. J. Cell Biol.*, 73, 182, 1997.
159. Amano, M., Mukai, H., Ono, Y., Chihara, K., Matsui, T., Hamajima, Y., Okawa, K., Iwamatsu, A., and Kaibuchi, K., Identification of a putative target for Rho as the serine-threonine kinase protein kinase N, *Science*, 271, 648, 1996.
160. Olson, M.F., Paterson, H.F., and Marshall, C.J., Signals from Ras and Rho GTPases interact to regulate expression of $p21^{Waf1/Cip1}$, *Nature*, 394, 295, 1998.
161. Ridley, A.J. and Hall, A., Signal transduction pathways regulating Rho-mediated stress fibre formation: Requirement for a tyrosine kinase, *EMBO J.*, 13, 2600, 1994.
162. Chong, L.D., Traynor-Kaplan, A., Bokoch, G.M., and Schwartz, M.A., The small GTP-binding protein Rho regulates a phosphatidylinositol 4-phosphate 5-kinase in mammalian cells, *Cell*, 79, 507, 1994.
163. Watanabe, N., Madaule, P., Reid, T., Ishizaki, T., Watanabe, G., Kakizuka, A., Saito, Y., Nakao, K., Jockusch, B.M., and Naumiya, S., p140mDia, a mammalian homolog of *Drosophila* diaphanous, is a target protein fro Rho small GTPase and is a ligand for profilin, *EMBO J.*, 16, 3044, 1997.
164. Mackay, D.J.G., Esch, F., Furthmayr, H., and Hall, A., Rho- and Rac-dependent assembly of focal adhesion complexes and actin filaments in permeabilized fibroblasts: an essential role for Ezrin/radixin/Moesin proteins, *J. Cell Biol.*, 138, 927, 1997.
165. Fensome, A., Whatmore, J., Morgan, C., Jones, D., and Cockcroft, S., ADP-ribosylation factor and Rho proteins mediate fMLP-dependent activation of phospholipase D in human neutrophils, *J. Biol. Chem.*, 273, 13157, 1998.
166. Stasia, M.-J., Jouan, A., Bourmeyster, N., Boquet, P., and Vignais, P.V., ADP-ribosylation of a small size GTP-binding protein in bovine neutrophils by the C3 exoenzyme of *Clostridium botulinum* and effect on the cell motility, *Biochem. Biophys. Res. Commun.*, 180, 615, 1991.
167. Koch, G., Norgauer, J., and Aktories, K., ADP-ribosylation of Rho by *Clostridium limosum* exoenzyme affects basal but not N-formyl-peptide-stimulated actin polymerization in human myeloid leukaemic (HL60) cells, *Biochem. J.*, 299, 775, 1994.
168. Barry, S.T., Flinn, H.M., Humphries, M.J., Critchley, D.R., and Ridley, A.J., Requirement for Rho in integrin signalling, *Cell Adhes. Commun.*, 4, 387, 1997.
169. Fujisawa, K., Madaule, P., Ishizaki, T., Watanabe, G., Bito, H., Saito, Y., Hall, A., and Narumiya, S., Different regions of Rho determine Rho-selective binding of different classes of Rho target molecules, *J. Biol. Chem.*, 273, 18943, 1998.
170. Hill, C.S., Wynne, J., and Treisman, R., The Rho family GTPases RhoA, Rac1, and CDC42Hs regulate transcriptional activation by SRF, *Cell*, 81, 1159, 1995.
171. Henning, S.W., Galandrini, R., Hall, A., and Cantrell, D.A., The GTPase Rho has a critical regulatory role in thymus development, *EMBO J.*, 16, 2397, 1997.
172. Genot, E., Cleverley, S., Henning, S., and Cantrell, D., Multiple p21ras effector pathways regulate nuclear factor of activated T cells, *EMBO J.*, 15, 3923, 1996.
173. Aullo, P., Giry, M., Olsnes, S., Popoff, M.R., Kocks, C., and Boquet, P., A chimeric toxin to study the role of the 21 kDa GTP binding protein rho in the control of actin microfilament assembly, *EMBO J.*, 12, 921, 1993.
174. Barth, H., Hofmann, F., Olenik, C., Just, I., and Aktories, K., The N-terminal part of the enzyme component (C2I) of the binary *Clostridium botulinum* C2 toxin interacts with the binding component C2II and functions as a carrier system for a Rho ADP-ribosylating C3-like fusion toxin, *Infect. Immun.*, 66, 1364, 1998.

175. Barth, H., Olenik, C., Sehr, P., Schmidt, G., Aktories, K., and Meyer, D.K., Neosynthesis and activation of Rho by *Escherichia coli* cytotoxic necrotizing factor (CNF1) reverse cytopathic effects of ADP-ribosylated Rho, *J. Biol. Chem.*, 274, 27407, 1999.
176. Hall, A., Small GTP-binding proteins and the regulation of the actin cytoskeleton, *Annu. Rev. Cell Biol.*, 10, 31, 1994.
177. Nobes, C.D. and Hall, A., Rho, Rac, and Cdc42 GTPases regulate the assembly of multimolecular focal complexes associated with actin stress fibers, lamellipodia, and filopodia, *Cell*, 81, 53, 1995.
178. Mackay, D.J.G., Nobes, C.D., and Hall, A., The Rho's progress: a potential role during neuritogenesis for the Rho family of GTPases, *Trends Neurosci.*, 18, 496, 1995.
179. Jalink, K., Van Corven, E.J., Hengeveld, T., Morii, N., Narumiya, S., and Moolenaar, W.H., Inhibition of lysophosphatidate- and thrombin-induced neurite retraction and neuronal cell rounding by ADP ribosylation of the small GTP-binding protein Rho, *J. Cell Biol.*, 126, 801, 1994.
180. Tigyi, G., Fischer, D.J., Sebök, A., Marshall, F., Dyer, D.L., and Miledi, R., Lysophosphatidic acid-induced neurite retraction in PC12 cells: neurite-protective effects of cyclic AMP signaling, *J. Neurochem.*, 66, 549, 1996.
181. Wahl, S., Barth, H., Ciossek, T., Aktories, K., and Mueller, B.K., Ephrin-A5 induces collapse of growth cones by activating Rho and Rho kinase, *J. Cell Biol.*, 149, 263, 2000.
182. Lamaze, C., Chuang, T.H., Terlecky, L.J., Bokoch, G.M., and Schmid, S.L., Regulation of receptor-mediated endocytosis by Rho and Rac, *Nature*, 382, 177, 1996.
183. Vögler, O., Krummenerl, P., Schmidt, M., Jakobs, K.H., and van Koppen, C.J., RhoA-sensitive trafficking of muscarinic acetylcholine receptors, *J. Pharmacol. Exp. Ther.*, 288, 36, 1999.
184. Schmalzing, G., Richter, H.P., Hansen, A., Schwarz, W., Just, I., and Aktories, K., Involvement of the GTP binding protein Rho in constitutive endocytosis in *Xenopus laevis* oocytes, *J. Cell Biol.*, 130, 1319, 1995.
185. Weernick, P.A.O., Schulte, P., Guo, Y., Wetzel, J., Amano, M., Kaibuchi, K., Haverland, S., Voβ, M., Schmidt, M., Mayr, G.W., and Jakobs, K.H., Stimulation of phosphatidylinositol-4-phosphate 5-kinase by Rho-kinase, *J. Biol. Chem.*, 275, 10168, 2000.
186. Schmidt, M., Voss, M., Weernink, P.A., Wetzel, J., Amano, M., Kaibuchi, K., and Jakobs, K.H., A role for Rho-kinase in Rho-controlled phospholipase D stimulation by the m3 muscarinic acetylcholine receptor, *J. Biol. Chem.*, 274, 14648, 1999.
187. Kuribara, H., Tago, K., Yokozeki, T., Sasaki, T., Takai, Y., Morii, N., Narumiya, S., Katada, T., and Kanaho, Y., Synergistic activation of rat brain phospholipase D by ADP-ribosylation factor and *rho*A p21, and its inhibition by *Clostridium botulinum* C3 exoenzyme, *J. Biol. Chem.*, 270, 25667, 1995.
188. Balboa, M.A. and Insel, P.A., Nuclear phospholipase D in Madin-Darby canine kidney cells — Guanosine 5′-O-(thiotriphosphate)-stimulated activation is mediated by RhoA and is downstream of protein kinase C, *J. Biol. Chem.*, 270, 29843, 1995.
189. Alberts, A.S., Geneste, O., and Treisman, R., Activation of SRF-regulated chromosomal templates by Rho-family GTPases requires a signal that also induces H4 hyperacetylation, *Cell*, 92, 475, 1998.
190. Sah, V.P., Hoshijima, M., Chien, K.R., and Brown, J.H., Rho is required for $G\alpha_q$ and α_1-adrenergic receptor signaling in cardiomyocytes — dissociation of Ras and Rho pathways, *J. Biol. Chem.*, 271, 31185, 1996.
191. Fromm, C., Coso, O.A., Montaner, S., Xu, N., and Gutkind, J.S., The small GTP-binding protein Rho links G protein-coupled receptors and $G\alpha_{12}$ to the serum response element and to cellular transformation, *Proc. Natl. Acad. Sci. U.S.A.*, 94, 10098, 1997.
192. Mao, J., Yuan, H., Xie, W., and Wu, D., Guanine nucleotide exchange factor GEF115 specifically mediates activation of Rho and serum response factor by the G protein α subunit Gα13, *Proc. Natl. Acad. Sci. U.S.A.*, 95, 12973, 1998.
193. Mao, J., Yuan, H., Xie, W., Simon, M.I., and Wu, D., Specific involvement of G proteins in regulation of serum response factor-mediated gene transcription by different receptors, *J. Biol. Chem.*, 273, 27118, 1998.
194. Lang, P., Guizani, L., Vitté-Mony, I., Stancou, R., Dorseuil, O., Gacon, G., and Bertoglio, J., ADP-ribosylation of the *ras*-related, GTP-binding protein RhoA inhibits lymphocyte-mediated cytotoxicity, *J. Biol. Chem.*, 267, 11677, 1992.
195. Laudanna, C., Campbell, J.J., and Butcher, E.C., Role of Rho in chemoattractant-activated leukocyte adhesion through integrins, *Science*, 271, 981, 1996.

196. Nemoto, E., Yu, Y.J., and Dennert, G., Cell surface ADP-Ribosyltransferase regulates lymphocyte function-associated molecule-1 (LFA-1) function in T cells, *J. Immunol.*, 157, 3341, 1996.
197. Moss, J., Zolkiewska, A., and Okazaki, I., ADP-ribosylarginine hydrolases and ADP-ribosyltransferases — partners in ADP-ribosylation cycles, *Adv. Exp. Med. Biol.*, 419, 25, 1997.
198. Laudanna, C., Campbell, J.J., and Butcher, E.C., Elevation of intracellular cAMP inhibits RhoA activation and integrin-dependent leukocyte adhesion induced by chemoattractants, *J. Biol. Chem.*, 272, 24141, 1997.
199. Stam, J.C., Michiels, F., Van der Kammen, R.A., Moolenaar, W.H., and Collard, J.G., Invasion of T-lymphoma cells: cooperation between Rho family GTPases and lysophospholipid receptor signaling, *EMBO J.*, 17, 4066, 1998.
200. Wojciak-Stothard, B., Entwistle, A., Garg, R., and Ridley, A.J., Regulation of TNF-α-induced reorganization of the actin cytoskeleton and cell-cell junctions by Rho, Rac, and Cdc42 in human endothelial cells, *J. Cell. Physiol.*, 176, 150, 1998.
201. Reif, K. and Cantrell, D.A., Networking Rho family GTPases in lymphocytes, *Immunity*, 8, 395, 1998.
202. Henning, S.W. and Cantrell, D.A., GTPases in antigen receptor signalling, *Curr. Opin. Immunol.*, 10, 322, 1998.
203. Vouret-Craviari, V., Boquet, P., Pouysségur, J., and Van Obberghen-Schilling, E., Regulation of the actin cytoskeleton by thrombin in human endothelial cells: role of Rho proteins in endothelial barrier function, *Mol. Biol. Cell*, 9, 2639, 1998.
204. Santos, M.F., McCormack, S.A., Guo, Z., Okolicany, J., Zheng, Y., Johnson, L.R., and Tigyi, G., Rho proteins play a critical role in cell migration during the early phase of mucosal restitution, *J. Clin. Invest.*, 100, 216, 1997.
205. Schmidt, G. and Aktories, K., Rho GTPase-activating toxins: cytotoxic necrotizing factors and dermonecrotic toxin, in *Regulators and Effectors of Small GTPases,* Balch, W.E., Der, C.J., and Hall, A., Eds., Academic Press, San Diego, 2000, 125.
206. Horiguchi, Y., *Escherichia coli* cytotoxic necrotizing factors and *Bordetella* dermonecrotic toxin: the dermonecrosis-inducing toxins activating Rho small GTPases, *Toxicon*, 39, 1619, 2001.
207. Goehring, U.-M., Schmidt, G., Pederson, K.J., Aktories, K., and Barbieri, J.T., The N-terminal domain of *Pseudomonas aeruginosa* exoenzyme S is a GTPase-activating protein for Rho GTPases, *J. Biol. Chem.*, 274, 36369, 1999.
208. von Pawel-Rammingen, U., Telepnev, M.V., Schmidt, G., Aktories, K., Wolf-Watz, H., and Rosqvist, R., GAP activity of the *Yersinia* YopE cytotoxin specifically targets the Rho pathway: a mechanism for disruption of actin microfilament structure, *Mol. Microbiol.*, 36, 737, 2000.
209. Shao, F., Merritt, P.M., Bao, Z., Innes, R.W., and Dixon, J.E., A *Yersinia* effector and a *Pseudomonas* avirulence protein define a family of cycteine proteases functioning in bacterial pathogenesis, *Cell*, 109, 575, 2002.
210. Fu, Y. and Galán, J.E., A *Salmonella* protein antagonizes Rac-1 and Cdc42 to mediate host-cell recovery after bacterial invasion, *Nature*, 401, 293, 1999.
211. Hardt, W.-D., Chen, L.-M., Schuebel, K.E., Bustelo, X.R., and Galán, J.E., S. typhimurium encodes an activator of Rho GTPases that induces membrane ruffling and nuclear responses in host cells, *Cell*, 93, 815, 1998.
212. Wilde, C., Chhatwal, G.S., and Aktories, K., C3stau, a new member of the family of C3-like ADP-ribosyltransferases, *Trends Microbiol.*, 10, 5, 2001.
213. Madden, J.C., Ruiz, N., and Caparon, M., Cytolysin-mediated translocation (CMT): a functional equivalent type III secretion in Gram-positive bacteria, *Cell*, 104, 143, 2001.
214. Vandekerckhove, J., Schering, B., Bärmann, M., and Aktories, K., *Clostridium perfringens* iota toxin ADP-ribosylates skeletal muscle actin in Arg-177, *FEBS Lett.*, 225, 48, 1987.
215. Vandekerckhove, J., Schering, B., Bärmann, M., and Aktories, K., Botulinum C2 toxin ADP-ribosylates cytoplasmic β/g-actin in arginine 177, *J. Biol. Chem.*, 263, 696, 1988.
216. Aktories, K. and Wegner, A., ADP-ribosylation of actin by clostridial toxins, *J. Cell Biol.*, 109, 1385, 1989.
217. Ohishi, I., Iwasaki, M., and Sakaguchi, G., Purification and characterization of two components of botulinum C2 toxin, *Infect. Immun.*, 30, 668, 1980.
218. Stiles, B.G. and Wilkens, T.D., Purification and characterization of *Clostridium perfringens* iota toxin: dependence on two nonlinked proteins for biological activity, *Infect. Immun.*, 54, 683, 1986.

219. Popoff, M.R., Milward, F.W., Bancillon, B., and Boquet, P., Purification of the Clostridium spiroforme binary toxin and activity of the toxin on HEp2 cells, *Infect. Immun.*, 57, 2462, 1989.
220. Perelle, S., Gibert, M., Bourlioux, P., Corthier, G., and Popoff, M.R., Production of a complete binary toxin (actin-specific ADP-ribosyltransferase) by *Clostridium difficile* CD196, *Infect. Immun.*, 65, 1402, 1997.
221. Montecucco, C. and Schiavo, G., Mechanism of action of tetanus and botulinum neurotoxins, *Mol. Microbiol.*, 13, 1, 1994.
222. Habermann, E. and Dreyer, F., Clostridial neurotoxins: handling and action at the cellular and molecular level, *Curr. Top. Microbiol. Immunol.*, 129, 93, 1986.
223. Rubin, E.J., Gill, D.M., Boquet, P., and Popoff, M.R., Functional modification of a 21-Kilodalton G protein when ADP-ribosylated by exoenzyme C3 of *Clostridium botulinum*, *Mol. Cell. Biol.*, 8, 418, 1988.
224. Nakamura, S., Serikawa, T., Yamakawa, K., Nishida, S., Kozaki, S., and Sakaguchi, G., Sporulation and C2 toxin production by *Clostridium botulinum* type C strains producing no C1 toxin, *Microbiol. Immunol.*, 22, 591, 1978.
225. Simpson, L.L., A comparison of the pharmacological properties of *Clostridium botulinum* type C1 and C2 toxins, *J. Pharmacol. Exp. Ther.*, 223, 695, 1982.
226. Ohishi, I. and Odagiri, Y., Histopathological effect of Botulinum C2 toxin on mouse intestines, *Infect. Immun.*, 43, 54, 1984.
227. Ohishi, I., Miyake, M., Ogura, H., and Nakamura, S., Cytopathic effect of botulinum C2 toxin on tissue-culture cells, *FEMS Microbiol. Lett.*, 23, 281, 1984.
228. Reuner, K.H., Presek, P., Boschek, C.B., and Aktories, K., Botulinum C2 toxin ADP-ribosylates actin and disorganizes the microfilament network in intact cells, *Eur. J. Cell Biol.*, 43, 134, 1987.
229. Aktories, K., Bärmann, M., Ohishi, I., Tsuyama, S., Jakobs, K.H., and Habermann, E., Botulinum C2 toxin ADP-ribosylates actin, *Nature*, 322, 390, 1986.
230. Ohishi, I., Activation of botulinum C2 toxin by trypsin, *Infect. Immun.*, 55, 1461, 1987.
231. Eckhardt, M., Barth, H., Blöcker, D., and Aktories, K., Binding of *Clostridium botulinum* C2 toxin to asparagine-linked complex and hybrid carbohydrates, *J. Biol. Chem.*, 275, 2328, 2000.
232. Barth, H., Blöcker, D., Behlke, J., Bergsma-Schutter, W., Brisson, A., Benz, R., and Aktories, K., Cellular uptake of *Clostridium botulinum* C2 toxin requires oligomerization and acidification, *J. Biol. Chem.*, 275, 18704, 2000.
233. Blöcker, D., Pohlmann, K., Haug, G., Bachmeyer, C., Benz, R., Aktories, K., and Barth, H., *Clostridium botulinum* C2 toxin: Low pH-induced pore formation is required for translocation of the enzyme component C2I into the cytosol of host cells, *J. Biol. Chem.*, 278, 37360, 2003.
234. Blöcker, D., Behlke, J., Aktories, K., and Barth, H., Cellular uptake of the binary *Clostridium perfringens* iota-toxin., *Infect. Immun.*, 69, 2980, 2001.
235. Haug, G., Leemhuis, J., Tiemann, D., Meyer, D.K., Aktories, K., and Barth, H., The host cell chaperone Hsp90 is essential for translocation of the binary *Clostridium botulinum* C2 toxin into the cytosol, *J. Biol. Chem.*, 274, 32266, 2003.
236. Simpson, L.L., Molecular basis for the pharmacological actions of *Clostridium botulinum* type C2 toxin, *J. Pharmacol. Exp. Ther.*, 230, 665, 1984.
237. Fujii, N., Kubota, T., Shirakawa, S., Kimura, K., Ohishi, I., Moriishi, K., Isogai, E., and Isogai, H., Characterization of component-I gene of botulinum C2 toxin and PCR detection of its gene in clostridial species, *Biochem. Biophys. Res. Commn.*, 220, 353, 1996.
238. Tsuge, H., Nagahama, M., Nishimura, H., Hisatsune, J., Sakaguchi, Y., Itogawa, Y., Katunuma, N., and Sakurai, J., Crystal structure and site-directed mutagenesis of enzymatic components from *Clostridium perfringens* iota-toxin, *J. Mol. Biol.*, 325, 471, 2003.
239. Li, M., Dyda, F., Benhar, I., Pastan, I., and Davies, D.R., Crystal structure of the catalytic domain of *Pseudomonas exotoxin* A complexed with a nicotinamide adenine dinucleotide analog: implications for the activation process and for ADP ribosylation, *Proc. Natl. Acad. Sci. U.S.A.*, 93, 6902, 1996.
240. Jung, M., Just, I., van Damme, J., Vandekerckhove, J., and Aktories, K., NAD-binding site of the C3-like ADP-ribosyltransferase from *Clostridium limosum*, *J. Biol. Chem.*, 268, 23215, 1993.
241. Carroll, S.F. and Collier, R.J., NAD binding site of diphtheria toxin: identification of a residue within the nicotinamide subsite by photochemical modification with NAD, *Proc. Natl. Acad. Sci. U.S.A.*, 81, 3307, 1984.

242. Domenighini, M., Pizza, M., and Rappuoli, R., Bacterial ADP-Ribosyltransferases, in *Bacterial Toxins and Virulence Factors in Disease,* Moss, J., Iglewski, B., Vaughan, M., and Tu, A.T., Eds., Marcel Dekker, Inc., New York, Basel, Hong Kong, 1995, 59.
243. Domenighini, M., Magagnoli, C., Pizza, M., and Rappuoli, R., Common features of the NAD-binding and catalytic site of ADP-ribosylating toxins, *Mol. Microbiol.*, 14, 41, 1994.
244. Takada, T., Iida, K., and Moss, J., Conservation of a common motif in enzymes catalyzing ADP-ribose transfer, *J. Biol. Chem.*, 270, 541, 1995.
245. Okazaki, I.J. and Moss, J., Common structure of the catalytic sites of mammalian and bacterial toxin ADP-ribosyltransferases, *Mol. Cell. Biochem.*, 138, 177, 1994.
246. Barth, H., Roebling, R., Fritz, M., and Aktories, K., The binary *Clostridium botulinum* C2 toxin as a protein delivery system, *J. Biol. Chem.*, 277, 5074, 2002.
247. Blöcker, D., Barth, H., Maier, E., Benz, R., Barbieri, J.T., and Aktories, K., The C-terminus of component C2II of *Clostridium botulinum* C2 toxin is essential for receptor binding, *Infect. Immun.*, 68, 4566, 2000.
248. Schmid, A., Benz, R., Just, I., and Aktories, K., Interaction of *Clostridium botulinum* C2 toxin with lipid bilayer membranes: formation of cation-selective channels and inhibition of channel function by chloroquine and peptides, *J. Biol. Chem.*, 269, No.24, 16706, 1994.
249. Leppla, S.H., Anthrax toxins, in *Bacterial Toxins and Virulence Factors in Disease,* Moss, J., Iglewski, B., Vaughan, M., and Tu, A.T., Eds., Marcel Dekker, New York, 1995, 543.
250. Melamed, I., Downey, G.P., Aktories, K., and Roifman, C.M., Microfilament assembly is required for antigen-receptor-mediated activation of human B lymphocytes, *J. Immunol. Meth.*, 147, 1139, 1991.
251. Duesbery, N.S., Webb, C.P., Leppla, S.H., Gordon, V.M., Klimpel, K.R., Copeland, T.D., Ahn, N.G., Oskarsson, M.K., Fukasawa, K., Paull, K.D., and Woude, G.F.V., Proteolytic inactivation of MAP-kinase-kinase by anthrax lethal factor, *Science*, 280, 734, 1998.
252. Leppla, S., Anthrax toxin edema factor: a bacterial adenylate cyclase that increases cyclic AMP concentrations in eukaryotic cells, *Proc. Natl. Acad. Sci. U.S.A.*, 79, 3162, 1982.
253. Milne, J.C., Furlong, D., Hanna, P.C., Wall, J.S., and Collier, R.J., Anthrax protective antigen forms oligomers during intoxication of mammalian cells, *J. Biol. Chem.*, 269, 20607, 1994.
254. Perentesis, J.P., Miller, S.P., and Bodley, J.W., Protein toxin inhibitors of protein synthesis, *BioFactors*, 3, 173, 1992.
255. Petosa, C., Collier, R.J., Klimpel, K.R., Leppla, S.H., and Liddingtom, R.C., Crystal structure of the anthrax toxin protective antigen, *Nature*, 385, 833, 1997.
256. Blaustein, R.O., Koehler, T.M., Collier, R.J., and Finkelstein, A., Anthrax toxin: channel-forming activity of protective antigen in planar phopholipid bilayers, *Proc. Natl. Acad. Sci. U.S.A.*, 86, 2209, 1989.
257. Mourez, M., Lacy, D.B., Cunningham, K., Legmann, R., Sellman, B.R., Mogridge, J., and Collier, R.J., 2001: a year of major advances in anthrax toxin research, *Trends Microbiol.*, 10, 287, 2002.
258. Collier, R.J. and Young, J.A., Anthrax toxin, *Annu. Rev. Cell Dev. Biol.*, 19, 45, 2003.
259. Bachmeyer, C., Benz, R., Barth, H., Aktories, K., Gibert, M., and Popoff, M., Interaction of *Clostridium botulinum* C2-toxin with lipid bilayer membranes and vero cells: inhibition of channel function by chloroquine and related compounds *in vitro* and intoxification *in vivo*, *FASEB J.*, 15, 1658, 2001.
260. Fritz, G., Schroeder, P., and Aktories, K., Isolation and characterization of a *Clostridium botulinum* C2 toxin-resistant cell line: evidence for possible involvement of the cellular C2II receptor in growth regulation, *Infect. Immun.*, 63, 2334, 1995.
261. Sandvig, K. and Van Deurs, B., Membrane traffic exploited by protein toxins, *Annu. Rev. Cell Dev. Biol.*, 18, 1, 2002.
262. Werner, G., Hagenmaier, H., Drautz, H., Baumgartner, A., and Zahner, H., Metabolic products of microorganisms. 224. Bafilomycins, a new group of macrolide antibiotics. Production, isolation, chemical structure and biological activity, *J. Antibiotics*, 37, 110, 1984.
263. Haug, G., Wilde, C., Leemhuis, J., Meyer, D.K., Aktories, K., and Barth, H., Cellular uptake of *Clostridium botulinum* C2 toxin: membrane translocation of a fusion toxin requires unfolding of its dihydrofolate reductase domain, *Biochemistry*, 42, 15284, 2003.
264. Haug, G., Aktories, K., and Barth, H., The host cell chaperone Hsp90 is necessary for cytotoxic action of the binary iota-like toxins, *Infect. Immun.*, 72, 3066, 2004.

265. Ratts, R., Zeng, H., Berg, E.A., Blue, C., McComb, M.E., Costello, C.E., Vanderspek, J.C., and Murphy, J.R., The cytosolic entry of diphtheria toxin catalytic domain requires a host cell cytosolic translocation factor complex, *J. Cell Biol.*, 160, 1139, 2003.
266. Popoff, M.R. and Boquet, P., *Clostridium spiroforme* toxin is a binary toxin which ADP-ribosylates cellular actin, *Biochem. Biophys. Res. Commun.*, 152, 1361, 1988.
267. Considine, R.V. and Simpson, L.L., Cellular and molecular actions of binary toxins possessing ADP-ribosyltransferase activity, *Toxicon*, 29, 913, 1991.
268. Perelle, S., Gibert, M., Boquet, P., and Popoff, M.R., Characterization of *Clostridium perfringens* iota-toxin genes and expression in *Escherichia coli*, *Infect. Immun.*, 61, 5147, 1993.
269. Borriello, S.P. and Carman, R.J., Association of iota-like toxin and *Clostridium spiroforme* with both spontaneous and antibiotic-associated diarrhea and colitis in rabbits, *J. Clin. Microbiol.*, 17, 414, 1983.
270. Carman, R.J., Van Tassell, R.L., and Wilkins, T.D., Production of iota toxin by *Clostridium spiroforme*: a requirement for divalent cations, *Veterin. Microbiol.*, 15, 115, 1987.
271. Simpson, L.L., Stiles, B.G., Zepeda, H., and Wilkins, T.D., Production by *Clostridium spiroforme* of an iotalike toxin that possesses mono(ADP-ribosyl)transferase activity: Identification of a novel class of ADP-ribosyltransferases, *Infect. Immun.*, 57, 255, 1989.
272. Popoff, M.R., Rubin, E.J., Gill, D.M., and Boquet, P., Actin-specific ADP-ribosyltransferase produced by a *Clostridium difficile* strain, *Infect. Immun.*, 56, 2299, 1988.
273. Stubbs, S., Rupnik, M., Gibert, M., Brazier, J., Duerden, B., and Popoff, M., Production of actin-specific ADP-ribosyltransferase (binary toxin) by strains of *Clostridium difficile*., *FEMS Microbiol. Lett.*, 186, 307, 2000.
274. Schering, B., Bärmann, M., Chhatwal, G.S., Geipel, U., and Aktories, K., ADP-ribosylation of skeletal muscle and non-muscle actin by *Clostridium perfringens* iota toxin, *Eur. J. Biochem.*, 171, 225, 1988.
275. Aktories, K., Ankenbauer, T., Schering, B., and Jakobs, K.H., ADP-ribosylation of platelet actin by botulinum C2 toxin, *Eur. J. Biochem.*, 161, 155, 1986.
276. Mauss, S., Koch, G., Kreye, V.A.W., and Aktories, K., Inhibition of the contraction of the isolated longitudinal muscle of the guinea-pig ileum by botulinum C2 toxin: evidence for a role of G/F-actin transition in smooth muscle contraction, *Naunyn-Schmiedeberg's Arch. Pharmacol.*, 340, 345, 1989.
277. Wegner, A. and Aktories, K., ADP-ribosylated actin caps the barbed ends of actin filaments, *J. Biol. Chem.*, 263, 13739, 1988.
278. Weigt, C., Just, I., Wegner, A., and Aktories, K., Nonmuscle actin ADP-ribosylated by botulinum C2 toxin caps actin filaments, *FEBS Lett.*, 246, 181, 1989.
279. Holmes, K.C., Popp, D., Gebhard, W., and Kabsch, W., Atomic model of the actin filament, *Nature*, 347, 44, 1990.
280. Kabsch, W., Mannherz, H.G., Suck, D., Pai, E.F., and Holmes, K.C., Atomic structure of the actin:DNase I complex, *Nature*, 347, 37, 1990.
281. Geipel, U., Just, I., Schering, B., Haas, D., and Aktories, K., ADP-ribosylation of actin causes increase in the rate of ATP exchange and inhibition of ATP hydrolysis, *Eur. J. Biochem.*, 179, 229, 1989.
282. Wegner, A., Aktories, K., Ditsch, A., Just, I., Schoepper, B., Selve, N., and Wille, M., Actin-gelsolin interaction, *Adv. Exp. Med. Biol.*, 358, 97, 1994.
283. Wille, M., Just, I., Wegner, A., and Aktories, K., ADP-ribosylation of the gelsolin-actin complex by clostridial toxins, *J. Biol. Chem.*, 267, 50, 1992.
284. Just, I., Wille, M., Chaponnier, C., and Aktories, K., Gelsolin-actin complex is target for ADP-ribosylation by *Clostridium botulinum* C2 toxin in intact human neutrophils, *Eur. J. Pharmacol. Mol. Pharmacol.*, 246, 293, 1993.
285. Ballweber, E., Galla, M., Aktories, K., Yeoh, S., Weeds, A.G., and Mannherz, H.G., Interaction of ADP-ribosylated actin with actin binding proteins, *FEBS Lett.*, 508, 131, 2001.
286. Al-Mohanna, F.A., Ohishi, I., and Hallett, M.B., Botulinum C2 toxin potentiates activation of the neutrophil oxidase. Further evidence of a role for actin polymerization, *FEBS Lett.*, 219, 40, 1987.
287. Norgauer, J., Kownatzki, E., Seifert, R., and Aktories, K., Botulinum C2 toxin ADP-ribosylates actin and enhances O_2^- production and secretion but inhibits migration of activated human neutrophils, *J. Clin. Invest.*, 82, 1376, 1988.
288. Grimminger, F., Sibelius, U., Aktories, K., Just, I., and Seeger, W., Suppression of cytoskeletal rearrangement in activated human neutrophils by botulinum C_2 toxin. Impact on cellular signal transduction, *J. Biol. Chem.*, 266, 19276, 1991.

289. Wenzel-Seifert, K., Lentzen, H., Aktories, K., and Seifert, R., Complex regulation of human neutrophil activation by actin filaments:Dihydrocytochalasin B and botulinum C2 toxin uncover the existence of multiple cation entry pathways, *J. Leukocyte Biol.*, 61, 703, 1997.
290. Norgauer, J., Just, I., Aktories, K., and Sklar, L.A., Influence of botulinum C2 toxin on F-actin and N-formyl peptide receptor dynamics in human neutrophils, *J. Cell Biol.*, 109, 1133, 1989.
291. Melamed, I., Stein, L., and Roifman, C.M., Epstein-Barr virus induces actin polymerization in human B cells, *J. Immunology*, 153, 1998, 1994.
292. Melamed, I., Turner, C.E., Aktories, K., Kaplan, D.R., and Gelfand, E.W., Nerve growth factor triggers microfilament assembly and paxillin phosphorylation in human B lymphocytes, *J. Exp. Med.*, 181, 1071, 1995.
293. Melamed, I., Franklin, R.A., and Gelfand, E.W., Microfilament assembly is required for anti-IgM dependent MAPK and p90rsk activation in human B lymphocytes, *Biochem. Biophys. Res. Commun.*, 209, 1102, 1995.
294. Ermert, L., Duncker, H.-R., Bruckner, H., Grimminger, F., Hansen, T., Rossig, R., Aktories, K., and Seeger, W., Ultrastructural changes of lung capillary endothelium in response to botulinum C2 toxin, *J. Appl. Physiol.*, 82, 382, 1997.
295. Melamed, I. and Gelfand, E.W., Microfilament assembly is involved in B-cell apoptosis., *Cell Immunol.*, 194, 136, 1999.
296. Matter, K., Dreyer, F., and Aktories, K., Actin involvement in exocytosis from PC12 cells: studies on the influence of botulinum C2 toxin on stimulated noradrenaline release, *J. Neurochem.*, 52, 370, 1989.
297. Gasman, S., Chasserot-Golaz, S., Popoff, M., Aunis, D., and Bader, M.-F., Involvement of Rho GTPases in calcium-regulated exocytosis from adrenal chromaffin cells, *J. Cell Sci.*, 112, 4763, 1999.
298. Wex, C.B.A., Koch, G., and Aktories, K., Effects of *Clostridium botulinum* C2 toxin-induced depolymerisation of actin on degranulation of suspended and attached mast cells, *Naunyn-Schmiedeberg's Arch. Pharmacol.*, 355, 319, 1997.
299. Prepens, U., Barth, H., Wilting, J., and Aktories, K., Influence of *Clostridium botulinum* C2 toxin on FcεRI-mediated secretion and tyrosine phosphorylation in RBL cells, *Naunyn-Schmiedeberg's Arch. Pharmacol.*, 357, 323, 1998.
300. Li, G., Rungger-Brändle, E., Just, I., Jonas, J.-C., Aktories, K., and Wollheim, C.B., Effect of disruption of actin filaments by *Clostridium botulinum* C2 toxin on insulin secretion in HIT-T15 cells and pancreatic islets, *Mol. Biol. Cell*, 5, 1199, 1994.
301. Considine, R.V., Simpson, L.L., and Sherwin, J.R., Botulinum C_2 toxin and steroid production in adrenal Y-1 cells: The role of microfilaments in the toxin-induced increase in steroid release, *J. Pharmacol. Exp. Ther.*, 260, 859, 1992.
302. Tsakiridis, T., Bergmann, A., Somwar, R., Taha, C., Aktories, K., Cruz, T.F., Klip, A., and Downey, G.P., Actin filaments facilitate insulin activation of the Src and collagen homologous/mitogen-activated protein kinase pathway leading to DNA synthesis and *c-fos* expression, *J. Biol. Chem.*, 273, 28322, 1998.
303. Valderrama, F., Duran, J.M., Babia, T., Barth, H., Renau-Piqueras, J., and Egea, G., Actin microfilaments facilitate the retrograde transport from the Golgi complex to the endoplasmic reticulum in mammalian cells, *Traffic*, 2, 717, 2001.
304. Schmid, G., Schürmann, A., Huppertz, C., Hofmann, F., Aktories, K., and Joost, H.-G., Inhibition of insulin-stimulated glucose transport in 3T3-L1 cells by *Clostridium difficile* toxin B, *Clostridium sordellii* lethal toxin, and *Clostridium botulinum* C2 toxin, *Naunyn-Schmiedeberg's Arch. Pharmacol.*, 357, 385, 1998.
305. Reuner, K.H., Schlegel, K., Just, I., Aktories, K., and Katz, N., Autoregulatory control of actin synthesis in cultured rat hepatocytes, *FEBS Lett.*, 286, 100, 1991.
306. Bershadsky, A.D., Glück, U., Denisenko, O.N., Sklyarova, T.V., Spector, I., and Ben-Ze'ev, A., The state of actin assembly regulates actin and vinculin expression by a feedback loop, *J. Cell Sci.*, 108, 1183, 1995.
307. Reuner, K.H., Van der Does, A., Dunker, P., Just, I., Aktories, K., and Katz, N., Microinjection of ADP-ribosylated actin inhibits actin synthesis in hepatocyte-hepatoma hybrid cells, *Biochemical J.*, 319, 843, 1996.

308. Barth, H., Klingler, M., Aktories, K., and Kinzel, V., *Clostridium botulinum* C2 Toxin delays entry into mitosis and activation of p34[cdc2] kinase and cdc25-C phosphatase in HeLa cells, *Infect. Immun.*, 67, 5083, 1999.

309. Verschueren, H., Van der Taelen, I., Dewit, J., De Braekeleer, J., De Baetelier, P., Aktories, K., and Just, I., Effects of *Clostridium botulinum* C2 toxin and cytochalasin D on *in vitro* invasiveness, motility and F-actin content of a murine T-lymphoma cell line, *Eur. J. Cell Biol.*, 66, 335, 1995.

310. Suttorp, N., Polley, M., Seybold, J., Schnittler, H., Seeger, W., Grimminger, F., and Aktories, K., Adenosine diphosphate-ribosylation of G-actin by botulinum C2 toxin increases endothelial permeability *in vitro*, *J. Clin. Invest.*, 87, 1575, 1991.

311. Sibelius, U., Hattar, K., Schenkel, A., Noll, T., Csernok, E., Gross, W.L., Mayet, W.J., Piper, H.M., Seeger, W., and Grimminger, F., Wegener's granulomatosis: anti-proteinase 3 antibodies are potent inductors of human endothelial cell signaling and leakage response, *J. Exp. Med.*, 187, 497, 1998.

312. Ermert, L., Brückner, H., Walmrath, D., Grimminger, F., Aktories, K., Suttorp, N., Duncker, H.-R., and Seeger, W., Role of endothelial cytoskeleton in high-permeability edema due to botulinum C_2 toxin in perfused rabbit lungs, *Am. J. Physiol. Lung Cell. Mol. Physiol.*, 268, L753, 1995.

313. Ermert, L., Rossig, R., Hansen, T., Schutte, H., Aktories, K., and Seeger, W., Differential role of actin in lung endothelial and epithelial barrier properties in perfused rabbit lungs, *Eur. Respir. J.*, 9, 93, 1996.

314. Schnittler, H.-J., Schneider, S.W., Raifer, H., Luo, F., Dieterich, P., Just, I., and Aktories, K., Role of actin filaments in endothelial cell-cell adhesion and membrane stability under fluid shear stress, *Pflügers Arch.-Eur. J. Physiol.*, 442, 675, 2001.

315. Nilius, B., Voets, T., Prenen, J., Barth, H., Aktories, K., Kaibuchi, K., Droogmans, G., and Eggermont, J., Role of Rho and Rho kinase in the activation of volume-regulated anion channels in bovine endothelial cells, *J. Physiol.*, 516, 67, 1999.

316. Vischer, U.M., Barth, H., and Wollheim, C.B., Regulated von Willebrand factor secretion is associated with agonist-specific patterns of cytoskeletal remodeling in cultured endothelial cell, *Arterioscler. Thromb. Vasc. Biol.*, 20, 883-891, 2000.

317. Sakr, S., Eddy, R., Barth, H., Xie, B., Greenberg, S., Maxfield, F.R., and Tabas, I., An *in-vitro* model for the uptake and degradation of lipoproteins by arterial-wall mocrophages during atheriogenesis, *J. Biol. Chem.*, 276, 37649, 2001.

318. Dreikhausen, U., Varga, G., Hofmann, F., Barth, H., Aktories, K., Resch, K., and Szamel, M., Regulation by rho family GTPases of IL-1 receptor induced signaling: C3-like chimeric toxin and *Clostridium difficile* toxin B inhibit signaling pathways involved in IL-2 gene expression, *Eur. J. Immmunol.*, 31, 1610, 2001.

319. Klussmann, E., Tamma, G., Lorenz, D., Wiesner, B., Maric, K., Hofmann, F., Aktories, K., Valenti, G., and Rosenthal, W., An inhibitory role of Rho in the vasopressin-mediated translocation of aquaporin-2 into cell membranes of renal principal cells, *J. Biol. Chem.*, 276, 20451, 2001.

320. Linseman, D.A., Laessig, T., Meintzer, M.K., McClure, M., Barth, H., Aktories, K., and Heidenreich, K.A., An essential role for Rac/Cdc42 GTPases in cerebellar granule neuron survival, *J. Biol. Chem.*, 276, 39123, 2001.

321. Hahn, A., Barth, H., Kress, M., Mertens, P.R., and Goppelt-Strübe, M., Role of Rac and Cdc42 in lysophosphatidic acid-mediated cyclooxygenase-2 gene expression, *Biochem. J.*, 362, 33, 2002.

322. Strey, A., Janning, A., Barth, H., and Gerke, V., Endothelial Rho signaling is required for monocyte transendothelial migration, *FEBS Lett.*, 517, 261, 2002.

323. Carton, I., Trouet, D., Hermans, D., Barth, H., Aktories, K., Droogmans, G., Jorgensen, N.K., Hoffmann, E.K., Nilius, B., and Eggermont, J., RhoA exerts a permissive effect on volume-regulated anion channels in vascular endothelial cells, *Am. J. Physiol. Cell Physiol.*, 283, C115, 2002.

324. Galle, J., Mameghani, A., Bolz, S.S., Gambaryan, S., Gorg, M., Quaschning, T., Raff, U., Barth, H., Seibold, S., Wanner, C., and Pohl, U., Oxidized LDL and its compound lysophosphatidylcholine potentiate AngII-induced vasoconstriction by stimulation of RhoA, *J. Am. Soc. Nephrol.*, 14, 1471, 2003.

325. Leemhuis, J., Boutillier, S., Barth, H., Feuerstein, T.J., Brock, C., Nürnberg, B., Aktories, K., and Meyer, D.K., Rho GTPases and Phosphoinositide 3-kinase organize formation of branched dendrites, *J. Biol. Chem.*, 279, 585, 2004.

20 Membrane Active Toxins

Richard W. Titball and Rodney K. Tweten

CONTENTS

20.1 INTRODUCTION

The clostridia are some of the most prolific producers of proteins that are able to act on eukaryotic cell membranes. Many of these proteins have served as prototypic molecules for basic studies on the modes of action, structures, and roles in disease of bacterial toxins. For example, pioneering work with the *Clostridium perfringens* α-toxin in the 1940s [1] showed for the first time that a bacterial toxin could also have an enzymatic mode of action (in this case a phospholipase C activity). However, while some clostridial toxins have been well characterized over the past 50 years, the basis of toxicity of others is only poorly understood, and it seems likely that there are membrane active toxins still awaiting discovery. This duality of an increasingly detailed knowledge of some toxins, and a lack of understanding of many others, make this field one that is both at the leading edge and still maturing. In this chapter, the properties of a range of membrane active toxins produced by the clostridia are reviewed. It is hoped that this will provide a basis for future studies by researchers.

20.2 PHOSPHOLIPASES C

Bacterial phospholipases can be grouped into A_2, C, and D class enzymes, according to the phospholipid bond that is targeted for hydrolysis (Figure 20.1). The bacterial phospholipases C have been further grouped according to substrate specificity into those enzymes with a preference for phosphatidylcholine, those that hydrolyze sphingomyelin, and those that are active toward phosphatidylinositol [2,3]. The clostridia have been reported to produce only phospholipases C and, more specifically, only phosphatidylcholine-preferring and phosphatidylinositol hydrolyzing enzymes [2,3]. Of these, the phosphatidylcholine-preferring phospholipases C are considered to be potent membrane active toxins and are considered in more detail in this chapter.

20.2.1 Sources and Properties

The action of *C. perfringens* α-toxin on egg yolk lipoprotein that has been incorporated into agar results in a zone of turbidity around colonies and is one of the tests used to identify *C. perfringens* in the laboratory [4]. Since the discovery that *C. perfringens* α-toxin was a phospholipase C [1], a range of clostridial species have been shown to produce similar enzymes, and genes encoding many of these proteins have been isolated and nucleotide sequenced (Table 20.1). In general, the enzymes that have been characterized are produced by clostridia of medical significance, and it seems likely that many as-yet-uncharacterized phospholipases C are produced by other clostridia.

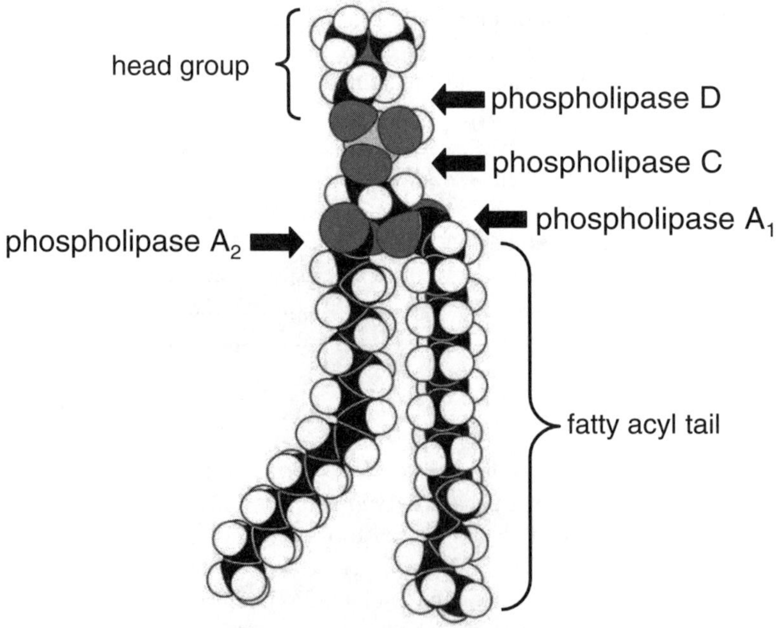

FIGURE 20.1 Phosphatidylcholine molecule showing the sites of cleavage by phospholipases A_1, A_2, C, or D. Other phospholipids have different head or tail groups.

TABLE 20.1
Phospholipases C Produced by the *Clostridia*

Enzyme	Gene	Substrate Specificity	Properties
C. perfringens α-toxin	Cloned [6]	PC, SM, PS [17]	Toxic, hemolytic [6,17]
Clostridium absonum PLC	Cloned [14]	PC, SM [14]	Toxic, hemolytic [14]
Clostridium barati PLC	Cloned	NR	ND
Clostridium bifermentans PLC	Cloned [24]	PC, SM [12]	Weakly toxic, weakly hemolytic [12]
Clostridium novyi β-toxin	Cloned [25]	PC, SM, PE, PS, PG [16]	Toxic, hemolytic [16]
Clostridium haemolyticum PLC	Cloned	PC, SM	Toxic, hemolytic
Clostridium sordellii PLC	Cloned	NR	NR

PC: phosphatidylcholine; SM: sphingomyelin; PS: phosphatidyl serine; PE: phosphatidylethanolamine; PI: phosphatidylinositol; PG: phosphatidylglycerol; NR: not reported.

A range of assays have been used to detect phospholipase C activity, the simplest of which involves measuring hydrolysis of egg yolk lecithin incorporated into agar plates [1,4,5]. Liquid-based microtiter tray assay have also been devised [6]. However, this assay suffers from the relative complexity of the substrate, which contains phospholipids in addition to phosphatidycholine. The relative insolubility of phospholipids in water makes any assay using monodispersed substrates difficult, though detergents have been used to solubilize phospholipids. The products of hydrolysis can then be identified after their separation from thin-layer chromatography [7]. In some cases, radiolabeled substrates have been used, and the excision of the products from thin-layer chromatography plates allows quantification of the products. To improve the solubility of the substrate, some workers have used phospholipid derivatives that either lack or have short fatty acyl tail groups [8,9]. For example, ρ-nitrophenolphosphorylcholine (NPPC), where the tail groups are replaced with ρ-nitrophenol, has the advantage that hydrolysis is accompanied by the release of the chromophore. However, these derivates do not accurately mimic phospholipid substrates, since it is known that fatty acyl chain lengths of between C6 and C14 are necessary for efficient phospholipid hydrolysis by the *B. cereus* PC-PLC and *C. perfringens* α-toxin, respectively [10,11]. Liposome-based assays allow the phospholipids to be presented to the enzyme in a form that more closely mimics the cell membrane, and the composition of the liposome can be selected to include different types of phospholipids and different phospholipid tail lengths. The measurement of the release of entrapped dyes such as carboxyfluorescein or calcein [11–13] allows the simple quantification of liposome lysis. Using a variety of these different assays, all of the characterized clostridial enzymes have been shown to hydrolyze phosphatidylcholine, albeit with different efficiencies. However, the enzymes show different activities toward phospholipids with alternative headgroups. For example, the *C. absonum* and *C. bifermentans* enzymes are apparently unable to hydrolyze phosphatidylserine [12,14], while the *C. novyi* enzyme is the only clostridial enzyme reported to be able to hydrolyze phosphatidylethanolamine [15,16].

A range of assays have been used to measure the cytotoxic and lethal activities of clostridial phospholipases C, ranging from the measurement of acute toxicity in small animals, such as mice, to cell-based assays. The measurement of lysis of erythrocytes provides one of the simplest cell-based assays [5,17,18]. Not all erythrocyte types are susceptible to lysis, and mouse, sheep, or rabbit cells have been used most frequently in this assay [6,18,19]. In the case of rabbit and sheep erythrocytes, complete lysis is only seen when the cells are first incubated with the phospholipase C at 37°C and then cooled to 4°C ("hot-cold" lysis). Cytoxicity assays using cells other than erythrocytes have been used less frequently [20–23].

The clostridial phospholipases that have been characterized all appear to be major exoproducts of the bacteria, and all possess typical signal sequences that would direct export of the protein across the cell membrane [6,14,24–26]. It is probably the fact that these proteins are found in abundance in culture fluids, and are therefore relatively easily isolated, that originally attracted the attention of researchers. However, the purity of enzymes used in many of the early studies is not certain — a problem that is compounded because other membrane active enzymes are also produced by these clostridia. Therefore, more recently, researchers have used purified recombinant proteins, produced in *Escherichia coli* [6,12,14,21,23,24,27–32].

20.2.2 Phylogenetic Relationships

The zincmetallophospholipases C show significant amino acid sequence identity, and residues that are known to play a key role in binding of the three zinc ions in *C. perfringens* α-toxin (Figure 20.2) are invariantly conserved. A phylogenetic tree (Figure 20.3), constructed from an alignment of the sequences of all of the zincmetallophospholipases C deposited with GenBank, reflects speciation with all of the *C. perfringens* α-toxin sequences clustered on one branch of the tree. There was no obvious clustering of strains isolated from cases of gas gangrene in humans, environmental isolates, or enteric isolates from calves, which are all represented in this phylogram. Similarly, there was no obvious linkage of strains that have been assigned to biotypes A to E [25,33]. Of the *C. perfringens* sequences, the α-toxin from *C. perfringens* strain SWCP showed the greatest divergence, and this might reflect the avian origin of this strain [32]. Whether other avian isolates of *C. perfringens* encode α-toxins that also show this level of sequence divergence awaits investigation. *C. bifermentans* and *C. sordellii* are known to be closely related [34] and although the phospholipase C sequences from these bacteria were clustered on a separate branch. The phospholipases C from *C. novyi*, *C. absonum,* and *C. haemolyticum* were also located on isolated branches in the phylogram.

20.2.3 Roles in Disease

Gas gangrene has been described as "the most fulminant necrotizing infection that affects humans" [35]. The disease is usually a consequence of a traumatic injury that allows both the ingress of the bacteria into the wound and damage to the blood supply that promotes the anoxic conditions required for bacterial growth [36]. Since the 1940s, it had been suggested that *C. perfringens* α-toxin played a major role in gas gangrene, but the difficulties of isolating protein free of the myriad of other protein toxins produced by this bacterium [37] have made proof of this suggestion difficult.

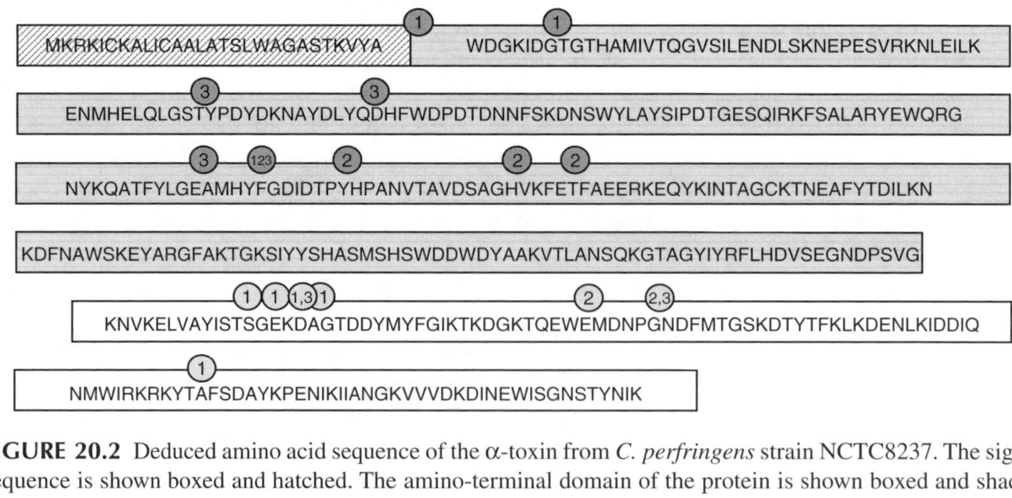

FIGURE 20.2 Deduced amino acid sequence of the α-toxin from *C. perfringens* strain NCTC8237. The signal sequence is shown boxed and hatched. The amino-terminal domain of the protein is shown boxed and shaded grey. The carboxy-terminal domain of the protein is shown boxed. Residues involved in the coordination of zinc ions 1, 2, or 3 in the amino-terminal domain, and residues involved in the coordination of calcium ions 1, 2, or 3 in the carboxy-terminal domain are indicated below the shaded circles.

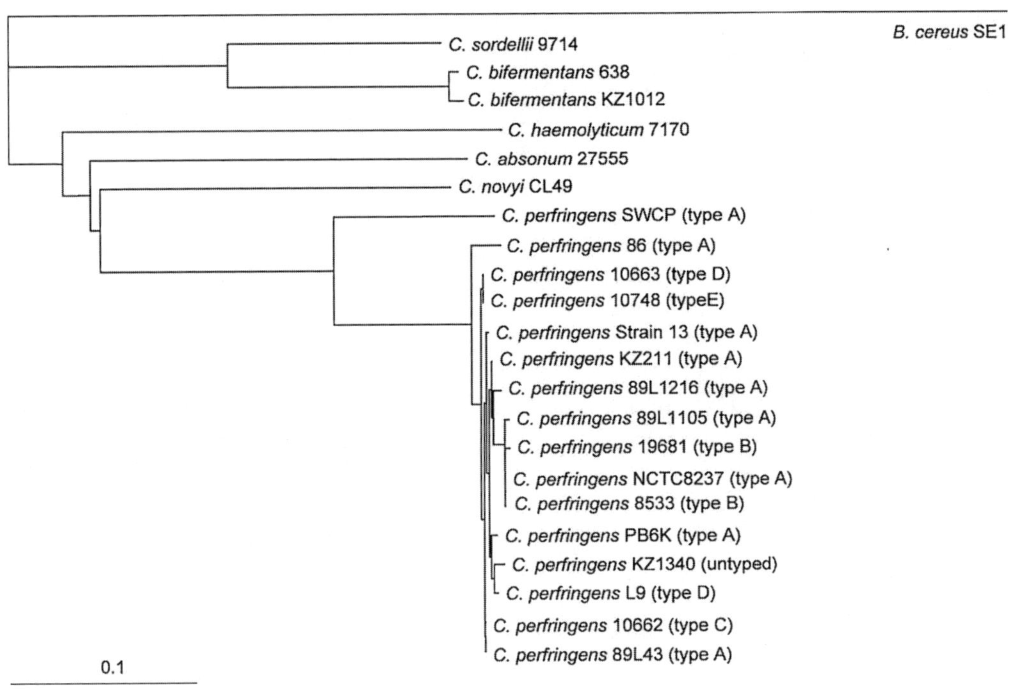

FIGURE 20.3 Phylogenetic tree for clostridial zincmetallophospholipases C, constructed using TreeDraw from a Clustal sequence alignment [240,241]. *Bacillis cereus* strain SE1 PC-phospholipase C (GenBank accession P09598); *C. sordellii* strain 9714 phospholipase C (GenBank accession BAB83263); *C. bifermentans* strain ATCC638 phospholipase C (GenBank accession AAD41623); *C. bifermentans* strain KZ1012 phospholipase C (GenBank accession BAB83265); *C. haemolyticum* strain 7170 beta-toxin (GenBank accession P59626); *C. absonum* strain ATCC 27555 phospholipase C (GenBank accession AAN78416); *C. novyi* strain CL49 α-toxin (GenBank accession BAA06851); *C. perfringens* strain SWCP α-toxin (GenBank accession AAF20094); *C. perfringens* strain 86 α-toxin (GenBank accession JQ0366); *C. perfringens* strain NCIB10663 α-toxin (GenBank accession BAA08721); *C. perfringens* strain NCIB10748 α-toxin (GenBank accession BAA06852); *C. perfringens* strain 13 α-toxin (GenBank accession AAA99193); *C. perfringens* strain KZ211 α-toxin (GenBank accession BAA06850); *C. perfringens* strain CER89L1216 α-toxin (GenBank accession AAA99195); *C. perfringens* strain CER89L1105 α-toxin (GenBank accession AAA99194); *C. perfringens* strain 19681 α-toxin (GenBank accession BAA06854); *C. perfringens* strain NCTC8237 α-toxin (GenBank accession A30565); *C. perfringens* strain NCTC8533 α-toxin; *C. perfringens* strain PB6KN5L7 α-toxin (GenBank accession BAA06849); *C. perfringens* strain KZ1340 α-toxin (GenBank accession BAA09944); *C. perfringens* strain L9 α-toxin (GenBank accession BAA08720); *C. perfringens* strain 10662 α-toxin (GenBank accession BAA01093); *C. perfringens* strain CER89L43 α-toxin (GenBank accession AAA99192).

Studies over the ensuing 30 years mainly involved assessing the toxic properties of the protein in animals or toward isolated cells. These studies all pointed toward the *C. perfringens* phospholipase C as the potent toxin found in culture supernatant fluid [1,38]. They also suggested that the enzyme was hemolytic, cytotoxic, dermonecrotic, and caused platelet aggregation [17,38–42]. However, none of these studies demonstrated unequivocally the role of α-toxin in disease. Even studies that showed that immunization with an α-toxoid provided protection against gas gangrene [43,44] are open to question, since there was always the possibility that immune responses to proteins other than α-toxin were responsible for the protection reported.

The first definitive proof that α-toxin played a role in gas gangrene was reported in 1993 [45], when immunization of mice with a recombinant polypeptide corresponding to the carboxy-terminal domain of the protein was shown to provide protection against experimental gas gangrene. Evidence for the precise role of the toxin in disease was later provided by Awad et al. [46], who showed that an allelic replacement mutant of *C. perfringens*, unable to produce α-toxin, was unable to cause disease in mice. Moreover, after challenge of mice with the α-toxin mutant, there was little or no evidence of infection, whereas animals challenged with the wild-type strain showed muscle necrosis around the site of challenge, as well as limb swelling and hematuria, and these animals eventually died [46]. It is now accepted that α-toxin plays a major role in the pathogenesis of gas gangrene caused by *C. perfringens*.

The evidence for the role of α-toxin in the pathogenesis of other diseases caused by *C. perfringens* type A is quite circumstantial. However, a range of diseases of domesticated livestock have been linked to the overgrowth of *C. perfringens* type A in the gut [47,48] and the consequent production of toxin. Probably the most often cited example of this is necrotic enteritis in farmed fowl. The incidence of this disease has increased markedly in some countries over the past few years [49] and appears to be linked to a reduction in the use of antibiotic growth promoters in the diet and the consequent overgrowth of *C. perfringens* type A in the gut [49,50]. The development of suitable animal models of this disease has not proved to be simple, and this has hampered studies on the etiology of the disease. Therefore, while it is generally accepted that α-toxin plays a key role in the pathogenesis of disease, it has yet to be proven. In support of this suggestion, one study has shown that the level of maternal antibodies to α-toxin correlated with reduced mortality from necrotic enteritis in chicks [49].

In other animals, such as calves [51], there has also been a suggestion that the overgrowth of *C. perfringens* type A in the gut and the production of α-toxin is linked to the development of disease. In this context, it may be significant that, in comparison with the α-toxin produced by a gas gangrene isolate, the α-toxin produced by a bovine enteric isolate of *C. perfringens* was more resistant to chymotrypsin [33]. This might be in accordance with the proposed role of this variant form of the toxin in enteric disease.

Compared with the *C. perfringens* α-toxin, the role of other clostridial phospholipases C in disease is much less well documented. However, many of the species that are known to produce these enzymes, including *C. novyi*, *C. bifermentans*, *C. sordellii*, and *C. histolyticum*, are occasionally associated with gas gangrene [43,52,53]. These enzymes are closely related to α-toxin at the primary amino acid sequence level (see below) and some cross-reactivity of antisera raised against α-toxin has been reported [53,54]. Although yet to be proven, it seems likely that these phospholipases C also play roles in the development of disease.

20.2.4 Molecular Architecture and Mechanism

More than 50 years after the isolation of *C. perfringens* α-toxin, the crystal structure of the protein produced by a bovine enteric isolate of the bacterium was reported [27]. This confirmed that the protein consisted of two domains, with the enzyme active site present in the amino-terminal domain (Figure 20.4). Three metal (zinc) ions were present in the active site cleft. The amino-terminal domain bore a striking resemblance to the structure of the nontoxic phosphatidylcholine preferring phospholipase C (PC-PLC) from *Bacillus cereus* [55]. This was not entirely unexpected, since significant amino acid sequence homology between the amino-terminal domain of α-toxin and the entire PC-PLC molecule had previously been reported [6,24]. However, some subtle differences were seen between these two structures. In particular, the structure of α-toxin reported [27] revealed two amino acids (Trp_{214} and Phe_{334}) with hydrophobic side chains that were exposed to the solvent phase (though these amino acids were not surface exposed in a later reported structure) [28]. As might be expected, the replacement of any of the zinc-coordinating amino acids with amino acids unable to perform this function (Table 20.2) markedly reduces the phospholipase C activity of α-toxin. The replacement of key residues believed to line the active site or play a role in substrate binding also markedly reduced phospholipase C activity. More significantly, these variant proteins are also devoid of hemolytic and, where tested, lethal activities. These findings indicate that the phospholipase C activity of α-toxin is directly related to toxicity.

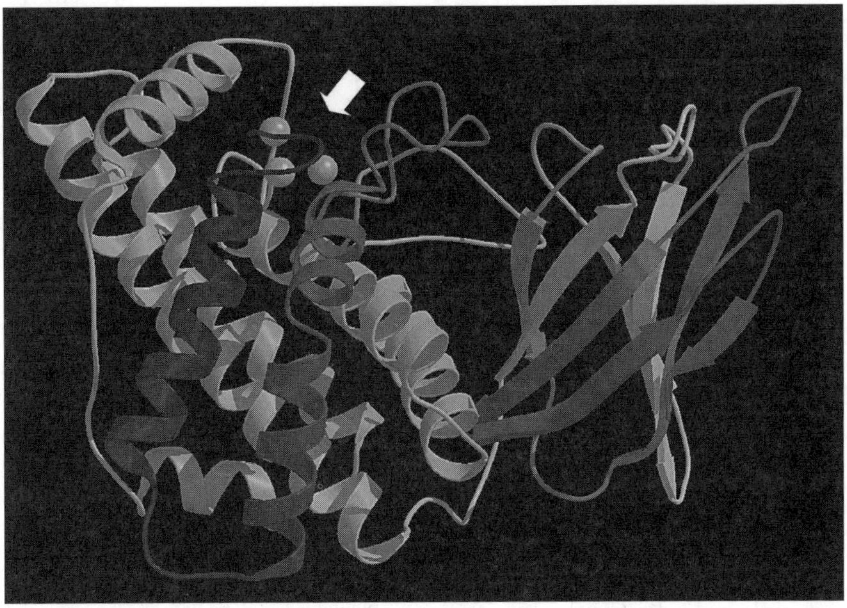

FIGURE 20.4 Crystal structure of α-toxin from *C. perfringens* strain CER89L43. Three zinc ions are shown in the putative active site cleft (arrowed).

The carboxy-terminal domain of α-toxin (which is lacking in the *B. cereus* PC-PLC enzyme) appeared to be structurally similar to the phospholipid-binding domains ("C2 domains") from a range of eukaryotic proteins, such as human pancreatic lipase, rat phosphatidylinositol phospholipase C, and synaptotagmin [27]. The deletion of the carboxy-terminal domain of α-toxin had previously been shown to abolish the hemolytic and lethal activities but not to affect the phospholipase C activity toward phosphatidylcholine [7,45]. Together, these observations suggested that the carboxy-terminal domain of α-toxin might play a role in the recognition of membrane phospholipids. A range of other studies support this suggestion. Calcium ions are required for the binding of α-toxin to phospholipid monolayers [56], and three partial calcium ion-binding sites have been identified in the carboxy-terminal domain [28]. The complete calcium ion-binding site could be created by the additional participation of other ligands, such as the phosphate groups of phospholipids. Variant forms of α-toxin, where these calcium-binding ligands have been replaced with other amino acids, show markedly reduced abilities to interact with phospholipids [21,23].

Together, these observations allow a model for the interaction of α-toxin with host cell membranes to be devised. The initial interaction is mediated by calcium ions that form molecular bridges between the carboxy-terminal domain of α-toxin and the phosphate groups of membrane phospholipids [27,28]. This binding event is accompanied by structural changes in α-toxin — hydrophobic amino acids such as Trp_{214} and Phe_{334} become surface exposed and then embedded in the hydrophobic core of the host cell membrane [30]. Finally, the binding of α-toxin to cell membranes appears to be accompanied by opening of the active site cleft (Figure 20.5) [30,57]. Experimental data also supports this suggestion. In an elegant study, phospholipid monolayers that were subjected to increasing lateral pressures were incubated with various phospholipases, including α-toxin and the *B. cereus* PC-PLC [58]. The maximum lateral pressure at which hydrolysis of the monolayer could be detected was much higher for α-toxin than PC-PLC (Figure 20.6). Significantly, the lateral pressure at which hydrolysis ceased when incubated with PC-PLC (31 dyne/cm) was lower than the typical lateral pressure in a cell membrane (33 dyne/cm). Therefore, overall, we can propose

TABLE 20.2
Amino Acid Substitutions in the Amino Terminal Domain of α-Toxin, Which Have Been Shown To Result in a Marked Reduction (<15% of Wild Type) of Phospholipase C and Hemolytic Activities

Amino Acid Variant	Proposed Function of Target Amino Acid
W_1S[a]	Zn^{++} binding
$H_{11}S$[a]	Zn^{++} binding
$H_{68}S$[a]	Zn^{++} binding
$H_{68}G$[b]	Zn^{++} binding
$H_{126}S$[a]	Zn^{++} binding
$H_{126}G$[b]	Zn^{++} binding
$H_{136}S$[a]	Zn^{++} binding
$H_{136}G$[b]	Zn^{++} binding
$H_{136}A$[b]	Zn^{++} binding
$H_{148}S$[a]	Zn^{++} binding
$H_{148}G$[b]	Zn^{++} binding
$H_{148}L$[b]	Zn^{++} binding
$E_{152}D$[c]	Zn^{++} binding
$E_{152}Q$[c]	Zn^{++} binding
$E_{152}G$[c]	Zn^{++} binding
$F_{69}C$[a]	Substrate binding
$E_{152}Q$[a]	Active site acidic pocket
$D_{56}N$[a]	Active site acidic pocket
$D_{56}N$[c]	Active site acidic pocket
$D_{56}E$[c]	Active site acidic pocket
$D_{56}G$[c]	Active site acidic pocket
$D_{56}S$[c]	Active site acidic pocket
$D_{130}N$[a]	Active site acidic pocket
$D_{130}E$[c]	Active site acidic pocket
$D_{130}G$[c]	Active site acidic pocket
$D_{130}A$[c]	Active site acidic pocket

[a] [237].
[b] [238].
[c] [239].

that the lack of toxicity of the *B. cereus* PC-PLC is explained by the absence of structural features that allow membrane phospholipid recognition.

20.2.5 Interaction with the Cell

The ability of the clostridial phospholipases to damage the host appears to be a consequence of the action of the enzyme on membrane phospholipids. The outer leaflet of the host cell membrane is generally rich in phosphatidylcholine and sphingomyelin [59], which are substrates for all of the characterized clostridial phospholipases. In the case of erythrocytes, the hydrolysis of membrane phospholipids results in hemolysis — easily measured by the release of hemoglobin. However, there are marked differences in the abilities of different clostridial enzymes to cause hemolysis. Although the carboxy-terminal domains of these different enzymes promote membrane binding, they perform this function with different efficiencies. For example, the replacement of the carboxy-terminal domain of α-toxin with the carboxy-terminal domain of the *C. bifermentans* phospholipase

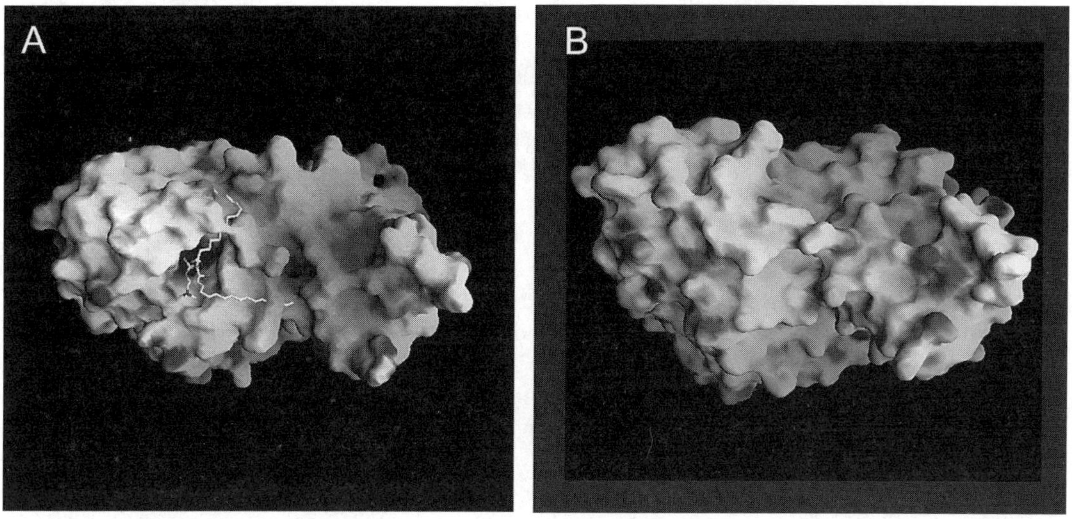

FIGURE 20.5 GRASP electrostatic surface potential surface of α-toxin. In panel A, a sphingomyelin molecule is shown with the head group in the active site cleft. In panel B, the surface potential of α-toxin with the active site cleft in a closed orientation is depicted.

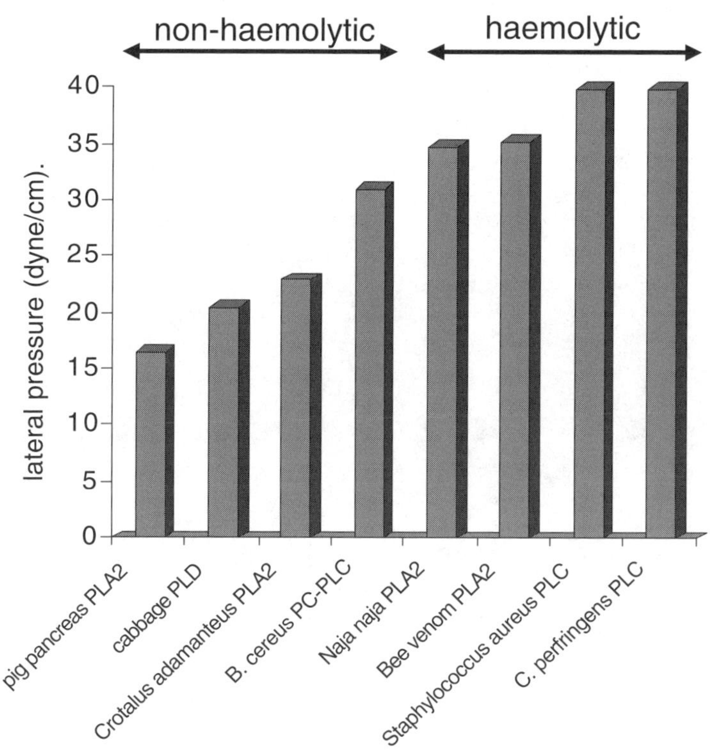

FIGURE 20.6 Maximum lateral pressure in a phospholipid monolayer at which hydrolysis by selected phospholipases C ceased [58]. PLA2 = phospholipase A_2; PLD = phospholipase D; PLC = phospholipase C.

C (a weakly hemolytic and weakly toxic protein) resulted in a reduction in hemolytic and toxic activity of the hybrid protein [12].

Additionally, part of the explanation for the different hemolytic (and lethal activities) of these enzymes is the difference in enzyme turnover rate. For example, the phospholipid turnover rate of *C. perfringens* α-toxin is higher than the turnover rate of *C. bifermentans* phospholipase C, and only the α-toxin is an efficient hemolysin [12]. The substrate specificity of the enzyme might also influence hemolytic activity — the hydrolysis of both phosphatidylcholine and sphingomyelin seems to be necessary for cell lysis. Enzymes that are only able to hydrolyze one substrate type are generally weakly hemolytic [5,60]. Indeed, a modified form of *C. perfringens* α-toxin that retained the ability to hydrolyze phosphatidylcholine, but had a reduced ability to hydrolyze sphingomyelin showed the concomitant loss of hemolytic activity [7].

Differences in the properties of the phospholipases are not the only factor influencing cytolytic activity. Cells are, of course, able to repair membrane damage, and the more efficiently they do this, the more resistant they are to lysis. However, surprisingly, it appears that the activation of the target cell metabolism can also contribute to cell lysis. The most convincing evidence for this comes from studies showing that endogenous (cell membrane) phospholipases C and D are activated in cells treated with *C. perfringens* α-toxin [61,62]. These activated enzymes appear to play a major role in auto-digestion of the host cell membrane, and the bacterial phospholipase C merely serves as an activator of the whole process.

While hemolysis assays provide one way of measuring the destruction of host cells, the role of this toxin in disease appears to be much more subtle. The low level hydrolysis of membrane phospholipids results in the generation of diacylglycerol, which then serves as a secondary messenger [63], activating a whole range of cellular pathways (Figure 20.7). Diacylglycerol can serve

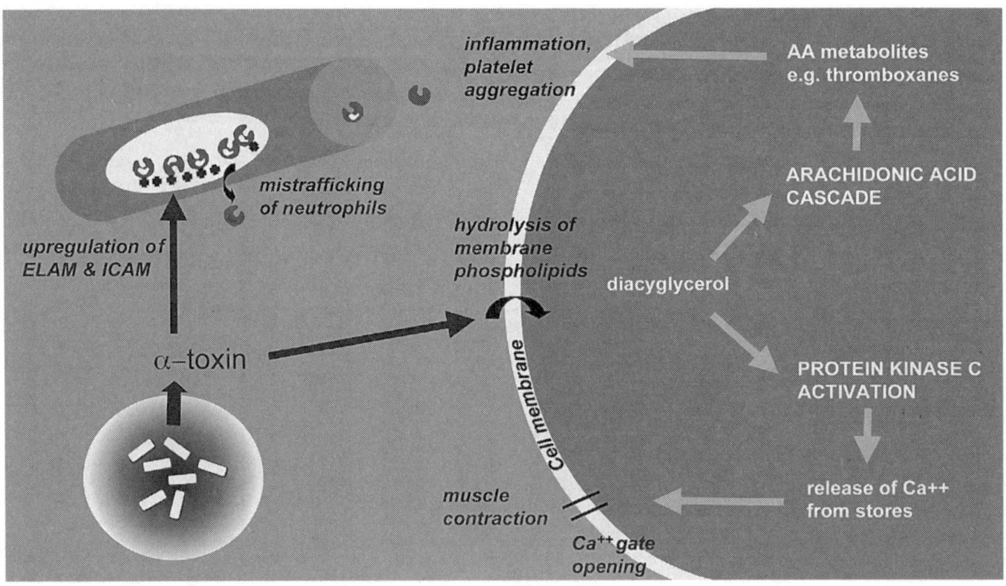

FIGURE 20.7 Cellular pathways activated by *C. perfringens* α-toxin. A focus of infection in soft tissues with bacteria producing α-toxin is shown; α-toxin, which diffuses away from the site of infection, causes the upregulation of ELAM and ICAM adherence markers on endothelial cells lining blood vessels, and the mistrafficking of neutrophils; α-toxin also interacts with other cell membranes, hydrolysing membrane phospholipids and releasing diacylglycerol. Diacyglycerol activates the arachidonic acid cascade after conversion by host diacylglycerol lipase (enzyme not shown). Diacyglycerol also activates protein kinase C, which is then able to modulate a range of cellular processes including the release of stored Ca^{++} ions.

as a substrate for the arachidonic acid cascade [64,65], with the resultant generation of leukotrienes, prostaglandins, or thromboxanes. Leukotriene C_4 and thromboxane A_2 appear to be especially produced in response to the exposure of cells to α-toxin [65]. Diacylglycerol is also able to activate protein kinase C [66], and there is some evidence of this response in cells treated with α-toxin [62,67]. One consequence of the activation of protein kinase C is the activation of endogenous membrane phospholipases, which, as noted above, contributes to membrane damage. The mobilization of calcium stores and opening of calcium gates might also be consequences of the activation of protein kinase C in cells. Overall, the activation of these pathways in cells treated with α-toxin would result in inflammatory responses, muscle contraction, and platelet aggregation.

What are the consequences of all this? The contraction of muscles exposed to α-toxin has been reported, and, in the case of muscles surrounding arteries, this would likely reduce the blood supply to infected tissues [65,68,69]. A further reduction in blood supply might be caused by the aggregation of platelets [39,41,42,70], which appears to be, at least in part, a consequence of the upregulation of gpIIbIIIa on the surface of platelets [71,72]. It is easy to envisage that, collectively, these events would reduce the blood supply to infected tissues, would promote anoxia, and would, therefore, allow the growth of *C. perfringens*. However, there is another important consequence of these events; for almost 100 years it has been known that tissues taken from the site of a myonecrotic *C. perfringens* infection are devoid of phagocytic cells [73]. The phagocytes (and especially neutrophils) that would normally become trafficked from the circulatory systems into intervascular spaces and to the site of infection become trapped within the blood vessels serving these tissues [46,74,75]. This appears to be a consequence of α-toxin mediated upregulation of adherence markers, such as intercellular adhesion molecule 1 (ICAM) and endothelial cell adherence marker 1 (ELAM 1) on the surface of these cells [76] and on the upregulation of platelet P-selectin and gpIIbIIIa [71,72,75]. Together, these molecules promote the binding of neutrophils to endothelial cells lining blood vessels and the formation of cellular aggregates that prevent phagocyte trafficking [75,77].

Death from gas gangrene is usually a consequence of profound shock and cardiac failure. Only rarely is large scale hemolysis seen in gas gangrene patients [78]. It seems that the α-toxin plays a role in these final events. The ability of α-toxin to reduce cardiac output appears to be a consequence of the reduced contractility of cardiac muscle [79,80], and it is possible that this is, in turn, related to the changes in intracellular calcium ion concentrations eluded to above.

20.2.6 Vaccines

Several early studies investigated the prophylactic potential of α-toxoids or the therapeutic value of antiserum raised against "pure" preparations of α-toxoid for the prevention or treatment of gas gangrene [43,53]. Many of the human trials with passively administered antisera were carried out during the first and second world wars. The results of these trials are generally inconclusive, but some reports clearly demonstrate the value of antisera, and it was generally accepted that antisera was of value in the prevention or treatment of gas gangrene [53]. For example, one study reports that the mean incubation period for the development of fulminant disease was increased from 33 hours to 68 hours when gas gangrene antisera was used prophylactically [81]. The greatest problem with the use of antisera given therapeutically might be the fact that the blood supply to infected tissues is invariably compromised, and this makes delivery of antibody to vulnerable tissues difficult. Additionally, it has been suggested that the generation of efficacious α-toxoids was not a straightforward task [53]. At least the latter problem has been resolved with the demonstration that a recombinant vaccine, derived from the carboxy-terminal domain of α-toxin, promotes high levels of antitoxic antibodies, and mice immunized with the vaccine were protected against and experimental gas gangrene infection. Such vaccines might have a utility for the prevention of disease in "at-risk" groups of humans [36], or for the generation of antisera for the treatment of disease.

Additionally, because α-toxin has been implicated in the etiology of a range of diseases in animals, an α-toxoid might be of value as a veterinary vaccine.

20.3 CHOLESTEROL-DEPENDENT CYTOLYSINS

20.3.1 Sources and Properties

The cholesterol-dependent cytolysins (CDCs) are the most widely distributed family of toxins currently known and are produced by at least 5 different genera of Gram-positive bacteria and nearly 25 different species [82]. Several CDCs have been identified in clostridial species (Table 20.3); however, it should be noted that only three clostridial CDCs, perfringolysin O [83–85], botulinolysin [86], and tetanolysin [87–89] have been purified and characterized. The CDCs are highly cytolytic, pore-forming toxins, and their activity is absolutely dependent on the presence of membrane cholesterol. Therefore, they are targeted to eukaryotic cells because of the cholesterol requirement. CDCs have been detected only in pathogenic bacterial species, although this does not preclude their presence in nonpathogenic bacteria. They form comparatively large pores that range in size from 250 to 300 Å and are sufficiently large to allow the passage of most proteins. The CDCs are responsible for the appearance of clear zone or β-hemolysis on blood agar plates.

The CDCs are expressed by the clostridia (as well as the other species) as highly soluble extracellular monomeric proteins that range in mass from 40 to 70 kDa. The sequenced clostridial CDC genes all exhibit a sec-dependent signal peptide that is presumably removed during secretion [90,91]. A *virR*/*virS* 2-component regulatory system has been shown to regulate *C. perfringens*-toxin (perfringolysin O or PFO) expression [92]. The expression of PFO was suggested to be positively regulated by the upstream *pfo*R gene [93]; however, recent studies have shown that an isogenic knockout of the *pfo*R gene did not affect PFO expression in *C. perfringens* [94]. Therefore, although *vir*R/*vir*S regulates PFO and other putative virulence genes in *C. perfringens,* apparently *pfo*R is not involved. The clostridial CDCs, as well as other CDCs, all appear to be easily expressed at high levels in recombinant forms when produced in *Escherichia coli* using a variety of inducible promoters and appear to fold properly whether secreted or expressed intracellularly in *E. coli* [91,95]. Purification of these toxins can be achieved relatively easily from their native sources by

TABLE 20.3
The Clostridial CDCs

CDC	Origin
Bifermentolysin****	*C. bifermentans*
Botulinolysin**	*C. botulinum*
Chauveolysin****	*C. chauvoei*
Histolyticolysin****	*C. histolyticum*
Novyilysin***	*C. novyi*
Perfringolysin*	*C. perfringens*
Septicolysin***	*C. septicum*
Sordellilysin****	*C. sordellii*
Tetanolysin*	*C. tetani*

* Gene sequence, purified protein
** Purified protein
*** Gene sequence
**** Uncharacterized

conventional chromatography; however, with the availability of modern molecular biological tools, it is much more efficient to clone and express these toxins in *E. coli* [95].

20.3.2 Phylogenetic Relationships

Fourteen CDCs have been sequenced from various bacterial species, and their primary structures exhibit a high degree of similarity with homologies that range from 40 to 70%. Of the clostridial CDCs, only *C. perfringens* [83], *C. tetani* [88], and *C. botulinum* [86] have been shown to definitively produce a CDC by its purification from the bacterial source. The primary structures of four CDCs have been determined from their gene sequence (Figure 20.8). Although the gene sequences for CDCs from *C. septicum* and *C. novyii* have been entered into the GenBank sequence database, it is unclear whether these organisms produce a CDC. Their sequences are surprisingly similar to that of *C. perfringens* CDC and perfringolysin O (PFO), both exhibiting nearly 98% sequence identity with the primary structure of PFO. The primary structure of tetanolysin clearly exhibits differences from the other three clostridial CDCs and only exhibits about 55% identity with PFO. These data suggest that the PFO type CDC was recently shared among *C. perfringens*, *C. novyi,* and *C. septicum,* since they exhibit little divergence in their primary structures.

A dendrogram generated from the primary structures of the sequenced CDCs shows that they appear to fall into three major branches (Figure 20.9). One branch includes the listerial CDCs that are the only pH-sensitive CDCs that exhibit significant cytolytic activity at pH 5.5 but little activity at neutral pH. Another branch includes the CDCs pyolysin and intermedilysin, which contain alanine instead of cysteine, and a proline for one of the tryptophan residues in the carboxy terminal undecapeptide peptide, which exhibits a highly conserved structure (ECTGLAWEWWR) and is a hallmark sequence in the CDC structure. All of the clostridial CDCs are found in the largest branch of this tree, which also includes the *Bacillus* CDCs and streptolysin O (SLO) from *Streptococcus pyogenes*.

It is clear that the CDCs have undergone divergent evolution, and, to some extent, this divergence might reflect a specific requirement of the pathogenic mechanism of the bacterium. Although structure-function differences that give rise to specific features that might enhance the compatibility of the CDC with a specific pathogenic mechanism are uncharacterized in the clostridia, there are examples of functional differences in CDCs from other genera that appear to be important to their pathogenic mechanisms. The *Streptococcus pneumoniae* CDC, pneumolysin, exhibits a complement binding activity that appears to be important in the mouse model for pneumonia [96] but not peritoneal infections [97]. Complement binding has not been shown to occur in other CDCs. The CDC listeriolysin O (LLO) has been shown to facilitate the escape of *Listeria monocytogenes*, a facultative intracellular parasite, from the phagosome of the macrophage. Unlike other CDCs, LLO is active at pH 5.5 but is largely inactive at neutral pH. Therefore, LLO is active in the phagosome containing the *L. monocytogenes* and is essential for its escape into the cytosol [98]. However, once the bacteria cell escapes into the macrophage cytoplasm, the activity of LLO must be eliminated to prevent lysis of the macrophage plasma membrane and allow the bacterium to successfully replicate within the protected environment of the macrophage cytoplasm [99,100]. The specific changes in the structure of the listerial CDCs that give rise to the pH sensitivity of the cytolytic mechanism remain unknown, although a single mutation in domain 4 of LLO effectively increases its activity at neutral pH and, therefore, reduces its ability to support intracellular replication of *L. monocytogenes* [99]. A PEST-like sequence is also present at the amino terminus of LLO and appears to be important in the regulation of its activity and might affect proteolytic breakdown of LLO once *L. monocytogenes* gains into the macrophage cytosol [101].

Other obvious structural differences that exist in the primary structure of CDCs include the lack of a sec-dependent signal peptide on pneumolysin and the presence of approximately 75 extra residues at the amino terminus of streptolysin O. The significance of these structural differences remains unknown, but why they exist remains a highly intriguing question.

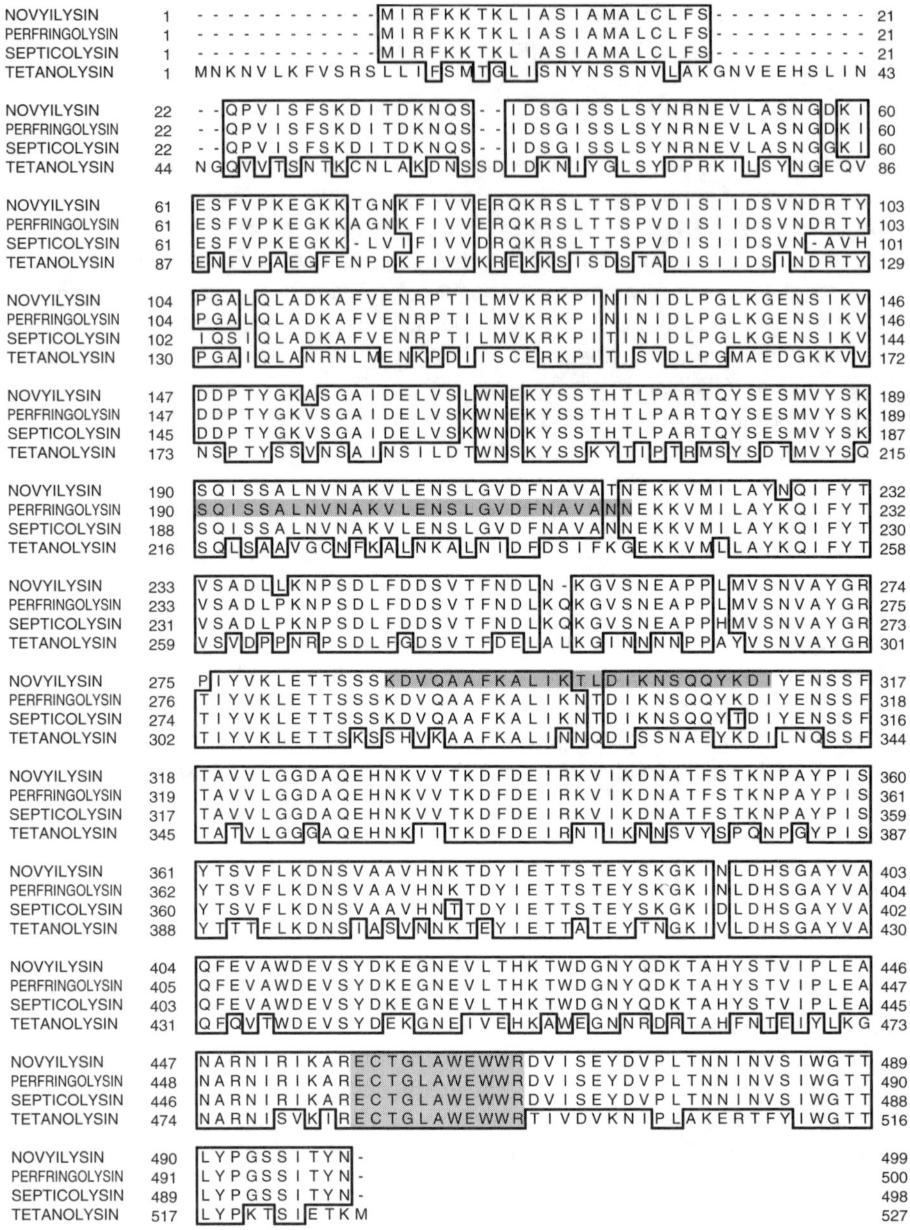

FIGURE 20.8 Alignment of the primary structures of sequenced clostridial CDCs. The primary structures of the four sequenced CDCs are shown. The carboxy terminal conserved undecapeptide sequence (ECTGAWEWWR) and the two regions in domain 3 that form the transmembrane β-hairpins 1 and 2 (residues 189–217 and 188–312, respectively, of PFO) of PFO are highlighted in grey. It should be noted that the sequences of novyilysin (novyilisin) and septicolysin (accession numbers AJ404676 and AJ276799, respectively) have been reported in GenBank, but have not been published.

20.3.3 Roles in Disease

The role of CDCs in clostridial disease remains largely unexplored. For most clostridial species, there is insufficient evidence to assess the contribution of the CDC to clostridial disease, although it appears that several clostridial pathogens produce a CDC. The widespread presence of CDCs in

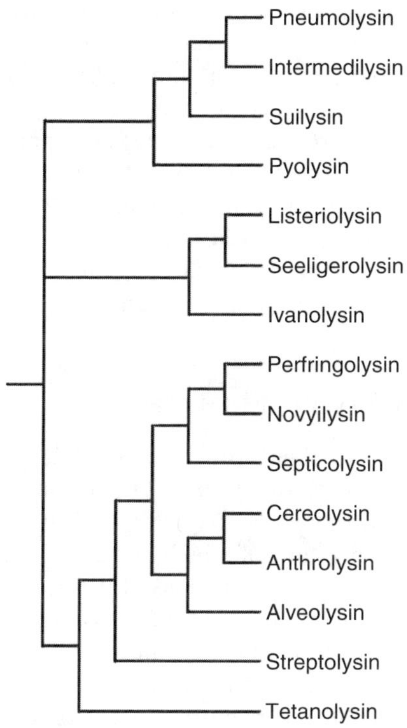

FIGURE 20.9 The phylogenetic relationships within the CDCs. The dendrogram showing the relationship of the DNA-derived primary structures of the CDCs was generated with the programs Clustalx [240,241] and Treeview X [242].

clostridial pathogens suggests that they are important to disease progression, but other than studies with perfringolysin O from *C. perfringens*, there is no available data on their roles in clostridial disease. Therefore, the discussion of the involvement of clostridial CDCs in disease will be restricted to the *C. perfringens* gas gangrene mouse model.

C. perfringens gas gangrene, or myonecrosis, is a fulminant infection that is defined by its restriction to the muscle. Typically, the infection is initiated when a traumatic tissue wound occurs in a major muscle and is contaminated by clostridial spores or vegetative cells. Contamination is most frequently from *C. perfringens* cells and spores present in the soil. Disease progression is rapid and is fatal if left untreated. Although massive muscle necrosis is typical, the victims of gas gangrene usually succumb to toxemia and shock. One hallmark of gas gangrene is the absence of polymorphonuclear lymphocytes (PMNs) from the necrotic tissue. It appears that PMNs exhibit leukostasis and do not penetrate into the infection. Therefore, the body's first line of immune defense is somehow circumvented by *C. perfringens*.

The available evidence suggests that PFO can play an important role in the inhibition of PMN migration into the necrotic tissue. Stevens and co-workers have found that PFO affects polymorphonuclear cell migration [102] and can suppress myocardial contractility [74]. Pure PFO can induce shock, and low concentrations can induce priming and degranulation of PMNs and upregulate the PMN-dependent adherence molecules integrin CD11/CD18. By using isogenic knockout mutants of *C. perfringens* that expressed PFO, α-toxin (PLC), or both, Awad et al. [103] showed that a PLC^-/PFO^- *C. perfringens* did not cause gas gangrene in the mouse model. The PLC^+PFO^- mutant exhibited most of the disease symptoms of wild type, but the progression of the disease was markedly reduced [46,103]. Mutants expressing just PFO did not succumb within the timeframe of the experiment (24 hours) and only exhibited a few of the symptoms associated with gas gangrene. Interestingly, histological examination of the infected tissue from the PLC^+PFO^- strain indicated that the loss of PFO also reduced the level of leukostasis associated with gas gangrene. Therefore, PFO appeared to specifically contribute to leukostasis. The $PLC^{+}PFO^{+}$ strain also exhibited a more rapid disease progression that more closely approximated that of the wild-type strain and showed that PLC and PFO functioned synergistically in the development of gas gangrene. Although it has been well established that the CDCs form a comparatively large pore on the membrane of susceptible cells and can affect several physiological parameters of various cells, how the CDC pore is used to induce leukostasis during myonecrosis remains unknown. Whether it is a direct lytic effect of the pore on the cell or a more subtle effect is also not clear.

One surprising finding of these studies is the fact that PFO did not appear to contribute significantly to tissue necrosis. This observation was unexpected since PFO is highly cytolytic to a variety of cell types.

20.3.4 Molecular Architecture and Mechanism

The crystal structure of PFO in its soluble form is the only CDC structure that has been solved to date [104] (Figure 20.10). Upon inspection of the PFO structure, it is immediately obvious that it is rich in β-sheet, a feature that is common among oligomerizing, pore-forming toxins [105–107]. The structure is divided into four domains of which domains 3 and 4 are directly involved in the interaction of PFO with the membrane. The functions of domains 1 and 2 remain unclear, but they likely undergo some molecular movement during the transition of domain 3 into the membrane. They may also be involved in the conformational coupling of domains 3 and 4, as described below.

The first step in the mechanism of PFO is membrane recognition, and the anchoring of the monomer to the membrane. It has been known for more than a decade that the carboxy terminal residues are involved with membrane recognition [108,109]. The conserved undecapeptide loop in domain 4 has been shown to penetrate the membrane [110,111] but does not penetrate it deeply [112]. More recently, it has been shown that in addition to the undecapeptide loop, the three other hydrophobic loops at the tip of the domain 4 β-sandwich insert into the bilayer surface and anchor PFO to the membrane [113]. The rest of domain 4 is apparently exposed to the aqueous milieu

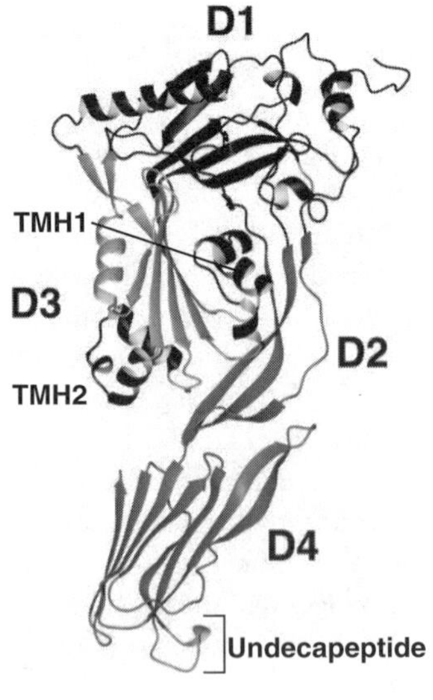

FIGURE 20.10 Ribbon representation of the PFO crystal structure showing domains 1-4 (D1-D4). Also shown are the locations of the two sets of α-helices that ultimately form the two transmembrane β-hairpins, TMH1 and TMH2, in D3, and the undecapeptide in D4. The ribbon structure was generated using MolMol [243] and PovRay (povray.org).

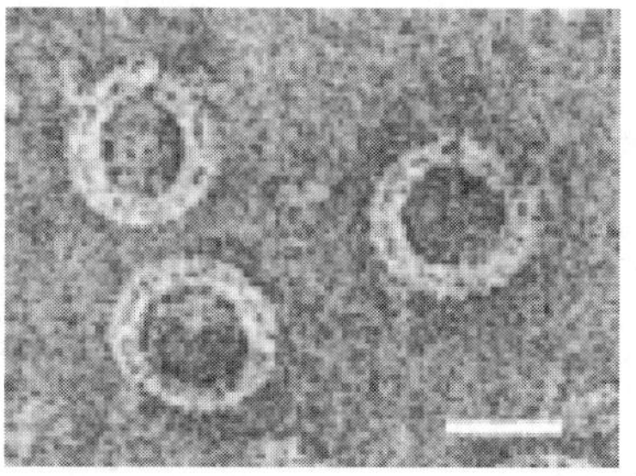

FIGURE 20.11 Electron micrograph of PFO membrane oligomers. Shown are several typical ring-shaped membrane oligomers of PFO. The bar represents 25 nm.

and does not contact either the membrane or neighboring monomers when oligomerized on the membrane surface. It is clear that the interaction of PFO with cholesterol occurs within domain 4 [114,115]. Since only the loops at the tip of the domain 4 β-sandwich interact with the membrane [113], presumably cholesterol interacts with one or more of these loops. It is tempting to speculate that the conserved undecapeptide interacts with cholesterol, since it is such a highly conserved feature of the CDCs. However, so far no direct evidence exists that this region is in contact with the cholesterol molecule, although mutations within the undecapeptide affect cell binding and interaction with cholesterol [116,117].

The bound monomers presumably diffuse laterally and interact with one another to form the large, ring-shaped membrane oligomer (Figure 20.11). The prepore oligomer is formed immediately prior to the formation of the oligomeric pore complex [118]. The prepore complex appears to be important to prepare the oligomer to subsequently convert to the pore complex by the insertion of the transmembrane β-barrel. The prepore must grow to a large size before it becomes insertion competent, since small pores do not appear to form early in the maturation process of the oligomeric complex [119].

The prepore is finally converted to the pore structure upon insertion of the transmembrane β-barrel. Although it is now clear that the tip of domain 4 anchors PFO to the membrane surface, it was discovered earlier that the transmembrane pore itself was formed from regions within domain 3. Domain 3 contains a core β-sheet, and two of the β-hairpins that form the core sheet have turns that are comprised of three short α-helices each (Figure 20.12). These α-helices have been shown to form the membrane-spanning β-barrel of PFO. They also have been shown to unfurl and form two extended amphipathic β-hairpins that cross the membrane bilayer [95,120]. Therefore, PFO is unique among pore-forming toxins in that there are two hairpins per monomer instead of one, as is found in other toxins that form a transmembrane β-barrel [107,121]. Also, the secondary structural transition of this region from α-helices to β-strands was unprecedented, and such changes in the secondary structure of proteins are rare [120]. Therefore, the final step in the PFO pore-forming mechanism (Figure 20.5) is the insertion of the two transmembrane β-hairpins of each monomer into the membrane. Since monomers in the oligomer are juxtaposed to one another, the β-hairpins of the neighboring monomers of the prepore complex interact to form the transmembrane β-barrel. Hence, the PFO transmembrane β-barrel may be comprised of up to 100 β-hairpins if 50 monomers comprise the membrane oligomer.

20.3.5 Interaction with the Cell

It is well known that cholesterol is essential to the cytolytic mechanism of the CDCs, and PFO has been shown to bind specifically to membrane cholesterol as well as the cholesterol-rich membrane raft domains [108,114,122–124]. Therefore, PFO appears to utilize cholesterol as a receptor; however, it is clear from studies with PFO and the *C. tetanii* CDC, tetanolysin, on artificial liposome

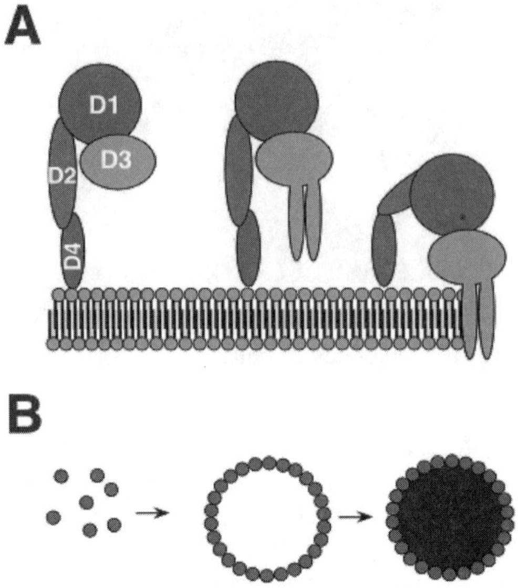

FIGURE 20.12 Model for the cytolytic mechanism of PFO. Shown in panel A are the transitions hypothesized for the various domains of PFO that occur in the monomer, prepore, and pore structures. Panel B shows the bird's-eye view of oligomer assembly. Pore formation only occurs as the domain 3 transmembrane. β-hairpins of each monomer in the prepore structure cooperatively insert to form the transmembrane β-barrel of the pore complex.

membranes, that the concentration of cholesterol required for binding is relatively high [89,112,125]. These data suggest that the presentation of cholesterol on the membrane, which may be influenced by its concentration, may also be important for PFO binding. However, there are other studies that suggest that although cholesterol is necessary to the mechanism of the CDCs, its role might not be to function only as a receptor. Jacobs et al. have shown that membrane cholesterol may function downstream of receptor binding, suggesting another role for the steroid [126]. Also, a recently discovered CDC from *Streptococcus intermedius*, intermedilysin (ILY), exhibits specificity for human cells having little or no effect on animal cells [127]. Although its mechanism appears to still require membrane cholesterol, it is difficult to envision how ILY maintains its specificity for human cells if cholesterol is the CDC receptor. Therefore, it is likely that the role of cholesterol in the mechanism of the CDCs is more complex than is currently understood.

As described above, during myonecrosis, it appears that PFO does not contribute significantly to the extensive tissue necrosis in the muscle, but does appear to exhibit an important contribution to the hallmark leukostasis observed during myonecrosis [128]. How PFO contributes to the leukostasis at the molecular level remains unclear; however, it is reasonable to assume that the PFO pore is utilized in this process. Recently, a discovery with the CDC from *Streptococcus pyogenes* (streptolysin O or SLO) showed that the CDC pore participated in a translocation process in which an NAD-glycohydrolase enzyme (SPN) was translocated from the bacterial cell into the cytoplasm of a keratinocyte [129]. In effect, the sec-dependent secretion system and the SLO pore were shown to be a functional analog of the type III secretion system found in various Gram-negative bacteria and is used to translocate cellular effectors directly into the cytoplasm of eukaryotic cells [130]. The SLO mediated introduction of the SPN into the keratinocyte had significant consequences for the eukaryotic cell, such that cell death was more rapid when SPN was tranlocated into the cytoplasm than if it was absent in an SPN^-/SLO^+ isogenic mutant of *S. pyogenes*. Therefore, it is possible that PFO and other CDCs can be used in the same way as is SLO, to facilitate the introduction of metabolic effector molecules into the eukaryotic cytosol. This possibility becomes even more intriguing for PFO in light of the recent revelation that the upstream *pfo*R gene that is juxtaposed to the PFO gene is not involved in the regulation of the PFO gene [94]. Furthermore, the *pfo*R protein appears to exhibit a high degree of similarity with the *fru*A gene, which is the fructose-specific enzyme IIC of the phosphotransferase system [131]. Hence, the intriguing possibility exists that the *pfo*R gene product could instead be an effector molecule that is translocated into eukaryotic cells similar to how the SPN of *S. pyogenes* is translocated into the cytosol of the keratinocyte by SLO.

20.3.6 Vaccines

Although there are reports of the CDCs being effective vaccines for various Gram-positive bacterial pathogens in animal models, to date there are no reports of their use as vaccines. Most clostridial animal vaccines likely contain a CDC since they are primarily derived from fermentation cultures and culture supernatants of various clostridial species. The contribution of the CDCs to protection in these complex vaccines remains unknown, but it is likely they contribute to the protection afforded by these vaccines. Other CDCs from nonclostridial sources that have been shown to be effective vaccines or vaccine components in experimental animal systems including suilysin from *Streptococcus suis* [132,133] and pneumolysin from *Streptococcus pneumoniae* [134].

20.4 AEROLYSIN-LIKE PORE FORMING TOXINS

20.4.1 Sources and Properties

The *C. perfringens* ε-toxin is produced only by types B and D strains [135] and appears to be plasmid encoded [136]. The *etx* gene encodes a 32.5 kD prototoxin, which is exported from the bacterial cell and then activated by proteolytic cleavage to yield a toxic protein of approximately 30.5 kD [137,138]. The protein can be readily purified from *C. perfringens* culture fluid [139] and, more recently, has been isolated from recombinant *E. coli* expressing the plasmid-cloned *etx* gene [140]. The toxicity of the prototoxin increases approximately 100-fold on proteolytic activation to the toxin [141] and for the activated toxins is typically reported as ranging from 50 to 100 ng/kg in mice [141,142]. A range of proteases, including trypsin [138,141], chymotrypsin [138], and the lambda-toxin, which is produced by some strains of *C. perfringens* [138,143], are able to activate the prototoxin. Activation was originally thought to involve proteolytic cleavage, which released a 13 or 14 amino-terminal peptide from the amino terminus of the protein [137,144,145]. More recently, proteolytic processing of the carboxy terminus of the protein has also been demonstrated [138]. The currently available evidence indicates that activation to the toxic form is dependent on proteolytic cleavage, which releases a 22 amino acid peptide from the carboxy terminus, whereas release of the amino-terminal peptide has no significant effect on toxicity [141].

The gene for *Clostridium septicum* α-toxin is chromosomal and encodes a 49.8 kDa protein. The α-toxin protein is secreted by *C. septicum* via a *sec*-dependent signal peptide [146,147]. The toxin can be purified from the culture supernatant of *C. septicum* [146], but the recombinant toxin can be expressed in *E. coli* and rapidly purified from the periplasmic fraction via a genetically engineered histidine tag at the carboxy terminus [148]. The LD_{50} of α-toxin for mice is 10 to 20 μg/kg body weight [146]. The secreted protein is inactive and requires the proteolytic cleavage of a 5 kDa carboxy-terminal propeptide to become cytolytically active [149], whereby it can lyse a variety of cell types. Cleavage of the propeptide can be accomplished *in vitro* with a variety of proteases that specifically cleave downstream of basic residues such as trypsin [149]. However, on the membrane of eukaryotic cells, it appears that furin or furin-like proteases may be the primary means of proteolytic activation [150]. Interestingly, although α-toxin is β-hemolytic, it does not appear to be activated on the surface of washed human erythrocytes, presumably because they lack an appropriate activating protease. Hemolysis can be induced by the addition of trypsin [149]. Even though erythrocytes are used as a common *in vitro* substrate to assay the cytolytic activity of α-toxin, it is unlikely that erythrocytes are a primary target of the toxin *in vivo* during *C. septicum* infections since they lack an activating protease.

20.4.2 Phylogenetic Relationships

The genes (*etxB* and *etxD*) encoding ε-toxin from a type B and a type D strain of *C. perfringens* have been nucleotide sequenced, and only two nucleotide differences were identified within the open reading frame, of which one results in an amino acid substitution [151]. The deduced amino acid sequence of ε-toxin does not reveal extensive homology with any other known bacterial toxin. However, there is 26% amino acid sequence similarity with the *Bacillus sphaericus* mosquitocidal toxins, which are also activated by proteolytic cleavage [152,153]. The significance of this sequence homology is not clear at this time. The most closely related toxin to *C. septicum* α-toxin is aerolysin from the Gram-negative pathogen *Aeromonas hydrophila* [154]. The similarity between these two toxins is based both in similarity of the primary structures (Figure 20.13) and molecular mechanisms. It has been shown that both toxins utilize GPI-anchored receptors [155–157], are activated by furin or furin-like enzymes [150,158], and each contains a carboxy-terminal propeptide [149,159]. The primary structures of each exhibit approximately 30% identity and 70% similarity [154] (Figure 20.13). It appears unlikely that these toxins have evolved by convergent evolution due to their similarities in structure and mechanism. Therefore, they are the only existing examples of a toxin gene that is found in both a Gram-negative and Gram-positive organism. It now appears that ε-toxin (discussed earlier) exhibits mechanistic similarities to both α-toxin and aerolysin. If

```
1     ...........LTNLEEGGY......ANHNNASSIK...................IFGYE
                 | .|::|..      .|:::| |:|                   |.| :
1     AEPVYPDQLRLFSLGQGVCGDKYRPVNREEAQSVKSNIVGMMGQWQISGLANGWVIMGPG

25    DN......................EDLKAKIIQDPE...............FIRNWANVA
       |                      ..|.| |.|.:               ||:  . :|
61    YNGEIKPGTASNTWCYPTNPVTGEIPTLSALDIPDGDEVDVQWRLVHDSANFIKPTSYLA

48    HSLGFGWCGGTANPNVGQGFEFKREVGAGGKVSYLLSARYNPNDPYASGYRAKDRLSMKI
      | ||::|.||. .. ||::.:..|: |.|      :  | | ||. ..|||. |: .:|:
121   HYLGYAWVGGNHSQYVGEDMDVTRD.GDG.....WVIRGN.NDGGCDGYRCGDKTAIKV

108   SNVRFVIDNDSIKLGTPKVKKLAPLNSASFDLINESKT.ESKLSKTFNYTTSKTVSKTDN
      ||. : :| ||:| |... .. . :...    :|:|.| :| .. |:.|.|... |||:.
173   SNFAYNLDPDSFKHGDVTQSDRQLVKTVVGWAVNDSDTPQSGYDVTLRYDTATNWSKTNT

167   FKFGEKIGVKTSFKVGLEAIADSKVETSFEFNAEQGWSNTNSTTETKQESTTYTATVSPQ
      : ::||:..|..|| .| : .     | |:|:.|:|:|.. |:...|.  |  .  :||.:.
233   YGLSEKVTTKNKFKWPLVGET....ELSIEIAANQSWASQNGGSTTTSLSQSVRPTVPAR

227   TKKRLFLDVLGSQIDIPYEGKIYMEYDIELMGFLRYTGNAREDHTEDRP....TVKLKFG
      .| .: :::. .:|..||| |  :.||:.| ||||:.|||: .|.::||    |. :
289   SKIPVKIELYKADISYPYEFKADVSYDLTLSGFLRWGGNAWYTHPDNRPNWNHTFVIGPY

283   KNGMSAEEHLKDLYSHKNINGYS.EWDWKWVDEKFGYLFKNSYDALTSRKLGGIIKGSFT
      |: |.   |      | |    |||.|. :. |.  ... | . |.: : |.|.|.
349   KDKASSIRYQWDKR...YIPGEVKWWDWNWTIQQNGLSTMQNNLARVLRPVRAGITGDFS

342   NINGTKIVIREGKEIPLP.DKKRRGKRSVDSLDARLQNEGIRIENIETQDVPGFRLN...
      . ..     |  | .:||: |.| |  ||||: :. |.|      ::.|::.|: :|
406   AESQFAGNIEIGAPVPLAADSKVRRARSVDGAGQGLRLEIP....LDAQELSGLGFNNVS

398   .SITYNDKKLILINNI
       |:|  ....
462   LSVTPAANQ.......
```

FIGURE 20.13 Alignment of the primary structures of α-toxin (upper sequence) and aerolysin (lower sequence). Vertical bars indicate identical residues, whereas two vertical dots represent conservative substitutions.

these toxins are related, then ε-toxin appears to be an even more distant relative of α-toxin than is aerolysin, since it exhibits only about 20% identity with either toxin, a level of identity that is not significantly different than these toxins exhibit with a randomized sequence.

20.4.3 Roles in Disease

C. perfringens types B and D are associated with a range of enterotoxemic diseases of domesticated livestock. For example, in lambs the disease caused by type B strains is called lamb dysentery, while diseases caused by type D strains are termed pulpy kidney [160]. These diseases are similar but not identical, and this reflects the overwhelming contribution of ε-toxin to disease caused by type D strains, whereas disease caused by type B strains involves both the ε- and β-toxins [160]. Disease caused by *C. perfringens* type D is often the result of a change from a poor to a rich diet, with the consequential overgrowth of the bacterium in the gut and the production of toxin [160]. The fatality rate is approaching 100%, and treatment of the disease is extremely difficult.

The symptoms of enterotoxemia caused by type D strains of *C. perfringens* are readily reproduced by administering toxin either by the i.g. or i.v. route into suspectible animals [161–163]. The ε-protoxin is produced in the gut of animals and activated by proteolysis. By an unknown mechanism, the toxin appears to cross the intestinal wall and enter the blood stream [164]. The toxin then appears to accumulate especially in the brain and in the kidneys [165,166]. With high doses of toxin, there are changes in the permeability of vascular endothelia in the brain within a few minutes [167–169]. As the disease progresses, serum proteins and, eventually, red cells leak from the vasculature, resulting in edema [170].

C. septicum causes a variety of diseases in both humans and animals, most of which are not well understood. In humans, it is a major cause of nontraumatic myonecrosis (gangrene) and necrotizing enterocolitis, infections that are often associated with a variety of predisposing disease syndromes such as colonic cancer, neutropenia, leukemia, and diabetes [171]. It is recommended that individuals that develop *C. septicum* infections should be examined for underlying carcinomas since it has been shown to be frequently associated with various cancers, especially colonic cancer [172]. More recently, *C. septicum* has been shown to be a superinfection in humans with *Escherichia coli* O157 hemolytic uremic syndrome (HUS) [173].

The most frequently reported and horrific disease caused by *C. septicum* is nontraumatic myonecrosis. The progression is rapid, often progressing from no clinically distinctive symptoms to complete necrosis of a major muscle by the organism within hours [174], and death usually ensues within 24 hours [171]. It is thought that the predisposing diseases result in direct (e.g., colonic tumors) or indirect (e.g., chemotherapy) damage to the colonic mucosa that allows the organism (or its spores) to gain access to the bloodstream from the intestinal tract. Presumably, if conditions are present in muscle tissue that allow the growth of *C. septicum,* the ensuing development of myonecrosis occurs. However, this route of disease development is largely conjectural since no animal model of this disease exists. Evidence that this route of infection occurs in humans is largely based on observation and anecdotal evidence. A high mortality rate, even with aggressive treatment, is due to the difficulty in early diagnosis of the infection and its rapid progression.

C. septicum is responsible for a variety of animal diseases including "braxy" in lambs, which occurs from the penetration of the abomasal lining, resulting in a fatal bacteremia. It also causes myonecrosis (i.e., "malignant edema") in cattle, usually from wound contamination. The frequency of myonecrosis in cattle is sufficiently high that it warrants vaccination of cattle herds.

The contribution of α-toxin to disease progression remains unclear, although it is likely that it plays a major role in disease progression and the usually fatal outcome of these infections. It is the only known lethal factor produced by *C. septicum,* with a LD_{50} in mice of about 10 μg/kg body weight [146]. Immunization of guinea pigs with formalin inactivated purified α-toxin provided complete protection against challenge with the organism [175].

20.4.4 Molecular Architecture and Mechanism

Little is know about the molecular architecture of ε-toxin, and although crystallization of the protein was reported in 1962 [142], the crystal structure of the toxin has not yet been reported. Early circular dichrosim studies indicated that the protein was predominantly in a beta-sheet conformation [176]. Chemical modification studies have indicated that one tryptophan [177], one histidine [178], one tyrosine [179], and three or four aspartic or glutamic acid residues [180] are essential for toxicity. More recently, site-directed mutants of the toxin have been tested for activity. Of the two histidine residues (H106 and H149) in the mature toxin, only the substitution of H149 with alanine resulted in a significant reduction in toxicity toward Madin-Darby Canine Kidney (MDCK) cells [181]. However, the single tryptophan residue in ε-toxin (W190) appears to play little or no role in toxicity [181].

The observation that edema is a common feature of animals suffering from *C. perfringens* enterotoxaemia led Worthington and Mülders [182] to suggest that the toxin is a permease. In support of this suggestion, the permeability of the blood–brain barrier was disrupted in mice treated with purified toxin, allowing the leakage of radiolabeled polyvinyl-pyrrolidone or human serum albumin into the brain [182]. It is only recently that any significant progress has been made in understanding the mode of action of ε-toxin. Epsilon-toxin is not hemolytic, and a major step forward in the study of the mode of action of ε-toxin was the finding that, unlike most cell lines, MDCK cells were highly sensitive to the action of the toxin [183,184]. Concentrations of ε-toxin as low as 8 ng/ml are reported to cause a significant reduction in the viability of cultured cells [183]. This cell line has been used to show that the cytotoxicity is temperature- and pH-dependent [185]. The binding of ε-toxin to MDCK cells appears to be correlated with the formation of a stable high molecular weight (155 kD) toxin-receptor complex [186,187]. By an unknown mechanism, the cells subsequently swell, show blebbing, and lyse [187,188].

The activated ε-toxin has been shown to bind with high affinity (K_Ds of 3.3 n*M*) to rat brain synaptosomes [189] and limited evidence suggests that the receptor for the toxin is a sialoglycoprotein [189,190]. After binding to either MDCK cells or rat synaptosomes, the toxin forms heptameric complexes within the cell membranes [141,191]. These heptamers appear to form preferentially within detergent-insoluble microdomains [191]. The molecular basis of toxicity of ε-toxin is not clear. The appearance of camp, glucose, catecholamines, adrenaline, and noradrenaline in the sera of intoxicated lambs is thought to be a consequence of brain edema and changes to the permeability of the blood–brain barrier.

As indicated above, *C. septicum* α-toxin exhibits sequence similarity with aerolysin. Based on the alignment of the primary structures of the two toxins (Figure 20.13), aerolysin contains an amino terminal region that is missing from α-toxin. A crystal structure for aerolysin has been determined, and it exhibits a high β-sheet content and a bilobal appearance with small and large lobes [105]. It is the large lobe of aerolysin that exhibits identity with α-toxin, and so it is possible that the 3D structure of α-toxin resembles the large lobe aerolysin (Figure 20.14). As described below, the small lobe of aerolysin appears to be responsible for the differences in receptor specificity of the two toxins.

Ballard et al. originally showed that α-toxin was a proteolytically activated, pore-forming toxin [146,149]. The α-toxin monomers were also found to oligomerize on the cellular membrane after proteolytic cleavage into a complex likely to be heptameric in size. The pore generated in the membrane by this complex was approximately 1.3 to 1.6 nm in diameter [149]. It was subsequently determined that the oligomer could form prior to the formation of the pore, suggesting that the pore-forming structures within the toxin did not penetrate the membrane until after the toxin had oligomerized into its membrane-bound oligomeric complex [192].

The 5 kDa carboxy-terminal propeptide was found to prevent oligomerization of α-toxin, unless it was proteolytic cleaved [149]. When the propeptide is cleaved, the activated toxin oligomerizes

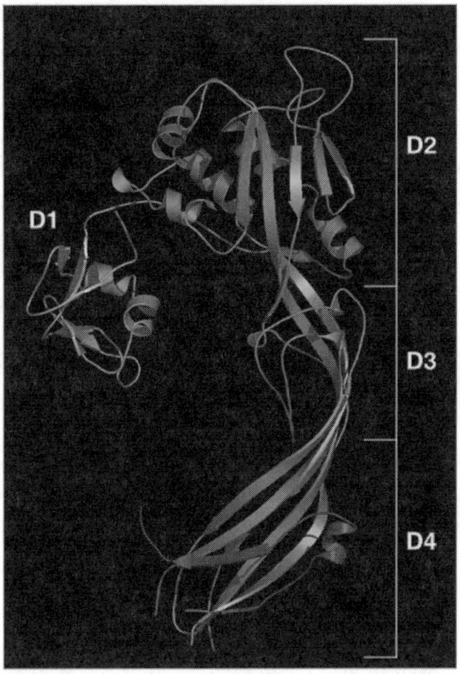

FIGURE 20.14 Molecular structure of aerolysin showing the small lobe (D1) and large lobe (D2–D4).

with other activated membrane-bound monomers of α-toxin into a heptameric complex. However, Sellman et al. [192] showed that the propeptide did not simply "fall off" the toxin after cleavage by protease. As suggested by the aerolysin crystal structure, there are many noncovalent interactions between the toxin and propeptide, suggesting that some additional force is necessary to displace the propeptide from the toxin after proteolysis. When the covalent bond between α-toxin and propeptide was cleaved by protease, the interaction of the monomers could displace the propeptide and, thus, form the oligomer [192]. However, if the toxin was not activated, then the propeptide could not be displaced, and, thus, it prevented oligomer formation. Furthermore, it was shown that the addition of excess synthetic propeptide to activated α-toxin could also prevent it from oligomerizing. This indicated that the excess propeptide could shift the equilibrium of the propeptide–toxin interaction such that the interaction of the activated monomers was not sufficient to displace the propeptide. They also found that if α-toxin was activated in solution, it would begin to form inactive solution oligomers. Therefore, proteolytic activation of α-toxin after it is bound to the cell surface is the optimal pathway for activation since activation prior to binding to the membrane results in the formation of an inactive oligomer and loss of activity. These characteristics showed that the propeptide functioned as an intramolecular chaperone sequence by preventing premature oligomerization (i.e., improper interactions) of the toxin until it was first bound to the membrane, where it could be activated and could oligomerize into an active complex [192].

C. septicum α-toxin recognizes and binds to a select group of GPI (glycosylphosphatidylinositol)-anchored proteins present on eukaryotic cells [193]. Aerolysin and α-toxin both utilize GPI-anchored proteins and recognize some of the same GPI-anchored receptors, but also appear to bind some different subsets of GPI-anchored receptors [193]. Aerolysin and α-toxin have been shown to recognize contactin and folate receptor on mammalian cells [155] and the GPI-anchored SAG1 and SAG3 proteins in the membrane of *Toxoplasma gondii* [194]. However, aerolysin binds Thy-1 and glycophorin, whereas α-toxin does not. Interestingly, α-toxin can be converted to an aerolysin-like molecule that binds the aerolysin-specific receptors by the fusion of the aerolysin small lobe (D1 in Figure 20.14) to the amino terminus of α-toxin [195]. These and other data [104,196–198] suggest that aerolysin can have two determinants for receptor, one in the small lobe and the second in the large lobe, whereas α-toxin appears to only have one determinant in the large lobe.

Once bound, α-toxin and aerolysin are activated by the membrane protease furin or furin-like enzymes [150,158]. Activation then allows the membrane-bound monomers to oligomerize into the membrane prepore complex that is then converted to the pore complex [192]. Although α-toxin appears to play a central role in the various diseases that result from infection with *C. septicum,* the specific molecular mechanisms by which α-toxin contributes to disease remains unclear. Injection of pure α-toxin results in the rapid death of mice, and α-toxin is the only lethal factor produced by cultures of *C. septicum* [146]. Only recently has it been shown that immunization with purified α-toxin can completely protect animals from experimental infections with *C. septicum* [175]. The formation of a pore in the membrane of a susceptible cell can result in cell lysis and death. Also, at sublytic levels of the toxin where there may be insufficient numbers of pores to cause cell lyses, it is possible that physiological perturbations of the cell may also be induced. Therefore, it is likely that α-toxin will have multiple effects on the infected host cells, but the dissection of these effects on the host will require additional studies.

20.4.5 Vaccines

Disease caused by *C. perfringens* type D is of great economic importance wherever animals are reared intensively. Vaccines containing ε-toxoid [199], which are prepared by treating *C. perfringens* culture fluid with formaldehyde, are widely used throughout the world to prevent disease in domesticated livestock. The vaccines are highly immunogenic [200] and appear to afford good protection against disease. As indicated above, a purified *C. septicum* α-toxoid has been shown to provide complete protection to guinea pigs challenged with a lethal number of *C. septicum* spores. Therefore, it is possible that α-toxin can be an effective vaccine for *C. septicum* disease, and it might be reasonable to utilize such a vaccine for at-risk human populations.

20.5 HLA-LIKE PORE-FORMING TOXINS

20.5.1 Sources and Properties

The α-toxin of *Staphylococcus aureus* is the prototypic of a class of pore-forming toxins that assemble from water soluble monomers into membrane-bound heptameric beta-barrel structures in target cell membranes [201–203]. The α-toxin is a potent cytolysin, and its activity toward a range of erythrocyte types is relatively easily measured in the laboratory [204]. On the basis of deduced amino acid sequence homology (Figure 20.15), it appears that several other bacterial toxins are likely to have similar pore-forming modes of action, including the β-toxin of *C. perfringens* [205]. β-toxin is produced only by type B and type C strains of the bacterium; indeed, production of this toxin is one of the criteria assigning strains into these biotypes. It is now known that the encoding gene (*cpb*) is located on a plasmid [136], and that the reported nucleotide sequences show little variation between type B and type C strains [205–207]. Although β-toxin can be isolated and purified from culture supernatant fluid [208], it is quite labile, and yields are often low. In part, this might reflect the importance of harvesting the cultures at the right time; the toxin is produced maximally during the logarithmic growth phase but levels decline rapidly as a result of proteolysis [209]. More recently, and to facilitate studies, recombinant β-toxin has been produced in *E. coli* [205] or in *Bacillus subtilis* [210]. Expression as a GST-fusion protein in *E. coli* allows simple purification of the protein [206], but removal of the GST moiety is required for full toxicity [206,211].

```
                           10        20        30        40        50        60        70        80        90       100
                            |         |         |         |         |         |         |         |         |         |
C.perfringens β-toxin  MKKKFISLVIVSSLLNGCLLSPTLVYANDIGKTTTITRNKTSDGYTIITQNDKQIISYQSVDSSSKNEDGFTASIDARFIDDKYSSEMTTLINLTGFMSS
S. aureus α-toxin      MKTRIVSSVTTTLLLGSILMNPVAGAADSDINIKTGTTDIGSN----TTVKTGDLVTYD-------KENGMHKKVFYSFIDDKNHNKKLLVIRTKGTIAG
                       **.:::* * .: **.. *:.*.   *:.  : .* * :  *:      * :  ::::*:        :*:*:  .:   *****  .:   :*. .* ::.

                          110       120       130       140       150       160       170       180       190       200
                            |         |         |         |         |         |         |         |         |         |
C.perfringens β-toxin  KKEDVIKKYNLHDVTNSTAINFPVRYSISILNESINENVKIVDSIPKNTISQKTVSNTMGYKIGGSIEIEEN-KPKASIESEYAESSTIEYVQPDFSTIQ
S. aureus α-toxin      -----QYRVYSEEGANKSGLAWPSAFKVQLQLP-DNEVAQISDYYPRNSIDTKEYMSTLTYGFNGNVTGDDTGKIGGLIGANVSIGHTLKYVQPDFKTIL
                            :   .: :*.:.: :*  :.:.:     ** .:* *  *:*:*. *   .*: * :.*.:  ::. *   . * :: : . *::******.**

                          210       220       230       240       250       260       270       280       290       300
                            |         |         |         |         |         |         |         |         |         |
C.perfringens β-toxin  TDHSTSKASWDTKFT---ETTRGNYNLKSNNPVYGNEMFMYGRYTNVPATENIIPDYQMSKLITGGLNPNMSVVLTAP--NGTEESIIKVKMERERNCYY
S. aureus α-toxin      ESPTDKKVGWKVIFNNMVNQNWGPYDRDSWNPVYGNQLFMKTRNGSMKAADNFLDPNKASSLLSSGFSPDFATVITMDRKASKQQTNIDVIYERVRDDYQ
                        . : .*..*.. *.    : . * *: .* ******::**  *   .: *::*::   : *.*::.*:.*:::.*:*      ..::: *.*  ** *: *

                          310       320       330       340
                            |         |         |         |
C.perfringens β-toxin  LNWNGANWVGQVYSRLAFDTPNVDSHIFTFKINWLTHKVTAI.
S. aureus α-toxin      LHWTSTNWKGTNTKDKWTDRSSER-----YKIDWEKEEMTN-.
                       *:*..:** *    .     * ..      :**:*  ..::*
```

FIGURE 20.15 Alignment of the deduced amino acid sequences of *C. perfringens* β-toxin and *S. aureus* α-toxin. Signal sequences are shown underlined. Histidine residues, which have previously been shown to be essential for the hemolytic activity of *S. aureus* α-toxin [244], are in bold type.

20.5.2 Roles in Disease

The LD_{50} dose of *C. perfringens* β-toxin is broadly similar to that of *S. aureus* α-toxin — less than 400 ng/kg, and making this one of the most potent toxins known [212]. The toxin is also active toward gastric mucosal surfaces, causing hemorrhagic necrosis and the destruction of intestinal villi [213], and this damage is similar to that seen in cases of necrotising enteritis caused by *C. perfringens* type C in humans and animals. Although necrotising enteritis mediated by β-toxin occurs in a range of animals including sheep, lambs, and calves [160], suckling pigs are especially susceptible to the disease, which is passed on from sows carrying *C. perfringens* [160,214–216]. In all cases, the disease is marked by hemorrhagic diarrhea, and the intestinal contents may contain up to 10^9 cfu of *C. perfringens* type C [160]. In some cases of the disease, there is neurological involvement. Death is the most likely outcome of the disease, and can occur so rapidly that in some parts of the U.K., the disease in sheep is referred to as "struck" because the animals appear to have died as the result of a lightning strike [160].

Disease in humans (which is often termed pig-bel) is linked to the prevalence of *C. perfringens* type C in pigs, and to the ingestion of contaminated pork [217,218]. However, a particular and unusual set of circumstances conspire to allow the development of disease. The toxin is especially sensitive to proteases [219], and in normal circumstances, the toxin would be rapidly destroyed. Cases of pig-bel arise when proteolytic activity in the gut is also markedly reduced, usually as a result of malnutrition or the ingestion of foods that contain protease inhibitors [219]. Sweet potatoes are especially implicated in the development of the disease [220]. This combination of factors, which leads to the development of the disease, is rare but does occur in parts of the world such as Papua New Guinea in Thailand, and in Vietnam [219,221,222]. Very rarely, the pig-bel is reported in individuals from developed countries, where underlying conditions, such as diabetes, can predispose individuals to disease [217].

20.5.3 Molecular Architecture and Mechanism

The precise mode of action of β-toxin is still not clear. The intravenous administration of toxin into rats results in an increase in blood pressure and a decrease in heart rate [223,224]. There is also the release of catecholamines [223], and it has also been suggested that the toxin acts on the autonomic nervous system, producing arterial constriction [224]. The amino acid sequence homology with *S. aureus* α-toxin would suggest that β-toxin is a pore-forming toxin, but, unlike α-toxin, β-toxin has not been reported to be hemolytic. However, there are some reports of cytotoxicity toward guinea pig monocytes [225] or Chinese hamster ovary cells [213], but many other cell types are apparently resistant to the toxin [213]. This issue of host cell specificity has also been highlighted by Steinthorsdottir et al., who showed that the toxin was able to form multimers in the cell membranes of human umbilical vein endothelial cells, but not in other cell types [226]. Multimer formation was associated with the release of arachidonic acid from these cells, but it is not clear whether this is a direct consequence of the activity of the toxin or merely the response of the cells to membrane perturbation.

The evidence that β-toxin is able to form pores in cell membranes comes from two studies. First, the release of inositol from human umbilical vein endothelial cells treated with β-toxin has been demonstrated, which would be consistent with pore formation [226]. More convincingly, the exposure of lipid bilayers to β-toxin resulted in a stepwise increased in their conductance, which was consistent with the formation of cation-selective pores [227]. The major types of channels formed by β-toxin were estimated to be 12.7 and 11.1 Å in diameter [227].

Other than that which can be inferred from the structure of *S. aureus* α-toxin, relatively little is known about the structure of the *C. perfringens* β-toxin. The *S. aureus* α-toxin assembles to form a mushroom-shaped heptamer [107,203], with a membrane-spanning channel of between 16 and

46 Å in diameter [228]. If the structure and mode of action of β-toxin are similar, then it may be able to form heptamers in target cell membranes. Early studies with purified β-toxin indicated a role for cysteine residues in toxicity, since oxidizing agents and reagents that react with sulfhydryl groups [such as 5,5′-dithio-bis(2-nitrobenzoic acid) (DTNB) and N-ethylmaleimide] were capable of abolishing toxicity [229]. However, more recent studies suggest that it is not the single cysteine residue, but rather the immediate molecular environment, that is critical for toxicity [210]. Replacement of the cysteine residue (Cys_{265}) with an amino acid with a bulky side chain (such as tyrosine or histidine) did reduce activity [211], possibly because of steric effects. Additionally, the replacement of amino acid flanking the cysteine residue, such as Tyr_{266}, Leu_{268}, or Trp_{275}, completely abolished lethality of the toxin [211]. To date, only one study, where Arg_{212} was changed to aspartate, has shown that a reduced ability to form pores in lipid bilayers is directly linked to a reduction in toxicity [227].

20.5.4 Vaccines

Several commercially available vaccines based on a β-toxoid are available for use in domesticated livestock, and immunization with these vaccines induces the production of neutralizing antibodies to the toxin [199]. For the prevention of disease in piglets, immunization of pregnant sows is a preferred option [215,230,231]. However, even in vaccinated populations, significant outbreaks of disease can occur [232,233]. These vaccines have occasionally been used in at-risk populations of humans [218]. For example, in Papua New Guinea, the incidence of pig-bel in Highland communities decreased by 80% after the introduction of vaccination [221]. While these vaccines are effective, they are not simple to produce because of the lability of the toxin. In the future, it is possible that the increasing knowledge of the molecular architecture of the toxin will allow genetic β-toxoids to be devised.

20.6 CONCLUSIONS

The clostridia are prolific producers of toxins that act on cell membranes, and this review has considered the main classes of membrane active toxins produced by the clostridia. However, there are some membrane active toxins that are less well-characterized and have not been discussed in this chapter. For example, the *C. perfringens* δ-toxin is a hemolytic protein [234] that shows a remarkable specificity for erythrocytes from even-toed ungulates (sheep, goats, and pigs). More recent studies have shown that the toxin binds to GM2 ganglioside receptors [235,236].

Although many of these membrane active toxins play important and, in some cases, critical roles in the pathogenesis of disease, the clostridia are most frequently found in the environment outside of animal hosts and especially in soil that is rich in organic matter. It seems likely that these proteins have evolved to support the lifestyle of the bacteria outside of mammalian hosts, rather than as virulence determinants per se. In the case of the phospholipases C, it is possible that these enzymes secreted from bacteria play a role in the scavenging of phosphate from decaying organic matter. The possible roles of the pore-forming toxins in nutrient acquisition are less easily explained.

The past decade has seen major advances in our understanding of the molecular basis of toxicity of several membrane active toxins produced by the clostridia. As our knowledge of the molecular biology of clostridial toxins has increased, so we have realized that these toxins are members of superfamilies of bacterial toxins. Often, the individual members of these superfamilies have very different properties. Comparisons of the deduced amino acid serquences and, more recently, the molecular structures is allowing us to address key questions on the way in which proteins in general interact with cell membranes. In the longer term, it might allow the properties of these toxins to be tailored to specific applications, for example in cancer therapy.

REFERENCES

1. MacFarlane, M.G. and Knight, B.C.J.G., The biochemistry of bacterial toxins I Lecithinase activity of *Cl welchii* toxins, 35, 884, 1941.
2. Titball, R.W., Membrane-damaging and cytotoxic phospholipases, in *The Comprehensive Sourcebook of Bacterial Protein Toxins*, Freer, J.H., Ed., Academic Press, London, 1999, 310.
3. Titball, R.W. and Rood, J.I., Bacterial phospholipases, in *Handbook of Experimental Pharmacology: Bacterial Protein Toxins*, Just, I., Ed., Springer-Verlag, Berlin, 2000, 529.
4. Gubash, S.M., Improved egg-yolk agar plate medium for the detection of Clostridial phospholipase-C activity, *Res. Microbiol.*, 142, 87, 1991.
5. Titball, R.W., Bacterial phospholipases-C, *Microbiol. Rev.*, 57, 347, 1993.
6. Titball, R.W., Hunter, S.E.C., Martin, K.L., Morris, B.C., Shuttleworth, A.D., Rubidge, T., Anderson, D.W., and Kelly, D.C., Molecular cloning and nucleotide sequence of the alpha-toxin (phospholipase-C) of *Clostridium perfringens*, *Infect. Immun.*, 57, 367, 1989.
7. Titball, R.W., Leslie, D.L., Harvey, S., and Kelly, D., Hemolytic and sphingomyelinase activities of *Clostridium perfringens* alpha-toxin are dependent on a domain homologous to that of an enzyme from the human arachidonic acid pathway, *Infect. Immun.*, 59, 1872, 1991.
8. Kurioka, S. and Matsuda, M., Phospholipase C assay using p-nitrophenolphosphorylcholine and its application to studying the metal and detergent requirement of the enzyme, *Anal. Biochem.*, 75, 281, 1976.
9. Synder, W.R., A continuous spectrophotometric assay for the *Bacillus cereus* phospholipase C using a thiophosphate analog: evaluation of assay conditions and chromogenic agents, *Anal. Biochem.*, 164, 199, 1987.
10. El-Sayed, M.Y., DeBose, C.D., Coury, L.A., and Roberts, M.F., Sensitivity of phospholipase C (*Bacillus cereus*) activity to phosphatidylcholine structural modifications, *Biochim. Biophys. Acta,* 837, 325, 1985.
11. Nagahama, M., Michiue, K., Kobayashi, K., and Sakurai, J., Effect of *Clostridium perfringens* alpha-toxin on phospholipid-cholesterol liposomes, *Jpn. J. Med. Sci. Biol.*, 49, 249, 1996.
12. Jepson, M., Howells, A., Bullifent, H.L., Bolgiano, B., Crane, D., Miller, J., Holley, J., Jayasekera, P., and Titball, R.W., Differences in the carboxy-terminal (putative phospholipid binding) domains of *Clostridium perfringens* and *Clostridium bifermentans* phospholipases C influence the hemolytic and lethal properties of these enzymes, *Infect. Immun.*, 67, 3297, 1999.
13. Lim, S.J., Koh, U.C., Lee, E.O., and Kim, C.K., Liposome-based assay for phospholipase C, *Bull. Korean Chem. Soc.*, 18, 761, 1997.
14. Clark, C., Briggs, D., Karasawa, T., Wang, X.M., Jayasekeera, P., Cole, A., Maegawa, T., Naylor, C.E., Moss, D.S., Miller, J., Nakamura, S., Basak, A., and Titball, R.W., *Clostridium absonum* α-toxin: new insights into Clostridial alpha-toxin substrate binding and specificity, *J. Mol. Biol.,* 333, 759, 2003.
15. Taguchi, R. and Ikezawa, H., Phospholipase C from *Clostridium novyi* type A.I., *Biochim. Biophys. Acta*, 409, 75, 1975.
16. Taguchi, R. and Ikezawa, H., Phosphatidyl inositol specific phospholipase C from *Clostridium novyi* Type A, *Arch. Biochem. Biophys.*, 186, 196, 1978.
17. Krug, E.L. and Kent, C., Phospholipase C from *Clostridium perfringens*: preparation and characterisation of homogenous enzyme, *Arch. Biochem. Biophys.*, 231, 400, 1984.
18. Ochi, S., Hashimoto, K., Nagahama, M.. and Sakurai, J., Hemolysis and phospholipid metabolism elicited by *Clostridium perfringens* alpha-toxin through activation of GTP-binding protein in rabbit erythrocytes, *Jpn. J. Med. Sci. Biol.*, 49, 266, 1996.
19. Takahashi, T., Sugahara, T.. and Ohsaka, A., Phospholipase C from *Clostridium perfringens*, *Methods Enzymol.*, 71, 710, 1981.
20. Flores-Diaz, M., Alape-Giron, A., Titball, R.W., Moos, M., Guillouard, I., Cole, S., Howells, A.M., von Eichel-Streiber, C., Florin, I.. and Thelestam, M., UDP-glucose deficiency causes hypersensitivity to the cytotoxic effect of *Clostridium perfringens* phospholipase C, *J. Biol. Chem.*, 273, 24433, 1998.
21. Jepson, M., Bullifent, H.L., Crane, D., Flores-Diaz, M., Alape-Giron, A., Jayasekeera, P., Lingard, B., Moss, D.. and Titball, R.W., Tyrosine 331 and phenylalanine 334 in *Clostridium perfringens* alpha-toxin are essential for cytotoxic activity, *FEBS Lett.*, 495, 172, 2001.

22. Titball, R.W., Fearn, A.M.. and Williamson, E.D., Biochemical and immunological properties of the C-terminal domain of the alpha-toxin of *Clostridium perfringens*, *FEMS Microbiol. Lett.*, 110, 45, 1993.
23. Walker, N., Holley, J., Naylor, C.E., Flores-Diaz, M., Alape-Giron, A., Carter, G., Carr, F.J., Thelestam, M., Keyte, M., Moss, D.S., Basak, A.K., Miller, J. and Titball, R.W., Identification of residues in the carboxy-terminal domain of *Clostridium perfringens* alpha-toxin (Phospholipase C) which are required for its biological activities, *Arch. Biochem. Biophys.*, 384, 24, 2000.
24. Tso, J.Y. and Siebel, C., Cloning and expression of the phospholipase-C gene from *Clostridium perfringens* and *Clostridium bifermentans*, *Infect. Immun.*, 57, 468, 1989.
25. Tsutsui, K., Minami, J., Matsushita, O., Katayama, S., Taniguchi, Y., Nakamura, S., Nishioka, M., and Okabe, A., Phylogenetic analysis of phospholipase-C genes from *Clostridium perfringens* Type-A to Type-E and *Clostridium novyi*, *J. Bacteriol.*, 177, 7164, 1995.
26. Karasawa, T., Wang, X.M., Maegawa, T., Michiwa, Y., Kita, H., Miwa, K., and Nakamura, S., *Clostridium sordellii* phospholipase C: gene cloning and comparison of enzymatic and biological activities with those of *Clostridium perfringens* and *Clostridium bifermentans* phospholipase C, *Infect. Immun.*, 71, 641, 2003.
27. Naylor, C.E., Eaton, J.T., Howells, A., Justin, N., Moss, D.S., Titball, R.W., and Basak, A.K., Structure of the key toxin in gas gangrene, *Nat. Struct. Biol.*, 5, 738, 1998.
28. Naylor, C.E., Jepson, M., Crane, D.T., Titball, R.W., Miller, J., Basak, A.K., and Bolgiano, B., Characterisation of the calcium-binding C-terminal domain of *Clostridium perfringens* alpha-toxin, *J. Mol. Biol.*, 294, 757, 1999.
29. Alape-Giron, A., Flores-Diaz, M., Guillouard, I., Naylor, C.E., Titball, R.W., Rucavado, A., Lomonte, B., Basak, A.K., Gutierrez, J.M., Cole, S.T., and Thelestam, M., Identification of residues critical for toxicity in *Clostridium perfringens* phospholipase C, the key toxin in gas gangrene, *Eur. J. Biochem.*, 267, 5191, 2000.
30. Eaton, J.T., Naylor, C.E., Howells, A.M., Moss, D.S., Titball, R.W., and Basak, A.K., Crystal structure of the *C. perfringens* alpha-toxin with the active site closed by a flexible loop region, *J. Mol. Biol.*, 319, 275, 2002.
31. Justin, N., Eaton, J., Naylor, C., Howells, A., Moss, D., Titball, R., and Basak, A., Structure function studies of *Clostridium perfringens* alpha-toxin; a gas gangrene causing protein, *J. Chem. Soc. Pak.*, 21, 239, 1999.
32. Justin, N., Walker, N., Bullifent, H.L., Songer, G., Bueschel, D.M., Jost, H., Naylor, C., Miller, J., Moss, D.S., Titball, R.W., and Basak, A.K., The first strain of *Clostridium perfringens* isolated from an avian source has an alpha-toxin with divergent structural and kinetic properties, *Biochemistry*, 41, 6253, 2002.
33. Ginter, A., Williamson, E.D., Dessy, F., Coppe, P., Bullifent, H., Howells, A., and Titball, R.W., Molecular variation between the alpha-toxins from the type strain (NCTC 8237) and clinical isolates of *Clostridium perfringens* associated with disease in man and animals, *Microbiology (U.K.)*, 142, 191, 1996.
34. Pons, J.L., Ageron, E., Mangeney, N., Tessedre, A.C., Niel, P., Leluan, G., and Grimont, P.A.D., Genetic relationships of *Clostridium bifermentans* and *Clostridium sordellii* analyzed by multilocus enzyme electrophoresis and DNA-DNA hybridization, *Syst. Appl. Microbiol.*, 20, 381, 1997.
35. Stevens, D.L. and Bryant, A.E., The role of clostridial toxins in the pathogenesis of gas gangrene, *Clin. Infect. Dis.*, 35, S93, 2002.
36. Titball, R.W. and Rood, J.I., *Clostridium perfringens*; Wound infections, in *Molecular Medical Microbiology*, Sussman, M., Ed., Academic Press, London, 2001, 1875.
37. Wadstrom, T., Advances in the purification of some bacterial protein toxins, in *Bacterial Toxins and Cell Membranes*, Wadstrom, T., Ed., Academic Press, New York, 1978, 9.
38. MacFarlane, M.G., Mechanisms of microbial pathogenicity, in *On the Biochemical Mechanisms of Action of Gas Gangrene Toxins*, Ollea, A.J., Ed., Cambridge University Press, Cambridge, 1955, 57.
39. Barzaghi, G., Cerletti, C., and Degaetano, G., Phospholipase C from *Clostridium perfringens* induces human platelet aggregation in plasma, *Thromb. Haemost.*, 59, 236, 1988.
40. Mitsui, K., Mitsui, N., and Hase, J., *Clostridium perfringens* exotoxins I. Purification and properties of the alpha-toxin, *Jpn. J. Exp. Med.*, 43, 65, 1973.

41. Ohsaka, A., Tsuchiya, M., Oshio, C., Miyaira, M., Suzuki, K., and Yamkawa, Y., Aggregation of platelets in the micro-circulation of the rat induced by the alpha-toxin (phospholipase C) of *Clostridium perfringens*, *Toxicon*, 16, 333, 1978.
42. Sugahara, T., Takahashi, T., Yamaya, S., and Ohsaka, A., *In vitro* aggregation of platelets induced by alpha-toxin (phospholipase C) of *Clostridium perfringens*, *Jpn. J. Med. Sci. Biol.*, 29, 255, 1977.
43. Boyd, N.A., Thomson, R.O., and Walker, P.D., The prevention of experimental *Clostridium novyi* and *Cl. perfringens* gas gangrene in high-velocity missile wounds by active immunisation, *J. Med. Microbiol.*, 5, 467, 1972.
44. Kameyama, S., Sato, H., and Murata, R., The role of a-toxin of *Clostridium perfringens* in experimental gas gangrene in guinea pigs, *Jpn. J. Med. Sci. Biol.*, 25, 200, 1975.
45. Williamson, E.D. and Titball, R.W., A genetically engineered vaccine against the alpha-toxin of *Clostridium perfringens* protects mice against experimental gas-gangrene, *Vaccine*, 11, 1253, 1993.
46. Awad, M.M., Bryant, A.E., Stevens, D.L., and Rood, J.I., Virulence studies on chromosomal alpha-toxin and theta-toxin mutants constructed by allelic exchange provide genetic evidence for the essential role of alpha-toxin in *Clostridium perfringens*-mediated gas-gangrene, *Mol. Microbiol.*, 15, 191, 1995.
47. Annett, C.B., Viste, J.R., Chirino-Trejo, M., and Classen, H.L., Necrotic enteritis: effect of barley, wheat and corn diets on proliferation of *Clostridium perfringens* type A, *Avian Pathol.*, 31, 599, 2002.
48. Craven, S.E., Colonization of the intestinal tract by *Clostridium perfringens* and fecal shedding in diet-stressed and unstressed broiler chickens, *Poult. Sci.*, 79, 843, 2000.
49. Heier, B.T., Lovland, A., Soleim, K.B., Kaldhusdal, M., and Jarp, J., A field study of naturally occurring specific antibodies against *Clostridium perfringens* alpha toxin in Norwegian broiler flocks, *Avian Dis.*, 45, 724, 2001.
50. Wierup, M., The Swedish experience of the 1986 year ban of antimicrobial growth promoters, with special reference to animal health, disease prevention, productivity, and usage of antimicrobials, *Microb. Drug Resist. Mechan. Epidemiol. Dis.*, 7, 183, 2001.
51. Manteca, C., Daube, G., Pirson, V., Limbourg, B., Kaeckenbeeck, A., and Mainil, J.G., Bacterial intestinal flora associated with enterotoxaemia in Belgian Blue calves, *Vet. Microbiol.*, 81, 21, 2001.
52. Cortinas, T.I., Mattar, M.A., and deGuzman, A.M.S., Alpha-toxin production by *Clostridium septicum* at different culture conditions, *Anaerobe*, 3, 199, 1997.
53. MacLennan, J.D., The histotoxic clostridial infections of man, *Bact. Revs.*, 26, 177, 1962.
54. Nakamura, S., Shimamura, T., Hayase, M., and Nishida, S., Numerical taxonomy of saccharolytic clostridia particularly Clostridium perfringens-like strains:descriptions of *Clostridium absonum* sp. n. and *Clostridium paraperfringens*, 23, 419, 1973.
55. Hough, E., Hansen, L.K., Birkness, B., Jynge, K., Hansen, S., Hordik, A., Little, C., Dodson, E., and Derewenda, Z., High resolution (1.5A) crystal structure of phospholipase C from *Bacillus cereus.*, *Nature*, 338, 357, 1989.
56. Moreau, H., Pieroni, G., Jolivetreynaud, C., Alouf, J.E., and Verger, R., A new kinetic approach for studying phospholipase-C (*Clostridium perfringens* alpha toxin) activity on phospholipid monolayers, *Biochemistry*, 27, 2319, 1988.
57. Titball, R.W., Naylor, C.E., Miller, J., Moss, D.S., and Basak, A.K., Opening of the active site of *Clostridium perfringens* alpha-toxin may be triggered by membrane binding, *Int. J. Med. Microbiol.*, 290, 357, 2000.
58. van Deenen, L.L.M., Demel, R.A., Guerts van Kessel, W.S.M., Kamp, H.H., Roelofsen, B., Verkleij, A.J., Wirtz, K.W.A., and Zwaal, R.F.A., Phospholipases and monolayers as tools in studies on membrane structure, in *The Structural Basis of Membrane Function*, Djavadi-Ohaniance, L., Ed., Academic Press, New York and London, 1976, 21.
59. Verkleij, A.J., Zwaal, R.F.A., Roelofsen, B., Comfurius, P., Kastelijn, D., and van Deenen, L.L.M., The asymmetric distribution of phospholipids in the human red cell membrane. A combined study using phospholipases and freeze-etch electron microscopy, 323, 178, 1973.
60. Gilmore, M.S., Cruzrodz, A.L., Leimeisterwachter, M., Kreft, J., and Goebel, W., A *Bacillus cereus* cytolytic determinant, cereolysin-AB, which comprises the phospholipase C and sphingomyelinase genes — nucleotide sequence and genetic linkage, *J. Bacteriol.*, 171, 744, 1989.
61. Sakurai, J., Ochi, S., and Tanaka, H., Evidence for coupling of *Clostridium perfringens* alpha-toxin induced hemolysis to stimulated phosphatidic acid formation in rabbit erythrocytes, *Infect. Immun.*, 61, 3711, 1993.

62. Sakurai, J., Ochi, S., and Tanaka, H., Regulation of *Clostridium perfringens* alpha-toxin activated phospholipase-C in rabbit erythrocyte membranes, *Infect. Immun.*, 62, 717, 1994.
63. Samuelsson, B., Leukotrienes: mediators of immediate hypersensitivity reactions and inflammation, *Science*, 220, 568, 1983.
64. Gustafson, C. and Tagesson, C., Phospholipase-C from *Clostridium perfringens* stimulates phospholipase-A2-mediated arachidonic acid release in cultured intestinal epithelial cells (Int-407), *Scand. J. Gastroenterol.*, 25, 363, 1990.
65. Fujii, Y. and Sakurai, J., Contraction of the rat isolated aorta caused by *Clostridium perfringens* alpha toxin (phospholipase-C) — evidence for the involvement of arachidonic acid metabolism, *Br. J. Pharmacol.*, 97, 119, 1989.
66. Berridge, M.J., Inositol triphosphate and diacylglycerol: two interacting secondary messengers, *Ann. Rev. Biochem.*, 56, 159, 1987.
67. Ochi, S., Miyawaki, T., Matsuda, H., Oda, M., Nagahama, M., and Sakurai, J., *Clostridium perfringens* alpha-toxin induces rabbit neutrophil adhesion, *Microbiology (U.K.)*, 148, 237, 2002.
68. Sakurai, J., Fujii, Y., and Shirotani, M., Contraction induced by *Clostridium perfringens* alpha toxin in the isolated rat ileum, *Toxicon*, 28, 411, 1990.
69. Sakurai, J., Tsuchiya, Y., Ochi, S., and Fujii, Y., Effect of *Clostridium perfringens* alpha-toxin on contraction of isolated guinea-pig diaphragm, *Microbiol. Immunol.*, 35, 481, 1991.
70. Hashizume, T., Sato, T., and Fujii, T., Sphingosine enhances platelet aggregation through an increase in phospholipase-C activity by a protein kinase-C-independent mechanism, *Biochem. J.*, 282, 243, 1992.
71. Bryant, A.E., Chen, R.Y.Z., Nagata, Y., Wang, Y., Lee, C.H., Finegold, S., Guth, P.H., and Stevens, D.L., Clostridial gas gangrene. II. Phospholipase C-induced activation of platelet gpIIbIIIa mediates vascular occlusion and myonecrosis in *Clostridium perfringens* gas gangrene, *J. Infect. Dis.*, 182, 808, 2000.
72. Bryant, A.E., Bayer, C.R., Hayes-Schroer, S.M., and Stevens, D.L., Activation of platelet gpIIbIIIa by phospholipase C from *Clostridium perfringens* involves store-operated calcium entry, *J. Infect. Dis.*, 187, 408, 2003.
73. McNee, J.W. and Dunn, J.S., The method of spread of gas gangrene into living muscle, 1, 727, 1917.
74. Bryant, A.E., Bergstrom, R., Zimmerman, G.A., Salyer, J.L., Hill, H.R., Tweten, R.K., Sato, H., and Stevens, D.L., *Clostridium perfringens* invasiveness is enhanced by effects of theta-toxin upon PMNL structure and function — the roles of leukocytotoxicity and expression of Cd11/Cd18 adherence glycoprotein, *FEMS Immunol. Med. Microbiol.*, 7, 321, 1993.
75. Bryant, A.E., Chen, R.Y.Z., Nagata, Y., Wang, Y., Lee, C.H., Finegold, S., Guth, P.H., and Stevens, D.L., Clostridial gas gangrene. I. Cellular and molecular mechanisms of microvascular dysfunction induced by exotoxins of *Clostridium perfringens*, *J. Infect. Dis.*, 182, 799, 2000.
76. Bryant, A. and Stevens, D., Phospholipase C and perfringolysin O from *Clostridium perfringens* upregulate endothelial cell adherence molecule 1 and intercellular leukocyte adherence molecule 1 expression and induce interleukin-8 synthesis in cultured human umbilical vein endothelial cells, *Infect. Immun.*, 64, 358, 1996.
77. Stevens, D.L., Tweten, R.K., Awad, M.M., Rood, J.I., and Bryant, A.E., Clostridial gas gangrene: evidence that alpha and theta toxins differentially modulate the immune response and induce acute tissue necrosis, *J. Infect. Dis.*, 176, 189, 1997.
78. Hungerland, E., Eiring, P., and Bultmann, B., Septicaemia with intravascular haemolysis caused by *Clostridium sordellii*, *Dtsch. Med. Wochenschr.*, 122, 1281, 1997.
79. Stevens, D.L. and Bryant, A.E., Pathogenesis of *Clostridium perfringens* infections: mechanisms and mediators of shock, *Clin. Infect. Dis.*, 25, S160, 1997.
80. Asmuth, D.M., Olson, R.D., Hackett, S.P., Bryant, A.E., Tweten, R.K., Tso, J.Y., Zollman, T., and Stevens, D.L., Effects of *Clostridium perfringens* recombinant and crude phospholipase C and theta-toxin on rabbit hemodynamic parameters, *J. Infect. Dis.*, 172, 1317, 1995.
81. MacLennan, J.D. and MacFarlane, M.G., The treatment of gas gangrene, *Brit. Med. J.*, II, 683, 1944.
82. Alouf, J.E., Cholesterol-binding cytolytic protein toxins, *Int. J. Med. Microbiol.*, 290, 351, 2000.
83. Hauschild, A.H., Lecroisey, A., and Alouf, J.E., Purification of *Clostridium perfringens* type C theta toxin, *Can. J. Microbiol.*, 19, 881, 1973.
84. Smyth, C.J., The identification and purification of theta-haemolysin (theta-toxin) of *Clostridium perfringens* type A, *J. Gen. Microbiol.*, 87, 219, 1975.

85. Yamakawa, Y., Ito, A., and Sato, H., Theta-toxin of *Clostridium perfringens*; purification and some properties, *Biochim. Biophys. Acta*, 494, 301, 1977.
86. Haque, A., Sugimoto, N., Horiguchi, Y., Okabe, T., Miyata, T., Iwanaga, S., and Matsuda, M., Production, purification, and characterization of botulinolysin, a thiol-activated hemolysin of *Clostridium botulinum*, *Infect. Immun.*, 60, 71, 1992.
87. Alving, C.R., Habig, W.H., Urban, K.A., and Hardegree, M.C., Cholesterol-dependent tetanolysin damage to liposomes, *Biochim. Biophys. Acta*, 551, 224, 1979.
88. Mitsui, N., Mitsui, K., and Hase, J., Purification and some properties of tetanolysin, *Microbiol. Immunol.*, 24, 575, 1980.
89. Rottem, S., Cole, R.M., Habig, W.H., Barile, M.F., and Hardegree, M.C., Structural characteristics of tetanolysin and its binding to lipid vesicles, *J. Bacteriol.*, 152, 888, 1982.
90. Tweten, R.K., Nucleotide sequence of the gene for perfringolysin O (theta-toxin) from *Clostridium perfringens*: significant homology with the genes for streptolysin O and pneumolysin, *Infect. Immun.*, 56, 3235, 1988.
91. Tweten, R.K., Cloning and expression in *Escherichia coli* of the perfringolysin O (theta-toxin) gene from *Clostridium perfringens* and characterization of the gene product, *Infect. Immun.*, 56, 3228, 1988.
92. Lyristis, M., Bryant, A.E., Sloan, J., Awad, M.M., Nisbet, I.T., Stevens, D.L., and Rood, J.I., Identification and molecular analysis of a locus that regulates extracellular toxin production in *Clostridium perfringens*, *Mol. Microbiol.*, 12, 761, 1994.
93. Shimizu, T., Okabe, A., Minami, J., and Hayashi, H., An upstream regulatory sequence stimulates expression of the perfringolysin O gene of *Clostridium perfringens*, *Infect. Immun.*, 59, 137, 1991.
94. Awad, M.M. and Rood, J.I., Perfringolysin O expression in *Clostridium perfringens* is independent of the upstream *pfoR* gene, *J. Bacteriol.*, 184, 2034, 2002.
95. Shepard, L.A., Heuck, A.P., Hamman, B.D., Rossjohn, J., Parker, M.W., Ryan, K.R., Johnson, A.E., and Tweten, R.K., Identification of a membrane-spanning domain of the thiol-activated pore-forming toxin *Clostridium perfringens* perfringolysin O: an alpha-helical to beta-sheet transition identified by fluorescence spectroscopy, *Biochemistry*, 37, 14563, 1998.
96. Rubins, J.B., Charboneau, D., Fasching, C., Berry, A.M., Paton, J.C., Alexander, J.E., Andrew, P.W., Mitchell, T.J., and Janoff, E.N., Distinct roles for pneumolysin's cytotoxic and complement activities in the pathogenesis of pneumococcal pneumonia, *Am. J. Respir. Crit. Care Med.*, 153, 1339, 1996.
97. Berry, A.M., Alexander, J.E., Mitchell, T.J., Andrew, P.W., Hansman, D., and Paton, J.C., Effect of defined point mutations in the pneumolysin gene on the virulence of *Streptococcus pneumoniae*, *Infect. Immun.*, 63, 1969, 1995.
98. Portnoy, D., Jacks, P.S., and Hinrichs, D., The role of hemolysin for intracellular growth of *Listeria monocytogenes*, *J. Exp. Med.*, 167, 1459, 1988.
99. Glomski, I.J., Gedde, M.M., Tsang, A.W., Swanson, J.A., and Portnoy, D.A., The *Listeria monocytogenes* hemolysin has an acidic pH optimum to compartmentalize activity and prevent damage to infected host cells, *J. Cell. Biol.*, 156, 1029, 2002.
100. Jones, S. and Portnoy, D.A., Characterization of *Listeria monocytogenes* pathogenesis in a strain expressing perfringolysin O in place of listeriolysin O, *Infect. Immun.*, 62, 5608, 1994.
101. Decatur, A.L. and Portnoy, D.A., A PEST-like sequence in listeriolysin O essential for *Listeria monocytogenes* pathogenicity, *Science*, 290, 992, 2000.
102. Stevens, D.L., Mitten, J., and Henry, C., Effects of alpha-toxins and theta-toxins from *Clostridium perfringens* on human polymorphonuclear leukocytes, *J. Infect. Dis.*, 156, 324, 1987.
103. Awad, M.M., Ellemor, D.M., Boyd, R.L., Emmins, J.J., and Rood, J.I., Synergistic effects of alpha-toxin and perfringolysin O in *Clostridium perfringens*-mediated gas gangrene, *Infect. Immun.*, 69, 7904, 2001.
104. Rossjohn, J., Feil, S.C., McKinstry, W.J., Tweten, R.K., and Parker, M.W., Structure of a cholesterol-binding thiol-activated cytolysin and a model of its membrane form, *Cell*, 89, 685, 1997.
105. Parker, M.W., Buckley, J.T., Postma, J.P.M., Tucker, A.D., Leonard, K., Pattus, F., and Tsernoglou, D., Structure of the *Aeromonas* toxin proaerolysin in its water-soluble and membrane-channel states, *Nature*, 367, 292, 1994.
106. Petosa, C., Collier, R.J., Klimpel, K.R., Leppla, S.H., and Liddington, R.C., Crystal structure of the anthrax toxin protective antigen, *Nature*, 385, 833, 1997.

107. Song, L.Z., Hobaugh, M.R., Shustak, C., Cheley, S., Bayley, H., and Gouaux, J.E., Structure of staphylococcal alpha-hemolysin, a heptameric transmembrane pore, *Science*, 274, 1859, 1996.
108. Iwamoto, M., Ohno-Iwashita, Y., and Ando, S., Effect of isolated C-terminal fragment of theta-toxin (perfringolysin-O) on toxin assembly and membrane lysis, *Eur. J. Biochem.*, 194, 25, 1990.
109. Tweten, R.K., Harris, R.W., and Sims, P.J., Isolation of a tryptic fragment from *Clostridium perfringens* theta toxin that contains sites for membrane-binding and self-aggregation, *J. Biol. Chem.*, 266, 12449, 1991.
110. Nakamura, M., Sekino, N., Iwamoto, M., and Ohnoiwashita, Y., Interaction of theta-toxin (perfringolysin-O), a cholesterol-binding cytolysin, with liposomal membranes — change in the aromatic side-chains upon binding and insertion, *Biochemistry*, 34, 6513, 1995.
111. Nakamura, M., Sekino-Suzuki, N., Mitsui, K., and Ohno-Iwashita, Y., Contribution of tryptophan residues to the structural changes in perfringolysin O during interaction with liposomal membranes, *J. Biochem. (Tokyo)*, 123, 1145, 1998.
112. Heuck, A.P., Hotze, E.M., Tweten, R.K., and Johnson, A.E., Mechanism of membrane insertion of a multimeric beta-barrel protein: perfringolysin O creates a pore using ordered and coupled conformational changes, *Mol. Cell.*, 6, 1233, 2000.
113. Ramachandran, R., Heuck, A.P., Tweten, R.K., and Johnson, A.E., Structural insights into the membrane-anchoring mechanism of a cholesterol-dependent cytolysin, *Nat. Struct. Biol.*, 9, 823, 2002.
114. Shimada, Y., Maruya, M., Iwashita, S., and Ohno-Iwashita, Y., The C-terminal domain of perfringolysin O is an essential cholesterol-binding unit targeting to cholesterol-rich microdomains, *Eur. J. Biochem.*, 269, 6195, 2002.
115. Shimada, Y., Nakamura, M., Naito, Y., Nomura, K., and Ohno-Iwashita, Y., C-terminal amino acid residues are required for the folding and cholesterol binding property of perfringolysin O, a pore-forming cytolysin, *J. Biol. Chem.*, 274, 18536, 1999.
116. Iwamoto, M., Ohno-Iwashita, Y., and Ando, S., Role of the essential thiol group in the thiol-activated cytolysin from *Clostridium perfringens*, *Eur. J. Biochem.*, 167, 425, 1987.
117. SekinoSuzuki, N., Nakamura, M., Mitsui, K.I., and OhnoIwashita, Y., Contribution of individual tryptophan residues to the structure and activity of theta-toxin (perfringolysin O), a cholesterol-binding cytolysin, *Eur. J. Biochem.*, 241, 941, 1996.
118. Shepard, L.A., Shatursky, O., Johnson, A.E., and Tweten, R.K., The mechanism of assembly and insertion of the membrane complex of the cholesterol-dependent cytolysin perfringolysin O: formation of a large prepore complex, *Biochemistry*, 39, 10284, 2000.
119. Heuck, A.P., Tweten, R.K., and Johnson, A.E., Assembly and topography of the prepore complex in cholesterol-dependent cytolysins, *J. Biol. Chem.*, preprint: May, 30, 2003.
120. Shatursky, O., Heuck, A.P., Shepard, L.A., Rossjohn, J., Parker, M.W., Johnson, A.E., and Tweten, R.K., The mechanism of membrane insertion for a cholesterol-dependent cytolysin: a novel paradigm for pore-forming toxins, *Cell*, 99, 293, 1999.
121. Benson, E.L., Huynh, P.D., Finkelstein, A., and Collier, R.J., Identification of residues lining the anthrax protective antigen channel, *Biochemistry*, 37, 3941, 1998.
122. Iwamoto, M., Morita, I., Fukuda, M., Murota, S., Ando, S., and Ohno-Iwashita, Y., A biotinylated perfringolysin O derivative: a new probe for detection of cell surface cholesterol, *Biochim. Biophys. Acta-Biomembr.*, 1327, 222, 1997.
123. Ohno-Iwashita, Y., Iwamoto, M., Mitsui, K., Ando, S., and Nagai, Y., Protease nicked θ-toxin of *Clostridium perfringens*, a new membrane probe with no cytolytic effect, reveals two classes of cholesterol as toxin-binding sites on sheep erythrocytes, *Eur. J. Biochem.*, 176, 95, 1988.
124. Waheed, A.A., Shimada, Y., Heijnen, H.F.G., Nakamura, M., Inomata, M., Hayashi, M., Iwashita, S., Slot, J.W., and Ohno-Iwashita, Y., Selective binding of perfringolysin O derivative to cholesterol-rich membrane microdomains (rafts), *Proc. Natl. Acad. Sci. U.S.A.*, 98, 4926, 2001.
125. Blumenthal, R. and Habig, W.H., Mechanism of tetanolysin-induced membrane damage: studies with black lipid membranes, *J. Bacteriol.*, 157, 321, 1984.
126. Jacobs, T., Darji, A., Frahm, N., Rohde, M., Wehland, J., Chakraborty, T., and Weiss, S., Listeriolysin O: cholesterol inhibits cytolysis but not binding to cellular membranes, *Mol. Microbiol.*, 28, 1081, 1998.
127. Nagamune, H., Ohnishi, C., Katsuura, A., Fushitani, K., Whiley, R.A., Tsuji, A., and Matsuda, Y., Intermedilysin, a novel cytotoxin specific for human cells secreted by *Streptococcus intermedius* UNS46 isolated from a human liver abscess, *Infect. Immun.*, 64, 3093, 1996.

128. Ellemor, D.M., Baird, R.N., Awad, M.M., Boyd, R.L., Rood, J.I., and Emmins, J.J., Use of genetically manipulated strains of *Clostridium perfringens* reveals that both alpha-toxin and theta-toxin are required for vascular leukostasis to occur in experimental gas gangrene, *Infect. Immun.*, 67, 4902, 1999.
129. Madden, J.C., Ruiz, N., and Caparon, M., Cytolysin-mediated translocation (CMT): a functional equivalent of type III secretion in gram-positive bacteria, *Cell*, 104, 143, 2001.
130. Blocker, A., Komoriya, K., and Aizawa, S., Type III secretion systems and bacterial flagella: insights into their function from structural similarities, *Proc. Natl. Acad. Sci. U.S.A.*, 100, 3027, 2003.
131. Prior, T.I. and Kornberg, H.L., Nucleotide sequence of *fruA*, the gene specifying enzyme IIfru of the phosphoenolpyruvate-dependent sugar phosphotransferase system in *Escherichia coli* K12, *J. Gen. Microbiol.*, 134 (Pt 10), 2757, 1988.
132. Jacobs, A.A., Loeffen, P.L., van den Berg, A.J., and Storm, P.K., Identification, purification, and characterization of a thiol-activated hemolysin (suilysin) of *Streptococcus suis*, *Infect. Immun.*, 62, 1742, 1994.
133. Jacobs, A.A., van den Berg, A.J., and Loeffen, P.L., Protection of experimentally infected pigs by suilysin, the thiol-activated haemolysin of *Streptococcus suis*, *Vet. Rec.*, 139, 225, 1996.
134. Alexander, J.E., Lock, R.A., Peeters, C.C.A.M., Poolman, J.T., Andrew, P.W., Mitchell, T.J., Hansman, D., and Paton, J.C., Immunization of mice with pneumolysin toroid confers a significant degree of protection against at least nine serotypes of *Streptococcus pneumoniae*, *Infect. Immun.*, 62, 5683, 1994.
135. McDonel, J.L., Toxins of *Clostridium perfringens* types A, B, C, D and E, in *Pharmacology of Bacterial Toxins*, Drews, J., Ed., Pergamon Press, Oxford, 1986, 477.
136. Katayama, S., Dupuy, B., Daube, G., China, B., and Cole, S.T., Genome mapping of *Clostridium perfringens* strains with I-CeuI shows many virulence genes to be plasmid-borne, *Mol. Gen. Genet.*, 251, 720, 1996.
137. Hunter, S.E.C., Clarke, I.N., Kelly, D.C., and Titball, R.W., Cloning and nucleotide sequencing of the *Clostridium perfringens* epsilon-toxin gene and its expression in *Escherichia coli*, *Infect. Immun.*, 60, 102, 1992.
138. Minami, J., Katayama, S., Matsushita, O., Matsushita, C., and Okabe, A., Lambda-toxin of *Clostridium perfringens* activates the precursor of epsilon-toxin by releasing its N- and C-terminal peptides, *Microbiol. Immunol.*, 41, 527, 1997.
139. Thompson, R.O., The fractionation of *Clostridium welchii* e-antigen on cellulose ion exchangers, *J. Gen. Microbiol.*, 31, 79, 1963.
140. Goswami, P.P., Rupa, P., Prihar, N.S., and Garg, L.C., Molecular cloning of *Clostridium perfringens* epsilon-toxin gene and its high level expression in *E. coli*, *Biochem. Biophys. Res. Commn.*, 226, 735, 1996.
141. Miyata, S., Matsushita, O., Minami, J., Katayama, S., Shimamoto, S., and Okabe, A., Cleavage of a C-terminal peptide is essential for heptamerization of *Clostridium perfringens* epsilon-toxin in the synaptosomal membrane, *J. Biol. Chem.*, 276, 13778, 2001.
142. Thompson, R.O., Crystalline epsilon prototoxin from *Clostridium welchii*, *Nature*, 193, 69, 1962.
143. Ghazi, F., Younan, M., Ardehali, M., and Muller, W., Differentiation of proteinase (Minor Toxin Lambda) in *Clostridium perfringens* strains from sheep and goats in Iran, *Dtsch. Tierarztl. Wochenschr.*, 104, 443, 1997.
144. Bhown, A.S. and Habeeb, A.F.S.A., Structural studies on e-prototoxin of *Clostridium perfringens* type D. Localisation of the site of tryptic scission necessary for activation to e-toxin, *Biochem. Biophys. Res. Commn.*, 78, 889, 1977.
145. Worthington, R.W. and Mülders, M.S., Physical changes in the epsilon prototoxin molecule of *Clostridium perfringens* during enzymatic activation, *Infect. Immun.*, 18, 549, 1977.
146. Ballard, J., Bryant, A., Stevens, D., and Tweten, R.K., Purification and characterization of the lethal toxin (alpha toxin) of *Clostridium septicum*, *Infect. Immun.*, 60, 784, 1992.
147. Ballard, J., Crabtree, J., Roe, B.A., and Tweten, R.K., The primary structure of *Clostridium septicum* alpha toxin exhibits similarity with *Aeromonas hydrophila* aerolysin, *Infect. Immun.*, 63, 340, 1995.
148. Sellman, B.R. and Tweten, R.K., The propeptide of *Clostridium septicum* alpha toxin functions as an intramolecular chaperone and is a potent inhibitor of alpha toxin-dependent cytolysis, *Mol. Microbiol.*, 25, 429, 1997.
149. Ballard, J., Sokolov, Y., Yuan, W.L., Kagan, B.L., and Tweten, R.K., Activation and mechanism of *Clostridium septicum* alpha-toxin, *Mol. Microbiol.*, 10, 627, 1993.

150. Gordon, V.M., Benz, R., Fujii, K., Leppla, S.H., and Tweten, R.K., *Clostridium septicum* alpha-toxin is proteolytically activated by furin, *Infect. Immun.*, 65, 4130, 1997.
151. Havard, H.L., Hunter, S.E.C., and Titball, R.W., Comparison of the nucleotide sequence and development of a PCR test for the epsilon toxin gene of *Clostridium perfringens* type-B and type-D, *FEMS Microbiol. Lett.*, 97, 77, 1992.
152. Liu, J.W., Porter, A.G., Wee, B.Y., and Thanabalu, T., New gene from nine *Bacillus sphaericus* strains encoding highly conserved 35.8-kilodalton mosquitocidal toxins, *Appl. Environ. Microbiol.*, 62, 2174, 1996.
153. Thanabalu, T. and Porter, A.G., A *Bacillus sphaericus* gene encoding a novel type of mosquitocidal toxin of 31.8 kDa, *Gene*, 170, 85, 1996.
154. Ballard, J., Crabtree, J., Roe, B.A., and Tweten, R.K., The primary structure of *Clostridium septicum* alpha-toxin exhibits similarity with that of *Aeromonas hydrophila* aerolysin, *Infect. Immun.*, 63, 340, 1995.
155. Gordon, V.M., Nelson, K.L., Buckley, J.T., Stevens, V.L., Tweten, R.K., Elwood, P.C., and Leppla, S.H., *Clostridium septicum* alpha toxin uses glycosylphosphatidylinositol-anchored protein receptors, *J. Biol. Chem.*, 274, 27274, 1999.
156. Cowell, S., Aschauer, W., Gruber, H.J., Nelson, K.L., and Buckley, J.T., The erythrocyte receptor for the channel-forming toxin aerolysin is a novel glycosylphosphatidylinositol-anchored protein, *Mol. Microbiol.*, 25, 343, 1997.
157. Nelson, K.L., Raja, S.M., and Buckley, J.T., The glycosylphosphatidylinositol-anchored surface glycoprotein Thy-1 is a receptor for the channel-forming toxin aerolysin, *J. Biol. Chem.*, 272, 12170, 1997.
158. Abrami, L., Fivaz, M., Decroly, E., Seidah, N.G., Jean, F., Thomas, G., Leppla, S.H., Buckley, J.T., and van der Goot, F.G., The pore-forming toxin proaerolysin is activated by furin, *J. Biol. Chem.*, 273, 32656, 1998.
159. Howard, S.P. and Buckley, J.T., Activation of the hole-forming toxin aerolysin by extracellular processing, *J. Bacteriol.*, 163, 336, 1985.
160. Songer, J.G., Clostridial diseases of animals, in *The Clostridia: Molecular Biology and Pathogenesis*, Titball, R.W., Ed., Academic Press, London, 1997, 153.
161. Pathak, D.C. and Parihar, N.S., Experimental production of enterotoxaemia in sheep, *Ind. J. Animal. Sci.*, 62, 915, 1992.
162. Uzal, F.A. and Kelly, W.R., Experimental *Clostridium perfringens* type D enterotoxemia in goats, *Vet. Pathol.*, 35, 132, 1998.
163. Uzal, F.A., Kelly, W.R., Morris, W.E., and Assis, R.A., Effects of intravenous injection of *Clostridium perfringens* type D epsilon toxin in calves, *J. Comp. Pathol.*, 126, 71, 2002.
164. Bullen, J.J. and Batty, I., The effect of *Clostridium welchii* type D culture filtrates on the permeability of the mouse intestine, *J. Pathol. Bacteriol.*, 71, 311, 1956.
165. Worthington, R.W., Mülders, M.S., and van Rensburg, J.J., *Clostridium perfringens* type D epsilon prototoxin. Some chemical, immunological and biological properties of a highly purified prototoxin, *Onderstepoort J. Vet. Res.*, 40, 143, 1973.
166. Nagahama, M. and Sakurai, J., Distribution of labeled *Clostridium perfringens* epsilon-toxin in mice, *Toxicon*, 29, 211, 1991.
167. Finnie, J.W., Blumbergs, P.C., and Manavis, J., Neuronal damage produced in rat brains by *Clostridium perfringens* type D epsilon toxin, *J. Comp. Pathol.*, 120, 415, 1999.
168. Finnie, J.W., Ultrastructural changes in the brain of mice given *Clostridium perfringens* type D epsilon toxin, *J. Comp. Pathol.*, 94, 445, 1984.
169. Buxton, D., Use of horseradish peroxidase to study the antagonism of *Clostridium welchii* (*Cl. perfringens*) type D epsilon toxin in mice by the formalised epsilon prototoxin, *J. Comp. Pathol.*, 86, 62, 1976.
170. Buxton, D. and Morgan, K.T., Studies on the lesions produced in the brains of colostrum-deprived lambs by *Clostridium welchii* (*Clostridium perfringens*) type D toxin, *J. Comp. Pathol.*, 86, 435, 1967.
171. Stevens, D.L., Musher, D.M., Watson, D.A., Eddy, H., Hamill, R.J., Gyorkey, F., Rosen, H., and Mader, J., Spontaneous, nontraumatic gangrene due to *Clostridium septicum*, *Rev. Infect. Dis.*, 12, 286, 1990.
172. Kirchner, J.T., *Clostridium septicum* infection beware of associated cancer, *Post. Med.*, 90, 157, 1991.

173. Barnham, M. and Weightman, N., *Clostridium septicum* infection and hemolytic uremic syndrome, *Emerg. Infect. Dis.*, 4, 321, 1998.
174. Price, C.I., Hollingsworth, A.B., and Tweten, R.K., A traumatic gas gangrene caused by *Clostridium septicum*. in *43rd Annual Meeting of the Southwestern Surgical Congress*, Las Vegas, NV, 1991.
175. Amimoto, K., Ohgitani, T., Sasaki, O., Oishi, E., Katayama, S., Isogai, M., and Ota, S., Protective effect of *Clostridium septicum* alpha-toxoid vaccine against challenge with spores in guinea pigs, *J. Vet. Med. Sci.*, 64, 67, 2002.
176. Habeeb, A.F.S.A., Lee, C.-L., and Atassi, M.Z., Conformational studies on modified proteins and peptides. VII. Conformation of e-prototoxin and e-toxin from *Clostridium perfringens*. Conformational changes assocaited with toxicity, *Biochim. Biophys. Acta*, 322, 245, 1973.
177. Sakurai, J. and Nagahama, M., Role of one tryptophan residue in the lethal activity of *Clostridium perfringens* epsilon toxin, *Biochem. Biophys. Res. Commn.*, 128, 760, 1985.
178. Sakurai, J. and Nagahama, M., Histidine residues in *Clostridium perfringens* epsilon toxin, *FEMS Microbiol. Lett.*, 41, 317, 1987.
179. Sakurai, J. and Nagahama, M., The inactivation of *Clostridium perfringens* epsilon toxin by treatment with tetranitromethane and N-acetylimidazole, *Toxicon*, 25, 279, 1987.
180. Sakurai, J. and Nagahama, M., Carboxyl Groups in *Clostridium perfringens* epsilon toxin, *Microb. Pathog.*, 3, 469, 1987.
181. Oyston, P.C.F., Payne, D.W., Havard, H.L., Williamson, E.D., and Titball, R.W., Production of a non-toxic site-directed mutant of *Clostridium perfringens* epsilon-toxin which induces protective immunity in mice, *Microbiology (U.K.)*, 144, 333, 1998.
182. Worthington, R.W. and Mülders, M.S.G., Effect of *Clostridium perfringens* epsilon toxin on the blood brain barrier of mice, *Onderstepoort J. Vet. Res.*, 42, 25, 1975.
183. Payne, D.W., Williamson, E.D., Havard, H., Modi, N., and Brown, J., Evaluation of a new cytotoxicity assay for *Clostridium perfringens* type D epsilon toxin, *FEMS Microbiol. Letts.*, 116, 161, 1994.
184. Knight, P.A., Queminet, J., Blanchard, J.H., and Tilleray, J.H., *In vitro* tests for the measurement of clostridial toxins, toxoids and antisera II. Titration of *Clostridium perfringens* toxins and antitoxins in cell culture, *Biologicals*, 18, 262, 1990.
185. Lindsay, C.D., Assessment of aspects of the toxicity of *Clostridium perfringens* epsilon-toxin using the MDCK cell line, *Hum. Exp. Toxicol.*, 15, 904, 1996.
186. Nagahama, M., Ochi, S., and Sakurai, J., Assembly of *Clostridium perfringens* epsilon-toxin on MDCK cell membrane, *J. Nat. Toxins*, 7, 291, 1998.
187. Petit, L., Gibert, M., Gillet, D., LaurentWinter, C., Boquet, P., and Popoff, M.R., *Clostridium perfringens* epsilon-toxin acts on MDCK cells by forming a large membrane complex, *J. Bacteriol.*, 179, 6480, 1997.
188. Donelli, G., Fiorentini, C., Matarrese, P., Falzano, L., Cardines, R., Mastrantonio, P., Payne, D.W., and Titball, R.W., Evidence for cytoskeletal changes secondary to plasma membrane functional alterations in the *in vitro* cell response to *Clostridium perfringens* epsilon-toxin, *Comp. Immunol. Microbiol. Infect. Dis.,* 26, 145–156, 2003.
189. Nagahama, M. and Sakurai, J., High-affinity binding of *Clostridium perfringens* epsilon-toxin to rat brain, *Infect. Immun.*, 60, 1237, 1992.
190. Payne, D., Williamson, E.D., and Titball, R.W., The *Clostridium perfringens* epsilon-toxin, *Rev. Med. Microbiol.*, 8, S28, 1997.
191. Miyata, S., Minami, J., Tamai, E., Matsushita, O., Shimamoto, S., and Okabe, A., *Clostridium perfringens* epsilon-toxin forms a heptameric pore within the detergent-insoluble microdomains of Madin-Darby canine kidney cells and rat synaptosomes, *J. Biol. Chem.*, 277, 39463, 2002.
192. Sellman, B.R., Kagan, B.L., and Tweten, R.K., Generation of a membrane-bound, oligomerized pre-pore complex is necessary for pore formation by *Clostridium septicum* alpha toxin, *Mol. Microbiol.*, 23, 551, 1997.
193. Gordon, A.E., Saadi, A.T., MacKenzie, D.A.C., Molony, N., James, V.S., Weir, D.M., Busuttil, A., and Blackwell, C.C., The protective effect of breast feeding in relation to sudden infant death syndrome (SIDS): III. Detection of IgA antibodies in human milk that bind to bacterial toxins implicated in SIDS, *FEMS Immunol. Med. Microbiol.*, 25, 175, 1999.
194. Wichroski, M.J., Melton, J.A., Donahue, C.G., Tweten, R.K., and Ward, G.E., *Clostridium septicum* alpha-toxin is active against the parasitic protozoan *Toxoplasma gondii* and targets members of the SAG family of glycosylphosphatidylinositol-anchored surface proteins, *Infect. Immun.*, 70, 4353, 2002.

195. Diep, D.B., Nelson, K.L., Lawrence, T.S., Sellman, B.R., Tweten, R.K., and Buckley, J.T., Expression and properties of an aerolysin — *Clostridium septicum* alpha toxin hybrid protein, *Mol. Microbiol.*, 31, 785, 1999.
196. Gruber, H.J., Wilmsen, H.U., Cowell, S., Schindler, H., and Buckley, J.T., Partial purification of the rat erythrocyte receptor for the channel-forming toxin aerolysin and reconstitution into planar lipid bilayers, *Molecular Microbiology*, 14, 1093, 1994.
197. Howard, S.P. and Buckley, J.T., Membrane glycoprotein receptor and hole-forming properties of a cytolytic protein, *Biochem.*, 21, 1662, 1982.
198. Green, M.J. and Buckley, J.T., Site-directed mutagenesis of the hole-forming toxin aerolysin: studies on the roles of histidines in receptor binding and oligomerization of the monomer, *Biochem.*, 29, 2177, 1990.
199. Lobato, F.C.F., Mora, E., Umehara, O., Assis, R.A., Martins, N.E., and Goncalves, L.C.B., Cattle and rabbits immune response against beta and epsilon *Clostridium perfringens* toxins induced by six commercial vaccines in Brazil, *Arq. Bras. Med. Vet. Zootec.*, 52, 313, 2000.
200. Uzal, F.A. and Kelly, W.R., Serum antibody responses to a *Clostridium perfringens* epsilon toxoid vaccine in goats, *Anaerobe*, 5, 287, 1999.
201. Menestrina, G., Dalla Serra, M., and Prevost, G., Mode of action of beta-barrel pore-forming toxins of the staphylococcal alpha-hemolysin family, *Toxicon*, 39, 1661, 2001.
202. Montoya, M. and Gouaux, E., Beta-barrel membrane protein folding and structure viewed through the lens of alpha-hemolysin, *Biochim. Biophys. Acta-Biomembr.*, 1609, 19, 2003.
203. Bhakdi, S., Walev, I., Palmer, M., and Valeva, A., Staphylococcal alpha-toxin, in *Bacterial Protein Toxins*, Just, I., Ed., Springer-Verlag, Berlin, 2000, 509.
204. Dinges, M.M., Orwin, P.M., and Schlievert, P.M., Exotoxins of *Staphylococcus aureus*, *Clin. Microbiol. Rev.*, 13, 16, 2000.
205. Hunter, S.E.C., Brown, J.E., Oyston, P.C.F., Sakurai, J., and Titball, R.W., Molecular genetic analysis of beta-toxin of *Clostridium perfringens* reveals sequence homology with alpha-toxin, gamma-toxin, and leukocidin of *Staphylococcus aureus*, *Infect. Immun.*, 61, 3958, 1993.
206. Steinporsdottir, V., Fridriksdottir, V., Gunnarsson, E., and Andresson, O.S., Expression and purification of *Clostridium perfringens* beta-toxin glutathione-S-transferase fusion protein, *FEMS Microbiol. Lett.*, 130, 273, 1995.
207. Goswami, P.P., Girish, K.S., Chaudhuri, P., Tiwari, V., Akare, S.J., and Harbola, P.C., Cloning and sequencing of beta toxin gene of *Clostridium perfringens* type C, *Indian J. Exp. Biol.*, 40, 109, 2002.
208. Sakurai, J. and Duncan, C.L., Purification of beta-toxin from *Clostridium perfringens* type C, *Infect. Immun.*, 18, 741, 1977.
209. Jansen, B.C., The beta-toxin of *Clostridum welchii* type B (Wilsdon), in relation to the production of a vaccine against lamb dysentery, *Onderstepoort J.*, 28, 495, 1961.
210. Steinthorsdottir, V., Fridriksdottir, V., Gunnarsson, E., and Andresson, O.S., Site-directed mutagenesis of *Clostridium perfringens* beta-toxin: expression of wild-type and mutant toxins in *Bacillus subtilis*, *FEMS Microbiol. Lett.*, 158, 17, 1998.
211. Nagahama, M., Kihara, A., Miyawaki, T., Mukai, M., Sakaguchi, Y., Ochi, S., and Sakurai, J., *Clostridium perfringens* beta-toxin is sensitive to thiol-group modification but does not require a thiol group for lethal activity, *Biochim. Biophys. Acta-Mol. Basis Dis.*, 1454, 97, 1999.
212. Gill, D.M., Bacterial toxins: a table of lethal amounts, *Microbiol. Revs.*, 46, 86, 1982.
213. Jolivet-Reynaud, C., Popoff, M.R., Vinit, M.-A., Ravisse, P., Moreau, H., and Alouf, J.E., Enteropathogenicity of *Clostridium perfringens* beta-toxin and other clostridial toxins, in *Bacterial Protein Toxins; 2nd European Workshop*, Falmagne, P., Ed., Gustav Fischer Verlag, Stuttgart, 1986, 145.
214. Gut, P., Luginbuhl, A., Nicolet, J., Boerlin, P., and Burnens, A.P., The necrotizing enteritis by *Clostridium perfringens* type C in piglets: II. Molecularepidemiological study, *Schweiz. Arch. Tierheilkd.*, 144, 275, 2002.
215. Selbitz, H.J., Schoepe, H., Springer, S., Neubauer, A., and Rosch, H., Immunoprophylaxis of clostridioses of piglets, *Tierarztl. Prax. Ausg. Grobtiere Nutztiere*, 28, 337, 2000.
216. Kohler, B., Necrotizing enteritis in suckling pigs (*Clostridium perfringens* type C enterotoxemia). II. Toxin formation, heat and drug resistance of *Clostridium perfringens* strains isolated from suckling pigs and broilers with necrotizing enteritis, *Arch. Exp. Veterinarmed.*, 32, 841, 1978.

217. Petrillo, T.M., Beck-Sague, C.M., Songer, J.G., Abramowsky, C., Fortenberry, J.D., Meacham, L., Dean, A.G., Lee, H., Bueschel, D.M., and Nesheim, S.R., Enteritis necroticans (Pigbel) in a diabetic child, *N. Engl. J. Med.*, 342, 1250, 2000.
218. Lawrence, G., The pathogenesis of enteritis necroticans, in *The Clostridia: Molecular Biology and Pathogenesis*, Titball, R.W., Ed., Academic Press, London, 1997, 197.
219. Lawrence, G. and Cooke, R., Experimental pigbel: the production and pathology of necrotizing enteritis due to *Clostridium welchii* type C in the guinea-pig, *Br. J. Exp. Pathol.*, 61, 261, 1980.
220. Murrell, T.G. and Walker, P.D., Pig-bel — a zoonosis?, *J. Trop. Med. Hyg.*, 81, 231, 1978.
221. Lawrence, G.W., Lehmann, D., Anian, G., Coakley, C.A., Saleu, G., Barker, M.J., and Davis, M.W., Impact of active immunisation against enteritis necroticans in Papua New Guinea, *Lancet*, 336, 1165, 1990.
222. Johnson, S., Echeverria, P., Taylor, D.N., Paul, S.R., Coninx, R., Sakurai, J., Eampokalap, B., Jimakorn, P., Cooke, R.A. and Lawrence, G.W., *Enteritis necroticans* among Khmer children at an evacuation site in Thailand, *Lancet*, 29, 496, 1987.
223. Sakurai, J., Fujii, Y., Matsuura, M., and Endo, K., Pharmacological effect of beta toxin of *Clostridium perfringens* type C on rats, *Microbiol. Immunol.*, 25, 423, 1981.
224. Sakurai, J., Fujii, Y., Dezaki, K., and Endo, K., Effect of *Clostridium perfringens* beta toxin on blood pressure of rats, *Microbiol. Immunol.*, 28, 23, 1984.
225. Allan, D., Cytotoxic effects produced by the beta-toxin of *Clostridium welchii* on guinea pig monocytes, *Immunol.*, 6, 3, 1963.
226. Steinthorsdottir, V., Halldorsson, H., and Andresson, O.S., *Clostridium perfringens* beta-toxin forms multimeric transmembrane pores in human endothelial cells, *Microb. Pathog.*, 28, 45, 2000.
227. Shatursky, O., Bayles, R., Rogers, M., Jost, B.H., Songer, J.G., and Tweten, R.K., *Clostridium perfringens* beta-toxin forms potential-dependent, cation-selective channels in lipid bilayers, *Infect. Immun.*, 68, 5546, 2000.
228. Gouaux, E., alpha-hemolysin from *Staphylococcus aureus*: An archetype of beta-barrel, channel-forming toxins, *J. Struct. Biol.*, 121, 110, 1998.
229. Sakurai, J., Fujii, Y., and Matsuura, M., Effect of oxidizing agents and sulfhydryl group reagents on beta toxin from *Clostridium perfringens* type C, *Microbiol. Immunol.*, 24, 595, 1980.
230. Matisheck, P.H. and McGinley, M., Colostral transfer of *Clostridium perfringens* type C beta antitoxin in swine, *Am. J. Vet. Res.*, 47, 1132, 1986.
231. Kelneric, Z., Naglic, T., and Udovicic, I., Prevention of necrotic enteritis in piglets by vaccination of pregnant gilts with a *Clostridium perfringens* type C and D bacterin-toxoid, *Vet. Med. (Praha)*, 41, 335, 1996.
232. Springer, S. and Selbitz, H.J., The control of necrotic enteritis in sucking piglets by means of a *Clostridium perfringens* toxoid vaccine, *FEMS Immunol. Med. Microbiol.*, 24, 333, 1999.
233. Luginbuhl, A., The necrotizing enteritis by *Clostridium perfringens* type C in piglets: practical observations, control and epidemiology, *Schweiz Arch Tierheilkd*, 144, 263, 2002.
234. Alouf, J.E. and Jolivetreynaud, C., Purification and characterization of *Clostridium perfringens* delta toxin, *Infect. Immun.*, 31, 536, 1981.
235. Jolivetreynaud, C., Hauttecoeur, B. and Alouf, J.E., Interaction of *Clostridium perfringens* delta toxin with erythrocyte and liposome membranes and relation with the specific binding to the ganglioside GM2, *Toxicon*, 27, 1113, 1989.
236. Jolivetreynaud, C., Launay, J.M., and Alouf, J.E., Damaging effects of *Clostridium perfringens* delta-toxin on blood platelets and their relevance to ganglioside GM2, *Arch. Biochem. Biophys.*, 262, 59, 1988.
237. Guillouard, I., Garnier, T., and Cole, S.T., Use of site-directed mutagenesis to probe structure-function relationships of alpha-toxin from *Clostridium perfringens*, *Infect. Immun.*, 64, 2440, 1996.
238. Nagahama, M., Ochi, S., Kobayashi, K., and Sakurai, J., The relationship between histidine residues and various biological activities of *Clostridium perfringens* alpha-toxin, *Abstr. Pap. Am. Chem. Soc.*, 209, 75, 1995.
239. Nagahama, M., Nakayama, T., Michiue, K., and Sakurai, J., Site-specific mutagenesis of *Clostridium perfringens* alpha-toxin: replacement of Asp-56, Asp-130, or Glu-152 causes loss of enzymatic and hemolytic activities, *Infect. Immun.*, 65, 3489, 1997.

240. Higgins, D.G. and Sharp, P.M., Clustal: a package for performing multiple sequence alignment on a microcomputer, *Gene*, 73, 237, 1988.
241. Higgins, D.G., Bleasby, A.J., and Fuchs, R., CLUSTAL V: improved software for multiple sequence alignment, *Comput. Appl. Biosci.*, 8, 189, 1992.
242. Page, R.D., TreeView: an application to display phylogenetic trees on personal computers, *Comput. Appl. Biosci.*, 12, 357, 1996.
243. Koradi, R., Billeter, M., and Wuthrich, K., MOLMOL: a program for display and analysis of macromolecular structures, *J. Mol. Graph.*, 14, 51, 1996.
244. Walker, B. and Bayley, H., Key residues for membrane binding, oligomerization, and pore forming activity of staphylococcal alpha-hemolysin identified by cysteine scanning mutagenesis and targeted chemical modification, *J. Biol. Chem.*, 270, 23065, 1995.

21 Clostridial Neurotoxins

Eric A. Johnson

CONTENTS

21.1 INTRODUCTION

Botulism and tetanus are among the most striking diseases known to humankind. Tetanus is characterized by spasticity, violent movements, and high mortality [1]. It is still endemic in various areas of the world, occurring most frequently in neonates following birth and causing thousands of deaths. Botulism is characterized as a flaccid paralysis progressing to suffocation and death in severe cases that do not receive adequate treatment [2]. Interestingly, both botulism and tetanus are caused by similar neurotoxins (NTs) with extraordinary neurospecificity and similar mechanisms of action [3–8]. The diseases are considered to be true toxemias, in which the toxins produced by the organisms are the responsible agents, and the bacteria are not directly involved in the spastic and paralytic conditions. The clostridial neurotoxins (CNTs), botulinum neurotoxin (BoNT) and tetanus toxin (TeNT), are produced by strains of neurotoxigenic clostridia.

During the past two decades, knowledge obtained regarding the biochemistry, structure, genetics, and the action of BoNTs and TeNT has led to these toxins becoming important tools in cell biology and has stimulated much interest in their basic nature and their actions on the human nervous system [3–8]. Studies on CNTs have provided insight into mechanisms by which nerves transmit information to other nerves and to nonneural tissues and organs [9]. Interest in BoNT has also been incited by awareness of its possible implementation in bioterrorism [10,11]. Yet the most remarkable discovery that has resulted from investigation of CNTs is the use of BoNTs as pharmaceuticals for the treatment of a myriad of neuronal and muscle disorders [12–16].

Research on the microbiology and genetics of neurotoxigenic clostridial organisms has lagged behind studies of the CNTs [17,18], but this area is gaining increased interest and research is advancing. The genome sequences of *C. tetani*, *C. botulinum*, and other *Clostridium* species, such as *Clostridium perfringens*, *Clostridium acetobutylicum*, and *Clostridium difficile,* have recently become available [19,20] or are near completion, mainly through efforts at the Sangre Centre. The available genome sequences will certainly provide valuable tools for investigation of the neurotoxigenic clostridia and their CNTs, and the important roles of these pathogens in causing disastrous diseases in humans and animals. Several contemporary books and excellent reviews are available describing the taxonomy, physiology, genetics, and diseases caused by the toxigenic clostridia, as well as the biochemistry, immunology, pharmacology, and medical uses of the CNTs [21–27]. This chapter describes the properties of *C. botulinum* and *C. tetani,* with an emphasis on microbiology and new findings on the organisms and their neurotoxins.

21.2 NEUROTOXIGENIC CLOSTRIDIA

21.2.1 The Clostridia — Prolific Producers of Toxins

One of the most interesting features of clostridial organisms is their formation of a wide diversity of toxins, and the clostridia produce more types of protein toxins than any other group of microorganisms [28–31]. More than 20 protein toxins have been identified from clostridia, some with very high potencies [28–30]. At least 15 species of *Clostridium* are known to produce protein toxins [29,30]. These agents include neurotoxins, lipases, lecithinases, haemolysins, enterotoxins, cytotoxins, collagenases, permeases, necrotizing toxins, proteinases, hyaluronidases, DNases, ADP-ribosyltransferases, neuraminidases, and some others that are simply known as "lethal toxins" [29–31]. Of specific importance, and in contrast to many other bacterial pathogens, *C. botulinum* and *C. tetani* cause true toxemias, in which the produced neurotoxin is responsible for the pathogenic effects. These neurotoxins may be elicited during infections, such as in wound infections by *C. tetani* and *C. botulinum*, or in infant botulism caused by *C. botulinum* and strains of *C. butyricum* and *C. baratii* [32,33]. Botulinum and tetanus neurotoxins both affect the nervous system, causing either a flaccid paralysis (botulism) or a spastic paralysis (tetanus).

21.2.2 Discovery of Botulism, Tetanus, and Neurotoxigenic Clostridia

In ancient times, an association was noted between wound infections and a striking disease characterized by spasticity, violent movements, and death [1,34]. Hippocrates and his contemporary

Aretaues described tetanus and its manisfestations about 30 centuries ago [1]. Galen implicated nerves in the disease and in motor function by showing that cutting a nerve in a patient with tetanus stopped the contractions and paralyzed the innervated muscles [1]. The etiology of tetanus remained unknown until the late 1800s, when Carle and Rattone [35,36] demonstrated that infected tissue from a diseased animal could transmit the disease to a healthy animal. Nicolaier [37] showed that inoculation of soil into mice, guinea pigs, or rabbits caused a syndrome that closely mimicked human tetanus. He noted the presence of long, thin bacilli in the wounds and pus of the inoculated animals and that bacilli were occasionally observed in the sciatic nerve sheath and spinal cord. *C. tetani* was first isolated in pure culture in 1889 by Kitasato in Berlin [38]. The isolation was facilitated by heating the pus at 80°C for 45 to 60 min to destroy vegetative organisms and by plating the culture on gelatin and incubating in a hydrogen atmosphere. Faber found that cell-free extracts of a culture caused tetanus-like symptoms when injected into laboratory animals, which gave evidence that a cell-free toxin (tetanus neurotoxin; often referred to as tetanospasmin) was responsible for the disease [39]. In 1890, Behring and Kitasato [40] discovered an immune response to inactive toxin (toxoid), and in the 1920s Glenny and Ramon independently discovered that the toxin inactivated with formalin when elicited to animals resulted in an immune response [41–43]. The prevention of tetanus intoxication by vaccination is one of the great discoveries of medicine. Vaccination has proved to be extremely effective in preventing tetanus intoxication in humans and animals.

Botulism was initially reported by Kerner in the 1700s from people who had consumed raw blood sausage long before the etiologic agent was isolated [44–46]. *C. botulinum* and properties of its neurotoxin were discovered between 1895 and 1897 in a remarkable investigation of the disease by Emile Pierre van Ermengem [47–48]. The investigations by van Ermengem and the isolation of *Bacillus botulinus* (*Clostridium botulinum*) is a classic study in the field of clostridial pathogens and their toxins. The association of botulism poisoning with various foods was also established [49–56]. The formation of resistant endospores by *C. botulinum* was of considerable significance to the control of food contamination and botulism. The implementation of measures to control *C. botulinum* and its toxins in foods provides the basis for many current food laws and regulations. In the mid-to-late 1900s, it was established that botulism also could occur from wound and intestinal infections in humans and in a wide variety of animals [21,32,57–61].

21.2.3 Isolation and Characterization of Neurotoxigenic Clostridia

The isolation and identification of neurotoxigenic clostridia present practical difficulties [61–66]. Clostridia are anaerobes, and culturing requires media of low redox potential and elimination of oxygen and toxic oxygen metabolites. Anaerobic jars, or preferably an anaerobic glove box, are needed for cultivation and manipulations. Prompt collection of foods or clinical samples and rapid analysis is necessary to prevent exposure to oxygen and reduction in viability of clostridial cultures. In general, *C. botulinum* and *C. tetani* are more sensitive to oxygen than certain other pathogenic clostridia such as *C. sporogenes* and *C. perfringens*. Oxygen sensitivity restricts the habitat of vegetative cells to anaerobic environments, though endospores can survive in aerobic environments. In most environmental, food, and fecal samples, clostridial endospores are often present and can be isolated using appropriate techniques [62–69]. Because of the resistance properties of spores, competitor vegetative organisms can be eliminated by heat or chemical treatments. Incubation of the source materials at 60 to 80°C for 10 to 20 min is usually sufficient to select for clostridial sporeformers. However, certain types of *C. botulinum* have heat-sensitive spores, and 50% alcohol treatment can facilitate the isolation of these cultures [30,50,62,66,67]. Mixed cultures are usually present in clinical specimens or foods, and oxygen-utilizing bacteria and fungi will utilize oxygen and provide an environment favorable for growth of clostridia. When clostridia are enriched in complex media from foods, fecal materials, or other complex sources, they can be outgrown by competitor organisms. Selective media have been useful for isolation of *C. botulinum* types A and B (group I) [62,66,67–69].

Difficulty might be encountered in obtaining and maintaining pure cultures of neurotoxigenic clostridia. Strains tend to show genetic instability and variation in phenotypic properties, including decrease in toxin production on repeated subculture in the laboratory [70–73]. Newly isolated cultures should be stored as pure cultures in cooked meat medium or a clear rich medium, such as TPGY, at –20 or –70°C. The ability to produce toxin by *C. tetani* and *C. botulinum* can decrease with storage and during subculture to levels that are unacceptable for toxin purification. Nontoxigenic variants can exhibit enhanced sporulation and changes in lysogeny [70–74]. Subculturing in cooked meat medium or veal infusion broth may help to maintain or restore toxin titers. Maintenance of high toxin producing cultures can require isolation of individual clones and testing for toxin production by assay in mice [13]. An immunologic method was developed for qualitative assessment of toxin production within individual colonies and for detection of mutants [75]. Several texts and reviews provide useful descriptions of materials and methods for culture and study of clostridia involved in disease, including *C. botulinum* and *C. tetani* and their characteristic neurotoxins [30,31,50,62,63–69,76,77].

21.2.4 Taxonomy of Neurotoxigenic Clostridia

The genus *Clostridium* has traditionally been classified as anaerobic rod-shaped, fermentative bacteria that form endospores and obtain energy for growth by fermentation of organic substrates [77,78]. *Clostridium* is a large and diverse group of prokaryotes. The ninth edition of *Bergey's Manual of Systematic Bacteriology* [77] presented 83 species in the genus *Clostridium*, and in a recent compilation, the number of *Clostridium* species extends to 177 [79]. Clostridia were formerly classified mainly by phenotypic properties [76,77], and more recently with emphasis of determining the 16S rRNA gene sequences [80,81]. Classification of Bacteria based on genetic and protein relatedness is useful for determining the evolutionary relationships of bacterial groups, but it may not be practical to use this approach for rapid identification of neurotoxigenic clostridia [82]. Most investigators will continue to classify neurotoxigenic clostridia in only a few species, based on the characteristics and medical importance of their toxins.

Initial presumptive identification of neurotoxigenic clostridia can include several of the following tests: (1) cell morphology, Gram stain reaction, and size and the shape and position of mature endospores in the mother cell; (2) biochemical reactions, including lecithinase, lipase, proteolysis, indole formation; (3) activity of metabolic pathways for utilization of specific carbohydrates as energy sources; (4) volatile fatty acid end products by gas chromatography or high pressure liquid chromatography; and (5) determination of neurotoxicity in mice with the use of proper controls [30,31,62,68,69,76]. For consistent identification, standardization of the growth conditions and the use of reference strains are recommended. Specific culture media are useful for determining phenotypic characteristics of *C. tetani* and *C. botulinum* [30,31,62,67,69,76,77]. Egg yolk agar is useful for detecting lecithinase and lipase enzymes secreted by *C. botulinum* and related organisms. Lecithinase (phospholipase C) is indicated by the appearance of an insoluble, whitish-opaque pearly layer surrounding the colonies in the agar. Lipase reaction appears as an iridescent sheen on the surface of the colonies. Like other phenotypic tests, many other nontoxigenic members of the genus *Clostridium* produce colonies with similar colony appearance. Many strains of *C. botulinum* and *C. tetani* are strongly proteolytic. Proteolysis is detected by digestion of chopped meat particles or by digestion of casein in skim-milk plates [30,62,67,69,76,77]. In all biochemical tests, it is desired to use positive controls to validate the method in use.

Identification of *C. botulinum* requires demonstration of the presence of BoNT [30,50,62,68,83]. Although not required for identification, it is also desired to detect TeNT in cultures in *C. tetani* [30,64,65]. Various methods for mouse bioassay and typing of BoNTs, including pitfalls in the techniques, have been described [31,62,68,69,83,84,85]. In recent years, the classification of clostridia has increasingly relied on molecular techniques, particularly sequencing of genes encoding rRNA or interspacial regions [80,81,86,87,88]. For definitive classification, the molecular analysis of the genes encoding rRNA, rRNA spacing region, or other genetic approaches should be complemented by phenotypic and toxin tests.

Epidemiologic analysis of botulism outbreaks can be assisted by typing methods for neurotoxigenic clostridia. Several methods have been used for genotyping, including restriction analysis of isolated genomic DNA, PCR, RFLP, RAPD, and pulsed field gel electrophoresis (PFGE) [89,90–92]. Of the various methods, most laboratories are using PFGE, as it appears to be most discriminating and provides consistent analyses. Determining the nucleotide sequence of regions of the gene encoding 16sRNA is also useful in taxonomic identification and strain discrimination within the neurotoxigenic clostridia. As the genome sequences and multitudes of genes become available, identification by genomic methods will become an important method for identification and characterization of *C. botulinum* and *C. tetani* strains.

The traditional neurotoxigenic clostridia, *C. tetani* and *C. botulinum*, were isolated more than 100 years ago. During the past two decades, strains of clostridia (*Clostridium baratii*, *Clostridium butyricum*) that produce BoNTs and have caused botulism in humans [17,30,31,33,62,92,93] have been isolated. These findings indicate that the genes encoding BoNTs and associated proteins of the toxin complexes can be laterally transferred to nonpathogenic clostridia [17]. *C. butyricum* and *C. baratii* toxigenic strains have mostly been isolated from cases of infant botulism, but recently they have also been isolated from diverse geographic environments and from poisonous food [92–99]. The isolation of neurotoxigenic *C. butyricum* and *C. baratii* suggest that the genes encoding the neurotoxin gene complex in *C. botulinum* can be laterally transferred to other clostridia [17,100].

21.3 PHENOTYPIC PROPERTIES OF NEUROTOXIGENIC CLOSTRIDIA

21.3.1 Habitat

Due to the formation of resistant endospores, clostridia are widely distributed in nature. The principal habitat of *C. botulinum* and *C. tetani* is soil [30], and spores have been found in soils worldwide [101]. *C. tetani* has been detected in human feces, but *C. botulinum* is rarely isolated from this source [30,102], with the exception of infants with botulism [32,33]. As a result of being predominant in soil, certain clostridia frequently occur in dust, soil-cultivated foods, milk, and sewage.

The geographic distribution of *C. botulinum* worldwide differs according to toxin type and physiology [21,30,60,101]. For example, proteolytic strains (group I) of *C. botulinum* type A occur commonly in the Western U.S., whereas proteolytic type B predominate in the Eastern U.S. Nonproteolytic type B (group II) predominate in Western Europe [101]. Type E is associated with aquatic habitats, including muds and brackish waters [60,101]. *C. botulinum*/*C. argentinense* producing type G neurotoxin has only been isolated from soils in Argentina [21,30,101]. The ecological factors that determine the distribution of *C. botulinum* spores in nature have not been elucidated but would be valuable in understanding the epidemiology of botulism, as its occurrence is associated with the incidence and type of spores in the environment.

21.3.2 Morphological Characteristics and Staining Reactions

The neurotoxigenic clostridia have several phenotypic features useful for identification and characterization. Of course, the most important characteristic is the production of CNTs, but other features are also important in their characterization. The spores in *C. botulinum* and *C. tetani* are usually broader than the vegetative organisms in which they are formed, imparting the characteristic spindle or "tennis-racket" morphology [30,31] (Figure 21.1). Young cultures (<8 h) of *C. botulinum* and *C. tetani* stain Gram-positive, while older cultures can be readily discolorized with ethanol and often appear Gram-negative. Most clostridia, including *C. botulinum* and *C. tetani,* lack a capsule. An S-layer has clearly been demonstrated in isolates of *C. difficile* [103] and has been

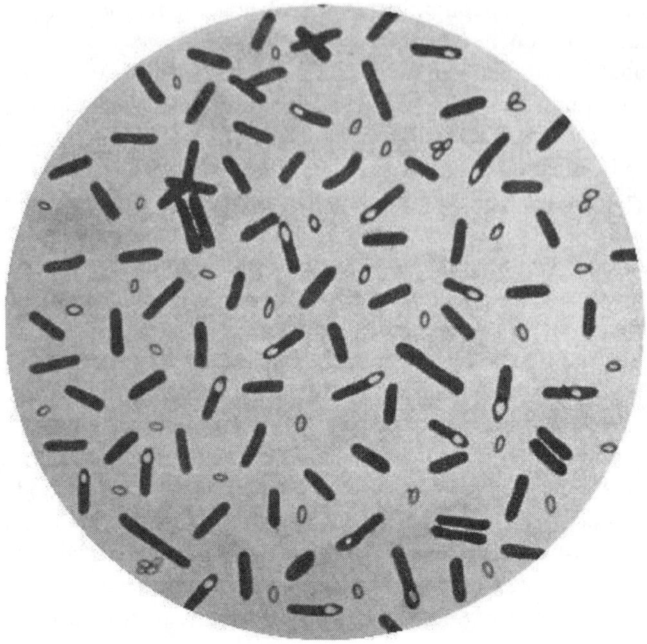

FIGURE 21.1 Typical morphology of *C. botulinum*. (From the CDC Public Health Image Library.)

reported for single strains of *C. botulinum* type E and *C. tetani* [104]. More research is needed to elucidate the cell surface structure of vegetative cells and spores of *C. botulinum* and *C. tetani*. Freshly isolated cultures of *C. botulinum* and *C. tetani* are motile by peritrichous flagella [30]. Young cultures (6 to 24 h) are commonly motile in fluid media. The occurrence of resistant endospores is a common feature of *C. botulinum* and *C. tetani,* and the spores are subterminal to terminal giving the cells a club-shape or tennis-racket appearance. Spores are easily visualized by phase contrast microscopy as bright, refractile oval bodies. Sporulation of neurotoxigenic clostridia has been reviewed [105,106] but relatively little is known of the mechanisms controlling sporulation, their resistance properties, and germination and outgrowth compared to *Bacillus* sp. Spores of *C. botulinum* and *C. tetani* are usually formed in chopped meat broth or agar after one to two weeks of incubation [30,67,76], and these species can also sporulate in clear media, such as trypticase-peptone-glucose-yeast extract (TPGY) broth. Sporulation in *C. botulinum* type B (group I) is stimulated by zinc [107], but other factors and the molecular mechanisms of sporulation in neurotoxigenic clostridia have not been elucidated. *C. botulinum* spores typically contain a characteristic outermost layer, termed the exposporium, with a structurally complex morphology [105]. The importance of spore resistance in food technology has been reviewed [105,108,109]. In addition to heat, clostridial endospores are resistant to oxygen, chemical agents, irradiation, and drying [105,109,110]. Spores can be inactivated with chlorine or high concentrations of hydrogen peroxide [110].

21.3.3 Plasmids

Plasmids are common in many strains of *C. botulinum* and *C. tetani* [30,70], but they have not been well studied and most have cryptic functions. Conjugative plasmids potentially of value in genetic analyses have not been detected [18], in contrast to *C. perfringens,* in which a conjugative plasmid has been isolated that has been very useful for genetic analyses [111]. Plasmids harbor the genes for TeNT and for BoNT/G [18,70,112,113]. In *C. tetani*, it was shown that 52 toxin-producing strains of *C. tetani* harbored a single large plasmid of about 75 mDa. It was subsequently shown by molecular analyses that the gene encoding tetanus neurotoxin resided on the plasmid [70]. The genes for type BoNT/G complex in *C. botulinum* type G (*C. argentinense*) are harbored on an 81 mDa plasmid [112,113]. The functions of these large plasmids in *C. tetani* and *C. botulinum* type G are not known, but they do not appear to have conjugative functions. A gene encoding a bacteriocin has been shown to reside on a plasmid in strains of *C. botulinum* types A and B [115].

21.3.4 Bacteriophages

Neurotoxigenic clostridia are known to contain a multitude of bacteriophages [70,100,116,117]. In *C. botulinum* types C and D, bacteriophages harbor genes for neurotoxins [70,116,117]. The bacteriophage carrier state in *C. botulinum* types C and D has been termed pseudolysogeny, referring to the observations that the prophage–bacterium relationship is unstable and nonlysogens occur frequently [70,100,116,117]. In this parasite–host relationship, populations of bacteriophage-infected and noninfected cells coexist in various ratios, depending on the conditions and phage susceptibilities of the nonlysogens. The relation of phages to toxin production has not been shown in *C. tetani* or serotypes of *C. botulinum* other than C and D [70,116,117]. Although strains of *C. botulinum* group I contain prophages, curing of phages with mitomycin C, other agents, or UV has not been correlated with loss of toxigenicity [70,100,116,117]. Certain strains of *C. botulinum* in group II lost their ability to produce neurotoxin following mitomycin C/acriflavine treatment [70]. However, these cured strains were not sensitive to phage infection under the conditions tested and toxigenicity. The inability to find indicator strains that can be infected and lead to a burst has hindered studies of the roles of phages in neurotoxigenic clostridia. Introduction of the transposon Tn*916* into *C. botulinum* strain 62A (type A) resulted in large deletions in three strains that were also affected in NT production [89,100,118]. These three strains had large deletions (~32 kb), and analysis showed that the deletions included the entire toxin gene complexes [89,118]. These data suggest that the BoNT toxin gene locus can be unstable and might be associated with mobile elements such as bacteriophages, of which deletion and loss could involve genome perturbation, such as transposon insertion. The concept that genes encoding the botulinum neurotoxin cluster can be transferred in nature is supported by the isolation of *C. butyricum* strains and *C. baratii* strains that produce neurotoxins E and F, respectively. However, analyses of genome structure in *C. botulinum* and *C. tetani* have not indicated that pathogenicity islands are present, which occur frequently in many other bacterial pathogens and have been associated with virulence gene transfer [119,120]. Indirect evidence has been presented that transfer of neurotoxigenicity can occur from *C. butyricum* to a nontoxigenic *C. botulinum* E-like strain, which may be mediated by either a plasmid or defective bacteriophage [121–123].

21.3.5 Bacteriocins

Bacteriocins are produced by a number of clostridia, including *C. botulinum* [70,115,124,125]. Recently, a bacteriocin designated boticin B was isolated and characterized from *C. botulinum* 213B [115]. DNA sequencing indicated that the boticin B structural gene, *btc*B, encodes a protein of 50 amino acids with a calculated molecular mass of 5138 Da. It had a narrow spectrum of activity against *C. botulinum* 62A and a *C. botulinum* strain 588 Ab of many strains examined. The gene for the bacteriocin from 213B resides on an 18.8 kb plasmid. This is the first reported bacteriocin to be isolated from a group I *C. botulinum* strain, and it could have significance in the inhibition of related clostridia in the intestine or in foods. Recently, strain *C. botulinum* ATCC 3502, whose genome has been sequenced, was shown to contain a ~20 kb plasmid that included a gene for a bacteriocin related to boticin B (http://www.sanger.ac.uk/Projects/C_botulinum).

21.4 CULTURAL DESCRIPTIONS OF THE SPECIES *C. BOTULINUM*, *C. TETANI* AND OTHER NEUROTOXIGENIC CLOSTRIDIA

21.4.1 *Clostridium tetani*

21.4.1.1 General Description of *C. tetani*

C. tetani is well recognized because it is the cause of the vivid disease known as lockjaw or tetanus [1,126]. The physiological properties of *C. tetani* have been described in detail [30,38,39,64,65,76,77]. The mol% GC content of the DNA is 25 to 26%. The bacilli are 0.5 to 1.7 × 2.1 to 18.1 μm and often possess terminal endospores giving sporulating cells a drumstick appearance. *C. tetani* is motile by peritrichous flagella. Optimal growth occurs at 37°C, and little or no growth occurs at 25 or 42°C. Agar cultures often form a film because of the organism's vigorous motility and swarming. *C. tetani* is similar culturally and biochemically to *C. cochlearium* and organisms identified as *C. tetanomorphum*. Nontoxigenic strains of *C. tetani* have been found that correlate well with toxigenic strains by phenotypic and DNA comparisons, and may have lost the plasmid bearing the gene encoding tetanus toxin. Since the clinical features of tetanus are so distinct and recognizable, isolation and laboratory investigation of cultures is rarely performed. Few organisms involved in tetanus have been thoroughly studied, and the strain in which the genome has been sequenced [19] is a nonsporing strain that has been used for toxoid production. The genome sequence has revealed several interesting properties of the organism including the presence of putative virulence factors and a unique metabolism based on sodium ion transport (see discussion below).

21.4.1.2 *C. tetani* Toxins

C. tetani produces two well-known pathogenic toxins, the tetanus neurotoxin (TeNT) (often referred to as tetanospasmin) and an oxygen-labile hemolysin referred to as tetanolysin [30,38,39,64,127]. In the genome analysis of *C. tetani*, Brüggeman et al. [19,128] described other putative virulence-related genes, including collagenase, fibronectin-binding protein, a putative PAF acetylhydrolase, a periplasmic immunogenic protein, myosin-crossreactive antigen, virulence factor MviN, internalin A homologue, and surface layer proteins. Many of these genes were proposed as virulence factors on the basis of their homology with virulence factors from other bacterial pathogens. Although animal models are available for study of wound infections by *C. tetani* [129], virulence factors have not been elucidated due to lack of genetic tools, particularly, directed mutational and gene replacement systems.

Unequivocal designation of pathogenic *C. tetani* requires demonstration of toxigenicity. Toxicity and neutralization tests are usually performed by intramuscular injection of 0.1 ml of culture supernatant into the rear leg muscle of a mouse; a second mouse is injected with a mixture of 0.1 ml of supernatant and 0.1 ml of tetanus antitoxin [30,64,65]. A positive culture will cause the development of a rigid paralysis in the injected muscle, with subsequent progression of the paralysis into other musculature regions of the body. Test animals should be euthanized as soon as a conclusive judgment can be made of paralytic signs. It is estimated that the human lethal intravenous or intramuscular dose of tetanus and botulinum neurotoxins are 0.2 to 0.5 ng per kg [13,130], and extreme caution is needed in handling these toxins. Unlike *C. botulinum*, TeNT is not active by the oral route, probably because it lacks protective proteins analogous to those in the BoNT toxin complexes [17,131].

21.4.2 *Clostridium botulinum*

21.4.2.1 General Description of *C. botulinum*

C. botulinum produces seven distinct serotypes of neurotoxins, A, B, C, D, E, F, and G [21,29,31,131]. Although the "species" differs widely in phenotypic and genetic relatedness, all

TABLE 21.1
Characteristics of the Six Known Clostridial Groups that Produce the Clostridial Neurotoxins

Neurotoxigenic Organism	Neurotoxins Formed	Location of Neurotoxin Gene	Non-Neurotoxigenic Equivalent Organism
C. botulinum group I (proteolytic)	A, B, F	Chromosome	*C. sporogenes*
C. botulinum group II (nonproteolytic)	B, E, F	Chromosome	No name given
C. botulinum group III	C, D	Bacteriophage	*C. novyi*
C. botulinum group IV (*C. argentinense*)	G	Plasmid	*C. subterminale*
C. baratii	F	Chromosome	All typical strains
C. butyricum	E	Chromosome	All typical strains
C. tetani	TeNT	Plasmid	*C. tetanomorphum* *C. cochlearium*

strains have the common property of producing this characteristic neurotoxic protein. With the exception of *C. botulinum* types C and D, strains of *C. botulinum* generally only produce one neurotoxin type [70,132]. There are reports, however, of strains that produce two toxin types, of which one is produced in much higher titers than the other, such as Af, Bf, and Ba [132], where the major toxin produced is designated by the uppercase letter.

Extreme caution is needed in working with the organism and its toxins [50,69,133]. The demonstration of botulinum toxicity by the traditional mouse bioassay is essential to identify toxigenic strains [62,68,85]. In addition to demonstration of neurotoxinogenicity, *C. botulinum* is also characterized by various cultural and physiological properties. *C. botulinum* was originally classified into three metabolic groups [134] but was expanded to four groups on discovery of *C. botulinum* type G (*C. argentinense*) [135,136]. Organisms in groups I–III are motile, produce lipase as demonstrated on egg yolk agar, liquefy gelatin, and ferment glucose. *C. botulinum* type G (*C. argentinense*, group IV) are proteolytic and nonsaccharolytic. Within the species *C. botulinum*, there exists considerable variation from strain to strain as shown by phenotypic and genotypic analyses [17]. The general properties of the groups I–IV of *C. botulinum* and the neurotoxigenic organisms *C. butyricum* producing E-like neurotoxin and *C. barati* producing F-like neurotoxin are presented in Table 21.1.

21.4.2.1.1 Description of Physiological Groups of C. botulinum

21.4.2.1.1.1 Group I: C. botulinum *type A, Proteolytic Strains of Types B and F*

The GC content was estimated to be 26 to 28 mol% [77], and the genome size of *C. botulinum* type A (strain 62A) was initially estimated as 4039 ± 40 kbp by PFGE [89]. The size of the genome of the "Hall" strain A (ATCC 3502) has been reported as 3,886,916 bp by genomic sequencing (http://www.sanger.ac.uk/Projects/C_botulinum). Although strains are commonly referred to by the source of isolation (e.g., Beans B, Beluga E) or by the person who isolated the strain (e.g., Hall A), strains with the same name designation can show marked differences in genetic composition and in toxin production [17,89]. Caution must be exercised in assuming that strains of the same name are identical. In fact, studies have shown considerable genetic variation in strains such as "Hall A" [89]. It would be valuable to have defined international reference strains for each serotype and group, ideally with their genome sequence available for comparative studies.

In TPGY broth, cells are straight to slightly curved rods, 0.6 to 1.4 × 3.0 to 20.2 μm, and are motile by peritrichous flagella [30,77]. The optimum temperature for growth of group I *C. botulinum* is 35 to 40°C, and the minimum and maximum are 10 and 48°C (these and other physiological parameters affecting growth in various environments have been reviewed) [30,138,139]. Most strains grow poorly below 12–15°C. The minimum pH for growth is 4.6 to 4.7 [140] and the maximum is 8 to 9. Spores are oval and subterminal and swell the rod. Sporulation is stimulated by zinc in the medium [107]. The incidence of spores in the environment shows an intriguing geographical distribution [21,31,101,142]. Nutritional growth factors required for *C. botulinum* types A and B are arginine and phenylalanine (high concentrations) and biosynthesis concentrations of methionine, leucine, valine, isoleucine, glycine, histidine, tryptophan, tyrosine, and the vitamins pyridoxamine, p-aminobenzoic acid, biotin, nicotinic acid, and thiamine [143]. The 10 required amino acids can be replaced by proteins by virtue of the organism's ability to synthesize proteases. Growth is stimulated by fermentable carbohydrates, including D-fructose, D-glucose, glycerol, maltose, pyruvate, sorbitol, sucrose, and starch [143]. Arginine is the organic nutrient required in highest concentrations by group I *C. botulinum* and strongly represses toxin and protease formation in *C. botulinum* types A and B [143–145].

21.4.2.1.1.2 Group II: C. botulinum *Type E and Nonproteolytic Types E, B, and F*

The "*Bacillus botulinus*" originally isolated by van Ermengem [47] was probably a strain of nonproteolytic *C. botulinum* type B (group II). Cells in TPGY are straight rods, 0.8 to 1.6 × 1.7 to 15.7 μm, and are motile by peritrichous flagella [30,76,77]. Spores are oval, subterminal, and generally swell the rod. Spores of type E have characteristic appendages and an exosporium, while spores of type F have an exosporium but lack appendages for strains examined.

Cultures ferment carbohydrates but are generally weak or non-proteolytic. Gelatin is liquefied, but casein or meat particles are not digested. *C. botulinum* type E requires glucose or another carbohydrate energy source for growth and does not utilize arginine or intact protein as a source of energy [143]. *C. botulinum* type E requires biosynthesis levels of seven amino acids (histidine, isoleucine, leucine, tryptophan, tyrosine, valine, and serine), adenine, and six vitamins (biotin, thiamine, pyridoxamine, folic acid, choline, and nicotinamide). This organism utilizes ammonia as a nitrogen source, although growth is stimulated by organic nitrogenous nutrients, especially glutamate and asparagine. Neurotoxin formation was strongly repressed by tryptophan in a minimal medium [146]. Unexpectedly, the nutritional requirements of nonproteolytic *C. botulinum* type B were similar to group I strains [143].

The optimum temperature for growth of group II *C. botulinum* is 25 to 30°C, and the minimum temperature for growth of types B and E is 3 to 4°C [147–150]. Most strains grow poorly below ~7°C. The minimum pH for growth is 5.0 and the maximum is ~8. The minimum A_w for growth of group II *C. botulinum* is 0.97, and growth is inhibited by 5% brine. Growth is also inhibited by bile. Spores of group II *C. botulinum* have relatively low heat resistance compared to group I [105,109,138,139]. Spore recovery can be enhanced by using lysozyme in the recovery medium, which substitutes for lytic enzymes that are denatured during heating [109].

21.4.2.1.1.3 Group III: Clostridium botulinum *Types C and D*

The mol% G + C content of the DNA is 26 to 28 [30,77]. Cells in TPGY broth appear as straight rods, 0.5 to 2.4 × 3.0 to 22.0 μm, and are motile by peritrichous flagella. Spores are oval, subterminal, and distend the cell. Holdeman and Brooks [134] recognized that toxicity was unstable in *C. botulinum* type C and that nontoxigenic strains were indistinguishable phenotypically from *C. novyi* type A. Cultures of *C. botulinum* type C and D exhibit an interesting cross-relation of bacteriophage sensitivity [61,70]. Bacteriophages determining toxin production can cross-infect the two serotypes, and strains cured of a specific bacteriophage no longer produced toxin [61,70]. Depending on the lineage, *C. botulinum* types C and D are closely related phenotypically and genotypically to *C. novyi* and *C. haemolyticum*.

Cultures within group III are saccharolytic and weakly nonproteolytic; strains can liquefy gelatin, and some can curdle casein, but they do not digest meat [134]. *C. botulinum* type D and *C. novyi* have the distinct feature among tested anaerobes of producing propionate as a major end product from lactate [76,134]. Strains of type C also catabolize glutamic acid, glycine, arginine, aspartic acid, and some strains convert threonine to propionate [151]. Growth of group III *C. botulinum* can occur between 15 and 47°C, and optimum growth occurs at 35 to 40°C. Most strains do not grow below 20–25°C. The optimum pH for growth is 6.6 to 7.3.

Production of BoNT/C_1 and BoNT/D neurotoxins is dependent on bacteriophage infection (reviewed in Eklund and Dowell [61]). The genes for C_1 and D neurotoxins are carried by pseudolysogenic bacteriophages [61,70,152,153]. The genes for C_1 neurotoxin and C_3 exoenzyme are associated with bacteriophages. Other genes encoding proteins in type C and D botulinum toxin complexes including hemagglutinin and nontoxin nonhemagglutinin are also associated with bacteriophages [61,70,122,153].

21.4.2.1.1.4 C. botulinum *Group IV:* C. argentinense *(sic* C. botulinum *Type G), Certain Strains of* C. subterminale *and* C. hastiforme *(Group IV)*

Group IV comprises organisms producing botulinum BoNT/G. *C. botulinum* type G was originally isolated by Giménez and Ciccarelli in 1969 from soils in Argentina [135,136,137]. A serologic relationship was demonstrated between *C. botulinum* type G and some strains of *C. subterminale* and *C. hastiforme* [154]. *C. subterminale* resembles *C. botulinum* type G genotypically and phenotypically, except for neurotoxin formation. Later, *Clostridium argentinense* (species novum) was proposed for organisms related to type G *C. botulinum* [155]. Although toxigenic strains of *C. argentinense* have only been isolated from Argentina, nontoxigenic strains of the species have been obtained from clinical specimens including wound samples and blood cultures.

The G + C content of the DNA of *C. botulinum* type G is 28 to 30 mol% [155], but the size of its genomic DNA has not been determined. Cells harbor a large plasmid of 81 mDa [112,113]. Cells in broth are straight to slightly curved rods, 0.5 to 2.0 μm wide and 1.6 to 9.4 μm long, motile by peritrichous flagella, and Gram-positive [135,136,154,155]. The organism has unusual phenotypic properties compared to other neurotoxigenic clostridia and is weakly proteolytic and asaccharolytic. The organism produces very low titers of type G botulinal toxin in culture [154,156,157]. The neurotoxin is activated by proteolysis, and endoproteinase lys C is generally preferred over trypsin for activation [156,157]. BoNT/G has a molecular size of 144,000 Da and a specific toxicity of ca. 1×10^7 LD50 per mg in mice [158]. The genes for *Clostridium botulinum* type G toxin complex and for a bacteriocin are on a plasmid of 81 kDa [112,113], but the plasmid has not been characterized for potential of conjugation and lateral transfer. Type G botulinum neurotoxin was lethal by oral, intravenous, or subcutaneous routes to several animals, including monkeys, guinea pigs, chickens, and mice [136]. As with the botulinum neurotoxin complexes from other serotypes, it was several hundred times less toxic by the oral route as compared to intravenous route of administration [136].

21.4.2.1.1.5 Neurotoxigenic Clostridia Demonstrated to Form Neurotoxins Similar to C. botulinum

C. butyricum has traditionally been considered to be nonpathogenic. However, two neurotoxigenic strains were originally isolated from cases of infant botulism in Rome, Italy [92,159,160]. Since the original isolation of neurotoxigenic *C. butyricum* from Rome, additional strains have been from intestinal botulism incidents, including cases associated with *C. difficile* colitis, and from foodborne botulism outbreaks in Italy, China, and India [92–95,97,98]. These data indicate that toxin gene transfer from *C. botulinum* type E to *C. butyricum* occurs at a higher frequency than originally suspected. Plasmids and bacteriophages have been detected in *C. butyricum* [121], and the gene for the neurotoxin may be associated with a plasmid [121,122] or a bacteriophage [123].

C. baratii has been isolated from human wounds and infections, rat feces, and from soil and certain vegetables [30,77]. Strains of *C. baratii* that produce type F botulinum neurotoxin have been isolated from infants with botulism and also from foodborne botulism cases [161–164]. The type E and F neurotoxins produced by *C. butyricum* and *C. baratii* have been partially characterized [165,166].

21.5 POTENCY AND ASSAY OF CLOSTRIDIAL NEUROTXINS (CNTs)

21.5.1 Potency of CNTs

The hallmark of neurotoxigenic clostridia is their production of a characteristic neurotoxin (NT) of extraordinary potency for humans and animals [3,10,13,130,167]. Botulinum neurotoxin (BoNT) and tetanus toxin (TeNT) are the most poisonous substances, and potencies have been estimated to be as little 0.1 to 1 ng per kg body weight by i.v. and i.m. injections [130]. However, experience with commercial preparations of botulinal neurotoxins in humans as well as nonhuman primate exposure studies have demonstrated that the intramuscular lethal dose is approximately 39 U/kg body weight for humans [167]. Assuming that a mouse LD50 is approximately 10 pg, then the i.m. lethal dose for a 70 kg human would be approximately 20 ng. Botulinum toxin is also extremely potent by the oral route, and ingestion of 10 to 70 μg has been estimated to be lethal for a 70 kg human (150 to 700 ng per kg) [10,168,169]. This remarkable potency is primarily due to their extraordinary neurospecificity and by their action as zinc-metalloenzymes, catalyzing the cleavage of neuronal substrates *in vivo* at exceedingly low concentrations of NT (10^{-12} *M*) [3,4,7].

21.5.2 Detection of BoNT and TeNT

The mouse bioassay is the most important laboratory test to identify *C. botulinum* and *C. tetani*, as well as to establish toxigenicity of novel neurotoxigenic strains such as *C. barati* and *C. butyricum* [13,30,62,68,85]. Neutralization of toxicity by tetanus antitoxin or type-specific botulinum antitoxins ensures positive identification of toxigenic strains [62]. Detection of neurotoxins is usually carried out on fecal specimens, blood (serum), suspect foods in botulism cases, and culture fluid following enrichment of the organism. Standard methods have been described [62,68,85]. Specimens of diluted culture supernatants are injected into mice after a portion is reacted with antitoxin. TeNT is injected into the hind leg or paw, and botulinum toxin is injected into the intraperitoneal cavity or into the tail vein in a manner analgous to that of TeNT [30,65].

Complications are often encountered in the mouse bioassay for BoNT, particularly deaths caused by nonbotulinum interfering agents or infectious agents. In assays for botulinum toxin, deaths occurring in less than 2 h (i.p.) or later than 48 h should be considered suspect, and the assay should be repeated [31,62]. Generally, it is possible to dilute out the interfering agent to an end point where the nonspecific deaths are eliminated. Occasionally, a sample might contain more than one serotype of botulinum toxin, other lethal toxins, or infectious agents, and it might be necessary to neutralize using more than one type of antitoxin or to use antibiotics to inhibit the infectious agent.

There has been considerable interest in developing assays as alternatives to the mouse bioassay. One of the primary thrusts has been in the area of immunological assays, particularly enzyme-linked immunosorbent assays (ELISA) and related assays, such as lateral-flow immunoassays and other platforms. Since BoNTs are zinc-metallopeptides with high specificity for their substrates, assays have been tested based on catalytic activity on the natural neuronal substrates combined with sensitive detection methods, including high-throughput fluorogenic assays and fluorescence resonance energy transfer (FRET) sensing systems [173,174]. A particularly exciting recent area of research has been the use of mass spectroscopy including high-resolution Maldi-Tof or electrospray to characterize the toxins and to detect reaction products from proteolytic cleavage of the neuronal substrates [175]. Recently, approaches have been directed toward development of biosensor devices including microfluidic and nanofluidic platforms [176]. The successful development of these advanced methods will require assay of BoNTs in complex matrices such as foods and clinical samples (blood, saliva, feces). A challenging assay for clostridial neurotoxins will be their detection in human or animal neuronal cytosol [177]. In the development of all new assays, the method should be controlled to the "gold standard" and time-tested mouse bioassay.

21.5.3 Safety Aspects of Working with CNTs

Biological safety precautions, security, and appropriate animal care is paramount for conducting research with the neurotoxigenic clostridia [69,133,178]. The laboratory should be designed according to recommended guidelines. A biosafety manual should be posted in the laboratory and contain emergency phone numbers and procedures for emergency response, spill control, and decontamination. All personnel should be trained in these procedures, and personal protection apparel should be worn in all experimental manipulations. In the U.S. and certain other countries, personnel who handle *C. botulinum* and BoNTs undergo immunization with a botulinum pentavalent toxoid [69]. When performing experiments where aerosols may be created (e.g., centrifugation), special precautions need to be taken. A class II or III biological safety cabinet or respiratory protection should be used during these procedures. The use of needles and syringes in bioassays also requires an extreme degree of caution. The Centers for Disease Control and Prevention (CDC) recommends Biosafety Level 3 primary containment and personnel precautions for facilities making production quantities of the toxin. Starting in 1997, *Clostridium botulinum* cultures and toxins have been included in a group of select agents whose transfer has been controlled by the CDC and the U.S. Department of Agriculture (USDA). To transfer these agents, both the person sending and the person receiving them must be registered with the CDC and USDA. Import and export permits must be attained for transfer of cultures in most areas of the world.

21.6 STRUCTURE AND ORGANIZATION OF BoNT TOXIN COMPLEXES

The genes encoding the protein components of the BoNT toxin complexes (progenitor complexes) reside as a cluster on the chromosome (BoNT/A/B/E/F), on bacteriophages (BoNT/C/D), or on plasmids (BoNT/G, TeNT) depending on the *C. botulinum* serotype [13,18,179]. Tetanus neurotoxin (TeNT) is believed to be synthesized as a single toxic protein without being associated with proteins or nucleic acids [3,4,13,18,180]. In contrast, BoNTs are produced in culture as molecular complexes consisting of BoNT, hemagglutinins (HA), nontoxic nonhemagglutinin (NTNH), possibly other uncharacterized proteins, and RNA [3,4,13,17,131,153,179,181–183]. In this chapter, I have used the "traditional" nomenclature for the genes and proteins known to exist in the toxin gene complexes, i.e., *bont* (BoNT), *ntnh* (NTNH), *ha* (HA), *botR* (BotR). orfX (OrfX) is the designation used for genes of unknown function or protein structure within the toxin complexes. The literature also contains the designations of, e.g., *p21* (P21), for genes encoding proteins of a given molecular mass (21 kDa). These gene and protein nomenclatures do not adhere to the systematic rules of genetic nomenclature. There has been constructive suggestions to use cnt as the gene designation, such as *cntA*, *cntB*, etc. for structural genes, and *cntR* for regulatory genes [18], but it has been difficult to use this nomenclature for other genes such as NTNH, HA, and uncharacterized orfs in the *C. botulinum* toxin complexes. Standard nomenclature should be adopted for the genes within the entire toxin complex.

The association of BoNT with other proteins was initially shown by ultracentrifugation in alkaline conditions and by the demonstration that the neurotoxic organisms and hemaglutinin activity could be dissociated [13]. Considerable research has been dedicated to understanding the genetic arrangement of the genes of the toxin complex, and in purifying and characterizing the protein components. A limited number of laboratories have purified and characterized all seven toxin complexes [13,17,31]. The toxin genes of the complex and associated proteins vary in structure among the various serotypes and strains (Figure 21.2). In addition to proteins, certain complexes have been shown to contain associated RNA, which is probably extracellular rRNA, which is abundant from the cells. The toxin complexes range in size from ~300 to 900 kDa depending on

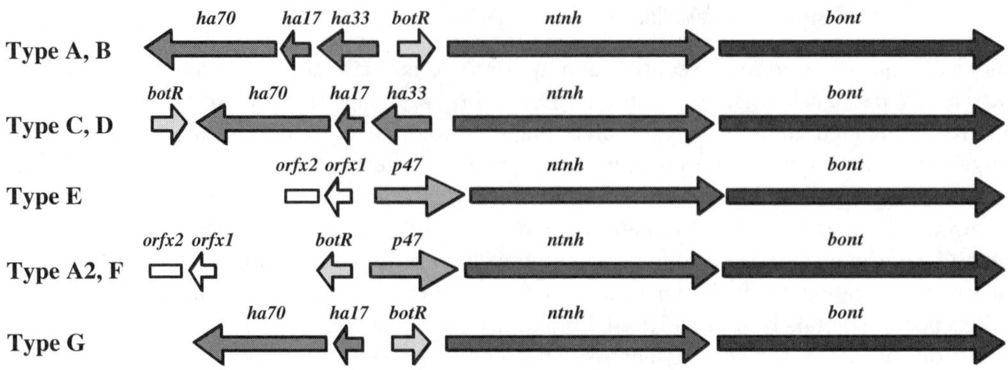

FIGURE 21.2 Structural representation of CNT toxin complexes. Adapted from Reference 17.

the serotype, strain, nutritional and environmental conditions of the culture, and method of isolation [13,17,31,131]. Nutritional and cultural conditions, including metal composition of the culture medium, presence of amino acids and peptides, pH and buffering capacity, temperature, and cell density have been reported to influence toxin complex production and structure [13,17].

Other than TeNT and the BoNT component of the toxin complexes which have catalytic activity on neuronal substrates [3,4], very little is known of the roles of the proteins and nucleic acid components of the BoNT complexes. The nontoxic proteins of the complexes are often discarded during purification of the BoNTs, and their structures and possible functions have not been studied in detail, although interest and study of their roles are increasing. It has been proposed, and certain evidence supports, that these nontoxic proteins have protective roles, and help to protect BoNT activity during oral ingestion [13,17,131,153], as well as during high dilution, which is necessary for compounding as a pharmaceutical [13,184]. The "adjuvant" proteins may also help slow diffusion of BoNT into neighboring tissues and resulting ptosis on injection as a pharmaceutical, but the evidence for this is equivocal [185]. BoNT dissociates from the nontoxic proteins in the lymph and in the blood at physiological pH [7,185,186] and in an unknown mechanism leaves the vasculature to reach motorneurons [7]. Some of the HA components have amino acid sequence homology with the region of *C. perfringens* enterotoxin that is important for insertion into target cells, and they also possess Arg-Gly-Asp (RGD) and Lys-Gly-Asp (KGD) sequences similar to regions of fibronectin, and thus may be involved in the association with target cells [5,227,239]. Studies have shown that HA components have affinity for galactose residues [181,187] and that gangliosides inhibited the hemagglutination reaction [188]. Thus, it appears that HA may bind to surface components on cells that contain sialic acid or D-galactose on the surface, but these proteins have not been identified. It is not clear [7,131,153] that the HA component has a role in the translocation of BoNT across the intestinal epithelium [7]. Structural studies have demonstrated that sugar residues bind to the H_c domain of TeNT [189]. Research is needed to more completely elucidate the biological roles of the proteins and nucleic acid components of the botulinum toxin complexes.

All of the BoNT genes analyzed to date reside in clusters, and the sole uniformity in gene arrangement among these clusters appears to be that the *ntnh* gene resides immediately upstream of *bont* (Figure 21.2). In several BoNT complexes, a four-component HA is also present, which is encoded by three genes in one operon. However, the presence of *ha* genes, the *botR* gene, and other genes encoding for unknown ORFS is variable depending on the cluster type. For example, *ha* is absent in subtype cluster A2, BoNT/E, and F clusters that have been analyzed, and ha17 and ha70 are the only *ha*s present in the BoNT/G cluster. In addition, our laboratory has recently found several genes encoding ORFS of unknown functions that appear to be associated with A2 clusters and are transcribed from a promoter within the traditional toxin gene cluster [183]. Thus, considerable variation has been found in the composition of the gene and protein components of the BoNT gene clusters [13,17,18,182,183,190,191], and these regions apparently undergo active recombination or translocation within regions in the chromosome. BoNT and TeNT gene clusters (except for BoNT/E) possess the positive regulatory genes, *botR* or *tetR*, that encode a protein of about 21 kDa. Certain strains of *C. botulinum* type A associated with infant botulism (A2 clusters) lack the *ha* genes [183,190,192], and certain strains contain silent genes for NTNH and BoNT that are not expressed due to various classes of mutations [17,193,194]. Characterization of a toxigenic *C. botulinum* and neurotoxigenic strains of *C. butyricum* showed the presence of additional ORFs (encoded by ORF-X1 and ORF-X2) in the type E, F, and A2 regions of the clusters [183,190]. These strains had a variable distribution of *orf22* or *p21* [*botR*], and *p47* [190]. The variation in the complement and organization of genes coding for the complexes, as well as the detection of silent or chimeric toxin gene complexes in numerous strains, indicates that considerable gene movement and possibly transposition or recombination occur within these regions in *C. botulinum* and toxigenic *C. butyricum*.

It has become apparent that a significant proportion of *C. botulinum* strains contain multiple toxin gene clusters [17,92,193]. Most experiments for the detection of secondary toxin gene clusters used PCR [17,92,193] and the clusters have not been mapped; it is not clear if they both reside on the chromosome and in close or distant locations. Many of these "secondary" clusters possess expressed genes as well as unexpressed (silent) genes of the complexes. In a detailed characterization of NCTC 2916, which carries an A cluster and a silent B cluster A(B), mapping of the genes by PFGE indicated that the silent and active toxin gene clusters were separated by 40 to 60 kb [194]. Interestingly, sequence analysis of NTNH adjacent to the defective BoNT/B gene was "chimeric," and the 5′ and 3′ regions of the gene showed high homology with the corresponding regions of the B NTNH gene, while the 471-amino acid sequence in the central region was identical to NTNH of type A [194]. A "mosaic" toxin gene was also detected in 3 strains of *C. botulinum* that consisted of parts of BoNT/C1 and BoNT/D [195]. Mosaic regions are more frequently being detected in clusters, particularly in the *ntnh* gene, and in one strain, the cluster contained sequences characteristic of ntnh/A and ntnh/B [195]. The chromosomal regions flanking the toxin gene complexes have not been well characterized and the genes are not known, other than a putative autolysin gene located downstream from the toxin gene cluster in *C. botulinum* NCTC 2916 [18]. Similar sequences were found in only a few other strains. Recent data from our laboratory indicate that there is considerable variation in the upstream and downstream regions in three type A *C. botulinum* strains being characterized [183]. Variations in these regions could be involved in differences noted in regulation of toxin production in different type A strains [13,17,182,183,196].

Most strains of *C. botulinum*, even if they contain more than one toxin gene cluster, only produce detectable levels of one serotype of BoNT. In early work on *C. botulinum* strain 62A, Tn*916* mutagenesis, PFGE and DNA hybridization analyses clearly proved that a single toxin gene cluster is chromosomally located [17,89]. The genome sequence obtained at the Sangre Centre has confirmed that the BoNT/A gene cluster is located on the chromosome. However, several strains of *C. botulinum* types A and B [A(b), A(f), B(a), B(f)] have been isolated that produce more than one serotype of toxin, usually in different quantities [17,98,132]. The cluster structures and genetic location are not known from many of these isolates. Production in certain strains is temperature-dependent, which appears to be an unusual trait for most group I *C. botulinum* strains. The regulation of expression of different toxin type genes from the same organisms would be expected to be complex and has not been elucidated.

21.7 REGULATION OF CNT EXPRESSION

Very little is known regarding the regulation of expression BoNT and TeNT. The production of various serotypes of botulinum toxin depends on the strain, medium, and conditions of culture.

Strains containing the BoNT genes on extrachromosomal elements often produce lower quantities of toxin than other serotypes. Production of BoNT in type C and D strains is notoriously unstable, and type G toxin is produced in very low quantities.

Regulation of BoNT production has been investigated in *C. botulinum* types A and B. Early nutritional studies in *C. tetani* and *C. botulinum* demonstrated that the expression of TeNT and BoNT are regulated. In pioneering studies, Mueller and colleagues showed that histidine diminished synthesis of TeNT, and that optimal synthesis was achieved by slowly allowing histidine metabolism [198–201]. This was accomplished by eliminating free histidine and supplying the amino acid in peptide form. Definitive studies were eventually performed in a chemically defined medium, and Mueller and colleagues at Harvard advocated the use of chemically defined media for production of antigens for vaccine production. In retrospect, the recommendation was prescient as evident by the concerns for potentially dangerous antigens, such as BSE in vaccines produced today. Recently, it was demonstrated that certain peptides derived by casein pancreatic digestion of hormones and other peptides enhanced the production of tetanus toxin [202].

Early studies showed also that certain peptones and casein digests, glucose, meat extracts, corn steep liquor, and other complex nutrients stimulated BoNT synthesis in *C. botulinum* type A [203]. BoNT is regulated by nitrogen sources, particularly arginine [13,17,144,145]. Our laboratory found that high levels of arginine repressed toxin formation and activation in *C. botulinum* Okra B, a group I strain with high proteolytic capabilities. At very high levels (5%), glucose also decreased expression to a lesser extent [13,17,144,145]. In type E *C. botulinum*, excess levels of tryptophan repressed synthesis of BoNT/E [146]. These studies demonstrated regulation by the nitrogen source, particularly arginine, and by high levels of glucose, but the molecular mechanisms governing this regulation have not been elucidated. These results imply that one function of NTs may be to obtain nutrients for growth. Sporulation in *C. botulinum* type B, which appears to be antagonistic to toxin production in certain strains, is enhanced by zinc in the medium [107]. This is interesting since zinc is present in the toxin and is also been shown to affect toxin complex assembly [13].

Although expression of BoNT and TeNT is clearly regulated by nutrition, the mechanisms of control are not known at the molecular level. Consequently, production of high titers of neurotoxins relies on use of a limited number of strains and empirically derived media and growth conditions. In type *C. botulinum* type A, BoNT synthesis onsets during late log phase and the corresponding mRNA is detected by Northern blots for only a few hours, and then synthesis subsides with higher onset of proteolytic activation to the dichain structure [196].

In addition to regulation indicated by nutrition, we found preliminary evidence for the involvement of a positive regulatory factor in BoNT production within the toxin gene complex [17,204]. The involvement of negative regulation was also supported by expression studies carried out in *C. botulinum* 62A and its nontoxigenic derivative (strain LNT01), which possesses transposon Tn*916* and is deleted for the entire toxin gene cluster [17,118,197]. These experiments were also the first demonstration of conjugative transfer and expression of cloned botulinal genes in *C. botulinum*. Further experiments are being performed to identify genes regulating BoNT synthesis. Popoff and collaborators have found intriguing results indicating BotR/A (previously called *orfX* or *orf21*) and TetR are positive regulators of toxin gene synthesis in *C. botulinum* and *C. tetani*, respectively [205–207]. The positive regulatory proteins, BotR and TetR, have characteristics of transcriptional regulators including basic pIs (10.4 and 9.3, respectively) and helix-turn-helix motif structures. BotR and TetR possess considerable homology (~50% identity) have 20 to 29% identity to some other regulatory proteins including UviA, a putative activator of bacteriocin synthesis in *C. perfringens*, and TxeR from *C. difficile*, a positive activator of toxins A and B [111,205–207]. BotR/A also has homology to BotR/C (previously called orf21) from *C. botulinum* type C [122]. Molecular studies indicate that BotR/A and TetR only activate transcription when RNA core polymerase is present. Available evidence indicates that BotR/A and TetR are related to the alternative sigma factor group in the σ^{70} family [208,209]. This subgroup also contains UviA and TcdR, which respectively regulate bacteriocin production in *C. perfringens* and toxins A and B in *C. difficile* [111,211–215]. These studies suggest a common regulatory mechanism governing toxin synthesis in toxigenic clostridia.

21.8 BIOCHEMISTRY, STRUCTURE, AND PHARMACOLOGY OF BoNT AND TeNT

"Crystalline" botulinum toxin was first precipitated from culture fluid in 1908 by Somer and later purified by Lamanna and colleagues [13]. Researchers in the U.S. and Japan demonstrated that the large (900 kDa) "crystalline" type A molecule could be separated into toxic and nontoxic moieties. Later, using centrifugation techniques or chromatography on ion exchange resins, the NT component was isolated [13,216]. Tetanus toxin was also isolated as a pure neurotoxic protein during this period [13].

Following the isolation of the toxin complex and demonstration of BoNTs and TeNT, studies increasingly focused on these components, and less research was conducted with nontoxic proteins of the BoNT complex or of proteins and other factors that might be associated with BoNT and TeNT. BoNTs and TeNT are produced as single chain molecules of ca. 150 kDa, which achieve their characteristic high toxicities of 10^7 to 10^8 MLD per mg (Table 21.2) by post-translational proteolytic cleavage within a narrow region of the polypeptide to form a dichain molecule comprised of a light (L_C; ~50,000 kDa) and heavy (H_C; ~100,000 kDa) chain linked by a disulfide bond (Figure 21.3) [3,181,217]. Most BoNTs produced by strains of *C. botulinum* type A achieve full

TABLE 21.2
Properties of Botulinum (BoNT) and Tetanus (TeNT) Neurotoxins

Toxin	Gene Location[a]	Representative Species Affected	Specific toxicity ($\times 10^8$)[b]	Lc Substrate	Peptide Bond Cleaved[c]
BoNT/A	C	Humans, chickens	1.05–1.86	SNAP-25	Gln-Arg
BoNT/B	C	Humans, cattle, horses	0.98–1.14	VAMP	Gln-Phe
BoNT/C_1	B	Birds, cattle, dogs, minks	0.88	Syntaxin	Lys-Ala
				SNAP-25	Arg-Ala
BoNT/D	B	Cattle	1.60	VAMP	Lys-Leu
BoNT/E	C	Humans, fish, aquatic birds	0.21–0.25	SNAP-25	Arg-Ile
BoNT/F	C	Humans	0.16–0.40	VAMP	Gln-Lys
BoNT/G	P	Unknown	0.1–0.3	VAMP	Ala-Ala
TeNT	P	Humans, cattle, horses, sheep, dogs, chickens, other animals		VAMP	Gln-Phe

[a] Gene location: C, chromosome; B, bacteriophage; P, plasmid. For genes with putative chromosomal locations, this location is inferred from PCR amplification of "chromosomal" DNA preparations, except for type A, in which toxin gene mutations have been mapped to the chromosome (see Lin and Johnson [17,89]).

[b] Specific toxicity refers to toxins activated by trypsinization when necessary for maximum toxicity. Toxicities are per milligram of protein. Most of the reported data is from Sugiyama [181]. Experiments in our laboratory and (Sugiyama [181]; Sakaguchi [131]) have shown that oral doses of type A or B toxin complexes in mice are 10 to 1000 times greater than the IP or IV lethal dose, depending on the size of the complex, diluent, and other factors.

[c] The specific peptide bond cleaved is shown [3]. However, the clostridial neurotoxins require a minimum peptide length of >20 aa residues and a characteristic substrate tertiary structure for catalytic activity [3].

activation during culture, but the degree of proteolytic modification depends on the medium composition, temperature, and time of incubation [181,218]. Proteolytic modification of the dichain toxin type A occurs in at least two steps: first, removal of the internal sequence Gly-Tyr-Asn-Lys-Ala-Leu-Asn-Asp-Leu-, and then cleaving the N-terminal Met-Ala at a slower rate, leaving Ala as the N-terminal residue [218,219]. The toxin can undergo further proteolytic modification during subsequent incubation and storage, including autocatalytic cleavage [220,221]. The quality and homogeneity of BoNTs can change during incubation, shipping, and storage, and conditions should be determined in the individual laboratories for maintenance of high quality toxins. The propensity of the toxin to degrade has presented difficulties in determinations of structure. The isolated L and H chains are particularly prone to proteolytic modification and other denaturation reactions, such as aggregation, as has been shown, particularly for type A [222,223]. The extent of autocatalysis and aggregation is dependent on the presence of certain protective substances such as glycerol, methionine, and urea (for separated chains). In nonproteolytic serotypes such as B and E, proteolytic modification appears to be minimal in toxin preparations in culture supernatants, and a high degree of activation (e.g., up to 100-fold) is attained by treatment with trypsin or other proteases [181,217,218]. When activating BoNTs with trypsin (including proteolytic serotypes such as BoNT/B and BoNT/F, which are only partially activated during culture), it is usually desirable to stop protease activity by addition of soybean trypsin inhibitor or another suitable inhibitor to stop unwanted secondary proteolytic modification. In certain serotypes, such as type G, optimal activation occurs with proteases other than trypsin, such as endoproteinase Lys C [156,157].

21.9 CATALYTIC ACTIVITIES OF CNTs

The molecular mechanisms by which botulinal and tetanus neurotoxins specifically inhibited neuroexocytosis for extended periods remained a mystery for many years. It had been postulated by various investigators that the CNTs might break down the actin cytoskeleton, but such an observation later appeared to be due to the presence of contaminating C3 toxin or other degradation enzymes in toxin preparations. Montecucco postulated early on that the CNTs might contain intracellular enzymatic activity, analogous to diphtheria toxin activity in cells [224]. The breakthrough in elucidation of function came about through cloning and sequencing of the genes encoding TeNT and the BoNTs and analysis of the corresponding amino acid sequences in the late 1980s and 1990s [3,4,8,18,192,225–227]. The analysis of the nucleotide and amino acid sequences revealed that the light chain contained a zinc-binding motif characteristic of a class of zinc-metalloproteases [3,4,8,225,228,229]. This discovery led to the remarkable finding that botulinum and tetanus neurotoxins selectively cleave key neuronal proteins involved in exocytosis of neurotransmitters and potentially other signalling molecules and chemokines [3,4,6,8,230–232]. Specifically, it was found that the L chains of TeNT and BoNT are metalloproteases that specifically cleave VAMP/synaptobrevin (TeNT, BoNT/B,/D,/F, and/G), SNAP-25 (BoNT/A,/C and/E), and syntaxin (BoNT/C) (Table 21.2). VAMP/synaptobrevin, SNAP-25, and syntaxin form a heterotrimeric complex (SNARE complex) that mediates the exocytosis of vesicles [232–233]. BoNTs belong to the thermolysin family of zinc proteases [234], but they are distinct because of their requirements for large substrates and apparently stringent substrate specificities. Despite the remarkable elucidation of the enzymatic properties of BoNTs and the identification of their neuronal substrates, many challenges still exist to understand the biochemical mechanisms and cellular biology of CNT recognition of substrates, including factors affecting cleavage of the substrates in neurons (various proteins enhance activity *in vitro* — are there analogous stimulatory factors in nerves?), eludication of the mechanisms for the extraordinarily long half-life of certain serotypes in nerves [235,236], and, most intriguingly, the evolution of BoNTs and TeNT as highly specific proteases for substrates within nerve cells.

21.10 STRUCTURE OF TETANUS AND BOTULINUM NEUROTOXINS

As expected, the gene and corresponding amino acid sequences of TeNT and BoNTs possess regions of homology that have been useful in the prediction of structure and function [3,18,225,227]. TeNT and BoNTs show high homology in the regions defining the catalytic active site, the translocation domain, and the two cysteine residues forming the disulfide bond connecting the H_C and L_C. The least degree of homology is in the carboxyl region of the H_C, which is involved in neurospecific binding. The binding domain appears to have sequences conferring the largest differences in antigenicity [227,237,238] and might be largely responsible for specification of serotype as determined by neutralization with polyclonal antibodies raised against BoNTs. This C-terminal subdomain differs markedly in surface exposed residues in types A and B [237,239,240]. Although the functional domains of the BoNTs have high homology, the presence of different structural motifs suggests that "domain swapping" occurred in their evolution [237,239]. The high homologies of amino acid sequences of CNT genes isolated from diverse clostridia suggests that a single ancestral gene was dispersed by lateral transfer, and six lineages of clostridial NT evolution have been proposed [226]. However, the recent observations of subtypes of toxins within a given serotype make the evolutionary lineage picture more complex. Nonetheless, these observations reinforce the hypothesis that genetic transfer of the toxin genes or regions of the genes to normally benign clostridia occurs in certain environments, and it will be fascinating to determine the mechanisms of toxin gene transfer [17,100].

After several years of effort, Stevens and colleagues obtained the structure of type A BoNT [239]. This was a considerable achievement and was affected by the complexity of the molecule and its tendency to be unstable and to degrade. It was essential to have high quality protein of uniform nature, particularly from the highly proteolytic type A toxin, in which autocatalysis is common [220,221]. The overall shape of BoNT/A is rectangular with dimensions of ~45 Å × 105 Å × 130 Å, and the molecule shows a linear arrangement of the three functional domains (Figure 21.3) [239]. The three functional domains have distinct structures, and the molecule contains an unusual loop that encircles the perimeter of the catalytic domain. The existence of this loop was

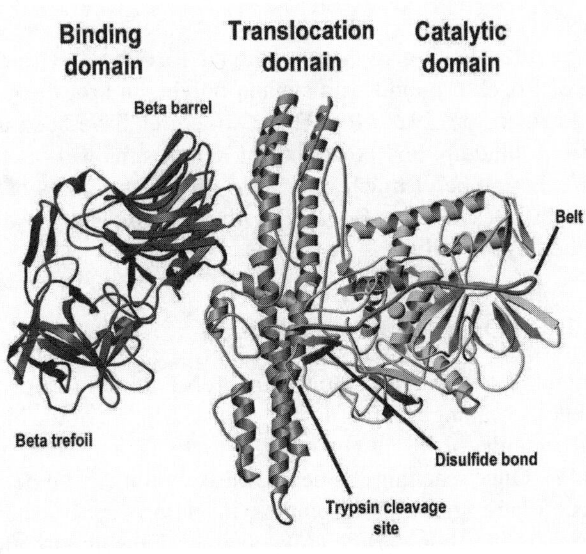

FIGURE 21.3 Structure of botulinum neurotoxin serotype A (BoNT/A). Figure kindly provided by Ray Stevens.

unexpected and presents a puzzle regarding catalysis, since it partly covers the catalytic active site. It is possible that the site opens on contact with the substrate (allosteric activation) or when the L_C is internalized within neurons. Structural analyses revealed that the ganglioside binding C-terminal subdomain has structural homology with proteins known to interact with sugars, such as the H_C fragment of TeNT, serum amyloid P, sialidase, various lectins, and the cryia and insecticidal ∂-endotoxin, which binds glycoproteins and creates channels in membranes. Thermolysin and leishmanolysin have high homology to the catalytic domain of BoNT/A, confirming that BoNT belongs to the thermolysin class of zinc proteases. The translocation domain is distinct in structure from bacterial pore-forming toxins and shows more resemblance to coiled-coil viral proteins such as HIV-1 gp41/GCN4, influenza hemagglutinin, and the MoMuLV TM fragment. BoNT/A appears to consist of a hybrid of varied structural motifs that may have evolved by combination of functional subunits to generate a highly toxic molecule. The structures of BoNT in various stages of intoxication, including oral ingestion, receptor binding, internalization and intraneuronal catalysis, have been elegantly discussed by Stevens and colleagues [4,239,240]. Stevens's group also obtained a low resolution electron density projection map by electron crystallography for the 900 kDa complex [241], and additional studies are needed to more accurately refine the structure of the large toxin complex. It is known that type A toxin complexes also exist in forms of approximately 500 kDa and 300 kDA, and it is likely that interconversion of these complexes occurs *in vitro*; thus, it may be difficult to isolate only a single form of the complex. A detailed structural analysis of the complex, including coverage of the nontoxic proteins, would provide considerable insight into the naturally produced complex, including increased understanding of oral toxicity and stability properties compared to lower molecular weight complexes and, particularly, the purified neurotoxin. Recently, the crystal structure of a truncated construct of BoNT/A L_C was obtained [242], and the structure appeared as a homodimer in the product-bound state. Interestingly, a four-stranded β-sheet at the center of the homodimer is formed between the F242 and V257 loops and is exchanged between the dimer partner. A large part of the dimer interface is inserted deeply into a cleft entering into the partner's active site. The loop was oriented in the reverse canonical direction than for other zinc proteases, such as thermolysin. It was proposed [242] that the BoNT/A binding mode and catalytic mechanism are substantially different from those proposed for BoNT/B [243] and perhaps other serotypes of BoNTs.

High-resolution structures have also been obtained for BoNT/B, the L_Cs of BoNT/A and BoNT/B, the structure of BoNT/B bound with synaptobrevin, and for the carboxy-terminus region of tetanus neurotoxin H chain [5,6,242–246]. These structures have been useful in understanding receptor binding, substrate binding, and poisoning of neurons, as well as for the rationale design and testing of inhibitors. It will be of much interest to obtain structures of BoNTs and TeNT that interact with different substrates, such as BoNT/C with syntaxin, and those that cleave at different peptide bonds within the CNT family.

21.11 CELLULAR BIOLOGY OF CNT ACTION

The mechanisms of neuronal poisoning by BoNT and TeNT initially involve entry into the vasculature and blood, whether by oral ingestion of BoNT and absorption of the intestinal tract, formation and translocation from wounds, or by inhalation [7,10]. BoNT and TeNT exit the vasculature by unknown mechanisms [7]. Once reaching the nerve, intoxication can be delineated into four major steps [5,6,8,9]: (a) specific and high affinity binding to motorneuron terminals; (b) entry into the endosomes or other intracellular compartments by poorly understood endocytotic processes; (c) trafficking of the L_C from intracellular compartments to the cytosol; and (d) cleavage of substrate proteins VAMP/synaptobrevin, SNAP-25, and syntaxin, culminating in blockage of endocytosis. The proposed general process is summarized in References 3, 4, and 7. The molecular mechanisms from exposure to neuronal poisoning are very complex, and much research is needed to further our understanding. Some of the outstanding questions include: (a) identification of the receptors for CNT binding; (b) elucidating the mechanisms of trafficking and the possible involvement of multiple/discrete pathways depending on the serotype; (c) multiple actions of the CNTs within neurons and; (d) the factors contributing to the long duration of action at activity at extraordinary low concentrations of CNTs in neurons ($<10^{-12}$ *M*). Relevant recent findings include the observation that synaptotagmin II, an intraluminal neuronal protein, appears to be involved in the receptor process for BoNT/B [247,248]. The trafficking mechanisms have also been receiving attention, and recently very intriguing data have been published on the possible mechanisms for chaperone involvement in trafficking and release from endosomes [249], as well as methods to follow pH-changed conformational changes during translocation through endosomes [250]. It is becoming clear that the neuronal environment and half-life play a role in the widely different durations of action that are observed for different serotypes of BoNTs and for TeNT [235,236]. Two general mechanisms have been debated — one being the control of the half-life of CNTs, and the other the regulation of neuronal function and synapse repair by products of CNT cleavage — and it is likely that both are important in controlling duration. VAMP/synaptobrevin cleavage by TeNT and BoNT is enhanced by acidic liposomes [251], and BoNTs of different serotypes appear to localize in different regions within the neuronal environment [252]. Since BoNTs and TeNT are zinc proteases, and zinc stimulates activity *in vitro*, it is possible that the metal environment of neurons affects activity, as it has been postulated to affect other neuronal diseases such as Alzheimer's disease and autism [253]. *In vivo,* BoNTs/A/B/E and TeNT undergo tyrosine phosphorylation, which affects their catalytic activity and stability [254,255], raising the intriguing question of whether BoNTs are interacting with a phosphorylation signaling pathway in mammalian cells. Besides being involved in regulating exocytosis, SNAP-25 and other neuronal proteins affect several functions in the nerve, including axonal growth, neurite sprouting, maturation, synaptogenesis, and G-protein regulation of presynaptic calcium channels, and other processes [5–7]. Research during the next few years will yield exciting results on the cellular biology of the CNTs, which will be important in understanding the mechanisms of intoxication.

21.12 BOTULINUM TOXIN AS A POTENTIAL BIOTERRORIST AGENT

Because botulinum toxin is the most poisonous substance known, and is active by the oral route, and potentially by inhalation, it has been considered as a potential bioterrorist agent [10,256]. Attempts have been made to develop or counteract BoNT preparations as weapons by several countries including Japan, Germany, Russia, Iraq, and the U.S. [10,257–260]. The earliest known use of botulinum neurotoxin was by General Shiro Ishii, head of the Japanese biological warfare command, who confessed to killing prisoners during Japan's occupation of Manchuria by feeding them cultures of *C. botulinum* [259]. The U.S. began intensive research into BoNTs and certain other toxins as potential weapons in 1944, when intelligence suggested that Germany had bombs containing the toxin [258]. Researchers at Fort Detrick and other camps produced large quantities of "crystalline" botulinum toxin and investigated its properties [13,16,258,260]. During the Gulf War in the 1990s, Iraq was reported to have produced 19,000 liters of concentrated botulinum toxin for impregnation of warheads [261,262]. There has been much debate on the conditions and skills needed to produce high quantities of botulinum toxin required for weaponization [10,13].

Botulinum toxin stability depends on its formulation (i.e., complexed with other proteins and other stabilizing molecules such as trehalose). A unique property of botulinum neurotoxin is its high oral toxicity, and a bioterrorist scenario could possibly involve administration of the toxin in food. Although poisoning can occur by aerosolization [7], research has suggested that administration by aerosolization would be limited by considerable technical difficulties [256]. Obviously, a bioterrorist event could result in a mass tragedy, and important measures to counteract the toxin on human exposure are available and in progress. Considerable research is being devoted to the development of rapid and sensitive sensing mechanisms for the toxin in the environment, foods, and in exposed individuals to enable a rapid medical response. Since botulinum toxin can inflict a paralysis of long duration, rapid detection and medical countermeasures would be of considerable value in preventing morbidity and mortality.

21.13 USE OF BOTULINUM TOXINS IN MEDICINE

Of the many outstanding properties and developments during the past two decades on CNTs, the most remarkable is the development of botulinum toxins for the treatment of human disease. The background and development of botulinum toxin as a pharmaceutical has been described [12–16,184], and it has become an extraordinarily effective drug for a myriad of disease disorders (Table 21.3). In common with many seminal discoveries, the development of botulinum toxin as a drug involved collaborations among scientists and physicians, creative experimentation, setbacks and frustration, more than a few twists of fate and good fortune, and extreme patience. The development of botulinum toxin as a drug involved productive collaborations between physicians and research scientists [13,16]. The guiding principle in the development of botulinum toxin as a drug followed the concept of "graded selective denervation." This concept had its roots in the works of early toxicologists and physiologists, and much of the seminal work is often attributed to Claude Bernard (1813–1878) in his studies with curare. He demonstrated that curare "kills rapidly without convulsions and renders the nerves inexcitable." [16,263,264]. While nerves become unexcitable after treatment with curare, the muscles still react on direct stimulation. He showed that the site of action of curare was motor nerve endings. Bernard further demonstrated that curare selectively inactivated the action of the motor nerves without impairing that of sensory nerves or the central nervous system. In essence, he demonstrated selective denervation with retention of voluntary muscle activity by a naturally occurring toxin.

TABLE 21.3
Pharmaceutical Uses of Botulinum Toxin in Medicine

Achalasia
Bruxism (TMJ)
Cosmetic (wrinkles, rhytides, facial asymmetry)
Debarking of dogs
Essential hyperhydrosis
Migraine headaches
Muscle contraction headaches
Myofascial pain
Myokimia
Nystagmus
Pain syndromes, e.g., back spasms
Painful rigidity
Pelvirectal spasms (anismus, vaginismus)
Protective ptosis
Selected inflammatory disorders
Spastic bladder
Spasticity
Strabismus
Stuttering
Tension headaches

Source: Compiled from various sources including [12–16].

The concept of using botulinum toxin as a selective chemodenervating agent was probably considered for many years by neurologists [11,12,13]. However, Drachman's fascinating studies of the effects of botulinum toxin in developing chicken embryos at Johns Hopkins University in Baltimore provided conclusive experimental proof that botulinum toxin A induced selective denervation, muscle weakening, and muscle atrophy [9]. The chick embryo model he employed was ideal for investigation of botulinum paralysis since respiratory gas exchange occurs by passive diffusion across the chorioallantoic membrane and large quantities of botulinum toxin could be administered without detrimental effects on respiration thus maintaining viability in the animal model. Drachman obtained "crystalline" toxin preparations from Ed Schantz, who was then at Fort Detrick, Maryland [13,16]. Following injection of enormous quantities (30 µg or approximately 30,000 units) of botulinum toxin, the thighs and legs of the chicks were markedly shrunken, devoid of muscle tissue, and showed signs of degeneration and fat accumulation [9]. The cardiac muscle of the chicks appeared normal, indicating that botulinum toxin did not have a generalized toxic effect on striated muscle but preferentially poisoned skeletal muscle. Drachman [9] proposed the characteristics of an "ideal (nerve) blocking agent" in terms of its: (a) *mode of action* in selectively blocking cholinergic transmission; (b) *specificity* in blocking only cholinergic transmission; (c) *reversibility* in not permanently impairing function or structure of nerve or muscle; (d) *generality of action* in blocking exocytosis from all motor neuron terminals of striated skeletal muscle; (e) *convenience of use* in requiring simple injection of a soluble toxin preparation into desired muscle regions; (f) *safety* if used with appropriate precautions and proper doses; and (g) *absence of systemic or CNS effects* [9,16].

The use of botulinum toxin type A as an injectable selective denervation and muscle-weakening agent in humans was investigated by Scott in concert with Drachman and, subsequently, with Schantz. Alan B. Scott (1932–present) primarily used macaque monkeys as a model in the late 1960s and late 1970s and in humans in the 1970s and 1980s. These experiments involved a unique collaboration with Edward J. Schantz (1908–present), and later with Schantz and myself in the 1980s and 1990s [13,16,184]. Scott's initial goal was to provide a pharmacologic alternative to surgery in patients with strabismus (crossed eyes) by injection of extraocular muscles with neurotoxic agents using the monkey model. Scott and other investigators had injected ethanol, phenol, or other drugs in attempts to achieve this goal, but these injections were extremely painful and were not of long duration. Remarkably, Scott was able to produce weakness in eye muscles of monkeys that lasted 2 to 8 months by injection of botulinum toxin. Scott also demonstrated that accurate injection of graded quantities of botulinum toxins altered ocular alignment. Dietary habits and activity appeared normal in the monkeys. Transient ptosis in the vicinity of the monkeys' eyes cleared within a few weeks. Three and a half months following injection, the EMG recorded from the injected muscle was of normal amplitude, and eye movement was also normal. In these remarkable experiments, permanent ocular alignment changes after temporary muscle paralysis was the clinical outcome without any need for incision or direct exposure of the muscle. The quantity of toxin needed to induce ocular alignment was approximately 700-fold less than the estimated lethal dose (~ 0.1 µg or ~28 U/kg) [167] for a 70 kg human, which has been estimated as 1 to 2 µg (1000–2000 units). Scott predicted that botulinum toxin injection was a suitable pharmacologic approach that could replace or augment existing methods of surgical correction of strabismus. Scott also predicted that botulinum toxin could be used as a denervating agent in humans for treatment of many other conditions characterized by overactive neuromuscular activity.

Following the 10- to 15-year studies with monkeys, the tools were in place for experimentation with humans. In response to a courageous request by Scott in the late 1970s, the National Institutes of Health granted him permission (in a similarly courageous response) to inject human adult volunteers suffering from strabismus. In 1980, he reported the results of 67 injections of botulinum A toxin for correction of strabismus. These studies established type A botulinum toxin injections as an effective nonsurgical alternative treatment for strabismus. Botulinum toxin type A appeared to be a specific, effective, and safe chemodenervating agent for weakening of muscles. After thousands of injections into patients, the Food and Drug Administration approved one batch of toxin produced in 1979 by Schantz at the Food Research Institute (batch 79-11) for use in the United States for strabismus, blepharospasm, and hemifacial spasm. Schantz (and later, in 1985, this author) prepared several batches of crystalline botulinum toxin using the Schantz strain (Hall A-*hyper*), proven fermentation conditions, and a purification procedure optimized at the University of Wisconsin Food Research Institute [13,16,17,184]. Schantz and I wrote sections of the Master Batch Record for FDA approval of batch 79-11. During this period, in the mid- to late 1980s, and extending into the late 1990s, botulinum toxin was increasingly being experimentally used for many other disorders, and this single batch of toxin (79-11; ~150 mg) eventually was used for more than 250,000 injections in humans. In the 1980s, Schantz and I prepared several other batches of type A botulinum toxin that were injected with excellent results in humans.

Scott [12,265–267] outlined the following principles and characteristics of botulinum toxin based on his monkey and human studies: (a) it showed no known focal or systemic effects apart from muscle paralysis; (b) it apparently did not elicit antibody production in the small doses that were used; (c) it diffused slowly out of the injected muscle region into adjacent muscles; (d) the toxin acted for several weeks to months; (e) and the paralytic intensity was strongly correlated with the dose injected.

As Scott [265] predicted in 1973, the utility of botulinum toxin A in medicine would greatly accelerate and expand to other neuromuscular disorders. Botulinum toxin type A has been effective in treatment of a myriad of disorders involving involuntary muscle contractions, dystonias, and spasticity in focal or segmental muscle regions. Somewhat unexpectedly, it has also been efficacious in treating pain syndromes and certain disorders associated with inflammation [268]. One of the first pain syndromes to be treated was myofascial pain [268]. Tension and migraine headaches have also been treated with successful results. The treatment of certain pain conditions suggests that it may affect autonomic nervous systems and nervous control of pain. Undoubtedly, the numbers of disorders treated with botulinum toxin will continue to grow, particularly as we learn more of its basic effects on nerve-muscle physiology.

21.14 CONCLUSIONS AND PERSPECTIVES

Outstanding advances have been achieved during the past three decades in the elucidation of the biochemistry, structure, and pharmacological mechanisms of botulinum and tetanus neurotoxins. The discovery that CNTs cleave neuronal substrates involved in transmitter vesicle transport and exocytosis in the presynaptic nerve terminal has contributed greatly to an understanding of the mechanisms of clostridial neurotoxins. These advances have certainly contributed to the remarkable success of botulinum toxin as a pharmacological agent for the treatment of a myriad of neuromuscular muscle diseases. Much information has also been gained regarding the genes encoding the toxins. Paradoxically, a basic understanding of the biology of *C. botulinum* and the mechanisms governing the lateral transfer of the CNT genes to other clostridial species have lagged behind the studies of CNT biochemistry, structure, and cellular biology. The development of genetic tools, as well as functional genomics, is expected to provide progress in this important area of the CNTs. Further knowledge of the properties of the neurotoxigenic clostridia and their neurotoxins should lead to improved pharmaceuticals for prevention and treatment of human disease.

ACKNOWLEDGMENTS

I am grateful for the many collaborations with colleagues in the CNT field, particularly to Ed Schantz for introducing me to many remarkable features of botulinum toxin. I also thank Mary

Whitmer for assistance with this chapter. Research in my laboratory has been supported by the NIH, USDA, FDA, the Food Research Institute, and the College of Agricultural and Life Sciences, University of Wisconsin – Madison.

REFERENCES

1. Bleck T.P., Tetanus: pathophysiology, management, and prophylaxis, *Disease-a-Month*, 37, 547–603, 1991.
2. Cherington, C., Clinical spectrum of botulism, *Muscle & Nerve*, 21, 701–710, 1998.
3. Montecucco, C. and Schiavo, G., Structure and function of tetanus and botulinum neurotoxins, *Quart. Rev. Biophys.*, 28, 423–472, 1995.
4. Schiavo, G. et al., Neurotoxins affecting neuroexocytosis, *Physiol. Rev.*, 80, 717–766, 2000.
5. Lalli, G. et al., The journey of tetanus and botulinum neurotoxins in neurons, *Trends Microbiol.*, 11, 431–437, 2003.
6. Turton, K., Chaddock, J.A., and Acharya, K.R., Botulinum and tetanus neurotoxins: structure, function, and therapeutic utility, *Trends Biochem. Sci.*, 27, 552–558, 2002.
7. Simpson, L.L., Identification of the major steps in botulinum toxin action, *Annu. Rev. Pharmacol. Toxicol.,* 44, 167–193, 2004.
8. Humeau, Y. et al., How botulinum and tetanus neurotoxins block neurotransmitter release, *Biochimie*, 82, 427–446, 2000.
9. Drachman, D.B., Botulinum toxin as a tool for research on the nervous system, in *Neuropoisons. Their Pathophysiological Actions. Vol. 1. Poisons of Animal Origin*, Simpson, L.L., Ed., Plenum Press, New York and London, 1971, 325–347.
10. Arnon S.S. et al., Botulinum toxin as a biological weapon. Medical and public health management, *JAMA*, 285, 1059–1070, 2001.
11. Caya, J.G., *Clostridium botulinum* and the ophthalmologist: a review of botulism, including biological warfare ramifications of botulinum toxin, *Survey of Ophthamol.*, 46, 25–34, 2001.
12. Scott, A.B., *Clostridial toxins as therapeutic agents*, Simpson L.L., Ed., Academic Press, San Diego, 1989, 399–412.
13. Schantz, E.J. and Johnson, E.A., Properties and use of botulinum toxin and other microbial neurotoxins in medicine, *Microbiol. Rev.*, 56, 80–92, 1992.
14. Jankovic, J. and Hallett, M., *Therapy with Botulinum Toxin,* 4th Edition, Marcel Dekker, New York, 1994.
15. Brin, M.F., Hallett, M., and Jankovic, J., Eds., *Scientific and Therapeutic Aspects of Botulinum Toxin*, Lippincott, Williams & Williams, Philadelphia, 2002.
16. Johnson, E.A., Clostridial toxins as therapeutic agents: benefits of nature's most toxic proteins, *Annu. Rev. Microbiol.*, 53, 551–575, 1999.
17. Johnson, E.A. and Bradshaw, M., *Clostridium botulinum*: A metabolic and cellular perspective, *Toxicon*, 39, 1703–1722, 2001.
18. Quinn, C.P. and Minton, N.P., Clostridial neurotoxins, in *Clostridia; Biotechnology and Medical Applications*, Bahl, H. and Dürre, P., Eds., Wiley-VCH, Weinheim, Germany, 2001, 211–250.
19. Brüggeman, H., et al., The genome sequence of *Clostridium tetani*, the causative agent of tetanus disease, *Proc. Natl. Acad. Sci. U.S.A.*, 100, 1316–1321, 2003.
20. Shimizu, T.K., et al., Complete genome sequence of *Clostridium perfringens*, an anaerobic flesh-eater, *Proc. Natl. Acad. Sci. U.S.A.*, 99, 996–1001, 2002.
21. Smith, L.D.S. and Sugiyama, H., *Botulism: The Organism, Its Toxins, the Disease*, 2nd Edition, Charles C. Thomas, Springfield, IL, 1988.
22. Simpson, L.L., *Botulinum Neurotoxin and Tetanus Toxin,* Academic Press, San Diego, 1989.
23. Bahl, H. and Dürre, P., Eds., *Clostridia, Biotechnology and Medical Applications,* Wiley-VCH, Weinheim, Germany, 2001.
24. Rood, J.L. et al., Eds., *The Clostridia, Molecular Biology and Pathogenesis,* Academic Press, San Diego, 1997.
25. Montecucco, C., Ed., *Clostridial Neurotoxins*, Springer, Berlin, 1995.

26. Alouf, J.E. and Freer, J.H., Eds., *The Comprehensive Sourcebook of Bacterial Toxins*, 2nd Edition, Academic Press, London, 1999.
27. Rappuoli, R. and Montecucco, C., Eds., *Guidebook to Protein Toxins and Their Use in Biology,* Oxford University Press, Oxford, 1997.
28. van Heyningen, W.E., *Bacterial Toxins*, Blackwell Scientific, Oxford, 1950.
29. Hatheway, C.L., Toxigenic clostridia, *Clin. Microbiol. Rev.*, 3, 67–98, 1990.
30. Hatheway, C.L. and Johnson, E.A., *Clostridium*: the spore-bearing anaerobes, in *Topley and Wilson's Microbiology and Microbial Infections, 9th Edition, Volume 2: Systematic Bacteriology,* Collier, L., Balows, A., and Sussman, M., Eds., Arnold, London, 1998, 731–782.
31. Johnson, E.A. and Goodnough, M.C., Botulism, in *Topley and Wilson's Microbiology and Microbial Infections, 9th Edition, Volume 3: Bacterial Infections*. Collier, L., Balows, A., and Sussman, M. Eds., Arnold, London, 1998, 723–741.
32. Arnon, S.S., Botulism as an intestinal toxemia, in *Infections of the Gastrointestinal Tract*, Blaser, M.J. et al., Eds., Raven Press, New York, 1995, 257–271.
33. Arnon, SS, Infant botulism, in *Textbook of Pediatric Infectious Diseases*, 5th Edition, Feigen, R.D., Cherry, J.D., Eds., W.B. Saunders, Philadelphia, 2004.
34. Cook, T.M., Protheroe, R.T., and Handel, J.M., Tetanus: a review of the literature, *Brit. J. Anasthesia*, 87, 477–487, 2001.
35. Carle, A. and Rattone, G., Studio experimentale sull' eziologia del tetano, *G. Acad. Med. Torino*, 32, 174, 1884.
36. Bytchenko, B., Microbiology of tetanus, in *Tetanus, Important New Concepts*, Veronesi, R., Ed., Excerpta Medica, Amsterdam, 1981, 28–39.
37. Nicolaier, A., Ueber infectiosen tetanus, *Dtsch. Med. Wochenschr.*, 10, 842–844, 1884.
38. Kitasato, S., Über den Tetanusbacillus, *Z. Hyg.*, 7, 224–234, 1889.
39. Finegold, S.M., Tetanus, in *Topley and Wilson's Microbiology and Microbial Infections, 9th Edition, Volume 3: Bacterial Infections*, Collier, L., Balows, A., and Sussman, M., Eds., Arnold, London, 1998, 693–722.
40. Behring, E. and Kitasato, S., Ueber das Zustandekommen der Diptherie-Immunität und der Tetanus-Immunität bei Thieren, *Dtsch. Med. Wochenschr.*, 16, 113–114, 1890.
41. Glenny, A.T., Hamp, A.G., and Stevens, M.F., Protection of horses against tetanus by active immunization with alum toxoid, *Vet. J.*, 88, 90, 1932.
42. Glenny, A.T. et al., Immunological notes, *J. Pathol. Bacteriol.*, 29, 31–40, 1926.
43. Ramon, G. and Lemétayer, E., Sur l'immunité antitétanique naturellement acquise chez quelques expèces de ruminants, *C. R. Soc. Biol.*, 116, 275, 1934.
44. Kerner, C.A.J., Vergiftung durch verdorbene Würste, *Tübinger Blättner für naturwissenshcaften und Arzneykunde*, 3, 1–45, 1817.
45. Kerner, C.A.J., Das Fettgift oder die Fettsäure und ihre Wirkungen auf den thierischen Organismus, ein Beytrag zur Untersuchung des in verdorgenen Wåursten giftig wirkenden *Stoffes*, Stuttgart und Tübingen, Cotta 1822.
46. Kessler, K.R. and Benecke, R., Botulinum toxin: from poison to remedy, *Neurotoxicology,* 18, 761–770, 1997.
47. van Ermengem, E., Ueber einen neuen anaeroben Bacillus and seine Beziehungen zum Botulismus, *Z. Hyg. Infektionskh.*, 26, 1–56, 1897. Abbreviated English translation, *Rev. Infect. Dis.*, 1, 701-719, 1979.
48. Devriese, P.P., On the discovery of *Clostridium botulinum*, *J. Hist. Neurosci.*, 8, 43–50, 1999.
49. Dewberry, E.B., *Food Poisoning. Its Nature, History and Causation Measures for Its Control*, Leonard Hill Limited, London, 1950.
50. CDC, Botulism in the United States, 1899–1977, in *Handbook for Epidemiologists, Clinicians, and Laboratory Workers,* U.S. Dept. Health, Education, Welfare, Public Health Service, Centers for Disease Control and Prevention, Atlanta, GA, 1979.
51. Kempner, W., Further contributions to the knowledge of meat poisoning. The antitoxin to botulism, *Z. Hyg. InfektKrankh*, 26, 481–500, 1897.
52. Leighton, G.R., *Botulism and Food Preservation (the Loch Maree Tragedy),* W. Collins & Sons, Ltd, Glasgow, U.K., 1923.
53. Meyer, K.F., The protective measures of the state of California against botulism, *J. Prevent. Med.*, 5, 261–293, 1931.

54. Meyer, K.F., The status of botulism as a world health problem, *Bull. Wld. Hlth. Org.*, 15281–15298, 1956.
55. Meyer, K.F., The rise and fall of botulism, *Calif. Med.*, 118, 63–64, 1973.
56. Meyer, K.F. and Eddie, B., Perspectives concerning botulism, *Zeitschr. f Hygiene, Bd.*, 133, S255–263, 1951.
57. Pickett, J. et al., Syndrome of botulism in infancy: clinical and electrophysiologic study, *N. Engl. J. Med.*, 295, 770–772, 1976.
58. Midura, T.F. and Arnon, S.S., Infant botulism: identification of *Clostridium botulinum* and its toxin in faeces, *Lancet*, ii, 934–936, 1976.
59. Davis, J.B., and Mattman, L.H., and Wiley, A.B., *Clostridium botulinum* in a fatal wound infection, *J. Amer. Med. Assoc.*, 146, 646–648, 1951.
60. Dolman, C.E., Botulism as a world health problem, *Botulism. Proceedings of a Symposium*, Lewis, H.H., and Casse, K. Jr., Eds., U.S. Dept. Health, Education, and Welfare. Cincinnati, OH, 1964, 5–30.
61. Eklund, M.W. and Dowell, V.R. Jr., *Avian Botulism: An International Perspective*, Thomas, C.C., Ed. Springfield, IL, 1987.
62. Hatheway, C.L., Botulism, in *Laboratory Diagnosis of Infectious Diseases: Principles and Practice*, Balows, A. et al., Eds. Springer-Verlag, New York, 1988, 111–133.
63. Johnson, E.A., Anaerobic Fermentations, in *Manual of Methods for Industrial Microbiology*, 2nd Edition, Demain, A.L. et al., Eds., ASM Press, Washington, DC, 1999.
64. Willis, A.T., *Clostridia of Wound Infection*, Butterworths, London, 1969.
65. Willis, A.T., *Anaerobic Bacteriology: Clinical and Laboratory Practice*, 3rd Edition, Butterworths & Co., London, 1977.
66. Dowell, V.R. Jr. and Dezfulian, M., Physiological characterization of *Clostridium botulinum* and development of practical isolation and identification procedures, in *Biomedical Aspects of Botulism*, Lewis, G.E.J., Ed., Academic Press, New York, 1981, 205–216.
67. Dowell, V.R. Jr. et al., *Media for Isolation, Characterization and Identification of Obligately Anaerobic Bacteria*, U.S. Department of Health and Human Services, Public Health Service, Centers for Disease Control, Atlanta, 1977.
68. Solomon, H.M. et al., *Clostridium botulinum* and its toxins, in *Compendium for the Microbiological Examination of Foods, 4th Edition*, Downes, F.P. and Ito, K., Eds., American Public Health Association, Washington, DC, 2001, 317–324.
69. CDC, Botulism in the United States, 1899–1996, in *Handbook for Epidemiologists, Clinicians, and Laboratory Workers*, Centers for Disease Control and Prevention, Atlanta, 1998.
70. Eklund, M.W., Poysky, F.T., and Habig, W.H., Bacteriophages and plasmids in *Clostridium botulinum* and *Clostridium tetani* and their relationship to production of toxins, in *Botulinum Neurotoxin and Tetanus Toxin*, Simpson, L.L., Ed., Academic Press, San Diego, 1989, 25–51.
71. Schantz, E.J. and Johnson, E.A., Properties and use of botulinum toxin and other microbial neurotoxins in medicine, *Microbiol. Rev.*, 56, 80–92, 1992.
72. Gunnison, J.B. and Meyer, K.F., The occurrence of nontoxic strains of *Cl. parabotulinum*, *J. Infect. Dis.*, 45, 77–86, 1929.
73. Gunnison, J.B. and Meyer, K.F., Cultural study of an international collection of *Clostridium botulinum* and *parabotulinum*, *J. Infect. Dis.*, 45, 119–134, 1929b.
74. Eklund, M.W., Poysky, F.T., and Habig, W.H., Bacteriophages and plasmids in *Clostridium botulinum* and *Clostridium tetani* and their relationship to production of toxins, in *Botulinum Neurotoxin and Tetanus Toxin*, Simpson, L.L, Ed., Academic Press, San Diego, 1989, 25–51.
75. Goodnough, M.C. et al., Colony immunoblot assay of botulinal toxin, *Appl. Environ. Microbiol.*, 59, 2339–2342, 1993.
76. Holdeman, L.V., Cato, E.P., and Moore, W.E.C., *Anaerobe Laboratory Manual*, 4th Edition, Department of Anaerobic Microbiology, Virginia Polytechnic Institute and State University, Blacksburg, 1977.
77. Cato, E.P., George, W.L., and Finegold, S.M., Genus *Clostridium*, in *Bergey's Manual of Systematic Bacteriology*, Sneath, P.H.A. et al., Eds., Williams & Wilkins, Baltimore, 1986, 1141–1200.
78. Cato, E.P. and Stackebrandt, E., Taxonomy and phylogeny, in *Clostridia*, Minton, N.P. and Clarke, D.J., Eds., Plenum Press, New York, 1989, 1–26.

79. DSMZ (Deutsche Sammlung von Mikroorganismen und Zellulturen/German Collection of Microorganisms and Cell Cultures), Bacterial nomenclature, Genus *Clostridium*, GmbH, Braunschweig, Germany, 2003.
80. Collins, M.D. et al., The phylogeny of the genus *Clostridium*: proposal of five new genera and eleven new species combinations, *Int. J. Syst. Bacteriol.*, 44, 812–826, 1994.
81. Finegold, S.M., Song. Y., and Liu, C., Taxonomy — General comments and update on taxonomy of clostridia and anaerobic cocci, *Anaerobe*, 8, 283–285, 2002.
82. Goodfellow, M. and O'Donnell, A.G., Roots of bacterial systematics, in *Handbook of New Bacteria Systematics*, Goodfellow, M. and O'Donnell, A.G., Eds., Academic Press, London, 1993, 3–54.
83. Giménez, D.F. and Giménez, J.A., The typing of botulinal neurotoxins, *Int. J. Food Microbiol.*, 27, 1–9, 1995.
84. Boroff, D. and Fleck, U., Statistical analysis of a rapid *in vivo* method for the titration of the toxin of *Clostridium botulinum*, *J. Bacteriol.* 92, 1580–1581, 1966.
85. Schantz, E.J. and Kautter, D.A., Standardized assay for *Clostridium botulinum* toxins, *J. Assoc. Off. Anal. Chem.*, 61, 96–99, 1978.
86. Olive, D. and Bean, P., Principles and applications for DNA-based typing of microbial organisms, *J. Clin. Microbiol.*, 35, 2160–2162, 1999.
87. Wilson, K.H., Molecular biology as a tool for taxonomy, *Clin. Infect. Dis.*, 20, S117–121, 1995.
88. Pourshaban, M. et al., Taxonomic identity of type E botulinum toxin-producing *Clostridium butyricum* strains by sequencing of a short 16S rDNA region, *FEMS Microbiol. Lett.*, 214, 119–125, 2002.
89. Lin, W.J. and Johnson, E.A., Genome analysis of *Clostridium botulinum* type A by pulsed-field gel electrophoresis, *Appl. Environ. Microbiol.*, 61, 4441–4447, 1995.
90. Hielm, S. et al., Genomic analysis of *Clostridium botulinum* group II by pulsed-field gel electrophoresis, *Appl. Environ. Microbiol.*, 64, 703–708, 1998.
91. Hielm, S. et al., Prevalence of *Clostridium botulinum* in Finnish trout farms: pulsed-field gel electrophoresis typing reveals extensive genetic diversity among type E isolates, *Appl. Environ. Microbiol.*, 64, 4161–4167, 1998.
92. Franciosa, G., Aureli, P., and Schechter, R., *Clostridium botulinum*, in *International Handbook of Foodborne Pathogens*, Miliotis, M.D. and Bier, J.W., Eds., Marcel Dekker, New York and Basel, 2003, 61–89.
93. Fenicia, L. et al., A case of infant botulism due to neurotoxigenic *Clostridium butyricum* type E associated with *Clostridium difficile* colitis, *Eur. J. Microbiol. Infect. Dis.*, 21, 736–738, 2002.
94. Fenicia, L. et al., Intestinal toxemia botulism in two young people caused by *Clostridium butyricum* type E, *Clin. Infect. Dis.*, 154, 207–211, 1999.
95. Chaudry, R. et al., Outbreak of suspected *Clostridium butyricum* botulism in India, *Emerg. Infect. Dis.*, 4, 506–507, 1998.
96. Harvey, S.M., Sturgeon, J., and Dassey, D.E., Botulism due to *Clostridium botulinum* type F toxin, *J. Clin. Microbiol.*, 40, 2260–2262, 2002.
97. Meng, X. et al., Isolation and characterization of neurotoxigenic *Clostridium butyricum* from soil in China, *J. Med. Microbiol.*, 48, 133–137, 1999.
98. Barash, J.R. and Arnon, S.S., Dual-toxin producing strain of *Clostridium botulinum* type Bf isolated from a California patient with infant botulism, *J. Clin. Microbiol.*, 42, 1713–1715, 2004.
99. Wang, X. et al., Genetic analysis of type E botulinum toxin-producing *Clostridium butyricum* strains, *Appl. Environ. Microbiol.*, 66, 4992–4997, 2000.
100. Johnson, E.A., Bacteriophages encoding botulinum and diphtheria toxin, in *Bacteriophages and Bacterial Pathogens,* Waldor, M., Friedman, A., and Adhya, S., Eds., ASM Press, Inc., Washington, DC, in press.
101. Dodds, K.L., *Clostridium botulinum* in the environment, in *Clostridium botulinum: Ecology and Control in Foods*, Hauschild, A.H.W. and Dodds, K.L., Eds., Marcel Dekker, New York, 1993, 105–117.
102. Easton, E.J. and Meyer, K.F., Occurrence of *Bacillus botulinus* in human and animal excreta, *J. Infect. Dis.*, 35, 207–212, 1924.
103. Calabi, E. and Fairweather, N., Patterns of sequence conservation in the S-layer proteins and related sequences in *Clostridium difficile*, *J. Bacteriol.,* 184(14), 3886–3897, 2002.

104. Takumi, K. et al., Characterization, self-assembly and reattachment of S layer from *Clostridium botulinum* type E saroma, *Tokushima J. of Exp. Med.,* 39(3–4), 101–107, 1992.
105. Setlow, P. and Johnson, E.A., Spores and their significance, in *Food Microbiology: Fundamentals and Frontiers*, Doyle, M.P., Beuchat, L.R., and Montville, T., Eds., ASM Press, Washington, DC, 2001.
106. Woods, D.R. and Jones, D.T., Physiological responses of *Bacteroides* and *Clostridium* strains to environmental stress, *Adv. Microbiol. Physiol.*, 28, 1–64, 1986.
107. Kihm, D. et al., Zinc stimulates sporulation of *Clostridium botulinum* 113B, *Curr. Microbiol.*, 17, 193–198, 1998.
108. Ito, K.A. et al., The thermal and germicidal resistance *of Clostridium botulinum* types A, B, and E spores, in *Proceedings of the First U.S.-Japan Conference on Toxic Micro-Organisms*, Herzberg, M., Ed., UJNR Joint Panels on Toxic Micro-Organisms and U.S. Dept. Interior, Washington, DC, 1968.
109. Lund, B.M. and Peck, M.W., *Clostridium botulinum*, in *The Microbiological Safety and Quality of Food*, Vol II, Lund, B.M., Baird-Parker, T.C., and Gould. G.W., Eds., Aspen Publishers, Inc., Gaithersburg, MD, 2000, 1057–1109.
110. Kim, J. and Foegeding, P.M., *Principles of Control: Ecology and Control in Foods,* Hauschild, A.H.W. and Dodds, K.L., Eds., Marcel Dekker, New York, 1993, 121–176.
111. Rood, J.I., Virulence genes of *Clostridium perfringens*, *Annu. Rev. Microbiol.*, 52, 333–360, 1998.
112. Zhou, Y., et al., The genes for the *Clostridium botulinum* type G toxin complex are on a plasmid, *Infect. Immun.*, 63, 2087–2209, 1995.
113. Eklund, M.W. et al., Evidence for plasmid-mediated toxin and bacteriocin production in *Clostridium botulinum* type G, *Appl. Environ. Microbiol.*, 54, 1405–1408, 1988.
114. Laird, W.J. et al., Plasmid-associated toxigenicity in *Clostridium tetani*, *J. Infect. Dis.*, 142, 623, 1980.
115. Dineen, S.S., Bradshaw, M., and Johnson, E.A., Cloning, nucleotide sequence, and expression of the gene encoding bacteriocin boticin B from *Clostridium botulinum* strain 213B, *Appl, Environ, Microbiol.*, 66, 5480–5483, 2000.
116. Adams, M.H., *Bacteriophages,* Wiley Interscience, New York, 1959.
117. Johnson, E.A., Extrachromosomal virulence determinants in the clostridia, in *Molecular Genetics and Pathogenesis of the Clostridia*, Rood, J. et al., Eds., Academic Press, London, 1997.
118. Johnson, E.A., et al., Characterization of neurotoxin mutants in *Clostridium botulinum* type A, *Clin. Infect. Dis.*, 25 (Suppl. 2), S168–170, 1997.
119. Canchaya, C. et al., Phages as agents of lateral gene transfer, *Curr. Opin. Microbiol.*, 6, 417–424, 2003.
120. Gentry-Weeks, P., Coburn, S., and Gilmore. M.S., Phages and other mobile virulence elements in Gram-positive pathogens, *Curr. Top. Microbiol. Immunol.,* 264, 79–94, 2002.
121. Hauser, D. et al., Plasmid localization of a type E botulinal neurotoxin gene homologue in toxigenic *Clostridium butyricum* strains, and absence of this gene in non-toxigenic *C. butyricum* strains, *FEMS Microbiol. Lett.*, 99, 252–256, 1992.
122. Hauser, D. et al., Botulinal toxin genes, clostridial neurotoxin homology and genetic transfer in *Clostridium botulinum*, *Toxicon*, 33, 515–526, 1995.
123. Zhou, Y., Sugiyama, H., and Johnson. E.A., Transfer of neurotoxigenicity from *Clostridium butyricum* to a nontoxigenic *Clostridium botulinum* type E-like strain, *Appl. Environ. Microbiol.*, 59, 3825–3831, 1993.
124. Kautter, D.A. et al., Antagonistic effect on *Clostridium botulinum* type E by organisms resembling it, *Appl. Microbiol.*, 14, 616–622, 1966.
125. Eklund, M.W. et al., Evidence for plasmid-mediated toxin and bacteriocin production in *Clostridium botulinum* type G, *Appl. Environ. Microbiol.*, 54, 1405–1408, 1988.
126. Adams, E.B., Lawrence, D.R., and Smith, J.W.G., *Tetanus*, Blackwell Scientific, Oxford, 1969.
127. Mitsui, N., Mitsui, K., and Hase, J., Purification and some properties of tetanolysin, *Microbiol. Immunol.*, 24, 575–584, 1980.
128. Brüggeman, H. and Gottschalk, G., Insights in metabolism and toxin production from the complete genome sequence of *Clostridium tetani, Anaerobe,* 10, 53–68, 2004.
129. Dezfulian, M., Animal models of botulism and tetanus, in *Botulinum Neurotoxin and Tetanus Toxin*, Simpson, L.L., Ed., Academic Press, San Diego, 1989, 335–350.
130. Gill, D.M., Bacterial toxins: a table of lethal amounts, *Microbiol. Rev.*, 46, 86–94, 1982.
131. Sakaguchi, G., *Clostridium botulinum* toxins, *Pharmacol. Ther.*, 19, 165–194, 1983.

132. Giménez, D.F. and Giménez, J.A., Serological subtypes of botulinal neurotoxins, in *Botulism and Tetanus Neurotoxins: Neurotransmission and Biomedical Aspects*, DasGupta, B.R., Ed., Plenum Press, New York, 1993.
133. Malizio, C.J., Goodnough, M.C., and Johnson, E.A., Purification of *Clostridium botulinum* type A neurotoxin, in *Bacterial Toxins: Methods and Protocols, Meth. Molec. Biol.,* Vol. 145, Holst, O., Ed., Humana Press Inc., Totowa, NJ, 2000, 27–39.
134. Holdeman, L.V. and Brooks, J.B., Variation among strains of *Clostridium botulinum* and related clostridia, in *Proceedings of the First U.S.-Japan Conference on Toxic Microorganisms*, Herzberg, M., Ed., U.S. Government Printing Office, Washington, DC, 1970, 278–286.
135. Giménez, D.F. and Ciccarelli, A.S., Another type of *Clostridium botulinum, Zentralbl Bakteriol I, Abt Orig A*, 215, 221–224, 1970.
136. Ciccarelli, A.S. et al., Cultural and physiological characterization of *Clostridium botulinum* type G and the susceptibility of certain animals to its toxin, *Appl. Environ. Microbiol.*, 34, 843–848, 1977.
137. Gimenez, D.F. and Ciccarelli, A.S. Studies on strain 84 of *Clostridium botulinum. Zentralblatt Bakteriol., Parasit., Infektionsk. Hyg. - 1 - Abt - Mediz.-Hyg. Bakteriol., Virusfors. Parasit. - Orig.*, 215(2), 212–220, 1970.
138. ICMSF (International Commission for Microbiological Safety of Foods), *Microorganisms in Foods 5, Characteristics of Microbial Pathogens*, Blackie Academic & Professional, London, 1996, 66–111.
139. Glass, K.A. and Johnson, E.A., Formulating low-acid foods for botulinal safety, in *Control of Foodborne Organisms,* Juneja, V.K. and Sofos, J.N., Eds., Marcel Dekker, New York, 2001, 323–350.
140. Townsend, C.T., Yee, L., and Mercer, W.A., Inhibition of the growth of *Clostridium botulinum* by acidification, *Food. Res.*, 19, 536–542, 1954.
141. Bell, C. and Kyriakides, A., *Clostridium botulinum*: *A Practical Approach to the Organism and Its Control in Foods,* Blackwell Science, Ltd., Oxford, 2000.
142. Popoff, M.R., Ecology of neurotoxigenic strains of clostridia, in *Clostridial Neurotoxins*, Montecucco, C., Ed., Springer, Berlin, 1995, 1–29.
143. Whitmer, M.E. and Johnson, E.A., Development of improved defined media for *Clostridium botulinum* serotypes A, B, and E, *Appl. Environ. Microbiol.*, 54, 753–759, 1988.
144. Patterson-Curtis, S.I. and Johnson, E.A., Roles of arginine in growth of *Clostridium botulinum* type B, *Appl. Environ. Microbiol.*, 58, 2334–2337, 1992.
145. Patterson-Curtis, S.I. and Johnson, E.A., Regulation of neurotoxin and protease formation in *Clostridium botulinum* Okra B and Hall A by arginine, *Appl. Environ. Microbiol.*, 55, 1544–1548, 1989.
146. Leyer, G.J. and Johnson, E.A., Repression of toxin production by tryptophan in *Clostridium botulinum* type E, *Arch. Microbiol.*, 154, 443–447, 1990.
147. Dolman, C.E., Growth and metabolic activities of *Clostridium botulinum* types in *Botulism,* in *Proc. of Symp., Cincinnati, Ohio, Jan. 1964,* Lewis, K.H. and Cassel, K. Jr., Eds., Public Health Service Publication No. 999-FP-1, PHS, Cincinnati, 1964, 43–68.
148. Eklund, M.W., Wieler, D.I., and Poysky, F.T., Outgrowth and toxin production of non-proteolytic type B *Clostridium botulinum* at 3.3 – 5.6 deg.C, *J. Bacteriol.*, 93, 1461–1462, 1967.
149. Ohye, D.F. and Scott, W.J., Studies on the physiology of *Clostridium botulinum* type E, *Aust. J. Biol. Sci.*, 10, 85–94, 1957.
150. Lynt, R.K., Kautter, D.A., and Solomon, H.M., Differences and similarities among proteolytic and nonproteolytic strains of *Clostridium botulinum* types A, B, E, and F: a review, *J. Food. Prot.*, 45, 466–474, 1982.
151. Mead, G.C., The amino acid-fermenting clostridia, *J. Gen. Microbiol.*, 67, 47–56, 1971.
152. Inoue, K. and Iida, H., Conversion of toxigenicity in *Clostridium botulinum* type C, *Jpn., J. Microbiol.*, 14, 87–89, 1970.
153. Oguma, K. et al., Structure and function of *Clostridium botulinum* progenitor toxin. *J. of Tox. - Tox. Rev.* 18(1), 17–34, 1999.
154. Glasby, C. and Hatheway, C.L. Flourescent antibody reagents for the identification of *Clostridium botulinum*, *J. of Clin. Microbiol.,*18, 1378–1383, 1983.
155. Suen, J.C. et al., *Clostridium argentinense*, sp. nov.: a genetically homogenous group composed of all strains of *Clostridium botulinum* toxin type G and some nontoxigenic strains previously identified as *Clostridium subterminale* or *Clostridium hastiforme*, *Int. J. Syst. Bacteriol.*, 38, 375–381, 1988.

156. Nukina, M. et al., Purification of *Clostridium botulinum* type G progenitor toxin, *Zbl. Bakt. Hyg. A.*, 268, 220–227, 1988.
157. Malizio, C.J., Sugiyama, H., and Johnson, E.A., unpublished data, 1996.
158. Schiavo, G. et al., Botulinum G neurotoxin cleaves VAMP/synaptobrevin at a single ala-ala peptide bond, *J. Biol. Chem.*, 269, 20213–20216, 1994.
159. Aureli, P. et al., Two cases of type E infant botulism caused by neurotoxigenic *Clostridium butyricum* in Italy, *J. Infect. Dis.*, 154, 207–211, 1986.
160. McCroskey, L. and Hatheway, C.L., et al., Characterization of an organism that produces type E botulinal toxin but which resembles *Clostridium butyricum* from the feces of an infant with type E botulism, *J. Clin. Microbiol.*, 23, 201–202, 1986.
161. Hall, J.D. et al., Isolation of an organism resembling *Clostridium barati* which produces type F botulinal toxin from an infant with botulism, *J. Clin. Microbiol.*, 21, 654–655, 1985.
162. McCroskey, L.M. et al., Type F botulism due to neurotoxigenic *Clostridium baratii* from an unknown source in an adult, *J. Clin. Microbiol.*, 29, 2618–2620, 1991.
163. Harvey, S.M., Sturgeon, J., and Dassey, D.E., Botulism due to *Clostridium botulinum* type F toxin, *J. Clin. Microbiol.*, 40, 2260–2262, 2002.
164. Paisley, J.W., Lauer, B.A., and Arnon, S.S., A second case of infant type F botulism caused by *Clostridium baratii*, *The Pediatr. Inf. Dis. J.*, 14, 912–914, 1995.
165. Giménez, J.A. and Sugiyama, H., Simplified purification method for *Clostridium botulinum* type E toxin, *Appl. and Environ. Microbiol.* 53, 2827–2830, 1987.
166. Giménez, J.A., Gimenez, M.A., and DasGupta, B.R., Characterization of the neurotoxin isolated from a *Clostridium baratii* strain implicated in infant botulism, *Infect. Immun.* 60, 518–522, 1992.
167. Scott, A.B. and Suzuki, D., Systemic toxicity of botulinum toxin by intramuscular injection in the monkey, *Movement Disord.*, 3, 333–335, 1988.
168. Herrero, B.A. et al., Experimental botulism in monkeys — a clinical pathological study, *Exp. Mol. Path.*, 6, 84–95, 1967.
169. Larson, A.E. and Johnson, E.A., Review of oral toxicity of botulinum toxin in primates, (Unpublished Review), 2004.
170. Ferreira, J.L. et al., Detection of botulinal neurotoxins A, B, E, and F by amplified enzyme-linked immunosorbent assay: collaborative study, *J. AOAC Int.*, 86(2), 314–331, 2003.
171. Ketema, F. et al., Assessment of the performance of a rapid, lateral flow assay for the detection of antibodies to HIV, *J. Acq. Imm. Deficien. Sys.*, 27, 63–70, 2001.
172. Sharma, S. K. et al., Evaluation of lateral flow *Clostridium botulinum* neurotoxin detection kits for food analysis, *Appl. Environ. Microbiol.*, in press, 2004.
173. Anne, C. et al., High-throughput fluorigenic assay for determination of botulinum type B neurotoxin protease activity, *Anal. Biochem.*, 291, 253–261, 2001.
174. Schmidt, J.J. and Stafford, R.G. Fluorigenic substrates for the protease activities of botulinum neurotoxins, serotypes A, B, and F., *Appl. Environ. Microbiol.*, 69(1), 297–303, 2003. [Erratum appears in *Appl. Environ. Microbiol.* 69, 3025, 2003.]
175. Barr, J., Characterization of botulinum neurotoxins by mass spectroscopy, personal communication, 2004.
176. Moorthy, J. et al., Microfluidic tectonics platform: a colorimetric, disposable botulinum toxin enzyme-linked immunosorbent assay system, *Electrophoresis,* 25, 1705–1713, 2004.
177. Liu, W. et al., Botulinum toxin type B micromechanosensor, *Proc. Natl. Acad. Sci. U.S.A.,* 100, 13621–13625, 2003.
178. Centers for Disease Control/National Institutes of Health, *Biosafety in Microbial and Biomedical Laboratories*, 4th Edition, U.S. Dept. of Health and Human Services, Washington, DC, 1993.
179. Johnson, E.A., Extrachromosomal virulence determinants in the clostridia, in *Molecular Genetics and Pathogenesis of the Clostridia*, Rood, J. et al., Eds., Academic Press, London, 1997.
180. Matsuda, M., The structure of tetanus toxin, in *Botulinum Neurotoxin and Tetanus Toxin*, Simpson, L.L., Ed., Academic Press, San Diego, 1989, 69–92.
181. Sugiyama, H., *Clostridium botulinum* neurotoxin, *Microbiol. Rev.*, 44, 419–448, 1980.
182. Dineen, S.S., Bradshaw M., and Johnson, E.A., Neurotoxin gene clusters in *Clostridium botulinum* type A strains: sequence comparison and evolutionary implications, *Curr. Microbiol.,* 46, 345–352, 2003.

183. Dineen, S.S. et al., Nucleotide sequence and transcriptional analysis of the type A2 neurotoxin gene cluster in *Clostridium botulinum, FEMS Microbiol. Lett.*, 235, 9–16, 2004.
184. Schantz, E.J. and Johnson, E.A., Botulinum toxin: the story of its development for the treatment of human disease, *Perspectives Biol. Med.,* 40, 317–327, 1997.
185. Diane, D.-S. et al., Intramuscular injection of ^{125}I-botulinum neurotoxin-complex versus^{125}I-botulinum-free neurotoxin: time course of tissue distribution, *Toxicon,* 42, 461–469, 2003.
186. Wohlfarth, K., Kampe, K., and Bigalke, H., Pharmacokinetic properties of different formulations of botulinum neurotoxin type A, *Mov. Disor.,* 19 Suppl 8, S65–67, 2004.
187. DasGupta, B.R. and Sugiyama, H., Inhibition of *Clostridium botulinum* types A and B hemagglutinins by sugars, *Can. J. Microbiol.,* 23(9), 1257–1260, 1977.
188. Kozaki, S. et al., Ganglioside GT1b as a complementary receptor component for *Clostridium botulinum* neurotoxins, *Microbial Path.*, 25, 91–99, 1998.
189. Emsley, P. et al., The structures of the H(C) fragment of tetanus toxin with carbohydrate subunit complexes provide insight into ganglioside binding, *J. Biol. Chem.,* 275(12), 8889–8894, 2000.
190. Kubota, T. et al., Gene arrangement in the upstream region of *Clostridium botulinum* type E and *Clostridium butyricum* BL 6340 progenitor toxin gene is different from that of other types, *FEMS Microbiol. Lett.*, 158, 215–221, 1998.
191. Raffestin, S. et al., Organization and regulation of the neurotoxin genes in *Clostridium botulinum* and *Clostridium tetani*, *Anaerobe,* 10, 93–100, 2004.
192. Popoff, M.R. and Marvaud, J.-C., Structural and genomic features of clostridial neurotoxins, in *The Comprehensive Sourcebook of Bacterial Protein Toxins*, Alouf, J.E. and Freer, J.H., Eds., Academic Press, London, 1999, 174–201.
193. Franciosa, G., Ferreira, J.L., and Hatheway, C.L., Detection of type A, B, and E botulism neurotoxin genes in *Clostridium botulinum* and other *Clostridium* species by PCR — evidence of unexpressed type B toxin genes in type A toxigenic organisms, *J. Clin. Microbiol.*, 32, 1911–1917, 1994.
194. Hutson, R.A. et al., Genetic characterization of *Clostridium botulinum* type A containing silent B neurotoxin gene sequences, *J. Biol. Chem.*, 271, 10786–10792, 1996.
195. Moriishi, K. et al., Mosaic structures of neurotoxins produced from *Clostridium botulinum* types C and D organisms, *Biochim. Biophys. Acta,* 1307, 123–126, 1996.
196. Bradshaw, M., Dineen, S.S., Maks, N.D., and Johnson, E.A., Regulation of neurotoxin complex expression in *Clostridium botulinum* strains 62A, Hall A-*hyper*, and NCTC 2916, *Anaerobe,* 10, 321–333, 2004.
197. Lin, W.-J. and Johnson, E.A., Transposon Tn*916* mutagenesis in *Clostridium botulinum*, *Appl. Environ. Microbiol.,* 57, 2946–2950, 1991.
198. Mueller, J.H. and Miller, P.A., Growth requirements of *Clostridium tetani*, *J. Bacteriol.*, 43, 763–772, 1942.
199. Mueller, J.H. and Miller, P.A., Production of tetanal toxin, *J. Immunol.*, 50, 377–384, 1945.
200. Mueller, J.H. and Miller, P.A., Unidentified nutrients in tetanus toxin production, *J. Bacteriol.*, 56, 219–233, 1948.
201. Mueller, J.H. and Miller, P.A., Essential role of histidine peptides in tetanus toxin production, *J. Biol. Chem.*, 223, 763–772, 1956.
202. Porfirio, Z. et al., Specific peptides of casein pancreatic digestion enhance the production of tetanus toxin, *J. Appl. Microbiol.,* 83, 678–684, 1997.
203. Lewis, K.H. and Hill, E.V., Practical media and control measures for producing highly toxic cultures of *Clostridium botulinum* type A, *J. Bacteriol.*, 53, 213–230, 1947.
204. Bradshaw, M., Goodnough, M.C., and Johnson, E.A., Conjugative transfer of the *E. coli-C. perfringens* shuttle vector pJIR1457 to *Clostridium botulinum* type A strains, *Plasmid*, 40, 233–237, 1998.
205. Marvaud, J.C. et al., TetR is a positive regulator of the tetanus toxin gene in *Clostridium tetani* and is homologous to botR, *Infect. Immun.* 66, 5698–5702, 1998.
206. Marvaud, J.C. et al., botR/A is a positive regulator of botulinum neurotoxin and associated non-toxin protein genes in *Clostridium botulinum* A, *Mol. Microbiol.*, 29,1009–1018, 1998.
207. Marvaud, J.C. et al., Regulation of the toxinogenesis in *Clostridium botulinum* and *Clostridium tetani*, *Biol. Cell*, 92, 455–457, 2000.
208. Wosten, M.M. et al., A signal transduction system that responds to extracellular iron, *Cell,* 103, 113–125, 2000.

209. Helman, J.D., The extracytoplasmic function (ECF) sigma factors, *Adv. Microbiol. Physiol.,* 46, 47–110, 2002.
210. Dupuy, B. and Sonenshein, A.L., Regulated transcription of *Clostridium difficile* toxin genes, *Mol. Microbiol.,* 27,107–120, 1998.
211. Garneir, T. and Cole, S.T., Studies of UV-inducible promoters from *Clostridium perfringens in vivo* and *in vitro, Mol. Microbiol.,* 2, 607–614, 1988.
212. Karlsson, S., Burman, L.G., and Akerlund, T., Suppression of toxin production in *Clostridium difficile* VPI10463 by amino acids, *Microbiol.,* 145, 1683–1693, 1999.
213. Karlsson, S. et al., Expression of *Clostridium difficile* toxins A and B and their sigma factor TcdD is controlled by temperature, *Infect. Immun.,* 71, 1784–1793, 2003.
214. Mani, N. and Dupuy, B., Regulation of toxin synthesis in *Clostridium difficile* by an alternative RNA polymerase sigma factor, *Proc. Natl. Acad. Sci. U.S.A.,* 98, 5844–5849, 2001.
215. Mani, N. et al., Environmental response and autoregulation of *Clostridium difficile* TxeR, a sigma factor for toxin gene expression, *J. Bacteriol.*, 184, 5971–5978, 2003.
216. DasGupta, B.R. and Boroff, D.A., Separation of toxin and hemagglutin from crystalline toxin of *Clostridium botulinum* type A by anion exchange chromatography and determination of their dimensions by gel filtration, *J. Biol. Chem.,* 243, 1065–1072, 1968.
217. DasGupta, B.R. and Sugiyama, H., A common subunit structure in *Clostridium botulinum* type A, B, and E toxins, *Biochem. Biophys. Res. Commun.,* 48, 108–112, 1971.
218. DasGupta, B.R. and Sugiyama, H., Molecular forms of neurotoxins in proteolytic *Clostridium botulinum* type B cultures, *Infect. Immun.,* 14, 680–686, 1976.
219. DasGupta, B.R. and Dekleva, M.L., Botulinum neurotoxin type A: sequence of amino acids at the N-terminus and around the nicking site, *Biochimie,* 72, 661–664, 1990.
220. DasGupta, B.R. and Foley, J. Jr., *C. botulinum* neurotoxin types A and E: isolated light chain breaks down into two fragments. Comparison of their amino acid sequences with tetanus neurotoxin, *Biochimie,* 71, 1193–1200, 1989.
221. Ahmed, S.A. et al., Enzymatic autocatalysis of botulinum A neurotoxin light chain, *J. Protein Chem.,* 3, 221–231, 2001.
222. Kurazano, H. et al., Minimal essential domains specifying toxicity of the light chains of tetanus toxin and botulinum neurotoxin type A, *J. Biol. Chem.,* 267, 14721–14729, 1992.
223. Baldwin, M.R. et al., The C-terminus of botulinum neurotoxin type A light chain contributes to solubility, catalysis and stability, *Protein Express. Purif.*, 37, 187–195, 2004.
224. Montecucco, C., On the enzymatic activity of tetanus toxin, *Toxicon,* 25, 1255–1257, 1987.
225. Niemann, H., Molecular biology of clostridial neurotoxins, in *A Sourcebook of Bacterial Protein Toxins*, Alouf, J.E. and Freer, J.H., Eds., Academic Press., London, 1991, 303–348.
226. Collins, M.D. and East, A.K., Phylogeny and taxonomy of the food-borne pathogen *Clostridium botulinum* and its neurtoxins, *J. Appl. Microbiol.,* 84, 5–17, 1998.
227. Lacy, D.B. and Stevens, R.C., Sequence homology and structural analysis of the clostridial neurotoxins, *J. Mol. Biol.*, 291, 1091–1104, 1999.
228. Schiavo, G. et al., Botulinum neurotoxins are zinc proteins, *J. Biol. Chem.*, 267, 23479–23483, 1992.
229. Wright, J.F. et al., Identification and partial characterization of a low affinity metal-binding site in the light chain of tetanus toxin, *J. Biol. Chem.*, 267, 9053–9058, 1992.
230. Rossetto, O., Tonello, F., and Montecucco, C., Proteases, in *Bacterial Protein Toxins,* Burns, D.L. et al., Eds., ASM Press, Washington, DC, 2003, 271–282.
231. Poulain, B. et al., Inhibition of neurotransmitter release by botulinum neurotoxins and tetanus toxin at *Aplysia* synapses: role of constituent chains, *J. Physiol.,* 84, 247–261, 1990.
232. Sutton, R.B. et al., Crystal structure of a SNARE complex involved in synaptic exocytosis at 2.4 A resolution, *Nature,* 395, 347–353, 1998.
233. Sollner, T.H., Regulated exocytosis and SNARE function, *Mol. Membrane Biol.*, 20, 209–220, 2003.
234. Matthews, B.W., Structural basis of the action of thermolysin and related zinc peptidases, *Acc. Chem. Res.,* 21, 333–340, 1988.
235. Keller, J.E., Cai, F., and Neale, E.A., Uptake of botulinum neurotoxin into cultured neurons, *Biochem,* 43, 526–532, 2004.

236. Foran, P.G. et al., Evaluation of the therapeutic usefulness of botulinum neurotoxin B, C1, E, and F compared with the long lasting type A. Basis for distinct durations of inhibition of exocytosis in central neurons, *J. Biol. Chem.,* 278, 1363–1371, 2003.
237. Chen, F., et al., Antibody mapping to domains of botulinum neurotoxin serotype A in the complexed and uncomplexed forms, *Infect. Immun.,* 65, 1626–1630, 1997.
238. Byrne, M.P. and Smith, L.A., Development of vaccines for prevention of botulism, *Biochimie*, 82, 955–966, 2000.
239. Lacy, D.B. et al., Crystal structure of botulinum neurotoxin type A and implications for toxicity, *Nat. Struct. Biol.,* 5, 898–902, 1998.
240. Hanson, M.I. and Stevens, R.C., Structural view of botulinum neurotoxin in numerous functional states, in *Scientific and Therapeutic Aspects of Botulinum Toxin,* Brin, M.F., Hallett, M., and Jankovic, J., Eds., Lippincott, Williams & Wilkins, Philadelphia, 2002.
241. Burkard, F. et al., Electron density projection map of the botulinum neurotoxin 900-kilodalton complex by electron crystallography, *J. Struct. Biol.,* 120, 78–84, 1997.
242. Segelke, B. et al., Crystal structure of *Clostridium botulinum* neurotoxin protease in a product-bound state: Evidence for noncanonical zinc protease activity, *Proc. Natl. Acad. Sci. U.S.A.,* 101, 6888–6893, 2004.
243. Swaminathan, S. and Eswaramoorthy, S., Structural analysis of the catalytic and binding sites of *Clostridium botulinum* neurotoxin B, *Nat. Struct. Biol.*, 7, 693–699, 2000.
244. Knapp, M., Segelke, B., and Rupp, B., The 1.61 Angstrom structure of the tetanus toxin ganglioside binding region: solved by MAD and MIR phase combination, *Am. Cryst. Assoc. Papers,* 25, 90, 1998.
245. Rupp, B. and Segelke, B., Questions about the structure of the botulinum neurotoxin B light chain in complex with a target peptide, *Nat. Struct. Biol.,* 8, 663–664, 2001.
246. Umland, T.C. et al., Structure of the receptor binding fragment H_C of tetanus neurotoxin, *Nat. Struct. Biol.,* 4, 788–792, 1997.
247. Nishiki, T. et al., Binding of botulinum type B neurotoxin to Chinese hamster ovary cells transfected with rat synaptotagmin II cDNA, *Neurosci. Lett.,* 208, 105–108, 1996.
248. Dong, M. et al., Synaptotagmins I and II mediate entry of botulinum neurotoxin B into cells, *J. Cell Biol.*, 162, 1293–1303, 2003.
249. Koriazova, L.K. and Montal, M., Translocation of botulinum neurotoxin light chain protease through the heavy chain channel, *Nat. Struct. Biol.,* 10, 13–18, 2003.
250. Puhar, A. et al., Comparison of the pH-induced conformational change of different clostridial neurotoxins, *Biochem. Biophys. Res. Commn.,* 319, 66–71, 2004.
251. Caccin, P. et al., VAMP/synaptobrevin cleavage by tetanus and botulinum neurotoxins is strongly enhanced by acidic liposomes, *FEBS Lett.,* 542, 132–136, 2003.
252. Fernandez-Salas, E. et al., Plasma membrane localization signals in the light chain of botulinum neurotoxin, *Proc. Natl. Acad. Sci. U.S.A.,* 101, 3208–3213, 2004.
253. Koh, J.Y., Zinc and disease of the brain, *Mol. Neurobiol.,* 24, 99–106, 2001.
254. Ferrer-Montiel, A.V. et al., Tyrosine phosphorylation modulates the activity of clostridial neurotoxins, *J. Biol. Chem.*, 271, 18322–18325, 1996.
255. Encinar, J.A. et al., Structural stabilization of botulinum neurotoxins by tyrosine phosphorylation, *FEBS Lett.,* 429, 78–82, 1998.
256. Patrick III, W.C., Analysis of botulinum toxin, type A, as a biological warfare threat, unpublished report, 1998.
257. Sugishima, M., Aum Shinrikyo and the Japanese law on bioterrorism, *Prehospital Disast. Med.*, 18, 179–183, 2003.
258. Broussard, L.A., Biological agents: weapons of warfare and bioterrorism, *Mol. Diag.,* 6, 323–333, 2001.
259. Atlas, R.M., Bioterrorism and biodefence research: changing the focus of microbiology, *Nat. Rev. Microbiol.,* 1, 70–74, 2003.
260. Covert, N.M., *Cutting Edge—A History of Fort Detrick, Maryland 1943–1993,* Public Affairs Office, Fort Detrick, 1994.
261. Alibek, K. and Handleman, S., *Biohazard,* Random House, New York, 1999.
262. Caya, J.G., *Clostridium botulinum* and the ophthalmologist: a review of botulism, including biological warfare ramifications of botulinum toxin, *Survey Ophthal.,* 46, 25–34, 2001.

263. Bernard, C., Analyse physiologique des propriétés des systèmes musculaire et nerveux au moyen du curare, *C. R. Acad. Sci. (Paris),* 43, 825–829, 1856.
264. Bernard, C., *An Introduction to the Study of Experimental Medicine*, translated by Green, H.C., The MacMillan Company, New York, 1927.
265. Scott, A.B., Rosenbaum, A., and Collins, C.C., Pharmacologic weakening of extraocular muscles, *Invest. Ophthal.*, 12, 924–927, 1973.
266. Scott, A.B., Botulinum toxin injection into extraocular muscles as an alternative to surgery, *Opthamol.*, 87, 1044–1049, 1980.
267. Scott, A.B., Botulinum toxin injection of eye muscles to correct strabismus, *Trans. Am. Opthal. Soc.*, 79, 734–740, 1981.
268. Borodic, G.E., Acquadro, M., and Johnson E.A., Botulinum toxin therapy for pain and inflammatory disorders: mechanisms and therapeutic effects, *Expert. Opin. Invest. Drugs,* 10, 1531–1544, 2001.

22 Clostridial Diseases in Domestic Animals

J. Glenn Songer

CONTENTS

22.1 NEUROTOXIGENIC CLOSTRIDIA

22.1.1 BOTULISM

Clostridium botulinum produces eight types of botulinum neurotoxin (BoNT) (Table 22.1) [1,2] that affect mammals, birds, and fish [3], but they have unique geographic distribution and species susceptibility patterns [3,4–8].

Botulism spores are ubiquitous in soil throughout the U.S. Type A spores are found in nearly one fifth of soils west of the Rocky Mountains [1] and are associated with rare cases of botulism in cattle in California, Utah, Idaho, Oregon, and Ohio [9].

Type B spores are responsible for more than 85% of cases of botulism in cattle and horses in North America [9,10]. The source is seldom identified in sporadic cases of equine botulism, but in outbreaks, it is typically associated with forage [11–16]. Contamination of grain by decomposing animal carcasses is rarely a factor [9], but high forage pH encourages germination of spores [17]. Hay stored in plastic bags is commonly associated with outbreaks in the U.S. and England.

Type C or D botulism is most commonly a consequence of feed contamination by a decomposing carcass [18,19] or transport of BoNT by ravens or crows feeding on a decomposing carcass [20,21]. Poultry litter containing chicken carcasses has frequently been associated with outbreaks of botulism when fed to beef cattle, including major outbreaks in Israel [22], Australia [23], and Europe [23–27].

TABLE 22.1
Host Predilection of Botulinum Toxin Serotypes

Toxin Type	Species Affected
A	Cattle, humans
B	Horses, cattle, humans
C	Birds, cattle
D	Cattle
E	Fish, fish-eating birds, humans
F	Humans
G	Humans

Type D botulism in South America and South Africa has been linked to pica, in which phosphorous-deficient cattle consume bones of comrades dead of botulism [25]. Type D botulism is rare in the U.S. [28], and types E, F, and G botulism apparently do not affect cattle or other domestic species [4,6,29].

Botulism occurs after ingestion of preformed BoNT or by dissemination of toxin from a wound or focus of clostridial multiplication in the gastrointestinal tract [8,30]. Horses are more sensitive to BoNT than cattle; preformed toxin may be degraded by rumen microbes in cattle [17,31] and absorbed proximal to the colon in horses.

Spores germinating in a wound can produce toxin and cause wound botulism [32]. Castration site infections are common in horses [32], as are umbilical hernia repairs and deep puncture wounds from injections [33]. Wound botulism also occurs in calves castrated by banding. Toxicoinfectious botulism in foals is often called "shaker foal syndrome" [30] and occurs following absorption of toxin from the gastrointestinal tract. Toxin can be detected in feces of about one-third of affected animals, and disease occurs up to about eight months of age [12,30]. In bovine, toxicoinfectious botulism, which occurs via the gastrointestinal tract, deaths may continue for a lengthy period after removal of contaminated feed, suggesting gastrointestinal residence.

The clinical course of mild bovine botulism is dose-dependent and ranges from 2 to 30 days. Cattle exposed to moderate amounts of BoNT in the field exhibit weakness for 24 to 48 h and recumbency for 2 to 3 days before death, with greater severity of symptoms in those that are physically more active. In experimental intoxications, clinical signs may be delayed more than a week after administration of small doses of BoNT to cattle, while large doses (e.g., 10^8 MLD_{50} units) can lead to recumbency and death in 24 h.

Smaller amounts of toxin (e.g., 10^3 IP MLD_{50} units) administered to horses cause less-severe signs, with onset over 3 to 10 days [9], which manifest as transient dysphagia and ultimate recovery. Peracute, rapidly progressive illness follows experimental inoculation with larger doses, and these same effects may be seen in animals exposed to massive amounts of toxin with forage.

Toxin is rarely detected in serum or plasma of acutely affected cattle and horses, but can be found frequently in birds, which are more resistant to the disease. Gross or histological lesions are usually absent, although inhalation pneumonia may occur due to an abnormal deglutition reflex. Affected animals are dysphagic and will usually have relatively empty gastrointestinal tracts. A diagnosis of botulism should be considered in afebrile, alert animals with progressive weakness and recumbency.

Cattle with mild clinical signs of botulism develop antibodies to BoNT approximately one month after recovery [34–36]. A type B toxoid is available in some countries [37,38], and protective antibody responses can usually be raised in horses in about three weeks [39]. There is no cattle-approved toxoid in the U.S., but type C and D toxoids are available in Australia to prevent disease in sheep and cattle. Annual revaccination is advocated in endemic areas.

A type C toxoid, approved for mink, is routinely used in horses, especially in endemic areas such as the southwestern U.S. Mares revaccinated before foaling should produce colostrum with adequate antibody titers to protect foals for 8 to 12 weeks [37]. Foals vaccinated as neonates develop antibodies, without apparent interference from dam-acquired passive antibodies [37].

22.1.2 Tetanus

Spores of *Clostridium tetani* are ubiquitous, but dogma regarding a preferential association with horse manure is likely without basis in fact. Nonuniform distribution of spores can affect incidence more than system of management.

Clostridium tetani often enters via traumatic wounds, including those from tail-docking and dehorning. Castration of lambs and calves via elastrator is a risk factor for infection, and adult cows can be infected by contamination of uterine prolapses. Growth in the rumen may also result in clinical tetanus. Horses are most often infected via contaminated wounds.

Spore germination is often facilitated by the presence of facultative anaerobes that reduce Eh in tissue. Bacterial multiplication and tetanospasmin (TeNT) production is enhanced by tetanolysin, a cholesterol-binding toxin that inhibits neutrophil and macrophage chemotaxis and causes local tissue necrosis. Spores can remain latent in healed wounds for 10 years or more.

Horses are protected by vaccination with tetanus toxoid, and annual revaccination is recommended [40]. Passive immunization can be useful in foals.

22.2 HISTOTOXIC CLOSTRIDIA

Common themes in infections by histotoxic clostridia include an endogenous or environmental source of the infecting organism, entry facilitated by trauma, local multiplication, extensive local and systemic damage, and rapid death; in all cases, the hallmark is impassioned toxinogenesis. Control by vaccination has decreased the incidence of many such infections.

22.2.1 *Clostridium perfringens*

Clostridium perfringens is the most important cause of clostridial disease in domestic animals (Table 22.2) [7,41], and most cases of myonecrosis are apparently caused by type A strains. Spores germinate and vegetative cells multiply in ischemic tissue, and infection spreads to healthy muscle, often with fatal results. Alpha (CPA) and theta (PFO) toxins play local and systemic synergistic roles in myonecrosis [42], although the entirety of our knowledge in this area comes from studies in mice. Specific immunization of domestic animals against infection by type A in North America is focused at present upon anti-alpha toxin immunity, and both toxoids and bacterin:toxoids are widely available elsewhere in the world. The protective effect of CPA-containing toxoids against

TABLE 22.2
Clostridia in Myonecrosis

Organism	Major Toxins	Diseases	Species Affected
C. perfringens type A	Alpha, theta	Gas gangrene, myonecrosis	All warm-blooded
C. septicum	Alpha	Abomasitis, malignant edema	Sheep, cattle
C. chauvoei	Alpha, beta	Blackleg	Sheep, cattle
C. novyi types A and B	Alpha, beta	Wound infections ("bighead"), infectious necrotic hepatitis	Sheep, goats
C. novyi type D	Beta	Bacillary hemoglobinuria	Cattle
C. sordellii	Hemolytic, lethal	Myonecrosis	Sheep, cattle

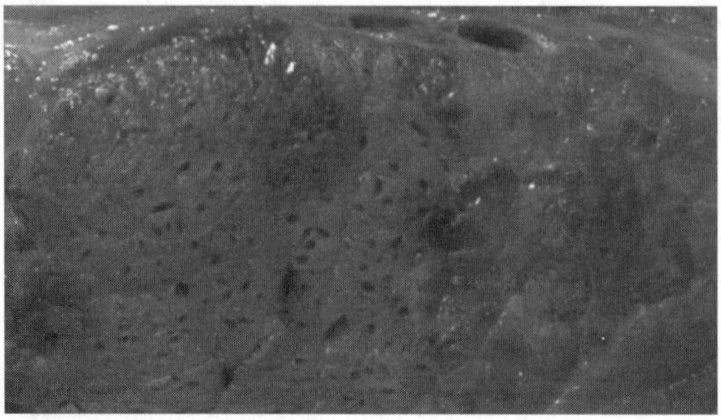

FIGURE 22.1 Bovine malignant edema due to *C. septicum* infection, with gas accumulation in muscle. (Figure courtesy of Raymond E. Reed.)

gas gangrene has long been known, and antibodies against native CPA and its C-terminus protect mice against challenge with toxin or multiple LD_{50} of spores [43,44].

22.2.2 Clostridium septicum

Clostridium septicum is found commonly in soil and feces [45–49], and it can enter hosts with one of the life stages of liver flukes [50,51]. It is a frequent post-mortem invader. Iatrogenic infections in horses are not uncommon [52–54], and ovine umbilical infections are a major source of loss.

The organism causes malignant edema (Figure 22.1), which has striking similarities to blackleg. Hemorrhage, edema, and necrosis develop as the infection spreads along muscular fascial planes. Tissues are initially painful and warm, with pitting edema, but become crepitant and cold, and the course is often <24 h. Pathogenesis is mediated largely by alpha toxin.

Prevention is preferred to treatment, given the fulminant clinical course. Antibodies to somatic and toxin antigens [55] provide lifelong immunity to malignant edema, although there are differences in immunogenicity by vaccine and by host species [56–58]. Death losses in feedlot cattle can be reduced by approximately 50% in vaccinates, with an estimated cost benefit of more than $10 per animal [59].

22.2.3 Clostridium chauvoei

Clostridium chauvoei causes blackleg [60,61] (Figure 22.2), an emphysematous, necrotizing myositis of ruminants and other domestic species (Table 22.2), which resembles malignant edema. Signs include high fever, anorexia, depression, and lameness, with crepitant lesions and sudden death. Lesions are often dry and emphysematous at the center, but edematous, hemorrhagic, and necrotic at the periphery, with little leukocytic infiltration.

The roles of alpha toxin, which is necrotizing, hemolytic, and lethal, and beta toxin, a DNase [62], remain undefined. Flagellar expression is associated with virulence, and phase variation occurs in motility and flagellation [63]; flagella are apparently protective antigens [64,65].

22.2.4 Clostridium novyi

Clostridium novyi type A causes wound infections, the hallmark of which is edema. "Bighead" of young rams is a rapidly spreading edema of the head, neck, and cranial thorax following clostridial invasion of subcutaneous tissues damaged by fighting [66]. Infectious necrotic hepatitis ("black disease") of sheep and cattle occurs when type B spores germinate in liver tissue damaged by fluke migration. Systemic effects with acute or peracute death follow dissemination of alpha toxin [67]. Cardio-, neuro-, histo-, and hepatotoxic effects result in edema, serosal effusion, and focal hepatic

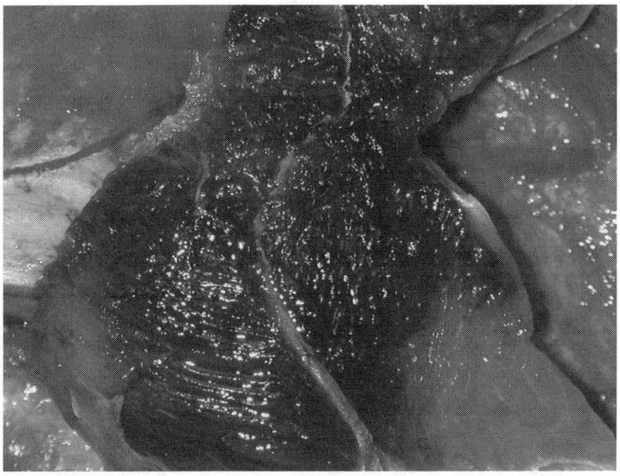

FIGURE 22.2 Bovine blackleg (*C. chauvoei* infection). (Figure courtesy of Gregory A. Bradley.)

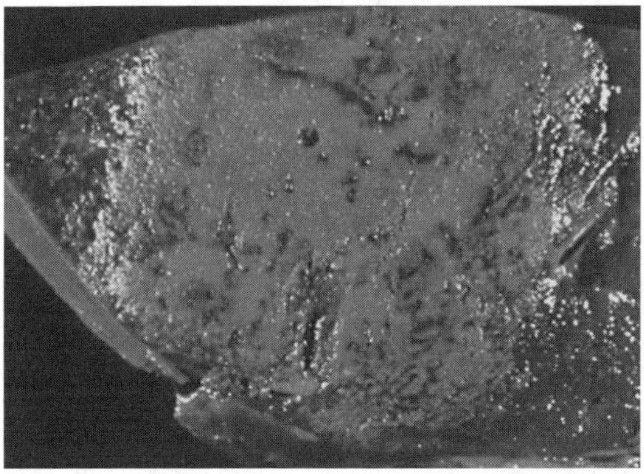

FIGURE 22.3 Liver infarct in *C. novyi* type D infection. (Figure courtesy of Raymond E. Reed.)

necrosis [67–70]. "Black disease" derives from the darkening of the underside of the skin due to venous congestion. *Clostridium novyi* type D (*C. haemolyticum*) causes bacillary hemoglobinuria of cattle (Figure 22.3) and other ruminants [71]. Differential production of alpha and beta toxins is the major phenotypic differentiating factor. The former, a "large clostridial cytotoxin" [72,73], is produced by strains of types A and B. Beta toxin, a phospholipase that is related to CPA of *C. perfringens*, is produced in small quantities by type B strains and in large amounts as the sole major toxin of type D strains.

Prevention of *C. novyi* infections can be achieved with bacterin:toxoids or toxoids. Second-generation vaccines can be based upon native or recombinant alpha [74] or beta toxoids.

22.2.5 *Clostridium sordellii*

Clostridium sordellii is an important cause of myonecrosis in domestic ruminants, with lesions similar to those in malignant edema or blackleg. Lethal and hemolytic toxins, which are central to pathogenesis, are antigenically and pathophysiologically similar to *Clostridium difficile* toxins B and A, respectively. Guinea pigs immunized with toxoids prepared from purified lethal toxin and partially purified hemolytic toxin are protected against spore challenge, and results of trials suggest that both toxins must be represented in the vaccine [75].

TABLE 22.3
Diseases Produced by Toxigenic Types of *C. perfringens*

Toxin Type	Major Diseases	Major Toxins
A	Food poisoning, poultry necrotic enteritis, lamb enterotoxemia, porcine neonatal necrotizing enterocolitis, bovine neonatal hemorrhagic enteritis	Alpha
B	Lamb dysentery, ovine chronic enteritis, bovine/equine hemorrhagic enteritis	Alpha, beta, epsilon
C	Fowl necrotic enteritis, hemorrhagic or necrotic enterotoxemia in piglets, lambs, calves, goats, foals, acute enterotoxemia ("struck") in adult sheep	Alpha, beta
D	Ovine enterotoxemia, caprine enterocolitis, bovine enterotoxemia (calves, possibly adults)	Alpha, epsilon
E	Bovine (possibly ovine) neonatal enterotoxemia	Alpha, iota

TABLE 22.4
Enteric Infections by Clostridia Other than *C. perfringens*

Organism	Virulence Factor	Diseases
C. septicum	Alpha (pore former)	Abomasitis (braxy) in sheep, calves
C. difficile	A (enterotoxin) B (cytotoxin)	Antibiotic-associated typhlocolitis (humans, rodents); equine neonatal hemorrhagic enterocolitis; porcine neonatal typhlocolitis
C. spiroforme	Iota toxin (ADP-ribosylating)	Rabbit diarrhea
C. colinum	None confirmed	Fowl ulcerative enteritis

22.3 ENTERIC CLOSTRIDIA

First among the enteric clostridia, in severity, economic impact, and overall incidence, are the various toxinotypes of *C. perfringens*. *C. difficile* has emerged as an important pathogen of horses and pigs, and other clostridia (*C. septicum*, *C. spiroforme*, and *C. colinum*) are common pathogens in some geographic areas (Table 22.3 and Table 22.4).

22.3.1 *Clostridium perfringens*

Production of the "major toxins" [CPA, beta (CPB), epsilon (ETX), and iota (ITX)] divides the species into five toxinotypes (Table 22.3). Strains of type A are widespread in the intestines of warm-blooded animals and in the environment [76]. Necrotic enteritis of domestic poultry is by type A strains, although type C strains are found occasionally [77,78]). Mild disease results in decreased rates of gain, but inappetence, anorexia, and diarrhea are common and can be part of a

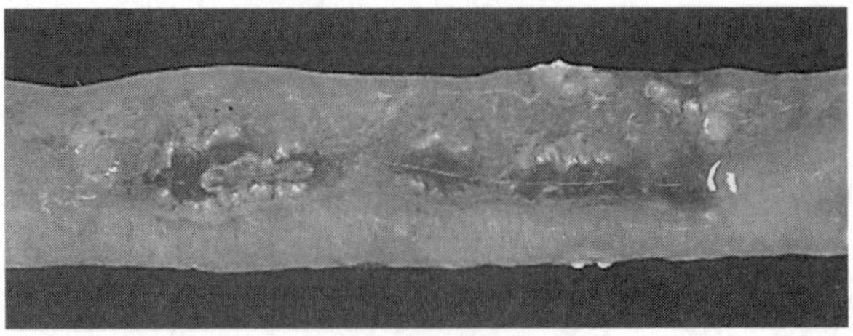

FIGURE 22.4 Mucosal necrosis in chick infected with *C. perfringens* type A. (Figure courtesy of J. Glenn Songer.)

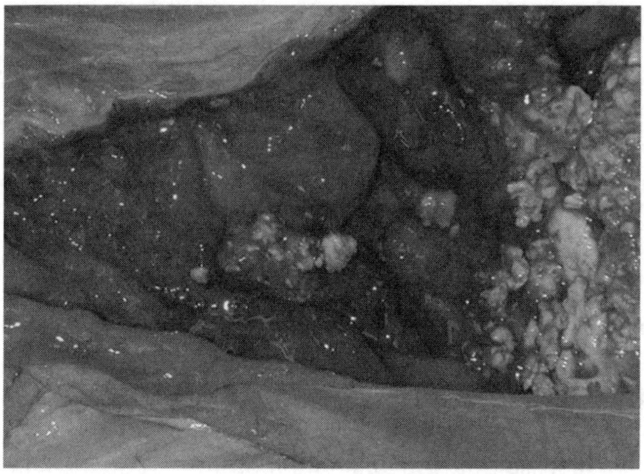

FIGURE 22.5 Hemorrhagic abomasitis in a one-week-old dairy calf infected with *C. perfringens* type A. (Figure courtesy of J. Glenn Songer.)

short clinical course culminating in death [77]. Jejunal and ileal necrosis (Figure 22.4) are characteristic (79). High-fiber litter, concurrent infection with coccidia [80], and composition of the diet can predispose to necrotic enteritis [81].

Type A infection of neonatal pigs takes the form of necrotizing enteritis with mild villous atrophy [82]. Jejunal and ileal lesions can be heavily colonized with *C. perfringens* [83], although it is common to find masses of organisms in the lumen. Disease has been reproduced in gnotobiotic colostrum-deprived and conventional pigs [82].

Neonatal foals develop hemorrhagic diarrhea in response to type A infection, and clinical signs can be as mild as watery-to-mucoid diarrhea or as severe as peracute death. Subserosal hemorrhage, diffuse mucosal necrosis, and hemorrhage of lamina propria and submucosa are common, and Gram-positive bacilli are associated with necrotic tissue [84].

Type A is a common cause of abomasal ulceration and tympany in beef calves [85]. Acute onset of abdominal tympany, colic, or death due to these infections is often mistaken for abomasal displacement or intestinal obstruction [86], but post-mortem findings frequently include abomasal distention and abomasitis (Figure 22.5), hemorrhage, and ulceration; type A is isolated in large numbers from affected tissues. Lamb enterotoxemia (yellow lamb disease), which occurs sporadically in the Pacific Northwest of the U.S. [87], is characterized by anemia, icterus, hemoglobinuria, and death after a brief clinical course.

Pathogenesis of type A enteric infections is poorly understood, but is likely to be multifactorial. Enterotoxemia in lambs and calves is compatible with the action of a hemolytic toxin (such as CPA) in circulation, with IV hemolysis and capillary damage, inflammation, platelet aggregation, shock, and cardiac effects [88]. CPA can contribute to intestinal mucosal necrosis in necrotic enteritis of poultry [89,90]. Type A strains cause enteropathy in neonatal pigs after substantial multiplication, sometimes with adherence to enterocytes. CPA alone causes mild enteritis and villous edema, with minimal damage to epithelium and blood vessels [91].

Another putative virulence attribute is the beta2 toxin (CPB2) [92–94]. *cpb2* has been associated with typical and atypical typhlocolitis in horses (~50% of affected animals) [93]. PCR screening revealed *cpb2* in 37.2% of field isolates [94]. Few isolates from normal pigs contain *cpb2*, but most from porcine neonatal enteritis (>90%) are positive. Most *cpb2*-positive porcine isolates (>96%) produce CPB2, but isolates from other species are less-commonly positive (50%). The specific role of CPB2 in pathogenesis is in question, but its strong association with enteric disease in pigs suggests that it is at least a marker of virulence.

Enterotoxigenic *C. perfringens* has been implicated in diarrheal disease in pigs [83,95,96]. Mucosal necrosis and villous atrophy are superficial [83,95]. Experimental infection causes signs ranging from creamy diarrhea and emaciation with low mortality to profuse, bloodstained diarrhea, enteritis, and death [97].

Type B strains cause lamb dysentery (Table 22.3). Disease occurs worldwide, but is relatively rare in North America. Infection commences early in life, with infection acquired from the dam or her environment. Nutrient spillover into small intestine in lambs suckling heavily lactating dams favors multiplication of the organism in the gut. Lambs lack appetite and have the tucked-up posture that suggests abdominal pain. Bloody diarrhea is followed shortly by recumbency, coma, and death in nearly 100% of cases. Extensive hemorrhage and ulceration of the small intestine is a common necropsy finding in lambs [98], although "pine" in older lambs, which takes the form of chronic abdominal pain without diarrhea, has less-severe lesions. Type B may also be associated with hemorrhagic enteritis in kids, calves, and foals. Pathogenesis is likely due to additive or synergistic effects of CPB and ETX.

Type C-associated disease is perhaps most common in newborn animals [99] (Table 22.3), but older animals can develop disease following alteration of flora by sudden dietary changes. Peracute disease in piglets is characterized by hemorrhagic diarrhea, sometimes with dysentery [100], and extensive emphysema and necrosis of mucosa, submucosa, and muscularis mucosa [101,102]. Case fatality rates approach 100%. The course in older piglets is not as acute, and the primary attributes of disease are nonbloody diarrhea and jejunal mucosal necrosis [100]. Similar disease is seen in neonatal calves [103], lambs [104], foals, and goats. Young sheep are subject to "struck," a form of enterotoxemia leading to rapid death, which simulates a lightning strike. Struck is initiated at sites of gastrointestinal mucosal damage, in which type C multiplies, causing local necrotic lesions (usually in abomasum and small intestine) and toxemia, but usually no dysentery or diarrhea.

Pathogenesis of type C infections is mediated principally by CPB, as judged indirectly by efficacy of vaccination with a CPB toxoid [105]. Damage to intestinal microvilli, mitochondria, and terminal capillaries occurs before the organism adheres to porcine jejunal mucosa [82,106]. Apices of jejunal villi are affected first, but this is followed by progressive, widespread mucosal necrosis [107]. Toxemia is probably the ultimate cause of death, which is often without premonitory signs [100]. CPB acts in jejunum in the absence of normal protease activity (e.g., early in the neonatal period or as a result of the action of protease inhibitors in colostrum) [100]. Increased capillary permeability caused by CPB may facilitate its uptake to circulation and promote systemic effects.

Type D infection manifests as enterotoxemia (also called sudden death or overeating), and affects calves and kids, but is most prevalent in young lambs (Table 22.3). Reports of disease in horses and adult cattle suggest the importance of ruling out type D infections in these species, but the consensus seems to be that disease is rare in these hosts. Enterotoxemia is associated with sudden changes to a rich diet (overeating) or from continuous feeding of a highly concentrated ration (perhaps with nutrient spillover to small intestine) [108]. Rapid multiplication of type D organisms is accompanied by production of ETX, which affects the CNS and other tissues, causing sudden death. Chronic intoxications often manifest as encephalomalacia (Figure 22.6).

The course is short and nearly 100% fatal [108]. Disease in sheep and calves does not always involve intestinal hemorrhage [109], but caprine infections present most commonly as chronic hemorrhagic enterocolitis [110]. Differential sensitivity of cell lines to ETX is perhaps an *in vitro* parallel [111]. The structure and mechanism of action of ETX have been investigated, but many aspects remain unexplored.

Immunization with ETX toxoids stimulates antitoxic immunity [112]. Anti-idiotypic monoclonal antibodies can also be used as immunogens [113], suggesting that neutralization can be achieved at a single epitope. Although sheep and goats respond equivalently to ETX toxoids [110] and toxemic deaths are averted in both species, goats still develop enterocolitis [110].

ETX is a candidate biological weapon, and is a Category B Select Agent. The attendant demand for higher security has introduced unique problems for veterinary biologics companies.

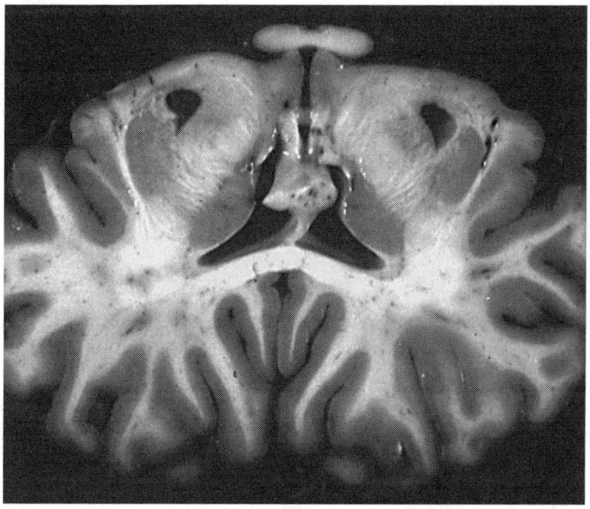

FIGURE 22.6 Ovine encephalomalacia associated with *C. perfringens* type D infection. (Figure courtesy of Raymond E. Reed.)

The unique feature of type E is its production of iota toxin (ITX) [114]. Iota enterotoxemia was reported more than 50 years ago in calves and lambs. Reports since that time have been mainly limited to calves [115,116].

In North America, type E strains are found solely in hemorrhagic enteritis (often with sudden death) in neonatal calves [117]. Gross lesions include abomasal and small intestinal edema, multifocal mucosal hemorrhage, acute inflammation, and submucosal edema. Morbidity in affected herds is typically 10%, and more than half of these cases are fatal. Type E isolates represent nearly 50% of all isolates from similar clinical cases in calves. Commercial toxoids offer no protection against type E infections.

Silent *cpe* sequences are highly conserved among these type E isolates, but contain nonsense and frameshift mutations and lack an initiation codon, promoters, and ribosome binding site [117].

Immunity to *Clostridium perfringens* is, as noted, mainly antitoxic. Equine hyperimmune antiserum administered early protects against infections by types B, C, and D for up to three weeks [118]. Active immunoprophylaxis is critical and is almost universally achieved by use of commercial toxoid vaccines [112]. More than 10-fold reductions in mortality are commonly achieved in piglets born to immunized sows [118]. The same applies to vaccination of ewes and cows against infection by types B, C, or D [112].

22.3.2 Clostridium septicum

Clostridium septicum is found in feces of domestic animals and humans and is a common postmortem invader in ruminants. Iatrogenic infections are most likely to occur in horses [53]. The best-known enteric infection by *C. septicum* is braxy in calves or lambs, in which the organism establishes in damaged abomasal epithelium and produces a fatal bacteremia [119]. Damage may result from ingestion of frozen feed [119–121]. Edema and hemorrhagic necrosis are seen in abomasal walls and proximal small intestine [121]. Alpha toxin is the key factor in pathogenesis [122,123]. Vascular endothelial cytotoxicity could result in loss of fluid from circulation and induction of shock. Prevention is preferred to therapy, and application of commercial bacterins and toxoids usually yields lifelong immunity [59]. Vaccination of mice with alpha toxoid partially protects from lethal infection [122].

22.3.3 *Clostridium difficile*

Human *C. difficile*-associated disease (CDAD) is responsible for one quarter of all cases of antibiotic-associated diarrhea [124–126], and clindamycin, ampicillin, amoxicillin, the cephalosporins, and many other antimicrobials are associated with its development [127]. Spores germinate in ileum, cecum, and colon [128], and vegetative cells fill empty niches and produce toxins. Disease presents as diarrhea, colitis, pseudomembranous colitis, or fulminant colitis [128].

CDAD is also important in domestic and laboratory animals. Disease occurs in antibiotic-treated hamsters [129] and guinea pigs. The organism is an important cause of diarrhea and fatal necrotizing enterocolitis in neonatal foals [130]. The clinical course is usually less than 24 h, with fatal outcome. *Clostridium difficile* has also become quite important as a cause of nosocomial, often-antimicrobial-associated diarrhea in adult horses.

CDAD is emerging as a cause of enteritis in neonatal pigs. More than one third of piglets submitted for diagnosis of enteritis in major swine-producing areas of the U.S. have uncomplicated CDAD; mixed infections occur in a further 20 to 25% [131,132]. Prevalence in a production system was 47.6% on a per-litter basis, and 90% of herds were infected.

Porcine CDAD occurs in piglets 1 to 7 days of age with a history of early onset scours and born to gilts or multiparous sows. Respiratory distress and sudden death (with hydrothorax or ascites) occur on occasion. Gross lesions usually include mesocolonic edema, and large intestines are often filled with pasty-to-watery yellowish feces. Extensive sampling in CDAD-affected herds has revealed that two thirds of litters and more than one third of individual pigs are toxin-positive in an infected barn. Piglets without enteric signs can be toxin-positive [133,134]. Focal suppuration in colonic lamina propria is the hallmark lesion, and colonic serosal and mesenteric edema and infiltration of mononuclear inflammatory cells and neutrophils in the edematous areas are common. Segmental erosion of colonic mucosal epithelium and volcano lesions (exudation of neutrophils and fibrin into the lumen) may be seen [131]. Pathogenesis of CDAD in domestic animals is likely mediated by the monomeric toxins A (TcdA, 308 kDa, an enterotoxin) and B (TcdB, 270 kDa, a cytotoxin).

Immunoprophylaxis of CDAD in domestic animals has not been studied, although precedent suggests that effective immunity will be antitoxic. Antibodies against TcdA prevent toxin binding in mouse and hamster models, eliminating secretion, inflammation, and clinical disease [135]. AntiTcdB antibodies also participate [136,137].

22.3.4 *Clostridium spiroforme*

Clostridium spiroforme causes iota enterotoxemia in rabbits and other laboratory rodents [138] (Table 22.4). Its distinct, loosely coiled, spiral form consists of uniform aggregation of individual semicircular cells joined end-to-end [139]. The organism produces a toxin that is cross-neutralized with *C. perfringens* iota toxin. Enterotoxemia can be produced with filtrates of cecal contents from affected rabbits.

Disease occurs spontaneously in weaned rabbits [140], but also upon destabilization of cecal microflora at weaning or by antimicrobial therapy. Diarrhea is evidenced by perineal staining, and the case fatality rate is high. Cecal dilation with watery contents is a remarkable feature, as are epithelial necrosis and pronounced inflammation of the lamina propria. Affected animals have toxin and high numbers of spores in cecal contents. Rabbits immunized with iota toxoid are protected against IP toxin challenge [141].

22.4 CONCLUSION

Many exciting findings regarding toxins and their molecular mechanisms of action, as well as development of genetic systems for some previously refractory species, provide opportunities for more extensive work on the interaction of these organisms with their hosts. Our knowledge in this area is embryonic for many species. Aspects of clostridial biology in the gut (e.g., quorum sensing) present opportunities for research, to the benefit of livestock production and basic understanding of the interaction of bacterial pathogens and hosts.

Diagnosis is also an important area for further research. Development and application of highly sensitive and serotype-specific assays for clostridial neurotoxins, perhaps based upon their metalloprotease activity, will provide the opportunity for greater understanding of the pathogenesis, epidemiology, and natural history of these intoxications. Availability of rapid and simple assays for detection of infections by histotoxic and enteric clostridia will provide better information for prevention and control strategies.

The veterinary biologics industry is seeking a new paradigm for preparation and delivery of immunoprophylactic products in the face of continuing concerns about undesirable post-vaccination effects. Delivery of recombinant proteins by conventional means or by *in vivo* expression from attenuated bacterial delivery systems is a focus of industry efforts.

ACKNOWLEDGMENTS

My work over the past 15 years has been generously supported by the U.S. Department of Agriculture, National Research Initiative, by the National Pork Board, and by commercial interests, including Alpharma, Boehringer-Ingelheim Vetmedica, Pfizer, Grand Laboratories, Novartis Animal Vaccines, Schering Plough Animal Health, and Colorado Serum Company.

REFERENCES

1. Smith, L.D.S., *Clostridium botulinum*: Characteristics and occurrence, *Rev. Inf. Dis.*, 1, 637–641, 1979.
2. Linial, M., Bacterial neurotoxins — a thousand years later, *Israel J. Med. Sci.,* 31, 591–595, 1995.
3. Smith, L.D.S., Botulism in Animals, in *Botulism: The Organism, Its Toxins, the Disease,* Thomas, C.C., Ed., Springfield, IL, 1977, 236.
4. Shapiro, R.L., Hatheway, C., and Swerdlow, D.L., Botulism in the United States: a clinical and epidemiologic review, *Ann. Int. Med.,* 129, 221–228, 1998.
5. Smith, G.R. and Milligan, R.A., *Clostridium botulinum* in soil on the site of the former metropolitan (Caledonian) cattle market, London, *J. Hyg. Camb.,* 83, 237–241, 1979.
6. Centers for Disease Control and Prevention: *Botulism in the United States, 1899–1996. Handbook for Epidemiologists, Clinicians, and Laboratory Workers,* Centers for Disease Control and Prevention, Atlanta, GA, 1998.
7. Hatheway, C.L., Toxigenic clostridia, *Clin. Microbiol. Rev.,* 3, 66–98, 1990.
8. Hatheway, C. L., Botulism: the present status of the disease, *Curr. Topics Microbiol. Immun.,* 195, 55–75, 1995.
9. Whitlock, R.H. and Williams, J.M., Botulism toxicosis of cattle, *Proc. Bovine Practitioner,* 32, 45–53, 1999.
10. Kelch, W.J. et al., Fatal *Clostridium botulinum* toxicosis in eleven Holstein cattle fed round bale barley haylage, *J. Vet. Diagn. Invest.,* 12, 453–455, 2000.
11. Kinde, H. et al., *Clostridium botulinum* type C intoxication associated with consumption of processed hay cubes in horses, *J. Am. Vet. Med. Assoc.,* 199, 742–746, 1991.
12. Whitlock, R.H., Botulism, *Vet. Clin. North Am. Equine Pract.,* 13, 107–128, 1997.
13. Wichtel, J.J., and Whitlock, R.H., Botulism associated with feeding alfalfa hay to horses, *J. Am. Vet. Med. Assoc.,* 199, 471–472, 1991.
14. Ricketts, S.W. et al., Thirteen cases of botulism in horses fed big bale silage, *Eq. Vet. J.,* 16, 515–518, 1984.
15. Franzen, P., Gustafsson, A., and Gunnardsson, A., Botulism in horses associated with feeding big bale silage, *Svensk Veterinar tidning*, 44, 555–559, 1992.

16. Kelly, A.R. et al., Outbreak of botulism in horses, *Eq. Vet. J.,* 16, 519–521, 1984.
17. Divers, T.J. et al., *Clostridium botulinum* type B toxicosis in a herd of cattle and a group of mules, *J. Am. Vet. Med. Assoc.,* 188, 382, 1986.
18. Enfors, E. et al., Outbreak of botulism caused by *Clostridium botulinum* type C in cattle, *Svensk. Vet.,* 27, 333, 1975.
19. Galey, F.D. et al., Type-C botulism in dairy cattle from feed contaminated with a dead cat, *J. Vet. Diagn. Invest.,* 12, 204–209, 2000.
20. Schoenbaum, M.A. et al., An outbreak of type C Botulism in 12 horses and a mule, *J. Am. Vet. Med. Assoc.,* 217, 365–368, 2000.
21. Whitlock, R.H. et al., Type C Botulism in New Mexico horses: an outbreak abstract; *40th Annual Proceedings American Association of Veterinary Laboratory Diagnosticians*, October 17–24, 1997.
22. Savoir, D., Mass outbreaks of botulism in ruminants associated with ingestion of feed containing poultry waste. III. Epidemiologic findings, *Refuah. Vet.,* 35, 105–108, 1978.
23. Trueman, K.F. et al., Suspected botulism in three intensively managed Australian cattle herds, *Vet. Record,* 130, 398–400, 1992.
24. Appleyard, W.T. and Mollison, A., Suspected bovine botulism associated with broiler litter waste, *Vet. Rec.,* 116, 522, 1985.
25. Dobereiner, J. et al., Epizootic botulism of cattle in Brazil, *Dtsch. Tierarztl. Wschr.,* 99, 188, 1992.
26. McLoughlin, M.F., McIlroy, S.G., and Neill, S.D., A major outbreak of botulism in cattle being fed ensiled poultry litter, *Vet. Rec.,* 122, 579–581, 1988.
27. Neill, S.D., McLoughlin, M.F., and McIlroy, S.G., Type C botulism in cattle being fed ensiled poultry litter, *Vet. Rec.,* 124, 558–560, 1989.
28. Whitlock, R.H. and Messick, J.B., Botulism in cattle and horses, in *Proc. 4th Ann. Am. College Vet. Internal Med.*, 1986, 57–62.
29. Sugiyama, H., Bott, T.L., and Foster, E.M., *Clostridium botulinum* type E in an island bay (Green Bay of Lake Michigan), in *Toxic Micro-Organisms,* Herzberg M., Ed., U.S. Dept. of Interior, 1970.
30. Swerczek, T.W., Toxicoinfectious botulism foals and adult horses, *J. Am. Vet. Med. Assoc.,* 176, 348–350, 1980.
31. Allison, M.J., Maloy, S.E., and Matson, R.R., Inactivation of *Clostridium botulinum* toxin by ruminal microbes from cattle and sheep, *Appl. Environ. Microbiol.,* 32, 685, 1976.
32. Bernard, W. et al., Botulism as a sequel to open castration in a horse, *J. Am. Vet. Med. Assoc.,* 191, 73, 1987.
33. Mitten, L.A. et al., Mechanical ventilation and management of botulism secondary to an injection abscess in an adult horse, *Equine Vet. J.,* 26, 420–423, 1994.
34. Thomas, R.J., Detection of *Clostridium botulinum* types C and D toxin by ELISA, *Aust. Vet. J.,* 68, 111–113, 1991.
35. Jubb, T.F., Ellis T.M., and Gregory, A.R., Diagnosis of botulism in cattle using ELISA to detect antibody to botulinum toxins, *Aust. Vet. J.,* 70, 226–227, 1993.
36. Main, D.C. and Gregory, A.R., Serological diagnosis of botulism in dairy cattle, *Aust. Vet. J.,* 73, 77, 1996.
37. Lewis, G.E., Kulinski, S., Fallon, E.H., Klyza, J.P., and Bryans, J.T., Evaluation of a *Clostridium botulinum* toxoid, type B, for the prevention of shaker foal syndrome, *Proc. Am. Assoc. Equine Pract.,* 27, 233, 1981.
38. Thomas, R.J., Rosenthal, D.V., and Rogers, R.J., A *Clostridium botulinum* type B vaccine for prevention of shaker foal syndrome, *Aust. Vet. J.,* 65, 79–80, 1988.
39. Crane, S.A. et al., *Clostridium botulinum* type-B toxoid for vaccination of adult horses, pregnant mares and foals: a study of vaccination protocols, in *Proc. 37th Conv. Am. Assoc. Equine Pract.,* 611, 1991.
40. Green, S.L. et al., Tetanus in the horse: a review of 20 cases (1970 to 1990), *J. Vet. Intern. Med.,* 8, 128–132, 1994.
41. Rood, J.I. and Cole, S.T., Molecular genetics and pathogenesis of *Clostridium perfringens, Microbiol. Rev.,* 55, 621–648, 1991.
42. Awad, M.M. et al., Construction and virulence testing of a collagenase mutant of *Clostridium perfringens, Microb. Pathog.,* 28, 107, 2000.
43. Titball, R.W., Fearn, A.M., and Williamson, E.D., Biological and immunological properties of the C-terminal domain of the alpha-toxin of *Clostridium perfringens*, *FEMS Microbiol. Lett.,* 110, 45–50, 1993.

44. Williamson, E.D. and Titball, R.W., A genetically engineered vaccine against the alpha-toxin of *Clostridium perfringens* protects mice against experimental gas gangrene, *Vaccine,* 11, 1253–1258, 1993.
45. Finegold, S.M., Sutter, V.L., and Mathisen, G.L., Normal indigenous intestinal flora, in *Human Intestinal Microflora in Health and Disease,* Hentges, D., Ed., Academic Press, New York, 1983, 3–31.
46. Kahn, C.M., Anaerobic spore-bearing bacteria of the human intestine in health and in certain diseases, *J. Infect. Dis.,* 35, 423–478, 1924.
47. Princewill, T.J.T., Sources of clostridial infections in Nigerian livestock and poultry, *Bull. Anim. Health Prod. Africa,* 33, 323–326, 1985.
48. Princewill, T.J.T., Agba, M.I., and Jemitola, S.O., Isolation of *Clostridium* species from paddock soil, *Bull. Anim. Health Prod. Africa,* 32, 263, 1984.
49. Princewill, T.J. and Oakley, C.L., Deoxyribonucleases and hyaluronidases of *Clostridium septicum* and *Clostridium chauvoei*. III. Relationship between the two organisms, *Med. Lab. Sci.,* 33, 110–118, 1976.
50. Gariev, B.G., Role of freshwater snails in the epidemiology of certain clostridial infections of sheep, *Veterinariya Moscow USSR,* 10, 39, 1986.
51. Petrov, Y.F. et al., Intestinal and hepatic microflora of sheep during fascioliasis, *Veterinariya Moscow USSR,* 2, 45–49, 1985.
52. Harwood, D.G., Apparent iatrogenic clostridial myositis in cattle, *Vet. Rec.,* 115, 412, 1984.
53. Mullaney, T.P., Brown, C.M., and Flint-Taylor, R., Clostridial myositis in horses following intramuscular administration of ivermectin, *Proc. Annu. Meet. Am. Assoc. Vet. Lab. Diagn.,* 27, 171–177, 1984.
54. van Nie, G.J., Liability, prostaglandins and Clostridium. Malignant oedema in a cow after injection of fenprostalene, *Tijdschrift voor Diergeneeskunde,* 109, 669–671, 1984.
55. Hjerpe, C.A., Clostridial disease vaccines, *Vet. Clin. North Amer. Food Anim. Pract.* 6, 222–234, 1990.
56. Green, D.S. et al., Injection site reactions and antibody responses in sheep and goats after the use of multivalent clostridial vaccines, *Vet. Rec.,* 120, 435–439, 1987.
57. Hjerpe, C.A., Clostridial disease vaccines, *Vet. Clin. North Amer. Food Animal Pract.,* 6, 222–234, 1990.
58. Morgan, K.L. et al., The injection site reaction and antibody response in sheep and goats following the use of multivalent clostridial vaccines, in *Proc. Workshop Goat Produc. Humid Tropics, University of Ife,* Smith, O.B. and Osman, H.G., Eds., 1987, 117–124.
59. Knott, G.K.L., Erwin, B.G., and Classick, L.G., Benefits of a clostridial vaccination program in feedlot cattle, *Vet. Med.,* 80, 95–97, 1985.
60. Burke, M.P. and Opeskin, K., Nontraumatic clostridial myonecrosis, *Am. J. Forensic Med. Pathol.,* 20, 158, 1999.
61. Kuhnert, P. et al., Phylogenetic positions of *Clostridium chauvoei* and *Clostridium septicum* based on 16S rRNA gene sequences, *Int. J. Syst. Bacteriol.,* 46, 1174–1176, 1996.
62. Ramachandran, S., Haemolytic activities of *Cl. chauvoei, Indian Vet. J.,* 46, 754–768, 1969.
63. Tamura, Y. et al., Reversible expression of motility and flagella in *Clostridium chauvoei* and their relationship to virulence, *Microbiology,* 141, 605–610, 1995.
64. Kojima, A. et al., Cloning and expression of a gene encoding the flagellin of *Clostridium chauvoei*, *Vet. Microbiol.,* 76, 359–372, 2000.
65. Verpoort, J.A., Joubert, F.J., and Jansen, B.C., Studies on the soluble antigen and haemolysin of *Clostridium chauvoei* strain 64, *S. Afr. J. Agric. Sci.,* 9, 153–172, 1966.
66. Sterne, M. and Batty, I., *Pathogenic Clostridia,* Butterworths, London, 1975.
67. Elder, J.M. and Miles, A.A., The action of the lethal toxins of gas gangrene clostridia on capillary permeability, *J. Pathol. Bacteriol.,* 74, 133, 1957.
68. Aikat, B.K. and Dible, J.H., The local and general effects of cultures and culture-filtrates of *Clostridium oedematiens, Cl. septicum, Cl. sporogenes,* and *Cl. histolyticum, J. Pathol. Bacteriol.,* 79, 227, 1960.
69. Cotran, R.S., Studies on inflammation. Ultrastructure of the prolonged vascular response induced by *Clostridium oedematiens* toxin, *Lab. Invest.,* 17, 39, 1979.
70. Rutter, J.M. and Collee, J.G., Studies on the soluble antigens of *Clostridium oedematiens* (*Cl. novyi*), *J. Med. Microbiol.,* 2, 395–417, 1967.
71. Bender, L.C. et al., Bacillary hemoglobinuria in a free-ranging elk calf, *J. Zoo. Wildl. Med.,* 30, 293, 1999.
72. Busch, C. et al., Characterization of the catalytic domain of *Clostridium novyi* alpha-toxin, *Infect. Immun.,* 68, 6378, 2001.
73. Selzer, J. et al., *Clostridium novyi* alpha-toxin-catalyzed incorporation of GlcNAc into Rho subfamily proteins, *J. Biol. Chem.,* 271, 25173–25177, 1996.

74. Amimoto, K. et al., The protective effect of *Clostridium novyi* type B alpha-toxoid against challenge with spores in guinea pigs, *J. Vet. Med. Sci.,* 60, 681, 1998.
75. Amimoto, K. et al., Protective effects of *Clostridium sordellii* LT and HT toxoids against challenge with spores in guinea pigs, *J. Vet. Med. Sci.,* 63, 879, 2001.
76. Smith, L. and. Williams, B.L., *The Pathogenic Anaerobic Bacteria, 3rd Edition,* Charles C. Thomas, Springfield, IL, 1984.
77. Long, J.R., Pettit, J.R., and Barnum, D.A., Necrotic enteritis in chickens. II Pathology and proposed pathogenesis, *Can. J. Comp. Med.,* 38, 467–474, 1974.
78. Prescott, J.F., The prevention of experimentally-induced necrotic enteritis in chickens by avoparcin, *Avian Dis.,* 23, 1072–1074, 1979.
79. Ratz, F., Szabo E., and Szentmiklossy, E. Observations on the pathomorphology of necrotic enteritis in chickens, *Magyar Allatorvosok Lapja,* 44, 133–140, 1989.
80. Al-Sheikhly, F. and Al-Saieg, A., Role of coccidia in the occurrence of necrotic enteritis of chickens, *Avian Dis.,* 24, 324, 1980.
81. Branton, S.L., Reece, F.N., and Hagler, W.M., Jr., Influence of a wheat diet on mortality of broiler chickens associated with necrotic enteritis, *Poult. Sci.*, 66, 1326, 1987.
82. Johannsen, U. et al., Studies into experimental *Clostridium perfringens* type A enterotoxaemia of suckled piglets: experimental provocation of the disease by *Clostridium perfringens* type A intoxication and infection, *Monatsh für Veterinaermed,* 48, 129–136, 1993.
83. Nabuurs, M.J.A. et al., Diarrhea in one to three week-old piglets associated with *Clostridium perfringens* type A, *Annales de Recherches Veterinaires,* 14, 408–411, 1983.
84. Bueschel, D.M. et al., Enterotoxigenic *Clostridium perfringens* type A necrotic enteritis in a foal, *J. Amer. Vet. Med. Assoc.*, 213, 1305, 1998.
85. Mills, K.W. et al., Laboratory findings associated with abomasal ulcers/tympany in range calves, *J. Vet. Diagn. Invest.,* 2, 208–212, 1990.
86. Roeder, B.L. et al., Isolation of *Clostridium perfringens* from neonatal calves with ruminal and abomasal tympany abomasitis and abomasal ulceration, *J. Am. Vet. Med. Assoc.,* 190, 1550–1555, 1987.
87. McGowan, G., Moulton, J.E., and Rood, S.E., Lamb losses associated with *Clostridium perfringens* type A, *J. Am. Vet. Med. Assoc.,* 133, 219–221, 1958.
88. Stevens, D.L. et al., Lethal effects and cardiovascular effects of purified a and q toxins, *J. Infect. Dis.,* 157, 272–279, 1988.
89. Köhler, B., Kolbach, S., and Meine, J. Untersuchungen zur nekrotischen enteritis der Hühner. 2. Microbiologische Aspekte, *Monatsh Veterinaermed,* 29, 385–391, 1974.
90. Fukata, T. et al., Influence of bacteria on *Clostridium perfringens* infections in young chickens, *Avian Dis.,* 35, 224–247, 1991.
91. Johannsen, U. et al., Experimental *Clostridium perfringens* type A enterotoxaemia in unweaned piglets, *Monatsh für Veterinarmed,* 48, 267–273, 1993.
92. Gibert M., Jolivet-Renaud, C., and Popoff, M.R., Beta2 toxin, a novel toxin produced by *Clostridium perfringens*, *Gene,* 203, 65–73, 1997.
93. Herholz C. et al., Prevalence of beta2-toxigenic *Clostridium perfringens* in horses with intestinal disorders, *J. Clin. Microbiol.,* 37, 358–361, 1999.
94. Bueschel, D.M. et al., Prevalence of *cpb2*, encoding beta2 toxin, in *Clostridium perfringens* field isolates: correlation of genotype with phenotype, *Vet. Microbiol.*, 94, 121, 2003.
95. Collins, J.E. et al., Diarrhea associated with *Clostridium perfringens* type A enterotoxin in neonatal pigs, *J. Vet. Diagn. Invest.,* 1, 351, 1989.
96. Estrada-Correa, A.E. and Taylor, D.J., Porcine *Clostridium perfringens* type A spores, enterotoxin, and antibody to enterotoxin, *Vet. Rec.,* 124, 606, 1989.
97. Olubunmi, P.A. and Taylor, D.J. *Clostridium perfringens* type A in enteric diseases of pigs, *Trop. Vet.,* 3, 28–33, 1985.
98. Frank, F.W., *Clostridium perfringens* type B from enterotoxemia in young ruminants, *Am. J. Vet. Res.,* 17, 492, 1956.
99. Songer, J.G., Clostridial enteric diseases of domestic animals, *Clin. Microbiol. Rev.,* 9, 216–234, 1996.
100. Niilo, L., *Clostridium perfringens* type C enterotoxemia, *Can. Vet. J.,* 29, 658–664, 1988.
101. Ohnuna, Y. et al., Necrotic enteritis due to *Clostridium perfringens* type C in newborn piglets, *J. Japan. Vet. Med. Assoc.,* 45, 738–741, 1992.

102. Okazaki, Y. et al., Mass outbreak of necrotic enteritis in newborn piglets caused by *Clostridium perfringens* type C on a pig farm, *J. Japan. Vet. Med. Assoc.,* 46, 214–217, 1993.
103. Griner, L.A. and Bracken, F.K., *Clostridium perfringens* (type C) in acute hemorrhagic enteritis of calves, *J. Am. Vet. Med. Assoc.,* 122, 99–102, 1953.
104. Griner, L.A. and Johnson, H.W., *Clostridium perfringens* type C in hemorrhagic enterotoxemia of lambs, *J. Am. Vet. Med. Assoc.,* 125, 125–127, 1954.
105. Lawrence, G.W. et al., Impact of active immunisation against enteritis necroticans in Papua New Guinea, *Lancet,* 336, 1165–1167, 1990.
106. Arbuckle, J.B.R., The attachment of *Clostridium welchii* (*Cl. perfringens*) type C to intestinal villi of pigs, *J. Pathol.,* 106, 65, 1972.
107. Kubo, M. and Watase, H., Electron microscopy of *Clostridium perfringens* in the intestine of neonatal pigs with necrotic enteritis, *Japan. J. Vet. Sci.,* 47, 497–501, 1985.
108. Uzal, F.A. and Kelly, W.R., Effects of the intravenous administration of *Clostridium perfringens* type D epsilon toxin on young goats and lambs, *J. Comp. Pathol.,* 116, 63–71, 1997.
109. Griner L.A., Aichelman, W.W., and Brown, G.D., *Clostridium perfringens* type D (Epsilon) enterotoxemia in Brown Swiss dairy calves, *J. Am. Vet. Med. Assoc.,* 129, 375–376, 1956.
110. Blackwell, T.E. et al., Differences in signs and lesions in sheep and goats with enterotoxemia induced by intraduodenal infusion of *Clostridium perfringens* type D, *Am. J. Vet. Res.,* 52, 1147, 1991.
111. Shortt, S.J., Titball, R.W, and Lindsay, C.D., An assessment of the *in vitro* toxicology of *Clostridium perfringens* type D epsilon-toxin in human and animal cells, *Hum. Exp. Toxicol.,* 19, 108–116, 2000.
112. Kennedy, K.K. et al., Vaccination of cattle and sheep with combined *Clostridium perfringens* type C and type D toxoid, *Am. J. Vet. Res.,* 38, 1515–1518, 1977.
113. Percival, D.A. et al., Anti-idiotype antibody-induced protection against *Clostridium perfringens* type D, *Infect. Immun.,* 58, 2487–2492, 1990.
114. Perelle, S. et al., Characterization of *Clostridium perfringens* iota-toxin genes and expression in *Escherichia coli*, *Infect. Immun.,* 61, 5147–5156, 1993.
115. Hart, B. and Hooper, P.T., Enterotoxemia of calves due to *Clostridium welchii* type E, *Aust. Vet. J.,* 43, 360–363, 1967.
116. Sterne, M. and Thomson, A., The isolation and identification of clostridia from pathological conditions of animals, *Bull. Off. Int. Epizoot.,* 59, 1487–1489, 1963.
117. Billington, S.J. et al., *Clostridium perfringens* type E animal enteritis isolates with highly conserved, silent enterotoxin gene sequences, *Infect. Immun.,* 66, 4531, 1998.
118. Ripley, P.H. and Gush, A.F., Immunisation schedule for the prevention of infectious necrotic enteritis caused by *Clostridium perfringens* type C in piglets, *Vet. Rec.,* 112, 201, 1983.
119. Saunders, G., Diagnosing braxy in calves and lambs, *Vet. Med.,* 81, 1050–1052, 1986.
120. Schamber, G.J., Berg, I.E., and Molesworth, J.R., Braxy or bradsot-like abomasitis caused by *Clostridium septicum* in a calf, *Can. Vet. J.,* 27, 194, 1986.
121. Ellis, T.M., Rowe, J.B., and Lloyd, J.M., Acute abomasitis due to *Clostridium septicum* infection in experimental sheep, *Aust. Vet. J.,* 60, 308, 1983.
122. Ballard, J. et al., Purification and characterization of the lethal toxin (alpha toxin) of *Clostridium septicum*, *Infect. Immun.,* 60, 784, 1992.
123. Ballard, J. et al., The primary structure of *Clostridium septicum* alpha toxin exhibits similarity with *Aeromonas hydrophila* aerolysin, *Infect. Immun.,* 63, 340, 1995.
124. Bartlett, J.G., Antibiotic-associated diarrhea, *Clin. Infect. Dis.,* 15, 573, 1992.
125. Bartlett, J.G. et al., Antibiotic-associated pseudomembranous colitis due to toxin-producing clostridia, *N. Engl. J. Med.,* 298, 531, 1978.
126. Borriello, S.P., and Wilcox, M.H., *Clostridium difficile* infections of the gut: the unanswered questions, *J. Antimicrobial. Chemotherapy,* 41 Suppl C, 67–69, 1998.
127. Johnson, S., Samore, M.H., and Farrow, K.A., Epidemics of diarrhea caused by a clindamycin-resistant strain of *Clostridium difficile* in four hospitals, *N. Engl. J. Med.* 341, 1645–1651, 1999.
128. Kelly, C.P., Pothoulakis, C., and LaMont, J.T., *Clostridium difficile* colitis, *N. Engl. J. Med.,* 27, 257–262, 1994
129. Libby, J.M., Jortner, B.S., and Wilkins, T.D., Effects of the two toxins of *Clostridium difficile* in antibiotic-associated cecitis in hamsters, *Infect. Immun.* 36, 822–829, 1982.

130. Jones, R.L. et al., Hemorrhagic necrotizing enterocolitis associated with *Clostridium difficile* infection in four foals, *J. Am. Vet. Med. Assoc.* 193, 76–79, 1988.
131. Songer, J.G. et al., Enteric infection of neonatal swine with *Clostridium difficile, Swine Health Produc.,* 8, 185–189, 2000.
132. Songer, J.G., A practical approach to clostridial infections, in *Emerging Infections and Molecular Pathogenesis,* Ann. Mtg. Am. Coll. Vet. Pathol., Salt Lake City, UT, December 2–5, 2001.
133. Yaeger, M., Funk, N., and Hoffman, L., A survey of agents associated with neonatal diarrhea in Iowa swine including *Clostridium difficile* and porcine reproductive and respiratory syndrome virus, *J. Vet. Diagn. Invest.,* 14, 281–287, 2002.
134. Waters, E.H. et al., Typhlocolitis caused by *Clostridium difficile* in suckling piglets, *J. Vet. Diagn. Invest.,* 10, 104–108, 1998.
135. Allo, M. et al., Prevention of clindamycin-induced colitis in hamsters by *Clostridium sordellii* antitoxin, *Gastroenterol.,* 76, 351, 1979.
136. Kink, J.A. and Williams, J.A., Antibodies to recombinant *Clostridium difficile* toxins A and B are an effective treatment and prevent relapse of *C. difficile*-associated disease in a hamster model of infection, *Infect. Immun.,* 66, 2018–2025, 1998.
137. Viscidi, R., Laughon, B.E., and Yolken, R., Serum antibody response to toxins A and B of *Clostridium difficile, J. Infect. Dis.,* 148, 93–100, 1983.
138. Carman, R.J. and Borriello, S.P., Infectious nature of *Clostridium spiroforme*-mediated rabbit enterotoxaemia, *Vet. Microbiol.,* 9, 497, 1984.
139. Borriello, S.P., Davies, H.A., and Carman, R.J., Cellular morphology of *Clostridium spiroforme, Vet. Microbiol.,* 11, 191, 1986.
140. Katz, L. et al., Experimental clindamycin-associated colitis in rabbits, *Gastroenterology,* 74, 246, 1978.
141. Ellis, T.M., Gregory, A.R., and Logue, G.D., Evaluation of a toxoid for protection of rabbits against enterotoxaemia experimentally induced by trypsin-activated supernatant of *Clostridium spiroforme, Vet. Microbiol.,* 28, 93, 1991.

Part IV

Regulatory Mechanisms

23 Two-Component Signal Transduction Systems in Clostridia

Jackie K. Cheung, Sheena McGowan, and Julian I. Rood

CONTENTS

23.1 INTRODUCTION

To survive and proliferate in constantly changing surroundings, bacteria have evolved regulatory systems that enable them to monitor and adapt to a diverse range of environmental conditions [1,2]. The most common adaptive response involves two-component signal transduction systems. The number of these regulatory systems identified in recent years has increased dramatically; they have been detected in Gram-positive and Gram-negative bacteria [2], as well as in some eukaryotes [3,4]. These regulatory networks generally consist of a membrane-associated sensor histidine kinase and its cognate cytoplasmic response regulator, which communicate by a His-Asp phosphorelay that usually culminates in the modulation of gene expression [5–7]. This regulatory cascade is activated upon the detection of the appropriate stimulus by the sensor histidine kinase. This event induces the protein to autophosphorylate, whereby the γ-phosphoryl group in ATP is transferred to a conserved His residue located in the cytoplasmic C-terminal domain. The N-terminal regulatory domain of the cognate response regulator then catalyzes the transfer of the phosphoryl group from the His residue to a conserved Asp residue, which activates the response regulator so it is able to bind to its target DNA and, subsequently, regulate gene expression [1,6,8].

This chapter will review what is known about two-component signal transduction systems in the clostridia, including data that have been revealed from the sequencing of a number of clostridial genomes. Since more is known about the VirS/R regulatory network from *Clostridium perfringens*, there will be an emphasis on the analysis of that two-component system.

23.2 THE VIRS/VIRR SYSTEM OF *CLOSTRIDIUM PERFRINGENS*

23.2.1 Identification of the VirS/R Two-Component Signal Transduction System

The isolation of *C. perfringens* mutants with an altered ability to produce extracellular toxins [9–11] implied that toxin production may be subject to global regulation. This early work provided the impetus for the identification of a two-component signal transduction system by two independent research laboratories [12,13]. Chemical mutagenesis of the transformable *C. perfringens* strain 13 led to the isolation of a mutant that could not produce perfringolysin O (θ-toxin), a cholesterol dependent cytolysin, and synthesised reduced amounts of α-toxin, collagenase (κ-toxin) and extracellular protease. A regulatory gene whose product showed sequence identity with the response regulator superfamily was identified and designated as *virR* [13]. At the same time, in our laboratory, Tn*916* mutagenesis was used to isolate a pleiotropic strain 13 mutant that had a similar toxin profile. These studies also led to the identification of the *virR* gene as well as a downstream overlapping gene, *virS*, which encoded a sensor histidine kinase. The Tn*916* insertion was located within the chromosomal *virS* gene [12]. Northern blot analysis of the *virR/virS* locus detected a single 2.1 kb mRNA molecule, indicating that these genes were co-transcribed [14]. Northern analysis and complementation of the *virR* and *virS* mutants with plasmid-encoded wild-type genes, confirmed that the VirS/R system positively regulated the transcription of the *pfoA* (encoding perfringolysin O), *plc* (encoding α-toxin) and *colA* (encoding collagenase) genes at the transcriptional level [14]. Subsequent studies have shown that this system also regulates the production of the extracellular protease α-clostripain [15].

Disruption of the *virRS* locus reduces both toxin levels and the virulence of the resultant strains [12]. In the *virS* mutant JIR4000, α-toxin, sialidase and total protease activities are significantly reduced, whereas no perfringolysin O activity is detected, implying that the VirS/R system has differential effects on the respective toxin structural genes [12]. Further analysis indicated that the transcription from the *pfoA* promoter was completely dependent upon the presence of a functional VirR protein, whereas transcription from the *plc* and *colA* promoters was only partially VirR dependent [14]. Virulence studies using the mouse myonecrosis model showed the VirS/R system regulates genes involved in virulence, since the *virS* mutant was less virulent for mice than the parent strain or a complemented derivative [12].

23.2.2 The VirS and VirR Proteins

23.2.2.1 VirS

Sequence analysis of the VirS sensor kinase indicates that it is most closely related to sensor kinases associated with Gram-positive quorum sensing systems [16]. Alignment of the VirS amino acid sequence with the AgrC, ComD, and PlnB sensor kinases from *Staphylococcus aureus*, *Streptococcus pneumonie,* and *Lactobacillus plantarum*, respectively, identified core kinase motifs including the H, N, and G2 boxes (Figure 23.1) [16]. Mutagenesis studies have shown that in VirS, these motifs are functional and that glutamate residues located within putative transmembrane domains were also required for VirS activity [17]. PhoA fusion studies showed that AgrC has six transmembrane helices [18]. Secondary structure predictions of the VirS protein indicate that it may also

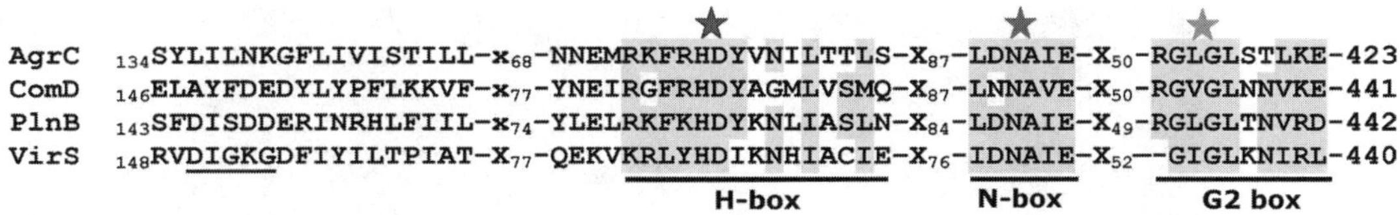

FIGURE 23.1 Sequence alignment of VirS with related sensors AgrC, ComD, and PlnB. Regions of sequence similarity are shaded in gray. The conserved histidine kinse motifs, the H, N, and G2 boxes, are indicated by the black lines under the sequence. The site of phosphorylation is indicated by the first star, the NA motif by the second star, and the GxGL box by the third star. The functional VirS glycine rich region is underlined.

have six transmembrane helices [12,17]. Interestingly, the AgrC/ComD/PlnB sensors do not appear to contain a typical or identifiable G1 box [16]. In the original analysis of VirS, a putative G1 box was identified as the DxGxG at position 154-158 [12] (Figure 23.1). Site-directed mutagenesis of the central glutamate residue (Glu-156) in this region indicated that this residue had an essential role in VirS function [17]. However, the putative G1 box appears to be located in a cytoplasmic loop between the fourth and fifth transmembrane segments [12], and in the alignment is located in a nonconserved region in these proteins [16]. This unusual location would limit the ability of these residues to interact with other conserved motifs, which, combined with the lack of conservation noted in the related proteins, argues against these residues comprising a G1 box.

23.2.2.2 The N-Terminal Domain of VirR

Sequence alignment of the N-terminal domain of VirR ($VirR_N$) with other response regulators revealed 16 to 25% amino acid sequence identity [12]. By comparison with the highly conserved N-terminal domains of other response regulators, potentially functional amino acids were identified. The D57 residue is the putative site of phosphorylation and the E8, D9, and K105 residues align to a conserved acidic pair and an invariant lysine, respectively. These residues should form the basic structure of a conserved phosphoacceptor pocket. The N-terminal receiver domain of VirR has been modeled with the program MODELLER [19] using the *E. coli* PhoB regulatory domain [20] as the template structure. Analysis of the model (Figure 23.2) demonstrated that $VirR_N$ shares the common doubly wound α/β structure that is characteristic of response regulator receiver domains. The region that corresponds to the active site of phosphorylation forms a pocket that contains the conserved residues, E8, D9, and D57, with the side-chain of K105 reaching back toward the pocket.

23.2.2.3 The C-Terminal Domain of VirR

Comparative sequence analysis of the 236 amino acid VirR protein revealed a C-terminal domain ($VirR_C$) that did not have close sequence similarity to any protein of known structure. Like most response regulators, VirR is a transcriptional activator. It does not, however, appear to contain the canonical helix-turn-helix (HTH) motif that is often associated with DNA binding proteins [21]. Consistent with the VirS sequence alignments, VirR has sequence identity with the response regulators AgrA, ComE, and PlnC [16].

In vivo and *in vitro* studies carried out using a deletion derivative, *virR*Δ2-125, which effectively removed the N-terminal receiver domain, indicated that the truncated VirRΔ2-125 protein was functional, providing evidence that the $VirR_C$ effector domain is capable of acting as an independent DNA binding domain [22]. Database searches identified two highly conserved regions in this domain: the FxRxHrS and the SKHR motifs [23]. Random mutagenesis identified one further residue, a functionally important methionine (M172), which is located outside the conserved FxRxHrS motif [22]. We have carried out extensive site-directed mutagenesis of these two motifs and have shown that amino acids R186, H188, and S190, which are key residues within the FxRxHrS

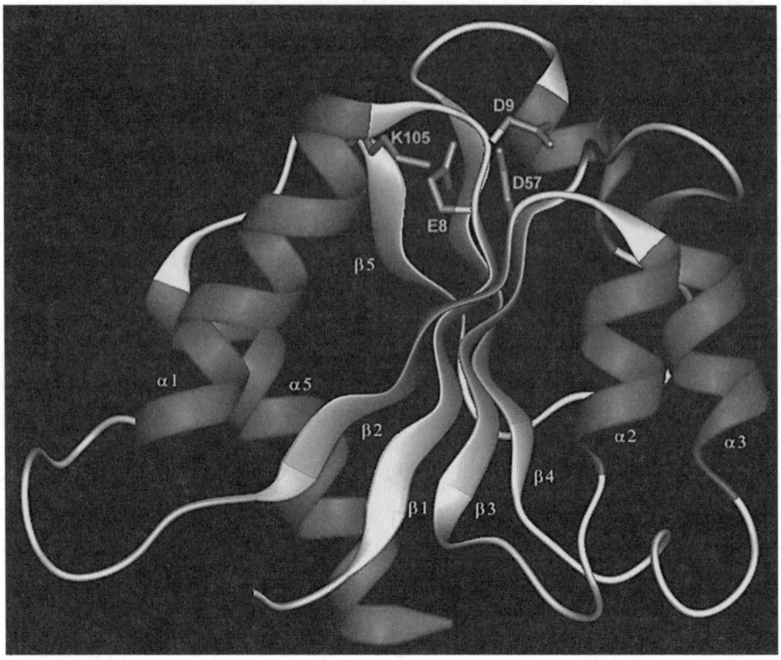

FIGURE 23.2 Proposed structure of $VirR_N$. The β-strands are shown as the ribbons in the middle, while the α-helices are represented by the outer coils. The side-chains of the proposed invariant residues E8, D9, D57, and K105 are shown at the top. This model was generated using the computer program MODELLER.

motif, and S216, K217, H218, and R219, which comprise the SKHR motif, are required for biological activity, since even conservative substitutions at these positions result in essentially nonfunctional proteins [22,23]. Seven mutant VirR proteins, including the random mutant M172V, were purified and shown to have a lower binding affinity for the VirR binding site, compared to the wild-type protein. It is concluded that the two VirR motifs are essential for proper DNA-protein interactions, and that the C-terminal domain constitutes a novel DNA binding domain [22,23].

An *in silico* study identified VirR as belonging to the newly described LytTR family of proteins [24], supporting the biological evidence obtained from our VirR studies. This study also identified the two conserved motifs and showed significant conservation of this region of the protein. These two regions were comparatively analyzed, but no conclusions could be drawn regarding their structural role and their involvement in the DNA binding domain [24].

Analysis of $VirR_C$ using a fold prediction server revealed closest similarity to the transcriptional regulator Mbp1, a CAP-like HTH DNA binding protein from *Saccharomyces cerevisiae* [23]. Comparative alignment of the C-terminal domains of VirR and Mbp1 indicated that the conserved 'RxHr' residues from the LytTR family, are predicted to align with DNA binding residues KxKR on helix-B of Mbp1. We have therefore proposed [23] that $VirR_C$ may contain a variation of the winged helix-turn-helix DNA binding domain found in Mbp1.

23.2.3 Direct Regulation of Virulence Associated Genes by VirS/R

Regulation of perfringolysin O production in *C. perfringens* was initially thought to be only partially controlled by VirR. The cloning and sequencing of the perfringolysin O structural gene, *pfoA,* led to the identification of a putative regulatory gene, *pfoR* [25]. The *pfoR* gene is located 591 bp upstream of *pfoA* and is transcribed in the same direction as *pfoA*, but the genes do not comprise an operon. The putative PfoR protein was reported to contain a HTH sequence motif [25] yet, when using the Dodd & Egan weight matrix [26], no HTH is detected. In a study carried out in *E. coli*, deletion of all or part of the *pfoR* gene was found to reduce perfringolysin O production [25]. This finding led to the hypothesis that PfoR had regulatory effects on *pfoA* expression. Subsequent studies in *C. perfringens*, however, have shown that deletion of all or part of the *pfoR* gene has no effect on perfringolysin O production [27]. More recent database searches suggest that PfoR is a transmembrane protein that has similarity to domains (COG1299) of a fructose-specific enzyme II component of a phosphoenolpyruvate dependent-phosphotransferase system that catalyzes the uptake and phosphorylation of fructose.

Sequence analysis of the *pfoA* gene region identified two imperfect direct repeats upstream of the *pfoA* promoter [14]. These repeats could not be identified in the upstream regions of the *colA* and *plc* genes [14]. Purification of a His-tagged VirR protein [28] and subsequent *in vitro* DNA binding analysis showed that the purified VirR protein was capable of binding to a 52 bp region of DNA that contained these two direct repeats [28]. In gel mobility shift experiments, two major protein-DNA complexes (CI and CII) were seen, with some higher order complexes also being observed. When nucleotides in these direct repeats were altered, VirR binding was severely reduced, indicating the critical role of the repeats, now referred to as VirR boxes, in VirR binding. VirR binds to the VirR boxes in a noncooperative manner; mutation of one VirR box does not affect binding to the other VirR box [28]. More recent work [72] indicates that both VirR boxes are required for biological function and that the spatial arrangement and the position of the VirR boxes in relation to the promoter is important for VirR-mediated activation of *pfoA*. This study also implicates other interactions, particularly with RNA polymerase, as being necessary for VirR-activated transcription of its target genes.

Gel mobility shift analysis of the upstream and internal regions of the *plc* and *colA* genes showed that purified VirR does not bind specifically to these regions [28]. It is clear from this work, and from previous observations [12–14,29], that regulation of α-toxin and collagenase production, unlike that of perfringolysin O, is regulated in an indirect manner by VirS/R and is not completely dependent upon the VirS/R system. Expression from the *plc* promoter has been previously shown to be temperature-dependent [29], while regulation of *colA* is partially dependent on VirR, but other as-yet-unknown regulatory factors must also be influencing its transcription. Recently, preliminary evidence suggested that the *virX* gene, which appears to encode a regulatory RNA molecule, may influence the expression of both *plc* and *colA* [30]. Further studies are required to confirm these findings.

The complete genome sequence of *C. perfringens* strain 13 was released in early 2002 [31]. Analysis of the genome identified five promoter regions that contained directly repeated variants of the VirR boxes (Figure 23.3). In addition to the *pfoA* gene, VirR boxes were located of two open reading frames of unknown function (CPE0845 & CPE0920), CPE0846, whose product has

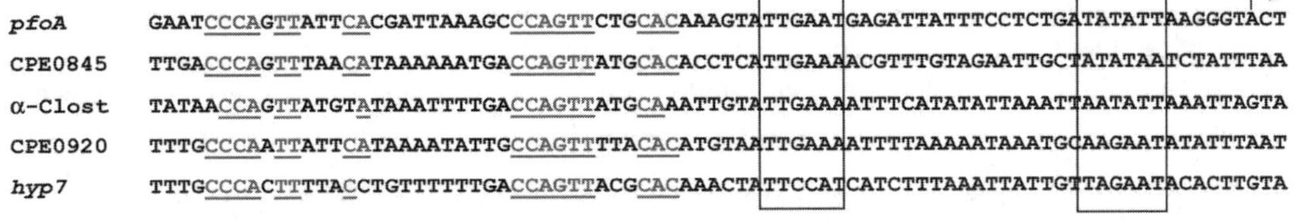

FIGURE 23.3 VirR boxes in the *C. perfringens* genome. VirR boxes are shown in gray and are underlined. The putative –35 and –10 sequences are boxed. The consensus VirR box motif is shown on the bottom. This figure is based on Figure 4 from Shimizu et al. [31].

sequence identity with the cysteine protease, clostripain, and a gene designated *hyp7* [31]. Recent experiments showed that purified VirR was capable of binding to all five of these VirR box regions and provided evidence that these genes were all directly regulated by VirR [72].

The *hyp7* gene was initially identified using a differential display method that aimed to identify VirS/R-regulated genes [32]. A plasmid library was constructed from chromosomal DNA from strain 13 and the plasmids were hybridized with cDNA probes prepared from RNA from strain 13 and its *virR* mutant, TS133. Hybridization analysis was used to identify which plasmids carried gene regions that were VirS/R regulated. A plasmid that carried the VirR-activated *hyp7* gene was identified and sequenced. Importantly, the *hyp7* gene product was shown to activate the transcription of both *plc* and *colA* [32]. Further investigations revealed that the active product was not the putative *hyp7*-encoded protein but a regulatory RNA molecule, named VR-RNA, that is encoded within the *hyp7* gene [33].

The 386 nt VR-RNA molecule is transcribed from a promoter that carries two directly repeated VirR boxes and is, therefore, likely to be directly activated by VirR. The transcriptional activation of *plc* and *colA* is subsequently mediated by VR-RNA [33] by an unknown mechanism that probably involves the formation of a VR-RNA-protein complex. Deletion analysis of VR-RNA revealed that the 3′ end of the molecule is required for transcriptional activation. This approximately 100 nt region of the molecule contains a putative rho-independent terminator, the presence of which is unexpected within the active region of a molecule involved in transcriptional activation.

VirR boxes were also identified upstream of the clostripain homologue, CPE0846. Clostripain is a trypsin-like cysteine protease produced by *C. histolyticum* that has stringent specificity for the carboxy-terminal side of arginine residues [34]. The CPE0846 gene product [31] has the putative active site (FDA<u>C</u>LMG) of clostripain [35] and has been, therefore, designated as α-clostripain [15]. Like the *C. histolyticum* enzyme, α-clostripain is susceptible to the trypsin inhibitors benzamidine, antipain, leupeptin, and is partially inhibited by EDTA [15]. Unusually, both proteases are resistant to the common cysteine protease inhibitor E-64 [15,36].

The identification of VirR boxes upstream of the α-clostripain gene [31], together with proteomic analysis [15], has confirmed earlier evidence for a VirR-regulated protease gene [12]. Northern blots of RNA from the *virR* mutant TS133 showed that no transcript was produced from the α-clostripain gene in this mutant, indicating that this gene is positively regulated at the transcriptional level by VirR [15]. In the same study, using protease inhibitors, it was shown that α-clostripain could degrade several secreted proteins. It was concluded that several proteins that had previously been identified as VirS/R-regulated, were actually being degraded by this VirS/R-regulated protease, emphasizing that secondary effects can compromise the interpretation of global proteomic studies [15].

23.2.4 Indirect Regulation by VirS/R

The VirS/R system indirectly regulates other genes by regulating the expression of VR-RNA. Several of these genes were identified by differential display [32]. With the exception of the *hyp7*/VR-RNA gene region, these VirS/R-dependent genes did not contain upstream VirR boxes, indicating that VirR did not directly regulate their transcription. These genes comprised the positively regulated *ptp* and *cpd* genes and a negatively regulated operon that contained four putative genes.

The product of the *ptp* gene has sequence similarity to eukaryotic protein tyrosine phosphatases (PTPases) [32]. PTPases are also found in several Gram-negative bacterial pathogens, including *Yersinia* spp., *Shigella* spp., *Escherichia coli,* and *Pseudomonas aeruginosa*. As tyrosine phosphorylation does not commonly occur in bacteria, it is likely that the proteins of these potentially intracellular pathogens have specifically evolved to modulate host cellular functions. In these bacteria, a novel type III protein secretion system has evolved to deliver effector proteins to the host intracellular niche [37], and PTPases have been implicated in this secretory pathway. Secretion of effectors initiates complex signal transduction events aimed at securing the microbe's ability to multiply and move to a new host cell [37]. Since, *C. perfringens* is a Gram-positive, extracellular pathogen, it is unlikely to use a PTPase in a similar manner to these bacteria. Perhaps this phosphatase reverses tyrosine phosphorylation within an internal signal transduction system. True prokaryotic tyrosine phosphorylation/dephosphorylation was first reported in the cyanobacterium, *Nostoc commune* [38]. The *C. perfringens* PTPase homologue has 26% sequence identity to the *N. commune* PTPase, IphP. The region of strongest identity surrounds the sequence characteristic of the PTPase active site, HCxAGxxR.

Northern blots have been used to determine *ptp* transcript levels in growing *C. perfringens* cultures [32]. Samples taken after two hours of growth showed that in the *virR* mutant, TS133, the *ptp* transcript was reduced in comparison to the wild-type strain, but was still present. By three hours, however, there was no discernible difference in transcript levels. Subsequent time points showed little transcript in either the wild-type or TS133 strains. These results indicate that the *ptp* gene is primarily transcribed in mid-exponential phase, in a partially VirS/R-dependent manner. In a more recent study [33], it was concluded that this VirS/R-dependence was VR-RNA-mediated, although these data require further confirmation.

A phosphodiesterase gene, *cpd*, was initially identified within the same cloned DNA fragment as the *ptp* gene [32]. The *cpd* gene product had similarity to CpdB from *E. coli* and the multifunctional phosphodiesterase, YfkN from *B. subtilis*. This gene was designated *cpdB,* as it was located immediately downstream as another apparent *cpd* gene, *cpdA* [31]. The *cpdA* gene is some 100 bp larger than *cpdB,* but has nearly 85% identity. Much like the upstream *ptp* gene, Northern analysis revealed that *cpdB* was only partially VirS/R-regulated at mid-exponential growth phase [32]. By contrast, proteomic analysis [15] indicated that *cpdA* was positively regulated by the VirS/R system but did not detect the regulatory effects on *cpdB* that were reported in the previous study [32]. It would appear that we do not have sufficient information to determine the precise role that VirR or VR-RNA plays in *cpd* regulation. It appears that there is partial transcriptional regulation of *cpdB,* but not enough to be detected by proteomics.

The final region that was identified by differential display [32] was a 4.9 kb mRNA transcript that encoded the *metB*, *cysK,* and *luxS* genes. The *metB* and *cysK* genes appeared to be involved in cysteine biosynthesis as their products had highest similarity to a cysteine synthase from *B. subtilis* and a cystathionine γ-synthase from *Helicobacter pylori,* respectively [32]. Chromosomal walking of the 5′ region identified a fourth gene, *ycgJ*, which had similarity to a gene of unknown function on the *B. subtilis* genome [39]. More recent searches have indicated that the YcgJ product has highest similarity to a S-adenosylmethionine (SAM)-dependent methyltransferase from *C. acetobutylicum*. Northern analysis revealed that all four genes were transcribed as an operon that was indirectly regulated by VirS/R [32,39], using VR-RNA as an intermediary repressor [33]. The downregulation of the two housekeeping genes, *metB* and *cysK,* was the first indication that the VirS/R system could regulate genes that were not involved in virulence [32].

Recent studies have indicated that VR-RNA is involved in the regulation of the β2-toxin gene (*cpb2*) and a possible collagen adhesin gene (*cna*) [40]. These genes were found to be differentially regulated, with VR-RNA activating the *cpb2* gene while repressing the *cna* gene. Interestingly, these genes are encoded on the native *C. perfringens* plasmid pCP13, indicating that the VirS/R system can control the expression of genes located on both the chromosome and plasmids [40].

23.2.5 The Role of Quorum Sensing

The identification of the *luxS* gene revealed that another regulatory network may operate in *C. perfringens*. Regulation of bacterial gene expression in response to cell density through cell-to-cell signaling, or quorum sensing, has been reported in a diverse range of bacteria. The LuxS protein from various bacterial species is involved in SAM metabolism and has been found to be required for the production of the universal cell-to-cell signaling molecule, the diffusable autoinducer 2 (AI-2) [41]. In some pathogenic bacteria, *luxS*-dependent signaling systems play an important role in the regulation of virulence factors, for example, hemolysin and cysteine protease production in *Streptococcus pyogenes* [42]. In a preliminary study, the *C. perfringens luxS* gene product was shown to be taken up by a *luxS* mutant and to enhance *pfoA* transcription levels [43]. The mechanism by which AI-2 regulates perfringolysin O activity in these strains remains unknown. It is possible that AI-2 is a signal to other regulatory factors, but the hemolytic nature of the *luxS* mutant indicates that AI-2 is not the extracellular signal responsible for VirS/R activation. In addition, autoinducers from Gram-positive bacteria are generally oligopeptides. By contrast, AI-2 has recently been shown to be a novel furanosyl borate diester [41]. Finally, in *C. perfringens,* the operon that contains the *luxS* gene appears to be principally involved in SAM modification and cysteine metabolism. Similarly, in *Borrelia burgdorferi, luxS* was found in an operon that was primarily involved in SAM metabolism, by converting toxic by-products to nontoxic intermediates and homocysteine [44]. It is possible that in *C. perfringens,* the primary role of the LuxS protein is the metabolism of SAM and that the production of AI-2 is secondary to this function.

The activation of VirS/R remains a key question for this regulatory pathway. In Gram-negative bacteria, autoinducers are known to be derivatives of N-acylhomoserine lactones that freely diffuse in and out of the cells and interact directly with intracellular regulatory proteins [45]. Autoinducers involved in quorum sensing in Gram-positive bacteria are generally small unique peptides that are actively secreted and subsequently function as ligands for signal receptors such as sensor histidine kinases [46]. For example, in the *agr* system of *Staphylococcus epidermidis,* the autoinducing peptide is only eight residues in length [47]. In systems similar to VirS/R, the gene encoding the precursor of the autoinducing peptide has been located immediately upstream of the sensor kinase gene. Sequence searches upstream and downstream of VirS have not identified any potential genes that may encode such a peptide.

Earlier studies [9,10,48] did demonstrate the production of a diffusible substance that could stimulate perfringolysin O production. Chemical mutagenesis was used to derive *C. perfringens* pleiotropic mutants that were unable to produce several extracellular toxins, including perfringolysin O [9]. It was found that when certain combinations of these mutants were cross-streaked on sheep blood agar, hemolysis could be restored. The mutants were classified into two groups (*a* and *b*) by their ability to complement in this way. It was found that hemolysis (complementation) would only occur when a group *b* mutant was cross-streaked with a group *a* mutant. It was concluded that the group *a* mutant and the parent *C. perfringens* strain were able to release a signal substance or substances that stimulated the production or release of perfringolysin O by group *b* cells. This diffusible product was termed "substance A" [9].

Substance A has been shown to be a very small molecule (<2 kDa) that is stable and active at pH 5.0 and 0°C. It is produced in exponential growth phase, but not in stationary phase [48]. This elusive molecule has been thought to be a small peptide that can function effectively as an autoinducer of the VirS/R system [49]. However, this extracellular signaling substance remains an enigma. The small peptide is highly labile, has never been purified [50], and even with genomic information, its coding region has not been identified. Its existence is not doubted, but its molecular nature and role in regulation remain to be determined.

The model of the VirS/R pathway first proposed [12] has been revised as each major piece of information has been published [28,33]. The most recent model (Figure 23.4) [33] implicates VR-RNA and an unknown RNA binding protein in the activation of all VirR-regulated genes, except *pfoA* and the other four genes that have VirR boxes, which are directly regulated by VirR. It is probable that the activation of the α- and κ-toxin structural genes involves an RNA binding protein and VR-RNA; however, the mechanism of regulation of the *ptp* and *cpd* genes remains unresolved. It is clear that there are two major pathways by which VirS/R regulates transcription. The first involves direct transcriptional activation resulting from the binding of VirR to the VirR boxes. The second involves the VirR-dependent VR-RNA activation pathway (Figure 23.4).

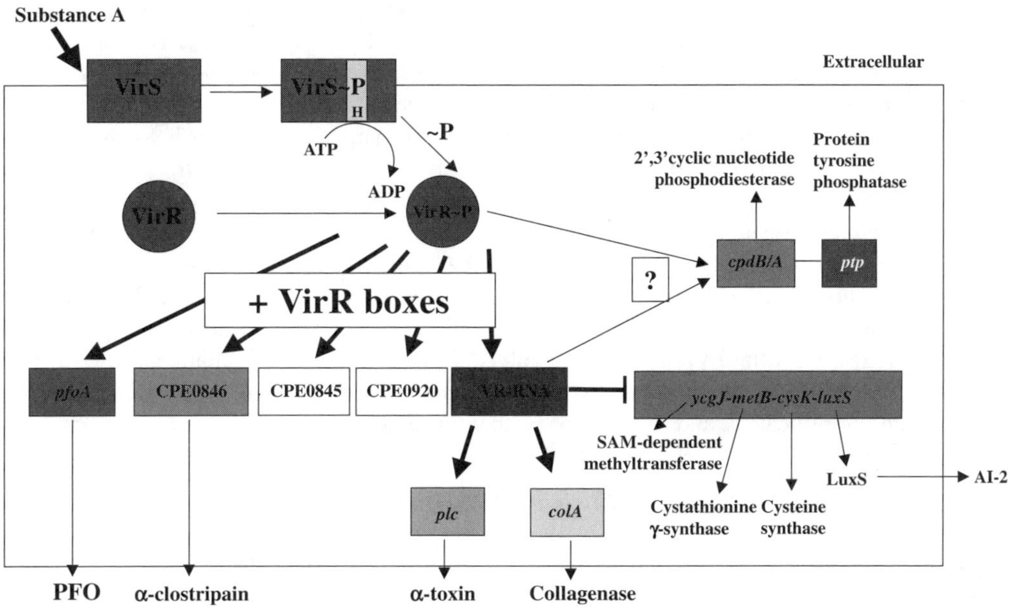

FIGURE 23.4 A model of the VirS/R regulatory pathway. Activation is shown by an arrow — a thick arrow for direct activation and thin arrow for partial control. Negative regulation is represented by a solid line. The VirS sensor kinase detects an unknown external signal, probably substance A. This stimulates autophosphorylation at His-255 in the cytoplasmic region of the protein. It is postulated that VirR then catalyzes the transfer of the phosphate group to the conserved Asp-57 in its N-terminal receiver domain. Phosphorylated VirR can then bind upstream of the *pfoA* gene and directly activate the production of perfringolysin O (PFO). Phosphorylated VirR also binds upstream of the sequences encoding CPE0846, producing the cysteine protease α-clostripain, VR-RNA, producing a regulatory RNA molecule and the hypothetical CPE0845 and CPE0920 ORFs. The regulatory VR-RNA molecule has been shown to activate the expression of *plc*, *colA* and to downregulate two cysteine biosynthesis genes (*metK* and *cysK*) and the *luxS* gene. There appears to be weak regulation of the *ptp* and the *cpd* genes, probably mediated through VR-RNA.

23.3 TWO-COMPONENT SYSTEMS IDENTIFIED FROM GENOME PROJECTS

In recent years, the complete sequences of three clostridial genomes have been published and several additional genomes are at various stages of completion. Analysis of these sequences has revealed that two-component signal transduction sequences are common in the clostridia, but has not revealed much information on their functional role. Examination of the sequence of *C. perfringens* strain 13 indicated that the *virRS* operon was one of 18 pairs of genetically linked two-component genes [31]. The products of one of these genome pairs, CPE0518 and CPE0519, have significant similarity to VirR and VirS, respectively, although their biological role is unknown. As discussed, in addition to *pfoA,* analysis of the genome sequence revealed four other genes with directly repeated VirR boxes located upstream of their putative promoters.

The solvent-producing species *Clostridium acetobutylicum* was the first clostridium to be completely sequenced, but its two-component signal transduction systems were not analyzed in detail [51]. Apart from *C. perfringens*, the only other published clostridial genome is that of *Clostridium tetani* [52]. In *C. tetani*, the tetanus toxin gene is carried on a 74 kb plasmid that also encodes a two-component signal transduction system (CTP21 and CTP22) of unknown function. The sequences of *Clostridium botulinum* and *Clostridium difficile* have been elucidated but have not yet been published (http://www.sanger.ac.uk/Projects/Microbes/). Analysis of the *C. difficile* genome sequence indicates that it contains at least 30 pairs of two-component signal transduction genes, with some of their products having similarity to VirS and VirR (K. Farrow and J. Rood, unpublished data). None of these genes have been analyzed at the functional level. Other workers have analyzed a series of two-component signal transduction genes that are linked to genes that encode subfamily 9 ABC transport systems in low G+C Gram-positive bacteria such as *Bacillus* and *Clostridium* [53]. This type of linkage is not found in other bacteria. In *B. subtilis,* there are five such systems, which all encode response regulators of the OmpR family and have a similar but not identical genetic and transcriptional organization. Six such linked gene clusters were detected in the *C. acetobutylicum* genome and 13 in *C. difficile*. The response regulators in most of these systems are clustered into two groups when subjected to phylogenetic analysis, and all of the related sensor histidine kinases appear to have only two N-terminal transmembrane domains. These data suggest that these systems have evolved from a single two-component system via an initial gene duplication event, followed by the independent recruitment of ABC transporter genes and subsequent amplification and divergence of the entire gene system.

23.4 OTHER CLOSTRIDIAL TWO-COMPONENT SYSTEMS

23.4.1 The KdpD/E Two-Component Signal Transduction System of *C. acetobutylicum*

The *kdpD* and *kdpE* genes were identified serendipitously from a *C. acetobutylicum* genomic library using a probe that was originally designed to identify the *sigE* sigma factor gene [54]. In *E. coli*, KdpFABC comprises a high-affinity K^+ uptake system that is expressed when osmotic pressure is reduced [55,56]. The *C. acetobutylicum kdp* gene region consists of *orfZ* (or *kdpZ*), *orfY* (or *kdpY*) and *kdpABCXDE* [57]. The products of *kdpD* and *kdpE* show similarity to sensor histidine kinases and response regulators, respectively, with the highest similarity to their *E. coli* counterparts. *In vitro* transcription/translation experiments showed that both genes were functional [54,57], and transcriptional analysis showed that they were arranged in an operon, with the transcriptional start point located 236 bp upstream of the *kdpD* start codon [54].

Hydropathy analysis indicated that the 101 kDa KdpD protein from *C. acetobutylicum* contains four putative transmembrane regions. Alignment of the amino acid sequences of this protein and KdpD from *E. coli* demonstrated that not only were these proteins similar in size and structure, but that they had 40% identity in their N-terminal regions, compared to 28% identity in their C-terminal domains [54]. Given that the N-terminal domain of *E. coli* KdpD has been proposed to be involved in signal recognition [58], it was suggested that these proteins recognized the same input signal [54]. One difference was the presence of a histidine-containing phosphotransfer (Hpt) domain in the clostridial KdpD protein, which is absent in the *E. coli* sensor. This domain was found to be almost identical to the Hpt domain of the *E. coli* ArcB sensor histidine kinase [59,60].

The *C. acetobutylicum* KdpE protein has the highest similarity to KdpE of *E. coli*, but is also very similar to other response regulators such as PhoB, OmpR, and CreB [54]. In *E. coli*, the KdpE protein has been found to bind upstream of the *kdp* promoter [61]. A similar sequence was identified upstream of the *orfZY/kdpA* promoter in *C. acetobutylicum* [57]. Gel mobility shift experiments carried out with overexpressed, tagged KdpE demonstrated that it was able to bind specifically to this region. Delimitation of the binding site showed that at 26 bp, the binding sequence was slightly longer than its *E. coli* counterpart [60]. Furthermore, unlike the *E. coli* protein, only phosphorylated KdpE from *C. acetobutylicum* was able to bind to DNA [60,61]. Analysis of the *C. acetobutylicum* genome revealed three other potential KdpE binding sites, albeit with several mismatched bases. Gel mobility shift experiments demonstrated that phosphorylated clostridial KdpE was not able to bind to these sequences, leading to the conclusion that this protein did not regulate additional genes [60]. The high binding specificity of the clostridial KdpE protein was further emphasized when it was shown that phosphorylated KdpE was not able to bind to the *E. coli kdp* binding site, despite the sequence similarity. By contrast, the *E. coli* KdpE protein not only bound to the *E. coli* DNA target, but also bound weakly to the clostridial target.

In *C. acetobutylicum*, expression of the *kdp* genes is induced by a reduction in the potassium concentration in the media [57]. At noninducing high potassium concentrations, the only transcript observed is that derived from the constitutive but low level expression of the *kdpDE* operon from the *kdpD* promoter [54]. However, when the *kdp* genes are induced, a highly unstable transcript that is longer than the transcript derived from the *kdpD* promoter is detected. This mRNA molecule represents transcription of the *kdpZYABCXDE* genes, with the resultant readthrough into the *kdpDE* operon increasing the level of *kdpDE* transcript [57]. It was suggested that KdpD and KdpE might be involved in the regulation of other systems besides *kdp* [57,60]. If so, this cross-regulation is unlikely to occur through KdpE since, as already discussed, this protein only binds to the *kdp* promoter region. Alternatively, KdpD may play a role in cross-regulation, due to the presence of the HPt domain [60]. Such domains are regarded as highly suited for cross-communication between different proteins [1], but whether KdpD actually acts as a regulatory hub between different pathways remains to be determined.

23.4.2 The Response Regulator Spo0A in the Clostridia

Sporulation has been most extensively studied in *B. subtilis,* and the most important protein in this process is the transcription factor, Spo0A. This protein is the response regulator of a complex phosphorelay cascade; once phosphorylated, it either represses or activates the expression of most early stationary phase and sporulation genes [62,63]. Homologues of the *spo0A* gene have been identified in various *Bacillus* species as well as a number of clostridia, including *C. pasteurianum, C. thermoaceticum, C. acetobutylicum, C. butyricum, C. innocuum* [64], *C. beijerinckii,* and *C. cellulolyticum* [65,66]. Alignment of the available sequences of the *Bacillus* and *Clostridium* Spo0A homologues revealed that these proteins could be divided into three domains [64]. The first domain comprised the first 120 amino acids and was similar to the phosphoacceptor domain of CheY and other response regulators. Both the aspartate residue that acts as the site of phosphorylation, and the lysine residue that is believed to be involved in the activation process, were conserved. The next domain of between 30 and 40 residues was not conserved and was proposed to be a flexible linker region. The final 110 amino acids, which were presumed to comprise the effector domain, could be further divided into three regions. Regions I and III were proposed to be involved in interdomain signaling, dimerization of dimeric or monomeric forms of the protein, or interactions with RNA polymerase that initiate transcription. Region II was found to be the most highly conserved area, with a possible HTH motif located in this segment [64]. It is presumed to be the DNA binding domain.

Similarly, alignment of the deduced amino acid sequences of Spo0A from *C. beijerinckii* and *C. cellulolyticum* with other sequences in the database [66], including those from *B. subtilis* and *C. acetobutylicum* [67], showed that the greatest similarity was in the C-terminal DNA binding domains. In particular, a 12-amino-acid motif (SRVERAIRHAIE), which encompassed a putative DNA recognition helix, was found to be highly conserved. It was therefore suggested that Spo0A recognized the same target DNA binding site in all bacilli and clostridia [66].

In *B. subtilis*, phosphorylated Spo0A directly regulates genes that contain one or more of the target binding sites in the upstream regulatory regions [63]. These binding sites are termed "0A boxes" and consist of a 7 bp sequence, TGNCGAA [68]. Similar binding sites were located upstream of several genes in various clostridial species. Possible 0A boxes were identified in the intergenic region between *spo0A* and *spoIVB* homologues in *C. pasteurianum* and *C. thermoaceticum* [64], upstream of genes that are either activated (*adc, ctfAB, adhE bdhA,* and *bdhB*) or repressed (*ptb, butK*) at the start of the solventogenesis process in *C. beijerinckii* and *C. acetobutylicum* [65,67], and in the promoter regions of the *spo0A* gene [67] and the *adhE* gene of the *sol* operon in *C. acetobutylicum* [69]. Gel mobility shift analysis carried out with the purified C-terminal domain of the *C. beijerinckii* and *B. subtilis* Spo0A proteins showed that both proteins were able to bind to the various 0A box targets, which included the promoter regions of the *adc* gene (encodes acetoacetate decarboxylase), the *ptb* gene (encodes phosphotransbutyrylase), and the control *B. subtilis abrB* gene. However, only the 0A boxes from the *B. subtilis abrB* promoter were protected by both proteins in DNase I footprinting experiments [66]. In *B. subtilis*, phosphorylation of Spo0A increases the affinity of the protein for the 0A boxes [70]. However, even phosphorylation of a purified clostridial Spo0A protein failed to protect the 0A boxes in the *adc* and *ptb* promoters in DNase I footprinting experiments [66]. These differences in 0A box binding were thought to be due to the presence of divergent residues at the *adc* and *ptb* promoters [66], which differ from the invariant *B. subtilis* promoters [68]. Analysis of mutated 0A box sequences in the *adc* and *ptb* promoters showed that boxes that conformed with the canonical 0A sequence were able to be footprinted, while those that diverged resulted in a reduction in binding affinity [66]. While the footprinting experiments strongly suggested that Spo0A was unable to bind specifically to the clostridial *adc* and *ptb* promoters *in vitro, in vivo* expression experiments showed that the Spo0A binding sites were functional in the native host. It was demonstrated that Spo0A in *C. beijerinckii* positively regulated the expression of *adc* during solventogenesis, while postexponential phase expression from the *ptb* promoter was downregulated [66] and that in *C. acetobutylicum*, Spo0A activated the expression of the *sol* operon [69].

The identification of putative 0A boxes upstream of genes involved in solventogenesis indicates that Spo0A may directly regulate more than just sporulation. The multifunctional role of Spo0A was demonstrated by inactivation of the *spo0A* gene by homologous recombination in *C. acetobutylicum* [64,67] and *C. beijerinckii* [66]. The mutants showed changes in colony morphology, a slower growth rate, and altered ability to synthesize solvents. The *C. beijerinckii spo0A* mutant produced similar levels of acetate, butyrate, and lactate compared to the wild-type strain. The production of butanol, however, was severely reduced, and the ability to produce acetone was completely abolished, indicating that the mutant was unable to switch from acidogenic to solventogenic metabolism at the end of exponential growth [66]. By contrast, the *C. acetobutylicum spo0A* mutant accumulated high levels of acetate and butyrate [67,71], but was still able to synthesize butanol and acetone, albeit at low levels [67,71]. These results were substantiated by Northern blot analysis, where the transcript level of the solvent genes was found to be decreased, but not eliminated. It was, therefore, suggested that in addition to Spo0A, other regulators may be involved in the expression of solvent genes in *C. acetobutylicum* [67].

Further evidence for the multifunctional nature of Spo0A was demonstrated by transcriptional analysis of the *C. acetobutylicum spo0A* mutant using a DNA microarray [71]. Most of the Spo0A regulated genes fell into two distinct clusters. First, genes that were differentially upregulated in the *spo0A* mutant and would therefore be directly or indirectly repressed by Spo0A in the wild-type strain; these genes included *abrB*, genes involved in motility and chemotaxis, and two glycosyltransferase genes. The second cluster consisted of genes that were differentially downregulated in the mutant, that is, they would normally be activated either directly or indirectly by Spo0A in the wild-type strain during stationary phase. These genes included the major *sol* locus genes (*aad, ctfAB,* and *adc*), butanol dehydrogenase genes (*bdhAB*), a number of sporulation genes (*sigF, spoIIAB, spoVS*, and *spoVAD*), and a gene encoding a sensor histidine kinase that has high similarity to the *B. subtilis* KinA and SpoIIJ sensor kinases. Also included in this cluster were genes associated with electron transport and sugar metabolism [71].

23.5 CONCLUDING REMARKS

With the exception of the well-studied VirS/R system from *C. perfringens*, the KdpD/E system from *C. acetobutylicum,* and Spo0A from the solventogenic clostridia, little is known about the role that two-component signal transduction systems play in the regulation of either virulence genes or genes involved in cellular metabolism. By inference from studies in other bacteria, and by taking into account the large number of linked two-component systems that are present on the genomes of the clostridia that have been sequenced, it is highly likely that these systems play a key role in the regulation of such processes. With the widespread availability of the genome sequences and the advent of microarray technology, we look forward over the next few years to a dramatic expansion of our knowledge of the functional role that these systems play in the clostridia.

ACKNOWLEDGMENTS

The research carried out in this laboratory was supported by grants from the Australian National Health and Medical Research Council. Sheena McGowan was supported by an Australian Postgraduate Award. We thank Dr. James Whisstock for generating the model of $VirR_N$.

REFERENCES

1. Stock, A., Robinson, V., and Goudreau, P., Two-component signal transduction, *Annu. Rev. Biochem.,* 69, 183, 2000.
2. West, A. and Stock, A., Histidine kinases and response regulator proteins in two-component signaling systems, *Trends Biochem. Sci.,* 26, 369, 2001.
3. Kennelly, P. and Potts, M., Fancy meeting you here! A fresh look at "prokaryotic" protein phosphorylation, *J. Bacteriol.,* 178, 4759, 1996.
4. Thomason, P. and Kay, R., Eukaryotic signal transduction via histidine-aspartate phosphorelay, *J. Cell. Sci.,* 113, 3141, 2000.
5. Goudreau, P. and Stock, A., Signal transduction in bacteria: molecular mechanisms of stimulus-response coupling, *Curr. Opin. Microbiol.,* 1, 160, 1998.
6. Robinson, V., Buckler, D., and Stock, A., A tale of two components: a novel kinase and a regulatory switch, *Nature Struct. Biol.,* 7, 626, 2000.
7. Hoch, J.A., Two-component and phosphorelay signal transduction, *Curr. Opin. Microbiol.,* 3, 165, 2000.
8. Stock, J.B., Surette, M.G., Levit, M., and Park, P., Two-component signal transduction systems: structure-function relationships and mechanisms of catalysis, in *Two-Component Signal Transduction,* Hoch, J.A. and Silhavy, T.J., Eds., American Society for Microbiology, Washington, DC,1995, 25.
9. Higashi, Y., Chazono, M., Inoue, K., Yanagase, Y., Amano, T., and Shimada, K., Complementation of θ-toxinogenicity between mutants of two groups of *Clostridium perfringens, Biken J.,* 16, 1, 1973.
10. Imagawa, T., Tatsuki, T., Higashi, Y., and Amano, T., Complementation characteristics of newly isolated mutants from two groups of strains of *Clostridium perfringens, Biken J.,* 24, 13, 1981.
11. Rood, J.I. and Wilkinson, R.G., Isolation and characterization of *Clostridium perfringens* mutants altered in both hemagglutinin and sialidase production, *J. Bacteriol.,* 123, 419, 1975.
12. Lyristis, M., Bryant, A.E., Sloan, J., Awad, M.M., Nisbet, I.T., Stevens, D.L., and Rood, J.I., Identification and molecular analysis of a locus that regulates extracellular toxin production in *Clostridium perfringens, Mol. Microbiol.,* 12, 761, 1994.
13. Shimizu, T., Ba-Thein, W., Tamaki, M., and Hayashi, H., The *virR* gene, a member of a class of two-component response regulators, regulates the production of perfringolysin O, collagenase, and hemagglutinin in *Clostridium perfringens, J. Bacteriol.,* 176, 1616, 1994.
14. Ba-Thein, W., Lyristis, M., Ohtani, K., Nisbet, I.T., Hayashi, H., Rood, J.I., and Shimizu, T., The *virR/virS* locus regulates the transcription of genes encoding extracellular toxin production in *Clostridium perfringens, J. Bacteriol.,* 178, 2514, 1996.
15. Shimizu, T., Shima, K., Yoshino, K., Yonezawa, K., and Hayashi, H., Proteome and transcriptome analysis of the virulence genes regulated by the VirR/VirS system in *Clostridium perfringens, J. Bacteriol.,* 184, 2587, 2002.
16. Pestova, E.V., Havarstein, L.S., and Morrison, D.A., Regulation of competence for genetic transformation in *Streptococcus pneumoniae* by an auto-induced peptide pheromone and a two-component regulatory system, *Mol. Microbiol.,* 21, 853, 1996.

17. Cheung, J. and Rood, J., Glutamate residues in the putative transmembrane region are required for the function of the VirS sensor histidine kinase from *Clostridium perfringens*, *Microbiology,* 146, 517, 2000.
18. Lina, G., Jarraud, S., Ji, G., Greenland, T., Pedraza, A., Etienne, J., Novick, R. P., and Vandenesch, F., Transmembrane topology and histidine protein kinase activity of AgrC, the agr signal receptor in *Staphylococcus aureus*, *Mol. Microbiol.,* 28, 655, 1998.
19. Sali, A. and Blundell, T.L., Comparative protein modelling by satisfaction of spatial restraints, *J. Mol. Biol.,* 234, 779, 1993.
20. Solà, M., Gomis-Rüth, F. X., Serrano, L., Gonzalez, A., and Coll, M., Three-dimensional crystal structure of the transcription factor PhoB receiver domain, *J. Mol. Biol.,* 285, 675, 1999.
21. Martinez-Hackert, E. and Stock, A.M., Structural relationships in the OmpR family of winged-helix transcription factors, *J. Mol. Biol.,* 269, 301, 1997.
22. McGowan, S., O'Connor, J.R., Cheung, J.K., and Rood, J.I., The SKHR motif is required for biological function of the VirR response regulator from *Clostridium perfringens*, *J. Bacteriol.,* 185, 6205, 2003.
23. McGowan, S., Lucet, I.S., Cheung, J.K., Awad, M.M., Whisstock, J.C., and Rood, J.I., The FxRxHrS Motif: a conserved region essential for DNA binding of the VirR response regulator from *Clostridium perfringens*, *J. Mol. Biol.,* 322, 997, 2002.
24. Nikolskaya, A.N. and Galperin, M.Y., A novel type of conserved DNA-binding domain in the transcriptional regulators of the AlgR/AgrA/LytR family, *Nucl. Acids Res.,* 30, 2453, 2002.
25. Shimizu, T., Okabe, A., Minami, J., and Hayashi, H., An upstream regulatory sequence stimulates expression of the perfringolysin O gene of *Clostridium perfringens*, *Infect. Immun.,* 59, 137, 1991.
26. Dodd, I. and Egan, J.B., Improved detection of helix-turn-helix DNA binding motifs in protein sequences, *Nucl. Acids Res.,* 18, 5019, 1990.
27. Awad, M.M. and Rood, J.I., Perfringolysin O expression in *Clostridium perfringens* is independent of the upstream *pfoR* gene, *J. Bacteriol.,* 184, 2034, 2002.
28. Cheung, J.K. and Rood, J.I., The VirR response regulator from *Clostridium perfringens* binds independently to two imperfect direct repeats located upstream of the *pfoA* promoter, *J. Bacteriol.,* 182, 57, 2000.
29. Katayama, S., Matsushita, O., Jung, C.-M., Minami, J., and Okabe, A., Promoter upstream bent DNA activates the transcription of the *Clostridium perfringens* phospholipase C gene in a low temperature-dependent manner, *EMBO J.,* 18, 3442, 1999.
30. Ohtani, K., Bhowmik, S.K., Hayashi, H., and Shimizu, T., Identification of a novel locus that regulates expression of toxin genes in *Clostridium perfringens*, *FEMS Microbiol. Lett.,* 209, 113, 2002.
31. Shimizu, T., Ohtani, K., Hirakawa, H., Ohshima, K., Yamashita, A., Shiba, T., Ogasawara, N., Hattori, M., Kuhara, S., and Hayashi, H., Complete genome sequence of *Clostridium perfringens*, an anaerobic flesh-eater, *Proc. Natl. Acad. Sci. U.S.A.,* 99, 996, 2002.
32. Banu, S., Ohtani, K., Yaguchi, H., Swe, T., Cole, S.T., Hayashi, H., and Shimizu, T., Identification of novel VirR/VirS-regulated genes in *Clostridium perfringens*, *Mol. Microbiol.,* 35, 854, 2000.
33. Shimizu, T., Yaguchi, H., Ohtani, K., Banu, S., and Hayashi, H., Clostridial VirR/VirS regulon involves a regulatory RNA molecule for expression of toxins, *Mol. Microbiol.,* 43, 257, 2002.
34. Rao, M.B., Tanksale, A.M., Ghatge, M.S., and Deshpande, V.V., Molecular and biotechnological aspects of microbial proteases, *Microbiol. Mol. Biol. Rev.,* 62, 597, 1998.
35. Gilles, A.M., De Wolf, A., and Keil, B., Amino-acid sequences of the active-site sulfhydryl peptide and other thiol peptides from the cysteine proteinase alpha-clostripain, *Eur. J. Biochem.,* 130, 473, 1983.
36. Gilles, A.M., Imhoff, J.M., and Keil, B., alpha-Clostripain. Chemical characterization, activity, and thiol content of the highly active form of clostripain, *J. Biol. Chem.,* 254, 1462, 1979.
37. Galan, J.E. and Collmer, A., Type III secretion machines: bacterial devices for protein delivery into host cells, *Science,* 284, 1322, 1999.
38. Potts, M., Sun, H., Mockaitis, K., Kennelly, P.J., Reed, D., and Tonks, N.K., A protein-tyrosine/serine phosphatase encoded by the genome of the cyanobacterium *Nostoc commune* UTEX 584, *J. Biol. Chem.,* 268, 7632, 1993.
39. Ohtani, K., Takamura, H., Yaguchi, H., Hayashi, H., and Shimizu, T., Genetic analysis of the *ycgJ-metB-cysK-ygaG* operon negatively regulated by the VirR/VirS system in *Clostridium perfringens*, *Microbiol. Immunol.,* 44, 525, 2000.

40. Ohtani, K., Kawsar, H.I., Okumura, K., Hayashi, H., and Shimizu, T., The VirR/VirS regulatory cascade affects transcription of plasmid-encoded putative virulence genes in *Clostridium perfringens* strain 13, *FEMS Microbiol. Lett.,* 222, 137, 2003.
41. Xavier, K.B. and Bassler, B.L., LuxS quorum sensing: more than just a numbers game, *Curr. Opin. Microbiol.,* 6, 191, 2003.
42. Lyon, W.R., Madden, J.C., Levin, J.C., Stein, J.L., and Caparon, M.G., Mutation of *luxS* affects growth and virulence factor expression in *Streptococcus pyogenes*, *Mol. Microbiol.,* 42, 145, 2001.
43. Ohtani, K., Hayashi, H., and Shimizu, T., The *luxS* gene is involved in cell-cell signalling for toxin production in *Clostridium perfringens*, *Mol. Microbiol.,* 44, 171, 2002.
44. Schauder, S., Shokat, K., Surette, M.G., and Bassler, B.L., The LuxS family of bacterial autoinducers: biosynthesis of a novel quorum-sensing signal molecule, *Mol. Microbiol.,* 41, 463, 2001.
45. Miller, M.B. and Bassler, B.L., Quorum sensing in bacteria, *Annu. Rev. Microbiol.,* 55, 165, 2001.
46. Novick, R.P., Regulation of pathogenicity in *Staphylococcus aureus* by peptide-based density-sensing system, in *Cell-Cell Signaling in Bacteria*, Dunny, G.M. and Winans, S.C., Eds., ASM Press, Washington, DC, 1999, 129.
47. Otto, M., Sussmuth, R., Jung, G., and Gotz, F., Structure of the pheromone peptide of the *Staphylococcus epidermidis agr* system, *FEBS Letters,* 424, 89, 1998.
48. Imagawa, T. and Higashi, Y., An activity which restores theta toxin activity in some theta toxin-deficient mutants of *Clostridium perfringens*, *Microbiol. Immunol.,* 36, 523, 1992.
49. Rood, J.I., Virulence genes of *Clostridium perfringens*, *Annu. Rev. Microbiol.,* 52, 333, 1998.
50. Shimizu, T., Okabe, A., and Rood, J.I., Regulation of toxin production in *C. perfringens*, in *The Clostridia: Molecular Biology and Pathogenesis*, Rood, J I., McClane, B.A., Songer, J.G., and Titball, R.W. Academic Press, London, 1997, 451.
51. Nölling, J., Breton, G., Omelchenko, M.V., Makarova, K.S., Zeng, Q., Gibson, R., Lee, H.M., Dubois, J., Qiu, D., Hitti, J., Wolf, Y.I., Tatusov, R.L., Sabathe, F., Doucette-Stamm, L., Soucaille, P., Daly, M.J., Bennett, G.N., Koonin, E.V., and Smith, D.R., Genome sequence and comparative analysis of the solvent-producing bacterium *Clostridium acetobutylicum*, *J. Bacteriol.,* 183, 4823, 2001.
52. Brüggemann, H., Bäumer, S., Fricke, W.F., Wiezer, A., Liesegang, H., Decker, I., Herzberg, C., Martinez-Arias, R., Merkl, R., Henne, A., and Gottschalk, G., The genome sequence of *Clostridium tetani*, the causative agent of tetanus disease, *Proc. Natl. Acad. Sci. U.S.A.,* 100, 1316, 2003.
53. Joseph, P., Fichant, G., Quentin, Y., and Denizot, F., Regulatory relationship of two-component and ABC transport systems and clustering of their genes in the *Bacillus/Clostridium* group, suggest a functional link between them, *J. Mol. Microbiol. Biotechnol.,* 4, 503, 2002.
54. Treuner-Lange, A. and Dürre, P., Molecular biological analysis of *kdpD/E*, a sensor histidine kinase/response regulator system in *Clostridium acetobutylicum*, *Anaerobe,* 2, 351, 1996.
55. Laimins, L.A., Rhoads, D.B., and Epstein, W., Osmotic control of *kdp* operon expression in *Escherichia coli*, *Proc. Natl. Acad. Sci. U.S.A.,* 78, 464, 1981.
56. Malli, R. and Epstein, W., Expression of the Kdp ATPase is consistent with regulation by turgor pressure, *J. Bacteriol.,* 180, 5102, 1998.
57. Treuner-Lange, A., Kuhn, A., and Dürre, P., The *kdp* system of *Clostridium acetobutylicum*: cloning, sequencing, and transcriptional regulation in response to potassium concentration, *J. Bacteriol.,* 179, 4501, 1997.
58. Zimmann, P., Puppe, W., and Altendorf, K., Membrane topology analysis of the sensor kinase KdpD of *Escherichia coli*, *J. Biol. Chem.,* 270, 28282, 1995.
59. Ishige, K., Nagasawa, S., Tokishita, S., and Mizuno, T., A novel device of bacterial signal transducers, *EMBO J.,* 13, 5195, 1994.
60. Behrens, M. and Dürre, P., KdpE of *Clostridium acetobutylicum* is a highly specific response regulator controlling only the expression of the *kdp* operon, *J. Mol. Microbiol. Biotechnol.,* 2, 45, 2000.
61. Sugiura, A., Nakashima, K., Tanaka, K., and T.M., Clarification of the structural and functional features of the osmoreglated *kdp* operon of *Escherichia coli*, *Mol. Microbiol.,* 6, 1769, 1992.
62. Sonenshein, A.L., Control of sporulation initiation in *Bacillus subtilis*, *Curr. Opin. Microbiol.,* 3, 561, 2000.
63. Hoch, J.A., *Spo0* genes, the phosphorelay, and the initiation of sporulation, in *Bacillus subtilis and Other Gram-Positive Bacteria. Biochemistry, Physiology, and Molecular Genetics*, Sonenshein, A.L., Hoch, J.A., and Losick, R., Eds., American Society for Microbiology, Washington, DC, 1993, 747.

64. Brown, D.P., Ganova-Raeva, L., Green, B.D., Wilkinson, S.R., Young, M., and Youngman, P., Characterization of *spo0A* homologues in diverse *Bacillus* and *Clostridium* species identifies a probable DNA-binding domain, *Mol. Microbiol.,* 14, 411, 1994.
65. Wilkinson, S.R., Young, D.I., Morris, J.G., and Young, M., Molecular genetics and the initiation of solventogenesis in *Clostridium beijerinckii* (formerly *Clostridium acetobutylicum*) NCINB 8052, *FEMS Microbiology Reviews,* 17, 275, 1995.
66. Ravagnani, A., Jennert, K.C.B., Steiner, S., Grünberg, R., Jefferies, J.R., Wilkinson, S.R., Young, D.I., Tidswell, E.C., Brown, D.P., Youngman, P., Morris, J.G., and Young, M., Spo0A directly controls the switch from acid to solvent production in solvent-forming clostridia, *Mol. Microbiol.,* 37, 1172, 2000.
67. Harris, L.M., Welker, N.E., and Papoutsakis, E.T., Northern, morphological, and fermentation analysis of Spo0A inactivation and overexpression in *Clostridium acetobutylicum* ATCC 824, *J. Bacteriol.,* 184, 3586, 2002.
68. Strauch, M., Webb, V., Spiegelman, G., and Hoch, J.A., The Spo0A protein of *Bacillus subtilis* is a repressor of the *abrB* gene, *Proc. Natl. Acad. Sci. U.S.A.,* 87, 1801, 1990.
69. Thormann, K., Feustel, L., Lorenz, K., Nakotte, S., and Dürre, P., Control of butanol formation in *Clostridium acetobutylicum* by transcriptional activation, *J. Bacteriol.,* 184, 1966, 2002.
70. Trach, K., Burbulys, D., Strauch, M., Wu, J.-J., Dhillon, N., Jonas, R., Hanstein, C., Kallio, P., Perego, M., Bird, T., Spiegelman, G., Fogher, C., and Hoch, J.A., Control of the initiation of sporulation in *Bacillus subtilis* by a phosphorelay, *Res. Microbiol.,* 142, 815, 1991.
71. Tomas, C.A., Alsaker, K.V., Bonarius, H.P.J., Hendriksen, W.T., Yang, H., Beamish, J.A., Paredes, C.J., and Papoutsakis, E.T., DNA array-based transcriptional analysis of asporogenous, nonsolventogenic *Clostridium acetobutylicum* strains SK01 and M5, *J. Bacteriol.,* 185, 4539, 2003.
72. Cheng, J.K., Dupuy, B., Deveson, D.S., and Rood, J.I., The spatial organization of the VirR boxes is critical for VirR-mediated expression of the perfringolysin O gene, *pfoA*, from *Clostridium perfringens*, *J. Bacteriol.,* 186, 3321–3330, 2004.

24 Comparative Genomic Analysis of Signal Transduction Proteins in Clostridia

Sabrina Doß, Carolin Gröger, Tina Knauber, David E. Whitworth, and Anke Treuner-Lange

CONTENTS

24.1 INTRODUCTION

Whole-genome analysis has energized research in the microbial field tremendously in the past decade. Since 1995, when The Institute for Genomic Research (TIGR) published the first complete genome sequence of *Haemophilus influenza* [1], 138 additional genome sequencing projects have been completed and 488 projects are underway (http://wit.integratedgenomics.com/GOLD/).

The methods of identifying genes in prokaryotic genomes have advanced in such a way that nearly all protein-coding regions can now be identified with confidence. Typically, about 40 to 60% of genes in a newly sequenced bacterial genome encode proteins that display significant sequence similarity to proteins or domains with known functions.

To date, three clostridial genome sequencing projects have been completed. The genome of *Clostridium acetobutylicum* was published in August 2001 [2], followed by the publications of the *C. perfringens* genome in January 2002 [3] and the *C. tetani* genome in February 2003 [4].

Clostridia are a diverse group of endospore-forming, Gram-positive, rod-shaped anaerobes. This group includes toxin-producing pathogens such as *C. difficile*, *C. botulinum*, *C. tetani,* and *C. perfringens* as well as terrestrial species such as *C. acetobutylicum*, *C. bifermentans*, and *C. propionicum*, which produce a variety of organic acids and solvents through fermentation.

The purpose of this chapter is to give an overview of the signal transduction proteins found in the completely annotated genomes of *C. acetobutylicum*, *C. perfringens,* and *C. tetani* E 88. Signal transduction systems represent a major mechanism by which microorganisms sense environmental stimuli and respond appropriately, usually through changes in gene expression. As the genome of a given organism reflects its lifestyle and physiology in addition to its phylogeny, significant differences in regard to signal transduction can be expected by a comparative genomic analysis of these three clostridial species.

24.2 THE GENOMES OF *C. ACETOBUTYLICUM, C. PERFRINGENS,* AND *C. TETANI*

The genomes and extrachromosomal plasmids of the three organisms differ significantly in size, which is reflected in the number of their annotated and assigned proteins (Table 24.1). A comparison of the *C. tetani* genome with the genomes of *C. perfringens* and the nonpathogenic *C. acetobutylicum* revealed 1506 open reading frames (ORFs), which were found in all three genomes [4]. Sequence similarity between orthologues from *C. acetobutylicum* and *B. subtilis* was greater than between *C. acetobutylicum* and Gram-negative bacteria [2]. Nevertheless, similarities were less pronounced than within the gamma-proteobacterial lineage [2].

The numbers of genes that have no homologues in the other two genomes differ significantly and do not correlate with genome size (1344 ORFs in the genome of *C. acetobutylicum*, 516 ORFs in the genome of *C. tetani*, and 228 ORFs in the genome of *C. perfringens*) [4]. The overlap of ORFs from just two of the three species is highest between *C. acetobutylicum* and *C. perfringens,* with 790 additional shared genes [4]. *C. tetani* and *C. perfringens* share 199 additional ORFs [4]. This group of potentially virulence-related genes encodes a high number of transport systems (see supporting information for Table 24.3 in reference [4]). The *C. tetani* genome has been shown to contain several genes encoding sodium ion-dependent systems, and a (V)-type ATPase, indicating active sodium ion-dependent bioenergetic processes in this organism [4]. A genomic comparison of *C. acetobutylicum* and *C. perfringens* has also shown that the former genome encodes several amino acid biosynthesis proteins that are not present in the latter, suggesting that *C. perfringens* depends on external amino acid sources [3].

TABLE 24.1
Comparison of Annotated Chromosomes and Plasmids

	C. acetobutylicum[a]	*C. perfringens*[b]	*C. tetani*[c]
Chromosome			
Size (bp)	3,940,880	3,031,430	2,799,251
GC content (%)	31[d]	28.6	28.6
rRNA operons	11	10	6
tRNA genes	73	96	54
Annotated proteins	3672	2660	2375
Proteins with assigned function	2861	2159	1957
Plasmid	pSol	pCP13	pE88
Size (bp)	192,000	54,310	74,082
GC content (%)	31[d]	25.5	24.5
Annotated proteins	178	63	61

[a] http://www.ncbi.nlm.nih.gov/genomes/chrom.cgi?gi=185&db=G.
[b] http://www.ncbi.nlm.nih.gov/genomes/framik.cgi?db=genome&gi=226.
[c] http://www.ncbi.nlm.nih.gov/genomes/framik.cgi?db=genome&gi=275.
[d] Analyzed by EMBOSS software (WWW.EMBOSS.ORG).

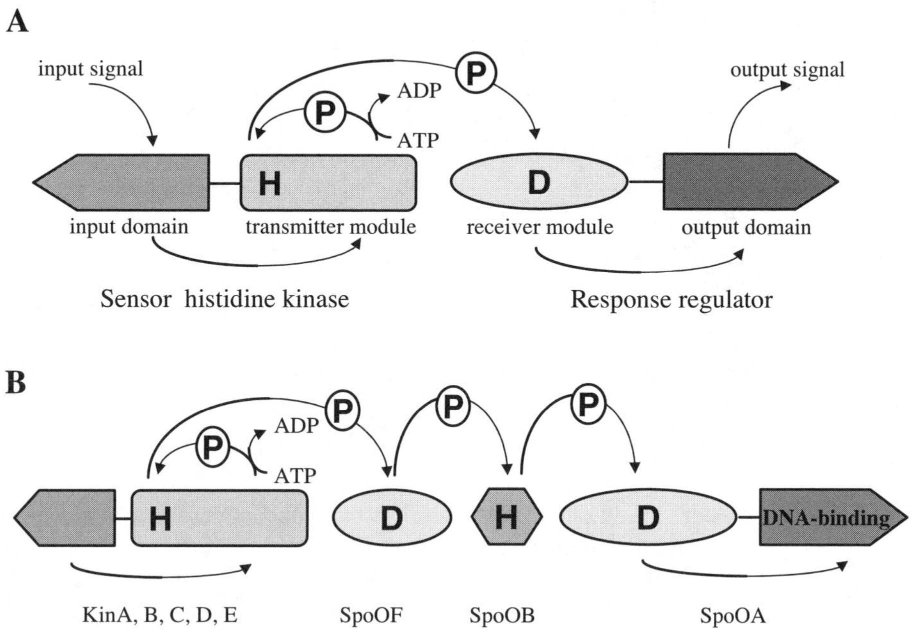

FIGURE 24.1 A typical two-component signal transduction system (A) and a multistep His-Asp phosphorelay (B). Upon receiving an appropriate input signal, the transmitter domain of a histidine kinase transmitter domain autophosphorylates a conserved histidine residue. The phosphate group is then passed to an aspartate residue on the partner response regulator receiver domain. Phosphorylation of the receiver domain leads to altered activity of the response regulator output domain (A). In the *B. subtilis* sporulation multistep phosphorelay (B), a phosphate group is sequentially transferred from the initial histidine kinases (KinA, B, C, D, E) to successive aspartate (Spo0F), histidine (Spo0B), and aspartate (Spo0A) residues. The final phosphoacceptor protein Spo0A has an output domain with DNA-binding activity, which is stimulated upon phosphorylation.

24.3 HIS-ASP SIGNAL TRANSDUCTION SYSTEMS

Bacteria are challenged with rapid and unexpected changes in nutrient and toxin levels, temperature, osmolarity, and acidity in their environment. In order to survive, cells need to respond to such changes by generating intracellular signals that lead to the modification of their metabolism, structure, and movements. Bacteria rely primarily on the histidine-aspartate signal-transduction system to adapt to such environmental changes. These systems consist of at least two different proteins, a histidine kinase and a response regulator. Such proteins comprise modular components called input sensing domains, output effector domains, transmitter and receiver domains (Figure 24.1A; [5,6]). As histidine kinases and response regulators need to interact in a highly specific manner and their genes are often coorganized, pairs of these proteins are called two-component systems. Nevertheless, these proteins can also be part of more complex signaling pathways known as multistep His-Asp phosphorelays, as illustrated by the example of the sporulation phosphorelay system of *Bacillus subtilis* (Figure 24.1B, [7–10]).

24.4 SER/THR SIGNAL TRANSDUCTION SYSTEMS

Reversible protein phosphorylation is a major mechanism for the control of cellular processes such as metabolism, transport, secretion, cell division, and signal transduction, and is predominantly performed by eukaryotic organisms. Nevertheless, there is a growing understanding of such systems

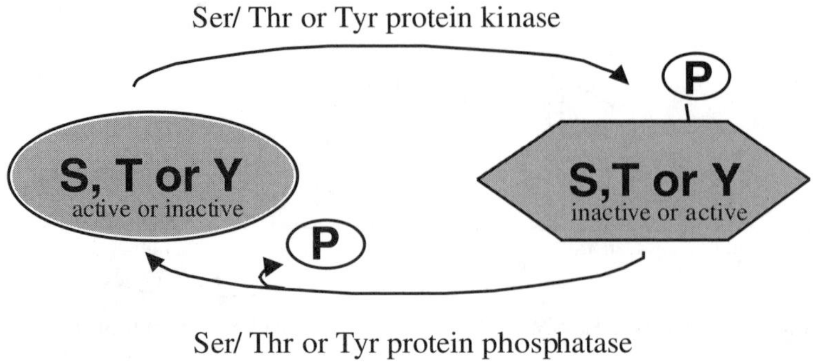

FIGURE 24.2 Reversible protein phosphorylation regulates the activity of a protein. A regulated protein can adopt two conformational states (one active, one inactive). Phosphorylation of the protein favors one conformation, while dephosphorylation favors the other conformation. Phosphorylation and dephosphorylation are mediated by regulated kinase and phosphatase activities, respectively.

in bacteria. These systems are based on the formation of stable phosphoesters of serine, threonine, or tyrosine. The phosphorylation or dephosphorylation of serine, threonine, and tyrosine residues triggers conformational changes in regulated proteins, thereby altering their biological properties. (Figure 24.2; [11,12]). The phosphorylation level and activity of a given protein is, therefore, a consequence of the activity of regulated protein kinases and opposing protein phosphatases acting on that protein.

24.5 GENOMIC COMPARISONS OF SIGNAL TRANSDUCTION SYSTEMS

In the clostridial genomes investigated, we found predominantly His-Asp signal transduction proteins and only a few examples of protein kinases and protein phosphatases (Table 24.2 and Table 24.3). The distribution of signal transduction genes through the clostridial genomes is depicted in Figure 24.3. The abbreviations used in Figure 24.3 correlate with the gene identifiers in Table 24.3.

Signal transduction proteins are more or less evenly distributed through the genomes, although there are some large regions that do not encode any signal transduction proteins. The clostridial plasmids encode very few signal transduction proteins, indicating that, generally, these proteins are crucial for survival of the organisms, whereas the plasmids might not be. Loss of the plasmid pSOL1 from *C. acetobutylicum*, which encodes proteins involved in solvent formation and sporulation, has been reported to coincide with loss of the capacity to produce acetone and butanol and to sporulate [13]. The pCP13 plasmid from *C. perfringens* predominantly encodes proteins of unknown function, proteins required for plasmid partitioning, and a few known virulence-associated proteins [3]. The pE88 plasmid from *C. tetani* is the only plasmid encoding a two-component system (CTP21/CTP22, TC26 in Figure 24.3). In addition, it harbors genes for the tetanus toxin (*tetX*) and its transcriptional regulator TetR. More than 50% of all pE88 ORFs are unique to *C. tetani* [4]. To our knowledge, none of these plasmids have been reported to be crucial for survival of the corresponding organism.

24.5.1 Clostridial Histidine Kinases

The highest number of histidine kinases was found in the genome of *C. acetobutylicum*. Compared to other microorganisms, the number of histidine kinases in *C. acetobutylicum* and *C. perfringens* is quite average in relation to genome size, whereas *C. tetani* possesses a relatively high number (Table 24.4).

Further details about two-components systems from *Escherichia coli, B. subtilis, Pseudomonas aeruginosa, Synechocystis* sp. PCC6803 and *Nostoc* sp. (PCC7120) can be found in references

TABLE 24.2
Number of Signal Transduction Proteins in the Clostridial Genomes

	C. acetobutylicum	*C. perfringens*	*C. tetani*
Chromosomally encoded:			
Histidine kinases:	37	27	31
Organized as two-component systems	28	18	25
As part of chemotaxis systems	2	—	1
Orphans	7[a]	9[a]	4[a]
Hybrid proteins with response regulator	—	—	1
Response regulators:	41	21	30
Organized as two-component systems	28	18	25
As part of chemotaxis systems	4	—	1
Orphans	9	3	3
Hybrid proteins with histidine kinase	—	—	1
Protein kinases[c]	5	5	4
Protein phosphatases[c]	7[b]	4	6
Plasmid encoded:			
Two-component systems	—	—	1
Orphan response regulators	1	—	—

[a] The genes for kinases CAC0317, CAC0323, CAC0863, CAC2730, CPE0951, CTC00159 are each organized with a response regulator gene; however, they are separated by 2 to 4 other genes.
[b] The protein CAC0035 has been annotated as an inactivated protein.
[c] The HPr proteins CAC1089, CPE1004, CTC01010 were counted as protein kinases as well as protein phosphatases (see text).

[14–18]. The genome of *Nostoc* sp. PCC7120 encodes a large number of histidine kinases and other signal transduction proteins, and it has been speculated that this number reflects a high degree of adaptability to changes in the environment [18]. Also, as *Synechocystis* sp. PCC 6803 lacks several of these proteins, they might be involved in heterocyst development or intercellular communication [18].

The 37 histidine or histidine-like kinases from *C. acetobutylicum* [2] belong to 6 different clusters of orthologous groups of proteins (COGs) [19]. The COG database is a tool for phylogenetic classification of proteins encoded in complete genomes. Each COG consists of orthologous proteins from at least three lineages. In 2001, the COG database contained 2791 COGs comprising 45,350 proteins from 30 bacterial genomes [19]. In 2003, the COG collection consisted of 138,458 proteins, which formed 4873 COGs and comprised 75% of the 185,505 (predicted) proteins encoded in 66 genomes of unicellular organisms [20]. This growing database and its defined clusters identify significant similarities and allow predictions of gene function.

The majority of clostridial histidine kinases belongs to COG 0642 and is predominantly transmembrane proteins. The membership of a given histidine kinase to one COG is predominantly based on the sequence and domain structure of the transmitter domain or because of another significant domain. For example, the protein KdpD is a prototypical histidine kinase like those of the 0642 cluster. However, its N-terminal input domain is so highly conserved that this protein and its orthologous proteins are grouped in their own cluster (COG 2205). Classification into different COGs can therefore show similarities with respect to input domains.

TABLE 24.3
Gene Identifiers of Signal Transduction Proteins

		C. acetobutylicum		*C. perfringens*		*C. tetani*	
TC1		CAC0080/CAC0081		CPE0119/CPE0120		CTC00189/CTC00191	
TC2		CAC0224/CAC0225		CPE0235/CPE0236		CTC00392/CTC00393	
TC3		CAC0239/CAC0240		CPE0457/CPE0458		CTC00411/CTC00412	
TC4		CAC0289/CAC0290		CPE0518/CPE0519		CTC00597/CTC00598	
TC5		CAC0371/CAC0372		CPE0531/CPE0532		CTC00628/CTC00628	
TC6		CAC0450/CAC0451		CPE0574/CPE0575		CTC00805/CTC00806	
TC7		CAC0524/CAC0525		CPE0840/CPE0841		CTC00848/CTC00849	
TC8		CAC0564/CAC0565		CPE0895/CPE0896		CTC00872/CTC00873	
TC9		CAC0653/CAC0654		CPE0927/CPE0928		CTC00934/CTC00935	
TC10		CAC0830/CAC0831		CPE1192/CPE1193		CTC00949/CTC00950	
TC11		CAC0864/CAC0865		CPE1500/CPE1501		CTC01130/CTC01131	
TC12		CAC1454/CAC1455		CPE1925/CPE1926		CTC01420/CTC01421	
TC13		CAC1506/CAC1507		CPE2087/CPE2088		CTC01481/CTC01482	
TC14		CAC1516/CAC1517		CPE2098/CPE2099		CTC01490/CTC01491	
TC15		CAC1552/CAC1553		CPE2239/CPE2240		CTC01523/CTC01524	
TC16		CAC1581/CAC1582		CPE2331/CPE2332		CTC01804/CTC01805	
TC17		CAC1700/CAC1701		CPE2363/CPE2364		CTC01818/CTC01819	
TC18		CAC2253/CAC2254		CPE2487/CPE2488		CTC01848/CTC01849	
TC19		CAC2434/CAC2435				CTC01905/CTC01906	
TC20		CAC2720/CAC2721				CTC01918/CTC01919	
TC21		CAC2759/CAC2760				CTC01951/CTC01953	
TC22		CAC2939/CAC2940				CTC01978/CTC01979	
TC23		CAC3219/CAC3220				CTC02155/CTC02156	
TC24		CAC3390/CAC3391				CTC02178/CTC02179	
TC25		CAC3429/CAC3430				CTC02322/CTC02323	
TC26		CAC3516/CAC3517				CTP22/CTP21	
TC27		CAC3662/CAC3663					
TC28		CAC3677/CAC3678					
H1	R1	CAC0317*	CAC0174	CPE0207	CPE0642	CTC00159*	CTC00163*
H2	R2	CAC0323*	CAC0321*	CPE0870	CPE0948*	CTC00456	CTC01569
H3	R3	CAC0437	CAC0585	CPE0951*	CPE1812	CTC00715	CTC01695
H4	R4	CAC0863*	CAC0860*	CPE0986		CTC00716	
H5	R5	CAC0903	CAC1233	CPE1316			
H6	R6	CAC2730*	CAC2071	CPE1512			
H7	R7	CAC3319	CAC2407	CPE1757			
	R8		CAC2735*	CPE1986			
	R9		CAC3286	CPE1987			
	R10		CAP0009				
HR1						CTC00084	
Che 1		CAC0117-CAC0122				CTC01729-CTC01735	
Che 2		CAC2218-CAC2222					
PP1		CAC0035		CPE1739		CTC00201 (SpoIIE)	
PP2		CAC0407		CPE1952		CTC00306	
PP3		CAC1727		CPE2473 (SpoIIE)		CTC00969	
PP4		CAC2787				CTC01224	

TABLE 24.3 (continued)
Gene Identifiers of Signal Transduction Proteins

	C. acetobutylicum	*C. perfringens*	*C. tetani*
PP5	CAC2881		CTC02045
PP6	CAC3205 (SpoIIE)		
PK1	CAC0404	CPE1334 (PrkA)	CTC01039 (PrkA)
PK2	CAC0579 (PrkA)	CPE1738 (PASTA)	CTC01225 (PASTA)
PK3	CAC1728 (PASTA)	CPE1990	CTC02248 (SpoIIAB)
PK4	CAC2307 (SpoIIAB)	CPE2049 (SpoIIAB)	
Hpr	CAC1089	CPE1004	CTC01010

TC = two component system; H = orphan histidine kinase; R = orphan response regulator; HR = hybrid protein comprising a histidine kinase and a response regulator domain; PK = protein kinase; PP = protein phosphatase. Hpr = bifunctional protein kinase/phosphatase; Che = chemotaxis gene clusters. * = proteins whose genes are in the vicinity of other orphan two-component system genes. The names of previously identified clostridial signal transduction proteins are given in parentheses.

The most abundant input domain found in the histidine kinases is HAMP domains, which are very common in prokaryotic signaling proteins and are a predicted structural element in homodimeric signaling proteins [21,22]. HAMP domains are located immediately inside the cytoplasmic membrane and are predicted to form two short amphipathic alpha-helices [23]. The HAMP domain is thought to be involved in transmitting conformational changes in periplasmic ligand-binding domains to cytoplasmic signaling domains [22]. As HAMP domains are not freely exchangeable between kinases, it has been concluded that the HAMP domain communicates to the transmitter domain that a specific ligand is bound [23]. This scenario might also apply to the clostridial histidine kinases, as more than two thirds of the HAMP-kinases contain transmembrane domains. Some kinases have been found to contain PAS domains. This domain can bind FAD or heme and is often found in histidine kinases that govern the activity of a transmitter domain in response to oxygen, carbon monoxide, or nitric oxide [21,24,25]. Only one clostridial histidine kinase was found to have a GAF domain (CAC2720). GAF domains have been shown to be involved in cGMP or photopigment binding in other proteins [21,26].

Phylogenetic studies and sequence comparisons have subdivided histidine kinases into clusters I–XI [27], families 1–11 [28], or types I–V [29]. Interested readers are encouraged to consult these reviews for further details. Here, we would like to explain an extended classification of Bilwes et al. [30] described by Koretke et al. [31]. Bilwes et al. [30] divided histidine kinases into two classes: class I contains all kinases except CheA and its homologues, which belong to class II. To show the differences between these two classes the graphical output of a SMART analysis of two clostridial histidine kinases is shown in Figure 24.4.

These two classes of proteins differ with respect to the domain within which the phosphorylated histidine residue resides. Class I proteins have transmitter domains, which comprise a HisKA domain in addition to a HATPase domain. The HisKA domain (dimerisation and phosphoacceptor domain of histidine kinases, also known as DHp) contains the phosphoacceptor histidine, while the HATPase domain has kinase activity. For class II proteins the HisKA domain is replaced by a HPT domain. In these cases, phosphate is transferred by the HATPase domain to a histidine residue within the HPT domain. Koretke et al. [31] suggested subdividing class I proteins into three subclasses based on the further path of phosphate transfer. Histidine kinases that transfer phosphate

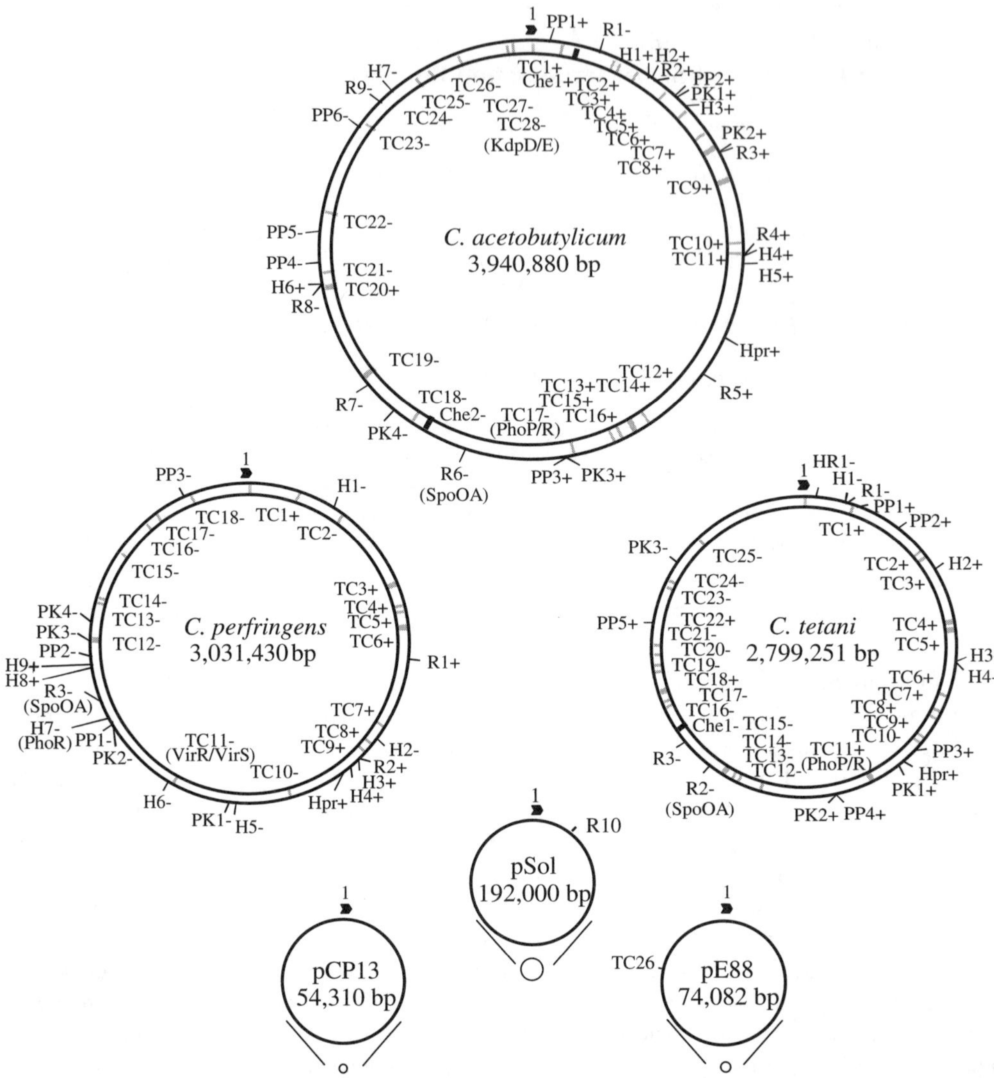

FIGURE 24.3 Distribution of signal transduction proteins on clostridial chromosomes and plasmids. See Table 24.3 for abbreviations of gene identifiers. +/– = strand of DNA on which the genes are encoded. Chromosomes have been drawn to the same scale, illustrating their differences in size. The sizes of the plasmids using the same scale are depicted under the corresponding plasmid map.

directly to a response regulator are grouped in subclass a. Kinases that transfer phosphate via a receiver domain and a second transmitter domain form subclass b. Subclass b includes kinases like KinA, B, C, D, and E, which are involved in the initiation of sporulation in *B. subtilis* and which act together with Spo0F, Spo0B, and Spo0A (Figure 24.1B, [7–9]). The second transmitter in this phosphorelay Spo0B (Figure 24.1B) is considered to be a degenerated transmitter [31]. Kinases that transfer phosphate via a receiver domain and then a HPT domain form subclass c. This class comprises a large number of hybrid kinases. Proteins of the latter two subclasses are parts of multistep phosphorelay systems.

We searched for further HPT domains in the genome of *C. acetobutylicum* to identify components of putative phosphorelay systems but could only find the two HPT domains of the two CheA kinases CAC2220 and CAC0118.

TABLE 24.4
Histidine Kinases in Sequenced Microbial Genomes

Organism	Histidine Kinases[A]	Chromosome (Mbp)[B]	A/B
E. coli[a]	28[15]	4.6[a]	6.1
Bacillus subtilis[b]	36[14]	4.2[b]	8.6
C. perfringens	27	3.0	9.0
C. acetobutylicum	37	3.9	9.5
P. aeruginosa[c]	63[17]	6.3[c]	10.0
C. tetani	31	2.8	11.1
Synechocystis sp. (PCC 6803)[d]	42[16]	3.6[d]	11.7
Nostoc sp. (PCC 7120)[e]	131[18]	6.4[e]	20.5

[a] http://www.ncbi.nlm.nih.gov/genomes/framik.cgi?db=genome&gi=115.
[b] http://www.ncbi.nlm.nih.gov/genomes/framik.cgi?db=genome&gi=27.
[c] http://www.ncbi.nlm.nih.gov/genomes/framik.cgi?db=genome&gi=163.
[d] http://www.ncbi.nlm.nih.gov/genomes/framik.cgi?db=genome&gi=112.
[e] http://www.ncbi.nlm.nih.gov/genomes/framik.cgi?db=genome&gi=217.

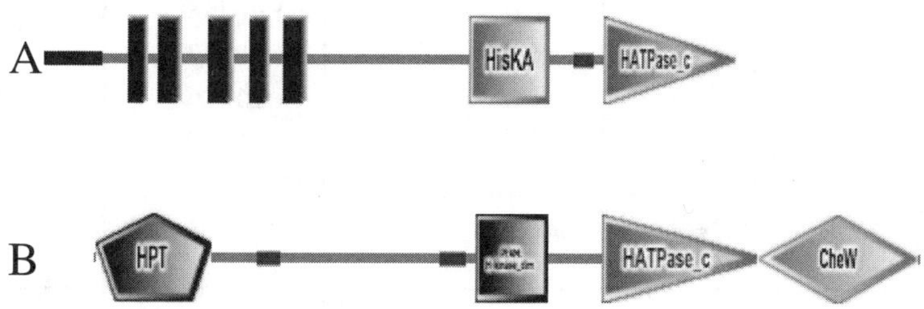

FIGURE 24.4 Domain structure of a class I (A, CAC3391) and a class II (B, CAC2220) kinase.

Several investigators have observed that clostridial Spo0A, a protein that seems to be exclusive to endospore-forming bacteria [32], plays an important role in regulating the initiation of sporulation, the accumulation of the storage polysaccharide granulose, and production of acetone and butanol [33,34]. As a DNA binding protein, Spo0A is thought to directly regulate expression of several genes. The identification of 0A-boxes in promoter regions of solventogenesis genes and the transcriptional consequences of modifying these boxes or by deleting Spo0A have been reported by Ravagnani et al. [35,36]. In *C. perfringens,* Spo0A has been shown to regulate not only sporulation, but also enterotoxin production [37]. Whereas Spo0A from *B. subtilis* is phosphorylated by a phosphorelay system to initiate sporulation (Figure 24.1B), the clostridial Spo0A protein might become phosphorylated directly by one or more histidine kinases. Besides Spo0A, none of the Spo0A-phosphorelay components from *Bacillus* can be found in the clostridial genomes [38]. A comparison of *spo0B* gene organization between two *Bacillus* and *Clostridium* species is given by Stragier [39].

As Spo0A is an orphan response regulator in the three clostridial genomes, it might be speculated that it is phosphorylated by an orphan histidine kinase. It might further be speculated that such a kinase would be conserved within clostridial species with significant similarity, particularly in the input domain. To demonstrate the different architectures of the clostridial histidine kinases, especially in the N-terminal input domain, we analyzed all of these proteins by SMART

SMART graphic	*C. acetobutylicum*	*C. perfringens*	*C. tetani*
COG 0642 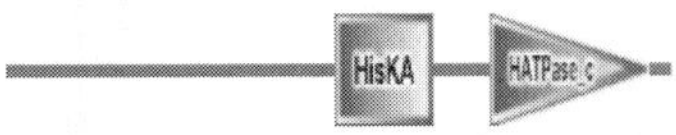	**CAC3319 (o)** CAC0863 (o*)		CTC00456 (o) CTC00715 (o)
			CTC00084#
			**CTC00716 (o)#**
	**CAC2730 (o*)** CAC0525	CPE0951 (o*)	CTC01804 CTC00159 (o*)
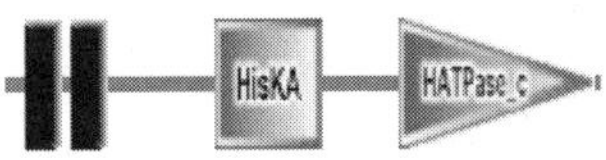	**CAC1517** CAC0372 CAC0290 CAC0654 CAC0225	CPE0841 CPE0458 CPE2363 CPE1986 CPE0120 CPE1512 (o)	CTC00393 CTC01818 CTC01918 CTP22
	**CAC3391**		
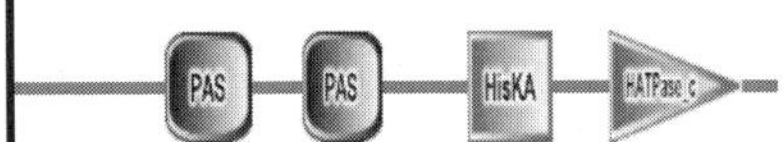	**CAC0437 (o)**	CPE0986 (o)	
	**CAC1701**	CPE1757 (o)	CTC01131

FIGURE 24.5 Domain organization of clostridial histidine kinases analyzed by SMART. The protein used for the representative SMART graphic is shown in bold. Proteins are sorted based on the COG classification of the kinases of *C. acetobutylicum*. The abbreviation (o) indicates orphan kinases; (o*) indicates that the kinase gene has a response regulator gene in the vicinity. As described in the text, we ran the program COGguess in the indicated cases (#). In some cases SMART predicted signal peptides in the proteins and, therefore, did not show some of the putative transmembrane domains. In these cases, we removed the first three amino acids of the protein sequences to get the graphical output showing all putative transmembrane domains.

[40] (http://smart.embl.de/) (Figure 24.5). As only the *C. acetobutylicum* kinases have been assigned to different COGs, we sorted Figure 24.5 to reflect that classification. Similar kinases from *C. perfringens* and *C. tetani* have been linked to the *C. acetobutylicum* kinases based on a similar domain architecture shown by SMART. In the case of poor similarities, we used COGguess (http://www-archbac.u-psud.fr/genomics/cog_guess.html) to evaluate our predictions. Additionally

SMART graphic	*C. acetobutylicum*	*C. perfringens*	*C. tetani*
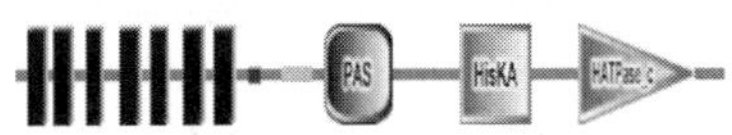	**CAC0903 (o)** CAC0323 (o)	CPE0207 (o) CPE1316 (o) CPE1987	
			**CTC01491#**
	**CAC2760** CAC3219 CAC0565 CAC2434 CAC0451 CAC0831 CAC1507	CPE2098 CPE0235 CPE2487 CPE1192	CTC02155 CTC01951 CTC00191 CTC00806 CTC00849 CTC00598 CTC00629 CTC01481 CTC01524
	**CAC1553** CAC0317 (o*) CAC3662 CAC3516	CPE1926 CPE2239 CPE2331	CTC00412 CTC00873 CTC01420 CTC01905
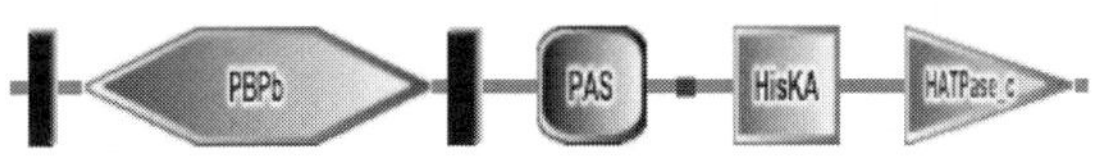			**CTC02323**

FIGURE 24.5 (continued.)

we performed ClustalW analysis [41,42] and ran Phylip Drawgram (PHYLIP program suite, J. Felsenstein, 1993, University of Washington, Seattle) to create the phylogenetic tree in Figure 24.6.

Figure 24.5 and Figure 24.6 illustrate a significant degree of similarity and dissimilarity. For example, in each genome, the predominant histidine kinases are transmembrane proteins with HAMP domains. In all three genomes, the ratio between cytoplasmic and transmembrane kinases is similar (CAC: 6/31; CPE: 4/26; CTC: 7/31), and HAMP and PAS domains are predominant. Similar pairs of histidine kinase and response regulators that are not immediately adjacent (o*) could be found in each genome. The corresponding kinases (CAC0863, CAC2730, CPE0951, CTC001599 are conserved and cluster together indicating that these systems might fulfill similar functions (orphan* cluster; Figure 24.6). Additionally, there is a significant clustering of the other orphan kinases (orphan cluster; Figure 24.6). Other groups of strongly conserved kinases found in all three genomes are the following: Cluster A: CAC1553, CPE2331, CTC00873; Cluster B: CAC3219, CPE2487, CTC00191; Cluster C: CAC2720, CPE0896, CTC02178.

However, there are also marked differences between proteins from the different genomes. For example, protein CTC02323 has a periplasmic substrate-binding domain and is only found in the *C. tetani* genome. In Gram-positive bacteria, such proteins are thought to work as receptors to

SMART graphic	*C. acetobutylicum*	*C. perfringens*	*C. tetani*
COG 4585 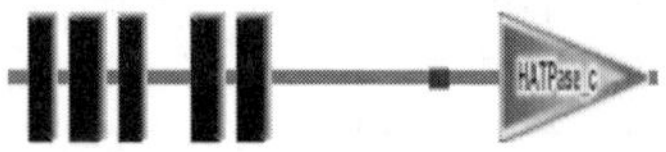	**CAC2253** CAC0239 CAC1454 CAC2940 CAC3430 CAC0864	CPE1500 CPE1519 CPE0870 (o)	
COG 3920 	**CAC2720**		
		CPE0896#	**CTC02178#**
COG2205 	**CAC3678**		
COG0643 	**CAC0118** CAC2220		CTC01731
COG 2972 	**CAC0080 (o)** CAC1582		CTC01978
		**CPE2088#**	CTC01848#
		**CPE0574**	
		**CPE0927**	CTC00934
			**CTC00949**

FIGURE 24.5 (continued.)

trigger or initiate translocation of a solute through the membrane (S). Among the clostridial genomes, the KdpD protein (CAC3678) is unique to *C. acetobutylicum* (see also Chapter 13). *C. tetani* is the only organism to possess a hybrid histidine kinase with a fused response regulator (CTC00084). As mentioned earlier, several hybrid proteins are class 1c proteins and act as part of phosphorelay systems. We could not detect any HPT-domains in clostridial proteins except for the HPT-containing CheA in the *C. tetani* genome. However, as most known HPT-domains were identified in Gram-negative bacteria, it cannot be ruled out that the clostridial genomes do contain genes encoding proteins with that function.

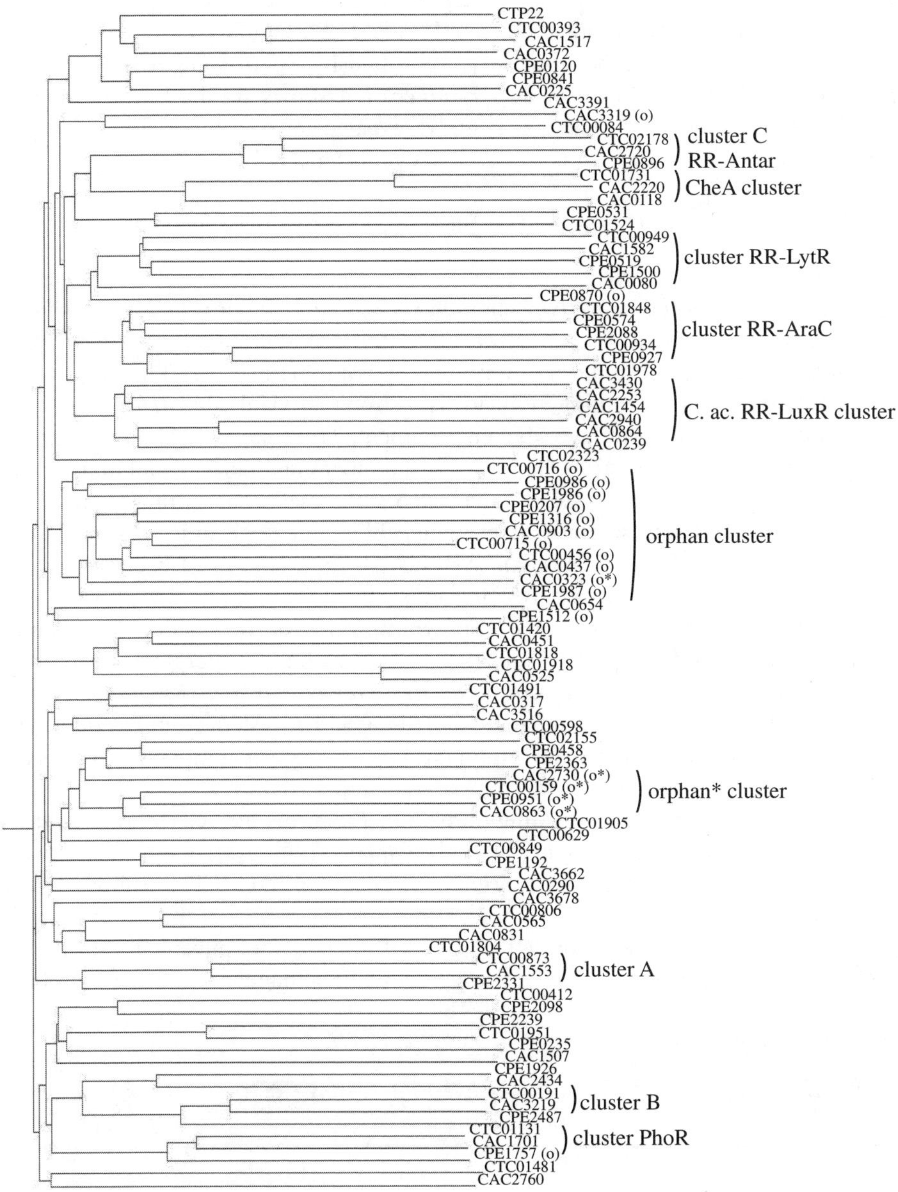

FIGURE 24.6 Phylogenetic tree of the clostridial histidine kinases using ClustalW and Phylip. The abbreviation (o) indicates orphan kinases; (o*) indicates that the kinase gene has a response regulator gene in the vicinity. The clusters are explained in the text.

Another interesting difference between the clostridial genomes regards the PhoB/R operon. Whereas *C. acetobutylicum* and *C. tetani* possess a PhoB/R two-component system (CAC1700/1701; CPE1130/1131; see also Chapter 14), in *C. perfringens* the PhoR-like protein (CPE1757) is an orphan histidine kinase.

In *C. perfringens,* next to the well-studied VirR/VirS (CPE1500/1501) two-component system, there is another VirR/VirS-like two-component system (CPE0518/CPE0519; see also Chapter 23); however, such a system could not be found in *C. tetani* or *C. acetobutylicum.* The most similar proteins to VirS in *C. tetani* and *C. acetobutylicum* (CAC1582; CTC00949; Figure 24.6) differ in their number of putative transmembrane helices (Figure 24.5). In a DNA array analysis, the kinase CAC3319 has been proposed to be positively controlled by Spo0A, either directly or indirectly [13]. That kinase has no homologue in *C. perfringens* but shows similarity to the kinase domain of the hybrid protein CTC0084. The same study revealed an overexpression of CAC3319 and CAC1701 (PhoR) in a strain lacking the pSOL1 megaplasmid [13].

24.5.2 Chemotaxis Systems, Specialized Signal Transduction Systems

As a nonmotile organism, *C. perfringens* lacks any chemotaxis system, whereas *C. tetani* contains one and *C. acetobutylicum* contains two chemotaxis systems (Figure 24.7). The *che* system of *C. tetani* resembles in composition and organization the second *che* system of *C. acetobutylicum*. Both *C. acetobutylicum* and *C. tetani* are motile and possess peritrichous flagella.

The capacity to control swimming behavior is a strong survival advantage as it allows bias of random swimming toward a more favorable environment. There are several conditions that constitute such an environment for bacteria. Changes of pH, temperature, nutrient and toxin concentration, etc. can be sensed by chemotaxis systems. The sensors, the methyl-accepting chemotaxis proteins (MCPs), are predominantly transmembrane proteins, although cytoplasmic versions also exist. The central sensory pathway of such chemotaxis systems is common to most bacteria.

Nevertheless, specific differences between such systems from *E. coli* and *B. subtilis* have been reported [43,44]. These differences are based on composition of the systems (for example, CheZ only in *E. coli*, CheC, CheD only in *B. subtilis*), as well as on the interplay of the proteins [43]. As chemotaxis systems of *Myxococcus xanthus, Helicobacter pylori, Rhodobacter sphaeroides* resemble the *B. subtilis*-type system, the well-studied paradigm system from *E. coli* might be more specific for γ-proteobacteria than originally thought. The clostridial chemotaxis systems also resemble the *B. subtilis*-type system, which will be briefly summarized [43].

All MCPs share a cytoplasmic signaling domain that controls the activity of a class 2 histidine kinase, the CheA protein (see also Figure 24.4). The small CheW protein, the linker protein, binds to the C-terminal domain of CheA and the signaling domain of the MCP, forming a ternary CheA-CheW-MCP complex. The signal that a chemoattractant has bound to the MCP is transduced to CheA causing increased autophosphorylation. CheA-P transfers its phosphate to the small single domain response regulator CheY. CheY-P can interact with the flagellar motor protein FliY, causing

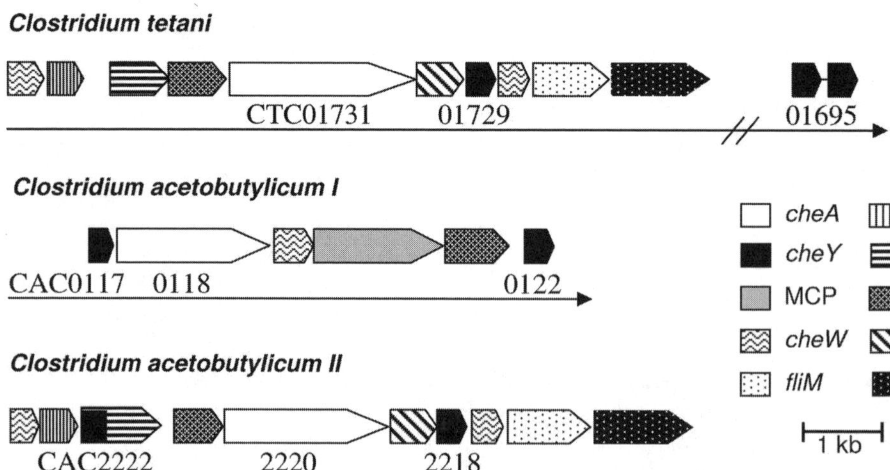

FIGURE 24.7 Organization of the clostridial chemotaxis operons. The different *che* genes are indicated in the figure. Gene identifier numbers are also shown for genes encoding histidine kinases or response regulators.

a change of the rotational direction of the flagellum resulting in swimming of the bacterium. In order to adapt to attractant or repellent stimuli, the *B. subtilis*-type chemotaxis system comprises further proteins. In addition to CheY, CheA-P can transfer phosphate to the response regulator CheB, which contains a C-terminal methyl esterase domain. Phospho-CheB is active and removes methyl groups from specific glutamate residues of the MCP. Addition or removal of an attractant causes rapid demethylation and slow remethylation of the MCP. CheR is a methyl transferase responsible for methylation of the MCPs. The changes in methylation state of the MCPs alter their signaling behavior to CheA and allow the MCPs to reset. Proteins CheC and CheD seem to coordinate methylation of the MCPs [43]. The chemotaxis system of *C. tetani* and the second system of *C. acetobutylicum* are coorganized with genes of the flagellar apparatus including *fliY*, indicating that these systems indeed control rotation of flagella (Figure 24.7). The differences between the CheB homologues seem to be based on a sequencing error. Assuming just one frame shift, the CheB protein from *C. tetani* becomes even more similar to CheB from *C. acetobutylicum*. Both systems encode all chemotaxis proteins mentioned above, except for the MCPs. The MCP genes are not co-organized with these two systems but, instead, are spread throughout the genome. However, the *che* 1 system of *C. acetobutylicum* contains a gene encoding a MCP. The *che* 1 system is not as complex as system 2 and lacks CheC, CheD, and CheB (Figure 24.7). Interestingly, the total number of annotated MCPs in the genome of *C. tetani* is 18 and is nearly doubled in *C. acetobutylicum*. Both numbers seem to be rather high compared to 5 MCPs in *E. coli* and 10 MCPs in *B. subtilis* [45]. Nevertheless, *Chromobacterium violaceum* has been reported to comprise 41 MCPs, probably transducing signals to three chemotaxis clusters [46].

24.5.3 Clostridial Response Regulators

Response regulators are characterized by a conserved domain of approximately 125 amino acids, known as a receiver domain (Figure 24.1). Receiver domains contain the conserved aspartate residue, which is involved in the phosphotransfer reaction. The receiver domain is typically linked to an effector or output domain, which usually has DNA binding activity. Based on these effector domains, response regulators fall into different classes [47]. The clostridial response regulators belong to 10 different COGs (Figure 24.8).

More than 50% of the clostridial response regulators belong to COG 0745, representing the OmpR-PhoB subgroup. These response regulators function as activators or repressors of σ70-dependent promoters. Where identified, DNA binding sites localize upstream of or within corresponding promoter regions. Interaction with the RNA-polymerase σ 70- or α–subunit has been reported for PhoB and OmpR [48,49]. The second most abundant clostridial response regulators belong to COG 0784, which includes a rather diverse group of response regulators comprising different chemotaxis-related proteins. *C. acetobutylicum* has four single domain CheY-like response regulators — three of them are part of the chemotaxis systems, while one is an orphan response regulator (CAC2407). CAC2407 could, therefore, be acting as a component of a phosphorelay system. All three Spo0A proteins belong to the same COG cluster as the single domain response regulators, and each organism has at least one additional Spo0A-like response regulator (CAC3286, CPE0532, CTC00848, CTC01523). Another group of response regulators belong to COG 2197, which contains proteins of the NarL/FixJ subgroup. NarL/FixJ family members also function as transcriptional activators of σ 70-dependent promoters. Interestingly, this type of response regulator is only found in *C. acetobutylicum*. All examples are part of two-component systems, and their corresponding histidine kinases cluster significantly (C. ac. RR-LuxR cluster; Figure 24.6). In all three clostridial genomes, two proteins of COG 3297 can be found. This cluster represents AlgR/AgrA/LytR-type response regulators. These regulators are reported to regulate production of virulence factors such as toxins, bacteriocins, and extracellular polysaccharides [50]. Again, the corresponding histidine kinases show significant similiarities (cluster RR-LytR; Figure 24.6). Every genome also encoded a protein with a receiver and ANTAR domain, grouped within COG 3707.

Smart graphic /COG	*C. acetobutylicum*	*C. perfringens*	*C. tetani*
COG 0745 	CAC0224, CAC0289, CAC0371, CAC0321 (o*), CAC0450, CAC0524, CAC0564, CAC0653, CAC0830, CAC0860 (o*), CAC1506, CAC1516, CAC1552, CAC1700, CAC2435, CAC2735 (o*), CAC2759, CAC3220, CAC3517, CAC3663, CAC3677	CPE0119, CPE0236, CPE0457, CPE0642 (o), CPE0840, CPE0948 (o*), CPE1193, CPE1925, CPE2099, CPE2240, CPE2332, CPE2364, CPE2488	CTC00163 (o*), CTC00189, CTC00392, CTC00411, CTC00597, CTC00628, CTC00805, CTC00872, CTC01130, CTC01421, CTC01482, CTC01490, CTC01805, CTC01819, CTC01906, CTC01919, CTC01953, CTC02156
COG 2197 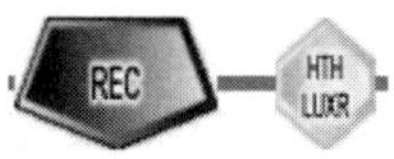	CAC0174 (o), CAC0240, CAC0865, CAC1455, CAC2254, CAC2939, CAC3429		
COG3297 	CAC0081, CAC1581	CPE0518, CPE1501	CTC00950, CTC01979
COG 3707 	CAC2721	CPE0895	CTC02179
COG 3947  	CAC3390 CAP0009		
COG 4753 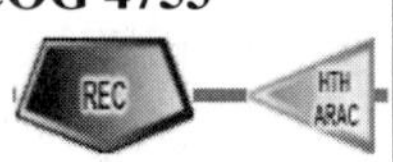		CPE0575, CPE0928, CPE2087	CTC00935, CTC01849

FIGURE 24.8 Domain organization of clostridial response regulators analyzed by SMART. Proteins are sorted based on the COG classification of the response regulators of *C. acetobutylicum*. The abbreviation (o) indicates orphan proteins; (o*) indicates that the response regulator gene has a kinase gene in the vicinity.

The ANTAR domain, an RNA binding domain, indicates a transcriptional antitermination mechanism for controlling gene expression [51]. These response regulators are organized as two-component systems, and the kinases are highly conserved (cluster C, RR-Antar; Figure 24.6).

Response regulators of COG 4753 can only be found in *C. perfringens* and *C. tetani.* Members of this AraC/XylS family are defined by a conserved stretch of 99 amino acids usually located in the C-terminal region containing all the elements required to bind DNA target sequences and to activate transcription from cognate promoters. Proteins of that family have three main regulatory functions in common: carbon metabolism, stress response, and pathogenesis [52]. An NtrC-like

Smart graphic / COG	*C. acetobutylicum*	*C. perfringens*	*C. tetani*
COG 0784			
	CAC0117 CAC0122 CAC22218 CAC2407 (o)		CTC01729
	CAC2071 (o) (Spo0A) CAC3286 (o)	CPE0532 CPE1812 (o) (Spo0A)	CTC00848 CTC01523 CTC01569 (o) (Spo0A)
	CAC0585 (o)		
	CAC1233 (o)		
			CTC01695 (o)
COG 2201 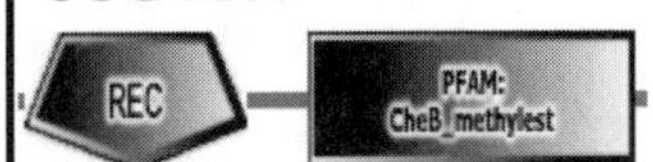	CAC2222		
COG 2204 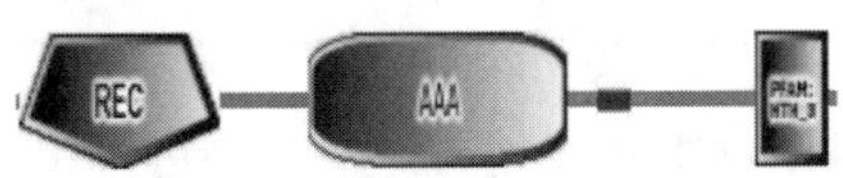			CTC02322
COG 0642 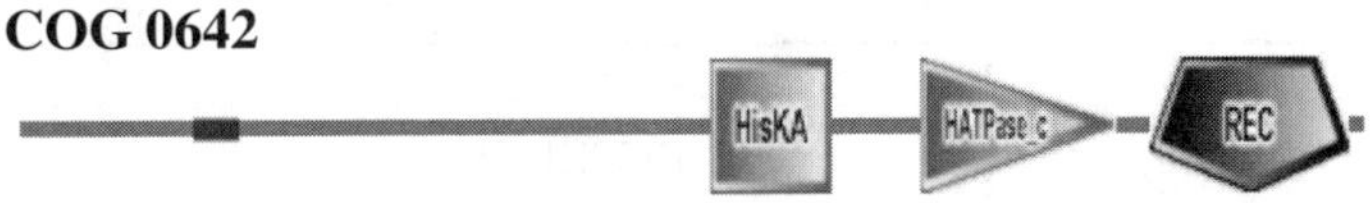			CTC00084

FIGURE 24.8 (continued.)

response regulator, a putative activator of a σ 54-dependent promoter, is exclusively found in *C. tetani* (COG 2204, CTC02322). That response regulator forms a two-component system with the histidine kinase CTC02323, which was also exclusive to *C. tetani*.

Using multisequence analysis Grebe and Stock [28] observed that not only do response regulators and histidine kinases fall into different classes, but that there is also a strong correlation between the classes of these proteins forming two-component systems [28]. We observed the same phenomenon with the clostridial histidine kinases and response regulators. An interesting regulatory relationship between two-component and ABC transport systems was described for the *Bacillus/Clostridium* group [53]. Several clusters between two-component and ABC transporter encoding genes have been found in the genomes of *B. halodurans*, *C. acetobutylicum,* and *C. difficile* [53]. We have observed the same clustering in the genomes of *C. perfringens* and *C. tetani.*

24.5.4 Protein Kinases and Protein Phosphatases

There are a very limited number of protein kinases and protein phosphatases in the three investigated clostridial genomes (Figure 24.3; Table 24.2 and Table 24.3).

A gene encoding a putative bifunctional HPr kinase/phosphorylase (HprK/P) exists in all three genomes. This protein catalyzes the phosphorylation and dephosphorylation of Ser-46 in HPr, a phosphocarrier protein of the phosphoenolpyruvate:carbohydrate phosphotransferase system (PTS) [54–56]. The phosphotransferase system and catabolite repression mechanism in clostridia have already been investigated [57–63], and analyses of HPr of *C. acetobutylicum* indicated that the mechanism of carbon catabolite repression might be similar to that of *B. subtilis* [58].

Another well-studied regulatory system based on protein phosphorylation controls activity of sigma factor F in *B. subtilis* [64,65]. Sigma factor F is a developmental regulator that is activated during the process of spore formation. Activation depends on the membrane-bound protein phosphatase SpoIIE, which dephosphorylates and activates the phosphoprotein SpoIIA. SpoIIA then causes release of the sigma factor F from the antisigma factor SpoIIAB. SpoIIAB is itself a protein kinase that catalyzes phosphorylation of SpoIIA. SpoIIE- and SpoIIAB-like proteins can be found in all three clostridial genomes (Table 24.3, Figure 24.3). Significant similarities between the sporulation sigma factors of *B. subtilis* and *C. acetobutylicum* were observed earlier [66].

In *B. subtilis* the PrkA protein, with distant homology to eukaryotic cAMP-dependent protein kinases, was shown to phosphorylate a 60-kDa protein at a serine residue [67]. Genes encoding PrkA-like proteins also have been identified in the clostridial genomes (CAC0579; CPE1334; CTC01039). One gene pair encoding a eukaryotic-type protein kinase and a PP2C-type phosphatase appear very conserved within each investigated clostridial genome (Figure 24.9). The protein kinases contain three PASTA domains in their C-terminal region (Figure 24.9). Similar putative protein kinases and PP2C-type phosphatases are also shown in Figure 24.9.

Protein kinase	Protein phosphatase
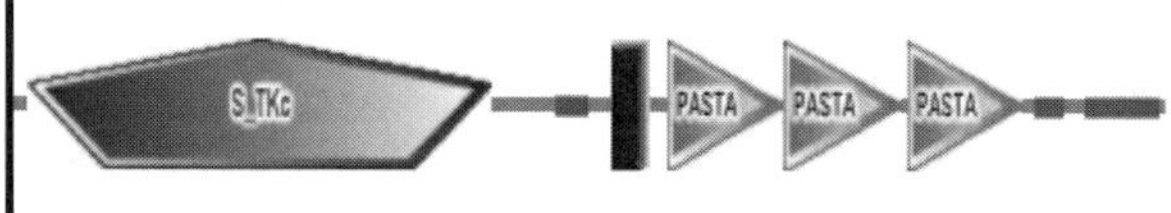	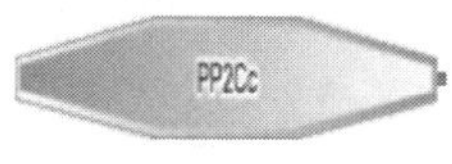
CAC1728 CPE1738 CTC01225	CAC1727 CPE1739 CTC01224
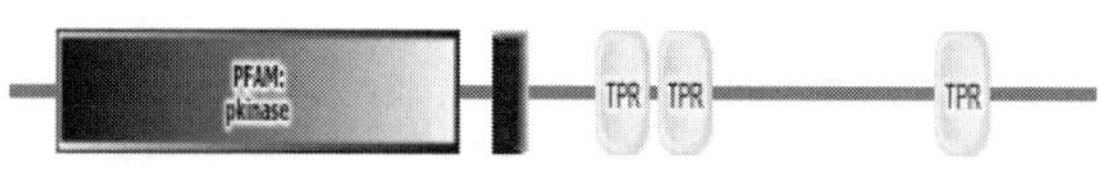	
CAC0404	CAC0407, CAC0035

CPE1990 CAC2400 (related, inactivated protein)	

FIGURE 24.9 Domain organization of clostridial protein kinases and protein phosphatases analyzed by SMART. The gene identifier numbers are given below the graphic.

The PASTA domain (for penicillin-binding protein and serine/threonine kinase associated domain) has been found in the high molecular weight penicillin-binding proteins and eukaryotic-like serine/threonine kinases of a range of pathogens, and these kinases have been suggested to be key regulators of cell-wall biosynthesis [68]. In the clostridial genomes, there are some further proteins annotated as protein phosphatases. Two of them have a hydrolase domain (CAC2787; CTC00969), two others look like low molecular weight protein tyrosine phosphatases (CAC2881; CTC00306), CTC0204 looks like an acid phosphatase, whereas CPE1952 does not resemble a tyrosine phosphatase. We did not find evidence for the presence of tyrosine kinases in the clostridial genomes, although it cannot totally be ruled out that the Ser/Thr kinases do not also phosphorylate tyrosine residues. These proteins need further analysis to show that they are indeed protein kinases and protein phosphatases. The existence of genes encoding putative FHA domains, a domain involved in phosphopeptide-specific protein-protein interaction, already has been reported for *C. acetobutylicum* [2]. Two of these genes (CAC0036, CAC0406) are coorganized with the already mentioned PP2C-type protein phosphatase genes CAC0035 and CAC0407.

24.6 CONCLUDING REMARKS

The comparative analysis of clostridial signal transduction proteins revealed marked similarities and dissimilarities among the three clostridial species. Nevertheless, we are aware that we have only compared what has already been recognized as signal transduction proteins during the annotation of the three clostridial genomes. Although we could identify additional proteins by Blast searches, our analysis might still be incomplete. For example, a new class of histidine kinases, the HWE kinases, has just been identified [69], which are not recognized as histidine kinases by SMART or PFAM [70]. Also, we did not include proteins comprising putative signaling domains like GGDEF or HD-GYP [21]. It will be very interesting to analyze further how all these proteins interact and constitute signaling networks in these different organisms.

REFERENCES

1. Fleischmann, R.D. et al., Whole-genome random sequencing and assembly *of Haemophilus influenzae* Rd, *Science,* 269, 496, 1995.
2. Nölling, J. et al., Genome sequence and comparative analysis of the solvent-producing bacterium *Clostridium acetobutylicum, J. Bacteriol.,* 183, 4823, 2001.
3. Shimizu, T. et al., Complete genome sequence of *Clostridium perfringens*, an anaerobic flesh-eater, *Proc. Natl. Acad. Sci. U.S.A.,* 99, 996, 2002.
4. Brüggemann, H. et al., The genome sequence of *Clostridium tetani*, the causative agent of tetanus disease, *Proc. Natl. Acad. Sci. U.S.A.,* 100, 1316, 2003.
5. Kofoid, E.C. and Parkinson, J.S., Transmitter and receiver modules in bacterial signaling proteins, *Proc. Natl. Acad. Sci. U.S.A.,* 85, 4981, 1988.
6. Parkinson, J.S. and Kofoid, E.C., Communication modules in bacterial signaling proteins, *Annu. Rev. Genet.,* 26, 71, 1992.
7. Hoch, J.A., Two-component and phosphorelay signal transduction, *Curr. Opin. Microbiol.,* 3, 165, 2000.
8. Varughese, K.I., Signal transmission and specificity in the sporulation phosphorelay of *Bacillus subtilis*, in *Histidine Kinases in Signal Transduction*, Inouye, M. and Dutta, R., Eds., Academic Press, San Diego, 2003, 203.
9. Phillips, Z.E. and Strauch, M.A., *Bacillus subtilis* sporulation and stationary phase gene expression, *Cell. Mol. Life Sci.,* 59, 392, 2002.
10. Varughese, K.I., Molecular recognition of bacterial phosphorelay proteins, *Curr. Opin. Microbiol.,* 5, 142, 2002.
11. Bakal, C.J. and Davies, J.E., No longer an exclusive club: eukaryotic signalling domains in bacteria, *Trends Cell. Biol.,* 10, 32, 2000.

12. Hunter, T., Protein kinases and phosphatases: the yin and yang of protein phosphorylation and signaling, *Cell,* 80, 225, 1995.
13. Tomas, C.A. et al., DNA array-based transcriptional analysis of asporogenous, nonsolventogenic *Clostridium acetobutylicum* strains SKO1 and M5, *J. Bacteriol.,* 185, 4539, 2003.
14. Fabret, C., Feher, V.A. and Hoch, J.A., Two-component signal transduction *in Bacillus subtilis*: how one organism sees its world, *J. Bacteriol.,* 181, 1975, 1999.
15. Mizuno, T., Compilation of all genes encoding two-component phosphotransfer signal transducers in the genome of *Escherichia coli, DNA Res.,* 4, 161, 1997.
16. Mizuno, T., Kaneko, T. and Tabata, S., Compilation of all genes encoding bacterial two-component signal transducers in the genome of the cyanobacterium, *Synechocystis sp.* strain PCC 6803, *DNA Res.,* 3, 407, 1996.
17. Rodrigue, A. et al., Two-component systems in *Pseudomonas aeruginosa*: why so many?, *Trends Microbiol.,* 8, 498, 2000.
18. Wang, L. et al., Genomic analysis of protein kinases, protein phosphatases and two-component regulatory systems of the cyanobacterium *Anabaena* sp. strain PCC 7120, *FEMS Microbiol. Lett.,* 217, 155, 2002.
19. Tatusov, R.L. et al., The COG database: new developments in phylogenetic classification of proteins from complete genomes, *Nucleic Acids Res.,* 29, 22, 2001.
20. Tatusov, R.L. et al., The COG database: an updated version includes eukaryotes, *BMC Bioinformatics,* 4, 41, 2003.
21. Galperin, M.Y., Nikolskaya, A.N. and Koonin, E.V., Novel domains of the prokaryotic two-component signal transduction systems, *FEMS Microbiol. Lett.,* 203, 11, 2001.
22. Aravind, L. and Ponting, C.P., The cytoplasmic helical linker domain of receptor histidine kinase and methyl-accepting proteins is common to many prokaryotic signalling proteins, *FEMS Microbiol. Lett.,* 176, 111, 1999.
23. Appleman, J.A., Chen, L.L. and Stewart, V., Probing conservation of HAMP linker structure and signal transduction mechanism through analysis of hybrid sensor kinases, *J. Bacteriol.,* 185, 4872, 2003.
24. Gilles-Gonzalez, M.A. and Gonzalez, G., Signal transduction by heme-containing PAS-domain proteins, *J. Appl. Physiol.,* 96, 774, 2004.
25. Taylor, B.L. and Zhulin, I.B., PAS domains: internal sensors of oxygen, redox potential, and light, *Microbiol. Mol. Biol. Rev.,* 63, 479, 1999.
26. Mutsuda, M. et al., Biochemical properties of CikA, an unusual phytochrome-like histidine protein kinase that resets the circadian clock in *Synechococcus elongatus* PCC 7942, *J. Biol. Chem.,* 278, 19102, 2003.
27. Koretke, K.K. et al., Evolution of two-component signal transduction, *Mol. Biol. Evol.,* 17, 1956, 2000.
28. Grebe, T.W. and Stock, J.B., The histidine protein kinase superfamily, *Adv. Microb. Physiol.,* 41, 139, 1999.
29. Kim, D. and Forst, S., Genomic analysis of the histidine kinase family in bacteria and archaea, *Microbiology,* 147, 1197–1212, 2001.
30. Bilwes, A.M. et al., Structure of CheA, a signal-transducing histidine kinase, *Cell,* 96, 131, 1999.
31. Koretke, K.K. et al., Molecular Evolution of histidine kinases, in *Histidine Kinases in Signal Transduction,* Inouye, M. and Dutta, R., Eds., Academic Press, San Diego, 2003, 483.
32. Brown, D.P. et al., Characterization of *spo0A* homologues in diverse *Bacillus* and *Clostridium* species identifies a probable DNA-binding domain, *Mol. Microbiol.,* 14, 411, 1994.
33. Wilkinson, S.R. et al., Molecular genetics and the initiation of solventogenesis in *Clostridium beijerinckii* (formerly *Clostridium acetobutylicum*) NCIMB 8052, *FEMS Microbiol. Rev.,* 17, 275, 1995.
34. Dürre, P. et al., Transcriptional regulation of solventogenesis in *Clostridium acetobutylicum, J. Mol. Microbiol. Biotechnol.,* 4, 295, 2002.
35. Ravagnani, A. et al., Spo0A directly controls the switch from acid to solvent production in solvent-forming clostridia, *Mol. Microbiol.,* 37, 1172, 2000.
36. Harris, L.M., Welker, N.E. and Papoutsakis, E.T., Northern, morphological, and fermentation analysis of *spo0A* inactivation and overexpression in *Clostridium acetobutylicum* ATCC 824, *J. Bacteriol.,* 184, 3586, 2002.

37. Huang, I.H. et al., Disruption of the gene (*spo0A*) encoding sporulation transcription factor blocks endospore formation and enterotoxin production in *enterotoxigenic Clostridium perfringens* type A, *FEMS Microbiol. Lett.,* 233, 233, 2004.
38. Stephenson, K. and Hoch, J.A., Evolution of signalling in the sporulation phosphorelay, *Mol. Microbiol.,* 46, 297, 2002.
39. Stragier, P., A gene odyssey: Exploring the genomes of endospore-forming bacteria, in *Bacillus subtilis and Its Closest Relatives: from Genes to Cells*, Sonnenshein, A.L., Ed., ASM Press, Washington, DC, 2002, 519.
40. Letunic, I. et al.SMART 4.0: towards genomic data integration, *Nucleic Acids Res.,* 32, 142, 2004.
41. Thompson, J.D., Higgins, D.G., and Gibson, T.J., CLUSTAL W: improving the sensitivity of progressive multiple sequence alignment through sequence weighting, position-specific gap penalties and weight matrix choice, *Nucleic Acids Res.,* 22, 4673, 1994.
42. Higgins, D.G., CLUSTAL V: multiple alignment of DNA and protein sequences, *Methods Mol. Biol.,* 25, 307, 1994.
43. Rao, C.V., Kirby, J.R., and Arkin, A.P., Design and diversity in bacterial chemotaxis: a comparative study in *Escherichia coli* and *Bacillus subtilis*, *PLoS Biol.,* 2, E49, 2004.
44. Bischoff, D.S. and Ordal, G.W., *Bacillus subtilis* chemotaxis: a deviation from the *Escherichia coli* paradigm, *Mol. Microbiol.,* 6, 23, 1992.
45. Armitage, J.P., Bacterial tactic responses, *Adv. Microb. Physiol.,* 41, 229, 1999.
46. Pereira, M. et al., Chemotaxis and flagellar genes of *Chromobacterium violaceum*, *Genet. Mol. Res.,* 3, 92, 2004.
47. Stock, A.M. and West, A.H., Response regulator proteins and their interactions with histidine protein kinases, in *Histidine Kinases in Signal Transduction*, Inouye, M. and Dutta, R., Eds., Academic Press, San Diego, 2003, 237.
48. Slauch, J.M., Russo, F.D. and Silhavy, T.J., Suppressor mutations in *rpoA* suggest that OmpR controls transcription by direct interaction with the alpha subunit of RNA polymerase, *J. Bacteriol.,* 173, 7501, 1991.
49. Makino, K. et al., DNA binding of PhoB and its interaction with RNA polymerase, *J. Mol. Biol.,* 259, 15, 1996.
50. Nikolskaya, A.N. and Galperin, M.Y., A novel type of conserved DNA-binding domain in the transcriptional regulators of the AlgR/AgrA/LytR family, *Nucleic Acids Res.,* 30, 2453, 2002.
51. Shu, C.J. and Zhulin, I.B., ANTAR: an RNA-binding domain in transcription antitermination regulatory proteins, *Trends Biochem. Sci.,* 27, 3, 2002.
52. Gallegos, M.T. et al., Arac/XylS family of transcriptional regulators, *Microbiol. Mol. Biol. Rev.,* 61, 393, 1997.
53. Joseph, P. et al., Regulatory relationship of two-component and ABC transport systems and clustering of their genes in the *Bacillus/Clostridium* group, suggest a functional link between them, *J. Mol. Microbiol. Biotechnol.,* 4, 503, 2002.
54. Vadeboncoeur, C., Frenette, M., and Lortie, L.A., Regulation of the *pts* operon in low G+C Gram-positive bacteria, *J. Mol. Microbiol. Biotechnol.,* 2, 483, 2000.
55. Dahl, M.K., CcpA-independent carbon catabolite repression in *Bacillus subtilis*, *J. Mol. Microbiol. Biotechnol.,* 4, 315–321, 2002.
56. Darbon, E. et al., Phosphotransfer functions mutated *Bacillus subtilis* HPr-like protein Crh carrying a histidine in the active site, *J. Mol. Microbiol. Biotechnol.,* 3, 439, 2001.
57. Tangney, M. and Mitchell, W.J., *Clostridium tetani* encodes a phosphocarrier protein, HPr, *Microbiology,* 150, 525, 2004.
58. Tangney, M. et al., Analysis of the elements of catabolite repression in *Clostridium acetobutylicum* ATCC 824, *J. Mol. Microbiol. Biotechnol.,* 6, 6, 2003.
59. Lee, J. and Blaschek, H.P., Glucose uptake in *Clostridium beijerinckii* NCIMB 8052 and the solvent-hyperproducing mutant BA101, *Appl. Environ. Microbiol.,* 67, 5025, 2001.
60. Reizer, J. et al., A hybrid response regulator possessing a PEP-dependent phosphorylation domain, *Microbiology,* 145, 987, 1999.
61. Brynestad, S. et al., A complex array of Hpr consensus DNA recognition sequences proximal to the enterotoxin gene in *Clostridium perfringens* type A, *Microbiology,* 140, 97, 1994.

62. Mitchell, W.J., Shaw, J.E., and Andrews, L., Properties of the glucose phosphotransferase system of *Clostridium acetobutylicum* NCIB 8052, *Appl. Environ. Microbiol.,* 57, 2534, 1991.
63. Reizer, J. et al., The phosphoenolpyruvate:sugar phosphotransferase system in gram-positive bacteria: properties, mechanism, and regulation, *Crit. Rev. Microbiol.,* 15, 297, 1988.
64. Carniol, K., Eichenberger, P., and Losick, R., A threshold mechanism governing activation of the developmental regulatory protein sigma F in *Bacillus subtilis, J. Biol. Chem.,* 279, 14860, 2004.
65. Arabolaza, A.L. et al., Characterization of a novel inhibitory feedback of the anti-anti-sigma SpoIIAA on Spo0A activation during development in *Bacillus subtilis, Mol. Microbiol.,* 47, 1251, 2003.
66. Sauer, U. et al., Sigma factor and sporulation genes in *Clostridium, FEMS Microbiol. Rev.,* 17, 331, 1995.
67. Fischer, C. et al., Cloning and characterization of the *Bacillus subtilis prkA* gene encoding a novel serine protein kinase, *Gene,* 168, 55, 1996.
68. Yeats, C., Finn, R.D., and Bateman, A., The PASTA domain: a beta-lactam-binding domain, *Trends Biochem. Sci.,* 27, 438, 2002.
69. Karniol, B. and Vierstra, R.D., The HWE histidine kinases, a new family of bacterial two-component sensor kinases with potentially diverse roles in environmental signaling, *J. Bacteriol.,* 186, 445, 2004.
70. Zhulin, I.B., Digging with experimental pick and computational shovel: a new addition to the histidine kinase superfamily, *J. Bacteriol.,* 186, 267, 2004.

25 Regulation of Catabolic Gene Systems

Martin Tangney and Wilfrid J. Mitchell

CONTENTS

25.1 INTRODUCTION

The clostridia, like other bacteria, possess the capacity to grow and survive under a variety of environmental conditions. They generally possess the capacity to metabolize a variety of different carbon sources, and they might have the ability to hydrolyze complex polymers (such as starch) into simple, more useful compounds (such as glucose or maltose) that may be transported directly into the cell to provide energy and carbon. This requires the synthesis of specific proteins. Although a bacterium can possess the capacity to produce such proteins, the precise protein content of the cell will vary depending on the environmental conditions in which it exists. This is true for clostridia; for example, *Clostridium acetobutylicum* can transport and metabolize sucrose, but intuitively it would appear inefficient to synthesize high levels of the enzymes required when this substrate is not available to the cell — and indeed this is the case, as the substrate is needed to induce enzyme synthesis [1].

It is important for bacteria to be able to sense their environment and generate the proteome that is best suited to particular needs. Furthermore, it is common for bacteria to differentially utilize carbon sources where more than one is available. Typically, the preferred carbon source is utilized first, and only when this is significantly depleted is the less favored substrate metabolized. Again, using the example of sucrose metabolism, when *C. acetobutylicum* is presented with a mixture of sucrose and the preferred substrate glucose, the enzymes required for sucrose uptake and metabolism are not produced until the available glucose has been used — despite the presence of the inducing substrate for these enzymes [1]. This is the phenomenon of carbon catabolite repression (CCR).

The ability to respond and adapt to changes in the environment is of fundamental importance to survival, and bacteria have evolved a variety of mechanisms for achieving this goal. The majority of such mechanisms in bacteria (including the clostridia) involve the regulation of transcription, whereby the expression of a particular gene can be switched on (induced) or off (repressed) in response to an environmental stimulus. This might involve a single gene (as is often the case for a specific extracelluar enzyme), or might require the coordinate regulation of a set of genes (as is often found for transport systems). Typically, where a number of proteins are required, the respective genes are found in an operon that is subject to regulation. This is best exemplified by operons that encode transport systems. Such systems encode the proteins necessary to translocate a substrate into the cell and can additionally encode enzymes specifically required to convert the translocated substrate into an intermediate that might enter central metabolism. As these proteins may only be required when the individual substrate is available, transcription of the operon is specifically increased in response to the presence of the relevant substrate in the environment.

An important transport mechanism in bacteria is the phosphoenolpyruvate (PEP) — dependent phosphotransferase system (PTS). Apart from its role in transport, the PTS plays a critical role in CCR, both in Gram-positive and Gram-negative bacteria (although the mechanism differs significantly between these two groups). PTS activity has been demonstrated for a variety of substrates in species of *Clostridium,* and both substrate-specific induction of PTS operons and CCR have been observed. The reader is referred to Chapter 8 for a detailed discussion of the PTS in clostridia, but in order to appreciate its role in gene regulation, it is necessary to summarize key aspects of the PTS here.

25.2 THE PHOSPHOTRANSFERASE SYSTEM

The PTS is an active transport mechanism in which the translocation of a specific substrate into the cell is coupled to its phosphorylation, thereby conserving energy and effectively trapping a phosphorylated derivative of the substrate within the cell. The energy for translocation is derived from hydrolysis of PEP to pyruvate. The phosphate from PEP is passed to the substrate by sequential transfer via two general PTS proteins, Enzyme I (EI) and HPr, which are common to (almost) all phosphotransferases in the cell, and a substrate-specific complex, enzyme II (EII), which confers specificity to the system. The central proteins EI and HPr are encoded by the genes *ptsI* and *ptsH,* which are typically found in an operon *ptsHI.* These proteins are very well conserved among bacteria and the key phosphorylation sites have been identified. In the PTS transport reaction, HPr is phosphorylated at a specific histidine residue, His-15. This is essential for PTS activity and the substitution of this residue results in a loss of PTS activity. However, in certain bacteria, HPr may also be phosphorylated at an alternative site (Ser-46) with significant consequences for gene regulation, as discussed later in this chapter.

The EII complex consists of three (or sometimes four) functional domains — IIA, IIB, and IIC (or IIC/IID) — two of which (IIA and IIB) are involved in the phosphoryl transfer chain. The IIA domain receives phosphate from phosphorylated HPr and, in turn, passes the phosphate to the

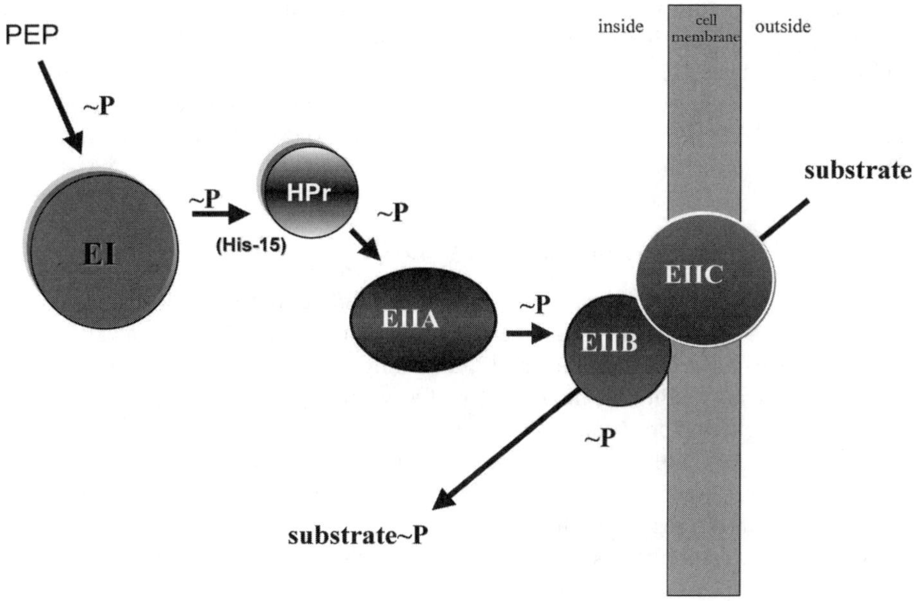

FIGURE 25.1 The bacterial phosphoenolpyruvate-dependent phosphotransferase system. The phosphoryl residue (~P) is transferred from PEP to the translocated substrate via the phosphoryl transfer chain, Enzyme I (EI) HPr and the Enzyme IIA and IIB domains of the multidomain enzyme II complex. Enzyme II is responsible for translocation of the substrate, via the IIC domain. The phosphorylation of HPr in the transport reaction is at residue His-15.

IIB domain, which is responsible for phosphorylating the incoming substrate. The EIIC domain is a transmembrane protein that is responsible for substrate translocation across the membrane. In the presence of the inducing substrate, the phosphate is passed to the substrate from PEP via the chain EI – HPr –EIIA –EIIB (Figure 25.1). Therefore, while sufficient substrate is available, the EIIB domain will predominantly exist in the dephosphorylated state. In contrast, when the inducer is absent, the EIIB domain is unable to donate its phosphate to the substrate and will, therefore, be in its phosphorylated state. As discussed later, this can have consequences for the regulation of gene expression.

The specificity of an individual PTS for its substrate is conferred by the EII complex, and, accordingly, a PTS gene system will at least encode the relevant EII. An EII can be a single protein encoded by a single gene; however, depending on domain nature and architecture, a specific EII complex might be encoded by more than one gene. This variation in gene arrangement is observed in clostridia; for example, the complete EII for sucrose in *C. acetobutylicum* is encoded by a single gene [1], while the EII for glucitol in *C. beijerinckii* is encoded by three distinct genes, each encoding separate components of the glucitol EII complex [2]. Therefore, solely to encode the EII, a PTS gene system may require more than one gene, necessitating coordinate regulation of expression. In addition to the possibility of multiple genes encoding the EII complex, there are often other indispensable genes associated with those encoding the PTS. The product of PTS transport is a phosphorylated substrate that usually requires further modification before it can enter central metabolism. It is, therefore, common for the genes encoding the substrate-specific EII to be in an operon with genes specifically required to modify the product of the PTS reaction. In the case of disaccharides, it is typical for the operon to encode a specific phosphohydrolase enzyme that cleaves the phosphorylated substrate into monosaccharide units. In *C. acetobutylicum,* for example, maltose is transported by a PTS, resulting in the accumulation of maltose 6-phosphate, which in turn is hydrolyzed to glucose and glucose 6-phosphate. This process requires two substrate specific activities — i.e., the EII for maltose and a maltose 6-phosphate hydrolase — the genes for which have been identified in an operon in this organism [3]. Depending on the nature of the substrate, there might also be additional genes in such operons encoding enzymes required for further processing of the hydrolysis products. The sucrose gene system in *C. acetobutylicum* is an example, where, in addition to genes encoding the sucrose EII and a sucrose 6-phosphate hydrolase, the operon also encodes a fructokinase [1]. In clostridia, as with other bacteria, PTS operons therefore appear to be complex multigene systems.

25.3 INDUCTION OF CATABOLIC OPERONS

The expression of catabolic gene systems that have been studied to date in clostridia is coordinately regulated in response to the presence of an inducing substrate. In other bacteria, a variety of different induction mechanisms have been characterized, including de-repression, activation, and antitermination. Such mechanisms require the involvement of a system-specific regulator (such as a repressor protein). Although regulator proteins act *in trans*, it is common for the gene encoding the system regulator to be associated with the target operon. Indeed, the regulator gene may be encoded as part of the target operon, with the regulator and substrate utilization genes transcribed in a single mRNA transcript (with the regulator subject to autoregulation).

The activity of the catabolic operon regulators is modulated in response to the availability of the inducer. This might be as a consequence of a direct interaction with the inducer (or, more likely, an intracellular metabolic derivative of the external inducer) as is often observed with repressor proteins. However, the regulator can also respond indirectly to availability of the inducer. The activity of certain types of regulators can be modulated by the PTS. In particular, regulators that possess conserved PTS regulation domains (PRDs) can be phosphorylated by PTS components to stimulate or inhibit regulator activity [4]. A PRD is a short region (approximately 100 amino acids) that includes two highly conserved histidine residues. The reversible phosphorylation of a PRD by the PTS occurs at the conserved histidine, modifying the activity of the regulator protein either positively or negatively. In fact, most regulator proteins possess two PRDs, and consequently, the PTS can in many cases exert both positive and negative control over the same regulator [4–6]. Such proteins are typically positive regulators, including transcription activators and an emerging family of antiterminator proteins [7].

As more and more operons are investigated in clostridia and with the availability of a growing number of completed genome sequences, it is apparent that there is a diversity of induction mechanisms employed in the genus.

25.3.1 Repressor-Mediated Regulation

The regulation of catabolic operons by means of a repressor protein is widespread among bacteria. The classic example is the *lac* operon in *Escherichia coli,* which is regulated by the LacI repressor. In the absence of the inducer substrate (lactose), the repressor binds to a specific operator sequence associated with the target *lac* promoter to prevent transcription of the operon. In the presence of external lactose, an internally generated effector molecule (allolactose) binds to and inactivates the repressor. In the absence of the functional repressor, transcription of the *lac* operon is free to proceed. This basic mechanism has been identified in many bacteria, where repressors similar to LacI have been characterized. These repressors belong to the LacI/GalR family and contain a characteristic helix-turn-helix DNA binding motif [8]. Although the *E. coli lac* operon remains the best studied of these systems, repressor mediated regulation has been well characterized for a number of catabolic systems in Gram-positive bacteria. Other families of repressor proteins have also been identified, such as the DeoR family, which includes proteins that negatively regulate the expression of gene systems including PTS operons in Gram-negative and Gram-positive bacteria [9,10]. It is not surprising, therefore, that repressor-mediated regulation of catabolic operons has also been described for species of *Clostridium.*

25.3.1.1 Sucrose Repressor

C. beijerinckii can grow on sucrose as a sole carbon source, using a pathway that involves a PTS transport mechanism, a sucrose 6-phosphate hydrolase, and a fructokinase [11]. The genes encoding the enzymes of this pathway were cloned in *E. coli* in a problematic cloning that required three stages to obtain the full operon [12]. The gene system was sequenced and found to constitute an operon in the order *scrARBK* (Figure 25.2); where *scrA* encodes a sucrose specific EII (EII^{Scr}), *scrB* encodes the sucrose 6-phosphate hydrolase, and *scrK* encodes the fructokinase. The *scrB* and *scrK* open reading frames partially overlap. The other gene in this system, *scrR*, encodes a repressor of the LacI/GalR family. The repressor contains a charactersitic N-terminal helix-turn-helix motif and shares a proposed ligand binding domain with the equivalent ScrR from *Staphylococcus xylosus* [13]. An obvious candidate for the role of effector ligand is the product of the PTS reaction, sucrose 6-phosphate, but this has not been investigated. RNA analysis confirmed that the *C. beijerinckii scr* genes are transcribed in an operon from a promoter located upstream of the first gene, *scrA*. Transcription of the operon is induced when the culture is grown on sucrose, but this is not detected when the culture is grown on other substrates including glucose, fructose, xylose, and maltose. A strain was constructed with a targeted integration to disrupt the *scrR* gene. As this gene lies downstream of the *scrA* gene it was possible to investigate the effects of the insertion in *scrR* on regulation of the operon by monitoring expression of *scrA*. In contrast to the parental strain, there was expression of the operon in the mutant strain under noninducing conditions — confirming the negative regulation of the operon by the ScrR repressor [12]. There was no evidence for an independent transcript containing the *scrR* gene, and, hence, given the gene order *scrARBK*, the authors concluded that the *scrR* gene must be subject to autoregulation. This is a common feature in the regulation of catabolic gene systems.

The complete genome of the pathogen *C. perfringens* has recently been published [14]. Very little is known about the PTS in this organism, but it is known to operate as glucose PTS activity has been demonstrated [15]. An analysis of the genome sequence reveals the presence of a number of putative PTS gene systems, although the designation of these gene systems is tentative, at best. Nevertheless, it is of interest that *C. perfringens* has a putative gene system equivalent to the

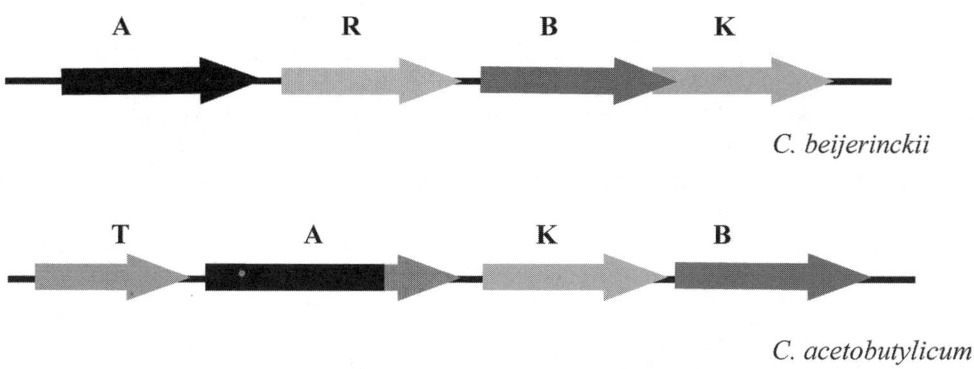

FIGURE 25.2 Sucrose PTS operons of *C. beijerinckii* and *C. acetobutylicum*. The equivalent gene systems encoding enzymes responsible for sucrose uptake and metabolism in *C. beijerinckii* and *C. acetobutylicum* are depicted — each encoding a PTS Enzyme II (*scrA*), a sucrose 6-phosphate hydrolase (*scrB*), and a fructokinase (*scrK*). The *C. acetobutylicum scrA* gene encodes an EIIA domain (shaded), which is not present in the *C. beijerinckii scrA*, while the gene order with respect to *scrB* and *scrK* is also reversed in the two species. Most strikingly, the operons encode very different regulatory proteins, with the *C. beijerinckii scrR* gene encoding a repressor protein, while the *C. acetobutylicum scrT* encodes a member of the BglG family of antiterminator proteins.

C. beijerinckii scr operon. There are four open reading frames in the same order, *scrARBK*, encoding proteins highly homologous to ScrA (72% identity; 83% similarity), ScrR (61% identity, 77% similarity), ScrB (50% identity; 73% similarity), and ScrK (63% identity; 78% similarity) of *C. beijerinckii*. As pointed out by Reid et al. [12], it is unusual for a fructokinase of this kind to be associated with a sucrose PTS gene system in other Gram-positive bacteria. Furthermore, the degree of conservation is such that even the overlap between *scrB* and *scrK* is conserved between the two species of *Clostridium*. The overall homology between the proteins and the conservation of gene order between the two species suggests that the putative *C. perfringens scr* operon is indeed a gene system for sucrose uptake and metabolism, which is under the control of the repressor protein ScrR.

25.3.1.2 Lactose Repressor

The uptake and metabolism of lactose by *C. acetobutylicum* ATCC 824 has been investigated at our laboratories. Lactose is transported by a PTS mechanism, and the product of the reaction is further metabolized by a phospho-β-galactosidase. The genes encoding these activities are found in an operon together with a gene encoding a repressor protein, which we have designated LacR. The lactose operon is induced by lactose and galactose (and repressed by glucose). Induction of the operon is believed to be negatively regulated by the activity of the LacR repressor, which is a member of the DeoR family of repressors. A detailed analysis of this operon will be presented elsewhere [16].

The induction of a lactose metabolic system has also been reported in *C. perfringens*, where lactose was shown to induce transcription of an operon containing a β-galactosidase gene, *pbg*, and a putative lactose permease — although, in this case, the uptake of lactose was not the subject of the investigation, and the regulatory mechanism was not addressed [17].

25.3.1.3 Fructose PTS

Substrate-induced fructose PTS activity has been demonstrated in a number of clostridia, including *C. beijerinckii,* where the product of the reaction is fructose 1-phosphate [18]. An examination of the *C. acetobutylicum* and *C. perfringens* genomes reveals the presence of homologous putative fructose PTS gene systems. In both of these organisms, the fructose EII complex is encoded in the domain order EIIABC, and the domains all share considerable homology, but the coding for these differs in the two species. In *C. perfringens* the complete EII is encoded by a single gene, while in *C. acetobutylicum,* the EIIA and IIBC domains are encoded by two tandem genes. These genes are preceded by a gene encoding a putative fructose 1-phosphate kinase. As with *C. beijerinckii*, we have evidence that the product of the PTS reaction in *C. acetobutylicum* is fructose 1-phosphate. The association of the EII fructose with a fructose 1-phosphate kinase allows for the conversion of this product to fructose 1, 6-bisphosphate, which can subsequently enter central metabolism.

The fructose operons in *C. acetobutylicum* and *C. perfringens* are each associated with a gene, *fruR*, which encodes a protein related to the DeoR family of transcriptional regulators. Of the two proteins, the *C. perfringens* FruR is the more closely related to other members of the DeoR family and shares 50% identity and 70% similarity with the FruR repressor of the fructose PTS operon in *B. subtilis* [19]. The putative *C. acetobutylicum* FruR protein only shares 36% identity and 57% similarity with the *C. perfringens* FruR but is equally related to DeoR regulator proteins from other organisms, including Gram-negative bacteria. Although fructose PTS activity in *C. perfringens* has not been investigated and the mechanism of induction of fructose PTS activity in *C. acetobutylicum* has not been characterized, we suggest that these related *fru* gene systems encode fructose PTS transport and metabolic pathways and that they are negatively regulated by a repressor protein of the DeoR family.

25.3.1.4 MalR

Repressor mediated regulation of catabolic gene systems other than PTS operons has also been observed in clostridia. A negatively regulated gene system has been studied in *C. butyricum*. The system is in the order *malRQP*, where *malQ* was shown experimentally to encode a 4-α-glucanotransferase and *malP* (by homology) encodes a maltodextrin phosphorylase [20]. The *malR* gene encodes a protein with significant homology to members of the LacI/GalR family, and it contains an N-terminal helix-turn-helix motif. When cloned in *E. coli*, the introduction of a frameshift mutation in *malR* resulted in a 10-fold increase in MalQ activity [21]. The natural inducing substrate for this gene system has not been identified nor have the effects of the frameshift in *malR* been investigated in the native background (due to the difficulties in genetically manipulating this organism). Nevertheless, the available evidence is consistent with the role of MalR as a repressor that negatively regulates expression of the *malQP* operon.

The *C. butyricum malQ* gene was used to probe a gene library from *C. beijerinckii* NCIMB 8052, and we identified a homologous gene in this organism. The clone was sequenced and found to encode a gene system also in the order *malRQP* [22]. As with *C. butyricum*, the natural inducing conditions for the *C. beijerinckii* gene system have not yet been identified — although it does not appear to be induced by maltose. There is indirect evidence that MalR negatively regulates expression of *malQP* in *C. beijerinckii*. Using reporter gene constructs, we have observed that the introduction of a frameshift mutation in *malR* results in a 10-fold increase in reporter activity in the native *C. beijerinckii* background — consistent with the designation of MalR as a repressor for this gene system [23]. It will be of interest to determine the natural inducing substrate for this gene system and to investigate induction in the native background.

25.3.1.5 Xylose Repressor

One of the best characterized repressor-regulated systems in Gram-positive bacteria is the xylose utilization system, which is found in a number of *Bacillus* species including *B. subtilis*, *B. megaterium*, and *B. licheniformis* [24–26]. Regulation of xylose utilization is negatively controlled by the XylR repressor, which is a member of a somewhat disparate family of repressors called the ROK family [27]. In the absence of xylose, XylR binds to a specific regulatory sequence in the target operons. The induction is strong and the repression is stringent, which has led to the exploitation of the elements of this regulatory mechanism as a tool for heterologous gene expression in *B. subtilis* [28].

It has been established that *C. acetobutylicum* is capable of growth on xylos [29]. It is, therefore, of interest that in a comparative analysis of pentose utilization systems in the *Bacillus/Clostridium* group, a putative XylR was identified in the *C. acetobutylicum* genome [27]. Although phylogenetic analysis revealed that the clostridial protein was only distantly related to the other XylR proteins in this group, it was proposed that XylR regulates xylose metabolism genes in this organism. A novel XylR regulatory sequence was proposed, based on the occurrence of a conserved sequence proximal to xylose metabolic gene systems. The proposed consensus for this new regulatory sequence matched the signal for another member of the ROK family, the NagC repressor from *E. coli* [27]. While the significance of the XylR repressor in *C. acetobutylicum* remains to be established, the possibility of exploiting this system for regulating expression of cloned genes, as has already been done so effectively in *Bacillus*, is an interesting prospect.

25.3.2 Antiterminator-Mediated Regulation

A mechanism of regulation that was first described in *E. coli*, and has since been identified for a variety of PTS operons in a number of Gram-positive bacteria, is that mediated by the BglG family of antiterminator proteins [7]. These proteins possess an RNA-binding domain that is involved in

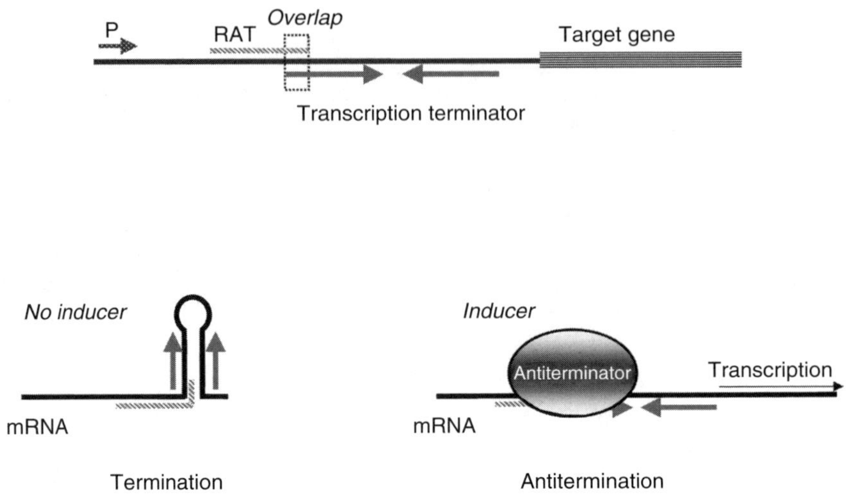

FIGURE 25.3 Regulation of transcription by BglG-type antiterminator proteins. Genes subject to regulation by antiterminator proteins of the BglG family are characteristically preceded by a transcription terminator that is partially overlapped by a ribonucleic antiterminator (RAT) sequence. Transcription is initiated constitutively at a promoter (P) upstream of these regulatory sequences. In the absence of inducer the transcription terminator structure forms and transcription of the subject gene is prevented. In the presence of the inducer, the antiterminator protein is activated and interacts with the RAT sequence to preclude the formation of the transcription terminator, thereby allowing transcription of the subject gene.

antitermination of transcription. Their activity is modulated (positively and negatively) at a PRD by components of the PTS. Genes or operons subject to regulation by a BglG type antiterminator are preceded by characteristic *cis*-acting sequences involved in regulation of transcription. The target gene (or operon) is preceded by a transcription terminator sequence, which in turn is partially overlapped at the 5′ end by a conserved sequence called a ribonucleic antiterminator (RAT). The RAT sequence has the potential to form a characteristic stem-loop secondary structure in the transcript. This structure is stabilized by an interaction with the associated antiterminator protein. As the RAT sequence overlaps the 5′ end of the terminator, the formation of the transcription terminator and the RAT structure are mutually exclusive. Transcription is initiated from a constitutive promoter that lies upstream of the RAT/terminator control region. Under noninducing conditions, the terminator forms — thereby preventing transcription into the regulated gene (or operon). Under inducing conditions, the associated antiterminator protein is activated to stabilize the RAT structure, thereby precluding the formation of the terminator and allowing transcription to proceed normally (Figure 25.3) [4,7,30].

This mechanism of gene regulation is associated with PTS gene systems and, in particular, those involved in transport and metabolism of β-glucosides such as the *bgl* operon in *E. coli* [31], which lends its name to this family of antiterminator proteins. In Gram-negative bacteria, BglG-type antiterminators appear to be exclusively associated with β-glucoside systems; however, these antiterminators are associated with a greater diversity of PTS substrates in Gram-positive bacteria (including sucrose, lactose, and glucose) [32]. A number of these proteins have been studied in detail, notably BglG from *E. coli* and the *B. subtilis* proteins SacT and LicT. Direct binding of these proteins to their target RAT sequences has been demonstrated *in vitro*. An N-terminal RNA binding domain has been identified, and it has been demonstrated that N-terminal fragments possess constitutive antitermination activity [7,33,34].

The regulation of antitermination activity is mediated by the two C-terminal PRDs found in these proteins. Typically, each PRD contains two histidine residues that may be phosphorylated by components of the PTS. Negative regulation of antiterminator activity is achieved by phosphorylation

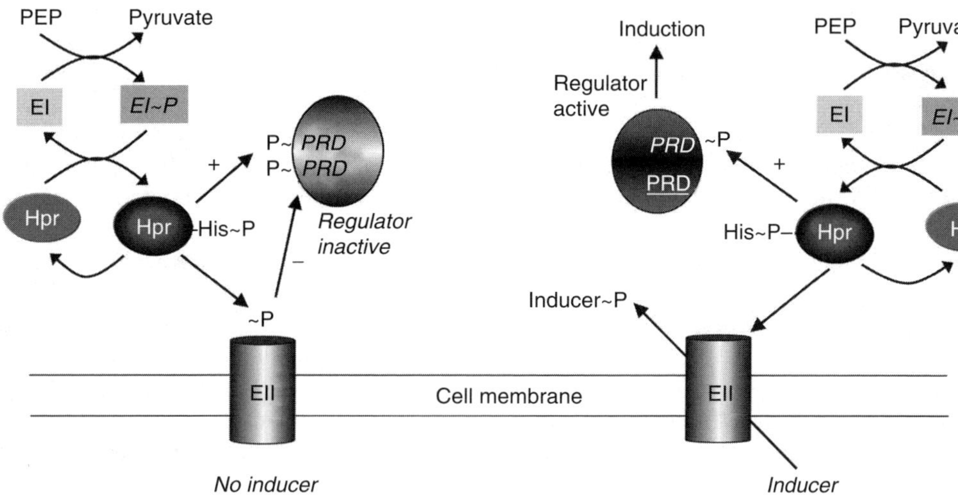

FIGURE 25.4 Modulation by phosphorylation of PRD containing regulator proteins. The diagram depicts the modulation of activity of a typical PRD-containing regulator (such as BglG-type antiterminator proteins), which may be positively phosphorylated (+) by HPr and negatively phosphorylated (-) by the cognate PTS EII. In the absence of inducer the PTS components accumulate in the phosphorylated state. Consequently, EII may mediate phosphorylation of the regulator at one of the PRDs (as may HPr at the other), and the regulator is in the inactive conformation. In the presence of inducer, phosphate is channeled through the PTS to the incoming substrate and the substrate specific EII is in the dephosphorylated state. The inhibiting phosphorylation of the regulator by the EII is therefore prevented by the preferential phosphorylation of the incoming substrate, while the regulator may additionally be positively phosphorylated by Hpr. Hence, in the presence of inducer, the regulator is in the activated state, thereby facilitating induction of the target gene system.

of a PRD mediated by the cognate EIIB of the substrate-specific EII complex for the PTS substrate. As mentioned previously, the availability of substrate is reflected in the phosphorylation state of the EIIB domain. In the absence of the inducer, the EIIB domain is predominately in its phosphorylated state, which allows it to phosphorylate and inactivate the antiterminator. However, in the presence of the inducer, the phosphate is siphoned away by the substrate during the PTS transport reaction. The EIIB is, therefore, predominately in its dephosphorylated state and is unable to phosphorylate the antiterminator, thereby relieving inhibition of the antiterminator and allowing transcription of the associated operon [4,6,32,35]. The substrate-specific EII complex therefore acts as the environmental sensor for substrate availability and, via the antiterminator, controls induction of the cognate gene system. Some (but not all) antiterminators can also be positively regulated via phosphorylation by components of the PTS in a reaction that is associated with CCR. This PEP-dependent phosphorylation is mediated by EI and HPr and is required for activation of antiterminator activity. In such antiterminators, when both PRDs are in the phosphorylated state, it is the EIIB-mediated negative regulation that is dominant and the antiterminator is inactive [4,6], as depicted in Figure 25.4.

25.3.2.1 The Sucrose Operon of *C. acetobutylicum* ATCC 824

Sucrose is transported and metabolized in *C. acetobutylicum* by the same sucrose-induced pathway identified in *C. beijerinckii*. The gene system encoding the coordinately induced EII^{scr} (*scrA*), sucrose 6-phosphate hydrolase (*scrB*) and fructokinase (*scrK*) has been identified [1]. The *C. acetobutylicum* enzymes, as expected, are very homologous to the *C. beijerinckii* proteins, although the *scrA* gene in *C. acetobutylicum* encodes an EIIA domain, which is absent in the *C. beijerinckii scrA*. The gene order is also reversed with respect to *scrK* and *scrB,* and there is no overlap between these two genes, as is seen in *C. beijerinckii* [12]. However, the most striking difference between the two systems is the absence of the repressor gene *scrR* in *C. acetobutylicum*. Instead, the operon is associated with an antiterminator, ScrT, of the BglG family — as is also found with two different sucrose PTS operons in *B. subtilis* and a sucrose PTS gene system in *B. stearothermophilus* [32,36].

The *C. acetobutylicum* sucrose operon is of the gene order *scrTAKB* (Figure 25.2). The ScrT protein is homologous to other known antiterminators, including the sucrose PTS system antiterminators found in *B. subtilis*, SacT, and SacY, as well as the *B. subtilis* LicT and the *E. coli* BglG [1,7]. ScrT also contains the four conserved histidine residues in the two PRDs of LicT and BglG, which are known to be phosphorylated by the components of the PTS [6,37,38]. An operon promoter was recognized upstream of *scrT,* and the genes are transcribed in a single transcript that is induced by sucrose [1]. The *scrT* and *scrA* genes are both immediately preceded by transcription terminators. The terminator preceding *scrT* is partially overlapped by a highly conserved RAT sequence, consistent with a BglG-type regulation. However, the terminator preceding *scrA* is partially overlapped by a very unusual RAT-like sequence. At 42 nt in length, the proposed RAT sequence is significantly longer than a typical RAT, and an alternative secondary structure has been postulated for this sequence [1]. In support of this proposal, a re-examination of the *arb* operon in *Erwinia chrysanthemi* [39] revealed that the novel clostridial RAT is not unique, as the *arb* operon has a homologous RAT sequence that can be folded to give a very similar secondary structure [1], as depicted in Figure 25.5.

We examined the *C. acetobutylicum* genome sequence for evidence of other PTS gene systems that may be regulated by a BglG antitermination mechanism. Two candidate systems were identified, on the basis that each of these putative PTS gene systems is preceded by a gene (in the same orientation) encoding a protein homologous to known antiterminator proteins, and in each case the relevant EII gene is preceded by a transcription terminator that is partially overlapped by a conserved RAT sequence, characteristic of this mechanism of gene regulation [40]. One of these systems is a putative glucose PTS and the other is a putative β-glucoside PTS. The regulation of β-glucoside systems by this mechanism is relatively common, while regulation of glucose PTS systems by this mechanism has been demonstrated in *B. subtilis* [41] and *Staphylococcus carnosus* [42]. We have confirmed PTS activity for glucose as well as the β-glucoside cellobiose in extracts of induced cultures of *C. acetobutylicum* [40]. There would, therefore, appear to be at least three PTS gene systems in *C. acetobutylicum* subject to the BglG-type antiterminator mechanism of regulation; namely sucrose, glucose, and β-glucoside — which, coincidentally or otherwise, are the same antiterminator-regulated substrate systems found in *B. subtilis*.

25.3.2.2 *C. longisporum abg* Operon

The first BglG-type antiterminator to be associated with a species of *Clostridium* was the *abg* operon in the rumen organism C. *longisporum* [43]. This encodes a PTS system for β-glucosides that includes a BglG-type antiterminator. There are three genes in the system *abgGFA*, where *abgG* encodes the antiterminator, *abgF* encodes a PTS EII, and *abgA* encodes a phosphoglucosidase. A typical RAT/terminator system was identified upstream of *abgF.* A second putative RAT/terminator system was identified upstream of *abgG,* although in this case the suggested RAT structure does not closely resemble other RATs. The gene system was not studied in the native background, but was cloned and investigated in *E. coli* by complementation analysis, whereby activity was ascribed to the EII and the phosphoglucosidase. A promoter was identified upstream of *abgG*, although the regulatory system did not appear to be functional in *E. coli* [43].

25.3.3 GlpP Antiterminator

Antitermination as mediated by the BglG family of regulators is not the only form of antitermination activity found in Gram-positive bacteria. In *B. subtilis,* glycerol uptake is by facilitated diffusion,

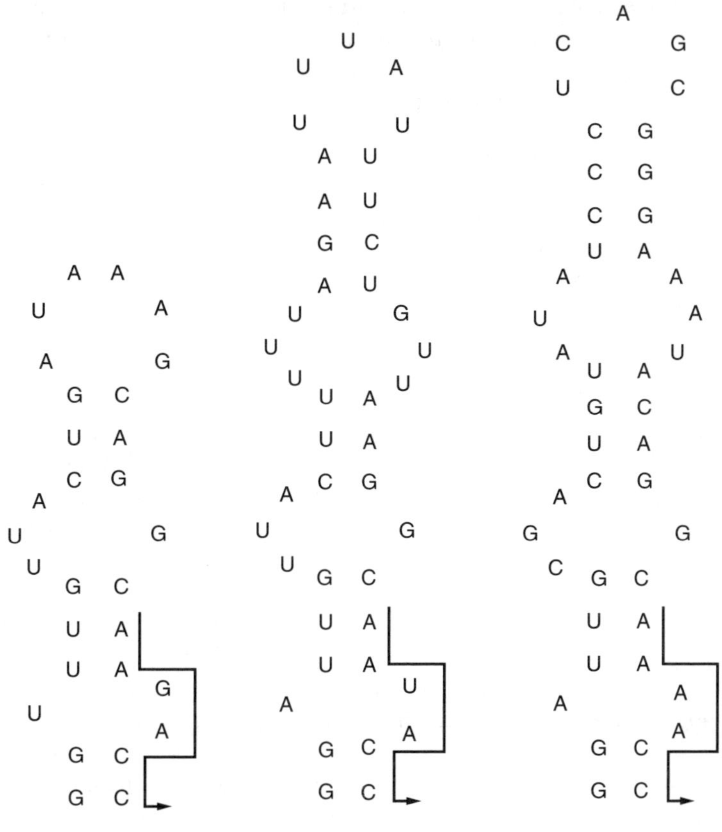

FIGURE 25.5 Ribonucleic antiterminator (RAT) structures. Depicted are the proposed secondary structures for the RAT sequences preceding the *C. acetobutylicum* genes (1) *scrT* and (2) *scrA*, and (3) the RAT located downstream of the *Erwinia chrysanthemi arbG* gene. The first of these (1) is a typical RAT structure, while the other two (2,3) are novel but very similar structures proposed for these unusually long RAT sequences.

with the translocated substrate being subsequently trapped within the cell via conversion to glycerol-phosphate by a glycerol kinase before subsequent metabolism. The *B. subtilis* GlpP antiterminator protein positively regulates the expression of the glycerol regulon, which encodes the genes involved in glycerol uptake and metabolism. Antiterminator activity is stimulated by glycerol-3-phosphate, which acts as the inducer for the system [7,44,45]. In addition to antitermination activity, it has also been reported that binding of glycerol-3-phosphate activated GlpP can stabilize its target mRNA [46].

A number of species of *Clostridium* can metabolize glycerol [47,48]; although, to the best of our knowledge, there have been no reports of characterization of a transport mechanism for glycerol in any of these organisms. An analysis of the respective genomes reveals the presence of equivalent genes encoding a glycerol uptake facilitator protein in *C. acetobutylicum* and *C. perfringens*. These genes are part of a gene system that appears to encode proteins involved in glycerol uptake and metabolism. In each organism, the genes are in the order *glpFPKA*, where, by homology to known proteins, *glpF* encodes the glycerol transport protein, *glpK* encodes a glycerol kinase, and *glpA* encodes a glycerol-3-phosphate dehydrogenase. If these gene systems do encode a glycerol uptake and metabolic pathway, then the identity of *glpP* genes within both of these systems is of interest, as by homology this gene encodes a GlpP antiterminator protein. Although the identity remains to be established experimentally, it appears that *glpP* encodes a glycerol 3-phosphate dependent transcription activator that represents another mechanism of gene regulation in clostridia.

25.3.4 Transcription Activator Proteins

In terms of control of expression of catabolic gene systems, transcription activators are proteins that bind to specific regulatory sequences (usually in the promoter region of the target gene system) and stimulate transcription. Activator proteins in Gram-positive bacteria are typically controlled indirectly by phosphorylation in response to substrate availability, although there are also activators that respond directly to the presence of the inducer. An example of the former class of activators has been described in *C. acetobutylicum.*

25.3.4.1 MtlR

The MtlR protein from *C. acetobutylicum* DSM 729 is encoded by, and thought to activate, the inducible mannitol PTS operon (*mtl*) in this organism [49]. There are four genes in the *mtl* operon, including *mtlR*, in the gene order *mtlARFD* where *mtlA* and *mtlF* encode the EIICB and IIA domains, respectively, of the PTS EII complex, and *mtlD* encodes a mannitol-1-phosphate dehydrogenase. As revealed by genome sequencing, an almost identical operon is present in *C. acetobutylicum* ATCC 824. The expression of the *mtl* operon has been investigated in *C. acetobutylicum* DSM 792 [49]. The genes are transcribed as a polycistronic mRNA, which is induced by mannitol by a mechanism thought to be mediated by the activator protein MtlR. As with other activators of this type, the *C. acetobutylicum* MtlR protein contains two PRDs as well as a C-terminal domain with significant homology to a PTS EIIA domain. The MtlR protein from *B. stearothermophilus* has been shown to be phosphorylated by components of the PTS both at the PRDs and at the EIIA domain. Phosphorylation of the MtlR-IIA domain is mediated by the mannitol IICB ($IICB^{mtl}$) and inhibits MtlR binding to DNA [50,51]. Therefore, in the absence of mannitol, $IICB^{mtl}$ phosphorylates MtlR, and it cannot activate transcription of the *mtl* operon. In contrast, in the presence of mannitol, the phosphate is channeled to the PTS transport reaction, and the MtlR-IIA domain is in its active dephosphorylated state. As with certain antiterminator proteins, MtlR can also be both positively and negatively phosphorylated by HPr at the PRDs — phosphorylation of PRD-I causing inhibition of DNA binding and PRD-II leading to activation of the protein [50,51].

25.3.4.2 Glucitol Transport

Glucitol uptake and metabolism in *C. beijerinckii* has been investigated, and a PTS transport system and associated metabolic enzymes were identified [2]. The gene system involved was cloned and sequenced. There are five genes required to encode the glucitol uptake and metabolic pathway, and, when expressed in *C. beijerinckii,* these genes are co-transcribed. The genes are coordinately induced by glucitol and subject to catabolite repression by glucose. In *E. coli,* there are two regulatory proteins, GutM and GutR, involved in the regulation of the equivalent *gut* operon [52]. The GutM protein is a transcription activator that is required for expression of the operon, while GutR is also involved in regulation and is similar to DeoR proteins. Since our initial publication, we have obtained additional sequence data from upstream of the published *C. beijerinckii gut* gene system and we have identified one complete upstream ORF that encodes a protein homologous to GutM, and an as-yet-incomplete ORF that appears to encode the C-terminal of GutR. If the complete gene does encode a GutR, then these two proteins, GutR and GutM, are equivalent to the previously characterized glucitol regulatory proteins GutR and GutM in the *E. coli gut* operon. The putative GutM and the available incomplete sequence of GutR are both most closely related to the equivalent proteins in *Streptococcus mutans* [53]. The gene order is also more closely related to the *S. mutans* operon, where, in contrast to the *E. coli* system, the regulatory genes precede the metabolic genes. While this conservation of regulatory regions is of interest, the sequence data for *gutR* is incomplete, and the precise roles of GutR and GutM in the regulation of the *C. beijerinckii gut* operon await further investigation.

25.3.5 Maltose PTS

Maltose uptake and metabolism has been characterized in *C. acetobutylicum,* and a regulated gene system has been identified, which includes a gene designated *malP*, encoding a PTS EII for maltose, and *malH,* which encodes a maltose 6-phosphate hydrolase [3]. Neither activity was found in cells grown on glucose or sucrose, but activity was detected in cells grown on maltose. The system is, therefore, clearly subject to regulation, but whether this represents induction by maltose, or catabolite repression of the system by both glucose and sucrose is not clear. The *malPH* gene system is also associated with an upstream ORF encoding a protein with low homology to putative transcriptional regulator proteins [3]. The putative protein has a predicted helix-turn-helix motif and is homologous to a protein designated YfiA encoded by the maltose PTS gene system in *B. subtilis* [54]. In *B. subtilis,* the *yfiA* gene lies between *malP* and the equivalent of *malH* — designated *malA* in this organism — in the gene order *malP-yfiA-malA*. Intriguingly, we have identified a homologous gene system in *C. perfringens*, which also encodes a YfiA homologue, in the gene order *malP-yfiA-malH*. Although these observations are consistent with the deduction that YfiA is a transcriptional regulator, the role (if any) of YfiA in the regulation of maltose metabolism in *C. acetobutylicum* and *C. perfringens* remains to be established.

25.4 CARBON CATABOLITE REPRESSION (CCR)

There are a variety of mechanisms by which bacteria effect CCR, not least of which is at the level of transcription. Irrespective of the mechanism by which induction of catabolic operons might be achieved, clostridia (like other bacteria) can exert an overriding mechanism of transcriptional control, whereby, in the presence of a preferred substrate (such as glucose), induction is by some means prevented. The functionally equivalent sucrose PTS operons in *C. acetobutylicum* and *C. beijerinckii,* for example, are induced by very different mechanisms (controlled by an antiterminator and a repressor, respectively), but both are subject to repression by glucose [1,12], while repression of transcription has also been demonstrated for the *mtl* operon in *C. acetobutylicum*, which is regulated by a multidomain PRD-containing regulatory protein [49]. Furthermore, it is not merely a matter of responding to the presence of glucose, as there is a hierarchy of preferred substrates such that the cell continually senses its environment and responds to the availability of whatever might be the most favored carbon source present. This necessitates an ability to react and adapt if a cell is to maximize its potential for survival in a changing environment. When one considers the diversity of induction mechanisms that can operate in clostridia, it would seem logical that some form of central CCR control system should evolve rather than a cell having to elicit a variety of independent CCR mechanisms to prevent the induction of operons controlled by unrelated regulatory mechanisms. Such a central mechanism has been proposed for a number of species of *Bacillus* as well as other low-GC Gram-positive bacteria.

25.4.1 Carbon Catabolite Repression in *Bacillus*

The PTS plays a pivotal role in CCR in low-GC Gram-positive bacteria, with the HPr protein being of particular importance. In contrast to Gram-negative bacteria, HPr in these bacteria may be phosphorylated not only by Enzyme I at His-15 in the PEP-dependent transport reaction, but also at Ser-46 by an ATP-dependent kinase. The latter reaction has a regulatory function and is the key to CCR in these bacteria. In fact, *B. subtilis* possess a second HPr-like protein, Crh, which is also phosphorylated at Ser-46 by HPr kinase and contributes to CCR, but which lacks His-15 and consequently is not involved in PTS transport [55]. In *B. subtilis,* HPr kinase activity is stimulated by fructose 1, 6-bisphosphate (FBP) — a glycolytic intermediate that is thought to serve as a signal for the availability of a preferred carbon source. A rapidly metabolized carbon source, such as glucose, will lead to high levels of glycolytic intermediates, including FBP. This stimulates HPr kinase activity, resulting in the phosphorylation of HPr at Ser-46 to form HPr(Ser~P). This reaction is reversible; in fact, HPr kinase is actually a bi-functional enzyme (kinase/phosphorylase) that can also dephosphorylate HPr(Ser~P) — a reaction that is stimulated by inorganic phosphate and forms pyrophosphate [56,57]. The kinase/phosphorylase enzyme, HprK/P, therefore acts as a direct sensor of the energetic state of the cell and responds accordingly with kinase or phosphorylase activity, as appropriate.

An immediate consequence of phosphorylation of HPr at Ser-46 is that the PEP-dependent EI-mediated phosphorylation of HPr in the PTS transport chain is inhibited, thereby reducing PTS activity in a general response to the energetic state of the cell [58]. This also has implications for operons subject to regulation by proteins that require activation via phosphorylation by HPr(His~P) at a PRD. Certain BglG-type antiterminators require phosphorylation by HPr(His~P) for activation of antiterminator activity [7] (as depicted in Figure 25.4), while transcriptional regulators such as MtlR also require this activation [50,51]. In the case of such antiterminators, when both the inducer and a repressor substrate are available, the negative phosphorylation of the antiterminator by the cognate EIIB~P may be relieved by phosphorylation of the substrate (as depicted in Figure 25.4), but the lack of positive phosphorylation of the antiterminator by HPr(His~P) results in continued repression of the target operon (i.e., CCR), despite the availability of the inducer substrate (i.e., the positive phosphorylation of the regulator in the presence of the inducer as depicted in Figure 25.4 does not occur when a repressing substrate is also present and the regulator remains inactive). Significantly, antiterminators such as the *B. subtilis* SacY, which are not subject to CCR, do not require the positive phosphorylation reaction for antiterminator activity, although SacY can be phosphorylated by HPr(His~P) *in vitro* [6]. Therefore, in the presence of inducer, this antiterminator can be activated independently of phosphorylation by HPr(His~P).

The phosphorylation of HPr by HPrK/P can also result in repression of catabolic gene systems via an interaction between HPr(Ser~P) and a specific co-repressor, the catabolite control protein CcpA [32,59]. The CcpA/HPr(Ser~P) regulation complex binds to a specific operator sequence called the catabolite responsive element, *cre*, which is typically located in the promoter region of catabolic systems that are subject to this mode of catabolite repression. The binding of the complex to a *cre* in such systems results in the prevention of transcription, as depicted in Figure 25.6.

The key effector proteins of this mechanism of CCR, namely, HPr, HPrK/P, and CcpA, together with target *cre* sequences and regulator proteins containing PRDs, have been identified in a variety of low-GC Gram-positive bacteria, and this model of CCR is either known or assumed to operate in these bacteria. The clostridia belong to the low-GC Gram-positive group of bacteria and recent work coupled with the sequencing of a number of clostridial genomes has provided valuable insight into CCR in clostridia.

25.4.2 HPr in Clostridia

The *ptsH* gene, encoding HPr, has recently been identified from the *C. acetobutylicum* genome sequence and cloned in *E. coli* [60]. The cloned clostridial gene complemented a *ptsH* mutant of *E. coli*, thereby confirming its identity as *ptsH*. An alignment of the deduced amino acid sequence of the *C. acetobutylicum* HPr with that of HPr sequences from other low-GC Gram-positive bacteria revealed that the clostridial protein shared considerable homology throughout its length and, unsurprisingly, was particularly well conserved in the region around His-15 (the site of phosphorylation for the PTS transport reaction). However, the sequence surrounding Ser-46 (the site of phosphorylation for the regulatory ATP-dependent kinase reaction), which is strictly conserved among other HPr sequences, was not well conserved in the *C. acetobutylicum* protein. Despite the relative lack of conservation in this region, *in vitro* phosphorylation experiments demonstrated that

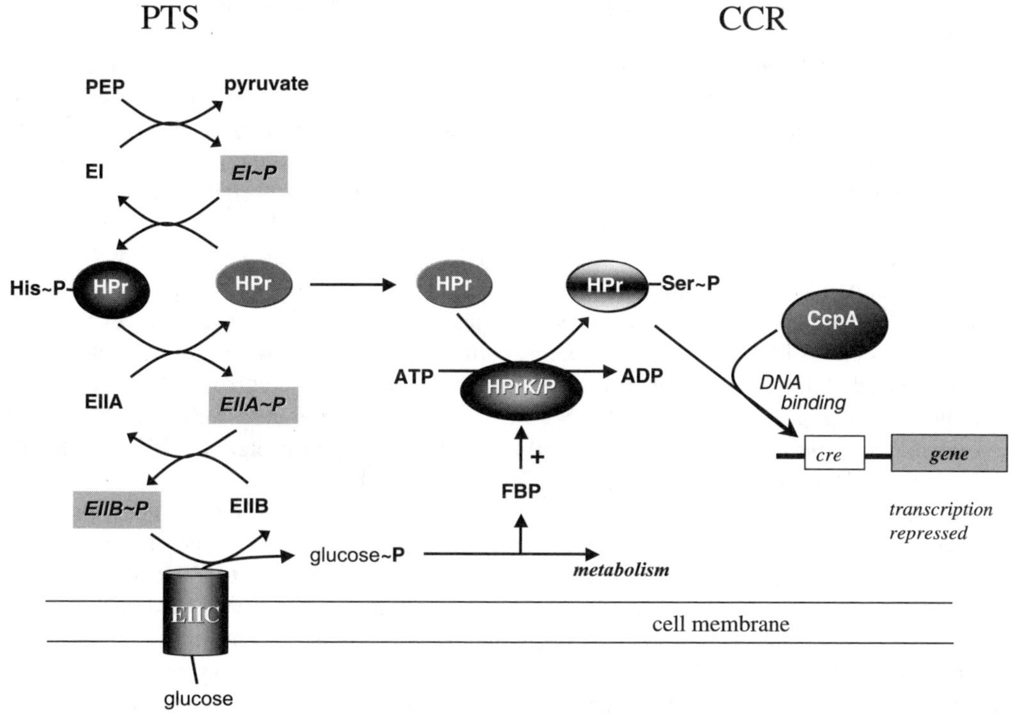

FIGURE 25.6 Model of HPr/CcpA mediated catabolite repression. The PTS is an important transport mechanism that also plays a critical role in this model for CCR in low GC-Gram positive bacteria. In the presence of a rapidly metabolized substrate, such as glucose, high levels of FBP are generated, which stimulates the kinase/phosphorylase enzyme HprK/P to phosphorylate HPr at residue Ser-46 (it is phosphorylated by EI at His-15 in the PTS phosphoryl transfer chain). The catabolite control protein CcpA can interact with HPr(Ser~P) to effect gene regulation by binding at specific regulatory sequences called catabolite responsive elements (CREs). In CCR, this results in repression of a target system, even when an inducer is present.

the *C. acetobutylicum* HPr protein is phosphorylated by an ATP-dependent kinase activity present in extracts of this organism. In *B. subtilis*, HPr kinase activity is stimulated by FBP, although this is not the case for every organism; for example, in a study of the purified HPrK/P from *S. salivarius,* it was discovered that FBP did not stimulate kinase activity, although it inhibited dephosphorylation of HPr(Ser~P) and protected kinase activity against inhibition by inorganic phosphate [61]. Nevertheless, *C. acetobutylicum* HPr kinase activity was stimulated by FBP. Purified HPr from *B. subtilis* was also phosphorylated by the clostridial HPr kinase in the presence of FBP, despite the differences in sequence surrounding the Ser-46 phosphorylation site in these two HPr proteins [60]. It is therefore clear that the key reaction of the *B. subtilis* model of CCR is functional in *C. acetobutylicum.*

To date, HPr kinase activity has only been demonstrated in one species of *Clostridium* (*C. acetobutylicum*); however, PTS activity has been observed in a variety of clostridia, and it is possible that HPr might also be involved in CCR in other clostridia. The availability of genome sequence data allows an investigation of this possibility. A putative *ptsH* gene is present in the completed *C. perfringens* genome, and recently, a *ptsH* has also been proposed for *C. tetani* [62]. Interestingly, in each of these three clostridial genomes, the *ptsH* gene is not linked to *ptsI,* and this would also seem to be the situation for *C. beijerinckii* [60]. That the *ptsH* and *ptsI* genes are not in an operon (*ptsHI*) is unusual but not unprecedented, as this lack of association has also been observed in a number of other bacteria including *Mycoplasma capricolum* [63] and *Streptomyces coelicolor* [64]. It will be of interest to determine whether this is a general characteristic of clostridial genomes.

The *C. acetobutylicum* HPr was the first clostridial HPr protein to be characterized at the sequence level. The Ser-46 region was found to be conserved (and functional), but an alignment revealed a number of deviations from a consensus found in 13 representative HPr proteins from 7 different genera of low-GC Gram-positive bacteria [60]. The observed deviations in this region of the *C. acetobutylicum* HPr are conserved in the HPr proteins from *C. perfringens* and *C. tetani* as shown in the alignment presented in Figure 25.7. In this alignment, the three deduced clostridial HPr sequences are compared with representative HPr sequences from six other low-GC Gram-positive genera. As can be seen, all of these proteins share considerable homology throughout their length, with the two phosphorylation regions being particularly well conserved. However, there are noticeable distinctions in the clostridial proteins, particularly in the Ser-46 region, and phylogenetic analysis reveals that these clostridial proteins form a cluster that is distinct from other Gram-positive HPr proteins (see Figure 8.4 in Chapter 8). As the region surrounding Ser-46 is functional in *C. acetobutylicum* and is almost identical in all three clostridial HPr proteins, it is likely that these closely related HPr proteins are substrates for HPr kinase.

Investigation of the completed *C. acetobutylicum*, *C. tetani,* and *C. perfringens* genomes reveals only a single gene encoding a HPr in these bacteria, while there is no evidence for an equivalent of the closely related HPr-like protein Crh, which is known to contribute to CCR in *Bacillus* [55]. In a number of *Bacillus* species, *crh* is present as part of a cryptic operon designated *yvc*. The equivalent of this cryptic gene system — but without *crh* — is also found in a number of other Gram-positive bacteria that lack *crh*, leading to the suggestion that Crh may perform a *Bacillus*-specific function [59]. Consistent with this proposal is the observation that a *yvc* equivalent appears to be present in *C. acetobutylicum,* but there is no evidence for *crH* in this operon or, indeed, elsewhere in the genome.

25.4.3 HPrK/P in Clostridia

The sensor for CCR in the *B. subtilis* model is the novel bifunctional enzyme HPrK/P. The HPrK/P proteins possess a signature sequence thought to be involved in the interaction with HPr [65] and a Walker A box that is involved in ATP binding [66]. The phosphorylation activity of HPrK/P has been demonstrated in *C. acetobutylicum,* and a putative *C. acetobutylicum* HPrK/P was identified from the genome sequence [60]. In a study of HPrK/P enzymes, the putative *C. acetobutylicum* HPrK/P sequence (which shares considerable homology with other HPrK/P sequences) was shown to possess both the HPrK/P signature sequence and the Walker A box [67]. A search of the *C. tetani* and *C. perfringens* genomes reveals the presence of putative *ptsK* genes encoding proteins with 71% identity/87% similarity and 63% identity/80% similarity, respectively, to the *C. acetobutylicum* HPrK/P protein. The *C. tetani* and *C. perfringens* proteins share 65% identity and 82% similarity. In a previous publication, we noted that the *C. acetobutylicum hprK* appears to be associated with a gene encoding a fructose bis-phosphatase, which is of considerable interest given the fact that HPr kinase activity in this organism is stimulated by FBP [60]. A similar gene is located upstream of the putative *ptsK* gene in *C. tetani*, but this is not the case in *C. perfringens*. Hence, the significance (if any) of the putative fructose bis-phosphatase remains to be established.

25.4.4 CcpA in Clostridia

The transcriptional regulator protein CcpA is a member of the LacI/GalR family and contains a typical helix-turn-helix DNA binding domain that facilitates binding at a specific *cre* target site. The role of CcpA in CCR is as a co-repressor, together with HPr, although in *B. subtilis* and other bacteria, CcpA also has an alternative function as an activator of transcription [32]. In CCR it appears that the level of CcpA in the cell is not affected by a repressing sugar (such as glucose),

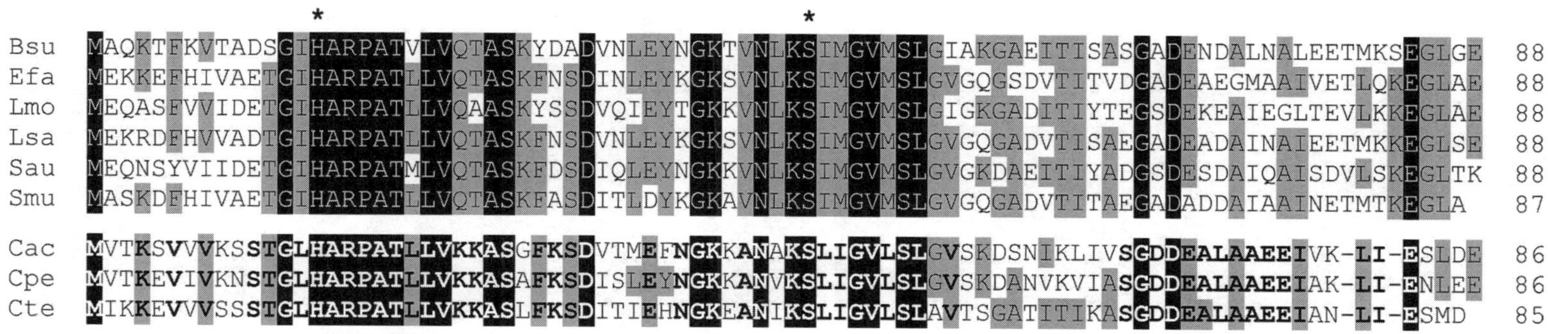

FIGURE 25.7 Multiple alignment of HPr sequences. The three identified clostridial HPr proteins are compared with HPr sequences from other low-GC Gram-positive bacteria. Residues that are identical in the clostridial proteins are shown in bold; residues that are identical in all sequences are within black boxes; conserved residues are shaded. The phosphorylation sites at His-15 and Ser-46 are indicated (*). Abbreviations (with accession numbers in brackets) are as follows: Bsu, *Bacillus subtilis* (A46238); Efa, *Enterococcus faecalis* (A25053); Lmo, *Listeria monocytogenes* (AF030824); Lsa, *Lactobacillus sakei* (AF172726); Sau, *Staphylococcus aureus* (X93205); Smu, *Streptococcus mutans* (A44562); Cac, *Clostridium acetobutylicum* (AY196477); Cpe, *C. perfringens* (NP_562585); Cte, *C. tetani* (DAA02130).

but its affinity for target *cre* sites is modulated in response to the availability of a preferred substrate. CcpA alone does not appear to bind well to *cre* sequences, and efficient binding requires interaction with the co-repressor HPr(Ser~P) to form a regulatory complex that effects repression of transcription at specific *cre* sequences [68,69] (Figure 25.6). The binding site for the co-repressor on CcpA has been identified, and this distinguishes CcpA from other members of the LacI/GalR family [70]. Studies of *in vitro* interactions between CcpA and different forms of HPr have demonstrated that CcpA can complex effectively with HPr(Ser~P), but not HPr(His~P) or unphosphorylated HPr [68,71]. Hence, it is HPr and HprK/P that sense the availability of substrate and affect CCR via direct interaction between the kinase-modulated HPr and the repressor protein CcpA.

A gene, designated *regA*, encoding a CcpA homologue, was cloned from a species of *Clostridium,* which at the time was believed to be *C. acetobutylicum* (strain P262) [72], although the species has subsequently been re-classified as *C. saccharobutylicum* [73]. The RegA protein is a member of the LacI/GalR family (with a typical DNA binding domain) and is homologous to CcpA proteins. Furthermore, when *regA* was expressed in a *B. subtilis ccpA* mutant, complementation of a CCR defect was observed, suggesting that RegA might be a CcpA regulator [72]. In a previous publication, we identified a CcpA from the *C. acetobutylicum* genome, which we found to share 62% identity with the *C. saccharobutylicum* RegA protein, while in a phylogenetic analysis both of these proteins were observed to cluster with other CcpA proteins [60]. Analysis of the *C. perfringens* and *C. tetani* genomes also reveals the presence of a putative CcpA in each organism, sharing 64% and 71% identity, respectively, with the *C. acetobutylicum* CcpA. All four of these putative CcpA proteins possess an N-terminal DNA binding domain.

The target for the activated HPr/CcpA repression complex is the *cre* operator sequence. The manipulation of a *cre* by mutagenesis can result in a reduced sensitivity to CCR [74,75]. A typical *cre* is a partially palindromic sequence, and a 14 bp consensus neucleotide sequence was originally proposed prior to the publication of the *B. subtilis* genome sequence [76]. A subsequent analysis of *cre* sequences in *B. subtilis* has provided fresh insight into the nature of this regulatory element, and a modified (longer) consensus has been proposed [77]. Consistent with the operation in clostridia of the HPr/CcpA mechanism of CCR, putative *cre* sites have been identified in the promoter region of almost all of the PTS gene systems that have been studied in species of *Clostridium.* A number of these sites overlap recognized –35 or –10 promoter sequences, a location that allows for effective repression of transcription. Consider, for example, the *mtl* operon in *C. acetobutylicum* DSM 792, encoding the mannitol PTS in this organism. Transcription analysis demonstrated that the *mtl* operon is induced by the substrate mannitol, but the transcript is not detected in cells grown in the presence of both mannitol and glucose, confirming catabolite repression of the system at the level of transcription. The promoter for this operon was mapped, and a putative *cre* was identified that overlaps the –35 region of the promoter [49]. On the other hand, the *cre* identified for the *scr* operon in *C. acetobutylicum*, which is induced by sucrose and repressed by glucose, overlaps the proposed –10 region of the promoter [1]. Similarly, *cre* sites have been proposed for other clostridial operons that are subject to CCR, but direct binding of the HPr(Ser~P)/CcpA regulatory complex to any of the clostridial *cre* sites has yet to be demonstrated experimentally.

25.4.5 Inducer Exclusion and Inducer Expulsion

It has long been established that species of clostridia exhibit CCR and may preferentially utilize one carbon source over another when presented with a mixture of different carbon sources. The available evidence suggests that HPr plays a central role in this process. All of the essential features for the *B. subtilis* model of CCR appear to be present in the completed genomes — even in the *C. tetani* genome, which would seem to encode only one PTS gene system. Nevertheless, further experimental investigation is required to gain a deeper understanding of gene regulation in this genus. Indeed, repression of transcription has only been demonstrated for a small number of clostridial gene systems, and evidence for CCR is frequently based on the demonstration of preferential uptake of sugars or the absence of specific enzyme activity when cultures are grown in the presence of both an inducer and a repressing substrate [78]. It is important to note that such observations can result not from the inhibition of transcription, but from inhibition of enzyme activity or an alternative mechanism of prevention of induction, such as inducer exclusion and inducer expulsion, both of which have been demonstrated in clostridia.

Inducer exclusion is a term that describes a process whereby the induction of a regulated catabolic gene system is prevented by means of disturbing the natural interaction between the inducer and its target — for example, by inhibiting the uptake of the inducer into the cell, or by preventing the formation of the internal inducing compound. Such inhibition can be demonstrated by monitoring the effects of a repressing substrate (such as glucose) on the uptake of a substrate in induced cells. In this manner, inducer exclusion has been demonstrated in a number of clostridia, although such experimentation does not reveal the mechanisms by which inducer exclusion is achieved. In a study of *C. pasteurianum*, competition for PEP between the glucose PTS and the sorbitol (glucitol) PTS was implicated as a means of effecting inducer exclusion of sorbitol in this organism [79].

A mechanism of inducer exclusion that has been well documented in a number of Gram-positive bacteria is the repression of glycerol uptake, as operated, for example, in *B. subtilis* [59,80]. As discussed earlier in this chapter, the inducer molecule for the glycerol transport system in *B. subtilis* is glycerol 3-phosphate, which is the product of glycerol kinase. However, glycerol kinase activity is stimulated by phosphorylation via HPr(His~P). Therefore, in the presence of both glucose and the inducing substrate (glycerol), the formation of the internal inducing compound (glycerol-3-phosphate) is reduced and CCR is effected by inducer exclusion. As a similar glycerol gene system appears to be present both in *C. acetobutylicum* and *C. perfringens,* it will be of interest to determine if this mechanism of CCR is also operational in these (and other) clostridia.

The phenomenon of inducer expulsion has been mostly studied in lactococci and streptococci and is a poorly understood energy-dependent process, whereby an accumulated substrate is expelled from the cell in the presence of a more favorable carbon source, thereby effecting CCR [81,82]. As the product of a PTS transport reaction is phosphorylated, inducer expulsion of a PTS substrate requires the dephosphorylation of the accumulated (phosphorylated) product of the PTS reaction prior to expulsion. An example of this phenomenon has been observed in *C. saccharobutylicum* P262 (formerly *C. acetobutylicum*), where the nonmetabolizable lactose analog thiomethylgalactoside (TMG) was found to be accumulated by a PTS mechanism in induced cultures, but was expelled from the cell into the medium in response to the presence of glucose [83]. The mechanism by which this is achieved is not understood; however, it was observed that glucose both inhibited uptake of TMG (inducer exclusion) and promoted inducer expulsion, while the nonmetabolizable glucose analog 2-deoxyglucose could inhibit the uptake of TMG (inducer exclusion), but did not promote inducer expulsion. It remains to be established how common this phenomenon is in the clostridia.

25.5 CONCLUSION

It is clear that clostridia share with other bacteria the ability to sense and respond to environmental conditions. The induction and repression of gene systems allow the cell to control its proteome, as it switches on or off a variety of genes to best suit the requirements of the cell at a particular moment. As our knowledge of gene regulation in clostridia continues to expand, it is obvious that this is a complex phenomenon, a full understanding of which will require considerable effort. A plethora of system-dependent mechanisms would appear to be employed to induce gene expression in these bacteria. However, it also seems likely that HPr plays a pivotal role in the repression of gene expression, although the investigation of the precise contribution to gene regulation of this multifunctional protein is still at an early stage. There is also a need to study in detail a greater number of individual gene systems and their associated regulators. Nevertheless, we are now entering an exciting period in the continued study of the genus *Clostridium* where rapid progress can be expected. Several genomes have already been sequenced, and, with more to be completed, the impact of genome sequence data on our understanding of these organisms is already apparent. This wealth of data, coupled with improvements in technology allowing detailed molecular investigation and proteomic analysis, should rapidly and significantly enhance our understanding of gene regulation in clostridia.

REFERENCES

1. Tangney, M. and Mitchell, W.J., Analysis of a catabolic operon for sucrose transport and metabolism in *Clostridium acetobutylicum* ATCC 824, *J. Mol. Microbiol. Biotechnol.*, 2, 71, 2000.
2. Tangney, M., et al., A gene system for glucitol transport and metabolism in *Clostridium beijerinckii* NCIMB 8052, *Appl. Env. Microbiol.*, 64, 1612, 1998.
3. Tangney, M., Winters, G.T., and Mitchell, W.J., Characterization of a maltose transport system in *Clostridium acetobutylicum* ATCC 824, *J. Ind. Microbiol. Biotechnol.*, 27, 298, 2001.
4. Stülke, J., et al., PRD — a protein domain involved in PTS-dependent induction and carbon catabolite repression of catabolic operons in bacteria, *Mol. Microbiol.*, 28, 865, 1998.
5. Tortosa, P., et al., Multiple phosphorylation of SacY, a *Bacillus subtilis* transcriptional antiterminator negatively controlled by the phosphotransferase system, *J. Biol. Chem.*, 272, 17230, 1997.
6. Tortosa, P., et al., Sites of positive and negative regulation in the *Bacillus subtilis* antiterminators LicT and SacY, *Mol. Microbiol.*, 41, 1381, 2001.
7. Rutberg, B., Antitermination of transcription of catabolic operons, *Mol. Microbiol.*, 23, 413, 1997.
8. Weickert, M.J. and Adhya, S., A family of bacterial regulators homologous to Gal and Lac repressors, *J. Biol. Chem.*, 267, 15869, 1992.
9. Mortensen, L., Dandanell, G., and Hammer, K., Purification and characterisation of the DeoR repressor of *Escherichia coli*, *EMBO J.*, 8, 325, 1989.
10. van Rooijen, R.J. and de Vos W.M., Molecular cloning, transcriptional analysis, and nucleotide sequence of *lacR*, a gene encoding the repressor of the lactose phosphotransferase system of *Lactococcus lactis*, *J. Biol. Chem.*, 265, 18499, 1990.
11. Tangney, M., et al., Sucrose transport and metabolism in *Clostridium beijerinckii* NCIMB 8052, *J. Appl. Microbiol.*, 84, 914, 1998.
12. Reid, S.J., Rafudeen, M.S., and Leat, N.G., The genes controlling sucrose utilization in *Clostridium beijerinckii* NCIMB 8052 constitute an operon, *Microbiol.*, 145, 1461, 1999.
13. Gering, M. and Brückner, R., Transcriptional regulation of the sucrase gene of *Staphylococcus xylosus* by the repressor ScrR, *J. Bacteriol.*, 178, 462, 1996.
14. Shimizu,T., et al., Complete genome sequence of *Clostridium perfringens*, an anaerobic flesh-eater, *Proc. Natl. Acad. Sci. U.S.A.*, 99, 996, 2002.
15. Groves, D.J. and Gronlund, A.F., Carbohydrate transport in *Clostridium perfringens* type A, *J. Bacteriol.*, 100, 1256, 1969.
16. Yu, Y. et al., Mechanism and regulation of lactose metabolism in *Clostridium acetobutylicum*, manuscript in preparation.
17. Kobayashi, T., et al., Transcriptional analysis of the β-galactosidase gene (*pbg*) in *Clostridium perfringens*, *FEMS Microbiol. Lett.*, 133, 65, 1995.
18. Mitchell, W.J., Carbohydrate uptake and utilization by *Clostridium beijerinckii* NCIMB 8052, *Anaerobe*, 2, 379, 1996.
19. Kunst, F., et al., The complete genome sequence of the Gram-positive bacterium *Bacillus subtilis*, *Nature,* 390, 249, 1997.
20. Goda, S.K., et al., Molecular analysis of a *Clostridium butyricum* NCIMB 7423 gene encoding 4-α-glucanotransferase and characterisation of recombinant enzyme produced in *Escherichia coli*, *Microbiology,* 143, 3287, 1997.
21. Goda, S.K., et al., Molecular analysis of the *malR* gene of *Clostridium butyricum* NCIMB 7423 gene a member of the LacI-GalR family of repressor proteins, *FEMS Microbiol. Lett.,* 165, 193, 1998.
22. Hu, D. et al., unpublished data.
23. Minton N., et al., unpublished data.

24. Dahl, M.K., Degenkolb, J. and Hillen, W., Transcription of the *xyl* operon is controlled in *Bacillus subtilis* by tandem overlapping operators spaced by four base-pairs, *J. Mol. Biol.*, 243, 413, 1994.
25. Schmiedel, D., et al., Regulation of expression, genetic organization and substrate specificity of xylose uptake in *Bacillus megaterium*, *Mol. Microbiol.*, 23, 1053, 1997.
26. Scheler, A. and Hillen, W., Regulation of xylose utilization in *Bacillus licheniformis*: Xyl repressor-*xyl*-operator interaction studied by DNA modification protection and interference, *Mol. Microbiol.*, 13, 505, 1994.
27. Rodionov, D.A., Mironov, A.A., and Gelfand, M.S., Transcriptional regulation of pentose utilisation systems in the *Bacillus/Clostridium* group of bacteria, *FEMS Microbiol. Lett.*, 205, 305, 2001.
28. Bhavsar, A.P., Zhao, X., and Brown, E.D., Development and characterization of a xylose-dependent system for expression of cloned genes in *Bacillus subtilis*: Conditional complementation of a teichoic acid mutant, *Appl. Env. Microbiol.*, 67 403, 2001.
29. Ounine, K., Petitdemange, H., and Gay, R., Regulation and butanol inhibition of D-xylose and D-glucose uptake in *Clostridium acetobutylicum, Appl. Env. Microbiol.*, 49, 874, 1985.
30. Aymerich, S., and Steinmetz, M., Specificity determinants and structural features in the RNA target of the bacterial antiterminator proteins of the BglG/SacY family, *Proc. Natl. Acad. Sci. U.S.A.*, 89, 10410, 1992.
31. Schnetz, K., Toloczyki, C., and Rak, B., Beta-glucoside (*bgl*) operon of *Escherichia coli* K12-nucleotide sequence, genetic organization, and possible evolutionary relationship to regulatory components of two *Bacillus subtilis* genes, *J. Bacteriol.*, 169, 2579, 1987.
32. Stülke, J. and Hillen, W., Regulation of carbon catabolism in *Bacillus* species, *Ann. Rev. Microbiol.*, 54, 849, 2000.
33. Manival, X., et al., From genetic to structural characterization of a new class of RNA-binding domain within the SacY/BglG family of antiterminator proteins, *EMBO J.,* 16, 5019, 1997.
34. Langbein, I., Bachem, S., and Stülke, J., Specific interaction of the RNA-binding domain of the *Bacillus subtilis* transcriptional antiterminator GlcT with its RNA target, RAT, *J. Mol. Biol.,* 293, 795, 1999.
35. Görke, B. and Rak, B., Catabolite control of *Escherichia coli* regulatory protein BglG activity by antagonistically acting phosphorylations, *EMBO J.,* 18, 3370, 1999.
36. Li, Y. and Ferenci, T., Gene organization and regulatory sequences in the sucrose utilisation cluster of *Bacillus stearothermophilus* NUB36, *Gene,* 195, 195, 1997.
37. Lindner, C., et al., Regulation of the activity of the *Bacillus subtilis* antiterminator LicT by multiple PEP-dependent, Enzyme I- and HPr-catalysed phosphorylation, *Mol. Microbiol.*, 31, 995, 1999.
38. Chen, Q., et al., BglF, the sensor of the *Escherichia coli bgl* system, uses the same site to phosphorylate both a sugar and a regulatory protein, *EMBO J.*, 16, 4617, 1997.
39. El Hassouni, M., et al., Nucleotide sequences of the *arb* genes, which control β-glucoside utilization in *Erwinia chrysanthemi* — comparison with the *Escherichia coli bgl* operon and evidence for a new β-glycohydrolase family including enzymes from eubacteria, archeabacteria, and humans, *J. Bacteriol.*, 174, 765, 1992.
40. Tangney, M. and Mitchell, W.J., unpublished data.
41. Stülke, J., et al., Induction of the *Bacillus subtilis ptsGHI* operon by glucose is controlled by a novel antiterminator, GlcT, *Mol. Microbiol.,* 25, 65, 1997.
42. Knezevic, I., et al., Regulation of the glucose-specific phosphotransferase system (PTS) of *Staphylococcus carnosus* by the antiterminator protein GlcT, *Microbiology*, 146, 2333, 2000.
43. Brown, G.D. and Thomson, J.A., Isolation and characterisation of an aryl-beta-D-glucoside uptake and utilisation system (*abg*) from the Gram-positive ruminal *Clostridium* species *C. longisporum*, *Mol. Gen. Genet.*, 257, 213, 1998.
44. Holmberg, C. and Rutberg, B., Expression of the gene encoding glycerol-3-phosphate dehydrogenase (*glpD*) in *Bacillus subtilis* is controlled by antitermination, *Mol. Microbiol.*, 5, 2891, 1991.
45. Glatz, E., Persson, M., and Rutberg, B., Antiterminator protein GlpP of *Bacillus subtilis* binds to *glpD* leader mRNA, *Microbiology,* 144**,** 449, 1998.
46. Glatz, E., et al., A dual role for the *Bacillus subtilis glpD* leader and the GlpP protein in the regulated expression of *glpD*: antitermination and control of mRNA stability, *Mol. Microbiol.*, 19, 319, 1996.
47. Forsberg, C.W., Production of 1, 3-propanediol from glycerol by *Clostridium acetobutylicum* and other *Clostridium* species, *Appl. Env. Microbiol.*, 53, 639, 1987.

48. Saint-Amans, S., et al., Regulation of carbon and electron flow in *Clostridium butyricum* VPI 3266 grown on glucose-glycerol mixtures, *J. Bacteriol.*, 183, 1748, 2001.
49. Behrens, S., Mitchell, W.J., and Bahl, H., Molecular analysis of the mannitol operon of *Clostridium acetobutylicum* encoding a phosphotransferase system and a putative PTS-modulated regulator, *Microbiol.*, 146, 75, 2001.
50. Henstra, S.A., et al., The *Bacillus stearothermophilus* mannitol regulator, MtlR, of the phosphotransferase system, *J. Biol. Chem.*, 274, 4754, 1999.
51. Henstra, S.A., Duurkens, R.H., and Robillard, G.T., Multiple phosphorylation events regulate the activity of the transcriptional regulator MtlR of the *Bacillus stearothermophilus* phosphoenolpyruvate-dependent mannitol phosphotransferase system, *J. Biol. Chem.*, 275, 7037, 2000.
52. Yamada, M. and Saier, M.H. Jr., Positive and negative regulators for glucitol (*gut*) operon expression in *Escherichia coli*, *J. Mol. Biol.*, 203, 569, 1988.
53. Boyd, D.A., et al., Identification of the operon for the sorbitol (glucitol) phosphoenolpyruvate:sugar phosphotransferase system in *Streptococcus mutans*, *Infect. Immun.*, 68, 925, 2000.
54. Yamamoto, H., et al., Determination of a 12 kb nucleotide sequence around the 768 region of the *Bacillus subtilis* chromosome, *Microbiology*, 142, 1417, 1996.
55. Galinier, A., et al., The *Bacillus subtilis crh* gene encodes a HPr-like protein involved in carbon catabolite repression, *Proc. Natl. Acad. Sci. U.S.A.*, 94, 8439, 1997.
56. Kravanja, M., et al., The *hprK* gene of *Enterococcus faecalis* encodes a novel bifunctional enzyme: the HPr kinase/phosphatase, *Mol. Microbiol.*, 31, 59, 1999.
57. Mijakovic, I., et al., Pyrophosphate-producing protein dephosphorylation by HPr kinase/phosphorylase: A relic of early life?, *Proc. Natl. Acad. Sci. U.S.A.*, 99, 13442, 2002.
58. Deutscher, J., et al., Bacterial phosphoenolpyruvate-dependent phosphotransferase system: P-ser-HPr and its possible regulatory function, *Biochem.,* 23, 4455, 1984.
59. Deutscher, J., Galinier, A., and Martin-Verstraete, I., Carbohydrate uptake and metabolism, in *Bacillus subtilis and Its Closest Relatives: From Genes to Cells*, Sonenshein, A.L. et al., Eds., ASM Press, Washington, DC, 2002, Chapter 11.
60. Tangney, M., et al., Analysis of the elements of catabolite repression in *Clostridium acetobutylicum* ATCC 824, *J. Mol. Microbiol. Biotechnol.*, 6, 6, 2003.
61. Frey, N., et al., The HPr(Ser) kinase of *Streptococcus salivarius*: a hexameric bifunctional enzyme controlled by glycolytic intermediates and inorganic phosphate, *FEMS Microbiol. Lett.*, 224, 67, 2003.
62. Tangney, M. and Mitchell, W.J., *Clostridium tetani* encodes a phosphocarrier protein HPr, *Microbiology*, 150, 525, 2004.
63. Zhu, P.P., et al., Unique monocistronic operon (*ptsH*) in *Mycoplasma capricolum* encoding the phosphocarrier protein, HPr, of the phosphoenolpyruvate: sugar phosphotransferase system. Cloning, sequencing, and characterization of *ptsH, J. Biol. Chem.*, 268, 26531, 1993.
64. Parche, S., Schmid, R., and Titgemeyer, F., The phosphotransferase system (PTS) of *Streptomyces coelicolor.* Identification and biochemical analysis of a histidine phosphocarrier protein HPr encoded by the gene *ptsH*, *Eur. J. Biochem.*, 256, 308, 1999.
65. Reizer, J. et al., A novel protein kinase that controls carbon catabolite repression in bacteria, *Mol. Microbiol.*, 27, 1157, 1998.
66. Walker, J.E., et al., Distantly related sequences in the α- and β-subunits of ATP synthase, myosin kinases and other ATP-requiring enzymes and a common nucleotide binding fold, *EMBO J.*, 1, 945, 1982.
67. Fieulaine, S., et al., X-ray structure of HPr kinase: a bacterial protein kinase with a P-loop nucleotide binding domain, *EMBO J.*, 20, 3917, 2001.
68. Jones, B.E., et al., Binding of the catabolite repressor protein CcpA to its DNA target is regulated by phosphorylation of its corepressor HPr, *J. Biol. Chem.*, 272, 26530, 1997.
69. Kraus, A., and Hillen, W., Analysis of CcpA mutations defective in carbon catabolite repression in *Bacillus megaterium*, *FEMS Microbiol. Lett.*, 153, 221, 1997.
70. Kraus A, et al., Identification of a co-repressor binding site in catabolite control protein CcpA, *Mol. Microbiol.*, 30, 955, 1998.
71. Deutscher, J., et al., Protein kinase-dependent HPr/CcpA interaction links glycolytic activity to carbon catabolite repression in gram-positive bacteria, *Mol. Microbiol.*, 15, 1049, 1995.

72. Davison, S.P., et al., A *Clostridium acetobutylicum* regulator gene (*regA*) affecting amylase production in *Bacillus subtilis*, *Microbiol.*, 141, 989, 1995.
73. Keis, S., Shaheen, R., and Jones, D.T., Emended descriptions of *Clostridium acetobutylicum* and *Clostridium beijerinckii*, and descriptions of *Clostridium saccharoperbutylacetonicum* sp. nov. and *Clostridium saccharobutylicum* sp. nov., *Int. J. Syst. Evol. Microbiol.*, 51, 2095, 2001.
74. Nicholson, W.L., et al., Catabolite repression-resistant mutations of the *Bacillus subtilis* alpha-amylase promoter affect transcription levels and are in an operator like sequence, *J. Mol. Biol.*, 198, 609, 1987.
75. Hueck, C.J., Hillen, W., and Saier, M.H., Analysis of a cis-active sequence mediating catabolite repression in gram-positive bacteria, *Res. Microbiol.*, 145, 503, 1994.
76. Weickert, M. and Chambliss, G., Site-directed mutagenesis of a catabolite repression operator sequence in *Bacillus subtilis*, *Proc. Natl. Acad. Sci. U.S.A.*, 87, 6238, 1990.
77. Miwa, Y., et al., Evaluation and charcterization of catabolite-responsive elements (cre) of *Bacillus subtilis*, *Nucl. Acids Res.*, 28, 1206, 2000.
78. Mitchell, W.J., Physiology of carbohydrate to solvent conversion by clostridia, *Adv. Mic. Physiol.*, 39, 31, 1998.
79. Roohi, M.S. and Mitchell, W.J., Regulation of sorbitol metabolism by glucose in *Clostridium pasteurianum*: a role for inducer exclusion, *J. Gen. Microbiol.*, 133, 2207, 1987.
80. Reizer, J., et al., Regulation of glycerol uptake by the phosphoenolpyruvate-sugar phosphotransferase system in *Bacillus subtilis, J. Bacteriol.*, 159, 243, 1984.
81. Thomson, J.D., Regulation of sugar transport and metabolism in lactic acid bacteria, *FEMS Microbiol. Rev.*, 46, 221, 1987.
82. Saier M.H., et al., Catabolite repression and inducer control in Gram-positive bacteria, *Microbiology*, 142, 217, 1996.
83. Diez-Gonzalez, F. and Russell, J.B., The regulation of thiomethylgalactoside transport in *Clostridium acetobutylicum* P262 by inducer exclusion and inducer expulsion mechanisms, *FEMS Microbiol. Lett.*, 136, 123, 1996.

26 RNA Polymerase and Alternative σ Factors

Abraham L. Sonenshein, Jeralyn D. Haraldsen, and Bruno Dupuy

CONTENTS

26.1 RNA POLYMERASE AND ALTERNATIVE σ FACTORS

The heart of the bacterial transcription machinery is DNA-dependent RNA polymerase and its associated σ factors. The core subunits of RNA polymerase, α, β, and β′, provide the catalytic activity of the enzyme, but lack the ability to distinguish promoter regions from other DNA sequences [1]. Thus, the σ subunits, which are sequence-specific DNA binding proteins, associate with the core polymerase and direct it to appropriate promoter sites [1]. σ-core interaction is transient; soon after the initiation of RNA synthesis, σ is usually released and the core (with or without other kinds of associated factors) then moves along the DNA generating its RNA product.

RNA polymerase holoenzyme (core plus σ) has been purified from *Clostridium acetobutylicum* [2], *Clostridium perfringens* [3], and *Clostridium difficile* [4,5]. In all cases, the purified enzyme was able to initiate transcription *in vitro* from appropriate promoter sites.

σ factors belong to two major families, reflecting in all likelihood their evolutionary history [6]. Members of the σ^{70} family, named after the principal σ factor of *Escherichia coli*, have an intrinsic DNA melting activity that allows them to open the promoter region and thereby facilitate transcription initiation without added proteins or ATP [6]. The σ^{54} family, named for the *E. coli* factor (also known as RpoN, NtrA, and GlnF) that directs transcription of certain nitrogen metabolism genes, has special requirements. σ^{54} is unable to bind tightly to promoter sites or melt them without the aid of auxiliary proteins and ATP [7]. Bacteria typically encode multiple σ^{70}-type factors, each having a unique promoter specificity. It is rare, however, for a bacterial species to have more than a single example of a σ^{54}-like factor.

The specificity of each σ factor for its cognate promoter sites resides principally in two domains, known as regions 2.4 and 4.2 [8]. Region 4.2 allows the σ factor, when bound to core polymerase, to make initial contact with the promoter region at a group of nucleotides located 30 to 38 bp upstream of the start point of transcription (the "-35 region"). When region 2.4 of the σ factor then makes an additional series of contacts with nucleotides located 6 to 13 bp upstream of the start point (the "-10 region"), the protein-DNA complex is stabilized as the DNA becomes locally melted. σ factors are distinguished from one another by variations in their 2.4 and 4.2 regions, as a result of which they recognize distinct and characteristic –10 and –35 regions. For the σ^{54} family, the conserved promoter sequences are located at positions –12 and –24.

In analyzing the state of knowledge about RNA polymerase and σ factors in *Clostridium* spp., we have relied as much as possible on published and unpublished experimental data, but many of our conclusions are based on comparisons by the authors of coding sequences currently available in databases. In most cases, those conclusions are not specifically referenced. Sequence comparisons carried out by the authors made use of the BLAST program and databases accessible through the Microbial Genomes Web site of the National Center for Biotechnology Information (http://www.ncbi.nlm.nih.gov/sutils/genom_table.cgi?). Alignments were made using the Clustal W program through the Web site of the Institut Pasteur (http://www.pasteur.fr/externe). At the time of writing, the *Clostridium* genome sequences available to us were those of *C. acetobutylicum* [9], *C. botulinum, C. difficile, C. perfringens* [10], *C. tetani* [11], and *C. thermocellum* (the last incomplete).

26.1.1 σ Factors Active during Growth

26.1.1.1 σ^A

Most genes expressed during exponential growth phase have promoter sites that are recognized by a particular member of the σ^{70} family. This factor, known as the principal or housekeeping σ, is called σ^{70} in *E. coli* and is known as σ^A in Gram-positive bacteria. All *Clostridium* spp. analyzed to date encode at least one close homologue of *Bacillus subtilis* σ^A, the founding member of this group [9–11] (Table 26.1). The *Clostridium* σ^A proteins are highly similar to each other, especially in regions 2.4 and 4.2 (Figure 26.1). Thus, it is very likely that all members of this subfamily recognize as preferred binding sites the consensus sequences established for *B. subtilis* promoters TTGACA [-35] and TATAAT [-10] [12].

As in *B. subtilis* [13], the *sigA* gene in *C. acetobutylicum* is located immediately downstream from and is apparently co-transcribed with *dnaE*, the gene for DNA primase [14]. Identical gene arrangements are found in the genomes of *C. perfringens, C. difficile, C. tetani, C. botulinum*, and *C. thermocellum*.

A second σ^A-like gene was found in *C. difficile*. It is 51% identical to *B. subtilis* σ^A and is almost identical in regions 2.4 and 4.2 (Figure 26.1). No function can be ascribed to the product of this gene at present.

26.1.1.2 σ^D

B. subtilis σ^D activates transcription of a large class of genes involved in chemotaxis, motility, and autolysis [15]. The *sigD* gene lies, in fact, within a large cluster of such genes [16]. σ^D-like proteins are encoded within the genomes of *C. acetobutylicum, C. difficile, C. botulinum*, and *C. thermocellum*, but not within the genomes of *C. tetani* and *C. perfringens* (Table 26.2). In the four positive cases, the *sigD* gene is surrounded by genes that potentially encode components of

TABLE 26.1
σ^A Homologues in *Clostridium* spp.[a]

	Bsu	Cac	Cbo	Cdi	Cpe	Cte	Cth
Bsu	100						
Cac	65	100					
Cbo	67	87	100				
Cdi	65	69	73	100			
Cpe	67	81	86	71	100		
Cte	67	83	86	72	83	100	
Cth	73	84	86	78	83	84	100

[a] Species are identified by a three-letter abbreviation. Bsu, *Bacillus subtilis*; Cac, *Clostridium acetobutylicum*; Cbo, *Clostridium botulinum*; Cdi, *Clostridium difficile*; Cpe, *Clostridium perfringens*; Cte, *Clostridium tetani*; Cth, *Clostridium thermocellum*. The values reported indicate percent identity of protein coding sequences after alignment using the Clustal W program.

	Region 2.4	Region 4.2
Bsu	IRQAITRAIADQARTIRIPVHM	TLEEVGKVFGVTRERIRQIEAKALRKLR
CaC	IRQAITRAIADQARTIRIPVHM	TLEEVGKEFNVTRERIRQIEAKALRKLR
Cbo	IRQAITRAIADQARTIRIPVHM	TLEEVGKEFNVTRERIRQIEAKALRKLR
Cdi	IRQAITRAIADQARTIRIPVHM	TLEEVGKEFDVTRERIRQIEAKALRKLR
Cdi*	IKQSITRYIDDCENTIRIPIHL	TLEQIGKVFGVTRERIRQIEAKAIRKLR
Cpe	IRQAITRAIADQARTIRIPVHM	TLEEVGKEFNVTRERIRQIEAKALRKLR
Cte	IRQAITRAIADQARTIRIPVHM	TLEEVGKEFNVTRERIRQIEAKALRKLR
Cth	IRQAITRAIADQARTIRIPVHM	TLEEVGKEFNVTRERIRQIEAKALRKLR

FIGURE 26.1 Sequences of DNA recognition domains of σ^A homologues in *Clostridium* spp. See Table 26.1 for species abbreviations. Cdi* indicates a second homologue of σ^A found in *C. difficile*.

TABLE 26.2
σ^D Homologues in *Clostridium* spp.[a]

Species	Bsu	Cac	Cbo	Cdi	Cth
Bsu	100				
Cac	34	100			
Cbo	35	72	100		
Cdi	31	50	48	100	
Cth	37	39	38	37	100
Cpe	ND[b]	ND	ND	ND	ND
Cte	ND	ND	ND	ND	ND

[a] See note to Table 26.1.
[b] ND = not detected.

TABLE 26.3
σ^L Homologues in *Clostridium* spp.[a]

Species	Bsu	Cac	Cbo	Cdi	Cpe	Cte	Cth
Bsu	100						
Cac	31	100					
Cbo	31	54	100				
Cdi	31	40	41	100			
Cpe	33	53	58	39	100		
Cte	33	53	60	39	53	100	
Cth	ND[b]	ND	ND	ND	ND	ND	ND

[a] See note to Table 26.1.
[b] ND = not detected in unfinished genome.

the chemotactic response, flagellar biosynthesis, and cell division regulation. The activity of σ^D is controlled by a small, secretable protein, called FlgM, in both Gram-negative and Gram-positive bacteria [17,18]. Homologues of *B. subtilis* FlgM are encoded in the genomes of *C. acetobutylicum, C. botulinum*, and *C. difficile* (19 to 31% identity). Surprisingly, a FlgM homologue was also found in *C. tetani*, which has no σ^D, but not in *C. thermocellum*, which does appear to encode σ^D.

26.1.1.3 σ^L

σ^L, the activator-dependent σ factor that is only distantly related to the σ^{70} family, has representatives encoded in all *Clostridium* species whose genomes have been completely sequenced (Table 26.3). At the time of this writing, the incomplete *C. thermocellum* genome had not yet revealed such a gene. A partial sequence for a *sigL*-like gene has been obtained from *Clostridium kluyveri* [19].

The *E. coli* version of this σ factor is primarily, but not exclusively, involved in transcription of genes whose products permit the utilization of various nitrogen sources and the biosynthesis of glutamine [20]. A broader role in metabolism has been ascribed to this type of σ factor in *B. subtilis*, where σ^L is implicated in utilization of arginine, branched chain keto acids, acetoin, and complex sugars [21]. There is no information at present about potential targets of σ^L in *Clostridium.*

26.1.2 σ Factors Activated by Stressful Conditions

26.1.2.1 σ^B

In *B. subtilis*, *Staphylococcus aureus,* and other Gram-positive bacteria, the σ^B factor activates transcription of many genes that are induced by exposure to acid, salt, ethanol, heat, or carbon starvation [22,23]. Of the *Clostridium* species whose genomes have been sequenced, only *C. difficile* appears to encode a σ^B-like factor (37% identity to *B. subtilis* σ^B). As in *B. subtilis*, the *sigB* gene of *C. difficile* is located immediately downstream of genes that encode an anti-σ factor (RsbW) and an anti-anti-σ (RsbV). In *B. subtilis*, another five genes involved in regulating σ^B activity reside in the same locus, but in *C. difficile* these other genes are absent. This arrangement implies that σ^B synthesis and activity in *C. difficile* are not subject to the same complex regulation as in *B. subtilis*.

The *rsbV-rsbW-sigB* gene cluster is homologous to *spoIIAA-spoIIAB-sigF*, the genes that encode the sporulation σ factor, σ^F, and its regulatory proteins (see below). We attribute to the putative *C. difficile* σ^B protein the functional role of σ^B based on its slightly greater sequence similarity to *B. subtilis* σ^B than to σ^F, especially in regions 2.4 and 4.2 (see Figure 26.2), and its location near *gyrA* rather than the *spoVA* operon. Without knowing the phenotype of a knockout mutant in the

```
                 Region 2.4                       Region 4.2

Bsu σB   IIGEIKRFLRDKTWSVHVPRRI     SQKETGDILGISQMHVSRLQRKAVKELR
Cdi σB   IVGEIKKYFRDKVWTLRVPRRI     TQSSIAKELGISQMTVSRLEKKVVEKLK

Bsu σF   IIGEIQRFIRDDG-TVKVSRSI     TQSEVAERLGISQVQVSRLEKKILKQIK
Cdi σF   ILGEIKRYLRDDG-MIKVSRSL     TQSEIGEMLNISQVQVSRIEKKVLSKLK
```

FIGURE 26.2 Sequences of DNA recognition domains of σ^B and σ^F homologues in *Clostridium* spp.

C. difficile sigB gene; however, we cannot rule out the possibility that *C. difficile* encodes two different and potentially redundant σ^F variants.

26.1.2.2 ECF Subfamily σ Factors

The extracytoplasmic function (ECF) σ factors, which form a distinct subfamily within the σ^{70} family, have several common features that distinguish them from all other σ factors [24]. First, in response to external signals, the ECF σ factors appear to control functions associated with the cell surface or transport. Second, they frequently recognize promoter elements with a conserved "AAC" motif in the –35 region, and in many cases the ECF σ factor is cotranscribed with a transmembrane anti-σ factor. The analysis of currently available genome sequences reveals that many bacteria contain one or more ECF σ factors and the number of the ECF σ factors is generally linked to its genome complexity: 2 in *E. coli*, 7 in *B. subtilis*, 10 in *Mycobacterium tuberculosis,* and 50 in *Streptomyces coelicolor* [25]. The ECF subfamily was first recognized when the *Streptomyces sigE* gene was cloned. In *S. coelicolor* [24], σ^E controls functions related to cell wall biogenesis. No obvious σ^E-like protein was found to be encoded in any *Clostridium* species whose genome has been completely or incompletely sequenced. The only ECF σ factor-like proteins found in the *Clostridium* genomes are potential orthologues and homologues of the *B. subtilis* ECF σ factors.

B. subtilis possesses seven ECF σ factors (σ^M, σ^V, σ^W, σ^X, σ^Y, σ^W, and YlaC) whose physiological functions are mostly unknown [26]. However, recent studies of three of them (σ^X, σ^W, and σ^M) indicated that they are involved in membrane and cell wall modifications in response to extracytoplasmic stresses (heat, oxidative stress, antimicrobial agents, and salts) [27–29]. Based on sequence comparisons of *B. subtilis* ECF σ factors and proteins encoded by *Clostridium* genomes, we could only find putative orthologues of *B. subtilis* σ^X, σ^W, and σ^V.

Of the *Clostridium* species sequenced, only *C. acetobutylicum* and *C. thermocellum* encode σ^X-like factors, which share 25% and 28% identity with *B. subtilis* σ^X, respectively ([30] and Table 26.4).

TABLE 26.4
σ^X Homologues in *Clostridium* spp.[a]

	Bsu	Cac	Cth
Bsu	100		
Cac	25	100	
Cbo	ND[b]	ND	ND
Cdi	ND	ND	ND
Cpe	ND	ND	ND
Cte	ND	ND	ND
Cth	28	21	100

[a] See note to Table 26.1.
[b] ND = not detected.

TABLE 26.5
σ^W Homologues in *Clostridium* spp.[a]

Species	Bsu	Cac	Cdi	Cpe	Cth
Bsu	100				
Cac	32	100			
Cdi	30	26	100		
Cpe	32	25	29	100	
Cth	40	35	26	30	100
Cbo	ND[b]	ND	ND	ND	ND
Cte	ND	ND	ND	ND	ND

[a] See note to Table 26.1.
[b] ND = not detected.

TABLE 26.6
σ^V Homologues in *Clostridium* spp.[a]

Species	Bsu	Cbo	Cdi	Cpe	Cte	Cth
Bsu	100					
Cac	ND[b]	ND	ND	ND	ND	ND
Cbo	31	100				
Cdi	57	36	100			
Cpe	34	42	35	100		
Cte	36	42	33	45	100	
Cth	55	36	52	33	38	100

[a] See note to Table 26.1.
[b] ND = not detected.

In *B. subtilis*, the gene immediately downstream of *sigX* encodes RsiX, a negative regulator of σ^X activity [31]. No such gene was found close to *sigX* or anywhere else in the *C. thermocellum* or *C. acetobutylicum* genomes, suggesting that σ^X-like proteins in *Clostridium* might be controlled by another mechanism. For instance, a two-component regulatory system (CseC/CseB) regulates the synthesis of the *S. coelicolor* ECF factors σ^E and BldN [32].

Proteins with 30 to 40% identity to *B. subtilis* σ^W are encoded within the genomes of *C. difficile, C. thermocellum, C. acetobutylicum,* and *C. perfringens,* but no such proteins are encoded within the genomes of *C. tetani* and *C. botulinum* (Table 26.5). In *B. subtilis*, an anti-σ factor gene *rsiW* (*ybbM*) is cotranscribed with *sigW* [33]. Only *C. perfringens* has a gene located downstream of the *sigW*-like gene that appears to be similar to a negative effector. Interestingly, there is a gene coding for a putative anti-σ^V in the *C. acetobutylicum* genome, but there is no *sigV* homologue. Perhaps the putative anti-σ factor is involved in controlling σ^X or σ^W in this bacterium. Finally, *B. subtilis* σ^V is the ECF σ factor that has representatives encoded in almost all of the *Clostridium* spp. (36 to 57% identity) (Table 26.6). In *B. subtilis*, a mutant strain has been generated for this σ factor but no function can yet be ascribed. Work in progress to define the *B. subtilis* σ^V regulon may be helpful in understanding the role of the best conserved ECF σ factor in *Clostridium*.

Other ECF σ factors putatively encoded within genomes of *Clostridium* species show weak but significant similarities to most of the *B. subtilis* ECF σ factors. In addition to σ^V and σ^W, *C. perfringens* [10] and *C. difficile* only encode one other ECF σ orthologue, whereas *C. botulinum* and *C. tetani* code for four and eight additional ECF σ factors, respectively.

σ^X

```
         Region 2.4                 Region 4.2

Bsu   IARHVAIDWFR--KQQTI   IQGYSIQETAKALRFSESKVKTTQHRGLKVLRK
Cac   IARNVLNDYFRGLKRHTL   GANLKNKEISELLGIKSNNVGIILFRTIKKLKS
Cth   ITLN-TIKSS—LKKKKL    IRDFSYKEIGKIMGISQEAAKMKVHRAKSILAK
```

σ^W

```
         Region 2.4                 Region 4.2

Bsu   IATNLTIDRIRKKKPD     IDELSLIEIGEILNIPVGTVKTRIHRGREALRKQ
Cac   ISVNTCLNFKRDNRGN     YEDLSYDEIASVLDMPIGTVKTKIYRAKNVLKKS
Cdi   IAKNKCKDFWKKKQPI     YHDLRFKDIAKIMESSVSVTKYRMSRALKKLELY
Cpe   ILINECNLILRRNKKT     FEDMSTKEISEMLNIAEGTVRSRLARAREKLRDF
Cth   VTTNACLDEIRKRKNK     IQGFTYEEIANIIKCPEGTVKSRINRARQALKKI
```

σ^V

```
         Region 2.4                 Region 4.2

Bsu   ILVRTAIDFLRKQKKIR    FEDLKLEEIAEITGENTNTVKTRLYRALKLMRIQ
Cbo   ILINSCITHINKRKRLI    FNDLTVPQIADIMDYPIGTVKTYIHKALKQLRIE
Cdi   ILVRTSIDFIRKNRKYN    FEDLKIEEVAIILDENVNTVKTRLYTALKKLKLK
Cpe   ILINVSIDFLKKKGKNE    FNDMKIKDISKVMEIPENTVKTYLRRAKQALGEV
Cte   ILINECITLLKKKNKIV    FQDLKISEISDVMQISDNTVKSHIRRGILSLKKS
Cth   IVVNTSLDFLRRQKRNV    FEDLKIDEIAEILNENVNTVKTRLYKSLEILRVK
```

FIGURE 26.3 Sequences of DNA recognition domains in ECF family σ factors in *Clostridium* spp.

Unlike other types of *Clostridium* σ factors, which are highly similar to each other, especially in the regions 2.4 and 4.2, similarities among the *Clostridium* ECF σ-like proteins are weaker and less conserved in these regions (Figure 26.3). Comparison of the *B. subtilis* σ^V orthologues and putative orthologues encoded in other *Bacillus* spp. shows similar diversity in protein sequences.

26.1.2.3 The TcdR Subfamily of σ Factors

The alternative σ factors differ sufficiently in sequence that it is sometimes difficult to identify them by standard sequence similarity searching methods. Even when the function has been proved biochemically, it is still a challenge to classify them within the σ^{70} family. Such a dilemma arose for TcdR, the σ factor for toxin gene expression in *C. difficile*. TcdR was formerly known as TxeR.

TcdR was first described as a positive regulator that greatly enhances expression from two toxin promoters (*tcdA* and *tcdB)* when expressed in *E. coli* [34]. If TcdR were a conventional positive regulator, one would have expected to see direct interaction between the protein and the regulatory regions of the toxin genes. No such interaction could be detected in gel mobility shift assays, suggesting that TcdR activates toxin gene expression by a different mechanism [4]. A Blast search of TcdR against protein databases revealed that TcdR is distantly related to *Bacillus* ECF σ factors. To test the ability of TcdR to act as a sigma factor, gel mobility shift assays were used to show that the mixture of TcdR and core RNA polymerase is able to bind to the promoter regions of the *tcdA, tcdB*, and *tcdR* genes [4]. *In vitro* run-off transcription reactions proved that TcdR activates specific transcription by core RNA polymerase from the toxin promoters [4]. Moreover, TcdR positively regulates its own expression in a manner analogous to its effect on the toxin promoters [5].

TABLE 26.7A
Sequence Similarities (% Identical Residues) of the Characterized Members within the TcdR Subfamily

σ Factors	TcdR	TetR	BotR	UviA
TcdR	100			
TetR	26	100		
BotR	24	67	100	
UviA	28	27	25	100

TABLE 26.7B
Sequence Similarities (% Identical Residues) to the TcdR σ Factor Family of Putative σ Factors Encoded on pE88 (CTP10 and CTP11) and pSOL1 (CAP1 and CAP2*)

σ Factors ORFs	TcdR	TetR	BotR	UviA
CTP10	28	21	21	24
CTP11	26	21	24	25
CAP1	21	23	24	26
CAP2	20	21	21	26

* CAP2 was previously named σ^Y [30].

Several other proteins required for transcription of *Clostridium* toxin or bacteriocin genes, including *C. tetani* TetR, *C. botulinum* BotR/A, and *C. perfringens* UviA, are related to TcdR ([35,36] and Table 26.7A). We and others have recently demonstrated that BotR/A, TetR, and UviA are respectively required, as σ factors, for the activation of the neurotoxin genes in *C. botulinum* and *C. tetani*, and the UV-inducible bacteriocin gene of *C. perfringens* [37,38]. These proteins appear to belong to the same family of alternative σ factors. In support of this idea, we have shown that they are functionally interchangeable. Expression of UviA *in trans* activates the *C. difficile tcdA* promoter and expression of TcdR activates the *C. perfringens bcn* promoter. These *in vivo* results were confirmed with *in vitro* transcription experiments. The presence of either TcdR or UviA permits RNA polymerase core enzyme to initiate transcription from the *bcn* and *tcdA* promoters [38]. Similar *in vivo* and *in vitro* functional complementarities have been seen for TetR and BotR/A and also for TcdR and BotR/A [36,38]. In addition, although the –10 regions are quite divergent, the –35 regions of the target gene promoters are nearly identical to each other, with a conserved "TTTACA" hexanucleotide motif. Thus, a common molecular mechanism, involving a new subgroup of the σ^{70} family of RNA polymerase σ factors, appears to control production of important toxins and a bacteriocin in several major pathogenic *Clostridium* species.

A search of all the complete and incomplete *Clostridium* genomes revealed additional homologues of the TcdR subfamily (Table 26.7B). In the solvent-producing strain ATCC 824 of *C. acetobutylicum*, the megaplasmid pSOL1 carries, in addition to genes involved in solvent formation, two genes coding for such σ factors. One of them (called σ^Y) was previously described to be phylogenetically related to *C. perfringens* UviA [30]. In *C. tetani*, the tetanus toxin-encoding plasmid pE88 harbors three genes that appear to code for members of the TcdR family, although one of them seems to be a pseudogene disrupted by a frameshift [11]. The roles that these additional factors might play in the control of solvent production or tetanus toxin synthesis remain to be determined. The fact that the genes for these factors, as well as for UviA and TetR, are carried on plasmids leads us to speculate that they may be derived from a common ancestor.

26.1.3 Sporulation-Associated σ Factors

The genera *Bacillus* and *Clostridium* are among the few that form metabolically dormant, environmentally resistant endospores. The process of spore formation has been studied in great detail in *B. subtilis*, as a result of which we know that a series of five σ factors directs RNA polymerase to transcribe large, distinct groups of genes at different stages of differentiation and in different intracellular compartments [39]. Sporulation is induced in dense populations of cells by limitation of certain key nutrients. A critical early event is the synthesis and activation of σ^H, which joins the core polymerase and directs it to certain key promoter sites, including the promoter of the *spo0A* gene. The Spo0A product is a transcription factor that positively or negatively regulates dozens of genes. Some of the products of these genes drive formation of a division septum near one pole of the cell. The forespore, the smaller of the two intracellular compartments thus formed, is destined to become the dormant spore; the larger, mother cell compartment plays an essential role in assembly of the spore but lyses when sporulation is complete. Other Spo0A-dependent genes encode the σ factors F and E, which are active in the forespore and mother cell compartments, respectively, after septation. At later stages of sporulation, σ^F and σ^E are replaced in their respective compartments by σ^G and σ^K. The occurrence in *Clostridium* spp. of genes resembling those that code for sporulation σ factors in *B. subtilis* has been the subject of previous reviews [40–42] and was noted in descriptions of completed genomes [9,10].

26.1.3.1 σ^H

σ^H is the earliest acting sporulation σ factor [39,43]. σ^H is essential for the initiation of sporulation because it controls the expression of *kinA,* a gene that encodes a histidine kinase that activates the Spo0A phosphorelay [44–46], the *spoIIA* operon, encoding the forespore-specific σ^F and its regulators [47], and high-level expression of *spo0A* [48]. σ^H is also thought to play a role in the switch from medial to polar cell division during sporulation, in part by recognizing the promoters of the *ftsA* and *minCD* genes [43,49,50].

Homologues of σ^H can be identified in all of the *Clostridium* spp. examined to date [14,51]. In all cases, with the possible exception of *C. thermocellum*, *sigH* is located directly upstream of a cluster of ribosomal proteins and the genes encoding the β and β′ subunits of RNA polymerase [41], as is true in *B. subtilis* as well [26]. (The region downstream of *sigH* in *C. thermocellum* could not be identified at the time of writing due to the incomplete nature of the genome.) The predicted σ^H proteins display 63 to 74% identity to σ^H from *B. subtilis* (Table 26.8), with nearly 100% conservation in regions 2.4 and 4.2 (Figure 26.4). Thus, σ^H from these organisms probably recognizes promoter DNA sequences that are very similar to those recognized by *B. subtilis* σ^H (–35, AGGAWWT, 12 to 14 bp spacer region, –10, RGAAT where W = A or T and R = A or G). In fact, consensus σ^H-like promoter sequences can be easily identified upstream of the *spoIIAA* gene in all of the *Clostridium* spp. examined (Figure 26.5). σ^H-dependent promoters have also been identified upstream of the *spo0A* genes of *C. acetobutylicum* [52] and *Clostridium pasteurianum* [53].

In *B. subtilis*, *sigH* is transcribed by both σ^A-containing and σ^H-containing forms of RNA polymerase [12]. Consensus σ^A promoter sequences can be identified upstream of the *sigH* ORF in all of the *Clostridium* spp. (Figure 26.5), but no σ^H-recognized promoter sequences are apparent.

TABLE 26.8
σ^H Homologues in *Clostridium* spp.[a]

	Bsu	Cac	Cbo	Cdi	Cpe	Cte	Cth
Bsu	100						
Cac	66	100					
Cbo	71	73	100				
Cdi	69	73	70	100			
Cpe	63	72	76	66	100		
Cte	67	73	83	66	72	100	
Cth	74	67	72	70	64	74	100

[a] See note to Table 26.1.

	Region 2.4	Region 4.2
Bsu	ITRQIITAIKTATRQKHIPLNS	DGRSYQEISDELNRHVKSIDNALQRVKRKLEK
Cac	ITRQIITAIKTATRQKHIPLNT	EGKSYQEIAKDLDRCSKSIDNALQRVKRKLEK
Cbo	VTRQIITAIKTATRQKHIPLNT	DGKSYQEIACDLDRHAKSIDNALQRVKRKLEK
Cdi	ITRQIITAIKTATRQKHIPLNS	NGKSYQFIADKLKRDVKSIDNALQRVKRKLEK
Cpe	VTRQIITAIKTATRQKHIPLNT	DGKSYQEIACDLDREAKSIDNALQRVKRKLEK
Cte	ITRQIITAIKTATRQKHIPLNT	DGKSYQEIACDLDRHAKSIDNALQRVKRKLEK
Cth	ITRQIITAIKTATRQKHIPLNS	QGKSYQEIAVELDRHVKSIDNALQRVKRKLEK

FIGURE 26.4 Sequences of DNA recognition domains in σ^H homologues in *Clostridium* spp.

26.1.3.2 σ^F

Since the *spoIIA* operon, which includes *sigF* (*spoIIAC*), the gene for σ^F, is expressed from a σ^H-dependent promoter [47], σ^F is synthesized in the predivisional cell. σ^F, however, is initially inactive due to the anti-σ activity of SpoIIAB, which is also encoded within the *spoIIA* operon [54–56]. Release of σ^F from interaction with SpoIIAB is regulated by the anti-σ factor antagonist SpoIIAA [57,58]. SpoIIAA, when in its active unphosphorylated form, competes with σ^F for interaction with SpoIIAB. The phosphorylation state of SpoIIAA is regulated by a phosphatase (SpoIIE) that resides in the polar septum [59–62], explaining in part the dependence of σ^F activity on the successful formation of the polar division septum.

Once activated, σ^F directs the transcription of early forespore-specific genes. It is also responsible for the appearance or activity of two additional developmental σ factors, the mother cell-specific factor σ^E and the forespore-specific factor σ^G. One of the gene products expressed by σ^F-RNA polymerase, SpoIIR, although present only in the forespore, acts as a signaling molecule that initiates the mother-cell program of sporulation gene transcription by activation of σ^E [63–65]. The continued development of the forespore compartment is achieved through σ^F-dependent transcription of *sigG* (*spoIIIG*) [66], encoding the later-acting forespore σ factor, σ^G.

σ^F homologues can be easily identified in all of the *Clostridium* species examined [41,51]. These proteins share 44 to 55% identity with *B. subtilis* σ^F (Table 26.9). Regions 2.4 and 4.2 of the predicted σ^F proteins are well conserved (Figure 26.6), indicating that similar σ^F promoter recognition sequences are utilized in these organisms (–35, WGCATA, 14 to 15 bp spacer region, –10, GGnRAYAMTW where W=A or T, R=A or G, and M=A or C). The identification of a putative σ^F-dependent promoter upstream of *sigG* (*spoIIIG*) in each of the *Clostridium* spp. analyzed illustrates this point (Figure 26.5).

In all cases, the putative *sigF* genes are located directly downstream of homologues of the regulators SpoIIAA and SpoIIAB [41]. A cluster of sporulation genes involved in dipicolinic acid

P*sigH*

```
σA consensus   TTGACA      (16-18)      TATAAT

Cac     GAAAATATTGACTTGAAAAATATTACTATTGTATAATTGTTTTGTGGGTTACATAGGGGGATATCAGTG
Cbo     ATTTACCTTGACGTTGTTATTAACACTAATGTATAATCAAATATGTGGTTATATTTATTTTGGGGGGGATACCTTG
Cdi     GAAAGCGTTGAAATTACTAGGTTCTTTAGACTATAATATACCTATAAATTTGTTTTATATATGGGGGAGATTTACATG
Cpe     GGTAATATTGACGAAGTCATATATAAACCTTATAATTTATTTGTAATTTATTGGATATTAGATTTTTATCTGGGGGTGTAGAGAATATG
Cte     AAATCCCTTGACGTATAAAAGAGATGTACGTTATAATTTAACTTGTAATGAGGGATTTTAGGGGGGATACTGTG
Cth     TTTATGCTTGACAATGTAAAGGAAACTTTTGTATAATCCCGATG
                -35      (17-18)       -10
```

P*spoIIAA-AB-sigF*

```
σH consensus          AGGAWWT   (12-14)   RGAAT

Cac     GTATTGTATAAAAAAGGAAATTGGATAAGTTTATAGAATAAATACTTAGTGCATTTGATTTATAATTTCAGGGGGAATTAATTATG
Cbo     ATAAAAGAAAAAGAAGGAAAATTATTCCATATGTAGAATAGTAATATTGATGATTTATAAAAAACATAAAATTGGGGGAATAGTATG
Cdi     TTAAATTTCATTGAAGGAATAAAAATATAATTATAGAATTGATTAAAAGTTACTATTTTTACAAAACTATTGAAGGAGGGATTTTTG
Cpe     GCTCTTGAAGATATAGGAAATAAAATTAGTGAAGAAATTTAATTTATTTTAGTATTATTTGACATAAATAATTAAAATATGTTAAAAT
Cte     ATAAAAATAAAAAGAGGAAAAAAATGTAATATGTAGAATAGTAAAAGAGAAGGACTTAGTAAAAACATAATGCAAAATGGGGGATTAATTTTG
Cth     ATATTCTTTGAAGGGGGAAATAATATTTTGAAGAATTAAGGTTAGGATAACTGACAGTGGAGGAGGATAAGCTTG
                        -35   (10-13)   -10
```

P*spoIIGA-sigE*

```
σA consensus  TTGACA      (16-18)         TATAAT

Cac     TATAAAACGACAAAATTTTAAACCATCAAGTAATATAATTATTTAAGAAAGGTGGAAGAAAATG
Cbo     TATGAATTGACAAATTTTTGAAATATTTAATTGTATAATAGTAAATAAATGGCAGTAGTTTAAGGGGGGAGATAACTATTG
Cdi     TCAAAAGTGACATAAAATTTTGTACATACTATCTTATAATATGTTTAAAGGTTTTAAGGGGGGATACATG
Cpe     TATATAATGACAAAATTTTAGAATAAGAAGCTTTATAATTATAAATAGGGAGTGAGTTGATTTATG
Cte     CAAAAAATAACAAAATTTTAAAATATTAAGATATATAATAATAAACAGGAGGGATTTTAATG
              -35        (21-22)         -10
```

P*sigG*

```
σF consensus    WGCATA  (14-15)   GGnRAYAMTW
σG consensus                     TGAATA     (17-18)     CATACTA

Cac     TTGTCTTTAGTATAAAATAGCTCCCCTGAGGGAATAATTTTTAATGCTATTATTTCAAAGGGGCTGATAAAGTG
Cbo     TTGTCCTTAGTATAAAATTAGATTCCTTTGGAAACAATTTAAATGCAATCATTCTGAAGGGACTGATGACTTTATG
Cdi     TCTTAAAAAGTATAAAAAAAATTGGAAATGTAAATAATTTCAATAGATGAATACTTTTGAGAGAGGACTGAAAATATATG
Cpe     GATTTAAGAGTATAAAATTAACCTCTACAGGTAATAATTAAAATGCGGAAGGGAATTACTTCGAAGGGGTTGATTTCAATTATG
Cte     TTTTATTATGTATAAAAGTCGGCTTCTAAGGAAATAATTATAATGCAATTATACTAAAGGGGCTGATTTACTTATG
Cth     TTTTTAGCAGTATAAAAATGACCTCCGTGGCAATAATGATACTGACTTTAAATTATTGCCGGGAGGTTTTTTCTATG
                -35     (14-15)        -10
                       σF              -35         (13- 18)       -10
                                                      σG
```

P*sigK*

```
σE consensus                                          GKCATATT    (13-15)    CATACAMT
σK consensus              AC    (16-17)    CATAnAnTA

Cdi     TGTTTTAAATTCTAAAAATATGCATATTATTTTATTAGATAACAGTCAACTTGGATAAGTCTCCTTCATATAATTTAAAG
Cth     GCCAAAACCATTTTGAAAAAGGGACTCTAATTTCTTGTTGTACTGGCACCTTAAACATTTTGGCTACATATATTTATAAG
Cpe     AAATGTTTATGAAGATAGAAAACCAGGAGGTGGAAATACTGCAGGTCCTATTTTCAGGGATATAGTTTTAGAACTTACAA
Cbo     AATCATATAAAAAAGCCTACATTATAATATATATTAAAAATATTGTCTTTTAATGCCTAATGGAAAATATGATTTTAAAT
Cte     TGATACTTAAAGAAAATTTATATAAAATAAGAACTAAAATTAGTCTCTCTGAATATTATAATAATAAGCACTATTTGGAG
Cac     TTAGAAAGTAACGAGAATACCTATAACAGCGTAGGATAAATAGTTTTAACCGTAATTGTGGTTAAACTTAAGTTAGTTTT
           -35    (13-18)      -10                      -35      (14-15)      -10
                    σK                                             σE
```

FIGURE 26.5 Consensus promoter sequences potentially recognized by sporulation σ factors in *Clostridium* spp.

transport into the developing spore, *spoVAA, AB, AC, AD*, and *AE*, is found downstream of *spoIIAC* in *B. subtilis* [26]. Three of these genes, *spoVAC, AD*, and *AE*, are also found at this locus in the *Clostridium* spp., but *spoVAA* and *AB* are conspicuously absent in all cases [41]. Although the chromosomal arrangement downstream of *spoIIAC* is conserved between *B. subtilis* and *Clostridium* spp., this is not the case for the region upstream of the *spoIIA* locus. In *C. acetobutylicum*,

TABLE 26.9
SpoIIAA, SpoIIAB, and σ^F Homologues in *Clostridium* spp.[a]

	Bsu	Cac	Cbo	Cdi	Cpe	Cte	Cth
			SpoIIAA				
Bsu	100						
Cac	32	100					
Cbo	35	56	100				
Cdi	43	49	46	100			
Cpe	42	56	60	48	100		
Cte	37	54	67	50	64	100	
Cth	40	45	42	42	51	39	100
			SpoIIAB				
Bsu	100						
Cac	54	100					
Cbo	54	71	100				
Cdi	55	63	56	100			
Cpe	57	68	70	65	100		
Cte	53	71	73	58	68	100	
Cth	54	62	60	53	58	58	100
			σ^F				
Bsu	100						
Cac	44	100					
Cbo	48	71	100				
Cdi	49	48	53	100			
Cpe	48	67	68	51	100		
Cte	48	66	74	54	66	100	
Cth	55	51	56	53	53	54	100

[a] See note to Table 26.1.

	Region 2.4	Region 4.2
Bsu	IIGEIQRFIRDDGTVKVSRSL	DQTQSEVAERLGISQVQVSRLEKKILKQIKVQM
Cac	IMGEIKRFLRDDGIIKVSRSV	DKTQTQVANMLGISQVQVSRIEKRILKAIREKL
Cbo	IMGEIKRFLRDDGIIKVSRSI	DKTQIEVAKQLGISQVQVSRIEKKVLKLMKEKL
Cdi	ILGEIKRYLRDDGMIKVSRSL	DMTQSEIGEMLNISQVQVSRIEKKVLSKLKEYI
Cpe	IIGEIKRFLRDDGIIKVSRST	DKTQIQVAKMLGISQVQVSRIEKKVLKIMKEKL
Cte	IMGEIKRFIRDDGIIKVSRSL	DKTQIEVAKMLGISQVQVSRIEKRVLKTMKEKL
Cth	IIGEIKRFIRDDGMIKVSRSL	EKTQVQIAKMLGISQVQVSRIEKKILEEIRKKI

FIGURE 26.6 Sequences of DNA recognition domains in σ^F homologues in *Clostridium* spp.

C. botulinum, and *C. tetani* the region upstream of *spoIIAC* encodes a number of proteins predicted to be involved in lipopolysaccharide biosynthesis, but there is no corresponding similarity in this region among the other organisms analyzed, including *B. subtilis*.

In *B. subtilis*, the *spoIIAA-IIAB-sigF* operon is transcribed from a σ^H-dependent promoter [47]. Strong consensus σ^H promoter sequences can easily be identified in the aligned promoter regions of the *Clostridium sigF* operons. Homologues of the regulatory proteins SpoIIAA and SpoIIAB are found upstream of *sigF* in all of the *Clostridium* spp. Putative homologues of SpoIIE can also be found in the *Clostridium* genomes [41], but they only share 23 to 28% identity with the *B. subtilis* phosphatase. Further analysis would be required to make a functional assignment.

26.1.3.3 σ^E

Although it is the first mother cell-specific σ factor, σ^E, like σ^F, accumulates in the predivisional cell because the gene encoding σ^E, *sigE* (*spoIIGB*), is expressed from a σ^A-dependent, Spo0A~P-dependent promoter [39,67–69]. Like σ^F, the activity of σ^E is inhibited until polar septation [70–72], but by a completely different mechanism. Expression of *sigE* results in the production of an inactive pro-σ factor, Pro–σ^E, containing a 27 amino acid N-terminal extension [71]. This pro-sequence must be cleaved by a protease, SpoIIGA, encoded by the gene directly upstream of *sigE* [70,73]. The proteolytic activation of Pro–σ^E is coupled to polar septum development because the activity of SpoIIGA is dependent on a forespore-generated signal. It is currently thought that this signal is transduced to SpoIIGA by SpoIIR, a protein produced in the forespore [64,65], but secreted across the forespore membrane [63]. SpoIIR then interacts with SpoIIGA, triggering its proteolytic activity and resulting in Pro-σ^E processing.

In *B. subtilis*, σ^E directs the transcription of many genes, some of which are involved in spore morphogenesis (primarily spore coat and cortex gene products) [74]. It is also required for the expression of the major mother cell transcription factor SpoIIID [75–77] and the later-acting mother cell-specific σ factor, σ^K [78].

Homologues of σ^E can be identified in all of the *Clostridium* genomes, with the exception of the unfinished *C. thermocellum* genome, where only a partial sequence can be identified [9,14,51,79,80]. The predicted σ^E proteins share 66 to 73% identity with *B. subtilis* σ^E (Table 26.10). As with the other σ factors discussed, regions 2.4 and 4.2 of the σ^E homologues are highly conserved (Figure 26.7), indicating that similar promoter sequences are recognized by σ^E in all of these organisms (consensus –35, GKCATATT, 13 to 15 bp spacer, –10, CATACAMT where K=G or T, and M=A or C). Reasonably good σ^E recognition sequences can be identified in the promoter regions of the *sigK* genes identified in the *Clostridium* spp. (Figure 26.5). *C. perfringens* σ^E is needed for transcription of the enterotoxin (*cpe*) gene [81]. In addition, a σ^E-dependent promoter has been identified upstream of the *spoIID* gene of *C. acetobutylicum* [40].

In all of the genomes analyzed, the *sigE* gene is found in the same three-gene clusters as in *B. subtilis*, i.e., directly downstream of *spoIIGA* and upstream of *sigG* (*spoIIIG*) [9,40,41,80,82]. Also, as in *B. subtilis*, the genes directly upstream of *spoIIGA* are usually *ftsAZ* [83]. *C. difficile*, however, appears to lack *ftsA* at this locus.

As expected, all of the *Clostridium sigE* genes appear to be transcribed by σ^A-containing RNA polymerase, as strong consensus σ^A –10 and –35 regions can be identified within the aligned promoter regions (Figure 26.5 and [83]). It is not easy to ascertain, however, whether *sigE* transcription is activated by Spo0A~P in these organisms because no 0A boxes are apparent. The deduced σ^E proteins from *Clostridium* all contain an N-terminal pro-sequence of approximately the same length (28 aa in *C. difficile*, 23 aa in all others) as that of *B. subtilis* (27 aa). Not surprisingly, as mentioned above, all of these organisms also contain a homologue of the protease SpoIIGA required to cleave this pro-sequence ([14,80] and Table 26.10). The SpoIIGA homologues in these organisms only share 20 to 25% identity to *B. subtilis* SpoIIGA, but can be confidently assigned as the cognate proteases based on their location directly upstream of *sigE* (*spoIIGB*). The SpoIIR signaling protein can also be identified in the genomes of all of the *Clostridium* spp. (34 to 38% identity with the *B. subtilis* protein) [41].

Introduction of the *C. acetobutylicum spoIIG* operon into a *sigE* null mutant strain of *B. subtilis* led to only partial restoration of sporulation because of an apparent defect in the timing of expression of the *Clostridium* operon [79].

TABLE 26.10
SpoIIGA and σE Homologues in *Clostridium* spp.[a]

	SpoIIGA					
	Bsu	**Cac**	**Cbo**	**Cdi**	**Cpe**	**Cte**
Bsu	100					
Cac	20	100				
Cbo	22	52	100			
Cdi	20	21	24	100		
Cpe	23	33	38	27	100	
Cte	25	45	43	21	33	100
Cth	ND[b]	ND	ND	ND	ND	ND

	σ^E						
	Bsu	**Cac**	**Cbo**	**Cdi**	**Cpe**	**Cte**	**Cth**
Bsu	100						
Cac	73	100					
Cbo	71	82	100				
Cdi	66	64	68	100			
Cpe	71	81	78	66	100		
Cte	72	79	80	68	79	100	
Cth[c]	(51)	(48)	(45)	(51)	(48)	(50)	(100)

[a] See note to Table 26.1.

[b] ND = not detected.

[c] Only partial sequence of the *C. thermocellum sigE* gene could be identified in the Microbial Genomes database at the time of writing. Values for percent identity were calculated based on the partial sequence.

	Region 2.4	Region 4.2
Bsu	IENEILMYLRRNNKIRS	EKTQKDVADMMGISQSYISRLEKRIIKRLRKEFNKMV
Cac	IENEILMYLRRNSKVKA	EKTQKEVADMLGISQSYISRLEKRIIKRLKKEINKML
Cbo	IENEILMYLRRNSKVKA	EKTQKEVADMLGISQSYISRLEKRIIKTLKKEINKMV
Cdi	IENEILMYLRKNNKKKT	EKTQKEVAGMLGISQSYISRLEKKIISRLQKEMKKFV
Cpe	IENEILMYLRRNSKVKA	EKTQKEVADMLGISQSYISRLEKKIIKRLKKEINKMI
Cte	IENEILMYLRRNNKIKS	EKTQKEVADMLGISQSYISRLEKRIIKRLKKEINKMI
Cth	IENEILMHIRSNKKTQS	GKTQREIAKMLGISRSYVSRIEKKAIKKLGKELKSES

FIGURE 26.7 Sequences of DNA recognition domains in σ^E homologues in *Clostridium* spp.

26.1.3.4 σ^G

σ^G is the final σ factor to act in the forespore compartment of the sporangium [34,84,85]. *sigG* (*spoIIIG*) is initially expressed from a σ^F-dependent promoter, but it also has a σ^G-dependent promoter [66,84,86]. It is thought that σ^G is sequestered by SpoIIAB, the same anti-σ factor known to regulate σ^F, but it is unclear how the proposed interaction of σ^G and SpoIIAB is relieved [87]. The activity of σ^G, however, is also dependent on prior developmental events. σ^G activity only occurs after complete engulfment of the forespore compartment by the mother cell [88]. Products of the mother cell *spoIIIA* operon can play a role in σ^G activation [87]. Once activated, σ^G directs

TABLE 26.11
σ^G Homologues in *Clostridium* spp.[a]

	Bsu	Cac	Cbo	Cdi	Cpe	Cte	Cth
Bsu	100						
Cac	74	100					
Cbo	74	85	100				
Cdi	68	74	77	100			
Cpe	75	87	86	75	100		
Cte	75	84	84	74	85	100	
Cth	74	76	80	73	79	76	100

[a] See note to Table 26.1.

	Region 2.4	Region 4.2
Bsu	IIGEIRRYLRDNNPIRVSRSL	GKTQMEVAEEIGISQAQVSRLEKAAIKQMNKNI
Cac	IIGEIRRYLRDNNSIRVSRSL	GRTQMEVADEIGISQAQVSRLEKTALKHMKKYV
Cbo	IIGEIRRYLRDNNSIRVSRSL	GRTQMEVADEIGISQAQVSRLEKTALKHMRKYV
Cdi	IIGEIRRYLRDNNPIRVSRSL	GRTQIEVADEIGISQAQVSRIEKNALKNMRKYV
Cpe	IIGEIRRYLRDNNTIRVSRSL	GRTQMEVADEIGISQAQVSRLEKIALKHMRKYV
Cte	IIGEIRRYLRDNNSIRVSRSL	GRTQMEVADEIGISQAQVSRLEKTALKHMRKFV
Cth	IIGEIRRYLRDNNAIRVSRSL	GRTQMEVAEEIGISQAQVSRLEKTALMHMRKYI

FIGURE 26.8 Sequences of DNA recognition domains in σ^G homologues in *Clostridium* spp.

transcription of late forespore-specific genes such as the small acid soluble proteins (SASPs) [85,86]. σ^G-containing RNAP also transcribes SpoIVB, a signaling protein involved in the activation of σ^K in the mother cell [89–91].

Clostridium σ^G homologues, identified in all of the genomes analyzed here, show 68 to 75% identity to *B. subtilis* σ^G [9,14,51,80] (Table 26.11). As with all of the sporulation σ factors discussed, regions 2.4 and 4.2 of the identified σ^G candidates are highly conserved (Figure 26.8). Thus, *Clostridium* σ^G is likely to recognize promoter sequences similar to the *B. subtilis*-derived consensus sequences (–35, TGAATA, 17 to 18 bp spacer region, –10, CATACTA). Consensus σ^G-like promoters have been identified upstream of *ssp* genes in *C. perfringens* [92,93].

The *sigG* gene is found downstream of the *spoIIGA-sigE* operon in all of the *Clostridium* genomes evaluated [9,14,41,51,80]. In most of the species, *sigG* is followed by a set of three or four open reading frames of unknown function and genes encoding a typical bacterial two-component system. In *C. botulinum* and *C. tetani*, gene synteny continues through a four-gene cluster encoding an ABC transporter. None of these genes are found in the region downstream of *sigG* in the *B. subtilis* genome. Interestingly, *C. difficile* shares gene synteny with the region downstream of *sigG* in *B. subtilis*, but also maintains an insertion of genes (different from those found in the other *Clostridium* species) not found in this region of the *B. subtilis* genome.

Identification of both putative σ^F and σ^G consensus promoter regions upstream of *sigG* [14,40,83] (Figure 26.5) supports the suggestion that in *Clostridium* spp. this gene is regulated in the same manner as in *B. subtilis*. As stated above, all of the *Clostridium* spp. contain homologues of SpoIIAB and all appear to contain the *spoIIIA* operon (*spoIIIAA-AH*) in full or in part [41]. The *C. acetobutylicum*, *C. difficile*, and *C. tetani* genomes contain homologues of SpoIIIAA-AH in a contiguous region of DNA, whereas *C. botulinum* appears to lack only *spoIIIAH*. *C. perfringens* has the entire operon, with the possible exception of *spoIIIAB*. At this time, incomplete sequence data make it difficult to know whether the *C. thermocellum* genome also contains the *spoIIIA* operon.

26.1.3.5 σ^K

In *B. subtilis*, σ^K directs the transcription of spore coat and cortex genes, as well as the late mother cell transcription factor GerE [39,94–96]. *sigK* is expressed from a promoter dependent on the previously activated mother cell σ factor, σ^E, [97] and the transcription factor SpoIIID [98]. As is the case for the other sporulation σ factors, the activity of *B. subtilis* σ^K is tightly regulated post-translationally. It is expressed as an inactive pro-σ factor [91,94,99]. Activation of Pro-σ^K requires the proteolytic processing of a 20 amino acid N-terminal peptide. This cleavage is performed by a membrane-localized protease, SpoIVFB, whose activity is triggered by a signaling protein, SpoIVB, produced in the forespore [100–103].

In *B. subtilis*, an additional event must occur to allow σ^K production. The gene encoding σ^K is disrupted by a large prophage-like element, termed *skin* (σ K intervening sequence) [104,105]. This disruption does not permanently destroy the ability of *skin*-containing strains of *B. subtilis* to sporulate because *skin* encodes a site-specific recombinase that is developmentally regulated [106,107]. Expression of the recombinase, specifically during sporulation, results in the precise excision of *skin* from the *sigK* ORF. Although the excision of *B. subtilis skin* was once predicted to play a critical role in the regulation of *sigK* expression or σ^K activity, it has since been shown that it is redundant to the proteolytic activation of Pro–σ^K, described above [108]. It has also been clear for some time that *skin* is not generally required for proper sporulation because all other spore-formers within the *Bacillus* and *Clostridium* genera that had been previously examined contain uninterrupted *sigK* genes [14,109–111].

All of the *Clostridium* genomes analyzed in this study contain homologues of σ^K, with identity to the *B. subtilis* σ factor ranging from 54 to 58% [14,51] (Table 26.12). As with the other sporulation σ factors, the regions of σ^K known to interact with promoter DNA are highly conserved in all of the identified proteins (Figure 26.9). Again, this indicates that σ^K in *Clostridium* should interact with promoter sequences similar to the consensus sequences (–35, AC, 16 to 17 bp spacer region, –10, CATAnAnTA) characteristic of *B. subtilis*. The *cpe* gene of *C. perfringens* has a σ^K-dependent promoter that corresponds to this *B. subtilis* consensus sequence (in addition to the σ^E-dependent sporulation promoter mentioned previously) [81].

In *C. acetobutylicum*, *C. botulinum*, *C. perfringens*, and *C. tetani,* the *sigK* gene is found upstream of the *spoIIG-sigG* locus, and separated from the other σ factor operons by three to five genes of unrelated function. In these organisms, *sigK* is preceded by a cluster of 7 to 9 genes whose products include a penicillin-binding protein, collagenase, SAM-dependent methyltransferase, and

TABLE 26.12
σ^K Homologues in *Clostridium* spp.[a]

	Bsu	Cac	Cbo	Cdi	Cpe	Cte	Cth
Bsu	100						
Cac	58	100					
Cbo	56	71	100				
Cdi[b]	55	63	61	100			
Cpe	54	65	71	59	100		
Cte[b]	57	66	73	62	65	100	
Cth	58	63	63	62	62	60	100

[a] See note to Table 26.1.

[b] Percent identity determined for predicted translation product after recombination to remove *skin*.

	Region 2.4	Region 4.2
Bsu	IENEILMHLRALKKTKK	EKTQREIAKELGISRSYVSRIEKRALMKMFHE
Cac	IENEILMLIRNNKKSKN	QPRNQREIAQILGISRSYVSRIEKRALKKLYR
Cbo	IENEILMLIRNNKKTKG	PRTQREIAGILNISRSYVSRIEKKALKKLYKE
Cdi	IENEILMNIRSNKKNKT	YKTQREIAQKLDISRSYVSRIEKKALKKLEKE
Cpe	IDNEILMFFRNTKKTKG	IKTQREIAGILGISRSYVSRIEKKALKKLNKE
Cte	IENEILMLIRNNKKTKG	IKTQREIAEILGISRSYVSRIEKRSLKKLCKE
Cth	IENEILMHIRSNKKTQS	GKTQREIAKMLGISRSYVSRIEKKAIKKLGKE

FIGURE 26.9 Sequences of DNA recognition domains in σ^K homologues in *Clostridium* spp.

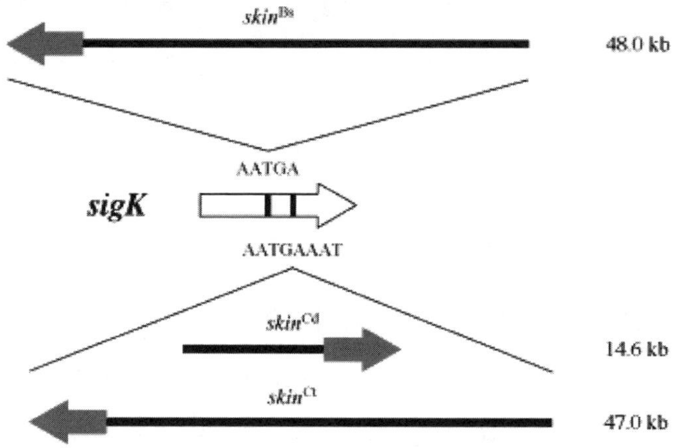

FIGURE 26.10 Location and orientation of *skin* elements in *B. subtilis, C. difficile,* and *C. tetani sigK* genes.

a predicted folate metabolism gene. The proximity of *sigE* and *sigG* in spore-formers might indicate that they arose by a gene duplication event followed by specialization of each sigma factor. The linkage of *sigK* to *sigE-sigG* in most of the *Clostridium* spp. provides additional support for such a model, since *sigK* and *sigE* show more similarity to each other than do *sigE* and *sigG*. In *C. difficile*, however, *sigK* appears to be located in a different region of the chromosome. Although four of the same genes found in the other *Clostridium* spp. are also adjacent to *sigK* in *C. difficile*, they are downstream of the gene, rather than upstream, and *C. difficile sigK* is not linked to the *spoIIG-sigG* locus. The chromosomal location of *C. difficile sigK* is also distinct from that of *B. subtilis sigK*.

Since the presence of the *skin* element in *B. subtilis* (designated *skin*^{Bs} for clarity) had, until recently, been the exception rather than the rule, it was surprising to find that the *sigK* genes of *C. difficile* and *C. tetani* are disrupted by different, but related, elements, termed *skin*^{Cd} and *skin*^{Ct} [51,112]. The *Clostridium* elements are integrated into exactly the same site within the *sigK* gene (AATGAAAT) but are in different orientations with respect to the encoded recombinase [51] (Figure 26.10). This site is close to but different from the site of *skin*^{Bs} integration in *B. subtilis*. The *skin* elements vary quite dramatically in size (Figure 26.10). *skin*^{Ct}, which is 47 kb, is similar in size to the 48 kb *skin*^{Bs}. In contrast, *skin*^{Cd} is only 14 kb. *skin*^{Ct}, like *skin*^{Bs}, appears to encode a temperate phage, whereas *skin*^{Cd}, if it ever corresponded to a prophage, appears to have undergone extensive genetic deterioration [51]. However, *skin*^{Cd} maintains the ability to excise from the chromosome during sporulation [51].

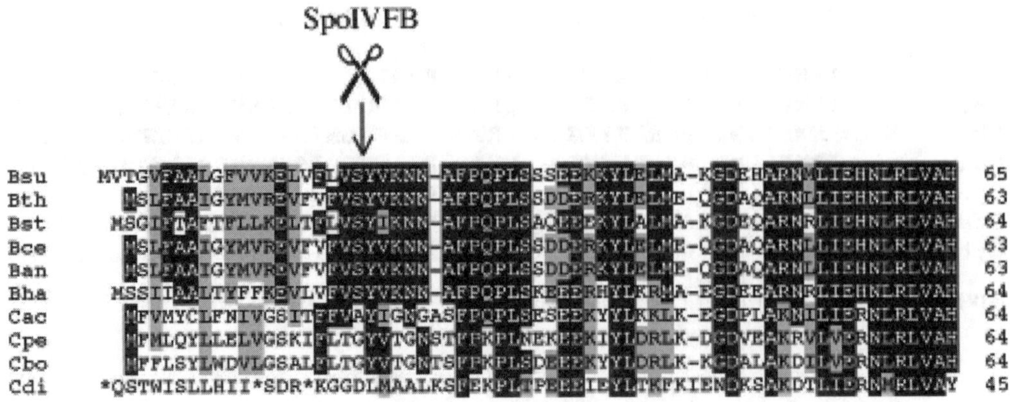

FIGURE 26.11 Alignment of N-terminal coding sequences of *sigK* genes in *Clostridium* spp. The scissors and arrow indicate the site at which proteolytic cleavage by SpoIVFB occurs in *B. subtilis* pro–σ^K. Translation of *C. difficile sigK* mRNA is thought to initiate with the amino acid sequence MAALKS. (Adapted with permission from Haraldsen and Sonenshein [51].)

The *sigK* gene in most *Clostridium* spp. encodes a pro-σ factor [41,51]. The pro-sequence varies in length from 19 to 21 amino acids. Homologues of SpoIVFB, the protease involved in Pro–σ^K activation, as well as regulators of its activity, SpoIVFA and BofA, can be identified in the corresponding genomes [41]. In addition, the signaling protein SpoIVB, produced in the forespore and required for activation of the SpoIVFB protease, can also be identified in all of these organisms [41].

Analysis of *C. difficile sigK* revealed a second surprise. The *sigK* gene in this organism encodes a mature σ factor, rather than an inactive pro-σ factor. The start codon aligns with the site of proteolytic activation in other Pro–σ^K sequences (Figure 26.11) [41,51]. Consistent with the absence of the pro-sequence, no homologues of SpoIVFB and its regulators can be identified in the *C. difficile* genome [41]. A homologue of the forespore signaling protein SpoIVB can be seen, however, suggesting that it has another role in sporulation in addition to Pro–σ^K activation. The absence of the pro-sequence in *C. difficile* also appears to make regulated *skin*^{Cd} excision a requirement for successful sporulation, since the regulatory redundancy observed in other spore-formers is eliminated. Isolates of *C. difficile* that naturally lack *skin*^{Cd} are defective for sporulation, whereas isolates that contain *skin*^{Cd} sporulate normally [51]. One would predict that strains of *C. tetani* lacking *skin*^{Ct}, if any exist, would not have a sporulation defect, since the *sigK* gene in this organism encodes a pro–σ factor. Thus, the *sigK* gene in *C. difficile* presents a novel situation.

ACKNOWLEDGMENTS

We thank the following institutions for making genome sequences available: for *B. subtilis*, the Institut Pasteur, Paris (http://genolist.pasteur.fr/SubtiList); for *C. acetobutylicum*, the DNA Data Bank of Japan (http://gib.genes.nig.ac.jp/) using sequence information provided by Genome Therapeutics Corp.; for *C. perfringens,* the University of Tsukuba, Japan; for *C. difficile* and *C. botulinum*, the Sanger Centre; for *C. tetani*, the University of Göttingen; and for *C. thermocellum*, the U.S. DOE Joint Genome Institute. Unpublished work from the authors' laboratories was supported by a research grant from the U.S. Public Health Service (GM42219) to ALS and by funds from the Institut Pasteur to B.D.

REFERENCES

1. Helmann, J.D. and Moran, C.P. Jr., RNA polymerase and σ factors, in *Bacillus subtilis and Its Closest Relatives: From Genes to Cells*, Sonenshein, A.L., Hoch, J.A., and Losick, R., Eds. ASM Press, Washington, DC, 2002, Chapter 21.
2. Pich, A. and Bahl, H., Purification and characterization of the DNA-dependent RNA polymerase from *Clostridium acetobutylicum, J. Bacteriol.,* 173, 2120, 1991.
3. Garnier, T. and Cole, S.T., Studies of UV-inducible promoters from *Clostridium perfringens in vivo* and *in vitro*, *Mol. Microbiol.*, 2, 607, 1988.
4. Mani, N. and Dupuy, B., Regulation of toxin synthesis in *Clostridium difficile* by an alternative RNA polymerase σ factor, *Proc. Natl. Acad. Sci. U.S.A.,* 98, 5844, 2001.
5. Mani, N. et al., Environmental response and autoregulation of *Clostridium difficile* TcdR, a sigma factor for toxin gene expression, *J. Bacteriol.,* 181, 5971–5978, 2002.
6. Lonetto, M., Gribskov, M., and Gross, C.A., The σ^{70} family: sequence conservation and evolutionary relationships, *J. Bacteriol.*, 174, 3843, 1992.
7. Shingler, V., Signal sensing by σ^{54}-dependent regulators: derepression as a control mechanism, *Mol. Microbiol.*, 19, 409, 1996.
8. Helmann, J.D. and Chamberlin, M.J., Structure and function of bacterial σ factors, *Ann. Rev. Biochem.*, 57, 839, 1988.
9. Nolling, J. et al., Genome sequence and comparative analysis of the solvent-producing bacterium *Clostridium acetobutylicum, J. Bacteriol.*, 183, 4823, 2001.
10. Shimizu, T. et al., Complete genome sequence of *Clostridium perfringens*, an anaerobic flesh-eater, *Proc. Natl. Acad. Sci. U.S.A.*, 99, 996, 2002.
11. Brüggemann, H. et al., The genome sequence of *Clostridium tetani*, the causative agent of tetanus disease, *Proc. Natl. Acad. Sci. U.S.A.*, 100, 1316, 2003.
12. Moran, C.P. Jr., RNA polymerase and transcription factors, in *Bacillus subtilis and Other Gram-Positive Bacteria: Biochemistry, Physiology, and Molecular Genetics*, Sonenshein, A.L., Hoch, J.A., and Losick, R., Eds., American Society for Microbiology, Washington, DC, 1993, Chapter 45.
13. Wang, L.F. and Doi, R.H., Nucleotide sequence and organization of *Bacillus subtilis* RNA polymerase major σ (σ^{43}) operon, *Nucleic Acids Res.*, 14, 4293, 1986.
14. Sauer, U. et al., Sporulation and primary σ factor homologous genes in *Clostridium acetobutylicum, J. Bacteriol.,* 176, 6572, 1994.
15. Marquez-Magana, L.M. and Chamberlin, M.J., Characterization of the *sigD* transcription unit of *Bacillus subtilis, J. Bacteriol.*, 176, 2427, 1994.
16. Chen, L. and Helmann, J.D., The *Bacillus subtilis* σ^{D}-dependent operon encoding the flagellar proteins FliD, FliS, and FliT, *J. Bacteriol.*, 176, 3093, 1994.
17. Hughes, K.T. et al., Sensing structural intermediates in bacterial flagellar assembly by export of a negative regulator, *Science*, 262, 1277, 1993.
18. Mirel, D.B., Lauer, P., and Chamberlin, M.C., Identification of flagellar synthesis regulatory and structural genes in a σ^{D}-dependent operon of *Bacillus subtilis, J. Bacteriol.*, 176, 4492, 1994.
19. Sohling, B. and Gottschalk, G., Molecular analysis of the anaerobic succinate dehydrogenase pathway in *Clostridium kluyveri, J. Bacteriol.*, 178, 871, 1996.
20. Fisher, S.H. and Débarbouillé, M., Nitrogen source utilization and its regulation, in *Bacillus subtilis and Its Closest Relatives: From Genes to Cells*, Sonenshein, A.L., Hoch, J.A., and Losick, R., Eds., ASM Press, Washington, DC, 2002, Chapter 14.
21. Magasanik, B., Regulation of nitrogen utilization, in *Escherichia coli and Salmonella: Cellular and Molecular Biology*, Neidhardt, F.C. et al., Eds., ASM Press, Washington, DC, 1996, Chapter 86.
22. Price, C.W., General stress response, in *Bacillus subtilis and Its Closest Relatives: From Genes to Cells*, Sonenshein, A.L., Hoch, J.A., and Losick, R., Eds., ASM Press, Washington, DC, 2002, Chapter 26.
23. Gertz, S. et al., Characterization of the σ^{B} regulon in *Staphylococcus aureus*, *J. Bacteriol.*, 182, 6983, 2000.
24. Lonetto, M.A. et al., Analysis of the *Streptomyces coelicolor sigE* gene reveals the existence of a subfamily of eubacterial RNA polymerase σ factor involved in the regulation of extracytoplasmic functions, *Proc. Natl. Acad. Sci. U.S.A.,* 91, 7573, 1994.

25. Helmann, J.D., The extracytoplasmic function (ECF) sigma factors, *Adv. Microb. Physiol.*, 46, 47, 2002.
26. Kunst, F. et al., The complete genome sequence of the Gram-positive bacterium *Bacillus subtilis*, *Nature*, 390, 246, 1997.
27. Huang, X. et al., The *Bacillus subtilis* σ^X protein is an extracytoplasmic function sigma factor contributing to the survival of high temperature stress, *J. Bacteriol.*, 179, 2915, 1997.
28. Huang, X. et al., Identification of target promoters for the *Bacillus subtilis* extracytoplasmic function σ factor, σ^W, *Mol. Microbiol.*, 31, 361, 1999.
29. Horsburgh, M.J. and Moir, A., Sigma M, an ECF RNA polymerase sigma factor of *Bacillus subtilis* 168, is essential for growth and survival in high concentration of salt, *Mol. Microbiol.*, 32, 41, 1999.
30. Behrens, S. et al., Identification of two genes encoding putative new members of the ECF subfamily of eubacterial RNA polymerase σ factors in *Clostridium acetobutylicum, J. Mol. Microbiol. Biotechnol.*, 2, 265, 2000.
31. Brutsche, S. and Braun, V., SigX of *Bacillus subtilis* replaces the ECF sigma factor FecI of *Escherichia coli* and is inhibited by RsiX, *Molec. Gen. Genet.*, 256, 416, 1997.
32. Paget, M. et al., The ECF sigma factors of *Streptomyces coelicolor* A3(2), in *Switches, Signals, Regulons and Cascades: Control of Bacterial Gene Expression,* Society for General Microbiology Symposium, Thomas, C.M. and Hodgson, D.A., Eds., Cambridge University Press, Cambridge, 2002, Volume No. 61.
33. Huang, X., Frederick, K.L., and Helmann, J.D., Promoter recognition by *Bacillus subtilis* σ^W: autoregulation and partial overlap with the σ^X regulon, *J. Bacteriol.*, 180, 3765–3770, 1998.
34. Moncrief, J.S., Barroso, L.A., and Wilkins, T.D., Positive regulation of *Clostridium difficile* toxins, *Infect. Immun.*, 65, 1105, 1997.
35. Marvaud, J.C. et al.,TetR is a positive regulator of the tetanus toxin gene in *Clostridium tetani* and is homologous to *botR*, *Infect. Immun.*, 66, 5698, 1998.
36. Marvaud, J.C. et al., *botR/A* is a positive regulator of botulinum neurotoxin and associated non-toxin protein genes in *Clostridium botulinum* A, *Mol. Microbiol.*, 29, 1009–1018, 1998.
37. Raffestin, S. et al., Bot R/A and TetR are alternative RNA polymerase sigma factors controlling expression of the neurotoxin and associated protein genes in *Clostridium botulinum* type A and *Clostridium tetani*, *Mol. Microbiol.*, 2004, in press.
38. Dupuy, B., et al., Transcription activation of a UV-inducible *Clostridium perfringens* bacteriocin gene by a novel σ factor, *Mol. Microbiol.*, 2005, in press.
39. Piggot, P.J. and Losick, R., Sporulation genes and intercompartmental regulation, in *Bacillus subtilis and Its Closest Relatives: From Genes to Cells*, Sonenshein, A.L., Hoch, J.A., and Losick, R., Eds., ASM Press, Washington, DC, 2002, Chapter 34.
40. Sauer, U. et al., σ factor and sporulation genes in *Clostridium*, *FEMS Microbiol. Rev.*, 17, 331, 1995.
41. Stragier, P., A gene odyssey: Exploring the genomes of endospore-forming bacteria, in *Bacillus subtilis and Its Closest Relatives: From Genes to Cells*, Sonenshein, A.L., Hoch, J.A., and Losick, R., Eds., ASM Press, Washington, DC, 2002, Chapter 35.
42. Dürre, P. and Hollergschwander, C., Initiation of endospore formation in *Clostridium acetobutylicum*, *Anaerobe,* 10, 69–74, 2004.
43. Britton, R. A. et al., Genome-wide analysis of the stationary-phase sigma factor (Sigma-H) regulon of *Bacillus subtilis*, *J. Bacteriol.*, 184, 4881, 2002.
44. Antoniewski, C., Savelli, B., and Stragier, P., The *spoIIJ* gene, which regulates early developmental steps in *Bacillus subtilis*, belongs to a class of environmentally responsive genes, *J. Bacteriol.*, 172, 86, 1990.
45. Hoch, J.A., Regulation of the phosphorelay and the initiation of sporulation in *Bacillus subtilis*, *Annu. Rev. Microbiol.*, 47, 441, 1993.
46. Burbulys, D., Trach, K.A., and Hoch, J.A., Initiation of sporulation in *B. subtilis* is controlled by a multicomponent phosphorelay, *Cell*, 64, 545, 1991.
47. Wu, J.J. et al., Transcription of the *Bacillus subtilis spoIIA* locus, *Gene*, 101, 113, 1991.
48. Predich, M., Nair, G., and Smith, I., *Bacillus subtilis* early sporulation genes *kinA*, *spo0F*, and *spo0A* are transcribed by the RNA polymerase containing sigma H, *J. Bacteriol.*, 174, 2771, 1992.
49. Gonzy-Tréboul, G., Karmazyn-Campelli, C., and Stragier, P., Developmental regulation of transcription of the *Bacillus subtilis ftsAZ* operon, *J. Mol. Biol.*, 224, 967, 1992.

50. Levin, P.A. and Losick, R., Transcription factor Spo0A switches the localization of the cell division protein FtsZ from a medial to a bipolar pattern in *Bacillus subtilis, Genes Dev.*, 10, 478, 1996.
51. Haraldsen, J.D. and Sonenshein, A.L., Efficient sporulation in *Clostridium difficile* requires disruption of the σ^K gene, *Mol. Microbiol.*, 48, 811, 2003.
52. Wilkinson, S.R. et al., Molecular genetics and the initiation of solventogenesis in *Clostridium beijerinckii* (formerly *Clostridium acetobutylicum*) NCIMB 8052, *FEMS Microbiol. Rev.*, 17, 275, 1995.
53. Brown, D.P. et al., Characterization of *spo0A* homologues in diverse *Bacillus* and *Clostridium* species identifies a probable DNA-binding domain, *Mol. Microbiol.*, 14, 411, 1994.
54. Duncan, L. and Losick, R., SpoIIAB is an anti-σ factor that binds to and inhibits transcription by regulatory protein s^F from *Bacillus subtilis*, *Proc. Natl. Acad. Sci. U.S.A.*, 90, 2325, 1993.
55. Kirchman, P. A. et al., Forespore-specific disappearance of the sigma-factor antagonist SpoIIAB: implications for its role in determination of cell fate in *Bacillus subtilis*, *Mol. Microbiol.*, 8, 663, 1993.
56. Min, K.T. et al., σ^F, the first compartment-specific transcription factor of *B. subtilis*, is regulated by an anti-σ factor that is also a protein kinase, *Cell*, 74, 735, 1993.
57. Duncan, L., Alper, S., and Losick, R., SpoIIAA governs the release of the cell-type specific transcription factor σ^F from its anti-sigma factor SpoIIAB, *J. Mol. Biol.*, 260, 147, 1996.
58. Diederich, B. et al., Role of interations between SpoIIAA and SpoIIAB in regulating cell-specific transcription factor σ^F of *Bacillus subtilis*, *Genes Dev.*, 8, 2653, 1994.
59. Arigoni, F. et al., SpoIIE governs the phosphorylation state of a protein regulating transcription factor σ^F during sporulation in *Bacillus subtilis*, *Proc. Natl. Acad. Sci. U.S.A.*, 93, 3238, 1996.
60. Barak, I. et al., Structure and function of the *Bacillus* SpoIIE protein and its localization to sites of sporulation septum assembly, *Mol. Microbiol.*, 19, 1047, 1996.
61. Arigoni, F. et al., Localization of a protein implicated in the establishment of cell type to sites of asymmetric division, *Science*, 270, 637, 1995.
62. Duncan, L. et al., Activation of cell-specific transcription by a serine phosphatase at the site of asymmetric division, *Science*, 270, 641, 1995.
63. Hofmeister, A.E.M. et al., Extracellular signal protein triggering the proteolytic activation of a developmental transcription factor in *B. subtilis*, *Cell*, 83, 219, 1995.
64. Karow, M.L., Glaser, P., and Piggot, P.J., Identification of a gene, *spoIIR*, that links the activation of σ^E to the transcriptional activity of σ^F during sporulation in *Bacillus subtilis*, *Proc. Natl. Acad. Sci. U.S.A.*, 92, 2012, 1995.
65. Londoño-Vallejo, J.A. and Stragier, P., Cell-cell signaling pathway activating a developmental transcription factor in *Bacillus subtilis*, *Genes Dev.*, 9, 503, 1995.
66. Sun, D., Cabrera-Martinez, R.M., and Setlow, P., Control of transcription of the *Bacillus subtilis spoIIIG* gene, which codes for the forespore-specific transcription factor σ^G, *J. Bacteriol.*, 173, 2977, 1991.
67. Gholamhoseinian, A. and Piggot, P.J., Timing of *spoII* gene expression relative to septum formation during sporulation of *Bacillus subtilis*, *J. Bacteriol.*, 171, 5747, 1989.
68. Kenney, T.J. et al., Genetic evidence that RNA polymerase associated with σ^A factor uses a sporulation-specific promoter in *Bacillus subtilis*, *Proc. Natl. Acad. Sci. U.S.A.*, 86, 9109, 1989.
69. Satola, S.W., Baldus, J.M., and Moran, C.P. Jr., Binding of Spo0A stimulates *spoIIG* promoter activity in *Bacillus subtilis*, *J. Bacteriol.*, 174, 1448, 1992.
70. Stragier, P., Bonamy, C., and Karmazyn-Campelli, C., Processing of a sporulation sigma factor in *Bacillus subtilis*: how morphological structure could control gene expression, *Cell*, 52, 697, 1988.
71. LaBell, T.L., Trempy, J.E., and Haldenwang, W.G., Sporulation-specific sigma factor σ^{29} of *Bacillus subtilis* is synthesized from a precursor protein, P^{31}, *Proc. Natl. Acad. Sci. U.S.A.*, 84, 1784, 1987.
72. Trempy, J.E., Morrison-Plummer, J., and Haldenwang, W.G., Synthesis of σ^{29}, an RNA polymerase specificity determinant, is a developmentally regulated event in *Bacillus subtilis*, *J. Bacteriol.*, 161, 340, 1985.
73. Jonas, R.M. et al., The *Bacillus subtilis spoIIG* operon encodes both σ^E and a gene necessary for σ^E activation, *J. Bacteriol.*, 170, 507, 1988.
74. Eichenberger, P. et al., The σ^E regulon and the identification of additional sporulation genes in *Bacillus subtilis, J. Mol. Biol.*, 327, 945, 2003.
75. Tatti, K.M., Jones, C.H., and Moran, C.P. Jr., Genetic evidence for interaction of σ^E with the *spoIIID* promoter in *Bacillus subtilis*, *J. Bacteriol.*, 173, 7828, 1991.

76. Stevens, C.M. and Errington, J., Differential gene expression during sporulation in *Bacillus subtilis*: structure and regulation of the *spoIIID* gene, *Mol. Microbiol.*, 4, 543, 1990.
77. Kunkel, B. et al., Temporal and spatial control of the mother-cell regulatory gene *spoIIID* of *Bacillus subtilis*, *Genes Dev.*, 3, 1735, 1989.
78. Kunkel, B. et al., The promoter for a sporulation gene in the *spoIVC* locus of *Bacillus subtilis* and its use in studies of temporal and spatial control of gene expression, *J. Bacteriol.*, 170, 3513, 1988.
79. Arcuri, E.F., Wiedmann, M., and Boor, K.J., Phylogeny and functional conservation of σ^E in endospore-forming bacteria, *Microbiology*, 146, 1593, 2000.
80. Wong, J., Sass, C., and Bennett, G.N., Sequence and arrangement of genes encoding σ factors in *Clostridium acetobutylicum* ATCC 824, *Gene*, 153, 89, 1995.
81. Zhao, Y. and Melville, S.B., Identification and characterization of sporulation-dependent promoters upstream of the enterotoxin gene (*cpe*) of *Clostridium pefringens*, *J. Bacteriol.*, 180, 136, 1998.
82. Wong, J. and Bennett, G.N., Recombination-induced variants of *Clostridium acetobutylicum* ATCC 824 with increased solvent production, *Curr. Microbiol.*, 32, 349, 1996.
83. Santagelo, J.D. et al., Sporulation and time course expression of σ-factor homologous genes in *Clostridium acetobutylicum*, *FEMS Microbiol. Lett.*, 161, 157, 1998.
84. Karmazyn-Campelli, C. et al., Tandem genes encoding σ-factors for consecutive steps of development in *Bacillus subtilis*, *Genes Dev.*, 3, 150, 1989.
85. Sun, D.X., Stragier, P., and Setlow, P., Identification of a new σ-factor involved in compartmentalized gene expression during sporulation of *Bacillus subtilis*, *Genes Dev.*, 3, 141, 1989.
86. Nicholson, W.L. et al., Promoter specificity of σ^G-containing RNA polymerase from sporulating cells of *Bacillus subtilis*: identification of a group of forespore-specific promoters, *J. Bacteriol.*, 171, 2708, 1989.
87. Kellner, E.M., Decatur, A., and Moran, C.P. Jr., Two-stage regulation of an anti-sigma factor determines developmental fate during bacterial endospore formation, *Mol. Microbiol.*, 21, 913, 1996.
88. Cutting, S., Panzer, S., and Losick, R., Regulatory studies on the promoter for a gene governing synthesis and assembly during development in *B. subtilis*, *Cell*, 62, 239, 1989.
89. Cutting, S. et al., Forespore-specific transcription of a gene in the signal transduction pathway that governs Pro-σ^K processing in *Bacillus subtilis*, *Genes Dev.*, 5, 456, 1991.
90. Cutting, S. et al., A forespore checkpoint for mother cell gene expression during development in *B. subtilis*, *Cell*, 62, 239, 1990.
91. Lu, S., Halberg, R., and Kroos, L., Processing of the mother-cell σ factor, σ^K, may depend on events occurring in the forespore during *Bacillus subtilis* development, *Proc. Natl. Acad. Sci. U.S.A.*, 87, 9722, 1990.
92. Cabrera-Martinez, R.M. and Setlow, P., Cloning and nucleotide sequence of three genes coding for small, acid-soluble proteins of *Clostridium perfringens* spores, *FEMS Microbiol. Lett.*, 61, 127, 1991.
93. Holck, A., Blom, H., and Granum, P.E., Cloning and sequencing of the genes encoding acid-soluble spore proteins from *Clostridium perfringens*, *Gene*, 91, 107, 1990.
94. Kroos, L., Kunkel, B., and Losick, R., Switch protein alters specificity of RNA polymerase containing a compartment-specific sigma factor, *Science*, 243, 526, 1989.
95. Zhang, J. et al., Regulation of the transcription of a cluster of *Bacillus subtilis* spore coat genes, *J. Mol. Biol.*, 240, 405, 1994.
96. Zheng, L. et al., Sporulation regulatory protein GerE from *Bacillus subtilis* binds to and can activate or repress transcription from promoters for mother-cell-specific genes, *J. Mol. Biol.*, 226, 1037, 1992.
97. Driks, A. and Losick, R., Compartmentalized expression of a gene under the control of sporulation transcription factor Sigma E in *Bacillus subtilis*, *Proc. Natl. Acad. Sci. U.S.A.*, 15, 9934, 1991.
98. Halberg, R. and Kroos, L., Sporulation regulatory protein SpoIIID from *Bacillus subtilis* activates and represses transcription by both mother-cell-specifc forms of RNA polymerase, *J. Mol. Biol.*, 243, 425, 1994.
99. Cutting, S. et al., A forespore checkpoint for mother cell gene expression during development in *B. subtilis*, *Cell*, 62, 239, 1990.
100. Rudner, D. Z. and Losick, R., A sporulation membrane protein tethers the Pro-σ^K processing enzyme to its inhibitor and dictates its subcellular localization, *Genes Dev.*, 16, 1007, 2002.
101. Green, D.H. and Cutting, S.M., Membrane topology of the *Bacillus subtilis* Pro-σ^K processing complex, *J. Bacteriol.*, 182, 278, 2000.

102. Yu, Y.T. and Kroos, L., Evidence that SpoIVFB is a novel type of membrane metalloprotease governing intercompartmental communication during *Bacillus subtilis* sporulation, *J. Bacteriol.*, 182, 3305, 2000.
103. Rudner, D.Z., Fawcett. P., and Losick, R., A family of membrane-embedded metalloproteases involved in regulated proteolysis of membrane-associated transcription factors, *Proc. Natl. Acad. Sci. U.S.A.*, 96, 14765, 1999.
104. Stragier, P. et al., Chromosomal rearrangement generating a composite gene for a developmental transcription factor, *Science*, 243, 507, 1989.
105. Takemaru, K. et al., Complete nucleotide sequence of a *skin* element excised by DNA rearrangement during sporulation in *Bacillus subtilis*, *Microbiology*, 141, 323, 1995.
106. Popham, D.L. and Stragier, P., Binding of the *Bacillus subtilis spoIVCA* product to the recombination sites of the element interrupting the σ^K-encoding gene, *Proc. Natl. Acad. Sci. U.S.A.*, 89, 5991, 1992.
107. Kunkel, B., Losick, R. and Stragier, P., The *Bacillus subtilis* gene for the development transcription factor Sigma K is generated by excision of a dispensable DNA element containing a sporulation recombinase gene, *Genes Dev.*, 4, 525, 1990.
108. Oke, V. and Losick, R., Multilevel regulation of the sporulation transcription factor σ^K in *Bacillus subtilis*, *J. Bacteriol.*, 175, 7341, 1993.
109. Takami, H. et al., Complete genome sequence of the alkaliphilic bacterium *Bacillus halodurans* and genomic sequence comparison with *Bacillus subtilis*, *Nucleic Acids Res.*, 28, 4317, 2000.
110. Vary, P.S., Prime time for *Bacillus megaterium*, *Microbiology*, 140, 1001, 1994.
111. Adams, L.F., Brown, K.L., and Whiteley, H.R., Molecular cloning and characterization of two genes encoding sigma factors that direct transcription from a *Bacillus thuringiensis* crystal protein gene promoter, *J. Bacteriol.*, 173, 3846, 1991.
112. Stragier, P., personal communication, 2002.

27 Transposable Genetic Elements of Clostridia

Dena Lyras and Julian I. Rood

CONTENTS

27.1 INTRODUCTION

This chapter will focus on describing the most recent data on transposable genetic elements of the clostridia. Earlier reviews can provide the reader with a more historical perspective of this research area [1–6]. Since most of these reviews were published, there have been significant advances in our knowledge and understanding of several of these elements, as well as the discovery of elements that were previously unknown.

We will highlight an unusual family of integrative mobilizable elements, the Tn*4451*/Tn*4453* family, which is found in *Clostridium perfringens* and *Clostridium difficile*, as well as the conjugative element Tn*5397* from *C. difficile*. The mechanism of transposition of these antibiotic resistance elements has been elucidated and is unusual in that the elements all rely upon a site-specific recombination process that is mediated by a large resolvase. Evidence implicating the involvement of a compound transposon, Tn*5565*, in the movement of the enterotoxin gene in *C. perfringens* will also be described. Finally, this chapter will describe group I introns and insertion sequences that have been discovered in the clostridia.

27.2 THE Tn*4451*/Tn*4453* FAMILY OF INTEGRATIVE MOBILIZABLE ELEMENTS

Chloramphenicol resistance in *C. perfringens* and *C. difficile* is often encoded by the closely related 6.3 kb elements Tn*4451* and Tn*4453*, respectively [7,8]. These unusual nonconjugative but mobilizable elements are grouped together as the Tn*4451/3* family of integrative mobilizable elements [6]. They contain six genes (Figure 27.1), one of which, *catP*, has been shown to encode chloramphenicol resistance [7,9]. Comparative analysis of Tn*4451* and Tn*4453a* has shown that although these elements contain identical *tnpV* and *catP* genes, there are many amino acid sequence differences between the products of the other genes [9]. Curiously, the region containing *catP* and part

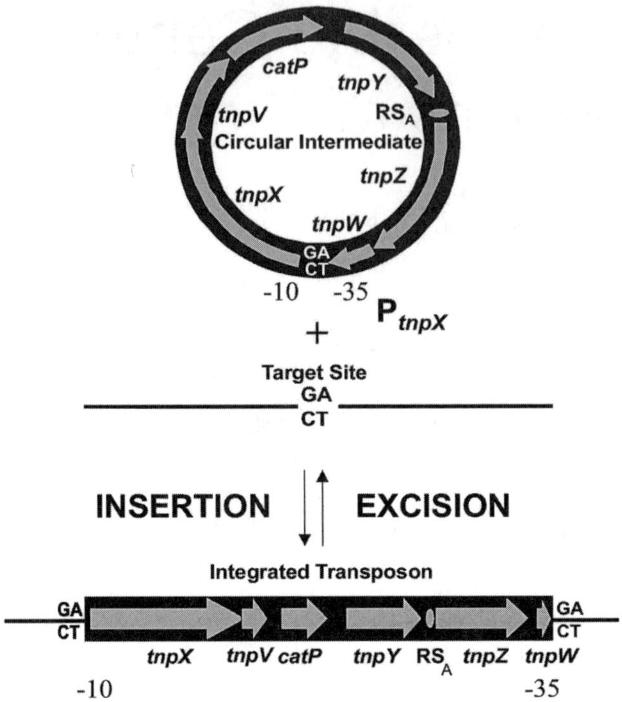

FIGURE 27.1 Excision and insertion model of the Tn*4451/3* family of elements. TnpX-mediated insertion of the circular intermediate is the reverse of the transposon excision event. Note the directly repeated GA residues and the location of the components of the *tnpX* promoter. (Based on Figure 1 from Adams et al. [6].)

of *tnpV* has been found in several *Neisseria meningitidis* isolates from different geographical locations and has been shown to be identical to that in the clostridia [10,11]. None of the other transposon-encoded genes were present in these neisserial isolates.

An unusual feature of this family of transposons is that they are capable of being mobilized into a recipient cell in the presence of another conjugative element, such as the IncP plasmid RP4 [8,12]. This process involves the transposon-determined *tnpZ* gene, which encodes a Mob protein, TnpZ, that mediates transposon mobilization by nicking the element at its RS$_A$ site (Figure 27.1). This site acts as an origin of transfer or *oriT* site and is located immediately upstream of *tnpZ* [12]. The presence of TnpZ and RS$_A$ allows Tn*4451/3*, or plasmids on which these elements are located, to be mobilized to recipient cells [9,12]. The presence of this mobilization module on these transposons has significant implications for intercellular movement and transposition, which will be discussed in more detail later in this chapter.

Transposition of the Tn*4451/3* family is very different from the movement of classical transposons because it is mediated by the large resolvase TnpX rather than by a DDE transposase. TnpX precisely excises the element, after a 2 bp staggered cut is made at terminal GA dinucleotides, to produce a circular molecule that is the integrative intermediate (Figure 27.1) [7,13]. The generation of the double-stranded circular molecule results in the formation of a strong promoter at the joint between the left and right ends of the linear element. This promoter leads to the expression of the *tnpX* gene (Figure 27.1) [9]. We have proposed that formation of this promoter is a transposon survival mechanism. Since the circular molecule does not encode its own replication, the formation of a *tnpX* promoter increases the probability of integration of the circular form [9]. This process is particularly important since the transposon carries the TnpZ/RS$_A$ mobilization module, which allows the nonreplicating circular form to be transferred into recipient cells. Whether this mobilization process does or does not occur, the circular intermediate must integrate into a replicating

DNA molecule (plasmid or chromosome) to survive. The presence of the promoter at the joint would ensure that TnpX is produced, particularly in naive recipient cells where no TnpX protein is present. The end result would be an increased likelihood of transposon survival and dissemination. Experimental data obtained in both conjugation and transformation experiments confirm that *tnpX* expression is occurring in the recipient cells [9]. Therefore, in contrast to most transposons, these elements have evolved several mechanisms by which they optimize their survival and dissemination; specifically, in addition to the normal transposition process, we refer to the TnpZ-mediated mobilization process and the expression of *tnpX* via the specific formation of a promoter in the circular intermediate [9].

The function of the *tnpV, tnpY* and *tnpW* genes is not known. The product of the *tnpV* gene has some similarity to the Xis protein from bacteriophage λ [7], and TnpY shows similarity to DNA repair proteins, such as RadA from *Archaeoglobus fulgidus* [14]. Recent studies, using *in vivo* deletion analysis of the transposon-encoded *tnpV, tnpY,* and *tnpW* genes in *E. coli*, showed that none of these genes are essential for excision or integration of this group of elements [14]. These results do not preclude the possibility that the proteins encoded by these genes might play a role in the native clostridial hosts or that the effect of deleting these genes in *E. coli* is too subtle to be detected using the current assay systems.

Genetic analysis using a recently developed *in vivo* insertion assay has now shown that TnpX is the only transposon-encoded protein required for the integration reaction [14], although it had been implied from previous experiments [9]. Significantly, the analysis of various TnpX mutants in this assay system showed that TnpX catalyzes insertion by a resolvase-like mechanism since the same resolvase catalytic residues responsible for excision were also found to be essential for insertion. Both genetic evidence [13] and data from an *in vivo* excision assay that excluded all other transposon-encoded proteins [14] have shown that TnpX is the only protein required for transposon excision. Our results show that TnpX can act on a DNA substrate encompassing 229 bp from the joint of the circular intermediate, 90 bp from the left or *tnpX* end and 139 bp from the right or *tnpW* end [14]. These studies, therefore, have provided good evidence that TnpX is responsible for both excision and insertion and that these reactions might be the reverse of each other. However, other experiments have shown that there are likely to be significant mechanistic differences between these two processes.

These results represent an important finding, since the only resolvases that mediate insertion as well as excision belong to the large resolvase group of site-specific recombinases — a group of proteins that are structurally quite distinct from other resolvase proteins [15]. TnpX is most closely related to the large resolvase TndX from the conjugative *C. difficile* transposon Tn*5397*, which is discussed in greater detail later in this chapter. Like TnpX, TndX is also the only transposon-encoded protein required *in vivo* for both excision and insertion [16]. The large resolvases constitute a very diverse range of proteins [15], many of which are encoded on bacteriophages or by integrative mobile resistance elements. Examples include the integrases from the *Mycobacterium tuberculosis* prophage-like element ΦRv1 and the temperate lactococcal bacteriophage TP901-1, as well as the CcrA and CcrB proteins from the SCC*mec* element from *Staphylococcus aureus*. Many of the phage-related large resolvases have been shown to require additional phage-encoded proteins for efficient excision *in vivo* [17,18]. There is, therefore, evidence that the transposon and phage-related large resolvase systems differ at the mechanistic level. Clearly, more work on large resolvases is required to gain a better understanding of the mechanism of action of this intriguing group of proteins.

The TnpX protein has been purified and, as already discussed, shown to catalyze excision *in vitro* in the absence of any other protein [14]. Further studies showed that supercoiled DNA was the preferred TnpX substrate, which is consistent with a classical resolvase reaction [19]. To date, TnpX-mediated insertion has not been demonstrated *in vitro*, suggesting that the optimal conditions for *in vitro* excision and insertion are different or that insertion might require a host-encoded factor [14].

The 707 amino acid TnpX protein can be divided into a number of discrete functional domains. The N-terminal domain (aa 1–100) has significant similarity to the small resolvase family, containing many of the conserved resolvase motifs [7,13]. Site-specific mutagenesis has shown that many of these motifs are essential for excision and insertion and, therefore, for the transposition process [7,8,13,14]. Deletion analysis has shown that the last 110 amino acids, which constitute a cysteine rich domain, are not essential *in vivo* for biological function, although this domain can play a subtle functional role [70]. *In vitro* binding studies on this truncated TnpX derivative confirmed that it was still capable of binding to DNA and of catalyzing excision. Other TnpX deletion derivatives, lacking 215 and 351 amino acids from the C-terminal of the protein, respectively, were not biologically functional and were not capable of binding to DNA. These studies, therefore, allowed a DNA binding domain to be identified between amino acids 493–597 [70].

We have recently carried out extensive DNA binding studies on purified TnpX, which exists as a dimer in solution. The binding region was found to be approximately 50 bp in size [70], which is similar in size to the binding regions found for other large resolvases, such as those encoded by ΦTP901-1 [20,21] and ΦC31 [22]. This target site is considerably smaller than that found for small resolvases, such as the *res* binding site for the Tn*3* resolvase, which is 120 bp in size [23]. Our studies have also shown that TnpX binds to the ends of the transposon and to the joint of the circular intermediate with very high affinity, but binds to its deletion and insertion sites with a considerably lower affinity. This is an unusual finding because it is in direct contrast to binding studies conducted on other large resolvases, such as those found on ΦC31 and ΦTP901-1. These studies showed that the binding affinity of the respective resolvase proteins for their target sites and phage-encoded components was very similar [20,22]. We have concluded that the resolvase-like excision and insertion reactions mediated by TnpX must be distinct processes. Since the same resolvase residues and catalytic reactions are involved in both excision and insertion [9,13,14], we suggest that the mechanistic differences must involve distinctive ways in which the resolvase-DNA synaptic complexes are formed and undergo conformational change [70]. TnpX is the first serine recombinase in which differential binding to its transposon and target sites has been demonstrated, although studies on this group of proteins are still in their early stages.

27.3 THE INTEGRATIVE CONJUGATIVE ELEMENT Tn*5397*

Early studies on tetracycline resistant isolates of *C. difficile* revealed that resistance was often transferable, in a plasmid independent manner [24,25]. Subsequent studies have focused on strain 630, which has the ability to transfer its tetracycline resistance by conjugation to genetically marked recipient strains of *C. difficile*, *Bacillus subtilis,* and *S. aureus*. The recipients have the ability to act as donors of tetracycline resistance in subsequent matings, indicating that transfer is mediated by a conjugative element. Subsequent experiments showed that transfer is mediated by a nonreplicative integrative conjugative element, Tn*5397*, that is related to the 18 kb conjugative transposon Tn*916* from *Enterococcus faecalis* [26,27]. Transfer of Tn*5397* has been demonstrated in a model of an oral biofilm, indicating that transfer is likely to occur in the environment of the human host [28].

Tn*5397* has now been sequenced and shown to have similar conjugation and tetracycline resistance modules to Tn*916* but a very different recombination cassette [29]. The element is 20,658 bp in length and has 21 potential open reading frames (ORFs). It consists of four distinct gene modules, which are potentially involved in conjugation, tetracycline resistance, gene regulation, or transposon insertion and excision. With the exception of the insertion/excision module, all of these gene regions are closely related to the equivalent regions from Tn*916* [30] and from a similar but defective element from *C. perfringens* (Figure 27.2) [29].

The conjugation region consists of ORFs 13 to 23 and is very closely related to Tn*916*. The major differences in Tn*5397* are the presence of a unique 180 bp of sequence at the very left end,

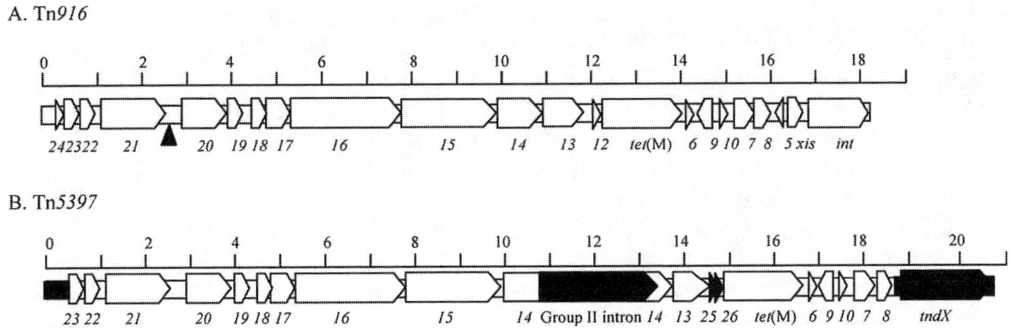

FIGURE 27.2 Schematic representation of the structural organization of (*A*) Tn*916* and (*B*) Tn*5397*. The top line shows the scale in kb. The arrows represent the open reading frames (ORFs); the point shows the direction of transcription. The ORFs have been named in accordance with the standard nomenclature for Tn*916*. The regions that are filled in Tn*5397* are unique to Tn*5397*. (Reproduced from Mullany et al. [36] with permission from the authors and publisher.)

the absence of the first seven nucleotides of ORF24 [29], and the presence of a group II intron in ORF14 [31]. The putative proteins encoded by ORF23 to ORF16 have 90 to 100% identity to their Tn*916* homologues; these proteins are involved in the conjugation process in Tn*916* and presumably in Tn*5397*. ORF15 has a 25 amino acid C-terminal truncation in Tn*5397*, and ORF14 has 87% identity. Finally, the *oriT* site that is present in Tn*916* is identical to that found in Tn*5397* [29]. These data all suggest that the conjugative transfer of Tn*5397* occurs by the same mechanism as that of Tn*916*. Unfortunately, little is known about the precise transfer mechanism.

Most unusually, ORF14 of Tn*5397* contains a group II intron [31,32]. Genes containing group II introns are transcribed normally and then the resultant RNA molecule may undergo a self-splicing process that precisely excises the intron-specific RNA, leading to the formation of a normal mRNA molecule that can subsequently be translated. In bacteria, group II introns are not only generally associated with mobile genetic elements, they are themselves mobile genetic elements and may encode a multidomain protein that has reverse transcriptase, endonuclease, and maturase activity [33,34]. Recent studies have shown that the intron located within ORF14 can undergo self-splicing *in vivo* and that mutation of the intron-encoded protein, although eliminating self-splicing, had no effect on conjugal transfer of the resultant Tn*5397* derivative [32]. Since the intron is inserted at a point 23 codons from the end of Tn*5397*, these results imply that the C-terminal 23 amino acids of ORF14 are not required for conjugal transfer.

The conjugation modules found in Tn*916* and Tn*5397* appear to be widely distributed in Gram-positive bacteria, with components being found in other integrative conjugative elements and in conjugative plasmids. Seven such modules appear to be present in the incomplete *C. difficile* strain 630 genome, although it is not known if any of these gene regions, other than Tn*5397*, represent functional conjugative elements [35].

The tetracycline resistance phenotype conferred by Tn*5397* is encoded by the *tet*(M) gene [36]. The resultant Tet(M) ribosomal protection protein has 90.5% amino acid sequence identify to Tet(M) from Tn*916* [29]. Interestingly, the Tet(M) protein from a defective Tn*916*-like conjugative transposon found in *C. perfringens* strain CW459 is much more closely related (99.7% identity) to the Tn*916* protein [29]. In Tn*916*, tetracycline resistance is induced by a transcriptional attenuation mechanism that involves the upstream ORF12 peptide [37]. Although ORF12 has been partially deleted in Tn*5397*, *tet*(M) is still inducible [29].

In Tn*916* ORFs 7, 8, and 9 are believed to play a regulatory role in the movement of the element [36,37]. ORF7 encodes a protein with similarity to the C-terminal DNA binding domain of RNA polymerase sigma factors, but has not been shown to be a functional sigma factor. ORF9 has a helix-turn-helix DNA binding domain that is commonly found in transcriptional regulators, but ORF8 has no significant similarity to any proteins with a verified biological function. ORFs 7 and 8 comprise an operon that appears to be repressed by the ORF9 product. In the presence of tetracycline, there is reduced expression of ORF9 and increased expression of ORFs 7 and 8, as well as longer transcripts that include the downstream *xis* and *int* genes. In the circular intermediate these transcripts may read through into the conjugation genes [37]. However, little is known about how this putative regulatory pathway is potentiated. All of these putative regulatory genes are present in Tn*5397,* but nothing is known about their potential role in gene regulation [29].

The excision module of Tn*5397* is different from that of Tn*916* [38] and more closely resembles the excision module of the integrative mobilizable element Tn*4451* [36]. Excision and insertion of Tn*916* is mediated by the products of the *int* and *xis* genes, which are located at the right end of the element. Catalysis involves the Int protein, a member of the integrase family of site-specific recombinases. At the right end of Tn*5397,* there is a single gene, *tndX*, that is essential for the excision and insertion of Tn*5397*, via the formation of a circular intermediate [16]. The TndX protein has been shown both *in vitro* and *in vivo* to catalyze the insertion and deletion of mini-transposon derivatives of Tn*5397* in the absence of any other transposon-encoded proteins [16]. TndX is a member of the large resolvase family of site-specific recombinases [15] and is closely related to the TnpX protein from Tn*4451* [6]. These enzymes excise Tn*5397* and Tn*4451*, respectively, to form circular intermediates and then catalyze the specific insertion of their respective target molecules into the genome. TndX and TnpX recognize different target sites and are not capable of catalyzing heterologous insertion or deletion events [38]. However, both enzymes make staggered 2 bp cuts on either side of a central GA dinucleotide and act via the transient formation of a phosphoserine linkage between the 5′ end of the digested DNA strand and the active site serine residue, which is located in the N-terminal resolvase domains of both proteins. Note that of the seven putative integrated conjugation modules detected in strain 630, four, including Tn*5397*, are associated with serine recombinases like TnpX and TndX, and three are associated with tyrosine recombinases (i.e., integrases) [35].

The ends of Tn*5397* and Tn*916* are not related [29,38], and these elements generally show different target site specificities in *C. difficile*. However, a *C. difficile* strain that has a copy of an element that appears indistinguishable from Tn*916* has been identified and shown to be inserted at the normal *C. difficile* insertion site for Tn*916* [39]. Tn*5397* only inserts into a single target site in *C. difficile*, which resembles the joint of the circular intermediate, but into different sites in *B. subtilis* [38]. Although the four *B. subtilis* target sites that were sequenced all had a central GA dinucleotide, they did not have any significant sequence similarity. These results suggest that Tn*5397* can insert into secondary sites if the primary recognition site is not present in the target genome [38].

27.4 THE MOBILIZABLE ELEMENT Tn*5398*

The *C. difficile* isolate that carries Tn*5397*, strain 630, also confers transferable resistance to macrolide-lincosamide antibiotics such as erythromycin, clindamycin, and lincomycin [40,41]. Resistance results from the production of a 23S rRNA methyltransferase encoded by one of two *erm*(B) genes that are located on the unusual genetic element Tn*5398*. This element is not really a transposon, since it does not contain any transposase or site-specific recombinase genes. However, despite the fact that it does not appear to carry any mobilization genes, it is mobilizable, at very low frequency, by either Tn*5397* or by an unknown chromosomal element in strain 630 [42].

Tn*5398* is 9.6 kb in size and contains several ORFs (Figure 27.3) [43]. Earlier studies in *C. perfringens* showed that the *erm*(B) gene carried by the mobilizable plasmid pIP402 had 1.3 kb directly repeated sequences on either side of the resistance gene; these sequences potentially

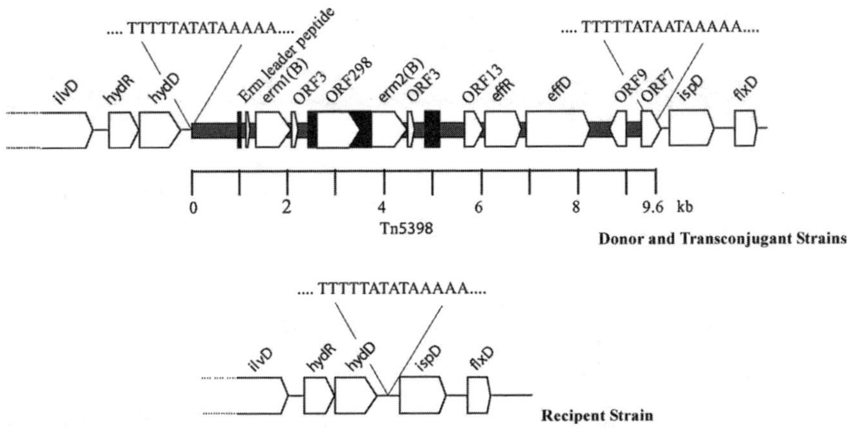

FIGURE 27.3 Genetic organization of Tn*5398*. Schematic representations of Tn*5398*, as observed in the donor and transconjugant strains, are shown to scale. ORFs and their direction of transcription, are represented as blocked arrows. The region encompassed by Tn*5398* is represented by a cross-hatched box and is further indicated by the scale below the diagram. Regions encompassing DR sequences are indicated by black boxes. The target site in the recipient strain CD37 is also shown. The location of each of the ends of the transposon and the target sequence is indicated. (Reproduced from Farrow et al. [42] with permission from the publisher.)

encoded a homologue of an ATPase that is a member of the Soj-like chromosomal partitioning or ParA plasmid replication family [44]. Similar sequences are found bordering *erm*(B) genes from other bacteria [44]. ParA-like proteins are ATPases that have helix-turn-helix DNA binding domains and function as transcriptional repressors [45]. Tn*5398* carries two identical *erm*(B) genes that are separated by a copy of the same directly repeated sequences, including the *parA* homologue ORF298 [46]. Furthermore, deleted versions of the direct repeats are found upstream and downstream of the *erm* genes [42]. These data suggest that these genes have all arisen from a common progenitor and have evolved by a combination of gene duplication and homologous recombination events. Other genes encoded by Tn*5398* include a putative efflux protein and its associated regulator and homologues of ORF13, ORF9, and ORF7 from Tn*916*. It does not appear to contain any gene regions with homology to the toxin A gene, as previously reported [41]. We have proposed that Tn*5398* is not capable of mediating either its own excision or conjugative transfer. We have suggested that mobilization involves the formation of a circular intermediate by either excision of the element, by an unknown resolvase or integrase enzyme, or by homologous recombination [6]. Mobilization to a suitable recipient strain is either mediated by Tn*5397* or by one of the other potentially conjugative integrated elements [35], by use of a putative *oriT* site located within Tn*5398*. Integration in the recipient is postulated to be the reverse of excision and might be the result of either site-specific or homologous recombination events. It is tempting to speculate that Tn*5398* is the derivative of a progenitor element that carried a site-specific recombinase gene and a conjugation module, but that both have been lost in its evolution. Further genetic and molecular analysis of other *erm*(B) positive strains of *C. difficile* might provide the evidence that is needed to verify this hypothesis.

Not all erythromycin-resistant strains of *C. difficile* carry two copies of the *erm*(B) gene. In many strains, resistance is mediated not by an *erm*(B) gene but by an as-yet-unknown mechanism [42,44,47,48]. Even within the *erm*(B) isolates, there is considerable genetic variation, with some strains only having one *erm*(B) gene, others having various derivatives of the directly repeated sequence [42,47], and several strains having two copies of the *erm*(B) gene, and Tn*5398*, as in strain 630 [42].

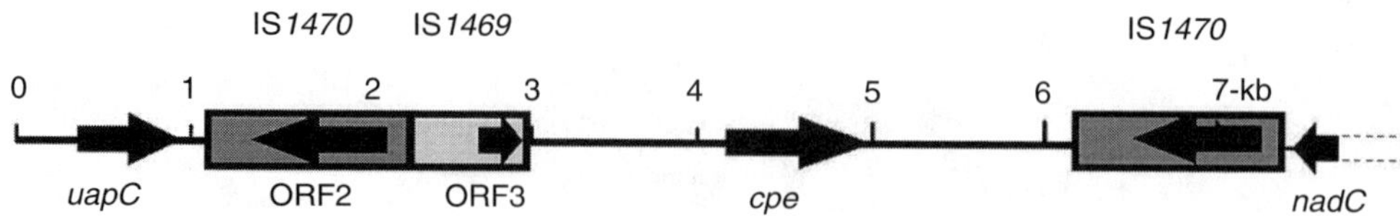

FIGURE 27.4 Genetic organization of the putative enterotoxin transposon Tn*5565*. The organization of the chromosomal *cpe* gene region is shown. ORFs are indicated by the arrows. The IS elements are shown by the gray boxes. The scale is shown in kb. (Reproduced from Rood [69] with permission from the publisher.)

27.5 THE PUTATIVE ENTEROTOXIN TRANSPOSON Tn*5565* FROM *C. PERFRINGENS*

The *C. perfringens cpe* gene encodes a sporulation-specific enterotoxin (CPE) that is responsible for human food poisoning, nonfoodborne diarrhea, and enterotoxemic diseases of animals. The *cpe* gene is chromosomally encoded in isolates from food poisoning outbreaks but is located on large plasmids in nonfoodborne and animal strains [49,50]. Although these plasmids have not been characterized in great detail, at least one of them has been shown to be conjugative [51].

In all CPE-positive strains, there is an insertion sequence, IS*1469*, located 1.3 kb upstream of the *cpe* gene [50]. In the chromosomal CPE-positive strains, the IS*1469-cpe* gene region is bordered by two copies of a different IS element, IS*1470* [52]. This 6.3 kb region, therefore, has the classical genetic organization of a compound transposon and has been designated as Tn*5565* (Figure 27.4). Neither Tn*5565*, IS*1470,* nor IS*1469* has been shown to move from one site in the genome to another. However, PCR evidence for the excision of Tn*5565* to form a circular molecule that could be a transposition intermediate has been reported [53]. In addition, although the region upstream of the *cpe* gene, including IS*1469*, is highly conserved in all CPE-positive strains, the downstream region is quite variable, with many strains carrying defective variants of IS*1470* [50]. These results are consistent with the hypothesis that the cpe^+ plasmids arose from the transposition of an element like Tn*5565* into a conjugative plasmid, downstream of a *dcm* gene, and that the IS*1470* elements were subsequently lost or underwent mutational or recombinational genetic drift. However, of the 13 CPE plasmids examined in this study, four carried a defective derivative of another insertion sequence, IS*1151*, located downstream of the *cpe* gene [50]. These results imply that the movement of the IS*1469-cpe* gene region might have occurred independently of IS*1470.* Further studies are required to determine if Tn*5565* represents a functional CPE transposon or whether the consistent presence of the IS*1470* elements simply provides evidence for the relatively clonal nature of the human food poisoning isolates.

27.6 GROUP I INTRONS

Group I introns are widely distributed and structurally conserved catalytic RNA molecules that are capable of self-splicing. That is, they are ribozymes responsible for their own excision from a pre-mRNA molecule and for the ligation of the resultant exons to produce a functional mRNA molecule. They often contain the genes that encode enzymes responsible for the endonuclease-mediated homing of the intron into the original target gene. In bacteria, group I introns are often found in phage or tRNA genes. Analysis of the genome sequence of *Clostridium botulinum* has led to the identification, within the tmRNA gene, of a small 287 nucleotide group I intron (Cb.ssrA) that has no internal ORF [54]. tmRNA or 10S RNA molecules can act as functional tRNA and mRNA homologues and are involved in the release of stalled mRNA molecules from ribosomes. This intron can be present in the same location in *Clostridium tetani* but is not found in *C. perfringens, C. difficile,* or *C. acetobutylicum* [54]. The only other nonphage encoded group I intron found in

Gram-positive bacteria is Cd*ISt1,* which was identified within the toxin A gene, *tcdA*, of *C. difficile* strain C34 [55].

Cd*ISt1* is 1975 nucleotides in size and, instead of encoding a typical endonuclease or maturase, includes two ORFs that encode putative transposases, TlpA and TlpB, which have similarity to transposases from IS*605*-like elements [55]. TlpA has similaity to the C-terminal end of a IS*200*-like transposase and is unlikely to be a functional enzyme. TlpB is a DDE-like transposase with similarity to the transposase TnpC from the *Deinococcus radiodurans* IS element IS*8301*. Cd*ISt1* appears to be a composite element derived from the fusion of a smaller group I intron and the 3′ truncated end of an IS*605*-like element. Evidence that Cd*ISt1* is properly spliced from the pre-mRNA molecule was obtained by PCR analysis and by experiments that showed that this strain produces a functional toxin A molecule of the correct molecular size. Finally, it was shown that Cd*ISt1* is widely distributed in *C. difficile* and is located at sites other than the *tcdA* gene, often at multiple sites within the same genome [55]. For example, there are at least 11 functional Cd*ISt1* elements in strain C34, all inserted downstream of a conserved sequence, TTGAT, and all are capable of self-splicing. However, strain 34 is the only *C. difficile* strain identified in which the element is inserted within *tcdA*.

27.7 INSERTION SEQUENCES

IS*1151* was originally found upstream of the plasmid-determined epsilon-toxin gene, *etx,* in a type D strain of *C. perfringens* and was the first complete IS element to be identified in the clostridia [56]. Variants of IS*1151* have also been found near the plasmid-determined beta-toxin gene, *cpb* [57], the iota-toxin genes, *iap* and *ibp* [58], and plasmid-determined *cpe* genes [49,50,58]. IS*1151* is a 1696 bp element that has 23 bp inverted repeats at each end and is a member of the IS*4* family [59]. The DDE transposase encoded by IS*1151* is most closely related to the transposase from the *Bacillus thuringiensis* transposon IS*231*, homologues of which are found on the toxin plasmid pXO1 from *Bacillus anthracis* [60]. The reason for this loose association of IS*231*-like elements with toxin genes is not known. Note that more recent studies in *C. perfringens* [50] indicate that IS*1151* is not as closely associated with the *cpe* gene as originally thought [49].

As already discussed, both IS*1469* and IS*1470* are associated with the *C. perfringens cpe* gene [50,52]. IS*1469* is a member of the IS*605* family of IS elements and is most closely related to IS*200* from *Salmonella typhimurium* [59]. By contrast, IS*1470* is related to the IS*30* family [59] and is more widely distributed in *C. perfringens*, with some strains carrying multiple copies that are not necessarily associated with the *cpe* gene [49,61]. Finally, it should be noted that part of a different element, which is related to IS*256*, has been reported downstream of the *etx* gene in a type B strain of *C. perfringens* [62].

Other clostridial species have also been shown to contain IS elements. At least nine copies of a 1764 bp IS*4*-like element, IS*Cb1*, have been identified in one strain of *Clostridium beijerinckii*, with multiple copies of this element being found in other *C. beijernickii* isolates [63]. IS*3*-like sequences have been identified downstream of the BoNT/A gene cluster in at least one strain of *C. botulinum* [64], and two IS elements have been detected in *Clostridium cellulolyticum* [65].

The *C. cellulolyticum cipC* gene encodes a scaffolding protein that is involved in the formation of a cellulosome complex. Two *cipC* mutants of *C. cellulolyticum* were shown to contain IS sequences. One mutant contained a 1292 bp clement, IS*Cce1*, which encoded a transposase with similarity to members of the IS*3* family. The other mutant had a second element, IS*Cce2* (1359 bp), inserted into the 3′ end of IS*Cce1* [65]. IS*Cce2* encoded a protein with similarity to the transposase from IS*256*. Additional experiments showed that there were multiple copies of both elements in the genome of the parent strain and that three copies of IS*Cce1* contained an IS*Cce2* insertion. In all, four of the seven copies of IS*Cce2* were inserted into IS*Cce1*, providing good evidence that the latter element contains an insertional hotspot for IS*Cce2*. Since these elements appear to be active in *C. cellulolyticum,* they have considerable potential for use in mutagenesis studies in the clostridia [65].

Finally, three complete clostridial genome sequences have been published [66–68], and several others are currently complete but not yet published. Analysis of the *C. acetobutylicum* genome sequence [66] revealed that there were only three transposase genes on the chromosome and one on a large plasmid. Only one of these genes appeared to be intact. Seven of the eleven transposase genes found on the *C. perfringens* strain 13 chromosome were intact [67], and 16 transposase genes, most of which appeared to be nonfunctional, were found on the *C. tetani* chromosome [68]. Two copies of a potential IS element were detected on pE88, the 74 kb toxin plasmid from the sequenced *C. tetani* strain. Although no functional studies have been carried out on these elements or any putative IS elements from the three sequenced strains, it does seem that they all have relatively few IS compared to other bacteria.

ACKNOWLEDGMENTS

The research carried out in this laboratory was supported by grants from the Australian National Health and Medical Research Council.

REFERENCES

1. Hächler, H. and Kayser, F.H., Genetics and molecular biology of antibiotic resistance in *Clostridium difficile*: general and specific overview, in *Genetics of Anaerobic Bacteria*, Sebald, M., Ed., Springer-Verlag, Berlin, 1992, 156.
2. Rood, J.I. and Cole, S.T., Molecular genetics and pathogenesis of *Clostridium perfringens*, *Microbiol. Rev.,* 55, 621, 1991.
3. Rood, J.I., Antibiotic resistance determinants of *Clostridium perfringens*, in *Genetics and Molecular Biology of Anaerobic Bacteria*, Sebald, M., Ed., Springer-Verlag, New York, 1993, 141.
4. Lyras, D. and Rood, J.I., Transposons and antibiotic resistance determinants from *Clostridium perfringens* and *Clostridium difficile*, in *The clostridia: Molcular Biology and Pathogenesis*, Rood, J., McClane, B., Songer, J., and Titball, R., Eds., Academic Press, London, 1997, 73.
5. Lyras, D. and Rood, J.I., Clostridial genetics, in *Gram-Positive Pathogens*, Fischetti, V.A., Novick, R.P., Ferretti, J.J., Portnoy, D.A., and Rood, J.I., Eds., American Society for Microbiology, Washington, DC, 2000, 529.
6. Adams, V., Lyras, D., Farrow, K., and Rood, J., The clostridial mobilisable transposons, *Cell. Mol. Life Sci.,* 59, 2033, 2002.
7. Bannam, T.L., Crellin, P.K., and Rood, J.I., Molecular genetics of the chloramphenicol-resistance transposon Tn*4451* from *Clostridium perfringens*: the TnpX site-specific recombinase excises a circular transposon molecule, *Mol. Microbiol.,* 16, 535, 1995.
8. Lyras, D., Storie, C., Huggins, A.S., Crellin, P.K., Bannam, T.L., and Rood, J.I., Chloramphenicol resistance in *Clostridium difficile* is encoded on Tn*4453* transposons that are closely related to Tn*4451* from *Clostridium perfringens*, *Antimicrob. Ag. Chemother.,* 42, 1563, 1998.
9. Lyras, D. and Rood, J., Transposition of Tn*4451* and Tn*4453* involves a circular intermediate that forms a promoter for the large resolvase, TnpX, *Mol. Microbiol.,* 38, 588, 2000.
10. Galimand, M., Gerbaud, G., Guibourdenche, M., Riou, J.Y., and Courvalin, P., High-level chloramphenicol resistance in *Neisseria meningitidis*, *New Eng. J. Med.,* 339, 868, 1998.
11. Shultz, T.R., Tapsall, J.W., White, P.A., Ryan, C.S., Lyras, D., Rood, J.I., Binotto, E., and Richardson, C.J., Chloramphenicol-resistant *Neisseria meningitidis* containing *catP* isolated in Australia, *J. Antimicrob. Chemother.,* 52, 856, 2003.
12. Crellin, P. and Rood, J., Tn*4451* from *Clostridium perfringens* is a mobilisable transposon that encodes the functional Mob protein, TnpZ., *Mol. Microbiol.,* 27, 631, 1998.

13. Crellin, P.K. and Rood, J.I., The resolvase/invertase domain of the site-specific recombinase TnpX is functional and recognizes a target sequence that resembles the junction of the circular form of the *Clostridium perfringens* transposon Tn*4451*, *J. Bacteriol.,* 179, 5148, 1997.
14. Lyras, D., Adams, V., Lucet, I., and Rood, J.I., The large resolvase TnpX is the only transposon-encoded protein required for transposition of the Tn*4451/3*-family of integrative mobilizable elements, *Mol. Microbiol.,* in press, 2004.
15. Smith, M.C. and Thorpe, H.M., Diversity in the serine recombinases, *Mol. Microbiol.,* 44, 299, 2002.
16. Wang, H. and Mullany, P., The large resolvase TndX is required and sufficient for integration and excision of derivatives of the novel conjugative transposon Tn*5397*, *J. Bacteriol.,* 182, 6577, 2000.
17. Bibb, L.A. and Hatfull, G.F., Integration and excision of the *Mycobacterium tuberculosis* prophage-like element, phiRv1, *Mol. Microbiol.,* 45, 1515, 2002.
18. Breüner, A., Bröndsted, L., and Hammer, K., Novel organization of genes involved in prophage excision identified in the temperate lactococcal bacteriophage TP901-1, *J. Bacteriol.,* 181, 7291, 1999.
19. Stark, W.M. and Boocock, M.R., Topological selectivity in site-specific recombination, in *Mobile Genetic Elements*, Sherratt, D.J., Ed., Oxford University Press, Oxford, 1995, 101.
20. Breüner, A., Bröndsted, L., and Hammer, K., Resolvase-like recombination performed by the TP901-1 integrase, *Microbiology,* 147, 2051, 2001.
21. Stoll, S.M., Ginsburg, D.S., and Calos, M.P., Phage TP901-1 site-specific integrase functions in human cells, *J. Bacteriol.,* 184, 3657, 2002.
22. Thorpe, H.M., Wilson, S.E., and Smith, M.C., Control of directionality in the site-specific recombination system of the streptomyces phage phiC31, *Mol. Microbiol.,* 38, 232, 2000.
23. Sherrat, D., Tn*3* and related transposable elements: site-specific recombination and transposition, in *Mobile DNA*, Berg, D.E. and Howe, M.M., Eds., American Society for Microbiology, Washington, DC, 1989, 163.
24. Smith, C.J., Markowitz, S.M., and Macrina, F.L., Transferable tetracycline resistance in *Clostridium difficile*, *Antimicrob. Ag. Chemother.,* 19, 997, 1981.
25. Wüst, J. and Hardegger, U., Transferable resistance to clindamycin, erythromycin, and tetracycline in *Clostridium difficile*, *Antimicrob. Ag. Chemother.,* 23, 784, 1983.
26. Hächler, H., Kayser, F.H., and Berger-Bächi, B., Homology of a transferable tetracycline resistance determinant of *Clostridium difficile* with *Streptococcus (Enterococcus) faecalis* transposon Tn*916*, *Antimicrob. Ag. Chemother.,* 31, 1033, 1987.
27. Mullany, P., Wilks, M., Lamb, I., Clayton, C., Wren, B., and Tabaqchali, S., Genetic analysis of a tetracycline resistance element from *Clostridium difficile* and its conjugal transfer to and from *Bacillus subtilis*, *J. Gen. Microbiol.,* 136, 1343, 1990.
28. Roberts, A.P., Pratten, J., Wilson, M., and Mullany, P., Transfer of a conjugative transposon, Tn*5397* in a model oral biofilm, *FEMS Microbiol. Lett.,* 177, 63, 1999.
29. Roberts, A.P., Johanesen, P.A., Lyras, D., Mullany, P., and Rood, J.I., Comparison of Tn*5397* from *Clostridium difficile*, Tn*916* from *Enterococcus faecalis* and the CW459*tet*(M) element from *Clostridium perfringens* shows that they have similar conjugation regions but different insertion and excision modules, *Microbiology,* 147, 1243, 2001.
30. Flannagan, S.E., Zitzow, L.A., Su, Y.A., and Clewell, D.B., Nucleotide sequence of the 18-kb conjugative transposon Tn*916* from *Enterococcus faecalis*, *Plasmid,* 32, 350, 1994.
31. Mullany, P., Pallen, M., Wilks, M., Stephen, J., and Tabaqchali, S., A group II intron in a conjugative transposon from the gram-positive bacterium, *Clostridium difficile*, *Gene,* 174, 145, 1996.
32. Roberts, A.P., Braun, V., von Eichel-Streiber, C., and Mullany, P., Demonstration that the group II intron from the clostridial conjugative transposon Tn*5397* undergoes splicing *In vivo*, *J. Bacteriol.,* 183, 1296, 2001.
33. Martinez-Abarca, F. and Toro, N., Group II introns in the bacterial world, *Mol. Microbiol.,* 38, 917, 2000.
34. Klein, J.R. and Dunny, G.M., Bacterial group II introns and their association with mobile genetic elements, *Front. Biosci.,* 7, d1843, 2002.
35. Burrus, V., Pavlovic, G., Decaris, B., and Guedon, G., The ICE*St1* element of *Streptococcus thermophilus* belongs to a large family of integrative and conjugative elements that exchange modules and change their specificity of integration, *Plasmid,* 48, 77, 2002.

36. Mullany, P., Roberts, A.P., and Wang, H., Mechanism of integration and excision in conjugative transposons, *Cell. Mol. Life Sci.,* 59, 2017, 2002.
37. Celli, J. and Trieu-Cuot, P., Circularization of Tn*916* is required for expression of the transposon-encoded transfer functions: characterization of long tetracycline-inducible transcripts reading through the attachment site, *Mol. Microbiol.,* 28, 103, 1998.
38. Wang, H., Roberts, A., Lyras, D., Rood, J., Wilks, M., and Mullany, P., Characterization of the ends and target sites of the novel conjugative transposon Tn*5397* from *Clostridium difficile*: excision and circularization is mediated by the large resolvase, TndX, *J. Bacteriol.,* 182, 3775, 2000.
39. Wang, H., Roberts, A.P., and Mullany, P., DNA sequence of the insertional hot spot of Tn*916* in the *Clostridium difficile* genome and discovery of a Tn*916*-like element in an environmental isolate integrated in the same hot spot, *FEMS Microbiol. Lett.,* 192, 15, 2000.
40. Hächler, H., Berger-Bächi, B., and Kayser, F.H., Genetic characterization of a *Clostridium difficile* erythromycin-clindamycin resistance determinant that is transferable to *Staphylococcus aureus*, *Antimicrob. Ag. Chemother.,* 31, 1039, 1987.
41. Mullany, P., Wilks, M., and Tabaqchali, S., Transfer of macrolide-lincosamide-streptogramin B (MLS) resistance in *Clostridium difficile* is linked to a gene homologous with toxin A and is mediated by a conjugative transposon, Tn*5398*, *J. Antimicrob. Chemother.,* 35, 305, 1995.
42. Farrow, K.A., Lyras, D., and Rood, J.I., Genomic analysis of the erythromycin resistance element Tn5398 from *Clostridium difficile*, *Microbiology,* 147, 2717, 2001.
43. Farrow, K.A., Lyras, D., Polekhina, G., Koutsis, K., Parker, M.W., and Rood, J.I., Identification of essential residues in the Erm(B) rRNA methyltransferase of *Clostridium perfringens*, *Antimicrob. Ag. Chemother.,* 46, 1253, 2002.
44. Berryman, D.I. and Rood, J.I., The closely related *ermB/AM* genes from *Clostridium perfringens*, *Enterococcus faecalis* (pAMβ1), and *Streptococcus agalactiae* (pIP501), are flanked by variants of a directly repeated sequence, *Antimicrob. Ag. Chemother.,* 39, 1830, 1995.
45. Gordon, G.S. and Wright, A., DNA segregation in bacteria, *Annu. Rev. Microbiol.,* 54, 681, 2000.
46. Farrow, K.A., Lyras, D., and Rood, J.I., The macrolide-lincosamide-streptogramin B resistance determinant from *Clostridium difficile* 630 contains two *erm*(B) genes, *Antimicrob. Ag. Chemother.,* 44, 411, 2000.
47. Spigaglia, P. and Mastrantonio, P., Analysis of macrolide-lincosamide-streptogramin B (MLS(B)) resistance determinant in strains of *Clostridium difficile*, *Microb. Drug Res.,* 8, 45, 2002.
48. Ackermann, G., Degner, A., Cohen, S.H., Silva, J. Jr., and Rodloff, A.C., Prevalence and association of macrolide-lincosamide-streptogramin B (MLSB) resistance with resistance to moxifloxacin in *Clostridium difficile*, *J. Antimicrob. Chemother.,* 51, 599, 2003.
49. Cornillot, E., Saint-Joanis, B., Daube, G., Katayama, S.-I., Granum, P.E., Canard, B., and Cole, S.T., The enterotoxin gene (*cpe*) of *Clostridium perfringens* can be chromosomal or plasmid-borne, *Mol. Microbiol.,* 15, 639, 1995.
50. Miyamoto, K., Chakrabarti, G., Morino, Y., and McClane, B.A., Organization of the plasmid *cpe* locus in *Clostridium perfringens* type A isolates, *Infect. Immun.,* 70, 4261, 2002.
51. Brynestad, S., Sarker, M.R., McClane, B.A., Granum, P.E., and Rood, J I., Enterotoxin plasmid from *Clostridium perfringens* is conjugative, *Infect Immun.,* 69, 3483, 2001.
52. Brynestad, S., Synstad, B., and Granum, P.E., The *Clostridium perfringens* enterotoxin gene is on a transposable genetic element in type A human food poisoning strains, *Microbiology,* 143, 2109, 1997.
53. Brynestad, S. and Granum, P.E., Evidence that Tn*5565*, which includes the enterotoxin gene in *Clostridium perfringens*, can have a circular form which may be a transposition intermediate, *FEMS Microbiol. Lett.,* 170, 281, 1999.
54. Williams, K.P., The tmRNA Website: invasion by an intron, *Nucl. Acids Res.,* 30, 179, 2002.
55. Braun, V., Mehlig, M., Moos, M., Rupnik, M., Kalt, B., Mahony, D.E., and von Eichel-Streiber, C., A chimeric ribozyme in *Clostridium difficile* combines features of group I introns and insertion elements, *Mol. Microbiol.,* 36, 1447, 2000.
56. Daube, G., Simon, P., and Kaeckenbeeck, A., IS*1151*, an IS-like element of *Clostridium perfringens*, *Nucl. Acids Res.,* 21, 352, 1993.
57. Katayama, S., Dupuy, B., Daube, G., China, B., and Cole, S.T., Genome mapping of *Clostridium perfringens* strains with I-*Ceu*I shows many virulence genes to be plasmid-borne, *Mol. Gen. Genet.,* 251, 720, 1996.

58. Billington, S.J., Wieckowski, E.U., Sarkar, M.R., Bueschel, D., Songer, J.G., and McClane, B.A., *Clostridium perfringens* type E animal enteritis isolates with highly conserved, silent enterotoxin gene sequences, *Infect. Immun.,* 66, 4531, 1998.
59. Mahillon, J. and Chandler, M., Insertion Sequences, *Microbiol. Mol. Biol. Rev.,* 62, 725, 1998.
60. Okinaka, R.T., Cloud, K., Hampton, O., Hoffmaster, A.R., Hill, K.K., Keim, P., Koehler, T.M., Lamke, G., Kumano, S., Mahillon, J., Manter, D., Martinez, Y., Ricke, D., Svensson, R., and Jackson, P.J., Sequence and organization of pXO1, the large *Bacillus anthracis* plasmid harboring the anthrax toxin genes, *J. Bacteriol.,* 181, 6509, 1999.
61. Brynestad, S., Iwanejko, L.A., Stewart, G.S.A.B., and Granum, P.E., A complex array of Hpr consensus DNA recognition sequences proximal to the enterotoxin gene in *Clostridium perfringens* type A, *Microbiology,* 140, 97, 1994.
62. Hunter, S.E.C., Clarke, I.N., Kelly, D.C., and Titball, R.W., Cloning and nucleotide sequencing of the *Clostridium perfringens* epsilon-toxin gene and its expression in *E. coli*, *Infect. Immun.,* 60, 102, 1992.
63. Liyanage, H., Holcroft, P., Evans, V.J., Keis, S., Wilkinson, S.R., Kashket, E.R., and Young, M., A new insertion sequence, IS*Cb1*, from *Clostridium beijernickii* NCIMB 8052, *J. Mol. Microbiol. Biotechnol.,* 2, 107, 2000.
64. Dineen, S.S., Bradshaw, M., and Johnson, E.A., Neurotoxin gene clusters in *Clostridium botulinum* type A strains: sequence comparison and evolutionary implications, *Curr. Microbiol.,* 46, 345, 2003.
65. Maamar, H., de Philip, P., Bélaich, J.P., and Tardif, C., IS*Cce1* and IS*Cce2*, two novel insertion sequences in *Clostridium cellulolyticum*, *J. Bacteriol.,* 185, 714, 2003.
66. Nölling, J., Breton, G., Omelchenko, M.V., Makarova, K.S., Zeng, Q., Gibson, R., Lee, H.M., Dubois, J., Qiu, D., Hitti, J., Wolf, Y.I., Tatusov, R.L., Sabathe, F., Doucette-Stamm, L., Soucaille, P., Daly, M.J., Bennett, G.N., Koonin, E.V., and Smith, D.R., Genome sequence and comparative analysis of the solvent-producing bacterium *Clostridium acetobutylicum*, *J. Bacteriol.,* 183, 4823, 2001.
67. Shimizu, T., Ohtani, K., Hirakawa, H., Ohshima, K., Yamashita, A., Shiba, T., Ogasawara, N., Hattori, M., Kuhara, S., and Hayashi, H., Complete genome sequence of *Clostridium perfringens*, an anaerobic flesh-eater, *Proc. Natl. Acad. Sci. U.S.A.,* 99, 996, 2002.
68. Brüggemann, H., Bäumer, S., Fricke, W.F., Wiezer, A., Liesegang, H., Decker, I., Herzberg, C., Martinez-Arias, R., Merkl, R., Henne, A., and Gottschalk, G., The genome sequence of *Clostridium tetani*, the causative agent of tetanus disease, *Proc. Natl. Acad. Sci. U.S.A.,* 100, 1316, 2003.
69. Rood, J.I., Virulence genes of *Clostridium perfringens*, *Annu. Rev. Microbiol.,* 52, 333, 1998.
70. Adams, V., Lucet, I.S., Lyras, D., and Rood, J.I., DNA binding properties of TnpX indicate that different synapses are formed in the excision and integration of the Tn4451 family, *Mol. Microbiol.,* 53, 1195–1207. 2004.

Part V

Complex Regulatory Networks

28 Sporulation (Morphology) of Clostridia

Ronald G. Labbé

CONTENTS

28.1 INTRODUCTION

Bacterial spores are the most heat resistant of all living organisms, having been first described by Robert Koch and Ferdinand Cohn in 1876 [1]. In 1933, Bayne-Jones and Petrilli [2] first presented photographic evidence that bacterial sporulation involved a series of morphological changes in which one cell gives rise to one spore. In the 1960s and 1970s, the development of cultural methods resulting in the synchronous sporulation of a population of cells and refinement of electron microscope techniques resulted in detailed knowledge of the fine structure of bacterial spores and the sporulation process.

It is generally recognized that nutrient deprivation is the trigger for bacterial endospore formation. An exception is *Clostridium acetobutylicum* in which the accumulation of acid end-products causes the inhibition of cell division and the initiation of solvent production and cell differentiation [3].

The vast majority of work describing the morphological events of the sporulation process has been done using *Bacillus* species, especially *Bacillus subtilis* and the *Bacillus cereus* group [4]. However, there are distinct similarities between the sporulation process of *Bacillus* and *Clostridium* species. In both cases, bacterial sporulation is subdivided into seven distinct morphological stages [5], with the last stage corresponding to the release of the mature spore from the mother cell.

28.1.1 Stages of Sporulation

Stage I: Axial filament formation. The DNA appears to change from a compact spherical shape to a single rope or axial thread (Figure 28.1). It is not clear if this process is directly related to sporulation, since this particular formation also occurs in situations involving cessation or reduced rate of DNA synthesis. In addition, this reconfiguration of the chromatin is reversible; axial filaments return to conventional patterns if the sporulation medium is replaced by fresh growth medium, and cells divide in the normal way.

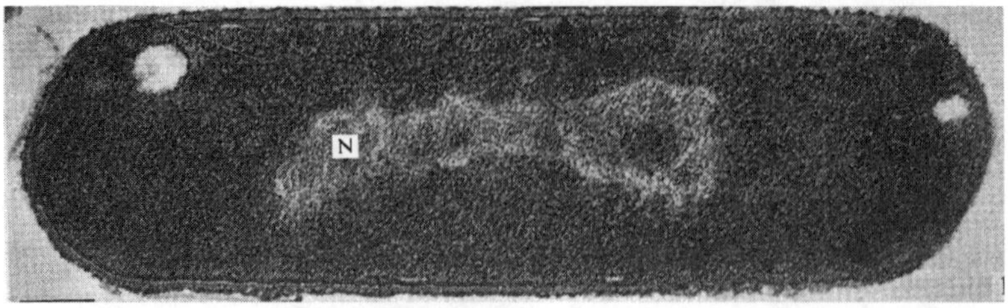

FIGURE 28.1 Axial filament formation during sporulation of *Clostridium pasteurianum.* (From Mackey and Morris, *J. Gen. Microbiol.,* 66, 1, 1971, with permission.)

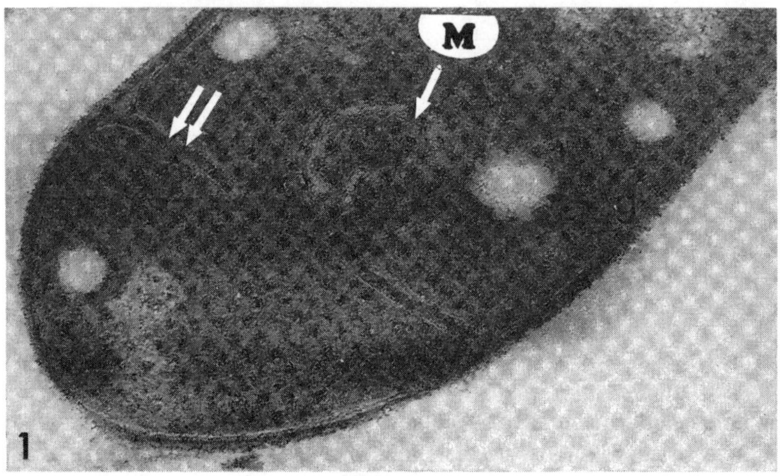

FIGURE 28.2 Asymmetric membrane formation during sporulation of *Clostridium butyricum* to form forespore with nearby mesosome [m]. (From Rousseau et al., *Ann. Inst. Pasteur,* 120, 23, 1971, with permission.)

Stage II: The first indication of the initiation of the sporulation process is the de novo formation of an asymmetrical double membrane [6] with little or no intervening peptidoglycan. Mesosomes are often seen associated with this stage [7,8] (Figure 28.2).

Stage III: The membrane produced during stage II grows around the smaller of the cells, the future spore protoplast (Figure 28.3). The leading edges of the encircling membrane fuse to form the characteristic forespore (Figure 28.4). The resulting two membranes possess opposite polarity, which affects nutrient transport [9]. At this stage, the developing spore is committed to the sporulation process.

Stage IV: The spore cortex next forms between the two membranes (Figure 28.5). This is composed of peptidoglycan, modified from that of vegetative cell wall [10–12]. This modified peptidoglycan includes repeating subunits of: a muramic lactam subunit without any attached amino acids; an alanine subunit with only a L-alanyl residue; and a tetrapeptide subunit bearing the sequence L-ala-D-glu-meso-DAP-D-ala. Unlike the vegetative cell wall, there is very little cross-linking between tetrapeptide chains.

Stage V: Although early pieces of coat protein are often seen at the beginning of peptidoglycan synthesis, the dramatic appearance of coat protein is associated with this stage. Coat protein, which accounts for half or more of all spore protein, develops outside the outer forespore membrane. The cystine-rich coat appears as discontinuous masses of dense coat material (Figure 28.6), which link together to form a continuous layer. Electron microscopic observations only reveal visible structure. Coat precursor synthesis undoubtedly begins early in the sporulation process, perhaps stage II [12]. It should be noted that

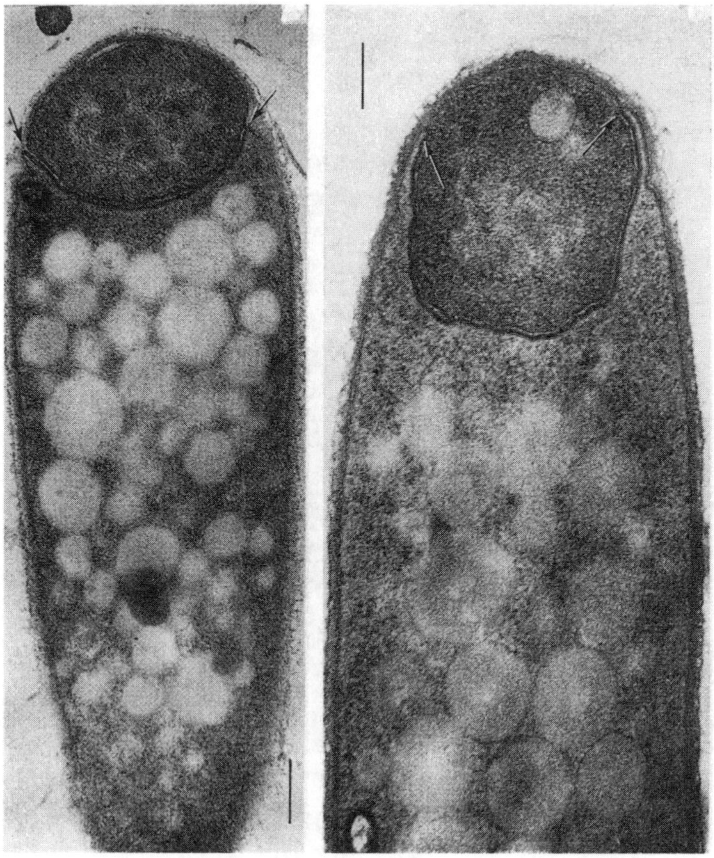

FIGURE 28.3 Engulfment of forespore by mother cell membrane in *Clostridium pasteurianum*; granulose bodies are readily apparent. (From Mackey and Morris, *J. Gen. Microbiol.*, 66, 1, 1971, with permission.)

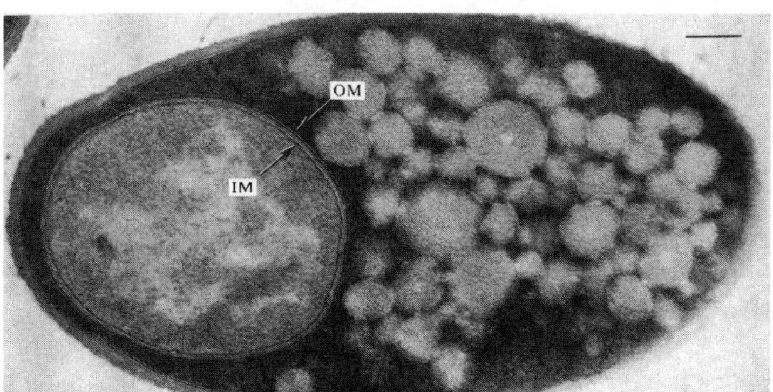

FIGURE 28.4 Completion of forespore of *C. pasteurianum* showing inner [IM] and outer [OM] membranes. (From Mackey and Morris, *J. Gen. Microbiol.*, 66, 1, 1971, with permission.)

in the case of at least one species, *Clostridium pasteurianum*, spore cortex is laid down later than the start of spore coat formation [6].

Stage VI: This stage is characterized by the maturation of the cortex and coat components. In electron micrographs, the mature cortex appear as a thick whitish envelope surrounding

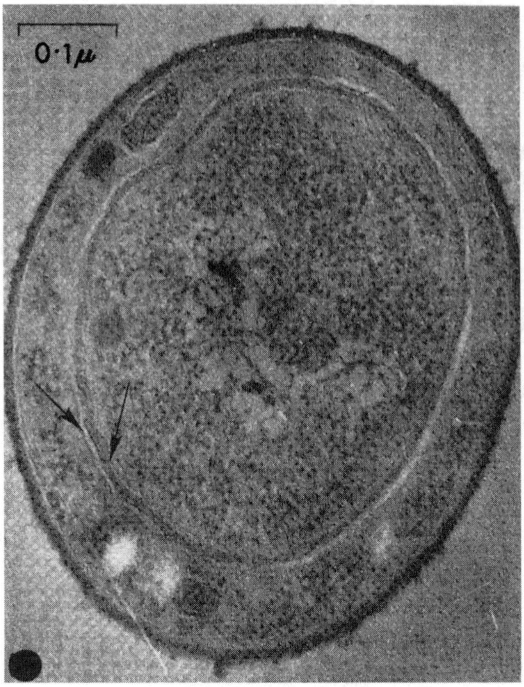

FIGURE 28.5 Beginning of spore cortical development between inner and outer membranes [arrows] in *Clostridium pectinovorum*. (From Fitz-James and Young, *The Bacterial Spore*, 39, 1969, with permission.)

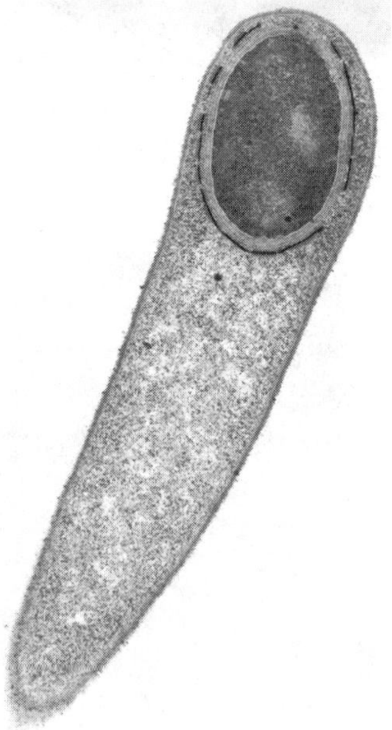

FIGURE 28.6 Appearance of coat fragments during sporulation of *Clostridium perfringens* (Labbé, R., unpublished data.)

the core [13] (Figure 28.7). The spore coat structure of those *Clostridium* species examined usually consist of one or more layers [13]. In some species, spore coat can also be seen as being composed of several lamellae (Figure 28.8). The spore lytic enzyme involved in

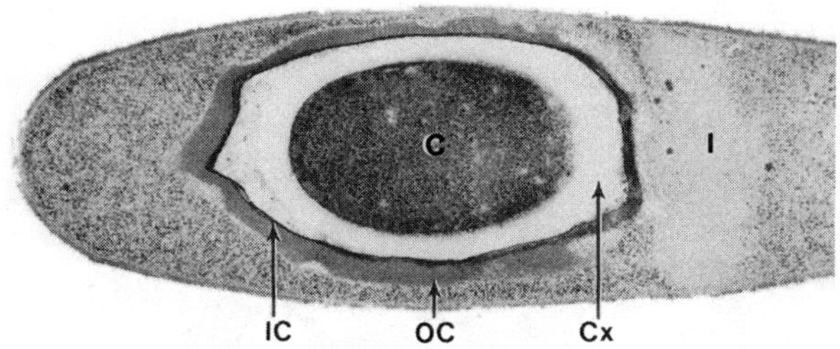

FIGURE 28.7 Maturation of sporulating cell of *Clostridium perfringens* showing core [c], cortex [cx], and inner [ic], and outer coat [oc] layers. An inclusion body (I) is also visible. (From Labbé and Rufner, *Can. J. Microbiol.*, 26, 1153, 1980, with permission.)

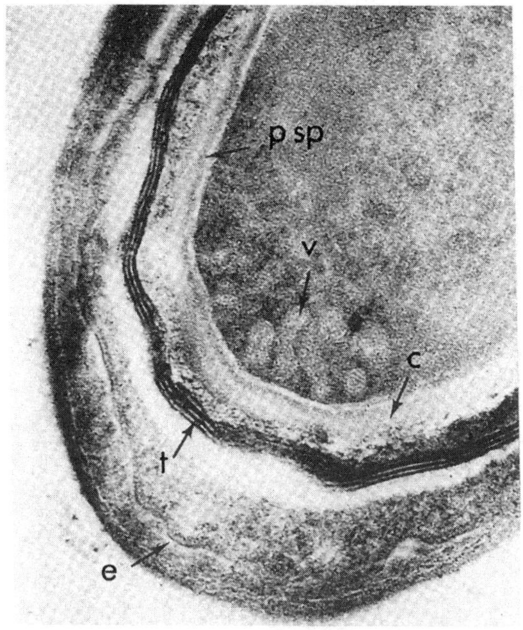

FIGURE 28.8 Sporulating cell of *Clostridium tyrobutyricum* showing coat structure [t] composed of several successive lamellae. (From Rousseau et al., *Ann. Inst. Pasteur,* 120, 33, 1971, with permission.)

the germination process [14] is believed to be located interior to the inner coat, close to its substrate, the spore cortex [15]. As viewed by phase contrast microscopy, the maturing spore takes on a refractile appearance (Figure 28.9), which is associated with the uptake of calcium, the synthesis of dipicolinic acid, and loss of water from the core. It is also at this stage that resistance to heat, solvents, and other environmental agents occurs.

Stage VII: A lytic enzyme acts to release the spore from the mother cell [16,17]. The contents of the sporangium are released as well, which, in the case of *C. perfringens*, includes an enterotoxin involved in foodborne illness [18].

28.1.2 Other Spore Structures

28.1.2.1 Exosporium

A loose fitting membrane-like structure is visible in thin sections of certain *Clostridium* sp. exterior to the spore coat (Figure 28.10) and can be selectively removed by mechanical shear. It is composed

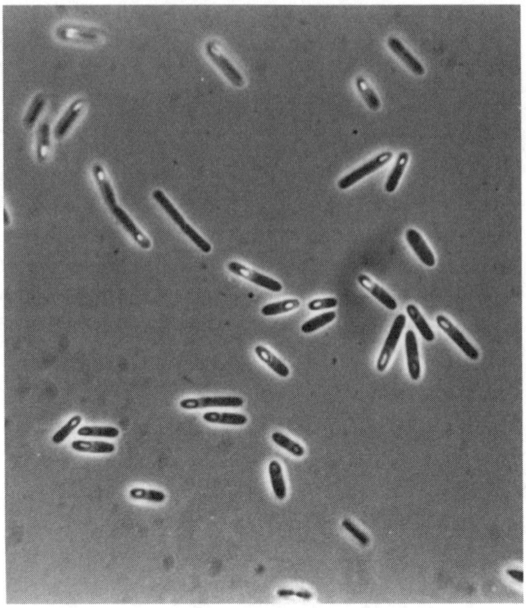

FIGURE 28.9 Phase contrast microscopy of sporulating cells of *C. perfringens*. (Labbé, R., unpublished data.)

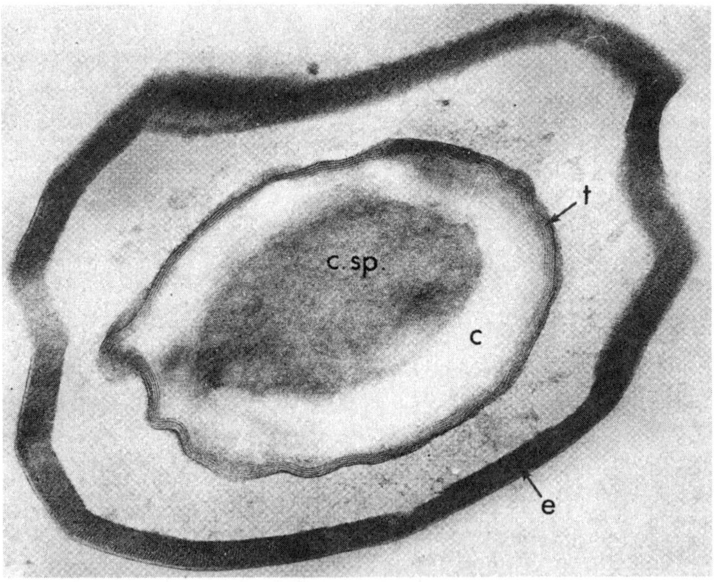

FIGURE 28.10 Free spores of *C. tyrobutyricum* showing thick exosporium [e] and layered coat [t]. (From Rousseau et al., *Ann. Inst. Pasteur,* 120, 33, 1971, with permission.)

of protein, polysaccharide, lipid, and phospholipid. By compartmentalizing and concentrating the coat subunits, the exosporium's function might be to ensure that coat fragments do not spontaneously assemble elsewhere in the cell. On the other hand, an exosporium has not been observed in at least one *Clostridium* sp., *C. perfringens* (Figure 28.7), as well as in certain *Bacillus* sp. [19], which questions the absolute necessity of this structure. Spores from which the exosporium has been removed retain both dormancy and germinability.

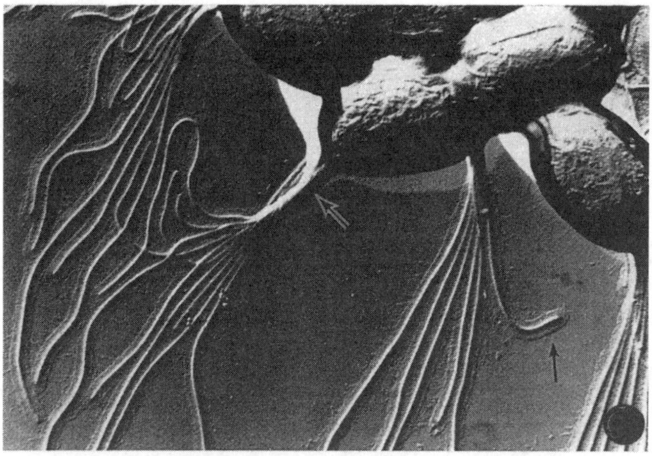

FIGURE 28.11 Appendages of *Clostridium bifermentans* attached to the spores by a common trunk [large arrow]; each appendage is surrounded by a layer of material [small arrow]. (From Pope et al., *J. Bacteriol.,* 94, 1206, 1967.)

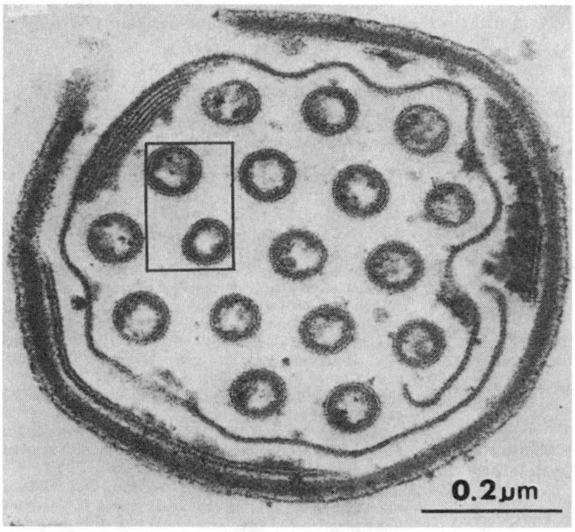

FIGURE 28.12 Cross-section of spore appendages of *C. bifermentans* showing their tubular structure. (From Samsonoff et al., *J. Bacteriol.,* 101, 1038, 1970, with permission.)

28.1.2.2 Appendages

A variety of spore appendages are associated with spores of *Clostridium* sp. [20–27]. These appear as long, tubular structures (Figure 28.11 and Figure 28.12). Other appendages have a ribbon-like structure extending from one or both spore ends (Figure 28.13) or capped, pin-like protrusions projecting from the spore surface (Figure 28.14). Their function remains obscure.

28.1.2.3 Inclusion Bodies

During sporulation, a number of anaerobes produce inclusion bodies and granules that are visible by phase contrast and electron microscopy. *C. bifermentans* produces parasporal inclusions lacking a crystalline structure [28,29]. These are toxic to mosquito larvae. Crystalline inclusions are produced by *C. perfringens* during sporulation (Figure 28.15) and are composed of closely packed parallel layers [30] (Figure 28.16), which can occupy a significant portion of the sporangium in

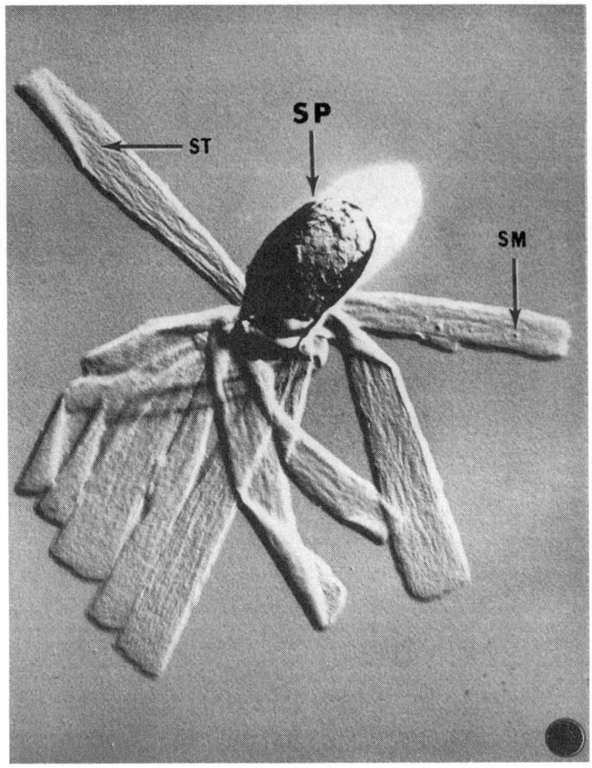

FIGURE 28.13 Free spore of *Clostridium* sp. N1 with ribbon-like appendages extending from one end; sp, spore; sm, smooth appendage; st, striated appendage. (From Rode et al., *J. Bacteriol.*, 93, 1160, 1967, with permission.)

FIGURE 28.14 Pin-like appendages of *C. bifermentans* showing spore shaft [s] capped by a lobed structure [c]. (From Yolton et al., *J. Bacteriol.*, 95, 231, 1968, with permission.)

high enterotoxin — producing strains [31]. These inclusions are composed of *C. perfringens* enterotoxin [CPE] subunits, perhaps representing a structural component where overproduced enterotoxin accumulates intracellularly [32,33]. CPE is known to be a sporulation-specific gene product [34].

Other inclusions are composed of carbohydrate or polyhydroxybutyrate (PHB). In the former case, concomitant with a decrease in intracellular glucose is the appearance of large granules composed of glucose subunits [35–37] prior to the end of exponential growth and early in the

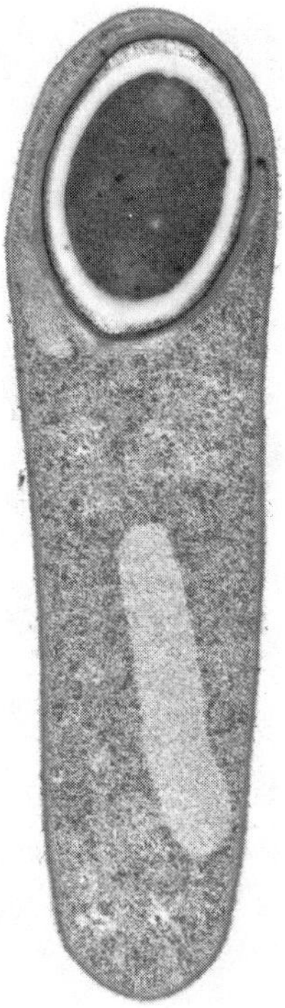

FIGURE 28.15 Sporulating cell of enterotoxigenic *C. perfringens* showing intracellular inclusion. (Labbé, R., unpublished data.)

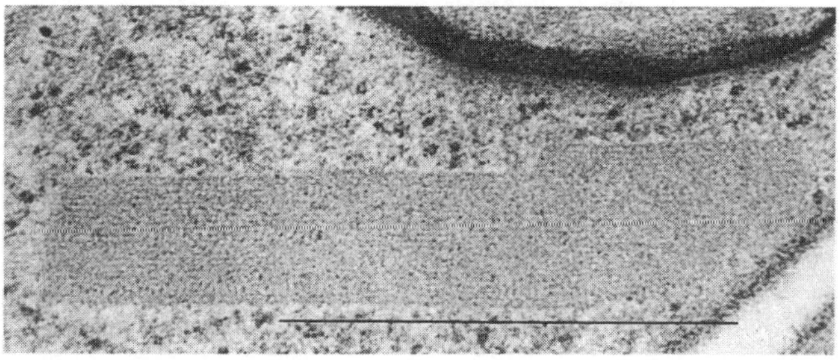

FIGURE 28.16 Subunit structure of *C. perfringens* inclusion body showing closely packed parallel layers. (From Duncan et al., *J. Bacteriol.*, 114, 845, 1973, with permission.)

sporulation process (Figure 28.17). These decrease in number during the last stages of sporulation. The synthesis of the inclusions seems to be a mechanism to lower the intracellular concentration of glucose to a level that will not inhibit sporulation. The disappearance of the carbohydrate bodies during the later stages of sporulation suggests that they may also provide endogenous carbon and energy for the synthesis of spore-specific components. A similar function has been attributed to PHB granules in the case of *Clostridium botulinum* [38]. Accumulation of PHB granules peaks in the stationary-phase cells of both sporogenic and asporogenic strains (Figure 28.18). Most of the PHB in the sporogenic strain is catabolized during the development of the spore, but in the asporogenic mutant it remains unchanged.

For a viable, biological entity, the resulting spores possess unusual resistance to physical and chemical agents. Their inactivation is the objective of many processing procedures in the food, pharmaceutical, and medical devices industries.

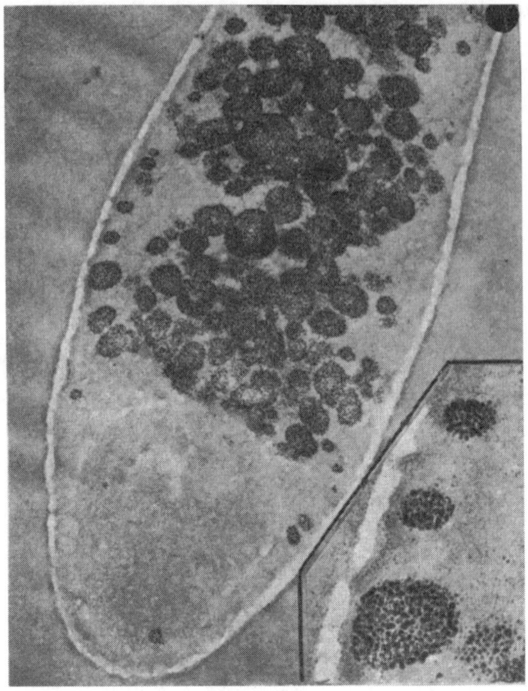

FIGURE 28.17 Sporulating cell of *C. butyricum* showing accumulation of polysaccharide granules inside mother cell cytoplasm. (From Bergere et al., *Ann. Microbiol.*, 126A, 295, 1975, with permission.)

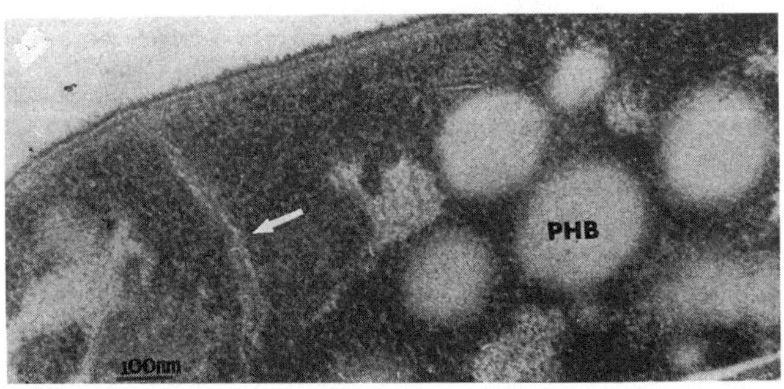

FIGURE 28.18 Sporulating cell of *C. botulinum* showing polyhydroxybutyrate (PHB) inclusions dispersed throughout mother cell cytoplasm; arrow, forespore septum. (From Emeruwa and Hawirko, *J. Bacteriol.*, 116, 989, 1973, with permission.)

REFERENCES

1. Brock, T., *Milestones in Microbiology*, American Society for Microbiology, Washington, DC, 1975.
2. Bayne-Jones, S. and Petrilli, A. Cytological changes during the formation of the endospore in *Bacillus megaterium*, *J. Bacteriol.*, 25, 261, 1933.
3. Long, S., Jones, D., and Woods, D., Initiation of solvent production, clostridial stage and endospore formation in *Clostridium acetobutylicum* P262, *Appl. Microbiol. Biotechnol.*, 20, 256, 1984.
4. Fitz-James, P. and Young, E., Morphology of sporulation, in *The Bacterial Spore*, Gould, G. and Hurst, A., Eds., Academic Press, New York, 1969, 39.
5. Ryter A., Étude morphologique de la sporulation chez *Bacillus subtilis, Ann. Inst. Pasteur,* 108, 40, 1965.

6. Mackey, B. and Morris, J., Ultrastructural changes during sporulation of *Clostridium pasteurianum, J. Gen. Microbiol.,* 66, 1, 1971.
7. Rousseau, M., Hermier, J., and Bergere, J.-L., Structure de certains *Clostridium* du groupe butyrique I. Sporulation de *Clostridium butyricum* et *Clostridium saccharobutyricum, Ann. Inst. Pasteur,* 120, 23, 1971.
8. Rousseau, M., Hermier, J., and Bergere, J.-L., Structure des certains *Clostridium* du groupe butyrique. II. Sporulation de *Clostridium tyrobutyricum, Ann. Inst. Pasteur,* 120, 33, 1971.
9. Freese, E. and Heinze, J., Metabolic and genetic control of bacterial sporulation. in *The Bacterial Spore*, Vol. 2, Hurst, A. and Gould, G., Eds., Academic Press, London, 1983, 101.
10. Leyh-Bouille, M. et al., LL-diaminopimelic acid containing peptidoglycan in walls of *Streptomyces* sp. and *Clostridium perfringens*, *Biochemistry*, 9, 2944, 1970.
11. Tang, S. and Labbe, R., Mode of action of *Clostridium perfringens* initiation protein [spore-lytic enzyme], *Ann. Inst. Pasteur/Microbiol.,* 138, 597, 1987.
12. Tipper, D. and Gauthier, J., Structure of the bacterial endospore, in *Spores V*, Halverson, H., Hanson, R., and Campbell, L., Eds., American Society for Microbiology, Washington, DC, 1972, 3.
13. Labbé, R. and Rufner, R., Ultrastruture of sporulating cells of *Clostridium perfringens* type A grown in the presence of raffinose, *Can. J. Microbiol.,* 26, 1153, 1980.
14. Gombas, D. and Labbé, R., Extraction of spore-lytic enzyme from *Clostridium perfringens* spores, *J. Gen. Microbiol.,* 126, 37, 1981.
15. Labbé, R., Reich, R., and Duncan, C., Alteration in ultrastructure and germination of *Clostridium perfringens* type A spores following extraction of spore coats, *Can. J. Microbiol.,* 24, 1526, 1978.
16. Labbé, R., Lu, H.P., and Tang, S., Extraction, purification and characterization of a peptidoglycan hydrolase from vegetative cells of *Clostridium perfringens* type A, *J. Gen. Appl. Microbiol.,* 36, 333, 1990.
17. Traci, P., Isolation and characterization of an autolysin present in spores, sporulating cells and vegetative cells of *Clostridium perfringens*, Ph.D. Thesis, University of Wisconsin, Madison, 1977.
18. Labbé, R. and Heredia, N., *Clostridium perfringens,* in *Guide to Foodborne Pathogens*, Labbé, R. and Garcia, S., Eds., John Wiley & Sons, New York, 2001, 133.
19. Lewis, J., *Dormancy in the Bacterial Spore*, Gould, G. and Hurst, A., Eds., Academic Press, New York, 39, 1969.
20. Hodgkiss, W., Ordal, Z., and Cann, D., The morphology and ultrastructure of the spore and exosporium of some clostridial species, *J. Gen. Microbiol.,* 47, 213, 1967.
21. Hodgkiss, W. and Ordal, Z., Morphology of the spore of some strains of *Clostridium botulinum* type E, *J. Bacteriol.,* 91, 2031, 1966.
22. Pope, L. and Rode, L., Spore fine structure in *Clostridium cochlearium, J. Bacteriol.,* 100, 994, 1969.
23. Pope, L., Yolton, D., and Rode, L., Appendages of *Clostridium bifermentans* spores, *J. Bacteriol.,* 94, 1206, 1967.
24. Rode, L., Crawford, M., and Williams, G., *Clostridium* spores with ribbon-like appendages, *J. Bacteriol.,* 93, 1160, 1967.
25. Rode, L. and Smith, L., Taxonomic implications of spore fine structure in *Clostridium bifermentans, J. Bacteriol.,* 105, 349, 1971.
26. Samsonoff, W., Hashimoto, T., and Conti, S., Ultrastructural changes associated with germination and outgrowth of an appendage-bearing Clostridial spore, *J. Bacteriol.,* 101, 1038, 1970.
27. Yolton, D. et al., Further electron microscope characterization of spore appendages of *Clostridium bifermentans*, *J. Bacteriol.,* 95, 231, 1968.
28. Charles, J.F. et al., *Clostridium bifermentans* serovar *Malaysia*: sporulation, biogenesis of inclusion bodies and larvicidal effect on mosquito, *Res. Microbiol.,* 141, 721, 1990.
29. De Barjac, H. et al., *Clostridium bifermentans* serovar *Malaysia*, une nouvelle bacterie anaerobie pathogene des larves de moustiques et de simulies, *C. R. Acad. Sci.* [Paris] [Ser. III], 310, 383, 1990.
30. Duncan, C., King, G., and Frieben, W., A paracrystalline inclusion formed during sporulation of enterotoxin-producing strains of *Clostridium perfringens* type A, *J. Bacteriol.,* 114, 845, 1973.
31. Garcia-Alvarado, J., Labbe, R., and Rodriguez, M., Sporulation and enterotoxin production by *Clostridium perfringens* type A at 37 and 43°C, *Appl. Environ. Microbiol.,* 58, 1411, 1992.
32. Loffler, A. and Labbe, R., Isolation of an inclusion body from sporulating, enterotoxin positive *Clostridium perfringens*, *FEMS Microbiol. Lett.*, 27, 143, 1985.

33. Loffler, A. and Labbe, R., Characterization of a parasporal inclusion body from sporulating, enterotoxin-positive *Clostridium perfringens* type A, *J. Bacteriol.,* 165, 542, 1986.
34. McClane, B., *Clostridium perfringens*, in *Food Microbiology, Fundamentals and Frontiers*, 2nd Edition, Doyle, M., Beuchat, L., and Montville, T., Eds., American Society for Microbiology Press, Washington, DC, 351, 2001.
35. Bergere, J., Rousseau, M., and Mercier, C., Polyoside intracellulaire implique dans la sporulation de *Clostridium butyricum* I. Cytologie, production et analyse enzymatique preliminaire, *Ann. Microbiol. [Inst. Pasteur]*, 126A, 295, 1975.
36. Laishley, E. et al., Isolation and morphology of native intracellular polyglucose granules from *Clostridium pasteurianum*, *Can. J. Microbiol.*, 19, 991, 1973.
37. Strasdine, G., Amylopectin accumulation in *Clostridium botulinum* type E, *Can. J. Microbiol.,* 14, 1059, 1968.
38. Emeruwa, A. and Hawirko, R., Poly-hydroxybutyrate metabolism during growth and sporulation of *Clostridium botulinum*, *J. Bacteriol.,* 116, 989, 1973.

29 Sporulation in Clostridia (Genetics)

Peter Dürre

CONTENTS

29.1 INTRODUCTION

Longing for survival is one of the cardinal biological principles. Many bacteria undergo transition to a resting state under unfavorable conditions, reducing metabolism to a minimum. However, their cells are still vulnerable by environmental impact. Few genera developed more sophisticated defense strategies. Cell differentiation leads to protected particles, better suited for long-term "bacterial hibernation". Examples are exospores produced by, e.g., *Streptomyces*, myxospores in myxobacteria, and cysts made by, e.g., *Azotobacter* and *Methylomonas*. These processes involve filament formation with subsequent septation and division, transformation of the complete progenitor cell, or a kind of a budding process [1–3]. Best protection against heat, loss of water, radiation, etc., however, is provided by endospores, the most resistant cell type known in all living organisms and produced, by e.g., *Bacillus* and *Clostridium* species. Endospore formation has so far almost only been observed in members of the Gram-positive bacterial phylum [4]. The model organism for investigating this process in detail is *Bacillus subtilis*, and numerous reviews have been published dealing with morphological, biochemical, and regulatory features occurring during this transition [5–13].

Due to significant morphological similarities of spore formation, it was generally assumed that the process in clostridia is identical to that in *Bacillus*. However, as detailed in this chapter, there are distinct regulatory differences between these two genera. In addition, isolation of members of novel bacterial genera within the Gram-positive phylum indicate that endospore formation might even provide more clues not thought of before. *Anaerobacter polyendosporus* and *Metabacterium polyspora* phylogenetically belonging to the *Clostridium* clusters, both form multiple spores (up to 5) inside their mother cells [14,15]. Here, an evolutionary link might exist to multiple internal offspring (daughter cells) within *Epulopiscium fishelsonii* [15], one of world's largest bacteria and also a member of the *Clostridium* clusters [16]. This underlines the importance of investigating in detail the endospore formation process in anaerobes. Clostridial spores and cell differentation are, however, not only fascinating objects for basic research, but meanwhile developed also into important tools for medical therapy, as detailed in Chapter 40.

29.2 MORPHOLOGY AND PHYSIOLOGY OF CLOSTRIDIAL ENDOSPORE FORMATION

The morphology of clostridial sporulation is described in detail in Chapter 28, and some reviews summarized the various developmental stages of and nutrient requirements for sporulation [17–20]. Therefore, only few complementary remarks will be added here.

Sporulation in *C. saccharobutylicum* starts by accumulation of the storage material granulose at the end of the exponential growth phase. Granulose is a high-molecular-weight polyglucan, containing only $\alpha(1\rightarrow4)$ linked D-glucopyranose subunits and serves as a carbon and energy source for spore formation and maturation [21]. Similar, partially branched polymers have been detected in *C. botulinum*, *C. butyricum*, and *C. pasteurianum* [22–26]. The accumulation leads to a swollen, cigar-like phenotype, referred to as "clostridial stage" and observed prior to onset of sporulation. This represents a major difference to the process in bacilli. Then, asymmetric septation of a progenitor cell takes place, yielding the larger mother cell and the smaller forespore. The original wall of the progenitor cell still encloses both compartments. The cytoplasmic membrane of the mother cell grows around the forespore, eventually completely engulfing it. The cortex, consisting of peptidoglycan and possibly also some protein, is synthesized between the two membranes. The spore coat, consisting of several protein layers, is formed outside the outer forespore membrane. After completion, lysis of the mother cell occurs, thus releasing the spore into the environment.

The genetic material inside the forespore, at least in *Bacillus subtilis*, has an A-helix conformation, the only naturally known example of this form. Small acid-soluble proteins (SASPs), highly conserved in bacilli and clostridia, are responsible for formation and maintenance of this condensed state of DNA [10,12,27]. During germination, they will serve as a nutrient source. Also, 3-phosphoglycerate is stored in large amounts in the spore and is used as a carbon and energy source in early germination [10].

29.3 ONSET OF SPORULATION

External signals are recognized and transmitted in *B. subtilis* via histidine kinases and phosphoryl transferases to initiate induction of sporulation-specific regulators, sigma factors, and structural proteins. In *Bacillus* species, nutrient limitation is recognized by at least five different kinases. After autophosphorylation, the phosphate is transferred via two other proteins to the master regulator of sporulation, Spo0A. The transmitters show typical features of bacterial two-component systems [28]. Spo0F is the first transmitter of the chain, also referred to as phosphorelay. This protein becomes phosphorylated by a kinase at an aspartate residue. The phosphate group is then transferred to a histidine residue in Spo0B. This protein resembles a typical phosphotransferase domain of two-component systems. The last link of the chain is Spo0A, which is phosphorylated at an aspartate residue and then acts as an ambiactive transcriptional regulator by repressing and activating its target genes by binding to specific recognition motifs, "0A boxes" [29]. The C-terminal domain of Spo0A is responsible for DNA binding and, thus, repression and activation. Its properties, however, are altered by the N terminus, when the protein becomes phosphorylated. This process causes the Spo0A monomer to form dimers, active in regulation [30]. Key targets for initiation of sporulation are the *spo0A* gene (positive autoregulation) and the *sigH* gene, encoding an alternative sigma factor, which is constitutively expressed at low level, but stimulated when sporulation is started (indirect action of Spo0A~P by inhibiting transcription of *abrB*, encoding another regulator) [31,32]. Spo0A~P is also responsible for expression of *sigF* and *sigE*, encoding sporulation-specific sigma factors (see Section 29.4). The decision for sporulation can be reversed by dephosphorylation of Spo0A~P by the phosphatase Spo0E.

The availability of clostridial genome sequences allowed to screen for the presence of phosphorelay components similar to those known from *Bacillus*. Surprisingly, neither *spo0F* nor *spo0B* genes could be detected in *C. acetobutylicum* and *C. difficile*, while *spo0A* is present [33]. The

TABLE 29.1
Similarities and Differences in Endospore Formation by *Bacillus* and *Clostridium* sp.

Feature	*Bacillus*	*Clostridium*
Induction by nutrient limitation	+	–
7 distinct morphological sporulation stages	+	+
Granulose formation	–	+
"Clostridial stage" cells	–	+
Presence of *spo0F*	+	–
Presence of *spo0B*	+	–
Presence of *spo0A*	+	+
Spo0A = master regulator	+	+
Presence of *sigH*	+	+
Presence of *sigF*	+	+
Presence of *sigE*	+	+
Presence of *sigG*	+	+
Presence of *sigK*	+	+
skin element *in sigK*	–/+ (*B. subtilis*)	–/+ (*C. difficile, C. tetani*)
Constitutive transcription of *spo0A* and *sigH* at constant level	–	+

same phenomenon is found in all other clostridia completely or partially sequenced [34]. Also, no homologues of significant similarity to the various *Bacillus* kinases interacting with Spo0F can be identified. Thus, the proteins phosphorylating Spo0A in clostridia are still unknown (compare also Chapter 24). The reason for this difference might be the underlying metabolic reactions leading to initiation of sporulation in the two genera. Endospore formation in *Bacillus* is caused by nutrient limitation, with different signals being fed into the transduction cascade at either Spo0F or Spo0B level. Thus, the bacteria can survive, until new carbon, nitrogen, and phosphorus sources become available. In clostridia, such as *C. acetobutylicum*, fermentation leads to production of acetate and butyrate, and these organic acids will diffuse in undissociated form across the cytoplasmic membrane at low pH. Since the intracellular pH of anaerobic bacteria is, in general, not constant but changes correspondingly to the external pH and is one unit higher [35], the undissociated acids will dissociate in the cytoplasm, leading to a collapse of the transmembrane pH gradient and, thus, to cell death, even in the presence of a surplus of nutrients. Clostridia, in a still unknown way, sense this dangerous condition and start sporulation (or buy time by solventogenesis, see Chapter 30 and Zickner et al. [36]). A summary of similarities and differences in sporulation by bacilli and clostridia is presented in Table 29.1.

Cloning and genome sequencing studies revealed that the presence of *spo0A* gene is always associated with the ability to form endospores. All nonsporulating bacteria investigated so far are devoid of such a gene [33,37]. Thus, presence of *spo0A* has now become a valuable taxonomic marker for attributing endospore-forming capability [38]. The respective clostridial gene, or part of it, has been subcloned from *C. acetobutylicum, C. beijerinckii, C. butyricum, C. innocuum, C. pasteurianum*, and *Moorella thermoacetica* [36]. In all clostridia, for which additional sequence information is available, the *spo0A* gene seems to be organized in a monocistronic operon. Interestingly, the gene located upstream always proved to be *spoIVB* (except for *C. difficile*), whereas the downstream DNA region shows considerable variation [33]. SpoIVB is a protein, which is involved in activation of the sporulation-specific sigma factor K (σ^K) and is associated with the membrane of the forespore [13]. Experimental evidence for the key role of Spo0A in clostridial sporulation was obtained by construction of mutants with an inactivated *spo0A* gene. Disruption of this gene in *C. beijerinckii* by multiple plasmid copy integration led to a strain lacking any apparent morphological evidence of sporulation [39]. In *C. acetobutylicum, spo0A* inactivation was obtained by double-crossover chromosomal integration. The resulting strain (SKO1) did neither transform into clostridial-stage cells nor endospores, but a large fraction of the cells formed filaments of connected rods [40]. Finally, disruption of *spo0A* in *C. perfringens* also abolished the ability to sporulate [41]. Thus, Spo0A represents the master control element of endospore formation in both bacilli and clostridia.

In *B. subtilis*, the monocistronic *spo0A* gene is controlled by tandem promoters [42]. During the exponential growth phase, a σ^A-dependent transcription initiation region allows a low-level formation of *spo0A* mRNA. At the transition to the stationary growth phase, when the onset of sporulation takes place, this distal promoter is repressed and a proximal, σ^H-dependent one is activated by binding of Spo0A~P to three 0A boxes (designated as promoter switching) [32,43]. In *C. beijerinckii*, also three 0A boxes are found upstream of *spo0A*, while in *C. acetobutylicum* and *C. difficile,* only one such motif has been detected [39]. In *C. acetobutylicum*, sequence inspection allows deduction of three putative promoters: P_1, the distal one, resembles a typical σ^A-dependent transcription initiation region (5′-TTGATT-17 bp-TAAAAT-3′) with a conserved T two bp upstream of the –10 region [44]. The 3′-T is located 171 bp upstream of the ATG start codon. P_2, the middle one, resembles a σ^H-dependent promoter (5′-AATGGAAAG-11 bp-ACAAAAT-3′, only 3 mismatches in total within the 16 nucleotides, but all in highly conserved residues [44]). Its 3′-T is located 147 bp upstream of the ATG start codon. P_3, the proximal one, shows a perfect match with the σ^H-consensus (5′-GAAGGATAA-11 bp-GTCGAAT-3′) [43]. Its 3′-T is located 63 bp upstream of the ATG start codon, and the –10 region overlaps with the 0A box [34]. Thus, the situation seems to be similar to that in *B. subtilis*. However, in contrast to *Bacillus*, Northern blots and RT-PCR experiments indicated a constant transcription level of *spo0A* throughout exponential and stationary growth phase [34,45]. Thus, if a promoter switching is occurring also in *C. acetobutylicum*, both promoter types (σ^A- and σ^H-dependent) should be of equal strength. In addition, it still must be verified, if indeed, two σ^H-dependent promoters are active upstream of *spo0A* in *C. acetobutylicum*. Reporter gene experiments using the *lucB* gene of *Photinus pyralis* [46] showed that, despite constant mRNA level, Spo0A is only synthesized during the exponential growth phase, indicating a post-transcriptional regulation [45]. The –10 region of the putative promoter P_3, and thus also the overlapping 0A box, is flanked by two imperfect palindromes. Disruption of the resulting stem-loop by targeted mutation of the upstream part led to increased and prolonged formation of the Spo0A protein, indicating a participation of this structure in regulation [45].

29.4 SPORULATION-SPECIFIC SIGMA FACTORS

In bacilli, after the asymmetric compartimentation into mother cell and forespore, σ^H and Spo0A~P activate transcription of σ^F in the forespore and Spo0A~P that of σ^E in the mother cell. A second set of sporulation-specific sigma factors (again, one in the forespore, σ^G, and one in the mother cell, σ^K) is produced by help of the former two. Thus, the cell achieves, in a timely, strictly coordinated sequence, by using respective sigma factor-dependent promoters, induction of genes and operons, the gene products of which are required for successful completion of spore formation. Regulatory patterns are different in the two compartments. σ^E and σ^K in the mother cell are translated as pro-proteins and become active only by specific proteolytic cleavage. On the other hand, σ^F and σ^G in the forespore are synthesized as active proteins, but blocked by binding to anti-sigma factors. This blockade is released if the anti-sigma factor itself is interacting with an anti-anti-sigma factor [5,6,8,9,11,13,47].

σ^H, in vegetative *B. subtilis*, is constitutively expressed at a low level. A burst in transcription at the initiation of sporulation is achieved by inhibition of transcription of the *abrB* gene, encoding a transition state regulator. This repression is mediated by Spo0A~P. σ^H and Spo0A~P then act together to express the *spoIIA* operon, encoding an anti-anti-sigma factor (SpoIIAA), an anti-sigma factor (SpoIIAB), and σ^F (encoded by the *spoIIAC* gene). SpoIIAB has binding sites for σ^F and either ATP or ADP. In the ATP state, the protein inactivates σ^F by binding. It also can inactivate SpoIIAA by phosphorylation. With bound ADP, SpoIIAB forms a complex with SpoIIAA, which is unable to bind σ^F. Active σ^F allows transcription of *spoIIGA* and *spoIIGB* in *B. subtilis*, which encode the protease required for pro-σ^E processing and σ^E itself. Directly downstream of this DNA region, *spoIIIG* is located, which encodes σ^G and is also controlled by a σ^F-dependent promoter [5,6,8,9,11,13,47].

This chromosomal arrangement is identical to the situation found in *C. acetobutylicum* by PCR-based cloning and RNA analysis, as well as the later genome sequencing [48–52]. In *C. acetobutylicum*, the *spoIIGB* gene is designated as *sigE* [48]. RT-PCR experiments confirmed the operon structure of *spoIIGA-sigE*, located directly downstream of *ftsZ*, and, using primer extension, single transcription start points were identified upstream of *spoIIGA* and *sigG*, respectively [51]. The deduced promoter upstream of *sigG* (corresponding to *spoIIIG* in *Bacillus*), represented by 5′-GTATA-15 bp-GGGAATAATT-3′, matches perfectly the σ^F *Bacillus* consensus sequence [44]. Upstream of *spoIIGA*, putative σ^A- and σ^H-dependent, overlapping promoter structures could be deduced [51], which, however, show significant deviation from the consensus motifs [44]. The σ^A-dependent promoter shows a perfect –10 region, but an unusually long distance (21 bp) to a putative –35 region with two mismatches (5′-ACGACA-3′) [44,51]. Similarly, the putative σ^H-dependent promoter has a reasonably well-conserved –10 region (only the highly conserved G residue being replaced by a T), but in a distance of 12 bp upstream, a –35 region is found in which 4 out of 9 nucleotides do not match, all of them belonging to the highly conserved group [44,51]. Thus, additional experimentation is required to identify the type of RNA polymerase holoenzyme, which is transcribing the *spoIIG* operon in *C. acetobutylicum*. A PCR-based amplification, cloning, and sequencing of *sigE* genes from *C. bifermentans, C. butyricum, C. perfringens*, and *C. sporogenes* confirmed the presence of this sigma factor in a variety of clostridial species and allowed construction of a phylogenetic tree of the respective protein. With the exception of SigE from *C. bifermentans*, a coherent cluster including the known *Bacillus* sequences was found [53]. However, complementation of a *sigE*-negative mutant of *B. subtilis* by the *sigE* gene from *C. acetobutylicum* was only possible at a low level, indicating a very specific interaction of sigma factor with its cognate RNA polymerase [53]. The genome sequencing projects completed for *C. acetobutylicum* [52], *C. perfringens* [54], and *C. tetani* [55], as well as the still incomplete sequence of *C. difficile* [56], all confirmed the presence of *sigF* genes. Also, experimental (Southern hybridization, [48]) and in silico data [52,54–56] provided evidence of *sigH* gene presence in clostridia. However, a major difference proved to be transcription level of *sigH* in *B. subtilis* and *C. acetobutylicum*. As mentioned before, in the former organism derepression of *spo0H* (encoding σ^H) by Spo0A~P-mediated *abrB* repression leads to a burst in transcript level. This is not the case in *C. acetobutylicum*. As also found for *spo0A*, the *sigH* gene is constitutively expressed at constant transcript level. The amount of *sigH* transcript was much lower than that of *spo0A* [34,45]. Although three putative 0A boxes could be identified upstream of *sigH* and additional four in the beginning of the structural gene, no binding of Spo0A~P to respective DNA fragments was observed [45]. Thus, caution should be used, when attributing regulatory features to in silico-determined 0A motifs.

The constitutive expression of *sigH* corresponds well to the possible role in transcription of the "macromolecular synthesis operon" in *C. acetobutylicum*, comprising *dnaE* and *sigA* — while in *B. subtilis* and *Escherichia coli*, *dnaE* is preceded by *orfp23* or *rpsU*, respectively [48,50,57]. In the latter two organisms, 5 σ^A-dependent promoters have been detected, and in *B. subtilis* two additional σ^H-dependent ones [44,50]. In *C. acetobutylicum*, a single transcription start has been determined by primer extension [51]. Deduction of putative promoters revealed an almost perfect match with the σ^H-consensus (only one mismatch in the conserved first residue in the –10 region) [44], while little homology can be found to a σ^A-consensus (5′-AGGATT-18 bp-TTAAAT-3′,

TABLE 29.2
Comparison of Specific Clostridial Promoter Sequences with *Bacillus* Consensus Motifs

<table>
<tr><th>Operon</th><th>Organism</th><th>Sigma Factor Required for Transcription</th><th>Nucleotide Sequence and Respective Consensus Motifs</th></tr>
<tr><td>spo0A, P₁</td><td>C. acetobutylicum</td><td></td><td>5′-TTGATT-14 bp- T--TAAAAT- 170 bp- ATG-3′</td></tr>
<tr><td></td><td></td><td></td><td>|||| | || |||</td></tr>
<tr><td></td><td></td><td>σ^A</td><td>TTGACA-13/15 bp-T--TATAAT</td></tr>
<tr><td>spo0A, P₂</td><td>C. acetobutylicum</td><td></td><td>5′-AATGGAAAG -11 bp- ACAAAAT- 146 bp- ATG-3′</td></tr>
<tr><td></td><td></td><td></td><td>|| ||||| ||| |||</td></tr>
<tr><td></td><td></td><td>σ^H</td><td>RnAGGAwWW-11/12 bp-RnnGAAT</td></tr>
<tr><td></td><td></td><td></td><td>|| |||||| ||| ||||</td></tr>
<tr><td>spo0A, P₃</td><td>C. acetobutylicum</td><td></td><td>5′-GAAGGATAA -11 bp- GTCGAAT- 62 bp- ATG-3′</td></tr>
<tr><td>sigG</td><td>C. acetobutylicum</td><td></td><td>5′-GTATA-15 bp-GGGAATAATT- 32 bp-GTG-3′</td></tr>
<tr><td></td><td></td><td></td><td>||||| |||||||||| |</td></tr>
<tr><td></td><td></td><td>σ^F</td><td>GywTA-15 bp-G gnr AnAnTw</td></tr>
<tr><td>spoIIGA</td><td>C. acetobutylicum</td><td></td><td>5′-ACGACA----TTA-12 bp-ATATAAT3-22 bp- ATG-3′</td></tr>
<tr><td></td><td></td><td></td><td>|||| ||||||</td></tr>
<tr><td></td><td></td><td>σ^A</td><td>TTGACA -16/18 bp- TATAAT</td></tr>
<tr><td></td><td></td><td></td><td>||| ||| |||</td></tr>
<tr><td></td><td></td><td>σ^H</td><td>RnAGGAwWW-12 bp-RnnGAAT</td></tr>
<tr><td></td><td></td><td>σ^A</td><td>TTGACA -17 bp- TATAAT</td></tr>
<tr><td></td><td></td><td></td><td>|| |||</td></tr>
<tr><td>dnaE-sigA</td><td>C. acetobutylicum</td><td></td><td>5′-AAAGGATTT- 18 bp- ATTAAAT- 184 bp- GTG-3′</td></tr>
<tr><td></td><td></td><td></td><td>||||||||| ||| |||</td></tr>
<tr><td></td><td></td><td>σ^H</td><td>RnAGGAwWW-11/12 bp-RnnGAAT</td></tr>
<tr><td>sigK</td><td>C. acetobutylicum</td><td></td><td>5′-ACA - 16 bp - AATAGTAT-3′</td></tr>
<tr><td></td><td></td><td></td><td>| | ||| ||</td></tr>
<tr><td></td><td></td><td>σ^E</td><td>Ata-16/18 bp- cATAcanT</td></tr>
<tr><td>cpe, P3</td><td>C. perfringens</td><td></td><td>5′-ATT - 16 bp - CATAAATG - 183 bp - ATG-3′</td></tr>
<tr><td></td><td></td><td></td><td>|| |||| ||</td></tr>
<tr><td></td><td></td><td>σ^E</td><td>ATa-16/18 bp- cATAcanT</td></tr>
<tr><td></td><td></td><td></td><td>|| ||| | ||</td></tr>
<tr><td>cpe, P2</td><td>C. perfringens</td><td></td><td>5′-ATT - 17 bp - AATACTTT - 111 bp - ATG-3′</td></tr>
<tr><td>cpe, P1</td><td>C. perfringens</td><td></td><td>5′-AC - 16 bp - CAAAATAA - 97 bp - ATG-3′</td></tr>
<tr><td></td><td></td><td></td><td>|| || ||||</td></tr>
<tr><td></td><td></td><td>σ^K</td><td>AC-16/18 bp- CATA nnnT</td></tr>
</table>

Data were taken from [44, 45, 51, 71]. N=A, C, G, T; R=A, G; W=A, T; Y=C, T.

resulting in a 50% mismatch [44]). Thus, it looks as if the macromolecular synthesis operon in *C. acetobutylicum*, whose gene products are required throughout the exponential phase, is transcribed by a σ^H-dependent RNA polymerase. Promoter comparisons are summarized in Table 29.2.

SigK in *B. subtilis* is encoded by the *spoIVCB* and *spoIVCA* loci, which are interrupted by a prophage-like element (*skin,* from *sigK* intervening sequence) encoding also a site-specific recombinase. *spoIVCB* represents the 5′-part with a σ^E-/σ^K-dependent promoter upstream. During sporulation, two regulatory events take place. Via recombination in the mother cell, the *skin* element is excised and an intact *sigK* gene is formed [13,33]. As already mentioned, the resulting gene product is proteolytically activated [58,59]. However, removal of *skin* did not affect sporulation capability, ruling out the recombination event as an essential regulatory mechanism [59]. Indeed, *B. anthracis*, *B. thuringiensis*, and *B. stearothermophilus* carry an intact *sigK* gene [33]. This is also true for *C. acetobutylicum,* as revealed by cloning and sequencing the *sigK* gene [48] and all other clostridia for which genome sequences are available. Two remarkable exceptions are *C. difficile* and *C. tetani* [55,60]. As in *Bacillus*, the *sigK* gene in *C. acetobutylicum* is transcribed from a σ^E-dependent promoter [51]. The sequence 5′-ACA-16 bp-AATAGTAT-3′ shows 4 mismatches out of 11 nucleotides, only 1 representing a highly conserved residue [44]. In *C. tetani*, the situation closely resembles *B. subtilis*. The *skin* element is of similar size (47 instead of 48 kbp) and orientated in the same way [55,58,60]. *C. difficile*, on the other hand, carries a smaller *skin* element (*skin*^{Cd} is only 14.6 kbp), which is divergently orientated (compared to *B. subtilis* and *C. tetani*), and the *sigK* gene lacks a sequence encoding an N-terminal propeptide, cleaved off for activation [60]. In accordance with this finding, two members of the proSigK processing pathway in *B. subtilis*, SpoIVFA and BofA are missing in this *Clostridium* [13,33]. In *C. difficile*, excision of *skin* from the chromosome in form of a circular molecule proved to be essential for sporulation, as absence of this element correlated with inability to sporulate [60].

In *B. subtilis*, the sporulation-specific sigma factors appear in a timely strictly coordinated sequence, namely σ^F, σ^E, σ^G, and σ^K. RNA analyses proved that this is also the case in *C. acetobutylicum* [51]. σ^E was induced before σ^G, and σ^K was the last sigma factor produced before mature endospores became microscopically detectable. At the time, when these experiments had been performed, a σ^F probe was not available [51].

Thus, it can be concluded that endospore formation is performed by a mechanistically similar, if not identical, process in both *Bacillus* and *Clostridium*, but that its regulation is differing in many aspects.

29.5 CONNECTION OF SPORULATION TO OTHER METABOLIC NETWORKS

In bacilli, Spo0A is not only the master regulator of sporulation, but controls, directly or indirectly, competence, motility, and production of extracellular hydrolases as well as synthesis of peptide antibiotics [61]. In clostridia, this transcription factor was also found to be involved in regulation of other metabolic networks.

Solventogenesis is a means by which some clostridia buy time to fight hostile environmental conditions, while initiating and completing the time-consuming process of endospore formation (see Chapter 30). Essential components of acetone and butanol synthesis are encoded by the *adc* and *sol* operons of *C. acetobutylicum* and *C. beijerinckii*. The *adc* operon of *C. acetobutylicum* is preceded by three 0A boxes to which purified and phosphorylated Spo0A indeed does bind, and Spo0A~P has been shown to have an important, although not exclusive, regulatory effect, as *gusA* and *lacZ* reporters still showed a residual regulatory activity in constructs with mutated or deleted 0A boxes. This was demonstrated by gel retardation, reporter gene constructs, and deletion mutants [39,45,62]. The same is true for the *sol* operon of that organism, the promoter of which is preceded by a single 0A box. Again, Spo0A~P proved to have an important, but not an exclusive, influence [39,63,64]. Sequences in the vicinity of the 0A boxes are not at all similar in the *adc* and *sol* operons. Thus, these data indicate that additional transcription factors act in concert with Spo0A, either phosphorylated or unphosphorylated, to induce solventogenesis and to repress acidogenesis.

A process, inextricably linked to sporulation, is the formation of enterotoxin in *Clostridium perfringens* (CPE) (compare Chapter 18 and Melville et al. and McClane and Rood [65,66]) and of C2 toxin in *Clostridium botulinum* (compare Chapter 17). CPE synthesis is the best-studied example of a sporulation-linked toxigenesis. The close connection of both processes was determined by mutant, protein, and RNA analyses [67–71]. The *cpe* gene is organized in a monocistronic operon, and transcription and translation only occur after the onset of sporulation. Three different promoters were identified by nested and internal deletions as well as reporter gene constructs. The distal and middle promoters (P3 and P2) revealed σ^E-dependent motifs, while the proximal P1 was suggested to be σ^K-dependent [71]. P3 shows only two mismatches in not highly conserved residues and one in a highly conserved residue, P2 three mismatches in not highly conserved residues, and P1 a single mismatch in a highly conserved residue [44,71]. This clearly represents the molecular basis for the connection to sporulation. In accordance with these data, a recent study reported that CPE production was blocked upon disruption of the *spo0A* gene [41]. Since Spo0A~P is required for σ^F and σ^E synthesis and these two sigma factors for σ^K formation, this result fits nicely into the currently accepted scheme.

ACKNOWLEDGMENTS

Work in my laboratory was supported by grants from the BMBF GenoMik project (Competence Network Göttingen) and the European Community (contract No. QLK3-CT-2001-01737).

REFERENCES

1. Whittenbury, R., Davies, S.L., and Davey, S.L., Exospores and cysts formed by methane-utilizing bacteria, *J. Gen. Microbiol.,* 61, 219, 1970.
2. Shimkets, L.J. and Brun, Y.V., Prokaryotic development: strategies to enhance survival, in *Prokaryotic Development*, Brun, Y.V. and Shimkets, L.J., Eds., American Society for Microbiology, Washington, DC, 2000, 1.
3. Chater K.F., Developmental decisions during sporulation in the aerial mycelium in *Streptomyces*, in *Prokaryotic Development,* Brun, Y.V. and Shimkets, L.J., Eds., American Society for Microbiology, Washington, DC, 2000, 33.
4. Sonenshein, A.L., Endospore-forming bacteria: an overview, in *Prokaryotic Development,* Brun, Y.V. and Shimkets, L.J., Eds., American Society for Microbiology, Washington, DC, 2000, 133.
5. Losick, R. and Stragier, P., Crisscross regulation of cell-type-specific gene expression during development in *Bacillus subtilis*, *Nature,* 355, 601, 1992.
6. Hoch, J.A., Regulation of the onset of the stationary phase and sporulation in *Bacillus subtilis*, *Adv. Microbial Physiol.,* 35, 111, 1993.
7. Lutkenhaus, J., Asymmetric septation in *Bacillus subtilis*, in *Regulation of Bacterial Differentiation*, Piggot, P., Moran, C.P. Jr., and Youngman, P., Eds., American Society for Microbiology, Washington, DC, 1994, 95.
8. Errington, J., Determination of cell fate in *Bacillus subtilis*, *Trends Genet.,* 12, 31, 1996.
9. Burkholder, W.F. and Grossman, A.D., Regulation of the initiation of endospore formation in *Bacillus subtilis*, in *Prokaryotic Development,* Brun, Y.V. and Shimkets, L.J., Eds., American Society for Microbiology, Washington, DC, 2000, 151.
10. Driks, A. and Setlow, P., Morphogenesis and properties of the bacterial spore, in *Prokaryotic Development,* Brun, Y.V. and Shimkets, L.J., Eds., American Society for Microbiology, Washington, DC, 2000, 191.
11. Levin, P.A. and Losick, R., Asymmetric division and cell fate during sporulation in *Bacillus subtilis*, in *Prokaryotic Development,* Brun, Y.V. and Shimkets, L.J., Eds., American Society for Microbiology, Washington, DC, 2000, 167.
12. Driks, A., Proteins of the spore core and coat, in *Bacillus subtilis and Its Closest Relatives. From Genes to Cells,* Sonenshein, A.L., Hoch, J.A., and Losick, R., American Society for Microbiology, Washington, DC, 2002, 527.

13. Piggot, P. and Losick, R., Sporulation genes and intercompartmental regulation, in *Bacillus subtilis and Its Closest Relatives. From Genes to Cells,* Sonenshein, A.L., Hoch, J.A., and Losick, R., Eds., American Society for Microbiology, Washington, DC, 2002, 483.
14. Siunov, A.V. et al., Phylogenetic status of *Anaerobacter polyendosporus*, an anaerobic, polysporic bacterium, *Int. J. Syst. Bacteriol.,* 49, 1119, 1999.
15. Angert, E.R., Brooks, A.E., and Pace, N.R., Phylogenetic analysis of *Metabacterium polyspora*: clues to the evolutionary origin of daughter cell production in *Epulopiscium* species, the largest bacteria, *J. Bacteriol.,* 178, 1451, 1996.
16. Angert, E.R., Clements, K.D., and Pace, N.R., The largest bacterium, *Nature,* 362, 239, 1993.
17. Labbé, R.G. and Shih, N.-J.R., Physiology of sporulation of clostridia, in *The Clostridia: Molecular Biology and Pathogenesis,* Rood, J.I. et al., Eds., Academic Press, Inc., San Diego, 1997, 21.
18. Mayer, F., *Cytology and Morphogenesis of Bacteria,* Gebrüder Borntraeger, Berlin-Stuttgart, Germany, 1986.
19. Woods, D.R. and Jones, D.T., Physiological responses of *Bacteroides* and *Clostridium* strains to environmental stress factors, *Adv. Microbial Physiol.,* 28, 1, 1986.
20. Bahl, H. and Dürre, P., Clostridia, in *Biotechnology,* 2nd Edition, Vol. 1, Sahm, H., Ed., VCH, Weinheim, Germany, 1993, 285.
21. Reysenbach, A.L. et al., Characterization, biosynthesis, and regulation of granulose in *Clostridium acetobutylicum, Appl. Environ. Microbiol.,* 52, 185, 1986.
22. Whyte, J.N.C. and Strasdine, G.A., An intracellular α-D-glucan from *Clostridium botulinum*, type E, *Carbohydr. Res.,* 25, 435, 1972.
23. Hobson, P.N. and Nasr, H., An amylopectin-type polysaccharide synthesized from sucrose by *C. butyricum, J. Chem. Soc.,* 1951, 1855, 1951.
24. Bergère, J.L., Rousseau, M., and Mercier, C., Polyoside intracellulaire impliqué dans la sporulation de *Clostridium butyricum.* I. Cytologie, production et analyse enzymatique préliminaire, *Ann. Microbiol. (Inst. Pasteur),* 126, 295, 1975.
25. Brown, R.G., Lindberg, B., and Laishley, E.J., Characterization of two reserve glucans from *Clostridium pasteurianum, Can. J. Microbiol.,* 21, 1136, 1975.
26. Darvil, A.G. et al., The intracellular reserve polysaccharide of *Clostridium pasteurianum, Can. J. Microbiol.,* 23, 947, 1977.
27. Mohr, S.C. et al., Binding of small acid-soluble spore proteins from *Bacillus subtilis* changes the conformation of DNA from B to A, *Proc. Natl. Acad. Sci. U.S.A.*, 88, 77, 1991.
28. Perego, M. and Hoch, J.A., Two-component systems, phosphorelays, and regulation of their activities by phosphatases, in *Bacillus subtilis and Its Closest Relatives. From Genes to Cells,* Sonenshein, A.L., Hoch, J.A., and Losick, R., Eds., American Society for Microbiology, Washington, DC, 2002, 473.
29. Spiegelman, G.B., Bird, T.H., and Voon, V., Transcription regulation by the *Bacillus subtilis* response regulator Spo0A, in *Two-Component Signal Transduction*, Hoch, J.A. and Silhavy, T.J., Eds., American Society for Microbiology, Washington, DC, 2002, 159.
30. Lewis, R.J. et al., Dimer formation and transcription activation in the sporulation response regulator Spo0A, *J. Mol. Biol.,* 316, 235, 2002.
31. Weir, J. et al., Regulation of *spo0H*, a gene coding for the *Bacillus subtilis* σ^H factor, *J. Bacteriol.,* 173, 521, 1991.
32. Strauch, M.A. et al., Spo0A activates and represses its own synthesis by binding at its dual promoters, *Biochimie,* 74, 619, 1992.
33. Stragier P., A gene odyssey: exploring the genomes of endospore-forming bacteria, in *Bacillus subtilis and Its Closest Relatives. From Genes to Cells,* Sonenshein, A.L., Hoch, J.A., and Losick, R., Eds., American Society for Microbiology, Washington, DC, 2002, 519.
34. Dürre, P., and Hollergschwandner, C., Initiation of endospore formation in *Clostridium acetobutylicum, Anaerobe,* 10, 69, 2004.
35. Dürre, P., Bahl, H., and Gottschalk, G., Membrane processes and product formation in anaerobes, in *Handbook on Anaerobic Fermentations*, Erickson, L.E. and Fung, D.Y.-C., Eds., Marcel Dekker, Inc., New York, 1988, 187.
36. Zickner, B. et al., Multiple levels of regulation of solventogenesis in *Clostridium acetobutylicum,* in *Regulatory Networks in Prokaryotes*, Dürre, P. and Friedrich, B., Eds., Horizon Scientific Press, Wymondham, U.K., 2003, 127.

37. Brown, D.P. et al., Characterization of *spo0A* homologues in diverse *Bacillus* and *Clostridium* species identifies a probable DNA-binding domain, *Mol. Microbiol.,* 14, 411, 1994.
38. Trotter, J.R., and Bishop, A.H., Phylogenetic analysis and confirmation of the endospore-forming nature of *Pasteuria penetrans* based on the *spo0A* gene, *FEMS Microbiol. Lett.,* 225, 249, 2003.
39. Ravagnani, A. et al., Spo0A directly controls the switch from acid to solvent production in solvent-forming clostridia, *Mol. Microbiol.,* 37, 1172, 2000.
40. Harris, L.M., Welker, N.E., and Papoutsakis, E.T., Northern, morphological, and fermentation analysis of *spo0A* inactivation and overexpression in *Clostridium acetobutylicum* ATCC 824, *J. Bacteriol.,* 184, 3586, 2002.
41. Huang, I.H. et al., Disruption of the gene (*spo0A*) encoding sporulation transcription factor blocks endospore formation and enterotoxin production in enterotoxigenic *Clostridium perfringens* type A, *FEMS Microbiol. Lett.,* 233, 233, 2004.
42. Chibazakura, T., Kawamura, F., and Takahashi, H., Differential regulation of *spo0A* transcription in *Bacillus subtilis*: glucose represses promoter switching at the initiation of sporulation, *J. Bacteriol.,* 173, 2625, 1991.
43. Strauch, M.A. et al., A positive feedback loop controls transcription of the *spo0F* gene, a component of the sporulation phosphorelay in *Bacillus subtilis, Mol. Microbiol.,* 7, 967, 1992.
44. Helmann, J.D., and Moran, C.P. Jr., RNA polymerase and sigma factors, in *Bacillus subtilis and Its Closest Relatives. From Genes to Cells,* Sonenshein, A.L., Hoch, J.A., and Losick, R., Eds., American Society for Microbiology, Washington, DC, 2002, 289.
45. Hollergschwandner, C., Unterschiede in der Regulation der Sporulation von *Clostridium acetobutylicum* und *Bacillus subtilis*, Ph.D. Thesis, University of Ulm, Germany, 2003.
46. Feustel, L., Nakotte, S., and Dürre, P., Characterization and development of two reporter gene systems for *Clostridium acetobutylicum, Appl. Environ. Microbiol.,* 70, 798, 2004.
47. Hoch, J.A., Control of cellular development in sporulating bacteria by the phosphorelay two-component signal transduction system, in *Two-Component Signal Transduction*, Hoch, J.A. and Silhavy, T.J., Eds., American Society for Microbiology, Washington, DC, 2002, 129.
48. Sauer, U. et al., Sporulation and primary sigma factor homologous genes in *Clostridium acetobutylicum, J. Bacteriol.,* 176, 6572, 1994.
49. Wong, J., Sass, C., and Bennett, G.N., Sequence and arrangement of genes encoding sigma factors in *Clostridium acetobutylicum* ATCC 824, *Gene,* 153, 89, 1995.
50. Sauer, U. et al., Sigma factor and sporulation genes in *Clostridium*, *FEMS Microbiol. Rev.,* 17, 331, 1995.
51. Santangelo, J.D. et al., Sporulation and time course expression of sigma-factor homologous genes in *Clostridium acetobutylicum, FEMS Microbiol. Lett.,* 161, 157, 1998.
52. Nölling, J. et al., Genome sequence and comparative analysis of the solvent-producing bacterium *Clostridium acetobutylicum, J. Bacteriol.,* 183, 4823, 2001.
53. Arcuri, E.F., Wiedmann, M., and Boor, K.J., Phylogeny and functional conservation of σ^E in endospore-forming bacteria, *Microbiol. U.K.,* 146, 1593, 2000.
54. Shimizu, T. et al., Complete genome sequence of *Clostridium perfringens*, an anaerobic flesh-eater, *Proc. Natl. Acad. Sci. U.S.A.,* 99, 996, 2002.
55. Brüggemann, H. et al., The genome sequence of *Clostridium tetani*, the causative agent of tetanus disease, *Proc. Natl. Acad. Sci. U.S.A.*, 100, 1316, 2003.
56. Genome sequence data of *Clostridium difficile* available at http://www.sanger.ac.uk/Projects/C_difficile
57. Lupski, J.R., Ruiz, A.A., and Godson, G.N., Promotion, termination, and anti-termination in the *rpsU-dnaG-rpoD* macromolecular synthesis operon of *E. coli* K-12, *Mol. Gen. Genet.,* 195, 391, 1991.
58. Stragier, P. et al., Chromosomal rearrangement generating a composite gene for a developmental transcription factor, *Science,* 243, 507, 1989.
59. Kunkel, B., Losick, R., and Stragier, P., The *Bacillus subtilis* gene for the developmental transcription factor σ^K is generated by excision of a dispensable DNA element containing a sporulation recombinase gene, *Genes Dev.,* 4, 525, 1990.
60. Haraldsen, J.D., and Sonenshein, A.L., Efficient sporulation in *Clostridium difficile* requires disruption of the σ^K gene, *Mol. Microbiol.,* 48, 811, 2003.
61. Strauch, M.A., and Hoch J.A., Transition state regulators: sentinels of *Bacillus subtilis* post-exponential gene expression, *Mol. Microbiol.,* 7, 337, 1993.

62. Böhringer, M., Molekularbiologische und enzymatische Untersuchungen zur Regulation des Gens der Acetacetat-Decarboxylase von *Clostridium acetobutylicum*, Ph.D. Thesis, University of Ulm, Germany, 2002.
63. Thormann, K. et al., Control of butanol formation in *Clostridium acetobutylicum* by transcriptional activation, *J. Bacteriol.,* 184, 1966, 2002.
64. Dürre, P. et al., Transcriptional regulation of solventogenesis in *Clostridium acetobutylicum*, *J. Mol. Microbiol. Biotechnol.,* 4, 295, 2002.
65. Melville, S.B., Collie, R.E., and McClane, B.A., Regulation of enterotoxin production in *Clostridium perfringens*, in *The Clostridia: Molecular Biology and Pathogenesis,* Rood, J.I. et al., Eds., Academic Press, Inc., San Diego, 1997, 471.
66. McClane, B.A. and Rood, J.I., Clostridial toxins involved in human enteric and histotoxic infections, in *Clostridia: Biotechnology and Medical Applications,* Bahl, H. and Dürre, P., Eds., Wiley-VCH Verlag GmbH, Weinheim, Germany, 2001, 169.
67. Duncan, C.L., Strong, D.H. and Sebald, M., Sporulation and enterotoxin production by mutants of *Clostridium perfringens*, *J. Bacteriol.,* 110, 378, 1972.
68. Melville, S.B., Labbé, R., and Sonenshein, A.L., Expression from the *Clostridium perfringens cpe* promoter in *Clostridium perfringens* and *Bacillus subtilis*, *Infect. Immun.,* 62, 5550, 1994.
69. Kokai-Kun, J.F. et al., Comparison of Western immunoblots and gene detection assays for identification of potentially enterotoxigenic isolates of *Clostridium perfringens*, *J. Clin. Microbiol.,* 32, 2533, 1994.
70. Czeczulin, J.R., Collie, R.E., and McClane, B.A., Regulated expression of *Clostridium perfringens* enterotoxin in naturally *cpe*-negative type A, B, and C isolates of *C. perfringens*, *Infect. Immun.,* 64, 3301, 1996.
71. Zhao, Y. and Melville, S.B., Identification and characterization of sporulation-dependent promoters upstream of the enterotoxin gene (*cpe*) of *Clostridium perfringens*, *J. Bacteriol.,* 180, 136, 1998.

30 Formation of Solvents in Clostridia

Peter Dürre

CONTENTS

30.1 INTRODUCTION

The solvents acetone, butanol, and 2-propanol (isopropanol) are bulk chemicals of major industrial importance and used for synthesis of a variety of products (Table 30.1). Commercial production is performed nowadays by chemical processes.

However, these compounds are also naturally formed by a number of clostridia. Major producers are *Clostridium acetobutylicum*, *C. beijerinckii*, *C. saccharobutylicum,* and *C. saccharoperbutylacetonicum*, whereas lower amounts are synthesized by *C. aurantibutyricum, C. butyricum, C. cadaveris, C. pasteurianum, C. puniceum, C. sporogenes, C. tetanomorphum*, and *C. thermosaccharolyticum* [for reviews see 1,2]. *C. felsineum* also forms butanol, and, according to recent 16S rRNA investigations, is closely related to *C. acetobutylicum* [3]. Older literature suffers partially from the fact that the National Collections of Industrial and Marine Bacteria (NCIMB), Aberdeen, U.K., listed strain NCIMB 8052 as the type strain of *C. acetobutylicum*, whereas in fact it represents a *C. beijerinckii*. Also, P262, a production strain of National Chemicals Products, South Africa, as well as N1-4 and N1-504 were originally referred to as *C. acetobutylicum*, but later found to represent taxonomically *C. saccharobutylicum* and *C. saccharoperbutylacetonicum*, respectively [4,5,6,7].

Biological butanol formation was first reported by the famous Louis Pasteur [8]. His *Vibrion butyrique* is unlikely to have represented a pure culture, but probably contained *Clostridium butyricum*, also being able to synthesize butanol under special conditions [1,9]. Pasteur's studies in this respect represent a biological landmark, since he was the first to demonstrate that life was possible without air (oxygen). Based on this observation, he coined the term "anaerobic" [10–12]. A number of other well-known microbiologists also worked on bacterial butanol formation, among them Martinus W. Beijerinck and Sergej N. Winogradsky [for a summary see 2]. However, Fitz was probably the first to isolate a pure culture of what he called *Bacillus butylicus* [13–16]. The term *Clostridium* was reintroduced later by Prazmowski [17], based on a previous designation by Trécul [18,19]. *Clostridium* stems from the Greek root "kloth" and means "small spindle" (for a

TABLE 30.1
Industrial Uses and Synthesis of Acetone, Butanol, and 2-Propanol

Acetone	Butanol	2-Propanol
Production of methacrylates and methyl isobutyl ketone (about 39% of the worldwide production), smokeless powder (cordite), solvent for adhesives, cellulose acetate, fats, lacquers, oils, paints, plastics, printing inks, resins, rubber cements, varnishes, waxes Use in wet and dry spinning, for extraction of compounds from animal and plant substances, and the preparation of vitamin intermediates Component of nail polish remover, paint, and varnish removers	Conversion to butyl acrylate, methacrylate, butyl glycol ethers butyl acetate for use in the coating industry Solvent for acid-curing lacquers and baking finishes Plasticizers for dispersions, plastics, and rubber mixes (such as butyl acrylate, benzylbutyl phthalate, and dibutyl phthalate) Use of esters for latex architectural paints in the housing market Solvent for dyes, e.g., in printing inks Extractant in the production of drugs and natural substances such as alkaloids, antibiotics, hormones, and vitamins Additive in polishes and cleaners, e.g., floor cleaners and stain removers Solubilizer in the textile industry, e.g., additive in spinning baths or carrier for coloring plastics Additive in de-icing fluids Anti-icing additive of gasoline Mobile phase in paper and thinlayer chromatography Humectant for cellulose nitrate Component in the manufacture of hydraulic fluids Diluent for formulating brake fluids suitable for use in passenger cars Feedstock for the production of the ore flotation agent butyl xanthate	Component in antifreeze mixtures, quick-drying oils, ink, hand and aftershave lotions, and in some healthcare personnel antiseptic handwash mixtures Solvent in perfume oils, and in extraction processes Intermediate or processing solvent in the synthesis of other chemicals (about 40% of the worldwide production) Coating solvent De-icing agent for liquid fuels (gasoline additive) Ingredient in liquid soaps and window cleaners One of the cheapest alcohols and has replaced ethanol for many uses
Commercial synthesis is mainly performed by catalytic dehydrogenation of 2-propanol	Commercial synthesis is mainly based on an oxo reaction with propylene, yielding butyraldehyde, which then is converted into 1-butanol by hydrogenation	Commercial synthesis is performed by dissolving propylene gas in sulfuric acid and then hydrolyzing the sulfate ester that is formed

more detailed report see Dürre [20]). A taxonomically valid description of *C. acetobutylicum*, however, was only published in 1926 [21]. More than 40 years after Pasteur's pioneering studies, the first report on the production of acetone in a microorganism was published. Schardinger found this compound to be formed by *Bacillus macerans* [22,23]. Just one year later, Pringsheim reported that 2-propanol (isopropanol) was a fermentation product by *C. americanum* [24,25]. Another bacterium synthesizing this alcohol is *Granulobacter butylicum*, originally isolated in 1893 by Beijerinck, as detected by Folpmers in 1920 [cited in 26]. *G. butylicum* was renamed first *C. butylicum,* and later *C. beijerinckii* [27].

30.2 HISTORY

The acetone-butanol fermentation had important biotechnological and political implications. Although the biological formation was known for several decades as described above, a biotechnological use was not initiated before 1913. In 1911 and 1912, Fernbach and Strange filed three patents [28–30] describing formation and recovery of acetone and higher alcohols (mainly butanol) from a mixed fermentation by "Thyrothrix tennis" and "butylic bacillus of Fitz." Their aim was to produce synthetic rubber from isoprene, dimethylbutadiene, or butadiene, which could be synthesized from the fermentation products amyl alcohol, acetone, or butanol, respectively. A production plant under the direction of Strange & Graham, Ltd. went into operation first at Rainham, U.K., and later at King's Lynn, U.K. [31]. The outbreak of World War I led to a dramatically increased demand for acetone, which was needed to manufacture smokeless ammunition (cordite). Weekly production averaged 970 pounds and was increased to 2340 pounds after the government took over the plant [31]. Except for the French bacteriologist Fernbach, Strange and Graham had also employed Perkins and Weizmann from Manchester University for their synthetic rubber project. Weizmann left the group in 1912, but continued research in this field and isolated a strain with superior properties from the standpoint of commercial fermentation [32]. A respective patent was issued to him in 1915 [33]. The Weizmann process, employing *C. acetobutylicum*, became the method of choice at King's Lynn and all other fermentation plants, later adapted from existing distilleries in Canada (Toronto) and the United States (Terre Haute) [31]. The enormously enhanced acetone yield unfortunately confirms the adage of the Greek philosopher Heraclitus: "War is father of all." Weizmann declined any honors by the British goverment, but raised the question of establishing a Jewish homeland in Palestine. This was substantiated by the Balfour Declaration in 1917 and the subsequent creation of the State of Israel. Weizmann, who meanwhile had changed his prename from Charles to Chaim, became the first president of Israel. Thus, few bacteria had a similarly important political impact as had *C. acetobutylicum*.

After the war had ended in 1918, there was no longer a high demand for acetone, and consequently, the production plants were closed. During the whole war, butanol had been considered a useless byproduct and was stored in large tanks. The situation changed when prohibition was introduced in the U.S. in 1920, meaning that no fusel oil or amyl acetate (from amyl alcohol) was available for industrial purposes [31]. On the other hand, Henry Ford's revolutionary new assembly-line methods for automobile production created an urgent need for a solvent of quick-drying nitrocellulose lacquers, which could be met by butyl acetate (for which butanol served as a feedstock) [2,32,34,35]. This led to reopening and additional construction of production plants in many parts of the world. The two major facilities in the U.S., at Terre Haute and Peoria, had a capacity of 148 fermenters with a volume of 50,000 gallons each (i.e., 189,270 l each and more than 28 million l in total) [31]. In 1945, two thirds of industrially used butanol were synthesized biologically [36]. Thus, the acetone-butanol fermentation is one of the largest biotechnological processes ever performed (exceeded in total volume only by ethanol fermentation).

During World War II, the focus shifted to acetone production again. However, after the war, rising substrate prices and lower sugar content in the molasses gave the petrochemical production of solvents an advantage over fermentation [34]. Consequently, biotechnological synthesis declined and was only continued (for a while) in countries that were cut off from international supplies for political or monetary reasons (e.g., Egypt, the Soviet Union, South Africa, China) [34,37]. Starting with the "oil crisis" in 1973, scientific interest in fermentation and product recovery, as well as in physiology, molecular biology, and cellular regulation of the bacteria involved, was renewed. Aims are the reintroduction of an economically competitive fermentation and the understanding of the underlying metabolic processes.

30.3 FERMENTATION

A typical feature of the clostridial acetone-butanol production is a biphasic fermentation, as shown with the model organism *C. acetobutylicum*. The bacteria first follow the typical butyric acid pathway, with acetate, butyrate, hydrogen, and carbon dioxide being the major products (Figure 30.1). Small amounts of acetoin can be formed, and under certain conditions lactate is produced. All this happens during the exponential growth phase. At the transition to stationary

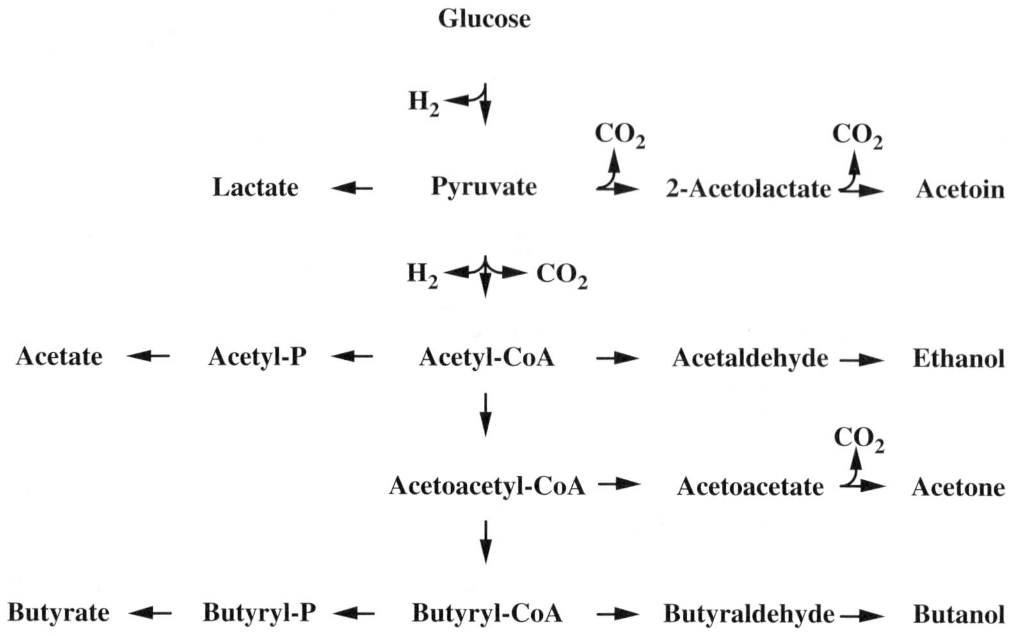

FIGURE 30.1 Catabolic pathways of acid and solvent formation in *Clostridium acetobutylicum.*

growth, formation of acids is decreased, and acetone and butanol become dominant products (or 2-propanol instead of acetone in some strains of *C. beijerinckii*). Part of the previously produced acids is taken up again and converted into solvents. As a consequence, the pH of the culture first decreases to between 4.5 and 4 and increases again, when solventogenesis starts. Concomitantly, endospore formation is initiated. As will be detailed later in this chapter, the regulatory networks of solventogenesis and sporulation are interconnected. A decreasing pH poses a severe threat for anaerobic bacteria. They are, in general, unable to maintain a constant internal pH, but rather keep the transmembrane pH gradient constant [38]. This is also true for *C. acetobutylicum* [39,40]. As a consequence, the internal pH will be approximately 1 pH unit more alkaline. At an external pH of 4.5 to 4, the butyrate and acetate formerly produced will be present as undissociated acids and, thus, be able to pass the cytoplasmic membrane via diffusion. Inside the cell, however, they will dissociate into salts and protons due to the higher pH. This way, the proton gradient across the membrane is destroyed, and the cell will die. Solventogenesis allows for fighting this deleterious effect. Conversion of acetate and butyrate into acetone and butanol leads to an increasing pH. On the other hand, the solvents produced also are toxic. The most dangerous compound is butanol. Abolition of the pH gradient across the membrane, inhibition of ATPase, release of intracellular metabolites, and inhibition of sugar uptake are caused by high concentrations of this alcohol [39, 41–44]. Thus, solventogenesis only serves as a kind of emergency reaction to let the cells buy time to complete endospore formation and, so, to guarantee long-time survival.

In Chapter 26, different commercial fermentation processes and product recovery methods have been detailed, so that only few remarks are given here. Fermentation has been performed as batch culture, fed-batch culture, and continuous culture (single- or two-stage chemostats) [2,35,45,46]. The latter will lead to enrichment of asporogenous mutants, as spores are constantly washed out, but without affecting solvent production [47]. In the past, product recovery has been achieved by distillation, which, however, is energetically unfavorable due to the high boiling point of butanol (117.7°C). Other possibilities are adsorption, gas stripping, liquid-liquid extraction, membrane evaporation, perstraction, pervaporation, or reverse osmosis [35]. Since membrane-based systems suffer from clogging and fouling, gas stripping might become the method of choice, although it does not allow complete removal of solvents from the fermentation broth.

30.4 ENZYMES

C. acetobutylicum usually feeds on starch and sugars. A number of respective enzymes have been purified and characterized, and the genome sequence (3.94 Mbp chromosome and 192 kbp megaplasmid pSOL1) provided additional valuable information [48]. Several potential amylase genes have been detected, but only two reports describe purification of an α-amylase from strain ATCC 824 [49,50]. The enzymes proved to have different pH optima, different temperature optima, showed a different response to Ca^{2+} ions, and were found to have different molecular masses. The enzyme purified by Paquet et al. [49], now referred to as AmyP [51], was determined to be monomeric with a size of 83 to 84 kDa, whereas the protein purified by Annous and Blaschek [50], has a molecular mass of 61 kDa. Genetic information is scarce. An α-amylase gene from *C. acetobutylicum* ATCC 824 has been cloned into *Escherichia coli* [52], and from nucleotide sequence analysis, a very unusual promoter sequence (spacing of 26 bp between the –35 and –10 regions) and a putative signal sequence have been proposed. The partially purified gene product was determined to possess a molecular mass of 60 to 65 [53]. However, the determined nucleotide sequence does not match any open reading frame from the genome sequence, which, however, revealed the coding sequence of a protein of 61 kDa with significant similarity (64%) to AmyP. Thus, the genetic location of the other amylase encoding sequence still remains obscure. *C. acetobutylicum* is unable to ferment cellulose, but can use cellobiose [54]. Interestingly, analysis of the genome sequence revealed the presence of a number of genes (most of them clustered), whose products were confidently identified as cellulosome components [48]. Single proteins proved to be functional when overexpressed in *Escherichia coli* or *C. acetobutylicum* [55,56]. In fact, an inactive cellulosome complex was detected [51]. The amount, however, was very low [57]. The reasons for the low production and the inability of *C. acetobutylicum* to degrade cellulose are still not understood. Since this polymer would be an economically attractive substrate for commercial fermentation, recent investigations focused on overexpression of heterologous and chimeric scaffoldins, the backbone of a functional cellulosome [57].

Sugars are taken up by phosphotransferase systems, as detailed in Chapter 8. While gluconate is fermented via a modified Entner-Doudoroff pathway [58], glucose degradation occurs by glycolysis (Embden-Meyerhof-Parnas pathway), according to the genome sequencing data [48]. Only the glyceraldehyde-3-phosphate dehydrogenase has been analyzed of the respective enzymes [59]. It has a native molecular mass of 160 kDa and consists of four identical subunits. In addition, a nonphosphorylating NADPH-dependent glyceraldehyde-3-phosphate dehydrogenase was recently purified [60]. The protein has a molecular mass of 50 kDa and also forms a homotetrameric enzyme. The final product of the glycolytic reactions, pyruvate, is converted by pyruvate:ferredoxin-oxidoreductase into acetyl-CoA, CO_2, and reduced ferredoxin. The enzyme is very oxygen-sensitive, contains iron, sulfur, and thiamine pyrophosphate, and consists of a yet-unknown number of 123 kDa subunits [61].

Several enzymes involved in acidogenesis have been purified and characterized. Lactate dehydrogenase of *C. acetobutylicum* consists of four identical subunits with a molecular mass of 36 kDa each, uses NADH as a coenzyme, and is specifically activated by fructose-1,6-diphosphate [62]. The enzyme from *C. saccharobutylicum* NRRL B643 is slightly larger (subunit size of 38 kDa), but shows the same activation feature [63]. An NADH-independent lactate dehydrogenase has been detected in *C. saccharobutylicum* (NC)P262. It is probably involved in lactate utilization and might be linked to an NADH-independent butyryl-CoA dehydrogenase [64,65]. Acetate is formed by action of phosphotransacetylase (Pta) and acetate kinase (Ack). Pta was only partially purified from *C. beijerinckii* and was found to possess a molecular mass of 56 to 57 kDa [66]. Ack of

C. acetobutylicum is a dimeric enzyme, consisting of identical subunits with a molecular mass of 43 kDa. The protein shows high specificity for the substrates ATP and acetate (formate and butyrate are not at all converted, propionate only with 2.3% of the activity obtained with acetate), and maximal activity requires Mn^{2+} as divalent cation [67]. The enzyme has also been purified from *C. saccharobutylicum* P262. It is a homodimer, consisting of subunits with a molecular mass of 42 kDa [68]. The C_4 compound butyrate is formed by successive action of thiolase (Thl), 3(or β)-hydroxybutyryl-CoA dehydrogenase (Hbd), crotonase or crotonyl-CoA hydratase (Crt or Cch), butyryl-CoA dehydrogenase (Bcd), phosphotransbutyrylase (Ptb), and butyrate kinase (Buk). ThlA of *C. acetobutylicum* is a homotetrameric enzyme, consisting of 44 kDa subunits [69]. A second thiolase is present in this organism (ThlB), but its physiological function is still unknown [70]. Hbd has only been purified from *C. beijerinckii* NRRL B593. The native molecular mass was determined to be 213 kDa, and it consists of identical subunits with a size of 31 kDa [71]. Crotonase is a homotetramer, formed by 40 kDa subunits. It converts only C_4- and C_6-enoyl CoA [72]. The strain used for purification has not been specified, but it might be *C. acetobutylicum* NRRL B528 [66,73,74]. Data on Bcd are scarce. The enzyme of *C. acetobutylicum* seems to be very oxygen-sensitive [75], and that of *C. saccharobutylicum* P262 is NADH-independent [65]. Phosphotransbutyrylase has been purified from both *C. acetobutylicum* and *C. beijerinckii* [76,77]. The sizes of the native enzymes differ considerably (264 vs. 205 kDa), while the subunit size is very similar (31 vs. 33 kDa). In both cases, the enzyme consists of identical subunits. Buk of *C. acetobutylicum* is a homodimer, consisting of two 39-kDa monomers. It is very specific for butyrate (with acetate, only 6% of the activity obtained with butyrate could be determined) [78]. A second butyrate kinase with a molecular mass of 43 kDa has been detected in *C. acetobutylicum*, but its physiological role is still unknown [79]. Finally, a NADH:rubredoxin-oxidoreductase of *C. acetobutylicum* has been characterized. It contains FAD and has a molecular mass of 41 kDa [80]. The physiological function is also still unknown. The enzyme seems to be only active during the acidogenic phase [81–83].

Most of the enzymes involved in solventogenesis have also been purified. Acetoacetyl-CoA: acetate/butyrate-coenzyme A transferase (CoA transferase or Ctf) catalyzes conversion of the previously produced acids into acyl CoA derivatives and acetoacetate. The latter compound is split into acetone and CO_2 by acetoacetate decarboxylase (Adc). In some *C. beijerinckii* strains, acetone is further reduced to 2-propanol by action of a primary/secondary alcohol dehydrogenase. Acetyl-CoA and butyryl-CoA are transformed into ethanol and butanol by a number of different aldehyde and alcohol dehydrogenases. CoA tranferase is a heterotetramer, consisting of two different subunits with 23 and 28 kDa in *C. acetobutylicum* ATCC 824 [84] and 22.7 and 23.7 kDa in strain DSM 792 [85,86]. The enzyme from *C. beijerinckii* looks very similar [66]. A conflicting report describes additional Ctf activity from a purified thiolase of *C. acetobutylicum* [87]. However, this has never been verified. Adc was extensively investigated more than 30 years ago [88,89]. The *C. acetobutylicum* enzyme is a dodecamer formed of identical subunits. The originally reported molecular masses of 340 kDa and 29 kDa [90] were later corrected to 330 kDa and 28 kDa [85,91]. The reaction mechanism of Adc is based on formation of a Schiff base between a lysine residue in the active site and acetoacetate [92]. The enzyme of *C. beijerinckii* strains NRRL B592 and B 593 is significantly smaller (200 to 230 kDa) [66]. Transposon mutagenesis and protein purification indicated that ethanol formation (which is constitutively produced during fermentation in only low amounts) is catalyzed by a specific acetaldehyde dehydrogenase that does not react with butyryl-CoA, and an NADPH-dependent alcohol dehydrogenase with a subunit size of 44 kDa [1,93,94]. The latter could be separated from the NADH-dependent activity by ultracentrifugation [95]. Another alcohol dehydrogenase (originally described as butanol dehydrogenase A or Bdh I [96]) might also be involved in ethanol formation (in addition to its serving as an electron sink in low butanol synthesis) [94,97,98]. BdhA consists of two identical subunits of 42 kDa, and the activity with acetaldehyde is half of that with butyraldehyde [96]. Butanol production in *C. acetobutylicum* is brought about by action of the bifunctional butyraldehyde/butanol dehydrogenase E (AdhE) and an additional butanol dehydrogenase (BdhB or BdhII) [86,97,99]. The physiological role of AdhE in butanol formation was shown by induction of its gene just prior to butanol synthesis [98], successful restoration of butanol production in a solvent-negative mutant upon transformation of the *adhE* (also called *aad* [99]) gene [100], and the drastic decrease of butanol formation when *adhE* was insertionally inactivated [101]. BdhB is a homodimer with a subunit size of 42 kDa and shows a 46-fold higher activity with butyraldehyde compared to acetaldehyde [96]. Separate butyraldehyde dehydrogenases have been isolated and characterized from *C. saccharobutylicum* [102] and *C. beijerinckii* [103]. They are homodimers with native molecular masses of 115 and 100 kDa and subunit sizes of 56 and 55 kDa, respectively. The primary/secondary alcohol dehydrogenase of *C. beijerinckii*, responsible for 2-propanol formation, consists of subunits with a molecular mass of 39 kDa [104]. *C. beijerinckii* NRRL B592, which does not produce 2-propanol, carries three other alcohol dehydrogenases with subunit sizes of 42 and 45 kDa and different specificities for either NADPH, or both NADH and NADPH [104]. Further details on the characteristics of all these enzymes can be found in a number of respective reviews [2,35,66,104–106].

30.5 GENES AND OPERONS

Gene cloning from *C. acetobutylicum* started in 1986 in David Woods' laboratory at the University of Cape Town, South Africa [reviewed in 107]. The first cloned genes encoding solventogenic enzymes were *ctfA/B* and *adc* [85,91,108]. All sequence data are available from the *C. acetobutylicum* genome [48]. Therefore, the following paragraphs will only deal with those publications that present additional and detailed material.

Of the genes encoding glycolytic enzymes, those for phosphofructokinase (*pfk*) and pyruvate kinase (*pyk*), as well as glyceraldehyde-3-phosphate dehydrogenase (*gap*), phosphoglycerate kinase (*pgk*), triosephosphate isomerase (*tpi*), and 2,3-phosphoglycerate-independent phosphoglycerate mutase (*pgm*(i)) have been cloned and analyzed. *pfk* forms an operon together with the adjacent gene, *pyk* [109]. The other genes mentioned are all clustered [59]. *gap*, *pgk*, and *tpi* form a common operon in this order, with transcription start points in front of *gap* and *tpi*. Transcript processing was also observed, indicating that the most abundant *gap* transcript is a stable degradation product of the full-length message. The downstream located *pgm*(i) forms, or is part of, a separate operon, controlled by its own promoter. Promoter sequences resembled those of σ^A-dependent start points, however, with unusually short spacing between the –35 and –10 regions (15 and 16 bp, respectively) [110].

A short report on cloning of a butyraldehyde dehydrogenase gene from *C. saccharobutylicum* NRRL B643 [111] later turned out as, in fact cloning, of the lactate dehydrogenase gene. The heterologously expressed enzyme was activated by fructose-1,6-diphosphate (see Section 30.4) [63]. The genes encoding phosphotransacetylase and acetate kinase form a common operon in the order *pta-ack* [112]. A transcription start point has been identified by primer extension upstream of *pta*. *thlA* is organized monocistronically and controlled by a typical σ^A-dependent promoter [70,113]. The two primer experiments yielded slightly varying start points (3-b difference). The gene of the second thiolase is located in an operon consisting of, at least, *thlR-thlB-thlC*. The gene product of *thlR* has significant similarity to transcriptional regulators of the TetR/AcrR family. ThlC, so far, only has homology to a protein of *C. beijerinckii*, whose gene is also located downstream of a thiolase gene. The *thlRBC* operon is transcribed from a vegetative sigma factor-dependent promoter, however, with an unusually long spacing (19 bp) [70]. The genes encoding 3-hydroxybutyryl-CoA dehydrogenase, crotonase, and butyryl-CoA dehydrogenase are clustered in *C. acetobutylicum*. Gene order was found to be *crt* (encoding crotonase) - *bcd* (encoding butyryl-CoA dehydrogenase) – *etfB* (encoding a gene product with homology to electron transfer flavoproteins) – *etfA* (encoding a gene product with homology to electron transfer flavoproteins) – *hbd* (encoding 3-hydroxybutyryl-CoA dehydrogenase) [114,115]. A putative promoter upstream of *crt* was deduced from primer extension expriments. Since no further transcription start points could be detected within 150 bp of the start codons of all remaining genes, it was concluded that *crt-bcd-etfB-etfA-hbd* form the *bcs* operon (butyryl-CoA synthesis) [115]. In *C. saccharobutylicum*, the *hbd* gene had also been subcloned. Downstream, in a distance of 354 bp, the *adh1* gene was found, encoding an NADPH-dependent alcohol (butanol) dehydrogenase. Both genes are part of a common transcription unit [116–118]. This is different in *C. acetobutylicum*, where a gene encoding a putative glycosylase is located in this region. *ptb* and *buk*, encoding phosphotransbutyrylase and butyrate kinase, are located adjacently on the chromosome of *C. acetobutylicum* [119]. They form a common operon and are transcribed from a start point 57 bp upstream of *ptb* [120]. The respective *ptb* promoter region has, meanwhile, become a valuable tool in clostridial molecular biology. Its strength is in the same range as the promoters of *thlA* and *adc,* and it is active throughout the acidogenic phase [121,122]. In *C. beijerinckii*, an identical arrangement of the two genes has been found, but their operon structure still has to be determined [123].

Almost all genes encoding solventogenic enzymes have been cloned and characterized. As already stated, *ctfA/B* and *adc* were the first representatives of this group [85,91,108]. The gene encoding the acetoacetate decarboxylase subunit is organized in a monocistronic operon [85,124]. The *adc* promoter shows a high similarity to the consensus motif of σ^A-dependent control regions (only 2 mismatches in the –35 region and a spacing of 18 bp [125]) and is one of the strongest among the ones controlling operons encoding solventogenic enzymes (1.3-fold higher than *bdhB* and 84-fold higher than *sol* [122]). A rho-independent transcription terminator consisting of a 28-bp stem-loop is located 6 bp downstream from the first of two consecutive UAA stop codons [85,124]. This termination structure is flanked on either side by a stretch of A or T residues and is, thus, functioning in both directions. On its other side, the *ctfB* gene is found, with convergent direction of transcription as compared to *adc*, representing the end of the *sol* operon. This transcription unit comprises genes encoding a butyraldehyde/butanol dehydrogenase and the CoA transferase in the order *orfL-adhE-ctfA-ctfB* [85,124]. *orfL* encodes a small peptide of still unknown function, which could be heterologously expressed in *E. coli* [126]. Originally, primer extension experiments led to the deduction of two promoters upstream of the *sol* operon. The distal one was designated P_1 or S_2, the proximal one P_2 or S_1. Stronger signals were observed from P_2 [86,99]. However, reporter gene studies unequivocally revealed that only P_1 represents a promoter, while P_2 obviously is an mRNA processing site [127]. *C. acetobutylicum* is the only solventogenic *Clostridium* known to carry an *adhE* gene. It is not present in *C. beijerinckii*, *C. saccharobutylicum*, and *C. saccharoperbutylacetonicum*. However, homologues have been detected in *Escherichia coli* [128], *Salmonella typhimurium* [129], *Lactococcus lactis* [130], *Streptococcus bovis* [131] as well as in the eukaryotes *Entamoeba histolytica* [132] and *Giardia lamblia* [133]. The bi- or even polyfunctional nature of the gene product (the *E. coli* enzyme has acetaldehyde and ethanol dehydrogenase, as well as pyruvate: formate-lyase deactivase activity [128,134]) led to speculations on convergent evolution from separate genes. Support for this assumption comes from the analysis of genes of succinate-semialdehyde dehydrogenase (*sucD*) and 4-hydroxybutyrate dehydrogenase (*4hbD*) from *C. kluyveri*, which are tandemly arranged (separated by just 36 nucleotides) and exhibit high homology to the 5′ and 3′ parts of the *adhE* gene of *C. acetobutylicum*, respectively [135]. Also, the C-terminus of AdhE proteins shows homology to group III alcohol dehydrogenases, which mostly are iron-activated, whereas the N-terminus is similar to the family of aldehyde: NADH-oxidoreductases [136,137]. The *sol* operon of *C. acetobutylicum* differs considerably from those of the other solventogenic clostridia (*C. beijerinckii, C. saccharobutylicum*, and *C. saccharoperbutylacetonicum*). The latter carry an aldehyde dehydrogenase gene (*ald*) instead of *adhE* and also include the *adc* gene [138,206].

Recently, a second *adhE* gene (*adhE2*) has been detected in *C. acetobutylicum*, representing the first case of an organism carrying two such genes [139]. Identification of this gene resulted from close inspection of the genome sequence. The gene product exhibits 66% identity to AdhE of *C. acetobutylicum* and was shown to express NADH-dependent butyraldehyde and butanol dehydrogenase activity. *adhE2* is organized in a monocistronic operon, and primer extension experiments revealed 2 signals upstream of the gene. Of the deduced promoters, only the distal one (S_2) showed convincing homology to σ^A-dependent control regions (perfect match), whereas the proximal one (S_1) showed 2 mismatches in each hexamer motif, as well as an unusually long spacing (20 bp) [139]. Thus, it is tempting to speculate that the regulatory mechanism is similar to the *sol* operon, with a single promoter (which would be S_2) and a mRNA processing site (which would be S_1). Further inspection of the sequence revealed a putative binding motif for the essential sporulation transcription factor Spo0A (5′-TGGCAAT-3′) just 35 bp upstream of S_2, supporting its assumed role as the proper promoter [140]. *adhE2* is not induced under either acidogenic or solventogenic conditions. Expression was only observed when special substrate combinations were used for culturing *C. acetobutylicum* [139]. A mixture of glucose and glycerol leads to a high $NADH/NAD^+$ ratio in the cells, which results in an "alcohologenic" fermentation with butanol and increased amounts of ethanol as products, but no acetone [105,141]. Such a situation can also be mimicked artificially by addition of redox dyes such as methyl viologen [142,143]. Indeed, RNA analyses performed after respective experiments provided the first indication for the presence of a second *adhE* gene [94,144].

The *adc*, *sol*, and *adhE2* operons of *C. acetobutylicum* are all located on the 192 kbp megaplasmid pSOL1 [48,145,146]. *adhE2* is separated from the *adc-sol* locus by approximately 47 kbp. In addition, two other genes encoding solventogenic enzymes have been found on the 3.94 Mbp chromosome. These two genes (*bdhA* and *bdhB*) encode butanol dehydrogenases and are organized in contiguous, monocistronic operons. Primer extension experiments allowed to deduce σ^A-dependent promoters, which, however, carry 4 and 3 mismatches, respectively. Transcription of both operons is stopped by typical rho-independent terminators [97,147]. Reporter gene studies revealed that the *bdhB* promoter is 17-fold stronger than that of *bdhA* and, thus, is in the range of the *adc* promoter (77% of the maximal strength observed from the *adc* promoter) [122].

A compilation of *C. acetobutylicum*'s known transcription units has recently been published [148].

30.6 REGULATION

First insight into regulation was obtained by following enzyme activities over the course of fermentation by *C. acetobutylicum*. Acetoacetate decarboxylase is induced at the transition from the exponential to the stationary growth phase [149]. The same is true for butyraldehyde and butanol dehydrogenase activity [95,150]. CoA transferase showed a steady increase over the whole growth period [150]. Activity of 3-hydroxybutyryl-CoA dehydrogenase and crotonase decreases rapidly at the end of the exponential growth phase, that of thiolase is less dramatic and stays constant at a lower level [75]. Phosphotransacetylase and acetate kinase are also active only during acidogenesis, similar to phosphotransbutyrylase and butyrate kinase. In contrast to the other enzymes, butyrate kinase still showed a significant activity at the end of culturing [75,150,151]. It is tempting to speculate that this activity might be attributed to the second butyrate kinase, mentioned before to be present in *C. acetobutylicum*, especially considering the genetic organization of *ptb* and *buk* genes (see above). The use of transcription and translation inhibitors (rifampicin and chloramphenicol) indicated that mRNA and protein synthesis were a requirement for increased enzyme activities [102,152,153], although conflicting conclusions have also been reported [154].

A more detailed analysis became possible by mRNA studies. The *adc* transcript can already be detected early during exponential growth but reaches maximal amounts in the stationary phase, and then abundance decreases [98,125]. This transient expression is not matched by enzyme activity, which is kept at a high level in stationary growth phase [125,149]. Proteome analysis, comparing acidogenic and solventogenic *C. acetobutylicum* cells, revealed that protein synthesis of Adc parallels the increase of transcript. It was surprising to identify by N-terminal sequencing two spots as acetoacetate decarboxylase [155], since only one such gene is present in the genome [48]. Both spots showed the same induction pattern [155]. Obviously, Adc is modified co- or post-translationally. Since it is the only solventogenic enzyme known to show such a feature and since it is the only solventogenic enzyme known to stay active even under aerobic conditions [125,149], the modification, the nature of which is still unknown, might be the means to stabilize catalytic activity. Transcriptional regulation involves the cardinal sporulation regulator Spo0A. This protein, in its phosphorylated form, is essential for initiation of endospore formation (see Chapter 29). Binding motifs for Spo0A~P (0A boxes) have been identified upstream of the *adc* promoter of *C. acetobutylicum.* Two 0A boxes (5′-TGCAGAA-3′ and 5′-TTTCGAA-3′) are located in a distance of 53 and 34 bp from the –35 region, while the third (in reverse orientation, 5′-TTCGAAA-3′ corresponding to TTTCGAA) is separated by 33 bp from the promoter's first hexamer motif [156]. *In vivo* relevance for regulation involving these motifs was verified by targeted mutation and gel retardation studies with purified and artificially phosphorylated Spo0A from *C. beijerinckii* [156] and *C. acetobutylicum* [157]. The involvement of the master sporulation regulator Spo0A in regulation of solventogenesis represents the controlling link between these two complex metabolic networks. However, a targeted mutation of the two upstream 0A boxes to 5′-TCGAGAG-3′ and 5′-TCTAGAG-3′, repectively [156], and even the complete removal of all three 0A boxes [158], resulted only in a reduced level of induction, while the underlying regulatory pattern still could be observed. This clearly points to the involvement of at least one additional transcription factor in *adc* regulation. The presence of such a second regulator would also nicely explain the phenomena that (i) solventogenesis and sporulation are not tightly connected [47]; (ii) under special growth conditions, the ratio of acetone and butanol formation can vary dramatically [159]; and (iii) in the alcohologenic fermentation, no acetone is produced at all without obvious effect on sporulation [105]. The external signal leading to *adc* induction is still not known. Several extra- and intracellular parameters such as concentration of undissociated acids [152,160,161], ATP/ADP ratio and NAD(P)H level [162], and levels of butyryl-phosphate and butyryl-CoA [39,163] have been proposed to serve this role, but unequivocal evidence has not yet been obtained. On the other hand, the degree of DNA supercoiling is dependent on all these factors. Indeed, *adc* transcription level is increased by 31% upon addition of novobiocin to exponentially growing cells (resulting in less negative supercoiling = relaxation of DNA) [164]. Thus, external (pH, acid and substrate concentration, even temperature) and internal conditions (as mentioned above) will change the topology of DNA, which, thus, could serve as a transcriptional sensor by allowing or restricting regulatory proteins to bind [165]. The transient expression of *adc* then might be explained by shutting down transcription from the *adc* promoter by activating the adjacent *sigY* promoter (separated by approximately 200 bp, but transcribed in the opposite direction) [166]. *sigY* encodes an alternative sigma factor of the ECF type with a still-unknown physiological function [167]. Such a transcription termination of *adc* would be in accordance with the "twin transcriptional-loop model" [166,168]. It implies late activation of the *sigY* promoter, and, indeed, a protein (still uncharacterized) was found to bind to this region and was detectable only late in the solventogenic phase [169]. The organization of the *adc* gene as a monocistronic operon is also worth comment. The seemingly illogical separation of *ctfA/B* and *adc* genes, the gene products of which are all needed for acetone biosynthesis, is based on our habit of looking at generalized metabolic pathways [94]. Under physiological conditions, however, only Ctf and AdhE are required for transformation of acetate and butyrate into butanol. Decarboxylation of acetone is only necessary if no sufficient reducing equivalents are available for butyryl-CoA and butyryl-phosphate reduction. Then, Adc must "pull" the thermodynamically unfavorable CoA transferase reaction (acetoacetyl-CoA + butyrate/acetate $\rightarrow$ acetoacetate + butyryl-/acetyl-CoA) by removal of acetoacetate [94]. This is not an essential requirement, as demonstrated both in solventogenic [159] and alcohologenic [105] fermentations with decreased acetone levels.

Northern blots and reporter gene experiments revealed that transcription of the *sol* operon is induced several hours before massive solvent production starts [98,122]. As already stated, *sol* is transcribed from only one promoter, P_1 [127]. Upstream of this control region, three imperfect repeats were detected [R1 (distal) - R2 - R3 (proximal)], of which R3 partially overlaps with a 0A box (only one reversed Spo0A binding motif was found 62 nucleotides upstream of P_1 [156]). Reporter gene studies revealed that R1 and R3 are essential for induction of the *sol* operon. While the effect of R3 might be due to its overlapping with the 0A box (a band shift of the *sol* operon promoter region including the 0A box could be demonstrated by using purified and *in vitro* phosphorylated Spo0A from *C. acetobutylicum* [157]), the comparatively long distance to R1 points to the involvement of another transcription factor [127]. Such a regulator might act in concert with Spo0A~P. Support for this hypothesis comes from a report describing that inactivation of the *spo0A* gene did reduce but did not completely abolish expression of the *sol* operon (and also of the *adc* operon) [170]. This putative regulator will certainly be different from the one obviously required for *adc* regulation, as no R1-R3 motifs are detectable upstream of *adc*, and regulation of butanol and acetone formation differs in some respects as evidenced by varying product concentrations (e.g., [159]). Since disruption of R1 resulted in loss of induction, the putative transcription factor possibly acts as an activator. A recent report, however, proposed a repressor to be encoded by the upstream gene (*orf5* [86] or, as designated in that publication, *solR*) [171]. This conclusion was based on *in silico* detection of a helix-turn-helix (HTH) motif catalyzing protein-DNA interaction, loss of solvent production upon overexpression of *orf5* (*solR*), and increased solvent production upon insertional inactivation of *orf5* (*solR*). A detailed investigation of the *orf5* gene product, however, revealed that the putative signal sequence could be experimentally verified by detecting the product at the extracellular side of the cytoplasmic membrane (which is in contrast to a transcriptional repressor) [172]. In addition, DNA binding could experimentally not be verified, and, on close inspection, the HTH motif turned out to be rather a tetratrico peptide repeat [173], assumed to be responsible for protein-protein interactions. A BLAST search revealed homology to O-linked UDP-N-acetylglucosamine-peptide: N-acetylglucosaminyl transferases (GlcNAc transferases) of archaea and eukaryotes. Indeed, deglycosylation activity of Orf5 could be experimentally verified [172]. Finally, overexpression of *orf5* did not lead to decreased solvent production as reported by Nair et al. [171], but to values similar to those obtained with the wild type [127]. Close inspection of the subcloning procedure used by Nair et al. [171] revealed that the regulatory part of the P_1 promoter of the *sol* operon (including both R1 and R3) had been erroneously subcloned with *orf5*. This part was responsible for the solvent-negative phenotype as verified experimentally by overexpression of this regulatory region [127]. As already mentioned, P_2 presumably represents an mRNA processing site. This conclusion was based on targeted mutations of the region between P_1 and the ribosome binding site, resulting in disruption of stem-loop structures at the start of the *adhE* gene and a thus altered secondary structure [127]. It is interesting to note significant similarities of *adhE* regulation in *C. acetobutylicum* and *E. coli*. In the latter, also, two transcription start points have been found, the second one representing a processing site by an RNase [174]. RNase III is obviously required for efficient translation initiation [174], while RNase G degrades the *adhE* transcript [175,176]. This enzyme probably recognizes the secondary structure rather than the primary sequence [177], which mimics the situation in *C. acetobutylicum*, where secondary structure of the mRNA is important for recognition of a processing RNase [127]. The region around the processing site P_2 might also have a negative role in *adhE* expression, as suggested from deletion studies [178].

The hypothesis that *adhE* genes stem from evolutionary gene fusions (see above) is further supported by the finding that the respective transcript of *C. acetobutylicum* yields two different translation products, the mature enzyme and the C-terminal alcohol dehydrogenase domain. This is probably due to the presence of a ribosome binding site sequence within the N-terminal butyraldehyde dehydrogenase domain of the structural gene in the correct distance to an ATG codon, representing the translational start of the C-terminal butanol dehydrogenase domain [127].

As already mentioned for *adc* expression, the signal for induction of the *sol* operon expression has not yet been identified. Cultural prerequisites are a surplus of carbon source, a pH below 4.3, limiting phosphate or sulfate concentrations, and high concentrations of acetate and butyrate [for reviews, see 2,35,166]. Of course, larger amounts of external acids can compensate for a higher pH (pointing to the influence of undissociated acids [160]). In general, the effects are, thus, interchangeable and additive (e.g., substrate and salt concentrations). Increasing temperature also leads to enhanced butanol production [179]. As mentioned before, all these parameters influence DNA topology directly. Relaxation of DNA by inhibition of DNA gyrase caused a dramatic increase in *sol* operon expression (300%) of exponentially growing *C. acetobutylicum* cells [164]. In addition, DNA isolated from *C. acetobutylicum* at various stages of growth was shown to become relaxed at the transition from acidogenesis to solventogenesis [180]. Thus, it seems that DNA topology represents a transcriptional sensor and, at least, participates in regulating the onset of solventogenesis [181].

Little is known about regulation of *adhE2*. A putative 0A box (5′-TGGCAAT-3′) is located 35 bp upstream of the putative proper promoter S_2 [140]. While this motif might represent the link to the sporulation network, no sequences resembling R1 or R3 of *adhE* are detectable. Thus, the putative transcription regulator involved in *sol* operon regulation probably does not play a role in *adhE2* expression. However, a putative FNR binding site was identified, located just 19 bp upstream of the predicted ribosome binding site of *adhE2* [139]. It might represent the motif for DNA-protein interaction of a regulator, responsible for redox state-dependent expression.

Similarly low is our knowledge on expression of the chromosomally located *bdhA* and *bdhB*. The gene product of the latter is essential for massive butanol production, as the respective transcript reaches the maximum level when *sol* operon expression already decreases [98]. Thus, the physiological role of BdhB seems to take over the task of production from AdhE (or is indeed responsible for massive production, while AdhE manages initiation of solventogenesis) [94]. It is still not known whether a separate butyraldehyde dehydrogenase is expressed after *sol* operon repression, or whether the N-terminal domain of AdhE remains active. The *bdhB* promoter is preceded by a 0A box (5′-TGTAGAA-3′) [156], in accordance with its expression late in exponential growth and its physiological function in butanol production. Reporter gene studies confirmed the Northern blots [122]. *bdhA*, on the other hand, seems to encode an alcohol (possibly both butanol and ethanol) dehydrogenase with the major function of removing reducing equivalents (electron sink). This gene, at least under certain conditions, is therefore constitutively expressed [98]. Contrary is the finding that upstream of the *bdhA* promoter two 0A boxes also were detected (5′-TGCAGAA-3′ and 5′-CGACGAA-3′) [156]. This raises the question of the physiological significance of *in silico* detected 0A boxes. Indeed, these must always be verified by experimental data, as 0A boxes upstream and within the *sigH* gene of *C. acetobutylicum* did not yield band shifts with Spo0A~P from *C. acetobutylicum* [157]. However, it must be stressed that interaction of Spo0A or Spo0A~P with regulatory regions can also lead to repression. This seems to be the case with the *ptb-buk* operon in *C. beijerinckii*, as this is transcriptionally shut down after the exponential growth phase and two 0A boxes have been detected upstream of the promoter [156]. In *C. difficile*, one 0A box also is associated with the *ptb* promoter, but no such motif has been detected upstream of *ptb-buk* in *C. acetobutylicum* [156].

A number of other operons are induced at the onset of solventogenesis. Proteome analysis led to identification of the *serCAXS* operon, encoding enzymes required for serine biosynthesis and incorporation into proteins (namely serine aminotransferase, 3-phosphoglycerate dehydrogenase, possibly 3-phosphoserine phosphatase, and seryl-tRNA synthetase). RNA analyses confirmed the operon structure, controlled by a σ^A-dependent promoter without obvious 0A boxes, and the transcriptional induction concomitantly with the *sol* operon [155]. Since no proteins with a high serine content are known to be involved in solventogenesis (or sporulation), the physiological basis for induction still remains mysterious. Another unexpected finding was that *tpi* transcript (see above) levels are enhanced by approximately 15% at the onset of solventogenesis, while the *gap* promoter is constitutively active [110]. Rather expected, on the other hand, is the induction of stress protein genes in parallel to solvent formation. Proteome and RNA analyses allowed to identify heat shock proteins such as DnaK, GroEL, and Hsp18, of which the latter is also modified co- or post-translationally [98,155,182–185], and a new putative stress protein, PdxY, possibly required for quenching reactive oxygen species [155]. An interesting case represents *thlA*. This operon is transcribed at high level during acidogenesis, then becomes repressed and induced again at the onset of solventogenesis [70]. No other operon of *C. acetobutylicum* is known with such a regulatory pattern. The *thlA* promoter, similar to the one controlling *ptb-buk*, has, meanwhile, become an important tool in clostridial recombinant DNA technology. The *thlRBC* operon also shows a transient, but low (approximately 3-fold) induction at initiation of solvent formation [70]. In 2003, another approach of *C. acetobutylicum* transcriptional analysis was launched. The group of Terry Papoutsakis developed the first DNA-microarray chips (compare Chapters 4 and 37). Since the results are detailed in those chapters, only few remarks will be made here. The cDNA arrays included 1019 genes from *C. acetobutylicum* (representing approximately 25% of the whole genome), among them 169 of 178 pSOL1 open reading frames (ORFS), 123 DNA replication and repair genes (90% of total such genes as identified by the genome annotation [48]), 97 cell division- and sporulation-related genes (92%), 85 carbohydrate or primary metabolism genes (31%), 67 energy conservation genes (52%), 63 membrane and cell envelope genes (36%), 48 lipid metabolism genes (80%), 42 motility and chemotaxis genes (39%), and all previously identified stress response genes (http://www.chem-eng.northwestern.edu/Faculty/papou.html). Examined in detail were solvent production [186], sporulation [187], stress response [188], *spo0A* overexpression [189], and butanol tolerance [190]. This led to a massive load of new data, not all of which are yet understood. As an example might serve overexpression of *adhE*, resulting in differential expression of 273 genes out of the 1019 tested. This surprising finding might be caused by titration of transcription factors or altered product concentrations that initiate a stress response [186]. It underlines that a real understanding of clostridial metabolism will still take many years.

30.7 METABOLIC ENGINEERING

This topic is also covered in detail in another chapter of this book (see Chapter 37). The major aim of metabolic engineering has always been to construct strains with improved solvent productivity. Almost all of these experiments have been performed with *C. acetobutylicum*. A summary of the achievements is given in Table 30.2.

The first attempts, about 10 years ago, were straightforward and focused on overexpression of *adhE* and inactivation of acid formation genes [99,100,191]. However, they did not lead to enhanced solvent production. This might have been due to testing conditions, as inactivation of the butyrate kinase gene (*buk*) and culturing at pH 5 or below led to solvent concentrations never observed before, namely 225 m*M* butanol and 76 m*M* acetone. Additional overexpression of *adhE* did not result in a further enhanced yield [193]. Similarly high butanol concentrations were obtained by inactivation of the deglycosylase/glycosylase-encoding *orf5* [192] (highest concentration of butanol reported to date) and overexpression of the chaperone-encoding *groESL* genes [188]. The latter approach, to increase solvent production by increased tolerance, was also the rationale for overexpression of the cyclopropane fatty acid synthase gene (*cfa*). However, although lipid composition was successfully altered and butanol resistance of the recombinant strain was higher compared to the wild type, butanol production was significantly lower [200]. Ethanol production could also be selectively increased to 190 m*M* by simultaneous inhibition of *ctfB* expression by asRNA and overexpression of *adhE* [186]. Even novel operons have meanwhile been constructed and expressed in either *E. coli* or *C. acetobutylicum*. A synthetic acetone production transcription unit consisted of *adc-ctfA-ctfB* (first generation) [196,198] and additional integration of *thlA* (second generation) [197]. In all these cases, however, it should be noted that the presence of plasmids in *C. acetobutylicum* cells causes a complex host response [188,195], and integration of the constructs into the genome would be desirable for industrial applications.

TABLE 30.2
Metabolic Engineering Using *Clostridium acetobutylicum* or Its Genes

Genes Inactivated or Repressed	Genes Overexpressed	Acetate	Butyrate	Acetone [m*M*]	Butanol	Ethanol	PHA*	Ref.
				Host *C. acetobutylicum*				
(Wild type)	(Wild type)	112	108	79	131	11	n.d.	[191]
		71	76	85	158	16	n.d.	[192]
adhE, ctfA/B	*adhE*	101	99	—	84	8	n.d.	[100]
pta	—	87	159	72	133	13	n.d.	[191]
buk	—	149	37	39	146	16	n.d.	[191]
		111	18	76	225	57	n.d.	[193]
buk	*adhE*	113	18	66	226	98	n.d.	[193]
orf5	—	84	74	97	197	29	n.d.	[192]
orf5	*adhE*	87	97	141	238	47	n.d.	[192]
ctfB	*adhE*	125	52	26	130	190	n.d.	[186]
—	*groESL*	80	70	148	231	21	n.d.	[188]
—	*adhE*			Same as wild type			n.d.	[99]
—	*buk, ptb*	60	30	80 (Plasmid effect reported)	140	10	n.d.	[194] [195]
—	*adc, ctfA, ctfB*	10	0.5	149	177	31	n.d.	[196]
				Host *E. coli*				
—	*thlA, adc, ctfA, ctfB*	23	—	93	—	3	n.d.	[197]
—	*adc, ctfA, ctfB* (acetoacetate added)	n.d.	—	8	—	n.d.	n.d.	[198]
—	*buk, ptb, phaE, phaC*						68.7	[199]

[a] PHA, poly(hydroxyalkanoic acid); amount given in % of cell dry weight.
n.d. = not determined.

Other approaches focused on converting *C. acetobutylicum* into a 2,3-butanediol- or isoamyl acetate producer. Since acetoin is a fermentation product by *C. acetobutylicum* (see Section 30.3.), it was attempted to reduce it in a recombinant strain to 2,3-butanediol, which is an effective antifreeze agent and can be used for production of butadiene and liquid fuel. The *Klebsiella pneumoniae* acetoin reductase gene (*ard*) was successfully subcloned and expressed in *C. acetobutylicum*. A specific activity of 0.48 U/mg protein could be observed in the recombinant [201]. However, this was not sufficient for 2,3-butanediol production with the low amount of acetoin produced during growth (3.6 m*M*). Only external addition of acetoin led to conversion of approximately 6.5% of that substrate to 2,3-butanediol [201]. Similarly, expression of a yeast alcohol acetyltransferase gene (*ATF2*) in *C. acetobutylicum* was hoped to result in production of the flavoring compound isoamyl acetate upon isoamyl alcohol addition. Unfortunately, the *C. acetobutylicum* construct produced a much lower amount of isoamyl acetate than respective *E. coli* recombinants. This was due to isoamyl acetate degradation during the later stage of growth (after 18 h) [202].

C. acetobutylicum genes (*ptb-buk*) have also been used together with the poly(hydroxyalkanoic acid) synthase-encoding genes *phaE* and *phaC* from *Thiocapsa pfennigii* to synthesize different poly(hydroxyalkanoic acids) from added hydroxyfatty acids in *E. coli* [199]. This was also possible *in vitro* by using the three purified enzymes [203].

Finally, a totally different approach to biotechnological butanol production should be mentioned. A research group from the Fraunhofer-Institute in Stuttgart, Germany, could show that several aerobic bacteria growing with butane as sole carbon and energy source oxidize this compound to butanol (95% 1-butanol, 5% 2-butanol), obviously by means of a monooxygenase [204; http://www.igb.fraunhofer.de/WWW/GF/Biokatalyse/dt/GFBK_233_Butanol.dt.html]. Since butane is a cheap substrate and can easily be obtained from natural gas, a recombinant strain overexpressing the monooxygenase and being solvent-tolerant could be an interesting alternative for biotechnological butanol production [205].

ACKNOWLEDGMENTS

Work in my laboratory was supported by grants from the Deutsche Forschungsgemeinschaft within the priority program Molekulare Analyse von Regulationsnetzwerken in Bakterien, the BMBF GenoMik project (Competence Network Göttingen), and the European Community (contract No. QLK3-CT-2001-01737).

REFERENCES

1. Sauer, U., Fischer, R., and Dürre, P., Solvent formation and its regulation in strictly anaerobic bacteria, *Curr. Top. Mol. Genet.,* 1, 337, 1993.
2. Dürre, P. and Bahl, H., Microbial production of acetone/butanol/isopropanol, in *Biotechnology,* Vol. 6, 2nd Edition, Roehr, M., Ed., VCH Verlagsgesellschaft mbH, Weinheim, Germany, 1996, 229.
3. Tamburini, E. et al., *Clostridium felsineum* and *Clostridium acetobutylicum* are two distinct species that are phylogenetically closely related, *Int. J. Syst. Evol. Microbiol.,* 51, 963, 2001.
4. Keis, S. et al., Taxonomy and phylogeny of industrial solvent-producing clostridia, *Int. J. Syst. Bacteriol.,* 45, 693, 1995.
5. Johnson, J.L. et al., Cultures of *"Clostridium acetobutylicum"* from various collections comprise *Clostridium acetobutylicum, Clostridium beijerinckii,* and two other distinct types based on DNA-DNA reassociation, *Int. J. Syst. Bacteriol.,* 45, 420, 1997.
6. Jones, D.T., Applied acetone-butanol fermentation, in *Clostridia. Biotechnological and Medical Applications,* Bahl, H. and Dürre, P., Eds., Wiley-VCH Verlag GmbH, Weinheim, Germany, 2001, 125.
7. Keis, S., Shaheen, R., and Jones, D.T., Emended descriptions of *Clostridium acetobutylicum* and *Clostridium beijerinckii,* and descriptions of *Clostridium saccharoperbutylacetonicum* sp. nov. and *Clostridium saccharobutylicum* sp. nov., *Int. J. Syst. Evol. Microbiol.,* 51, 2095, 2001.
8. Pasteur, L., Quelques résultats nouveaux relatifs aux fermentations acétique et butyrique, *Bull. Soc. Chim. Paris,* May 1862, 52, 1862.
9. Zoutberg, G.R. et al., Solvent production by an aggregate-forming variant of *Clostridium butyricum, Appl. Microbiol. Biotechnol.,* 32, 22, 1989.
10. Pasteur, L., Animacules infusoires vivant sans gaz oxygène libre et déterminant des fermentations, *C.R. Hebd. Séances Acad. Sci.,* 52, 344, 1861.
11. Pasteur, L., Expériences et vues nouvelles sur la nature des fermentations, *C.R. Hebd. Séances Acad. Sci.,* 52, 1260, 1861.
12. Pasteur, L., Recherches sur la putréfaction, *C.R. Hebd. Séances Acad. Sci.,* 56, 1189, 1863.
13. Fitz, A., Ueber die Gährung des Glycerins, *Ber. Dtsch. Chem. Ges.,* 9, 1348, 1876.
14. Fitz, A., Ueber Schizomyceten-Gährungen II [Glycerin, Mannit, Stärke, Dextrin], *Ber. Dtsch. Chem. Ges.,* 10, 276, 1877.
15. Fitz, A., Ueber Schizomyceten-Gährungen III, *Ber. Dtsch. Chem. Ges.,* 11, 42, 1878.
16. Fitz, A., Ueber Schizomyceten-Gährungen. VII. Mittheilung, *Ber. Dtsch. Chem. Ges.,* 15, 867, 1882.
17. Prazmowski, A., Untersuchungen über die Entwickelungsgeschichte und Fermentwirkung einiger Bacterien-Arten, Ph.D. Thesis, University of Leipzig, Germany, 1880.

18. Trécul, A., Production de plantules amylifères dans les cellules végétales pendant la putréfaction. Chlorophylle cristallisée, *C.R. Hebd. Séances Acad. Sci.,* 61, 432, 1865.
19. Trécul, A., Matière amylacée et cryptogames amylifères dans les vaisseaux du latex de plusieurs apocynées (Sur les *Amylobacter*), *Ann. Sci. Nat. Bot. 5ième Série,* 7, 204, 1867.
20. Dürre, P., From Pandora's box to cornucopia: clostridia — a historical perspective, in *Clostridia. Biotechnology and Medical Applications*, Bahl, H. and Dürre, P., Eds., Wiley-VCH Verlag GmbH, Weinheim, Germany, 2001, 1.
21. McCoy, E. et al., A cultural study of the acetone butyl alcohol organism, *J. Infect. Dis.,* 39, 457, 1926.
22. Schardinger, F., Azetongärung, *Wiener Klin. Wochenschr.,* 17/8, 207, 1904.
23. Schardinger, F., *Bacillus macerans*, ein Aceton bildender Rottebacillus, *Zbl. Bakteriol. Parasitenkd. Infektionskrankh., II. Abt.,* 4, 772, 1905.
24. Pringsheim, H.H., Ueber den Ursprung des Fuselöls und eine Alkohole bildende Bakterienform, *Cbl. Bakteriol. Parasitenkd. Infektionskrankh., II. Abt.,* 15, 300, 1906.
25. Pringsheim, H., Ueber ein Stickstoff assimilierendes *Clostridium*, *Cbl. Bakteriol. Parasitenkd. Infektionskrankh., II. Abt.,* 15, 795, 1906.
26. Osburn, O.L. and Werkman, C.H., Utilization of agricultural wastes. II. Influence of nitrogenous substrate on production of butyl and isopropyl alcohols by *Clostridium butylicum*, *Ind. Eng. Chem.,* 27, 416, 1935.
27. George, H.A. et al., Acetone, isopropanol, and butanol production by *Clostridium beijerinckii* (syn. *Clostridium butylicum*) and *Clostridium aurantibutyricum*, *Appl. Environ. Microbiol.,* 45, 1160, 1983.
28. Fernbach, A. and Strange, E.H., Improvements in the manufacture of products of fermentation, *Br. Patent,* 15, 203, 1911.
29. Fernbach, A. and Strange, E.H., Improvements in the manufacture of higher alcohols, *Br. Patent,* 15, 204, 1911.
30. Fernbach, A. and Strange, E.H., Improvements connected with fermentation processes for the production of acetone, and higher alcohols, from starch, sugars, and other carbohydrate materials, *Br. Patent,* 21, 073, 1912.
31. Gabriel, C.L., Butanol fermentation process, *Ind. Eng. Chem.,* 28, 1063, 1928.
32. Gabriel, C.L. and Crawford, F.M., Development of the butyl-acetonic fermentation industry, *Ind. Eng. Chem.,* 30, 1163, 1930.
33. Weizmann, C., Improvements in the bacterial fermentation of carbohydrates and in bacterial cultures for the same, *Br. Patent,* 4845, 1915.
34. Jones, D.T. and Woods, D.R., Acetone-butanol fermentation revisited, *Microbiol. Rev.,* 50, 484, 1986.
35. Dürre, P., New insights and novel developments in clostridial acetone/butanol/isopropanol fermentation, *Appl. Microbiol. Biotechnol.,* 49, 639, 1998.
36. Rose, A.H., *Industrial Microbiology,* Butterworths, London, 1961, 160.
37. Santangelo, J.D. and Dürre, P., Microbial production of acetone and butanol: Can history be repeated? *Chimica oggi/Chemistry today,* 14, 29, 1996.
38. Dürre, P., Bahl, H., and Gottschalk, G., Membrane processes and product formation in anaerobes, in *Handbook on Anaerobic Fermentations*, Erickson, L.E. and Fung, D.Y.-C., Eds., Marcel Dekker, Inc., New York, 1988, 187.
39. Gottwald, M. and Gottschalk, G., The internal pH of *Clostridium acetobutylicum* and its effect on the shift from acid to solvent formation, *Arch. Microbiol.,* 143, 42, 1985.
40. Huang, L., Gibbins, L.N., and Forsberg, C.W., Transmembrane pH gradient and membrane potential in *Clostridium acetobutylicum* during growth under acetogenic and solventogenic conditions, *Appl. Environ. Microbiol.,* 50, 1043, 1985.
41. Moreira, A.R., Ulmer, D.C., and Linden, J.C., Butanol toxicity in the butylic fermentation, *Biotechnol. Bioeng. Symp.,* 11, 567, 1981.
42. Moreira, A.R., Acetone-butanol fermentation, in *Organic Chemicals from Biomass*, Wise, D.L., Ed., Benjamin Cummings Publ. Co. Inc., Menlo Park, CA, 1983, 385.
43. Bowles, L.K. and Ellefson, W.L., Effects of butanol on *Clostridium acetobutylicum*, *Appl. Environ. Microbiol.,* 50, 1165, 1985.
44. Hutkins, R.W. and Kashket, E.R., Phosphotransferase activity in *Clostridium acetobutylicum* from acidogenic to solventogenic phases of growth, *Appl. Environ. Microbiol.,* 51, 1121, 1986.

45. Chin, H.-L., Chen, Z.-S., and Chou, C.P., Fedbatch operation using *Clostridium acetobutylicum* suspension culture as biocatalyst for enhancing hydrogen production, *Biotechnol. Prog.*, 19, 383, 2003.
46. Ezeji, T.C., Qureshi, N., and Blaschek, H.P., Acetone butanol ethanol (ABE) production from concentrated substrate: reduction in substrate inhibition by fed-batch technique and product inhibition by gas stripping, *Appl. Microbiol. Biotechnol.*, 63, 653, 2004.
47. Meinecke, B., Bahl, H., and Gottschalk, G., Selection of an asporogenous strain of *Clostridium acetobutylicum* in continuous culture under phosphate limitation, *Appl. Environ. Microbiol.*, 48, 1064, 1984.
48. Nölling, J. et al., Genome sequence and comparative analysis of the solvent-producing bacterium *Clostridium acetobutylicum*, *J. Bacteriol.*, 183, 4823, 2001.
49. Paquet, V. et al., Purification and characterization of the extracellular α-amylase from *Clostridium acetobutylicum* ATCC 824, *Appl. Environ. Microbiol.*, 57, 212, 1991.
50. Annous, B.A. and Blaschek, H.P., Isolation and characterization of α-amylase derived from starch-grown *Clostridium acetobutylicum* ATCC 824, *J. Ind. Microbiol.*, 13, 10, 1994.
51. Sabathé, F., Bélaich, A., and Soucaille, P., Characterization of the cellulolytic complex (cellulosome) of *Clostridium acetobutylicum*, *FEMS Microbiol. Lett.*, 217, 15, 2002.
52. Verhasselt, P. et al., Cloning and expression of a *Clostridium acetobutylicum* α-amylase gene in *Escherichia coli*, *FEMS Microbiol. Lett.*, 59, 135, 1989.
53. Verhasselt, P. and Vanderleyden, J., Cloning and expression of *Clostridium acetobutylicum* genes involved in carbohydrate utilization, in *Genetics and Molecular Biology of Anaerobic Bacteria*, Sebald, M., Ed., Springer-Verlag New York, Inc., New York, 1993, 301.
54. Lee, S.F., Forsberg, C.W., and Gibbins, L.N., Cellulolytic activity of *Clostridium acetobutylicum*, *Appl. Environ. Microbiol.*, 50, 220, 1985.
55. Lopéz-Contreras, A.M. et al., Production by *Clostridium acetobutylicum* ATCC 824 of CelG, a cellulosomal glycoside hydrolase belonging to family 9, *Appl. Environ. Microbiol.*, 69, 869, 2003.
56. Sabathé, F. and Soucaille, P., Characterization of the CipA scaffolding protein and *in vivo* production of a minicellulosome in *Clostridium acetobutylicum*, *J. Bacteriol.*, 185, 1092, 2003.
57. Perret, S. et al., Production of heterologous and chimeric scaffoldins by *Clostridium acetobutylicum* ATCC 824, *J. Bacteriol.*, 186, 253, 2004.
58. Andreesen, J.R. and Gottschalk, G., The occurrence of a modified Entner-Doudoroff pathway in *Clostridium aceticum*, *Arch. Microbiol.*, 69, 160, 1969.
59. Schreiber, W. and Dürre, P., The glyceraldehyde-3-phosphate dehydrogenase of *Clostridium acetobutylicum*: isolation and purification of the enzyme, and sequencing and localization of the *gap* gene within a cluster of other glycolytic genes, *Microbiol. U.K.*, 145, 1839, 1999.
60. Iddar, A. et al., Expression, purification, and characterization of recombinant nonphosphorylating NADP-dependent glyceraldehyde-3-phosphate dehydrogenase from *Clostridium acetobutylicum*, *Protein Expression Pur.*, 25, 519, 2002.
61. Meinecke, B., Bertram, J., and Gottschalk, G., Purification and characterization of the pyruvate-ferredoxin oxidoreductase from *Clostridium acetobutylicum*, *Arch. Microbiol.*, 152, 244, 1989.
62. Freier, D. and Gottschalk, G., L(+)-lactate dehydrogenase of *Clostridium acetobutylicum* is activated by fructose-1,6-bisphosphate, *FEMS Microbiol. Lett.*, 43, 229, 1987.
63. Contag, P.R., Williams, M.G., and Rogers, P., Cloning of a lactate dehydrogenase gene from *Clostridium acetobutylicum* B643 and expression in *Escherichia coli*, *Appl. Environ. Microbiol.*, 56, 3760, 1990.
64. Diez-Gonzalez, F., Russell, J.B., and Hunter, J.B., The role of an NAD-independent lactate dehydrogenase and acetate in the utilization of lactate by *Clostridium acetobutylicum* P262, *Arch. Microbiol.*, 164, 36, 1995.
65. Diez-Gonzalez, F., Russell, J.B., and Hunter, J.B., NAD-independent lactate and butyryl-CoA dehydrogenases of *Clostridium acetobutylicum* P262, *Curr. Microbiol.*, 34, 162, 1997.
66. Chen, J.-S., Properties of acid- and solvent-forming enzymes of clostridia, in *The Clostridia and Biotechnology*, Woods, D.R., Ed., Butterworth-Heinemann, Stoneham, MA, 1993, 51.
67. Winzer, K., Lorenz, K., and Dürre, P., Acetate kinase from *Clostridium acetobutylicum*: a highly specific enzyme that is actively transcribed during acidogenesis and solventogenesis, *Microbiol. U.K.*, 143, 3279, 1997.

68. Diez-Gonzalez, F., Russell, J.B., and Hunter, J.B., The acetate kinase of *Clostridium acetobutylicum* strain P262, *Arch. Microbiol.,* 166, 418, 1997.
69. Wiesenborn, D.P., Rudolph, F.B., and Papoutsakis, E.T., Thiolase from *Clostridium acetobutylicum* ATCC 824 and its role in the synthesis of acids and solvents, *Appl. Environ. Microbiol.,* 54, 2717, 1988.
70. Winzer, K. et al., Differential regulation of two thiolase genes from *Clostridium acetobutylicum* DSM 792, *J. Mol. Microbiol. Biotechnol.,* 2, 531, 2000.
71. Colby, G.D. and Chen, J.-S., Purification and properties of 3-hydroxybutyryl-coenzyme A dehydrogenase from *Clostridium beijerinckii* ("*Clostridium butylicum*") NRRL B593, *Appl. Environ. Microbiol.,* 58, 3297, 1992.
72. Waterson, R.M. et al., Purification and characterization of crotonase from *Clostridium acetobutylicum, J. Biol. Chem.,* 247, 5266, 1972.
73. Jones, D.T. and Keis, S., Origins and relationships of industrial solvent-producing clostridial strains, *FEMS Microbiol. Rev.,* 17, 223, 1995.
74. Johnson, J.L. and Chen, J.-S., Taxonomic relationships among strains of *Clostridium acetobutylicum* and other phenotypically similar organisms, *FEMS Microbiol. Rev.,* 17, 233, 1995.
75. Hartmanis, M.G.N. and Gatenbeck, S., Intermediary metabolism in *Clostridium acetobutylicum*: levels of enzymes involved in the formation of acetate and butyrate, *Appl. Environ. Microbiol.,* 47, 1277, 1984.
76. Wiesenborn, D.P., Rudolph, F.B., and Papoutsakis, E.T., Phosphotransbutyrylase from *Clostridium acetobutylicum* ATCC 824 and its role in acidogenesis, *Appl. Environ. Microbiol.,* 55, 317, 1989.
77. Thompson, D.K. and Chen, J.-S., Purification and properties of an acetoacetyl coenzyme A-reacting phosphotransbutyrylase from *Clostridium beijerinckii* ("*Clostridium butylicum*") NRRL B593, *Appl. Environ. Microbiol.,* 56, 607, 1990.
78. Hartmanis, M.G.N., Butyrate kinase from *Clostridium acetobutylicum, J. Biol. Chem.,* 262, 617, 1987.
79. Huang, K.-X et al., Identification and characterization of a second butyrate kinase from *Clostridium acetobutylicum* ATCC 824, *J. Mol. Microbiol. Biotechnol.,* 2, 33, 2000.
80. Petitdemange, H. et al., Isolation and properties of reduced nicotinamide adenine dinucleotide-rubredoxin oxidoreductase of *Clostridium acetobutylicum, Biochem. Biophys. Res. Commun.,* 4, 1258, 1979.
81. Marczak, R. et al., Regulation of the biosynthesis of NADH-rubredoxin oxidoreductase in *Clostridium acetobutylicum, Curr. Microbiol.,* 10, 165, 1984.
82. Marczak, R. et al., Differential levels of ferredoxin and rubredoxin in *Clostridium acetobutylicum, Biochimie,* 67, 241, 1985.
83. Ballongue, J. et al., Rôle de l'acétate et du butyrate dans l'induction de la NADH: rubrédoxine oxydoréductase chez *Clostridium acetobutylicum, Biochimie,* 68, 575, 1986.
84. Wiesenborn D.P., Rudolph, F.B., and Papoutsakis, E.T., Coenzyme A transferase from *Clostridium acetobutylicum* ATCC 824 and its role in the uptake of acids, *Appl. Environ. Microbiol.,* 55, 323, 1989.
85. Gerischer, U. and Dürre, P., Cloning, sequencing, and molecular analysis of the acetoacetate decarboxylase gene region from *Clostridium acetobutylicum, J. Bacteriol.,* 172, 6907, 1990.
86. Fischer, R. J., Helms, J., and Dürre, P., Cloning, sequencing, and molecular analysis of the *sol* operon of *Clostridium acetobutylicum*, a chromosomal locus involved in solventogenesis, *J. Bacteriol.,* 175, 6959, 1993.
87. Nakamura, T., Igarashi, Y., and Kodama, T., Acetoacetyl-CoA transferase and thiolase reactions are catalyzed by a single enzyme in *Clostridium acetobutylicum* ATCC824, *Agr. Biol. Chem.,* 54, 267, 1990.
88. Westheimer, F.H., Acetoacetate decarboxylase from *Clostridium acetobutylicum, Methods Enzymol.,* 14, 231, 1969.
89. Fridovich, I., Acetoacetate decarboxylase, in *The Enzymes*, Vol. 6, Boyer, P.D., Ed., Academic Press, New York, 1972, 255.
90. Tagaki, W. and Westheimer, F.H., Acetoacetate decarboxylase. The molecular weight of the enzyme and subunits, *Biochem.,* 7, 895, 1968.
91. Petersen, D.J. and Bennett, G.N., Purification of acetoacetate decarboxylase from *Clostridium acetobutylicum* ATCC 824 and cloning of the acetoacetate decarboxylase gene in *Escherichia coli, Appl. Environ. Microbiol.,* 56, 3491, 1990.
92. Warren, S., Zerner, B., and Westheimer, F.H., Acetoacetate decarboxylase. Identification of lysine at the active site, *Biochem.,* 5, 817, 1966.

93. Bertram, J., Kuhn, A., and Dürre, P., Tn*916*-induced mutants of *Clostridium acetobutylicum* defective in regulation of solvent formation, *Arch. Microbiol.,* 153, 373, 1990.
94. Dürre, P. et al., Solventogenic enzymes of *Clostridium acetobutylicum*: catalytic properties, genetic organization, and transcriptional regulation, *FEMS Microbiol. Rev.,* 17, 251, 1995.
95. Dürre, P. et al., Enzymatic investigations on butanol dehydrogenase and butyraldehyde dehydrogenase in extracts of *Clostridium acetobutylicum, Appl. Microbiol. Biotechnol.,* 26, 268, 1987.
96. Welch, R.W., Rudolph, F.B., and Papoutsakis, E.T., Purification and characterization of the NADH-dependent butanol dehydrogenase from *Clostridium acetobutylicum* (ATCC 824), *Arch. Biochem. Biophys.,* 273, 309, 1989.
97. Petersen, D.J. et al., Molecular cloning of an alcohol (butanol) dehydrogenase gene cluster from *Clostridium acetobutylicum* ATCC 824, *J. Bacteriol.,* 173, 1831, 1991.
98. Sauer, U. and Dürre, P., Differential induction of genes related to solvent formation during the shift from acidogenesis to solventogenesis in continuous culture of *Clostridium acetobutylicum, FEMS Microbiol. Lett.,* 125, 115, 1995.
99. Nair, R.V., Bennett, G.N., and Papoutsakis, E.T., Molecular characterization of an aldehyde/alcohol dehydrogenase gene from *Clostridium acetobutylicum* ATCC 824, *J. Bacteriol.*, 176, 871, 1994.
100. Nair, R.V. and Papoutsakis, E.T., Expression of plasmid-encoded *aad* in *Clostridium acetobutylicum* M5 restores vigorous butanol production, *J. Bacteriol.,* 176, 5843, 1994.
101. Green, E.M. and Bennett, G.N., Inactivation of an aldehyde/alcohol dehydrogenase gene from *Clostridium acetobutylicum* ATCC 824, *Appl. Biochem. Biotechnol.,* 57/58, 213, 1996.
102. Palosaari, N. and Rogers, P., Purification and properties of the inducible coenzyme A-linked butyraldehyde dehydrogenase from *Clostridium acetobutylicum, J. Bacteriol.,* 170, 2971, 1988.
103. Yan, R.-T. and Chen, J.-S., Coenzyme A-acylating aldehyde dehydrogenase from *Clostridium beijerinckii* NRRL B592, *Appl. Environ. Microbiol.,* 56, 2591, 1990.
104. Chen, J.-S., Alcohol dehydrogenase: multiplicity and relatedness in the solvent-producing clostridia, *FEMS Microbiol. Rev.,* 17, 263, 1995.
105. Girbal, L. and Soucaille, P., Regulation of solvent production in *Clostridium acetobutylicum, Trends Biotechnol.,* 16, 11, 1998.
106. Mitchell, W.J., Physiology of carbohydrate to solvent conversion by clostridia, *Adv. Microbial Physiol.,* 39, 31, 1998.
107. Young, M., Minton, N.P., and Staudenbauer, W.L., Recent advances in the genetics of the clostridia, *FEMS Microbiol. Rev.,* 63, 301, 1989.
108. Cary, J.W. et al., Cloning and expression of *Clostridium acetobutylicum* ATCC 824 acetoacetyl-coenzyme A: acetate/butyrate: coenzyme A-transferase in *Escherichia coli, Appl. Environ. Microbiol.,* 56, 1576, 1990.
109. Belouski, E., Watson, D.E., and Bennett, G.N., Cloning, sequence, and expression of the phosphofructokinase gene of *Clostridium acetobutylicum* ATCC 824 in *Escherichia coli, Curr. Microbiol.,* 37, 17, 1998.
110. Schreiber, W. and Dürre, P., Differential expression of genes within the *gap* operon of *Clostridium acetobutylicum, Anaerobe*, 6, 291, 2000.
111. Contag, P.R. and Rogers, P., The cloning and expression of the *Clostridium acetobutylicum* B643 butyraldehyde dehydrogenase by complementation of an *Escherichia coli* aldehyde dehydrogenase deficient mutant, *Abstracts 88th Ann. Meeting ASM*, Miami Beach, May 8–13, 1988, 169, H-146.
112. Boynton, Z.L. et al., Cloning, sequencing, and expression of genes encoding phosphotransacetylase and acetate kinase from *Clostridium acetobutylicum* ATCC 824, *Appl. Environ. Microbiol.,* 62, 2758, 1996.
113. Stim-Herndon, K.P., Petersen, D.J., and Bennett, G.N., Characterization of an acetyl-CoA acetyltransferase (thiolase) gene from *Clostridium acetobutylicum* ATCC 824, *Gene,* 154, 81, 1995.
114. Bennett, G.N. and Rudolph, F.B., The central metabolic pathway from acetyl-CoA to butyryl-CoA in *Clostridium acetobutylicum, FEMS Microbiol. Rev.,* 17, 241, 1995.
115. Boynton, Z.L. et al., Cloning, sequencing, and expression of clustered genes encoding β-hydroxybutyryl-coenzyme A (CoA) dehydrogenase, crotonase, and butyryl-CoA dehydrogenase from *Clostridium acetobutylicum* ATCC 824, *J. Bacteriol.,* 178, 3015, 1996.
116. Woods, D.R. and Reid, S.J., Regulation of nitrogen metabolism, starch utilisation and the *β-hbd-adh1* gene cluster in *Clostridium acetobutylicum, FEMS Microbiol. Rev.,* 17, 299, 1995.

117. Youngleson, J.S., Jones, D.T., and Woods, D.R., Homology between hydroxybutyryl and hydroxyacyl coenzyme A dehydrogenase enzymes from *Clostridium acetobutylicum* fermentation and vertebrate fatty acid β-oxidation pathways, *J. Bacteriol.,* 171, 6800, 1989.
118. Youngleson, J.S. et al., Molecular analysis and nucleotide sequence of the *adh1* gene encoding an NADPH-dependent butanol dehydrogenase in the Gram-positive anaerobe *Clostridium acetobutylicum, Gene,* 78, 355, 1989.
119. Cary, J.W. et al., Cloning and expression of *Clostridium acetobutylicum* phosphotransbutyrylase and butyrate kinase genes in *Escherichia coli, J. Bacteriol.,* 170, 4613, 1988.
120. Walter, K.A. et al., Sequence and arrangement of two genes of the butyrate-synthesis pathway of *Clostridium acetobutylicum* ATCC 824, *Gene,* 134, 107, 1993.
121. Tummala, S.B., Welker, N.E., and Papoutsakis, E.T., Development and characterization of a gene expression reporter system for *Clostridium acetobutylicum* ATCC 824, *Appl. Environ. Microbiol.,* 65, 3793, 1999.
122. Feustel, L., Nakotte, S., and Dürre, P., Characterization and development of two reporter gene systems for *Clostridium acetobutylicum, Appl. Environ. Microbiol.,* 70, 798, 2004.
123. Oultram, J.D. et al., Cloning and sequence analysis of the genes encoding phosphotransbutyrylase and butyrate kinase from *Clostridium acetobutylicum* NCIMB 8052, *Gene,* 131, 107, 1993.
124. Petersen, D.J. et al., Sequence and arrangement of genes encoding enzymes of the acetone-production pathway of *Clostridium acetobutylicum* ATCC 824, *Gene,* 123, 93, 1993.
125. Gerischer, U. and Dürre, P., mRNA analysis of the *adc* gene region of *Clostridium acetobutylicum* during the shift to solventogenesis, *J. Bacteriol.,* 174, 426, 1992.
126. Lorenz, K., Gezielte Mutagenese des *sol*-Operons und physiologische Untersuchungen zur enzymatischen Aktivität der Aldehyd/Alkohol-Dehydrogenase E aus *Clostridium acetobutylicum*, Ph.D. Thesis, University of Ulm, Germany, 1997.
127. Thormann, K. et al., Control of butanol formation in *Clostridium acetobutylicum* by transcriptional activation, *J. Bacteriol.,* 184, 1966, 2002.
128. Goodlove, P.E. et al., Cloning and sequence analysis of the fermentative alcohol-dehydrogenase-encoding gene of *Escherichia coli, Gene,* 85, 209, 1989.
129. Dailly, Y.P., Bunch, P., and Clark, D.P., Comparison of the fermentative alcohol dehydrogenases of *Salmonella typhimurium* and *Escherichia coli, Microbios,* 103, 179, 2000.
130. Arnau, J. et al., Cloning of the *Lactococcus lactis adhE* gene, encoding a multifunctional alcohol dehydrogenase, by complementation of a fermentative mutant of *Escherichia coli, J. Bacteriol.,* 180, 3049, 1998.
131. Asanuma, N., Yoshii, T., and Hino, T., Molecular characteristics and transcription of the gene encoding a multifunctional alcohol dehydrogenase in relation to the deactivation of pyruvate formate-lyase in the ruminal bacterium *Streptococcus bovis, Arch. Microbiol.,* 181, 122, 2004.
132. Yang, W. et al., *Entamoeba histolytica* has an alcohol dehydrogenase homologous to the multifunctional *adhE* gene product of *Escherichia coli, Mol. Biochem. Parasitol.,* 64, 253, 1994.
133. Sánchez, L.B., Aldehyde dehydrogenase (CoA-acetylating) and the mechanism of ethanol formation in the amitochondriate protist, *Giardia lamblia, Arch. Biochem. Biophys.,* 354, 57, 1998.
134. Kessler, D., Leibrecht, I., and Knappe, J., Pyruvate: formate-lyase-deactivase and acetyl-CoA reductase activities of *Escherichia coli* reside on a polymeric operon particle encoded by *adhE, FEBS Lett.,* 281, 59, 1991.
135. Söhling, B. and Gottschalk, G., Molecular analysis of the anaerobic succinate degradation pathway in *Clostridium kluyveri, J. Bacteriol.,* 178, 871, 1996.
136. Reid, M.F. and Fewson, C.A., Molecular characterization of microbial alcohol dehydrogenases, *CRC Crit. Rev. Microbiol.,* 20, 13, 1994.
137. Membrillo-Hernández, J. et al., Evolution of the *adhE* gene product of *Escherichia coli* from a functional reductase to a dehydrogenase, *J. Biol. Chem.,* 275, 33869, 2000.
138. Toth, J., Ismaiel, A.A., and Chen, J.-S., The *ald* gene, encoding a coenzyme A-acylating aldehyde dehydrogenase, distinguishes *Clostridium beijerinckii* and two other solvent-producing clostridia from *Clostridium acetobutylicum, Appl. Environ. Microbiol.,* 65, 4973, 1999.
139. Fontaine, L. et al., Molecular characterization and transcriptional analysis of *adhE2*, the gene encoding the NADH-dependent aldehyde/alcohol dehydrogenase responsible for butanol production in alcohologenic cultures of *Clostridium acetobutylicum* ATCC 824, *J. Bacteriol.,* 184, 821, 2002.

140. Dürre, P., Solventogenesis by clostridia, in *Strict and Facultative Anaerobes: Medical and Environmental Aspects,* Nakano, M.M. and Zuber, P., Eds., Horizon Scientific Press, Wymondham, U.K., 2004, 329.
141. Girbal, L. et al., Regulation of metabolic shifts in *Clostridium acetobutylicum* ATCC 824, *FEMS Microbiol. Rev.,* 17, 287, 1995.
142. Rao, G. and Mutharasan, R., Alcohol production by *Clostridium acetobutylicum* induced by methyl viologen, *Biotechnol. Lett.,* 8, 893, 1986.
143. Rao, G. and Mutharasan, R., Altered electron flow in continuous cultures of *Clostridium acetobutylicum* induced by viologen dyes, *Appl. Environ. Microbiol.,* 53, 1232, 1987.
144. Zickner, B. et al., Multiple levels of regulation of solventogenesis in *Clostridium acetobutylicum*, in *Regulatory Networks in Prokaryotes*, Dürre, P. and Friedrich, B., Eds., Horizon Scientific Press, Wymondham, U.K., 2003, 127.
145. Cornillot, E. and Soucaille, P., Solvent-forming genes in clostridia, *Nature,* 380, 489, 1996.
146. Cornillot, E. et al., The genes for butanol and acetone formation in *Clostridium acetobutylicum* ATCC 824 reside on a large plasmid whose loss leads to degeneration of the strain, *J. Bacteriol.,* 179, 5442, 1997.
147. Walter, K.A., Bennett, G.N., and Papoutsakis, E.T., Molecular characterization of two *Clostridium acetobutylicum* ATCC 824 butanol dehydrogenase isozyme genes, *J. Bacteriol.,* 174, 7149, 1992.
148. Paredes, C.J., Rigoutsos, I., and Papoutsakis, E.T., Transcriptional organization of the *Clostridium acetobutylicum* genome, *Nucl. Acids Res.,* 32, 1973, 2004.
149. Davies, R., Studies on the acetone-butyl alcohol fermentation. 2. Intermediates in the fermentation of glucose by *Cl. acetobutylicum.* 3. Potassium as an essential factor in the fermentation of maize meal by *Cl. acetobutylicum*, *Biochem. J.,* 36, 582, 1942.
150. Andersch, W., Bahl, H., and Gottschalk, G., Level of enzymes involved in acetate, butyrate, acetone and butanol formation by *Clostridium acetobutylicum*, *Eur. J. Appl. Microbiol. Biotechnol.,* 18, 327, 1983.
151. Hartmanis, M.G.N., Klason, T., and Gatenbeck, S., Uptake and activation of acetate and butyrate in *Clostridium acetobutylicum*, *Appl. Microbiol. Biotechnol.,* 20, 66, 1984.
152. Ballongue, J. et al., Induction of acetoacetate decarboxylase in *Clostridium acetobutylicum*, *FEMS Microbiol. Lett.,* 29, 273, 1985.
153. Welch, R.W. et al., Effects of rifampicin and chloramphenicol on product and enzyme levels of the acid- and solvent-producing pathways of *Clostridium acetobutylicum* (ATCC 824), *Enzyme Microb. Technol.,* 14, 277, 1992.
154. Hüsemann, M.H.W. and Papoutsakis, E.T., Comparison of *in vivo* and *in vitro* enzyme activities in continuous and batch fermentations of *Clostridium acetobutylicum*, *Appl. Microbiol. Biotechnol.,* 30, 585, 1989.
155. Schaffer, S. et al., Changes in protein synthesis and identification of proteins specifically induced during solventogenesis in *Clostridium acetobutylicum*, *Electrophoresis,* 23, 110, 2002.
156. Ravagnani, A. et al., Spo0A directly controls the switch from acid to solvent production in solvent-forming clostridia, *Mol. Microbiol.,* 37, 1172, 2000.
157. Hollergschwandner, C., Unterschiede in der Regulation der Sporulation von *Clostridium acetobutylicum* und *Bacillus subtilis*, Ph.D. Thesis, University of Ulm, Germany, 2003.
158. Böhringer, M., Molekularbiologische und enzymatische Untersuchungen zur Regulation des Gens der Acetacetat-Decarboxylase von *Clostridium acetobutylicum*, Ph.D. Thesis, University of Ulm, Germany, 2002.
159. Bahl, H. et al., Nutritional factors affecting the ratio of solvents produced by *Clostridium acetobutylicum*, *Appl. Environ. Microbiol.,* 52, 169, 1986.
160. Holt, R.A., Stephens, G.M., and Morris, J.G., Production of solvents by *Clostridium acetobutylicum* cultures maintained at neutral pH, *Appl. Environ. Microbiol.,* 48, 1166, 1984.
161. Monot, F., Engasser, J.-M., and Petitdemange, H., Regulation of acetone butanol production in batch and continuous cultures of *Clostridium acetobutylicum*, *Biotechnol. Bioeng. Symp.*, 13, 207, 1983.
162. Grupe, H. and Gottschalk, G., Physiological events in *Clostridium acetobutylicum* during the shift from acidogenesis to solventogenesis in continuous culture and presentation of a model for shift induction, *Appl. Environ. Microbiol.,* 58, 3896, 1992.
163. Boynton, Z.L., Bennett, G.N., and Rudolph, F.B., Intracellular concentrations of coenzyme A and its derivatives from *Clostridium acetobutylicum* ATCC 824 and their roles in enzyme regulation, *Appl. Environ. Microbiol.,* 60, 39, 1994.

164. Ullmann, S., Kuhn, A., and Dürre, P., DNA topology and gene expression in *Clostridium acetobutylicum*: implications for the regulation of solventogenesis, *Biotechnol. Lett.*, 18, 1413, 1996.
165. Wang, J.Y. and Syvanen, M., DNA twist as a transcriptional sensor for environmental changes, *Mol. Microbiol.*, 6, 1861, 1992.
166. Dürre, P. et al., Transcriptional regulation of solventogenesis in *Clostridium acetobutylicum*, *J. Mol. Microbiol. Biotechnol.*, 4, 295, 2002.
167. Behrens, S. et al., Identification of two genes encoding putative new members of the ECF subfamily of eubacterial RNA polymerase sigma factors in *Clostridium acetobutylicum*, *J. Mol. Microbiol. Biotechnol.*, 2, 265, 2000.
168. Liu, L.F. and Wang, J.C., Supercoiling of the DNA template during transcription, *Proc. Natl. Acad. Sci. U.S.A.*, 84, 7024, 1987.
169. Nakotte, S., Funktionelle Analyse von Regulationsstrukturen essentieller Gene für die Lösungsmittelbildung in *Clostridium acetobutylicum*, Ph.D. Thesis, University of Ulm, Germany, 1998.
170. Harris, L.M., Welker, N.E., and Papoutsakis, E.T., Northern, morphological, and fermentation analysis of *spo0A* inactivation and overexpression in *Clostridium acetobutylicum* ATCC 824, *J. Bacteriol.*, 184, 3586, 2002.
171. Nair, R.V. et al., Regulation of the *sol* locus genes for butanol and acetone formation in *Clostridium acetobutylicum* ATCC 824 by a putative transcriptional repressor, *J. Bacteriol.*, 181, 319, 1999.
172. Thormann, K. and Dürre, P., Orf5/SolR: a transcriptional repressor of the *sol* operon of *Clostridium acetobutylicum*?, *J. Ind. Microbiol. Biotechnol.*, 27, 307, 2001.
173. Lamb, J.R., Tugendreich, S., and Hieter, P., Tetratrico peptide repeat interactions: to TPR or not to TPR?, *Trends Biochem. Sci.*, 20, 257, 1995.
174. Aristarkhov, A. et al., Translation of the *adhE* transcript to produce ethanol dehydrogenase requires RNase III cleavage in *Escherichia coli*, *J. Bacteriol.*, 178, 4327, 1996.
175. Umitsuki, G. et al., Involvement of RNase G in *in vivo* mRNA metabolism in *Escherichia coli*, *Genes Cells*, 6, 403, 2001.
176. Wachi, M. et al., A novel RNase G mutant that is defective in degradation of *adhE* mRNA but proficient in the processing of 16S rRNA precursor, *Biochem. Biophys. Res. Commn.*, 289, 1301, 2001.
177. Tock, M.R. et al., The CafA protein required for 5′-maturation of 16S rRNA is a 5′-end-dependent ribonuclease that has context-dependent broad sequence specificity, *J. Biol. Chem.*, 275, 8726, 2000.
178. Scotcher, M.C. et al., Sequences affecting the regulation of solvent production in *Clostridium acetobutylicum*, *J. Ind. Microbiol. Biotechnol.*, 30, 414, 2003.
179. Bahl, H., Kontinuierliche Aceton-Butanol-Gärung durch *Clostridium acetobutylicum*. Ph.D. Thesis, University of Göttingen, Germany, 1983.
180. Wong, J. and Bennett, G.N., The effect of novobiocin on solvent production by *Clostridium acetobutylicum*, *J. Ind. Microbiol.*, 16, 354, 1996.
181. Ullmann, S. and Dürre, P., Changes in DNA topology are involved in triggering the onset of solventogenesis in *Clostridium acetobutylicum*, *Recent Res. Dev. Microbiol.*, 2, 281, 1998.
182. Terracciano, J.S., Rapoport, E., and Kashket, E.R., Stress- and growth phase-associated proteins of *Clostridium acetobutylicum*, *Appl. Environ. Microbiol.*, 54, 1989, 1988.
183. Pich, A., Narberhaus, F., and Bahl, H., Induction of heat shock proteins during initiation of solvent formation in *Clostridium acetobutylicum*, *Appl. Microbiol. Biotechnol.*, 33, 697, 1990.
184. Sauer, U. and Dürre, P., Sequence and molecular characterization of a DNA region encoding a small heat shock protein of *Clostridium acetobutylicum*, *J. Bacteriol.*, 175, 3394, 1993.
185. Bahl, H. et al., Expression of heat shock genes in *Clostridium acetobutylicum*, *FEMS Microbiol. Rev.*, 17, 341, 1995.
186. Tummala, S.B., Junne, S.G., and Papoutsakis, E.T., Antisense RNA downregulation of coenzyme A transferase combined with alcohol-aldehyde dehydrogenase overexpression leads to predominantly alcohologenic *Clostridium acetobutylicum* fermentations, *J. Bacteriol.*, 185, 3644, 2003.
187. Tomas, C.A., et al., DNA array-based transcriptional analysis of asporogenous, nonsolventogenic *Clostridium acetobutylicum* strains SKO1 and M5, *J. Bacteriol.*, 185, 4539, 2003.
188. Tomas, C.A., Welker, N.E., and Papoutsakis, E.T., Overexpression of *groESL* in *Clostridium acetobutylicum* results in increased solvent production and tolerance, prolonged metabolism, and changes in the cell's transcriptional program, *Appl. Environ. Microbiol.*, 69, 4951, 2003.

189. Alsaker, K.V., Spitzer, T.R., and Papoutsakis, E.T., Transcriptional analysis of *spo0A* overexpression in *Clostridium acetobutylicum* and its effect on the cell's response to butanol stress, *J. Bacteriol.*, 186, 1959, 2004.
190. Tomas, C.A., Beamish, J., and Papoutsakis, E.T., Transcriptional analysis of butanol stress and tolerance in *Clostridium acetobutylicum*, *J. Bacteriol.*, 186, 2006, 2004.
191. Green, E.M., et al., Genetic manipulation of acid formation pathways by gene inactivation in *Clostridium acetobutylicum* ATCC 824, *Microbiol. U.K.*, 142, 2079, 1996.
192. Harris, L.M., et al., Fermentation characterization and flux analysis of recombinant strains of *Clostridium acetobutylicum* with an inactivated *solR* gene, *J. Ind. Microbiol. Biotechnol.*, 27, 322, 2001.
193. Harris, L.M., et al., Characterization of recombinant strains of the *Clostridium acetobutylicum* butyrate kinase inactivation mutant: need for new phenomenological models for solventogenesis and butanol inhibition?, *Biotechnol. Bioeng.*, 67, 1, 2000.
194. Walter, K.A., Mermelstein, L.D., and Papoutsakis, E.T., Studies of recombinant *Clostridium acetobutylicum* with increased dosages of butyrate formation genes, *Ann. New York Acad. Sci.*, 721, 69, 1994.
195. Walter, K.A., Mermelstein, L.D., and Papoutsakis, E.T., Host-plasmid interactions in recombinant strains of *Clostridium acetobutylicum* ATCC 824, *FEMS Microbiol. Lett.*, 123, 335, 1994.
196. Mermelstein, L.D., et al., Metabolic engineering of *Clostridium acetobutylicum* ATCC 824 for increased solvent production by enhancement of acetone formation enzyme activities using a synthetic acetone operon, *Biotechnol. Bioeng.*, 42, 1053, 1993.
197. Bermejo, L.L., Welker, N.E., and Papoutsakis, E.T., Expression of *Clostridium acetobutylicum* ATCC 824 genes in *Escherichia coli* for acetone production and acetate detoxification, *Appl. Environ. Microbiol.*, 64, 1079, 1998.
198. Guillot, J.-P.K.M., Konstruktion und Expression artifizieller Transkriptionseinheiten zur Lösungsmittelsynthese in *Escherichia coli*, Diploma Thesis, University of Ulm, Germany, 1997.
199. Liu, S.-J. and Steinbüchel, A., A novel genetically engineered pathway for synthesis of poly(hydroxyalkanoic acids) in *Escherichia coli*, *Appl. Environ. Microbiol.*, 66, 739, 2000.
200. Zhao, Y. et al., Expression of a cloned cyclopropane fatty acid synthase gene reduces solvent formation in *Clostridium acetobutylicum* ATCC 824, *Appl. Environ. Microbiol.*, 69, 2831, 2003.
201. Wardwell, S.A. et al., Expression of the *Klebsiella pneumoniae* CG21 acetoin reductase gene in *Clostridium acetobutylicum* ATCC 824, *J. Ind. Microbiol. Biotechnol.*, 27, 220, 2001.
202. Horton, C.E. et al., Heterologous expression of the *Saccharomyces cerevisiae* alcohol acetyltransferase genes in *Clostridium acetobutylicum* and *Escherichia coli* for the production of isoamyl acetate, *J. Ind. Microbiol. Biotechnol.*, 30, 427, 2003.
203. Liu, S.-J. and Steinbüchel, A., Exploitation of butyrate kinase and phosphotransbutyrylase from *Clostridium acetobutylicum* for the *in vitro* biosynthesis of poly(hydroxyalkanoic acid), *Appl. Microbiol. Biotechnol.*, 53, 545, 2000.
204. Strandberg, L. et al., Biotechnological production of 1-butanol, *ASM Meeting on Biodegradation, Biotransformation, and Biocatalysis*, San Juan, Puerto Rico, October 2001, B3.
205. Duda, M. et al., Verfahren zur Oxidation von Kohlenwasserstoffen unter Verwendung von Mikroorganismen, patent application pending, 2000.
206. Jones, D.T., presentation at the 9th International Symposium on the Genetics of Industrial Microorganisms (GIM), July 1–5, 2002, Gyeongju, Korea.

Part VI

Special Groups and Ecosystems

31 Bacteriophages of *Clostridium*

David T. Jones

CONTENTS

31.1 INTRODUCTION

Bacteriophages (phages) constitute the largest viral group, and virtually all bacterial species that have been investigated have proved to be susceptible to infection by one or more phages [1,2]. Phages are abundant and widely distributed in natural environments and can exist as both free virions or as integrated prophages in a host bacterium. The study of phages has tended to go hand in hand with studies of the various genera and species of host bacteria. From the beginnings of molecular biology to present times, phage systems have proved to be useful for investigating the molecular mechanisms underpinning genetics [3]. Phages have proved to be useful experimental systems, as they are convenient to maintain and relatively easy to manipulate. In addition to their use in fundamental research, phages have also been extensively studied in a number of bacterial groups with respect to their practical significance and applications, particularly for those bacterial species of medical and industrial importance.

Susceptibility to infection by phages has been reported for many of the well-studied species belonging to the genus *Clostridium.* The exception to this appears to be for thermophilic species, where little evidence for the presence of phages has been recorded so far [4]. When compared with studies undertaken on the phages infecting many groups of well-characterized aerobic bacteria, phages infecting *Clostridium* species have, in general, been far less extensively characterized. A number of the early studies were directed mainly at the identification and classification of different morphological phage types [4–6]. Most of the research undertaken with clostridial phages has been directed toward finding solutions to practical problems. In almost all cases, these investigations have either focused on species of pathogenic clostridia or species with industrial applications.

Investigations undertaken on phages infecting pathogenic species have been directed toward addressing specific challenges in medical microbiology. These have included attempts to develop phage-typing for species with epidemiological importance, the elucidation of the role of phages in toxigenic conversion of pathogenic strains, and their use as therapeutic agents.

Following the initial discovery of bacterial viruses, their potential for use as therapeutic agents for controlling bacterial diseases was soon recognized. However, although various phages were subjected to assessment as control agents, these early attempts were uniformly unsuccessful [7]. During World War II, Russian workers isolated phages active against *Clostridium histolyticum, Clostridium oedematiens, Clostridium perfringens, Clostridium putrificus,* and *Clostridium septicum.* Phages that caused lysis of 96% of *C. perfringens* strains tested were reported and were apparently used with success in the treatment of gas gangrene in humans [5]. In general, the narrow host range of most phages coupled with problems of restriction/modification, resistance, and immunogenicity have continued to mitigate against the successful use of bacteriophage therapy [7]. Recently, there has been some resurgence of interest in the potential for using phages as therapeutic agents for treating bacterial diarrhea, as well as for use as control agents in aquatic environments such as water treatment plants and fish farms [7,8].

A number of studies involving phages infecting pathogenic clostridia were undertaken to evaluate the feasibility of utilizing phage typing for strain identification. Most work has been conducted on the phages of *C. perfringens,* but phages infecting other clostridial species associated with gas gangrene, such as *C. histolyticum, Clostridium novyi, C. oedematiens, C. putrificus, C. septicum,* and *Clostridium sporogenes,* were also investigated. For detailed coverage of these early studies, readers should consult the comprehensive review on this topic by Mahony [5]. The feasibility of developing phage typing systems for some of the other major groups of pathogenic clostridia such as *Clostridium botulinum* and *Clostridium tetani* have also been investigated with limited success [4,5].

A number of studies undertaken on the major groups of pathogenic clostridia have demonstrated that many strains harbor temperate phages [10]. In a series of pioneering studies, toxin production in some groups of *C. botulinum* was shown to be encoded for by temperate phages [9,10]. Although some phenotypic effects of temperate phages have been demonstrated in *C. perfringens* strains, changes in toxin production in *C. perfringens* infected with a temperate phage has not been demonstrated [11]. Cowles [12] also examined toxin production in *C. tetani* without being able to demonstrate that bacteriophages were involved in toxin production.

The earliest report of phages infecting solvent-producing clostridial species utilized for the industrial production of acetone and butanol dates back to 1923 [13]. Over the following 40-year period, the fermentation process continued to operate as the major route for the commercial production of solvents. During this time, sporadic reports of ongoing problems with phage contamination were found in the literature [14,15]. A series of devastating phage attacks, which blighted the industrial fermentation process in Japan was comprehensively reviewed by Ogata and Hongo [6]. More recently, phage infections occurring in the acetone-butanol fermentation process operated in South Africa during the 40-year post-WWII period have also been documented [15,16].

Much of the work that was carried out on clostridial phages predates the era of modern molecular biology. The earliest studies relied on the assessment of properties such as filterability, immunogenicity, and plaque formation. The advent of the electron microscope in the early 1950s enabled the morphology on the virus particles to be determined and described. However, in the majority of studies undertaken during the 1950s and 1960s, no attempts were made to determine the composition of the phage genomes, although these are generally assumed to consist of double-stranded DNA (dsDNA). Due to the practical focus of many of these studies, little is know about the genomic characteristics of most clostridial phages at the molecular level. Most of the fundamental information relating to the characterization of clostridial phages has been generated as a spinoff from more applied studies.

31.2 MORPHOTYPES AND CLASSIFICATION

Phages can differ significantly with respect to their morphology and genomic characteristics. Phages are grouped according to the nature of their genomic material, which can include dsDNA, ssDNA, dsRNA, and ssRNA, and are classified and named according to the system of virus taxonomy contained in the *Seventh Report of the International Committee on the Taxonomy of Viruses* [17]. Phages containing dsDNA are by far the most abundant and are divided into nine families. Phages with genomes consisting of ssDNA include three families and phages containing either dsRNA or ssRNA are comprised of a single family each. The different families of phages each exhibit distinguishable morphotypes associated with the virions. For detailed information on the classification and nomenclature of phages, refer to the report on virus taxonomy [17].

In a survey reported by Ackermann [1], the frequency of phage morphotypes infecting all types of bacteria were determined. More than 4,500 phages were included, infecting around 140 genera of bacteria. Of these, 96% consisted of dsDNA phages with a tailed structure belonging to the order *Caudovirales* [1]. The remaining 4% of phages fall into 11 families that exhibit cubic, filamentous, or pleomorphic morphology that infect various groups of archea, mycoplasmas, and bacteria. Examples of these uncommon phage types are rare among Gram-positive bacteria with low guanine plus cytosine (G+C) ratios [18]. The majority of phages that have been characterized consist of those with tailed morphotypes assigned to three families. These are: the *Myoviridae*, consisting of tailed phages with contractile tails; the *Siphoviridae*, consisting of phages with long noncontractile tails; and the *Podoviridae*, consisting of phages with short tails. Of the 96% of phages surveyed with tailed morphotypes, 24% were characterized as *Myoviridae*, 62% as *Siphoviridae,* and 14% as *Podoviridae* [1]. The survey conducted by Ackermann lists 170 phages reported to infect various species of *Clostridium*. Of these, 92 (54%) were classified as members of the *Myoviridae*, 57 (34%) as members of the *Siphoviridae*, and 21 (12%) as members of the *Podoviridae* [1].

Clostridial phages belonging to the *Myoviridae* consist of phages with isometric heads and tails with contractile sheaths. The tail structures often exhibit complex appendages including base plates, spikes, and tail fibers. Head sizes have been reported to range from around 40 nm up to 170 nm, with tail structures varying in length from around 100 nm to almost 400 nm [5]. Examples of clostridial phages belonging to the *Myoviridae* include phages infecting pathogenic species *C. botulinum* [9,19–21], *Clostridium difficile* [22], *C. novyi* [23], *C. perfringens* [5], *C. sporogenes* [24], *Clostridium sordellii* [25], and *C. tetani* [26,27]. Among the solvent–producing clostridial species, phages belonging to the *Myoviridae* have been reported to infect *Clostridium acetobutylicum* [28], *Clostridium beijerinckii* [15], *Clostridium saccharoperbutylacetonicum* [29,30], and a temperate phage infecting an unidentified solvent-producing *Clostridium* sp. [31].

The majority of clostridial phages belonging to the *Siphoviridae* typically exhibit isometric heads with a relatively long flexible, noncontractile tail, which may or may not have tail fibers or other structures at the distal end. The dimensions of individual phages exhibit a considerable range in size, with heads reported to range from around 60 nm to more than 100 nm, with filamentous tails ranging in length from around 120 nm to up to 400 nm [5]. Some phages exhibit simple, tubular, noncontractile tails. One common morphotype typically has distinctly striated flexible tails, often with an end plate present. A second common morphotype exhibits long flexible tail cores, typically from 6 to 10 nm in diameter, that can be partially surrounded by a sheath varying between 16 and 20 nm in diameter. The sheaths can be of varying length and are usually positioned adjacent to the phage head.

Phages exhibiting the uncommon elongated prolate head structure have been reported to infect strains classified as belonging to the *Myoviridae C. sporogenes* [24,32] and *C. botulinum* [33]. It has been suggested that these phages, bearing a strong resemblance to those infecting strains in the *Caryophanon, Staphylococcus,* and *Streptococcus* genera, might reflect a common origin [4,33]. Clostridial phages belonging to the *Myoviridae* include phages infecting pathogenic species such as *C. botulinum* [19,21], *C. difficile* [22], *C. histolyticum* [34], *C. novyi* [23], *C. perfringens* [5,35–37], *C. sporogenes* [24], and *C. tetani* [26,27]. Phages classified as *Myoviridae* have also been reported to infect solvent-producing species, including *C. beijerinckii* [15] and *C. saccharoperbutylacetonicum* [29].

Clostridial phages belonging to the *Podoviridae* have, in general, somewhat smaller isometric heads than the other two groups of tailed phages and are characterized by a short noncontractile tail, which often carries complex appendages, such as a base plate, spikes, and tail fibers. The majority of phages belonging to this group have isometric heads in the range of 30 to 40 nm or 50 to 60 nm in diameter. The tail structures commonly range between 10 to 40 nm in length, with the majority being around 30 nm in length. Phages exhibiting this morphotype often exhibit an upper collar or disk at the junction between the head and the tail. The distal portion of the tail can be wider than the upper portion. In most examples of phages with this morphotype, there is also a lower platelike structure or collar situated below the upper disk to which 4 to 12 appendages might be attached, with the core projecting beyond the base plate. Evidence suggests that this complex tail structure plays some role in attachment to the host cell and injection of phage DNA into the host [6]. Phages belonging to the *Podoviridae* have been reported in *C. perfringens* [5,38], *C. saccharoperbutylacetonicum* [30], and *C. beijerinckii* [15].

For those species of *Clostridium* that have been most extensively investigated, the occurrence of a variety of tailed phages have been described, which include examples of both virulent and temperate phages belonging to the three families. The range of phage morphotypes infecting a particular bacterial species is, in general, a poor indicator of relatedness above the species level [4].

A ssDNA phage-like structure has also been reported to occur in a solvent-producing strain of *C. beijerinckii* (formerly classified as *C. acetobutylicum*) [39]. The defective, filamentous virus-like particle designated CAK1 was released from cell cultures of *C. beijerinckii* NCIMB 6444 in the absence of cell lysis. The entity consisted of a ssDNA molecule of 6.6 kb complexed with a coat protein. The CAK1 particle was reported to be structurally similar to the filamentous phage M13, and the CAK1 DNA was found to be homologous to the DNA from phage M13mp9 [39].

31.3 GENOMIC CHARACTERIZATION

Data on genome structure and replicative properties are only available for a small number of well-studied phage types [17]. Among the phages that have been characterized, those belonging to the *Podoviridae* are reported to have a single linear genome varying in size from 19–44 kb, those belonging to the *Myoviridae* have a single linear genome varying in size from 39–169 kb, and those belonging to the *Siphoviridae* have a single linear genome varying in size from 22–121 kb. The phage genomes that have been sequenced encode from 27 to more than 200 genes [17]. Those clostridial phages with tailed morphotypes with genomes that have been fully or partially characterized are consistent with these more general findings.

The nucleic acid sequences of six phages involved in the conversion of *C. botulinum* to type C and D toxin production were characterized. These phages were shown to contain dsDNA, with genome sizes from 110 kb to 150 kb, and the existence of the structural genes for toxin production was confirmed by hybridization tests [40]. The completed nucleotide sequence of the 33.5 kb genome of a temperate phage belonging to the *Siphoviridae* isolated from a lysogenic *C. perfringens* strain has been determined [41]. The genome size of two phages, belonging to the *Podoviridae,* isolated from infections of solvent-producing clostridia, was investigated. The morphology of the two phages was very similar and the genomes consisting of linear dsDNA were of a similar size, with the genome of the phages infecting *C. saccharoperbutylacetonicum* being 17.7 kb and that infecting *C. beijerinckii* being 17.2 kb [42]. These genome sizes are among the smallest reported for any phage. The genome of a large temperate phage infecting an industrial strain of *C. beijerinckii,* belonging to the *Siphoviridae,* was also characterized [43]. This phage was shown to exist as an extrachromosomal element within the host strain, and the linear dsDNA genome had a size of 215 kb, making it among the largest phage genomes to be reported.

The majority of phages have G+C ratios that are close to those exhibited by the host strains they infect [17]. The G+C content of clostridial host strains and their phages are among the lowest known and fall within the range of 25 to 40% [44]. Most clostridial phages that have been investigated have G+C ratios that are similar to those exhibited by the host strains [6]. However, in some instances, the G+C ratio of the phage DNA has been shown to exceed that of the host bacteria. Examples of such cases are: a phage with a G+C ratio of 39.5% infecting a *C. sporogenes* host strain with a G+C ratio of 33.2% [45], and a phage with a G+C ratio of 35.2% infecting a *C. saccharoperbutylacetonicum* host strain with a G+C ratio of 31% [30]. There appear to be no reports of clostridial phages containing unusual nucleotide bases.

31.4 PHAGE STABILITY

Reports dealing directly with the determination of stability of clostridial phages are sparse. However, a perusal of the literature suggests that, in general, phages infecting *Clostridium* species do not appear to exhibit any significant differences with respect to stability, compared with phages infecting more extensively studied aerobic species. No instances of free phage particles being sensitive to exposure to air appear to have been reported, and the loss of viability in phage lysates appears to occur at a similar rate irrespective of whether lysates are stored under aerobic or anaerobic conditions (unpublished results). As might be anticipated, exposure of the host clostridial strain to aeration can have a marked effect on phage infection and multiplication. Studies on the effect of aeration on phage production showed that aeration of the culture at the time of infection, or at any time during the latent period, greatly prolonged the latent period and prevented phage multiplication. The inhibition was reversible with the return to anaerobic conditions [5,46]. Aeration at the end of the latent period had no marked effect on phage production, and aeration itself was shown to have no adverse effect upon phage viability.

Little comprehensive data is available regarding the thermostability of clostridial phages. In studies that have been conducted, various groups of phages remained viable for around 10 min at 60°C; however, the rate of phage inactivation differed according to the type of phage and the nature of the suspending medium [5,6]. The pH of the suspending medium along with the presence of divalent cations, polyvalent cations, and proteins can all influence thermostability. With respect to pH stability, most phages investigated were stable in the range of pH 5 to 8 but lost viability when held in more acidic or alkaline conditions [6].

31.5 PROPAGATION AND ENUMERATION

Methods that have been employed for the study of phages affecting *Clostridium* species are essentially the same as those used for aerobic species. The main differences relate to modifications required as a consequence of work with obligate anaerobes. Early studies involving clostridial phages were hampered by a lack of satisfactory methods for enumerating the phages coupled with difficulties of working with obligate anaerobes [6]. Those early studies relied on liquid dilution assay methods that only indicated the rough order of magnitude of the phage titre [47]. Modifications of the standard double-layer method were first applied to studies of phages affecting solvent-producing clostridia by Kinoshita and Teramoto [48]. Similar techniques were applied to the study of phages infecting *C. botulinum* [10,20], *C. perfringens* [35,37], and *C. sporogenes* [45].

With most clostridial phage host systems, there is no difficulty in using the standard agar plate soft overlay technique provided anaerobic cultivation facilities are available. For all but the strictest anaerobic species, anaerobic growth cabinets are not a requirement. For stringent anaerobes, all procedures need to be carried out in an anaerobic glove box. For most species, the procedures can be performed under aerobic conditions on the laboratory bench. Appropriate dilution series of phage suspensions are added to molten soft overlay medium containing the mixture of indicator cells and phage and then are poured onto the surface of suitable agar plates and allowed to set. Host cell growth and plaque formation are not normally affected by exposure to air, provided exposure times do not exceed 20 to 30 min before transfer to anaerobic conditions. The anaerobic cultural conditions required for the incubation of the plates can be provided by the use of standard anaerobic jars or cabinets. The evacuation of jars or chamber interchanges can, however, cause problems with the detachment of soft overlays. Additional problems with the disruption of surface overlays can arise from active rapid growth of the host strain, resulting in the evolution of gases that can cause cracking of the overlay and distortion through the formation of gas bubbles [6].

Factors affecting the formation and size of plaques include the number of cells, the age of cells, the concentration of agar, the thickness of the upper overlay, the composition and pH of the culture medium, and the temperature and period of incubation. Trials under different conditions might be required to achieve optimum plaque formation. The propagation of phages is usually optimum under conditions that allow good growth of the host organisms. In most instances, healthy actively growing host cells in mid-exponential phase produce the best lawns. Some clostridial species can be particularly problematic, as they do not produce even lawns on the agar plates under any conditions so that plaque formation is poor or nonexistent [6]. Assay or propagation techniques that need to be undertaken in liquid culture, such as those used for obtaining one-step growth curves and burst size of the phage, are best undertaken in an anaerobic glove box. If this item of equipment is not available, it might be possible to use liquid cultures bubbled under nitrogen [6].

For the production of large-volume high-titre phage lysates, either liquid or solid culture media can be used. Bulk liquid culture techniques are simpler but are less reliable. Host cells need to be at an optimal density and growth stage, and the infecting phage has to be added at the optimum multiplicity of infection to obtain complete lysis of liquid cultures. These conditions can often be difficult to determine and achieve. Solid culture techniques involve seeding agar plates with mixtures of phage and host cultures to achieve confluent lysis. Plates can be inoculated by spreading or by flooding and subsequent draining of the inoculum. After incubation, the plates are normally harvested by irrigating with an appropriate sterile phage buffer and scraping with a glass spreader or similar implement. Standard methods for the preparation and purification of phages can be accessed in various practical handbooks, such as *Methods for General and Molecular Bacteriology* [49].

31.6 GROWTH CYCLES

The growth cycle of virulent clostridial phages involves the processes of phage adsorption, replication, and assembly, followed by liberation by cell lysis. Phage adsorption can be affected by pH, temperature, and other culture conditions and in some instances requires the presence of divalent cations. The majority of phages adsorb to the cell wall by attachment using tail structures located at the distal end of the tail. Most, if not all, tailed phages seem to have lytic enzymes such as a lysozyme and holin associated with the phage tail structure capable of breaching the cell wall to enable the phage DNA to be injected directly into the host cell [17]. With a high multiplicity of infection, almost complete lysis of the cell culture can occur soon after infection.

The latent period following infection with virulent clostridial phages has been reported to vary widely, ranging between 20 and 120 min, with the majority exhibiting a latent period of around 45 min [5]. The latent periods of temperate phages following induction with UV irradiation or other mutagenic agents have been reported to vary between 20 and 50 min [5,37,50]. A close relationship between host cell doubling times and phage development can be anticipated, with phages inoculated into rapidly growing host strains cultivated under optimal conditions exhibiting shorter latent periods than phages developing in slow-growing species or cultured under suboptimal conditions. In a study of phage development in a temperate phage with a latent period of 45 min, Mahony and Easterbrook [36] reported that head forms became visible intracellularly around 30 min post-infection and tails were visible by 35 min, with full assembly being completed around 40 min post-infection. The typical phage burst size produced by infection with a single phage particle appears to vary widely between different clostridial phage types, with burst sizes reported to range from as low as 20 to almost 2000 [5]. Phage-induced lytic enzymes activated during the final stages of phage maturation normally result in cellular lysis from within, enabling the release of the mature phage particles by hydrolysis of the bacterial cell wall and capsular polysaccharide.

31.7 HOST RANGE AND PHAGE TYPING

In general, phages tend to exhibit a narrow host range specificity, with a particular phage type usually only being capable of infecting a limited number of closely related bacterial strains. However, there are some exceptions, with certain phages infecting *Bacillus* and *Streptomyces* species reported to have quite wide host ranges [2,6,18].

Information available in the literature suggests that, for clostridial phages that have been investigated, narrow host range specificity is the norm. In a number of investigations on virulent phages capable of lysing various *C. perfringens* strains, investigators observed that the host ranges of the individual phages were quite restricted [5,51]. In some cases, more than half of the strains tested were resistant to any phage, and the temperate phages included only infected strains of the same type that harbored the phage. Although phage typing studies of limited epidemiological value were undertaken, the restricted host range of the phages tested have led to the conclusion that a general phage typing scheme for *C. perfringens* was not feasible [5]. Similar conclusions were reached with respect to phage host range and the development of phage typing schemes for other species of pathogenic clostridia of epidemiological importance. Studies investigating the host range of phages infecting *C. botulinum* [5], *C. histolyticum* [34,52], *C. novyi* [23], and *C. sporogenes* [45] suggested that the host range of the phages tested were too restricted for use in routine typing applications.

Similar findings have been reported with respect to the host range of phages infecting various species of industrial solvent-producing clostridia, where phage infection appears to have been quite strain specific [6,53]. The phages infecting industrial strains of *Clostridium madisonii* (subsequently reclassified as *C. beijerinckii*) and *C. acetobutylicum* were found to exhibit narrow host ranges and only infect some strains of the same species with similar characteristics and fermentation ability [28,54]. The host ranges of a large number of phages infecting *C. saccharoperbutylacetonicum,* isolated from contaminated industrial fermentations in Japan, were investigated by Hongo and Murata [29]. All of these phages were found to be infectious only to derivatives of *C. saccharoperbutylacetonicum* industrial strains, indicating very narrow host ranges. Seven phages were used for phage typing in a taxonomic study of more than 50 strains belonging to four species of solvent-producing clostridia [55]. All of these phages exhibited restricted host ranges that were confined to the infection of strains belonging to a single genomic DNA fingerprint group. The use of phage typing has now largely been superseded by molecular identification techniques, such as genomic DNA fingerprinting and 16S rRNA sequencing.

31.8 RESTRICTION AND MODIFICATION SYSTEMS

Bacterial restriction and modification systems are known to provide an effective defense system against the entry of foreign DNA into bacterial host cells [56]. A bacterium encoding for a restriction endonuclease and modification methylase has the capacity to recognize the invading DNA as foreign and degrade it via the action of restriction endonucleases. However, the associated modification methylases can also provide opportunities for the adaptation of a phage population so that it will no longer be constrained by the action of restriction endonucleases [57].

Host controlled restriction/modification systems have been reported to occur in a number of different clostridial species [58,59]. Richards et al. [58] prepared crude extracts from eleven strains of *Clostridium* and screened them for restriction endonuclease activity, using phage DNA as a substrate. *Clostridium pasteurianum* ATCC 6013 was shown to contain an isoschizomer of *Tha*I (*Fnu*DII) [5′-CGCG-3′]. *Clostridium thermohydrosulfuricum* DSM 568 contained an isoschizomer of *Mbo*I (*Sau*3A) [5′-GATC-3′] that was inactive on dam-methylated substrates. The chromosomal DNA of this strain was dam-methylated, which prevented self-digestion [58]. In addition, isoschizomers of *Mbo*I were found in *C. pasteurianum*, isoschizomers of *Bcl*I and *Eco*RI in *Clostridium thermocellum* ATCC 27405, and *Cfo*I in *Clostridium formicroaceticum* ATCC 23439 [60,61].

Experiments involving phage infection conducted on sporulation mutants of *C. perfringens* implied that the wild-type strain might carry a gene coding for a defective restriction endonuclease, which could mutate to an active form [62]. Based on the findings of this study, the authors postulated the existence of a restriction/modification system in this strain. The susceptibility of DNA to cleavage by certain restriction endonucleases [63], along with difficulties encountered in conducting conjugation and transformation studies [64,65], has also provided convincing evidence for the presence of restriction/modification systems in a number of strains of *C. perfringens*. Rood and Cole [66] have reported that preliminary studies have identified a restriction endonuclease that is an isoschizomer of *Mbo*I.

The type strain of *C. acetobutylicum* has also been shown to contain a potent restriction/modification system. This restriction endonuclease can act as a potential barrier to genetic transformation, which cuts GC-rich DNA fragments with a very high frequency [59]. The presence of this restriction/modification system has been postulated as the main reason why gene transformation difficulties have been encountered with plasmids containing *E. coli*-plasmid fragments [67].

31.9 LYSOGENY AND PSEUDOLYSOGENY

Phages are classified as virulent or temperate according to their relationship with the host bacterium. Temperate phages are capable of establishing a long-term stable relationship with a host bacterium that prevents lysis of the host cell and ensures that all progeny cells carry the phage [68]. The temperate phage genome is maintained as a prophage in the host bacterial cell, which is replicated along with the host chromosome and transferred to the daughter cells. To ensure distribution to both progeny cells at each cell division, the prophage DNA must replicate at the same rate as the host chromosomal DNA. Bacterial strains harboring a prophage that has the potential to produce phage particles are referred to as lysogens. Two types of prophage relationships have been reported. The first type includes most of the extensively studied temperate phages, whose genomes are integrated into the bacterial chromosome. Integration can occur either by a site-specific recombination mechanism between the phage DNA and the bacterial chromosomal DNA or by transposition [69]. A second type of prophage is exemplified by phages P1 or P4 that rarely become inserted into the host chromosome, but exist as extrachromosomal plasmids that replicate independently within the bacterial cell [70]. For the maintenance of the lysogenic state, the initial effect of the phage infection must be either nondetrimental to cell survival — or reversible. The survival of the host cell and the maintenance of the prophage rely on the interaction between phage and host, which ensures that most of the phage-specific proteins, especially those involved in late function, such as the production of lysozyme and the virion structural proteins, must be suppressed to prevent phage particle assembly cell lysis. However, the lysogenic state cannot be maintained in every situation. Under certain conditions, it is possible for the prophage to revert to the vegetative cycle and initiate a lytic replicative process. This conversion normally comes about through the inactivation of a repressor system, which can occur spontaneously or be induced by exposure to some mutagenic agents, such as ultraviolet radiation or Mitomycin C. When a temperate phage enters the lytic state, it follows a very similar process to that of a virulent phage [71].

There are numerous reports relating to the occurrence of temperate phages in various strains belonging to the specific toxin types of *C. botulinum,* which may either be released spontaneously or after induction following treatment with UV-irradiation, Mitomycin C, or other mutagenic agents [5,9,10,19–21,72–74]. The majority of studies involving temperate phages have been focused on the elucidation of the role of temperate phages in the conversion of host strains to toxin production. Studies relating to the toxigenic conversion of *C. botulinum* strains are considered in more detail in Section 31.12.

Lysis of *C. tetani* cultures following treatment with Mitomycin C or UV irradiation was reported by Prescott and Altenbern [27], but they were unable to demonstrate plaque formation from induced cultures. Plaque formation was demonstrated from 11 of 26 toxin-producing strains of *C. tetani* investigated by Roseman and Richardson [75]. A temperate phage isolated from one of three lysogenic strains of *C. histolyticum* was partially characterized by Guélin et al. [34].

The occurrence of temperate phages in *Clostridium* has been most comprehensively studied in strains of *C. perfringens* belonging to the various toxin types. A number of investigations have reported the occurrence of strains harboring temperate phages, and some of these lysogenic systems in *C. perfringens* have been partially characterized [5,11,35–37,50,51]. In these various studies, the frequency of lysogeny was reported to vary between 13% and 43% in the strains tested. When the findings of these studies are considered together, a cumulative total of around 400 strains are reported to have been investigated. Of these, approximately 80 strains were reported to be lysogenic given an overall average frequency of around 20% for the occurrence of lysogeny among *C. perfringens* strains. In some instances, lysogenic strains were shown to be inducible by treatment with UV light and Mitomycin C. A high proportion were not inducible and produced phage spontaneously. The morphotypes exhibited by temperate phages included examples belonging to the *Podoviridae*, *Myoviridae,* and *Siphoviridae*. Individual phage types exhibited considerable structural variation in tail structure and head size, which can range from 30 to 90 nm in diameter. The conclusions reached in a number of these reports with respect to host range specificity strongly support the view that these temperate phages are restricted to a very narrow host range. In virtually all of the phages investigated, infection was restricted to closely related strains that did not harbor the prophage genome. Since many of the *C. perfringens* strains were not susceptible to any of these phages, it was concluded that these *C. perfringens* temperate phages had little value for epidemiological typing schemes [5].

More recent studies on *C. perfringens* temperate phages have involved the mapping of the chromosomal attachment sites of two lysogenic phages previously designated as Ø29 and Ø59. Comparison wild-type and lysogenic derivatives of strains of the host strain were undertaken using pulsed-field gel electrophoresis. The results demonstrated that the Ø29 site was near the *Mlu*I site at 1.0 Mb and the Ø59 attachment site was located near the *Mlu*I site at 2.9 Mb on the chromosomal map [76]. The Ø29 site was located within a 180-kb *Fsp*I fragment, which also contained the *nan*H gene, which encodes neuraminidase (sialidase) production [77]. The Ø59 site was located near the *atp*D gene, which codes for ATP synthase [76]. The nucleotide sequence of the genome of another UV inducible temperate phage designated Ø3626 isolated from a lysogenic *C. perfringens* strain has been determined [41]. The phage was characterized by electron microscopy, protein profiling, and host range determination and was shown to be a member of the *Siphoviridae,* with an isometric capsid of 55 nm in diameter and tail of 170 nm in length. The genome consisted of a linear double-stranded DNA molecule with 3′-protruding cohesive ends consisting of 9 residues. Integration of the viral genome into the bacterial chromosome was shown to occur at the *att*B attachment site, located within the 3′ end of a *gua*A homologue. This essential housekeeping gene was shown to be functionally independent of the integration status due to reconstitution of its terminal codons by the phage sequence. The Ø3626 phage attachment site *att*P was located in a noncoding region immediately downstream of *int*. The nucleotide sequence of the genome consisting of 33,507 bp was determined. Fifty open reading frames were identified, organized within 3 main life-cycle-specific gene clusters. The genes involved in lytic development occurred in the opposite orientation to those involved in the control region for lysogeny. Functions were assigned to 19 gene products, based upon bioinformatic analyses, N-terminal amino acid sequencing, or experimental evidence. These included structural components, DNA-packaging proteins, a dual lysis system, a putative lysogeny switch, and proteins involved in replication, recombination, and modification of phage DNA. The identification of phage encoded genes for a putative sporulation-dependent sigma factor and transcription regulator was taken as indicative of a possible interaction of the Ø3626 phage with initiation of sporulation in *C. perfringens* [41].

Compared with the pathogenic clostridia, there are only sparse reports of the occurrence of temperate phage among the various species of solvent-producing *Clostridium*. Lysogenic strains have been reported in *C. acetobutylicum* [48] and an unidentified solvent-producing species [31]. A large temperate bacteriophage, designated CMX, infecting a strain of *C. madisonii* (subsequently reclassified as *C. beijerinckii*) was isolated from an abnormal industrial acetone butanol fermentation [78–80]. The genome of this phage, belonging to the *Siphoviridae,* consisted of a 215 kb linear dsDNA genome. The inducible prophage was shown to exist as a circular extrachromosomal element maintained within the host strain in low copy number [43].

In some instances, certain phages, such as the *B. subtilis* phages PBS1 and SP10, can resemble temperate phages by way of their ability to establish a relatively long-term relationship with the host strain, which mimics normal lysogeny, and results in the production of turbid plaques, as do most temperate phages [81]. However, treatment of pseudo-lysogenic cultures with phage-specific antiserum led to loss of phage, indicating that continuous reinfection is essential to maintain the prophage state [82]. It has been observed that little, if any, of the prophage DNA is integrated into the host chromosome, and the exponentially growing cultures release very high titre of free phage [81]. Phages of this type are not considered to be true temperate phages, and the association of these phages with the host cell is considered to represent a "pseudolysogenic" or "carrier state" [81].

Hongo and Murata [29] reported on the isolation of a number of closely related phages from abnormal butanol fermentation broth capable of infecting *C.saccharoperbutylacetonicum*. The majority of the isolates were virulent phages, but concurrently they also isolated a variant that appeared to be a temperate phage, which they designated HMT. All of these related phages exhibited the short tail characteristics of the *Podoviridae* and exhibited no significant difference with respect to the physical and biochemical properties of phages including serology, morphology, host range, DNA base composition, and other properties. The HMT phage was found to be capable of forming a lysogenic relationship with the host strain, but the frequency of lysogenization was low (around 0.1%). A notable feature of the HMT-lysogenized strain was the spontaneous release of phage particles at a very high titre (up to 10^{10} pfu/ml) in thc culturc [31]. These workers indicated that the phage genome was not integrated into the host chromosome but existed in the cytoplasm. The origin of phage HMT was not clear, but it was suggested that the virulent phages might have been mutants derived from HMT phages. Detailed information relating to a wide range of studies undertaken on this group of phages is contained in the comprehensive review by Ogata and Hongo [6]. Pseudolysogeny has also been reported to occur in a phage infecting *C. botulinum* [10]. This is covered in more depth in Section 31.12.

31.10 DEFECTIVE PHAGES AND PHAGE-LIKE PARTICLES

The ability of bacterial cells to produce noninfective phage-like particles is a widespread occurrence found in many species [68,83]. Particles morphologically similar to phages, or phage-like components such as structures resembling empty phage heads or phage tails, have been reported to occur in a variety of Gram-positive and Gram-negative bacteria. These particles appear to be unable to produce productive phage infections; although, in a number of instances, these particles have been shown to possess lytic or bacteriocin-like activity. These phage-like particles are frequently referred to as defective phages [84]. In common with many temperate bacteriophages, the defective phages or phage-like particles are produced either spontaneously or induced following treatment of the cells with agents that induce the SOS response, such as UV irradiation, Mitomycin C, or other mutagenic agents [85]. The particles can be readily observed by electron microscopy of culture supernatants in which a producer strain has been grown.

A number of species of clostridia have been reported to produce phage-like particles [5,19,21,25,27]. A survey of the production of phage-like structures in 34 strains belonging to 16 species of *Clostridium* was undertaken by Beatriz et al. [32]. A wide variety of phage or phage-like particles were observed to be produced by 20 of the strains belonging to 14 clostridial species. These ranged in morphology from apparently intact phage virions to rod-shaped structures, sheaths, filaments, and fibrils. Some of the particles were only produced after induction with mutagenic agents, while other strains exhibited spontaneous production. Growth inhibition and bacteriocin-like activity associated with different species of *Clostridium* were also investigated. Of the 34 strains tested, 25 were found to exhibit some bacteriocin-like activity that inhibited the growth of one or more of the other strains tested. However, with one exception, cell-free filtrates were found to be inactive. One *Clostridium bifermentans* strain was shown to inhibit the growth of 23 of the 34 strains investigated. A range of similar structures was observed to be produced by a number of solvent-producing strains included in the taxonomic studies undertaken on four species of solvent-producing clostridia [55] (unpublished results).

Bacteriocin-like activity associated with particulate structures has been investigated in more depth in a number of clostridial species. These appear to exhibit a morphological resemblance to the high molecular weight pyocins produced by *Pseudomonas aeruginosa* that are postulated to be encoded for by a defective prophage within the bacterial chromosome [86]. Similar killer particles have also been reported to occur in a number of Gram-positive species including *Actinomyces, Bacillus, Streptococcus,* and *Streptomyces* [33]. These phage-like killer particles are able to adsorb to, and kill, sensitive cells without multiplying in them. These particles are generally referred to as particulate bacteriocins. The amount of bacteriocin produced can usually be greatly increased by treatment with inductive agents.

A particulate bacteriocin, designated boticin P, produced by a nontoxigenic type E strain of *C. botulinum* was investigated by Lau et al. [87]. It was postulated that these particles were mutants of a normal phage that consisted of a heat and pH sensitive tail-like structure with a contracted sheath. Dolman and Chang [19] also reported the presence of a killer particle in *C. botulinum,* similar to the PBSX killer particle in *Bacillus,* that consisted of a phage-like entity comprised of a small head and a long, thick contractile tail exhibiting striations and a base plate. This defective phage-related killer particle was referred to as a high molecular weight bacteriocin.

Two phage tail-like particles, designated as clostocins O and M, were produced in the Mitomycin C or UV-induced lysates of *C. saccharoperbutylacetonicum* [88]. The particles resembled a sheathed phage tail, but no equivalent head-like structures were observed. These particles did not contain DNA and were incapable of propagating themselves, but were capable of causing lysis of sensitive clostridial strains. The particles occurred in two forms, one being described as the intact sheathed form and the other as the contracted inactive form [89]. Some of the intact forms were observed to be converted into the contracted form during purification [89]. When the intact particles adsorbed to the cell surface of sensitive host strains, they were observed to convert into the contracted form [88]. The particles did not contain DNA and consisted mainly of protein with a small amount (2 to 3%) of RNA [89]. It was shown the RNA was not involved in the killing action of the particle, and it was suggested that it might be involved in the assembly of the particle [89]. It was concluded that the lysis of mature clostocin O-producing cells was due to the cell-wall hydrolysis, which was controlled by the clostocin O lysin. Clostocin O was found to kill only clostocin M-producing strains of *C. saccharoperbutylacetonicum*, and clostocin M only killed clostocin O-producing strains [88]. The killing mechanism was not determined, but it was reported that clostocin O acted on the mRNA synthesis step as its first target [90]. Defective phage-like particles resembling PBSX particles from *Bacillus* have also been observed to be produced in Mitomycin C-induced cultures of some strains of solvent-producing *C. beijerincki* [43]. Lysis caused by defective phage or phage-like particles have narrow activity spectra, but the host range specificity is still broader than that of virulent phages, as the particles can cause lysis in other species of *Clostridium* [6,32].

The evidence available indicates that many clostridial species, in common with other Gram-positive genera, are capable of releasing a wide variety of phage-like particles under appropriate conditions. In some instances where intact phage particles containing phage DNA are produced, these might represent viable temperate phages for which suitable host indicator strains have not yet been discovered. In other instances, these might represent defective phage particles that are incapable of infection, self-replication, and propagation. A number of clostridial strains have been observed to produce both head structures and tail structures, but intact phage particles consisting of fully assembled heads and tails are not observed. Where head-like structures are present, these can be empty, or they can appear to contain genomic material. Tail-like components constitute the most commonly observed particulate structures. These often take the form of headless rod-shaped tails with visible striations present. These particles can display the form of an uncontracted tail sheath, or of a contracted tail sheath, or a combination of both. Less commonly encountered are particulate structures that display the form of long narrow flexible filaments or wider flexible sheaths of varying length, which show structural similarities to the core and sheath structures of phages with flexible noncontractile tails. In some instances, finer filamentous forms with either a straight or a curly appearance, reminiscent of phage tail attachment fibers, might also be observed. In some instances, these particulate structures have been demonstrated to have lytic activity and are capable of acting as killer particles or particulate bacteriocins; in other instances, bacteriocin-like activity was not demonstrated. The particles can either be produced spontaneously or after exposure to an inducing agent. What limited experimental evidence exists supports the interpretation that these various particulate structures are encoded for by genes situated on localized regions of the host cell chromosome that exhibit the features in common with prophages or defective prophage remnants.

31.11 USE OF PHAGES AS GENETIC TOOLS

Phages have provided useful tools for the study of genetics in a number of well-studied Gram-negative and Gram-positive bacterial species. A large number of virulent and temperate phages infecting *B. subtilis* have been extensively characterized, and a number of these have proved to be useful for achieving either generalized or specialized transduction. Generalized transduction mediated by a pseudolysogenic flagellatrophic phage PBS1 that has a genome in excess of 200 kb has provided a useful tool for the course level positioning of genetic markers on genomic maps. Another useful tool for specialized transduction in *B. subtilis* is the large 130 kb temperate phage SPβ. In addition, two different temperate phages, SPβ and Ø105, have been developed for use as cloning vectors in *B. subtilis*. Both vector systems enable the construction of recombinant specialized transducing phages that can be used for the cloning of genes by complementation. The techniques for preparing lysates of these phages and information relating to standard genetic applications for which these phages can be utilized can be accessed from a number of technical handbooks, such as *Methods for General and Molecular Bacteriology* [49].

Although gene transfer and cloning systems have been developed for a number of *Clostridium* species involved in solvent production, as well as those of medical importance, phage transduction and cloning techniques, for gene characterization and gene transfer, have not been developed for any clostridial species so far [59]. Although phages of *C. perfringens* have been quite extensively studied over a number of years, there are no reports of the transduction of either chromosomal or plasmid-determined genes. Attempts to obtain transduction of erythromycin resistance using temperate phages were unsuccessful [35]. An indicator strain of *C. perfringens*, designated strain 13, capable of being lysogenized by the temperate phage S9 was investigated by Mahoney and Kalz [37]. Although this phage recipient is readily transformable due to an apparent lack of a restriction system, little attempt has been made to use transduction as a means of strain construction in derivatives of this strain [65]. Phage resistance has been used as a marker for the genetic analysis of transconjugants [91], but studies of phage genetics have, in general, proved to be unrewarding and played little role in the development of *C. perfringens* genetics [66]. Among the solvent-producing clostridia, phage DNA has been used as a marker for the development of protoplast transformation techniques [92]. Phages have also been used in the construction of a shuttle vector for transfection of *C. acetobutylicum* [59].

The occurrence of phenotypic changes in host strains, mediated by phage genomes acquired by infection of a temperate phage, have been investigated in *C. perfringens.* Structural changes in *C. perfringens* cells infected with a short-tailed bacteriophage originally isolated by Vieu et al. [38] were first reported by Bradley and Hoeniger [93]. Lysogenization of a derivative of strain 13 by the temperate phage S9, along with lysogenized and cured derivatives of the original lysogenic parent strain, was reported to result in the production of an increased number of heat-resistant spores and a shortening of the sporulation process in strains harboring the prophage [11]. Additional studies were undertaken on the UV-inducible bacteriophage S9 [37] and on a second phage isolated in the same laboratory [36]. These findings were interpreted as an example of phage-mediated phenotypic conversion involving phage encoded sporulation genes in *C. perfringens,* but confirmation of this conclusion would require further in-depth molecular studies. Attempts to correlate the presence of particular phages with toxin production in *C. perfringens* have been unsuccessful [4].

31.12 TOXIGENIC PHAGE CONVERSION

Phages are commonly associated with *C. botulinum* strains and, in some cases, have been shown to be implicated in toxin production. Inoue and Iida [9] were the first to report the induction of phages from strains of *C. botulinum* types A–F using UV irradiation or Mitomycin C treatment. They grouped the phages into three morphotypes that associated with various toxin types and suggested that the toxigenicity of *C. botulinum* types C and D were dependent on the presence of phages. Subsequently, they demonstrated that when Mitomycin C-induced toxigenic cultures of *C. botulinum* type C were mixed with a nontoxigenic mutant of the same strain, the nontoxigenic strains were converted to toxin production [94]. They also showed that lysates prepared from a type C strain could convert a nontoxigenic type D strain into a type C toxin-producing strain and that toxigenicity was concomitantly lost with the curing of the phage infection [95]. The loss and reacquisition of toxigenicity by curing and reinfection with a lysogenizing phage have also been reported by other workers [10]. Eklund et al. [10] also reported the existence of two phages carried by *C. botulinum* type C, of which only one exhibited toxin converting ability. They demonstrated that toxin-producing strains lost toxigenicity after treatment with phage antisera. Further studies demonstrated the toxigenicity of *C. botulinum* types C and D and *C. novyi* types A and B depended on the presence of phages, and the association between phages and toxigenicity in type D strains was also confirmed [72]. Further evidence for phage conversion was provided by demonstrating that lysates from already converted strains were able to convert nontoxic strains of types C and D [96]. Eklund and Poysky [20] demonstrated nontoxigenic types C and D strains could be converted by infection with phages from toxigenic types C and D and postulated that toxigenic types C and D might lose their prophages in their natural environment and reinfection could convert strains to either type C or D toxin production. It was also demonstrated that some strains of *C. botulinum* could lose prophages and still produce toxins, indicating that in these strains either prophages remain uninduced or not all toxin production in *C. botulinum* was phage controlled.

Eklund et al. [20,73] also demonstrated the alpha toxin production in *C. novyi* types A and B was also governed by the presence of phages. When type A strains lost their phage, the resulting cured nontoxigenic type A organisms more closely resembled *C. botulinum* types C and D than the other types of *C. novyi*. They also demonstrated that a nontoxigenic mutant of *C. botulinum* type C could be converted not only to toxigenic type C or D by respective phages, but also to *C. novyi* type A by infection with a phage from that strain. These findings confirm the close genotypic and phenotypic relationship between *C. botulinum* types C and D, and *C. novyi* types A–C, reported by Johnson and Francis [97].

Hariharan and Mitchell [98] reported that the toxigenic type C strains they examined were infected with one or two phage types and that the phage-host relationship was characterized by a stable lysogenic type of association. In a study of stability of toxigenicity in *C. botulinum* types C and D, Oguma [99] showed that most converted strains lost toxigenicity upon serial transfer with or without phage antiserum — or when the bacterial spores are heated at 70°C for 20 min. This indicated that the prophage was not integrated into the host chromosome but existed in a carrier state, indicating a pseudolysogenic relationship rather than a true lysogenic conversion. He also suggested that nonconverting mutant phages might have arisen that could render the host strain phage-resistant but nontoxigenic, thus explaining the loss of toxigenicity commonly observed in some type C and D strains of *C. botulinum*. Oguma et al. [100] also demonstrated that hemagglutinin production in *C. botulinum* types C and D also appeared to be controlled by phage infection and that this property could be transmitted concomitantly or separately from toxin production.

The genomic characteristics of the converting phages were characterized and the existence of the structural genes for the toxins was confirmed by hybridization tests [40]. The five converting phages and one nonconverting phage that were investigated were shown to contain dsDNA, with a genome size ranging from 110 kb to 150 kb. The absence of true lysogenization was confirmed by demonstrating that gene-specific probes for the toxin gene hybridized only with phage DNA, and not host chromosomal DNA. The location of toxin genes on the phage genome was established via recombinant techniques, and the nucleotide sequences encoding for the C1 and D neurotoxins were determined [101,102].

A direct relationship between phage and toxigenicity has only been established for *C. botulinum* types C and D strains belonging to group III. Phages induced from group I and II *C. botulinum* strains have been shown to exist in true lysogenic association. As a consequence, phage-sensitive variants are seldom isolated. The phage-sensitive variants that have been investigated were found to retain the ability to produce toxins, suggesting that these lysogenic phages were not involved in toxigenic conversion [103]. However, there is some evidence that suggests that botulinum toxin production might be associated with lysogenic prophages in some clostridial species [104]. In a study undertaken on toxigenic *Clostridium butyricum* strains, the position of the CNT toxin gene was located in the chromosomes of the toxigenic strains *C. butyricum* ATCC 5839 and 5521 using a gene-specific probe. The CNT toxin gene was not detected on the chromosome of the nontoxigenic strain ATCC 19398 [105]. Both the toxigenic and nontoxigenic strains were shown to harbor prophages, which had similar genome sizes of around 34 kb, but proved to be unrelated [105]. The finding that both the phage probe and the toxin probe appeared to hybridize to the same chromosomal fragment suggested there could be some common linkage of these elements on the host chromosome. This suggestion was supported by the finding that toxin production could be transferred from *C. butyricum* ATCC 5839 to a nontoxigenic *C. botulinum* strain by a phage-mediated process. Transfer only occurred when the recipient was incubated in a phage lysate derived from the donor supplemented with a filtrate of a helper strain, indicating that the phage particles produced by ATCC 5839 were defective and only become infective through the participation of the helper strain. Subsequent investigations revealed that, although purified phage DNA was able to act as a template for the PCR mediated amplification of a fragment encoding the toxin gene, the product produced did not hybridize to a toxin gene probe, suggesting that only a minority of the purified phage genomes contained the toxin gene [105]. It has been proposed that juxtaposition of the prophage and toxin genes on the host chromosome could have given rise to phage derivatives carrying fragments of host chromosomes carrying toxin genes, through a process of imperfect excision, similar to specialized transduction [104]. Indirect evidence for the involvement of phages in neurotoxin gene transmission has also come from the analysis of DNA regions in the vicinity of the botulinum toxin locus [104].

Historically, toxicity and pathogenicity were utilized as key characteristics in the classification and taxonomy of the pathogenic and toxigenic clostridia. One of the most obvious examples was the emphasis placed on toxin types in the classification of *C. botulinum*, where the characteristics of the species were defined by its ability to produce neurotoxins. This approach led to some major anomalies with respect to classification and nomenclature, from a phylogenetic perspective. It was subsequently shown that, in some instances, strains belonging to the same genotype produced different toxins; in other instances, the same toxin was shown to be produced by strains belonging to different genotypes. This observation was elucidated by the discovery of toxigenic phage conversion. This has led to the outcome of the different types of *C. botulinum* strains being placed into three taxonomic clusters that are unrelated phylogenetically [4].

31.13 OCCURRENCE FREQUENCY AND CAUSES OF PHAGE INFECTIONS IN THE INDUSTRIAL AB FERMENTATION

From the beginning of the 1920s to the end of the 1950s, the industrial Acetone-Butanol (AB) fermentation process utilizing solvent-producing clostridia flourished in the U.S. and Britain, serving as the major route for the commercial production of solvents [14]. The AB fermentation process was subsequently established in many other countries and continued to operate in Japan until the early 1960s, in South Africa until the early 1980s, and in China until recent times. Wherever the process was established, contamination by phages emerged as a serious problem for the fermentation industry [6]. Based on reports spanning more than 60 years, during the period that the AB fermentation was operated in the Western world, it can be concluded that no matter how effective the plant hygiene and operating practices were, sporadic phage infections could not be prevented [15,106,107]. Although infections by phages were identified as a major problem in the AB fermentation, little detailed information relating to specific phage infections was ever published in the scientific literature. The main reason for this was that almost all of the work carried out on phage infections was undertaken by commercial companies and was considered to be commercially sensitive information. There are numerous general references to the effects of phage infections in the AB fermentation contained in reviews dealing with the AB fermentation process [6,53,106–109]. The incidence and effects of phage infections occurring in the industrial AB fermentation processes operated in the U.S., Puerto Rico, Japan, and South Africa were detailed in a recent review [15].

The first occurrence of phage contamination in the AB fermentation was reported to have occurred in 1923 in an industrial fermentation plant operated by the Commercial Solvents Corporation in the U.S. [13]. This company later undertook extensive research into the problem of phage contamination, which resulted in the development of patented immunized strains that were resistant to phage infection. Problems with phage contamination were also reported to have affected an industrial AB fermentation process operated in Puerto Rico during the early 1940s [79]. At least four different infections occurred involving serologically distinct phages in a newly established process producing solvents from molasses, utilizing a patented strain designated as *C. madisonii* (now reclassified as *C. beijerinckii)* [55]. A series of immunized strains were generated to counter the phage contamination, but each new immunized strain developed proved susceptible to a new phage infection [78–80].

The AB fermentation operated in Japan during the World War II encountered problems with abnormal fermentations at several factories during the early 1940s. The problems with sluggish fermentations were subsequently shown to have been caused by phage infections [28]. A new solvent-producing clostridial species isolated from soil capable of producing high levels of butanol was employed in the industrial fermentation process to give more favorable butanol ratios [29].

These new clostridial strains were patented under the name *C. saccharoperbutylacetonicum*. However, a series of phage infections occurred during the industrial application of these strains that detracted from their success. Twelve separate phage infections occurred within a period of about a year and large number of phages were isolated, which were characterized as virulent phages belonging to the *Podoviridae, Myoviridae,* and *Siphoviridae.*

An industrial AB fermentation process was operated by National Chemical Products in South Africa over a period of 46 years. During this time, four confirmed phage infections occurred, and another two suspected but unconfirmed phage infections were recorded. The interval between phage infections ranged from 2 years up to 16 years. A range of virulent and pseudo-lysogenic phages were recovered from these infections that were identified as belonging to the *Podoviridae, Myoviridae,* and *Siphoviridae*. The two earliest infections took place during the period when the fermentation process was undergoing a conversion from using maize to molasses as the fermentation substrate. Another phage contamination incident occurred after one of the factory fermenters exploded, and a later infection resulted from switching from the use of carbon dioxide to unfiltered nitrogen as the gas used to produce anaerobic conditions while the fermenters were being filled [15,16].

From the records that are available, it would appear that the AB fermentation process was particularly vulnerable to phage contamination at the time that new industrial plants were being established, or when new production strains were being introduced. Other infections were associated with chance mishaps and errors or changes in operating procedures [15,16].

31.14 SYMPTOMS AND DIAGNOSIS OF PHAGE INFECTIONS IN THE INDUSTRIAL AB FERMENTATION

The typical symptoms produced by a phage infection in the industrial acetone-butanol fermentation process are sluggish fermentations with extended fermentation times, decreased solvent concentrations and yields, and alterations in solvent ratios. Phage contamination is also usually accompanied by changes in the cell morphology which can include loss of motility, cell wall changes, cell lysis, protoplast formation, and, on occasion, the production of elongated filaments. Changes in cell morphology usually occurred concurrently with a marked reduction in cell numbers. The effects produced by a phage infection can vary significantly, depending on the specific circumstances associated with the particular infection. A heavy infection by a virulent phage with a short latent period and large burst size can produce very rapid and severe symptoms. Such infections can result in the fermentation coming to a complete stop, with the majority of cells undergoing complete lysis within a few hours. When an infection occurs early on in the fermentation process, either during inoculation or through contamination of the starter culture, the infection is usually widespread often affecting all fermenters. Such infections typically lead to stalled fermentations, which can sometimes start up again after a delay of one or two days. Fermentations occurring after such delayed starts almost always exhibit protracted fermentation times and produce poor solvent concentrations and yields. Phage infections that occur at a later stage in the fermentation process, or those resulting from infections by lysogenic or pseudo-lysogenic phages, can often produce less marked and widespread effects. These typically result in incomplete fermentations, characterized by a reduction in gas evolution, high acid levels, and the presence of unfermented substrate resulting in the mash remaining dark colored. Symptoms associated with low level persistent contamination by a phage can be less obvious with more protracted symptoms that can be much harder to diagnose. Such chronic infections can prove particularly difficult to eradicate from infected fermentation plants. The composition of the fermentation substrate, and the physical and chemical conditions in the fermenters, can also exert an effect, but in general the symptom produced by phage infections are similar, regardless of whether the infections occurred in starch- or sugar-based fermentation substrates. A definitive diagnosis of a phage infection is normally achieved by utilizing standard liquid or agar plate phage assay techniques involving the transmission of phage infection to an indicator strain, coupled with examination using an electron microscope [15,16].

31.15 PREVENTION AND TREATMENT OF PHAGE INFECTIONS IN THE INDUSTRIAL AB FERMENTATION

The standard industrial batch fermentation process was run under sterile conditions using pure cultures. This required the sterilization of the fermentation substrate, the water, and the anaerobic make-up gas. The design of the fermentation plant was crucial in minimizing opportunities for phage contamination to occur, with adequate provision for steam sterilization of the fermenters, lines, valves, and other items of the plant, coupled with facilities for acid or alkali treatment when required. It was also critical that operating procedures were adopted that minimized the likelihood of such infections occurring and that a high standard of factory hygiene was maintained [15,16].

A widespread approach for making provisions to deal with the effects of phage infections was the maintenance of a battery of strains that were resistant to a range of different phages [53,106–109]. The development of phage-resistant mutant strains was referred to as phage immunization. The method of producing immunized strains normally involved the serial transfer and reinfection of cells surviving exposure to phage infection. Between 10 and 18 selection cycles were normally undertaken, coupled with exposure of surviving spores to heat sterilization treatment. However, in many instances, immunized strains produced lower solvent concentrations and yields and exhibited altered solvent ratio compared with the original parent strain [53]. A further drawback associated with the production of immunized strains was that it was not possible to immunize a single strain against a range of different phages. In addition, some resistant strains proved to be lysogens that resulted in persistent temperate phage infections [109].

When a phage infection did occur, it was standard practice to endeavor to eliminate all of the contaminated material, followed by rigorous cleaning, disinfection, and sterilization. This often required closure of the entire plant while decontamination was undertaken. Common liquid and gaseous disinfectants were used in the plant to control contamination such as fumigation with formalin vapor, washing and spraying with quaternary ammonium compounds, and disinfection of fermenters, lines, and valves with a caustic or formalin treatment, followed by steam sterilization [15,16].

Various methods were attempted to salvage phage-infected industrial fermentation substrates, including treatment with antibiotics or detergents and the addition of chelating agents to remove divalent cations known to be essential for infection with some phages. However, no successful method for recovering phage-contaminated fermentations was ever developed [6].

REFERENCES

1. Ackermann, H.-W., Frequency of morphological phage descriptions in 1995, *Arch. Virol.*, 141, 209, 1996.
2. Ackermann, H.-W., Tailed Bacteriophages: The Order *Caudovirales*, *Adv. Virus Res.*, 51, 101, 1998.
3. Douglas, J., *Bacteriophages*, Chapman & Hall, London, 1975.
4. Cato, E.P. and Stackebrandt, E., Taxonomy and Phylogeny, in *Clostridia*, Vol. 3, Minton, N.P. and Clarke, D.J., Eds., Plenum Press, New York, 1989, 1.
5. Mahony, D.E., Bacteriocins, bacteriophage and other epidemiological typing methods for the genus *Clostridium*, in *Methods in Microbiology*, Vol. 13, Bergan, T. and Norris, J.R., Eds., Academic Press, New York, 1979, 1.
6. Ogata, S. and Hongo, M., Bacteriophages of the genus *Clostridium*, *Adv. Appl. Microbiol.*, 25, 241, 1979.
7. Summers, W.C., Bacteriophage therapy, *Ann. Rev. Microbiol.*, 55, 437, 2001.

8. Carson, C F. and Riley, T.V., Non-antibiotic therapies for infectious diseases, *Commun. Dis. Intell.*, 27, S144, 2003.
9. Inoue, K. and Iida, H., Bacteriophages of *Clostridium botulinum*, *J. Virol.*, 2, 537, 1968.
10. Eklund, M.W., Poysky, F.T., Reed, S.M., and Smith, C.A., Bacteriophage and the toxigenicity of *Clostridium botulinum* type C, *Science*, 172, 480, 1971.
11. Stewart, A.W. and Johnson, M.G., Increased numbers of heat-resistant spores produced by two strains of *Clostridium perfringens* bearing temperate phage s9, *J. Gen. Microbiol.*, 103, 45, 1977.
12. Cowles, P.B., A bacteriophage for *Clostridium tetani*, *J. Bacteriol.*, 27, 163, 1934.
13. Gabriel, C.L., Butanol fermentation process, *Ind. Eng. Chem.*, 20, 1063, 1928.
14. Jones, D.T. and Woods, D.R., Acetone-Butanol fermentation revisited, *Microbiol. Rev.*, 50, 484, 1986.
15. Jones, D.T., Shirley, M., Wu, X., and Keis, S., Bacteriophage infections in the industrial acetone butanol (AB) fermentation process, *J. Mol. Microbiol. Biotechnol.*, 2, 21, 2000.
16. Jones, D.T., Applied Acetone-Butanol Fermentation, in *Clostridia; Biotechnology and Biomedical Applications*, Bahl, H. and Dürre, P., Eds., Wiley-VCH, Weinheim, 2001, 125.
17. van Regenmortel, M.H.V., Fauquet, C.M., Bishop, D.H.L., Carstens, E.B., Estes, M.K., Lemon, S.M., Maniloff, J., Mayo, M.A., McGeoch, D.J., Pringle, C.R., and Wickner, R.B., *Virus Taxonomy: Seventh Report of the International Committee on Taxonomy of Viruses*, Academic Press, San Diego, 2000.
18. Zahler, S.A., Temperate Bacteriophages, in *Bacillus subtilis and Other Gram Positive Bacteria.*, Sonenshein, A.L., Hoch, J.A., and Losick, R., Eds., American Society for Microbiology, Washington, DC, 1993, 831.
19. Dolman, C.E. and Chang, E., Bacteriophages of *Clostridium botulinum*, *Can. J. Microbiol.*, 18, 67, 1972.
20. Eklund, M.W., Poysky, F.T., Meyers, J.A., and Pelroy, G.A., Interspecies conversion of *Clostridium botulinum* type C to *Clostridium novyi* type A by bacteriophage, *Science*, 186, 456, 1974.
21. Eklund, M.W., Poysky, F.T., and Boatman, E.S., Bacteriophages of *Clostridium botulinum* types A, B, E, and F and nontoxigenic strains resembling type E, *J. Virol.*, 3, 270, 1969.
22. Mahony, D.E., Bell, P.D., and Easterbrook, K.B., Two bacteriophages of *Clostridium difficile*, *J. Clin. Microbiol.*, 21, 251, 1985.
23. Schallehn, G. and Lenz, W., *Clostridium novyi* type A bacteriophages, *Zbl. Bakt. Abt. Orig. I*, 232, 100, 1975.
24. Betz, J.V., Some properties of bacteriophages active on the obligate anaerobe *Clostridium sporogenes*, *Virology*, 36, 9, 1968.
25. Shimamura, T., Nakamura, S., Hayase, M., and Nishida, S., Mitomycin-induced lysis of *Clostridium sordellii*, *J. Med. Microbiol.*, 7, 277, 1974.
26. Prescott, L.M. and Altenbern, R.A., Detection of bacteriophages from two strains of *Clostridium tetani*, *J. Virol.*, 1, 1085, 1967a.
27. Prescott, L.M. and Altenbern, R.A., Inducible lysis of *Clostridium tetani*, *J. Bacteriol.*, 93, 1220, 1967.
28. Kinoshita, S. and Teramoto, K., Studies on the bacteriophages for *Clostridium acetobutylicum.* Part 6. (1) Classification of strains by phage. (2) On the relationship between sluggish fermentation and phage action, *J. Agri. Chem. Jpn.*, 29, 955, 1955.
29. Hongo, M. and Murata, A., Bacteriophages of *Clostridium saccharoperbutylacetonicum.* Part 1. Some characteristics of the 12 phages obtained from the abnormally fermented broths, *Agri. Biol. Chem.*, 29, 1135, 1965.
30. Ogata, S., Nagao, N., Hidaka, Z., and Hongo, M., Bacteriophages of *Clostridium saccharoperbutylacetonicum.* Part XVII. The structure of phage HM2, *Agr. Biol. Chem.*, 33, 1541, 1969.
31. Hongo, M., Murata, A., and Ogata, S., Bacteriophages of *Clostridium saccharoperbutylacetonicum.* Part XVI. Isolation and some characters of a temperate phage, *Agr. Biol. Chem.*, 33, 337, 1969.
32. Beatriz, M.N., Gil, F., and Castillo, F.J., Growth inhibition activity and bacteriophage and bacteriocin like particles associated with different species of *Clostridium*, *Can. J. Microbiol.*, 27, 216, 1981.
33. Reanney, D.C. and Ackermann, H.-W., Comparative biology and evolution of bacteriophages, in *Advances in Virus Research*, Vol. 27, Lauffes, M.A., Bang, F.B., Gavamorosch, K., and Smith, K.G., Eds., Academic Press, New York, 1982, 205.
34. Guélin, A.M., Beerens, H., and Petitprez, A., An anaerobic bacteriophage active on *Clostridium histolyticum*, *Ann. Inst. Pasteur (Paris)*, 111, 141, 1966.
35. Grant, R.B. and Riemann, H.P., Temperate phages of *Clostridium perfringens* type C, *Can. J. Microbiol.*, 22, 603, 1976.

36. Mahony, D.E. and Easterbrook, K.B., Intracellular development of a bacteriophage of *Clostridium perfringens*, *Can. J. Microbiol.*, 16, 983, 1970.
37. Mahony, D.E. and Kalz, G.G., A temperate bacteriophage of *Clostridium perfringens*, *Can. J. Microbiol.*, 14, 1085, 1968.
38. Vieu, J.F., Guélin, A.M., and Dauguet, C., Morphology of the *Welchia perfringens* bacteriophage 80, *Ann. Inst. Pasteur (Paris)*, 109, 157, 1965.
39. Kim, A.Y. and Blaschek, H.P., Isolation and characterization of a filamentous viruslike particle from *Clostridium acetobutylicum* NCIB 6444, *J. Bacteriol.*, 173, 530, 1991.
40. Fujii, N., Oguma, K., Yokosawa, N., Kimura, K., and Tsuzuki, K., Characterisation of bacteriophage nucleic acids obtained from *Clostridium botulinum* types C and D, *Appl. Environ. Microbiol.*, 54, 69, 1988.
41. Zimmer, M., Scherer, S., and Loessner, M.J., Genomic analysis of *Clostridium perfringens* bacteriophage ϕ3626, which integrates into *guaA* and possibly affects sporulation, *J. Bacteriol.*, 184, 4359, 2002.
42. Keis, S., Characterization of two bacteriophages infecting *Clostridium acetobutylicum*, Department of Microbiology, University of Otago, Dunedin, 1992.
43. Wu, X., Characterization of a large temperate bacteriophage of *Clostridium beijerinckii*, Department of Microbiology, University of Otago, Dunedin, 1998.
44. Ackermann, H.W., Classification of *Bacillus* and *Clostridium* bacteriophages, *Path. Biol.*, 22, 909, 1974.
45. Betz, J.V. and Anderson, K.E., Isolation and characterization of bacteriophages active on *Clostridium sporogenes*, *J. Bacteriol.*, 87, 408, 1964.
46. Gàspàr, G., Studies of *Clostridium perfringens* phage. II. The latent period of its growth curve, *Acta Microbiol. Acad. Sci. Hung*, 7, 269, 1960.
47. Gold, W. and Watson, D.W., Studies on the bacteriophage infection cycle, *J. Bacteriol.*, 59, 17, 1950.
48. Kinoshita, S. and Teramoto, K., Studies on bacteriophages for *Clostridium acetobutylicum.* Part 2. On the method of assay of phage titer by plaque count method, *J. Agri. Chem. Jpn.*, 29, 436, 1955.
49. Gerhardt, P., Murray, R.G.E., Wood, W.A., and Krieg, N.R., *Methods for General and Molecular Bacteriology*, American Society for Microbiology, Washington, DC, 1994.
50. Ionesco, H., Wolff, A., and Sebald, M., Production induite de bactériocine et d'un bactériophage par la souche BP6K-N5 de *Clostridium perfringens*, *Ann. Inst. Pasteur (Paris)*, 125B, 335, 1974.
51. Smith, H.W., The bacteriophages of *Clostridium perfringens*, *J. Gen. Microbiol.*, 21, 622, 1959.
52. Popovitch, M. and Sebald, M., Bacteriophages of *Clostridium histolyticum*. Isolation and lysotype, *Bull. Off. Inst. Epiz.*, 67, 1183, 1967.
53. Beesch, S.C., Acetone-Butanol Fermentation of Sugars, *Ind. Eng. Chem.*, 44, 1677, 1952.
54. Sylvester, J.C., Studies on bacteriophage affecting an acetone-butanol anaerobe, in *Agricultural Bacteriology*, University of Wisconsin, Madison, 1943.
55. Keis, S., Bennett, C.F., Ward, V.K., and Jones, D.T., Taxonomy and phylogeny of industrial solvent-producing clostridia, *Int. J. Syst. Bacteriol.*, 54, 936, 1995.
56. Hill, C., Bacteriophage and bacteriophage resistance in lactic acid bacteria, *FEMS Microbiol. Rev.*, 12, 87, 1993.
57. Klaenhammer, T.R., Interaction of bacteriophages with lactic Streptococci, *Adv. Appl. Microbiol.*, 30, 1, 1984.
58. Richards, D.F., Linnett, P.E., Ouitram, J.D., and Young, M., Restriction endonucleases in *C. pasteurianum* ATCC 6013 and *C. thermohydrosulfuricum* DSM 568, *J. Gen. Microbiol.*, 134, 3151, 1988.
59. Young, M., Minton, N.P., and Staudenbauer, W.L., Recent advances in the genetics of the Clostridia, *FEMS Microbiol. Rev.*, 63, 301, 1989.
60. Kessler, C. and Holtke, J., Specificity of restriction endonucleases and methylases — a review, *Gene*, 47, 1, 1986.
61. Roberts, R.J., Restriction enzymes and their isoschizomers, *Nucleic Acids Res.*, 15, 189, 1987.
62. Dreyfus-Fourcade, M., Sebald, M., and Zavadova, M., Restriction et modification du phage par *Clostridium perfringens* NCTC 8798 et ses mutants, *Ann. Inst. Pasteur (Paris)*, 122, 1117, 1972.
63. Garnier, T. and Cole, S.T., Characterization of a bacteriocinogenic plasmid from *Clostridium perfringens* and molecular genetic analysis of the bacteriocin-encoding gene, *J. Bacteriol.*, 168, 1189, 1986.
64. Rood, J.I., Transferable tetracycline resistance in *Clostridium perfringens* strains of porcine origin, *Can. J. Microbiol.*, 29, 1241, 1983.

65. Scott, P.T. and Rood, J.I., Electroporation-mediated transformation of lysostaphin-treated *Clostridium perfringens*, *Gene*, 82, 327, 1989.
66. Rood, J.I. and Cole, S.T., Molecular genetics and pathogenesis of *Clostridium perfringens*, *Microbiol. Rev.*, 55, 621, 1991.
67. Papoutsakis, E.T. and Bennett, G. N., Cloning, structure and expression of acid and solvent pathway genes of *Clostridium acetobutylicum*, in *The Clostridia and Biotechnology*, Woods, D.R., Ed., Butterworth-Heinemann, Stoneham, MA, 1993, 157.
68. Birge, A.E., *Bacterial and Bacteriophage Genetics*, Springer-Verlag, New York, 1988.
69. Harshey, R.M., Phage Mu, in *The Bacteriophage*, Calendar, R., Ed., Plenum Press, New York, 1988, 193.
70. Yarmolinsky, M.B. and Sternerg, N., Bacteriophage P1, in *The Bacteriophage*, Calendar, R., Ed., Plenum Press, New York, 1988, 291.
71. Harshey, A.D., *Bacteriophage Lambda*, Cold Spring Harbor Laboratory, New York, 1971.
72. Eklund, M.W., Poysky, F.T., and Reed, S.M., Bacteriophage and the toxigenicity of *Clostridium botulinum* type D, *Nature New. Biol.*, 235, 16, 1972.
73. Eklund, M.W., Poysky, F.T., Peterson, M.E., and Meyers, J.A., Relationship of bacteriophages to alpha toxin production in *Clostridium novyi* types A and B, *Infect. Immun.*, 14, 793, 1976.
74. Iida, H., Oguma, K., and Inoue, K., Phage-conversion of toxigenicity in *Clostridium botulinum* types C and D, *Jpn. J. Med. Sci. Biol.*, 27, 101, 1974.
75. Roseman, D. and Richardson, R.L., Isolation of bacteriophage for *Clostridium tetani*, *J. Virol.*, 3, 350, 1969.
76. Canard, B. and Cole, S.T., Genome organization of the anaerobic pathogen *Clostridium perfringens*, *Proc. Natl. Acad. Sci. U.S.A.*, 86, 6676, 1989.
77. Roggentin, P., Rothe, B., Lottspeich, F., and Schauer, R., Cloning and sequencing of a *Clostridium perfringens* sialidase gene, *FEBS Lett.*, 238, 31, 1988.
78. McCoy, E., Industrial importance of bacteriophage, *J. Bacteriol.*, 45, 75, 1943.
79. McCoy, E., McDaniel, L.E., and Sylvester, J.C., Bacteriophage in a butyl fermentation plant, *J. Bacteriol.*, 47, 443, 1944.
80. Peri, B.A. and Watson, D.W., Studies on host-virus relationships: Lysogenesis in a strain of *Clostridium madisonii*, *Proc. Meet. Soc. Am. Bact.*, 1, 26, 1948.
81. Hempill, H.E. and Whiteley, H.R., Bacteriophage of *Bacillus subtilis.*, *Bacteriol. Rev.*, 39, 257, 1975.
82. Kawakami, M. and Landman, O. E., Nature of the carrier state of bacteriophage SP-10 in *Bacillus subtilis*, *J. Bacteriol.*, 90, 1575, 1968.
83. Reeves, P., The concept of bacteriocins, *Inf. Krankheiten und Hygiene*, 244, 78, 1979.
84. Garro, A.J., Leffert, H., and Marmur, J., Genetic mapping of a defective bacteriophage on the chromosome of *Bacillus subtilis*, *J. Virol.*, 6, 340, 1970.
85. Tomizawa, J. and Ogawa, T., Effect of ultraviolet irradiation on bacteriophage Lambda immunity, *J. Mol. Biol.*, 23, 247, 1967.
86. Bradley, D.E., Ultrastructure of bacteriophages and bacteriocins, *Bacteriol. Rev.*, 31, 230, 1967.
87. Lau, A.H.S., Hawirko, R.Z., and Chow, C.T., Purification and properties of boticin P produced by *Clostridium botulinum*, *Can. J. Microbiol.*, 20, 385, 1974.
88. Ogata, S., Mihara, O., Ikeda, Y., and Hongo, M., Bacteriocins of nonpathogenic *Clostridium* species. 3. Inducible phage tail-like particles of *Clostridium saccharoperbutylacetonicum* and its related strains, *Agr. Biol. Chem.*, 36, 1413, 1972.
89. Ogata, S., Kato, F., Myanaga, M., Hay, A.S., Hayashida, S., and Hongo, M., Purification and some properties of an inducible bacteriocin, clostocin O of *Clostridium saccharoperbutylacetomicum*, *J. Gen. Appl. Microbiol.*, 24, 223, 1978.
90. Kato, F., Ogata, S., and Hongo, M., Lytic action of inducible bacteriocin clostocin O, *Agr. Biol. Chem.*, 40, 1101, 1976.
91. Brèfort, G., Magot, M., Ionesco, H., and Sebald, M., Characterization and transferability of *Clostridium perfringens* plasmids, *Plasmid*, 1, 52, 1977.
92. Reid, S.J., Alcock, E.R., Jones, D.T., and Woods, D.R., Transformation of *Clostridium acetobutylicum* protoplasts with bacteriophage DNA, *Appl. Environ. Microbiol.*, 45, 305, 1983.
93. Bradley, D.E. and Hoeniger, J.F.M., Structural changes in cells of *Clostridium perfringens* infected with short-tailed bacteriophage, *Can. J. Microbiol.*, 17, 397, 1971.

94. Inoue, K. and Iida, H., Conversion of toxigenicity in *Clostridium botulinum* type C, *Jpn. J. Microbiol.*, 14, 87, 1970.
95. Inoue, K. and Iida, H., Phage-conversion of toxigenicity in *Clostridium botulinum* types C and D, *Jap. J. Med. Sci. Biol.*, 24, 53, 1971.
96. Oguma, K., Iida, H., and Ioune, K., Bacteriophage and toxigenicity in *Clostridium botulinum:* an additional evidence for phage conversion, *Jpn. J. Microbiol.*, 17, 425, 1973.
97. Johnson, J.L. and Francis, B.S., Taxonomy of the clostridia: Ribosomal ribonucleic acid homologies among the species, *J. Gen. Microbiol.*, 88, 229, 1975.
98. Hariharan, H. and Mitchell, W. R., Observations on bacteriophages of *Clostridium botulinum* type C isolates from different sources and the role of certain phages in toxigenicity, *Appl. Environ. Microbiol.*, 32, 145, 1976.
99. Oguma, K., The stability of toxigenicity in *Clostridium botulinum* type C and D, *J. Gen. Microbiol.*, 92, 67, 1976.
100. Oguma, K., Iida, H., and Shiozaki, M., Phage conversion to haemagglutinin production in *Clostridium botulinum* Type C and D, *Infect. Immun.*, 14, 597, 1976.
101. Binz, T., Kurazono, H., Popoff, M.R., Eklund, M.W., Sakaguchi, G., Kozaki, S., Krieglstein, K., Henschen, A., Gill, D.M., and Niemann, H., Nucleotide sequence of the gene encoding *Clostridium botulinum* neurotoxin type D, *Nucleic Acids Res.*, 18, 5556, 1990.
102. Hauser, D., Eklund, M.W., Kurazono, H., Binz, T., Niemann, H., Gill, D.M., Boquet, P., and Popoff, M.R., Nucleotide sequence of *Clostridium botulinum* C1 neurotoxin, *Nucleic Acids Res.*, 18, 4924, 1990.
103. Eklund, M.W., Poysky, F.T., and Habig, W.H., Bacteriophages and plasmids in *Clostridium botulinum* and *Clostridium tetani* and their relationships to production of toxins, in *Botulinum neurotoxin and tetanus toxin*, Simpson, L.L., Ed., Academic Press, San Diego, 1989, 26.
104. Quinn, C.P. and Minton, N.P., Clostridial neurotoxins, in *Clostridia: Biotechnology and Medical Applications*, Bahl, H. and Dürre, P., Eds., Wiley-VCH, Weinheim, Germany, 2001, 211.
105. Zhou, Y., Sugiyama, H., and Johnson, E.A., Transfer of neurotoxigenicity from *Clostridium butyricum* to a nontoxigenic *Clostridium botulinum* type E-like strain, *Appl. Environ. Microbiol.,* 59, 3825, 1993.
106. Hastings, J.J., Development of the Fermentation Industries in Great Britian, in *Advances in Applied Microbiology*, Vol. 14, Perlman, D., Ed., Academic Press, New York, 1971, 1.
107. Walton, M.T. and Martin, J.L., Production of Butanol-Acetone by Fermentation, in *Microbial Technology*, Vol. 1, Peppler, H.J. and Perlman, D., Ed., Academic Press, New York, 1979, 187.
108. McCutchan, W.N. and Hickey, J.R., The Butanol-acetone fermentations, *Ind. Fer.*, 1, 347, 1954.
109. Ryden, R., Development of anaerobic fermentation processes: Acetone-butanol, in *Biochemical Engineering*, Steel, R., Ed., Heywood, London, 1958, 125.

32 Acetogenic Clostridia

Harold L. Drake and Kirsten Küsel

CONTENTS

32.1 INTRODUCTION TO ACETOGENIC CLOSTRIDIA

For the purposes of this chapter, acetogenic clostridia are defined as species of the genus *Clostridium* that utilize the acetyl-CoA pathway for the reductive synthesis of acetyl-CoA from CO_2. It must be noted that the genus *Clostridium* was restructured in the mid-1990s [32]. This restructuring is not without consequence relative to the content of this chapter; this point will be addressed later. The acetyl-CoA pathway is a terminal electron-accepting process and is also termed the Wood-Ljungdahl pathway in recognition of the two biochemists Harland G. Wood and Lars G. Ljungdahl, who were responsible for elucidating most of its enzymological features. Although the focus of this chapter is on acetogens in the genus *Clostridium*, 20 other bacterial genera also contain acetogens, and the reader is directed to recent reviews that describe these other genera and outline the more far-reaching aspects of acetogenesis and enzymological features of the acetyl-CoA pathway [42,53,134,154,169]. Although emphasis will be placed on acetogenic clostridia, much of the information in this chapter applies to all acetogenic genera. Purinolytic clostridia (e.g., *Clostridium acidiurici* and *Clostridium cylindrosporum*) will not be considered in this chapter because the CO_2-fixing process by which they synthesize acetate (i.e., the glycine pathway) is different than that used by acetogens (see Andreesen [5] for an excellent review on the physiology and enzymology of purinolytic clostridia).

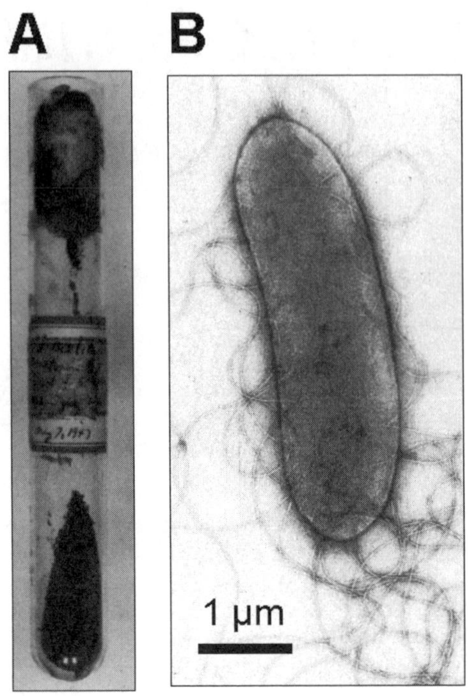

FIGURE 32.1 (A) Tube containing dried soil and spores of the first acetogen to be isolated, *Clostridium aceticum*. (B) Electron micrograph of peritrichously flagellated cell of *C. aceticum*. (Used with permission from Braun et al. [23].)

32.2 HISTORICAL PERSPECTIVES

Acetogenesis was first reported in 1932, when it was observed that acetate was synthesized from H_2-CO_2 by unknown microorganisms in wastewater [57]. The first acetogen was isolated shortly thereafter and was named *Clostridium aceticum* [207–209]. Thus, the first known acetogen was a clostridial species. This mesophilic bacterium had the ability to catalyze the following reaction, which at that time was considered a biological novelty:

$$4\ H_2 + 2\ CO_2 \rightarrow CH_3COOH + 2\ H_2O \tag{32.1}$$

As history would have it, the original culture of *C. aceticum* was lost until an old culture tube (dated May 7, 1947 [Figure 32.1A]) containing dried spores was discovered several decades later. The spores were revived, and *C. aceticum* was reisolated and newly described (Figure 32.1B) (see Braun et al. [23] for a detailed description of the reisolation; see also Adamse [2] and Gottschalk and Braun [72]). Although relatively little research has been conducted with *C. aceticum*, the organism has special importance for being not only the first acetogen isolated, but for also being the first acetogen shown to catalyze the conversion of H_2-CO_2 to acetate, which has since become a primary hallmark of acetogens.

The thermophilic bacterium *Clostridium thermoaceticum* was discovered a few years after the isolation of *C. aceticum* [58] and, likewise, catalyzed a novel reaction, the near stoichiometric conversion of glucose to acetate:

$$C_6H_{12}O_6 \rightarrow 3\ CH_3COOH \tag{32.2}$$

As will be described below, reactions 32.1 and 32.2 have a common underlying feature: the reduction of CO_2 to acetate via the acetyl-CoA pathway. Although *C. thermoaceticum* was originally characterized as an obligate heterotroph, it was shown to be capable of autotrophic growth on H_2-CO_2 nearly five decades after its isolation [37]. *C. thermoaceticum* was reclassified as *Moorella thermoacetica* in the mid-1990s when the genus *Clostridium* was taxonomically restructured (based on phylogenetic assessment of the 16S rRNA gene sequences of its members). Although *M. thermoacetica* is no longer classified as a species of *Clostridium*, some of its features will be described in this chapter for four reasons: (i) it was originally classified as a species of the genus *Clostridium*; (ii) the organism is still sometimes referred to as *C. thermoaceticum* in the literature; (iii) it was the model acetogen used to resolve the acetyl-CoA pathway [39,214] and, therefore, constitutes the most historically important acetogen relative to understanding the biochemistry of acetogenesis (indeed, the first ^{14}C-tracer experiments in biology were performed with this organism in studies that demonstrated the autotrophic nature of the acetyl-CoA pathway [11,93]); and (iv) it displays an enormous metabolic versatility [39] and continues to be an important model acetogen in laboratory studies. Nonetheless, this organism will be referred to hereafter as *M. thermoacetica*.

Acetogens were viewed as somewhat obscure microorganisms for many decades following their discovery in the 1930s, and the microbiology of acetogens drew little interest until it was realized that both the acetyl-CoA pathway and acetogens were widely distributed in nature. Though beyond the scope of this chapter, it should be mentioned that nonacetogenic bacteria (e.g., sulfate-reducing bacteria) and members of the domain *Archaea* (e.g., methanogens) use metabolic pathways that are very similar to the acetyl-CoA pathway for either the assimilation of CO_2 (i.e., carbon) into biomass or the oxidation of acetate [64,65]; thus, the acetyl-CoA pathway is not restricted to acetogens and has broad biological significance [49].

32.3 PHYLOGENY AND TAXONOMIC FEATURES

All known acetogens are members of the domain *Bacteria*, and the acetogenic clostridia are members of the spore-forming low G+C Gram-positive bacteria. Based on the phylogenetic analysis of the 16S rRNA genes of acetogens, they do not form a phylogenetically exclusive group, i.e., acetogenic species are often dispersed in bacterial genera that also contain nonacetogenic species [51,53,193]. The genus *Clostridium* is no exception. For example, the closest phylogenetic relative of the acetogen *Clostridium formicoaceticum* (the third acetogen isolated [4]) is the nonacetogen *Clostridium felsineum* (99.3% 16S rRNA gene sequence similarity). In addition, clostridial species that were not originally described as acetogens were later shown to be acetogens, i.e., capable of forming acetate via the acetyl-CoA pathway. For example, *Clostridium coccoides* strain 1410 and *Clostridium scatologenes* strain SL1 were isolated as acetogens, yet the type strains of these species were not known to be acetogens. A reassessment of the type strains of these two species demonstrated that they, too, are acetogens [95,121]. Certain acetogens lose their acetogenic capabilities after prolonged laboratory cultivation. For example, *Clostridium glycolicum* strain RD-1 and strain 22 are acetogens, yet the type strain of *C. glycolicum* does not grow acetogenically [122,157]. Thus, the ability to engage the acetyl-CoA pathway is not always a stable metabolic feature. In this context, it is interesting to note that the pathogenic type strain of *Clostridium difficile* has not been characterized as an acetogen, yet *C. difficile* strain AA1 was shown to be an acetogen [30,78,176]. Collectively, these results demonstrate that many clostridia that are not known to be acetogens might, in fact, be acetogens.

Approximately 100 different acetogenic species have been isolated from extremely diverse habitats, and these species are dispersed in 21 genera [53]. The acetogens belonging to the genus *Clostridium* are outlined in Table 32.1. Acetogens that were originally classified as clostridial species, then later reclassified, are outlined in Table 32.2. Although acetogenic species of the genus *Clostridium* have been isolated from a variety of habitats, all are mesophiles. However, acetogens

TABLE 32.1
Acetogenic Clostridia[a]

Acetogen	Source of Isolate	Deposited As	16S rRNA Gene Seq. Accession No.	Ref.
Clostridium aceticum	Soil	DSM 1496^T	Y18183	23,207[b]
"Clostridium autoethanogenum" (?)	Rabbit feces	DSM 10061	Y18178	1
Clostridium coccoides	Mice feces, human feces	DSM 935^T	M59090	96,98
Clostridium difficile AA1	Rumen, newborn lamb	DSM 12056	AJ310756	176
Clostridium formicoaceticum	Sewage	DSM 92^T	X77836	4
Clostridium glycolicum 22	Sewage	ATCC 29797 (22)	AJ291746	157
Clostridium glycolicum RD-1	Seagrass roots	DSM 13865 (RD-1)		122
Clostridium ljungdahlii	Chicken waste	ATCC 49587^T	M59097	10,192
Clostridium magnum	Freshwater sediment	DSM 2767^T	X77835	179
Clostridium mayombei	Soil-feeding termite, gut	DSM 6539^T	M62421	97
Clostridium methoxybenzovorans	Olive oil mill wastewater	DSM 12182^T	AF067965	148
Clostridium scatologenes	Soil, coal mine pond sediment	DSM 757^T DSM 12750 (SL1)	M59104 Y18813	121,201
Clostridium ultunense	Swine manure digester	DSM 10521^T	Z69293	182
Clostridium sp. CV-AA1	Sewage sludge	n.d.[c]		3
Clostridium sp. M5a3	Human feces	n.d.		15,129,130
Clostridium sp. F5a15	Human feces	n.d.		15,129,130
Clostridium sp. Ag4f2	Human feces	n.d.		15
Clostridium sp. TLN2	Human feces	n.d.		15

[a] The species of *Clostridium* listed appear to use the acetyl-CoA pathway for the synthesis of acetate and growth (modified from Drake et al. [53]). A question mark after the name of an organism indicates that the acetogenic nature of the organism is not certain. An organism not having a validated name is enclosed in quotation marks. Type strains are marked with a "T" adjacent to the deposition number.

[b] See also Adamse 1980 [2].

[c] n.d., not deposited.

TABLE 32.2
Reclassified Clostridial Acetogens[a]

Original Name	Current Name	Source of Isolate	Deposited As	16S rRNA Gene Seq. Access. No.	Ref.
Clostridium fervidus (?)	*Caloramator fervidus*	Hot spring	ATCC 43204^T	L091287	158
Clostridium pfennigii	*Oxobacter pfennigii*	Rumen, steer	DSM 3222^T	X77838	112
Clostridium thermoaceticum	*Moorella thermoacetica*	Horse manure	ATCC 35608^T	M59121	58
Clostridium thermoautotrophicum	*Moorella thermoautotrophica*	Hot spring	ATCC 33924^T	L09168	204

[a] See Collins et al. 1994 [32]. A question mark after a name of an organism indicates that the acetogenic nature of the organism is not certain.

that are capable of psychrophilic or thermophilic growth occur in other genera. For example, the reclassified acetogen *M. thermoacetica* is a thermophile [58].

Three metabolic features that are characteristic for acetogens are: (i) the use of chemolithoautotrophic substrates (H_2-CO_2 or CO-CO_2) as sole sources of carbon and energy under anoxic conditions; (ii) the capacity to convert certain sugars in near stoichiometry to acetate; and (iii) the ability to *O*-demethylate methoxylated aromatic compounds and metabolize the *O*-methyl group via the acetyl-CoA pathway. Many clostridial acetogens, including the first acetogen isolated, *C. aceticum*, possess all three of these capabilities. These metabolic features can be taken advantage of to selectively enrich acetogens and are useful when taxonomically describing a new acetogen.

A phylogenetic evaluation of acetyl-CoA synthase, a centrally important enzyme in the acetyl-CoA pathway (see Section 32.4.1), indicates that the different organisms (e.g., acetogens and methanogens) that have this enzyme (or enzymes that are biochemically and evolutionarily related to it) likely had a common ancestor [134]. Although the matter is still open to debate, there is good reason to think that an acetyl-CoA synthase-dependent pathway constituted the first autotrophic process on earth [134,214].

32.4 PHYSIOLOGY OF ACETOGENIC CLOSTRIDIA

Acetogens can utilize numerous electron donors and also numerous electron acceptors. Their diverse metabolic capabilities make them unlike the more metabolically limited methanogens, a group of anaerobes that might compete with acetogens in certain habitats. At the onset of this section, one should note that the metabolic capabilities of acetogens vary from one acetogen to the next and that very few acetogens have been extensively characterized. Thus, the current information base is too limited to describe the complete physiological features of each clostridial acetogen. This section will, therefore, highlight the general metabolic features of acetogens.

32.4.1 CO_2 as Terminal Electron Acceptor

The acetyl-CoA pathway is a terminal electron-accepting process and is composed of two reductive branches, both of which reduce CO_2 (Figure 32.2). The methyl branch of the pathway reduces CO_2 to the methyl level via several tetrahydrofolate-dependent reactions, and the carbonyl branch reduces CO_2 to the carbonyl level via the enzyme acetyl-CoA synthase. The acetyl-CoA formed by the pathway is a catabolic precursor for acetate and an anabolic precursor for biomass. Pyruvate ferredoxin oxidoreductase, which is important for the oxidation of pyruvate when the pathway functions under heterotrophic conditions [48,49,214], is also important to the assimilation of acetyl-CoA during autotrophic growth [66]. The standard redox potential of the CO_2/acetate half-cell reaction approximates –290 mV. This redox potential is one of the most negative redox potentials in biology relative to its being an energy-conserving terminal electron accepting process.

All of the enzymes of the acetyl-CoA pathway have been purified and characterized. A description of these enzymes is beyond the scope of this chapter, and the reader is directed to several reviews that highlight the numerous historical studies that led to the resolution of the enzymological features of the pathway [42,49,53,167,214]. Additional references will serve as a gateway to some of the more recent enzymological studies on the acetyl-CoA pathway [9,21,40,134,135,145,146,166,168–170]. Given the importance of acetyl-CoA synthase in the pathway, it is of particular note that recent studies on the crystal structure of the enzyme from *M. thermoacetica* have shown that the alpha subunits of acetyl-CoA synthase display both closed and open conformations and that the active form of the enzyme has an Ni-Ni-[Fe_4-S_4] cluster at the active site [40]. Copper, once thought to be a part of this nickel-containing metal cluster [47], is an inhibitor of acetyl-CoA synthase and not a component of the catalytically active enzyme [21,40]. A nickel insertase has recently been identified that is involved in the biosynthesis of acetyl-CoA synthase [140].

When acetogens are cultivated on hexoses, the main theoretical function of the acetyl-CoA pathway is the recycling (i.e., oxidation) of reduced electron carriers (e.g., $NADH^+$ or reduced ferredoxin). As illustrated in Figure 32.3, 4 ATP are produced by substrate-level phosphorylation (ATP_{SLP}) when a hexose (e.g., glucose) is converted stoichiometrically to 3 acetates. Thus, acetogens conserve more energy by substrate-level phosphorylation than ethanol, lactate, and butyrate fermenters.

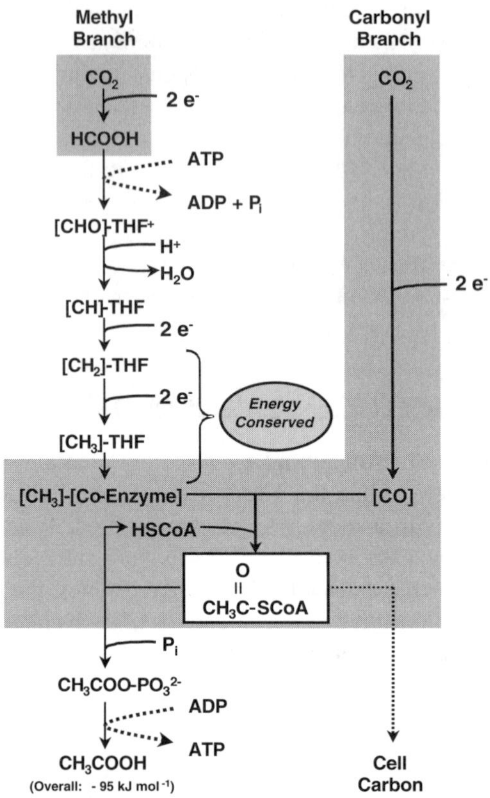

FIGURE 32.2 The acetyl-CoA "Wood-Ljungdahl" pathway. The two enzymes responsible for the initial reduction of CO_2 are formate dehydrogenase (small shaded box at the top of the Methyl Branch) and acetyl-CoA synthase (shaded area of the Carbonyl Branch; note that acetyl-CoA synthase catalyzes both the reduction of CO_2 and also the synthesis of acetyl-CoA). Abbreviations: THF, tetrahydrofolate; HSCoA, coenzyme A; P_i, inorganic phosphate; e^-, reducing equivalent; Co-Enzyme, corrinoid enzyme. Brackets indicate that a particular C1 unit is bound to a cofactor or structurally associated with an enzyme.

The "homoacetogenic conversion" of hexoses to acetate (reaction 32.2) is often cited as a property of acetogens, but is seldom realized in laboratory studies, mainly because: (i) most acetogens can form other products in addition to acetate from hexoses, (ii) carbon recoveries are low, and (iii) acetogens can utilize terminal electron acceptors that route reductant away from the acetyl-CoA pathway. Thus, homoacetogenesis is a conditional rather than absolute property of acetogens. Under chemolithoautotrophic conditions (i.e., when growth is at the expense of H_2-CO_2), the pathway must serve both anabolic and catabolic functions, and acetate tends to be the sole catabolic product under such conditions.

Because CO_2 is utilized as a terminal electron acceptor, exogenous CO_2 can be required for the oxidation of reduced electron carriers and optimal growth with certain substrates. For example, acetogenesis at the expense of highly reduced one-carbon substrates (e.g., methanol or methoxy groups of aromatic compounds) is strictly dependent on exogenous CO_2 [86,144]), and exogenous CO_2 can likewise be required for the optimal acetogenic utilization of hexoses and CO [4,178]. Carbonic anhydrase, which catalyzes the following reversible reaction, is highly active in acetogens [24]:

$$CO_2 + H_2O \leftrightarrow HCO_3^- + H^+ \quad (32.3)$$

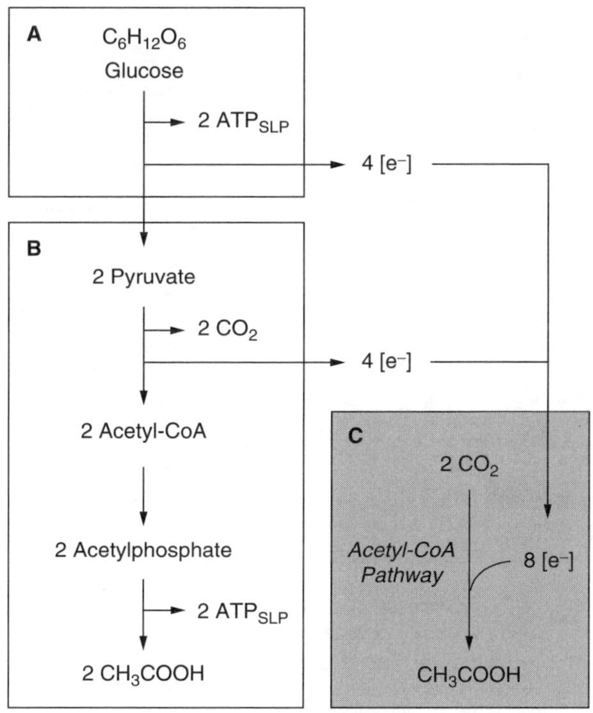

FIGURE 32.3 Homoacetogenic conversion of glucose to acetate. Box A: Conversion of glucose to pyruvate via glycolysis. Box B: Oxidation and decarboxylation of pyruvate; the acetyl-CoA produced from pyruvate is converted to acetate by phosphotransacetylase and acetate kinase. Box C (shaded): The acetyl-CoA pathway. The majority of the CO_2 that is reduced in the acetyl-CoA pathway is apparently derived from exogenous CO_2 rather than the CO_2 produced via the decarboxylation of pyruvate. Abbreviations: ATP_{SLP}, ATP that is produced by substrate-level phosphorylation; $[e^-]$, reducing equivalent. (Modified from Drake [49].)

Given the need for CO_2 in the acetyl-CoA pathway, one physiological function of carbonic anhydrase in acetogens might be to increase intracellular levels of CO_2.

32.4.2 Conservation of Energy

Clostridial acetogens conserve energy by both substrate-level phosphorylation and chemiosmotic mechanisms. Substrates that yield glycolytic intermediates or acetyl-CoA can be expected to yield ATP_{SLP} (Figure 32.3). Depending on the conditions of growth, both substrate-level phosphorylation and chemiosmotic processes can be utilized simultaneously for the conservation of energy. The net production of ATP_{SLP} does not occur in the acetyl-CoA pathway (Figure 32.2) (one ATP is consumed when formate is converted to formyltetrahydrofolate, and one ATP is formed by acetate kinase in the terminal reaction of the pathway). Thus, energy is only conserved by a chemiosmotic process during growth under chemolithoautotrophic conditions. As illustrated in Figure 32.2, the reduction of methylenetetrahydrofolate to methyltetrahydrofolate and subsequent transmethylation reactions are thermodynamically very favorable and thought to be coupled to the conservation of energy [44,45,53,154,211].

Two different chemiosmotic mechanisms are utilized by acetogens for conserving energy. One mechanism generates a membranous proton gradient that can subsequently be used to drive energy-dependent processes, such as active transport and the synthesis of ATP by proton-dependent ATPase [139] (Figure 32.4A). These acetogens include those that have membranous electron transport systems or certain membrane-associated oxidoreductases (e.g., hydrogenase) [41,44,50,89,94,139].

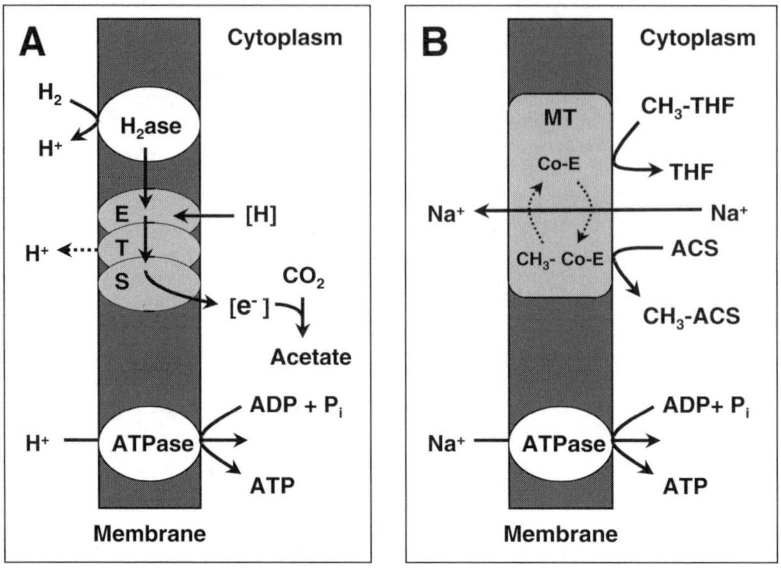

FIGURE 32.4 Mechanisms for the chemiosmotic conservation of energy by acetogens. Abbreviations: H_2ase, hydrogenase; ETS, electron transport system; e⁻, reducing equivalent; ATPase, ATP synthase; THF, tetrahydrofolate; Co-E, corrinoid enzyme; MT, methyltransferase; ACS, acetyl-CoA synthase. (Modified from Drake et al. [53].)

Those acetogens that have membranous electron-transport systems and proton-dependent ATPases include *M. thermoacetica* and *Sporomusa sphaeroides*.

The other chemiosmotic mechanism generates a membranous sodium gradient that can likewise be used to drive energy-dependent processes, such as the synthesis of ATP by sodium-dependent ATPase [153,154] (Figure 32.4B). The methyltransferase-mediated step in the acetyl-CoA pathway appears to facilitate the translocation of sodium ions [154]. Under certain conditions, some acetogens require sodium for growth, motility, and the optimal synthesis of acetate [67,82,152,217]. For example, sodium is required for the autotrophic growth, but not for organotrophic growth, of *Thermoanaerobacter kivui*; in contrast, sodium is not required for either the autotrophic or organotrophic growth of *M. thermoacetica* [217]. *Acetobacterium woodii* and *T. kivui* are examples of acetogens that use sodium translocation and sodium-dependent ATPases for the conservation of energy. Sodium-proton antiporters can be used in the formation of electrochemical gradients by acetogens and may be important to their ability to conserve energy [139,154,194,217]. The reduction of aromatic acrylates (see Section 32.4.4) by certain acetogens is coupled to the synthesis of ATP and is also a sodium-dependent process [79,91,198].

The capacity of an acetogen to conserve energy varies from one acetogen to the next. Thus, acetogens might display different growth efficiencies when grown on the same substrate. For example, the cell yields of *M. thermoacetica* on H_2 and glucose are half those of *T. kivui* on these same substrates [37]. It seems likely that the different mechanisms used by acetogens to conserve energy, as well as the engagement of substrate-induced differential electron transport systems [94], might account for some of the differences in growth efficiencies of acetogens.

32.4.3 Diverse Electron Donors

Acetogens can utilize a wide range of electron donors and electron acceptors (Figure 32.5A), and have various metabolic mechanisms that allow them to cope with O_2 and the toxic effects of oxic conditions. These diverse metabolic capabilities likely contribute to their competitiveness in nature and will be outlined in the following three sections. The capacity of acetogens to utilize diverse redox couples enables them to form junction points in different biological cycles (Figure 32.5B).

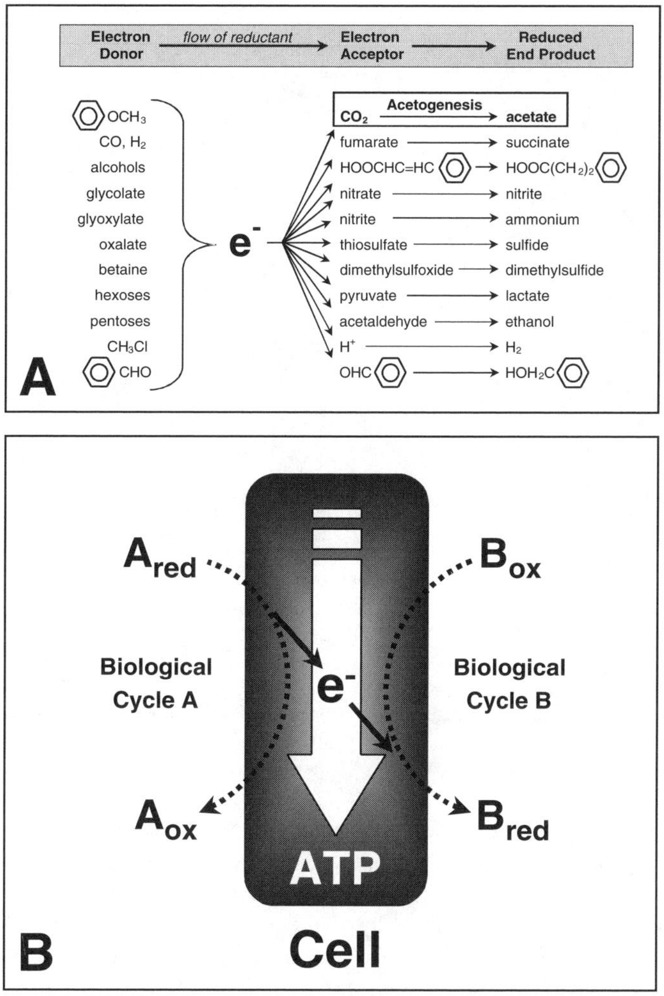

FIGURE 32.5 (A) Composite of the diverse types of electron donors and electron acceptors that can be utilized by acetogens. Abbreviation: e^-, reducing equivalent. (Modified from Drake et al. [50].) (B) Intercycle coupling (see text).

For example, H_2-CO_2-dependent acetogenesis (reaction 32.1 [most acetogens]) links the hydrogen and carbon cycles, and the dissimilation of nitrate via the oxidation of organic carbon (e.g., *M. thermoacetica*) links the nitrogen and carbon cycles. This redox coupling provides the cell with a means of conserving energy and forms a basis for intercycle coupling at the ecosystem level (Figure 32.5B).

CO, H_2, carbohydrates, alcohols, carboxylic acids, dicarboxylic acids, aldehydes, substituent groups of various aromatic compounds, and certain halogenated substrates are examples of substrates that acetogens can oxidize. Although the use of large polymers does not appear to be a general property of the acetogens isolated to date, the recently described acetogen *Bryantella formatexigens* (strain 1–52 [212]), isolated from human feces, utilized amorphous (cabbage) cellulose when the organism was first isolated [213]. The organism lost this ability upon prolonged cultivation. Nonetheless, this finding is significant, as it demonstrates that certain acetogens may indeed be capable of degrading cellulose or similar polymers. This possibility is strengthened by the recent description of a cellulose-degrading bacterium that appears to be a species of *M. thermoacetica* [100].

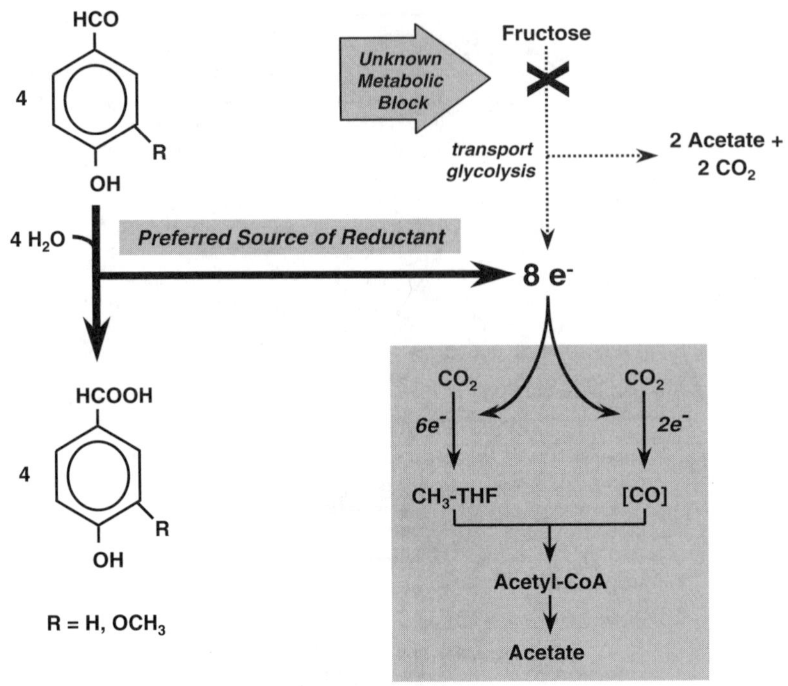

FIGURE 32.6 Metabolic scheme illustrating the preferred flow of reductant from aromatic aldehydes to the acetyl-CoA pathway by the acetogen *Clostridium formicoaceticum*. Abbreviations: see Figure 32.3 caption. (Modified from Drake and Küsel [52].)

Acetogens are specialists in terms of their ability to use substituent groups of aromatic compounds, and their use of such substituent groups can have significant effects on how they metabolize sources of carbon and reductant. For example, the aldehyde group of certain aromatic compounds (e.g., 4-hydroxybenzaldehyde) are oxidized by *C. formicoaceticum*, and the reductant derived from this oxidation is used in preference to fructose-derived reductant for the reduction of CO_2 to acetate [52,59,69]. Thus, the redox coupling between the oxidation of aldehyde substituent groups and the reductive acetyl-CoA pathway inhibits the cell's ability to use fructose, which otherwise is a preferred substrate (Figure 32.6). Likewise, the ability of *M. thermoacetica* to decarboxylate carboxylated aromatic compounds provides the cell with growth-supportive CO_2 equivalents that are used in the acetyl-CoA pathway for the synthesis of both acetate and cell carbon [86,87].

Many acetogens, including many clostridial acetogens, utilize methoxyl groups of aromatic compounds (e.g., vanillate) [7,38,60,61]. The capacity to selectively utilize methoxyl groups of aromatic compounds appears to be a specialized feature of acetogens and might be of importance to acetogens in certain habitats. A nonclostridial acetogen ("*Acetobacterium dehalogenans*") grows methylotrophically on methyl chloride (reaction IV) and illustrates that this capability might occur in other acetogenic species [197].

$$4\ CH_3Cl + 2\ CO_2 + 2\ H_2O \rightarrow 3\ CH_3COOH + 4\ HCl \qquad (32.4)$$

Figure 32.7 illustrates how methyl groups are metabolized by acetogens. The demethylase system utilized (i) is induced by the substrate that is demethylated, (ii) can display broad specificity for several methoxylated substrates, and (iii) might require ATP for maintaining corrinoid proteins in an active form [14,38,61,77,99,102,110,155,216].

Although acetogens are specialized in utilizing various substituent groups of aromatic compounds, it does not appear that the full degradation of aromatic compounds is a widespread

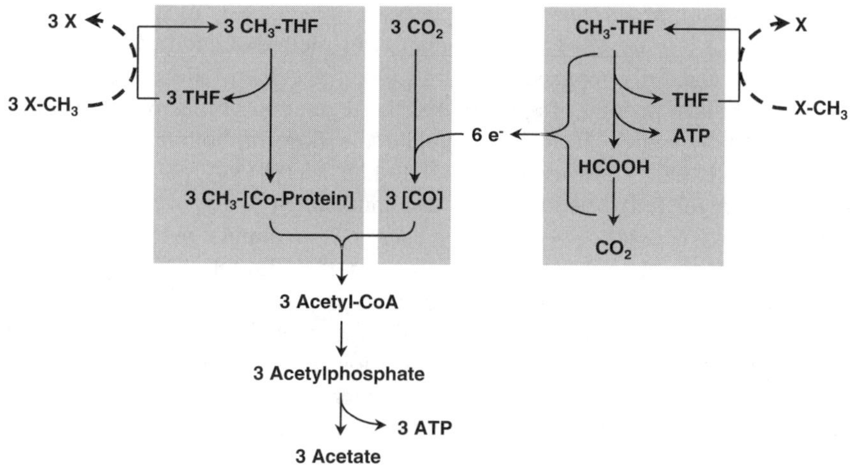

FIGURE 32.7 Metabolic scheme illustrating hypothetical routes by which methyl groups from methoxylated aromatic compounds (e.g., vanillate) or methyl chloride can be utilized by acetogens. Abbreviations: see Figure 32.3 caption. (Modified from Drake et al. [53].)

capability of acetogens. However, the acetogen *H. foetida* degrades aromatic rings [8,109,133], and it must be assumed that future studies will likely identify other acetogens with such capabilities. The capacity of acetogens to utilize alternative terminal electron acceptors, i.e., electron acceptors whose half-cell reactions are poised at redox potentials more positive than that of the CO_2/acetate half-cell reaction, increases this likelihood.

32.4.4 Diverse Terminal Electron Acceptors

Because the acetyl-CoA pathway is not the only terminal electron-accepting process that acetogens can utilize (Figure 32.5A), the adaptation strategies of acetogens are not solely determined by their ability to engage the acetyl-CoA pathway. A few generalizations can be made about the ability of acetogens to reduce different terminal electron acceptors:

a. Almost all known acetogens can utilize more than one terminal electron acceptor.
b. The efficiency by which acetogens oxidize a particular substrate is dependent on the type of terminal electron acceptor that is utilized. For example, the oxidation of H_2 and H_2-dependent growth can be significantly improved when an alternative terminal electron acceptor (e.g., aromatic acrylates or nitrate) is utilized [35,63,186]. Likewise, *M. thermoacetica* cannot use ethanol and propanol as acetogenic substrates, but readily oxidizes both alcohols when nitrate is utilized as a terminal electron acceptor [63].
c. An acetogen can utilize other terminal electron acceptors, even when exogenous CO_2 is readily available. For example, *C. formicoaceticum* can simultaneously utilize fumarate and CO_2 as terminal electron acceptors when methanol is oxidized [144]. Likewise, *R. productus* reduces phenylacrylates to phenylpropionates or pyruvate to lactate (both phenylpropionates and lactate are end-products) concomitant to the reduction of CO_2 to acetate [150,151]. In contrast, the reduction of CO_2 to acetate can occur preferentially. For example, *C. aceticum* preferentially utilizes CO_2 instead of fumarate as a terminal electron acceptor when H_2 is oxidized [144].
d. The use of an alternative electron acceptor can have regulatory effects on the acetyl-CoA pathway. For example, when an acrylate group of certain aromatic compounds (e.g., caffeate) is used as a terminal electron acceptor by certain acetogens, the reduction of CO_2 to acetate can be repressed [198]. Likewise, the dissimilation of nitrate to nitrite

and ammonium by *M. thermoacetica* represses the synthesis of a membranous *b*-type cytochrome that is essential for the formation of methyltetrahydrofolate in the acetyl-CoA pathway and fully represses the ability of the cell to reduce CO_2 to acetate and biomass [51,63]. In general, it appears that the regulation of electron transport systems is one means by which acetogens regulate the flow of both carbon and reductant [63,94,187,188]. Other processes central to the acetyl-CoA pathway might also be affected (i.e., regulated) when nitrate is dissimilated [6].

e. Some reductive processes do not conserve energy. For example, reductive dehalogenation by acetogens does not appear to conserve energy [56,62,80,85,195,197]. However, secondary coupling that is linked to reductive processes can be indirectly coupled to growth and the conservation of energy. For example, the reduction of protons to H_2 does not appear to be directly coupled to the conservation of energy, but the ability to produce H_2 (i.e., reduce protons) can be important for interspecies transfer of H_2 and the conservation of energy under certain *in situ* conditions [210].

32.4.5 Cellular Responses to O_2

Although acetogens have been termed strict anaerobes, they colonize habitats that receive periodic influxes of O_2 and must, therefore, be able to cope with unstable redox conditions (see Section 32.6). It is therefore not surprising that acetogens can both tolerate and consume O_2 [18,51,52,101,122]. There are three ways that acetogens cope with oxic conditions:

a. Close trophic interaction between acetogens and O_2-consuming microaerophiles might enhance both the survival and activity of acetogens in habitats with low levels of O_2. In one such case that has been documented, *M. thermoacetica* and the fermentative microaerophile *Thermicanus aegyptius* grow commensally on oligosaccharides (Figure 32.8)

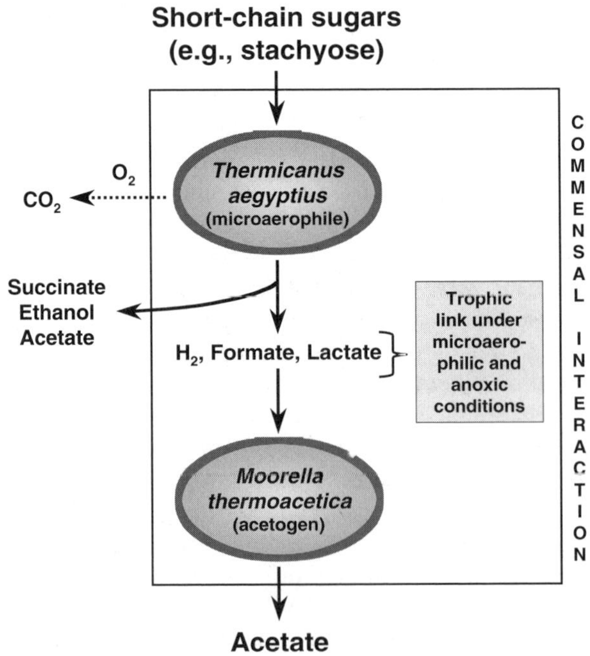

FIGURE 32.8 Scheme illustrating the commensal interaction of the acetogen *Moorella thermoacetica* and the facultative microaerophile *Thermicanus aegyptius*. (Modified from Gößner et al. [71].)

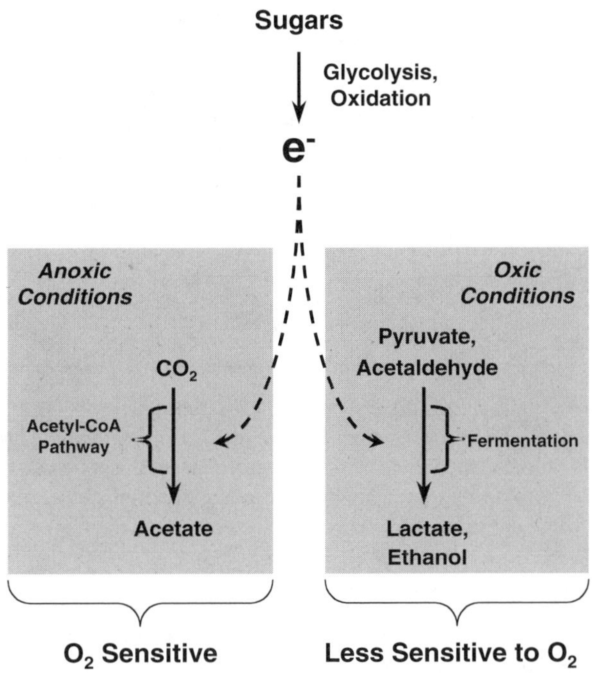

FIGURE 32.9 Scheme illustrating how oxic conditions can result in a redirection of electron flow away from acetogenesis and toward classic fermentation processes. (Modified from Küsel et al. [122].)

[71]. The fermentative microaerophile not only consumes O_2, and thus, in theory, would reduce the level of incoming O_2 in microzones inhabited by the commensally associated acetogen, but also produces products (e.g., lactate and H_2) that are utilized as substrates by the acetogen for acetogenesis.

b. As outlined above, acetogens can use alternative terminal electron acceptors for recycling reduced electron carriers. In general, these alternative catabolic processes are less sensitive to O_2 than the acetyl-CoA pathway and also occur at higher redox potentials than the CO_2/acetate half-cell reaction. Thus, by utilizing alternative terminal electron acceptors, acetogens can metabolically bypass the acetyl-CoA pathway for recycling reduced electron carriers under more oxidative conditions, i.e., conditions that would otherwise be deleterious to their survival and growth. The acetogen *Clostridium glycolicum* RD-1 (isolated from seagrass roots) is aerotolerant and can tolerate up to 4% O_2 in the headspace of agitated cultures [122]. Under more oxic conditions, acetogenesis is repressed or inhibited, and a combined lactate-ethanol fermentation is used for growth on sugars. Recently, an acetogenic species has been cultivated under aerated conditions [124]. Under these conditions, homolactate fermentation serves as the main energy-conserving process. Thus, certain acetogens can tolerate oxic conditions by metabolizing sugars via a fermentation pathway that is less sensitive to O_2 than is the acetyl-CoA pathway (Figure 32.9).
c. The following oxidative stress enzymes have been detected in acetogens at either the activity or gene level: NADH-oxidase, peroxidase, superoxide dismutase, rubredoxin oxidoreductase, and rubrerythrin. Superoxide dismutase forms O_2, and it would not seem to be very rational for a "strict anaerobe" to have such an enzyme. Thus, it is not surprising that only one acetogen, *C. glycolicum* RD1, an acetogen that displays a very high tolerance to O_2 (above), has been found to have this enzyme [122]. *C. glycolicum* RD1 (Figure 32.10) also contains peroxidase and NADH-oxidase. *Clostridium magnum* and other acetogens

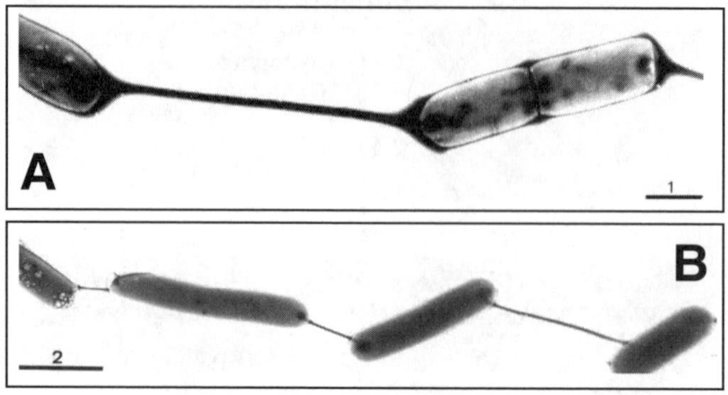

FIGURE 32.10 Electron micrographs of (A) the acetogen *Clostridium glycolicum* RD1 (used with permission from Küsel et al. [122]) and (B) the nitrogen-fixing bacterium *Clostridium akagii* [116] (used with permission from Drake et al. [53]). Both of these organisms have connecting filaments that might provide cells with a means of remaining in close proximity to one another for either structural or communication purposes under certain *in situ* conditions. Bars are in micrometers.

also have peroxidase and NADH-oxidase, but appear to lack superoxide dismutase [101]. Catalase, which also forms O_2, has thus far only been detected in three acetogens, i.e., two species of *Sporomusa* [18] and one species of *Acetonema* [96]; thus, catalase does not appear to be an oxidative stress enzyme in most acetogens. Rubrerythrin and rubredoxin oxidoreductase are oxidative stress enzymes in certain sulfate-reducing bacteria and reductively consume hydrogen peroxide and superoxide, respectively [142]. Genes for similar proteins have been detected in *M. thermoacetica* [43]. A flavoprotein/rubredoxin combination functions as an NADH:O_2 oxidoreductase in *M. thermoacetica* [190]. These collective findings demonstrate that acetogens have the capacity to reductively remove O_2 and its toxic side products peroxide and superoxide.

32.5 APPLICATIONS

Given their diversity of metabolic processes, acetogens might be considered attractive candidates for certain biotechnological applications. However, a commercial-scale application of an acetogen has not been reported. Nonetheless, numerous investigations have assessed their potential use, and this section will briefly highlight some of these studies. (For additional information, see [27,53,141,205,206].)

Production of acetic acid. During homoacetogenesis, acetogens convert hexoses stoichiometrically to three molecules of acetate (reaction 32.2). Because acetic acid is important in the chemical industry (global production in 2001 approximated 10^{10} kg), acetogens have been evaluated for their ability to produce acetate at the commercial level. Although homoacetogenesis can be viewed as more efficient than the two-stage vinegar process that requires two organisms and yields two molecules of acetic acid per molecule hexose, two limitations of homoacetogenesis have prevented the commercialization of the process: (i) acetogens are inhibited by high concentrations of acetate and (ii) acetogens do not grow under acidic conditions (i.e., at pHs lower than 4 [note: most acetogens cannot grow below pH 5]). Thus, no known acetogen produces acetic acid at concentrations required (i.e., 50 g acetic acid per liter) for the process to be commercially (i.e., economically) feasible [206]. Mutants with increased tolerance to acidic acid and acidic conditions grow poorly [185,206], mainly because acetic acid and protons destroy the proton motive force (ΔpH) and transmembrane electrical potentials (Δ) of acetogens [13]. Despite the fact that certain strains of acetogens have been obtained that have improved capabilities to synthesize acetate, these collective problems have prevented the use of acetogens for the commercial production of calcium-magnesium acetate as an environmentally safe road de-icer [31,137,138,205,206]. The recent demonstration that certain acetogens are capable of converting cellulose to glucose [100,213] might spark renewed interest in this area. Immobilization of acetogens might also enhance their effectiveness in certain biotechnological applications [191].

Bioconversion of synthesis gas. Indirect liquefaction of coal yields H_2, CO, and CO_2, which are collectively referred to as synthesis gas. These gases are converted to acetate by acetogens, and, as in the case of hexose-coupled homoacetogenesis, acetogens have been evaluated for possible use in the bioconversion of synthesis gas to acetic acid [74]. The acetogen *Clostridium ljungdahlii*, "*Clostridium autoethanogenum*," and other acetogens produce ethanol from synthesis gas [1,28,74,163,192], and other acetogens can form butyrate and *n*-butanol from the components of synthesis gas [75,215]. Despite these metabolic potentials, commercialization of such a process has not been reported.

Bioremediation. Acetogens contribute significantly to the turnover of organic matter in methanogenic bioreactors and landfills [12,90,147,206]. However, because acetogenic processes are tightly linked in the complex foodwebs of such systems, their *in situ* activities are masked and remain poorly resolved. Thus, the bioremediation potentials of acetogens have not been extensively examined. Although certain acetogens dehalogenate anthropogenic compounds [56,62,197], the *in situ* importance of this metabolic potential is unresolved. For the most part, acetogens do not appear to be capable of degrading aromatic rings. The acetogen *Holophaga foetida* is a noted exception [133]. The acetyl-CoA synthase of acetogens transforms 2,4,6-trinitrotoluene (TNT) [88,164]; however, application of this catalytic potential has not been reported. *M. thermoacetica* sequesters the heavy metal cadmium [36]; whether acetogens can be used in the remediation of materials contaminated with heavy metals remains an open question.

Fine chemicals. The potential to commercially produce corrinoids (i.e., vitamin B_{12}), cysteine, acetate kinase, and a variety of other fine chemicals with acetogens has been evaluated [54,55,92,103,128,183,206]; however, commercial-scale production has not been reported.

32.6 ECOLOGICAL PERSPECTIVES OF CLOSTRIDIAL ACETOGENS

By virtue of their ability to use a wide range of electron donors and electron acceptors, acetogens likely have numerous trophic links to other organisms under *in situ* conditions. However, the magnitudes of these links are, for the most part, not resolved. Likewise, the use of diverse electron acceptors by acetogens and their ability to cope with oxic conditions indicate that acetogens can accommodate a wide range of redox conditions. Nonetheless, their *in situ* activities under different redox conditions are also unresolved. Several clostridial acetogens (*C. aceticum*, *C. formicoaceticum*, and *Clostridium magnum*) fix N_2 [19,20,30]. However, the *in situ* importance of this ability is not known. Such disparities are reason why one must describe the ecology of acetogens with proper qualification. Nonetheless, it is now quite clear that acetogens, including clostridial acetogens, occur in very diverse habitats, and that their ecological functions are complex and not restricted to a single metabolic feature. This section will outline a few of the more important ecological features of clostridial and other acetogens.

32.6.1 *In situ* Competitiveness and Vigor

In anoxic habitats, acetogens compete with fermentative microbiota for compounds that are derived from the initial breakdown of biological polymers (e.g., cellulose and lignin) [147]. Under certain conditions, acetogens are quite competitive, while under other conditions they are less so. For example, acetogenesis appears to be the most competitive anaerobic glucose-consuming process in anoxic paddy soils and certain fresh water sediments [111]. In contrast, even though H_2-CO_2-dependent acetogenesis is one of the more important physiological features of acetogens, at the concentrations of reactants and products that are often found in anoxic environments, H_2-CO_2-dependent methanogenesis is more favorable than H_2-CO_2-dependent acetogenesis [35]. The minimum concentration of H_2 required for its cellular uptake (i.e., termed the "H_2 threshold") decreases as the redox potential of the electron-accepting half-cell reaction increases [34]. Because the CO_2/acetate half-cell reaction (–290 mV) is one of the most negative terminal electron-accepting reactions in biology (the half-cell reactions for methanogenesis and sulfate reduction have more positive redox potentials), the H_2 threshold of acetogens in pure cultures is higher than that of other H_2-consuming anaerobes [35]; thus, at low concentrations of H_2, acetogens are generally considered to be uncompetitive when H_2-dependent acetogenesis is their only possible energy-conserving process.

Numerous factors contribute to the competitiveness and durability of acetogens. The spores of *M. thermoacetica* have a decimal reduction time (i.e., the time required to achieve a 90% kill) of up to 111 minutes at 121°C and are some of the most heat resistance spores known [29]. Mixotrophic conditions and co-substrate utilization can increase the ability of acetogens to compete for H_2 and other substrates [136,161]. Furthermore, low pH and low temperature do not favor hydrogenotrophic methanogenesis and acetogens may become more competitive under such conditions [33,156,162,218]. In addition, the *in situ* success of acetogens can be enhanced by their close association with H_2-producing microbes, as is the case in certain termite-gut ecosystems [126]. Transient amounts of O_2 can also influence competition between acetogens and other anaerobes. For example, methanogens appear to be more sensitive to O_2 than acetogens; thus, acetogens are more active and more numerous than methanogens and sulfate reducers in habitats that are subject to transient fluctuations of O_2 and drying (e.g., soil and litter [117,118,120,159,160,171,200]). The ability of acetogens to remain competitive in such habitats is due in part to their ability to reductively remove traces of O_2 and its toxic by-products, and commensal association with facultative microaerophiles or other O_2-consuming microbiota (see Section 32.4.5).

32.6.2 Ubiquitous Nature

Most acetogens have thus far been obtained from permanently anoxic habitats (e.g., sediments, sewage, and gastrointestinal tracts); however, they are also easily isolated from numerous other habitats (e.g., hypersaline environments, aquifers, oxic soils, and plant roots). Their physiological diversities likely contribute to their seemingly ubiquitous nature, and one can easily conclude that they will continue to be discovered in more and more environments. This section highlights a few of the habitats from which they have been isolated.

32.6.2.1 Gastrointestinal Tracts

Mammalian gastrointestinal tracts are colonized by acetogens [132,143,165,212,213]. Part of the daily production of acetate in the human intestine can be attributed to the activity of acetogens, as evidenced by the synthesis of [^{13}C]acetate from $^{13}CO_2$ [125] and the formation of [1,2-^{14}C]acetate from [3,4-^{14}C]glucose [149] by human fecal suspensions. Cultured numbers of H_2-utilizing fecal acetogens of non-CH_4-excreting animals are higher than those of CH_4-excreting animals [46,165]. Such correlations indicate that H_2-dependent acetogenesis is important in gastrointestinal ecosystems that are poorly colonized by methanogens [15,46]. Some of the H_2-utilizing acetogens isolated from human feces occur within clostridial cluster XIVa [16,17,32,95,129,212] (Table 32.1).

Acetogens are plentiful in the rumen, and cultured values range from 10^6 to 10^9 per gram ruminal content [113,131,143,173–176] (cultured methanogens in the rumen approximate 10^8 to 10^9 per gram ruminal content [131]). The cultured numbers of acetogens from the rumen of steers maintained on a high-forage diet are higher than the cultured numbers of ruminal methanogens, suggesting that diet affects the relative distribution of acetogens and methanogens in the rumen [131]. Numerous acetogens, including *C. difficile* AA1 (Table 32.1), have been isolated from the rumen [73,113,143,173–176). However, the ecology of acetogens (i.e., the magnitude of their *in situ* activities) in ruminal ecosystems remains poorly resolved.

Cellulolytic protozoa and bacteria symbiotically achieve a nearly homoacetogenic decomposition of wood polysaccharides in the hindgut of the wood-feeding "lower" termites [22]. The protozoa convert cellulose to acetate, H_2, and CO_2, and acetogens subsequently convert H_2-CO_2 to acetate [25]. Microbially produced acetate can provide up to 100% of the energy requirement of wood-feeding termites [26]. H_2-consuming acetogenic spirochetes (e.g., *Treponema* sp., strain ZAS-X) in the central region of the hindgut (where most of the H_2 is consumed) are often attached to the H_2-producing protozoa; this symbiosis provides concentrations of H_2 well above the known H_2-threshold values for acetogens [126]. Acetogens in the posterior hindgut might utilize substrates other than H_2 [181,196]. Although the *in situ* involvement of clostridial acetogens in the termite gut has not been resolved, *Clostridium mayombei* was isolated from the gut of a soil-feeding termite [97].

32.6.2.2 Water-Saturated Habitats

Acetogens are ubiquitous in the sediments of water-saturated habitats. In such habitats, acetogens might have to compete for H_2. Although the dissimilation of sulfate and methanogenesis appear to be more effective than acetogenesis as terminal electron accepting processes for the oxidation of low concentrations of H_2 [35], acetogens appear to be competitive for H_2 in sediments under mildly acidic or low temperature conditions [33,156,162,184,218]. *Clostridium scatologenes* SL1 is an H_2-utilizing acetogen isolated from acidic sediments and can grow at pH 4 [121]. Sodium-proton antiporters [194] might enable certain acetogens to cope with broad variations in pH [180]. Psychrotolerant acetogens have been isolated [33,107,108], thus underscoring the potential importance of acetogenesis in low temperature sediments. Although the partial pressure of H_2 in pore waters might be lower than that needed for H_2-dependent acetogenesis, the *in situ* partial pressure of H_2 might be more optimal for acetogenesis if acetogens were closely associated with H_2-producing organisms.

The flow of carbon and reductant shifts from sulfate reduction to methanogenesis when sulfate becomes depleted in marine sediments. During this transition of processes, the production of [^{14}C]-acetate from $^{14}CO_2$ occurs at rates comparable to those of methanogenesis or sulfate reduction during their respective period of dominance [84]. Cultured numbers of H_2-utilizing acetogens and acetate-utilizing sulfate reducers from a marine sediment were not that dissimilar, approximating 10^4 and 10^5 cells per g wet wt. sediment, respectively [119]. Acetogens also occur in hypersaline habitats [218–220], subsurface aquifers [81,104–106,114], and oceanic sediments [202,203]. Thus, it is very likely that the acetogenic reduction of CO_2 occurs in a broad range of water-logged habitats.

32.6.2.3 Rhizosphere

Relatively little is known about the ability of anaerobic bacteria to colonize the rhizosphere [76,83]. Although the rhizospheres of salt marsh vegetation and rice is generally regarded as anoxic, a fluctuating gradient of O_2 is generated around roots via the transport of O_2 that is produced by oxygenic photosynthesis in the plant [68,172]. Thus, anaerobes experience periods of elevated O_2 tension in rhizospheres. Seagrass rhizosphere has higher numbers of acetogens than unvegetated soil, acetogenic *O*-demethylation activity is tightly associated with seagrass roots, and seagrass root thin sections hybridize with ^{33}P-labeled acetogen probes, indicating that seagrass rhizophere and roots are colonized by acetogens [119]. An H_2-utilizing acetogen, *Clostridium glycolicum* RD1, obtained from the highest, growth-positive dilution of a seagrass root most-probable number series displays a high tolerance to O_2 [122] (see Section 32.4.5). Acetogens that are associated with salt marsh plant roots might also display a high tolerance to O_2 [124]. Analysis of formyltetrahydrofolate synthetase sequences from salt marsh plant roots indicates that such roots are colonized by diverse acetogenic genera, including the genus *Clostridium* [127]. Acetogens also occur in flooded paddy soils, suggesting that acetogenesis might occur in the rhizophere of rice [33,177].

32.6.2.4 Aerated Soils

Anoxic microzones occur in aerated soils [189,199], and soils form enormous amounts of acetate from endogenous organic matter under anoxic conditions (up to 15 g C-acetate per kg dry wt. soil) [117,160,200]. The acetate-forming microbiota of soils is complex, but acetogens are a part of this microbiota [118,171]. Supplemental H_2, CO, and ethanol are converted to acetate by acetogens that colonize soils, and acetogenic activities of soils are relatively stable to fluxes of O_2 [115,117,120,200]. H_2-utilizing acetogens are a dominant group of the cultured anaerobes in soil or litter and approximate 10^4 to 10^5 cells per g of dry soil or litter [120,159]. *M. thermoacetica* has a wide geographical distribution in well-drained, high temperature soils [70,71]. The consumption of acetate that is formed anaerobically appears to be linked to oxidative processes, such as O_2-dependent respiration, denitrification, or the reduction of Fe(III) [117,120,123,200]. Thus, acetogens and other acetate-producing bacteria in anoxic microzones of soils appear to be involved in the anaerobic production of acetate, which subsequently serves as a trophic link to oxidative microbial populations.

It is worth recalling that the first acetogen isolated, *C. aceticum*, was isolated from ditch mud soil [207]. It is likely that this habitat was also subject to periods of dryness and aeration, thus underscoring not only the occurrence of acetogens in such habitats, but also the ability of acetogens to metabolically cope with unstable water activities and fluctuating redox conditions. Indeed, the durability of acetogens is well exemplified in the revival of *C. aceticum* after decades of dessication (Figure 32.1).

32.7 CONCLUSIONS

Clostridial acetogens are a metabolically robust group of bacteria. Although this chapter has mostly highlighted features of special relevance to acetogenic species of the genus *Clostridium*, the literature on the 21 genera of acetogens has reached enormous proportions. This large body of information attests to not only their diverse capabilities in the laboratory, but also to their abundance and importance in nature as well. One of the challenges of future studies will be to more closely resolve the *in situ* activities of acetogens and thus determine both the magnitude and importance of their trophic links in the environment.

ACKNOWLEDGMENTS

The authors thank Anita Gößner for her excellence in keeping the laboratory acetogens healthy and happy, Georg Acker for electron microscopy of isolates, Meyer Wolin and Terry Miller for providing unpublished information, and Volker Müller for helpful discussions. Current support for the authors' laboratory is derived in part from funds from the German Research Society (DFG) and the German Ministry of Education, Research and Technology (BMBF), and is gratefully acknowledged.

REFERENCES

1. Abrini, J., Naveau, H., and Nyns, E.-J. *Clostridium autoethanogenum*, sp. nov., an anaerobic bacterium that produces ethanol from carbon monoxide, *Arch. Microbiol.,* 161, 345–351, 1994.
2. Adamse, A.D., New isolation of *Clostridium aceticum* (Wieringa), *Antonie van Leeuwenhoek,* 46, 523–531, 1980.
3. Adamse, A.D. and Velzeboer, C.T.M., Features of a *Clostridium,* strain CV-AA1, an obligatory anaerobic bacterium producing acetic acid from methanol, *Antonie van Leeuwenhoek,* 48, 305–313, 1982.
4. Andreesen, J.R., Gottschalk, G., and Schlegel, H.G, *Clostridium formicoaceticum* nov. spec. isolation, description and distinction from *C. aceticum* and *C. thermoaceticum, Arch. Microbiol.,* 72, 154–174, 1970.
5. Andreesen, J.R., Acetate via glycine: a different form of acetogenesis, in *Acetogenesis,* Drake, H.L., Ed., Chapman & Hall, New York, 1994, 568–629.
6. Arendsen, A.F., Soliman, M.Q., and Ragsdale, S.W., Nitrate-dependent regulation of acetate biosynthesis and nitrate respiration by *Clostridium thermoaceticum, J. Bacteriol.,* 181, 1489–1495, 1999.
7. Bache, R. and Pfennig, N., Selective isolation of *Acetobacterium woodii* on methoxylated aromatic acids and determination of growth yields, *Arch. Microbiol.,* 130, 255–261, 1981.
8. Bak, F., Finster, K., and Rothfub, F., Formation of dimethylsulfide and methanethiol from methoxylated aromatic compounds and inorganic sulfide by newly isolated anaerobic bacteria, *Arch. Microbiol.,* 157, 529–534, 1992.
9. Banerjee, R. and Ragsdale, S.W., The many faces of vitamin B_{12}: catalysis by cobalamin-dependent enzymes, *Ann. Rev. Biochem.,* 72, 209–247, 2003.
10. Barik, S., Prieto, Harrison, S.B., Clausen, E.C., and Gaddy, J.L., Biological production of alcohols from coal through indirect liquefaction, *Appl. Biochem. Biotech.,* 18, 363–378, 1988.
11. Barker, H.A. and Kamen, M.D., Carbon dioxide utilization in the synthesis of acetic acid by *Clostridium thermoaceticum, Proc. Natl. Acad. Sci. U.S.A.,* 31, 219–225, 1945.
12. Barlaz, M.A., Microbial studies of landfills and anaerobic refuse decomposition, in: *Manual of Environmental Microbiology,* Hurst, C.J., Ed.,. ASM Press, Washington, DC, 1997, 541–555.
13. Baronofsky, J.J., Schreurs, W.J.A., and Kashket, E.R., Uncoupling by acetic acid limits growth of and acetogenesis by *Clostridium thermoaceticum, Appl. Environ. Microbiol.,* 48, 1134–1139, 1984.
14. Berman, M.H. and Frazer, A.C., Importance of tetrahydrofolate and ATP in the anaerobic O-demethylation reaction for phenylmethylethers, *Appl. Environ. Microbiol.,* 58, 925–931, 1992.
15. Bernalier, A., Lelait, M., Rochet, V., Grivet, J.-P., Gibson, G.R., and Durand, M. Acetogenesis from H_2 and CO_2 by methane- and non-methane-producing human colonic bacterial communities, *FEMS Microbiol. Ecol.,* 19, 193–202, 1996.
16. Bernalier, A., Rochet, V., Leclerc, M., Doré, J., and Pochart, P., Diversity of H_2/CO_2-utilizing acetogenic bacteria from feces of non-methane-producing humans, *Curr. Microbiol.,* 33, 94–99, 1996.
17. Bernalier, A., Willems, A., Leclerc, M., Rochet, V., and Collins, M.D., *Ruminococcus hydrogenotrophicus* sp. nov., a new H_2/CO_2-utilizing acetogenic bacterium isolated from human feces, *Arch. Microbiol.,* 166, 176–183, 1996.
18. Boga, H.I. and Brune, A., Hydrogen-dependent oxygen reduction by homoacetogenic bacteria isolated from termite guts, *Appl. Environ. Microbiol.,* 69, 779–786, 2003.
19. Bogdahn, M., Andreesen, J.R., and Kleiner, D., Pathways and regulation of N_2, ammonium and glutamate assimilation by *Clostridium formicoaceticum, Arch. Microbiol.,* 134, 167–169, 1983.
20. Bomar, M., Hippe, H., and Schink, B., Lithotrophic growth and hydrogen metabolism by *Clostridium magnum, FEMS Microbiol. Lett.,* 83, 347–350, 1991.
21. Bramlett, M.R., Tan, X., and Lindahl, P.A., Inactivation of acetyl-CoA synthase/carbon monoxide dehydrogenase by copper, *J. Am. Chem. Soc.,* 125, 9316–9317, 2003.
22. Brauman, A., Kane, M.D., Labat, M., and Breznak, J.A., Genesis of acetate and methane by gut bacteria of nutritionally diverse termites, *Science,* 257, 1384–1387, 1992.
23. Braun, M., Mayer, F., and Gottschalk, G., *Clostridium aceticum* (Wieringa), a microorganism producing acetic acid from molecular hydrogen and carbon dioxide, *Arch. Microbiol.,* 128, 288–293, 1981.

24. Braus-Stromeyer, S.A., Schnappauf, G., Braus, G.H., Gößner, A.S., and Drake, H.L., Carbonic anhydrase in *Acetobacterium woodii* and other acetogenic bacteria, *J. Bacteriol.,* 179, 7197–7200, 1997.
25. Breznak, J.A. and Switzer, J.M., Acetate synthesis from H_2 plus CO_2 by termite gut microbes, *Appl. Environ. Microbiol.,* 52, 623–630, 1986.
26. Breznak, J.A., Acetogenesis from carbon dioxide in termite guts, in *Acetogenesis*, Drake, H.L., Ed., Chapman & Hall, New York, 1994, 303–330.
27. Busche, R.M., Extractive fermentation of acetic acid: economic tradeoff between yield of *Clostridium* and concentration of *Acetobacter, Appl. Biochem. Biotechnol.,* 28/29, 605–621, 1991.
28. Buschhorn, H., Dürre, P., and Gottschalk, G., Production and utilization of ethanol by the homoacetogen *Acetobacterium woodii, Appl. Environ. Microbiol.,* 55, 1835–1840, 1989.
29. Byrer, D.E., Rainey, F.A., and Wiegel, J., Novel strains of *Moorella thermoacetica* form unusually heat-resistant spores, *Arch. Microbiol.,* 174, 334–339, 2000.
30. Cato, E.P., George, W.L., and Finegold, S.M., Genus *Clostridium* Prazmowski 1880. in *Bergey's Manual of Systematic Bacteriology,* Vol. 2, Sneath, P.H.A., Ed., Williams & Wilkins, Baltimore, 1986, 1141–1200.
31. Cheryan, M., Parekh, S., Shah, M., and Witjitra, K., Production of acetic acid by *Clostridium thermoaceticum, Adv. Appl. Microbiol.,* 43, 1–33, 1997.
32. Collins, M.D., Lawson, P.A., Willems, A., Cordoba, J.J., Fernandez-Garayzabal, J., Garcia, P., Cai, J., Hippe, H., and Farrow, J.A.E., The phylogeny of the genus *Clostridium*: Proposal of five new genera and eleven new species combinations, *Int. J. Syst. Bacteriol.,* 44, 812–826, 1994.
33. Conrad, R., Bak, F., Seitz, H.J., Thebrath, B., Mayer, H.P., and Schütz, H., Hydrogen turnover by psychrotrophic homoacetogenic and mesophilic methanogenic bacteria in anoxic paddy soil and lake sediment, *FEMS Microbiol. Ecol.,* 62, 285–294, 1989.
34. Conrad, R., Soil microorganisms as controllers of atmospheric trace gases (H_2, CO, CH_4, OCS, N_2O, and NO), *Microbiol. Reviews,* 60, 609–640, 1996.
35. Cord-Ruwisch, R., Seitz, H.-J., and Conrad, R., The capacity of hydrogenotrophic anaerobic bacteria to compete for traces of hydrogen depends on the redox potential of the terminal electron acceptor, *Arch. Microbiol.,* 149, 350–357, 1988.
36. Cunningham, D.P. and Lundie, L.L. Jr., Precipitation of cadmium by *Clostridium thermoaceticum, Appl. Environ. Microbiol.,* 59, 7–14, 1993.
37. Daniel, S.L., Hsu, T., Dean, S.I., and Drake, H.L., Characterization of the H_2- and CO-dependent chemolithotrophic potentials of the acetogens *Clostridium thermoaceticum* and *Acetogenium kivui, J. Bacteriol.,* 172, 4464–4471, 1990.
38. Daniel, S.L., Keith, E.S., Yang, H., Lin, Y-S., and Drake, H.L., Utilization of methoxylated aromatic compounds by the acetogen *Clostridium thermoaceticum:* expression and specificity of the CO-dependent O-demethylating activity, *Biochem. Biophys. Res. Commn.,* 180, 416–422, 1991.
39. Daniel, S.L. and Drake, H.L., Physiology of the thermophilic acetogen *Moorella thermoacetica, Research Microbiol.,* 155, 869–883, 2004.
40. Darnault, C., Volberg, A., Kim, E.J., Legrand, P., Vernède, X., Lindahl, P.A., and Fontecilla-Camps, J.C., Ni-Zn-[Fe_4-S_4] and Ni-Ni-[Fe_4-S_4] clusters in closed and open α subunits of acetylCoA synthase/carbon monoxide dehydrogenase, *Nature Structural Biology,* 10, 271–279, 2003.
41. Das, A., Ivey, D.M., and Ljungdahl, L.G., Purification and reconstitution into proteoliposomes of the F_1F_0 ATP synthase from the obligately anaerobic Gram-positive bacterium *Clostridium thermoautotrophicum, J. Bacteriol.,* 179, 1714–1720, 1997.
42. Das, A. and Ljungdahl, L.G., Acetogenesis and acetogenic bacteria, in Lederberg, J., Ed., *Encyclopedia of Microbiology,* 2nd Edition, Vol. 1, Academic Press, San Diego, 2000, 18–27.
43. Das, A., Coulter, E.D., Kurtz, D.M. Jr., and Ljungdahl, L.G., Five-gene cluster in *Clostridium thermoaceticum* consisting of two divergent operons encoding rubredoxin oxidoreductase — rubredoxin and rubrerythrin-type flavodoxin — high-molecular-weight rubredoxin, *J. Bacteriol.,* 183, 1560–1567, 2001.
44. Das, A. and Ljungdahl, L.G., Electron-transport system in acetogens, in Ljungdahl, L.G., Adams, M., Barton, L., Ferry, J.G., and Johnson, M., Eds., *Biochemistry and Physiology of Anaerobic Bacteria,* Springer-Verlag, New York, 2003, 191–204.
45. Diekert, G. and Wohlfarth, G., Energetics of acetogenesis from C_1 units, in *Acetogenesis*, Drake, H.L., Ed., Chapman & Hall, New York, 1994, 157–179.

46. Doré, J., Pochart, P., Bernalier, A., Goderel, I., Morvan, B., and Rambaud, J.C., Enumeration of H_2-utilizing methanogenic archaea, acetogenic and sulfate-reducing bacteria from human feces, *FEMS Microbiol. Ecol.*, 17, 279–284, 1995.
47. Doukov, T.I., Iverson, T.M., Sevavalli, J., Ragsdale, S.W., and Drennan, C.L., Ni-Fe-Cu center in a bifunctional carbon monoxide dehydrogenase/acetyl-CoA synthase, *Science,* 298, 567–572, 2002.
48. Drake, H.L., Hu, S.-I., and Wood, H.G., Purification of five components from *Clostridium thermoaceticum* which catalyze synthesis of acetate from pyruvate and methyltetrahydrofolate: properties of phosphotransacetylase, *J. Biol. Chem.,* 255, 7174–7180, 1981.
49. Drake, H.L., Acetogenesis, acetogenic bacteria, and the acetyl-CoA "Wood-Ljungdahl" pathway: past and current perspectives, in *Acetogenesis*, Drake, H.L., Ed., Chapman & Hall, New York, 1994, 3–60.
50. Drake, H.L., Daniel, S. L., Küsel, K., Matthies, C., Kuhner, C., and Braus-Stromeyer, S., Acetogenic bacteria: what are the *in situ* consequences of their diverse metabolic versatilities? *BioFactors,* 6, 13–24, 1997.
51. Drake, H.L., Küsel, K., and Matthies, C., Ecological consequences of the phylogenetic and physiological diversities of acetogens, *Antonie van Leeuwenhoek,* 81, 203–213, 2002.
52. Drake, H.L. and Küsel, K., How the diverse physiological potentials of acetogens determine their *in situ* realities, in *Biochemistry and Physiology of Anaerobic Bacteria*, Ljungdahl, L.G., Adams, M., Barton, L., Ferry, J.G., and Johnson, M., Eds., 171–190, Springer-Verlag, New York, 2003.
53. Drake, H.L., Küsel, K., and Matthies, C., Acetogenic prokaryotes, in *The Prokaryotes,* Dworkin, M., et al., Eds., 3rd ed.: An Evolving Electronic Resource for the Microbiological Community, release 3.17, August 2004, Springer, New York, http://Springeronline.com, 2004.
54. Eck, R. and Simon, H., Preparation of (*S*)-2-substituted succinates by stereospecific reductions of fumarate and derivatives with resting cells of *Clostridium formicoaceticum, Tetrahedron,* 50, 13631–13640, 1994.
55. Eck, R. and Simon, H., Preparation of both enantiomers of malic and citramalic acid and other hydroxysuccinic acid derivatives by stereospecific hydrations of *cis* and *trans* 2-butene-1,4-dioic acids with resting cells of *Clostridium formicoaceticum, Tetrahedron,* 50, 13641–13654, 1994.
56. Egli, C., Tschan, T., Scholtz, R., Cook, A.M., and Leisinger, T., Transformation of tetrachloromethane to dichloromethane and carbon dioxide by *Acetobacterium woodii, Appl. Environ. Microbiol.,* 54, 2819–2824, 1988.
57. Fischer, F., Lieske, R., and Winzer, K., Biologische Gasreaktionen. II. Über die Bildung von Essigsäure bei der biologischen Umsetzung von Kohlenoxyd und Kohlensäure mit Wasserstoff zu Methan, *Biochem. Z.,* 245, 2–12, 1932.
58. Fontaine, F.E., Peterson, W.H., McCoy, E., Johnson, M.J., and Ritter, G.J., A new type of glucose fermentation by *Clostridium thermoaceticum* n. sp., *J. Bacteriol.,* 43, 701–715, 1942.
59. Frank, C., Schwarz, U., Matthies, C., and Drake, H.L., Metabolism of aromatic aldehydes as co-substrates by the acetogen *Clostridium formicoaceticum, Arch. Microbiol.,* 170, 427–434, 1998.
60. Frazer, A.C. and Young, L.Y., A Gram-negative anaerobic bacterium that utilizes O-methyl substituents of aromatic acids, *Appl. Environ. Microbiol.,* 49, 1345–1347, 1985.
61. Frazer, A.C., O-demethylation and other transformations of aromatic compounds by acetogenic bacteria, in *Acetogenesis*, Drake, H.L., Ed., Chapman & Hall, New York, 1994, 445–483.
62. Freedman, D.L. and Gosset, J.M., Biodegradation of dichloromethane and its utilization as a growth substrate under methanogenic conditions, *Appl. Environ. Microbiol.,* 57, 2847–2857, 1991.
63. Fröstl, J.M., Seifritz, C., and Drake, H.L., Effect of nitrate on the autotrophic metabolism of the acetogens *Clostridium thermoautotrophicum* and *Clostridium thermoaceticum, J. Bacteriol.,* 178, 4597–4603, 1996.
64. Fuchs, G., CO_2 fixation in acetogenic bacteria: variations on a theme, *FEMS Microbiol. Rev.,* 39, 181–213, 1986.
65. Fuchs, G., Variations of the acetyl-CoA pathway in diversely related microorganisms that are not acetogens, in *Acetogenesis*, Drake, H.L., Ed., Chapman & Hall, New York, 1994, 507–520.
66. Furdui, C. and Ragsdale, S.W., The role of pyruvate ferredoxin oxidoreductase in pyruvate synthesis during autotrophic growth by the Wood-Ljungdahl pathway, *J. Biol. Chem.,* 275, 28494–28499, 2000.
67. Geerligs, G., Schönheit, P., and Diekert, G., Sodium dependent acetate formation from CO_2 in *Peptostreptococcus productus* (strain Marburg), *FEMS Microbiol. Lett.,* 57, 253–258, 1989.

68. Gilbert, B. and Frenzel, P., Methanotrophic bacteria in the rhizosphere of rice microcosms and their effect on porewater methane concentration and methane emission, *Biol. Fertil. Soils,* 20, 93–100, 1995.
69. Gößner, A., Daniel, S.L., and Drake, H.L., Acetogenesis coupled to the oxidation of aromatic aldehyde groups, *Arch. Microbiol.,* 161, 126–131, 1994.
70. Gößner, A. and Drake, H.L., Characterization of a new thermophilic acetogen (PT-1) isolated from aggregated Kansas prairie soil, *Abstr. Ann. Meet. Am. Soc. for Microbiol.,* 401, Abstr. N-122, 1997.
71. Gößner, A., Devereux, R., Ohnemüller, N., Acker, G., Stackebrandt, E., and Drake, H.L., *Thermicanus aegyptius* gen. nov., sp. nov., isolated from oxic soil, a facultative microaerophile that grows commensally with the thermophilic acetogen *Moorella thermoacetica, Appl. Environ. Microbiol.,* 65, 5124–5133, 1999.
72. Gottschalk, G. and Braun, M., Revival of the name *Clostridium aceticum, Int. J. System. Bacteriol.,* 31, 476, 1981.
73. Greening, R.C. and Leedle, J.A.Z., Enrichment and isolation of *Acetitomaculum ruminis*, gen. nov., sp. nov.: acetogenic bacteria from the bovine rumen, *Arch. Microbiol.,* 151, 399–406, 1989.
74. Grethlein, A.J. and Jain, M.K., Bioprocessing of coal-derived synthesis gases by anaerobic bacteria, *TIBTECH,* 10, 418–423, 1992.
75. Grethlein, A.J., Worden, R.M., Jain, M.K., and Datta, R., Evidence for production of *n*-butanol from carbon monoxide by *Butyribacterium methylotrophicum, J. Ferment. Bioengineer.,* 72, 58–60, 1991.
76. Groβkopf, R., Stubner, S., and Liesack, W., Novel euryarchaeotal lineages detected on rice roots and in the anoxic bulk soil of flooded rice microcosms, *Appl. Environ. Microbiol.,* 64, 4983–4989, 1998.
77. Häggblom, M.M., Berman, M.H., Frazer, A.C., and Young, L.Y., Anaerobic O-demethylation of chlorinated guaiacols by *Acetobacterium woodii* and *Eubacterium limosum, Biodegradation,* 4, 107–114, 1993.
78. Hall, I.C. and O'Toole, E., Intestinal florain newborn infants with a description of a new pathogenic anaerobe, *Bacillus difficilis, Am. J. Dis. Child,* 49, 390–402, 1935.
79. Hansen, B., Bokranz, M., Schönheit, P., and Kröger, A., ATP formation coupled to caffeate reduction by H_2 in *Acetobacterium woodii* Nzva16, *Arch. Microbiol.,* 150, 447–451, 1988.
80. Hashsham, S.A. and Freedman, D.L., Enhanced biotransformation of carbon tetrachloride by *Acetobacterium woodii* upon addition of hydroxocobalamin and fructose, *Appl. Environ. Microbiol.,* 65, 4537–4542, 1999.
81. Haveman, S.A. and Pedersen, K., Distribution of culturable microorganisms in Fennoscandian Shield groundwater, *FEMS Microbiol. Ecol.,* 39, 129–137, 2002.
82. Heise, R., Müller, V., and Gottschalk, G., Sodium dependence of acetate formation by the acetogenic bacterium *Acetobacterium woodii, J. Bacteriol.,* 171, 5473–5478, 1989.
83. Hines, M.E., Evans, R.S., Sharak Genthner, B.R., Willis, S.G., Friedman, S., Rooney-Varga, J.N., and Devereux, R., Molecular phylogenetic and biogeochemical studies of sulfate-reducing bacteria in the rhizosphere of *Spartina alterniflora, Appl. Environ. Microbiol.,* 65, 2209–2216, 1999.
84. Hoehler, T.M., Albert, D.B., Alperin, M.J., and Martens, C.S., Acetogenesis from CO_2 in an anoxic marine sediment, *Limnol. Oceanogr.,* 44, 662–667, 1999.
85. Holliger, C. and Schraa, G., Physiological meaning and potential for application of reductive dechlorination by anaerobic bacteria, *FEMS Microbiol. Rev.,* 15, 297–305, 1994.
86. Hsu, T., Daniel, S.L., Lux, M.F., and Drake, H.L., Biotransformations of carboxylated aromatic compounds by the acetogen *Clostridium thermoaceticum*: generation of growth-supportive CO_2 equivalents under CO_2-limited conditions, *J. Bacteriol.,* 172, 212–217, 1990.
87. Hsu, T., Lux, M.F., and Drake, H.L., Expression of an aromatic-dependent decarboxylase which provides growth-essential CO_2 equivalents for the acetogenic (Wood) pathway of *Clostridium thermoaceticum, J. Bacteriol.,* 172, 5901–5907, 1990.
88. Huang, S., Lindahl, P.A., Wang, C., Bennett, G.N., Rudolph, F.B., and Hughes, J.B., 2,4,6-trinitrotoluene reduction by carbon monoxide dehydrogenase from *Clostridium thermoaceticum, Appl. Environ. Microbiol.,* 66, 1474–1478, 2000.
89. Hugenholtz, J., Ivey, D.M., and Ljungdahl, L.G., Carbon monoxide-driven electron transport in *Clostridium thermoautotrophicum* membranes, *J. Bacteriol.,* 169, 5845–5847, 1987.
90. Ibba, M. and Fynn, G.H., Two stage methanogenesis of glucose by *Acetogenium kivui* and acetoclastic methanogenic sp., *Biotechnol. Lett.,* 13, 671–676, 1991.

91. Imkamp, F. and Müller, V., Chemiosmotic energy conservation with Na^+ as the coupling ion during hydrogen-dependent caffeate reduction by *Acetobacterium woodii, J. Bacteriol.,* 184, 1947–1951, 2002.
92. Inoue, K., Kageyama, S., Miki, K., Morinaga, T, Kamagata, Y., Nakamura, K., and Mikami, E., Vitamin B_{12} Production by *Acetobacterium* sp. and its tetrachloromethane-resistant mutants, *J. Ferment. Bioengineer.,* 73, 76–78, 1992.
93. Kamen, M.D., The early history of carbon-14, *J. Chem. Ed.,* 40, 234–242, 1963.
94. Kamlage, B., Boelter, A., and Blaut, M., Spectroscopic and potentiometric characterization of cytochromes in two *Sporomusa* species and their expression during growth on selected substrates, *Arch. Microbiol.,* 159, 189–196, 1993.
95. Kamlage, B., Gruhl, B., and Blaut, M., Isolation and characterization of two new homoacetogenic hydrogen-utilizing bacteria from the human intestinal tract that are closely related to *Clostridium coccoides, Appl. Environ. Microbiol.,* 63, 1732–1738, 1997.
96. Kane, M.D. and Breznak, J.A., *Acetonema longum* gen. nov. sp. nov., an H_2/CO_2 acetogenic bacterium from the termite, *Pterotermes occidentis, Arch. Microbiol.,* 156, 91–98, 1991.
97. Kane, M.D., Brauman, A., and Breznak, J.A., *Clostridium mayombei* sp. nov., an H_2/CO_2 acetogenic bacterium from the gut of the African soil-feeding termite, *Cubitermes speciosus, Arch. Microbiol.,* 156, 99–104, 1991.
98. Kaneuchi, C., Benno, Y., and Mitsuoka, T., *Clostridium coccoides*, a new species from the feces of mice, *Int. J. Syst. Bacteriol.,* 26, 482–486, 1976.
99. Kappler, O., Janssen, P.H., Kreft, J.-U., and Schink, B., Effects of alternative methyl group acceptors on the growth energetics of the *O*-demethylating anaerobe *Holophaga foetida, Microbiology,* 143, 1105–1114, 1997.
100. Karita, S., Nakayama, K., Goto M., Sakka, K., Kim, W.-J., and Ogawa, S., A novel cellulolytic, anaerobic, and thermophilic bacterium, *Moorella* sp. strain F21, *Biosci. Biotechnol. Biochem.,* 67, 183–185, 2003.
101. Karnholz, A., Küsel, K., Gößner, A., Schramm, A., and Drake, H.L., Tolerance and metabolic response of acetogenic bacteria toward oxygen, *Appl. Environ. Microbiol.,* 68, 1005–1009, 2002.
102. Kaufmann, F., Wohlfarth, G., and Diekert, G., *O*-demethylase from *Acetobacterium dehalogenans*, substrate specificity and function of the participating proteins, *Eur. J. Biochem.,* 253, 706–711, 1998.
103. Koesnandar, Nishio, N., Yamamoto, A., and Nagai, S., Enzymatic reduction of cystine into cysteine by cell-free extract of *Clostridium thermoaceticum, J. Ferment. Bioengineer.,* 72, 11–14, 1991.
104. Kotelnikova, S. and Pedersen, K., Evidence for methanogenic *Archaea* and homoacetogenic *Bacteria* in deep granitic rock aquifers, *FEMS Microbiol. Rev.,* 20, 339–349, 1997.
105. Kotelnikova, S. and Pedersen, K., Distribution and activity of methanogens in deep granitic aquifers at Äspö Hard Rock Laboratory, Sweden, *FEMS Microbiol. Ecol.,* 26,121–134, 1998.
106. Kotelnikova, S., Microbial production and oxidation of methane in deep subsurface, *Earth Sci. Rev.,* 58, 367–395, 2002.
107. Kotsyurbenko, O.R., Simankova, M.V., Nozhevnikova, A.N., Zhilina, T.N., Bolotina, N.P., Lysenko, A.M., and Osipov, G.A., New species of psychrophilic acetogens: *Acetobacterium bakii* sp. nov., *A. paludosum* sp. nov., *A. fimetarium* sp. nov., *Arch. Microbiol.,* 163, 29–34, 1995.
108. Kotsyurbenko, O.R., Nozhevnikova, A.N., Soloviova, T.I., and Zavarin, G.A., Methanogenesis at low temperatures by microflora of tundra wetland soil, *Antonie van Leeuwenhoek,* 69, 75–86, 1996.
109. Kreft, J.-U. and Schink, B., Demethylation and degradation of phenylmethylethers by the sulfide-methylating homoacetogenic bacterium strain TMBS 4, *Arch. Microbiol.,* 159, 308–315, 1993.
110. Kreft, J.-U. and Schink, B., Specificity of *O*-demethylation in extracts of the homoacetogenic *Holophaga foetida* and demethylation kinetics measured by a coupled photometric assay, *Arch. Microbiol.,* 167, 363–368, 1997.
111. Krumböck, M. and Conrad, R., Metabolism of position-labelled glucose in anoxic methanogenic paddy soil and lake sediment, *FEMS Microbiol. Ecol.,* 85, 247–256, 1991.
112. Krumholz, L.R. and Bryant, M.P., *Clostridium pfennigii* sp. nov. uses methoxyl groups of monobenzenoids and produces butyrate, *Int. J. Sys. Bacteriol.,* 35, 454–456, 1985.
113. Krumholz, L.R. and Bryant, M.P., *Syntrophococcus sucromutans* sp. nov. gen. nov. uses carbohydrates as electron donors and formate, methoxymonobenzenoids or *Methanobrevibacter* as electron acceptor systems, *Arch. Microbiol.,* 143, 313–318, 1986.

114. Krumholz, L.R., Harris, S.H., Tay, S.T., and Suflita, S.M., Characterization of two subsurface H_2-utilizing bacteria, *Desulfomicrobium hypogeium* sp. nov. and *Acetobacterium psammolithicum* sp. nov., and their ecological roles, *Appl. Environ. Microbiol.,* 65, 2300–2306, 1999.
115. Kuhner, C.H., Frank, C., Grießhammer, A., Schmittroth, M., Acker, G., Gößner, A., and Drake, H.L., *Sporomusa silvacetica* sp. nov., an acetogenic bacterium isolated from aggregated forest soil, *Int. J. Syst. Bacteriol.,* 47, 352–358, 1997.
116. Kuhner, C.H., Matthies, C., Acker, G., Schmittroth, M., Gößner A., and Drake, H.L., *Clostridium akagii* sp. nov. and *Clostridium acidisoli* sp. nov.: acid-tolerant, N_2-fixing clostridia isolated from acidic forest soil and litter, *Int. J. System. Evol. Microbiol.,* 50, 873–881, 2000.
117. Küsel, K. and Drake, H.L., Effects of environmental parameters on the formation and turnover of acetate by forest soils, *Appl. Environ. Microbiol.,* 61, 3667–3675, 1995.
118. Küsel, K. and Drake, H.L., Anaerobic capacities of leaf litter, *Appl. Environ. Microbiol.,* 62, 4216–4219, 1996.
119. Küsel, K., Pinkart, H.C., Drake, H.L., and Devereux, R., Acetogenic and sulfate-reducing bacteria inhabiting the rhizoplane and deep cortex cells of the sea grass *Halodule wrightii, Appl. Environ. Microbiol.,* 6, 5117–5123, 1999.
120. Küsel, K., Wagner, C., and Drake, H.L., Enumeration and metabolic product profiles of the anaerobic microflora in the mineral soil and litter of a beech forest, *FEMS Microbiol. Ecol.,* 29, 91–103, 1999.
121. Küsel, K., Dorsch, T., Acker, G., Stackebrandt, E., and Drake, H.L., *Clostridium scatologenes* strain SL1 isolated as an acetogenic bacterium from acidic sediments, *Int. J. Syst. Evol. Microbiol.,* 50, 537–546, 2000.
122. Küsel, K., Karnholz, A., Trinkwalter, T., Devereux, R., Acker, G., and Drake, H.L., Physiological ecology of *Clostridium glycolicum* RD-1, an aerotolerant acetogen isolated from sea grass roots, *Appl. Environ. Microbiol.,* 67, 4734–4741, 2001.
123. Küsel, K., Wagner, C., Trinkwalter, T., Gößner, A.S., Bäumler, R., and Drake, H.L., Microbial reduction of Fe(III) and turnover of acetate in Hawaiian soils, *FEMS Microbiol. Ecology,* 40, 73–81, 2002.
124. Küsel, K., Gößner, A., Lovell, C.R., and Drake, H.L., Ecophysiology of an aerotolerant acetogen, *Sporomusa* ST-1, isolated from *Juncus* roots, *Abstr. Ann. Meet. Am. Soc. for Microbiol.,* Abstract Q-375, 582, 2003.
125. Lajoie, S.F., Bank, S., Miller, T.L., and Wolin, M.J., Acetate production from hydrogen and [^{13}C]carbon dioxide by the microflora of human feces, *Appl. Environ. Microbiol.,* 54, 2723–2727, 1988.
126. Leadbetter, J.R., Schmidt, T.M., Graber, J.R., and Breznak, J.A., Acetogenesis from H_2 plus CO_2 by sprirochetes from termite guts, *Science,* 283, 686–689, 1999.
127. Leaphart, A.B., Friez, M.J., and Lovell, C.R., Formyltetrahydrofolate synthetase sequences from salt marsh plant roots reveal a diversity of acetogenic bacteria and other bacterial functional groups, *Appl. Environ. Microbiol.,* 69, 693–696, 2003.
128. Lebloas, P., Loubiere, P., and Lindley, N.D., Use of unicarbon substrate mixtures to modify carbon flux improves vitamin B_{12} production with the acetogenic methylotroph *Eubacterium limosum, Biotech. Lett.,* 16, 129–132, 1994.
129. Leclerc, M., Bernalier, A., Donadille, G., and Lelait, M., H_2/CO_2 metabolism in acetogenic bacteria isolted from the human colon, *Anaerobe,* 3, 307–315, 1997.
130. Leclerc, M., Bernalier, A., Lelait, M., and Grivet. J.-P., ^{13}C-NMR study of glucose and pyruvate catabolism in four acetogenic species isolated from the human colon, *FEMS Microbiol. Lett.,* 146, 199–204, 1997.
131. Leedle, J.A.Z. and Greening, R.C., Postprandial changes in methanogenic and acidogenic bacteria in the rumens of steers fed high- or low-forage diets once daily, *Appl. Environ. Microbiol.,* 54, 502–506, 1988.
132. Leedle, J.A.Z., Lotrario, J., Hovermale, J., and Craig, A.M., Forestomach anaerobic microflora of the bowhead whale (*Balaena mysticetus*), *Abstr. Ann. Meet. Am. Soc. Microbiol.,* Abstr. N-8, 334, 1995.
133. Liesack, W., Bak, F., Kreft, J.-U., and Stackebrandt, E., *Holophaga foetida* gen. nov., sp. nov., a new, homoacetogenic bacterium degrading methoxylated aromatic compounds, *Arch. Microbiol.,* 162, 85–89, 1994.
134. Lindahl, P.A. and Chang, B., The evolution of acetyl-CoA synthase: origins of life and evolution of the biosphere, *Orig. Life Evol. Biosph.,* 31, 403–434, 2001.

135. Lindahl, P.A., The Ni-containing carbon monoxide dehydrogenase family: light at the end of the tunnel?, *Biochemistry (Mosc.),* 41, 2097–2105, 2002.
136. Liu, S. and Suflita, J.M., H_2/CO_2-dependent anaerobic *O*-demethylation activity in subsurface sediments and by an isolated bacterium, *Appl. Environ. Microbiol.,* 59, 1325–1331, 1993.
137. Ljungdahl, L.G., Carreira, L.H., Garrison, R.J., Rabek, N.E., and Wiegel, J., Comparison of three thermophilic acetogenic bacteria for production of calcium magnesium acetate, *Biotechnol. Bioeng. Symp.,* 15, 207–223, 1985.
138. Ljungdahl, L.G., Hugenholtz, J., and Wiegel, J., Acetogenic and acid-producing clostridia, in *Clostridia*, Minton, N.P. and Clarke, D.J., Eds., Plenum Press, New York, 1989, 145–191.
139. Ljungdahl, L.G., The acetyl-CoA pathway and the chemiosmotic generation of ATP during acetogenesis, in *Acetogenesis*, Drake, H.L., Ed., Chapman & Hall, New York, 1994, 63–87.
140. Loke, H.K. and Lindahl, P.A., Identification and preliminary characterization of AcsF, a putative Ni-insertase used in the biosynthesis of acetyl-CoA synthase from *Clostridium thermoaceticum,* J. *Inorg. Biochem.,* 93, 33–40, 2003.
141. Lowe, A., Jain, M.K., and Zeikus, J.G., Biology, ecology, and biotechnological applications of anaerobic bacteria adapted to envionmental stresses in temperature, pH, salinity, or substrates, *Microbiol. Revs.,* 57, 451–509, 1993.
142. Lumppio, H.L., Shenvi, N.V., Summers, A.O., Voordrouw, G., and Kurtz, D.M. Jr., Rubrerythrin and rubredoxin oxidoreductase in *Desulfovibrio vulgaris*: a novel oxidative stress protection system, *J. Bacteriol.,* 183, 101–108, 2001.
143. Mackie, R.I. and Bryant, M.P., Acetogenesis and the rumen: syntrophic relationships, in *Acetogenesis,* Drake, H.L., Ed., Chapman & Hall, New York, 1994, 331–364.
144. Matthies, C., Freiberger, A., and Drake, H.L., Fumarate dissimilation and differential reductant flow by *Clostridium formicoaceticum* and *Clostridium aceticum, Arch. Microbiol.,* 160, 273–278, 1993.
145. Maynard, E.L. and Lindahl, P.A., Evidence of a molecular tunnel connecting the active sites for CO_2 reduction and acetyl-CoA synthesis in acetyl-CoA synthase from *Clostridium thermoaceticum, J. Am. Chem. Soc.,* 121, 9221–9222, 1999.
146. Maynard, E.L. and Lindahl, P.A., Catalytic coupling of the active sites in acetyl-CoA synthase, a bifunctional CO-channeling enzyme, *Biochemistry,* 40, 13262–13267, 2001.
147. McInerney, M.J. and Bryant, M.P., Basic principles of bioconversions in anaerobic digestion and methanogenesis, in *Biomass Conversion Processes for Energy and Fuels*, Sofer, S.S. and Zaborsky, O.R., Eds., Plenum Press, New York, 1981, 277–296.
148. Mechichi, T., Labat, M., Patel, B.K.C., Woo, T.H.S., Thomas, P., and Garcia, J.-L. *Clostridium methoxybenzovorans* sp. nov., a new aromatic *o*-demethylating homoacetogen from an olive mill wastewater treatment digester, *Int. J. Syst. Bacteriol.,* 49, 1201–1209, 1999.
149. Miller, T. and Wolin, M.J., Pathways of acetate, propionate, and butyrate formation by the human fecal microbial flora, *Appl. Environ. Microbiol.,* 62,1589–1592, 1996.
150. Misoph, M., Daniel, S.L., and Drake, H.L., Bidirectional usage of ferulate by the acetogen *Peptostreptococcus productus* U-1: CO_2 and aromatic acrylate groups as competing electron acceptors, *Microbiology,* 142, 1983–1988, 1996.
151. Misoph, M. and Drake, H.L., Effect of CO_2 on the fermentation capacities of the acetogen *Peptostreptococcus productus* U-1, *J. Bacteriol.,* 178, 3140–3145, 1996.
152. Müller, V. and Bowien, S., Differential effects of sodium ions on motility in the homoacetogenic bacteria *Acetobacterium woodii* and *Sporomusa sphaeroides, Arch. Microbiol.,* 164, 363–369, 1995.
153. Müller, V., Aufurth, S., and Rahlfs, S., The Na^+-cycle in *Acetobacterium woodii*: identification and characterization of a Na^+-translocating F_1F_0-ATPase with a mixed oligomer of 8 and 16-kDa proteolipids, *Biochim. Biophys. Acta*, 1505, 108–120, 2001.
154. Müller, V., Energy conservation in acetogenic bacteria, *Appl. Environ. Microbiol.,* 69, 6345–6353, 2003.
155. Naidu, D. and Ragsdale, S.W., Characterization of a three-component vanillate O-demethylase from *Moorella thermoacetica, J. Bacteriol.,* 183, 3276–3281, 2001.
156. Nozhevnikova, A., Kotsyurbenko, O.R., and Simankova, M.V., Acetogenesis at low temperature, in *Acetogenesis,* Drake, H.L., Ed., Chapman & Hall, New York, 1994, 416–431.
157. Ohwaki, K. and Hungate, R.E., Hydrogen utilization by clostridia in sewage sludge, *Appl. Environm. Microbiol.,* 33, 1270–1274, 1977.

158. Patel, B.K.C., Monk, C., Littleworth, H., Morgan, H.W., and Daniel, R.M., *Clostridium fervidus* sp. nov., a new chemoorganotrophic acetogenic thermophile, *Int. J. Sys. Bacteriol.,* 37, 123–126, 1987.
159. Peters, V. and Conrad, R., Methanogenic and other strictly anaerobic bacteria in desert soil and other oxic soils, *Appl. Environ. Microbiol.,* 61, 1673–1676, 1995.
160. Peters, V. and Conrad, R., Sequential reduction processes and initiation of CH_4 production upon flooding of oxic upland soils, *Soil Biol. Biochem.,* 28, 371–382, 1996.
161. Peters, V., Janssen, P.H., and Conrad, R., Efficiency of hydrogen utilization during unitrophic and mixotrophic growth of *Acetobacterium woodii* on hydrogen and lactate in the chemostat, *FEMS Microbiol. Ecol.,* 26, 317–324, 1998.
162. Phelps, T.J. and Zeikus, J.G., Influence of pH on terminal carbon metabolism in anoxic sediments from a mildly acidic lake, *Appl. Environ. Microbiol.,* 48, 1088–1095, 1984.
163. Phillips, J.R., Clausen, E.C., and Gaddy, J.L., Synthesis gas as substrate for the biological production of fuels and chemicals, *Appl. Biochem. Biotech.,* 45/46, 145–157, 1994.
164. Preuss, A., Fimpel, J., and Diekert, G., Anaerobic transformation of 2,4,6-trinitrotoluene (TNT), *Arch. Microbiol.,* 159, 345–353, 1993.
165. Prins, R.A. and Lankhorst, A., Synthesis of acetate from CO_2 in the cecum of some rodents, *FEMS Microbiol. Lett.,* 1, 255–258, 1977.
166. Radfar, R., Shin, R., Sheldrick, G.M., Minor, W., Lovell, C.R., Odom, J.D., Dunlap, R.B., and Lebioda, L., The crystal structure of N10-formyltetrahydrofolate synthetase from *Moorella thermoacetica, Biochemistry (Mosc),* 39, 3920–3926, 2000.
167. Ragsdale, S.W., The Eastern and Western branches of the Wood/Ljungdahl pathway: how the East and West were won, *BioFactors,* 6, 3–11, 1997.
168. Ragsdale, S.W., Nickel containing CO dehydrogenases and hydrogenases, in *Enzyme-Catalyzed Electron and Radical Transfer,* Vol. 35, Holzenburg, A. and Scrutton, N., Eds., Plenum Press, New York, 2000, 487–518.
169. Ragsdale, S.W., Anaerobic one-carbon catalysis, in *Encyclopedia of Catalysis,* Horvath, I.T., Iglesia, E., Klein, M.T., Lercher, J.A., Russell, A.J., and Stiefel, E.I., Eds., John Wiley & Sons, Inc., New York, 2003, 665–695.
170. Ragsdale, S.W., Pyruvate:ferredoxin oxidoreductase and its radical intermediate, *Chem. Rev.,* 103, 2333–2346, 2003.
171. Reith, F., Drake, H. L., and Küsel, K., Anaerobic activities of bacteria and fungi in moderately acidic conifer and leaf litter, *FEMS Microbiol. Ecol.,* 41, 27–35, 2002.
172. Revsbech, N.P., Pedersen, O., Reichardt, W., and Briones, A., Microsensor analysis of oxygen and pH in the rice rhizosphere under field and laboratory conditions, *Biol. Fertil. Soils,* 29, 379–385, 1999.
173. Rieu-Lesme, F., Fonty, G., and Doré, J., Isolation and characterization of a new hydrogen-utilizing bacterium from the rumen, *FEMS Microbiol. Lett.,* 125, 77–82, 1995.
174. Rieu-Lesme, F., Dauga, C., Morvan, B., Bouvet, O.M.M., Grimont, P.A.D., and Doré, J., Acetogenic coccoid spore-forming bacteria isolated from the rumen, *Res. Microbiol.,* 147, 753–764, 1996.
175. Rieu-Lesme, F., Morvan, B., Collins, M.D., Fonty, G., and Willems, A., A new H_2/CO_2-using acetogenic bacterium from the rumen: description of *Ruminococcus schinkii* sp. nov., *FEMS Microbiol. Lett.,* 140, 281–286, 1996.
176. Rieu-Lesme, F., Dauga, C., Fonty, G., and Doré, J., Isolation from the rumen of a new acetogenic bacterium phylogenetically closely related to *Clostridium difficile, Anaerobe,* 4, 89–94, 1998.
177. Rosencrantz, D., Rainey, F.A., and Janssen, P.H., Culturable populations of *Sporomusa* spp. and *Desulfovibrio* spp. in the anoxic bulk soil of flooded rice microcosms, *Appl. Environ. Microbiol.,* 65, 3526–3533, 1999.
178. Savage, M.D., Wu, Z., Daniel, S.L., Lundie, L.L. Jr., and Drake, H.L., Carbon monoxide-dependent chemolithotrophic growth of *Clostridium thermoautotrophicum, Appl. Environ. Microbiol.,* 53, 1902–1906, 1987.
179. Schink, B., *Clostridium magnum* sp. nov., a non-autotrophic homoacetogenic bacterium, *Arch. Microbiol.,* 137, 250–255, 1984.
180. Schink, B., Diversity, ecology, and isolation of acetogenic bacteria, in *Acetogenesis,* Drake, H.L., Ed., Chapman & Hall, New York, 1994, 197–235.

181. Schmitt-Wagner, D. and Brune, A. Hydrogen profiles and localization of methanogenic activities in the highly compartmentalized hindgut of soil-feeding higher termites (*Cubitermes* spp.), *Appl. Environ. Microbiol.,* 65, 4490–4496, 1999.
182. Schnürer, A., Schink, B., and Svensson, B.H., *Clostridium ultunense* sp. nov., a mesophilic bacterium oxidizing acetate in syntrophic association with a hydrogenotrophic methanogenic bacterium, *Int. J. Syst. Bacteriol.,* 46, 1145–1152, 1996.
183. Schulz, M., Leichmann, H., Günther, H., and Simon, H., Electromicrobial regeneration of pyridine nucleotides and other preparative redox transformations with *Clostridium thermoaceticum, Appl. Microbiol. Biotechnol.,* 42, 916–922, 1995.
184. Schulz, S. and Conrad, R., Influence of temperature on pathways to methane production in the permanently cold profundal sediment of Lake Constance, *FEMS Microbiol. Ecol.,* 20, 1–14, 1996.
185. Schwartz, R.D. and Keller, F.A. Jr., Isolation of a strain of *Clostridium thermoaceticum* capable of growth and acetic acid production at pH 4.5, *Appl. Environ. Microbiol.,* 43, 117–123, 1982.
186. Seifritz, C., Daniel, S.L., Gößner, A., and Drake, H.L., Nitrate as a preferred electron sink for the acetogen *Clostridium thermoaceticum, J. Bacteriol.,* 175, 8008–8013, 1993.
187. Seifritz, C., Fröstl, J.M., Drake, H.L., and Daniel, S.L., Influence of nitrate on oxalate- and glyoxylate-dependent growth and acetogenesis by *Moorella thermoacetica, Arch. Microbiol.,* 178, 457–464, 2002.
188. Seifritz, C., Drake, H.L., and Daniel, S.L., Nitrite as an energy-conserving electron sink for the acetogenic bacterium *Moorella thermoacetica, Curr. Microbiol.,* 46, 329–333, 2003.
189. Sexstone, A.J., Revsbech, N.P., Parkin, T.B., and Tiedje, J.M., Direct measurement of oxygen profiles and denitrification rates in soil aggregates, *Soil Sci. Soc. Am. J.,* 49, 645–651, 1985.
190. Silaghi-Dumitrescu, R., Coulter, E.D., Das, A., Ljungdahl, L.G., Jameson, G.N.L., Huynh, B.H., and Kurtz, D.M. Jr., A flavodiiron protein and high molecular weight rubredoxin from *Moorella thermoacetica* with nitric oxide reductase activity, *Biochemistry,* 42, 2806–2815, 2003.
191. Talabardon, M., Schwitzguébel, J.-P., Péringer, P., and Yang, S.-T., Acetic acid production from lactose by an anaerobic thermoophilic coculture immobilized in a fibrous-bed bioreactor, *Biotechnol. Prog.,* 16, 1008–1017, 2000.
192. Tanner, R.S., Miller, L.M., and Yang, D., *Clostridium ljungdahlii* sp. nov., and acetogenic species in clostridial rRNA homology group I, *Int. J. Sys. Bacteriol.,* 43, 232–236, 1993.
193. Tanner, R.S. and Woese, C.R., A phylogenetic assessment of the acetogens, in *Acetogenesis*, Drake H.L., Ed., Chapman & Hall, New York, 1994, 254–269.
194. Terracciano, J.S., Schreurs, W.J.A., and Kashket, E.R., Membrane H^+ conductance of *Clostridium thermoaceticum* and *Clostridium acetobutylicum*: evidence for electrogenic Na^+/H^+ antiport in *Clostridium thermoaceticum, Appl. Environ. Microbiol.,* 53, 782–786, 1987.
195. Terzenbach, D.P. and Blaut, M., Transformation of tetrachloroethylene by homoacetogenic bacteria, *FEMS Microbiol. Lett.,* 123, 213–218, 1994.
196. Tholen, A. and Brune, A., Localization and in situ activities of homoacetogenic bacteria in the highly compartmentalized hindgut of soil-feeding higher termites (*Cubitermes* spp.), *Appl. Environ. Microbiol.,* 65, 4497–4505, 1999.
197. Traunecker, J, Preub, A., and Diekert, G., Isolation and characterization of a methyl cloride utilizing strictly anaerobic bacterium, *Arch. Microbiol.,* 156, 416–421, 1991.
198. Tschech, A. and Pfennig, N., Growth yield increase linked to caffeate reduction in *Acetobacterium woodii, Arch. Microbiol.,* 137, 163–167, 1984.
199. van der Lee, G.E.M., de Winder, B., Bouten, W., and Tietema, A., Anoxic microsites in douglas fir litter, *Soil Biol. Biochem.,* 31, 1295–1301, 1999.
200. Wagner, C., Grießhammer, A., and Drake, H.L., Acetogenic capacities and the anaerobic turnover of carbon in a Kansas prairie soil, *Appl. Environ. Microbiol.,* 62, 494–500, 1996.
201. Weinberg, M. and Ginsbourg, B., *Données récentes sur les microbes anaérobies et leur role en pathologie*, Masson et Cie, Paris, 1927, 1–291.
202. Wellsbury, P., Goodman, K., Barth, T., Cragg, B.A., Barnes, S.P., and Parkes, R.J., Deep marine biosphere fuelled by increasing organic matter availability during burial and heating, *Nature,* 388, 573–576, 1997.
203. Wellsbury, P., Mather, I., and Parkes, R.J., Geomicrobiology of deep, low organic carbon sediments in the Woodlark Basin, Pacific Ocean, *FEMS Microbiol. Ecol.,* 42, 59–70, 2002.

204. Wiegel, J., Braun, M., and Gottschalk, G., *Clostridium thermoautotrophicum* species novum, a thermophile producing acetate from molecular hydrogen and carbon dioxide, *Curr. Microbiol.,* 5, 255–260, 1981.
205. Wiegel, J., Carreira, L.H., Garrison, R.J., Robek, N.E., and Ljungdahl, L.G., Calcium magnesium acetate (CMA) manufacture from glucose by fermentation with thermophilic homoacetogenic bacteria, in *Calcium Magnesium Acetate*, Wise, D.L., Levendis, Y., and Metghalchi, M., Eds., Elsevier, Amsterdam, 1990, 359–416.
206. Wiegel, J., Acetate and the potential of homoacetogenic bacteria for industrial applications, in *Acetogenesis*, Drake, H.L., Ed., Chapman & Hall, New York, 1994, 484–504.
207. Wieringa, K.T., Over het verdwijnen van waterstof en koolzuur onder anaerobe voorwaarden, *Antonie van Leeuwenhoek,* 3, 263–273, 1936.
208. Wieringa, K.T., The formation of acetic acid from carbon dioxide and hydrogen by anaerobic spore-forming bacteria, *Antonie van Leeuwenhoek, J. Microbiol. Seriol.,* 6, 251–262, 1939–1940.
209. Wieringa, K.T., Über die Bildung von Essigsäure aus Kohlensäure und Wasserstoff durch anaerobe Bazillen, *Brennstoff-Chemie,* 22, 161–164, 1941.
210. Winter, J.U. and Wolfe, R.S., Methane formation from fructose by syntrophic associations of *Acetobacterium woodii* and different strains of methanogens, *Arch. Microbiol.,* 124, 73, 1980.
211. Wohlfarth, G. and Diekert, G., Thermodynamics of methylenetetrahydrofolate reduction to methyltetrahydrofolate and its implications for the energy metabolism of homoacetogenic bacteria, *Arch. Microbiol.,* 155, 378–381, 1991.
212. Wolin, M.J. and Miller, T.L., Acetogenesis from CO_2 in the human colonic ecosystem, in *Acetogenesis,* Drake, H.L., Ed., Chapman & Hall, New York, 1994, 365–385.
213. Wolin, M.J., Miller, T.L., Collins, M.D., and Lawson, P.A., Formate-dependent growth and homoacetogenic fermentation by a bacterium from human feces: description of *Bryantella formatexigens* gen. nov., sp. nov., *Appl. Environ. Microbiol.,* 69, 6321–6326, 2003.
214. Wood, H.G. and Ljungdahl, L.G., Autotrophic character of the acetogenic bacteria, in *Variations in Autotrophic Life*, Shively, J.M. and Barton, L.L., Eds., Academic Press, San Diego, 1991, 201–250.
215. Worden, R.M., Grethlein, A.J., Zeikus, J.G., and Datta, R., Butyrate production from carbon monoxide by *Butyribacterium methylotrophicum, Appl. Biochem. Biotech.,* 20/21, 687–698, 1989.
216. Wu, Z., Daniel, S.L., and Drake, H.L., Characterization of a CO-dependent O-demethylating enzyme system from the acetogen *Clostridium thermoaceticum, J. Bacteriol.,* 170, 5747–5750, 1988.
217. Yang, H. and Drake, H.L., Differential effects of sodium on hydrogen- and glucose-dependent growth of the acetogenic bacterium *Acetogenium kivui, Appl. Environ. Microbiol.,* 56, 81–86, 1990.
218. Zavarzin, G.A., Zhilina, T.N., and Pusheva, M.A., Halophilic acetogenic bacteria, in *Acetogenesis*, Drake, H.L., Ed., Chapman & Hall, New York, 1994, 432–444.
219. Zhilina, T.N., Zavarzin, G.A., Detkova, E.N., and Rainey, F.A., *Natroniella acetigena* gen. nov. sp. nov., an extremely halophilic, homoacetogenic bacterium: a new member of *Haloanaerobiales, Curr. Microbiol.,* 32, 320–326, 1996.
220. Zhilina, T.N., Detkova, E.N., Rainey, F.A., Osipov, G.A., Lysenko, A.M., Kostrikina, N.A., and Zavarzin, G.A., *Natronoincola histidinovorans* gen. nov., sp. nov., a new alkaliphilic acetogenic anaerobe, *Curr. Microbiol.,* 37, 177–185, 1998.

33 Halophilic Clostridia

Aharon Oren

CONTENTS

33.1 INTRODUCTION

In view of the wealth of information on the genus *Clostridium* and its relatives, it is surprising to realize how little is known on the occurrence of endospore-forming anaerobic members of the low G+C group of the *Firmicutes* in marine and hypersaline environments. Out of the around 150 species of the genus *Clostridium* with standing in the bacterial nomenclature, no more than four show a marked requirement for salt. Information on the distribution of these species in the marine environment is virtually lacking.

Most of the relatively few marine studies that mention the clostridia at all deal with the monitoring of non-halophilic species, notably with the occurrence of *Clostridium perfringens* as an indicator of pollution of the sediments from terrestrial sources, e.g., to monitor the distribution of sewage outfall in coastal marine environments. Many species of the genus *Clostridium* have thus been isolated from marine sediments. However, it is more than likely that these organisms had survived in the saline environment as dormant endospores. Studies on the possible activity and growth of "terrestrial" clostridia in the sea appear to be lacking altogether, and little is known about the longevity of their endospores in marine sediments.

In this chapter, the term "clostridia" is taken in the broad sense of obligatory anaerobic, fermentative or homoacetogenic representatives of the low G+C group of the *Firmicutes*. A number of moderately thermophilic fermentative organisms have been isolated from submarine hydrothermal vent areas that phylogenetically cluster within the *Clostridium* group. These have been described as representatives of newly created genera. Some of these produce endospores; in others, sporulation was not observed. These marine organisms, closely related to some of the well-known terrestrial and freshwater clostridia, will be included in this review.

The genus *Clostridium* contains one species that was isolated from a hypersaline environment, namely *Clostridium halophilum*. This organism grows optimally at a salt concentration about twice that of seawater, and tolerates at least 10% salt [1]. Another isolate, *Thermohalobacter berrensis*, which grows in the same salinity range and which was never shown to produce endospores, has thermophilic properties and is closely related to some *Clostridium* species [2]. However, it appears that the niche of fermentative saccharolytic and proteolytic anaerobes adapted to life at high salt concentrations is mostly occupied not by members of the Clostridiales but by another order of low G+C *Firmicutes*: the Halanaerobiales, families Halanaerobiaceae and Halobacteroidaceae. The last named family contains many endospore-forming representatives. The order Halanaerobiales consists entirely of halophiles, and its representatives can be found in hypersaline environments worldwide. This group has been reviewed several times in the past [3–8]. A short overview of the properties of the members of this order is given in this chapter, including the most recent additions to the growing list of isolates belonging to this group.

33.2 NON-HALOPHILIC CLOSTRIDIA IN MARINE SEDIMENTS

Although quantitative information is scarce, there is little doubt that endospores of non-halophilic clostridia can survive for prolonged periods in marine sediments. A laboratory simulation experiment using model marine sediments showed that the viability of *C. perfringens* spores was not affected during a 28-day incubation [9]. Recovery of elevated levels of viable spores of *C. perfringens* from different depths within the sediment of Long Island Sound was used as proof that sewage pollution there has persisted for more than a century [10]. *C. perfringens* is, therefore, sometimes used as an indicator organism to monitor fecal pollution of marine environments in the present and in the past.

Many additional non-halophilic species of *Clostridium* have been recovered from marine sediments since 1938, when Hotchkiss published what was probably one of the first reports on the isolation of clostridia from marine sediments [11]. The most abundant species found in Puget Sound (WA) sediments collected from depths between 10 to 240 m were *C. perfringens*, *C. bifermentans*, and *C. novyi*. Additional isolates were identified as *C. barati*, *C. botulinum*, *C. butyricum*, *C. cadaveris*, *C. difficile*, *C. fallax*, *C. ghonii*, *C. kluyveri*, *C. mangenotii*, *C. oroticum*, *C. paraputrificum*, *C. sphenoides*, *C. sporogenes*, *C. sticklandii*, and *C. subterminale* [12,13]. The authors stated that these species undoubtedly reached the sea through sewage, food processing waste discharge, and land drainage by seepage or through rivers. A number of psychrophilic *Clostridium* isolates obtained from Puget Sound sediments grew at 12°C, and these have partially been characterized. They did not require salt for growth [12].

A comparative study of the relatively oligotrophic sediments of the western equatorial Atlantic and the Caribbean, as compared to the organic matter-rich sediments from the upwelling areas of the continental shelf of Peru, showed that the richer sediments of the Peru current contained more *Clostridium* species, especially saccharolytic ones such as *C. novyi* and *C. perfringens*. Further abundant species were *C. sporogenes*, *C. sticklandii*, and *C. subterminale*. It was concluded that (1) proteolytic clostridia might be an essential member of marine sediments, both in productive and in nonproductive areas; (2) populations of primarily saccharolytic clostridia can maintain themselves only when the organic content of the sediments is relatively high and is being constantly replenished; and (3) more species of clostridia can be found in environments with a high content of organic matter [14]. Sediment samples from the North Sea primarily yielded *C. bifermentans*, *C. perfringens*, and *C. sordellii* [15]. *C. botulinum* has also been recovered from marine sediments [16–18].

None of the studies cited above addressed to what extent the cultures recovered were derived from cells actively growing in the saline environment or grew from resting spores that germinated upon transfer to standard low-salt media. It can be tacitly assumed that most isolates probably emerged from dormant endospores.

Clostridium endospores can also resist exposure to hypersaline conditions; viable clostridia belonging to nonhalophilic species have even be recovered from sediments of the Dead Sea. The first report on the isolation of living microorganisms from the Dead Sea dates from the last decade of the 19th century, and deals with the occurrence of clostridia causing tetanus and gas gangrene within the mud [19]. The study does not provide in-depth information about the mode of sample collection, use of aseptic techniques used during handling, etc. However, an attempt to reproduce the work 100 years later largely confirmed the earlier findings [20].

33.3 SALT-REQUIRING CLOSTRIDIA AND RELATED BACTERIA IN THE MARINE ENVIRONMENT

Among the approximately 150 named species of the genus *Clostridium,* there are only four isolates that have been obtained from the marine environment and that specifically require salt for growth: *C. oceanicum*, *C. litorale*, *C. halophilum*, and *C. grantii*. Their properties are summarized in the upper rows of Table 33.1.

Clostridium oceanicum was isolated from coastal sediments off Peru, Ecuador, Brazil, and the Caribbean [21]. It grows optimally at a salinity similar to that of seawater. However, it does not have a specific salt requirement, and good growth is also achieved in freshwater-based media. Cells are rod-shaped, and cells can show development of endospores at both ends, without any obvious division plane (Figure 33.1E). The organism was used as a test object in a study in which novel preparation techniques for scanning electron microscopy were developed [22].

Clostridium litorale (Figure 33.1C, c and d) was isolated from marine mud collected from the North Sea shore on Jadebusen and Sylt, northern Germany. This organism grows at salt concentrations between 0.2 and 6%, with an optimum at 1%. Cells stain Gram negative. The organism does not ferment carbohydrates but grows in complex media by fermentation, forming mainly acetate and butyrate, or by means of the Stickland reaction, a reaction in which different amino acids (glycine, alanine, leucine, isoleucine, valine, phenylalanine) or molecular hydrogen serve as electron donor with glycine betaine as electron acceptor. Glycine betaine is converted to trimethylamine and acetate. Sarcosine is fermented using alanine as electron donor [1].

Most isolates of *Clostridium halophilum* (Figure 33.1C, a and b), an organism that grows between 1.5 and 10% salt with an optimum at 6%, have been obtained from environments with a salinity exceeding that of seawater — hypersaline sediments from the Maldives Islands and from Solar Lake, Sinai, Egypt. However, the species has also been recovered from marine sediments [1].

Clostridium grantii was isolated from the gut of mullets. It is only very slightly halophilic, growing optimally at 0.1% and tolerating up to 1% salt. It ferments different carbohydrates, including alginate to acetate, ethanol, formate, and CO_2. Hydrogen is not evolved during the fermentation [23].

It is surprising that beyond the above-mentioned taxonomic species descriptions, little work has been done toward a further characterization of the properties of these few marine clostridia. Their mechanisms of adaptations to life in the marine environment have never been investigated. Information on the distribution of these species in marine environments worldwide is lacking as well. However, a few biochemical studies have been performed on the enzymes involved in the Stickland reaction in *C. litorale*. The glycine reductase consists of two selenocysteine-containing proteins, P-A (159 amino acids) and the glycine-specific selenoprotein P-B (437 amino acids). The two genes encoding the components of the complex, *grdA* and *grdB*, both contain a TGA codon in-frame to code for selenocysteine [24,25]. The thioredoxins (and thioredoxin reductase of *C. litorale*) have been purified and characterized as well [26].

The deep-sea hydrothermal vents present at several locations on the ocean bottom have recently yielded a number of interesting new species of *Clostridium* and phylogenetically related organisms. These share a requirement for seawater salinities, and they all have thermophilic properties.

Clostridium caminithermale belongs to cluster XI of the *Clostridium* group, as defined by Collins et al. [27]. Phylogenetically, its closest relative is *Clostridium halophilum* [28]. It was isolated from a hydrothermal vent located at a depth of 800 m in the North Atlantic Ocean ridge. It grows optimally at 3% salt and requires at least 1.2% salt for growth. Its temperature optimum is 45°C, and it grows up to 58°C. It produces subterminal oval spores. *C. caminithermale* ferments carbohydrates to acetate, CO_2, and H_2, with, in addition, some propionate and butyrate. Succinate is converted to propionate. Fermentation of arginine, cysteine, glycine, proline, and tyrosine mainly yields acetate. Glutamate is degraded to propionate; from histidine, propionate and formate are formed; lysine is fermented to acetate and butyrate; methionine to propionate and acetate; isoleucine

TABLE 33.1
Clostridia and Other Fermentative Endospore-Forming Bacteria from Marine and Hypersaline Environments and Their Properties (Order Clostridiales)

Name	Nomenclatural and Taxonomic Comments	Habitat	Salt Optimum and Concentration Range for Growth	Temperature Optimum and Range for Growth	pH Optimum and Range for Growth	Metabolic Properties[a]	Production of Endospores/Production of Gas Vacuoles	DNA G+C Content (mol%)	Type Strain	Ref.
Clostridium oceanicum		Marine sediments	4% (0 to <8%)	30 to 37°C (5 to <45°C)	6.5 to 8.6	Sugars fermented to Ac, IBu, Bu, IVa, ICa, Et-OH, Pr-OH	+/–	27–28	ATCC 3082	21
Clostridium litorale	Belongs to cluster XI of the *Clostridium* subphylum	Seashore, north Germany	1% (2 to 6%)	28°C (17 to 38.5°C)	7.3 (6.5 to 8.4)	Sugars fermented to Ac, Bu Performs the Stickland reaction	+/–	26.1	DSM 5388	1
Clostridium halophilum	Originally named *Clostridium halophilum*; belongs to cluster XI of the *Clostridium* subphylum	Hypersaline sediments, the Maldives Islands, Solar Lake (Egypt), marine sediments	6% (1.5 to >10%)	41°C (18 to 49°C)	7.4 (6 to 8)	Sugars fermented to Ac, Et-OH, H_2, CO_2; performs the Stickland reaction	+/–	26.9	DSM 5387	1
Clostridium grantii	Belongs to cluster I of the *Clostridium* subphylum	Mullet guts	0.1% (0.1 to 1%)	30°C optimum	6.5 optimum	Sugars fermented to Ac, Et-OH, Fo, CO_2; H_2 not produced	+/–	30.2	DSM 8605	23

Clostridium caminithermale	Belongs to cluster XI of the *Clostridium* subphylum	Hydrothermal chimney, North-Atlantic Ocean Ridge	3% (1.2 to 5.5%)	45°C (20 to 58°C)	6.6 (5.8 to 8.2)	Sugars fermented to Ac, Pr, Bu, H_2, CO_2; performs the Stickland reaction	+/–	33.1	DSM 15212	28
Caminicella sporogenes	Belongs to cluster XI of the *Clostridium* subphylum	Hydrothermal chimney, East-Pacific Rise	2.5 to 3% (2 to 6%)	53 to 60°C (45 to 65°C)	7.5 to 8 (4.5 to 8)	Ferments sugars and other substrates to Bu, Et-OH, Ac, Fo, Ala	+/–	24.2	DSM 14501	29
Tepidibacter thalassicus	Belongs to cluster XI of the *Clostridium* subphylum	Hydrothermal chimney, East-Pacific Rise	1.5 to 6%	50°C (33 to 60°C)	6.5 to 6.8 (4.8 to 8.5)	Ferments proteins and starch to Et-OH, Ac, H_2, CO_2	+/–	24	DSM 15285	30
Caloranaerobacter azorensis	Belongs to cluster XII of the *Clostridium* subphylum	Hydrothermal chimney, Mid-Atlantic Ridge	3% (1 to 10%)	65°C (45 to 65°C)	7 (5.5 to 9)	Ferments sugars and other substrates to Ac, IVa, Ala	+/–	27	DSM 13643	31
Tindallia magadiensis	Belongs to cluster XI of the *Clostridium* subphylum; originally named *Tindallia magadii*	Lake Magadi, Kenya	0.5 to 1 *M* NaCl or $NaHCO_3$; no growth without Na^+	37°C (19 to 47°C)	8.5 (7.5 to 10.5)	Ferments amino acids (arginine, ornithine) to Ac, Pr, Fo, IVa, H_2, NH_3	+/–	37.6	DSM 10318	34
Anoxynatronum sibiricum	Belongs to cluster XI of the *Clostridium* subphylum	Lake Nizhnee Beloye, Siberia	0.25 to 0.86 *M* Na^+ (0.076 to 1.27 *M*)	35°C (25 to 41°C)	9.1 (7.1 to 10.1)	Ferments sugars, sugar alcohols and certain amino acids to Ac, Et-OH	+/–	48.4	DSM 15060	35
Thermohalobacter berrensis	Belongs to cluster XII of the *Clostridium* subphylum	Saltern, South of France	5% (2 to 15%)	65°C (45 to 70°C)	7 (5.2 to 8.8)	Ac, Et-OH, H_2, CO_2	+/–	33	CNCM 105955	2

[a] Fo = Formate; Ac = Acetate; Pr = Propionate, Bu = Butyrate; IBu = Isobutyrate; IVa = Isovalerate; ICa = Isocaproate; Et-OH = Ethanol; Pr-OH = Propanol; Ala = Alanine.

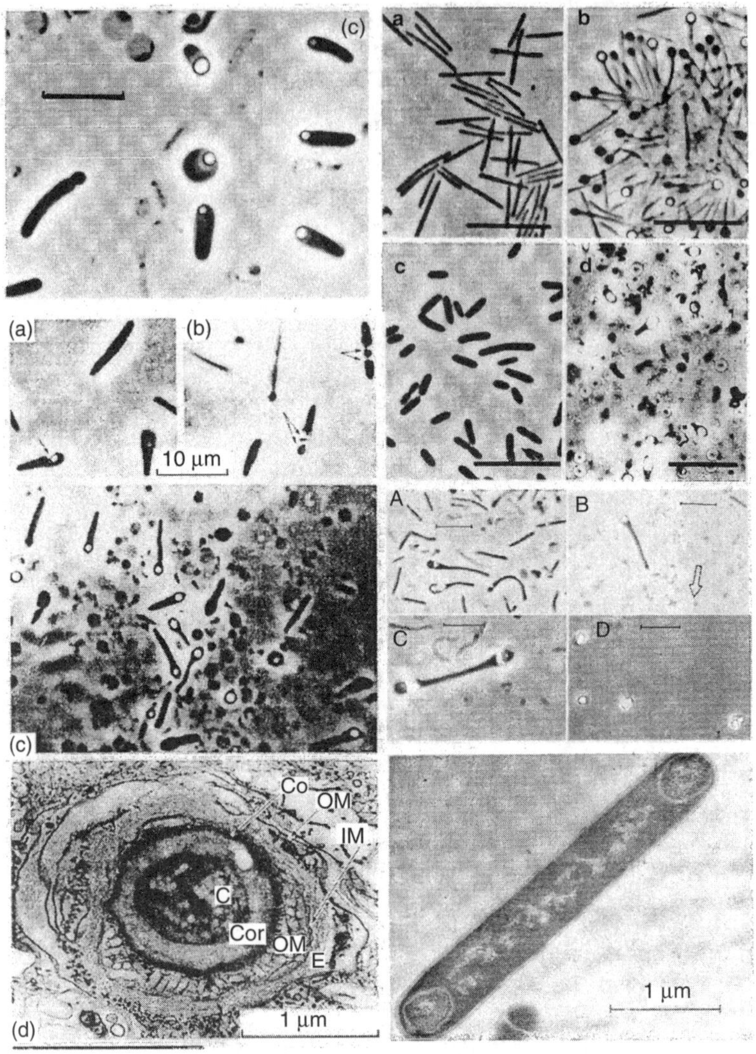

FIGURE 33.1 Micrographs of halophilic endospore-forming anaerobic bacteria. (A) Sporulating cells of *Orenia salinaria* at different stages of sporulation. Bar = 10 μm. (From Mouné, S. et al., *Int. J. Syst. Evol. Microbiol.,* 50, 721, 2000 [56]. With permission.) (B) *Orenia sivashensis*. (a, b, c) Phase contrast micrographs; the arrows indicate gas vacuoles. Bar = 10 μm. (d) An ultrathin section of a mature spore. C, core; Co, coat; E, exosporium, IM, the exosporium inner membrane; OM, the exosporium multilayered outer membrane. Bar = 1 μm. (From Zhilina, T.N. et al., *Microbiology,* 68, 452, 1999 [78]. With permission.) (C) Vegetative cells (a, c) and sporulating cells (b, d) of *Clostridium halophilum* (a, b) and *Clostridium litorale* (c, d). Bar = 10 μm. (From Fendrich, C., Hippe, H., and Gottschalk, G., *Arch. Microbiol.,* 154, 127, 1990 [1]. With permission.) (D) *Sporohalobacter lortetii*: (a) Different stages of development of endospores and gas vacuoles; (b) a sporulating cell with gas vacuoles at both ends; (c) a cell with gas vacuoles and beginning endospore formation at both ends; (d) mature endospores. Bar = 5 μm. (From Oren, A., *Arch. Microbiol.,* 136, 42, 1983 [57]. With permission.) (E) A sporulating cell of *Clostridium oceanicum* with developing endospores at both ends. (From Smith, L.D.S., *J. Bacteriol.,* 103, 811, 1970 [21]. With permission.)

to methyl-2-butyrate; and leucine to isovalerate. The organism can also perform the Stickland reaction, using isoleucine as electron donor and methionine or glycine betaine as electron acceptor [28].

Three more fermentative thermophilic isolates closely related to the genus *Clostridium* have been described from similar environments as representatives of new genera: The endospore-producing

Caminicella sporogenes and *Tepidibacter thalassicus* were isolated from hydrothermal vents of the East Pacific Rise. Phylogenetically, both group with cluster XI of the clostridia [29,30]. *Caminicella* grows optimally at 2.5 to 3% salt, has its optimal temperature at 53 to 60°C, and tolerates temperatures up to 65°C. It ferments sugars and amino acids to butyrate, acetate, ethanol, and other products. *Tepidibacter* has its growth optimum at 1.5 to 6% salt and 50°C and tolerates up to 60°C. It ferments a range of sugars and amino acids to ethanol, acetate, CO_2, and H_2. *Caloranaerobacter azorensis*, an organism that phylogenetically belongs to cluster XII of the *Clostridium* group, was never shown to produce endospores. It was obtained from an Atlantic ridge hydrothermal vent. It grows at salt concentrations between 1 and 10%, with an optimum at 3%, and at temperatures up to 65°C. Sugars and other substrates are fermented to acetate, isovalerate, and, in certain cases, alanine [31].

33.4 MEMBERS OF THE ORDER CLOSTRIDIALES IN HYPERSALINE ENVIRONMENTS

Relatively few representatives of the genus *Clostridium* and related members of the order Clostridiales have been isolated from environments in which the salinity exceeds that of seawater.

Clostridium halophilum, a species also found in marine sediments (see above), was first isolated from anoxic hypersaline sediments of the Maldives Islands and from Solar Lake, Sinai, Egypt. Its salt optimum (6%) is significantly above seawater concentration, and growth was observed up to 10% and higher. Salt concentrations of at least 1.5% are required for growth. *C. halophilum* is, thus, a true halophile. It ferments a range of simple sugars to acetate, ethanol, and H_2. In addition, it can grow by means of the Stickland reaction, using H_2, glycine, alanine, leucine, isoleucine, valine, phenylalanine, and histidine as electron donors and glycine betaine as electron acceptor [1].

The sediments of a saltern pond on the Mediterranean coast of France yielded *Thermohalobacter berrensis*, a nonsporulating thermophilic (growing at up to 65°C) halophile (optimum growth being obtained at 5% salt, while tolerating up to 15%). Phylogenetically, this organism is a member of cluster XII of the *Clostridium* group. It ferments different monosaccharides, disaccharides, and starch to acetate, ethanol, H_2, and CO_2 [2].

Bottom sediments of saline alkaline lakes are an especially interesting habitat for fermentative members of the *Firmicutes*. The East African rift valley of Kenya and Tanzania contains a great number of soda lakes (pH of 10 to 11 and higher) with salinities increasing from slightly saline to saturation. The relatively low salinity lakes Elmenteita and Bogoria yielded a number of haloalkaliphilic fermentative isolates that phylogenetically cluster with *Clostridium* group XI. Phenotypically, these isolates are quite diverse. They ferment a variety of simple sugars or amino acids to acetate and propionate or buyrate, and tolerate salt concentrations up to 4 to 12% [32,33]. The isolates still await a full taxonomic characterization.

From the sediments of the salt-saturated Lake Magadi, Kenya, three haloalkaliphilic strains were obtained that phylogenetically are positioned between clusters VIII and IX of the *Clostridium* group, forming a separate, well-defined group with *C. thermaceticum* and *C. thermautotrophicum* (cluster VI) as closest relatives. The strains grow optimally at pH 9.5, require at least 12 to 16% salt, and tolerate salt concentrations up to 25%. They ferment a range of sugars, and excrete mainly isovaleric acid with smaller amounts of isobutyric acid and acetic acid as end products [32,33]. A different haloalkaliphilic anaerobe isolated from Lake Magadi is *Tindallia magadiensis*. This alkaliphile grows optimally at pH 8.5, and requires salt (3 to 6% NaCl or even better equivalent concentrations of $NaHCO_3$). Phylogenetically, it belongs to *Clostridium* group XI. Its preferred substrates are arginine and compounds of the ornithine cycle. Citrate and pyruvate are also metabolized. Fermentation products are mainly acetate, with minor amounts of formate, propionate, and isovalerate. Hydrogen is also produced. Moreover, the organism reduces dimethylsulfoxide to dimethylsulfide and Fe(III) to Fe(II) [34]. Another alkaliphilic Na-requiring anaerobe belonging to

Clostridium group XI, related to *Tindallia*, is *Anoxynatronum sibiricum*, isolated from a Siberian soda lake. It requires at least 76 mM Na^+ and grows up to 1.27 M Na^+. Chloride is not required for growth. It is an aerotolerant aerobe that ferments sugars, sugar alcohols, and some amino acids. Per mol glucose fermented, 2.5 mol acetate is formed, together with minor amounts of ethanol. This organism also reduces Fe(III) to Fe(II). Formation of endospores was never observed [35].

33.5 THE ORDER HALANAEROBIALES: HALOPHILIC ENDOSPORE-FORMING AND NONENDOSPORE-FORMING LOW G+C *FIRMICUTES* FROM HYPERSALINE ENVIRONMENTS

Although, as described above, there are a few representatives of the *Clostridium* group that grow at salinities above that of seawater, the niche of anaerobic fermentative bacteria in hypersaline sediments is not primarily occupied by members of the Clostridiales. Instead, the related order of the Halanaerobiales, which belongs as well to the low G+C *Firmicutes*, harbors most of the halophilic fermentative bacteria isolated thus far. This order consists of two families, the Halanaerobiaceae and the Halobacteroidaceae [36,37]. The first report of the isolation of a member of the order was probably the description of the "Bacteroides halosmophilus," isolated from salted anchovies [38]. Unfortunately, this isolate has not been preserved. Table 33.2 presents an overview of the species characterized in these two families. The table also provides information on the many nomenclature changes that have been proposed over the years for many representatives of the order [37,39]. The properties of this group have been reviewed several times in the past [3–8].

All known members of the Halanaerobiales are strictly anaerobic and moderately halophilic. NaCl concentrations between 3 and 20% are required for optimal growth, and no growth is observed below 2 to 10 NaCl, depending on the species. The most halophilic species known is *Halanaerobium lacusrosei*, which grows between 6 and 34% NaCl, with an optimum at 20% [40]. Alkaliphilic and thermophilic representatives are also known.

Species belonging to the order Halanaerobiales are abundant in any hypersaline anaerobic environment in which simple organic compounds, such as sugars and amino acids, are available. They have been isolated from thalassohaline hypersaline lakes, such as the Great Salt Lake, Utah [41–43], and the Salton Sea, California [44–46], hypersaline lakes and lagoons in the Crimea [47–52] and Senegal [40,53], and saltern evaporation ponds in California [54] and France [55,56]. They have also been recovered from lakes with an athalassohaline ionic composition, such as the Dead Sea [57–60] and the alkaline hypersaline lakes Magadi, Kenya [45,61,62], and Big Soda Lake, Nevada [45,46], and from a hot hypersaline lake in Tunisia [63]. A number of interesting species have been isolated from brines associated with oil wells and petroleum reservoirs [64–69]. *Halanaerobium* strains growing between 5 and 34% salt have been recovered from the boundary layer between the oxic seawater and the anoxic hypersaline brines of Kebrit Deep, an 84-deep brine layer located at the bottom of the Red Sea at a depth of 1.5 km. In addition, 16S rDNA sequences affiliated with the Halanaerobiales have been amplified from this interesting environment [70]. Finally, they can be present in salted fermented foods [71,72].

Phylogenetically, the Halanaerobiales form a coherent cluster close to the bifurcation point that separates the Actinomycetes subphylum and the Bacillus/Clostridium subphylum. Their 16S rRNA genes contain signature nucleotides that have been defined as characteristic of members of the Bacillus/Clostridium subphylum, while they lack the Actinomycetes-specific nucleotides [37,73–75]. The deep branching justifies classification in a separate order [37]. The phylogenetic affiliation of *Halanaerobium praevalens*, the type species of the type genus of the Halanaerobiales, with the *Bacillus/Clostridium* group was confirmed by a comparison of the amino acid sequences of the ribosomal A-protein [76]. Physiologically, the group is also coherent to the extent that no known aerobes and no nonhalophiles cluster within the order.

TABLE 33.2
Halophilic Fermentative Endospore-Forming and Non-Endospore-Forming Bacteria of the Order Halanaerobiales, Families Halanaerobiaceae and Halobacteroidaceae

Name[a]	Nomenclatural and Taxonomic Comments	Habitat	Salt Optimum and Concentration Range for Growth	Temperature Optimum and Range for Growth	pH Optimum and Range for Growth[b]	Metabolic Properties[c]	Production of Endospores/ Production of Gas Vacuoles	DNA G+C Content (mol%)	Type Strain	Ref.
				Members of the Halanaerobiaceae						
Halanaerobium praevalens	Originally named *Haloanaerobium praevalens*	Sediment, Great Salt Lake, Utah	13% (2 to 30%)	37°C (5 to 50°C)	7 to 7.4 (6 to 9)	Ferments sugars to Ac, Bu, Pr, H_2, CO_2	–/–	27–28	DSM 2228	43
Halanaerobium alcaliphilum	Originally named *Haloanaerobium alkaliphilum*	Sediment, Great Salt Lake, Utah	10% (2.5 to 25%)	37 to 40°C (25 to 50°C)	6.7 to 7 (5.4 to 8)	Ferments sugars to Ac, Bu, La, H_2, CO_2	–/–	31	DSM 8275	41
Halanaerobium acetethylicum	Basonym: *Halobacteroides acetoethylicus*	Offshore oil rig, Gulf of Mexico	10% (5 to 22%)	34°C (15 to 45°C)	6.3 to 7.4 (5.4 to 8)	Ferments sugars to Ac, H_2, CO_2	–/–	32	DSM 3532	69
Halanaerobium salsuginis	Originally named *Haloanaerobium salsugo*	Petroleum reservoir fluid, Oklahoma	9% (6 to 24%)	40°C (22 to 51°C)	6.1 (5.6 to 8)	Ferments sugars to Ac, Et-OH, H_2, CO_2	–/–	34	ATCC 51327	66
Halanaerobium saccharolyticum subsp. *saccharolyticum*	Basonym: *Haloincola saccharolyticus* (*saccharolytica*)	Lake Sivash, Crimea	10% (3 to 30%)	37 to 40°C (15 to 47°C)	7.5 (6 to 8)	Ferments sugars to Ac, H_2, CO_2	–/–	31.3	DSM 6643	77
Halanaerobium saccharolyticum subsp. *senegalensis*	Basonym: *Haloincola saccharolyticus* (subsp. *senegalensis*)	Lake Retba, Senegal	7.5 to 12.5% (5 to 25%)	40°C (20 to 47°C)	7 (6.3 to 8.7)	Ferments sugars to Ac, H_2, CO_2	–/–	31.7	DSM 7379	53

TABLE 33.2 (continued)
Halophilic Fermentative Endospore-Forming and Non-Endospore-Forming Bacteria of the Order Halanaerobiales, Families Halanaerobiaceae and Halobacteroidaceae

Name[a]	Nomenclatural and Taxonomic Comments	Habitat	Salt Optimum and Concentration Range for Growth	Temperature Optimum and Range for Growth	pH Optimum and Range for Growth[b]	Metabolic Properties[c]	Production of Endospores/ Production of Gas Vacuoles	DNA G+C Content (mol%)	Type Strain	Ref.
Halanaerobium congolense	Originally named *Haloanaerobium congolense*	Offshore oil well, Congo	10% (4 to 24%)	42°C (20 to 45°C)	7 (6.3 to 8.5)	Ferments sugars to Ac, H_2, CO_2	–/–	34	DSM 11287	68
Halanaerobium lacusrosei	Originally described as *Haloanaerobium lacusroseus*	Lake Retba, Senegal	18 to 20% (7.5 to 34%)	40°C (20 to 50°C)	7 (6.3 to 8.5)	Ferments sugars to Ac, Et-OH, H_2, CO_2	–/–	32	DSM 10165	40
Halanaerobium kushneri	Originally described as *Haloanaerobium kushneri*	Petroleum reservoir, Oklahoma	12% (9 to 18%)	30°C (20 to 45°C)	6.5 to 7.5 (6 to 8)	Ferments sugars to Ac, Et-OH, Fo, H_2, CO_2	–/–	34.1	ATCC 700103	67
Halanaerobium fermentans	Originally described as *Haloanaerobium fermentans*	Fermenting salted puffer fish ovaries	10% (7 to 25%)	35°C (15 to 45°C)	7.5 (6 to 9)	Ferments sugars to Ac, Et-OH, Fo, La, H_2, CO_2	–/–	33.3	JCM 10494	71
Halocella cellulosilytica	Originally described as *Halocella cellulolytica*	Lake Sivash, Crimea	15% (5 to 20%)	30°C (20 to 50°C)	7 (5.5 to 8.5)	Ferments sugars to Ac, Et-OH, La, H_2, CO_2	–/–	29	DSM 7362	47
Halothermothrix orenii		Hypersaline lake sediment, Tunisia	10% (4 to 20%)	60°C (45 to 68°C)	6.5 to 7	Ferments sugars to Ac, Et-OH, H_2, CO_2	–/–	39.6	OCM 544	63

Members of the Halobacteroidaceae										
Halobacteroides halobius		Dead Sea sediments	9 to 15% (7 to 19%)	37 to 42°C (30 to 47°C)	N/A		–/–	30.7	ATCC 35273	58
Halobacteroides elegans	Originally described as a strain of *Halobacteroides halobius*	Lake Sivash, Crimea	10 to 15% (10 to 30%)	40°C (28 to 47°C)	7 (6.5 to 8)	Ferments sugars to Ac, Et-OH, H_2, CO_2	+/–	30.5	DSM 6639	52
Acetohalobium arabaticum		Lake Sivash, Crimea	15 to 18% (10 to 25%)	38 to 40°C (up to 47°C)	7.4 to 8 (5.8 to 8.4)	Ac	–/–	33.6	DSM 5501	49
Halanaerobacter chitinivorans	Originally described as *Haloanaerobacter chitinovorans*	Saltern, California	12 to 18% (3 to 30%)	30 to 45°C (23 to 50°C)	7 (range NR)	Ferments sugars to Ac, IBu, H_2, CO_2; performs Stickland reaction	–/–	34.8	OCG 229	54
Halanaerobacter lacunarum	Basonym: *Halobacteroides lacunaris*	Lake Chokrak, Kerech Peninsula	15 to 18% (10 to 30%)	35 to 40°C (25 to 52°C)	6.5 to 7 (6 to 8)	Ferments sugars to Ac, Et-OH, H_2, CO_2	–/–	32.4	DSM 6640	101
Halanaerobacter salinarius	Originally described as *Haloanaerobacter salinarius*	Saltern pond, France	14 to 15% (5 to 30%)	45°C (10 to 50°C)	7.4 to 7.8 (5.5 to 8.5)	Ferments sugars to Ac, Et-OH, Pr, Fo, H_2, CO_2; performs Stickland reaction	–/–	31.6	DSM 12146	55
Sporohalobacter lortetii	Basonym: *Clostridium lortetii*	Dead Sea sediments	8 to 9% (4 to 15%)	37 to 45°C (up to 55°C)	N/A	Ferments amino acids to Ac, Pr, Bu, IBu, IVa	+/+	31.5	ATCC 36069	57
Orenia marismortui	Basonym: *Sporohalobacter marismortui*	Dead Sea sediments	3 to 12% (3 to 18%)	36 to 45°C (25 to 50°C)	N/A	Ferments sugars to Ac, Et-OH, Fo, Bu, H_2, CO_2	+/–	29.6	ATCC 35420	59
Orenia sivashensis		Hypersaline lagoon, Lake Sivash, Crimea	7 to 10% (5 to 25%)	40 to 45°C (up to 50°C)	6.3 to 6.6 (5.5 to 7.8)	Ferments sugars to Et-OH, Ac, Fo, Bu, H_2, CO_2	+/+	28.6	DSM 12596	78

TABLE 33.2 (continued)
Halophilic Fermentative Endospore-Forming and Non-Endospore-Forming Bacteria of the Order Halanaerobiales, Families Halanaerobiaceae and Halobacteroidaceae

Name[a]	Nomenclatural and Taxonomic Comments	Habitat	Salt Optimum and Concentration Range for Growth	Temperature Optimum and Range for Growth	pH Optimum and Range for Growth[b]	Metabolic Properties[c]	Production of Endospores/Production of Gas Vacuoles	DNA G+C Content (mol%)	Type Strain	Ref.
Orenia salinaria		Saltern, South of France	5 to 10% (10 to 25%)	40 to 45°C (up to 50°C)	7.2 to 7.4 (5.5 to 8.5)	Ferments sugars to Et-OH, Fo, Ac, La, H_2, CO_2	+/–	33.7	ATCC 700911	56
Natroniella acetigena		Bottom mud, Lake Magadi, Kenya	12 to 18% (10 to 26%)	37°C (up to 42°C)	9.7 to 10 (7.7 to 10.3)	Homoacetogen, grows on La, Et-OH, Py, Glu, Pr	+/–	31.9	DSM 9952	61
Halonatronum saccharophilum		Coastal lagoon mud, Lake Magadi, Kenya	7 to 12% (3 to 17%)	36 to 55°C (up to 60°C)	8 to 8.5 (7.7 to 10.3)	Ferments sugars to Fo, Ac, Et-OH, H_2, CO_2	+/–	34.4	DSM 13868	62
Selenihalanaerobacter shriftii		Dead Sea sediment	21% (10 to 24%)	38°C (16 to 42°C)	7.2 (5.3 to 8.9)	Grows by anaerobic respiration on organic electron donors (glycerol, glucose), using selenate or nitrate as electron acceptors; Fermentative growth not observed	–/–	31.2	ATCC BAA-73	100

[a] For details on the nomenclature, see also Rainey et al. [37], Oren [39], and Trüper and de'Clari [112].
[b] N/A = No data available.
[c] Fo = Formate; Ac = Acetate; Pr = Propionate; Bu = Butyrate; IBu = Isobutyrate; IVa = Isovalerate; La = Lactate; Py = Pyruvate; Et-OH = Ethanol; Glu = Glutamate.

Most species have DNA of low G+C contents, ranging between 27 and 36.9 mol%. The thermophilic *Halothermothrix orenii* (family Halanaerobiaceae) has the highest value at 39.6 mol%. All species stain Gram negative, and a typical Gram-negative type cell wall with an outer membrane and a pronounced periplasmic space is often seen in electron micrographs of thin sections. Meso-diaminopimelic acid was detected in the peptidoglycan of *Halanaerobium saccharolyticum* subsp. *saccharolyticum* [77].

Formation of endospores has never been reported in any member of the Halanaerobiaceae, but heat-resistant endospores are produced by a number of species of Halobacteroidaceae. These include *Orenia salinaria* [56] (Figure 33.1A), *Orenia sivashensis* [78] (Figure 33.1B), *Sporohalobacter lortetii* [57] (Figure 33.1D), *Halobacteroides elegans* [52], *Orenia marismortui* [59], and *Halonatronum saccharophilum* [62]. *Natroniella acetigena* produces spores only infrequently [61]. When initially isolated, *Acetohalobium arabaticum* produced spores, but sporulation was lost during subsequent transfers [79]. Growth on solid media or in nutrient-poor liquid media can stimulate sporulation in certain species [57,59]. In *Sporohalobacter lortetii,* cells developing endospores at both poles were occasionally observed [58] (Figure 33.1D, c); a similar phenomenon has been documented in *Clostridium oceanicum* [21] (Figure 33.1E).

Hydrolysis of the D-isomer of *N′*-benzoyl-arginine-*p*-nitroanilide (BAPA) can be used as a phenotypic test, the result of which may be correlated with the phylogenetic position of the Halanaerobiales within the phylum of the Gram-positive bacteria and with the ability to form endospores. Three out of four representatives of the Halanaerobiales (*Halobacteroides halobius*, *Orenia marismortui*, and *Halanaerobium praevalens*, the last belonging to the family Halanaerobiaceae, in which endospore formation has never yet been observed) hydrolyzed D-BAPA, while L-BAPA was not hydrolyzed [80].

Presence of heat-resistant endospores can be exploited to selectively enrich *Halobacteroides*-like bacteria. Pasteurization of hypersaline mud samples obtained from a sulfur spring on the shore of the Dead Sea and from a saltern evaporation pond (salt concentrations between 5.8 and 15.3%) for 10 to 20 min at 80 to 100°C, followed by inoculation of suitable anaerobic high-salt media, yielded growth of *Halobacteroides*-like long, slender fermentative bacteria [81].

Two species synthesize gas vesicles adjacent to the endospores: *Sporohalobacter lortetii* [57] (Figure 33.1D) and *Orenia sivashensis* (Figure 33.1B) [78]. Vegetative cells are devoid of gas vesicles. A thin section through a mature endospore of *Orenia sivashensis* (Figure 33.1B, d) suggests the gas vesicles to be located between the spore cortex and the exosporium. Production of gas vesicles adjacent to endospores has also been reported for certain soil clostridia [82–84] and for the sulfate reducer *Desulfotomaculum acetoxidans* [85]. It has been speculated that these gas vesicles might aid in the dispersion of the mature endospores [86].

Three members of the Halanaerobiales (*Halobacteroides halobius*, *Sporohalobacter lortetii*, and *Orenia marismortui*) have been recovered from the relatively low pH environment of the sediments of the Dead Sea, whose brines contain around 1.9 *M* $MgCl_2$, 1.7 *M* NaCl, 0.4 *M* $CaCl_2$, and 0.2 *M* KCl, and that have a pH of around 6. *Sporohalobacter lortetii* (basonym *Clostridium lortetii*) [57] was named after M.L. Lortet, who, in the last decade of the 19th century, had isolated (nonhalophilic) clostridia from Dead Sea mud [19].

Lake Magadi, an alkaline (pH 10.2) hypersaline lake in Kenya, was shown to harbor a varied anaerobic community, including cellulolytic, proteolytic, saccharolytic, and homoacetogenic bacteria [87]. From this interesting environment, two obligatory alkaliphilic representatives of the family Halobacteroidaceae have been characterized: *Natroniella acetigena* and *Halonatronum saccharophilum* [61,62]. *Natroniella* grows between pH 8.1 and 10.7 and 10 to 26% salt, with an optimum at pH 9.7 to 10 and 12 to 15% salt. It is a homoacetogen that ferments few substrates only (lactate, pyruvate, glutamate, ethanol, propanol) and forms acetate as the sole product [61]. *Halonatronum* is an endospore-forming alkaliphilic and halophilic sugar fermenter with moderately thermophilic properties. Optimal growth is obtained at 7 to 12% salt and pH 8 to 8.5; salt concentrations between 3 and 17% and pH values between 7.7 and 10.3 support growth. High concentrations of Na_2CO_3 or $NaHCO_3$ are specifically required; no growth was observed in high NaCl media buffered with serine at the optimal pH (8 to 8.4) in the absence of carbonates [62]. Enzymatic assays showed that carbohydrate fermentation follows the Embden-Meyerhof pathway. A low activity of methylviologen-dependent formate dehydrogenase was detected as well [88].

Halothermothrix orenii, a member of the Halanaerobiaceae isolated from Chott El Guettar, a warm saline lake in Tunisia, is a thermophile that grows optimally at 60°C and tolerates up to 68°C [63]. The α-amylase of this interesting organism has been cloned and expressed in *Escherichia coli* and characterized. The enzyme is optimally active at 65°C in 5% NaCl and retains significant activity in 25% NaCl. The protein does not display an unusually high content of acidic amino acids [89]. The protein has been crystallized [90], and the elucidation of its three-dimensional structure is in progress. A vector library of 3360 clones of average 3.5 kb has been prepared from *H. orenii* DNA, and many genes have been identified. Acidic amino acids were not present in high excess in the proteins of this organisms [91]. Another member of the Halanaerobiales with moderately thermophilic properties is *Halonatronum saccharophilum*, which grows at temperatures up to 60°C [62].

Most members of the Halanaerobiales obtain their energy by fermenting simple sugars. An active transport system for glucose has been characterized in a number of *Halobacteroides* strains [92,93]. Carbohydrate fermentation is the dominant type of dissimilatory metabolism observed thus far in the family Halanaerobiaceae. *Halanaerobium alcaliphilum* was reported to ferment glycine betaine as well, with the formation of acetate and trimethylamine [41]. Fermentation products from sugars typically include acetate, carbon dioxide, and hydrogen. Some species produce, in addition, formate, propionate, butyrate, or lactate. Ethanol is also a common fermentation product. However, *Halanaerobium congolense* does not form ethanol [68]. Starch is hydrolyzed by a variety of species, while other polysaccharides, such as chitin and cellulose, can be degraded by specialized organisms such as *Halanaerobacter chitinivorans* and *Halocella cellulosilytica*, respectively [47,54,94,95].

Several members of the Halanaerobiales ferment glycerol, trehalose, and glucosylglycerol, all compounds synthesized by halophilic and hatolerant microorganisms and accumulated intracellularly as high concentrations to be used as osmotic stabilizers. Such compounds may, therefore, be available in large amounts in microbial mats that develop in many hypersaline environments. Glycerol, produced as compatible solute by the halophilic green alga *Dunaliella*, is fermented by *Halanaerobium saccharolyticum* [53,77], and by *Halanaerobium lacusrosei* [40]. Glycerol oxidation by anaerobic halophiles was markedly improved through interspecies hydrogen transfer when the glycerol fermenters were grown in co-culture with H_2-consuming sulfate-reducing bacteria [40]. *Halanaerobium saccharolyticum* was isolated from a benthic cyanobacterial mat from a hypersaline lagoon in the Crimea, dominated by *Microcoleus chthonoplastes*. Its ability to use glucosylglycerol, the compatible solute produced by *Microcoleus*, might therefore be of importance in its natural habitat. *Halanaerobium saccharolyticum* also degrades trehalose, another osmotic solute produced by many cyanobacteria [77].

The metabolic diversity within the family Halobacteroidaceae is much greater than within the Halanaerobiaceae. In addition to sugar fermenters, there are species that ferment amino acids, either alone or using the Stickland reaction. *Sporohalobacter lortetii* is probably an amino acid fermenter; glucose added to the medium was used only after other, more easily fermentable, substrates were exhausted [57]. *Halanaerobacter salinarius* and *Halanaerobacter chitinivorans* use the Stickland reaction, oxidizing serine while using glycine betaine as electron acceptor with the formation of acetate, trimethylamine, CO_2, and NH_3 [55].

The neutrophilic *Acetohalobium arabaticum* and the alkaliphilic *Natroniella acetigena* are homoacetogens, which are unable to use carbohydrates, but grow chemoheterotrophically on substrates such as lactate, ethanol, pyruvate, glutamate, propanol, and glycine betaine. *Acetohalobium arabaticum* also grows chemolithotrophically on $H_2 + CO_2$ or on CO, as a methylotroph on trimethylamine with formation of acetate and NH_3, and as an organotroph on formate, trimethylamine, glycine betaine, lactate, pyruvate, histidine, aspartate, glutamate, or asparagine [48,49,96]. A flavoprotein has been identified in the electron transport system used during autotrophic growth on $H_2 + CO_2$; NAD, quinones, and cytochromes are not involved [97,98]. An Na^+ potential generated via a Na^+/H^+ antiport system is essential for the functioning of the cell, and monensin, an inhibitor of Na^+/H^+ antiport and Na^+-dependent ATP synthesis, was found inhibitory for growth [97].

Natroniella acetigena cannot grow as a chemolithotroph. Lactate, ethanol, pyruvate, glutamate, and propanol are used as substrates for energy generation in a homoacetogenic metabolism [61]. At least part of the acetate is produced via the acetyl-CoA pathway involving CO dehydrogenase. Also, here, the energy metabolism relies on the generation of a transmembrane electrochemical Na^+ gradient mediated by Na^+/H^+ antiporter activity [99].

The family Halobacteroidaceae contains one representative that does not appear to use a fermentative metabolism, but generates its energy by anaerobic respiration instead — *Selenihalanaerobacter shriftii* isolated from bottom sediment of the Dead Sea. It was obtained from an anaerobic enrichment culture that contained diluted Dead Sea water, glycerol, and selenate [100]. Glycerol and glucose are oxidized to a mixture of acetate and CO_2 by anaerobic respiration, using nitrate, trimethylamine *N*-oxide, or selenate as electron acceptor. Selenate is reduced to a mixture of selenite and elemental selenium. *S. shriftii* is the only member of the Halanaerobiales known to possess cytochromes. No cytochromes were detected in *Halanaerobium acetethylicum* [69], *Halobacteroides halobius* [58], *Sporohalobacter lortetii* [57], *Orenia marismortui* [59], and *Acetohalobium arabaticum* [97].

Several *Halanaerobium* species can use oxidized sulfur compounds as electron acceptors or electron sinks. *Halanaerobium congolense* uses thiosulfate or elemental sulfur as electron acceptors. Addition of thiosulfate stimulated carbohydrate utilization and enhanced growth rates. The growth yield was increased 6-fold and 3-fold following addition of thiosulfate or sulfur, respectively. Presence of thiosulfate also alleviated growth inhibition by accumulating hydrogen [68]. Reduction of elemental sulfur to sulfide was also reported in *Halanaerobium saccharolyticum* subsp. *saccharolyticum* [77], in *Halanaerobacter lacunarum* [101], in *Halobacteroides elegans* [52,102], and in *Halonatronum saccharophilum* [62]. Thiosulfate reduction was also observed in *Orenia marismortui* [59]. *Acetohalobium arabaticum* slowly reduces elemental sulfur to sulfide, but addition of sulfur did not enhance growth rates [79,103]. *Halanaerobium praevalens* and *Orenia marismortui* reduce nitrosubstituted aromatic compounds (nitrobenzene, nitrophenols, 2,4-dinitrophenol, and 2,4-dinitroaniline) to the corresponding amino derivatives [104,105].

Several species can use methanethiol as sulfur source for assimilation. These include *Halanaerobium saccharolyticum*, *Halobacteroides halobius*, *Halobacteroides elegans*, and *Halanaerobacter lacunarum* [101–103]. *Orenia salinaria* was reported to grow on N_2 as assimilatory nitrogen source [56].

In contrast to most other halophilic representatives of the domain Bacteria, which use organic solutes such as glycine betaine and ectoine to provide osmotic balance of their cytoplasm with the surrounding medium, the Halanaerobiales appear to accumulate inorganic salts intracellularly. High concentrations of Na^+, K^+, and Cl^- have been measured inside the cells of *Halanaerobium praevalens*, *Halanaerobium acetethylicum*, and *Halobacteroides halobius* [106–108], high enough to be approximately isotonic with the medium. X-ray microanalysis with the transmission electron microscope showed that K^+ was the major cation in exponentially growing cells of *Halanaerobium praevalens*, accounting for 70% of the cation sum. In stationary-phase cells, a great variability among individual cells was found in the intracellular Na^+ and K^+ concentrations [107]. Some species might also rely on organic solutes to provide osmotic stabilization. Glycine betaine was found intracellularly in quite high concentrations in *Orenia salinaria* when yeast extract (which contains glycine betaine) was included in the growth medium. Intracellular concentrations were estimated at 1.9 to 2.2 μmol/mg protein, values that correspond with concentrations of around 0.6 to 0.8 *M*, assuming the intracellular volume to be between 2.5 and 3 μl/mg protein [56].

In agreement with the high intracellular ionic concentrations present in the Halanaerobiales, those intracellular enzymes tested all function well in the presence of high salt. These include glyceraldehyde-3-phosphate dehydrogenase, NAD-linked alcohol dehydrogenase, pyruvate dehydrogenase, and methyl viologen-linked hydrogenase of *Halanaerobium acetethylicum*, the fatty acid synthetase complex of *Halanaerobium praevalens*, and hydrogenase and CO dehydrogenase of *Acetohalobium arabaticum*. Many of these enzymes function better at molar concentrations of salts than in a salt-free medium [79,108,109].

As expected for organisms that maintain salt-dependent and salt-resistant enzymes, the bulk cellular protein of *Halanaerobium praevalens*, *Halobacteroides halobius*, and *Sporohalobacter lortetii* has a high content of acidic amino acids [106]. However, no especially high content of acidic amino acids was detected in the ribosomal A-protein of *Halanaerobium praevalens* [76]. Likewise, the proteins of the thermophilic *Halothermothrix orenii* are not especially acidic, as proven by a partial genomic sequence analysis [90].

The Halanaerobiales have found a number of biotechnological applications. They can be involved in the production of certain salted fermented foods. *Halanaerobium fermentans* was isolated from "fugunoko nukaduke," a traditional Japanese food prepared by salting puffer fish ovaries for at least 6 months, whereafter they are fermented for several years with rice bran and fish sauce [71]. *Halanaerobium* species can also be involved in the fermentation of herrings in Sweden ("surströmming") [72]. Other *Halanaerobium* species can be used in microbially enhanced oil recovery by plugging of porous reservoirs and by anaerobically metabolizing nutrients with the production of gases, biosurfactants, and polymers [64]. Lastly, it has been proposed that anaerobic halophilic bacteria can be used in the industrial fermentation of complex organic matter for the production of organic solvents [110,111].

REFERENCES

1. Fendrich, C., Hippe, H., and Gottschalk, G., *Clostridium halophilium* sp. nov., and *C. litorale* sp. nov., an obligate halophilic and a marine species degrading betaine in the Stickland reaction, *Arch. Microbiol.*, 154, 127, 1990.
2. Cayol, J.-L. et al., *Thermohalobacter berrensis* gen. nov., sp. nov., a thermophilic, strictly halophilic bacterium from a solar saltern, *Int. J. Syst. Evol. Microbiol.*, 50, 559, 2000.
3. Oren, A., The ecology and taxonomy of anaerobic halophilic eubacteria, *FEMS Microbiol. Rev.*, 39, 23, 1986.
4. Oren, A., Anaerobic degradation of organic compounds in hypersaline environments: possibilities and limitations, in *Bioprocessing and Biotreatment of Coal*, Wise, D.L., Ed., Marcel Dekker, New York, 1990, 155.
5. Oren, A., The genera *Haloanaerobium*, *Halobacteroides*, and *Sporohalobacter*, in *The Prokaryotes, A Handbook on the Biology of Bacteria: Ecophysiology, Isolation, Identification, Applications, 2nd ed., Vol. II.*, Balows, A. et al., Eds., Springer-Verlag, New York, 1992, 1893.
6. Oren, A., Anaerobic heterotrophic bacteria growing at extremely high salt concentrations, in *Trends in Microbial Ecology*, Guerrero, R. and Pedrós-Alió, C., Eds., Spanish Society for Microbiology, Madrid, 1993, 539.
7. Oren, A., The order Haloanaerobiales, in *The Prokaryotes: An Evolving Electronic Resource for the Microbiological Community*, 3rd ed., Dworkin, M., Falkow, S., Rosenberg, E., Schleifer, K.-H., and Stackebrandt, E., Eds., release 3.2, July 25, 2001. Springer-Verlag, New York, 2001. http://link.springer-ny.com/link/service/books/10125.
8. Ollivier, B. et al., Anaerobic bacteria from hypersaline environments, *Microbiol. Rev.*, 58, 27, 1994.
9. Davies, C.M. et al., Survival of fecal microorganisms in marine and freshwater sediments, *Appl. Environ. Microbiol.*, 61, 1888, 1995.
10. Brink, M.R.B., Mecray, E.L., and Galvin, E.L., *Clostridium perfringens* in Long Island Sound sediments: an urban sedimentary record, *J. Coastal Res.*, 16, 591, 2000.
11. Hotchkiss, M., The presence in the sea of bacteria belonging to the genus *Clostridium*, *J. Bacteriol.*, 36, 254, 1938.
12. Finne, G. and Matches, J.R., Low-temperature-growing clostridia from marine sediments, *Can. J. Microbiol.*, 20, 1639, 1974.

13. Matches, J.R. and Liston, J., Mesophilic clostridia in Puget Sound, *Can. J. Microbiol.,* 20, 1, 1974.
14. Smith, L.D.S., The clostridial flora from marine sediment from a productive and from a non-productive area, *Can. J. Microbiol.,* 14, 1301, 1968.
15. Davies, J.A., *Clostridia from North Sea sediments*, Ph.D. Thesis, University of Aberdeen, U.K., 1967.
16. Cann, D.C. et al., The incidence of *Clostridium botulinum* type E in fish and bottom deposits in the North Sea and off the coast of Scandinavia, *J. Appl. Bacteriol.,* 28, 426, 1965.
17. Carroll, B.J. et al., Presence of *Clostridium botulinum* in the Gulf of Venezuela and the Gulf of Darien, *Appl. Microbiol.,* 14, 837, 1966.
18. Ward, B.Q. et al., Survey of the U.S. Gulf coast for the presence of *Clostridium botulinum*, *Appl. Microbiol.,* 15, 629, 1967.
19. Lortet, M.L., Researches on the pathogenic microbes of the mud of the Dead Sea, *Pales. Expl. Fund,* 48, 1892 (Microbes pathogènes des vases de la Mer Morte, *Lyon Medical,* 67, 519, 1891).
20. Oren, A., Tetanus bacteria and other pathogens in the Dead Sea?, *Salinet,* 6, 84, 1991.
21. Smith, L.D.S., *Clostridium oceanicum* sp. n., a sporeforming anaerobic isolate from marine sediments, *J. Bacteriol.,* 103, 811, 1970.
22. Hernández, F. and Guillén, R., Microwave processing for scanning electron microscopy, *Eur. J. Morphol.,* 38, 109, 2000.
23. Mountfort, D.O. et al., *Clostridium grantii* sp. nov., a new obligately anaerobic, alginolytic bacterium isolated from mullet gut, *Arch. Microbiol.,* 162, 173, 1994.
24. Kreimer, S. and Andreesen, J.R., Glycine reductase of *Clostridium litorale*. Cloning, sequencing, and molecular analysis of the *grdAB* operon that contains two in-frame TGA codons for selenium incorporation, *Eur. J. Biochem.,* 234, 192, 1995.
25. Kreimer, S., Söhling, B., and Andreesen, J.R., Two closely linked genes encoding thioredoxin and thioredoxin reductase in *Clostridium litorale*, *Arch. Microbiol.,* 168, 328, 1997.
26. Harms, C., Meyer, M.A., and Andreesen, J.R., Fast purification of thioredoxin reductases and of thioredoxins with an unusual redox-active centre from anaerobic, amino-acid-utilizing bacteria, *Microbiology U.K.,* 144, 793, 1998.
27. Collins, M.D. et al., The phylogeny of the genus *Clostridium*: proposal of five new genera and eleven new species combinations, *Int. J. Syst. Bacteriol.,* 44, 812, 1994.
28. Brisbarre, N. et al., *Clostridium caminithermale* sp. nov., a slightly halophilic and moderately thermophilic bacterium isolated from an Atlantic deep-sea hydrothermal chimney, *Int. J. Syst. Evol. Microbiol.,* 53, 1043, 2003.
29. Alain, K. et al., *Caminicella sporogenes* gen. nov., sp. nov., a novel thermophilic spore-forming bacterium isolated from an East-Pacific Rise hydrothermal vent, *Int. J. Syst. Evol. Microbiol.,* 52, 1621, 2002.
30. Slobodkin, A.I. et al., *Tepidibacter thalassicus* gen. nov., sp. nov., a novel moderately thermophilic, anaerobic, fermentative bacterium from a deep-sea hydrothermal vent, *Int. J. Syst. Evol. Microbiol.,* 53, 1131, 2003.
31. Wery, N. et al., *Caloranaerobacter azorensis* gen. nov., an anaerobic thermophilic bacterium isolated from a deep-sea hydrothermal vent, *Int. J. Syst. Evol. Microbiol.,* 51, 1789, 2001.
32. Jones, B.E. and Grant, W.D., Microbial diversity and ecology of the soda lakes of East Africa, in *Microbial Biosystems: New Frontiers, Proceedings of the 8th International Symposium on Microbial Ecology*, Bell, C.R., Brylinsky, M., and Johnson-Green, P., Eds., Atlantic Canada Society for Microbial Ecology, Halifax, 1999.
33. Jones, B.E. et al., Microbial diversity of soda lakes, *Extremophiles,* 2, 191, 1998.
34. Kevbrin, V.V. et al., *Tindallia magadii* gen. nov., sp. nov.: an alkaliphilic anaerobic ammonifier from soda lake deposits, *Curr. Microbiol.,* 37, 94, 1988.
35. Garnova, E.S. et al., *Anoxynatronum sibiricum* gen. nov., sp. nov. alkaliphilic saccharolytic anaerobe from cellulolytic community of Nizhnee Beloe (Transbaikal region), *Extremophiles*, 7, 213, 2003.
36. Oren, A., Paster, B.J., and Woese, C.R., *Haloanaerobiaceae*: a new family of moderately halophilic, obligatory anaerobic bacteria, *Syst. Appl. Microbiol.,* 5, 71, 1984.
37. Rainey, F.A. et al., The taxonomic status of the fermentative halophilic anaerobic bacteria: description of Halobacteriales ord. nov., Halobacteroidaceae fam. nov., *Orenia* gen. nov. and further taxonomic rearrangements at the genus and species level, *Anaerobe,* 1, 185, 1995.

38. Baumgartner, J.G., The salt limits and thermal stability of a new species of anaerobic halophile, *Food Res.*, 2, 321, 1937.
39. Oren, A., Change of the names *Haloanaerobiales*, *Haloanaerobiaceae*, and *Haloanaerobium* to *Halanaerobiales*, *Halanaerobiaceae*, and *Halanaerobium*, respectively, and further nomenclature changes within the order *Haloanaerobiales*, *Int. J. Syst. Evol. Microbiol.*, 50, 2229, 2000.
40. Cayol, J.-L. et al., *Haloanaerobium lacusroseus* sp. nov., an extremely halophilic fermentative bacterium from the sediments of a hypersaline lake, *Int. J. Syst. Bacteriol.*, 45, 790, 1995.
41. Tsai, C.-R. et al., *Haloanaerobium alcaliphilum* sp. nov., an anaerobic moderate halophile from the sediments of Great Salt Lake, Utah, *Int. J. Syst. Bacteriol.*, 45, 301, 1995.
42. Zeikus, J.G., Metabolic communication between biodegradative populations in nature, in *Microbes in Their Natural Environments. 34th Symp., Soc. Gen. Microbiol.*, Slater, J.H., Whittenbury, R., and Wimpenny, J.W.T., Eds., Cambridge University Press, Cambridge, 1983, 423.
43. Zeikus, J.G. et al., Isolation and description of *Haloanaerobium praevalens* gen. nov. and sp. nov., an obligately anaerobic halophile common to Great Salt Lake sediments, *Curr. Microbiol.*, 9, 225, 1983.
44. Shiba, H., Anaerobic halophiles, in *Superbugs. Microorganisms in Extreme Environments*, Horikoshi, K. and Grant, W.D., Eds., Japan Scientific Societies Press, Tokyo/Springer-Verlag, Berlin, 1991, 191.
45. Shiba, H. and Horikoshi, K., Isolation and characterization of novel anaerobic, halophilic eubacteria from hypersaline environments of western America and Kenya, in *Proceedings of the FEMS Symposium on the Microbiology of Extreme Environments and Its Biotechnological Potential*, Portugal, 1988, 371.
46. Shiba, H., Yamamoto, H., and Horikoshi, K., Isolation of strictly anaerobic halophiles from the aerobic surface sediments of hypersaline environments in California and Nevada, *FEMS Microbiol. Lett.*, 57, 191, 1989.
47. Simankova, M.V. et al., *Halocella cellulolytica* gen. nov., sp. nov., a new obligately anaerobic, halophilic, cellulolytic bacterium, *Syst. Appl. Microbiol.*, 16, 385, 1993.
48. Zhilina, T.N. and Zavarzin, G.A., Extremely halophilic, methylotrophic, anaerobic bacteria, *FEMS Microbiol. Rev.*, 87, 315, 1990.
49. Zhilina, T.N. and Zavarzin, G.A., A new extremely halophilic homoacetogenic bacterium *Acetohalobium arabaticum* gen. nov., sp. nov., *Dokl. Akad. Nauk. SSSR*, 311, 745, 1990 (in Russian).
50. Zhilina, T.N. and Zavarzin, G.A., Anaerobic bacteria participating in organic matter destruction in halophilic cyanobacterial community, *J. Obshey. Biol.*, 52, 302, 1991 (in Russian).
51. Zhilina, T.N. et al., Isolation of saccharolytic anaerobes from a halophilic cyanobacterial mat, *Microbiology (English translation of Mikrobiologiya)*, 60, 101, 1991.
52. Zhilina, T.N. et al., Reclassification of *Halobacteroides halobius* Z-7287 on the basis of phylogenetic analysis as a new species *Halobacteroides elegans* sp. nov., *Microbiology (English translation of Mikrobiologiya)*, 66, 97, 1997.
53. Cayol, J.-L. et al., *Haloincola saccharolytica* subsp. *senegalensis* subsp. nov., isolated from the sediments of a hypersaline lake, and emended description of *Haloincola saccharolytica*, *Int. J. Syst. Bacteriol.*, 44, 805, 1994.
54. Liaw, H.J. and Mah, R.A., Isolation and characterization of *Haloanaerobacter chitinovorans* gen. nov., sp. nov., a halophilic, anaerobic, chitinolytic bacterium from a solar saltern, *Appl. Environ. Microbiol.*, 58, 260, 1992.
55. Mouné, S. et al., *Haloanaerobacter salinarius* sp. nov., a novel halophilic fermentative bacterium that reduces glycine-betaine to trimethylamine with hydrogen or serine as electron donors; emendation of the genus *Haloanaerobacter*, *Int. J. Syst. Bacteriol.*, 49, 103, 1999.
56. Mouné, S. et al., *Orenia salinaria* sp. nov., a fermentative bacterium isolated from anaerobic sediments of Mediterranean salterns, *Int. J. Syst. Evol. Microbiol.*, 50, 721, 2000.
57. Oren, A., *Clostridium lortetii* sp. nov., a halophilic obligatory anaerobic bacterium producing endospores with attached gas vacuoles, *Arch. Microbiol.*, 136, 42, 1983.
58. Oren, A. et al., *Halobacteroides halobius* gen. nov., sp. nov., a moderately halophilic anaerobic bacterium from the bottom sediments of the Dead Sea, *Syst. Appl. Microbiol.*, 5, 58, 1984.
59. Oren, A., Pohla, H., and Stackebrandt, E., Transfer of *Clostridium lortetii* to a new genus *Sporohalobacter* gen. nov. as *Sporohalobacter lortetii* comb. nov., and description of *Sporohalobacter marismortui* sp. nov., *Syst. Appl. Microbiol.*, 9, 239, 1987.

60. Oren, A., Photosynthetic and heterotrophic benthic bacterial communities of a hypersaline sulfur spring on the shore of the Dead Sea, in *Microbial Mats. Physiological Ecology of Benthic Microbial Communities*, Cohen, Y. and Rosenberg, E., Eds., American Society for Microbiology, Washington, DC, 1989, 64.
61. Zhilina, T.N. et al., *Natroniella acetigena* gen. nov. sp. nov., an extremely haloalkaliphilic, homoacetogenic bacterium: a new member of *Haloanaerobiales*, *Curr. Microbiol.,* 32, 320, 1996.
62. Zhilina, T.N. et al., *Halonatronum saccharophilum* gen. nov., sp. nov.: a new haloalkaliphilic bacterium of the order *Haloanaerobiales* from Lake Magadi, *Microbiology (English translation of Mikrobiologiya),* 70, 64, 2001.
63. Cayol, J.-L. et al., Isolation and characterization of *Halothermothrix orenii* gen. nov., sp. nov., a halophilic, thermophilic, fermentative, strictly anaerobic bacterium, *Int. J. Syst. Bacteriol.,* 44, 534, 1994.
64. Bhupathiraju, V.K. et al., Isolation and characterization of novel halophilic anaerobic bacteria from oil field brines, *Dev. Petrol. Sci.,* 31, 132, 1991.
65. Bhupathiraju, V.K., McInerney, M.J., and Knapp, R.M., Pretest studies for a microbially enhanced oil recovery field pilot in a hypersaline oil reservoir, *Geomicrobiol. J.,* 11, 19, 1993.
66. Bhupathiraju, V.K. et al., *Haloanaerobium salsugo* sp. nov., a moderately halophilic, anaerobic bacterium from a subterranean brine, *Int. J. Syst. Bacteriol.,* 44, 565, 1994.
67. Bhupathiraju, V.K. et al., *Haloanaerobium kushneri* sp. nov., an obligately halophilic, anaerobic bacterium from an oil brine, *Int. J. Syst. Bacteriol.,* 49, 953, 1999.
68. Ravot, G. et al., *Haloanaerobium congolense* sp. nov., an anaerobic, moderately halophilic, thiosulfate- and sulfur-reducing bacterium from an African oil field, *FEMS Microbiol. Lett.,* 147, 81, 1997.
69. Rengpipat, S., Langworthy, T.A., and Zeikus, J.G., *Halobacteroides acetoethylicus* sp. nov., a new obligately anaerobic halophile isolated from deep subsurface hypersaline environments, *Syst. Appl. Microbiol.,* 11, 28, 1988.
70. Eder, W. et al., Microbial diversity of the brine-seawater interface of the Kebrit Deep, Red Sea, studied via 16S rRNA gene sequences and cultivation methods, *Appl. Environ. Microbiol.,* 67, 3077, 2001.
71. Kobayashi, T., Kimura, B., and Fujii, T., *Haloanaerobium fermentans* sp. nov., a strictly anaerobic, fermentative halophile isolated from fermented puffer fish ovaries, *Int. J. Syst. Evol. Microbiol.,* 50, 1621, 2000.
72. Kobayashi, T., Kimura, B., and Fujii, T., Strictly anaerobic halophiles isolated from canned Swedish fermented herrings (Surströmming), *Int. J. Food Microbiol.,* 54, 81, 2000.
73. Patel, B.K.C. et al., Reevaluating the classification of *Halobacteroides* and *Haloanaerobacter* species based on sequence comparisons of the 16S ribosomal RNA gene, *FEMS Microbiol. Lett.,* 134, 115, 1995.
74. Tourova, T.P., The role of DNA-DNA hybridization and 16S rRNA gene sequencing in solving taxonomic problems by the example of the order Haloanaerobiales, *Microbiology (English translation of Mikrobiologiya),* 69, 741, 2000.
75. Tourova, T.P. et al., Phylogenetic study of haloanaerobic bacteria by 16S ribosomal RNA sequence analysis, *Syst. Appl. Microbiol.,* 18, 189, 1995.
76. Matheson, A.T. et al., The primary structure of the ribosomal A-protein (L12) from the halophilic eubacterium *Haloanaerobium praevalens*, *Biochimie,* 69, 1013, 1987.
77. Zhilina, T.N. et al., Ecology, physiology and taxonomy studies on a new taxon of *Haloanaerobiaceae*, *Haloincola saccharolytica* gen. nov., sp. nov., *Syst. Appl. Microbiol.,* 15, 275, 1992.
78. Zhilina, T.N. et al., *Orenia sivashensis* sp. nov., a new moderately halophilic anaerobic bacterium from Lake Sivash lagoons, *Microbiology (English translation of Mikrobiologiya),* 68, 452, 1999.
79. Zavarzin, G.A., Zhilina, T.N., and Pusheva, M.A., Halophilic acetogenic bacteria, in *Acetogenesis*, Drake, H.L., Ed., Chapman & Hall, New York, 1994, 432.
80. Oren, A., Gofshtein-Gandman, L.V., and Keynan, A., Hydrolysis of Nα benzoyl-D-arginine-p-nitroanilide by members of the *Haloanaerobiaceae*: additional evidence that *Haloanaerobium praevalens* is related to endospore forming bacteria, *FEMS Microbiol. Lett.,* 58, 5, 1989.
81. Oren, A., A procedure for the selective enrichment of *Halobacteroides halobius* and related bacteria from anaerobic hypersaline sediments, *FEMS Microbiol. Lett.,* 42, 201, 1987.
82. Krasil'nikov, N.A. and Duda, V.I., Ultrastructure of caps on spores of anaerobic bacteria, *Dokl. Akad. Nauk. SSR,* 179, 970, 1968 (in Russian).

83. Krasil'nikov, N.A., Duda, V.I., and Pivovarov, G.E., Characteristics of the cell structure of soil anaerobic bacteria forming vesicular caps on their spores, *Mikrobiologiya,* 40, 681, 1971 (in Russian).
84. Krasil'nikov, N.A., Pivovarov, G.E., and Duda, V.I., Physiological properties of anaerobic soil bacteria which form vesicular caps on their spores, *Mikrobiologiya,* 40, 896, 1971 (in Russian).
85. Widdel, F. and Pfennig, N., Sporulation and further nutritional characteristics of *Desulfotomaculum acetoxidans*, *Arch. Microbiol.,* 129, 401, 1981.
86. Duda, V.I. and Makar'eva, E.D., Morphogenesis and function of gas caps on spores of anaerobic bacteria of the genus *Clostridium*, *Mikrobiologiya,* 46, 689, 1977 (in Russian).
87. Zhilina, T.A. and Zavarzin, G.A., Alkaliphilic anaerobic community at pH 10, *Curr. Microbiol.,* 29, 109, 1994.
88. Garnova, E.S. and Krasil'nikova, E.N., Carbohydrate metabolism of the saccharolytic alkaliphilic anaerobes *Halonatronum saccharophilum*, *Amphibacillus fermentum*, and *Amphibacillus tropicus*, *Microbiology (English translation of Mikrobiologiya),* 72, 558, 2003.
89. Mijts, B.N. and Patel, B.K.C., Cloning, sequencing and expression of an α-amylase gene, *amyA*, from the thermophilic halophile *Halothermothrix orenii* and purification and biochemical characterization of the recombinant enzyme, *Microbiology U.K.,* 148, 2343, 2002.
90. Li, N. et al., Crystallization of an α-amylase, AmyA, from the thermophilic halophile *Halothermothrix orenii*, *Acta Crystallogr.,* D58, 2125, 2002.
91. Mijts, B.N. and Patel, B.K.C., Random sequence analysis of genomic DNA of an anaerobic, thermophilic, halophilic bacterium, *Halothermothrix orenii*, *Extremophiles,* 5, 61, 2001.
92. Senyushkin, A.A., Severina, L.O., and Zhilina, T.N., Influence of the environmental conditions on glucose transport in cells of halophilic anaerobic eubacteria of the genus *Halobacteroides*, *Microbiology (English translation of Mikrobiologiya),* 60, 545, 1992.
93. Severina, L.O., Senyushkin, A.A., and Zhilina, T.N., Glucose uptake systems in halophilic anaerobic eubacteria of the genus *Halobacteroides*, *Microbiology (English translation of Mikrobiologiya),* 61, 237, 1992.
94. Bolobova, A.V., Simankova, M.C., and Markovich, N.A., Cellulase complex of a new halophilic bacterium *Halocella cellulolytica*. *Microbiology (English translation of Mikrobiologiya),* 61, 557, 1992.
95. Siman'kova, M.V. and Zavarzin, G.A., Anaerobic decomposition of cellulose in lake Sivash and hypersaline lagoons of Arabat spit, *Microbiology (English translation of Mikrobiologiya),* 61, 193, 1992.
96. Kevbrin, V.V., Zhilina, T.N., and Zavarzin, G.A., Physiology of the halophilic homoacetic bacterium *Acetohalobium arabaticum*, *Microbiology (English translation of Mikrobiologiya),* 64, 134, 1995.
97. Pusheva, M.A. and Detkova, E.N., Bioenergetic aspects of acetogenesis on various substrates by the extremely halophilic acetogenic bacterium *Acetohalobium arabaticum*. *Microbiology (English translation of Mikrobiologiya),* 65, 516, 1996.
98. Pusheva, M.A. et al., Properties of periplasmatic hydrogenase of *Acetohalobium arabaticum*, an extremely halophilic homoacetogenic bacterium, *Microbiology (English translation of Mikrobiologiya),* 61, 653, 1992.
99. Pusheva, M.A., Bioenergetics of acetogenesis in the extremely alkaliphilic homoacetogenic bacteria *Natroniella acetigena* and *Natronoincola histidinovorans*, *Microbiology (English translation of Mikrobiologiya),* 68, 568, 1999.
100. Switzer Blum, J. et al., *Selenihalanaerobacter shriftii* gen. nov., sp. nov., a halophilic anaerobe from Dead Sea sediments that respires selenate, *Arch. Microbiol.,* 175, 208, 2001.
101. Zhilina, T.N., *Halobacteroides lacunaris* sp. nov., new saccharolytic, anaerobic, extremely halophilic organism from the lagoon-like hypersaline lake Chokrak, *Microbiology (English translation of Mikrobiologiya),* 60, 495, 1992.
102. Kevbrin, V.V. and Zavarzin, G.A., Methanethiol utilization and sulfur reduction by anaerobic halophilic saccharolytic bacteria, *Curr. Microbiol.,* 24, 247, 1992.
103. Kevbrin, V.V. and Zavarzin, G.A., Effect of sulfur compounds on the growth of the halophilic homoacetogenic bacterium *Acetohalobium arabaticum*, *Microbiology (English translation of Mikrobiologiya),* 61, 563, 1992.

104. Oren, A., Gurevich, P., and Henis, Y., Reduction of nitrosubstituted aromatic compounds by the halophilic eubacteria *Haloanaerobium praevalens* and *Sporohalobacter marismortui*, *Appl. Environ. Microbiol.,* 57, 3367, 1991.
105. Oren, A. et al., Microbial degradation of pollutants at high salt concentrations, *Biodegradation,* 3, 387, 1992.
106. Oren, A., Intracellular salt concentrations of the anaerobic halophilic eubacteria *Haloanaerobium praevalens* and *Halobacteroides halobius*, *Can. J. Microbiol.,* 32, 4, 1986.
107. Oren, A., Heldal, M., and Norland, S., X-ray microanalysis of intracellular ions in the anaerobic halophilic eubacterium *Haloanaerobium praevalens*, *Can. J. Microbiol.,* 43, 588, 1997.
108. Rengpipat, S., Lowe, S.E., and Zeikus, J.G., Effect of extreme salt concentrations on the physiology and biochemistry of *Halobacteroides acetoethylicus*, *J. Bacteriol.,* 170, 3065, 1988.
109. Oren, A. and Gurevich, P., The fatty acid synthetase complex of *Haloanaerobium praevalens* is not inhibited by salt, *FEMS Microbiol. Lett.,* 108, 287, 1993.
110. Lowe, S.E., Jain, M.K., and Zeikus, J.G., Biology, ecology, and biotechnological applications of anaerobic bacteria adapted to environmental stresses in temperature, pH, salinity, or substrates, *Microbiol. Rev.,* 57, 451, 1993.
111. Wise, D.L., Meeting report — First international workshop on biogasification and biorefining of Texas lignite, *Res. Conserv.,* 15, 229, 1984.
112. Trüper, H.G. and de'Clari, L., Taxonomic note: erratum and correction of further specific epithets formed as substantives (nouns) "in apposition," *Int. J. Syst. Bacteriol.,* 48, 615, 1998.

34 Syntrophism among Clostridiales

Caroline M. Plugge and Alfons J.M. Stams

CONTENTS

34.1 OUTLINE

The ability to transfer electrons to a partner organism is an important metabolic feature associated with many physiologically diverse microorganisms. This trait is usually referred to as syntrophism. Syntrophism is a special type of symbiosis between two microorganisms, where the growth of one organism is dependent or highly improved by growth factors, nutrients, or substrates, either removed or provided by the partner organism. Especially among anaerobic microorganisms, cooperation of several metabolic types of bacteria in the food chain is important. Knowledge of the occurrence, diversity, and distribution of these syntrophic consortia is steadily increasing [1,2,3]. Also among Clostridiales, many species can grow in cooperation with syntrophic partners. In this chapter, we review important interspecies electron transfer-based syntrophic conversions and describe the microorganisms involved. Some of the organisms are capable of syntrophic growth on a variety of compounds. In this case, we describe the best-studied conversion. Detailed information on the biochemistry and bioenergetics of syntrophic organisms can be found elsewhere [2,3]. There are probably many other members of the Clostridiales with the ability to grow syntrophically. However, such capacities are only rarely checked for.

34.2 INTRODUCTION

34.2.1 The Concept of Syntrophism in the Anaerobic Food Chain

Complete conversion of high molecular weight organic matter (such as cellulose, polysaccharides, and protein) to methane and carbon dioxide is catalyzed by the combined action of different metabolic groups of bacteria, including fermentative microorganisms, acetogenic bacteria, and methanogenic archaea [1,3,4,5]. The first group in the food chain consists of a variety of fermentative bacteria (Figure 34.1). They excrete hydrolytic enzymes that degrade the biopolymers to their corresponding monomeric substances. Subsequently, these compounds are fermented to reduced organic compounds (short-chain fatty acids, alcohols, lactate, and alanine), and to compounds that can be used directly by methanogenic archaea (acetate, formate, CO_2, and H_2). The reduced compounds are further oxidized to acetate, formate, CO_2, and H_2 by acetogenic bacteria, the second trophic group. Ultimately, a third group, which consists of methanogenic archaea, metabolizes acetate, formate, and CO_2, plus H_2, to form CH_4. These groups of microorganisms have strong metabolic interactions; the H_2 removal by the methanogens enables a faster metabolism of the fermentative and acetogenic bacteria. Especially the acetogenic bacteria are strongly dependent on a low hydrogen partial pressure to catabolize their substrates. In extreme cases, the acetogenic bacterium depends entirely on cooperation with a methanogen to metabolize.

The first example of a syntrophic conversion was described by Bryant and co-workers [6]. They discovered that the previously isolated pure culture of an ethanol-oxidizing methanogen, *Methanobacterium omelianskii* [7,8], consisted of two microorganisms: one organism termed the "S-organism," which oxidized ethanol to acetate and H_2 and a second organism, *Methanobacterium bryantii*, which converted H_2 and CO_2 to CH_4. The "S-organism" was found to be strictly dependent on the presence of *M. bryantii* for growth.

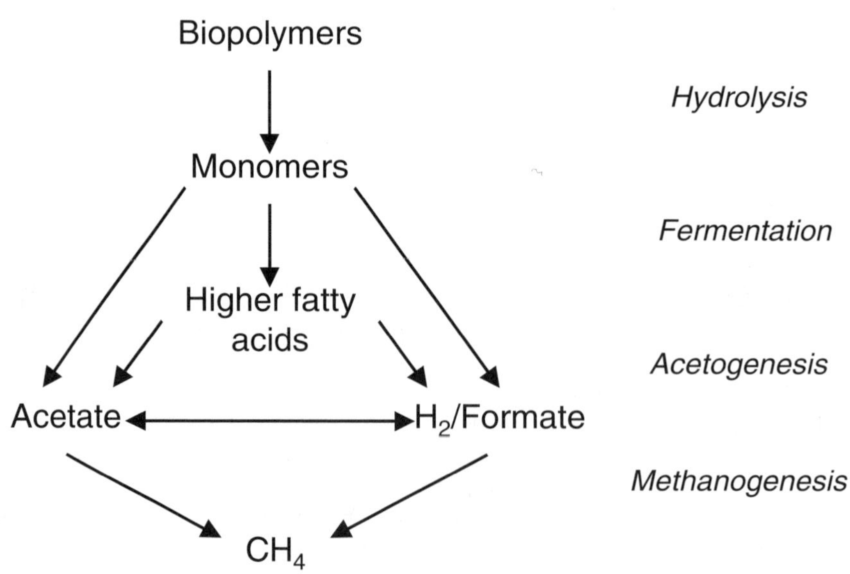

FIGURE 34.1 Reaction scheme for the anaerobic degradation of organic matter to methane. (Adapted from Gujer and Zehnder [98].)

The dependency can be explained using the changes in Gibbs free energy ($\Delta G^{0\prime}$) for the oxidation of ethanol to H_2, CO_2, and acetate. Under defined standard conditions with gases at 10^5 Pa pressure, 1 *M* of products/substrates, a pH of 7, and 298 K the Gibbs free energy value for ethanol oxidation is positive: +9.6 kJ/reaction (Table 34.2). This indicates that the reaction cannot take place, let alone that any microbe can gain energy from this oxidation. However, the Gibbs free energy becomes negative when the hydrogen partial pressure (pH_2) is kept low. This first example of interspecies hydrogen transfer is characteristic for the way organic matter is degraded under methanogenic conditions.

The majority of the acetogens in the anaerobic food web that oxidize, e.g., butyrate, propionate, or ethanol, cannot grow without the presence of a methanogen, or another hydrogen scavenger, which keeps the pH_2 at a low level, typically 1 to 100 Pa. These acetogens form obligate syntrophic associations with methanogens. In addition, cocultivation of fermentative hydrogen-producing bacteria with methanogens often results in a shift in product formation. Several textbook examples are known to illustrate this phenomenon [9,10]. When the fermentative bacteria *Clostridium thermocellum* or *Ruminococcus albus* are cultivated in pure culture on glucose acetate, hydrogen and reduced end-products (lactate and ethanol) are formed (Figure 34.2).

At a high hydrogen partial pressure, the bacteria are no longer able to form hydrogen through NADH or ferredoxin oxidation. These components are common redox mediators in these bacteria. The redox couples NAD/NADH and $Fd_{(ox)}/Fd_{(red)}$ are higher than the redox couple H^+/H_2 (Table 34.1). This means that the electron transfer from the electron mediators to protons is energetically unfavorable. Only when the hydrogen partial pressure is kept low, the reactions can occur. The effect of the substrate and product concentration on the Gibbs free energy change is given by:

$$\Delta G' = \Delta G^{0\prime} + RT \ln \left(\left[\text{products}\right]/\left[\text{reactants}\right]\right) \tag{34.1}$$

$\Delta G'$ is the change in Gibbs free energy under conditions where substrate and product concentrations do not equal 1 *M* or 10^5 Pa. $\Delta G^{0\prime}$ is the change in Gibbs' free energy under standard conditions and R is the gas constant (8.31 kJ * mol^{-1}* K^{-1}). $\Delta G^{0\prime}$ can be calculated from the standard redox couples according to:

$$\Delta G^{0\prime} = -n * F * \Delta E^{0\prime} \tag{34.2}$$

where n is the number of electrons transferred; F is the Faraday constant, and $\Delta E^{0\prime}$ is the difference between the redox couples. In the case of Fd and NADH oxidation coupled to proton reduction, the reactions are as follows:

$$2\text{ Fd (red)} + 2\text{ H}^+ \rightarrow 2\text{ Fd (ox)} + \text{H}_2 \quad \Delta G^{0\prime} = +3.1\text{ kJ/mole} \tag{34.3}$$

$$\text{NADH} + \text{H}^+ \rightarrow \text{NAD}^+ + \text{H}_2 \quad \Delta G^{0\prime} = +18.1\text{ kJ/mole} \tag{34.4}$$

Below a partial pressure of H_2 of 100 Pa, it becomes energetically favorable to convert NADH to NAD plus H_2.

The reducing equivalents formed by *Clostridium thermocellum* growing on glucose are disposed directly upon acetyl-CoA or pyruvate to form ethanol and lactate, respectively (Figure 34.2). When these bacteria are cultivated in the presence of *Methanothermobacter thermoautotrophicus*, a hydrogen-consuming organism, less of these reduced products is formed, because ferredoxin and NADH oxidation coupled to proton reduction can proceed. Consequently, more acetate, and coupled to this, more ATP is formed.

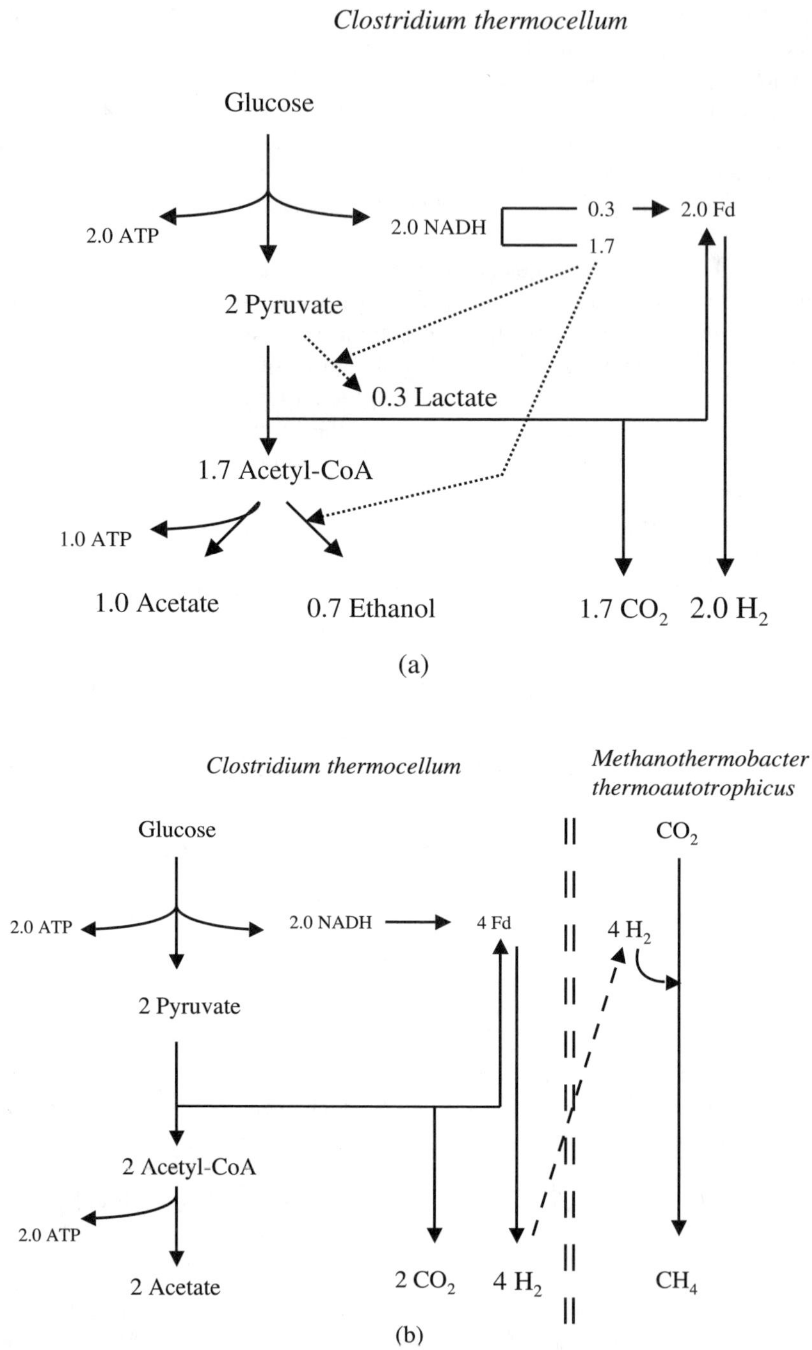

FIGURE 34.2 Fermentation of glucose by *Clostridium thermocellum* in pure culture (a) and in coculture with *Methanothermobacter thermoautotrophicus* (b).

34.2.2 Ecology and Distribution of Syntrophic Prokaryotes

Syntrophic organisms are widespread in nature and human-made systems. They occur in many ecosystems and play a significant role in the anaerobic mineralization of organic matter. There is hardly any transformation process in an anoxic environment in which syntrophic interactions do

TABLE 34.1
Redox Potentials of Some Redox Couples Involved in Interspecies Electron Transfer

Redox Couple	$E^{0\prime}$ (mV)
CO_2/formate	–432
H^+/H_2	–414
Fd (ox)/Fd (red)	–398
$NAD(P)^+/NAD(P)H$	–320
$FAD/FADH_2$	–220

not take place. Syntrophic organisms are most often found in ecosystems with high biomass densities, such as anoxic lake sediments and anaerobic bioreactors, since they are dependent on the physical presence of their partner organism. An important factor in syntrophic interactions is the distance between the hydrogen-producing and the hydrogen-consuming organism. Schink and Thauer [11] describe that the diffusion distances for the transfer of metabolites should be as short as possible. The diffusion of hydrogen from producer to consumer can be described by a simple equation: $Flux_{H2} = -A*D* (c_2 - c_1)* d^{-1}$ $mol*sec^{-1}$ (A is the total surface area of the hydrogen producer ($4\pi r^2 \times$ number of hydrogen producers), D is the diffusion constant for hydrogen ($4.9*10^{-5}$ cm^2*sec^{-1} at 298 K), c is the concentration of hydrogen in water and d is the distance between producer and consumer). Aggregation of bacteria from different metabolic groups enables a faster degradation of complex organic material. This is the case in p.e. high rate anaerobic bioreactors, where the biomass is immobilized in granules (Figure 34.3). The upflow anaerobic sludge blanket (UASB) process has been widely used

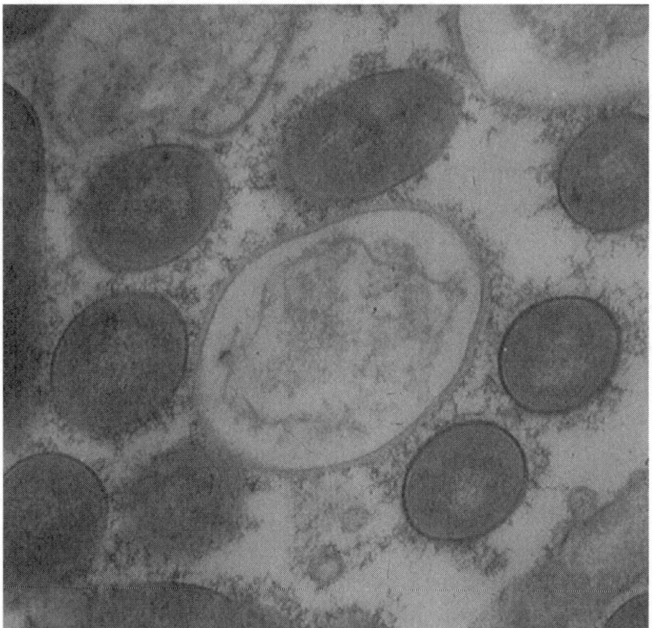

FIGURE 34.3 Transmission electron micrograph of an ultrathin section of a granule from an anaerobic bioreactor, showing a syntrophic bacterium in the center, surrounded by methanogenic archaea. (The image was provided by Tim Grotenhuis.)

over the past decades mainly for the treatment of medium- and high-strength organic wastewaters. The granulation of sludge is a very interesting phenomenon. Granulation forms dense spherical aggregates, in which several metabolic groups of microbes reside for complete mineralization of organic matters. In addition, within the granules, close interactions among the microorganisms linked to the food web in anaerobic organic degradation take place, including syntrophic interactions.

Furthermore, in more extreme environments such as haloalkaliphilic soda lakes, hydrogen-forming microbes have been identified [12], suggesting the potential presence of syntrophic processes in these environments. Zhilina and Zavarzin [13] speculate on the presence of syntrophic conversions in soda lakes, based on missing links in the overall degradation scheme of organic matter. In hyperthermophilic habitats, the significance of syntrophic conversions has been the subject of a limited number of studies. The growth of *Thermotoga maritima* in pure culture is inhibited by the hydrogen it generates [14]. This has also been demonstrated for *Pyrococcus furiosis* [15,16]. Muralidharan and co-workers [17] describe that a coculture of *Thermotoga maritima* and *Methanococcus jannaschii* had a significant higher yield when growing on glucose than the sum of the yields of pure cultures of the same strains. Presumably in their natural environments, mixed communities of these organisms optimize the production and consumption of molecular hydrogen.

34.2.3 Taxonomy of Syntrophic Organisms from the Clostridiales Group

Based on their 16S rRNA gene sequence similarities, the species belonging to the low GC Gram-positive Clostridiales are not tightly clustered phylogenetically, but are dispersed throughout the whole group. The trait to transfer electrons to a syntrophic partner is not restricted to a narrow group within the low GC Gram-positive bacteria (e.g., many different obligate and facultative syntrophic organisms within taxonomically different groups can perform fermentative degradation of amino acids). On the other hand, the trait to oxidize butyrate is restricted to a phylogenetically closely related group of organisms. Nevertheless, it can hardly be assumed that syntrophy is associated with a single evolutionary trait.

34.3 SYNTROPHIC DEGRADATION OF ORGANIC COMPOUNDS

34.3.1 Acetate Oxidation

Acetate oxidation to CO_2 and H_2 is energetically one of the most unfavorable reactions. It requires the input of 104.6 kJ/mole acetate (Table 34.2). Only in the presence of hydrogen scavengers are microorganisms able to perform this reaction. Metabolically, this syntrophic association consists of two reactions that were originally proposed by Barker [18]. In the first reaction, acetate is oxidized to form hydrogen and CO_2, which are, in the second reaction, converted to CH_4.

To date, two syntrophic acetate oxidizing Clostridiales growing in association with hydrogenotrophic methanogens are known. The first description was strain AOR, which was a Gram-negative thermophilic acetate oxidizer isolated from a methanogenic reactor in coculture with a *Methanobacterium* sp. [19]. The isolate was later found to be a homoacetogen, which forms acetate from hydrogen and CO_2 in pure culture, whereas acetate oxidation occurs in coculture with the methanogen [20,21,22]. The phylogenetic position of the thermophilic strain AOR is not known. Moreover, the isolate has not been deposited in any culture collection and is no longer available. The first Gram-positive acetate-degrading syntroph described was *Clostridium ultunense*, a mesophile, which oxidizes acetate in the presence of hydrogenotrophic methanogens [23,24]. The second species is *Thermoacetogenium phaeum* [25], a thermophilic strain that has been isolated from an anaerobic reactor. One of the features that distinguishes *T. phaeum* from the other two strains is that it can oxidize acetate in pure culture by using sulfate as the electron acceptor. This is also a property of thermophilic sulfate-reducing bacteria belonging to the genus *Desulfotomaculum*. *T. phaeum* is phylogenetically affiliated with the Gram-positive low G-C group, which also includes

TABLE 34.2
Hydrogenogenic Reactions Involved in Syntrophic Conversions Performed by Clostridiales and the Gibbs Free Energy Changes[a]

Reaction	$\Delta G^{0\prime}$ kJ/Reaction
$Acetate^- + 4\ H_2O \rightarrow 4\ H_2 + 2\ HCO_3^- + H^+$	+104.6
$Propionate^- + 3\ H_2O \rightarrow 3\ H_2 + acetate^- + HCO_3^- + H^+$	+76.1
$Benzoate^- + 7\ H_2O \rightarrow 3\ H_2 + 3\ acetate^- + HCO_3^- + H^+$	+58.9
$Butyrate^- + 2\ H_2O \rightarrow 2\ H_2 + 2\ acetate^- + H^+$	+48.1
$Methanol + 2\ H_2O \rightarrow 3\ H_2 + HCO_3^- + H^+$	+23.5
$Glycolate^- + 3\ H_2O \rightarrow 3\ H_2 + 2\ HCO_3^- + H^+$	+19.3
$Ethanol + H_2O \rightarrow 2\ H_2 + acetate^- + H^+$	+9.6
$Glutamate^- + 4\ H_2O \rightarrow 2\ H_2 + propionate^- + 2\ HCO_3^- + NH_4^+ + H^+$	–5.8
$Glutamate^- + 3\ H_2O \rightarrow H_2 + 2\ acetate^- + HCO_3^- + H^+ + NH_4^+$	–33.9
$Glucose + 4\ H_2O \rightarrow 4\ H_2 + 2\ acetate^- + 2\ HCO_3^- + 4\ H^+$	–206.3

[a] Data from Thauer et al. [99].

Desulfotomaculum species. Biochemical studies on the mechanism of interspecies electron transfer in syntrophic acetate oxidation by *T. phaeum* revealed that not only hydrogen, but also formate was involved [25]. Moreover, indications for syntrophic acetate conversions have been found in lake sediments [26,27].

34.3.2 Propionate Oxidation

The degradation of propionate under anaerobic conditions is energetically very unfavorable (Table 34.2). Under standard conditions, the formation of acetate, CO_2, and H_2 requires the input of 76.1 kJ/mole propionate. In methanogenic environments, syntrophic consortia are required to degrade propionate. Although many well-studied syntrophic propionate-degrading organisms are Gram negative [28,29,30,31], there are a few syntrophic propionate oxidizers discovered recently that belong to the Clostridiales. *Desulfotomaculum thermobenzoicum* subsp. *thermosyntrophicum* was the first Gram-positive propionate oxidizer described [32]. This organism is a moderate thermophilic propionate-oxidizing syntroph isolated from an anaerobic methanogenic reactor. It degrades propionate to acetate CO_2 and H_2 via the methylmalonyl-CoA pathway, a pathway used by the majority of propionate-degrading microorganisms [29,33,34]. Recently, the mesophilic *Pelotomaculum schinkii* [35,36] and the thermophilic *Pelotomaculum thermopropionicum* [37] were described. They metabolize propionate in a similar fashion as *D. thermobenzoicum* subsp. *thermosyntrophicum*. In an axenic culture, the thermophilic propionate oxidizers could grow on several substrates, such as pyruvate and fumarate. Moreover, *D. thermobenzoicum* subsp. *thermosyntrophicum* is able to couple propionate oxidation to sulfate reduction. Substrates like pyruvate, fumarate, or propionate plus fumarate have been used to obtain axenic culture of these syntrophs. They bypass the energetically unfavorable steps in propionate oxidation [29,38]. *Pelotomaculum schinkii*, however, could not be obtained until today in a pure culture, nor could it grow on any other compound than propionate. This makes this organism the first true obligate syntrophic organism.

34.3.3 Butyrate and Long-Chain Fatty Acids

The oxidation of butyrate and long-chain fatty acids (C4–C18) is performed by members belonging to the family of Syntrophomonadaceae. The Syntrophomonadaceae are part of the phylum Clostridiales with low G + C content. The trait to oxidize butyrate can be found in a limited number of genera:

Thermosyntropha, Syntrophomonas, Syntrophospora, Pelospora, and *Syntrophothermus* [39]. The first butyrate-oxidizing organism isolated is *Syntrophomonas wolfei* [40].

The members of this family are found in environments where no terminal electron acceptor is present and methanogenesis is the predominant process. Moreover, members of the Syntrophomonadaceae grow in obligate association with methanogens, since the degradation of fatty acids to acetate and hydrogen is endergonic under standard conditions (Table 34.2). The methanogens keep the hydrogen partial pressure lower than 10 Pa. Hydrogen partial pressures and acetate concentrations in active sewage sludge and various sediments [41] are in the range of 1–10 Pa and 10–100 μ*M*, respectively. When these organisms were first isolated, no mechanism other than syntrophy was known for the growth of these organisms [42]. However, in the late 1980s, it was found that these obligate syntrophs were capable of crotonate dismutation [43]. This metabolic feature made it possible to grow the organisms in pure culture and study their biochemistry. *Pelospora glutarica* cannot grow on crotonate, but grows on succinate and glutarate in pure culture [44]. Crotonate is metabolized by dismutation; one molecule of crotonate is oxidized to 2 acetate molecules and a second crotonate molecule is reduced to butyrate [45]. The use of crotonate bypasses the energetically unfavorable step of butyryl-CoA to crotonyl-CoA.

The pathway of butyrate oxidation in syntrophic butyrate-oxidizing bacteria has been elucidated to a certain extent for *Syntrophomonas wolfei* [46]. The pathway involves a fatty acid β-oxidation via acetoacetyl CoA.

Syntrophic oxidation of long-chain fatty acids from lipid hydrolysis probably proceeds via β-oxidation with concomitant release of electrons as hydrogen via reversed electron transport, analogous to butyrate oxidation.

34.3.4 Alcohol Oxidation

34.3.4.1 Ethanol Oxidation

Syntrophism was demonstrated first in an ethanol oxidizing co-culture [6], as described in the introduction. The case of "*Methanobacillus omelianskii*" is *the* classical example of interspecies hydrogen transfer. After description of the cooperative nature of this process, the original S-strain was lost, but other syntrophically ethanol-oxidizing Clostridiales have been isolated, such as *Thermoanaerobium brockii* [47], *Clostridium magnum* [48], and *Clostridium* EE121 [49].

C. magnum and *Clostridium* EE121 could ferment ethanol to acetate using bicarbonate as an external electron acceptor [49,50]. The stoichiometry of ethanol degradation showed that a pure culture of EE121 converted 1 mole of ethanol to 1.5 moles of acetate. One mole of acetate was directly formed by oxidation of ethanol and another 0.5 mole was formed from the reduction of one mole bicarbonate. When strain EE121 was cultivated in coculture with a methanogen, a shift in product formation was found. Since homoacetogens are less efficient hydrogen scavengers than methanogens, all reducing equivalents were used for methanogenesis, which resulted in the production of 1 mole of acetate and 0.5 mole of methane from one mole of ethanol.

34.3.4.2 Methanol Oxidation

Interspecies hydrogen transfer in cocultures with methanol as a substrate has been observed in a limited number of studies. Davidova and Stams [51] report methanol conversion by a thermophilic consortium. However, the identity of the microorganisms in the coculture remains unclear. Methanol was converted by a coculture of the thermophilic strains *Desulfotomaculum kutznetzovii* and *Methanothermobacter thermoautotrophicus* strain ΔH [52]. Balk and co-workers [53] describe a thermophilic homoacetogenic bacterium, *Moorella mulderi,* that can grow on a variety of compounds. A coculture of *M. mulderi* and a methanogen converts methanol to methane and CO_2.

34.3.5 Sugar Oxidation

Saccharolytic clostridia are ubiquitous in nature and have been isolated from a large variety of ecosystems. They play an important role in the initial breakdown of polymeric plant material using exoenzymes such as pectinases, cellulases, and xylanases. They grow in most environments when the oxygen partial pressure is sufficiently low. Most species require very few growth factors and, by their ability to form spores, they survive longer periods of low nutrient availability.

Saccharolytic clostridia grow on a large variety of carbohydrates; producing a mixture of acids as reduced end-products. Most species convert glucose to acetate according to the following reaction:

$$C_6H_{12}O_6 + 2\ H_2O \rightarrow 2\ CH_3COOH + 2\ CO_2 + 4\ H_2 \qquad (34.5)$$

Acetate and hydrogen are the major fermentation products formed by saccharolytic clostridia. However, the amounts produced are usually lower than predicted from the reaction equation. Part of the reducing equivalents produced is used for the production of secondary fermentation products such as ethanol, butyrate, and lactate. Although the fermentative bacteria can generally grow without a syntrophic partner, additional energy is obtained from acetate production. When these saccharolytic bacteria are cultivated in the presence of electron scavengers, such as methanogens, acetate production becomes more predominant and one additional ATP is produced from acetyl-CoA (Figure 34.2).

Syntrophococcus sucromutans was isolated from rumen contents as the predominant utilizer of methoxylated aromatics [54]. However, it requires the presence of sugar as electron donor to convert the methoxylated compound. It can use a variety of sugars as electron donor, but is depending on the presence of hydrogen-utilizing organisms as an external electron acceptor. Furthermore, it can use formate and methoxybenzenoids as external electron acceptor in pure culture. Although *S. sucromutans* stains Gram negative, its phylogenetic position is within a Gram-positive phylum that contains Gram-negative organisms with a Gram-positive cell wall structure.

Newly isolated saccharolytic clostridia are not often tested for their ability to transfer electrons to syntrophic partners. However, when they are growing in environments with high biomass density such as biofilms and sludge granules it can be speculated that these bacteria metabolize sugars with the help of syntrophic partners. This allows them to grow with a maximum energy yield.

34.3.6 Amino Acid Oxidation

Amino acids, the hydrolysis products of proteins, can serve as carbon and energy sources for many types of anaerobic bacteria. Utilization of amino acids has traditionally been used as a classification characteristic for clostridia [55,56]. Mixtures of amino acids are often degraded by coupled fermentation of pairs of amino acids via the Stickland reaction [5]. In methanogenic environments, methanogens can act as scavengers of reducing equivalents in the oxidation of amino acids, taking over the role of the reductive part of the Stickland reaction. Nagase and Matsuo [57] observed that in mixed methanogenic communities, the degradation of alanine, valine, and leucine was inhibited when methanogens were inhibited. Nanninga and Gottschal [58] could stimulate the degradation of these amino acids by addition of hydrogen-scavenging sulfate-reducers. These early observations indicated that syntrophic interactions play a significant role in the degradation of amino acids. Some amino acids are degraded by the same organism, either in pure culture or in coculture with methanogens; however, methanogens can cause a shift in metabolism resulting in a changed product formation pattern and an increased growth rate and growth yield. *Clostridium sporogenes* [59,60], *Eubacterium acidaminophilum* [61], *Acidaminobacter hydrogenoformans* [62], *Aminomonas paucivorans* [63], *Aminobacterium colombiense,* and *Aminobacterium mobile* [64,65], as well as strain

PA-1 [66] are mesophilic Clostridiales that degrade one or more amino acids in syntrophic association with hydrogen consumers. Moderately thermophilic Clostridiales include *Caloramator coolhaasii* [67], *Caloramator proteoclasticus* [68,69], *Thermanaerovibrio acidaminovorans* [70,71], *Gelria glutamica* [72], and *Clostridium* P2 [73,74]. There are probably many other Clostridiales with the ability to grow syntrophically with certain amino acids and to release reducing equivalents as molecular hydrogen.

34.3.6.1 Glutamate Oxidation

The effect of methanogens on the metabolism of amino acid-fermenting anaerobes has been studied best with glutamate. Glutamate fermentation is carried out by a variety of anaerobes, including a number of *Clostridium* species, *Peptostreptococcus asaccharolyticus,* and *Acidaminococcus fermentans* [75]. These microorganisms ferment glutamate to acetate and butyrate by either the β-methylaspartate or the hydroxyglutarate pathway [76]. In this fermentation, reducing equivalents formed in the oxidation of glutamate to acetate are disposed of either partly or completely by reductive formation of butyrate from acetate (acetyl-CoA). *Anaeromusa acidaminophila* ferments glutamate to acetate plus propionate [70,77]. In this bacterium, reducing equivalents are disposed of by reduction of pyruvate to propionate.

In the last decade, several bacteria have been isolated that, during growth on glutamate, release reducing equivalents exclusively as hydrogen, both in the formation of acetate and the formation of propionate. Microorganisms that ferment glutamate to acetate only include *C. coolhaasii* [67] and *C. proteoclasticus* [69]. Only propionate is formed from glutamate by *Aminobacterium colombiense* [64] and *Gelria glutamica* [72]. *A. hydrogenoformans* [62,78], *T. acidaminovorans* [70,71], and *Aminomonas paucivorans* [63] form both acetate and propionate from glutamate.

34.3.6.2 Other Amino Acids

Acidaminobacter hydrogenoformans, *Aminomonas paucivorans,* and *T. acidaminovorans* can grow with histidine, ornithine, arginine, lysine, and threonine when cocultured with hydrogenotrophic bacteria [62,70,71]. Remarkably, *T. acidaminovorans* and *A. paucivorans* grow in pure culture with arginine, forming citrulline or ornithine as products. In this conversion, carbamyl phosphate is formed as an intermediate that is further converted to yield ATP [79]. Ornithine and citrulline could be degraded when these bacteria were cocultured with a hydrogenotrophic methanogen. It is not clear which pathway is used for ornithine degradation and which metabolic step is hampering growth of these bacteria in pure culture.

A. hydrogenoformans degrades threonine and lysine in syntrophic association with a hydrogen-consuming anaerobe. Acetate is the main product of the degradation of these two amino acids, and large amounts of hydrogen are formed. In pure culture, these compounds do not support growth. The stoichiometry of the fermentation is not exactly known, and the pathways involved in these fermentations have not been resolved yet.

34.3.7 Oxidation of Aromatics

Aromatic, polyaromatic, and chlorinated aromatic compounds are widely used in industries, and these chemicals are often discharged in industrial wastewater. The toxicity of these compounds requires that they are removed from the waste streams. However, under anoxic conditions these components are very persistent and hard to degrade. Degradation of benzoate and related compounds has been found in anaerobic methanogenic bioreactors [80,81,82,83]. The anaerobic methanogenic degradation of benzoate occurs mainly via interspecies hydrogen transfer, since the degradation requires the input of 58.9 kJ/mole benzoate under standard conditions (Table 34.2). Certain anaerobes can utilize benzoate fermentatively in pure culture [32,84]. Some aspects of anaerobic benzoate degradation have been studied, especially the biochemistry and bioenergetics [2,84,85]. However, little is known about the species diversity and ecology of the bacteria that anaerobically degrade benzoate, due to the difficulties of growing these bacteria in pure or defined cultures. Three Gram-negative species, belonging to the genus *Syntrophus,* have been described to date, and only one species belonging to the Low GC Gram-positive lineage *Sporotomaculum syntrophicum* [86]. *S. syntrophicum* degrades benzoate to acetate and CO_2, with the intermediate formation of butyrate. The strain was not capable of growing on compounds other than benzoate. In pure culture, it was capable of crotonate fermentation. Crotonate was fermented to acetate and butyrate, similar to *Syntrophomonas* species [43].

34.3.8 Miscellaneous Compounds

Glycolate is an important excretion product of algal cells. It is syntrophically oxidized by homoacetogenic and methanogenic cocultures to 2 CO_2, and hydrogen is the electron carrier between the fermenting bacterium *Syntrophobotulus glycolicus* and a hydrogen-oxidizing partner bacterium [87,88]. The primary conversion is an endergonic process under standard conditions (Table 34.2) and needs coupling to a hydrogen scavenger. In the degradation pathway, the oxidation of glycolate to glyoxylate is the critical oxidation step [87].

A special situation is the fermentative conversion of acetone to methane and CO_2, which is performed by syntrophic consortia as well. In this case, acetate is the only intermediate between both partners, as illustrated by the following equations:

$$CH_3COCH_3 + HCO_3^- \rightarrow 2\ CH_3COO^- + H+ \qquad \Delta G^{0\prime} = -25.8\ \text{kJ/mol} \qquad (34.6)$$

$$2\ CH_3COO^- + 2\ H_2O \rightarrow 2\ CH_4 + 2\ HCO_3^- \qquad \Delta G^{0\prime} = -71.8\ \text{kJ/mol} \qquad (34.7)$$

Overall reaction:

$$CH_3COCH_3 + H_2O \rightarrow 2\ CH_4 + HCO_3^- \qquad \Delta G^{0\prime} = -97.6\ \text{kJ/mol} \qquad (34.8)$$

Although all partial reactions are exergonic under standard conditions, the primary fermenting bacterium depends on the methanogenic partner, and acetone degradation in the mixed culture is inhibited in the presence of acetylene as an inhibitor of methanogens [89]. Experiments with the primary acetone-fermenting bacterium in dialysis cultures revealed that acetate accumulation inhibited growth and acetone degradation at concentrations higher than 10 m*M* [90]. Since acetone metabolism by these bacteria starts with an endergonic carboxylation reaction, this might be the amount of energy they need to invest into this primary substrate activation reaction. Unfortunately, the acetone-fermenting bacterium was never obtained in pure culture and, hence, detailed studies were never performed.

34.4 FUTURE PERSPECTIVES

Over the past 35 years, after the discovery of Bryant and co-workers [6], syntrophic populations within anaerobic methanogenic ecosystems have been intensively studied, mainly with conventional techniques such as methane-producing activity, most probable number (MPN)-counting, and conventional culturing techniques. The internal architecture of anaerobic sludge granules has been also surveyed with methods such as scanning electron microscopy and immunohistochemical techniques [91,92]. Nevertheless, only a fraction of the whole microbial community structure has been clarified so far; this is largely attributed to the limitations of the conventional techniques used. An extra complicating factor is that hydrogen-producing bacteria require the presence of methanogens.

Recent cultivation-independent molecular approaches have overcome some drawbacks of these techniques, and small subunit ribosome DNA (rDNA) and rRNA-based molecular techniques are good tools for microbial community structure analysis. In recent years, these molecular techniques have been applied to methanogenic bioreactors, and several new findings have been reported [93,94]. The molecular-based microbial community structure analyses revealed biodiversity and detailed spatial organization of microbes of interest in sludge granules, and conventional cultivation techniques together with molecular techniques could enlarge our fundamental knowledge of the functions of significant syntrophic populations in bioreactors.

The drawback of rDNA approaches is that phylogeny cannot be linked to physiology and function. Interest in establishing the function and identity of syntrophs without their cultivation has seen a recent rapid increase in the use of many techniques. The use of ^{13}C-labeled substrates unraveled the function of syntrophs in complex ecosystems by identifying labeled metabolites and metabolic pathways used by the targeted organisms under conditions that approach those *in situ* [95].

Several recent publications describe the use of "heavy" ^{13}C-labeled growth substrates to link microbial function with identity and phylogeny *via* selective recovery of "heavy" ^{13}C-labeled RNA [96,97]. The use of this technique in combination with those of classical and molecular microbiology might offer clearer insight into the identity and metabolic function of syntrophs in the environment.

ACKNOWLEDGMENTS

This work was supported by the Research Council for Earth and Life Sciences (ALW) with financial aid from the Netherlands Organization for Scientific Research (NWO).

REFERENCES

1. Stams, A.J.M., Metabolic interactions between anaerobic bacteria in methanogenic environments, *Antonie van Leeuwenhoek,* 66, 271, 1994.
2. Schink, B., Energetics of syntrophic cooperations in methanogenic degradation, *Microbiol. Mol. Biol. Rev.,* 61, 262, 1997.
3. Schink, B. and Stams, A. J. M., Syntrophism among prokaryotes, in *The Prokaryotes*, (Electronic Third Edition), Dworkin, M., Schleifer, K.-H., and Stackebrandt, E., Eds., Springer-Verlag, New York, 2001, without page numbers.
4. Bryant, M.P. et al., *Methanobacillus omelianskii*, a symbiotic association of two species of bacteria, *Arch. Microbiol.*, 59, 20, 1967.
5. McInerney, M.J., Anaerobic hydrolysis and fermentation of fats and proteins, in *Biology of Anaerobic Microorganisms*, Zehnder, A.J.B., Ed., John Wiley & Sons, New York, 1988, Chapter 8.
6. Bryant, M.P., Microbial methane production — theoretical aspects, *J. Anim. Sci.*, 48, 193, 1979.
7. Barker, H.A., Studies upon the methane fermentation. IV. The isolation and culture of *Methanobacterium omelianskii, Antonie van Leeuwenhoek*, 6, 201, 1940.
8. Stadtman, T.C. and Barker, H.A., Studies on methane formation. VII. Tracer studies. *Arch. Biochem.*, 21, 256, 1949.
9. Ianotti, E.L., et al., Glucose fermentation products of *Ruminococcus albus* in continuous culture with *Vibrio succinogenes*, changes caused by inter species hydrogen transfer of H_2, *J. Bacteriol.,* 114, 1231, 1973.
10. Weimer, P.J. and Zeikus, J.G., Fermentation of cellulose and cellobiose by *Clostridium thermocellum* in the absence and presence of *Methanobacterium thermoautotrophicum, Appl. Environ. Microbiol.*, 33, 289, 1977.
11. Schink, B. and Thauer, R.K., Energetics of syntrophic methane formation and the influence of aggregation, in *Granular Anaerobic Sludge; Microbiology and Technology,* G. Lettinga et al., Eds., Pudoc, Wageningen, 1988, 5–17.

12. Kevbrin, V.V. et al., *Tindallia magadii* gen. nov., sp. nov., an alkaliphilic anaerobic ammonifier from soda lake deposits, *Curr. Microbiol.*, 37, 94, 1998.
13. Zhilina, T.N. and Zavarzin, G.A., Alkaliphilic anaerobes from soda lakes, in *Proceedings of 1st Fems Conference Ljubliana*, Slovenia, 2003, 181.
14. Huber, R., et al., *Thermotoga maritima* sp. nov. represents a new genus of unique extremely thermophilic eubacteria growing up to 90°C, *Arch. Microbiol.*, 144, 324, 1986.
15. Fiala, G. and Stetter, K.O., *Pyrococcus furiosus* sp. nov. represents a novel genus of marine heterotrophic archaebacteria growing optimally at 100°C, *Arch. Microbiol.*, 145, 56, 1986.
16. Kengen, S.W.M. and Stams, A.J.M., Formation of L-alanine as a reduced end product in carbohydrate fermentation by the hyperthermophilic archaeon *Pyrococcus furiosis, Arch. Microbiol.*, 161, 168, 1994.
17. Muralidharan, V. et al., Hydrogen transfer between methanogens and fermentative heterotrophs in hyperthermophilic cocultures, *Biotechnol. Bioeng.*, 56, 268, 1997.
18. Barker, H.A., On the biochemistry of methane formation, *Arch. Microbiol.*, 6, 404, 1936.
19. Zinder, S.H. and Koch, M., Non-aceticlastic methanogenesis from acetate: acetate oxidation by a thermophilic syntrophic coculture, *Arch. Microbiol.*, 138, 263, 1984.
20. Lee, M.J. and Zinder, S.H., Isolation and characterization of a thermophilic bacterium which oxidizes acetate in syntrophic association with a methanogen and which grows acetogenically on H2-CO_2, *Appl. Environ. Microbiol.*, 54, 124, 1988.
21. Lee, M.J. and Zinder, S.H., Hydrogen partial pressures in a thermophilic acetate-oxidizing methanogenic cocultures, *Appl. Environ. Microbiol.*, 54, 1457, 1988.
22. Lee, M.J. and Zinder, S.H., Carbon monoxide pathway enzyme activities in a thermophilic anaerobic bacterium grown acetogenically and in a syntrophic acetate-oxidizing coculture, *Arch. Microbiol.*, 150, 513, 1988.
23. Schnürer, A., Houwen, F.P., and Svensson, B.H., Mesophilic syntrophic acetate oxidation during methane formation by a triculture at high ammonium concentration, *Arch. Microbiol.*, 162, 70, 1994.
24. Schnürer, A., Schink, B., and Svensson, B.H., *Clostridium ultunense* sp. nov., a mesophilic bacterium oxidizing acetate in syntrophic association with a hydrogenotrophic methanogenic bacterium, *Int. J. Syst. Bacteriol.*, 46, 1145, 1996.
25. Hattori, S. et al., *Thermacetogenium phaeum* gen. nov., sp. nov., a strictly anaerobic, thermophilic, syntrophic acetate-oxidizing bacterium, *Int. J. Syst. Evol. Microbiol.,* 50, 1601, 2000.
26. Nuesslein, B. et al., Evidence for anaerobic syntrophic acetate oxidation during methane production in the profundal sediment of subtropical Lake Kinneret (Israel), *Environ. Microbiol.*, 3, 460, 2001.
27. Nuesslein, B. et al., Stable isotope biogeochemistry of methane formation in profundal sediments of Lake Kinneret (Israel), *Limnol. Oceanogr.*, 48, 1439, 2003.
28. Boone, D.R. and Bryant, M.P., Propionate-degrading bacterium, *Syntrophobacter wolinii* sp. nov. gen. nov., from methanogenic ecosystems, *Appl. Environ. Microbiol.*, 40, 626, 1980.
29. Wallrabenstein, C., Hauschild, E., and Schink, B., *Syntrophobacter pfennigii* sp. nov., new syntrophically propionate-oxidizing anaerobe growing in pure culture with propionate and sulfate, *Arch. Microbiol.*, 164, 346, 1995.
30. Harmsen, H.J.M. et al., *Syntrophobacter fumaroxidans* sp. nov., a syntrophic propionate-degrading sulfate-reducing bacterium, *Int. J. Syst. Bacteriol.*, 48, 1383, 1998.
31. Liu, Y. et al., Characterization of the anaerobic propionate-degrading syntrophs *Smithella propionica* gen. nov., sp. nov. and *Syntrophobacter wolini, Int. J. Syst. Bacteriol.*, 49, 545, 1999.
32. Plugge C.M., Balk, M,. and Stams, A.J.M., *Desulfotomaculum thermobenzoicum* subsp. *thermosyntrophicum* sp. nov., a thermophilic syntrophic propionate-oxidising spore-forming bacterium, *Int. J. Syst. Evol. Microbiol.,* 52, 391, 2002.
33. Houwen, F.P. et al., Enzymatic evidence for involvement of the methylmalonyl-CoA pathway in propionate oxidation by *Syntrophobacter wolinii, Arch. Microbiol.*, 155, 52, 1990.
34. Plugge, C.M., Dijkema, C., and Stams, A.J.M., Acetyl-CoA cleavage pathway in a syntrophic propionate-oxidizing bacterium growing on fumarate in the absence of methanogens, *FEMS Microbiol. Letters,* 110, 71, 1993.
35. Harmsen, H.J.M. et al., Phylogenetic analysis of two syntrophic propionate-oxidizing bacteria in enrichment cultures, *Syst. Appl. Microbiol.*, 18, 67, 1995.
36. DeBok, F.A.M. et al., *Pelotomaculum schinkii* sp. nov., an obligate syntrophic propionate-oxidizing spore-forming bacterium isolated in co-culture with *Methanospirillum hungatei*, Submitted for publication.

37. Imachi, H. et al., *Pelotomaculum thermopropionicum* gen. nov., sp. nov., an anaerobic, thermophilic, syntrophic propionate-oxidizing bacterium, *Int. J. Syst. Evol. Microbiol.*, 52, 1729, 2002.
38. Stams, A.J.M. et al., Growth of syntrophic propionate-oxidizing bacteria with fumarate in the absence of methanogenic bacteria, *Appl. Environ. Microbiol.,* 59, 1114, 1993.
39. Sobieraj, M. and Boone D.R., Syntrophomonadacaea, in *The Prokaryotes*, (Electronic Third Edition), Dworkin, M., Schleifer, K.-H., and Stackebrandt, E., Eds., Springer-Verlag, New York, 2001, without page numbers.
40. McInerney, M.J. et al., *Syntrophomonas wolfei* gen. nov. sp. nov., an anaerobic, syntrophic, fatty acid-oxidizing bacterium, *Appl. Environ. Microbiol.*, 41, 1029, 1981.
41. Oremland, R.S., Biogeochemistry of methane production, in *Biology of Anaerobic Microorganisms*, Zehnder, A.J.B., Ed.,. John Wiley & Sons, New York, Chapter 13.
42. McInerney, M.J., Bryant, M.P., and Pfennig, N., Anaerobic bacterium that degrades fatty acids in syntrophic association with methanogens, *Arch. Microbiol.*, 122, 129, 1979.
43. Beaty, P.S. and McInerney, M.J., Growth of *Syntrophomonas wolfei* in pure cultures on crotonate, *Arch. Microbiol.,* 147, 389, 1987.
44. Matthies, C. et al., *Pelospora glutarica* gen. nov., sp. nov., a glutarate-fermenting, strictly anaerobic, spore-forming bacterium, *Int. J. Syst. Evol. Microbiol.*, 50, 645, 2000.
45. McInerney, M.J. and Wofford, N.Q., Enzymes involved in crotonate metabolism in *Syntrophomonas wolfei, Arch. Microbiol.*, 158, 344, 1992.
46. Wofford, N.Q., Beaty, P.S., and McInerney, M.J., Preparation of cell-free extracts and the enzymes involved in fatty acid metabolism in *Syntrophomonas wolfei*, *J. Bacteriol.*, 167, 179, 1986.
47. Ben-Bassat, A., Lamed, R., and Zeikus, J.G., Ethanol production by thermophilic bacteria: metabolic control of end product formation in *Thermoanaerobium brockii, J. Bacteriol.*, 146, 192, 1981.
48. Schink, B., *Clostridium magnum* sp. nov., a non-autotrophic homoacetogenic bacterium, *Arch. Microbiol.,* 137, 250, 1984.
49. Plugge, C.M., Grotenhuis, J.T.C., and Stams, A.J.M., Isolation and characterization of an ethanol-degrading anaerobe from methanogenic granular sludge, in *Microbiology and Biochemistry of Strict Anaerobes Involved in Inter Species Hydrogen Transfer,* Belaich, J.P., Bruschi, M., and Garcia, J.L., Eds., Plenum Publishers Corporation, New York, 1990, 439.
50. Bomar, M., Hippe, H., and Schink, B., Lithotrophic growth and hydrogen metabolism by *Clostridium magnum*, *FEMS Microbiol. Lett.*, 83, 347, 1991.
51. Davidova, I.A. and Stams, A.J.M., Sulfate reduction with methanol by a thermophilic consortium obtained from a methanogenic reactor, *Appl. Microbiol. Biotechnol.*, 46, 297, 1996.
52. Goorissen, H.P., *Thermophilic methanol utilization by sulfate reducing bacteria*. Ph.D. Thesis, University of Groningen, the Netherlands, 2002, Chapter 4.
53. Balk, M. et al., Methanol utilization by a novel thermophilic homoacetogenic bacterium *Moorella mulderi* sp. nov., isolated from a bioreactor, *Arch. Microbiol.*, 179, 315, 2003.
54. Krumholz, L.R. and Bryant, M.P., *Syntrophococcus sucromutans* sp. nov., gen. nov., uses carbohydrates as electron donors and formate, methoxybenzenoids or *Methanobrevibacter* as electron acceptor systems, *Arch. Microbiol.*, 143, 313, 1986.
55. Mead, G.C., The amino acid-fermenting clostridia, *J. Gen. Microbiol.*, 67, 47, 1971.
56. Elsden, S.R. and Hilton, M.G., Amino acid utilization in Clostridial taxonomy, *Arch. Microbiol.,* 137, 1979.
57. Nagase, M. and Matsuo, T., Interaction between amino-acid degrading bacteria and methanogenic bacteria in anaerobic digestion, *Biotechnol. Bioeng*., 24, 2227, 1982.
58. Nanninga, H.J. and Gottschal, J.C., Amino acid fermentation and hydrogen transfer in mixed cultures, *FEMS Microbiol. Ecol.*, 31, 261, 1985.
59. Wildenauer, F.X. and Winter, J., Fermentation of isoleucine and arginine by pure and syntrophic cultures of *Clostridium sporogenes*, *FEMS Microbiol. Ecol.*, 38, 373, 1986.
60. Winter, J., Schindler, F., and Wildenauer, F. X., Fermentation of alanine and glycine by pure and syntrophic cultures of *Clostridium sporogenes, FEMS Microbiol. Ecol.*, 45, 153, 1987.
61. Zindel, U. et al., *Eubacterium acidaminophilum* sp. nov., a versatile amino acid-degrading anaerobe producing or utilizing H_2 or formate. Description and enzymatic studies, *Arch. Microbiol.*, 150, 254, 1988.

62. Stams, A.J.M. and Hansen, T.A., Fermentation of glutamate and other compounds by *Acidaminobacter hydrogenoformans* gen. nov., sp. nov., an obligate anaerobe isolated from black mud. Studies with pure cultures and mixed cultures with sulfate-reducing and methanogenic bacteria, *Arch. Microbiol.*, 137, 329, 1984.
63. Baena, S. et al., *Aminomonas paucivorans* gen. nov., sp. nov., a mesophilic, anaerobic, amino-acid-utilizing bacterium, *Int. J. Syst. Bacteriol.,* 49, 975, 1999.
64. Baena, S. et al., *Aminobacterium colombiense*, gen. nov., sp. nov., an amino acid degrading anaerobe isolated from anaerobic sludge, *Anaerobe,* 4, 241, 1998.
65. Baena, S. et al., *Aminobacterium mobile*, sp. nov., a new anaerobic amino-acid-degrading bacterium, *Int. J. Syst. Evol. Microbiol.*, 50, 259, 2000.
66. Barik, S., Brulla, W.J., and Bryant, M.P., PA-1, a versatile anaerobe obtained in pure culture, catabolizes benzenoids and other compounds in syntrophy with hydrogenotrophs, and P-2 plus *Wolinella* sp. degrades benzenoids, *Appl. Environ. Microbiol.*, 50, 304, 1985.
67. Plugge, C.M., Zoetendal, E.G., and Stams, A.J.M., *Caloramator coolhaasii*, sp. nov. a glutamate-degrading, moderately thermophilic anaerobe, *Int. J. Syst. Bacteriol.*, 50, 1155, 2000.
68. Tarlera, S. et al., *Caloramator proteoclasticus* sp. nov., a new moderately thermophilic anaerobic proteolytic bacterium, *Int. J. Syst. Bacteriol.*, 47, 651, 1997.
69. Tarlera, S. and Stams, A.J.M., Degradation of proteins and amino acids by *Caloramator proteoclasticus* in pure culture and in coculture with *Methanobacterium thermoautotrophicum* Z245, *Appl. Microbiol. Biotechnol.*, 53, 133, 1999.
70. Baena, S. et al., Phylogenetic relationships of three amino-acid-utilizing anaerobes, *Selenomonas acidaminovorans*, *'Selenomonas acidaminophila'* and *Eubacterium acidaminophilum*, as inferred from partial 16S rDNA nucleotide sequences and proposal of *Thermanaerovibrio acidaminovorans* gen. nov., comb. nov. and *Anaeromusa acidaminophila* gen. nov., comb. nov., *Int. J. Syst. Bacteriol.*, 49, 969, 1999.
71. Cheng, G. et al., *Selenomonas acidaminovorans* sp. nov., a versatile thermophilic proton-reducing anaerobe able to grow by decarboxylation of succinate to propionate, *Arch. Microbiol.*, 157, 169, 1992.
72. Plugge, C.M., et al., *Gelria glutamica*, gen. nov., sp. nov., a thermophilic obligate syntrophic glutamate-degrading anaerobe. *Int. J. Syst. Evol. Microbiol.*, 52, 401, 2002.
73. Ørlygsson, J., Houwen, F.P., and Svensson, B.H., Anaerobic degradation of protein and the role of methane formation in steady state thermophilic enrichment cultures, *Swedish J. Agric. Res.*, 23, 45, 1993.
74. Ørlygsson, J., *The role of interspecies hydrogen transfer on thermophilic protein and amino acid metabolism*, Ph.D. Thesis, Swedish University of Agricultural Sciences, Uppsala, 1994.
75. Gottschalk, G., *Bacterial Metabolism*, Second Edition, Springer-Verlag, New York, 1986.
76. Buckel, W. and Barker, H.A., Two pathways of glutamate fermentation by anaerobic bacteria, *J. Bacteriol.*, 117, 1248, 1974.
77. Nanninga, H.J., Drent, W.J., and Gottschal, J.C., Fermentation of glutamate by *Selenomonas acidaminophila* sp. nov., *Arch. Microbiol.*, 147, 152, 1987.
78. Meijer, W.G., Nienhuis-Kuiper, M.E., and Hansen, T.A., Fermentative bacteria from estuarine mud: phylogenetic position of *Acidaminobacter hydrogenoformans* and description of a new type of Gram-negative, propionigenic bacterium as *Propionibacter pelophilus* gen. nov., sp. nov., *Int. J. Syst. Bacteriol.*, 49, 1039, 1999.
79. Plugge C.M. and Stams A.J.M., Arginine catabolism by *Thermanaerovibrio acidaminovorans, FEMS Microbiol. Lett.,* 195, 259, 2001.
80. Macarie, H. and Guyot, J.P., Inhibition of the methanogenic fermentation of p-toluic acid (4-methylbenzoic acid) by acetate, *Appl. Microbiol. Biotechnol.*, 38, 398, 1992.
81. Cheng, S.S., Ho, C.T., and Wu, J.H., Pilot study of UASB process treating PTA manufacturing wastewater, *Water Sci. Technol.,* 36, 73, 1997.
82. Kleerebezem, R. et al., Anaerobic pretreatment of petrochemical effluents: terephthalic acid wastewater, *Water Sci. Technol.,* 36, 237, 1997.
83. Wu, J.-H. et al., Characterization of microbial consortia in a terephthalate-degrading anaerobic granular sludge system, *Microbiology,* 147, 373, 2001.

84. Elshahed, M.S. et al., Metabolism of benzoate, cyclohex-1-ene carboxylate, and cyclohexane carboxylate by "*Syntrophus aciditrophicus*" strain SB in syntrophic association with H_2-using microorganisms, *Appl. Environ. Microbiol.*, 67, 1728, 2001.
85. Jackson, B.E. and McInerney, M.J., Anaerobic microbial metabolism can proceed close to thermodynamic limits, *Nature*, 415, 454, 2002.
86. Qiu, Y.L. et al., *Sporotomaculum syntrophicum* sp. nov., a novel anaerobic, syntrophic benzoate-degrading bacterium isolated from methanogenic sludge treating wastewater from terephthalate manufacturing, *Arch. Microbiol.*, 179, 242, 2003.
87. Friedrich, M., Laderer, U., and Schink, B., Fermentative degradation of glycolic acid by defined syntrophic cocultures, *Arch. Microbiol.*, 156, 398, 1991.
88. Friedrich, M. et al., Phylogenetic position of *Desulfofustis glycolicus* gen. nov. sp. nov. and *Syntrophobotulus glycolicus* gen. nov. sp. nov., two strict anaerobes growing with glycolic acid, *Int. J. Syst. Bacteriol.*, 46, 1065, 1996.
89. Platen, H. and Schink, B., Methanogenic degradation of acetone by an enrichment culture, *Arch. Microbiol.*, 149, 136, 1987.
90. Platen, H., Janssen, P.H., and Schink, B., Fermentative degradation of acetone by an enrichment culture in membrane-separated culture devices and in cell suspensions, *FEMS Microbiol. Lett.*, 122, 27, 1994.
91. Grotenhuis, J.T.C. et al., Bacteriological composition and structure of granular sludge adapted to different substrates, *Appl. Environ. Microbiol.*, 57, 1942, 1991.
92. Koorneef, E. et al., Methanogens revealed immunologically in granules from five upflow anaerobic sludge blanket (UASB) bioreactors grown on different substrates, *FEMS Microbiol. Ecology*, 73, 225, 1992.
93. Harmsen, H.J.M. et al., Detection and localization of syntrophic propionate-oxidizing bacteria in granular sludge by in situ hybridization using 16S rRNA-based oligonucleotide probes, *Appl. Environ. Microbiol.*, 62, 1656, 1996.
94. Sekiguchi, Y. et al., Fluorescence *in situ* hybridization using 16S rRNA-targeted oligonucleotides reveals localization of methanogens and selected uncultured bacteria in mesophilic and thermophilic sludge granules, *Appl. Environ. Microbiol.*, 65, 1280, 1999.
95. Stams, A.J.M. et al., Contribution of ^{13}C-NMR spectroscopy to the elucidation of pathways of propionate formation and degradation in methanogenic environments, *Biodegradation*, 9, 463, 1998.
96. Manefield, M. et al., RNA stable isotope probing, a novel means of linking microbial community function to phylogeny, *Appl. Environ. Microbiol.*, 68, 5367, 2002.
97. Radajewski, S., McDonald, I.R., and Murrell, J.C., Stable-isotope probing of nucleic acids: a window to the function of uncultured microorganisms, *Curr. Opinion. Biotechnol.*, 14, 298, 2003.
98. Gujer, W. and Zehnder, A.J.B., Conversion processes in anaerobic digestion, *Wat. Sci. Technol.*, 15, 127, 1982.
99. Thauer, R.K., Jungermann, K., and Decker, K., Energy conservation in chemotrophic anaerobic bacteria, *Bacteriol. Rev.*, 41, 100, 1977.

35 Ecology and Activity of Clostridia in the Intestine of Mammals

Sergey R. Konstantinov, Hauke Smidt, and Antoon D.L. Akkermans

CONTENTS

35.1 INTRODUCTION

The importance of intestinal anaerobic clostridia in many aspects of mammalian host health and performance, including provision of nutrients, stimulation of immune response, protection from pathogens, and production and metabolism of toxins, is becoming increasingly apparent. There is, therefore, a great need for both the accurate identification and enumeration of the intestinal clostridia populations in intestinal ecosystems and a functional assessment of their *in situ* activities in the gut.

The gastrointestinal (GI) tracts of mammals including ruminant (e.g., cattle, sheep) and non-ruminant animals (e.g., humans, horses, pigs, mice) contain vast communities of anaerobic micro-organisms. Molecular biological studies involving polymerase chain reaction (PCR) recovery of 16S rRNA gene sequences from GI tract samples have highlighted the extreme diversity of the microbiota in gut ecosystems. The results of such studies (on cattle, humans, horses, rodents, pigs) have shown that the majority of recovered sequences are phylogenetically affiliated to several known groups of anaerobic bacteria, namely the low% G+C Gram-positive bacteria, including the genus *Clostridium*. The taxon *Clostridium* consists of a highly diverse group of spore-forming anaerobes and is subdivided into various subgroups that need further reclassification. In this chapter, we will

TABLE 35.1
Cultivated Bacterial Species Within the *Clostridium Coccoides* Group[a]

Clostridium aminophilum
Clostridium aminovalericum
Clostridium celerecrescens
Clostridium clostridiiforme
Clostridium coccoides
Clostridium herbivorans
Clostridium nexile
Clostridium oroticum
Clostridium polysaccharolyticum
Clostridium populeti
Clostridium sphenoides
Clostridium symbiosum
Clostridium xylanolyticum

[a] The *Clostridium* species are listed alphabetically, and the data are summarized from the studies of Finegold et al. [48] and Tannock [43]. For a detailed phylogenetic analysis, see Collins et al. [14].

use the group in a broad sense, including groups that already have been recognized as "related to clostridia." A combination of direct sequence analysis of 16S rRNA genes, group-specific dot-blot hybridization and fluorescent *in situ* hybridization has suggested the occurrence of a large number of yet-uncultured *Clostridium* species within the mammalian gut. A significant fraction of them was affiliated with two phylogenetically distinct groups, namely the *Clostridium coccoides* group and the *Clostridium leptum* subgroup [1–6]. In contrast, there are only a few intestinal bacterial species within the *Clostridium coccoides* group cultivated (Table 35.1). This chapter is aimed to unveil the recent progress in the field of molecular detection and characterization of clostridia in the mammalian GI tract. In addition, the role of some commensal (nonpathogenic) clostridia in the creation and maintenance of a healthy mammalian GI tract ecosystem will be highlighted.

35.2 ECOLOGY OF CLOSTRIDIA IN THE MAMMALIAN INTESTINE

35.2.1 Ruminants

The rumen is a complex, open ecosystem that harbors a dense and competitive mixture of populations of obligatory anaerobic microorganisms, including fungi, protozoa, bacteria, and archaea. The anaerobic part of the system has been subjected to intensive studies over the past more than 40 years. A number of studies have been published describing the isolation and identification of a wide variety of aerobic, facultative, and anaerobic bacterial strains from ruminants of different ages, health conditions, geographic locations, and diets [7,8]. Twenty-two species of ruminant bacteria were estimated to predominate in the rumen [9]. However, considering that only a small fraction of the total microbial diversity in natural ecosystems can be recovered by cultivation-based methods [10], it is not clear whether these descriptions accurately reflect bacterial species distribution in the rumen. In recent studies, a comparative DNA sequence analysis of 16S rRNA genes was undertaken to further characterization of bacterial communities in the rumen fluid of dairy cattle [11–13]. The phylogenetic analysis unveiled that 55% of the sequences were similar to low G+C Gram-positive bacteria related to the genus *Clostridium*, the majority of which fall into clusters XIVa and XIVb as defined by Collins et al. [14]. For 38% of the sequences, the similarity with database sequences was in the range of 90 to 98%, whereas for the remaining clones analyzed in these studies, the similarity was less than 90%. The rRNA gene sequence analysis data suggested a vast uncultured diversity of the genus *Clostridium* that has not yet been isolated.

35.2.2 Rodents

Mice and rats have often been used as models to study nutrition and immune responses, and as a consequence, many studies have been made on the composition of the microbiota in the GI tract of mice. Initially, these studies used conventional cultivation methods [15,16], but more recently, applied molecular techniques, such as 16S rRNA gene sequence analysis [5] have been used. Several sequences of clostridial species have been found, most of them belonging to the *Eubacterium rectale-Clostridium coccoides* group (e.g., *C. clostridiiformes, C. celerecrescens, Ruminococcus schinkii*). Special attention was given to the group of uncultured segmented filamentous bacteria (SFB), which have been found attached to the intestine of a wide range of vertebrates, and even invertebrates. Besides in rat and mouse, SFB were found in frog, toad, duck, guinea pig, zebra, dog, cat, sheep, pig, quail, chicken, turkey, and horse [17]. Since SFB morphotypes have also been observed in evertebrates, it has been concluded that this group is ubiquitous in the animal kingdom. The phylogenetic position was determined by cloning and sequencing of 16S rRNA genes amplified from ileum tissue from monoassociated mice [18]. Further molecular studies were performed with rat, chicken [19], and trout [20], and the provisional name "*Candidatus Arthromitus*" was proposed for this distinct group of clostridia [19]. SFB are absent during the suckling period of mice and rats, but colonize the ileum after weaning and disappear several months after weaning. In immunodeficient (scid/scid) mice, however, they remained visible for several months [21]. These observations were confirmed by 16S rRNA gene-targeted denaturing gradient gel electrophoresis (DGGE) fingerprinting studies with immunocompetent and immunodeficient mice from the same litter (van Schieveen, de Vries, Snel, and Akkermans, unpublished results).

SFB are usually regarded as nonpathogenic commensal bacteria, although in birds the occurrence of SFB was related with diarrhea [22]. In rodents, SFB have been suggested to be active in stimulation of the gut mucosal immune system [23,24]. The mechanism of the stimulation of SFB to the mucosal immune cells has not been well defined. Yamauchi and Snel [25] proposed enterocyte phagocytosis and processing of attached SFB to be the mechanism of immune cell presentation in broiler chicks. However, in a recent study on SFB colonization of the ileum of 4-to-5-week-old pigs, a direct mechanism of SFB presentation for mucosal immune stimulation is suggested [26].

Insight into the quantitative significance of these microbes is obtained by the development of group-specific fluorescent probes against members of the *Eubacterium rectale-Clostridium coccoides* group [1] and SFB [18].

35.2.3 Horses

So far, little systematic studies have been published on the composition of the microbiota in the hindgut of horses. Most studies are related to intestinal diseases such as acute laminitis as a result of fermentation of carbohydrates in the hindgut. Fermentation of an excess of carbohydrates was associated with higher numbers of streptococci and lactobacilli [27], and no reports are available on the involvement of clostridia. In a recent comparative study of the intestinal microbiota of healthy horses and those suffering from equine grass sickness (EGS), it was shown that fecal samples of the latter group contained 10-fold (in acute EGS, even 100-fold) higher numbers of clostridia than in healthy animals [28]). Toxin producing *Clostridium* species have often been found as causes of death of equine. In a recent study of Hunter et al. [29], *Clostridium botulinum* type C was found to be associated with equine sickness, while the occurrence of *C. difficile* was found to be associated with enterocolitis and lactose intolerance in a foal [30].

35.2.4 Pigs

The majority of the microorganisms isolated from the fecal and colonic microbiota of adult swine were Gram-positive obligate anaerobe and microaerophilic microorganisms [31]. Among the anaerobe isolates estimated by plate counting, the genus *Clostridium* was found to comprise up to 2% of the viable counts in adult pigs. While the fermentative conversion of ingested feed by individual members of the genus *Clostridium* has been demonstrated, others such as *Clostridium perfringens* were cultured from the intestinal mucus layer of neonatal piglets [32]. Recent application of 16S rRNA gene-targeted culture-independent detection techniques have allowed for a more comprehensive assessment of the *Clostridium* diversity in the pig GI tract, although this approach might be biased as well [33]. Phylogenetic analysis of the 16S rRNA sequences retrieved from pig intestine has revealed that a significant proportion of the bacterial species colonizing the intestine of pigs, and in particular the ones related to the genus *Clostridium*, have not been characterized [4,6]. In unison with these studies, the application of molecular fingerprinting techniques such as DGGE has shown that the predominant fecal bacteria of weaning piglets were similar to sequences derived from the *Clostridium coccoides* group and the *Clostridium leptum* subgroup [34]. The latter two groups were recovered both from the feces and from samples of the colonic lumen, colonic wall, and cecal lumen from a pig [4], and their phylogenetic analysis showed that the sequences fell into *Clostridium* cluster XIVa [14]. In a recent comprehensive study based on fast sequence output of 16S rRNA genes, the phylogenetic diversity of the pig intestinal bacteria was assessed [6]. Pigs representing a variety of diets, ages, and herd health status were analyzed, and a library was constructed comprising more than 4000 cloned 16S rRNA gene sequences. A large fraction (29%) of the low-G+C Gram-positive phylotypes was affiliated with the *Clostridium coccoides* group and the *Clostridium leptum* subgroup, which contain many of the classic intestinal bacteria. However, most of the *Clostridium* phylotypes showed an average similarity of 92.2% to cultured intestinal or ruminal strains. The results documented that the intestinal microbial community is very complex and that the majority of the *Clostridium* species colonizing the gastrointestinal tract in pigs have not been characterized. These findings, based on molecular analyses, have also raised the question on the functional metabolic properties of the hitherto uncultured *Clostridium* phylotypes with only limited similarity to sequences present in the databases. Their phylogenetic inferences can be estimated from the dendrograms, but the extrapolation of functional properties from well-characterized cultured strains to distantly related phylotypes might not be justified [6]. Part of the low-similarity phylotypes might represent novel species and higher taxonomic entities. Other phylotypes might already have been cultured and characterized, but their 16S rRNA sequence is not available in the databases. Recently, it has been shown that for some of the large clusters of *Prevotella* and *Bacteroides* detected by molecular methods, cultured strains are indeed available [35]. A similar situation might indeed be the case of *Clostridium* clusters comprising nonpathogenic anaerobic bacteria that have not received much attention throughout the history of microbiology.

35.2.5 Humans

35.2.5.1 Development in Time

Under normal conditions, babies are born without bacteria or viruses in the body. However, during and after the delivery, microbes from the mother and their environment invade the body and start to colonize the skin, the mouth, the stomach, and the intestine. First colonizers of the intestine often are facultative anaerobes, such as *Escherichia coli*, but soon anaerobes from the *Clostridium* group are occupying the intestine [36,37]. 16S rRNA gene analysis of fecal samples of babies revealed the presence of *C. disporicum* and *C. perfringens*. This early presence of *Clostridium* species was also described in previous studies based on conventional plate counting [38,39]. In normal, breastfed babies *Bifidobacterium* spp. are the dominant bacteria, comprising up to 80% of the total community. During the first weeks, a rapid change of microbial populations occur with major increase in the diversity after weaning. At later stages, the composition of the microbial community stabilizes and becomes more or less characteristic of each individual [40,41]. Of particular interest is the occurrence of sequences of the uncultured *Ruminococcus obeum*-like bacteria in human feces [2,40,41]. It belongs to the *Clostridium coccoides-Eubacterium rectale* group, a group of dominant bacteria in the human feces. Flow cytometric analysis with 16S rRNA-specific fluorescent probes demonstrated that the group comprised about 15% of the total microbiota, while the *R. obeum* as uncultured bacterium comprised approximately 2.5% of the total bacterial community in feces of adult humans [42].

35.2.5.2 Spatial Distribution

For practical reasons, most studies on the GI tract microbiota were based on analysis of fecal samples. Samples from different parts of the GI tract usually were taken from sites that are relatively easy to reach by biopsy. Other areas were only sampled in patients or dead subjects, which complicates the interpretation of the data. Biopsy studies have demonstrated that there are major differences in the composition and the diversity of the microbiota in the GI tract. The small intestine is usually occupied by a consortium of bifidobacteria and other anaerobes, including clostridia, while the large intestine is more diverse and mainly occupied strictly by anaerobes.

Clostridium species, such as *C. cadaveris, C. clostridioforme, C. innocuum, C. paraputrificum, C. perfringens, C. ramosum,* and *C. tertium* are often found in human feces and can occur in numbers of up to 10^9 per gram fresh weight of feces, which is less than 1% of the total microbiota [43]. The occurrence of cultured usually proteolytic clostridia often has been related to disease and toxin production. Saccharolytic clostridia usually are nonpathogenic and are involved in the breakdown of polysaccharides. Despite these results derived from cultivation assays, the data obtained with molecular methods clearly demonstrate that the majority of the clostridia belong to previously uncultured bacteria from which we only know the 16S rRNA gene sequence. It is evident that pure cultures are needed in order to unravel the physiological role of these bacteria.

35.3 ACTIVITY OF CLOSTRIDIA IN THE MAMMALIAN INTESTINE

The fermentation in single-stomached animals, including humans, is being recognized as having important implications for maintenance of host health [44]. The GI tract fermentation occurs mainly in the large intestine due to the longer transit time there and the high density of the residential microbiota. The principal physiological functions of the healthy colon in the monogastric animals include the salvage of energy and nutrients through its symbiotic relationship with the GI tract microbiota. Whereas the time taken for the intestinal contents to travel the length of the small intestine is only 2 to 4 h in humans, the large bowel transit time is normally 20 to 80 h, so there is ample time for development and activity of the microbiota, including the species belonging to the clostridia. As numerically predominant anaerobes that occur in the intestine, most of the clostridia found in the intestine are both saccharolytic and amino acid-fermenting bacteria. Although digestion by host enzymes and absorption from the small intestine are very effective, there is still a constant supply of nutrients to the large intestine in the form of undigested dietary components, host enzymes, and desquamated gut mucosal cells. The absence of glucose means that inducible enzymes are produced by the microbiota, which facilitate utilization of indigestible nutrients, such as resistant starch [45,46].

35.3.1 Carbohydrate Fermentation

Bacteria in the colon metabolize available carbohydrate to obtain energy for their own growth and maintenance (including motility, enzyme synthesis, maintenance of ionic and osmotic gradients, and active ion transport). Such an anaerobic fermentation has been summarized by the following general equation [47]:

$$57.5\ C_6H_{12}O_6 + 45\ H_2O = 65\ \text{acetate} + 20\ \text{propionate} + 15\ \text{n-butyrate} + 140\ H_2 + 95\ CO_2 + 288\ \text{ATP}.$$

A general molar ratio of the acids produced is 1 acetate:0·31 propionate:0·23 n-butyrate, though the proportions can and do vary, depending on the type of substrate available, the composition of the anaerobic flora, and the prevailing pH. However, the equation shows that 73% of the C units and 68% of the energy value of fermented carbohydrate can reappear in the form of completely metabolizable volatile fatty acids (VFAs).

Fermentation of carbohydrates by clostridia in the large intestine is part of the global anaerobic fermentation in the gut and results in the following principal metabolic products: acetate, propionate, n-butyrate, lactate, and ethanol. Well-studied solvent-producing bacteria such as *C. beijerinckii* and *C. butyricum* were isolated from human intestine after an extensive survey [48]. The biochemical pathways used for the conversion of carbohydrates into hydrogen, carbon dioxide, VFA, and solvents by these clostridial species have been extensively studied and characterized [49,50].

The colon rapidly absorbs free VFA, thus conserving energy and reducing the osmotic load. The bacterial VFA are absorbed in the colon along a concentration gradient from lumen to serosa without evidence of active transport. VFA are an important fuel for large intestinal colonocytes, with butyric acid being the most important reviewed [44,51]. Disregarding regional differences along the length of the colon, epithelial cells prepared from the entire length of the colon revealed a preference of metabolic fuels in the following order: VFA, ketone bodies, amino acids, and glucose. This is in contrast to the cells of the small intestine, where the preference for metabolic fuels is: glutamine, glucose, and ketones in ruminants. These differences are not surprising given that the cells of both regions have undoubtedly adapted to the fuels available. For example, while glucose is potentially available from the intestinal blood supply, it is unlikely that significant amounts of glucose will reach the large intestine via the gut lumen. Further, the physiological significance of the different VFA will be highlighted.

Acetate is an essential metabolic fuel in ruminant animals because all glucose reaching the rumen is fermented by the resident bacteria. In humans, its importance is less certain [52]. Blood acetate is derived primary from the gut. Acetate is metabolized by the skeletal and cardiac muscle and brain. Patients who have no large intestine have very low blood acetate levels [51].

Propionate metabolism has been extensively studied in ruminants, where it is a major glucose precursor [53]. Much less is known about its role in humans. Propionate can be found in the portal blood, although some can be metabolized by the colonic epithelium and can be a differentiating factor, but with less power than n-butyrate. Propionate supplement was found to lower the concentration of the cholesterol in rats and pigs. In humans, the propionate had no effect, or slightly increased the serum cholesterol in short studies [51,52].

N-butyrate is rapidly metabolized by the colonic epithelium, where it is involved in the regulation of the cell growth and differentiation. In addition, in a recent *in vivo* study with weaning piglets, it was found that increasing butyrate concentration in the feed led to increased numbers of lactobacilli, while *E. coli* decreased, in all sections of the GI tract. Consequently, the authors concluded that there was a selective antimicrobial effect of n-butyrate. They classified the n-butyrate sensitivity of microbes, as follows: *Clostridium acetobutylicum, E. coli, Streptococcus cremoris, Lactococcus lactis* and *L. cremoris,* and *Salmonella* spp. [44].

35.3.2 Amino Acid Fermentation

Many of the intestinal clostridial species like *C. perfringens* and *C. bifermentans* are both saccharolytic and strongly proteolytic. In the colon, as carbohydrate sources (starch and other fermentable carbohydrates) become depleted due to fermentation, carbohydrate:N of the caecum decreases and the fermentation shifts toward proteolysis. Hence, some of the intestinal VFA might originate from polypeptides, which are apparently the major source of the mainly branched-chain VFA (isobutyrate, valerate, and isovalerate) formed by the metabolism of branched-chain amino acids such as valine, leucine, and isoleucine. Unlike carbohydrate metabolism, many of the products of protein breakdown are toxic to the host. A proteolytic fermentation can lead to the formation of such metabolites as NH_3, amines, volatile phenols, and indoles, which are toxic and found only in small amounts in the healthy colon. In fact, deamination of amino acids from both dietary and endogenous protein is the main source of NH_3 in the colon [44]. NH_3 generated in the colon readily passes across the gut wall, thereby gaining access to other tissues of the body. NH_3 can disturb the development of the mucosa of the intestine. Furthermore, NH_3 produced and absorbed must be excreted as urea with an energy cost of about 7% of total energy expenditure in single-stomached as well as in ruminant animals. Subacute levels of NH_3 as low as 10 m*M* can alter the morphology, influence metabolism of intestinal cells, and can potentially be involved in the initiation of cancer in humans [54].

Studies on the formation of phenolic and indolic compounds by intestinal clostridia have demonstrated that both p-cresol and skatole are formed in pure culture [55]. Both compounds are involved in a variety of disease states in humans and in animals [56]. As with the formation of phenols and indoles, many different *Clostridium* spp. are known to produce amines [57]. A number of amines produced in the colon, such as histamine, cadaverine, tyramine, and putrescine, are pharmacologically active and can affect a variety of body functions [54,58].

Recently, it has been hypothesized that the production of potentially harmful compounds, such as tryptophan catabolism to skatole, would be reduced in the presence of a carbohydrate source in the colon. The presence of fermentable carbohydrate can, therefore, lead to an increased demand for amino acids by microbiota, as indicated by reduced protein catabolism in the hindgut and, hence, lower NH_3 concentrations [44].

35.4 CONCLUDING REMARKS

Studying the interactions of our intestinal clostridia and their mammalian hosts represents an experimental challenge. Recent high throughput analysis using DNA microarray technology unveiled the intestinal transcriptional responses to colonization of germ-free mice with *Bacteroides thetaiotamicron,* a prominent member of the normal murine and human intestinal microbiota [59]. The results showed that this commensal strain was able to modulate expression of host genes participating in diverse and fundamental physiological functions, including nutrient absorption, mucosal barrier fortification, xenobiotic metabolism, angiogenesis, and postnatal intestinal maturation. The identified changes in gene expression highlight how host physiology can be affected by changes in the composition of the indigenous microbiota. Further assessment of the *in vivo* gene expression and activity of a specific *Clostridium* species in a complex ecosystem will unravel the enigma of their colonization of and persistence within the mammalian gut.

ACKNOWLEDGMENTS

This work has been carried out with financial support of the European Community specific RTD program Quality of Life and Management of Living Resources research projects: HEALTHYPIGUT (QLK5-LT2000-00522).

REFERENCES

1. Franks, A. H. et al., Variation of the bacterial population in human feces measured by fluorescent in situ hybridization with group-specific 16S rRNA-targeted oligonucleotide probes, *Appl. Environ. Microbiol.*, 64, 3336, 1998.
2. Suau, A. et al., Direct analysis of genes encoding 16S rRNA from complex communities reveals many novel molecular species within the human gut, *Appl. Environ. Microbiol.*, 65, 4799, 1999.
3. Sghir, A. et al., Quantification of bacterial groups within the human fecal flora by oligonucleotide probe hybridisation, *Appl. Environ. Microbiol.*, 66, 2263, 2000.
4. Pryde, S.E. et al., Molecular analysis of the microbial diversity present in the colonic wall, colonic lumen, and cecal lumen of a pig, *Appl. Environ. Microbiol.*, 65, 5372, 1999.
5. Salzman, N.H. et al., Analysis of 16S libraries of mouse gastrointestinal microflora reveals a large new group of mouse intestinal bacteria, *Microbiology*, 148, 3651, 2002.
6. Leser, T.D. et al., Culture-independent analysis of gut bacteria: the pig gastrointestinal tract microbiota revisited, *Appl. Environ. Microbiol.*, 68, 673, 2002.
7. Briant, M.P., Bacterial species in the rumen, *Bacteriol. Rev.*, 23, 125, 1959.
8. Stevens, C.E. and Hume, I.D., Contributions of microbes in vertebrate gastrointestinal tract to production and conservation of nutrients, *Physiol. Rev.*, 78, 393, 1998.
9. Krause, D.O. and Russell, J.B., How many rumen bacteria are there?, *J. Dairy Sci.*, 79, 1467, 1996.
10. Amann, R.I. and Ludwig, W., Ribosomal RNA-targeted nucleic acid probes for studies in microbial ecology, *FEMS Microbiol. Rev.*, 24, 555, 2000.
11. Tajima, K. et al., Rumen bacterial diversity as determined by sequence analysis of 16S rDNA libraries, *FEMS Microbiol. Ecol.*, 29, 159, 1999.
12. Tajima, K. et al., Rumen bacterial community transition during adaptation to high-grain diet, *Anaerobe*, 6, 273, 2000.
13. Whitford, M.F. et al., Phylogenetic analysis of rumen bacteria by comparative sequence analysis of cloned 16S rRNA genes, *Anaerobe*, 4, 153, 1998.
14. Collins, M.D. et al., The phylogeny of the genus *Clostridium*: proposal of five new genera and eleven new species combinations, *Int. J. Syst. Bacteriol.*, 44, 812, 1994.
15. Schaedler, R.W., Dubos, R., and Costello, R., The development of the bacterial flora in the gastrointestinal tract of mice, *J. Exp. Med.*, 122, 59, 1965.
16. Dubos, R., et al., Indigenous, normal, and autochthonous flora of the gastrointestinal tract, *J. Exp. Med.*, 67, 1965.
17. Klaasen, H.L., et al., Intestinal, segmented, filamentous bacteria, *FEMS Microbiol. Rev.*, 88, 165, 1992.
18. Snel, J. et al., Phylogenetic characterization of *Clostridium* related segmented filamentous bacteria in mice on 16S ribosomal RNA analysis, *System. Appl. Microbiol.*, 17, 172, 1994.
19. Snel, J. et al., Comparison of 16S rRNA sequences of segmented filamentous bacteria isolated from mice, rats, and chicken and proposal of "*Candidatus Arthromitus*," *Int. J. Syst. Bacteriol.*, 780, 1995.
20. Urdaci, M.C., Regnault, B., and Grimont, P.A.D., Identification by in situ hybridization of segmented filamentous bacteria in the intestine of diarrheic trout (*Oncorhynchus mykiss*), *Res. Microbiol.*, 152, 67, 2001.
21. Jiang, H.Q., Bos, N.A., and Cebra, J.J., Timing, localization, and persistence of colonization by segmented filamentous bacteria in the neonate mouse gut depend on immune status of mothers and pups, *Infect. Immun.*, 69, 3611, 2001.
22. Morishita, T., Lam, K., and McCapes, R.H., Isolation of two filamentous bacteria associated with enteritis in turkey poults, *J. Poultry Sci.*, 203, 1992.
23. Klaasen, H.L., et al., Apathogenic, intestinal, segmented, filamentous bacteria stimulate the mucosal immune system of mice, *Infect. Immun.*, 61, 303, 1993.
24. Umesaki, Y., et al., Differential roles of segmented filamentous bacteria and clostridia in development of the intestinal immune system, *Infect. Immun.*, 67, 3504, 1999.
25. Yamauchi, K.E. and Snel, J., Transmission electron microscopic demonstration of phagocytosis and intracellular processing of segmented filamentous bacteria by intestinal epithelial cells of the chick ileum, *Infect. Immun.*, 68, 6496, 2000.
26. Meyerholz, D.K., Stabel, T.J., and Cheville, N.F., Segmented filamentous bacteria interact with intraepithelial mononuclear cells, *Infect. Immun.*, 70, 3277, 2002.

27. Bailey, S.R., et al., Identification of equine cecal bacteria producing amines in an *in vitro* model of carbohydrate overload, *Appl. Environ. Microbiol.*, 69, 2087, 2003.
28. Garret, L.A., Brown, R., and Poxton, I.R., A comparative study of the intestinal microbiota of healthy horses and those suffering from equine grass sickness, *Vet. Microbiol.*, 87, 81, 2002.
29. Hunter, L.C., Miller, J.K., and Poxton, I.R., The association of *Clostridium botulinum* type C with equine grass sickness: a toxicoinfection?, *Equine Vet. J.*, 31, 492, 1999.
30. Weese, J.S., Straempfli, H.R., and Prescott, J.F., A prospective study of the roles of *Clostridium difficile* and enterotoxigenic *Clostridium perfingens* in equine diarrhoea, *Equine Vet. J.*, 33, 403, 2001.
31. Stewart, C.S., Microorganisms in hindgut fermemtors, in *Gastrointestinal Microbiology*, Vol. 2, R.I. Mackie, White, B.A., and Isaacson, R.E., Eds., Chapman & Hall, New York, 1997, 142.
32. Deplancke, B. et al., Selective growth of mucolytic bacteria including *Clostridium perfringens* in a neonatal piglet model of total parenteral nutrition, *Am. J. Clin. Nutr.*, 76, 1117, 2002.
33. von Wintzingerode, F., Gößel, U.B., and Stackebrandt, E., Determination of microbial diversity in environmental samples: pitfalls of PCR-based rRNA analysis, *FEMS Microbiol. Review*, 21, 213, 1997.
34. Konstantinov, S.R. et al., Effect of fermentable carbohydrates on faecal bacterial communities as revealed by DGGE analysis of 16S rDNA, *FEMS Microbiol. Ecol.*, 43, 225, 2003.
35. Ramsak, A. et al., Unravelling the genetic diversity of ruminal bacteria belonging to the CFB phylum, *FEMS Microbiol. Ecol.*, 33, 69, 2000.
36. Favier, C.F. et al., Molecular monitoring of succession of bacterial communities in human neonates, *Appl. Environ. Microbiol.*, 68, 219, 2002.
37. Favier, C.F., Lallès, P.J., and Sève, B., Intestinal bacteriological changes in piglets after weaning onto a diet based on wheat, *Reproduction Nutrition Development*, 42, 18, 2002.
38. Cooperstock, M. et al., Influence of age, sex, and diet on asymptomatic colonization of infants with *Clostridium difficile*, *J. Clin. Microbiol.*, 17, 830, 1983.
39. Benno, Y., Sawada, K., and Misuoka, T., The intestinal microflora of infants: composition of fecal flora in breast-fed and bottle-fed infants, *Microbiol. Immunol.*, 28, 975, 1984.
40. Zoetendal, E.G., Akkermans, A.D.L., and deVos, W.M., Temperature gradient gel electrophoresis analysis of 16S rRNA from human fecal samples reveals stable and host-specific communities of active bacteria, *Appl. Environ. Microbiol.*, 64, 3854, 1998.
41. Zoetendal, E.G., Akkermans, A.D.L., and de Vos, W.M., Molecular characterization of microbial communities in the GI-tract based on 16S rRNA sequence diversity, in *New Approaches for the Generation and Analysis of Microbial Fingerprints*, L. Dijkshoorn, Towner, K.J., and Struelens, M., Eds., Elsevier, Amsterdam, 2001, 267.
42. Zoetendal, E.G. et al., Mucosa-associated bacteria in the human gastrointestinal tract are uniformly distributed along the colon and differ from the community recovered from feces, *Appl. Environ. Microbiol.*, 68, 3401, 2002.
43. Tannock, G.W., *Normal microflora. An introduction to microbes inhabiting the human body*, Chapman & Hall, New York, 1995.
44. Williams, B.A., Verstegen, M.W.A., and Tamminga, S., Fermentation in the large intestine of single-stomached animals and its relationship to animal health, *Nutrition Research Reviews*, 14, 207, 2001.
45. Bugaut, M., Occurrence, absorption and metabolism of short chain fatty acids in the digestive tract of mammals, *Comp. Biochem. Physiol.*, 86, 439, 1987.
46. Bugaut, M. and Bentéjac, M., Biological effects of short-chain fatty acids in nonruminant mammals, *Annual Reviews in Nutrition*, 13, 217, 1993.
47. Ewing, W.N. and Cole, D.J.A., *The Living Gut. An Introduction to Micro-Organisms in Nutrition*, Dungannon, Ireland: Context, 1994.
48. Finegold, M.S. et al., Fecal microflora in seventh day adventists, *Am. J. Clin. Nutr.*, 30, 1781, 1977.
49. Jones, D.T. and Woods, D.R., Acetone-butanol fermentation revisited, *Microbiol. Rev.*, 50, 484, 1986.
50. Dürre, P. and Bahl, H., Microbial production of acetone/butanol/isopropanol, in *Products of Primary Metabolism.*, Vol. 6, M. Roehr, Ed., VCH, Weinheim, Germany, 1996, 229.
51. Cummings, J.H. and Macfarlane, G.T., Gastrointestinal effects of prebiotics, *Br. J. Nutr.*, 87 Suppl. 2, S145, 2002.
52. Cummings, J.H., Short chain fatty acids, in *Human Colonic Bacteria Role in Nutrition, Physiology and Pathology*, Vol. 1, Gibson, G.R. and Macfarlane, G.T., Eds., CRC Press, Boca Raton, FL, 1995, 101.

53. Bergman, E.N., Production and utilization of metabolites by the alimentary tract as measured in portal and hepatic blood, in *Digestion and Metabolism in the Ruminants*, McDonald, I.W. and Wagner, A.C.I., Eds., University of New England, Sydney, Australia, 1975, 295.
54. Macfarlane, G.T. and Gibson, G.R., Metabolic activities of the normal colonic flora, in *Human Health: the Contribution of Microorganisms*, Vol. 1, Stewart, A.W. and Gibson, G.R., Eds., Springer-Verlag, New York, Great Britain, 1994, 17.
55. Elsden, S.R., Hilton, M.G., and Waller, J.M., The end products of the metabolism of aromatic amino acids by clostridia, *Arch. Microbiol.*, 107, 283, 1976.
56. Yokohama, M.T. et al., The effect of antibiotics in the weanling pig diet on growth and the excretions of volatile phenolic and aromatic bacteral metabolites, *Am. J. Clin. Nutr.*, 35, 1417, 1982.
57. Andreesen, J.R., Bahl, H., and Gottschalk, G., Introduction to the physiology and biochemistry of the genus *Clostridia*, in *Clostridia, Biotechnology Handbooks 3*, Minton, N.P. and Clarke, D.J., Eds., Plenum Press, New York, 1989, 27.
58. Macfarlane, G.T., et al., Contribution of the microflora to proteolysis in the human large intestine, *J. Appl. Bacteriol.*, 64, 37, 1988.
59. Hooper, L. and Gordon, J.I., Commensal host-bacterial relationships in the gut, *Science*, 292, 1115, 2001.

Part VII

Novel Developments and Applications

36 Industrially Relevant Fermentations

Thaddeus C. Ezeji, Nasib Qureshi, and Hans P. Blaschek

CONTENTS

36.1 INTRODUCTION

Acetone-butanol (AB) fermentation has a long history as an industrially significant fermentation. The original observation that bacteria produce butanol was made by Pasteur in 1861. Interest in commercializing the process came about in 1909 in England, primarily as a means to obtain butadiene as a raw material for the production of synthetic rubber. In 1912, the father of the AB fermentation process, Dr. Chaim Weizmann, studied the fermentation and isolated a bacterium called *Clostridium acetobutylicum*, which was able to ferment starchy grains to acetone, butanol, and ethanol. In 1915, a patent was issued to Dr. Weizmann for the process, which was subsequently transferred to Canada in 1916. In 1918, a butanol-acetone plant was built in Terre Haute, Indiana. Following World War I, Dupont developed lacquers for the automobile industry that required butyl acetate as a solvent. To supply butanol for conversion to butyl acetate, the acetone-butanol unit in Terre Haute was operated by the Commercial Solvents Corporation. Following expiration of the patent in 1936, AB plants were built throughout the U.S., and from 1916 to the late 1930s, the AB fermentation process was the major process for producing butanol. There were large facilities in the Midwestern U.S., including a facility in Peoria, Illinois, which had ninety-six 50,000 gallon fermentors.

Because of unfavorable economics and competition from the petroleum industry, AB fermentation was discontinued in the early 1960s in the U.S. Following the uncertainty of supply associated with the oil embargo of the early 1970s, the finite nature of petrochemical feedstocks, and increasing costs, there was renewed interest in AB fermentation. Research programs throughout the 1980s and 1990s were focused on the development of systems for genetic manipulation of the solventogenic clostridia, with an eye toward the development of strains with improved fermentation characteristics. It became clear that an improved and economical AB process would require overcoming a major limitation associated with this fermentation, namely the low solvent concentration in the fermentation broth. The cost of the substrate and the energy required for distillative recovery of butanol from a dilute product stream were the major costs having an impact on the economics of the AB fermentation.

Economic modeling studies were carried out on AB fermentation. A two-pronged approach involving modification of the existing cultures, combined with improved downstream recovery of product, was believed to be necessary to make AB fermentation economically viable. We report herein some recent advances in AB fermentation involving the hyper-butanol producing *C. beijerinckii* BA101 strain used in combination with improved downstream product recovery systems. The results obtained suggest that the biological approach for butanol production can be competitive with the petrochemical route for butanol production. This statement is based on the use of starch or starch by-products as substrate. The use of cellulosic substrates will likely improve the economics of AB fermentation even further.

36.2 ACETONE-BUTANOL (AB) FERMENTATION

36.2.1 Substrate and Nutrient Requirements

Species making up the solventogenic clostridia have fairly simple substrate and nutrient requirements and are able to utilize a variety of substrates containing carbohydrates (pentose or hexose sugars), proteins, and mineral salts. Generally, a mixture of organic and inorganic nitrogen sources is required for good growth and solvent production by *C. acetobutylicum* [1] and *C. beijerinckii* [2]. Organic nitrogen sources such as yeast extract provide essential amino acids and growth factors that promote good growth. It was suggested that a low carbon/nitrogen ratio or nitrogen-limited fermentation is required to achieve high levels of solvent production [1]. In a study examining the effect of nitrogen in batch and continuous culture, Lai and Traxler [3] observed that butanol production was reduced at carbon/nitrogen ratios above 7.3. Monot and Engasser [4] demonstrated that it was possible for *C. acetobutylicum* to produce solvent in a nitrogen-limited chemostat, which was maintained at low pH and operated at a low dilution rate. In general, the literature suggests that an optimal nutrient concentration in the fermentation medium is important for production of high levels of solvent. Most of the studies examining acetone-butanol production have employed complex media containing growth supplements such as yeast extract, tryptone, vitamins, and minerals. However, the cost of these supplementary ingredients might not be economical for industrial processes and commercialization. Corn steep water (CSW), a by-product of the corn wet-milling industry that contains a rich complement of important nutrients such as amino acids, vitamins, nitrogen, and minerals was determined to be a good substitute for the more expensive organic-nitrogen-source yeast extract [5,6,7]. This finding is important as it impacts the economics of the acetone-butanol fermentation. However, while CSW is a rich source of growth promoters, it can contain growth inhibitors such as lactic acid, sulfites, phytic acid, and heavy metals (Zn, Cu) and their salts, which must be diluted to reduce their inhibitory effect before CSW can be used as nutrient source for fermentation [6].

The effect of carbon limitation on the onset and maintenance of solvent production in batch, fed-batch, and continuous fermentations is well known. The presence of a limited amount of carbon results in, predominantly, acids being produced [8,9]. The shift to solvent production can occur in cultures in which a large excess of sugar is present, as catabolite repression of solvent-producing enzymes apparently does not occur [4,10]. The presence of an excess of available carbon has been associated with high levels of solvent production. It has been our experience that the cost of substrate (carbon source) has the greatest impact on the cost of producing butanol [11]. Over the years, various substrates such as corn, molasses, potatoes, jerusalem artichokes, cheese whey, apple pomace, algal biomass, etc. have been examined for use in acetone-butanol production. Of the listed substrates, only corn and molasses have been used for the industrial production of acetone-butanol, although there has been a report that sweet potatoes are being used in China to produce acetone-butanol. Efforts are currently underway to examine the use of starch-based waste packing peanuts [12], agricultural waste such as rotten apples and pears [12], soy molasses [13], and corn fiber [14] for acetone-butanol production, with the objectives of reducing the cost of fermentation and utilization of agricultural waste. However, the use of these alternative substrates for the acetone-butanol fermentation can have an impact on the butanol/acetone ratio, solvent yield, and associated solvent productivities. For example, Maddox [15] reported butanol/acetone ratios as high as 10:1 when whey filtrate was the substrate compared to ratios of 3:1 when glucose was the substrate. Bahl et al. [16] demonstrated that nutritional factors affected the ratio of solvents produced by *C. acetobutylicum*.

In general, *Clostridium* spp. require a low redox potential for cellular metabolism and growth [17]. Since the conversion of pyruvate to acetyl-CoA in the solventogenic clostridia involves a ferredoxin oxidoreductase iron-sulfur protein that accepts and donates electrons at very low potential [18,19], the maintenance of a low redox potential is important for the acetone-butanol fermentation. This makes iron an important mineral supplement in clostridia media formulation. In related studies carried out by Parekh et al. [6], it was observed that the addition of $FeSO_4.7H_2O$ to a 6% glucose-1.6% solids CSW medium in a batch culture resulted in a 26% increase in butanol concentration, and the ratio of butanol to acetone was higher than when the culture was grown in the absence of added ferrous ions. However, the addition of manganese and magnesium ions apparently does not have an impact on solvent production. In addition to an increase in solvent production, ferrous ions were found to have a dramatic effect on the butanol/acetone ratio [16]. Furthermore, under iron limitation in a fermentation carried out by *C. acetobutylicum*, lactate instead of butyrate and acetate is the predominant product at a pH greater than 5 [10,19].

36.2.2 Factors Affecting the Fermentation

36.2.2.1 pH

The ability of cells to maintain an intracellular pH value near neutrality during the fermentation of sugars is subject to stress imposed by the fermentation products [20]. The extent to which cells can cope with the stress is very important, since any change in intracellular pH will lead to alteration of cellular functions [20,21]. Due to the sensitivity of the solventogenic clostridia to variations in the environment, the pH of the medium has been recognized to exert a significant influence on the acetone-butanol fermentation.

The typical acetone-butanol batch fermentation carried out by *C. acetobutylicum* or *C. beijerinckii* is characterized by two distinct phases. The first, or acidogenic, phase involves the rapid formation of acetic and butyric acids. These acids are secreted into the medium, causing the pH of the medium to decrease to a low value. The second phase (solventogenic) commences after the "pH breakpoint" has been reached during which acids are re-assimilated and butanol and acetone become the major fermentation products. The assumption has been that a low pH is a prerequisite for solvent production [10,22]. *C. acetobutylicum* cultures maintained at relatively high pH produced more acid and less solvent than cultures maintained at low pH. It has been demonstrated that *C. beijerinckii* is capable of producing acetone-butanol in a medium in which the pH is maintained at pH 7 throughout the fermentation, although such production is dependent on supplementation of the medium with acetate and butyrate at high concentrations [23]. The fact that high concentrations of acetate and butyrate are required to trigger solventogenesis at pH 7 focuses attention on the effect of the undissociated forms of these acids on the cell culture. Several related parameters such as acidic pH, high weak acid concentrations, and high cell densities appear to be involved in triggering solvent production [24].

The onset of acetone-butanol production is normally associated with a decrease in the pH of the medium, which is linked to the accumulation of acid end-products, which are in undissociated form [10]. In poorly buffered media, the pH can decrease to below 4.5 during the early part of fermentation due to sudden termination of solventogenesis, and a phenomenon known as "acid crash" may occur. The acid crash is associated with a rapid termination of solventogenesis after the switch has occurred and when the combined concentration of undissociated acetic and butyric acids exceeds a critical threshold value in the fermentation broth [25]. The acid crash should not be confused with culture degeneration, which occurs as a result of an apparent failure in switching from the acidogenic to the solventogenic phase and takes place over a period of time as might occur during continuous fermentation. Based on this information, increasing the buffering capacity of the growth medium can be a simple method for achieving a high concentration of the less toxic butyrate ion before inhibitory levels of undissociated butyric acid are reached. This would allow for increased growth and carbohydrate utilization, as well as provide a greater amount of butyrate to serve as a precursor for butanol production [26].

In industrial acetone-butanol production, pH has been shown to play an important role in solvent production. During the acetone-butanol fermentation by *C. beijerinckii* BA101 in P2 medium [27], the fermentation was initiated by growing the culture in P2 medium with a pH value near neutrality and then allowing the pH to fall to the optimum value (pH 5.0 to 5.5), which triggers solventogenesis. *C. beijerinckii* BA101, when grown in P2 medium, is able to maintain the optimum pH value for acetone-butanol production after the onset of solventogenesis and requires no external pH adjustment [28,29].

36.2.2.2 Sugar Concentration

Following a change in the environment during bacterial culturing, a lag phase is frequently observed during which cells increase in mass but do not divide. In other words, there is no change in cell number. The increase in cell mass and the duration of the lag phase depends on the clostridial spp. and the composition of the culture medium. Substrate inhibition is a commonly known problem in solvent fermentations. A high sugar concentration (160 g/l) was found to be toxic to *C. beijerinckii* BA101 [29]. *C. beijerinckii* BA101 differs in osmotolerance when compared to the *C. acetobutylicum* P262 industrial strain [30], which is able to tolerate more than 250 g/L lactose present in whey permeate [31]. Qureshi and Blaschek [32], during their evaluation of the effect of different initial sugar concentrations on growth of *C. beijerinckii* BA101, demonstrated that the maximum cell concentration (1.74 g/L dry weight) was achieved at a glucose concentration of 60 to 102 g/L. Cell growth was severely inhibited above 158 g/L glucose, while the lag phases for cell growth at glucose concentrations of 60 and 158 g/L were 6 and 23 h, respectively. The effect of a high initial sugar concentration within the tolerable limits of the microorganism appears to have a more profound effect on the lag phase duration of the culture than on solvent production (Table 36.1). Having survived the high initial osmotic pressure, the cells appear to be able to carry on with their normal metabolism and produce acetone-butanol. However, our experience with the *C. beijerinckii* BA101 fermentation of industrial liquefied starch has shown that the effects of the initial sugar concentration on acetone-butanol production during the batch fermentation process varies with the composition and presence of inhibitors in the medium. The presence of inhibitors lowers the sugar tolerance level of *C. beijerinckii* BA101. Therefore, during operation of fermentation processes,

TABLE 36.1
Effect of Different Initial Sugar Concentrations on Lag Phase, Solvent Production and Cell Concentration Produced by *Clostridium beijerinckii* BA101

Sugar Concentration (g/l)	Lag Phase (h)	Solvent Concentration (g/l)	Cell Concentration (g/l)	Ref.
60, 102, 158, 200	6, 15, 23, 100	24.2, 23.0, 21.9, 1.7	1.7, 1.7, 1.6, 0.35	[28,32]
5, 25, 60, 100, 162	3, 3, 3, 6, 18	1.7, 8.1, 18.3, 16.7, 14.6	1.9, 2.7, 3.1, 2.8, 2.1	[74]

attention must be paid to the concentration of inhibitors that are present in the medium in order to determine the appropriate initial sugar concentration to be employed.

36.2.2.3 Acetate in Feed

It is well known that solvent-producing clostridia will lose the ability to produce solvents following repeated subculture or continuous cultivation — a phenomenon known as degeneration. The addition of sodium acetate to the fermentation medium was found to prevent degeneration in *C. beijerinckii* NCIMB 8052 and *C. beijerinckii* BA101 [33]. Studies have shown that in *C. acetobutylicum*, a 210-kb plasmid (pSOL1) encoding solventogenic genes (*ctfA*, *ctfB*, and *adhE/aad*) is lost during the degeneration process [34,35]. *C. acetobutylicum* grown in phosphate-limited continuous culture did not lose pSOL1 and, therefore, remained stable. Although no indigenous plasmids are found in *C. beijerinckii* NCIMB 8052, which has a 6.7-Mbp single circular chromosome [36], the microorganism is known to degenerate following a period of repeated subculture or continuous cultivation [37]. Chen and Blaschek [38] evaluated the effect of acetate on the molecular and physiological aspects of *C. beijerinckii* NCIMB 8052 solvent production and strain degeneration. It was shown that *C. beijerinckii* NCIMB 8052 grown in modified P2 medium without added acetate produced large amounts of acids and little solvent, consistent with it being a degenerated culture. The addition of 20 m*M* sodium acetate was able to stabilize solvent production by *C. beijerinckii* NCIMB 8052 and to maintain the cell density, while growth decreased rapidly in the absence of additional acetate [33,38]. Different concentrations of sodium acetate have been used during batch fermentations by *C. beijerinckii* NCIMB 8052. The highest butanol concentrations observed were 0.6, 5.3, and 13.9 g/l for cultures grown in medium containing 0, 20, and 60 m*M* sodium acetate, respectively [38]. When grown in medium containing higher concentration of sodium acetate, the culture also utilized more glucose.

RNA and enzyme analyses showed that coenzyme A (CoA) transferase was highly expressed and has higher activity in *C. beijerinckii* NCIMB 8052 grown in medium containing added sodium acetate than when the culture was grown in the absence of sodium acetate. CoA transferase is an important enzyme responsible for acid reassimilation in solventogenic *Clostridium* spp. In addition to CoA transferase, *C. beijerinckii* NCIMB 8052 demonstrated higher acetate kinase- and butyrate kinase-specific activity when the culture was grown in medium containing added acetate [38]. The increase in solvent production by *C. beijerinckii* NCIMB 8052 grown in medium containing higher concentrations of added acetate might be related to acetate assimilation, which increases the carbohydrate utilization efficiency of the culture [33]. Therefore, acid uptake by CoA transferase would result in lower intracellular acetyl-CoA, which can cause the rate of glycolysis to increase. The higher specific activity of acetate kinase- and butyrate kinase enzymes exhibited by *C. beijerinckii* grown in medium containing high concentrations of sodium acetate appears to have contributed to an increase in acid reassimilation, which permits the direct utilization and conversion of the added acetate to solvent.

36.2.2.4 Solvent (AB) Toxicity

An important problem associated with the typical acetone-butanol fermentation is product inhibition due to solvent toxicity, and although butyric acid is more toxic than butanol, its concentration during a normal solvent-producing fermentation remains low. During the acetone-butanol fermentation, solventogenic clostridia are able to carry out cellular functions until the concentration of the solvent reaches inhibitory levels at around 20 g/liter. Above this level, further cellular metabolism ceases [10]. Butanol toxicity has limited the concentration of sugar solutions for fermentation and resulted in low reactor productivity, low acetone-butanol yield, and a low final acetone-butanol concentration. The low final acetone-butanol concentration results in increased recovery costs, the need for large-capacity fermentors, and, ultimately, is a major limitation in the industrial production of solvents by the solventogenic clostridia.

Alcohols (including ethanol and butanol) are known to damage the structure of cell membranes and, thereby, inhibit cellular functions. The ability of butanol to inhibit cellular functions is directly related to the impact it has on the fluidity of the cell membrane [39]. The autolytic degradation of solvent-producing cells of *C. acetobutylicum* P262 was linked to a toxic level of butanol, which was suggested to be involved in the triggering of the release of a cell-free autolysin during the solventogenic phase [10,40,41].

In a typical acetone-butanol batch fermentation, acetone-butanol concentrations in excess of 20 g/liter over a period of 40 to 60 h are difficult to achieve [42]. It has been suggested that an increase to 22 to 28 g/liter would be required to make biological synthesis economically competitive once again [42,43]. In order to achieve this acetone-butanol concentration level, it has been further suggested that increasing the tolerance of solvent-producing clostridia cells to butanol might result in the production of higher concentrations of solvents. To date, however, this dream has not been realized. Solvent toxicity still remains an important impediment to the commercialization of biologically produced acetone-butanol. Therefore, to solve this problem, *in situ*/online butanol removal currently appears to be the most viable path to follow. A variety of alternative methods, including membrane-based systems, such as pervaporation [44], perstraction [45], and reverse osmosis [46], as well as adsorption [47], liquid-liquid extraction [48], and gas stripping [29,49] have been examined. The application of some of these techniques results in the use of a concentrated sugar solution in the fermentor [50], and a reduction in butanol inhibition and high sugar utilization [30], thereby reducing volumes of the process streams. In such systems, up to 100% utilization of the sugar available in the feed has been demonstrated [29].

36.3 RECENT TECHNOLOGIES FOR ACETONE-BUTANOL (AB) RECOVERY: INDUSTRIALLY RELEVANT PROCESSES

Early industrial production of acetone-butanol (AB) used batch fermentation processes, which suffered from severe product inhibition. The traditional product recovery technique for the AB fermentation process is distillation, which is carried out at the end of the process. The substrate concentration in the batch is low, and the concentration of butanol in the fermentation broth is limited due to toxicity of butanol to the culture. This approach results in high volume process streams, including waste water streams, leading to high recovery and waste disposal costs. We believe that to improve the product yield and reactor productivity, to bring about a reduction in process streams, and improve the economics of AB production, it will be necessary to use integrated AB fermentation technologies. Over the last two decades, considerable effort has been directed toward the development of efficient AB fermentation processes that are integrated with *in situ* product recovery. The simultaneous fermentation and product recovery approach appears to have solved most of the early problems associated with the AB fermentation, such as product inhibition, poor reactor productivity, and high volume process streams encountered in the traditional AB batch fermentation process. Integrated fermentation-product recovery techniques such as pervaporation, perstraction, and reverse osmosis, as well as adsorption, liquid-liquid extraction, and gas stripping, have been examined by various investigators. In recent years, the principles and the adaptability of these simultaneous fermentation and recovery techniques to AB fermentation have been investigated in the authors' laboratory at the University of Illinois, Urbana. Gas stripping and pervaporation appear to be the most promising of the *in situ* AB fermentation and recovery techniques, but in terms of cost effective industrial application, gas stripping appears to be the most promising.

36.3.1 Gas Stripping

Gas stripping has been described as one of the most important techniques for removing butanol from the fermentation broth [51,52]. Gas stripping allows for selective removal of volatiles from fermentation broth and uses no membrane [29]. Ennis et al. [53] were the first to investigate the application of gas stripping in butanol fermentation using *C. acetobutylicum*. The volatile properties of the AB permit easy product removal by gas stripping. Gas can be sparged into the fermentor through a rotating fermentor shaft, and volatiles can be condensed and recovered from the condenser. During gas stripping, stringent anaerobic conditions were maintained using oxygen-free nitrogen gas.

In Figure 36.1a, a typical batch or fed-batch fermentation process integrated with an *in situ* gas stripping product-removal system is shown. A batch fermentation (33 to 35°C) by *C. beijerinckii* BA101 was allowed to proceed up to 15 to 36 h. Depending on the initial concentration of the substrate when the AB concentrations approached 3 to 4 g/liter, gas stripping was initiated by bubbling nitrogen or fermentation gas (CO_2 and H_2) through the fermentation broth and recycling these gases through the system using a twin-head peristaltic pump [29,54]. The AB vapors were cooled to –2°C in a condenser using ethylene glycol (50% vol/vol) circulated at a flow rate of 600 ml/min through the condenser, and the stripped AB was pumped into the solvent collector [29]. In this system, both the fermentation and gas-stripping process take place inside the fermentor vessel. A separate gas stripper can be used in which the fermentation broth is circulated over a stripping column and the stripping gas is introduced at the bottom of the column as shown in Figure 36.1b.

The gas-stripping process has a number of advantages over other removal processes, for example, it is simple and inexpensive to operate and does not suffer from fouling or clogging due to the accumulation of biomass [29]. Ezeji et al. [29], working on batch ABE fermentation and *in situ* recovery by gas stripping, demonstrated that incorporation of gas stripping during fermentation resulted in a decrease in fermentation time, a high glucose utilization rate, and complete utilization of glucose and acids. This work demonstrated that solvent removal by gas stripping promotes acid utilization and conversion to solvent. Furthermore, due to decreased inhibition, an elevated cell concentration was observed in the fermentor, leading to increased reactor productivity. It is interesting to note that gas stripping does not remove acids from the fermentor. In addition to making the use of concentrated sugar solution possible and improving the glucose utilization rate in such an integrated fermentation, a concentrated AB stream is obtained that requires less energy for further recovery.

36.3.2 Pervaporation

Pervaporation is a membrane-based process that is used to remove solvents from fermentation broth selectively by using a membrane [52,55]. The volatiles diffuse through a solid membrane leaving behind dissolved solids (nutrients and sugars), and microbial cells. The concentration of solvents across the membrane depends upon membrane selectivity, which is a function of feed solvent concentration and membrane composition. Pervaporation has been used by Groot et al. [44], Larrayoz and Puigjaner [56], and Qureshi and Blaschek [28] in batch butanol fermentation processes. Pervaporation has also been used for the removal of butanol from the fermentation broth

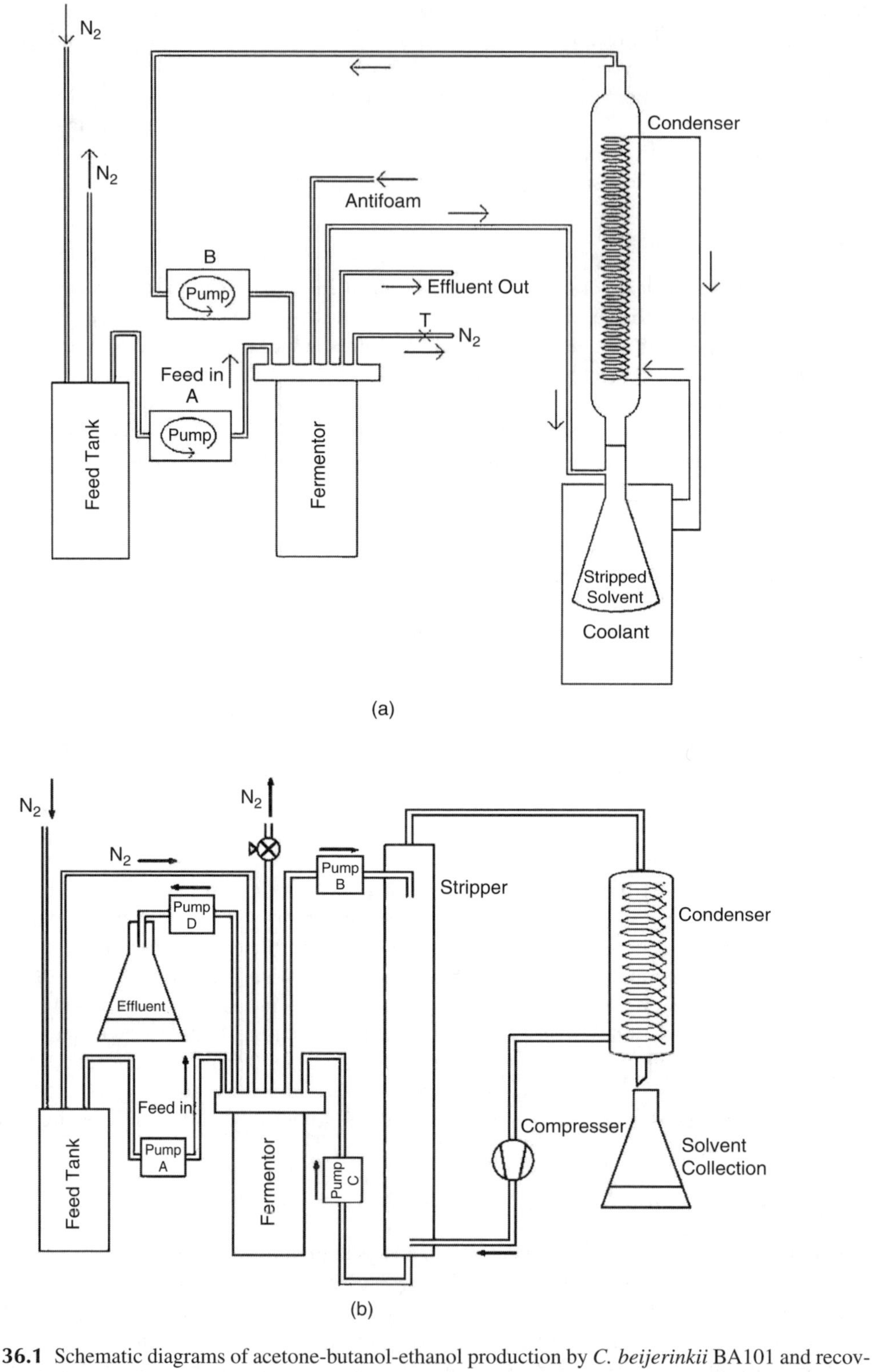

FIGURE 36.1 Schematic diagrams of acetone-butanol-ethanol production by *C. beijerinkii* BA101 and recovery by gas stripping. (a) Gas stripping process without a separate stripping column; (b) gas stripping process with a separate stripping column.

in fed-batch bioreactors [57,58]. In the fed-batch bioreactors, a concentrated sugar solution was used to reduce the process stream volume. Qureshi et al. [45] and Friedl et al. [59] used a polypropylene membrane through which acids diffused at high acid concentration in the fermentation broth. It is desirable that acids (butanol fermentation reaction intermediates) do not diffuse through the membrane. The retention of these intermediates results in an improved acetone-butanol yield.

The use of a high selectivity liquid membrane, has been described by Matsumura et al. [60]. The membrane contained oleyl alcohol and was supported on a microporous polypropylene 25-mm-thick flat sheet. The butanol diffused through the membrane and demonstrated a selectivity of 180 as compared to a silicone membrane selectivity of approximately 45. The authors estimated that if this pervaporation membrane was used as a pretreatment process for butanol separation, the energy requirements would be only 10% that of conventional distillation. It should be noted that the membrane was not stable.

In order to develop a stable and highly selective superior membrane, silicalite was included in a silicone membrane [61]. The selectivity of the silicone-silicalite membrane was improved from approximately 40 to 209. The membrane was used for butanol recovery from both butanol model solutions and fermentation broths. A comparison of various membranes suggested that this membrane might be superior to other available membranes [27].

Recovery of butanol by pervaporation has received considerable attention. Use of pervaporation results in an increased rate of solvent production, a higher rate of sugar utilization, and recovery and partial concentration of butanol and acetone. Pervaporation does not harm the culture [56,62,63]. Although several membrane types have been studied [silicone (polydimethylsiloxane, PDMS), polypropylene, liquid (oleyl alcohol and polypropylene), polytetrafluoroethylene, and silicone membranes filled with silicalite [63]; for the removal of butanol from fermentation broth, silicone-based membranes are still considered to be the best available.

36.4 ADVANCED AB FERMENTATION: SIMULTANEOUS FERMENTATION AND RECOVERY

Acetone-butanol production using the clostridia has been widely studied, especially following the 1973 oil crisis, which led to renewed interest in the AB fermentation as well as investigation into product recovery. Although the energy problem does not seem as urgent today as it did 30 years ago, there has been recent concern throughout the world about the diminishing oil reserves and the need for alternative energy approaches. Butanol can be used as fuel or a fuel extender, which has several advantages over other fuels produced by fermentation. Some of these advantages include high energy content, miscibility with gasoline and diesel fuel, and a low vapor pressure [28,29]. At present, the concentration of butanol obtained during a nonintegrated batch, fed-batch, or continuous fermentation is too low to be economical due to the high energy cost that is required to recover butanol from the fermentation broth. However, the possible reintroduction of a large-scale AB fermentation process appears increasingly feasible. This is due in part to recent advances in our understanding of the genetics and physiology of solvent production by the clostridia, improved downstream recovery processes, and economic and environmental considerations [12,29,32,43,54,64]. In recent years, considerable progress has been made toward the development of efficient integrated fermentation processes (especially gas stripping and pervaporation) involving simultaneous bioconversion and AB removal to avoid the inhibitory effect of butanol on the culture during fermentation [28–29,54]. These processes have allowed for the use of concentrated sugar solutions (up to 500 g/liter) and the production of high concentration AB streams [29,54]. The development of these processes has brought AB fermentation closer to commercialization. A comparison of reactor productivity, sugar utilization, solvent yield, and recovered AB concentration in the presence and absence of gas stripping is shown in Table 36.2.

TABLE 36.2
Effect of *in situ* Product Recovery by Gas Stripping on AB Fermentation Parameters

Control Fermentation Parameters			Fermentation Parameters with Gas Stripping					
ABE Yield (gg^{-1})	Productivity ($gl^{-1}h^{-1}$)	Sugar Utilization (gl^{-1})	ABE Yield (gg^{-1})	Productivity ($gl^{-1}h^{-1}$)	Sugar Utilization (gl^{-1})	Recovered ABE Concentration (gl^{-1})	Comments	Ref.
0.40	0.29	44.6	0.47	0.60	161.7	45–92	Sugar batch fermentation	[29]
0.26	0.07	31.0	0.35	0.32	199.0	20–120	Lactose batch fermentation	[30]
0.22	0.39	29.0	0.27[a]	0.31	60	75.9	Whey batch fermentation	[53]
0.39	0.29	45.4	0.47	1.16	500.1	40–172	Fed-batch fermentation	[54]

[a] Low value due to loss of solvents.

36.4.1 Batch Culture Systems

The general characteristics of batch culture systems are well known. Cultural conditions undergo continual change, the cell population increases, and the substrate concentration decreases. Usually, the time required to carry out a batch AB process is 48 to 90 h, depending on the culture and the type of substrate being used. Early in the 20th century, industrial AB production by *C. acetobutylicum* flourished. AB fermentation was carried out in a batch process, using 50,000- to 200,000-gallon fermentors, which lacked mechanical agitation systems [10]. The final concentration of solvents produced ranged from 12 to 20 g/liter, which were separated from the fermentation broth by distillation. Batch technology was the process of choice due to the ease of operation and low requirements for complete sterilization of the substrate. However, the batch process as practiced at that time does not stand the economic competitiveness test of the 21st century. The successful development of a commercially viable clostridial batch fermentation would be enhanced if the fermentation could be integrated with product recovery.

Recently, fermentation and gas stripping were integrated in order to achieve a high productivity and a concentrated AB stream. In a batch gas stripping integrated process, *C. beijerinckii* BA101 produced 75.9 g/liter total solvents when compared to a nonintegrated batch reactor where 17.7 g/liter total solvents was produced [29]. In this process, a sugar concentration as high as 161.7 g/liter was used, as compared to 59.2 g/liter in a nonintegrated batch reactor. The AB concentration in the recovered stream was 45 to 92 g/liter, depending on the concentration in the fermentation broth. Furthermore, AB productivities and yield were improved up to 200 and 18%, respectively, when compared to nonintegrated batch fermentations. In a similar investigation involving fermentation integrated with gas stripping, Groot et al. [49] demonstrated that product inhibition by butanol resulted in the termination of fermentation activity after 80 h. At that time, the butanol concentration was 8.8 g/liter and the glucose consumption was 37 g/liter. In fermentations employing gas stripping for product recovery, the butanol concentration in the broth was lowered, and a reduction in product inhibition enabled the continuation of substrate consumption, resulting in complete utilization of substrate [49]. It should be noted that during gas stripping, acids are not removed from the fermentation broth.

36.4.2 Fed-Batch Culture Systems

Fed-batch fermentation is an industrial technique in which the bioreactor is initiated with a relatively low substrate concentration in order to reduce substrate inhibition. This may or may not be a growth-limiting substrate; as the substrate is consumed, it is replaced by adding a concentrated substrate solution at a low rate, while keeping the substrate concentration in the reactor below the toxic level [54,57,58]. The major objectives of fed-batch culture systems in a classical fermentation are to improve the overall reactor productivity and to increase the final product concentration in the fermentation broth. The production of a highly concentrated product stream will reduce the cost of product recovery, which is a major factor in the commercial success of biologically produced AB.

Fed-batch fermentation is advantageous in cases where an initial high substrate concentration is toxic to the culture and a high product concentration in the fermentation broth is desired. However, when fed-batch is applied to systems that involve the production of a toxic compound, a product-removal technique must be applied in combination with the fed-batch fermentation process. Gas stripping has been successfully used to remove solvents from fed-batch fermentations [54]. The substrate and butanol toxicity can be kept below the inhibitory levels by feeding the reactor at a slow and controlled rate, while the product-removal technique is applied simultaneously to remove the AB being produced. The application of these two techniques solves two toxicity problems, namely substrate inhibition and butanol toxicity. In the integrated fed-batch fermentation and gas stripping recovery system, a total of 500 g glucose was used to produce 232.8 g solvents (77.7 g acetone, 151.7 g butanol, 3.4 g ethanol) in a 1 liter culture broth, and solvent productivities were improved by 400% over that of the control batch fermentation productivities (Table 36.2).

36.4.3 Continuous Culture Systems

For many bacterial physiological investigations, it is desirable to grow the cells over an extended period of time, during which the cultural conditions are kept constant and the cells grow at a fairly constant exponential rate. The continuous culture system can be classified into four types, namely: (a) chemostat, in which the feed rate is set and cell growth is limited by a selected nutrient; (b) auxostat, in which a growth dependent parameter (pH, cell density, products, dissolved oxygen concentration, etc.) is kept constant by adjusting the feeding rate of the medium; (c) multistage, in which the maximum cell growth, acid production, solvent production, etc. may occur in separate bioreactors; and (d) cell recycle, in which the continuous fermentation is operated at a dilution rate higher than the maximum specific growth rate and the concentrations of the biomass and fermentation products are increased by recycling using either centrifugation, immobilization, sedimentation, or filtration techniques.

The continuous culture system has been an important research tool for the determination of parameters responsible for the changes in the physiology and activity of solvent-producing clostridia during growth. An industrial fermentation process that can be operated in a continuous mode might provide some advantages over a batch process. For example, only one batch of inoculum culture would be needed over the AB production period, the volume of the fermentor can be reduced without reduction in production capacity, and the time necessary for cleaning and sterilization of the equipment would be drastically reduced. Due to the complexity of AB fermentation, especially the production of fluctuating levels of solvents and the problems of culture stability, the use of single stage continuous processes for the industrial production of AB does not appear to be feasible at this time. It is well known that the final concentration of solvents in a batch culture system rarely exceeds 20 g/liter. The final concentration in a typical continuous culture system is even lower when compared to that achieved in a batch culture system. The longevity of the fermentation by solventogenic clostridia is reduced at high butanol concentrations, and as a consequence, the culture must be maintained in a state of simultaneous growth and solvent production [65,66]. For example, in a single stage continuous reactor, Fick et al. [67] maintained a stable continuous culture of *C. acetobutylicum* for one month at a productivity of 0.75 g/lh and at a 13g/l solvent concentration. Higher overall productivity can be obtained in a single stage continuous reactor, but this usually occurs at the expense of reduced substrate utilization and volumetric productivity.

To be industrially competitive, a continuous culture process should yield a high final AB concentration similar to that obtainable with the batch culture process. The reactor effluent is difficult to recycle due to the presence of butanol, which is toxic. To recycle reactor effluent and improve sugar utilization and volumetric productivity, it is desirable to remove butanol from the effluent prior to recycling. During continuous ABE fermentation by *C. acetobutylicum*, increases in lactose utilization, solvent concentration, and productivity were achieved when using in-line product removal by gas stripping [68]. Solvent levels in the fermentation broth were drastically reduced due to gas stripping, which decreases the toxicity and redirects the equilibrium toward solvent production.

36.5 CURRENT ECONOMIC SCENARIOS

During the past 13 years, a number of developments have been made at the University of Illinois in the production of butanol from agricultural resources. These developments include:

1. Development of hyper-butanol producing strain of *C. beijerinckii* BA101 [69]
2. Optimization of media and growth conditions for enhanced butanol production in batch reactors [6,7]
3. Immobilized cell continuous fermentation and pervaporative recovery (ICCFPR) [70]
4. Continuous fermentation and gas stripping recovery (CFGSR) [71,73]
5. Batch fermentation and distillative recovery (BFDR) [72]
6. Batch fermentation and pervaporative recovery (BFPR) [28]
7. Fed-batch fermentation and pervaporative recovery (FBFPR) [58]
8. Fed-batch fermentation and gas stripping recovery (FFGSR) [54]

It is believed that the further development of these processes will lead to commercialization of the AB process. With the aim of identifying the most energy-efficient process, economic evaluations of some of these processes were performed. It was assumed that the fermentation facility would be an extension of an existing wet corn milling facility located in the Midwestern United States. For such a situation, the cost of butanol was estimated to be \$0.55 kg^{-1} for a BFDR process [72]. During this exercise, it was determined that ABE yield and cost of substrate are the two most influential factors that affect butanol cost. An additional factor that affects the cost of butanol is by-product credit [73].

By-product credit was estimated to be approximately 3.7 times that of the cost of butanol [73]. Based on this by-product credit, butanol cost for BFPR, FBFPR, and ICCFPR was estimated to be \$0.14–0.39, \$0.12–0.37, and \$0.11–0.36 kg^{-1}, respectively. However, these amounts are based on a number of assumptions, some of which have been outlined previously [72,73]. At the time of writing the report by Qureshi and Blaschek [73], the butanol price was reported to be \$1.21 kg^{-1} (Chemical Marketing Reporter; October 16, 2000). Since these economic evaluations were carried out, two other integrated processes (CFGSR and FFGSR) have been developed. Although an economic evaluation of these processes has not yet been performed, they are considered to be superior to BFPR, FBFPR, and ICCFPR processes. In processes 3 and 4 (ICCFPR and CFGSR) and 7 and 8 (FBFPR and FFGSR), a concentrated sugar feed stream is used to reduce the energy requirements for fermentation and downstream processing.

Considering the recent developments in fermentation, upstream, and downstream processing, we believe that butanol production by fermentation is close to commercialization. As of this writing, scale-up of the AB fermentation is being attempted at one of the largest corn processing facilities in the world, located in the Midwestern United States. It is anticipated that corn steep liquor, a by-product of the wet corn milling process, would be utilized as a nutrient source.

36.6 CONCLUSIONS

Acetone-butanol fermentation is a relatively difficult process for large-scale industrial application due to butanol toxicity to the producing culture and strain degeneration. The potential for incorporating in-line product recovery processes during AB fermentation has generated considerable interest. Further development of clostridial strains that can produce and tolerate higher levels of butanol during fermentation would be ideal. We feel that, given the prevailing circumstances, AB fermentation can be run economically on an industrial scale if integrated with product-removal techniques such as gas stripping or pervaporation. By employing these approaches, the butanol toxicity to the culture and the cost of product recovery are drastically reduced.

REFERENCES

1. Madihah, M.S., Ariff, A.B., Sahaid, K.M., Suraini, A.A., and Karim, M.I.A., Direct fermentation of gelatinized sago starch to acetone-butanol-ethanol by *Clostridium acetobutylicum*, *World J. Microbiol. Biotechnol.*, 17, 567, 2001.
2. Nimcevic, D., Schuster, M., and Gapes, J.R., Solvent production by *Clostridium beijerinckii* NRRL B592 growing on different potato media, *Appl. Microbiol. Biotechnol.*, 50, 426, 1998.
3. Lai, M.C. and Traxler, R.W., A coupled-stage continuous fermentation for solvents production by *Clostridium acetobutylicum*, *Enzyme Microbial Technol.*, 16, 1021, 1994.
4. Monot, F. and Engasser, J.M., Production of acetone and butanol by batch and continuous culture of *Clostridium acetobutylicum* under nitrogen limitation, *Biotechnol. Lett.*, 5, 213, 1983.
5. Hull, S., Yang, B.Y., Venzke, D., Kulhavy, K., and Montgomery, R., Composition of corn steep water during steeping, *J. Agric. Food Chem.*, 44, 1857, 1996.
6. Parekh, M., Formanek, J., and Blaschek, H.P., Development of a cost-effective glucose-corn steep medium for production of butanol by *Clostridium beijerinckii*, *J. Ind. Microbiol. Biotechnol.*, 21, 187, 1998.
7. Parekh, M., Formanek, J., and Blaschek, H.P., Pilot-scale production of butanol by *Clostridium beijerinckii* BA101 using a low-cost fermentation medium based on corn steep water, *Appl. Microbiol. Biotechnol.*, 51, 152, 1999.
8. Campos, E.J, Qureshi, N., and Blaschek, H.P., Production of acetone butanol ethanol from degermed corn using *Clostridium beijerinckii* BA101, *J. Appl. Biochem. Biotechnol.*, 98, 553, 2002.
9. Qureshi, N. and Maddox, I.S., Continuous solvent production from whey permeate using cells of *Clostridium acetobutylicum* immobilized by adsorption onto bonechar, *Enzyme Microbial Technol.*,9, 668, 1987.
10. Jones, D.T. and Woods, D.R., Acetone-butanol fermentation revisited, *Microbial Reviews,* 50, 484, 1986.
11. Qureshi, N. and Blaschek, H.P., ABE production from corn: a recent economic evaluation, *J. Ind. Microbiol. Biotechnol.*, 27, 292, 2001.
12. Jesse, T.W., Ezeji, T.C., Qureshi, N., and Blaschek, H.P., Production of butanol from starch-based waste packing peanuts and agricultural waste, *J. Ind. Microbiol. Biotechnol.*, 29, 117, 2002.
13. Qureshi, N., Lolas, A., and Blaschek, H.P., Soy molasses as fermentation substrate for production of butanol using *Clostridium beijerinckii* BA101, *J. Ind. Microbiol. Biotechnol.*, 26, 290, 2001.
14. Qureshi, N., Ebener, J., Ezeji, T.C., Dien, B., Cotta, M.A., and Blaschek, H.P., Production of butanol from corn fiber: hydrolysis and fermentation using *Clostridium beijerinckii,* The Twenty-Fifth Symposium on Biotechnology for Fuels and Chemicals. Breckenridge, CO, May 4–7, 2003.
15. Maddox, I.S., Production of n-butanol from whey filtrate using *Clostridium acetobutylicum* NCIB2951, *Biotechnol. Lett.*, 2, 493, 1980.
16. Bahl, H., Gottwald, M., Kuhn, A., Rale, V., Andersch, W., and Gottschalk, G., Nutritional factors affecting the ratio of solvents produced by *Clostridium acetobutylicum, Appl. Environ. Microbiol.*, 52, 169, 1986.
17. Douglas, F., Hambleton, R., and Rigby, G.J., An investigation of the oxidation-reduction potential and of the effect of oxygen on the germination and outgrowth of *Clostridium butyricum* spores, using platinum electrodes, *J. Bacteriol.*, 36, 625, 1973.

18. Kim, J., Bajpai, R., and Iannotti, E.L., Redox potential in acetone-butanol fermentations, *Appl. Biochem. Biotechnol.*, 18, 175, 1988.
19. Peguin, S., Goma, G., Delorme, P., and Soucaille, P., Metabolic flexibility of *Clostridium acetobutylicum* in response to methyl viologen addition, *Appl. Microbiol. Biotechnol.,* 42, 611, 1994.
20. Huang, L., Forsberg, C.W., and Gibbins, L.N., Influence of external pH and fermentation products on *Clostridium acetobutylicum* intracellular pH and cellular distribution of fermentation products, *Appl. Environ. Microbiol.*, 51, 1230, 1986.
21. Konings, W.N. and Veldkamp, H., Energy transduction and solute transport mechanisms in relation to environments occupied by microorganism, *Symp. Soc. Gen. Microbiol.*, 34, 153, 1983.
22. Kim, B.H., Bellows, P., Datta, R., and Zeikus, J.G., Control of carbon and electron flow in *C. acetobutylicum* fermentations: utilization of carbon monoxide to inhibit hydrogen production and to enhance butanol yields, *Appl. Environ. Microbiol.,* 48, 764, 1984.
23. Holt, R.A., Stevens, G.M., and Morris, J.G., Production of solvents by *Clostridium acetobutylicum* cultures maintained at neutral pH, *Appl. Environ. Microbiol.,* 48, 1166, 1984.
24. George, H.A. and Chen, J.-S., Acidic conditions are not obligatory for onset on butanol formation by *Clostridium beijerinckii* (synonym. *C. butylicum*), *Appl. Environ. Microbiol.*, 46, 321, 1983.
25. Maddox, I.S., Steiner, E., Hirsch, S., Wessner, S., Gutierrez, N.A., Gapes, J.R., and Schuster, K.C., The cause of "acid crash" and "acidogenic fermentations" during the batch acetone-butanol-ethanol (ABE-) fermentation process, *J. Mol. Microbiol. Biotechnol.,* 2, 95, 2000.
26. Bryant, D.L. and Blaschek, H.P., Buffering as a means for increasing growth and butanol production by *Clostridium acetobutylicum, J. Ind. Microbiol.*, 3, 49, 1988.
27. Qureshi, N. and Blaschek, H.P., Butanol recovery from model solutions/fermentation broth by pervaporation: evaluation of membrane performance, *Biomass Bioenergy,* 17, 175, 1999.
28. Qureshi, N. and Blaschek, H.P., Production of acetone butanol ethanol (ABE) by a hyper-producing mutant strain of *Clostridium beijerinckii* BA101 and recovery by pervaporation, *Biotechnol. Prog.*, 15, 594, 1999.
29. Ezeji, T.C., Qureshi, N., and Blaschek, H.P., Production of butanol by *Clostridium beijerinckii* BA101 and *in-situ* recovery by gas stripping, *World J. Microbiol. Biotechnol.,* 19, 595, 2003.
30. Maddox, I.S., Qureshi, N., and Roberts-Thomson, K., Production of acetone butanol ethanol from concentrated substrates using *Clostridium acetobutylicum* in an integrated fermentation-product removal process, *Process Biochem.,* 30, 209, 1995.
31. Qureshi, N. and Maddox, I.S., Production of butanol from concentrated lactose/whey permeate using *Clostridium acetobutylicum* and removal by perstraction. Biotechnology Conference for Fuels and Chemicals, Breckenridge, CO, May 4–7, 2003, 25.
32. Qureshi, N. and Blaschek, H.P., Recent advances in ABE fermentation: hyper-butanol producing *Clostridium beijerinckii* BA101, *J. Ind. Microbiol. Biotechnol.,* 27, 287, 2001.
33. Chen, C.-K and Blaschek, H.P., Acetate enhances solvent production and prevents degeneration in *Clostridium beijerinckii* BA101, *Appl. Microbiol. Biotechnol.,* 52, 170, 1999.
34. Stim-Herndon, K.P., Nair, R.V., Papoutsakis, E.T., and Bennett, G.B., Analysis of degenerate variants of *Clostridium acetobutylicum* ATCC 824, *Anaerobe,* 2, 11, 1996.
35. Cornillot, E., Nair, R.V., Papoutsakis, E.T., and Soucaille, P., The genes for butanol and acetone formation in *Clostridium acetobutylicum* ATCC 824 reside on a large plasmid whose loss leads to degeneration of the strain, *J. Bacteriol.*, 179, 5442, 1997.
36. Wilkinson, S.R. and Young, D.I., Physical map of the *Clostridium beijerinckii* (formerly *Clostridium acetobutylicum*) NCIMB 8052 chromosome, *J. Bacteriol.*, 177, 439, 1995.
37. Kashket, E.R. and Cao, Z.-Y., Isolation of a degeneration-resistant mutant of *Clostridium acetobutylicum* NCIMB 8052, *J. Bacteriol.*, 59, 4198, 1993.
38. Chen, C.-K. and Blaschek, H.P., Effect of acetate on molecular and physiological aspects of *Clostridium beijerinckii* NCIMB 8052 solvent production and strain degeneration, *Appl. Environ. Microbiol.*, 65, 499, 1999.
39. Baer, S.H., Blaschek, H.P., and Smith, T.L., Effect of butanol challenge and temperature on lipid composition and membrane fluidity of butanol-tolerant *Clostridium acetobutylicum, Appl. Environ. Microbiol.*, 53, 2854, 1987.
40. Van der Westhuizen, A., Jones, D.T., and Woods, D.R., Autolytic activity and butanol tolerance of *Clostridium acetobutylicum, Appl. Environ. Microbiol.*, 44, 1277, 1982.

41. Barber, J.M., Robb, F.T., Webster, J.R., and Woods, D.R., Bacteriocin production by *Clostridium acetobutylicum* in industrial fermentation process, *Appl. Environ. Microbiol.*, 37, 433, 1979.
42. Woods, D.R., The genetic engineering of microbial solvent production, *Trends Biotechnol.,* 13, 259, 1995.
43. Dürre, P., New insights and novel developments in clostridial acetone/butanol/isopropanol fermentation, *Appl. Microbiol. Biotechnol.*, 49, 639, 1998.
44. Groot, W.J., van den Oever, C.E., and Kossen, N.W.F., Pervaporation for simultaneous product recovery in the butanol/isopropanol batch fermentation, *Biotechnol. Lett.*, 6, 709, 1984.
45. Qureshi, N., Maddox, I.S., and Friedl, A., Application of continuous substrate feeding to the ABE fermentation: relief of product inhibition using extraction, perstraction, stripping, and pervaporation, *Biotechnol. Prog.*, 8, 382, 1992.
46. Garcia, A., Lanotti, E.L., and Fischer, J.L., Butanol fermentation liquor production and separation by reverse osmosis, *Biotechnol. Bioeng.*, 28, 785, 1986.
47. Nielson, L., Larsson, M., Holst, O., and Mattiasson, B., Adsorbents for extractive bioconversion applied to the acetone butanol fermentation, *Appl. Microbiol. Biotechnol.*, 28, 335, 1988.
48. Evans, P.J. and Wang, H.W., Enhancement of butanol fermentation by *Clostridium acetobutylicum* in the presence of decanol-oleyl alcohol mixed extractants, *Appl. Environ. Microbiol.*, 54, 1662, 1988.
49. Groot, W.J., van der Lans, R.G.J.M., and Luyben, K.Ch.A.M., Batch and continuous butanol fermentation with free cells: integration with product recovery by gas stripping, *Appl. Microbiol. Biotechnol.*, 32, 305, 1989.
50. Qureshi, N. and Blaschek, H.P., Recovery of butanol from fermentation broth by gas stripping, *Renewable Energy*, 22, 557, 2001.
51. Maddox, I.S., The acetone-butanol-ethanol fermentation: recent progress in technology, *Biotechnol. Genetic Eng. Reviews*, 7, 189, 1988.
52. Groot, W.J., van der Lans, R.G.J.M., and Luyben, K.Ch.A.M., Technologies for butanol recovery integrated with fermentations, *Process Biochem.*, 27, 61, 1992.
53. Ennis, B.M., Marshall, C.T., Maddox, I.S., and Paterson, A.H.J., Continuous product recovery by in situ gas stripping/condensation during solvent production from whey permeate using *Clostridium acetobutylicum, Biotechnol. Lett.*, 8, 725, 1986.
54. Ezeji, T.C., Qureshi, N., and Blaschek, H.P., Acetone-butanol-ethanol (ABE) production from concentrated substrate: reduction in substrate inhibition by fed-batch technique and product inhibition by gas stripping, *Appl. Microbiol. Biotechnol.,* 63, 653, 2004.
55. Maddox, I.S., The acetone-butanol-ethanol fermentation: recent progress in technology, *Biotechnol. Genetic Eng. Reviews*, 7, 190, 1989.
56. Larrayoz, M.A. and Puigjaner, L., Study of butanol extraction through pervaporation in acetobutylic fermentation, *Biotechnol. Bioeng.,* 30, 692, 1987.
57. Qureshi, N. and Blaschek, H.P., Butanol production using hyper-butanol producing mutant strain of *Clostridium beijerinckii* BA101 and recovery by pervaporation, *Appl. Biochem. Biotechnol.,* 84, 225, 2000.
58. Qureshi, N., Meagher, M.M., Huang, J., and Hutkins, R.W., Acetone butanol ethanol (ABE) recovery by pervaporation using silicalite-silicone composite membrane from fed-batch reactor of *Clostridium acetobutylicum, J. Membrane Sci.*, 187, 93, 2002.
59. Friedl, A., Qureshi, N., and Maddox, I.S., Continuous acetone-butanol-ethanol (ABE) fermentation using immobilized cells of *Clostridium acetobutylicum* in a packed bed reactor and integration with product removal by pervaporation, *Biotechnol. Bioeng*., 38, 518, 1991.
60. Matsumura, M., Kataoka, H., Sueki, M., and Araki, K., Energy saving effect of pervaporation using oleyl alcohol liquid membrane in butanol purification, *Bioproc. Eng.,* 3, 93, 1988.
61. Qureshi, N., Meagher, M.M., and Hutkins, R.W., Recovery of butanol from model solutions and fermentation broth using a silicalite/silicone membrane, *J. Membrane Sci.*, 158, 115, 1999.
62. Groot, W.J., Den, M.C.H., Baart de la Faille, T., van der Lans, R.G.J.M., and Luyben, K.Ch.A.M., Integration of pervaporation and continuous butanol fermentation with immobilized cells. I: Experimental results, *Chem. Eng. J.*, 46, B1, 1991.
63. Qureshi, N., Recovery of alcohol fuels using selective membranes by pervaporation, Ph.D. Dissertation, University of Nebraska, Lincoln, 1997.
64. Lee, J., Mitchell, W.J., and Blaschek, H.P., Glucose uptake in *Clostridium beijerinckii* NCIMB 8052 and the solvent-hyperproducing mutant BA101, *Appl. Environ. Microbiol.*, 67, 5025, 2001.

65. Monot, F., Engasser, J.M., and Petitdemange, H., Regulation of acetone butanol production in batch and continuous cultures of *Clostridium acetobutylicum, Biotechnol. Bioeng. Symp.,* 13, 207, 1983.
66. Godin, C. and Engasser, J.M., Two-stage continuous fermentation of *Clostridium acetobutylicum*: effects of pH and dilution rate, *Appl. Microbiol. Biotechnol.,* 33, 269, 1990.
67. Fick, M., Pierrot, P., and Engasser, J.M., Optimal conditions for long-term stability of acetone-butanol production by continuous culture of *Clostridium acetobutylicum, Biotechnol. Lett.,* 7, 503, 1985.
68. Ennis, B.M., Qureshi, N., and Maddox, I.S., In-line toxic product removal during solvent production by continuous fermentation using immobilized *Clostridium acetobutylicum, Enzyme Microb. Technol.*, 9, 672, 1987.
69. Annous, B.A. and Blaschek, H.P., Isolation and characterization of *Clostridium acetobutylicum* mutants with enhanced amylolytic activity, *Appl. Environ. Microbiol.*, 57, 2544, 1991.
70. Lienhardt, J., Schripsema, J., Qureshi N., and Blaschek, H.P., Acetone butanol ethanol (ABE) production in an immobilized cell reactor of *Clostridium beijerinckii* BA101 and recovery by pervaporation, *Appl. Biochem. Biotechnol.,* 98, 591, 2002.
71. Ezeji, T.C., Qureshi, N., and Blaschek, H.P., Continuous production of butanol using *Clostridium beijerinckii* BA101 and recovery by gas stripping. *Clostridium* 2002: The Seventh International Conference and Workshop on Regulation of Metabolism, Genetics and Development of the Solvent and Acid Forming Clostridia. Rostock, Germany, September 19–21, 2002.
72. Qureshi, N. and Blaschek, H.P., Economics of butanol fermentation using hyper-butanol producing *Clostridium beijerinckii* BA101, *Trans IChemE*, 78, Part C, 139, 2000.
73. Qureshi, N. and Blaschek, H.P., Evaluation of recent advances in butanol fermentation, upstream, and downstream processing, *Bioprocess Biosystems Eng.,* 24, 219, 2001.
74. Ezeji, T.C., Qureshi, N., and Blaschek, H.P., Unpublished data, 2002.

37 Metabolic Engineering of Solventogenic Clostridia

Christopher A. Tomas, Seshu B. Tummala, and Eleftherios T. Papoutsakis

CONTENTS

37.1 INTRODUCTION

Metabolic engineering (ME) can be defined as "directed modification of cellular metabolism and properties through the introduction, deletion and/or modification of metabolic pathways using recombinant DNA and other molecular biological techniques," which is a slight modification of the original definition by Bailey [1]. Recombinant DNA technology has enabled metabolic pathway modification using targeted genetic modifications, in addition to the traditional mutagenesis and selection approach. Since the mid-1980s, numerous examples of metabolic engineering have been reported [2]. It was originally thought that it would be a straightforward task to change the metabolic fluxes in a desired fashion by simply introducing the genes for a new pathway or amplifying or deleting the genes of other metabolic pathways. However, this has not been the case. This is because cells have evolved to use complex schemes to regulate gene expression and enzyme levels in order to economize energy and biosynthetic resources. Thus, modification of cellular traits might require changes in numerous genes that are loosely related to the pathways for a particular cellular trait. For example, it has been recently reported that for the 3-PG *Escherichia coli* based DuPont/Genencor Process more than 60 genetic modifications proved necessary to generate an industrially significant strain [3]. In addition to maximizing the flux for a desirable product, improved robustness and prolonged productivity of the biocatalyst (the cells) under realistic bioprocessing conditions is an equally important ME goal. Therefore, ME strategies that increase the ability of cells to withstand "stressful" bioprocessing conditions (such as accumulation of toxic products, byproducts, or substrates as encountered in most applications) without loss of productivity, must also be pursued. Identification of key target genes, and implementation of genetic modification make it imperative that advanced high-throughput investigative tools (such as DNA microarrays), as well as crucial genetic tools and strategies (such as easy and precise gene inactivation and screening technologies) are developed and effectively employed.

Metabolically engineered solventogenic clostridia can potentially lead to industrial processes for production of butanol, butyric acid, acetone, butanediol, propanol, acetoin, polysaccharides, vitamins, and enzymes or for biotransformations [4–7]. A major advantage of solventogenic clostridia is their ability to utilize an unusually large variety of substrates: mono-, oligo-, and polysaccharides, including many pentoses and hexoses [4–6] and, as such, to utilize biomass hydrolysates. Clostridia also grow under a low redox potential [4,6,8], which enables them to undertake a variety of stereospecific reductions yielding chiral products that are difficult to synthesize chemically. In addition, these and related clostridial species can degrade a number of toxic chemicals and have a large potential for bioremediation applications [9–12]. The list of toxic chemicals includes various aromatic compounds and their derivatives (including chlorinated aromatics such as PCP) [13–15], natural and synthetic polyesters [16], derivatized triazines [17], toxic fuel components [18], and mercury compounds [19].

ME in clostridia was reviewed and discussed in substantial detail in a 1999 book chapter [20]. Thus, this chapter is predominantly focusing on work performed since, with only a brief overview of the earlier work. A large focus is now placed on the use of DNA microarrays as an investigative tool and of the antisense RNA technology as a ME tool.

37.2 METHODS FOR METABOLIC ENGINEERING

37.2.1 Gene Overexpression

The first attempts to overexpress genes in *C. acetobutylicum* utilized *B. subtilis-C. acetobutylicum* shuttle vectors that overexpressed the *adc* (CAP0165) and *ptb* (CAC3076) genes of *C. acetobutylicum* and have been extensively reviewed [21,22]. Since cloning with *B. subtilis-C. acetobutylicum* shuttle vectors was more cumbersome than working with *E. coli-C. acetobutylicum* shuttle vectors, a method for easily transforming *E. coli-C. acetobutylicum* shuttle vectors into *C. acetobutylicum* was sought after. Due to the existence of a restriction endonuclease (*Cac*824I) in *C. acetobutylicum* that recognizes a commonly found sequence in *E. coli* plasmid DNA [21], transformation of *C. acetobutylicum* was difficult with *E. coli-C. acetobutylicum* shuttle vectors. This issue was resolved when Mermelstein and Papoutsakis developed a methylation protocol that overcame the restriction system in *C. acetobutylicum* by methylating the desired *E. coli-C. acetobutylicum* shuttle vector with the ϕ3T I methyltransferase (encoded by *B. subtilis* phage ϕ3T) prior to transformation with *C. acetobutylicum* [23]. Many *E. coli-C. acetobutylicum* shuttle vectors have since been developed for gene overexpression as well as gene inactivation in *C. acetobutylicum* [20].

The latest series of *E. coli-C. acetobutylicum* shuttle vectors used for gene overexpression in *C. acetobutylicum* are derived from the pSOS series of plasmids (pSOS94, 95, and 96) [24–27]. These plasmids were originally developed to overexpress the genes of the acetone formation pathway using different promoters [*ptb* – pSOS94, *thl* (CAC2873) – pSOS95, and *adc* – pSOS96]. Due to convenient restriction sites, these plasmids can be digested with *Bam*HI and *Ehe*I such that the acetone formation pathway genes can easily be removed without disrupting the essential portions of the shuttle vector (i.e., the antibiotic resistance genes and origins of replication for both *E. coli* and *C. acetobutylicum,* as well as promoter and transcriptional terminator sequences). By inserting the desired gene (in the sense orientation) downstream of the promoter of a digested pSOS series vector, one can tailor the gene expression of the desired gene depending on the choice of pSOS series shuttle vector used for overexpression. Shuttle vectors derived from the pSOS series have been used to drive the expression of heat-shock genes and numerous antisense RNA molecules in *C. acetobutylicum* [24–26]. With the development of several gene expression reporter systems in *C. acetobutylicum* [28,29], new promoter sequences for gene overexpression can be discovered and inserted into the pSOS series or other *E. coli-C. acetobutylicum* shuttle vectors to produce an improved generation of expression vectors.

37.2.2 Gene Inactivation

The task of targeting genes for inactivation remains a daunting one despite many years of efforts from several laboratories. Almost all reported gene knockouts were created using nonreplicating or suicide plasmids, but unfortunately the method is not very reproducible. Wilkinson and Young were the first to use nonreplicative vectors to create gene disruptions by homologous recombination in clostridia [30]. Using a conjugation method in *Clostridium beijerinckii* NCIMB 8052, the *gutD* (sorbitol dehydrogenase structural gene) and *spo0A* were inactivated. The precise nature of the genetic alteration has not been determined. Green and Bennett used electroporation to introduce nonreplicative plasmids containing internal gene fragments, which allowed for recombination between the cloned gene fragment and the homologous chromosomal gene copy, thereby disrupting the chromosomal gene with one or more copies of the plasmid [31,32]. Using this technique, recombinant strains of *C. acetobutylicum* ATCC 824 with inactivated *buk* (CAC3075), *pta* (CAC1742), *aad/adhE1* (CAP0162), and *solR/orf5* (CAP0161) genes were constructed [31–33]. Low recombination frequencies (less than one transformant per μg plasmid DNA) and the lack of reproducibility and ease of these methods made it necessary that an alternative approach be sought.

The use of replicative plasmids for gene inactivation theoretically provides for higher recombination efficiencies due to the prolonged (i.e., over several generations) proximity of the plasmid to the chromosome. Replicative plasmids used for targeted gene inactivation contain an origin of replication functional in the host strain, a DNA fragment homologous to the target gene, and an antibiotic resistance marker for selection of the plasmid after transformation. The plasmid pLHKO, a chloramphenicol resistant vector containing a multiple cloning site and a pIM13 Gram-positive replicon, is one such plasmid [34]. An internal fragment from the targeted gene is cloned into the multiple cloning site. A second antibiotic resistance marker such as MLS^r (providing resistance to erythromycin) can then be cloned into the center of the internal gene fragment, resulting in a plasmid capable of producing a double crossover integration at the target site. The strain carrying the pLHKO derivative is grown on plates without antibiotic selection and is replica plated daily onto fresh plates. Plating onto antibiotic-free plates is performed to cure the host of the pLHKO-derived plasmid. After several days, the culture is replica plated onto antibiotic selection plates (erythromycin only) and bacteria that have incorporated the MLS^r marker into the chromosome are isolated. The precise nature of the genetic alteration is determined using PCR amplification of the chromosomal region, followed by sequencing. This method was used to inactivate the *spoOA* (CAC2071) gene (discussed in detail next).

37.2.3 DNA-Array Analysis

With few exceptions [35,36], DNA-array applications in prokaryotic systems have focused mostly on transcriptional profiling of *E. coli* and *B. subtilis*. Some relevant *E. coli* studies include transcriptional analysis of growth on rich vs. minimal media [37], growth on acetate [38], transitional growth [39], stress response during high cell density fermentations [40], quorum-sensing mutants [41], and H_2O_2 stress [42]. Novel uses of arrays in *E. coli* include parallel gene-trait mapping and operon structure prediction [43,44]. Microarray analysis of Gram-positive organisms has been primarily limited to *B. subtilis* [45–47], and has only been recently reported for any clostridia [48]. With the development of a working cDNA-array protocol in *C. acetobutylicum*, the use of DNA-arrays in other clostridia is likely to expand. DNA-array analysis has already been used to analyze a spo0A (CAC2071) knockout and strain degeneration [48], the effects of *ctfA/B* (CAP0163/CAP0164) downregulation by asRNA [with and without *aad/adhE1* (CAP0162) overexpression] [49], *groESL* overexpression and host-plasmid interactions [26], and the effects of butanol stress [50,51], several of which are discussed below.

37.2.4 Proteome Analysis

Proteome analysis examines the interrelationships between proteins within a system as a means to understand the cell's pattern of functioning genes through the analysis of the final products of genes: proteins. That pattern is dependent on many factors, such as the genetic state, which can allow determination of the effect of a particular mutation on the cell physiology and protein pattern; the growth state, which can show the effects of environmental parameters on the cell status; the post-translational protein modifications, which may activate, inactivate, stabilize or mark for degradation a particular protein to suit the cell's needs. Many proteins can be present in alternate forms at the same time within a cell. Proteome profiles may or may not correlate with the gene expression pattern as measured by DNA-array analyses, depending on the respective RNA and protein stabilities and the differential efficiencies with which proteins are translated. Part of the appeal of proteome analysis is that mass spectrometry is able to provide precise molecular weights of proteolytically cleaved peptides, which are subsequently used to probe a database to identify the protein with high accuracy. A number of studies in *B. subtilis* have validated such an approach, and these include several studies on the organism's stress response [52,53]. While protein separation by 2-D electrophoresis, followed by analysis of spots by mass spectrometry (e.g., Maldi-TOF) [54], remains the most widely applied proteome analysis method, advances in separation methods, capillary chromatography/mass spectrometry combinations, and high throughput systems have allowed more complete analyses. These have been applied to proteome analysis of bacterial pathogens among other systems [55–59]. Proteome studies in solventogenic clostridia are limited to one recent *C. acetobutylicum* report in which changes in protein synthesis were examined during the switch from acidogenesis to solventogenesis [60], and an earlier study illustrating that a number of proteins are phosphorylated in response to stress [61]. Schaffer et al. compared protein synthesis in acidogenic and solventogenic *C. acetobutylicum* cells using two-dimensional gel electrophoresis to illustrate the dramatic change in protein patterns associated with the switch to solventogenesis [60]. Of 130 proteins analyzed, they found that expression of 52 proteins increased more than twofold, while expression of 34 proteins decreased by at least one half in solventogenic cells vs. acidogenic cells. Included in the set of proteins with higher expression were acetoacetate decarboxylase [AADC (CAP0165)], three stress proteins [DnaK (CAC1282), GroEL (CAC2703), and Hsp18 (CAC3714)], two serine biosynthesis enzymes, and a seryl-tRNA synthetase. Another research group identified also GroES (CAC2704) as being induced during the switch to solventogenesis, in addition to confirming the AADC, GroEL, and Hsp18 expression patterns [62]. Novel proteins defined in the genome annotation as being unique to clostridia by having a clostridial hydrophobic W-containing (ChW) repeat were also found to be induced in the solvent stage (corresponding to genes CAC1532 and CAC2584) [62]. The DNA regions upstream from some ChW genes were examined, and putative spo0A box sequences were identified. The region upstream of those containing the putative promoter has been cloned, and studies using reporter plasmids to better assess the expression pattern are planned. Acetoacetate decarboxylase was the only protein identified that is directly involved in solventogenesis. Two proteins (acetoacetate decarboxylase and Hsp18) were found at two positions on 2-dimensional gels, differing in pI values but having identical molecular masses [60]. It is possible that the appearance of a second spot is a result of phosphorylation, acetylation, deamindation, adenylation, or amino acid substitution. A separate study also found triosephosphate isomerase, one of the major proteins of glycolysis, at two distinct positions [62]. The type of modification and its role are not known, but the revelation of this new area might suggest an expansion of work on protein modification in bacterial systems.

37.3 METABOLIC ENGINEERING APPLICATIONS

37.3.1 Metabolic Engineering of Acid and Solvent Formation Pathways

37.3.1.1 Gene Overexpression of Solvent Pathways

After development of an effective transformation procedure of *C. acetobutylicum* [23], gene overexpression to modify solvent formation pathways became a practical task. The first pathway to be overexpressed in *C. acetobutylicum* was the acetone formation pathway. Mermelstein et al. created a vector that contained a synthetic operon (called the "*ace* operon") in which the three homologous acetone-formation genes [*adc* (CAP0165), *ctfA* (CAP0163), and *ctfB* (CAP0164)] are transcribed from the *adc* promoter [63]. This vector was used to develop strain *C. acetobutylicum* ATCC 824 (pFNK6). In pH-controlled fermentations of 824 (pFNK6), acetone-formation enzyme activities were expressed earlier and led to earlier induction of acetone formation compared to the parental strain. Furthermore, relative to the parental strain in pH 5.5 controlled fermentations, 824 (pFNK6) produced 95, 37, and 90% higher final concentrations of acetone, butanol, and ethanol, respectively, and a 22-fold lower mass of residual carboxylic acids.

Soon after, Nair et al. cloned and overexpressed *aad* (CAP1062; also known as *adhE1*) [64] in *C. acetobutylicum* [65]. Despite dramatic increases in NAD-specific BYDH and ACDH and NADH-specific BDH and EDH activities in bioreactor experiments with the *aad/adhE1* overexpressing strain [*C. acetobutylicum* ATCC 824 (pCAAD)] compared to the plasmid control strain, final butanol and acetone concentrations were actually higher in the plasmid control strain. However, the higher ethanol-to-acetone and butanol-to-acetone ratios in the *aad* overexpressing strain indicate that 824 (pCAAD) produces alcohols with improved selectivity compared with the plasmid control and parental strains.

Genes of the acetone formation pathway of *C. acetobutylicum* have also been used to metabolically engineer *E. coli* to enhance acetone production and reduce detrimental acetate accumulation. Bermejo et al. developed vector pACT that contained a synthetic acetone operon (*ace4*) composed of four *C. acetobutylicum* ATCC 824 genes [*adc*, *ctfA*, *ctfB*, and *thl* (CAC2873)] under the control of the *thl* promoter [66]. In bioreactor experiments with an *E. coli* strain containing pACT [*E. coli* ATCC 11303 (pACT)], final concentrations of acetone (125 to 154 m*M*) were shown to be equal to or higher than those produced by wild-type *C. acetobutylicum* fermentations.

37.3.1.2 Gene Inactivation of Acid Pathways

In hopes of redirecting carbon flow from acid formation to solvent formation, the acid formation pathway genes *buk* (CAC3075) and *pta* (CAC1742) were first targeted for gene inactivation via plasmid integration by a Campbell-like mechanism [31]. As expected, the strain containing a mutation in the *buk* gene (named PJC4BK) exhibited drastic differences in BK enzyme activities (fivefold lower) and solvent production (10% more butanol, and 50% less acetone) in pH 4.5 controlled fermentations compared to the parental strain. Despite sevenfold lower PTA activity, PJC4PTA did not exhibit any drastic differences in solvent production compared to the parental strain.

In 1999, strain PJC4BK was revisited, and fermentations of this strain at pH ≥5.0 fermentations were performed [67]. Under these fermentation conditions, PJC4BK was shown to be a solvent superproducer, producing 225 m*M* butanol, 76 m*M* acetone, and 57 m*M* ethanol. To further enhance solvent production in this strain, a plasmid that contained the *aad/adhE1* (CAP0162) gene (pTAAD) was introduced into PJC4BK. Overexpression of *aad* in PJC4BK resulted in similar acetone and butanol levels but a drastically higher final ethanol concentration (98 m*M*). Metabolic flux analyses of these strains showed a decrease in butyrate formation fluxes by up to 75% and an increase in acetate formation fluxes of up to 100% during early growth compared to wild-type fermentations. In addition, the mean specific butanol and ethanol formation fluxes in early growth increased significantly in these recombinant strains, up to 300 and 400%, respectively, compared to wild-type.

Onset of solvent production in PJC4BK pH ≥5.0 fermentations occurred during the exponential growth phase when the culture optical density was very low and when total and undissociated butyric acid levels were <1 m*M*. This was in sharp contrast to the classical model correlating the end of exponential phase and high undissociated butyric acid concentrations (17 m*M*) [68,69] with the switch from acidogenesis to solventogenesis. Thus, a new phenomenological model to explain solvent formation was needed. Results from this study [67], as well as several studies in *E. coli* that suggest acyl phosphates can act as regulatory agents involved in two-component signal transduction systems [70,71], suggest that the accumulation of butyryl-P, and not butyrate itself, might play a regulatory role and trigger solvent production in *C. acetobutylicum*.

37.3.1.3 Metabolic Engineering of Acid and Solvent Formation Pathways Utilizing Antisense RNA Strategies

Although not widely exploited in prokaryotic systems, antisense RNA (asRNA) is a potent and flexible tool for manipulating microbial metabolism. Desai and Papoutsakis examined the effectiveness of asRNA strategies for the metabolic engineering of *C. acetobutylicum* [24]. They developed two different asRNA molecules against two genes involved in the primary metabolic network of *C. acetobutylicum* ATCC 824 (Figure 37.1). One molecule was directed toward the mRNA of the butyrate kinase (*buk*; CAC3075) gene, while the other one was directed toward the mRNA of the phosphotransbutyrylase (*ptb*; CAC3076) gene. The strain of *C. acetobutylicum* ATCC 824 containing the *buk*-asRNA [824(pRD4)] exhibited 85% lower butyrate kinase (BK) activity than the control strain. This strain also resulted in 50 and 35% higher final concentrations of acetone and butanol, respectively, than the control strain. In addition, the strain of *C. acetobutylicum* ATCC 824 containing the *ptb*-asRNA molecule [824(pRD1)] exhibited 70% lower phosphotransbutyrylase activity but 96 and 75% lower final acetone and butanol concentrations, respectively, than the control. Such drastic differences in solvent production due to downregulation of enzymes in the same pathway appear to be illogical. Nonetheless, these results are consistent with the hypothesis that butyryl-P is a crucial intermediate involved in the initiation of solventogenesis and are discussed further in Section 37.3.2.2 [67,72].

Tummala et al. performed a series of experiments using antisense RNA for the metabolic engineering of the acetone-formation pathway in *C. acetobutylicum* [25]. They investigated the effect of antisense RNA (asRNA) on downregulation of the enzymes in the acetone-formation pathway [acetoacetate decarboxylase (AADC) and CoA-transferase (CoAT)] to increase the butanol to acetone ratio of *Clostridium acetobutylicum* ATCC 824. First, they generated three strains, that contained plasmids that produce asRNA of various lengths against the AADC (*adc*; CAP0165) transcript. Despite significant downregulation of AADC in these strains, there were no concomitant effects on acetone formation. These findings suggested that AADC was not limiting acetone formation and, thus, they targeted next the CoAT and developed a strain [824(pCTFB1AS)] that

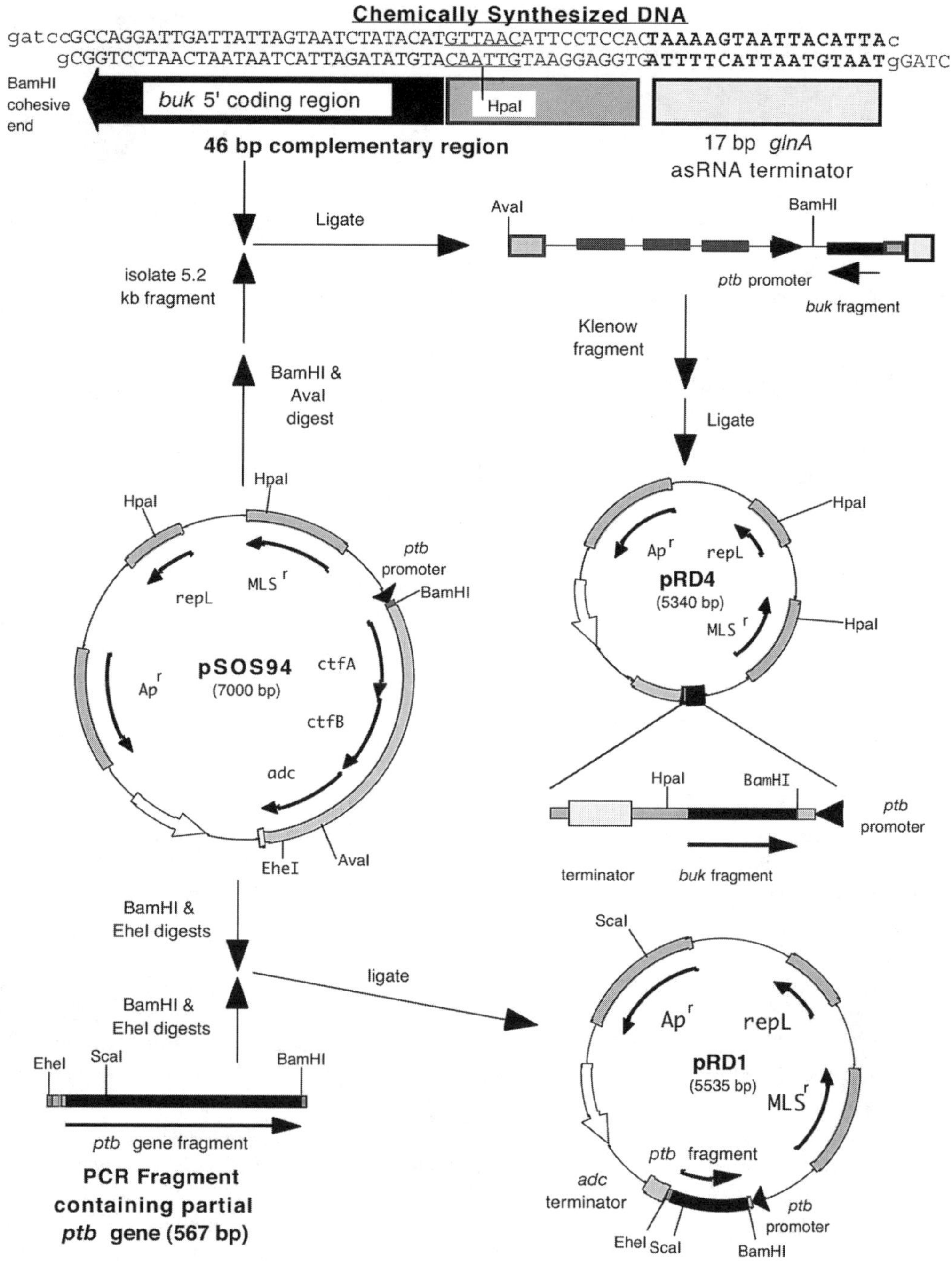

FIGURE 37.1 Construction of plasmids pRD4 and pRD1 [24]. For each plasmid, the locations and directions of transription of relevant genes are indicated (arrows). Relevant restriction sites are shown. Abbreviations: Ap^r, ampicillin resistance gene; MLS^r, macrolide-lincosamide-streptogramin resistance gene; repL, gene required for replication in Gram-positive organisms.

expresses an asRNA to downregulate the CoAT subunit B gene (CAP0164). Compared to the plasmid control strain, this strain produced substantially lower levels of acetone and butanol, and Western blot analyses showed significantly lower levels of both CoAT subunits. These results showed that CoAT is the rate-limiting enzyme in the acetone formation pathway.

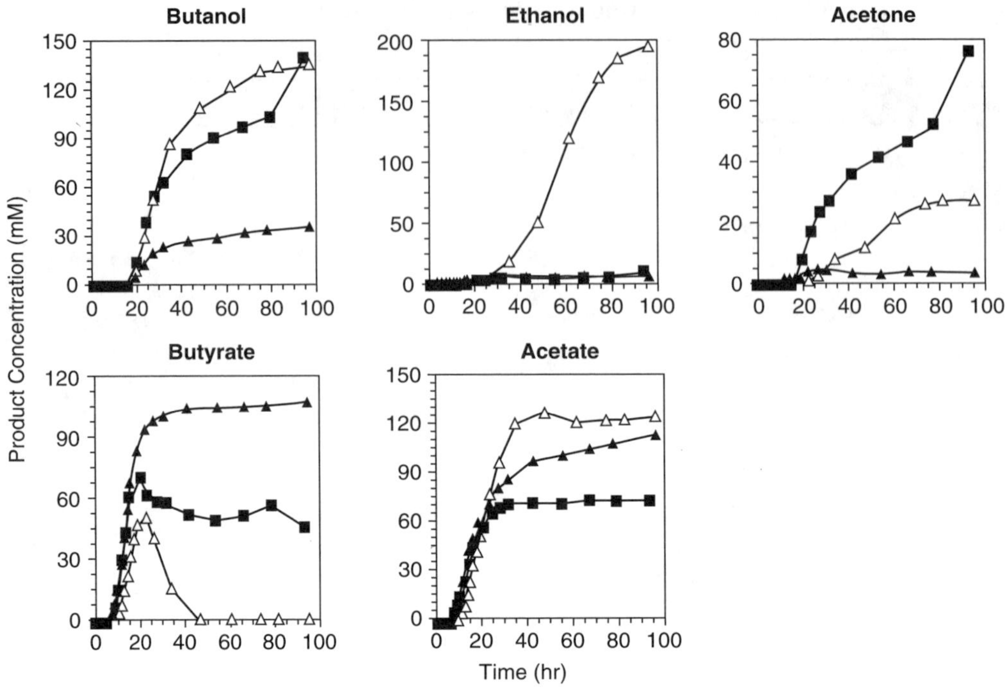

FIGURE 37.2 Fermentation kinetics of 824(pSOS95del), 824(pCTFB1AS), and 824(pAADB1) [49]. Acid and solvent profiles. The name of each profile in 824(pSOS95del) (■), 824(pCTFB1AS) (▲), and 824(pAADB1) (Δ) is indicated above each graph.

Interestingly, butanol levels in CoAT-asRNA expressing strains were also significantly decreased, suggesting that expression of the gene responsible for butanol production, alcohol-aldehyde dehydrogenase (*aad/adhE1*; CAP0162), might also be altered by CoAT-asRNA. This is not surprising in view of the fact that the *aad* resides on the same polycistronic message as *ctfA* and *ctfB*. These results suggested that the asRNA in these strains is resulting in partial degradation of the tricistronic mRNA.

To overcome this likely degradation of *aad* mRNA in 824(pCTFB1AS), they hypothesized that overexpressing *aad/adhE1*, while still suppressing acetone formation with the asRNA produced by pCTFB1AS (*ctfb1*-asRNA), would increase butanol production substantially compared to 824(pCTFB1AS) and, thereby, increase the butanol/acetone ratio of *C. acetobutylicum*. Plasmid pAADB1 for the overexpression of the alcohol-aldehyde dehydrogenase (*aad*) gene and downregulation of the coenzyme A transferase using antisense RNA (*ctfb1*-asRNA) was employed in order to increase the butanol to acetone ratio of *Clostridium acetobutylicum* ATCC 824 fermentations (Figure 37.2) [49]. Compared to strain 824(pCTFB1AS), 824(pAADB1) fermentations exhibited two profound differences. First, butanol levels were approximately 2.8-fold higher in 824(pAADB1) and restored back to plasmid control levels, thus supporting the hypothesis that asRNA downregulation of *ctfB* leads to degradation of the whole *aad-ctfA-ctfB* transcript. Second, ethanol titers in 824(pAADB1) were approximately 23-fold higher and the highest (approximately 200 m*M*) ever reported in *C. acetobutylicum*. Western blot analysis confirmed that CoAT was downregulated in 824(pAADB1) at nearly the same levels as in strain 824(pCTFB1AS).

To assess the global impact of *ctfb1*-asRNA expression with and without overexpression of *aad* on the transcriptome, Tummala et al. used DNA microarrays for a comparative analysis of strains 824(pCTFB1AS) and 824(pSOS95del), and 824(pCTFB1AS) and 824(pAADB1) [49]. DNA microarray analysis showed that expression patterns of the CoA transferase genes (*ctfA* and *ctfB*) and *aad* were consistent with the overexpression of *aad* and asRNA downregulation of *ctfB* in these strains. DNA-array based transcriptional analysis also revealed that the large changes in product concentrations (and notably butanol concentration) due to *ctfb1*-asRNA expression alone and in combination with *aad* overexpression resulted in dramatic changes of the cellular transcriptome. Cluster analysis and gene expression patterns of established and putative operons involved in stress response, motility, sporulation and fatty-acid biosynthesis indicate that these simple genetic changes dramatically alter the cellular programs of *C. acetobutylicum*. Comparisons of gene expression and flux analysis data pointed out new possible flux-controlling steps and suggested unknown regulatory mechanisms, including a potential feedback mechanism involving *adc* and the potential role of the putative ferredoxin oxidoreductase gene in alcohol formation.

37.3.2 Metabolic Engineering Using Regulatory Genes and Global Cellular Programs

37.3.2.1 The *sol* Operon

The *sol* operon is one of the first to be transcribed at the onset of solventogenesis [64] and includes the genes [aad/*adhE1* (CAP0162) and *ctfA/B* (CAP0163 and CAP0164)] essential for solvent formation. Several conclusions have been drawn regarding the control of this important solvent formation operon. First, DNA topology has been shown to play a critical role in the induction of the *sol* operon. An increase in transcription was observed after inhibition of DNA gyrases, which results in DNA relaxation (less negative supercoiling) [73]. This is further supported by the observation that the DNA of *C. acetobutylicum* is less negatively supercoiled at the onset of solventogenesis [74]. Second, insertional inactivation of the gene originally named solR (CAC0161) directly upstream of aad/*adhE1* was found to result in increased solvent production [33]. Further analysis of the inactivated gene product revealed that it (1) is localized at the extracellular side of the cytoplasmic membrane, (2) is involved in glycosylation/deglycosylation reactions, and (3) contains a tetratrico peptide repeat motif for protein-protein interactions [75]. Thormann et al. [75] concluded that the inactivated gene was not acting as a repressor, but rather is regulated through transcriptional activation. The presence of a Spo0A binding site suggests a role for Spo0A and is supported by lack of reporter activity upon disruption of the Spo0A-binding motif. In addition, three imperfect repeats were found preceding the promoter of the *sol* operon, the results of which suggest a possible second activator that might work in concert with Spo0A. Additional analysis of the disrupted region between the integrated plasmid (pOX1) and the *sol* operon in one of the mutant strains (mutant H) was subsequently performed to further clarify the nature of the observed effect [62]. The promoter activity of sequences inserted adjacent to *adhE* during the construction of mutant H was evaluated by isolating several segments of the region upstream of *adhE* in mutant H by PCR amplification. These experiments (which were designed to detect if there is another complicating effect on expression of downstream genes occurring due to the nature of sequences introduced during the disruption of *orf5*) did not suggest a direct effect of the inserted pOX1 plasmid. A complementation experiment was done with pCO1, bearing the *solR* gene and some surrounding DNA, and, as expected, this resulted in decreased levels of solvents. A thiamphenicol resistant derivative of the plasmid was prepared and tested to see if it would complement mutants H and B. Solvent production was reduced in strains carrying this plasmid. To further investigate the cause of the complementation, derivatives that had deletions within the *solR* coding region were tested for their ability to restore solvent production in SolRH and SolRB to wild-type levels. The *solR* deleted plasmids also showed complementation, indicating that the region on the plasmid responsible for complementation was not *solR* but a short fragment between *solR* and the *sol* operon. The promoter region upstream of *adhE* was examined to identify sites required for induction of the promoter during solvent phase. Several promoter variants were placed on a chloramphenicol acetyl-transferase reporter plasmid and assayed under acid phase and solvent phase growth. It was determined that the presence of a spo0A box sequence (required for Spo0A binding) alone is not sufficient for full induction and that a region beyond this site seems to be required. It is in this region that the repeat sequences observed by Thormann et al. reside [75]. These results are consistent with the titering model proposed for the action of plasmids containing this region without a functional promoter. A possible explanation for this is that the added DNA affects the overall conformation of DNA in the region, thus making the binding sites more suitable for interaction with the factors stimulating transcription.

37.3.2.2 Butyrate Kinase Inactivation Mutant

Inactivation of the butyrate kinase gene (*buk*; CAC3075) in *C. acetobutylicum* had profound global effects on primary metabolism. Butyrate-formation fluxes in strain PJC4BK were 65 to 75% lower than in wild-type *C. acetobutylicum* ATCC 824 [31,67]. A reduction in butyrate-formation fluxes represents a probable metabolic burden as a result of an associated reduction in ATP production. Cellular demands for ATP in PJC4BK are likely met by increased acetate production, which is also responsible for ATP production. Strain PJC4BK was a solvent superproducer under low pH conditions (pH 5.0) with butanol, acetone, and ethanol levels reaching 225 m*M* (16.7 g/l), 76 m*M* (4.4 g/l), and 57 m*M* (2.6 g/l), respectively. Overexpression of the *aad/adhE1* gene (CAP0162) in PJC4BK resulted in a doubling of the ethanol titers, but did not significantly increase butanol production. This suggests that AdhE1 is not a limiting factor for butanol production in PJC4BK under low pH fermentation conditions and is further evidence that AdhE1 plays a significant role in ethanol formation [65]. Butanol production in PJC4BK began during the early exponential phase of cell growth (A600 <0.5), in sharp contrast to wild-type fermentations where the onset of solvent production occurs near the end of the exponential phase. The low levels (0.4 m*M*) of undissociated butyric acid (UBA) in the PJC4BK cultures during the switch to solventogenesis did not conform to the classical model, which required levels above 6 to 18 m*M* UBA [68,69]. This suggested that the accumulation of a metabolic precursor of butyrate, such as butyryl CoA or butyryl-P, might be responsible for triggering solvent production. AsRNA studies aimed at downregulating the phosphotransbutyrylase (PTB) and butyrate kinase (BK) enzymes of the butyrate formation pathway also suggest that butyryl-P is responsible for triggering solventogenesis [24]. Downregulation of BK and PTB led to drastic differences in solvent formation with downregulation of BK resulting in increased solvent production and downregulation of PTB resulting in much lower solvent production. Since PTB is responsible for butyryl-P formation, downregulation of PTB results in lower butyryl-P levels. Furthermore, downregulation of BK, which converts butyryl-P to butyrate, results in an accumulation of butyryl-P. Thus, the correlation between high levels of butyryl-P and increased solvent production in the strain with lower BK levels, in conjunction with the low levels of butyryl-P and decreased solvent production in the strain with lower PTB levels, further suggests that butyryl-P plays a key regulatory role in the triggering of solventogenesis.

37.3.2.3 *spoOA* Inactivation and Overexpression in *C. acetobutylicum* ATCC 824

The *spo0A* gene (CAC2071) was cloned, and two recombinant strains were generated, a *spo0A* inactivation strain (SKO1) and a *spo0A* overexpression strain [824(pMPSOA)] [34]. SKO1 was developed by targeted gene inactivaction using a replicative plasmid capable of double crossover chromosomal integration. SKO1 was severely deficient in solvent formation: only 2 m*M* of acetone and 14 m*M* of butanol were produced compared to the 93 m*M* of acetone and 171 m*M* of butanol produced by the parental strain. After 72 hours of growth on solid media, SKO1 formed long filaments of rod-shaped cells that failed to septate properly. SKO1 cells never achieved the swollen clostridial form typical of wild-type and did not form endospores. Additionally, mRNA levels produced during SKO1 fermentations showed that the *spo0A* transcript was not made in SKO1 and showed limited transcription of two solvent formation operons (*aad-ctfA-ctfB* and *adc*) in the absence of Spo0A. Product formation patterns in strain 824(pMSPOA) were largely unaltered compared to the control strain [824 (pIMP1)]. However, microscopy indicated that spore formation occurred earlier due to *spo0A* overexpression. Expression of the key solvent formation genes (*aad-ctfA-ctfB* and *adc*) and two sporulation-specific genes [*sigE* (CAC1695) and *sigG* (CAC1696)] was observed earlier in 824(pMSPOA) than in the plasmid control. These data support the hypothesis that Spo0A is a transcriptional regulator that positively controls sporulation and solvent production.

37.3.2.4 Transcriptional Analysis of *spo0A* Overexpression

Transcriptional analysis of 824(pMSPOA) against its plasmid control 824(pIMP1) confirmed that *spo0A* (CAC2071) overexpression leads to earlier induction of solventogenic gene expression [51]. In addition, transcripts of several fatty acid metabolism genes, heat shock proteins, and the Fts family of cell division proteins were differentially expressed, suggesting roles in sporulation and the solvent stress response. However, 824(pMSPOA) showed significant concerted downregulation of many glycolytic genes during stationary phase; metabolic flux analysis showed that the carbon flux through this pathway was reduced in 824(pMSPOA) by 38 to 42% during stationary phase. Although these genes are probably not directly regulated by *spo0A*, this might be an indirect result of sporulating cells with reduced energy requirements. Also, the *spo0A*-overexpressing strain showed only nominal changes in transcription of motility (flagella) and gene regulators [*abrB* (most likely CAC3647, but also could be CAC0310 or CAC1941), *sigF* (CAC2306)], genes whose expression was significantly altered in SKO1.

37.3.2.5 Transcriptional Analysis of Asporogenous, Nonsolventogenic *C. acetobutylicum* Strains SKO1 and M5

The large-scale transcriptional program of strains SKO1 and M5 relative to the parent strain (WT) was examined using DNA arrays [48]. Glass DNA arrays containing a select set of 1019 genes covering more than 25% of the whole genome were designed, constructed and validated for data reliability. Strain SKO1, with an inactivated *spo0A* (CAC2071) gene displays an asporogenous, filamentous, and largely deficient solventogenic phenotype (described above) [34]. SKO1 displays downregulation of all solvent formation genes, *sigF* (CAC2306), carbohydrate-metabolism genes (similar to genes expressed as part of the stationary-phase response in *B. subtilis*), but also several electron transport genes. A major cluster of genes upregulated in SKO1 includes the genes from the major chemotaxis and motility operon, *abrB* (most likely CAC3647, but also could be CAC0310 or CAC1941), and glycosylation genes. Stress genes were mostly downregulated in SK01 as well. Strain M5 displays an asporogenous and nonsolventogenic phenotype as a result of the megaplasmid pSOL1 loss [76]. Consistent with that, M5 displays a downregulation of all pSOL1 genes expressed in WT. Notable among other genes expressed higher in WT compared to M5 were *sigF* (CAC2306), several two-component histidine kinases, *spo0A* (CAC2071), *cheA* (CAC2220), *cheC* (CAC2219), many stress response genes, *fts* family genes, DNA topoisomerases, and central-carbon metabolism genes. Genes expressed higher in M5 include electron transport genes (but different from those downregulated in SKO1), and several motility and chemotaxis genes. Most of these expression patterns were consistent with phenotypic characteristics. Several of these expression patterns are new or different from what is known in *B. subtilis* and can be used to test a number of functional-genomic hypotheses. Furthermore, the relationship between gene expression and fluxes of the primary metabolism in strain SK01 vs. 824 was examined. While solvent and butyrate fluxes show good agreement with key corresponding genes (e.g, for alcohol formation the key genes are those that correspond to AAD, BDH I and II), the glycolytic and acetate fluxes are not in agreement with gene expression patterns for several possible reasons, some of which can be elucidated by proteome analysis.

37.3.2.6 GroESL Heat Shock Proteins (HSPs)

824(pGROE1) was created in an effort to assess the effects of *groESL* overexpression on the metabolism and solvent production in *C. acetobutylicum* [26]. Overexpression of the *groESL* operon was verified at both the transcriptional level (DNA arrays and Q-RT-PCR) and at the protein level (Western analysis). Overexpression resulted in increased final solvent titers relative to both the WT (40% higher) and the plasmid control strain 824(pSOS95del) (33% higher). Increased acetone production in 824(pGROE1) relative to 824(pSOS95del) was achieved primarily through increased acetate uptake, rather than butyrate uptake. Increased acetone production in 824(pGROE1) relative to the control strain was observed despite the fact that AADC and CoAT protein levels were not significantly higher during the stationary phase. Increased levels of GroES and GroEL can hold the AADC and CoAT proteins in a more active state, resulting in increased acetone production. Alternatively, overexpression of *groESL* can increase the level of metabolic intermediates available for acetone formation by stabilizing the proteins involved in their formation. Increased butanol titers similar to those observed in 824(pGROE1) have been achieved through overexpression of the alcohol-aldehyde dehydrogenase gene (*aad*; CAP0162) [67,77]. However, *aad* is likely too large (96 kD) to be stabilized by the GroESL machinery, which has a size limit of approximately 56 kD in most bacterial species [78]. This suggests that increased solvent production is a result of a broader stabilization of the biosynthetic machinery. This is supported by the fact that cells overexpressing *groESL* were better able to grow in the presence of butanol (Figure 37.3). In addition, analysis of specific *in vivo* fluxes related to carbon metabolism (rGLY1, rGLY2, and rHYD) and solvent formation (rBUOH and rACTN) revealed that overall metabolic activity in 824(pGROE1) is significantly higher relative to 824(pSOS95del) (Figure 37.4). Increased levels of GroES and GroEL can also help to overcome the stress associated with the presence of a plasmid, resulting in initially higher cell densities and a motility/chemotaxis gene expression pattern that is more like the WT strain.

Several conclusions with regard to the regulation of HSPs in *C. acetobutylicum* were able to be drawn. First, a CIRCE element was found upstream of the *hsp90* start site, and its transcriptional expression pattern is identical to that of the *dnaKJ* operon in this and several previous studies [26,49]. Therefore, *hsp90* is likely part of the CIRCE heat shock regulon. Second, the decrease in expression of the *dnaKJ* and *hsp*90 operon genes at the transcriptional level, along with decreased DnaK protein levels, suggests that *groESL* acts as a major modulator of the CIRCE heat shock regulon, similar to *B. subtilis* [79]. The *dnaKJ* and *groESL* operons in *C. acetobutylicum* have the same organization as in *B. subtilis*, including a HrcA homologue (CAC1280) and CIRCE elements.

37.4 CONCLUDING REMARKS

Although a large number of crucial genetic and ME tools (transformation protocols, expression vectors, flux analysis, gene inactivation, reporter systems, antisense RNA (asRNA) technology, as well as many interesting strains, have been developed for solventogenic clostridia, the available ME toolbox in clostridia remains underdeveloped when compared to other well-studied organisms such as *E. coli*, *B. subtilis*, and *S. cerevisiae*. However, the recently completed genome sequences of *C. acetobutylicum* and those of a few clostridial pathogens offer great new opportunities to advance the goals of ME in clostridia for industrial, bioremediation, and medical applications [80]. The ability to apply postgenomic tools and analyses to advance our understanding of crucial cellular programs and processes in clostridia and to develop ME tools and strategies that would have been unthinkable even a few years ago is now feasible and will be an essential prerequisite for fast progress. For example, a desirable end-product of ME research would be to generate nonsporulating strains with deregulated solvent production in order to increase the productivity of the cells, further increase solvent yields and titers, and make possible the use of continuous processes for solvent

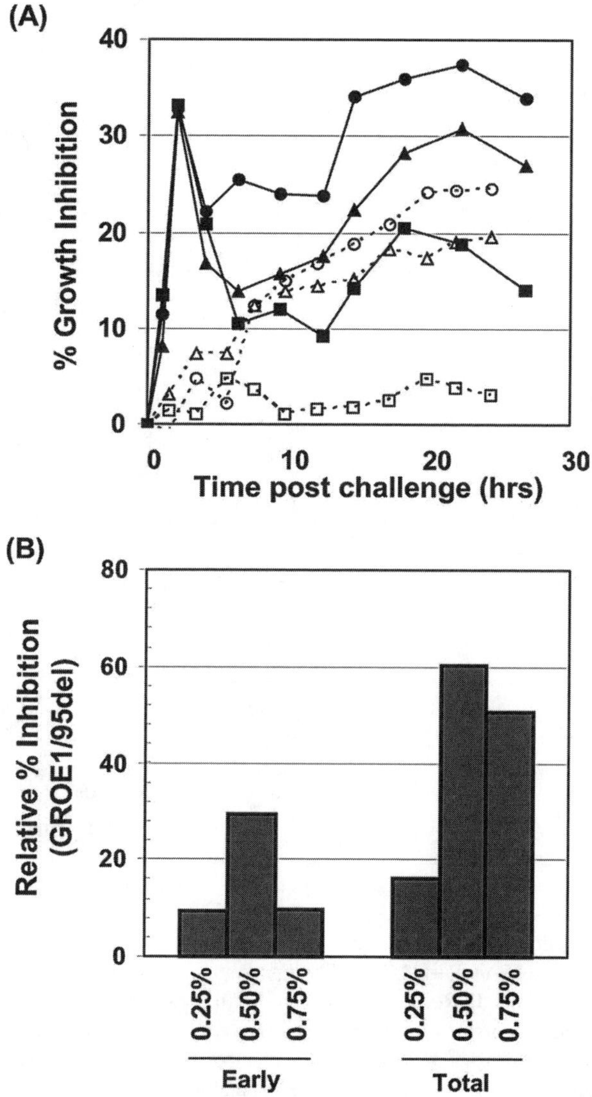

FIGURE 37.3 Butanol challenge experiments [26]. (A) Growth inhibition profiles for 824(pSOS95del) (solid symbols) and 824(pGROE1) (open symbols) at various levels of butanol challenge: squares, 0.25%(v/v); triangles, 0.50%(v/v); circles, 0.75%(v/v). (B) Relative inhibition of 824(pGROE1) to 824(pSOS95del) for the early (0 to 6 h) and total course of butanol challenge.

production. The use of DNA microarrays, proteome analysis, and a "systems biology" approach for the analysis of cellular programs, phenotypes (solventogenesis and degeneration), and recombinant strains will be crucial in achieving this ME objective. ME tools that need to be developed include an efficient recombination system for chromosomal integration and gene deletion, the development of regulatable promoters, improved transformation protocols, and master strains. Also, a detailed molecular understanding of small (e.g., solventogenesis, pathogenesis) and large (e.g., sporulation, stress response, motility/chemotaxis) programs will greatly facilitate ME applications. Furthermore, development of computational and bioinformatic tools to annotate the various genomes in terms of operon organization will prove invaluable for functional genomic studies as preludes to ME applications. In addition, the development of genetic-network models to understand

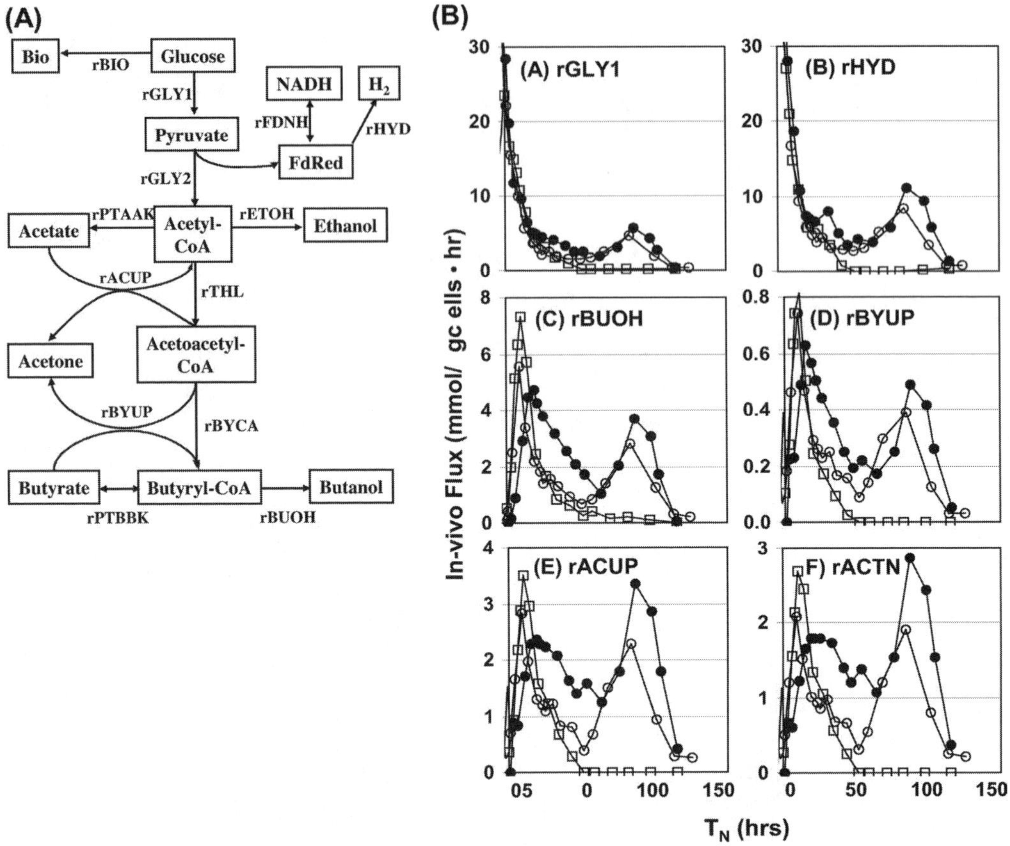

FIGURE 37.4 *C. acetobutylicum* primary metabolic pathways (A) and corresponding *in vivo* fluxes and time course profiles of several metabolic fluxes in cultures of WT (-□-), 824(pSOS95del) (-○-), and 824(pGROE1) (-●-) strains (B) [26]. Profiles shown represent the average of 2 (for recombinant strains) or 3 (for WT) fermentations. Note: rACTN = rBYUP + rACUP. TN represents normalized time such that $T_N = 0$ h at $A_{600} = 1.0$. Abbreviations: Bio, biomass; FdRed, reduced ferrodoxin; rACUP, acetate uptake flux; rBIO, biomass formation flux; rBUOH, butanol formation flux; rBYCA, butyryl-CoA formation flux; rBYUP, butyrate uptake flux; rETOH, ethanol formation flux; rFDNH and rHYD, electron transport fluxes; rGLY1 and rGLY2, pyruvate and acetyl-CoA formation fluxes, respectively; rPTAAK, acetate formation flux; rPTBBK, butyrate formation flux; rTHL, acetoacetyl-CoA formation flux.

the interconnections of the various metabolic and regulatory networks and programs of the cell will be useful to examine possible relationships between the cellular transcriptome and the metabolic flux network, which is another important objective of ME. While it is clear that the potential of ME in clostridia is great, one should not underestimate the magnitude of the effort necessary to make progress, and the need for cooperation and coordination among various groups and laboratories in this, still rather small, research community must be emphasized. In fact, it could be argued that substantial growth in the clostridial community would be necessary if fast progress in the genetics and ME of clostridia is to be achieved.

ACKNOWLEDGMENTS

This work was supported by grants BES-9911231 and BES-9905669 from the National Science Foundation and grant R828562 from the Environmental Protection Agency.

LIST OF ABBREVIATIONS

aad	alcohol-aldehyde dehydrogenase gene (same as *adhE1*)
AADC	acetoacetate decarboxylase
ACDH	acetaldehyde dehydrogenase
adc	acetoacetate decarboxylase gene
adhE1	alcohol-aldehyde dehydrogenase gene (same as *aad*)
asRNA	antisense RNA
BDH	butanol dehydrogenase
BK	butyrate kinase
buk	butyrate kinase gene
butyryl-P	butyryl-phosphate
BYDH	butyraldehyde dehydrogenase
CIRCE	controlling inverted repeat of chaperone expression
CoAT	coenzyme A transferase
ctfA	coenzyme A transferase subunit A gene
ctfB	coenzyme A transferase subunit B gene
EDH	ethanol dehydrogenase
gutD	sorbitol dehydrogenase structural gene
HSPs	heat shock proteins
ME	metabolic engineering
MLS^r	macrolide-lincosamide-streptogramin B resistance gene
NADH/NAD	nicotinamide adenine dinucleotide
pta	phosphotransacetylase gene
PTB	phosphotransbutyrylase
ptb	phosphotransbutyrylase gene
solR/orf5	*sol* locus repressor gene
thl	thiolase gene
UBA	undissociated butyric acid
WT	wild-type

REFERENCES

1. Bailey, J.E., Toward a science of metabolic engineering, *Science,* 252, 1668, 1991.
2. Lee, S.L., Papoutsakis, E.T., *Metabolic Engineering,* Marcel Dekker, New York, 1999.
3. Soucaille, P. and Emptage, M.H., Metabolic pathway engineering for the production of 1,3-propanediol from glucose, *Abstracts of Papers of the American Chemical Society,* 224, U220, 2002.
4. Jones, D.T. and Woods, D.R., Acetone-butanol fermentation revisited, *Microbiological Reviews,* 50, 484, 1986.
5. McNeil, B., The acetone butanol fermentation, *Adv. Appl. Microbiol.,* 31, 61, 1986.
6. Rogers, P., Genetics and biochemistry of *Clostridium* relevant to development of fermentation processes, *Adv. Appl. Microbiol.,* 31, 1, 1986.
7. Young, M. et al., Genetic exchange and prospects for cloning in Clostridia, in *Bacillus Molecular Genetics and Biotechnology Applications*, Ganeson, A.T. and Hoch, J.A., Eds., Academic Press, Orlando, 1986, 259.
8. Kim, B.H. et al., Control of carbon and electron flow in *Clostridium acetobutylicum* fermentation: Utilization of carbon monoxide to inhibit hydrogen production and to enhance butanol yields, *Applied and Environmental Microbiology,* 48, 764, 1984.
9. Hughes, J.B. et al., Reduction of 2,4,6-trinitrotoluene by *Clostridium acetobutylicum* through hydroxylamino-nitrotoluene intermediates, *Environmental Toxicology and Chemistry,* 17 (3), 343, 1998.
10. Francis, A.J. and Dodge, C.J., Dissolution of Ferrites by *Clostridum* sp., *Geomicrobiology Journal* 9 (1), 27, 1991.

11. Francis, A.J. et al., XPS and Xanes Studies of Uranium Reduction by *Clostridium* sp., *Environmental Science & Technology,* 28 (4), 636, 1994.
12. Spain, J.C., Biodegradation of nitroaromatic compounds, *Annu. Rev. Microbiol.,* 49, 523, 1995.
13. Gibson, J. and Harwood, C.S., Metabolic diversity in aromatic compound utilization by anaerobic microbes, *Annu. Rev. Microbiol.,* 56, 345, 2002.
14. Hughes, J.B., Wang, C.Y. and Zhang, C.L., Anaerobic biotransformation of 2,4-dinitrotoluene and 2,6-dinitrotoluene by *Clostridium acetobutylicum*: a pathway through dihydroxylamino intermediates, *Environmental Science & Technology,* 33 (7), 1065, 1999.
15. Tartakovsky, B. et al., Enhanced selection of an anaerobic pentachlorophenol-degrading consortium, *Biotechnol. Bioeng.,* 73 (6), 476, 2001.
16. Abou-Zeid, D.M., Muller, R.J., and Deckwer, W.D., Degradation of natural and synthetic polyesters under anaerobic conditions, *J. Biotechnol.,* 86 (2), 113, 2001.
17. Zhang, C.L. and Hughes, J.B., Biodegradation pathways of hexahydro-1,3,5-trinitro-1,3,5-triazine (RDX) by *Clostridium acetobutylicum* cell-free extract, *Chemosphere,* 50 (5), 665, 2003.
18. Powell, S. et al., Degradation of 2-nitrodiphenylamine, a component of Otto Fuel II, by *Clostridium* spp., *Anaerobe,* 4 (2), 95, 1998.
19. Narita, M. et al., Identification and characterization of anaerobic mercury-resistant bacteria isolated from mercury-polluted sediment, *Water Science and Technology,* 42 (3–4), 109, 2000.
20. Tummala, S.B. et al., Genetic tools for solventogenic Clostridia, in *Clostridia: Biotechnology and Medical Applications*, Bahl, H. and Dürre, P., Eds., John Wiley & Sons, New York, 2001, 105.
21. Mermelstein, L.D. et al., Expression of cloned homologous fermentative genes in *Clostridium acetobutylicum* ATCC 824, *Biotechnology (NY),* 10 (2), 190, 1992.
22. Papoutsakis, E.T. and Bennett, G.N., Cloning, structure, and expression of acid and solvent pathway genes of *Clostridium acetobutylicum*, in *The Clostridia and Biotechnology*, Woods, D.R., Ed., Butterworth-Heinemann, Stoneham, MA, 1993, 157.
23. Mermelstein, L.D. and Papoutsakis, E.T., In vivo methylation in *Escherichia coli* by the *Bacillus subtilis* phage phi 3T I methyltransferase to protect plasmids from restriction upon transformation of *Clostridium acetobutylicum* ATCC 824, *Appl. Environ. Microbiol.,* 59 (4), 107710, 1993.
24. Desai, R.P. and Papoutsakis, E.T., Antisense RNA strategies for metabolic engineering of *Clostridium acetobutylicum*, *Appl. Environ. Microbiol.,* 65 (3), 936, 1999.
25. Tummala, S.B., Welker, N.E., and Papoutsakis, E.T., Design of antisense RNA constructs for downregulation of the acetone formation pathway of *Clostridium acetobutylicum*, *J. Bacteriol.,* 185 (6), 1923, 2003.
26. Tomas, C.A., Welker, N.E., and Papoutsakis, E.T., Overexpression of groESL in Clostridium acetobutylicum results in increased solvent production and tolerance, prolonged metabolism, and changes in the cell's transcriptional program, *Appl. Environ. Microbiol.,* 69 (8), 4951, 2003.
27. Soucaille, P. and Papoutsakis, E.T., Personal communication, 1995.
28. Tummala, S. and Papoutsakis, E.T., Development and characterization of a gene-expression reporter system for *Clostridium acetobutylicum* ATCC 824, *Appl. Environ. Microbiol.,* 65, in press, 1999.
29. Girbal, L. et al., Development of a sensitive gene expression reporter system and an inducible promoter-repressor system for *Clostridium acetobutylicum*, *Appl. Environ. Microbiol.,* 69 (8), 4985, 2003.
30. Wilkinson, S.R. and Young, M., Targeted integration of genes into the *Clostridium acetobutylicum* chromosome, *Microbiology,* 140, 89, 1994.
31. Green, E.M. et al., Genetic manipulation of acid formation pathways by gene inactivation in *Clostridium acetobutylicum* ATCC 824, *Microbiol. U.K.,* 142, 2079, 1996.
32. Green, E.M. and Bennett, G.N., Inactivation of an aldehyde/alcohol dehydrogenase gene from *Clostridium acetobutylicum* ATCC 824, *Appl. Biochem. Biotechnol.,* 57–58, 213, 1996.
33. Nair, R.V. et al., Regulation of the *sol* locus genes for butanol and acetone formation in *Clostridium acetobutylicum* ATCC 824 by a putative transcriptional repressor, *J. Bacteriol.,* 181, 319, 1999.
34. Harris, L.M., Welker, N.E., and Papoutsakis, E.T., Northern, morphological, and fermentation analysis of *spo0A* inactivation and overexpression in *Clostridium acetobutylicum* ATCC 824, *J. Bacteriol.,* 184 (13), 3586, 2002.
35. Gill, R.T. et al., Genome-wide dynamic transcriptional profiling of the light-to-dark transition in *Synechocystis* sp. strain PCC 6803, *J. Bacteriol.,* 184 (13), 3671, 2002.

36. Muffler, A. et al., Genome-wide transcription profiling of *Corynebacterium glutamicum* after heat shock and during growth on acetate and glucose, *J. Biotechnol.,* 98 (2–3), 255, 2002.
37. Tao, H. et al., Functional genomics: expression analysis of *Escherichia coli* growing on minimal and rich media, *J. Bacteriol.,* 181 (20), 6425, 1999.
38. Oh, M.K. et al., Global expression profiling of acetate-grown *Escherichia coli, J. Biol. Chem.,* 277 (15), 13175, 2002.
39. Chang, D.E., Smalley, D.J., and Conway, T., Gene expression profiling of *Escherichia coli* growth transitions: an expanded stringent response model, *Molecular Microbiology,* 45 (2), 289, 2002.
40. Gill, R.T. et al., Genomic analysis of high-cell-density recombinant *Escherichia coli* fermentation and "cell conditioning" for improved recombinant protein yield, *Biotechnol. Bioeng.,* 72 (1), 85, 2001.
41. DeLisa, M.P. et al., DNA microarray-based identification of genes controlled by autoinducer 2-stimulated quorum sensing in *Escherichia coli, J. Bacteriol.,* 183 (18), 5239, 2001.
42. Zheng, M. et al., DNA microarray-mediated transcriptional profiling of the *Escherichia coli* response to hydrogen peroxide, *J. Bacteriol.,* 183 (15), 4562, 2001.
43. Gill, R.T. et al., Genome-wide screening for trait conferring genes using DNA microarrays, *Proc. Natl. Acad. Sci. U.S.A.,* 99 (10), 7033, 2002.
44. Sabatti, C. et al., Co-expression pattern from DNA microarray experiments as a tool for operon prediction, *Nucleic Acids Research,* 30 (13), 2886, 2002.
45. Britton, R.A. et al., Genome-wide analysis of the stationary-phase sigma factor (sigma-H) regulon of *Bacillus subtilis, J. Bacteriol.,* 184 (17), 4881, 2002.
46. Fawcett, P. et al., The transcriptional profile of early to middle sporulation in *Bacillus subtilis, Proc. Natl. Acad. Sci. U.S.A.,* 97 (14), 8063, 2000.
47. Petersohn, A. et al., Global analysis of the general stress response of *Bacillus subtilis, J. Bacteriol.,* 183 (19), 5617, 2001.
48. Tomas, C.A. et al., DNA array-based transcriptional analysis of asporogenous, nonsolventogenic *Clostridium acetobutylicum* Strains SKO1 and M5, *J. Bacteriol.,* 185 (15), 4539, 2003.
49. Tummala, S.B., Junne, S.G., and Papoutsakis, E.T., Antisense RNA downregulation of coenzyme A transferase combined with alcohol-aldehyde dehydrogenase overexpression leads to predominantly alcohologenic *Clostridium acetobutylicum* fermentations, *J. Bacteriol.,* 185 (12), 3644, 2003.
50. Tomas, C.A. and Papoutsakis, E.T., Transcriptional analysis of butanol stress and tolerance in *Clostridium acetobutylicum, J. Bacteriol.*, 2003 (submitted).
51. Alsaker, K.V., Spitzer, T.R., and Papoutsakis, E.T., Transcriptional analysis of *spo0A* overexpression in *Clostridium acetobutylicum* and its effect on the cell's response to butanol stress, *App. Environ. Microb.*, 2003 (submitted).
52. Bernhardt, J. et al., Specific and general stress proteins in *Bacillus subtilis* — a two-dimensional protein electrophoresis study, *Microbiology,* 143 (3), 999, 1997.
53. Movahedi, S. and Waites, W., A two-dimensional protein gel electrophoresis study of the heat stress response of *Bacillus subtilis* cells during sporulation, *J. Bacteriol.,* 182 (17), 4758, 2000.
54. Lay, J.O. Jr., MALDI-TOF mass spectrometry of bacteria, *Mass. Spectrom. Rev.,* 20 (4), 172, 2001.
55. Cordwell, S.J., Nouwens, A.S., and Walsh, B.J., Comparative proteomics of bacterial pathogens, *Proteomics,* 1 (4), 461, 2001.
56. Graham, M.R. et al., Toward a genome-scale understanding of group A Streptococcus pathogenesis, *Curr. Opin. Microbiol.,* 4 (1), 65, 2001.
57. Bjorkholm, B.M. et al., Genomics and proteomics converge on *Helicobacter pylori, Curr. Opin. Microbiol.,* 4 (3), 237, 2001.
58. Bumann, D., Meyer, T.F., and Jungblut, P.R., Proteome analysis of the common human pathogen *Helicobacter pylori, Proteomics,* 1 (4), 473, 2001.
59. Chakravarti, D.N. et al., Mining genomes and mapping proteomes: identification and characterization of protein subunit vaccines, *Dev. Biol.,* 103, 81, 2000.
60. Schaffer, S. et al., Changes in protein synthesis and identification of proteins specifically induced during solventogenesis in *Clostridium acetobutylicum, Electrophoresis,* 23 (1), 110, 2002.
61. Balodimos, I.A., Rapaport, E., and Kashket, E.R., Protein phosphorylation in response to stress in *Clostridium acetobutylicum, Appl. Environ. Microbiol.,* 56 (7), 2170, 1990.
62. Bennett, G.N., Personal communication, 2003.

63. Mermelstein, L.D. et al., Metabolic engineering of *Clostridium acetobutylicum* Atcc-824 for increased solvent production by enhancement of acetone formation enzyme-activities using a synthetic acetone operon, *Biotechnol. Bioeng.,* 42 (9), 1053, 1993.
64. Dürre, P. et al., Solventogenic enzymes of *Clostridium acetobutylicum*: catalytic properties, genetic organization, and transcriptional regulation, *FEMS Microbiol. Rev.,* 17 (3), 251, 1995.
65. Nair, R.V., Bennett, G.N., and Papoutsakis, E.T., Molecular characterization of an aldehyde/alcohol dehydrogenase gene from *Clostridium acetobutylicum* ATCC 824, *J. Bacteriol.,* 176 (3), 871, 1994.
66. Bermejo, L.L., Welker, N.E., and Papoutsakis, E.T., Expression of *Clostridium acetobutylicum* ATCC 824 genes in *Escherichia coli* for acetone production and acetate detoxification, *Appl. Environ. Microbiol.,* 64 (3), 1079, 1998.
67. Harris, L.M. et al., Characterization of recombinant strains of the *Clostridium acetobutylicum* butyrate kinase inactivation mutant: need for new phenomenological models for solventogenesis and butanol inhibition?, *Biotechnol. Bioeng.,* 67 (1), 1, 2000.
68. Husemann, M.H.W. and Papoutsakis, E.T., Solventogenesis in *Clostridium acetobutylicum* fermentations related to carboxylic-acid and proton concentrations, *Biotechnol. Bioeng.,* 32 (7), 843, 1988.
69. Terracciano, J.S. and Kashket, E.R., Intracellular conditions required for the initiation of solvent production by *Clostridium acetobutylicum, Appl. Environ. Microbiol.,* 52, 86, 1986.
70. Feng, J. et al., Role of phosphorylated metabolic intermediates in the regulation of glutamine synthetase synthesis in *Escherichia coli, J. Bacteriol.,* 174 (19), 6061, 1992.
71. McCleary, W.R., Stock, J.B., and Ninfa, A.J., Is acetyl phosphate a global signal in *Escherichia coli*?, *J. Bacteriol.,* 175 (10), 2793, 1993.
72. Desai, R.P. et al., Metabolic flux analysis elucidates the importance of the acid-formation pathways in regulating solvent production by *Clostridium acetobutylicum, Metab. Eng.,* 1 (3), 206, 1999.
73. Ullmann, S., Kuhn, A., and Dürre, P., DNA topology and gene expression in *Clostridium acetobutylicum*: Implications for the regulation of solventogenesis, *Biotechnol. Lett.,* 18 (12), 1413, 1996.
74. Wong, J. and Bennett, G.N., The effect of novobiocin on solvent production by *Clostridium acetobutylicum, J. Ind. Microbiol.,* 16 (6), 354, 1996.
75. Thormann, K. et al., Control of butanol formation in *Clostridium acetobutylicum* by transcriptional activation, *J. Bacteriol.,* 184 (7), 1966, 2002.
76. Cornillot, E. et al., The genes for butanol and acetone formation in *Clostridium acetobutylicum* ATCC 824 reside on a large plasmid whose loss leads to degeneration of the strain, *J. Bacteriol.,* 179 (17), 5442, 1997.
77. Harris, L.M. et al., Fermentation characterization and flux analysis of recombinant strains of *Clostridium acetobutylicum* with an inactivated *solR* gene, *J. Ind. Microbiol. Biotech.,* 27 (5), 322, 2001.
78. Ranson, N.A., White, H.E., and Saibil, H.R., Chaperonins, *Biochem. J.,* 333 (2), 233, 1998.
79. Mogk, A. et al., The GroE chaperonin machine is a major modulator of the CIRCE heat shock regulon of *Bacillus subtilis, Embo. J.,* 16 (15), 4579, 1997.
80. Nolling, J. et al., Genome sequence and comparative analysis of the solvent-producing bacterium *Clostridium acetobutylicum, J. Bacteriol.,* 183 (16), 4823, 2001.

38 Biodegradation of Hazardous Materials by Clostridia

Farrukh Ahmad, Joseph B. Hughes, and George N. Bennett

CONTENTS

38.1 INTRODUCTION

The metabolic diversity of clostridial organisms has been the subject of many reviews and is covered elsewhere in this series. Clostridia are generally classified as Gram-positive endospore-forming obligate anaerobes that are incapable of the dissimilatory reduction of sulfate [1–3]. With close to 100 species [2,3], the genus *Clostridium* is one of the largest genera among prokaryotes [1]. Clostridia are ubiquitous in nature due to their fermentative diversity and due to their ability to form resistant spores [4,5]. They have been isolated from soils, decomposing biological materials, and from the lower gut of mammals. In addition, they have been widely studied because some proteolytic species (e.g., *C. tetani*) produce strong toxins and because other species (e.g., *C. acetobutylicum*) are of industrial importance due to their solvent-producing ability. Clostridia are known to possess strong "nitroreductase" properties that have been extensively investigated [6–12]. Clostridia have been isolated from an anaerobic consortium that was used to transform nitroaromatic compounds in the presence of a suitable carbohydrate carbon source [13,14]. Therefore, saccharolytic clostridia, typically nonpathogenic organisms, are especially well suited for the study of biotransformations of TNT in anaerobic engineered systems.

Considering the widespread existence of clostridium in anaerobic environments and the importance of biodegradation of chemicals in subsurface environments, there has been a relatively modest effort to exploit or analyze biodegradation by this group of organisms. The contribution of these organisms in communities where biodegradation takes place has been noted, and, in many cases, the effect is through positive interactions beneficial to other organisms of the community. In a number of cases, the diverse metabolism of clostridial species extends to direct transformation of the hazardous or recalcitrant material. The action of clostridial species on a number of specific types of molecules will be addressed.

Clostridial species have been used as components of bioremediation processes for a variety of substances. Among those are nitroaromatics, chlorinated compounds, and toxic metals. The review will consider the studies of clostridial reactions and species involved in these processes and discuss their role and potential in remediation. A number of reviews of clostridial action in degradation of nitro compound explosives have been published [15–20]. Effects of the fermentative clostridia in reduction of metals [uranium (VI)] as well as the more common sulfate-reducing bacteria [21] have been noted. Of particular interest are recent descriptions of association of clostridium in sediments degrading chlorinated hydrocarbons such as PCE (tetrachloroethylene) (*C. bifermentans* DPH-1 [22]) and PCB (*C. hydroxybenzoicum* [23]). Recent reports of clostridial action on dyes (*C. paraputrificum* [24,25]), cinnamic acid (*C. glycolicum* [26]), methylglyoxal [27], and polyhydroxybutyrate derivatives [28] further enhance the potential for application of clostridial organisms in environmental bioprocesses of industrial interest. On the other hand, concerns have been raised about the presence of pathogenic clostridia and spores of these organisms in solid waste landfills [29,30] and in biocomposting processes [31].

38.2 REACTIONS WITH CHLORINATED COMPOUNDS

Chlorinated solvents are widely used in industrial operations and have become public health hazards in many locations. Bioremediation and monitoring of such compounds have generated public and governmental interest. Most studies have focused on the degradation of perchloroethylene (tetrachloroethylene, PCE) or chlorinated cyclic compounds (e.g., PCB). There are a number of ways hologenated compounds can be degraded aerobically or anaerobically, and the basic biochemical processes have been reviewed [32–34]. The action of microbes in attenuation of chlorinated compounds has gained attention as a cost-effective alternative, and a recent review addresses this point [35]. Organisms generally degrade PCE to dichloroethylene, which is more difficult to metabolize but is still of toxic concern, although it can be degraded by aerobic organisms such as *Rhodococcus rhodochrons* or *Nitrosomonas europaea*. However, there are reports of a dehalogenase from *Dehaloococcoides ethenogenes* [36] that converts PCE to ethylene.

The biotransformation of chlorinated compounds by clostridial species was reported by Galli and McCarty [37,38]. Recent studies have been conducted with *C. bifermentans* DPH-1, and cell-free extracts were shown to be active on PCE and mainly degraded the tetrachloro-compound to cis-1,2-dichloroethylene. The dehalogenase activity was inhibited by cyanide and stimulated by NADH [39]. This work was continued with the report of a purified enzyme from *C. bifermentans* DPH-1 [22], which catalyzed the conversion of PCE to cis-1,2-dichloroethylene with trichloroethylene as an intermediate. The enzyme had a pH optimum of 7.5 and a K_m of 12 μM for PCE. The enzyme had a lower V_{max} (73 nmol/mg protein) than other dehalogenases and was unstable during purification. The enzyme PceC showed activity on a broad spectrum of aliphatic chlorinated compounds. N-terminal sequence analysis on the homodimeric protein allowed oligonucleotides to be constructed for isolating the gene. A gene, *pceC*, was cloned, but the encoded protein (approximately 360 aa) did not show homology with other known dehalogenases. Some homology of *pceC* to open reading frames in some other bacterial species are found. The environmental role of this enzyme and the presence in other clostridial species are still undetermined.

A number of clostridial strains were isolated that could degrade para- or meta-PCB (polychlorinated biphenyl) [23]. Since these toxic materials contaminate sediment, the degradation by anaerobes such as clostridia would be significant. The analyses of 16s RNA clones from the PCB-degrading bacteria showed clusters with the phylogeny of *C. hydroxybenzocium* strains, and para-PCB degraders were the most common. Other strains were related to *C. beijerinckii* (Otu-A), *C. intestinalis* (Otu-D), and *C. thermolactium* (Otu-J) and were mostly active on meta-PCB. A few clostridial isolates could act on both para- and meta-PCB. Recently, some strains have been reclassified as *Sedimentibacter hydroxybenzoicus* [40].

The reductive debromination of (α-bromoisovaleryl)urea to (3-methylbutyryl)urea has been observed with extracts from *C. sporogenes* [41]. The reaction seemed to involve the reduction of flavin cofactors, and debromination occurred by using the reduced flavins as an electron donor for a dehalogenase type enzyme. These reactions by intestinal microbes could have an effect on the metabolism of halogenated compounds in the animal gut.

Besides these actions of clostridia directly on chlorinated compounds, clostridial species contribute to the degradation of chlorinated compounds by their production of hydrogen. As described later, hydrogen production is a common means of release of reducing equivalents generated during glycolytic processes in clostridial fermentations. Hydrogen stimulates biotransformation of halogenated compounds by providing a source of reductant. The ability of hydrogen to readily diffuse away from the clostridial component of the natural environment ecosystem and to the dehalogenating organisms acts to aid the energetics of both species.

38.3 REACTIONS WITH TNT AND RELATED NITRO COMPOUNDS

Due to the synthesis and processing of trinitrotoluene (TNT), this nitroaromatic compound contaminates the surface and subsurface soil near the surface of many military facilities [42]. TNT contamination in easily accessible surficial soils allows treatment by either *in situ* or *ex situ* engineered biological systems. The removal and incineration of such soils have been employed at some sites, but is quite expensive compared to bioremediation. Many sites are also contaminated by related nitro compounds. These include compounds such as the relatively polar dinitrotoluene (DNT) impurities of TNT. The explosive compound, 1,3,5-trinitrohexahydro-1,3,5-triazine (RDX), has also been found in shallow groundwater aquifers, since it does not bind to soils as readily as TNT-related compounds [43–45].

The nitro groups of TNT are electrophilic in nature and, thus, they are prone to reductive attack, even under moderate oxidative conditions [46,47]. In order to complete reduction of a nitro group to an amino group, a six-electron transfer is required. The reduction pathway proceeds via the potentially stable two- and four-electron transfer intermediates of nitroso and hydroxylamino groups, respectively. Highly unstable anion radicals can be obtained from odd number electron transfer reactions, but these are rarely observed in natural systems [48]. Although the intermediates are thermodynamically stable compared to nitro groups, nitroso and hydroxylamino groups are rather chemically reactive. Such intermediates can undergo a variety of reactions, including rapid abiotic condensation reactions, to form toxic azoxy compounds [48,49]. The complete reduction of polynitroaromatics, such as TNT, becomes progressively slower with the reduction of each nitro group as less electron withdrawing groups are present. The complete reduction of TNT to 2,4,6-triaminotoluene (TAT) has been observed only under strictly anaerobic conditions in bacterial systems [46].

Recent efforts focusing on the cleanup of contamination from herbicides [50] and explosives [12,13] have brought attention to anaerobic systems because of their ability to reduce polynitroaromatics. A substantial amount of research has been conducted with mixed anaerobic cultures. Such natural cultures have yielded two commercially implemented biological processes for TNT treatment. Both processes attain complete reduction of TNT, but have been less employed for more mechanistic studies of the biochemistry employed by the organisms involved. A clostridial species was recently isolated from a commercially available anaerobic consortium capable of reducing TNT [14]. This observation has stimulated research in the ability of clostridia to reduce nitro-containing compounds. Clostridia have long been known to possess enzymes capable of rapid nitro group reduction. These enzymes have the general name "nitroreductases," but many different proteins can carry out this reaction. The rates of nitro group reduction achieved in studies with pure clostridial cultures, cell-free extracts, and purified enzymes appear significantly faster than those reported for commercially available anaerobic consortia, if an equal degree of reduction is considered. On the other hand, the complete reduction of polynitroaromatics by clostridial enzymes has been inconclusive. The purified systems analyzed so far have shown conversion to the hydroxylamino compounds.

38.3.1 Basis of Related Clostridial Metabolism

This section provides only a brief summary of clostridial metabolic pathways and processes. There are many extensive reviews of related topics: acid production [51]; substrate utilization [52]; solvent production [53–59]; proteolytic and purinolytic fermentations [1]; and acetate production [51]. Readers can turn to these publications and other chapters in this book for detailed discussions on specific topics. The purpose of this section is to provide some background material on the broader metabolic capabilities of clostridia that are relevant to nitroreductase activity.

Clostridia lack an electron-transport phosphorylation system and are limited to substrate-level phosphorylation for energy generation [4]. Consequently, fermentative organisms, including clostridia, must rely on a high throughput of substrate for generation of energy for active growth. As a result, anaerobic fermenters tend to overproduce reducing power in the form of reduced carriers (e.g., NADH or reduced ferredoxin) [4,5]. While a portion is utilized for biosynthesis in the form of NADPH [58], the reoxidation of the reduced carriers is essential because the carriers are present in limited quantities within a cell.

In pure cultures of clostridia, excess reducing power can be dissipated by the evolution of hydrogen gas via the ferredoxin/hydrogenase system, the formation of a larger high-energy intermediate such as butyryl-CoA derived from acetyl-CoA, or the generation of reduced fermentation end-products such as solvents ethanol, butanol, acetone, or isopropyl alcohol. In saccharolytic fermentations, the reduction of an unrelated electrophilic compound, such as TNT, can provide an additional means for reoxidizing reduced electron carriers. A dye or added substrate can also be used as an acceptor to generate a reduced compound.

Clostridia demonstrate considerable fermentative diversity by utilizing a wide variety of substrates and are commonly classified by the type of substrate they ferment. They are generally divided into four categories: (1) saccharolytic (e.g., *C. perfringens, C. acetobutylicum*); (2) proteolytic; (3) both saccharolytic and proteolytic; and (4) possessing specialized metabolism (e.g., *C. acidiurici* ferments purines and *C. kluyveri* can ferment ethanol, propanol, and succinate). The fermentation pathways usually proceed through common high-energy intermediates such as acetyl-CoA and butyryl-CoA and lead to formation of acids or solvents [5]. Clostridia are also grouped by the type of fermentation end products they produce, e.g., solventogenic (alcohol and acetone producers, *C. acetobutylicum*), acidogenic (organic acid producers, *C. pasteurianum, C. clostridiiforme*), and acetogenic (acetate producers, *C. thermoaceticum*). Some saccharolytic clostridia, such as *C. acetobutylicum,* are acidogenic but undergo a switch in their metabolism to solventogenesis during the course of typical batch culture and have been used industrially for the production of solvents. The main organisms used in studies of nitro reduction have been *C. acetobutylicum*, *C. bifermentans*, *C. pasteurianum*, *C. kluyveri*, *C. clostridiiforme*, *C. perfringens*, and *C. paraputrificum*.

Some clostridia are capable of autotrophic metabolism [51,60]. Examples of such species are *C. aceticum* and *C. thermoaceticum*. Such organisms can derive carbon for cell growth from the reduction of carbon dioxide via carbon monoxide to form acetate. A key enzyme in the autotrophic pathway is carbon monoxide dehydrogenase (CODH). The reducing power for the pathway is directly generated via the uptake of molecular hydrogen by hydrogenase or derived from reduced ferredoxins formed during the fermentation of an electron donor [51].

38.3.2 Clostridial Enzymes Involved in the Transformation of Nitro Groups

Interest in the "nitroreductase" capability of clostridial enzymes has derived from several observations. One was the study of uncouplers of phosphorylation by 2,4-dinitrophenol (DNP) [6]. It was discovered that DNP could undergo reduction in ferredoxin-rich systems such as chloroplasts of photosynthetic plants where reduced ferredoxin is formed [61–63]. del Campo et al. [9] explored reactions that could reduce DNP and found that DNP could be effectively reduced to 2,4-diaminophenol by H_2 in the presence of the hydrogenase/ferredoxin system of *C. pasteurianum.* The H_2 uptake stoichiometry indicated a direct correlation. Another interest in nitro reduction came from medical researchers attempting to halt protein synthesis of a rapidly growing culture of clostridia. The researchers were unable to easily block protein synthesis by chloramphenicol [11], a broad-spectrum antibiotic often used with aerobic organisms. This seemed to conflict with an earlier report [64] that chloramphenicol, a nitroaromatic compound, was effective against a wide variety of clostridia. After this report, the nitro reduction of this antibiotic by clostridia alerted the profession to a potential mechanism for the antibiotic resistance. The finding was of medical importance due to the pathogenic nature of certain clostridia. Other nitro compounds, such as metronidazole, have been used for antibacterial agents against anaerobic organisms, since reduction activates the compound to a toxic form [65,66]. This is a common treatment for *C. difficile* infections [67]. With regard to the focus of this article, attempts to remediate soil contaminated with di- and polynitroaromatics have prompted further interest in anaerobic biotransformations, since nitroaromatics undergo complete reduction only under strictly anaerobic conditions [46]. For this reason, the enzyme systems of clostridia are being investigated for their ability to rapidly transform compounds such as TNT to potentially less harmful products.

The identity of the enzymes involved has been a major focus of research. It was considered that the reduction of nitro groups by clostridia was due to the gratuitous activity of well-defined oxidoreductases. Such proteins are capable of reducing electron carriers with an extremely low mid-point redox potential and include hydrogenase, pyruvate-ferredoxin oxidoreductase, and NADH-ferredoxin oxidoreductase. Table 38.1 presents a compilation of studies that have utilized clostridial crude cell extracts or purified cell components to reduce the nitro groups of various compounds. The clostridial enzymes conclusively identified as generating cellular nitroreductase activity, namely hydrogenase and carbon monoxide dehydrogenase (CODH), are different from the classical type I (oxygen insensitive) and type II (oxygen sensitive) nitroreductases purified from aerobic and facultative bacteria. The classical bacterial nitroreductases found in those organisms are flavin mononucleotide (FMN), requiring flavoproteins that operate at a suitably high mid-point redox. Such a flavoprotein can accept electrons from either NADH or NADPH. A more detailed review of classical bacterial nitroreductases is presented by Bryant and McElroy [71]. In contrast, clostridial hydrogenase and CODH contain highly reducing iron-sulfur redox centers at their active sites that allow transfer of electrons to electron carriers at a redox value near that of the hydrogen electrode. Recent research also suggests that other transformations of nitro compounds such as the rearrangement of hydroxylamino compounds to aminophenols [71] can also be catalyzed by clostridial enzymes. The reactions mediated by such enzymes could potentially divert the partially reduced intermediates of nitroaromatic compounds away from the reduction pathway catalyzed by nitroreductases, if they were present at a sufficient level in the cell.

TABLE 38.1
Reduction of Compounds Containing Nitro Substituents By Crude Cell Extracts or Purified Components From Clostridia

Species	Biological Component			Electron Donor/Reductant	Electron Carrier	pH	Starting Compound	Intermediates	Final Product	Ref.
	Extract	Enzyme	Electron Carrier							
C. acetobutylicum	X			PYR	MV/FD	7.0	Chloramphenicol	NR	NR	[11] (*)
	X			PYR	MV	7.0	2/3-NP	NR	NR	
	X			PYR	MV	7.0	2/3/4-NBA	NR	NR	
	X			PYR	MV	7.0	2NBD	NR	NR	
	X			PYR	MV	7.0	NF	NR	NR	
	X			PYR	MV	7.0	3NAn	NR	NR	
	X			PYR	MV	7.0	24DNP	NR	NR	
	X			H_2	—	7.8	TNT	DHA6NT	Aminophenol	[68]
				H_2	—	7.8	24DNT, 26DNT	HANTS, HAATS	pHATS	[69]
C. kluyveri	X			H_2/NADH	—	8.5/11.5	pNB	—	pHAB	[6,7]
		X		H_2	FD	8.5	pNB	—	pHAB	
C. pasteuranium		X		H_2	FD/FMN/BV	8.0	24DNP	NR	NR (6e⁻/NO_2)	[9] (*, **)
	X			H_2	FD	7.5	TNT	4HADNT	TAT	[10]
			X	Dithionite	FD/MV	7.0	Metronidazole	NR	NR (4e⁻/NO_2)	[70] (*, **)
		X		H_2	FD/FLV	8.0	Metronidazole	NR	NR	[8] (*)
	X			H_2	FD	8.5	pNB	—	pHAB	[6,7]
		X		H_2	MV	8.0	DA6NT	—	DA6HAT	[12]
C. sporogenes	X			H_2	FD	8.5	pNB	—	pHAB	[7]
C. thermoaceticum		X		CO	MV	8.0	DA6NT	—	DA6HAT	[12]

Abbreviations

Electron Carriers: FD = Ferredoxin; FLV = Flavodoxin; FMN = Flavin mononucleotide; BV = Benzyl viologen; MV = Methyl viologen.

Electron Donors: PYR = Pyruvate; H_2 = Hydrogen gas; NADH = Reduced nicotinamide adenine dinucleotide; CO = Carbon monoxide.

Compounds: DA6HAT = 2,4-diamino-6-hydroxylaminotoluene; DAP = Diaminophenol; DNP = Dinitrophenol; 24DNT = 2,4-dinitrotoluene; 26DNT = 2,6-dinitrotoluene; HAATs =hydroxylaminoaminotoluenes; HANTs = hydroxylaminonitrotoluenes; 4HADNT = 4-hydroxylamino2,6-dinitrotoluene; NAn = Nitroaniline; NBA = Nitrobenzoic acid; NBD = Nitrobenzaldehyde; NF = Nitrofurantoin; NP = Nitrophenol; pHAB = para-Aminobenzoate; pNB = para-Nitrobenzoate; TAT = 2,4,6-triaminotoluene; TNT = 2,4,6-trinitrotoluene.

Other: NR = Not reported; * = Loss of reactant monitored; ** = Hydrogen uptake monitored.

Reprinted from Ahmad and Hughes [19].

38.3.2.1 The Hydrogenase/Ferredoxin System

Clostridial hydrogenases and ferredoxins contain highly reducing inorganic iron-sulfur centers that participate in electron transfer reactions at extremely low redox potentials [73]. In clostridia, the iron hydrogenases are believed to contain iron-sulfur clusters of the general form [4Fe-4S] [74,75]. Iron and nickel-iron hydrogenases are known, and the structures and phylogeny have been reviewed [76,77]. The clostridial hydrogenases are usually "bidirectional" [73,74]. This means that they can catalyze the transfer of electrons to and from an electron carrier:

$$\text{(Electron Carrier)}_{\text{reduced}} + 2\text{H}^+ \xrightleftharpoons[]{\substack{\text{Bidirectional}\\\text{Hydrogenase}}} \text{H}_2 + \text{(Electron Carrier)}_{\text{oxidized}}$$

One key factor in dealing with hydrogenases is that exposure to oxygen inactivates hydrogenase, whereas exposure to carbon monoxide reversibly inhibits hydrogenase activity [74]. Certain hydrogenases and mutant enzymes can sustain longer exposure to oxygen before inactivation.

In obligate anaerobes, hydrogenases interact with an electron carrier that has a suitably low midpoint redox potential. In clostridia, the most abundant electron carrier with a low midpoint potential is ferredoxin. Clostridial ferredoxins have been well studied and generally possess two [4Fe-4S] clusters that can each cycle between +3 and +2 oxidation states, thereby allowing each ferredoxin molecule to transfer one or two electrons [73,78]. The midpoint redox potentials of clostridial ferredoxins are in the range of –390 to –434 mV [73], based on the particular species from which they are extracted. In typical sugar fermentations, ferredoxins can be reduced by oxidoreductases involved in the pyruvate phosphoroclastic reaction (pyruvate:ferredoxin oxidoreductase) [79] or directly by NADH via the enzymatic action of NADH-ferredoxin oxidoreductase. *In vitro,* electron transfer with hydrogenase can be maintained if the ferredoxins are replaced with other biological or nonbiological electron carriers such as flavodoxins and redox dyes, respectively [9,74]. Together, hydrogenase and ferredoxin play a critical part in clostridial fermentative metabolism by using this system to dissipate excess reducing power by reducing protons to produce hydrogen gas. Other species of clostridia, such as *C. thermoaceticum,* use the system to acquire reducing power that is used in the autotrophic synthesis of acetyl-CoA from CO_2.

Clostridia exhibit an interesting regulatory mechanism under iron-limiting conditions. Under such conditions, the iron-containing redox factor ferredoxin is not formed, and the iron of existing ferredoxin molecules is recycled for use in the production of essential enzymes [80], while small non-iron-containing flavoproteins, flavodoxins, are induced [81]. Flavodoxins require FMN as cofactor, lack iron and labile sulfide, and have a sufficiently low mid-point redox potential to substitute for ferredoxins as electron carriers and can interact in processes where ferredoxins are normally employed. It should be noted that flavodoxins are single electron carriers. Therefore, under conditions where the electron carrier might change, particularly under iron limiting conditions, different reduction kinetics can appear, and this factor can help to explain some of the variation in findings regarding the reduction of TNT by clostridia. This could be a factor where other materials bind iron or limit iron availability in various mixed media or natural bioremediation environments.

The hydrogenase/ferredoxin system was first observed to reduce nitro groups by del Campo et al. [9]. They reduced DNP with purified hydrogenase and ferredoxin from *C. pasteurianum* using hydrogen gas as the electron donor. They also showed that spinach ferredoxin, benzyl viologen, and FMN were suitable replacements for the clostridial ferredoxin electron carrier. O'Brien and Morris [11] used crude cell-free extracts and pyruvate as the electron donor to expand these findings to include various other nitroaromatic compounds, including the antibiotic chloramphenicol. In addition, using ferredoxin-free extracts, they showed that ferredoxin or a similar electron carrier was essential for effective nitro reduction. They also demonstrated the inability of FMN, FAD, NADH, and NADPH to interact directly with the hydrogenase/ferredoxin system, contradicting some of the earlier findings [9]. The reduction of chloramphenicol has medical significance, as the reduction limits the antibacterial effectiveness. The toxicity of chloramphenicol is known and has been related to the reduction and release of NO by metabolism of the drug [82].

Lindmark and Müller [70] were able to partially reduce the antibiotic metronidazole, a nitroimidazole (2-methyl-5-nitroimidazole-1-ethanol), with electron carriers that had been previously reduced using dithionite. The rates of the nonenzymatic reduction with only reduced carriers were much slower than rates with the hydrogenase/electron carrier system. Building on the earlier studies, Chen and Blanchard [8] developed a hydrogenase-linked electron carrier assay using the easily observed reduction of metronidazole with the H_2/hydrogenase/ferredoxin system. They went on to note that this system was more sensitive to the concentration of electron carrier as compared to that of the electron donor and suggested that levels of the carrier might also play a role in activity in whole cells if they fluctuate appreciably in a sensitive range.

The mechanism of reduction of metronidazole by hydrogenase was further studied [83–86] with the idea that the dye or metronidazole might react with the hydrogenase at a position where reaction is stimulated by interactions with an electron donor (ferredoxin or other dye). The presence of multiple Fe-S centers could allow such events. Susceptibility to metronidazole was also used in the cloning of redox genes from clostridia [87].

The first experiment that utilized the H_2/hydrogenase/ferredoxin system to reduce TNT was reported by McCormick et al. [10]. They reported the formation of 2,4,6-triaminotoluene (TAT) via 4-hydroxylamino-2,6-dinitrotoluene (4HADNT) using ferredoxin-rich cell extracts of *C. pasteurianum*. The study advanced the detail of study and introduced methods for monitoring the formation of the partially reduced hydroxylamino intermediates. The separation method was thin layer chromatography followed by the application of various dyes. Earlier studies had only monitored the loss of nitroaromatic reactants and analyzed the stoichiometric consumption of electron donor. Prediction of reduced products based on hydrogen uptake assays can lead to an overestimation of the degree of reduction, especially when other ferredoxin-oxidizing enzymes (e.g., clostridial NADH-ferredoxin oxidoreductase) are present.

Since accumulation of hydroxylamino products was observed in two separate studies conducted with purified hydrogenase, it appears that the hydrogenase/ferredoxin system is inefficient for the complete reduction of nitro substituents to amino substituents. Angermaier [6,7] studied the ability of purified components of the H_2/hydrogenase (*C. kluyveri*)/ferredoxin (*C. sp. La1*) system to partially reduce *para*-nitrobenzoate (pNB) to *para*-hydroxylaminobenzoate (pHAB). In these experiments, a dianion radical was detected using electron paramagnetic resonance (EPR) spectroscopy. The presence of this radical intermediate confirmed that odd-numbered electron reductions can occur in these systems [48]. They also obtained similar results with cell extracts of *C. kluyveri*, indicating this was a major process in cells. However, they noted that during H_2 uptake studies with clostridial cell cultures, there was a slow uptake of the third mole of hydrogen. This would be needed for the complete reduction of the nitro group to an amino substituent. The accumulation of the hydroxylamino compound led Angermaier [6,7] to conclude that in the H_2/hydrogenase/ferredoxin system, the pHAB product must be a poor substrate for the reoxidation of the reduced ferredoxin. A similar finding by Preuss, Fimpel, and Diekert [12] discovered that 2,4-diamino-6-hydroxylaminotoluene (DAHAT) accumulated as the final product when 2,4-diamino-6-nitrotoluene (DANT) was incubated with purified hydrogenase from *C. pasteurianum,* and methyl viologen was used as the electron carrier. In the same study, they reported the complete reduction of DANT to TAT with intact cells of *C. pasteurianum*. Based on these findings, Preuss, Fimpel, and Diekert [12] suggested that a second enzyme might exist that was involved in the complete reduction of the nitro group of DANT to an amino group.

Hughes et al. [72] performed experiments with cell extracts of *C. acetobutylicum*, and showed partial reduction of the nitroaromatic compound TNT. Hughes et al. [72] provided rigorous proof of the accumulation of an aminophenolic product (2-hydroxylamino-4-amino-5-hydroxy-6-nitrotoluene) formed from TNT via a partially reduced 2,4-dihydroxylamino-6-nitrotoluene (DHANT). The results suggested that DHANT underwent a Bamberger rearrangement [17,18,48,88,89] to form the aminophenolic product. A similar biologically mediated rearrangement has been reported with hydroxylaminobenzene in aerobic cultures of *Pseudomonas pseudoalcaligenes* [17,18,88]. In *P. pseudoalcaligenes*, unlike those in clostridia, the reduction of the nitro group followed by the Bamberger rearrangement was part of a productive catabolic sequence providing nutrients for the cell. It should be noted, however, that aminophenol formation was not detected when clostridial cell extracts were incubated with the related compounds 2,4-dinitrotoluene (DNT) and 2,6-DNT [69], indicating a difference in chemical properties of the hydroxyamino between these molecules with two functional groups vs. the three of TNT. The difference could also be partially due to the specificity of possible proteins involved.

There has been less study of the degradation of another explosive, RDX (hexahydro-1,3,5-trinitro-1,3,5-triazine), by clostridia. Regan and Crawford [14] reported transformation of RDX by *C. bifermentans*. A study by Bhushan et al. [90] showed transformation of RDX by N-denitration with a diaphorase from *Clostridium kluyveri*. Using NADH as the electron donor, the products formaldehyde, ammonium, nitrite, and nitrous oxide were found. The reaction seemed to proceed through a methylene dinitramine, giving rise to one nitrite per RDX molecule. Subsequent ring cleavage and decomposition then occurred. A mechanism involving FMN in the redox from NADH resulting in an RDX anion radical was proposed. Work reported by Zhang and Hughes [91] examined the biodegradation of RDX by *C. acetobutylicum*. Biodegradation was observed using whole cells and H_2 as the electron donor. A number of compounds were observed including nitroso, hydroxylamino, and amino derivatives.

Isolation of an enzyme from *C. acetobutylicum* based on TNT reducing activity indicated the Fe-hydrogenase of *C. acetobutylicum* was the enzyme involved [92]. *In vivo* experiments, where the activity of the Fe-hydrogenase was altered by genetic means, also correlated the activity of the hydrogenase with the TNT reduction activity in cells [92].

The presence of hydroxylamino products has repeatedly been reported during the reduction of nitro compounds with clostridial hydrogenase/ferredoxin systems. Difficulties encountered with complete reduction of nitroaromatic compounds by the hydrogenase/ferredoxin system can facilitate the occurrence of competing biologically mediated reactions, such as Bamberger rearrangements, that can form products amenable to aerobic mineralization by other organisms [17,18,88].

38.3.2.2 Other Clostridial Enzymes Catalyzing Reduction of Nitro Groups

In studies with nitroaromatic compounds, two other clostridial enzymes have been implicated for their "nitroreductase" activity. In a study that focused mainly on the nitroreductase activity of the hydrogenase/ferredoxin system, Angermaier et al. [6] made the unusual observation that the nitroreductase activity of cell extracts of *C. kluyveri* was NADH-dependent and was at a maximum pH of 11.5. These conditions are not those expected for a hydrogenase reaction, since hydrogenase cannot interact directly with the NAD^+/NADH couple and hydrogenase typically has a much lower optimal pH [74]. Several enzymes were ruled out as contributors by performing multiple enzymatic assays on fractions collected from the separation of the extract on a DEAE-Sepharose column. They proposed that the enzyme responsible for the nitroreductase activity was butyryl-CoA dehydrogenase, an enzyme that participates in the formation of butyryl-CoA from crotonyl-CoA in butyric acid clostridia. It now seems likely this enzyme utilizes an electron-transfer flavoprotein as redox coupling agent [93]. It should be noted that the authors did not rule out NADH-ferredoxin oxidoreductase, another ferredoxin reducing enzyme common in clostridia.

Preuss, Fimpel, and Diekert [12] reported results corresponding to their work with the H_2/hydrogenase/methyl viologen system when they employed the CO/CO-dehydrogenase/methyl viologen system to reduce DANT. An accumulation of DAHAT was also observed in the CO/CODH/methyl viologen system. The CO-dehydrogenase used in those studies had been partially purified from *C. thermoaceticum*. A more recent study employing purified CODH from this organism showed that an electron carrier was unnecessary to produce dihydroxylaminonitrotoluenes from TNT [94] when CO was used as the electron donor. The highly reducing redox center of CO-DH contains nickel, iron and [4Fe-4S] clusters [51]. The CO electron donor directly forms a complex with the nickel in the redox center. The reaction catalyzed by this complex is a key step to the homoacetogenic pathway of *C. thermoaceticum.*

Incubation of nitroaromatic compounds with purified clostridial enzymes, in the presence of a carrier to produce a source of electrons, has resulted in the rapid accumulation of hydroxylamino compounds. The initial reduction of nitroaromatic compounds seems to be a nonspecific property of enzymes involved in the normal fermentative and homoacetogenic metabolic pathways of clostridia. However, the complete reduction of a nitro group to an amino substituent might involve other factors besides the "nitroreductase" proteins studied to date. There are many oxidoreductase-like proteins encoded in the clostridial genome, and contribution of each to specific reactions has not been completely defined.

The availability of the complete genomes of several clostridia (*C. acetobutylicum* [95], *C. perfringens* [96], *C. tetani* [97]) indicates the presence of other genes encoding enzymes that could play a role in nitroreduction. For example, in the genome of *C. acetobutylicum*, eight genes are found with some homology with other nitroreductases. In addition to the Fe-hydrogenase, genes are found encoding a Ni-Fe hydrogenase and numerous small Fe-S redox proteins [95]. While the Fe-hydrogenase seems the major contributor in *C. acetobutylicum* [92], the possible contributions of other enzymes under some circumstances have not been examined.

38.3.3 Studies of Transformation of TNT by Clostridial Cells

O'Brien and Morris [11] reported the nitro group of chloramphenicol could be reduced by actively growing cultures of *C. acetobutylicum* using pyruvate as the electron donor. McCormick et al. [10] expanded the findings to include the reduction of TNT in the presence of a pure culture of *C. pasteurianum* and used hydrogen as the electron donor. The above findings implicated ferredoxin-reducing enzymes in nitroreductase activity and prompted the exploration of various practical applications. Such applications included the biocatalysis of optically active aliphatic hydroxylamines to their corresponding amines [98], and the treatment of soil contaminated with explosive compounds. The recent isolation and characterization of *Clostridium* species as the predominant active microorganism in a commercially available anaerobic consortium that effectively reduced levels in TNT- and RDX-contaminated soil [14] has generated an interest in remediation studies with pure clostridial cultures.

38.3.3.1 Mixed Cultures and Composting

Quite a number of studies have been made on the use of compost techniques to treat nitro-compound contaminated soils. Clostridia are considered as major active species in the composting method. A summary of the processes is presented in Table 38.2. Based on this information with mixed or natural cultures, a variety of studies utilizing different clostridial and other species, nutrients (synthetic media/suspensions vs. complex media/active cultures), growth phases (active cells vs. resting cells), and nitro compounds have been conducted. Table 38.3 provides a listing of such pure culture biotransformation studies. An evaluation of trends evident among the studies follows.

38.3.3.2 Cometabolic Nitro Reduction by Clostridial Cultures

Clostridia have diverse nutritional requirements for growth (e.g., salts, trace metals, vitamins, amino acids, etc.), and some species can grow on defined media. In laboratory experiments, such requirements are most conveniently met by complex media such as peptone-yeast extract-glucose (PYG), chopped meat-carbohydrate (CMC), or other extract-enriched media [1]. These factors have made it difficult to conduct studies [115] on the role of nitroaromatic compounds as a sole source of carbon or nitrogen. Attempts to simplify the culture media in transformation studies involving

TABLE 38.2
Examples of Pilot-Scale Composting Treatments of Soil Contaminated with Explosives

Treatment	Treatment Time (days)	Temperature Range °C	Amount of Soil (dry wt%) and origin	Concentration of Explosive Initial/Final (mg/kg dry wt)	Reduction of Extractable Explosives (%)	Quantity of Treated Soil	Ref.
In-Vessel Static Pile							
Hay/horse feed compost	47	39–67	11.5	TNT: 46,178/1,716	96	470 kg	[99]
			LAAP sediment	ADNT: 27.5/170	+618		
				RDX: 3,580/600	83		
				HMX: 575/161.5	79		
Sewage sludge-wood chips	47	40–53	15.7	TNT: 39,236/10,446	73	504 kg	
			LAAP sediment	ADNT: 1,153/1,005	13		
				RDX: 6,698/2,471	63		
				HMX: 962/914	5		
Horse manure	56	26–71	11.8	TNT: 31,021/138	99.6	260 kg	
			LAAP sediment	ADNT: 3,000/0	100		
				RDX: 7,400/445	94		
				HMX: 2,800/532	81		
Static Pile							
Alfalfa/manure/horse feed	153	35 (mesophilic)	3	TNT: 11,187/50	99.6	4400 kg	[100]
			LAAP sediment	RDX: 4,630/242	94.8		
				HMX: 643/84	86.8		
Alfalfa/manure/horse feed	153	55 (thermophilic)	3 (wt/wt)	TNT: 11,840/3.0	99.9		
			LAAP sediment	RDX: 5,293/45	99.1		
				HMX: 739/26	96.5		
Sawdust/apple/chicken manure/potatoes	90	50 (thermophilic)	10	TNT: 4,984/200	95.9	2.3 m^3	[100–102]
			UMDA soil	RDX: 1,008/542	46.2		
				HMX: 180/12	21.3		

Continued

TABLE 38.2 (continued)
Examples of Pilot-Scale Composting Treatments of Soil Contaminated with Explosives

Treatment	Treatment Time (days)	Temperature Range °C	Amount of Soil (dry wt%) and origin	Concentration of Explosive Initial/Final (mg/kg dry wt)	Reduction of Extractable Explosives (%)	Quantity of Treated Soil	Ref.
Mechanically Agitated In-Vessel Composting (MAIV)							
Horse manure/buffalo manure/alfalfa/horse feed	44	No data	10 UMDA soil	TNT: 3,452/90 RDX: 1,011/104 HMX: 169/120	97.4 89.7 28.8	2.3 m^3	[100–102]
Horse manure/buffalo manure/alfalfa/horse feed	44	No data	10 UMDA soil	TNT: 3,126/5 RDX: 574/3 HMX: 199/6	99.8 99.3 94.9	2.3 m^3	
Sawdust/apple/potato/cow manure	44	No data	25 UMDA soil	TNT: 5,208/14 RDX: 597/18 HMX: 161/51	99.7 97.0 68.0	2.3 m^3	
Windrow Composting							
Sawdust/apple/potato/cow manure with forced aeration	40	Thermophilic	30 UMDA soil	TNT: 1,869/4 RDX: 1,069/8 HMX: 175/47	99.8 99.3 73.1	23.0 m^3	[100–102]
Unaerated				TNT: 1,038/20 RDX: 944/2 HMX: 159/5	99.7 99.8 96.9	23.0 m^3	
Full scale				TNT: 296/2.3 RDX: 290/1.2 HMX: 26.9/11.0	99.2 99.6 59.1		
Composition unknown	86	Thermophilic	64 soil from "Tanne"	TNT: 699/>0.5 No metabolites	100	70 tons	[103]
Dynamic Pile (Anaerobic/Aerobic Composition)							
Byproduct of food production	180	20–65	80 soil from "Tanne"	TNT: 917/1.3 Negligible amounts of transformation products	99.9	64 tons	[104–106]

Note: UMDA: Umatilla Military Depot Activity, Oregon; LAAP: Louisiana Army Ammunition Plant, Louisiana; "Tanne": former ammunition plant in Germany.

Reprinted from Bruns-Nagel et al. [20].

TABLE 38.3
Reduction of Nitroaromatic Compounds By Intact Cells of Clostridia

Clostridial Species	Nitroaromatic or Intermediate Transformed	Reference (listed chronologically)
C. acetobutylicum	Chloramphenicol	[11]
	TNT	[107]
	TNT, DA6NT	[108]
	TNT, DHA6NT	[68]
	TNT	[72]
	24DNT,26DNT	[69]
C. bifermentans	TNT, RDX	[14]
	TNT	[109]
	TNT, DA6NT, TAT	[110]
	TNT	[111]
	TNT, DA6NT	[108]
C. clostridiiforme	1NPy, pNBA, 1,3DNPy, 1,6DNPy	[112]
C. kluyveri	pNB	[6]
	pNB	[7]
C. leptum	1NPy, pNBA, 1,3DNPy, 1,6DNPy	[112]
C. paraputrificum	1NPy, pNBA, 1,3DNPy, 1,6DNPy	[112]
C. pasteuranium	TNT	[10]
	pNB	[6]
	pNB	[7]
	TNT, DA6NT	[12]
	pNP, mNP, 2,4DNP, pNB, pNA	[113]
C. perfringens	6NC	[114]
C. sordellii	TNT, DA6NT	[108]
C. sp.	1NPy, pNBA, 1,3DNPy, 1,6DNPy	[112]
C. sporogenes	pNB	[7]
	TNT, DA6NT	[108]
C. sp. *W1*	pNP, mNP, 2,4DNP, pNB, pNA	[113]
C. thermoaceticum	TNT, DA6NT	[12]

Compound Abbreviations
DA6NT = 2,4-diamino-6-nitrotoluene; DHA6NT = 2,4-dihydroxylamino-6-nitrotoluene; DAP = Diaminophenol; 2,4DNP = 2,4-dinitrophenol; 1,3DNPy = 1,3-dinitropyrene; 1,6DNPy = 1,6-dinitropyrene; 24DNT = 2,4-dinitrotoluene; 26DNT = 2,6-dinitrotoluene; mNP = meta-nitrophenol; 6NC = 6-nitrochrysene; pNA = para-nitroaniline; pNB = para-nitrobenzoate; pNBA = para-nitrobenzoic acid; pNP = para-nitrophenol; pNPy = para-nitropyrene; RDX = Research Department Explosive (Hexahydro-1,3,5-trinitro-s-triazine); TAT = 2,4,6-triaminotoluene; TNT = 2,4,6-trinitrotoluene.

Reprinted from Ahmad and Hughes [19].

nitroaromatic compounds have inevitably led to drastic reductions in growth [68,108,110,111]. Furthermore, the slow growth has been accompanied by correspondingly slower rates of reduction, and the isolated products are previously analyzed partial reduction intermediates derived from the parent nitro compound [68,111]. This is particularly noted when resting cells with low acidogenic activity [111] or synthetic media supplemented with inadequate carbon sources [110,111] have been used. With these considerations, the focus has been on using other methods to determine if their metabolism is strictly cometabolic or if some metabolites are consumed productively.

Transformation studies performed with ^{14}C-labeled nitroaromatic compounds failed to show the generation of significant quantities of radiolabeled carbon dioxide, indicating that parent nitroaromatic compounds are not mineralized in the usual sense by pure clostridial cultures [68,107,109,110]. The use of ^{14}C-labeled material also allows effective analysis of the fate of the carbon of the nitro compound to be followed and the overall mass balance can be traced. Mass balances performed in radiolabeled studies have shown the majority of the radioactivity resides in solution [107,109], suggesting that components of the nitroaromatic compounds are not incorporated into biomass. The question of the fate of this released material has been less studied. Studies of whether it can be further converted by other chemical processes or other organisms are difficult due to the many possible reactions of derivatives with culture components and soils.

Since ferredoxin-reducing enzymes can stimulate nitroreduction, and these are continually present in growing cells, it is logical that clostridia carry out at least the initial reductions of nitroaromatic compounds without any prior acclimation [11,107]. The absence of an acclimation phase, together with the lack of mineralization evidence, shows that cometabolism of nitroaromatic compounds under fermentative conditions is the main, if not the only, process. Although the reduction of nitroaromatic compounds provides no nutritional benefit to clostridia, the reduction can serve to reoxidize reduced electron carriers needed for energy production via substrate-level formulation. Due to the amounts of nitro compounds compared to the amounts of sugars metabolized, it seems that this would be a minor alternative recycling pathway under most conditions. It is noted that in the presence of quantities of some electron carrier dyes, an early shift away from hydrogen production to solvent production occurs.

38.3.3.3 Acidogenic Cultures in the Reduction of Nitro Groups

Studies of the reduction of nitroaromatic compounds have shown growing cultures of fermentative acidogenic clostridia are most effective [68,107–110,116]. Such organisms and growth conditions are active in dissipating excess reducing power generated from glycolytic reactions by employing one or more of the three reaction pathways. The first of these dissipation pathways during clostridial fermentations is the reduction of protons via the ferredoxin/hydrogenase system to liberate hydrogen gas. Under acidogenic conditions, the generation of molecular hydrogen serves as the means of maintaining the redox balance produced by formation of a large quantity of partially oxidized fermentation products. Reduced ferredoxin is generated by either pyruvate-ferredoxin oxidoreductase or NADH-ferredoxin oxidoreductase. The second pathway is common to all butyric acid forming clostridia. In this pathway, butyryl-CoA is generated from acetyl-CoA via several reducing reactions. Butyric acid clostridia can utilize the hydrogen-producing and butyrate-producing pathways simultaneously as indicated, and high levels of hydrogenase activity exist under acidogenic conditions [1,58,116]. The third pathway is present in species that can induce enzymes capable of producing solvents or "solventogenesis." This pathway is usually active after sufficient levels of acids have accumulated. A common example of such a species is the acetone-, ethanol-, and butanol-producing species, *C. acetobutylicum*, which undergoes a switch in its metabolism from producing acetate and butyrate to producing the more reduced solvents. Although it has been widely shown that the onset of solventogenesis occurs in batch cultures of *C. acetobutylicum* during late exponential growth phase, the exact mechanism for the transformation is unclear. However, low extracellular pH, the difference between intracellular and extracellular pH, and intracellular undissociated butyric acid concentration have been cited as important factors in the sensing/response process that shifts the cell toward solventogenesis under non-nutrient limiting conditions [57,58,118–127]. Morphological changes have also been associated with the onset of solventogenesis and include cell size and lack of motility [58,128]. It should also be noted that hydrogen production and solvent production pathways for dissipating reducing power are mutually exclusive; a concomitant drop in hydrogenase activity occurs with the onset of solventogenesis [1,58,59,117], and if hydrogenase activity is inhibited prematurely, early solvent formation occurs.

One study that attempted to identify a pathway or cell state associated with nitroreduction was reported by Khan et al. [107]. When solventogenesis was induced by carbon monoxide, a common inhibitor for hydrogenase activity, TNT was reduced to hydroxylamino compounds much more slowly than it is converted in fully acidogenic controls. A few researchers have attempted nitroaromatic compound reduction with stationary phase or resting cells [107,111]. Such experiments have not demonstrated reduction of TNT beyond the monohydroxylamino stage. The results suggested that enzymes involved in solvent production and the acetyl-CoA to butyryl-CoA pathway are not very effective in the reduction of nitroaromatic groups, thereby confirming the importance of hydrogenase and acidogenic enzymes in nitroaromatic reductions.

38.3.3.4 Incomplete Nitro Reduction

Recent studies of the reduction of nitroaromatic compounds [68,72,107,109–111,116], especially those employing radiolabeled TNT in the presence of active clostridial cultures [68,107,109–111], have demonstrated the formation of considerable quantities of polar products other than TAT. The polar nature was inferred by chromatography elution position on HPLC columns and partitioning during extractions. In one study, the polar product formed was identified as an aminophenol [72] formed by a Bamberger-type rearrangement [47,48,89] of a characterized hydroxylamino intermediate (the dihydroxylamino derivative) of TNT. In an earlier study with active cultures, a reasonable mass balance could not be established between the aminonitrotoluene reduction intermediates and the completely reduced TAT product [10]. The result suggested the potential existence of alternative pathways.

Studies that have revealed aminophenol formation have been performed with active clostridial cultures grown in batch systems without pH control. Since the pH in such cultures drops substantially over time due to acidogenic activity, it is difficult to determine whether the Bamberger rearrangement is biologically mediated, or is nonbiological in nature but stimulated by a chemical reaction in the spent culture broth because of the low pH [47,89]. However, more recent studies conducted with extracts prepared from acetogenic cells at neutral pH have duplicated the results obtained from growing cultures [72], suggesting that the Bamberger rearrangement is enzyme catalyzed. Further evidence supporting this finding is the fact that rearrangement cannot be reproduced from arylhydroxylamino starting compounds in the absence of cell extract or active cell cultures [72].

Although the formation of polar products other than TAT has been observed in a number of studies, TAT has not been detected as an intermediate in such studies [68,72,107]. Incubations of cell cultures with TAT have not produced any other polar products [110], suggesting that TAT is a biologically dead-end product. Similar results have been obtained with mixed anaerobic cultures [129]. It can be inferred from these findings that the formation of aminophenols is a biologically mediated pathway that can compete with the complete reduction of TNT to TAT. Further studies are needed to determine the factors controlling the degree of TNT reduction and the extent of competing reaction pathways, such as aminophenol formation.

38.4 PRACTICAL APPLICATIONS AND LIMITATIONS

Bioremediation processes based on clostridial metabolism require additional research to better understand the fate of the contaminant and predictability of process performance. Several areas comprising fundamental (e.g., analytical methods, biochemical, microbiological, or toxicological studies) as well as applied engineering considerations are under current study. A brief discussion of specific areas follows.

38.4.1 Analysis of Products

Biotransformation studies with TNT have relied heavily on reverse phase high performance liquid chromatographic (RP-HPLC) separations followed by analysis of the UV spectra of separated components [13,50,107,113,130]. With the recent discovery of polar products, including aminophenols in clostridial systems, new methods should be developed using analytical systems [e.g., capillary electrophoresis (CE) followed by MS compound confirmation] that are more suitable for non-UV-absorbing or polar compounds. The inherent instability of hydroxylamino intermediates also requires better analytical methods for quantification. The hydroxylamino intermediates can be confused with their corresponding amino analogs due to their similar UV spectra but generally elute slightly earlier than the corresponding amino compounds on RP-HPLC columns. Pre-column derivatization procedures used in conjunction with RP-HPLC/diode array detection and improved elution regimens with temperature control might help to resolve this problem by stabilizing the hydroxylamino intermediates and giving them distinct UV spectra. Improved analytical methods, together with radiolabeled studies, can more completely determine the true fate of nitro-compound contaminants in bioremediation studies and at *in situ* sites.

38.4.2 Pathways and Enzymes

The enzymes and reactions of certain species of clostridia have been extensively studied, but relatively little is known about their potential for the biotransformation of xenobiotic compounds. Conventional research conducted with the hydrogenase/electron carrier system has yielded incomplete reduction of polynitroaromatic compounds such as TNT. The influx of genomic information on various clostridia, the additional clostridial species being studied, and the better appreciation of interactions among bacterial species in mixed consortia will allow more complete models of metabolism of these compounds to be developed.

38.4.3 Microbiology Considerations

Microbiological studies are needed to correlate the findings of clostridial pure culture and enzyme studies and the mixed culture studies typical of compost treatment. Current anaerobic mixed culture processes rely on saccharolytic fermentation conditions to reduce the polynitroaromatic compounds [13,47,50,116,130]. Under such conditions, the acidogenic organisms maintain a redox balance by the production of molecular hydrogen. Solventogenic organisms can divert some electron flow away from reduced ferredoxin reduced solvent compounds. Therefore, studies employing protein or molecular biology methods such as enzyme assays, antibody detection, and DNA phylogenetic probing for enumeration of specific organisms capable of acidogenic fermentation would be useful for determining the potential nitroreductase activity of an undefined anaerobic consortium. Such information would allow better prediction of bioremediation options and be useful in long-term monitoring efforts.

Research conducted with clostridial cultures and extracts has demonstrated the rapid nitroreductase activity of these organisms. Therefore, efforts to enhance the competitive position of these organisms in bioremediation sites would be useful. Since clostridia are the only fermentative endospore-forming anaerobes, they can be enriched by supplying a suitable fermentation substrate under anaerobic conditions or by supplying an excess of the fermentation substrate to produce anaerobic conditions. The development of field-scale methods for the selection and enrichment of clostridia in mixed anaerobic cultures could help develop startup conditions for *in situ* bioremediation applications. Development of an inoculum rich in clostridial endospores could be used for bioaugmentation.

38.4.4 Toxicology of Intermediates and Products

Although the toxicity and mutagenicity of TNT have been investigated due to widespread contamination at military sites [131,132], limited information is available on the toxicity and mutagenicity of amino and hydroxylamino intermediates of TNT. The mutagenicity of aminonitrotoluenes has been demonstrated [133]. The inherent unstable nature of the hydroxylamino intermediates in the

TABLE 38.4
Comparison of Toxicological Data of Different Compost Treatments of Explosive-Contaminated Soil

	Percent Reduction from Start to End of Composting				
	Windrow		Static Pile		MAIV*
Toxicological Test	**Non-Aerated 30% Soil**	**Aerated 30% Soil**	**7% Soil**	**10% Soil**	**25% Soil**
Mutagenicity					
Ames test					
TA-98	99.7	99.2	88.2	83.6	97.1
TA-100	97.9	97.5	99.0	87.2	97.8
Acute toxicity toward *C. dubia*					
Survival	92	91	>80	93	>80
Reproduction	87	91	>86	>93	>97

* MAIV: mechanically agitated in-vessel composting.

Reprinted from Bruns-Nagel et al. [20].

environment makes them unsuitable for use in long-term chronic toxicity tests. However, when *C. acetobutylicum* cell extracts are incubated with TNT, the mutagenicity of the solution is maximum when the 2,4-dihydroxylamino-6-nitrotoluene concentration is maximal [134,135]. Similar mutagenic effects of hydroxylamino derivatives of DNT have been observed [135]. The finding is in agreement with the fact that 2,6-dinitrotoluene is highly mutagenic via covalent binding to DNA and proteins after activation by acetylation or sulfation of its hydroxylamino intermediate in rat liver [136]. Hydroxylamino intermediates can also engage in abiotic condensation reactions with nitroso intermediates to form highly toxic azoxy compounds [48,49]. Aminophenolic products show only a fraction of the mutagenicity displayed by the hydroxylamino intermediates of TNT [134]. The low level of toxicity observed is probably due to the instability of such compounds under conditions of the test. This evidence suggests that the goal of any anaerobic remediation process should be to carry on TNT treatment past the initial hydroxylamino intermediates. In regard to this issue, toxicological studies of the products of compost bioremediation have been undertaken. These studies from the work of Griest et al. [137,138] (summarized in Table 38.4) have shown a considerable decrease in toxicity during the treatment process.

38.4.5 Features of a Useful Process

The selection of a practical process for the biodegradation of nitroaromatic compounds by anaerobic fermenters such as clostridia must take into account two crucial factors for the operation of the process: maintaining acidogenic activity and the limiting of sorptive interactions of biodegradation intermediates with soil. Currently available mixed culture processes have not optimized these parameters.

Acidogenic cultures could be maintained more easily in continuous flow systems because such systems provide a means of removal of the acidified fermentation end-products from the culture. Also, when clostridia capable of solvent production are involved, it becomes essential to avoid the low pH conducive to solventogenesis.

The interactions of nitroaromatic compounds and their amino intermediates with soil have been vigorously investigated in recent years [139–145]. Nitroaromatic compounds undergo sorption via cation exchange interactions at siloxane groups of soil mineral media [139,143], whereas amino compounds covalently bind to natural organic matter in soil [141,144]. The strong interactions of hydroxyamino compounds with DNA and proteins in mutagenicity studies suggest that their sorption behavior with soil components might be similar to that of related amino compounds. A recent study of the reaction of partially reduced TNT with natural organic matter [146] indicated a reaction between nitroso derivatives and peat humic acid, and thiol groups and oxygen were implicated in the process.

In addition to the above considerations, the incorporation of an aerobic stage after the anaerobic treatment is another system that should be further investigated. Aerobic cultures of *Pseudomonas pseudoalcaligenes* catalyze the ring cleavage of an aminophenol formed from the rearrangement of hydroxylaminobenzene [17,18,78]. The incorporation of such an aerobic stage to cleave aminophenol-derived compounds formed in clostridial systems could offer a pathway to the mineralization of nitroaromatic contaminants.

38.5 CONCLUSION AND PERSPECTIVES

This review points out some of the biodegradation processes of clostridia that are relevant for alleviating pollution problems in environmental situations. There is information indicating a role for these organisms in degradation of hazardous chemicals, and work has primarily focused on their role in the degradation of nitroaromatic compounds. The study of the impact of these common soil organisms on hazardous chemicals and their role in supporting the functioning of microbial communities capable of degradation is at an early stage. Much is yet to be learned about their metabolic potential on a variety of compounds and the nature of interactions of clostridial species with other soil organisms.

ACKNOWLEDGMENTS

This material is based upon work supported by, or in part by, the U.S. Army Research Laboratory and the U.S. Army Research Office under grant No. DAAD 19-01-1-0524 and DOD Army W911NF-04-1-0179. Support was also obtained from the Strategic Environmental Restoration Development Program (SERDP). A grant from the Robert A. Welch Foundation (C1268) is gratefully acknowledged.

REFERENCES

1. Anderseen, J.R., Bahl, H., and Gottschalk, G., Introduction to the physiology and biochemistry of the genus *Clostridium*, in *Clostridia (Biotechnology Handbooks, Vol. 3)*, Minton, N.P. and Clarke, D.J., Eds., Plenum Press, New York, 1989.
2. Cato, E.P., George, W.L., and Finegold, S.M., Genus Clostridium, in *Bergey's Manual of Systematic Bacteriology Vol. 2*, Sneath, P.H.A. et al., Eds., Williams & Wilkins, Baltimore, 1986, 1141.
3. Cato, E.P. and Stackenbrandt, E., Taxonomy and phylogeny, in *Clostridia (Biotechnology Handbooks, Vol. 3)*, Minton, N.P. and Clarke, D.J., Eds., Plenum Press, New York, 1989.
4. Gottschalk, G., *Bacterial Metabolism*, Springer-Verlag, New York, 1979, 182.
5. Holland, K.T., Knapp, J.S., and Shoesmith, J.G., *Anaerobic Bacteria*, Blackie & Son Limited, Glasgow, 1987, 42 and 97.
6. Angermaier, L., Hein, F., and Simon, H., Investigations on the reduction of aliphatic and aromatic nitro compounds by Clostridium species and enzyme systems, in *Biology of Inorganic Nitrogen and Sulfur*, Springer, Berlin, 1981, 266.
7. Angermaier, L. and Simon, H., On the reduction of aliphatic and aromatic nitro compounds by Clostridia, the role of ferredoxin and its stabilization, *Hoppe-Seyler's Z. Physiol. Chem.*, 364, 961, 1983.
8. Chen, J.S. and Blanchard, D.K., A simple hydrogenase-linked assay for ferredoxin and flavodoxin, *Anal. Biochem.*, 93, 216, 1979.

9. del Campo, F.F. et al., Ferredoxin and the dark and light reduction of dinitrophenol, *Biochem. Biophys. Res. Commun.*, 22, 547, 1966.
10. McCormick, N.G., Feeherry, F.E., and Levinson, H.S., Microbial transformation of 2,4,6-nitrotoluene and other nitroaromatic compounds, *Appl. Environ. Microbiol.*, 31, 949, 1976.
11. O'Brien, R.W. and Morris, J.T., The ferredoxin-dependent reduction of chloramphenicol by *Clostridium acetobutylicum*, *J. Gen. Microbiol.*, 67, 265, 1971.
12. Preuss, A., Fimpel, J., and Diekert, G., Anaerobic transformation of 2,4,6-trinitrotoluene (TNT), *Arch. Microbiol.*, 159, 345, 1993.
13. Funk, S.B. et al., Initial-phase optimization for bioremediation of munition compound-contaminated soils, *Appl. Environ. Microbiol.*, 59, 2171, 1993.
14. Regan, K.M. and Crawford, R.L., Characterization of *Clostridium bifermentans* and its biotransformation of 2,4,6-trinitrotoluene and 1,3,5-triaza-1,3,5-trinitrocyclohexane (RDX), *Biotechnol. Lett.*, 16, 1081, 1994.
15. Rosser, S.J. et al., Microbial transformation of explosives, *Adv. Appl. Microbiol.* 49, 1, 2001.
16. Esteve-Nunez, A., Caballero, A., and Ramos, J.L., Biological degradation of 2,4,6-nitrotoluene, *Microbiol. Mol. Biol. Rev.*, 65, 335, 2001.
17. Spain, J.C., Bacterial degradation of nitroaromatic compounds under aerobic conditions, in *Biodegradation of Nitroaromatic Compounds*, Spain, J.C., Ed., Plenum Press, New York, 1995.
18. Spain, J.C., Biodegradation of nitroaromatic compounds, *Annu. Rev. Microbiol.*, 49, 523, 1995.
19. Ahmad, F. and Hughes, J.B., Anaerobic transformation of TNT by Clostridium, in *Biodegradation of Nitroaromatic Compounds and Explosives*, Spain, J.C., Hughes, J.B., and Knackmuss, H.J., Eds., Lewis Publishers, Boca Raton, FL, 2000, 185.
20. Bruns-Nagel, D. et al., Composting (humification) of nitroaromatic compounds, in *Biodegradation of Nitroaromatic Compounds and Explosives*, Spain, J.C., Hughes, J.B., and Knackmuss, H.J., Eds., Lewis Publishers, Boca Raton, FL, 2000, 357.
21. Suzuki, Y. et al., Microbial populations stimulated for hexavalent uranium reduction in uranium mine sediment, *Appl. Environ. Microbiol.*, 69, 1337, 2003.
22. Okeke, B.C. et al., Purification, cloning, and sequencing of an enzyme mediating the reductive dechlorination of tetrachloroethylene (PCE) from *Clostridium bifermentans* DPH-1, *Can. J. Microbiol.*, 47, 448, 2001.
23. Hou, L. and Dutta, S.K., Phylogenetic characterization of several para- and meta-PCB dechlorinating Clostridium species: 16s rDNA sequence analyses, *Lett. Appl. Microbiol.*, 30, 38, 2000.
24. Moir, D., Masson, S., and Chu, I., Structure-activity relationship study on the bioreduction of azo dyes by *Clostridium paraputrificum*, *Environ. Toxicol. Chem.*, 20, 479, 2001.
25. Rafii, F. and Cerniglia, C.E., Comparison of the azoreductase and nitroreductase from *Clostridium perfringens*, *Appl. Environ. Microbiol.*, 59, 1731, 1993.
26. Chamkha, M. et al., Isolation of a cinnamic acid-metabolizing *Clostridium glycolicum* strain from oil mill wastewaters and emendation of the species description, *Int. J. Syst. Evol. Microbiol.*, 51, 2049, 2001.
27. Liyanage, H. et al., *Clostridium beijerinckii* and *Clostridium difficile* detoxify methylglyoxal by a novel mechanism involving glycerol dehydrogenase, *Appl. Environ. Microbiol.*, 67, 2004, 2001.
28. Abou-Zeid, D.M., Muller, R.J., and Deckwer, W.D., Degradation of natural and synthetic polyesters under anaerobic conditions, *J. Biotechnol.,* 86, 113, 2001.
29. Bujoczek, G. et al., Co-processing of organic fraction of municipal solid waste and primary sludge — stabilization and disinfection, *Environ. Technol.*, 23, 227, 2002.
30. Ortiz, N.E. and Smith, G.R., Landfill sites, botulism and gulls, *Epidemiol. Infect.*, 112, 385, 1994.
31. Bohnel, H. and Lube, K., *Clostridium botulinum* and bio-compost. A contribution to the analysis of potential health hazards caused by bio-waste recycling, *J. Vet. Med. B. Infect. Dis. Vet. Public Health*, 47, 785, 2000.
32. Fetzner, S. and Lingens, F., Bacterial dehalogenases: biochemistry, genetics, and biotechnological applications, *Microbial Rev.*, 58, 641, 1994.
33. Fetzner, S., Bacterial dehalogenation, *Appl. Microbiol. Biotechnol.*, 50, 633, 1998.
34. Janssen, D.B., Oppentocht, J.E., and Poelarends G.J., Microbial dehalogenation, *Curr. Opin. Biotechnol.*, 12, 254, 2001.
35. Lee, M.D., Odom, J.M., and Buchanan, R.J. Jr., New perspectives on microbial dehalogenation of chlorinated solvents: insights from the field, *Ann. Rev. Microbiol.*, 52, 523, 1998.

36. Maymo-Gatell, X. et al., Isolation of a bacterium that reductively dechlorinates tetrachloroethene to ethene, *Science*, 276, 1568, 1997.
37. Galli, R. and McCarty, P.L., Biotransformation of 1,1,1-trichloroethane, trichloromethane, and tetrachloromethane by a Clostridium sp., *Appl. Environ. Microbiol.*, 55, 837, 1989.
38. Galli, R. and McCarty, P.L., Kinetics of biotransformation of 1,1,1-trichloroethane by Clostridium sp. strain TCAIIB, *Appl. Environ. Microbiol.*, 55, 845, 1989.
39. Chang, Y.C. et al., In vitro dehalogenation of tetrachloroethylene (PCE) by cell-free extracts of *Clostridium bifermentans* DPH-1, *Bioresour. Technol.*, 78, 141, 2001.
40. Breitenstein, A. et al., Reclassification of *Clostridium hydroxybenzoicum* as *Sedimentibacter hydroxybenzoicus* gen. nov. comb. Nov., and description of *Sedimentibacter saalensis* sp. nov., *Inst. J. Syst. Evol. Microbiol.*, 52, 801, 2002.
41. Kitamura, S. et al., Reductive debromination of (alpha-bromoiso-valeryl)urea by intestinal bacteria, *J. Pharm. Pharmacol.*, 51, 79, 1999.
42. Noyes, R., *Chemical Weapons Destruction and Explosive Waste/Unexploded Ordinance Remediation*, Noyes Publications, Park Ridge, NJ, 1996, 102.
43. Bradley, P.M. et al., Microbial transformation of nitroaromatics in surface soils and aquifer materials, *Appl. Environ. Microbiol.*, 60, 2170, 1994.
44. Bradley, P.M. et al., Potential for intrinsic bioremediation of a DNT-contaminated aquifer, *Ground Water*, 35, 12, 1997.
45. Spanggord, R.J. et al., Environmental fate studies on certain munition wastewater constituents, Final Report Phase I: Literature Review, prepared by SRI International for the U.S. Army Medical Research and Development Command, Fort Detrick, MD, Contract No. DAMD17-78-C-8061, 1980.
46. Preuss, A. and Rieger, P.G., Anaerobic transformation of 2,4,6-trinitrotoluene, in *Biodegradation of Nitroaromatic Compounds*, Spain, J.C., Ed., Plenum Press, New York, 1995.
47. Rieger, P.G. and Knackmuss, H.J., Basic knowledge and perspectives on biodegradation of 2,4,6-trinitrotoluene and related nitroaromatic compounds in soil, in *Biodegradation of Nitroaromatic Compounds*, Spain, J.C., Ed., Plenum Press, New York, 1995.
48. Corbett, M.D. and Corbett, B.R., Bioorganic chemistry of the arylhydroxylamine and nitrosoarene functional groups, in *Biodegradation of Nitroaromatic Compounds*, Spain, J.C., Ed., Plenum Press, New York, 1995.
49. Kaplan, D.L. and Kaplan, A.M., Thermophilic biotransformation of 2,4,6-trinitrotoluene under simulated composting conditions, *Appl. Environ. Microbiol.*, 44, 747, 1982.
50. Kaake, R.H. et al., Bioremediation of soils contaminated with the herbicide 2-sec-butyl-4,6-dinitrophenol (dinoseb), *Appl. Environ. Microbiol.*, 58, 1683, 1992.
51. Ljungdahl, L.G., Hugenholtz, J., and Wiegel, J., Acetogenic and acid-producing clostridia, in *Clostridia (Biotechnology Handbooks, Vol. 3)*, Minton, N.P. and Clarke, D.J., Eds., Plenum Press, New York, 1989.
52. Mitchell, W.J., Physiology of carbohydrate to solvent conversion by clostridia, *Adv. Microb. Physiol.*, 39, 31, 1998.
53. Dürre, P. et al., Solventogenic enzymes of *Clostridium acetobutylicum*: catalytic properties, genetic organization, and transcriptional regulation, *FEMS Microbiol. Rev.*, 17, 251, 1995.
54. Dürre, P., New insights and novel developments in clostridial acetone/butanol/isopropanol, *Appl. Microbiol. Biotechnol.*, 49, 639, 1998.
55. Bahl, H. and Dürre, P., *Clostridia — Biotechnology and Medical Applications*, Wiley-VCH, Weinheim, Germany, 2001.
56. Dürre, P. et al., Transcriptional regulation of solventogenesis in *Clostridium acetobutylicum*, *J. Mol. Microbiol. Biotechnol.*, 4, 295, 2002.
57. Bennett, G.N. and Petersen, D.J., Cloning and expression of *Clostridium acetobutylicum* genes involved in solvent production, in *Genetics and Molecular Biology of Anaerobic Bacteria*, Sebald, M., Ed., Springer-Verlag, New York, 1993.
58. Jones, D.T. and Woods, D.R., Acetone-butanol fermentation revisited, *Microbiol. Rev.*, 50, 484, 1986.
59. Jones, D.T. and Woods, D.R., Solvent production, in *Clostridia (Biotechnology Handbooks, Vol. 3)*, Minton, N.P. and Clarke, D.J., Eds., Plenum Press, New York, 1989.
60. Ragsdale, S.W. and Wood, H.G., Acetate biosynthesis by acetogenic bacteria, *J. Biol. Chem.*, 260, 3970, 1985.

61. Arnon, D.I., Ferredoxin and photosynthesis, *Science*, 149, 1460, 1965.
62. Losada, M. et al., Light and dark reduction of nitrate in reconstituted chloroplast system, *Biochim. Biophys. Acta*, 109, 86, 1965.
63. Wessels, J.S.C., Mechanism of the reduction of organic nitro compounds by chloroplasts, *Biochim. Biophys. Acta*, 109, 357, 1965,
64. Kiser, J.S. et al., Chemotherapy of experimental clostridial infections, *J. Infect. Dis.*, 28, 76, 1950.
65. Rosenblatt, J.E. and Edson, R.S., Symposium on antimicrobial agents. Metronidazole, *Mayo Clin. Proc.*, 62, 1013, 1987.
66. Freeman, C.D., Klutman, N.E., and Lamp, K.C., Metronidazole. A therapeutic review and update, *Drugs*, 54, 679, 1997.
67. Stoddart, B. and Wilcox, M.H., *Clostridium difficile*, *Curr. Opin. Infect. Dis.*, 15, 513, 2002.
68. Hughes, J.B. et al., Reduction of 2,4,6-trinitrotoluene by *Clostridium acetobutylicum* through hydroxylamino intermediates, *Environ. Toxicol. Chem.*, 17, 343, 1998.
69. Hughes, J.B., Wang, C.Y., and Zhang, C., Anaerobic biotransformation of 2,4-dinitrotoluene and 2,6-dinitrotoluene by *Clostridium acetobutylicum*: a pathway through dihydroxylamino intermediates, *Environ. Sci. Technol.*, 33, 1065, 1999.
70. Lindmark, D.G. and Müller, M., Antitrichomonad action, mutagenicity, and reduction of metronidazole and other nitroimidazoles, *Antimicrob. Agents Chemother.*, 10, 476, 1976.
71. Bryant, C. and McElroy, W.D., Nitroreductases, in *Chemistry and Biochemistry of Flavoproteins*, Vol. II, Muller, F., Ed., CRC Press, Boca Raton, FL, 1991.
72. Hughes, J.B. et al., Bamberger rearrangement during TNT metabolism by *Clostridium acetobutylicum*, *Environ. Sci. Technol.*, 32, 494, 1998.
73. Yoch, D.C. and Carithers, R.P., Bacterial iron-sulfur proteins, *Microbial Rev.*, 43, 384, 1979.
74. Adams, M.W., Mortenson, L.E., and Chen, J.S., Hydrogenase, *Biochim. Biophys. Acta*, 594, 105, 1981.
75. Beinert, H., Recent developments in the field of iron-sulfur proteins, *FASEB J.*, 4, 2483, 1990.
76. Wu, L.-F. and Mandrand, M.A., Microbial hydrogenases: primary structure, classifications, signatures, and phylogeny, *FEMS Microbiol. Rev.*, 104, 243, 1993.
77. Horner, D.S. et al., Iron hydrogenase — ancient enzymes in modern eukaryotes, *Trends Biochem. Sci.*, 27, 148, 2002.
78. Mortenson, L.E. and Nakos, G., Bacterial ferredoxins and/or iron-sulfur proteins as electron carriers, in *Iron-sulfur Proteins*, Vol. 1, Lovenberg, W., Ed., Academic Press, New York, 1973.
79. Rabinowitz, J.C., The *Clostridium pasteurianum* ferredoxin gene, in *Genetics and Molecular Biology of Anaerobic Bacteria*, Sebald, M., Ed., Springer-Verlag, New York, 1993.
80. Schonheit, P., Brandis, A., and Thauer, R.K., Ferredoxin degradation in growing *Clostridium pasteurianum* during periods of iron deprivation, *Arch. Microbiol.*, 120, 73, 1979.
81. Ludwig, M.L. and Luschinsky, C.L., Structure and redox properties of clostridial flavodoxin, in *Chemistry and Biochemistry of Flavoenzymes*, Vol. III, Muller, F., Ed., CRC Press, Boca Raton, FL, 1991.
82. Holt, D.E. and Bajoria, R., The role of nitro-reduction and nitric oxide in the toxicity of chloramphenicol, *Hum. Exp. Toxicol.*, 18, 111, 1999.
83. Church, D.L., Rubin, H.R., and Laishley, E.J., Reduction of 2-, 4- and 5-nitroimidazole drugs by hydrogenase 1 in *Clostridium pasteurianum*, *J. Antimicrob. Chemother.*, 25, 15, 1990.
84. Lockerby, D.L., Rubin, H.R., and Laishley, E.J., Role of the phosphooclastic reaction of *Clostridium pasteurianum* in the reduction of metronidazole, *Antimicrob. Agents Chemother.*, 27, 863, 1985.
85. Church, D.L., Rubin, H.R., and Laishley, E.J., Role of hydrogenase 1 of *Clostridium pasteurianum* in the reduction of metronidazole, *Biochem. Pharm.*, 37, 1525, 1988.
86. Church, D.L. and Laishley, E.J., Reduction of metronidazole by hydrogenase from Clostridia, *Anaerobe*, 1, 81, 1995.
87. Santangelo, J.D., Jones, D.T., and Woods, D.R., Metronidazole activation and isolation of *Clostridium acetobutylicum* electron transport genes, *J. Bact.*, 173, 1088, 1991.
88. Nishino, S.F. and Spain, J.C., Degradation of nitrobenzene by a *Pseudomonas pseudoalcaligenes*, *Appl. Environ. Microbiol.*, 59, 2520, 1993.
89. Shine, H.J., The rearrangements of phenylhydroxylamines, in *Aromatic Rearrangements*, Elsevier, Amsterdam, 1967, 182.
90. Bhushan, B. et al., Diaphorase catalyzed biotransformation of RDX via N-denitration mechanism, *Biochem. Biophys. Res. Commun.*, 296, 779, 2002.

91. Zhang, C. and Hughes, J.B., Biodegradation pathways of hexahydro-1,3,5-trinitro-1,3,5-triazine (RDX) by *Clostridium acetobutylicum* cell-free extract, *Chemosphere*, 50, 665, 2003.
92. Watrous, M.M. et al., 2,4,6-Trinitrotoluene reduction by an Fe-only hydrogenase in *Clostridium acetobutylicum*, *Appl. Environ. Microbiol.*, 69, 1542, 2003.
93. Boynton, Z.L., Bennett, G.N., and Rudolph, F.B., Cloning, sequencing, and expression of clustered genes encoding beta-hydroxybutyryl-coenzyme A (CoA) dehydrogenase, crotonase, and butyryl-CoA dehydrogenase from *Clostridium acetobutylicum* ATCC 824, *J. Bacteriol.*, 178, 3015, 1996.
94. Huang, S., et al., 2,4,6-Trinitrotoluene (TNT) reduction by carbon monoxide dehydrogenase from *Clostridium thermoaceticum* kinetic characterization and inhibitor study, *Appl. Environ. Microbiol.*, 66, 1474, 2000.
95. Nolling, J. et al., Genome sequence and comparative analysis of the solvent-producing bacterium *Clostridium acetobutylicum*, *J. Bacteriol.*, 183, 4823, 2001.
96. Shimizu, T. et al., Complete genome sequence of *Clostridium perfringens*, an anaerobic flesh-eater, *Proc. Natl. Acad. Sci. U.S.A.*, 99, 996, 2002.
97. Bruggemann, H. et al., The genome sequence of *Clostridium tetani*, the causative agent of tetanus disease, *Proc. Natl. Acad. Sci. U.S.A.*, 100, 1316, 2003.
98. Braun, H. et al., Microbial reduction of n-allylhydroxylamines to n-allylamines using Clostridia, *Tetrahedron*, 47, 3329, 1991.
99. Doyle, R.C. et al., Composting explosives/organics contaminates soils, U.S. Army Toxic and Hazardous Materials Agency, Aberdeen Proving Ground, MD, NTIS-Publication No. ADA 169994, 1986.
100. Craig, H.D. et al., Bioremediation of explosives-contaminated soils: a status review, in 10th Annual Conference on Hazardous Waste Research, Manhattan, KS, 1995, 168–179.
101. Keehan, K., Erfahrungen in den Vereinigten Staaten bei der Sanierung von sprengst-offkontaminierten Standorten, Oral presentation at Expertengespräch "Rüstung-saltlasten aus der Produktion und Verarbeitung von Sprengstoff," Marburg, Germany, September 1992.
102. Track, T and Shenk, D., Recherche über das Abbau-/Transformationsverhalten von 2,4,6-Trinitrotoluol, 2,4- & 2,6-Dinitrotoluol sowie den Entwicklungsstand von Sanierungstechnologien für Sprengstoffkontaminationen, Johannes Gutenberg-Universität, Mainz, Germany, unpublished, 1998.
103. Fischer, D. and Walter, U., *TNT-Hunifizierung — Das Umweltschutz Nord-Verfahren, Werk Tanne — Erste Ergebnisse der Bodenbehandlung*, Umweltschutz Nord GmbH & Co., Germany, 1999.
104. Dohnalik-Droste, A. and Winterberg, R., Das Plambeck ContraCom-Verfahren zur Sanierung TNT-kontaminierter Böden, in *Biotechnologie im Unweltschutz: Bioremediation:Entwiclung — Anwendung — Perspektiven*, Heiden, S. et al., Eds., Erich Schmidt Verlag, Berlin, 1999, 281.
105. Winterberg, R., Maβstabsgerechte Erprobung biologischer Verfahren zur Sanierung von Rüstungsaltlasten: Dynamisches Beeverfahren, p. L1-L12, in Verbundvorhaben Langzeit- und Remobilisierungverfahren von Schadstoffen, Statusseminar 22 und 23, in Bremen, Umweltbundesamt, Germany, 1998.
106. Winterberg, R. and Dohnalik-Droste, A., Technikumsversuche zur Optimierung und zum scale up der Sanierungsverfahren, in Verbundvorhaben, Biologische Sanierung von Rüstungsaltlasten. Tagungsband zum 4, Statusseminar in Clausthal-Zellerfeld, CD edited by the German Federal Ministry of Education, Science, Research and Technology (BMBF), 1998.
107. Khan, T.A., Bhadra, R., and Hughes, J., Anaerobic transformation of 2,4,6-TNT and related nitroaromatic compounds by *Clostridium acetobutylicum*, *J. Ind. Microbiol. Biotechnol.*, 18, 198, 1997.
108. Ederer, M.M., Lewis, T.A., and Crawford, R.L., 2,4,6-Trinitrotoluene (TNT) transformation by clostridia isolated from a munition-fed bioreactor: comparison with non-adapted bacteria, *J. Ind. Microbiol. Biotechnol.*, 18, 82, 1997.
109. Shin, C.Y. and Crawford, D.L., Biodegradation of trinitrotoluene (TNT) by a strain of *Clostridium bifermentans*, in *Bioaugmentation for Site Remediation*, Hinchee, R.W., Fredrickson, J., and Alleman, B.C., Eds., Battelle Press, Columbus, OH, 1995.
110. Lewis, T.A. et al., Products of anaerobic 2,4,6-trinitrotoluene (TNT) transformation by *Clostridium bifermentans*, *Appl. Environ. Microbiol.*, 62, 4669, 1996.
111. Shin, C.Y., Lewis, T.A., and Crawford, D.L., 2,4,6-Trinitrotoluene (TNT) biodegradation by *Clostridium bifermentans*, in *In situ and On-site Bioremediation, volume 2*, Battelle Press, Columbus, OH, 1997.

112. Rafii, F. et al., Reduction of nitroaromatic compounds by anaerobic bacteria isolated from the human gastrointestinal tract, *Appl. Environ. Microbiol.*, 57, 962, 1991.
113. Lewis, T.A. et al., Microbial transformation of 2,4,6-trinitrotoluene, *J. Ind. Microbiol. Biotechnol.*, 18, 89, 1997.
114. Manning, B.W. et al., Metabolism of 6-nitrochrysene by intestinal microflora, *Appl. Environ. Microbiol.*, 54, 197, 1988.
115. Bruhn, C., Lenke, H., and Knackmuss, H.J., Nitrosubstituted aromatic compounds as nitrogen source for bacteria, *Appl. Environ. Microbiol.*, 53, 208, 1987.
116. Gorontzy, T. et al., Transformation of the explosive TNT by clostridia, in *In situ and On-Site Bioremediation, Volume 2*, Battelle Press, Columbus, OH, 1997.
117. Hartmanis, M.G.N. and Gatenbeck, S., Intermediary metabolism in *Clostridium acetobutylicum*: levels of enzymes involved in the formation of acetate and butyrate, *Appl. Environ. Microbiol.,* 47, 1277, 1984.
118. Martin, J.R., et al., Effects of acetic and butyric acids on solvent production by *Clostridium acetobutylicum*, *Biotechnol. Lett.*, 5, 89, 1983.
119. Gottshal, J.C. and Morris, J.G., The induction of acetone and butanol production in cultures of *Clostridium acetobutylicum* by elevated concentrations of acetate and butyrate, *FEMS Microbiol. Lett.*, 12, 385, 1981.
120. Monot, F., Engasser, J.M., and Petitdemange, H., Influence of pH and undissociated butyric acid on the production of acetone and butanol in batch culture of *Clostridium acetobutylicum*, *Appl. Microbiol. Biotechnol.*, 19, 422, 1984.
121. Gottwald, M.A., Hoppe, H., and Gottschalk, G., The internal pH of *Clostridium acetobutylicum* and its effect on the shift from acid to solvent formation, *Arch. Microbiol.*, 143, 42, 1984.
122. Terracciano, J.S. and Kashket, E.R., Intracellular conditions required for initiation of solvent production by *Clostridium acetobutylicum*, *Appl. Environ. Microbiol.*, 52, 86, 1986.
123. Huesemann, M.H.W. and Papoutsakis, E.T., Solventogenesis in *Clostridium acetobutylicum* fermentations related to carboxylic acid and proton concentrations, *Biotechnol. Bioeng.*, 32, 843, 1988.
124. Huang, L., Gibbins, L.N., and Forsberg, C.W., Transmembrane pH gradient and membrane potential in *Clostridium acetobutylicum* during growth under acetogenic and solventogenic conditions, *Appl. Environ. Microbiol.*, 50, 1043, 1985.
125. Huang, L., Forsberg, C.W., and Gibbins, L.N., Influence of external pH and fermentation products on *Clostridium acetobutylicum* intracellular pH and cellular distribution of fermentation products, *Appl. Environ. Microbiol.*, 51, 1230, 1986.
126. Rogers, P. and Gottschalk, G., Biochemistry and regulation of acid and solvent production in Clostridia, in *The Clostridia and Biotechnology*, Woods, D.R., Ed., Butterworth-Heinemann, Stoneham, MA, 1993, Chapter 2.
127. Grupe, H. and Gottschalk, G., Physiological events in *Clostridium acetobutylicum* and its effect from acidogenesis to solventogenesis in continuous culture and presentation of a model for shift induction, *Appl. Envrion. Microbiol.*, 58, 3896, 1992.
128. Jones, D.T. et al., Solvent production and morphological changes in *Clostridium acetobutylicum*, *Appl. Environ. Microbiol.*, 43, 1434, 1982.
129. Hawari, J. et al., Characterization of metabolites in the biotransformation of 2,4,6-trinitrotoluene with anaerobic sludge: role of triaminotoluene, *Appl. Environ. Microbiol.*, 64, 2200, 1998.
130. Gorontzy, T., Küver, J., and Blotevogel, K.-H., Microbial transformation of nitroaromatic compounds under anaerobic conditions, *J. Gen. Microbiol.*, 139, 1331, 1993.
131. Tan, E.L., et al., Mutagenicity of trinitrotoluene and its metabolites formed during composting, *J. Toxicol. Environ. Health*, 36, 165, 1992.
132. Won, W.D., di Salvo, L.H., and Ng, J., Toxicity and mutagenicity of 2,4,6-trinitrotoluene and its microbial metabolites, *Appl. Environ. Microbiol.*, 31, 576, 1976.
133. Spanggord, R.J. et al., Mutagenicity in *Salmonella typhimurium* and structure activity relationships of waste water components emanating from the manufacture of trinitrotoluene, *Environ. Mutagen.*, 4, 163, 1982.
134. Padda, R.S. et al., Mutagenicity of trinitrotoluene and its metabolites formed during anaerobic degradation by *Clostridium acetobutylicum* ATCC 824, *Environ. Toxicol. Chem.*, 19, 2871, 2000.

135. Padda, R.S. et al., Mutagenicity of nitroaromatic degradation compounds, *Environ Toxicol. Chem.*, in press.
136. Parkinson, A., Biotransformation of xenobiotics, in *Casarett & Doull's Toxicology: The Basic Science of Poisons*, Klaassen, C.D., Ed., McGraw-Hill, New York, 1996.
137. Griest, W.H. et al., Chemical and toxicological testing of composted explosives-contaminated soil, *Environ. Toxicol. Chem.*, 12, 1105, 1993.
138. Griest, W.H. et al., Chemical characterization and toxicological testing of windrow composts from explosives-contaminated sediments, *Environ. Toxicol. Chem.*, 14, 51, 1995.
139. Daun, G. et al., Biological treatment of TNT-contaminated soil. 1. Anaerobic cometabolic reduction and interaction of TNT and metabolites with soil components, *Environ. Sci. Technol.*, 32, 1956, 1998.
140. Drzyzga, O. et al., Incorporation of ^{14}C-labeled 2,4,6-trinitrotoluene metabolites in different soil fractions after anaerobic and anaerobic-aerobic treatment of soil/molasses mixtures, *Environ. Sci. Technol.*, 32, 3529, 1998.
141. Elovitz, M.S. and Weber, E.J., Sediment-mediated reduction of 2,4,6-trinitrotoluene and fate of the resulting (poly)amines, *Environ. Sci. Technol.,* 33, 2617, 1999.
142. Haderlein, S.B. and Schwarzenbach, R.P., Adsorption of substituted nitrobenzenes and nitrophenols to mineral surfaces, *Environ. Sci. Technol.*, 27, 316, 1993.
143. Haderlein, S.B., Weissmahr, K.W., and Schwarzenbach, R.P., Specific adsorption of nitroaromatic explosives and pesticides to clay minerals, *Environ. Sci. Technol.*, 30, 612, 1996.
144. Knicker, H. et al., Characterization of 15N-TNT residues after an anaerobic/aerobic treatment of soil/molasses mixtures by 15N NMR spectroscopy. 1. Determination and optimization of relevant NMR spectroscopic parameters, *Environ. Sci. Technol.*, 33, 343, 1999.
145. Shen, C.F. et al., Fate of explosives and their metabolites in bioslurry treatment processes, *Biodegradation*, 8, 339, 1998.
146. Ahmad, F. and Hughes, J.B., Reactivity of partially reduced arylhydroxylamine and nitrosoarene metabolites of 2,4,6-trinitrotoluene (NT) toward biomass and humic acids, *Environ. Sci. Technol.*, 36, 4370, 2002.

39 Clostridial Collagenase in Wound Repair

David W. Brett

CONTENTS

39.1 WOUND REPAIR

Normal wound repair is a coordinated and predictable series of cellular and biochemical events, generally characterized by the following phases: hemostasis, inflammation, granulation, and remodeling. It requires complex and dynamic processes that restore cellular structures and tissue layers by an intricate interplay among a variety of cell types, endogenous chemoattractants and growth factors, fibrous proteins, proteinases, and angiogenic factors. This multifaceted and coordinated series of events includes chemotaxis, phagocytosis, neocollagenesis, collagen degradation, and collagen remodeling. In addition, angiogenesis, epithelialization, and the production of new glycosaminoglycans (GAGs) and proteoglycans are vital to the wound healing milieu. The culmination of these biological processes results in the replacement of normal skin structures with fibroblastic mediated scar tissue. However, certain pathophysiologic conditions and metabolic factors can alter this course of events, resulting in chronic, nonhealing wounds [1–3].

Chronic wounds can be caused by or characterized by one or more of the following: repeated trauma, local tissue ischemia, necrotic tissue, heavy bacterial burden, and tissue breakdown. Chronic wounds are typically characterized by being "frozen" in one of the aforementioned phases of wound healing, typically the inflammatory phase. In this instance, inflammatory cells, microphages, and neutrophiles are overstimulated, resulting in an overproduction of certain pro-inflammatory cytokines, such as TNFα and IL-1. As these polypeptides are chemical messengers activating their target cells (such as fibroblasts to product matrix metalloproteinases — MMPs), an overproduction of pro-inflammatory cytokines ultimately results in an overproduction of MMPs. As MMPs degrade wound matrix proteins, it becomes very difficult for a functional extracellular matrix (ECM) to develop. Without a functional ECM, the wound cannot heal, as many cells involved in the healing process depend upon this matrix protein "scaffold" for their migration and proliferation.

39.2 WOUND BED PREPARATION

The fundamental definition of wound bed preparation (WBP) is "the process of preparing the wound bed to promote optimal healing." This is accomplished by removing the barriers to healing, such as: necrosis, bacteria, exudates and encouraging keratinocyte migration from the wound margin. Removal of the necrotic tissue is essential to reduce the bacterial burden in the wound, which in turn decreases the amount of exudate produced.

The goal of WBP is to accelerate endogenous healing or to facilitate the effectiveness of other therapeutic measures. A successful outcome is a healed re-epithelialized wound, with well-vascularized granulation tissue and no signs of local infection, such as drainage, cellulitus, or odor.

Optimal healing requires that the wound bed be both *clean* and *functional*. *Clean* refers to an absence of the necrotic tissue that often covers the receptor sites of cells in the wound bed. A clean site permits cytokines or growth factors to travel through the wound environment and bind to the cell surface receptors on another cell. This binding of growth factors or cytokines to the extracellular receptor site causes specific chemical reactions to occur at the intracellular domain and leads to a biological response by the cell [4].

Many of the newer technologies, such as topically applied growth factors (PDGF-Regranex®, Ortho-McNeil, Raritan, NJ) or bioengineered tissue (Dermagraft®, Smith and Nephew, Largo, FL), require that the wound bed be free of devitalized tissue.

Functional suggests a favorable environment where cellular healing can occur. Often, even after the wound is clean and a topical growth factor or bioengineered tissue has been applied, healing fails to occur. Recent research attributes this to very high levels of inflammatory cytokines and proteases, coupled with a relatively narrow range of growth factors [1,5,6].

Research also confirms that the resident cells in chronic wounds may be senescent, phenotypically altered, and unable to perform their intended function [5,6]. Though WBP is a holistic approach to wound healing taking into account such things as the nutritional status of the patient, the mental and emotional state, as well as many other aspects, there are four key components of the WBP model: (1) The removal of necrotic tissue, (2) managing the bioburden, (3) managing wound exudates and (4) the stimulation of the epidermal margin. Of this "quartet", we will explore the removal of necrotic tissue as the focus of this chapter [7].

39.3 TYPES OF DEBRIDEMENT

Why debride?

- Enhance wound assessment
- Decrease potential for infection
- Establish earlier wound bed coverage
- Optimize wound healing

39.3.1 Surgical Debridement

In this selective method of debridement, the clinician uses instruments to remove the necrotic tissue from the wound. While lasers have been used, most clinicians employ scissors, scalpels, and forceps to accomplish surgical or sharp debridement. Individual skill plays a large role in the effectiveness of this treatment. Knowledge of underlying anatomy and proficiency with surgical instruments are essential. Surgical debridement should be considered when large amounts of necrotic tissue must be removed quickly. If the patient can tolerate this intervention, and the procedure is performed by a skilled practitioner, surgical debridement is beneficial.

39.3.2 Mechanical Debridement

In this nonselective method of debridement, the clinician uses force to remove the necrotic tissue. This can include the use of irrigation fluids delivered by pulsed lavage, the agitation of water during whirlpool therapy, or the removal of tissue via wet-to-dry dressings. Mechanical debridement is considered nonselective in that viable and nonviable tissues are both removed during the process.

39.3.3 Autolytic Debridement

In this selective method of debridement, the clinician uses technologies such as occlusive or semiocclusive moisture-retentive dressings (e.g., hydrocolloids, transparent films, amorphous hydrogels, hydrogel sheets). The dressing is placed over the wound, thus allowing the natural wound fluids to soften eschar and the endogenous proteolytic enzymes to digest tissue.

39.3.4 Enzymatic Debridement

Enzymatic debridement uses topical enzymes to remove necrotic tissue by digesting and dissolving the devitalized tissue in the wound bed. Enzymatic debriding agents are either *selective* for nonviable tissue only or *nonselective* for both viable and nonviable tissue. In addition, enzymes are either *specific* relative to the target substrate (e.g., collagen, fibrin, elastin), or *nonspecific,* digesting a variety of different proteins or substrates. The appropriate class of enzyme one employs is dependent upon the tissue composition in the wound bed, the patient's condition, and mode of action (MoA) specific to the products being considered. For instance, various proteolytic enzymes have been employed (papain, ficin, streptokinase, streptodornase, trypsin-chymotrypsin, etc.) for the debridement of wounds. However, they have had only limited success because they *do not* attack native collagen [8,9].

39.3.4.1 Historical Overview of Enzymatic Debriders

Over the years, various proteolytic enzymes have been employed (papain, ficin, streptokinase, streptodornase, trypsin-chymotrypsin, collagenase, etc.) for the debridement of wounds. This section provides an overview of these various enzymes.

The concept of using proteolytic enzymes to digest dead tissue as an adjunct to the management of dirty, infected wounds is an old one, probably stemming from the observation of the natives of tropical countries where the papain-rich latex obtained by scratching the skin of the green fruit of the papaw tree (*Carica papaya*) has long been used for the treatment of eczema, warts, ulcers, and other types of foul sores [8]. It is also known that in addition to applying papain-rich latex to a wound, the wounds were at times exposed to urine, and then wrapped in green leaves from the same plant. This is interesting as these three naturally occurring materials contain the chemical compounds papain, urea and chlorophyllin, respectively. Urea is a component of mammalian urine, and chlorophyllin is a derivative of chlorophyll. Papain, urea, and chlorophyllin are the active ingredients of one of the most well-known enzymatic debriders in topical use today.

Before the turn of thc 20th century, literature on the use of papaya latex preparation for treating sloughing ulcers, removing impacted cerumen, and dissolving diphtheritic membranes became available [10–12]. More recently, it has been found that the major insoluble constituents of inflammatory exudates, fibrin and desoxyribonucleoprotein derived from the nuclei of dead degenerating cells, could be rapidly lysed by the local application of a mixture of enzymes obtained from the secretory products of certain strains of hemolytic streptococci. The major constituents of this enzyme mixture, streptokinase (an activator of plasminogen, the naturally occurring precursor of a proteolytic and fibrinolytic enzyme of human plasma) and streptodornase (streptococcal desoxyribonuclease), provided the basis for an enzymatic debridement [13–19].

Specific enzyme preparations that have been used or are in current use for purposes of local debridement include the following.

39.3.4.1.1 Microbe-Derived Enzymes

- **Sutilain:** A water-soluble mixture of enzymes derived from the bacteria *Bacillis subtilis* that is relatively nonspecific in its action and is capable of breaking down a variety of necrotic tissue types within an optimal pH range of 6.0 to 6.8.
- **Collagenase:** A water-soluble enzyme that specifically attacks and breaks down undenatured (natural) collagen. In actuality, collagenase is known to affect denatured collagen as well. Collagenase is commercially obtained from bacterial *(Clostridia histolyticum)* sources. Collagenase is active over a pH range of 6 to 8. Bacterial collagenase, although a zinc metalloenzyme that uses calcium, bears little or no structural relationship to mammalian collagenase. Bacterial collagenase rapidly attacks and degrades human collagen into small peptides. The commercially available collagenase is made up of proteolytic enzymes that break collagen into small peptides of differing molecular weights, most of which are tripeptides. Six collagenases have been identified in the culture filtrate of *C. histolyticum,* and all have been purified to homogeneity. These collagenases uniquely cleave the interstitial collagens and exhibit both endopeptidase and tripeptidyl-carboxypeptidase activities. The combined activity of endo- and tripeptidyl-C-peptidase makes these enzymes ideally suited for rapid collagen degradation.
- **Streptokinase-streptodornase mixtures:** This preparation is only partially purified and contains a number of other streptococcal enzymes, such as a ribonuclease, hyaluronase, nucleotidase, and nucleosidase, all of which can contribute to the effects observed. The enzyme mixture is essentially free of streptolysin and streptococcol proteinase. It does not contain any proteolytic enzymes in the conventional sense. The mixture contains enzymes that act upon nonprotein substrates; much of its virtue lies in its content of streptodornase, which rapidly reduces the viscosity of purulent exudates. It is unclear whether streptokinase requires another humoral factor (proactivator) to activate human plasminogen. Plasmin, the proteolytic enzyme formed from the latter precursor, is active at neutral pH and, though distinct from trypsin, resembles it in many respects (pH optima, types of links split, etc.). The major attribute of streptokinase lies in its special fibrin-dissolving properties. In contrast to the rapid inhibition of proteolytic enzymes by naturally occurring humoral antiproteolytic substances, streptokinase is inactivated at a relatively slow rate (except in the presence of an excess of a specific antibody, antistreptokinase) [8]. Streptokinase-streptodornase preparations are the agents of choice for liquefying clotted blood, loculated effusions, and purulent exudate in closed body cavities. A significant incidence of pyrogenic [pyogenic] and inflammatory reactions to the locally administered enzyme mixture has limited its usefulness, since the therapeutic procedure can be complicated by the patient's discomfort and the need for frequent and repeated drainage [8].
- **Streptodornase (streptococcal desoxyribonuclease):** Acts directly upon deoxyribonucleic acid (DNA), rapidly depolymerizing this highly complex substance into smaller units [16,18,20]. The activity of streptodornase is enhanced by the presence of Mg^{2+} or other divalent metallic ions and inhibited by the presence of substances, such as citrate, which form complexes with the metallic cofactor (i.e., chelating agents) [16].
- **Fungal** proteases have also been employed as topically applied enzymatic debriders.

39.3.4.1.2 Animal-Derived Enzymes

- **Fibrinolysin:** Commercially obtained from bovine plasma, then activated by chloroform, it specifically attacks and breaks down the fibrin component of blood clots and fibrinous exudates.
- **Desoxyribonuclease:** Obtained from bovine pancreatic tissue, it acts specifically on the nucleoprotein components (DNA) of purulent exudates.
- **Trypsin:** Crystalline trypsin preparations of bovine pancreatic origin are currently used. Trypsin directly hydrolyzes a large number of naturally occurring proteins. It is thought not to affect living cells or require any cofactors, and its action on denatured proteins is usually more extensive than on native proteins. Trypsin has advantages over streptokinase for surface wound debridement since it does not require additional factors for its action, acts upon a greater number of proteins than plasmin, and degrades them more extensively [8].
- **Chymotrypsin:** This preparation is of bovine pancreatic origin and is the other major proteolytic enzyme of the pancreas. Pancreatic enzymes are usually standardized in terms of their proteolytic activity. Though chymotrypsin acts upon different bonds in proteins than does trypsin or plasmin, its spectrum of activity on whole proteins is somewhat similar to that of trypsin [8].
- **Hyaluronidase:** Another common animal-derived enzyme used for topical enzymatic debridement.

39.3.4.1.3 Plant-Derived Enzymes

- **Bromelain:** A water-soluble enzyme derived from the stem or fruit of the pineapple plant, bromelain is actually a mixture of proteolytic enzymes that is also reported to be effective in breaking down a variety of different necrotic tissue substrates over a fairly wide pH range (5.5 to 8.5). It should be noted that cases of anaphylactic shock have been reported with enzymes derived from the pineapple plant, as well as with other plant-derived enzymes. Papain dust inhalation resulted in anaphylactic shock in one reported case [21,22].
- **Papain:** As previously mentioned, papain is a latex protein obtained from the skin and green fruit of the papaw tree (*Carica papaya*). Papain acts upon a wide variety of proteins; its activity can be considerably enhanced by the addition of cysteine or other reducing agents or by protein denaturants, such as urea [8]. Indeed, without the presence of urea, papain displays very little proteolytic activity. The enzyme's activity is optimal over a pH range of 6 to 7. It has been stated that at low pH, papain is capable of digesting collagen. Though papain preparations have been used occasionally in acetic acid solutions to digest collagenous tissue, the success of this method has not been established. Other literature sources have described papain as having no effect on collagen. Papain has been cited [20] as attacking cysteine-containing proteins *only*. As collagen contains no cysteine and urea has not been shown to denature collagen, a papain-urea system is likely to have no effect on collagen. In 1957, J. Miller showed that papain-urea lacks the ability to degrade native collagen [23].
- **Ficin:** Another plant-derived enzyme found in the latex or sap of fig trees.

The history of enzymatic debriders has been a turbulent one, with only two enzymatic systems currently used widely, (i.e., bacterial collagenase and papain-urea). One reason for this turbulent history might be related to an enzyme's ability to degrade collagen. Howes et al. [24] and Rao et al. [25] have demonstrated that necrotic tissue is anchored to the wound surface by strands of undenatured collagen.

Until these fibers are severed, debridement cannot take place, granulation is slowed, and, thus, no supportive base is available for proper epithelialization. Consequently, the wound fails to heal. Over the years various proteolytic enzymes have been employed (papain, ficin, streptokinase, streptodornase, fibrinolyin/desoxynucleic acid, and trypsin-chymotrypsin) for the debridement of wounds. Trypsin and chymotrypsin have been used to remove necrotic tissue from dermal wounds, but neither of these enzymes demonstrates any activity against native collagen. Streptokinase and streptodornase have been used topically for wound debridement, but these enzymes do not attack collagen at all, nor do other enzymes, which they activate. The SH enzymes, papain, ficin, and bromelain which are derived from plants, are relatively unspecific and have been shown to have proteolytic activity and to attack collagen under certain conditions. In short, an enzymatic debrider with the ability to degrade nonviable collagen, while having no detrimental effects on viable tissue, is desirable, yet prior to the introduction of bacterial collagenase, none of the enzymes being used as topical wound debriders possessed such characteristics.

39.4 MODE OF ACTION OF BACTERIAL COLLAGENASE

Over the years, various proteolytic enzymes have been employed (papain, ficin, streptokinase, streptodornase, trypsin-chymotrypsin, etc.) for the debridement of wounds. They have had only limited success because they *do not* attack native collagen [8,9]. We will now focus on the MoA of bacterial collagenase.

The catalytic mechanism of metalloproteinases, such as collagenase, leads to the formation of a noncovalent tetrahedral intermediate after the attack of a zinc-bound water molecule on the carbonyl group of the scissile bond. This intermediate is further decomposed by transfer of a glutamic acid proton to the leaving group. The mechanism is hydrolysis [26] (see Figure 39.1).

In vitro, human types I and III collagen, extracted and purified from placental tissue, have been digested by incubation with bacterial collagenase. After analysis on Superex-30 gel sieve chromatography, the breakdown products were shown to be of the size of di- and tripeptides. The collagen-derived peptides were added to rat fibroblast culture to evaluate the effects of these breakdown products on cell proliferation and biosynthetic activity. By means of the neutral red test, stimulated cell proliferation was demonstrated when collagen breakdown products, at a concentration of 5 to 50 ng/mL of medium, were added. It would appear that bacterial collagenase has the ability to rapidly digest collagen and promote healing.

Bacterial collagenase is produced via a fermentation reaction orchestrated by *Clostridium histolyticum*. In processing the enzyme, bacterial cells are washed clean of the nutrient media via repeated rinses in aqueous media. Once free of the media, the cells are then lysed, fractured, filtered, and dried. What results is a "crude" form of collagenase, which can contain a very low percentage of actual enzymes, the bulk of the dried material being composed of bacterial cell debris. This "crude" material can be used in topically applied enzymatic deriding agents as an enzymatic agent as is or can be purified to specific collagenases of varying molecular weights. Six collagenases have been identified in the culture filtrate of *C. histolyticum,* and all have been purified to homogeneity. These collagenases uniquely cleave the interstitial collagens and exhibit both endopeptidase and tripeptidylcarboxypeptidase activities. The combined activity of endo- and tripeptidyl-C-peptidase makes these enzymes ideally suited for rapid collagen degradation.

In contrast, mammalian collagenases act differently by cleaving interstitial collagen at a single locus within the triple helical structure, giving rise to two large fragments, TC_A and TC_B [27,28]. These portions of the helix are then attacked by other nonspecific proteases, released by connective tissue cells, to be further degraded into small peptides [28].

It is believed that because necrotic tissue does not contain viable cells, bacterial collagenase treatment is not simply augmentation therapy but, rather, provides an essential biochemical activity

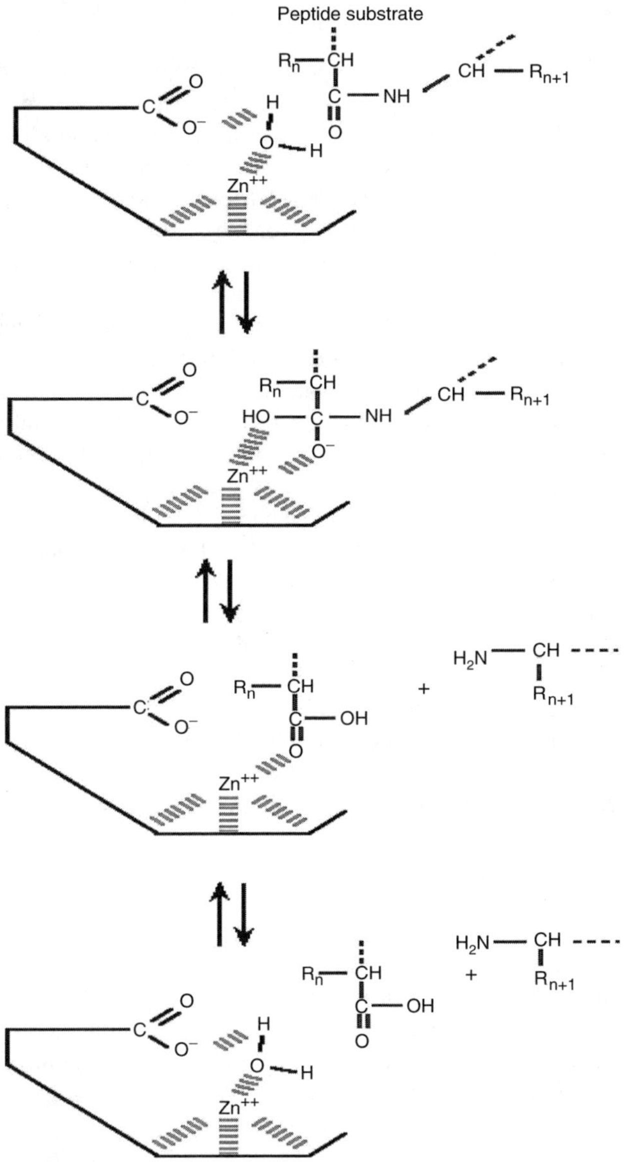

FIGURE 39.1 Catalytic mechanism of metallo proteinases.

to areas that are incapable of producing proteases. Without such treatment, slower, less effective healing would be predicted [30].

It has been shown in diabetic ulcers that necrotic tissue is anchored to the wound surface by a layer of perpendicular strands of undenatured collagen [31]. Collagenases, by definition, are enzymes capable of solubilizing fibrous collagen by peptide bond cleavage under physiologic pH and temperature conditions. Thus, collagen attacks not only necrotic tissue, but also fibers of undenatured collagen. It is suspected that the fibers of undenatured collagen anchor the eschar plug to the wound bed. With use of collagenase, the entire plug could be released and the remaining anchoring fibers would be removed. This would tend to lead to a cleaner and more thoroughly prepared wound bed. In support of this concept, it has been documented that collagenase treatments result in a decrease in visible scarring [31,32].

Since necrotic tissue is an important local cause of failure of the wound to heal, it would seem obvious that such binding must be severed so that debridement and healing can occur. Otherwise, granulation is slowed and no supportive base is available for epithelialization. Collagenase is irreversibly inactivated in a low-pH environment. It functions best in the pH range of 6 to 8 and temperatures below 56°C. Chelating agents (EDTA, citric acid, sodium citrate, etc.) also inactivate the enzyme by interacting with Ca^{2+} ions and Zn^{2+} ions, essential constituents of the structure and function of collagenase. Not all references state that calcium ions are part of the structure of collagenase, however, it is now known that Ca^{2+} ions play a role in stabilizing the 3^0 structure of the protein. It is well known that Zn^{2+} is located in the active site and is necessary for enzymatic activity. Collagenase hydrolyzes the peptide bonds in collagen. It does not attack other proteins such as hemoglobin or fibrin, important components in the formation of granulation tissue. In addition, collagenase does not attack growth factors, tissue inhibitors of metalloproteinase (TIMPs), and other critical components of the wound repair cycle [31]. Collagenase has been reported as an effective agent for the debridement of thermal burns. Although any protease would thoroughly digest degraded matter in the center of the burn eschar, only collagenase would effectively attack necrotic edges of the eschar, including the perpendicular fibers of undenatured collagen. As previously mentioned, these perpendicular collagen fibers anchor the eschar plug to the wound bed, and their removal is likely to be key for optimal wound bed preparation and potentially less scarring.

Collagenase has also been found to be useful in the debridement of third-degree burns. Otteman and Stahlgren compared the lytic effects of a number of enzymes on experimentally induced burns. The enzymes studied included streptokinase-streptodornase, trypsin-chymotrypsin, papain, ficin, desoxyribonuclease-fibrinolysin, and collagenase. Of these enzymes, only collagenase and papain were more than 90% effective in the digestion of wound debris and necrotic material [33].

As a result of the rather extensive investigation into the MoA collagenase systems, the following summary can be made:

Endogenous/mammalian collagenase:

1. Collagenase cleaves the triple stranded helix at a single point, (Gly^{775}–Ile^{776} in the α_1 chain; Gly^{775}–Leu^{776} in α_2 chain), which is located approximately three fourths of the distance from the N-terminus of the collagen molecule.
2. Results in two fragments: a 225-kDa fragment (TC_A) and a 75-kDa fragment (TC_B).
3. Fragments denature spontaneously into randomly coiled gelatin peptides.
4. Gelatin peptides are attacked by a variety of enzymes (less specific), including the gelatinases.

Figure 39.2 is a depiction of this MoA:

Bacterial collagenase:

1. Cleaves bonds of nonviable collagen
2. Results in small fragments
3. Fragments are degraded by less specific proteases

The MoA is very similar to that of mammalian collagenase, with a few important distinctions. The bacterial collagenase does not cleave the triple helical collagen in a single place, but attacks at many different points. Anywhere a Gly-X-Y (where X = proline and Y= hydroxyproline) exists, it is believed bacterial collagenase can attack [30].

This explains the more rapid degradation of collagen via bacterial collagenase when compared to endogenous/mammalian collagenase. In wound debridement, bacterial collagenase migrates to the base of the eschar, where it degrades the strands of undenatured collagen fibers, which hold

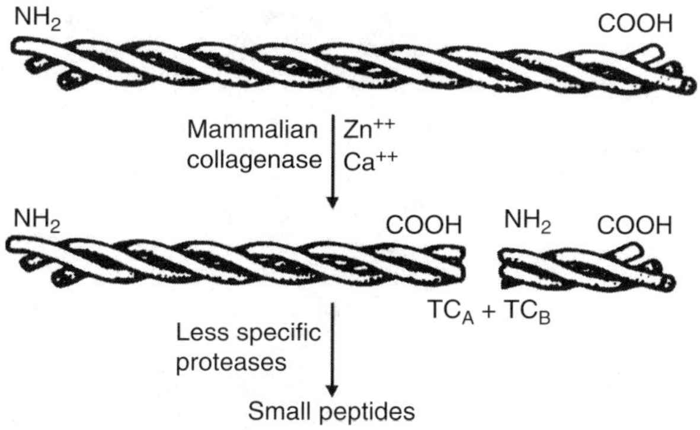

FIGURE 39.2 Action of mammalian collagenase.

the eschar plug to the wound bed. For this reason, it is believed that bacterial collagenase works from the "bottom-up."

The following is a proposed view of how a debriding agent (collagenase) might have multiple beneficial effects on wound bed preparation and healing: (A) The necrotic plug is being removed by the enzymatic action; (B) the collagen bundles are being cut at the necrotic-viable tissue interphase; (C) rapid epithelialization that might occur either indirectly (as necrotic tissue is removed) or through the direct effect of collagenase on keratinocyte migration; (D) the wound, now free of necrotic tissue, has good granulation tissue and is mostly epithelialized [5].

Figure 39.3 and Figure 39.4 are depictions of these events.

Several clinical studies have provided insight into the MoA of various enzymatic systems. Mainly, the issue of substrate selectivity vs. substrate nonselectivity is brought out via clinical

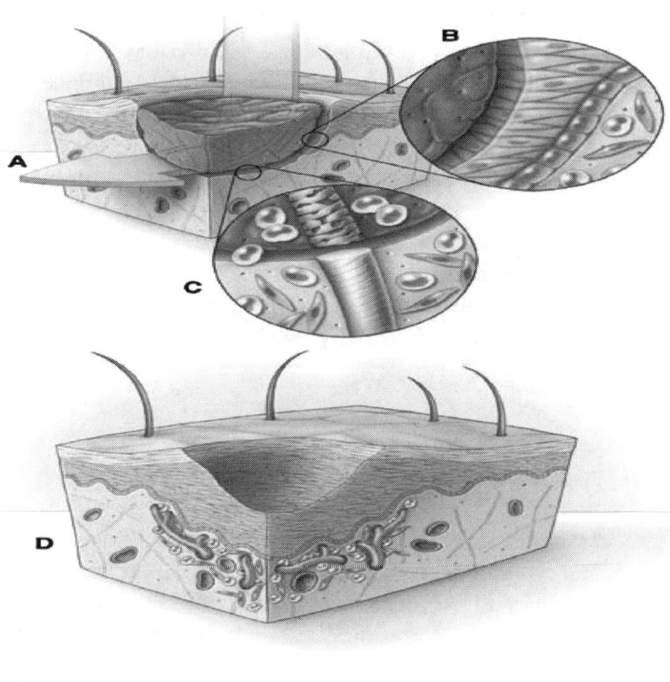

FIGURE 39.3 Proposed view of how a debriding agent (collagenase) might have multiple beneficial effects on WBP and healing.

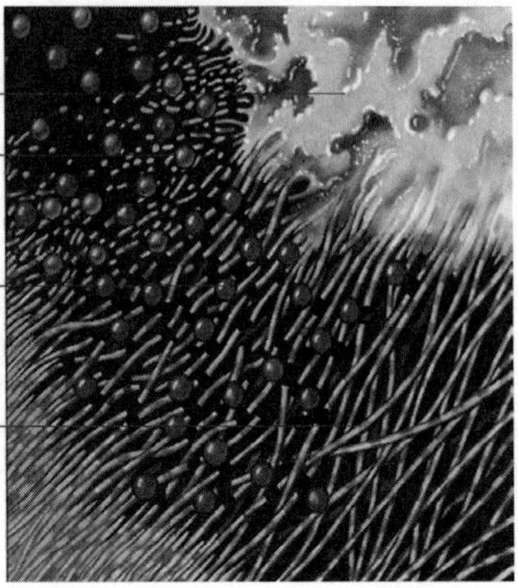

FIGURE 39.4 Clostridial collagenase attacks at the necrotic-viable tissue interface.

studies. To date, side effects such as inflammation, blistering, etc. have not been reported in treatments using more specific enzymes, such as bacterial collagenase. In fact, it is well reported [29–31,34–39] that collagenase itself or the breakdown components associated with collagenase (bacterial or endogenous) activity actually promote chemotaxis and healing and do not promote or prolong an inflammatory response. More recent *in vitro* work seems to indicate that collagenase might play a role in promoting angiogenesis.

In a study of the effect of various enzymatic debriders available on the market, on the biological activity of recombinant human PDGF-BB (rhPDGF-BB), several interesting and explainable results were obtained. These explanations are yet another example of how the differences between substrate-specific and substrate nonspecific enzymatic systems manifest themselves. The commercially available enzymes studied are as follows: clostridial collagenase, trypsin, papain-urea, and plasmin. The biological activity of rhPDGF-BB, expressed as the rate of rhPDGF-BB–stimulated DNA synthesis, was measured in an *in vitro* Swiss 3T3 mouse fibroblast system after incubation at 37°C for 2, 24, and 48 hours. Clostridial collagenase did not affect the biological activity of rhPDGF-BB. In the presence of trypsin, the biological activity of rhPDGF-BB was completely lost after incubation for 2 hours. Biological activity of rhPDGF-BB in the presence of papain was sustained, but partial loss in the presence of papain-urea, suggesting that the combination of papain-urea might interfere with the biological activity of rhPDGF-BB. Similarly, there was a complete loss of biological activity of rhPDGF-BB after 2 and 24 hours of incubation with plasmin [40].

In this study, the substrate specificity of collagenase and the lack thereof of the other enzymes tested is demonstrated, as are the potential effects of the use of substrate nonspecific enzymes. In summary, of the enzymes tested, clostridial collagenase was the only enzyme that did not affect *in vitro* biological activity of rhPDGF-BB.

39.5 EFFECTS OF ENZYMATIC DEBRIDERS ON THE PROCESS OF WOUND REPAIR

The healing and repair of wounds is a complex biological process that proceeds at an optimal rate under normal environmental conditions. However, the process can be retarded by systemic factors such as circulatory deficiencies and by local factors such as bacterial infections, surface debris, and necrotic matter. Debilitation, poor nutritional status, hyperglycemia, uremia, and devitalizing disease states can further impede healing. Howes et al. [24] and Rao et al. [25] have demonstrated that necrotic tissue is anchored to the wound surface by strands of undenatured collagen. Until these fibers are severed, debridement cannot take place, granulation is slowed, and, thus, no supportive base is available for proper epithelialization. Consequently, the wound fails to heal.

Normal cutaneous wound healing involves an orderly progression of events to reestablish the integrity of the injured tissue. The initial injury starts an orderly series of independent, yet separate, responses that include re-epithelialization and epithelial proliferation, inflammation, angiogenesis, fibroplasia, matrix accumulation, and, eventually, resolution of the wound. During each stage in the process, proteinases are needed to remove or remodel extracellular matrix components in both the epithelial and interstitial compartments, thereby accommodating cell migration and tissue repair [41]. Although extracellular matrix proteins can be degraded by various proteinases, fibrillar type I collagen, the most abundant protein in the dermis, is resistant to degradation by most enzymes [35].

Over the years, various proteolytic enzymes have been employed (papain, ficin, streptokinase, streptodornase, trypsin-chymotrypsin, etc.) for the debridement of wounds. They have had only limited success because these agents do not attack native collagen. Howes and Boxer demonstrated that collagenase could hydrolyze native collagen and, thereby, facilitate rapid debridement and healing of ulcers [24,32].

39.5.1 Function of Collagen in Wound Healing

Previously, collagens were thought to function only as structural support; however, it is now evident that collagen and collagen-derived fragments control many cellular functions, including cell shape and differentiation [42,43], migration [44], and synthesis of a number of proteins. A study of the expression of type I collagen in wound healing in the rat demonstrated a strict temporal and spatial response to injury. Six days after wounding, collagen expression is seen throughout the wound bed, but it becomes confined to the upper layers of the dermis as healing proceeds [45]. Later, collagen production is completely turned off throughout the dermal compartment — with the exception of a few subdermal fibroblasts [46]. A new extracellular matrix is synthesized and deposited during the early stages of wound repair, and this synthesis is turned off in the later stages of healing. This carefully controlled pattern of dermal matrix production suggests that collagen might influence cell phenotype and behavior during healing. By growing dermal fibroblasts in a preformed three-dimensional collagen lattice, the metabolism of the cells is reprogrammed to promote the synthesis of structural proteins, including various collagens and proteoglycans, as well as matrix-degrading proteinases and cytoskeletal proteins similar to the production of the molecules in dermal fibroblasts in an actual wound [47].

Of interest, results have shown that the downregulation of collagen synthesis is controlled by multiple cellular mechanisms. The production of most structural proteins is controlled by regulating gene transcription, the process by which genetic information is transcribed into mRNA — which, in turn, is translated into protein. Since protein synthesis is the end result of separate intracellular processes involving different molecules (DNA, mRNA, and protein), the amount of a protein produced can be controlled at any point during its synthesis. It has been found that downregulation of collagen synthesis in fibroblasts is controlled by both a decrease in procollagen α-chain gene transcription and the reduced stability of its corresponding mRNA [48,49]. By using dual mechanisms, cells can rapidly and effectively decrease their production of collagen [37]. It could be hypothesized that destroying TIMP, DNA, mRNA, etc. could result in errors in the downregulation of collagen synthesis and promote scarring and also slow the healing of wounds. When applied to a wound, nonspecific enzymes could have a negative effect on these various regulatory macromelocules.

Findings suggest that cell contact with precise extracellular matrix molecules influences cell behavior by regulating the quantity and quality of matrix deposition and degradation as well as cell differentiation, migration, and proliferation. These effects are likely mediated by specific cell surface receptors called integrins, composed of α and β subunits of two noncovalently linked polypeptide chains, which together confer their specificity for particular extracellular matrix proteins.

With regard to healing, regulation of matrix-degrading enzymes might be an important step in early wound debridement and removal of necrotic tissue. These observations emphasize that cellular functions are regulated by extracellular matrix. The information provided by extracellular matrix macromolecules is processed and transduced into the cells by specialized cell surface receptors, the integrins [50]. These are heterodimers of one α- and β-chain, which assemble noncovalently to bind a specific extracellular matrix protein. Evidence demonstrates that the $\alpha 2\beta 1$ receptor plays a major function in contraction of wounds [51,52], migration of epithelial cells [53], and induction of matrix-degrading collagenase, whereas collagen deposition appears to involve the $\alpha 1\beta 1$ receptor [37]. Cell surface receptors, being proteins with a tertiary structure, could be affected by compounds such as urea found in some of the nonspecific topical enzymatic debridement formulations.

The remodeling of connective tissue is mediated by serine proteinases and metalloproteinases, although other enzyme families have important functions as well [54].

The serine proteinases (chymotrypsin, trypsin, elastin, and subtilisin) are potent enzymes with broad catalytic specificity and are readily available when needed. In contrast, the metalloproteinases have more defined substrate specificity and are generally produced on demand. Metalloproteinases are released and participate in normal regulated tissue processes such as wound repair, morphogenesis during development, and involution [55], but they can be overproduced and destructive under prolonged inflammatory conditions [30].

39.5.2 Metalloproteinases

Metalloproteinases play a major role in wound repair. The basal epidermis is a prominent source of metalloproteinases, especially type I collagenase. Findings suggest that collagenase performs an essential function in normal wound repair. Interstitial collagenase is produced by basal keratinocytes in wounded skin. It has often been assumed that this enzyme is produced primarily by fibroblasts, macrophages, and other cells within the granulation tissue. Collagenase was not produced in dermal cells in samples of either acute human wounds or healthy skin [56]. In this way, viable collagen is not attacked by collagenase.

The degradative activity of dermal and epidermal collagenase may be involved in distinct healing processes. Since epidermal repair is common to all wounds and is ongoing when healing is most apparent, it is reasonable to believe that collagenase is expressed by migrating keratinocytes. Basal keratinocytes normally rest on a basement membrane composed of various forms of laminin, entactin, proteoglycans, and type IV collagen [57].

In response to wounding, keratinocytes migrate from the edge of the wound, under a provisional matrix of fibrin and fibronectin, and over the dermis, which includes structural macromolecules such as type I collagen, microfibrils, and elastin, which are distinct from the macromolecules of the basement membrane. Loss of contact with the basement membrane and establishment of new cell-matrix interaction with components of the dermal and provisional matrices can be a critical determinant that controls keratinocyte phenotype and induces collagenase production [30].

Collagenase production is induced in isolated human basal keratinocytes grown on a surface coated with type I collagen, which is the most abundant component of dermal matrix. In contrast, components of the basement membrane and other proteins of the interstitial matrix do not affect collagenase expression. The inductive effect of collagen is dependent on its native fibrillar structure. Although keratinocytes will adhere to denatured collagen (gelatin), collagenase production is not turned on in response to this substrate [58]. Keratinocytes have been known to recognize and migrate on type I collagen substratum, resulting in enhanced collagenase production [59]. Collectively, these studies demonstrate a key role for type I collagen in initiating keratinocyte collagenase synthesis in the epithelial response to wounding [30].

Pericellular degradation of the extracellular matrix is achieved via precise cell-matrix interactions in which cells recognize and adhere to specific extracellular matrix molecules. If the cell were activated for tissue remodeling, the appropriate proteolytic enzyme would be released into pericellular compartments where degradation would occur. Excess and "used" enzymes would be released into the tissue space and rapidly neutralized by specific inhibitors. An example of the specific inhibitors mentioned here are TIMPs. Keratinocytes producing collagenase are not in contact with the basement membrane. Most collagenase-positive keratinocytes are not in contact with the basement membrane, but rather migrate over the dermal matrix. The signal for collagenase mRNA diminishes rapidly in areas with intact basal lamina.

39.5.3 Role of Metalloproteinases in Wound Repair

The invariant and prominent production of interstitial collagenase by basal keratinocytes in both acute and chronic wounds indicates that this metalloproteinase plays a critical and required role in re-epithelialization rather than in dermal remodeling [30]. Table 39.1 denotes the source and frequency of production of a few metalloproteinases involved in wound repair.

TABLE 39.1
The Source and Frequency of Production of Metalloproteinases Involved in Wound Repair

Enzyme	Keratinocytes (%)	Fibroblasts (%)	Macrophages (%)
Collagenase	100	80	80
Stromelysin-1	80	90	90
Stromelysin-2	50	0	0
92-kDa gelatinase	0	0	0
72-kDa gelatinase	0	0	0
Matrilysin	0	0	0

Topical collagenase, derived from bacteria *(Clostridium hystolyticim)*, debrides and, thereby, promotes wound repair. Interstitial collagenase is produced by various cell types in response to inflammation. Human interstitial collagenase acts primarily on fibrillar collagens, such as types I and III, which are abundant in the dermis [30].

Interstitial collagenase catalyzes a single cleavage in collagen fibers, which in turn allows these proteins to be completely degraded by any other proteinase.

During wound healing, interstitial collagenase is produced by migrating keratinocytes and is released at the basal surface; it is believed that this metalloproteinase acts on denatured collagen fibers in intimate contact with the cells and within protected pericellular spaces. This process occurs within a viable part of the wound bed.

Bacterial collagenase rapidly attacks, degrading human collagen into small peptides. Human types I and III collagen, extracted and purified from placental tissue, then digested by incubation with bacterial collagenase, produce breakdown products the size of di- and tripeptides. These oligopeptides, upon addition to rat fibroblast culture, stimulate cell proliferation. Degradation products might be responsible for chemotactic phenomena. Breakdown products obtained by the digestion of collagen with bacterial collagenase have been found to affect chemotactic migration of fibroblasts and blood cells. The presence of collagen breakdown products at concentrations ranging from 5 to 50 ng/mL stimulates cell growth. The human collagen peptides obtained after bacterial collagenase digestion were able to stimulate cell proliferation in a dose-dependent relationship at a concentration range of 5 to 50 ng/mL [29].

Macrophages and precursor monocytes, which are key promoters in the wound healing process, demonstrate enhanced chemotaxis upon contact with collagen-derived peptides from bacterial collagenase digestion. It has also been demonstrated that collagen oligopeptides obtained both by chemical cleavage and by bacterial collagenase hydrolysis exert chemotactic effects on human blood monocytes [39]. As previously noted, specific cells (i.e., basal keratinocytes) secrete collagenase to denature the fibrillar collagen into gelatin, which they can move across. Perhaps the di- and tripeptides produced via bacterial collagenase activity perform a similar function (i.e., a more navigable surface), or perhaps they are recognized by the relevant integrins, resulting in elevated activity of macrophages and monocytes.

The action of proteolytic enzymes on the macromolecular constituents of the extracellular matrix gives rise to many peptides during wound healing. These degradation products seem to have a chemotactic effect on the recruitment of other cells, such as fibroblasts, mononuclear cells, and neutrophils. Again, this may be an indication that the di- and tripeptides are recognized by the relevant integrins resulting in elevated activity of these cells. Given its rapid hydrolytic activity, bacterial collagenase at wound sites can significantly enhance both the debridement and healing of wounds. In addition, the increased production of many breakdown products could contribute to chemotactic activity for connective tissue cell and blood cell recruitment in the wound area [29].

Misregulation of proteolysis is a suspected source of connective tissue damage in several diseases, such as emphysema and arthritis. The balance of proteinases and inhibitors is essential for the repair and maintenance of a normal extracellular matrix, and it can be disrupted in favor of proteolysis by the action of cytokines encountered at the wound site. Proteinase inhibitor complexes such as those composed of neutrophil elastase and α_1-proteinase inhibitor can be inflammatory mediators. Thus, interfering with the proteinase inhibitor balance in favor of inhibition might help maintain a neutrophil presence well after the neutrophil phase of inflammation should already have been resolved [60]. Also, TIMPs have been shown to inhibit angiogenesis and cytokine activation. Many cytokines important to wound healing can increase or decrease proteinase transcription and translation [61,62]. An interesting point here is that cytokines have a tertiary structure, which could be negatively affected by denaturants such as urea.

The process of wound healing is, in part, orchestrated by the release of and response to peptide mediators, such as growth factors. Growth factors, a group of peptides, have been shown to exert remarkable biological activities upon cells in culture. These hormone-like molecules interact with specific cell surface receptors to orchestrate the process of tissue repair. Upon initial injury, these macromolecules are released by platelets, leukocytes, and local tissue stores to stimulate the proliferation of new connective tissue, blood vessels, and epithelium, and their migration to the injury site. Inflammation advances to become tissue replacement. Local expression of these growth factors controls the rate and extent of repair. Growth factors can be released from the extracellular matrix during its turnover, and they can be sequestered by newly deposited matrix [63].

In chronic ulcers, poorly healing wounds, and severe burns, protease production is poorly regulated. More important, there is an excessive buildup of insoluble connective tissue proteins — in addition to collagen and necrotic tissues that block cell migration and impair healing. Because necrotic tissue does not contain viable cells, collagenase treatment is not simply augmentation therapy, providing essential biochemical activity to areas incapable of producing their proteases. Without such treatment, slower, less effective healing would be expected [30].

The enzymatic debriding agents available to us for clinical use are fibrinolytic, nonspecific proteolytics, or collagenolytics. It has been demonstrated that papain fails to discriminate between necrotic and normal tissue [9]. Additionally, it was shown that enzymes like papain and ficin do not digest collagen at a significant rate. Denaturing agents such as urea must be incorporated in formulations of these enzymes in order for them to attack collagen [31]. Other literature cites papain-urea systems as having no effect on collagen under any circumstances [23]. However, with the addition of strong reducing agents or detergents, papain can be activated toward collage [64]. The downside is that adding strong reducing agents or detergents to the wound bed would likely be detrimental to the healing process.

As previously mentioned, collagenases, by definition, are enzymes capable of solubilizing fibrous collagen by peptide bond cleavage under physiologic conditions of pH and temperature. Thus, collagen attacks not only necrotic tissue but also fibers of undenatured collagen [31]. This is an interesting point, as it provides insight into the action of collagenase and how it could potentially remove all residual collagen, including the undenatured collagen, which is suspected of anchoring the eschar "plug" to the wound bed [24,25]. This process would tend to clean and prepare the wound bed for healing. Collagenase hydrolyzes the peptide bonds in collagen. It does not attack other proteins such as hemoglobin and fibrin, important components in the formation of granulation tissue [31]. In addition, collagen does not attack growth factors and (in theory) TIMPs, nor other critical components of the wound repair cycle, based on the MoA. However, other enzymatic systems, such as papain-urea (in theory, based on the MoA) have the potential to attack these critical components.

In a study evaluating the use of *C. histolyticum* for the debridement of dermal ulcers and decubiti, the time required for complete debridement of the treated lesions averaged 10.5 days. Debridement was followed by gradual granulation and epithelialization, which generally proceeded at a faster rate than was expected in chronic dermal lesions of the type under treatment. Until the anchoring undenatured collagen fibers are severed, thereby allowing the necrotic plug to be removed, debridement cannot take place, granulation is slowed, and no supportive base is available for epithelialization [65]. Even though any protease would thoroughly digest degraded matter in the center of the burn eschar, only collagenase would effectively attack the necrotic edges of the eschar, including the perpendicular fibers of undenatured collagen [66]. In an animal model, investigating full thickness hot water burns, lesions treated with collagenase ointment showed increasing areas of debridement and were 95 to 100% free of necrotic material at the end of 4 days [67].

In a study of bacterial collagenase used on dermal ulcers and decubiti, healing occurred within 2 to 4 weeks in 11 of 17 patients, but complete epithelialization occurred only after about 10 weeks of treatment on 3 ulcers in a patient with advanced arteriosclerosis. An increased rate of healing might have been due to the presence of a granulation-stimulation factor in the collagenase ointment, which remains to be identified. This could be chemotaxis due to breakdown products. A markedly reduced incidence of hypertrophic scarring was also noted with the use of collagenase preparations. This reduced scarring might be related to the fact that epithelialization of the ulcer treated with collagenase proceeds from a completely debrided base [32].

Based on *in vitro* observations and a growing awareness that collagenase plays a pivotal role during the response to injury, further studies using clostridial collagenase to foster matrix remodeling during wound healing are warranted [34]. During the events that lead to cutaneous wound repair, the epidermal keratinocytes and vascular epithelial cells migrate and proliferate in an extracellular matrix-dependent fashion [68,69]. Keratinocyte locomotion, which is an essential component of the process, begins very early on during the wound healing response, as cells from the basal layer spread and move in from the wound margins [68–72]. The basement membrane glycoproteins — e.g., collagens, fibronectin, and vitronectin — play important roles in regulating keratinocyte adhesion and spreading [34,72–75]. Not only are these cell-matrix interactions essential for epidermal remodeling, but integrin-mediated endothelial signaling plays a pivotal role in regulating angiogenesis [69,76,77].

During re-epithelialization, matrix turnover is largely facilitated by the metalloproteinases, a family of enzymes that has diverse substrate specificity consistent with the dynamics of extracellular tissue remodeling during wound healing [73,78–80]. The interstitial collagenases, which cleave types I, II, III, and X collagen, act in concert with the stromelysins and matrilysins, enzymes that possess a broader substrate specificity and attack such matrix molecules as fibronectin, laminin, and the proteoglycans. Within the basement membrane, collagens IV and V, as well as elastin, are attacked by the 72-kDa and 92-kDa gelatinases. Several studies have documented the modulatory role that cytokines and TIMPs play in controlling these matrix-degrading enzymes. Recent results from wound healing studies have revealed that specific metalloproteinases (e.g., stromelysins) are uniquely associated with keratinocytes engaged in the migratory response to injury [78,79].

Results suggest that cell-matrix interactions modulate metalloproteinase expression. Results also support the notion that the keratinocyte phenotype can be modulated when cells contact the dermal matrix during wound healing [80,81]. Because of the demonstrable role that matrix metalloproteinases play during wound healing processes, researchers have become increasingly interested in understanding whether active remodeling of the collagenous subcellular matrix is necessary for potentiating the cellular responses to injury [34]. During endothelial cell proliferation and migratory responses to injury, using an *in vitro* wound healing model, the authors reported that collagenase treatment of the subendothelial matrix potentiated postinjury motility sevenfold over control cultures [69,82].

In vitro studies of clostridial collagenase have indicated that collagenase digestion of the subkeratinocyte matrix induces cell proliferation and migration several fold over controls. Addition of collagenase to the culture media of keratinocytes migrating or proliferating on collagenase-treated matrices significantly amplifies the collagenase-potentiated state [34].

The addition of bacterial collagenase to healing wounds could actually accelerate wound healing and complement the naturally occurring process of *in vivo* tissue repair. Clostridial collagenase enhances human keratinocyte proliferation *in vitro* and stimulates keratinocyte migration. Collagenase potentiates human keratinocyte postinjury migration *in vitro*. The matrix modulates the wound healing responses [34].

There has been a growing awareness that the extracellular matrix plays a pivotal role in regulating the cellular responses to injury [68,69,76,82,83]. There have been several studies demonstrating that degradation of the extracellular environment (i.e., the provisional matrix, as well as the basement membrane) is a required step in order for keratinocyte migration, as well as new blood vessel formation, to ensue during reparative processes. Initially, keratinocytes begin their migratory response within a provisional matrix derived from the plasma constituents fibrin and fibronectin. Furthermore, it has been established that when the basement membrane has been interrupted, as in the case of deep puncture wounds at the edge of ulcers in pyogenic granuloma, keratinocytes contact the underlying stromal (dermal) matrix, which contains interstitial collagens (I, III) [65,69]. It is of interest that under these conditions, keratinocyte production of interstitial collagenase is strongly induced [81]. These observations suggest that keratinocytes involved in re-epithelialization processes take on a collagenolytic phenotype. This is consistent with research that reveals that the keratinocyte's response to injury (migration and proliferation) is markedly enhanced when cells are allowed to recover on matrices treated with collagenase. The fact that cell-derived collagenolytic events can be markedly amplified *in vitro* if clostridial collagenase is added to the culture media strongly suggests that its incorporation into healing wounds (when the basement membrane is still interrupted) may potentiate tissue repair *in vivo* [34].

Matrix remodeling during tissue repair is dependent upon the action of several MMPs [78,79]. An indirect association between 92-kD type IV collagenase and interstitial collagenase has been documented via the TIMPs [84], and, perhaps even more interestingly, TGF-α has been shown to exert a negative influence on interstitial collagenase activity while stimulating the production of human keratinocyte type IV collagenase [85,86]. While this relationship suggests that the association of cytokines with the fibrillar components of the matrix is important in modulating wound healing responses, it also suggests that the local interactions between cells and extracellular components dictate wound healing rates [34]. If TGF-α inactivates interstitial collagenase while stimulating type IV collagenase production, we might predict a restricted or retarded keratinocyte migratory response to injury when the basement membrane is disrupted, in which case keratinocytes contact interstitial rather than basement membrane collagens. The fact that both keratinocyte migration and proliferation can be accelerated by clostridial collagenase treatment of the subcellular matrix or by its inclusion in the culture media lends credence to this notion [34].

Integrins mediate cell-matrix interaction during wound healing. Several reports support the fact that wound healing is dependent upon the association of epidermal and dermal cells with their associated matrix components [34,76,87]. In the mid-1980s, it was demonstrated that the matrix could modulate endothelial cell motility after injury *in vitro* [73]. While the glycoproteins, fibronectin, and interstitial collagen were comparable to one another in regard to their migration-promoting effects, basement membrane collagen was most effective in retarding migration after injury. It was deduced that the transmembrane signaling events that occur via the integrins were responsible for stimulating cytoskeletal reorganization and stress fiber formation [34]. These events, in turn, stimulate increased adhesion and decreased locomotion [76,82,88]. If the subcellular collagenous matrix with metalloproteinases were modified, the postinjury migration rate could be markedly elevated [69].

Support for the hypothesis that matrix remodeling actively influences wound healing (by stimulating the release of matrix-degrading enzymes) is also derived from studies marking the role of anti-α_5 integrin antibodies of fibronectin fragments in stimulating the release of stromelysin from stromal fibroblasts [71]. Moreover, based on earlier published reports on keratinocyte integrin expression during wound healing [71,73,89,90] together with recent results, it is predicted that integrin-mediated signals are likely to play a pivotal role in regulating the collagenase-induced alterations seen during the keratinocyte migratory and active responses to injury [34].

The use of collagenase in the treatment of pressure ulcers has been documented. Collagenase is a proteolytic enzyme involved in the process of healing of ulcers and wounds; as such, it acts as a debriding agent that stimulates the formation of granulation tissue. Necrotic and devitalized tissue covering the wound delays healing and is an adequate medium for organism multiplication, with consequences that can range from simple bacterial colonization to hematogenous spreading and severe sepsis. Collagenase is released from granulocytes and macrophages into the extracellular spaces and is involved in the natural process of debridement during the inflammatory phase. The major source of collagenase is the basal keratinocytes. The enzyme specifically splits collagen into two moieties, allowing other, unspecific proteases to complete collagen degradation [89].

Collagenase has the ability to dissolve desmosomes, enabling the cells to migrate in a matrix of fibronectin. The enzyme also fosters the migration of fibroblasts and epithelial cells of keratinocytes. Cleavage products resulting from collagen degradation stimulate fibroblast proliferation. Finally, during the maturation phase, collagenase plays an important role in the healing process by loosening necrotic tissue and enhancing its digestion by neutrophils and macrophages, stimulating fibroblast proliferation, and fostering remodeling of collagen fibers. Topical application of exogenous collagenase confers clear clinical benefits in cases in which endogenous collagen synthesis is decreased, as in the elderly, in diabetic patients, in patients with malnutrition, or in those treated with corticosteroids [90].

39.6 CONCLUSION

Wound healing is a complex biological process that proceeds in a timely and orderly fashion under normal environmental conditions. However, the process can be impaired by a variety of both systemic and local factors, resulting in a chronic wound. In managing these wounds, the concept of WBP can be used as a road map to healing. The quartet — the removal of necrotic tissue, bacterial management, exudate management, and the stimulation of the epidermal margin — is key in treating these wound types. In this chapter, we have focused on the removal of necrotic tissue. In the case of chronic wound, the wound often requires "help" in removing necrotic tissue. In many instances, the use of a topically applied enzymatic debrider is appropriate.

Over the years, various proteolytic enzymes have been employed for the debridement of wounds. They have had limited success because they *do not* attack native collagen. Howes et al. [24] and Rao et al. [25] have demonstrated that necrotic tissue is anchored to the wound surface by strands of undenatured collagen. Until these fibers are severed, debridement cannot take place, granulation is slowed, and, thus, no supportive base is available for proper epithelialization. Consequently, the wound fails to heal. In this chapter papain-urea, endogenous collagenase, and bacterial collagenase have been reviewed. Other enzymes, such as ficin, sutilains, bromelain, trypsin, and chymotrypsin, have been touched on as well. Most notably, the MoA of bacterial collagenase has been detailed and compared to endogenous collagenase. Although the mechanisms of action and the substrates upon which various enzymes act are interesting from an academic standpoint, they more importantly provide an insight into their effects on the wound bed and the subsequent healing process. We have learned that bacterial collagenase breaks collagen down into oligopeptides, of 2 to 3 amino acids in length. These di- and tripeptides have been shown to promote chemotaxis of fibroblasts in the dermis and keratinocytes in the epidermis. Collagenase is also involved in angiogenesis, resulting in well-vascularized granulation beds and, ultimately, wound healing. We have also seen that due to the selective nature of bacterial collagenase in that it does not attack viable tissue or other compounds, such as growth factors, it is key to the healing process. The use of bacterial collagenase is an ideal agent for preparing a clean wound bed, free of necrotic tissue and nonviable collagen. As a result, the quality of healing is improved. A decrease in scarring has been associated with its use, equating to a closed wound with greater tensile strength and a reduced chance of recurrence. All of these properties of bacterial collagenase make it a very useful tool in the management of wounds and, ultimately, wound healing.

REFERENCES

1. Staiano-Coico, L., Higgins, P.J., and Schwartz, S.B. et al., Wound fluids: a reflection of the state of healing, *Ostomy Wound Manage,* 46(suppl. 1A), 85S–93S, 2000.
2. Lazarus, G.S., Cooper, D.M., and Knighton, D.R. et al., Definitions and guidelines for assessment for wounds and evaluation of healing, *Arch. Dermatol.,* 130, 489–493, 1994.
3. Shultz, G. and Mast, B., Molecular analysis of the environment of healing and chronic wounds: cytokines, proteases and growth factors, *Wounds,* 10 (suppl. 6F), 1F–9F, 1998.
4. Whalen, G.F. and Zetter, B.R., Angiogenesis, in *Wound Healing Biochemical and Clinical Aspects,* Cohen, I.K., Dregelmann, R.F., and Lindblad, W.J., Eds., W.B. Saunders Company, Philadelphia, PA, 1992, 77–95.
5. Falanga, V., Wound bed preparation and the role of enzymes: a case for multiple actions of therapeutic agents, *Wounds,* 14(2), 47–57, 2002.
6. Falanga, V., Classifications for wound bed preparation and stimulation of chronic wounds, *Wound Repair and Regen.,* 8(5), 347–352, 2000.
7. Schultz, G.S. et al., Wound bed preparation: a systematic approach to wound management, *Wound Repair and Regeneration,* 11, 1–28, 2003.
8. Sherry, S. and Fletcher, A.P., Clinical pharmacology and therapeutics. Proteolytic enzymes: a therapeutic evaluation, 1(2), 202–226, 1960.
9. Mandl, I., *Collagenase,* Rev. 1, Gordon & Breach, Science Publishers, New York, 1972.
10. Modder, E.E., On the Medicinal Uses of *Carica Papaya* (Lin.), Ceylon MJ, 1887–1888, 115.
11. Johnston, R. McK., Notes on the use of papain in ear diseases, Edinburgh, MJ, 35, 621–624, 1890.
12. Jacobi, A., Note on Papayotin, *Therap. Gaz.,* 2, 145–147, 1886.
13. Tillett, W.S. and Sherry, S., Effect I patients of streptococcal desoxyribonuclease in fibrinous, purulent and sanguinous pleural exudations, *J. Clin. Invest.,* 28, 173–190, 1949.
14. Sherry, S., Tillett, W.S., and Christensen, L.R., Presence and significance of desoxyribose nucleoprotein in the purulent pleural exudates of patients, *Proc. Soc. Exper. Biol. Med.,* 68, 179–184,1948.

15. Tillett, W.S., Sherry, S., and Christensen, L.R., Streptococcal desoxyribonuclease (streptodornase): significance in lysis of purulent exudates and production by strains of hemolytic streptococci, *Proc. Soc. Exper. Biol. Med.,* 68, 184–188, 1948.
16. Sherry, S., Johnson, A., and Tillett, W.S., The action of Streptococcal desoxyribonuclease (streptodornase), *in vitro* and on purulent pleural exudations of patients, *J. Clin. Invest.,* 28, 1094–1104, 1949.
17. Johnson, A.J., Cytological studies in association with local injections of streptokinase-streptodornase into patients, *J. Clin. Invest.,* 29, 1376–1386, 1950.
18. Sherry, S. and Goeller, J.P., The extent of the enzymatic degredation of desoxyribonucleic acid (DNA) in purulent exudates by streptodornase, *J. Clin. Invest.,* 29, 1588–1594, 1950.
19. Sherry, S., The action of streptococcol desoxyribonuclease (streptodornase) on purulent exudates, *Tr. New York Acad. Sc.,* 14, 138–140, 1952.
20. McCarty, M.I., The occurrence of nucleases in culture filtrates of group A hemolytic streptococci, *J. Exper. Med.,* 88, 181–188, 1948.
21. Freye, H.B., Papain anaphylaxis: a case report, *Allergy Proc.,* 9(5), 571–574, 1988.
22. Quinones, D., Alonso, S., Lopez, R., Sanchez, I., Rodriquez, F., and Fernadez, L., Contact urticaria, rhinoconjunctivitis and branchial asthma from occupational use of papain. Secci6n de Alergologia, Hospital Universito "Marques de Valdecilla" Santander, Cantabria, Spain. Allegro Immunopathol., (Madr) September–October, 1999.
23. Miller, J.M. and Howard, F., The interaction of papain, urea and water-soluble chlorophyllin in a proteolytic ointment for infected wounds, *Surgery,* 43(6), 939–948, 1957.
24. Howes, E.L., Mandl, I., Zaffuto, S., and Ackerman, W., The use of *Clostridium histolyticum* enzymes in the treatment of experimental 3rd degree burns, *Surg. Gyn. Obst.,* 109, 177, 1959.
25. Rao, D.B., Sane, P.G., and Georgiev, E.L., Collagenase in the treatment of dermal and decubitus ulcers, *J. Am. Geriatr.,* 23(1), 22–30, 1975.
26. Moreau, T., *Introduction to Proteases,* Prolysis, University of Tours, France, 1999–2000, 1–3.
27. Mignatti, P., Welgus, H.G., and Rifkin, D.B., Role of degradative enzymes in wound healing, in *The Molecular and Cellular Biology of Wound Repair,* Clarke, R.A.F. and Henson, P.M., Eds., Plenum Press, New York and London, 1988, 497–524.
28. Welgus, G.H., Jeffrey, J.J., and Eisen, A.Z., The collagen substrate specificity of human skin fibroblast collagenases, *J. Biol. Chem.,* 256, 9511–9515, 1981.
29. Cortivo, R., Radice, M., Brun, P., and Abatangelo, G., Biological activity of human collagen breakdown products on fibroblasts, *WOUNDS — A Compendium of Clinical Research and Practice,* Vol. 7, Supplement A, September/October 1995, 38A–44A.
30. Parks, W.C., The production, role, and regulation of matrix metalloproteinsases in the healing epidermis, *WOUNDS — A Compendium of Clinical Research and Practice,* Vol. 7, Supplement A, September/October 1995, 23A–33A.
31. Altman, M.I., Goldstein, L., and Horowitz, S., Collagenase-an adjunct to healing trophic ulcerations in the diabetic patient, *J. Am. Podiatric. Assoc.,* 68(1), 11–15, 1978.
32. Boxer, A.M., Gottesman, N., Bernstein, H., and Mandl, I., Debridement of dermal ulcers and decubiti with collagenase, *Geriatrics,* 24, 75–86, 1969.
33. Ottman, M.G. and Stahlgren, L.H., A laboratory method of the quantitative measurements of the lysis of burn eschars in animals by chemical debriding agents, *Surg. Forum,* 13, 41, 1962.
34. Herman, I.M., Stimulation of human keratinocyte migration and proliferation *in vitro*: insights into the cellular responses to injury and wound healing, *Wounds — A Compendium of Clinical Research and Practice,* 8(2), 33–41, 1996.
35. Pilcher, B.K., Dumin, B.D., Sudbeck, S.M., Krane, H.G., Parks, W., and Parks, W.C., The activity of collagnese-1 is required for keratinocyte migration on a type 1 collagen matrix, *J. Cell. Biol.,* 137(6), 1445–1457, 1997.
36. Silverstein, P., Ruzicha, F.J., Helmkamp, G.M., Lincoln, R.A., and Mason, A.D., In-vitro evaluation of enzymatic debridement of burn wound eschar, *Surgery,* 73(1), 15–22, 1973.
37. Krieg, T., Collagen in the healing wound, *Wounds — A Compendium of Clinical Research and Practice,* Vol. 7, Supplement A, September/October 1995, 5A–12A.
38. Postlethwaite, A.E., Seyer, J.M., and Kang, A.H., Chemotactic attraction of human fibroblasts to type I, II and III collagens and collagen-derived peptides, *Proc. Natl. Acad. Sci. U.S.A.,* 75, 871–875, 1978.

39. Postlethwaite, A.E. and Kang, A.H., Collagen and collagen peptide-induced chemotaxis of human blood mononcytes, *J. Exp. Med.*, 143, 1299–1307, 1976;
40. Gosiewska, A., Yi, C.-F., Brown, L., and Geesin, J.C., *The Effect of Enzyme Debriders on the Biological Activity of Recombinant Human Platelet-Derived Growth Factor-BB (rhPDGF-BB), the Active Agent of Regranex Gel,* J&J Wound Healing Technology Resource Center, Skillman, NJ, 1996.
41. Mignatti, P., Welgus, H.G., Rifkin, D.B., and Parks, W.C., Proteinases and tissue remodeling, in *The Molecular and Cellular Biology of Wound Repair,* 2nd ed., Clark, R.A.F., Ed., Plenum Press, New York, 1996, 427–474.
42. Montesano, R., Orci, L., and Vasselli, F., *In vitro* rapid organization of endothelial cells into capillary-like networks is promoted by collagen matrices, *J. Cell. Biol.*, 97, 1648–1652, 1983.
43. Madri, J.A. and Marx, M., Matrix composition, organization and soluble factors: modulation of microvascular cell differentiation *in vitro, Kidney Int.*, 41, 560–565, 1992.
44. Albini, A. and Dadelmann-Grill, B.C., Collagenolytic cleavage products of collagen type I as chemoatractants for human dermal fibroblasts, *Eur. J. Cell. Biol.,* 36, 104–107, 1985.
45. Scharffetter, K. and Stolz, W., Lankat-buttgereit B et al., Localization of collagen α1(I) gene expression during wound healing *in situ* hybridization, *J. Invest. Dermatol.*, 93, 405–412, 1989.
46. Oono, T., Specks, U., and Eckes, B., et al., Expression of type VI collagen mRNA during wound healing, *J. Invest. Dermatol.*, 100, 329–334, 1993.
47. Bell, E., Ivarsson, B., and Merill, C., Production of a tissue-like tructure by contract of collagen latices by human fibroblasts of different proliferative potential *in vitro, Proc. Natl. Acad. Sci. U.S.A.,* 76, 1274–1278, 1979.
48. Mauch, C., Hatamochi, A., Scharggetter, K., and Krieg, T., Regulation of collagen synthesis in fibroblasts within a 3-D collagen gel, *Exp. Cell. Res.*, 1787, 493–503, 1988.
49. Eckes, B., Mauch, C., Huppe, G., and Krieg, T., Downregulation of collagen synthesis in fibroblasts within dimensional and posttranscriptional mechanisms, *FEBS Lett.,* 250, 301–305, 1989.
50. Hynes, R.O., Integrins versatility, modulation and signalling in cell adhesion, *Cell,* 69, 11–25, 1992.
51. Klein, C.E., Dressel, D., and Steinmayer, T., et al., Integrin a α21 is unregulated in fibroblasts and highly aggressive melanoma cells in 3-D collagen lattices and mediates the reorganization of collage in I fibrils, *J. Cells Biol.,* 115, 1427–1436, 1991.
52. Schiro, J.A., Chan, B.M.C., and Roswit, W.T., et al., Integrin α21 (VLA-2) mediates reorganization and contraction of collagen matrices by human cells, *Cells,* 67, 403–410, 1991.
53. Scharffetter-Kochanek, K., Klein, C.E., and Heinen, G., et al., Migration of human keratinocyte cell line HaCaT to interstitial collagen type I is mediated by α21 integrin receptor, *J. Invest. Dermatol.,* 98, 3–11, 1992.
54. Werb, Z., Proteinases and matrix degradation, in *Textbook of Rheumatology,* Kelley, W.N., Harris, E.D. Jr., and Ruddy, S., Eds., Saunders, Philadelphia, 1989, 300–326.
55. Matrisian, L.M., The matrix-degrading metalloproteinases, *BioEssays,* 14, 455–463, 1992.
56. Jeffrey, J., Metalloproteinases and tissue turnover, *Wounds — A Compendium of Clinical Research and Practice*, Vol. 7, Supplement A, September/October 1995, 13A–22A.
57. Stenn, K.S. and Malhorta, R., Epithelialization, in *Wound Healing: Biochemical and Clinical Aspects,* Cohen, I.K., Diegelmann, R.F., and Lindblad, W.J., Eds., WB Saunders Company, Philadelphia, 1992, 115–127.
58. Sudbeck, B.D., Welgus, H., Parks, W.C., and Pentland, A.P., Collagen-mediated induction of collenase by basal keratinocytes involves distinct intracellular pathways, *J. Biol. Chem.,* 269, 30022–30029, 1994.
59. Petersen, M.J., Woodley, D.T., Stricklin, G.P., and O'Keefe, E.J., Enhanced synthesis of collagenase by human keratinocytes cultured on type I or type IV collagenase, *J. Invest. Dermatol.,* 94, 341–346, 1990.
60. Banda, M.J., *Wounds — A Compendium of Clinical Research and Practice,* 7SA, 45A–52A, 1995.
61. McNaul, K.L., Chartrain, N., and Lark, M., et al., Discoordinate expression of stromelysin, collagenase and TIMP-1 in rheumatoid human synovial fibroblasts. Synergistic effects of IL-1 and TNF-alpha on stromelysin expression, *J. Biol. Chem.,* 265, 17238–17245, 1990.
62. Edwards, D.R., Murphy, G., and Reynolds, J.J., et al., TGF-beta modulate the expression of collagene and metalloproteinase inhibitor, *EMBO J.,* 1899–1904, 1987.
63. Davidson, J.M., *Wounds — A Compendium of Clinical Research and Practice,* 7SA, 53A–63A, 1995.

64. Miller, J.M. and Goldman, B., Preliminary and short report — the digestion of collagen, *J. Invest. Dermatol.,* 30, 217–219, 1958.
65. Hoover, N.W. and Ivins, J.C., Wound debridement, *Arch. Surg.,* 79, 701, 1959.
66. Connell, J.R., Bowe, J.J., Del Guercio, L., and Rousselot, L.M., Evaluation of present day concepts in the treatment of the severely burned patient, *Amer. J. Surg.,* 93, 694, 1957.
67. Linder, J.S., Personal communication.
68. Grinnell, F., Tod, K.-I., and Takashima, A., Activation keratinocyte fibronectin receptro funtion during cutaneous wound healing, *J. Cell. Sci. Su.,* 8, 199–209, 1987.
69. Herman, I.M., Molecular mechanisms regulating vascular endothelial cell motile response to injury, *J. Cardio. Pharm.,* 22, 25–36, 1993.
70. Chen, J.D., Helmhold, M., Kim, J.P., Wynn, R., and Woodley, D.T., Human keratinocytes synthesizes high levels of collagenase that degrades basement membrane type IV collagen, *Dermatology,* 188, 6–12, 1994.
71. Kim, J.P., Zhang, K., Chen, J.D., Wynn, K.C., Kramer, B., and Woodley, D.T., Mechanisms of human keratinocyte migration on fibronectin: unique roles of RGD and integrins, *J. Cell. Physiol.,* 151, 443–450, 1992.
72. Brown, C., Stenn, K.S., Falk, R.J., Woodley, D.T., and O'Keefe, E.J., Vitronectin: effects on keratinocyte motility and inhibition of collagen-induced motility, *J. Inv. Dermatol.,* 96, 724–728, 1991.
73. Gailit, J., Welch, M.P., and Clark, R.A.F., TGF-beta 1 stimulates expression of keratinocyte integrins during re-epithelization of cutaneous wounds, *J. Inv. Dermatol.,* 103, 221–227, 1994.
74. Kim, J.P., Zhang, K., Chen, J.D., Kramer, B., and Woodley, D.T., Vitronectin-driven keratinocyte locomotion mediated by the $\alpha v\beta 5$ integrin receptor, *J. Bio. Chem.,* 269, 26926–26932, 1994.
75. Herman, I.M., ECM-cytoskeletal interactions, *Tissue & Cell,* 19, 1–19, 1987.
76. Marikovsky, M., Breuing, K., Liu, P.Y., Erikson, Higashiyama, S., Farber, P., Abrahm, J., and Klagsby, M., Appearance of eparne-binding EGF-1 growth factor in wound fluid as a response to injury.
77. Kim, J.P., Zhang, K., Chen, J.D., Wynn, K.C., Kramer, B., and Woodley, D.T., Mechanisms of human keratinocyte migration on fibronectin: unique roles of RGD and integrins, *PNSA,* 90, 3889–3893, 1993.
78. Saarialho-Kere, U.K., Chang, E.S., Welgus, H.G., and Parks, W.C., Interstitial collagenase is expressed by migrating kertinocytes at the epidermal wound edge, *J. Clin. Invest.,* 90, 1952–1957, 1992.
79. Saarialho-Kere, U.K., Pentland, A., Birkedal-Hansen, H., Parks, W.C., and Welgus, H.G., Distinct population of basal keratinocytes express stromelysin-1 and -2 in chronic wounds, *J. Clin. Invest.,* 94, 79–88, 1994.
80. Sarret, Y., Woodley, D.T., Goldberg, G.S., Kronberg, A., and Wynn, K.C., Constitutive synthesis of a 92 kDa keratinocyte-derived type IV collagenase enhanced by type I collagen and decreased by type IV collagen matrices, *J. Inv. Dermatol.,* 99, 836–841, 1992.
81. Saarialho-Kere, U.K., Kovacs, S.O., and Pentland, A.P., et al., Cell-matrix interactions modulate interstitial collagenase expression by human keratinocytes actively involve in wound healing, *J. Clin. Invest.,* 92, 2858–2866, 1993.
82. Young, W.C. and Herman, I.M., ECM modulation of endothelial cell shape and motility following injury *in vitro, J. Cell. Sci.,* 73, 19–32, 1985.
83. Grinnell, F., The activated keratinocyte: up regulation of cell adhesion and migration during wound healing, *J. Trauma,* 30, S144–149, 1990.
84. Goldberg, G.I., Strongin, A., Collier, I.E., Genrich, L.T., and Marmer, B.L., Interaction of the 92kDa type IV collagenase w/the TIMPs prevents dimerization, complex formation with interstitial collagenase and activation of the proenzyme with stromelysin, *J. Biol. Chem.,* 267, 4583–4591, 1992.
85. Overall, C.M., Wrana, J.L., and Sodek, J., Independent regulation of collagenase, 72kDa progelatinase and metalloendopeptidase inhibitor expression in human fibroblasts by TGF-beta, *J. Biol. Chem.,* 264, 1860–1869, 1989.
86. Salo, T., Lyons, J.G., Ranemtulla, F., Birkedal-Hansen, H., and Latjava, H., TGF-beta up regulates human keratinocytes, *J. Biol. Chem.,* 226, 11436–11441, 1991.
87. Larjava, H., Lyons, J.G., Salo, T., Makela, M., Miovisto, L., Birkedal-Hansen, H., Akiyama, S.K., Yamada, K.M., and Heina, J., Anti-integrin antibodies induce type IV collagenase expression in keratinocytes, *J. Cell. Physiol.,* 157, 190–200, 1993.
88. Yost, J.C. and Herman, I.M., Substratum-induced stress fiber assembly in vascular endothelial cells during spreading *in vitro, J. Cell. Sci.,* 95, 507–520, 1990.

89. Wang, N. and Ingber, D.E., Control of cytoskeletal mechanisms by ECM, cell shape and mechanical tension, *Biophysical J.,* 66, 2181–2189, 1994.
90. Hebda, P.A., Kevin, J., and Dohnar, J.E., Evaluation of efficacy of enzymatic debriding agents for removal of necrotic tissue and promotion of healing in porcine skin wounds, *Wounds — A Compendium of Clinical Research and Practice,* Vol. 10, 5, 83–96, 1998.

40 Clostridia As Production Systems for Prokaryotic and Eukaryotic Proteins of Therapeutic Value in Tumor Treatment

Lieve Van Mellaert, Jan Theys, Oliver Pennington, Sofie Barbé, Sandra Nuyts, William Landuyt, Philippe Lambin, Nigel P. Minton, and Jozef Anné

CONTENTS

40.1 INTRODUCTION

As strict anaerobic bacteria, clostridia can be exploited for the treatment of cancer. This is because, upon spore administration to tumor-bearing organisms, *Clostridium* specifically colonizes the hypoxic/necrotic region of solid tumors, which is highly resistant to radio- and chemotherapy and, hence, with a negative outcome for tumor control. In this respect, recombinant clostridia expressing and eventually secreting therapeutic proteins within the tumor microenvironment have potential in tumor treatment, as they can transfer proteins of therapeutic interest to a critical region of the tumor. The delivery of therapeutic proteins to the hypoxic/necrotic region of the tumor may, therefore, allow its control from inside. Moreover, the *Clostridium*-mediated protein transfer provides several advantages compared to the classical gene therapy for cancer. Tumor cells do not need to be genetically modified, evading possible cancerogenic events caused, for instance, by random insertion of foreign DNA into the genome. In addition, the bacterial transfer system can be regarded as safe, since targeted gene expression can be stopped at any time by elimination of the tumor-colonizing bacteria through administration of suitable antibiotics. In combination with the conventional tumor treatment modalities like radio- and chemotherapy, it may lead to the increase of the therapeutic ratio of cancer.

The use of clostridia as protein transfer systems, however, demands the availability of genetic tools including well-established transformation protocols, suitable vectors, and a battery of regulatory sequences for establishment of expression and secretion of the proteins of interest. In this chapter, we will review the genetic elements currently in use in order to achieve heterologous gene expression and, eventually, protein secretion in recombinant clostridia. Since the final aim of the *Clostridium* recombinant protein production system described in this chapter is situated in the field of tumor control, the heterologous genes of interest expressed within this bacterial transfer system are encoding therapeutic agents. Mainly, they belong to two major groups: prodrug-converting enzymes and cytotoxic proteins. The heterologous protein yield so far obtained in recombinant clostridia will be discussed and related to the *in vivo* evaluation of the clostridial protein transfer system to tumors. Finally, with respect to the improvement of this promising modality in cancer treatment, future approaches will be discussed.

40.2 TUMOR COLONIZATION BY CLOSTRIDIAL SPORES

It has been over 50 years since the therapeutic potential of injecting spore suspensions of *Clostridium histolyticum* directly into the transplanted sarcomas of mice was tested [1]. Following germination, vegetative growth of the organism resulted in significant tumor lysis and, in those animals co-treated with penicillin and antitoxin, extended their survival compared to tumor-bearing mice that were not injected with clostridial spores. Subsequently, it became apparent that direct injection was not necessary and that tumor colonization was readily obtained following intravenous administration of spores [2]. Moreover, these experiments served to illustrate the exquisite selectivity of the delivery system, as the spores used were derived from *Clostridium tetani*. Thus, only tumor-bearing mice were found to succumb to tetanus poisoning some 48 hours after injection, a consequence of the elaboration of tetanus toxin by vegetative cells growing in the hypoxic center of tumors. This was in stark contrast to nontumor-bearing mice. These remained entirely unaffected by the same spore treatment, as the bacterium was incapable of establishing itself in the absence of appropriate anaerobic conditions.

Following these original observations, over many years, the potential of the system was analyzed for treating solid tumors in both animals and humans, including clinical trials experiments (see Minton et al. [3] for review). Overall, these studies showed that clostridial spore treatment frequently led to the destruction of large parts of the tumor ("oncolysis"). However, an outer viable rim invariably remained, from which tumor regrowth frequently occurred. Despite this, it was clear that the ability of clostridial spores to germinate in the hypoxic regions of solid tumors represented a highly specific delivery system. Accordingly, the advent of gene transfer systems (see below) has rekindled interest in the use of clostridial spores in cancer therapy [4]. This is because it is anticipated that these organisms can now be engineered to produce other therapeutic agents, which in combination with oncolysis, might overcome previous deficiencies and bring about effective tumor regression.

In the early years, the majority of studies had involved the use of a proteolytic clostridial strain specifically isolated for this purpose by Möse and co-workers (designated *Clostridium butyricum* M-55, later *Clostridium oncolyticum,* and now reclassified as *Clostridium sporogenes* ATCC 13732), which they had demonstrated to be benign by injecting themselves [3]. When interest in using clostridial spores was revived in the early 1990s, a major driver was to improve the efficacy of the strains utilized through their endowment with genes encoding therapeutic proteins (see next). Initial attempts to introduce plasmid into strain M-55 were, however, unsuccessful. As a consequence, the effectiveness of genetically amenable saccharolytic strains, such as *Clostridium acetobutylicum* and *Clostridium beijerinckii,* at colonizing tumors was reassessed [5]. Thus, EMT6 tumor-bearing mice were injected intravenously with 10^8 spores (0.1 ml of a saline suspension containing 10^9 spores per ml) of *C. beijerinckii* NCIMB 8052 (formerly *C. acetobutylicum*). High numbers of vegetative rods in tumor tissue were shown to be present some 24 h later [5]. Similar findings were obtained in subsequent studies with four other saccharolytic clostridial strains [6]. In this case, however, *C. acetobutylicum* was found to more effectively colonize tumors than the *C. beijerinckii* strain (ATCC 17778) that was tested. In all of these studies, histochemical staining confirmed that vegetative bacterial cells were confined to the hypoxic and acellular necrotic areas of the tumor, and were not present in the oxygenated regions of the tumor or in other tissues, such as the heart, kidney, liver, lung, and spleen.

It was apparent, however, that the saccharolytic clostridia employed failed to colonize tumors to the levels of population density achieved using *C. sporogenes* M-55, e.g., the bacterial cell numbers in tumors colonized by *C. beijerinckii* NCIMB 8052 were two orders of magnitude lower than comparable tumors colonized by strain M-55 [7]. Moreover, the benefits of a more aggressive colonizer have subsequently been shown in a study that utilized a nontoxinogenic, proteolytic strain of *C. novyi* [8].

40.3 *CLOSTRIDIUM* CLONING VECTORS

For many years, the lack of suitable molecular DNA techniques for *Clostridium* spp. hampered the genetic analyses and rational biotechnological exploitation of this diverse group of microorganisms. This bottleneck was removed during the 1980s, when the first host vector systems were developed in order to unravel the toxicity of pathogenic clostridia such as *C. perfringens* and to enhance the exploitation of industrially important solventogenic *Clostridium* spp., including *C. acetobutylicum.* The successful deployment of clostridia in cancer therapy is also reliant on the availability of efficient cloning tools in order to stably introduce heterologous genes into the clostridial host cell. Since only the nonpathogenic clostridial species are suitable for use in tumor therapy, developments made in pathogenic species will not be discussed in this chapter.

The shuttle vectors used thus far to transform *Clostridium* spp., are based on several different plasmid replication regions, derived either from plasmids isolated from *Clostridium* spp. or from broad host-range plasmids commonly used in other Gram-positive bacteria. For example, the minimal replicon of pCB102 originating from *C. butyricum* [9] forms the basis of a number of frequently used *E. coli-Clostridium* shuttle vectors, which can be stably maintained in several *Clostridium* species. Other clostridial shuttle vectors developed for solventogenic clostridia, are pCP1 from *C. paraputrificum* [10], pCS86 from *C. acetobutylicum* strain no. 86 [11], and pDM11 from *C. acetobutylicum* NCIB 6443 [12]. Nonclostridial replicons are derived mainly from *Enterococcus faecalis* pAMβ1 [13] and *Bacillus subtilis* pIM13 [14]. The majority of shuttle vectors in use also contain a Gram-negative replication origin. This provides the facility to undertake necessary cloning steps in *E. coli,* a more effective and efficient cloning host. This replicon is invariably that of the plasmid ColE1.

While PEG-mediated protoplast transformation was initially employed to introduce these shuttle vectors into clostridia [e.g.,15], this method has now been superseded by the development of procedures based on electroporation. This latter technology avoids the tedious task of protoplast production and the difficulties encountered with cell wall regeneration implicit in PEG-mediated protoplast transformation. A number of electrotransformation protocols have been devised for the different *Clostridium* strains under investigation by the scientific community [7,16,17,18].

In addition to transformation procedures, shuttle vectors can also be transferred into the clostridial host by conjugation. This was achieved by using a conjugative, broad host-range enterococcal plasmid, in particular pAMβ1 [13], or by incorporating into already existing shuttle vectors the *cis*-acting origin of conjugative DNA transfer (*oriT*) from the extremely broad host-range Gram-negative IncP plasmid, RK2 [19]. The other conjugation functions are provided *in trans* by the donor cell.

The erythromycin resistance gene (*erm*) from pAMβ1 is the most widely used selectable marker to screen *Clostridium* transformants. Other useful markers are the kanamycin resistance (*aphA-3*) gene of Tn*1545*, tetracycline resistance genes from clostridial origin (*tetP*) or Tn*916* (*tetM*), and chloramphenicol acetyl transferase (*cat*) gene (in combination with thiamphenicol — not chloramphenicol — as selective compound) [19,20,21]. The dramatic increase of multidrug resistant bacteria in both community and hospital settings, however, banishes the use of antibiotic resistance markers in clinical applications. Therefore, new strategies for introducing heterologous genes into clostridia that avoid the use of selectable markers are under development. One of the options is the use of integrative phage-based vectors, in which heterologous genes are stably replicated with the bacterial genome.

Studies on segregational and structural stability of distinct shuttle vectors revealed that most vectors are rather stably maintained in *Clostridium*. It was previously suggested that pCB101-based vectors would be more susceptible to structural instabilities because of the rolling circle replication mechanism, resulting in the presence of significant quantities of recombination-sensitive ssDNA [22]. However, this seemed not to be the case [19,23]. On the other hand, pAMβ1-based plasmids (e.g., pMTL500E) that do not generate measurable ssDNA, could be stably maintained within *C. beijerinckii* [23] but appeared less stable in *C. acetobutylicum* [10,24], indicating the involvement of strain-specific factors. Segregational — but not structural — instability of pIM13-based vectors was observed within *C. acetobutylicum* NI-4081, which could be avoided in the mutant strain NI-4082 [25]. The mutation, which appeared to be specific for pIM13 replication, was related to the multimerization of the pIM13 plasmids within the respective strains. Other strain-specific elements shown to be major barriers to efficiently transform *Clostridium* spp. are host restriction/modification systems. For example, the restriction endonuclease *Cac*824I present in *C. acetobutylicum* ATCC 824 inhibited the use of *E. coli/Clostridium* shuttle vectors in this strain due to the large number of *Cac*824I sites in the *E. coli* plasmid portion. However, by means of simultaneous vector replication and expression of the *Bacillus subtilis* phage ϕ3tI methyltransferase cloned on a compatible plasmid, the shuttle vector was *in vivo* methylated in an *E. coli* host and could, as a consequence, be introduced into *C. acetobutylicum* ATCC 824 without problems [26].

For other industrially important bacteria, bacteriophage DNA has often been used as base for vector constructs. In spite of the availability of a variety of bacteriophages isolated from different *Clostridium* species [27,28], only the filamentous virus-like particle CAK1 isolated from *C. beijerinckii* NCIMB 6444 was further developed to phage-based vector systems, including phagemids pCAK1 and pYL102E [29,30]. The latter was evaluated by expressing the heterologous *Thermoanaerobacterium polysaccharolyticum manA* gene and the *C. acetobutylicum* DSM 792 *sol* operon [30].

40.4 REGULATORY SEQUENCES USED IN EXPRESSION/SECRETION SYSTEMS

Renewed interest in clostridia as biomass converters and solvent-producing cell factories has led to the genetic characterization of their biochemical pathways. Consequently, regulatory sequences promoting specific gene expression and protein secretion, together with the development of new genetic tools to discover alternative regulatory sequences, have become available.

40.4.1 PROMOTERS

Knowledge of the structural features of clostridial promoters and other regulatory sequences for gene expression has consistently lagged behind their more illustrious aerobic counterparts, *Bacillus*

	-35		-10		Ref.
P*eglA*	AAAAA TTATTA	ATGTAAAAATATACTAAG	TATAGA	ATATTTA	[35]
P*closI*	AATAA TGTAAA	ACTTTAAATAATAACTCT	TATAAT	GGTTTTT	[36]
P*fd*	CACTT TTAAAA	AGTTTAAAAACATGA	TACAAT	AAGTTAT	[37]
P*glnA*-1	GTTAT TTTTTA	TATTTATCAATTTGATAAA	TATAAG	CTTTAAT	[39]
P*glnA*-2	AAAAT TAAAGT	TTTTACCACAAAAATACTTAC	TTTTTT	AAAAAAT	[39]
P*glnA*-3	ATGAA TTGAGC	AAAGGCGTTCAATAGATGGTA	TACCGC	TTCTAT	[39]
P*glnA*-4	AAAAA TCTATT	GATTTGTTGAAACGG	TGGTAT	ATAATT	[39]
Consensus clostridia	**TTGA**C**A**	(17 bp)	**TA**TA**AT**		[10]

FIGURE 40.1 Putative promoter sequences of the β-galactosidase (*eglA*), the clostripain (*closI*), the ferredoxin (*fd*), and the glutamine synthetase (*glnA*) genes used within clostridial expression vectors. They are compared to the consensus promoter sequence recognized by the vegetative RNA polymerase holoenzyme. The –35 and –10 regions are underlined; boldface characters within the consensus sequence indicate the highly conserved nucleotides.

spp. The recent availability of genome sequence of various *Clostridium* spp., including *C. acetobutylicum* ATCC 824 [31], is now redressing this balance.

Like other Gram-positive bacteria, *Clostridium* species encode a number of σ factors (17 predicted σ factors within *C. acetobutylicum*) differently expressed during different growth phases and changing environmental conditions. Alternative σ factors change promoter recognition by RNA polymerases and, as such, allow coordinate transcription of different gene sets for a variety of bacterial responses. Furthermore, via *in silico* analyses of the *C. acetobutylicum* genome sequence [31], numerous predicted specific transcriptional regulators have been found (e.g., members of the AcrR/TetR, the LysR and the LacI family). Notwithstanding these *in silico* data, only a subset of promoters has been analyzed in detail, in particular the promoters involved in solventogenic processes, biomass-converting pathways, and heat shock response [32,33,34]. Although these promoters are commonly recognized by the vegetative σ factor containing RNA polymerase holoenzyme, their activity is not constitutive but controlled by regulatory proteins, activators, or repressors. For that reason, most of the characterized promoters are not suitable for the application in anti-tumor therapy. As a consequence, only a limited number of promoters have been utilized to promote therapeutic protein production in nonpathogenic *Clostridium*, namely the promoters preceding the encoding regions of *C. acetobutylicum* β-1,4-endoglucanase [35], *C. histolyticum* clostripain [36], and *C. pasteurianum* ferredoxin [37]. These promoters all resemble the promoter consensus sequence (Figure 40.1) recognized by the major form of *Clostridium* RNA polymerase, which shows homology with those for other bacteria [10].

In a further search for additional promoters that efficiently promote expression of the envisaged therapeutic genes, versatile reporter systems initially developed to study promoter regulation in clostridia have been and will be valuable tools. These systems contain a promoterless reporter gene of which the expression can easily be quantified. Reporter systems constructed for *C. acetobutylicum* ATCC 824 and *C. beijerinckii* NCIMB 8052 carry the promoterless β-galactosidase (*lacZ*) gene of *Thermoanaerobacterium thermosulfurogenes* EM1 (formerly *Clostridium thermosulfurogenes* EM1) and the promoterless β-1,4-endoglucanase (*eglA*) gene from *C. acetobutylicum* P262, respectively [38,39]. Using the latter system, it was demonstrated that the promoter region of the *C. beijerinckii* glutamine synthetase (*glnA*) gene directed an 11-fold higher β-1,4-endoglucanase activity than the native *eglA* promoter [39]. Further analysis of the *glnA* promoter region revealed 4 separate promoters (Figure 40.1). This observation resulted in the selection and current evaluation of the *glnA* promoter region for heterologous gene expression in the clostridial transfer system.

40.4.2 Radiation-Induced Promoters

In the approach to combat tumor growth with recombinant clostridia, a further specificity was sought. Besides the spatial control realized by the tumor-related hypoxia/necrosis, temporal control might be desirable. As such, the therapeutic gene would only be expressed both in a specific region and under defined conditions. The latter could be achieved by placing the heterologous gene under the control of a radiation-induced promoter. This would ensure its expression by the clostridial host solely in the irradiated tumor tissue. Consequently, the specificity of the envisaged *Clostridium*-mediated transfer system would be further increased, avoiding unwanted gene expression in non-tumor hypoxic environments like abscesses or infarcted tissues. An additional advantage of this combination is that the anti-tumor effect of radiotherapy impaired in hypoxic conditions could be compensated by a therapeutic or radiation-sensitizing agent expressed within this radioresistant hypoxic/necrotic tumor region.

Radiation-mediated gene activation or "genetic radiotherapy" has already been investigated in several experimental gene therapy strategies using radiation-inducible promoters/enhancers from eukaryotic origin, particularly the *egr1* promoter/enhancer region [40,41,42]. Since these elements are obviously not functional in clostridia, the existence of radiation-induced genes in *Clostridium* was examined.

Bacteria are endowed with several regulatory systems that are differentially activated in response to a variety of DNA-damaging agents, those of *Escherichia coli* being the most intensively studied [43,44]. The *E. coli* SOS regulon includes more than 30 unlinked genes and is regulated by the products of the *recA* and *lexA* genes, which also belong to this regulon. The LexA protein is a negatively acting transcriptional regulator and specifically binds to a consensus sequence called, in *E. coli* and other Gram-negative bacteria, the SOS box [45]. In Gram-positive bacteria (e.g., *Bacillus subtilis* and *Mycobacterium* spp.), the binding site for LexA, designated Cheo box [46], consists of a region of 12 bp with dyad symmetry (5′-GAAC-N_4-GTTC-3′) but without homology to the Gram-negative SOS box. When the DNA is not affected, LexA (DinR in Gram-positives) represses the SOS repair mechanism by binding the SOS or Cheo box. However, upon DNA damage, the RecA protein acquires an active conformation and binds single-stranded DNA regions produced by DNA damage-mediated inhibition of replication, through which the protein gains apoprotease activity. Because of this activation, RecA promotes the autocatalytic cleavage of LexA, which triggers the induction of the SOS genes. After repair of DNA, the single-stranded regions disappear and the RecA protein is no longer activated, resulting in an increase in the LexA repressor level, which in turn inhibits SOS gene expression [43]. In this context, the *C. acetobutylicum* counterparts of genes involved in the DNA repair system were isolated to assess their radiation inducibility.

Northern blot hybridization experiments showed that *C. acetobutylicum* DSM 792 *recA* as well as *recN* were radiation-induced at a dose of 2 Gy [47]. Since this dose is the standard dose per fraction used in classical radiation treatment schedules, promoters induced at this dose level look promising for further clinical application.

Following the location of *recA* and *recN* on the *C. acetobutylicum* genome sequence [31], the preceding DNA encompassing the promoter regions could be isolated using PCR techniques. Analysis of these nucleotide sequences in more detail (Figure 40.2) revealed that both promoters were accompanied by a Cheo box consensus sequence. The isolated *C. acetobutylicum recA* and *recN* promoters were then tested for their radiation inducibility by cloning the promoter regions in a reporter system carrying the promoterless *lacZ* gene of *T. thermosulfurogenes* EM1 (see above). Recombinant clostridia containing the respective reporter plasmids were irradiated with different doses. Culture samples taken at different time intervals were subsequently analyzed for β-galactosidase activity. Irradiation of cultures with a 2-Gy dose resulted in an increase of enzyme activity, but the observed induction was significant only for the *recA* promoter (Figure 40.3). This can be explained by the 50- to 100-fold lower activity of the *recN* promoter compared to the *recA* promoter, of which β-galactosidase activity was barely detectable [47].

		Ref.
PrecA	TATATTGACAAATTAATAAATTACTATATAATTATATGTATA**GAA**CAAAT**GTTC**GAGAGA -35 -10	[48]
PrecN	TAGAT**GAA**TAGCAG**GTTC**AATAGCAGGCGATAACACTATTTTTATACTTTCTAGAAGTGA -35 -10	

FIGURE 40.2 Promoter regions of *C. acetobutylicum* DSM 792 *recA* and *recN* genes with indication of the putative –35/–10 regions (underlined) and of the accompanying Cheo box (boxed). The conserved bases of the Cheo consensus sequence are in boldface.

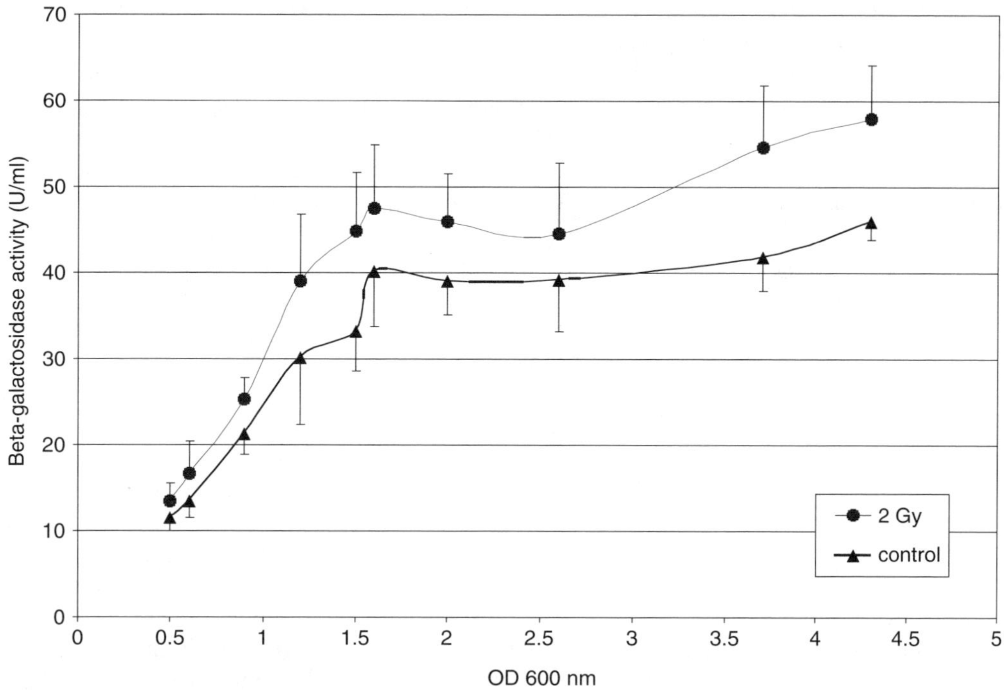

FIGURE 40.3 β-galactosidase activity of irradiated *versus* nonirradiated *C. acetobutylicum* DSM 792 containing pIMPrecAlacZ in function of growth. Results are the mean of four independent experiments. Vertical bars represent the standard errors of the mean. (From Nuyts et al. [47], with permission.)

By deleting the 12-bp Cheo box from the *recA* promoter region, it was demonstrated that only this genetic element is responsible for the radiation responsiveness of the promoter [48]. Exploiting the knowledge that the Cheo box can function outside its native sequence environment, any constitutive promoter might be converted to a radiation-responsive element. Proof of principle was delivered by introducing a Cheo box consensus sequence in the proximity of the strong, constitutive *eglA* promoter. Recombinant clostridia containing this mutated promoter produced up to 2.4 times more heterologous protein (i.e., mTNFα) following irradiation with a dose of 2 Gy, compared to the nonirradiated conditions. In contrast, the wild-type promoter resulted in the production of equal amounts of heterologous protein in irradiated and nonirradiated samples (see also Section 40.5.2). Furthermore, insertion of an extra Cheo box in the *recA* or of one Cheo box in the *eglA* promoter region decreased the basal activity of the tested promoters because of the higher repression levels under noninducing conditions [48]. It is, therefore, reasonable to assume that the addition of two or more Cheo boxes might further decrease basal activity of the promoter, thus providing the desired conditional control of therapeutic gene expression in tumors.

40.4.3 Signal Sequences

In most instances, it is desirable that the clostridial cells that colonize the tumor secrete their recombinant payload (cytokines or prodrug-converting enzymes) specifically into the local tumor microenvironment. This can be achieved when the gene of interest is preceded by a signal sequence upstream and in frame with the ORF encoding the therapeutic protein. The signal sequence encodes a signal peptide, which is an NH_2-terminal extension of secretory proteins, required for the translocation-competent state of the precursor, its targeting to the membrane, and the initiation of translocation across the membrane. The membrane-associated signal peptide is cleaved off by the signal peptidase during, or shortly after, translocation.

In *Clostridium,* being Gram-positive, proteins translocated across the cytoplasmic membrane are released directly in the extracellular environment. Consequently, when growing in a tumor, the secreted proteins are released into the tumor microenvironment. For that reason, when the extracellular presence of the therapeutic agent is required to obtain anti-tumor activity, the use of *Clostridium* is preferred compared to attenuated *Salmonella* Typhimurium, an alternative bacterial protein transfer system in anticancer therapy [49,50]. The presence of a periplasm in the latter Gram-negative strain makes extracellular protein secretion more complex.

As far as is known, clostridial signal peptides show the common features of a typical signal peptide with a positively charged n-region containing one or more basic amino acids, followed by a central hydrophobic h-region and a c-region, which comprises the signal peptidase cleavage site. In order to develop *C. acetobutylicum* to an efficient delivery system that expresses and secretes relevant proteins, expression/secretion vectors based on clostridial regulatory sequences were developed. In first instance, the signal sequences from the *C. acetobutylicum* P262 endo-$\beta_{1,4}$-glucanase (*eglA*) gene [35] and from the *C. histolyticum* clostripain gene [36] were chosen to support heterologous protein secretion.

The *eglA* gene of *C. acetobutylicum* P262 encodes a precursor of 448 amino acids starting with a 38-aa long signal peptide. This signal peptide consists of a relatively long n-region containing five positively charged Lys residues followed by a core of hydrophobic amino acids and, finally, a stretch of more polar amino acid residues at its C-terminal end (Figure 40.4). As often found in signal peptides, a β-turn-promoting residue (i.e., Gly) is present at the border of the h- and c-region.

Clostripain is a heterodimeric cysteine endopeptidase originally isolated from *C. histolyticum* culture filtrates. The putative signal peptide (Figure 40.4) spans 27 amino acids and contains 3 positively charged residues at position +3 to +5, a central hydrophobic core segment, and a c-region with the signal peptidase recognition site. Signal peptidase cleavage occurs presumably between Ala_{27} and Asn_{28}. Since clostripain expression under control of its own regulatory elements yielded high amounts of secreted active protease in the Gram-positive host cell *Bacillus subtilis* [51], the use of the signal peptide appeared to be promising for heterologous protein secretion in *Clostridium.*

```
                                                                                  Ref.
SP-EglA          MFSKIKKINFFKKTFS  FLIAVVMMLFTVL  GTNTYKAEA ↓ ATT                [35]
                     + + +    + +

SP-Clostripain           MLRRKVST  LLMTALITTSFL   NSKPVYA   ↓ NPV                [36]
                           + + +
                 ________________  _____________  _________   _____
                         n-region       h-region   c-region   mature protein
```

FIGURE 40.4 Amino acid sequence of EglA and clostripain signal peptides (SP) used to promote heterologous protein secretion in *Clostridium.* The characteristic n-, h-, and c-regions are indicated (+, positively charged residue within the n-region; ↓, signal peptidase cleavage site).

40.5 PROTEINS OF INTEREST FOR TUMOR THERAPY

Proteins currently considered to have an anti-tumor effect when delivered to the hypoxic/necrotic tumor region by bacterial delivery systems belong to two major groups of proteins: prodrug-converting enzymes and cytotoxic proteins. The first attempt to introduce a therapeutic gene in *Clostridium* was reported by Schlechte and Elbe [52]. They transformed the oncolytic *Clostridium butyricum* M-55 (*Clostridium sporogenes* ATCC 13732) with an *E. coli*-derived gene coding for Colicin 3, a bacteriocin with cancerostatic properties. The data with regard to the recombinant M-55 strain, however, were not convincing.

Below, we will briefly review some results obtained with prokaryotic and eukaryotic proteins expressed in *Clostridium*, which might be potentially useful tools in the *Clostridium*-mediated tumor treatment.

40.5.1 Prodrug-Converting Enzymes

The concept of directed enzyme prodrug therapy (DEPT) was first described by Bagshawe [53]. By targeting prodrug-converting enzymes specifically to the tumor, the administration of relevant nontoxic prodrugs leads to the production of active drugs only at the tumor site, while sparing normal tissue from damage. In addition, through a "bystander effect" neighboring tumor cells will also be killed by the activated prodrug. The bystander phenomenon can be defined as an extension of the killing effects of the active drug to untransfected neighboring cells. This implies that even if only 5 to 10% of the target cells are genetically modified and express the therapeutic gene, tumor eradication might still be achieved. Since the first report on DEPT, several vector types have been used to target the enzyme to the tumor, including viruses (VDEPT), and monoclonal or recombinant antibodies (ADEPT). More recently, polymers have been employed (PDEPT) in a novel two-step anti-tumor approach that uses a combination of a polymeric prodrug and polymer-enzyme conjugate to generate a cytotoxic drug rapidly and selectively at the tumor site [54]. In gene-directed enzyme prodrug therapy (GDEPT), the enzyme is targeted to tumor cells using viral or nonviral vectors, followed by integration of the enzyme-encoding sequence in the genome of the tumor cells. Since recombinant DNA techniques became available for *Clostridium*, this organism was also evaluated as vector in directed enzyme prodrug therapy (CDEPT) [4].

Out of several enzyme/prodrug combinations proposed for cancer therapy [55], the following three have been tested so far using the CDEPT approach:

1. Cytosine deaminase (CDase), encoded by the *E. coli codA* gene, converts the nontoxic 5-fluorocytosine to the toxic 5-fluorouracil, which can then be converted to either 5-fluorouridine triphosphate or 5-fluorodeoxyuridine monophosphate. These highly reactive products inhibit RNA-mediated protein synthesis and DNA synthesis, respectively [56].
2. Bacterial nitroreductase activates CB 1954 (5-aziridino-2,4-dinitrobenzamide) to a bifunctional alkylating agent, which causes DNA interstrand cross-links [57].
3. *Pseudomonas*-derived carboxypeptidase G2 hydrolyzes aromatic N-substituted glutamates with formation of benzoic acid, phenol, and aniline mustards. The benefits of the latter released toxins are the cell phase-independent cytotoxicity and the one-step activation, avoiding the potential problem of rate limitation in subsequent metabolism of the product [58].

As already mentioned, improved tumor control might be attained by the combination with alternative treatment modalities. This might also be the case for CDEPT since, for example, combination of radiotherapy with 5-FU produced from 5-FC by CDase might result in beneficial anti-tumor effects. The CDase/5-FC molecular chemotherapy has been shown to enhance the effects of radiotherapy in animal models of cholangiocarcinoma [59].

In 1996, Fox et al. [23] reported on *C. beijerinckii* transformants producing significant levels of active CDase. Upon transformation of *C. beijerinckii* with the *E. coli codA* gene, which was cloned in a pCB102-based shuttle vector under control of the constitutive *C. pasteurianum* ferrodoxin promoter, an enzymatic activity of over 3000 pmol of 5-FC converted to 5-FU per 10^9 cells per min could be measured within cell lysate, as well as in culture filtrates [23]. Addition of supernatant from transformed bacteria to EMT6 carcinoma cells resulted in a 500-fold enhancement of 5-FC cytotoxicity, which is comparable to the increase of toxicity described in other studies in which the *codA* gene was transfected into mammalian cells. In a similar way, *C. beijerinckii* was transformed with the *E. coli* nitroreductase gene. Supernatants of the nitroreductase-producing *C. beijerinckii* strain increased the EMT6 tumor cell killing by conversion of the prodrug CB 1954 22-fold, compared to media taken from nonrecombinant clostridia cultures [5]. Very recently, Liu et al. [7] described the successful transformation of *C. sporogenes* with a recombinant plasmid promoting intracellular production of CDase. Adding cell extract of this recombinant strain to SCCVII cancer cells increased the cytotoxicity of 5-FC by a factor 10^4.

The fact that in some of these experiments cytotoxic activity was measured in the supernatant of the transformed clostridial strains, even in the absence of a signal sequence, indicates that cell lysis occurred, thereby releasing the recombinant protein into the medium. An option to ensure that the therapeutic protein would be released in the tumor microenvironment without the necessity for bacterial cell lysis is to fuse a signal sequence at the 5′ end of the therapeutic gene. For this purpose, plasmid pKNT19closcodA was constructed and introduced in *C. acetobutylicum*. This shuttle vector contains the *codA* gene placed under control of the clostripain *closI* promoter and fused to the *closI* signal sequence for the secretion of CDase [60]. After introduction of pKNT19closcodA into *C. acetobutylicum*, CDase production was monitored. The 52-kDa CDase protein could be detected via Western blotting both in culture filtrates and cell lysates prepared from early-logarithmic growth phase samples, thereby confirming the functionality of the *closI* expression and secretion signals in the recombinant strain. In cell lysates of *C. acetobutylicum* [pKNT19closcodA], maximum activity was reached in the stationary growth phase (OD_{600nm}~1.5), and this activity level was prolonged during the 20-h follow-up period. The CDase activity yielded about 1000 pmol 5-FC converted to 5-FU/min/ml cell lysate (derived from circa 10^8 cells). In supernatants of the corresponding culture, the maximum activity (i.e., circa 700 pmol 5-FC converted to 5-FU/min/ml supernatant) was reached in the early-log growth phase. At later growth stages, a decrease in extracellular CDase activity was observed. This might be caused by the acidification of the medium resulting in denaturation of CDase or by the proteolytic breakdown of the enzyme as a consequence of proteases secreted by the host strain [60].

Besides CDase and nitroreductase, the prodrug-converting enzyme carboxypeptidase G2 derived from *Pseudomonas* was envisaged to be secreted by *Clostridium* [4]. However, overproduction of this enzyme could not be achieved in the clostridial host. A possible reason for this failure might have been the high-GC content of the pseudomonad gene (70% G+C) causing DNA instability and inefficient transcription/translation in the low-GC background of the clostridial host.

40.5.2 Proteins with Cytotoxic Activity

The anti-tumor activity of a protein can result either from a direct cytotoxic effect of the specific protein or from an indirect effect. Some proteins in the tumor microenvironment can induce a cascade of reactions finally resulting in cell death. On the other hand, other proteins can enhance the efficacy of a chemotherapeutic drug or of radiotherapy, thereby improving the therapeutic outcome. A common problem seen with the systemic administration of therapeutic proteins causing tumor cell death is the toxicity limiting severely their clinical applications. To avoid systemic side effects, the local administration of therapeutic proteins or the local expression of their encoding sequences in the tumor microenvironment is, therefore, advantageous.

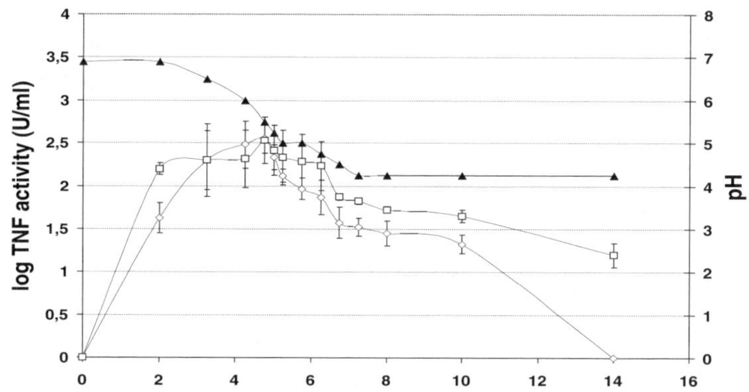

FIGURE 40.5 Amount of biologically active mTNFα in supernatant (□) and lysates (◇) of *C. acetobutylicum* DSM 792 containing pIMP1eglATNF and evolution of pH (▲) in function of growth. Error bars represent standard deviations. (From Theys et al. [24], with permission.)

Among several potential proteins reported to be effective in tumor treatment [61,62] in first instance tumor necrosis factor α (TNFα) was chosen for expression in *Clostridium* [24]. This homotrimeric compound composed of 17 kDa monomers is secreted by different mammalian cells in response to invasive stimuli (bacterial, viral, fungal, parasitic, or neoplastic). Although acknowledged as possessing great potential in cancer therapy, TNFα exhibits systemic toxicity at very low doses. It, therefore, represents an attractive therapeutic agent to be evaluated in the clostridial protein transfer approach. TNFα has a direct cytotoxic effect on some tumor cells while it is cytostatic on others, it can induce both necrotic and apoptotic forms of cell death, it can activate T cell-mediated immunity and it causes destruction of the tumor microvasculature, resulting in thrombosis and necrosis. Preclinical studies demonstrated the anti-tumor capacity of TNFα [63]. Moreover, TNFα have been shown to act synergistically or additively in combination with radiotherapy [40]. Additionally, synergistic anti-tumor effects between TNFα and 5-FU, and between TNFα and IL2 or IFNα-A/D, have been reported [64,65]. Recombinant clostridia producing a combination of therapeutic proteins with synergistic effects represents an attractive option.

With the aim to express and secrete mTNFα in *C. acetobutylicum*, mTNF cDNA was placed under the control of *eglA* or *closI* regulatory sequences and the resulting expression/secretion cassettes were cloned in versatile shuttle vectors [24]. Lysates and supernatants from cultures of recombinant *C. acetobutylicum* containing one of those vectors were analyzed by Western blotting. Cell lysates contained both the precursor and processed form of mTNFα, 21 and 17 kDa in size, respectively, while in supernatants only the processed form could be detected. Addition of lysates or supernatants of recombinant strains to WEHI164 clone 13 cells resulted in cytotoxicity. Biologically active mTNFα was produced at levels of up to $10^{5.5}$ U/ml lysate and circa 10^4 U/ml supernatant (Figure 40.5), representing concentrations likely to bring about anti-tumor effects on sensitive cells. Under conditions used for *in vitro* growth of the recombinant clostridia, a decrease in mTNF activity as a function of growth phase was observed, which might be explained by mTNF instability at pH values below 5.5.

mTNFα expression and secretion in recombinant *C. acetobutylicum* strains containing the *eglA*/mTNFα fusion gene under control of the radiation-inducible *recA* in place of the constitutive *eglA* promoter proved the functionality of the *recA* promoter in these recombinants. Moreover, mTNFα expression was shown to be induced by irradiation of the *Clostridium* cultures with a 2-Gy dose. After radiation induction, mTNFα in supernatant yielded approximately 880 U/ml, which is lower than the values obtained using the *eglA* promoter, but probably still sufficiently high to result in an *in vivo* cytotoxic effect on human tumor cells [66,67]. The insertion of a second Cheo box (see Section 40.4.2) in the *recA* promoter region resulted in a decrease of the basal *recA* promoter activity. Upon 2 Gy irradiation, secreted mTNFα concentrations of about 200 U/ml culture supernatant were measured [48]. Further investigations on the radiation inducibility of heterologous gene expression in *Clostridium* are currently ongoing in order to develop a clostridial expression system in which minimum basal gene expression is combined with maximum radiation-induced production of the protein of interest.

In conclusion, these results clearly demonstrate that genetically engineered *Clostridium* strains can express and secrete therapeutic proteins of interest from both prokaryotic and eukaryotic origin.

40.6 *IN VIVO* EVALUATION OF RECOMBINANT CLOSTRIDIA EXPRESSING PROTEINS USEFUL IN TUMOR THERAPY

The ultimate steps in the evaluation of the *Clostridium*-mediated protein transfer approach are to assess the effectiveness of the constructed recombinant strains to deliver the heterologous proteins specifically to the tumor site and to evaluate their efficacy in tumor control.

In this respect, the first results published described that after injection of spores of recombinant *C. beijerinckii* expressing the *E. coli* nitroreductase into tumor-bearing mice, this protein was detected solely in the tumors. This was evidenced by Western blot analysis of tumor and normal tissue homogenates prepared up to 5 days after administration of recombinant spores to EMT6 tumor-bearing mice [5]. More recently, tumor-specific expression of CDase was demonstrated in rat tumors following intravenous injection of genetically engineered spores of *C. acetobutylicum* [60]. Upon administration of recombinant *C. acetobutylicum* expressing *codA*, CDase activity was detected in 50 to 70% of the tested tumor homogenates, while neither control homogenates nor normal tissues (i.e., liver, spleen) showed CDase activity. Moreover, therapeutic dose intensities can be increased using a vascular-targeting agent, Combretastatin A-4 phosphate [60]. Combretastatin A-4 phosphate (CombreAp; OXiGENE, Lund, Sweden; presently: Watertown, MA) induces an acute vascular shutdown and necrosis within 1 to 3 days, without obvious side effects [68,69]. As demonstrated [70], these effects result in improved tumor colonization by the *Clostridium* strains. By treatment of tumor-bearing rats with CDase-producing clostridia and simultaneously with nontoxic doses (25 mg/kg) of CombreAp, CDase activity was not only detected in 100% of the analyzed tumors, but conversion efficiency of 5-FC to 5-FU was also 4-fold increased in tumor homogenates from animals concomitantly treated with CombreAp compared to tumor samples taken from animals solely receiving recombinant clostridia [60].

Other experiments demonstrated tumor-specific expression of mTNFα upon injection of recombinant mTNFα-producing clostridia to tumor-bearing rats. ELISA measurements revealed significantly higher mTNFα levels in tumor homogenates from animals treated with recombinant clostridia as compared to homogenates from animals injected with nonrecombinant clostridial spores or from sham-treated animals [Theys, unpublished data]. The slight difference in mTNFα concentrations measured in these two control groups also indicated that injection of recombinant clostridia only resulted in a very moderate inflammatory reaction. Taken together, these studies have established the proof of principle that potentially any heterologous (therapeutic) protein of choice can be expressed exclusively in the tumor microenvironment following systemic administration of recombinant clostridial spores.

Of course, determination of the productivity of recombinant bacteria expressing anticancer therapeutics *in vivo* is only a first step to the evaluation of an anti-tumor response following their administration. In that context, a pilot study was performed to determine potential anti-tumor effects using CDase-producing *C. acetobutylicum* in nu/nu mice with subcutaneously transplanted HT29 tumors [Theys, unpublished data]. The data obtained showed there was a tumor growth delay within the follow-up period, especially in the animals bearing larger tumors (and, hence, more hypoxia/necrosis) at the start of the treatment. This anti-tumor effect, observed in the absence of an adjuvant therapy, is highly encouraging, since many chemotherapeutics administered at concentrations that are useful in combination with other therapies often exhibit little or no effect when administered alone. Nevertheless, the levels of 5-FU produced should at least be sufficient for significant improvement in combination with radiotherapy. In a theoretical study, it was concluded that sufficient 5-FU should be available if conversion efficiency from 5-FC is as low as 1 to 3%. Sensitization enhancement ratios of 1.1 to 1.2 should be achievable with daily 2-Gy fractions [71]. Based on the recent studies, it is believed that this radiation-sensitization enhancement is attainable by means of the *Clostridium*-mediated transfer of CDase to the tumor.

The expected anti-tumor effect, however, could not be observed *in vivo* by using genetically engineered *C. beijerinckii,* producing CDase or nitroreductase upon systemic injection of the relevant prodrug. This lack of measurable effect on tumor growth was explained by the insufficient amounts of therapeutic proteins correlated to the relatively low number of metabolically active recombinant clostridia in the tumor [7]. This suggested that the colonization capacity of the *Clostridium* strain used is critical.

In this respect, *Clostridium sporogenes* has superior colonization capacities compared to other clostridia used for tumor colonization. Whereas *C. sporogenes* produces up to 2×10^8 bacteria per gram of tumor, *C. beijerinckii*, for example, only produces up to 10^6 colony forming units/g. Consequently, *C. sporogenes* has been suggested to be an excellent host in the *Clostridium*-mediated tumor therapy approach. However, this promising host remained, for a long time, nontransformable. A breakthrough has been made with the recently reported, successful transformation of this strain [7]. After injection of tumor-bearing mice with spores of recombinant *C. sporogenes* producing *E. coli* CDase, systemic delivery of the prodrug 5-FC into these mice resulted in greater tumor growth delay than could be achieved by the maximum tolerated doses of the 5-FU drug.

In conclusion, the results of *in vivo* experiments support the expectations about the *Clostridium*-mediated protein transfer strategy to solid tumors. Because of the absence of toxicity, its highly selective growth in tumors, and its potential beneficial effect when combined with conventional anticancer therapies, this bacterial protein transfer system opens new perspectives for future tumor therapy.

40.7 CONCLUSIONS AND PERSPECTIVES

The advent of recombinant DNA techniques for clostridia has provided the means to engineer selected species to be able to express and secrete a variety of proteins of prokaryotic as well as of eukaryotic origin. Although such production strains could be of interest for different purposes, a principal focus has been for the purposes of delivering therapeutic proteins to the tumor environment.

New developments in anticancer therapies are important because of the limitations of the conventional therapies, including surgical removal and radio- and chemotherapy. Therefore, considerable effort has been directed toward gene therapy approaches, which aim to introduce suicide genes or genes that restore the action of tumor suppressor proteins in tumor cells or that have an immunomodulatory action. Besides classical vector systems based on different viruses and nonviral systems such as liposomes and molecular conjugates, within the past decade, bacteria including *Clostridium*, *Bifidobacterium,* and attenuated *Salmonella* Typhimurium have been more intensively investigated as potential vectors. Bacterial vectors offer several advantages compared to other delivery vectors. These include their high tumor specificity, removal of the need to bring about tumor cell transduction, and the easy removal of bacteria upon administration of appropriate antibiotics — a critical issue in those cases where foreign DNA is integrated within the tumor genome, i.e., as in VDEPT. Upon successful transformation of *Clostridium* with a therapeutic gene, it has been demonstrated that genetically engineered clostridia can specifically grow in the tumor microenvironment. Consequently, nonpathogenic clostridia represent a promising delivery system and might form an alternative for protein delivery to the hypoxic/necrotic region of solid tumors.

Despite promising preclinical results, clinical trials using recombinant clostridia have yet to be initiated. Data, to date, indicate that the administration of clostridial spores to humans is a fundamentally safe procedure. However, undeserved doubts over safety aspects are still an important obstacle to overcome. Another reason retarding the clinical application is the current use of antibiotic resistance markers in the recombinant constructs generated to date, which comes into conflict with the measures taken to combat the increasing multidrug resistance of bacterial pathogens. The development of integrative cloning vectors for clostridia will be helpful in this respect. Additionally, there is still scope for improvement in the yields of recombinant therapeutic protein produced, and the range of proteins to be expressed needs to be extended.

Yield improvement can be attempted at different levels, including transcription, translation, or secretion. Transcription can be manipulated, e.g., by using more efficient promoters. The availability of the complete genome sequence of different *Clostridium* strains and of suitable promoter probe vectors developed for *Clostridium* spp. will be helpful in this respect. The combination of promoter elements taken from different promoter regions or the optimization of the selected promoters by means of site-directed mutagenesis might further increase the heterologous gene transcription, possibly increasing the final protein yield. Improved translation can be obtained by the use of an optimized ribosome binding site, which positively influences mRNA translation and, thus, the final protein production. Furthermore, the low GC content of clostridial genomes — and, consequently, the strongly biased codon usage to AU-rich codons [10] — has to be considered. It may negatively influence mRNA translation and stability of heterologous mRNA that contains multiple minor codons. Adaptation of minor codons within the specific gene to the codon usage of the host could have a positive effect. Also at the secretion level, yield improvement can be expected when using different signal peptides. From other bacterial production systems, it is known that different signal peptides can have a distinct effect on the secretion efficiency of a particular protein, and so far only two signal sequences have been tested. Therefore, the identification of additional signal sequences, either based on the available *Clostridium* genome sequences, or experimentally identified, might be helpful.

Other factors that might influence the final production levels are, for instance, the presence of cytoplasmic or extracellular proteases leading to a proteolytic breakdown of the protein of interest. In that case, the construction of protease-deficient strains needs to be considered. Additionally, compared to some other bacterial host systems, *Clostridium* is hardly characterized regarding proteins involved in the secretion process, for example, chaperone proteins, folding catalysts, and the secretion machinery itself. A better insight into these processes and into the specific interaction between the relevant proteins will certainly open new perspectives to improve production levels of this bacterial host.

Alternative means to improve the efficacy of the *Clostridium*-mediated tumor therapy could rely upon the construction of recombinant strains that express and secrete two distinct proteins with synergistic anti-tumor action or on the combination of the clostridial gene delivery system with other tumor treatment modalities. Tumor colonization by clostridial spores was obviously improved in animal models that received simultaneously a vascular targeting compound. By affecting the neovasculature of tumors, relatively higher levels of necrosis might be induced within the tumor, creating better conditions for *Clostridium* growth. Finally, the combination of radiotherapy and the *Clostridium*-mediated gene delivery system in which the heterologous gene is placed under control of a radiation-inducible promoter holds promises for future applications by improving the tumor specificity of the therapeutic produced by the tumor-colonizing clostridia.

In conclusion, an extended arsenal still needs to be tested to fully realize the potential of the *Clostridium* delivery system. In this context, available animal tumor-bearing models will prove invaluable in the assessment of anti-tumor efficacy. It is anticipated that all of these improvements will enable the clinical application of recombinant clostridia in tumor therapy, extending the usefulness of this fascinating genus to humankind.

REFERENCES

1. Parker, R.C. et al., Effect of histolyticus infection and toxin on transplantable mouse tumours, *Proc. Soc. Exp. Biol. Med.,* 66, 46, 1947.
2. Malmgren, R.A. and Flanigan, C.C., Localization of the vegetative form of *Clostridium tetani* in mouse tumours following intravenous spore administration, *Cancer Res.,* 15, 473, 1955.
3. Minton, N.P. et al., Clostridia in cancer therapy, in *Clostridia — Biotechnology and Medical Applications,* Bahl, H. and Dürre, P., Eds., Wiley-VCH, Weinheim, Germany, 2001, 251.
4. Minton, N.P. et al., Chemotherapeutic tumour targeting using clostridial spores, *FEMS Microbiol. Rev.,* 17, 357, 1995.
5. Lemmon, M.J. et al., Anaerobic bacteria as a gene delivery system that is controlled by the tumor microenvironment, *Gene Ther.,* 4, 791, 1997.
6. Lambin, P. et al., Colonisation of the *Clostridium* in the body is restricted to hypoxic and necrotic areas of tumours, *Anaerobe,* 4, 183, 1998.
7. Liu, S.-C. et al., Anticancer efficacy of systemically delivered anaerobic bacteria as gene therapy vectors targeting tumor hypoxia/necrosis, *Gene Ther.,* 9, 291, 2002.
8. Dang, L.H. et al., Combination bacteriolytic therapy for the treatment of experimental tumors, *Proc. Natl. Acad. Sci. U.S.A.,* 98, 15155, 2001.
9. Minton, N.P. and Morris, J.G., Isolation and partial characterization of three cryptic plasmids from strains of *Clostridium butyricum, J. Gen. Microbiol.,* 127, 325, 1981.
10. Young, M., Minton, N.P., and Staudenbauer, W.L., Recent advances in the genetics of the clostridia, *FEMS Microbiol. Rev.,* 63, 301, 1989.
11. Yoshino, S. et al., Construction of shuttle vector plasmid between *Clostridium acetobutylicum* and *Escherichia coli, Agric. Biol. Chem.,* 54, 437, 1990.
12. Blaschek, H.P. and White, B.A., Genetic systems development in the clostridia, *FEMS Microbiol. Rev.,* 17, 349, 1995.
13. Oultram, J.D. and Young, M., Conjugal transfer of plasmid pAMβ1 from *Streptococcus lactis* and *Bacillus subtilis* to *Clostridium acetobutylicum, FEMS Microbiol. Lett.,* 27, 129, 1985.
14. Truffaut, N., Hubert, J., and Reysset, G., Construction of shuttle vectors useful for transforming *Clostridium acetobutylicum, FEMS Microbiol. Lett.,* 58, 15, 1989.
15. Reysset, G. et al., Transfection and transformation of *Clostridium acetobutylicum* strain NI-4081 protoplasts, *Biotechnol. Tech.,* 2, 199, 1988.
16. Birrer, G.A., Chesbro, W.R., and Zsigray, R.M., Electro-transformation of *Clostridium beijerinckii* NRRL B-592 with shuttle plasmid pHR106 and recombinant derivatives, *Appl. Microbiol. Biotechnol.,* 41, 32, 1994.
17. Nakotte, S. et al., Electroporation of, plasmid isolation from and plasmid conservation in *Clostridium acetobutylicum* DSM 792, *Appl. Microbiol. Biotechnol.,* 50, 564, 1998.
18. Tyurin, M. et al., Electrotransformation of *Clostridium acetobutylicum* ATCC 824 using high-voltage radio frequency modulated square pulses, *J. Appl. Microbiol.,* 88, 220, 2000.
19. Williams, R., Young, D.I., and Young, M., Conjugative plasmid transfer from *Escherichia coli* to *Clostridium acetobutylicum, J. Gen. Microbiol.,* 136, 819, 1990.
20. Green, E.M. and Bennett, G.N., Genetic manipulation of acid and solvent formation in *Clostridium acetobutylicum* ATCC 824, *Biotechnol. Bioeng.,* 58, 215, 1998.
21. Bertram, J. and Dürre, P., Conjugal transfer and expression of streptococcal transposons in *Clostridium acetobutylicum, Arch. Microbiol,* 151, 551, 1989.
22. Minton, N.P. and Oultram, J.D., Host:vector systems for gene cloning in *Clostridium, Microbiol. Sci.,* 10, 310, 1988.
23. Fox, M.E. et al., Anaerobic bacteria as a delivery system for cancer gene therapy: *in vitro* activation of 5-fluorocytosine by genetically engineered clostridia, *Gene Ther.,* 3, 173, 1996.
24. Theys, J. et al., Stable *Escherichia coli-Clostridium acetobutylicum* shuttle vector for secretion of murine tumor necrosis factor alpha, *Appl. Environ. Microbiol.,* 65, 4295, 1999.
25. Azeddoug, H., Hubert, J., and Reysset, G., Stable inheritance of shuttle vectors based on plasmid pIM13 in a mutant strain of *Clostridium acetobutylicum, J. Gen. Microbiol.,* 138, 1371, 1992.

26. Mermelstein, L.D. and Papoutsakis, E.T., In vivo methylation in *Escherichia coli* by the *Bacillus subtilis* phage ϕ3T I methyltransferase to protect plasmids from restriction upon transformation of *Clostridium acetobutylicum* ATCC 824, *Appl. Environ. Microbiol.*, 59, 1077, 1993.
27. Ogata, S. and Hongo, M., Bacteriophages of the genus *Clostridium*, *Adv. Appl. Microbiol.*, 25, 241, 1979.
28. Nieves, B.M., Gil, F., and Castillo, F.J., Growth inhibition activity and bacteriophage and bacteriocin-like particles associated with different species of *Clostridium*, *Can. J. Microbiol.*, 27, 216, 1981.
29. Kim, A.Y. and Blaschek, H.P., Construction and characterization of a phage-plasmid hybrid (phagemid), pCAK1, containing the replicative form of virus-like particle CAK1 isolated from *Clostridium acetobutylicum* NCIB 6444, *J. Bacteriol.*, 175, 3838, 1993.
30. Li, Y. and Blaschek, H.P., Molecular characterization and utilization of the CAK1 filamentous viruslike particle derived from *Clostridium beijerinckii*, *J. Ind. Microbiol. Biotechnol.*, 28, 118, 2002.
31. Nölling, J. et al., Genome sequence and comparative analysis of the solvent-producing bacterium *Clostridium acetobutylicum*, *J. Bacteriol.*, 183, 4823, 2001.
32. Dürre, P. et al., Solventogenic enzymes of *Clostridium acetobutylicum*: catalytic properties, genetic organization and transcriptional regulation, *FEMS Microbiol. Rev.*, 17, 251, 1995.
33. Han, S.O. et al., Transcription of *Clostridium cellulovorans* cellulosomal cellulose and hemicellulase genes, *J. Bacteriol.*, 185, 2520, 2003.
34. Bahl, H. et al., Expression of heat shock genes in *Clostridium acetobutylicum*, *FEMS Microbiol. Rev.*, 17, 341, 1995.
35. Zappe, H. et al., Structure of an endo-β-1,4-glucanase gene from *Clostridium acetobutylicum* P262 showing homology with endoglucanase genes from *Bacillus* spp., *Appl. Environ. Microbiol.*, 54, 1289, 1988.
36. Dargatz, H. et al., The heterodimeric protease clostripain from *Clostridium histolyticum* is encoded by a single gene, *Mol. Gen. Genet.*, 240, 140, 1993.
37. Meyer, J., Cloning and sequencing of the gene encoding the [2Fe-2S] ferredoxin from *Clostridium pasteurianum*, *Biochem. Biophys. Acta*, 1174, 108, 1993.
38. Tummala, S.B., Welker, N.E., and Papoutsakis, E.T., Development and characterization of a gene expression reporter system for *Clostridium acetobutylicum* ATCC 824, *Appl. Environ. Microbiol.*, 65, 3793, 1999.
39. Quixley, K.W.M. and Reid, S.J., Construction of a reporter gene vector for *Clostridium beijerinckii* using *Clostridium* endoglucanase gene, *J. Mol. Microbiol. Biotechnol.*, 2, 53, 2000.
40. Hallahan, D.E. et al., Spatial and temporal control of gene therapy using ionizing radiation, *Nature Med.*, 1, 786, 1995.
41. Seung, L.P. et al., Genetic radiotherapy overcomes tumor resistance to cytotoxic agents, *Cancer Res.*, 55, 5561, 1995.
42. Marples, B. et al., Development of synthetic promoters for radiation-mediated gene therapy, *Gene Ther.*, 7, 511, 2000.
43. Friedberg, E.C., Walker, G.C., and Siede, W. *DNA Repair and Mutagenesis*, ASM Press, Washington, DC, 1995.
44. Koch, W.H. and Woodgate, R., The SOS response, in *DNA Damage and Repair: DNA Repair in Prokaryotes and Lower Eukaryotes,* Nickoloff, J.A. and Hoekstra, M.F., Eds., Humana Press, Towata, NJ, 1998, 107.
45. Walker, G.C., Mutagenesis and inducible responses to deoxyribonucleic acid damage in *Escherichia coli*, *Microbiol. Rev.*, 48, 60, 1984.
46. Cheo, D.L., Bayles, K.W., and Yasbin, R.E., Cloning and characterization of DNA damage-inducible promoter regions from *Bacillus subtilis*, *J. Bacteriol.*, 173, 1696, 1991.
47. Nuyts, S. et al., The use of radio-induced bacterial promoters in anaerobic conditions: a means to control gene expression in *Clostridium*-mediated therapy for cancer, *Radiation Res.*, 155, 716, 2001.
48. Nuyts, S. et al., Insertion or deletion of the Cheo box modifies radiation inducibility of *Clostridium* promoters, *Appl. Environ. Microbiol.*, 67, 4464, 2001.
49. Pawelek, J.M., Low, K.B., and Bermudes, D., Tumor-targeted *Salmonella* as a novel anticancer vector, *Cancer Res.*, 57, 4537, 1997.
50. Mei, S. et al., Optimization of tumor-targeted gene delivery by engineered attenuated *Salmonella typhimurium*, *Anticancer Res.*, 22, 3261, 2002.

51. Witte, V. et al., Heterologous expression of the clostripain gene from *Clostridium histolyticum* in *Escherichia coli* and *Bacillus subtilis*: maturation of the clostripain precursor is coupled with self-activation, *Microbiology*, 140, 1175, 1994.
52. Schlechte, H. and Elbe, B., Recombinant plasmid DNA variation of *Clostridium oncolyticum* — model experiments of cancerostatic gene transfer, *Zbl. Bakt. Hyg. A*, 268, 347, 1988.
53. Bagshawe, K.D., Antibody directed enzymes revive anti-cancer prodrug concept, *Br. J. Cancer*, 56, 531, 1987.
54. Satchi-Fainaro, R. et al., PDEPT: polymer-directed enzyme prodrug therapy. 2. HPMA copolymer-beta-lactamase and HPMA copolymer-C-Dox as a model combination, *Bioconjug. Chem.*, 14, 797, 2003.
55. Connors, T.A., The choice of prodrugs for gene directed enzyme prodrug therapy of cancer, *Gene Ther.*, 2, 702, 1995.
56. Martin, D.S., Purine and pyrimidine biochemistry, and some relevant clinical and preclinical cancer chemotherapy research, in *Metabolism and Action of Anti-cancer Drugs*, Powis, G. and Prough, R.A., Eds., Taylor & Francis, London, 1987, 91.
57. Knox, R.J., Friedlos, F. and Boland, M.P., The bioactivation of CB 1954 and its use as a prodrug in antibody-directed enzyme prodrug therapy (ADEPT), *Cancer Metastasis Rev.*, 12, 195, 1993.
58. Friedlos, F. et al., Three new prodrugs for suicide gene therapy using carboxypeptidase G2 elicit bystander efficacy in two xenograft models, *Cancer Res.*, 62, 1724, 2002.
59. Pederson, L.C. et al., Molecular chemotherapy combined with radiation therapy enhances killing of cholangiocarcinoma cells *in vitro* and *in vivo*, *Cancer Res.*, 57, 4325, 1997.
60. Theys, J., et al., Specific targeting of cytosine deaminase to solid tumors by engineered *Clostridium acetobutylicum, Cancer Gene Ther.*, 8, 294, 2001.
61. Parmiani, G. et al., Cytokines in cancer therapy, *Immunology Lett.,* 74, 41, 2000.
62. Balkwill, F., Tumor necrosis factor or tumor promoting factor?, *Cytokine & Growth Factor Rev.*, 13, 135, 2002.
63. Mueller, H., Tumor necrosis factor as an antineoplastic agent: pitfalls and promises, *Cell. Mol. Life Sci.*, 54, 1291, 1998.
64. McIntosh, J.K. et al., Combination cytokine immunotherapy with tumor necrosis factor α, interleukin 2, and interferon and its synergistic antitumor effects in mice, *Cancer Res.*, 49, 1408, 1989.
65. Li, Y. et al., Combined effect of 5-fluorouracil and recombinant tumor necrosis factor against human gastric carcinoma cell lines, *Anticancer Res.*, 16, 1771, 1996.
66. Nuyts, S. et al., Radio-responsive *recA* promoter significantly increases TNFα production in recombinant clostridia after 2 Gy irradiation, *Gene Therapy*, 8, 1197, 2001.
67. Hallahan, D.E. et al., The interaction between recombinant human tumor necrosis factor and radiation in 13 human tumor cell lines, *Int. J. Radiat. Oncol. Biol. Phys.*, 19, 69, 1990.
68. Landuyt, W. et al., Vascular targeting of solid tumours: a major “inverse” volume-response relationship following combretastatin A-4 phosphate treatment of rat rhabdomyosarcomas, *Eur. J. Cancer*, 36, 1833, 2000.
69. Tozer, G.M. et al., The biology of the combretastatins as tumour vascular targeting agents, *Int. J. Exp. Pathol.*, 83, 21, 2002.
70. Theys, J. et al., Improvement of *Clostridium* tumour targeting vectors evaluated in rat rhabdomyosarcomas, *FEMS Immunol. Med. Microbiol.*, 30, 37, 2001.
71. Lambin, P. et al., The potential therapeutic gain of radiation-associated gene therapy with the suicide gene cytosine deaminase, *Int. J. Radiat. Biol.*, 76, 285, 2000.

Index

D

E

F

G

H

I

K

L

M

N

O

P

Q

R

S

T